**INTERNATIONAL SOCIETY
FOR ROCK MECHANICS**

**SOCIÉTÉ INTERNATIONALE
DE MÉCANIQUE DES ROCHES**

**INTERNATIONALE GESELLSCHAFT
FÜR FELSMECHANIK**

**International Congress
on Rock Mechanics**

**Congrès international
de mécanique des roches**

**Internationaler Kongress
über Felsmechanik**

PROCEEDINGS / COMPTES-RENDUS / BERICHTE

VOLUME / TOME / BAND 2

MELBOURNE / 1983

Proceedings Fifth Congress of the International Society for Rock Mechanics

Comptes-rendus Cinquième congrès de la Société Internationale de Mécanique des Roches

Berichte Fünfter Kongress der Internationalen Gesellschaft für Felsmechanik

MELBOURNE (AUSTRALIA) 1983

Rock Mechanics for Resource Development, Mining and Civil Engineering

La mécanique des roches en rapport avec l'exploitation de ressources naturelles, l'industrie minière et le génie civil

Felsmechanik für den Aufschluss von Bodenschätzen, Bergbau und Tiefbau

VOLUME 2 Theme C Deep Underground Excavations
Theme D Rock Dynamics
Theme E Special Topics in Rock Mechanics

TOME 2 Thème C Excavation a grande profondeur
Thème D Dynamique des roches
Thème E Aspects particuliers de la mécanique des roches

BAND 2 Thema C Tiefe unterirdische Hohlräume und Abbaue
Thema D Felsdynamik
Thema E Sonderthemen der Felsmechanik

A. A. BALKEMA / ROTTERDAM / 1983
and Australian Geomechanics Society
et Société Australienne de Géomécanique
und Australische Gesellschaft für Geomechanik

For the complete set of three volumes, ISBN 90 6191 236 9
For volume 1, ISBN 90 6191 237 7
For volume 2, ISBN 90 6191 238 5
For volume 3, ISBN 90 6191 239 3

© 1983 by the authors concerned

Published by A.A. Balkema, P.O. Box 1675, Rotterdam, Netherlands
Distributed in U.S.A. and Canada by M.B.S., 99 Main Street, Salem, NH 03079

Printed in Australia by Brown Prior Anderson Pty Ltd
5 Evans Street Burwood Victoria Australia 3125

THEMES VOLUME 2

Theme C Deep Underground Excavations
1 – Mining excavations and mining methods including caving
2 – Permanent underground excavations including tunnels, power stations and storage caverns
3 – Coal mining including ground control and gas outbursts
4 – Prediction, control and measurement of subsidence
5 – Nuclear waste disposal and thermal behaviour of rocks

Theme D Rock Dynamics
1 – Drilling and blasting
2 – Crushing and grinding
3 – Petroleum reservoir behaviour and in situ fracture methods for resource development

Theme E Special Topics in Rock Mechanics
1 – Fracture and flow of the earth's crust, including tectonic stresses
2 – Numerical modelling of rock behavior
3 – Future developments and directions in rock mechanics

THÈMES TOME 2

Thème C Excavation à Grande Profondeur
1 – Excavation Minière et Méthodes minières, y compris l'ouvrage par éboulement
2 – Excavations souterraine permanente, y compris les tunnels, les centrales et les galeries de stockage
3 – Les houillères, y compris le contrôle en surface et les coups de grisou
4 – Prédiction, contrôle et mesure d'affaissements
5 – Traitement des déchets nucléaires et comportement thermique des roches

Thème D Dynamique des Roches
1 – Forage et sautage
2 – Concassage et broyage
3 – Comportement des reservoirs de pétrole et méthodes de fracture in situ pour l'exploitation des ressources naturelles

Thème E Aspects Particuliers de la Mécanique des Roches
1 – Fracture et écoulement de l'écorce terrestre, y compris les contraintes tectoniques
2 – Modelage numérique de comportement des roches
3 – Les voles futures en mécanique des roches

THEMEN BAND 2

Thema C Tiefe unterirdische Hohlräume und Abbaue
1 – Bergbauliche Hohlräume und Bergbaumethoden, einschliesslich Bruchbau
2 – Longlebige Grubenbaue, einschliesslich Tunnel, Kraftwerke und unterlrdische Speicherhohlräume
3 – Kohlenbergbau, einschliesslich Gebirgsbeherrschung und Gasausbrüchen
4 – Voraussage, Kontrolle und Vermessung von Absenkungen
5 – Beseitigung radioaktiven Abfalls und thermisches Verhalten von Felsen

Thema D Felsdynamik
1 – Bohr- und Sprengarbeiten
2 – Zerkleinerung und Vermahlung
3 – Verhalten von Erdölbecken und in-situ-Aufbrüche zur Erschliessung von Bodenschaft

Thema E Sonderthemen der Felsmechanik
1 – Bruch und Kriechen der Erdkruste, einschliesslich tektonischer Spannungen
2 – Erstellung numerischer Modelle des Felsverhaltens
3 – Zukünftige Entwicklungen und Entwicklungsrichtungen auf dem Gebiet der Felsmechanik

Themes and contents of the volumes
Thèmes et contenu des tomes
Themen und Inhalt der Bände

Volume 1
- Theme A   Site Exploration and Evaluation
- Theme B   Surface and Near-surface Excavations

Volume 2
- Theme C   Deep Underground Excavations
- Theme D   Rock Dynamics
- Theme E   Special Topics in Rock Mechanics

Volume 3
Reports of General Reporters and Discussion

Tome 1
- Thème A   Exploration et evaluation in situ
- Thème B   Excavation en surface et à faible profondeur

Tome 2
- Thème C   Excavation à grande profondeur
- Thème D   Dynamique des roches
- Thème E   Aspects particuliers de la mécanique des roches

Tome 3
Rapports des rapporteurs généraux et discussion

Band 1
- Thema A   Untersuchung und Beurteilung des Betriebspunktes
- Thema B   Übertägige und oberflächennahe Felsbauwerke

Band 2
- Thema C   Tiefe unterirdische Hohlräume und Abbaue
- Thema D   Felsdynamik
- Thema E   Sonderthemen der Felsmechanik

Band 3
Berichte der Hauptreferenten und Diskussionen

Foreword

The 5th International Congress of the International Society for Rock Mechanics was held in Melbourne, Australia in April 1983. The overall theme for the Congress was:

"Rock Mechanics for Resource Development,
Mining and Civil Engineering"

This theme was chosen having regard to new developments and applications of rock mechanics throughout the world, and particular interests in rock engineering in Australia.

Each national group of the Society in the fifty-five countries participating in ISRM was responsible for the selection and review of the papers from their country. The papers have been reproduced in these volumes from the manuscripts as received.

The first two volumes of the proceedings contain the papers received in each of the five main themes of the Congress.

The third volume of the proceedings is to contain the General Reports, together with reports of discussions presented by Congress participants, and reports of the Opening and Closing Ceremonies. These General Reports, prepared by internationally recognised specialists, represent an international survey of the state of the art in rock mechanics at the time of the Congress.

The International Society for Rock Mechanics is represented in Australia by the Australian Geomechanics Society, which is jointly sponsored by the mining and engineering communities in Australia represented by the learned bodies, The Australasian Institute of Mining and Metallurgy, and The Institution of Engineers, Australia.

The Australian Geomechanics Society also represents in Australia the International Society for Soil Mechanics and Foundation Engineering, and the International Association of Engineering Geologists. In Australia we have found it professionally valuable for all those having an interest in geomechanics—mining engineers, civil engineers and geologists—to share joint learned activities in geomechanics on a national and state level.

The Australian Geomechanics Society was honoured to be invited by our international colleagues in rock mechanics to host the 5th International Congress in Australia in 1983. The arrangements for the Congress were in the hands of an Organising Committee of the Australian Geomechanics Society, which was supported very strongly indeed by the resources of The Australasian Institute of Mining and Metallurgy, who kindly accepted the invitation to provide the secretariat for the Congress.

Professor H. G. Poulos,
Chairman,
Australian Geomechanics Society

Professor L. A. Endersbee, AO
Congress Chairman

April 1983

Avant-propos

Le 5ᵉ Congrès International de la Société de Mécanique des Roches s'est tenu à Melbourne, Australie, en avril 1983, et a eu pour thème général:

"Rôle de la Mécanique des Roches dans l'Exploitation des Ressources Naturelles, l'Industrie Minière et le Génie Civil"

Ce thème a été choisi pour sa pertinence aux innovations et aux applications récentes en mécanique des roches à travers le monde et en rapport avec les intérêts spécifiques en mécanique des roches en Australie.

Chaque groupe national de la Société dans les cinquante-cinq pays membres de la SIMR s'est chargé de selectionner et revoir les communications soumises par leurs groupements nationaux. Ces communications sont reproduites dans les volumes telles qu'elles ont été reçues.

Les deux premiers volumes des comptes rendus regroupent les communications ayant trait à chacun des thèmes majeurs du Congrès.

Le troisième volume doit comporter les Rapports Généraux, ensemble avec des rapports de discussions des participants au Congrès ainsi que des rapports des Cérémonies d'Ouverture et de Fermeture du Congrès. Ces Rapports Généraux, préparés par des spécialistes mondialement connus, représentent un exposé international du stade d'avancement de la mécanique des roches au moment du Congrès.

La Société Internationale de Mécanique des Roches est représentée en Australie par la Société Australienne de Géomécanique qui est sous l'égide conjointe des communautés minière et d'ingénierie qui sont elles-mêmes représentées par l'Institut Australasien des Mines et de la Métallurgie et l'Institut d'Ingénieurs d'Australie.

La Société de Géomécanique Australienne représente également la Société Internationale de Mécanique des Sols et du Génie des Fondations de même que l'Association Internationale des Géologues-Ingénieurs. En Australie, nous avons trouvé qu'au niveau national, et au niveau des Etats, la mise en commun des activités de recherche en géomécanique par les ingénieurs des mines, les ingénieurs du génie civil et les géologues peut être d'une aide professionnelle précieuse.

La Société Australienne de Géomécanique est sensible à l'honneur d'avoir été invitée par nos collègues internationaux d'être la Société hôte du 5ᵉ Congrès International en Australie, en 1983.

L'organisation du Congrès a été le fait d'un Comité Organisateur fourni par la Société Australienne de Géomécanique. Ce Comité a été pleinement soutenu par l'Institut Australasien des Mines et de la Métallurgie qui a eu l'amabilité d'accepter l'invitation d'assurer les travaux de Secrétariat du Congrès.

Le Professeur H. G. Poulos
Président
Société Australienne de Géomécanique

Le Professeur L. A. Endersbee, AO
Président du Congrès

Avril 1983

Vorwort

Der 5. Internationale Kongress der Internationalen Gesellschaft für Felsmechanik fand in Melbourne, Australien, im April 1983 statt. Das Generalthema des Kongresses lautete:

"Felsmechanik für den Aufschluss von Bodenschätzen, Bergbau und Tiefbau"

Dieses Thema wurde nicht nur mit Bezug auf neue Entwicklungen und Anwendungen auf dem Gebiet der Felsmechanik auf der ganzen Welt, sondern auch insbesondere mit Bezug auf das Felsingenieurwesen in Australien gewählt.

Jede Nationalgruppe der Gesellschaft in den fünfundfünfzig Mitgliedsländern der IGFM war für die Auswahl und Rezension der Beiträge aus dem eigenen Lande verantwortlich, Beiträge, welche in diesen Bänden in der Originalfassung abgedruckt worden sind.

Die ersten zwei Bände des Tätigkeitsberichtes enthalten die Beiträge zu jedem der fünf Hauptthemen des Kongresses.

Der dritte Band des Tätigkeitsberichtes wird die Allgemein-Berichte sowie Einzelheiten der Diskussionen der Kongressteilnehmer als auch Berichte über die Eröffnungs- und Abschluss-Sitzungen enthalten. Diese von international anerkannten Fachleuten verfassten Allgemein-Berichte bieten einen Überblick über den Entwicklungsstand auf der ganzen Welt auf dem Gebiet der Felsmechanik zur Zeit des Kongresses.

Die Internationale Gesellschaft für Felsmechanik wird in Australien von der Australischen Gesellschaft für Geomechanik vertreten, welche gemeinsam von dem australischen Bergbau- und Ingenieurwesen durch die Fachvereinigungen "Australisches Institut für Bergbau und Metallurgie" und "Gesellschaft Australischer Ingenieure" unterstützt wird.

Die Australische Gesellschaft für Geomechanik ist gleichzeitig die australische Mitgliedsvereinigung der Internationalen Gesellschaft für Bodenmechanik und für das Fundament-Ingenieurwesen, als auch der Internationalen Vereinigung für Ingenieurgeologie. In Australien hat es sich für sämtliche an der Geomechanik interessierten Fachleute, wie z.B. Bergwerks- und Bauingenieure und Geologen, als zweckmässig erwiesen, ihre beruflichen Interessen in der Geomechanik gemeinschaftlich auf nationaler und auch auf regionaler Grundlage zu verfolgen.

Der Australischen Gesellschaft für Geomechanik wurde die Ehre zuteil, von ihren internationalen Kollegen auf dem Gebiet der Felsmechanik eingeladen zu werden, den 5. Internationalen Kongress im Jahre 1983 in Australien abzuhalten. Die Kongress-Vorbereitungen wurden von einem Organisationskomitee der Australischen Gesellschaft für Geomechanik getroffen. Dieses genas die volle Unterstützung des "Australischen Instituts für Bergbau und Metallurgie", welches freundlicherweise auch das Sekretariat für den Kongress zur Verfügung stellte.

Professor H. G. Poulos
Vorsitzender
Australische Gesellschaft für Geomechanik

Professor L. A. Endersbee, AO
Vorsitzender des Kongresses

April 1983

Organization of the fifth ISRM Congress

CONGRESS ORGANISING COMMITTEE

W. E. Bamford (Chairman of Committee)
Professor L. A. Endersbee, AO (Congress Chairman)
Dr. J. R. Barrett
Dr. A. G. Bennet
W. J. Cuming
Professor I. B. Donald
J. R. Enever
Dr. R. S. Evans
Dr. I. W. Johnston
Dr. M. Kurzeme
Professor H. G. Poulos
W. M. G. Regan
E. D. J. Stewart
W. E. Vance (Executive Secretary)
Mrs Judy Webber (Secretary to the Committee)

ADVISORY COMMITTEE

President of the ISRM: W. Wittke
Secretary General of the ISRM: A. Silverio

Vice-President for
- Africa: A. Chaoui
- Asia: M. Yoshida
- Australasia: W. E. Bamford
- Europe: S. Uriel Romero
- North America: T. C. Atchison
- South America: O. Moretto

GENERAL REPORTERS

Theme A
  Professor David H. Stapledon, South Australian Institute of Technology, Adelaide, Australia
  Dr. Peter Rissler of Ruhrtalsperrenverein, Essen, West Germany

Theme B
  Professor Richard E. Goodman, University of California, Berkeley, U.S.A.
  Professor Klaus W. John, Ruhruniversität, Bochum, West Germany

Theme C
  Professor Charles Fairhurst and Dr. Barry H. G. Brady, University of Minnesota, Minnesota, U.S.A.
  Professor Yoshio Hiramatsu, Kyoto University, Kyoto, Japan

Theme D
  Dr. Per Anders Persson, Director, Research and Development, Nitro Nobel AB, Sweden
  Roger Holmberg, President, Swedish Detonic Research Foundation Stockholm, Sweden

Theme E
  Professor François G. Cornet, University of Pierre and Marie Curie, Paris, France

Organisation du cinquième Congrès de la SIMR

COMITE D'ORGANISATION DU CONGRES

W. E. Bamford (Président du Comité)
Professeur L. A. Endersbee, AO (Président du Congrès)
Dr. J. R. Barrett
Dr. A. G. Bennet
W. J. Cuming
Professeur I. B. Donald
J. R. Enever
Dr. R. S. Evans

Dr. I. W. Johnston
Dr. M. Kurzeme
Professeur H. G. Poulos
W. M. G. Regan
E. D. J. Stewart
W. E. Vance (Secrétaire exécutif)
Mme Judy Webber (Secrétaire du comité)

COMITE CONSULTATIF

Président de la SIMR: W. Wittke
Secrétaire Général de la SIMR: A. Silverio

Vice-Président pour
l'Afrique: A. Chaoui
l'Asie: M. Yoshida
l'Australasie: W. E. Bamford

l'Europe: S. Uriel Romero
l'Amérique du Nord: T. C. Atchison
l'Amérique du Sud: O. Moretto

RAPPORTEURS GÉNÉRAUX

Thème A

Professeur David H. Stapledon, Institut de Technologie de l'Australie du Sud, Adelaïde, Australie
Dr. Peter Rissler, Ruhrtalsperrenverein, Essen, République Fédérale d'Allemagne

Thème B

Professeur Richard E. Goodman, Université de Californie, Berkeley, E.U.
Professeur Klaus W. John, Université de la Ruhr, Bochum, République Fédérale d'Allemagne

Thème C

Professeur Charles Fairhurst et Dr. Barry H. G. Brady, Université de Minnesota, E.U.
Professeur Yoshio Hiramatsu, Université de Kyoto, Japan

Thème D

Dr. Pers Anders Persson, Directeur, Recherche et Développement, Nitro Nobel AB, Suède
Roger Holmberg, Président, Fondation Suédoise pour la Recherche Détonique, Stockholm, Suède

Thème E

Professeur François G. Cornet, Université de Pierre et Marie Curie, Paris, France

Organisation des fünften Kongresses der IGFM

ORGANISATIONSKOMITEE DES KONGRESSES

W. E. Bamford (Vorsitzender des Komitees)
Professor L. A. Endersbee, AO (Vorsitzender des Kongresses)
Dr. J. R. Barrett
Dr. A. G. Bennet
W. J. Cuming
Professor I. B. Donald
J. R. Enever
Dr. R. S. Evans

Dr. I. W. Johnston
Dr. M. Kurzeme
Professor H. G. Poulos
W. M. G. Regan
E. D. J. Stewart
W. E. Vance (Geschäftsführer)
Frau Judy Webber (Schriftführerin des Komitees)

BERATENDES KOMITEE

Der Präsident der IGFM: W. Wittke
Der Generalsekretär der IGFM: A. Silverio

Vizepräsident für
Afrika: A. Chaoui
Asien: M. Yoshida
Australien und Ozeanien: W. E. Bamford

Europa: S. Uriel Romero
Nordamerika: T. C. Atchison
Südamerika: O. Moretto

HAUPTREFERENTEN

Thema A

Professor David H. Stapledon, South Australian Institute of Technology, Adelaide, Australien
Dr. Peter Rissler des Ruhrtalsperrenvereins, Essen, Westdeutschland

Thema B

Professor Richard E. Goodman, University of California, Berkeley, U.S.A.
Professor Klaus W. John, Ruhruniversität, Bochum, Westdeutschland

Thema C

Professor Charles Fairhurst und Dr. Barry H. G. Brady, University of Minnesota, Minnesota, U.S.A.
Professor Yoshio Hiramatsu, Kyoto Universität, Kyoto, Japan

Thema D

Dr. Per Anders Persson, Direktor, Forschung und Entwicklung, Nitro Nobel AB, Schweden
Roger Holmberg, Präsident der Swedish Detonic Research Foundation, Stockholm, Schweden

Thema E

Professor François G. Cornet, Université de Pierre et Marie Curie, Paris, Frankreich

Contents / Contenu / Inhalt
Volume / Tome / Band 2

C Theme / Thème / Thema **Deep Underground Excavations**
Excavation à Grande profondeur
Tiefe unterirdische Hohlräume und Abbaue

1 Mining excavations and mining methods including caving
Excavation Minière et Méthodes minières, y compris l'ouvrage par éboulement
Bergbauliche Hohlräume und Bergbaumethoden, einschliesslich Bruchbau

C. L. de Jongh J. W. Klokow	THE USE OF A SEISMIC NETWORK AS A MANAGEMENT TOOL Un réseau sismique comme aide à l'exploitation des mines Die Anwendung eines seismischen Netzwerkes als Hilfsmittel der Betriebsführung	D1
M. Wallner	STANDSICHERHEITSBERECHNUNGEN FÜR DIE PFEILERDIMENSIONIERUNG IM SALZBERGBAU Stability calculations concerning a room and pillar design in rock salt Calculs de stabilité pour les dimensions de chambres et piliers dans les mines de sel gemme	D9
Dr. Thomas F. Herbst	DIE ANWENDUNG DER ANKERSICHERUNG IM BERGBAU Application of anchor securing in mines Utilisation des ancrages au rocher dans les mines	D17
G. Borm	ANALYSIS OF CREEP AND RELAXATION AROUND AN UNDERGROUND OPENING IN ROCK SALT Etude du fluage et de la relaxation autour d'une cavité percée dans un massif de sel gemme Analyse der Hohlraumkonvergenz und Gebirgsrelaxation im Steinsalzgebirge	D23
S. Bywater R. Cowling B. N. Black	STRESS MEASUREMENT AND ANALYSIS FOR MINE PLANNING Mesures de tension et analyse en projets miniers Spannungsmessungen und Analyse für den Entwurf von Bergwerken	D29
P. G. Fuller	THE POTENTIAL FOR CABLE SUPPORT OF OPEN STOPES Le potentiel des réseaux de câbles dans le cas d'exploitation en gradins dans les mines à ciel ouvert Das Potential einer Felsbewehrung in Kammerbruchbau	D39
G. Beer J. L. Meek R. Cowling	PREDICTION OF THE BEHAVIOUR OF SHALE HANGING WALLS IN DEEP UNDERGROUND EXCAVATIONS Prédiction du comportement des toits schisteux en excavations profondes Vorhersage des Verhaltens von Hangendem im tiefliegenden Ausbau in Schiefer	D45

M. A. Coulthard J. M. Crotty M. W. Fabjanczyk	COMPARISON OF FIELD MEASUREMENTS AND NUMERICAL ANALYSES OF A MAJOR FAULT IN A MINE PILLAR Comparaison de mesures in situ et d'analyses numériques d'une faille importante dans un pilier de mine Vergleich von Feldmessungen und numerischen Analysen einer grösseren Störung in einem Grubenpfeiler	D53
C. H. Page A. Haines G. S. Esterhuizen	GEOTECHNICAL INVESTIGATION – DESIGN CRITERIA FOR A ROOM AND PILLAR FLUORSPAR MINE Une investigation géotechnique: Critères de conception pour une exploitation en chambres et piliers d'une mine de fluorine Geotechnische Untersuchung – Entwurfskriterien für eine Flusspatgrube im Kammer- und Pfeilerbau	D61
T. J. Kotze	STRATA CONTROL PROBLEMS RESULTING FROM MINING AT SHALLOW DEPTH IN THE HARD ROCK MEASURES OF THE SOUTH AFRICAN BUSHVELD COMPLEX Problèmes de contrôle des toits en exploitations minières peu profondes dans des strates de roches dures du "Bushveld Complex" dans la République d'Afrique du Sud Probleme der Schichtungskontrolle als Folge niedriger Teufe in hartem Gestein in den Gruben des Bushveldes in Südafrika	D67
G. Crea A. Lembo-Fazio R. Ribacchi G. Pantaleone	GEOMECHANICAL INVESTIGATIONS FOR THE DEVELOPMENT OF THE CAMPIANO MINE Investigations géomécaniques pour le développement de la mine de Campiano Geomechanische Studien zum Bergbauplan in Campiano	D77
Chunting Liao Zhaoxian Shi	IN-SITU STRESS MEASUREMENTS AND THEIR APPLICATION TO ENGINEERING DESIGN IN THE JINCHUAN MINE Mesures de contraintes in situ et applications au projet d'ingénierie de la mine au Jinchuan Die in situ Spannungsmessungen und ihre Anwendung zur Grubenplanung im Jinchuan Bergbaubezirk	D87
C. Chambon J. Arcamone J. P. Josien J. P. Piguet	EVALUATION DE LA STABILITE D'EXPLOITATIONS PAR CHAMBRES ET PILIERS ET COMPORTEMENT DU TOIT Estimation of the stability of room and pillar workings and roof behaviour Bestimmung der Standfestigkeit im Kammerpfeilerbau und das Verhalten des Hangenden	D91
Ph. Weber	UNE APPROCHE ENERGETIQUE DE LA STABILITE D'EXPLOITATIONS PAR CHAMBRES ET PILIERS ABANDONNES, TENANT COMPTE DU COMPORTEMENT POST-RUPTURE DES PILIERS An energy approach to room and pillar exploitations based on post-failure behaviour of pillars Eine energetische Methode zur Stabilität verlassener Kammerpfeilerbauten, gegründet auf dem Verhalten der Pfeiler nach dem Bruch	D97
P. S. Sarkka	THE INTERACTIVE DIMENSIONING OF A CROWN PILLAR IN THE RAUTUVAARA MINE La planification interactive des dimensions d'un pilier de toit dans la mine de Rautuvaara Die interaktive Dimensionierung eines Firstenpfeilers in der Grube Rautuvaara	D101
Yun-Mei Lin	DISPLACEMENT REGULARITY IN DEEP OPENINGS NEAR AN OREBODY Régularité du déplacement des galeries profondes situées près de masses minérales Typische Verschiebungen in tiefgelegenen Strecken in der Nähe von Erzkörpern	D109

I. M. Petukhov A. M. Linkov	THEORETICAL PRINCIPLES AND FUNDAMENTALS OF ROCK BURST PREDICTION AND CONTROL	D113
	Les fondements théoriques de la prédiction et méthodes de la prévention des coups de charge	
	Theoretische Grundlagen der Prognose und Verhütung von Gebirgsschlägen	
N. A. Filatov V. D. Beliakov A. V. Dokukin V. F. Trumbachev O. K. Slavin	PHOTOMECHANICAL PRACTICE OF STUDYING CONDITIONS OF SOLID ROCK NEAR UNDERGROUND WORKINGS DRIVEN AT GREAT DEPTH	D121
	La pratique des recherches photomécaniques appliquée aux massifs de roches à proximité de galeries de mines à grande profondeur	
	Praxis der photomechanischen Untersuchungen von Bergmassivzuständen in der Nähe von Untertageräumen in grosser Teufe	
S. Vujec D. Zeljkovic	STOPE STABILITY IN UNDERGROUND BAUXITE EXPLOITATION	D129
	Stabilité des zones d'abattage en exploitation souterraine de bauxite	
	Stabilität von Untertagehohlräumen im Bauxit-Bergbau	
P. Ramirez Oyanguren P. Carpenter F. J. Alfageme J. M. Canto	MATHEMATICAL MODEL OF A CUT AND FILL MINE	D135
	Modèle mathématique d'une mine en coupe et remblayage	
	Mathematisches Modell eines Cut-and-Fill Bergwerkes	
Eduardo Bolivar, B. Guillermo Krstulovic, L.	GEOTECHNICAL STUDY OF THE STABILITY OF THE SANTA CLARA STOPE AT THE EL SOLDADO MINE IN CHILE	D141
	Etude technique de la stabilité de la banquette Santa Clara à la mine de El Soldado au Chili	
	Geotechnische Stabilitätsstudie des Santa Clara Abbauortes in dem El Soldado Bergwerk in Chile	
A. G. Paşamehmetoglu T. Y. Irfan N. Bölükbaşi A. Bilgin A. Özgenoglu C. Karpuz	AN INVESTIGATION INTO THE ROCK MECHANICS ASPECTS OF SUBLEVEL OPEN STOPE MINING	D151
	Recherche en mécanique des roches. Un aspect de l'exploitation par chambres à sous étage	
	Untersuchung der felsmechanischen Aspekte eines Abbaus mit Teilsohlenstollen	

2 Permanent underground excavations including tunnels, power stations and storage caverns
Excavations souterraine permanente, y compris les tunnels, les centrales et les galeries de stockage
Longlebige Grubenbaue, einschliesslich Tunnel, Kraftwerke und unterirdische Speicherhohlräume

E. Gartung P. Bauernfeind	CONSTRUCTION OF ADJACENT TUNNEL TUBES BY N.A.T.M.	D163
	Construction de tunnels adjacents	
	Ausführung eng nebeneinander liegender Tunnel	
Walter Wittke Jose Luis Soria	EXPLORATION, DESIGN AND EXCAVATION OF THE POWERHOUS CAVERN ESTANGENTO SALLENTE IN SPAIN	D167
	Exploration, dimensions et excavation de la galerie de la centrale souterraine de Estangento Sallente en Espagne	
	Versuchsprogramm, Entwurf und Bau der Maschinenkaverne Estangento Sallente in Spanien	
W. D. Ortlepp	CONSIDERATIONS IN THE DESIGN OF SUPPORT FOR DEEP HARD-ROCK TUNNELS	D179
	Eléments de la conception de soutènements de tunnels profonds en roche dure	
	Überlegungen zur Planung der Stützmassnahmen tiefliegender Tunnel	

Manuel Romana Davor Simic	DESIGN OF TUNNELS IN SWELLING MARLS Conception de tunnels dans les marnes gonflantes Entwurf eines Tunnels in blähenden Mergeln	D189
Ryoji Kobayashi Norimasa Kubo Koji Matsuki	CONCRETE SUPPORT COMBINED WITH OUTER AND INNER STEEL SETS IN THE HEAVY PRESSURE ZONE Supports en béton combinés avec cadres métalliques internes et externes dans la zone de hautes pressions Ein mit äusseren und inneren Stahlausbauten versehener Betonausbau im sehr druckhaften Gebirge	D195
Satoshi Hibino Mutsumi Motojima Tadashi Kanagawa	BEHAVIOUR OF ROCKS AROUND LARGE CAVERNS DURING EXCAVATION Comportement de roches autour d'une grande caverne pendant l'excavation Das Verhalten des Gebirges in grossen Untertagehohlräumen während des Ausbruchs	D199
T. Saito K. Tsukada E. Inami H. Inoma Y. Ito	STUDY ON ROCKBURSTS AT THE FACE OF A DEEP TUNNEL, THE KAN-ETSU TUNNEL IN JAPAN BEING AN EXAMPLE Etude de la chute de roches sur le front de taille d'un tunnel profond – cas du tunnel de Kan-Etsu, Japon – Untersuchung über Gebirgsschläge and der Ortsbrust eines tiefliegenden Tunnels, dargestellt am Beispiel des Kan-Etsu Tunnels in Japan	D203
Manuel Romana Samuel Estefania	FIELD INVENTORY OF TUNNELS FOR CLASSIFICATION PURPOSES Inventaire de tunnels aux fins de classification Das Inventar von Geländebeobachtungen in Tunneln zum Zwecke einer Klassifikation	D207
Dominique Fourmaintraux	PERFORMANCES DES TUNNELIERS AU ROCHER: RESULTATS ACTUELS, PREVISION ET CONTROLE Performances of TBM in rocks: Current results, forecasting and control Das Verhalten von Vollschnittmaschinen im Gebirge: Aktuelle Ergebnisse, Voraussage und Kontrolle	D211
Nobuaki Kondo Masayoshi Yamashita	BEHAVIOUR OF ROCK AROUND THE OKUYOSHINO UNDERGROUND POWERHOUSE Comportement de roches autor de la centrale souterraine d'Okuyoshino Gebirgsverhalten des Okuyoshino Kavernenkraftwerkes	D217
P. Berest J. M. Morisseau J. M. Noe G. Souquet	ETANCHEITE DES STOCKAGES SOUTERRAINS DE G.P.L. EN GALERIES NON RECOUVERTES Containment of unlined caverns used for L.P.G. storage Dichtheit von unterirdischen Kavernenspeichern ohne Innenschale	D221
P. Berest D. Nguyen, Minh	COMPORTEMENT MECANIQUE DES CAVITES PROFONDES DE STOCKAGE D'HYDROCARBURES DANS LE SEL Mechanical behaviour of deep salt caverns for the storage of hydrocarbons Mechanisches Verhalten tiefliegender Hohlräume zwecks Kohlenwasserstoffspeicherung im Salzgebirge	D227
D. Nguyen Minh P. Burest J. Bergues	ANALYSE DU COMPORTEMENT DIFFERE DES OUVRAGES SOUTERRAINS Analysis of the time-dependent behaviour of underground works Untersuchung des Zeitverhaltens unterirdischer Hohlräume	D233

S-M Tijani G. Vouille B. Hugout	LE SEL GEMME EN TANT QUE LIQUIDE VISQUEUX Rock salt as a viscous liquid Steinsalz als viskose Flüssigkeit	D241
Thierry Doucerain	ETUDE EXPERIMENTALE D'UNE CONDUITE FORCEE BLINDEE SOUTERRAINE SOUS TRES FORTE PRESSION Experimental study of an underground penstock under very high pressure In situ Versuche mittels einer unterirdischen Druckrohrleitung unter sehr hohen Drücken	D247
K. H. O. Saari R. E. Goodman	ANALYSIS OF SQUEEZING SEAMS IN ROCK Analyse de filons compressés dans de la roche Analyse quetschender Schichten	D253
Prof. Dr. ir Tan Tjong Kie	SWELLING ROCKS AND THE STABILITY OF TUNNELS Roches gonflantes et stabilité des tunnels Drückendes Gebirge und die Standfestigkeit von Tunneln	D261
Keshan Zhu Xiesheng Lie	PRESSURE TESTS IN ROCK CHAMBERS Essai en galeries rocheuses sous pression interne Druckkammerversuche in Felskavernen	D267
Shiwei Bai Weishen Zhu Kejun Wang	SOME ROCK MECHANICS PROBLEMS RELATED TO A LARGE UNDERGROUND POWER STATION IN A REGION WITH HIGH ROCK STRESS Quelques problèmes de mécanique des roches relatifs à une centrale souterraine importante dans une région à fort niveau de contrainte des roches Einige felsmechanische Probleme beim Bau einer Kraftwerkskaverne im Gebirge mit hohen Spannungen	D271
Bernhard Maidl	NEW SUPPORTING METHODS AND MATERIALS FOR THE LINING OF TUNNELS AND SHAFTS IN MINING Méthodes nouvelles et matériaux nouveaux pour le soutènement définitif des tunnels et des puits de mine Neue Verfahren und Sicherungsmaterialien zur Auskleidung von Tunneln und Schächten im Bergbau	D275
G. Innerhofer H. Loacker	DER MASCHINELLE AUSBRUCH DES 21 KM LANGEN WALGAUSTOLLENS Excavation of the 21 km Walgautunnel by a full-face tunnelling machine Percement du tunnel de Walgau (21 km de long) avec un tunnelier de pleine section	D281
B. Bonapace	TESTS AND MEASUREMENTS FOR THE PRESSURE TUNNELS AND SHAFTS OF THE SELLRAIN-SILZ HYDROELECTRIC POWER SCHEME WITH EXTREMELY HIGH HEAD Essais et mesures pour les galeries d'amenée et les conduites forcées en charge extrême de l'aménagement hydroélectrique de Sellrain-Silz Versuche und Messungen für die hochbeanspruchten Druckstollen und Schächte der Werksgruppe Sellrain-Silz der TIWAG	D287
Giancarlo Ceriani Gunnar Nord	MAJES PROJECT: GEOLOGICAL CONDITIONS AND CONSTRUCTION Projet Majes: Conditions géologiques et construction Projekt Majes: Geologische Verhältnisse und Bauausführung	D293
S. G. A. Bergman H. Stille	ROCK BURST PROBLEMS IN A 2.6 MILLION m^3 UNDERGROUND CRUDE OIL STORAGE IN GRANITE Problèmes d'éclatement de roche dans un entrepôt souterrain de 2,6 millions de m^3 creusé dans du granite et destiné au stockage de pétrole brut Abplatzungsprobleme in einer 2,6 Millionen m^3 grossen Granitkaverne für Rohöl	D301

B. J. Holmgren	TUNNEL LININGS OF STEEL FIBRE REINFORCED SHOTCRETE	D311
	Revêtement de tunnels en béton projeté à fibres d'acier	
	Tunnelauskleidungen aus Stahlfaserspritzbeton	
P. F. R. Altounyan P. D. Shelton Wang Hao	SHAFT LINING PRESSURES DURING SINKING THROUGH DEEP AQUIFER ROCKS	D315
	Pressions exercées sur le revêtement des puits lors du forage à travers des roches aquifères profondes	
	Druckbelastungen des Schachtausbaus beim Abteufen durch wasserführende Gesteinsschichten in grosser Teufe	
N. S. Bulychev N. N. Fotiyeva Yu. A. Veksler S. K. Tutanov G. A. Katkov D. I. Kolin E. L. Kokosadze S. A. Chesnokov	LINING DESIGN FOR PERMANENT WORKINGS AND TUNNELS	D323
	Calcul des excavations définitives et des galeries	
	Dimensionierung des Ausbaues in permanenten Hohlraumbauten und Tunneln	
P. K. Kaiser S. Maloney N. R. Morgenstern	TIME-DEPENDENT BEHAVIOUR OF TUNNELS IN HIGHLY STRESSED ROCK	D329
	Déformation en fonction du temps des tunnels en roche surchargée	
	Das zeitabhängige Verhalten von Tunneln im überbeanspruchten Gebirge	
Ivan Vrkljan Prof. Ervin Nonveiller Antun Szavits-Nossan Zvonimir Lisac Ivan Viseć	CONTROLLED SINKING OF AN OPEN END CAISSON IN WEAK ROCK	D337
	Forage contrôlé d'un puits dans une roche fragile	
	Absenken eines grossen Brunnens in weichem Fels	

3 Coal mining including ground control and gas outbursts
Les houillères, y compris le contrôle en surface et les coups de grisou
Kohlenbergbau, einschliessich Gebirgsbeherrschung und Gasausbrüchen

J. Bergues J. Grolier J. C. Soula P. Travert	INTERPRETATION DE LA STRUCTURE DU BASSIN HOUILLER DE ST-ELOY, D'APRES LES RESULTATS D'ESSAIS MECANIQUES ET LES MODELES TECTONIQUES	E1
	Structural interpretation of the St. Eloy coal basin from mechanical tests and tectonic models	
	Interpretierung der Geologie des St. Eloy Kohlebeckens mit Hilfe von Ergebnissen aus mechanischen Versuchen und von tektonischen Modellen	
Y. Hiramatsu T. Saito N. Oda	STUDIES ON THE MECHANISM OF GAS AND COAL BURSTS IN JAPANESE COAL MINES	E7
	Etudes du mécanisme de coup de grisou dans les mines de houille au Japon	
	Untersuchung über den Mechanismus der plötzlichen Kohle- und Gasausbrüche in japanischen Kohlebergwerken	

Authors	Title	Page
K. Fukuda Y. Ishijima S. Kinoshita	STRATA CONTROL IN DEEP COAL MINES IN HOKKAIDO – IN-SITU MONITORING AND INTERPRETATION OF STRESS CHANGES IN COAL SEAMS Contrôle des couches dans les mines souterraines profondes à Hokkaido – Surveillance in situ et interprétation du changement de contrainte dans les couches de charbon Flözbeherrschung in den tiefen Kohlebergwerken in Hokkaido – Feldmessung und Analyse der Beanspruchungsänderungen in Kohleflözen	E11
J. Abad B. Celada E. Chacon V. Gutierrez E. Hidalgo	APPLICATION OF GEOMECHANICAL CLASSIFICATION TO PREDICT THE CONVERGENCE OF COAL MINE GALLERIES AND TO DESIGN THEIR SUPPORTS Applications des classifications géomécaniques pour prévoir les convergences des galeries de mines de charbon et pour calculer leur soutènement Anwendung der geomechanischen Klassifizierung zur Vorabschätzung der Konvergenz und zum Entwurf des Ausbaus von Strecken in Kohlegruben	E15
Fritz Schuermann	PLANMÄSSIGE ÜBERWACHUNG WICHTIGER GRUBENRÄUME IM STEINKOHLENBERGBAU MIT FERNÜBERTRAGUNG DER MESSWERTE Systematic monitoring of important cavities in hard coal mines with remote data transmission Contrôle régulier d'ouvrages souterrains importants dans les mines de charbon comprenant la télé-transmission de données de mesure	E21
Z. T. Bieniawski	NEW DESIGN APPROACH FOR ROOM-AND-PILLAR COAL MINES IN THE U.S.A. Une nouvelle méthode de dimensionnement pour abattage par chambres et dépilage dans les houillères aux Etats Unis Eine neue Dimensionierungsmethode für Kammer- und Pfeilerbau im amerikanischen Kohlenbergbau	E27
Paul H. Lu, Ph.D.	STABILITY EVALUATION OF RETREATING LONGWALL CHAIN PILLARS WITH REGRESSIVE INTEGRITY FACTORS Evaluation de la stabilité des chaînes-piliers dans les longwalls au moyen de facteurs d'intégrité régressifs Bewertung der Stabilität von Strebkettenpfeilern mittels regressiver Integritätsfaktoren	E37
S. A. Khristianovich R. L. Salganik	SEVERAL BASIC ASPECTS OF THE FORMING OF SUDDEN OUTBURSTS OF COAL (ROCK) AND GAS Quelques questions fondamentales concernant la formation de dégagements instantanés dans les mines de charbon Einige Grundfragen zum Entstehen plötzlicher Kohle- (Gesteins-) und Gasausbrüche	E41
R. N. Gupta I. W. Farmer	STRATA DEFORMATION AND SUPPORT PERFORMANCE AT A LONGWALL COAL FACE Déformation des bancs du toit et rendement obtenu dans une longue taille de charbon Gebirgsverformung und Verhalten des Schildes im Streb	E51
L. J. Wardle J. R. Enever	APPLICATION OF THE DISPLACEMENT DISCONTINUITY METHOD TO THE PLANNING OF COAL MINE LAYOUTS Application de la méthode de déplacement discontinu à la planification des traçages de houillères Anwendung der Verschiebungs-Diskontinuitäts-Methode auf die Planung von Kohlegruben	E61
R. A. Yeates J. R. Enever B. K. Hebblewhite	INVESTIGATIONS PRIOR TO THE INTRODUCTION OF LONGWALL MINING Recherche avant la mise en oeuvre de l'exploitation minière par "Longwall" Untersuchungen vor Einführung des Strebbaues	E71

J. Hanes R. D. Lama J. Shepherd	RESEARCH INTO THE PHENOMENON OF OUTBURSTS OF COAL AND GAS IN SOME AUSTRALIAN COLLIERIES	E79
	Recherche sur les dégagements instantanés dans les mines de charbon australiennes	
	Forschung über das Phänomen der Kohle- und Gasausbrüche in einigen australischen Kohlegruben	
M. M. Singh R. A. Cummings	PREDICTING MOISTURE-INDUCED DETERIORATION OF SHALES	E87
	Prédiction de la détérioration des schistes argileux provoquée par l'humidité	
	Die Vorhersage der feuchtigkeitsbedingten Entfestigung von Schiefer	
J. Carrasco	INSTRUMENTATION SYSTEM FOR THE MEASUREMENT AND RECORDING OF TRANSIENT GEODYNAMIC PHENOMENA	E97
	Système d'enregistrement et de mesure des phénomènes géodynamiques transitoires	
	Instrumentierung zur Messung und Aufzeichnung kurzzeitiger geodynamischer Phänomene	

4 Prediction, control and measurement of subsidence
Prédiction, contrôle et mesure d'affaissements
Voraussage, Kontrolle und Vermessung von Absenkungen

T. R. Stacey H. A. D. Kirsten B. L. Wiid	MAJOR DEVELOPMENT ON AN UNDERMINED SITE IN A CENTRAL CITY AREA	E101
	Constructions importantes sur d'anciennes mines situées sous un centre urbain	
	Grossbauten in einem innerstädtischen, vom Untertagebau beeinflussten Baugebiet	
M. D. G. Salamon	LINEAR MODELS FOR PREDICTING SURFACE SUBSIDENCE	E107
	Modèles linéaires pour prédiction d'affaissements de surface	
	Lineare Modelle zur Voraussage von Oberflächensetzungen	
J. Arcamone R. Poirot	EXPLOITATIONS DE SURFACE AU-DESSUS D'EXPLOITATIONS SOUTERRAINES – PROBLEMES D'INTERACTION	E115
	Open-pit mining over underground extraction – Interactive problems	
	Tagebau über Untertagebauten – Probleme der gegenseitigen Beeinflussung	
P. A. Mikula G. E. Holt	PREDICTION OF MINE SUBSIDENCE IN EASTERN AUSTRALIA BY MATHEMATICAL MODELLING	E119
	La prédiction par modélisation mathématique des affaissements provoqués par l'extraction de charbon en Australie orientale	
	Über die mathematische Vorraussagung von Bodensenkung infolge Kohlegewinnung in Ostaustralien	
J. M. Galvin, Ph.D.	THE DEVELOPMENT OF THE THEORY OF DOLERITE SILL BEHAVIOUR	E127
	Le développement de la théorie du comportement d'un sill de dolerite	
	Entwicklung einer Theorie über das Verhalten von Doleritschichten	
Dr. G. Petrasovits	ANALYSIS OF SURFACE SUBSIDENCES CAUSED BY TWIN TUNNELS BUILT IN COHESIVE SOILS	E133
	Analyse des affaissements de surface au-dessus des tunnels doubles construits dans un sol cohérent	
	Analyse der Oberflächensenkungen, hervorgerufen durch in bindigen Böden gebaute Doppeltunnel	

5 Nuclear waste disposal and thermal behaviour of rocks
Traitement des dechets nucleaires et comportement thermique des roches
Beseitigung radioaktiven Abfalls und thermisches Verhalten von Felsen

S. Ehara M. Terada T. Yanagidani	THERMAL PROPERTIES OF STRESSED ROCKS Propriétés thermiques des roches sous compression Thermische Eigenschaften von belasteten Gesteinen	E137
G. Barla N. Innaurato G. Pantaleoni	HEAT TRANSFER IN THE ROCK MASS AROUND MINE OPENINGS Diffusion de la chaleur dans la roche autour des vides miniers Wärmeaustausch in der Umgebung bergmännischer Felshohlräume	E141
Michio Kuriyagawa Isao Matsunaga Tsutomu Yamaguchi	AN IN SITU DETERMINATION OF THE THERMAL CONDUCTIVITY OF GRANITIC ROCK Détermination in situ de la conductivité thermique de la roche granitique In situ Bestimmung der Wärmeleitfähigkeit eines granitischen Gneises	E147
K. Röshoff, Ph.D. O. Stephansson H. Larsson, Civ.Eng. R. Stanfors, Ph.D. K. Eriksson, Civ.Eng.	CLAB – AN INTERMEDIATE STORAGE FOR SPENT NUCLEAR FUEL IN SWEDEN Le CLAB – Un stockage intermédiaire de combustibe nucléaire usé en Suède CLAB – Vorläufige Verwahrung von verbrauchtem Kernbrennstoff in Schweden	E151
J. A. Hudson	UK ROCK MECHANICS RESEARCH FOR RADIOACTIVE WASTE DISPOSAL Recherches en cours au Royaume Uni sur la mécanique des roches en ce qui concerne l'entreposage des déchets radioactifs Felsmechanikforschungen in Grossbritannien für die Endlagerung radioaktiver Abfallstoffe	E161
R. A. Wagner, M. C. Loken H. Y. Tammemagi	THERMOMECHANICAL ROOM REGION ANALYSIS OF FOUR POTENTIAL NUCLEAR WASTE REPOSITORY SITES IN SALT Analyse thermomécanique de la zone entourant une chambre à quatre sites éventuels pour les résidus nucléaires en gisements de sel Thermomechanical Analyse der Einlagerungsstrecke für radioaktiven Abfall an vier potentiellen Endlagerstandorten im Salz	E167
Sten Bjurström, Dr.Eng. Juri Martna Göran Rehbinder, Dr. Eng. Kennert Röshoff, Ph.D.	STABILITY OF A ROCK OPENING SUBJECTED TO PULSATING TEMPERATURE Stabilité d'une caverne rocheuse soumise à des températures pulsatiles Die Stabilität eines Bergraumes, der pulsierenden Temperaturen ausgesetzt ist	E173

D Theme / Thème / Thema
Rock Dynamics
Dynamique des Roches
Felsdynamik

1 Drilling and blasting
Forage et sautage
Bohr- und Sprengarbeiten

E. Mikura	KLASSIFIZIERUNG DER BOHRBARKEIT DES GEBIRGES DURCH IN SITU- UND LABORUNTERSUCHUNGEN Classification of the drillability of a rock mass by in situ and laboratory testing Classification de la forabilité des massifs rocheux par des essais in situ et en laboratoire	E181

P. N. Calder A. Bauer	PRE-SPLIT BLAST DESIGN FOR OPEN-PIT AND UNDERGROUND MINES Tirage en deux temps pour l'exploitation à ciel ouvert et exploitation au fond Entwurf zum pre-split Sprengen für Tage- und Untertagebau	E185
M. Hisatake S. Sakurai T. Ito Y. Kobayashi	ANALYTICAL CONTRIBUTION TO TUNNEL BEHAVIOUR CAUSED BY BLASTING Contribution analytique sur le comportement des tunnels suite à l'abattage Ein analytischer Beitrag zum Verhalten eines Tunnels beim Sprengen	E191
A. Mouraz Miranda F. Mello Mendes	DRILLABILITY AND DRILLING METHODS Forabilité et méthodes de forage Bohrbarkeit und Bohrmethoden	E195
F. Pechalat Y. Lefin	UTILISATION DES JETS D'EAU A HAUTE PRESSION DANS LES TRAVAUX MINIERS The use of high-pressure water jets in mining operations Einsatz von Hochdruckwasserstrahlen im Grubenbetrieb	E201
R. J. Fowell O. Tecen	STUDIES IN WATER JET ASSISTED DRAG TOOL ROCK EXCAVATION Etudes sur l'excavation par outils à dragline secondés par jet d'eau Untersuchungen über das Zerspanen von Gesteinen mit Radialmeissel und Hochdruck—Wasserstrahl	E207
C. K. McKenzie P. D. Forbes G. E. LeJuge I. H. Lewis P. A. Lilly J. D. Lilly	LIMIT BLAST DESIGN EVALUATION Evaluation du modèle de charge explosive limite Überlegungen zum Entwurf von begrenztem Sprengen	E215
K. Nishi T. Kokusho Y. Esashi	DYNAMIC SHEAR MODULUS AND DAMPING RATIO OF ROCKS FOR A WIDE CONFINING PRESSURE RANGE Modules de cisaillement dynamique et rapports d'amortissement de roches soumis à une grande variation de pression reserrants de confinement Dynamischer Schermodulus und Dämpfungsfaktor für dreiachsig-beanspruchte Gesteine	E223
Takashige Haga Jiro Saito	A NEW CRUSHING TEST FOR ROCK EXCAVATION PROPERTIES Un nouvel essai de concassage appliqué aux propriétés d'excavation en roches Eine neue Quetschprüfung zur Ermittlung der Aushubeigenschaften von Gesteinen	E227
R. Holmberg K. Maki W. Hustrulid H. Sellden	BLAST DAMAGE AND STRESS MEASUREMENTS IN THE LKAB-MALMBERGET FABIAN OREBODY Dommages causés par une charge explosive et détermination de contraintes dans l'amas métallifère "Fabian" de LKAB-Malmberget Sprengauflockerung und Spannungsmessungen im Erzkörper "Fabian" der LKAB-Malmberget Grube	E231

2 Crushing and grinding
Concassage et broyage
Zerkleinerung und Vermahlung

Mossaid Al-Hussaini	EFFECT OF PARTICLE SIZE AND STRAIN ON THE STRENGTH OF CRUSHED ROCK Influence de la dimension et de la déformation des particules sur la résistance des roches broyées Einfluss der Partikelgrösse und der Dehnung auf die Festigkeit zerkleinerten Gesteins	E239

Lineu Azuaga Ayres Da Silva Wildor Theodoro Hennies	A NEW METHOD FOR PARTICLE SHAPE DETERMINATION Une nouvelle technique pour la détermination de la forme des particules Ein neues Verfahren zur Kornformbestimmung	E245

3 Petroleum reservoir behaviour and in situ fracture methods for resource development
Comportement des reservoirs de pétrole et méthodes de fracture in situ pour l'exploitation des ressources naturelles
Verhalten von Erdölbecken und in-situ-Aufbrüche zur Erschliessung von Bodenschätzen

Rolf K. Bratli, Per Horsrud Rasmus Risnes	ROCK MECHANICS APPLIED TO THE REGION NEAR A WELLBORE Mécanique des roches appliqué à proximité d'un puits Gesteinsmechanik, angewandt auf die Bohrstellenumgebung	F1
H. Gross	COMMUNICATING ELEMENTS FOR THE SIMULATION OF CRACK GROWING PROCESSES IN ROCKMASSES Eléments en action réciproque pour simulation de la propagation des fissures dans la roche Kommunizierende Elemente zur Simulation von Risswachstumsprozessen im Gebirge	F19
J. P. Bruggeman J. P. Coyette E. Lousberg J. F. Thimus	SIMULATION NUMERIQUE DE L'ECOULEMENT DE L'EAU DANS UNE VEINE DE CHARBON SUITE A DES ESSAIS DE FRACTURE ENTRE DEUX FORAGES Numerical simulation of water flow in a coal vein by cracking tests between two boreholes Numerische Simulation des Wasserflusses in einer Kohlenschicht auf Grundlage von Bruchversuchen zwischen zwei Bohrungen	F25
Anders Carlsson Tommy Olsson	ROCK STRESS INFLUENCE ON WATER FLOW IN FRACTURES Effect de la contrainte des roches sur l'écoulement des eaux dans les fractures Der Einfluss von Spannungen im Felsen auf die Wasserströmung in Klüften	F31
T. W. Thompson K. E. Gray P. N. Jogi	THE TIME-DEPENDENT BEHAVIOUR OF RESERVOIR ROCKS IN RELATION TO FLUID PRODUCTION Comportement des roches d'un gisement géothermique par rapport à la production des fluides, et en fonction du temps Das zeitliche Verhalten von Speichergesteinen bei fortschreitender Förderung	F35
L. W. Teufel N. R. Warpinski	IN-SITU STRESS VARIATIONS AND HYDRAULIC FRACTURE PROPAGATION IN LAYERED ROCK — OBSERVATIONS FROM A MINEBACK EXPERIMENT Variations des contraintes in situ et propagation des fractures hydrauliques dans une roche litée: Observations effectuées à partir d'une galerie creusée après coup Imhomogene in situ Spannungszustände und die Ausdehnung eines hydraulisch erzeugten Bruches in geschichteten Gebirge — Beobachtungen eines Versuches vor Ort	F43

Y. I. Protasov I. F. Oksanich P. S. Mironov	DEVELOPMENT AND ESTIMATION OF ROCK BREAKING METHODS	F49
	Le développement de la technologie et du calcul d'abattage des roches	
	Die Entwicklung der Verfahren und der Berechnung zum Lösen des Gebirges	

E Theme / Thème / Thema
Special Topics in Rock Mechanics
Aspects Particuliers de la Mécanique des Roches
Sonderthemen der Felsmechanik

1 Fracture and flow of the earth's crust, including tectonic stresses
Fracture et écoulement de l'écorce terrestre, y compris les contraintes tectoniques
Bruch und Kriechen der Erdkruste, einschliesslich tektonischer Spannungen

T. Borg N. Krauland	THE APPLICATION OF THE FINITE ELEMENT MODEL OF THE NÄSLIDEN MINE TO THE PREDICTION OF FUTURE MINING CONDITIONS	F55
	Application du modèle en éléments finis établi pour la mine de Näsliden à la prévision des conditions futures d'exploitation à cette mine	
	Anwendung eines Modells mit finiten Elementen für das Bergwerk Näsliden zur Vorhersage künftiger Bergbaubedingungen	
W. Rahn	UNTERSUCHUNG MÖGLICHER AUSWERTUNGSFEHLER BEI DER INTERPRETATION VON IN-SITU SPANNUNGSMESSUNGEN IN ANISOTROPEN GESTEINEN	F63
	Analysis of Potential Errors of Interpretation of In-Situ Stress Measurements in Anisotropic Rocks	
	Etude des erreurs potentielles d'interprétation des contraintes primaires en roches anisotropes	
K. Balthasar E. Wenz	DIE BESTIMMUNG DES VOLLSTÄNDIGEN SPANNUNGSZUSTANDES IM GEBIRGE MIT DER KOMPENSATIONSMETHODE	F69
	The determination of the complete state of stress in rock with the flat jack method	
	La détermination des contraintes dans les massifs rocheux par la méthode du vérin plat	
J. L. Hernandez Enrile	FAULT MECHANISM IN THE TOLEDO SHEAR ZONE IN SPAIN	F75
	Mécanisme des failles dans la zone de cisaillement de Tolédo en Espagne	
	Mechanismus einer geologischen Störung in der Scherzone von Toledo in Spanien	
José Loureiro Pinto José Gabriel Charrua-Graça	DETERMINATION OF THE STATE OF STRESS OF ROCK MASSES BY THE SMALL FLAT JACK (SFJ) METHOD	F79
	Détermination de l'état de contrainte des massifs rocheux par la méthode des petits vérins plats (SFJ)	
	Bestimmung des Gebirgsspannungszustandes mit der Methode der kleinen Druckkissen (SFJ)	
T. Kanagawa H. Komada M. Hayashi	MEASUREMENTS OF TECTONIC STRESSES, STRAIN RATES RELATED TO ACTIVE FAULTS AND OBSERVED EARTHQUAKES AROUND LARGE CAVERNS	F85
	Mesure de contraintes tectoniques, progression de cisaillement géodésique des failles actives, et observations de tremblements de terre aux alentours d'une centrale électrique souterraine	
	Messungen tektonischer Spannungen, Dehnungsgeschwindigkeiten an aktiven Störungen und von Erdbeben nahe grosser Kraftwerkskavernen	

J. Deramond P. Sirieys J. C. Soula	MECANISMES DE DEFORMATION DE L'ECORCE TERRESTRE – STRUCTURES ET ANISOTROPIE INDUITES	F89
	Mechanisms of the deformation of the Earth's crust – Induced structures and anisotropy	
	Mechanismen der Deformation der Erdkruste – Induzierte Strukturen und Anisotropie	
Prof. Dr. ir Tan Tjong Kie Kang Wen Fa	TIME-DEPENDENT DILATANCY PRIOR TO ROCK FAILURE AND EARTHQUAKES	F95
	Dilatation en fonction du temps avant la fracture de roches et les tremblements de terre	
	Zeitabhängige Dilatanz vor dem Felsbruch und vor Erdbeben	
P. J. Huergo	EVALUATION DES CONTRAINTES NATURELLES DANS LES COUCHES SUPERIEURES D'UN MASSIF SCHISTEUX	F103
	Assessment of in situ stresses in the upper layers of a schistous bedrock	
	Abschätzung der natürlichen Spannungen in den oberen Schichten eines Schieferuntergrundes	
J. Martna R. Hiltscher K. Ingevald	GEOLOGY AND ROCK STRESSES IN DEEP BOREHOLES AT FORSMARK IN SWEDEN	F111
	Géologie et contraintes des roches dans les trous de forage en profondeur à Forsmark en Suède	
	Geologie und Gebirgsspannungen in tiefen Bohrungen in Forsmark in Schweden	
N-K. Ren J.-C. Roegiers	DIFFERENTIAL STRAIN CURVE ANALYSIS – A NEW METHOD FOR DETERMINING THE PRE-EXISTING IN-SITU STRESS STATE FROM ROCK CORE MEASUREMENTS	F117
	Analyse de la courbe des déformations différentielles – Une nouvelle méthode de détermination des contraintes in-situ à partir de mesures sur carottes	
	Differentialanalyse von Dehnungskurven – Eine neue Methode zur Bestimmung des Spannungszustandes in situ durch Messungen an Gesteinskernen	
O. Stephansson	ROCK STRESS MEASUREMENT BY SLEEVE FRACTURING	F129
	Mesure par la technique du manchon fissurant des contraintes subies par une roche	
	Gebirgsdrucksmessung durch Manschetten-Bruchbelastung	
S. V. Kuznetsov D. M. Bronnikov I. A. Parabuchev V. D. Parphenov I. T. Aitmatov G. A. Markov	THE STATE OF STRESS IN ROCK AND METHODS OF ITS DETERMINATION	F139
	L'état de contrainte dans les masses rocheuses et méthodes de sa détermination	
	Der Spannungszustand des Gebirges und die Methoden seiner Bestimmung	

2 Numerical modelling of rock behaviour
Modelage numérique de comportement des roches
Erstellung numerischer Modelle des Felsverhaltens

I. Carol E. E. Alonso	A NEW JOINT ELEMENT FOR THE ANALYSIS OF FRACTURED ROCK	F147
	Un nouvel élément sur les joints pour l'analyse des massifs rocheux fissurés	
	Ein neues Kluftelement für die Analyse von geklüftetem Fels	
Antonio P. Cunha	ANALYSIS OF ADVANCING TUNNELS IN ROCK	F153
	Comportement des tunnels au voisinage du front de taille	
	Analytische Untersuchungen zum Gebirgsverhalten beim Tunnelvortrieb	

A. P. Peirce J. A. Ryder	**EXTENDED BOUNDARY ELEMENT METHODS IN THE MODELLING OF BRITTLE ROCK BEHAVIOUR**	F159
	Extension des méthodes d'éléments limite pour modélisation du comportement de roches dures et fragiles	
	Erweiterte "boundary element" Methode zur Modellierung von Sprödbruchverhalten von Gebirge	
L. R. Sousa	**THREE-DIMENSIONAL ANALYSIS OF LARGE UNDERGROUND POWER STATIONS**	F169
	Analyse tridimensionelle de grandes cavernes pour des usines hydroélectriques	
	Dreidimensionale Analyse grosser unterirdischer Kraftwerke	
L. Rochet	**MODELES NUMERIQUES D'ANALYSE A LA RUPTURE AVEC DILATANCE**	F175
	Numerical models for failure analysis with dilatancy	
	Numerische Modelle zur Bruchanalyse bei Dilatanz	
A. Guenot M. Panet	**ETUDE NUMERIQUE D'OUVRAGE EN MASSIF ROCHEUX A STRUCTURE PLANAIRE**	F181
	Numerical analysis of rock masses with planar structure	
	Numerische Analyse von geschichtetem Gebirge	
Jia-Shou Zhuo Yin-Tang Wang	**NON-LINEAR ANALYSIS OF ROCK FOUNDATIONS WITH SOFT INTERFACES**	F187
	L'analyse non-linéaire de fondations en roche avec interfaces molles	
	Nichtlineare Analyse von Felsgründungen mit weichen Zwischenschichten	
J. A. Quiblier K. Ngokwey	**MECANIQUE DES ROCHES ET MODELES MATHEMATIQUES APPLIQUES A LA GEOLOGIE**	F191
	Rock mechanics and mathematical models applied to geology	
	Gesteinsmechanik und auf Geologie angewandte mathematische Modelle	
Lin Dezhang Liu Baoshen	**BOUNDARY ELEMENT METHOD FOR LINEAR VISCO-ELASTIC STRESS ANALYSIS IN A ROCK MASS AND ITS APPLICATION IN ROCK ENGINEERING**	F195
	Méthode des éléments limites appliquée à des massifs rocheux viscoélastiques linéaires — analyse des contraintes et son application en géotechnique	
	Granzelementmethode zur Spannungsanalyse für lineare Viskoelastizität des Gebirges und ihre Anwendung im Felsbau	
S. D. Priest A. Samaniego	**A MODEL FOR THE ANALYSIS OF DISCONTINUITY CHARACTERISTICS IN TWO DIMENSIONS**	F199
	L'analyse des caractéristiques de discontinuités en deux dimensions	
	Ein ebenes Computermodell für die Untersuchung der Klufteigenschaften	
C. S. Desai I. M. Eitani C. Haycocks	**AN APPLICATION OF FINITE ELEMENT PROCEDURE FOR UNDERGROUND STRUCTURES WITH NON-LINEAR MATERIALS AND JOINTS**	F209
	Application de la méthode des éléments finis à des structures souterraines avec des matériaux non-linéaires et des joints	
	Die Anwendung einer Finiten Element Prozedur im Hohlraumbau mit nichtlinearen Materialien und Klüften	
G. N. Pande C. M. Gerrard	**THE BEHAVIOUR OF REINFORCED JOINTED ROCK MASSES UNDER VARIOUS SIMPLE LOADING STATES**	F217
	Le comportement de masses rocheuses jointes et armées sous divers chargements simples	
	Das Verhalten eines armierten, geklüfteten Felsgebirges unter verschiedenen einfachen Belastungszuständen	

P. M. Warburton	APPLICATIONS OF A NEW COMPUTER MODEL FOR RECONSTRUCTING BLOCKY ROCK GEOMETRY – ANALYSING SINGLE BLOCK STABILITY AND IDENTIFYING KEYSTONES	F225
	Applications d'un nouveau modèle informatique pour la réconstitution de la géométrie des rochers en blocs – l'analyse de la stabilité des blocs simples et la détermination des clefs de voûtes	
	Anwendungen eines neuen Rechnermodells zur Nachahmung der Blockgesteinsgeometrie, zur Analyse der Stabilität einfacher Blöcke sowie zur Identifizierung von Schluss-Steinen	
Manuel Casteleiro Eugenio Oñate Antonio Huerba Jordi Roig Eduardo Alonso	THREE-DIMENSIONAL ANALYSIS OF NO TENSION MATERIALS	F231
	Analyse tridimensionelle des matériaux "no traction"	
	Dreidimensionale Analyse für Materialien ohne Zugfestigkeit	
P. Loven J. Oksanen K. Äikäs	ELEMENT METHODS IN PLANNING OF MINE OPENINGS IN HIGHLY STRESSED PRECAMBRIAN BEDROCK	F237
	Die Element-Methoden in der Planung von Grubenhohlraumen in stark beanspruchten präkambrischen Gesteinen	
	Methodes a elements dans la planification d'excavations minieres en roches precambriennes soumises a de fortes contraintes	

3 Future developments and directions in rock mechanics
Les voies futures en mécanique des roches
Zukünftige Entwicklungen und Entwicklungsrichtungen auf dem Gebiet der Felsmechanik

Lubomir Siska	GEGENWÄRTIGER STAND UND TENDENZEN IN DER ENTWICKLUNG DER GESTEINSMECHANIK IN DER TSCHECHOSLOWAKEI	F243
	The present state and tendencies in the development of rock mechanics in Czechoslovakia	
	L'état actuel et tendance du développement futur de la mécanique des roches en Tchécoslovaquie	
Mei Jian-Yun Fu Bing-Jun Kang Wen-Fa	THE DEVELOPMENT AND CURRENT STATE OF ROCK MECHANICS IN CHINA	F251
	Développement et situation actuelle de la mécanique des roches en Chine	
	Die Entwicklung und der gegenwärtige Stand der Felsmechanik in China	
M. Gálos P. Kertész	RESULTS OF ROCK MECHANICS IN HUNGARY – AN ENGINEERING GEOLOGICAL MODEL OF ROCKS	F259
	Résultats dans la mécanique des roches en Hongrie – Un modèle géotechnique des roches	
	Fortschritte der Felsmechanik in Ungarn – Ein ingenieur – geologisches Gesteinsmodell	

Index to Authors
Index des Auteurs
Inhaltsverzeichnis nach Schriftstellern

Volume / Tome / Band 2

ABAD, J.	E15	EHARA, S.	E137	
AIKAS, K.	F237	EITANI, I.M.	F209	
AITMATOV, I.T.	F139	ENEVER, J.R.	E61	E71
ALFAGEME, F.J.	D135	ENRILE, J.L.H.	F75	
AL-HUSSAINI, M.	E239	ERIKSSON, K.	E151	
ALONSO, E.E.	F147 F231	ESASHI, Y.	E223	
ALTOUNYAN, P.F.R.	D315	ESTEFANIA, S.	D207	
ARCAMONE, J.	D91 E115	ESTERHUIZEN, G.S.	D61	
BAI, S.	D271	FABJANCZYK, M.W.	D53	
BALTHASAR, K.	F69	FARMER, I.W.	E51	
BAOSHEN, L.	F195	FILATOV, N.A.	D121	
BARLA, G.	E141	FORBES, P.D.	E215	
BAUER, A.	E185	FOTIYEVA, N.N.	D323	
BAUERNFEIND, P.	D163	FOURMAINTRAUX, D.	D211	
BEER, G.	D45	FOWELL, R.J.	E207	
BELIAKOV, V.D.	D121	FU, B.J.	F251	
BEREST, P.	D221 D227 D233	FUKUDA, K.	E11	
BERGMAN, S.G.A.	D301	FULLER, P.G.	D39	
BERGUES, J.	D233 E1			
BIENIAWSKI, Z.T.	E27	GALOS, M.	F259	
BILGIN, A.	D151	GALVIN, J.M.	E127	
BJURSTROM, S.	E173	GARTUNG, E.	D163	
BLACK, B.N.	D29	GERRARD, C.M.	F217	
BOLIVAR, B.E.	D141	GOODMAN, R.E.	D253	
BOLUKBASI, N.	D151	GRAY, K.E.	F35	
BONAPACE, B.	D287	GROLIER, J.	E1	
BORG, T.	F55	GROSS, H.	F19	
BORM, G.	D23	GUENOT, A.	F181	
BRATLI, R.K.	F1	GUPTA, R.N.	E51	
BRONNIKOV, D.M.	F139	GUTIERREZ, V.	E15	
BRUGGEMAN, J.P.	F25			
BULYCHEV, N.S.	D323	HAGA, T.	E227	
BYWATER, S.	D29	HAINES, A.	D61	
		HANES, J.	E79	
CALDER, P.N.	E185	HAO, W.	D315	
CANTO, J.M.	D135	HAYASHI, M.	F85	
CARLSSON, A.	F31	HAYCOCKS, C.	F209	
CAROL, I.	F147	HEBBLEWHITE, B.K.	E71	
CARPENTER, P.	D135	HENNIES, W.T.	E245	
CARRASCO, J.	E97	HERBST, T.F.	D17	
CASTELEIRO, M.	F231	HIBINO, S.	D199	
CELADA, B.	E15	HIDALGO, E.	E15	
CERIANI, G.	D293	HILTSCHER, R.	F111	
CHACON, E.	E15	HIRAMATSU, Y.	E7	
CHAMBON, C.	D91	HISATAKE, M.	E191	
CHARRUA-GRACA, J.G.	F79	HOLMBERG, R.	E231	
CHESNOKOV, S.A.	D323	HOLMGREN, B.J.	D311	
COULTHARD, M.A.	D53	HOLT, G.E.	E119	
COWLING, R.	D29 D45	HORSRUD, P.	F1	
COYETTE, J.P.	F25	HUDSON, J.A.	E161	
CREA, G.	D77	HUERBA, A.	F231	
CROTTY, J.M.	D53	HUERGO, P.J.	F103	
CUMMINGS, R.A.	E87	HUGOUT, B.	D241	
CUNHA, A.P.	F153	HUSTRULID, W.	E231	
DA SILVA, L.A.A.	E245	INAMI, E.	D203	
DE JONGH, C.L.	D1	INGEVALD, K.	F111	
DERAMOND, J.	F89	INNAURATO, N.	E141	
DESAI, C.S.	F209	INNERHOFER, G.	D281	
DEZHANG, L.	F195	INOMA, H.	D203	
DOKUKIN, A.V.	D121	IRFAN, T.Y.	D151	
DOUCERAIN, T.	D247	ISHIJIMA, Y.	E11	

ITO, T.	E191		ODA, N.	E7	
ITO, Y.	D203		OKSANEN, J.	F237	
			OKSANICH, I.F.	F49	
JOGI, P.N.	F35		OLSSON, T.	F31	
JOSIEN, J.P.	D91		ONATE, E.	F231	
			ORTLEPP, W.D.	D179	
KAISER, P.K.	D329		OYANGUREN, P.R.	D135	
KANAGAWA, T.	D199	F85	OZGENOGLU, A.	D151	
KANG, W.F.	F95	F251			
KARPUZ, C.	D151		PAGE, C.H.	D61	
KATKOV, G.A.	D323		PANDE, G.N.	F217	
KERTESZ, P.	F259		PANET, M.	F181	
KHRISTIANOVICH, S.A.	E41		PANTALEONE, G.	D77	
KIE, ir T.T.	F95		PANTALEONI, G.	E141	
KINOSHITA, S.	E11		PARABUCHEV, I.A.	F139	
KIRSTEN, H.A.D.	E101		PARPHENOV, V.D.	F139	
KLOKOW, J.W.	D1		PASAMEHMETOGLU, A.G.	D151	
KOBAYASHI, R.	D195		PECHALAT, F.	E201	
KOBAYASHI, Y.	E191		PEIRCE, A.P.	F159	
KOKOSADZE, E.L.	D323		PETRASOVITS, G.	E133	
KOKUSHO, T.	E223		PETUKHOV, I.M.	D113	
KOLIN, D.I.	D323		PIGUET, J.P.	D91	
KOMADA, H.	F85		PINTO, J.L.	F79	
KONDO, N.	D217		POIROT, R.	E115	
KOTZE, T.J.	D67		PRIEST, S.D.	F199	
KRAULAND, N.	F55		PROTASOV, Y.I.	F49	
KRSTULOVIC, G.L.	D141				
KUBO, N.	D195		QUIBLIER, J.A.	F191	
KURIYAGAWA, M.	E147				
KUZNETSOV, S.V.	F139		RAHN, W.	F63	
			REHBINDER, G.	E173	
LAMA, R.D.	E79		REN, N-K.	F117	
LARSSON, H.	E151		RIBACCHI, R.	D77	
LEFIN, Y.	E201		RIE, ir T.T.	D261	
LeJUGE, G.E.	E215		RISNES, R.	F1	
LEMBO-FAZIO, A.	D77		ROCHET, L.	F175	
LEWIS, I.H.	E215		ROEGIERS, J-C.	F117	
LIAO, G.	D87		ROIG, J.	F231	
LIE, X.	D267		ROMANA, M.	D189	D207
LILLY, J.D.	E215		ROSHOFF, K.	E151	E173
LILLY, P.A.	E215		RYDER, J.A.	F159	
LIN, Y-M.	D109				
LINKOV, A.M.	D113		SAARI, K.H.O.	D253	
LISAC, Z.	D337		SAITO, J.	E227	
LOACKER, H.	D281		SAITO, T.	D203	E7
LOKEN, M.C.	E167		SAKURAI, S.	E191	
LOUSBERG, E.	F25		SALAMON, M.D.G.	E107	
LOVEN, P.	F237		SALGANIK, R.L.	E41	
LU, P.H.	E37		SAMANIEGO, A.	F199	
			SARKKA, P.S.	D101	
MAIDI, B.	D275		SCHUERMANN, F.	E21	
MAKI, K.	E231		SELLDEN, H.	E231	
MALONEY, S.	D329		SHEPHERD, J.	E79	
MARKOV, G.A.	F139		SHELTON, P.D.	D315	
MARTNA, J.	E173	F111	SHI, Z.	D87	
MATSUKI, K.	D195		SIMIC, D.	D189	
MATSUNAGA, I.	E147		SINGH, M.M.	E87	
McKENZIE, C.K.	E215		SIRIEYS, P.	F89	
MEEK, J.L.	D45		SISKA, L.	F243	
MEI, J-Y.	F251		SLAVIN, O.K.	D121	
MENDES, F.M.	E195		SORIA, J.L.	D167	
MIKULA, P.A.	E119		SOULA, J.C.	E1	F89
MIKURA, E.	E181		SOUQUET, G.	D221	
MIRANDA, A.M.	E195		SOUSA, L.R.	F169	
MIRONOV, P.S.	F49		STACEY, T.R.	E101	
MORGENSTERN, N.R.	D329		STANFORS, R.	E151	
MORISSEAU, J.M.	D221		STEPHANSSON, O.	E151	F129
MOTOJIMA, M.	D199		STILLE, H.	D301	
			SZAVITS-NOSSAN, A.	D337	
NGOKWEY, K.	F191				
NGUYEN, D.M.	D227	D233	TAMMEMAGI, H.Y.	E167	
NISHI, K.	E223		TECEN, O.	E207	
NOE, J.M.	D221		TERADA, M.	E137	
NONVEILLER, E.	D337		TEUFEL, L.W.	F43	
NORD, G.	D293		THIMUS, J.F.	F25	

THOMPSON, T.W.	F35	WARDLE, L.J.	E61
TIJANI, S-M.	D241	WARPINSKI, N.R.	F43
TRAVERT, P.	E1	WEBER, Ph.	D97
TRUMBACHEV, V.F.	D121	WENZ, E.	F69
TSUKADA, K.	D203	WIID, B.L.	E101
TUTANOV, S.K.	D323	WITTKE, W.	D167
VEKSLER, Yu.A.	D323		
VISEC, I.	D337	YAMAGUCHI, T.	E147
VOUILLE, G.	D241	YAMASHITA, M.	D217
VRKLJAN, I.	D337	YANAGIDANI, T.	E137
VUJEC, S.	D129	YEATES, R.A.	E71
WALLNER, M.	D9		
WAGNER, R.A.	E167	ZELJKOVIC, D.	D129
WANG, K.	D271	ZHU, K.	D267
WANG, Y.T.	F187	ZHU, W.	D271
WARBURTON, P.M.	F225	ZHUO, J.S.	F187

THE USE OF A SEISMIC NETWORK AS A MANAGEMENT TOOL
Un réseau sismique comme aide à l'exploitation des mines
Die Anwendung eines seismischen Netzwerkes als Hilfsmittel der Betriebsführung

C. L. de Jongh
Assistant Consulting Engineer, Gold Fields of S.A. Ltd., Johannesburg, Republic of South Africa

J. W. Klokow
Chief Rock Mechanics Engineer, West Driefontein Gold Mine, Carletonville, Republic of South Africa

SYNOPSIS
The seismic network currently being installed at the Kloof, East Driefontein, West Driefontein and Doornfontein gold mines of the Gold Fields of South Africa Group, on the West Witwatersrand in South Africa is probably the biggest of its kind in the world. The inclusion of two more mines is being considered. The four mines concerned experience an annual average of 200 serious rock bursts which affect production and the safety of personnel. The network will be used to provide management with immediate intelligence regarding the occurrence and location of seismic incidents. It will also be used to identify areas with high seismic activity and for fundamental scientific studies.

RESUME
L'installation d'un réseau d'appareils pour l'enregistrement de données sismiques est actuellement en cours aux mines d'or de la compagnie minière "Gold Fields of South Africa" à Kloof, East Driefontein, West Driefontein et Doornfontein, à l'ouest du Witwatersrand. Ce réseau est probablement le plus considérable de ce genre dans le monde. L'inclusion au réseau de deux autres mines est envisagée. Les quatre mines surmentionnées ont une moyenne de 200 graves secousses sismiques par an, ce qui affecte la production et la sécurité du personnel. Le réseau d'appareils servira à donner à la direction des mines des renseignements immédiats, sur la fréquence et la location de ces secousses sismiques, il servira aussi à identifier les endroits de haute activité sismique et servira pour des études fondamentales scientifiques.

ZUSAMMENFASSUNG
Das größte und modernste seismische Netzwerk der Welt wird gegenwärtig auf den zum Goldfields of South Africa Konzern gehörigen Goldbergwerken Kloof, East Driefontein, West Driefontein und Doornfontein installiert. Diese Bergwerke befinden sich im westlichen Teil des Witwatersrand von Südafrika. Die Einbeziehung von zwei weiteren Bergwerken wird erwogen. Die vier oben-angeführten Betriebe werden pro Jahr durchschnittlich von 200 schwereren Gebirgsschlägen heimgesucht, wobei Produktion und Grubensicherheit beeinträchtigt werden. Das Netzwerk wird vor allem dazu dienen, der Betriebsführung sofortige Informationen über Auftreten und Lokalisierung seismischer Schwingungen anzuzeigen. Desweiteren wird das Netzwerk zur Erkundung hoher seismischer Aktivität zur Grundlagenforschung eingesetzt werden.

1.0 INTRODUCTION

Presently gold mines in the Republic of South Africa are mining at depths down to 3 500m below surface and operations as deep as 4 000m below surface are being planned. Diamond drilling has shown that the ore bodies (reefs) extend to beyond 4 000m below surface along the West Wits Line. (See locality plan figure 1.1).

All the gold reefs are near-parallel tabular ore bodies. On the West Wits Line up to three economic reefs are being mined, namely the Ventersdorp Contact Reef, Middelvlei (Main) Reef and Carbon Leader Reef (See Fig. 1.2). On average, the reefs dip between 20 and 35 degrees and stoping widths average 1,5m.

These narrow tabular reefs are mined using both scattered and longwall open stoping methods (See figures 1.3). Scattered mining is more suitable for extracting reefs which are extensively faulted or disturbed by dykes, or have limited pay zones, but has the following disadvantages :-

(a) Numerous highly stressed isolated pillars (remnants) are formed which increase the probability of violent rock failure (Rock bursts).

(b) Predeveloped haulages are damaged by high abutment stresses, especially at depth, when stoping takes place across them.

Accordingly, scattered mining is generally not suitable for mining at depth, and long-walling is employed. This system creates less remnants and haulages do not become highly stressed since they are developed behind the advancing stope faces. Another advantage is that it generates a more uniform increase in stress with increasing mined out spans. However, high stresses still develop because of the large spans and the formation of a final remnant between two adjacent long-walls. Consequently, there is still a rock burst problem associated with long-wall mining.

Seismic events are everyday occurrences on the mines of the West Wits Line, but only a small percentage cause serious damage to underground excavations. Due to the high frequency of events even this small number constitutes a safety hazard and causes production losses. For example, on the West Driefontein Gold Mine 106 damaging rock bursts were reported during 1981.

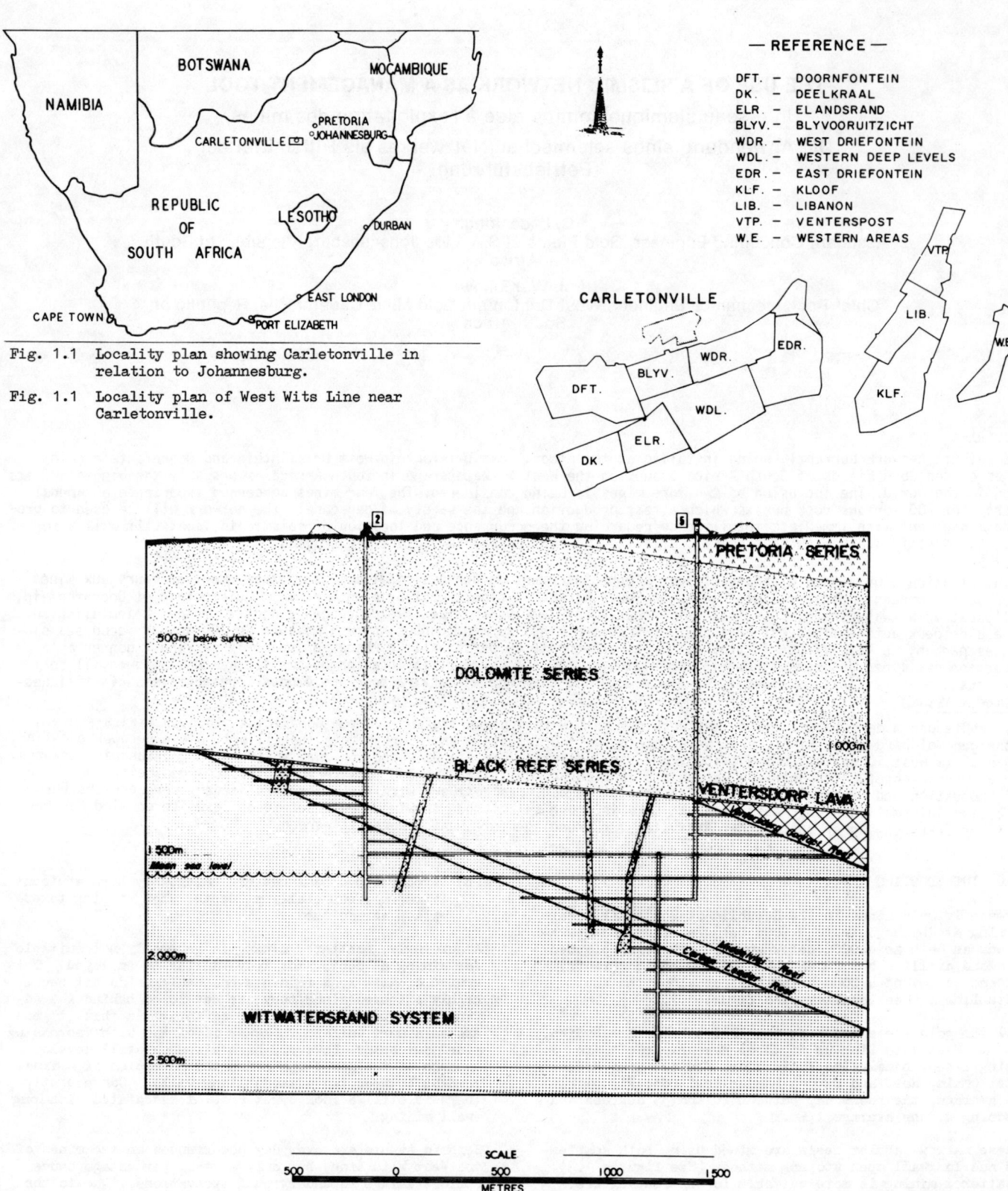

Fig. 1.1 Locality plan showing Carletonville in relation to Johannesburg.

Fig. 1.1 Locality plan of West Wits Line near Carletonville.

Fig. 1.2 Typical geological section through a mine on the West Wits Line.

2.0 ROCK BURSTS

2.1 Reducing the Incidence of Rock Bursts

In the past, mining engineers had to rely on experience and empirical data in planning mining layouts and sequences. It became clear, however, that a more scientific approach was needed in mine planning in order to reduce the rock burst hazard.

Currently computer models based on the theory of elasticity are used as an aid to determine and evaluate mining strategies which result in reduced stress levels and seismic activity.

Statistical data has shown that a large number of events are associated with geological discontinuities such as faults and dykes. These seismically active discontinuities can be stabilised by leaving adjacent reef pillars of sufficient dimensions.

Mining induced stresses can be controlled by restricting the elastic movement of the rock mass into the voids created, Salamon and Wagner (1979). This can partially be achieved by sacrificing parts of the orebody as stabilising pillars. Another method is to use a suitable back-fill. Back filling with dewatered slime is presently being tested and evaluated under South African conditions.

While the use of pillars has undoubtedly led to a reduction in both the incidence and severity of rock bursts the engineering problem which remains is one of deciding optimum pillar sizes such that a minimum of ore is lost. Data collected from the seismic network will assist in setting the required parameters.

2.2 Minimising the Effects of Rock Bursts

Rock bursts cannot be totally eliminated in deep level gold mining on the Witwatersrand and therefore it is necessary to employ suitable support methods to reduce their effects.

To reduce production losses and increase the safety of personnel, the support must be capable of maintaining the integrity of the fractured hangingwall as effectively as possible under rock burst conditions.

The most common stope supports used in gold mines are timber packs, sandwich packs (timber packs incorporating concrete bricks), cement-sand-grout filled packs, pipe-sticks, (mine poles partly sheathed in steel sleeves) and rapid-yielding hydraulic props. Of these, sandwich packs and rapid-yielding props have proved to be the most effective in reducing the damaging effects of rock bursts on the Gold Fields mines.

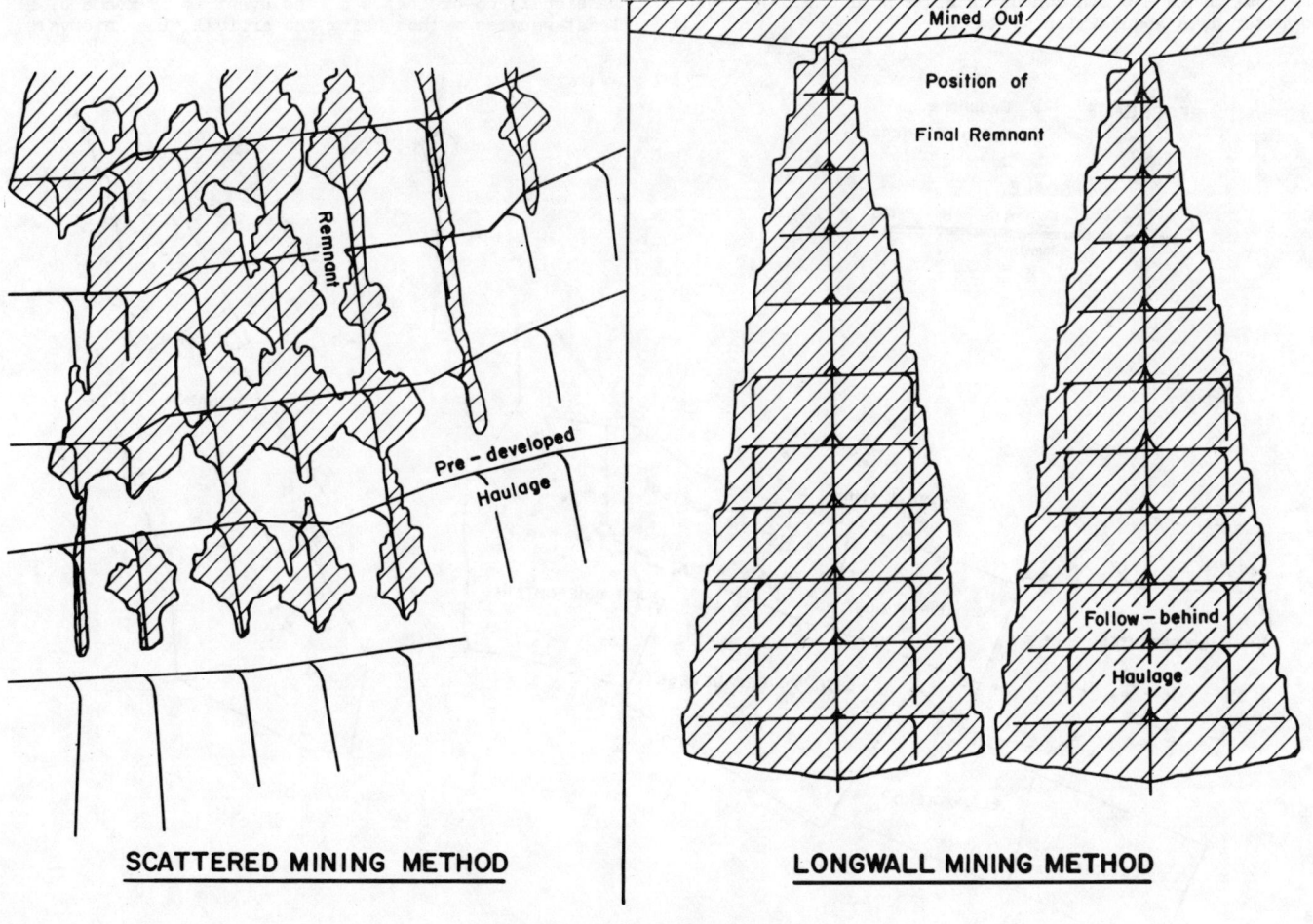

Fig. 1.3 Scattered and longwall mining methods.

3.0 DESCRIPTION OF NETWORK

3.1 General

The system will monitor seismic activity from a central station, covering an area of approximately 250 km². To meet the objectives of the system it is necessary to obtain a high degree of accuracy in the location of events This is dependent on the density of monitoring stations (geophones). It was therefore decided to install at least 90 geophones. There are other systems which monitor much larger areas with less geophones, but as far as can be ascertained the Gold Fields system will be the only one monitoring events over a relatively large area using a large concentration of fixed stations. (See figure 3.1).

3.2 Technical

Geophones (4,5 Hz) are grouted into a hole, drilled from a haulage. Each of these is connected by a single pair cable to a nearby logarithmic amplifier and frequency modulator, powered by a 12V battery source. The geophone signals are logarithmically amplified and frequency modulated onto a carrier frequency of 2,5kHz and transmitted to a 16-bit mini computer on surface via multi-core cables of which there will be some 400 km on the four mines. (See diagram in figure 3.2). Each of the four mines has its own cable network and minicomputer data acquisition system.

Each data acquisition system consists of demodulators, filters, a triggering unit and a 16-bit computer with built-in analog to digital converters. The function of the 16-bit computer at each mine is to receive data from up to 32 geophones on that mine, scan the signal and either accept or reject it on the basis of the triggering criteria used. Each geophone channel is sampled every milli-second. The function of the triggering unit is to determine whether a valid seismic event has occurred. This means that the event must be registered by at least four geophones. If the answer is positive, the data is written into a disc file on the 16-bit computer in a form suitable for transmission, via a UHF radio link, to a centrally situated 32-bit computer at the West Driefontein mine office. Only data from geophones which are close to those which caused the trigger, are selected for transmission.

The 32-bit computer enquires at milli-second intervals at each of the four 16-bit computers whether new data has been stored. If the answer is positive, the data is transmitted to the 32-bit computer and written into a disc file in a form suitable for further processing.

On the central computer, seismograms are plotted on an interactive graphics terminal for selection of the relative compressive (p) and shear (s) wave arrival times at each geophone. Calculation of the hypocentre (3 dimensional) co-ordinates of the event is by means of a least-squares method using the arrival time intervals.

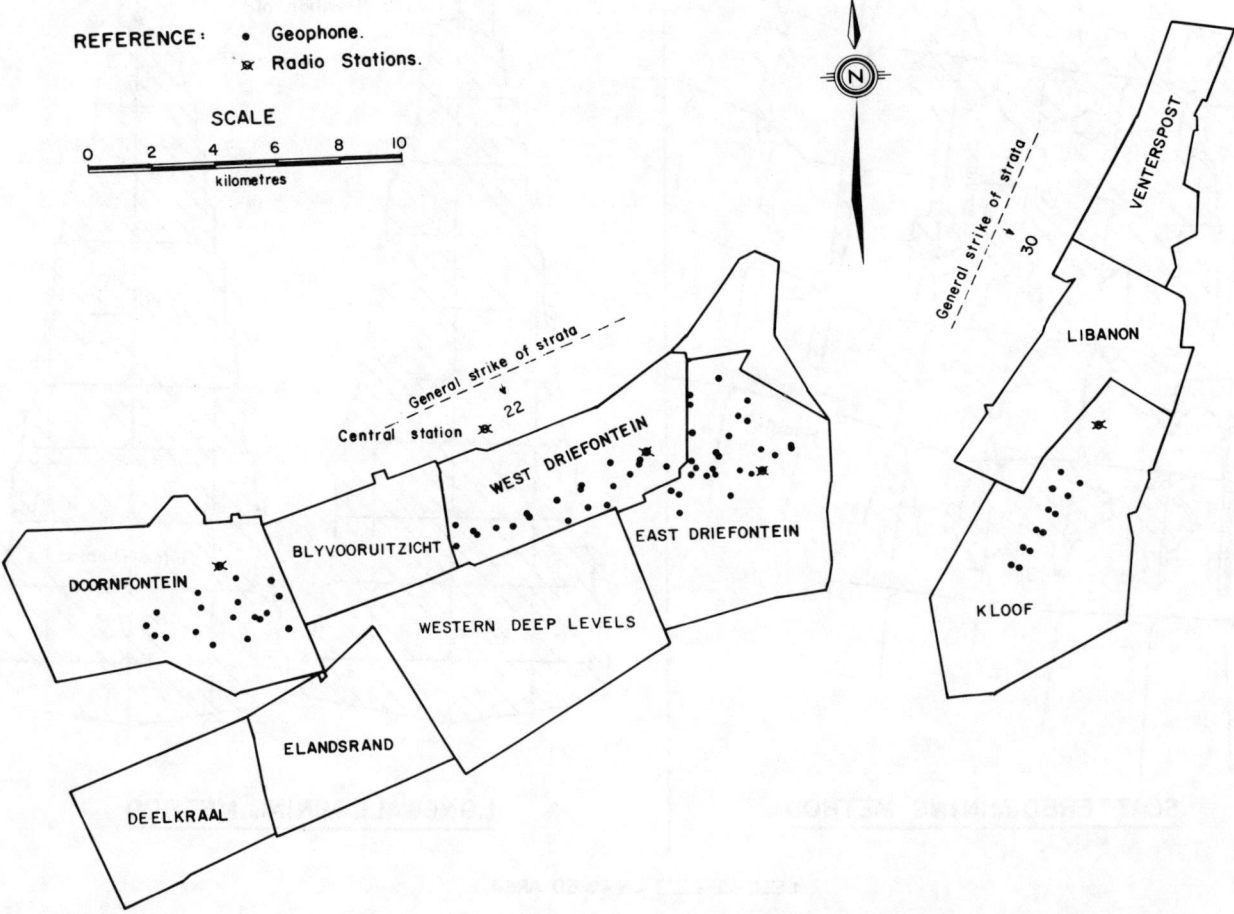

Fig. 3.1 Plan showing location of geophones and radio transmitters on Gold Fields mines.

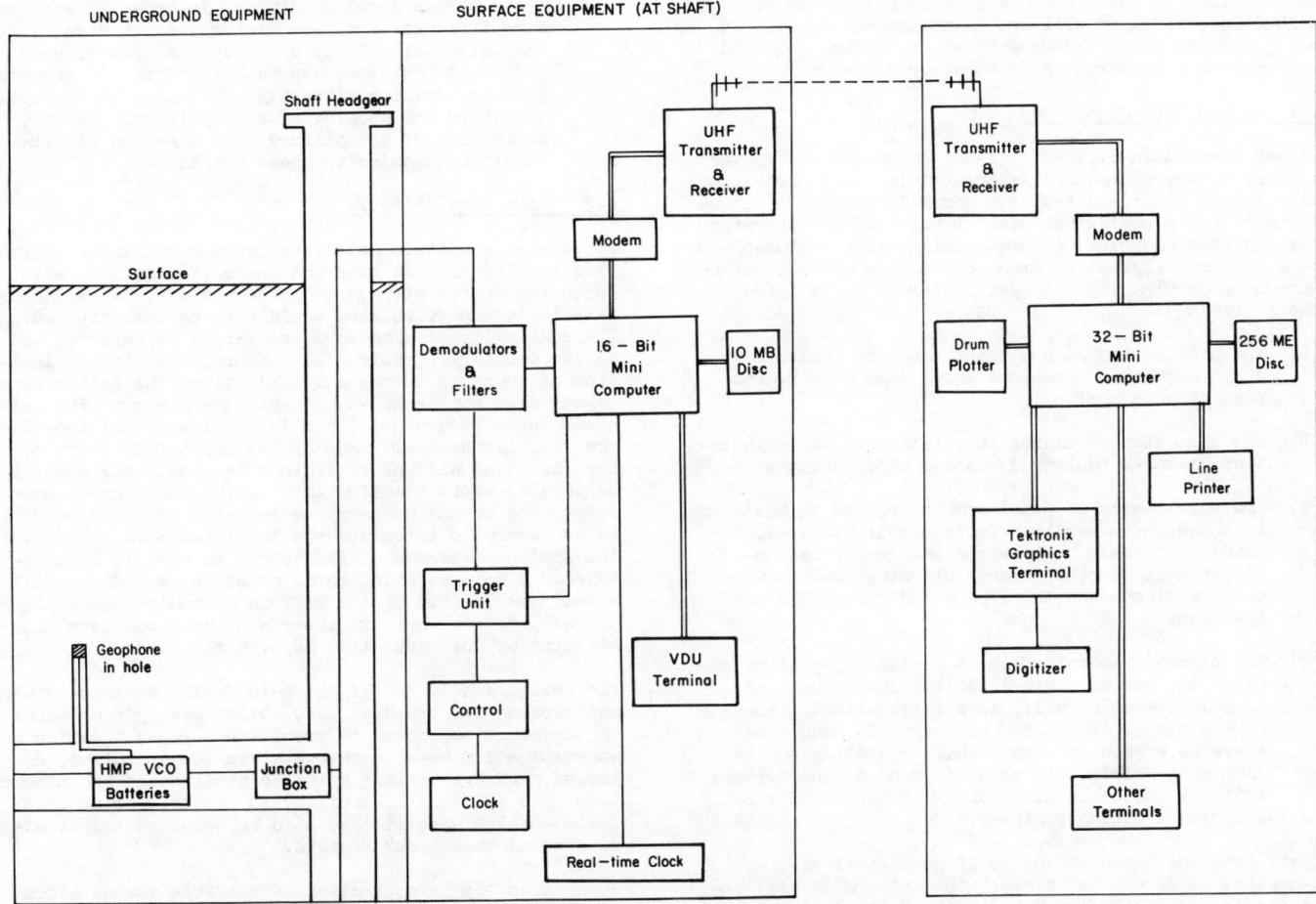

Fig. 3.2 Diagrammatic presentation of seismic installations and processing equipment.

Once co-ordinates have been calculated they are transmitted back to the four mines via the radio link. All data captured on the central computer are stored on disc or magnetic tape for further study.

4.0 OBJECTIVES OF THE SYSTEM

4.1 Immediate Objectives

Mine management has been concerned for many years with reducing injuries, fatalities and production losses resulting from seismic events. Due to mining at ever-increasing depths, it can be expected that these incidents will increase unless effective measures are taken to reduce them.

It can be appreciated that, due to the restricted space available in the mining of narrow tabular orebodies, rock falls resulting from seismic events during the shift frequently injure people and often disrupt the ventilation to the areas involved. In many cases sympathetic seismic events occur immediately after the first one and may result in more than one area being subjected to rockfalls. It is essential therefore, that the areas affected are identified as quickly as possible to enable management to institute rescue operations without delay.

Many seismic events do not result in tremors on surface, and may pass unnoticed. Even when tremors are felt on surface, the location of the event is unknown as it can originate from any of a number of adjacent mines. (See plan, fig. 3.1). Mine managements are at present hampered in their efforts to initiate rescue operations timeously in the absence of a rapid locating system. The situation can be aggravated by underground communication difficulties. On a big gold mine it is very difficult to maintain an effective underground telephone network due to the large number of tunnels and mining areas involved. The communications network could also be damaged by the rockfalls. This is particularly true in a situation where a section of the mine is cut off from the rest of the mine after a rock burst.

A particularly difficult problem arises when seismic events occur immediately before blasting at the end of the shift. Without a system capable of locating these events rapidly, management will not be aware of their location and extent. In the past, blasting operations have been stopped when a tremor of some severity has been felt on surface in case miners are trapped and could be exposed together with rescue teams to the danger of toxic fumes. In many cases it materialised that this was unnecessary, because the events did not affect the area where the blast was stopped, and resulted in unnecessary production losses.

With the aid of a fast and accurate locating system, it will be possible to ascertain when a rockburst has occurred whether the safety of personnel may be involved. It may only be necessary then to stop those blasting operations in the affected area which may result in people being exposed to fumes and which could trigger additional rockfalls in that area. A seismic network as described in this paper will enable management to make an early and far better assessment of the action required when potentially damaging seismic events occur.

4.2 Medium Term Objectives

It has been shown by Heunis (1976) and Hagan (1981) that a seismic network can fairly accurately locate seismically active dykes and faults. Normally these are extensions of known geological discontinuities, and therefore readily recognisable. Accurate delineation of these and other seismically active areas could have a major influence on mine planning. In particular it would affect the following :-

(a) The positioning of new shafts and the location of large service excavations which form an integral part of such shafts.

(b) The direction of mining in relation to the orientation of known faults, dykes and major joint sets.

(c) The positioning of final remnant blocks of ore which in a deep mine are invariably highly stressed, depending on their size, shape and geological environment. Correct advance planning would have a major influence on their safe and successful extraction.

(d) The planned removal of shaft protection pillars towards the end of a mine's life. The success of such an operation will, to a large extent, depend on the stress levels and geology. In most cases there is a need to keep a shaft operating for as long as possible. Close monitoring of the seismic activity in such a diminishing pillar would greatly assist the mining engineer.

Apart from the known influence of geological discontinuities on deep level mining, the network is also expected to provide information on the following :

(i) The effect of using tailings or rockfill in order to reduce stresses by reducing elastic closure. For a meaningful and proper evaluation of these benefits, a seismic network could, over a period of time provide valuable guidance.

(ii) Confirmation of deductions from seismic data, as collected by the Chamber of Mines of South Africa Research Organisation, which indicates that most events cluster around the reef plane, with equal distribution above and below. This is the case when the hanging and the footwall consist of quartzite. Some of the mines in the Gold Fields Group mine a reef where the rock formation in the hanging consists of hard massive lava, and the footwall is softer quartzite or shales. It will be useful to know if the event distribution is the same in this case. This information can then be used for the siting of connecting tunnels and their support requirements.

(iii) Checking on the practice, which for some years now has been to use reef pillars alongside and parallel to faults and dykes, to improve their stability. The width of most of these pillars varies between 10m and 20m. Results to date indicate a very high degree of success for this method. With the computer programmes at present available to the Rock Mechanics Engineer, accurate seismic information would allow optimisation of the rib sizes.

(iv) Testing the Chamber of Mines Research Organisation's recommendation that stabilising pillars be used in deep level gold mines in order to achieve a real reduction in seismic activity. Mines of the Gold Fields Group using such a system, have indicated that this approach is correct. There is however, still a diversity of opinion on the optimum width and spacing of such pillars. Seismic monitoring of the pillars over a period of time should help to solve these problems.

4.3 Long Term Objectives

In general a seismic network's basic function is to provide the location in time and space of a seismic event. Depending on the structure of the network, other facets of seismic energy release could also be investigated. The complexity of a network can vary considerably, and so too can the objectives of the project. If the location of an event is the only objective, the entire seismogram need not be known and only the minimum of detail needs to be recorded. For a three dimensional location the relative arrival times of the event from four recording sites will be sufficient to enable the event to be located without ambiguity. On the other hand, to assess the amount of energy associated with the event it is necessary to integrate the total vibration. For this the entire waveforms of the vibration need to be monitored. A more detailed scrutiny of the waveforms will reveal the details of the failure mechanism and rupture process, facets that are important for a complete understanding of the generation of events.

The design and layout of the Gold Fields seismic network anticipated its broadest use, whilst giving prominence to the prime objective of providing a rapid location of an event and mine-wide coverage was the objective, detailed study of certain specific areas could be informative. To this end flexibility in the triggering criteria enables specific areas to be selected and treated as special "research" regions.

Apart from the investigation of possible stress migrations, other facets associated with the dynamic process have not been researched fully and the network design anticipates some form of study along the lines of dynamic stress assessment, together with seismic moment and magnitude analysis. Yet other areas of research and practical significance are the total volumes of rock involved in the failure and how the stress distribution within the mine relates to the seismicity energy distribution.

Laws of attenuation pertaining to velocity, displacement and acceleration are of importance in predicting damage in areas adjacent to failure and need further investigation. These laws could affect future engineering design criteria for shafts, hydraulic props and support systems.

Another important aspect of network study is the relationship of the slope magnitude frequency distribution to the stress changes and pattern. This technique may in future have important predictive aspects associated with it.

5.0 RESULTS TO DATE

The West Driefontein section of the network came into operation at the end of July 1982. The whole system is expected to be in full operation by the end of June 1983.

Indications at this stage are that the location of seismic events is already being determined within 50m of their hypocentres. We are, however, aiming to reduce this to 20m. It may be necessary to do a calibration blast for this purpose. Time requirement to locate an event is currently in the order of 12 minutes. It is expected that this interval will be reduced to about 4 minutes with improved radio communication and minor changes in the computer software.

The Management of West Driefontein Mine has welcomed the introduction of the seismic network and is already making wide use of it as a management tool. We believe similar responses will be forthcoming from the rest of the mines once they are connected to the network.

6.0 REFERENCES

Heunis, R. (1976) Rock Bursts and the search for an early warning system : S.A. Mining and Engineering Journal, December 1976.

Hagan, T.O. Micro-seismic events : An aid in studies of deep-mine rock structures : S.A. Institute of Mining and Metallurgy, November 1981, 319-320.

Salamon, M.D.G. and Wagner, H. (1979). The role of stabilising pillars in the alleviation of rock burst hazard in deep mines : Chamber of Mines of South Africa Research Organisation, Research Report No. 27/79.

STANDSICHERHEITSBERECHNUNGEN FÜR DIE PFEILERDIMENSIONIERUNG IM SALZBERGBAU

Stability calculations concerning a room and pillar design in rock salt

Calculs de stabilité pour les dimensions de chambres et piliers dans les mines de sel gemme

M. Wallner
Bundesanstalt für Geowissenschaften und Rohstoffe, Hannover, Bundesrepublik Deutschland

ZUSAMMENFASSUNG

Eine optimale Ausnutzung der Gebirgstragfähigkeit im Kammer-Pfeilerbau des Salzbergbaus stellt erhöhte Anforderungen an die Pfeilerdimensionierung und die Abbauführung. Ein praxisgeregelter Standsicherheitsnachweis wird sich neben langjährigen bergmännischen Erfahrungen sowie umfangreichen Messungen in situ und im Labor zur Pfeilertragfähigkeit auch auf rechnerische Untersuchungen abstützen. Es wird das speziell für die Berechnung untertägiger Hohlraumstrukturen im Salzgestein entwickelte FEM-Rechenprogramm ANSALT vorgestellt, das die nichtlinearen zeit- und temperaturabhängigen Stoffeigenschaften der Salzgesteine entsprechend dem derzeitigen Kenntnisstand berücksichtigt. Anhand ausgewählter Beispiele wird gezeigt, welchen Beitrag derartige Berechnungen zur Standsicherheitsanalyse liefern können.

SYNOPSIS

Deep salt mining and excavation at high extraction rates require safe design methods which allow for an acceptable utilization of the bearing capacity of the rock mass. The procedure to prove the structural stability will be based on practical experience as well as on laboratory and in situ measurements. Apart from that, computations concerning the stress/strain analysis will help to derive site-specific design criteria. A newly developed FEM computer code, ANSALT, is presented, which includes a model of the constitution of rock salt appropriate to the current knowledge of salt mechanics. Some calculated examples are discussed to demonstrate the usefulness of such stability computations.

RESUME

Des procédés de conception fiables permettant une utilisation optimale des qualités de soutènement de la roche s'imposent pour l'exploitation du sel gemme à des taux de défruitement élevés dans les mines profondes. La stabilité structurale des masses rocheuses peut être évaluée en s'appuyant sur une longue expérience minière et sur des mesures de solidité des piliers effectuées in situ et en laboratoire, et dont les données auront été traitées sur ordinateur. Un programme FEM-ANSALT, spécialement conçu pour le calcul de structures de cavités souterraines en formations salifères, tient compte des propriétés non-linéaires qui sont fonction du temps et de la température des roches salifères, et cela conformément aux connaissances actuelles en la matière. On montre, à l'aide d'exemples sélectionnés, dans quelle mesure de tels calculs sont utiles à l'analyse.

1. EINLEITUNG

Die notwendige Versorgung unserer Wirtschaft mit dem Rohstoff Kali und Steinsalz führt dazu, daß der Bergbau zunehmend in größere Teufen vorstößt und bestrebt ist, unter optimaler Ausnutzung der Gebirgstragfähigkeit, verlustarm abzubauen. Ein sicherer und zugleich wirtschaftlicher Abbau erfordert eine auf gebirgsmechanischer Grundlage aufbauende Bergwerksdimensionierung. Die Nichtbeachtung oder Fehleinschätzung gebirgsmechanischer Gegebenheiten kann katastrophale Folgen haben, was durch verschiedene Spannungs- oder Gebirgsschläge im Kalibergbau der flachen Lagerung belegt werden kann (Grimm, Pforr 1961).

Im Folgenden sollen ausgehend von einer zusammenfassenden Darstellung zum Festigkeits- und Verformungsverhalten von Salzgestein einige mit dem FEM-Rechenprogramm ANSALT durchgeführte Berechnungen zur Pfeilerdimensionierung diskutiert werden, die geeignet sind, rechnerisch zur Absicherung von praktischen Dimensionierungsregeln beizutragen.

2. MATERIALVERHALTEN DER SALZGESTEINE

In der Bundesanstalt für Geowissenschaften und Rohstoffe (BGR) wurden in den letzten Jahren umfangreiche Untersuchungen zum mechanischen Verhalten von Salzgesteinen durchgeführt. Ziel der Laboruntersuchungen war es, ein umfassendes Bild über das spannungs- und temperaturabhängige rheologische Verhalten der Salzgesteine zu erhalten und physikalisch begründete Stoffgleichungen abzuleiten.

Über einzelne Ergebnisse der Untersuchungen ist bereits mehrfach berichtet worden (Langer 1979, Wallner et al 1979, Wallner 1981, Staupendahl et al 1982). Einige typische Versuchsergebnisse zur Abgrenzung bruchloser Verformungen gegenüber dem Bruch sind in Abb. 1 dargestellt.

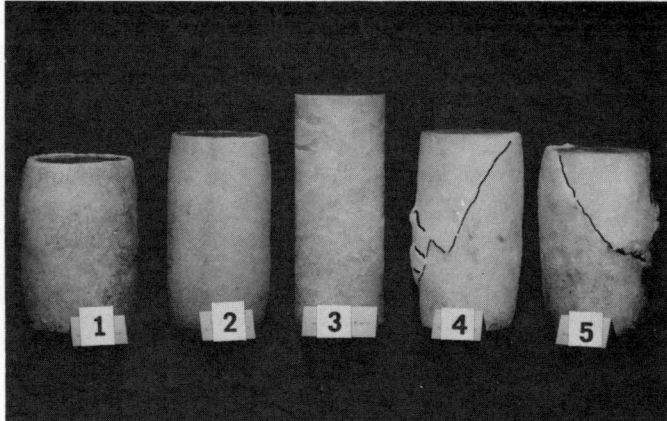

Abb. 1: Triaxial geprüfte Steinsalzproben, Asse-Liniensalz Na3β
1. $\sigma_3 = 5{,}0$ MPa, $\dot{\varepsilon} = 8 \times 10^{-7}$ (s^{-1})
2. $\sigma_3 = 2{,}5$ MPa, $\dot{\varepsilon} = 8 \times 10^{-7}$ (s^{-1})
3. ungeprüfte Probe
4. $\sigma_3 = 2{,}5$ MPa, $\dot{\varepsilon} = 8 \times 10^{-5}$ (s^{-1})
5. $\sigma_3 = 5{,}0$ MPa, $\dot{\varepsilon} = 8 \times 10^{-5}$ (s^{-1})

Die Proben 4 und 5 zeigen einen typischen Schubbruch. Die Proben 1 und 2 wurden bei gleichen Manteldrücken, aber einer um den Faktor 100 geringeren axialen Verformungsgeschwindigkeit geprüft. Sie weisen trotz großer Verformungen keine Brucherscheinungen auf.

Formänderungsverhalten

Aus den umfangreichen Laborversuchen wurde ein pragmatisches Stoffmodell für Steinsalz entwickelt. Das Gesamtdehnungsinkrement $\dot{\varepsilon}_{ges}$ setzt sich zusammen aus:
1. einem elastischen Verzerrungsinkrement $\dot{\varepsilon}^{el}$
2. einem Kriechinkrement $\dot{\varepsilon}^{cr}$ (zeitabhängige irreversible Gestaltänderungen)
3. einem Bruchverformungsinkrement $\dot{\varepsilon}^f$ (irreversible, geschwindigkeitsunabhängige Formänderungen mit Ausbildung diskret verteilter Bruchflächen und Dilatation)

$$\dot{\varepsilon}_{ges} = \dot{\varepsilon}^{el} + \dot{\varepsilon}^{cr} \mid \dot{\varepsilon}^f \quad (1)$$

$$\dot{\varepsilon}^{el}_{ij} = -\frac{\nu}{E} \dot{\sigma}_{kk} \delta_{ij} + \frac{1+\nu}{E} \dot{\sigma}_{ij} + \alpha_t \Delta \vartheta \delta_{ij} \quad (2)$$

$$\dot{\varepsilon}^{cr}_{ij} = \frac{3}{2} \frac{\dot{\varepsilon}^{cr}_{eff}}{\sigma^{cr}_{eff}} s_{ij} \quad (3)$$

$$\dot{\varepsilon}^{cr}_{eff} = \sqrt{\frac{2}{3} \dot{\varepsilon}_{ij} \dot{\varepsilon}_{ij}} \quad (4)$$

$$\sigma_{eff} = \sqrt{\frac{3}{2} s_{ij} s_{ij}} \quad (5)$$

$$\dot{\varepsilon}^{cr}_{eff} = \sum_{i=1}^{3} \dot{\varepsilon}^{cr}_{eff} \, i(S, \sigma_{eff}, \vartheta) \quad (6)$$

$$\dot{\varepsilon}^{cr}_{eff} 1 = A_1 e^{\frac{-Q_1}{R\vartheta}} \left(\frac{\sigma_{eff}}{G}\right)^{n_1} \quad (7)$$

$$\dot{\varepsilon}^{cr}_{eff} 2 = A_2 e^{\frac{-Q_2}{R\vartheta}} \left(\frac{\sigma_{eff}}{G}\right)^{n_2} \quad (8)$$

$$\dot{\varepsilon}^{cr}_{eff} 3 = 2(B_1 e^{\frac{-Q_1}{R\vartheta}} + B_2 e^{\frac{-Q_2}{R\vartheta}}) \sinh(D < \frac{\sigma_{eff} - \sigma^0_{eff}}{G} >) \quad (9)$$

$$\dot{\varepsilon}^f_{ij} = \dot{\lambda} \frac{\delta f}{\delta \sigma_{ij}} \quad (10)$$

$$f = \tau_0 - C \sigma_0^\alpha \quad (11)$$

Abb. 2: Materialmodell für Steinsalz

Die konstitutiven Gleichungen für das Materialmodell sind in Abb. 2 zusammengefaßt.

Für das Kriechverzerrungsinkrement $\dot{\varepsilon}^{cr}$ ist nur stationäres Kriechen als wesentlicher Anteil der Langzeit-Kriechdeformation berücksichtigt. Die mathematische Formulierung basiert auf einem von Munson and Dawson (1979) entwickelten Ansatz für das Kriechen als Summe dreier unterschiedlicher Kriechmechanismen:

- Versetzungsklettern bei niedrigen Spannungen und hohen Temperaturen
- Übergangsversetzungsmechanismus bei Lagerstättentemperaturen und -spannungen
- Versetzungsgleiten bei hohen Spannungen

Im Falle der Überschreitung des Kriteriums für die Langzeitfestigkeit wird anstatt Kriechen ein plastisches Bruchverzerrungsinkrement berücksichtigt.

Festigkeitsverhalten

Spannungs- bzw verformungsgeregelte ein- und mehraxiale Versuche zur Bruchfestigkeit ergaben zusammenfassend folgende, auch aus der Literatur (Dreyer 1967, Menzel et al 1978, Böhnel et al 1981) bekannte Ergebnisse:

- Die Festigkeit hängt vom Mineralbestand, der Mineralverteilung und dem Gefüge ab.
- Salzgesteine besitzen eine sehr geringe Zugfestigkeit
- Die Druckfestigkeit nimmt mit steigendem isotropen Druck zu.
- Bei einaxialer Beanspruchung besitzt Steinsalz eine ausgeprägte Festigkeitsanisotropie, deren Einfluß mit zunehmendem isotropen Druck wieder verschwindet.

Darüber hinaus hängt die Bruchfestigkeit entscheidend auch von der Beanspruchungsgeschwindigkeit ab (Abb. 3).

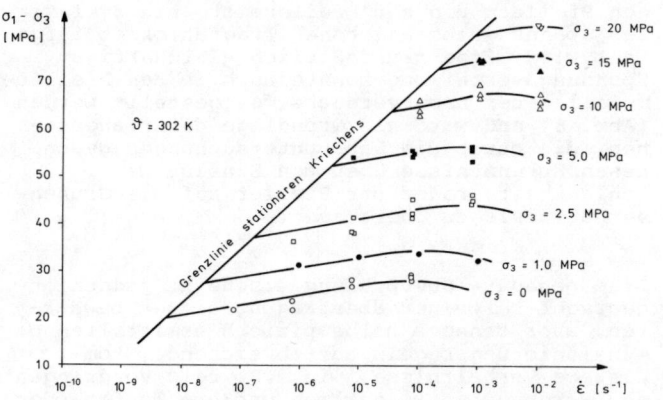

Abb. 3: Abhängigkeit der Bruchfestigkeit von der Verzerrungsgeschwindigkeit Asse Liniensalz Na3ß

Aus Abb. 3 geht hervor, daß - abhängig von der Größe des Manteldruckes σ_3 - Brucherscheinungen erst bei Überschreitung einer bestimmten Größe der Verzerrungsgeschwindigkeit auftreten. Als Grenzlinie ergibt sich die für Versetzungsgleiten ermittelte Bezeichnung für stationäres Kriechen. Bei gleichem σ_3 steigt die Festigkeit mit zunehmendem $\dot{\varepsilon}$ bis zu einem Höchstwert an und fällt dann wieder ab. Auf dieses Ergebnis wurde auch von Passaris (1982) hingewiesen.

Der durch höhere Verformungsgeschwindigkeiten hervorgerufene Wiederabfall der Festigkeit verbietet es nach Ansicht des Verfassers, die Zunahme der Festigkeit mit zunehmender Verformungsgeschwindigkeit in praktischen Fällen zu berücksichtigen.

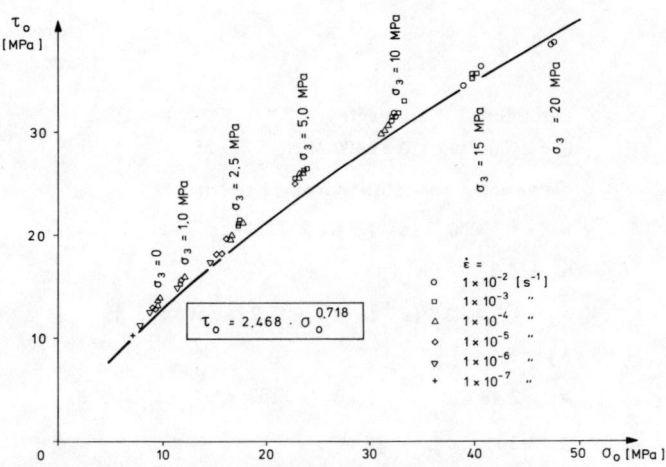

Abb. 4: Kurve der Langzeitfestigkeit

Die Schnittpunkte der Bruchzustandslinien mit der Grenzlinie für stationäres Kriechen ergeben die Langzeitbruchfestigkeit, wie sie in Abb. 4 dargestellt ist.

Als Kriterium für die Tragfähigkeit einer Laborprobe und entsprechend auch eines Grubenpfeilers wird das in Abb. 5 dargestellte Diagramm vorgeschlagen. Neben der Langzeitbruchfestigkeit sind die bei einer bestimmten stationären Kriechgeschwindigkeit bruchlos auf nehmbaren Spannungen (horizontale Geraden) eingetragen. Die Tragfähigkeit wird danach einerseits durch die Langzeitbruchfestigkeit und andererseits durch eine zulässige Konvergenz bzw. örtlich vorhandene Pfeilerdeformation begrenzt.

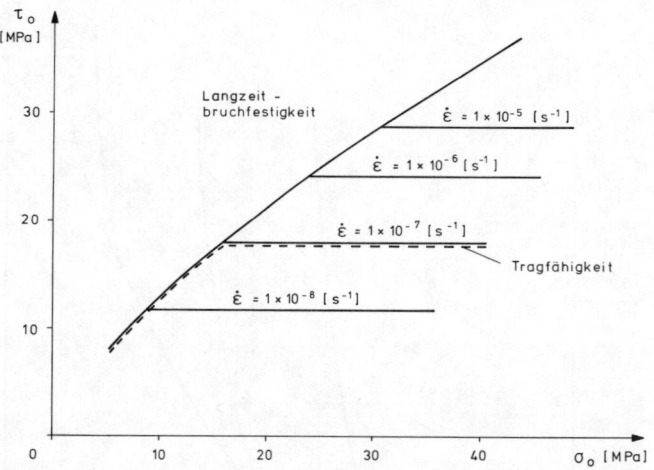

Abb. 5 Kriterium der Tragfähigkeit

Kriechbruch

Aus Untersuchungen zur Bruchfestigkeit von Salzgesteinen geht hervor, daß unterhalb einer bestimmten Grenzbeanspruchung Salzgestein unbegrenzt tragfähig ist und sich beliebig bruchlos verformt (Knoll 1973, Wallner 1981). Versuche, die klären sollten, unter welchen Bedingungen Kriechbruch zu erwarten ist oder nicht, ergaben, daß ein Versagen der Proben durch Scherbruch bei einer bestimmten Konfiguration der Zustandsgrößen Spannung, Verformungsgeschwindigkeit und Temperatur eintritt.

In Abb. 6, einem für die Darstellung der Ergebnisse von Kriechversuchen üblichen $\dot{\varepsilon}/\sigma_1-\sigma_3$ Diagramm, sind die Versuchsspuren dreier unterschiedlicher Experimente eingetragen, die unter vergleichbaren Versuchsbedingungen alle zum gleichen Bruchpunkt führen. Das Phänomen des Kriechbruches unter konstanter Beanspruchung kann danach zurückgeführt werden auf den Effekt einer begrenzten Zunahme der Festigkeit bei höheren Verformungsgeschwindigkeiten. In der primären Kriechphase wird ausgehend von einem Niveau höherer Festigkeit ein Zustand erreicht, der zum Bruch führt, ohne daß sich ein stationäres Gleichgewicht (sekundäres Kriechen) einstellen kann.

3. FEM PROGRAMMSYSTEM ANSALT

Das Rechenprogramm ANSALT wurde von der BGR in Zusammenarbeit mit Control Data GmbH, Hamburg, zur Berechnung thermomechanischer Beanspruchungen von untertägigen Hohlraumstrukturen im Salzgestein entwickelt (Wallner, Wulf 1982). Neben der Implementation des zuvor erläuterten

Stoffmodells für Salzgestein und verschiedener Strategien zur Lösung nichtlinearer Spannungs-Verformungsprobleme, enthält das Programm geeignete Rechentechniken, die eine realistische Erfassung gebirgsmechanischer Vorgänge (Ausbruch, Verfüllung, gebirgsmechanischer Ausbau) und die Simulation thermisch-mechanischer Wechselwirkungen erlauben.

ler bei großflächigem Abbaufeld die volle Überlagerungslast zu tragen haben. Die besondere Tragwirkung der hochbeanspruchten Pfeiler ergibt sich dadurch, daß bei hinreichend plattigen Pfeilern ein zur Pfeilermitte hin dreiaxialer Spannungszustand hoher Tragfähigkeit aufgebaut wird. Eine grundsätzlich gleichartige Spannungsverteilung konnte auch in den Pfeilermodellen der Laborversuche festgestellt werden (Abb. 8) und wird als Grundlage dafür angesehen, die durch die Laboruntersuchungen gewonnenen Erkenntnisse über den Einfluß des Schlankheitsgrades der Pfeiler auf die Grubenverhältnisse zu übertragen.

Eine genauere Überprüfung erscheint jedoch angebracht bei einer Übertragung dieser bewährten, aber dennoch halbempirisch ermittelten Dimensionierungsregeln auf abweichende, komplexere Verhältnisse, wie z.B. beim Vordringen des Bergbaus in wesentlich größere Teufen oder in Gebiete mit ungünstigen Lagerungsverhältnissen.

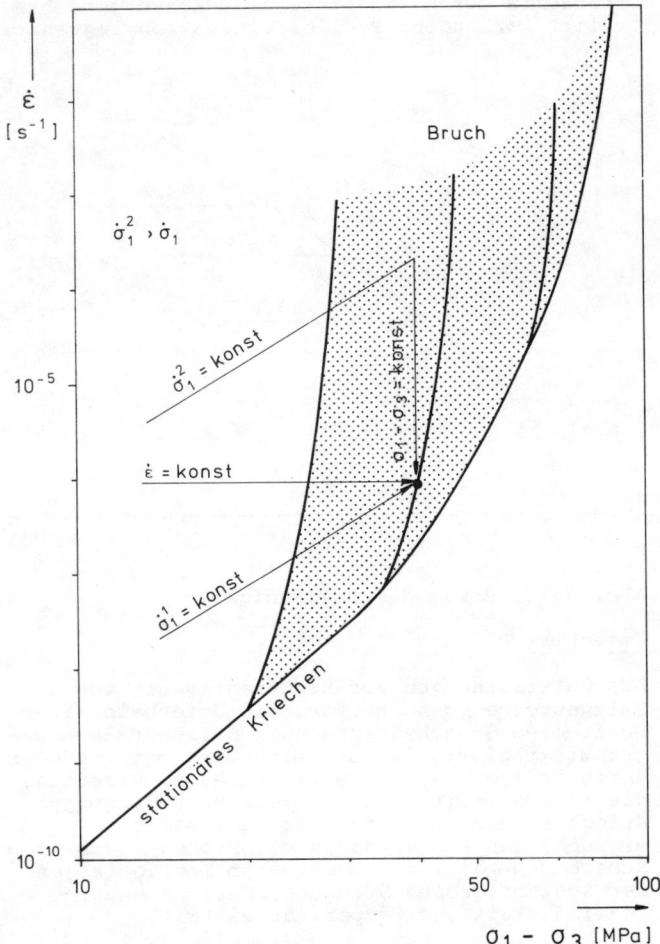

Abb. 6: Kriechbruchmodell

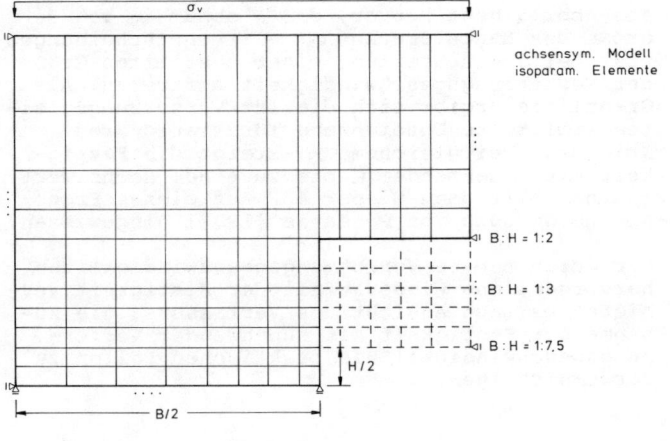

elastische Kennwerte:

$E = 24000$ MPa, $G = 9600$ MPa, $\nu = 0.25$

Kennwerte für stationäres Kriechen:

$A_1 = 1.2 \times 10^{22}$ [s^{-1}], $Q_1 = 27$ [kcal mol^{-1}]

$n_1 = 5.5$

$A_2 = 1.7 \times 10^{14}$ [s^{-1}], $Q_2 = 12.9$ [kcal mol^{-1}]

$n_2 = 5.0$

$B_1 = 2.15 \times 10^7$ [s^{-1}], $B_2 = 6.67 \times 10^{-1}$ [s^{-1}]

$D = 2300$, $\sigma^o_{eff} = 20$ MPa

Bruchkennwerte:

$C = 2.468$, $\alpha = 0.718$

Abb. 7: Tragfähigkeit eines Abbaupfeilers, FE-Netz, mechanische Kennwerte

4. PFEILERTRAGFÄHIGKEIT

Als Grundlage für die sichere Gestaltung eines versatzlosen Abbaus von carnallitischen Kaliflözen im Werra-Kalirevier wurden von Uhlenbecker (1971, 1978) Regeln für die Pfeilerdimensionierung angegeben, die sich in der Praxis bewährt haben. Diese Richtwerte für die Pfeilerdimensionierung wurden aufgrund langjähriger bergmännischer Erfahrungen vor Ort, umfangreicher untertägiger Messungen und deren Korrelation mit grubenähnlichen Großmodellversuchen im Labor gewonnen.

Durch markscheiderische Messungen im Grubengebäude konnte nachgewiesen werden, daß die Pfei-

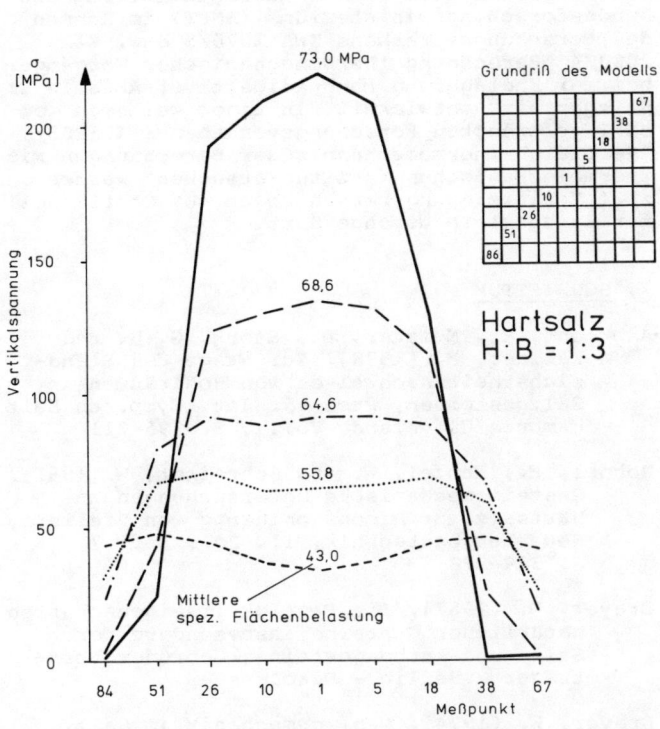

Abb. 8: Meßergebnisse zur Verteilung der Vertikalspannungen in einem Pfeilermodell (nach Uhlenbecker)

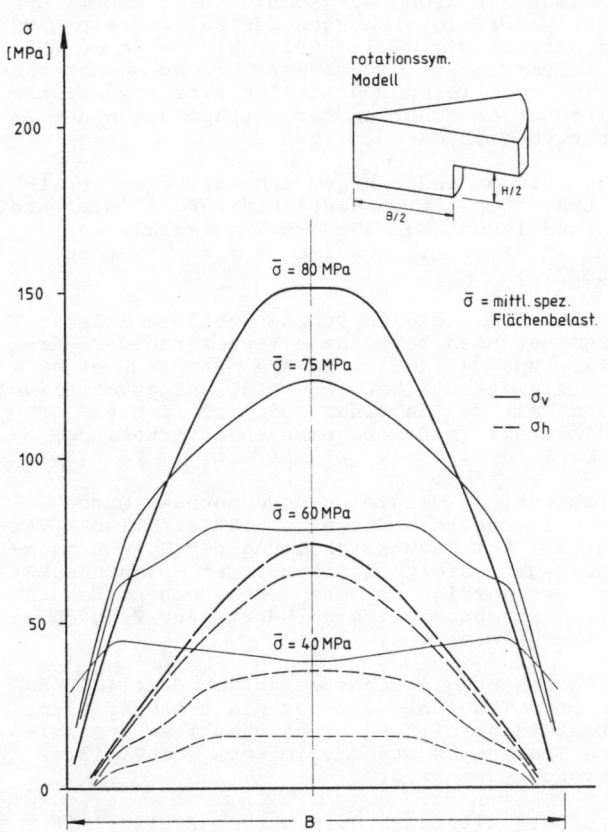

Abb. 9: Berechnete Spannungsverteilung in dem Pfeiler H:B = 1:3

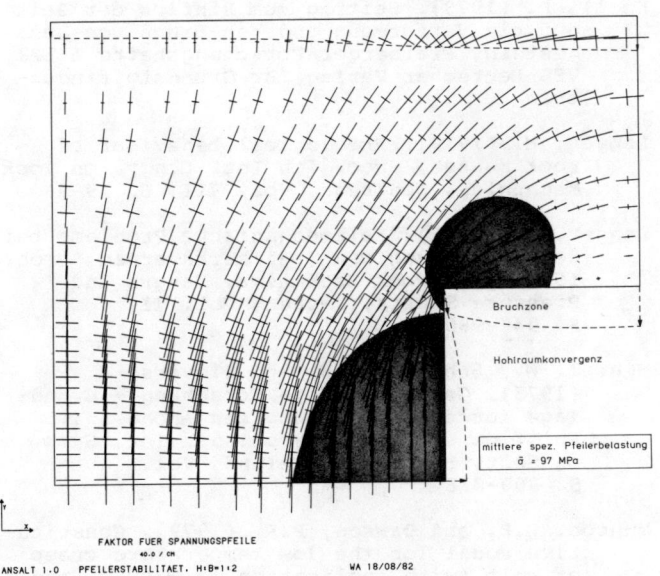

Abb. 10: Hauptspannungsverteilung und Bruchzonen, Pfeiler H:B = 1:2

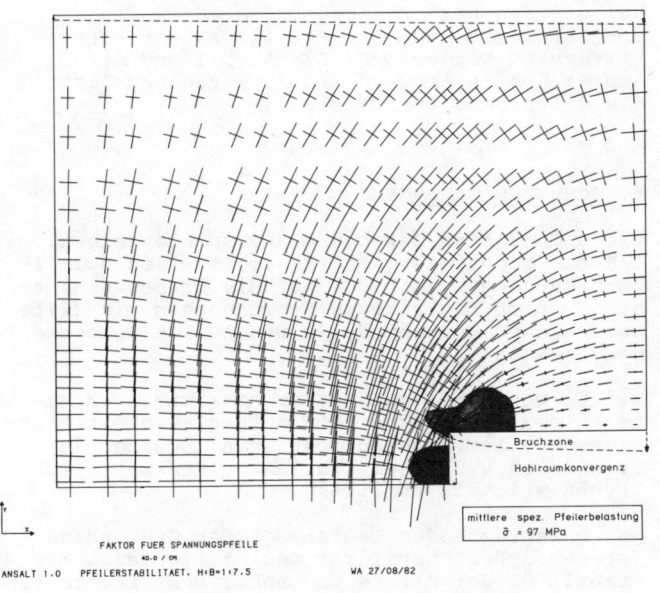

Abb. 11: Hauptspannungsverteilung und Bruchzonen, Pfeiler H:B = 1:7,5

Numerische Berechnungen zur Pfeilertragfähigkeit auf der Grundlage konsistenter konstitutiver Materialgleichungen für Salzgestein geben uns hierbei die Möglichkeit, die im Labor und im Grubengebäude gemessenen Verformungen besser zu interpretieren und stellen eine wesentliche Hilfe zur Absicherung der Extrapolation der Laborergebnisse dar.

Dies soll anhand einiger Ergebnisse zur Pfeilertragfähigkeit verdeutlicht werden, die unter Zugrundelegung der für Asse-Liniensalz bestimmten Materialkennwerte berechnet wurden (Abb.7).

In Abb. 9 ist die zu repräsentativen Belastungszuständen berechnete Verteilung der Vertikal- und der Horizontalspannung im Pfeiler dargestellt. Die berechnete Spannungsverteilung stimmt mit der in einem Modellpfeiler experimentell im Labor gemessenen Verteilung (Abb. 8) qualitativ gut überein.

Durch mehrere Vergleichsberechnungen wurde nachgewiesen, daß dieser charakteristische Verlauf nur bei Berücksichtigung der Spannungsumlagerungen infolge Bruch und der Spannungsumlagerung infolge Kriechen entsprechend dem in Abb. 5 veranschaulichten Gesetz zur Tragfähigkeit erhalten wird.

Die Berechnungsergebnisse zeigen deutlich, daß die hohe Tragfähigkeit des plattigen Pfeilers durch die Ausbildung eines günstigen dreiaxialen Spannungszustandes im Kern des Pfeilers hervorgerufen wird.

In einer weiteren Studie werden Pfeiler mit unterschiedlichem Schlankheitsgrad untersucht. In einigen Berechnungen, in denen zur Verdeutlichung der Brucheffekte der zeitabhängig günstige Einfluß von Spannungsumlagerung durch Kriechen vernachlässigt wurde, zeigt sich, daß neben der Stabilität des Pfeilers bei höheren Belastungen entsprechend größeren Teufenlagen des Bergwerks auch die Stabilität der Firste betrachtet werden muß. Die Ausbildung der Bruchzonen in Abb. 10 und 11 veranschaulicht die Gefährdung der Firste.

5. SCHLUSSFOLGERUNGEN

Die angeführten Berechnungsbeispiele zeigen, daß die Übertragung der Ergebnisse von Modelluntersuchungen im Labor auf die Grubenverhältnisse durch begleitende Berechnungen mit Hilfe numerischer Rechenmodelle wesentlich verbessert werden kann.

Bei Verwendung konsistenter konstitutiver Materialmodelle läßt sich die Tragfähigkeit plattiger Pfeiler auch auf der Grundlage der in triaxialen Versuchen bestimmter Kennwerte zur Bruchfestigkeit ermitteln.

Bei entsprechender Teufenlage der Grubenbaue ist neben der Stabilität der Pfeiler auch die Stabilität der Firste und Sohle der Strecke von Bedeutung, die ihrerseits Rückwirkungen auf die Gesamttragwirkung haben kann. Zur Zeit werden weitere Berechnungen zur Erfassung der Langzeitstabilität und zur Erfassung großräumiger abbaudynamischer Vorgänge durchgeführt.

6. ANMERKUNGEN

Das Programmsystem ANSALT wurde im Auftrag des Bundesforschungsministeriums (BMFT) im Rahmen des Forschungsvorhabens KWA 2070/8 bzw. KWA 2090/4 "Berechnung thermomechanischer Vorgänge bei der Endlagerung hochradioaktiver Abfälle im Salzgestein" entwickelt. In einem weiteren vom BMFT geförderten Forschungsvorhaben KWA 5203/7 "Vergleich thermomechannischer Berechnungen mit thermomechanischen in situ Versuchen" werden z.Zt Verifizierungsberechnungen zum Stoffmodell für Salzgestein durchgeführt.

7. SCHRIFTTUM

Albrecht, H., Meister, D., Stork, G.-H. und Wallner, M. (1978). Zur Frage des Standsicherheitsnachweises von Hohlräumen in Salzgesteinen, Proc. 5. Int. Symp. on Salt Hamburg/Cleveland, Vol.I, S. 195-211

Böhnel, H., Menzel, W. und Schreiner, W.(1981). Gesteinsmechanische Untersuchungen an Hartsalz zur Dimensionierung von Pfeilern, Neue Bergbautechnik, 11. Jg., Heft 7, S. 374-378

Dreyer, W. (1967). Die Festigkeitseigenschaften natürlicher Gesteine insbesondere der Salz- und Karbongesteine, Gebrüder Bornträger - Berlin - Nikolassee

Dreyer, W. (1974). Gebirgsmechanik im Salz, Ferdinand Enke Verlag, Stuttgart

Grimm, W. und Pforr, H. (1961). Gebirgsschläge im Kalibergbau unter Berücksichtigung von Erfahrungen des Kohlen- und Erzbergbaus, Freiberger Forschungshefte A 173, Akademie Verlag Berlin

Knoll, P. (1973). Beitrag zum Einfluß der Zeit auf die Verformung und den Bruch von Salzgestein, Freiberger Forschungshefte A 528, VEB Deutscher Verlag für Grundstoffindustrie

Langer, M. (1979). Rheological behaviour of rock masses, Proc. 4th Int. Congr. on Rock Mechanics, Montreux, Vol. III, S. 29-96

Langer, M. (1982). Felsmechanische Probleme bei der Errichtung von Speicherkavernen, Proc. ISRM Symp. Rock Mechanics. Cavern and Pressure Shafts, Aachen, Vol. II, S. 947-960

Menzel, W., Schreiner, W. and Sievers, J. (1978). Geomechanische Forschung - Grundlage für die Gestaltung des Abbaues im Kaliflöz Thüringen, Proc. 5th Int. Symp. on Salt, Hamburg/Cleveland, Vol. I, S. 309-318

Munson, D.E. and Dawson, P.R. (1979). Constitutive model for the low temperature creep of salt (with application to WIPP), Sand-79-1853, Sandia National Laboratories, Albuquerque NM

Passaris, E.K.S. (1982). Fatigue characteristics of rock salt with reference to underground storage caverns, Proc. ISRM Symp. Rock Mechanics: Caverns and Pressure Safts, Aachen, Vol. II, S. 983-989

Staupendahl, G., Gessler, K. und Wallner, M. (1982). Zusammenfassende Darstellung von Versuchsergebnissen zum spannungs- und temperaturabhängigen Festigkeits- und Verformungsverhalten von Steinsalzen, Proc. ISRM Symp. Rock Mechanics: Cavern and Pressure Shafts, Aachen, Vol. III, in print

Uhlenbecker, F.W. (1971). Gebirgsmechanische Untersuchungen auf dem Kaliwerk Hattorf (Werra-Revier), Kali und Steinsalz Bd. 5 Heft 10, S. 345-359

Uhlenbecker, F.W. (1978). Neuere Forschungsergebnisse in der Gebirgsmechanik im Hinblick auf den Abbau von carnallititschen Kaliflözen, Proc. 5th Int. Symp. on Salt, Hamburg/Cleveland, Vol. I, S. 413-422

Wallner, M., Caninenberg, C. und Gonther, H. (1979). Ermittlung zeit- und temperaturabhängiger mechanischer Kennwerte von Salzgesteinen, Proc. 4th Int. Congr. Rock Mechanics, Montreux, Vol. 1, S. 313-318

Wallner, M. (1981a). Critical examination of conditions for ductile fracture of rock salt, Proc. OECD/NEA Workshop on Near Field Phenomena in Geologic Repositories for Radioactive Waste, Seattle p. 243-253

Wallner, M. (1981b). Analysis of thermomechanical problems related to the storage of heat producing radioactive waste in rock salt, Proc. 1st Conf. on the Mechanical Behavior of Salt, The Pennsylvania State University

Wallner, M. and Wulf, A. (1982). Thermomechanical calculations concerning the design of a radioactive waste repository in rock salt, Proc. ISRM Symp. Rock Mechanics: Cavern and Pressure Shafts, Aachen, Vol. II, S. 1003-1012

DIE ANWENDUNG DER ANKERSICHERUNG IM BERGBAU
Application of anchor securing in mines
Utilisation des ancrages au rocher dans les mines

Dr. Thomas F. Herbst
Dyckerhoff and Widmann AG, München, Germany

ZUSAMMENFASSUNG

Die Verwendung der Anker als Teil des Tragmodells "Bewehrtes Gebirge" wird für den Bergbau von immer größerer Bedeutung, je größer die Abbauteufen und je größer die Ausbruchsflächen, bedingt durch moderne Technologien, werden. Die Rolle und Wirkung der Anker als Gebirgsbewehrung mit Blickrichtung auf noch bestehende Kenntnislücken wird diskutiert.

SYNOPSIS

The application of anchors in mines as an element in the reinforced rock support system is gaining in importance as mines become deeper, and the cross sections of excavations increase due to the requirements of new technologies. The role and effect of the anchors as part of the rock reinforcement is discussed with special emphasis on areas where theoretical knowledge is not sufficiently far advanced.

RESUME

L'utilisation des tirants et des boulons d'ancrage en tant qu'éléments de systèmes de soutènement statique du rocher armé prend de plus en plus d'importance pour les mines au fur et à mesure qu'augmente la profondeur des gisements et que croissent les sections d'excavation, et cela grâce à de nouvelles technologies. Le rôle et la fonction des tirants et boulons en ce qui concerne le renforcement du rocher sont discutés en tenant compte des secteurs où les connaissances techniques présentent encore des lacunes.

EINLEITUNG

Die Anwendung von Ankern im Bergbau ist so alt oder sogar älter als die im Tunnel- und Kavernenbau. Anker wurden lange Zeit nur als Fixierungselemente für Installationen, für lose Felsblöcke und als Mittel der Ausbruchflächensicherung betrachtet und sollten einen äquivalenten Ersatz für Türstock oder Bogenausbau bilden. Daraus entstanden in den verschiedenen Regionen und Bergwerken die unterschiedlichsten Ankertypen, die nach Material und Bauart variieren. Die Philosophie der Neuen Österreichischen Tunnelbauweise (Müller-Salzburg 1978) zeigte aber, daß der systematisch angewendete Anker wesentlich mehr Möglichkeiten bietet, was durch viele imposante Bauwerke des Untertagebaus nachgewiesen wurde. Zweck dieses Beitrags ist es mit Blickrichtung auf den Bergbau, auf die Einflußgrößen der Anker einzugehen, damit die mit Ankersicherungen gemachten Erfahrungen übertragbar werden.

1. UNTERSCHIEDE ZWISCHEN TUNNELBAU UND BERGBAU

Eine der Schlüsselfragen ist, inwieweit Tunnel- und Bergbauprobleme miteinander vergleichbar sind und wo sie sich grundsätzlich unterscheiden. Nachfolgend werden mögliche, aber nicht notwendige Unterschiede aufgeführt. Sie sollten aber im Auge behalten werden, wenn Erfahrungen aus beiden Bereichen miteinander verglichen werden.

Die Hohlraumauffahrung des Bauwesens hat fast immer permanenten Charakter. Spannungen im Gebirge und Stützmaßnahmen müssen daher zu einem verformungsfreien Gleichgewicht gelangen. Wo immer möglich wird man Zonen hoher Gebirgsspannungen, niedriger Gebirgsfestigkeit oder großen Wasserandranges ausweichen. Die Position der Lagerstätten im Bergbau läßt dazu nur beschränkten Spielraum. Dafür ist die Lebensdauer der Strecken vielfach begrenzt. Verformungen werden bis zu betrieblich bedingten Grenzwerten hingenommen.

Umfangreiche felsmechanische Voruntersuchungen mit Probestollen sowie laufende Ausbruchsüberwachung, zum Vergleich von Annahmen und vorgefundenen Werten, gehören zum Instrumentarium der Auffahrung von Hohlräumen für dauerhafte Nutzung. Allein wirtschaftliche Gesichtspunkte verhindern beim Bergbau zu aufwendige Untersuchungen, da jede Strecke ja nur Hilfsmaßnahmen auf dem Wege der Gewinnung der Bodenschätze ist. Durch die zeitlich begrenzte Nutzungsdauer vor allem von Abbaustrecken steht viel stärker der Gedanke eines temporären Gleichgewichtes im Vordergrund, bei dem eine gewisse Verformungsgeschwindigkeit nicht überschritten werden darf.

Die Überlagerungshöhen bzw. Tiefenstufen sind bei den Bergwerken eine Größenordnung höher, als bei den Tunneln des Bauwesens, wenn man einmal von den Basistunneln in den Gebirgsregionen absieht.

Im Bergbau ist durch die starke Durchörterung und durch den Abbau der Lagerstätten teilweise mit sehr unklaren Spannungsverhältnissen zu rechnen, die sich durch die abbaubedingten Bewegungen noch laufend verändern können.

Die Auffahrquerschnitte sind im Bergbau bisher wesentlich kleiner als im Bauwesen. Selten erreichten bisher Strecken die Querschnitte von großen Straßentunneln. Dadurch gibt es Platzbeschränkungen beim Einbau von Ankern, was einer Mechanisierung hinderlich ist.

2. AUFGABEN UND MÖGLICHKEITEN DER ANKERSICHERUNG

2.1 Anwendungsbereiche

Die klassischen Aufgaben der Ankersicherung stehen nach wie vor im Mittelpunkt: Die Aufhängung von Leitungen, von Masch-endrahtgewebe zur Sicherung gegen Fall von losem Gestein, die Konsolidierung der Ausbruchsoberfläche. Solange die Ankersicherung nur unter diesen traditionellen Gesichtspunkten gesehen wird, kann das Verständnis für die Wirkungen eines "bewehrten Gebirges" jedoch nicht geweckt werden. Hier aber liegen die großen Möglichkeiten der Ankersicherung. Werden die Gebirgsdrücke durch größere Tiefenlagen höher, und steigen die Spannungen wegen der größer werdenden Auffahrquerschnitte um den Ausbruch herum, so gibt es kaum noch wirtschaftliche Alternativen zum Ankerausbau (Fig.1). Durch ihn allein können im druckhaften Gebirge Traggewölbe gebildet werden, die den Hohlraum sichern.

Obwohl der Türstock oder Bogenausbau ein höheres Gefühl der Sicherheit für den Mineur vermittelt, zeigt die Unfallstatistik, daß der Ankerausbau sicherer ist. Die frühzeitige Warnung durch Verformungen am Ankerkopf erlaubt es, Nachankerungen rechtzeitig durchzuführen. Dadurch daß Anker im Gebirge verbleiben, entfällt auch das Gefahrenmoment des Raubens. Durch die Tiefenwirkung von Ankern kann sich der Bruchmechanismus im Gebirge nicht mehr frei entfalten. Diese Sicherheitsaspekte spielen um so mehr eine Rolle, je industrialisierter der Bergbau betrieben wird.

2.2 Tragmodelle

Sollen die Anker wirkungsvoll eingesetzt werden, so muß das Tragmodell stimmen.
Dieses erfordert die Kenntnis der vor dem Auffahren der Öffnung vorhandenen Spannungen po und Spannungsrichtungen, sowie die Wirkung des Traggewölbes, das sich um den Ausbruch einstellt. Von Gewölbe wird hier deshalb gesprochen, da in den seltensten Fällen die Gebirgseigenschaften zusammen mit den Ankern das Modell eines Biegebalkens zulassen, und wegen der fast immer vorhandenen Klüfte nur Druck und Schubkräfte übertragbar sind. Dabei können jedoch hohe Horizontalspannungen die Wirkung eines vorgespannten Trägers ergeben.

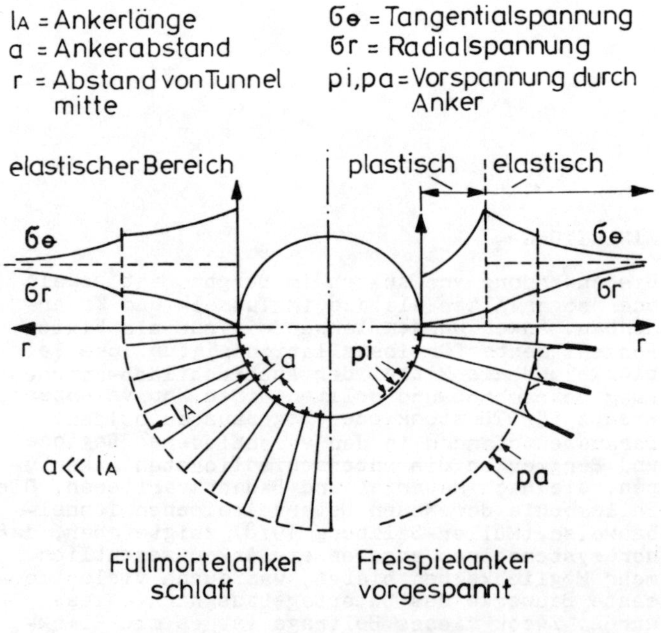

Fig. 1 Verwendung von vorgespannten Freispielankern in einer Schachtglocke

l_A = Ankerlänge
a = Ankerabstand
r = Abstand von Tunnelmitte
σ_θ = Tangentialspannung
σ_r = Radialspannung
p_i, p_a = Vorspannung durch Anker

Füllmörtelanker schlaff   Freispielanker vorgespannt

Fig. 2 Wirkung von Systemankerungen mit Füllmörtelankern oder vorgespannten Freispielankern

Obwohl die Wirkung des Traggewölbes unbestritten ist, gibt es verschiedene Modelle, wie die Systemanker dort wirksam werden. Daraus resultieren unterschiedliche Ankerdichten, Ankerkräfte und Ankerlängen. Natau und Leischnitz (1978) sehen in einem durch dichte Bewehrung hochfest gemachten Tragring begrenzter Stärke die wirksamste Lösung. Die kurzen Anker kommen dabei den Einbaubedingungen vor allem im Bergbau entgegen. Egger (1978) sieht in einem Kontinuum, das Hohlraum, eine darum liegende plastische Zone und anschließend eine elastische Zone umfaßt, den geeigneten Ansatz für die Ankerdimensionierung (Fig.2). Das Materialgesetz, das Post- Failure Festigkeit mit einschließt, berücksichtigt die elastischen und plastischen Verformungen des Gebirges, die Ankerkräfte gehen entweder als Ausbauwiderstand pi am Ausbruchsrand bei vorgespannten Freispielankern ein, auf ganze Länge vermörtelte Anker gehen als zusätzliche Kohäsion ein. Beide Modelle, obwohl schwer harmonisierbar, sind Teile eines Gesamttragmodells, bei einem steht die Steuerung der Gebirgsfestigkeit am Ausbruchsrand, beim anderen die Gesamtbeanspruchung des Gebirges durch die Hohlraumöffnung im Vordergrund.

2.3 Wirkung der Systemanker

Die Wirkung des Ankers bei einer systematischen Ankerung liegt nun nicht darin, direkt die auf den Hohlraum wirkenden Kräfte abzufangen, sondern die Materialfestigkeit des das Gewölbe bildenden Gebirges zu beeinflussen und zu erhöhen. Durch die Hohlraumauffahrung ist ja die senkrecht zum Ausbruchsrand wirkende Spannung dort auf null reduziert worden, sodaß hier eine niedrigere einachsige Druckfestigkeit vorhanden ist. Die Wiederherstellung des 3achsigen Spannungszustandes ist daher die vornehmliche Aufgabe der Systemankerung. Je nach Ausbildung können die Anker sowohl als äußere Kräfte bei Vorspannankern als auch als innere Kräfte bei schlaffen vermörtelten Ankern eingesetzt werden.

Es ist nun entscheidend, ob das Gebirge noch seine ursprünglichen Festigkeitseigenschaften beibehalten hat, wenn der Anker eingebaut und wirksam wird, oder ob es nur mehr die nach dem Bruch verbleibende Restfestigkeit hat. Der Einfluß der Gefügeauflockerung auf die Tragfähigkeit läßt sich auch aus Fig.3 ersehen. Die wesentlich niedriger verlaufende Einhüllende eines gestörten Gefüges ergibt bei gleichem Ankerausbauwiderstand erheblich niedrigere Festigkeiten.

Je weiter die Entfestigung ins Gebirge fortgeschritten ist, desto größer wird die Gewölbestärke und die Grundlänge der Stützlinie. Eine zu weitgehende Entfestigung birgt die Gefahr, daß die Verformungen nicht mehr in den Griff zu bekommen sind. Mit auf volle Länge vermörtelten Ankern (Füllmörtelankern), die vor Wirksamwerden der Ausbruchsspannungen gesetzt wurden, sind sehr gute Erfahrungen gemacht worden. Dies zeigt den bedeutenden Einfluß der Auflockerung auf den Gefügewiderstand des Gebirges.

Die bekannte Relation zwischen Ausbauwiderstand pi und der in den Hohlraum gerichteten Verformung u_i, die die Entlastung des Ausbauwiderstandes zeigt, darf nicht zu dem Schluß führen, daß der starke Abfall zur Restfestigkeit des Gebirges im plastischen Bereich generell erwünscht ist, da der Entspannung die Festigkeitsabminderung entgegensteht.

2.4 Beanspruchungsarten der Anker

Bei der Systemverankerung lassen sich 2 wesentliche Aufgaben und daraus resultierend mögliche Beanspruchungen der Anker im Gebirge feststellen:

a) Verhinderung von Verformungen im Gebirge, die im wesentlichen in Ankerachsrichtung verlaufen. Hierzu gehören das Ausknicken von ausbruchparallelen Schichtpaketen bei schichtparalleler Belastung, das Herabfallen von Blöcken aus der Firste

b) Verhinderung von Verformungen im Gebirge, die nach anfänglich achsialer Belastung den Anker auf Scherung und Biegung beanspruchen.

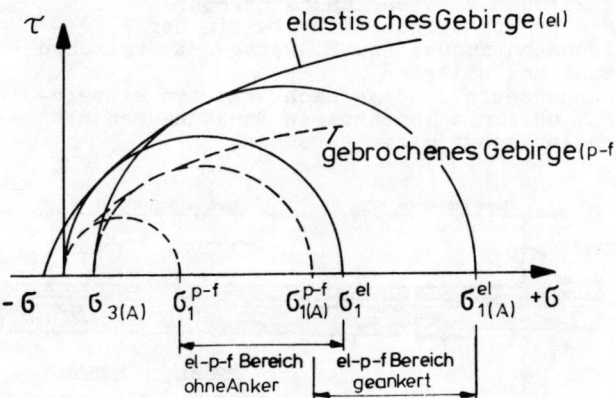

Fig.3 Schematische Darstellung des Einflusses des Gefügezustandes auf die Druckfestigkeit des Gebirges

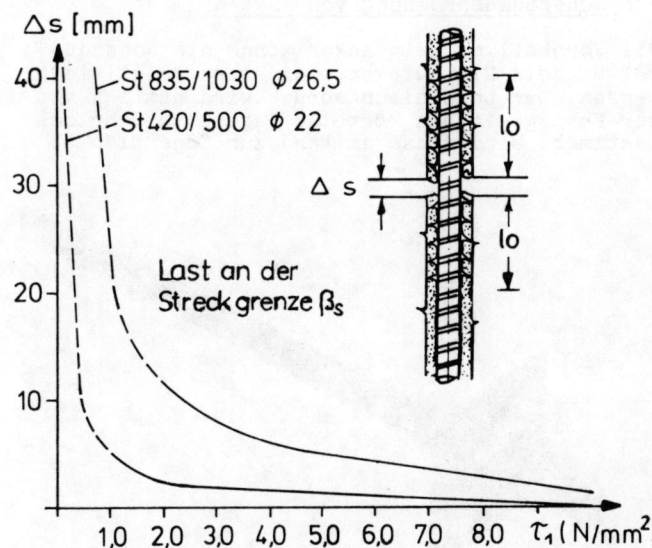

Fig. 4 Einfluß des Haftverbundes zwischen Stahl und Gebirge auf die Weite der Trennflächenöffnung

Dazu gehören:
Schubbruchsicherung entlang der Trennflächen,
Spaltbrüche an den Ulmen,
Gleitkeile in die Hohlräume hinein.

Diese Unterscheidung ist dann nicht zu machen, wenn durch Vorspannung ein verformungsarmes Gebirgsgefüge erzwungen wird oder durch einen entsprechenden Querbewegungsraum im Bohrloch Scherbeanspruchungen des Ankers verhindert werden.

2.5 Haftverbund der Anker

Das Zusammenwirken zwischen Gebirge und schlaffen Anker setzt wie im Stahlbeton einen guten Haftverbund voraus. Können zu große Spaltweiten entstehen, sodaß die Verzahnung in den Trennflächen geschwächt oder aufgehoben wird, so wird der Anker bei Schubbruchsicherung auf Abscheren bzw. Verbiegen beansprucht. Hiermit setzt ein Mechanismus größerer Verformungen ein. Die Abhängigkeit der theoretischen Trennflächenöffnung vom Haftverbund zwischen Stahl und Gebirge läßt sich aus Fig.4 ablesen. Die Haftspannung τ_1 muß, wenn Spaltrisse verhindert oder klein gehalten werden sollen, groß sein. Es eignen sich daher für die Systemankerung mit nicht vorgespannten, vermörtelten Felsankern im besonderen Stähle mit starker Rippung und niedriger Stahlgüte. Auch darf die Verbundspannung nicht in einer zum Bohrloch konzentrischen Bruchfläche durch Gebirgsauflockerung stark herabgesetzt sein.

Leider ist noch wenig bekannt über die Verbundfestigkeit in den verschiedenen Post-Failure Stadien eines mit Ankern bewehrten Gebirges. Die Gefügeauflockerung kann dabei je nach Zeitpunkt des Wirksamwerdens der Anker vor, während oder nach dem Ankereinbau eintreten. Trennflächen mit unterschiedlichen Winkeln zur Hauptbeanspruchungsrichtung erschweren dabei die Erfassung der Zusammenhänge.

2.6 Scherbeanspruchung von Ankern

Die Vorstellung, ein Anker könne als Scherdübel wie ein Niet wirken, muß stark relativiert werden. Der Lochreibungsdruck wird nämlich von der Festigkeit des Mörtelauflagers im Bohrloch bestimmt. Wird diese am Rand zur Scherfläche überschritten, so setzt dort fortschreitend eine plastische Verbiegung mit Zugbeanspruchung des Ankerzuggliedes mit gleichlaufend zunehmender Gebirgsverformung ein. Die mögliche Scherkraft des Stahles kann daher nicht genützt werden. Müssen von den Ankern aus dem Gebirge Scherkräfte aufgenommen werden, so empfiehlt sich eine Vielzahl kleiner Stabdurchmesser mehr als wenige große Stabdurchmesser. Die Stähle müssen insgesamt eine hohe Gleichmaßdehnung haben.

2.7 Ankersicherungen mit kontrollierter Verformungsgeschwindigkeit

Im Bergbau werden als Besonderheit Überlegungen angestellt, den Ausbauwiderstand nur in solcher Größe zu halten, daß die Verformungsgeschwindigkeit kontrollierbar bleibt. Dabei wird für einen vorgegebenen Zeitraum eine Verformung des Ausbruchrandes zugelassen, die die üblichen Dehnungseigenschaften der Anker weit überschreitet. Anker die hierfür eingesetzt werden und als Rutschanker bezeichnet werden, müssen lastabhängig nachgeben. Können die Gebirgsverformungsmechanismen nicht wirklichkeitsnahe erfaßt werden, besteht die Gefahr, daß das Verformungsverhalten des Gebirges durch die Anker nicht optimal beeinflußt wird.

3. AUSBILDUNG UND WIRKUNGSWEISE DER ANKER

3.1 Korrosionsschutz

Da die Ausbildung der Ankerteile (Zugglied, Ankerkopf u. bergseitige Verankerung) als bekannt vorausgesetzt werden kann, (z. B. Fig. 5) soll hier nur der Korrosionsschutz kurz gestreift werden. Er besteht bei Temporärankern aus einfachem Korrosionsschutz, d. h.
- Zementmörtel oder Kunstharzmörtel zwischen Zugglied und Bohrloch
- bei Freispielankern: Hüllrohr in der freien Stahllänge

bei Dauerankern aus doppeltem Korrosionsschutz (Fig. 6) d. h.
- in der Verankerungsstrecke: geripptes Kunststoffrohr aus Hart-PVC, PE, PP mit Zementmörtel
- in der freien Stahllänge: wie oben mit zusätzlichem glatten Kunststoffrohr oder Kunststoffrohr mit plastischer Korrosionsschutzmasse (z. B. Denso-Jet) zwischen Stahl und Hüllrohr

Die Ankerköpfe sind je nach Ankertyp einbetoniert, oder mit abnehmbaren Schutzhauben mit Korrosionsschutzmasse versehen.

Fig. 5 Stabanker mit Gewindestahl

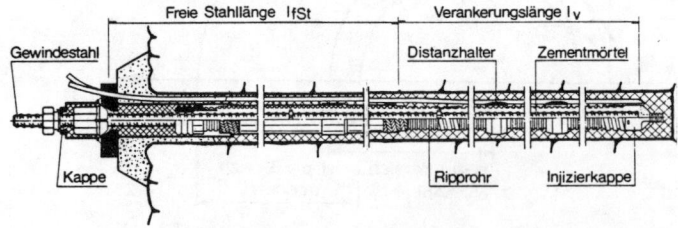

Fig. 6 Dauer-Einstabanker mit doppeltem Korrosionsschutz DYWIDAG

Diese anerkannten Regeln für den Korrosionsschutz beziehen sich auf die Verwendung von Ankern mit großen Ankerkräften oder Verwendung von Ankern mit Spannstählen.

3.2 Wirkungsweise der Anker

Die in Fig.7 dargestellten Wirkungsweisen unterscheiden zwischen

	dauerhaft elastisch	blockiert
Vorgespannt	Freispielanker z.B. Verpreßanker	Freispielanker nachträglich blockiert z.B. Felsanker
Schlaff	nicht vorgespannter Freispielanker z.B. Spreizdübel-, Kunstharzklebeanker	Füllmörtelanker z.B. SN-Anker, Kunstharzklebeanker

Fig. 7 Schema der Ankerwirkungsweisen

3.2.1 Vorgespannte und schlaffe Anker

Die Vorspannung nimmt die Längsdehnung zur Aktivierung der Kraftaufnahme vorweg. Mit Vorspannankern wird ein verformungsarmes Tragsystem geschaffen, was sich vor allem in hartem geklüftetem Gebirge anbietet. Die Vorspannung erfordert eine Freispielstrecke, die den Gebirgsbereich, der mit einer Zusatzspannung beaufschlagt werden soll, beschreibt. Innerhalb eines Kraftverteilungsbereiches wirkt die Vorspannung als σ_3-Spannung oder Ausbauwiderstand. Es werden hierfür vornehmlich Spannstähle als Zugglieder verwendet, da sie am unempfindlichsten hinsichtlich Spannungsverlust bei einer Verschiebung des Ankerkopfes sind.

Schlaffe Anker erhalten ihre Last durch die Dehnung. Um die Verformungen gering zu halten, werden niedrigere Stahlgüten mit größeren Querschnitten oder eine größere Anzahl von Ankern verwendet, letzteres ist oft im Interesse einer guten Oberflächensicherung.

3.2.2 Freispielanker und blockierte Anker

Ist die elastische Länge eines Freispielankers immer wirksam, so führen starke Kluftweitenveränderungen an einer Stelle meist nur zu unbedeutenden Spannungsänderungen im Zugglied. Wird der Anker nach dem Vorspannen in der freien Stahllänge durch Verpressen blockiert, d. h. mit dem Gebirge verbunden, so können lokale Kluftweitenvergrößerungen zur örtlichen Überbeanspruchung führen. Blockierungen sollten immer erst nach dem Abklingen aller Gebirgsverformungen vorgenommen werden, wenn Spannstähle und vorgespannte Anker verwendet werden.

3.2.3 Füllmörtelanker

Der Anker wird auf gesamter Länge mit dem Bohrloch durch Kunstharz- oder Zementmörtel verbunden. Die Last wird durch die Gebirgsverformung aufgebracht. Bei hohem Haftverbund zwischen Zugglied und Gebirge wird eine gleichmäßige und feine Rißverteilung erzwungen. Die Belastung des Zuggliedes folgt den Gebirgsverformungen. Ansonsten gilt das für schlaffe Anker aufgeführte.

Ohne daß der gesamte Anker gespannt wird, kann die Ankerplatte gegen das Auflager gespannt werden um die Gebirgsoberfläche etwas unter Spannung zu setzen. Da keine Elastizität die Vorspannung aufrecht erhalten kann, ist es müßig über Blockierkräfte zu diskutieren. Ein Anziehen mit einem Schlagschrauber zur Herstellung eines Kraftschlusses ist ausreichend.

3.2.4 Einfluß auf das Gebirge

Die aus den Bauarten resultierenden Unterschiede sind auch Steuerungselemente für die Ausbildung des Stützgewölbes. In einem Gebirge mit klaren Trennflächenscharen können durch geeignete Anordnung von Vorspannankern und schlaffen Ankern verschiedener Länge, Kräfte und Dichte, verformbare und starre Bereiche geschaffen werden. Die Stützlinie wölbt sich dabei im allgemeinen der Last entgegen, starre Zonen ziehen sie an den Ausbruchsrand, nachgiebige Zonen verlagern sie ins Berginnere.

SCHLUSSBEMERKUNG

Nicht überall lassen sich ankeroptimale Bedingungen schaffen, da geologische und betriebstechnische Gesichtspunkte berücksichtigt werden müssen. Macht man sich aber bewußt, daß die Ankersicherung zu einer Schlüsseltechnik im Bergbau werden kann, so stehen noch viele Optimierungen und Änderungen der bisherigen Gepflogenheiten im Hinblick auf die Möglichkeiten der Ankersicherung im Bergbau am Horizont.

LITERATUR:

Egger P., (1978)
Neue Gesichtspunkt bei Tunnelankerungen in: Grundlagen und Anwendung der Felsmechanik
Trans. Tech. Publications, Clausthal.

Herbst Th., Kern G., (1979)
Felsankerungen im Hohlraumbau und deren Dauerbeständigkeit
Proc. 4. Int. Konf. für Felsmechanik, 1.Band, S. 441 - 449, Montreux.

Müller-Salzburg, L. (1978)
Der Felsbau, 3. Band Tunnelbau
Enke-Verlag, Stuttgart.

Natau O., Leichnitz W., (1978)
Die Verbundwirkung Systemankerung - Gebirge in: Grund-lagen und Anwendung der Felsmechanik
Trans. Tech. Publications, Clausthal.

ANALYSIS OF CREEP AND RELAXATION AROUND AN UNDERGROUND OPENING IN ROCK SALT

Etude du fluage et de la relaxation autour d'une cavité percée dans un massif de sel gemme

Analyse der Hohlraumkonvergenz und Gebirgsrelaxation im Steinsalzgebirge

G. Borm
Lehrstuhl für Felsmechanik, Universität Karlsruhe, Bundesrepublik Deutschland

ZUSAMMENFASSUNG

Das rheologische Gebirgsverhalten in der Umgebung unausgebauter oder ausgebauter Hohlräume in Steinsalzstöcken wird analysiert. Als Materialmodell ist ein nichtlinearer, verallgemeinerter Maxwell-Körper angenommen, bei dem die viskose Deformationsrate einem Potenzansatz der Spannungsdifferenzen folgt. Besonders berücksichtigt ist die Wechselwirkung von Kriechen und Relaxation im Steinsalzgebirge. Die Gebirgsentspannungsvorgänge lassen sich mit Hilfe nichtlinearer viskoelastischer Stoffansätze realistisch erfassen. Es zeigt sich, daß die zirkumpolaren Gebirgsspannungen mit wachsender Stoßtiefe einem nichtmonotonen Verlauf mit auffallender Ähnlichkeit zu der bekannten Schutzhüllenausbildung in elastoplastischen Berechnungen folgen. Die dort verwendete Fließgrenze wird durch eine geschwindigkeitsabhängige Spannungsbedingung ersetzt. Die maximal aufnehmbare Differenzspannung ist umso größer, je höher die Kriechgeschwindigkeit ist. Als Folge der Spannungsrelaxation nimmt die Hohlraumkonvergenz monoton mit wachsender Zeit ab, so daß eine scheinbare Dehnungsverfestigung des Steinsalzes vorgetäuscht wird. Die chronische Konvergenz der Hohlraumwände ist bei Annahme kontinuierlichen Materialverhaltens und bei konstanten äußeren Bedingungen durch eine Trend-Analyse prognostizierbar.

SYNOPSIS

The rheological behaviour of salt rock around underground cavities is studied by analytical and numerical calculations. As a material model a nonlinear generalized Maxwell body is assumed, where the rate of the viscous deformation is a power function of the effective stress. Special emphasis is directed to the interaction of creep and relaxation in the ground. Stress relaxation is calculated for nonlinear visco-elastic rock materials. The time dependent maximum circumferential stresses form a temporary screen around the opening, similar to the nonmonotonous distribution of maximum tangential stresses in elastoplastic tunnel analyses. The flow condition depends on the rate of deformation. As a result of stress relaxation, the rates of cavity wall convergences decrease monotonously with time, indicating apparently strain hardening material properties. The chronical convergences of the cavity walls can be predicted by trend analysis, if the material deforms continuously, and if the boundary conditions are independent of time.

RESUME

L'objet de cette étude est l'analyse du comportement rhéologique du sel gemme aux abords d'une cavité libre ou en soutènement. Le comportement du matériau sera approché par un modèle de Maxwell généralisé, non-linéaire. Pour ce dernier, le taux de déformation visqueux s'exprime comme puissance de la différence des contraintes. On considère en particulier l'influence réciproque du fluage et de la relaxation dans un massif de sel. Le processus de déchargement en contraintes du massif se laisse approcher de façon réaliste par une loi non-linéaire visco-plastique. On constate que, si la profondeur de la cavité augmente, les contraintes tangentielles en coordonnées cylindriques suivent un chemin non monotone. Celui-ci ressemble d'une manière frappante à la formation de voûtes dans les calculs élasto-plastiques. Le critère d'écoulement, qui y est utilisé, est remplacé ici par une condition sur la contrainte en fonction du temps. La différence des contraintes, qui peut être absorbée au maximum, est d'autant plus grande, que la vitesse de fluage est importante. Comme conséquence de la relaxation en contraintes, il résulte que la convergence de la cavité décroit avec le temps, de telle sorte qu'on simule un écrouissage à froid du sel. La convergence, avec le temps, des cavités peut être prévue par une analyse de tendance, sous l'hypothèse d'un comportement continu du matériau et pour des conditions limites constantes.

1. EINFÜHRUNG

Zur numerischen Analyse der rheologischen Deformationen und Spannungsumlagerungen eignet sich die Methode der finiten Elemente, die sich durch große Allgemeinheit auszeichnet. Der betrachtete Bereich des als kontinuierlich angenommenen Gebirges wird in eine endliche Anzahl endlich kleiner Teilbereiche zerlegt, die man finite Elemente nennt. Fig. 1 zeigt z. B. ein finites Ring-Element.

Die Feinteilung der Finite Element Struktur richtet sich nach den Gradienten des zu erwartenden Geschwindigkeitsfeldes. Unter der Annahme, daß die Feldfunktionen innerhalb der einzelnen Elemente hinreichend genau durch lineare oder polynome Ansätze interpoliert werden können, genügt die Kenntnis der Knotenpunktgeschwindigkeiten im Element-Innern, um hieraus die Element-Dehnungs- und Spannungsraten zu approximieren.

Nichtlineare Stoffgesetze lassen sich bei der Methode der finiten Elemente dadurch berücksichtigen, daß man die Berechnung in inkrementellen, kleinen Lastschritten mit abschnittsweise linearisierten Stoffgesetzen durchführt. Bei der Methode der Anfangslasten z. B. werden die nichtelastischen Geschwindigkeitsanteile mittels fiktiver Ersatzkräfte auf der Basis eines ideal elastischen Stoffgesetzes ermittelt.

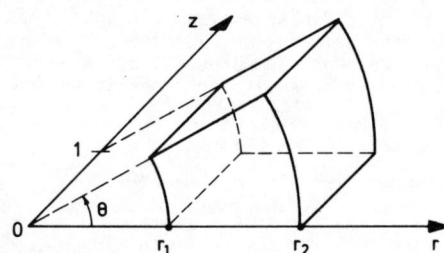

Fig. 1 Finites Ring-Element

2. BERECHNUNGSMODELL

Als Berechnungsmodell werde z. B. ein horizontaler Schnitt durch ein Bohrloch betrachtet (Fig. 2). Das umgebende Salzgebirge wird mit eindimensionalen Lamé'schen Ring-Elementen unter Annahme eines ebenen Formänderungszustandes simuliert. Am Außenrand der Scheibenstruktur wirkt die (als isotrop angenommene) primäre Gebirgsspannung σ_h, die sich im einfachsten Fall aus der Gesteinswichte γ und der Teufe h berechnet. Das Randwertproblem ist mit einem programmierbaren Taschenrechner lösbar.

Die rheologische Strukturbewegungsgleichung lautet für das gezeigte Modell

$$\underline{K} \int_{t_o}^{t_a} \underline{v} \, dt = \int_{t_o}^{t_a} \underline{\dot{F}} \, dt \qquad (1)$$

Darin sind

$\underline{K}$ die Struktursteifigkeitsmatrix,

$\underline{v}$ der Knotenpunktgeschwindigkeitsvektor und

$\underline{\dot{F}}$ der Knotenpunktlastratenvektor,

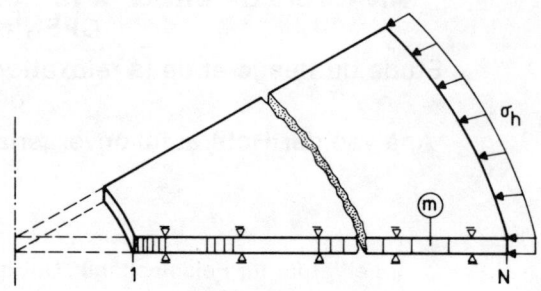

Fig. 2 Finitisierte Scheibenstruktur für das Hohlraummodell im isotropen Gebirgsspannungsfeld

die sich zusammensetzen aus

$$\underline{K} = \sum_{m=1}^{N-1} \underline{K}^{(m)} \qquad (2a)$$

$$\underline{v} = (v_1, v_2, \ldots, v_N)^T \qquad (2b)$$

$$\underline{\dot{F}} = (\dot{F}_1, \dot{F}_2, \ldots, \dot{F}_N)^T \qquad (2c)$$

(N = Anzahl der Knotenpunkte)

Die elastischen Elementsteifigkeiten $\underline{K}^{(m)}$ sind für das Lamé'sche Ring-Element gegeben durch

$$\underline{K}^{(m)} = \frac{2G}{1-2\nu} \frac{1}{1-x^2} \begin{bmatrix} 1-2\nu+x^2 & 2\nu x-2x \\ 2\nu x-2x & 1-2\nu x^2+x^2 \end{bmatrix} \qquad (3)$$

mit

$$x \doteq r_1/r_2 \qquad (4)$$

G = Kirchhoff'scher Schubmodul

ν = Poisson'sche Querkontraktionszahl

Die elastischen Verformungsraten $\dot{\varepsilon}_r$ und $\dot{\varepsilon}_\theta$ ermitteln sich im geometrischen Mittelpunkt $\tilde{r}$ des Ring-Elementes

$$\tilde{r} = \sqrt{r_1 r_2} \qquad (5)$$

aus den Knotenpunktgeschwindigkeiten v_1 und v_2 zu

$$\begin{bmatrix} \dot{\varepsilon}_r \\ \dot{\varepsilon}_\theta \end{bmatrix} = \frac{x}{r_1(1-x)} \begin{bmatrix} -1 & 1 \\ b & b \end{bmatrix} \begin{bmatrix} v_1 \\ v_2 \end{bmatrix} \qquad (6a)$$

mit

$$b = (1-x)/(1+x) \qquad (6b)$$

Die elastischen Spannungsraten $\underline{\dot{\sigma}}$ folgen aus den elastischen Deformationsraten über das Hooke'sche Gesetz.

3. STOFFGESETZ

Als Stoffgesetz wird der Ansatz eines nichtlinearen Maxwell Körpers angenommen, bei dem sich die inkrementellen Dehnungsraten $\underline{\dot{\varepsilon}}$ additiv aus einem elastischen Anteil $\underline{\dot{\varepsilon}}^{ela}$ und einem viskosen Anteil $\underline{\dot{\varepsilon}}^{vis}$ zusammensetzen:

$$\underline{\dot{\varepsilon}} = \begin{bmatrix} \dot{\varepsilon}_r \\ \dot{\varepsilon}_\theta \end{bmatrix} = \begin{bmatrix} \dot{\varepsilon}_r \\ \dot{\varepsilon}_\theta \end{bmatrix}^{ela} + \begin{bmatrix} \dot{\varepsilon}_r \\ \dot{\varepsilon}_\theta \end{bmatrix}^{vis} \tag{7}$$

(a) Elastische Spannungsraten

$$\underline{\dot{\sigma}}^{ela} \equiv \begin{bmatrix} \dot{\sigma}_r \\ \dot{\sigma}_\theta \end{bmatrix}^{ela} = \frac{2G}{1-2\nu} \begin{bmatrix} 1-\nu & \nu \\ \nu & 1-\nu \end{bmatrix} \begin{bmatrix} \dot{\varepsilon}_r \\ \dot{\varepsilon}_\theta \end{bmatrix}^{ela} \tag{8}$$

(b) Viskose Deformationsraten

Umfangreiche experimentelle Untersuchungen zum rheologischen Verhalten von natürlichem, polykristallinem Steinsalz sind von Heard (1979) und von Wallner (1979) durchgeführt worden. Dabei handelte es sich um Zylinderdruckversuche mit unterschiedlichen Temperaturen und veränderlichen Manteldrücken. Die Auswertungen ergaben im Beanspruchungsbereich der Polygonisierung das stationäre Kriechgesetz

$$\dot{\varepsilon}^{vis} = A \exp(-E/R_B T) \sigma^\alpha \tag{9}$$

Darin sind A, E und α Meßgrößen, R_B ist die universelle Gaskonstante, T ist die absolute Temperatur im Prüfkörper, und σ ist die Hauptspannungsdifferenz.

Beim axialsymmetrischen, ebenen Formänderungszustand ist die zu σ äquivalente Hauptdifferenz $\bar{\sigma}$ gegeben durch

$$\bar{\sigma} = |\sigma_\theta - \sigma_r| \tag{10}$$

4. RHEOLOGISCHES BERECHNUNGSVERFAHREN

Der viskose Anteil der Deformationsrate wird nach der Anfangslastmethode mithilfe fiktiver Ersatzlastraten $\underline{\dot{F}}^{vis}$ auf der Basis eines elastischen Stoffgesetzes ermittelt. Diese Ersatzlastraten berechnen sich bei einer aus jeweils ähnlichen Lamé'schen Ring-Elementen aufgebauten Modellstruktur zu

$$\underline{\dot{F}}^{vis} = \sum_{m=1}^{N-1} \underline{\dot{F}}^{vis}_{(m)} \quad \text{wobei} \tag{17a}$$

$$\underline{\dot{F}}^{vis}_{(m)} \equiv \begin{bmatrix} \dot{F}_m \\ \dot{F}_{m+1} \end{bmatrix}^{vis} = \frac{r_1}{2x^m} \begin{bmatrix} -1-x & 1-x \\ 1+x & 1-x \end{bmatrix} \begin{bmatrix} \dot{\sigma}_r^{(m)} \\ \dot{\sigma}_\theta^{(m)} \end{bmatrix}^{vis} \tag{17b}$$

mit $x = r_m/r_{m+1} = r_1/r_2$ (18)
N = Anzahl der Knotenpunkte

Symbolisch schreiben wir für Gl.(17a)

$$\underline{\dot{F}}^{vis} = \sum_{m=1}^{N-1} \underline{C}_{(m)} \underline{\dot{\sigma}}^{vis}_{(m)} \tag{17c}$$

Die viskosen Ersatzspannungsraten $\underline{\dot{\sigma}}^{vis}_{(m)}$ folgen aus den viskosen Deformationsraten $\underline{\dot{\varepsilon}}^{vis}_{(m)}$ über das elastische Gesetz

$$\underline{\dot{\sigma}}^{vis}_{(m)} = 2G \underline{\dot{\varepsilon}}^{vis}_{(m)} \tag{19}$$

Die viskosen Deformationsraten sind bei den einzelnen Ring-Elementen gegeben durch die Gln.(9) und (10)

$$\begin{bmatrix} \dot{\varepsilon}_r^{(m)} \\ \dot{\varepsilon}_\theta^{(m)} \end{bmatrix}^{vis} = \dot{\varepsilon}_o \left| \frac{\sigma_\theta^{(m)} - \sigma_r^{(m)}}{\sigma_o} \right|^\alpha \begin{bmatrix} 1 \\ -1 \end{bmatrix} \tag{20}$$

mit $\dot{\varepsilon}_o = A \exp(-E/R_B T) \sigma_o^\alpha$ (21)

σ_o = Referenzspannung

Das Ziel der rheologischen Berechnungen ist eine quantitative Beschreibung der zeitlichen Entwicklung von Verformungs- und Spannungsfeldern. Zu diesem Zweck ist die Integration der Gl.(1) nach der Zeit erforderlich. Ausgangspunkt der inkrementellen Berechnung ist ein Gleichgewichtszustand zu einer Zeit t, für den die Spannungen $\sigma(t)$, die Dehnungen $\varepsilon(t)$ und die Verschiebungen $\underline{u}(t)$ bekannt sind. Die Bestimmungsgleichung für das Inkrement des Knotenpunktverschiebungsvektors $\Delta\underline{u}$ lautet dann näherungsweise

$$\underline{K}\Delta\underline{u} = \underline{K}\,\underline{v}\,\Delta t = \Delta\underline{F} + \Delta\underline{F}^{vis} \tag{22a}$$

mit

$$\Delta\underline{F} = \underline{F}(t+\Delta t) - \underline{F}(t) \tag{22b}$$

und

$$\Delta\underline{F}^{vis} = 2G\dot{\varepsilon}_o \Delta t \sum_{m=1}^{N-1} \underline{C}_{(m)} \begin{bmatrix} 1 \\ -1 \end{bmatrix} \left| \frac{\sigma_\theta^{(m)} - \sigma_r^{(m)}}{\sigma_o} \right|^\alpha \tag{22c}$$

Die Knotenpunktverschiebungen $\underline{u}(t)$, die Elementverformungen $\underline{\varepsilon}(t)$ und -spannungen $\underline{\sigma}(t)$ folgen durch numerische Integration der Knotenpunktgeschwindigkeiten $\underline{v}$, der Elementverformungsraten $\underline{\dot{\varepsilon}}$ und -spannungsraten $\underline{\dot{\sigma}}$ nach der Zeit t.

5. KRIECHKONVERGENZEN UND GEBIRGSRELAXATION

Die numerische Berechnung verschiedener Modelle von Bohrlöchern, Schächten und Tunneln im tiefen Salinar ergibt für die zeitabhängigen Stoßkonvergenzen und -entspannungen ein einheitliches Bild bei beliebiger Wahl der Ausbruchsradien R, des Schubmoduls G, der Gebirgstemperatur T, der primären Gebirgsspannung σ_h und der Referenzkriechrate $\dot{\varepsilon}_o$. Der Exponent in Gl.(9) sei durch $\alpha = 5.0$ festgelegt (Wallner et al., 1979), und die Referenzspannung σ_o sei gleich der primären Gebirgsspannung σ_h gewählt. Ferner führen wir die folgenden dimensionsfreien Darstellungen der Zeit t, der Stoßkonvergenz u(t) und der tangentialen Gebirgsspannung $\sigma_\theta(t)$ am Ausbruchsrand $r = R$ ein (Rolnik, 1979):

$$t^* = 8(G/\sigma_o)\dot{e}_o t \qquad (23)$$

$$u^* = 2(G/\sigma_o)(u/R) - 1 \qquad (24)$$

$$\sigma_\theta^* = \sigma_\theta/\sigma_o \qquad (25)$$

Dann läßt sich die Stoßkonvergenz u^* in Abhängigkeit von der Zeit t^* mit guter Näherung durch die Parabel

$$u^* = \tfrac{1}{2}\sqrt{t^*} \qquad (26)$$

approximieren. Die Konvergenzgeschwindigkeit nimmt monoton mit der Zeit ab, erreicht aber auch nach langer Zeit t^* nicht den Wert Null (Fig. 3)

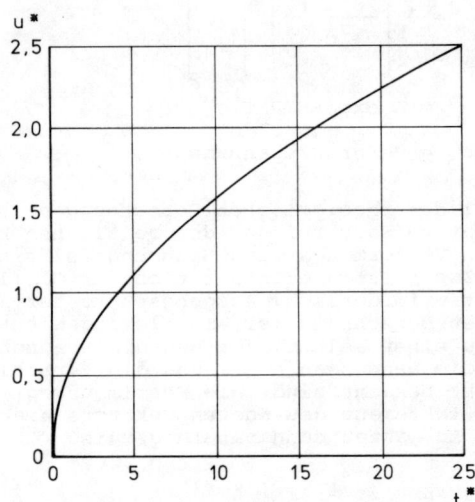

Fig. 3 Zeitabhängige Stoßkonvergenz beim zylindrischen Hohlraum im Salinar

Die monotone Abnahme der Konvergenzrate mit der Zeit hat ihre Ursache in der Relaxation der deviatorischen Gebirgsspannungen (Fig. 4), wobei Kriechen und Relaxation sich wechselseitig beeinflussen. Die Relaxation, die beim unausgebauten Hohlraum vor allem die Umfangsspannungen σ_θ^* betrifft, endet mit dem Verschwinden der Deviatorspannungen, also mit dem Einstellen eines isotropen Spannungszustandes in der Hohlraumumgebung. Dieses ist i. a. mit dem vollständigen Schließen des Hohlraumes nach langer Standzeit verbunden.

Die auf σ_o bezogenen tangentialen und radialen Gebirgsspannungen des betrachteten Modelles sind in Abhängigkeit vom bezogenen radialen Abstand r/R für verschiedene Zeitstufen (aufsteigende Reihenfolge der Numerierung) in Fig. 5 dargestellt. Kurvenparameter ist die Zeit t^*. Die Spannungsrelaxation findet in einem zeitabhängigen Gradientenfeld des mittleren Druckes statt.

Die tangentialen Gebirgsspannungen folgen mit wachsendem Abstand vom Bohrlochrand einem nichtmonotonen Verlauf. Es stellt sich eine auffallende Ähnlichkeit zu der bekannten Schutzhüllenausbildung in elasto-plastischen Berechnungen heraus. Die dort verwendete Elastizitätsgrenze ist hier durch die geschwindigkeitsabhängige Spannungsbedingung Gl.(9) ersetzt worden. Die maximal zulässige Differenzspannung σ ist umso größer, je höher die Kriechgeschwindigkeit $\dot{e}$ ist. Das Ergebnis entspricht den experimentellen Beobachtungen an natürlichen Salzgesteinen und läßt sich sinngemäß auch auf Hohlräume allgemeinerer Geometrie im Salinar übertragen (Borm, 1980).

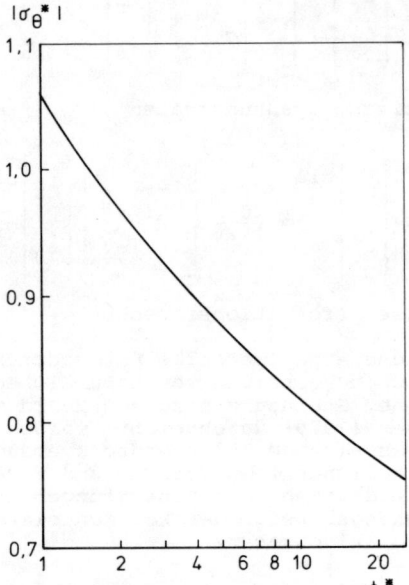

Fig. 4 Relaxation der tangentialen Gebirgsspannungen am Ausbruchsrand

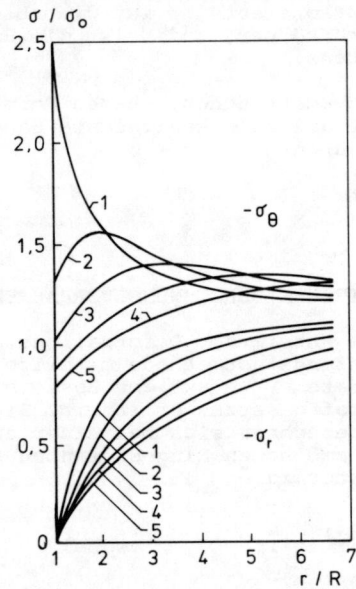

Fig. 5 Bezogene tangentiale und radiale Spannungen in der Hohlraumumgebung in Abhängigkeit von der bezogenen Stoßtiefe r/R und vom Zeitparameter t^* (Tab. 1)

Tab. 1: t^*-Werte in Fig. 5

Kurve Nr.	1	2	3	4	5
t^* =	0	0.2	2	10	30

6. SCHLUSSBEMERKUNGEN

Die vorstehenden Untersuchungen wurden für inkompressibles, bruchloses Materialverhalten und für isotherme Umgebungstemperaturen durchgeführt. Temperaturänderungen lassen sich in einfacher Weise berücksichtigen, doch erfordert Wärmetransport eine besondere Behandlung.

Die Parameter für das nichtlinear viskose Kriechgesetz von Wallner et al. (1979) wurden an relativ kleinen Laborproben ermittelt und entsprechen nicht unbedingt den Gebirgsparametern in situ, wie auch u. a. die Konvergenzmessungen von Dreyer (1974) an Laborprüfkörpern und an Großbohrlöchern gezeigt haben. Bei Kenntnis des zeitlichen Konvergenzverlaufes in situ für ein gewisses Zeitintervall ist eine Modifikation der Stoff-Eingangsparameter in Gl.(9) sinnvoll.

Das scheinbare Verfestigungsverhalten des Salzgebirges, das sich durch die zeitlich monoton abnehmenden Konvergenzraten andeutet, ist eine Folge der Spannungsrelaxation. Wirkliches Dehnungs-Verfestigungsverhalten ist für Steinsalz bisher nicht bekannt. Aus diesem Grunde kommt den Relaxationsuntersuchungen an Steinsalzproben eine ganz besondere Bedeutung zu.

Herrn Prof. Dr.-Ing. O. Natau, Herrn Dipl.-Math. H. Groß und Herrn Dr.-Ing. H.-B. Mühlhaus danke ich für zahlreiche klärende Diskussionen und der Deutschen Forschungsgemeinschaft (DFG) für die finanzielle Förderung der numerischen Untersuchungen.

7. LITERATUR

Borm, G. (1980). Zur Analyse chronischer Gebirgsdeformationen beim Felshohlraumbau. Veröff. Inst. f. Bodenmech. u. Felsmech., Nr. 88, Universität Karlsruhe.

Dreyer, W. (1974). Gebirgsmechanik im Salz. Enke Verlag, Stuttgart.

Heard, H. C. (1972). Steady state flow in polycrystalline halite at pressures of 2 kilobars. In: Flow and fracture of rocks, Amer. Geophys. Union, Washington D. C.

Lurje, A. T. (1963). Räumliche Probleme der Elastizitätstheorie. Akademie Verlag, Berlin.

Rolnik, H. (1982). Evolution dans le temps de parois d'une cavité cylindrique dans un massif de sel gemme. Vertieferarbeit, Lehrstuhl für Felsmechanik, Universität Karlsruhe.

Wallner, M., C. Caninenberg und H. Gonther (1979) Ermittlung zeit- und temperaturabhängiger mechanischer Kennwerte von Salzgesteinen. Proc. 4. Congr. ISRM, Montreux, Vol. 1, 313- 318.

Address of the author:

Dr. rer. nat. habil. G. Borm
Lehrstuhl für Felsmechanik
Universität Karlsruhe
7500 Karlsruhe 1
Federal Republic of Germany

STRESS MEASUREMENT AND ANALYSIS FOR MINE PLANNING
Mesures de tension et analyse en projets miniers
Spannungsmessungen und Analyse für den Entwurf von Bergwerken

S. Bywater
Rock Mechanics Research Engineer

R. Cowling
Senior Research Engineer

B. N. Black
Senior Mining Engineer — Planning

(All from Mount Isa Mines Limited, Queensland, Australia)

SYNOPSIS

During open stoping between 13 and 15 levels of the Lead area at Mount Isa Mines, ground behaviour has been extensively monitored, including regular stress measurement. In parallel, computer based stress analysis techniques have been developed to permit modelling of stress redistribution resulting from the excavation of many stopes in multiple, tabular orebodies. One computer program, based on the displacement discontinuity method, has been validated by back analyses and parameter studies. It is now being used to examine the design and scheduling of open stopes in current and future mining areas.

RESUME

Lors de l'exploitation par piliers abandonnés entre les étages 13 et 15 de la zone à plomb des mines de Mount Isa, on a réalisé une étude poussée des terrains en y incorporant des mesures régulières des contraintes. En même temps, on a développé des techniques d'analyse de contrainte sur ordinateur afin d'en réaliser un modèle de rédistribution. La rédistribution des contraintes résulte de l'excavation de nombreux chantiers d'abattage dans de multiples amas minéralisés tabulaires. Un programme ordinateur basé sur la méthode des déformations discontinues a été confirmé par analyse ultérieure du comportement des paramètres. Ce programme est actuellement utilisé pour étudier la conception et l'exploitation par piliers abandonnés dans les zones minières déjà en exploitation ou qui le seront dans l'avenir.

ZUSAMMENFASSUNG

Bei dem Kammerabbau des Bleivorkommens zwischen Etagen 13 und 15 im Bergwerk Mount Isa wurde das Verhalten des Bodens sorgfältig überwacht, einschließlich laufenden Spannungsmessungen. Zur gleichen Zeit wurden elektronische Rechenmethoden zur Spannungsanalyse entwickelt, um die Spannungsumlagerung, die vom Ausbruch vieler Hohlräume in einer Anzahl von tafelförmigen Erzkörpern resultiert, voraussagen zu können. Ein elektronisches Rechenprogramm, das auf der Methode der Verformungs-Diskontinuität beruht, wurde durch Rückrechnung und Parameterstudien überprüft. Es wird jetzt dazu verwendet, die Bemessung und den Entwurf von versatzlosen Abbaufirsten in derzeitigen und zukünftigen Ausbaugebieten zu prüfen.

INTRODUCTION

Open stoping, with associated pillar recovery, in several steep-dipping, close-spaced, tabular orebodies accounts for between 60% and 70% of Mount Isa Mines annual lead-zinc-silver ore production of 3.3 million tonnes. At any time, as many as twelve stopes and pillars are in various stages of development, production or filling. Stopes are typically 30 m x 25 m in plan and 100 m high. At present the lower level for the majority of mining is 15 level, 730 m below surface. Over the next five years production from 19 level, 960 m below surface, will become increasingly important. In some areas the open stopes are planned to be 150 m to 200 m high.

The remaining annual ore production is achieved by cut-and-fill methods, either from narrow extensions of, or from separate orebodies adjacent and parallel to, the open-stoped areas.

Of major importance in successful attainment of annual production targets is the scheduling of the numerous open and cut-and-fill stopes. Metal grades vary within and between orebodies which imposes constraints on the order in which stopes can be extracted. In addition pre-mining, major principal stresses between 15 level and 19 level are from 30 to 35 MN/m^2. A viable, successful schedule is one which achieves metal targets whilst accounting for the complex stress redistributions created by concurrent extraction of multiple stopes in several orebodies.

Detailed descriptions of mining methods in the Lead orebodies have been prepared by Goddard (1977), (1981).

This paper describes rock mechanics input to the examination of previous and present stoping areas, with the objective of predicting ground behaviour critical to mining operations in future areas. Particularly important is the interaction between the Mine Planning and Rock Mechanics functions. Part of the role of the latter is the testing and validation of a psuedo-three-dimensional stress analysis program and numerous stress determinations using the CSIRO triaxial cell, Worotniki, (1976). Their role in the examination of open stoping sequences forms the main part of the paper.

GEOLOGICAL SETTING

Only a brief summary of the geology of Isa Mine is presented here. Detailed description is provided by Mathias, (1976). The majority of lead-zinc-silver orebodies, hereafter referred to as lead orebodies, lie mainly to the North of a central shaft complex. The Urquhart Shale Formation contains all known economic mineralisation and consists of three major subdivisions:

- massive, recrystalised dolomitic and siliceous breccia, known locally as "silica-dolomite", and host to the copper ore,

- well bedded dolomitic, pyritic and tuffaceous shale, which is host to the lead ore, and

- barren, dolomitic, tuffaceous and carbonaceous shale and silstone.

Bedding in the shale and enclosed lenses of silica-dolomite strikes North-South and dips 65° to the West. Lead orebodies are disposed in a roughly en-echelon pattern down-dip and along strike and commonly interfinger with the silica-dolomite at their lower and southern extremities, Figures 1 and 2. They

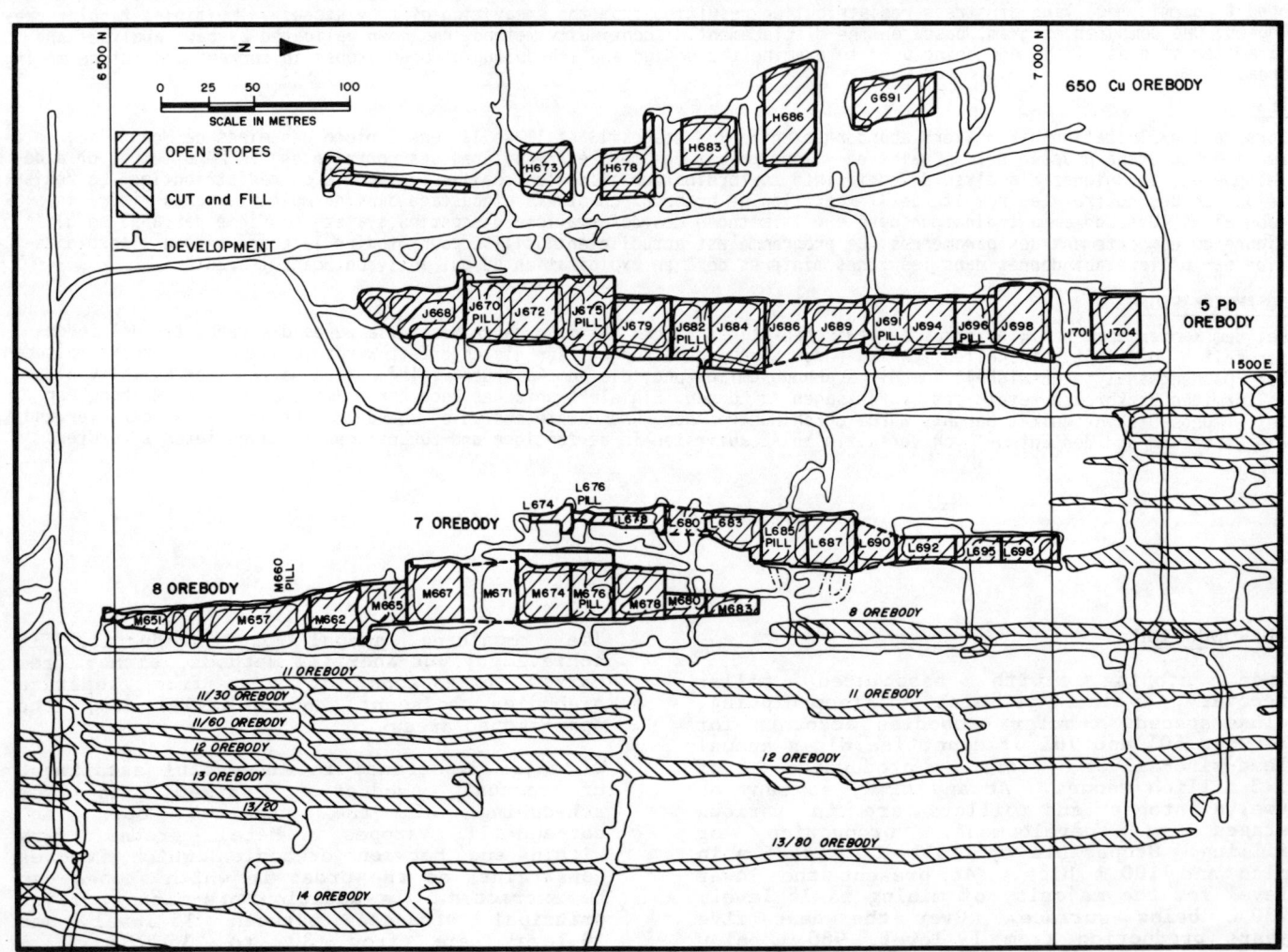

FIG. 1 - PLAN OF NORTHERN PART OF MOUNT ISA MINES

are naturally divided into two groups: Black Star orebodies to the hangingwall (West) and Racecourse orebodies to the footwall (East).

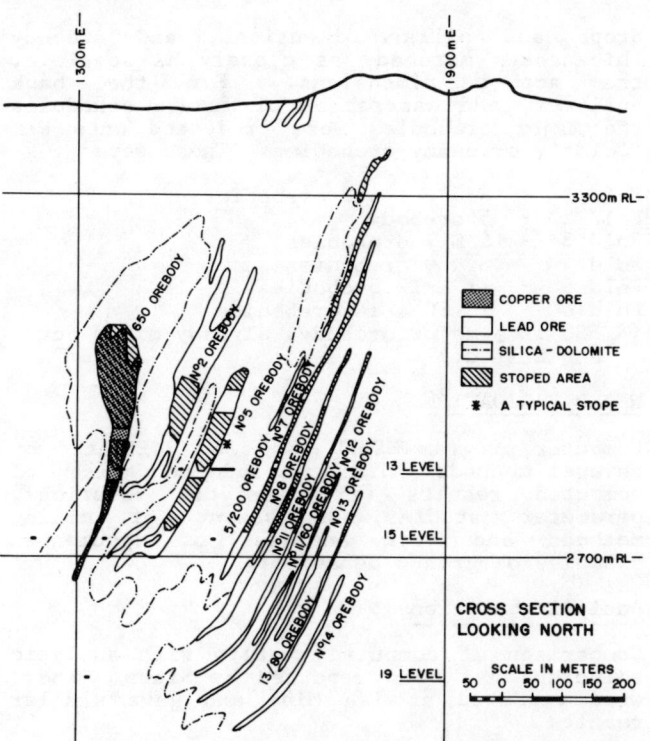

FIG. 2 - CROSS SECTION THROUGH LEAD OREBODIES

The most important geological features in the shales are:

- bedding plane breaks, defined by a thin film of carbonaceous material,
- fractures striking North-South, and dipping about 60° East,
- two sets of orthogonal extension fractures with orientations normal to bedding, and
- conjugate shear fractures.

MINING METHODS

Generally, orebodies wider than 10 m are extracted by open stoping and those of less than 10 m by cut-and-fill methods. An open stope is created by enlarging a cut-off raise, at the side of the stope, by sequential blasting of drill-holes. The slot so created is used as an expansion void into which the remainder of the stope is blasted, either in single or multiple vertical slices, Goddard, (1981). Typical open stopes are full width of the orebody, 30 m along strike, and 100 m high and contain up to 220 000 t. Depending on the sequence in a particular orebody, pillars containing up to 150 000 tonnes and 20 m along strike are massfired into adjacent, empty stopes. The resulting overhanging inclined wall, or hangingwall, can be as much as 50 m along strike. Naturally, these massfires cause rapid redistribution of stress which can result in one or a combination of the following:

- damage to nearby development and pillars, and
- initiation of significant hangingwall collapse.

For a discussion on hangingwall behaviour and modelling see Beer, (1983).

Open stoping has been extensively used for extracting the Black Star orebodies, due to their greater width and more competent ore and hangingwall. The major Racecourse orebodies that have been partially recovered by open-stoping are 7 and 8 orebodies. This has been aided by the presence of silica-dolomite within the hangingwall of some of the stopes.

Originally, minimal permanent pillars were left between stoping compartments. However, the current objective is to extract 100% of the ore along strike. Semi-permanent pillars are designed to be large enough to enable their safe recovery between stopes filled with cemented hydraulic fill (Leahy, 1978). All stopes are filled after completion of extraction, the type of fill being dependent on the subsequent mining sequence.

MINE PLANNING AND ROCK MECHANICS

The bases of mine plans are two and five year production schedules. Although not explicitly defined on the schedules, they have to take account of such varied factors as tonnes and grade, ventilation, equipment availability, filling and expected ground behaviour. These factors are assessed by discussion with the various specialist service groups such as Geology, Ventilation and Rock Mechanics.

One of the overall objectives of the Isa Mine rock mechanics programme is to develop models (empirical, physical or numerical), which account for the major components of observed ground behaviour. These models are then to be used to assess the likely ground behaviour arising from future mining strategies. Today, physical models are no longer used, and, whilst empirical models are still the major contribution in day-to-day operations, much effort has gone into investigating, developing and applying appropriate stress analyses techniques.

Development of empirical models, and the data for validating numerical models, is founded on rock mechanics monitoring programmes in various mining areas. A typical rock mechanics programme includes the following instruments and techniques:

- rod and wire multi-point extensometers,
- re-surveys of development,
- absolute stress measurements using the CSIRO triaxial cell,

- stress change monitoring in pillars and abutment areas using the CSIRO triaxial cell, and more recently IRAD vibrating wire stressmeters, and

- observation, with photographs. (Mosaics of photographs are made at various stages of extraction, and have proved invaluable in identifying overbreak in, and collapse around, stopes.)

The above programmes are supported by mine site laboratories, which perform tests on rock, fill and ground support, and computer facilities for recording and analysing data.

Stress measurement is an essential component of the rock mechanics programmes. In excess of 40 absolute stress determinations have been made in the northern section of the mine over the last five years. Results have proven to be reliable, in themselves, and correlate well with observed and computed behaviour.

Fourteen absolute stress measurements have been used to determine the virgin stress profile. The results indicate a linear relationship between σ_2 and σ_3 and depth below surface. It is uncertain whether σ_1 has a bi-linear or curvilinear variation with depth, but its orientation is normal to bedding, Alexander, (1981).

STRESS ANALYSES

Modelling of multiple excavations in several tabular orebodies with the combined objectives of realistic input and output, and acceptable preparation and run times proved intractable until a psuedo-three-dimensional program, NFOLD, based on the displacement discontinuity method, was developed, Sinha (1979).

The program has as input the geometry of the various orebodies, elastic properties of the orebodies and intervening rock medium, and stress field components, (σx, σy, σz, τxy, τyz, τzx) as a function of depth. Output consists of plots (either line printer or drum plotter) of normal stress, dip and strike shear stress, and convergence, dip and strike ride displacements in the planes of the orebodies. Post processing programs are also used to determine principal stresses and absolute displacements at points off the planes of the orebodies, and to carry out detailed re-analysis in smaller areas of the original problem.

In the analyses which are reported in the remainder of this paper, the following inputs have been used:

Orebody and intervening rock -

Modulus   80 GN/m^2
Poisson's ratio 0.2

Stress field -

σx = $(10.0 + .026z)$MN/m^2 (East-West)
σy = $(7.0 + 0.012z)$MN/m^2 (North-South)
σz = $(0.0 + 0.028z)$MN/m^2 (Vertical)
τzx = $(-6.0 + 0.001z)$MN/m^2

where z = depth in metres below surface.

Stope and pillar dimensions, and orebody thicknesses matched, as closely as possible, the actual dimensions. For the back analyses, and assessment of future schedules the major orebodies were projected onto six 'folds', or dummy orebodies. These were:

Fold 1 - 650*, 1 & 2 orebodies
Fold 2 - 5 orebody
Fold 3 - 6 & 7 orebodies
Fold 4 - 8 & 9 orebodies
Fold 5 - 11 & 12 orebodies
Fold 6 - 13/80 & 14 orebodies
(* 650 is a major orebody, already mined out and filled.)

NFOLD VALIDATION

Computer program NFOLD has been validated by several methods. These include comparison of computed results with analytic solutions, parameter studies, assessment of mining methods and back analyses of documented examples of ground behaviour.

Analytic Solution

Comparison of computed results with analytic solutions has been reported by Sinha. These were repeated at Isa Mine and gave similar results.

Parameter Studies

As a final step to assessing the applicability of the technique for the examination of mining geometries, numerous parameter studies were carried out. Figure 3 shows the results of two of the studies. These results could also have been obtained by use of three-dimensional boundary element or finite element techniques. However, to expand the exercise to multiple, parallel orebodies, with these techniques, would be a prohibitive exercise both in terms of preparation and computer costs.

The above discussion has demonstrated that it is possible to model simplistic mining geometries and obtain realistic results. To be useful as a tool with which to examine real mining geometries, the next stage of the validation was to carry out a series of back analyses. In these analyses an attempt was made to model all of the excavations in the lead area. Figure 4 depicts the geometries of 5, 7 and 8 orebodies. Similar geometries had to be prepared for the other orebodies.

BACK ANALYSES

Numerous back analyses have been undertaken at key stages of extraction in several orebodies. These analyses have shown good correlation with ground behaviour measured and observed underground. Two examples are presented below.

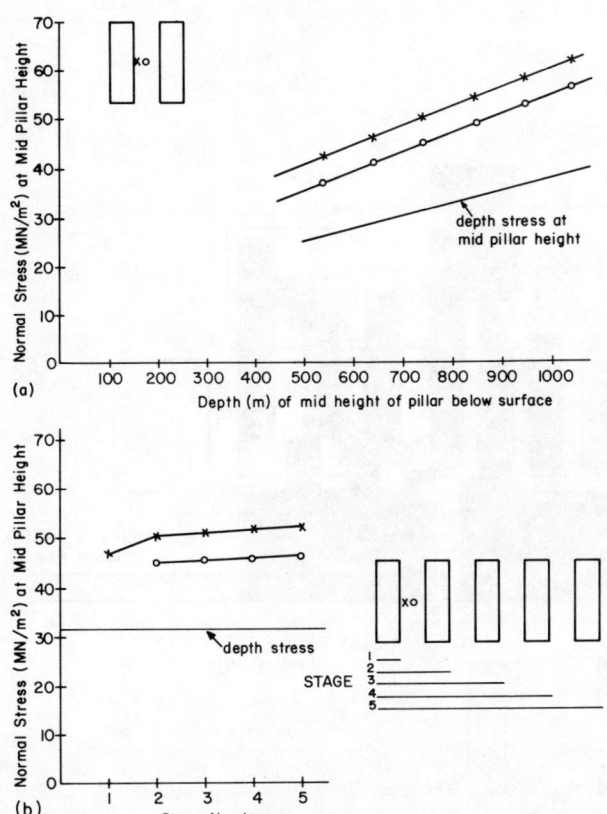

FIG. 3 - PARAMETER STUDIES

L685 Triplet, 7 Orebody (Figure 5)

This part of 7 orebody was extracted by the 'triplet' method of open stoping. The method is based on a compartment consisting of two stopes and an intervening pillar. One stope is extracted and filled with cemented fill, the second stope extracted and the pillar massblasted into the void created by the second stope. When this ore is extracted the remaining void is also filled with cemented fill. In this case L683 and L687 are the stopes, and L685 is the pillar. L680 and L690, respectively to the North and South, are semi-permanent pillars (Figure 5a).

Two analyses were undertaken: the first was the geometry after the extraction of L683 stope and the second after the massfire of L685 pillar. (All geometry changes in other orebodies were also incorporated in the analyses.)

Stage 1, (Figure 5b) after the extraction of L683 stope, indicates normal stress levels greater than 70 MN/m² in the crown pillar region of the triplet to the North (L692-5-8). These high stress levels were confirmed by absolute stress measurements and observations on 13 level prior to and during the extraction of L695 triplet. Crown pillar failures occurred in both L692 and L698 stopes due to high stresses as stoping commenced. Absolute stress measurements on 13 and 14 levels in L695 pillar, prior to the extraction of L695 triplet are given below.

	Principal Stress MN/m²		Dip/Dip Direction
13 Level	σ_1	70	25°/098°
	σ_2	21	50°/335°
	σ_3	18	30°/203°
14 Level	σ_1	32	26°/110°
	σ_2	27	64°/285°
	σ_3	18	02°/019°

The massfire of L685 pillar resulted in severe damage to development on 13 level and failure of L685-7 crown pillar, leading to fill dilution from previous cut-and-fill stopes above. Severe spalling in development and cracking occurred in L690 pillar on 13 level. Large cracks developed in the hangingwall of L685 and L687. Rock noise was audible for up to three weeks after the massfire. Damage to 14C sub level was less severe, generally evident as spalling in development. Damage to 14 level was negligible.

Northern 5 Orebody (Figure 6)

The 'triplet' method was also adopted for the extraction of 5 orebody between 13 and 15 levels. J686 and J696 were designed as semi-permanent pillars, but because of the success of the method in the area and favourable ground conditions, a decision was made to fire J691 into J694, and when these were empty, to then fire J696 into the void. The extensions of J691 and J696 above 13 level were retained to maintain hangingwall stability and prevent failure through to the stopes above. J704 has now been extracted to 13 level. J701 pillar will be fired into the J701 void, followed by their extensions above 13 level.

NFOLD analyses are presented (Figure 6b-6e) at key stages in the extraction of northern 5 orebody. The stages are:

1. prior to J691, 2. prior to J696,
3. prior to J704, 4. post J704.

Stage 1

An absolute stress determination was made on 14 level in J691 pillar immediately after the extraction of J694 slot.

	Principal Stress MN/m²	Dip/Dip Direction
σ_1	29	17°/088°
σ_2	23	71°/243°
σ_3	8	07°/000°

This result cannot be directly compared with Figure 6b, as the slot accounted for only 5 m (or one eighth) of the stope strike length indicated in the figure. Prior to slot extraction NFOLD analyses predicted normal stress levels of mid to high thirties in the area of the stress measurement. What the analyses do clearly indicate is the influence

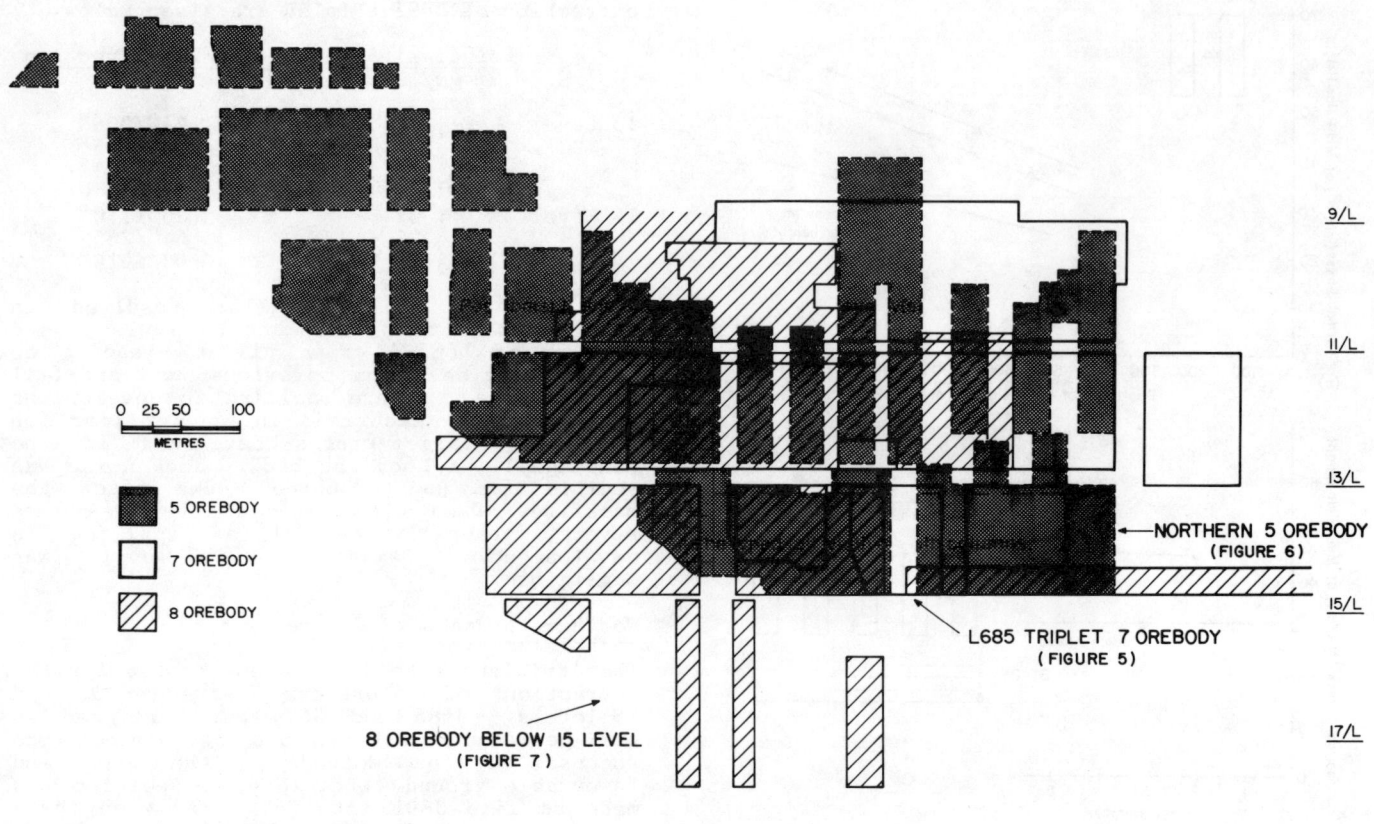

FIG. 4 - LONG SECTION OF 5, 7 AND 8 OREBODIES

of the excavations in the 650 copper orebody, in the hangingwall (Figure 2). The upper part of J691, above 14 level, is in a stress shadow, whilst below 14 level there is a stress concentration created by the stresses passing under the 650 orebody.

When J694 was fully extracted and prior to J691 extraction, spalling from side walls was apparent in crosscuts on 14 level through J691 and J696 pillars. On 13 level and 14C sublevel there was no evidence of spalling.

Stage 2

Following massfiring of J691 into J694 (Figure 6c), absolute stress measurements were made in J696 pillar on 13 and 14 levels. The 14 level results, below, indicated increased, and high, stress levels.

Principal Stress MN/m^2	Dip/Dip Direction
σ_1 68	39°/093°
σ_2 17	48°/299°
σ_3 10	13°/194°

The resolved normal component is about 65 MN/m^2 and is in good agreement with the computed value at that location. Results of the absolute stress measurements on 13 level were inconsistent with respect to principal stress directions, but were consistently low in magnitude, thus confirming the influence of the 650 orebody. Spalling continued on 14 level crosscut through J696 pillar.

Stage 3

J696 was massfired into the J691-4 void (Figure 6d). NFOLD analyses predicted a significant increase in the normal stesses at J701, which now formed the extreme abutment of northern 5 orebody. This was evidenced underground as spalling in the crosscut through J701 on 14 level. Development through the pillar on the levels above showed little change, again supported by the low stress levels predicted by the analyses.

Stage 4

J704 stope was extracted to 13 level (Figure 6e), with the cut-off adjacent to J701. Significant damage, (extensive

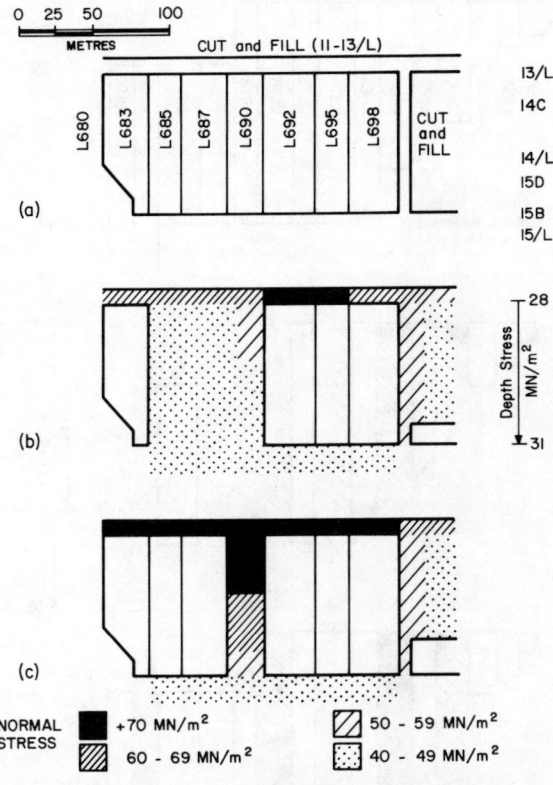

FIG. 5 - L685 TRIPLET, 7 OREBODY

slabbing and spalling, from the side walls) resulted to the crosscut through J701 on 14 level. Minor damage was also evident in the northern abutment of J704 on 14 level.

Throughout the above described extraction sequence J686 semi-permanent pillar showed significant evidence of high stress. As there is no development through this pillar on 14 level or 14C sublevel it was not possible to confirm this by stress measurement. However, buckling and high loads on rockbolt plates were apparent in the footwall access drive on both levels.

PLANNING GUIDELINES

Previous numerical and empirical studies (Brady, 1977) concluded that the mass strength of the shales was between 90 and 95 MN/m^2. Following an approach similar to that reported by Herget, (1976), size effect assessment of laboratory tests of rock cores indicated a lower limit for rock mass strength of between 60 and 80 MN/m^2.

Assessment of the results of back analyses provides further guidelines with which to examine the results of analyses of alternative mining schedules. Areas experiencing predicted normal stresses of 70 to 80 MN/m^2, generally exhibit spalling and re-adjustment on structures. Where stresses exceed 80 MN/m^2, and conditions permit (i.e. the stope below, or to the side of the highly stressed rock, is empty), local and regional failures, resulting in collapse

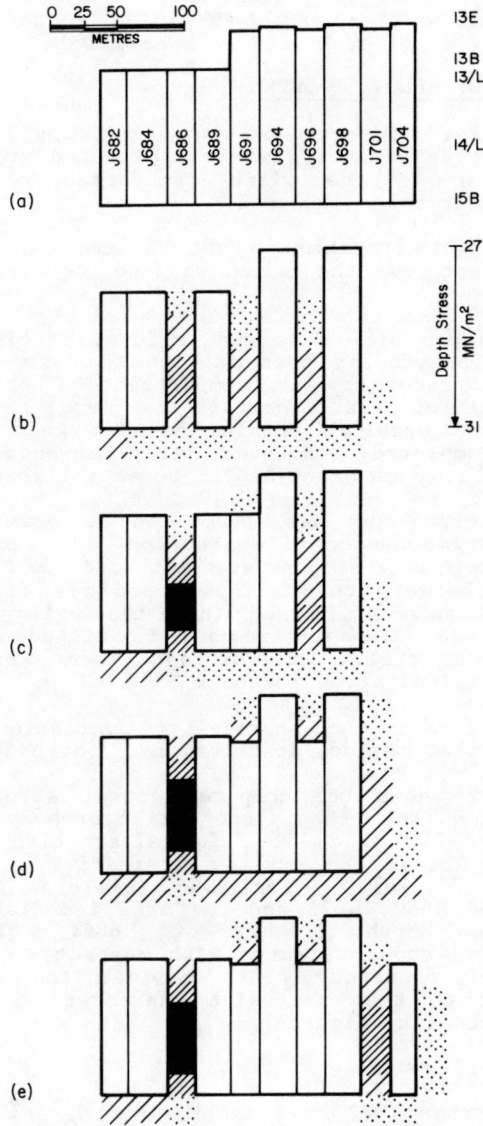

FIG. 6 - NORTHERN 5 OREBODY

have occurred. While the absolute magnitude of the normal stress is important, sudden large increases (greater than 10 MN/m^2) can also result in adverse ground conditions at stress levels less than 70 MN/m^2. Other important effects include the influence of stress shadows in orebodies which are close to each other, and stress concentrations projected from the plane of one orebody onto the plane of other orebodies. These latter two effects appear to be restricted to distances within 80 to 100 m.

With the guidelines established from the above considerations, NFOLD was used to examine alternative sequences of stoping in the Lead orebodies for the Years 1981/82 to 1984/85. Two alternative sequences were analysed, with the mining geometry at

December 1980 as the datum. For both alternatives the expected mining geometries at December of each year were analysed.

8 OREBODY BELOW 15 LEVEL (Figure 7)

Of major importance in examining the alternative schedules was the expected ground behaviour in the plane of 8 orebody, in particular below 15 level.

The first extraction below 15 level in the North end of the mine was to come from 8 orebody.

For reasons of access and regional stability it had previously been decided to carry out initial recovery in 'compartments' between regional pillars. The pillars were to be recovered once all mining between them had been completed. However, the sequence of stoping in each panel had to be established.

Alternative one was based on a primary/secondary sequence of extraction. In such a sequence, two stopes are extracted with a pillar between them. The stopes are filled with cemented fill and then the pillar is recovered. The pillar void is filled with uncemented fill. The Portland Cement saving for a typical stope is about $300 000-00.

Alternative two was basically a variation of the triplet method, described for 7 orebody.

Figure 7 shows the computed normal stresses predicted for the plane of 8 orebody at December 1982 and 1983, for alternative one (Figure 7b, 7c), and alternative two (Figure 7d, 7e). Regional pillars are labelled K660, K675 and the area immediately to the North (right) of K688. Their locations appear strange with respect to the size of 8 orebody, but are situated with respect to the en-echelon character of all the orebodies, Figure 1.

Alternative One

By December 1982 only small, local areas are predicted to have normal stresses in excess of 70 MN/m^2 (Figure 7b). By December 1983 (Figure 7c), K665 and K686 pillars, and K675 regional pillar have been isolated and all are predicted to experience significant areas in excess of 70 MN/m^2. As the pillars (K665 and K686) would be isolated to the North and South by cemented fill, the recovery of both would require that a cut-off slot be developed in a very highly stressed region.

Alternative Two

Stress levels at December 1982 are similar to those of alternative one, at the same stage. The stress situation by December 1983 is, however, much more promising. K678 stope has the potential for being recovered by use of a side slot against K680, as has K688 to the North of the same panel. Recovery of K663 could perhaps best be achieved by including part of it in K665 stope, and leaving the remainder with the K660 regional pillar for later recovery.

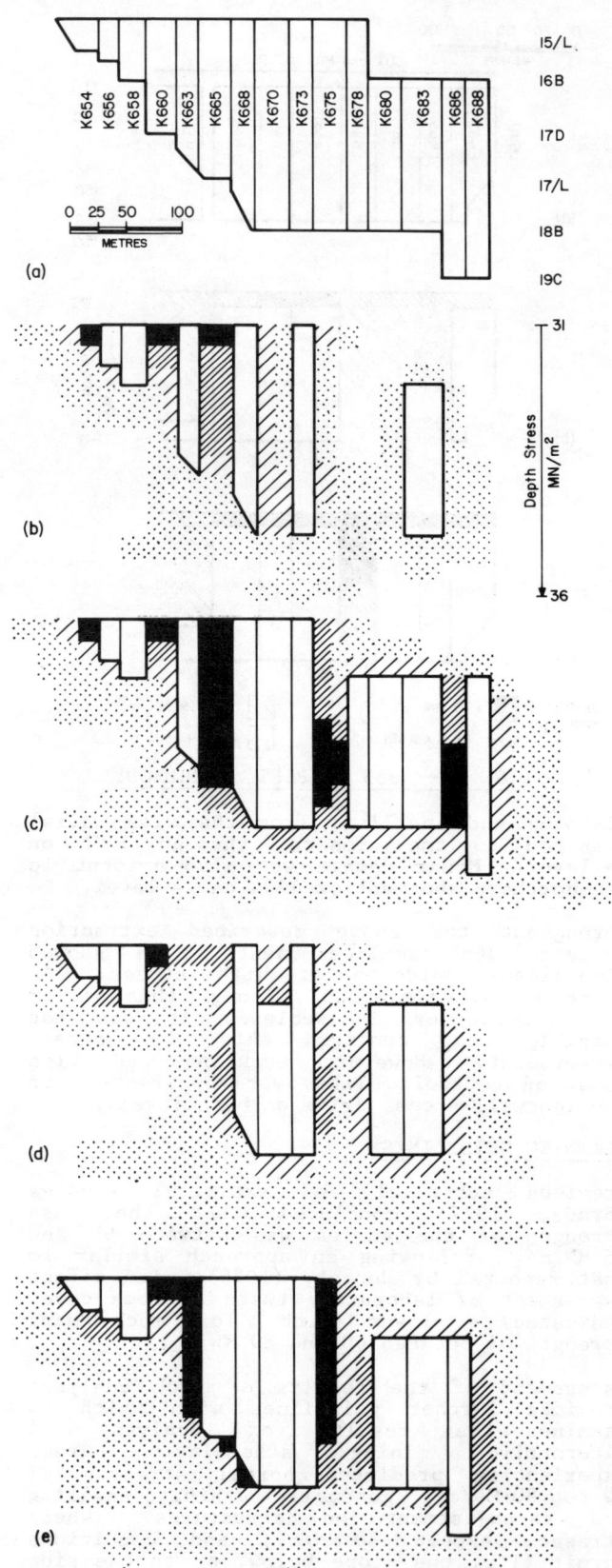

FIG. 7 8 OREBODY BELOW 15 LEVEL

Initial experience in the area has supported the general picture of stress distribution predicted by the analyses. Further interaction between Mine Planning and Rock Mechanics has improved the extraction schedule. In particular, detailed analyses at individual stope stages, instead of at annual intervals, has assisted in finalising the schedule.

CONCLUSIONS

The roles of stress measurement and the displacement discontinuity method of stress analyses in examining mine production schedules has been demonstrated. Elastic analyses were used throughout, and good agreement was achieved between measured and computed stress values for a variety of mining geometries.

Although good agreement has been obtained between model and prototype, further work is required to assess the suitability of the technique to the examination of cut-and-fill stopes. Non-linear models of rock behaviour are coded in the program, and these are currently being tested; initially by parameter studies and eventually by back analyses.

It must be emphasized that rock mechanics assessment is only one stage in the attainment of a viable mine production schedule. If such a thing as an ideal, rock mechanics, production schedule could be established, it is unlikely that other factors, such as tonnes and grade, could also be attained. Compromise is necessary.

Mine Planning is inherently an iterative process, but one which can be greatly assisted by the application of suitable stress analysis techniques.

ACKNOWLEDGEMENTS

The permission of the management of Mount Isa Mines Limited to prepare this paper is acknowledged. Numerous colleagues at Mount Isa Mines were involved in preparing and assessing the schedules and back analyses.

REFERENCES

Alexander, E.G. and Fabjanczyk, M.W. (1981). Extraction design using open stopes for pillar recovery in the 1100 orebody at Mount Isa.: International Conference on Caving and Sub Level Stoping, Denver.

Beer, G. et al (1983). Prediction of the behaviour of shale hangingwalls of deep, underground excavations: 5th ISRM Conference, Melbourne, Australia.

Brady, B.H.G. (1977). An analysis of rock behaviour in an experimental stoping block at the Mount Isa Mine, Queensland, Australia: International Journal of Rock Mechanics and Mining Science, Vol 14, p 59.

Goddard, I.A. and Bridges, M.C. (1977). Development in lead-zinc mining methods at Mount Isa, Australia: Lead/Zinc Update, Society of Mining Engineers of the A.I.M.E., New York.

Goddard, I.A. (1981). The development of open stoping in lead orebodies at Mount Isa Mines Limited: International Conference on Caving and Sub Level Stoping, Denver.

Herget, G. and Unrug, K. (1976). In situ rock strength from triaxial testing: International Journal of Rock Mechanics and Mining Science, Vol. 13, p 209.

Leahy, F.J. and Cowling, R. (1978). Stope fill development at Mount Isa: 12th Canadian Rock Mechanics Symposium, Sudbury, Ontario.

Mathias, B.V. and Clark, G.J. (1976). Mount Isa copper and silver-lead-zinc orebodies - Isa and Hilton Mines 'Economic Geology of Australia and Papua New Guinea', Vol. 1 Metals: Australian Institute of Mining and Metallurgy.

Sinha, K.P. (1979). Displacement discontinuity technique for analysing stresses and displacements due to mining in seam deposits: Ph.D. Thesis, University of Minnesota.

Worotnicki, G. and Walton, R.J. (1976). Triaxial "Hollow inclusion" gauges for determination of rock stress in situ. I.S.R.M. Symposium - Investigation of stress in rock - advances in stress measurement: Institution of Engineers, Australia, Publication No. 7614, Supplement, 1976, p 1.

THE POTENTIAL FOR CABLE SUPPORT OF OPEN STOPES

Le potentiel des réseaux de câbles dans le cas d'exploitation en gradins dans les mines à ciel ouvert

Das Potential einer Felsbewehrung in Kammerbruchbau

P. G. Fuller
Principal, Mining Research Associates, Melbourne, Australia

SYNOPSIS

Cable support technology has been transferred from cut and fill mining to open stoping with only minor modification and refinement. The experience with cable support of walls and crowns of open stopes has been reviewed and, to date, the method can only be regarded as marginally successful. A simplified analysis of a cable supported hangingwall demonstrates a need for increasing the cable density compared to current practice to improve hangingwall stability. This should form the basis for future trials with the support method to minimize hangingwall dilution.

RESUME

La technique du câble support a été transférée des mines du type creusage/remplissage à l'exploitation en gradins à ciel ouvert, avec seulement quelques transformations et un raffinement mineurs. L'expérience du support des parois et des couronnes à l'aide d'un réseau de câbles dans le cas des mines en gradins à ciel ouvert a été revisée et, jusqu'à présent, la méthode n'était considérée que marginalement bonne. Une analyse simplifiée d'une paroi maintenue par un réseau de câbles démontre le besoin d'augmenter la densité du câblage en comparaison avec la méthode courante pour augmenter la stabilité de la paroi. Ceci devrait constituer la base d'essais futurs en utilisant la méthode du support pour minimiser l'écroulement des parois.

ZUSAMMENFASSUNG

Die Felsbewehrungstechnologie aus dem Cut-and-Fill Bergbau ist mit nur geringfügigen Änderungen und Verbesserungen in den Kammerbruchbau übertragen worden. Die Erfahrung, die dabei mit Felsbewehrungen in Firsten und Stoß gemacht worden sind, werden besprochen, mit dem Ergebnis, daß diese Methode bislang nur als bedingt erfolgreich anzusehen ist. Eine einfache Analyse eines felsbewehrten Hangenden zeigt die Notwendigkeit auf, den Abstand der Bewehrungsstränge zu verringern. Dies könnte eine neue Basis für kommende Versuche sein, in denen die Verunreinigungen des gebrochenen Erzes durch Nachfall aus dem Hangenden zu verringern versucht wird.

1. INTRODUCTION

Cable support is a method with the potential to improve rock mass stability in underground mines by pre-reinforcing ground ahead of existing working levels. It usually involves installing long, steel cables in a regular array from access development within a stope or from development in country rock near the stope. Practical constraints of limited headroom in the development mean that the steel must be either a flexible continuous length or a series of rigid, shorter lengths coupled together. Both approaches have been tried but flexible cables are now used almost exclusively. In the past, cables have been either pre-tensioned or left untensioned and fully bonded with cement grout. Current practice in most cases is to leave the steel untensioned.

Some of the earliest trials of the method were reported by Gramoli (1975) in which pre-tensioned hoist ropes were used to reinforce unstable backs of open stopes in a Canadian mine. The method developed concurrently in Australia to improve back stability in cut and fill stopes. Clifford (1974) and Palmer et al (1976) describe the early Australian developments and highlight the substantial improvement in back stability achieved within a variety of ground conditions.

Attempts to quantify the performance of cables in a cut and fill stope by measuring the strain at various points along their length were outlined by Fuller (1981). Measurements indicated that the steel became tensioned immediately above a newly formed stope back and at localized zones further into the rock mass. The local tensions have been interpreted as a response of the cable to shear and/or dilation at the intersection of the cables with geological structures. Cables close to the back become loaded due to stress relaxation and rock movement into the opening. More recently, Stheeman (1981) reports similar developmental work on cable supports in South Africa which resulted in a cable design method based on supporting the weight of rock beneath a parabolic arch in the back. While this approach may be appropriate in some conditions, the low cable strains reported by Fuller (1981) indicate that cables serve to reinforce the rock mass so that it becomes partially self stabilizing.

During the late 1970's, many underground metalliferous mines in Australia changed from cut and fill to the long-hole open stope mining method. In open stopes, both wall and back exposures are typically much larger than in cut and fill and the success of this bulk mining method is largely dependent on controlling the amount of dilution from back and wall failures. Cable supports were an obvious choice to try to control dilution and their use in open stopes soon became widespread.

This paper briefly reviews the Australian experience with cable support of open stopes and examines, with a simplified theoretical approach, the potential for cables to support a large, down dip exposure in a layered hangingwall.

2. EXPERIENCE WITH CABLE SUPPORT OF OPEN STOPES

Dimensions of open stopes vary widely according to ground conditions, continuity of the orebody and the blast hole drilling capabilities. In all cases, stopes are planned to minimize the amount of development required to bring a stope into production. Development is normally limited to a top sill and a series of drawpoints into a bottom sill, some 50m below the top sill. Where ground conditions are such that wall exposures can be increased, extra development is normally provided within the orebody for drilling access.

The result of this concept of mining large stopes with minimal development is that the access available for installing cable support is restricted. The options available can be broadly summarized by sketches shown in Figs. 1 to 4. In Fig. 1 where the height of the stoping block is less than 30m, it is possible to achieve an even distribution of cables at the designed position of the hangingwall, provided a strike drive is developed at the sub-level horizon on the hangingwall side.

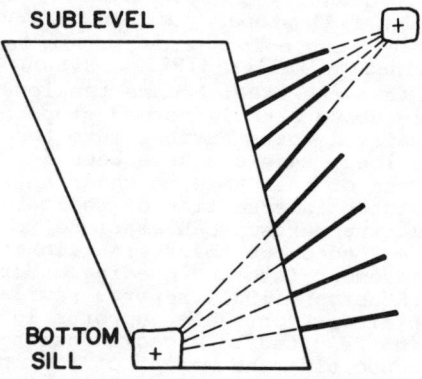

Fig. 1 Uniform cable distribution in the hangingwall

In higher stopes with an intermediate drilling level (see Fig. 2) access for cable support installation is limited to the hangingwall drill drives. Clearly, the support coverage is localized, particularly near the hangingwall. The intended function of this layout is to divide the hangingwall into a number of smaller unsupported but stable spans.

Where significant overbreak of a barren hangingwall is anticipated and the value of the ore is high, additional hangingwall development may be justified for cable support installation. Costs for extra development are approximately A$800/m and cable support costs are A$250/m along strike, so operators need to be assured of a substantial reduction in overbreak before undertaking the extra development.

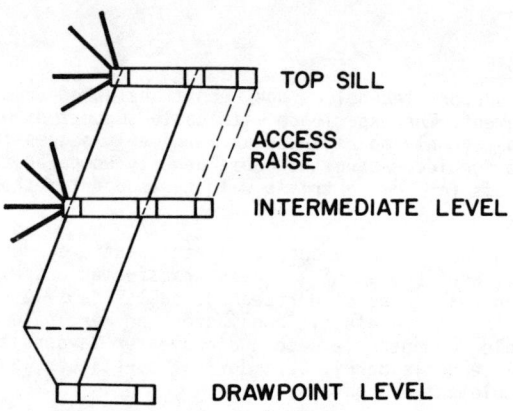

Fig. 2 Localized support with cables installed from hangingwall drill drives

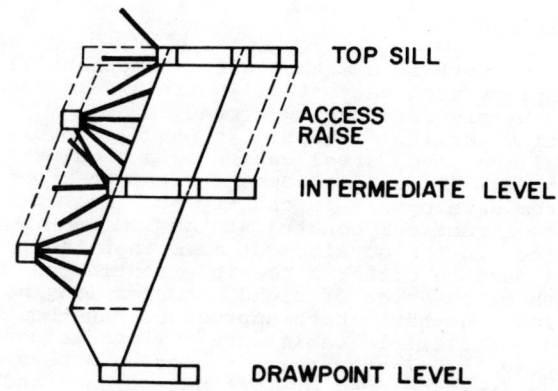

Fig. 3 Cable support from specially driven access and hangingwall drill drives

Reinforcing crown pillars or backs of open stopes is normally less of a problem because depending on stope width one, two or three strike drives are developed as drilling access. This geometry allows an even cable density to be achieved in the centre of the crown but at the stope back the coverage is localized (see Fig. 4).

Most cable support patterns in open stope walls and crown are in rings of radiating holes which are repeated at regular intervals along the stope. In Australian practice, rings are spaced at between 1.5 and 3m but in other countries, spacings up to 10m have been tried. Cable orientations in each ring are often chosen on an empirical basis but in a few cases some attempts have been made for cables to intersect structures known from past experience to cause ground instability.

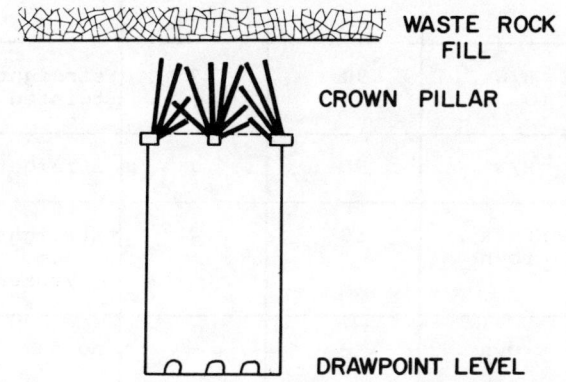

Fig. 4 Crown pillar reinforcement from top sill development

The Australian experience with cable support around open stopes has been widely variable. Fabjanczyk (1982) has summarized the experience on a mine wide basis at ten Australian mines. At four of these, individual stopes completed before December 1981 have been treated as case studies for this paper to examine the cable support performance in more detail. Case study data are summarized in Table 1 and highlight the low success rate of cables stabilizing layered and blocky ground; particularly in stope hangingwalls. In these conditions, both the localized cable patterns and the even density of cables have been unsuccessful in preventing failure. Steel failure occurred in isolated cases but in the majority of failures, the steel strands were left protruding from hangingwalls either twisted or straight. The twisted, curled shape seems consistent with a failure mechanism in which blocks slide off the steel by failure of the grout - steel bond.

The straight cables are more difficult to understand because they show no signs of having been loaded in shear or tension during the failure. Only two rock failure mechanisms seem consistent with the observed behaviour; these are:

- rock failure between cables and/or along the axis of cable drill holes to allow blocks to break away from cables, or

- rock failure through cables which were not adequately grouted due to some deficiency in grouting technique.

While there have been practical problems with grouting cables in the past, it seems unlikely that grouting practice would be deficient in all four mines. Hence, on the available evidence, rock failure from between and around cables seems the more likely explanation for the observed straight cables.

The case studies show that in massive ground, few failures have occurred where cable supports have been used. Because the studies did not cover unsupported walls or crowns in massive ground, it is not possible to draw any definite conclusions on the role of cable supports in such situations. Clearly, as more open stopes are developed in massive ground and operators assess the stability of unsupported walls, any stabilizing influence of cable supports in such conditions will become more clear.

The implications of the failures recorded in Table 1 in terms of mining economics are very significant. On a volume basis, the failures represent stope dilutions ranging from 8 to 23% but their economic effect depends on whether the dilution has significant grade. In most cases, a loss of profit occurs when dilution is mined and hence there is a substantial economic attraction to use cable supports.

To date, only minimal development work has been undertaken on the mechanics of supporting open stope crowns and walls with cable arrays. There is a vast difference between cut and fill and open stope mining because, in open stopes:

- areas of ground exposed are large,

- minor failures can be tolerated, and

- support density; including rock bolts and cables, is small.

In light of these points, the lack of success in applying cable support methods to open stoping is not surprising.

It is obvious from the review and the economic significance of reducing hangingwall dilution without additional development that the concept of "localized" cable support of hangingwalls requires more detailed study. The simplest geometry to examine is a hangingwall with continuous structure parallel to the wall.

3. CABLE SUPPORT OF LAYERED HANGINGWALLS - A THEORETICAL APPROACH

The simplest means of assessing hangingwall stability in layered rock is to assume that the wall behaves as a beam rather than a plate. This represents a worst case which may approximate conditions at midspan, along strike. The intention in this theoretical analysis is treat the rock layers as homogeneous and elastic and to examine a worst case situation in terms of geometry,

Table 1. Statistics on cable supported open stopes in four Australian mines.

CASE No	STOPE DIMENSIONS (m) LEN.	WID.	HT	DIP (deg)	DEPTH BELOW SURFACE* (m)	CABLE SUPPORTED ZONE+	GROUND CONDITIONS IN SUPPORTED ZONE	OBSERVED PERFORMANCE			
								OVERBREAK LOCATION	OVERBREAK AREA• (%)	AVERAGE FAILURE THICKNESS (m)	CABLE CONDITIONS
1	29	25	95	60	670	H/W local	layered	H/W	25	8	70% broken
2	16	20	30	76	420	H/W uniform & crown	blocky	H/W & crown	80	5	straight
3	29	28	95	65	670	H/W local	layered	H/W	90	7	straight & twisted
4	23	28	95	65	670	H/W local	layered	H/W	90	7	straight
5	34	20	95	65	670	H/W local & crown	layered & massive	H/W & crown	50	4	straight - some broken
6	31	11	100	67	670	H/W local	layered & massive	crown	-	-	no failure
7	29	25	140	65	650	H/W local	massive	nil	nil	nil	no failure
8	20	25	100	65	670	H/W local	massive	nil	nil	nil	no failure
9	43	46	97	68	210	back	blocky	back	50	10	back failed above cables
10	100	28	90	75	590	crown	blocky	nil	nil	nil	no failure
11	55	15	53	73	370	crown	blocky	crown	50	3	straight
12	50	45	36	90	460	crown	massive	crown	90	6	straight

NOTES:

* Depth below surface is taken from mid height of the stope

+ H/W local signifies localized support coverage whereas H/W uniform indicates that cables were evenly distributed in the hangingwall

• Overbreak area is expressed as a percentage of the total wall or crown area

Data have been collated from studies at Mount Isa Mines; CSA Mine; Cobar; Rosebery Mine and Mount Charlotte.

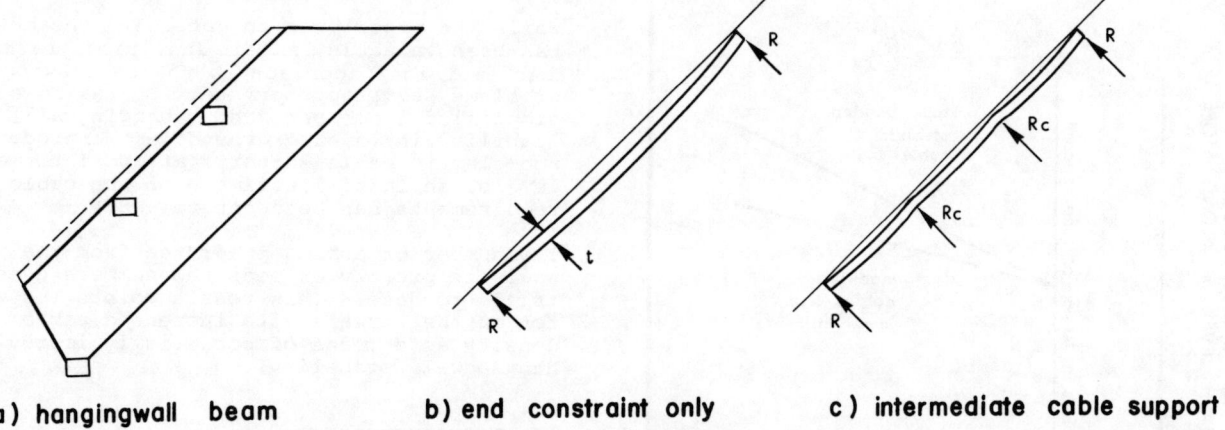

Fig. 5. Beam simulation of a hangingwall supported at two intermediate points.

end contraints and applied load on the beam to check the viability of stabilizing it with cable supports. A 45° dipping hangingwall is considered with a layer of thickness t being detached from the country rock (see Figure 5a). Under a component of self weight loading, the beam will deflect into the stope. Maximum deflection will result if the upper and lower ends are simply supported as shown in Fig. 5b.

Cable supports can be simulated by reactions Rc at the two intermediate points in Fig. 5c, but when a beam is supported at more than two points it becomes statically indeterminate. Because cables are passive supports, some beam deflection is required at the intermediate points before cable reactions develop. Of the several methods available to solve for the beam reactions, the Three Moment Equation (Timoshenko and Young (1965)) has the advantage that it can readily allow for deflection at the support points.

The following parameters have been assumed in the analysis:

- uniformly distributed beam loading based on a rock specific gravity of 4.0.

- maximum allowable tensile stress in the rock = 15×10^6 N/m^2

- $E_{rock} = 60 \times 10^9$ N/m^2

- overall beam length = 90m

- cable support reactions based on pullout test curves for various bonded lengths of twin 15.2mm ⌀ strand cables in Fig. 6.

Results of the analysis are illustrated in Fig. 7 and show that beams less than 1m thick exceed the allowable tensile stress in the rock at the support points and indications are that they cannot be stabilized with cables. Even assuming that the support stiffness could be reduced by a cable design change, thin beams still fail in tension. Beams thicker than 1m can, in theory, be stablized with a minimum of four cables per m along strike at each intermediate level.

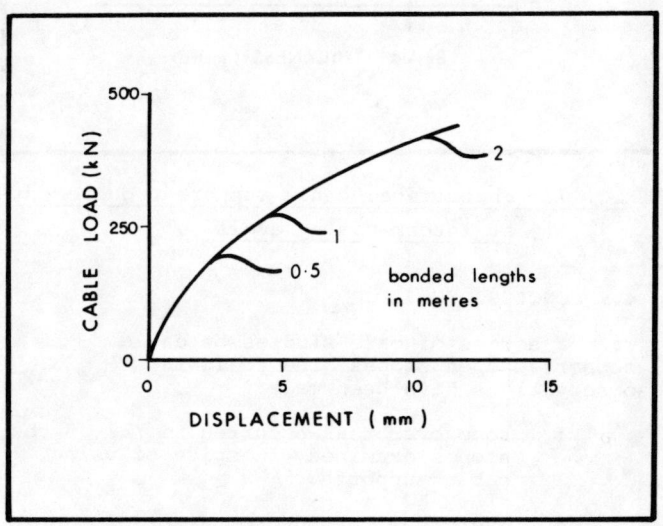

Fig. 6. Pullout test data for twin 15.2mm ⌀ strand cable.

Fig. 7 indicates that the number of cable supports could be marginally reduced if they were designed to be less stiff by debonding but, as this usually involves some increase in cable cost, any economic benefits of this approach are likely to be insignificant. Alternatively the improvement in stability in these situations is negligible.

The results in Fig. 7 show that the cable density required to maintain stability greatly exceeds that tried to date in any Australian operation (Table 1). However, the analysis was conceived to represent a worst case. In open stopes with steeper walls and where the specific gravity of the rock is lower, fewer cables than indicated by Fig. 7 would probably be required. The most rewarding aspect of the analysis is that it gives some confidence that an increase in the density of cables placed from development at intermittent height up the walls may lead to a dramatic reduction in hangingwall overbreak.

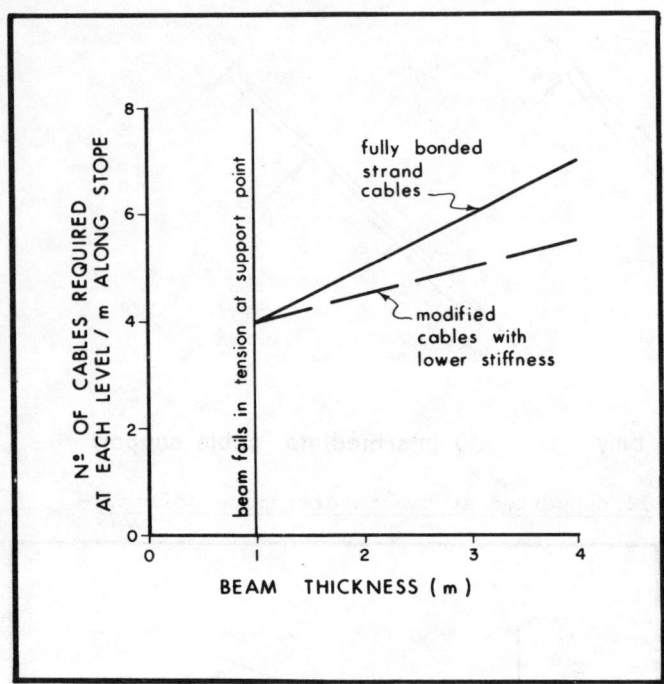

Fig. 7. Theoretical cable support requirements at intermediate levels.

4. CONCLUSIONS

From a series of case studies on cable supported open stopes, the following observations have been made:

- some overbreak occurred in 75% of the stopes examined, in spite of using cable supports.

- overbreak ranged from 8 to 23% by volume.

- both localized and even cable distributions were unsuccessful in limiting dilution caused by hangingwall failures in blocky and layered ground.

- steel failure has been uncommon and when stope wall or crown failures occur, cables often remain with either a straight or twisted and curled shape.

This experience with cable supports in open stopes is in marked contrast with that in cut and fill stopes and indications are that the cable support technology developed previously is not directly transferable to open stopes. Since overbreak in open stopes normally makes mining economics less attractive, there is a considerable incentive to improve the performance of cable supports of open stopes.

While the experience to date with the method has been valuable, a more analytical rather than empirical approach to the cable support problems seems appropriate. By taking a simplified beam approach to hangingwall stability in layered ground, and introducing non-elastic cable supports at two intermediate levels, an initial estimate of the cable requirements has been determined.

The number of cables predicted from the analysis greatly exceeds the numbers used in trials to date. This result points the way for further trials with increased cable density as a means of potentially improving hangingwall stability.

5. ACKNOWLEDGEMENT

This work forms part of an AMIRA Ltd research project entitled: "Cable Support for Long Hole Open Stopes". Support from AMIRA and from mining companies sponsoring the project is gratefully appreciated.

6. REFERENCES

Clifford, R.L. (1974). Long rockbolt support at New Broken Hill Consolidated Limited: Proc. Aust. Inst. Min. Metall. (251), 21-26.

Fabjanczyk, M. (1982). Review of ground support practice in Australian Underground Metalliferous mines: Proc. Annual Conf. Aust. Inst. Min. Metall., 337-349, Melbourne.

Fuller, P.G. (1981). Pre-reinforcement of cut and fill stopes: Proc. Application of Rock Mechanics to Cut and Fill Mining; Inst. Min. Metall., 55-62.

Gramoli, L. (1975). Tensioned cable rock anchorage at Geco division of Noranda Mines Ltd: Proc. CIM 1st Underground Operators Conf., Sudbury.

Palmer, W.T., Bailey, S.G., and Fuller, P.G. (1976). Experience with pre-placed supports in timber and cut and fill stopes: In Influence of Excavation Design and Ground Support on Underground Mining Efficiency and Costs; Symposium Wollongong, 1976 (Parkville, Vict., Australian Mineral Industries Research Association, 1976), 45-71.

Stheeman, W.H. (1982). A practical solution to cable bolting problems at the Tsumeb Mine: CIM Bulletin (75), 838, 65-77.

Timoshenko, S.P., and Young, D.H. (1965). Theory of Structures. 2nd Edition. 629pp. New York: McGraw-Hill.

PREDICTION OF THE BEHAVIOUR OF SHALE HANGING WALLS IN DEEP UNDERGROUND EXCAVATIONS

Prédiction du comportement des toits schisteux en excavations profondes

Vorhersage des Verhaltens von Hangendem im tiefliegenden Ausbau in Schiefer

G. Beer
J. L. Meek
Department of Civil Eng., University of Queensland, Brisbane, Australia

R. Cowling
Mining Research Dept., Mount Isa Mines, Mount Isa, Australia

SYNOPSIS

The paper describes results of investigations into the behaviour of hanging-walls of deep underground excavations in shale rock. The influence of various parameters which affect their deformation behaviour such as frequency of bedding planes, stress field and geometry is studied using a numerical model. The aim of the research is to isolate important factors which contribute to the overbreak of the hanging-wall rock.

RESUME

Cette communication donne les résultats de recherches sur le comportement des toits d'excavations profondes en roche schisteuse. En se servant d'une modèle numérique, on examine l'influence des divers paramètres, tels que la fréquence des discontinuités, le champ et la géométrie des contraintes, qui affectent les déformations. Cette recherche a pour but d'isoler les facteurs majeurs qui contribuent aux ruptures des roches du toit.

ZUSAMMENFASSUNG

In diesem Beitrag werden Resultate der Forschung über das Verhalten von Hangendem in tiefliegendem Ausbruch in Schiefer präsentiert. Der Einfluß von verschiedenen Parametern, die das Verformungsverhalten beeinflussen, wie die Kluftfrequenz, das Spannungsfeld und die Geometrie, wird anhand eines numerischen Modells untersucht. Das Ziel der Forschung ist es, wichtige Faktoren, die zum Überbruch des hangenden Felses beitragen, zu bestimmen.

INTRODUCTION

The stability of hangingwalls of deep underground excavations is of particular importance in mining. In cut-and-fill mining methods, there is the potential danger of men working in the mining area, or stope. In open stoping, hangingwall collapse may lead to dilution of the ore and, in extreme cases, to loss of production.

This paper describes results of investigations carried out at the Department of Civil Engineering at the University of Queensland and sponsored by Mount Isa Mines Limited. The paper concentrates on the behaviour of the inclined wall, or hangingwall, formed during the recovery, by the open stoping method, of lead-zinc-silver ores from well bedded shale rocks.

Open stoping is a non-entry form of mining and is described with reference to Figure 1 (Goddard 1981). Initially a cut-off raise is bored the full height of the stope. This cut-off raise, typically 1.8m diameter, is then expanded by firing along holes parallel to the raise, until a cut-off slot is formed. The remainder of the ore is then blasted into the slot, either in single vertical slices, or several slices at a time (mass blasting). All ore is removed through drawpoints at the bottom of the stope. Open stopes vary in size, and on present production schedules, are typically 30m x 30m x 110m high or 200m high. In excavating the stope, two inclined walls are formed at the orebody limits and the over hanging rock is called the hangingwall This wall may or may not be supported by arrays of cable dowels.

The geology of the Urquhart Shale formation is described in detail by Mathias, (1976). Two zones occur, one being host to the copper ore while the other is host to the lead-zinc-silver ore. The latter is composed of well bedded, dolomitic and pyritic shales and silt stones striking north-south and dipping 65° to the west. The most important geological structures in the shale are:

- bedding - plane breaks
- fractures which strike north-south, and dip about 60° to the east

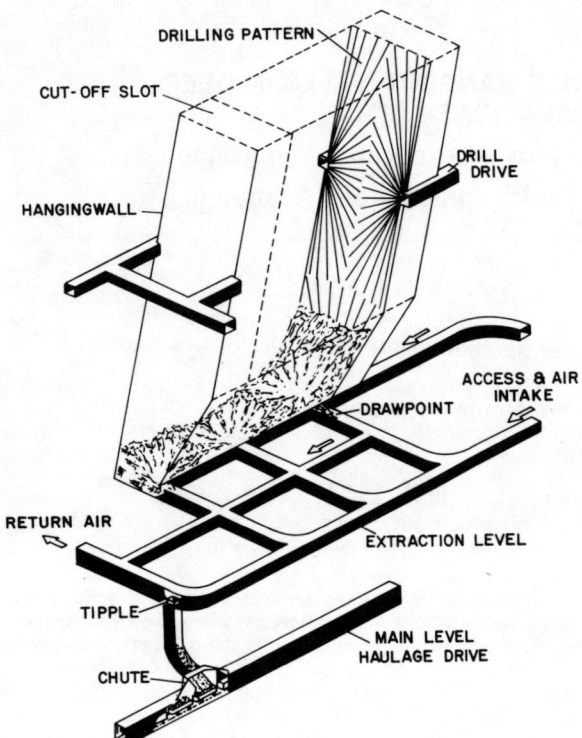

FIGURE 1. Schematic of lead open stope

- two sets of orthogonal extension fractures with orientations normal to bedding

A hypothesis on the behaviour of the shale hangingwall permits selection of suitable numerical models and parameters to be investigated. The numerical model used, in this particular study however, is also applicable to other mining methods.

Hypothesis of Hanging Wall Behaviour in Bedded Rock

Based on underground observations and considerations of statics and dynamics, an hypothesis of hangingwall behaviour has been developed. This hypothesis is then used to select suitable numerical models and parameters to be investigated.

Consider an excavation being made in a prestressed rock (In this case σ_1 is the major principal virgin stress and is perpendicular to the hangingwall Lee 1980). When such an excavation is made, the hanging wall and foot-wall will deform due to the release of stress at the free surface. The displacements shown in Figure 2 will take place. The rock material will be completely destressed in the direction perpendicular to the hangingwall except at the abutments where a pressure arch forms. Due to the dynamic effects of blasting and gravity, tensile stresses perpendicular to bedding are introduced to the hanging wall rock. These tensions will cause the formation of micro-cracks along planes of weakness (bedding planes with lowest cohesion etc.). Also, because of the high shear stress near the abutments, slip on bedding planes, which have the lowest angle of friction, will occur (Lee and Bridges, 1980). This slip will cause the micro-cracks to open and initiate the formation of individual rock beams or plates (Figure 2b). Depending on the thickness of the plates, frequency of cross fracture, presence and location of ground support and the stope dimensions, these plates are either stable or will collapse. Collapse will progress inwards and a stable equilibrium is reached because of either an increase in plate thickness or a decrease in span (Figure 2c). This is indeed what is observed after failure underground.

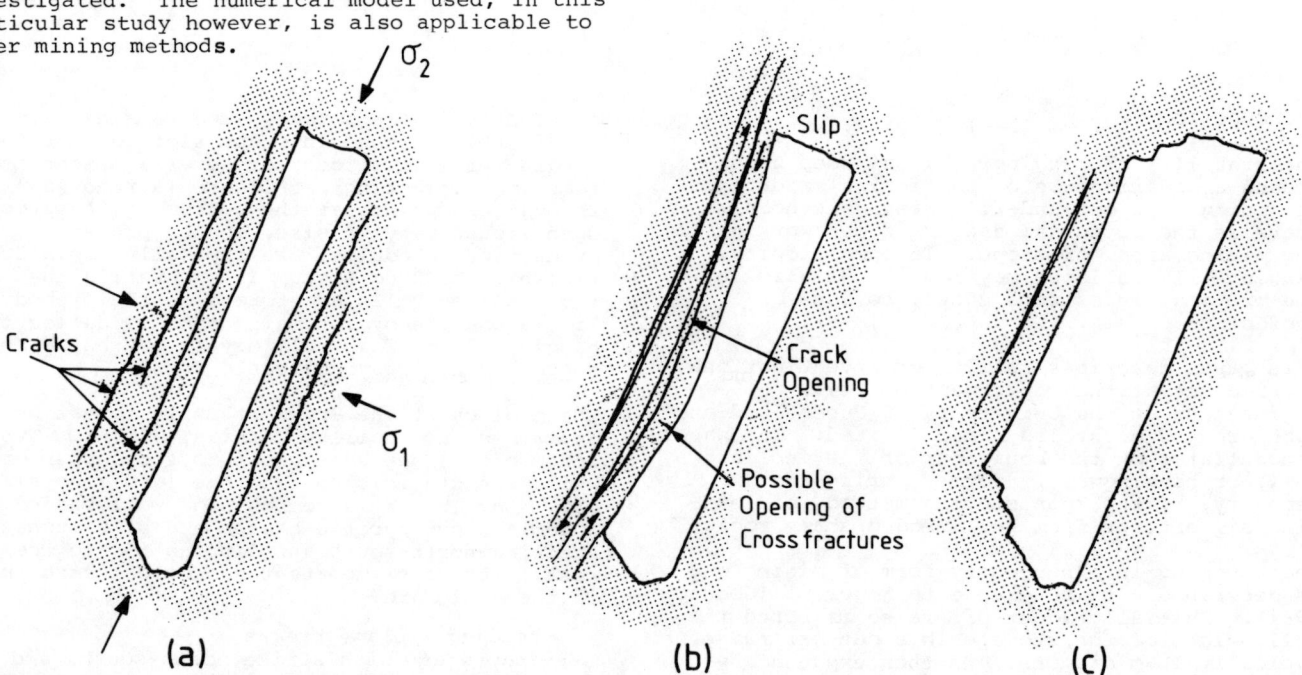

FIGURE 2. The three stages of hanging wall failure

From this hypothesis the following important parameters in the analysis will emerge:

- Frequency and spacing of bedding planes and cross fractures; their frictional and cohesive properties.
- Geometry of excavation (span, presence of drives undercutting the hanging wall etc.)
- Premining stress field
- Ground support, location and size of cable dowels etc.

NUMERICAL MODELS

Deterministic numerical models can be used for parameter studies to help determine the important parameters and the way they influence the behaviour of hanging walls. Various numerical methods are available. They fall into three main categories:

- Continuum models (Finite Element, Boundary Element)
- Discontinuum models (Distinct Element Method by Cundall)
- Simplified limit Equilibrium methods

Continuum models essentially assume that the displacements are continuous everywhere. They either use local displacement functions throughout the continuum as in the Finite Element method (Zienkiewicz 1978) or on the boundary only as in the Boundary Element method (Brebbia 1974). The advantage of the Boundary Element method is that it can model unbounded problems in underground mining. The Finite Element method, on the other hand, is well suited to model material failure and sequential excavation and can be extended to large displacements/strains. The effect of discontinuous behaviour, when the spacing of the discontinuities is small, can also be modelled reasonably accurately using a multi-laminate model (Zienkiewicz 1977). Both methods are applicable to plane-strain and three-dimensional situations. Extensions which can be made to continuous models, in particular the Finite Element method, include Joint Elements, contacts at nodes of Elements and the treatment of large rigid body displacements.

The feature of the discontinuous models or blocky models (Cundall 1971) is that they treat the rock mass as an assembly of rigid blocks. Contacts between blocks can occur anywhere on the block surfaces and the method can handle edge-edge, corner-edge and corner-corner contacts. At present, the method is restricted to plane strain conditions. Dynamic relaxation techniques are used to analyse the system of blocks and there is no limitation on the magnitude of the displacements. The blocky model has been applied to the analysis of hanging by Stewart (1981). The results so far are disappointing in that they have given behaviour which does not agree with in situ observations. The main problems are in the modelling of the far field (i.e. artificial and unrealistic boundary conditions were introduced only a few blocks away from the free surface) and in the assumption that the blocks are undeformable. Extensions which have been made recently by Cundall (1978) include the implementation of block deformability and block failure (splitting of blocks). However, these new developments at the time of writing have not been tested on practical problems. Work is currently underway by Sofianos (1982) on combination of the Boundary Element method and the blocky model to enable the representation of the far field. Extension to three dimensions is required to allow the realistic analysis of hanging wall panels. To the authors best knowledge, the extension of the blocky model to three-dimensions has not been attempted.

Simplified equilibrium models were used well before the advent of the computer and modern numerical methods. The most important of these are the wedge failure theories (Obert 1967) and the "voussoir" beam theory by Evans (1940). An advantage of these techniques is that they allow determination of the stability of an assembly of rock blocks and predict collapse conditions with a minimum of computational effort. Recently, techniques have been developed to predict the collapse conditions for square and rectangular plates consisting of a no-tension material (Beer 1982a).

For the analysis of the hanging walls the following procedure has been adopted:

1. For the first stage of failure, that is, the initiation of cracks, slip and opening on bedding planes and the formation of blocks small displacement continuum models (Finite Element and Boundary Element) are used. The Finite Element method is modified to allow the modelling of discontinuous behaviour along weak bedding planes and is coupled with the Boundary Element method to allow the exact modelling of the response of the far field (Beer 1982b). This first stage of hanging wall failure (termed "primary failure") is restricted to small displacements (order of magnitude less than the thickness of the hanging wall plates). This is termed phase one of the investigation.

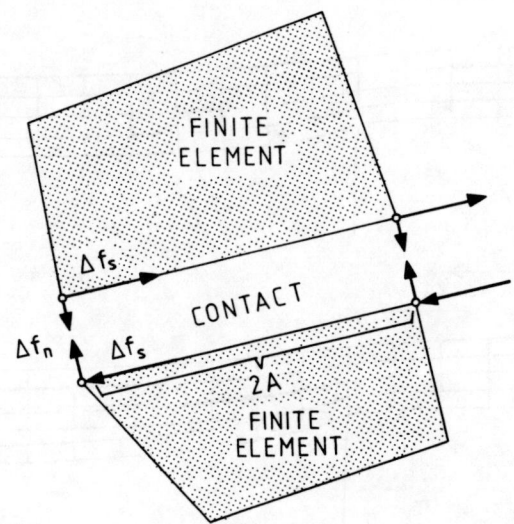

FIGURE 3. Definition of contact forces

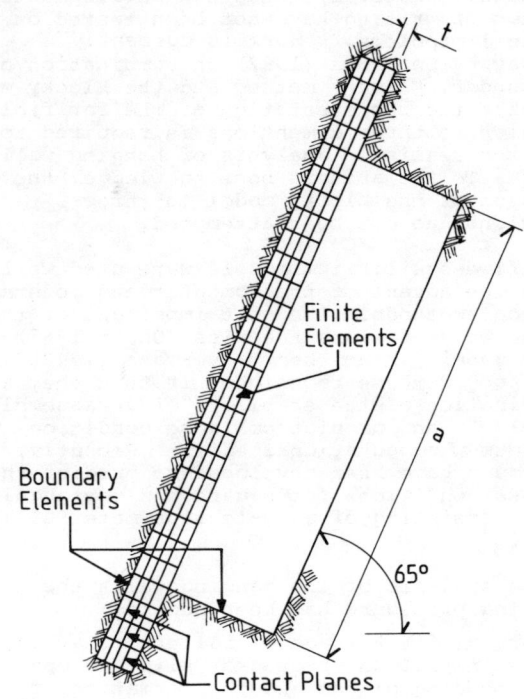

FIGURE 4. Discretisation of idealised stope

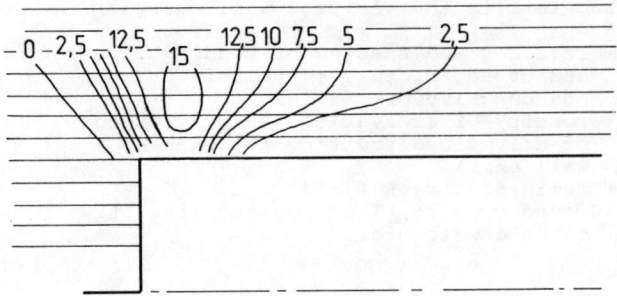

FIGURE 5. Contours of excessive shear stress

The Discontinuous Finite Element Model

In this model, the Finite Element Method is modified to incorporate discontinuous behaviour along cracks

The method described herein does not use Joint Elements but cracks are regarded as boundary interactions between Finite Elements at the nodes. The reason for not choosing the joint elements developed by Goodmann (1968) is that these are still based on a continuum formulation. With these joints for example, a relative rotation of adjacent Elements is associated with strain energy in the contact plane. However, if this deformation corresponds to a crack opening, no strain energy should be generated. Also, perhaps because of this, problems have been experienced with these Elements when modelling mining excavations (Jonasson 1980).

The theoretical background of the discontinuous Finite Element model will be described. Similar models have been used recently by others to analyse piles and fault zones (Frank, Sweet 1982). The boundary interactions (Fig. 3) are assigned stiffness in the directions normal and parallel to the contact. The force displacement relationships at each node are:

2. The second stage of failure can involve large displacements (i.e. in the same order of magnitude as the plate thickness). This stage either leads to the collapse of hanging wall blocks or to a stable equilibrium condition. Investigation of these techniques will constitute phase two of the study. Only the results of phase one are reported in this paper.

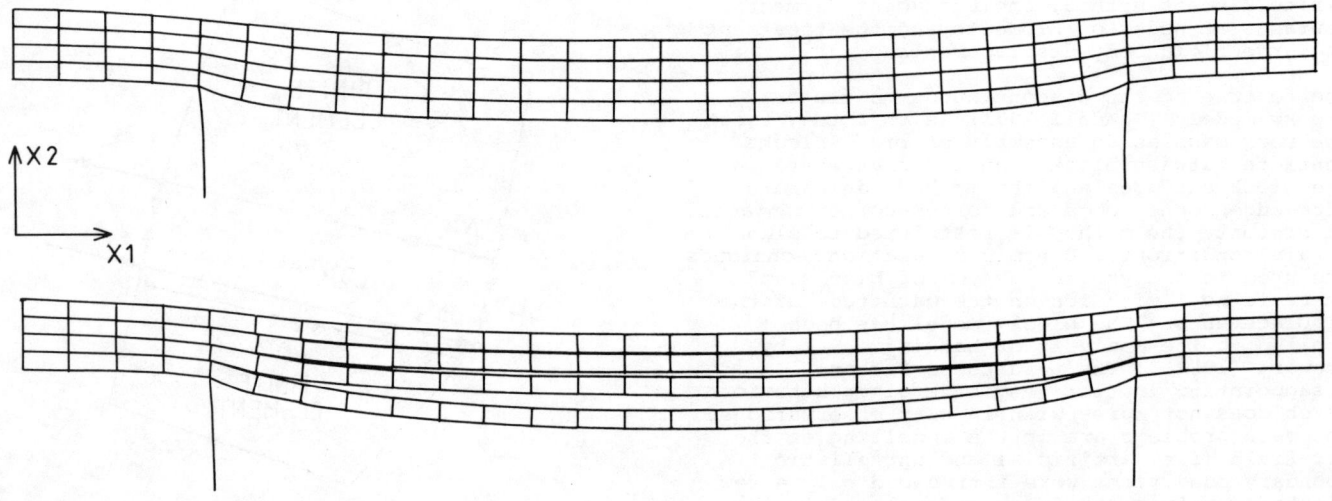

FIGURE 6. Deflection of hanging wall before and after slip on bedding takes place

$$\Delta f_N = k_N \Delta d_N \quad \text{Normal interaction} \quad (1a)$$

$$\Delta f_S = k_S \Delta d_S \quad \text{Shear interaction} \quad (1b)$$

In the above, Δf_N is an increment in contact force normal to the joint, k_N is the normal stiffness and Δd_N is the increment in relative displacement. Similarly, Δf_S, k_S and Δd_S are the corresponding values in the tangential direction. Two yield conditions are enforced on the contact:

Yield condition 1 (no tension)
$$F_1 = \sigma_N - T < 0 \quad (2)$$

where $\sigma_N = f_N/A$ is the normal stress (A is shown in Fig. 3 for a 2-D problem) and T is the allowable tensile stress.

Yield condition 2 (friction law)
$$F_2 = |\tau| + \sigma_N \tan\phi - c < 0 \quad (3)$$

where $\tau = f_S/A$ is the shear stress, ϕ is the angle of friction and c the cohesion of the contact. When either of the above conditions is violated, the stress by which the yield condition is exceeded must be redistributed. The analysis proceeds as follows:
First, an elastic analysis is carried out with high iteration stiffness and the contact forces obtained at each point. Now three possibilities exist:

Case 1: $F_1 < 0$ and $F_2 < 0$ contact is rigid (4a)

Case 2: $F_1 > 0$ contact is broken (4b)

Case 3: $F_1 < 0$ but $F_2 > 0$ contact is slipping (4c)

In case 2 set
$$k_N = k_S = 0 \quad (5)$$

and apply the residual forces
$$\Delta f_S^R = -f_S \quad (6a)$$
$$\Delta f_N^R = -\sigma_N \quad (6b)$$

In case 3 set
$$k_S = 0 \quad (7)$$

and
$$\Delta f_S^R = -\text{sign}(f_S) \cdot F_2 \cdot A \quad (8)$$

The problem is then reanalysed with only the residual forces applied. The results are added to the elastic stresses and the yield conditions checked again with the new contact stresses. The process continues until the yield conditions are satisfied everywhere. Convergence is generally within a few iterations.

Strain softening contact behaviour can be handled by reducing the angle of friction and cohesion after failure and it is also possible to consider dilatant behaviour of contact planes (i.e. effect of asperities in the contact plane).

Sample Analysis

A sample analysis was carried out to verify that the model behaves as intended. An idealised cross section of a typical open stope in the Racecourse area of Mount Isa was analysed. Figure 4 shows the idealised cross section. Based on a series of stress measurements, the virgin stress field was assumed to be:

$$\sigma_1 = 0.066 \, D \quad \text{MPa} \quad \text{(for D < 500m)}$$

$$\sigma_1 = 33 + 0.01 \, D \quad \text{MPa} \quad \text{(for D > 500m)}$$

and

$$\sigma_2 = \sigma_3 = 7 + 0.017 \, D \quad \text{MPa}$$

where D is the depth below surface in metres.

The following figures show some results of the analysis of a stope which has been excavated at a depth of 600m in virgin ground. The spacing of weak bedding planes ($\phi = 10°$, c = 0, T = 0) has been assumed to be 2m and the properties of intact rock are E = 80,000 MPa and ν = 0.3. Figure shows contours of the yield function F_2 and this essentially indicates the amount of shear stress to be redistributed. Figure 6 shows the magnified displacements of the hangingwall rock after excavation before and after slip has taken place on the bedding planes. The analyses show that the opening of hanging wall cracks is due to bedding plane slip. This means that for this example, the redistribution of the shear stress energy is more important than the tension energy as typified by 'no tension' analyses.

Parameter Studies

A series of analyses were carried out using the simplified model to study the effect of various parameters on the deformation behaviour of shale hanging walls. The following values of displacements were obtained from the model:

1. maximum absolute displacement of hangingwall in direction normal to bedding with reference to the virgin state. (deflection)
2. maximum relative normal displacement of contact surfaces (crack opening)
3. maximum relative shear displacement of contact surfaces (slip)

Some of the results are shown in the following figures:-

(1) Depth from ground surface
In Figure 7, the influence of the depth of the stope on closure is plotted together with the assumed virgin stress field. Values of displacement increase about proportional with σ_1, (stress perpendicular to bedding).

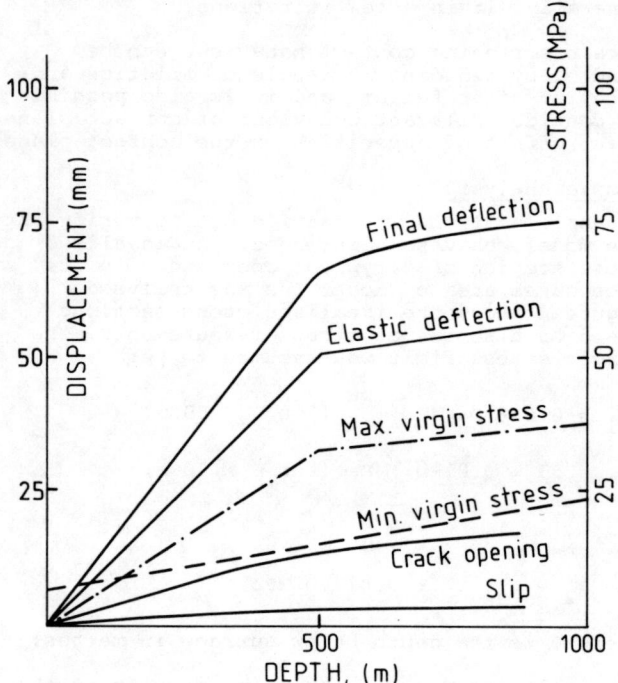

FIGURE 7. Influence of depth on hanging wall displacements.

2. **Frequency of weak bedding planes**

The plot in figure 8 shows a significant increase in hanging wall displacement and opening of bedding planes as spacing decreases. The maximum amount of slip on a bedding plane on the other hand decreases with increasing bedding plane frequency.

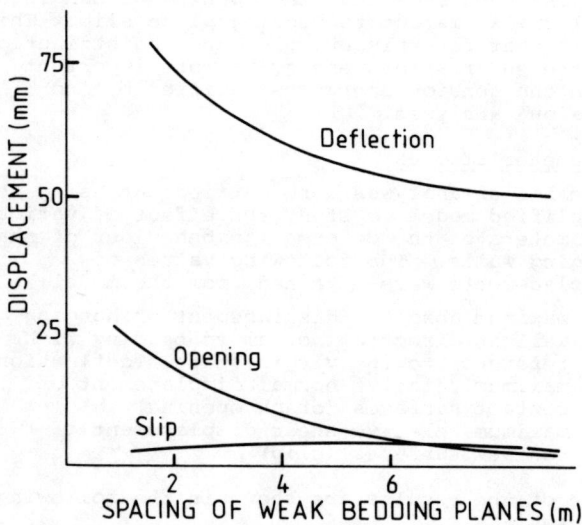

FIGURE 8. Influence of bedding plane spacing

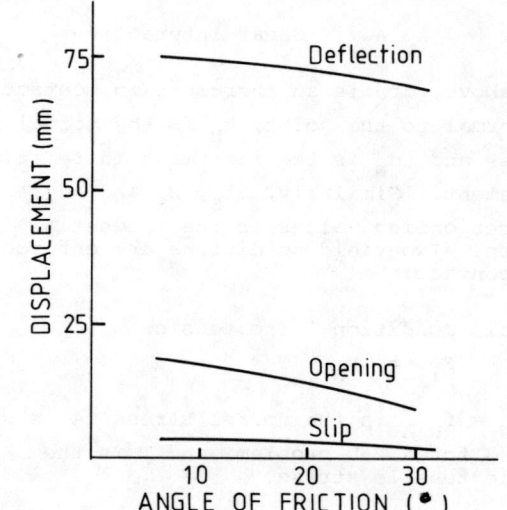

FIGURE 9. Influence of angle of friction at weak bedding planes.

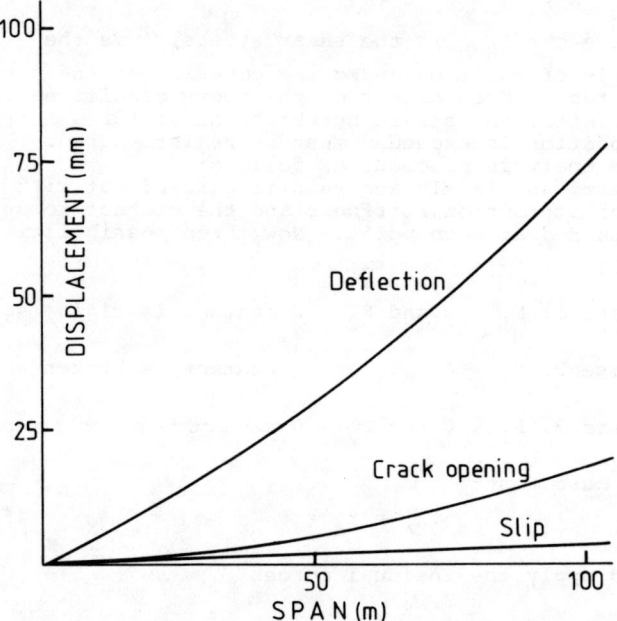

FIGURE 10. Influence of span.

3. **Angle of bedding plane friction**

The effect is shown in Figure 9. A decrease in the friction angle results in a small increase of displacements. In the range considered, cohesion was found not to be a major influence.

4. **Span of stope**

The influence of span 'a' is shown in figure 10. There is a major effect on crack opening because the deflection of the hanging wall plates increases rapidly with span.

At the time of writing this paper, the effect of frequency of cross fractures is being studied. The following conclusions are made from the parameter study carried out so far:

1. Opening of hanging wall cracks is influenced by slip on bedding planes near the abutments.
2. The major parameter isolated so far which affects the opening of cracks and deflection is the spacing of weak bedding planes and span of the opening.
3. For the stress field at Mount Isa, hanging-wall deflections and opening of cracks increase approximately linearly with depth.
4. Within the range considered the frictional properties of weak bedding planes are not a major parameter.

CONCLUSIONS

In this paper, an attempt has been made to predict the deformation behaviour of shale hanging walls using a numerical model. The model characteristics and the parameters to be studied were selected through observations in-situ and consideration in statics and dynamics. The conclusions of the parameter study reported here are only preliminary since the study of the effect of the initiation and opening of cross fractures has not been completed at the time of writing. In addition, the present numerical model is restricted to small displacements and as such can not predict the stability or collapse of these hangingwalls. Thus, phase 2 of the study will concentrate on the stability of the cracked hangingwall. Several avenues such as large displacement Finite Element, Blocky Model and limit equilibrium methods will be explored.

The numerical model used is not restricted to plane-strain analyses and three-dimensional analyses are being carried out at the time of writing. The authors feel that the consideration of the three-dimensional state of stress is essential when hanging wall geometry such as encountered at Mount Isa is considered.

The deterministic analyses will finally lead to a statistical treatment and enable the prediction of probability of failure as a function of statistically defined parameters.

ACKNOWLEDGEMENT

The work presented here has been carried out as part of a research project entitled "Computer based excavation design techniques for multiple orebodies in jointed rock" which has been sponsored by Mount Isa Mines Ltd., Australia.

REFERENCES

Beer, G. and Meek, J.L. (1982a). Design curves for Roof and Hanging Walls in Bedded Rock Based on "voussoir" Beam and Plate Solutions. Trans. Inst. Min. & Met. (Section A: Mining Industry), 91, A18-22.

Beer, G. and Meek, J.L. (1982b). Efficient Analysis of Problems in Geomechanics. Proc. 4th Int. Conf. on Num. Methods in Geomech. (1), 5-13, Edmonton.

Brebbia, C.A. (1974). The Boundary Element Method for Engineers, London: Pentech Press.

Cundall, P.A. (1971). A Computer Model for Simulating Progressive, Large Scale Movements in Blocky Rock Systems. Symp. of I.S.R.M., Nancy, France.

Evans, W.H. (1940). The Strength of Undermined Strata. Trans. Instn. Min. & Met., 50, 475-532.

Frank, R. et. al. (1982). Numerical Analysis of Contacts in Geomechanics. Proc. 4th Int. Conf. Num. Methods in Geomech., Edmonton, Canada, Vol. 1, 37-45.

Goddard, I.A. (1981). The Development of Open Stoping in Lead Orebodies at Mount Isa Mines Limited. in Design and Operation of Caving and Sub level Stoping Mines, Society of Mining Engineers, A.I.M.E.

Goodman, R.E. et al (1968). A Model for the Mechanics of Jointed Rock. J. Soil. Mech. Found. Div. ASCE, 94, SM3, 637-659.

Jonasson P. (1980). Shear Behaviour of Joint Elements in the BEFEM Code. Proc. Conf. on Applications of Rock Mechanics to Cut-and-Fill Mining, I.M.M., London.

Lee, M.F. and Bridges, M.C. (1980). Rock Mechanics of Crown Pillars Between Cut-and-Fill Slopes at The Mount Isa Mine. Proc. Conf. on Application of Rock Mechanics to Cut-and-Fill Mining, I.M.M., London.

Mathias, B.V. and Clark, G.J. (1976). Mount Isa Copper and Silver-Lead-Zinc Orebodies - Isa and Hilton Mines. in Economic Geology of Australia and Papua New Guinea, The Aust. Inst. of Min. & Met. 1976.

Obert, L. and Duvall, W.I. (1967). Rock Mechanics and the Design of Structures in Rock, Wiley New York.

Sofianos, A. (1982). Ph.D. Dissertation, Royal School of Mines, Imperial College, in preparation.

Stewart, I.J. (1981). Numerical and Physical Modelling of Underground Excavations in Discontinuous Rock. Ph.D. Dissertation, Royal School of Mines, London.

Sweet, J. et. al. (1982). Instantaneous Strain Release in Faulted Media. Proc. 4th Int. Conf. Num. Methods in Geomechanics, Edmonton, Canada, Vol. 2, 495-502.

Zienkiewicz, O.C. and Pande, G.N. (1977). Time Dependent Multi-laminate Model of Rocks - A Numerical Study of Deformation and Failure of Rock Masses. Int. Jnl. Num. Anal. Meth. Geomech. 1, 219-247.

Zienkiewicz, O.C. (1978). The Finite Element Method - Third Edition, London: Wiley.

COMPARISON OF FIELD MEASUREMENTS AND NUMERICAL ANALYSES OF A MAJOR FAULT IN A MINE PILLAR

Comparaison de mesures in situ et d'analyses numériques d'une faille importante dans un pilier de mine

Vergleich von Feldmessungen und numerischen Analysen einer grösseren Störung in einem Grubenpfeiler

M. A. Coulthard
Principal Research Scientist, CSIRO, Division of Applied Geomechanics, Mount Waverley, Australia

J. M. Crotty
M. W. Fabjanczyk
Experimental Officers, CSIRO, Division of Applied Geomechanics, Mount Waverley, Australia

SYNOPSIS

A pillar created during mining of the 1100 orebody at Mount Isa Mine has been observed in the field and studied, using finite element and boundary element stress analyses. Monitoring showed that, in the early stages of mining, a fault transecting the pillar limited its longitudinal load bearing capacity; at later stages, substantial shear along the fault and yielding in adjacent rock were observed. The essential features of this behaviour have been reproduced in numerical analyses which allow yielding in both fault and the rock mass. General implications of this study for future mining elsewhere in the orebody are outlined.

RESUME

Un pilier créé au cours de l'exploitation minière de l'amas de minerai 1100 à Mount Isa Mine a été observé in situ et a fait l'objet d'une étude utilisant les analyses de résistance d'élément limite et d'élément frontier. Le contrôle a fait ressortir qu'au cours de la première période d'exploitation, une faille recoupant le pilier transversalement avait limité sa capacité de charge longitudinale; à un stade ultérieur, un cisaillement substantiel dans le sens de la faille et un fléchissement de la roche adjacente ont été observés. Les caractéristiques essentielles de ce comportement ont été traduites en analyses numériques qui tiennent compte du fléchissement à la fois de la faille et de la masse rocheuse. Les conséquences générales de cette étude pour l'exploitation future dans le corps minéralisé sont indiquées.

ZUSAMMENFASSUNG

Ein im Laufe des Abbaues gebildeter Pfeiler im Erzlager 1100 der Mount Isa Gruben wurde vor Ort beobachtet und mit Hilfe der Finite Element und Boundary Element Methoden untersucht. Die Beobachtung zeigte, daß eine den Pfeiler durchsetzende Störung dessen Längslast-Tragfähigkeit zu Beginn des Abbaues begrenzte, und daß später bedeutende Schubkräfte entlang dieses Sprunges und Zusammenbrüche im umliegenden Gebirge eintraten. Die wichtigsten Kennzeichen dieses Verhaltens wurden in numerischen Analysen reproduziert, die das plastische Fließen sowohl der Störung als auch der Gebirgsmasse berücksichtigten. Die allgemeine Bedeutung dieser Untersuchung für den zukünftigen Abbau an anderen Orten des Erzlagers wird skizziert.

1. INTRODUCTION

The 1100 orebody at Mount Isa Mine, Queensland, Australia, is a massive copper orebody which is being mined using open stope methods along with the placement of cemented rock fill to enable complete extraction of the ore. In the course of mining, details of which are given in Section 2, substantial transverse pillars are created. The presence of several major faults which pass through the orebody has been found to affect the behaviour of some of the transverse pillars, and will affect others as mining proceeds. Results of laboratory testing of rock and fault samples from the orebody are also given in Section 2.

Because of the implications of fault movement for the mining strategy being employed, the development of stresses within the P519 transverse pillar, and movements on the T53 fault which transects it, were carefully monitored. These field observations are outlined in Section 3.

Numerical stress analyses have now been performed in an attempt to understand the observed behaviour of P519 pillar and to elucidate the factors which govern the mechanics of a faulted pillar in the orebody. The boundary element and finite element models which have been used are described in Section 4 and, in Section 5, the results obtained are presented and discussed.

Finally, in Section 6, the results of this study are assessed. The relative merits of the two stress analysis methods for analysing this problem are discussed, and the implications of the results for future mining elsewhere in the 1100 orebody are outlined.

2. MINING ENVIRONMENT

The 1100 orebody at Mount Isa is being mined by a series of primary open stopes on an irregular grid, with pillar recovery taking place between cemented rock fill (Alexander and

Fabjanczyk, 1981). Figure 1 shows the location of the P519 pillar in relation to other stoping and Figure 2 the layout of primary stopes and pillars in a cross-section of the stoping block immediately to the north.

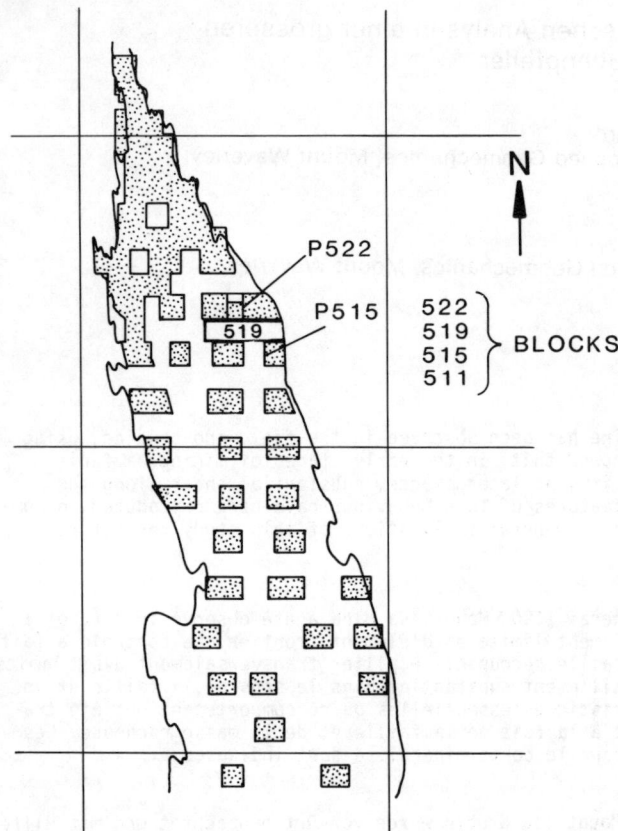

Fig. 1. Location of P519 pillar in 1100 orebody

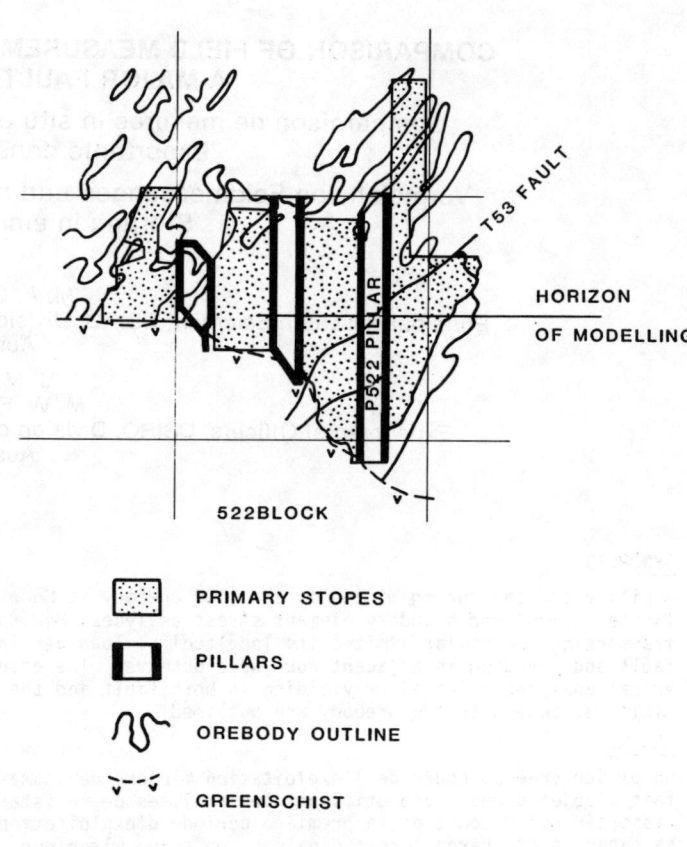

Fig. 2. Primary stopes and pillars in vertical cross-section

2.1 Geology

The P519 pillar is located in a large siliceous and dolomitic breccia mass with an irregular upper, and uniform eastern boundary formed by a shale sequence dipping at 65° to the west (Mathias & Clark, 1975). The main structure in the pillar is the T53 fault which cuts the pillar as shown in Figures 2 and 3. The fault consists of several branches up to 3 m thick with graphite and quartz infilling.

2.2 Monitoring

The monitoring of the P519 pillar was carried out as part of a comprehensive program covering the whole orebody. The program consisted of repeated stress measurements in transverse pillars using the CSIRO soft inclusion cell (Worotnicki & Walton, 1976); monitoring of fault movement using orthogonal multipoint extensometers which gave absolute shear and normal displacement; routine survey of development; and visual inspection. Transverse pillar dilation was also studied using multipoint extensometers and borehole television cameras.

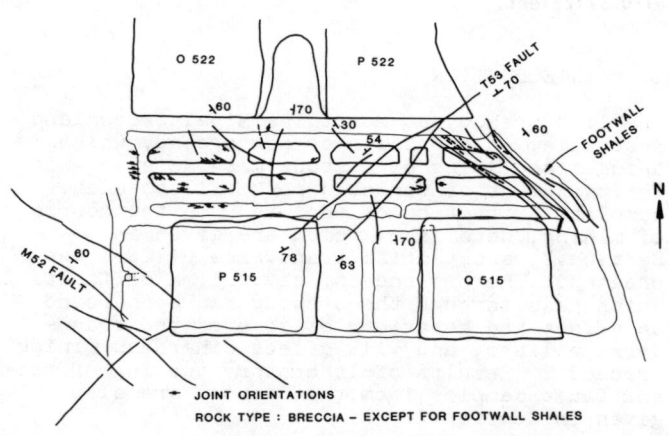

Fig. 3. Structures in vicinity of P519 pillar on horizon of modelling

2.3 Rock properties

The rock properties used in the numerical analyses were derived from extensive laboratory tests on samples from throughout the orebody and from further determination of elastic constants using bi-axial compression tests of overcored rock containing the CSIRO soft inclusion cell, as obtained in the stress measurement program. The properties are shown in Table I.

Table I. Material properties

ROCK MASS

Young's modulus	65	GPa
Poisson's ratio	0.20	
Cohesion	20	MPa
Internal friction angle	35°	
Effective tensile strength (est.)	10	MPa

FAULT ZONES

Normal stiffness (est.)	10	GPa/m
Tangential stiffness (est.)	1-5	GPa/m
Cohesion	0.3	MPa
Friction angle	22°	
Tensile strength	0	

Stresses measured in the pillar at the commencement of movement on the pillar fault were resolved on to the fault plane. The angle of friction for the fault was then obtained by back-calculation, assuming the fault was at limiting equilibrium. The value obtained is in agreement with triaxial tests carried out on joints with similar infilling.

2.4 Stress field

The virgin stress field affecting the pillar was obtained from a series of premining measurements, the depth-stress relationship being shown in Figure 4. Further information was obtained from measurements in this and surrounding pillars. In the plane of modelling (see Figure 2), the principal stresses were approximately 22 MPa N-S and 36 MPa E-W.

3. OBSERVED GROUND BEHAVIOUR

The formation of the P519 pillar can be divided into three stages: mining of the primary stopes and the successive extraction of the two connecting rib pillars (see Figure 5). Between each of these 'stages' the mined stopes were filled with cemented rock fill, to provide regional and local ground support.

3.1 Stage 1

At the time of mining of the primaries, only limited mining had taken place to the north and west, with the 515 block stopes (Figure 1) forming the southern abutment of the excavations. As the T53 fault did not transect the sections of the transverse pillar so formed, the east-west stresses through these could be expected to reach up to twice the premining stress levels, i.e. about 70 MPa. This level of stress would be sufficient to cause axial cracking adjacent to the north and south walls of the stopes, which was later observed with borehole television camera inspection.

3.2 Stage 2

The P522 pillar was mined when primary stoping had extended 700 m to the south and limited pillar recovery had taken place to the north. Dilation measurement had shown only 3 mm movement over the 35 m width of P519 pillar adjacent

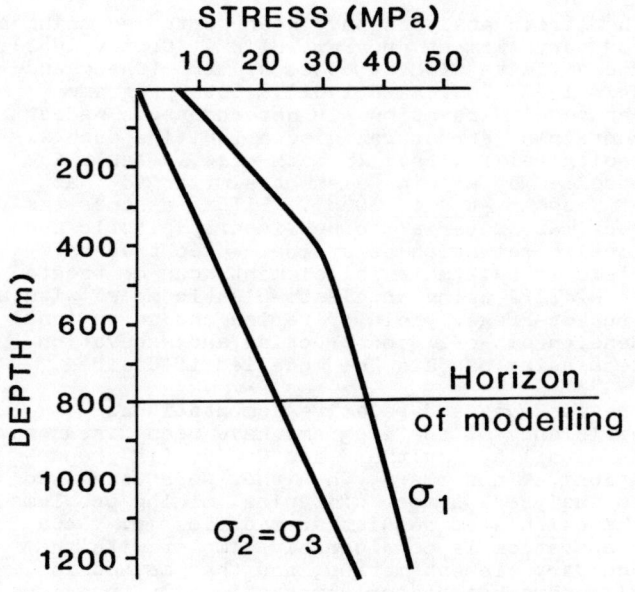

Fig. 4. Variation with depth of premining stress field

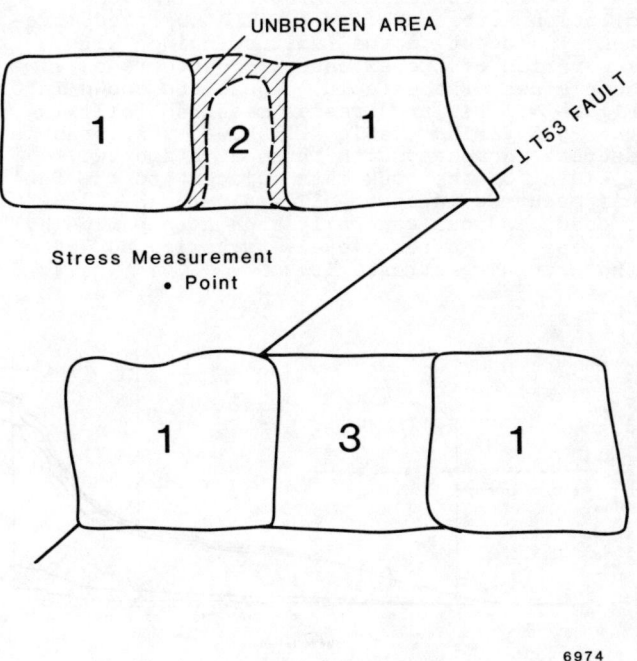

Fig. 5. Mining sequence for formation of the P519 pillar

to the extracted stope. Although fault monitoring did not occur in this pillar, monitoring in the P511 pillar to the south with a similar geometry showed only small movement, about 8 mm at a similar stage of extraction. More extensive development at a later date confirmed this as well as indicating some yielding on the north and south edges of the pillar as a result of east-west cracking. Pillar stresses measured at this time are given in Table II.

Table II. Stress measured in P519 pillar
(See Figure 5)

	AFTER STAGE 2	AFTER STAGE 3
Principal stresses (MPa)		
σ_1	53 ± 7	30 ± 1
σ_2	33 ± 4	19 ± 1
σ_3	16 ± 3	7 ± 2
Orientations (deg.) dip/dip bearing		
σ_1	11/112	74/283
σ_2	79/310	16/92
σ_3	3/203	3/183

3.3 Stage 3

With the extraction of the P515 pillar, movement was observed on the T53 fault zone as well as yielding of the rock mass adjacent to the fault, shown by opening of pre-existing structural features subperpendicular to the fault plane. Measurement of the fault movement in the P511 pillar at a similar stage of extraction showed 25 mm of horizontal shear as well as a similar amount of vertical movement with limited dilation across the fault. The observed movements adjacent to the fault coincided with separation of the extensometer anchors on both sides of the fault. These are shown in Figure 6. Pillar stresses measured following the P515 firing (Table II) showed a substantial decrease compared with those at stage two. Yielding of the rock mass adjacent to the fault has occurred in other pillars through the orebody and has generally been accompanied by decreases of stress levels to as low as 30% of the premining stress field.

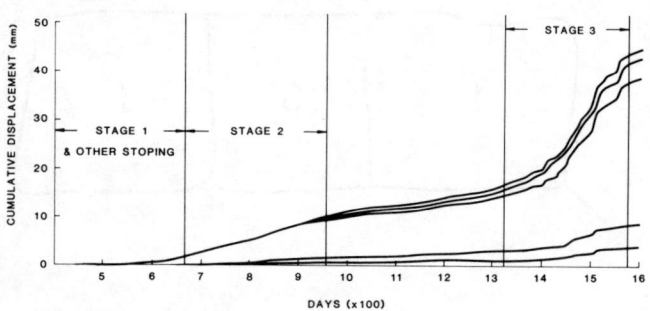

Fig. 6. Extensometer measurements across T53 fault in P511 pillar

4. NUMERICAL MODELLING

Two-dimensional stress analyses of a horizontal section have been performed (see Figures 2 and 3). This plane strain approximation would be excellent for an individual stope, and should also be satisfactory for the system considered here. The premining stress field is largely consistent with a plane strain approximation also, in that one principal stress is essentially vertical and the stresses do not vary greatly over the central region of the 1100 orebody. The final major source of potential deviation from plane strain is the dip of the T53 and other faults (see Figure 3). However, it has been shown by Coulthard and Crotty (in preparation) that a joint which dips at an angle β relative to the plane of a plane strain analysis can be represented by an equivalent joint, dipping normal to that plane. The effective parameters for an equivalent joint (upper case characters) are given in terms of those for the actual joint (lower case) by:

$$K_N = k_N \cdot \cos \beta \cdot (1 + \frac{k_T}{k_N} \cdot \tan^2 \beta)$$

$$K_T = k_T / \cos \beta$$

$$C = c / \cos \beta$$

$$\tan \Phi = \tan \phi / [\cos \beta (1 + \frac{k_T}{k_N} \cdot \tan^2 \beta)]$$

where k_N and k_T are respectively the normal and tangential stiffnesses of the joint, c is the cohesion and ϕ the angle of friction used to define its shear strength. This equivalent joint formulation has been used in our analyses to account, in part, for the dip of the major faults.

The stress analyses have been performed using a boundary element program, BITEMJ (Crotty 1982), and a finite element code, NTJTEP2 (Chang and Nair 1973, Coulthard 1982). Both programs can model excavations in heterogeneous media containing structural discontinuities such as geological faults. In each case, a fault is modelled by a joint element similar to that of Goodman et al. (1968). BITEMJ assumes the rock mass material to be linearly elastic and models excavation as a single-step process. Yield in bulk material elements can be treated in NTJTEP2 using an elasto-plastic model with a Drucker-Prager yield criterion and an optional tension cut-off; construction and excavation sequencing can also be modelled if desired.

The accuracy and relative computational efficiency of the programs have been discussed in detail by Coulthard and Crotty (in preparation). In summary, when they were each used to analyse a number of typical mining problems, the calculated results agreed closely. Data preparation is considerably simpler with the boundary element method, and that method is also somewhat faster computationally, particularly for nonlinear iterations. The principal advantage of the finite element method is its greater versatility.

5. RESULTS OF STRESS ANALYSES

The full transverse pillar was created via the mining stages shown in Figure 5. The T53 fault was included in the meshes (see Figures 7 and 8), as well as another fault to represent the effect of the faulting in the SW corner of Figure 3. In most analyses, the remnants of P515 pillar (see Figure 5) were ignored because only thin (<3 m) diaphragms of rock remained on the P519 pillar sides of the stope.

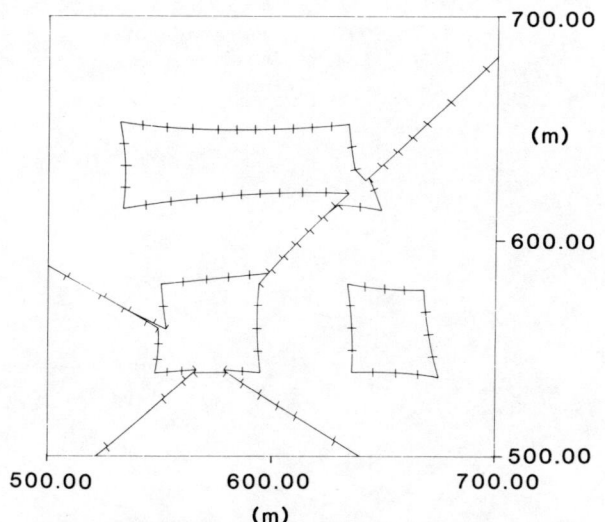

Displacement scale: ⊔ = 0.1m

6976

Fig. 7. Deformation of boundary element mesh after stage 2; case (1) in Table III

The values for the material properties used in these analyses are listed in Table I. The two programs use different sign conventions. In the results presented here, compressive stress has been taken as positive.

Premining stresses were assumed to be constant across the horizontal section represented by the meshes; in most cases, the principal stresses were taken to be as in Section 2.4. As these principal stress directions are not aligned with either of the faults, non-zero initial shear stresses will be calculated for the joint elements. In each program, these may be set to zero to allow for the possible dissipation over geological time of shear stress in an inactive fault.

5.1 Boundary element method

Analyses of the system after each of the three stages of the mining sequence were performed with BITEMJ. Each stage was modelled separately as a single-step excavation.

The boundary element mesh was equivalent to the finite element discretisation around boundaries and along faults (see Figures 7 and 8).

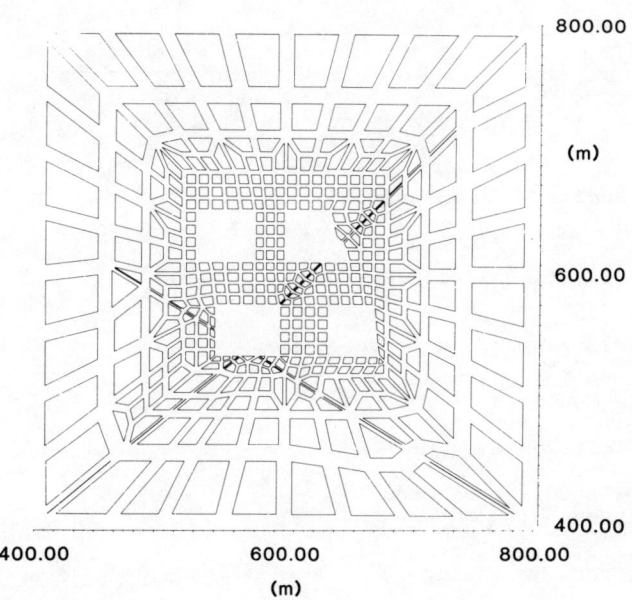

Fig. 8. Inner region of finite element mesh

Wide variations in the assumed magnitude and orientation of premining stresses caused little qualitative change in the stresses calculated near the measurement point in the P519 pillar (see Figure 5). When the faults were included, the calculated minor principal stress was usually tensile, with a compressive maximum principal stress inclined at an angle of around 15° south of east. Without the faults, these stresses were generally more compressive and their orientation varied much more, indicating that yielding of the T53 fault tends to dampen out any variation due to the premining stresses. It thus appears that only the overall magnitude of the initial E-W stress field is of importance in determining the behaviour of the transverse pillar.

Further runs including only that part of T53 fault which lies within P519 pillar altered the pillar and fault stresses and displacements by less than 5%.

The inclusion of additional stopes to the west and south-west, in analyses without faults, also yielded only small changes in the calculated behaviour of the transverse pillar.

Modelling only partial extraction of P515 rib pillar (see Figure 5) resulted in somewhat more compressive pillar principal stresses, aligned more closely with the analysis axes.

Some of the boundary element predictions of stresses at the measurement point in P519 pillar and movement on T53 fault where it transects that pillar, are presented in Table III; movements on the mesh for case (1) are shown in Figure 7.

Table III. Comparison of field measurements and calculations

| | STRESSES IN P519 PILLAR (MPa) | | | | | | SHEAR MOVEMENT ON T53 FAULT (mm) | |
| | STAGE 2 | | | STAGE 3 | | | STAGE 2 | STAGE 3 |
	σ_{EW}	σ_{NS}	τ	σ_{EW}	σ_{NS}	τ		
Field measurements	47	21	-13	19	7	-1	8 (in P511 pillar)	25
Boundary element method*								
(1) $k_T=k_N/10$ and zero initial shear stress in faults	40.4	0.5	-7.2	28.8	0.1	0.0	65	127
(2) $k_T=k_N/10$ and non-zero initial shear stress in faults.	23.8	0.3	-12.1	10.0	-0.3	-3.6	158	236
Finite element method*								
(3) full rock mass strength; $k_T=k_N/10$ and zero initial shear stress in faults	43.0	3.0	-4.5	34.8	3.0	1.2	37	77
(4) as for (3), except $k_T=k_N/2$	46.4	2.9	-3.2	42.4	3.0	2.6	16	35
(5) as for (3), except non-zero initial shear stress in faults	30.8	2.9	-8.2	18.9	2.8	-1.4	99	162
(6) as for (3) except reduced rock mass strength	43.8	6.4	-3.1	35.3	5.2	2.5	34	71

*Note: Boundary element analyses use single-step excavation process; finite element results are for sequential excavations.

At the first stage of mining, little movement in the pillar fault was predicted. Substantial shear movement occurs at stage 2, and this approximately doubles when the second rib pillar is extracted. The results agree qualitatively with the observations, but are about an order of magnitude larger than the measured values in P511 pillar. The results show the importance of the assumed pre-mining shear stresses in the faults. Using the same initial stress field for the faults and the rock mass raises the amount of shear stress which must be shed from overstressed joint elements during the mining sequence, and leads to large tangential displacements in T53 fault. Zero initial shear stresses in the faults correspondingly produce less fault slip and higher longitudinal stresses in P519 pillar. The amount of movement for a given excess shear stress depends upon the fault stiffness, as shown by the finite element results (3) and (4) in Table III.

Yielding on the faults thus limits the pillar longitudinal stress, as observed, but the ratio of the measured principal stresses (3:1 compression at stage 2) is not approached in any of these analyses, which consistently predict a tensile or only slightly compressive minor stress.

5.2 Finite element method

As boundary element analyses were not able to reproduce the measured stresses in P519 pillar, additional finite element analyses were performed. These enabled investigation of the effects of sequential excavation of the primary stopes and rib pillars, the placement of cemented fill, and non-linear behaviour within the rock mass.

The inner region of the finite element mesh is shown in Figure 8; the full mesh contained another three outer rings of elements, and covered a region 1200 m square. As boundary element analyses including additional stoping did not yield significantly different results, the concentration here on the stopes immediately adjacent to P519 pillar is not expected to introduce any substantial errors.

The excavation sequence was modelled in the three stages outlined above. The placement of fill was modelled by applying an average horizontal stress to the stope walls (about 0.4 MPa is appropriate for these stope dimensions; Barrett et al., 1978) and then the properties of the elements within stopes were altered to represent the stiffness and strength of cured cemented rock fill (Dight and Cowling, 1979). The vertical shear interaction between fill and rock must be ignored in this plane strain analysis.

Potential yielding in the rock mass was allowed for by using the parameters from Table I in the Drucker-Prager shear strength criterion with tension cut-off.

A large number of finite element analyses were performed, and the calculated stresses in P519 pillar and movements on T53 fault were compared with the measured values. Some of these results are included in Table III. The main conclusions which follow from the analyses are:

    i) The pressures from, and stiffness of, cemented fill are too low to have any significant effect on rock mass stresses in this model.
    ii) Increased shear displacement on T53 fault across P519 pillar results in a reduced longitudinal (EW) stress component in that pillar, e.g. compare, in Table III, the effects of the initial shear stress or shear stiffness assumed for the faults.
    iii) When the rock mass strength parameters in Table I are used, no shear or tensile rock yield is calculated at any stage of mining, i.e. apart from modelling sequential excavation, the finite element analyses are equivalent to the above boundary element runs.
    iv) The calculated shear displacements on T53 fault at stage 2 are reduced by 30 - 50% by sequential excavation, and the EW stress in the pillar is correspondingly higher.
    v) Omission of the dip corrections to joint parameters has very little effect on the calculated response of P519 pillar.
    vi) At stage 2 of mining, a significant N-S compressive component of stress near the stress measurement point is calculated only in analyses in which the rock mass is assigned reduced strength values (cohesion of 5 MPa, tensile strength of 1 MPa).

Stress vectors for the last mentioned case are shown in Figure 9.

The finite element analyses also reproduce only the qualitative aspects of observed behaviour, in that the E-W stress in the transverse pillar is reduced when further fault movement occurs in stage 3 of mining. Allowance for limited strength of the rock mass here shows a tendency for a compressive component of stress to develop across the transverse pillar, in agreement with measurements. Further, the finite element results suggest that the placement of cemented fill and nonlinear effects due to sequential excavation are unlikely to significantly alter the calculated stresses. There remain, then, some substantial differences between measured and calculated behaviour of P519 pillar.

Although the dip of faults has been allowed for in the analyses, other three-dimensional effects which have been ignored may be partly responsible for these differences. In particular, out-of-plane shear displacement on the fault was observed to be as large as the in-plane movement which is considered here. Uncertainty in the stiffness parameters for the faults and in the premining stress field may also contribute.

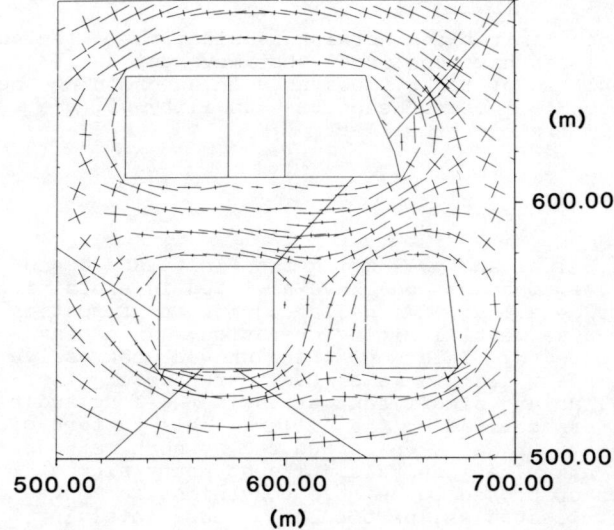

(a) stage 2

Stress scale: ⊔ = 30 MPa

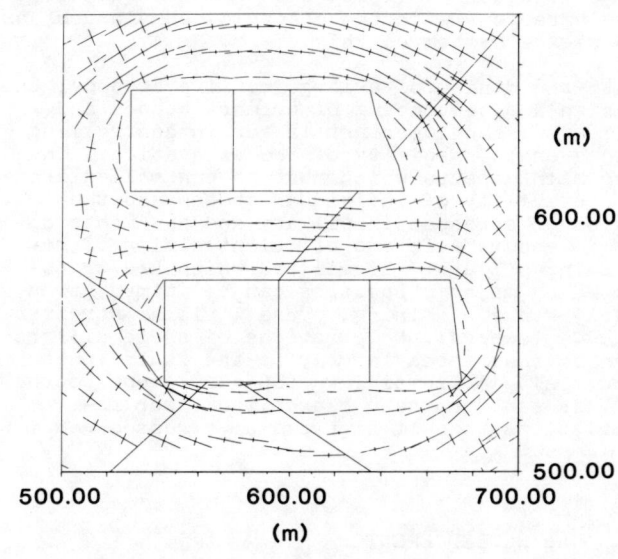

(b) stage 3

Stress scale: ⊔ = 30 MPa

Fig. 9. Stresses calculated in finite element analysis; case (6) in Table III

The observed movements and longitudinal stress could perhaps be understood better in terms of a more realistic joint material model which included a higher initial strength and shear stiffness and, after yield, lower residual values. The large transverse stress measured at stage 2 seems more difficult to explain.

These limitations notwithstanding, the numerical analyses have confirmed that movement on T53 fault is of major importance in determining the stresses in, and hence the stability of, the P519 transverse pillar.

6. CONCLUSIONS

Numerical analyses using the finite and boundary element methods have assisted in explaining the behaviour of a pillar during its formation in a sequential mining operation. As mining proceeded, field monitoring showed progressive movement along a major fault transecting the pillar and significant drops in the longitudinal stress carried by the pillar. This pattern of behaviour has been reproduced by both methods of numerical analysis. The boundary element method proved to be more convenient for parametric studies, particularly those involving mesh geometry changes. Agreement with the measured ratio of the pillar principal stresses was improved with the finite element model by allowing yield in the pillar rock mass as well as along the fault, but considerable differences remained.

It is suggested that three-dimensional effects and/or a more realistic joint constitutive model will need to be included before numerical modelling can be expected to produce quantitative agreement with the observed behaviour of as complex a system as this.

With the mining of a large massive orebody, the design and sequencing of primary stopes and pillars is an important factor in controlling the overall stability of the excavation. The two main aspects which must be controlled are the stability of individual stopes and the level of stresses in pillars and abutments so as to ensure safe and efficient mining of remaining pillars. If pillars which are transected by major structures can be formed at an early stage of mining, overall pillar stresses can be lowered. Deformations of these pillars can be restricted largely to the fault zones and mining of remaining stopes adjacent to the pillar can then take place with fewer restrictions than if high pillar stresses were present.

7. ACKNOWLEDGEMENTS

The field work for this study was performed by one of the authors, M.W. Fabjanczyk, when he was Monitoring Geologist with Mount Isa Mines Ltd. The permission of Mount Isa Mines to publish these results is appreciated. The numerical analyses have been performed as part of a joint project between CSIRO and the Australian Mineral Industries Research Association Limited; financial support from AMIRA is gratefully acknowledged. R.H. MacKinnon has assisted considerably with the numerical analyses, particularly with mesh preparation and plotting of results.

8. REFERENCES

Alexander, E.A. and Fabjanczyk, M.W. (1981). Extraction design using open stopes for pillar recovery in the 1100 orebody at Mount Isa. Symp. 'Design and Operation of Caving and Sublevel Stoping Mines', Denver.

Barrett, J.R., Coulthard, M.A. and Dight, P.M. (1978). Determination of fill stability. 12th Canadian Rock Mech. Symp., 'Mining with Backfill', Sudbury, Can.I.M.M. Spec. Pub. 19;85-91.

Chang, C.Y. and Nair, K. (1973). Development and application of theoretical methods for evaluating stability of openings in rock. Report to U.S. Bureau of Mines on Contract No. HO 220038.

Coulthard, M.A. (1982). Plane strain nonlinear finite element program NTJTEP2 - Modifications and corrections for use in mining geomechanics. CSIRO Aust., Div. Appl. Geomech., Technical Report No. 129.

Coulthard, M.A. and Crotty, J.M. (In preparation). Comparison of two-dimensional finite element and boundary element stress analyses of excavations in rock masses with structural discontinuities.

Coulthard, M.A. and Crotty, J.M. (in preparation). Effective stiffness and strength parameters for plane strain analysis of a dipping joint.

Crotty, J.M. (1982). User's manual for program BITEMJ - Two-dimensional stress analysis for piecewise homogeneous solids with structural discontinuities, CSIRO Aust., Div. Appl. Geomech., Geomechanics Computer Program No. 5.

Dight, P.M. and Cowling, R. (1979). Determination of material parameters in cemented fill. Proc. 4th Congress Int. Soc. Rock Mechanics, Montreux, 1;353-360.

Goodman, R.E., Taylor, R.L. and Brekke, T.L. (1968). A model for the mechanics of jointed rock. J. Soil Mech. Found. Div., A.S.C.E., 94;637-659.

Mathias, B.V. and Clark, E.J. (1975). Copper and Silver-lead-zinc orebodies - Mount Isa and Hilton Mines. Economic Geology of Australia and Papua New Guinea. Vol. 1, Metals, pp. 351-372, Aust. Instit. of Mining and Metallurgy.

Worotnicki, G. and Walton, R.J. (1976). Triaxial hollow inclusion gauges for determination of rock stresses _in situ_. Supplement to ISRM Symp. 'Investigations of Stress in Rock - Advances in Stress Measurement' Inst. Engr. Aust. National Conf. Publ. No. 76/4:1-8.

GEOTECHNICAL INVESTIGATION – DESIGN CRITERIA FOR A ROOM AND PILLAR FLUORSPAR MINE

Une investigation géotechnique: Critères de conception pour une exploitation en chambres et piliers d'une mine de fluorine

Geotechnische Untersuchung – Entwurfskriterien für eine Flusspatgrube im Kammer- und Pfeilerbau

C. H. Page
A. Haines
G. S. Esterhuizen
Steffen, Robertson and Kirsten, Consulting Engineers, Johannesburg, South Africa

SYNOPSIS

A geotechnical investigation is described in which the rock mass was classified from geotechnical logging of surface outcrops and cores and limited exploration development. Critical structures were identified and their influence on pillar stability was assessed using numerical techniques. The ore zone varied from 2.5 m to 20 m thick at depths of 60 m to 100 m below surface. The objective of the analysis was to obtain basic design criteria from which the extraction percentages could be estimated so that further financial evaluation of the deposit could be carried out.

RESUME

On décrit une investigation géotechnique au cours de laquelle une masse rocheuse a été classée à partir d'une étude géotechnique des affleurements de surface, de carottes et de développements exploratoires limités. Les structures critiques ont été identifiés et leur influence sur les piliers a été évaluée par des techniques numériques. L'épaisseur de la zone de minerai varie de 2,5 m à 20 m pour une profondeur de 60 m à 100 m. Le but de l'étude a été d'obtenir des critères de conception fondamentaux à partir desquelles les pourcentages d'extractions puissent être estimés, ce qui permettra de réaliser une évaluation financière complémentaires du gisement.

ZUSAMMENFASSUNG

Eine geotechnische Untersuchung wird beschrieben, in der das Gebirge durch die geotechnische Aufnahme von Gesteinsaufschlüssen und von Bohrkernen und durch begrenzte Erkundungen untertage klassifiziert wurde. Kritische Gefüge wurden festgestellt und deren Einfluß auf die Stabilität der Pfeiler mittels numerischer Methoden bestimmt. Die Erzzone variierte von 2,5 bis zu 20 m Mächtigkeit bei einer Teufe von 60 bis 100 m. Das Ziel der Analyse war es, grundlegende Entwurfsparameter zu gewinnen, die Erzhöffigkeit abzuschätzen und weitere Kostenanschläge der Erzlagerstätte auszuführen.

1 INTRODUCTION

The geotechnical feasibility investigation described in this paper was carried out for an underground fluorspar deposit situated in the Marico district of the Western Transvaal. The objectives of the study were to:

- assess the rock mass classification from limited rock testing and geotechnical logging of available surfaces and cores

- identification of potential critical structures and their frequency of occurrence

- establishment of basic design criteria for a room and pillar mining system and from these criteria produce an estimate of initial extraction ratios and support requirements

Data for the survey were available from the following sources:

- eight geotechnically logged vertical core intersections

- two geotechnically logged angled core intersections

- shaft to the footwall contact (for bulk sample)

- 8 m by 8 m by 2,5 m high excavation off shaft bottom

Due to the paucity of data, the investigation was primarily concerned with establishing lower limits for ore recovery. These lower limits would be used in a financial evaluation in the knowledge that when further underground development took place, further information would enable a more optimistic assessment of the potential recovery.

The investigation attempted to classify the rock mass in terms of structure, strength and using empirical classification techniques. From this analysis, a sensitivity approach was adopted in order to qualify the potential response of the rock mass to a room and pillar system.

1.1 Geology

The fluorspar deposits occur within the upper part of the Dolomite Series of the Transvaal System which is overlain unconformably by the cherts, shales and quartzites of the Pretoria System Series of the same system. The entire sequence has been metamorphosed by the later intrusion of the Bushveld Igneous Complex which has subsequently been eroded. A representative lithological profile based on available borehole information is shown in Figure 1.

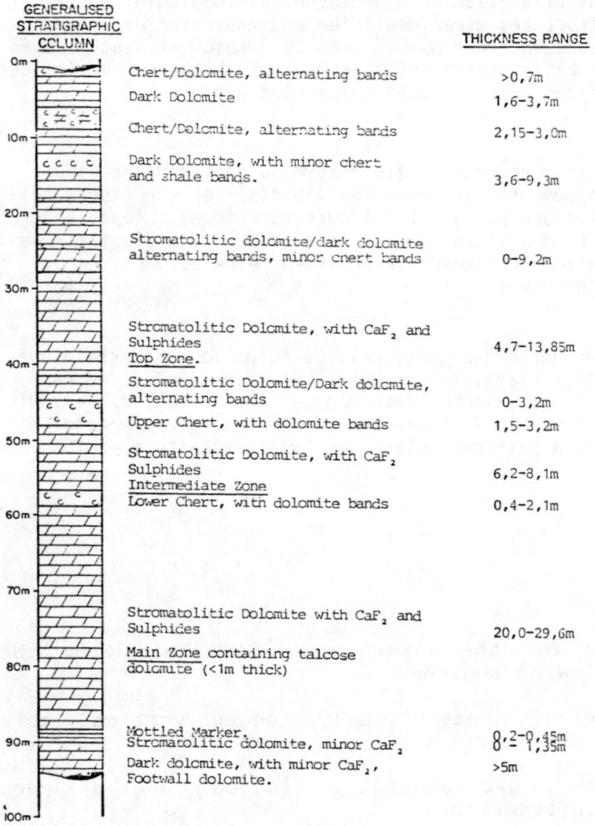

Fig 1 Lithological profile

The stratiform mineralisation of the deposit is of the replacement type with the fluorspar concentrating along the more permeable zones in the dolomites, such as algal mats and stromatolitic structures, vugs, cavities and joints. In this way, the stratiform mineralisation can also be very erratic and contains much barren or low grade dolomite.

The dip of the ore zone varies from 0° to 15° and the thickness of the economic zones varies from 2,5 m to 20 m. The footwall of the economic zones is situated within 60 m to 100 m of the surface.

2 ROCK MASS CHARACTERISATION

The assessment of the rock mass characteristics was derived from analysis of:

- aerial photographs for an initial structural analysis
- engineering geological mapping of surface outcrops
- structural logging of vertical and inclined boreholes
- engineering geological mapping of the shaft and test room
- field and laboratory testing of selected rock samples
- statistical analysis and interpretation of collected data for the NGI - Tunnelling Index (Barton, 1974)

2.1 Engineering geological mapping

Engineering geological mapping of the quartzite and chert outcrops and of the dolomite exposures underground was carried out. The properties recorded for each surface were:

- rock type
- rock hardness
- structure of discontinuity
- dip angle and dip direction
- continuity, both down dip and along strike
- infilling type and thickness
- surface roughness by recording amplitudes and base length of asperities
- water seepage
- surface compressive strength by use of the Schmidt hammer

The discontinuity orientation was plotted on an equal area polar projection and contoured to identify the major discontinuity sets. Figure 2 was derived from the mapping of the shaft and trial room. It was possible to make a direct correlation between the range of bearings of the surface features and the underground features. This enabled prediction to be made concerning the continuity of the various sets.

The underground development, although extremely limited, was of the utmost value. It illustrated the true interrelation of the identified joint systems and enabled conclusions to be drawn regarding the critical structures. The relative proportions of

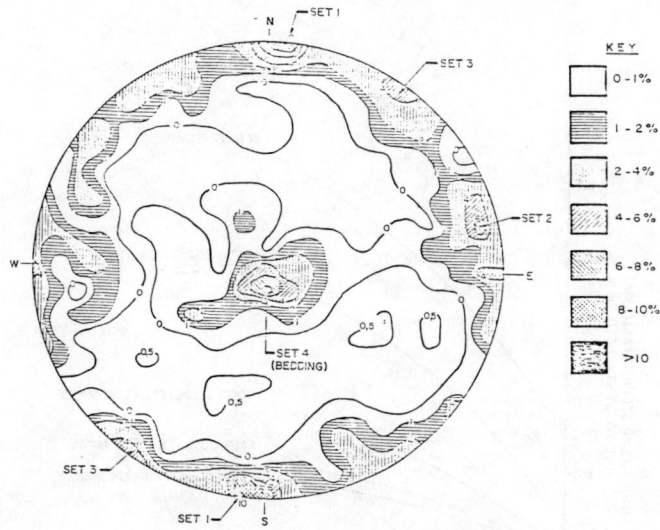

Fig 2 Distribution of discontinuities on equal area polar projection

discontinuity infilling and attitude are given in the table below.

Table I Relative proportions of discontinuity infilling

Dip angle °	Infilling Type (%)					
	A	B	C	D	E	F
0 - 4	66		34			
5 - 9	37	15	37	11		
10 - 14	36	3	44	15	2	
15 - 19	38	5	28	24	5	
20 - 24	30	22	22	17	9	
25 - 29	58	8	34			
30 - 34	44	19	25		12	
35 - 39	56	11	22		11	
40 - 44	56	11	22	11		
45 - 49	40		50			10
50 - 54		70	20			10
55 - 59	20		40	20	20	
60 - 64			75	50		
65 - 69		40	20	40		
70 - 74			20	80		
75 - 79			28	57	15	
80 - 84	9	18	27	37		9
85 - 90		11	44	22	5	18

where :

Type A = Polished pyritic graphite
Type B = Pyritic talc, often slickensided
Type C = Pyritic fluorspar
Type D = Chlorite calcite, often slickensided
Type E = Iron oxide staining
Type F = Open, often infilled with pyritic green clay, water bearing

The most significant result from the mapping of the trial underground room was the occurrence of green pyrithic clay infilled joints, F in the table above, associated with Set 1, Figure 2. The continuity of this joint set was up to 155 m.

Generally, the joint sets are subvertical, but it was evident that the clay infilled joint could occur at shallower dips. In this case it would affect both the stability of the rooms and the pillars.

2.2 Rock mass classification

A rock mass may be classified in a generalised manner in order to empirically assess the average behaviour of the rock mass when surrounding excavations of various dimensions. The result from such a classification can be expressed as a maximum unsupported span that will remain stable for a given length of time. The most commonly used system is the NGI - Tunnelling Index (Barton, 1974) which relates the following indices:

RQD - rock quality designation
J_n - joint set number
J_r - joint roughness number
J_a - joint alteration number
J_w - joint water reduction factor
SRF - stress reduction factor

in the form of :

$$Q = \frac{RQD}{J_n} \cdot \frac{J_r}{J_a} \cdot \frac{J_w}{SRF} \quad \ldots\ldots\ldots(1)$$

Levels of support installation for various excavation spans can be empirically derived from the 'Q' value.

The 'Q' value for the above analysius was 6,7 which indicates a maximum unsupported span of 10,7 m for a permanent mine opening. A second analysis was done in which it was assumed that the joint set 1 always contained a clay infilling. In this case, the maximum unsupported span became 3,2 m for a permanent opening and 10 m for a temporary opening.

This result illustrated the importance of further underground development to establish the frequency of clay infilling. The difficulty being that, with the drilling, the clay tends to be eroded by the drilling fluid.

2.3 Material testing

The core from one borehole was tested in the laboratory and the results used as a base case for comparison across the site using point load index testing. It was found that the individual strengths were related to the occurrence of fluorspar and the metamorphic mineral talc. The value of uniaxial compressive strength varied from 40 MPa for a talcose dolomite band in the upper part of the ore zone to 220 MPa for low fluorspar content dolomite towards the footwall of the ore zone.

The identification of a weak talcose band was a significant result for pillar design since the band was found to be up to 1 m in thickness.

3 PILLAR DESIGN

No underground extraction of the fluorspar has

taken place and therefore no prior knowledge of pillar behaviour exists. Pillar design was therefore based on a recent design method (Hoek, 1980) in which expressions were derived in terms of the rock mass classification that can be used to describe the rock mass behaviour under load. The criteria of behaviour was based on:

$$\frac{\sigma_1}{\sigma_c} = \frac{\sigma_3}{\sigma_c} + \sqrt{m\frac{\sigma_3}{\sigma_c} + s} \quad \ldots\ldots\ldots\ldots\ldots (2)$$

where σ_1 = major principal stress at failure

σ_3 = minor principal stess

σ_c = intact rock compressive strength

m and s = constants which depend on the properties of the rock mass and its structure and are given by the rock mass classification

With the above, Hoek (1980) determined a set of approximate relationships within broad categories of rock mass classification and then generated strength/stress contours for different pillar shapes. By following this procedure, (which is given in detail, Hoek 1980) an approximate design curve could be derived for the fluorspar ore zone.

It will be noted from Figure 3 that the pillar strength depends on the intact rock strength, but at this site it was apparent that the pillars could be made up of more than one strength rock. Since this was at the preliminary feasibility stage, it was concluded that a lower strength value comprising the talcose dolomite should be used. It could be argued that as the talcose dolomite layer remains the same thickness regardless of pillar height then the pillar strength should remain constant. However, the mode of pillar failure could be affected by the jointing, as illustrated in Figure 4. The freedom of movement of the defined blocks becomes greater as the pillar becomes taller and therefore the tendency to unravel due to the squeezing action of the talcose dolomite would also be greater. Therefore, it was concluded that consideration of pillar height remained valid.

In the previous section the presence of the thick clay infilling in joint set 1 was identified. The inclined geotechnical drilling with the information from the trial room indicated an average spacing of 12 m for this infilling. Although the jointing was predominantly vertical, there was an association of shallow angle joints with this joint set. The presence of a clay infilled joint within the pillar would severely modify its behaviour.

3.1 Pillar strength modification

Five pillar shapes were postulated as shown in Figure 5. The significant of each pillar type in terms of reduction of load-carrying capacity was assessed by numerical stress analysis techniques. The stress results were analysed using equation (2) with m = 0,7 and s = 0,004 for a good quality rock mass, to delineate overstressed zones within the pillars. Typical

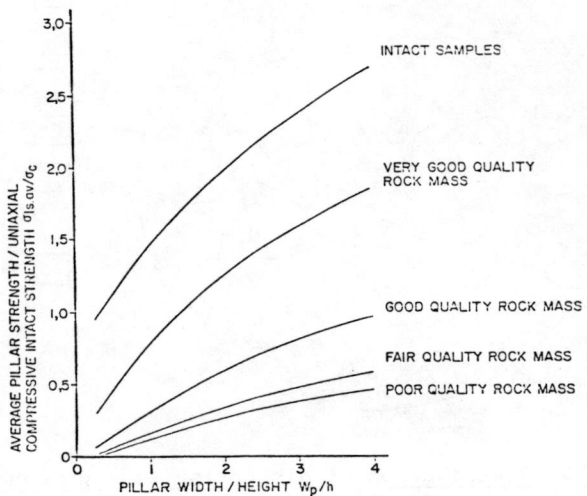

Fig 3 Pillar design curves modified for the rock mass rock quality

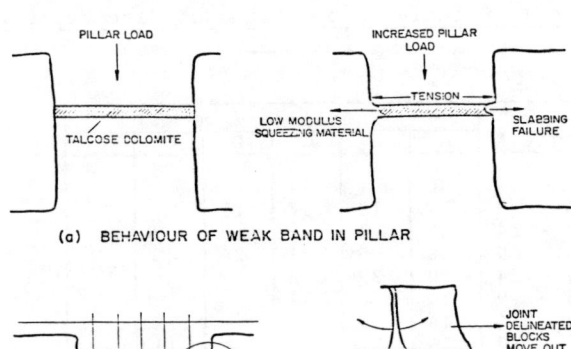

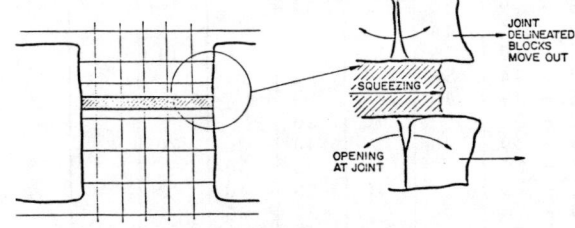

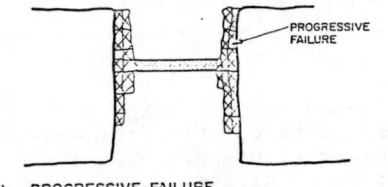

Fig 4 Pillar failure with weak band and vertical jointing

results are shown in Figure 6. From this analysis a subjective assessment of the various pillar types in terms of strength reduction was made with values shown in Figure 5.

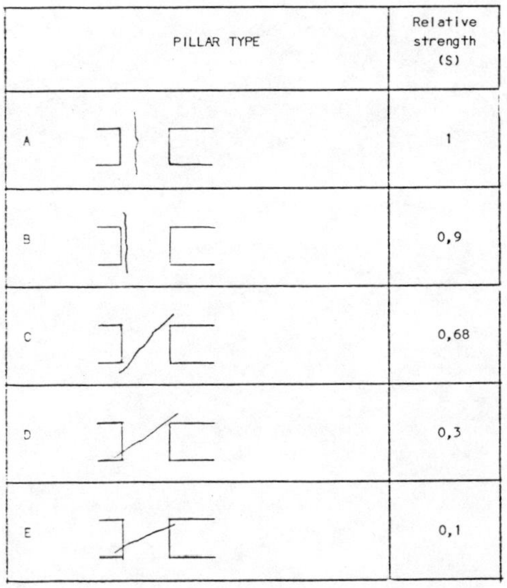

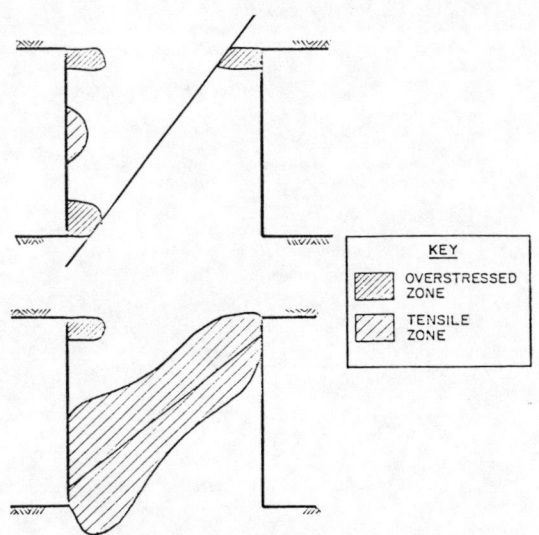

Fig 5 Pillar types and relative strength

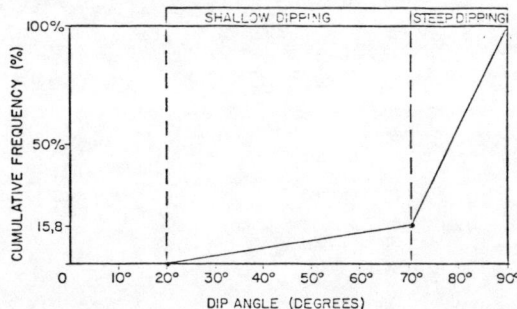

Seam height	Weighted average strength reduction factor
5	0,95
10	0,92
15	0,87
20	0,83

Fig 7 Simplified cumulative and frequency distribution of clay-filled joint dip angles

A design factor of safety of 1,6 was used for the final pillar design. First, the pillar strength was derived from Figure 3 for a range of orebody thicknesses and depths below surface. These values were then adjusted for the presence of the clay-infilled joint. Final design values derived from this process for a depth below surface of 100 m are given in the table below.

Seam thickness (m)	Extraction Limit (%)
5	75
10	64
15	55,5
20	50,5

4 DISCUSSION

A cursory assessment of the rock mass would have resulted in far higher extraction ratios than shown in the above table. It was only by following a structured investigation that the two structural weaknesses, the talcose band and clay-infilled joint were identified. It was then possible to carry out a sensitivity analysis for their effect on pillar strength. Nevertheless, it was concluded that, in general, the results were conservative. However, conservatism was appropriate at that stage of the investigation.

Further work would have entailed a detailed analysis of the decline that was to be sunk. This detailed analysis would have concentrated on gathering further data regarding the extent and thickness of both the talcose band and the clay infilling of joints.

REFERENCES

Barton, N, Lien, R and Lund, J (1974). Engineering classification of rock masses for the design of tunnel support: R Mech (6), 4, 189-236.

Hoek, E and Brown, E T (1980). Underground Excavation in Rock. 527pp. London : Inst Min Metall.

Fig 6 Typical numerical analysis results for joint effects

The occurrence of the above pillar types was analysed probabalistically based on the joint spacing data for the clay infilled joint, Figure 7. The final evaluation is shown in the table below. It will be noted that as the seam height increases then the potential for the joint to 'daylight' in a pillar wall also increases. Therefore, the effect of the structural deficiency is more pronounced.

STRATA CONTROL PROBLEMS RESULTING FROM MINING AT SHALLOW DEPTH IN THE HARD ROCK MEASURES OF THE SOUTH AFRICAN BUSHVELD COMPLEX

Problèmes de contrôle des toits en exploitations minières peu profondes dans des strates de roches dures du "Bushveld Complex" dans la République d'Afrique du Sud

Probleme der Schichtungskontrolle als Folge niedriger Teufe in hartem Gestein in den Gruben des Bushveldes in Südafrika

T. J. Kotze
Assistant Group Rock Mechanics Engineer, Fellow of the S.A.I.M.M., Associate Member of the A.M.M.S.A., General Mining Union Corporation Limited, Republic of South Africa

SYNOPSIS

Since the start of mining operations in 1968, strata control problems at Impala Platinum Limited ranged between small blocks falling out between joints, huge rock lenses sliding down along curved joints, and massive plug-like failures extending right to the surface, causing damage to surface structures. The paper describes the arguments which led to the adoption of an elaborate support system comprising continuous dip pillars plus small dip pillars in between. Local support requirements led to the development of the cluster-stick pack and are stimulating investigations into more effective support measures such as the sand-cell and the mini-grout base pack.

RESUME

Dès le début des opérations d'exploitation minière en 1968 à Impala Platinum Limited, il y a eu des problèmes de contrôle du toit: éboulement de petits blocs entre des fissures, glissement d'énormes blocs lenticulés le long de fissures courbes et affaissements massifs remontant jusqu'à la surface pour y provoquer des dégâts aux structures. Cette communication examine le raisonnement qui a abouté à l'adoption d'un système compliqué de soutènement comprenant des piliers continus avec de petits piliers intermédiaires. Les besoins locaux particuliers de soutènement ont entraîné l'élaboration du système de "cluster stick pack" (faisceau de baguettes) et incitent à rechercher des moyens de soutènement plus efficaces, tels que des disques de sable comprimé et le "mini grout base pack" (pile de bois remplie de lait de ciment).

ZUSAMMENFASSUNG

Seit der Inbetriebnahme des Bergwerkes "Impala Platinum Limited" im Jahre 1968 wurden gebirgsmechanische Probleme wahrgenommen. Diese reichten von lokalen Hangendausbrüchen, Bewegungen von Gesteinsblöcken entlang Trennfugen bis zu regionalen Verbrüchen, die sich bis zur Tagesoberfläche erstreckten. Der Artikel beschreibt die Überlegungen, welche zur Einführung eines pfeilerartigen Ausbausystems bestehend aus kontinuierlichen Barrierepfeilern und internen Kleinpfeilern besteht. Lokale Ausbauanforderungen führten zur Entwicklung von "Cluster-stick" Holzkästen und dem Einsatz von steifen Sandzellen und Betonausbaukästen. Die diesen Entwicklungen zugrundeliegenden Überlegungen werden diskutiert.

1. INTRODUCTION.

1.1 General description of the mining area.

This paper deals only with the strata control problems experienced on Impala Platinum Limited which controls four producing mines starting with Bafokeng North in the north followed by Bafokeng South, Wildebeestfontein North and Wildebeestfontein South towards the South. Mining and initial processing is conducted on a large area ± (10700 ha) belonging partly to the Bafokeng Tribe and the State in the independent Republic of Bophuthatswana which lies 16 km north-west of Rustenburg, a large town in the Transvaal province of the R.S.A.

The Merensky reef which is the main source of the platinum group metals outcrops in the south-westerly region of the lease area (fig. 1) and dips at about 9 degrees in a north-easterly direction to a depth of about 1000 m below surface at the opposite boundary.

1.2 A brief geological description.

The Merensky reef consists of a dark green pyroxenite overlying a thin 25 mm chromitite seam which is underlain by varying thicknesses of coarse pegmatoid.
Although historically the igneous hangingwall strata used to be described as massive, these measures are in fact extremely discontinuous. The immediate hangingwall consists of alternating layers of mottled and spotted anorthosite (fig. 2) with a distinct horizontal weakness occurring in the form of a layer of pyroxenite (Bastard Merensky reef) 9 m above the reef. The footwall below the reef consists of alternating layers of norite and anorthosite.

Horizontal stratification within these layers is present and although in many instances the contacts occur as gradational features, they can also be sharply defined with no cohesion across them. Bedding contacts within the pyroxenite beam have also been observed and as can be seen in figure 3 these contact planes are often un-

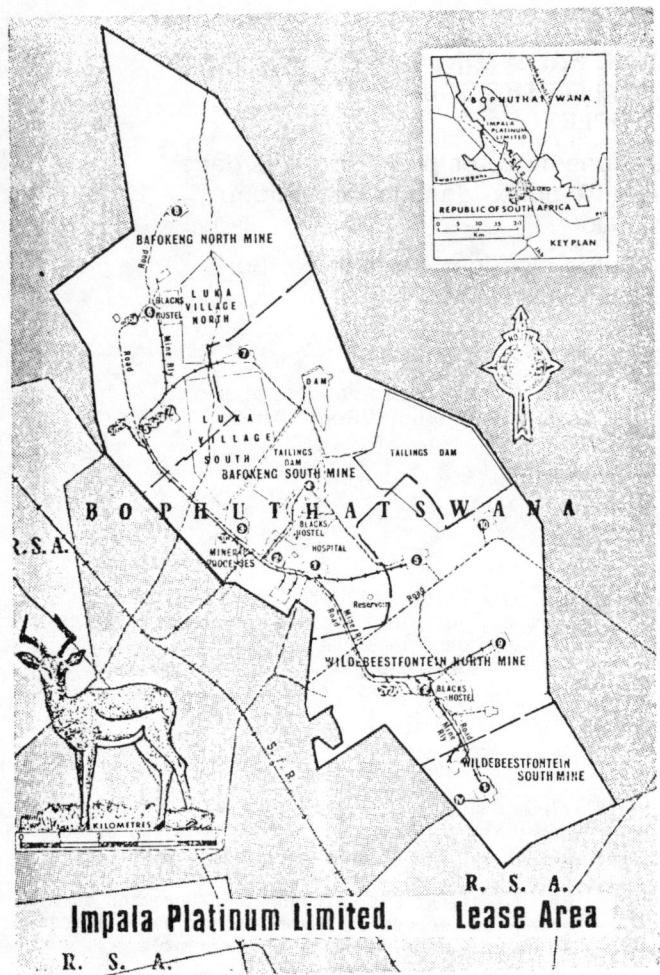

Fig. 1 General surface plan of the lease area.

SOIL	
GABBRO	25 m
MOTTLED ANORTHOSITE (N/W 5)	40 m
SPOTTED ANORTHOSITE (H/W 4)	10 m
MOTTLED ANORTHOSITE (H/W 3)	10 m
SPOTTED ANORTHOSITE (H/W 2)	1 m
NORITE (H/W 1)	5 m
BASTARD MERENSKY REEF (PYROXENITE)	3 m
MOTTLED ANORTHOSITE (M3)	4 m
SPOTTED ANORTHOSITE (M2)	5 m
MERENSKY REEF	
NORITE (F/W 1)	3-5 m
NORITE (F/W 2)	1 m
NORITE (F/W 3)	8 m
ANORTHOSITE (F/W 4)	1,5 m
NORITE (F/W 5)	3 m
ANORTHOSITE (F/W 6)	8 m
NORITE (F/W 7)	35-50 m

Fig. 2 Section showing the geological succession above and below the Merensky reef.

dulating, resulting in feather edges. Intense jointing is, however, the main feature which makes the hangingwall discontinuous.

Apart from features such as slump structures, which are called "potholes" locally, the other geological discontinuities which have played dominant roles in creating instability are faults and dykes. Because the continuity and spacing of these features largely control the incidence and magnitude of strata control problems, they have been studied in detail.

1.3 Mining method.

Since the stoping width is of the order of a metre, many of the well known Witwatersrand Gold Mining techniques are used to exploit the ore.

Crosscuts are developed at 45 m to 55 m vertical intervals from inclined or vertical shafts to a position immediately below the reef horizon. Footwall haulages are driven from these crosscuts on strike at vertical distances of 14 m below the reef. Ore-passes and travelling ways are developed to the reef from the footwall drives to connect with a series of raise-winze connections spaced at 120 m intervals.

Figure 3.

The stope faces are advanced on strike utilizing a series of strike tracks spaced at 8 m intervals on dip. The blasted ore is handloaded into one ton stope cars which are pushed manually on mono rails along the strike tracks and dumped into the dip scraper gulleys. Using double drum winches the ore is scraped over grizzlies feeding the ore-passes through which the ore is fed via hand operated chutes into five ton trucks operating in the footwall drives.

Support of workings at Impala is provided in three ways.

1.3.1 Below a depth of 100 m a primary support system comprising 20 m wide dip barrier pillars spaced at regular intervals. (Fig. 4)

1.3.2 A secondary support system comprising 5 m square pillars spaced at 32 m intervals on dip and 30 m intervals on strike. (Fig. 4)

1.3.3 A tertiary system comprising timber support which includes cluster-stick packs spaced at 4 m centres on dip and strike and timber props spaced at 2 m centres on dip and strike.

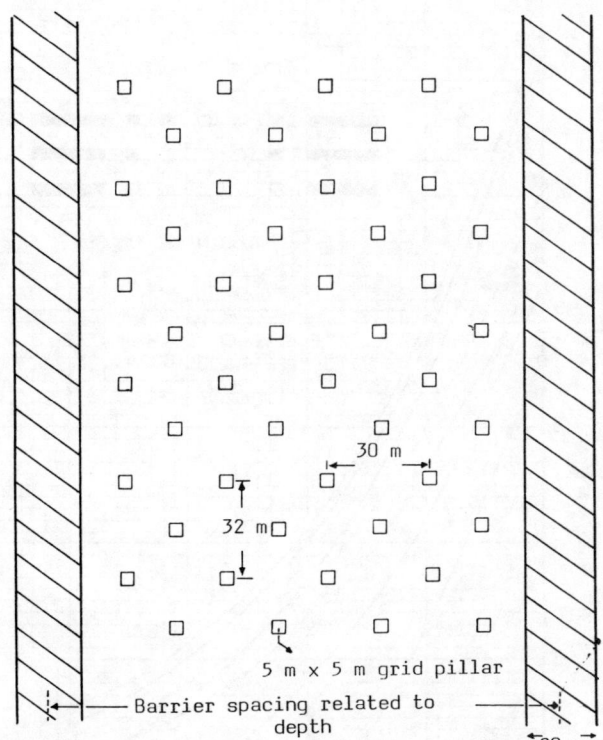

Fig. 4 Sketch showing pillar support pattern.

2. HISTORICAL REVIEW OF STRATA CONTROL PROBLEMS

Stoping operations on Impala commenced towards the end of 1968, employing conventional gold mine support elements, namely, timber mat packs and sticks. Soon afterwards in 1969, hangingwall failures occurred in two stopes. The failures occurred overnight and in each case a massive block of hangingwall sheared off at the stope abutments, its upper and lower limits being determined by faults and joints. From available information it was estimated that the failures affected the overlying 9 m of strata up to the Bastard Merensky reef horizon, which is known to have no cohesive strength. Although the depth of the workings was only 40 m and the span in both cases 90 m, no surface subsidence could be detected. Since the failures indicated that traditional timber support was incapable of supporting these massive blocks of hanging without precompression, it was decided to support the workings initially with small pillars (3 m x 3 m) spaced at 32 m x 30 m centres, supplemented by timber props at 2 m centres.

The timber props functioned adequately except for the deeper sections where the appearance of flattish curved joints (figure 5) caused the formation of unstable rock lenses which the timber props could not support without failure, usually by toppling. In the absence of any suitable substitute, matpack support was re-introduced.

Figure 5

The pillar system succeeded in keeping the stopes open until 1974 when a massive collapse of hanging occurred at Bafokeng South. A large area (600 m x 900 m) was involved at a depth of 160 m below surface. The hangingwall failure affected surface structures on this occasion, causing damage to the mine hospital. Underground observations revealed that convergences were in some cases as much as 300 mm despite the presence of the pillars. The fall itself was bounded on the up dip and the down dip extremities by a fault and a dyke respectively.

The above incident was soon followed by two further collapses at Bafokeng North and again the surface was affected. Re-assessing the situation it became apparent that in the areas where the bulk of the mining had been carried out in the past four years, losses of ground in the form of potholes had limited the actual extraction to 92 percent thereby masking the true strength of the pillars. In the stope which collapsed there were no potholes present and hence the actual extraction achieved was close to 99 percent, which meant that the pillars started failing due to overloading. Other factors such as the presence of faults, excessive spans and limited depth changed the previously stiff hangingwall loading system to a 'soft' system resulting in the violent failure of the pillars and the associated collapse of the hangingwall.

For obvious reasons the need for improving the support system became very urgent at that stage. In view of the limited strength of small pillars it was decided to opt for a system of barrier or rib pillars as regional support of the hangingwall and to retain the system of small pillars as local support of the potentially unstable hangingwall between the barriers. To improve local strata control a 'fifth' pillar was introduced in the centre of the existing 4 pillar grid system and the pillar dimensions were increased to 5 m x 5 m resulting in an overall extraction of 88 percent at a depth of 600 m. In the absence of any significant elastic convergence, local support was further improved by the development and introduction of the cluster-stick pack which will be discussed later in this paper.

3. SUPPORT SYSTEMS

3.1 Primary support systems (Barrier Pillars).

3.1.1 Barrier pillar strength.

No suitable formula for estimating the strength of a barrier pillar exists and hence barrier pillars are all 20 m wide. This width satisfies the criterion in use in South Africa namely that a long pillar with a width 10 times the stoping width will be indestructable and should on the other hand also limit the stress build-up to levels which will prevent the pillars from punching the hangingwall or footwall.

Attempts to monitor the stress build-up in barrier pillars with the C.S.I.R. doorstopper and with strain wires have not been surprisingly low. Visually the pillars are not causing foundation failures and petroscope studies of boreholes drilled into the barriers, reveal no failure propagation.

3.1.2 Barrier spacing.

In the case of Impala it was decided that barriers should be spaced at distances which will ensure that the hangingwall loading system remained stiff irrespective of whether small pillars are present or not.

3.1.2.1 In order to arrive at this "safe" mining span there are a number of options open. Figure 6 represents the results of a two-dimensional computer study where the removal of the small pillars (simulating failure) between a system of barriers was modelled. The stiffness of the hangingwall at each pillar being removed was calculated and plotted for varying barrier spacings and depths. The results are reasonably predictable and underline the fact that for decreasing depth-to-span ratios, the mine stiffness also decreases. If these stiffness values are compared with the post-failure stiffness of pillar material it can be concluded that depth-to-span ratios in excess of one should be in order. Unfortunately this type of investigation assumes elastic behaviour of the hangingwall throughout and does not take into account the effect of movements of large blocks of rock along dykes and faults, on the hangingwall stiffness.

3.1.2.2 If the disruptive affects of faults are to be included, the failure of stability of the hangingwall between two barriers can be assessed by assuming a Mohr-Coulomb failure condition for the two vertical planes next to the barriers which are subjected to the highest shear stresses. If proper estimates of cohesion and internal friction angle are available, suitable spacings can be calculated. Using this approach a depth/span ratio approaching 2 appeared to be safe and was in fact the norm aimed for in determining barrier pillar spacings on Impala.

3.1.2.3 Figure 7 is a sketch showing the tensile zones around a stope with depth/span ratios of 2 and 1 respectively. The difference is quite dramatic with the tensile zone reaching to surface in the latter case and it follows that depth/span ratios of 2

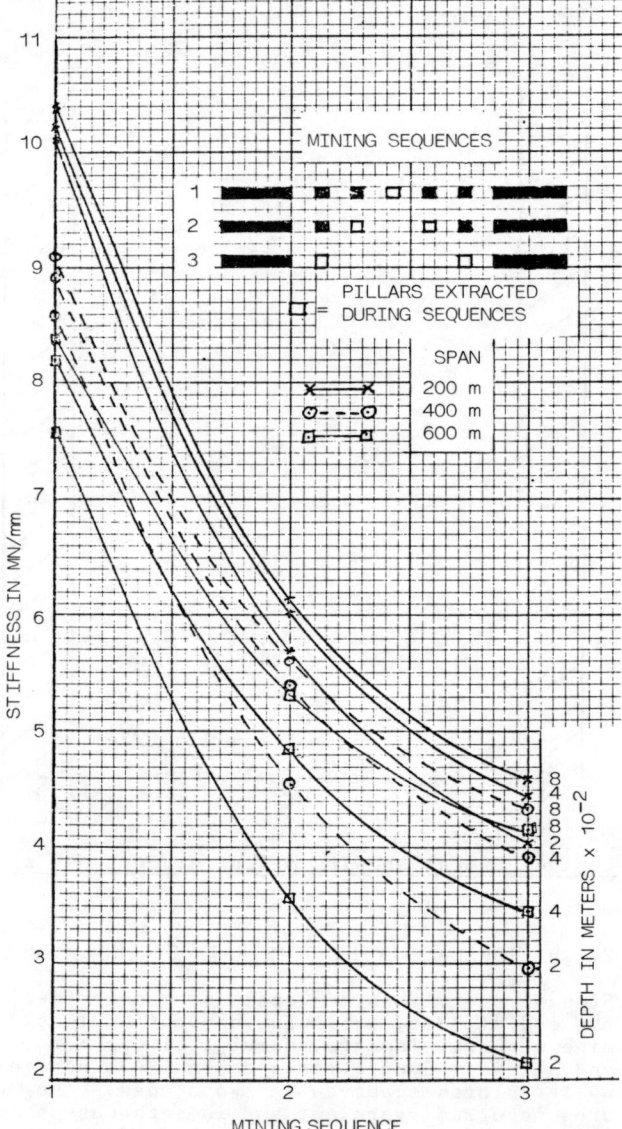

Fig. 6 Internal stiffness for varying stages of pillar failure.

are preferable if stability of the hanging is to be ensured.

3.1.2.4 Table I summarizes the elastic convergences at varying distances from the stope face for varying depth and barrier spacing. It can be seen that at a depth of 1600 m and a span of 800 m (H/L = 2) that the maximum elastic convergence at the centre of the span approaches 1 metre. This implies that, for the case where no small pillars are being left and a stope width approaching 1 metre, the need for leaving barrier pillars disappears when the depth of the working exceeds 1600 metres.

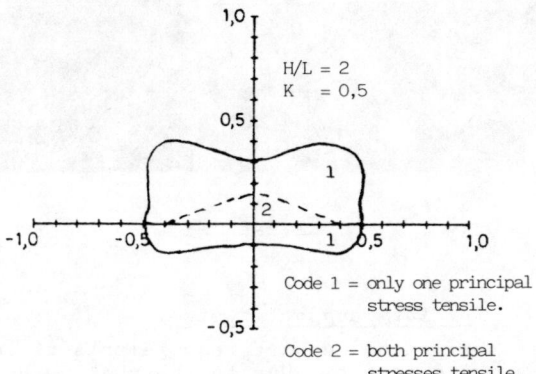

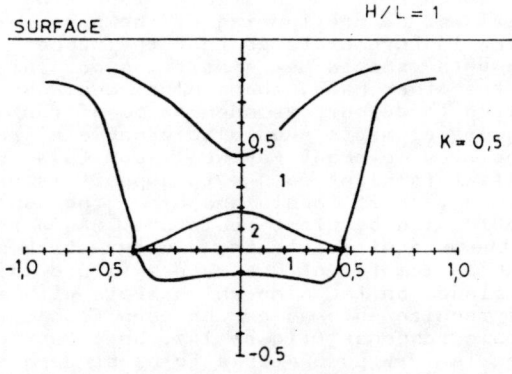

Fig. 7 Diagram showing tensile zones surrounding a stope with depth/span ratios of 2 and 1.

3.2 Secondary support system (Small Pillars).

3.2.1 Pillar strength.

Despite the fact that pillars have been studied in detail no reliable formula for strength has been developed. Although the possibility of using the well known coal pillar strength formula of Salamon is very attractive and has in fact been applied to arrive at an estimate of expected pillar strength, the pillar material is in fact so variable and jointed that the reliable application of such a formula is ruled out. A pillar may in addition consist of as many as three different rock types with compressive strengths varying between 190MPa and 40MPa. Underground observations confirm that pillar stiffness varies tremendously and that the total support load is never shared evenly between the pillars.

Table I Elastic convergence at varying distances from stope face for varying depth and barrier spacing.

Depth(m)	Barrier Spacing	Elastic Convergence in metres		
H	S(m)	1/8 Span from face	1/4 Span	1/2 Span
600	400	0,12	0,157	0,181
800	400	0,16	0,209	0,241
1000	500	0,249	0,327	0,377
1400	700	0,489	0,640	0,739
1600	800	0,638	0,856	0,965

3.2.2 Mode of pillar failure.

The mode of pillar failure is further completely determined by factors such as joint frequency, presence of horizontal weaknesses, softness of the chromitite seam and whether the contacts between the different layers are frozen or not. If for example a pillar is intensely jointed it tends to act as an assemblage of loose blocks as opposed to a solid unit with a tendency for the individual blocks to slide outwards against each other (see figs. 8 and 9) when the pillar is subjected to an excessive load.

Figure 8.

Figure 9.

Figure 10 illustrates the tearing influence of a soft thin layer of chromitite being squeezed out on the adjoining rock layers.

Figure 10.

Figure 11 on the other hand demonstrates the difference in failure when a pillar is made up of two rock types, one strong (lower layer) and one weak (upper layer).

Figure 11.

Figure 12 illustrates the more classic brittle stress failure with the associated outward buckling of the slabs.

Figure 12.

3.3 Tertiary support system.

Tertiary stope support requirements at Impala are quite severe. Due to the shallow depth allied to the pillar support system, elastic convergence is insignificant resulting in little or no pre-loading of the passive support which in turn dictates that the support elements should be as stiff as possible. On the other hand, the support could be required to support as much as 5m of rock and support elements should be capable of resisting ride movements as well. Due to a number of factors blast damage to support is extremely severe at Impala and hence the support should also be blast resistant. As a result of these requirements the cluster-stick pack with a diameter of 0,6 m (see figure 13) was developed on the mine which meets with most of the requirements as can be seen from the load-strain response (figure 14). Unfortunately this pack is very susceptible to blast damage if it is not well constructed with the result that support close to the working face is often lacking.

To overcome this drawback underground trials are at present underway to assess the blast resistance and general performance of the mini grout base pack (figure 15) and the sand-cell (figure 16) since both these support elements have extremely good support characteristics as shown in figure 14.

Since the strike spacing of support elements such as cluster-stick packs dictates that the last row of support can be as far as 6 m from the stope face, temporary support in the form of mechanical steel jacks or timber props without headboards are installed next to working face for the protection of persons working there.

Figure 13.

Figure 15.

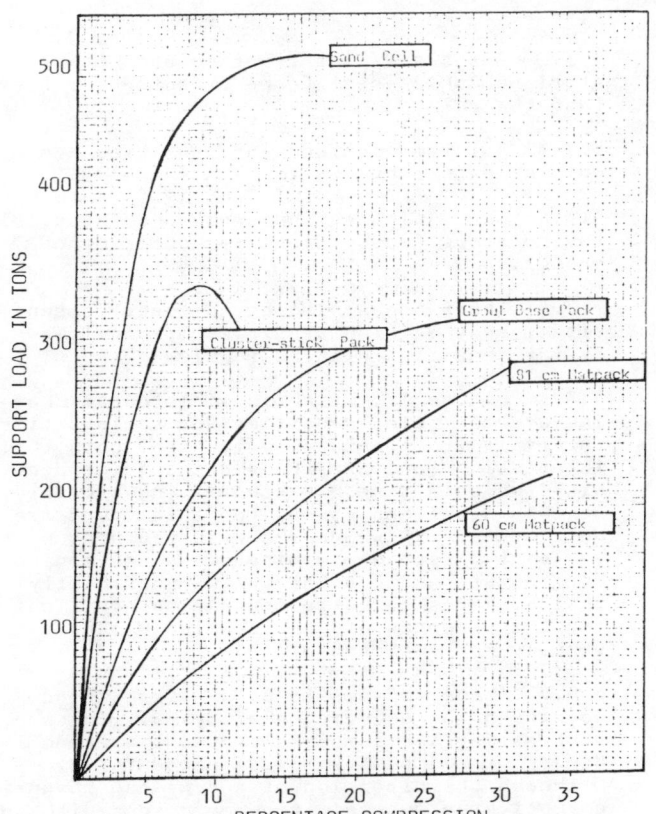

Fig. 14 The load deformation response of different packs in use and on trial.

Figure 16.

D 73

4. CASE STUDY EXPERIMENT WITH AN OPEN STOPE

As the cutting of small pillars is a notoriously time consuming operation productivity tends to be lower in pillar stopes. To overcome this drawback an open stope experiment was conducted to assess whether conventional support could cope with the fracture zone, if the open span of the stope was limited to 1/4 of the depth.

The experimental stope, a section of which had to be stopped prematurely because of dangerous strata conditions, revealed that this approach would be feasible provided that the quality of the stope support could be improved upon. It further stands to reason that as a result of the depth/span restriction, this approach would only be economically viable in the deeper (i.e. below 600 m) sections of the mine.

Convergence measurements confirmed that non-elastic movements were severe in the open stope, 300 mm as opposed to 70 mm in a 400 m span pillar supported stope. Although no flat curved joints resulting in 'domes' were encountered in the open stope, it was found that pillars had more success in preventing the lower beam from separating from the upper hangingwall measures.

In the open stope extensive separation occurred across the Bastard Merensky reef whereas in pillar supported stopes this separation is fairly insignificant.

In the Impala mining environment it may be concluded that in pillar supported stopes, the stability of the immediate skin of the stope hangingwall is not as sensitive to the quality of the stope support as is the case in an open stope.

5. JOINT SURVEYS

5.1 Objective.

Observations made in the collapsed workings underlined the dominant role which geological discontinuities played in initiating falls and collapses. Because data, such as frequency and direction of faults and dykes dictate the orientation of barrier pillars and information relating to the density and orientation of joints could alter local support requirements from stope to stope, it was decided to carry out surveys on a routine basis. Consequently a team of joint observers (total strength at present is 26) was assembled and trained. Joint surveys commenced early in 1976 and are still carried out. The results of the surveys are now being used as a management tool to determine where support should be intensified.

5.2 Routine joint survey procedures.

Each working stope is surveyed once a month but in order to make this practically feasible it was decided to carry out strike surveys at 50 m dip intervals only. This method has two limitations namely the stoped areas are not completely surveyed and the survey is heavily biased towards locating dip joints. The joint surveys embrace the following:

5.2.1 Fixing the position of the tape.

5.2.2 Locating the position of the joints where they cross the tape.

5.2.3 Measuring the angle at which they cross the tape.

5.2.4 Measuring the dip angle of the joint.

5.2.5 Estimating whether the joint is shorter or longer than 10 metres.

5.2.6 Noting the presence of gouge on the joint walls.

Based on the survey measurements the following parameters are calculated.

5.2.7 _Joint density_: This is a number which expresses the average number of joints per metre of hangingwall for the total length of survey. Joint density trends are highlighted by drawing contours of joint densities on mine plans.

5.2.8 _Joint direction_: By plotting the individual joints on stereonets using the lower hemisphere equal angle stereographic projection, the joints can be grouped together and presented as joint sets.

5.3 Results.

General conclusions are as follows:

5.3.1 Generally 3 joint sets can be identified although this does not rule out the occurrence of only one set or as many as five sets.

5.3.2 Joint sets intersecting each other at right angles seem to be exceptions rather than the rule.

5.3.3 Joint directions and densities are not influenced by faults.

5.3.4 One joint set appears to be parallel to the stope faces which are mostly mined breast.

5.3.5 There appears to be no specific density trends as high densities are followed by low densities with no regular pattern.

5.3.6 Figure 17, which is a histogram showing average joint densities for various time intervals from January 1977, highlights the interesting fact that the average joint density for each mine has been rising steadily. At this stage the causes for this trend are not entirely known but observed features such as joint direction, mining spans and mining configuration all point to the possibility that higher shear stresses may be highlighting the presence of joints which were previously not noticeable.

5.3.7 There appears to be no fixed trend governing the occurrence of dykes. In some stopes they are completely absent whereas in others quite a number may be encountered. Although the direction of a dyke can change abruptly the general trend coincides with the direction of one of the joint sets. This supports the accepted theory that the dykes were intruded subsequent to the formation of

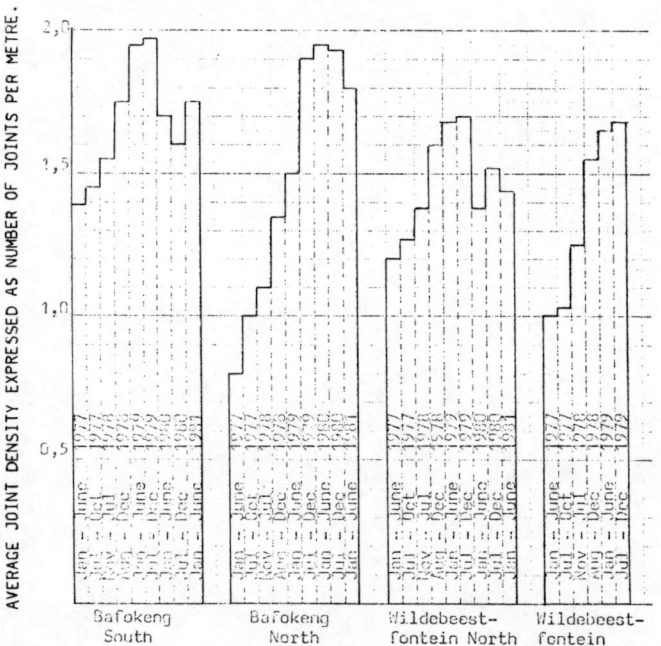

Fig. 17 A histogram showing the variation of the average joint density per mine with time.

joints as the molten dyke material would have sought existing paths of weakness, i.e. joint planes, to intrude.

5.3.8 At Impala faults are often most disruptive as far as stability is concerned since they extend right to surface and are extensive on strike but fortunately, faults are not abundant in the Bushveld Complex. On Wildebeestfontein North only one fault has thus far been encountered whilst on Bafokeng North there are six which thus far represents the upper limit as far as density is concerned. Generally the faults strike in directions tending to be parallel to the reef strike which confirms that the barrier pillars are orientated correctly.

5.4 Future improvements.

Investigations are currently being carried out to improve joint surveys. Instead of the former 50 m intervals, strike surveys are now being conducted on every stope track coupled to a dip survey along the stope face. In an endeavour to speed up surveys (since the new procedures involve much more work) orientations of the joints are not measured anymore and only the number of joints crossing the survey tapes are counted.

Since average joint density is a parameter which has a number of disadvantages, an alternative parameter such as the number of joint intersections is being investigated. A statistical investigation is in progress to determine whether a reliable correlation exists between joint density and the expectation of a dangerous situation.

6. CONCLUDING REMARKS

The paper describes some of the findings resulting from the research that has been carried out at Impala over the past 14 years. It should be stressed that most of the work was directed at finding quick solutions to the most pressing problems which were confronting the mines expecially in the early stages of exploitation.

It should also be pointed out that in the late sixties very little was known in South Africa about strata control problems in hard rock mining at shallow depth and hence there were no ready-made solutions to all the strata control problems which were experienced at Impala. Often new problems occurred before previous ones could be resolved with the result that some of the solutions were not elegant textbook material but were strictly tailored to meet the practical demands.

In retrospect this can be described as a success story for applied rock mechanics research as a support system has been evolved which has prevented further instability whilst also ensuring a high degree of extraction and hence profitability.

The routine joint surveys which are conducted monthly in all working stopes is certainly a novel feature and has succeeded in making production personnel extremely aware of the presence of joints and other discontinuities. Further work is being carried out to assess whether the results can be used to arrive at more meaningful predictions.

In conclusion it should be said that as mining proceeds deeper i.e. below the 1000 m depth mark the small pillar system will have to be replaced. Already seismicity is a daily occurrence in certain areas with tremors noticeable on surface. Pillars have exploded in some of the back areas and research is underway to determine when and how the small pillars will be phased out.

ACKNOWLEDGEMENT

The author wishes to thank the Consulting Engineer of Impala Platinum Limited for granting permission to submit this paper.

REFERENCES

Salamon M.D.G. A method of designing bord and pillar workings. J.S.A.I.M.M. November 1967.

GEOMECHANICAL INVESTIGATIONS FOR THE DEVELOPMENT OF THE CAMPIANO MINE

Investigations géomécaniques pour le développement de la mine de Campiano

Geomechanische Studien zum Bergbauplan in Campiano

G. Crea
Italian Bureau of Mines, Rome, Italy

A. Lembo-Fazio
Research Associate, Institute of Mining, Faculty of Engineering, University of Rome

R. Ribacchi
Professor of Rock Mechanics, Faculty of Engineering, University of Rome

G. Pantaleone
Solmine S.p.A., Boccheggiano (Grosseto), Italy

SYNOPSIS

In the Campiano mine (Tuscany), a pyrite and mixed sulphide lode, ranging in thickness from 10 to 40 m and with a 45° dip, will be mined. Access to the deposit has been provided by 2 shafts and a decline tunnel that reaches a depth of 700 m below ground surface. A sublevel stoping method will be adopted and the stopes will be subsequently backfilled. This paper discusses the laboratory and in situ mechanical characteristics of the rock formations, the support systems and the static behaviour of the excavations, with special reference to the situation following an accident which caused partial flooding for a few months. Geomechanical analyses of various stoping schemes have provided preliminary indications as to the stability of the stopes and on the influence of the exploitation on the shafts and other service excavations.

RESUME

Dans la mine de Campiano on va exploiter un filon de pyrite et divers sulfures épais de 10-20 m et avec une inclinaison de 45°. Parmi les travaux préparatoires il y a une galerie hélicoïdale et deux puits qui arrivent à une profondeur de 700 m au dessous de la surface. L'exploitation sera effectuée par chambres vides avec sous-niveaux qui seront ensuite partiellement remblayées. On décrit les propriétés géomécaniques des roches qui sont intéressées par les ouvrages et le comportement statique des excavations en se rapportant plus particulièrement au cas d'une venue d'eau qui a inondé les excavations pendant quelques mois. On a aussi exécuté des analyses géomécaniques qui ont donné des indications préliminaires sur la stabilité des chantiers d'abattage et sur l'influence de l'exploitation vis-à-vis les puits et les autres ouvrages préparatoires.

ZUSAMMENFASSUNG

In Campiano (Toskana) wird ein Sulfyde-Erzgang im Durchmesser schwankend zwischen 10 und 40 Metern und bis zu einer Neigung von 45° abgebaut. Der Zugang zu dem Lager wird durch zwei Schächte und eine Spiralrampe gewährt, die eine Tiefe von 700 Meter unter der Oberfläche erreichen. Es wird eine Teilsohlenbau-Methode angewandt, und die Abbauräume werden mit Fremdversatz nachgefüllt. Es wird hier der Einfluß der geomechanischen Eigenschaften der verschiedenen durchquerten Formationen auf das statische Verhalten der Vorbereitungswerke und der zukünftigen Abbauwerke untersucht. Die Untersuchungen der verschiedenen Abbauschemas haben vorbereitende Hinweise zur Standsicherheit der Bergwerke und Einfluß auf die Abbauwerke in Bezug auf die Hilfsbaue und Schachtarbeiten erteilt.

1. INTRODUCTION

Sulphide ore deposits are widespread throughout southwestern Tuscany ("Metalliferous Hills"). The former Boccheggiano mine exploited a pyrite and chalcopyrite lode down to a depth of 220 m where the mining activities were discontinued because of the decrease in Cu content. In the early seventies deep exploration drillings discovered, at a depth of 500 m below the surface, a down-dipping thick lode (Campiano orebody) consisting of massive pyrite with concentrations of mixed sulphides. So far the orebody has been fully explored for about 500 m in strike and about 220 m in depth, but the prospective reserves are expected to cover a much larger area. The lode has a dip of 45° and a thickness varying between 10 and 40 m.

The development system of the mine consists of two main shafts, a spiral decline tunnel connected by crosscuts to the shafts, a number of ventilation raises, three main level galeries along the strike, ancillary ramps and a cavern for the underground precrushing of the ore (Fig.1). The excavation of the decline tunnel, having an inclination of about 20%, made the sinking of the shafts easier, and besides, during the exploitation phase the mining machines will have direct access to the stopes.

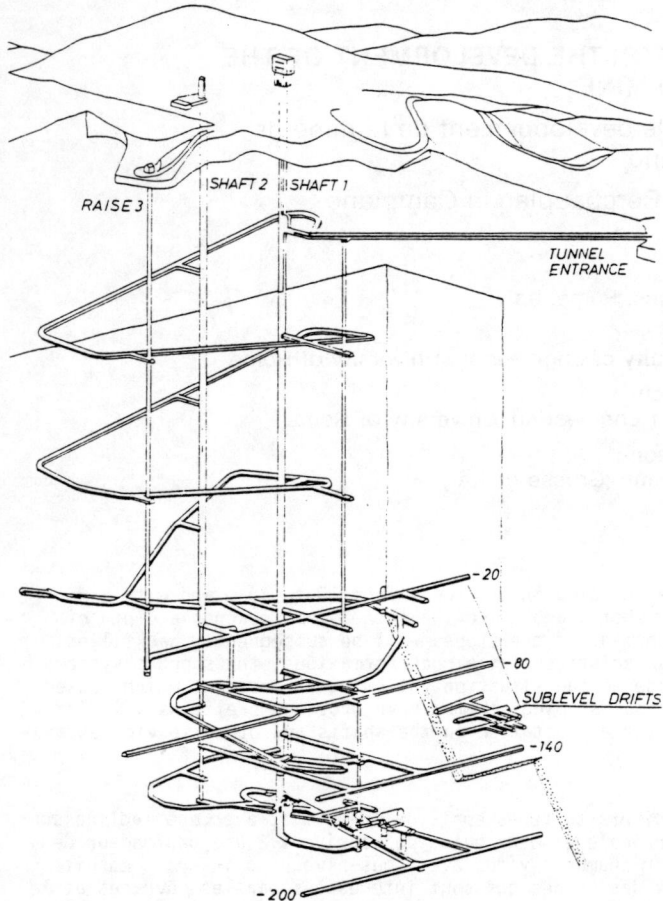

Fig. 1 - Sketch of the development works for the Campiano mine.

The tunnel was excavated full face by conventional drilling and blasting. Only in the first stretch of 370 m was a boom-type Paurat roadheader tried, but it did not provide satisfactory results because of the presence of hard layers at the face. For the sinking of the shafts, preholes, 0.28 m in diameter were drilled from the surface by means of a Raise Borer Ingersol RBM7. They were subsequently reached from the tunnel by means of crosscuts and reamed upwards to a diameter of 2.4m. The final section of the shaft was obtained by sinking downwards with the conventional blasting technique; muck was disposed of through the central hole and the decline tunnel.

The characteristics of the development works are summarized in Table 1.

The exploitation will be carried out with a sublevel mining method; the stopes will be backfilled with cemented rockfill obtained from a surface quarry. The works will roughly progress from the two extremities of the explored area towards the central zone where the shafts and the decline tunnel are located.

A major difficulty for the mining activities will be represented by the high original rock temperature (up to 75° at a depth of 700 m) and by the high conductivity of the orebody.

Table 1   Excavation Techniques

Decline Tunnel
Length	3550 m
Width	6 m
Height	4.5 m
Cross-section	25 m^2
Holes per round	73
Advance per round	2.2 m
Specific charge	2.64 kg/m^3
Contour holes spacing	0.5 m

Shafts (final enlargement)
Diameter	6.1 m
Holes per round	70
Advance per round	2.16 m
Specific charge	1.59 kg/m^3

2. GEOLOGICAL SITUATION

In southwestern Tuscany a basement of Paleozoic rocks and miogeosynclinal deposits of the Tuscan units were overlain by eugeosynclinal allochthonous formations ("Liguride Complex") during the Alpine orogenetic phase (late Cenozoic).

In the Campiano area, (Fig.2) the basement rocks are represented by phyllites with anhydritic intercalations ("Boccheggiano Phyllites") which are considered of Silurian-Devoniano age (BAGNOLI et al.,1979). The miogeosynclinal deposits are represented in depth by anhydrites of the Triassic age which, in the outcrops, have been turned into vuggy limestones. Finally, the "Liguride Complex" is represented here by the "Palombini shales" formation, of Cretassic age, consisting of shales interbedded with competent layers of siliceous limestones ("palombini").

The deformations of the basement rocks determined by pre-alpine orogenic phases are so far largely unknown. The Alpine orogenesis determined compressive structures (nappes and overthrusts). However, its latest phases led to distension structures, marked by direct faults having mainly a NW-SW strike and in some cases a SW-NE strike.

The main structural feature of the Campiano site is one such fault, known as the Boccheggiano fault, which has a throw of about 1000m. Together with its subsidiary fractures, this fault may have been the route along which the ore-bearing fluids ascended. The orebody was formed through a selective replacement of the evaporites with pyrite at an earlier stage and subsequently with mixed sulphides and magnetite (VIGHI,1972).

3. GEOMECHANICAL PROPERTIES OF THE CAMPIANO ROCK

3.1 Palombini Shales

This formation was crossed by the decline tunnel down to a depth of 500 m (Fig. 2); for a detailed description of its characteristics, reference can be made to a previous paper (BERRY et al.,1977), therefore only the most important data are recalled here.

Its main component (70%) is a mudrock; lithic layers of siliceous limestones and (more rarely) sandstones are also present. In most outcrops the formation is rhythmically and regularly stratified with layers some 0.1-0.5 m in thickness. On the contrary, throughout the excavation of the

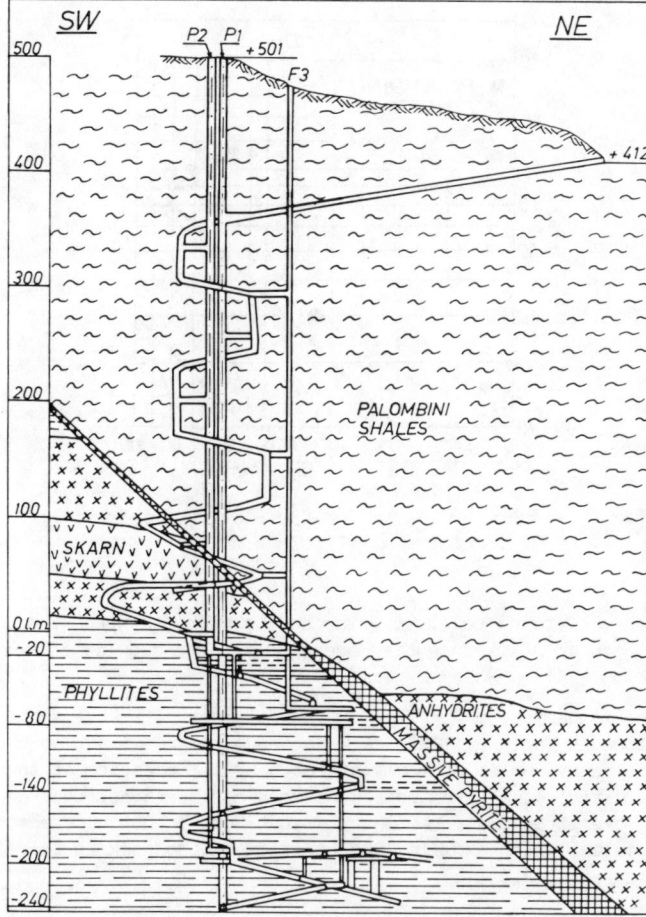

Fig. 2 - Geological situation of the Campiano orebody.

Fig. 3 - Typical disarranged structure of the "Palombini shales" at the face of the tunnel.

tunnel and of the shafts, the formation always showed a brecciated and disarranged structure which very likely was originated by complex tectonic deformations (Fig.3). The striking difference between the ordered structure of the natural outcrops and the disarranged structure which appears to be prevalent in the bulk of the formation, has been observed also in similar Italian formations (MANFREDINI et al., 1981) and may be attributed to the selective effects of weathering and erosion on the two types of material.

The mudrocks of the "Palombini shales" formation can be distinguished into three lithological types:

Shale A It is formed mainly of clay minerals (85%) having a strong preferred orientation which induces a marked fissility; the clay minerals are illite (75%) and chlorite (25%). Besides the fissility planes, weakness surfaces perpendicular to the latter are also present. Shale A is the typical mudrock forming the ordered sequences of the "Palombini shales" formation.

Shale B Its mineralogical composition is virtually identical to that of the previous lithotype; however, it tends to foliate according to curved surfaces (sometimes slickensided) forming flakes or scales. Often a network of calcite veinlets is present. The rock does not break down when immersed into water; it is however sensitive to degradation caused by the alternation of wetting and drying cycles (slake durability index Id_2=90%.

Shale B is by far the prevailing pelitic component of the disarranged or brecciated domains of the Palombini shale formation. It obviously derives from Lithotype A following diffused shear deformations; the formation of strong cohesive bonds both among the scales of the pelitic material and also between the latter and the lithic fragments.

Shale C It consists of a breccia having a pelitic matrix with fragments of shale B and of "palombini". The mineralogical composition of the pelitic matrix differs from that of lithotype A and B because of the presence of kaolinite (10-20%); it could derive from deep alterations of the two previous lithotypes. This material quickly disintegrates in water. (Slake durability index Id_2=65%.

The predominant lithologic type involved by the excavation is shale B, and so the mechanical tests were carried out mainly on this material.

The results are summarized in Table 2. The low porosity indicates the intense diagenetic actions which affected the original clay deposit. In the uniaxial compression tests, failure was markedly brittle and it occurred along pre-existing weakness surfaces. In the triaxial compression tests the formation of a single shear plane was generally observed; in some cases the residual strength (or better the strength for large strains) was determined. The triaxial strength values (Fig.4) can be represented by means of a linear Coulomb law; however, the intercept on that axis is somewhat higher than the true uniaxial strength, as is common for many types of rock.

The strength and modulus values of the shale, even if low (Fig.5), turn out to be higher than the typical values of many similar Italian "scaly" shales (AGI 1979); this could

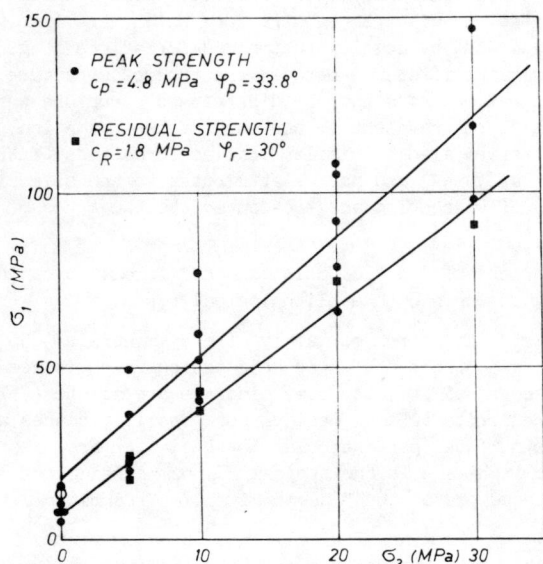

Fig. 4 - Triaxial strength of the B-shale; the Coulomb line was determined by a regression analysis excluding the results of the uniaxial samples.

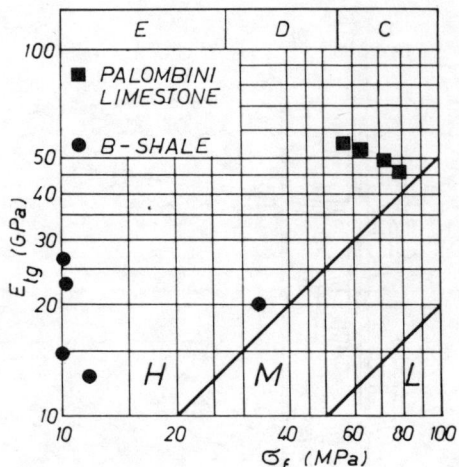

Fig. 5 - Deere's diagram for the rock materials of the "Palombini shales" formation.

derive from the high degree of diagenesis to which the rock was subjected.

The "Palombini limestones" are formed by an aggregate of calcite crystals of about $3-4\mu$ in size and of larger quartz granule ($30-40\mu$); the calcite content varies between 40% and 70%. A typical feature of the "Palombini" is the presence of plane calcitic veins and veinlets between 10 and 0.1 mm in thickness which are mostly perpendicular to the bedding (Fig.3).

The mechanical properties of the Palombini limestones are shown in Table 2. This rock is tough and resistant, but its behaviour is strongly affected by the presence of the calcitic veins which represent weakness surfaces; they lower the strength of the material but they do not reduce its modulus, which explains the position of the rock in Deere's diagram (Fig.5).

TABELLA 2 - Geomechanical properties of the Campiano rocks.

	B - SHALE			LIMESTONE			PHYLLITE			PHYLLITE			PHYLLITIC BRECCIA			ANYDRITE			SKARN			MASSIVE PYRITE		
	N	$\bar{x}$	v	N	$\bar{x}$	v	N	$\bar{x}$	v	N	$\bar{x}$	v	N	$\bar{x}$	v	N	$\bar{x}$	v	N	$\bar{x}$	v	N	$\bar{x}$	v
DRY DENSITY ϱ_d (t/m³)	25	2.69	1.7	12	2.71	1.9	65	2.76	2.8	-	-	-	4	2.88	25.	40	2.92	1.3	21	3.35	7.	72	4.37	7.9
MATRIX DENSITY ϱ_r (t/m³)	6	2.75	1.9	-	-	-	6	2.78	3.6	-	-	-	-	-	-	-	-	-	-	-	-	8	4.47	7.4
POROSITY n (%)	21	1.84	54.	-	-	-	10	.84	76.	-	-	-	4	.76	43.	7	.5	37.	9	1.4	69.	26	2.7	-
SEISMIC VELOCITY * v_l (km/sec)	17	3.8	26.	8	5.02	18.	4	3.4	22.	12	5.41	4.6	4	4.98	59.	12	4.55	12.	10	5.35	12.	13	5.04	20.
COMPRESSIVE STRENGTH σ_f (MPa)	7	13.1	62.	4	68.	-	5	79.	20.	6	17.	65.	3	100.	-	10	76.	14.	8	137.	19.	23	102.	23.
TENSILE STRENGTH σ_t (MPa)	4	2.0	-	8	6.9	10.	4	4.9	70.	9	2.4	23.	4	6.5	-	25	5.2	38.	20	13.2	39.	55	5.6	42.
YOUNG'S MODULUS * E_{sec} (GPa)	6	18.5	30.	4	49.	-	5	4.8	50.	6	16.3	37.	3	64.7	-	8	36.5	50.	6	70.	30.	17	63.	88.
POISSON'S RATIO ν_{sec}	4	.14	-	-	-	-	5	.02	90.	6	.11	87.	3	.2	-	4	.11	-	6	.23	34.	17	.2	78.
COHESION c (MPa)	-	4.8	-	-	14.6	-	-	2.35	-	-	9.	-	-	-	-	-	28.	-	-	26.	-	-	31.	-
ANGLE OF FRICTION φ (°)	-	33.8	-	-	43.	-	-	35.6	-	-	33.	-	-	-	-	-	37.	-	-	45.3	-	-	42.8	-

N number of samples, x mean values, v variation coefficient(%), * harmonic mean

As regards the characteristics of the rock mass, it must be pointed out that the rigid layers are strongly bound to the shale portions. Except for limited areas, extended discontinuity surfaces or joints are not frequent and therefore the characteristics of the rock mass may be considered as being approximately isotropic and comparable to those of the shale B material.

This is confirmed by the results of the seismic refraction surveys which indicate the seismic velocity of the rock mass as varying between 4.5 and 4.7 km/s. These values closely correspond to those determined on laboratory samples of the B shale at confining pressures greater than 10 MPa.

3.2 Boccheggiano Phyllites

The formation consists of silty-pelitic sediments that were transformed into sericitic and chloritic phyllites by a weak epizonal metamorphism. The rock has a strongly oriented texture and a marked fissility. Locally, evaporitic seams or evaporitic-phyllitic rithmites are also present within the formation.

Due to tectonic movements along the fault, the footwall rock has been turned, for a thickness of a few metres, into a breccia which is strongly cemented by quartz and other epigenetic minerals; the original anisotropy of the rock was therefore greatly reduced.

The results of the laboratory tests are summarized in Table 2 where the loading conditions, respectively perpendicular ($\perp$) and parallel ($//$) to the schistosity, are also indicated.

The results of the triaxial compression test (Fig.6) were analysed on the basis of a Mohr-Coulomb strength law and on the basis of a power law; the latter gives a better approximation. The influence of the orientation of the specimens with respect to the direction of the load is felt more by the "cohesion" parameters.

Deere's diagram (Fig.7) shows a high scatter of the strength and of the deformability values which can be partially caused by the disturbance induced by coring and by sampling. The same factor is very likely responsible for the opening of microfissures lying along the schistosity planes which caused a decrease in the minor principal modulus and a strong dependence of the latter on the applied stress. The moduli of the in situ formation are therefore probably higher and the elastic anisotropy lower as compared with the values obtained in laboratory tests, as was observed in many similar Italian schistous formations (RIBACCHI, 1980).

The mechanical characteristics of the phyllitic breccia (Table 2) are generally better than those of the phyllite formation because of the cementing actions of the epigenetic materials.

As regard the in situ characteristics, the Boccheggiano phyllites are crossed by three well-clustered sets of joints (Fig. 8). The first one corresponds to the schistosity plane having a southward dip and an inclination of about 10°; the joint surfaces are smooth, slightly wawy and lustrous; in most cases the joints found on the walls of the excavation are really pre-existing weakness sur-

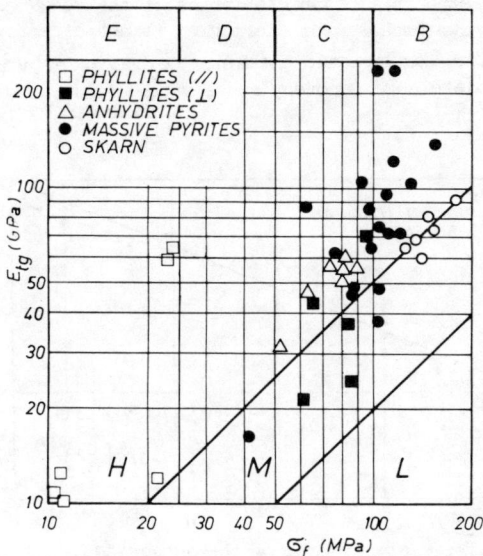

Fig. 7 - Deere's diagram for the rock materials of the Campiano mine.

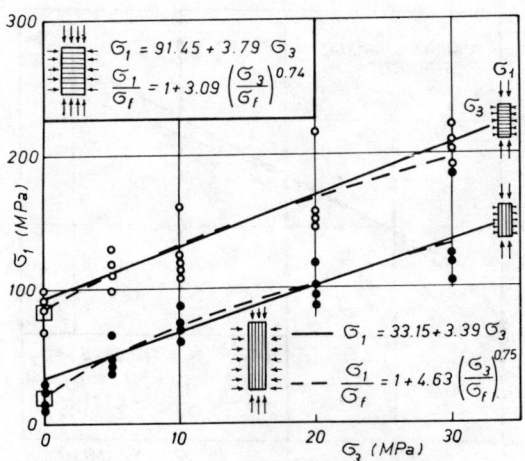

Fig. 6 - Triaxial strength of phyllites.

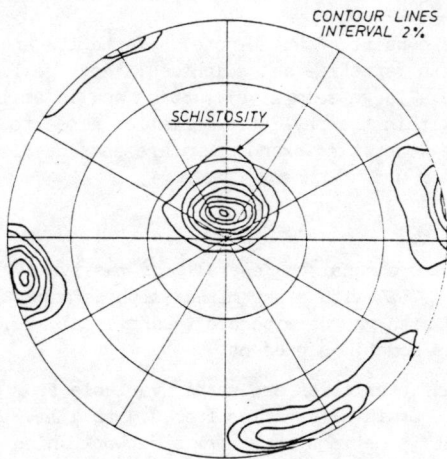

Fig. 8 - Orientation of the joints in the "Boccheggiano phyllites". Schmidt equiareal projection, lower hemisphere.

faces opened up by the effects of blasting. The other two
sets have a subvertical orientation and a mean spacing
of about 0.40 m; the joints are plane and extensive and
the surfaces show a thin quartz coating. It is interesting to mark that the orientations of the joints are not
related to that of the Boccheggiano fault, except in the
zone just adjacent to the lode.

3.3 Anhydrites

The evaporite formation which forms the hanging wall of
the lode consists almost exclusively of anhydrites having
a saccaroid texture and (average) grain-size of a few
tenths of a millimetre; limestone and dolomite intercalations are scanty. The experimental data summarized in
Table 2 and in Figs. 7 and 9 show that the scatter of the
geomechanical properties is low; this can be attributed
to the mineralogical and structural homogeneity of the
rock. The structural characteristics of the rock mass are
not well known because this formation is not directly involved in the excavations; the mining exploratory drillings
show a complete core recovery and a wide-spacing of the
joints.

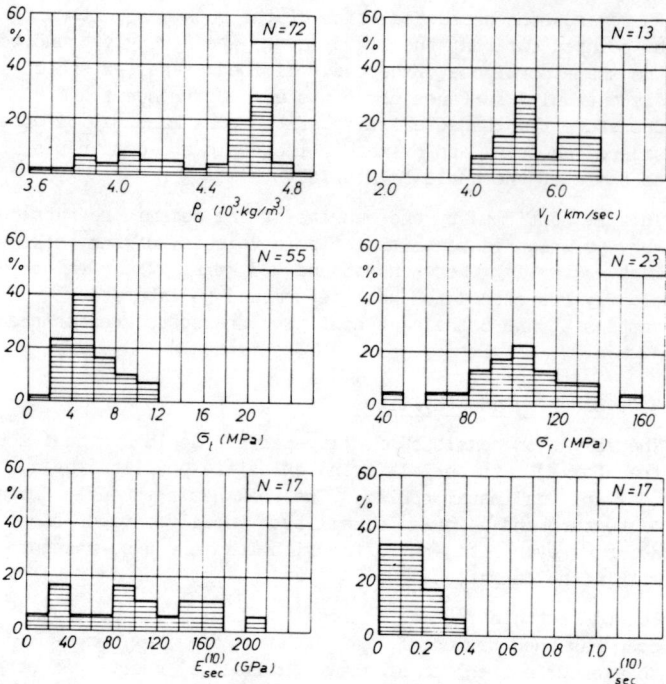

Fig. 10 - Histogram of the mechanical properties of the massive pyrite.

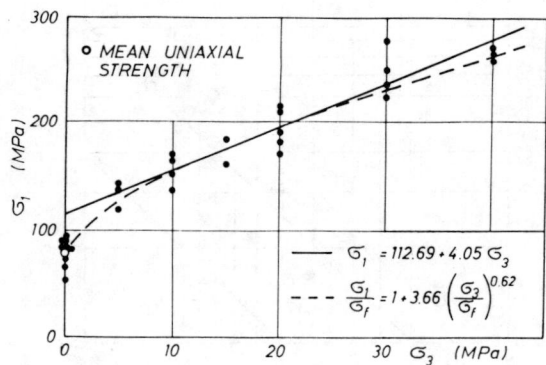

Fig. 9 - Triaxial strength of the anhydrites.

3.4 Skarn Rocks

Skarn rocks consist of an epidote, hederbegite and garnets
combined with magnetite and sulphides; they are found at
Campiano in a large seam at el. +80 m and in small masses
scattered within the phyllitic formation close to the footwall of the lode. The skarn rocks are characterized by
high strength and toughness (Table 2).

3.5 Pyrite Ore

The Campiano ore consists generally of massive pyrite (mean
content over 75%) with other mixed sulphide minerals (Pb,
Zn, Cu) and with gangue minerals (quartz, calcite, skarn);
magnetite is sometimes present.

Textural characteristics are rather variable from one site
to another. Grain-size varies from 0.1 to 1 mm. Fractures
are frequent; sometimes they form a network which divides
the rock up into elements of about 10 mm which are more or
less cemented by quartz.

The laboratory tests (Table 2 and Figs. 7,10,11) show a con-

siderable scatter which can be ascribed to the variability
of the pyrite grade and of the textural characteristics.
The influence of these factors can be appreciated by comparing the values obtained in the tests with those of a
pure pyrite mineral: density about 5.1 Mg/m^3, Young's
modulus 3 05 GPa, and seismic velocity 7.7 km/s. The
triaxial tests were interpreted both by means of a linear
and a power law as for the other rocks (Fig.11), the latter
gives a better fit to the experimental data.

The results of a structural survey is shown in the contoured pole plot of Fig.12. Joint attitudes are scattered but
three main sets can be identified, of which one closely
corresponds to the orientation of the Boccheggiano fault
The joints are often irregular, not very extensive and have
a spacing of about 0.3-1.0 m.

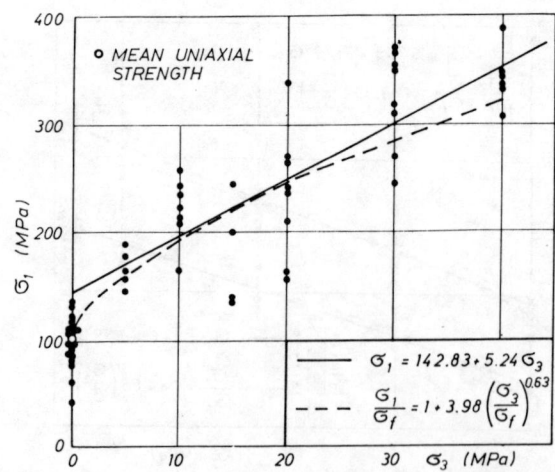

Fig. 11 - Triaxial strength of the massive pyrite.

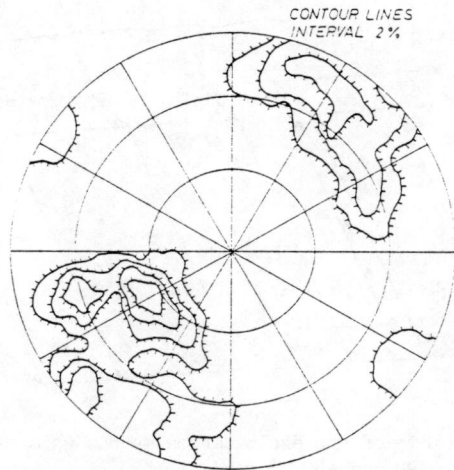

Fig. 12 - Orientation of the joints in the massive pyrites Schmidt equiareal projection, lower hemisphere.

4. SUPPORT BEHAVIOUR

4.1 "Palombini Shales" Formation

In the stretch of tunnel (2150 m long) within the "Palombini shale" formation, the support consists of sprayed concrete, 5 to 10 mm thick, whose main objective is that of preventing fretting and alteration of the shale; usually the shotcrete was applied every three rounds in order to maximise advance rate. At the cross-cuts with the shafts the support system was supplemented by resin-grouted bolts, 1.8 m long, and by hanging straps.

Diffused fracturing of the shotcrete occurred throughout most of the tunnel at depths greater than 180 m below ground surface. In most cases the fractures were hair-thin and they started to appear some 10 days after excavation, progressing initially at a high rate and then at a lower rate for a few months. The formation of these fractures was not taken as being detrimental to the stability of the tunnel; in fact convergence measurements started after their formation, and taken over a period of 1 year or more, showed that further convergence of the section was generally less than 1mm.

However, in a few more severely stressed sections, showing large fractures and the development of wedges of unstable rock, supplementary reinforcement by means of bolts, steel arches and other layers of shotcrete had to be applied. Generally, these stretches corresponded to zones of faults concentration.

As for the shafts, the final lining is formed by a 100x100x 4 mm weld mesh and by layers of shotcrete for a total thickness of about 250 mm; the minor raises are not lined at all.

So long as the tunnel ran through the "Palombini shale" formation water inflow into the workings was negligible, notwithstanding the high piezometric level. However, when the decline tunnel had to cross the lode zone, the water inflow began to increase, until a blasting round at el. +39 exposed an open fracture discharging water at the unexpected rate of 0.12 m^3/s which exceeded the capacity of the available pumps; the face had to be abandoned, the water flooded the workings and the shafts up to el. +310.

The lowering of the piezometric level required the pumping of 850000 m^3 of water; it was thereafter maintained with a rate of pumping of 0.03 m^3/s. The rate of pumping from the lowest level of the mine is about 0.01 m^3/s.

Most of the tunnel and shaft did not suffer any damage because of the flooding. In most cases local failures or falls of blocks occurred in sections where distress had already been observed before the accident; stabilization was easily obtained by means of grouted bolts and steel arches. Globally about 12% of the decline tunnel was supported by means of steel arches.

The major failure in the decline tunnel occurred at el. +218, where the tunnel crosses a subvertical fault with a set of subsidiary smaller faults. The caving extended for about 15 m along the tunnel and for a height of 10 m over the tunnel roof. Excavation was resumed by removing the rubble from the tunnel just to provide minimum access to the cavity formed by the collapse; a concrete slab some 3 m thick was built by means of a shotcrete gun and, as the rubble beneath the slab was gradually removed, 1.2 m spaced steel arches were installed. Subsequently, the void above the concrete slab was filled with rubble and shotcrete.

The same fault caused another large collapse at el.+236m involving the walls of shaft 1, which at that time had a diameter of 2.4 m. The caved material filled the shaft down to the crosscut at el. +103m. The zone involved by the collapse was identified by means of drillings and was found to be 25 m high and 12 m wide at the most. During the widening of the shaft to its final diameter of 6.1 m, the already caved and blasted material could be, without problems, drawn off from the base of the shaft at the intersection with the crosscut. The remaining void outside the final section of the shaft was filled with shotcrete.

Behaviour over time of the rock mass around the tunnel inside the shale formation was studied by "miniseismic" refraction surveys repeated over a 3-year period at various depths along the tunnel (BRIZZOLARI, 1981). Each miniseismic alignment was 18 m long and it was implemented by 7 electromagnetic geophones.

The results indicate the presence around the tunnel of a layer with a seismic velocity of 2.0-3.5 km/s as compared with the 4.5-4.75 km/s of the surrounding undisturbed rock. The thickness and the seismic characteristics of this layer, which probably corresponds to the plastic zone around the tunnel, are not significantly influenced by the height of the overburden nor by the position of the bases, whether on the wall or on the roof of the tunnel. The extent of the plastic zone initially shows a slight increase with time, but after about 2 years it stabilizes around values

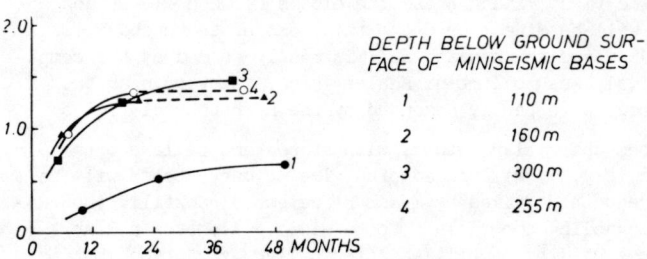

Fig.13 - Variation with time of the thickness of the plastic zone around the tunnel. Flooding and draining of tunnel occurred in between the last two observations at bases 3 and 4.

of about 1.3 m. Only for the first miniseismic base located in the stretch excavated with the roadheader, was the thickness of the yielded zone markedly smaller (<0.5 m). It must be pointed out that the flooding and the subsequent draining of the tunnel in the deepest measurement sites did not appear to cause any change in the extent and in the characteristics of the yielded layer.

A back analysis of tunnel behaviour on the basis of the extent of the yielded zone (BERRY et al.,1977) shows that the peak cohesion of the rock mass should be comparatively high with respect to that of the rock material (shale B), being of the order of 35-60% of the latter; this is in accordance with the forecast made on the basis of the appearance of the formation at the excavation face.

4.2 Boccheggiano Phyllites

A support consisting of resin-grouted bolts, ϕ 30 mm, 2.4 m long, and of a layer of shotcrete, 50-100 mm thick, was adopted. In general, three bolts per section were used, and a 2-m spacing between the sections.

A weldmesh (100x100x4 mm) was used at the crosscuts with the shafts or in limited stretches of heavily fractured rock.

The formation shows a marked rheologic behaviour which is particularly intense in the deeper levels of the mine. In this case the delayed deformations, even if they do not jeopardize the overall stability of the excavation, they do cause intense fracturing in the shotcrete layer which, after a 2-3 month period has to be removed. Fresh shotcrete and, if necessary further bolt reinforcement are subsequently applied.

The cavern for the underground precrushing plant is 8 m wide, 12 m long and 20 m high; the support of the roof and the walls consist of resin-grouted bolts, ϕ 22 mm and 2.40 m long, placed at a density of 1 per m^2, and a 100 millimetres-thick reinforced shotcrete lining. The diam. of the bolts used here is smaller than that of the bolts used in the decline tunnel because a bolting machine could not be used in this case.

4.3 Pyrite Formation

The drifts within the pyrite mass are stable without any need for support.

5. STABILITY OF THE EXPLOITATION STOPES

In the first mining phase the blocks between el.-20 and -80 will be mined; here the thickness of the orebody is 10-15 m and the hanging wall is mostly formed by the comparitively weak palombini shales which, however, near the orebody were partly cemented by the hydrothermal fluids.

A room and pillar method, with stopes and pillars both 60 m wide (Fig.14) will be adopted. The primary stopes will be subsequently filled by means of cemented rockfill, thus enablingable the pillars to be mined. Also the pillar stopes will be backfilled after pillar extraction, but in this case by using loose rockfill (cemented rockfill could be possibly used only in the lower part of the fill pillar). The fill will be placed in the stope with pneumatic packing machines, a technique which is currently used in cut and

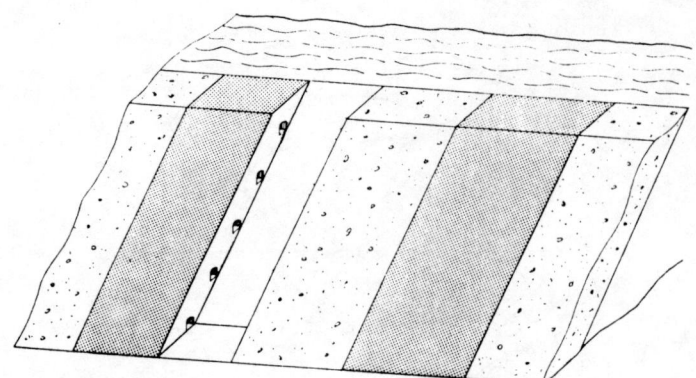

Fig.14 - Scheme of the exploitation method used between el.-20 and -80.

fill mining practice in other mines in Tuscany. Both the primary blocks and the pillars will be mined by means of sub-level drifts at a vertical distance of 12 m.

A similar method will be used for the underlying levels (below el.-80), but the vertical distance between the sub-level drifts will be certainly increased.

The purpose of the fill is to provide regional support in order to avoid excessive deformation in the surrounding rock which may damage the service excavations and cause excessive subsidence at the ground surface.

The filling of the primary stopes is intended to be a structural support for the walls of the lode during pillar recovery, thus providing a free-standing face adjacent to the exploited block; for this reason the backfill of the primary stopes must be cemented. Given the width of the lode and its inclination, a maximum unsupported vertical height of about 20 m is expected. Limit equilibrium analyses indicate that only a small cohesion of the fill (about 0.1 MPa) is needed to ensure stability of the face. However, the deformation of the lode walls during recovery of the ore pillars will induce compressive stresses in the fill pillar which may exceed the peak strength of the material; the fill must maintain some cohesion also in these conditions, to avoid the collapse of the face and dilution of the ore.

To cut down on the transportation costs, it would be preferable to use as aggregate material quarried from the "Palombini" shales formation.

Preliminary tests were carried out to evaluate whether the mechanical properties of this type of fill would allow it to be applied as support for the voids in spite of the low strength of the aggregate and the weak cohesion bond between the cement and the surface of the mudstone.

For testing, cylindrical samples, ϕ 150 mm, were cored from 2 m blocks prepared with aggregates of appropriately dimensioned grain-sizes. Compression tests were also carried out in 0.25 m cubical samples.

The results of the tests (Fig.15) showed a strong influence of the cement content on the strength and deformability; however, even for a cement content of 100-130 kg/m^3 (i.e. values which will be probably adopted at the

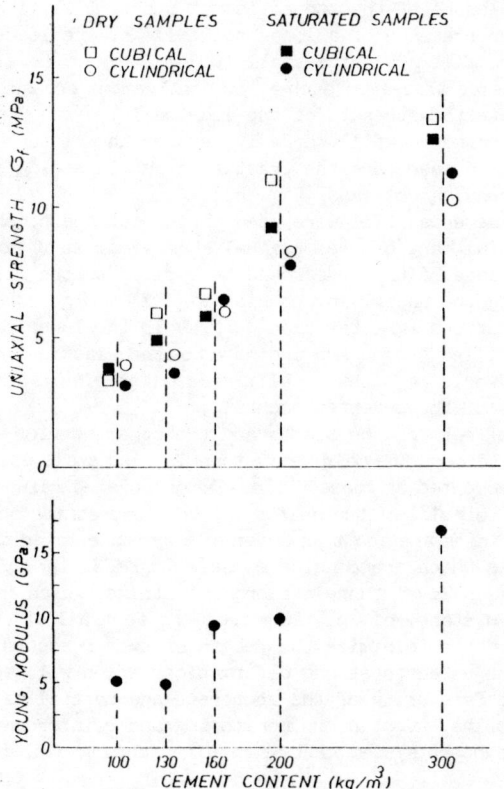

Fig. 15 - Mean uniaxial strength and Young modulus (secant values at a stress level of 2.5 MPa) of cemented rockfill.

outset of exploitation), the strength and rigidity of the fill appear to be satisfactory. No significant decay due to softening of the mudrock was observed after three months of immersion in water.

In uniaxial compression tests, the samples failed in a brittle manner with axial cleavage fractures; in the material with the lowest cement content, collapse of honeycomb structures was also observed.

A number of triaxial tests were carried out in cylindrical samples, ⌀ 70 mm, even though the results cannot be considered as very significant because of the small diameter of the samples as compared with the dimensions of the fill elements. In any case, the samples always showed a "plastic" behaviour, without a clearly defined peak strength. This kind of behaviour is obviously favourable to the stability of the fill pillars in the mine.

A number of triaxial tests were carried out in cylindrical samples, ⌀ 70 mm, even though the results cannot be considered as very significant because of the small diameter of the samples as compared with the dimensions of the fill elements. Any way, the samples always showed a "plastic" behaviour, without a clearly defined peak strength. This kind of behaviour is obviously favourable to the stability of the fill pillars in the mine.

Expected difficulties in the application of the cemented fill as described in the foregoing are:
- possible clogging of the packing machine with clay formed by the alteration and degradation of the mudstone aggregates;
- the segregation of the fill components during stowing and in particular the percolation of water and cement that would result in a fill having a variable cementing degree. This could be partly controlled by means of set accelerators; however, it must be pointed out that a greater cementation in the lower part of the fill would provide a more resistant roof for the mining of the underlying blocks;
- the instability behaviour of the fill pillar faces and bases when subjected to the stress concentration caused by the convergence of the walls of the lode during the exploitation of respectively the adjacent and underlying ore pillars.

The importance of these problems and the possible methods to overcome them can be assessed only in the field at the onset of exploitation.

To evaluate the influence of the fill characteristics on the regional stability and on the deformation of the service excavations during the exploitation of the lode, several preliminary numerical analyses were carried out by using the displacement-discontinuity technique (CROUCH, 1979).

In these analyses the following values of the geomechanical parameters were assumed:
- host rock modulus: 20 GPa
- orebody modulus: 40 GPa
- vertical original stress: corresponding to the overburden
- ratio between horizontal and vertical stress: 0.7.

For the cemented rockfill, a modulus of 1 GPa - somewhat lower than the values determined with the laboratory tests - was assumed. This reduction was suggested by the consideration that in situ, the packing and the cementation degree will be much more variable than in laboratory samples.

The loose rockfill was assigned an edometric modulus which increases with the applied stress according to the relationship: $\sigma = 2.4 \cdot \varepsilon + 120.0 \cdot \varepsilon^2$.

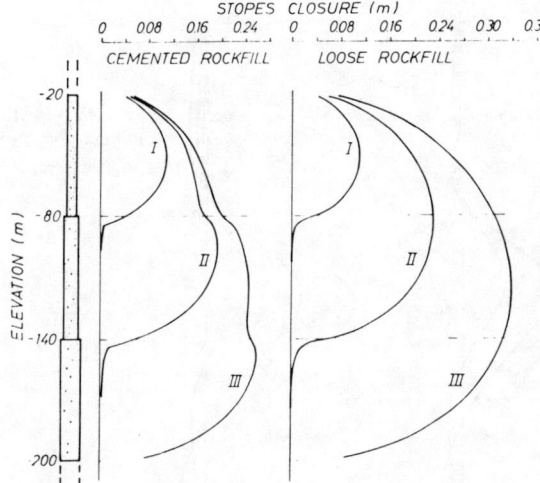

Fig. 16 - Influence of the type of fill on the closure of the stopes. I, II and III refers to the conditions existing after the exploitation of the first (-20 ÷ -80m.), second (-80 ÷ -140m.), and third (-140 ÷ -200m.) levels.

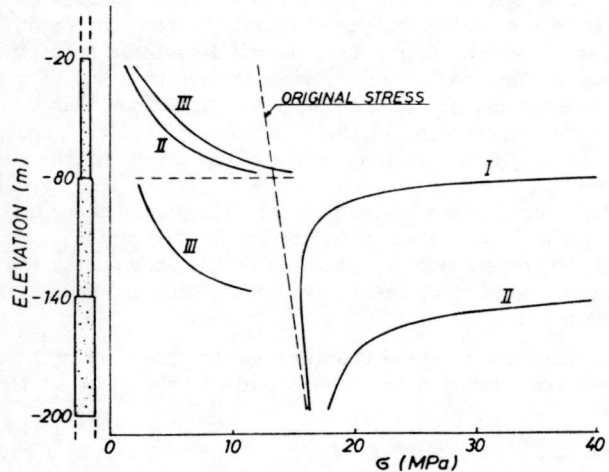

Fig. 17 - Stresses acting on the cemented rockfill and on the lode at various exploitation stages.

This estimate is based on the results of tests in various types of fills by means of an edometric cell of large dimensions (FRASSONI et al.,1982).

The thickness of the orebody was assumed equal to 10, 15 and 20 m in the first (I), second (II) and third (III) levels respectively. The analyses were carried out for plane strain conditions assuming that the fill (either loose or cemented) is placed inside the stopes after the exploitation of each level. This simplification obviously leads to an overestimation of the convergences and of the deformations in the area surrounding the exploitation and to an underestimation of the stresses to which the fill is subjected.

The following indications were drawn from the preliminary analyses:

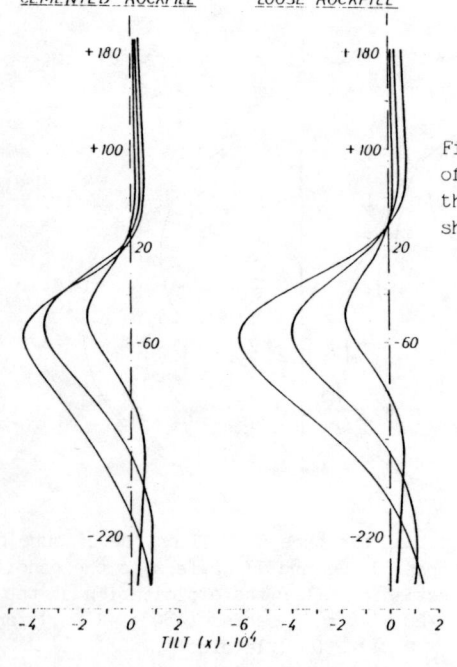

Fig. 18 - Influence of type of fill on the deformation of shaft 1.

- the closure of the stopes is not substantially modified by the presence of a loose rockfill and the loads applied to the fill are quite small (0.1-0.2 MPa); however, it is deemed that even such a low confinement pressure would improve the strength of the lode walls;
- a cemented rockfill markedly reduces the closure of the lode and therefore the fracturing and loosening of the surrounding rockmass (Fig.16);
- if a cemented fill were used throughout the block, a substantial part of the original stress component normal to the plane of the lode would be applied to the fill when mining the underlying panels (Fig.17). High stress concentrations (greater than 10 MPa) in the lower part of the filled stopes are indicated by the elastic model but they would be substantially reduced by plstic deformations of the cemented rockfill;
- the effects on the shafts and the other service excavations (except the main level tunnels) are small using either the cemented or loose fills. The calculated values of the tilt (Fig.18) of the shafts and the compressive strain in the lining are about an order of magnitude lower than the values which are considered safe (DAEMEN, 1972);
- the main level tunnels along the strike, which are placed at a distance of only 20 m from the footwall of the lode in order to minimize the length of the crosscuts, will be subjected to strong deformations which will certainly cause fracturing of the shotcrete and partial failure of the bolts. Continuous monitoring and reinforcement of the support system will be required during exploitation.

REFERENCES

A.G.I.(1979). Some Italian experiences on the mechanical characterization of structurally complex formations: 4th Congr.Int.Soc.Rock Mech.,(1),827-846, Montreux.

Bagnoli G., Gianelli G., Puxeddu M., Rau A., Squarci P., Tongiorgi M.(1979). A tentative stratigraphic reconstruction of the Tuscan paleozoic basement: Mem. Soc.Geol.It.,(20), 99-116.

Berry P., Brizzolari E., Pantaleoni G., Ribacchi R., Sciotti M., (1977). Static behaviour of a tunnel excavated in a complex mudrock formation: "The Geotechnics of Structurally Complex Formations", (1), 77-94, Capri.

Brizzolari E.(1981) Miniseismic investigations in tunnels - Methodology and results: Geoexploration, (18), 259-267.

Crouch S.L.(1979). Computer simulation of mining in faulted ground. J.South.Afr.Inst.Min.Metall.,(79), 159-173.

Daemen J.J.K.(1972). The effect of protective pillars on the deformation of mine shafts: Rock Mech.,(4), 89-113.

Frassoni A., Hegg U., Rossi P.P.(1982). Large-scale laboratory tests for the mechanical characterization of granular materials for embankment dams: XIV Congr. des Grands Barrages, (4), 727-751, Rio de Janeiro.

Manfredini G., Martinetti S., Ribacchi R., Santoro V.M., Sciotti M., Silvestri T.(1981). An earthflow in the Sinni valley (Italy): X I.C.S.M.F.E.,(3), 457-462, Stoccolma.

Ribacchi R.(1980). Caratteristiche meccaniche degli ammassi rocciosi: XIV Conv.Naz.di Geotecnica,(2), 28-31, Firenze.

Vighi L.(1972). Il nuovo giacimento di pirite e di solfuri misti di Campiano presso Boccheggiano nella Maremma Toscana: Boll. Ass. Min. Sub., (VIII), 1-2, 117-137.

ACKNOWLEDGEMENTS

The work reported in this paper was partially supported by the Commission of the European Communities within the framework of the European Mining Technology Program.

IN-SITU STRESS MEASUREMENTS AND THEIR APPLICATION TO ENGINEERING DESIGN IN THE JINCHUAN MINE

Mesures de contraintes in situ et applications au projet d'ingénierie de la mine au Jinchuan

Die in situ Spannungsmessungen und ihre Anwendung zur Grubenplanung im Jinchuan Bergbaubezirk

Chunting Liao
Engineer, Institute of Geomechanics, Chinese Academy of Geological Science, Beijing, China

Zhaoxian Shi
Engineer, Seismo-geological Expedition, National Seismological Bureau, Beijing, China

SYNOPSIS

This paper presents the results of rock stress measurements and characteristics of ground stresses and, combining with engineering geological investigation and theoretical analysis, points out the main causes of drift deformation and damage and provides precautions against them. In addition, some suggestions are made for the design of the mine.

RESUME

Cet article présente les résultats de mesures de contraintes in situ et les caractéristiques de contraintes souterraines d'un district minier au Jinchuan. On indiques les causes essentielles des déformations et des destructions de galeries et leur prévention par l'investigation géotechnique et des analyses théoriques. De plus, il présente des propositions pour le projet de développement de l'exploitation de la mine.

ZUSAMMENFASSUNG

In der vorliegenden Arbeit werden hauptsächlich die Ergebnisse der in situ Spannungsmessungen und die Eigenschaft der Grundspannung im Jinchuan Bergbaubezirk dargestellt. In Verbundung mit der Untersuchung der Ingenieurgeologie und der theoretischen Analyse werden die wesentlichen Ursachen der Deformation der Strecke im Bergbaubezirk und der Zerstörung, sowie ihre Verhütungsmaßnahmen aufgestellt. Außerdem werden Vorschläge hinsichtlich der weiteren Planung der Grube gemacht.

1. INTRODUCTION

The Jinchuan mine is situated where the Longshoushan uplift adjoins the chaoshui Basin in western Gansu Province. It is a large copper and nickel deposit, ranking second in reserves in the world. During the construction and mining of the mine, the drifts were strongly deformed and severely damaged. To solve this problem, we have made stress measurements and engineering geological investigation in the mine district.

2. OUTLINE OF GEOLOGY AND TECTONICS IN THE MINE DISTRICT

The outline of the geological structure of the mine district is shown in Figure 1.

The major faults in the mine district are numbered as F_1, F_{16}, F_{16-1}, F_8 and F_{17}. The faults F_1 and F_{16} are compressive and have an important bearing on the characteristics of engineering geology of the whole mine district.

In the whole mine district, faults, joints, and interlayer sliding planes, etc. are well developed, of which the WNW-, E-W-and NNW-trending faults are all compressive with generally wide and uncemented crushed zones. These factors greatly affect the engineering geological conditions in the district.

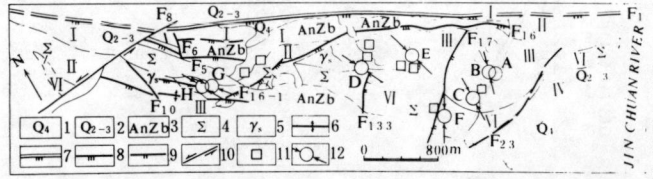

1 - Holocene; 2 - Middle-upper pleistocene; 3 - Presinian; 4 - Ultrabasic rock; 5 - Yellowish pink granite; 6 - Anticlinal axis; 7 - Regional fault; 8 - Compressive fault; 9 - Tensile fault; 10 - Strike-slip fault; 11 - Shaft; 12 - Stress measuring point; I - migmatite; II - Schist and gneiss; III - Marble; IV - Banded migmatite; V - Granite; VI - Ore-bearing ultrabasic rook.

Fig. 1 Sketch map showing the structure of the mine and the locations of stress measurements

3. RESULTS OF IN-SITU STRESS MEASUREMENTS

The magnetostriction method was used in stress measurements. Field measurement results and laboratory tests indicate that the values obtained are fairly consistent and reliable, with discrepancies usually less than 10% for stress magnitudes and less than 3° for stress directions (Wang, 1981).

3.1 Results of measurements

Measurements were made at different structural positions and depths and in different rocks. Their locations are shown in Figure 1 and the results in Table I.

Table I Results of ground stress measurements

Measuring point	Lithology	Depth below surface (m)	Max. principal stress			Intermediate principal stress			Min. principal stress			Time of measuring	Remark
			Value (MN/m²)	Direction	Dip	Value (MN/m²)	Direction	Dip	Value (MN/m²)	Direction	Dip		
A	Marble	20	2.4						2.3			1977	
B	Marble	44	4.2	N20°E					3.5			1978	
C	Marble	375	19.8	N 3°E					10.8			1975	Hanging wall, about 50m away from the F_{16} fault
D	Granite	480	24.5	N25°W					15.4			1980	
E	Marble	460	50.0	N13°W	∠6°SE	33.4	N76°E	∠6°NE	28.2	S63°E	∠81°NW	1976	Foot wall, about 80m away from the F_{17} fault
F	Very rich ore	480	32.0	N32°E	∠6°SW	21.4	S43°E	∠67°NW	20.6	N60°W	∠22°SE	1978	
G	Rich ore	240	34.4	N42°W	∠39°NW	21.1	N48°E	horizontal	2.6	S41°E	∠51°SE	1977	
H	Marble	120	16.8	N28°W	∠57°SE	12.1	S35°W	∠16°NE	5.8	S63°E	∠28°NW	1979	

3.2 Characteristics of the ground stresses of the mine district

Based on the results of stress measurements some conclusions are drawn as follows:

1) The stresses operating in the mine district are predominantly horizontal, compressive stresses which agree with the results of the survey of engineering geology and drift deformation carried out in the mine.

2) The maximum principal stress near the surface is about $3MN/m^2$, which is very close to the results obtained in North China (Wang, 1981). The value of stress increases with depth, and the maximum principal stress is generally $30MN/m^2$ or so at 200-300m depth. Figure 2a shows the variation of horizontal principal stresses with depth.

In a definite range of depth the mean horizontal stress is higher than the vertical, and the variation with depth of the ratio between it and the vertical stress is shown in Fig.2b, which is basically consistent with the statistical results of ground stress measurements in other regions (Brown, 1978).

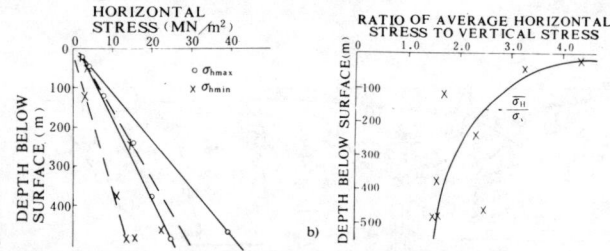

Fig. 2 a) Variation of horizontal principal stresses with depth
b) Variation with depth of the ratio between the mean horizontal stress and the vertical stress

And it can also be seen that horizontal stress is predominant to a depth of at least within 1000m below the surface.

3) Stresses measured at different structural positions vary in magnitude, direction and dip, suggesting that the stress distribution is related to geological structures.

Generally, owing to the limitations of measurements from a point, it is necessary to analyse statistically large quantities of field data. In order to get a knowledge of the stress field of the mine district, model experiment, finite element analysis and engineering geological investigation should follow.

4. RELATION BETWEEN DRIFT DEFORMATION AND STATE OF STRESS AND PRECAUTIONARY MEASURES PROPOSED

An upright-wall arch section has been used for most of the drifts in the mine. This is supported by precast concrete lining and grouting. The drifts deform mainly by halfway cracking of the upright wall, swelling of the bottom or strong distortion of the whole section (Fig.3, Fig.5 and Fig.6b).

 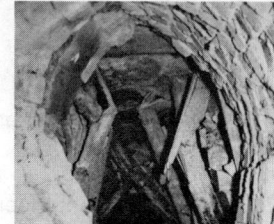

Fig. 3 Deformation types of drifts. a) Severe upswelling of the floor; b) Strong deformation

One of the main causes of strong deformation and severe damage of drift is the action of the stresses in the mine. The deposit is buried at depth where stresses are high.

4.1 Relation between deformation and direction of drift

Since the state of stress around a drift is related to the direction of the drift, its stability will vary with its direction. For example, the drift at Level 1250 passing through the fault F_{16-1} at an acute angle to the fault and almost normal to the maximum principal stress has been strongly deformed and repeatedly repaired since its completion, while the drift at Level 1200, being nearly normal to the fault and approximately parallel to the maximum principal stress, is much more stable (Fig.4).

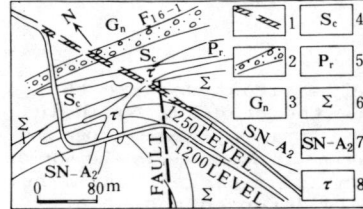

1 - Strongly deformed drift; 2 - Fault shattered zone; 3 - Gneiss; 4 - Graphite-chlorite schist; 5 - Pegmatite granite; 6 - Ultrabasic rock; 7 - Poor ore body; 8 - Plagiaplite

Fig. 4 Deformation of the two drifts at Lv Level 1200 and Level 1250 respectively

The stresses imposed upon the drift from all sides will be small and homogeneous if the drift is chosen to be parallel to the maximum principal stress. Therefore, a choice of the optimum direction of drift axis from an overall consideration of the measured direction of maximum principal stress, engineering geology and construction facilities is one of the effective measures to reduce drift deformation.

4.2 Relation between deformation and geometric shape of the cross-section of drift

Under the dominance of horizontal stresses, it is easy to cause fracturing of the side walls and swelling of the floor. Theoretically, a drift of flat elliptical cross-section is more reasonable, but it is difficult for construction. With the advantages of an elliptical cross-section as a basis, a quasi-elliptical cross-section of drift suited to the stress state of the mine has however been devised.

Fig. 5

In order to determine the feasibility of such a cross-section, a drift more than 300m long was purposely constructed. It has proved very successful during a period of several years. The side walls of the adjoining drift with an upright wall arch became swelled, while the quasi-elliptical drift remained unchanged (Fig.5). The axial ratio of the quasi-ellipse is best determined by the measured stresses.

4.3 Relation between drift deformation and engineering geological condition

It was found that the deformed and damaged drifts are mostly located where the engineering geological conditions are the worst. The presence of swelling rocks may also facilitate the deformation of drift.

For example, a N40° W-trending drift segment was strongly deformed on the side close to the fault when passing over the N60°W fault whereas the opposite side remained almost intact, and a roof collapse occurred where the fault meets the drift. This suggests a close relationship between the drift deformation and the geological structures (Fig.6).

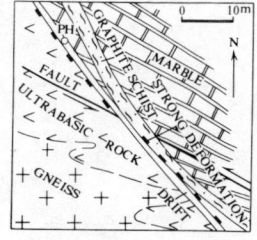 

Fig. 6 (a) Sketch plan showing the relation between the deformation of drift and the distribution of faults; PH: location of deformation (b)

The nearer the drift is to the fault, the more intense is the deformation, and the higher the deformation rate. Observations of the deformation have been made at two points of a drift at Level 1250 which are 220m and 70m away from fault F_{16-1} respectively. The deformation is 6cm for the first month at the first point and up to 23cm at the second.

4.4 Influence of technological level and quality of construction on stability of drift

The technological level and quality of construction has a great influence on the stability of a drift. When construction is carried out in broken rock formations, support measures should be taken simultaneously with digging in order to strictly control the initial displacement and reduce to a minimum the disturbance of in-situ rocks. Furthermore, the dilation effect should be reduced as much as possible, the shape and dimensions of drift cross-section controlled strictly, and a second time support used. These measures are favourable to the stability of drift and have proved effective in practice.

Deformation and damage of a drift is the result of a combination of various interacted factors.

5. STRESS STATE AND DISIGN OF STOPE

The results of ground stress measurements can provide basis not only for the reasonable design of a drift, but for the correct selection of mining methods and the safety in mine production. Suggestions for stope designing are made as follows:

1) The direction of the long axis of the stope should be selected approximately consistent with the direction of the maximum shear stress.

2) The span of the stopes should be determined in accordance with the concrete engineering geological conditions and the manner of stress action in individual mine blocks. In the case of a roof protection, an increase of the span of a stope is more favourable for its stability.

3) An arch shape of stope roof is considered preferable for the present state of stress of the mine. The same is true for the envelope of the stope roofs at the same level along the strike of the ore body.

4) Based on the study of the mechanism of the effect of filling materials, use of low-strength filling materials is suggested.

6. CONCLUSIONS

We have discussed in the above the relation between the action of stress and drift deformation from a viewpoint of rock mechanics on the basis of stress measurements and engineering geological survey. Measures for the prevention and control of drift deformation and damage are accordingly proposed. These measures have proved feasible and successful. In addition, a stope design has also been suggested.

In addition to the measurements of absolute stress, it is necessary to monitor the change and accumulation of stress during mining especially in stress concentrated areas so as to prevent the mine from rock burst and roof collapse.

In the disign of an underground construction project, a comprehensive consideration of the results of ground stress measurements, rock mechanical test, engineering geological investigation and finite element analysis is important for the upgrading of design and the stability of the project.

We are grateful to professor Chen Qingxuan for his guidance in the preparation of this paper.

REFERENCES

Brown, E. T. And Hoek, E. (1978). Technical Note Trends in Relationships between Measured In-situ Stresses and Depth. Int. J. Rock Mech. Min. Sci. Vol.15, pp. 211-215.

Wang Lianjie et al. (1981). Principles and Applications of Ground Stress Measurement. 48-142. Beijing.

EVALUATION DE LA STABILITE D'EXPLOITATIONS PAR CHAMBRES ET PILIERS ET COMPORTEMENT DU TOIT

Estimation of the stability of room and pillar workings and roof behaviour

Bestimmung der Standfestigkeit im Kammerpfeilerbau und das Verhalten des Hangenden

C. Chambon
Professeur à l'Ecole des Mines Laboratoire de Mécanique des Terrains
ENSMIM Nancy France

J. Arcamone
J. P. Josien
J. P. Piguet
Ingénieurs au CERCHAR Laboratoire de Mécanique des Terrains
ENSMIM Nancy France

RESUME
Comme le montre l'analyse de cas d'effondrements observés, la stabilité des exploitations par chambres et piliers ne se réduit pas à la pérennité des piliers. Les études de stabilité passent donc nécessairement par la détermination des éléments critiques dans la structure: le toit, le mur ou les piliers. Une méthode d'approche exposée dans cet article consiste à représenter l'ensemble de la mine (pilier, mur, recouvrement) au moyen des modèles numériques par éléments finis. Ces modèles peuvent évidemment être construits de manière à respecter fidèlement les caractéristiques réelles des terrains. Mais la priorité est donnée à la validation du modèle qu'on cherche à caler sur les mesures in situ de contraintes, de déformations et de déplacements ainsi que sur des observations plus qualitatives du comportement réel (dégradations locales, ou instabilité générale). Pour tenir compte de ces données d'expérience, on montre sur des exemples, comment on peut préférer adopter en définitive dans les calculs des propriétés ou des caractéristiques géométriques et géomécaniques équivalentes.

SYNOPSIS
The analyis of real cases of collapse shows that stability of room-and-pillar mines does not depend only upon everlasting pillars. Stability studies must also involve the determination of critical factors in a particular structure, which implies rock mechanics analysis of roof, floor, pillars and limits of mining. An approach presented in the paper is based on the representation of the whole mine under study (pillar, floor, overburden) by numerical finite element models. The models can of course be constructed so as to comply with the characteristics of strata. But priority is given to validation of the model which should be adjusted in accordance with in situ measurements of stress, deformation and displacement, and in accordance with more qualitative observations of effective behaviour.

ZUSAMMENFASSUNG
Wie aus der Analyse von reellen Bruchfällen hervorgeht, hängt die Stabilität der Kammer- und Pfeilerbauten nicht nur von zeitlich unveränderlichen Pfeilern ab. Bei Standfestigkeitsuntersuchungen sind auch die kritischen Bestandteile zu bestimmen: Hangendes, Liegendes und Pfeiler. Bei dem im Vortrag beschriebenen Verfahren wird der ganze Grubenau (Pfeiler, Deckgebirge) mit Hilfe von Modellen nach der Finite-Elemente Methode dargestellt. Das Modell wird mit besonderer Rücksicht auf die wirklichen Grundzüge der Gebirgsarten gebaut. Den Vorrang aber hat immer die Gültigmachung des Modells, das an in situ Messungen angepaßt werden soll, d.h. an Spannungs-, Verformungs- und Verschiebungsmeßwerte, sowie an mehr qualitative Beobachtungen über das reelle Verhalten (örtliche Zerstörungen, allgemeine Unbeständigkeit). Um solche Erfahrungsdaten zu berücksichtigen, wird an Hand von Beispielen gezeigt, wie bei Berechnungen es schließlich vorteilhaft sein kann, angepaßte äquivalente geomechanische oder geometrische Eigenschaften oder Merkmale einzusetzen.

1. INTRODUCTION

Comme le montre l'analyse de cas d'effondrements observés (cf tableau 1) la stabilité des exploitations par chambres et piliers ne se réduit pas à la perennité des piliers. Dans un certain nombre de situations, il faut incriminer le comportement du toit ou du mur immédiats sur la stabilité desquels, des piliers trop résistants et trop rigides peuvent même jouer un rôle néfaste (par poinçonnement par exemple). Dans d'autres cas, c'est dans les propriétés du recouvrement qu'il faut chercher les causes premières de l'instabilité (exemple des effondrements spontanés).

C'est pourquoi, les études qui traitent ces problèmes doivent nécessairement passer par la détermination préalable des éléments critiques dans l'ensemble de la structure (pilier, toit, mur). En l'illustrant par deux exemples, nous proposons une méthodologie qui s'efforce d'apporter une solution.

2. METHODE D'IDENTIFICATION DES ELEMENTS CRITIQUES DANS LA STRUCTURE

Le développement des moyens modernes de calculs nous permet aujourd'hui d'aborder ce problème en représentant l'ensemble de l'exploitation (piliers, chambres, mur, recouvrement) à l'aide de modèles numériques par Eléments Finis.

Il est en effet possible aujourd'hui de respecter fidèlement non seulement :
- la géométrie dans ses trois dimensions
- la lithologie et les caractéristiques mécaniques des terrains
- les conditions aux limites connues (déplacements imposés, contraintes gravitationnelles ou d'origine tectonique..)
mais encore
- les discontinuités représentées par des éléments-joints caractérisés par leurs raideurs tangentielles et normales, et par leur aptitude à se décoller ou à glisser

DATE	SITE	EXPLOITAT.	RECOUVRE-MENT(m)	PENDAGE	CHAMBRES m	PILIERS m	DEFRUITEMENT (calculé)(*)	OUVERTURE m	RESISTANCE bars	VICTIMES (**)	CAUSES INVOQUEES
1871,77,83,91	Bougival	Craie	40 à 50		4,3	6,5	57 %	7	20 à 30(sat)	0	Piliers imbibés d'eau
1873	Varangéville	Sel	146	1,2 %	8	6	83 %	5,5	332 à 461	2T, 17B	Poinçonnement des marnes du mur
1879	Fuveau	Lignite	∼ 400		10	10	75 %	6		0	Rupture des piliers
1880	Vieux Ports	Pierre à chaux	0 à 50		6	6	75 %	6	45	26T,8B au jour	Rupture des piliers
1883	Villainen-la-Carelle	Calcaire	7 à 30	∼ 0	6	6	> 90 %	2 à 3,5	67 (sec) 55 (sat)	0	
1885	Chancelade	Calcaire	≤ 69		7 à 9	5	85 à 90 %	6 à 8	82,6	13T(fond)	Effondrement subit
1892	Mery/Oire	Calcaire grossier	20 à 35		Tournés		élevé		90 à 130	0	Taux de défruitement élevé
1899	Pantin	Gypse (abandon.)	46 à 48	12à15%	8 à 9	3 x 4	(92 %)	5 à 6	∼ 100	3T(jour)	Toit marneux désagrégé par nappes d'eau
1902	Terres Rouges	Fer	121,5				72,5 %	2couches		0	Charges dues au recouvrement
1904	La Fourchette (Montvernier)	Gypse	40 à 50	élevé						2 T	Eaux d'infiltration
1905	Livry Gargan	Gypse	10	∼ 0	4 à 5	5	70 à 75 %	6	100	0	Faille affectant le gisement (infiltrations)
1905	Roche Noire (St Jean de Maurienne)	Gypse			10 à 14	4 à 5	90 à 95 %	8 à 10		0	Dimensions des chambres. Présence de crevasses. Eaux d'infiltration d'un torrent
1908	Limons	Calcaire	8 à 25	5 %	3	1	94 %	1,65	35(sat)	0	Défruitement exagéré
1911	La Boulloy (Emeville)	Calcaire	22		6 à 7	4,5	85 à 88 %	3 à 4	100	0	Enfoncement des piliers dans les sables du mur. Infiltrations d'eau
1913, 32,37	Orsonnette	Calcaire	40	∼ 0	5	5	75 %	2couches	125(sat)	0	
1919	Rochonvillers	Fer	170		6 à 7	6	60 à 65 %	6	259à354	24 T	Défruitement trop élevé quartiers voisins dépilés non effondrés
1920	Roncourt	Fer					65 %			0	
1922	Seyssins	Pierre à ciment	100	29°	5	4	75 %	4 à 5	619 (sec) 308(sat)	0	Délitement des piliers sous l'action de l'air humide
1926, 49	Pont-du-Château	Calcaire bitumin.	10 à 80		5	4	70 à 80 %	3 à 8	95	0	Piliers renversés, hauteur atteignant 10 et 12 m
1927	Port Marly	Craie (abandonnée)			4	6 x 7	62 %	7		0	
1928	Ivry en Montagne	Gypse	24			∅ 8	82 %	5 à 5,35	138 à 252	0	Taux de défruitement non respecté
1928	Rancié	Fer		70°			"inconsidéré"			3T,4B	Masses supérieures instables
1929	Mormoiron	Ocre	35	30à60°	2	6 x 8		2		3 T	Reprise du quartier d'exploit. après abandon
1929, 50, 61	Soumont	Fer	170à250°	30°		5 x 12	80 à 85 %	3		0	Poinçonnage des épontes, schistes tendres de 4 ans
1931	St Remy/Orne	Fer	<200	0à90°	5	2,5	90 %	2,5		0	
1932	Ste Marie-aux-Chênes	Fer	153			10 x 11,°	65 %	5		0	Charges dues au recouvrement et tensions résultant des zones voisines dépilées
1940	Auboué	Fer	121à134			10 x 11,°	65 à 73 %	2 couches		0	Flambement des piliers (sur haut. 8à10m) car délitage de l'intercalaire marneux
1943	Louvenciennes	Craie	50		5	7	66 %	8	20à30(sat)	0	Rupture de piliers en mauvais état, barrés de sillons argileux
1943, 61, 62, 64	Plateau de Vitry	Gypse (abandonnée)	35		5	6 à 7	60 à 70 %			0	Délitage des piliers
1953	Nanterre	Calcaire (abandonnée)	5 à 6		8 à 9	5x4 à 6	jusqu'à 90 %		105	1 B	Fatigue des piliers (infilt., gel, ébranlement)
1953	Houilles	Calcaire (abandonnée)	7		2,5 à 3,5		90 %	2 à 3		0	Piliers trop faibles disposés sans ordre
1954	Moyeuvre	Fer	146,5		7		> 71,5 %	7,5		1 B	Taux de défruitement trop élevé et influence du front de dépilage
1956	Grenoque	Lignite	∼ 500	8 à 10°		5x10 à 12		2,5		0	
1957	Gargas	Ocre (abandonnée)	20		3 à 2,5	6	65 %	3 à 6		0	
1959	Roncourt	Fer	140				75 %	2 couches (1,8 et 2,8)		5T,10B	Affaiss. des piliers sous le poids du recouvrement
1959, 62	Port-Maron (Vaux/Seine)	Gypse	73 à 75		7,5	6,5	80 %		63 à 150	0	Fissuration de la planche du toit. Poinçonnage du mur
1960, 63	Puy Notre Dame	Tuffeau (abandonnée)	10 à 25	0	6 à 9	5	50 %	3 à 4		0	Affaiblissement progressif des piliers
1961	Pierrefitte	Plomb		20°	6 à 10	2 à 3	> 80 %	4 à 10	110(sec) 27(sat)	0	Présence de failles et de cassures
1961	Clamart et Issy-les-Moulineaux	Craie et calcaire grossier (abandonnée)	48		4	7 x 40 et 6 x 6	≤ 65 %	2 couches (6à7 et 5) int. de 3	217,5à8 au jour		Infiltrations (pluies abondantes)
1964	Azay-le-Rideau	Tuffeau (abandonnée)	10		6 à 8,5	2 x 5 x	85à92 %	5 à 7			Piliers de dimensions insuffisantes. Dégel.
1964	Champagnole	Calcaire	160à170	1,5 %	8		90 %	4,5	120 à 595	6T, 1 au jour	Mise en charge hydraulique

TABLEAU I - EFFONDREMENTS SURVENUS DANS DES EXPLOITATIONS FRANCAISES A PILIERS ABANDONNES
(d'après une enquête de la Direction des Mines du 14/10/1964) ; Circulaire DM/H n° 252
du 23 avril 1965 - Direction des Mines, Service hygiène et sécurité minière
(*) Taux fournis par les rapports, les valeurs entre parenthèses ont été calculées à partir des dimensions des chambres et des piliers
(**) T = tués   B = Blessés

- les lois de comportement complexes (élasto-plasticité avec écrouissage positif ou radoucissement dans le domaine plastique, visco-élasticité etc...)

Toutefois, lorsqu'il s'agit de modéliser une mine tout entière, la mise en oeuvre de ces modèles devient rapidement lourde et très onéreuse. D'autant que leur validation ne pouvant s'envisager qu'en confrontant leurs résultats avec ceux de mesures ou d'observations qualitatives, effectuées in situ, on est amené à faire varier les valeurs des paramètres les plus mal connus ou ceux dont l'influence sur les résultats risque d'être la plus sensible.

Il faut ajouter que le nombre de ces paramètres et surtout des combinaisons possibles entre leurs modalités augmente très vite avec le degré de sophistication du modèle.

Cela conduit à la réalisation de calculs si nombreux que le coût total est rapidement déraisonnable. En cherchant à le réduire on se condamne à une validation très superficielle du modèle construit, ce qui peut être dangereux.

L'approche proposée est plus pragmatique et les deux exemples suivants en montrent l'efficacité. Dans un cas, il s'agit d'une mine de sel pour laquelle l'utilisation de lois de comportement du type élasto-viscoplastique serait naturellement justifié. Dans l'autre cas, le rôle, à l'évidence déterminant, de petites discontinuités naturelles aurait pu inciter à les prendre en compte systématiquement dans le modèle.

Cependant, dans les deux exemples, le recours à des modèles élastiques plus rustiques mais plus maniables, a permis, moyennant une confrontation soignée avec les données recueillies in situ, de formuler des conclusions opérationnelles.

3. PREMIER EXEMPLE

Une mine de sel gemme au nord-est de la France est exploitée depuis le 19ème siècle grâce à une méthode par chambres et piliers abandonnés. Le secteur Nord de la mine, qui fit l'objet de plusieurs études de stabilité [1] est défruité à 75 % sur une hauteur de 4,50 m et sous un recouvrement moyen de 180 m. Celui-ci est constitué d'une alternance de sel et de niveaux marneux, puis de grès et de marnes peu résistants jusqu'à proximité de la surface où on remarque un banc beaucoup plus dur, quoique hétérogène et d'épaisseur variable : la dolomie de Beaumont.

En première approximation, l'application du calcul très simple de l'aire tributaire conduit à penser que les piliers ont largement dépassé leur limite élastique, pour autant que celle-ci ait un sens dans le cas du sel. On constate en outre que même les piliers les plus anciens, découpés depuis un siècle, continuent à se déformer avec une vitesse de l'ordre de 0,4 mm/m par an, en moyenne. Dans le quartier Nord-Ouest (zone exploitée sur environ 400 x 450 m) où les piliers carrés ont 10 m de côté et les chambres 10 m de large, on n'observe pas au fond de dégradations importantes. En revanche, dans le quartier Nord-Est, plus récent, exploité sur 1200 m x 700 m avec piliers carrés de 15 m de côté et chambres de 15 m, on constate :

- un soufflage spectaculaire de la planche de sel laissée au mur de la chambre, décollée des marnes sous-jacentes.

- une fissuration des coins des piliers avec formation d'écailles sur 0,5 m à 1 m de profondeur.

- des chutes de placages au toit des chambres, de 10 à 30 cm d'épaisseur décollées sur toute la largeur le long de minces lits marneux.

Pour tenter d'identifier les éléments critiques dans la structure, l'ensemble de l'exploitation, y compris le mur et le recouvrement jusqu'à la surface ont été modélisés [2][3] dans un plan vertical perpendiculaire à la plus grande dimension de la surface exploitée, sous l'hypothèse des déformations planes et d'un comportement élastique linéaire des terrains.

La troisième dimension est prise en compte, soit en introduisant des chambres plus larges que dans la réalité, soit en attribuant aux terrains sus-jacents une densité sur-évaluée, de façon à retrouver une contrainte moyenne verticale sur les piliers correspondant à celle que permet de calculer l'aire tributaire, pour un taux de défruitement de 75 %.

Les grandes déformations mesurées au fond sont retrouvées par le modèle en affectant aux piliers un module élastique sensiblement plus faible (500 MPa) que celui qui résulte des essais de compression en laboratoire (variables de 3000 à 20 000 MPa selon la procédure d'essai employée). La difficulté à déterminer des caractéristiques mécaniques sûres pour le sel, donne en l'occurence une plus grande liberté pour choisir des valeurs équivalentes compatibles avec les mesures in situ. C'est ainsi que notre modèle permet de calculer des convergences au bord des piliers et des affaissements à la surface du même ordre de grandeur que les valeurs mesurées sur le site depuis plus de dix ans et extrapolées pour obtenir les mouvements cumulés depuis l'origine de la mine.

L'analyse des états de contraintes calculés également autour des chambres rend par ailleurs remarquablement bien compte des dégradations réelles observées (soufflage du mur, écaillages et chutes de placages) (figure 1).

Les contraintes verticales et horizontales ont été mesurées dans certains piliers et dans le bord ferme par la méthode de surcarottage[7][8]. Les résultats ne sont pas en contradiction avec ceux du calcul mais ils sont rendus très imprécis par suite d'une difficulté technique ayant nécessité une importante correction sur les valeurs trouvées : il s'agit du discage des carottes dû aux contraintes d'origine thermique engendrées par le carottage à sec dans le sel.

Enfin, en ce qui concerne les caractéristiques géomécaniques du recouvrement, inspirées au départ de quelques données recueillies sur des sondages, elles sont en définitive choisies de manière à caler le modèle sur le comportement observé au-dessus des cavités exploitées par dissolution du sel dans le voisinage de la mine (dans les mêmes conditions par conséquent).. On connaît en effet, par expérience, sur des exploitations voisines par dissolution, les dimensions approximativement atteintes

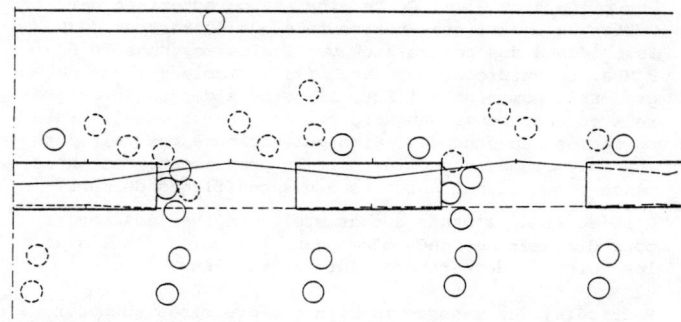

Figure 1 : Modélisation des piliers, chambres de 15 m
Répartition des ruptures, déformée du toit, déplacement du mur.

Pour les calculs : Rc(sel) = 20MPa   Rt(sel) = 2 MPa
Echelle du dessin 1/250è - des déplacements 1/50è
Les points encerclés sont en rupture.

par ces cavités lorsqu'elles se foudroient ainsi que la forme prise par le foudroyage. On retrouve ce comportement sur le modèle en y introduisant ces cavités dont on fait varier le diamètre (calcul en symétrie de révolution) comme le montre la figure 2.

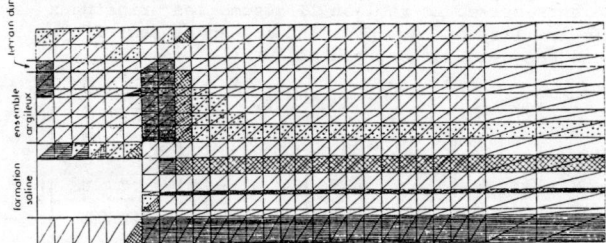

Figure 2 : Modélisation, cavité lessivée ∅ 180 m, zones de rupture.

▓ ruine
▬ rupture probable à court terme
▨ rupture probable à moyen terme
▦ rupture probable à long terme.

Finalement, la réponse du modèle ainsi construit, et appliqué au problème de la mine, montre que la flexion du toit doit engendrer des ruptures par traction au-dessus du centre de la zone exploitée avant tout autre désordre. Une série de calculs où on a fait varier l'épaisseur et la raideur relative de certains bancs du recouvrement (comme la dolomie de Beaumont) a montré que la localisation et l'intensité de ces ruptures dépendent de la lithologie du toit mais ne modifient pas qualitativement notre conclusion.

Des ruptures du toit au droit des bords fermes ne pourraient intervenir qu'avec une flèche du toit d'amplitude beaucoup plus importante que celle enregistrée jusqu'à présent, c'est-à-dire bien après que les ruptures par traction se soient produites au centre de l'exploitation.

4. DEUXIEME EXEMPLE

La mine qui fait l'objet du deuxième cas d'application est une exploitation de charbon située aux Etats-Unis dans les Appalaches. Le recouvrement est globalement très résistant et comprend plusieurs couches de charbon. Une seule, d'une épaisseur de 120 cm sera exploitée à une profondeur de 400 m environ par longues tailles foudroyées. Actuellement, les traçages des voies d'accès sont en cours : 5 voies parallèles de 6 m de large séparées par des piliers de 21 m. Cette mine est caractérisée par de fortes contraintes horizontales de l'ordre de 20MPa associées à des contraintes verticales moyennes de 6 à 8 MPa. La nature du toit est très variable : formé de grés très homogène à l'Est, il passe à des schistes gréseux et charbonneux au Nord par des variations latérales de faciès peu étendues. Bien que boulonné, ce toit est instable et des chutes de toit se produisent. Ces phénomènes sont souvent associés à des soufflages du mur.

L'objectif de l'étude a consisté à proposer des moyens pour diminuer ces phénomènes dans les voies d'accès et les voies de desserte des futures tailles.

Pour cela, des mesures in situ ont été mises en oeuvre :

- variations de contraintes par capsules pressiométriques dans les piliers
- expansions dans le toit et les piliers
- relevés des soufflages

Une caractérisation géomécanique du toit et du mur a été réalisée. Les comportements élastique et plastique du charbon ont été soigneusement étudiés. Cette caractérisation a permis de distinguer les principaux terrains et leurs caractéristiques moyennes.

Pour comprendre les mécanismes qui induisent chutes de toit et soufflages du mur, une étude sur modèles numériques basés sur la méthode des éléments finis [4] a été entreprise. Le tableau II résume les principaux calculs. Ils ont tous été conduits en faisant l'hypothèse des déformations planes.

Tableau II : Bilan des principaux calculs effectués pour l'étude de la mine de charbon.

Le modèle élastique global (figure 3) représente une coupe perpendiculaire à l'avancement des voies. D'après ces calculs, l'ensemble du massif demeure dans l'état élastique. Seuls les piliers sont fracturés au bord sur une épaisseur de 1 m environ. Ce résultat est validé par la comparaison des contraintes et des convergences mesurées et calculées entre lesquelles la concordance est satisfaisante.

Ce modèle montre, en outre, que les poutres du toit et du mur sont partiellement séparées du terrain sus et sous-jacent au niveau des voies. Cela expliquerait les fortes convergences mesurées in situ dans les zones lithologiquement hétérogènes au voisinage de la couche. Des calculs en élastoplasticité [5] ont été effectués sur un maillage très fin des poutres du mur et du toit. Ils ont confirmé les résultats du modèle global en montrant qu'une poutre est instable dès qu'elle a une épaisseur inférieure à 30 cm. Un phénomène analogue a été constaté in situ.

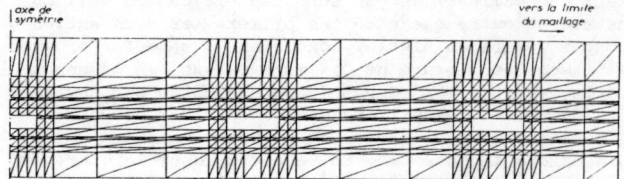

Figure 3 : Maillage utilisé pour les calculs en élasticité

Des calculs complémentaires en élastoplasticité sur un modèle de piliers ont permis de montrer que, jusqu'au niveau de la couche de charbon immédiatement surincombante la déformation du toit est très continue si celui-ci est formé de terrains épais et homogènes : les convergences calculées sont voisines de celles mesurées dans le quartier Est qui présente un toit très régulier. En outre, les expansions mesurées et calculées au bord du pilier sont voisines, ce qui tend à confirmer que l'épaisseur plastifiée au bord du pilier est de 1 mètre environ.

En conclusion, les fortes convergences sont dues aux hétérogénéités du toit et du mur qui provoquent, en raison des fortes contraintes horizontales, l'individualisation de bancs détachés du reste du massif. Les poutres ainsi formées sont instables car elles sont soumises au poinçonnement intense des piliers, attesté par leur faible fracturation au bord. De nombreuses modifications de l'exploitation ont été testées à partir du modèle élastique (cf tableau II). Elles ont été infructueuses car le pilier est resté élastique et, dans ce cas, les voies ne s'influencent pas mutuellement.

Pour améliorer l'exploitation et diminuer le soufflage et les chutes de toit, il a donc été recommandé d'imposer un comportement plastique aux bords de manière à diminuer le poinçonnement des piliers et l'influence des fortes contraintes horizontales. Cette modification de l'état du pilier peut être obtenue par des trous de détente ou par une diminution de leur largeur de 21 m à 15 m. Cette mesure a été essayée ; elle donne de bons résultats et le pilier demeure stable.

CONCLUSION

L'avènement des grands ordinateurs modernes permet d'envisager des calculs très complets pour apprécier le comportement de structures de géométrie complexe et dont les matériaux constitutifs peuvent obéir à des lois rhéologiques variées. C'est le cas des exploitations minières à faible et moyenne profondeur utilisant la méthode des chambres et piliers.

Mais cette situation obligera à définir une doctrine pour l'emploi de ces modèles afin d'éviter :
- soit le danger de formuler des conclusions insuffisamment validées par l'expérience
- soit de s'écarter coûteusement des réalités physiques mesurables par excès de sophistication.

A travers cet article une contribution à la réflexion sur ces problèmes a été recherchée. Une méthodologie d'approche des études de stabilité dans les exploitations par chambres et piliers a été proposée.

BIBLIOGRAPHIE

[1] TINCELIN, FINE et al (1972)
Etude de la stabilité de la mine de Varangéville
St Nicolas France
Ecole Nationale Supérieure des Mines de Paris. Centre de Mécanique des roches de Fontainebleau.

[2] NIANGOULA A. (septembre 1981)
Contribution à l'étude de la stabilité d'une ancienne exploitation de sel gemme par chambres et piliers abandonnés. Application au cas de la mine de sel de Varangéville.
Thèse docteur-ingénieur - Institut National Polytechnique de Lorraine. Ecole des Mines de Nancy.

[3] DEJEAN M., NIANGOULA A, et al (1981)
Etude de la stabilité des anciens travaux de la mine de sel de Varangéville.
Rapport établi par le Cerchar pour la "Commission des Recherches scientifiques sur la sécurité dans les mines et carrières.

[4] DEJEAN M., ARCAMONE J. (1978)
La méthode des éléments finis en élasticité. Applications au programme ELFI3F.
Laboratoire de Mécanique des Terrains - Rapport interne CERCHAR

[5] D.R.J. OWEN, E. HINTON (1981)
Finite Elements in Plasticity - Theory and Practice.
Pineridge Press United Swansea U.K.

[6] JOSIEN JP. (1977)
Surveillance de la stabilité d'une exploitation par des mesures de déformations - choix d'une méthode d'alarme.
Annales des Mines.

[7] HELAL H. (Mars 1982)
Etude et développement d'une méthode de mesure des contraintes par surcarottage. Applications à l'étude de stabilité d'ouvrages souterrains.
Thèse Docteur-Ingénieur INPL Ecole des Mines de Nancy.

[8] BONNECHERE F. (1971)
Contribution à la détermination de l'état de contrainte des massifs rocheux.
Thèse Science devant l'Université de Liège.

[9] SALAMON M.D.G., A.H. MONRO : A study of the Strengh of Coal Pillars, Journal of the South African Institute of Mining and Metallurgy, sept. 1967

[10] SALAMON M.D.G.
A method of designing Bord and Pillar Working, Journal of the South African Institute of Mining and Metallurgy - sept. 1967

[11] HUSTRULID W.A
A review of coal pillar Strengh Formulas - Rock Mechanics n° 8 (1976)

[12] HOLLAND G.T
The strengh of Coal in Mine Pillar - Proceedings of the 6th Symposium on Rock Mechanics. University of Missouri (1964).

[13] Roy S. CLELAND, K.H. SINGH
Development of post pillar mining at Falcombridge Nickel Mines Limited-the canadian Mining and Metallurgical (CIM) avril 1973.

UNE APPROCHE ENERGETIQUE DE LA STABILITE D'EXPLOITATIONS PAR CHAMBRES ET PILIERS ABANDONNES, TENANT COMPTE DU COMPORTEMENT POST-RUPTURE DES PILIERS

An energy approach to room and pillar exploitations based on post-failure behaviour of pillars

Eine energetische Methode zur Stabilität verlassener Kammerpfeilerbauten, gegründet auf dem Verhalten der Pfeiler nach dem Bruch

Ph. Weber
Professeur, Ecole Nationale Supérieure des Techniques Industrielles et des Mines d'Alès
France

RESUME
On propose une approche énergétique de la stabilité d'une exploitation souterraine par chambres et piliers, en schématisant le comportement mécanique des piliers par un modèle radoucissant à quatre paramètres: module d'élasticité, résistance à la rupture, module d'écrouissage négatif et résistance résiduelle. Un calcul énergétique conduit à la définition des conditions de stabilité globale de la structure: il apparaît ainsi un taux de défruitement critique, correspondant à une instabilité globale de l'exploitation. Le phénomène ainsi analysé doit constituer l'un des mécanismes initiateurs plausibles des effondrements spontanés.

SYNOPSIS
An energy approach to room and pillar exploitations is proposed by modelling the mechanical behaviour of pillars by a four-parameter model: Young's modulus, compressive strength, negative yield modulus and residual strength. Calculating the mechanical energy of the system leads to the definition of global stability conditions. When a critical extraction ratio is reached, the result is a global instability of the whole exploitation. This phenomenon probably constitutes one of the likely initiating mechanisms of rock bursts.

ZUSAMMENFASSUNG
Eine energetische Methode zur Stabilität eines Kammerpfeilerbaus wird vorgeschlagen. Das mechanische Verhalten der Pfeiler wird in einem Modell mit vier Parametern, Elastizitätsmodul, Bruchfestigkeit, negativer Kalthärtungsmodul und Restfestigkeit dargestellt. Eine energetische Rechnung führt zu einer Bestimmung der Bedingungen für die gesamte Stabilität. Es ergibt sich eine kritische Lagerstättenausbeutung, die einer Unbeständigkeit des gesamten Grubenbaus entspricht. Das untersuchte Phänomen könnte ein möglicher Mechanismus für das Entstehen von Bergschlägen sein.

1. INTRODUCTION

Les approches classiques de la stabilité d'exploitations par chambres et piliers font généralement référence à la notion de limite élastique de la roche constitutive des piliers, et supposent que la contrainte moyenne sur les piliers les plus chargés ne saurait excéder cette limite ; ces approches sont souvent contredites par l'expérience qui montre que certaines exploitations sont stables, malgré une fracturation manifeste des piliers, laissant supposer qu'ils subissent des déformations largement supérieures à leur déformation de première rupture.

Dans l'approche proposée ici, le comportement des piliers est schématisé par un modèle élastoplastique radoucissant classique à quatre paramètres E, R, m, r (fig. 1). Le reste de l'exploitation (toit immédiat et terrains susjacents) est supposé rester en équilibre élastique. Moyennant une hypothèse simplificatrice sur la déformée du toit de l'exploitation, un calcul fournit l'expression explicite de l'énergie mécanique totale du système. Il apparaît alors un taux de défruitement critique (τc) qui, lorsqu'il est atteint par les travaux d'exploitation, confère au modèle une instabilité globale.

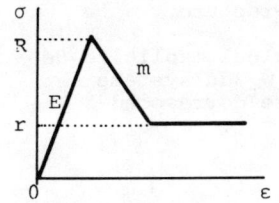

E module de Young
R résistance en compression
m module d'écrouissage négatif
r résistance résiduelle

Figure 1   Modélisation du comportement des piliers

.../...

2. HYPOTHESES DE BASE

Le modèle proposé est plan et défini par les grandeurs suivantes (fig. 2) :

- 2 L largeur d'exploitation,
- H   puissance de la couche,
- h   épaisseur du banc constitutif du toit immédiat,
- p   pression des terrains de recouvrement,
- E   module d'élasticité de la roche,
- τ taux de défruitement de l'exploitation.

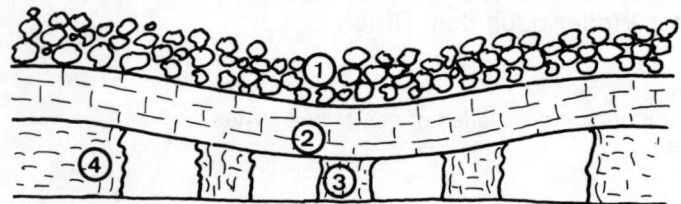

Figure 2 Schéma d'une exploitation par chambres et piliers

1. Terrains sus-jacents
2. Toit immédiat
3. Piliers
4. Bords fermes

Dans le modèle schématisé figure 3, la couche exploitée est assimilée à une couche fictive continue de module $\overline{E} = E(1-\tau)$

Lors de la mise en charge par la pression p, le modèle se déforme et l'hypothèse fondamentale onsiste à supposer que, quel que soit l'état de déformation des piliers, la flèche v(x) du toit, assimilé à une poutre mince, est de la forme : (fig. 3)

$$v(x) = \frac{V}{2}(1+\cos\frac{\pi x}{L}) \quad (1)$$

Cette expression de la déformée est cinématiquement admissible ; la constante V=v(o) représente la flèche maximale du toit, au centre de l'exploitation, et sa donnée définit l'état de déformation de tout le modèle. Deux configurations peuvent a priori se présenter :

. la flèche maximale V induit une déformation maximale des piliers, inférieure à leur déformation de rupture : $V < \varepsilon r.H$, l'ensemble du système reste donc élastique.

. la flèche maximale V induit une déformation maximale des piliers, supérieure à leur déformation de rupture : $V > \varepsilon r.H$, une partie des piliers de la zone centrale est alors en état de déformation post-rupture.

Avec l'hypothèse (1), le calcul explicite de l'énergie mécanique totale W_t du système est alors possible ; l'équilibre correspond à la flèche V qui assure $dW_t/dV = 0$.

.../...

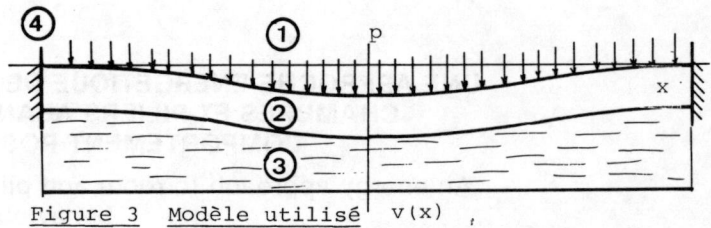

Figure 3 Modèle utilisé   v(x)

1. Pression p des terrains sus-jacents
2. Poutre mince bi-encastrée
3. Couche fictive équivalente à la couche
4. Encastrements

3. L'APPROCHE ENERGETIQUE

L'énergie mécanique totale W_t du système est la somme de 3 termes $W_t = W_{pot} + W_{el} + W_{pil}$ avec

W_{pot} = potentiel des forces de pression,
W_{el} = énergie élastique de la poutre constitutive du toit immédiat,
W_{pil} = travail de déformation de la couche fictive "équivalente" aux piliers.

On démontre que, compte tenu de (1),

$W_{pot} = - pVL$

$W_{el} = \frac{\pi^4}{8} \frac{EIV^2}{L}$ (I = $h^3/12$, inertie du toit)

$W_{pil} = \int_{-L}^{+L} dx \int_{0}^{v(x)} f(u)\, du$

f(u) étant obtenue par transformation d'axes à partir de la courbe de la figure 1.

L'expression de W_{pil}, quoique complexe, peut être calculée explicitement.

4. EXEMPLES

Les exemples qui suivent ont pour but de mettre en lumière l'influence des paramètres post-rupture des piliers sur la stabilité d'une exploitation définie par les données géomécaniques suivantes :

Largeur de l'exploitation : 2L = 100 m
Profondeur 300 m   : p = 7,5 MPa
Puissance exploitée   : H = 4,5 m
Epaisseur du banc du toit
         immédiat   : h = 10 m
Module de la roche   : E = 20 000 MPa
Résistance en compression : R = 20 MPa, (d'où $\varepsilon_r = 10^{-3}$, le tassement de rupture des piliers est alors égal à 0,45 cm.

4.1 - Premier exemple

On a fixé m = 0,5 E et r = 0 (forte pente post-rupture et résistance résiduelle nulle). Les courbes $W_t(V)$ tracées pour différentes valeurs du taux de défruitement (fig. 4) présentent toutes un minimum correspondant à la position d'équilibre du système ; l'abcisse de ce minimum est égale à la flèche réelle du toit, au centre de l'exploitation.

. le fait marquant est qu'il apparaît un taux de défruitement critique τ_c tel que pour $\tau < \tau_c$ = 53 %, les configurations d'équilibre correspondent à des flèches V faibles (V<0,52cm) et donc à un soutien du toit par les piliers ; pour $\tau > \tau_c$, les nouvelles configurations d'équilibre correspondent à des flèches très importantes (V =116 cm) ; pour des telles valeurs, et compte tenu de l'hypothèse sur le comportement post-rupture de la roche, les piliers ont "volé en éclats" et ne participent plus à la stabilité de l'édifice. Il apparaît donc (fig.5) une discontinuité de la fonction V(τ) pour $\tau = \tau_c$.

. On note également que pour τ compris entre 51 % et 53 %, il apparaît un domaine d'équilibre dans lequel les piliers de la zone centrale de l'exploitation sont en équilibre post-rupture (V compris entre 0,45 et 0,52 cm).

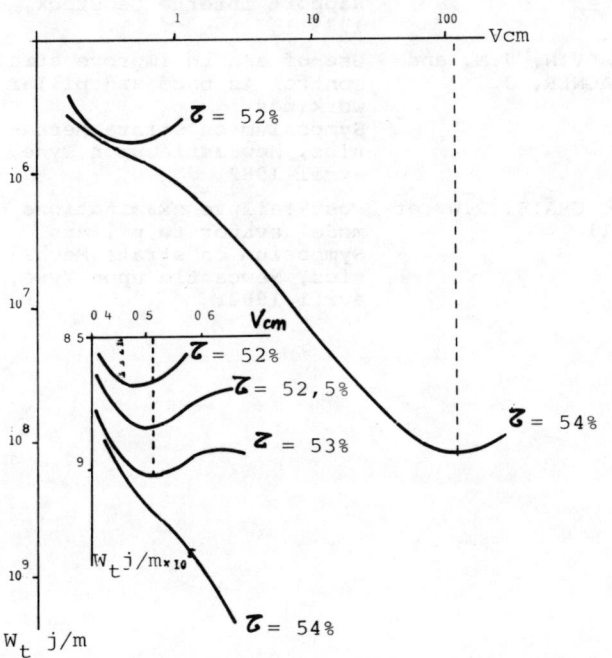

Figure 4 Influence de τ sur les fonctions $W_t(V)$

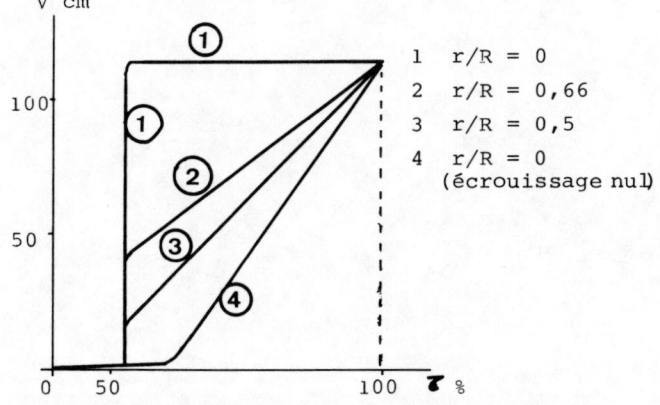

Figure 5 Influence de r/R sur les fonctions V(τ), pour m = 0,5 E

4.2 - Influence de la résistance résiduelle r

On a représenté, sur la figure 5, les fonctions V(τ) pour quatre valeurs du rapport r/R. Le fait le plus important concerne l'influence du rapport r/R sur la discontinuité de la flèche lorsque $\tau = \tau_c$: l'ampleur de la discontinuité diminue lorsque r/R augmente, la fonction V(τ) devenant continue lorsque r/R = 1 (roche élasto-plastique sans écrouissage).

r/R	Amplitude de la discontinuité pour $\tau = \tau_c$, cm
0	116
0,5	42
0,66	18
1	0

4.2 - Influence du module d'écrouissage m (r=0)

La valeur du module d'écrouissage m a une influence sur le taux de défruitement critique et sur l'amplitude de la discontinuité pour $\tau = \tau_c$.

m	∞	0,5	0,01	0
τ_c	51%	53%	61%	(*)
amplitude de la discontinuité pour $\tau = \tau_c$, cm	116	116	116	0

(*) non défini car V(τ) est continu.

5. CONCLUSION

L'approche présentée a pour objet de préciser les conditions de stabilité d'une exploitation souterraine par chambres et piliers abandonnés, tenant compte d'un éventuel comportement post-rupture des piliers. Le modèle souffre évidemment de plusieurs limitations :

- restriction à des configurations bidimentionnelles planes,
- modélisation du toit par une poutre élastique, quelle que soit l'amplitude du déplacement,
- prise en compte de la rupture du système uniquement au niveau des piliers,
- hypothèses outrageusement simplificatrices concernant le comportement post-rupture des piliers,
- enfin hypothèse restrictive sur la déformée du toit de l'exploitation.

Malgré ces limitations, la démarche et le modèle proposés permettent d'aborder certains aspects de la stabilité d'une exploitation, mettant en oeuvre une formulation énergétique globale, et tenant compte du comportement post-rupture des piliers.

L'intérêt majeur du modèle réside probablement dans le fait qu'il prend en compte l'exploitation dans son ensemble, substituant ainsi à l'approche analytique et locale classique, une approche globale du phénomène de stabilité.

Le résultat essentiel est la mise en évidence d'un taux de défruitement critique qui, lorsqu'il est atteint par les travaux d'exploitation, confère à l'édifice une instabilité se traduisant par un basculement et une descente violente du toit, sollicitant les piliers par des déformations très importantes.

L'amplitude de la descente du toit est fonction des caractéristiques post-rupture, module d'écrouissage négatif et résistance résiduelle, prises en compte au niveau des piliers.

Dans la réalité des travaux souterrains, une telle discontinuité doit s'accompagner d'altérations profondes des terrains sus-jacents, qui doivent coopérer à l'effondrement.

La sensibilité du modèle aux caractéristiques post-rupture montre tout l'intérêt qu'il y a à renforcer les piliers, par remblayage ou boulonnage : ce renforcement, en améliorant les caractéristiques de portance des piliers dans leur phase de rupture, doit augmenter la valeur du taux de défruitement critique et par conséquent améliorer la sécurité de l'édifice.

Le mécanisme qui vient d'être ainsi sommairement analysé constitue sans doute un "mécanisme initiateur plausible" des effondrements spontanés des exploitations par chambres et piliers abandonnés.

BIBLIOGRAPHIE

GERMAIN, P. — Cours de Mécanique des Milieux Continus, Masson et Cie, 1973

TINCELIN, E. et SINOU, J. — Effondrements brutaux et généralisés. Coups de toit. Revue de l'Industrie Minérale, Avril 1962.

LABASSE, H. — Les pressions de terrains dans les carrières souterraines ; coups de toit et coups de charge. Revue de l'Industrie Minérale, août 1973

MAURY, V. — Effondrements spontanés, synthèse d'observations et possibilité de mécanisme initiateur par mise en charge hydraulique. Rapport interne Géostock, 1979

GALVIN, J.M. and WAGNER, J. — Use of ash to improve strata control in bord and pillar workings. Symposium on strata Mechanics, Newcastle upon Tyne, avril 1982

Mc CRAIE, R.W. et all — Post-failure examinations of model evaporite pillars. Symposium on strata Mechanics, Newcastle upon Tyne, avril 1982.

THE INTERACTIVE DIMENSIONING OF A CROWN PILLAR IN THE RAUTUVAARA MINE

La planification interactive des dimensions d'un pilier de toit dans la mine de Rautuvaara

Die interaktive Dimensionierung eines Firstenpfeilers in der Grube Rautuvaara

P. S. Sarkka
Senior Fellow, Helsinki University of Technology, Department of Mining and Metallurgy,
SF-02150 Espoo 15, Finland

SYNOPSIS

The Rautuvaara iron ore mine is located about 120 km north of the Arctic Circle. It produces annually ca. 1.25 Mt of iron ore. The shape and size of a +200-level crown pillar was determined by computing the loading-deformation curve caused by country rock on the pillar with the finite-element method. Comparable loading-deformation curves for different pillar shapes were determined with small-scale model tests in a servo-controlled testing machine. The equilibrium points for pillar-country rock systems were determined by matching these curves appropriately scaled. The final dimensioning decisions, stoping experiences and results of control measurements during stoping are also presented.

RESUME

La mine de minerai de fer de Rautuvaara est située à environ 120 km au nord du cercle arctique. L'extraction annuelle de minerai est près de 1,25 Mt. La forme et les dimensions des piliers du niveau +200 mètres ont été déterminées à l'aide de la méthode des éléments finis. Le calcul de la courbe contrainte-déformation pour des formes variées de piliers a été défini par des essais de modèles en miniature à l'aide d'une machine d'essai asservie. Les points d'équilibre du système pilier-roches encaissantes ont été définis par la comparaison de ces courbes réalisées à des échelles appropriées. Les décisions finales des dimensions, les expériences d'abattage et les résultats de mesures contrôlées pendant l'abattage sont également presentés.

ZUSAMMENFASSUNG

Die Eisenerzgrube Rautuvaara liegt etwa 120 km nördlich vom Polarkreis. Die jährliche Förderung liegt bei ca. 1,25 Mt Eisenerz. Mit der Finite-Element Methode wurden unter Einbeziehung der von den Nebengesteinen über dem Pfeiler verursachten Belastungs-Deformationskurve, Form und Dimension eines Firstenpfeilers der +200-Sohle festgelegt. Die Bestimmung vergleichbarer Belastungs-Deformationskurven für verschiedene Pfeilerformen erfolgte in Modellversuchen kleinen Maßstabs in einer servo-kontrollierten Materialprüfungsmaschine. Die Gleichgewichtspunkte des Pfeiler-Nebengestein-Systems wurden durch Vergleich dieser Kurven, bei entsprechender Änderung der Maßstäbe, bestimmt. Die endgültigen Dimensionierungsbeschlüsse, Abbauerfahrungen und Resultate der Kontrollmessungen während des Abbaus werden ebenfalls vorgelegt.

1. GENERAL

1.1 Rautuvaara mine

Rautuvaara iron ore mine of Rautaruukki Oy is located in western Finnish Lapland near to the Swedish border, at a latitude of 67°30', some 120 km north of the Arctic Circle (Fig. 1).

The deposit was first located in 1956. The decision to develop it to a mine was made in 1970 and full production was obtained in 1976. Today the mine has an annual hoist of 1.25 Mt of magnetite iron ore, which covers about 25 % of the Finnish iron ore consumption.

The orebodies occur in two groups (Fig. 2), roughly 1 km apart. They lie in a 100 m wide zone of quarz-feldspar schists, skarns and amphibolites. The schist zone is surrounded by monzonitic and dioritic plutonic rocks. Skarn is the host rock of the ore.

The schists and the orebodies strike NE-SW and dip 80° SE, the orebodies in addition having long axis which plunges 50^g SW. The average width of the two largest orebodies, Green and Blue, is 15 m and their lengths are 150 m and

250 m respectively. These orebodies have been proved to persist to a depth of 475 m.

The hanging-wall rocks are mostly diorite, with skarn and quartz-feldspar schists in places. Shear zones are also present in the hanging-wall in some locations. The footwall consists of amphibolites or skarn (Fig. 2b).

The ore reserves in the Rautuvaara area are currently estimated as 16 Mt of proved reserves, averaging 45 % Fe, with a further inferred tonnage of 18 Mt.

The initial mining at Rautuvaara was by open-pit methods on the outcropping Green and Blue orebodies, two pits being taken down to the +35 m and +20 m levels respectively.

In the upper levels of the NE group of orebodies, immediately under the original open pits, extraction was carried out by sublevel caving. Caving did not seem to occur naturally, and this caused difficulties in winter when the mine was exposed to arctic winter temperatures through an opening to the surface.

Due to the temperature difficulties and high waste rock dilution sublevel caving was gradually abandoned, being phased out altogether by the end of 1980.

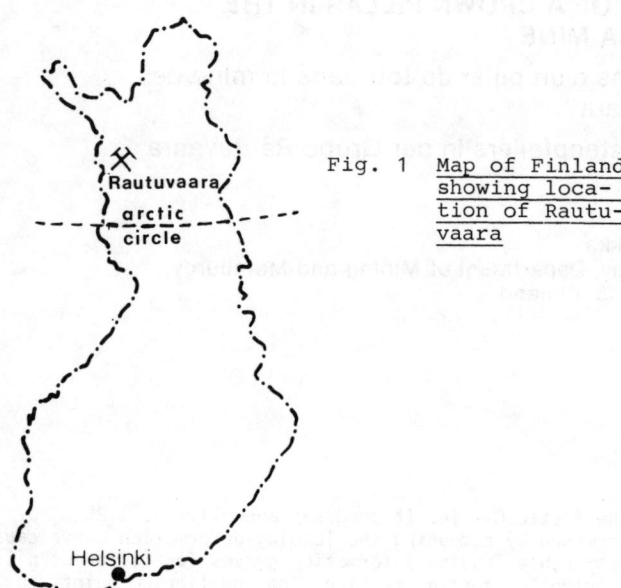

Fig. 1 Map of Finland showing location of Rautuvaara

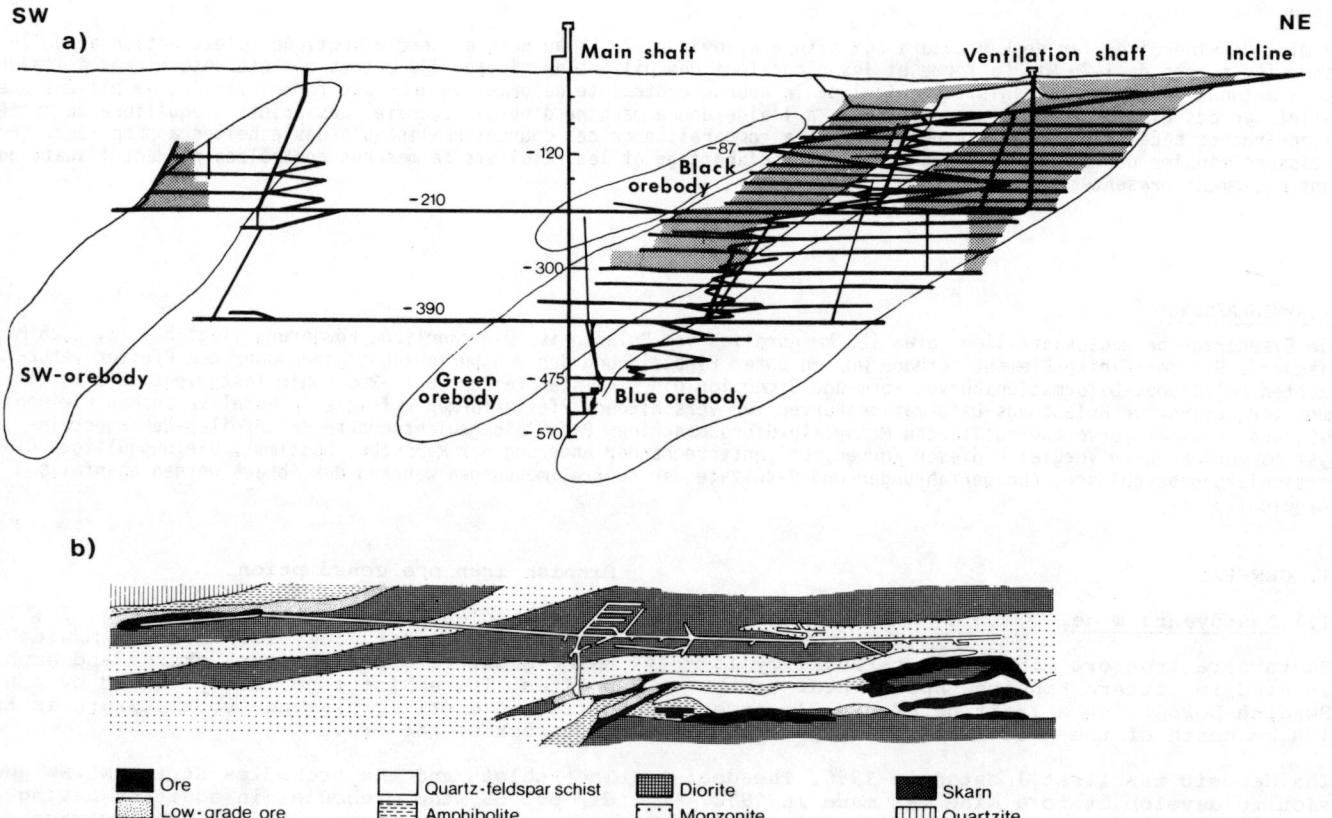

Fig. 2 a) Longitudinal section through the Rautuvaara mine, b) Geological map of Rautuvaara mine at the +210 m level

Currently all ore extraction is by sublevel stoping using a sublevel interval of 25 m, sometimes 30 m. Stopes are typically 150...200 m long, 100 ...200 m high and 7...23 m wide. Most production is currently coming from the +300 to +210 m levels in the Green and Blue orebodies and from between the +210 and +120 m levels in the SW-orebody.

1.2 The +200 m level crown pillar

In the NE district the stoping method used above the +200 m level was sublevel caving. This had to be abandoned due to the undeveloped caving of country rocks. In places the stopes still are open from the +200 m level to the surface, a height of ca 200 m.

The stopes above and under the +200 level are opposite to each other at a distance of about 400 m altogether. The facts favoring a crown pillar between the stopes are:
- without an isolating pillar surface air is able to enter freely to sublevels under the +200 m level. This causes inconvenience especially in winter when the temperature can temporarily fall down to -40° C. The drifts and drill holes freeze and fog is frequently met.
- a possible caving in above the +200 m level would harass loading in stopes under this level by increased waste rock dilution.

On the other hand, the pillar will contain iron ore about 20 000 t/m of thickness, i.e. totally about 250 000 t of ore. This can possibly, but not evidently, be stoped in a later stage. If a pillar is left, there are all reasons to optimize the thickness of it.

The cross-section of the pillar is to be found in Fig. 4. The upper border of the pillar has been formed as a loading drift of longitudinal sublevel caving in an earlier phase of stoping. There is no access to the area and the drift is probably filled with rock.

The lower border of the pillar can be chosen quite freely, however, in the limits the existing drifts allow to the drilling geometry and to the economy of stoping.

The upper border of the pillar can be supposed broken at least to the depth of 2 m from the planned stoping limit in connection with sublevel caving (Vuolio 1980).

With cautious blasting the broken zone at the lower border can be limited to about 1 m.

1.3 Pillar dimensioning

The conventional pillar design is based on the dimensioning of pillars to resist the complete load concentrated on them. This is a safe method but especially with single or few pillars it leads to overestimated pillars (Hoek and Brown, 1980). Besides, these methods are mainly aimed for the pillars in horizontal or near to horizontal seams and not easily applicable to steep stopes.

A newer approach to the problem is the concept of stiffness of rock, concerning both the pillar and the surroundings. First work on the subject

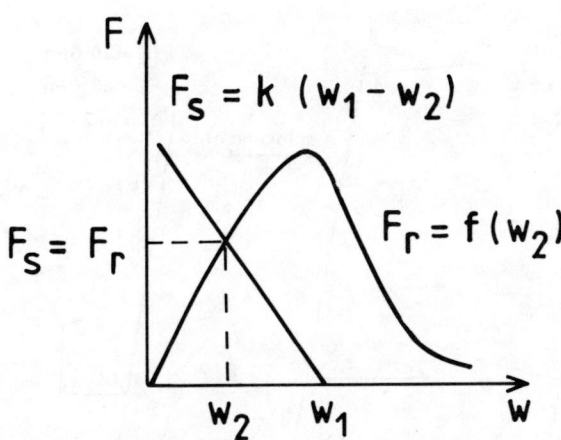

Fig. 3 The concept of stiffness in pillar dimensioning

was performed by Salamon (1967, 1970) and Starfield and Fairhurst (1968). They concentrated to the calculation of coal pillar-coal mine stiffness systems and to the dimensioning of room-and pillar systems respectively. The latter performed laboratory experiments, too, but only on the elastic area of the specimen. Some later theoretical work is described by Brady (1979) and Petukhov and Linkov (1979).

The concept of the use of stiffness in pillar dimensioning can be presented in Fig. 3. If a pillar is left in a mine opening, around which stresses (either regional and/or gravitational) influence, the opening deforms until it reaches a stable situation. If the rock mass remains intact, this deformation can be represented by the curve F_s in Fig. 3. Due to this deformation the pillar in the opening deforms, too. This deformation is represented by the curve F_r in Fig. 3. The intersection points of these curves are the equilibrium points of the system.

2 INTERACTIVE DIMENSIONING

The concept of interactive dimensioning is based on the fact that the deformations and stresses around an opening in rock can be calculated today quite reliably with different element methods, at least in elastic-brittle conditions of rock mass. This gives us the stiffness of rock mass for the studied geometry.

The stiffness of the pillar is not so easy to obtain. The main difficulty lies in the determination of the post-failural deformation and stiffness. This can be experimentally related, however, to small-scale tests performed with the same rock material and the same W/H ratios as in the real pillar. Especially with W/H ratios smaller than 1 the main governing factors are the structure and the characteristics of the rock matrix and the homogeneity of the rock.

2.1 Finite-element calculations

The interactive planning of the +200 m level crown pillar was started in 1980, when the decision to change finally over to sublevel stoping

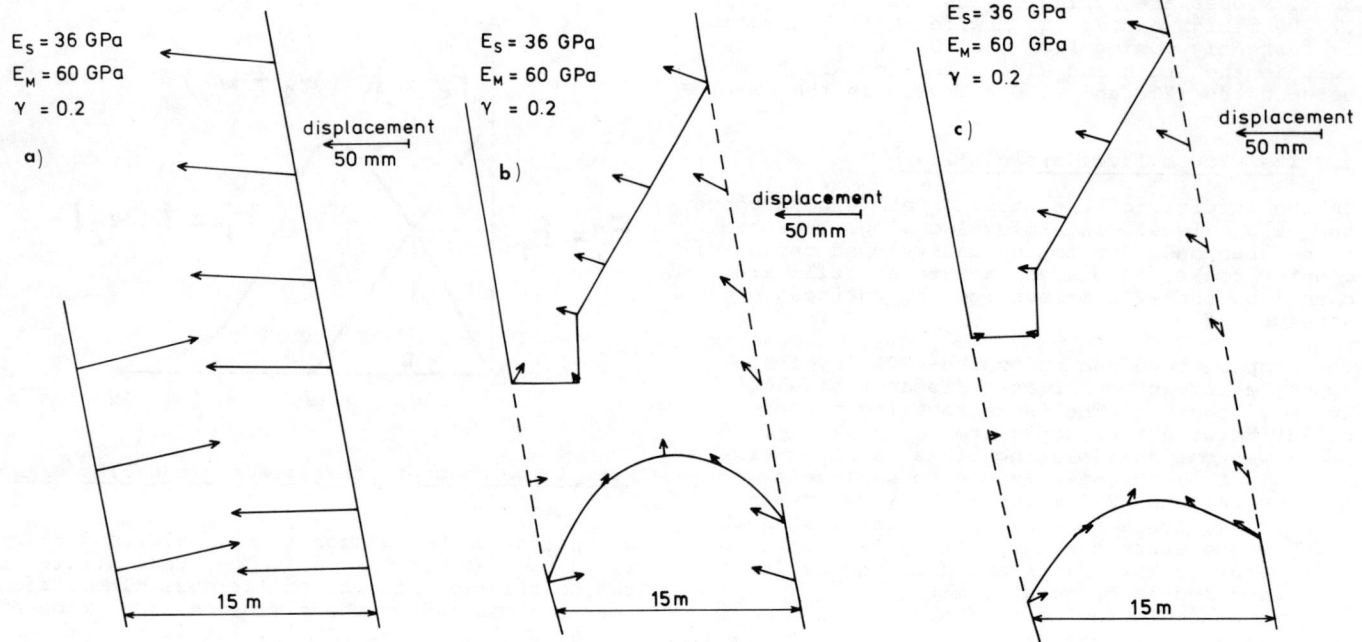

Fig. 4 The displacements of ore-country rock contacts with different pillar thicknesses. $E_{ore} = 60$ MPa, $E_{country\ rock} = 36$ MPa. a) no pillar, b) pillar thickness 6 m, c) pillar thickness 11.5 m

was made. The rock mass characteristics had been determined earlier (Saari and Ylinen, 1981) and their averages were used in the finite-element calculations of displacements in ore contacts. The calculations were performed with a simple two-dimensional program ELE (Aalto, 1975).

The material was assumed to be completely elastic. The differences between the country rock characteristics were so small that the computation could be done with only two materials, ore and country rock.

The horizontal stress was about 10 MPa and vertical stress equaled to the overburden pressure, i.e. 6 MPa. The resulting displacements are presented in Fig. 4.

The state of stress in the pillar was calculated by numerically integrating the stress of single elements at the thinnest point of the pillar. From this data the loading-compression curves caused by the country rock for different pillar thicknesses were constructed (Fig. 5). These curves suppose that the country rock behaves elastically.

2.2 Small-scale tests

The loading-compression curves for pillars with different thicknesses were obtained by small-scale tests in a servo-controlled compression machine. The details of testing are presented in Särkkä (1978).

The samples were drilled from the pillar area in the direction of the pillar. From the ∅ 32 mm drill cores specimens were made with L/D ratios of 1.15, 1.45, 1.95 and 3.00. These represented the pillar widths of 13 m, 10.5 m, 7.5 m and 5 m, respectively. There were 3 specimens of each size. The obtained force-compression curves are presented in Fig. 6. The shadowed area represents the variation of the curves.

The curves are scaled up to the pillar size and completed with the comparable loading curves of the country rock-pillar system from Fig. 5.

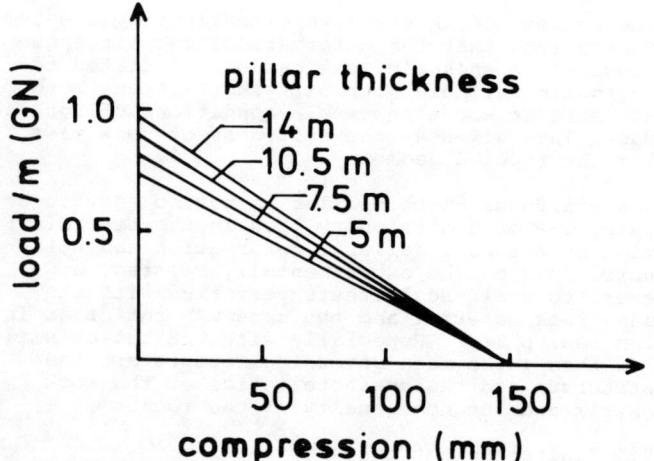

Fig. 5 The loading curves caused by country rock to different pillar thicknesses

D 104

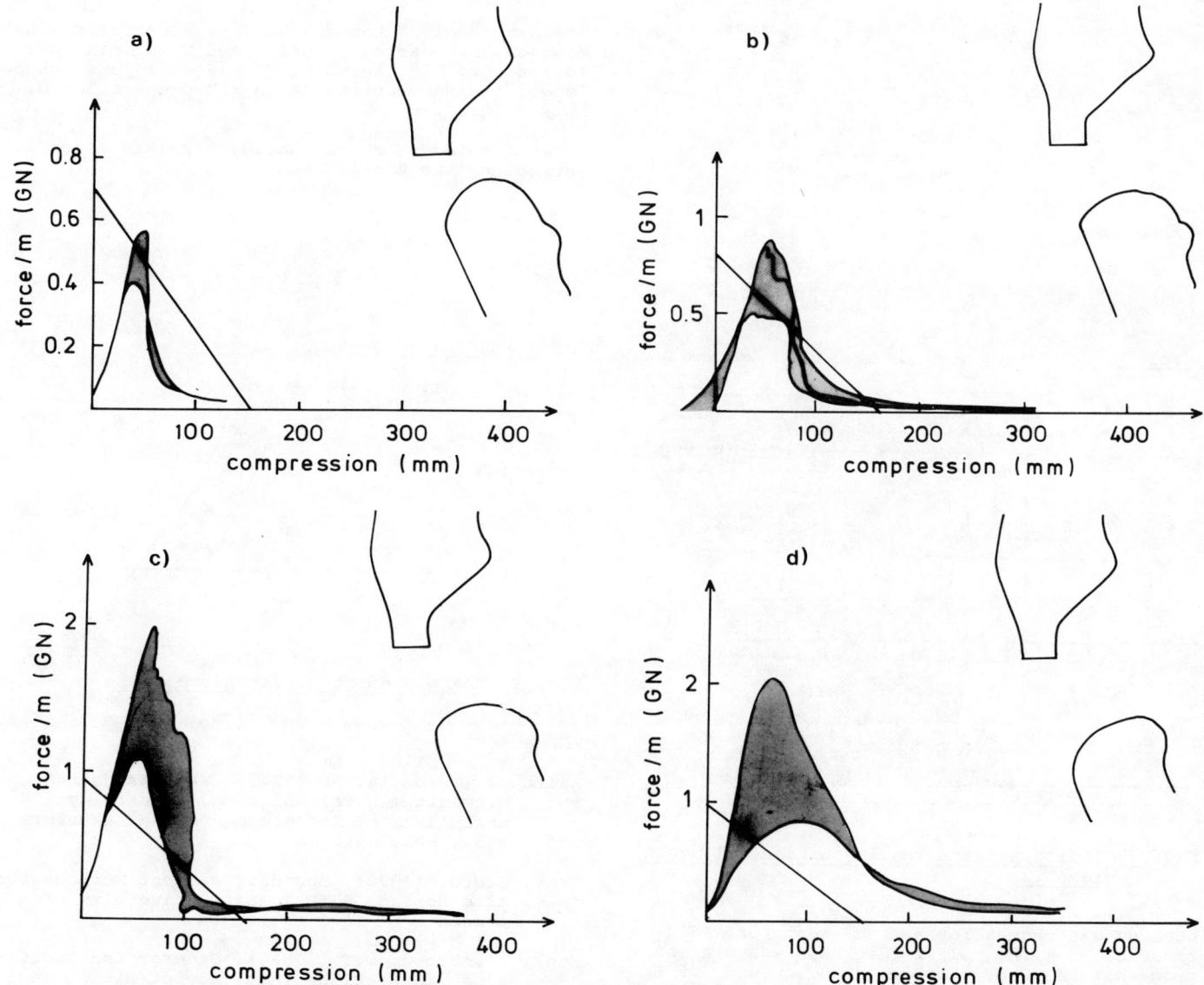

Fig. 6 The loading compression curves of the pillar-country rock system with different pillar thicknesses. a) pillar thickness 5 m, b) pillar thickness 7.5 m, c) pillar thickness 10.5 m and d) pillar thickness 13 m

2.3 Matching

From Fig. 6 can be seen that if the pillar thickness is only 5 m, it has a probability of only 30 % to stay intact, i.e. quite surely at least some part of the pillar will cave in.

With the following pillar thickness, 7.5 m, the probability has risen to 80 % for the pillar to stay, but this means again that a part of the pillar would cave in.

First the pillar with a thickness of 10.5 m would stay 100 % up, supposed that there would not exist any crushed zones in the pillar.

In the thickest part, 13 m, even the minor crushed zones would not cause failure, this due to the bigger stiffness in the pillar than in the country rock mass.

With pillar thickness here is meant the "solid" thickness of the pillar. In the dimensioning the broken zones caused by blasting on both sides of the pillar have to be added to this.

2.4 Decisions

The final dimensioning decisions were made in the beginning of 1981. The mine was aware of the results and of the fact that the minimum thickness for a stable pillar was about 14 m including the broken zones.

However, it was decided to test the results in the Blue orebody (Fig. 7). The NE part of the

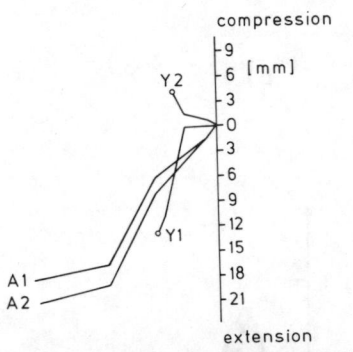

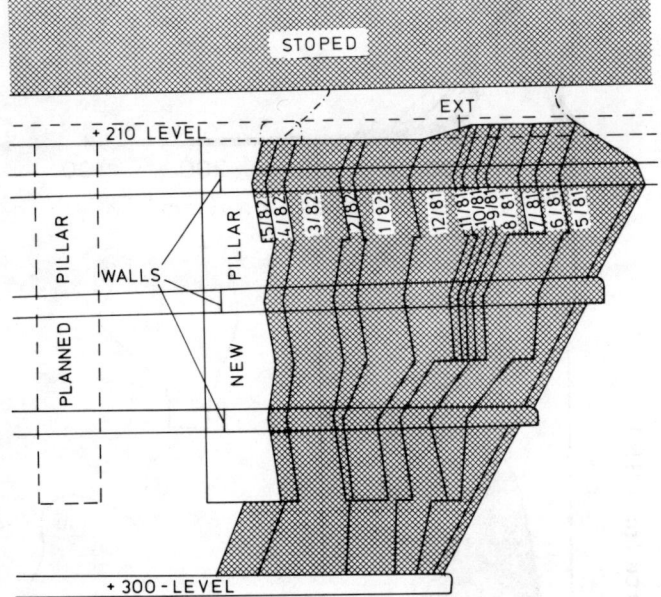

Fig. 7 Stoping and the results of control measurements

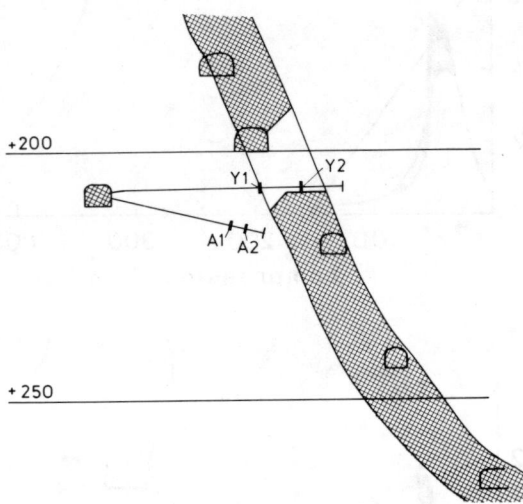

Fig. 8 The placement of extensometers

ACKNOWLEDGEMENTS

The personnel of Rautaruukki Oy, especially mine manager, Dr. Aarre Juopperi, and chief mine geologist, Mr. Ole Lindholm, MSc (Geol.), are acknowledged their co-operation in the preparation of this paper.

I am indebted to Mr. A Öhberg, MSc (Min.) for correcting the English text.

REFERENCES

Aalto, J., Promila, A. (1975). Yksinkertainen elementtimenetelmäohjelma. Rakenteiden mekaniikka 8. Rakenteiden mekaniikkaseura. 3-34, Helsinki.

Bray, B.H.G. (1979). Boundary element methods for mine design. Ph.D.Thesis, University of London.

Hoek, E., Brown, E.T. (1980). Underground Excavations in Rock.The Institute of Mining and Metallurgy. 527 pp, London.

Petukhov, I.M., Linkov, A.M. (1979). The theory of post-failure deformations and the problem of stability in rock mechanics. Int.J. Rock Mech.Min. Sci. (16), 2, 57-76.

Saari, K., Ylinen, A. (1980). The Stability Calculations of the Rautuvaara mine. Conference on Application of Rock Mechanics to Cut-and-Fill Mining. Vol. 1. 517-525, Luleå.

Salamon, M.D.G. (1970). Stability, instability and design of pillar workings. Int.J.Rock Mech.Min. Sci. (7), 6, 613-631.

Salamon, M.D.G. (1967). A method of designing bord and pillar workings. J. S African Inst.Min. Metall., 68-78.

orebody was chosen for a test area. For a distance of 20 m the pillar thickness was 8 m and from that on 10 m, which meant 5.5 m and 7.5 m in the model tests respectively. Two two-pint rod extensometers were placed in the pillar and in the footwall (Fig. 8).

3 FOLLOW-UP

When stoping progressed, the pillar was showing high horizontal compression. When stoping had passed the thinner part of the pillar, the pillar deformation had obtained the elastic limit which was calculated to be about 20 mm. In the same time the extensometers in the pillar were lost due to caving of the lower part of the pillar.

In May 1982 the whole pillar caved down (dotted lines in Fig. 7). The place of the intermediate pillar and the walls built into the sublevels were changed. In the new design the crown pillar will be down to the first sublevel, ca 20 m, and it is planned to recover in a later stage of stoping.

Starfield, A.M., Fairhurst, C. (1968). How high-speed computers advance design of practical mine pillar systems. Engineering and Mining Journal (169), 5, 78-84.

Särkkä, P. (1978). The Failure Behaviour of Some Finnish Mine Rocks in Uniaxial Compression. Ph.D.Thesis, Helsinki University of Technology.

Vuolio, R. (1980). Louhintaräjäytysten suunnittelu ja suorittaminen. Suomen Maanrakentajien Keskusliitto r.y. 188 pp. Helsinki.

DISPLACEMENT REGULARITY IN DEEP OPENINGS NEAR AN OREBODY

Régularité du déplacement des galeries profondes situées près de masses minérales

Typische Verschiebungen in tiefgelegenen Strecken in der Nähe von Erzkörpern

Yun-Mei Lin
Director of Rock Mechanics Laboratory, Northeast Institute of Technology, Shenyang, China

SYNOPSIS

The regularity with which mining excavation affects neighbouring orebodies is very important in metal mines, particularly those of considerable depth. Based on observations in dangerous locations the regularity is described with which rock pressure occurs. Moreover there is a report on deformation measurements carried out at a depth of 1000 m over a period exceeding 2 years, On the basis of these field observations, the modelling work and analytical investigations it will be shown that the regularity of the displacements and failure mechanisms which occur typically near orebody openings, can be explained by using the stability theory of plates.

RESUME

Le processus d'exploitation a une influence importante sur les galeries situées près de gisements, surtout dans le cas de mines profondes. Cette communication précise la régularité du mouvement de pression des terrains observé lors d'études dans des zones dangereuses. D'autre part on donne les résultats de deux années de mesures du déplacement de la roche alentour de galeries creusées à une profondeur de mille mètres. La recherche sur modèle et l'étude théorique réalisée à partir des données obtenues démontrent que la théorie d'instabilité des plaques peut expliquer la régularité du déplacement et le mécanisme de rupture qui se produisent dans les galeries près de gisements.

ZUSAMMENFASSUNG

Die Regelmäßigkeit, mit der der Abbau benachbarte Erzkörper beeinflußt, ist für Erzgruben von großer Bedeutung, insbesondere bei Gruben großer Teufe. Ausgehend von Beobachtungen in gefährlichen Grubenbereichen wird die Regelmäßigkeit beschrieben, mit der der Gebirgsdruck auftritt. Ferner wird über Deformationsmessungen von über zwei Jahren Dauer in 1000 m teufen Gruben berichtet. Aufbauend auf diesen Geländebeobachtungen, Modellversuchen und analytischen Untersuchungen wird gezeigt, daß die Regelmäßigkeit der Verschiebungen und Bruchmechanismen, die typischerweise in Strecken in der Nähe von Erzkörpern auftreten, mit Hilfe der Stabilitätsanalyse von Platten erklärt werden kann.

DISPLACEMENT OF SURROUNDING ROCK IN OPENINGS NEAR OREBODY

As shown in Fig.1, the angle of dip of orebody is between 80 and 85. Its hanging wall is composed of firm white marble and footwall of interstratified band of thin marble, slaty phyllite, etc. The measured openings lie in the footwall, 980 meters below the ground surface. The measuring points of displacement in both sides are arranged at four different regions A,B, C and D. In each group of measuring points, three measuring rods, 0.5m, 1.5m and 2m long, are contained. The layout of points is shown in Fig.1. The main results of measuring are as follows:

a) An increase of displacement is closely correlated with excavating process and regularly varied with its productive operation. Fig. 2 is the representative of hundreds displacement-time curves, which clearly expresses the variation of displacement velocity at different stages.

b) When the sublevel is not worked, a chamber can only obviously influence on the opening within that chamber range. In Fig.3, it is shown that group B which is 10 m from chamber is slightly influenced by stoping, whereas group A situated beyond 60 m is almost not influenced. Practical data of group A indicate that under certain conditions the variation of displacement of surround rocks in openings against time is negligible. Indeed, u-t curve reflects the relationship between displacement and productive factors under the present conditions.

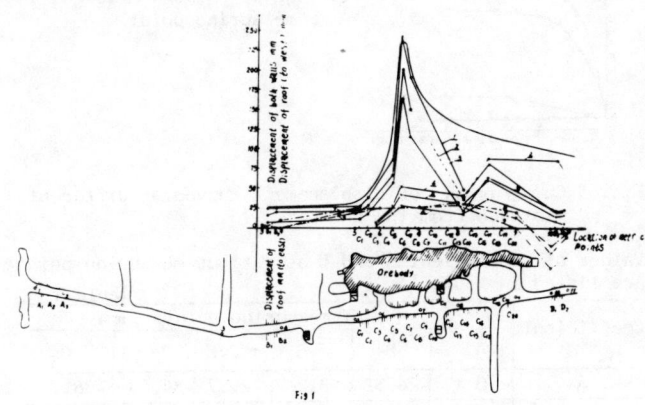

Fig.1 The layout of measuring points and distributed curves of displacement magnitude along strike

Through regression analysis, empirical equation* of displacement between overcut and storage stages is found as follows:

$$U + A = B \ln(t_1 + t_2) \quad (1)$$

* Coefficient of correlatability = 0.92 ~ 0.98

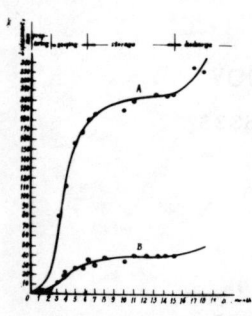

A -- 0.5m measuring rod.
during undercutting -- 11.5 mm/month
during begining of over-cut --- 62 mm/month
during finishing of overcutting --- 20 mm/month
during storage of overcutting --- 2 mm/month
during begining of greatly discharging -- 15 mm/month
B -- 1.5m measuring rod

Fig. 2 Total displacement variation curve of both two sides at measuring rods group C_6.

Or it may be written as:

$$U + A = \frac{K\gamma H}{D} \ln(\frac{h}{v_1} + t_2) \qquad (2)$$

Where: U - displacement measured at the begining of overcut period, mm;
A - coefficient correlated to displacement produced at surrounding rock in opening at the period of exploitation cutting;
k - coefficient of stress concentration;
γ - unit weight of rock;
H - depth of opening from ground surface;
$D = \frac{Eb^3}{12(1-v^2)}$ --- flexible rigidity of rock slice, where b --- mean thickness of rock slice;
h - stoping height
v_1 - stoping velocity
$t_1 = \frac{h}{v_1}$ --- time required from begining of overcut, minute;
t_2 - time required for storage, minute;

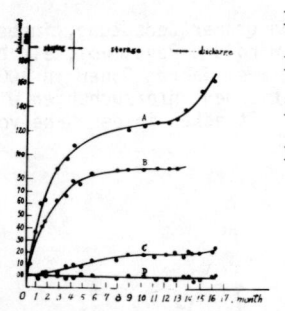

I -- group C_7, 0.5m long measuring point
II -- group C_{17}, 0.5 m long measuring point
II -- group B_2, 0.5 m long measuring point
IV -- group A_3, 0.5 m long measuring point

Fig. 3 Comparasion of displacement curves at different distance from the chamber.

Values of Coefficient A and B at various measuring point are listed in table I

Table I

Coefficient	No. of measuring points, mm					
	A	B_1	B_2	C_1	C_2	C_5
A	0	-26.52	-31.66	-29.7	-34.53	-38.7
B	0	7.35	8.48	8.92	10.84	17.45
	C_6	C_7	C_{14}	C_{17}	D_1	D_2
A	-77.2	-56.1	-25.1	-25.9	-125.1	-103.36
B	45	32.1	10.33	9.27	35.84	29.44

From equation (2) it is easily understood that to raise stoping velocity can reduce displacement of surrounding rock and facilitate the maintaining of stope opening.

c). The regularity of variation of displacement of both sides along axial direction is shown in fig. 1. This figure and the coefficient of table 1 indicate that the peak value of two sides displacement sum occurs at the location of group C_6, lying in south of stope centre and the value at opposite ends are obviously reduced, especially at the south end. Thus it can be seen that the max. value of bearing pressure approximately occurs at the centre of chamber.

d) The displacement regularity of measuring points on roof is that during overcutting the points on roof of section C within chamber length generally move upwords and towards the direction of stope and after overcutting the points gradually subside. This fact above proves that moving direction of opening will submit to the tendency of stope wall displacing towards larger excavated space. Except each side of opening move toward its own space, also the whole opening flex towards stope. (Fig. 1. dotted line)

Basing on the practical data obtained from horizontal displacement of points No 3~9 on roof, we gain regression equation as:

$$W = a + b\sin\frac{\pi x}{l} \qquad (3)$$

Where: W --- horizontal displacement of roof, mm;
x --- distance from south end of chamber, m;
l --- length of chamber along strike, m; l = 50
a and b --- coefficient of points equation, a = 11.4~17.7, b = 11 ~ 20;

This equation explains that displacement of roof is simplex trigonometric function of $\frac{x}{l}$. To solve the stability problem of rectangular plate simply supported at both ends in the principal direction of subjecting pressure, equation of flexure plant is

$$\frac{\partial^4 w}{\partial y^4} + 2\frac{\partial^4 w}{\partial x^2 \partial y^2} + \frac{\partial^4 w}{\partial x^4} + \frac{P}{D}\frac{\partial^2 w}{\partial y^2} = 0$$

When y/h is constant, we get the solution as follows:

$$W = B \sin\frac{\pi x}{l} \ldots \qquad (4)$$

The similarity of equations (3) and (4) indicates that the displacement and failure of steeply inclined stope wall should be explained as unstability problem of the plate system, subjected to pressure in the principal direction.

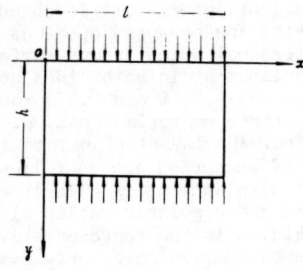

Fig. 4. Scheme of simply supported rectangular plate subjected to pressure in the principal direction

REGULARITY OF GROUND PRESSURE APPEARED IN OPENINGS NEAR OREBODY DURING SUBLEVEL

a) In sublevel stoping the critical time of ground pressure activation against openings near orebody is at the time of pillar falling down. Before this time openings near orebody may be maintained by different supporting means. After it they will partially or entirely be destroyed, namely no methods can be used to maintain them. In addition, the most apparent characteristic is a slipcrack line tens of meter long along strike appeared on the roof and floor of opening. Between cracks a fall exists (Fig. 5 and 6). This phenomenon can be used as a precise signal which predicts the approaching of staged pillar failure.

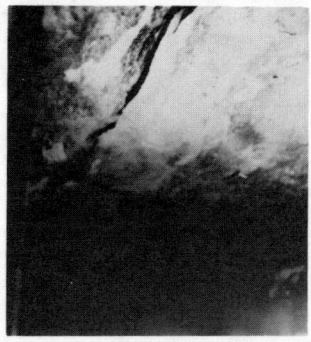 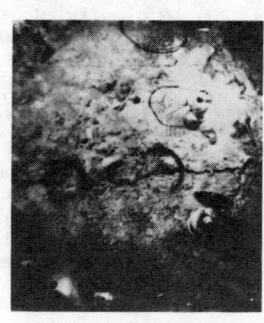

Fig.5. Strike slipcrack on the top of opening. located 920 m below surface

Fig.6. Strike slipcrack on the bottom of opening. located 920 m below ground surface

b) It is very interesting that the form of final failured opening is similar to that of unfailured opening. Depending on different factors such as thickness of orebody, structural features of rock mass and distance from orebody, it varies regularly.
For example:
(1) Near thin ore vein (2 -3m)
When the surrounding rock in opening is of laminar structure, degree of failure of both sides is similar; when the surrounding rock is of massive structure, a interesting phenomenon appears, that is the wall near by the stope is entirely destroyed, whereas the opposite wall (away from the stope) is still intact (Fig. 7).

Fig. 7. Opening in massive rock, one wall is destroyed and the other is intact.

(2) Near thick ore vein
No matter what the surrounding rock structure is, the failure reaches up to the roof and makes opening entirely choking up.
(3) Near the intersection of level excavated along the vein and cross level:
A special kind of large scale fall in the form of "pull out" often occurs in the foof (Fig. 8). This made of falling occurs only within the width of roof. Although the visible fracture occurs inside the roof, there is still lack of any relevant fracture on the floor. This explains that the fracture is not due to the result of complete rock slip, its probable reason is that the horizontal displacement of west wall towards stope is greater than that of east wall (firm wall). In this way a tensile stress takes place inside the roof strata and makes it to be opened. This analysis above has been proved by model study.

Fig.8 Fracture of roof at intersection

(4) The direction of movement of whole surrounding rock is towards emptied stope without exception. As in Fig. 9 the left side is chamber, and rock stratum of right wall is pushed towards the opening, while the left wall (near stope) contracts into the face of rock. Fig. 10 illustrates the front of opening is emptied stope, theoreby a stepped multi-state settlement appears.

Fig.9. Horizontal movement   Fig.10. Vertical movement

(5) After the whole rupture of pillar, the effect of end begins to become apparently. Not like in this sublevel stoping, chamber not only effects the level within the length range of this chamber, but also severely extends the level outside the chamber length range. The influencing degree is weakened gradually as the distance from stope is increased. Comparison of Fig. 11 to Fig. 12 may clearly see the change of the end influence.

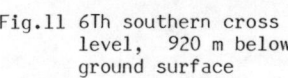

Fig.11 6Th southern cross level, 920 m below ground surface

Fig.12 8Th souther cross level.(50m from 6th southern cross level)

These results obtained from the partially or entirely failure openings are of great value to reader, because such investigation was carried out at dangerous places and similar findings had not been reported in reference literature prior to this paper.

MODEL STUDIES

Physical model studies are mainly concerned with the influence of excavation on displacement characteristic of opening near orebody. The dimension of plane strain model is 42×48mm², using bi-axial loading apparatus, loading value is measured by electrical resistance strain gauge which is adhered on steel bearing beam and calculated by a previous calibrated strain-pressure curve. Deformation and displacement of model are determined by photographic comparision with a large size of projector. The geometerical constant of similarity of model α is 1:100. clay, tailings, plaster and sand are used to make models and mica powder is used to simulate the weakness plane. Before excavation, firstly a certain vertical and horizontal load is applied to model, stope and level is constructed by hand drill and manual cutting. According to the result of field investigation, the ratio value of horizontal load and vertical load is defined to be $\lambda = \frac{1}{5}$. The results of laboratory test are consistent with that of field measurement. In model the similar displacement regularities of surrounding rock, such as direction of opening roof displacement initially is towards the stope and later backwards it; displacement velocity is consistent with productive process, are not only reproductive, but also obtain data of displacement around the opening at different times, involving initial deformation immediately after driving. These are useful for the analysis of displacement and failure mechanism of opening near orebody. For example, Fig. 13 is curves of location variation of surrounding rock at 1 cm position from wall face, illustrating the phenomenon of anti-symmetrical flexibility truly takes place at wall rock, therefore it is suitable to express the failure mechanism of bulging fracture of wall (Fig. 14) by the theory of unstability plate system by pressure.

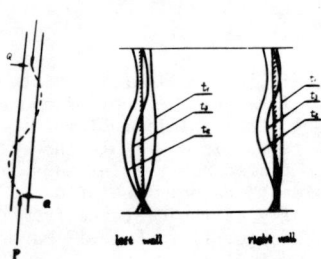 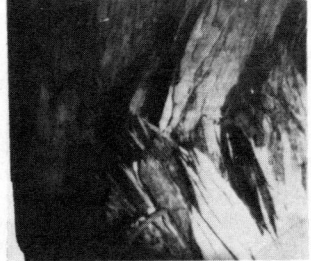

Fig.13 curves of variation of position of wall face

Fig.14 Typical phenomenon of bulging fracture

Similarly in modelling stope we may also see such failure phenomenon of unstability (Fig. 15)

Fig. 15 Type of modelling stope failure

Therefore, on the basis of stability theory of plates, we may use the following equation to determine the limited height of opening and chamber h or the critical bearing pressure Pc.
a) The plate wall of level may be simplified as plane strain problem:

$$Pc = \frac{4\pi^2 \cdot D}{h^2} Fs \qquad (5)$$

b) It may be simplified as simply supported rectangular plate and mainly subjected to the vertical pressure from the stope wall:

$$Pc = \frac{\pi^2 D}{h^2} \left(m + \frac{1}{m} \cdot \frac{h^2}{l^2} \right) Fs \qquad (6)$$

Where:
Fs -- safety factor, taking Fs = 0.5 0.6
m -- ratio value depending on the length and the height of chamber.
Values of m may be taken according to the value of $\frac{h}{l}$, when the value of $\frac{h}{l}$ is greater, may take $m \approx \frac{h}{l}$.

When the cap pillar and intermediate pillar fell down it corresponds the value of h and l increased, the capability of wall to subject the critical loading abruptly decrease and is led to losing of stability and collapse within greater range. At same time it will result in movement along strata plane, and the appearance of a slip crack. Therefore it is reasonable to use the strike slipcrack line as a signal, which predicts the arrival of staged pillar failure.

Moreover from equation (6) it is known that decrease of span l may result in the increase of Pc which is several times that of Pc produced by decreasing the height h of stope. Therefore, when stability of chamber is insufficient, the main method is to shorten the chamber length, i.e. increasing the intermediate pillar, but not decreasing the stope height to raise stability.

Conclusions:

1. In the stoping process an increase of displacement is closely correlated with excavating process. During stages of overcutting and discharging displacement abruptly increases. During these periods it is necessary to intentify the maintaining of opening and shorten the production cycle to improve the ground pressure state of opening.
2. In sublevel stoping the critical time of ground pressure activation against openings near orebody is at the time of pillar falling down. Pillar collapse within greater range can lead to fully destroy the openings. So the service life of opening near orebody depends on the time of pillar collapse.
3. The appearance of slipcrack along strike can be used as the predict of collapse of staged pillar.
4. The type of final failure of opening is still dependent on distance from emptied stope, thickness of orebody and structural feature of rock mass.
5. Using the stability theory of anisotropic plates to explain the mechanism of wall failure in steeply inclined orebody and to calculate the limit load of wall is reasonable. But in this respect, a large amount of theoretical investigation work is required to carry out.

Acknowledgement
Many thanks are expressed to Mr. Hong Yuan-zhen, Hau chun-hua and other person, who have participated with the authers in this research working and helped in every time wanted.

References
Lin Yun-mei et. al Report of Field Investigation of Mines (unpublished)
Xu Zhi-ren Mechanics of Elasticity

THEORETICAL PRINCIPLES AND FUNDAMENTALS OF ROCK BURST PREDICTION AND CONTROL

Les fondements théoriques de la prédiction et méthodes de la prévention des coups de charge

Theoretische Grundlagen der Prognose und Verhütung von Gebirgsschlägen

I. M. Petukhov
Head of the Department of Rock Bursts, All-Union Research Institute of Mine Geomechanics and Surveying (VNIMI), Leningrad, USSR

A. M. Linkov
Senior Scientist of the Department of Rock Bursts, All-Union Research Institute of Mine Geomechanics and Surveying (VNIMI), Leningrad, USSR

SYNOPSIS

The present state of the rock burst problem in the USSR is briefly discussed. Necessary and sufficient conditions for the occurrence of rock bursts are reviewed. The main statements of the authors' theory of rock bursts are presented in their connection with the practical applications. Suggestions for the future development of the rock burst theory are made.

RESUME

On examine brièvement l'état actuel des connaissances en ce qui concerne le problème des coups de charge en URSS et passe en revue les conditions nécessaires et suffisantes pour que ce phénomène se produise. Les principales données de la théorie des coups de charge élaborée par les auteurs sont présentées en relation avec les applications pratiques et on évoque des développements futurs possibles en la matière.

ZUSAMMENFASSUNG

Es wird der gegenwärtige Stand der Untersuchungen über die Gebirgsschläge in der UdSSR diskutiert. Die notwendigen und hinreichenden Bedingungen für die Entstehung dieser dynamischen Erscheinungen werden behandelt. Die hauptsächlichen Grundsätze der von den Autoren entwickelten Theorie der Gebirgsschläge werden in Verbindung mit ihren praktischen Anwendungen betrachtet. Vorschläge zur weiteren Entwicklung der Theorie von Gebirgsschlägen werden unterbreitet.

The rock burst problem in the world mining practice arose about two hundred years ago and during recent fifty years it became particularly acute for many mining regions of the world. In the USSR the first rock bursts occurred in Kizel coalfield in the late 1940s. Later on, the burstprone conditions were ascertained in a number of coalfields. The investigations on the rock burst problem are conducted under the scientific and methodical supervision of the All-Union Research Institute of Mine Geomechanics and Surveying (VNIMI). In the early 1950s, on the basis of these investigations the important conclusions were drawn; they were concerned with the nature of rock bursts, their classification by intensity and place of occurrence (loading conditions), significance of properties of the material to be fractured and its surrounding rocks, energy balance (Avershin, 1955; Petukhov, 1954 and 1957), which formed the basis for the rock burst prediction and prevention methods and allowed to solve successfully the rock burst problem in the coal mines of the USSR (Bich, 1972; Petukhov, 1972).

At present, the deep mining associated with the further development of the extractive industry acutely poses the problem of the burst hazard of not only the coalfields but also the most of ore deposits. The necessity of the early and fast solution of this problem increases the role of the theoretical generalization that enables us to develop the practical recommendations beforehand - not waiting till the rock bursts will occur constantly at new depths and in new ore deposits and give, of course, not only the additional scientific information but also, first of all, the information on the economical losses and victims.

The current state of the rock burst mechanics, which is characterized by the transition from the qualitative description to the completed quantitative theory, serves as a basis for such generalization. The given report presents the review of this generalization carried out to give a simple and complete idea of the theoretical aspects of the rock burst problem, wherever possible. And to make the understanding easier and fix an attention on the essence of the problems, the authors deliberately avoided using the complex formulae and calculations as well as describing the various mathematical methods of solving the problems being arisen. At the same time it should be borne in mind that all sections of the modern theory given below are well provided mathematically. The data on these formal aspects of the problem are given in references.

Revealed in the early 1950s (Petukhov,1954 and 1957) an important analogy between the violent fracture of a rock specimen in a conventional soft testing machine and the dynamic fracture of rock in a colliery or mine serves as a clue that helps to understand the nature of rock bursts and the contents and the structure of the rock burst theory. In this analogy the element being fractured around a working acts as a rock specimen and rocks surrounding the element - as a testing machine. The analogy clearly reveals the necessary and sufficient conditions of rock burst occurrence taking into account the rock mechanical behaviour data obtained owing to designing and utilizing the special rigid testing machine that enables us to investigate rock properties after the limit strength (post-failure characteristics).

To begin with, it is necessary to give close attention to the difference between what is known as s o f t loading and r i g i d loading (Cook,1965). With this aim in view, it is expedient to consider the "ideally soft" loading and "ideally rigid" one. Direct application of load to a specimen under test is an example of "ideally soft" loading, Fig. 1a. In this case load P on the specimen is exactly equal to load being applied however much the specimen under the action of fixed force is deformed. These are the conditions of the specified load. Rigidity C of such a testing machine is equal to zero.

With the "ideally rigid" loading, displacement u is controlled rather than load P at a specimen boundary. To clearly illustrate, it is possible to imagine a screw-type testing machine made of very rigid materials, Fig. 1b. In the limit, its rigidity C is equal to infinity.

When testing the rock specimen b e f o r e i t s l i m i t s t r e n g t h, the load-displacement relationship P(u) is the same for any mode of loading as shown by the corresponding uniform ascending parts of curves OP_m in fig.1. At the same time when maximum load P_m is attained, the pattern of deformation in an "ideally soft" testing machine and "ideally rigid" one is quite different.

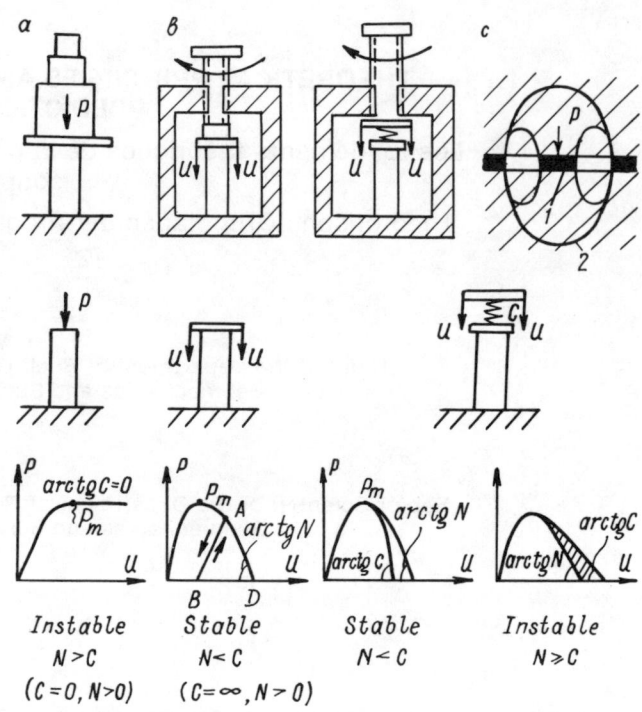

Fig.1. Stability dependence on loading types

Recording is interrupted in the soft testing machine because acting force remains invariable and specimen resistance diminishes. Process of failure is that of instable nature and if acting force is sufficiently great, the greater energy is released too.

Failure is observed to be dynamic and specimen fragments are scattered at high speed. By its nature it is a rock burst reproduced in the laboratory.

The post-failure deformations are quite different in a rigid testing machine. In this case, the displacement control provides exact equality of applied load and specimen resistance. As a result, the descending (post-failure) part of the curve (part P_mD in fig. 1) is obtained. The system "testing machine - specimen" remains stable, macroscopic dynamic effects are not observed. The specimen may be unloaded at any moment of post-failure deformation (part AB in fig. 1) and conserves its cohesiveness under unloaded conditions. It is possible to subject the specimen to deformation once again (part BA in fig. 1) and continue the post-failure test (part AD) from point A. The slope of drop N of the post-failure curve defines "rigidity" of a specimen while softening. "Rigidity" N may be converted to the m a t e r i a l characteristic by means of the transition to stresses (by dividing forces P by specimen cross-sectional area S) and to strains (by dividing displacements u by specimen height l). This important characteristic $M = Nl/S$ is called a drop modulus and it defines the rock properties after the limit strength. As for the graphical representation the drop modulus is well similar to the modulus of elasticity but in contrast to the latter it is shown as a slope of the descending part of the curve and not of the ascending one.

In practice, actual loading (specimen in a true testing machine, pillar or seam edge) never can be either ideally soft or, the more so, ideally rigid. To clearly illustrate, one can imagine that loading is carried out by means of a spring having rigidity C (Fig. 1c) and here $0 < C < \infty$. Before attaining maximum load P_m, the deformation, as with the ideal loading cases having been considered, is not dependent on relationship between rigidities C of the outside system (loading system) and element N being deformed. However, when the maximum load P_m is attained, this relationship assumes a decisive importance. If the rigidity of the testing machine (surrounding rocks) is more than that of a specimen (pillar, deposit edge), i.e. $C > N$, the displacements are controlled and "rigid" loading is carried out and deforming proceeds calmly in accordance with the descending part of the curve P(u). Otherwise, when $C < N$, the process is that of uncontrollable and instable nature. It continues spontaneously and dynamically till the complete failure and energy contained in the area between patterns of the testing machine and specimen, is released (hatched area in fig. 1). Transition of this energy to the kinetic energy of the flying fragments takes place and if this energy is sufficiently great the process appears to be a violent dynamic phenomenon or, in practice, to be a rock burst.

As mentioned above, in situ, the element of rock mass around the working and being deformed after the limit strength acts as a rock specimen and surrounding it rocks in stress play the role of the testing machine. Thus, pillar I yield (right upper diagram in fig.1) causes the release of energy from rocks located within outline 2.

In the early 1950s (Petukhov, 1954) a working hypothesis on rock burst nature and mechanism was developed on the basis of these statements. In conformity with this hypothesis the complete system "coal (ore) being fractured - surrounding rocks" participates in the preparation and manifestation of a rock burst. And the rock burst as well as the violent failure of the specimen in the testing machine was considered as a brittle failure of the element stressed to a limit load; this failure occurs when the absorbed energy in the mass being destructed exceeds the released energy from the surrounding elastic regions and the excess energy is transformed into the kinetic energy of the flying fragments and into the oscillations of the rock parts having not been destructed. The hypothesis included all principal statements of the modern rock burst theory and served as a basis for determining the composition of the combined rock burst investigations and for practical solving the problems of rock burst prediction and prevention.

The anology with the violent nature of the specimen fracture in the soft testing machine and a rock burst distinctly reveals three conditions of the dynamic fracture:

(i) condition of the limit load - load P on the element being fractured (specimen, pillar, seam edge) should reach the maximum value P_m;

(ii) condition of instability - rigidity N of the element being deformed should be greater than rigidity C of the loading system

$$N > C$$

or, that is to say, the energy inflow $(-\Delta \partial)$ should exceed the consumption W_p for fracture

$$-\Delta \partial > W_p ;$$

(iii) energy condition - level of energy to be released should be sufficiently high that the loss of stability might be considered as a violent dynamic phenomenon.

These three conditions completely determine the structure and contents of the rock burst theory and its applications (see Table).

Table. Conditions of dynamic destruction and sections of rock burst theory

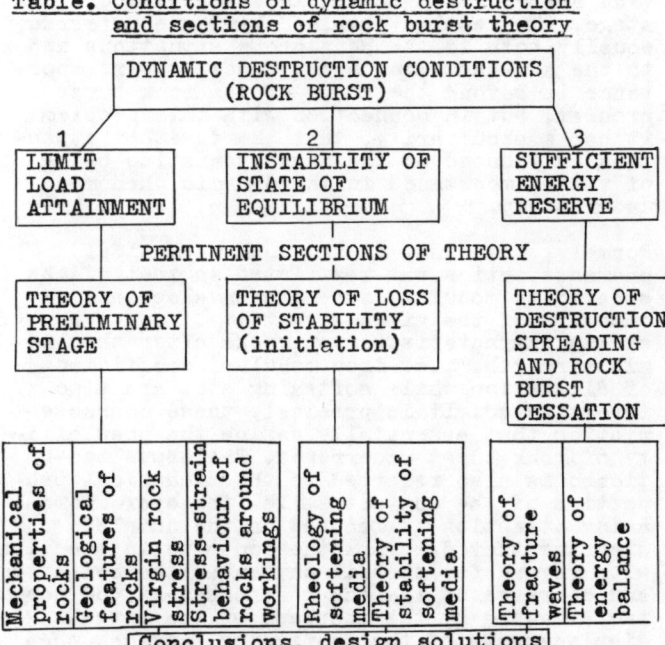

The investigation on the preliminary stage of the dynamic phenomenon, i.e. mechanical properties of rocks, geological features and the virgin rock stress, stress-strain behaviour of rocks around the workings, energy stores in state of equilibrium, is required by the first-mentioned condition. The elaboration of the loss-of-stability theory of the elements under limit stress, i.e. initiation of the rock bursts, is dictated by the second-mentioned condition. It presupposes the development of the mechanics of the post-failure deformations (after the limit strength) occurring in the rock mass and at the contacts of blocks and layers and of the pertinent theory of stability of such softening media as well. The importance of studying the features of the spreading and cessation of the fracture ensues from the third-mentioned condition that is taken into account in the theories of the destruction waves and energy balance. All the above mentioned sections are used for the development of the applied part of the theory which includes the substantiation of the methods of the rock burst prediction and prevention.

The extensive studies have been performed in accordance with each of the above mentioned sections and the important results have been obtained. Their detailed description is beyond the scope of the present report. Nevertheless, it seems to be useful to describe, on the whole, their contents, present state and tendency of their further development.

Theory of the preliminary stage

Studying the mechanical properties of rocks, geological features, the virgin rock stress and the in-situ state of stress opening driving are included in the investigation of this stage. The results obtained here are referred equally both to the burstprone situations and to the non-hazardous ones. Hence, their importance is beyond the scope of the rock burst problem, but in connection with this problem it has a peculiarity, that the foremost attention is focused on the characteristics being of vital importance in the dynamic phenomena under study.

Formely, the study of the rock mechanical characteristics was restricted to that of the elasticity modulus, strength envelope and parameters of the visco-elasticity. Now, the strain characteristics of rocks after the limit strength, i.e. drop modulus, coefficient of dilatation while softening etc. are also investigated. It is precisely these characteristics that essentially define the possibility of rock burst occurrence. The above mentioned is also referred to the mechanical properties of the contacts, i.e. the ever increasing attention is focused on the complete diagrams that describe the contacts interaction with regard for the dilatation at the contacts and decrease of the shear stresses at the contacts caused by the increase of the mutual displacements of the contacting surfaces when the maximum shear load is attained.

While analyzing the geological features of the deposits in connection with the rock burst problem, the attention is focused on the two points: on the one hand, the influence of these features on the mechanical properties and, on the other hand, - on the stresses fields in the massif. Thus, studying the geology is closely connected with two adjacent sections of the rock burst theory.

Obtaining the reliable data on the virgin rock stress is mainly an experimental task, because the possibility of the exact determination of the stresses by means of calculation methods of solid mechanics is extremely limited by the lack of the non-loaded state from which one would estimate the displacements, by the initiation of the sresses, directly in the course of the massif formation, by the complex tectonic processes having been occurred during the various geological epochs. The experimental methods or the combinations of the calculation and experimental methods are more relevant. The great amount of data on the virgin rock stress is experimentally obtained for ore deposits. These data are used for laying-out efficiently the mine fields in accordance with the direction and correlation of the main stresses, for determining the optimal cross-sections of the workings, for elaborating the recommendations on the methods of mining, opening driving and supporting. The data are also used as an initial information required for the subsequent calculations of the stress state in the course of mining, the stability and energy being released. Some of the important features of the stress distribution dependent on the relief, non-uniformity, anisotropy and other rock characteristics are studied on the basis of the comparatively simple solid models too.

At present, for solving the rock burst problem, study of the mechanical properties, geological features and the virgin rock stress is combining together in one section which is called as a dynamic deposits districting.

The stress-strain rock mass behaviour is changed after the opening driving. A new state of equilibrium is also investigated by means of the combination of the experimental methods, modelling, analytical and numerical methods which mutually supplement each other. At the same time the great attention is paid to the task schematization based on the generalization of extensive experimental studies and observation data. The mathematical methods of solving the problems, many of that have been realized as the operating and systematically used computer programs, are provided for the schemes having been developed. Thus, the programmes based on the finite elements methods, boundary integral equations, expansion in series and quadratic programming were developed and are successfully for solving the problems which are extremely complicated in respect to the calculations. Having been developed in VNIMI, the package of the programmes (Petukhov, 1981) enables us to investigate the influence of the various geological and mining factors on the stress-strain behaviour of rocks, the rock burst hazard, boundaries of the zones being protected and ones of the high rock pressure. Among these factors are the depth of occurrence, dip angle, thickness of a seam being mined, layered structure of the massif, puncturing action of the pillars, pressure of the undermined rocks on the floor of the working, three-dimensional outline of the workings, their mutual influence etc. The summary of the main results are, for instance, given in (Petukhov, Linkov et al.,1976) where many other original works are mentioned too.

The similar computations are not restricted by the research problems, they gain the industrial nature due to the perfection of the methods of the acquisition and preparation of the initial data to be set in the computer and due to the outcome of the computations which are necessary for the enhineering personnel; the computation results are given with the help of the graph plotters in a convenient graphic form (in particular, it is provided plotting the isobars around the workings, the boundaries of the zones protected against the rock bursts and zones with high rock pressure). Such graphs may be used as the documents specifying the procedure and methods of working-out the deposits.

Loss-of-stability theory

Having determined the stress-strain behaviour of the elements of the massif around the workings, it is also necessary to estimate the stability of this state of equilibrium. Here, the progress, also, depends, to a considerable extent, upon the advancement in the experimental area and comprehension of the mechanical tests results. The introduction of the rigid testing machine in practice has culminated in obtaining the post-failure characteristics of rocks and was a new stage in the development of the researches on the rock burst theory. It was stimulated by the coherent understanding of the above mentioned relation of the post-failure deformations to the stability problem, and the latter - to the rock burst problem (Blake, 1972; Cook, 1965; Crouch, Fairhurst, 1974; Petukhov, Linkov, 1974 and 1975; Salamon, 1970 and 1974). However, both the theoretical and practical development of the promising prerequisite was initially that of the completely quantative character or was restricted to the simpliest cases of the substitution of the rocks around the working, which are deformed after the limit strength, by some equivalent specimen being subjected to the uniaxial compression. The great possibilities of the post-failure diagrams having been obtained were not fully used because they were not viewed as the graphical representation of the peculiar constitutive equations which close the system of the equations of equilibrium and deformations simultaneity. In the last few years such view was formed directly in the course of the development of the rock burst theory. In conformity with this approach, the investigation is carried out among the usual solid mechanics problems and it enables us to make use of the wide experience in the development of the mechanics of the hardening and softening media.

The essence of this new approach is that the important rock burst theory sections under discussion have become the part of the general theory of stability under fracture. Thus, one closes the mechanical description of a series "stress-strain behaviour - loss-of-stability - vigorous release of energy" which is a sequence of the events during the rock burst. The earlier theoretical developments are, therewith, combined within the scope of the general theory of the rock bursts at the coal, ore and non-metalliferous deposits. The differences has been found to be applied only to the specific values of the parameters and their combinations which characterize the hazard rather than the nature of the phenomena being described. Such parameters and their combinations themselves and their critical values are revealed to develop the method of rock bursts prediction and prevention. It was impossible to achieve this result without the development of the new rheological models, methods of calculation of the stress-strain behaviour and stability which would consider new data on the mechanical properties of rocks; these data become available after introduction of the rigid testing machines in practice. In the above mentioned sections of the rock burst theory a significant progress was made towards the development of the deformation and increment descriptions of the post-failure deformations of rocks; the formulation of the constitutive equations for the rough contacts taking into account the possibilities of dilatation and softening at the contacts; obtaining the new equations and methods for solving the problems for the systems of blocks and layers taking into consideration the softening and complicated conditions of the interaction at the contacts; the development of the theory of stability of rocks surrounding the openings based on the most important for rocks physical non-linearity which is reflected by the post-failure diagrams of rocks themselves and interaction of their contacts; obtaining the theoretical indices and criteria of the hazard and safety (Linkov, 1977 and 1981; Petukhov, Linkov, 1974, 1975, 1976 and 1979; Petukhov, Linkov et al., 1981; Salamon, 1970 and 1974).

Theory of destruction spreading and cessation of dynamic phenomenon

Material being destructed is involved into motion when its stability is lost. In typical occurrence of rock bursts, the material is separated into the fragments which are ejected onto the direction of the mined-out space. The elements that are directly adjacent to the opening is understood to be moved in the first place, the elements following them etc. Therefore, it is expedient to consider the fracture wave front for the typical rock bursts: the material maintains its cohesion on one side of the fracture wave front and on another side of it the material is represented by the elements being not interconnected. Usually, the fracture waves during bursting are investigated on the base of the laws of conservation of mass, impulse and energy at the leap surface, i.e. on the fracture wave front, and it gives the relations determining the possibility of the destruction spreading (Petukhov, Linkov et al., 1976). These relations serve to determine the parameters of the effective means of controlling the bursting. They include the important rock burst energy characteristics - the energy to be released and one to be absorbed. The determination of these characteristics forms the important section of the general theory - theory of the energy balance, which characterizes the energy background of the dynamic phenomena taking place. The energy balance during bursting is given by the following formula:

$$W_M + (-\Delta\vartheta) = W_P + W_K + W_Б + W_C$$

The left part of this equation characterizes the energy to be released during bursting and the right one - the energy consumption. The energy to be released is a sum of part W_M having been contained in the material to be destructed and part $(-\Delta\vartheta)$ coming from the surrounding rocks. It is spent for destruction W_P and imparting the kinetic energy to the fragments of the material having been destructed W_K. The residual amount of the energy is absorbed by the surrounding rocks near the place of destruction W_6 and comparatively small amount of energy (less than 10%) leaves the neighbouring zone in the form of the seismic oscillations W_C. From the theory, it follows that the extremely significant characteristic of the energy level and, at the same time, the degree of stability is given by the energy coming from the surrounding rocks. To calculate this energy, the mathematic methods were elaborated that enables us to compare the various mining situations with the possible intensity of bursting (Cook, Hoek et al.,1966; Petukhov, 1972; Petukhov, Linkov, 1974 and 1976; Salamon, 1970 and 1974). It is distinctly revealed that the well-known rock burst classification by the place of occurrence (Petukhov, 1954, 1957 and 1972), in facts, arranges the situations in a class by the energy criteria and degree of stability. Concurrent with the quantitative assesements of energy, such classification favours the development of mining.

The rock burst ceases due to the different reasons, the main ones are: fracture wave entering the area where its spreading is difficult, e.g. due to rock inhomogeneity; appearance of the upthrust produced by the material having been destructed; reduction of rock mass volume before the destruction front, e.g. in pillars; formation of a stable destruction boundary.

Theory applied aspects

All the preceeding sections of the rock burst theory contain the statements for predicting and controlling these dynamic phenomena.

In conformity with three conditions of a rock burst occurrence one can, practically, solve whether the rock burst hazard is available having answered three questions: (i) whether the maximum (limit) load on the element being destructed is obtained in stress-starin behaviour under investigation; (ii) whether this behaviour is instable; (iii) whether a sufficiently high level of energy is available to assume the loss of stability as a hazardous dynamic manifestation. If the answers are positive the rock burst hazard is available. And if only one of the answers is negative there is no hazard. Consequently, on the one hand, if it has been ascertained that all three conditions are satified, the hazard is predicted and, on the other hand, it is sufficient to put an end to any of these conditions for rock burst preventing. This circumstance serves as a basis for the methods of the potentional and current prediction and prevention of bursting. Let us, briefly, consider the relation of these methods to the theory.

As it is evident from the foregoing explanations, the rocks are neither hazardous nor safe by themselves. They became like that only under the particular loading conditions. The element may be calmly deformed if the environment is "rigid" and it may be scattered while soft loading. The behaviour of the element being deformed and of the system outside to this element (surrounding rocks, in the laboratory - a testing machine) is a factor of particular importance. Therefore, it impossible, strictly speaking, to pose the burst hazard problem without taking into account the loading conditions. If, in the coal and ore mines, these loading conditions, i.e. the characteristics of the "testing machines", are changed over a wide range, the problem on the potential rock burst prediction would be actually meaningless. Nevertheless, the practice and the theory prove that the number of the loading conditions being considerably different, so to speak, "the types of the testing machines", are actually limited by several main types. They, as a whole, meet the requirements of the rock burst classification by the place of occurrence. Having demanded that there will be no rock burst under any of these loading conditions one can obtain the characteristics of the potential burst hazard which are formulated o n l y in the terms of p r o p e r t i e s of rock under investigation. It is precisely the restriction of the number and parameters of the "testing machines" having been actually realized in the practice that enables us to introduce the characteristics of the potential burst hazard which are good borne out by their use in the coal and ore mines.

A wide variety of the brittleness factors is proposed for comparing the potential burst hazard. Their mutual relations and facts of the burst hazard were investigated and it was found that, as a whole, the available statements are in line with the data obtained in practice. It refers, to a great extent, to the factor $\lambda = M/E$, directly proposed on the basis of the theory of stability, that represent the relation of the drop modulus to the modulus of elasticity (critical value is equal to unit) and to the brittleness factor K_ε that is correlated to factor λ and determined from the fraction of the elastic deformations when loading maximum up to 80% and with the critical value equal to 0.7.

Compared to the estimations of the potential burst hazard, the current local prediction has one additional peculiarity which provides, under the theoretical conditions of stability, the use of not only the material properties but also the factors that determine the outside system. The complete set of the parameters which characterize the burst hazard under the particular mining situations is involved and relative (correlative in nature) estimations of the hazard are deprived of the equality being usual for the potential rock burst prediction, and values which are directly included in the theoretical conditions have gained advantage.

However, with this method, the great difficulties arise if it is used not only for the general estimation of the mining situation but for the detailed current prediction which is carried out, for instance, at each step of face advance. The difficulties are caused by the fact that carrying out the measurements and calculations under the conditions of the fast changes of the geological and mining situation is a complex and expensive affair. Thus, to determine the large size cube strength or its drop modulus required for the direct application of the conditions of instability is a laborious question and its solution is likely justified for expeditious controlling the rapidly changing mining situation.

One should again focus his attention to the aproximate methods that are only indirectly related to the values and combinations of the values being included in the theoretical conditions due to the variability of the rock properties being common in practice and owing to the difficulties in determining them. Having been adequately adapted to the conditions of the particular deposits, these inderect proximate methods have become a very suitable practical way of current rock burst prediction. Thus, the destructibility under the acting loads is integrally evaluated by bore meals output (in coal), by core samples division into convex and concave disks (in sandstone and salt), by seismo-acoustic activity. By developing and applying such simplified prediction methods one should, however, clearly understand their links with the theoretical conditions. On the one hand, it allows to imagine the limits of these methods applicability and, on the other hand, to make the best of their possibilities.

Let us also note that, as a rule, it is unreal to raise a problem on the prediction of the exact time and place of the rock burst occurrence because the random nature of the changes of geological and mining factors eliminates the possibility of the strictly determinated approach to predicting the current burst hazard, i.e. it is evaluated only in the terms of the theory of probability. It is likely that one can pose a problem on p r o b a b i l i t y of a rock burst at the particular place. The effectiveness of this method is, to some extent, problematical owing to the difficulties of collecting, processing and interpretating the pertinent information. However, at present, there is a tendency to develop the systems of rock mass behaviour controlling which can have a decisive effect on the success of the probability approach to the current rock burst prediction.

As to the rock burst prevention methods it is necessary, first of all, to pay attention to the general guiding principles of safe mining that are well known in the USSR and given in the acting instructions. They, completely, result from the theory too. Thus, it is recommended to reduce the rock pressure by means of rational mine laying-out taking into consideration the calender plan which provides the systematic working-out preventing the formation of districts with high stress concentrations; to use obligatory advance protective seams working-out which provides successive working-out the hazardous seams of coal and ore without using the local measures of rock burst prevention; to reduce or to eliminate, at all, development in advance of mining, particularly, the productive one; to eliminate the retreating and following one another faces; to use mining without pillars, wherever possible,etc.

The theory also provides the quantitative analysis of the various means of controlling the bursting. The calculation of the protected zones boundaries is the most striking example of such quantitative evaluation. These results have been covered in detail in a number of the monographies, for instance, (Petukhov,Linkov et al.,1976) and there is no need to repeat them. The other quantitative applications of the theory may be the calculations of the protective zones size while carrying out the local means on reducing the rock burst incidence, as well as the calculations of the pillar systems and excavation supports with regard for rock burst hazard.

It seems to us that the q u a l i t a t i v e relation of the recommended means to the theoretical assumptions being presented is rather apparent for dwelling on the various effective means of reducing the incidence of bursting. However, it should be noted that the q u a n t i t a t i v e investigation of this relation is far from being exhausted despite all the successes achieved.

Conclusions

Development of the theory and its applications should be, as before, conducted by combining the in-situ, laboratory, analytical and numerical investigations.

While studying the virgin rock stress it is expedient to increase the degree of the comprehensive and reliable stress determination. For this purpose it is necessary to improve the theoretical and methodical basis and to develop the dynamic deposit districting that combine the structural geology, geomorphology, geophysics and geomechanics methods. Therewith, it should be borne in mind that the main task is to give the particular conclusions and recommendations for mining.

While studying the stress-strain behaviour of rocks surrounding an opening it is expedient to perfect further the methods of acquisition of the initial data on the mechanical properties of rocks and rock mass behaviour as well as to improve the calculation schemes. It is advisable to turn to initial data obtaining by means of the geophysical methods, to complete the formation of the blocky rock mass mechanics as a wholly completed section of the rock mechanics. To do this, firstly, it is necessary to improve the scientific and methodical level of experimental methods development, particularly the geophysical ones, in their connection with the general statements of the rock burst theory and, secondly, to develop, in a short time, and to use systematically the computer programs for calculating the blocky rock mass behaviour for the post-failure deformations of rock and softening at the contacts. As yet, there is no complete qualitative correlation of all practical methods with the theory and this fact makes difficulties in unifying them as well as decreases their reliability and complicates answering the number of questions which remain unsolved or solved slowly because of the difficulties of the empirical analysis. Among these problems one can mention the problem on the effect and degree of the effect of the local inhomogeneity as well as the loading speed on rock mass strength.

The problem of the development of the automatic systems of controlling rock mass behaviours and bursting remains the central problem. All the prerequisites for developing and implementing such systems are already there but it requires the energetic and purposeful activity to realize them. It seems that this activity acceleration requires some organization decisions, i.e. their centralization and proper planning, scientific and technical provision are desired.

There is also a number of the special applied problems being worthy of notice. Thus, taking into consideration the increase of mining depth and geological peculiarities of a number of deposits it is necessary to investigate systematically the yielding pillars behaviour with due regard for bursting. Special investigations and development of the methodical instructions on calculating the excavation supports with allowance for bursting are of use.

The development of the unified theory of the dynamics and gas dynamics phenomena in the underground coal and ore mines will also be essential and useful contribution to the science and practice. It is dictated by the availability of the sufficiently developed theories of the particular kinds of the phenomena having in common a whole number of the principle sections (stress virgin rock, rock mass behaviour around the openings, rheology of the blocky and softening media, theory of instability etc). The unified theory development enables one to unify, to the maximum extent, the specification requirements, the methods of rock bursts and outbursts prediction and prevention. Finally, it will contribute to the safety augmentation and effectiveness of mining.

References

Avershin,S.G. (1955).Rock bursts (in Russian). 235 pp. Moscow: Ugletekhizdat.

Bich,J.A. (1972). Rock bursts and methods of their prediction (in Russian). 101 pp. Moscow: CNIEIugol.

Blake,W. (1972).Rock bursts mechanics. Col. Sch. Mines Q. (67), 1-64.

Cook,N.G.W. (1965).A note on rockbursts considered as a problem of stability. J.S.Afr.Inst.Min. & Metall.(65), 8, 437-446.

Cook,N.G.W.,Hoek,E. et al. (1966).Rock mechanics applied to the study of rockbursts. J.S.Afr.Inst.Min. & Metall. (66),May, 435-528.

Crouch,S.L.,Fairhurst,C. (1974).Mechanics of coal mine bumps. Trans.Am.Inst.Min. Engrs. (256), 4 , 317-323.

Linkov,A.M. (1977).On the conditions of stability in mechanics of destruction. Proc. Acad.Sci. USSR (233), 1 , 45-48.

Linkov,A.M. (1977).On allowance for the postfailure deformations in solving rock mechanics problems. (in Russian). Trudy VNIMI (103),71-76.

Linkov,A.M. (1981).Stability and stressed behaviour of elastic blocks. (in Russian).Trudy VNIMI,"Borba s gornjimi udarami", 8-11.

Petukhov,I.M. (1954).Behaviour of rocks and coal in Kizel burstprone coalfield mines (in Russian). 23 pp. Author's abstr.thesis Deg.Cand.Sc.-(Tech.), Leningrad Mining Inst.

Petukhov,I.M. (1957).Rock bursts in Kizel coalfield mines (in Russian), 143 pp. Perm: Perm Publ.

Petukhov,I.M. (1972).Coal mine bumps (in Russian), 229 pp. Moscow:Nedra.

Petukhov,I.M. (1981).Methodical instructions on use of programmes for calculating and graphical construction of stresses in rock mass around minings (in Russian), 51 pp. Leningrad: VNIMI.

Petukhov,I.M.,Linkov,A.M. (1974).On the energy criterion of stability in the rock burst theory (in Russian). Trudy VNIMI (91), 182-184.

Petukhov,I.M.,Linkov,A.M. (1975).On estimations of material tendency for violent failure (in Russian). Trudy VNIMI (95), 97-102.

Petukhov,I.M.,Linkov,A.M. et al. (1976).The theory of protective seams (in Russian), 224 pp. Moscow: Nedra.

Petukhov,I.M.,Linkov,I.M. (1979).The theory of post-failure deformations and the problem of stability in rock mechnics. Int.J.Rock Mech.Min.Sci. & Geomech.Abstr. (16), 2, 57-76.

Petukhov,I.M.,Linkov,A.M.,Rabota,A.N.(1981). On solution of discretized problems of rock mechanics with allowance for softening and unloading (in Russian),Fiz.-tekhn. problemy razrabotki poleznykh iskopajemykh, (3),26-33.

Salamon,M.D.G. (1970). Stability,instability and design of pillar workings. Int.J. Rock Mech. Min.Sci. (7), 6, 613-631.

Salamon,M.D.G. (1974).Rock mechanics of underground excavations: 3-d Congress Int. Rock Mech., (1),B,951-1099.

PHOTOMECHANICAL PRACTICE OF STUDYING CONDITIONS OF SOLID ROCK NEAR UNDERGROUND WORKINGS DRIVEN AT GREAT DEPTH

La pratique des recherches photomécaniques appliquée aux massifs de roches à proximité de galeries de mines à grande profondeur

Praxis der photomechanischen Untersuchungen von Bergmassivzuständen in der Nähe von Untertageräumen in grosser Teufe

N. A. Filatov
Doctor of Technical Sciences

V. D. Beliakov
Candidate of Technical Sciences
VNIMI, Leningrad, USSR

A. V. Dokukin
Corresponding Member of USSR Academy of Sciences

V. F. Trumbachev
Doctor of Technical Sciences

O. K. Slavin
Candidate of Technical Sciences
IGD named after A. A. Skochinsky, Moscow, USSR

SYNOPSIS

Selections of parameters required for the stability of underground workings, interaction of supports with solid mass at various static and dynamic effects is substantiated on the basis of the photo-mechanical analysis of models simulating mining operations for coal extraction.

RESUME

A partir d'une analyse photomécanique de modèles simulant l'exploitation de charbonnages, les auteurs justifient le choix de paramètres de stabilité des galeries et du comportement de soutènement et du massif vis à vis de divers phénomènes tant statiques que dynamiques.

ZUSAMMENFASSUNG

Auf der Grundlage der photomechanischen Untersuchung von Modellen, die Kohlebergbaue wiedergeben, wird die Auswahl von Einflußgrößen zur Standfestigkeit von bergmännischen Hohlräumen begründet. Ferner wird die Auswahl von Einflußgrößen zur Wechselwirkung von Ausbau und Bergmassiv unter verschiedenen statischen und dynamischen Einwirkungen begründet.

1. INTRODUCTION

Considerable advances have been made in development of theoretical and applied trends of the mining science and its fundamental field - mining geomechanics (Dokukin, 1981). But along with this, problems connected nowadays with wide transition of mining operations to deeper levels and with development of new mineral deposits require more extended studies and there arise some new rather important and difficult problems.

Optimization of main industrial processes in the mining industry is possible only on the basis of obligatory knowledge of laws of rock pressure manifestation, which determine causality of their relation with the medium and influencing factors determined by the general mechanical laws applied to the complexes of conditions existing in underground workings. With transition to deeper levels rock pressure rises, danger of sudden coal and gas outbursts increases and problems of underground working control and maintenance become more complicated. It becomes necessary to improve techno - logy of mining operations, methods and technique of mineral extraction, and control of solid rock condition. To control dynamic phenomena a new relief method, i.e. powerful camouflette-shock blasting which induces influence of stress waves on the solid rock and stirring of coal seam under stressed condition, is used nowadays.

Photomechanical methods are effectively used in the USSR (VNIMI, 1975) for studying conditions of solid rock near underground workings and establishing laws of rock pressure appearing in the workings including those driven at great depth.

Current role of photomechanical methods in applied aspects of mining is characterized by improved methodical regulations and technique of simulation, on the one hand, and by considerably widened range of more sophystical solvable problems which reflect particularities of a real solid rock and variety of static and dynamic loads acting on the solid rock, on the other hand.

As a rule the photomechanical studies cope with the whole complex of the mining engineering problems connected with the deposit opening and its development. The general scheme of stages and sequence of mining operations is expressed by the fact that in any deposit the shafts (which suffer no influence and effects of other workings at the first stage of their operation) are sinked in the first turn and then other workings are driven, i.e. permanent workings, development workings and, finally, stopes, which are able to exert pronounced influence on the stress-strained state of the region wherein the opening workings and haulageways have been driven.

Solutions of the above mentioned problems are given in the present paper as typical examples based on photomechanics (at static and dynamic effects) and attention is paid to new trends of studies or to such trends which are not yet studied in details but are important for the mining theory and practice.

2. MODERN STOCK OF PHOTOMECHANICAL METHODS AND TECHNIQUE.

In recent time the photomechanical methods have been used in the mining art mainly for studying static-elastic deformation of the tested samples (Trumbachev, Molodsev, 1963).

Nowadays the methods of experimental solution of problems on plasticity, creepage, rock failure as well as various dynamic problems on rock mechanics have been developed and successfully used (Trumbachev, Slavin, 1975; Filatov, Beliakov, 1979; Chesin, 1976). And the studies are carried out with due consideration of inhomogeneity of solid rock, availability of weak surfaces and fractures in it, tectonical faults in the region (VNIMI, 1978, Gzovsky, 1975).

Recently the stock of the methods for photomechanical studies of solid rock state using two-dimensional models is enriched by an effective purely optical method of determining (distinguishing) normal stresses. A picture of interference bands, which characterize distribution of difference of main normal stresses (or maximum tangential stresses) in a model cannot be interpreted with an accuracy required for determination of all components of the stress tensor. Therefore the optical method of determination (distinguishing) of stresses in two-dimensional photoelastic models developed in the USSR is the most effective and the simpliest of the now existing methods.

This method is based on two theoretical conceptions:

1. In loaded two-dimensional model which is under the conditions of the plane stressed state, there are no stresses in the planes of the third degree, perpendicular to the plate plane (negligibly small as compared to the stresses available in the model plane).

2. The stresses acting on a model or on a section of this model in the direction of raying cannot be registered and exert no influence upon the interference band picture. Let us cut sections perpendicular to the model plane from a two-dimensional model with a band picture "frozen" in it, and subject these sections to radiographic test in polarization optical set. The picture of the interference bands observed in these sections will characterize the stresses acting in the tested model along the direction of these sections.

Basing on the above studies the possibilities of simulating initial conditions in the models have been widened and hence the experimental studies become more reliable and nearer to natural conditions in some aspects.

In some institutions laboratory facilities such as electronic and optical instruments, polarization interference and golographic sets including those fitted with high-speed instrumentation cameras have been modernized. To automize the optical measurements and their processing micro-computers with displays and graph plotters are generally used. Corresponding technique of establishing and testing two-dimensional and three-dimensional models at simple and complicated kinds of loading have been developed. Materials based on epoxy resins, epoxy gels (IGD named after Skochinsky, 1970), polystyrene (VNIMI, 1975), polyurethane elastomers (Krovopusk et al., 1979) are widely used as optically sensitive model materials.

3. PROBLEMS ON STRESS-STRAINED STATE OF ROCKS AND WORKING SUPPORTS AT STATIC ACTIONS.

3.1. General.

In modern practice many problems on determination of conditions of stability retaining and loss in exposed workings have been solved by photomechanical methods using two-dimensional models. The problems connected with determination of working support operating conditions and the nature of the support interaction with rock are covered to a lesser degree. Some concrete methodical conceptions of solving sophystical spatial thin-walled structures with realization of affinor-geometrical similarity are described in the paper of Slavin et al. (1980) as well as in the VNIMI paper (1982).

A shaft junction with horizontal workings and adjacent rooms calls for careful study of stress-strained state.

3.2. Examples of problem experimental studies.

The nature of the ring monolithic support operation of shaft crossing a high-dipping weak seam has been studied using a three-dimensional models, the cross-section of which is given in Fig. 1. The model is made of epoxy resin ED-16. The model has been loaded with hydrostatic pressure of P=0.5 MPa. Ratio of elasticity modulus of main rock and weak zone is equal to 20.

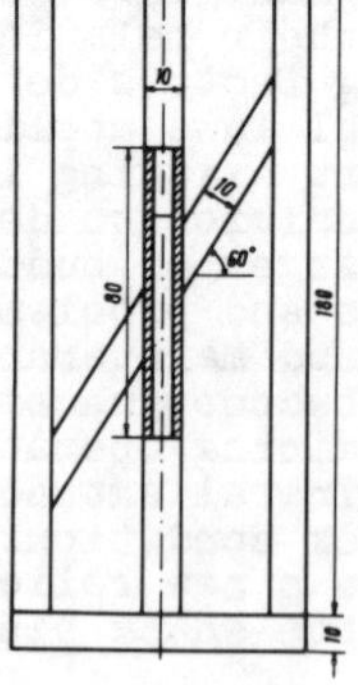

Fig. 1

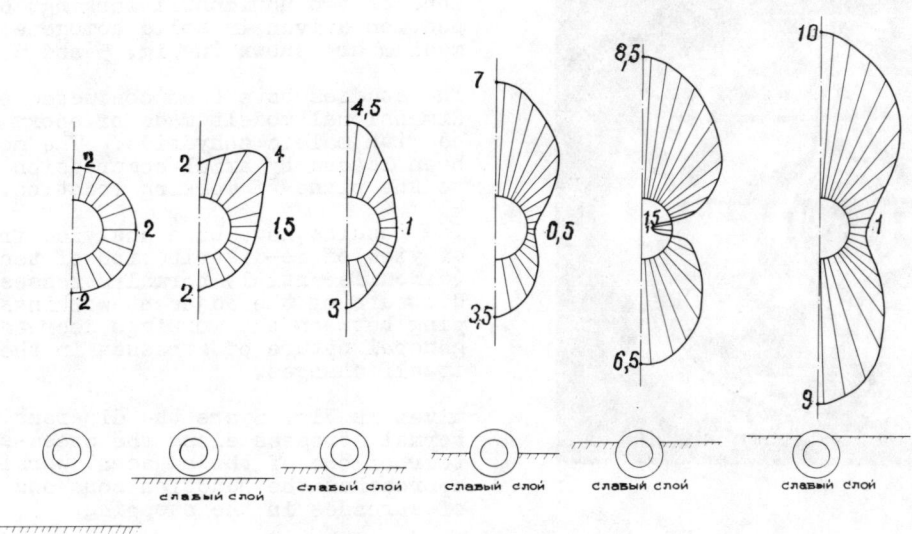

Fig. 2

Given in Fig. 2 are the diagrams of distribution of tangential normal stresses along the inner contour of the support in various cross-sections in relation to a weak zone. The analysis of the diagrams shows that as the shaft approaches the zone of junction with weak seam, the loads on the shaft support increases, the irregularity of their distribution along the outer contour of the support cross-section is strongly pronounced and hence the support suffers the action of bending moments of opposite signs. The working conditions of the support in the zone where the shaft crosses the weak seam at high dip become more complicated as compared to the section outside the zone of the weak seam influence.

The example of studying stress-strained state of bedded solid mass (homogeneous layers) in the zone of shaft junction with a horizontal working is given in Fig. 3.

Shown in Fig. 4 as an example are the pictures of the interference bands in the sections of the similar bedded model which has been subject to hydrostatic pressure. Given in Fig. 4 are the following sections: (a) vertical section in adjacent workings axes plane; (b) horizontal section directly above the adjacent horizontal working; (c) horizontal section in the axis plane of the adjacent horizontal working.

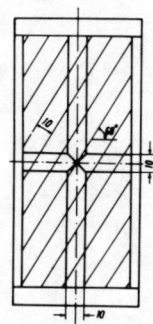

Fig. 3

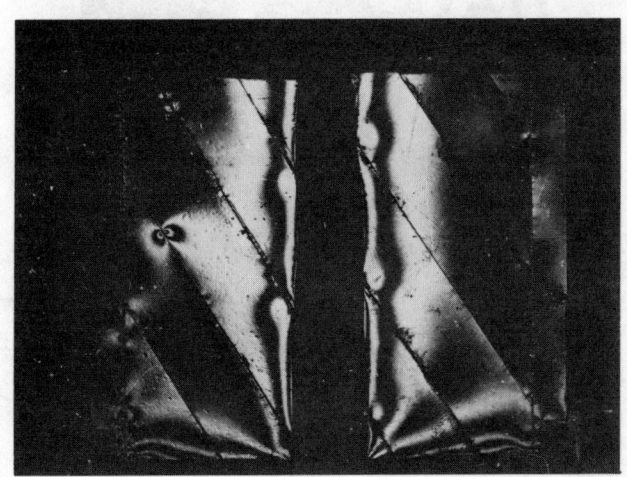

Fig. 4a

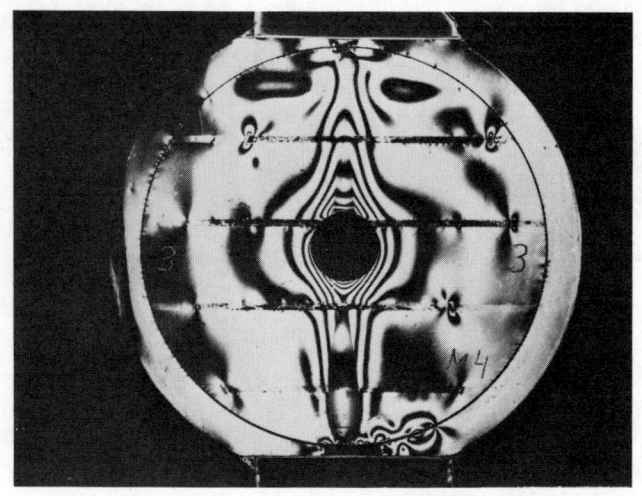

Fig. 4b

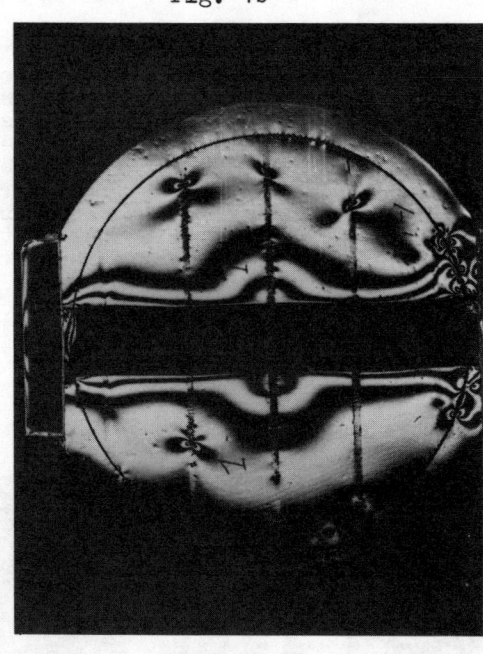

Fig. 4c

Results of studies on evaluation of stress-strained state of solid rock in the junction zone of two horizontal workings of round cross-section driven in solid homogeneous elastic medium are shown in Fig. 5 and 6.

The studies have been conducted on three-dimensional models made of epoxy resin hardened with maleic anhydride. The models have been loaded by axial compression perpendicular to the plane of working location.

The results have been analyzed from the point of view of re-distribution of tangential (circumferential) normal stresses along the contours of the adjacent workings as the stopping between the workings decreases and the general nature of stresses in the stopping itself changes.

Given in Fig. 6 are the diagrams of tangential normal stresses along the cross-section contour of one of the adjacent workings as it approaches the junction zone and the diagram of stresses in the stopping.

The analysis of the results shows that compression stresses along the contour of the adjacent workings (horizontal cross-section) decrease with approaching the junction zone. The tensile contour stresses (vertical cross-section) increase with approaching the junction zone, and they reaches their maximum at the moment when the cross-section of the adjacent workings takes a form of an ellips as compared to the values of the tensile stresses along the contour of a single working suffering the influence of the same field of forces.

The efficiency of the purely optical method of stresses distinguishing may be demonstrated by an example of determining normal stresses in the zone of bearing pressure near the stoping face when rock bedding is available in the roof. The model diagram and location of sections in the model is given in Fig. 7.

Fig. 8a represents a picture of interference bands in the tested model (distribution of maximum tangential stresses).

Given in Fig. 8b are the charts of normal stresses in the zone of bearing pressure plotted using the methods described above.

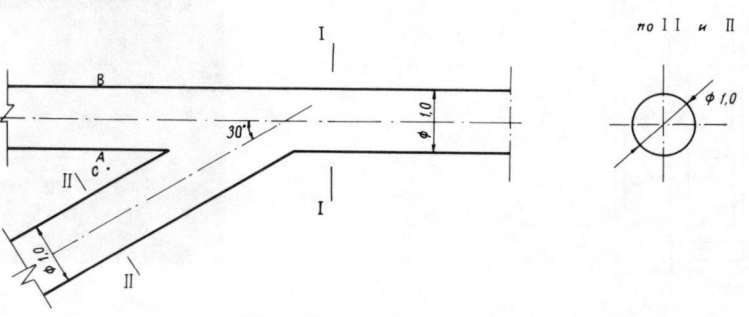

Fig. 5

D 124

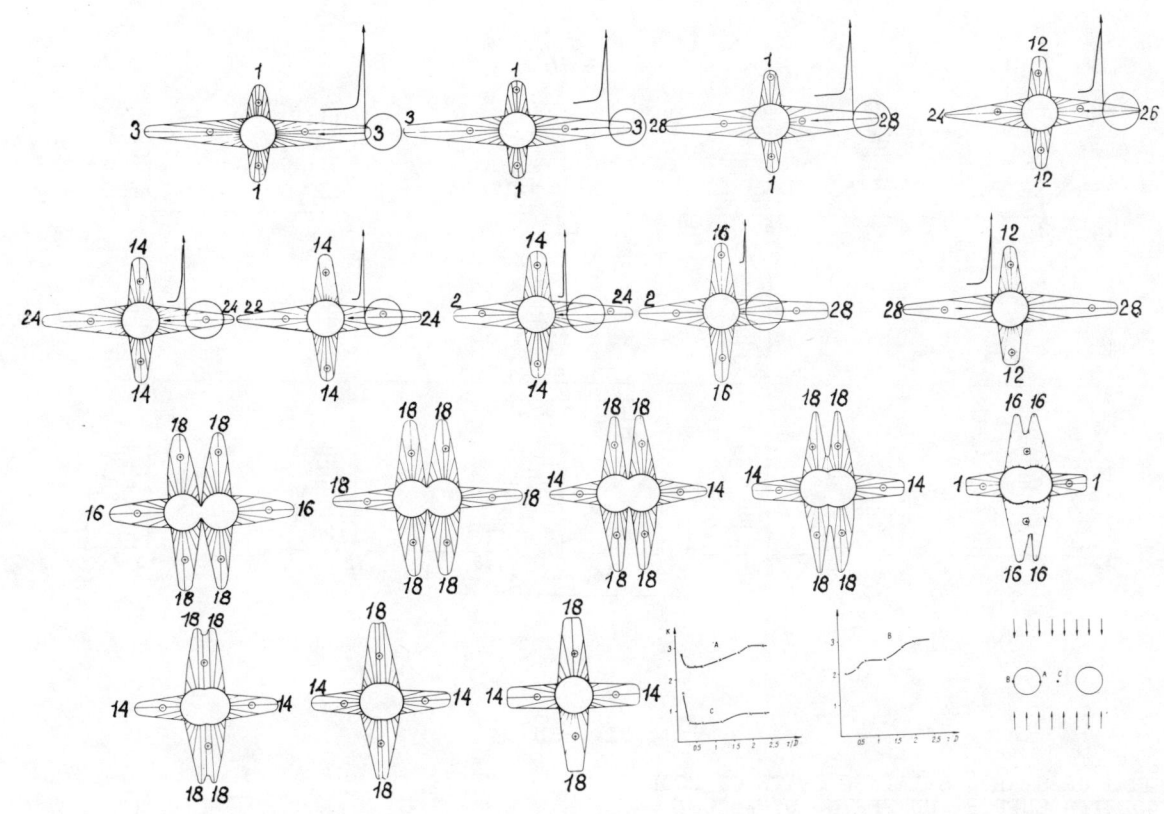

Fig. 6

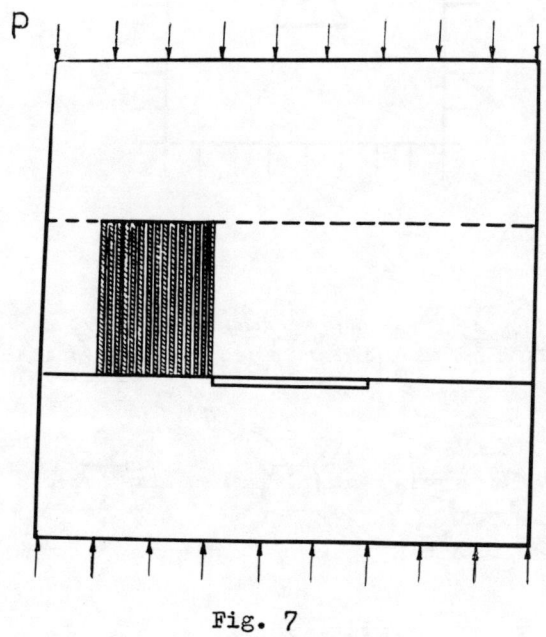

Fig. 7

Fig. 8a

D 125

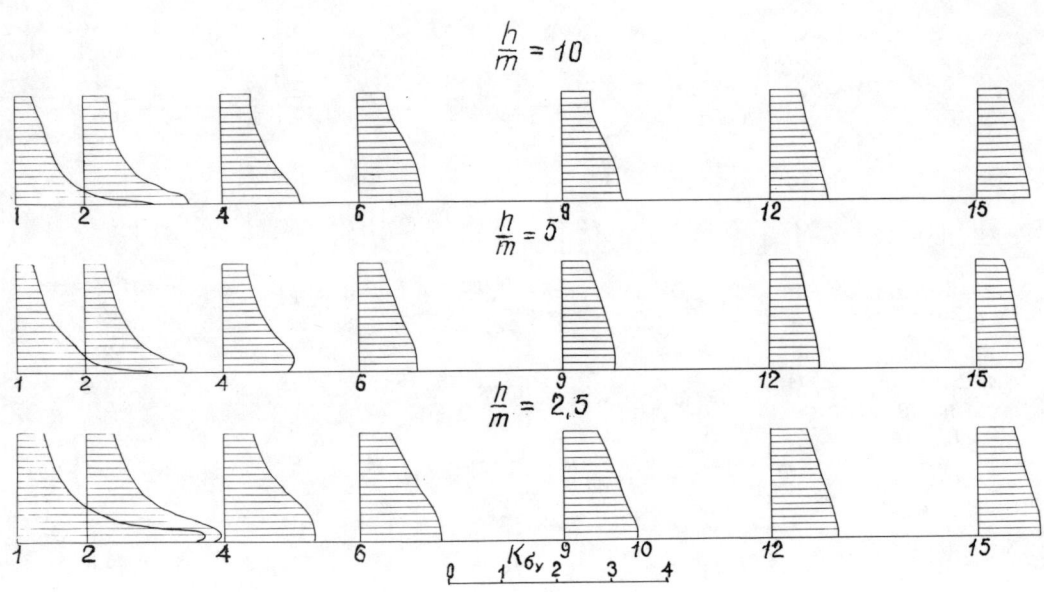

Fig. 8b

4. PROBLEMS ON STRESS-STRAINED STATE OF ROCK AND WORKING SUPPORT SUFFERING DYNAMIC LOADS.

4.1. General.

Study of dynamic problems of mining geomechanics makes it possible to determine dynamic concentration of stresses around underground workings of different purposes with account of factors influencing their stability, to determine fields of wave stresses and shifting in lump-heterogeneous, bedded, discrete disturbed media at supported or not supported underground workings.

The results of experimental solutions of such problems at combined action of static and dynamic loads have shown the satisfactory agreement with the theory, and this gives the reason to consider some simplifying assumptions inevitable at simulation to be quite allowable, exerting insignificant influence upon the nature of the obtained relations (Beliakov, Filatov, 1980).

4.2. Examples of experimental solutions of problems.

Combined action of static and dynamic loads upon an unsupported working of trapezoidal cross-section has been studied using two-dimensional models of polyurethane material (Fig.9a).

A high-speed polarization-dynamic instrumentation camera operating in the modes of slot scanning and frame photography (2.5 mln. frames

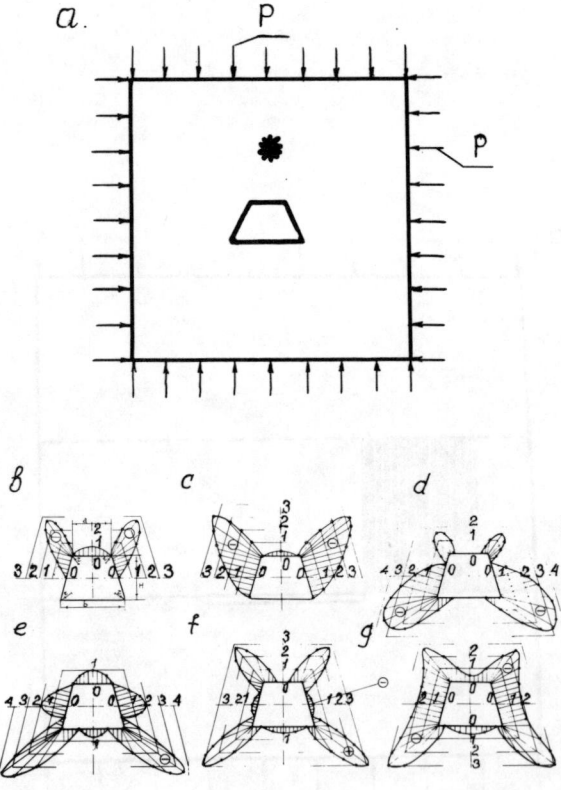

Fig. 9

per second) has been used for studies. Interpretation of the experimental data gives the possibility to plot the curves (Figs. 9b-9f) of main dynamic stresses σ_1 (x, y, t) along the working cross-section contour within the whole time range of action, as well as the time envelope curve (Fig. 9g) of maximum dynamic stresses.

The curves of static (1 and 2) and maximum dynamic (3 and 4) coefficients of stress concentration have been determined on this basis (Fig. 10). The following data are represented in Fig. 10: 1 - horizontal load p /4; 2 - vertical uniformly distributed load p, 3 and 4 - dipping of P-wave at an angle $\pi/4$ and parallel to the working roof respectively.

These studies have revealed general nature of longitudinal wave difraction at pulse actions which appear to be considerably dependent upon its length. When long waves (of seismic type) interact with the working at $H/\lambda = 0.04-0.25$ (where H - hole height, λ - wave length) the coefficinets of stress dynamic concentration in the lower corners of the trapezoidal workings are 15-20% higher than the static coefficients and their distribution along the contour is near to the static solution for the field of stresses around a cylindrical hole, caused by a linear source.

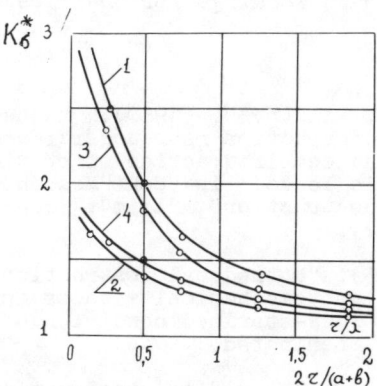

Fig. 11

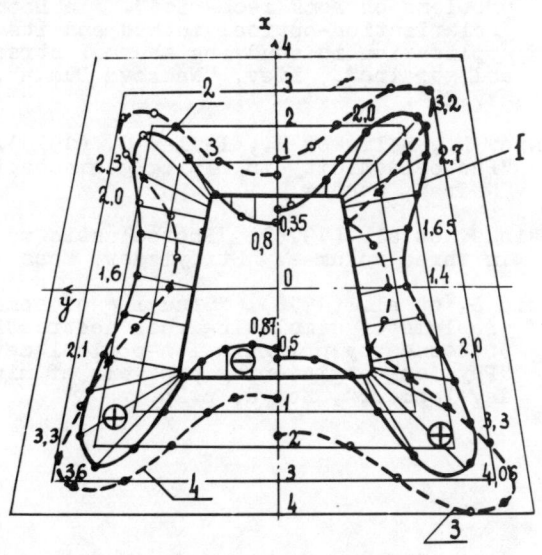

Fig. 10

At stress wave difraction in a trapezoidal working the concentration of dynamic stresses depends to a great degree upon the radius r of the working corner zone curvature, i.e. of the roof and particularly of the floor, as well as upon the working location in relation to the incident wave front. The results of the studies are given in Fig. 11 in relative units, where a and b are geometrical dimensions of the working cross-section. The Figure shows that maximum coefficients of dynamic stress concentration appear in corner zones when wave dipping is $\alpha = \pi/4$ to the working symmetry axis.

Analysis of relationships $K_\sigma^* = f[2r/(a+b)]$ shows that the concentration coefficient increases as the geometrical parameter (a+b)2r decreases. In the result of studying stress concentration for $\alpha = \pi/4$ and $\alpha = 0$, i.e. two extreme cases of dip angle , which characterizes inclination of the longitudinal wave front in respect to the working symmetry axis it is felt that at occasional location of the working in relation to the wave front the values K_σ^* are located within the determined extreme values (Fig. 11). The analysis of the curves $K_\sigma^* = (r/\lambda)$ shows that the coefficient of dynamic stress concentration does not depend upon the relative size of the working at $\alpha = \pi/4$ and (a+b)/2r = 1.2, which is stipulated only by a value of the radius of the working corner zone. Thus, within the longwave field the value K_σ^* depends mainly upon the working shape and to a lesser degree from the corner zone radius. Within a short-wave range the dynamic stress concentration is determined mainly by a local curvature of the working contour, but only by its geometry.

At $r/\lambda = 0.25$ the maximum values of K_σ^* for a trapezoidal working under the action of transversal S wave are respectively 30 and 60% higher as compared to the action of longitudinal P wave at $\alpha = \pi/4$ and $\alpha = 0$.

5. CONCLUSION.

Simultaneously with further development of theoretical aspects of simulation of static states, slow and quick processes based on photomechanics it is feasible to search for the conditions at which the mechanical properties of the model materials become similar as far as it is possible to real rock properties at a preset accuracy and reliability of the planned experiments in combination with other methods of simulation.

Establishing and studying of models in which the methods of photomechanics are combined with the method of equivalent materials is felt to be promising and that will make it possible to determine the criteria and indices of rock and support strain and to recommend to the industry the most rational and economical measures of supporting and maintenance of underground workings for their required life.

REFERENCES:

Beliakov V, Filatov N. (1980). Mechanism of rock destruction near underground workings at combined action of static and dynamic loads. In Book "Mechanics of rock destruction", "Ilim", Frunze, 321-324.

VNIMI (1975). "Methodical instructions on using photomechanical methods for studying stress-starined condition of rock", VNIMI, Leningrad.

VNIMI (1978). "Methodical instructions on studying dislocated and faulty solid rocks on models", VNIMI, Leningrad.

VNIMI (1982). "Methodical recommendations for making vertical working supports models of optically sensitive materials and their testing", VNIMI, Leningrad

Gzovsky M. (1975). "Fundamentals of tectonic physics", "Nauka", Moscow.

Dokukin A. (1981). "Problems of mining science", "Nauka", Moscow.

Dokukin A., Trumbachev V. et al. (1979). "Simulation of problems on rock mechanics by photomechanical methods using golography". In Book "Mining pressure and methods of its control", "Ilim", Frunze, 5-15.

Dokukin A. "Propagation of stress waves in bedded viscoelastic solid rock". In Book "Transactions of IGD names after Skochinsky", Moscow.

IGD named after Skochinsky (1970). "Methodical instructions on production of low-modulus optically sensitive materials (epoxy gels) for polarization-optical method of stress studies". IGD named after Skochinsky, Moscow.

Krovopusk O. et al. (1979). "Optically sensitive materials based on polyurethanes and epoxy polyurethanes for studying great deformations". Book "Transactions of VIII All-Union Conference on photoelastic method". Academy of Sciences of Estonian Soviet Republic, Tallin, 163--165.

Slavin O., Trumbachev V, Tarabasov N. "Simulation in strength calculation of thin-walled machine-building structures by photomechanical methods". In Book "Strength calculations". Collected papers, No. 21, Machinostroenie, Moscow, 53-68.

Trumbachev V, Molodsev L. (1963). "Optical method for studying stress state of work around underground workings of round cross-section". Academy of Sciences, Moscow, USSR.

Trumbachev V, Slavin O. (1975). "Simulation of solid rock by photomechanical methods". IGD named after Skochonsky, Moscow.

Trumbachev V., Katkov G. (1976). "Polarization-optical method of studying stresses in problems on rock mechanics". In Book "Polarization-optical method and its application in studying thermal stresses and strains". Kiev, "Naukova Dumka", 217-224.

Filatov N., Beliakov V., Ievlev G. (1975). "Photoelasticity in mining geomechanics". "Nedra", Moscow.

Chesin G. et al. (1975). "Photoelastic method". In three volumes. Stroyizdat, Moscow.

Chesin G. et al. (1976). "Studying of some problems of simulating rock destruction processes by polarization-optical method". "Physical engineering problems of mineral development", No. 6, p.23.

STOPE STABILITY IN UNDERGROUND BAUXITE EXPLOITATION
Stabilité des zones d'abattage en exploitation souterraine de bauxite
Stabilität von Untertagehohlräumen im Bauxit-Bergbau

S. Vujec
Prof. dr. Ing., Mining-geological-oil, faculty of Zagreb University

D. Zeljkovic
Mr. Sci. Ing., Manager, Bauxite mine, Jajce, Yugoslavia

SYNOPSIS

Underground exploitation of bauxite is very special in comparison with other ore reserves due to technological and mining-geological conditions. Dilution of ore is not allowed. The hanging wall of studied deposits contains widely spread seams ranging from breccia and limestone to marl and clay. Application of the sublevel method causes — depending on the hanging wall — either immediate roof caving or the formation of open spaces up to 2800 m² (horizontal cross section). The stability limit of open spaces was determined by means of observation in situ and calculation of stresses and strains (FEM). The final aim of the study is a modification of the mining method for the purpose of increasing safety and productivity.

RESUME

L'exploitation souterraine de bauxite est très particulier en comparaison avec les autres gîtes métallifères car à cause de contraintes technologiques et minières la dilution n'est pas permise. Le toit des gîtes étudiées contient une large variété de lits allant de la brèche et des roches calcaires jusqu'à la marne et la glaise. L'application de la méthode en sous-étages provoque, suivant les types de toits, soit l'éboulement immédiat, soit la formation d'espaces ouvertes allant jusqu'à 2.800 mètres carrês. La limite de stabilité des espaces ouvertes a été déterminée à l'aide d'observations in situ et des calculs de tensions et de contraintes (Méthode des éléments finis). Cette étude vise la modification des méthodes minières en vue d'améliorer le niveau de sécurité et du rendement.

ZUSAMMENFASSUNG

Die Untertagegewinnung von Bauxit ist im Vergleich zu anderen Erzlagerstätten sehr eigenartig, was auf besondere technologische und montangeologische Bedingungen zurückzuführen ist. Das Hangende der studierten Lagerstätten ist verschiedenartig ausgebildet — von Breccie und Kalkstein bis zu Mergel und Ton. Die Anwendung des Zwischensohlenbruchbaus ist von der Stabilität des Hangenden abhängig und fordert entweder den Firstenbruch oder die Entstehung offener Räume bis 2800 m². Die maximale Spannweite der offenen Räume wurde mittels Beobachtungen in situ und Rechnens der Spannungen und Verformungen bestimmt (Finite Element Methode). Der Endzweck der Studie ist eine Abänderung der Gewinnungsmethode mit dem Ziele der Sicherheits- und Produktivitätssteigerung.

INTRODUCTION

Underground exploitation of bauxite comprises a small part of the total production of bauxite in the world and is done in a few countries amongst Yugoslavia is included. In comparison with other mineral ores special demands are imposed on this exploitation because dilution is not allowed and bauxite is included amongst inexpensive raw materials. The sublevel method is mainly used in Yugoslav mines with underground exploitation. In ore deposits with hanging walls where caving is accompanied by mining, ore losses amount up to 50%, and in those with firm roof, empty spaces are formed behind the stope which even with application of loading machines with remote control makes exploitation insufficiently safe. In that case losses during exploitation are less than 10%. For this reason, investigation work has been performed in several Yugoslav mines for the purpose of finding out a method of exploitation which eliminate the above mentioned deficiencies. The text deals with the results of investigation work performed at the Poljane location of the Jajce bauxite mine.

The investigation is performed in three stages:

- collection of data on the rock massif,
- definition of conditions of caving of the roof,
- modification of exploitation method.

The first and the second stages were performed with the aim of explaining and understanding the caving process and the stability limits of open spaces under geotechnically complex conditions, and the third stage which is in the phase of elaboration ought to result in safer and more effective exploitation methods.

GEOLOGICAL CONDITIONS

Bauxite deposits at the Poljane location are of the Senonian age. The lenticular ore bodies have the form of irregular lenses up to several hundred thousand tons dimensions. Bauxite seam thickness ranges from several up to 40 metres.

The bauxite floor consists of limestone on the surface with sharp pyramids frequently penetrating deep into the bauxite bodies. The layer thickness ranges from several centimetres up to 2 metres with numerous fissures filled with clay.

The roof of the ore bodies is characterised by frequent lateral and vertical alteration. The most frequent petrographic members in the roof are:

1. breccia, conglomerates, calcarenites,
2. limestones, mostly calcarenites,
3. marly limestones or lime marls,
4. marls.

The immediate bauxite roof is mostly comprised of roughly-clastic rocks with irregular lumps. The upper roof contains limestone with transition to marl, while the upper-most roof contains marl. Thickness of individual lithomembers ranges from 10 to 30 metres. Locally this roof course is interrupted, thus layered calcarenite or lime marls can be found within the direct bauxite roof. The general incline of the roof deposits is about 17°. A characteristic geological deposits profile is shown in Fig. 1.

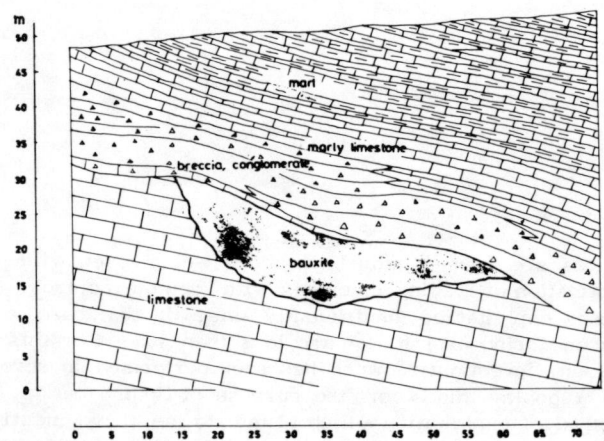

Fig. 1 Geological profile (after Sakač)

GEOTECHNICAL ROCK CHARACTERISTICS

Discontinuity of the roof, the floor and the ore body is particularly expressive. From the point of view of excavation space stability, three families of discontinuity are of particular significance.

Layered roof planes represent the most strongly pronounced discontinuity family. The layer thicknesses vary in individual deposits. Conglomerates, breccias and occasionally limestone have layer thickness of up to 2 metres. Marly limestones and marls are most frequently thinly layered, with the layer thicknesses of several centimetres up to 1 metre. The other two families of fissures are in mutual, almost vertical relation, and they are perpendicular to the layer plane. These fissures are most frequently filled with clayey filling material. It is a frequent occurrence that these fissures overlap along the vertical direction through several layers but it has also been noted that there have been overlappings down to the depth of 60 metres from the surface.

Thus regular parallelopipeds are formed in the roof which naturally in the tensile stress zones, fall down into the excavated space. Contoure diagram (a projection onto the upper hemisphere) of the main discontinuity systems is shown in Fig. 2. Mapping of the discontinuities was performed on the surface, on the core during drilling and during pit excavation works.

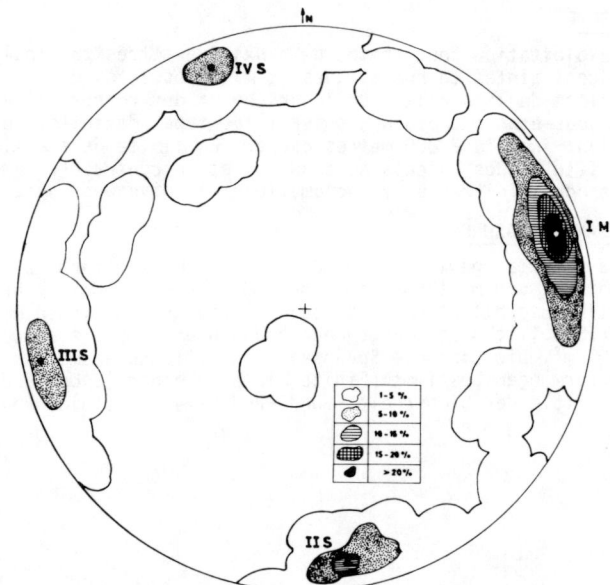

Fig. 2 Equal area joint net (des. I. Tomašić)

Laboratory testing of samples taken from pit works and from cores resulted in certain average values shows in Table 1.

EXPLOITATION METHODS

Underground bauxite mining methods have developed in such a way as to fulfill two basic conditions: that the production should be as inexpensive as possible (due to low price of bauxite) and dilution of ore can amount to maximum 1% since in the technological processing,

TABLE 1

Type of rock	Density ρ kg/m^3	Uniaxial compressive strength MN/m^2	Module of elasticity E MN/m^2	Poisson's ratio ν	Cohesion c MN/m^2	Internal friction angle φ °
Marl	2700	58-70	5000-7000	0.35	8.0	45-67
Marly limestone	2700	58-90	5000-8000	0.30	8-13	45-55
Limestones	2700	70-100	6000-8000	0.35	8-13	45-67
Breccias	2700	90-100	7000-8000	0.30	8-16	45-60
Conglomerates	2700	90-100	7000-8000	0.30	8-16	45-60
Floor limestones	2700	50-75	7000-1000	0.26	8-18	45-67
Bauxite	2900	20-50	7000	0.12	5-7	50-60

content of $CaCO_3$ should not exceed 1.5% and the ore most frequently contains about 0.5% of $CaCO_3$. Ore dressing has not yet yeilded good results. In countries where underground exploitation of bauxite is practised depending on geological conditions, several methods of exploitation have been developed. In Yugoslav mines with underground bauxite exploitation, the method of sublevel caving is mainly used. The ore deposit is divided into levels 6 to 10 metres high, provided that main level gallery as a rule is built along the roof contact and the cross excavation galleries have the position shown in Fig. 3. Gallery dimensions depend on applied mechanization. In majority of mines, loading and transport to the vertical chutes is performed by means of selfpropelled machines driven by compressed air or diesel oil. Depending on the geomechanical properties, at application of this excavation method two cases occur in practice: In the first case immediatele following ore blasting, the roof caves in. This happens in cases when thinly layered and cracked materials are present in the roof. With regard to mixing of the barren rocks and ore, recovery of ore body in that case amounts up to about 50%, thus this variant of sublevel caving is unsatisfactory.

In the second case the roof does not cave in after ore excavation thus the method of sublevel caving changes into the excavation method with open stopes. This case occurrs when there are thickly layered conglomerates and breccias in the roof which is the most frequent occurrence in the observed mine. Ore losses in that case amount to less than 10%. Even when remote control loading machines are used (worker do not enter the open space), this method from the point of view of safety is not satisfactory due to possible greater uncontrolled cavings of the roof.

DETERMINATION OF THE OPEN SPACE STABILITY

Determination of the open space stability was performed through observation in situ and through calculation of stress and deformation distribution by application of the finite element method.

Calculation of stress and deformation distribution was performed by means of finite elements method on a model for two dimension state of stress, which was

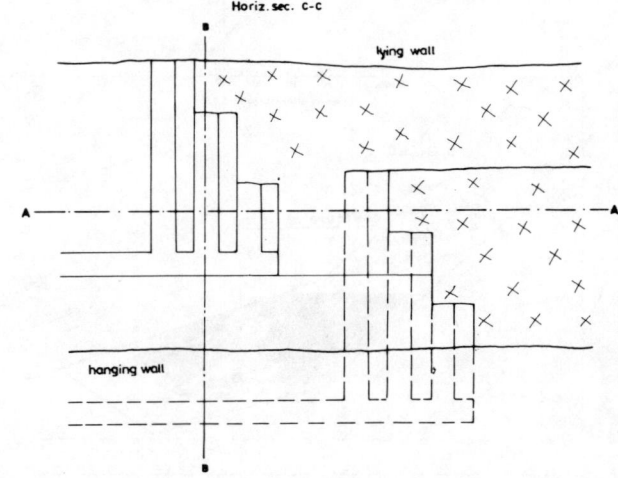
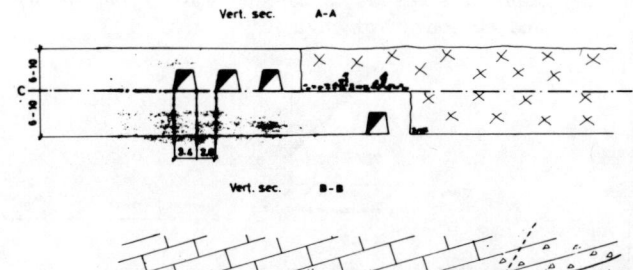
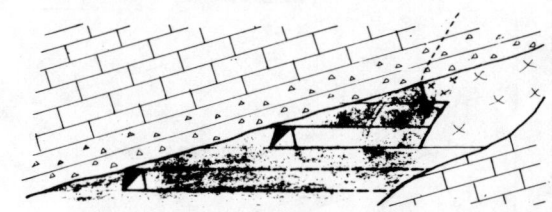

Fig. 3 Sublevel caving method

formed as a several-layered model (maximum number of used materials was 7) on which it was possible to alter the dimensions of the open space arbitrarily, as well as the depth and the physical-mechanical properties of individual layers in the roof.

Various open space widths up to 100 m were studied as well as the depth of bauxite deposits down to 150 m. Thus, somewhat greater values were included that the real ones in practice.

The model consisted of 320 to 543 elements and of 326 to 480 nodal points. Three zones of stress appear in the open space roof: the compression zone ($\sigma_1 \leq \sigma_2 \leq 0$) the tension zone ($\sigma_1 \geq \sigma_2 \geq 0$), and the zone of compressive and tensile stresses ($\sigma_1 \leq 0 \leq \sigma_2$). From the point of wiev of undiscontinual roof stability critical stresses are maximal compressive on the open space supports and maximal tensile stress in the part of the open space span. Limitation of the roof tension zone and of the critical points are shown for 40 and 60 m spans in Figs. 4 and 5.

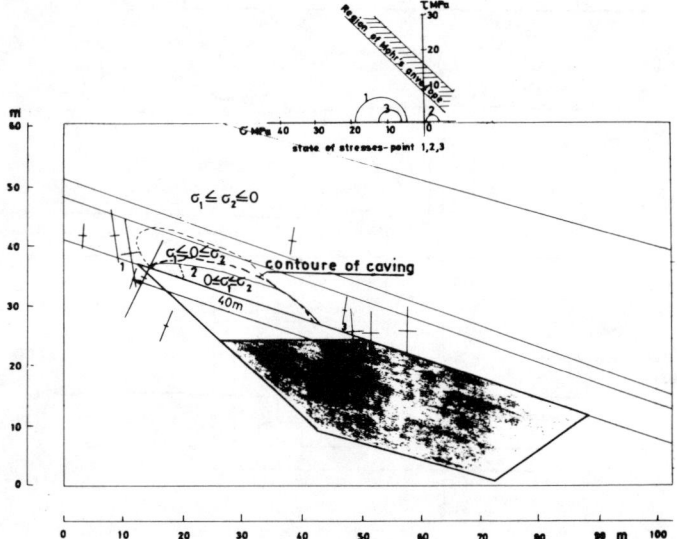

Fig. 4 State of stresses in hanging wall (span 40 n) and contoure of caving

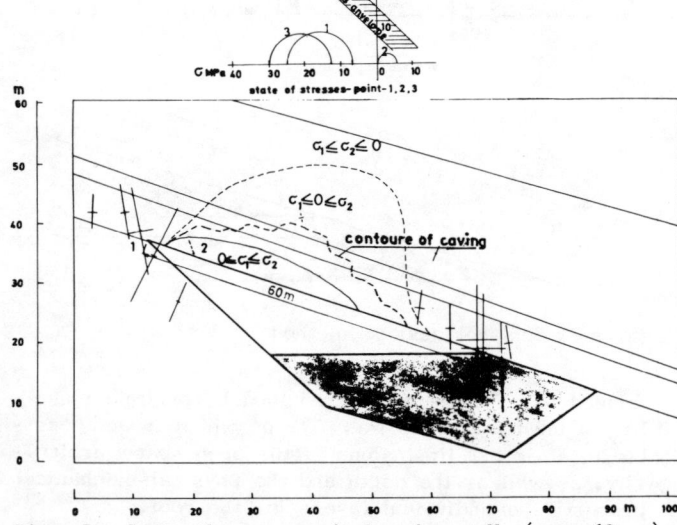

Fig. 5 State of stresses in hanging wall (span 60 m) and contoure of caving

The data for diagram $\sigma - \tau$ have been determined through laboratory investigation as the region of Mohr's envelope with regard to the wider area of cohesion value and the angle of internal friction. The diagrams and the data obtained from calculations indicate that compressive stresses are not critical for spans up to 100 m, however tensile stresses reach the critical value for spans of about 60 m.

Roof with vertical and subvertical discontinuities as shown in Fig. 2 does not have tensile strength because of which individual parallelopipeds fall out of the roof as soon as they find themselves in the tension zone.

The change of the stress zone with the depth was interesting for some ore deposits. Contoures of the tensile and compressive stresses are drawn in Fig. 6 for the depths of 30, 70 and 120 m and the open space span of 40 m. It is obvious that the zone where $\sigma_1 \leq 0 \leq \sigma_2$ decreases with the depth.

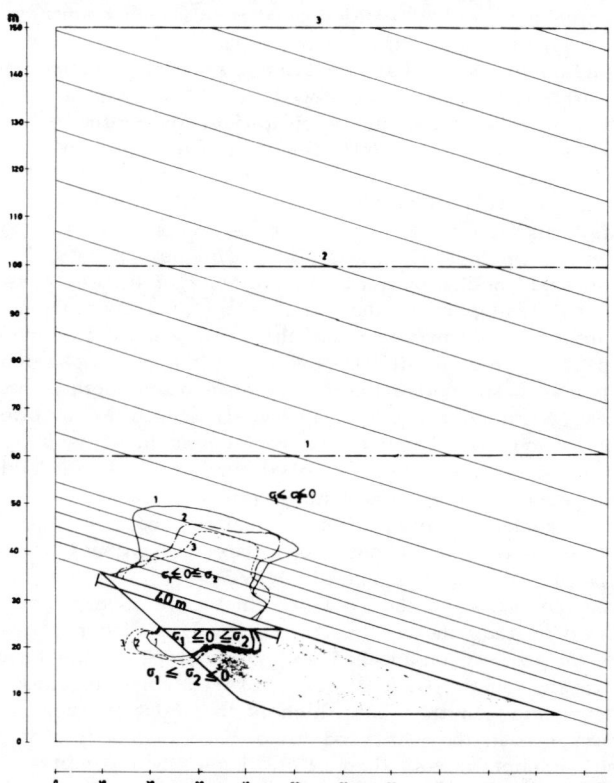

Fig. 6 State of stresses for different depths
Depths of: 1 - 30 m; 2 - 70 m; 3 - 120 m

Surveying of open spaces in situ was performed on 11 ore bodies whose basic dimensions are shown in Table 2.

The limits of the tension and compression zone could in several cases be traced according to the behaviour of discontinuities with filling material. At small span while the roof is in the compression zone clay is squeezed out of the fissure. Together with the increase in span when a tension zone appears the

TABLE 2

Ore body no.	Reserves t·10³	Area of ore body m²·10³	Open space max. before caving m²	Open space max. permanent m²	Ore depth m
1	320	18	1000		20
2	220	22	2000		30
3	175	6	2800		30
4	83	3.4	2250		100
5	37	2		2550	130
6	16	5		900	100
7	99	10	2600		77
8	274	4			
9	112	3			
10	310	16		2800	75
11	330	17		2000	42

ceasure of squeezing out of clay from the discontinuity is visible.

Observation in situ have revealed that in roof containing fissures, falling out of the roof occured in a span somewhat smaller than that which corresponds with the span where the tension zone appeared.

In the vertical section caving was halted between the zones of the tensile and compressive stresses. Figs. 4 and 5 show the shapes of caved space of 40 and 60 m spans.

In ore deposits in which during mapping of the first level no fissures were observed, spans reached approximately values that had been obtained through calculations. Cavings were in that case of larger dimensions. For conglomerations and breccias whose characteristics are given in Table 1 this occured at spans of about 60 m which is the largest observed open space for this materials at the depth of about 100 m.

Ore bodies were of variying ground plane shapes. For elongated ore bodies vertically cut onto the longer axis, results of stress analysis and in situ observations can be compared with sufficient safety while in regular ore bodies the influence of space distribution is considerable.

MODIFICATION OF THE EXCAVATION METHOD

Starting from the above described geological-mining conditions of exploitation and from the knowledge gained through the study of caving conditions for various types of ore bodies three excavation methods have been proposed. This phase of investigation work is in the course of execution.

1. For ore bodies with a span less than 30 m application of anchors in the roof has been proposed. The ore would be mined by means of sublevel caving with sublevels 10 m high (Fig. 7a). Anchor application should ensure formation of a self-supporting roof.

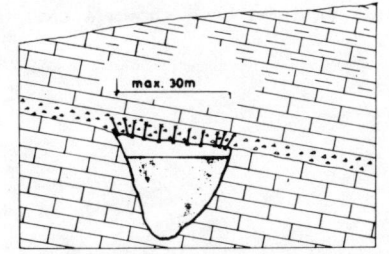
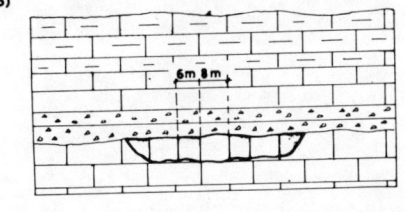
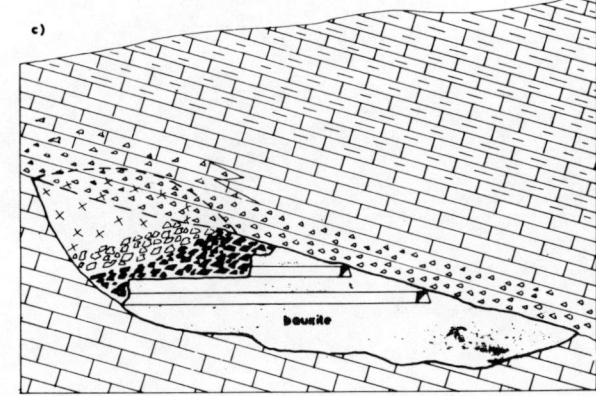

Fig. 7 Modified methods of exploitation
a) Open stope, b) Room and pillar, c) Sublevel caving

2. For ore bodies which are only several metres thick, and they spread for over 30 metres, application of a room and pillar method has been envisaged. For the beginning of work, the width of pillars is to be 6 m and the chamber width of 8 m (Fig. 7b). With the

application of anchors and for this variant the recovery could be increased.

3. For ore bodies of great thickness and of large span the method with sublevel caving and a protective layer of ore in the roof has been forseen (Fig. 7c). The ore layer has the purpose of protection from sudden caving of the roof.

For all the three forseen variants designed values will be checked by detailed in situ measuring and investigation of the stability in several ore bodies during production as well as by control mathematical analyses.

CONCLUSION

With the aim of improving of the underground bauxite exploitation method, calculation of strain and deformation distribution was performed by means of application of the finite element method and by means of observations of the stope stability in situ, after collection of data. Results obtained have shown a satisfactory matching of data and were used for designing of new excavation methods.

Bibliography:

Gudheus, G. (1977). Finite Elements in Geomechanics. London: John Wiley and Sons.

Hatzigiannelis, G.J. (1978). Underground Mining Methods and Mechanization in the Otavi-Sila Bauxite Mines. Proc. 4th. Int. Congress of ICSOBA (1), 297-311, Athens.

Standinger, J. (1969). Some Problems of Technology in Underground Bauxite Mining. Proc. 2nd. Int. Symposium of ICSOBA (2), 253-261. Budapest.

Vujec, S., Perić, B., and Zeljković D. (1978). Methods of Underground Bauxite Exploitation. Proc. 4th. Int. Congress of ICSOBA (2), 951-964, Athens.

MATHEMATICAL MODEL OF A CUT AND FILL MINE
Modèle mathématique d'une mine en coupe et remblayage
Mathematisches Modell eines Cut-and-Fill Bergwerkes

P. Ramirez Oyanguren
P. Carpenter
F. J. Alfageme
Dames & Moore, Iberia, S.A.

J. M. Canto
Minas de Almagrera, S.A.

SYNOPSIS

The design of cut and fill mines using the post-pillar technique is difficult to achieve theoretically, given that the small pillars quickly plastify and deform, distributing their load to less deformed and stressed parts of the rock mass. The simulation of the process of controlled deformation and transfer stress during excavation requires the incorporation into the mathematical model of the following features: joints, mechanical properties of the rock mass in the plastic state, in-situ stresses in the orebody prior to excavation, the process of excavating the ore and the progressive failure of the pillars and the rock mass surrounding the excavation.

RESUME

L'avant-projet d'une mine en coupe et remblayage dans laquelle on utilise la technique de "post-pillar" est difficile à réaliser théoriquement parce que les petits piliers se déforment et se plastifient rapidement, distribuant leur charge à d'autres zone du massif rocheux qui se trouvent moins déformées et moins chargées. La simulation du processus de déformation contrôlée et de transfert de la charge pendant l'excavation demande l'incorporation dans le modèle mathématique des aspects suivants: discontinuité, propriétés mécaniques du massif rocheux dans l'état plastique, tension naturelle dans le gisement avant l'ouverture de la mine, processus d'excavation du minerai et rupture progressive des piliers et du massif rocheux autour de la mine.

ZUSAMMENFASSUNG

Der Entwurf eines cut-and-fill Bergwerkes bei Verwendung brechender Pfeiler ist theoretisch schwer zu verwirklichen, denn die kleinen Pfeiler plastifizieren und deformieren sich schnell und verlagern ihre Last auf weniger deformierte und beanspruchte Teile des Gebirges. Die Simulation des Prozesses der kontrollierten Deformation und Übertragung der Belastung während des Felsaushubs bedingt die Einführung eines mathematischen Modelles entsprechend folgenden Aspekten: Klüfte, mechanische Eigenschaften der Felsmasse in plastischem Zustand, in-situ Beanspruchung der Erzschicht vor dem Aushub, Prozeß der Erzgewinnung und progressiver Bruch der Pfeiler und der umgebenden Felsmasse.

INTRODUCTION

The cut and fill mine of which mathematical modelling is described in this paper is the Sotiel Mine. This mine is located in the province of Huelva, Spain. The ore is pyrite and complex sulfides.

Based on the characteristics of the deposit, it was decided to mine it between the 600 and 700 metres levels (the surface is at about the 1000 metres level) by the mechanized cut and fill method with post pillars. The mine will be about 250 metres long (on the strike) and 85 metres wide on a horizontal section. The dip of the ore-body is 45º and its thickness 60 metres.

In the design that has been modelled, the only support to the back is provided by square post-pillars 13 metres wide and 12 metres apart. The height of the cut is three metres and the distance between the fill (dry) and the back varies between a maximum of seven and a minimum of four metres.

GEOLOGICAL MODEL

At the floor of the ore-body there is a shale band (intermediate shale) which lies on top of another shale bed (black shale) and underneath it there are volcanic tuffs. At the roof there is either intermediate shale or volcanic cinders. The deposit is of Carboniferous age and has been strongly affected by the Hercinic movements. The contacts between the pyrite and the intermediate shale or cinders, and between the intermediate shale and the black shale are net although in some places there is evidence of displacements. All the other contacts are not well defined.

GEOTECHNICAL MODEL

A geotechnical quality survey, based on the Q value (Barton, Lien and Lunde, 1974), was conducted on 2,400 metres of galleries on the 700 horizon. It was found that the average Q ratings range from 15 (good) in the massive sulfides to 0,5 (very poor) for faulted intermediate shale.

The samples for laboratory testing were obtained from borehole cores. Density, uniaxial compressive strength, tensile strength, Young's modulus, Poisson's ratio, cohesion and friction were determined for each lithology. From these laboratory values the properties of the rock masses were estimated based on the author's experience and the rock mechanics literature. The properties of the fill were taken directly from published results. The geotechnical properties of the rock masses and the fill are summarized in Table I.

Only the net and well defined lithological contacts have been taken as geotechnical discontinuities. Their shear strength and shear and normal stiffness were estimated using a well known approach (Barton and Choubey, 1977). For all the contacts, a friction angle of 30º, a shear stiffness of 50 MPa/m and a normal one of 250 MPa/m has been assumed.

Field stresses were measured at two locations within the pyrite in the 700 horizon, using triaxial cells and the overcoring technique. But the results were not completely conclusive possibly because it was not possible to achieve a good bond between the pyrite and the cell. The measurements show that most probably the major principal stress direction is nearly horizontal and at right angles to the strike of the deposit. Its magnitude is about three times the level of the theoretical overburden vertical pressure. The determination of the last stress vector is uncertain. The best fit values for this vector, which is horizontal and parallel to the strike of the formation, ranges from cero to tensile.

MATHEMATICAL MODEL

The modelling of a cut and fill mine where the post-pillar technique is used, is difficult to achieve theoretically, given that the small pillars quickly plastify and deform, distributing their load to less deformed and stressed parts of the rock mass. Pillar yield is accompanied by a relaxation of the stresses in the immediate back area of the slopes, which gives to the formation of a pressure arch. If the support provided by the fill restrained post-pillars is not enough for the development of a pressure arch above the ore-body over the mining interval, the planned layout is unworkable. The purpose of the mathematical modelling should therefore be to calculate the stress distribution and the deformations around the excavation.

The simulation of the process of controlled deformation and transfer of stress during the excavation of the Sotiel ore-body, requires the incorporation into the mathematical model of the following features: joints, mechanical properties of the rock mass before and after failure and a failure criteria, in-situ stresses in the ore-body prior to excavation, excavation process of the ore, and progressive failure of the pillars and the rock mass surrounding the excavation.

There is not in our knowledge, a three dimensional program that could handle this problem taking into account all the above mentioned features. We have therefore looked for a two-dimensional program and decided to use SAGE.

ROCK TYPE	DENSITY g/cm^3	ROCK MASS		FAILED ROCK MASS		RESIDUAL STRENGTH		FAILURE CRITERIA (MOHR-COULOMB)	
		G MPa	K MPa	G MPa	K MPa	C MPa	∅ grados	C MPa	∅ grados
CINDERS	2,90	3.030	7.400	270	7.400	0,60	28	6,0	33
INTERMEDIATE SHALES	3,20	2.300	4.500	200	4.500	0,50	33	5,0	38
PYRITE	4,70	13.500	23.600	1.100	26.300	0,44	50	4,4	55
BLACK SHALE	2,70	2.400	3.800	200	3.800	0,20	33	2,0	38
TUFFS	2,90	5.900	11.400	500	11.400	0,70	33	7,0	38

TABLE I - ROCK MASS PROPERTIES

This is a finite difference computer program written by Geognosis Limited for Dames & Moore that has been developed based on early work by Cundall (1976).

SAGE can model the mechanical behaviour of rock subjected to an external static load and the behaviour of an arbitrary structure connected to the rock at various points, the discontinuities are treated as structural elements. Even though SAGE is designed to model static problems, it works by making many small steps in time, so that the final displacements and stresses are obtained when equilibrium is reached after some time has elapsed. Progressive failure can therefore be represented naturally, since the process corresponds closely to what happens physically.

For plane-strain problems such as this one, the grid is to be regarded as slice through the site, this means that the square post-pillars are modelled as long rectangular pillars. To compensate for their increase in length the width of the pillars has been reduced from 13 to 7 metres and the stopes widened accordingly.

The mechanical behaviour of the five mentioned rock masses under load, has been described by the Mohr-Coulomb constitutive law, with different shear and bulk modulus before and after failure and with a residual strength, as shown in Table I.

In-situ stresses can be approximated from the lithostatic pressure and an estimate of coefficient of lateral pressure, $K = 3$ in this case. For problems with complex geometry such as this one, it is very likely that the in-situ stresses plus the body forces due to gravity, that have been introduced in the model as externally-applied loads, will not be in equilibrium. It will therefore be necessary to do a preliminary program run under this combination of forces alone, a process that it is known as "gravity" run.

Excavation of each cut has been simulated by changing the properties of the ore by those of the fill, and by setting the stresses to cero in the excavated grid zone. Due to limitations in the number of grid zones, it has only been possible to model five cuts among the 33 that will actually be done.

For each cut the stresses at the centre of each cell and the displacements of the grid points have been calculated. In Figures 1 and 2 the stresses and the plastified zones after the fourth and fifth cuts respectively, are shown. The model predictions for ground behaviour in and around the mining area are consistent with what might be expected to occur for that particular layout.

The model indicated mining could progress from the 600 to 700 metres horizon, but with signs of approaching instability with the fourth stage of mining. The peak stress in the pressure arch around the excavation increased by 10 MPa over the uniform values recorded in the first three mining stages. A further 7 MPa increase came with the much smaller fifth mining stage which did little to increase the overall mining span.

An analysis of displacements in the immediate hanging wall shale is also interesting. Displacements over the stope backs in the ore were small up to the end of the fourth cut. On the fifth cut, displacements over the stopes backs away from the hanging wall contact increased significantly.

CONCLUSIONS

The mathematical modelling shows that the support provided by the fill restrained post-pillars is not enough for the development of a pressure arch above the mining interval and therefore, two barrier pillars should be left.

The model (SAGE) has very well simulated the process of controlled yield of the post-pillars and transfer of stress during the excavation of the geotechnically complex Sotiel ore-body as is being mined by cut and fill. The model predictions are consistent with what might be expected to occur for that particular layout, which makes SAGE a very good tool for analysing these mining situations.

ACKNOWLEDGEMENTS

We are very grateful to P. Oliver from Incotech for his ideas and suggestions.

REFERENCES

Barton et al. (1974). Engineering Classification of Rock Masses for the Design of Tunnel Support. Rock Mechanics, no. 6.

Barton and Choubey (1977). The Shear Strength of Rock Joints in Theory and Practice. Rock Mechanics, December 1977.

Cundall (1976). Explicit finite difference Methods in Geomechanics. 2nd International Conference on Numerical Methods in Geomechanics. Blacksburg, Virginia, 2-6 June.

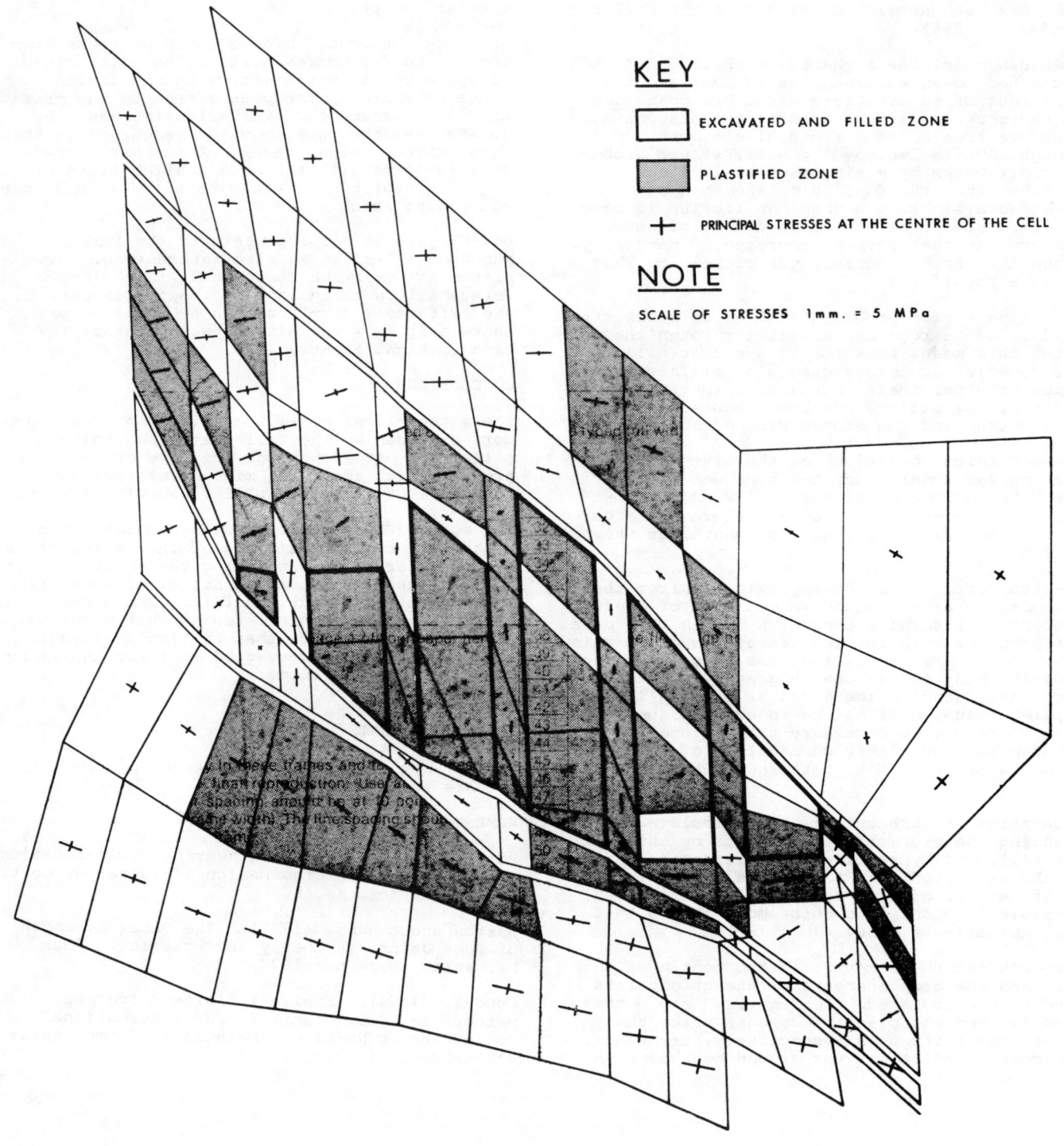

FIGURE 1
PRINCIPAL STRESSES INDUCED BY THE FOURTH CUT
SCALE 1:1.000

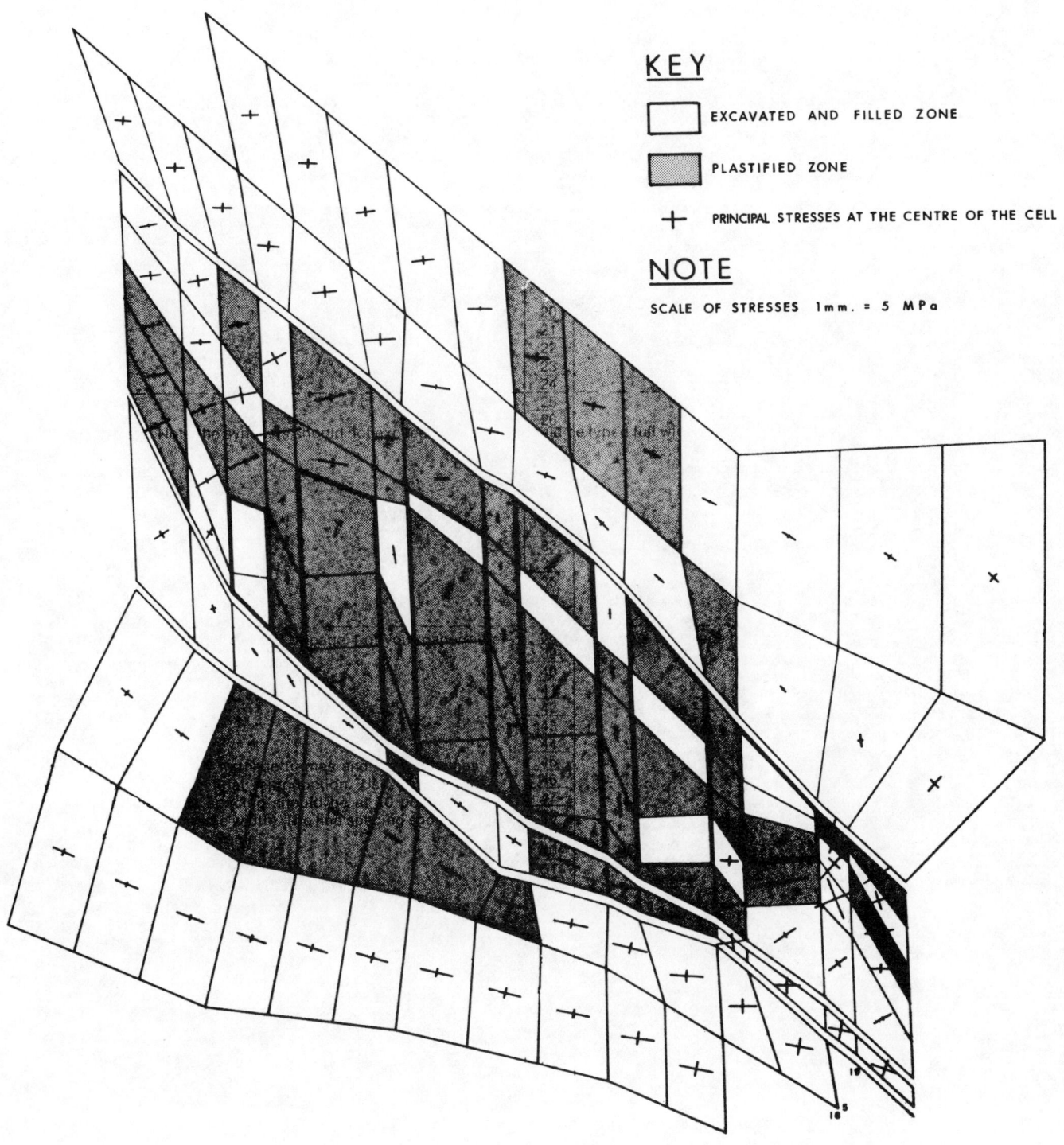

FIGURE 2
PRINCIPAL STRESSES INDUCED BY THE FIFTH CUT
SCALE 1:1.000

GEOTECHNICAL STUDY OF THE STABILITY OF THE SANTA CLARA STOPE AT THE EL SOLDADO MINE IN CHILE

Etude technique de la stabilité de la banquette Santa Clara à la mine de El Soldado au Chili

Geotechnische Stabilitätsstudie des Santa Clara Abbauortes in dem El Soldado Bergwerk in Chile

Eduardo Bolivar, B.
Asst. Technical Manager, Cía. Minera Disputada de Las Condes, Chile

Guillermo Krstulovic, L.
Head of Geomechanics Division, Mining Department, University of Chile, Santiago, Chile

SYNOPSIS

The sub-level stoping mining system used at the El Soldado copper mine in Chile has produced a large open cavity underground in a sector named Santa Clara. This opening is approximately 300 m long, 100 m wide and 150 m high. The cavity has not collapsed nor broken through to the surface, thus putting into question the subsequent mining in the stope and nearby. Consequently, the geomechanics conditions in the stope were studied in a three-part stability analysis: evaluation of the in-situ rock quality; simulation of geological and tectonic conditions of the stope; implementation of an instrumentation program in the field.

RESUME

Le méthode d'exploitation minière, le "sub-level stoping" (chambres à sous-étage) dans la mine de cuivre de El Soldado au Chili a produit une grande chambre ouverte souterraine, dans le sector connu sous le nom de Santa Clara. La cavité a, à peu près, 300 m de longueur, 100 m de largeur et 150 m de hauteur. Cette ouverture ne s'est effondrée ni s'est cassée jusqu'à la surface, ainsi remettant en question les conditions d'exploitation dans les secteurs environnants. Par conséquent les conditions géomécaniques de secteur ont été étudiées, dans une analyse de stabilité, divisés en trois parties, à savoir, évaluation de la qualité de la roche in situ; simulation des conditions géologiques et tectoniques dans le secteur; établissement d'instrumentation sur le terrain.

ZUSAMMENFASSUNG

Das in dem El Soldado Kupferbergwerk (Chile) angewandte "sub-level stoping" Abbauverfahren hat in dem Santa Clara Sektor einen großen bergmännischen Hohlraum hinterlassen. Die ungefähren Ausmaße des Hohlraumes sind 300 m Länge, 100 m Weite und 150 m Höhe. Da dieser hohle Abbauraum weder eingefallen noch zur Oberfläche durchgebrochen ist, wird der Abbau von anliegenden Weitungen in Frage gestellt. Folglich wurden die geomechanischen Bedingungen der Weitungen mittels einer dreifachen Stabilitätsstudie untersucht: Begutachtung der Steinfestigkeit im unverritzten Gebirge, Simulierung von geologischen und tektonischen Bedingungen in Weitungen, Einsatz von Meßgeräten im Felde.

INTRODUCTION

Cía. Minera Disputada de Las Condes' underground mining system at its El Soldado copper mine in Chile has produced a large, open cavity underground in a sector named Santa Clara. As this cavity has not collapsed nor broken through to the surface, various doubts have existed regarding future mining activities in the Santa Clara stope itself or in the other nearby sectors.

Consequently, a study of the existing geomechanical conditions in the Santa Clara sector was carried out to define in quantitative terms the stability of the roof and rocks surrounding the present underground cavity. Follow-up investigations have studied the operational alternatives for rational exploitation of the Santa Clara orebody.

The analysis method employed in this study was determined by the quality and quantity of geotechnical information available. Briefly, the following geotechnical background information was considered:

- Geometric and topographic information on the Santa Clara sector, provided by the El Soldado Geology Department. This was accompanied by a geologic interpretation of cross-and plan sections.

- Rock quality information, obtained from conventional mechanical competence tests on laboratory samples and geoseismic zoning in the field.

- Tectonic information, i.e., on local stresses in rock surrounding the Santa Clara stope. This information was obtained for two typical sectors of this stope by the overcoring technique.

The conclusions and recommendations of this study result from methematical simulation of the geometric, topographic and tectonic conditions and mechanical competence of the rock existing in the Santa Clara sector. These conditions have been examined in the laboratory and in field investigations. Thus, the information available for the mathematical simulation of the stope should be considered as optimal and limited only by the natural restrictions on the processes involved in obtaining this information.

REGARDING THE INFORMATION AVAILABLE FOR ANALYSIS

Geology and Topography

As shown in Figure 1, the Santa Clara stope is located under a minor gorge which descends from the principal mountain range in the El Soldado area. The mine site is located 150 km north of Santiago in the Andean Cordillera at an elevation of 1100 m.

Figure 2 is a section showing the more significant mine workings surrounding the Santa Clara stope. The stope itself is an open cavity approximately 300 meters long (N70°W-S70°E) by an average 100 meters wide. Vertically, the estimated 150 meters of open space extends from the abandoned production level it cuts through at 845 m elevation to below the +175 level of the mine (probable elevation, 1000 m). Drifts on the +175 Level into the stope roof bridge rock give access for geotechnical investigation.

The geology of the problem sector is controlled principally by major faults, which limit the extension of the copper orebody. In the stope sector, the Santa Clara fault on the south and the San Jose fault on the north are significant. Both dip subvertically and have a series of minor branches trending towards the interior of the body.

The principal rock types of the Santa Clara sector are andesites, porphyries, breccias and tuffs. Cross -and plan- sections of the sector show a confused distribution of these rocks. These sections suggest that the Santa Clara ore body host rocks were emplaced as semi-layered flows of andesite rock, followed by the intercalation of breccias and porphyries at depth.

This arrangement of rocks is interrupted by the

Santa Clara and San Jose fault systems, which follow the strike of the gorge on the surface. The Santa Clara stope has been developed in this failure sector in the central core zone without extending to the peripheral rock.

In-Situ Fracture Pattern

Apart from studies of the major faults mentioned above, the Santa Clara rock body was investigated for clusterings of minor structures. A detailed map of these minor structures was obtained from mappings of drifts surrounding the stope. The structures thus identified were processed statistically by using stereographic projections obtained with the PATCH computer program. Stereonet techniques have been amply reported in the literature (1) and will not be extensively treated in this paper. Briefly we can summarize that the stereographic representation of failures or fractures arranged in a spatial pattern can be projected onto horizontal equatorial planes, referenced to mine coordinates (north, south, etc.).

For each structure or failure plane identified in terms of dip and strike in the field, a vector pole may be assigned perpendicular to the failure plane. This pole may be represented in a stereonet equatorial plane, as shown in Figure 3. The occurrence of a large amount of poles in any stereonet sector implies the existence of a cluster or preferred failure tendency.

For Santa Clara, four clusters have been detected. A statistical count of the distribution of these failures indicates that 70% of these structures are located in the predominant sector represented by P1 and P2. The orientations of the four dominant Santa Clara clusters are as follow:

Cluster Plane	Dip	Strike
P1	N 31° E	72° W
P2	N 53° E	86° W
P3	N 30° W	71° W
P4	N 61° W	82° W

At a later date, this information will be the subject of a separate analysis to quantify its influence on the stability of the Santa Clara roof bridge.

Tectonism

Tectonism refers to the existence of an in-situ stress field of a magnitude and orientation different from the values attributable to the effect of gravity. For Santa Clara, the existence of an irregular topography with high cliffs in the vicinity has suggested (in consideration of tectonism evidence from elsewhere in the Chilean Andean range) the use of the overcoring technique to reconnoiter the local tectonism. This technique for in-situ stress measurement, developed by the U.S. Bureau of Mines, and its recent modifications have been amply reported in the literature and will not be treated separately in this report (2).

Measurements were made from two sites selected for overcoring in the sectors surrounding the Santa Clara stope. The first was made from an access drift in the roof bridge on the +175 Level, as shown in Figure 2. The second measurement, towards the northern side of the cave parimeter being investigates, was made on the +125 Level, i.e., at an intermediate height for the stope, as also shown in Figure 2.

The results of these measurements are presented in Figure 4. We have determined from these values that the rock surrounding Santa Clara is subjected to significant stresses, principally oriented subhorizontally and striking E-W. The values obtained from these two overcorings (in the Santa Clara roof sector and walls) perfectly agree with those expected from the predominant topography in the area being investigated, and also agree with the orientations measured for in-situ fatigue direction investigations made in other regions of Chile.

The magnitude and orientation of the tectonism detected at Santa Clara largely controls the stability conditions for the Santa Clara stope, thus constituting a ground factor of high priority in the stope stability analysis.

Mechanical Properties of the Rocks (In-Situ and at the Laboratory)

The mechanical proporties of the rock material

involved in the stability of the Santa Clara stope were investigates in two data collection campaigns.

The first campaign was designed to verify the competence of typical, predominant rock volumes of Santa Clara. For this, core samples were chosen by the geologic description mentioned earlier in this paper. These selected samples were laboratory tested to obtain background information on the deformability of the sound rock, i. e., cohesion, friction, tensile strenght; deformation, Young's, Poisson's modules. Additionally, the samples were systematically investigates with non-destructive tests, i.e., pulse velocity and dynamic modules.

Of course, all the above information was obtained from sound core samples having evidently superior quality than that of the average in-situ rock. Thus, the following step was designed to relate these values to those of the more-deteriorated in-situ rock, examined in the second data collection campaing.

For the Santa Clara sector, the range of deterioration was obtained from a series of geoseismic profiles using the "petit seismic" technique (3). A total of 15 geoseismic profiles were made of the rock surrounding Santa Clara according to the following distribution: six profiles of the surface topography (P1 through P6), five from the access drift through the stope roof bridge (S1 through S5), and four of the Santa Clara lower level (floor level - S6 through S9). From these profiles, the in-situ rock Vp velocities and corresponding dynamic properties were obtained.

From this information, the values or quality indexes of the in-situ rock have been obtained, thus establishing the degree of deterioration suffered by this rock with respect to the supposedly sound laboratory rock. Table 1 gives the P wave values obtained for typical rocks in the laboratory and from the surface and underground geoseismic profiles. Additionally, Table 1 shows the percentage of deterioration of Vp values experienced by the in-situ rocks.

The quality indexes mentioned above were used to estimate the deterioration of in-situ rock parameters: cohesion, friction, tensile strength and deformation modules. These in-situ values are shown in Table 2.

SANTA CLARA STOPE STABILITY ANALYSIS

In accordance with the quantity of information available for this investigation program, two types of stability analyses have been employed. The first deals with the evaluation of the quality of in-situ rock, considering the deterioration which would have occurred as a result of the mining of Santa Clara. The second deals with a simulation of the previous and present geometric, geologic and tectonic conditions of Santa Clara.

The following sections contain abrief description of the technology used for each of these analyses.

Stability of the Roof Bridge According to Stereonet

As indicated previously in this paper, the stability of massive rock bodies is principally controlled by the existing fractures or faults of major dimensions and/or by fracture tendencies and clusters, which are a product of coincident minor structures. For this reason, the evolution and present geometry of Santa Clara could have been favored or restricted by the existence of these fractures and clusters. The influence of these clusters on the Santa Clara stope roof bridge can be established by using Stereonet techniques.

As stated previously in this paper, at least four clusters are predominant in the rock being investigated. These rocks containing fracture tendencies are susceptible to caving and/or collapsing in a gradual way until the failure planes (i.e., the weakness vector from the intersection of these planes) open up as a free face. In Santa Clara, all four cluster planes intercept each other along various lines, which define the dips and strikes of many others weakness vectors.

Among these weakness vectors, the most significant results are from the clusters P1 and P2 in Figure 3. These two clusters account for 70% of all fractures investigates at Santa Clara. Moreover, these weakness vectors have the lowest strike angles of all cases investigated.

To evaluate the influence of all these weakness vectors on the present stability of the Santa Clara stope roof bridge, we must assume a simplified model of this rock volume. In Figure 5, this roof bridge has been represented in a modular way as a beam of triaxial prism. The caving of this beam would be caused by fault planes which cross it (i.e., its interception vectors). These fault planes can be estimated starting from the critical angles β and α. Thus, all weakness vectors of strike angle smaller than these critical angles originate and die out on the rock boundary of the roof bridge. On the other hand, all weakness vectors of strike angle larger than the critical angles will daylight to the free face of the roof bridge and contribute to its caving.

In Figure 6, a stereonet representation has been made of the critical angles controlling this stability along axes A and B of the roof bridge. In this figure, the solid line I represents the structure of the roof bridge. Note in this figure that the weakness vectors (Sij) are contained within the rock volume investigated and thus cause the daylighting of the intercepting lines of all the rock weakness contained in the area covered by the figure. This configuration will produce the caving of the roof bridge until the dimensions of the model beam (along axis h) are reduced to position II. In this final position, weakness vector S12 (with 70% of all fracture structures) will not daylight to the cavity, and, thus the possibility of caving will be significantly reduced.

Figure 6 considers the hypothetical case for the Santa Clara roof bridge as deduced from the position II stereographic representation. In this figure, the Santa Clara roof has a dome shaped geometry, whose typical dimensions could be: lenght, 50 meters; width, 30 meters; thickness, 55 meters.

In summary, the following conclusions are drawn from the stereographic geometric interpretation analysis of the influence of minor structures on the stability of the Santa Clara stope roof bridge:

1. Minor structures have preferred orientation along two principal directions which group 70% of all known fractures and faults. These clusters correspond to typical weakness planes that dip N 31° E to N 53° E and strike 72° W to 86° W.

2. The interception line, or weakness vector for these planes, strikes approcimatelt S 60° W, 62° W. This orientation of weakness produces the tendency towards a natural dome configuration for the roof cavity. The slope angles of this dome would be 15° to 18°, depending on the axes of the stope.

3. The conclusion deduced from the above data is that the minor structures of Santa Clara have not contributed significantly to the caving of the stope roof bridge; on the contrary, they have favored the forming of a dome shaped roof, by nature a very stable geometry.

Mechanical Deterioration of the Rock and In-Situ Effects of Tectonism.

The rock parameter values determined by investigation of the in-situ rock of Santa Clara show that invariably the rigidity and mechanical competence of this rock has deteriorated, especially when compared with material preselected for laboratory analysis. Most of this deterioration process is the result of the redistribution of in-situ stresses, producing deformation and cracks in the rock body. As stated earlier in this paper, the Santa Clara Stope rock material is subjected to a tectonic stress field of singular magnitude, oriented E-W at an acute angle to the major axis of the stope cavity.

Regarding our assimilation of the above field data in this stability analysis, we refer to a recent publication reporting investigations of the influence on the block caving method of the involved cavity's orientation with respect to the local tectonic field (4). The results of this research, expressed in terms of "caving indexes" for rock subjected to caving (see Table 3) suggest that the following conclusions are applicable to the Santa Clara case:

1. Horizontal tectonic stresses drastically reduce the caving of underground cavities as compared with identical cavities subjected to the effect of gravity alone.

2. Underground openings oriented with their major axes parallel to the direction of the maximum stress produce greater caving than do openings whose major axes are oriented at an angle to the maximum stress.

3. Conclusions 1 and 2 suggest that the magnitude and orientation of the local tectonism at the Santa Clara has a favorable influence on the stability of the roof bridge over the stope. Additionally, the quality of the rock material in this stope yields insignificant caving indexes.

Stability Analysis by Simulation Model

Considering the quality and quantity of the geotechnical information available for the Santa Clara stope stability evaluation, use was made of a two-dimensional mathematical simulation model for the analysis. The principal reasons for this decision are the following:

1. The availability of two dimensional simulation techniques capable of reproducing the more relevant geotechnical parameters of the Santa Clara configuration.

2. The geologic background information is principally in plan sections, thus giving maximun sensitivity for establishing the influence of the major faults cutting across the Santa Clara stope.

3. A two-dimensial profile simulation produces critical level results for a three-dimensional reality. Thus, although the geometric, geologic and tectonic configuration of Santa Clara is actually three dimensional, the difference with the two dimensional representation is not significant because, in the majority of cases, the third dimension contributes to improve rock support. This consideration places us on the safe side with the results.

For these reasons, a two-dimensional finite element simulation computer program was implemented for this investigation. The original computer program (5) allows the simulating of the excavation ot the cave in sequential steps.

For the analysis, we selected two cross-sections, oriented N - S and E - W, and one plan section at 890 meters elevation. These three sections approximatelt represent the most critical dimensions of the stope. (Figure 7 shows a typical finite element model used in the analysos.).

The simulation of Santa Clara stability conditions was made along the following guidelines:

- The present geometry of the Santa Clara stope is the end result of stope exploitation in sequencial stages. The simulation analysis **considered** four stages of mining, from the beginning up to the present stope geometry. This procedure resulted as being especially interesting since the historical mining problems of Santa Clara are well documented (6). Thus, the results of the simulation could be corroborated with historical information.

- The simulation models were designed to investigate the stability of the roof bridge with consideration given also to the exploitation of a peripheral rock body named "Caseron 53NW", as shown in Figure 8. The total tonnage to be mined from the stope is approximately 140,000 tonnes. Thus, the Caseron 53NW stope dimensions would probably be 70 meters long by 25 meters wide by possibly 40 meters high. A protection pillar with an average 10 meters width will be left between this new stope and the Santa Clara stope. The geologic and geotechnical competence of this protection pillar has been reported separately (7).

In this analysis, we have defined the "stability" concept by adopting the criterion of comparing the mechanical competence of the rocks (as discussed earlier in this paper) with the stress conditions that could result, as simulated for the rock by the finite element model. In the analysis process, for each element of the simulation grid the computer assigned a safety factor defined according to the Mohr-Coulomb failure criterion, graphically illustrated in Figure 8. Thus, when any safety factor resulted as less than one, then the rock area simulated by the "element" was considered unstable. On the other hand, if the safety factor was higher than one, then the area was considered stable. Furthermore, given that rocks are generally very sensitive to tensile stresses, those sectors subjected to tension were interpreted as rocks with fracture forming tendencies.

The simulation program also provided information on the deformation undergone by the rock surrounding the stope. This deformation could be corroborated by appropiate instrumentation in the field.

The more significant conclusions of the simulation analysis are the following:

1. The results of simulating sequential exploitation of Santa Clara coincide with the historical knowledge of caving in the stope. These results also show that the major faults considered in the orebody do not contribute significantly in the caving and collapsing of the stope. On the contrary, most of the caving was produced by stress concentrations, which fractured the rock by tension and shearing. This collapsing process continued until the present dome shape of the roof bridge was achieved.

2. In the walls surrounding the stope, some stress sectors have developed; these should constitute open craks in these rocks. Of all the major faults considered in the simulation, only the southern portion of the Santa Clara fault show signs of being worked on in a growing magnitude. Notwithstanding these observations, the Santa Clara roof bridge remains equally stable. The safety factors curves for the entire rock volume on a typical section are shown in Figure 9.

3. The deformation experienced by the rock surrounding the stope is shown in Figure 9. In general, the results obtained suggest that all free faces have displaced towards the interior (including a minor uplifting of the stope floor). In conjunction with the surface topographic displacement, this could be evaluated in a relative way by observing that the maximum deformation are to the south and/or as an uplifting of the southern sector relative to the northern sector.

INSTRUMENTATION PROGRAM

As a result of the conclusions presented in this paper, two recommendations were made: The exploitation of "Caseron 53NW" and the implementation of an instrumentation program designed to detect incipient changes in the present stability of the entire stope.

The instrumentation selection was based on the following operational needs and long range objectives:

- The recording in Santa Clara rock masses of rock stability parameters, such as interior deformations and/or convergence of walls or exposed surfaces. Additionally, the recording of variations in stresses in the rock.
- Field implementation of equipment systems with operating facilities available at the mine so that mine personnel may be trained to gather this data in the future.

At the El Soldado, the equipment implemented needs to provide coverage for all the volume of rock susceptible to being affected by future mining activities in and around Santa Clara. A large variety of equipment partially or fully complies with these requirements. Among these, extensometers deserve special mention.

In principle, extensometers are mechanical devices for the precision determination of incipient deformations in affected rock, mainly when large volumes of rock are involved. Basically, all of these devices are installed in holes (drill holes for the most part), leaving anchor points fixed at different depths in the hole. A comparison mechanism is used to check periodically the relative distance between these points, considered as fixed, and a reference head located at the mouth of the hole. With this arrangement, variations in these relative distances are detected, i.e, when movements of anchors are brought about by cracks in movement. For our case, the type selected was a Multiple Position Borehole Extensometer with eight anchors (MPBX-8).

Additionally, stressmeters were recommended for recording rock stresses. These instruments are installed in small diameter drill holes located inside the volume of rock for which the load variation measurements are to be made. The most reliable stressmeter is the type making use of the vibrating wire principle to detect load variations. This instrument is field installed in such a way that the equipment remains solidly attached to the rock in the hole in which it is housed. With this arrangement, the loads on the rock are transmitted directly to the stressmeter and to the wire it contains. Thus, variations in load and stress activity in the rock will cause changes in the length of the wire and its vibration frequency in response to pulses from a field recorder unit. When used with pre-established correlations, this equipment makes it possible to detect rock load variations as small as one kilogram.

Surface Equipment

Figure 10 gives a plan-section view of the upper stope roof rock features and the estimated stope contour. Considering the abrupt topography over the stope sector, it was decided that the minimun surface instrumentation should consist of four MPBX-8 extensometers, ideally to be installed at each of the four Santa Clara roof boundaries. This ideal arrangement could not be acomplished for the following reasons:

1. The geometric arrangement of major faults suggested that more effective monotoring could be achieved towards the eastern boundary of the stope. Also, the simulation analysis suggested that the first signs of roof instability could be expected towards

the eastern slope of the surface topography (see Photos 1 and 2).
2. The nature of the overlying topography made access difficult for drilling the holes in which the MPBX units would be placed. Access to the NE and SE boundary areas could be more readily obtained.

Based on this considerations, the four MPBX units recommended were installed along the NE-E-SE contour of the Santa Clara stope in a radial arrangement as shown in Figure 10. In this fashion, each instrument has a range of influence covering approximately 0.5 hectares.

Underground Equipment

The stressmeters were installed underground to identify any significant change in the stresses on the rock investigated, thus enabling the early detection of fracturing in these rocks. Consequently, the stressmeters were installed mainly in the pillars and walls of the new Caseron 53NW stope adjoining Santa Clara, In the field to date, only ten stressmeters have been installed: five of these on the blasting level, in in the moddle sector of the pillar separating the 53 NW stope from the Santa Clara boundary and five in the middle sector of the pillars on the haulage level under the 53 NW stope. All of these instruments are shown schematically in Figure 11.

Premilimary Conclusions from the Instrumentation

In spite of the reduced number of instruments emplaced on the surface and underground, some significant results already have been obtained. The most relevant conclusions are as follow:

1. The drilling of the four holes required for the MPBX extensometers has led to a better estimate of the Santa Clara roof contour since the drill holes intercepted the stope cavity. The total hole lenghts fluctuated between 108 and 138 meters, drilled at a 70° angle. Thus, the thickness of the roof bridging over the stope is less than 60 meters, i.e., in line with the 55 meters estimated from the stereographic analys.
2. The RQD data from these drill holes was compared with the "petit seismic" data obtained previously. With this analysis, the empirical formula for relating laboratory and field seismic velocities with RQD values was corroborated, as shown in the following table:

Empirical Formula:

$$(\%)RQD = \left(\frac{\text{Field Seismic Velocity } V_p}{\text{Laboratory Seismic Velocity } V_p}\right)^2$$

V_p at Drill Hole Extensometer N°	Estimated RQD Empirical Formula	Average RQD from Drill Hole
1	38	35
2	30	25
3	42	50
4	79	80

The above data is for the upper sector of the surface topography, or the maximum depth reached by petit seismic tecniques (i.e., 20 meters). Notwithstanding these upper sector RQD results, the lower sector with the Santa Clara stope roof rock yielded higher than normal rock quality indexes.

3. The field recordings obtained to date from the MPBX extensometers indicate the absolute stability of the Santa Clara stope roof in regards to the new excavations in the 53 NW stope. Likewise, the rock enveloping the 53 NW stope has shown only minor signs of load increases. This minor stress concentration is observed principally in the rock pillars on the haulage level, approximately 12 meters below the blasting level, where five other stressmeters are monitoring the stress concentration while the level is blasted out. These five stressmeters are being continuously reinstalled in a retreating pattern as the rock is blasted, as shown in Figure 11. Despite the above monitoring, no significant stress changes have been observed on these stressmeters, thus suggesting that the rock of the drifts on the blasting level is safe.

4. Obviously, the partial results presented here cannot be taken as holding for the future mining of additional stopes surrounding Santa Clara. Due to this, a continuous geotechnical program is underway to investigate mining alternatives for the exploitation of the Santa Clara Orebody.

REFERENCES

1. Yevenes Israel; Krstulović Guillermo - Programa de Computación para Análisis Estadístico de Cluster. RI-75-1 Universidad de Chile Depto. Minas. Septiembre 1975.

2. Hooker Verne; Bickel David - Overcoring Equipment and Techniques USed in Rock Stress Determination. U.S. Bureau of Mines. Information Circular 8618. 1974

3. Araneda Manuel - Informe de Resultados de Estudios Geosísmicos en Santa Clara. Cía. Minera Disputada de Las Condes. Enero 1981.

4. Krstulović Guillermo - Influence of Tectonic Stresses on the Caving Process. 4° International Congress ISRM. 1979.

5. Kulhway Fred - Analysis of Underground Openings in Rock by Finite Element Methods. U.S. Bureau of Mines. Contract Report S-69-8. 1973.

6. Glavić Marcelo - Cía. Minera Disputada de Las Condes, El Soldado Mine. Personal Communication. 1981.

7. Krstulović Guillermo - Análisis Estructural para Explotación de Caserón 53 NW. Cía. Minera Disputada de Las Condes, Mina El Soldado. Internal Report. 1981.

TABLE N°1 - VP WAVES VALUES FOR TYPICAL ROCKS VALUES OBTAINED IN THE LABORATORY AND FROM THE SURFACE AND UNDERGROUND GEOSEISMIC PROFILES

ROCK TYPE	VP WAVES VELOCITY (m/s)		
	LABORATORY	SURFACE	UNDERGROUND
PORFIDO	5300.	3000-3500	4000-4200
BRECHAS	5400.	2700-3700	2000-3500
ANDESITA	6000.	4700.	4000-4600

PERCENTAGE OF DETERIORATION OF VP VALUES EXPERIENCED BY THE IN-SITU ROCKS

ROCK TYPE	DETERIORATION (%)	
	SURFACE	UNDERGROUND
PORFIDO	60.	70-75
BRECHAS	50-65	35-65
ANDESITAS	75.	65-75

TABLE 2.- ROCK PARAMETERS FROM LABORATORY TESTS AND IN-SITU VALUES ADOPTED IN THE FINITE ELEMENT SIMULATION MODEL

ROCK TYPE	ROCK PARAMETERS			
	Cohesion (Ton/m^2)	Friccion (°)	Tensil Strength (Ton/m^2)	Deformation Modulus (Ton/m^2)
Data From Laboratory				
PORFIDO	460	50	300	6.850.000
BRECHAS	600	35-45	650	6.650.000
ANDESITA	600	45-50	700	9.200.000
Data Adopted For Simulation				
PORFIDO	280	40	100	3.500.000
BRECHAS	330	30	100	2.700.000
ANDESITA	480	36	100	4.000.000

TABLE 3.- CAVABILITY INDES FOR DIFFERENT VALUES OF ROCK PARAMETERS

BLOCK WITH SQUARE SECTION

GRAVITATIONAL LOAD

	ANGLE OF INTERNAL FRICTION								
	45°			55°			65°		
	TENSILE STRENGTH (psi)								
COHESION (psi)	0	1.40	7.00	0	1.40	7.00	0	1.40	7.00
1750	1.47	0.05	0.00	1.45	0.05	0.00	1.51	0.05	0.00
350	2.36	0.15	0.00	2.51	0.18	0.02	2.93	0.22	0.10
0		2.45			5.70			4.24	

TECTONIC LOAD

	ANGLE OF INTERNAL FRICTION								
	45°			55°			65°		
	TENSILE STRENGTH (psi)								
COHESION (psi)	0	1.40	7.00	0	1.40	7.00	0	1.40	7.00
1750	0.80	0.05	0.00	0.51	0.05	0.00	0.81	0.05	0.00
350	1.00	0.09	0.03	0.92	0.09	0.04	1.03	0.10	0.08
0		3.54			2.26			1.57	

BLOCK WITH RECTANGULAR SECTION

GRAVITATIONAL LOAD

	ANGLE OF INTERNAL FRICTION								
	45°			55°			65°		
	TENSILE STRENGTH (psi)								
COHESION (psi)	0	1.40	7.00	0	1.40	7.00	0	1.40	7.00
1750	3.35	0.47	0.00	3.37	0.47	3.38	3.38	0.49	0.00
350	3.75	0.69	0.25	3.70	0.71	0.34	3.84	0.78	0.81
0		9.72			7.66			5.52	

TECTONIC STRESSES ACTING ON Y DIRECTION

COHESION (psi)	0	1.40	7.00	0	1.40	7.00	0	1.40	7.00
1750	2.39	0.17	0.00	2.43	0.17	0.00	2.47	0.19	0.00
350	2.48	0.29	0.07	2.90	0.30	0.12	2.95	0.34	0.22
0		7.52			5.17			4.04	

TECTONIC STRESSES ACTING ON X DIRECTION

COHESION (psi)	0	1.40	7.00	0	1.40	7.00	0	1.40	7.00
1750	1.44	0.17	0.00	1.45	0.17	0.00	1.49	0.18	0.00
350	1.71	0.29	0.17	1.74	0.28	0.20	1.77	0.30	0.20
0		4.94			3.10			2.30	

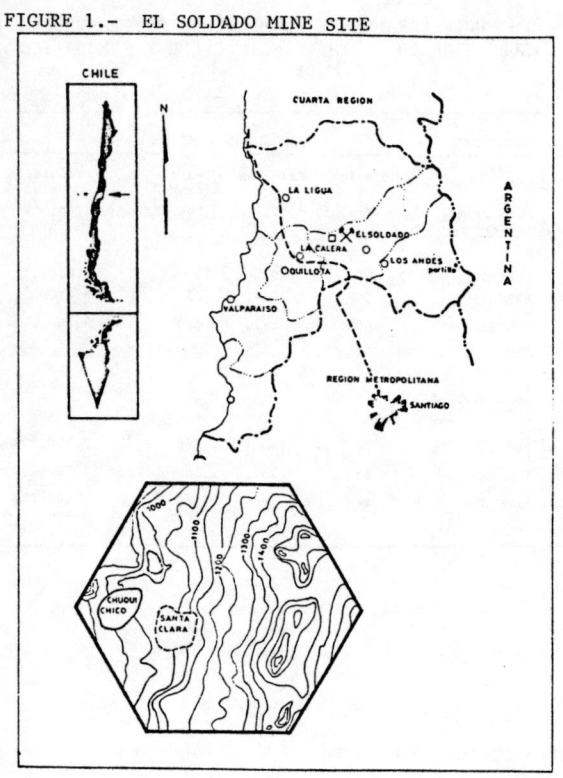

FIGURE 1.- EL SOLDADO MINE SITE

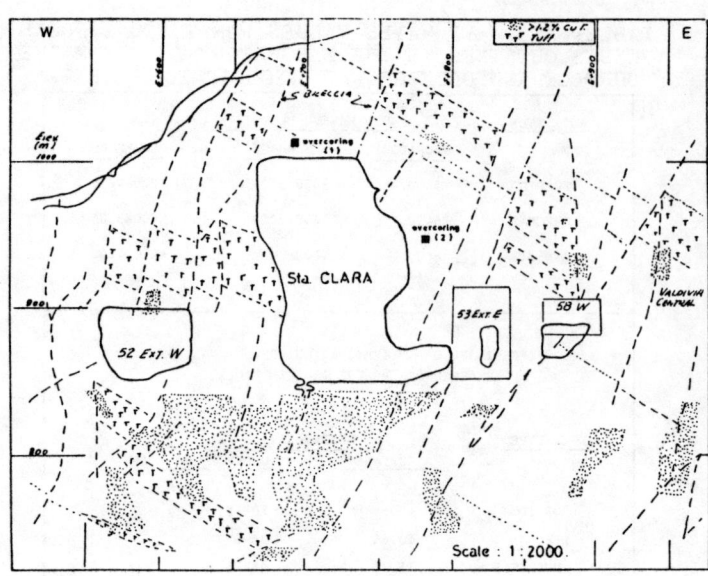

FIGURE 2.- MINE WORKING SURROUNDING SANTA CLARA STOPE

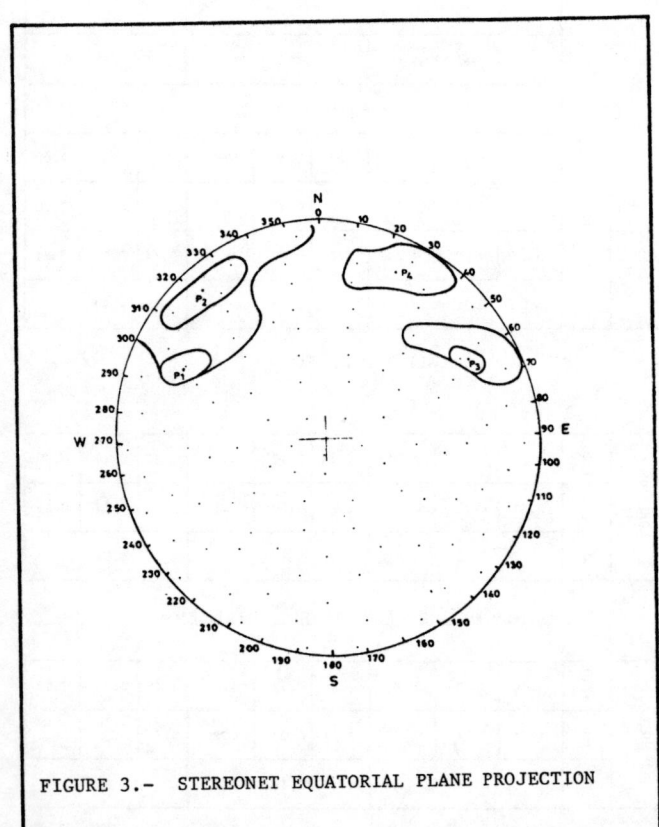

FIGURE 3.- STEREONET EQUATORIAL PLANE PROJECTION

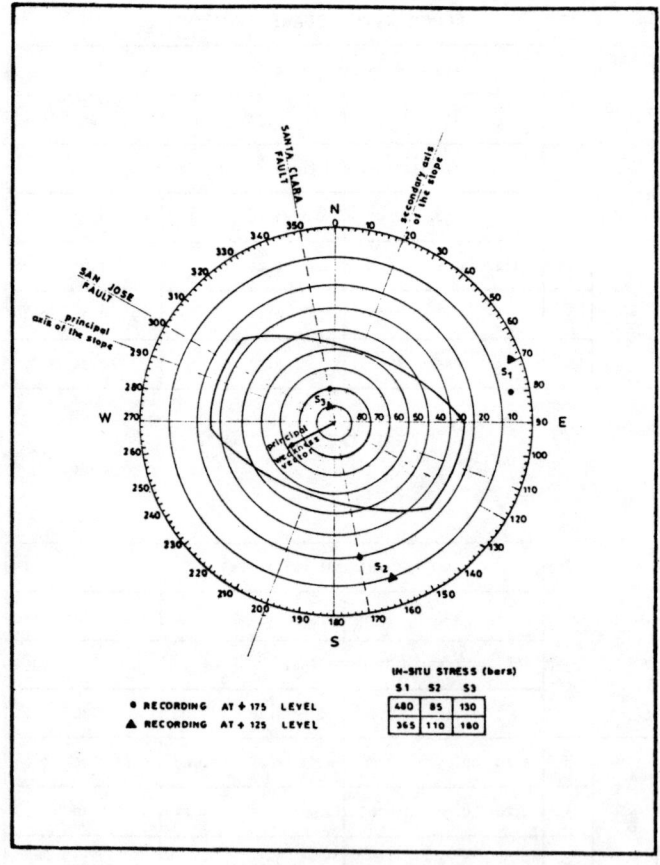

FIGURE 4.- IN-SITU STRESS PATTERN AT EL SOLDADO SANTA CLARA STOPE

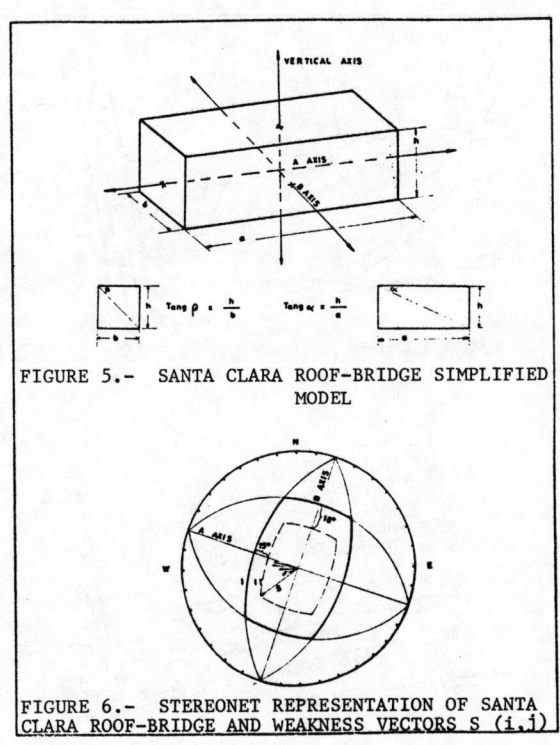

FIGURE 5.- SANTA CLARA ROOF-BRIDGE SIMPLIFIED MODEL

FIGURE 6.- STEREONET REPRESENTATION OF SANTA CLARA ROOF-BRIDGE AND WEAKNESS VECTORS S (i,j)

FIGURE 7.- FINITE ELEMENT MODEL FOR SANTA CLARA STOPE

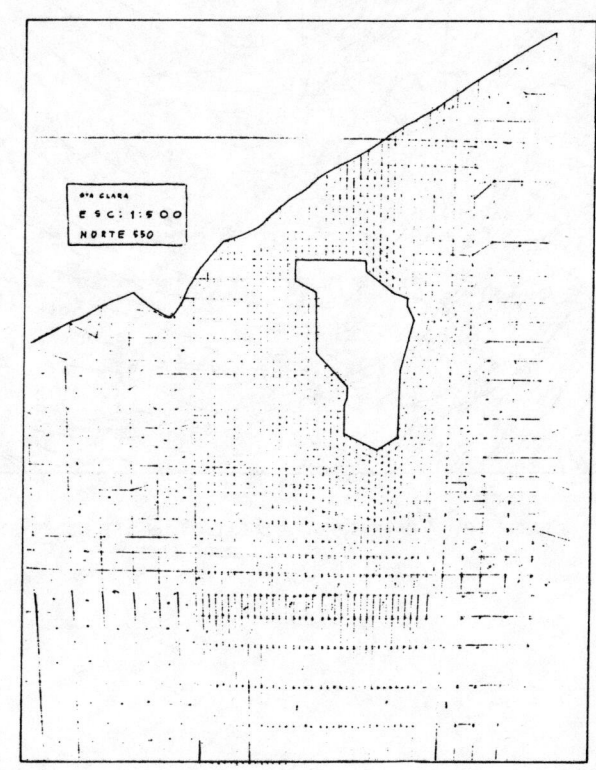

FIGURE 8.- MOHR-COULUMB FAILURE CRITERION

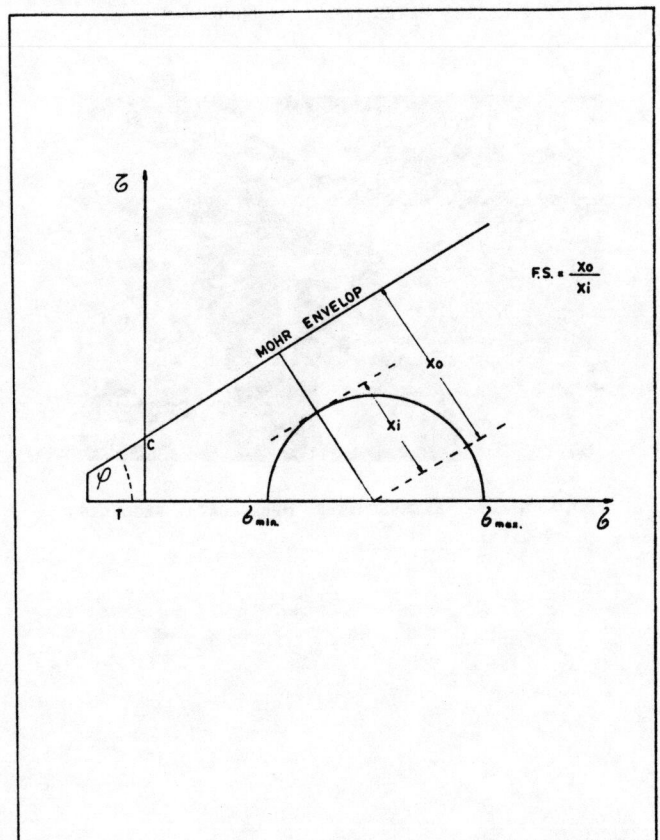

FIGURE 9.- SAFETY FACTOR CURVES FOR TYPICAL SECTION

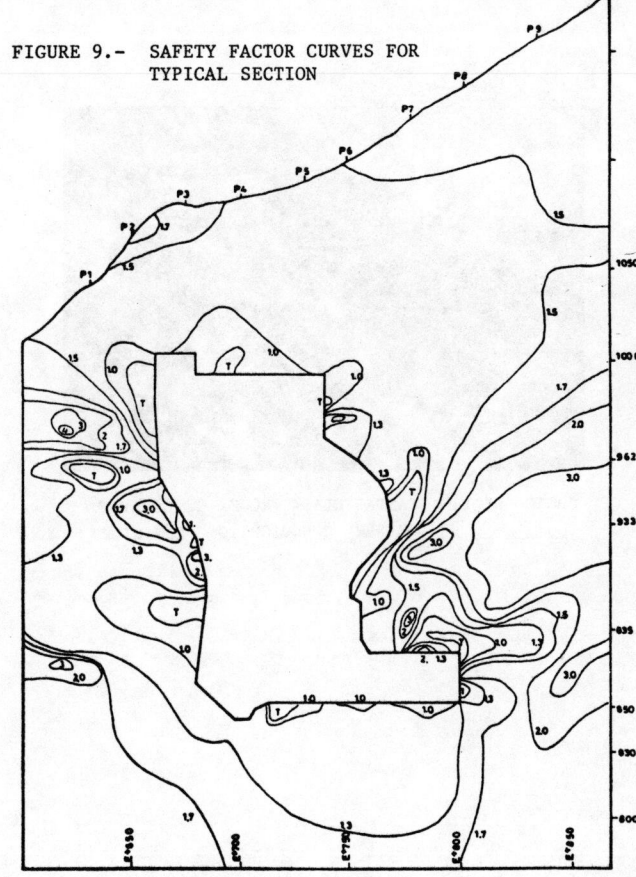

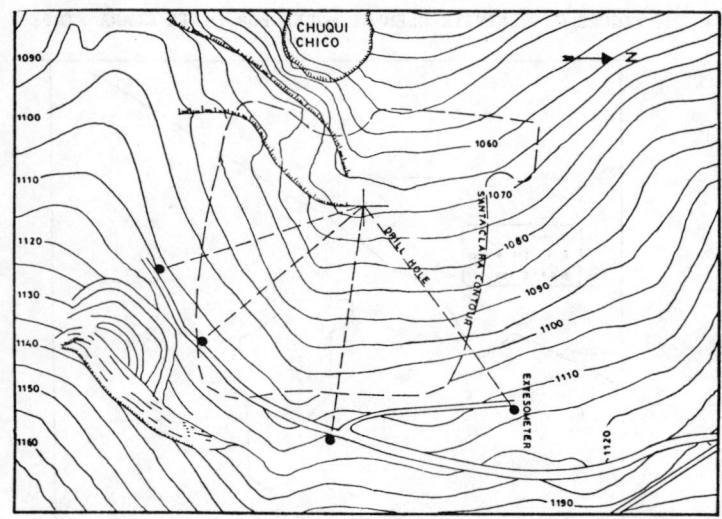

FIGURE 10.- EXTENSOMETER LOCATION ON THE UPPER STOPE ROOF ROCK

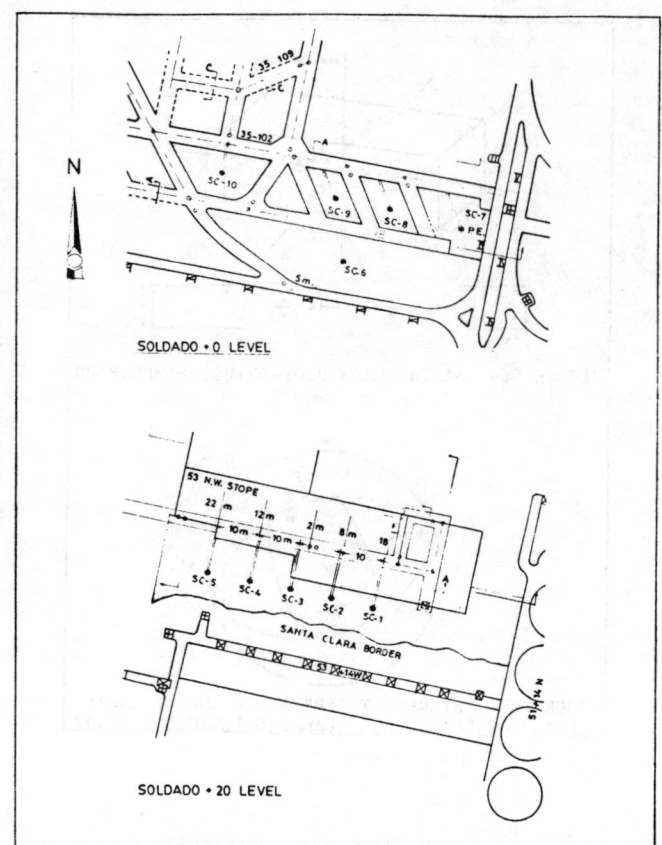

FIGURE 11.- STRESSMETER LOCATION UNDERGROUND

PHOTO N° 1.- SANTA CLARA ROOF TOPOGRAPHY NEARBY CHUQUICHICO OPEN STOPE

PHOTO N° 2.- EXTENSOMETER MPBX DATA RECORDING

AN INVESTIGATION INTO THE ROCK MECHANICS ASPECTS OF SUBLEVEL OPEN STOPE MINING

Recherche en mécanique des roches. Un aspect de l'exploitation par chambres à sous étage

Untersuchung der felsmechanischen Aspekte eines Abbaus mit Teilsohlenstollen

A. G. Paşamehmetoglu
T. Y. Irfan
N. Bölükbaşi
A. Bilgin
A. Özgenoglu
C. Karpuz

Mining Engineering Department, Middle East Technical University, Ankara, Turkey

SYNOPSIS

Rock mechanics investigations are carried out to evaluate the sublevel open stope mining method used at the tungsten mine in Bursa (Turkey) with respect to the stability of stopes, rib pillars and its various galleries. Although the orebody is massive in nature, it shows various rock mass strength characteristics. Major geological patterns and their strength characteristics are extensively studied together with the physical and mechanical properties of the ore-bearing rock (scarn) and the wall rocks. Using different rock mass classification systems, a careful evaluation of rock mass is established. Back analyses are carried out. The stability of stopes and pillars is studied. Design procedures and measures to be taken are suggested.

RESUME

Des recherches en mécaniqe de roches ont été faites pour évaluer la méthode d'exploitation par chambres à sous étage, employée à la mine de Wolfram à Bursa-Uluda, du point de vue de la stabilité des chantiers de travail, des piliers et des galeries. Bienque la formation soit massive, elle montre des résistances mécaniques différentes. Ses principaux aspects géologiques ainsi que ses résistances caractéristiques ont été largement étudiées. De même les propriétés physiques et mécaniques de la roche contenant le minerai (skarn) et les roches formant le toit et le mur ont été bien étudiées. En utilisant de différents systèmes de classification des roches massives, une évaluation soigneuse des roches a été établie. Une analyse ultérieure a été effectuée. Des procédés pour la réalisation du projet et les précautions à prendre ont été suggerés.

ZUSAMMENFASSUNG

Felsmechanische Untersuchungen werden durchgeführt, um den im Wolfram-Bergbau in Bursa (Türkei) angewandten Abbau mit Teilsohlenstrecken hinsichtlich der Stabilität der Kammern, des Langfrontabbaupfeilers und der diversen Strecken auszuwerten. Obwohl das Erz in der Natur als massiver Körper auftritt, zeigt das Gebirge verschiedene Festigkeitseigenschaften. Die geologischen Hauptformen und ihre Festigkeitseigenheiten sowie die physikalischen und mechanischen Eigenschaften des erzführenden Gesteins (skarn) und des Nebengesteins wurden ausführlich untersucht. Eine sorgfältige Auswertung des Gebirges unter Verwendung verschiedener Gebirgsklassifizierungssysteme wurde unternommen und Rückanalysen wurden durchgeführt. Die Stabilität von Kammern und Pfeilern wurde untersucht. Entwurfsprozeduren und Sicherheitsmaßnahmen werden vorgeschlagen.

1. INTRODUCTION

Uludağ tungsten deposits in Bursa, Turkey are located over 2100 meters near the peak of Uludağ Mountain and have been exploited by open pit and underground mining. Primary underground production comes from sublevel open stopes of full orebody height, which will subsequently be filled to aid pillar recovery (Fig.1). At present production from three primary sublevel open stopes is completed and at the fourth one it is near to completion. The stopes and pillars between them are 15 meters in width. The height and length of the stopes are approximately 30 meters and 100 meters respectively.

The first part of continuing rock mechanics research initiated at this mine concerns with the

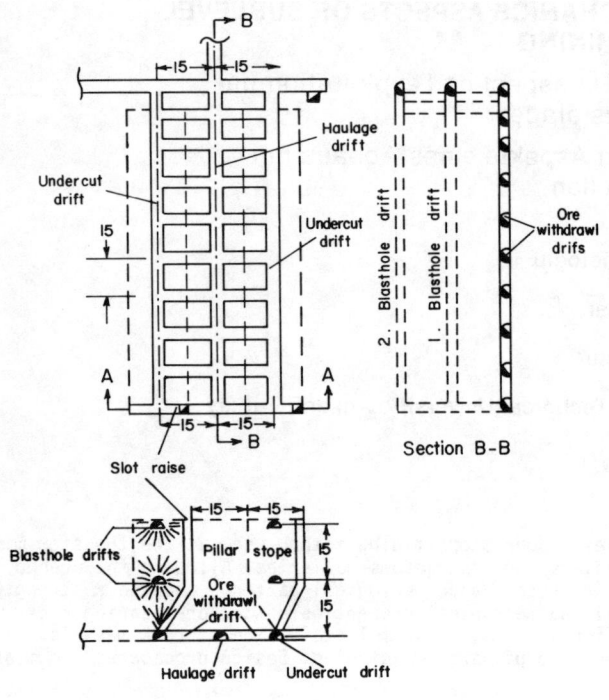

Figure 1 Sublevel stoping method as applied at Uludağ tungsten mine

re-evaluation of the sublevel open stope mining method with respect to stability of stopes, rib pillars and its various galleries and drifts. Here, outcome of first part of this extensive rock mechanics investigations will be presented.

2. GEOLOGY OF TUNGSTEN DEPOSITS

2.1 Geological setting

Uludağ tungsten mine is geologically situated in the scarn rocks between marbles of Paleozoic-aged Metamorphic Series and the younger granodioritic batholith. The Metamorphic Series which include amphibolites, gneisses and marbles at the base with mica-schists, phyllites and semi-marbles at the top, and the overlying Permian-Carboniferous limestones and clastic rocks have been folded into a gigantic anticline during the Alpine orogeny accompanied by the intrusion of a batholith (Ketin 1947, Pınar-Lahn 1954, Ronner 1954, Kaaden 1958, Öztunalı 1967). Continued tectonism and vertical uplift has subjected the Uludağ massif to intense weathering and erosion in Neogene and Recent times resulting in the exposure of granodiorite, marble and ore-bearing scarn in the vicinity of the summit region (İnan 1979).

Tunsten bearing minerals are mainly found in scarn which is formed from marbles by the contact metamorphism of granodioritc intrusion. Two types of scarn have been recognised (İnan 1979):

1. Endoscarn is formed as a thin band within the magmatic mass of the contact zone and represents metasomatically altered granodiorite ("granitic scarn").

2. Exoscarn is formed within the marbles by contact metamorphism process and is generally dark green or reddish brown (garnet rich) in color with clinopyroxene, garnet, quartz, epidote and calcite as the main constituents. It also contains scheelite, wolframite, chlorite, tremolite, pyrite, chalcopyrite, fluorite, sphalerite, magnetite, hematite, molibdenite and bismuthite (Kaaden 1954).

Marbles are white, coarsely crystalline, jointed and contain relict bedding structures. These rocks are poor in ore and tungsten-bearing scheelite mineral is only present in thin vein formations. Granodiorite is a grey, medium-grained biotite-rich rock with plagioclase (40-45%), alkali feldspar (13-15%), quartz, biotite and muscovite as the main constituents. Fine grained aplitic dykes and granodioritic porphyrites cut marbles, amphibolites and gneisses in the mine area and have been extensively altered within the scarn.

2.2 Scarn formation and mineralization

Uludağ scarn belt is formed by the intrusion of granodioritic batholith into Palaeozoic metamorphic rocks and subsequent contact metamorphism of the marbles under the control of regional structural elements. Uludağ scarns are developed at pressure-temperature conditions of albite-epidote hornfels and hornblende hornfels facies of contact metamorphism below $620^{\circ}C$ (İnan 1979). There is very little tungsten mineralization during this phase. In the hydrothermal phase following the scarn formation tungsten rich hydrothermal-pneumatolytic solutions from the igneous intrusion have acted along the previously formed structural discontinuities and deposited ore mineral scheelite and various other minerals in them. Tungsten mineral scheelite is generally found at the granodiorite-scarn, marble-scarn contacts, along discontinuities within scarn and in areas of sheared, intensely fractured rocks. Endoscarn formation is related to diffusion process during contact metamorphism (İnan 1979). Granodiorite has been extensively altered (kaolinized) as a result of hydrothermal alteration and more recent weathering.

2.3 Structure

The location of scarn sheets is controlled by the regional structural elements in the mine area. Metamorphism has been controlled by the geologic structure and as a result scarn is formed as sheets generally parallel to the bedding planes in the marbles and along fracture zones (Fig.2). Away from the granodiorite contact scarn zones thin away and disappear. Commercial scheelite mineralization is particularly developed in the brecciated contact zone and in the joints, fractures, shear zones within scarn and also along scarn-marble contacts. Scarn-granodiorite contact zone is generally sheared, crushed and brecciated due to the forcible intrusion of the batholith and the tectonism. Shear zones are also present within granodiorite, scarn and marbles. A number of these zones cross the proposed ore storage chamber presenting stability problems in the underground mine (Paşamehmetoğlu et al 1982).

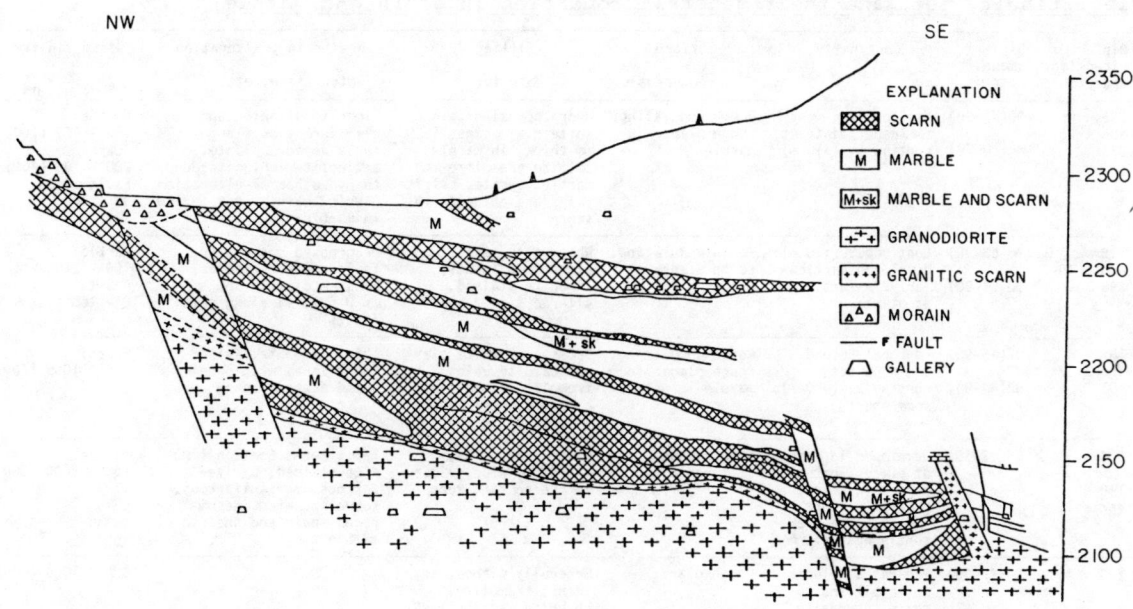

Figure.2 Uludag tungsten mine geological section in a NW - SE direction

Thickness of individual scarn sheets vary between 0 and 75 m and the total thickness of scarn reaches 250 m. Scarn sheets show a general dip of 15°-35° towards NE and the scarn-granodiorite contact dips towards SE. Relatively new NE-SW striking faults which cut the ore-bearing scarn and marbles are present in the area (Ketin 1947).

3. GEOTECHNICAL INVESTIGATIONS

3.1 Geotechnical mapping

Extensive engineering geological field and laboratory studies have been undertaken in the underground galleries excavated at various levels and the ore production chambers of Uludag tungsten mine to determine the rock mass characteristics of the ore-bearing scarn and the wall rocks (marble and granodiorite) and to re-evaluate the stability of the mine in the light of the presently used Sublevel Stoping mining method. Engineering geological investigations have been carried out in accordance with the rock mass description schemes recommended by Geological Society Working Party (1972), and ISRM (1978) and geomechanical rock mass classification systems proposed by Bieniawski(1974, 1979) and Barton et. al. (1974). Various rock material and rock mass properties such as orientation, spacing, persistence, roughness, strength, aperture, filling, seepage, number of sets of discontinuities, weathering and alteration state of the discontinuities and the rock mass, groundwater, etc. have been described in standard terms suggested by the above mentioned classification systems.

Table.I Discontinuity sets and their general properties for granodiorite

Discontinuity No	Type	Dip direction, degrees	Dip amount, degrees	Continuity	Filling material	Weathering alteration	Discontinuity spacing, mm
1	Joint	083 and ± 10 263	90(75-90)	The most dominant joint set; continuous (>5 m)	Intense quartz and quartz-pyrite veins with decomposed soil (kaolinite and quartz) zones; occasional calcite veins	Decomposed soil zones of varying thickness along some joints, intense at places	250 - 500 (300) fractured areas: 200
2	Joint	354 and ± 15 184	82(70-90) 90(70-90)	Second dominant joint set; generally continuous (>5 m)	Less frequent quartz, quartz-pyrite veins; kaolinite and quartz in altered joints	Less frequent decomposed soil zones; weathering increases towards the surface	500 - 1250 (700)
3	Joint	295 ± 7	60(50-70)	Developed at places particularly close to the contact	Occasional quartz-pyrite veins	Occasional decomposed soil along some joints	300
4	Joint	135 ± 7	65(60-70)	Locally well developed	Occasional quartz-pyrite veins		450
Shear zones	NNE-SSW striking, 1 - 20 m thick, sheared, epidotized, kaolinized weak rock and soil zones with pegmatitic quartz and pyrite veins and pockets; rich in scheelite.						

Table.II Discontinuity sets and their general properties in scarn and marble

Discontinuity No.	Type	Dip direction, degrees	Dip amount, degrees	Continuity	Surface roughness	Filling Material	Weathering, alteration water state, etc.	Discontinuity spacing, mm
1 a	Joint	093. and ± 20 273	90(75-90) 90(70-90)	Regional the most dominant joint set; continuous (>10 m)	Rough undulating; rough planar in marble	Hydrothermal-origin white clay veins, 1-20 mm thick, in marble; calcite, tremolite in marble, pyrite, calcite veins and chlorite in scarn	Limonite staining and clay formation along fault zones; epidote, actinolite, chlorite due to hydrothermal alteration. Dry in scarn, minor inflow in marble	Marble: 150 - 500 (250) Scarn: 200 - 700 (400) Fault zones: 30 - 100
1 b	Joint, fault	118 and ± 15 298	90(68-90) 90(74-90)	Continuous (>10 m); developed particularly in marble at places	Rough undulating; rough planar at places	White clay veins along some joints; limonite stained, altered soil zones	Decomposed scarn along fault and shear zones. Minor-medium inflow (< 0.5 lt/s) along some faults	Marble: 100 - 250 Faults: 0 - 100
2	Joint	181 and ± 12 001	82(65-90) 88(84-90)	Regional. Second dominant set; continuous (>10 m); some short	Rough undulating; rough planar in marble	Occasional white clay and calcite veins; tremolite in marble, chlorite, pyrite, epidote in scarn	Thin, limonite stained fault zones with decomposed/sheared scarn	Scarn: 250 - 1000 (400)
3	Joint, "Bedding" fault	000 to 110 variable	19(12-30)	Developed in marble at places; not apparent in scarn; generally short (<5 m), some are continuous (>10 m)	Smooth undulating; to rough planar	White clay veins, 1 - 40 mm wide, close to the contact zone	Fault zones contain limonite stained, pyriteferrous, manganiferrous, scheelite rich, decomposed scarn and sheared marble	Marble: 150 - 2000 (300)
4	Joint	223 ± 10	53(45-65)	Present at places in marble and scarn; generally short (<5 m)	Rough planar	Generally without any veins; tremolite, chlorite, actinolite on joint surfaces		300 - 1000 (500)
5	Joint, fault	134 and ± 6 314	90(80-90)	Continuous (>10 m)	Rough or irregular undulating	Limonite staining, clay, calcite, pyrite veins in faults	Very closely fractured marble and limonite stained, decomposed scarn	Faults: 0 - 100

3.2 Discontinuities

Dip directions and dips of discontinuities measured in various already opened galeries and chambers have been plotted on equal area Schmidt net and the dominant discontinuity sets have been determined. The results of discontinuity analyses are shown in Table I for granodiorite and Table II for scarn and marble. The most dominant joint sets in granodiorite have strikes N07W (No.1) and N84E (No.2) and are vertical or near vertical. The other two sets have average dip directions and dips of $295°/60°$ (No.3) and $135°/65°$ (No.4). Decomposed soil zones of varying thickness which are formed as a result of hydrothermal alteration and weathering, are mainly seen around quartz-pyrite veins along N07W striking joint set. These kaolinized soil zones are even seen in areas where granodiorite is least altered. Other structural elements in granodiorite are NNE-SSW striking, 1-20 m thick shear zones and NE-SW striking young faults with throws less than 30 m. Roof collapses and spalling along these shear zones during excavation and later resulted in heavy timber and steel set supporting.

Regionally developed N-S striking discontinuity set No.1 is the most dominant joint set in marble and scarn, and has two concentrations at 093 (or 273)/$90°$ (No.1a) and 118 (or 298)/ $90°$ with dips changing between $65°$ and $90°$(Table II). This set is well developed both in marble and scarn, and seen running parallel to the long axes of the ore production chambers and the main transport galleries since these were opened in approximately N-S direction in the mine. Faults are generally developed along 1b discontinuity set. White clay, calcite, pyrite and tremolite veins of hydrothermal origin, closely spaced at places, are also common in this direction. The second dominant set is an E-W striking steeply dipping joint set giving a concentration at 181/$82°$; this set cuts the production chambers and the transport galleries at right angles. Clay, calcite and pyrite veins and faults are scarcer in this direction.

The discontinuity set No.3 represents the relict bedding structure of original limestones before metamorphism. At depth this set is only locally developed in marble, but close to the ground surface and on the surface due to the solution and stress release bedding planes are well developed. The dip direction changes between 000 and $110°$ in relation to the main structure of the anticlinal fold and dips generally at $12°$ to $35°$. White clay veins are also common along the bedding planes particularly close to the contact zone. The discontinuity set No.4 is developed locally in marble, not very distinct in scarn and shows variation in dip direction and dip amount. A fifth set in marble and scarn (No.5) is present and represents NE-SW vertical or near vertical faults and fault zones. These zones are up to 6m wide, limonite stained and are composed of clay infilled, closely fractured, sheared marble and/or decomposed, crushed scarn. Vertical karstic solution holes reaching up to 3m in diameter are present along some of these faults with appreciable amount of groundwater flow. The faults have throws of 10-40m and cut granodiorite as well as scarn and marble (Figure 2).

NE-SW to N-S striking faults are rarer in the east and south-east of the mine where the present production is going on with an average spacing of 30-50m; in the centre areas, south and north

Table.III Geomechanical classification of granodiorite at Uludag tungsten mine

Rock type	Mass alteration grade	Description	Bieniawski (1979) Calculated (possible) range of RMR values	Rock mass class	Barton (1974) Calculated (possible) range of Q values	Rock mass class
Fresh granodiorite	GI	Grey, medium-grained muscovite-biolite-granodiorite, moderately widely spaced joints	73 - 95 (73 - 95)	Very good rock	15 - 50 (15 - 50)	Very good good rock
Slightly altered granodiorite	GII	Slight discoloration on the joint surfaces				
Moderately altered granodiorite	GIIIi	Occasional decomposed (kaolinized) zones. Soil <10%	42 - 62 (44 - 84)	Fair-good rock	2.5 - 7.5 (0.7 - 50)	Good-poor rock
	GIIIii	Frequent decomposed zones Soil: 10 - 50%	30 - 45 (12 - 46)	Fair-good rock	0.3 - 0.5 (0.01 - 1.5)	Very poor Extremely poor rock
Highly altered granodiorite	GIV	Decomposed soil >50%	10 - 20 (3 - 35)	Poor-very poor rock	0.03 - 0.07 (0.003 - 0.10)	Extremely poor rock
Completely altered granodiorite	GV	Completely decomposed to soil (kaolinite and quartz)	<10	Very poor rock	<0.001	Exceptionally poor rock
Metasomatically altered granodiorite (granitic scarn)	GIII -- GIV	Kaolinized, epidotized, pyriteferrous	8 - 33	Poor-very poor rock	0.005 - 0.20	Extremely poor rock
Sheared granodiorite		Sheared, kaolinized weak rock and soil mixture with closely spaced fractures	<10	Very poor rock	<0.01	Exceptionally poor rock

of the mine the spacing decreases to 20-30 m whereas along the contact region and south-west of the mine the spacing decreases to 5-15 m. Occasionally, 4-7 m thick fault zones with low dips of $20^{\circ}-35^{\circ}$ and dip directions of N-NE are present along scarn marble contacts and represent the shear zones in the marble formed during the intrusion of granodiorite. Later following the scarn formation and mineralization, these faults moved again during the continued tectonism. Another major fracture zone, 7-15 m wide, is seen south of the mine. This zone extends in an E-W direction and crosses the southern end of the ore production chambers. Towards west, near the granodiorite contact it is occupied by highly altered granodiorite "tongue".

3.3 Weathering and alteration

Granodiorite has been extensively altered by hydrothermal solutions and weathered by atmospheric agents, and as a result it has changed to soil-rock mixtures or soil in many places. Altered soil zones which are mainly composed of quartz and kaolinized feldspars, are generally developed in N-S direction. Granodiorite shows increased alteration in areas of intense quartz-pyrite veining and fault zones, and the effect of weathering increases towards the surface.

Granodiorite has been classified according to mass alteration grade(Table III)and geotechnical mapping has been carried out in terms of alteration grades (Paşamehmetoglu et. al. 1982). One of the problems encountered in the galleries excavated in granodiorite is the rapid weatherability. Short after the excavation, soil zones in contact with the atmosphere and water soften and failure occurs in the form of spalling and rock falls resulting in the enlargement of these galleries if left unsupported.

Hydrothermal solutions from the magma which deposited ore minerals along discontinuities in scarn have also altered these rocks along the fracture systems and chloritization, epidotization are the common phonemena. Marble is generally fresh but shows limonite staining, clay formation and solution channels along fault zones.

3.4 Rock mass classification

Two classification systems recently proposed, namely Bieniawski's (1973,1979) Rock Mass Rating, RMR, and Barton et. al.'s (1974) Tunnelling Quality Index, Q, have been used in this investigation because they include sufficient information to provide a realistic assessment of the factors which influence the stability of an underground excavation (Hoek and Brown 1980). RMR and Q values have been determined for selected localities in galleries excavated in granodiorite, marble and scarn, and the present support used in the mine at each locality is also noted. Various rock types encountered in the underground mine have been classified in terms of RMR and Q values (Table III and IV). RQD values which are used in calculating the RMR and Q values have been compiled and classified from the borehole records.

3.4.1 Assessment of the mine in terms of geomechanical rock classes

Fresh and slightly altered granodiorite, GI-II, are classified as very good rock according to Bieniawski (1979) classification and very good-good rock according to Barton (1974) classification. However, this sort of granodiorite is not present in the already opened galleries of the mine. The best quality granodiorite encountered in the mine is a moderately altered rock, $GIII_i$, occuring as 10-30 m thick zones with occasional kaolinized soil zones of varying thickness, 1-500 mm, and the soil amount less than 10% giving a rock mass class of fair-good rock (RMR or Q-system). This type of rock is seen along the main transport gallery at 2120 m level and is without any support. Moderately altered granodiorite containing 10-50 percent decomposed soil is classed as poor rock (RMR) or very poor rock(Q); temporary galleries, 3-4 m in diameter,

Table.IV Geomechanical classification of scarn and marble at Uludag tungsten mine

Rock Type	Description	Bieniawski (1979) Calculated (possible) range of RMR values	Rock mass class	Barton (1974) Calculated (possible) range of Q values	Rock mass class
Marble	White, coarsely crystalline, jointed marble with thin white clay, calcite and scarn veins at places	52 - 74 (52 - 77)	Good-fair rock	5.8 - 8.4 (1.3 - 35)	Good-fair rock
Fractured Marble	Limonite stained, closely spaced fractures with clay veins	33 - 52 (28 - 63)	Fair-poor rock	0.91 - 4.44 (0.12 - 7.5)	Fair-poor rock
Marble and Scarn	Marble and scarn mixtures with scarn veins; joint surfaces may be kaolinized and contain chlorite, tremolite, etc...	36 - 58 (36 - 63)	Fair-poor rock	0.75 - 7.50 (0.41 - 7.50)	Good-fair rock
Scarn	"Fresh" scarn with occasional marble blocks. Poorly pyriteferrous; joint surfaces contain chlorite, epidote, actinolite, etc...	58 - 63 (46 - 77)	Good-fair rock	60 - 15.0 (1.9 - 15.0)	Good-fair rock
Slightly Decomposed Scarn	Pyriteferrous scarn with pockets or zones of decomposed soil; calcite and pyrite-veined	32 - 44 (31 - 60)	Fair-poor rock	1.45 - 4.2 (0.37 - 7.5)	Fair-very poor rock
Marble and Decomposed Scarn	Marble with decomposed scarn veins; intensely veined marble; fault with decomposed scarn	22 - 36 (14 - 58)	Poor rock	0.30 - 0.75 (0.01 - 0.94)	Very poor - extremely poor rock
Decomposed scarn; contact scarn	Highly decomposed scarn; very closely fractured scarn; epidotized scarn; pyriteferrous scarn	8 - 20 (5 - 30)	Poor-very poor-rock	<0.2 (0.002 - 0.4)	Very poor - exceptionally poor rock
Fault and shear zones	0.5 - 7 m thick fault and shear zones with crushed, sheared marble/decomposed scarn	8 - 20 (3 - 32)	Poor-very poor rock	0.02 - 0.38 (0.002 - 0.4)	Very poor - exceptionally poor rock

in this type of rock mass are supported with very close to 2m spaced timber whereas permanent galleries, 3.5-5 m in diameter, are supported with 1.5-2 m spaced timber strengthened with 0.5-1 m spaced steel sets at places. Highly to completely altered and metasomatically altered granodiorite (endoscarn) where decomposed soil amount is over 50 percent give very low RMR, poor-very poor rock, and Q values, very poor-extremely poor rock; the galleries are supported by very closely spaced timber supports and/or steel sets. Roof collapses occur along the shear zones in granodiorite, very poor or exceptionally poor rock.

Marble with no clay veins and "fresh" scarn are classified as good-fair rock (Table IV); 2.5-4 m diameter galleries and the ore production chambers I and II (Figure 3) opened in both types of rocks have been standing unsupported, some

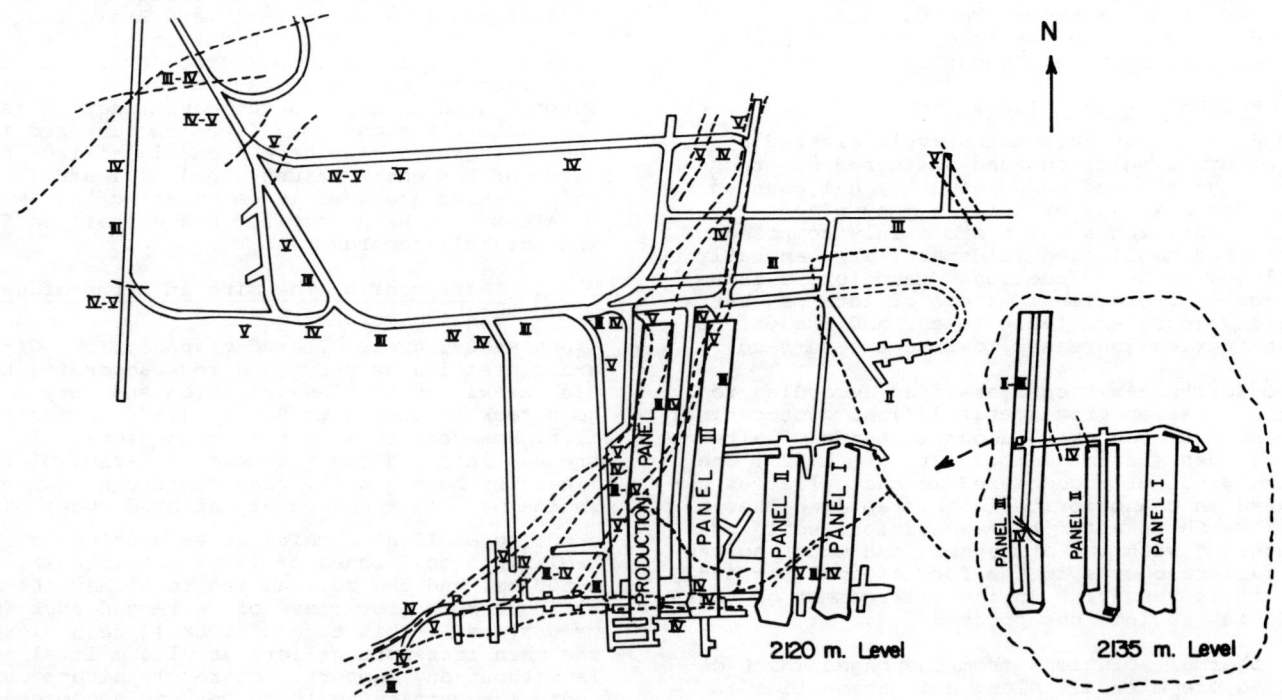

Figure.3 Geomechanics classification (Bieniawski's) map of 2120 and 2135 meter levels

since 1972. Marble with closely spaced discontinuities and clay veins, marble and scarn mixtures and slightly decomposed scarn are all classed as fair-poor rock in both classifications, and the galleries are generally unsupported except along fault zones or in areas of closely spaced clay veins. Marble with decomposed scarn zones gives lower RMR and Q values and is classified as poor rock (Bieniawski) or very poor-extremely poor rock (Barton); 1-2 m spaced timber supports used in most galleries. Highly decomposed scarn, scarn along contact zone, epidotized, pyriteferous scarn and the fault zones are in the poor-very poor rock (RMR) class or very poor-exceptionally poor rock (Q) class, the galleries in these zones are supported with heavy timber and/or closely spaced steel sets.

The following generalizations may be made in assessing the rock mass qualities in different parts of the mine, and the south-eastern part where production chambers I and II are, generally good-fair quality marble and scarn are present except at the souhtern edge of the chambers where E-W running fracture zone contains poor-very poor rock. Moving west towards the granodiorite-scarn contact fair-poor rock with NNE-SSW to NE-SW striking very poor rock zones are present at both 2120 m and 2150 m main levels. The contact zone and northern part of the mine which consist of decomposed scarn, granitic scarn, fault and shear zones are of poor-very poor quality rock. At the extreme south where epidotized, kaolinized granodiorite tongue occupies a previous fracture zone, the rocks are of very poor-extremely poor rock mass. In the granodiorite, centre and west of the mine at 2120 m level and the north, north-west of the mine at 2150 m level, consist of generally poor quality rock with zones of very poor-extremely poor rock along NE-SW extending shear belts and zones of fair rock. West of the mine poor-very poor quality rock (granodiorite and granitic scarn) is overlain by poor-fair quality marble and scarn with very poor quality altered scarn and fracture zones.

A final note in the application of the geomechanical rock mass classifications to Uludag tungsten mine is that the long and short axes of the ore production chambers and the majority of the underground galleries are opened in NNE-SSW and E-W directions, i.e. either in the direction of or slightly oblique to the dominant discontinuities and the fault zones. If the main transport galleries and the long axes of the production chambers were opened in a NW-SE direction, i.e. oblique to the joint set No.1, the geomechanical values of the rock mass would have improved significantly in each case (see Tables III and IV for possible range of values of each classification).

4. STABILITY ANALYSIS FOR THE GALLERIES

Since, load-haul-dump cars are planned to be used by mine management for ore-withdrawal, the size of the some of the main galleries and future ore-withdrawal and service drifts is to be increased to 14 m^2 cross-section. To estimate the support requirements for them, back analysis for the existing galleries at various sections of the mine is carried out. That is, support system proposed by Bieniawski's (1979) and Barton et. al.'s (1974) rock mass classification systems are compared with the applied supports.

Under the light of this comparison, it is found that Bieniawski's classification is rather conservative for this case. This is not surprising, considering that Bieniawski (1980) himself stated the need for some modifications to rock mass geomechanic classification for mining tunnels. On the other hand, the support system

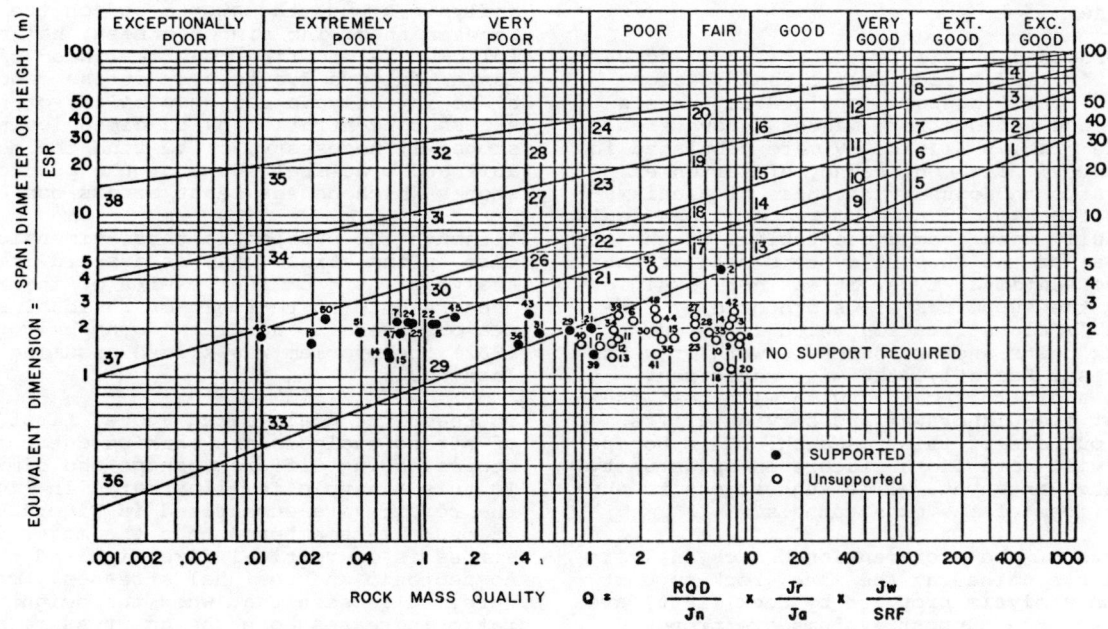

Figure.4 Case records incorporated to Barton's excavation support chart

Table.V Actual supports used and support recommendations for tunnels in granodiorite

Actual Support Used and Basis for Support Recommendation	Slightly Altered Granodiorite	Moderately Altered Granodiorite	Highly Altered Granodiorite	Completely Altered Granodiorite	Metasomatically Altered Granodiorite (Granitic Scarn)	Sheared Granodiorite
Bieniawski's Geomechanics Classification (Bieniawski, 1979)	Systematic bolts 3 m long, spaced 1.5 - 2 m in crown and walls with mesh in crown and shotcrete in crown 50 - 100 mm and in sidewalls 30 mm thick	Systematic bolts 3 - 4 m long, spaced 1 - 1.5 m with wire mesh and shotcrete in crown 100 - 150 mm, in sidewalls 100 mm and light ribs spaced 1.5 m where required	Systematic bolts 4 - 5 m long, spaced 1 - 1.5 m with wire mesh and shotcrete in crown 150 - 200 mm, in sidewalls 150 mm and on face 50 mm thick and medium to heavy ribs spaced 0.75 m with steel lagging and forepoling if required	Same as "Highly Altered Granodiorite"	Same as "Highly Altered Granodiorite"	Same as "Highly Altered Granodiorite"
Barton, Lien and Lunde's Tunnelling Quality Index (Barten et al. 1974)	No support	Untensioned grouted bolts at 1 m centres and 50 mm shotcrete with mesh	Tensioned bolts at 1 m centres and 50 - 100 mm shotcrete with mesh	Out of Barton's support recommendation range	Tensioned rock bolts at 1 m centres and 50 - 100 mm shotcrete with mesh	Tensioned rock bolts at 0.5 - 1 m centres and 100 - 200 mm shotcrete with mesh
Actual support used	Unsupported	Temporary roadways: timber supports at 0 - 2 m centres and/or steel ribs at 1 - 2 m centres. Main roads: steel ribs usually at 1.5 - 2 m centres, occasionally 0.5 - 1 m centres	Temporary roadways: timber supports at 0 - 1 m centres and/or steel ribs at 0.5 - 2 m centres. Main roads: steel ribs at 0.3 - 2 m centres (sometimes with timber supports)	Main road: Concrete or thick shotcrete	Steel ribs at 0.25 - 0.5 m centres	Temporary roadways: closely spaced timber supports at 0 - 2 m centres. Main roads: closely spaced steel ribs
Rock Support Interaction Analysis (Hoek, 1980)	No support (19 mm, 1 - 1.5 long grouted bolts at 2 m centres if required)	25 mm, 3 m long bolts at 1.2 m centres with mesh or 25 mm shotcrete or light section ribs at 1.5 - 2 m centres with mesh	50 mm shotcrete and 25 mm, 4 m long resin tensioned, grouted bolts at 1 m centres with mesh or medium section ribs at 1.5 m centres with mesh	75 - 100 mm shotcrete and 25 mm, 4 m long grouted bolts at 1 m centres with mesh or medium section steel ribs at 0.5 - 0.75 m centres with mesh	50 mm shotcrete and 25 mm, 4 m long resin tensioned, grouted bolts at 1 m centres with mesh or medium section steel ribs at 1.5 m centres with mesh	75 - 100 mm shotcrete and 25 mm, 4 m long resin grouted bolts at 1 m centres with mesh or medium section steel ribs at 0.5 - 0.75 m centres with mesh

proposed by Barton et. al.'s Q-index is more consistent with the applied support systems, especially for unsupported sections, This fact could be seen clearly from Fig.4. Out of 28 unsupported cases only one fell within the "support required" zone whereas out of 20 supported cases only two cases fell within "no support required" zone.

Supports proposed by Bieniawski's (1979) and Barton et. al.'s (1974) rock mass classification systems and actual supports applied at tunnels driven through various rock masses encountered in the mine (Tables III and IV) are tabulated in Tables V and VI. The consistency of Barton et. al.'s Q-system and conservativeness of Bieniawski's classification to unsupported sections could also be easily seen from these tables. On the other hand, since the support systems used at the mine are conventional type, it was not possible to compare the support systems proposed by Barton et. al.'s classification which is based on mainly rock bolts and/or shotcrete with or without wire mesh. Recently, the mine management started to apply rockbolting with wire mesh especially at ore-withdrawal and service drifts after the outcome of this research. It is hoped that this will give in the future an opportunity to check the Barton et. al.'s suggestions to supported sections of the mine tunnels.

In the ligth of the aforementioned back analysis and experience gained at the mine, rock-support interaction analysis proposed by Hoek (1980) are conducted and the support systems are proposed (Tables V and VI). Together with support recommendations, monitoring of deformations around galleries are also suggested.

5. STRESS ANALYSIS

Finite element method is used for the stress analysis around the stopes and on the pillars between them. The aim of stress analysis was as follows: a) to find the likelihood of rock material failure due to over stress concentration; b) to investigate the possibility of reducing stress concentration by re-dimensioning of stopes if necessary; c) to find the most suitable and practical shape of the production stopes which causes least stress concentration.

Although the problem is three-dimensional in nature, plane strain case is assumed in stress analysis. As from the results of insitu stress measurements no tectonic or residual stresses are observed in the area (Paşamehmetoglu et. al. 1982) the problem was solved assuming gravity loading.

To reach the goals cited above, first, a number of stress analysis are carried out for single stopes having different shape and dimensions. To this a sample is illustrated in Figure.5 and, the results are summarized in Figure.6. Here, it must be remembered that the major insitu stress is in vertical direction and there were no tectonic and residual stresses. From the figure, it is seen that when the height to width ratio increases both the compressive and tensile stress concentrations decrease. Consequences of this means that for a given width as the height

Table.VI Actual supports used and support recommendations for tunnels in marble and scarn

Actual Support used and Basis for Support Recommendation	Marble or Scarn	Fractured Marble or Marble and Scarn	Slightly Decomposed Scarn	Marble and Decomposed Scarn	Decomposed Scarn, Contact Scarn	Fault and Shear Zones
Bieniawski's Geomechanics Classification (Bieniawski, 1979)	Locally bolts in crown 3 m long, spaced 2.5 m with occasional mesh and 50 mm shotcrete in crown where required	Systemetic bolts 3 m long, spaced 1.5 - 2 m in crown and walls with mesh in crown and shotcrete in crown 50 - 100 mm and in sidewalls 30 mm thick	Systemetic bolts 3 m long, spaced 1.5 m in crown and walls with mesh in crown and shotcrete in crown 50 - 100 mm and in sidewalls 30 mm thick	Systematic bolts 3 - 4 m long, spaced 1 - 1.5 m with wire mesh and shotcrete in crown 100-150 mm, in sidewalls 100 mm thick and light ribs spaced 1.5 m where required	Systematic bolts 4 - 5 m long, spaced 1 - 1.5 m with wire mesh and shotcrete in crown 150 - 200 mm, in sidewalls 150 mm and on face 50 mm thick and medium to heavy ribs spaced 0.75 m with steel lagging and forepoling if required	Same as "Decomposed Scarn"
Barton, Lien and Lunde's Tunnelling Quality Index Barton et al.,1974)	No support	No support (At weaker zones or main roadways untensioned grouted bolts at 1 m centres or 25 - 30 mm shotcrete if required)	No support	Untensioned grouted bolts at 1 m centres and 50 mm shotcrete with mesh	Tensioned rock bolts at 1 m centres and 50 - 100 mm shotcrete with mesh	Tensioned rock bolts at 1 m centres and 50 - 100 mm shotcrete with mesh
Actual support used	Unsupported	Unsupported (Fault zones and zones with closely spaced fractures and thick clay veins are timber supported at 1 - 1.5 m centres)	Unsupported (Fault and highly decomposed zones are supported with timber ribs at 1 - 1.5 m centres)	Timber supports at 1 - 2 m centres	Temporary roadways: Usually timber supports at 1 - 2 m centres (occasionally very tightly spaced at 0 - 1 m centres). Main roadways: Steel ribs at 1 - 2 m centres with mesh	Timber supports at 0.5 - 2 m centres
Rock support Interaction Analysis (Hoek, 1980)	No support	No support (19 mm, 1 - 1.5 m long grouted bolts at 1.5 m centres if required)	No support (19 mm, 1 - 1.5 m long grouted bolts at 1.5 m centres if required)	25 mm, 3 m long bolts at 1 m centres with mesh or 25 mm shotcrete or light ribs at 1.5 - 2 m centres with.mesh	50 mm shotcrete and 25 mm, 4 m long resin tensioned, grouted bolts at 1 m centres with mesh or medium section steel ribs at 1.5 m centres with mesh	75 - 100 mm shotcrete and 25 mm, 4 m long grouted bolts at 1 m centres with mesh or medium section steel ribs at 0.5 - 0.75 m centres with mesh

of the stope increases stress concentration would be better as far as the stability is concerned.

Secondly, a series of stress analysis are conducted to evaluate the stability of rib pillars and stopes, a sample of which is given in Fig.7. From this figure major and minor principal stress concentrations are seen. The shaded areas show the tensile stress regions. The summary of these analysis is presented in Figure 8 in terms of average pillar stress at the

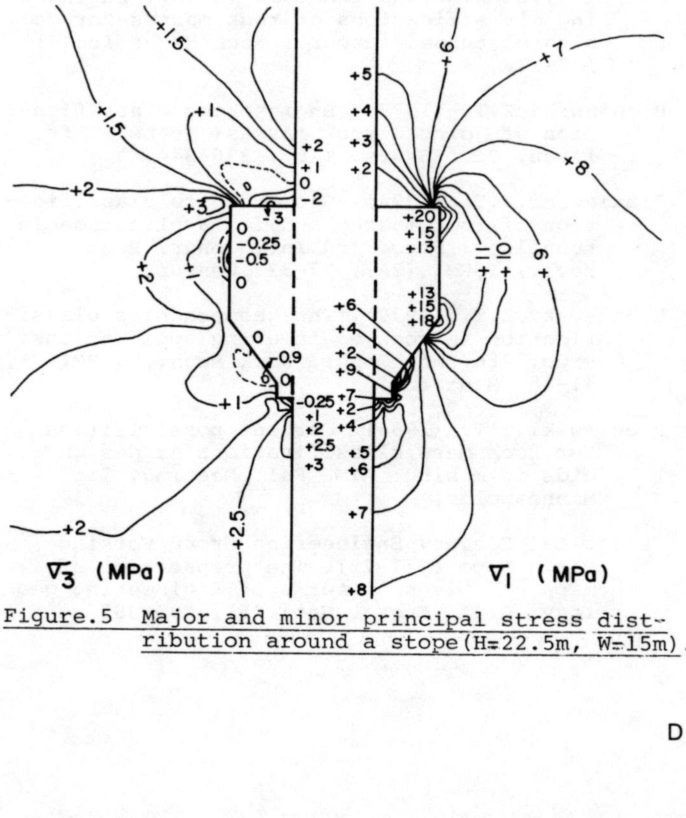

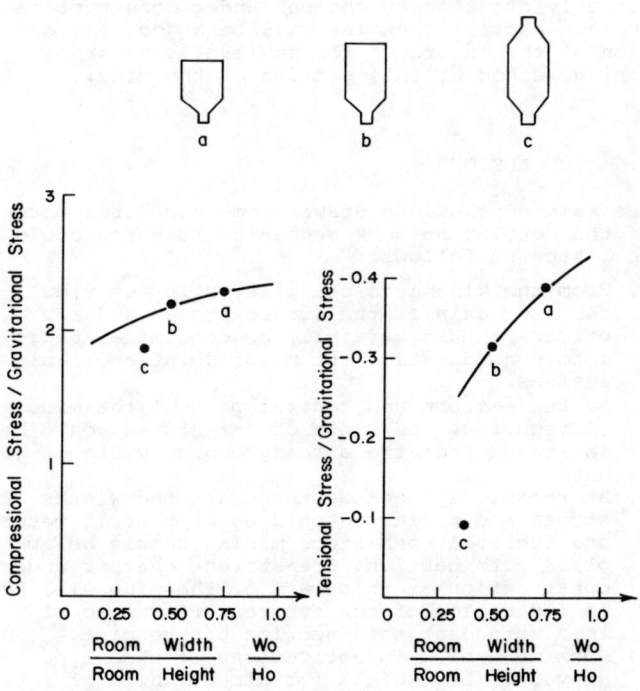

Figure.5 Major and minor principal stress distribution around a stope(H=22.5m, W=15m).

Figure.6 Compressional and tensional stress concentrations w.r. to room/pillar width.

D 159

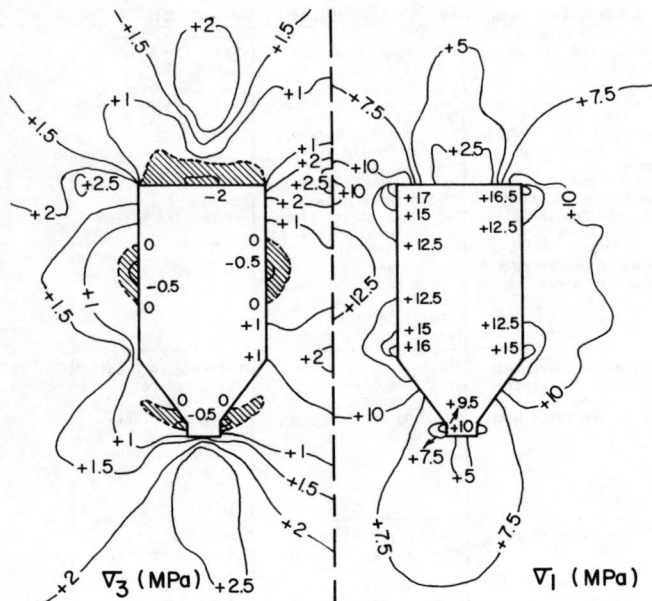

∇_3 (MPa) ∇_1 (MPa)

Figure.7 Major and minor principal stress distribution around stopes and pillar between them.

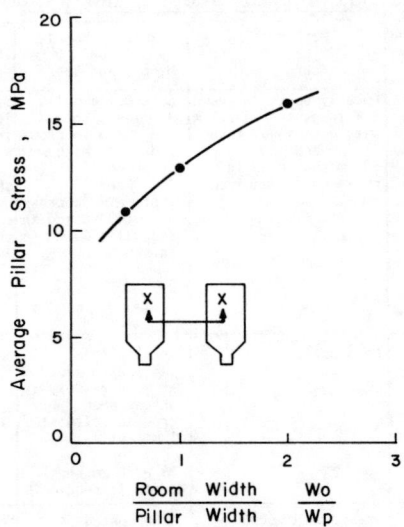

Figure.8 Average pillar stress against room/pillar width.

mid-height of the pillar as a function of stope to pillar width. When these stress concentrations and the strength of the orebody are considered, it would be said that, if it is desired, the width of the stopes could be easily increased at the east and central parts of the mine, from the already chosen width of 15 meters provided that stope width to pillar width ratio is not more than two. Whereas at the weak west part of the mine already chosen pillar and stope widths of 15 meters each could be applied with caution. Actually ore-storage chamber under construction at this part of the mine will be a good indication of the future of the sublevel open stope mining method at this section of the mine.

6. CONCLUSIONS

The main conclusions drawn from this first phase of the continuing rock mechanics research could be listed as follows:

1. From the kinematic stability point of view the long axis of the future stopes and rib pillars should carefully be determined taking into consideration the major joint and fault systems.
2. At the eastern and central part of the mine, if required, the width of the stopes could be increased from the already chosen width of 15 meters.
3. At relatively weak western part the widths of stopes and pillars should be kept at 15 meters and sublevel open stope mining should be applied with caution. Ore-storage chamber under construction at this part of the mine will be an indication of the future performance of this method at this section of the mine.
4. Rock mechanics investigations of this type proved to be usefull for dimensioning of stopes and pillars and for assessment of the stability of mines. If studies such as described in this paper are increased the knowledge thus accumulated will contribute significantly to operational planning and the design of future mines.

7. ACKNOWLEDGEMENTS

The authors express appreciation to ETİBANK for financial and practical support given to this project and for their permission to publish this paper. Any opinions expressed are those of the authors and not necessarily those of ETİBANK.

8. REFERENCES

Barton,N.,Lien,R.,and Lunde,J. (1974). Engineering classifications of rock masses for design of tunnel support: Rock Mechanics(6), 189-236.

Bieniawski,Z.T. (1973). Engineering classification of jointed rock masses: Trans.S.Afr. Instu. Civ. Engrs. (15), 335-344.

Bieniawski,Z.T. (1974). Geomechanics classification of rock masses and its application in tunneling: Proc. 3rd Int. Congr. Rock Mech., ISRM, (2A), 27-32, Denver.

Bieniawski,Z.T. (1979). The geomechanics classification in rock engineering applications: Proc. 4th Int. Congr. Rock Mech., ISRM,(2), 41-48, Montreux.

Bieniawski,Z.T. (1980). Current possibilities for rock mass classifications as design aids in mining: AIME Fall Meeting, 7pp, Minneapolis.

Geological Society Engineering Group Working Party Report (1972). The preparation of maps and plans in terms of Engineering geology: Q.Jl. Engng. Geol.(5), 295-382.

Hoek,E. (1980). An emprical strength criterion and its use in designing slopes and tunnels in heavily jointed weathered rock: Proc. 6th Souhteast Asian Confr. Soil Mech.(2), 111-158, Taipeh.

Hoek,E. and Brown,E.T. (1980). Underground excavations in rock. 572pp. London: The Institution of Mining and Metallurgy.

İnan,K. (1979). Petrogenesis and geochemistry of Uludag scarn belt (in Turkish): I.T.U. Faculty of Mining Engng.,Istanbul.

International Society for Rock Mechanics Commission on Standardization of Laboratory and Field Tests (1978). Suggested methods for the quantitative description of discontinuities in rock masses: Int. J. Rock Mech. Min. Sci. and Geomech. Abstr. (15), 319-368.

Kaaden,G.V. (1954). Geologic report on Uludag scheelite deposit (in Turkish): M.T.A. Institute (2202), Ankara.

Kaaden, G.V. (1958). On the genesis and mineralization of the tungsten deposit, Uludag, Province of Bursa, Turkey: Bull. Min. Res. Expl. Inst. Turkey (50), 33-42.

Ketin,I. (1947). About the tectonics of Uludag massif(in Turkish): Istalbul Geological Society Bull.(1).

Öztunalı,O.(1967). Petrology and geochronology of Uludag (Northwest Anatolia) and Egrigoz (West Anatolia) massifs (in Turkish): I.U. Faculty of Science, Istanbul.

Paşamehmetoglu,A.G.,Irfan,T.Y.,Bilgin,A., and others (1982). A rock mechanics investigation for the stability of an underground ore storage chamber: will appear in Proc. ISRM Sym. Rock Mechanics Related to Caverns and Pressure Shafts, Aachen.

Pınar,N., and Lahn,E. (1954) La position tectonique de l'anatolie dans le systeem orogenique: 19. Congr. Geol. Inst. Compt. Rend. (15), 18, 172-180, Algeria.

Ronner,F. (1954). Geologie und tektonik der wolframlagerstaette Uludag (Bursa) und depennaehemen umgebung: M.T.A. Institute /2203), Ankara.

CONSTRUCTION OF ADJACENT TUNNEL TUBES BY N.A.T.M.
Construction de tunnels adjacents
Ausführung eng nebeneinander liegender Tunnel

E. Gartung
Chief geotechnical engineer, LGA Bayern, Nuremberg, FR Germany
P. Bauernfeind
Managing director, Subway construction dept. City of Nuremberg, FR Germany

SYNOPSIS

In the soft sedmientary rocks of Nuremberg in Germany the New Austrian Tunnelling Method has proved to be very economical. So in the case of the subway station Schweinau it was employed in lieu of the originally designed open cut and cover construction method, because it turned out to be less costly. The station consists of twin tunnels spaced so closely, that a pillar of rock of only 36 cm was left between the two large underground openings of 8.5 m width. The excavation could be carried out safely according to detailed specifications for the construction sequence, based on parametric structural analyses and monitored by a measurement program in the field.

RESUME

La construction de tunnels par la Nouvelle Méthode Autrichienne est très économique dans les roches tendres de Nuremberg. On l'a appliquée lors de la réalisation d'une excavation ouverte pour la station de Schweinau qui consiste en deux tunnels adjacents chacun de 8,5 m de diamètre. Ces deux tunnels sont si rapprochés l'un de l'autre qu'il ne reste qu'un pilier de roche de 36 cm d'épaisseur entre eux. Les analyses des paramètres structuraux obtenus lors des programmes de terrain en ont permis l'excavation dans de bonnes conditions de sécurité en suivant les recommandations sur l'ordre d'exécution des travaux.

ZUSAMMENFASSUNG

Die Neue Österreichische Tunnelbauweise hat sich in den Sedimentgesteinen geringer Festigkeit in Nürnberg als sehr kostengünstig erwiesen. Daher wurde sie beim Bau des U-Bahnhofs Schweinau anstelle der offenen Tunnelbauweise angewandt. Der Bahnhof besteht aus zwei 8,5 m weiten Tunnelröhren, die durch einen Felspfeiler von 36 cm Dicke getrennt sind. Parametrische Standsicherheitsberechnungen, Messungen und sorgfältige Vorbereitungen ermöglichen eine sichere Ausführung.

1. INTRODUCTION

Most of the stations of the subway system in the City of Nuremberg in West Germany were constructed in open pits according to the cut and cover method. So far, only 2 stations had been excavated underground by the application of the New Austrian Tunneling Method. In one case, this had been necessary, because downtown traffic and city life did not permit the excavation of a large open pit. In the other case, the station was situated below a freeway which could not be relocated, so underground construction became imperative because of traffic requirements. Both stations consisted of twin tunnel tubes 8,5 m wide with a pillar of rock of 2,5 m in between. They were constructed in relatively soft Keuper sandstone under shallow rock cover, supported by shotcrete during the stages of construction. A water tight inner reinforced concrete shell was installed as permanent lining.

Recently, the third subway station, Nuremberg - Schweinau, was executed according to the New Austrian Tunneling Method.

In this case, open cut and cover construction had been proposed by the City authorities and was the basis for the design of the track and platform alignments. However, with several years of experience in tunneling in soft Keuper sandstone, the contractors submitted the alternative proposal to execute the construction of the station by underground excavation. Since this alternative proposal was offered at much lower cost than the open pit, the City of Nuremberg accepted it. Apart from the savings, there was the benefit of gaining experience with tunneling under extreme boundary conditions. This aspect is very important for the future, because it becomes more and more difficult to open large construction sites in cities. Since in Nuremberg several subway stations will have to be built in the same geological strata as the one reported here, they will most likely be designed for execution according to the New Austrian Tunneling Method, which turned out to be more economical and much more favourable under environmental and traffic aspects, than open cut and cover construction.

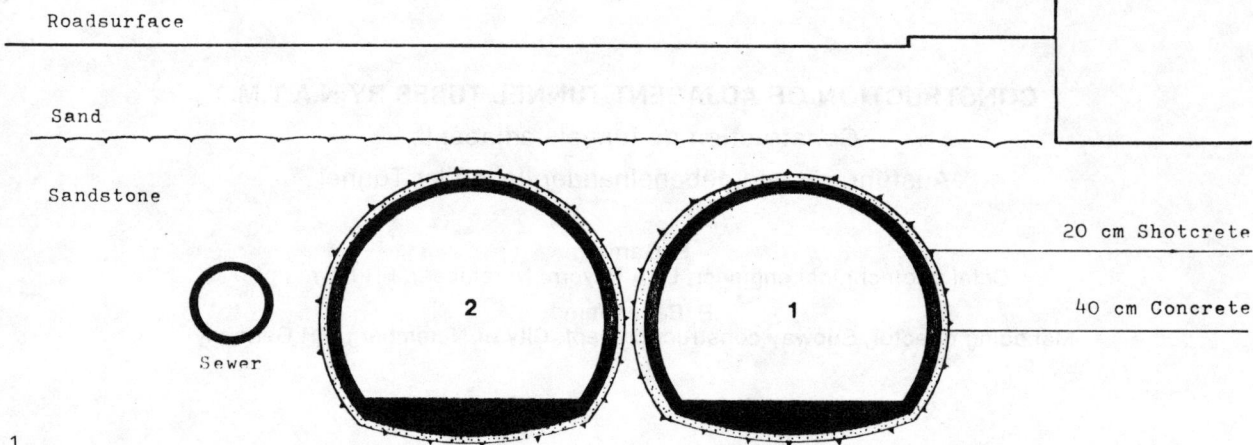

fig. 1
Cross section of the subway station Nuremberg - Schweinau

2. DESIGN OF SUBWAY STATION

The tracks and the platforms had been designed under the assumption that the subway station would have been constructed in a large open excavation. So, they were placed close to the ground surface, and a center spacing of 11,5 m was chosen for the tracks. When the underground construction method was selected, this configuration of tracks and platforms, which could not be modified any more, lead to extreme boundary conditions. The rock cover above the crown of the tunnels became very thin, and the twin tubes almost touched each other. These circumstances called for stability analyses and very careful preparations of the construction work.

The ground exploration prior to the design had indicated, that the rock cover above the tunnels consisted of 0,6 to 1,6 m of sandstone below 2,0 m of silty, slightly cemented sand in the weathered zone. The properties of the soft Keuper sandstone had been studied before in detail and described by Smoltczyk and Gartung.

Since no water had to be anticipated in the crown area of the tunnels, and the geological exploration had shown, that the sandstone had relatively uniform strength properties and was essentially free from dangerous discontinuities, the thin cover of competent sandstone did not involve too much of a risk. In order to control the stability of the crown area, the following measures seemed to be adequate.

- Bench following the top heading of the excavation at a short distance (about 3 m)
- Preserving the rock strength by excavation with heading machines
- Small advances (maximum 1,2 m)
- Light steel ribs (at 0,8 m centers)
- Immediate sealing of rock surfaces by shotcrete
- Early placement of the shotcrete shell
- Particular caution of the personnel at the site

The distance between the tunnel sections was so small that the remaining rock pillar had a theoretical minimum thickness of 24 to 36 cm. Obviously, it was necessary to install some kind of structural members, because the thin rock pillar alone could not carry any load. For this purpose, the contractor proposed to first excavate one tunnel and complete it with the installation of the inner reinforced concrete shell, which would then support the opening during the subsequent excavation of the adjacent tunnel. There was no doubt, that this concept would provide adequate safety during the construction of the tunnels. However, it was difficult to determine the stresses in the first, relatively stiff inner concrete shell during the phase of excavation of the second underground opening. So the dimensioning of the steel reinforcement of the concrete shells appeared to be a problem, and more importantly, it was not sure, whether the formation of cracks and leaks could safely be excluded and the required watertightness of the concrete shell could be warrented.

Therefore, the designers were contemplating, whether it would be possible, to avoid stressing of the inner concrete shell due to the excavation process. The aim was, to reinforce the rock in such a way, that, in spite of the thin rock cover and in spite of the weakening of the rock between the tunnel sections due to excavation, the strength of the rock could be improved such, that the loads due to the excavation process could be assigned to the rock and its reinforcement. Following the principles of the New Austrian Tunneling Method, a concept was developed to advance both tunnels simultaneously, with a lag between both in longitudinal direction, providing support by shotcrete, steel reinforcement and rock bolts. So, both tunnels could be excavated along their full length and the watertight inner concrete shell be placed afterwards. Besides the fact, that the inner concrete shells would not be stressed due to excavation, this construction sequence had the advantages of being less time consuming than that proposed by the contractor and clearly distinguishing between the processes of tunnel excavation and subsequent concreting. This construction method was more economical, but it required very cautious execution by experienced personnel.

3. EVALUATION OF STABILITY

Finite element analyses were carried out for the evaluation of the load bearing behavior of the rock and the support system. Since a three dimensional study would have become too expensive in this case, plane strain analyses for simplified conditions were performed. The finite element mesh with 1500 degrees of freedom included a sewer section next to the tunnels which had to be constructed just prior to the excavation of the tunnels. The sequence of analysis was as follows:

Step 1: Determination of initial state of stress

Step 2: Excavation of sewer section without support

There were no stability problems with the small unlined sewer tunnel.

Step 3: Starting from step 2, the excavation of tunnel section 2 was numerically simulated, taking the concrete liner of the sewer and the shotcrete of the excavated tunnel section into account

Step 4: Starting from the results of step 3, the excavation of tunnel section 1 was simulated, taking the shotcrete shells of both tunnels into account

This analysis applied to construction conditions where the distance between the faces of the first and second tube would be so large, that the stress rearrangements due to the excavation of the first tunnel section would have been completed before the second excavation would reach the analysed section. Under these conditions the shotcrete shell of the first excavated tunnel experienced larger stresses than that one of the section excavated subsequently.

In order to determine the influence of the distance between both tunnel faces in the stresses of the support system, the hypothetical case was analysed, that both tunnel faces were excavated simultaneously in the same plane section.

Step 5: Starting from the results of step 2, the excavation of both tunnels was simulated taking both shotcrete shells into account.

It was noticed, that the distribution of normal forces and bending moments in both shotcrete shells was symmetrical with respect to the center line between both tunnels and that the maximum stresses were smaller than according to step 4 of the analysis. So it seemed to be advisable for the execution of the job, to keep the distance between the two tunnel faces as small as possible.

The results of the analyses until step 5 indicated, that the stresses in the rock and in the shotcrete shells did not reach the material strength. However, it was assumed in steps 4 and 5, that the narrow rock pillar between the twin tubes remained intact and provided confinement to the shotcrete. Since the validity of this assumption seemed not sufficiently secure, the effect of possible yielding of the thin rock pillar was studied by another step of the finite element analysis.

Step 6: Starting from step 3 in which the first tunnel section had been excavated, the rock pillar between both tunnels was eliminated over a height of 2 m before the second tunnel was excavated

Step 7: Analysis of step 6 was repeated with a height of the yielding part of the rock pillar of 4 m

The analysis showed, that it was possible to reinforce the shotcrete shell sufficiently to cope with yielding of the rock pillar over a height of 2 m. But if it would yield over 4 m, the system would no longer remain stable.

The sequence of steps of the parametric analysis elucidated the relevant factors with respect to the stability of the twin tube system during construction. Consequently it was advisable to keep the distance between both tunnel faces small, and it was crucial to maintain the rock strength particularly in the zone between both tunnels by very smooth and careful excavation techniques. Furthermore, it was important that the shotcrete was placed quickly, and gained early sufficient strength to resist the relatively high compressive stresses. The thickness of the shotcrete shell of the first excavated tunnel was increased to 30 cm next to the adjacent opening and reinforced for bending moments which might be associated with possible local yielding of the thin rock pillar.

4. CONSTRUCTION OF THE TUNNELS

The twin tunnels were excavated simultaneously with a distance between the tunnel faces of 6 m to 10 m. So the shotcrete of the tube which was constructed first, had at least the age of 7 days and sufficient strength for the critical condition, when the second tunnel was excavated, and the confinement of the first shotcrete shell was reduced locally for a limited time.

The excavation was done by heading machines using the heading and bench method with a distance of 3 to 4 rib spacings of 80 cm between the faces of heading and bench. The maximum unsupported excavated length was 1,2 m. Special attention was given to cutting a very precise tunnel profile. In order to avoid any damage to the shotcrete shell, the last 30 cm next to the shotcrete shell were excavated using hand tools.

Immediately after the excavation, all fresh rock surfaces were sealed with 3 cm of shotcrete. Then the steel ribs were installed, and the steel mesh reinforcement was placed with adequate overlays. The shotcrete shell which had a regular thickness of 20 cm was executed with a thickness of 30 cm in the first excavated tube at the side, where the adjacent tunnel was to follow soon.

Rock zones around the tunnels with concentrations of deviatoric stresses were reinforced by steel rock bolts of 3 m length and 22 mm diameter. Where both tunnel tubes were almost touching each other, the thin rock pillar was confined additionally by prestressed short steel bolts.

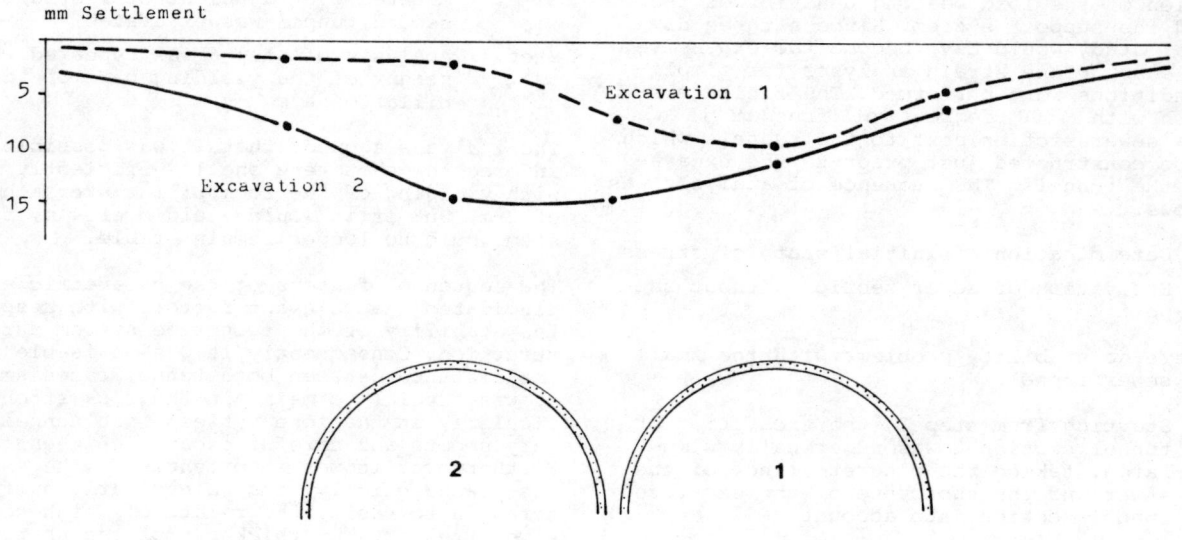

fig. 2   Settlement profile

While the first advancing tunnel could be excavated from the starting pit to the open terminating pit without interruption, the excavation of the second tunnel was stopped 10 m before the terminating pit was reached, and the last 10 m were advanced in the inverse direction from the terminating towards the starting pit.

5. MEASUREMENTS

The stability of the underground openings was continuously checked by measurements during construction. The geotechnical analyses had indicated, that the shotcrete shell of the first advancing tunnel was most unfavourably stressed when the adjacent opening was excavated, and before it was supported by shotcrete. So it had to be proved by measurements, that the safety margin was sufficient during this stage of construction.

For this purpose, the deformations of the tunnel cross section were measured, and the stresses in the shotcrete shell were monitored by Gloetzl-pressure cells. The deformation measurements did not indicate any development of critical conditions, and the stress readings showed the anticipated magnitudes.

The pressure cells which recorded the contact pressure between the rock and the shotcrete shell showed stable values shortly after installation and the advance of the first tunnel. They indicated a considerable increase in contact pressure of about 23 %, when the adjacent tunnel excavation reached the monitored cross section.

A sharp increase in tangential stresses in the shotcrete shell of the first advanced tunnel was observed, when the second excavation reached the monitored cross section. The increase in compression amounted to 57 to 61 % at the side wall of the adjacent opening, 26 to 49 % at the crown and 56 % at the spring line of the other side of the first excavated tunnel. The compressive stresses observed in the shotcrete shell of the tunnel which was excavated second, agreed well with those measured in the same cross section in the tunnel excavated first before the second excavation reached the monitored cross section. These observations were qualitatively in agreement with the plane strain finite element analyses.

The observed vertical displacements at the ground surface clearly showed the influence of the construction sequence.

6. CONCLUSIONS

Stability analyses, careful preparations and cautious excavation of the tunnels controlled by adequate simple measurements facilitated the safe construction of adjacent subway station tunnels in a sandstone under difficult boundary conditions.

REFERENCES

Smoltczyk, U., Gartung, E.: Geotechnical properties of a soft Keuper sandstone, Proc. 4th Int. Conf. ISRM Montreux, 639 - 644.

EXPLORATION, DESIGN AND EXCAVATION OF THE POWERHOUS CAVERN ESTANGENTO SALLENTE IN SPAIN

Exploration, dimensions et excavation de la galerie de la centrale souterraine de Estangento Sallente en Espagne

Versuchsprogramm, Entwurf und Bau der Maschinenkaverne Estangento Sallente in Spanien

Walter Wittke
Prof. Dr.-Ing. W. Wittke, Beratende Ingenieure für Grundbau und Felsbau GmbH, Aachen—Germany

Jose Luis Soria
Director de Obra, PSP Estangento Sallente FECSA-OPTO. Construccion y Produccion Hidraulica y Termica, Barcelona—Spain

SYNOPSIS

The pumped storage plant Estangento Sallente, which at present is under construction, is located in the Pyrenees in the province of Cataluna in Spain. The main structure is the 20 m wide, 38 m high and 90 m long powerhouse cavern, which is located in schist. The overburden amounts to 100 m. An extensive exploration and a rock mechanical testing program as well as stability analyses using the Finite Element Method have been carried out. The selected support of the vault consists of reinforced shotcrete and bolting and additional prestressed anchors in places. Parallel to the excavation of the vault an extensive monitoring program was performed. A result of the geological mapping of the vault is the unexpected local occurrence of discontinuities running parallel to the cavern axis and revealing additional stability analyses of the cavern walls.

RESUME

La centrale de pompage Estangento Sallente, actuellement en construction, est située dans les Pyrénées, dans la province de Catalogne, en Espagne. La caverne de l'usine souterraine est la structure principale et a une largeur de 20 m, une hauteur d'environ 38 m et une longueur d'environ 90 m. Elle se trouve dans des schistes cristallins et a un terrain de recouvrement d'une épaisseur d'environ 100 m. Un important programme d'investigations a été exécuté pour la reconnaissance des lieux des points de vue de la géologie et de la mécanique des roches. La stabilité de la cavité a été réalisée au moyen de la méthode des éléments finis et a conduit au blindage de la calotte à l'aide de béton armé projeté et de boulons d'ancrage. De plus, des tirants précontraints ont été prévus localement. Pour le contrôle de stabilité, des mesures des déformations, des contraintes et des forces d'ancrage ont été prises parallèlement au creusement de la calotte. La cartographie géologique de la surface excavée montre l'apparition locale inattendue de surfaces de discontinuité se développant parallèlement à l'axe de la caverne. Pour cette raison des analyses de stabilité complémentaires des murs de la caverne se sont avérées nécessaires.

ZUSAMMENFASSUNG

Das im Bau befindliche Pumpspeicherwerk Estangento Sallente liegt in der spanischen Provinz Catalunien in den Pyrenäen. Das Hauptbauwerk is die 20 m breite, 38 m hohe und 90 m lange Maschinenkaverne, die in kristallinen Schiefern liegt und eine Überdeckung von 100 m besitzt. Zur Erkundung der geologischen und felsmechanischen Verhältnisse wurde ein umfangreiches Untersuchungsprogramm durchgeführt. Die Standsicherheit des Hohlraums wurde nach der Methode der Finiten Elemente untersucht und führte zu einer Sicherung der Kalotte mittels armiertem Spritzbeton und Felsnägeln. Örtlich wurden zusätzlich Vorspannanker vorgesehen. Parallel zum Ausbruch der Kalotte wurden Verformungs-, Spannungs- und Ankerkraftmessungen zur Überwachung der Standsicherheit durchgeführt. Die geologische Kartierung der Ausbruchlaibung ergab, daß örtlich Trennflächen auftreten, die parallel zur Kavernenachse streichen und vorher nicht erwartet worden waren. Aus diesem Grunde wurden ergänzende Standsicherheitsuntersuchungen der Kavernenwände notwendig.

1. INTRODUCTION

The underground powerhouse of the Pumped Storage Plant Estangento Sallente in the Pyrenees in Spain is still under construction. In the following paper after a brief description of the project the results of the geological and rock mechanical exploration are presented. Further the parameters adopted for the stability analyses, which were performed by the Finite Element Method, are described, The selected support consists of reinforced shotcrete, bolting and prestressed anchors. Also the measurements of the displacements, the stresses in the shotcrete and the prestressing loads of the anchors, which were performed during excavation of the vault are outlined in the paper. The detailed geological mapping performed during the excavation of the vault have led to the necessity of a modification of the assumptions adopted during the design phase and consequently revealed additional stability analyses of the cavern walls, which are also briefly described.

The authors like to acknowledge the support of the Managing Direction of Fuerzas Electrica de Cataluna given towards the publication of this paper. In particular we would like to thank Dr. Riverola and Dr. Veiga for their particular interest in this publication. We also wish to thank all the engineers on the site for their cooperation.

2. PROJECT

The pumped storage plant Estangento Sallente, which is located in the Pyrenees in the Province of Cataluna in Spain, is equipped with four units with a total capacity of 415 MW (Fig. 1).

Fig. 1   Location of the Hydroelectric Scheme Estangento Sallente, Spain

The upper reservoir is the natural lake Estangento with a capacity of 3 hm^3 (Fig. 2 - 5). This lake is connected with a series of other natural lakes by transmission tunnels. The lower reservoir of 5.4 hm^3 capacity is created by a rock fill dam with an asphalt concrete membrane on the upstream face. The height of this dam, which was designed by INYSA, Consulting Engineers, Madrid, is 50 m and the length of the crest is 350 m.

The head between the upper and lower reservoirs is approximately 400 m. The two pressure shafts with an internal diameter of 3.50 m are steel lined and inclined at 51°. The discharge capacity amounts to 125 m^3/s. The excavation is performed by a tunnel boring machine (Fig. 3 - 5).

The underground powerhouse cavern has a maximum height of 37.5 m and an overburden of approximately 100 m (Fig. 3). The width is 20 m. The axis of the 89 m long opening runs perpendicularly to the axes of the pressure shafts, thus revealing the shortest possible length of the pressure shafts (Fig. 2). More detailed information on the project is given in the references (FECSA, 1980).

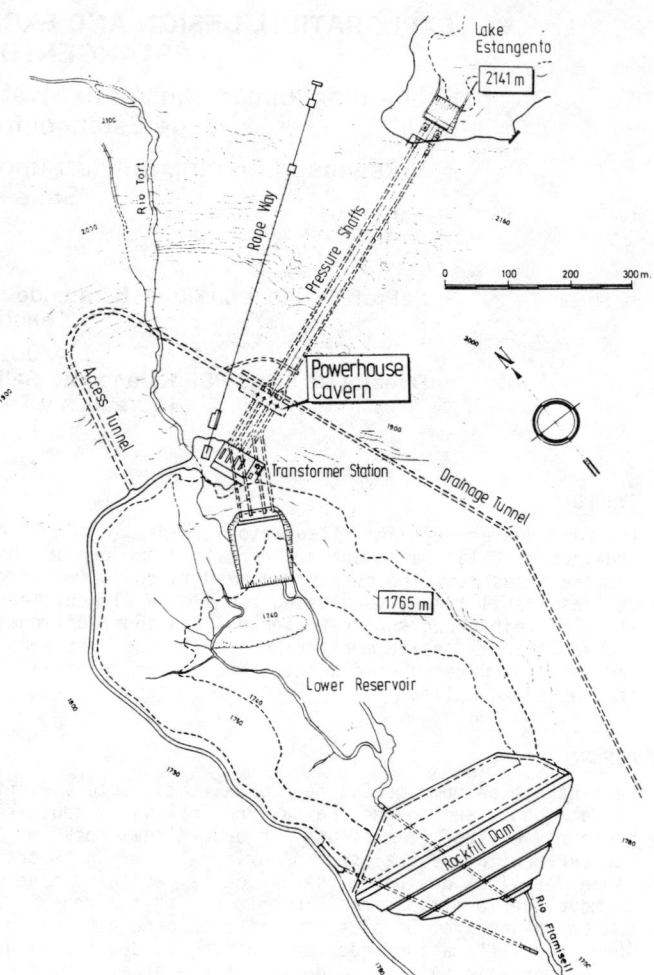

Fig. 2   General Plan of the Project

3. GEOLOGY OF THE AREA

The rock in the area of the Hydroelectric Project Estangento Sallente belongs to the massif at the south end of "la Maladeta" granite bathylith in the "Cordillera de los Pirineos". This bathylith has been created by an igneous process by intrusions, the lithological composition of which is biotitic-granodioritic. This intrusion has created a thermical metamorphism in the area of the transition to the surrounding rocks, the so called Cambio Ordovidic material, consisting of quartz-phyllites, phyllites, quartz-schists and schists.

The granite bathylith consists of uniform, medium grained granodiorite, which contains numerous quartz and aplite veins. The transition to the surrounding rocks, which extends over a few hundreds of meters to some kilometers, consists of a diorite with dark grey colour and the so-called corneans. The corneans are compact, dark coloured rocks, with a high degree of metamorphism but no oriented texture, because the schistosity of the original rock vanished due to the mentioned metamorphism.

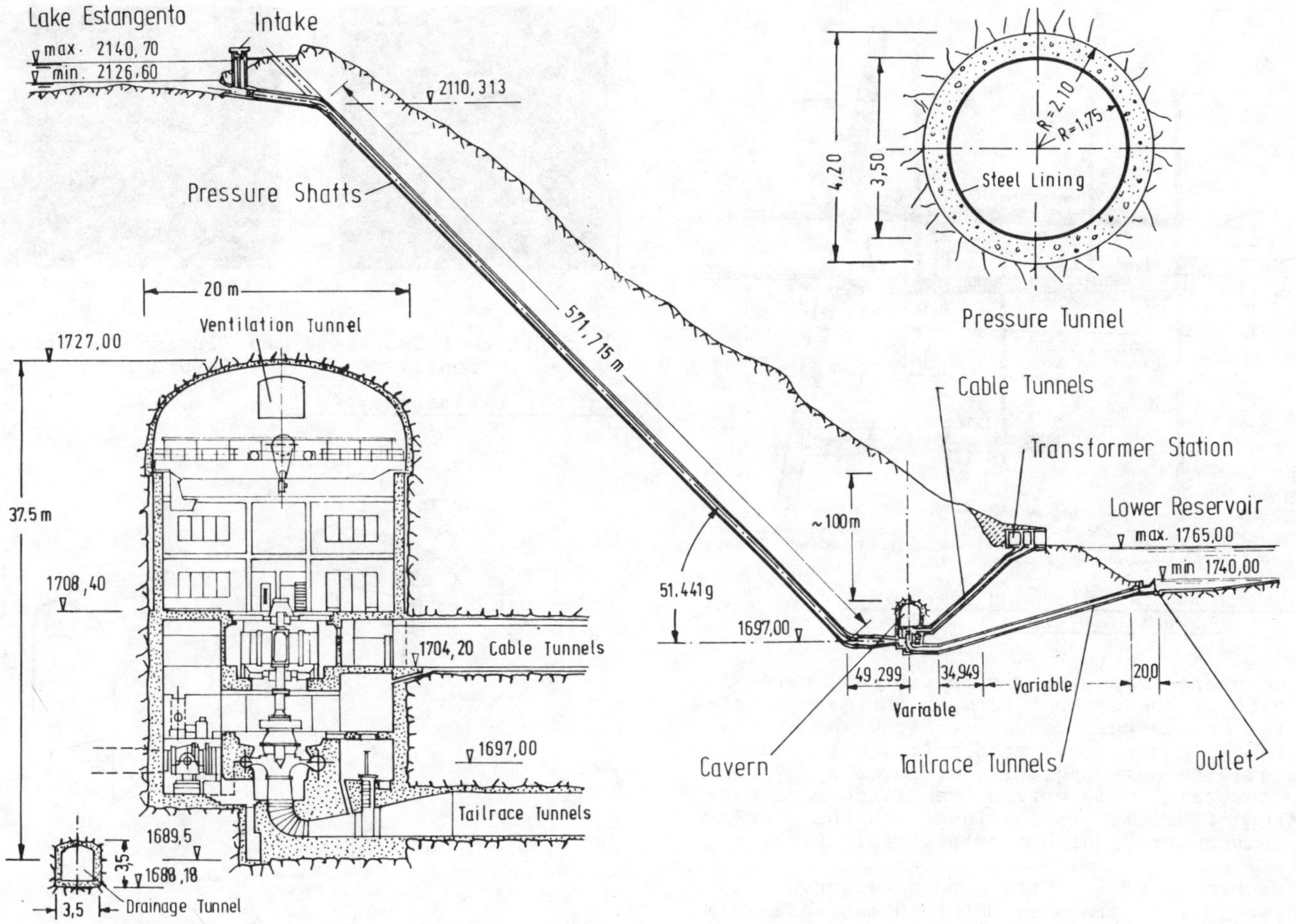

Fig. 3   Section of Project and Powerhouse

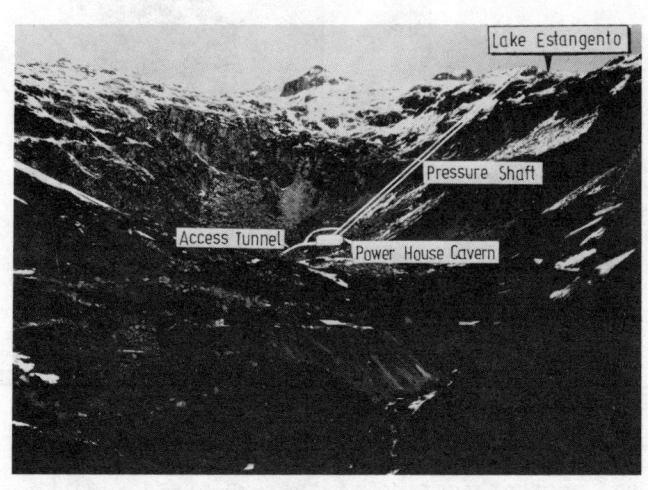

Fig. 4   View of the Project Site

Fig. 5   Aerial View of the Project Site

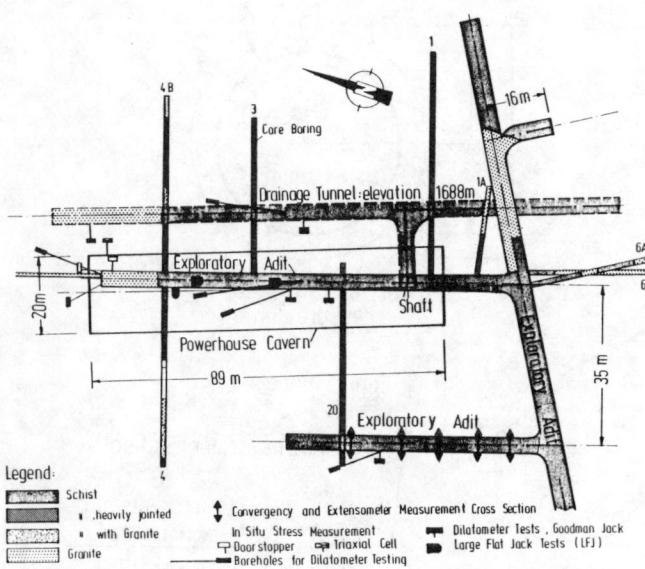

Fig. 6 Plan of Adits and Boreholes, Elevation 1723 m

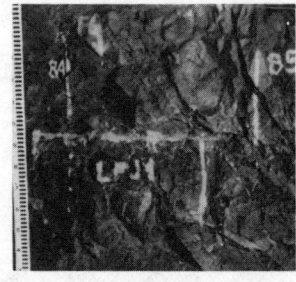
a) Location of LFJ-Test in Schist

b) Placing Doorstopper in Granite

Fig. 7 Exploratory Adit

The underground powerhouse cavern is located in the area of the contact between the granodiorite and the metamorphic rock. The north end of the cavern is located in granodiorite, which subsequently is named "granite" (Fig. 6 and 7). Also the relatively narrow transition zone consisting of diorite is combined with the granite and subsequently not separately dealt with.

The remaining part of the powerhouse cavern is located in the corneans, which subsequently are named "Schist" (Fig. 6 and 7). The transition between the rock types is a slightly curved plane (Fig. 8), which diagonally intersects the cavern. The contact between the rock types is welded and does not have the form of a discontinuity.

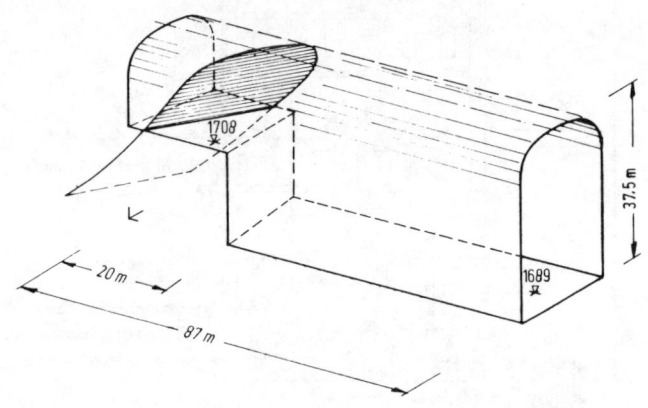

Fig. 8 Transition from Schist to Granite

As can be seen from Fig. 6, granite rock is also present beyond the transition zone shown in Fig. 8. This is due to branches of the intrusion and / or tectonic movements which took place after the granite intrusion.

4. EXPLORATION AND TESTING
4.1 RESULTS OF ADITS AND BOREHOLES

In addition to the geological mapping of outcrops at the adjacent slopes an inclined exploratory adit was excavated, the portal of which is located in the area of the outlet structure of the tailrace tunnels (Fig. 2). The direction of the axis of this adit in plan is N 60° E and it is located south of the powerhouse cavern (Fig. 6). In the area of the cavern an adit branches off. This branch approximately follows the axis of the cavern and is located within the cross section of the vault (Fig. 9). During operation it will be used as a ventilation tunnel (Fig. 3). From this section of the exploratory adit a series of core borings, which are located in sections running perpendicularly to its axis were drilled horizontally, vertically and inclined for exploration of the rock mass

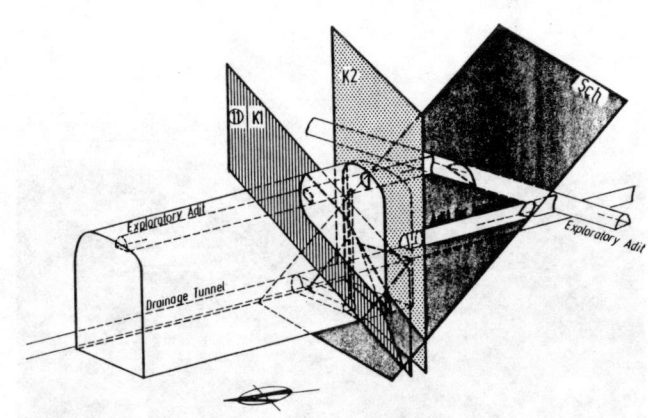

Fig. 9 Master Joints in the Area of the Cavern

adjacent to the cavern. Some of these borings are shown in Fig. 6 (1, 2D, 3, 4 and 4B). 35 m west another section of the exploratory adit branches off. This section of the adit was mainly used for rock mechanical measurements and testing (Fig. 6).

Furthermore a tunnel for drainage of the rock

in the powerhouse area was excavated at an early phase of the project (Fig. 2, 6, 9). This tunnel, which runs at the elevation of the invert of the cavern, also served the exploration of the rock conditions and is used for access to the powerhouse, if the access road to the project is blocked due to snow fall. In a later phase of the project also the main access tunnel to the cavern served for additional explorations (Fig. 2).

The explorations lead to the distribution of rock types in the powerhouse area, already described in paragraph 3. Accordingly the north end of the cavern is located in "Granite", whereas the major part of the opening has to be excavated in the so-called "Schist" (Fig. 6 - 8).

Though the granite proved to be far less jointed and thus far more competent than the schist, it was decided not to relocate the cavern into the granite, which could have been done, by shifting it to the north. This decision was justified because from the overall point of view of the project, as e.g. the alignment of the pressure shafts, the manifolds and the tailrace tunnels, the selected location of the cavern and the direction of its axis is advantageous.

A detailed geological mapping of the orientation, spacing and extent of all discontinuities exposed in the exploratory adit and the drainage tunnel in the area of the powerhouse was performed. A total of 1068 discontinuities were mapped and are represented in Fig. 10 with regards to their strike and dip. It is obvious that the families of discontinuities Sch and K1 occur most frequently. On the average they dip at 45° and intersect the cavern axis at 45° and 50° respectively. Far less frequent are the discontinuities of family K2, which are vertical and intersect the cavern axis at 40°.

Mainly parallel to the discontinuities Sch and K1 so-called master joints, which are water bearing and filled with cohesive material over larger extent, occur. Locally also master joints parallel to K2 occured. Furthermore some vertical master joints with clay fillings, running perpendicularly to the cavern axis, were mapped. They belong to a joint family K4, which has a rather low frequency and thus did not show up in the overall Schmidt's net of Fig. 10.

Because these master joints extended over larger areas the mapping was evaluated with regards to the existence of large rock mass wedges bounded by the cavern walls and discontinuities of this type, as is shown for an example at the southern end wall of the cavern in Fig. 9.

4.2 LABORATORY AND FIELD TESTING

In the western branch of the exploratory adit cross-sections for convergency and extensometer measurements were installed to monitor the displacements of the rock mass resulting from excavation (Fig. 6). Further in the section of the adit located within the cavern vault in-situ stress measurements using the triaxial cell (Rocha et al., 1969) and the doorstopper (Leeman, 1971) measurement technique as well as LFJ-tests (Rocha et al., 1970) were performed (Fig. 6, 7). Also dilatometer tests (Rocha et al.,

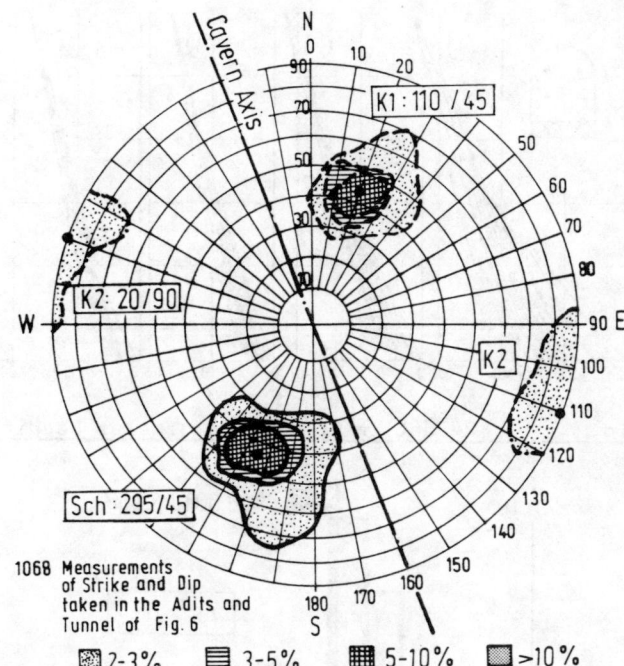

Fig. 10   Schmidt's Net

1966) and tests with the Goodman Jack (Goodman et al., 1968) were performed adjacent to the walls of the adits and in boreholes drilled from the adits (Fig. 6). Finally core samples were taken from the drillholes and tested in the laboratory. Since the major part of the cavern will be located in schists, testing was concentrated on this rock type. Thus within this paper only results on schist will reported.

From laboratory tests the unconfined compressive strength resulted to $\bar{\sigma}_{uc} \approx 1500$ kp/cm², thus revealing a rather high strength of the intact schist. The readings of the four deformeters of one jack of a flat jack test (LFJ) in jointed schist reveal, with one exception, a practically linear elastic stress-strain behaviour of the rock (Fig. 11). There is neither a difference between the 1st loading and the un- and reloading cycle, nor is there any difference between the results of tests with different jack orientation, thus proving that the rock is practically isotropic. From the histogram of the measured Young's moduli a mean value of $\bar{E} = 190\,000$ kp/cm² is obtained (Fig. 12). The same order of magnitude of the Young's moduli resulted from LFJ-tests performed in granite.

A marked difference between the 1st and the un- and reloading cycles resulted from the dilatometer tests. Also a much wider range of scatter of the single values resulted from these tests, as was expected due to the comparatively small size of the probe (Fig. 13).

Probably the LFJ-tests reveal the more reliable results.

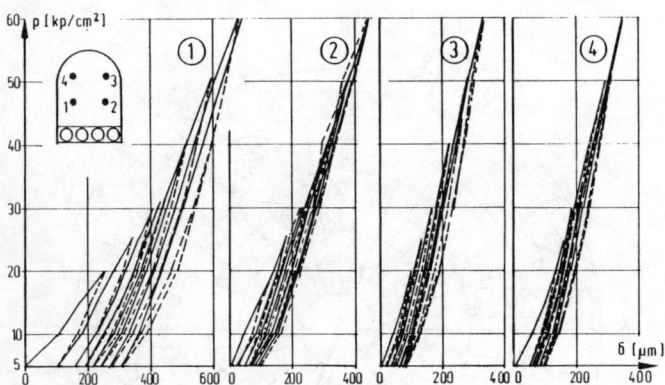

Fig. 11 LFJ-Test in Schist, Deformeter Reading

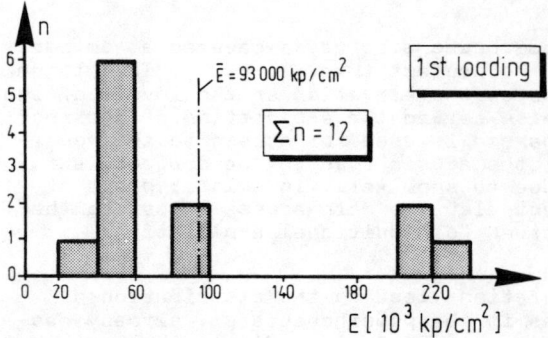

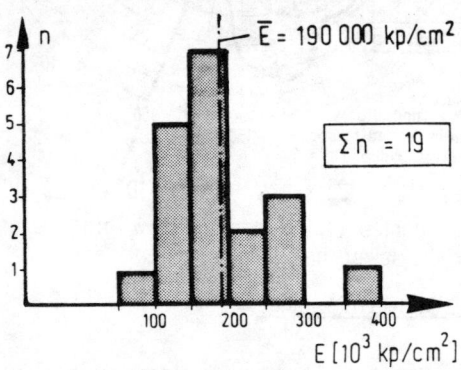

Fig. 12 Frequency Distribution of Young's Moduli, LFJ-Test in Schist

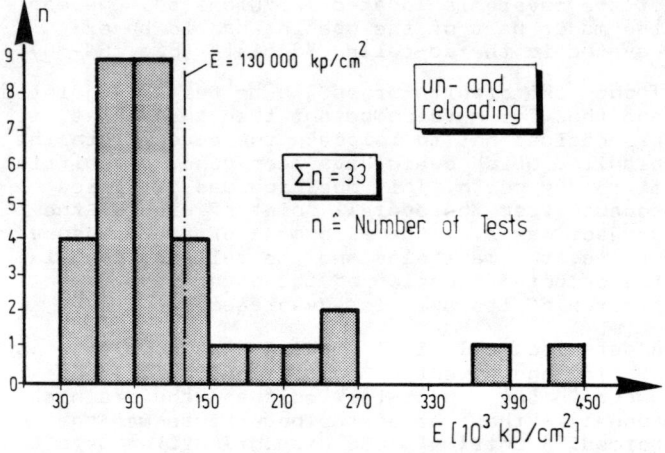

Fig. 13 Frequency Distribution of Young's Moduli, Dilatometer Test in Schist

From the two measurements performed with the triaxial cell a maximum normal principal stress of $\sigma_1 \sim 90 - 100$ kp/cm² horizontal and perpendicular to the cavern axis resulted (Fig. 14). The intermediate principal stress runs approximately parallel to the cavern axis, whereas the minimum normal principal stress is approximately vertical and equal to the weight of the overburden within the cavern area (Fig. 14).

Because of this rather unusual measurement result of a horizontal stress equal to approximately three times the weight of the overburden, the LFJ-tests were also used for measurement of the in-situ stresses. This was done by measuring the jack pressure, which compensated for the displacements of the walls adjacent to the jack due to cutting the corresponding slot. Since the jacks in the various tests were oriented horizontally and vertically as well as parallel and perpendicular to the cavern axis, normal stresses parallel and perpendicular to the cavern axis, the latter in horizontal and vertical direction, could be measured (Fig. 14). The measurements resulted in horizontal stresses of 40 kp/cm², which are considerably smaller than those measured by triaxial cells. Here also the vertical stress approximately equals the weight of the overburden (Fig. 14).

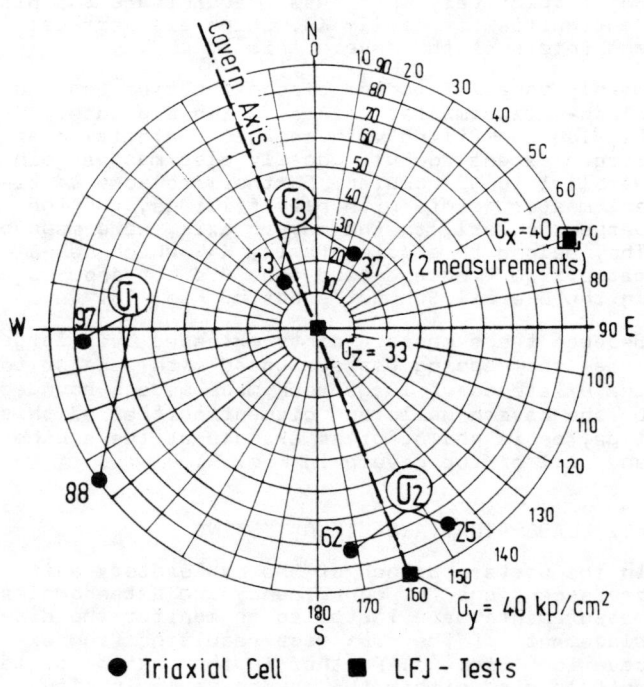

Fig. 14 In-Situ Stresses, measured by Triaxial Cells and LFJ-Tests

Because of the larger size of the flat-jacks it is expected, that the stresses measured by this technique are more reliable than those obtained from the triaxial cells, especially since only two tests have been performed with the triaxial cell. Therefore the interpretation of the measurements in paragraph 4.3 and the stability analyses in chapter 5 are based on the in-situ stresses measured in the LFJ-tests. It will have to be proven by monitoring during excavation of the cavern wether this assumption is realistic.

4.3 INTERPRETATION OF THE IN-SITU STRESS MEASUREMENTS

In-situ stresses are the stresses existing in the rock mass prior to excavation of the planned structure, in this case the powerhouse cavern. They result from dead weight and are thus influenced by the unit weight of the rock, the topography and - if the in-situ state is elastic - by the elastic constants of the rock mass. Furthermore tectonic and other stresses may be superimposed on the stresses resulting from the dead weight.

The measured stresses contain all influencing effects and do not enable a separation which is however required for a stability analysis by means of the FE-method. Therefore an interpretation of the measurement results was performed by calculating the stresses resulting from dead weight only. This was done by an elastic FE-calculation with the program outlined in references (Semprich, 1980) and (Pierau, 1981) for the 3D calculative section A, presented in Fig. 15, which contains the powerhouse area, the lake Estangento and the surrounding massif.
The Young's modulus was assumed to be $E = 150\ 000\ kp/cm^2$, the Poisson's ratio to be $\nu = 0,28$ and the unit weight to be $\gamma = 2,8\ t/m^3$ (Fig. 15). The selected displacement boundary conditions are also presented in Fig. 15. The stresses resulting from this calculation for the area of the powerhouse are plotted in Fig. 16. A comparison with those measured in the LFJ-tests shows that they have approximately the same orientation but different absolute values.

The differential stresses between measured and calculated values are also presented in Fig. 16. They should be at least approximately equal to the tectonic stresses existing in the area of the cavern.

From these analyses displacements boundary conditions were evaluated for the calculative section B (Fig. 15), which was used for the stability analyses of the cavern, in such a way, that the measured in-situ stresses resulted for this section for the undisturbed case, i.e. prior to excavation of the cavern (Fig. 17).

5. STABILITY ANALYSES

Also the subdivision of calculative section B into Finite Elements can be seen in Fig. 17. Further section C, for which the results of the analyses are subsequently presented, is specially marked. The same elastic constants and unit weight were used as before.

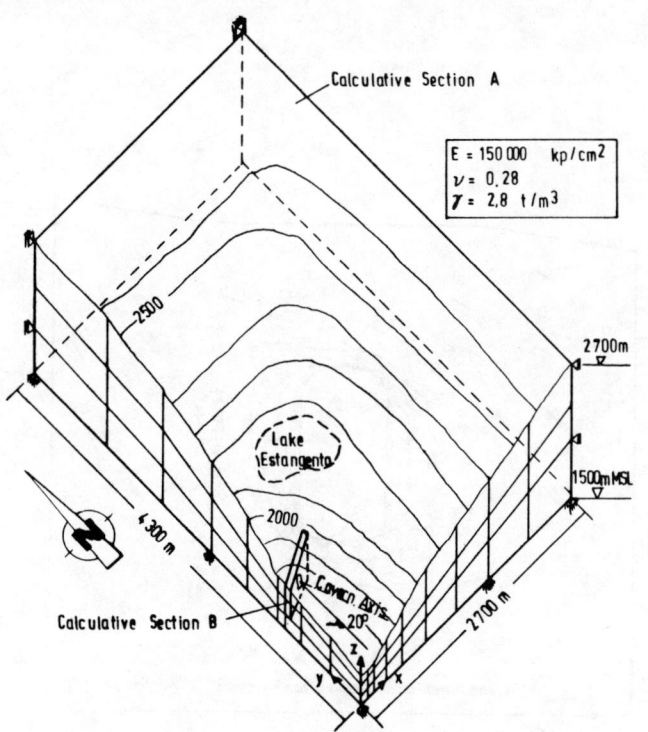

Fig. 15 Calculative Section for Evaluation of Stresses in the Slope due to Self Weight

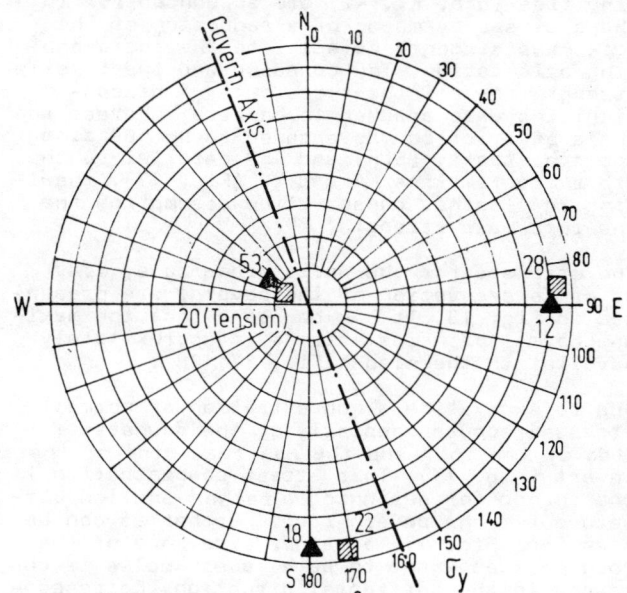

Fig. 16 Stresses resulting from Self Weight and evaluated Tectonic Stresses in the Area of the Cavern

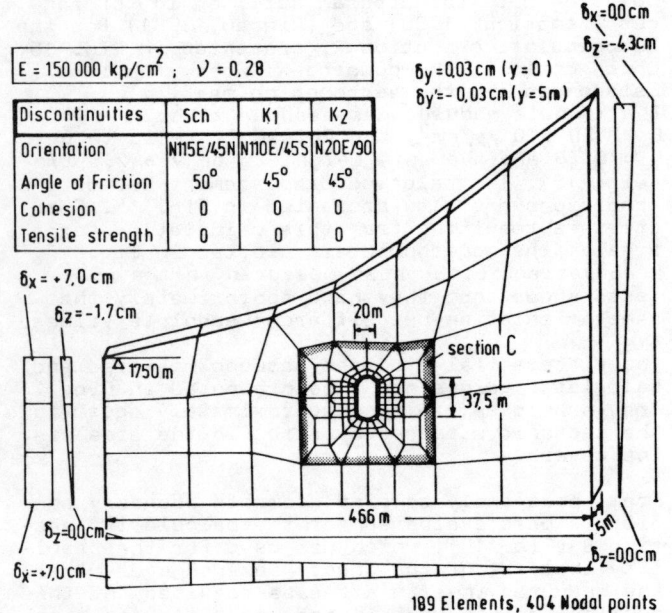

Fig. 17   Calculative Section B, FE-Mesh and Displacement Boundary Conditions (δ_x, δ_y, δ_z)

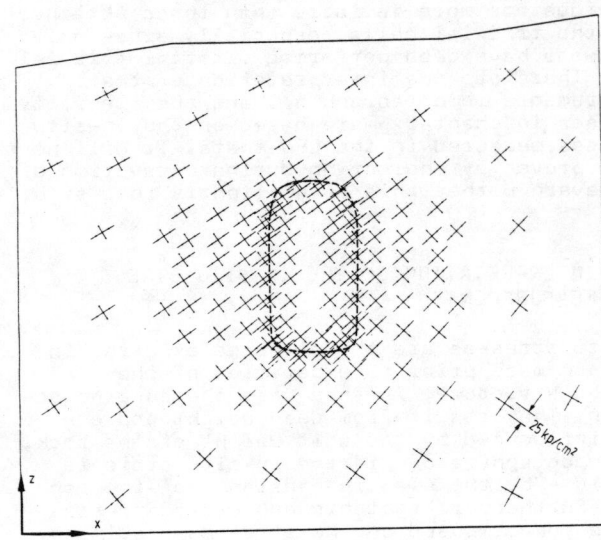

Fig. 18   Calculated In-Situ State of Stress due to Self Weight and Tectonic Stresses (Section C, Fig. 17)

In addition the three families of discontinuities (Sch, K1, K2) are accounted for in the analyses by means of a reduction of the rock mass strength parallel to the corresponding orientation. The cohesion and the tensile strength for all the three sets of discontinuities were assumed to be zero, whereas the angle of friction was assumed to be 50° along the schistosity (Sch) and 45° parallel to the other two families K1 and K2 (Fig. 17), resulting in a rather conservative assumption for the rock mass strength.

The stresses for the undisturbed case, i.e. prior to excavation of the cavern, are presented in Fig. 18. It can be seen, that the maximum normal principal stress is approximately parallel to the slope (Fig. 17, 18).

Due to excavation a concentration of normal stresses occurs specially at the downstream side of the roof and the upstream side of the invert (Fig. 19). This stress concentration is the reason for designing a rather shallow curvature for the cavern roof. Further as can be seen from Fig. 19, a rather wide zone of the rock mass adjacent to both cavern walls is unloaded in the horizontal direction. Correspondingly the horizontal displacements of the walls are high in comparison to the settlement at the roof, the latter being close to zero (Fig. 20).

In addition the horizontal displacements reach relatively far beyond both the cavern walls (Fig. 20). It can however be stated that the FE-analyses revealed the stability of the cavern on the basis of a rather small amount of support.

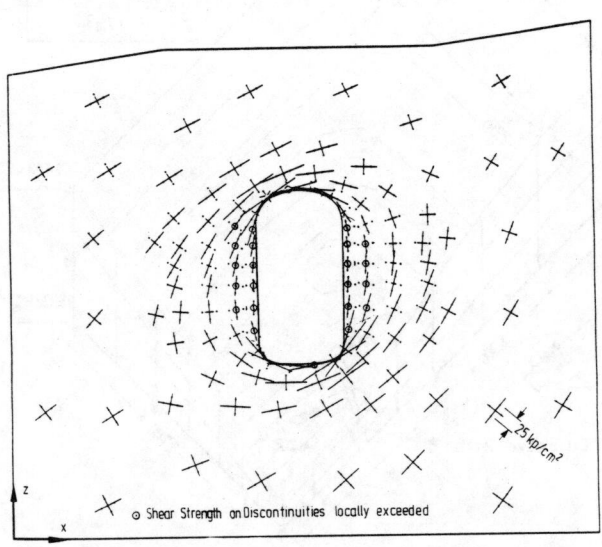

Fig. 19   Stress Redistribution due to Excavation (Section C, Fig. 17)

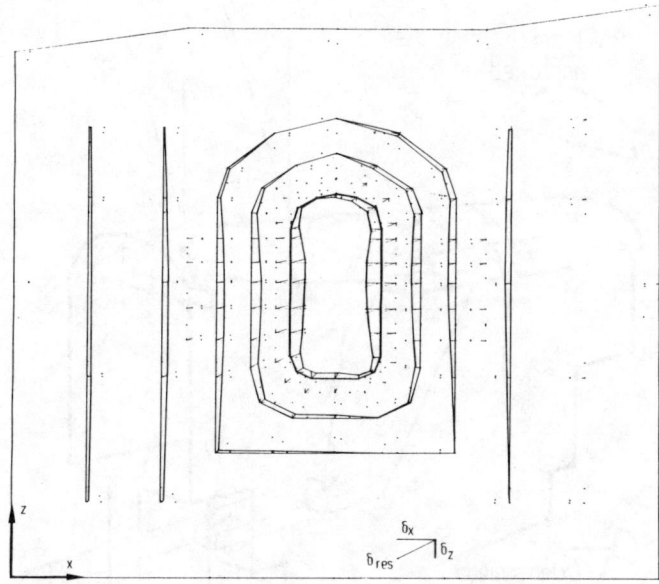

Fig. 20   Displacements due to Excavation
(Section C, Fig. 17)

6. EXCAVATION OF THE CAVERN VAULT
6.1 EXCAVATION AND INSTALLATION OF SUPPORT

The selected systematic support consists of 25 cm thick shotcrete reinforced by two layers of square wire mesh and a systematic grid of 5 m long bolts with a spacing of 1,5 m x 1,5 m (Fig. 21). Locally for the support of larger rock mass wedges 15 m and 20 m long prestressed anchors with a capacity of 63 tons were planned.

The installation of this support in the area of the vault is shown in Fig. 22. From these photographs it can also be seen that the surface of the excavation is rather uneven, this being due to the intensive jointing of the schist.

 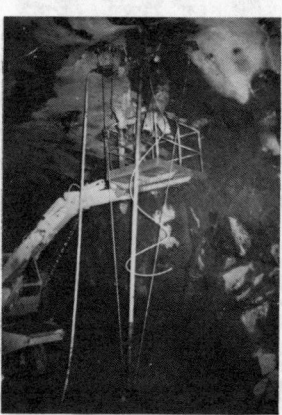

a) Shotcrete and Bolts   b) Prestressed Anchor

Fig. 22   Installation of Support

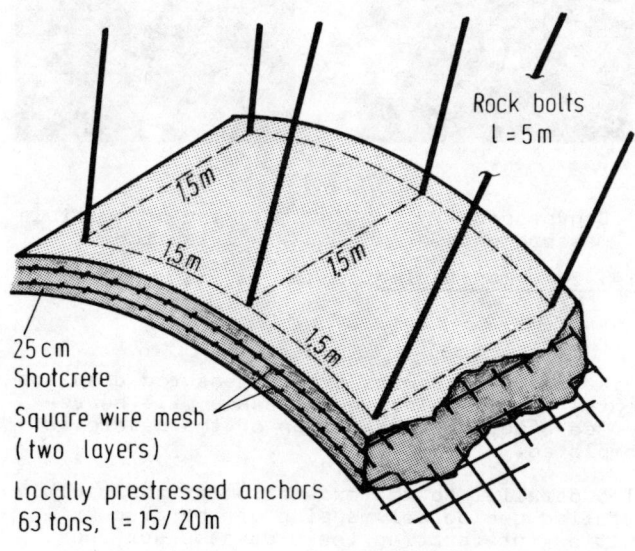

Fig. 21   Support of the Vault

The excavation of the vault was performed in three steps. In a first step the middle section of the vault was excavated by enlargement of the exploratory adit to a size of approximately 7 m x 7 m. This so-called Phase I can be seen on the right-hand side of Fig. 23. Parallel however, after a certain delay, the vault was enlarged to its complete width at the western side wall, (Phase II, Fig. 23, left-hand side) and in a third step the vault was completely excavated by removing an approximately 7 m wide section at the eastern side of the cavern.

The complete vault has now been successfully excavated (Fig. 24) and the lower part of the cavern is currently being excavated in benches.

6.2 MONITORING

Parallel to the excavation an extensive measuring program for monitoring the stability of the opening during and after construction is being carried out.

During the excavation of the vault readings from four convergency measurement cross-sections were taken (Fig. 25). These convergency measurement cross-sections were equipped with vertical, horizontal and diagonal measurement stretches which were installed parallel to excavation (Fig. 26 a). Further multiple extensometers were installed from the exploratory adits and the drainage tunnel respectively prior to the excavation (Fig. 25, 26 b).

Fig. 23 Excavation of Vault in three Phases

Fig. 24 Excavated Vault

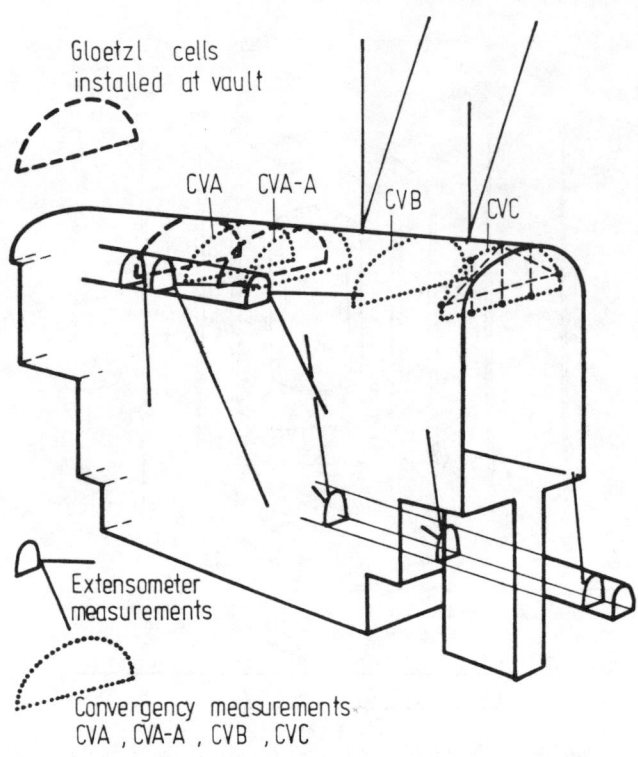

Fig. 25 Measuring Program for Monitoring

a) Convergency Measurements

b) Head of multiple Extensometer

Fig. 26 Monitoring

In addition a number of the installed prestressed anchors were equipped with measuring heads to monitor any increase of the prestressing load as excavation proceeded (Fig. 27a). Gloetzel cells for measurement of the tangential stresses in the shotcrete, which are expected to develop especially due to excavation of the benches, were installed in two cross-sections (Fig. 25, 27b).

As expected the displacements resulting from excavation of the vault were rather small, since as mentioned vertical displacements of the roof because of the high horizontal stresses are small and horizontal displacements are only expected to develope parallel to the excavation of the benches. Accordingly also no unusual large increase of the prestressing loads of the anchors was observed. The tangential stresses in the shotcrete showed however already measurable values.

More detailed results of the measured displacements, anchor loads and stresses will be reported after the excavation of the cavern is completed.

Also parallel to the excavation of the vault a detailed geological mapping of the discontinuities intersecting the excavated surfaces was performed. The result of the statistical evaluation of the measured angles of strike and dip is presented in Fig. 28. It was proven that the discontinuities of families K1 and

a) Measuring Head on Prestressed Anchor

b) Pressure Cell

Fig. 27   Monitoring

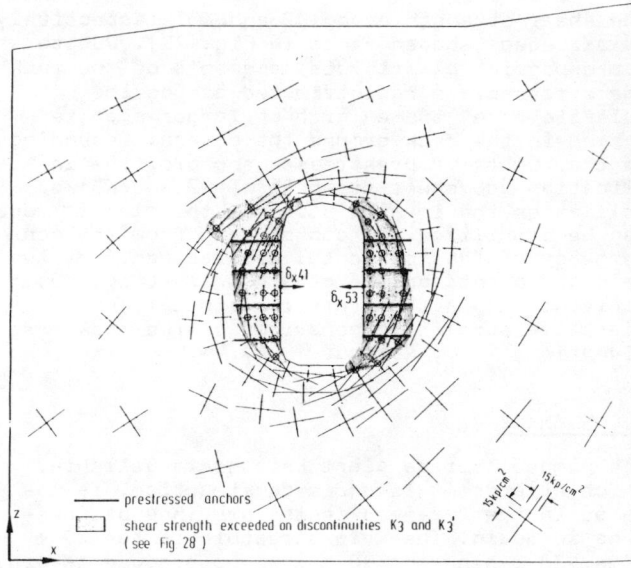

Fig. 29   Stress Redistribution influenced by Discontinuities K3 and K3'

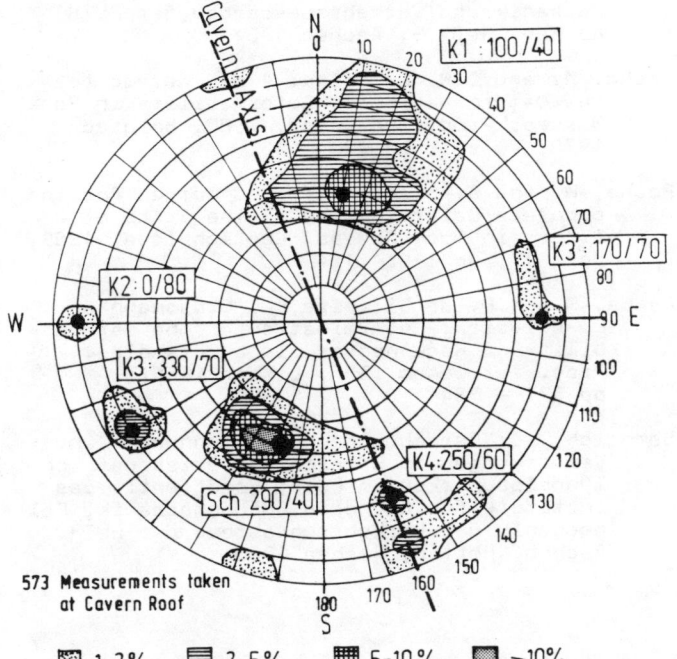

Fig. 28   Strike and Dip of Discontinuities measured during Excavation of Vault

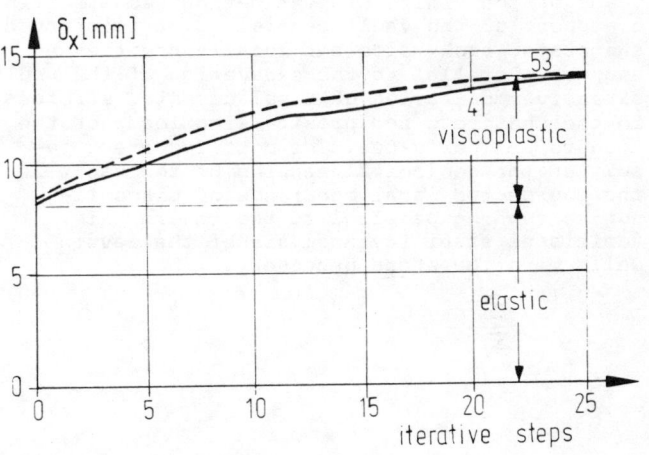

Fig. 30   Horizontal Component of Displacement of Nodal Points 41 and 53, Fig. 29

Sch are the most frequent. In addition discontinuities of families K2 and K4 were found. An unexpected result was however the local occurence of the two families of discontinuities K3 and K3' which run parallel to the axis of the cavern and dip steeply. Because of the latter discontinuities the number of prestressed anchors in the area of the vault had to be locally increased in order to fix rock mass wedges bounded by the discontinuities of families K1, Sch, K3 and K3' above the vault.

Also the stability of the walls obviously is unfavourably influenced by these discontinuities wherever they occur. Therefore additional stability analyses of the powerhouse walls accounting for discontinuities K3 and K3' were performed.

7. STABILITY ANALYSES OF THE CAVERN WALLS

In these stability analyses the inclination and the shear strengths of the discontinuities K3 and K3' were varied. Further the influence of a variation of the in-situ stresses and the number of prestressed anchors installed in the walls was investigated within parametric studies.

Here the result of only one calculation will be shown. As can be seen from Fig. 29, adjacent to the side walls extensive zones occur in which the shear strength along K3 and K3' respectively is exceeded (shaded areas in Fig. 29). Due to the corresponding plastic displacements of the rock, the stresses are redistributed around the "plastic zone" and an arch of larger span is formed in the rock around the cavern. Depending on the number of prestressed anchors, the inclination and shear strength of K3 and K3' as well as on the in-situ stresses the plastic zone can be stabilized, as can be seen from the convergency of the horizontal displacements of two selected points on the cavern walls (Fig. 30), achieved within the iterative simulation of the elastic viscoplastic behaviour of the rock mass (Semprich, 1980; Pierau, 1981).

8. SUMMARY

The pumped storage plant Estangento Sallente, which at present is under construction, is located in the Pyrenees in the province of Cataluna in Spain. The main structure is the 20 m wide, 38 m high and 90 m long powerhouse cavern, which is located in schist and has an overburden of ca. 100 m. An extensive exploration and rock mechanical testing program has been carried out, consisting of exploratory adits and boreholes, in-situ stress measurements and deformability tests. Stability analyses have been performed by means of the Finite Element Method resulting in a support of the vault consisting of reinforced shotcrete, rock bolts and locally prestressed anchors. Parallel to the excavation of the vault extensive monitoring of displacements, stresses in the shotcrete and prestressing loads of the anchors was performed. The most important result of the geological mapping of the vault is the unexpected local occurence of discontinuities running parallel to the cavern axis. Additional stability analysis of the cavern walls were therefore necessary.

9. REFERENCES

FECSA (Fuerzas Electricas de Cataluna S.A.) Proyecto de Construccion de la Central Hidraulica de Bombeo Reversible Estangento - Sallente en el Rio Flamisell - Lerida; Barcelona, Julio 1980

Goodmann, R.E., T.K. Van and F.E. Heuzé: Measurement of Rock Deformability in Boreholes, Proc. 10th Symp. Rock Mech., Austin, Texas, 1968

Leeman, E.R.: The C.S.I.R. Doorstopper and triaxial Rock Stress Measuring Instruments, Rock Mech., 1971, Vol. 3, pp. 25 - 50

Pierau, B.: Tunnelbemessung unter Berücksichtigung der räumlichen Spannungs- Verformungszustände an der Ortsbrust, Veröffentl. des Instituts für Grundbau, Bodenmechanik, Felsmechanik und Verkehrswasserbau der RWTH Aachen, Heft 9, Aachen 1981

Rocha, M. and J.N. Da Silva: A new Method for the Determination of Deformability in Rock Masses, Proc. 2nd Congr. ISRM, Belgrad 1970, Vol. I, pp 423 - 437

Rocha, M. and A. Silvério: A new Method for the complete Determination of the State of Stress in Rock Masses, Géotechnique, 1969, Vol. 19, pp 116 - 132

Rocha, M., A.F. da Silveira, N. Grossmann and E. Oliveira: Determination of the Deformability of Rock Masses along Boreholes, Proc. 1st Congr. ISRM, Lisbon 1966, pp 697 - 704

Semprich, S.: Berechnungen der Spannungen und Verformungen im Bereich der Ortsbrust von Tunnelbauwerken im Fels; Veröffentl. des Instituts für Grundbau, Bodenmechanik, Felsmechanik und Verkehrswasserbau der RWTH Aachen, Heft 8, Aachen 1980

CONSIDERATIONS IN THE DESIGN OF SUPPORT FOR DEEP HARD-ROCK TUNNELS

Eléments de la conception de soutènements de tunnels profonds en roche dure

Überlegungen zur Planung der Stützmassnahmen tiefliegender Tunnel

W. D. Ortlepp
Group Rock Mechanics Engineer, Rand Mines Limited, Johannesburg, South Africa

SYNOPSIS

An examination of the basic rationale for the use of rock-reinforcement methods in tunnels subject to high stress or rockburst damage, emphasizes the importance of yielding of the support elements. The concept of 'critical bond length' is proposed as a determinant of whether fully-grouted tendons will rupture or yield. Experimentally-determined characteristics of support elements are related to actual observations of tunnel damage to confirm the relevance and importance of these ideas.

RESUME

Un examen des raisons fondamentales de l'utilisation de soutènements dans le tunnels soumis à de grandes pressions ou à des coups de terrain a montré que les propriétés élastiques de ces soutènements sont primordiales. Le concept de "longueur critique de scellement" est proposé pour la détermination des critères de rupture ou de déformation des boulons d'ancrage. La relation trouvée entre les caractéristiques des éléments de soutènement déterminées expérimentalement et les observations in situ des dommages dans les tunnels a confirmé l'intérêt de ce concept.

ZUSAMMENFASSUNG

Eine Untersuchung primärer Stützmaßnahmen für Tunnel mit hohen Gebirgsdrücken und großer Gebirgsschlaghäufigkeit betont die Wichtigkeit von nachgiebigen Ausbaumethoden. Das Konzept der "kritischen Haftlänge" wird als Entscheidungsfaktor ob vollverklebte Anker als steifer oder nachgiebiger Ausbau wirken, vorgeschlagen. Versuchsergebnisse werden mit in situ Beobachtungen in verbrochenen Tunnelstrecken verglichen, um die Bedeutung und Wichtigkeit dieses Konzepts zu bestätigen.

INTRODUCTION

Rockbolting has been used for a long time as temporary or supplementary support in rock tunnels. About fifteen years ago the South African gold-mining industry first considered using steel tendons systematically as a complete method of tunnel support to supplant the traditional systems based on steel arches or timber sets. Recently, end-anchored rockstuds have almost entirely given way to fully-grouted reinforcing bar or looped wire rope as the preferred form of tendon.

Usually the tunnel walls are fractured to a greater or lesser degree because of depth or induced stresses. Often this fractured rock must be contained by means of a fabric of wire mesh backed-up by wire rope laced between the tendons. On one large mine a total length of over 2000m of tunnel is supported in this manner each month at a cost of R110/linear metre. Some 98 000m of wire rope strand is required for tendons and lacing and 20 000m^2 of wire mesh for the fabric.

Variations in detail that have evolved to meet differing local conditions have been empirically developed. While the empirical approach has mostly worked well it has become evident that a more careful consideration of system requirements is necessary for the extreme conditions at very great depths or, more particularly, where severe rockbursts occur.

This paper examines some of the premises implicit in the development of this type of support and presents some quantitative performance characteristics of the main components of the system.

2. TUNNEL SUPPORT ELEMENTS

In the South African gold-mining industry the rock-reinforcing system of tunnel support is usually, somewhat loosely, referred to as "wire-mesh support". Although there are some variations in detail the essential components of the system are, in order of importance, the tendons, the lacing and the wire-mesh.

A brief description of the most common types of these elements is necessary before their functional requirements are considered in detail.

2.1 Tendons

The length of tendons and their spacing patterns tend to be based on broad, empirical guidelines[1]. Since most main tunnels have typical cross-sectional dimensions of 3 to 3,5m, the most common length for tendons is 2,5 to 3m and the spacing between tendons is typically 1 to 2m. The tendons generally range from 12mm to 20mm in diameter.

'Full-column' grouting is almost invariably used as the means of providing anchorage. A thick cement slurry is used with wire rope tendons and, usually, resin capsules or cementitious cartridges with reinforcing bar.

Where very severe increases in field stress are anticipated and the tunnel is particularly important, pre-stressed rock anchors up to 7,5m long are used to supplement the normal 2,5m grouted tendons — Figure 1. These may be fully-grouted after tensioning or have a "free" length able to extend elastically.

Figure 1 : Intensively supported haulage. 2,5m grouted tendons and lacing are of de-stranded hoist rope. Pre-stressed anchors are 7,5m long and 890kN capacity.

2.2 Lacing

The type and size of wire rope that is employed for lacing through the looped ends of the tendons is determined largely by practical considerations and availability, rather than by design. Most mines find that single strands from large discarded hoist ropes can be used satisfactorily. While these are naturally far cheaper than new rope there is not always a sufficient quantity available. Where new rope has to be obtained scraper rope of 6 x 7 construction is usually chosen. The diameter of rope or strand used for lacing can thus vary from 9mm to 16mm.

The lacing method is to some extent determined by the density and position of the grouted tendons but a diamond-pattern is normally intended — Figure 1. Thus, typically, two ropes cross over through each tendon loop or 'eye' to form rhomboidal 'windows' about 1m wide.

2.3 Mesh

Although it represents the largest materials cost item in the system, there has been less enquiry into the effectiveness of wire mesh than into any of the other components. With few exceptions, diamond mesh is chosen for tunnels and weld-mesh for larger chambers. The diamond mesh is usually made of 3,2 or 4,0mm diameter wire with apertures ranging from 50mm to 100mm. The maximum strength of wire from which this type of mesh can be manufactured is 450MPa. The ease of handling and installation which results from the limpness of diamond mesh are important considerations in the relatively confined space of the tunnel.

Although the stiffness of weld mesh makes it somewhat difficult to mould to rough surfaces, it has the important advantage in larger excavations that its large apertures allow better shortcreting. Weld-mesh can be made from wire of greater tensile strength than most other mesh types.

3. DESIGN RATIONALE

Apart from a restricted zone of fractured rock surrounding most excavations, the rock mass in South African gold mines behaves as an elastic continuum. Thus the use of well-established digitally-computed analyses permits reliable calculation of the field stresses induced by the main stoping excavations at the site of any existing or proposed tunnels[2]. Empirical criteria for determining tunnel support requirements in terms of these field stresses have long been in use[1].

Since the strength behaviour of the fractured rock immediately surrounding the tunnel is not sufficiently well known, no attempt has been made to determine the corresponding support requirements in a rigorous or quantitative manner. However the broad rationale for supporting a highly-stressed tunnel has been expressed in qualitative terms[3]. Basically it recognises that the support must preserve the integrity of the fractured rock surrounding the tunnel as far as possible. To do this it must restrict the relative displacements between individual blocks or fragments while accommodating the inevitable bulk movement of the fractured mass. Essentially the aim is to utilize the potentially large residual strength of the failed rock material to enable it to support itself.

In the simplest terms, then, the requirements of the support elements are that they should:

i) be stiff so that they act as soon as possible, but

ii) retain the ability to yield through appreciable displacements at load values less than critical.

Under the dynamic loading conditions of a severe rockburst there is the additional requirement that the yield should be able to take place at high strain rates[3].

In recent analyses by Wagner[4] more quantitative estimates of these requirements have been given. He shows that the hypocentral distance for severe tunnel damage resulting from a rockburst of $M_L = 4,0$ could be decreased from about 140m to 60m by providing the support tendons with the ability to yield an additional 40mm after the yield point of (say) 100kN was exceeded.

The essential correctness of these findings is supported by actual examples where rockburst damage in tunnels has been relatively restricted even in the close vicinity of very large seismic events. A careful consideration of the probable damage mechanism obtaining in such circumstances can lead to important further conclusions regarding the yield requirement.

4. THE MECHANISM OF DAMAGE

In most deep gold mines the dominant mode of damage is slabbing and buckling of the tunnel walls. This results from the development of fractures parallel to these surfaces and to the direction of the maximum principal stress - photo 2. While they are often closely-spaced near the surface, the distance between fractures must increase with increased depth into the surrounding rock — Figure 3. Beyond some depth which is typically 0,4 to 0,7 times the height of the sidewall, the rockmass is intact apart from natural discontinuities such as joints. The frequency and depth of penetration of the fractures is probably closely related to the intensity of field stress and the shape of the tunnel.

Although there are instances where shear dislocation must occur, generally the tendons are subject to dominantly tensile loading by the dilatation and bulking associated with the formation of the fractures and the buckling of the resulting slabs — Figure 3. It is not possible, within practical and economic limits, to prevent these deformations[5] so the tendon must accommodate the resulting strain. In the case of a conventional end-anchored rock-stud the strain will be uniformly distributed but, provided there is no slip of the anchor, the total amount of elongation will be limited to 20 to 50mm after which the stud will fracture.

Clearly the distribution of strain along a fully-grouted tendon will be discontinuous and the dominant effect will be localized <u>at</u> fractures rather than <u>between</u> fractures. The strain concentration will be particularly pronounced where buckling or bulk displacement of a thick slab takes place.

The resulting effect on the steel tendon will be either an abrupt rupture of the element itself or a breakdown of the bond between the steel and the grout.

The stability of the tunnel could be crucially dependent on which of these two possibilities takes place. In a quasi-static situation it is conceivable that massive sidewall slabs could remain locked in position even if failure of some tendons had occurred deep within the supported walls. However, it is more likely that extra load would be thrown on the neighbouring tendons and these would fail, leading to a general collapse.

Figure 2 : Fracturing in sidewall of crosscut 3400m B.S.

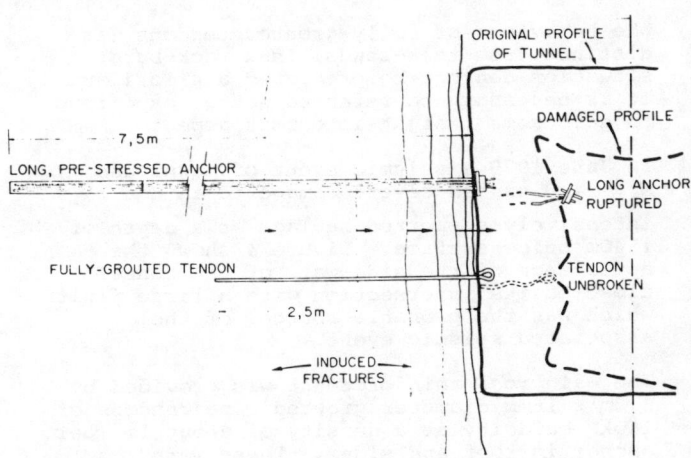

Figure 3 : Diagrammatic sketch of fracture spacing in tunnel sidewall.

In a dynamic stress situation, if failure of the bond occurred, very considerable frictional forces would still exist to prevent individual blocks of rock sliding freely along the tendons. Moreover the mesh and lacing attached to the tendon ends would prevent a final collapse provided they did not tear or break.

On the other hand it is almost certain that a total collapse would result if rupture of the tendons occurred rather than failure of the bond. The likelihood of this occurring obviously needs to be carefully considered particularly in the light of the known tendency of most steels to suffer brittle failure at high strain rates.

The interaction of the transient stress waves, arising from a large seismic event, with the complex system of fractures and fractured blocks surrounding a highly-stressed tunnel is obviously complicated and, at present, almost entirely unknown. Often the seismic source is located at some distance from the tunnel but, in some of the most severe instances, it would appear that the source region actually intersects the tunnel[6] or is closely parallel to it. There is even some suggestion that the source surface could envelop the tunnel.

Using the relationship suggested by McGarr, Wagner has shown for the simple case, that ground velocities in excess of 10m/s can occur at hypocentral distances of as much as 50m in the case of the larger seismic events. It is also possible that the reflection of the strain wave from the fracture surfaces can cause a doubling of the ground velocity, and a corresponding increase in the strain rate imposed on the support elements. The reality of intolerably high strain rates has been confirmed in many instances where conventional end-anchored rock-studs have been observed to have failed in a brittle fashion with no visible indication of ductile elongation. Specially-designed yielding rock-bolts have been shown[3] to be able to withstand high strain rates but these have not been used in practice.

The behaviour of fully-grouted tendons (as distinct from rock-studs) in a rock-burst situation can not be predicted a priori and it is necessary to refer to actual experience to gain some insight into this aspect.

In June 1975 a seismic event of magnitude $M_L = 4,0$ caused severe damage to an intensively-supported haulage at a depth of 1540m below surface. Figure 4 shows the appearance of the side-wall of the tunnel close to its intersection with a large fault which was the probable source[6] of the associated seismic event.

The main rock reinforcement was provided by 2,5m x 16mm diameter grouted rope tendons of 160kN capacity at a density of about 1m² per anchor in roof and sides. These were supplemented with 1300kN 6-strand, pre-stressed 7,5m long rock anchors at a density of approximately 8m² per anchor in the side-

Figure 4 : Rockburst damage in intensively-supported haulage, $M_L = 4,0$.

walls. The overall resistance was thus about 320kN/m².

The 3,2mm by 65mm aperture galvanized diamond mesh was backed by lacing of 16mm scraper rope with a breaking strength of about 180kN.

A careful examination of the scene of damage showed several significant features. Across the intersection of the fault with the haulage the damage was well contained although the west wall had bulged inwards by at least 1m in places — Figure 4. The mesh and lacing were stretched but intact and none of the 2,5m tendons appeared to have failed although several of the 7,5m long pre-stressed anchors had fractured or slipped at the collar.

Some important inferences can be drawn from interpretations based on these observations although these must be somewhat speculative.

The spacing between the fractures that existed, or were suddenly created, along the length of the 2,5m grouted ropes was sufficiently close that nowhere could a bond strength develop large enough to break the tendon. Thus the tendons remained as friction-locked ties or 'ligaments' to maintain some coherence in a bulging, buckling, moving mass of fractured fragments. The total displacement of this mass far exceeded the limited elongation available in the 7,5m pre-stressed anchors which were long enough to extend well beyond the fractured region and so were anchored in the stationary remote rock mass. Obviously the loads generated by the moving fractured rock were far in excess of the unit strength of the anchors and failure was inevitable.

A consideration of the likely extent of the complete collapse that almost certainly would have occurred if the 2,5m grouted ropes had also failed, suggests strongly that the most important requirement of such tendons is that they should be protected from rupture even if this means that the bond strength must be deliberately decreased.

If this were done sliding of the rock fragments along the tendons would become relatively easier and the mesh and lacing would be subjected to greater loads. A determination of their required characteristics thus becomes more important in the total design of the support system.

To better demonstrate the importance of this concept of protecting the tendons and promoting yield or compliance and also to gain some quantitative understanding of the factors that would lead to its attainment, a series of experiments have been performed.

5. LOAD-DEFORMATION CHARACTERISTICS OF SUPPORT ELEMENTS.

5.1 Tendons

Following the reasoning in section 4 based on the observations and interpretation of Figures 2 and 3, it was felt that a properly simulated test of the tendons would require that the traction be transmitted from the surrounding rock through the grout annulus to the interface between the grout and the steel tendon.

Thick walled steel tubing of 60mm O.D. and 39mm I.D. with slightly rough inner and outer surfaces was used to represent the surrounding rock. With the same grout-injection equipment used underground suitable lengths of tubing were filled with the standard paste of portland cement and various types of tendons inserted. The steel tube could be sawn part-way through at several positions along its length to simulate various spacings of fracture. Where the ends of the tube were gripped in the jaws of the large tensile-test machine in the laboratory, the wall-thickness was sufficient to preclude any spurious clamping effect on the tendon.

As the tube was tensioned some elongation would occur and the tendon contained within would be subjected to more-or-less uniform strain as might be expected with the initial dilatation of the stressed side-wall of a tunnel.

When fracture of the tube occurred at the pre-determined positions the elongation would be translated into localized strain at those portions of the tendon exposed across the fractures, in just the same way as must occur across the joint or fracture openings in the tunnel side. The subsequent load-elongation behaviour would depend on:-

i) The strength and type of tendon,
ii) The strength and nature of the grout/tendon bond, and
iii) on the length of bond subjected to the applied traction.

The behaviour of six types of fully-grouted tendon and three end-anchored tendons are compared in Figure 5. Curves (4a),(5),(6),(7) and (8) were obtained from tests using the thick-walled steel tube with one centrally-located weakening cut. Curve (3) describes the characteristics of a fully-grouted smooth steel bar joining two sawn blocks of real rock[7].

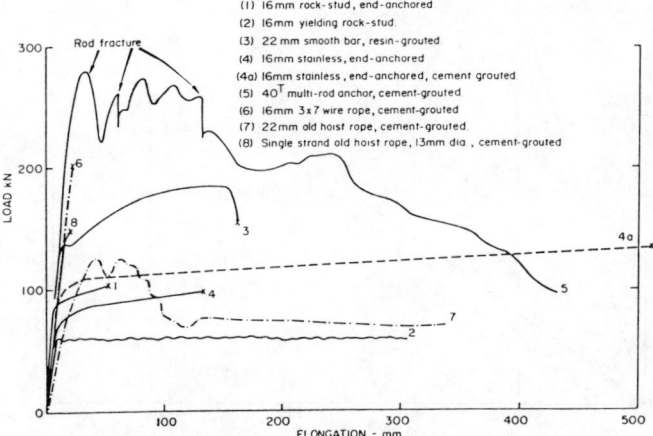

Figure 5 : Characteristics of various tendons.

The expectation, based on the discussion in section 4, that there would be <u>either</u> a breakdown of the bond between tendon and grout <u>or</u> that the tendon would fracture, was confirmed.

The effect of bond-breakdown is best illustrated in the case of the 22mm rope (7), which maintained a resistance of about 115kN until 80mm of fracture-opening had occurred. Then failure of the bond proceeded rapidly until complete at about 120mm of opening. Finally a frictional force that slowly declined from 75kN to 66kN remained as a residual resistance to further opening of the fracture.

Clear illustrations of the alternative mode of failure viz. rupture of the tendon, are provided by (6) and (8). The special grouting rope (6) rapidly developed a load of 200kN. Because of the very strong purchase afforded by the deeply-convoluted 3-strand helical surface, failure of the bond could not occur and the few millimetres of fracture-opening represented an intolerable strain in the very short length of rope over which it was localized. A cleaned single strand of old hoist rope (8) behaved in almost identical fashion, failing at a slightly lower load of 150kN.

Tendon (5) showed an intermediate type of behaviour because it consisted of six individual 7mm diameter rods. Three of the rods ruptured at values of fracture-opening of 24,50 and 120mm. The remaining three pulled out of the grout against a frictional resistance which rapidly decreased from 220kN to 100kN.

The criterion for determining which mode of failure will occur is that the separation between fractures traversed by the tendon should be less than the critical bond length L_C. The critical bond length is defined as that length which develops a total bond strength just greater than the tensile strength of the tendon. Actual measurements of L_C for tendons (5) and (7) were made by determining the force necessary to break

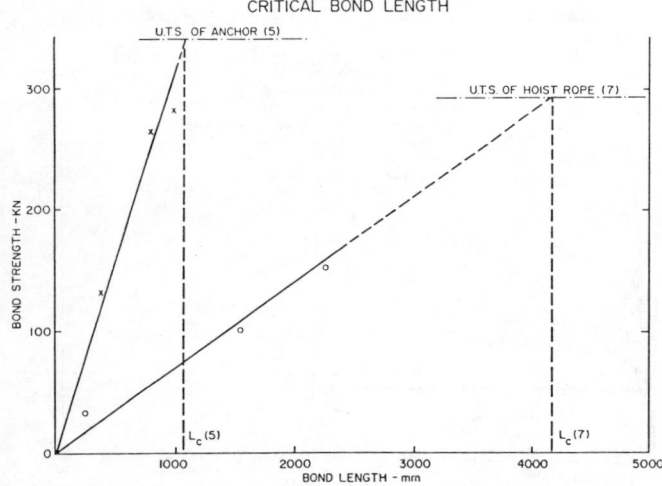

Figure 6 : Determination of critical bond length.

the bond of successively increasing lengths of thick-walled tube containing the grouted tendon. A plot of these bond strengths against bond length — Figure 6 — showed a linear relationship in each case which, extrapolated to the respective value of tendon strength, provided an estimate of L_C.

A value of L_C = 1000mm was obtained for the multi-rod anchor and its accuracy was confirmed by the fact that rupture of three of the rods did occur in test (5) where the tested length was 1m.

A further interesting effect that the enveloping grout annulus has on smooth, solid bar was strikingly evident in the behaviour of fully-grouted tendons (3) and (4a) compared with their ungrouted counterparts (1) and (4). Although the separation was made to occur at a single simulated fracture as before, the resulting strain concentration did not cause the localized 'necking' and 'cup-and-cone' fracture that is typical of ductile metals. Instead the 'necking' and accompanying increased elongation, progressed along the length of the bar causing a distributed rather than a localized plastic yield of the steel.

It is not known whether this process would occur with the less-ductile high-tensile steel normally used for tendons or whether it would constitute a significant yielding mechanism in practice.

5.2 Lacing

Because it is very difficult to tension the lacing properly there is a fair amount of slack in the 'fabric' of most tunnel support in practice. This slack is probably the main reason that failure of the lacing very seldom occurs even in the event of substantial rockburst damage. When failure does occur it is likely to be at the junction of the lacing cross-over with the tendon loop.

Two tests were carried out using the apparatus shown in Figure 7 where tension is applied between the tendon loop and the

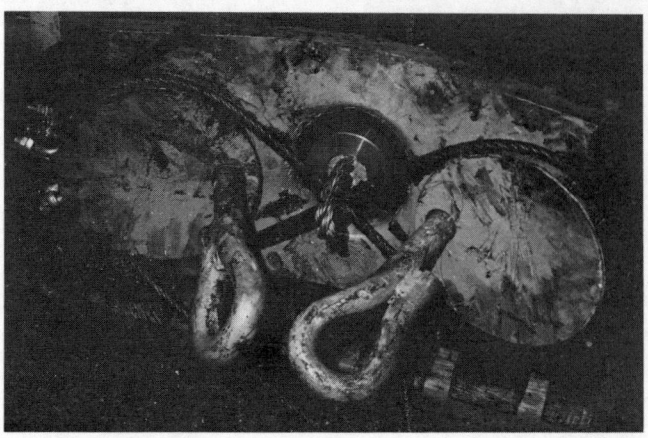

Figure 7 : Arrangement for testing lacing/loop.

two eye-bolts of the saddle arrangement.

The loop of 3 x 7 construction, 16mm diameter (tendon (6) in Figure 5) failed at 165kN when 13mm rope was used as lacing. With rope of 10mm diameter, failure occurred in the lacing at a load of 52kN.

5.3 Mesh

The device for testing the mesh consisted of a heavy steel framework forming a square window of 1,1m side with a peripheral clamping arrangement that enabled various panels of mesh to be stretched across the opening — Figure 8. Load was applied centrally through an articulated arrangement of four steel plates covering an area of 0,42m x 0,52m.

Details of the several types of mesh for which reliable and representative results were obtained, are given in Table I. The load-deflection characteristics are shown in Figure 9.

Figure 8 : Framework for testing wire mesh.

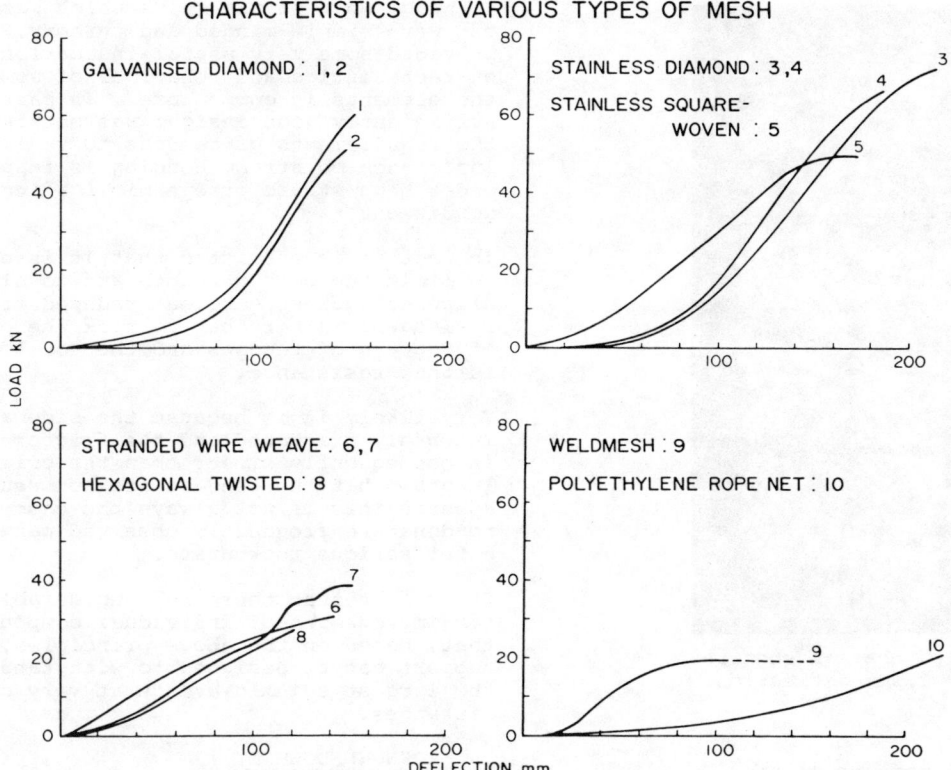

Figure 9 : Characteristics of various types of mesh.

The different types of mesh are listed in Table I in order of decreasing total load at failure. Obviously this does not represent a meaningful merit ranking as the physical dimensions vary widely and thus the mass of material involved is very different. This effect was taken into account by normalizing the failure load with respect to the total cross-sectional area of the wires or strands in a unit area of mesh, to produce the 'nominal index of relative strength' in the seventh column of Table I.

In some cases where the mesh aperture was large the load was actually applied to relatively few strands. This influence was recognized by normalizing the failure load with respect to the number of strands actually subjected to direct load, giving the 'loaded index of relative strength' in column 8 of the table.

Failure of the strands invariably occurred at a mesh intersection or 'cross-over'. Here different degrees of stress-concentration and impairment of material properties co-existed, depending on whether the cross-over was linked, woven, welded or knotted. Obviously the original strength of the material and its specific yield behaviour would have a strong influence on the 'cross-over' strength. To assess the merit of a particular mesh type in terms of its efficiency of utilizing the potential strength of the strand, it was necessary finally to normalize with respect to the material strength - column 9 in Table I.

Viewed in this light, the inter-linking construction of diamond mesh causes the least impairment of the potential strength of the wire. However it cannot be made of high-tensile wire or with apertures larger than 100mm.

The woven, and the stranded-wire weave types of mesh do not have any limitations in regard to aperture size or strength of wire and can, although less efficiently, produce higher strength mesh by using high-tensile wire.

5.4 Shotcrete

In terms of the rationale (i) in section 3 it is advantageous to have the 'fabric' as stiff as possible provided that its yielding properties are not impaired. The only effective way of stiffening wire mesh is to couple it to the rock surface by means of shotcrete.

In order to gain some estimate of the effective stiffness and yieldability of shotcreted wire mesh, an in-situ test was performed. An inflatable rubber envelope of $0,41m^2$ effective area was sandwiched between the tunnel wall and a 120mm thick layer of shotcrete applied over 75mm x 4,0mm diamond mesh. The rubber envelope was inflated hydraulically by means of a small hand pump while the deflection was measured by means of dial gauges — Figure 10. The load-deflection characteristic is shown in Figure 11. The first crack appeared after 0,3mm deflection, and the surface appeared substantially fractured after 2,5mm of bulging — Figure 10. However a final deflection of 25mm, at the centre of a base length of about 1250mm, was achieved without any failure of wire mesh or fall of fragments from the fractured shotcrete shell.

D 185

Figure 10 : In-situ test of shotcrete, after 2,5mm deflection.

Although most of the tests which were carried out were simple-minded and cursory, they provided some very useful indications of the characteristics and behaviour of some of the elements in common use. In particular, a sufficiently good insight was obtained into the requirements of tendons to realize that insistence on strong bonding is inappropriate under high stress or dynamic loading conditions.

In fact it is submitted that it is necessary to limit the bond strength and to allow slipping against somewhat reduced frictional resistance rather than to risk the possibility of rupture of tendons and the total loss of further resistance.

Very likely it is because the sidewalls are intensely fractured and the fracture spacing is consequently closer than the critical bond length, that the tendons are not usually broken. However this is not always the case and failed tendons are frequently observed particularly after serious rockbursts.

It is felt that there is considerable scope for improvement of individual components and that, based on the above principles, tunnel support can be designed to withstand all but the largest seismic events at very close focal distances.

6. CONCLUSIONS

An attempt has been made to relate conditions observed in failed tunnels under severe stress conditions to the measured characteristics of the various elements involved in their support.

7. ACKNOWLEDGEMENT

The permission of Rand Mines Limited, for the publication of this paper is gratefully acknowledged.

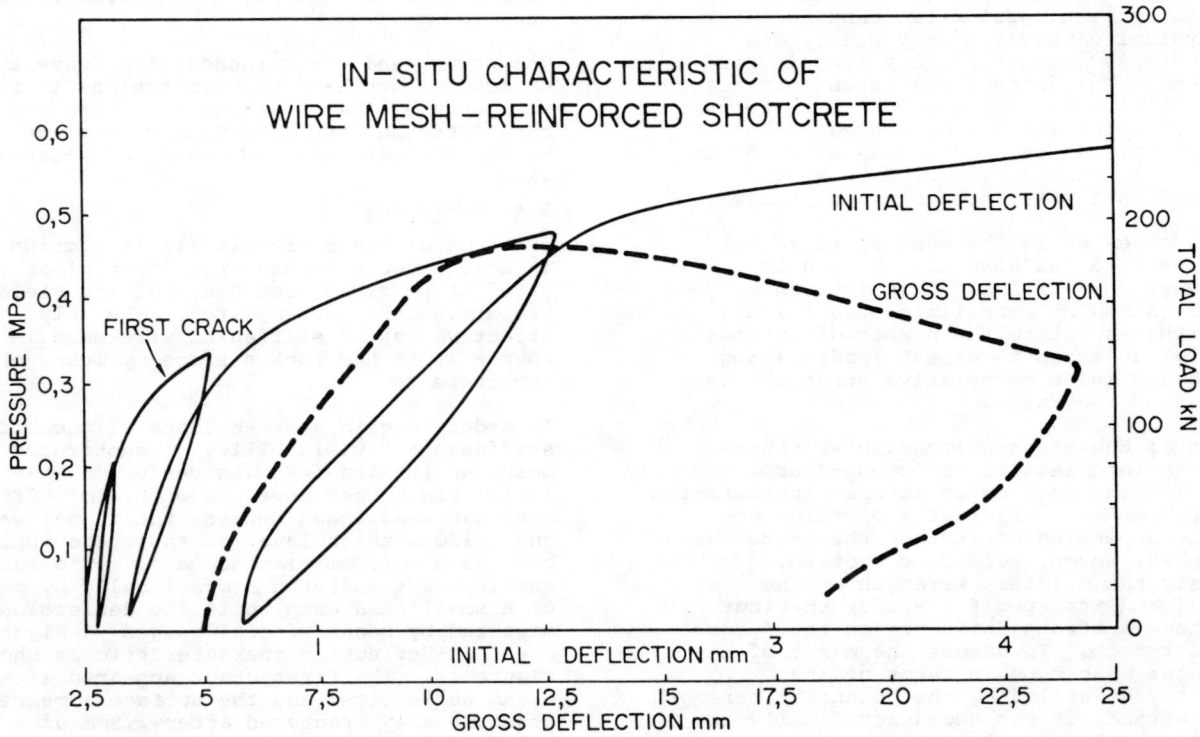

Figure 11 : Load-deflection characteristic of mesh-reinforced shotcrete.

TABLE I - SPECIFIC STRENGTH OF VARIOUS TYPES OF MESH

MESH TYPE	GRAPH NO.	DIMENSIONS mm (STRENGTH) MPa	CROSS SECTIONAL AREA (mm^2)		LOAD AT FAILURE kN	RELATIVE STRENGTH INDEX		
			Nominal[1]	Loaded[2]		Nominal[3]	Loaded[4]	Normalized[5]
DIAMOND STAINLESS (type 316)	3	2,5 x 55 (650)	123	49	71,3	58	145	22,3
(type 304)	4	2,5 x 65 (650)	103	41	66,8 / 64,1	65 / 62	162 / 156	24,9 / 24,0
DIAMOND MESH, GALVANIZED	-	3,2 x 55 (450)	195	86	60,0	31	70	15,5
	1	3,2 x 50	218	86	58,8 / 57,7	27 / 26	68 / 67	15,1 / 14,8
	2	4,0 x 105 (450)	163	75	51,0 / 44,5	31 / 27	68 / 59	15,1 / 13,1
	-	4,0 x 100	176	75	50,0	28	67	14,8
SQUARE WOVEN, STAINLESS (type 304)	5	2,3 x 60 (1000)	69	30	51,8 / 50,0	75 / 72	173 / 166	17,3 / 16,6
STRANDED WIRE WEAVE	7	3,2x125; 2,2x110 (1200)	73	40	38,0	52	95	7,9
	6	3,2x110; 2,5x120 (850)	81	39	30,0	37	76	8,9
	-	3,2x100; 2,0x75 (850)	81	31	23,0	21	73	8,6
	-	3,2x125; 2,5x110 (850; 1200)	88	39	16,0	18	41	4,8
HEXAGONAL TWISTED	8	3,0 x 120 (450)	145	43	28,0	19	64	14,3
	-	2,7 x 100 (450)	150	42	25,0	17	59	13,2
	-	2,7 x 120	119	36	20,0	17	56	12,5
WELD-MESH	9	3,2 x 100 (485)	78	39	20,0 / 19,0	26 / 24	51 / 49	11,4 / 10,8
POLYETHYLENE ROPE NET	10	10 x 100	310	125	60,6	19	48	-
	-	8 x 150	140	55	14,0	10	25	-

NOTE:
1. Cross-sectional area of strand multiplied by number of strands per linear metre.
2. Cross-sectional area of strand multiplied by number of strands passing under the short side of central loaded area.
3. Failure load x 100 divided by nominal cross-section area.
4. Failure load x 100 divided by loaded cross-section area.
5. Loaded relative strength x 100 divided by strength of wire.

8. REFERENCES

1. Ortlepp, W.D., More O'Ferrall, R.C. and Wilson, J.W. Support Methods in Tunnels. Symp. on Strata Control & Rockburst Problems of the S.A. Goldfields. A.M.M. of S.A. 1972-73. Johannesburg 1975. pp. 167-195.

2. Cook, N.G.W. Siting of Mine Tunnels and Factors Affecting their Layout & Design. Symp. Strata Control & Rockbursts Problems of S.A. Goldfields. A.M.M. of S.A. 1972-73. Johannesburg 1975. pp.199-215.

3. Ortlepp, W.D. An Empirical Determination of the Effectiveness of Rockbolt Support under Impulse Loading. Proc. Int. Symp. Large Permanent Underground Openings. ed. T.L. Brekke & F.A. Jorstad, Universitetsforlaget, Oslo, 1970. pp. 197.205.

4. Wagner, H. Support Requirements for Rockburst Conditions. Proc. First Int. Symp. on Seismicity in Mines. Johannesburg 1982.

5. Goodman, R.E. & Ewoldsen, H.M. A Design Approach for Rockbolt Reinforcement in Underground Galleries. Proc. Int. Symp. Large Permanent Underground Openings. ed. T.L. Brekke & F.A. Jorstad, Universitetsforlaget, Oslo, 1970. pp. 181-195.

6. Ortlepp, W.D. Rockbursts in South African Gold Mines: A Phenomenological View. Proc. First Int. Symp. on Seismicity in Mines. Johannesburg 1982.

7. Pells, P.J.N. The Behaviour of Fully Bonded Rockbolts. Proc. Third Congress of Int. Soc. Rock Mechanics. Denver 1974. pp. 1212-1217.

DESIGN OF TUNNELS IN SWELLING MARLS
Conception de tunnels dans les marnes gonflantes
Entwurf eines Tunnels in blähenden Mergeln

Manuel Romana
Professor of Geotechnical Engineering, Polytechnical Univ., Valencia, Spain

Davor Simic
Civil Engineer, INTECSA, Spain

SYNOPSIS

The problem of the tunnel excavation in a swelling rock is frequently encountered in many places. Expansive formations include many different rocks such as anhydrite, marls, shales and so forth. Mineralogical analyses and geotechnical property tests can be used to predict swelling behaviour, and special tests carried out in oedometer cells to quantify the swelling deformation of a given rock. These parameters can be used to compute the forces acting on the lining. This paper studies the expansive properties of some marly rocks in Algeria and Spain correlating them with some geological and mineralogical factors. The method of calculation of forces on the lining is also described, showing some representative results.

RESUME

Il s'agit d'un problème qu'on rencontre souvent. Les formations gonflantes comprennent plusieurs sortes de roches tels que l'anhydrite, les marnes et les ardoises. Pour mettre en relief le caractère gonflant, on peut employer des analyses minéralogiques et des essais dans des cellulles oedométriques. Ces paramètres peuvent être utilisés pour définir les forces agissant sur le revêtement. Ce rapport étudie les propriétés gonflantes de quelques roches marneuses en Algérie et en Espagne en liaison avec quelques facteurs géologiques et minéralogiques. On présente également la méthode de calcul des efforts sur le revêtement avec quelques résultats représentatifs.

ZUSAMMENFASSUNG

Die Konstruktion von Tunneln in blähendem Fels ist ein häufiges Problem. Von Laboratoriumsversuchen kann man die Material-Kennwerte erhalten. Dann ist es möglich, die Spannungen an der Auskleidung zu berechnen. In diesem Beitrag werden geotechnische Kennwerte für Mergel von Algerien und Spanien mit geologischen und mineralogischen Faktoren verbunden. Es wird die Berechnungsmethode für die Spannungen an der Auskleidung beschrieben.

1. INTRODUCTION

Within the works of the second track of the railway line Martorell-Castellbisbal (Barcelona, Spain), the Spanish Railway Company (RENFE) is constructing a 850 m long tunnel, with a maximum rock cover of 100 m. The survey of ground conditions established that aproximately one half of the tunnel alignment is bored through Miocene claystones and marls, with fine interbeddings of dolomies, limestones and gypsum; being the remaining half bored through Ordovician shales. The Miocene strata, which are discordantly bedded against the Ordovician substratum, dip gently (less than abour 15°) parallel to the tunnel axis. Figure 1 shows a geological profile along the tunnel alignment. Preliminary surveys showed the expansive properties of Miocene marls, so it was a mayor concern during the design stage to account for the forces on the lining due to swelling phenomena.

An also important tunnelling design involving swelling rock was studied by the Authors in Algeria for the Bou-Roumi hydraulic scheme. It consists of the design of three tunnels for the diversion of the waters of the rivers Oued Chiffa, Oued Djer and Oued Harbil (see figure nr. 1). The three tunnels are 12,5 Km, 3 Km and 4,5 Km long, with maximum rock cover of 500 m, 200 m and 80 m respectively. This geological province is located south of the Telliennes Chains, with Cretaceous and Miocene materials. The Cretaceous rocks are schists and calc-schists and the Miocene rocks are a thick series of marls with a basal level of limestones and sandstones. The Cretaceous formation is the northern bank of a great anticlinorium and the Miocene materials rest discordantly on it with a gentle northward dip. The Miocene marls show swelling behaviour, and are the subject of study in this paper.

2. GEOTECHNICAL PROPERTIES OF THE SWELLING MARLS

2.1 Properties of the matrix

In both cases, the swelling rock belongs to the lithological group of marls. However, there are significant differences between the rocks of the two projects. Plates nr. 1 and nr. 2 show the main geotechnical characteristics of these

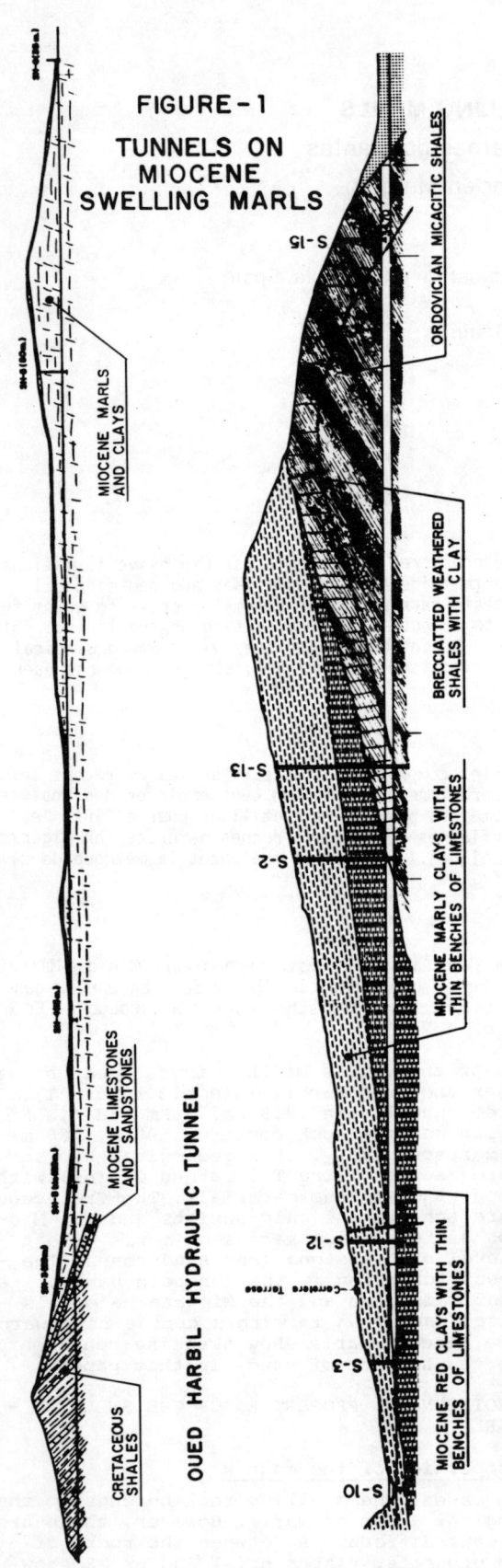

FIGURE - 1
TUNNELS ON MIOCENE SWELLING MARLS

OUED HARBIL HYDRAULIC TUNNEL

MARTORELL RAILWAY TUNNEL
GEOLOGICAL CROSS SECTION

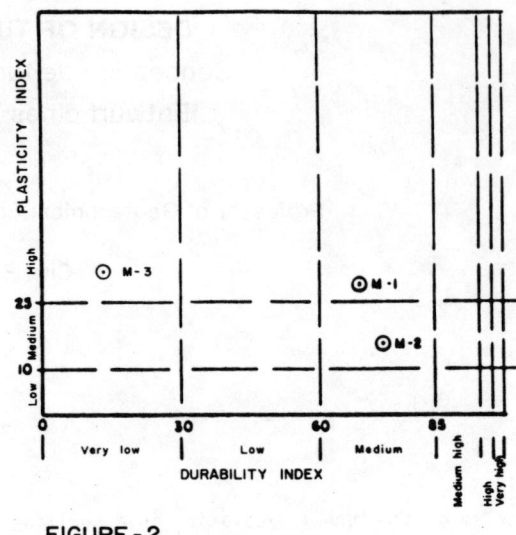

FIGURE - 2
MIOCENE SWELLING MARLS
GAMBLE CRITERIUM OF ROCK DURABILITY

M-1 } MARTORELL RAILWAY TUNNEL
M-2
M-3 OUED HARBIL HYDRAULIC TUNNEL

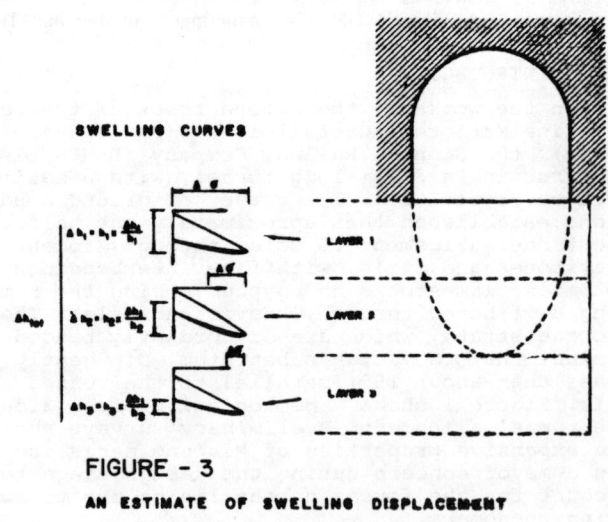

FIGURE - 3
AN ESTIMATE OF SWELLING DISPLACEMENT

rocks. It can be seen that the Algerian marl contains less carbonates and is much more plastic than the Spanish one. From the mineralogical standpoint, the latter has less proportion of clayey minerals, only a 20% against the 50% of the former. In both cases, a major clay-mineral constituent is montmorillonite.

Compression laboratory tests show a great dispersion in both rocks, partly because of some calcitic or dolomitic interbedding in very thin layers which appear somewhat randomly.

2.2 Joints of the rock mass

As both marls have a high amount of clay, they behave more or less plastic under the tectonic stress fields. Therefore faults are neat, with smooth surfaces filled with calcite or, in the Spanish project, with gypsum. No breccia materials are found along those faults. Stratification is slightly apparent, except for the case of sandstone layers interbedded in the Algerian formation. Joints, that are well developed in the more calcareous materials, don't seem to break the marly formation. From this study it is concluded that discontinuities are not significant for the behaviour of the marls.

3. BEHAVIOUR OF THE MIOCENE MARLS TO WATER ACTION.

A major group of rock materials, commonly featuring a high clay content, have a tendency to swell, soften and disintegrate when subjected to quick wetting or drying. To predict such mechanical behaviour, there are different kinds of tests suggested by the ISRM (1972), that give a simple way of identifying and assigning a property index to each material. Given the high clay content of the Miocene marls, and the detection of montmorillonite as a constituent, some ISRM suggested test were carried out.

3.1 Slake durability test

This test measures the strength of the rock to the weakening and disintegration when subjected to two standard cycles of drying and wetting. The test result is the weight of dry sample in the end of the process, as percentage of the initial weigh. Figure nr. 2 shows the classification of the Miocene marls according to the Gamble criterium, which combines the durability index with the plasticiy index.

3.2 Swelling index

This test measures the axial swelling deformation under a constant axial load, when an intact rock sample with radial restraint is saturated. Some results obtained from samples of Miocene marls, tested in oedometer cells, are shown in Plate nr. 2.

4. Analysis of the forces against the lining of a tunnel bored in a swelling rock

The swelling behaviour of some rocks has yielded major damage in the lining of many tunnels (Schillinger, 1970; Groß, 1972 ; Einstein et al, 1975; Wittke, 1978).

From the observation of such failures, it has been concluded that a great part of deformation takes place as invert heave. The inward movement of the lower abutment part is usually not due to lateral swelling of the rock but invert heave, as clear voids are found in that zone between the lining and the surrounding rock (Einstein et al, 1975).

Basically, the swelling process is switched on by the tensional changes in the rock mass due to tunnel excavation. In the vicinity of the opening, the rock follows an unloading stress path yielding to water absoption from the surrounding mass and leading to swelling deformation which in some cases is very noticeable in the form of invert heave or lining failure.

To evaluate the lining forces due to the expansive character of Miocene marls, the authors have followed the method proposed by Wittke and Rißler (1976), which tries to adjust the model for calculation to the real swelling phenomenon: the rock swells when subjected to an unloading stress path (given the necessary amount of water to switch on the process). As this method takes into account the rock-lining interaction, it is more realistic than the simplified procedures that estimate the forces of the lining when subject to a loading equal to the swelling pressure of the rock.

4.1 Input data of swelling deformation

The deformability of the rock due to swelling is studied by Huder et al (1970), who suggest a standard test to quantify the expansive deformation. This test has been widely mentioned in the literature(Einstein et al, 1975; Wittke et al, 1976; Gysel, 1977) due to its simplicity. An intact sample of swelling rock is put in an oedometer cell and subject to a cycle of loading - unloading - reloading with its natural moisture content. In this situation, it is saturated and subjected to an unloading path in succesive steps measuring the axial deformation.

This methodology has been applied to different representative samples of the Miocene marls. In figure nr. 4 it is shown the loading paths vs. deformation in a semi-logarithmic scale, where it can be appreciated a linear law relating the unloading and the swelling deformation:

$$\Delta \varepsilon_z = K \log \frac{P}{P_o}$$

where

$\Delta \varepsilon_z$ = unit vertical expansion, when the axial stress decreases from P_o down to P.

K = coeficient depending on rock sample. From figure nr. 4 , it can be seen that in the case of the Algerian Miocene marl, there is a progressively increasing K - value with proximity of the sample to the centre of the sedimentary bassin. Plate 2 shows the average values adopted in computations.

PLATE 1: INDEX PROPERTIES AND MINERALOGY OF TWO SWELLING ROCKS

Project	Dry Density (T/m3)	Moisture content (%)	Liquid Limit	Plasticity Index	Carbonate content (5)	Mineralogy	
						Quartz + carbonates	Clayey Minerals
Martorell Railway tunnel	2,40	4,5	34	15	40	on total: 80%	on 20%: Illite: 60 Montmorillonite: 30 Kaolinite: 10
Oueds Hydraulic tunnels	2,27	6	54	30	25	on total 50%	on 50%: Interstratified Illite+Montmorillonite:75 Kaolinite: 25

PLATE 2: LABORATORY STRENGTH AND DEFORMABILITY OF TWO SWELLING ROCKS

Project:	Unconfined compression strength (MPa)	Ratio Elasticity Modulus / Compression Strength	Poisson Modulus
Martorell Railway tunnel	1,5 to 16	100 to 200	0,15
Oueds Hydraulic tunnels	8,5 (average)	110 (average)	0,25

PLATE 2

SWELLING DEFORMATION INDEX OF MIOCENE MARLS

	INITIAL Moisture %	LOAD OF SATURATION (MPa)	FINAL MOISTURE	SWELLING DEFORMATION %
Martorell Miocene Marl	5,7 / 7,3	0,3 / 0,6	12 / 13	0,3 / 0,5
Oued Harbil Miocene Marl	12,7 / 12,6 / 25,1 / 11,6	0,01 / 0,16 / 0,22 / 0,8	13,5 / 15,5 / 29,7 / 15,1	0,12 / 2,45 / 7,8 / 9,4

PLATE 2

SWELLING MODULS OF MIOCENE MARLS

$$\Delta\varepsilon_z = K \log \frac{P}{P_o}$$

SAMPLES		AVERAGE K - VALUES
MARTORELL RAILWAY TUNNEL		$26,5 \cdot 10^{-3}$
OUEDS HYDRAULIC TUNNELS	DERIVATION DJER	$6,00 \cdot 10^{-3}$
	DERIVATION HARBIL - FIRST STRETCH	$8,00 \cdot 10^{-3}$
	DERIVATION HARBIL CENTER OF SEDIMENTARY BASSIN	$46 \cdot 10^{-3}$

The aforementioned law is strictly valid for oedometric stress paths (lateral restraint). A generalisation for tridimensional states of stress is proposed by Wittke, writing the equation above in terms of variation of sample volume and the first stress invariant. As the authors' analysis is carried out in a bidimensional model of plane deformation, it is more adequate to find the elasticity modulus from the oedometer state from the expression:

$$E = \frac{\Delta P}{\Delta\varepsilon z} \left(1 - \frac{2 K_o^2}{1 + K_o}\right)$$

where:

E = secant elasticity modulus in the step from P_o to $P_o - \Delta P$

K_o = coeficient of pressure

This modulus is valid only for the areas of the rock mass where there is an unloading stress path. For the zones of the rock where there is a loading path, it is assumed a behaviour defined by the elasticity modulus obtained from the unconfined compression tests. This assumption is justified by the fact of the massive character of the Miocene marl formation.

4.2 Method of calculation

If it is assumed that the oedometer stress conditions hold for the in-situ rock, it would be quite simple to calculate the swelling deformations (Einstein et al, 1975). The stress difference between the primary state and after the excava-

FIGURE-4
SWELLING BEHAVIOUR OF MARLS
RESULTS OF EXPANSION TESTS IN OEDOMETER CELLS

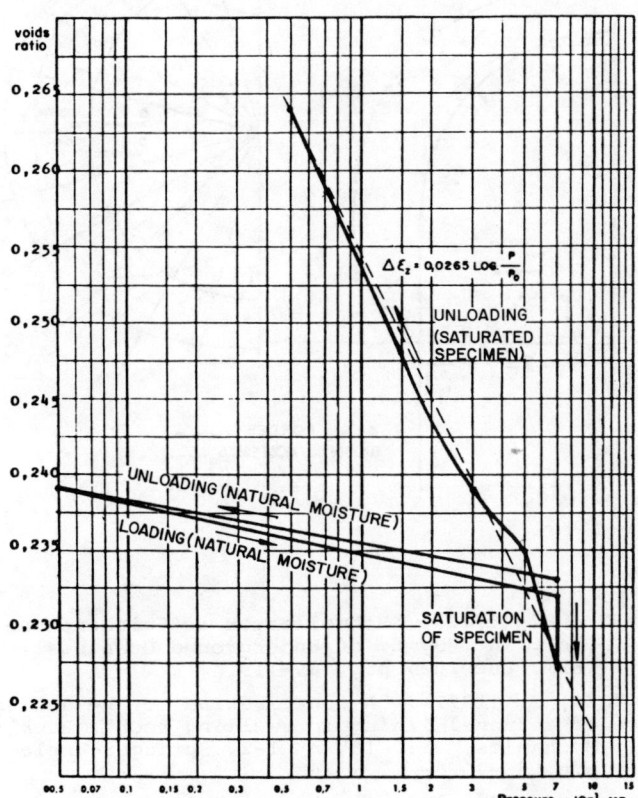
RAILWAY TUNNEL - MARTORELL (SPAIN)

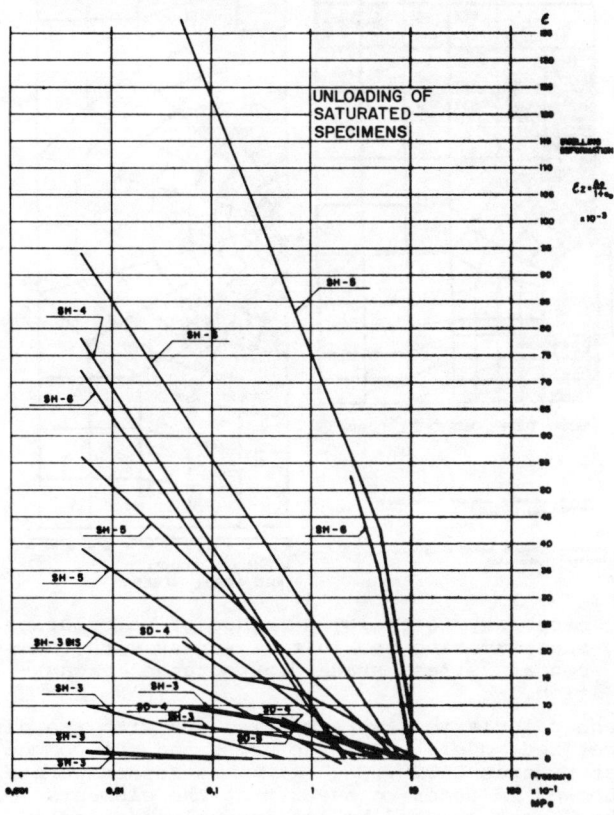
HYDRAULIC TUNNELS - OUEDS (ALGERIA)

tion corresponds to a swelling deformation in the diagram (see figure nr. 3) which, multiplied by the thickness of the i-layer, h_i, gives the displacement of this layer (see figure nr. 3). However, this simplified procedure involves gross simplification in assuming uniaxial behaviour and taking no account of the rock-lining interaction. Gysel (1977) presents a calculation method based on a volumetric swelling factor obtained from the Huder-Amberg test, and the model of a circular tunnel in an elastic space. This method is formally similar to the one proposed by Wittke and Rißler (1976).

The model employed by the authors in the case of Martorell railway tunnel and Oueds hydraulic tunnels, involves the simulation of tunnel construction within the rock mass taking into account the fact that the stress variation will yield a swelling behaviour of those zones of rock that unload, whereas the zones where the rock suffers a loading process, the assigned moduli correspond to the intact sample.

To carry out such simulation, a method of calculation is needed that allows for the rock-lining interaction and for different moduli for different rock zones. The finite element method was chosen because fulfills the above needs. The computer program employed was SSTIP, developed by the Berkeley University. This program features the additional advantage of considering different moduli according to stress levels, which is particularly useful in this study as the unloading modulus varies logarithmically with the stress.

With the above assumptions, the computation method was as follows:

a) In a first estimate, the rock zones which load and unload due to the tunnel excavation are established. This was done by means of the approximate method of considering the elastic stress state around a circular tunnel (Savim, 1961).

b) A discretization of the rock mass and tunnel lining is carried out, using isoparametric finite elements for the rock and beam elements for the lining. The simulation is made in terms of plane deformation. Figure nr. 5 shows the meshes employed in computations.

c) As the tunnel excavation is a variation of geometry of an elastic space subject to a given stress field (gravitatory loads), it is

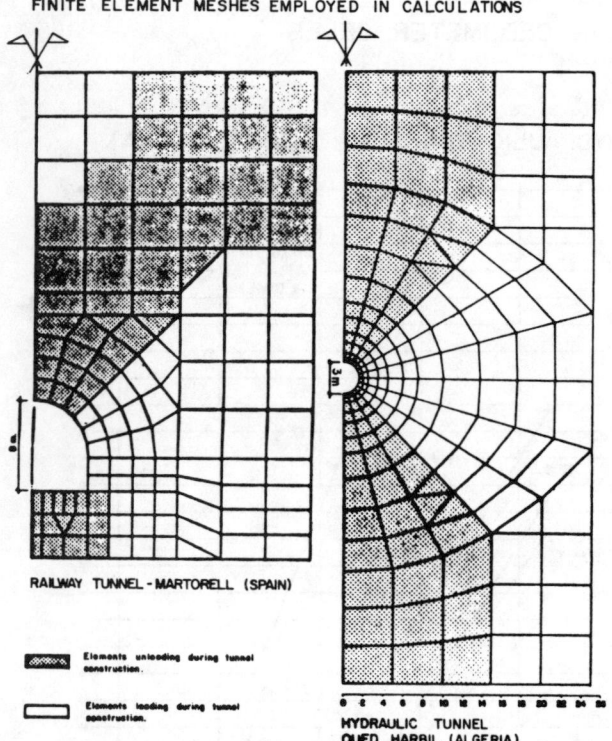

FIGURE - 5
FINITE ELEMENT MESHES EMPLOYED IN CALCULATIONS

RAILWAY TUNNEL - MARTORELL (SPAIN)

HYDRAULIC TUNNEL
OUED HARBIL (ALGERIA)

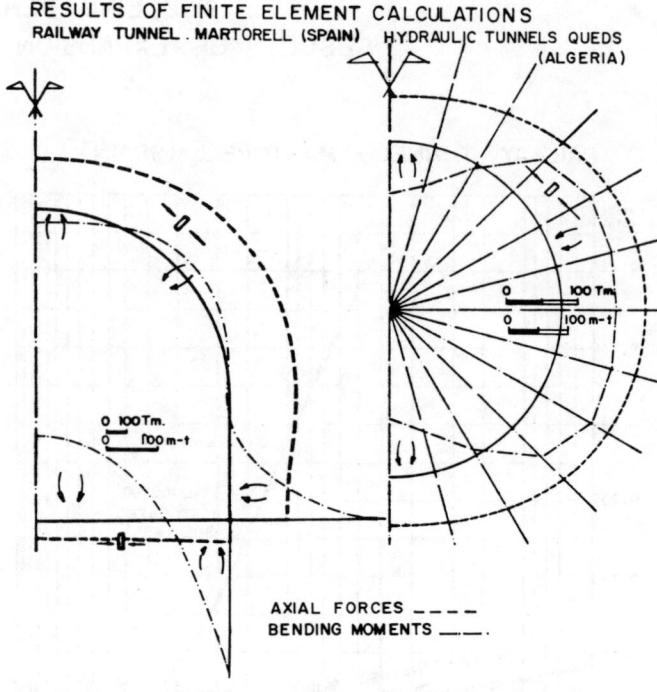

FIGURE - 6
RESULTS OF FINITE ELEMENT CALCULATIONS
RAILWAY TUNNEL. MARTORELL (SPAIN)   HYDRAULIC TUNNELS QUEDS (ALGERIA)

AXIAL FORCES -----
BENDING MOMENTS -----

necessary to employ a method of calculation superimposing the initial stress with the secondary stress state due to tunnel excavation.

Such calculations were carried out with the discretized models of figure nr. 5 and the computer program SSTIP. The output stresses were checked in order to asses that the elements - which were assumed to lead an unloading path, effectively followed it. Some iterations were needed to adjust all the elements.

4.3 Results of the computations

Figure nr. 6 show the axial forces and bending moments obtained from the computations. It can be seen that overstressed sections of lining in the invert of horseshoe section are obtained, whereas the circular section works much better.

5. ACKNOWLEDGEMENTS

The authors wish to thank the permission of Spanish railway system, RENFE and the Algerian Hydraulic Direction, DGIH to publish the results of the investigations. The help of INTECSA is also acknowledged.

6. BIBLIOGRAPHY

EINSTEIN, H.H.; BISCHOFF, N. (1975). "Design of tunnels in Swelling rock" 16 th Symp. on Rock Mechanics. Minneapolis. ISRM.

GAMBLE, J.C. (1971). "Durability-Plasticity Classification of shales and other argillaceous rocks". Ph. D. Thesis, Univ. of Illinois. Citado en ISRM (1972).

GROB, H. (1972): "Schwelldruck in Belchentunnel" Proc. Int. Symp. on Underground Openings. - ISRM. Lucerna, p.p. 99-119.

GYSEL, M. (1977): "A contribution to the design of a tunnel lining in swelling rock". Rock - Mechanics, vol. 10, n° 1-2. Springer-Verlag, p.p. 55-71.

ISRM, (1972): "Suggested methods for determining water content, porosity, density, absorption and related properties and slake durability - index properties". Comision on standardization of laboratory and field tests. Lisboa.

PORTILLO, E.; ROMANA, M., CEDRUN, G. (1981): - "El uso de las clasificaciones geotécnicas - en el túnel ferroviario de Martorell". SUIS. Madrid.

SAVIM, G.N. (1961): "Stress concentrations - around holes". Pergamon Press.

SCHILLINGER, G. (1970): "Die Feldsdrücke im - Gipskeuper beim Bau des Belchentunnels". Publication n° 80 Schweiz Ges. für Bodenmechanik und Foundationstechnik, p.p. 31-36.

WITTKE, W. (1978): "Fundamentals for the design and Construction of tunnels located in swelling rock and their use during construction of the Turning Loop of the Subway Struttgart. Univ. of Aachen.

WITTKE, W.; RIBLER, P. (1976): "Dimensioning the lining of underground openings in swelling - rock applying the finite element method". - Univ. of Aachen, p.p. 7-47.

CONCRETE SUPPORT COMBINED WITH OUTER AND INNER STEEL SETS IN THE HEAVY PRESSURE ZONE

Supports en béton combinés avec cadres métalliques internes et externes dans la zone de hautes pressions

Ein mit äusseren und inneren Stahlausbauten versehener Betonausbau im sehr druckhaften Gebirge

Ryoji Kobayashi
Professor, Tohoku University, Sendai, Japan

Norimasa Kubo
General Manager, Dowa Mining Co. Ltd., Ohdate, Japan

Koji Matsuki
Research Associate, Tohoku University, Sendai, Japan

SYNOPSIS

At the Matsumine mine in Japan, a tougher concrete support combined with outer steel sets and inner steel rings was developed in order to establish a permanent support system in the heavy pressure zone instead of repeating repairs of the drifts. To make clear the mechanism of this support, in situ measurements on the strains both in the inner rings and in the concrete lining were carried out for more than a year at the mine, and the effects of support parameters such as the thickness of the concrete lining and the spacing of steel sets are analyzed with the Finite Element Method.

RESUME

A la mine de Matsumine au Japon, on a mis au point un support plus solide en béton et l'a combiné avec des cadres métalliques externes et des anneaux de soutènement internes de manière à établir un dispositif de soutien permanent dans la zone de hautes pressions pour éviter la réparation continuelle des dérives. Afin de clarifier le mécanisme de ce support, on a mesuré in situ pendant plus d'un an des efforts de contrainte subies par les anneaux métalliques de soutènement et le revêtement de béton. Les effets des paramètres de soutien tels que l'épaisseur du revêtement en béton et l'espacement des cadres métalliques sont analysés à l'aide de la méthode des éléments finis.

ZUSAMMENFASSUNG

Im Matsumine-Bergwerk in Japan wurde ein mit äußeren Stahlausbauten und inneren Stahlringen verbundener starker Betonausbau entwickelt, um ein dauerhaftes Ausbausystem in der schweren Druckzone zu schaffen, statt wiederholte Reparaturen der Strecke in Kauf zu nehmen. Um den Mechanismus des Ausbausystems zu klären, wurden in diesem Bergwerk in mehr als einem Jahr die Verformungen in den inneren Stahlringen und in der Beton-Auskleidung gemessen und die Wirkung der Ausbauparameter, wie die Dicke der Beton-Auskleidung und der Abstand der Stahlausbauten, mit endlichen Element Methoden analysiert.

1. INTRODUCTION

A large portion of non-ferrous metals in Japan has been produced by the black ore mines.
Since black ore deposits lie in the vicinity of the weak rock formations of mudstones or argillized tuffs, the drifts nearby the deposits are exposed to the heavy rock pressure which makes heavy steel rings useless in a few months. Therefore, overcoming the difficulties such as the road closure has been an important problem to reduce the maintenence cost of the drifts, especially in the main haulage levels.

A tougher support system was developed at Matsumine mine, Dowa Mining Co. Ltd., in 1975 in order to establish a permanent support system instead of repeating repairs of the drifts and has been employed successfully. In this support, the drifts are usually driven with steel ribs at first and after installing steel rings inside, concrete is filled in the surrounding voids. In the heaviest pressure zone, the outer steel ribs are replaced to the steel rings.

To make clear the mechanism of this support and contribute to the rational design of the support system, in-situ measurements on the strains both in the inner rings and in the concrete lining were carried out at the mine for more than a year, and the effects of the support parameters such as the thickness of the concrete lining and the spacing of the steel sets are analized with finite element method for axi-symmetric cases.

2. IN-SITU MEASURMENT

In-situ measurements were carried out at two sites of 195m level of the mine. Both sites are in the heavy pressure zone and the depth is about 260m from the surface. One is a main haulage drift in the argillized tuff zone with-

Table I. Mechanical and physical properties of the argillized tuff.

Apparent Specific Gravity	Water Content (%)	Porosity (%)	
2.14-2.32	11.6-16.3	27.0-39.3	
P-wave Velocity (km/sec)	Tensile Strength (MPa)		
1.82-2.25	0.44-1.20		
Young's Modulus (MPa)	Maximum Differential Stress (MPa)		
487-1656	(Confining Pressure (MPa))		
Poisson's Ratio	(0)	(2.94)	(5.88)
0.09-0.23	1.66-8.58	3.32-9.25	6.24-9.28

in the pyrite zone and the other is an access drift in the more or less argillized tuff zone.

Several blocks of the argillized tuff were picked up from the access drift and cylindrical rock specimens were prepared in a dry way keeping the water content the same as in-situ. Mechanical and physical properties of the rock sample are summarized in Table I. The viscosity of the dashpot in a Maxwell element in Burgers model obtained in the uniaxial compressive creep tests ranges from 10^7 to 10^9 MPa sec. These data show that the argillized tuff is not only liable to flow, but also soft and weak compared to the overburden pressure as the competence factor ranges from 0.27 to 1.35.

The cross sections of the drifts at measured sites are shown in Fig.1, where (A) is the main haulage drift and (B) the access drift. The diameter of the drift is 2.7m for both sites. The dimension of I-cross section for both the outer ribs and inner rings is 125mm × 100mm and the spacing of the steel sets is 60cm for both sites. At the access drift, the remarkable closure of the drift before pouring concrete made the concrete lining thinner. Yield stress of the steel is 421MPa and the uniaxial compressive strength of the concrete is 18.6MPa at four weeks of curing. Measured points in the inner rings are shown by open circles in Fig.1. The strains of two steel rings were measured at the main haulage drift and three rings at the access drift. Fig.2 shows the positions of the strain gauges glued to the steel ring. Six gauges are glued to one side; therefore, twelve gauges are used for a measured point. Strain gauges are waterproof types and they were covered with steel boxes filled with epoxy resin as a protector. As for the strain measurement in the concrete lining, two strain gauges of buried type were buried in advance in a block of mortar, which was, then, buried in the concrete lining at the center between the measured points of the steel rings. Two strain gauges in the mortar block are perpendicular each other; one is in the circumferential direction and the other parallel to the drift axis.

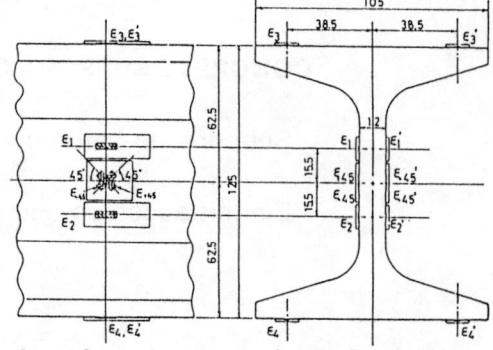

Fig.2. The positions of the strain gauges

Two blocks of mortar were buried at the main haulage drift and eight blocks at the access drift. Individual strain is measured one by one using dummy gauges in steel pipes buried in the concrete lining.

Bending moment in the plane of the steel ring M_z, bending moment in the lateral direction M_y, axial force N, shear force Q and torsional moment T were considered as internal forces in the steel rings. The relationships between the strains in Fig.2 and the internal forces are linear in the elastic region and there are more measured values than the unknowns. In this case, optimum internal forces are determined with least squares method, which has an advantage that the optimum values could be obtained if a few strain gauges should become useless because of the corrosion. The confidence intervals of the optimum values were calculated for every measurement to estimate the accuracy of the measurements and to detect the deteriorated strain gauges.

Figs.3 and 4 show the axial forces and the bending moments in the plane of the steel ring with time after pouring concrete at the access drift, respectively. Another internal forces are too small as to be neglected. The symbols such as "2R" in these figures represent the measured points. For example, "2R" represents the second and right-hand side measured point, and C1, C2 and C3 show the time when the concrete lining was constructed ahead of the measured point as the heading proceeded. The bending moments are positive and the axial forces are large at the right-hand side, while the bending moments are negative and the axial forces are small at the left-hand side. This suggests that the principal rock pressure is not vertical, but rather deviated from the vertical line to the left as diagrammatically shown in Fig.5. Although this principal rock pressure causes non-uniform change of the axial force when the stress in the rock mass is redistributed accompanied with heading or construction of concrete lining ahead of the measured point, the axial forces in the inner rings begin to increase steadily at both sides of the drift after a few months as seen in Fig.3. This phenomenon, which is found also at the main haulage drift related to the repair of the drift, seems to be caused by the hydrostatic secondary rock pressure created by the reaction of this rigid support. The rates of the maximum compressive stress at this stage range from 0.02 to 0.04 Kg/mm²/day at the main haulage drift and from 0.04 to 0.06 Kg/mm²/day at the access drift. The maximum compressive stresses show no symptoms of decrease through measure-

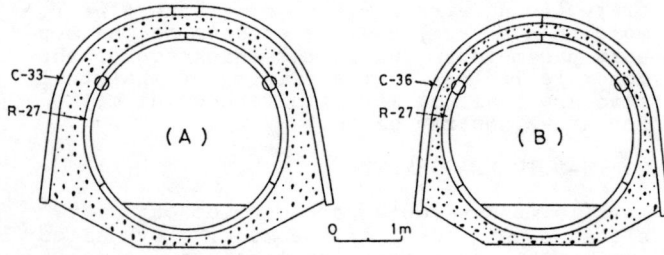

Fig.1. The cross sections of the drifts at the measured sites.

ments. Fig.6 shows the strains in the concrete lining with time at the main haulage drift. P and Q represent the directions of the strain gauges; P is parallel to the drift axis and Q in the circumferential direction. This figure clearly shows the hardening process of the concrete lining; the mortor block is, at the beginning, subjected to the hydrostatic pressure corresponding to the weight of the concrete, then, the compressive strains gradually decrease as the concrete hardens and, finally, the strains in the mortor block begin to increase as the concrete lining begins to carry the loads (see arrows in the figure). About 250 days after pouring the concrete, the strains began to decrease especially at the left-hand side as a few fractures initiated in the concrete lining. On the other hand, at the access drift, considerable fractures were initiated in the early stage due to the heavier rock pressure. However, in the concrete support in combination with outer ribs and inner rings, the fracture of the concrete lining does not directly result in the failure of the whole support in contrast with the conventional concrete support because the inner rings carry the load instead of the fractured concrete lining. Moreover, as mentioned previously, both the torsional moment and the bending moment in the lateral direction are still very small after more than 450 days. This means that the load bearing capacity of the inner rings is increased by the concrete lining restraining the lateral deformation of the inner ring even if the concrete lining is fractured. Consequently, it can be said that, cooperating each other, the inner rings and the concrete lining increase the load carrying capacity of the whole support.

3. FINITE ELEMENT ANALYSIS ON THE EFFECT OF SUPPORT PARAMETERS

The model for the axi-symmetric finite element analysis is shown in Fig.7. The unit hydrostatic pressure is applied to the outer boundary of the model so that the stress in the support may be expressed as the ratio to the applied pressure p. The displacements along the boundary vertical to the drift axis are fixed in the direction of the drift axis because of the symmetry. The dimension of I-cross section of the steel ring and the elastic constants of the materials are the same as those in-situ. Although the actual rock pressure is not always uniform, the fundamentals on the concrete support combined with the steel sets are obtained with this simple model. Some cases are analyzed including double rings and outer or inner rings alone. As the stress state in the support is three-dimensional, the maximum shear stress τ_{max} is adopted as the representative

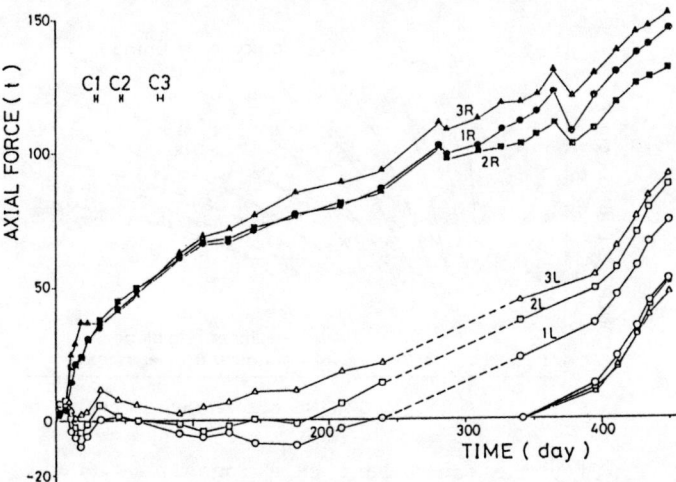

Fig.3. The axial forces in the steel rings with time in days (access drift).

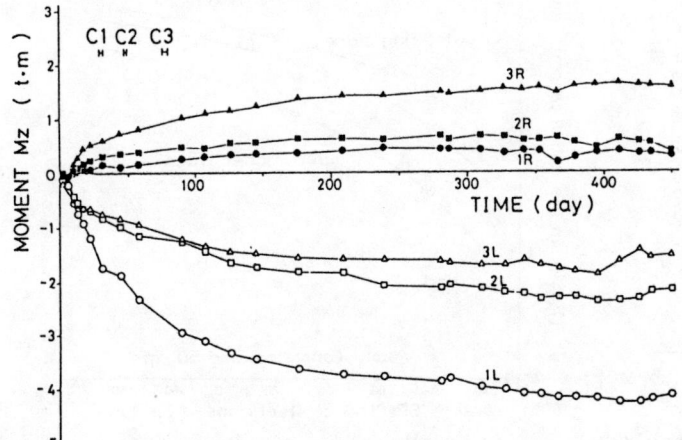

Fig.4. The bending moments in the plane of the steel ring with time in days (access drift).

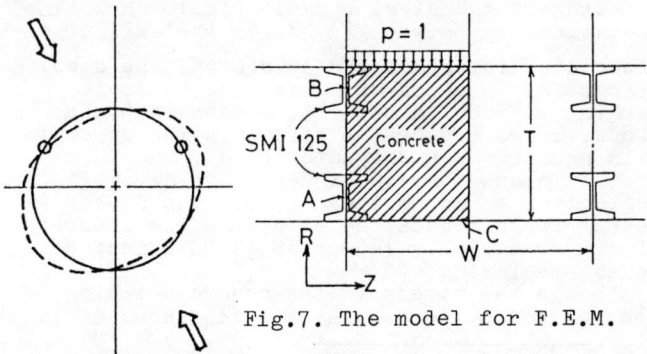

Fig.5. The direction of the principal rock pressure.

Fig.7. The model for F.E.M.

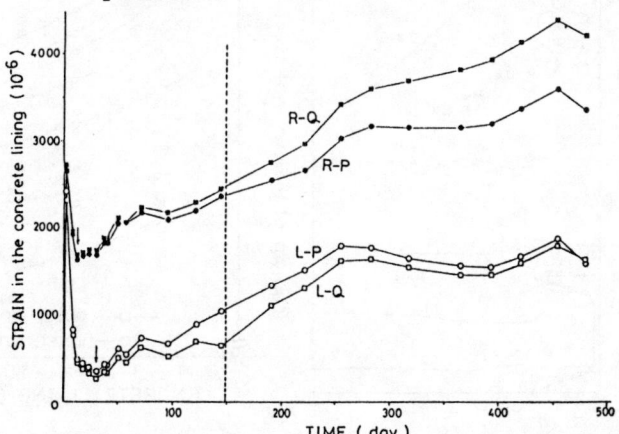

Fig.6. The strains in the concrete lining with time in days (main haulage drift).

stress. Fig.8 shows an example of the stress distribution in the concrete lining in the case of double steel rings, where the spacing of the steel rings is 60cm and the concrete lining is 50cm thick. The stress is not uniform in the concrete lining, but stress concentration occurs around the flanges of the I-cross section. Compared with the concrete lining, the stress state in the steel rings are rather homogeneous where the maximum shear stress is more than 25 times as large as that in the concrete lining corresponding to the ratio of Young's moduli. In order to estimate the effect of the support parameters, τ_{max} at the center of the steel ring (points A and B in Fig.7) is considered and τ_{max} in the wall of concrete lining nearly at the center between steel rings (point C in Fig.7) is extrapolated.

The relationships between τ_{max}/p at the points A and B in the steel rings and the thickness of the concrete lining for the case of the double steel rings are shown in Fig.9 where W is the spacing of the steel rings. τ_{max} is larger in the inner ring than in the outer ring although the difference between them becomes smaller as the concrete lining becomes thinner. The effect of the thickness of the concrete lining is very small if the spacing of the steel rings is small. Similarly, τ_{max}/p at the point C in the concrete wall decreases with the thickness of the conceret lining and increases with the spacing of the steel rings. Fig.10 shows the relationship between τ_{max}/p in the wall of the concrete lining and the spacing of the steel rings for three cases, where the concrete lining is 50cm thick. It is seen that the inner rings reduce the stress in the concrete lining more than the outer rings, and this result means that the inner rings are more effective on supporting loads. The double steel rings reduce the stress in the concrete lining remarkably superposing the effect of each steel ring.

Although the stress in the concrete lining can be reduced by increasing the thickness of the concrete lining or decreasing the spacing of the steel rings as previously mentioned, it may be more economical to decrease the spacing of the steel rings rather than to increase the thickness of the concrete lining. Fig.11 shows the equivalent relationships between the spacing of the steel rings and the thickness of the concrete lining, which gives the same stress concentration in the wall of the concrete lining for both the case of double steel rings and that of inner rings alone. This curve is very steep if the allowable stress concentration is small as in the heavy pressure zone. This means that less reduction of the spacing of the steel rings can decrease more the thickness of the concrete lining, and this may be very important to design the support parameters.

4. CONCLUSION

As described above, it has been proved that in the concrete support in combination with outer and inner steel sets, the interaction between the inner rings and the concrete lining is very effective on supporting loads, and that the inner rings play more important role than the outer steel sets.

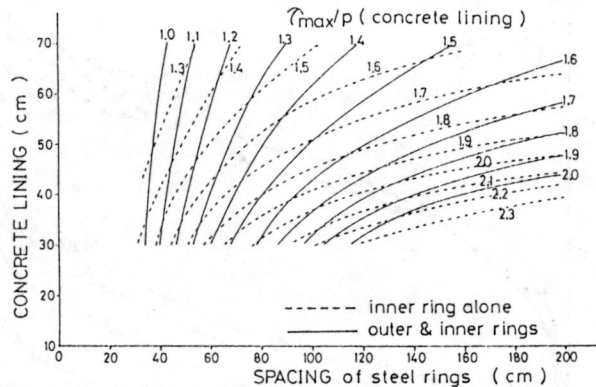

Fig.11. The equivalent relationship between the thickness of the concrete lining and the spacing of the steel rings.

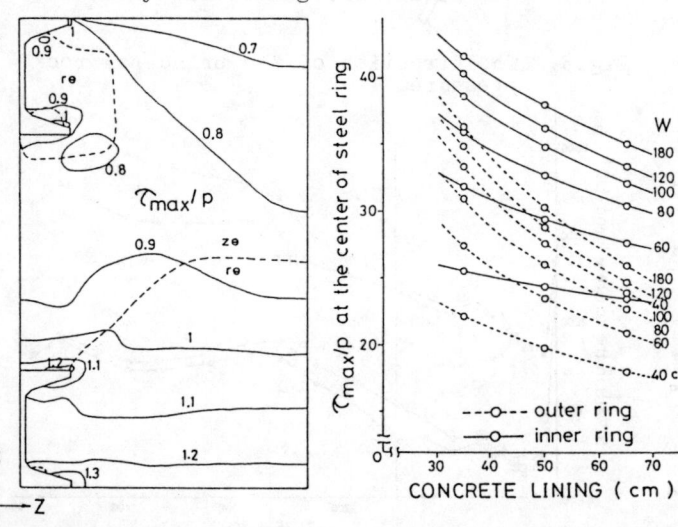

Fig.8. An example of the stress distribution in the concrete lining.

Fig.9. The effect of the concrete lining on τ_{max}/p at the center of the steel ring.

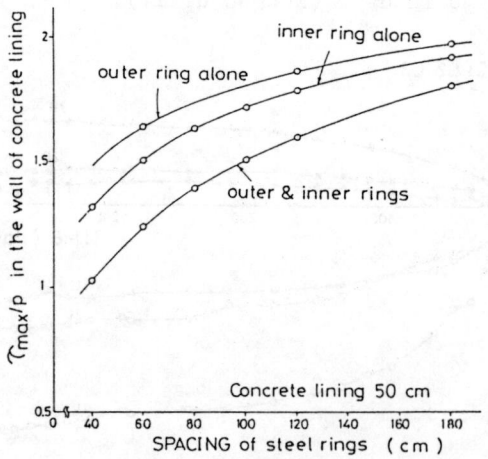

Fig.10. The comparison of τ_{max}/p in the concrete wall among three cases.

BEHAVIOUR OF ROCKS AROUND LARGE CAVERNS DURING EXCAVATION
Comportement de roches autour d'une grande caverne pendant l'excavation
Das Verhalten des Gebirges in grossen Untertagehohlräumen während des Ausbruchs

Satoshi Hibino
Manager, Fellow research engineer

Mutsumi Motojima
Tadashi Kanagawa
Senior research engineers
Central Research Institute of Electric Power Industry, Abiko-shi, Chiba-ken, Japan

SYNOPSIS
This paper presents characteristic features of rock behaviour during excavation obtained through field measurements at many sites for underground power station caverns. Some findings are as follows: (1) Joint openings play a great role in rock displacement. (2) Stresses in arched concrete linings do not depend on horizontal stress of ground pressure, but on the horizontal displacements of cavern walls. (3) The degraded states of relaxed zones are observed by means of the elastic wave velocity test, the Lugeon test, and the bore-hole television method.

RESUME
Cette communication présente les caractéristiques de comportement des roches lors d'excavations et obtenues par des mesures in situ relevées dans de nombreuses galeries de centrales souterraines. En voici quelques résultats: 1) L'ouverture des joints joue un rôle important dans le déplacement des roches. 2) L'effort dans le revêtement de béton de l'arc ne dépend pas de l'effort horizontal de la pression du sol mais du déplacement horizontal de la paroi de la galerie. 3) L'état d'altération des zones de décompression est observé par essais de vélocité des ondes élastiques de Lugeon et par introduction d'une caméra de télévision dans les sondages.

ZUSAMMENFASSUNG
Diese Abhandlung präsentiert die charakteristischen Eigenschaften des Felsverhaltens während des Ausbruchs, die durch in-situ Messungen an vielen Baustellen für Kavernen für unterirdische Kraftwerke erhalten wurden. Nachfolgend werden einige dieser Charakteristiken angeführt: (1) Kluftöffnungen spielen eine große Rolle bei der Gebirgsverformung. (2) Die Spannungen in den Betonauskleidungen hängen nicht von der horizontalen Spannung des primären Spannungsfeldes, sondern von der horizontalen Verdrängung der Kavernenwände ab. (3) Der degradierte Zustand der entspannten Zonen wurde durch elastische Wellengeschwindigkeitsprüfung, Lugeon-Prüfung und Bohrlochfernsehmethode nachgewiesen.

1. INTRODUCTION

A lot of large caverns for underground power stations have been excavated in Japan. The authors have been developing one method of numerical analysis to render these excavation sites as safe as possible (Hayashi et al., 1970). Comparison between forecasting by the method and measured results has proved the method to be fairly reliable (Hibino et al., 1977).

Presented in this paper are the characteristics of rock behaviour obtained through the measurements:

(1) the different characteristics of displacements in ceiling rocks and cavern walls,

(2) the typical behaviour of joint openings,

(3) the dependency of stresses in an arched concrete lining on convergences of cavern walls,

(4) the degradation of relaxed zones.

2. INITIAL CONDITIONS OF THE CAVERNS

Geological conditions, magnitudes of ground pressure, sizes of the caverns and the other conditions relating to the stability of caverns are listed in Table I.

The sizes of the caverns are nearly the same, about 50 m in height and 25 m in width. The vertical stresses of ground pressure range from 1.9 to 10.8 MPa, which means that the overlying rock depths are 80~480 m.

At each site the following in situ tests have been carried out to obtain mechanical properties of rock foundation. These include jack tests for deformability E_0, rock shearing tests for shearing strength τ_0, creep tests for creep factor α, and stress relief method (overcoring method) for ground pressure.

In advance of excavation, forecasting of rock behaviour around cavities had been performed with the numerical analysis method developed by the authors. During the excavation works, measurements of rock behaviour around the caverns were performed for the purpose of safety during construction and to ensure the validity of the numerical analysis. The main kinds of measurements were as follows: deformations of rock foundation (subsidences of ceiling rocks and horizontal displacements of walls of the caverns), stresses in arched concrete linings, and variation of stresses in the strands of reinforcement.

Table I Outline of the caverns and the initial conditions

Project		Kind of rock mass	Size of cavern (m)			Ground pressure (MPa)			Properties of rock mass[1]		
			Height	Width	Length	σ_h	σ_v	τ	E_0 (GPa)	τ_0 (MPa)	α
1 Kisenyama	(1969)	Shale, sandstone, chert	49.6	25.6	60.4	1.3	3.9	—	6 ~ 12	1.5 ~ 2.9	0.16
2 Niikappu	(1971)	Schalstein	44.6	19.6	52	3.3	4.4	0.7	24	2.4	0.16
3 Okutataragi	(1973)	Rhyolite, tuff, diabase	49	25	127	5.8	6.5		5 ~ 10	3.9 ~ 4.9	0.05
4 Nabara	(1974)	Granite	47	25	86	7.2	6.3	1.1	3 ~ 9	0.5 ~ 1.5	0.2
5 Oohira	(1975)	Sandstone, slate	45	23	83	5.7	7.8	1.6	10 ~ 29	1.0 ~ 2.5	0.17
6 Shintakase	(1975)	Granodiorite, diorite	56	28	166	2.0	5.9	2.2	14/7[2]	1.3/0.3[2]	1
7 Okuyoshino	(1976)	Shale, sandstone	42	20	160	6.6	6.9	2.3	13/6	2.0/0.8	0.3
8 Okuyahagi	(1978)	Granite	48.5	23.2	102	7.4	10.8	1.0	15/7	2.9/1.2	0.8
9 Numazawa No. 2	(1979)	Rhyolite	47.6	26	96.5	4.5	3.4	0.8	10	1.4	0.1
10 Tanbara	(1979)	Conglomerate	49.4	26.6	116	4.5	7.0	1.9	16 ~ 20	2.4 ~ 2.9	0.4
11 Arimine	(1979)	Granite	20.8	14.6	30	1.3	1.9	—	4/2	1.7/1.3	0.4
12 Honkawa	(1980)	Black shist	47.4	26.3	98	5.4	7.1	0.9	12/8	2.5/1.3	0.7

1) E_0: Deformability, τ_0: Cohesive strength, α: Creep coefficient   2) Anisotropy

3. CHARACTERISTIC DISPLACEMENTS OF CAVERNS DURING EXCAVATION

Displacements are usually measured with rock displacement meters. In order to detect the sizes of relaxed zones, the distribution of displacements should be cleared so that the measuring lengths of the displacement meters, which are fixed at the excavation surface, will often be 1, 3, 5, 10, 15 m, etc.

3-1 Subsidences of ceiling rock masses

A typical example of time history for subsidence in ceiling rocks is shown in Fig. 1 (Hibino et al., 1979). Observed results at the other sites have the same tendencies as those of Fig. 1. Fig. 2 gives the distributions of subsidences at the completion of excavation at several sites. From these Figs., the following findings are clear:

(1) Almost all of the total subsidence was generated at the stage of the arch part excavation, and the following excavation of the main part did not increase any subsidence; on the contrary, there was a tendency toward decreasing of the settlement. In Fig. 1 the results of the forecast by the numerical analysis are also shown, and rather good agreement between the measured and calculated settlement was found in both total values and time history.

(2) The largest values of the total measured subsidences occurred in a limited part of the rock surface in a region of only 0 ~ 5 m (Fig. 2). The calculated displacement distributions, however, showed rather large settlements in the deeper rock at a distance of more than 5 m from the surface. This discrepancy derives from the inelastic behaviour of the rock masses. Displacement of rock mass due to excavation consists of an elastic displacement and a non-elastic displacement. The latter one originates from joint openings etc., which were not fully considered in the numerical analysis at the present stage.

3-2 Displacements of walls

Characteristics of wall displacement are summarized below:

(1) The horizontal displacements of the walls increased continuously during each stage of the main part excavation, but did not increase after the completion of the cavern (Fig. 3).

(2) The wall displacements depended largely on the kinds of rock masses. In the case of the jointy type rock, the maximum displacements took a rather large value of 40 ~ 60 mm, which originated in joint openings. In the case of non-jointy type rock, the displacements were not so large. Fig. 4 shows one example of a granite site of the former case (Motojima et al., 1981-a): At the final excavation stage the magnitude of the relative displacement between points A and B (8 m length) reached 45.8 mm, which meant more than 5×10^{-3} strain was experienced. Such large strain could not be possible without joint openings.

4. STRESSES IN ARCHED CONCRETE LININGS

Stresses in arched concrete linings are evaluated by measurement with reinforcement strain meters. For estimation of stresses σ_c in concrete, the following equation is available:

$$\sigma_c = \sigma_s \times k \times E_c/E_s \qquad (1)$$

where σ_s is the stress in reinforcement, E_c and E_s are moduli of elasticity of concrete and reinforcement respectively. k is a corrective coefficient for shrinking and creeping of concrete material and has the value 0.68.

Fig. 5 shows the time history of reinforcement stresses σ_s. The excavation was finished at the beginning of Oct. 1976. The decrease in stress after that showed only seasonal variation. Figs. 6 and 7 show the relationships between the lining stress σ_c and the convergences or the lengths of caverns (Motojima et al., 1981-b). Typical features of lining stresses are as follows:

(1) Stresses in the lining did not depend on the overburden depth nor on the initial horizontal ground pressure (figures in parentheses in Fig. 6 show the values), but on the convergences of the caverns. In the case of jointy rock, wall displacements were, in general, larger than those of the non-jointy rock, so that the stresses in the linings were relatively larger.

(2) The stresses in the linings were found to depend linearly on the lengths of the caverns (Fig. 7). If the length of a cavern exceeded the width by two times, the convergence of a cavern would become nearly

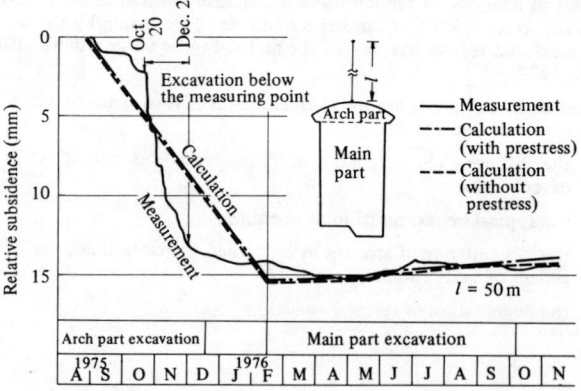

Fig. 1 Time history of the subsidence of the ceiling rock (project No. 7)

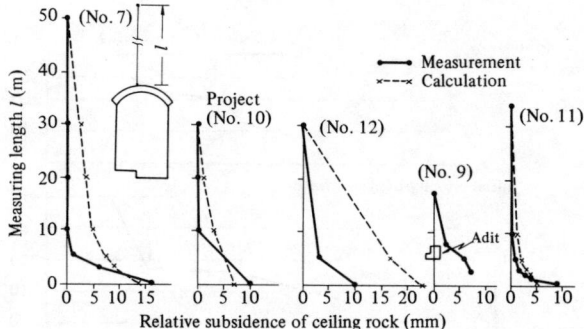

Fig. 2 Distribution of the subsidences of the ceiling rocks

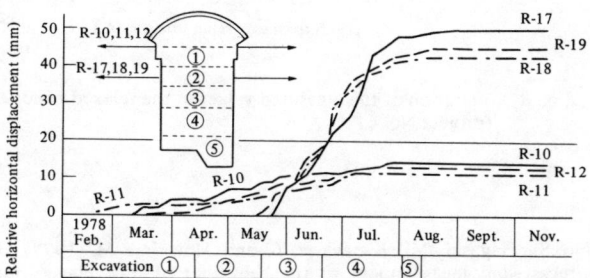

Fig. 3 Measured relative horizontal displacements of the wall (project No. 8)

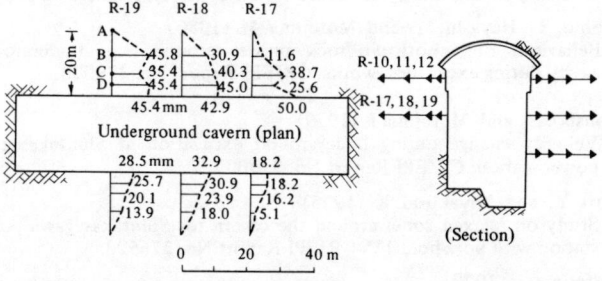

Fig. 4 Distribution of the measured horizontal displacements at the final excavation stage (Project No. 8)

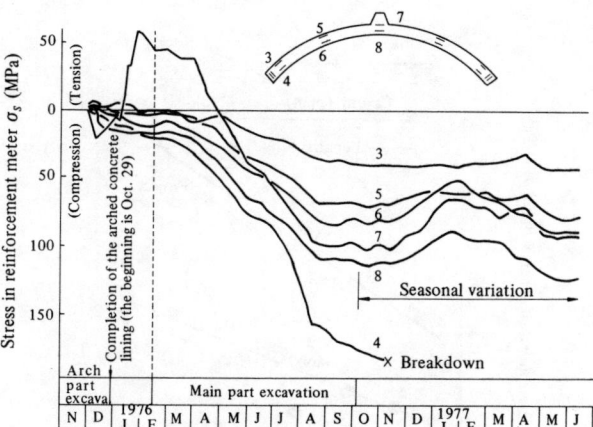

Fig. 5 Time history of the stresses in the arched concrete lining (project No. 7)

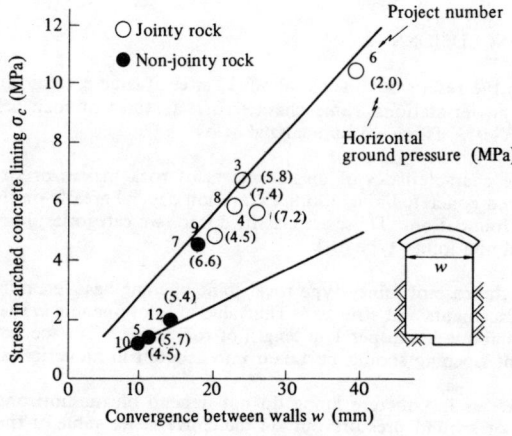

Fig. 6 Relationship between the stresses and the convergences

constant in the case of continuous rock, and the stresses would not be expected to become larger. However, the results were different from the estimation mentioned above. One of the reasons would be the discontinuity of the rock masses. That is, the wall rocks behave as a rock plate with joint openings.

5. CHARACTERISTIC PROPERTIES OF RELAXED ZONES

Rock foundation is partly degraded by stress concentration or blasting during excavation works. The moduli of deformation and strength of rock masses become small in those regions, which we tentatively called "relaxed zones." In order to estimate properties of relaxed zones, a few kinds of measurement were carried out at the site of project No. 6. Those included the elastic wave velocity test, the Lugeon test and the joint opening measurement with a bore-hole TV camera. Fig. 8 summarizes the results (Hibino et al., 1980):

(1) The Lugeon value L became logarithmically larger by 500 times the original value L_0 at a distance of nearly 5 m from the excavation surface (Motojima, 1979).

(2) The velocity of the elastic wave V_p decreased by 40% (Honsho et al., 1979).

(3) Accumulative joint opening totalled 20 ~ 27 mm, which meant about 1 mm joint opening per 1 m length of rock mass (Hori et al., 1975).

In Fig. 8 two singular points are seen, one is located about 11 ~ 12 m distant from the rock surface and the other point about 6 ~ 7 m distant. At the 1st point the rate of increase decreases. Inversely, at the 2nd point the rate increases greatly. The first singular point probably comes from the effects of reinforcing strands, which have 1.18 MN prestressing and are 15 m in length including a 5 m fixing part. The 2nd singular point is due to the effects of blasting, because the point arose just after blasting was done.

D 201

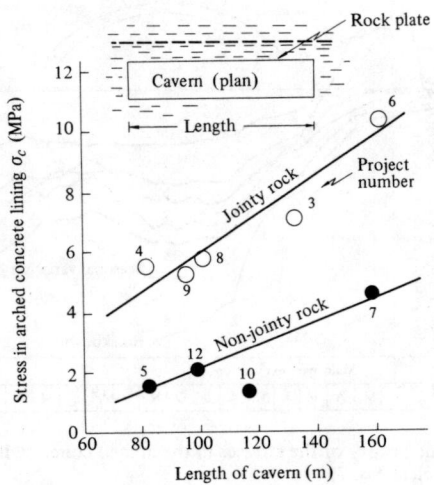

Fig. 7 Relationship between the stresses and the lengths

Fig. 8 Variation of the measured values in the relaxed zone (project No. 6)

6. CONCLUSIONS

Through the results obtained at about 12 sites of large caverns for underground power stations, some characteristic features of rock behaviour become clear. Those are summarized below:

(1) The characteristics of displacements of rock masses or stresses in the arched concrete lining during excavation depend greatly on the types of rock foundation. Those are classified into two categories; jointy type rock and non-jointy type rock.

(2) In the case of jointy type rock, joint opening has a great influence on displacements and stresses. The value of joint opening in the granite site was about 1 mm per 1 m length of rock mass. The mechanism of this joint opening should be taken into account in numerical analyses.

(3) Stresses in concrete lining do not depend on the horizontal component of ground pressure but on the convergence value of the cavern walls.

(4) In the relaxed zones, properties of rock mass were degraded, and changes in the Lugeon value and in the velocity of the elastic wave were observed.

REFERENCES

Hayashi, M. and Hibino, S. (1970)
Visco-plastic analysis on progressive relaxation of underground excavation works, Proc. of the 2nd Conf. on Rock Mechanics, Vol. 2, No. 4-25.

Hibino, S., Hayashi, M., Kanagawa, T. and Motojima, M. (1977)
Forecast and measurement of the behaviour of rock masses during underground excavation works, Proc. of the Int. Symp. on Field Measurement on Rock Mechanics.

Hibino, S. and Motojima, M. (1979)
Forecast and measurements on behaviour of anisotropic rock masses around a large underground cavity, Central Research Institute of Electric Power Industry (in abbreviation CRIEPI), Report No. 378029.

Hibino, S., Hayashi, M. and Motojima, M. (1980)
Behaviour of anisotropic rock masses around large underground cavity during excavation works, CRIEPI Report No. 379028.

Honsho, S. and Motojima I. (1979)
Velocity change during underground excavation at Shintakasegawa power station, CRIEPI Report No. 379003.

Hori, Y. and Miyakoshi, K. (1975)
Study on relaxed zones around the cavern for Shintakasegawa power station with bore-hole TV, CRIEPI Report No. 376528.

Motojima, I. (1979)
Study on the permeability change of rock mass due to underground excavation, Proc. of the 12th Interior Symp. on Rock Mechanics.

Motojima, M. and Hibino, S. (1981-a)
Behaviour of jointy granitic rock mass around the underground cavity during excavation work, CRIEPI Report No. 380046.

Motojima, M. and Hibino, S. (1981-b)
Behaviour of schisty rock mass around the underground cavity during excavation works, CRIEPI Report No. 38014.

STUDY ON ROCKBURSTS AT THE FACE OF A DEEP TUNNEL, THE KAN-ETSU TUNNEL IN JAPAN BEING AN EXAMPLE

Etude de la chute de roches sur le front de taille d'un tunnel profond — cas du tunnel de Kan-Etsu, Japon —

Untersuchung über Gebirgsschläge and der Ortsbrust eines tiefliegenden Tunnels, dargestellt am Beispiel des Kan-Etsu Tunnels in Japan

T. Saito
K. Tsukada
Dept. of Mineral Science and Technology, Kyoto University, Kyoto, Japan

E. Inami
H. Inoma
Y. Ito
Japan Highway Public Corporation, Tokyo, Japan

SYNOPSIS

In the Kan-Etsu Tunnel, which is one of the deepest expressway tunnels in Japan, rockbursts occurred mostly at the tunnel face. In order to prevent them, the authors carried out several investigations, including the measurements of initial rock stresses and the change of rock stress ahead of the advancing face. It was found that the areas of the core disking, observed at the coreboring before the excavation, fairly corresponded to those of rockbursts. Through these investigations and observations, the mechanism of rockbursts at the tunnel face is discussed.

RESUME

Dans le tunnel Kan-Etsu, l'un des tunnels d'autoroute les plus profonds, des chutes de roches ont eu lieu en particulier au front de la taille. Afin d'éviter ce phénomène, certaines investigations telles que mesures de contrainte initiale de roches et mesures de changement de contrainte de roches en avant du front, ont été faites. Il s'est avéré que les zones de disquage de carotte observées lors due carottage réalisé avant l'excavation correspondent bien a celles des chutes de roches. En se basant sur ces investigations et observations, on examine le mécanisme de chute des roches sur le front de taille.

ZUSAMMENFASSUNG

Im Kan-Etsu Tunnel, der einer der tiefsten Autobahntunnel in Japan ist, traten Gebirgsschläge zumeist an der Tunnelortsbrust auf. Um dieses Phänomen zu verhindern, wurden mehrere Untersuchungen durchgeführt, einschließlich Messung der geologischen Spannungen und der Spannungsänderungen vor der Ortsbrust. Hierbei wurde festgestellt, daß Bereiche, in denen der Bohrkern diskenförmig zerbrach, ziemlich genau denen der Gebirgsschläge entsprechen. Anhand dieser Untersuchungen und Beobachtungen wird der Mechanismus von Gebirgsschlägen an der Tunnelfront erörtert.

1. Introduction

Kan-Etsu Tunnel is 10,885m-long expressway tunnel, under the heavy overburden over 1000m in height, located at the place where Kan-Etsu Expressway passes throgh Tanigawa-Range which is one of the steepest mountain ranges in Japan. The construction of this tunnel started in summer, 1977. The excavation of the sub-tunnel (20m^2 section) and the main tunnel (86m^2 section), which are parallel each other at a distance of 30m, were finished in Feb. 1981 and in Feb. 1982, respectively.

Under Tanigawa-Range there are three railway tunnels in operation already (Dai-shimizu Tunnel, New Shimizu Tunnel and Shimizu Tunnel). In the excavation of every tunnel, they experienced the occurrence of rocknoises and rockbursts. Therefore it had been expected that rockbursts would occur at the excavation of Kan-Etsu Tunnel. As expected, over the area of 1100m length rockbursts occurred intermittently. Most of them were not so large in fracture zones, the same as a spalling. However they occurred mostly at the tunnel face rather than at the side wall, which is different from past experiences, and were very serious for the excavation works. Fortunately they caused no severe accident but interrupted construction work frequently.

In order to prevent the rockbursts and to secure the safty working, the authors carried out some investigations in the field. Through these results, the mechanism of rockbursts and the countermeasures against them are discussed from the point of view of rock mechanics.

2. Features of Rockbursts in Kan-Etsu Tunnel

The rock masses surrounding Kan-Etsu Tunnel are mainly consist of quartz diorite and hornfels (See in Fig.3), and include no apparent faults. In the region of quartz diorite, the part containing the regular joints and the relatively massive part appear alternately at the intervals varied from about 20m to 130m, while hornfels contains many fine joints. The average uniaxial compressive strength of quartz diorite is 230MPa

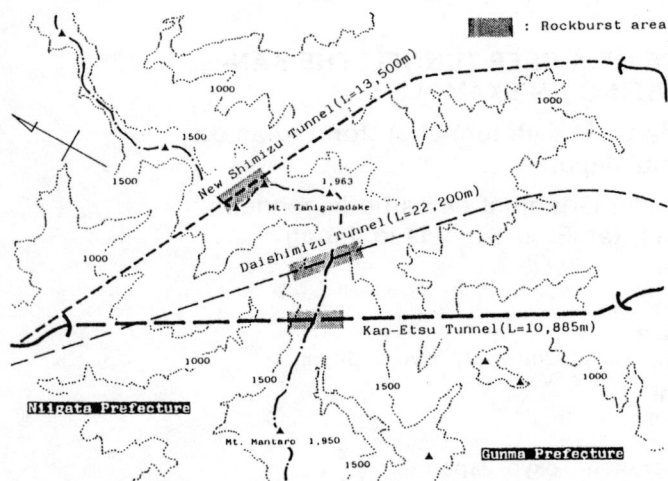

Fig. 1 Rockburst area of three tunnels under the Tanigawa-Range

Table 1 Initial rock stresses at the three points along the Kan-Etsu Tunnel
$\sigma_1, \sigma_2, \sigma_3$: principal stress
σ_v : vertical stress
σ_{h1} : max. horizontal stress

	No.1 point	No.2 point	No.3 point
rock	diorite	diorite	diorite & hornfels
overburden	260 m	960 m	920 m
σ_1	14.6 MPa	22.9 MPa	31.6 MPa
σ_2	6.3	10.7	22.2
σ_3	5.9	7.5	6.0
σ_v	6.2 MPa	16.4 MPa	31.3 MPa
σ_{h1}	14.6	17.1	22.5

and that of hornfels is 310MPa. Kan-Etsu Tunnel had been excavated at the rate of about 100m per month by the full face method. The largest rockburst in this tunnel occurred in quartz diorite at a distance of 4,327m from the north entrance. Thereafter the intermittent occurrence of rockbursts had been observed until the bed rock changed into hornfels completely.

Fig.1 shows the region in which rockbursts occurred frequently in the three neighboring tunnels. The fact that every region lies under the same ridge is interested, suggesting that the area under this ridge has the potentiality causing the rockbursts. Rockbursts in Kan-Etsu Tunnel occurred mostly at the face as shown in Fig.2 which illustrates the fracture zone of the largest rockburst and can be classified into some degrees of fractures, such as bursts or spallings of rock, cracks on the face without a spalling and only rocknoises.

Some other features of rockbursts in Kan-Etsu Tunnel can be pointed out as follows.
(1) It is assumed that the occurrence is gently related with joints. The rockbursts are apt to occur in the uniform and dry rock masses which have several closed joint sets with the regular orientation, while not in the rock masses which have many fine joints, or in the massive rock which has few joints.
(2) It seems that the workings, such as blasting, chopping and drilling, provoke rockbursts at the face.
(3) The sizes of broken rock pieces are various, but the shapes of them are generally flat plate.
(4) They have experienced rockbursts in the main tunnel more frequently than in the sub-tunnel.
(5) The rockbursts occurred only in quartz diorite under the heavy overburden over 750m.

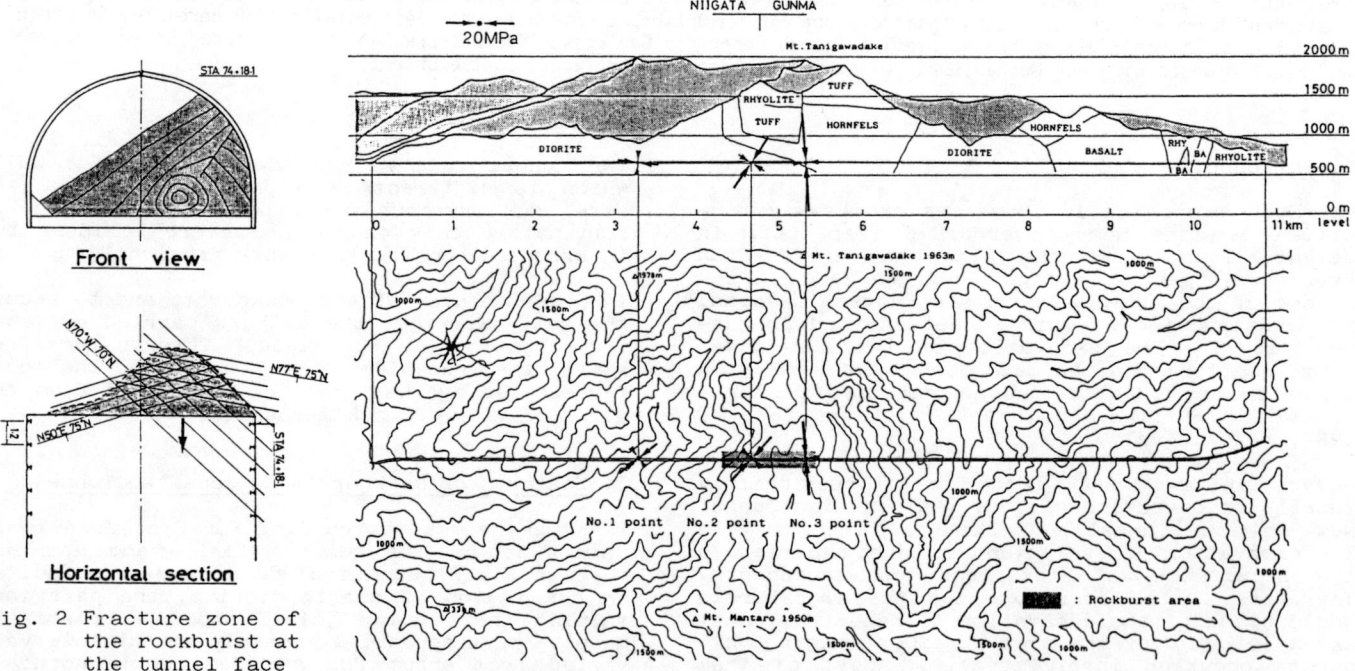

Fig. 2 Fracture zone of the rockburst at the tunnel face

Fig. 3 Measuring points, direction and magnitude of initial rock stresses and the topography

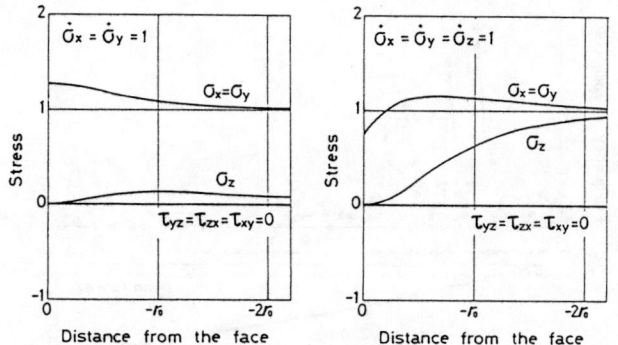

Fig. 4 Elastic stress distributions ahead of the tunnel face by unit loads
y : vertical axis, z : tunnel axis

3. Initial Rock Stresses

One of the factors which cause the rockburst seems to be the states of stress around the tunnel face. Therefore the authors tried to measure the initial rock stresses at three points along the sub-tunnel around the rockburst area.

The measuring method is a kind of the stress relief technique, using the 8 elements moulded gauge bonded on the bottom of a borehole, like the door stopper type. Table 1 shows the measured values of initial principal stresses and Fig.3 illustrates the direction and magnitude of rock stresses in the vertical and horizontal plane on the topographical map. It is found

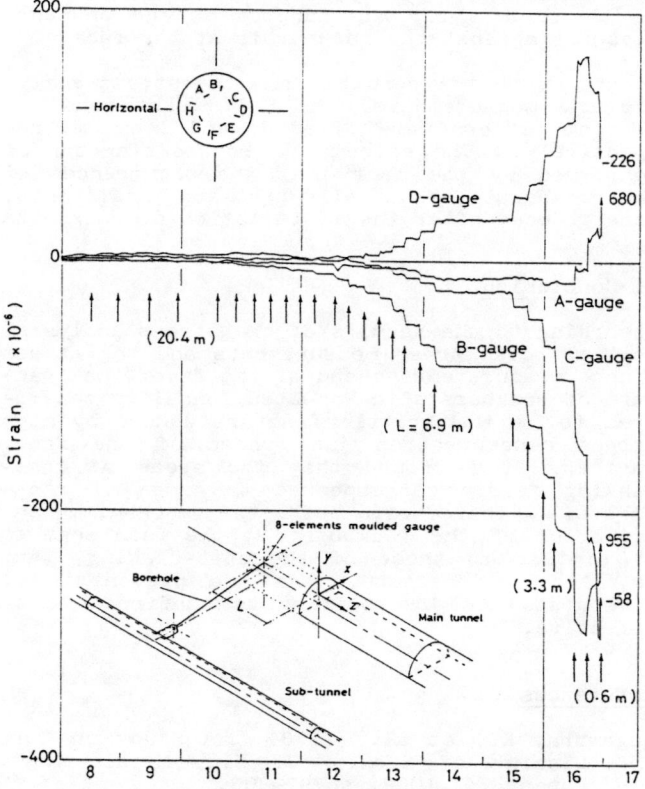

Fig. 5 Strain variations ahead of the face with the advance of the face

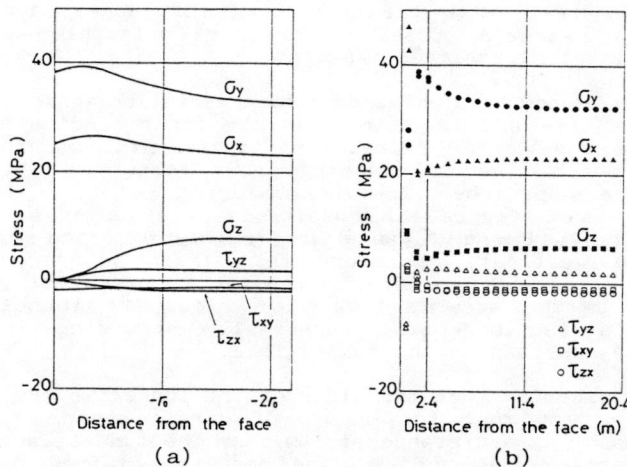

Fig. 6 Stress distribution ahead of the face in the rockburst area (a): calculated from initial rock stresses (b): calculated from measured strain variation

that the directions of rock pressure seems to fairly correspond to ones expected by the topography. On the magnitude, the values of No.2 and No.3 point, which belong to the rockburst area, are more than that of No.1 point, and amount to about 20MPa in the horizontal. In any case, it is considered that the occurrence and the appearance of rockburst deeply depend upon the direction and magnitude of intial rock stresses.

4. Elastic Stress States around the Tunnel Face

It is important to know the states of stress around the tunnel face caused by rock pressure to consider the mechanism of rockbursts. The elastic stress states around the tunnel face were analyzed by means of FEM applied to axisymmetric elastic body under asymmetric loads, in which the tunnel structure was expressed as the corresponding cylindrical cavity.

Fig.4 shows the examples of the stress distributions ahead of the tunnel face along the tunnel axis, in case of Poisson's ratio $\nu = 0.15$. At a distance of the tunnel diameter $2r_0$ away from the face, the states of stress are found to nearly come to the applied initial states. It is interested that when $\overset{*}{\sigma}_z = 0$, the peak of stress concentration is at the tunnel face, while, when $\overset{*}{\sigma}_z = 1$, it moves a little to the inner part from the face. The maximum stress concentration factor around the face is in the range of 1.16-1.32, which is generally lower than one at the side wall.

5. Stress Measurements ahead of the Face

5.1 The Stress Variation with the Advance of the Face

To clarify the actual stress states at the tunnel face in the rockburst area, the following measurements were carried out. Drilling the borehole of 40m length from the sub-tunnel which was going ahead of the main tunnel, directed to the front of the advancing face of the main tunnel, and setting the 8 elements moulded gauge on the bottom of the borehole, the changes of

D 205

the strain with the advance of the tunnel face were measured, based on the same principle as initial stress measurements.

Fig.5 shows the obtained strain variation against the time and the arrows in the figure indicate the time of blasting. The incremental strain on each time of blasting grows larger as the face approaches to the measuring point. In the same figure, the positions of 8 gauges are shown. B and C gauge are roughly directed to the vertical, therefore the strains measured by these gauges are increasing compressively, while the strain of D gauge along the tunnel axis is increasing tensionally according to the approach of the tunnel face.

Considering that the width of the tunnel is 18m, the fact that the change of strain begins to appear at a distance of 16m from the tunnel face, fairly corresponds to the results obtained by the elastic stress analysis. Futher, it is found that even at the interval between each blasting the strains increase gradually when the tunnel face approaches to the measuring within a distance of 4.2m-3.3m. It is interested that these deformations are seem to be due to the time dependent characteristics of rock or the redistribution of stresses.

5.2 The Stress States ahead of the Face in the Rockbursts Area

Fig.6(a) shows the stress distribution ahead of the face obtained from the elastic stress analysis using measured initial rock stresses, and Fig.6(b) shows the same stress distribution calculated from the measured strain variations and initial stresses. Comparing these two figures, it is found that these two results fairly coincide with each other, and therefore, the behavior of rock masses around the face is not plastic, but elastic as far as the point very close to the surface. The fact that the fracture initiated at the point more close to the face than 1.5m in this measurement, suggests that high stress concentration can appear on the surface in case of more competent rock.

The fracture of rockbursts is considered to be the brittle fracture caused by high stress concentration just ahead of the face, which value seems to be more than 40MPa, about 20% of the uniaxial compressive strength, in case of Kan-Etsu Tunnel.

6. Mechanism of Rockburst at the Tunnel Face

The correlation between occurrences of the gas and rock outburst in coal mines and the core-disking phenomena, has been pointed out. Then, the occurrences of core-disking were observed at the core boring before the excavation of the main tunnel.

Fig.7 shows these observations in comparison with the events of rockburst at the excavation of the main tunnel. Core-disking was observed at several points which fairly corresponded to the areas where rockbursts frequently occurred.

The fracture of core-disking is known to be the tensile fracture along the boring axis caused by the high compressive rock stress perpendicular to the axis, which amounts to more than about four times of the tensile strength.

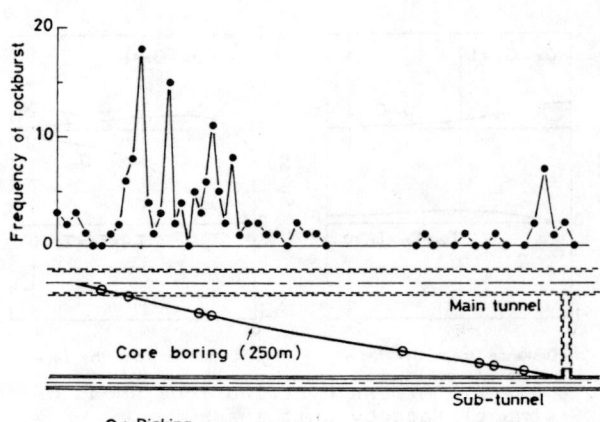

Fig. 7 Events of rockburst and core-disking

Considering the appearance of rockbursts and initial rock stresses in the rockburst area, the authors assume that both the mechanism and criterion of the rockburst at the face seem to be similar to that of core-disking.

7. The Prevention of Rockbursts by Rockbolting

The tensile fracture near the surface, mentioned above, can be prevented by a slight confining pressure, obtained with such as rockbolting or shotcrete. In fact, it seems to be due to the rockbolting that few rockbursts occurred at the side wall in Kan-Etsu Tunnel. Therefore, in this tunnel, the rockbolting on the face as well as the side wall, was adopted as the countermeasures against the rockbursts at the face.

According to the results from the stress analysis and measurements, the bolt length and the bolting pattern were fixed 3m and 2m x 2m, respectively. The effects of rockbolting can be confirmed by the fact that the occurrences of the rockbursts, especially the large scale ones, were reduced after the installation of rockbolts.

8. Conclusion

According to the results of the stress analysis, initial rock stress measurements and the stress change measurements ahead of the face, the fracture of rockbursts in Kan-Etsu Tunnel is considered to be the brittle fracture caused by high stress concentration just ahead of the face. Further, it is found that the areas of core-disking fairly corresponded to those of rockbursts. Therefore, both the mechanism and criterion of the rockburst at the face seem to be similar to those of the core-disking, that is, tensile fracture. As the prevention of rockbursts, the rockbolting is confirmed to be effective.

References

Sugawara, K., et al. (1978). A Study on Core Discing of Rock: J. Min. Metal. Inst. Japan, 94, 1089, pp797-803.
Saito, T. and Sato, K. (1981). Gas and Rock Bursts in Horonai Coal Mine: Proc. Fall Meeting of MMIJ, E2.

FIELD INVENTORY OF TUNNELS FOR CLASSIFICATION PURPOSES
Inventaire de tunnels aux fins de classification
Das Inventar von Geländebeobachtungen in Tunneln zum Zwecke einer Klassifikation

Manuel Romana
Professor of Geotechnical Engineering, Polytechnical Univ. of Valencia, Spain

Samuel Estefania
Civil T. Eng., Geologist, INTECSA, Spain

SYNOPSIS

This paper presents a field inventory of old existing railway tunnels in a mountain line crossing the divide between the Spanish Central Region and the North Coast. The condition of the tunnels was assessed by means of four very simple factors. The inventory helped to define geotechnical zones and indicated the importance of regional residual tectonic stresses.

RESUME

Ce rapport présente un inventaire des tunnels de chemin de fer actuels dans un tronçon de montagne qui traverse la ligne de partage entre le plateau Central Espagnol et la côte nord. L'état des tunnels est représenté par 4 facteurs simples. L'inventaire permet de définir des zones géotechniques et met en relief l'importance des efforts tectoniques résiduels.

ZUSAMMENFASSUNG

Es wird ein Feldinventar von alten bestehenden Bahntunneln der Linie von Central Spanien zur Nordküste vorgestellt. Die Tunnel wurden mit vier einfachen Faktoren beschrieben. Das Inventar ist von großer Hilfe für die Bestimmung von geotechnischen Zonen gewesen und zeigte die Bedeutung tektonischer Restspannungen.

1.- INTRODUCTION OF THE PROBLEM

On the section La Robla-Pola de Lena, the present Madrid - Oviedo - Gijón - Railway makes its way through the Cantabrian Cordillera divide from the Meseta, at almost 1000 m average altitude the coastline.

The existing railroad, which is about 80 Km long has more tha 80 tunnels - some rough gradients and a great number of small radius curves slow speeds and, heavy traffic, make this section one of the main bottlenecks of the Spanish railways. (Fig. 1) The Spanish National Railways (RENFE), has decided to build a new railroad, with more traffic - capability at higher speed.
To select the best layout, 42 solutions were examined though a wide strip of land. These solutions were compared by a multi-criterion programme taking - into account, among other aspects, the geology, the geotechnical factors, the costs, the operation, the maintenance, the protection of the environment, etc.

As part of the studies, a field inventory was established of several tunnels of the railroad, in order to observe - their actual state and try to find a - relationship with the geology of the - area.
The present paper describe the inventory method and the achieved results.

2.- GEOLOGY OF THE AREA

Most of the area is Palaeozoica material included in the heavily folded and - thrust substrate of the south branch of the socalled Asturian Joint. The - lithological groups are very variable - and the surface outcrops are large and irregular. The deep structure of these materials is very complex and has not - been stablished. The most important - structural accidents are the Pajares - Overthrust Fold and, less important, - the Villamarín Overthrust Fold. (Fig.2).

The remaining part of the area under study comprises materials included in - the largest productive coal in ---- Asturias. There are many field layers showing alternative heterogeneous and - different materials. Their folding - structure is looser and their fracture discontinuities are of a less important tectonic intensity.

Rain and snow cause intense erosive - processes There is a number of gravity slides and some fractures of glacial - origin.

3.- TUNEL INVENTORY

From the geological data, an important primary state of stress, due to tectonic foldind, could be forecast. As the -- solutions to be compared included a big

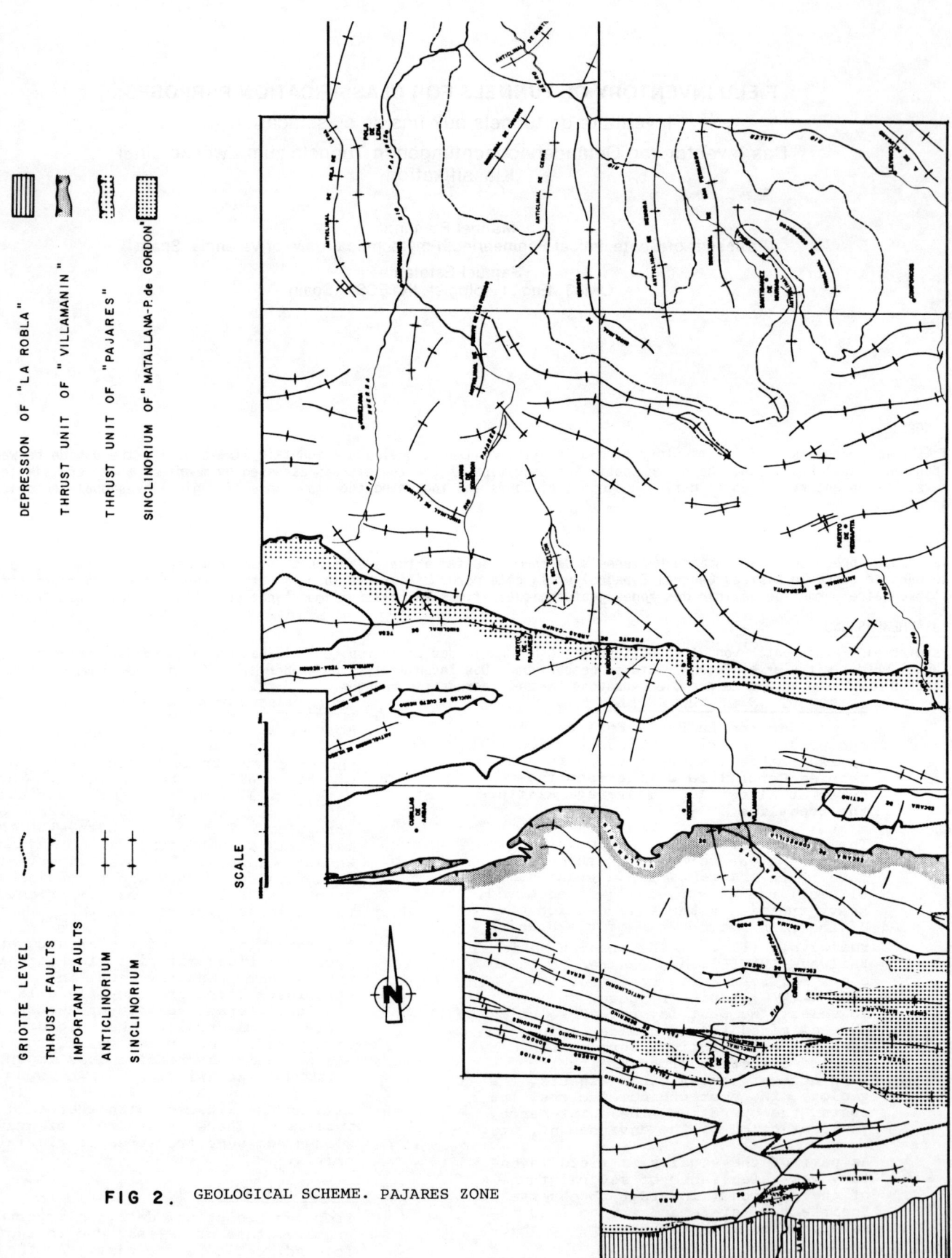

FIG 2. GEOLOGICAL SCHEME. PAJARES ZONE

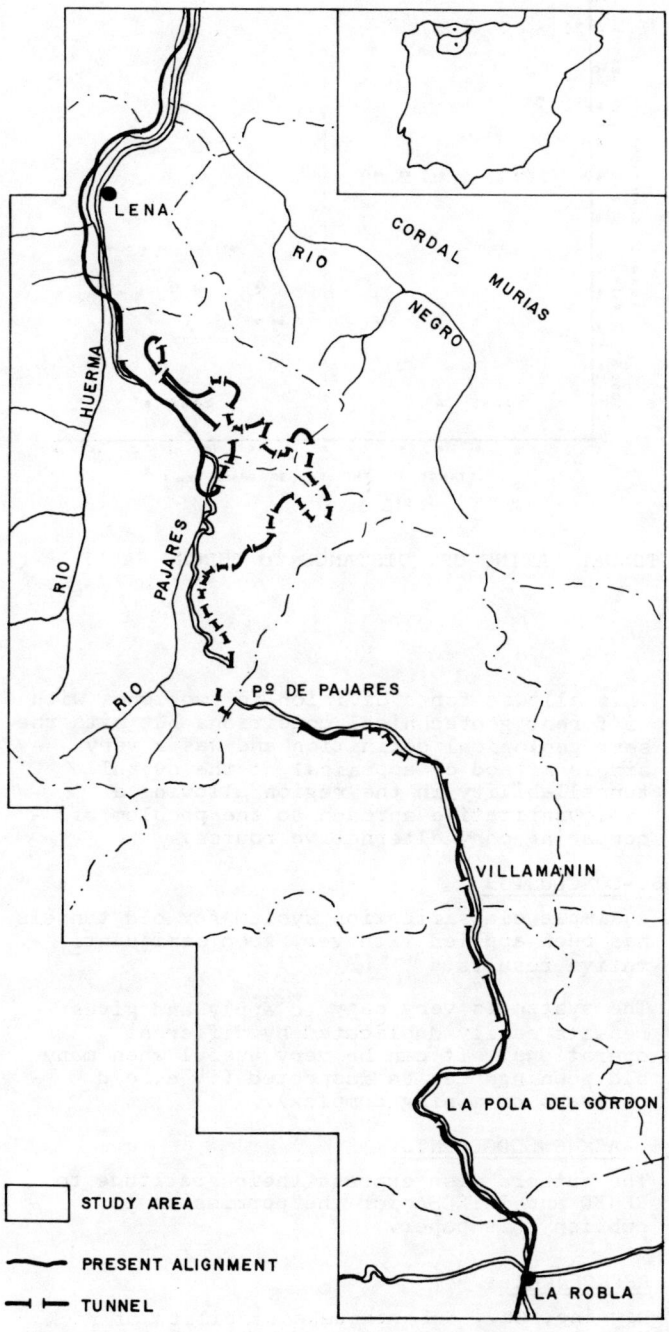

FIG. 1 EXISTING RAILWAY LINE

STUDY AREA
PRESENT ALIGNMENT
TUNNEL

number of tunnels it was necessary to appraise the effect of tectonic stresses and complex geologic conditions to many tunnels.

For this purpose a field inventory of existing tunnels was done with particular attention being gave to tunnels located on the northerm side of the Pajares Overthrust Fold.

Field slope inventories , proposed by HOEK (1970) , and used by several authors , have proven to be a practical method for estimating the rock mass strength properties. Inventories have been used by Dowding (1979) to study sismic damages or tunnels.

Although rock mass classification methods (BIENAWSKI, BARTON etc.) give good estimations for excavation and support, they are not useful to evaluate lining conditions and slow damage in tunnels.

In order to standardise, as much as possible, the observations, the following factors were taken into consideration in the studied tunnels:

. presence of water

. deformations

. state of lining

. state of portals

These factors were given a numerical appraisal of 0 to 3, 0 being good and 3 bad. Thus we achieved, for each tunnel, a value representing its condition. The values corresponding to each item were taken as follows:

Presence of Water		Deformations	
. Dry	0	. None	0
. Light drops	1	. Slight	1
. Average drops	2	. Only on mountain side	2
. Abundant drops	3	. Serious	3

Lining		Portals	
. No damages	0	. No damages	0
. Slight damages	1	. Slight damages	1
. Average damages	2	. Average damages	2
. Heavy damages	3	. Heavy damages	3

The next page table summarizes the value of each factor for the studied tunnels.

TUNNEL	WATER	STRAINS	LINING	PORTALS	TOTAL
11	3	1	3	1	8
12	3	2	2	2	9
13	2	2	2	1	7
14	3	3	3	2	11
15	3	3	3	2	11
16	3	2	2	2	9
17	2	1	1	1	5
18	3	3	2	3	11
19	2	2	2	1	7
20	2	1	1	1	5
21	1	1	2	2	6
22	2	2	2	2	8
23	3	1	1	2	7
24	3	2	2	1	8
25	2	1	1	3	7
26-2	2	2	2	1	7
57	2	1	1	1	5
58	1	2	1	1	5
59	0	1	2	1	4
60	1	1	1	2	5
61	0	1	2	2	5
62	0	1	1	1	3
63	0	2	1	1	4
67	1	2	1	1	5
68	1	1	1	1	4
69	0	1	1	1	3
71	0	1	1	1	3
72	0	1	1	1	3

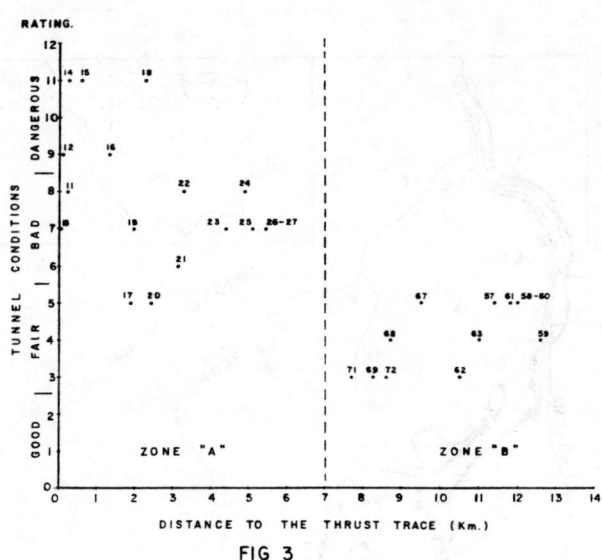

FIG 3

TUNNEL RATING US. DISTANCE TO THRUST FAULT

4.- RESULTS

The classification system is very simple to apply, but gives an adequate rating for old tunnels.

Rating	Tunnel characteristics
0-2	Good. No problems
3-5	Fine. Need some repairs
6-8	Bad. Need repairs
9-12	Dangerous

From the experience gathered during this aplication, the system seems very consistent. Different operators gave the same or very similar ratings without dispersion. But maybe the factors are not fully independent. Strong correlation seems to appear between factors 1 (water inrush), 2 (strains) and 3 (lining conditions).

When the total rating for the 28 tunnels was compared with the distance of each tunnel to the Pajares Overthrust Fold (see Fig. 3), a clear pattern was established between damages and distance. All tunnels (12) at more than 7 Km. were in "Fine" conditions. Tunnels located at 6 Km. or less, were in "Dangerous" (5), "Bad" (9) or "Fine" (2). Therefore, the existence of important residual stresses is clear.

This allowed for a division in two zones with different geotechnical conditions but with the same geological definition and was a very simple method of appraisal of the overall tunnellability in the region allowing a semiquantitative aproach to the problem of comparing many alternative routes.

5.- CONCLUSION

A simple classification system for old tunnels has been applied with very good semiquantitative results.

The system is very easy to apply and gives results easily duplicated by different operations. It can be very useful when many old openings can be inspected (i,e, old railways on mining complex).

6.- ACKNOWLEDGEMENTS

The authors wish express their gratitude to RENFE and INTECSA for the permission to publish this paper.

REFERENCES:

-Dowding, Ch., - Earthquake stability of rock Tunnels - Tunnels and Tunnelling, June 1979.

-Hoek, E.,(1970) - Estimating the stability of excavated slopes in opencast mines. Extract from Transaction. Section A of the Institution of Mining du Metallugy, Volume 79, London.

-Martínez Alvarez, J.A.,(1981) - Estudio geológico - geotécnico del paso por ferrocarril de la Divisoria Astur-Leonesa.

PERFORMANCES DES TUNNELIERS AU ROCHER: RESULTATS ACTUELS, PREVISION ET CONTROLE

Performances of TBM in rocks: Current results, forecasting and control

Das Verhalten von Vollschnittmaschinen im Gebirge: Aktuelle Ergebnisse, Voraussage und Kontrolle

Dominique Fourmaintraux
Laboratoire Central des Ponts et Chaussées, Paris — France

RESUME
La réussite d'un creusement de galerie avec un tunnelier tient d'une part à une reconnaissance suffisante fournissant les éléments d'une prévision pertinente et de propositions réalistes, et d'autre part à un suivi en temps réel approfondi en cours de travaux; une performance optimale peut ainsi être atteinte et les bases d'une rémunération réaliste posées.

SYNOPSIS
Success of tunnelling with a TRM depends on the one hand on an adequate site investigation which provides the elements for sound forecasting and realistic proposals, and on the other hand on a thorough-going real-time follow up during operations.

ZUSAMMENFASSUNG
Eine erfolgreiche Anwendung von Vollschnittmaschinen im Tunnelbau hängt von zweierlei ab: Von einer hinreichend genauen Erkundung, die gute Voraussagen und realistische Vorschläge ermöglichen soll, und von einer genauen Erfassung der aktuellen Leistungswerte beim Einsatz.

INTRODUCTION

L'objectif principal lors de l'utilisation d'un tunnelier est de parvenir à une performance à l'avancement élevée et régulière dans des terrains de caractéristiques et comportement variés. Des résultats non satisfaisants sont à imputer à une insuffisance des connaissances du maître d'oeuvre, de l'entrepreneur ou du constructeur en ce qui concerne les réponses aux trois questions suivantes :

1. Quelles seront les conditions, concernant les roches et le massif rocheux, rencontrées lors de l'avancement?

2. De quels équipements devra être doté le tunnelier pour surmonter ces conditions?

3. Comment organiser les opérations pour optimiser le processus d'avancement et le contrôler en fonction des conditions rencontrées?

PERFORMANCES ACTUELLES DES TUNNELIERS AU ROCHER

Où en sommes-nous actuellement? Le tableau I fournit un panorama résumé d'environ 100 comptes-rendus ou analyses de chantiers mécanisés au cours des quinze dernières années. L'évolution des résultats s'établit ainsi (tab.II).

Période	Performances en m/jour		Nombre de cas étudiées
	Moyenne	maxi	
avant 1969	20 à 36 m/j	65 m/j	8
1970-1972	5,5	20	3
1973-1978	12,2	27	21
1978-1979	14,7	26	8
1980	13,2	34,4	14
1981 (82)	18	46,4	16

Tableau II - Evolution des performances des tunneliers (voir tableau I en fin)

L'analyse détaillée fait apparaitre une volonté continue d'élargir le domaine d'utilisation du tunnelier à la fois vers les roches plus résistantes et plus abrasives et vers les massifs rocheux plus fracturés et hétérogènes plus difficiles nécessitant la mise en place immédiate d'un soutènement. Deux exemples volontairement extrêmes illustrent bien cette évolution : en 1966, le tunnel OSO est foré dans les schistes peu résistants, homogènes et réguliers du Nouveau Mexique avec un avancement moyen de 60 m/j (120 m/j pour le meilleur) sur plus de 6 km; mais il fallut 200 jours pour franchir 100 m de formations glaciaires à blocaux, en renonçant à l'usage du tunnelier, d'où une performance moyenne de 21 m/jour. En 1981-82, la galerie hydro-électrique de Wölla est forée dans un diamètre comparable, dans des gneiss et schistes cristallins résistants, à travers un massif tectoniquement mouvementé, aquifère, des Alpes autrichiennes à la vitesse moyenne de 34 m/jour (63 m/j pour le meilleur) et 932 m/mois . La régularité de la performance, qui en est l'aspect le plus remarquable est à attribuer d'après KARNELO et BUECHI(1982) à la coopération étroite et constante pour le suivi et l'organisation des opérations entre l'entrepreneur, le maître d'oeuvre, le constructeur, le géologue et le géotechnicien.

RECONNAISSANCE ET PREVISION

L'avancement du tunnelier dépend de la forabilité de la roche (comment les molettes pénêtreront-elles la roche?) qui déterminent la vitesse de foration ou de pénétration instantanée (en m/h), et de la forabilité du massif (comment le tunnelier sera-t-il opérationel dans la galerie compte tenu du comportement du massif?). La combinaison de ces deux éléments mène à l'obtention d'une vitesse d'avancement de l'ensemble du processus d'excavation qui s'exprime le plus pratiquement en m/jour.

La reconnaissance doit être menée de façon à fournir les paramètres contrôlant ces deux éléments et permettre ainsi une prévision.

FORABILITE DE LA ROCHE

La seule considération de la résistance en compression uniaxiale σ_c est reconnue par tous les auteurs comme insuffisante. On retiendra au moins le couple des valeurs de la résistance et de la continuité de la roche, (Nishimatsu et Ikeda - 1981, Fourmaintraux - 1974, Gripp - 1982) : résistance en compression σ_c ou en traction indirecte σ_{tb}, à partir de la célérité des ondes P Indice de Continuité IC (Fourmaintraux, 1974). Le rôle des discontinuités de la matrice rocheuse a été largement démontré par Wanner et Aeberli (1979), Fourmaintraux, (1979, 1980), Kutter et Sanio (1982). Ces données permettront au constructeur de dimensionner le tunnelier (poussée, couple, espacement des molettes). Les tunneliers actuels développent des poussées de plus de 25 t. (et jusqu'à 40 t.) sur des molettes de Ø 40 cm (en 1970, respectivement 7 à 10 t. et Ø 30 cm). Léonard (1981) donne comme "règle du pouce" une charge par molette exprimée en pounds (lbs)) au moins égale au double de la résistance en compression uniaxiale exprimée en PSI.

CONSOMMATION D'ENERGIE

La foration au tunnelier peut s'exprimer par le bilan énergétique du nombre de kwh (ou de MJ) nécessaires pour excaver 1m³ de roche : cette quantité d'énergie est *spécifique à la fois de la roche forée mais aussi de la machine de foration;* en particulier la rigidité du système de foration des tunneliers peut être très différente et Hignetts (1982) a montré qu'elle a un effet significatif sur l'efficacité du travail de la molette, donc sur l'énergie spécifique ES. Le tableau III fournit les valeurs de l'énergie spécifique ES déterminées in-situ sur quelques chantiers du tableau I. Elle varie de 5 à 35 kwh/m³ (18 à 126 MJ/m³). En laboratoire, Lindqvist (1982) obtient un ordre de grandeur comparable (18 kwh/m³ = 65 MJ/m³ dans un granite).

USURE DES MOLETTES

L'usure des disques modifie leur profil et réduit l'efficacité de leur action : l'outil doit être interchangé et reconditionné. Les coûts directs en fourniture et indirects en temps d'arrêt, peuvent prendre des proportions importantes (tableau III) : de 1000 m³ à moins de 40 m³ par molette. Des consommations rédhibitoires ont dans certains cas entraîné l'abandon du tunnelier et toujours des difficultés de règlement. La détermination de l'abrasivité des roches traversées est indispensable : elle doit utiliser des méthodes d'essai adaptées mettant en oeuvre simultanément la nature des minéraux et la cohésion entre les minéraux et la roche (Fourmaintraux, 1978) ou les combinant (Blindheim, 1970, 1979). Ces essais de "dureté" superficielle par rayure ou empreinte ne sont pas adaptés à la structure et au grain des roches.

PREVISION

Les conditions les plus favorables pour la foration sont *"une roche homogène, de résistance moyenne, d'un degré de fissuration élevé à moyen et d'abrasivité faible"* : elles définissent la classe de forabilité de la roche la meilleure; à partir des valeurs prises par ces paramètres, il est possible d'établir des classes (1 à 6) pour des conditions de plus en plus défavorables, en regard desquelles, pour un type de tunnelier donné (diamètre, puissance, vitesse de rotation, couple) comme le proposent Beckmann et Simmons (1982), on portera des gammes de vitesse de foration instantanée décroissantes (figure 1).

FORABILITE DU MASSIF ROCHEUX

Les conditions favorables à l'évolution du tunnelier sont (Deere, 1981) "un massif régulier, faiblement discontinu, sans faille ni accidents, homogène non ou peu altéré, dans un état de contraintes modérément élevé par rapport à la résistance moyenne des roches traversées et où les venues d'eau sont faibles à nulles". Le coefficient d'utilisation C.U., rapport du temps de foration au temps total, tend alors vers sa valeur limite, égale au coefficient de disponibilité de la machine. Celle-ci peut être estimée vers 70 à 75 %, par comparaison à la valeur jugée satisfaisante de 80 % pour un ripper D-9 travaillant et entretenu en surface, dans des conditions plus favorables que le souterrain. Le tableau I indique des valeurs maximales du C.U. de 70 %, et une moyenne autour de 45 à 50 %; le minimum annoncé est 11; six classes de forabilité du massif rocheux (de I à VI) peuvent ainsi être graduées en coefficient d'utilisation décroissant de < 65 % à < 10 % et combinées aux classes de forabilité de la roche (fig. 1).

PREVISION

Comment estimer et prévoir dans laquelle de ces six classes se trouvera placé le tunnelier? L'instabilité des terrains n'est pas le seul élément susceptible de rendre un tunnelier moins opérationnel. Robbins (1982) rassemble les conditions défavorables au tunnelier en cinq catégories:

1. Formations et chutes de blocs, dièdres et écailles rocheux en toit, en parement et à front,

2. Instabilités du front de taille; convergences importantes de la galerie; écaillages plus ou moins violents,

3. Détériorations des parements par délitage; gonflements ou dissolutions, entraînant des instabilités,

4. Venues d'eau "exceptionnelles" d'autant plus catastrophiques qu'imprévues, telle rencontre d'une zone broyée ou faille en charge, d'un contact tectonique, de cavernes karstiques (Merrit, 1981, Marin, 1980, - Berry et Fink, 1979). Une description soignée et complète des cadres géologique, hydrogéologique et des caractéristiques du massif rocheux ("Basic Géotechnical Description" de la S.I.M.R., 1976; Fourmaintraux et al., AFTES, 1978) fournira les éléments pour une analyse systématique des conditions en galerie et une détermination des équipements et accessoires dont le tunnelier devra être doté pour les surmonter au mieux.

De cette façon, des propositions raisonnées de performances à l'avancement et de coûts peuvent être établies et comparées.

SUIVI ET CONTROLE DES PERFORMANCES

Le creusement au tunnelier doit être considéré comme un processus industriel qu'il faut suivre en continu pour en évaluer l'efficacité, déceler les causes de pertes de rendement et y remédier. Un suivi efficace commence par la collecte systématique poste par poste des temps consacrés à huit activités et sept quantités essentielles (d'après Hulshizer, 1981).

<u>Activités</u> :

A 1 - Foration (avance du tunnelier)

A 2 - Foration de sondages de reconnaissance à l'avancement

A 3 - Inspection et changements de molettes

A 4 - Pose de soutènement

A 5 - Réparation du convoyeur principal

A 6 - Réparation d'équipements du tunnelier (Indisponibilité)

A 7 - Réparation du système d'évacuation des déblais en aval du tunnelier

A 8 - Autres pannes mineures

<u>Quantités</u> :

Q 1 - Longueur forée par le tunnelier

Q 2 - Poussée moyenne utilisée

Q 3 - Couple moyen développé

Q 4 - Longueur forée en reconnaissance

Q 5 - Nombre de boulons, cintres et autres soutènements posés...

Q 6 - Débit d'eau en galerie dans la zone d'avancement

Q 7 - Nombre de molettes changées

Ces données sont immédiatement introduites après chaque poste dans un système de traitement informatique qui éditera pratiquement en temps réel une série de graphiques permettant de replacer les résultats du poste dans les performances quotidiennes et hebdomadaires du processus, par exemple :

- un graphe donnant les variations jour par jour de la moyenne journalière des activités n° A1 à A9 et des quantités Q.1 à Q.7

- un autre, en donnant les moyennes et les variations, semaine après semaine, depuis le début des travaux

- un graphe donnant les variations des activités réparties en trois groupes : Foration (A1); soutènement (A2+A3+A4); Pannes (A5+A6+A7+A8).

- des graphes donnant par semaine les variations des performances quotidiennes ou hebdomadaires (longueur forée, vitesses de foration et d'avancement du tunnelier, coefficient d'utilisation).

Des traitement complémentaires comme par exemple l'établissement d'une moyenne mobile sur les quatre dernières semaines permettent des projections ou des comparaisons avec la semaine en cours ainsi que la mise en évidence de tendances de certaines activités ou quantités. Il est possible ainsi de déceler un fonctionnement différent plus ou moins insatisfaisant du tunnelier ou de certains équipements correspondant soit à une mauvaise mise en oeuvre, un déréglage, une usure ou une dégradation. L'analyse du type d'activités ou quantités en cause, de leurs variations permet de localiser cet élément ou de détecter une évolution réelle, prévue ou imprévue, des conditions de forabilité tant de la roche que du massif rocheux. Il convient alors de contrôler ces conditions réelles rencontrées avant de les mettre en cause dans la variation de performance du tunnelier. Quelques mesures simples pourraient être réalisées systématiquement sur le rocher ou sur le massif au droit du tunnelier au cours de l'avancement (célérité d'ondes P; carottage; nombre de fractures par mètre; venues d'eaux) et leurs variations visualisées.

Le traitement informatique de toutes ces données permet de disposer des graphes ci-dessus pour une analyse immédiate , pour les conversations entreprises-maître d'oeuvre et les décisions des responsables du chantier. Ces graphes peuvent également constituer une vigie décelant une dégradation progressive des conditions à l'approche de zones difficiles. Enfin, ce suivi fournit les éléments pour une comparaison réaliste entre les conditions et performances prévues et réelles, base d'une rémunération plus équitable. Beckmann et Simmons proposent pour cette dernière une solution originale et intéressante du type contrat à objectif avec incitation qui partage les surcoûts dûs aux conditions plus difficiles mais aussi les économies dues aux conditions plus faciles.(Beckmann et Simmons, 1982).

CONCLUSIONS

Une reconnaissance correcte est adaptée à une analyse en termes de classes de FORABILITE DE LA ROCHE et classes de FORABILITE DU MASSIF. Ces deux aspects sont gradués par l'entrepreneur associé au constructeur respectivement en vitesse de foration (m/h) et coefficient d'utilisation (C.U. %) correspondant au type de tunnelier choisi et combinés en termes de performance journalière (m/jour) et en coût par ml. Un suivi détaillé en temps réel peut être réalisé grâce à un traitement informatisé. Il constitue le meilleur tableau de bord pour conduire l'avancement, juger l'adéquation des prévisions et contrôler pas à pas les bases d'une rémunération réaliste.

REFERENCES BIBLIOGRAPHIQUES

BUECHI et KARNELO, 1982. Pre-investigation and geological follow-up of a TBM project in Austria, Pr. ISRM Symp., Aachen

NISHIDA et al., 1982. Rock mechanical viewpoint on excavation of pressure tunnel by TBM, Pr. ISRM Symp., Aachen (also in Tunneling'82)

FOURMAINTRAUX, 1974. Methode de quantification des discontinuités des roches et massif rocheux; leurs influences sur les performances d'un TBM, Pr. 3° Cong. Int., SIMR, Denver

FOURMAINTRAUX, 1975. Les discontinuités des roches : une méthode pour leur évaluation in-situ et leur effet sur la foration, Pr., 4° Cong. Int. SIMR, Montreux

FOURMAINTRAUX et COMES, 1980. Creusement mécanique des roches au tunnelier. Approche géologique et géotechnique, Revue Ind. Min., fév.80

GRIPP, 1982. Etude prévisionnelle de l'abattage mécanique dans différentes roches. Thèse Doc., Ing., Ecole Nat. Sup. des Mines de Paris

WANNER et AEBERLI, 1979. TBM performance in jointed rock Pr. 4 th Int. Cong. ISMR, Montreux

KUTTER et SANIO, 1982. Die Beeinflussung der Vortriebgeschwindigkeit von Vollschnittmaschnen mit Diskelmeisseln durch die Gebirgsanisotropie, Pr. ISRM symp., Aachen

LEONARD, 1981. TBM's use, Where are we? What can we do about it? Pr. RETC San Francisco

HIGNETTS, 1980. Specific Energy for mechanical cutting of chalk, T & Tunneling, Janv. 80

HIGNETTS & TEMPORAL, 1982. Effect of cutting rig rigidity on spécific energy, T & Tunneling, Janv. 82

LOUIS, 1979. In Progres récents de l'abattage mécanisé, Communication du CFMR au 4° Cong. Int. SIMR, Montreux

LINDQVIST, 1982. Energy consumption in disc cutting of hard rock, Tunneling'82

FOURMAINTRAUX et MASSIEU, 1978. Détermination de l'abrasivité des roches, Pr. 3° Cong. Int. AIGI, Madrid

BLINDHEIM,1979, Drillability predictions in hard rock unneling, Tunnelling'79

BECKMANN & SIMMONS. TBM Payment on basis of actual rock quality effect, Tunneling '82 (1982)

DEFRE, 1981. Adverse geology and TBM problems, Pr. RETC San Fransisco

ROBBINS, 1982. The application of TBM to bad conditions Pr. ISMR Symp., Aachen

MARIN, 1978. Le franchissement des accidents géologiques Tunnels et Ouvrages Souterrains, Sept. Oct. 78

BERRY & FINK, 1979. 28 km of rock tunnel for Kielder water scheme, Tunneling'79

GUERTIN & FLANAGAN, 1982. Effect of artesian aquifer on feasability of Buffalo LRRT Project, Tunneling'82

Int. Soc., of Rock Mechanic, 1981. Basic geotechnical Description of rock masses., Int., J., Rock Mech., & Min., Sc., 1981

FOURMAINTRAUX et al, 1978. Description des massifs rocheux utile aux travaux souterrains; Recommandation AFTES, Tunnels et Ouvrages Souterrains, 28, 1978

HULSHIZER et al., 1981. Production expérience and computerized evaluations of the Seabrook tunnel excavation, Pr. RETC, San Francisco

CHANTIER (Job location)	Date	Const. (1)	Ø (m)	AVANCE PAR JOUR, mètre (daily advance rate, meter)			C.U.	Longueur	Roche	Soutènement (2)
				moyenne (m/j)	maxi (m/j)	mini (m/j)				
TRAVAUX PUBLICS (Publics Works)										
KERCHOFF 2 (Californie)	1982	RO	7,3	19			55	6500		
GRAND-MAISON Vaujany (F)	1982	RO	7,7	15				(3500)	Gneiss	+
WOLLA (AUS)	1982	JA	3,5	34	63	3,7	60	6700	Gneiss	+
S-BISSORTE (F)	1981	WI	3,3	7,6 à 14	-	-	-	-	Grès dur	0 à +
MILWAUKEE HC	1981	JA	2,27	11	(30)	-				+
" NE 1	1981	JA	2,6	15,6	42	-		1725		+
CHICAGO (T.A.R.P.)	1981	JA	9,1	24	45	-		2200	Dolomie	0 à +
	1981	JA	6,7	-	52	-			Dolomie	0 à +
	1981	RO	6,7	-	45	-			Dolomie	0 à +
	1981	RO	5,5	-	48	-			Dolomie	0 à +
LEMONT (Chicago)	1981	JA	2	22			57	5350	Dolomie	0 à +
WOODSIDE (Maryland)	1981	JA	2,2	13			43	350	Gneiss	0 à +
MARYLAND W80	1981	JA	2,3	7			71	3900	Gneiss	0 à +
MILWAUKEE NE 2	1981	JA	2,6	12,2			11	2700	Calc.	0
				36			33			+ +
ATLANTA (Three rivers)	1981	JA	3,2	24,4			63	8200	Grès/gneiss	0 à +
BI COUNTY E Maryland	1981	JA	3,8	18			49	4600	Gneiss	0 à +
" W	1981	JA	3,8	16,75			51	550	Gneiss	0 à +
CHIOTAS (I)	1980	RO	2,5	8,5	30		34	1000	Granite	+
SEABROOK NPS (MASS.)	1980	RO	6,7	10,6	32,3	5	24	10500	Diorite	+ à ++
" "	1980	RO	6,7	10	29	4,8	30	10500	Diorite	+ à ++
BODENDORF (RFA)	1980	RO	3,5	26	45					
BUFFALO LRRT (NY)	79/80	RO	5,7	17,8	50,9	-	34	2800	Dolomie/Calc.	0 à +
" "	"	RO	5,7	20	51,2	-	32	2800	"	
" "	"	RO	5,7	14,8	39	-	39	2800	"	
" "	"	JA	5,7	15,4	39	-	43	2800	"	
ABATEMARCO (I)	1980	PRI	3,5	11 à 13						
MANHATTAN 3 AV T5	1980	RO	6,7	6,28	19,3	1,3		445	Schist.raides	
" T6	1980	RO	6,7	8,10	21,6			450	" "	
GUBRIST (CH)	1980	WI	11,5	9,8	18	-	37	6600	Grès	+ à ++
MONTREAL	79/80	JA	4	13,4	38	1,5	36	7700	Schiste	
BARMEN-Wal (AUS)	1978	DE	3,1	20	30	-	-	7700	Schist./Grès	
GRAND MAISON (P) 30° (F)	1979	WI	3,3	12,1	18	1,3	40	1460	Schist./Gneiss	0 à +
" (P) 30° (F)	1980	WI	3,3	13,6	30	1,3	40	1460	" "	0 à +
" DP (F)	1978	WI	3,3	18	30	1,3	50	850	Gneiss	0
ROCHESTER (NY)G.R.I.S.										
G.R.I.S. Tunnel	76/78	RO	5,7	13	22	2,5	47	2500	Dolomie	0
LYON Crémallière (F)	1978	WI	3,0	3,2	10	0,7		230	Granite	+ +
CRESPERA GEMMO (I)	1978	RO		(1,9)	(32)	1,5 à 2	(60)		"	
MONACO	1978	RO	3,3	18,5	40	1 (Man)			Calc.	0 à +
KIELDER (G.B.)	1976	DE	3,5	-	42,5	3,5 à 0,7	21	1600	Grès, Shales	+ à ++
"	1976	DE	3,5	-	30,6	(Man)	27	1600	" "	"
"	1976	RO	3,5	-	47	(Man)	27	1600	" "	"
ALBI (I)	76/78	RO	3,3	13	-	3 à 0			Granite	0 à ++
ORICHELLA (I)	76/78	RO	4,3	10 à 15	-	4			"	"
BRAMEFARINE (F)	1976	RO	8,1	9,16	20 à 25	1,5	55	3850	Schistes	+
BELLEDONNE (F)	74/75	WI	5,8	10 à 16	14 à 35	1 à 0	60	10000	Granite	0
CERN (F-CH)(45)	1976	RO	4,8	20,6	39	-	24		Grès tendre	0 à +
PARIS-RER (F)	1976	RO	7,0	13	37	-			Calc.tendre	+ +
BUSNAU (RFA)	1974	DE	2,8	14,5	32	-			Grès tendre	+
SUVIANA - BRASIMONE (I)	1974	RO	6,6	14	32	7			Grès,Shales	0 à +
ECHAILLON - (F)	1973	WI	5,8	10 à 13	16 à 34	1		4700	Gran./Schist.	0 à +
LA COCHE (F)	72/75	RO	3	10		0(D+B)		5000	Calc./Schist.	0 à +
YOUGOSLAVIE	19	DE	2,3	6,8	19,1			8500		
"	1975	RO	7,1	10,5	19,3			6200		
"	à	RO	7	5	14,2			2100		
"	1977	SC.4	7,2	8	14,2			1700		
"		SC.4	7,2	11	15,3			1700		
MANHATTAN	1973	JA	3,0	15	30					0 à +
ALBULA (I)	1972	DE	3,8	17	32				Calc./Schist.	0 à +
DRENSTEINFURT (RFA)	68/73	DE	2,3	14	19,7			700	Calc.tendres	
KOHLFURTH (RFA)	1973	DE	2,3	17,1	32			1435	Grès/Schistes	0 à +
HEILIGEN HAUS (RFA)	à	DE	2,8	19,3	37,4			1273	" "	"
SCHWELME (RFA)	1968	RO	4,0	6,0	23,4			2542	Calc./dolomie	0 à +

TABLEAU I

(to be continued)

CHANTIER (Job location)	Date	Const. (1)	Ø (m)	AVANCE PAR JOUR, mètre (daily advance rate, meter)			C.U.	Longueur	Roche	Soutènement (2)
				moyenne (m/j)	maxi (m/j)	mini (m/j)				
NAST TUNNEL (US)	1970	WI	3,0	5,3	22	0(D+B)	< 57	2500	Granite	
EMOSSON (CH)	1968	WI	2,9	5,3	15,8	-		1140	"	
AVEROLE (F)	1966	(RO)	2,2	2,0	(9)		24	1010	Schist.+ m	
LECHEYLAS (F)	1967	(RO)	2,2	-	10 à 15		46	553	Schistes	
PORT HURON (CAN)	68/70	LAW	5,6	34	64			9640	Schistes	
RIVER MTS (US)	68/69	JA	3,6	22,5 à 33	89			5180	Roche tendre	0
STARVATION (US)	1967	RO	2,8	13,2	40			1660	Grès tendre	0
NEW - MEXICO Navajo	1967	RO	6,0	11 à 15	29 à 49				Grès tendre	0 (à +)
" " Water Hollow	68/70	RO	4,0	13 à 29	55				Grès/schistes	0 (à +)
BLANCO (US)	65/67	RO	3,2	33 à 47	88			11500	Schistes	0
AZOTEA (US)	65/68	RO	3,8	16,8	22 à 46			20000	Grès/Schistes	0
OSO (US)	1966	RO	3,25	21,5 à 60	123	manuel 1,3		7840	Sch./glaciair.	0 (à +)
MANCHE (Beaumont)	1886	Beaum.	2,14	-	24,8			810	Craie	0
HOUILLERES (Collieries)										
GOTTELBORN (RFA)	1981	DE	6,0	15	21	6,0	30	2260	Grès/Houiller	+ à ++
SAARBERG WK	1980	DE	6,0	13	18	-	-		Grès/Houiller	+ à ++
NIEDERHEIM (RFA)	1979	DE	6,0	13	41				Houiller	+ à ++
"	1977	DE	6,0	14,4	32				"	
"	1976	DE	6,0	16,2	34				"	
"	1975	DE	6,0	8,4	21				"	
"	1973	DE	6,0	12,3	25				"	
RHEINLAND BERG WK (RFA)	(1980) (1973)	DE	6,0	12,2à(13) 16,2	42,5			5 tronçons 2600 à 2900	"	+ à ++
DORTMUND (Arge Vic.)	1977	DE	6,1	13	31				"	+ à ++
DAWDSON Colliery (Test)	1975	THY		13 à 18				tests		
MINES (Gold mine)										
FREE STATE GEDULD	1979	RO	3,3	4,4	19				Quartzite	
MACHINES PARTICULIERES										
NEUFCHATEL (CH)	1974		8	12,8			52à63	115	Calcaire	
INNSBRUCK	à	Mini	12	17,8			32à46	515	Cal.Dolomie	
SYDNEY	1976	Full Facer	14	23			50à69	600	Grès dur	
ROCHESTER (NY)	1973	Atlas Copco	8,3	15			-	210	Calc.dolomie	
WASHINGTON (DC)	à		11,5	19			-	235	Schistes/gneiss	
QUINCY (IL)	1980	Ø circulaire equivalent: env. 2,4 m	11,5	36			-	2380	Calc./schistes	
MONTREAL (QU)			3,8	6,8			-	150	Calcaire	
BUFFALO (NY)			17,9	26			72	440	Calc.dolomie	
MONTREAL (QUE)			5	-			60	135	granitique	
SILVERTON (CO)			3,75	8			34	538		
TORONTO (ONT)	à		8,1	13,5			48	180	Shale	
WADWORTH (OM)			5	14,4			21	140	Shale	
COLUMBIA (MO)			8,2	12			75	115	Métamorphique	
PEWAUKEE			7,7	12,3			55	270	Calc.silex	
LONGUEIL (QUE)			17,8	35,3			55	685	Shale/granite	
PALUEL (F)	1981	ZOKOR	5	6 à 7	15,5			900	Craie	Inj.

POUR COMPARAISON : quelques exemples de TBM avec bouclier en terrain tendre (to compare : tunneling jobs involved TBM with shield in soft ground)

THUNDERBAY (ONT)	1978	LOVAT	2,5	-	50				Argile	VS
Bouclier à bentonite (TBM with bentonite shield)										
NAGOYA (J)	1979	Hydros.		5 à 6	-				Silt	VS
WARRINGTON (GB)	1976	Nuttal	2,9	-	13				Argile	VS
ANTWERPEN (B)	1978 à 1981	Wet F	6,4	6,5 à 17,5				4000 (1400à 310)	Sable/argile	VS
BERLIN (Spandau)	1981	Wet F	6,4	8,5				1150	Sable/blocs	VS

Note (1) Constructeurs : JA = JARVA; RO = ROBBINS; WI = WIRTH, DE = DEMAG; THY = THYSSEN; PRI = PRIESTLEY
BOUY = BOUYGUES; SCU = SCHÄFIR et URBACH; W et F = WAYSE et FREITAG; LAW = LAWRENCE

Note (2) Soutènement : 0 = pas de soutènement; + = soutènement non systématique
++ = soutènement systématique, VS = voussoirs préfabriqués (précast linings)
Inj. = injections à l'avancement

TABLEAU I (suite...)   (Tab. I continued)

Roche Classe \ MASSIF Classe / C.U. % / V_{for} m/h	I <65	II 55	III 45	IV 30	V 20	VI 10
1 >5	78,7	66,6	54,5	36,3	24,2	12,1
2 4 à 5	64,4	54,5	44,6	29,7	19,8	9,9
3 3 à 4	50,0	42,4	34,7	23,1	15,4	7,7
4 2 à 3	35,8	30,3	24,8	16,5	11,0	5,5
5 1 à 2	21,5	18,2	14,9	9,9	6,6	3,3
6 0,5 à 1	10,7	9,0	7,4	5,0	3,3	1,7

Avancement en m/j

V_{for} : Vitesse de foration instantanée en m/h (instant pénétration, m/h)

C.U. : Coefficient d'utilisation du tunnelier en % (rapport du temps de foration au temps total) (TBM utilization)

Les graduations ne sont données qu'à fin d'exemple (cas d'un tunnelier de Ø 3m dans des roches de résistantes à tendres) d'après Beeckmann et Simmons (1981)

Figure 1 - Proposition d'une matrice d'avancement du tunnelier combinant les classes de forabilité des roches et de forabilité du massif (Advance matrice for TBM drive; combine "Rocks Boreability" and "Rock Mass Boreability" classes)

(1) Site ( ) voir note	(2) E.S. kWh/m^3 (MJ/m^3)	(3) U m^3 par molette	(4) Roche
Oslo (5)	–	15	diabase
La rouzille (5)	–	30 (à 8)	granite
Gd Maison D.P.	18 (65)	31	gneiss
Gd Maison P.I.	19,4 (70)	38,5	gneiss
" Vaujany	–	45	gneiss
Chiotas	27 (97)	48	granite/gneiss
Belledonne	16 (57,5)	55	Sch.Crist. "
Shimogo (6)	–	166	Grès/Rhyolite
Montréal	10 (36)	154 à 200	Schistes
Oslo (1)	–	200	Schistes
Gd Maison P.I.	9,2 (33)	200	Schistes
Wölla	25 (90)	240	Gneiss/schistes
Echaillon	14 (50,4)	300	Schistes
Damas (5)	–	300	Conglomérats
Bramefarine	8 (29)	320	Schistes
Chambery (5)	70 (252)	700	Calc.marneux
Paris (RER)	5 (18)	1000	Calc.tendre
Hignetts (7) (1980)	2,2 (8)	–	Craie
Lindqvist (8)	18 (65)	–	Granite

(1) Job Site; (2) Energie spécifique (Specific Energy); (3) Usure (Wear Tool); en m^3 foré par outil usé (bored m^3/tool used; (4): (Rock type); (5): TBM Bouygues Ø 3m (C. Louis, 1979); (6) TBM Wirth Ø 3,3/5,8 (Nishida, 1982) (7) en labo; TBM échelle réduite, à Pic (in Lab; pilot scale TBM with Picks; (8) en Lab. banc.linéaire à molettes (in lab., linéar testing rig with discs.)

Tableau III - Consommation d'énergie et consommation d'outil

BEHAVIOUR OF ROCK AROUND THE OKUYOSHINO UNDERGROUND POWERHOUSE

Comportement de roches autor de la centrale souterraine d'Okuyoshino

Gebirgsverhalten des Okuyoshino Kavernenkraftwerkes

Nobuaki Kondo
Masayoshi Yamashita
The Kansai Electric Power Co., Inc.

SYNOPSIS

Prior to the construction works of the Okuyoshino underground powerhouse, the authors estimated the behaviour of bedrock around the powerhouse cavity. During the cavity excavation, the behaviour of rock and the stress of concrete lining were measured by the instruments installed for safety during the work. By comparing the calculated values with the measured ones, the reliability of our estimating method could be verified successfully.

RESUME

Avant les travaux de construction de la centrale souterraine d'Okuyushino, les auteurs ont évalué le comportement des roches en place autour de la cavité centrale. Pendant l'excavation de la cavité, le comportement des roches et l'effort dans le revêtement de béton ont été mesurés par les instruments posés sur le chantier par souci de sûreté. En comparant les valeurs calculées et mesurées on a pu vérifier de façon satisfaisante la crédibilité de cette méthode.

ZUSAMMENFASSUNG

Vor der Bauausführung beurteilen die Autoren das mechanische Verhalten des Gebirges des Okuyoshino Kavernenkraftwerkes. Während der Aushubarbeiten wurde das Verformungsverhalten des Gebirges beobachtet, und die Spannungen in der Betonauskleidung wurden gemessen. Der Vergleich von berechneten und gemessenen Werten zeigt, daß die angewandte Methode zuverlässig und erfolgreich war.

1. INTRODUCTION

The Okuyoshino power plant is a pure pumped-storage power plant which has the maximum power of 1,206 MW under a net head 505 m and is located in Nara Prefecture in central Japan. The powerhouse (machine hall) is constructed approximately 180 m underground and houses six generating units in the cavity of 20.5 m width, 41.6 m height and 157.8 m length. (see Fig. 1)

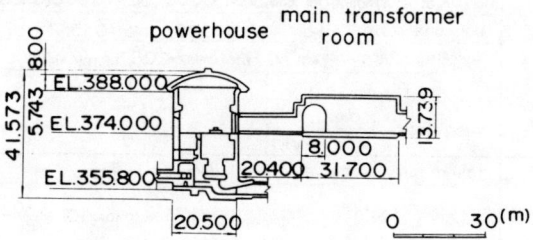

Fig. 1 Longitudinal section of powerhouse

2. GEOLOGICAL ASPECTS

The powerhouse site terrain is composed mainly of shale, and partially of sandstone and alternations of sandstone and shale. (see Fig. 2) The rock in the vicinity of the powerhouse cavity is relatively hard.

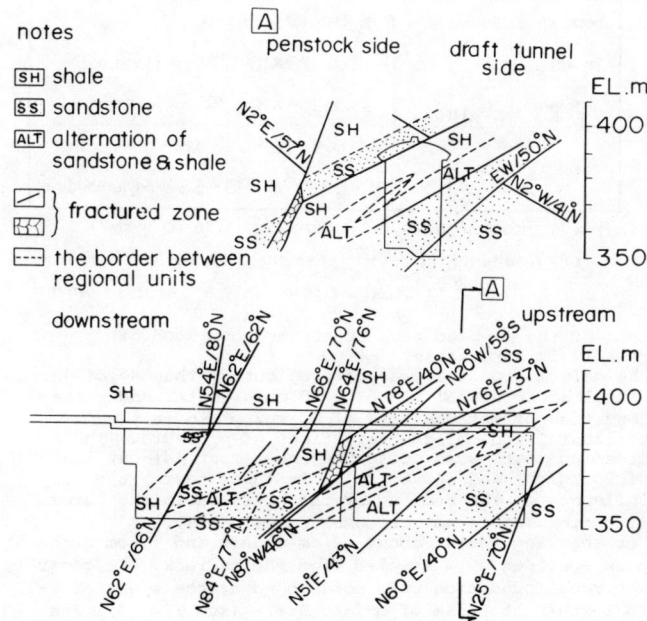

Fig. 2 Geological cross section of powerhouse

The bedding plane is EW/40°N in dip and strike, and intersects with a longitudinal axis of the powerhouse at the angle of approximately 45°. The planes of both joints and fractured zones point to almost the same direction as that of bedding. The bedrock stratifies and gives the considerable anisotropy in shearing strength.

3. PROPERTIES OF ROCK

In order to seize the rock properties required for the design and excavation of the powerhouse cavity, deformation, creeping and shearing tests were carried out by EL.430 test adit. The sampling spots for rock foundation tests are shown in Fig. 3, and the results of the tests are given in Table 1.

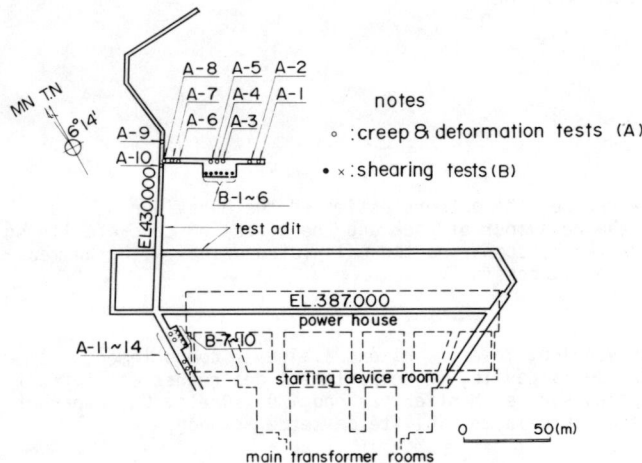

Fig. 3 Sites of rock foundation tests

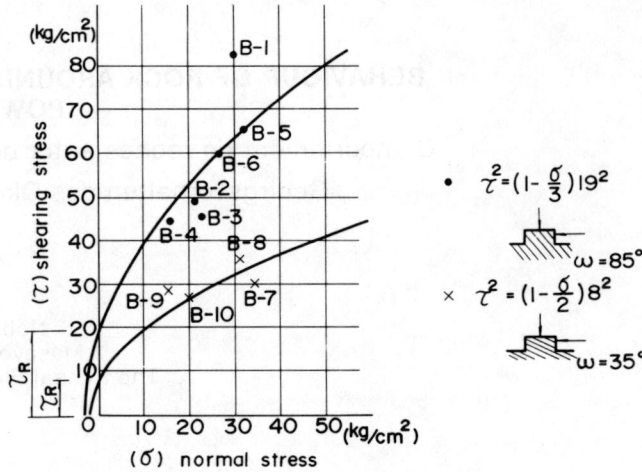

Fig. 4 Results of shearing tests

The bedding plane forms with the direction of the maximum compressive stress the angles of $\omega = 35°$ at $\times$ - mark spots and 85° at the ● - mark ones, on the average. This difference in the shearing stress values of the rock foundation is assumed to be due to that in such angles as mentioned above.

4. ESTIMATION FOR ROCK DEFORMATION

Before the design of the powerhouse, the behavior of bedrock in the vicinity of the powerhouse was analyzed in the Central Research Institute of Electric Power Industry, Japan (HIBINO S., MOTOJIMA M., 1979). In this analysis the bedrock deformation was calculated by non-linear step by step analysis (ANEXCRIEPI) in use of F.E.M. Table 11 showes the initial geo-stresses, initial values of rock properties and relax rules used in this analysis.

Table 1 Preperties of rock

	results of measurement
shearing strength	$\tau_R = 8 \sim 19$ kg/cm²
modulus of deformation ※1)	$D = 38 \sim 129 \times 10^3$ kg/cm² (loading)
	$D = 43 \sim 156 \times 10^3$ kg/cm² (unloading)
creep factor ※2)	$\alpha = 0.14 \sim 0.20$, $\beta = 0.4 \sim 1.6$ 1/day (loading)
	$\alpha = 0.21 \sim 0.37$, $\beta = 0.3 \sim 5.8$ 1/day (unloading)

※1) secant modulus of deformation (0 to 70 kg/cm²)
※2) results of 140 kg/cm² loading, $\varepsilon_c = \alpha \cdot \varepsilon_e (1 - e^{-\beta t})$
     ε_c : creep strain, ε_e : elastic strain

Most of the sampled rock specimens consisted of relatively hard shale.
The deformation test was carried out by the use of jacks with loading discal plate of 30 cm diameter under the condition of maximum load of 80 kg/cm² in vertical, horizontal and oblique directions of pressure. The creeping tests were performed under four (4)-day load of 70 kg/cm² and 140 kg/cm². In these tests, the influence on the bedding by the pressure in the three directions could not be observed evidently.
For shearing tests, about 60 cm square and 10 cm high rock specimen were sampled from the bedrock. In general, the rock foundation were more sound at the spots of B-7 to 10 than at those of B-1 to B-6. (see Fig. 3) The results of the shearing tests are plotted in Fig. 4, and the sampled foundation rocks marked with $\times$ are lower in shearing strength than the ones marked with ● .

Table 11 Initial conditions and relax rules

Initial geo-stresses	horizontal component σ_{xo} (kg/cm²)	−64.0
	vertical component σ_{yo} (")	−68.0
	shearing component τ_{xyo} (")	−23.0
	maximum principal stress σ_2 (")	−90.0
	minimum principal stress σ_1 (")	−42.0
	inclination of σ_2 from virtical axis	42
Initial values of rock properties	modulus of deformation D_o (kg/cm²)	$D_{o1} = 130,000$, $D_{o2} = 78,000 (0.6 D_{o1})$
	poisson's ratio ν_o	$\nu_{o1} = 0.25$, $\nu_{o2} = 0.15$
	shearing strength τ_{RO} (kg/cm²)	$\tau_{RO}\,\omega=90°=20.8$, $\tau_{RO}\,min=10.4$
	σ_1/τ_R	$0.1102 \sim 0.2326$
	creep factor α_o	0.3
	β_o (1/day)	1.0
Relax rules	failure intrinsic curve	$(\tau/\tau_R)^2 = 1 - \sigma/\sigma_1$
	relax factor	$R = (2.72 \times d_{min})/(\sigma_1 - \frac{\sigma_1 + \sigma_2}{2})$
	modulus of deformation D	$R \geq 1 : D = D_o$ $1 > R > 0 : D = R^{1/2} \times D_o$ $0 \geq R : D_{min} = 20,000$ kg/cm²
	poisson's ratio ν	$D > 0.25 D_o$ $\nu = 0.45 - (0.45 - \nu_o) \times \frac{1}{0.75 D_o} \cdot \frac{D - 0.25}{}$ $D \leq 0.25 D_o$ $\nu = 0.45$
	shearing strength τ_R	$\tau_R = \frac{D}{D_o} \tau_{RO}$
	creep factor α, β	$\alpha = \alpha_o$ (const.), $\beta = \beta_o$ (const.)

Since the initial geo-stresses vary in accordance with the elevation of bedrock, the values of these geo-stresses at the powerhouse site were calculated by F.E.M. Subsequently, it was confirmed that the calculated values were nearly equal to the initial geo-stress value obtained by the over-coreing method at the test adit.

The modulus of deformation of bedrock, obtained by the measured values, was estimated at $D_2/D_1 = 0.5$ in this analysis to take account of the anisotropy of bedrock (D_1: modulus of deformation in the direction parallel to the bedding plane, D_2: the one in the direction perpendicular to the bedding plane). For our estimation, $D_{02}/D_{01} = 0.6$ was adopted from the fact that the direction of the bedding plane was not at right angles to the two-dimensional cross section of the powerhouse in calculation.

The Poission's ratios were derived from practical values for bedrocks.

The shearing strength values were determined on the measured ones. The results of the tests showed the anisotropy in shearing strength. For this reason, the shearing strength values were transformed in accordance with the angle of ω. In this connection, the details are given in the report (HAYASHI M., KANAGAWA T., HIBINO S., MOTOJIMA M., KITAHARA Y., 1979).

But in our estimation the shearing strength (τ_{R0}) was modified because the bedding plane is not perpendicular to the cross-section of the powerhouse, and creep factors were take in from the measured values.

Our calculation for the excavation of the powerhouse cavity was made in thirteen (13) steps consisting of eight (8) for excavating works and five (5) for prestressing by rock bolts and strands.

According to the resuts of this calculation, the values at the completion of the powerhouse excavation were estimated at 10 to 61 kg/cm^2 in arch concrete lining stresses, 29 mm in maximum horizontal displacement of bedrock and 7 to 8 m in extent of relaxed zone.

The relaxed zone is defined as the loosened zone with Poisson's ratio of more than 0.45. Fig. 5 gives the distribution of horizontal displacements of bedrock after excavating of powerhouse cavity.

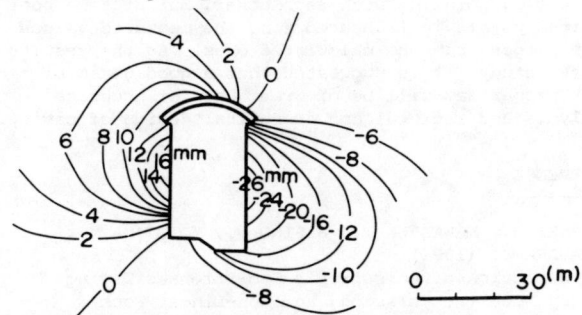

Fig. 5 Horizontal displacement of bedrock

5. REINFORCEMENT OF BEDROCK

For the relaxed zones obtained by our analysis, the reinforcement by prestressing strands (bars) was designed to prevent the collapse along a circular arc enveloping the relaxed zones themselves.
Only the angles of internal friction in the bedrock were taken into account. Fig. 6 shows the standard arrangement of PC steel strands (bars).

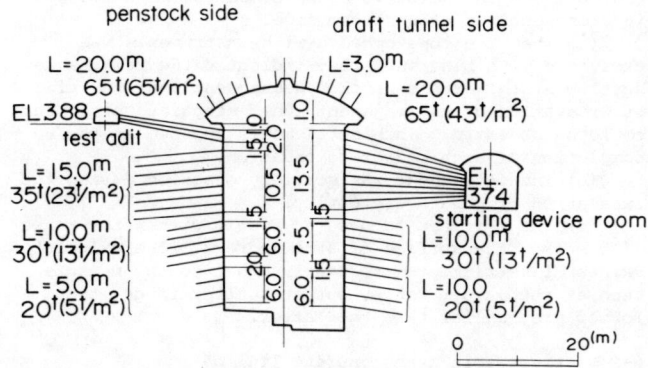

Fig. 6 PC steel strands (bars) arrangement

6. MEASUREMENT

A number of measuring instruments were set for safety in excavation work of the powerhouse cavity as shown in Fig. 7.

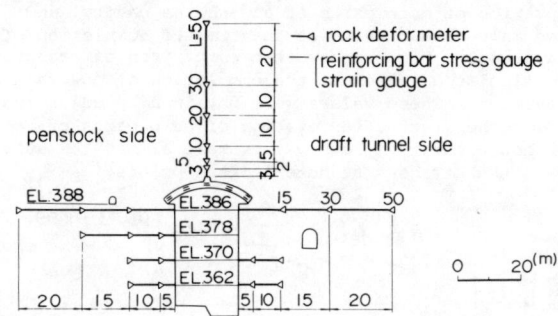

Fig. 7 Arrangement of measuring instruments

6-1 Displacement of arch bedrock

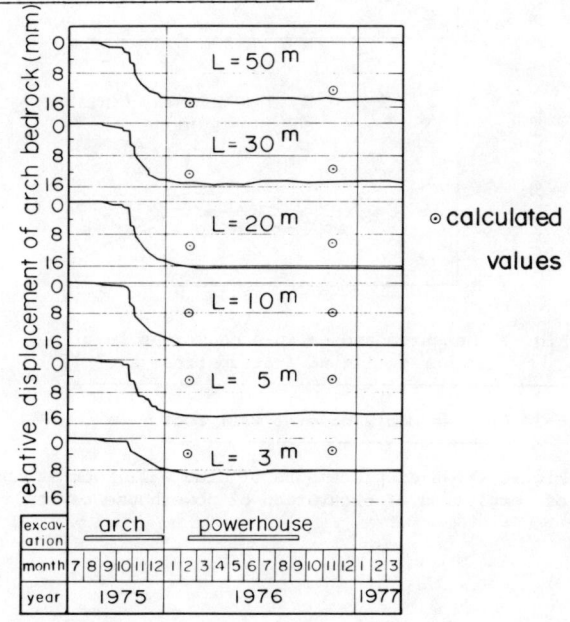

Fig. 8 Relative displacement of arch bedrock

Fig. 8 shows the relative displacement of arch bedrock in cross-section of upstream side.

(a) When the zone embeded with instruments was excavated with blasting, they indicated the remarkable settlements of arch bedrock. Subsequently at the time of excavation of the adjacent zone, stepwise settlements could be observed. Such settlements stopped at the completion of arch bedrock excavation.

(b) The settlement was scarcely observed during the excavation of powerhouse cavity.

(c) The settlements occured mostly within the range of 5 m from the surface of arch bedrock. The calculated values of settlements are nearly equal to the measured ones at the larger range, but at smaller range, the former are smaller than the latter.

6-2 Stresses in arch concrete lining

On the basis of the stresses in arch concrete lining at one month after setting the instruments, they were all compressive at the time of completion of powerhouse excavation ; 50 to 130 kg/cm^2 at the crown of arch, 20 to 40 kg/cm^2 at the abuttment. Most of the stresses ranged from 20 to 70 kg/cm^2. Stresses were hardly observed during the excavation of arch segment, but increased in value in accordance with the progress of excavation of main parts of powerhouse cavity, and ended in constant value at the time of completion of excavation. Fig. 9 shows the comparison of measured and calculated valves at the completion of powerhouse excavation. These values were obtained by adjusting to zero at the start of excavation of main parts of powerhouse cavity. Both values were almost the same expect the stresses at some limited points.

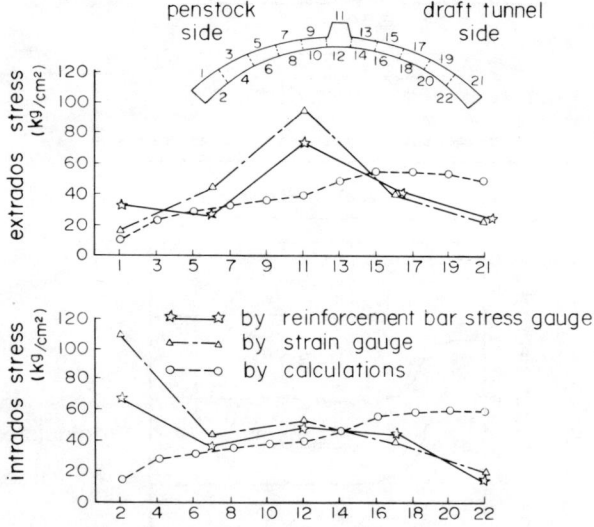

Fig. 9 Measured and calculated stress in arch concrete lining (at upstream side)

6-3 Displacements of side wall rock

Fig.10 shows displacements of side wall rock at the time of completion of excavation of powerhouse cavity.

Within the range of 50 m depth, the measured values were equal to calculated ones expect one point where displacement was supposed to be limitedly affected by the poor rock properties. Within 15 m depth, (a) at EL.386 m, measured values were somewhat smaller than calculated ones on the penstock side, and vi·ce ver·sa on the draft tunnel side.

That is assumed to be due to the influence produced on the draft tunnel side by the excavation of main transformer rooms and starting device room. (b) The both values were almost the same below EL.378 m.

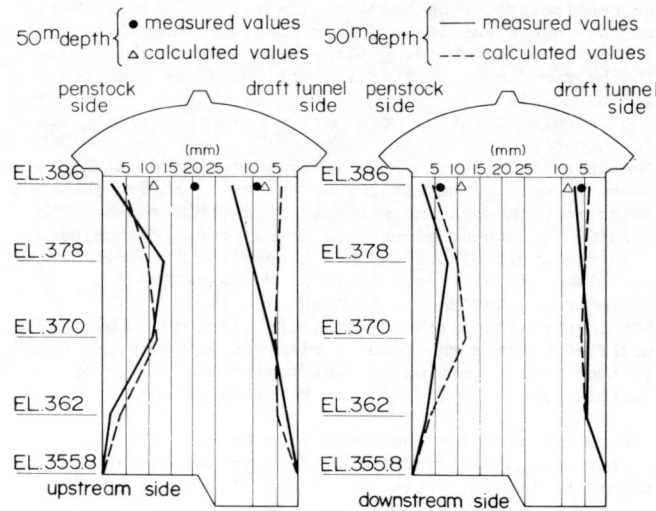

Fig. 10 Horizontal displacement of side wall rock

7. CONCLUSION

The behavior of the bedrock in the vicinity of the Okuyoshino underground powerhouse is described with the calculated and the measured values. Both values, on the whole, coincide with each other, but at some poor limited points in fractured zone the measured values were larger than the calculated ones. As the results of the study, it is suggested that a good grasp of rock properties will be of great use for accurate analysis and rational and economical design of cavity.

REFERENCE

HAYASHI M., KANAGAWA T., HIBINO S., MOTOJIMA M., KITAHARA Y. (1979)
  Detection of Anisotropic Geo-Stresses Trying by Acoustic Emission, and Non-Linear Rock Mechanics on Large Excavating Caverns. Proc. 4th I.S.R.M., 211-218

HIBINO S., MOTOJIMA M. (1979)
  Forecast and Measurements on Behavior of Anisotropic Rock Masses Around a Large Undergoud Cavity - Effects of Pre-Stressing Bars - . The Report (No.378558) of the Central Research of Electric Power Industry, Japan.

ETANCHEITE DES STOCKAGES SOUTERRAINS DE G.P.L. EN GALERIES NON RECOUVERTES
Containment of unlined caverns used for L.P.G. storage
Dichtheit von unterirdischen Kavernenspeichern ohne Innenschale

P. Berest
Laboratoire de Mécanique des Solides — Ecole Polytechnique, 91128 Palaiseau — France

J. M. Morisseau
J. M. Noe
G. Souquet
Géostock, Tour Aurore Cédex 5, 92080 Paris La Défense — France

RESUME

On examine le problème de l'étanchéité des galeries de stockage de GPL non recouvertes sous la nappe. On discute la notion de gradient hydraulique et on introduit les notions de pression d'entrée du gaz et de facteur de forme.

SYNOPSIS

This paper deals with the containment of unlined caverns used for compressed gas storage and located in water-bearing rock formations. The difference of pressure between the surrounding aquifer and the stored products required to ensure the containment is analysed; the hydraulic gradient method is examined and two new ideas, the threshold pressure of the gas and the form factor are introduced.

ZUSAMMENFASSUNG

Der Beitrag befaßt sich mit der Bestimmung des Problems der Dichtheit von unterirdischen Kavernenspeichern für flüssiges Gas, ohne Innenschale unter dem Grundwasserspiegel. Der Druckunterschied zwischen dem Grundwasserspiegel und den Speichern, welcher eine Bedingung for die Dichtheit ist, wird erklärt. Zwei neue Begriffe werden eingeführt: der Einführungsgasdruck und der Formfaktor.

INTRODUCTION

Les gaz de pétrole liquéfiés (G.P.L.) ou butane et propane, occupent une part croissante du marché international des produits pétroliers. La capacité de stockage doit suivre cette évolution. La solution du stockage souterrain est très souvent la meilleure : elle est plus sûre ; son emprise en surface est sensiblement plus faible ; enfin elle est moins coûteuse lorsque les volumes à stocker sont de quelques dizaines de milliers de mètres cubes, ou même moins lorsque les conditions géologiques sont favorables. Elle pose néanmoins un ensemble de problèmes, spécifiques à l'environnement souterrain, qui font l'objet du présent article.

CARACTERISTIQUES PHYSIQUES DES G.P.L.

Les G.P.L. peuvent être assez commodément liquéfiés, comme le montrent les figures 1 et 2. Cette propriété permet de réduire par un facteur d'environ 300 leur volume ; elle distingue l'économie de leur emploi de celles des alcanes plus légers comme le méthane (gaz naturel). La liquéfaction peut être obtenue par réfrigération (point R des figures 1 et 2) ou par compression (point S des figures 1 et 2).

- La réfrigération est le plus souvent utilisée dans les réservoirs aériens ; le stockage souterrain à l'état réfrigéré peut aussi être envisagé pour le propane ; l'expérience récente de Schelle a confirmé que ce type de stockage pouvait être envisagé tant en roche tendre qu'en roche dure, sous réserve que la technique mise en oeuvre soit parfaitement maitrisée (Boulanger et Luyten, 1982)
- Le stockage sous forme liquide obtenue par compression est la solution la plus classique dans les réservoirs souterrains et s'applique tant au butane qu'au propane.

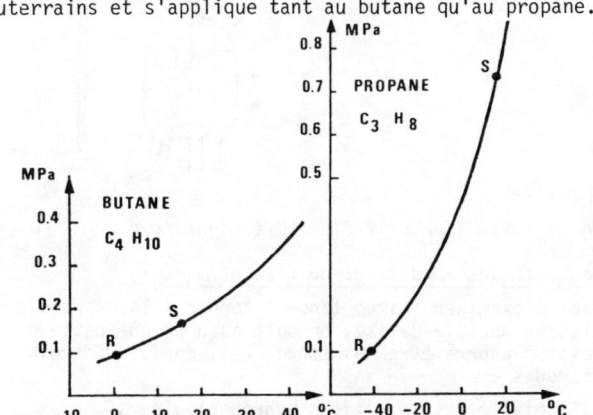

Fig.1.2-Diagramme équilibre liquide-vapeur (produits purs)

STOCKAGES A ETANCHEITE PASSIVE

On peut assurer en principe l'étanchéité d'un stockage souterrain de G.P.L. en l'implantant dans un massif réputé imperméable ; les problèmes de fuite sont alors circonscrits au voisinage des cuvelages de liaison avec la surface. Le stockage en cavités lessivées dans le sel gemme donne ainsi de très bons résultats, car les propriétés d'imperméabilité de cette roche sont excellents du fait de sa faculté à "s'autocicatriser" ; toutefois, il faut bien entendu disposer d'un gisement de sel. Dans d'autres matériaux, les déboires sont fréquents car la perméabilité globale "in situ" peut être très supérieure à celle que laissait prévoir la reconnaissance géotechnique ; les investissements consentis dans le creusement des galeries sont ainsi à la merci d'aléas géologiques qu'il est très difficile de corriger. On verra plus loin que la technique de "l'étanchéité dynamique" est beaucoup plus flexible : les exigences relatives aux qualités naturelles de la roche sont moins draconiennes et il est possible dans une assez large mesure de remédier à une situation moins favorable que prévu.

Principe du stockage à étanchéité dynamique

On peut estimer à plusieurs centaines le nombre de stockages qui utilisent cette technique. Son principe n'est pas d'interdire les mouvements de fluide dans le massif, comme dans les techniques précédentes, mais de provoquer ces mouvements pour empêcher la migration des hydrocarbures. Ceux-ci sont stockés dans des galeries à la pression d'équilibre entre les phases gazeuse et liquide, compte-tenu de la température de fond ; la profondeur des galeries sous la nappe doit être suffisante pour que l'eau du massif s'écoule vers la galerie en interdisant le mouvement inverse des produits vers le massif. Cette profondeur doit donc être au moins égale à la hauteur d'eau correspondant à la pression des produits, comme le montre la figure 3 ; on verra plus loin que cette condition n'est pas en fait suffisante. Enfin, la venue d'eau dans les galeries qui résulte de l'écoulement doit être compensée par un soutirage régulier pour que le stockage ne se remplisse pas d'eau.

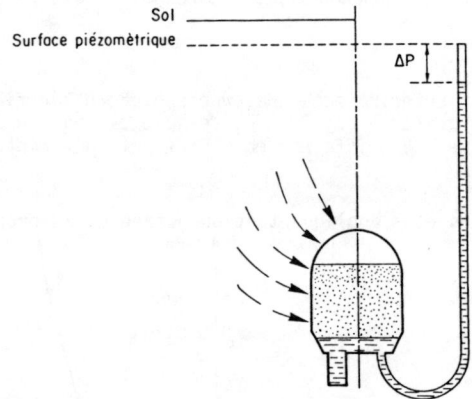

Fig. 3 - *Principe de l'étanchéité dynamique*

Stabilité des conditions de l'écoulement

Avant d'examiner l'importance à donner à la dépression existant dans la cavité, on doit assurer que cette dépression pourra être maintenue. Ceci conduit à trois remarques :

a- Stabilité des conditions hydrogéologiques

Le niveau piézométrique minimal de la nappe doit à l'évidence rester suffisant tant vis à vis des circonstances naturelles (période de sécheresse par exemple), que de la drainance assurée par la cavité elle-même. De ce dernier point de vue, il est souhaitable que la réalimentation de la nappe soit bonne ; on peut craindre alors que le débit d'exhaure soit fort, ce qui exigera un pompage coûteux.

La circonstance la plus favorable est que le massif entourant la galerie soit assez peu perméable, mais en communication avec un niveau supérieur beaucoup plus perméable : ainsi, la réalimentation est satisfaisante et le débit reste modéré. Cette configuration peut exister à l'état naturel ; sinon, on peut s'en approcher artificiellement au moyen d'un "rideau d'eau": une galerie horizontale est creusée au-dessus du stockage lui-même; elle est en communication par exemple avec un puits vertical qui permet d'en contrôler le niveau piézométrique. Un faisceau de sondages horizontaux rayonne depuis cette galerie ; le stockage est ainsi recouvert par un "rideau" que l'on alimente facilement pour y maintenir, si besoin, une hauteur d'eau constante.

A titre d'exemple, les stockages de Donges et Lavéra en France comportent un rideau d'eau ; le stockage du Vexin réalisé dans la craie turonienne n'en comporte pas en raison de sa bonne réalimentation et de la présence d'un horizon détritique à la base de la craie sénonienne qui joue de manière naturelle le même rôle qu'un rideau d'eau.

Quelle que soit la configuration, il est nécessaire que l'évolution de la surface libre de la nappe soit régulièrement suivie : on réalise un tel suivi au moyen de piézomètres : il s'agit simplement de puits ouverts dans lesquels on lit directement le niveau de l'eau.

Le problème de la conception d'un système de surveillance piézométrique est l'un des plus délicats que pose la réalisation d'un stockage à étanchéité dynamique et déborde le cadre du présent article.

b- Stabilité de la pression interne

La pression des produits stockés doit de même rester stable, vis à vis d'incidents tels qu'une panne des pompes d'exhaure ou une augmentation imprévue du débit d'exhaure. C'est la coexistence des phases gazeuse et liquide qui assure cette stabilité : la montée du niveau d'eau se traduit par une réduction progressive du volume de la phase gazeuse sans modification sensible de la pression tant, du moins, que le processus est assez lent (ce qui serait vraisemblablement toujours le cas).

c- Stabilité de la géométrie des cavités

La stabilité des conditions géométriques est tout aussi importante. On a vu -et ce point sera précisé par la suite- que la distance entre le point le plus haut de la cavité et la surface libre de la nappe était déterminante pour que l'écoulement s'opère dans le sens désiré. Si le toit de la cavité est instable, les premiers bancs du toit vont chuter et la cloche ainsi formée va se remplir de gaz. Le point le plus haut de la cavité n'est ainsi plus le même, et la condition d'étanchéité peut être violée (figure 4).

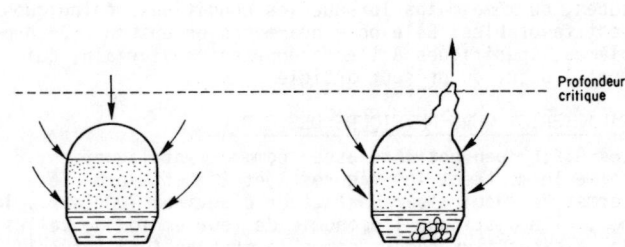

Fig. 4 - *Effet d'une rupture du toit sur l'étanchéité*

Approches théoriques pour les critères d'étanchéité dynamique

La condition d'étanchéité dynamique est en principe très simple à établir mathématiquement : il suffit de résoudre les équations de l'écoulement de l'eau en milieu poreux et de vérifier que l'écoulement de l'eau est, en tout point de la paroi des galeries, dirigé vers l'intérieur de la cavité. Par exemple pour un milieu perméable homogène et isotrope :

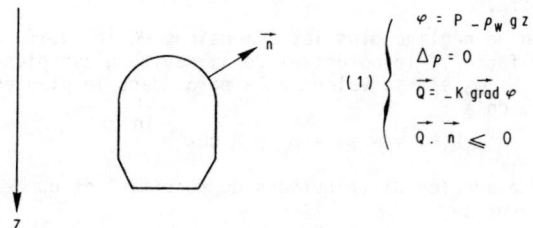

Fig. 5 - *Notations et Système Fondamental*

On peut regrouper les travaux théoriques effectués à propos de cette solution en deux groupes :

- L'un s'intéresse à deux phénomènes physiques qui ne sont pas pris en compte par le système (1) : la faible densité des produits stockés peut rendre instable la solution de ce système, à moins qu'on puisse montrer qu'un début de migration du gaz serait de manière naturelle refoulé dans la cavité ; cette analyse repose sur la notion de "gradient hydraulique" ; par contre, les forces de capillarité à l'interface eau-gaz donnent une marge de sécurité que l'on désigne par la notion de "pression d'entrée de l'eau".
- L'autre discute la validité même du système (1) ; d'une part l'incertitude sur les propriétés hydrogéologiques réelles du massif doivent être prises en compte ; d'autre part la solution de ce système n'est correcte que si les conditions à la limite de l'écoulement sont soigneusement analysées, en raison de caractéristiques particulières à l'écoulement de l'eau vers une galerie en gaz sous pression que l'on désigne par le terme "d'effet de forme".

1- Approche par le gradient hydraulique

Si l'on conçoit assez bien qu'une cavité en dépression par rapport à la nappe environnante "attire" l'eau du massif, on peut se demander si les produits stockés ne peuvent pas néanmoins remonter "à contre courant" vers la surface, en raison de leur faible densité. Le problème se pose principalement bien sûr pour la phase gazeuse des produits.

L'analyse de ce problème a été développée en particulier par le Professeur Aberg (1977) et ses résultats ont servi à dimensionner des stockages à étanchéité dynamique, en particulier dans les pays scandinaves. On peut résumer cette analyse par une formule très simple : on note ρ_w et ρ_g les masses volumiques respectives de l'eau et de la phase gazeuse des produits stockés ; si la pression de l'eau, en un point à la profondeur z sous la surface de la nappe, est notée P on appelle classiquement "potentiel" la quantité $\varphi = P - \rho_w g z$.

L'équation fondamentale de la dynamique, appliquée à une particule de gaz qui aurait commencé à pénétrer dans le massif, indique que cette particule est refoulée vers la cavité si la condition suivante est vérifiée :

(2) $\qquad |\overrightarrow{\text{grad}}\, \varphi| > (\rho_w - \rho_g)\, g$

Cette condition serait bien entendu violée si le fluide était immobile ($\overrightarrow{\text{grad}}\, \varphi = \vec{0}$) ; elle serait très difficile à vérifier pour un stockage qui posséderait deux dimensions horizontales non négligeables devant sa profondeur ; elle peut être vérifiée pour une galerie de faible diamètre, en raison du resserrement des équipotentielles de l'écoulement au voisinage de la galerie qui est propre à cette géométrie. Dans beaucoup de cas, la condition ne peut être assurée que par l'adjonction d'un rideau d'eau qui augmente le gradient en rapprochant de la galerie une surface à potentiel imposé.

2- Commentaires sur la méthode du gradient hydraulique

Le mérite de l'analyse précédente réside dans sa grande simplicité. Elle présente pourtant un défaut important qui réside dans le "télescopage" de deux niveaux d'approche du problème de l'écoulement d'eau dans un massif. Il est en effet incorrect d'utiliser simultanément des points de vue macroscopique et microscopique :
- soit on calcule le potentiel dans le voisinage de la galerie au moyen des équations des écoulements en milieu poreux ; on doit alors écrire au lieu de l'équation fondamentale de la dynamique les équations de Darcy pour un écoulement diphasique ; comme ces équations tiennent compte des frottements liés à l'écoulement et de l'existence d'une pression capillaire qui se développe à l'interface eau/gaz, on obtiendra des conditions de refoulement sensiblement différentes de celles résumées par la formule(2).
- soit on se place dans un "chenal" particulier dont la largeur est grande vis à vis des dimensions moyennes des pores de la roche constitutive du massif (ce chenal représente par exemple une faille) ; mais alors ce chenal est le lieu d'un écoulement préférentiel de l'eau et le calcul effectué au moyen des équations de Darcy dans l'ensemble du massif ne renseigne aucunement sur la distribution du potentiel dans ce chenal particulier.

3- Approche par la pression d'entrée du gaz

L'approche précédente supposait que la pénétration du gaz dans le massif était déjà initiée. En fait, il existe un obstacle à une telle pénétration, même en l'absence d'écoulement de l'eau, qui tient à l'existence des forces capillaires. Supposons la pression de l'eau dans le massif égale à celle du gaz dans la cavité de part et d'autre d'un élément de surface de la paroi. On peut alors penser que, en raison de sa faible densité, le gaz va pénétrer le milieu et remonter vers la surface, puisqu'il n'y a plus d'écoulement d'eau dans l'autre sens. Même dans ces conditions défavorables une telle évolution n'est pas certaine en raison de la moindre mouillabilité de la roche au gaz par rapport à l'eau : il pourra se former des ménisques à l'interface gaz-eau qui s'appuieront sur les grains qui constituent la roche (figure 6) : dans des conditions favorables même un excès de pression du gaz sur l'eau, s'il reste modéré, ne suffira pas à faire déplacer l'eau par le gaz. On appelle "pression d'entrée du gaz" la valeur de l'excès de pression pour laquelle s'initie un tel déplacement ; il faut la distinguer de la pression capillaire, notion utilisée dans le cas d'un écoulement biphasique déjà initié dans un milieu poreux.

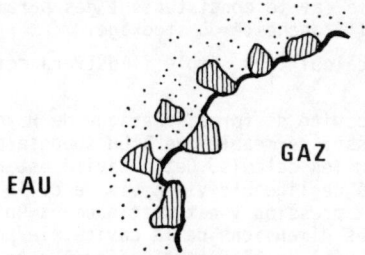

Fig. 6 - *Formation de ménisques à la paroi*

Notons que l'existence d'une "pression d'entrée du gaz" est le phénomène qui explique l'étanchéité des couvertures, par exemple argileuses, qui protègent les stockages de gaz naturel en aquifère.

Ainsi, si la pression d'entrée du gaz dans le massif excède la pression déterminée par une colonne d'eau de même hauteur que la caverne, l'équilibre du système et l'absence de tout mouvement de fluide dans un sens ou dans l'autre est possible. Ce phénomène est observable dans des mines noyées puis réouvertes : on remarque que de l'air a pu rester piégé dans des "cloches" verticales après l'invasion de la mine par l'eau. Malheureusement il n'est pas question de vérifier cette propriété sur les cavités réelles, car un test risquerait de provoquer un dénoyage partiel du voisinage des galeries ; elle ne constitue donc qu'une sécurité ultime virtuelle, donc l'existence apparaitra dans de nombreux cas extrêmement probable.

4- Incertitude sur les paramètres de l'écoulement

La connaissance des paramètres hydrogéologiques est souvent incertaine : d'une part ces paramètres présentent presque toujours une grande variabilité dans l'espace ; d'autre part les mesures effectuées sur échantillons ou in situ ne sont représentatifs que d'une très faible proportion du volume de roches intéressé par l'écoulement ; enfin, la présence de discontinuités (failles, fissures, diaclases) qu'il est souvent difficile de bien reconnaître peut induire de grandes différences entre l'écoulement réel et l'écoulement estimé par le calcul.

Ainsi, la période de creusement et d'essais des galeries permet une vérification en vraie grandeur du système d'hypothèses retenues particulièrement utile.

Les calculs doivent donc toujours être affectés de marges de sécurité suffisantes qui tiennent compte de ces incertitudes. Une bonne méthode consiste à effectuer divers calculs en modifiant les valeurs de certains paramètres afin de tester la sensibilité des résultats du calcul ; on peut l'appliquer utilement à un cas particulier de stockage. Une méthode d'intérêt plus général consiste à apprécier pour un schéma simple d'écoulement vers une galerie l'influence de divers phénomènes envisageables tels que : existence d'un hydrodynamisme naturel, anisotropie des propriétés hydrogéologiques, présence d'une faille ou d'un système de failles, succession de couches aux valeurs de perméabilité contrastées afin de disposer d'un "atlas" des différentes configurations de perturbations possibles aux schémas classiques d'écoulement.

5- Approche par le calcul de l'écoulement : "Effet de forme"

L'analyse par le "gradient hydraulique" contenant comme on l'a vu une certaine part d'arbitraire, il a été décidé pour les stockages actuellement réalisés en France, et en accord avec les autorités administratives chargées du contrôle de ces ouvrages, de vérifier la condition résumée par la formule (2) à titre conservatoire, mais de porter l'effort d'analyse sur le calcul exact de l'écoulement à l'échelle de l'ensemble de la cavité. Une analyse précise a alors permis de mettre en évidence que la forme et les dimensions de la cavité constituaient des paramètres essentiels pour l'étanchéité du stockage.

Un exemple de calcul très simple illustrera cette influence.

Imaginons une cavité de forme sphérique de rayon R, creusée dans un massif perméable que l'on supposera infini pour simplifier les calculs. Cette cavité est remplie de gaz, de densité négligeable vis à vis de celle de l'eau, de sorte que la pression y est pratiquement uniforme. Si l'on néglige les dimensions de la cavité, le potentiel de la cavité est aussi pratiquement uniforme ; on le note $-\varphi_0$ en choisissant par convention que le potentiel dans le massif est nul à l'infini ; alors les distributions du potentiel et du débit sont :

$$\varphi = - \frac{\varphi_0 R}{r}$$

$$\vec{q} = - K \frac{\varphi_0 R}{r^2} \vec{n}$$

où $\vec{n}$ est le vecteur unitaire radial ; en tout point de la paroi de la galerie le débit est négatif :

$$\vec{q}.\vec{n} = - K \frac{\varphi_0}{R^2}$$

Autrement dit, l'écoulement se fait bien du massif vers la cavité.

Si l'on ne néglige plus les dimensions de la cavité ("effet de forme"), le potentiel de la cavité n'est plus uniforme. Si φ_0 est sa valeur à la paroi dans le plan équatorial, on a :

$$\varphi(R,\theta) = - \varphi_0 - \rho_w g R \cos \theta$$

De sorte que les distributions du potentiel et du débit s'écrivent :

$$\varphi(r,\theta) = - \frac{\varphi_0 R}{r} - \rho_w g \frac{R^3 \cos \theta}{r^2}$$

$$\vec{q}(r,\theta) = - K(\varphi_0 \frac{R}{r^2} + 2 \rho_w g \frac{R^3 \cos\theta}{r^3})\vec{n} - K \rho_w g \frac{R^3}{r^3} \sin \theta \vec{p}.$$

Le débit n'est négatif en tout point de la paroi que si :

$$\frac{\varphi_0}{R} - 2 \rho_w g > 0$$

On voit qu'une analyse précise conduit à des conditions plus sévères que les conditions "intuitives" : une cavité de stockage de propane à la pression $P_i = 0,77$ MPa dont le rayon est 10 mètres, doit avoir son centre implanté à 97 mètres au moins sous la surface libre de la nappe, et non 87 mètres comme l'indiquent certains auteurs. De plus, la condition fournie est plus sévère que celle donnée par la règle du gradient hydraulique appliquée sans tenir compte de "l'effet de forme" qui donnerait :

$$\frac{\varphi_0}{R} > (\rho_w - \rho_g) g$$

La figure 7 donne l'allure des lignes de courant pour une cavité suffisamment profonde (centre à 107 mètres) et pour une cavité insuffisamment profonde (centre à 96 mètres) ; dans ce dernier cas certaines lignes de courant sortent de la cavité ; la solution présentée n'a plus de sens physique puisque l'équation de Darcy doit être écrite pour du gaz lorsqu'un fluide sort de la cavité.

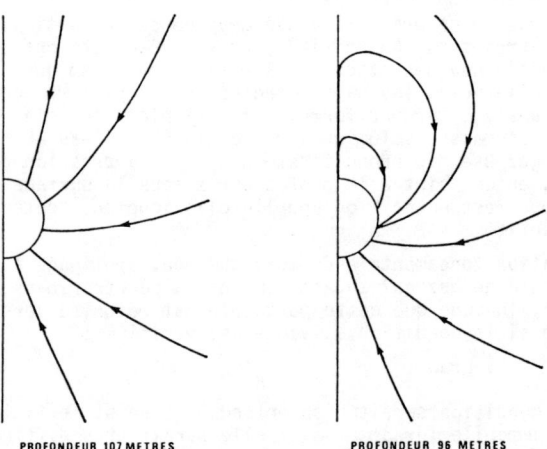

PROFONDEUR 107 METRES   PROFONDEUR 96 METRES

Fig. 7 - *Effet d'une profondeur insuffisante*

6- Généralisation de la méthode de "l'effet de forme"

L'exemple précédent concerne une géométrie très particulière, qui n'est pas applicable aux galeries de stockage réelles : il a en fait pour but d'illustrer dans un cas simple le principe de "l'effet de forme".

La conception des ouvrages, et spécialement la détermination de leur forme et de leur profondeur, implique une approche théorique similaire, mais néanmoins plus complexe car devant faire intervenir la géométrie exacte des cavernes, les rideaux d'eau, s'il y a lieu, ainsi que d'éventuelles inhomogénéités des propriétés pétrophysiques du massif.

Une telle démarche conduit à la détermination d'un terme que l'on peut dénommer "paramètre de forme", qui joue donc un rôle prépondérant dans l'application du critère d'étanchéité.

REFERENCES

A. BOULANGER, W. LUYTEN, "*Underground storage of liquefied gas at low temperature*", Comptes Rendus du Congrès GASTECH, Paris, Octobre 1982.

B. ABERG, "*Prevention of gaz leakage from unlined reservoirs in rock*", RockStore 1977, Proc. of the First Int. Symp., Stockholm, Pages 339 à 413.

H. KOMADA, K. NAKAGAWA, Y. KITAHISA, "*Study on seepage flow through rock mass surrounding caverns for petroleum storage*", RockStore 1980, "Subsurface space", Proc. of the Int. Symp., Stockholm, pages 303 à 310.

W.F. BAWDEN, H.C. ROEGIERS, "*Two phase flow through rock fractures*", Rockstore 1980, "Subsurface space", Proc. of the Int. Symp., Sockholm, Pages 563 à 570.

G. FONTAN, "*Underground Storage and Environmental Protection*", RockStore 1980, "Subsurface Space", Proc. of the Int. Symp., Stockholm, pages 273 à 278.

P. BEREST, "*Note sur la profondeur minimale d'un stockage à étanchéité dynamique*", Rapport interne du Laboratoire de Mécanique des Solides, Ecole Polytechnique, Août 1981.

J. GOGUEL, "*Géologie de l'Environnement*", Editions Masson, 1980, Pages 160 à 161.

P. BEREST, E. LEDOUX, B. TILLIE, "*Etanchéité des stockages d'hydrocarbures liquéfiés en galeries non revêtues dans un milieu aquifère*", Revue de l'Institut Français du Pétrole, Vol. 37, n° 3, Mai-Juin 1982.

Numéro de Septembre 1981 de "GAZ D'AUJOURD'HUI", Pages 293 à 319.

OIL and GAS JOURNAL, "*Containment method reduces underground storage leakage*", 23 Juin 1980.

Brevet d'invention "*procédé de réalisation de stockage souterrain de fluide*", Société Géostock.

COMPORTEMENT MECANIQUE DES CAVITES PROFONDES DE STOCKAGE D'HYDROCARBURES DANS LE SEL
Mechanical behaviour of deep salt caverns for the storage of hydrocarbons
Mechanisches Verhalten tiefliegender Hohlräume zwecks Kohlenwasserstoffspeicherung im Salzgebirge

P. Berest
D. Nguyen, Minh
Laboratoire de Mécanique des Solides École Polytechnique – ENSMP – ENPC – CNRS
91128 Palaiseau – France

RESUME

L'importance majeure des effets différés explique de nombreuses erreurs faites dans l'utilisation des essais de laboratoire au calcul des ouvrages souterrains dans le sel. En utilisant de nombreuses données in-situ, on montre qu'un modèle viscoplastique simple rend compte de l'essentiel des observations. L'intérêt et les limites de ce modèle sont discutés.

SYNOPSIS

On account of the extremely important delayed effects in the rheological behaviour of rock salt, the interpretation of laboratory experiments has led to many errors in the calculation of underground caverns. On the basis of a large number of available in situ data it can be shown that a viscoplastic model makes it possible to predict most in situ observations.

ZUSAMMENFASSUNG

Aufgrund der äußerst bedeutsamen Verzögerungswirkungen im Steinsalz hat das Interpretieren von Laborversuchen bei der Berechnung unterirdischer Bauwerke zu vielfachen Irrungen geführt. Anhand zahlreicher Meßwerte wird gezeigt, wie ein einfaches, viskoplastisches Modell es erlaubt, die meisten in situ Beobachtungen vorherzusagen. Die Brauchbarkeit und Zuverlässigkeit dieses Modells werden untersucht.

INTRODUCTION

Le calcul d'un ouvrage souterrain suit en général une démarche simple qui est d'ailleurs commune au calcul de n'importe quel ouvrage : le comportement des matériaux constitutifs est analysé en laboratoire sur échantillons ; un modèle rhéologique est retenu ; en prenant en compte ce modèle, la géométrie du problème et les chargements imposés, on peut alors calculer, par exemple au moyen d'un programme implanté sur ordinateur, l'ensemble des grandeurs mécaniques intéressantes (déplacements et contraintes) et éventuellement les comparer à un ensemble de "critères" qui indiquent si l'ouvrage sera stable ou non.

Un exemple des difficultés que soulève cette démarche est fourni par le calcul des ouvrages souterrains dans le sel gemme, mines ou stockages souterrains. L'expérience prouve que ces ouvrages continuent souvent à se déformer de manière importante plusieurs dizaines d'années après leur ouverture ; ces faits témoignent qu'une partie du comportement du matériau, très importante dans la pratique, est affectée par des temps de réponse très longs. Ainsi, les expériences de laboratoire sur échantillons, dont la durée n'excède souvent pas quelques heures, négligent par nature une partie substantielle de la rhéologie réelle ; leur interprétation imprudente peut conduire à des résultats désastreux ; des exemples sont décrits dans Baar (1977).

Un progrès sensible est apparu depuis plusieurs années avec un allongement substantiel de la durée des expériences et une plus grande précaution dans la mise en oeuvre et l'interprétation des résultats. Des difficultés demeurent néanmoins, de sorte qu'il est intéressant de prendre le problème "à l'envers", c'est-à-dire de tenter de définir la rhéologie du matériau uniquement à partir des données disponibles relatives au comportement des ouvrages réels.

Il est clair qu'il ne s'agit pas de négliger l'apport des expériences de laboratoire, qui demeure essentiel à une interprétation complète des phénomènes ; mais cet apport est provisoirement mis de côté en l'attente d'une synthèse qui paraît encore très difficile.

UTILISATION DES DONNEES "IN-SITU"

L'utilisation des données qui proviennent de la mesure du comportement des ouvrages réels soulève peut être encore plus de difficultés de principe que l'utilisation des résultats d'expériences en laboratoire :

- pour comparer le comportement d'ouvrages réalisés dans des sites différents, on est conduit à postuler de façon arbitraire une relative uniformité du comportement rhéologique du sel d'un site à l'autre ;

- les données "in-situ" sont le plus souvent de qualité médiocre, en particulier pour les cavités de stockage d'hydrocarbures, dans lesquelles les mesures sont peu précises et difficiles ;

- enfin, au contraire des essais en laboratoire dans lesquels on peut se placer dans des conditions opératoires simples, les données sont relatives au comportement de cavités de géométrie complexe, soumises à des chargements variables, dans des massifs composés de couches de terrain aux propriétés très diverses. Ainsi, doit-on traiter un "problème inverse" en essayant de retrouver le comportement élémentaire du matériau à travers sa réponse à des sollicitations complexes. Il est clair que ce problème n'admet pas une solution unique et que l'on sera conduit à faire largement appel à l'intuition pour simplifier les données, négliger certains facteurs du comportement, ou se laisser guider dans le choix des modèles rhéologiques par le principe de simplicité qui consiste à retenir, s'ils sont suffisants, les modèles les plus rudimentaires même s'ils paraissent contradictoires avec la complexité que révèlent par ailleurs les expériences de laboratoire ; l'absence de préjugé en cette matière constituant précisément la règle du jeu.

COMPORTEMENT DES CAVITES REELLES : EFFETS DE LA PROFONDEUR ET DE LA PRESSION INTERIEURE

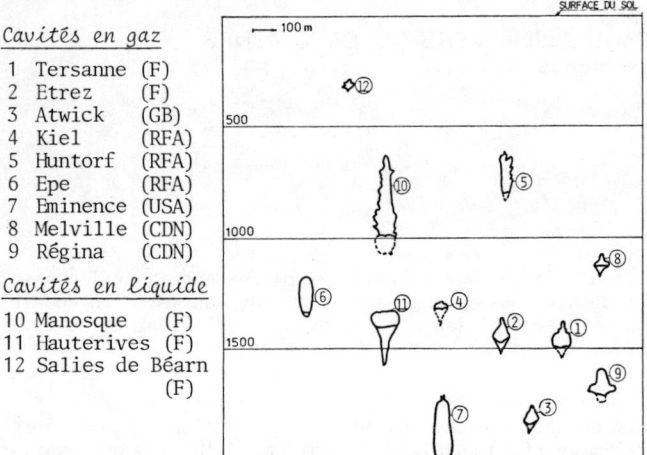

Fig. 1 : *Profondeur et forme de diverses cavités.*

1 - *Définition d'un indicateur simple* :

Précisons d'abord que nous nous intéressons aux seules cavités profondes, dont le rapport entre la profondeur et la plus grande dimension est supérieur à cinq.

a- On ne dispose que de peu de mesures de contraintes in-situ dans le sel, en général déduites d'opérations de fracturation hydraulique. Elles autorisent à estimer que l'état de contrainte naturel est en général peu différent de l'état isotrope qui résulte du poids des terrains surincombants, dont la masse volumique est de l'ordre de 2300 kg/m³ :
(1) $\sigma_{ij} = -P\delta_{ij}$, $P = 0.023\, H$

où H est la profondeur (en mètres) et P la pression (en Mega Pascal).

b- Le régime des pressions intérieures dépend très fortement de la nature des produits stockés :

. dans le cas des produits liquides ou liquéfiés, le sondage est équipé d'un tube central rempli de saumure depuis le fond jusqu'à la surface, de sorte que tout mouvement de produits soit compensé par un mouvement inverse de saumure : la cavité et le sondage sont toujours pleins de liquide, quel que soit le niveau du stock d'hydrocarbures. La pression intérieure résulte simplement de la masse volumique de la saumure (1200 kg/m³) et de la profondeur :
(2) $P_i = 0.012\, H$;

. dans le cas du gaz naturel : il n'y a pas de compensation par la saumure : la pression varie donc dans une fourchette assez large. On fixe toutefois une valeur maximale (pour ne pas compromettre l'étanchéité). En France, Gaz de France a retenu pour ses cavités, profondes d'environ 1500 mètres, les règles suivantes :
(3) $8\,\text{MPa} \leq P_i \leq 0.0166\, H$.

c- L'indicateur le plus simple de l'intensité maximale des efforts subis par la cavité est fourni par la différence entre la pression géostatique (formule (1)) et la pression intérieure minimale (formule (2) ou (3) suivant le cas). La figure 2 donne la valeur de cet indicateur, en fonction de la profondeur, pour différentes cavités (les chiffres renvoient à la légende de la figure 1).

Les points représentant les stockages de liquide sont alignés suivant une droite dont l'équation se déduit des formules (1) et (2) :
(4) $P - P_i = (0.023 - 0.012) H = 0.011\, H$.

Par contre, les stockages de gaz sont représentés par des segments de droite, puisque leur pression intérieure varie dans une certaine plage : le point le plus haut du segment correspond à la valeur minimale de la pression intérieure.

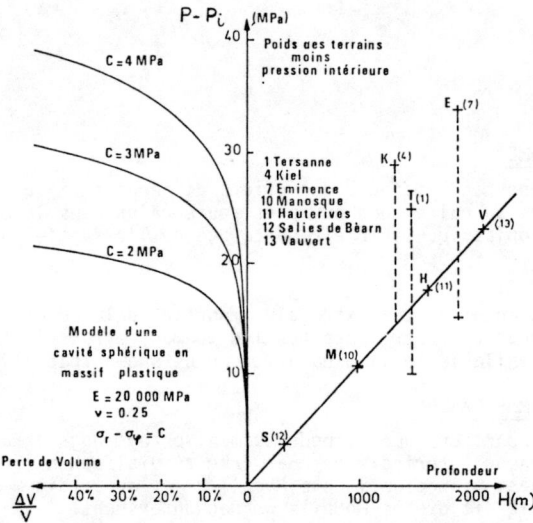

Fig. 2 : *Indicateur de l'intensité de la sollicitation.*

2 - *Fluage des cavités* :

La mesure de la forme de la cavité, et a fortiori de son évolution, est moins commode que dans d'autres ouvrages souterrains. La meilleure est l'échométrie par sonar, bien que sa précision ne soit pas inférieure à quelques pour-cent ; elle n'est utilisable que pour les cavités remplies de liquide. Pour les stockages de gaz naturel, l'équation d'état du gaz permet de déduire le volume global de la connaissance de la pression, de la température et du stock en place ; mais les incertitudes sont importantes. On peut les lever en partie par des méthodes de traçage par d'autres gaz (Boucly et Legreneur, (1979)). L'erreur possible reste assez large ; on peut néanmoins dégager quelques conclusions :

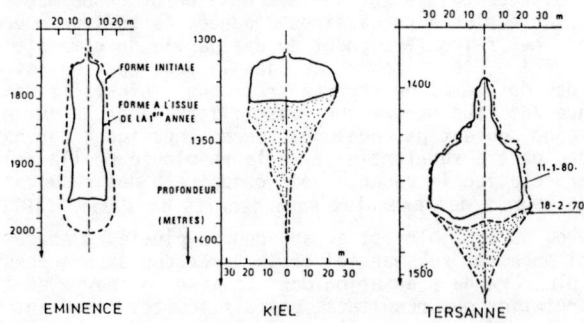

Fig. 3 : *Cavités présentant de fortes pertes de volume.*

a- Aucun mouvement de la surface du sol n'a été noté, bien que certaines cavités soient en opération depuis des dizaines d'années.

b- Les stockages de produits liquides ou liquéfiés sont affectés d'une faible convergence (inférieure à 5% du volume initial).

c- Plusieurs stockages de gaz naturel, ou plus précisément des stockages opérés à très faible pression minimale, ont présenté des dommages plus importants (voir fig. 3) :

. *Eminence Salt Dome, USA* : la cavité était située entre 1725 mètres et 1965 mètres ; la pression intérieure varia entre 7 MPa et 28 MPa, avec plusieurs cycles chaque année. Après deux ans, le fond avait remonté de 36 mètres et la perte globale de volume était de 40% environ (Allen, (1974)).

. *Kiel, RFA* : cette cavité expérimentale était située entre 1300 mètres et 1400 mètres. A l'issue du lessivage, la saumure a été pompée pour vider le sondage, de sorte que la pression intérieure a été pratiquement annulée. Puis, on a laissé la saumure remonter dans le cuvelage. Quarante cinq jours après l'essai, la perte de volume était estimée à 7500m³ (pour un volume initial de 68000 m³, dont 28000 m³ d'insolubles foisonnés). Cinq mois plus tard, une perte additionnelle de volume de 1900 m³ était mesurée (Kühne, Röhr, Sasse, (1973)).

. *Tersanne, France* : deux cavités, lessivées entre 1400 mètres et 1500 mètres ont été en opération pendant dix ans ; la pression intérieure varia entre 8 MPa et 22 MPa avec un ou deux cycles annuels. A l'issue de cette période, la perte de volume était de l'ordre de 25% à 30% (Boucly, Legreneur, (1980)).

Trois conclusions peuvent être dégagées de cet ensemble d'observations :

a- les pertes de volume ont été beaucoup plus importantes que prévues pour ces trois cavités (ou groupes de cavités);

b- les déformations ont été étalées sur plusieurs mois ou années ;

c- les grandes convergences ont été observées quand la valeur de l'indicateur suggéré à la fin du paragraphe précédent excède 20 MPa (fig. 2).

ELASTOPLASTICITE

L'indicateur $P - P_i$ de l'intensité des efforts appliqués apparaît ainsi jouer un rôle majeur. Il faut noter que les pertes de volume ne lui sont toutefois pas proportionnelles, puisqu'elles sont faibles lorsque l'indicateur est en dessous du seuil de 20 MPa. Ces faits suggèrent que le comportement est non linéaire.

Un modèle de comportement non linéaire particulièrement simple est fourni par le modèle élastoplastique de Tresca ou Von Misès. Soient E le module de Young, C la cohésion, ν le coefficient de Poisson. La perte de volume (comptée négativement) d'une cavité de forme sphérique peut alors être exprimée en fonction de $P - P_i$:

$$Q = \frac{3}{4C}(P - P_i)$$
$$(5) \quad \frac{1}{3}\frac{\Delta V}{V} = \frac{4C}{3E}\left\{(1 - 2\nu)Q - \frac{3(1-\nu)}{2}\exp(Q-1)\right\}$$

Si des valeurs vraisemblables des paramètres élastiques sont choisies, par exemple $E = 2\ 10^4$ MPa et $\nu = 0.25$, ce modèle très simple rend bien compte des données in-situ, comme en témoigne la partie gauche de la figure 2, pourvu que la valeur retenue de la cohésion soit assez faible (un peu moins de 3 MPa) : le seuil de 20 MPa pour la différence $P - P_i$ permet en effet, alors, de séparer les cavités dont la perte de volume est inférieure à 5%, donc pratiquement indiscernable, des cavités à très forte convergence qui occupent le haut du diagramme.

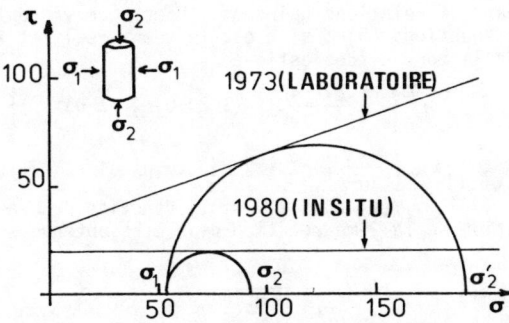

Fig. 4 : *Comparaison entre les valeurs du critère plastique déterminées in-situ et en laboratoire.*

Il est intéressant de revenir sur les données de laboratoire : dans le cas de Tersanne, par exemple, un ensemble très complet d'expériences, menées par des équipes différentes, avaient été rassemblées par Pottier (1969) au moment de la création du stockage : elles avaient permis de définir un critère de Mohr Coulomb (voir fig. 4) qui s'est avéré a posteriori incapable de rendre compte des observations effectuées.

Le défaut majeur du modèle proposé dans ce paragraphe réside dans l'absence d'influence propre du temps : alors que les observations montrent que la convergence s'accroît progressivement, le modèle plastique suppose que la convergence finale est obtenue dès l'application de la charge ou, plus précisément, que la convergence finale est bien celle observée au moment de la mesure -alors que rien ne permet d'affirmer que pour la même charge appliquée la perte de volume ne serait pas encore supérieure si la mesure avait été effectuée plus tard-. L'utilisation du cadre de la viscoplasticité permet d'approfondir cette question.

VISCOPLASTICITE

L'analyse plastique présentée précédemment laisse en suspens deux questions : d'une part, la convergence est-elle destinée à se poursuivre jusqu'à fermeture totale ? d'autre part, la période et l'amplitude des cycles ont-elles une influence sur cette convergence, une fois admis le rôle manifeste de l'intensité de la pression intérieure ? La logique de la méthode proposée conduit à chercher d'abord dans l'expérience la réponse à ces questions. Mais, il est vite apparu nécessaire de disposer d'un modèle de calcul simple, analogue au modèle utilisé plus haut, afin de pouvoir tester les hypothèses que suggèrent les observations. Plus précisément, il était souhaitable de disposer d'une solution explicite du problème de l'évolution d'une cavité sphérique soumise à une pression intérieure variable. Ce problème avait été résolu dans des cas de chargement très simple ou par des méthodes numériques, par exemple Aufaure (1975), Tijani (1978), Wierzbicki (1963). Une solution complète a pu être établie, dont on trouvera un exposé détaillé dans Bérest et Nguyen Minh Duc, (1981).

Ceci posé, le choix des paramètres du comportement viscoplastique pose un problème difficile : les données de laboratoire ne peuvent être utilisées et les données "in-situ" deviennent insuffisantes quand le modèle se complique. Dans le cas le plus simple, celui du modèle viscoplastique de Bingham, il faut opérer un choix des trois constantes (C, η, E) : cohésion, viscosité, module d'élasticité, (le coefficient de Poisson a moins d'importance, et on le fixe conventionnellement à 0,25).

A cette fin, on dispose par exemple dans le cas de Tersanne de la donnée suivante : la perte de volume est d'environ 30% après dix ans. On peut, pour simplifier encore, estimer que la cavité a été soumise pendant toute cette période à sa pression minimale. L'évolution du système est

D 229

régie par des relations qui sont l'homologue viscoplastique des équations (5) : si t est le temps réel, et x le rayon de la zone viscoplastique :

$$(6) \quad Q = \frac{3(P-P_i)}{4C} \quad \tau = \frac{Et}{2\eta(1-\nu)} \quad \begin{cases} \frac{1}{3}\frac{\Delta V}{V} = \frac{4C}{3E}\left\{(1-2\nu)Q - \frac{3}{2}(1-\nu) x^3\right\} \\ \frac{dQ}{d\tau} + Q - 1 = \frac{dx^3}{d\tau} + \log x^3 \end{cases}$$

On note $Q = (1+s)Y(t)$ où $Y(t)$ est la fonction de Heaviside, la solution de la 2ème relation peut être obtenue sous la forme :

$$(7) \quad \tau' = \int_{s}^{x^3} \frac{d\zeta}{s - \log\zeta} .$$

Si on se donne $t = 10$ ans et $\Delta V/V = -30\%$, une relation différentielle peut être obtenue entre C et η lorsque E est fixé. La surface représentative dans l'espace (E, C, η) des valeurs possibles des paramètres est donnée sur la figure 5.

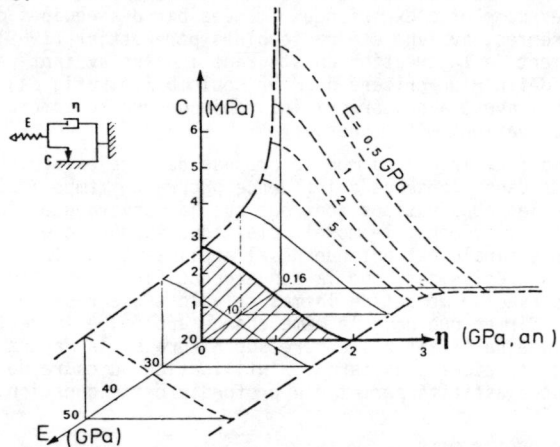

Fig. 5 : *Valeurs des paramètres qui rendent compte d'une perte de volume de 30% après dix ans.*

Plusieurs auteurs (Clerc-Renaud et Dubois, (1978), Boucly, communication personnelle) ont déduit d'essais de compressibilité in situ que le module d'élasticité "instantanée" était de l'ordre de $E = 10000$ MPa à $E = 30000$ MPa. On voit qu'à l'intérieur de ces limites l'influence de E n'est pas énorme (figure 5) ; on a donc choisi la valeur moyenne de 20000 MPa.

La courbe soulignée de la figure 5, qui lie C à η, montre alors que :

- soit on choisit une cohésion élevée ; on se rapproche du comportement plastique parfait ; au-delà des 10 ans la perte additionnelle de volume sera petite.

- soit la cohésion est basse, la constante de viscosité élevée : le sel se comporte comme un fluide de Newton ; la convergence continuera pendant des siècles jusqu'à la fermeture totale.

La tendance actuelle semble de retenir des valeurs moyennes : une cohésion de 1,5 MPa à 2 MPa et une constante de viscosité de 1000 MPa par an (voir par exemple Boucly et Legreneur, (1979)). La convergence finale serait alors très importante, quoique obtenue progressivement.

VISCOPLASTICITE OU VISCOELASTICITE NON LINEAIRE ?

On a retenu jusqu'ici un comportement viscoplastique, mais il est clair qu'une infinité de modèles rhéologiques pourraient convenir, du moins si on se limite au seul aspect mathématique du problème qui est celui de l'ajustement d'un certain nombre de constantes. Il n'est pas inutile de rappeler que plus on introduit de constantes dans un modèle, "meilleure" sera sa capacité à rendre compte des faits déjà connus, mais que son pouvoir prédictif n'en sera éventuellement pas amélioré, de sorte que le progrès sera illusoire.

Un modèle alternatif simple est fourni par la viscoélasticité non linéaire (du type Maxwell) :

$$\dot{\varepsilon} = A \left(\frac{\sigma}{\sigma_0}\right)^n ,$$

où n est une constante de l'ordre de 3 à 5 (Hardy, (1982), Langer, (1979), Waversick, (1981)).

Il permet également de rendre compte des faits observés sur les cavités ; toutefois, il induit des conséquences à long terme très différentes : du fait de l'absence de seuil les cavités creusées dans un tel matériau doivent à terme se refermer complètement, alors que dans le modèle viscoplastique une situation stable est possible si la cavité n'est pas de grande hauteur. Un argument en faveur de ce dernier modèle peut être trouvé dans certaines constatations géologiques : l'existence de nombreux gisements de sel en couche, très peu déformés après des dizaines de millions d'années, paraît peu compatible avec l'hypothèse d'un comportement de liquide (non Newtonien). Au contraire l'existence de dômes de sel prouve que le seuil plastique, s'il existe, est très faible (Goguel, (1975)). Cette argumentation reste néanmoins fragile ; en tout état de cause, notre modèle viscoplastique est sans doute également trop fruste ; par exemple la dépendance linéaire entre vitesse de déformation et charge appliquée rend compte de faits tel que l'influence du taux de défruitement sur la convergence de piliers dans des mines de sel exploitées par la méthode des chambres et piliers abandonnés.

AUTRES PARAMETRES DU COMPORTEMENT

Dans les paragraphes précédents, la pression interne et la pression géostatique (en fait, la différence entre ces deux quantités) sont apparues comme des paramètres très significatifs. D'autres facteurs jouent manifestement un rôle dans la stabilité de la cavité. Malheureusement, ils ne permettent le plus souvent qu'une analyse qualitative ; aussi beaucoup des idées suivantes restent assez conjecturelles.

Effets d'une pression interne variable :

Nous avons vu que les variations de pression sont négligeables dans un stockage de produits liquides et importantes dans un stockage de gaz naturel : à Tersanne par exemple, les limites sont de 8 MPa et 22 MPa. La période qui sépare le passage entre deux pressions extrêmes dépend du site considéré : elle est de l'ordre d'une année pour un stockage de modulation de consommation comme Tersanne, d'un mois pour un stockage proche d'un champ de production de gaz comme Eminence Salt Dome, et même d'un jour pour un stockage d'air comprimé comme celui de Huntorf (Boucly, Legreneur (1980), Allen (1974), Crotogino, Quast (1980)). Toutefois, le nombre de cas ayant été étudiés reste petit vis à vis de la variété des sollicitations possibles, qui doivent être caractérisées par la période, la pression moyenne et l'amplitude des cycles. On peut simplement affirmer que, contrairement aux idées avancées par certains auteurs, le caractère cyclique des sollicitations n'est pas l'élément explicatif principal de l'intensité de la convergence : les cas de Kiel et Vauvert (Bérest, (1979)) montrent qu'une cavité peut considérablement fluer sous une charge constante ou de variation monotone ; a contrario la cavité de Huntorf soumise à des cycles d'amplitude faible mais très rapides ne semble pas avoir connu de désordres particuliers. Les calculs présentés dans D. Nguyen Minh, P. Bérest (1981) conduisent à des conclusions théoriques analogues.

Forme de la cavité :

On a pu remarquer que les trois cavités qui ont connu un large fluage, (Kiel, Tersanne, Eminence) ont des formes très différentes. Le toit plat, de grande portée, de la cavité expérimentale de Kiel est tombé sur plusieurs mètres d'épaisseur : rien de semblable n'est survenu à Eminence

Salt Dome ou à Tersanne, dont le toit était de forme "pointue". Des ingénieurs de la Société Texas Eastern nous ont indiqué qu'une de leurs cavités, situées dans le Nord Est des USA, avait à l'origine un toit plat de 150 mètres de diamètre ; ce toit s'est effondré dans sa partie centrale. Au contraire, une cavité du stockage stratégique fédéral américain, sur le site de West Hackberry, a pour section horizontale un cercle de 250 mètres de diamètre et son toit est pratiquement plat. Cette cavité paraît constituer l'exemple de ce qu'il ne faut pas faire : elle n'a pourtant connu aucun problème particulier.

Pour analyser ces différences, il est peut-être nécessaire de distinguer les stockages en dômes de sel des stockages en couches salifères : dans ce dernier cas, le massif offre de nombreuses discontinuités lithologiques qui découpent dans un toit trop étendu des plaques horizontales qui supportent difficilement leur propre poids. Les instabilités provoquées par ce type de mécanisme peuvent aller de pair avec un comportement global viscoplastique (continu) à une échelle plus large : une analogie peut être trouvée avec certaines mines de sel, dans lesquelles un fluage généralisé se superpose au jeu des discontinuités naturelles dans le voisinage des cavités ; ces dernières peuvent provoquer des manifestations locales spectaculaires (soufflage du mur, écaillage des parements, chute du toit).

L'effet de la gravité :

L'effet propre de la gravité est rarement pris en compte dans le calcul des ouvrages souterrains profonds ; il est en effet négligeable, dès que la hauteur de l'ouvrage est petite devant sa profondeur. Les cavités de stockage, dont la hauteur peut être de plusieurs centaines de mètres, soulèvent donc de manière originale ce problème.

On a constaté que la cavité d'Eminence Salt Dome avait vu son fond remonter de manière considérable, de sorte qu'une part importante de la perte globale de volume était due au seul mouvement du fond. Ce phénomène apparaît également très nettement sur une section verticale de la cavité n° 2 de Tersanne (fig. 4), mais dans ce dernier cas les particularités de la géométrie peuvent suggérer des interprétations spécifiques. Un exemple plus net est donné par Thoms et Mogharrebi, (1979), qui ont effectué des mesures sur deux sondages de neuf pouces dont le découvert dans le sel était à 1500 mètres de profondeur. 90 jours après le forage l'un des découverts avait perdu 1,75 pouces au voisinage du fond, mais la perte de diamètre diminuait lorsqu'on se rapprochait du sabot.

Une explication de ces phénomènes (mais ce n'est pas la seule) peut être trouvée dans l'effet spécifique de la gravité. Cet effet est négligé dans les modèles simples présentés dans les paragraphes précédents, qui supposent que les produits stockés ont même densité que les terrains. Cette approximation est particulièrement inexacte pour les stockages de produits gazeux.

On peut tenter d'estimer l'erreur ainsi commise sur un exemple très simple : on suppose que le sel se comporte comme un fluide incompressible (très) visqueux. La vitesse de déplacement radial à la paroi, soit v pour une cavité sphérique, se déduit alors de la même vitesse, soit v_o, calculée en négligeant l'effet de la gravité par la formule (voir Bérest et Nguyen M.D., (1981)) :

$$(8) \quad v = v_o \left(1 + \frac{4}{3} \frac{\rho g R}{P - P_i} \cos \theta \right).$$

Pour une cavité assez grande, la différence est sensible entre la partie haute et la partie basse : à la convergence générale centripète se superpose un déplacement vers le haut, analogue à celui d'une bulle d'air dans de l'eau. Cet effet serait plus sensible si l'on retenait un modèle viscoélastique non linéaire ; mais le modèle viscoplastique fournit un résultat analogue. On montre en effet que si le seuil viscoplastique est petit devant le poids d'une colonne de terrains de même hauteur que la cavité, la cavité est nécessairement instable dans sa forme actuelle, de sorte que le mouvement du sel vers le fond de la cavité doit être particulièrement important.

Effet de la température :

Une explication complémentaire du phénomène précédent peut être trouvée dans les différences de température liées au gradient géothermique. De nombreux auteurs ont montré que la vitesse de déformation du sel sous une même charge dépendait fortement de la température (par exemple, Baar, (1977), Waversick, (1978), Langer, (1979), Vouille,(1981)). Si l'on divise l'espace par le plan équatorial de la cavité, la partie du demi-espace inférieur affectée de déformation viscoplastique est en moyenne plus chaude que son symétrique supérieur : les vitesses y sont donc plus grandes.

Plus généralement, cet effet doit être pris en compte lors de la comparaison de sites exploités à des profondeurs différentes. La valeur de 1000 MPa par an pour la constante de viscosité est adaptée à l'étude de cavités profondes (de l'ordre de 1500 mètres). Cette valeur apparaît un peu trop pessimiste lorsqu'elle est appliquée, par exemple, à des cavités moins profondes du stockage de pétrole de Manosque. Un exemple plus caractéristique encore est donné par la mine de sel de Varangeville (France). Un quartier de cette mine a été exploité, il y a un siècle environ, par une méthode de chambres et piliers : la mine est sous deux cent mètres de recouvrement, et le taux de défruitement est voisin de 75% ; la déformation verticale de ces vieux piliers a été mesurée avec un grand soin depuis une dizaine d'années ; la valeur moyenne est de 5.10^{-4} par an. Niangoula (1981) a montré que les conditions d'application de la théorie de l'aire tributaire paraissaient au moins en première approximation remplies. En appliquant alors le modèle viscoplastique de Bingham :

$$(9) \quad \dot{\varepsilon} = \frac{1}{\eta} (\sigma - 2C) \qquad \sigma \simeq 18 \text{ MPa} \qquad \dot{\varepsilon} \simeq 5.10^{-4} \text{ par an}$$

on trouve des valeurs des constantes C et η beaucoup plus grandes que pour les cavités souterraines profondes, par un ordre de grandeur de dix au moins : ce fait peut être expliqué par une température assez basse (de l'ordre de 18°C dans la mine de Varangeville, au lieu de 65°C. dans une cavité de Tersanne) qui détermine des vitesses de fluage beaucoup plus modérées.

REFERENCES

K. ALLEN, "ANALYSIS ...", 1974, (CITÉ PAR BAAR, 1977).
M. AUFFAURE, "ÉTUDE ...", JOURNAL DE MÉCANIQUE, VOL. 14, N° 2, PP. 221-235, 1975.
C.A. BAAR, "APPLIED ...", I, ELSEVIER SCIENTIFIC PUBLICATION COMPANY,1977.
P. BÉREST, E. LEDOUX, B. LEGAIT, G. DE MARSILY, "EFFETS ...", C.R. DU 4È CONGRÈS DE LA S.I.M.R., MONTREUX, 1979.
P. BÉREST, D. NGUYEN MINH, "STABILITY ...", INT. SYMP. ON WEAK ROCKS, SEPT. 21-24, 1981, TOKYO, JAPON.
P. BÉREST, "STABILITÉ ...", REVUE FRANCAISE DE GÉOTECHNIQUE, N° 16, AOUT 1981.
PH. BOUCLY, J. LEGRENEUR, "HYDROCARBON ...", ROCK STORE, STOCKHOLM, 1980.
K.M. BORCHERT, H. KLAPPERICH, T. RICHTER, B. WALZ, "BERECHNUNGEN ...", C.R. DU 4È CONGRÈS DE LA S.I.M.R., MONTREUX, 1979.
A. CLERC-RENAUD, D. DUBOIS, "EXPLOITATION ...", V SYMPOSIUM ON SALT, HAMBOURG, 1978.
F. CROTOGINO, P. QUAST, "COMPRESSED ...", ROCK STORE, STOCKHOLM, 1980.
J. GOGUEL, TRAITÉ DE TECTONIQUE, 1965, EDITIONS MASSON.
H.R. HARDY JR, "BASIC ...", ISRM SYMPOSIUM, AACHEN, 1982.
G. KUHNE, H,U. ROHR, W. SASSE, "KIEL ...", 12TH WORLD GAZ CONFERENCE, NICE, 1973.
M. LANGER, "RHEOLOGICAL ...", C.R. DU 4È CONGRÈS DE LA S.I.M.R., MONTREUX, 1979.
K.H. LUX, R.B. ROKAHR, "SOME REMARKS ...", ROCK STORE, STOCKHOLM, 1980.
A. NIANGOULA, "CONTRIBUTION ...", THÈSE DE DOCTEUR-INGÉNIEUR, INSTITUT NATIONAL POLYTECHNIQUE DE LORRAINE, NANCY, 08-09-1981.
D. NGUYEN MINH, P. BÉREST, "MODÉLISATION ...", REVUE FRANCAISE DE GÉOTECHNIQUE, N° 16, AOUT 1981.
D. NGUYEN MINH, P. BÉREST, "RESPONSE ...", INT. JOUR. SOL. & STRUC., 1982, (TO BE PUBLISHED).
POTTIER, COMMUNICATION AU COMITÉ FRANCAIS DE MÉCANIQUE DES ROCHES, REVUE DE L'INDUSTRIE MINÉRALE, NUMÉRO SPÉCIAL DU 15 AVRIL 1974.
H.U. ROHR, "MECHANICAL ...", IV SYMPOSIUM ON SALT, 1974.
R.L. THOMS, M. MOGHAREBBI, "BOREHOLE ...", 20TH US SYMPOSIUM ON ROCK MECHANICS, AUSTIN, TEXAS, JUNE 4-6 1979.
R.L. THOMS, M. NATHANY, R. GEHLE, "LOW FREQUENCY ...", VOL. 2, ROCK STORE, STOCKHOLM, 1980.
S.M. TIJANI, "RÉSOLUTION ...", THÈSE DE DOCTEUR-INGÉNIEUR, UNIVERSITÉ DE PARIS VI, 1978.
G. VOUILLE, J. FINE, S.M. TIJANI, P. BOUCLY, "DÉTERMINATION ...", C.R. DU 4È CONGRÈS DE LA S.I.M.R., MONTREUX, 1979.
W.R. WAVERSIK, "RECENT ...", FIRST CONFERENCE ON THE MECHANICAL BEHAVIOR OF SALT, 1981.
T. WIERZBICKI, "A THICK-WALLED ...", ARCHIVUM MECHANIKI STOSOWANEJ, VOL. 2, N° 15, PP. 297-307, 1963.

ANALYSE DU COMPORTEMENT DIFFERE DES OUVRAGES SOUTERRAINS
Analysis of the time-dependent behaviour of underground works
Untersuchung des Zeitverhaltens unterirdischer Hohlräume

D. Nguyen Minh
P. Burest
J. Bergues
Laboratoire de Mécanique des Solides, Ecole Polytechnique — ENSM — ENPC — CNRS 91128
Palaiseau — France

RESUME

Les auteurs mettent en évidence deux aspects du comportement différé des ouvrages souterrains dans les roches fragiles et résistantes et dans les roches molles et plastiques. On montre que la prise en compte du comportement radoucissant dans le cas des roches fragiles peut expliquer l'importance des phénomènes différés que l'on rencontre parfois dans les ouvrages souterrains. On montre également que pour les revêtements rigides des tunnels creusés dans les roches molles, on a une disparition de la zone viscoplastique.

SYNOPSIS

The authors show two aspects of the time-dependent behaviour of structures in either resistant-brittle or soft-plastic rocks. In the case of hard rocks it is shown that strain softening can account for observed retarded deformations in existing structures. Finally it is shown that in the case of a rigid lining of cavities inside soft rocks the viscoplastic zone disappears.

ZUSAMMENFASSUNG

Zwei Aspekte des zeitabhängigen Verhaltens von Untertagebauten in sprödem oder weich-plastischem Fels werden beschrieben. Gezeigt wird, daß die Berücksichtigung des weicheren Verhaltens von sprödem Fels die Bedeutung von Verzögerungserscheinungen erklären kann, welche manchmal im Untertagebau vorkommen. Außerdem wird gezeigt, daß bei steifen Tunnelverkleidungen in weichem Fels die viskoplastische Zone verschwindet.

INTRODUCTION

On sait depuis longtemps que les ouvrages souterrains sont soumis à des effets différés s'étalant parfois sur plusieurs mois, voire plusieurs années. Il est clair, sur une telle échelle des temps, que l'avancement du front de taille n'a plus d'action, et qu'il s'agit bien du comportement rhéologique du massif encaissant (Panet, 1979).
Les conditions géotechniques pour lesquelles ces phénomènes ont lieu peuvent être très différentes. Très souvent il s'agit de tunnels peu profonds dans des roches molles telles que les argiles plastiques (Habib et al. 1965). A l'autre extrême, on peut trouver des tunnels profonds (1000 à 2000 m) dans des roches dures et fragiles. Les phénomènes différés peuvent être limités comme dans le cas de la galerie de Belledone (gneiss, granite) (Plichon, 1979), ou peut-être le tunnel routier du Mont Blanc dans la protogine qui a donné lieu, cependant, à des phénomènes de décompression violents, (Panet, 1976) ; dans d'autres cas ces phénomènes différés peuvent être importants que ce soit par la pression exercée, mesurée sur le revêtement au cours du temps : tunnel ferroviaire du Mont Cenis dans les calcschistes, (Panet, 1976), ou tunnel de l'Arlberg (gneiss, micaschiste) (Pacher, 1979) ou par l'évolution des convergences mesurées dans le tunnel autoroutier du Fréjus (Lunardi, 1980).

Il est clair que la connaissance du comportement différé du massif rocheux revêt une grande importance pour le calcul du revêtement. Les modèles viscoélastiques classiques ne semblent pas convenir, dès lors que la limite d'écoulement ou de rupture du matériau est excédée. La formulation viscoplastique est plus adaptée.

Nous exposons dans cet article deux aspects remarquables du comportement différé des ouvrages souterrains en liaison avec les cas extrêmes décrits ci-dessus :

a) roches fragiles : on montre expérimentalement et théoriquement que la viscosité affectant le radoucissement des roches se traduit, sur une géométrie complexe, par des constantes de temps de fluage inhabituellement grandes dans l'optique plus classique d'un modèle viscoplastique parfait.

b) roches molles plastiques : la raideur relative du soutènement peut conditionner la nature du phénomène, et en particulier occasionner une disparition complète de la zone viscoplastique.

I - SCHEMATISATION D'UN TUNNEL

Dans ce qui suit on adopte quelques hypothèses simplificatrices pour schématiser le problème du tunnel, dans le but de mettre en évidence certains aspects du comportement ; on néglige en particulier l'effet de la forme du tunnel et le gradient de la pesanteur.

La géométrie admet une symétrie cylindrique : le tunnel est circulaire, de rayon intérieur unité, et de rayon extérieur ρ. La pression extérieure P est la pression géostatique ; la pression intérieure P_i englobe une pression de soutènement fictive due à la proximité du front de taille, P_i^f, et une pression de revêtement P_i^r.

$$P_i = P_i^f + P_i^r \qquad (1)$$

L'histoire de P_i sera précisée dans le texte ; elle conduit de toute façon à une contraction du tunnel.

La loi d'écoulement est indépendante de la pression moyenne, (critère de type "Tresca"), ce qui permet de ramener le problème à l'étude du cylindre soumis à la seule contrainte normale $\sigma_i(t)$ sur sa paroi interne, l'état initial étant l'état naturel :

$$\sigma_i(t) = P - P_i(t) \qquad (2)$$

Cette simplification importante n'est certainement pas valable pour toutes les roches qui possèdent en général un angle de frottement interne Les résultats des calculs sont donc plutôt qualitatifs, mais ils permettent une discussion complète du fait des solutions explicites obtenues.

II - MISE EN EVIDENCE EXPERIMENTALE DU COMPORTEMENT DIFFERE DES ROCHES FRAGILES

Depuis plus de vingt ans, on sait d'après les expériences de fluage uniaxial, que les roches fragiles, admettent un comportement différé, (Morlier, 1966), qu'on attribue essentiellement à la fissuration (Scholz, 1968) ; mais les expérimentateurs reconnaissent que la part de la déformation différée est négligeable devant les déformations instantanées du matériau, d'autant plus que la courbe de fluage classique à trois phases est très délicate à obtenir, puisque, sans transition très nette selon le niveau du déviateur, on avait soit une stabilisation rapide des déformations, soit une rupture (Waversik, 1974, Bergues et al, 1982). Les essais plus récents de relaxation dans la phase après rupture, utilisant les presses asservies, (Peng, 1972, Bergues, 1982, Morgenstern, 1981), ont permis de constater que le radoucissement de la roche était affecté de viscosité, laquelle était responsable de la phase de fluage tertiaire ; plusieurs modèles de comportement ont ainsi été proposés, (Kaiser et Morgenstern, 1981, Nguyen Minh et al. 1981).

Il n'était cependant pas certain que cette viscosité puisse rendre compte des observations in-situ, puisque les constantes de temps de relaxation ne dépassaient pas la journée -sinon l'heure- alors que par exemple les convergences du tunnel du Fréjus ont duré plusieurs mois. Une expérience sur maquette susceptible de représenter un modèle réduit de tunnel permett ait d'avoir une part d'explication du phénomène.

Il s'agit d'un essai sur tube épais de roche soumis à une pression extérieure. Le montage est conçu de telle sorte qu'il est possible de faire soit des essais de fluage (pression extérieure constante), soit des essais de relaxation (déformation intérieure constante), (Nguyen Minh et Bergues, 1982).

Les tests sont effectués avec deux roches fragiles (marbre de St Beat et craie sèche) ; si l'essai de relaxation donne des effets différés de l'ordre de quelques heures, la viscosité apparente et la ductilité du tube en fluage sont par contre nettement supérieures à celles des essais uniaxiaux, comme le montre le tableau comparatif ci-dessous donnant les valeurs des déformations différées pour des éprouvettes soumises à un chargement monoaxial et pour des tubes.

	Déformations stabilisées sur éprouvettes chargées à 90% de la rupture	Fluage sur tube
craie	$0,005 \times 10^{-2}$	$1,5 \times 10^{-2}$
marbre	$0,015 \times 10^{-2}$	$0,1 \times 10^{-2}$

Le fluage a pu être poursuivi sur plusieurs semaines et les trois phases ont pu être obtenues,

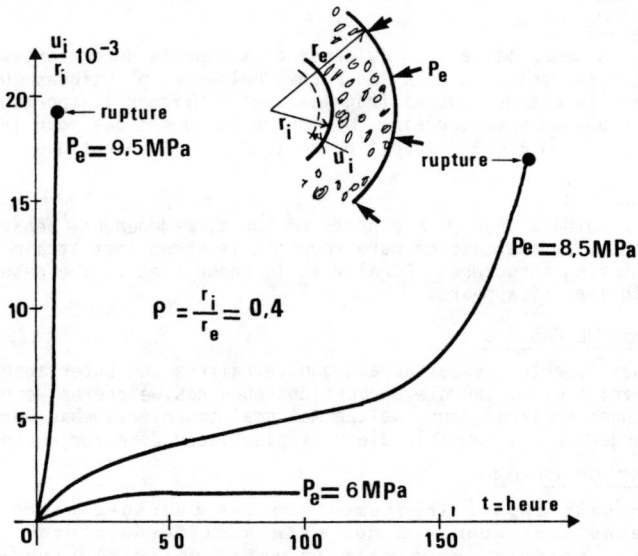

Fig 1 Fluage d'un tube de craie seche

Or il faut bien voir que c'est le fluage qui est le plus proche des sollicitations réelles dans un tunnel en roche dure, puisque la rigidité relative du revêtement est alors plutôt faible.

Ce résultat remarquable permettrait d'envisager une méthodologie expérimentale pour caractériser le comportement différé des roches fragiles utilisant un état de contraintes proches de celui régnant in situ. Il était auparavant nécessaire d'interpréter les résultats au moyen d'un modèle de calcul.

III - COMPORTEMENT DIFFERE DES TUNNELS EN ROCHES FRAGILES

III.1.- Modèle de comportement

On utilise un modèle de comportement élastoviscoplastique représenté par le groupement rhéologique à trois éléments, (fig.2), discuté par ailleurs dans (Nguyen Minh 1981). La traduction à trois dimensions de ce modèle est donnée en annexe ; on a envisagé deux modèles d'amortisseurs différents :

- "Modèle I" : la viscosité η est limitée à la seule phase radoucissante et est nulle au-delà ; ceci repose sur l'hypothèse selon laquelle pour certains matériaux fragiles la phase résiduelle n'intéresse que le frottement entre les fragments rompus du matériau.

- "Modèle II" : la viscosité η reste constante.

Il est possible qu'un comportement réaliste se situe entre ces deux modèles. Quelques courbes de réponse sous chargement uniaxial sont données sur la fig.3. On remarque que le fluage uniaxial est uniquement accéléré ; les phases primaires et secondaires ont été négligées.

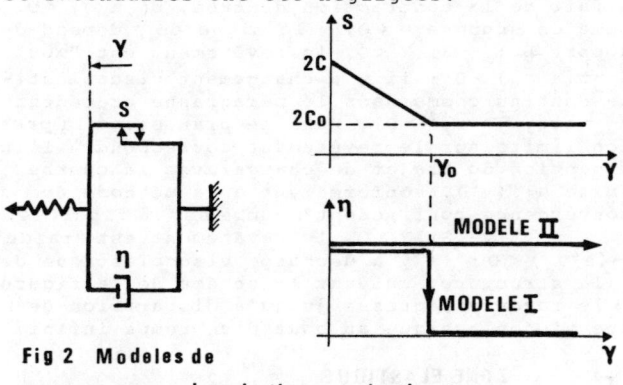

Fig 2 Modeles de comportement visçoplastique radoucissant

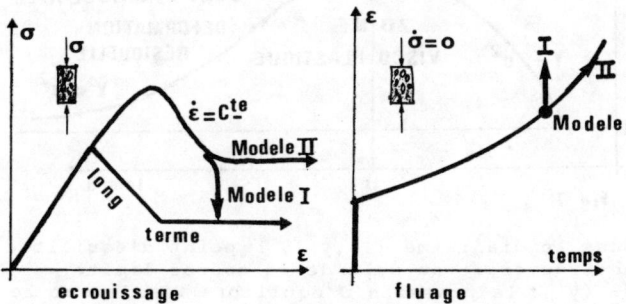

Fig 3 Reponse des modeles sous chargement uniaxial

On admet dans le calcul que le matériau est incompressible, ce qui entraîne que le coefficient de Poisson ν égale 0,5. Cette hypothèse permet de relier directement le déplacement radial u au rayon de la zone viscoplastique y :

$$\frac{u}{r} = -\frac{2C}{E'}\frac{y^2}{r^2} \qquad (3)$$

C étant la cohésion initiale du matériau, $E' = \frac{4}{3} E$ le module d'Young, équivalent en déformation plane ($\nu = 0,5$).

Cette formule montre que la contraction du tube entraîne un chargement viscoplastique continu ($\dot{y}^2 > 0$).

Remarque : l'hypothèse $\nu = 0,5$ est légitime dans la mesure où, comme en élastoplasticité, son poids est relativement faible lors du chargement. Nous reviendrons plus précisément sur la portée de cette simplification au §IV.

III.2.- Résultats et discussions

La solution du problème, pour une histoire quelconque du chargement $\sigma_i(\tau)$ est exposée en annexe 2. Dans le plan σ_i-y^2 (y : rayon de la frontière viscoplastique), nous avons tracé les chargements particuliers envisagés dans les applications consistant à imposer une relaxation ou un fluage du tube après un chargement élastique instantané jusqu'au niveau initial σ_i^0.

On montre (Nguyen Minh, 1982) que la relaxation se stabilise toujours sur la courbe de réponse limite $\sigma_i^\infty(y^2)$ du tube, obtenue par la solution élastoplastique sous-jacente au modèle, (voir annexe 2), avec une constante de temps d'un ordre de grandeur comparable à celle de la relaxation uniaxiale.

On sait, (Nguyen Minh, Berest, 1979), que pour un matériau radoucissant, la courbe limite $\sigma_i^\infty(y^2)$ admet toujours un maximum lorsque le rayon extérieur du tube n'est pas trop grand ; selon que le niveau σ_i^0 est inférieur au maximum de $\sigma_i^\infty(y^2)$, on montre (Nguyen Minh, 1982) que le fluage sera asymptotique ou admettra les trois phases comme il a été constaté expérimentalement. Les constantes de temps du phénomène n'ont ici plus aucun lien apparent avec la constante de temps de relaxation uniaxiale ; si le niveau σ_i^0 est très proche du maximum de $\sigma_i^\infty(y^2)$, on peut avoir un phénomène de très longue durée.

Si on se réfère, maintenant, au problème du tunnel revêtu, le chargement élastique instantané correspond à une annulation brutale de la pression fictive due au front de taille P_i^f. La relaxation simule un revêtement infiniment rigide posé immédiatement et le fluage un revêtement infiniment souple.

Plus généralement, la figure 4 représente le chemin suivi par le chargement lorsqu'on pose immédiatement un revêtement élastique de raideur K ; l'existence d'un fluage à trois phases est alors conditionnée par la possibilité de rencontre entre ce trajet de charge et la courbe $\sigma_i^\infty(y^2)$.

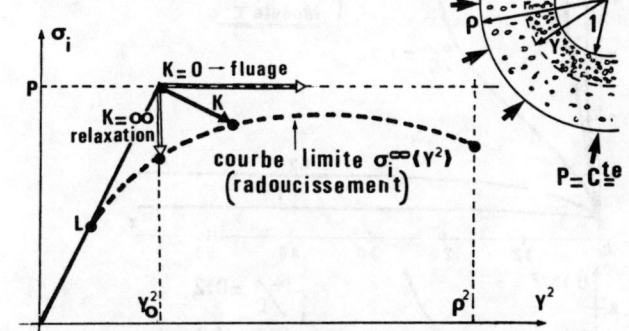

Fig 4 Trajets de charge suivant les differentes rigidites du revetement

IV - COMPORTEMENT DIFFERE DES TUNNELS EN ROCHES MOLLES

Dans les roches molles il est souvent suffisant de négliger l'écrouissage E_2. Avec les mêmes hypothèses simplificatrices précédentes, on est ramené à la seule équation différentielle (4), (voir Annexe 2). Il convient de remarquer ici que la roche étant molle, la raideur relative du revêtement n'est plus négligeable ; on montre alors que l'hypothèse d'incompressibilité précédemment faite risque de masquer un phénomène nouveau : la possibilité d'une décharge viscoplastique. On s'intéresse ici à la valeur finale de la pression sur le revêtement.

Les courbes des figures 5 et 6 illustrent les effets des différents paramètres sur la réponse des modèles, avec $C_o/C = 0,2$ et $\rho = 6$.

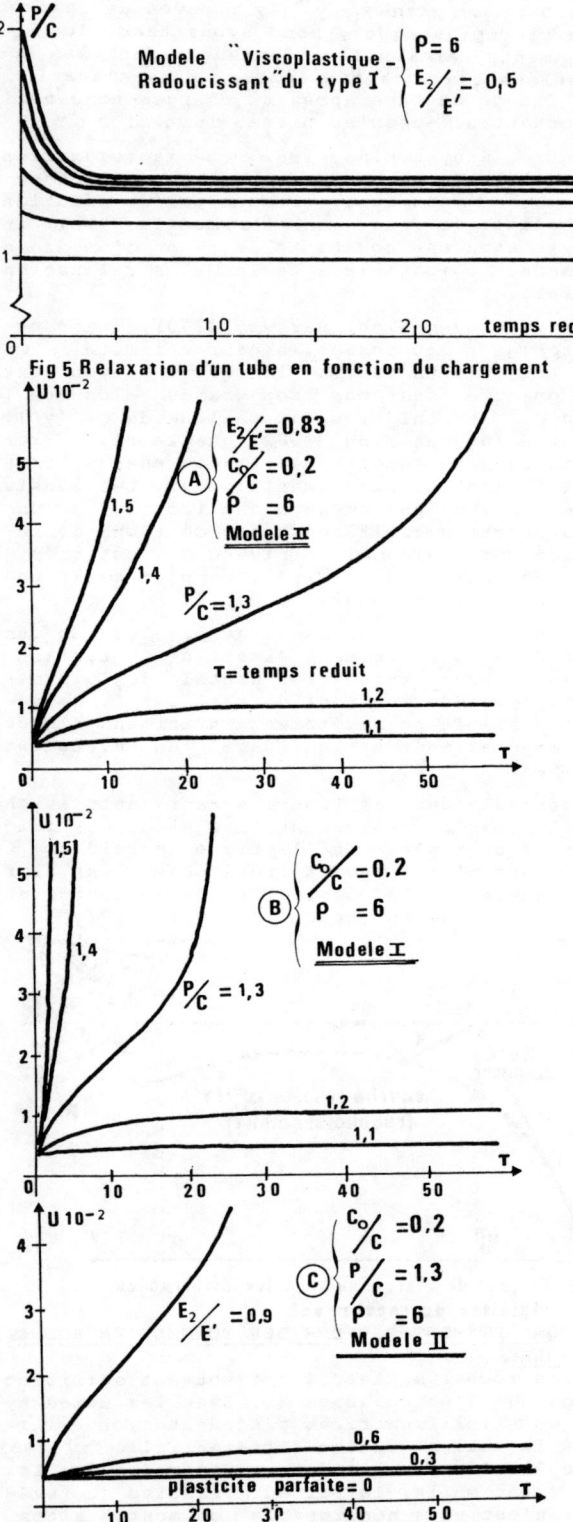

Fig 5 Relaxation d'un tube en fonction du chargement

Fig 6 Fluage d'un tube viscoplastique radoucissant
_A_ Effet de la pression _ Modele II
_B_ " _ Modele I
_C_ Effet du radoucissement E_2

Appelons y la frontière de la zone viscoplastique, qui peut éventuellement régresser ; notons e la valeur maximale jamais atteinte par y durant son évolution, et σ_e la contrainte radiale en e. Le coefficient de Poisson étant quelconque, la formule (3) est remplacée par :

$$E' \frac{u}{r} = \frac{(1-2\nu)}{1-\nu}\sigma_r - 2\frac{e^2\sigma_e}{r^2}$$

après la pose du revêtement, supposé élastique, de raideur K, on tire de la formule précédente, écrite en r = 1 :

$$\gamma \dot{\sigma}_i = \frac{d}{dt}(e^2\sigma_e) \qquad \gamma = \frac{1 - \frac{1-2\nu}{1-\nu}\frac{K}{E'}}{-2\frac{K}{E'}}$$

Du fait de la contraction du tube, on a $\dot{\sigma}_i < 0$. Comme on suppose $\nu \neq 0,5$, le signe de γ dépend du rapport $\frac{K}{E}$: - Si $\gamma < 0$, le revêtement est "mou", et $\frac{d}{dt}(e^2\sigma_e) > 0$: il y a chargement viscoplastique continu comme dans le paragraphe précédent (σ_e = cte, $\dot{y}^2 = \dot{e}^2 > 0$). Dans le plan σ_i, y^2 la pression limite sur le revêtement correspond à l'intersection du trajet de charge avec la courbe limite $\sigma_i^\infty(y^2)$, conformément à la méthode de "convergence-confinement", comme sur la figure 4.
- Si $\gamma > 0$, le revêtement est "raide" $\frac{d}{dt}(e^2\sigma_e) < 0$: il y a décharge viscoplastique de la structure, suivant le schéma de la figure 7 : le rayon y régresse jusqu'à disparition de la zone viscoplastique au bout d'un temps infini.

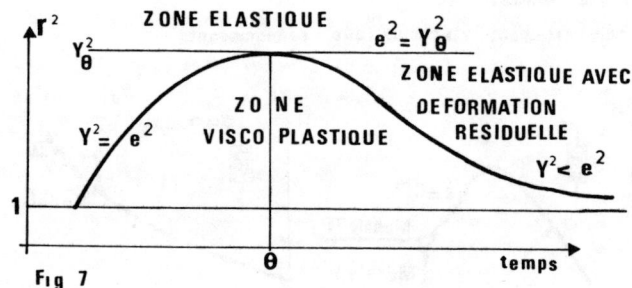

Fig 7

Dans le diagramme (σ_i, y^2), le point d'équilibre se trouverait au-delà de la courbe limite $\sigma_i^\infty(y^2)$; la pression d'équilibre finale sur le revêtement est donc supérieure à ce qu'aurait prévu le schéma de la figure 4, (quoique d'assez peu dans les applications vraisemblables).
Les abaques de la figure 8 donnent, pour une pression géostatique réduite P_∞/C, et différents instants de pose θ, les valeurs des pressions de revêtement en fonction du facteur de rigidité γ : le milieu est supposé infini ($\rho = \infty$). Les détails du calcul sont exposés dans (Berest et al., 1982).

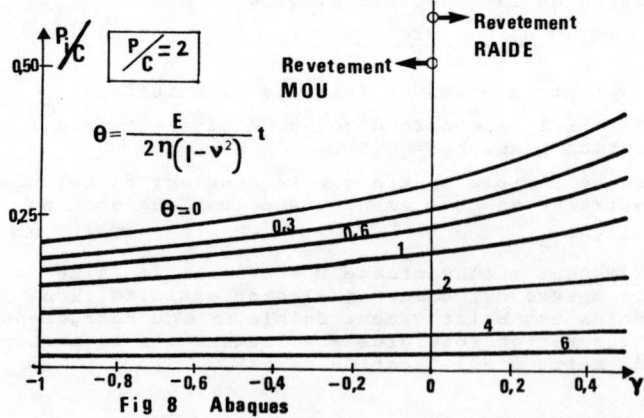

Fig 8 Abaques

Bien que le coefficient de Poisson soit un paramètre assez mal défini en général, on peut penser que de nombreux ouvrages sont susceptibles de répondre au deuxième type de comportement, dès lors que le module d'Young du massif est plus faible que le coefficient de rigidité K du soutènement ; c'est le cas des exemples consignés sur le tableau ci-dessous tirés de la littérature (Gaudin et Ricard, 1974, Habib et al., 1965), si on estime que K est de l'ordre de 900 MPa, pour un anneau en béton dont le rapport de l'épaisseur au rayon est de 0,06 et dont le module d'Young à long terme est de 16000 MPa.

	P/C^*	$E(MPa)$	Roches
Tunnel ferroviaire (Monaco)	~ 1 à 2	200 à 300	marne grise raide
Emissaire Sud (Paris)	~ 5	100	argile plastique (Sparnacien)
Métro (Marseille)	~ 2	50 à 400	marne (Stampien)
Tunnel (Menton)	< 1	500 à 3000	grès tendre inclusions marneuses

(*) dans l'hypothèse $\varphi = 0$.

CONCLUSIONS

Nous avons essayé de dégager deux aspects fondamentaux du comportement différé des ouvrages souterrains, en limitant l'étude à des cas extrêmes de comportement : roches dures et fragiles, roches molles et plastiques. Il est probable que dans la plupart des cas usuels, le comportement du massif rocheux inclue ces deux aspects dans des proportions variables : que ce soit en massifs homogènes (marnes très raides, charbon, sel gemme peut-être...), ou en massifs hétérogènes affectés de discontinuités ou composés de roches de différentes natures.

Les modèles de calculs élastoviscoplastiques présentés dans ces cas élémentaires peuvent alors être adaptés sans trop de difficultés à chaque problème particulier et conduire à une meilleure analyse des phénomènes.

ANNEXE 1
Modèle Elastoviscoplastique Radoucissant

Le patin "radoucissant" du modèle rhéologique à trois éléments (fig.2) est traduit ici par un critère de Tresca dont la cohésion $C(\gamma)$ diminue avec le paramètre d'écrouissage isotrope γ, qui est ici la déformation plastique équivalente :

$$\gamma = \frac{1}{2} \int_0^t \sum_1^3 |\dot{\varepsilon}_i^{vP}| dt \qquad \dot{\varepsilon}_i^{vP} = \text{vitesse de déformation viscoplastique principale.}$$

Soit :
$$F(\sigma_1, \sigma_3) \equiv \sigma_1 - \sigma_3 - 2C(\gamma)$$
$$\text{et} \quad C(\gamma) = C_0 + (C - C_0) <1 - \frac{\gamma}{\gamma_0}>$$

C est la cohésion initiale du matériau
C_0 est la cohésion résiduelle
<> signifie "valeur positive" de l'expression
σ_1, σ_3 : contraintes principales ordonnées :

$\sigma_1 > \sigma_2 > \sigma_3$ (la compression est négative)

γ_0 : paramètre positif qui traduit l'intensité du radoucissement ; on le remplace dans les calculs par le paramètre :
$$E_2 = \frac{2(C - C_0)}{\gamma_0}.$$

On admet alors que les vitesses de déformation viscoplastique dérivent d'un potentiel Ω :
$$\dot{\varepsilon}_i^{vP} = \frac{\partial \Omega}{\partial \sigma_i} \qquad i = 1,2,3 \text{ avec :}$$

$$\Omega = \frac{1}{2\eta} <F(\sigma_1, \sigma_3)>^2 \quad \text{si} \quad \sigma_1 > \sigma_2 > \sigma_3$$

$$\Omega = \frac{1}{2\eta} \left\{ \lambda <F(\sigma_1, \sigma_3)>^2 + (1-\lambda) <F(\sigma_2, \sigma_3)>^2 \right\}$$

si on a un régime d'arête, avec $\sigma_1 = \sigma_2 > \sigma_3$ par exemple ; $\lambda \in [0,1]$ est un multiplicateur à déterminer.

η est une constante de viscosité ; on considère ici deux modèles de viscosité différents :

Modèle I : $\eta = $ cte si $\gamma \in [0, \gamma_0]$; $\eta = 0$ si $\gamma > \gamma_0$.
Modèle II : $\eta = $ cte $\forall \gamma$.

Les vitesses de déformation totales s'écrivent alors :
$$\dot{\varepsilon}_i = \dot{\varepsilon}_i^e + \dot{\varepsilon}_i^{vP}$$

$\dot{\varepsilon}_i^e$ sont les vitesses de déformation élastique linéaire (E module d'Young ; ν coefficient de Poisson).

<u>Nota</u> : de la même façon on pourra prendre en compte le frottement interne du matériau avec un critère de Coulomb radoucissant.

ANNEXE 2
Contraction d'un tube épais élastoviscoplastique radoucissant

1.- Notations

Appelons :

$E' = \frac{E}{1 - \nu^2} = \frac{4}{3} E$: module d'Young équivalent

C : cohésion initiale du matériau
C_0 : cohésion résiduelle

$E_2 = \frac{2(C - C_0)}{\gamma_0}$: intensité du radoucissement (cf. Annexe 1)

γ_0 : valeur du paramètre d'écrouissage pour lequel apparaît la plasticité résiduelle

$$\alpha = \frac{E'}{E' - E_2} \quad ; \quad \beta = \frac{2C}{E' - E_2}.$$

Dans ce qui suit, on utilise un temps réduit :
$$(1) \qquad \tau = \frac{(E' - E_2)}{\eta} t.$$

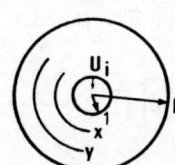

Le symbole $(\dot{})$ désigne la dérivation par rapport à τ. u est le déplacement radial ; u_i est la valeur de u sur la paroi intérieure ; y désigne la frontière de la zone viscoplastique ; x < y est la frontière de la zone en phase plastique résiduelle.

Ces notations sont valables dans le cas d'une charge viscoplastique continue de la structure, ce qui est le cas ici.

2.- Réponse à long terme

Appelons :

$$(2) \begin{cases} \phi_1(y^2) \equiv C\left[\alpha(1 + \text{Log } y^2) + y^2(1 - \alpha - \frac{1}{\rho^2})\right] \\ \phi_2(x^2, y^2) \equiv C_o \text{Log } x^2 + C\left[\alpha(1 + \text{Log}\frac{y^2}{x^2}) + y^2(\frac{1-\alpha}{x^2} - \frac{1}{\rho^2})\right] \end{cases}$$

Soit $\sigma_i^\infty(y^2)$ la réponse $\sigma_i(y^2)$ à long terme, de nature élastoplastique, de la structure, sous un chargement infiniment lent.
$y^2 = y_1^2 = \frac{\gamma_o}{\beta}$ étant la valeur d'apparition de la zone plastique résiduelle, on a :

$$(3) \begin{array}{l} \text{Si } y^2 < \inf(y_1^2, \rho^2) \Rightarrow \sigma_i^\infty(y^2) = \phi_1(y^2) \\ \text{Si } y_1^2 \leq y^2 \leq \rho^2 \Rightarrow \begin{cases} \sigma_i^\infty(y^2) = \phi_2(x^2, y^2) \\ \text{avec } y^2/x^2 = y_1^2 \end{cases} \end{array}$$

3.- Réponse différée

a) Cas où la zone plastique résiduelle n'est pas apparue.

Le problème est régi par l'équation différentielle :

$$(4) \qquad \sigma_i + \dot\sigma_i = \dot{y}^2 C(1 - \frac{1}{\rho^2}) + \phi_1(y^2)$$

b) Cas où la zone plastique résiduelle est apparue.

Il faut ici distinguer les solutions relatives à chacun des deux modèles de viscosité envisagés. Le modèle I a été exposé et discuté (Nguyen Minh 1982). La solution du modèle II s'obtient de façon tout à fait analogue, nous nous contenterons donc d'en exposer les résultats. Ces modèles satisfont à un système d'équations différentielles du premier ordre, dont une des équations est commune, qui exprime que le paramètre d'écrouissage γ atteint γ_o sur la frontière x :

$$(5) \qquad \dot{x^2}/x^2 = - \frac{y_1^2 - y^2/x^2}{y_1^2 - \exp{-(\tau - \tau_o(x^2))}}$$

où $\tau_o(x^2)$ est l'instant où le point $r = x$ entre en viscoplasticité.
Pour le modèle I, les inconnues x et y satisfont aux deux équations (5) et :

$$(6) \qquad \dot\sigma_i + \sigma_i = C\left[\frac{\dot{x^2}}{x^2}(y_1^2 - y^2/x^2) + \dot{y}^2(\frac{1}{x^2} - \frac{1}{\rho^2})\right] + \phi_2(x^2, y^2).$$

Quant au modèle II, on a trois inconnues : x, y, et σ_x (contrainte radiale sur la frontière x), qui satisfont au système d'équations (5) et aux deux suivantes :

$$\dot\sigma_i + \alpha\sigma_i = (\alpha-1)(\sigma_x + C_o \text{Log } x^2) + C\dot{y}^2(1 - \frac{1}{x^2}) + \phi_2(x^2, y^2)$$

$$\dot\sigma_i + \alpha\sigma_i = \dot\sigma_x + \alpha(\sigma_x + C_o \text{Log } x^2) + C\left[\dot{y}^2(1 - \frac{1}{x^2}) + \frac{\dot{x^2}}{x^2}(\frac{y^2}{x^2} - y_1^2 + \frac{C_o}{C})\right]$$

Remarque : pour le modèle I, l'annulation brutale de la viscosité au passage de la frontière x vers la zone plastique résiduelle introduit une discontinuité de la déformation plastique et des quantités $\frac{\partial \sigma_r}{\partial r}$ et $\frac{\partial \sigma_r}{\partial \tau}$, qui se traduit dans les courbes d'évolution de fluage par une augmentation momentanée de la vitesse de déformation. Ceci n'est pas le cas pour le modèle II.

ANNEXE 3

Valeur de la pression limite sur le revêtement dans le cas d'un matériau viscoplastique de Bingham

On distingue deux cas :

- $\gamma < 0$: revêtement "mou" -

La valeur limite du rayon plastique y_F est donnée par :

$$\frac{y_F^2 - y_\theta^2}{\gamma} - 1 - \text{Log } y_F^2 + P/C = 0$$

y_θ^2 étant la valeur du rayon y à l'instant θ de la pose du revêtement. Alors, la pression sur le revêtement vaut :

$$P_r^F = P - \frac{C}{\gamma}(y_F^2 - y_\theta^2).$$

- $\gamma > 0$: revêtement "raide" -

La pression limite sur le revêtement vaut :

$$P_r^F = P - \left\{I(y_\theta^2) + J(P/C) \exp{-(\theta/\gamma)}\right\} \times C$$

où I et J sont fonctions de y^2 et satisfont aux équations différentielles :

$$\begin{cases} \frac{dJ}{dy^2} - (1 - \frac{1}{\gamma}) \frac{J}{y^2 - 1 - \text{Log } y^2} + \frac{1}{\gamma} = 0 \\ J(1) = 0 \end{cases}$$

$$\begin{cases} \frac{dI}{dy^2} + \frac{I}{\gamma(P - 1 - \text{Log } y^2)} + \frac{1}{\gamma} = 0 \\ I(P/C) = 0 \end{cases}$$

BIBLIOGRAPHIE

Berest, P., Nguyen Minh, D. (1982), *Modèle viscoplastique pour le comportement d'un tunnel revêtu*. A paraître dans Revue Française de Géotechnique.

Bergues, J., Nguyen Minh, D. (1982), *Applications des concepts de comportement après la rupture des roches à l'analyse de la convergence différée dans les galeries souterraines*. Rapport D.G.R.S.T., n°79-7-1359.

Gaudin, M., Ricard, M. (1974), *Sollicitations différées des revêtements de tunnels creusés dans les marnes*. Revue de l'Industrie Minérale, Cahier n°6 CFMR, pp.3-12.

Habib, P., Bernède, J., Carpentier, L. (1965), *Résultats des mesures de contraintes effectuées dans divers souterrains en France*. Annales de l'ITBTP n°210, juin, pp. 825-834.

Kaiser, P.K., Morgenstern, N.R. (1981, *Phenomenological model for rock with time-dependent strength*. Int. J. Rock. Mech. Min. Sci & Geomech Abst., vol.19, pp. 153-165.

Lunardi, P. (1980), *Application de la mécanique des roches aux tunnels : cas des tunnels du Gran Sasso et du Fréjus*. Revue Française de Géotechnique, n°12, pp. 5-43.

Morlier, P. (1966), *Le fluage des roches*. Annales de l'ITBTP, n°217, pp. 91-111.

Nguyen Minh,D. (1982), *The contraction of hollow circular cylinders with time-dependent strain softening behaviour - Application to supported tunnels*. A paraître dans Int.J.Rock.Mech. & Min. Sc.

Nguyen Minh,D., Berest,P. (1979), *Etude de la stabilité des cavités souterraines avec un modèle de comportement élastoplastique radoucissant*. Proc. 4th Int. Cong. Rock. Mech., SIMR, Montreux vol.1.

Nguyen Minh,D., Berest,P., Bergues,J., Habib,P. (1981), *Strain softening behaviour of rocks in engineering practice*. C.R. Congrès Int. Mech. Roches, Tokyo.

Panet,M. (1976), *La mécanique des roches appliquée aux ouvrages du génie civil*. Association des Anciens Elèves ENPC.

Panet,M. (1979), *Les déformations différées dans les ouvrages souterrains*. C.R. 4ème Congrès SIMR Montreux, pp. 291-301.

Pacher,F. (1979), Intervention à la Journée d'Etudes de l'AFTES, Paris, octobre 1978. Tunnels et Ouvrages Souterrains, n°32, mars-avril, pp. 128-129.

Peng, Podnieks,E.R. (1972), *Relaxation and the behavior of failed rock*. Int.J.Rock.Mech.Min.Sc. vol.9, pp. 699-712.

Plichon,J.N. (1979), *Mesure de l'épaisseur de la zone décomprimée dans un souterrain à forte couverture*. AFTES, Paris 1978, Tunnels et Ouvrages Souterrains, n°32, pp. 132-134.

Scholz,C.H. (1968), *Mechanism of creep in brittle rock*. Journal of Geophys. Res., vol.73, n°10 pp. 3295-3302.

Waversik,W.R. (1974), *Time-dependent behaviour of rock in compression*. C.R. 3ème Congrès de la SIMR, Denver, vol.II, tome A, pp. 357-363.

LE SEL GEMME EN TANT QUE LIQUIDE VISQUEUX
Rock salt as a viscous liquid
Steinsalz als viskose Flüssigkeit

S-M Tijani
G. Vouille
Centre de Mécanique des Roches de l'Ecole Nationale Supérleure des Mines de Paris
35, rue Saint Honoré — 77305 Fontainebleau

B. Hugout
Département d'Etudes et Techniques Nouvelles de Gaz de France
Département Réservoirs Souterrains Courcellor 1
1-3 Rue Arthur L'Adwig 92531 Levallois Perret — France

RESUME

Des campagnes d'essais récentes ont permis de comprendre que certains paramètres mécaniques du sel gemme, notamment sa cohésion, avaient été jusqu'à présent très surestimés. On montre que les résultats des essais de fluage peuvent s'interpréter en considérant le sel comme un matériau viscoplastique à cohésion nulle obéissant à la loi rhéologique de J. LEMAITRE. La très grande viscosité de ce "liquide" explique que l'analyse des résultats des essais de courte durée (compression et relaxation) ait conduit à attribuer au sel une cohésion qui n'est qu'apparente. Le dimensionnement des édifices souterrains conçus dans un tel matériau doit impérativement faire intervenir la notion de durée d'utilisation ainsi que celle de déformation maximale admissible.

SYNOPSIS

Recent mechanical tests have proved that some mechanical parameters of rock salt, especially its cohesion, had been up to now greatly overestimated. We show that creep test results may be interpreted by assuming that rock salt is a viscoplastic material with null cohesion, obeying J. LEMAITRE's rheological law. The great viscosity of this "liquid" explains that an apparent cohesion may be derived from the results of tests of short duration (uniaxial compressive tests or relaxation tests). When designing an underground opening in such a material one must take into account how long the opening will be used for human purposes and how large the allowed strains are.

ZUSAMMENFASSUNG

Neuere Daten zeigen, daß gewisse Parameter des Steinsalzes, insbesondere seine Kohäsion, bis jetzt weit überschätzt wurden. Es wird gezeigt, wie sich die Ergebnisse der Kriechversuche erklären lassen, indem man das Salz als kohäsionsloses viskoplastisches Material ansieht, das dem rheologischen Gesetz von J. LEMAITRE gehorcht. Die sehr große Viskosität dieser Flüssigkeit erklärt, daß die Ergebnisse der kurzfristigen Versuche (Kompression und Spannungsrelaxation) dazu geführt haben, daß man dem Salz eine Kohäsion zuschreibt. Wenn man einen unterirdischen Hohlraum in diesem Material vorsieht, muß man unbedingt die vorgesehene Verwendungsfrist und die maximal erlaubte Verformung in Betracht ziehen.

I / INTRODUCTION

La rhéologie du sel gemme a fait l'objet de nombreuses études expérimentales au Centre de Mécanique des Roches. Ces études, financées principalement par Gaz de France, ont eu pour but une définition aussi précise et aussi complète que possible de la loi de comportement mécanique du sel gemme afin de l'utiliser dans des codes numériques de calcul permettant de dimensionner les excavations souterraines réalisées dans des couches salines et de prévoir leur évolution dans le temps.

La première application concrète des résultats des essais réalisés sur le sel gemme a eu lieu en 1970 lors du calcul de stabilité d'un réseau de cavités de stockage du gaz qui devaient être creusées par dissolution dans une couche saline à 1500 m de profondeur dans le site de TERSANNE (FRANCE). Ce calcul qui devait fournir la distance minimale à prévoir entre les cavités, la pression minimale du gaz stocké et la perte maximale que subirait le volume des cavités, a été effectué en admettant que le sel de TERSANNE était un matériau élastoplastique parfait défini principalement par un critère de Coulomb avec une cohésion C de 3 MPa et un angle de frottement ∅ de 5 degrés. Le dimensionnement du réseau de cavités de TERSANNE et leur pression minimale d'exploitation ont été déterminés pour que les zones plastiques apparaissant autour de chaque cavité ne se rejoignent pas. La perte maximale prévue pour le volume a été de 2.5 pour cent. Une fois les cavités réalisées, Gaz de France a procédé à des mesures in-situ utilisant diverses techniques modernes (thermométrie, comptage hydrogène, écholog ...) qui ont montré que certaines cavités de TERSANNE, au bout de 10 ans d'exploitation, ont perdu près de 30% de leur volume. L'écart notable entre prévision (2.5%) et réalité (30%) a déclenché une remise en question des résultats acquis concernant la rhéologie du sel gemme et a été le point de départ de nouvelles études expérimentales qui ont montré que les esssais classiques de compression ne permettent pas une détermination précise du domaine d'élasticité d'un matériau tel que le sel gemme.

L'utilisation d'essais durant lesquels la longueur est maintenue constante (relaxation) a permis de mieux connaître la frontière plastique en tant qu'ensemble de points représentant des états d'équilibre limite qui se sont avérés obéir à un critère de Tresca (angle de frottement nul) pour lequel la cohésion C a été initialement estimée à 3 MPa dans le cas du sel de TERSANNE. Cependant, lorsque l'on a réalisé les mêmes essais de relaxation en appliquant le confinement à une vitesse plus faible que précédemment, la cohésion obtenue a été de l'ordre de 2 MPa.

Ainsi, qu'il s'agisse des essais classiques (compression) ou des nouveaux essais (relaxation), les résultats obtenus dépendent fortement de la vitesse de sollicitation utilisée lors de l'essai : le sel est un matériau dont le comportement mécanique s'accompagne de phénomènes de viscosité loin d'être négligeables et que l'on a essayé de quantifier à partir d'essais expérimentaux de longue durée.

II / ESSAIS DE FLUAGE

1. Résultats bruts

Indépendemment de la provenance du sel étudié (Etrez et Tersanne en FRANCE, Kirkuk en IRAK), de la température à laquelle a été réalisé l'essai de fluage (291, 308, 328, 343 ou 353 degrés Kelvin), du confinement (0, 8, ou 20 MPa) et du signe des variations de la longueur de l'échantillon (contraction ou extension), l'évolution dans le temps de la valeur absolue de la déformation longitudinale e s'effectue à <u>vitesse positive décroissante, quasi-infinie au temps zéro</u> (les pressions sont appliquées instantanément puis maintenues constantes durant l'essai) (voir les six graphiques)

Par ailleurs, les résultats des essais de fluage sont caractérisés par <u>l'absence de déformations différées lorsque le déviateur est nul</u> (le déviateur est égal à la valeur absolue de la différence des deux pressions latérale et axiale) et lorsque le confinement est nul (fluage monoaxial) et <u>uniquement dans ce cas</u>, l'apparition d'une phase d'accélération souvent appelée fluage tertiaire (changement de la concavité de la courbe) qui se manifeste après une durée plus ou moins longue et se termine par la rupture de l'échantillon.

Pour une même durée de fluage, la déformation différée e^v est une fonction croissante du déviateur σ et de la température T et elle est peu sensible à la valeur du confinement P dès que celui-ci est différent de zéro (fluage triaxial).

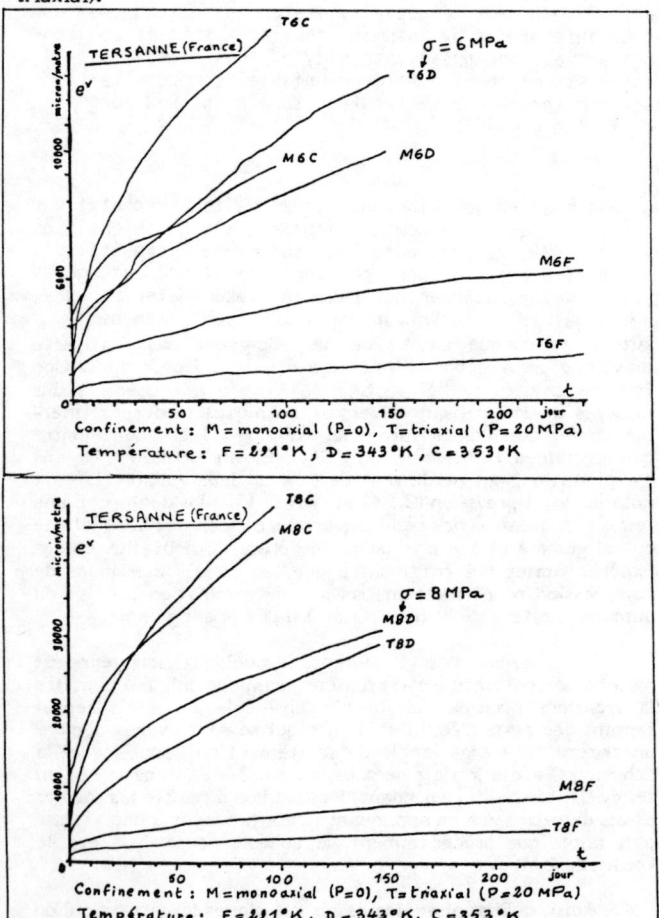

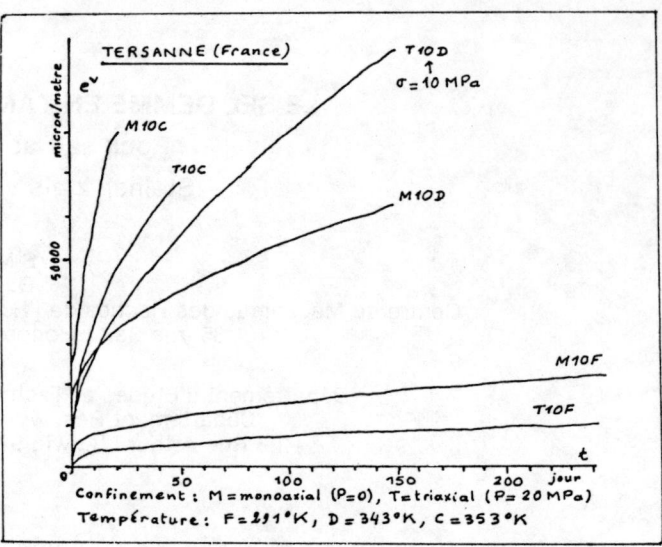

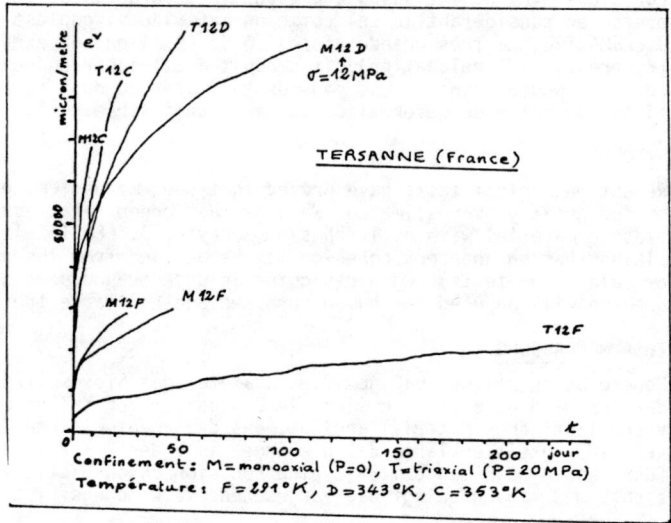

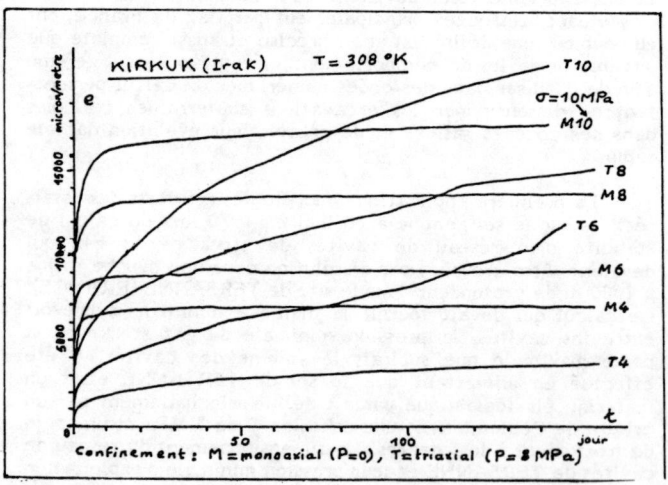

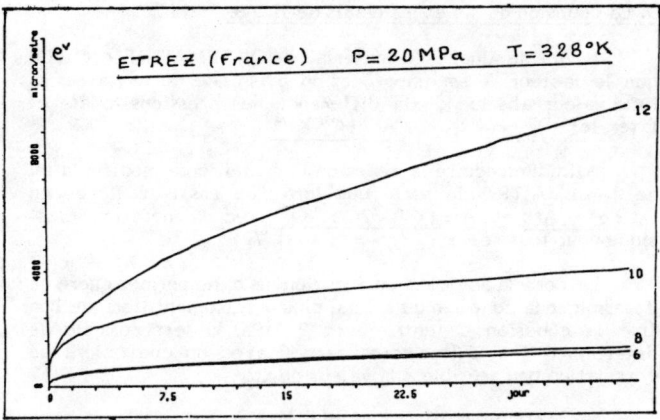

2. Interprétation des résultats des essais de fluage

Le but d'une telle interprétation est non pas d'obtenir une loi de fluage (déformation en fonction des contraintes et du temps) mais d'aboutir à une loi de comportement (loi rhéologique intrinsèque fournissant les contraintes à tout instant t en fonction de l'historique des déformations jusqu'à l'instant t).

Une longue recherche a conduit à conclure que parmi les modèles rhéologiques classiques celui qui rend le mieux compte des résultats des essais de fluage est le modèle de J. LEMAITRE dont la réponse lorsque les contraintes sont constantes est une loi puissance du temps : $e^v = At^\alpha$ où t est le temps mesuré en jour, e^v représente la valeur absolue de la déformation longitudinale différée en micron par mètre, A est une fonction des contraintes et de la température régnant durant l'essai de fluage et α est un nombre compris entre 0 et 1 caractérisant le matériau.

Il est certes ambitieux de vouloir qu'une loi aussi simple puisse représenter avec une bonne précision les résultats d'une vingtaine d'essais de fluage (sel de TERSANNE). En effet, les ajustements (au sens des moindres carrés) courbe par courbe ont fourni des paramètres α relativement différents d'un essai à l'autre, et l'on aurait pu garder ces ajustements en affirmant que le modèle rhéologique du sel est tel que α soit une fonction de la température et des contraintes, fonction que l'on aurait été incapable de préciser par manque de données expérimentales.

Mais le but étant d'obtenir la loi de comportement la plus simple (nombre minimal de caractéristiques mécaniques) dont les prévisions s'écartent le moins possible des mesures, non seulement les ajustements ont été refaits en imposant au paramètre α d'être intrinsèque au matériau mais de plus, les valeurs obtenues pour les facteurs A ont été corrélées aux contraintes en admettant que celles-ci n'interviennent que par leur déviateur (A est une fonction du déviateur pour une température donnée).

Il est à signaler cependant que l'hypothèse " A ne dépend pas du confinement P " n'est valable que dans une certaine plage de valeurs de P. En particulier, si l'on utilise les valeurs des paramètres mécaniques déduites des essais de fluage sous confinement (Cf. tableaux et figures ci-après) les déformations que l'on prédirait pour les essais de fluage en compression simple seraient notablement inférieures à celles réellement constatées dans le cas du sel de TERSANNE.

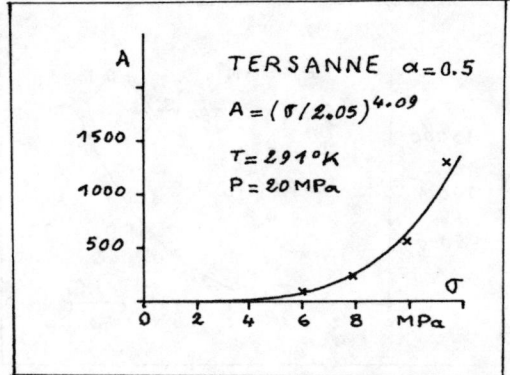

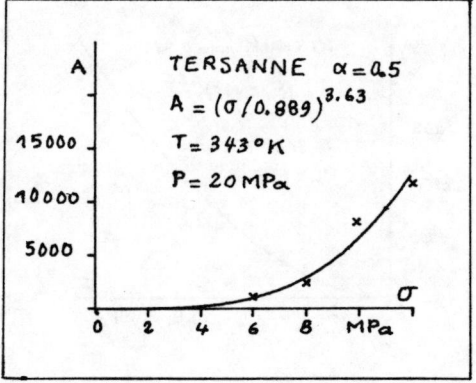

D 243

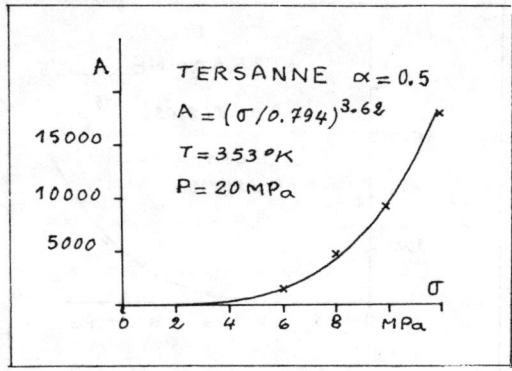

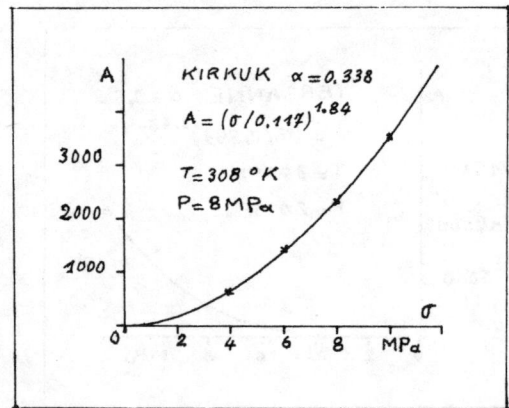

3. La cohésion du sel et les essais de fluage

La loi de fluage du matériau de J. LEMAITRE est telle que le facteur A est une fonction puissance du déviateur. (σ = valeur absolue de la différence des pressions axiale et latérale) : $A = (\sigma/K)^\beta$

Afin d'introduire la cohésion C du sel, on a modifié la loi de J. LEMAITRE de sorte que lors d'un essai de fluage on ait : $e^v = At^\alpha$ et $A = (<\sigma - 2C>/K)^\beta$ avec la notation classique pour tout réel x : $<x> = (x + |x|)/2$

La corrélation de A en fonction de σ ne permet guère de déterminer la cohésion du sel ou, plus précisément, lorsque l'on fixe la cohésion C (entre 0 et 3 MPa), il est possible de déterminer K et β (moindres carrés) avec un coefficient de corrélation peu sensible à la valeur de C.

En particulier, il est possible, sans accroître l'écart entre prévisions et mesures, de conserver le modèle de J. LEMAITRE (cohésion nulle) pour interpréter les résultats des essais de fluage du sel gemme.

4. Matériau de J. LEMAITRE

L'énoncé exact de la loi de comportement du matériau de J. LEMAITRE postule la décomposition du tenseur de déformation $\tilde{\varepsilon}$ en partie élastique $\tilde{\varepsilon}^e$ et partie visqueuse $\tilde{\varepsilon}^v$.

La partie élastique $\tilde{\varepsilon}^e$ est reliée au tenseur de contrainte $\tilde{\sigma}$ par une loi d'élasticité isotrope :
$$E \tilde{\varepsilon}^e = (1+\nu) \tilde{\sigma} - \nu (\operatorname{tr}\tilde{\sigma}) \tilde{1}$$
où $\operatorname{tr}\tilde{\sigma} = \sigma_{11} + \sigma_{22} + \sigma_{33}$ et $\tilde{1}$ est le tenseur unité.

La partie visqueuse $\tilde{\varepsilon}^v$ a une vitesse $\dot{\tilde{\varepsilon}}^v$ parallèle et de même sens que le tenseur $\tilde{\sigma}' = \tilde{\sigma} - (\operatorname{tr}\tilde{\sigma}/3)\tilde{1}$ (déviateur de $\tilde{\sigma}$) et plus précisément on a :
$$\sigma = \sqrt{3(\sum_i \sum_j \sigma'^2_{ij})/2} \text{ et } \dot{\tilde{\varepsilon}}^v = (3 d\xi^\alpha/dt/2/\sigma)\tilde{\sigma}'$$
où ξ est une variable interne fonction non décroissante du temps dont la vitesse est : $\dot{\xi} = (\sigma/K)^{\beta/\alpha}$.

Les 5 caractéristiques mécaniques du matériau E, α, β, ν et K sont supposées connues chaque fois que la température est fixée.

En particulier, lors d'essais mécaniques sur des éprouvettes cylindriques soumises à des pressions axiale Q et latérale P, on a : $\sigma = |Q - P|$, $e^v = \xi^\alpha$ (valeur absolue de la déformation longitudinale visqueuse en micron par mètre) et $\dot{\xi} = (\sigma/K)^{\beta/\alpha}$. Et lorsqu'il s'agit d'essais de fluage, on a : $\xi = (\sigma/K)^{\beta/\alpha} t$ et $e^v = (\sigma/K)^\beta t^\alpha$.

III / LE SEL EN TANT QUE LIQUIDE VISQUEUX

Le matériau défini au paragraphe précédent est siège de déformations non élastiques dès lors que le tenseur contrainte s'écarte de l'état isotrope. Par ailleurs, en phase de relaxation, les contraintes évoluent vers un état isotrope (déviateur nul). Il s'agit donc bien d'un matériau visqueux (viscoélasticité non linéaire).

Par conséquent, affirmer que le sel gemme obéit à la loi de J. LEMAITRE, c'est affirmer que le sel est un liquide visqueux particulier.

Cependant, la preuve indiscutable de cette affirmation est loin d'être établie d'une part parce que la confrontation entre les résultats des essais de fluage et le modèle ajusté est d'une précision parfois médiocre et d'autre part parce que des essais en cours (sollicitations cycliques) mettent en évidence certains phénomènes dont ne rend pas compte même qualitativement le modèle de Lemaître

Le but ici est de montrer que l'hypothèse de la nullité de la cohésion du sel gemme ne peut pas être rejetée a priori car elle n'est en contradiction avec aucun des résultats des divers essais mécaniques réalisés au laboratoire sur cette roche.

1. Essais classiques

Un essai classique de compression consiste à soumettre une éprouvette à une pression axiale augmentant à vitesse quasi-constante. Durant l'essai, réalisé sous confinement constant et éventuellement nul, on mesure la diminution relative de la longueur l de l'échantillon que l'on exprime généralement en micron par mètre.

La courbe représentant le déviateur σ en fonction de la déformation e est généralement interprétée en admettant que pour $\sigma \leq 2C$ le matériau est élastique linéaire (portion de droite). On détermine ainsi, en plus du module d'Young, la limite élastique 2C donc la cohésion C du matériau qui, en tant que paramètre intrinsèque, devrait être indépendant de la vitesse de mise en charge.

Or, on a observé que dans le cas du sel gemme, la vitesse $v = d\sigma/dt$ a une forte influence sur l'ordonnée 2C du point où la courbe effort-déformation quitte sa tangente à l'origine, phénomène que le modèle rhéologique défini ci-dessus permet d'interpréter.

En posant $E^* = 10^{-6} E$ où E est le module d'Young du matériau et E^* est la pente à l'origine de la courbe σ = fonction de e,

il vient : $e = \sigma/E^* + \xi^\alpha$ et $\dot{\xi} = (\sigma/K)^{\beta/\alpha}$

d'où : $v\,d\xi/d\sigma = (\sigma/K)^{\beta/\alpha} \longrightarrow \xi = (\alpha K/v/(\alpha+\beta))(\sigma/K)^{\beta/\alpha+1}$

Par conséquent : $e = \sigma/E^* + \alpha^\alpha (\alpha+\beta)^{-\alpha} v^{-\alpha} K^{-\beta} \sigma^{\alpha+\beta}$

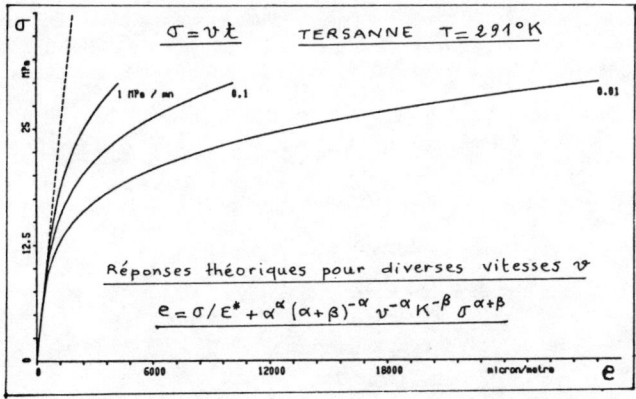

La figure ci-dessus représentant des courbes effort-déformation théoriques pour diverses valeurs de la vitesse de mise en charge permet d'illustrer l'influence de cette vitesse sur la limite élastique pratique (S = 2C) que l'on peut définir par convention comme étant la valeur de σ pour laquelle la déformation non-élastique ($e - \sigma/E^*$) est une fraction $f\sigma/E^*$ de la déformation élastique, f étant un coefficient inférieur à 1 fixé a priori.

$$\alpha^\alpha (\alpha+\beta)^{-\alpha} v^{-\alpha} K^{-\beta} S^{\alpha+\beta} = f S/E^*$$

En posant : $\gamma = \alpha/(\alpha+\beta-1)$

et $R = f^{\gamma/\alpha} E^{*-\beta/\alpha} \alpha^{-\gamma} (\alpha+\beta)^\gamma K^{\beta\gamma/\alpha}$

il vient $S = R v^\gamma$.

Si l'on adopte pour f la valeur de 0.01 (L'abscisse du point d'ordonnée S sur la courbe effort-déformation s'écarte de 1 pour cent de l'abscisse du point correspondant sur la tangente à l'origine), on obtient :

$$0.14 \leq \gamma \leq 0.16 \text{ et } 2.8 \leq R \leq 7$$

En particulier, sachant que les essais classiques de compression sont généralement réalisés à température ambiante ($S = 7 v^{0.14}$) et à des vitesses v variant entre 0.1 et 1 MPa par minute, la cohésion pratique obtenue à partir de tels essais varierait théoriquement de 2.5 à 3.5 MPa ce qui est compatible avec la valeur de 3 MPa obtenue initialement.

2. Essais de relaxation

En utilisant le modèle de J. LEMAITRE avec les paramètres obtenus pour le sel de TERSANNE à température ambiante, on montre aisément que lorsqu'un échantillon est soumis spontanément à un déviateur σ^0 à l'instant t = 0 et lorsque durant l'essai l'on maintient constante la nouvelle longueur de l'éprouvette (relaxation), le déviateur σ décroît en fonction du temps, très rapidement dans les premières minutes puis sa vitesse de décroissance s'affaiblit notablement et l'on constate que pour les durées d'essai usuelles le déviateur est loin d'être nul (Cf. figure). Ceci explique les fortes "cohésions" déduites des premiers essais de relaxation et met en évidence les limitations du Dispositif de Relaxation Biaxiale et Isotherme (DRBI) lorsqu'il s'agit de l'utiliser pour déterminer la cohésion de matériaux tels que le sel, à moins d'envisager des essais dont la durée se mesurerait en mois et non en jour.

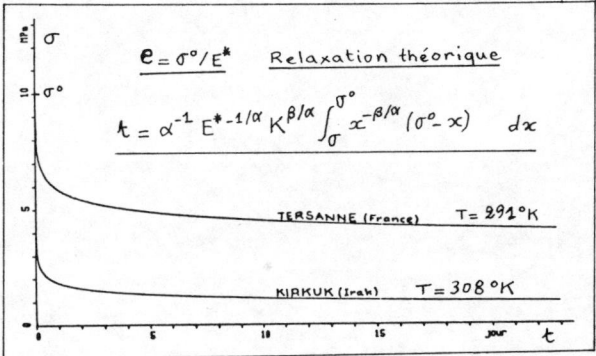

IV / CONCLUSION

Tant que l'on n'avait pas de doute sur l'existence d'une cohésion du sel gemme, on abordait l'étude mécanique des ouvrages conçus dans des couches salines en effectuant des calculs en élastoplasticité, le sel étant assimilé à un matériau élastoplastique parfait obéissant à un critère de Tresca. Lorsqu'un tel calcul convergeait (les zones plastiques autour des excavations ne se rejoignant pas), on pouvait affirmer que l'ouvrage évoluerait vers un état asymptotique caractérisé par le maximum de déformation et de plastification.

Ainsi les résultats d'un calcul élastoplastique (état asymptotique) permettaient de déterminer le taux maximal de défruitement d'une mine de sel, la pression minimale d'exploitation d'un réseau de cavités de stockage ...

Cependant, les résultats d'expériences réalisées au laboratoire et certaines observations in-situ laissent penser que la cohésion du sel est nulle et qu'il se comporte mécaniquement comme un liquide visqueux.

Grâce à des codes numériques adéquats on pourra toujours simuler l'évolution dans le temps d'excavations réalisées dans le sel considéré comme un matériau viscoélastique non linéaire et déterminer à chaque instant la répartition des contraintes et des déformations dans le massif rocheux.

En revanche, les critères de dimensionnement ne pourront plus faire intervenir la notion de zone plastique puisque dès qu'est constituée la mine ou la cavité de stockage, la "zone plastique" s'étend jusqu'à l'infini, tout le massif se mettant en mouvement pour combler les vides formés. Il semble donc qu'il faille faire intervenir conjointement la notion de déformation maximale admissible (pour des raisons techniques ou économiques) et celle de durée d'utilisation de l'ouvrage, pour obtenir, cas par cas, un critère de dimensionnement qui soit compatible avec le comportement mécanique du sel considéré comme un liquide visqueux.

REFERENCES

Boucly, P. (1981).
   Comportement mécanique des cavités dans le sel. Etat actuel de nos connaissances. Congrès. Association Technique de l'Industrie du Gaz de France. 15-18 septembre 1981, Biarritz.

Fine, J., Tijani, S-M., et Vouille, G. (1980).
   Nouveau dispositif expérimental pour l'étude de la rhéologie des roches. Publication. Revue Française de Géothechnique, n°8 23-26. Paris.

de Grenier, F., Tijani, S-M., et Vouille, G. (1981).
   Experimental determination of the rheological behavior of Tersanne rock salt. First Conference on the mechanical behavior of salt, 9-11 november 1981. The Pennsylvania State University.

Lemaître, J. (1970).
   Sur la détermination des lois de comportement des matériaux élastoviscoplastiques. Thèse. Publication ONERA n°135, Paris.

Wawersik, W.R. et Preece, D.S., (1981).
   Creep testing of salt. Procedure, problems and suggestions. First Conference on the mechanical behavior of salt, 9-11 november 1981. The Pennsylvania State University.

ETUDE EXPERIMENTALE D'UNE CONDUITE FORCEE BLINDEE SOUTERRAINE SOUS TRES FORTE PRESSION

Experimental study of an underground penstock under very high pressure

In situ Versuche mittels einer unterirdischen Druckrohrleitung unter sehr hohen Drücken

Thierry Doucerain
Ingénieur Direction de l'Equipement E.D.F.

RESUME

L'essai de conduite forcée bloquée au rocher réalisé sur le site de Super-Bissorte permet d'étudier le comportement du massif rocheux autour d'une galerie circulaire soumise à une pression intérieure. Un modèle analytique fondé sur les lois de l'élasticité a été utilisé pour la détermination des modules d'élasticité du rocher. Il a permis de calculer:
- dans les zones voisines de la galerie des valeurs voisines des mesures réalisées au vérin à plaque de charge,
- dans le massif sain des valeurs élevées mais confirmées par les vitesses sismiques.

SYNOPSIS

The test performed on the penstock anchored to the rock at the Super-Bissorte site makes it possible to study the behaviour of the rock mass around a circular tunnel subjected to an inner pressure. An analytical model based on elasticity rules was used in order to determine the elastic moduli of the rock. In particular, this model served to calculate:
- values close to those measured with a loading plate jack in the areas near the tunnel,
- high values confirmed by the seismic velocities in the sound rock mass.

ZUSAMMENFASSUNG

Ein Versuch mit einer im Felsgestein verankerten Druckleitung der Super-Bissorte-Anlage gestattet es, das Verhalten des Gebirges zu untersuchen, das einen kreisförmigen, einem inneren Druck ausgesetzten Stollen umschließt. Zur Bestimmung der Elastizitätsmoduli des Felsgesteins wurde ein auf den Gesetzen der Elastizität beruhendes analytisches Modell benutzt. Anhand der Anordnung wurden die folgenden Daten berechnet:
- für die Randzonen des Stollens Werte, die in der Nähe der Messungen liegen, die mit dem Plattendruckversuch durchgeführt wurden,
- für das unversehrte Gebirge hohe Werte, die durch seismische Geschwindigkeitsmessungen bestätigt wurden.

INTRODUCTION

L'étude du dimensionnement des conduites forcées du projet de SUPER-BISSORTE situé en Savoie dans la Vallée de l'Arc a montré la nécessité pour les aménagements de haute chute en cours de réalisation ou d'étude de revoir certaines hypothèses de calcul du blindage des conduites forcées.

La pression de fonctionnement des conduites d'amenée (14,1 MPa) conduit avec les méthodes habituelles, à des blindages de 5 cm d'épaisseur entraînant d'une part un coût élevé et d'autre part des difficultés technologiques dans la mise en place des viroles (soudure en particulier).

Il faut donc élaborer une méthode de calcul cernant de plus près la réalité et prenant en compte le complexe tôle + béton de blocage + rocher dans son ensemble. A cet effet, un modèle mathématique (qui utilise la méthode par éléments finis) a été mis au point par E.S.I. (Engineering System International).

.../...

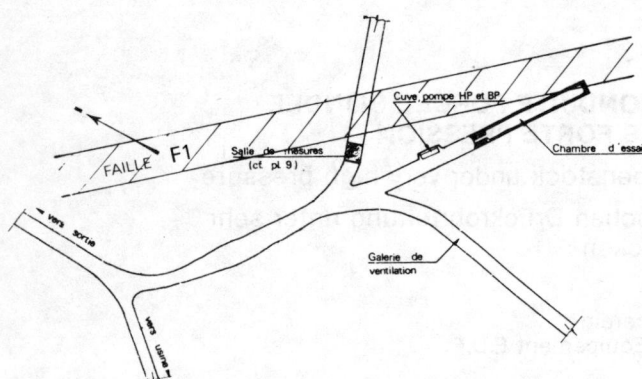

Fig. 1 : Implantation de la caverne d'essai

Pour caler ce modèle et vérifier sa validité, un essai en vraie grandeur a été réalisé sur un tronçon de conduite bloquée au rocher, implantée sur le site de SUPER-BISSORTE. Ce tronçon, long d'une vingtaine de mètres, a été équipé d'un dispositif d'auscultation de l'ensemble tôle-béton-rocher.
Cet essai déborde largement des limites strictes de la mécanique des roches. Il présente cependant de ce point de vue, un intérêt non négligeable.
L'essai conduit à étudier le comportement du rocher autour d'une excavation soumise à une pression intérieure. La mesure des déplacements du rocher à différentes profondeurs permet en particulier d'estimer, par ajustement sur des modèles analytiques simples fondés sur les lois de l'élasticité, les valeurs du module de déformation.

1. LES CONDITIONS DE L'ESSAI (voir détail en réf. 1)

1.1. Choix et implantation de la caverne

Le tronçon de galerie destiné à l'essai a été implanté sur l'un des rameaux de la galerie de reconnaissance de SUPER-BISSORTE, à proximité d'une importante faille (voir fig. 1).

1.2. Description

Il s'agit d'une galerie de 3,20 m de diamètre longue de 30 m. La virole en acier a une longueur de 20 m et un diamètre intérieur de 2 m. Son blocage est assuré par un anneau de béton de 0,60 m d'épaisseur.

1.3. Dispositif d'auscultation

L'auscultation de l'acier et du béton est décrite précisément en (1). Dans le rocher, les appareils utilisés sont des cannes extensométriques ancrées en fond de forage, munis de capteurs inductifs ; ceux-ci mesurent des déplacements relatifs avec une precision de ± 5 microns.
Deux sections ont été auscultées. Elles sont équipés de 4 cannes ancrées à 4 m et munis de 4 capteurs. La section 1 comprend une canne ancrée à 40 m et munie de 8 capteurs (voir fig. 2).

1.4. Déroulement

L'essai a comporté cinq cycles (à 5, 10, 15, 20, 25 MPa) comportant chacun :
- deux phases de mise en charge et décharge progressives
- six matraquages (charges et décharges rapides)
- un palier de fluage.

2. LES RESULTATS DEJA CONNUS (Réf. 1)

La part de pression prise par le rocher est, sur les deux sections, supérieure au dernier cycle, à 60 % de la pression intérieure.
Ce résultat permet d'estimer la pression en MPa s'exerçant sur les parements de la galerie (voir tableau ci-dessous).

Cycle (MPa)	5	10	15	20	25
Section 1	1,1	2,7	5,1	7,1	8,5
Section 2	1,7	3,7	6,0	8,2	10,0

3. ROCHER

3.1. Géologie

La caverne est entièrement située dans des niveaux gréseux.

3.2. Propriétés mécaniques

3.2.1. Les essais de laboratoire

Les mesures mettent en évidence une anisotropie de comportement (voir sur le tableau ci-après : les valeurs moyennes de Rc, Rtb, modules et vitesse).

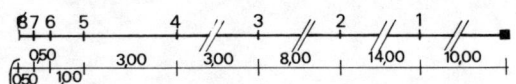

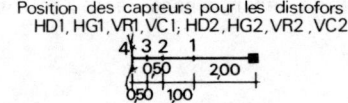

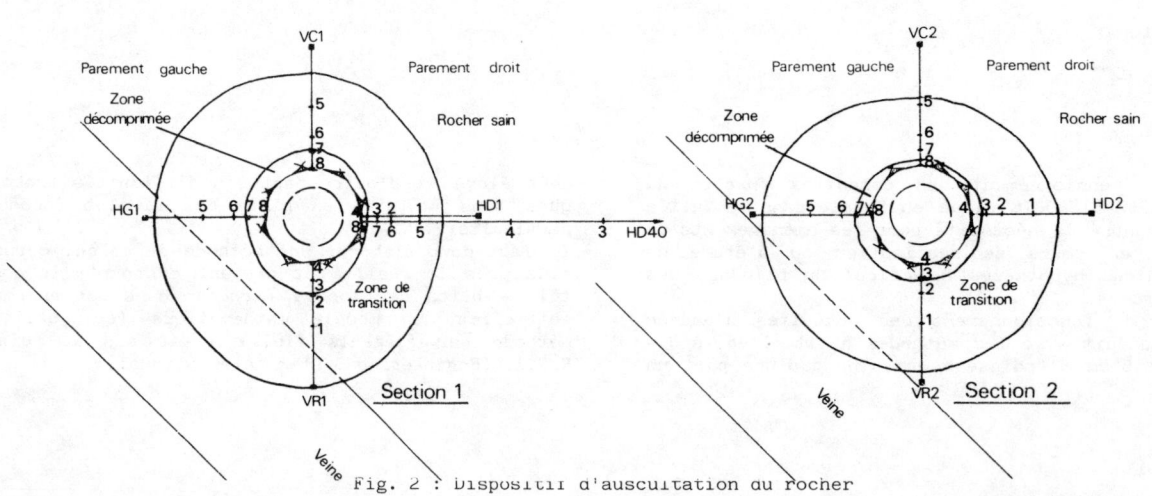

Fig. 2 : Dispositif d'auscultation au rocher

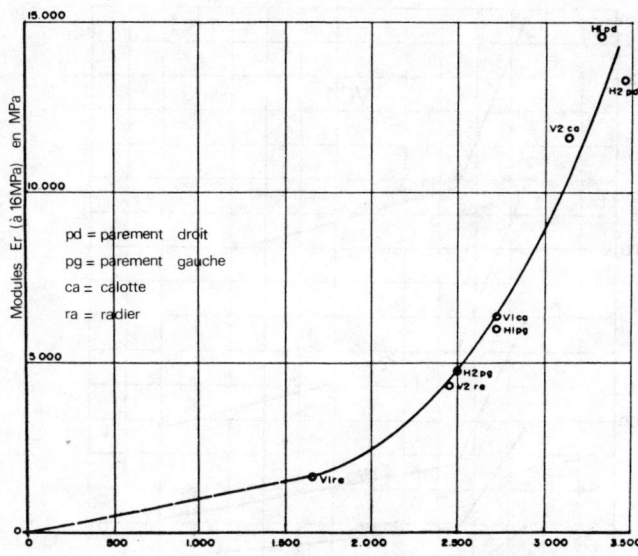

Fig. 3 : Corrélation module - vitesses sismiques

	⊥ Stratification	// Stratification
RC (MPa)	101,80	175,75
Rt	12,6	3,75
Vitesse longitudinale m/s	3 500	4 260
Module global (MPa)	26 200	36 800
Module réversible	30 900	41 900

3.2. Les reconnaissance in-situ

Elles ont compris :
- des mesures sismiques et microsismiques à la paroi,
- deux campagnes de mesures de contraintes au vérin plat (août 1977 et mai 1981),
- des mesures de module au vérin à plaque de charge dans deux sections correspondant aux sections auscultées au cours de l'essai.

a) La sismique a mis en évidence trois zones distinctes autour de l'excavation (voir fig. 2) :
  - zone décomprimée (vitesse comprise entre 2 000 et 3 000 m/s) plus étendue en parement gauche qu'en parement droit,
  - zone de transition : 4 300 à 4 500 m/s en parement droit et 3 500 à 3 700 m/s pour le parement gauche,
  - zone saine : 5 500 m/s. Cette valeur est très élevée et peut s'expliquer par la présence de contraintes élevées.

b) Il existe une corrélation étroite entre les valeurs de module (vérin à plaque de charge) et les vitesses sismiques (mesurées par microsismique) (voir fig. 3).

c) Les modules les plus faibles sont obtenus sur les zones les plus proches de la faille F1 (radier et parement gauche).

d) Les deux campagnes de mesures de contrainte ont montré une diminution des contraintes horizontale et longitudinale (voir tableau ci-dessous).

Date	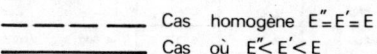		
Août 1977	11,1	34,0	6,2
Mai 1982	10,2	3,8	3,7

La galerie présente donc un phénomène d'évolution différée du matériau. Une étude numérique a permis de montrer que la redistribution des contraintes observées résultait d'une diminution importante de la cohésion due à l'aggravation de la microfissuration au voisinage de l'excavation.

Si la galerie était rigoureusement circulaire, excavée dans un milieu élastique orthotrope où les directions de contraintes principales sont horizontale et verticale, les contraintes verticale et horizontale initiales seraient de l'ordre de 170 et de 100 bars. La contrainte verticale serait donc très supérieure au poids de la couverture rocheuse (300 m).

4. ANALYSE DU COMPORTEMENT MECANIQUE DU ROCHER AUTOUR DE L'EXCAVATION

4.1. Choix du modèle analytique (voir fig. 4)

Le modèle proposé découle du schéma proposé par la sismique à la paroi. Il considère une galerie circulaire de rayon r_0 soumise à une pression intérieure radiale P creusée dans un massif rocheux de module d'Young initial E. Autour de l'excavation on trouve une zone décomprimée (module E'') d'épaisseur $r_1 - r_0$ et une zone de transition (module E') d'épaisseur $r_2 - r_1$.
Le comportement est supposé élastique. Le déplacement u à la distance r du centre de l'excavation s'écrit :

$$u = \frac{P}{E}(1+\nu)\, r_0^2 \left[\alpha r + \frac{\beta}{r}\right]$$

α et β ont dans chacune des zones les valeurs suivantes :

	α	β
Zone saine $r > r_2$	0	1
Zone de transition $r_2 > r > r_1$	$\frac{1}{r_2^2}\left[1 - \frac{E}{E'}\right]$	$\frac{E}{E'}$
Zone décomprimée $r_1 > r > r_0$	$\frac{1}{r_2^2}\left[1 - \frac{E}{E'}\right] + \frac{1}{r_1^2}\left[\frac{E}{E'} - \frac{E}{E''}\right]$	$\frac{E}{E''}$

4.2. Application

La méthode consiste pour chaque appareil, à tracer la courbe $u \times r$ en fonction de r (voir fig. 4), ce qui

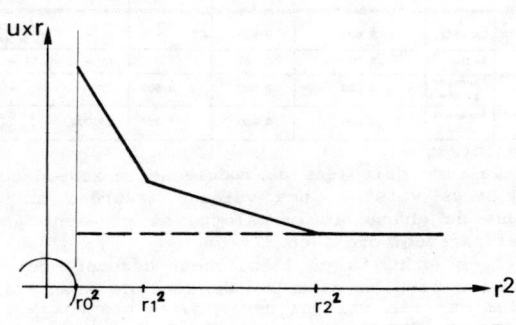

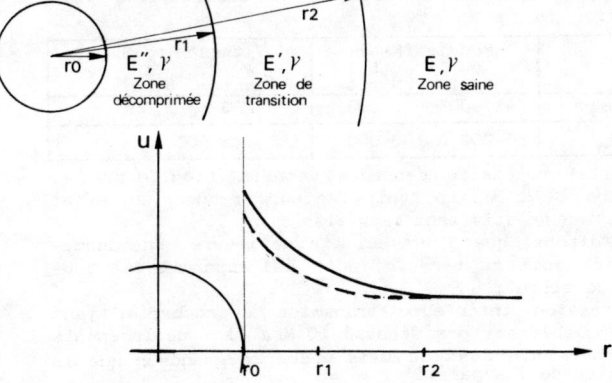

Fig. 4 : Description du modèle choisi

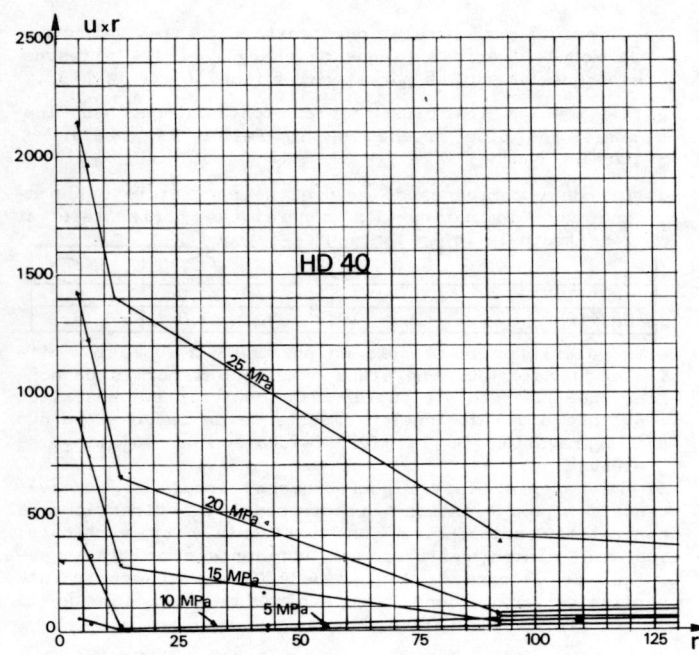

Fig. 5 : Distofor HD 40

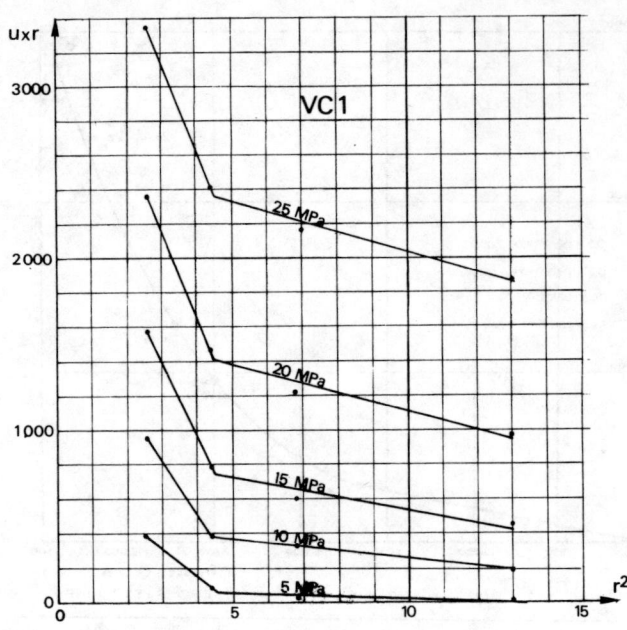

Fig. 6 : Distofor VC 1

permet d'estimer :
- E, E', E" à l'aide du distofor de 40 m,
- E' et E" grâce aux distofors de 4 m.

L'estimation de E" ne peut cependant être envisagé systématiquement. Elle nécessite en effet la présence au sein de la zone décomprimée d'au moins deux mesures.
Cette condition n'est réalisée, ni en parement droit (sur S1 et S2) ni en calotte sur S2 : la zone décomprimée y est très peu épaisse (voir fig. 2).

4.3. Exemples

Les figures 5 et 6 donnent le tracé des courbes u x r en fonction de r pour le distofor HD 40 (ancrée à 40 m) et le distofor VC1 (en calotte section 1).
On voit que le modèle s'applique de manière satisfaisante sur le distofor VC1 où on peut calculer E", E' et E. HD 40 ne permet, comme on pouvait s'y attendre, que de calculer E' et E.

5. RESULTATS - DISCUSSION

5.1. Comportement du massif sain

Les déplacements mesurés du rocher à 16 m de profondeur correspondent à des modules d'élasticité variant de 50 000 à 100 000 MPa, soit des valeurs 2 à 3 fois supérieures aux résultats des essais en laboratoire.
Cette différence est confirmée par les mesures de vitesses du son réalisées in-situ et en laboratoire (voir tableau).

	Module (MPa) $\parallel$ Strat	Module (MPa) $\perp$ Strat	Vitesse (m/s) $\parallel$ Strat	Vitesse (m/s) $\perp$ Strat
Laboratoire	41 900	31 000	4 260	3 500
In-situ	50 000 à 100 000		5 500	

Ce résultat nécessite cependant confirmation. Dans les conditions de l'essai, trois facteurs rendent en effet l'estimation relativement imprécise :
- on ne dispose que d'un seul axe de mesure. Une anomalie liée soit au terrain, soit à l'appareil est toujours possible ;
- la pression intérieure transmise au rocher n'ayant vraisemblablement pas dépassé 10 MPa, les déplacements du rocher sain sont du même ordre de grandeur que la précision de l'appareil ;
- la valeur de la pression transmise au rocher est en toute rigueur inconnue.

En première analyse, la différence observée entre les mesures sur éprouvettes et in-situ pourrait résulter d'un champ de contraintes élevées que les mesures de contraintes au vérin plat ont laissé entrevoir.

5.2. Comportement du massif au voisinage de l'excavation

Le modèle analytique utilisé, a permis en premier lieu de confirmer un certain nombre de données qualitatives.
- en parement droit, la zone décomprimée est très peu marquée (épaisseur faible ou module relativement élevé) ;
- en parement gauche, on retrouve la présence d'une zone plus déformable qui est probablement la faille ;
- l'épaisseur de la zone décomprimée, mise en évidence par les mesures, est de 0,50 m, de qui est tout-à-fait compatible avec la géophysique.

Le tableau suivant rassemble les valeurs :
- des modules mesurés par vérin à plaque de charge (module de la zone décomprimée),
- des vitesses mesurées par sismique à la paroi,
- des modules calculés.

		Module mesuré par essai au vérin	Vitesse mesurée (m/s) ZD	Vitesse mesurée (m/s) ZT	Modules calculés ZD	Modules calculés ZT
Section 1	Calotte	5 600	2 730	-	4 500	16 500
	Radier	1 200	1 650	-	-	-
	Parement gauche	5 200	2 730	3 600	4 500	7 000
	Parement droit	13 300	3 450	4 300	-	6 000 à 12 000
Section 2	Calotte	9 600	3 150	-	-	6 000 à 9 000
	Radier	3 700	2 450	-	-	-
	Parement gauche	3 500	2 500	3 600	-	-
	Parement droit	12 000	3 450	4 300	13 000	14 000 à 18 000

On constate que :
- les valeurs calculées du module de la zone décomprimée sont très voisines des valeurs mesurées au vérin à plaque de charge (voir calotte et parement gauche en S1 et parement droit en S2) ;
- dans les endroits où l'épaisseur décomprimée est faible, les modules calculés de zone de transition sont également très voisins des modules mesurés au vérin à plaque de charge (parement droit S1 et S2 et calotte en S2).

5.3. Conclusion

Le dépouillement des mesures de l'essai et des reconnaissances préalables a permis de montrer une corrélation étroite entre :
- les valeurs des vitesses du son et des modules statiques (en laboratoire et in-situ) ;
- les valeurs de modules statiques mesurées au vérin et calculées par le modèle analytique.

Le modèle analytique, malgré sa simplicité, permet donc de décrire la massif rocheux de manière satisfaisante.
Cette conclusion permet d'une part de considérer le comportement du rocher autour d'une galerie circulaire soumise à une pression intérieure comme élastique et d'autre part de justifier les valeurs élevées calculées pour le module du massif sain (50 000 à 100 000 MPa).

6. REFERENCE   A. MARTINET et T. DOUCERAIN, 1982

1 Etude du comportement d'une conduite forcée souterraine bloquée au rocher par la réalisation d'un essai en caverne.
Symposium International de Mécanique des Roches - AIX-LA-CHAPELLE 1982, Balkema éditeur

ANALYSIS OF SQUEEZING SEAMS IN ROCK
Analyse de filons compressés dans de la roche
Analyse quetschender Schichten

K. H. O. Saari
Senior Research Engineer, Technical Research Centre of Finland
R. E. Goodman
Professor of Geological Engineering, University of California, Berkeley

SYNOPSIS

Squeezing of seams intersecting tunnels and shafts in rock is treated as elasto-plastic deformation of the material. Approximate analytical solutions are given for elastic, rigid-plastic and elastic/rigid-plastic material models. Numerical methods are applied for non-dilatant, cohesive and dilatant, frictional material models. The parameters for the models were determined in standard laboratory tests. Results of the numerical analyses are compared to the measurements done in laboratory model tests.

RESUME

La compression des filons intersectant tunnels et puits en roche est traitée comme déformation élastoplastique du matériau. Des solutions analytiques approchées sont données pour des matériaux élastiques, plastiques rigides et élastiques/plastiques rigides. Des méthodes numériques sont appliquées à des modèles en matériaux non-dilatants, cohésifs et dilatants. Les paramètres pour les modèles ont été déterminés par des essais en laboratoire normaux. Les résultats des analyses numériques sont comparés avec les mesures faites dans des essais de modèle en laboratoire.

ZUSAMMENFASSUNG

Die Quetschung einzelner Schichtlagen, die von Tunneln oder Schächten durchschnitten werden, wird als elastisch-plastische Deformation des Gesteins behandelt. Für elastische, starr-plastische und elastische/starr-plastische Stoffgesetze werden analytische Näherungslösungen gegeben. Für bindige, nichtdilatierende und rollige, dilatierende Materialien, deren Stoffparameter mittels Laboruntersuchungen bestimmt wurden, sind numerische Methoden entwickelt worden. Die Resultate der numerischen Lösungen werden mit Meßergebnissen von Modellversuchen verglichen.

1. INTRODUCTION

Terzaghi's soft ground classification for tunneling purposes (Terzaghi 1950) defines squeezing ground as a ground which advances into the tunnel from all sides without any signs of fracturing. All materials may squeeze because an opening made in a stressed medium tends to shrink. Squeezing is notable in soft or medium clays and argillaceous rocks.

The development of deformations is called squeezing in underground construction when the displacements of the opening perimeter are large. Although no definitive values are agreed upon to call the ground behavior squeezing, a tunnel closure of more than one percent can be considered squeezing deformation. Laboratory tests on squeezing materials show that plastic yielding must take place to produce the strain observed in squeezing ground because a large part of the deformations is permanent.

In the present paper squeezing is treated as elastic-plastic deformation of rock or soil around underground openings. Analytical solutions exist which take into account the viscous properties of geologic materials in tunnel design. Most of the presented solutions are based on visco-elastic models (Goodman, 1980). Recently, analytical (Nonaka, 1978 and 1980; Zhu, 1981) and numerical (Gioda, 1982) solutions have been presented for visco-plastic material models. All analytical visco-elastic and visco-plastic solutions known at this point are for plane strain conditions and hydrostatic stress fields, which assumptions make the problem geometrically one dimensional.

2. ANALYTICAL SOLUTIONS

Approximate elastic and rigid-plastic solutions are developed for a layer with a circular opening, figure 1, Saari (1982). The material of the layer is assumed isotropic and loading is symmetric. Internal pressure inside the opening is denoted by p_i. The lining of the tunnel or shaft can be represented by an elastic spring with spring constant S. The boundary tractions at the parallel faces of the layer are the axial stress σ_z and shear stress τ_{rz} both of which are functions of radius as are the other two stress components σ_r and $\sigma_{\theta\theta}$. The stresses in the layer also depend on the axial dimension z. If the layer is thin compared to the radius of the opening an approximate solution can be found when the average values of stresses in the axial direction are used.

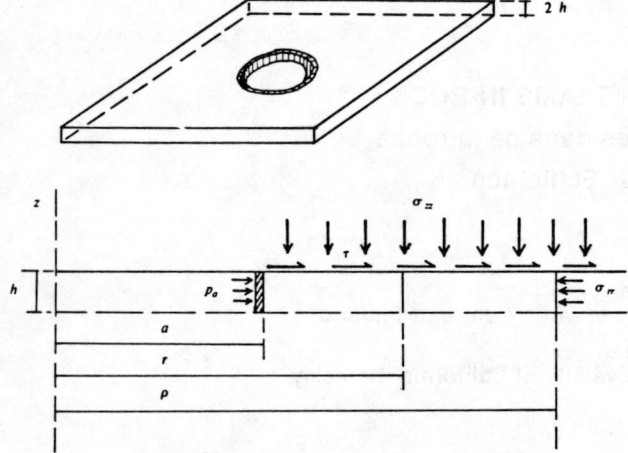

Figure 1. Boundary tractions of a squeezing layer with a circular hole.

If σ_{rr} and $\sigma_{\theta\theta}$ are assumed to be independent of z the average equilibrium condition is

$$\frac{1}{h}\int_0^h [\frac{\partial \sigma_{rr}}{\partial r} + \frac{\partial \tau_{rz}}{\partial z} + \frac{\sigma_{rr} - \sigma_{\theta\theta}}{r}]dz = \frac{\partial \sigma_{rr}}{\partial r} + \frac{\tau}{h} + \frac{\sigma_{rr} - \sigma_{\theta\theta}}{r} = 0 \quad (1)$$

2.1. Elastic Solutions

The axial displacement of the parallel boundaries in relation to the center plane of the layer is Δw. Because the surrounding rock is assumed to be rigid the axial displacement is constant in the radial direction and the strains are

$$\epsilon_{zz} = -\frac{\Delta w}{h}$$
$$\epsilon_{\theta\theta} = \frac{u}{r} \quad (2)$$
$$\epsilon_{rr} = \frac{\partial u}{\partial r}$$

The equilibrium equation then becomes

$$(\lambda + 2G)(\frac{\partial^2 u}{\partial r^2} + \frac{1}{r}\frac{\partial u}{\partial r} - \frac{u}{r^2}) + \frac{\tau}{h} = 0 \quad (3)$$

2.1.1. Cohesive Interface

If the interface between the layer and rock is assumed frictionless and the maximum shear stress is purely cohesive and constant, $\tau = \tau_0$, the differential equation (3) has the general solution

$$u(r) = A_1 r + A_2 \frac{1}{r} - \frac{\tau_0}{3h(\lambda + 2G)}r^2 \quad (4)$$

The boundary conditions for equation (4) are given by the condition

$$\sigma_{rr}(a) = S\, u(a) + p_i \quad (5)$$

at the surface of the opening and by the conditions

$$u(\rho) = 0 \qquad \frac{\partial u(\rho)}{\partial r} = 0 \quad (6)$$

at the distance ρ where no radial displacement is assumed. In an exact solution ρ would be infinitely far from the opening. In the approximate case ρ is the distance where no slip at the interface takes place. The constants A_1, and A_2 are determined from equations (4) - (6) and an equation for ρ is obtained and thus radial displacement and stresses become known, figure 2.

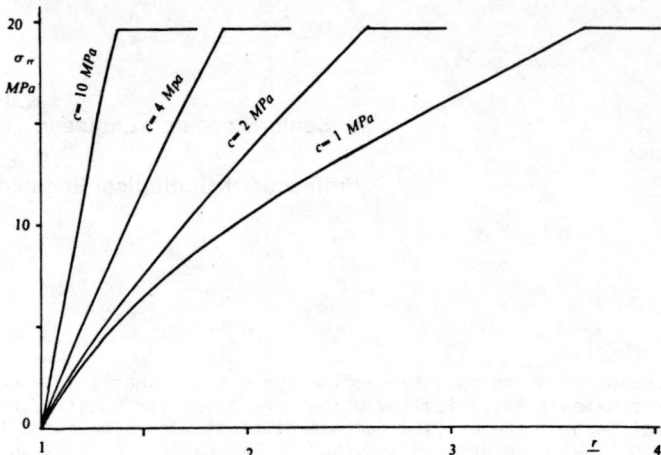

Figure 2. Influence of interface cohesion on stress distribution around a circular hole in elastic layer.

2.1.2. Frictional Interface

The shear stress at a frictional interface is with μ = coefficient of friction, $\tau = \mu \sigma_{zz}$

In this case the equation of equilibrium becomes

$$(\lambda + 2G)(\frac{\partial^2 u}{\partial r^2} + \frac{1}{r}\frac{\partial u}{\partial r} - \frac{u}{r^2} \mu \frac{\Delta w}{h^2}) + \lambda \frac{\mu}{h}(\frac{\partial u}{\partial r} + \frac{u}{r}) = 0 \quad (7)$$

Equation (7) has the general solution

$$u(r) = Dr + A_1 \frac{1}{r} + A_2(1 + \frac{1}{Ar})e^{-A\rho} \quad (8)$$

A transcendental equation for ρ is obtained from which ρ can be solved iteratively. Once ρ is known the radial displacement and stress distribution are known, figure 3.

2.2. Rigid-Plastic Solution

An analytical solution can be found for the circular hole with a thin layer confined between two rigid surfaces for a material with von Mises-type yield function and plastic potential. The initial volume of the plastic region between radii r and R is

$$V_0 = \pi (R^2 - r^2)h \quad (9)$$

The volume of the plastically deformed region is

$$V_1 = \pi [R^2 - (r+u)^2](h - \Delta w) \quad (10)$$

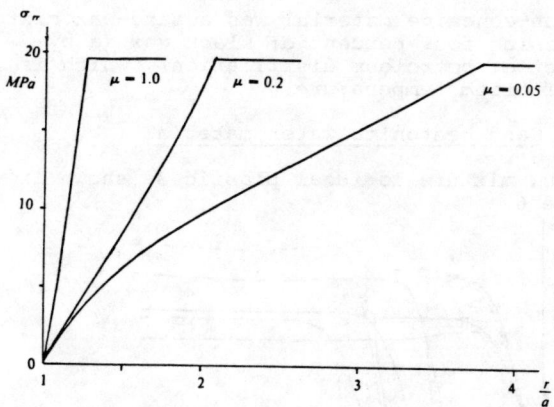

Figure 3. Influence of interface friction on stress distribution around a circular hole in elastic layer.

If there is no volume change $V_1 = V_0$ and for small displacements,

$$u(r) = -\frac{R^2 - r^2}{2r}\frac{\Delta w}{h} \quad (11)$$

It is assumed that the shear stresses are negligible and the stresses $\sigma_{rr}, \sigma_{\theta\theta}, \sigma_{zz}$ are the principal stresses. Then the von Mises yield function is

$$f(\sigma_{ij}) = \left\{\frac{1}{6}[(\sigma_{rr} - \sigma_{\theta\theta})^2 + (\sigma_{\theta\theta} - \sigma_{zz})^2 + (\sigma_{zz} - \sigma_{rr})^2]\right\}^{1/2} - k \quad (12)$$

The plastic strain increments are

$$\dot{\epsilon}^p_{ij} = \Lambda \frac{\partial f(\sigma_{ij})}{\partial \sigma_{ij}}$$

$$\dot{\epsilon}_{rr} = \Lambda \frac{\sigma_{rr} - \sigma_m}{2\sqrt{J_2}} = \frac{R^2 + r^2}{2r^2}\frac{\Delta w}{h}$$

$$\dot{\epsilon}_{\theta\theta} = \Lambda \frac{\sigma_{\theta\theta} - \sigma_m}{2\sqrt{J_2}} = -\frac{R^2 - r^2}{2r^2}\frac{\Delta w}{h} \quad (13)$$

$$\dot{\epsilon}_{zz} = \Lambda \frac{\sigma_{zz} - \sigma_m}{2\sqrt{J_2}} = -\frac{\Delta w}{h}$$

where $\sigma_m = \frac{1}{3}(\sigma_{rr} + \sigma_{\theta\theta} + \sigma_{zz})$. Equations (12) and (13) yield

$$\Lambda = \frac{\Delta w}{h}\frac{1}{r^2}(R^4 + 3r^4)^{1/2}\frac{\sqrt{J_2}}{k} \quad (14)$$

Substituting Λ into equations (13) gives

$$\sigma_{rr} - \sigma_{\theta\theta} = \frac{2kR^2}{(R^4 + 3r^4)^{1/2}} = 0 \quad (15)$$

The equation of equilibrium is thus

$$\frac{\partial \sigma_{rr}}{\partial r} + \frac{\tau_0}{h} + \frac{2kR^2}{r(R^4 + 3r^4)^{1/2}} = 0 \quad (16)$$

Equation (16) can be integrated giving

$$\sigma_{rr} = k \ln \frac{R^2 + \sqrt{R^4 + 3r^4}}{R^2 + \sqrt{R^4 + 3a^4}} + S\,u(a) + p_a - \frac{\tau_0}{h}(r - a) \quad (17)$$

The radius R of the plastic region can be determined from the condition that at the boundary of the plastic region the radial stress must be continuous, figure 4.

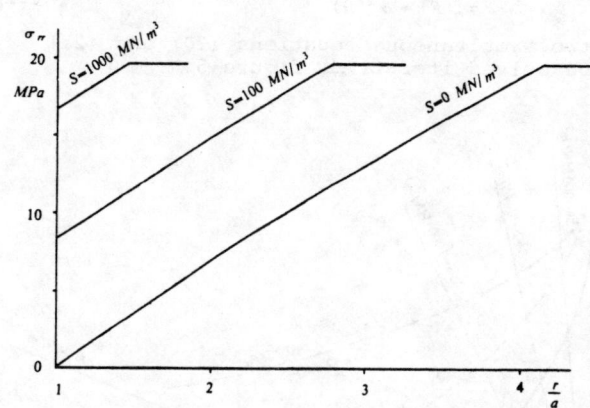

Figure 4. Influence of lining stiffness on radial stress distribution around circular opening in rigid-plastic layer. k=2.6 MPa, h=0.2a, τ_0 =1.0 MPa.

2.3. Elastic/Rigid-Plastic Solution

A refinement to elastic and rigid-plastic solutions is found in the form of an elastic/rigid-plastic solution where elastic deformations are assumed negligble in the plastic region. The plastic incompressibility requirement leads to equation (18) for the radial displacement in the plastic region

$$u(r) = \frac{R\,u(R)}{r^2} - \frac{R^2 - r^2}{2r}\frac{\Delta w}{h} \quad (18)$$

The radial stress in the plastic region is found to be

$$\sigma^p_{rr}(r) = kR(R - 2u(R)\frac{h}{\Delta w})\frac{1}{\alpha^2} \ln\left[\frac{(\alpha^2 + \sqrt{\alpha^4 + 3r^4})a^2}{(\alpha^2 + \sqrt{\alpha^4 + 3a^4})R^2}\right]$$

$$-\frac{\tau_0}{h}(r-a) + Su(a) + p_a \quad (19)$$

Both the radius of the plastic region, R, and the radius of the elastic region, ρ, are now unknown. One equation for their solution can be obtained from the requirement that at the elastic-plastic boundary, r=R, both the elastic and plastic stresses have to satisfy the yield condition (12) and an equation in terms of the radial displacement, u(r), is obtained

$$(\frac{\partial u}{\partial r} + \frac{u}{r})^2 - 3\frac{\partial u}{\partial r}\frac{u}{r} + \frac{\Delta w}{h}(\frac{\partial u}{\partial r} + \frac{u}{r}) + (\frac{\Delta w}{h})^2 - \frac{3k^2}{4G^2} = 0 \quad (20)$$

where u(r) is given by equation (18). The second equation is given by the requirement that the radial stress must be continuous at the elastic-plastic boundary r=R

$$\sigma_r^p(R) = \sigma_r^e(R) \quad (21)$$

The two simultaneous equations (20) and (21) can be solved iteratively figure 5.

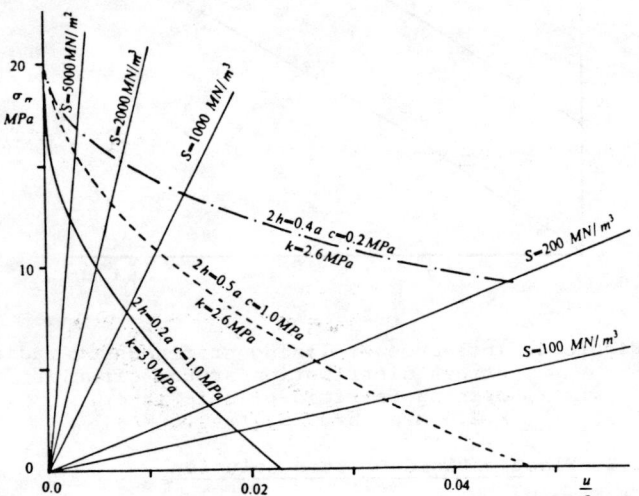

Figure 5. Characteristic lines of a circular hole in elastic/rigid-plastic layer.

3. MODEL TESTS AND NUMERICAL MODELS

3.1. Model tests

In the theoretical treatment of the problem it was found that plastic yielding of a layer is observed as squeezing type displacement of the opening perimeter. It was found necessary to verify the analytical and numerical results in laboratory tests.

The behavior of squeezing layers was studied in the laboratory using sand-bentonite-water and sand-wax mixtures as model materials. The layer was consolidated under a known axial stress and the excavation was simulated by removing a steel cylinder from the opening. The thickness of the layer was not allowed to change after the initial consolidation.

A mixture of five weight units of sand, two units of bentonite and 2.8 units of water was selected as the non-frictional model material.

The non-cohesive material was a sand-wax mixture containing four percent of slack wax (a by-product of petroleum distillation), which is soft in room temperature.

3.2. Sand bentonite water material

The SBW mixture is ideal plastic as shown in figure 6.

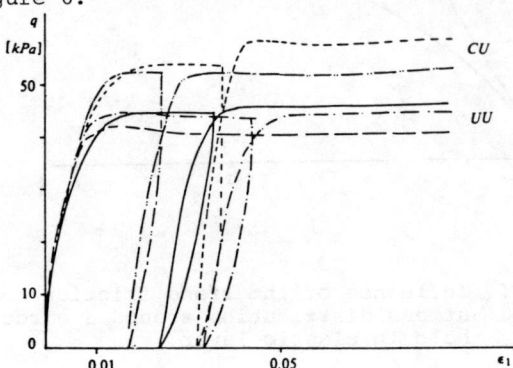

Figure 6. Axial stress-strain plots of triaxial tests with sand-bentonite-water material.

The volume compressibility of the SBW mixture is shown in figure 7, which shows results of two test done in the model test apparatus.

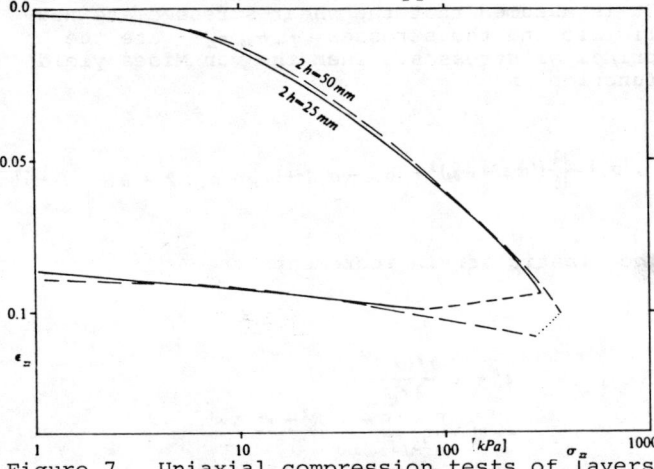

Figure 7. Uniaxial compression tests of layers 550 mm in diameter.

The stress paths of the undrained unconsolidated tests of SBW samples are shown in figure 8, where $p' = \frac{\sigma_1' + 2\sigma_3'}{3}$, $q = \sqrt{\frac{3}{2} s_{ij} s_{ij}}$

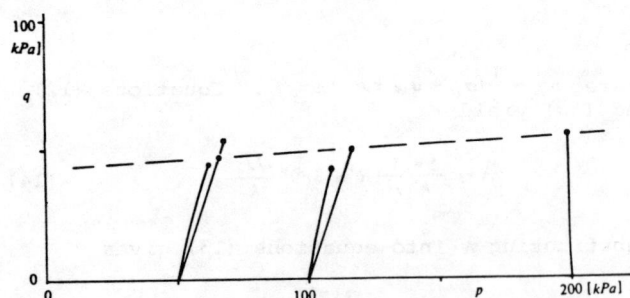

Figure 8. Stress paths of unconsolidated undrained SBW samples.

The material is partially saturated and thus the effective stresses are close to the total stresses.

The non-linear elastic properties of the material were modeled assuming the elastic component of the volumetric strain increment to be

$$\dot{\epsilon}_v^e = \frac{\Delta v}{v} = -\kappa \frac{\Delta p}{v p} = \frac{\Delta p}{B} \quad (22)$$

where v = specific volume = 1+e
  κ = slope of the unloading curve in (v, ln p)-plot, figure 9.
  B = tangent bulk modulus

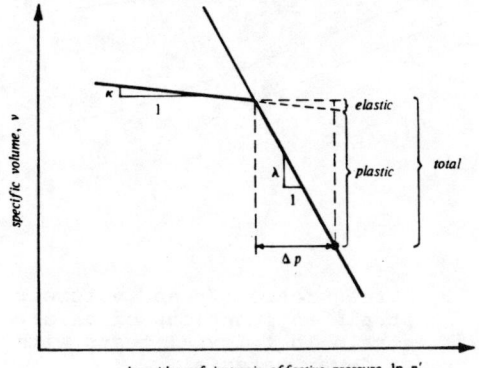

Figure 9. Elastic and plastic strains in uniaxial compression test.

The value of the Young's modulus was assumed constant and equal to the initial slope of (σ_1, ϵ_1)-plots.

Plastic yielding of the SBW material was modeled using the modified Cam Clay model (Chang and Duncan 1977), figure 10.

$$f_1(\sigma_{ij}) = p - p_r + \frac{q^2}{M^2(p - p_r)} - \exp(\frac{1+e}{\lambda + \kappa} \epsilon_v) \quad (23)$$

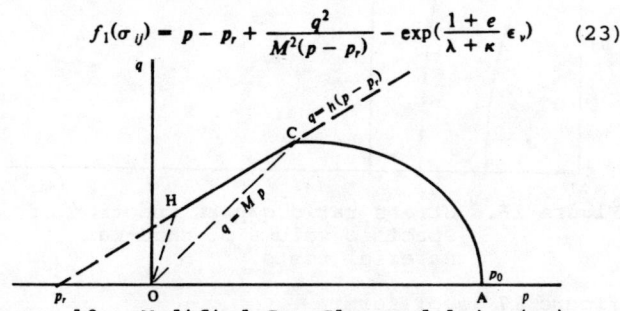

Figure 10. Modified Cam Clay model in (pq) coordinate system.

The elasto-plastic material model predicts accurately the instantaneous displacements as can be seen in figure 11, where measured and calculated radial displacements are compared. Time dependent analysis of the displacements would require a visco-plastic material model (Saari 1982).

3.3. Sand-Wax Material

Axial stress-strain plots of triaxial tests with sand-wax material are depicted in figure 12.

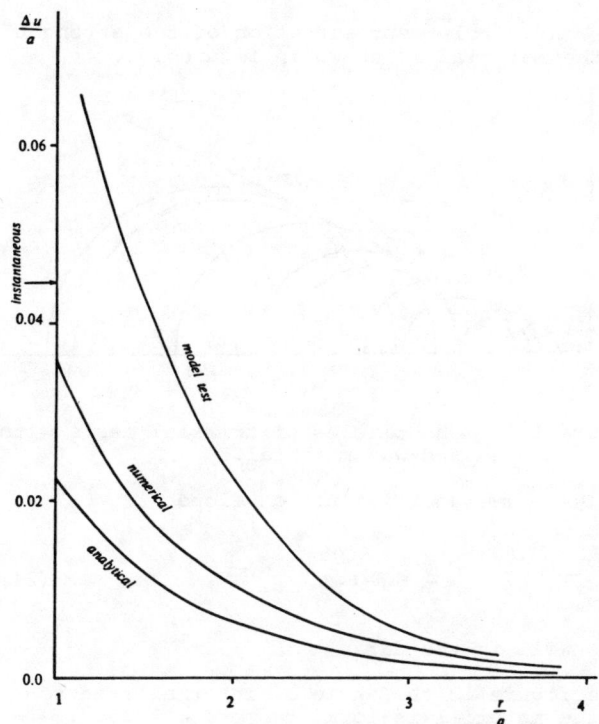

Figure 11. Measured and calculated radial displacements around a circular opening in SBW layer.

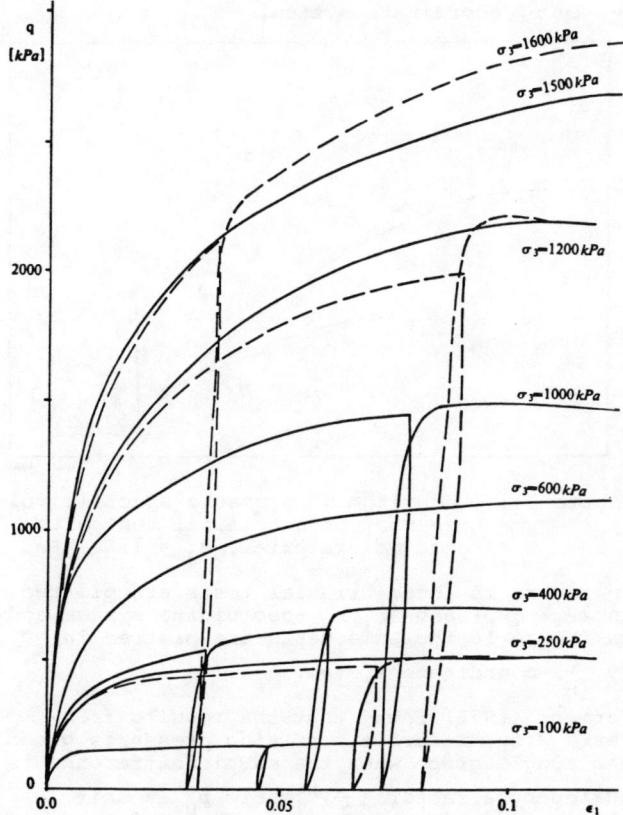

Figure 12. Stress-strain plots of sand-wax material.

The Mohr circle representation of the strength of the material is shown in figure 13.

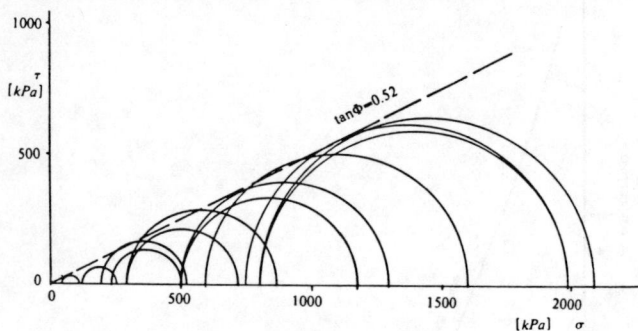

Figure 13. Mohr circles of triaxial tests with sand-wax material.

The Drucker-Prager failure envelope

$$\frac{q}{\sqrt{3}} + \alpha p - k = 0 \quad (24)$$

fits well in the data.

From figure 13 the angle of internal friction of the material is formd to be $\phi = 27.5°$. The Drucker-Prager function parameters are $\alpha = 0.63$, $k = 0.0$.

In figure 14 one of the tests is plotted in $(v, \ln p)$ coordinate system.

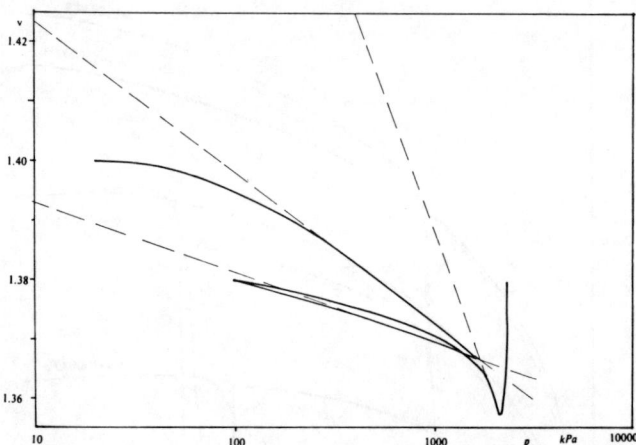

Figure 14. Logarithmic pressure-specific volume plot of a triaxial test with sand wax material, $\sigma_3 = 1500$ kPa.

In figure 15 three triaxial tests are plotted in $(\varepsilon_1, q/p)$ and $(\varepsilon_1, \varepsilon_v)$-coordinate system and in figure 16 the same tests are plotted in $(v, \frac{q}{p})$-coordinate system.

Vermeer (1978) found that the results from tests with different confining pressures give one single graph when the strain difference is scaled by a factor $(\frac{p_0}{p})^\beta$ where p_0 is unit stress and β is a constant. The sand-wax material needs an additional scaling factor for the stress ratio $\frac{q}{p}$ to produce nearly identical curves for tests with confining pressures ranging between 100 and 1500 kPa. The graph in

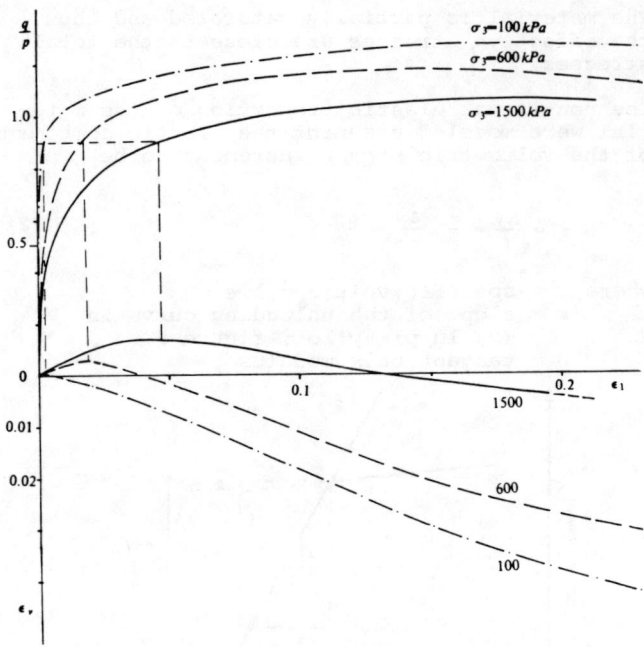

Figure 15. Stress ratio q/p and volumetric strain as functions of axial strain in triaxial tests with sand-wax material.

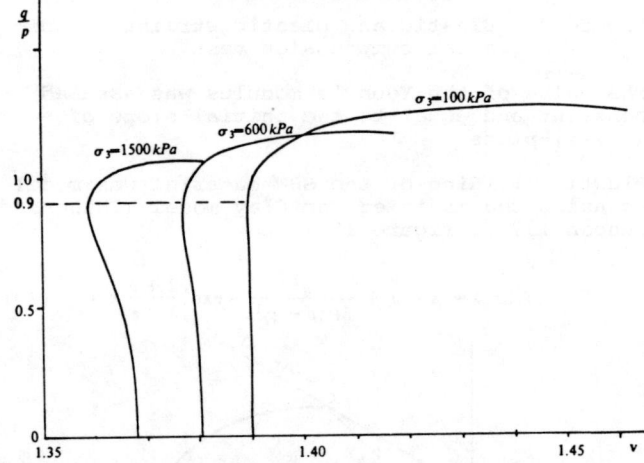

Figure 16. Stress ratio q/p as function of specific volume of sand-wax material tests.

figure 17 is of form

$$\frac{q}{p^{1-\alpha}} = a \, \bar{e}_s^p \frac{p_0^{\alpha+\beta}}{H q p^\beta + \bar{e}_s^p p_0^\beta} \quad (25)$$

where $\bar{e}_s^p = \varepsilon_1^p - \varepsilon_3^p$ for triaxial test and the values of the parameters are

a = 1.25
H = 0.000012
α = 0.070
β = 0.23

The graphs for constant values of $\bar{e}_s^p$ in the (p-q) coordinate system are plotted in figure 18.

The inverse relationship of equation (26) is

$$f_2(\sigma_{ij}) = \frac{q^2 H}{ap\left(\frac{p_0}{p}\right)^{\alpha+\beta} - q} - \bar{e}_s^p \qquad (26)$$

Prevost and Höeg (1975) proposed a double hardening model for soils. Their model consisted of a horizontal line and an inclined line with negative slope in the (p-q)-plane.
Based on large body of previous work, Vermeer's (1978) article specifically and the laboratory results it was found that the yielding of the sand-wax material and possibly other frictional materials can be modeled by a double hardening model where the two surfaces are of the form

$$\begin{aligned} f_1(\sigma_{ij},\epsilon_v) &= F_1(\sigma_{ij}) - \Gamma_1(\epsilon_v) = 0 \\ f_2(\sigma_{ij},\epsilon_s) &= F_2(\sigma_{ij}) - \Gamma_2(\bar{e}_s^p) = 0 \end{aligned} \qquad (27)$$

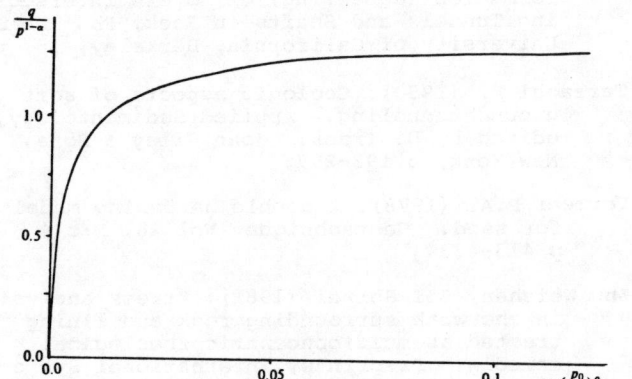

Figure 17. Normalized hyperbolic yield function, equation (25).

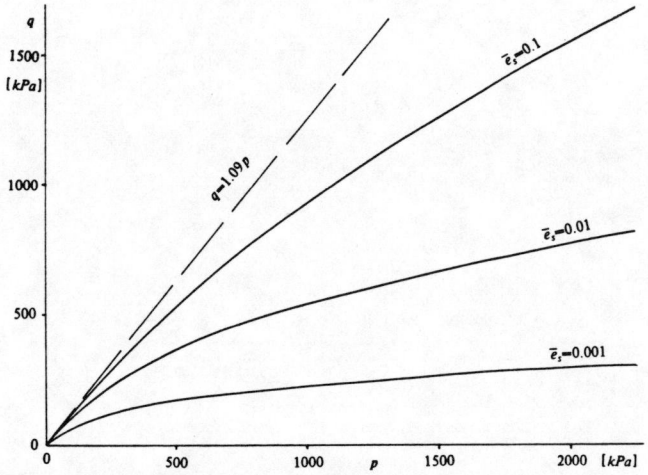

Figure 18. Hyperbolic yield function in (p,q)-plane.

Particular formulations for equations (27) are equations (23) and (26). A likely choice for the plastic potentials to be used with these functions are the functions themselves. This implies an associated flow rule.

In figure 19 the measured and calculated radial displacements are compared in a layer 2h = 50 mm thick and with a circular hole of diameter 2a = 114 mm. The numerical model was for the unsupported case. The bolt support in the model consisted of 32 wooden bolts of length ℓ = 50 mm installed equally spaced in two cross sections in the layer.

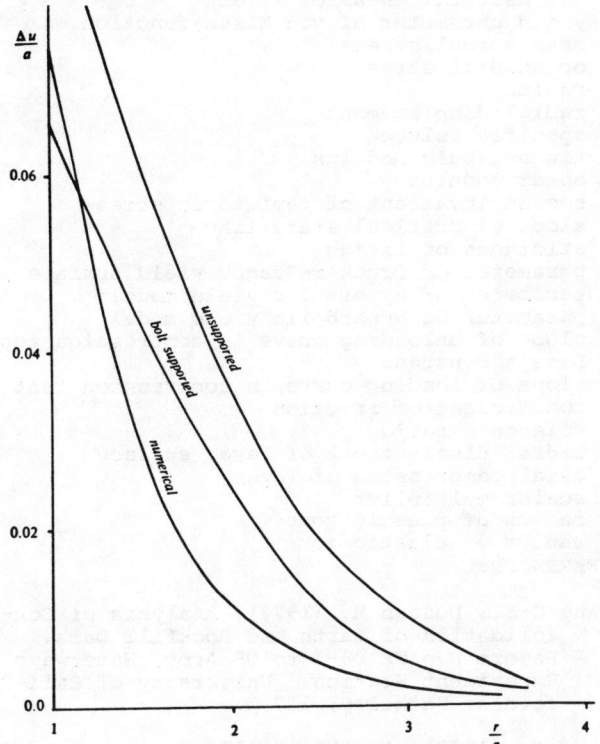

Figure 19. Radial displacements around a circular hole in sand-wax layer.

The modified Cam Clay model was applied to the sand-wax material and the resulting displacements were only one half of those of the two surface model, Saari (1982).

4. SUMMARY

The numerical methods used to analyze squeezing of layers around circular openings predict accurately the time-independent deformations of cohesive non-dilatant materials.

The displacement values of the analytical solutions are lower than the measured and numerical values because the stress-dependency of the elastic constants was not taken into account.

The two surface yield model gives smaller displacements than the model tests. The differencies are caused by the application of small deformation theory and by the model itself.
The main reason for the large measured displacements is believed to be the failure of the material at the surface of the opening which was not included in the model.

List of symbols

- a radius of opening
- a parameter of two surface yield function
- c cohesion
- e void ratio
- e_s equivalent shear strain
- $f(\sigma_{ij})$ yield surface
- h half thickness of layer
- k yield parameter of von Mises function
- p mean normal stress
- q octahedral stress
- r radius
- u radial displacement
- v specific volume
- B tangent bulk modulus
- G shear modulus
- J_2 second invariant of deviatoric stress
- M slope of critical state line
- S stiffness of liming
- α parameter of Drucker-Prager yield surface
- α parameter of hyperbolic yield model
- β parameter of hyperbolic yield model
- κ slope of unloading curve in compression test
- λ Lame's constant
- λ slope of loading curve in compression test
- μ coefficient of friction
- ν Poisson's ratio
- Δu radial displacement of layer surface
- Δw axial compression of layer
- Λ scalar multiplier
- R radius of plastic zone
- ρ radius of elastic zone

REFERENCES:

Chang C-S., Duncan M. (1977): Analysis of Consolidation of Earth and Rockfill Dams. Report N:o TE 88-3 to US Army, Waterways Experiment Station. University of California, Berkeley, 127 p.

Gioda G. (1982): On the non-linear 'squeezing' effects around circular tunnels. International Journal of Numerical and Analytical Methods in Geomechanics, Vol 6, N:o 1, p 21-46.

Goodman R.E. (1980): Introduction to Rock Mechanics. John Wiley and Sons. 478 p.

Nonaka T. (1978): An elasto-visco-plastic analysis for spherically and cylindrically symmetric problems. Ingenieur-Archiv, Vol 7 p 27-33.

Nonaka T. (1981): A time-independent analysis for the final state of an elasto-visco-plastic medium with internal cavities. International Journal of Solids and Structures, Vol 17. N:o 10 pp. 961-967.

Prévost J-H., Höeg K. (1975): Effective stress-strain-strength model for soils. Journal of the Geotechnical Engineering Division, ASCE, Vol 101, N:o GT3, pp 259-278.

Saari K.H.O. (1982): Analysis of Plastic Deformation (Squeezing) of Layers Intersecting Tunnels and Shafts in Rock, PhD Thesis, University of California, Berkeley, 183 p.

Terzaghi K. (1950): Geologic aspects of soft ground tunneling. Applied Sedimentology, edited by D. Trask. John Wiley & Sons, New York, p 193-209.

Vermeer P.A. (1978): A double hardening model for sand. Geotechnique, Vol 28, N:o 4 p 413-433.

Zhu Weishen, Bai Shiwei (1981): Stress analysis in the weak surrounding rock and lining treated as multiconcentric rheological memedia. Preprints, International Symposium on Weak Rock, Tokyo, Theme 3. pp 176-180.

SWELLING ROCKS AND THE STABILITY OF TUNNELS
Roches gonflantes et stabilité des tunnels
Drückendes Gebirge und die Standfestigkeit von Tunneln

Prof. Dr. ir Tan Tjong Kie
Wen Xuan Mei Res. Assoc., Institute of Geophysics, Academia Sinica, Beijing, China

SYNOPSIS

A characterization of swelling rocks is being presented. The mechanism of the swelling is described by means of the:
1) Physico-chemical effects, 2) Rheological effects: creep, the elastic recovery and dilatancy. The importance of mineralogical and physico-chemical tests is stressed. Some rheological testing methods are suggested. An analysis is given of the routine swelling test. Constitutive equations for dilatancy behaviour in two dimensions and a calculation of the dilatant zone around a tunnel are described.

RESUME

Cette communication donne une description de roches gonflantes. Le mécanisme de gonflement est attribué: 1) aux effets physico-chimiques, 2) aux effets rhéologiques — le fluage, la contraction élastique et la dilatation. Le plus souvent il y a une combinaison des deux phénomènes. On souligne l'importance des essais minéralogiques et physico-chimiques et propose quelques tests rhéologiques. Par ailleurs, on explique la marche à suivre pour l'analyse des résultats d'essais d'expansion et les relations de comportement dilatants bi-dimensionnels, ainsi qu'un calcul de la zone dilatante autour du tunnel.

ZUSAMMENFASSUNG

Diese Arbeit gibt eine Charakterisierung drückender Gesteine. Der Mechanismus der Schwellung wird beschrieben an:
1) Physikalisch Chemischen Effekten 2) Rheologischen Effekten: das Kriechen, die Elastische Nachwirkung und Dilatanz. Gewöhnlich arbeiten diese Effekte zusammen. Die Wichtigkeit von mineralogischen und physikalisch-chemischen Versuchen ist hervorgehoben. Empfehlungen zur Durchführung rheologischer Versuche werden gegeben und eine Basis für die Analyse von Schwellversuchen wird vorgestellt. Ausgangsgleichungen für das Dilatanz-Verhalten in zwei Dimensionen und eine Berechnung der dilatanten Zone um die Tunnel werden angegeben.

Part I. SWELLING ROCK, CHARACTERIZATION AND TESTING METHODS*

The basic concept is accepted that a rock mass is a rock structure with discontinuity planes and that the deformation and strength of this mass are governed by the rheological properties of the discontinuities and its fillings (joints, interbedding clayey layers etc.) Due to the processes of rockgenesis and tectonic history further inherent properties of the rock mass are (1) presence of cracks also in the intact rocks (2) internal stresses (3) heterogeneity and (4) anisotropy. Swelling problems must be analysed on the basis of these

* A major portion of Part I has been submitted to the ISRM Commission for Swelling Rocks.

fundamental concepts.

The inwards motions of the boundary surface of tunnels and excavations, which in many cases are accompanied by the cracking of the linings and the structure of powerhouses, have been attributed by several authors to the causative swelling of rocks. Swelling is generally considered as a time dependent volume increase, which is related to mineralogical composition and physico-chemical effects. Such effects are very clear in the case of potentially swelling rocks, mudrocks, slates, marls, anhydrite rocks and sandstones. However, in general, time-dependent inwards motion is the complex result of the mutually strengthening interaction of two main factors: 1.Physico-Chemical swelling which is a (generally anisotropic) volume increase 2.Mechanical processes as the recovery of the rock-layers after (partial) stress-relief and the volume dilatancy which is a volume increase due to deviatoric stresses exceeding a certain upper yield-value. The mechanical effects are generally accompanied by void formation due to the opening of cracks and formation of new cracks and gradual disintegration.

A Physico-mathematical description of the time-dependent dilatancy in three dimensions is given recently **by the first author (Tan, Kang 1980,1983).**

A: CHARACTERIZATION OF SWELLING ROCKS

A1. Phenomenological definition

Swelling is the time dependent volume increase of a rockmass (i.e. an overall volume increase due to the swelling of the material of the intact rock, the material within the discontinuities and the separation of the discontinuity surfaces). Inwards motion of the surface boundaries of cavities, tunnels, excavations is the integrated result of the combination of volumetric strains (swelling) and shear strains; so inwards motion is only partly due to swelling.

A2. Underlying mechanisms of swelling

Swelling may be due to the following mechanisms:

A2.1 Physico-chemical effects:

1) Water is taken up by clay-minerals of the montmorillonite-chlorite-illite groups;
2) Anhydrite-gypsum transformations due to hydration Further similar transformations leading to volume increase can be included.
3) As these minerals are not homogeneously distributed, expanding kernels will be formed within the rock-mass (intact rock and materials of discontinuities), creating extra deviatoric or tensile stresses in their surroundings, which in turn will lead to fissures (voids) formation. So the physico-chemical swelling and the kernels will be accompanied by a mechanical swelling and deterioration of the structure.
4) Temperature Changes
   1. Freezing: dilatation due to water-ice transformation, which as a result of heterogeneity also may lead to fissuring; then thawing will result into a decrease in strength properties.
   2. Temperature increase;

A2.2 Rheological effects:

1) Time-dependent dilatancy
Dilatancy is the volume increase, increasing with the time, due to void formation under the action of deviatoric stresses. From triaxial and torsion experiments it is known that in the beginning reversible volume changes occur; but that this is followed by a volume increase as soon as the deviatoric stresses exceed an upper yield value (called f_3). The volume increase can be observed after the stress differences exceed a critical limit f^{**}, which for many rocks is larger than 50% of the strength σ_f (see Table 1).

2) Squeezing is an engineering description for the time dependent plasto-viscous flow (large finite strains) of rock materials subjected to deviatoric stresses far exceeding their strengths. It is a dangerous type of inwards motion. Due to the extravagant non-compatible straining and mutual sliding of the grains and splitting of grains such a squeezing process is always accompanied by void formation or swelling and increasing structural disintegration.

3) Time dependent-elastic recovery after unloading

In nature a rock mass is always subjected to initial stressing due to:
   a. Current tectonic stress field;
   b. Locked in stresses created during rock genesis and tectonic history;
   c. Gravity field (overburden, topology);
   d. Temperature field;
   e. Waterpressure.

So in general the rock mass is subjected to a 3 dimensional total stress field and absorbs a certain amount of elastic energy. Local and partial unloading of the rock mass due to excavations will be followed by and elastic recovery, which is observed as inwards motion of the walls of tunnels after excavation and upheavals of their bottoms increasing with the time. Such strain release processes (3-dimensional) will be accompanied by the opening of fissures and the formation of new cracks due to the volume expansion of heterogeneously distributed kernels, which before was not possible due to the high compressional stresses.

The time dependent process of elastic strain recovery i.e. decrease of the current strain, which is due to unloading must not be confused with creep. Creep is the increase of the strain with the time due to loading.

The strain tensor has 6 components. Usually we make a subdivision of the strain tensor into the hydrostatic strain tensor and the deviatoric strain tensor. The hydrostatic strain tensor gives the volume changes, including swelling whereas the deviatoric strain tensor describes changes in shapes; **dilatant strains can also cause volume deformations.**

A2.3 Combination of physico-chemical and mechanical effects

In practice a combination of physico-chemical and mechanical effects always occurs. Crucial is increasing void formation; thereby the specific surface which is exposed to physico-chemical action due to the changing stress field is increased. The consequent expansion of the kernels in turn create an increasing number of cracks, and further branching and coalescence of these cracks with steadily increasing exposed surfaces. Shearing stresses may break cementing bounds within clay aggregates. Thereby largely increasing surface areas are exposed; in this respect the determination of the specific surface can give valuable indications.

Infiltration of water particles in this newly created fissures is now possible due to capillary effects. Furthermore in regions above the ground water migration of water vapour may be possible due to thermo-osmosis under thermal gradients.

The following sequence is active:
  Formation and opening of fissures and voids due to dilatancy under deviatoric stresses exceeding the upper yield value and elastic recovery→migration of water particles→physico-chemical expansion→creation of heterogeneous extra intergranular stress field→decrease of rigidity parameters→larger deformation under changing stress field combination→more fissures and voids→more exposed surfaces→larger physico-chemical activity→further decrease of rigidity, more intensive **time effects** maintaining a certain or longer time→long term failure of tunnel or slope.

B: TESTING

Testing is a powerful tool to get the necessary information of the condition and the future behaviour of the rock mass before and after construction of the engineering structure. The following tests, which are in current use in China, will now be described for reference:
Three types of tests can be considered:

B1. Exploratory tests

This type of tests must be combined with geological explorations and can be carried out in for instance exploration tunnels and boreholes to get some information of rock properties in different places in the rock mass. These types of tests must be simple, quick and cheap such that large numbers of tests can be carried out by a semi-skilled field laboratory staff very conveniently and give an adequate and overall information of the condition of the rock mass.

The following tests can be considered:

1. Permeability measurements in the field and the change of permeability after elapse of swelling time;
2. Strength-testing in portable equipments; uni-axial strength testing is convenient, but direct shear testing is preferable in layered and foliated mud rock layers, intercalations and bedding planes;
3. Durabilty test by determining the ratio of the strengths in fresh samples and after long exposure to weathering; furthermore the time of complete or partial disintegration after immersion of the sample in water is an useful indicator.

On the basis of the above informations in combination with geological data and a proper design analysis we can plan two important types of tests: laboratory tests and field tests and monitoring.

B2. Laboratory testing

Laboratory experiments can be performed in order to get information of a. Mineralogical, physico-chemical and mechanical factors underlying swelling in particular and rock motion in general; b. To determine the parameters needed for design and computation. The following tests may be useful:

Mineralogical Analysis:

- X- ray determination
- Scanning electron microscopy
- Differential thermal analysis
- Microscopic analysis of grains and textures

Physico-Chemical Analysis:

- Base exchange capacities
- Specific surface determinations
- Chemical analysis, especially of cementing and coating agents, which play a role in the mechanical behaviour.

The above analysis can be carried out on different granular fractions in the natural state and after drying and grounding. The influence of P_H and ions in the infiltrating ground water must be studied, since a change in the ion occupation may alter the mechanical properties on the long term.

Mechanical Tests:

Case records show that cracking of linings occurs in the course of time, it may be weeks to many tens of years. The rheology of rocks is now understood to be very important and the first session of the IV Congress in Montreux 1979 was devoted to rheological problems.

In the case of physico-chemical swelling with gradual deterioration the rheological parameters must be considered as function of the physico-chemical swelling.

Rheological tests can be carried out on fresh and disturbed samples and on samples after this physico-chemical swelling are "completed" are very useful. The results must be compared.

The following tests can be recommended:

1. Swelling determinations during percolation of natural groundwater or water plus electrolytes at different P_H.
2. Swelling pressure determinations at constant volumes and waterpercolation
3. Permeability determinations simultaneously in the above tests.
4. Creep tests (Triaxial or Direct shear) at constant stresses, including volume dilatancy
5. Relaxation test (Triaxial or Direct shear) at constant strains (Tan 1981)
6. Shear strength tests (Triaxial or Direct Shear, Uniaxial); including volume dilatancy

Thereby shearing series of tests to study creep at constant stresses and or relaxation tests at constant strains are enlightening and most straight forward for analysis. In the most elementary case short duration shearing tests can be carried out to measure strength parameters for fresh samples and samples after subjection to long term swelling. The tests 4,5 and 6 must be carried out on samples of "intact" rocks, and the material in the discontinuities.

B3. Field Tests:

Field tests are indicated where laboratory testing on small samples can not give sufficiently representative data due to the complexity of the rock structure. The samples must have such dimensions that the parameters are adequately representative and can be used for computational analysis.

Due to design requirements the following tests are in use:

1. Swelling tests;
2. Permeability tests in boreholes in tunnels;
3. Shear testing on intercalations, bedding planes etc;
4. Deformation tests.

Laboratory and especially field tests are expensive and therefore crucial is a correct diagnosis of the complex problem, coupled to efficient and economical methods of testing. A proper choice can be made of the experiments recommended above.

C. ANALYSIS OF SWELLING TESTS

Swelling tests in the oedometer-type swelling apparatus, whereby the vertical swelling pressure is observed at various constant deformations give valuable informations shown in Fig.1.

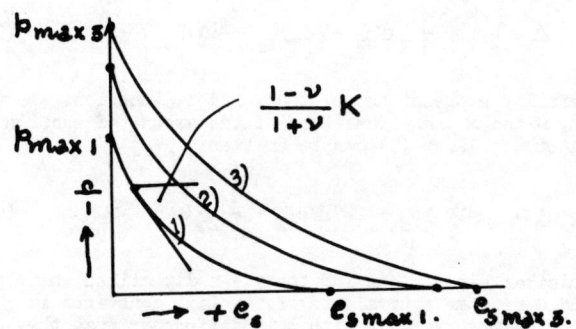

Fig.1 Swelling curves.

Curve 1) represent the swelling characteristics of "fresh" rocks
Curve 2) the same for disrupted samples
Curve 3) shows the characteristics for man-made samples; for this purpose the rock material is finely ground, dried and then moulded into a sample.

So Curve 3) shows an "upper bound" for physico-chemical swelling; $e_{s\ max3}$ (for $p = 0$) (free swelling), the maximal pressure p_{max3} (for $e_s = 0$) are good indications for the physico-chemical properties of the swelling substance. The curves can roughly be described by the following equation (compression is negative):

$$p = p_{max} + Ae^{\delta} \quad (1)$$

Whereby p_{max}, A and δ depend upon two factors:

1. degree of disruption of rock-structure due to mechanical causes of heterogeneous swelling;
2. Potentiality of swelling substance within the rock.

For example in curve 1) ⟨fresh sample⟩ the swelling pressure will not exceed p_{max1} i.e. for pressures exceeding p_{max} no swelling is possible; furthermore free swelling will not exceed $e_{s\ max1}$. Only then, when the internal structure is disrupted and the specific area of the exposed surface is increased then curve 1) merges into curve 2) and higher values of p_{max2} and $e_{s\ max2}$ can be found.

So swelling is a combined action of physico-chemical and mechanical processes.

We will now present an attempt to describe the swelling process.
The swelling strain $e_s = e_{sx} + e_{sy} + e_{sz}$ will now be written:

$$e_s = 3\alpha\Theta \; ; \quad \Theta = \Theta_o \exp(\varphi/RT) \quad (2)$$

Whereby α = parameter dependent on the actual specific area of the exposed surface involved in the swelling process. It is a measure for mechanical factors as crack formation due to tensile stresses and dilatancy.
Θ = parameter which accounts for the swelling activity of the substance as clay minerals; φ = activation energy, R = gas constant, T = absolute temperature. Then the stress strain relationships can be written in the following incremental form;

$$\Delta e_{sx} = -\frac{1}{E}\left\{\Delta\sigma_{sx} - \nu\Delta\sigma_{sy} - \nu\Delta\sigma_{sz}\right\} + \alpha\Theta \quad (3)$$

and similar expressions for the y and z direction, whereby: E, ν and α, are functions of the amount of suction of water e_s. From (3) can be written:

$$\Delta\sigma_{sx} = \lambda\Delta e_s + 2G\Delta e_{sx} - \frac{\alpha E}{1-2\nu}\Theta \quad (4)$$

and similar expressions for the y - z directions whereby λ and G are the equivalents of the Lamé constants in elasticity theory, only with this difference that they depend on the swelling e_s.

For the routine swelling tests we get. from (4)

$$\sigma_x = p = \frac{1-\nu}{1+\nu}Ke_{sx} - \alpha K\Theta \quad (5)$$

whereby

$$K = E/(1 - 2\nu)$$

For simplicity we can take ν a constant then the variation of K with e_s can be found from the family of swelling curves:
Thus:

$$\frac{1-\nu}{1+\nu}K = -\frac{dp}{de_{sx}} \quad (Fig.1); \quad p_{max} = -\alpha K\Theta$$

as it can be found from unconfined expansion tests in the swelling apparatus.

The magnitude of $\alpha\Theta$ depends on the actual distribution of the swelling within the mass, which changes with the time as a result of diffusion and suction of watermolecules. The governing equations apparently are of the non-linear diffusion type.

Part II STABILITY PROBLEMS OF TUNNELS IN POTENTIALLY SWELLING ROCKS:

IIA COMPLICATED CASE HISTORIES

The inwards motion of swelling rocks in tunnels which increases with the time may be a serious problem on which the attention of many engineers is focused. The bottom upheaval especially in railway tunnels may assume inacceptable proportions. The result is that the traffic must be suspended and the tunnel repaired, which is a considerable economic loss in tunnels with a length of a few km; such cases have been encountered in mountaineous regions, where the overburden may amount to 500 m and higher and horizontal tectonic stresses are considerable. The rocks often are folded layered mudstone and silty sandstone with a high content of potential clay minerals as montmorillonites, illites and chlorites. The fissures after excavation may be open but they are generally filled with a muddy substance with high swelling potentials.

Usually the side wall and roof lining is constructed from concrete blocks of ca 50 cm thickness cemented together; but no bottom slab is provided. This type of lining seems to be sufficient in more competent rocks but has given rise to many troubles in potentially swelling rocks. In one of the tunnels the bottom upheaval amounted to 290 mm, three years after the end of its completion. Then it was brought up to profile and a bottom arch was constructed. Five years hereafter the bottom upheaval showed an additional increase of 170 mm; so the total upheaval reached 460 mm.

In all those problems the following common factors may be summarised.
1) rocks are of the potentially swelling type, time effects play an important role.
2) the drainage of underground water was unefficient.
3) fissured are opened and new cracks are formed due to dilatant Volume-increase and tensile stresses.
4) the lining was not efficient to help strengthening the surrounding rock; in many cases even bottom slabs were not provided.
5) rock properties: mineralogy, physico-chemical, rheological properties were not studied adequately.
6) residual tectonic stresses play an important role.

The deterioration of the structure due the formation, opening of fissures and the suction of water, which result into a decrease of the overall rigidity and strength of the surrounding rock mass is a basic factor

in the stability of tunnels in potentially swelling rocks. Hence our remedy must be
1) to reduce this opening of fissures
2) to reduce or better prevent water from moving into those capillaries.

IIB. SIMPLIFIED CONSTITUTIVE EQUATIONS

In order to take efficient measures we first must have an idea of the thickness of the deteriorated annulus around the tunnel. For this purpose we will make use of the constitutive equations for the deformation, creep and dilatancy and swelling of rocks. For plane strain a great simplification can be made by assuming that the intermediate principal stress σ_2 plays no significant role in anelastic deformations. For the present we will not consider time-effects of creep and dilatancy and study only instantaneouces deformations.

The equations for the Dilatancy given in the equations (1),(2),(3),(4),(5) (Tan 1983) can be simplified for Finite Element Programming of plane strain problems when the influence of the intermediate principal stress σ_2 can be neglected for dilatant deformations. Then we get:

$$p = I_1 = (\sigma_x + \sigma_y)/2 = (\sigma_1 + \sigma_3)/2$$

$$\sigma_{oct} = \sqrt{J_2} = \left[\left(\frac{\sigma_x - \sigma_y}{2}\right)^2 + \tau_{xy}^2\right]^{\frac{1}{2}} = \frac{\sigma_1 - \sigma_3}{2}$$

$$e_d = e_{dx} + e_{dy}$$

$$e_{dx} = \left(\frac{\sigma_1 - \sigma_3}{2f_3}\right)^n \left[D^* + C\frac{\sigma_x - \sigma_y}{\sigma_1 - \sigma_3}\right] \quad (6)$$

$$e_{dy} = \left(\frac{\sigma_1 - \sigma_3}{2f_3}\right)^n \left[D^* + C\frac{\sigma_y - \sigma_x}{\sigma_1 - \sigma_3}\right] \quad (7)$$

$$\gamma_{xy} = 2e_{xy} = 4C\left(\frac{\sigma_1 - \sigma_3}{2f_3}\right)^n \frac{\tau_{xy}}{\sigma_1 - \sigma_3} \quad (8)$$

$$e_d = 2D^*\left(\frac{\sigma_1 - \sigma_3}{2f_3}\right)^n \quad (9)$$

IIC. FINITE ELEMENT ANALYSIS

In order to study the dilatant zone, stress-and displacement fields around a tunnel with B=6m, H=9m, we have carried out some Finite Element computations for the instanteous elastic and dilatant state, taking:
$D^* = 10^{-4}$; $C = 0.6 D^*$; $n = 5$; $f_3 = 4-6-8$ MPa; $E = 10^4$ MPa, $\nu = 0.4$. Two examples are studied: Horizontal Tectonic stress $\sigma_T = 0$ and $\sigma_T = 10$ Mpa. The overburden is 500 m. The results which are due to the second author are shown in Fig.2A and 2B respectively.

The following conclusions can be made:
1) Increase in overburden increases the dilatant zone and inwards motion of the walls; when the tectonic stress is zero or constant.
2) Increase in tectonic horizontal stress σ_T increase the dilatant zone and inwards motion of the roof and bottom for constant overburden. σ_T creates large stress concentrations in the roof and bottom parts of the tunnel.
3) Increase in yield value f_3 will result in decrease of the dilatant zone.
4) The computed inwards motions which are in the order of many cm represent the instantaneous part of the displacement directly after excavation. The time-dependent parts of the displacements due to creep and dilatancy will increase with the time, after the instantaneous stage, are unconsidered above.
5) Water will move into the capillaries and absorbed to the potentially swelling kernels. Then the displacement will increase considerably.

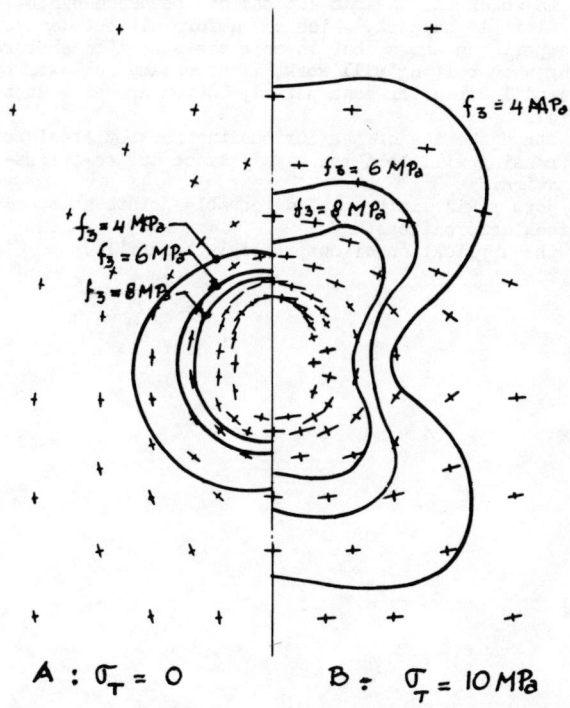

Fig.2 Dilatant zone, around tunnels
A. horizontal tectonic stress $\sigma_T = 0$;
B. horizontal tectonic stress $\sigma_T = 10$ Mpa;

Hence in mountaineous regions where horizontal tectonic stresses are active, the bottom upheavals have been found to be very troublesome. It can go on despite a lapse of more than 20 years.

CONCLUSION

In this report some basic factors influencing the complicated process of dilatancy and physico-chemical swelling is discussed. As in every branch of Rock mechanics a correct judgment based on field exploration, experience, experiments and numerical estimation is necessary.

Due to the high stress levels a large amount of elastic energy is stored in the rocks. This energy accumulated in an annular zone with an average dimension of three time the dimension of the tunnel,will be liberated after excavation; this process is known in Rheology as the elastic recovery, which is increasing with the time.

Hence it is not wise to place a rigid lining or support system in direct contact with the rock immediately after excavation as it will be subjected to considerable forces and will be liable to excessive deformation or failure
On the other hand any cracking in the rock will occur due to dilatancy, swelling, and lateral creep deformation due to concentrated normal stresses acting tangentially to the tunnel periphery. On the long duration this deterioratic processes may lead to tunnel instability; a striking example is the excessive upheaval of the bottom of tunnels without bottom arch, which are still increasing despite a period of 20 years. Therefore the lining must be so designed as to accommodate rock displacements in the order of 10cm and even more for potentially swelling rocks. There are several alternative ways in which the inwards motions can be accommodated.

1) flexible support, which can deform without damage; our experience shows that in some cases a 15cm shotcrete lining with bolting **will work**, if necessary followed by a second layer after some period; bottom arches must be applied.
2) strengthening of the surrounding rock by steel rods; prestressing will lost its significance due to stress-relaxation.
3) more rigid lining with deformable joints as to allow for rock deformations.
4) the application of compressible back fill.
5) an effective drainage system is essential, and sealing by grouting can be applied, wherever possible.
Of course combinations of the above measures can be adopted.

On the whole it is a matter of economic efficiency. It may be cheaper to construct an efficient but more expensive lining than to repair many times and to stop the traffic.

REFERENCES

Tan Tjong Kie and Kang Wen Fa (1980).
  Locked in stresses, creep and dilatancy of rocks, and constitutive equations, Rock Mechanics 13, pp 5-22

Tan Tjong Kie and Kang Wen Fa (1983).
  Time dependent dilatancy prior to rockfailure and earthquakes. Proc. this Congress.

Tan Tjong Kie and Li Ke Ri (1981).
  Relaxation and creep properties of thin interbedded clayey seams and their fundamental role in the stability of dams. Proc. Int. Symp. Weak Rock, Tokyo pp 369-374.

PRESSURE TESTS IN ROCK CHAMBERS
Essai en galeries rocheuses sous pression interne
Druckkammerversuche in Felskavernen

Keshan Zhu
Professor in Civil Engineering, Chongqing Inst. of Architecture and Eng. Chongqing,
Sichuan, China

Xiesheng Lie
Vice-President and Deputy Chief Engineer, Chengdu Design Inst. of Hydroelectric Power,
Chengdu, Sichuan, China

SYNOPSIS

The successful performance of a test on a rock cavern lined with thin steel plate under internal water pressure revealed the complex nature of the interaction between linings and the surrounding rock. Since then, quite a number of pressure tests have been carried out in different geological formations in China. This paper presents some critical reviews on the findings from three tests both in lined and unlined rock chambers in grano-diorite and diorite and concludes that there would be little difference in the lining capacity and deformation behaviours between reinforced concrete linings and shotcreting if good care is taken to avoid cold joints.

RESUME

Des essais en galeries rocheuses garnies de plaques d'acier ont montré la complexité de l'interaction entre la garniture et les roches à l'entour. Depuis lors, de nombreux tests en pression interne sous différentes formations géologiques ont été faits en Chine. On présente les résultats obtenus pour trois essais en chambres rocheuses dans une granodiorite avec ou sans garniture, et on en tire la conclusion qu'il y a peu de différence de résistance à la fissuration et le comportement de déformation entre la garniture en béton armé et celle du béton projeté, pourvu qu'on prenne garde à éviter les joints.

ZUSAMMENFASSUNG

Es gelang uns, beim Versuch an ausgekleideten Felskavernen, deren innere Schicht aus Stahlblech besteht, die komplizierte gegenseitige Wirkung zwischen Ausbau und Felskörper festzustellen. Danach wurde in China eine Reihe von Innendruckversuchen unter verschiedenen geologischen Bedingungen durchgeführt. In diesem Artikel die Ergebnisse der Innendruckversuche an zwei Kavernen, eine mit, die andere ohne Ausbau in grano-dioritischen und dioritischen Felskörpern bewertet. Durch geeignete Maßnahmen können nahezu gleiche Risslasten und Verformungseigenschaften bei Verkleidung aus Stahlbeton und aus Spritzbeton erzielt werden.

1. MOTIVATION

In the early sixties, in order to evaluate the interaction of steel penstocks with the surrounding rock mass, a test chamber was excavated in broken basalt with an overburden of only about 45m thick. The general size of the pieces was 5-20 cm with occasional ones over 40-50 cm. There were four main joint sets exposed on the excavated surface, all of them dipped steeply and met the chamber axis at rather small angles. The deformation modulus of the rock mass by the plate-bearing test was 1.01-1.59 GPa. The in-situ rock stresses were 0.59-0.89 MPa vertically and 0.86-1.28 MPa horizontally.

The excavated opening was roughly 3.5 m round and 25 m long. The lining consisted of a 12 mm thick welded steel pipe, 2.3 m in diameter and 10.14 m in length, and two thicker steel conic shells capping the pipe. There were privided several stiffeners along the pipe. After backfilling with concrete and all possible gaps grouted, a series of pressure tests were run. The maximum pressure attained was 13,63 MPa due to insufficient pumping capacity. Still, it was possible to maintain at this for four hours. Then a constant pressure of 11.77 MPa was maintained for 32 days without significant creeping. In all the loading and unloading cycles, The penstock behaved well and even under the maximum pressure, only portion of the thin wall yielded as evidenced by strain readings as well as several local depressions found near the pipe bottom after dewatering. Further investigations showed that this was probably due to the loose backfilling there.

Since then, quite a number of rock chamber tests have been carried out in a variety of geological formations. Particular interest has been concentrated on reinforced concrete linings, shotcreting, as well as unlined tunnels. This paper presents some critical reviews on the findings from such three chamber tests in more or less similar geological formations of grano-diolite and diorite where severe deformation of the rock and cracking of linings occurred in an underground hydro-electric powerhouse due to excavation as described elsewhere (Lee, 1979, 1980)

2. GEOLOGICAL CONDITIONS OF CHAMBERS

The underground hydroelectric power plant is anout 800 m north of the regional fault and about 250 m south of a small techtonic fault. The shotcreted rock chamber No. 1 was located in the connection gallery between the transportation drift and tailrace gallery, about 200 m form the river and 50 m downstream from the plant. The mountain slope rises at about 40 degrees to reach the range height of some 600 m. The rock mass is mainly of the relatively fresh granodiorite intrusion, with frequent fissures and small faults. See Fig. 1.

Test chamber No. 2 was unlined. It was a part of an old exploration gallery of 70 m long, the last 10 m was used for the test. The rock mass exposed in the gallery was coarsegrained blocky granodiorite with very few and no faults. See Fig. 2.

Test chamber No. 3 was lined with single and double layer reinforced concrete. It was a part of extension of another exploration gallery. The rock mass exposed was mainly of fresh fine-grained diorites and diorite porphyry. There were usually through-going cracks along the contact faces. The joints were 0.1-0.3 cm wide at an average spacing of 0.5 m and only half of them were filled. There was a small fault, 250/65 to 235/80. The shear zone was 2-10 cm wide and filled with borken rock. See Fig. 3.

The deformation modulus of granodiorite was 0.98 GPa and that of diorite was 0.98-1.37 GPa. The compression strength of the former was 157 MPa and that of the latter 167 MPa. The in-sith rock stress near the corner surface of a heavy rock pillar was about 17 MPa while that on the side wall surface was 7.6 MPa vertically and 3.2 horizontally.

3. TEST RESULTS

The open end of the chamber was encased in rock with bulk-headed concrete and grouted to prevent leakage. As a rule, diameter extensometers intersecting at 45 degrees were installed. A few piezometers and strain gages were installed near the rock surface. During all the loading and unloading cycles, gage readings were carefully taken.

3.1. Shotcreted chamber No. 1

The excavated profile was quite irregular as can be seen from Fig. 1c though the cross section was roughly circular. The diameter varied from 4 m at the open end to 2.5 m and the test length was 14 m. A rigidity coefficient of 3-4 was assigned to it.

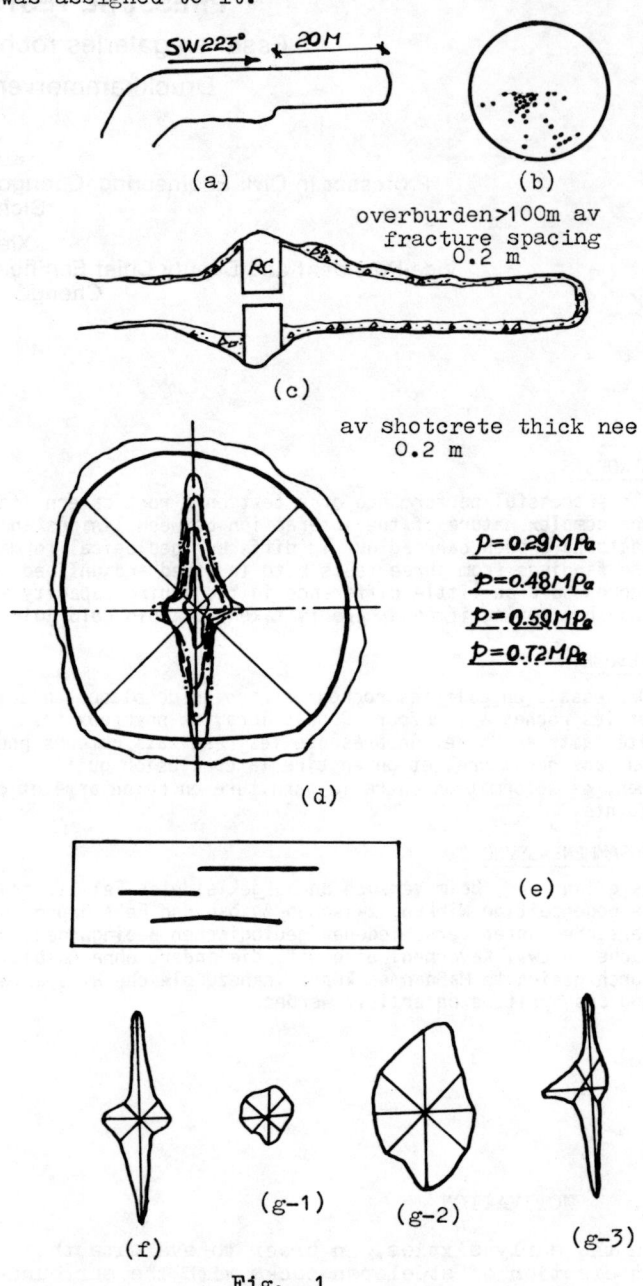

Fig. 1

Fig. 1 Shotcreted chamber No.1
(a) plan; (b) stereographic plot (lower hemisphere) of normals to fracture planes; (c) profile; (d) cross section and measured deformation; (e) crack pattern; (f) measured deformation averaged at p - 9.8 KPa; (g) computed deformation at p - 9.8 KPa: g1-elastic homogemeous and isotropic rock and lining; g2 - transversely isotropic rock; g3 - nonhomogeneous transversely isotropic rock and lining.

Test results are shown in Fig. 1d, e. The cracking load appeared to be around 0.49 MPa and the maximum pressure attained was 0.75 MPa. The deformation pattern could not be explained alone by the rock joint system. Analysis suggested tentatively that there should be some cold joints in the shotcrete.

3.2 Unlined chamber No. 2

The cross section was roughly of a square with side length of 2 m. The test length was 10 m and a rigidity coefficient of 7-8 was assigned. The maximum pressure attained was 1.96 MPa. Except loadings in small increments, there were 18 loading and unloading cycles from 0-1.96-0 MPa and then maintained at the maximum pressure for three days. Inspection of the unlined chamber after dewatering showed that only small rockfalls occurred and the filling materials washed out leaving a thin layer on the chamber bottom.

overburden 55-70m average fracture spacing 2.5 m

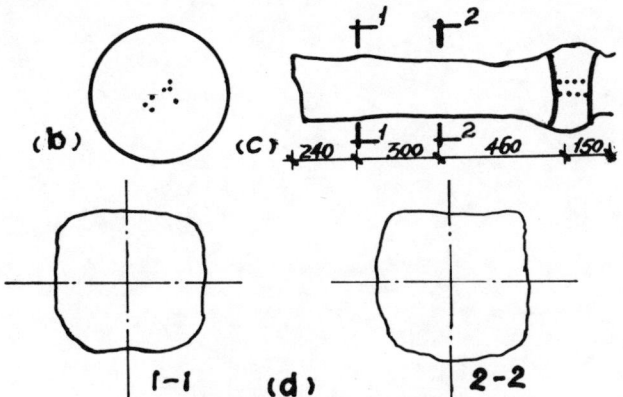

Fig. 2 Unlined chamber No. 2
(b), (c), (d) Ibid

3.3 R.C. lined chamber No. 3

The lined circular chamber had an inside diameter of 2.5 m. The test length was 15 m and the rigidity coefficient was 4. A layer of 5-12 mm round rebars per m was extended along the whole length, but an outer layer of the same extended only for the first 8 m. The maximum pressure attained 1.18 MPa and the cracking load appeared to be 0.58 MPa for the single layered section and 0.88 MPa for the double layered section.

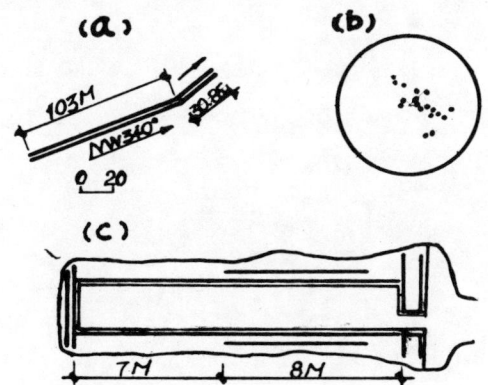

lining aver thickness 0.5-0.6 m
overburden 110-13- m av. fracture spacing 0.5m
single double-layered rebars

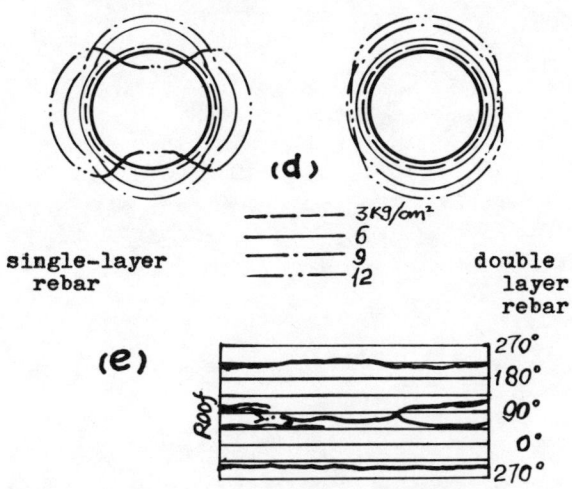

Fig. 3 R.C. lined chamber No. 3
(a), (b), (c), (d), (e) Ibid

4. CONCLUSION

By comparison test results of No. 1 and No. 3 it may be concluded that there would be little difference in the cracking capcacity and deformation behavior if good care is taken to avoid cold joints. The low permeability of shotcrete is to be noted. In all cases, the maximum pressure attained was limited not by the overall strength of the rock but the pumping capacity.

Acknowledgements

The authors are grateful to Messrs. Du Yunzhong, Duan Lozhai, Wang Chenlian of the Chengdu Design Inst of Hydroelectric Power and Associate Prof. Bai Shaolian, Dr. Zhu Jialin, Messrs Wu Yinsi and Zhu Mei for their assistance in the preparation of the manuscript.

References:

Li Xiesen and others (1979). Cracking in an underground hydroelectric powerhouse due to excavation. A Technical report, Chongqing Inst of Arch and Eng.
Li Xiesen and others (1980). Analysis and Discussion on stress and deformation of the surrounding rock in the underground hydroelectric power plant of the Fisherman's Creek I, Chinese Journal of Geotechnical Engineering, V2, N2.

SOME ROCK MECHANICS PROBLEMS RELATED TO A LARGE UNDERGROUND POWER STATION IN A REGION WITH HIGH ROCK STRESS

Quelques problèmes de mécanique des roches relatifs à une centrale souterraine importante dans une région à fort niveau de contrainte des roches

Einige felsmechanische Probleme beim Bau einer Kraftwerkskaverne im Gebirge mit hohen Spannungen

Shiwei Bai
Assistant Professor

Weishen Zhu
Associate Professor

Kejun Wang
Assistant Professor
Institute of Rock and Soil Mechanics, Academia Sinica, Wuhan, China

SYNOPSIS

The phenomenon of disk-shaped fractures of cores and of rock burst are described. This is followed by a presentation of stress measurements carried out in the river bed and in galleries. Fracture mechanisms of rock and stress distributions surrounding the power house are analysed by the Finite Element Method. Furthermore there will be a report of the results of tests gained from rock samples tested in a servo-controlled press using the acoustic emission registration technique.

RESUME

On décrit respectivement les phénomènes de carottes disques et des éclatements rocheux qui ont lieu dans un site de barrage. Ensuite on présente une série de résultats de mesures de contrainte obtenus dans le lit et les galeries. Le mécanisme de la fracture de roches et la distribution de contrainte autour de l'usine souterraine sont analysés par la méthode d'élément finis. On décrit les résultats d'essais qui ont été réalisés sur certains échantillons mis en pression au moyen d'une machine rigide à l'aide de servo-asservissement et étudiés grâce à la technique d'émission acoustique.

ZUSAMMENFASSUNG

Das Auftreten scheibenförmigen Zerbrechens von Bohrkernen und Gebirgsschlägen wird beschrieben. Danach werden Ergebnisse von Spannungsmessungen, die im Flußbett und in Stollen vorgenommen wurden, präsentiert. Bruchmechanismen des Gebirges und Spannungsverteilungen um die Kraftwerkskaverne werden mit Hilfe der Finite-Element-Methode analysiert. Des weiteren werden Versuchsergebnisse mitgeteilt, die an Gesteinsproben mit einer servo-kontrollierten Prüfpresse bei Anwendung der AE-Aufzeichnungstechnik gewonnen wurden.

The Ertan Hydropower Station will be located in the remote mountain and gorge region of the lower reaches of the Yalong River in south-west China. Along the banks of the dam site are high steep mountains (about 400-500m) with an average slope of 30 to 40 degrees. An arch dam with a height of 240m will be built in this site, while a large underground power house will probably be arranged on the left bank near the dam abutment.

1. THE PHENOMENA OF THE ROCK FRACTURE OBSERVED AT THE EXPLORATON STAGE

The dam site is located on the Gonghe Fault Block. This block is situated in the west side of the middle segment of Sichuan-Yunnan Structure Band with a south-north strike. The structure fracture inside the Fault Block is slight and the rock mass is hard and intact. The rock strata mainly consist of basalt(β) and deuterogenously intrusive syenite (ξ).
In order to make engineering geology conditions clear, nearlz 100 prospecting boreholes with a total length over 10,000m ere drilled. As a result, core ruptures presenting disk forms were found in many boreholes. The picture (Fig.1) shows the core failure appearance. These rock disks have even thickness with average h of $d/4 < h < d/3$ (d is diameter of a core). Fracture planes are fresh and rough. The top surfaces of the disks are concave and the bottom convex. The thickness is generally proportional to the diameter. It is unexpected, that there are rock disk phenomena in 40 boreholes among 48 ones located in the river bed. Rupture phenomena took place zonally and alternatively at various depths in every borehole. The distribution probability of 332 fracture bands of 54 boreholes along the altitude of the river bed is shown in Fig.2(Shi 1979). From the figure, it can be seen that the highest fracture probability appeared at the altitudes between 930-975m. This band is 20-40m just underneath the surface of rock base. According to the results of a series of experiments and analyses, it can be defined that the stress concetration in bottom of core and partial unloasing during the drilling process in high level stress region are the main reason of

Fig.1

disk phenomena. Also from Fig.2, it can be seen clearly that greatly dense fracture bands are gathered in the stress concentration region of valley. An initial stress value of fracture rock disks was obtained successfully in later stress relief measurements of deep boreholes in the river bed. The max. and min. principal stresses were 650 and 291 kg/cm² respectively (Bai 1982). Rock burst occured many times in prospect and testing galleries on the left bank as well. Among these, the axis of branched gallery No. 3 is approximately perpendicular to the direction of max. principal stress of the stress field. It was originally planned to cut a few rock blocks with a plane of 40 X 40 cm² for shear tests. However, rock burst was occurring when the cutting depths of these blocks reached about 10cm. There were two forms of rock burst. One is scaly rock flakes with a thickness of 1-8mm bursting on the top of a block; the other is the block rooted up. The rents were fresh, fractures accompanied by noises. The edges of the flakes were thicker, the center thinner, and concave facing down. Rock disk phenomena occured many times also at the stress concentration regions, when stress relief measurements were performed in the river bed and prospective galleries of the dam area.

Fig.2

2. The results of stress measurements

To make a thorough investigation of the initial stress around the dam, a lot of in situ measurements has been done by means of overcoring method, including stress measurement of three dimensions and two dimensions methods at 15 points (Fig.3) as well as the measurements in the two deep perpendicular boreholes at the river bed with depths 59.4m and 53m respectively (Fig.4).

Point	σ_1 kg/cm²
3	254
4	188
5	195
6	222
10	256
7	96
8	384
9	296
12	304
11	303
1	245
2	169
13	314
14	210
15	213

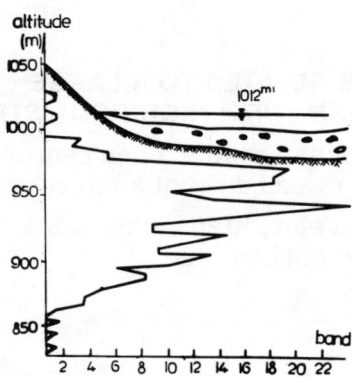

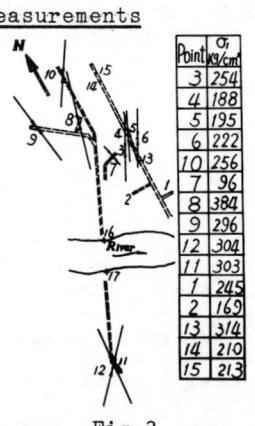

Fig.3

Fig.3 shows that the direction of max. principal stress is NE 11°-46° at most of the points except points 7 and 9, where the orientation of stress depends on local topography and change of properties of rock mass. Points 16 and 17 show the directions of max. principal stresses to be NE 12°-50° at the boreholes with a depth more than 30m. The statistical average of the above mentioned results is about NE 30°, which is just perpendicular to the direction of river with strike NE 60°. Therefore, it can be seen that the stress direction in rock mass near the slope is controlled to a great extent by topography. The stress magnitude is also shown in Fig.3. The max. principal stress in undisturbed syenite is about 200kg/cm², but about 300kg/cm² in basalt. The main cause of the difference is due to the different rock mechanical properties at various points. The rock mass in different geological regions has different conditions of accumulating stress. In general, the rock with higher elastic modulus would have higher stress. The max. principal stress intersects horizontal plane at small angle, which shows that there is horizontal tectonic stress and its effect is much more than that of gravity.

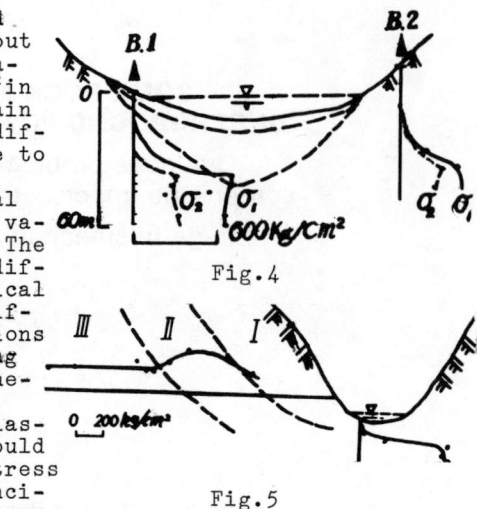

Fig.4

Fig.5

There is serious stress concentration in the bottom of river bed. The max. principal stress is over 650kg/cm². It is obviously shown in Fig.5 that stress distribution in the part of bank slopes may be divided into three areas: I, stress relaxation area, II, stress concentration area, III, stress stable area.
The plane linear elastic finite element analysis has been completed in order to form the whole outline of stress field. The calculation model is shown as Fig.6. It is a well-known fact that horizontal stress increases linearly with the depth though there has been no mature theory on the stress distribution law in surface layer of crust. Therefore effect of gravity and laterally applied triangular load increasing with depth were taken into account in the finite element analysis.
Let $\sigma x = KrH$, where K is coefficient of lateral pressure for determining the grade of triangular load.
Table 1 shows the comparison of the calculated stresses with the corresponding measured stresses. Results of inverse calculation have good agreement with the measurement results in all four different positions. The whole outline of the stress field of dam site is delineated clearly by the analysis.

Fig.6

Knot	σ_1	σ_2	β	Point	σ_1	σ_2	β
28	135		35	6	96		31°
21	198		25	5	195		57°
93	262	30	25	1	245	29	31°
157	659		7	16	659		

Tab.1

3. Prediction of stability of underground power station during excavation

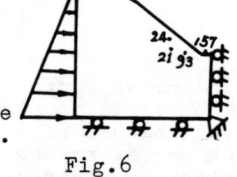

3.1. FEM analysis

The main power house A, which is planned to have a size of about 63m.(h) X 27.5m.(w) X 240m.(l), and the main transformer chamber B and the pressure regulator chamber C are arranged as in Fig.7. Taking account of safetz of the buildings in such a high stress region, at first, two dimentional FEM analyses have been done; and three dimentional nonlinear analysis will be made later. It can be considered that the results of elastic analyses are accurate to a certain degree because of higher hardness and fairly ideal elasticity of confining rock mass and high stres-

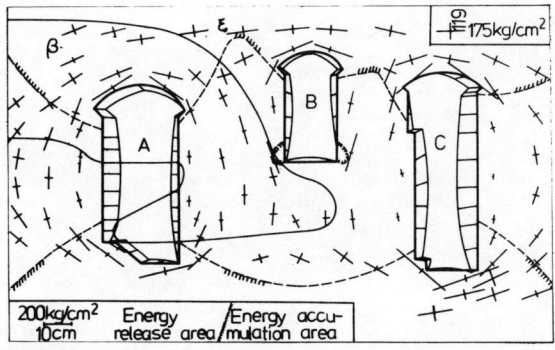

Fig.7

ses making the joints arounding caverns close tightly. The values of most redistributive stresses forming a stress-reduction region whthin the areas surrounded by three caverns are lower than the initial stresses for the horizontal stresses are higher (up to 175 kg/cm²) than vertical one. All the side walls of three houses have lager inward deformation, especially outside walls of two side ones. As shown in Fig.7, the side walls of three chambers have a large deformation inwards them. The max. displacement in the middle of the side walls of the two side chambers might reach up to 6.7cm and 7.6cm, when introducing $E=20 \times 10^4 kg/cm^2$ for syenite(ξ) and $E=16 \times 10^4 kg/cm^2$ for basalt(β), while the max, periphic stress with value up to 400-500kg/cm² takes place in the vault and in the bottom parts of surrounding rock; these values of displacements are expected to be larger if the effects of joints are taken into account and stress concentration are most serious at the top and bottom parts of the regulator chamber owing to its larger ratio of height to width. It also can be seen that the elastic strain energy concentration regions are basically located also in these parts of rock; and the max. value takes place in vault and bottom of regulator room while the magnitude is over 6-7 times as large as initial one. Therefore the side walls should be protected from overdeformation and vaults and bottom parts should be protected from shear failure and rock burst. Because of increasing stress accumulation due to gradual excavation from top to bottom and because of prior supporting of the vaults in general, the probability of failure occuring in vault is expected to be lower than that occurring in bottom. In summary, more attention should be paid to the safety of bottoms.

FEM analysis was also made to rock burst phenomena occurring in cutting testing blocks in branch gallery 3 mentioned above. The results indicated that sudden rock failure had resulted from high lateral stresses with a tension stress more than 25 kg/cm² at the root of the blocks.

3.2. Research on mechanics properties of rock

A series of laboratory tests have been done to determine why rock there behave as such a marked brittle failure and almost every syenite sample made intensive burst sounds with flying out broken piece when failure occurred,in which the tests performed on a rigid machine were monitored by AE and the results are shown in Fig. 8 and 9, It can be seen that the properties of the curve of the dry sample are quite different from those of the sample saturated with water. The former has a trength about one third larger than the latter, the latter presents yield when the pressure reaches 60% of the former (beyond point p in Fig. 9). It can be seen, from these curves, that suddenness character of AE takes place for the former while the failure of latter occurs resulting from continual small fractures and AE count rates are far below the former's, which shows that rock becomes greatly softened after saturation with water to reduce suddenness of failure.

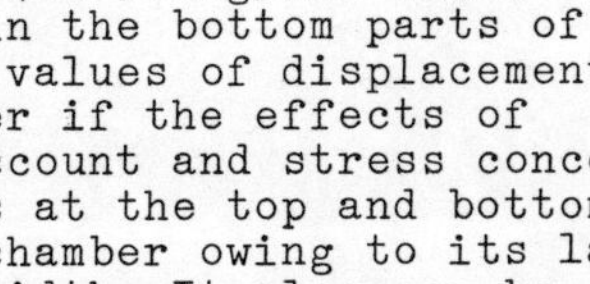

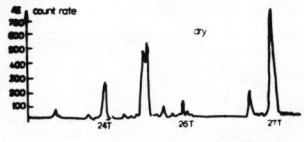

Fig.8a

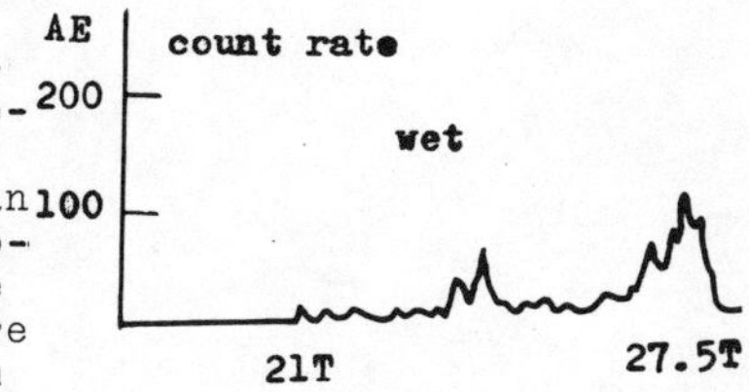

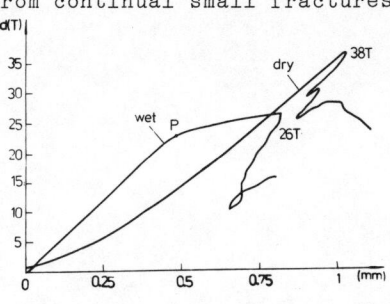

Fig.8b

Fig.9

Conclusions

1. Phenomena of disk cores and rock bursts take place mainly at stress concentration regions; the evident brittleness of syesnite helps to form this kind of failure.

2. There exists rather high earth stress and its distribution is intensively affected by landforms.

3. The areas with great accumulation of strain energy are situated near the parts of the vaults and the bottoms of underground buildings; more attention should be focused on rock fracture at bottoms during excavation.

4. Syenite becomes evicently softened after saturation with water, with the help of which rock burst phenomena would be reduced.

Acknowledgements

Many data in situ were offered by Institute of Power Survey and Design,Chengdu. Some of tests were made by Mr. Nie Shifeng and Miss Lin Zhuoying and assistance in writing this paper was made by Mr. Zhu Zuoduo and Miss Shi Baozhen and others.

References

Bai Shiwei et al. (1982). Stress measurement of rock mass in situ and the law of stress distribution in a large dam site: 23rd U.S. Symposium on Rock Mechanics, Berkeley.

Shi Jinliang. (1979). The brittle fracture of rocks in region of high earth stress: Internal report of Institute of Power Survey and Design, Chengdu.

NEW SUPPORTING METHODS AND MATERIALS FOR THE LINING OF TUNNELS AND SHAFTS IN MINING

Méthodes nouvelles et matériaux nouveaux pour le soutènement définitif des tunnels et des puits de mine

Neue Verfahren und Sicherungsmaterialien zur Auskleidung von Tunneln und Schächten im Bergbau

Bernhard Maidl
Prof. and Chief of Department for Construction Methods and Construction Management,
Ruhr-Universität Bochum, Bochum, F.R.G.

SYNOPSIS

The methods of support for the construction of shafts for the excavation and maintenance of underground mines are undergoing changes. These developments are encouraged in particular by the experience gained with new materials for tunnel support. Firstly, the paper deals with the general requirements and conditions, such as working zone, rock behaviour, excavation influences etc., and then describes the requirements of supports depending on the excavation methods. With respect to this, supporting methods and materials will be described and experiences gained in establishing underground construction will be reported.

RESUME

Les méthodes de soutènement définitif pour la construction des puits de mine évoluent en cours de réalisation des travaux. En particulier, les résultats obtenus grâce aux nouveaux matériaux de soutènement favorisent cette évolution. Cette communication commence par examiner les exigences des travaux miniers et les conditions rencontrées, telles que: zones d'exploitation, comportement des roches, influences d'excavation, etc., et ensuite discute les servitudes du soutènement définitif en fonction des méthodes d'excavation. C'est sur cette base, ainsi que sur celle de l'expérience obtenue lors de l'exécution des travaux au fond, qu'on décrit les méthodes et les matériaux de soutènement.

ZUSAMMENFASSUNG

Die Sicherungsverfahren für den Ausbau von Schächten für die Erschließung und Unterhaltung von untertägigen Bergwerken unterliegen einer Wandlung. Insbesondere die Erfahrungen mit neuen Sicherungsmaterialien aus dem Tunnelbau geben dieser Entwicklung wesentliche Impulse. Der Beitrag befaßt sich zunächst mit den äußeren Anforderungen und Bedingungen, wie Nutzungsort, Gebirgsverhalten, Abbaueinflüsse etc. und geht dann auf die Anforderungen der Sicherung in Abhängigkeit von den Abbauverfahren ein. Auf dieser Grundlage werden Sicherungsverfahren und Sicherungsmaterialien behandelt und über Erfahrungen bei der Ausführung von Untertagebauten wird berichtet.

1. DRIVING OF TUNNELS IN GERMAN COAL MINES

The conventional supporting of mining excavations consists of a retaining structure, for example, arches with mesh reinforcements and rock backfill and it is so constructed that at first it does not come into contact with the surrounding rock. Therefore, it is not considered as a load carrying structure. The support gets loaded only when deformation occur in rock. Due to this deformations the existing pressure in rock is released. Hence the surrounding rock is not capable of handling the loads as a combined system together with the retaining structure. So the support has to be designed to carry the substantial amount of the rock pressure.

As the excavation proceeds to greater depths the steel contents of the tunnel lining increases steadily corresponding to each metre depth of excavation. The supporting measures are, however, often brought at a comparatively later stage when loosening and deformations have already commenced. This happening can be prevented only by expensive measures such as deeply penetrated rock grouting.

In most of the cases tunnels which are constructed using conventional methods are to be relined after a lapse of time for safety and to maintain operation ability. Therefore, the costs of construction (including these repairs) becomes more.

The fundamental ideas and principles of the New Austrian Tunnelling Method (NATM) are being applied on a large scale for the first time in German coal mining as a part of a research programme. Albers, Jagsch (1981).

The main principle of this construction method is to exploit as far as possible the existing carrying capacity of the rock. Immediately after excavation a supporting layer is constructed thereby the pressure of the rock is not disturbed so that the bearing capacity of the rock can be fully utilized.

The driven exit mine air section is located on the shaft plant "NORDSTERN" at a depth of about 1,100 m in the 12th bottom. The cross section consists of clay slates of relatively low strength with thin coal layers in between.

Within the project area four different seams for getting the coal were established from 1960 to 1977 for this work. The lowest getting zones were lying about 120 m above the driven section. In some cases, the ribs were lying massed one upon the other. High rock pressures were to be expected in this area, as the rock was considerably stressed due to tectonic reasons.

The driven cross section of the tunnel was the result of a compromise solution wherein the rock mechanical aspects and the requirements of driving were considered in the analysis (see Fig.1).

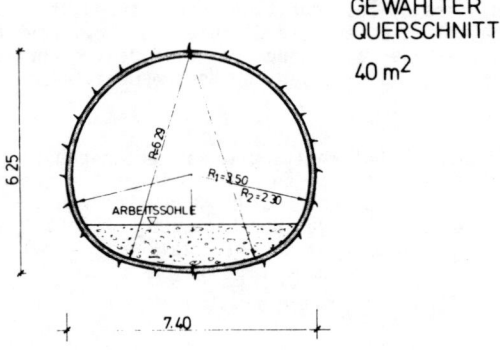

Fig. 1 Cross section of tunnel on the 12th seam. Albers, Jagsch (1981)

For the driving work the following machines were put into operation:
- an electrohydraulic drill hole and anchor wagon on a caterpillar truck
- two dieseldriven tractor showels with a bucket payload capacity of 2 cubic metres
- a pneumatic lifted floor elevator on a caterpillar truck and
- a central concrete mixing plant for dry shotcrete process.

The excavation is performed by blasting in full face excavation with length of rounds of maximum 3.30 m. Experiences in execution show that because of the expected rock and pressure conditions a support of shotcrete shell and anchoring system seems to be more suitable than the shotcrete shell and steel arches system.

The disadvantage of using steel arches lies in that at higher loads they try to break away from shotcrete shell due to improper bond and loose the capacity to carry the loads. On the contrary, a prestressed anchoring system is more flexible and is therefore in a position to allow rock deformations to a certain degree and to increase the bearing capacity of the rock.

Summing up it can be stated that by the support materials used and by construction according to NATM the deformations of underground excavations are kept small even in great depths of coal mining and stabilization of rock and support lining is achieved. These deformations would be in conventional construction method only imaginable with an expensive, rigid lining.

The bearing capacity of the rock is maintained in this method and therefore the rock is in a better position to handle the dynamic stresses due to subsequent underground working, than in usual supports in mining. Damages cannot always be avoided, but they can be recognized at an earlier stage by means of a measuring programme. This allows the strengthening work in the support lining.

Due to better utilization of the surrounding rock this method tends to economical solution for construction and maintenance of underground excavations.

2. DRIVING OF SHAFT BULGES WITH SINGLE SHELL AND DOUBLE SHELL SUPPORT SYSTEMS

Till now for a support lining of shaft bulges brickwork or structures of steel arches with mesh reinforcements were used. It was typical in all permanent linings that the lining could only be installed after completion of the total excavation. Moreover, higher expenditures were necessary for shaping the profile brickwork and the steel fitting.

For the projects which shall be described the construction method was chosen that makes it possible to construct the permanent lining after each step of excavation. The supporting of underground excavation by shotcrete and anchoring system is flexible and can satisfy the requirements of the driving and the static analysis.

2.1 Shaft bulge GENERAL BLUMENTHAL (1978), a single shell support system

As a rule, a single shell support lining is not appropriate to handle expected rock deformations without damages. A double shell would be able to withstand these rock movements and would moreover conform with the safety regulations which would normally apply to complex shaft installations in mining zones.

After technical consultations with the local mining authority for shaft bulge GENERAL BLUMENTHAL a single lining system was approved. This decision was influenced both by the results of investigations and by previous mining operations in similar depth, which had shown that the rock was strong and had resisted convergence movements. The approved shaft bulge construction, by contrast, is now installed in the form of a

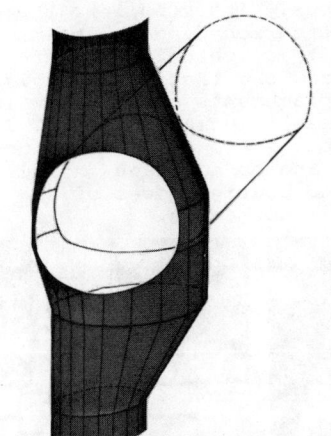

Fig. 2 Idealized geometry of mine shaft bulge for calculation purposes

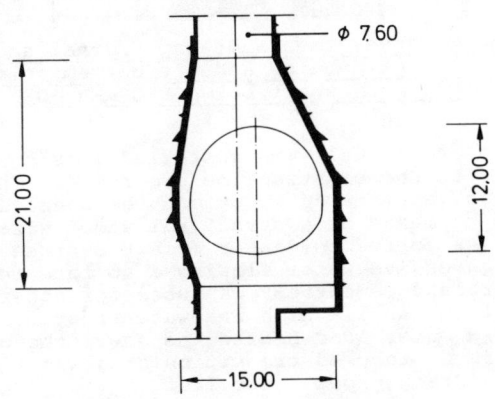

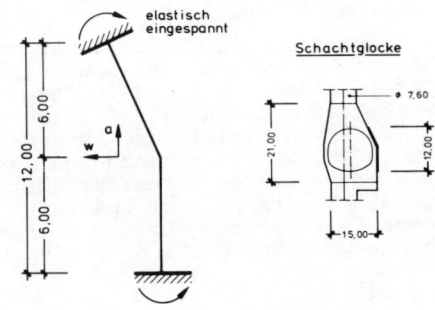

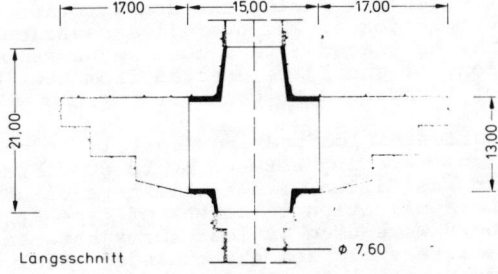

Fig. 3 Sectional area of mine shaft bulge and degree of fixing assumption as well as mine shaft bulge with lateral extensions

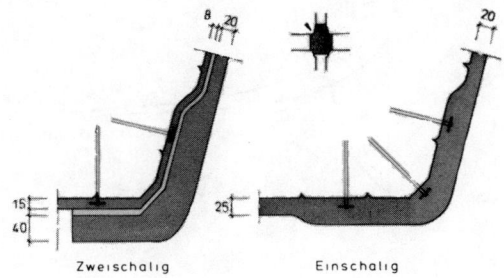

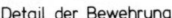

Detail der Bewehrung

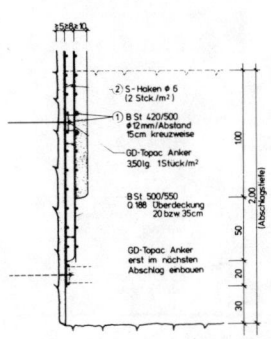

Fig. 4 Support lining of a double shell (left) and a single shell (right) support system as well as details of support installation showing consecutive layers of sprayed concrete and overlap of steel bars

single shell lining of between 200 and 250 mm thickness of sprayed concrete. At the transition edges between the shaft bulge and the lateral extensions, the reinforced concrete thickness was doubled to 400 mm and attached to the rock at intervals of 1 m by means of 5 m anchors.

The other anchors in the support system are 3.5 m long and were installed in a pattern of one anchor per square metre - a safety precaution requested by the mining authorities. The stipulated anchor capacity was 150 kN per anchor according to Maidl (1980).

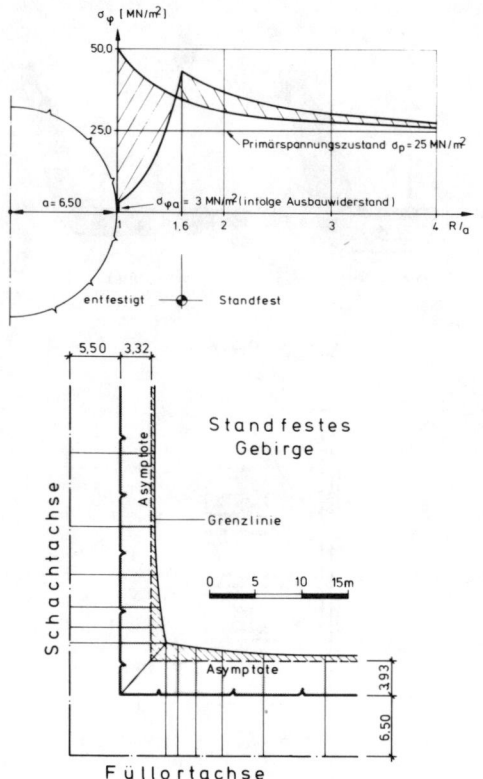

Fig. 5 Stress distribution and boundary line of loosening zones

2.2 Shaft bulge PROSPER 10 (1979) constructed in a double shell system

Temporary support

Rock bolting and shotcrete shell are at first the temporary support of the newly exposed rock of underground excavations and will become subsequently load bearing components of the outer shell as the permanent support.

Immediately after excavation a shotcrete layer of 5 mm thickness is applied. Following this, rock bolts, 2 metres long with a diameter of 24 mm are installed in a planning grid of 1 m on the bottom of boreholes, bonded by high-strength adhesive.

For anchorage of the rock a well adjustable type with free bolt ends is used which is installed by sectors in a planning grid of 2 metres. As they have to penetrate through the calculated loosening zone and the boundary of this zone into the rigid rock their length and gradients are different. Single anchoring bolts of system DYWIDAG are used with a diameter of 32 mm and 650 kN test load in boreholes of 115 mm corresponding to German standards DIN 4125 "Erd- und Felsanker" (Ground and Rock Anchors) as permanent tensioned anchors. Preliminary the tests have shown that with due regard to most unfavourable conditions the resin-bonded part of the bolts in the rigid rock should have the length of 2.5 m. The free adjustable part of the bolt is protected against corrosion by an encasing tube, previously inserted into the borehole. The annular space surrounding the bolt in the tube is filled with a bentonite-cement-mix of less strength. Provided that the bolt is placed centrically, displacements of layers which may occur will not result in shearing off. The reason is that the bolts are allowed to incline to a certain extent and to maintain in this way to a large extent the bearing power of the anchorage despite of disturbances.

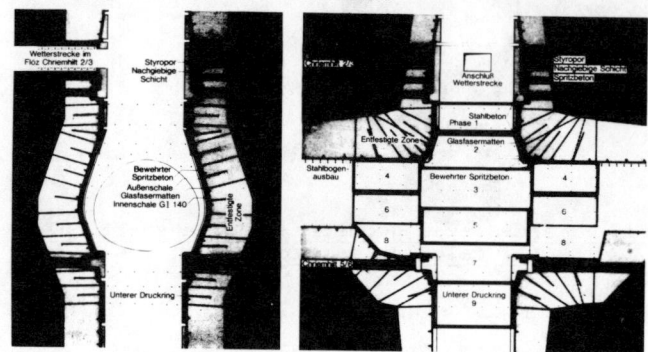

Fig. 6 Cross-section and longitudinal section of shaft bulge and pit bottom on floor 636 m of shaft bulge PROSPER 10

The shotcrete shell was installed in stages according to the construction progresses. The wall thickness amounts to 10 cm and the shell is reinforced by means of a structural steel wire mesh Q 188. As corrosion danger exists because of very aggressive water flowing a sulfate resisting Portland-cement 45F was used for construction of the shell which was watertight. In order to avoid water load behind the shell the upper part of the tapered transition area was equipped with 16 drain pipes for relief.

Flexible intermediate shell

The flexible intermediate shell was 80 mm thick and consisted of fiber mattresses which corresponds to German standards DIN 4120 "Brandverhalten von Baustoffen und Bauteilen" (Fire resistance of construction materials and members) The used material "ISOVER" is siliconbased and therefore has a high degree of waterrepellent characteristics. For the drainage of the water coming from the upper part of the shaft openings were provided below the load bearing retaining ring.

The inner shell

From the statical point of view the inner shell was not necessary during the construction stage. After completion of the overall excavation this shell may be placed as a second sequence of construction. It should be erected from the lowest point of foundation to the roof.

A prefabricated loadbearing scaffold of sections GI-140 in a spacing between 50 to 80 cm was used. Concrete was filled with a thickness of 28 cm with the application of cement of type PZ 45-HS. SZ-members were used as lost forms between the intermediate shell and the expanded steel mesh reinforcements at the inner side. Both these formworks were connected to each other. The SZ-elements were protected against corrosion by shotcrete of 2 cm thickness.

Ketteler, H., Hörning, W, Edeling, H. (1980)

REFERENCES

Albers, H., Jagsch, D. (1981)
Auffahrung einer Strecke nach der "Neuen Österreichischen Tunnelbauweise" auf der Schachtanlage Nordstern: Proc. Rock Mechanics (suppl. 11) 127-137, Salzburg

Edeling, H., Maidl, B.
Ausbautechniken im Tunnelbau und ihre mögliche Anwendung beim Abbau der Ortsbrust

Edeling, H., Maidl, B. (1980)
Tunnelling support methods and their possible application to machine rock face in coal mining: Proc. Eurotunnel '80, 119-129, Basel

Ketteler, H., Hörning, W., Edeling, H. et al. (1980)
Zweischaliger Ausbau für eine Schachtglocke mit Füllort im Schacht Prosper 10: Tunnelbautaschenbuch 1980 (116), 877-883, Essen

Maidl, B., Rapp, R. (1979)
Weiterentwicklung der Tunnelbauverfahren durch Stahlfaserbeton: Proc. 4. Kongreß der Internationalen Gesellschaft für Felsmechanik, (1), 485-493, Montreux

Maidl, B., Edeling, H. (1980)
Schacht Prosper 10, Füllort 636 m-Sohle: Firmen-Mitteilungen der Gewerkschaft Walter, 11-17

Maidl, B. (1980)
Supporting Construction of a Shaft Bulge by the Shotcrete and Anchoring System: Tunnelling and Geology, 18-29, London

Maidl, B. (1981)
Bauen unter Tage: Skriptum der Ruhr-Universität Bochum

Maidl, B. (1982)
Konstruktion und Bauverfahren im Tunnelbau: Volume 1 of this book will be published in the near future by Verlag Glückauf, Essen

DER MASCHINELLE AUSBRUCH DES 21 KM LANGEN WALGAUSTOLLENS
Excavation of the 21 km Walgautunnel by a full-face tunnelling machine
Percement du tunnel de Walgau (21 km de long) avec un tunnelier de pleine section

G. Innerhofer
H. Loacker
Leiter der Projektierungsabteilung — Chefgeologe Vorarlberger Illwerke AG
Bregenz — Österreich

ZUSAMMENFASSUNG

Der Walbaustollen ist Anlageteil der Wasserkraftanlage Walgauwerk in Vorarlberg im Westen Österreichs. Der Stollen ist 21 km lang und durchfährt im Gebirgsstock des Rätikons einen großen Teil der tektonischen Stockwerke der Nordalpen. Dank des lebhaften Wechsels von Gesteinsarten unterschiedlicher tektonischer Beanspruchung konnten reiche Erfahrungen gesammelt werden.

SYNOPSIS

The Walgau tunnel forms part of the hydro-electric power station Walgauwerk at Vorarlberg in Western Austria. The tunnel has a length of 21 km and in the Rätikon massif it passes through a major part of the tectonic layers of the Northern Alps. The many changes of rock types of varying tectonic strains made it possible to collect valuable experiences.

RESUME

Le tunnel de Walgau fait partie de la centrale hydro-électrique à Vorarlberg à l'ouest de l'Autriche. Ce tunnel a une longueur de 21 km et traverse, dans le massif de Rätikon, une grande partie des strates tectoniques des Alpes du Nord. La grande variété des roches de divers types tectoniques rencontrées lors de la réalisation des travaux a permis d'obtenir des données précieuses.

1. BAUGEOLOGISCHE VORAUSSETZUNGEN

Im Rätikon sind die Nördlichen Kalkalpen auf die Flyschzone aufgeschoben und ihrerseits in mehrere Teilschollen zerlegt. Sie grenzen im Südosten an die Glimmerschiefer und Phyllitgneise des Kristallins. Sie sind aus Kalken, Dolomiten, Mergeln, Tonschiefern, Quarziten und Rauhwacken, Gesteine die vom Perm bis in die Oberkreide reichen, aufgebaut. In den obertriadischen sehr stark gestörten Raibler-Schichten kommen neben den vorerwähnten Gesteinen auch ausgedehnte Gips- und Anhydritlager vor.

Der Flysch besteht aus einer intensiven Wechsellagerung von Kalk- und Sandsteinbänken mit Mergeln und Tonschiefern. Er ist stark verfaltet, die Hartgesteine sind geklüftet, die Mergel und Tonschiefer feinblättrig tektonisiert.

Die in der Überschiebungsfläche entstandene Quetschzone ist auch zwischen die einzelnen Schollen eingepreßt. In dichten, feinblättrig tektonisierten Tongesteinen schwimmen Reste der Hartgesteine, zum Teil als tektonische Gerölle. Die Mächtigkeit dieser Quetschzonen schwankt zwischen wenigen und mehreren 100 Metern.

Die stark teilbeweglichen Gips- und Anhydriteinlagerungen führten auch in den Raibler-Schichten zu schichtparallelen Bewegungen. Hiebei wurde der Dolomit als spröde Komponente zum Teil staubfein zerrieben, zum Teil zu bruchschotterartigen Lockermassen zertrümmert. Im Hangenden der großen Anhydritlager ist der Hauptdolomit durch Gipslösung allenthalben aufgelockert. Dies führt zu bedeutenden schichtparallelen Hohlräumen, die in Summe bis über 10 m betragen und mit Lehm und Schluff gefüllt sind.

2. HYDROGEOLOGISCHE VORAUSSETZUNGEN

Die tektonische und petrographische Vielfalt führt zu ebenso lebhaft wechselnden Gebirgswasserverhältnissen. Die tonreichen Gesteine sind bei der Verformung durch die Gebirgsbildung vorwiegend dicht geblieben. Die Hartgesteine aber sind in den Muldenkernen und Störzonen zerbrochen und nunmehr wasserführend. Der Zertrümmerungsgrad ist am höchsten in den spröden Dolomiten. Die Anhydritlager in den Raibler-Schichten sind randlich vergipst und von Karstschloten durchzogen. Auch die Kalke neigen zu Verkarstung. Da der alte Talboden wesentlich tiefer als die heutige Verschotterung liegt, reichen die Karstsysteme bis unter das Stollenniveau. Konzentrierte Bergwasserströme haben sich in den Muldenkernen oberhalb der wasserstauenden dichten Gesteine und in den Karstgesteinen entwickelt. Diese speichern bedeutende Wassermassen.

3. GEOLOGISCHE UND HYDROGEOLOGISCHE VERHÄLTNISSE

Das gute Erfassen der wechselhaften Voraussetzungen im Projektierungsstadium war für das Gelingen des Stollenbauvorhabens Voraussetzung. Nach baugeologischer Überarbeitung und Ergänzung der vorliegenden Detailkartierung wurden in verschiedenen Ebenen geologische Vertikal- und Horizontalschnitte ausgearbeitet und diese zu einem räumlichen Modell vergittert (Fig. 1). Neben dem Gebirgsbau stellen sich in diesem auch die hydrogeologischen Verhältnisse klar dar. Durch Kartierung der Quellaustritte und Messung der Schüttungen konnten diese Aussagen erhärtet werden.

Fig. 1 Geologisches Modell

Unklar blieben die Gebirgswasserverhältnisse im Bereich des vorgesehenen Fensterstollens. Stark schwankende Quellen deuten dort auf eine Verkarstung. Wegen der hohen Überlagerung war es wirtschaftlich nicht möglich Bohrungen bis zur Stollentrasse auszuführen. Zur Klärung wurde in der Achse des Fensterstollens ein Sondierstollen mit einer Robbins-Vollschnittmaschine, Ø 3,90 m, ausgeführt. In diesem wurden im Schutze einer dichten Mergelzone die Plattenkalke, an deren Ausbiß im Talboden die Karstquellen liegen, angebohrt und der Bergwasserdruck gemessen. Dieser Sondierstollen gab weiters Aufschluß über Gesteinsgrenzen und im besonderen über die Fräsbarkeit des Gebirges, d.h. über dessen technologisches Verhalten bei mechanischem Vortrieb.

Weiters wurden von allen zu durchörternden Gesteinsarten, teils in bestehenden Hohlraumbauten, teils an der Oberfläche, Proben entnommen und diese geomechanisch, besonders hinsichtlich der Druck- und Scherfestigkeit und des Quarzgehaltes untersucht.

4. TECHNISCHE LÖSUNG

4.1 Linienführung des Stollens

Der Stollen ist als Druckstollen ausgebildet und überwindet dank seiner Länge den größten Teil der Fallhöhe des Kraftwerkes. Die anschließende Falleitung ist nur mehr kurz. Bei der Festlegung der Linienführung wurde versucht, einen Kompromiß zwischen nachstehenden Forderungen zu finden:

- die stollenbautechnisch ungünstigen Zonen auf kürzestem Weg zu queren,

- zu hohe Überlagerungen und damit Gebirgsdruckerscheinungen zu vermeiden,

- den Stollen so tief in den Berg zu legen, daß der Gebirgswasserdruck über dem Betriebsinnendruck des Stollens liegt.

Diese letzte Forderung wurde erhoben, um Abdichtungsmaßnahmen zu vermeiden. Wo sie erfüllt ist, sind nur mehr statisch erforderliche Auskleidungen vorgesehen.

Der Stollen wurde in 2 Baulose zu je rd. 10 km geteilt. Im Hinblick auf den zu erwartenden großen Bergwasserandrang kam nur ein Vortrieb im Steigenden mit möglichst großem Gefälle in Frage. Aus fördertechnischen Gründen war dieses mit etwa 10 ‰ zu begrenzen. Die Überlegungen führten zu dem in Fig. 2 dargestellten sägezahnartigen Längsschnitt mit dem Fensterstollen Bürs im Tiefpunkt und dem Entlüftungsschacht Alvier im Hochpunkt. Die Lösung hat den weiteren Vorteil, daß die schwierigste Zone beidseits des Entlüftungsschachtes mit der Arosazone und den Raibler-Schichten nahe dem Fensterstollen liegt und ohne Zeitdruck ausgebrochen werden kann.

4.2 Vortriebsmethode

Der konventionelle und maschinelle Vortrieb wurden eingehend miteinander verglichen und schließlich bei nur geringem Preisvorteil für den maschinellen Vortrieb entschieden. Dem hiefür höher zu bewertenden Risiko der unterschiedlichen geologischen Voraussetzungen und der Wassereinbrüche stand die größere zu erwartende Vortriebsleistung in günstigen Zonen gegenüber. Auch bei Eintreten von Unterbrechungen war insgesamt eine kürzere Bauzeit als bei konventionellem Vortrieb zu erwarten. In jedem Baulos wurde vom jeweiligen Auftragnehmer eine Robbins-Vollschnittmaschine, Ausbruchdurchmesser 6,25 m, eingesetzt. In Bürs wurde zuerst vom Fensterstollen in Richtung Hochpunkt Alvier vorgetrieben, mit der Zielsetzung, nach Erreichen der Quetschzone die Maschine abzubauen und sie vom Fensterstollen in die entgegengesetzte Richtung nach Rodund einzusetzen. Diese Zone und vor allem die folgenden Raibler-Schichten sollten dann konventionell ausgebrochen werden.

Auch in den im Valkastielstollen zu erwartenden Störungen war nicht auszuschließen, daß der mechanische Vortrieb bereichsweise seine Grenzen erreicht und daß über Umfahrungsstollen konventionell ausgebrochen werden muß.

Für die immerhin kühne Entscheidung für maschinellen Vortrieb war jedoch Voraussetzung, daß die eingesetzten Vortriebsmaschinen und Nachlaufbetriebe den gegebenen Voraussetzungen gerecht werden und nachstehende Forderungen erfüllen:

- Die Sohle ist schon unmittelbar hinter dem Bohrkopf durch Sohltübbinge, die einen großen Wassergraben enthalten, zu sichern.

- Unmittelbar hinter dem Bohrkopf muß der Einbau von Stahlringen, Stahlbögen und Ankern möglich sein. In besonderen Fällen ist hier auch schon Spritzbeton aufzubringen.

- Spritzbeton muß großflächig von einer Bühne möglichst nahe der Vortriebsbrust aufgebracht werden können.

4.3 Stollenprofile, Sicherung und Auskleidung

In Fig. 3 sind die wesentlichen Stollenprofile dargestellt. Je nach dem zu erwartenden Wasserandrang werden Tübbinge mit verschieden tiefen Wassergräben verwendet. Der größte vermag bei 10 ‰ Neigung 500 l/s und bei Überflutung bis Schienenoberkante 1.500 l/s Wasser abzuführen.

Standfest:

Das Profil im Hartgestein bleibt fallweise unverkleidet.

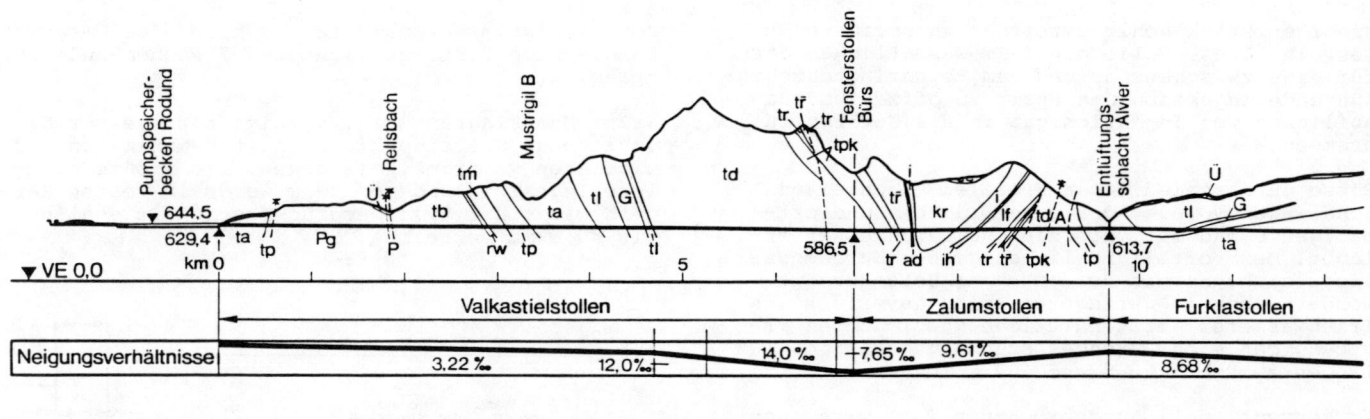

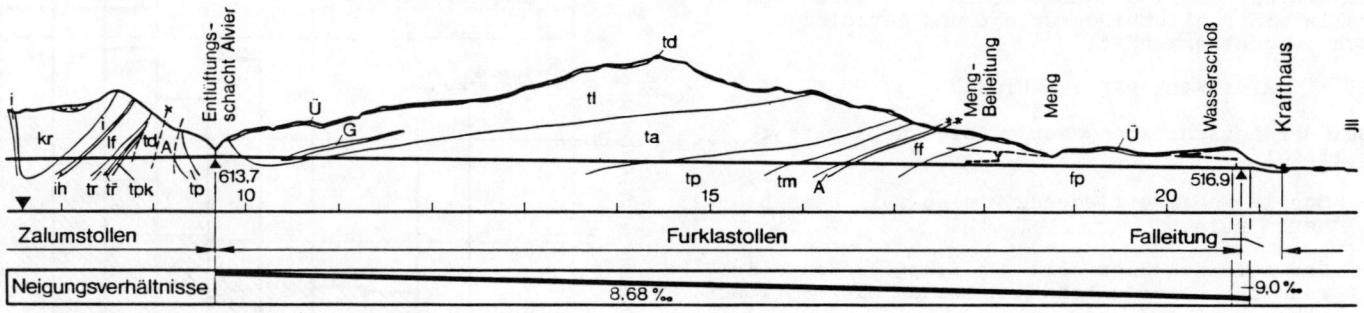

≡≡≡	Rutschung in den Kreideschiefern
Ü	Überlagerung (Hangschutt, Bergsturz, Moräne und Verbauungsschotter)
---*	Störzone, Deckengrenze

Penninikum Vorarlberger Flysch
- ff Fanola Serie
- fp Plänkner-Brücke Serie

Unterostalpin
- A Arosa-Zone

Oberostalpin Lechtaldecke und Basisschollen
- kr   Kreideschiefer
- i   Aptychenkalke
- ih   Radiolarite und rote Kalkmergel
- lf   Liasfleckenmergel
- ad   Adnether Schichten
- tŕ   Oberrhätischer Riffkalk
- tr   Kössener Schichten
- tpk Plattenkalk
- td   Hauptdolomit
- tl   Raibler Schichten
- G   Raibler Gips
- ta   Arlbergschichten
- tp   Partnachschichten
- tm   Muschelkalk
- rw   Reichenhaller Rauhwacken
- tb   Buntsandstein
- P   Paläozoische Schiefer
- Pg   Pyllitgneis und Glimmerschiefer

Fig. 2 Walgaustollen - Längenschnitt

Mit Betonauskleidung:

Das Profil ist durch Stahlringe, wo erforderlich in Verbindung mit Baustahlgitter, Streckmetall, Verzugsblechen und Spritzbeton gesichert. Es erhält eine durchgehende Betonauskleidung.

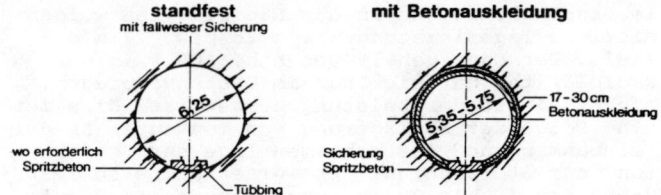

Fig. 3 Regelprofile

Eine besondere Behandlung erschien für die Anhydritstrecken erforderlich. Grundsätzliche physikalische Überlegungen ergaben, daß der bei Wasseraufnahme des Anhydrit entstehende Quelldruck durch den Ringdruck einer Auskleidung kaum zu beherrschen ist. Weiters setzt die Umwandlung von Anhydrit im Gips die Gesteinsfestigkeit herab. Dies würde im Bereich hoher Überlagerungen zu einer unerwünschten zusätzlichen Belastung der Auskleidung führen. Es wurde daher entschieden die Anhydritstrecken durch eine dichte Auskleidung vor dem Stollenwasser zu schützen. Aufbauend auf die günstigen diesbezüglichen Erfahrungen beim Rotenbergstollen des Kraftwerkes Langenegg wurde eine Foliendichtung gewählt.

Der Sohltübbing wird in zwei Teilen ausgeführt. Der Oberteil mit dem Wassergraben kann angehoben und die Folie auf den Unterteil verlegt und verschweißt werden. Nach Wiedereinsetzen des Oberteiles wird der Innenring betoniert und die Folie und der Betonring nach dem Prinzip des Kernringverfahrens durch eine Mörtelinjektion auf eine gewisse Vorspannung gebracht.

Nur in einzelnen kurzen Abschnitten, in denen bestehende Wasserwege das Gebirge so stark entwässern, daß der Bergwasserspiegel unter dem Betriebsdruck liegt, sind weitere Dichtungsmaßnahmen vorgesehen. Auch diese sind sehr variabel und richten sich nach den wechselnden Erfordernissen und den felsmechanischen Voraussetzungen. Sie reichen von Manschetteninjektionen als einfachstes Mittel in festen nur wenig wasserwegigen Kalken bis zu einer Folienabdichtung im nachgiebigen stark hohlraumreichen wasserführenden Gestein. Diese Folie dient im wesentlichen dazu für eine zwischen ihr und dem Betonring durchzuführende Injektion den Spalt zu öffnen und das Abfließen von Injektionsgut in das Gebirge zu bremsen.

Diese unterschiedlichen Profile setzen einen Schalwagen voraus, der einen raschen und oftmaligen Umbau auf verschiedene Durchmesser erlaubt. Der Vorteil möglichst großer Durchmesser liegt nicht nur in der geringeren Betonkubatur, sondern auch im geringeren Reibungsverlust des Triebwassers. Wirtschaftliche Überlegungen ergeben, daß sich Durchmesseränderungen schon für Längen zwischen 50 m und 100 m lohnen.

Im konventionell ausgebrochenen Teil wird ebenfalls ein Sohltübbing verlegt und derselbe Schalwagen verwendet.

5. PRINZIPIEN DER VERGÜTUNG

Bei der Ausschreibung wurde folgende Strategie verfolgt:

- Das im Gebirge liegende Risiko soll beim Bauherrn sein.

- Der Auftragnehmer soll möglichst großen Spielraum für seine unternehmerischen Dispositionen haben.

- Der Positionsaufbau muß die ständige Anpassung der Sicherungsmethoden an das Gebirgsverhalten ermöglichen.

- Dem Auftragnehmer sollen jene Leistungen vergütet werden, die er tatsächlich erbringt. Das Gebirge wird nicht nach felsmechanischen Parametern klassifiziert, sondern nach den jeweils erforderlichen Stützmaßnahmen. Diese beeinflußen die Vortriebsleistung und somit die Einheitspreise. Für jede Gebirgsklasse gibt es nach Menge abgestufte Zuschläge für Wassererschwernisse.

In Bereichen, in denen die aus Gesteinsbeschaffenheit, Gebirgsdruck und Bergwasserandrang resultierenden Schwierigkeiten so groß sind, daß die Vortriebsleistung nicht mehr kalkulierbar ist, werden Ausbruch und Sicherung in Regie vergütet.

Es konnten damit die komplexen Verhältnisse des Walgaustollens gut erfaßt werden. Selbstverständlich zeigt die Praxis, daß die Natur mehr Phantasie hat als der sie beschreibende Mensch, so daß laufende Gespräche zwischen den Vertragspartnern erforderlich werden. Die auf gegenseitiges Vertrauen begründete Bereitschaft hiezu ist, wie überall, auch hier Grundlage für ein gutes Gelingen des Bauwerkes.

6. ERFAHRUNGEN BEI AUSBRUCH UND SICHERUNG DES STOLLENS

6.1 Die beiden Vortriebsmaschinen wurden zum Jahreswechsel 1980/81 angeliefert, in vorbereiteten Kavernen montiert und konnten noch im Kalenderjahr 1981 einen Vortrieb von 5.634 m bzw. 5.923 m, insgesamt von 11.557 m, leisten. Dieses günstige Ergebnis gewinnt mit den schwierigen durchfahrenen Zonen an Bedeutung. Weiters war die Maschine im Baulos Bürs, nach Beendigung des Vortriebes im Zalumstollen während des Umsetzens in die neue Vortriebsrichtung 7 Wochen außer Betrieb.

6.2 Das Diagramm Fig. 4 zeigt für die verschiedenen Gesteine die erzielten maximalen und mittleren Vortriebsleistungen. Die größte mittlere Tagesleistung von 32 m wurde in festen Mergeln mit einer Zylinderdruckfestigkeit von 60 bis 80 N/mm² erreicht.

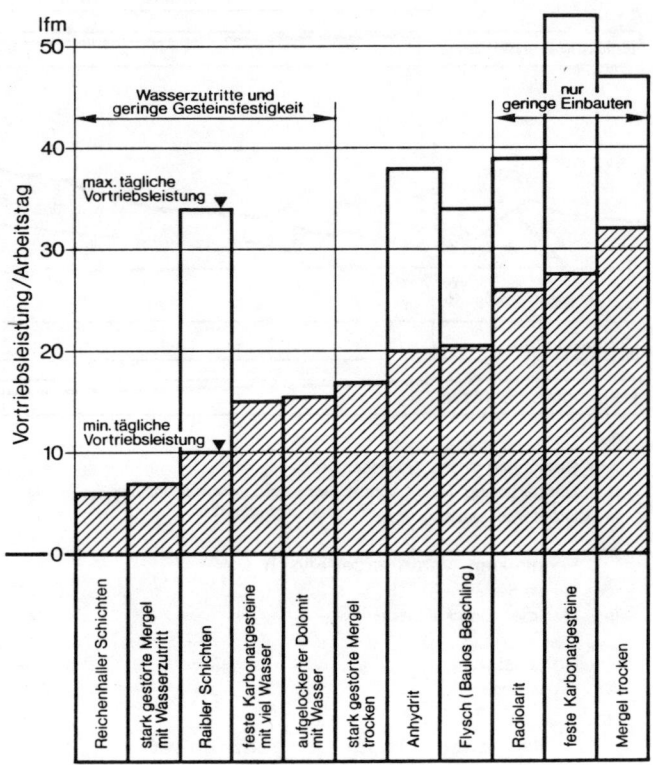

Fig. 4 Vortriebsleistungen im Baulos Bürs

Im Hauptdolomit und in den Radiolariten wurden mittlere Tagesleistungen von 28 bzw. 27 m erzielt. Der Quarzgehalt von nahe 100 % in den Radiolariten hat sich nur im Diskenverbrauch, nicht aber in der Leistung ausgewirkt. Dies ist eine Folge der Einlagerung von Tonhäuten in der Feinbankung und der schrägen Lage der Schichtung zur Stollenbrust. Es wurden Chips in der Größe von 10 bis 20 cm ausgebrochen. Im Hauptdolomit wurde in einem Monat 1 km aufgefahren und fallweise mit Firstbögen gesichert. Im Anhydrit, in den gestörten Mergeln und in der Quetschzone wurden bei durchgehendem Stahlringausbau mit Spritzbetonsicherung 20 bzw. 17 m pro Tag erzielt. Im leicht wasserführenden stark aufgelockerten Hauptdolomit im Hangenden des Anhydrits sank die Leistung auf 15 m pro Tag. In den stark wechselnden Raibler-Schichten schwankt die Leistung zwischen 2 bis 4 m und 34 m und erreichte im Mittel 10 m pro Tag.

6.3 Als weitere besondere Leistung darf die Beherrschung von Schwierigkeiten verschiedener Art gewertet werden:

6.31 Im Valkastielstollen sind an einer großen Störung Raibler-Schichten in die Arlberg-Schichten eingeschuppt. Unter hohem Gebirgswasserdruck steht ein feinkörnig zertrümmerter Dolomit an. Nach Durchörterung des abdichtenden Gesteinspaketes floß der Dolomitgrus breiartig in den Stollen aus. Hiebei entstand eine mehrere Meter hohe Kaverne, die bis 16 m vor den Bohrkopf reichte. Der Materialeinbruch betrug 450 m³, der Wasserandrang ca. 40 l/s. Nach Einbringen von Pumpbeton in den oberhalb der Firste gelegenen Hohlraum, wurde vor dem Bohrkopf ein Pilotstollen mit anschließendem händischen Vollausbruch und schweren Sicherungen bis in das feste Gestein vorgetrieben. Eine Betonsohle schützte den Fräskopf vor Absinken. Die Aufarbeitung dieses Verbruches dauerte einen Monat.

6.32 Im Liegenden des wasserführenden Muschelkalkes waren die an sich dichten Reichenhaller Rauhwacken stark durchfeuchtet. Beim Auffahren dieser Zone verklemmten die breiartig aufgeweichten Rauhwacken den Bohrkopf. Es wurde über dem Bohrkopf ein Stollen vorgetrieben und von diesem aus die aufgelockerte Zone mit Polyuretanschaum und Zement injiziert. Anschließend wurde der Firststollen bis in die trockene Rauhwacke weiter vorgetrieben und ausbetoniert. Im Schutze des damit geschaffenen Firstbalkens konnte wieder maschinell vorgetrieben werden. Die Aufarbeitung dieses Verbruches dauerte 17 Tage.

6.33 Wassereinbrüche bis 500 l/s an Karstklüften konnten ohne nennenswerte Verzögerungen beherrscht werden.

6.34 Durchörtern einer 9 m breiten Gipskarstzone in 11 Tagen. Diese Zone besteht aus skelettartig ausgewaschenem Gips, der mit weichem gelblichem Lehm gefüllt ist. Bei einem Wassereinbruch von 1.600 l/s wurden rd. 3.000 m³ Feststoffe ausgespült.

6.35 Im konventionell vorgetriebenen Teil des Stollens (siehe nächstes Kapitel) wurde im Hangenden einer geringmächtigen Anhydritlage ein Gipskarstsystem angefahren. Der Wasserandrang in den Stollen betrug 400 l/s. Der im fallenden Trum liegende Stollen wurde eingestaut; ca. 1.100 m³ Material wurden eingeschwemmt. Nach Abpumpen des Wassers und Ausräumen des Materials trat 10 Tage später ein neuerlicher Wassereinbruch ein, der im Mittel 2.000 l/s und stoßweise bis 8.000 l/s schüttete. Hiebei wurden 3.000 m³ Material eingeschwemmt und der Stollen neuerlich überflutet. Die Ursache dieses zweiten Wassereinbruches ist wahrscheinlich der Durchbruch eines benachbarten Karstsystems. Durch die Materialeinschwemmung und Verstürzungen in den Karsthohlräumen dichtete sich dieses Karstsystem gegenüber dem Stollen wieder teilweise ab. Dies ließ den Gebirgswasserdruck neuerlich auf 8 bar ansteigen. Dieser konnte erst in zahlreichen Entwässerungsbohrungen abgebaut werden. Um der stark verstürzten und gestörten Zone auszuweichen, wurde die Stollentrasse um 40 m verlegt. Nach Injektionen konnte dort im unverstürzten Bereich diese Gipskarstzone ohne größere Schwierigkeiten durchörtert werden.

7. GRENZEN DES MASCHINELLEN AUSBRUCHES

Die Raibler-Schichten, wie vorerwähnt, wurden schon in der Planung als das für den Stollenbau problemreichste Gestein erkannt. In den Staubdolomiten und im hohlraumreichen Dolomitgrus spült starker Wasserandrang beim Drehen des Bohrkopfes große kavernenartige Hohlräume vor und neben dem Bohrkopf aus. Diese müssen noch vor dem Setzen der Stahlringe und dem Verlegen der Sohlstübbinge mit Beton verfüllt werden, um der Maschine und den Gripperplatten ein Widerlager zu bieten. Das Schwemmgut hinter dem Bohrkopf muß zum Teil händisch geladen werden. Auch in festeren Bereichen zerfällt der gestörte Dolomit beim Fräsen in so feinen Grus, daß dieser ausgeschwemmt wird. Dennoch erwies sich auch hier bei Tagesleistungen zwischen 2 und 6 m der maschinelle Vortrieb noch als wirtschaftlich.

Im fallenden Trum wenige Meter jenseits des Hochpunktes wurde schließlich bei einem Wasserandrang von 200 l/s an der Brust die Grenze des wirtschaftlich möglichen maschinellen Vortriebes erreicht und auf konventionellen Vortrieb umgestellt. Die Dolomite sind dort teilweise so entfestigt, daß auch keine Sprenglösung möglich war. Im Vortrieb mit Getriebezimmerung mußte die Brust mit Spritzbeton abgesperrt werden.

8. VERHALTEN DES GEBIRGES UND FELSMECHANISCHE DEUTUNG

Im Diagramm, Fig. 5, sind die maximalen Überlagerungshöhen eingetragen, die in den verschiedenen Gesteinszonen wirksam sind. Hiebei ist zu beachten, daß die Belastung durch den ansteigenden Bergkamm erhöht wird. Weiters ist die Grenzlinie für beginnende Bruchzonen, die sich aus der Scheibentheorie unter der Annahme von $\sigma_v / \sigma_h = 2$ ergibt, ausgewiesen. Zufolge der tiefen, steilen Taleinschnitte war zu vermuten, daß die horizontalen Spannungen kleiner als die vertikalen sind. Tatsächlich ist oberhalb dieser Linie durchwegs Bruch der Felslaibung eingetreten. Die Lage und Gestalt der Bruchzonen spiegeln deutlich die Hauptspannungsrichtung, die meist nicht senkrecht steht, sondern etwas in Richtung der ansteigenden Bergflanken geneigt ist, wieder. Im darunterliegenden Bereich bis auf halbe Höhe sind fallweise Klüfte und Schichtflächen, je nach deren Lage zur Stollenachse, leicht aufgerissen. Bei geringen Überlagerungen sind keine Brucherscheinungen aufgetreten. Bei vielen Gesteinen, besonders beim homogenen Anhydrit ist die Abhängigkeit des Verhaltens von der Überlagerungshöhe augenfällig.

Während die einfache Bruchbetrachtung nach der Scheibentheorie in der Natur bestätigt wurde, ist dies bei der rechnerischen Abschätzung der Abhängigkeit von Ausbauwiderstand und erforderlicher Deformation weit weniger der Fall. Die Deformation der gebrochenen Bereiche wurden nach üblicher Weise ermittelt, wobei das Traggewölbe vom Ausbauwiderstand ausgehend, unter Berücksichtigung des kritischen Hauptspannungsverhältnisses zwischen Radial- und Tangentialspannung aufgebaut wird. Nach dieser Betrachtung führen hohe Reibungswinkel im post failure - Verhalten zu geringen Deformationen. Die Reibungswinkel sind bei großen Überlagerungshöhen wichtiger als die Druckfestigkeit vor dem Bruch. In den Quetsch- und Störzonen und weichen Mergeln mit Reibungswinkeln von nur wenig über 10° waren bei den verfügbaren Ausbauwiderständen, die im schwersten Ausbau mit 200 N/cm² beschränkt sind, Verformungen zu erwarten, die weit über dem Deformationsvermögen der Stahlringe liegen. Da auch für die Erhöhung des Ausbauwiderstandes praktische wie wirtschaftliche Grenzen gesetzt sind, wurde der Einbau von elastischen Zwischenschichten in den Stahlringen und in Spritzbetonschlitzen sowie ein späteres Ausbessern einer bereichsweisen deformierten oder zerbrochenen Vorausverkleidung erwogen. Tatsächlich konnte aber mit dem starren Ausbau überall das Auslangen gefunden werden. Mit nur oberflächlichen nicht tiefgreifenden Bruchzonen ist offensichtlich sehr bald ein günstiges Spannungsbild gefunden worden. Überraschend ist, daß nirgends echte Druckerscheinungen aufgetreten sind. Auch in den ungünstigen Zonen wurde der Bohrkopf nicht verklemmt. Voraussetzung hiefür war allerdings ein Vortrieb bei Vermeidung von Stillständen.

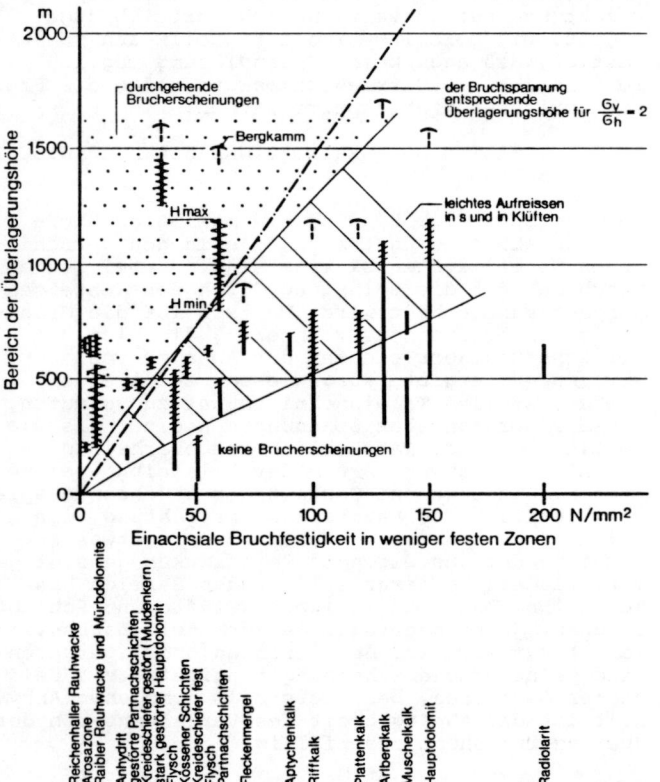

Fig. 5 Verhalten der durchörterten Gesteinsarten bei verschiedenen Überlagerungshöhen

Sohlhebungen sind nirgends aufgetreten. Konvergenzmessungen zeigen nur Deformationen im cm-Bereich.

Als Gründe für dieses günstige Verhalten werden angenommen:

Die verformbaren Zonen im Gebirge hängen sich an den starren auf. Dies gilt vor allem für die steil stehenden Quetschzonen. Da diese Aufhängung im Valkastielstollen offensichtlich auch bei einigen 100 m mächtigen Zonen gegeben ist, darf angenommen werden, daß sich diese im Gebirge schon vor dessen Durchörterung vollzogen hat. Diese Aufhängung wird voraussichtlich durch die Absenkung des Bergwasserdruckes beim Vortrieb noch erhöht.

Die erforderliche Deformation des Ausbaues wird durch die schon vor dem Bohrkopf und in dessen Bereich stattfindende Verformung reduziert. Weiters ermöglichen die Stahlringe Brucherscheinungen in den Zwischenräumen, wobei aber die Deformationen im Schutze des Ausbaues "geordnet" ablaufen.

Durch die geschlossenen Ringe und die schweren seitlich eingespritzten Sohltübbinge wird der Ringschluß rasch erreicht. Ein entscheidender Vorteil des maschinellen Vortriebes, nämlich das Gebirge nicht zu stören, wird voll gewahrt und damit einer wesentlichen Forderung der NÖT nachgekommen. Besonders im wenig festen Gebirge mit nicht elastischem Verhalten, in dem das Superpositionsgesetz nicht gilt, beeinflußt die Spannungsgeschichte das Materialverhalten bestimmend. Im Gebirgsinnern, wo das Gestein sein volles inneres Gleichgewicht gefunden hat, ist das durch Scherfestigkeit und Verformungsmodul ausgedrückte Wohlverhalten am größten. Jede Veränderung bedeutet somit eine Minderung von Festigkeit und Modul.

9. ZUSAMMENFASSUNG

Der Walgaustollen durchfährt außerordentlich wechselndes Gebirge mit stollenbautechnisch schwierigen Zonen. Gute geologische und planerische Vorarbeit, eine wohl aufeinander abgestimmte Einrichtung und hochqualifiziertes Personal ließen diese Schwierigkeiten beherrschen. Die erzielten Leistungen sind befriedigend.

Die besonders kennzeichnenden, auch den Prinzipien der NÖT entsprechenden Merkmale der angewandten Methode beruhen darin, daß der Stollenausbruch im geschlossenen Ring sofort gesichert wird und Verformungen möglichst vermieden werden. Die Sohle wird durch Tübbinge vor Beanspruchung geschützt. Nur die grundsätzlichen Entscheidungen werden bei der vorgängigen Planung, alle weiteren aber laufend bei der Ausführung vor Ort gefällt. Voraussetzung hiefür ist die Bereitschaft täglich oder täglich mehrmals die Ausbauart zu wechseln. Diese wohl auch der österreichischen Mentalität entsprechende Auffassung steht im Widerspruch zu anderer, die einen wirtschaftlichen Erfolg nur im Durchziehen einer möglichst einheitlichen, großzügigen Lösung sieht.

Auch bei großen Überlagerungshöhen bis zu 1.500 m und in wenig festem Fels hat der Gebirgsdruck zu keinen nennenswerten Verzögerungen geführt. Das Gleichgewicht konnte mit kleineren Ausbauwiderständen und Deformationen als rechnerisch abgeschätzt gefunden werden. Bedeutende Schwierigkeiten hingegen ergaben sich durch Zusammentreffen von wenig festem Gebirge und großem Wasserandrang. Nicht entlasteter Bergwasserdruck hat diese entscheidend erhöht. Die stützende Wirkung des schweren Bohrkopfes hat offensichtlich größere Materialeinbrüche begrenzt.

TESTS AND MEASUREMENTS FOR THE PRESSURE TUNNELS AND SHAFTS OF THE SELLRAIN-SILZ HYDROELECTRIC POWER SCHEME WITH EXTREMELY HIGH HEAD

Essais et mesures pour les galeries d'amenée et les conduites forcées en charge extrême de l'aménagement hydroélectrique de Sellrain-Silz

Versuche und Messungen für die hochbeanspruchten Druckstollen und Schächte der Werksgruppe Sellrain-Silz der TIWAG

B. Bonapace
Head of Planning and Civil Engineering Department, Tiroler Wasserkraftwerke AG (TIWAG)
Innsbruck, Austria

SYNOPSIS

A report is presented on the determination of the rock mechanics for the concrete lining (prestressed by interface grouting) of the upper stage shaft (800 m head) and the steel lining of the lower stage shaft (1500 m head), as well as on the test chamber for a concrete lining with plastic sheeting.

RESUME

Compte-rendu des recherches en mécanique des roches pour le revêtement en béton (précontraint par injection) du puits de Kühtai (P_{max} = 800 m) et pour le blindage du puits forcé de Silz (P_{max} = 1500 m) ainsi que des essais de pression pour un revêtement étanche en béton avec feuille de plastique.

ZUSAMMENFASSUNG

Bericht über die Ermittlung der felsmechanischen Grundlagen für die durch Injektionen vorgespannte Betonauskleidung des Oberstufenschachtes (800 m Druck) und die Panzerung des Unterstufenschachtes (1500 m Druck), sowie über den Druckkammerversuch für eine Betonauskleidung mit PVC-Dichtfolie.

1. OVERALL PROJECT DESCRIPTION

The 780-MW Sellrain-Silz Power Plant built by the Tiroler Wasserkraftwerke (TIWAG) makes use of a total head of 1678 m in two stages (Fig.1). The PSS Kühtai in the upper stage powers the water from the 60 million-m3 Finstertal seasonal reservoir through a 1800-m-long pressure shaft without surge chamber to the 400-m-lower Längental equalizing reservoir (capacity 3 million m3). The two reversible units with pump turbines in the Kühtai Shaft Power Plant are designed for a maximum capacity of 285 MW using turbines and 250 MW using pumps.

The lower stage is a typical high head power plant, characterized by its total head of 1257m. From the Längental Reservoir, the water runs through a 4600-m-long headrace tunnel with throttled twin surge chamber and a 2400-m-long steel-lined pressure shaft to the Silz Power House and powers two generating sets with six-nozzle Pelton turbines having a total capacity of 495 MW.

For both the execution of the prestressed con-

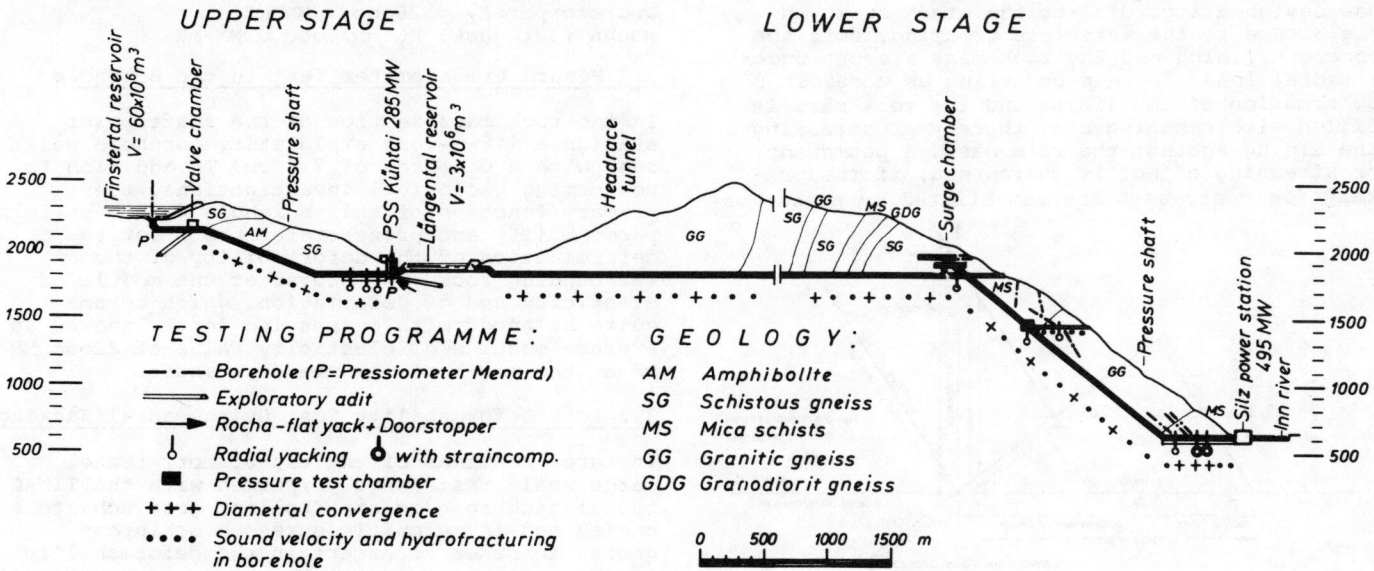

Fig.1: Cross section showing geology and rock engineering investigations

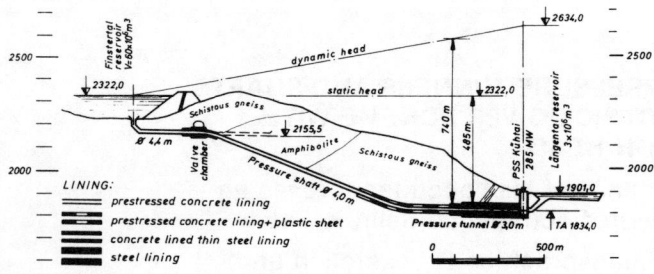

Fig.2: Cross section, Kühtai upper stage

crete lining with and without plastic sheeting as well as for the design of the steel lining required to share the internal pressure with the rock in the heavily loaded shafts of the two power stages, comprehensive investigations, examinations and rock mechanical test were necessary.

2. THE PRESSURE SHAFT FOR KÜHTAI'S UPPER STAGE

The pressure shaft for the PSS Kühtai was constructed in variable geology including schistous gneiss and amphibolite of the Ötz Valley Cristalline Series and is situated so far underground that it was possible to apply a prestressed concrete lining in a large portion of the shaft (Fig.2). The schistous gneiss was partially disturbed in the lower elbow, for which reason a waterproof membrane of PVC sheeting was used along 4oo m. The transition to the 27o-m-long steel-lined portion adjoining the Kühtai Shaft Power Station is bridged by a 35-m section with thin-walled steel lining and a concrete inner lining. The particular operating conditions of the pump turbines result in an exceptionally large increase in dynamic pressure up to 60% of the static pressure in the shaft. This put extreme loads of up to max. 485 mlc static head and max. 74o mlc dynamic head on the concrete lining. In order to prevent the concrete lining from cracking under this internal pressure, it was prestressed by interface grouting at high pressure using the TIWAG procedure (LAUFFER 1968). The prestressed concrete lining was designed according to the graph shown in Fig.3. Due to the interface grouting, both the concrete lining and the rock mass are put under a radial load. The gap occurring as a result of deformation of the lining and the rock mass is filled with cement grout, thereby prestressing the lining against the rock mass. A permanent prestressing effect is guaranteed, if the primary field stresses are remobilized by prestress grouting under sufficient rock overburden in order to resist the internal pressure. For this reason it was attempted in the planning stage to measure the in-situ stresses and the load lines of the rock mass in an exploratory tunnel in the level section. The 12oo-m-long section of the pressure shaft up to the valve chamber was driven mechanically by TBM (PIRCHER 198o). The even circular profile provides ideal conditions for prestress grouting of the concrete lining.

3. ROCK MECHANICAL MEASUREMENTS AND INVESTIGATIONS FOR THE UPPER STAGE PRESSURE SHAFT

In the course of exploration work, bore cores were taken from the surface and an exploratory tunnel was driven in the upper and lower level headrace tunnel sections, which were used for geologic and petrographic evaluation of the rock mass as well as for quantitative description of the discontinuities by the geologist. In the following, a description is given of the comprehensive rock mechanical measuring program performed to evaluate the rock mass properties, which provided the basis for selection of the shaft's lining (points 3.1 to 3.4).
After excavation of the shaft, additional procedures were applied along its entire length for the purpose of relative measurement of the rock mass properties (points 3.4 to 3.7).

3.1 In-Situ Stress Measurements

At the lower end of the pressure shaft, in the area of the lowest overburden, attempts were made in the perimeter of the exploratory tunnel to determine the in-situ stress conditions in the rock mass by means of the Rocha flat jack stress relief test combined with doorstopper tests. The results are scattered widely with the values generally higher than the vertical stress calculated from the overburden $\sigma_V = \gamma_r \cdot h = 2.5$ MN/m2. Mean values from the doorstopper measurements:
$\sigma_H = 3.3$ MN/m2, $\sigma_V = 4.1$ MN/m2
Rocha flat jack: $\sigma_H = 4.0$ MN/m2.
By measuring the stress relief during the flat jack pressure test, the modulus of elasticity was also determined.
Doorstopper: $E_r = 26000$. MN/m2
Rocha flat jack: $E_r = 16000$. MN/m2.

3.2 Menard Pressiometer Test in the Borehole

In the rock mass section of the shaft power station a 115-m-long exploratory borehole was sunk with a diameter of 7.6 cm. In addition to performing geological investigations, the primary function of this borehole was to perform permeability and pressiometer tests for the determination of the deformability of the surrounding rock. The course of the moduli of elasticity and of deformation, which became quite balanced with increasing depth, showed an average modulus of elasticity value of 22ooo MN/m2 at the mouth of the pressure shaft.

3.3 Rock Deformability Test Using Radial Jacking

In three sections of the exploratory tunnel large-scale tests were performed with the TIWAG radial jack in good amphibolite, good schistous gneiss and in mylonitic stressed schistous gneiss in order to ascertain the deformability of the rock mass under internal pressure (Fig.4).

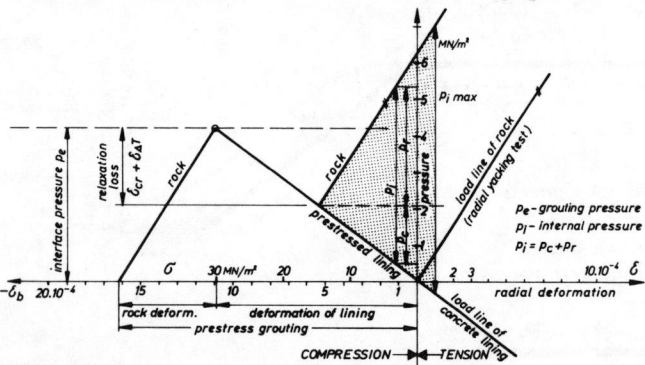

Fig.3: Design chart for concrete lining

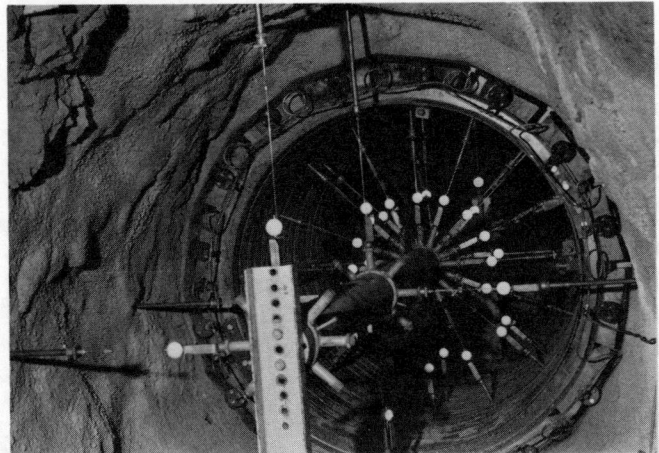

Fig.4: Radial jacking test chamber

The load lines of the rock (Fig.5) measured directly with this procedure were taken into consideration during design of the concrete and steel linings of the pressure shaft as rock mass' contribution (pr) to sharing internal pressure.

3.4 Laboratory Sample Testing

The uniaxial compressive strength, the modulus of elasticity and, using a direct shear box, the shear strength of various amphibolite and schistous gneiss core samples were measured at a 90° and 45° angle as well as parallel to the sample's bedding.

If a comparison is made of the modulus of elasticity values obtained through various procedures, it is seen that the values for samples of the same size correspond well and that they drop with an increase in the measurement area. The importance of the large-scale in-situ tests as the basis for design becomes clear.

Modulus of Elasticity (Mean Values): MN/m2
Doorstopper (borehole Ø 7.6 cm) ... E_r=26000
Pressiometer (borehole Ø 7.6 cm)... E_r=22000
Laboratory Sample (Ø 10 cm) .... E_r=25000
Rocha Flat Jack Test (100x125 cm).. E_r=16000
Radial Jacking Test (Ø 300 cm) .... E_r=11500

3.5 Measuring Diametral Convergence

An attempt was made to systematically measure the convergence of the cavity after excavation. The measurement bolts were installed as close as possible to the working face (1.2 m behind it for blasting driving and 1.5 m behind it for mechanical driving), in order to cover the major part of the stress relief's path. The measurement results were compared with the results of the seismic measurements and, just as these, show a strong dependency on rock type, degree of decomposition and on the range of loosening due to blasting or drilling (Fig.6).

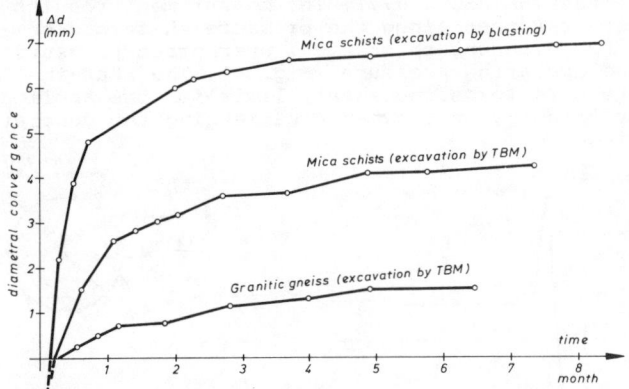

Fig.6: Diametral convergence measurement

3.6 Seismic Velocity Measurements in Borehole

In order to be able to transpose along the shaft's alignment the deformation values obtained with radial jacking tests in the exploratory tunnel, 35-mm radial drillholes were made along the wall every 10 m, in which the wave velocity was measured to a depth of 4 m. After these seismic velocity values were calibrated in the radial jack's measurement area, a correlation to the deformation modulus values was established. As a further result, the thickness of the loosened zone can be seen from the diagrams which reaches at least 1 m in depth for blasting driving and hardly 1/2 m in depth for the more careful mechanical driving by TBM (Fig. 7).

3.7 Hydrofracturing in the Borehole

Permeability tests at high pressure to fracturing of the rock were systematically performed in the boreholes for the seismic measurements to obtain information on the following:
a) permeability of the rock mass
b) critical pressure for fracturing the rock mass during prestress grouting and service
c) least rock mass stresses σ_{min} in the vicinity of the shaft, as a safety limit for maintaining permanent prestressing effect and the rock mass' contribution.

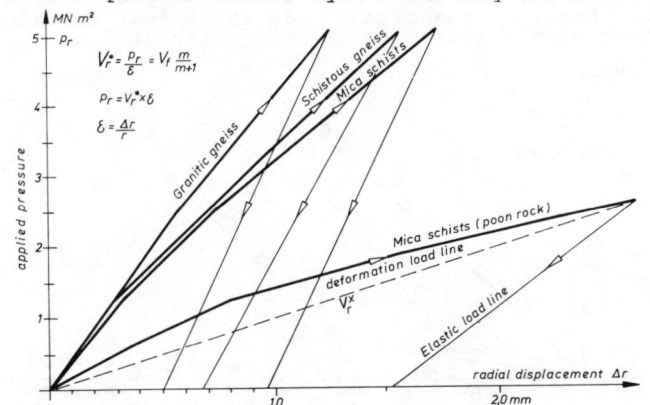

Fig.5: Radial jacking test, rock load lines

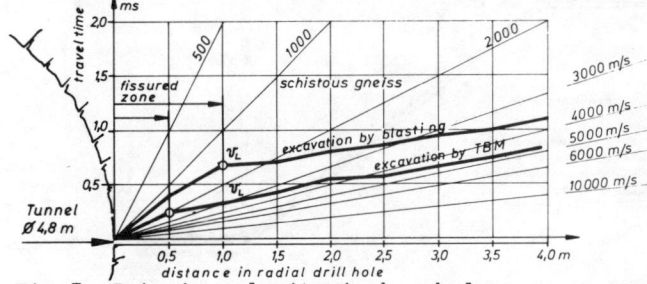

Fig.7: Seismic velocity in boreholes

Each borehole was sealed outside the disturbed zone at a depth of 1.0 m to 1.5 m and pressurized to the full depth of 4.0 m. First a Lugeon test was performed under low water pressure and then the pressure was increased until the rock fractured. Hydraulic fractures develop along the path of least resistance, i.e. in a plane perpendicular to the direction of the least in-situ compressive strength. The pressure needed to keep the fracture open (shut-in pressure) corresponds to the smallest in-situ rock mass stress (HAIMSON 1977). The measurement results were recorded along the pressure shaft and compared with the theoretical overburden pressure and operating pressure (Fig. 8). The shut-in pressure forms the safety limit for the applied grouting pressure when prestressing the concrete lining.

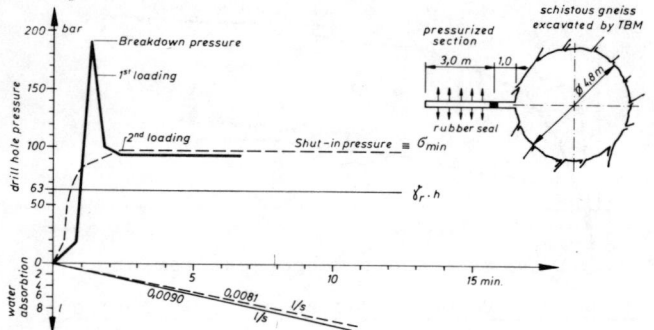

Fig.8: Hydraulic fracture operation

4. LOWER STAGE PRESSURE SHAFT

Due to its extremely high static head in the lower stage, namely 1257 m, and its dynamic pressure head of 1500 m, the Silz Pressure Shaft is under the greatest stress in the whole world. It is steel lined along its entire length (Fig.9). The shaft was situated so that on the one hand the overburden pressure would exceed the portion of internal pressure to be absorbed by the rock mass and on the other hand so that the water pressure in the rock mass would not provide too much external water pressure. Nevertheless, in the upper section the steel lining was protected against buckling by means of a concrete lining. The 500-m-long level section was first driven as an exploratory tunnel and later expanded. The 1900-m-long inclined shaft was mechanically driven upwards at a slope of 80%. The shaft ist situated in geologically varying formations of cristalline series. The upper and level sections are in mica schists, the lower shaft section in ganitic gneiss. For economic reasons, the steel lining was designed to absorb the high internal

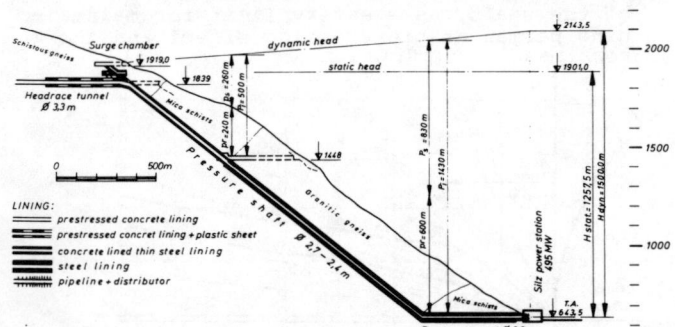

Fig.9: Cross section, Silz pressure shaft

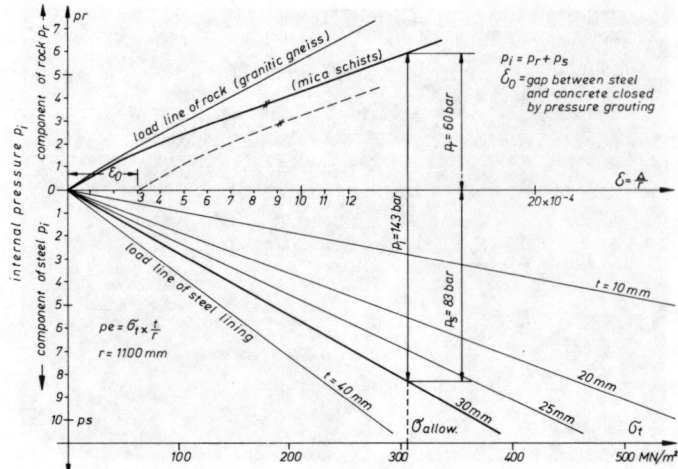

Fig.10: Design chart for steel lining

pressure while making good use of the rock mass' contribution to the procedure already applied by the TIWAG at the Kaunertal Power Plant (LAUFFER, SEEBER, 1961 and 1966). With the load lines of the various rock types directly obtained from the radial jacking tests it was possible to dimension the steel lining with the graph (Fig.10). Here, the rock load plotted upwards in dependency on the radial strain is the load line of the rock or the portion of internal pressure (pr) absorbed by the rock. That portion of the internal pressure is plotted downward, which is absorbed by the steel lining (ps).

5. ROCK MECHANICAL MEASUREMENTS AND INVESTIGATIONS FOR THE LOWER STAGE PRESSURE SHAFT

Already several years before construction was started, core samples were taken, exploratory tunnels driven and the geological and hydrogeological conditions of the slope site of the pressure shaft investigated (Fig.1). In order to evaluate the rock properties, a measurement program similar to that described under point 3 was carried out with its main point of focus being to determine the rock mass' contribution under internal pressure in the various rock zones - as the basis for designing the steel lining. For this purpose six measurement sections were laid out for the radial jack, three in the exploratory tunnel of the level area and two in the exploratory tunnel of the center adit tunnel, all of these in mica schists and granitic gneiss and one in the schistous gneiss of the headrace tunnel adit. In these test chambers the seismic measurements were calibrated and

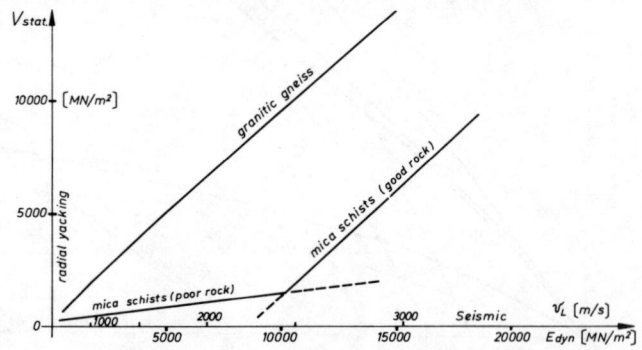

Fig.11: Correlation Eseismic / Vstatic

brought in correlation with the moduli of deformation (Fig.11). After this, using the borehole seismism the moduli of rock deformation were given along the shaft alignment and the design of the steel lining was checked.

5.1 Radial Jack Testing with Compensation of the Excavation Convergence

In the level section's exploratory tunnel the thin-layered mica schist offered poor prerequisites for measuring the in-situ rock mass stresses according to the usual methods. Therefore, an attempt was made to compensate the deformations that excavation had caused by pressing them back with the radial jack. The necessary pressure would correspond to the overburden pressure. These attempts did not run satisfactorily, however should be mentioned nonetheless.

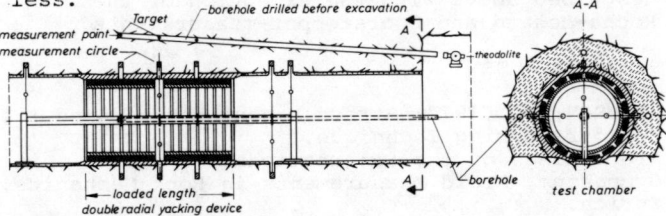

Fig.12: Convergence compensation test chamber

Before excavating the test chambers (Fig. 12), 8-m-long borings were made in the roof and the side wall in good and poor mica schists of the two twin measurement sections with the radial jack. Pipes with measurement markers were installed in these boreholes which were geodetically measured during tunnel driving and then under stress with the radial jack. In good rock the convergences were small and the redeformation caused by radial jacking tests up to 5o bars pressure with the geodetic method could no longer be measured with sufficient accuracy. Radial deformations up to 4o mm occurred in the poor mica schist due to excavation of the exploratory tunnel. In radial jacking attempts at 4o bars pressure these could not be brought back to normal. In the measurement circuit for the radial jack, deformations up to 12 mm were measured, which however decrease by one power of ten at the measurement markers 1 m deeper.

Overburden at this point was approximately 17om.

6. TEST CHAMBER

A prototype trial section of the concretelined pressure tunnels sealed with plastic membrane was constructed in the form of a test chamber with 2.o m internal diameter and 25 m of length in an exploratory adit of the lower stage (Fig. 13). The objectives of the test program were to determine the in-situ casting of the lining with PVC sheeting, the behavior during prestress grouting with the adapted TIWAG system and the short- and long-term operating behavior of lining and rock bedding under internal pressure conditions up to 5o bars. The test chamber was located in an area representative of the weaker rock conditions of the waterway where a radial jacking test had previously been performed to determine the deformability of the rock mass under internal loading. The prototype lining system consists, from inside to outside, of the following layers:
- concrete lining
- waterproof 3-mm thick PVC sheeting
- plastic felt (a loosely woven geotextile to protect the PVC sheet from the roughness of the rock or shotcrete surface)
- thin plastic sheet (to prevent the grout from intruding into the plastic felt when the circumferential joint between rock and concrete lining is pressure grouted).

The felt is fixed to the rock or shotcrete surface using a special PVC roundel. The sheet is then fixed to the roundels by heat welding. To allow grout injection through the sheet, special injection collars are welded to the sheet. In order to control grout distribution, rings or spirals of plastic felt are attached to the shotcrete prior to placement of the sheeting. As can be seen from the test program (Fig. 14), in addition to installing the plastic sheet system, its waterproofness was tested in an external pressure test. Then, the lining was prestressed using high-pressure grouting and tested for permeability at 5o bars and long-term loading. The extensive instrumentation in the pressure chamber during prestress grouting and the permeability test supplied information on the lining's behavior. The plastic sheet that was slightly damaged during installation was made practically watertight by grouting; the

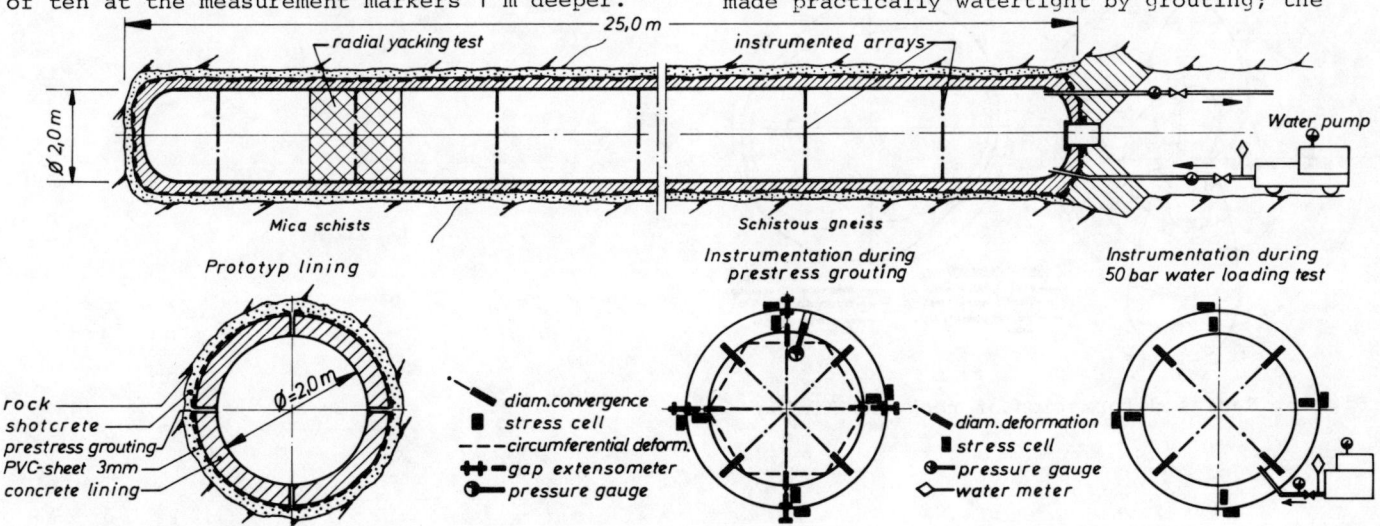

Fig.13: Test chamber layout and instrumentation arrangement

50-bar internal pressure test showed a water loss of only 0.8 l/min.=5.10^{-3} l/min/m2. Figure 15 shows the radial deformations caused by prestress grouting and the rock mass' contribution during the permeability test and, as a comparison, the radial rock deformations during the first load application by means of the radial jack. Using water records, the volumetric increase in the chamber under internal pressure was determined and the mean modulus of elasticity calculated, which corresponds well to the mean modulus of elasticity of the radial measurement section. This lining system was installed at the Sellrain-Silz Power Plant in the pressure shaft and the tailrace shafts of the upper stage and in the lower stage's headrace tunnel and surge chamber along a total tunnel length of 1400 m.

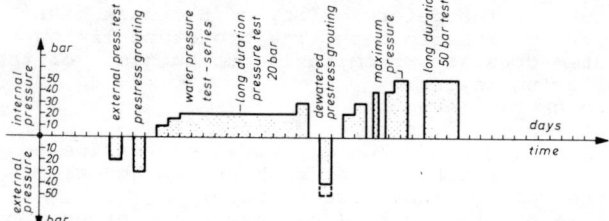

Fig.14: Test chamber program

7. SUMMARY

Before the Sellrain-Silz Power Plant was put into operation, the waterways of the upper and lower stages were subject to pressure test to check the design and execution of the lining. The pressure shaft of the upper stage was subject to the maximum static pressure from the valve chamber to the power plant for one month. The water loss for the 700-m-long, purely concrete section was only 1.5 l/s after reaching the original seapage level. Diameter measurements during the pressure tests showed that the behavior of concrete lining and rock mass correlated well with predictions based on elastic theory and proved that the lining would remain crackfree in service.
No longitudinal cracks were observed during subsequent inspection.

In the pressure test, the steel lining of the pressure shaft in the lower stage was put under the full dynamic pressure of 150 bars. Extensive stress measurements of the steel lining confirmed that the stress on the steel does not exceed the allowable and also confirms the economic design of the lining making good use of the prevailing rock mass' contribution. The good results in the pressure test and satisfactory operation of the pressure shafts and tunnels over two years are proof that the extensive, sometimes simple, methods were an effective means of systematically establishing the rock mass behavior under internal pressure.

8. ACKNOWLEDGEMENTS

The entire planning for the power plant was performed by the TIWAG as well as the tests described above with the exception of the Rocha-Menard-and Doorstopper-measurements.

9. REFERENCES

HAIMSON,B.C.,1977: Stress Measurements using the Hydrofracturing Technique.
Proccedings of the international
Symposium, Field Measurements in Rock Mechanics, Zurich.

LAUFFER,H.,SEEBER,G.,1961: Design and Control of Linings of Pressure Tunnels and Shafts based on Measurements of the Deformability of the Rock.
7th ICOLD-Congress Rome.

LAUFFER,H., SEEBER,G.,1966: Measurements of deformability of rock mass using the TIWAG-radial-jack and control with stress measurements on the steel lining of the pressure shaft of Kaunertal Power Scheme. (German).
1st ISRM-Congress, Lisbon.

LAUFFER,H.,1968: Prestressing grouting for pressure tunnels (German).
Der Bauingenieur 7.

PIRCHER,W.,1980:Experience with Mechanised Tunnelling for the Sellrain-Silz Hydropower Scheme (German).
Rock Mechanics, Suppl. 10.

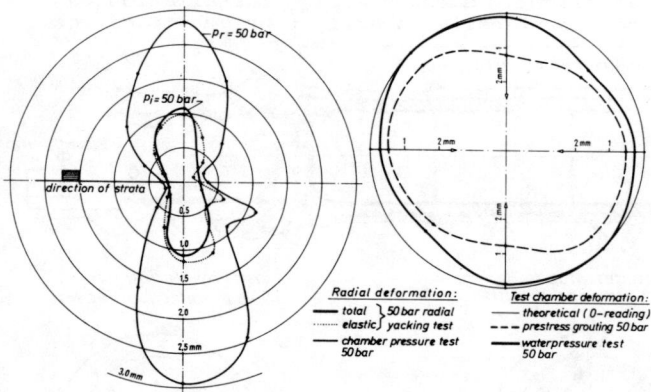

Fig.15: Radial deformation of rock and lining

MAJES PROJECT: GEOLOGICAL CONDITIONS AND CONSTRUCTION
Projet Majes: Conditions géologiques et construction
Projekt Majes: Geologische Verhältnisse und Bauausführung

Giancarlo Ceriani
MSc. – Geologist, ELC-Electroconsult, Italy

Gunnar Nord
MSc. Civil Engineering, Skanska Cementgjuteriet, Sweden

SYNOPSIS

Since the mid 70's 100 km of tunnels and 40 km of canals were constructed high up in the Andes in Southern Peru. The pre-investigation was carried out on a low-budget basis but turned out to be a good instrument for the planning and cost estimate. The pre-investigation program and the consequences for design and construction are presented by describing and analyzing the actual conditions for some tunnel stretches of special interest.

RESUME

Depuis le milieu des années 70, 100 km de tunnels et 40 km de canaux ont été construits à des altitudes élevées dans les Andes au sud du Pérou. La pré-investigation, bien que faite sur la base d'un budget restreint, s'est avérée très utile pour le planning et l'estimation des coûts. La pré-investigation et les effets de celle-ci sur les travaux de conception et de construction sont décrits en présentant et analysant les conditions réelles de quelques zones limitées des tunnels.

ZUSAMMENFASSUNG

In der zweiten Hälfte der siebziger Jahre und in den achtziger Jahren wurden 100 km Tunnel und 40 km Kanäle in den Hoch-Anden Südperus gebaut. Die Voruntersuchung wurde im Rahmen eines möglichst begrenzten Budgets durchgeführt, erwies sich aber dennoch als eine gute Unterlage für die Planung und Kostenberechnung. Die Voruntersuchung und deren Auswirkung in Bezug auf Konstruktion und Bau werden erläutert, und die gegebenen Verhältnisse in einigen begrenzten Tunnelstrecken werden beschrieben und analysiert.

1. GENERAL DESCRIPTION OF THE PROJECT

1.1 The purpose

The daringly conceived and impressive Majes Project, located close to Arequipa in Southern Peru, is at present under construction.

The purpose of the project is to divert the water from the Colca River to the coastal desert plains by means of tunnels and canals, for irrigation and hydroelectric power.

1.2 Layout of the Project

The scheme of the entire project has been designed by the Italian consulting firm ELC - Electroconsult of Milan, who are also in charge of the supervision of the construction in co-operation with two Peruvian companies.

Due to its complexity, the construction of the project has been divided into two stages.

The first stage includes:

a) the Condoroma Dam, 5 million m^3;

b) the Colca - Huambo Upper Conveyance with 88 km of tunnels with a cross section of approx. 12 m^2, and 13 km of canals;

c) The Majes Lower Conveyance with 11 km of tunnels with a cross section of 8 m^2 and 10 km of canals;

d) The Distribution Canal Network, for the irrigation of 23,000 ha.

This stage is ready for operation, except for the Condoroma Dam due to be completed at the end of 1984. The Majes Consortium (MACON), an International Joint Venture of five contractors sponsored by Skanska of Sweden is responsible for the construction of this stage.

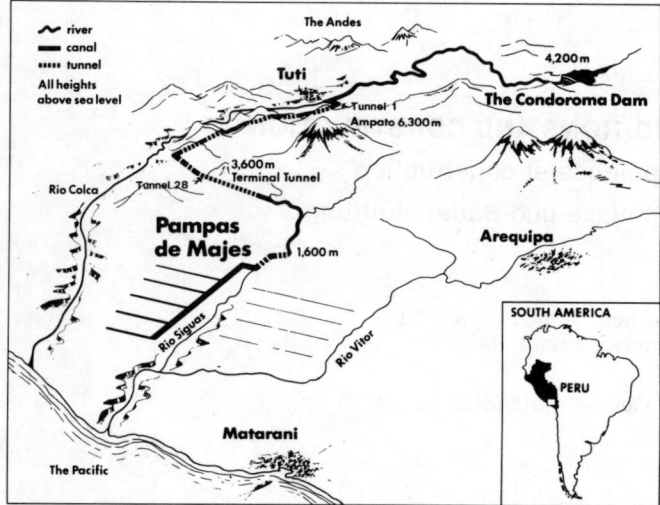

Figure 1   Layout of the Majes Project

The second stage comprises:

a) The Angostura Dam, 3 million m^3;

b) Tunnels and canals, 33 km;

c) The Distribution Canal Network for the irrigation of the remaining 34,000 ha;

d) Two hydroelectric power plants with a total installed capacity of 656 MW.

1.3 Time - Cost - Financing - Benefits

The construction period for the first stage was stipulated at 5 years, but will be 7 years.

However, the first cultivation of the desert plain has already started in 1982, and the whole project will be completed by the 1990s.

The cost for the first stage is estimated at about USD 700 million, while the cost of the entire project is estimated at about USD 2,000 million.

The project is financed to about 50 % by local Peruvian funds, and the balance by the Majes Consortium by means of a medium-term loan.

The project will affect the economy of the entire region of Southern Peru and also the income and welfare of future generations.

2. GENERAL DESIGN CONSIDERATIONS AT THE DESIGN STAGE

2.1 Geological conditions

The tunnels are to be driven through greatly varying ground conditions, mainly volcanic formations with a very complex structure derived from the Quaternary tectonic activity in the Andes, which has affected the Mesozoic metamorphic basement in the Project area.

The rock types with more favourable characteristics for tunnel work are: sound andesite and basalt, volcanic breccia, ignimbrite, porphyritic diorite, limestone and quartzite.

Less favourable conditions will be encountered in weathered volcanic rock, or intensively fractured limestone and quartzite, in slate and shale, sandstone, slightly cemented conglomerate and consolidated mud-flow. The tunnels will also pass through colluvial deposits, soft sedimentary rock and very weathered volcanic and metamorphic rock.

The worst tunnelling conditions will be encountered in volcanic rock deeply altered by residual hydrothermal activity with secondary clay products giving the ground swelling and squeezing characteristics, and in volcanic sand and ash deposits, dry and with water.

2.2 Design - decision model

The design of the Majes Project implies a progressive implementation of the basic design criteria in order to match the original philosophy with construction reality.

In the decade between the drawing up of the Contract Documents (1964-1967) and the start of construction (1975), the general situation has

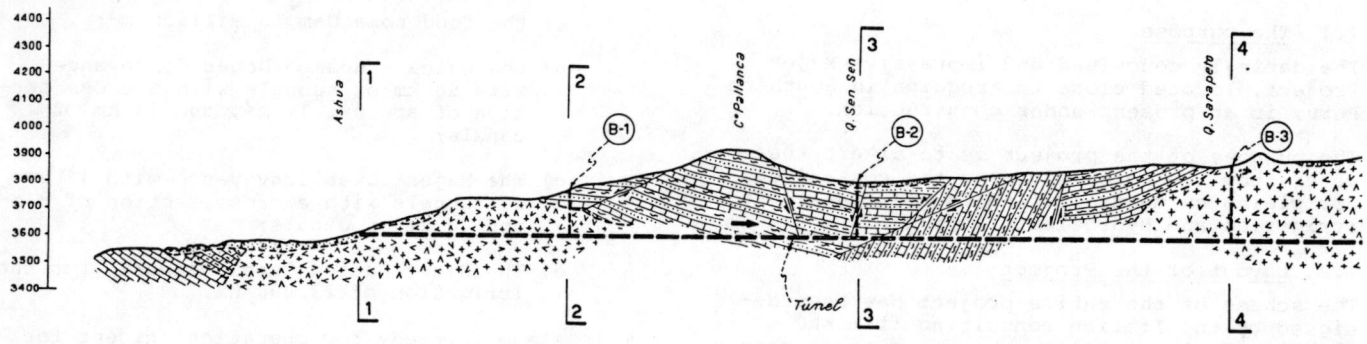

Figure 2

Geological section of Terminal Tunnel as a result of the preinvestigations.

evolved in such a way that serious reconsideration of the bases of the previous design had to be made by the Designer, taking into account for example:

a) the increased cost of canal construction by a factor of 2-3

b) the very limited availability of a skilled workforce

c) the favourable evolution of modern techniques for underground works

d) the increased cost of land expropriation even in hillside terrain which has been intensively cultivated for centuries

e) and of paramount importance - the safety of the Project in respect of high seismic risk, slope stability and costly maintenance of the canals

2.3 Contractual Conditions

The Majes Project presents working conditions and situations with many unforeseen and unquantified aspects which require the application of a "Target Value" type of contract. This type of contract guarantees a proper equilibrium between earnings and losses to both the Contractor and the Client, by evaluating the balance between the "estimated cost" and the "actual cost" of all the works.

The Target Value type of contract is flexible enough to absorb project modifications smoothly without the difficulties these would cause with a standard "Unit Price" type of contract.

2.4 Hazards

The geological investigations indicate the possibility of three main types of hazards in the tunnel excavations: fault zones, water inflows and high temperatures.

Most of the tunnels will cross mountain ridges affected by faults irregularly oriented, in which millenary erosion processes have created valleys and gorges with streams and small rivers. These spots will represent special hazards for tunnel driving with water-clogged, fractured and shear zones, which will require the application of adequate excavation and support techniques.

There are many contacts between permeable and impermeable geologic formations which may originate conditions in which water pockets are formed.

The Colca - Huambo Conveyance will also be influenced in some areas by a residual hydrothermal activity of volcanic origin. Thermal water (50 - 60°C) flows out close to the area through which the last km of tunnel No 1-2 will pass, and high temperatures and warm water are to be expected in the vicinity of this part. Outside evidence of some mineralisation in the mountains through which tunnels No 6-7 and No 10 will pass indicates the possibility of encountering special rock conditions somewhere inside.

3. THE PREINVESTIGATION PROGRAM

3.1 The system and philosphy

During the 60's and 70's, geological and geotechnical studies were carried out over the whole project area. The entire route of the tunnels and canals was carefully studied by photogeology and detailed surface geological mapping with special attention to inlet and outlet areas of the proposed tunnels where exploratory adits were excavated. The canal route was investigated by 35 full-size canal open cuts, 10 trenches and 30 pits, in order to get the soil-rock data for future excavation. The longest tunnel called Terminal Tunnel (14.9 km) was also investigated by 5 rotary drill borings (750 m in all) at both inlet and outlet areas, plus an almost 100 m long adit at the inlet.

The philosophy of the investigation program was based on economy and good sense. Money was not available at that time (1964-1967) for a complete investigation of the large area covered by the project routes. The expensive exploratory drilling holes had to be kept to a minimum and concentrated on the dam sites, with the exception only of the longest and more problematic tunnel.

The Designer had as best he could to utilise the photogeology interpretation, the field geology and geotechnics, extrapolating the results with good scientific sense and experience for the final design of the scheme.

In 1975, when the Contractor MACON started the works, more money was available for some extra investigations for the better adjustment of the project design to the construction phase. Almost 100 km of seismic refraction profiles were executed, mainly on the canal routes and tunnel portals, the most accessible areas, which also contained the majority of variable and unknown aspects where construction was concerned.

3.2 The results

The preinvestigation data have provided the following results:

a) from a <u>design</u> point of view, they have proved to be an efficient tool to calibrate the entire tunnels - canals route in order to avoid as far as possible any major constructional and long term functioning - maintenance problems

b) from a <u>construction</u> point of view, they allowed the Designer to give the Contractor a reasonable estimation of the different kinds of rock and their distribution, tunnel by tunnel. In practice the rock types to be tunnelled and their estimated quantities were as follows:

<u>A rock - high-quality rock</u>: sound volcanic rock, such as andesite, basalt, breccia and ignimbrite (volcanic tuff) - porphyritic diorite, limestone, quartzite - about 55 %.

<u>B rock - medium-quality rock</u>: slate, shale, sandstone, conglomerate, consolidated mud flow - type A rock with

alteration - about 30 %.

C rock - low-quality rock: colluvial deposits, soft sedimentary rock. Type A and B rock very altered - about 14 %.

D rock - difficult rock: deeply altered rock. Volcanic sand and ash deposits. Fault zones. About 1 %.

4. PLANNING OF THE EXCAVATION PROCEDURE

At an early stage in the planning it was settled that the tunnel driving had to be done by the drill and blast method. The prognoses from the preinvestigation were not very accurate and the drill and blast method offered great flexibility to cope with varying and unknown conditions. It was not considered worthwhile to use a roadheader due to the fact that the dominating rock types could not be excavated economically by such equipment.

The cross section for excavation was settled for 18 m². Most of the tunnels were planned for a first 500 metres of excavation using rubber-wheeled equipment, due to the need of material for the activities at the tunnel portals. Pneumatically-powered rock drills were chosen and mounted on a railbound Atlas Copco tunnel drill rig in order to achieve as little overbreak as possible by accurate alignment of the drillrods. The smooth blasting technique was applied in order to limit damage to roofs and walls and to minimise overbreak and support. The trim holes were closely spaced and charged with a low-velocity explosive Maconite.

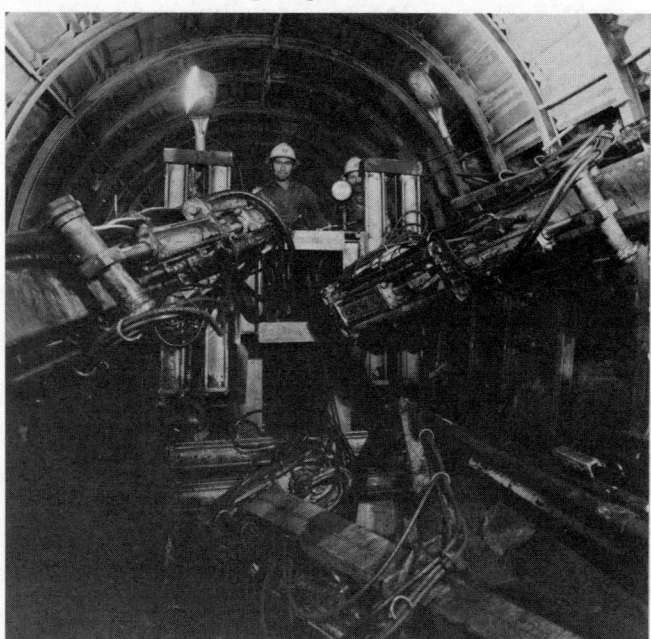

Figure 3   Drillrig with booms for accurate alignment of drillrods.

In order to work out practical guidelines for the support work, the prognosis was extended by assumed support measures and stand-up time, see table below. The distribution of the various rock conditions was specified for each tunnel.

This information made it possible to plan the input of labour, plant and materials and to determine the time schedule.

Table 1

Rock type	Temporary support during excavation	Estimated stand-up time for unsupported rock
A High quality rock	None (shotcrete 1"-2" in limited areas and only when the final lining was delayed)	Months or a year
B Medium quality rock	Shotcrete 1"-3" occasionally with mesh Ø 5 100 x 100 mm and rockbolts	More than 1 day
C Low quality rock	Steel arches 4"x4" and 5"x5" at 0.75-2.0 m spacing, shotcrete and occasionally lagging and reinforcement bars	Hours taking into account shorter rounds
D Difficult rock	Steel arches, inv. struts, forepoling, pregrouting, concrete prelining	Less than hours

The various rock classes and support measures

Table 2

Tunnel	A rock	B rock	C rock
1 - 2	73 %	22 %	5 %
10	70 %	15 %	15 %
15	50 %	30 %	20 %

Examples of specification of various rock classes for each tunnel.

5. DESCRIPTION OF THE EXCAVATION PROCEDURE

The results of the preinvestigation could not be taken as a detailed specification of the various types of support measures along the tunnel alignment. For each individual tunnel the estimated total quantities of various support classes were specified. The construction team had to be prepared to carry out any kind of support throughout the tunnels. To be successful under such conditions it was necessary to have experienced construction personnel on site, authorised to take decisions. A proper geological follow-up had to be done in order to take advantage of the experience gained. An open dialogue between Engineer and Contractor was, under these conditions especially important.

Excavation in A and B rock

When excavating in A and B rock, no great difference was observed in the average driving speed. Sometimes other factors like the skill of the crew, and the condition of the equipment had a greater impact on the excavation rate than the rock conditions. It can be stated that the highest rate of advance was achieved in the fresh andesite, volcanic breccia, tuff and ignimbrite with only minor seepage of water. It is natural that fresh and sparsely jointed rock will give the best driving conditions no matter what the type of rock is. The average speed was between 200 and 250 m per month. There were not many alterations in the drilling and charging pattern with respect to the rock conditions.

As can be seen from the table below, changes in the distribution of various rock types were encountered, but had in fact very little impact on the tunnel driving. It can thus be stated that the input of preinvestigation work in this case was very well balanced.

Figure 4   A-type rock

Excavation in C rock

When excavating in C rock, the length of the rounds was shortened to 1/2 to 1/3 of the length used in A and B rock in order to achieve a stand-up time long enough to allow for putting up steel sets.

Certainly there were situations where bolts would have worked out as well as steel ribs and shotcrete. But bolts were very sparsely used. The decision to use steel sets was made at the planning stage and might have been influenced by the traditional and American way of thinking.

Even during the excavation procedure it might be difficult to determine the correct rock conditions. Tunnel 10 exemplifies this problem: 500 metres of this tunnel were classified as A and B rock but changed to class C rock during 1/2 to 1 year's time. The normal excavation procedure was followed and good advance rates were achieved, just under 200 metres per month. The rock was andesite and volcanic breccia which looked more or less fresh and no more than 1-2 layers of shotcrete were considered necessary as temporary support. This support was also adequate during the whole excavation. The tunnel was then virtually abandoned for 1/2 year until the lining operation was due to start.

When preparations for lining got under way, it was observed that the shotcrete had started to peel off partly in large chunks and the bolts holding pipes and ducts had started to loosen. Even larger fall-ins had taken place. The rock had simply decayed behind the shotcrete and there was no adhesion at all between shotcrete and rock in places. Squeezing or swelling, however, were not observed.

Figure 5   C-type-rock support

Consequently the tunnel had to be repaired before the lining operation. New layers of shotcrete were applied. This operation was partially unsuccessful due to the fact that the shotcrete seemed to accelerate the decaying of the rock. To solve this problem the rock was drained. This method included an invert slab so that water could not find its way up behind the shotcrete layer by capillary forces. Once the rock was dry the decaying process stopped and lining could start.

As a point of interest it can be mentioned that pyrite mineralisation was observed within the andesite.

Table 3: Advances for the Upper Conveyance Colca-Huambo, 88 km of tunnel excavation

Type of Rock	Planned	
	Rock Type % of length	Advance m/month
A	55	200
B	30	120
C	14	60
D	1	-

Type of Rock	Actual	
	Rock Type % of length	Advance m/month
A	36.8	200
B	41.1	180
C	19.8	90
D	2.3	40

Note that the planned distribution on various rock types is based on 61 km of tunnels.

Excavation in D rock

Excavation in D rock will be illustrated by two examples.

In all, 400 metres of altered rock causing swelling or squeezing were met with in the total of 100 km of tunnels excavated in the entire project. The first stretch, 80 m occurred in Tunnel 6 and the second stretch, 320 m, in Tunnel 10.

The potential for squeezing ground was not discovered during the preinvestigation and consequently no excavation in squeezing ground outside fault zones was anticipated. When stability problems arose, they came as a surprise.

When the excavation entered the altered andesite in Tunnel 6, the squeezing effect was not immediately observed. It was, however, clear that the tunnel face had entered into altered rock with slickenside on the joint surfaces. Normal support with shotcrete and steel sets was applied as the first support step.

As the deformation of the first support was as much as 0.5 m at springline, it was obvious that it would not be stable much longer. Then the second support step was taken by applying heavier steel sets 6" x 6" spaced at 0.6 m and invert struts. But even this support turned out to be partially insufficient. A third support step was then carried out in the form of a complete temporary concrete prelining outside the final lining. The excavation rate with this 3-step support was as low as 22 m per month.

The alteration of both the andesite and volcanic breccia was so far advanced that pieces of rock could be broken off very easily by hand although most of the original structure was intact.

The question then arises as to how the construction procedure could have benefited if better information had been available. This question is best answered by looking at what happened when similar conditions were met with in Tunnel 10 later on.

In Tunnel 10, when the tunnel heading entered into altered volcanic breccia heavy steel sets were immediately used in an enlarged tunnel section to provide for both squeezing/swelling and adequate space for a temporary concrete lining. First shotcrete was applied and then steel sets with invert struts were erected plus a steel blocking between the steel and the shotcrete with 70 cm spacing. This was a method that had been used at a late stage in Tunnel 6 and had turned out to be successful. When these support measures became insufficient to limit the deformations, the prelining was applied and carried out at 3-5 metre intervals. By using this two-step method, the rate of advance was double the rate in Tunnel 6.

Figure 6   Prelining in tunnel 10

To identify these very limited stretches of squeezing ground in a preinvestigation is generally very costly. However, with the excavation completed, and knowing what happened it can be stated that the conditions could have been foreseeable taking into account what is said under the heading hazards about mineralisations on the slope surface.

Figure 7   Excavation in D-rock (Volcanic ash) with forepoling.

The hazards of heavy water inflow and high temperature

As mentioned earlier, high temperatures were anticipated in the last part of Tunnel 1-2 in conjunction with heavy water inflows. This prediction turned out to be accurate and caused a slow down in the driving speed of the tunnel. At 4 points water inflows of up to 50 l/sec were recorded and the temperature went up to 45°C. These conditions did not interrupt the excavation but reduced the driving speed by 25-30%, mainly due to water inflow.

High temperatures of more than 50°C and heavy inflows of water occurred in Tunnels 6 and 10 but not at the same chainages. In this case the obstacles were not predicted, and caused a slow-down in the driving speed of about 30%.

The last 900 m of the 5.5 km long Tunnel No 10 cut a shallow water table, which originated from a widespread water filtration totalling 200 l/sec constant flow. The water has been collected and diverted out of the tunnel and is used to solve an irrigation problem in two villages.

The 15 km long Terminal Tunnel, mainly driven through limestone, hit 10 sources of water giving inflows varying from as much as 600 l/sec to 100 l/sec, the mountain water table being completely drained.

Large water inflows were anticipated in the Terminal Tunnel and therefore it had been designed with inclined drifts from both ends.

Site follow-up

The excavation fronts were checked daily by two survey teams (Contractor and Supervision) and by a geology team (Supervision).

The surveyors had constantly to check the tunnel alignment and the excavated sections in order to define the over-under excavations and the future volumes of the final lining concrete.

The geology team had three main duties to perform, i.e.:

a) detailed geological mapping of the tunnel excavations on logs at a scale of 1:200. Special attention was paid to the rock types, their geotechnical characteristics such as discontinuities, fractures, fissure-filling materials, shear zones etc, and to any water presence.

b) classification of the rock according to the four rock types established, both for contractual and technical purposes.

c) immediate discussion with the Contractor in order to reach mutual agreement on the most adequate type of support to be installed.

This ambitious follow-up served three purposes:

1. as an input in the learning process for better excavation and support as the work proceeded, and to provide an improved prognosis of conditions ahead.

2. to document what is hidden behind the final lining in the event that any additional work has to be done in the tunnel.

3. feedback for future projects for the parties involved.

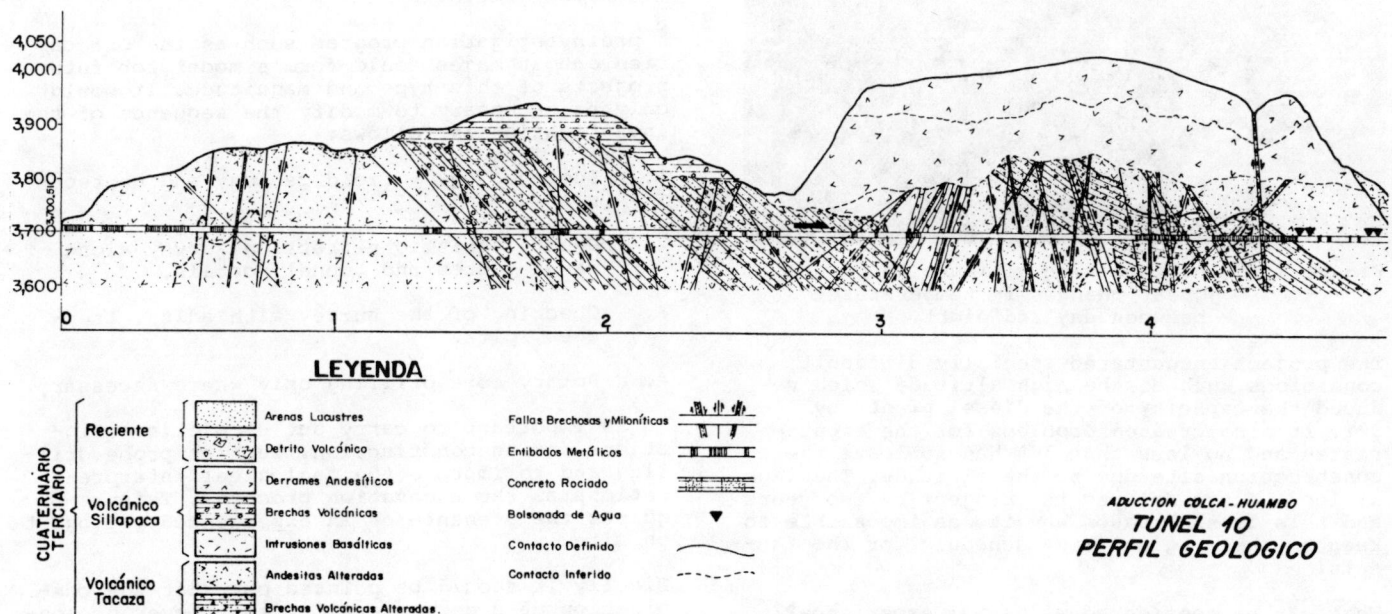

Figure 8   Documentation of the geological follow-up presented in a very condensed form.

6. WHAT CAN BE LEARNT FROM THE MAJES PROJECT

First some words about the limitation of the experience gained.

We are dealing with small tunnel driving, although some shorter tunnels with larger cross sections of up to 60 m^2 were also constructed.

There were a lot of tunnel headings available to work at as most activities were going on at 12 headings at the same time. This meant greater freedom than usual to reorganise the construction work if major problems arose to which immediate solutions could not be found.

Most of the tunnels were lateral (parallell to hillside) and only two passed under a mountain ridge. The main water table in the lateral ones was below the tunnel level which meant that large water inflow was limited to minor stretches where local groundwater was present.

The vertical overburden of rock was generally moderate but the tunnels were rather close to the hillside and this could have caused an unfavourable stress situation with rockburst and spalling. However, this did not occur.

The impression is that no high residual tectonic stresses existed in the tunnel areas. Most of the rock seemed to be released.

Figure 9   At this high altitude there are great changes in temperature between day and night.

The project encountered specially difficult conditions such as the high altitude which reduced the capacity of the diesel plant by 30%. It also created problems for the expatriates and no less than 10% had to leave the construction site due to the altitude. The lack of local funds delayed the project by two years and this is the reason why it was impossible to keep to the original time schedule for the tunnels.

What can be considered as useful experience?

It is most important to have the water situation well under control. In some cases it is necessary to check water conditions ahead of the tunnel face. Hydrothermal activity has created problems. In retrospect it can be stated that these effects could have been discovered when cuts for the access roads to the tunnels were constructed. The magnitude of the hydrothermal effect is however very difficult to foresee and the correct counteraction can only be learnt by trial and error during the construction process.

When excavating in dry conditions, do not just accept rock surfaces which look good, because leakage of water at later chainage may cause soaking in the dry sections. Therefore carry out simple testing of slaking and swelling.

When making waterways in areas with seismic activity, favour tunnels instead of canals and limit the amount of connections between tunnels and canals, due to scree hazard.

It is advisable, in a situation like this when the well-known NATM is difficult to apply due to its sophistication, to be prepared to put in a stiff temporary lining i swelling and squeezing ground.

How could the preinvestigation program have been improved? The seismic investigation of the tunnel portals and the canals gave good results. One mistake was made in the interpretation but it is not likely that a systematic drilling exploration at the portals is the right solution. Core drilling equipment shall be available on site during construction work.

One of the major experiences gained is that it is most important to focus the investigation efforts on the difficult rock conditions. In this case difficult rock was 2.3% of the total tunnel length but was 5-10 times more cost- and time-consuming. The effort to localize the difficult rock conditions should match that ratio.

A preinvestigation program such as the one carried out at Majes could form a model for future projects of this type and magnitude. It would only be necessary to modify the sequence of the investigation as follows:

1. Photogeology - field geology and geotechnics

2. General seismic refraction survey along canal routes and tunnel portals.

3. Checking of the survey with adits, trenches, pits.

4. Rotary core drilling only where necessary.

It is important to carry out simple investigations during construction, such as probe drilling and to improve the geological interpretations as the excavation proceeds. This requires the presence of an experienced geologist on site.

Finally it should be pointed out that the combination of a good overall view, clever geological judgement and good production experience has given good results. A daily and open dialogue between the Engineer and the Contractor is, however, a prerequisite for success.

ROCK BURST PROBLEMS IN A 2.6 MILLION m³ UNDERGROUND CRUDE OIL STORAGE IN GRANITE

Problèmes d'éclatement de roche dans un entrepôt souterrain de 2,6 millions de m³ creusé dans du granite et destiné au stockage de pétrole brut

Abplatzungsprobleme in einer 2,6 Millionen m³ grossen Granitkaverne für Rohöl

S. G. A. Bergman
Dr.Eng., Consultant, Stocksund, Sweden

H. Stille
Dr.Eng., Ass. Prof., Dept. of Soil and Rock Mechanics, Royal Institute of Technology, Stockholm, Sweden

SYNOPSIS

At Brofjorden in the western part of Sweden a 2 600 000 m³ crude oil depot has been excavated in very good granite. Three parallel twin-tunnels with an average length of 700 m have a span of 20 m and a height of 30 m. Although similar tunnels had been excavated nearby without any manifest difficulties, in this case rock burst phenomena from the roofs came to have a serious influence on the tunnelling operations. The events are briefly described including in situ stress measurements, various support methods, convergence measurements and laboratory tests. The contracted roof support method with spot bolting and a thin shotcrete layer had to be changed to wire mesh protection, which in turn necessitated a re-design of the storage operation system to meet the altered risk for gas explosions. FEM-calculations made after the excavations are compared with measurements and other experiences.

RESUME

Dans un excellent granite à Brofjorden dans l'Ouest de la Suède on a creusé un dépôt souterrain pour pétrole brut ayant une capacité de 2 600 000 m³. Il s'agit de trois galeries jumelles et parallèles longues de 700 m en moyenne, 20 m de potrtée et 30 m de hauteur. Des galeries similaires ont été percées à proximité sans présenter de difficultés évidentes, mais dans le cas présent un éclatement de roche dans la voûte a affecté considérablement les opérations de creusement des galeries. Nous décrivons brièvement les manifestations de ce phénomène, y compris les mesures de tension prélevées in situ, diverses méthodes de soutènement, des mesures de convergence et des tests en laboratoire. Nous avons abandonné la méthode de soutènement contracté de voûte avec boulonnage ponctuel et injection d'une mince couche de béton au profit d'un dispositif de protection en treillis, ceci nécessitant à son tour une conception nouvelle du système des opérations d'entreposage afin de faire face aux risques de coups de grisou. On compare les calculs FEM post-excavation avec des mesures prélevées et avec d'autres expériences.

ZUSAMMENFASSUNG

In der Nähe des Brofjorden an der schwedischen Westküste ist ein 2,6 Million Kubikmeter großer Rohöl-Lagerraum aus festgefügtem Fels (Granit) ausgebrochen worden. Drei parallele Doppelstollen haben eine durchschnittliche Länge von 700 m, eine Höhe von 30 m und eine Breite von 20 m. Obwohl nicht weit entfernt ähnliche Hohlräume ohne nennenswerte Schwierigkeiten ausgebrochen wurden, kam es im Firstbereich der hier beschriebenen Stollen zu Abplatzerscheinungen, die sich schwerwiegend auf die Vortriebsarbeiten auswirkten. Der Bericht bringt eine Kurzbeschreibung dieser Vorgänge sowie eine Darstellung der Messung von Spannungen im Ursprungszustand, der verschiedenen Abstützungsverfahren, der Konvergenzmessungen und der Laboruntersuchungen. Von der ursprünglich vorgesehenen Absicherungsmethode mit Punktbolzen und einer dünnen Spritzbetonauskleidung mußte abgegangen und statt dieser ein Drahtgeflechtausbau eingesetzt werden. Als Folge davon mußte — um der Gefahr von Gasexplosionen entgegenzuwirken — ein neues Betriebssystem für die Lagerung erarbeitet werden. Nach den Ausbrucharbeiten erstellte FEM-Berechnungen werden Meßergebnissen und gewonnenen Erfahrungswerten gegenübergestellt.

1. GEOLOGY

The bedrock at the plant site is only to a small extent covered by earth and consists of granite with slight incrustations of pegmatite. Seismic profiles (8) and core drillings (10) showed a very good rock mass quality - RQD = 0.9-1.0.

Three more or less perpendicular sets of joints - two vertical and one horizontal - dominated. A few steep fault zones with thin clay filled joints traversed the tunnels at favourable angles. Water loss tests indicated a very tight rock mass at tunnel level about 60-90 m below ground level.

The rock mass can be described as a very good or good rock according to Bieniewski's classification system as follows:

		Points
Rock strength:	high	12
RQD-value:	90%	20
Joint spacing:	0.2-2 m	10-15
Joint roughness:	rough	25-30
Water conditions:	damp	10
Point orientation:	favourable	-2
	RMR-value	75-80

2. LAY-OUT AND EXCAVATION TECHNIQUE

The general lay-out of the Brofjorden plant is shown in Fig. 1. The storage units consist of three parallel twin-tunnels with the storage volumes of 750 000, 850 000 and 1 000 000 m^3, respectively. The tunnel lengths vary between 600 and 950 m. The excavated masses have been taken out through 65 m^2 transport tunnels between the storage twin tunnels with adits at the three excavation levels to the respective storage tunnel, Fig. 1.

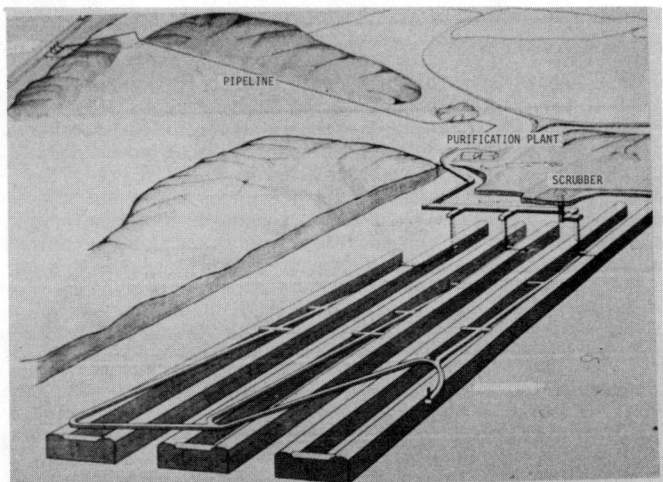

Fig. 1 Sketch of the plant showing the access tunnels to the three excavation levels of the Twin tunnels

The storage tunnel dimensions are shown in Fig. 2. The excavation method was a 7 m high top heading, a 8 m high horizontal drilling bench and a 15 m high vertical drilling bench, Fig. 2. The height of the roof arch was 5 m giving an arch span ratio of 0.25.

The owner, the Swedish National Board of Economic Defence, stipulated in the contract that the tunnel roofs - including the ends of spot bolts for roof support - should be covered by 25 mm shotcrete. The contractor was John Mattson Building Co.

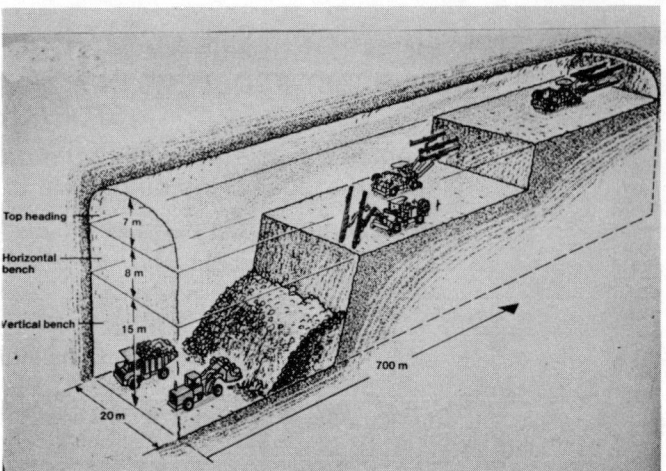

Fig. 2 Sketch of excavation method with top heading, horizontal drilling bench and vertical drilling bench

3. ROCK BURST PHENOMENA

3.1 Top heading excavation

Spalling rock began to occur in the roof of the transportation tunnels after a few hundred meters of excavation. Slices of rock with a thickness varying between 2-15 cm were spalled in roof areas, which had been scaled and sometimes bolted. In some cases crack noises were reported as accompanying the spalling. Bolts with washers, shotcrete and in some areas wire mesh gave adequate protection.

This indication that there might be rock burst trouble ahead caused the Owner to order in situ stress measurements in a drillhole placed centrally in the site area. The measurements were made by the Swedish State Power Boards, Ingevald et al., 1977, with their method, Hiltscher et al., 1979, at a level about 5 m above the roofs of the planned storage tunnels. The measured horizontal in situ stresses are shown in Fig. 3. Some peculiarities in the individual values obtained at the 4 measuring points, e.g. vertical tensile stresses at 60 m depth, indicated the possibility of residual stresses. The average values, Fig. 3, however, indicated horizontal in situ compression stress components of 12-15 MPa perpendicular to the tunnels and 6-9 MPa along the tunnels.

When the top headings were opened and excavation proceeded, spalling of thin rock slices and/or shotcrete occurred occasionally, usually most frequent in the region 20-50 m behind the headings.

Although many regions of Sweden according to successive in situ stress measurements seem to hold fairly high horizontal compressive stresses, rock burst phenomena are rare in tunnelling and the Swedish experience is fairly limited. In our neighbour country Norway, on the other hand, these phenomena have been studied extensively. Russenes, 1974, had summarized the Norwegian experience as shown in Fig. 4. The liability for spalling to occur

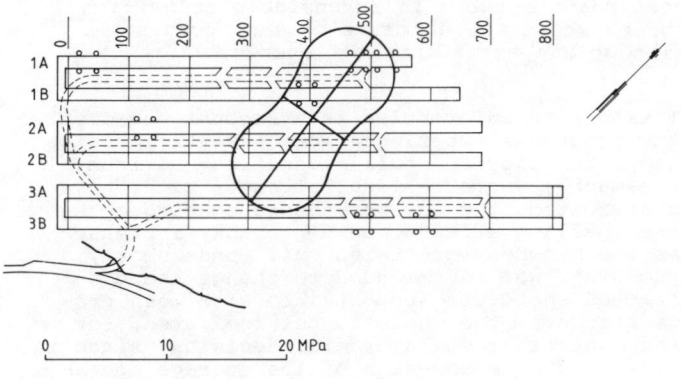

Fig. 3 Horizontal in situ compression stresses measured at storage tunnel top level. Positions for convergence measurements are shown.

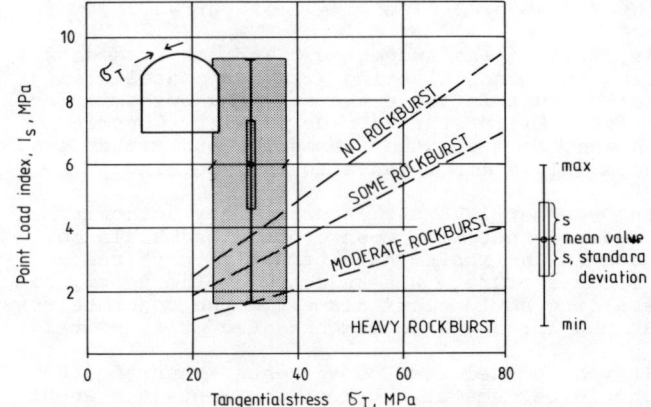

Fig. 4 Diagram of rock burst risks according to Russenes (1974). Results from point load tests near the rock surface, see Fig. 7, and calculated tangential stresses are also presented.

depends on the tangential compressive stress σ_T at the tunnel boundary and the point load index I_s of the rock.

The Owner and the Contractor decided to put up an Advisory Board constituted by independent consultants. This Board followed the excavation development, suggested various rock mechanics investigations and advised about supporting methods.

In March 1978 when about 20% of the top headings were excavated it became evident that the contracted support with spot bolting and 25 mm shotcrete did not give adequate protection against rock spalling. Estimates based on elastic behaviour indicated that the tangential compressive stresses at the tunnel boundaries might be at most about twice the in situ horizontal stress, i.e. $\sigma_T \approx 25-35$ MPa, at the flank tunnels 1A and 3B, Fig. 3. The point load index of the granite was estimated to $I_s = 13-22$ MPa. Thus, according to the Norwegian experiences as given in Fig. 4 there should be an ample margin of safety against rock burst, even if a considerable reduction of the σ_T-value as pertaining to the rock mass is allowed for. So, although the Brofjorden granite did not care to behave according to experience, there was at least hope that the spalling would be mild.

In considering the situation the Advisory Board concluded that the gross-stability of the tunnel roofs was good due to the in situ compression stresses, and that the fairly high height:span ratio of the roof arches was advantageous since it concentrated the rock spalling areas to the fairly flat roof section at the arch crest. The sloping, often step-formed roof parts along the tunnel walls were expediently supported by bolts in a "flexible" way that reduced spalling effects, see Fig. 5.

The Owner did not want to change the contracted roof support with shotcrete. The Board therefore recommended that support against spalling effects be made using 1 m long tensioned expansion bolts (Farex) with 200x200x10 mm washers in a systematic pattern with spacing 1.5 m. Bolting should follow the heading closely and shotcrete be applied within 15 days. The roof should be scaled immediately before shotcreting. Fig. 5 gives an example of the bolting pattern, where some expansion bolts near a blocky abutment have been replaced with long grouted spot bolts.

The Board also recommended that the blasting program should be revised in order to find out if a more cautious blasting might be possible.

Tunnel parts (about 20%), where the contracted support had been used, were investigated with scaling rods. Areas with boom sound reflections were supplemented with 1 m Farex bolts.

When the recommendations given by the Board had been inaugurated into the working cycle, the top heading excavation could be carried through safely.

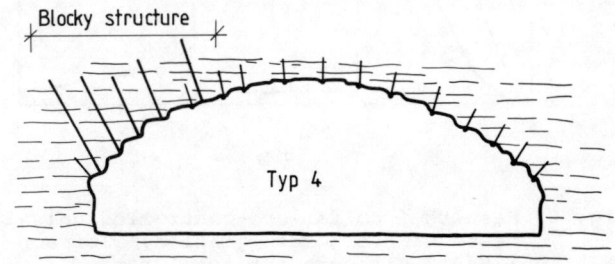

Fig. 5 Examples of roof reinforcement. The short anchor bolts are supplemented with grouted long bars for structural purposes.

3.2 Upper bench excavation

In November 1978 upper bench excavation, Fig. 2, began in some tunnels. Fairly soon some spalling of rock slices + adhering shotcrete began from roof parts between the bolts. The spalling covered about 3-5% of the roof surface.

At first it was suspected that rock fragments from the bench blasting with horizontal drillholes might initiate the spalling by hammering effect. Field studies made by Nitro Consult showed, however, that hammering and vibrations were only a minor cause to the spalling.

In December 1978 the labour safety authority required that wire mesh protection should be used in the roofs. An alternative with reinforced shotcrete was rejected by the Board, since spalling could occur also from the concrete cover as had already been demonstrated at the site.

It was decided that convergence measurements should be made at the roof abutments and about 1 m above the lower bench top in some tunnels. The measurement sections are shown in Fig. 3. Further, the technical characteristics of the Brofjorden granite as regards post-failure behaviour (ratio of brittleness) should be investigated. The results from these investigations are given in Section 4.

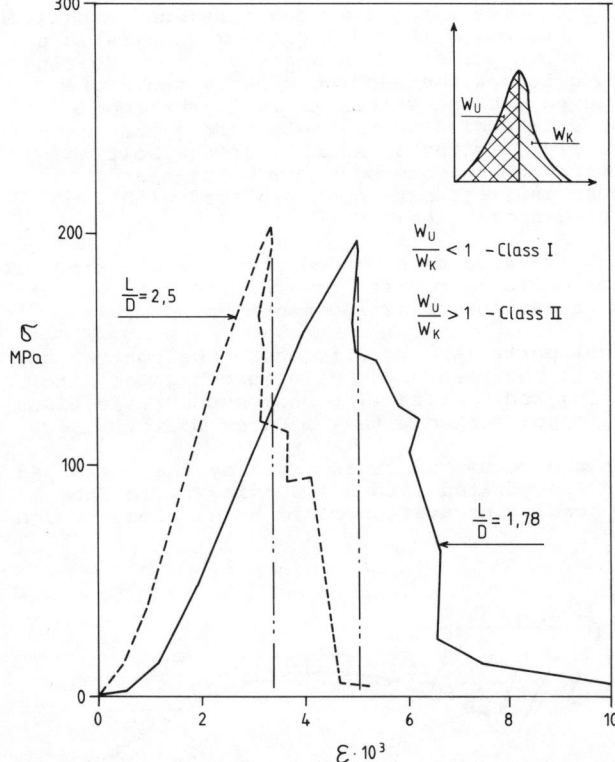

Fig. 6 Results from failure-controlled uniaxial tests of Brofjorden granite, Olofsson (1979). Although the total ratio of brittleness is <1 indicating "tough" behaviour, there is a tendency to behave "brittle" immediately after the initiation of failure.

The laboratory tests on post-failure behaviour indicated a ratio of brittleness of 0.65-0.8, which would mean that the Brofjorden granite is not especially brittle. However, a close inspection of the stress-strain curves in Fig. 6 shows that right after the ultimate stress has been reached there is an unstable transition phase, where the deformation can increase without any extra input of energy.

Taking this information into account and also the practical experience won, that the granite had a tendency to spall along fresh horizontal cleavages, where residual stresses might have contributed, the Advisory Board concluded that the spalling effects would probably increase as the benches were taken out. Consequently, the Owner was recommended to change the contracted shotcrete support into wire mesh protection over the whole tunnel roof area. For the Owner this was a crucial decision, since it must lead to a redesign of the storage operation system using inert gas to avoid the risk for gas explosions.

However, in March 1979 it was decided accordingly that all roofs be protected with wire mesh 50x50x3 mm anchored in 2 m long grouted bolts with triangular 200 mm washers. The bolts were put systematically with spacing 2.0x1.5 m. All further shotcreting with the exception of reinforced shotcrete in weak zones was discarded. The decision was taken at a time when almost all wire meshes could be put up from the top heading floor, Fig. 2.

3.3 Lower bench excavation

The wire meshes satisfied their protective function very well also during the excavation of the 15 m lower bench. The content of spalled material in the wire meshes was estimated by ocular inspection, and it could be stated that the main spalling occurred in the vicinity of the excavation headings, thus manifesting the stress redistributions also shown in the convergence measurements. The total amount of spalled materia varied considerably from section to section. Since the rock quality seemed to vary very little, we conclude that the many different cave geometries, which occurred during the excavation operations, have induced varying stresses in different tunnel sections. However, there are also indications that the in situ stresses may have varied in different parts of the rock mass.

When spalled materia assembled in the wire meshes, it turned out that there was no factual basis for estimating when tapping of the meshes was necessary. Opinions differed quite a lot. It is a both dangerous and costly operation to tap and repair a wire mesh at a height of 30 m. A simple fullscale test, where 4 sections of wire mesh were loaded with rock material dropped from a height of 0.5-0.75 m showed that every section could be loaded with at least 600 kg without failure within 3 weeks. Only a few sections were tapped after this test.

3.4 Cave-ins

Two cave-ins occurred from the roof in tunnel 3B, Fig. 3. In this tunnel the rock mass quality turned out to be considerably lower than in the other tunnels. Both cave-ins were clearly caused by the compression stresses in the rock mass and were accompanied by fully audible cracks and noise both before, under and some time after the cave-in.

The first cave-in occurred at section 170-190 m, Fig. 3. A shallow wedge-shaped strip of rock-depth a little more than one meter - which had been anchored by grouted bolts, was crushed and fell down when the upper bench excavation had just passed section 190. It was a classical wedge-formed "key-stone" that failed. Contributory cause to the failure was probably the fact, that an extra communication tunnel had been taken up between tunnels 3A and 3B at top heading level in this area, causing an extra stress concentration.

The other cave-in occurred between the sections 550 and 580, Fig. 3. A major joint with a dip of 60° intersected the roof about 6 m from and parallel to the outer wall. When the lower bench excavation had reached section 590 (proceeding SW, Fig. 3) a slip occurred along this joint and a considerable mass of rock fell down notwithstanding the fact that the area had been supported by bolts and 80 mm reinforced shotcrete. After the cave-in the stresses had been equalized and the roof profile was stabilized with grouted bolts and reinforces shotcrete.

A more detailed description has been given by Bergman and Johnsson, 1981.

4. ROCK MECHANICS INVESTIGATIONS

4.1 In situ stress measurements

During the last stage of the excavations the Owner planned to build a similar underground storage in a granite monolite a few hundred meters from the object here treated. In this connection in situ stress measurements have been performed in 3 drillholes at the planned site at tunnel roof level - 5 local measurements in every drillhole, Ingevald et al., 1980. The average horizontal compression stresses were evaluated to about 17 MPa NW-SE and about 8.5 MPa NE-SW, which is in fair correspondence with the results shown in Fig. 3. Notable is that also here the local vertical stress components were tensile stresses in 9 out of 15 points, which gave the evaluated average vertical stress as ±0 MPa as compared to a theoretical gravity compression stress about 1.6 MPa.

Thus, these results seem to support the conclusions drawn from the first stress measurements that there might be a considerable local influence from residual and/or block-bound stresses in the Brofjorden granite.

4.2 Rock post-failure characteristics

Laboratory investigations were made at Luleå University on granite cores from Brofjorden with the following results:

Young's modulus	$E = 54$ GPa
Poisson's ratio	$\nu = 0.13$
Uniaxial compressive strength	$\sigma_c = 200$ MPa
Tensile strength in bending	$\sigma_t = 9.3 - 18.8$ MPa
Point load index	$Is_{(50)} = 8.7$ MPa

Failure-controlled uniaxial compression tests were also made in order to determine the ratio of brittleness as defined by Waversik, 1968. The results as reported by Olofsson (1979) are shown in Fig. 6 and indicate a fairly low ratio of brittleness $W_u:W_k = 0.65-0.8$. However, right after the ultimate strength there is an unstable transition phase where the deformation can increase without extra input of energy, i.e. where the material behaves "brittly".

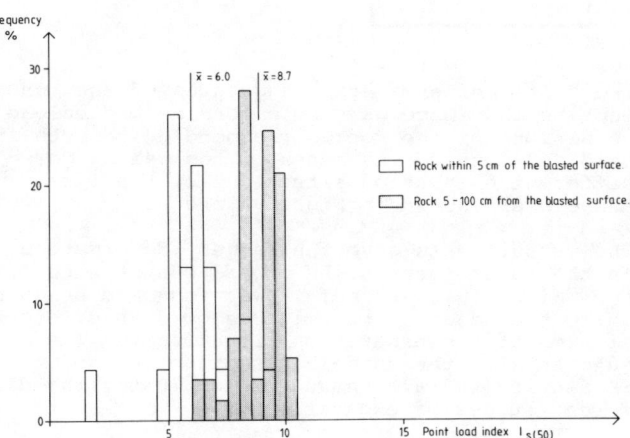

Fig. 7 Point load tests on rock from Brofjorden

Small cores were drilled from spalled rock slabs and from the wall of the top heading. It was found that close to the surface, up to a distance of around 5 cm, the point load index was 6.0 MPa with a coefficient of variation of 0.24 (23 tests). Deeper into the wall, from 5 cm up to 80 cm, the point load index was 8.7 with a coefficient of variation of 0.11 (57 tests), see also Fig. 7. These results indicate that a decreased strength close to the surface may have occurred.

4.3 Convergence measurements

The convergence measurements were performed in order to get information of the magnitude of the rock deformation and also to get an indication if unstability would occur.

Seven measurement sections were installed, Fig. 3. The deformations were measured at the roof abutments and 1 m above the lower bench and at each level in two separate measurement distances, 20 m apart, in every section, see Fig. 8. The results from three of the sections - tunnel 1 A section 045 and 462 and tunnel 1B section 403 - are shown here.

The results were influenced by the following:

o Instrumentation accuracy
o Local geological conditions
o Excavation sequences

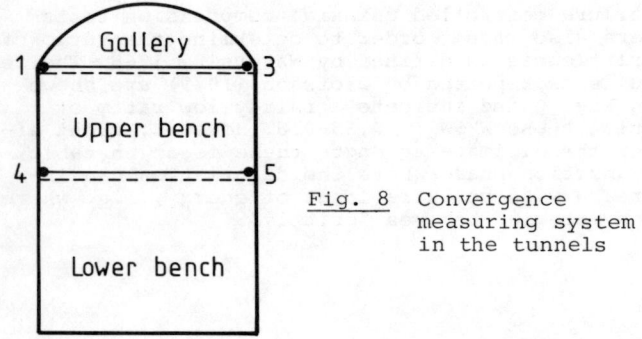

Fig. 8 Convergence measuring system in the tunnels

The different excavation sequences of the tunnels in time and space gave both loading and unloading conditions of the rock corresponding to both inward and outward movements of the walls. The different excavation situations of the three tunnels are shown in Figs. 9a and b.

The measurements gave the largest deformations in the outer tunnels. An inward convergence of the walls of the tunnel 1A was found to be 40 mm while the adjacent tunnel 1B showed an divergence of 9 mm at the excavation situation IV. The measurements in the tunnel 1A section 462 and tunnel 1B section 403 are shown in Fig. 10 for the different stages of excavation.

Unfortunately no measurement was performed after the excavation was completed.

Indications of unstability by studying deformation versus time could only be obtained during periods of no excavation work in the surrounding tunnels. This could only be performed in a few cases. An example from tunnel 1B section 403 is shown in Fig. 11. No indication of unstability of the walls could be seen from measurements.

The differences of the convergences measured in the two parallel measurement distances were normally around 1 mm. In some cases, however, the differences were much bigger, up to 10 mm, and no obvious geological difference in rock structure could be seen, see Fig. 12. This is important to remember when interpreting the measurements and at comparison with calculations. The difference between a separate measured value and a calculated one can therefore be several millimeters and still imply that a good agreement has been achieved.

5. FEM-CALCULATIONS

After the storage was completed finite element calculations were performed to compare the obtained experiences with theoretical calculations.

The calculations were done with JOBFEM, a program developed at the Department of Soil and Rock Mechanics, KTH (Fredriksson, 1982). Both elastic and plastic conditions were assumed.

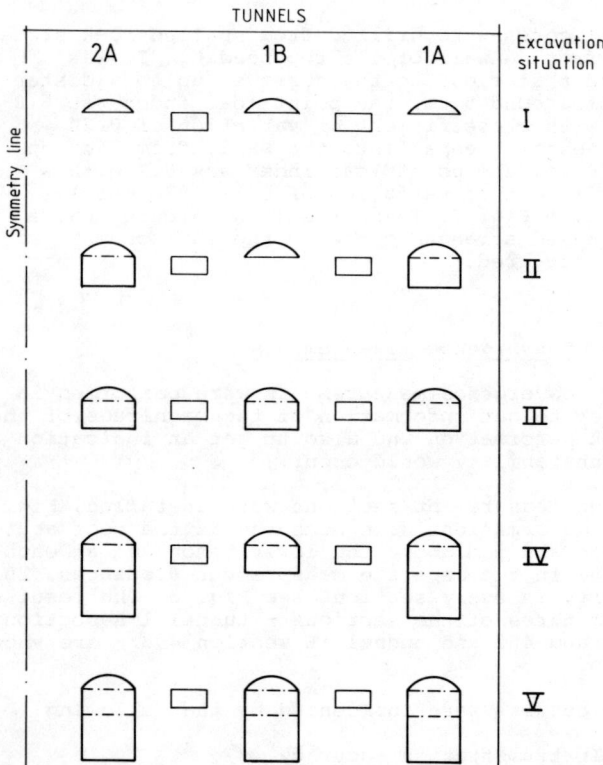

Fig. 9a Excavation stages in tunnel 1A section 462 and tunnel 1B section 403

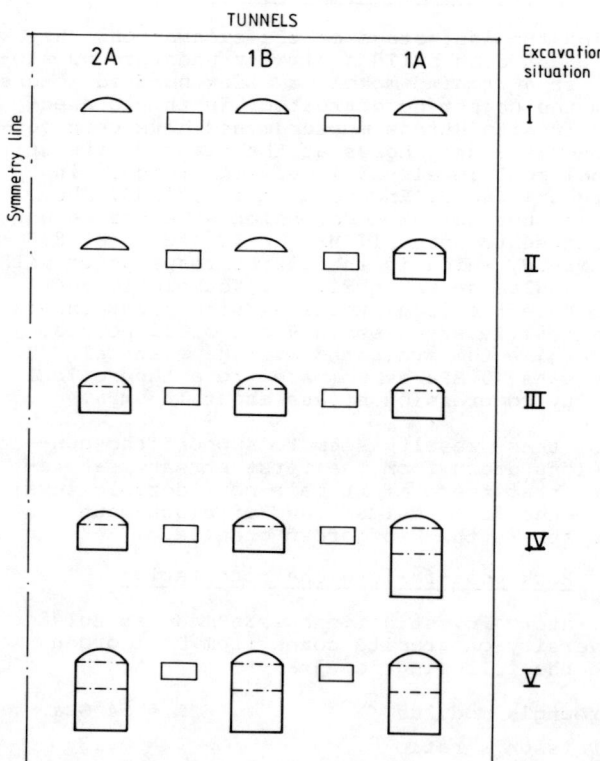

Fig. 9b Excavation stages in tunnel 1A section 045

The properties of the rock mass were first estimated. The deformation modulus was estimated by Bieniewski's formula

$$E_m = 2\ RMR - 100$$

where RMR is the rock mass rating. With a RMR value of 75-85 the E_m value can be calculated to 50-70 GPa. Since the modulus of small specimens was found to be 54 GPa, the smaller value was chosen.

The rock mass strength was roughly estimated with Mohr-Coulomb's failure criterion, c = 2.5 MPa, ρ = 45°, corresponding to an uniaxial compressive strength of 12 MPa, Stille, Groth and Fredriksson, 1982.

The parameters used in the calculation are shown in Table 1.

The results of the calculations expressed as the deformation during each excavation step are shown in Table 2 a, b and c. In the tables the corresponding measured values are also shown.

The measured deformations are in general 3 to 10 times bigger than the calculated values. The deformation pattern, however, is the same with inward or outward movements depending on the excavation steps.

Table 1. Parameters used in the FEM-calculations

Horizontal initial stress field,	σ_κ	14 MP
Deformation modulus,	E_m	50 GPa
Poisson's ratio,	ν	0.1
Density	ρ	2.65 t/m³
Cohesion	c	2.5 MPa
Friction angle,	ϕ	45°
Dilatancy angle,	ψ	45°
Tensile strength,	σ_t	0 MPa
Residual cohesion,	c_r	2.5 MPa

Part of the element mesh around the tunnels is shown in Fig. 13.

The maximum tangential stresses in the middle of the roof varied between 17 to 35 MPa depending of the excavation situation and tunnel for the elastic cases which is up to 2.5 times higher than the initial horizontal stress field, cf. Fig. 4.

In the plastic case a plastic zone about 1-2 m deep was developed in the centre of the roofs. The largest yielding occurred in the abutments and in the pillars between the rooms and the transport tunnels. This zone could at some excavation steps go through the whole pillars.

A more complete presentation of the calculations including stress and deformation graphs can be found in Qvarnström (1982).

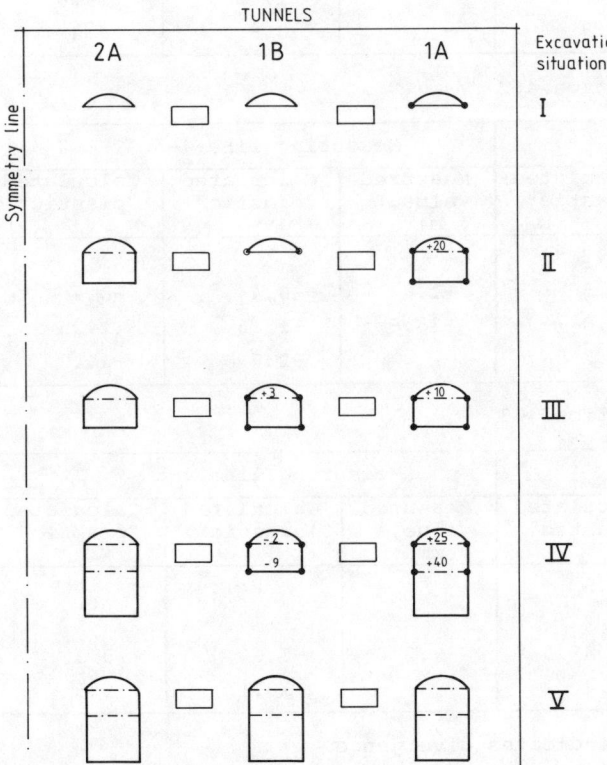

(+ implies an inward movement of the walls)

Fig. 10 Measurements in tunnel 1A section 462 and tunnel 1B section 403

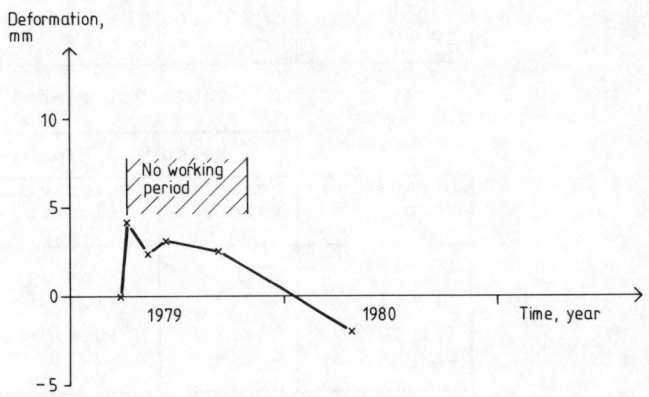

Fig. 11 Deformation versus time for measurements distance 1-3 from section 403 in tunnel 1B

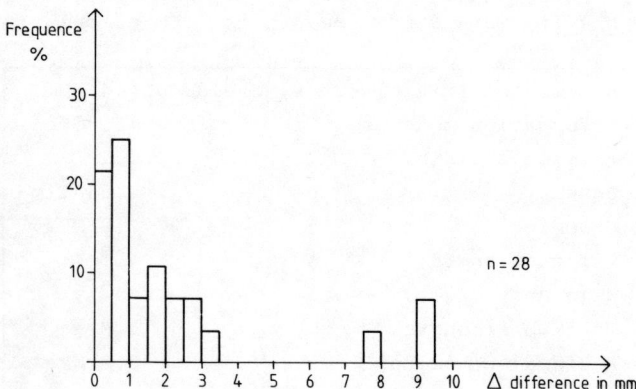

Fig. 12 Difference Δ in measured convergence at the two 20 m apart distances at every measurement sections

Table 2a: Tunnel 1A section 045

	Measuring line 1-3			Measuring line 4-5		
Excavation step	Measured value, mm	Calculated "elastic"	Calculated "plastic"	Measured value, mm	Calculated "elastic"	Calculated "plastic"
I -II	+2.4	+4.0	+5.9	-	-	-
II -III	-7.4	-0.6	-0.7	-8.3	-0.5	-0.6
III-IV	-	+4.4	+5.8	-	+11.2	+14.6
IV -V	-	-1.8	-1.9	-	-3.6	-4.4

Table 2b: Tunnel 1A section 462

	Measuring line 1-3			Measuring line 4-5		
Excavation step	Measured value, mm	Calculated "elastic"	Calculated "plastic"	Measured value, mm	Calculated "elastic"	Calculated "plastic"
I -II	+20.4	+3.8	+4.2	-	-	-
II -III	-10.3	-0.5	-0.5	+1.0	-0.4	-0.4
III-IV	+15.2	+3.8	+5.8	+38.6	+12.5	+14.5
IV -V	-	-1.3	-1.2	-	-2.7	-2.4

Table 2c: Tunnel 1B section 403

	Measuring line 1-3			Measuring line 4-5		
Excavation step	Measured value, mm	Calculated "elastic"	Calculated "plastic"	Measured value, mm	Calculated "elastic"	Calculated "plastic"
I -II	-	-0.5	-0.9	-	-	-
II -III	+2.7	+1.6	+3.7	-	-	-
III-IV	-4.7	+0.4	-2.0	-9.2	-3.6	-3.1
IV -V	-	-0.3	+3.4	-	+10.1	+9.5

+ indicates convergence   - indicates divergence

6. CONCLUSIONS

As can be seen from Fig. 4 the measured scatter of point load index I_s as well as of calculated tangential stress σ_T defines a certain probability of rock burst phenomena, which is in accordance with the real experience in the Brofjorden plant.

However, it should be noted that point load indices taken from virgin rock may lead to an interpretation of the rock burst tendency that is too optimistic.

Only fairly mild rock burst can be supported by shotcrete in combination with bolts. Moderate or heavy rock burst may destroy even reinforced shotcrete and cause fall of concrete fragments.

The FEM-calculations indicate that in the Brofjorden plant the possibilities to reduce the risk for rock burst by adapted excavation sequences were fairly small.

It seems evident that the rock burst phenomena are strongly influenced by factors like the rock mass structure, the rock petrography and residual or block-bound stresses. The general conclusions that can be drawn from the Brofjorden case are therefore limited.

REFERENCES

Bergman, S.G.A. and Johansson, N.E. (1981). Rock burst problems in ÖEF 600 m² tunnels in Brofjorden, Sweden. Swedish Rock Mech. Res. Found., Rock Mech. Meeting Stockholm, 121-147 (in Swedish).

Fredriksson, A. (1982). Analysis of geotechnical problems with finite element Methods (in Swedish). Dept. of Soil and Rock Mechanics KTH, Stockholm.

Hiltscher, R., Martna, J. and Strindell, L. (1979). The measurement of triaxial rock stresses in deep boreholes and the use of rock stress measurements in the design and construction of rock openings: Proc. 4th Int. Congr. ISRM, Vol. 2, 227-234, Montreux.

Ingevald, Kj. and Strindell, L. (1979). Brofjorden. In situ stress measurements at plant 3904: Swed. State Power Board, rep. BTH-SV-L-566 (in Swedish).

Ingevald, Kj. and Strindell, L. (1980). Brofjorden II. In situ stress measurements at plant 3904 C: Swed. State Power Board, rep. BTH-SV-L-602 (in Swedish).

Olofsson, T. (1979). Determination of parameters for Brofjorden granite: Report 1979-03-11, Rock Mech. Dept., University of Luleå, Sweden (in Swedish).

Qvarnström, T. (1982). FEM-analysis of the rock burst problems in Brofjorden project. Dept. of Soil and Rock Mechanics KTH, Stockholm (in Swedish).

Russenes, B.F. (1974). Rock burst analysis for tunnels in mountain slopes: Geologisk Institutt, Trondheim (in Norwegian).

Stille, H., Groth, T. and Fredriksson, A. (1982). FEM-analysis of rock mechanical problems with JOBFEM. BeFo 307:1/82. Stockholm (in Swedish).

Waversik, W.R. (1968). Detailed analysis of rock failure in laboratory compression tests: Ph.D. Thesis. University of Minnesota, Minneapolis, USA.

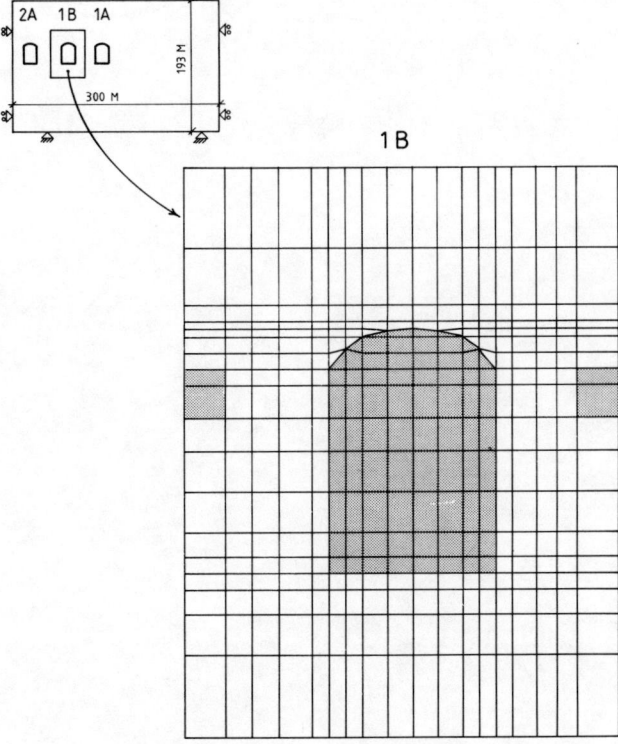

Fig. 13 Part of the element mesh around tunnel 1B

TUNNEL LININGS OF STEEL FIBRE REINFORCED SHOTCRETE
Revêtement de tunnels en béton projeté à fibres d'acier
Tunnelauskleidungen aus Stahlfaserspritzbeton

B. J. Holmgren
Research Scientist of the Swedish Rock Engineering Research Foundation, Stockholm,
Sweden and Senior Lecturer in Structural Mechanics and Engineering, Royal Institute of
Technology, Stockholm, Sweden

SYNOPSIS

Experimental investigations of the strength and deformability of steel fibre reinforced shotcrete linings are presented. The tests show that there are several advantages associated with the use of high strength steel fibres with hooks in shotcrete linings:
(i) These fibres slip under a certain resistance in the matrix and the failure becomes very plastic both in bending and in shear.
(ii) Advanced support systems utilizing the interaction of rock bolts and shotcrete are simplified in the design when steel fibre reinforcement is used.
(iii) The fibres prevent the breaking up of the shotcrete, and the risk of downfall is decreased.

RESUME

La résistance et les déformations de béton projeté à fibres d'acier ont fait l'objet de recherches expérimentales. Les essais mettent en évidence plusieurs avantages que présentent l'emploi, dans les revêtements de béton projeté, les fibres en acier de haute résistance terminées par des crochets.
(i) Ces fibres glissent dans la matrice en développant une certaine résistance et la rupture devient très plastique aussi bien en flexion qu'en cisaillement.
(ii) L'emploi de revêtements armés de fibres d'acier simplifie la conception de systèmes de soutènement basés sur l'action réciproque d'ancrages et de béton projeté.
(iii) Les fibres empêchent la désagrégation du béton projeté et diminuent le risque d'effondrement.

ZUSAMMENFASSUNG

Experimentelle Untersuchungen über die Stärke und Biegsamkeit des Stahlfaserspritzbetons liegen vor. Bei Tunnelauskleidungen aus Spritzbeton unter Verwendung von Fasern aus hochfestem Stahl mit gebogenen Enden, erweisen die Experimente mehrere Vorteile:
(i) Diese Fasern gleiten im Material unter gewissem Widerstand und resultieren in ausgesprochen plastischen Biege- und Schubbrüchen.
(ii) Der Entwurf schwieriger Konstruktionen aus Spritzbeton und Felsankern wird vereinfacht bei Verwendung von stahlfaserbewehrtem Spritzbeton.
(iii) Die Fasern verhindern das Zerbrechen des Betons und reduzieren das Einsturzrisiko.

1. INTRODUCTION

In 1973 a joint project between the Swedish Rock Engineering Research Foundation and the Royal Swedish Fortifications Administration started. Its aim was to get better knowledge of the strengthening function of a shotcrete lining when applied to a hard rock.

The investigation started with large scale tests on the failure mechanism and strength of shotcreted jointed rock (Holmgren, 1979). It was then continued with laboratory tests on the tensile adhesion of shotcrete to different rock types (Hahn, 1979).

It was found that the adhesion between the rock and the shotcrete plays a determining rôle for the strength of the lining unless other support is arranged e.g. systematic rock bolting. In Holmgren, 1979 there are described tests on bolt supported reinforced shotcrete linings. These results served as a basis for a standardized design for bolt supported shotcrete linings in military caverns. This solution is also used in an underground storage for radioactive waste.

There are many advantages with this design:
(i) The strength is independent of the obtained adhesion strength.
(ii) The strength is possible to calculate according to principles for concrete plates on columns.
(iii) A systematic rock bolting may be required for other reasons, too.
(iv) End anchored rock bolts with rolled threads interacting with reinforced shotcrete seems to be the most promising support in squeezing rock.

The disadvantages are:
(i) The installation of mesh reinforcement is very costly and time consuming.
(ii) The irregularities of the rock surface make it difficult to shotcrete a reinforced lining.
(iii) The probability that the reinforcement fits the distribution of the bending moments in the lining is very small.
(iv) It is impossible to reinforce a shotcrete lining against shear failure using conventional reinforcement.

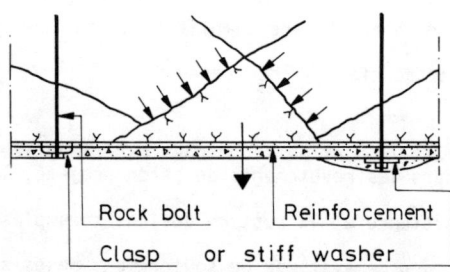

Fig. 1 Design principles for bolt supported reinforced shotcrete linings.

By using fibre reinforcement instead of a conventional one most of the above mentioned difficulties disappear. The most important remaining one is the cost because of fibres being an expensive material.

2. TEST PROGRAM

2.1 General remarks

The punching of a single block through a shotcrete lining which is supported by rock bolts was also studied by Holmgren, 1979.

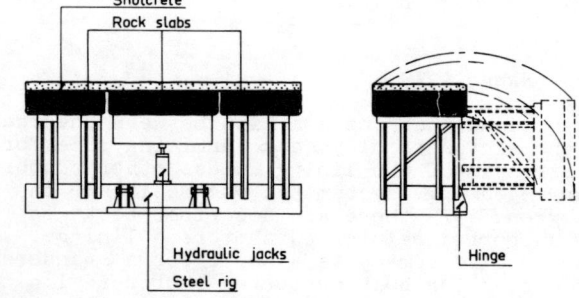

Fig. 2 Test rig

The tests were performed in a rig consisting of three large granite blocks on a steel stand.

The shotcrete was produced according to the wet mix method and the fibres were added at the nozzle by means of a so called Fibre Feeder which is patented by the Swedish company Besab.

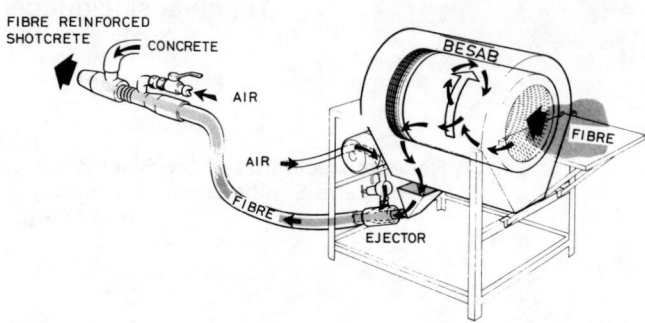

Fig. 3 Equipment used for the production of steel fibre reinforced shotcrete

The fibres were hooked steel fibres Bekært Lz 35x.35 and Lz 45x.35 with a failure strength of about 1200 MPa.

Standard cement and 0 - 4 mm sand were used for the concrete.

Most test specimens were cured for eight days.

2.2 Details about the test specimens

The test program consisted of four parts:
(i) Large scale tests on possible designs of a fibre reinforced shotcrete lining supported by rock bolts. There were tested stiff washers of two sizes at the end of the bolt, flexible washers and also bolts without washers.
(ii) Large scale tests for the comparison of mesh reinforced shotcrete and fibre reinforced shotcrete. There were tested meshes of cold tensioned steel and meshes of mild steel.
(iii) A few ad hoc tests in large scale on extremely thin fibre reinforced linings. (These tests were not finished at the time of writing this paper).
(iv) Small scale shear tests on cracked fibre reinforced specimens. (At the time of writing this paper these tests were performed but not the analytical treatment).

3. TEST RESULTS

3.1 General remarks

In connection with all tests control tests were shotcreted. 15 cm standard cubes were also cast.

The compressive strength of sawn out 8 cm cubes and cast 15 cm cubes respectively was 60 MPa and 40 MPa. It was very little affected by the presence of fibres. The bending tensile strength of beams 8x8x80 cm was about 6 MPa without fibres and 13 MPa at best with fibres. These values of course scatter a lot.

The registrations of the main tests are illustrated in diagrams showing the average displacement of the loading block as a function of the force which acts at each joint.

As a reference to older tests there are curves marked A2, A4 and A5 in the diagrams. These tests are reported in Holmgren, 1979.

Shortly expressed A2 and A4 correspond to reinforced shotcrete of to-day's practice and A5 to a somewhat more heavily reinforced shotcrete lining.

kinds of tests are shown in the same diagram.

The tests showed that even a rather small amount of fibres (about 0.5 % when long fibres are used) gives better rock support than when conventional mesh reinforcement is used. They also showed that a large amount of fibres give the same support strength as a conventional heavy reinforcement. As expected 45 mm fibres gave better results than 35 mm fibres. By using an extremely small washer (ø 60 mm) a local failure around the washer was caused. This failure was not a brittle one, however, but very plastic.

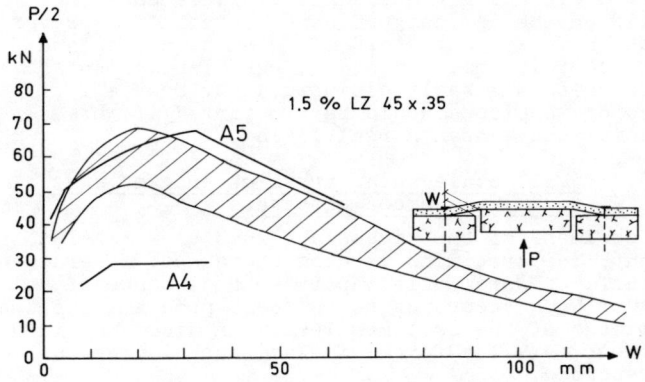

Fig. 5 Dashed area represents force-displacement curves for bolt supported, steel fibre reinforced shotcrete layers. Amount of fibre 1.5 % by volume, length of fibre 45 mm.

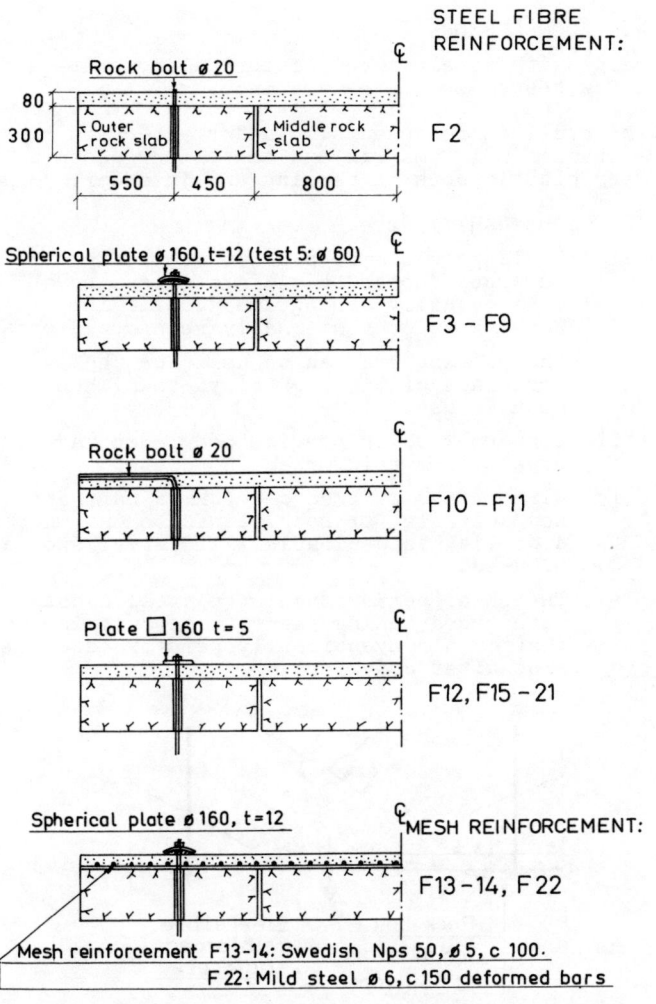

Fig. 4 Specimens for large scale tests on bolt supported shotcrete linings

3.2 Large scale tests on steel fibre reinforced shotcrete layers interacting with rock bolts

3.21 Bolts with washers

Tests were made with three amounts of fibre: 0.5, 1.5 and 2.0 % by volume. The tests showed that there was no particular difference between stiff washers and flexible ones. The reason for this is that fibre reinforced concrete is not sensitive to the stress concentrations which occur in the vicinity of the rock bolt under the flexible washer. Therefore results from both

3.3 Large scale tests on mesh reinforced shotcrete layers interacting with rock bolts and stiff washers

3.31 Mesh made of cold-tensioned steel

Test specimens F13 and F14 were made using a welded mesh with a strength of 500 MPa at 0.2 % permanent elongation. The strength is achieved by cold-tensioning. This mesh unfortunately has been very common in Swedish shotcrete works.

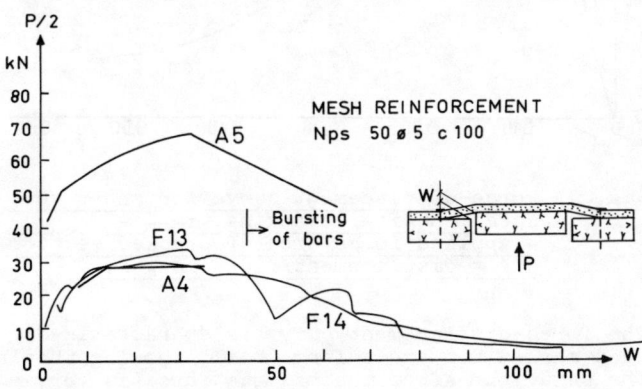

Fig. 6 Force-displacement curves for bolt supported, mesh reinforced shotcrete layers. Cold-tensioned mesh.

These specimens showed a lower carrying capacity than the fibre reinforced ones. Moreover the bars of the mesh were torn off at about 50 mm displacement of the movable block.

3.32 Reinforcement mesh made of mild steel

Specimen F22 was made using a welded annealed mesh with a yield strength of 220 MPa.

The strength of this specimen was very low because of the low steel strength and the small steel content. When the test results have been adjusted for this it is possible to compare F22 with F13 and F14. At small displacements the difference is insignificant. In test F22, however, the bars were not torn off even at a displacement of 150 mm. At these large displacements the concrete was badly disintegrated. Downfall of concrete pieces would have occurred if this should have been a real lining.

3.4 Small scale shear tests on cracked steel fibre reinforced shotcrete

Specimens of the dimension 4x5x8 cm were sawn out. The specimens were then cracked. The cracked surfaces were pulled apart to a distance of 1, 2 or 5 mm according to the test program. The two pieces of the specimen were then displaced in a shear box. The force-displacement diagram was recorded.

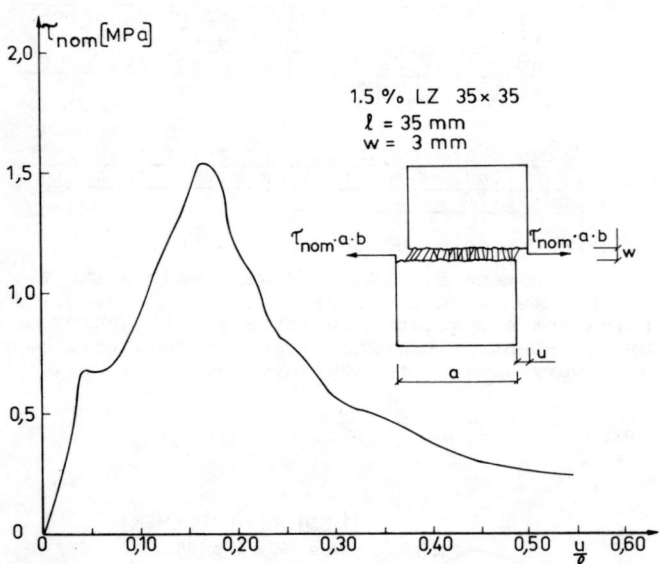

Fig. 7 Force-displacement curve for shear test specimen after cracking. τ_{nom} = = applied force/area of cracked section. u/ℓ = displacement/fibre length.

The force-displacement curve is characterized by a marked tip and a long "tail" showing that the behaviour after the maximum force is very plastic with a permanently decreasing resistance.

A shotcrete lining with a bar reinforcement at the center which is subjected to shear displacements will get a spalling failure. This will take place at a small displacement and the resistant force will be small, too. A steel fibre reinforced lining might have a shear strength in the cracked stage which is five times the strength of a conventionally reinforced one. Furthermore it has the advantage of the great plasticity after the maximum load. Since many loading cases are of the type forced deformations this plasticity in shear is of the utmost importance.

Tests in a large scale on this subject should have to be made before any definite conclusions can be drawn.

4. CONCLUSIONS

Tests with steel fibre reinforced shotcrete and a few tests with mesh were made.

The steel fibres used slip under a certain resistance in the matrix and the lining becomes very plastic both in bending and in shear.

The main results were:
  (i) It is possible to produce steel fibre reinforced shotcrete linings which are at least equally strong and ductile in bending as conventionally reinforced ones.
 (ii) Cheaper washers can be used for the anchorage of the rock bolts than when mesh is used.
(iii) Bent rock bolts provide poor anchorage in steel fibre reinforced shotcrete.
 (iv) Wire meshes with cold-deformed bars are not suitable for rock reinforcement where ductility is desirable. Mild steel should be used.
  (v) The shear performance of a steel fibre reinforced lining seems to be superior to that of a conventionally reinforced one also after cracking.

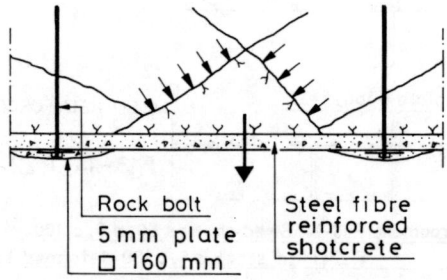

Fig. 8 Design principle for bolt supported steel fibre reinforced shotcrete lining. The dimensions are to be varied with the rock conditions.

5. REFERENCES

Hahn, T., Holmgren, J. "Adhesion of shotcrete to various types of rock surfaces", Proc. of the fourth International Congress on Rock Mechanics, Montreux, 1979.

Holmgren, J. "Punch-loaded shotcrete linings on hard rock", Publ. no. 7:2/79, Swedish Rock Engineering Research Foundation, Stockholm, 1979.

SHAFT LINING PRESSURES DURING SINKING THROUGH DEEP AQUIFER ROCKS

Pressions exercées sur le revêtement des puits lors du forage à travers des roches aquifères profondes

Druckbelastungen des Schachtausbaus beim Abteufen durch wasserführende Gesteinsschichten in grosser Teufe

P. F. R. Altounyan
P. D. Shelton
Wang Hao
University of Newcastle upon Tyne, Newcastle upon Tyne, U.K.

SYNOPSIS

Total and hydrostatic stresses and lining strains and temperatures were measured during the construction of cast-in-place plain concrete linings through saturated triassic and Coal Measures sandstones at depths up to 655 m at Whitemoor, North Selby and Riccall Mines, Yorkshire, U.K. After final grouting of the lining, radial hydrostatic stresses were close to the full hydrostatic head, and geostatic stresses were found to be negligible.

RESUME

Les efforts globaux et hydrostatiques et les contraintes sur le revêtement ainsi que les températures ont été mesurés pendant la construction d'un revêtement de 1,2 m d'épaisseur et de 7,315 m de diamètre interne, réalisé en béton lors du percement d'un banc de grès carbonifère saturé d'eau à une profondeur de 655 m à la mine de Whitemoor, North Selby et Riccall dans le Yorkshire (Royaume Uni). Après la coulée du béton on a mesuré les efforts hydrostatiques radiaux qui ont été pratiquement équivalents à la pression hydrostatique. L'effort géostatique a été pratiquement négligeable.

ZUSAMMENFASSUNG

Hydrostatische und Gesamtdruckbelastung sowie Formänderung und Temperaturen der Schachtauskleidung wurden beim Einbringen eines Ortbetonausbaus mit einer Nenndicke von 1,2 m und einem Innendurchmesser von 7,315 m im Bergwerk Riccall, Yorkshire, G.B., beim Durchteufen einer wassergesättigten Sandsteinschicht im flözführenden Karbon in 655 m Teufe gemessen. Nach Abschluß der Kontaktinjektion wurden radiale hydrostatische Drücke entsprechend des vollen hydrostatischen Drucks gemessen.

1. INTRODUCTION

The principles of design of plain concrete shaft linings are well documented (see for instance Auld 1979, 1982 and Bell, 1982). Difficulties in both design and construction can occur where aquifer rocks are encountered at depth.

In the case of lining design, a limiting lining thickness of 1200mm can normally be constructed in a conventional 7.315m (24 ft) internal diameter shaft. Using elastic methods of calculation and the strongest available concrete this can be shown to safely support a hydrostatic head of 650m (Bell, 1982). Using plastic methods of calculation (Auld, 1982) this head can be increased to below 1000m. However, the lower factor of safety inherent in limit state design, requires more detailed information on the relative contribution of radial earth pressures and hydrostatic pressures on the lining.

In the case of construction, difficulties are always encountered where high hydrostatic pressures and rock permeabilities allow ingress of significant quantities of water. These can be controlled by grouting in fissured rock and by groundwater freezing in near surface rocks where flow through pores inhibits grouting. However in deep aquifer rocks where flow is through porespace, and particularly where these exist in thin layers, considerable problems may be encountered in controlling groundwater. Under these conditions both freezing and grouting may prove difficult and drainage may prove to be the most satisfactory method of groundwater control.

During sinking of ten shafts for the Selby project in U.K., most types of aquifer rock have been encountered, and most methods of groundwater control have been used during construction. As part of an experimental programme aimed at improving the design of shaft linings, the opportunity was taken to observe hydrostatic and geostatic lining pressures in several aquifer rocks.

2. GEOLOGY OF THE AREA

The geology of the area in which the shafts have been sunk is reasonably regular and comprises Permo-Trias strata unconformably overlying the Carboniferous Coal Measures. Strata dip at about 1 in 40 to the East. At the surface there are about 20m of glacial clays and silts overlying the Bunter Sandstone which is between 200 and 300m thick depending on the shaft location. Below the Bunter Sandstone are the Upper Permian Marls (30 - 45m thick and including anhydrite layers), the Upper Magnesian Limestone (a flaggy limestone 20 - 30m thick), the Middle Permian Marls (30- 50m thick including gypsum and halite layers), the Lower Magnesian Limestone (60 - 120m thick), the Lower Permian Marl (1 - 2m thick), the Basal Permian Sands (up to 9m thick) and the Coal Measures.

The Coal Measures comprise a typical British irregular cyclic sequence of seatearths, sandstones shales and coal seams. Some of the sandstones contain water in significant quantities.

The major aquifer rocks are however, the Bunter Sandstone and to a lesser extent the Lower Magnesian Limestone, and water from these is controlled respectively by groundwater freezing and fissure grouting. The Bunter Sandstone has uniformly high permeability and porosity. The Upper and Lower Magnesian limestone have lower and more variable permeability. Typical packer permeability inflow borehole test results are given in Table 1.

Depth(m)	Formation	Permeability Coefficient (m/s)	Estimate inflow per 10m shaft (1/min)
42-52	Bunter Sst	5.01×10^{-6}	175
131-144	Bunter Sst	1.80×10^{-6}	205
201-211	Bunter Sst	2.10×10^{-6}	340
283-301	Up. Mag. Lst	zero	0
344-358	L. Mag. Lst	6.7×10^{-6}	20
356-372	L. Mag. Lst	8.7×10^{-6}	25
372-390	L. Mag. Lst	1.14×10^{-7}	35
387-405	L. Mag. Lst	1.15×10^{-7}	40
405-424	L. Mag. Lst	1.26×10^{-7}	40
424-432	L. Mag. Lst & Basal Permian Sands	3.6×10^{-7}	115

Table 1. Results of pump-in packer permeability tests - Selby Project.

Note. The average horizontal permeability coefficient of intact borehole specimens of Bunter Sandstone tested in the laboratory was 8.1×10^{-6} m/s and of Magnesian Limestone 4.5×10^{-11}. The average porosity of intact borehole specimens of Bunter Sandstone was 34.4%.

3. INSTRUMENTATION

The design of an instrumentation system to measure radial and hydrostatic stresses acting on a shaft lining and subsequent strains in the lining has been described by Altounyan (1982). The system utilised vibrating wire strain and temperature gauges and vibrating wire transducers for pressure measurement. The use of a single transducer facilitated multiplexing of signals at source to allow monitoring through a single shaft cable.

The basic elements of the system which are illustrated in Figures 1 and 2 comprised: (a) 300mm diameter Soil Instruments Ltd. pressure cells with high lateral stiffness (> 200 GN/m^2) for radial stress; (b) 38mm diameter by 240mm long porous pot piezometers with deformable diaphragm installed in boreholes and plugged with bentonite (c) 140mm long vibrating wire strain gauges installed free or pre-cast into briquettes with concrete lining; (d) 65mm vibrating wire temperature gauges installed in the lining and the shaft wall. These were used in a subsidiary experiment (see Altounyan et al 1982) to monitor thermal gradients in the concrete.

The multiplexer which behaves as a remote controlled rotary switch allows gauges to be monitored through a four core shaft cable from a surface mounted vibrating wire gauge comparator controlled by a micro-processor which also acts as a data logger.

4. WHITEMOOR NO. 2 SHAFT

The shafts at Whitemoor Mine have a planned depth of 920m. The first 20m of shaft is through 20m of glacial drift overlying 260m of Bunter Sandstone. The surface level is at 7m AOD and the groundwater level is approximately 10m below surface level. Core recovery during drilling through the Bunter Sandstone was low confirming a general picture of weak high porosity sandstone in which the major part of the water flow was through porespace. Additional average borehole geotechnical properties to those in Table 1 were:

Uniaxial compressive strength (saturated)	9.8 MN/m^2
" (saturated frozen)	36.2 MN/m^2
Uniaxial deformation modulus (saturated)	5.5 GN/m^2
" (saturated frozen)	7.5 GN/m^2
Coefficient of internal friction	0.50
Dry unit weight	1.84 kN/m^2
Specific gravity	2.67

The full depth of the Bunter Sandstone was frozen through freeze tubes of average spacing 687mm on a 14m diameter ring. Average steady state freeze tube temperature was -30°C. Nominal shaft finished diameter was 7.32m with a nominal lining thickness of 0.6m although this increased to 1.5m with overbreak in places.

Several sites were chosen for instrumentation, the general arrangement for that at a depth of 232m being illustrated in Figure 1. Specimen temperature/time curves are included in Figure 3. These illustrate the temperature in the frozen ground and the concrete during hydration immediately after casting. Although not strictly speaking the subject of the paper, the results are interesting insofar as they demonstrate a significant thaw zone in the concrete during hydration which lasts at gauge 11, 50mm into the concrete, for about 14 days. This allows adequate time for efficient hydration of the concrete.

Accelerated freezing of the concrete started 15 days after the pour (Day zero in Figure 3) with a reduction in freeze temperature to -15°C and rose in stages to 0°C. The first major breach in the ice wall occurred after 209 days (Table 1) and lining stresses which exhibited a minor total stress component after re-freezing of the ice wall - gradually rose to hydrostatic levels.

Table 1. Total stress and hydrostatic pressure readings - Whitemoor No. 2 Shaft.

Time after concrete pour (days)	Total Stress- pressure cells MN/m^2			Hydrostatic pressure - piezometers MPa		
	I	II	III	I	II	III
15	0.37	0.40	0.33	0	0	0
200	0.40	0.42	0.36	0.03	0.05	0.04
209	0.67	0.66	0.63	0.50	0.53	0.56
217	0.68	0.69	0.65	0.48	0.38	0.49
300	2.20	2.11	2.17	2.22	2.20	2.21

From data on total stress and hydrostatic pressures obtained from pressure cells and piezometers in Table 1, the following points may be noted:

(i) There is a close correlation between piezometric and total stresses. The average piezometric stress for three gauges after 300 days was 2.1 MN/m^2. This compares with a theoretical piezometric pressure of 2.18 MN/m^2 if the ground water level is assumed 10m below ground surface. There was no noticeable leakage of groundwater through the lining. The average total stress is 2.16 MN/m^2, indicating zero contribution to lining stresses from the

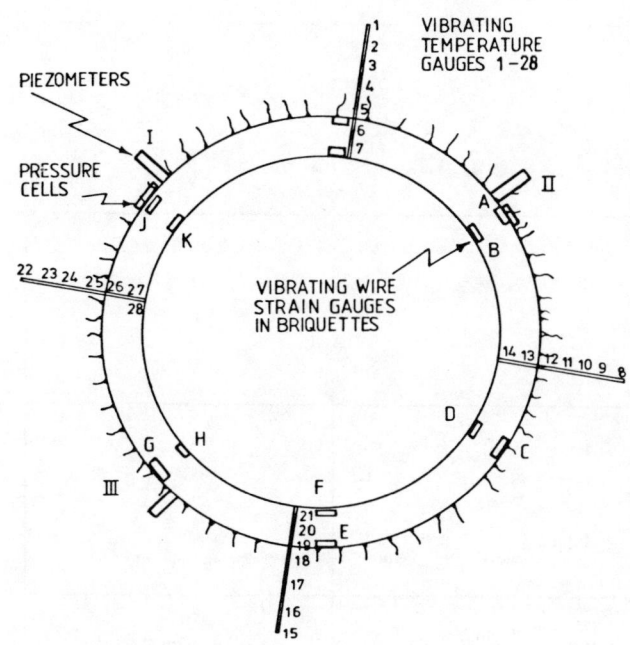

Figure 1. General arrangements of instrumentation at a depth of 232m at Whitemoor No. 2 Shaft in the frozen Bunter Sandstone.

backwall rock - even though this was a relatively weak rock. This confirms previous observations by Altounyan and Farmer (1981) of the very low strata disturbance caused by groundwater freezing.

(ii) The lining strains can be shown to exactly relate to the hydrostatic stresses in accordance with Lame's theorem, and a computed deformation modulus of 33.6 GN/m^2 compares with the design modulus range 45 MN/m^2 concrete of 27-38 GN/m^2.

5. NORTH SELBY NO. 2 SHAFT

The shafts at North Selby Mine have a planned depth to inset of 1000m and pass through 20m of glacial deposits and Bunter Sandstone to a depth of 250m. Shaft dimensions, concrete and freezing layouts were identical to those at Whitemoor Shafts. Geotechnical and hydro-geological data were similar at the two shaft locations. A section of the shaft was instrumented in a similar way to those at Whitemoor at a depth of 240m. The results in Table 3 are similar to those in Table 2 indicating:

(a) a small total stress component following refreezing of the ice wall at 14 days

(b) a hydrostatic pressure similar to the calculated piezometric pressure (2.35 MPa) at the instrumented level

(c) a total stress similar to or lower than the hydrostatic stress

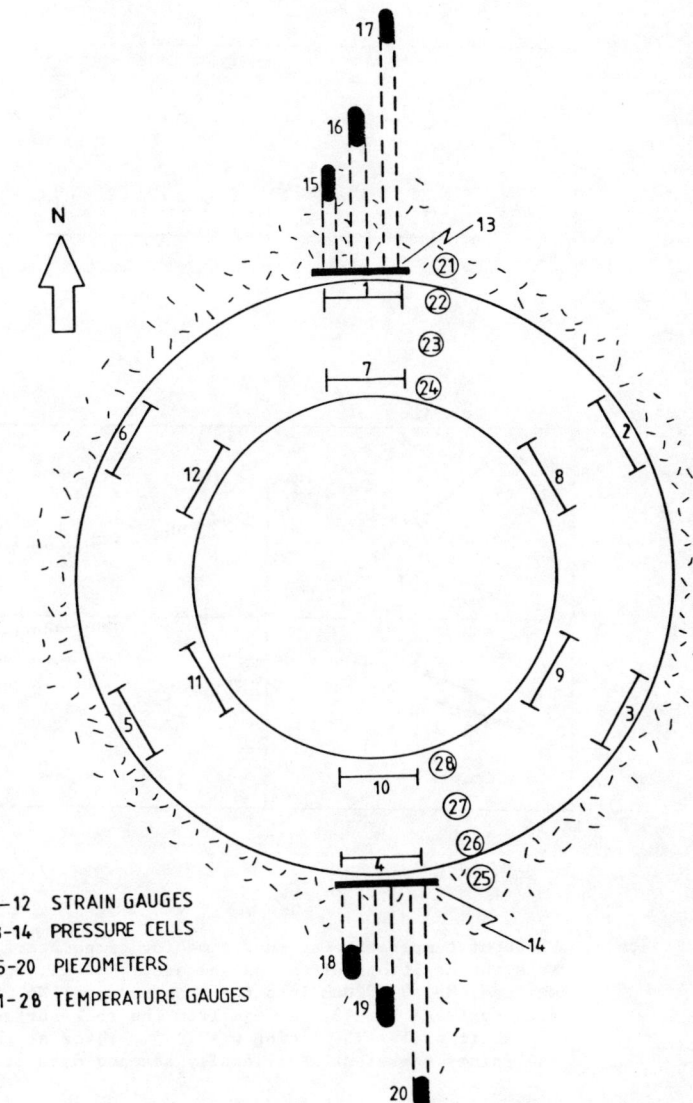

Figure 2. General arrangement of instrumentation at a depth of 665m in the Wooley Edge Sandstone at Riccall No. 2 shaft. Lining thicknesses are:

Position	Thickness (mm)
N	1665
NE	1645
SE	1805
S	1790
SW	1735
NW	1855

Table 3. Total stress and hydrostatic pressure readings - North Selby No. 2 shaft - Depth 234m.

Time After Pour (days)	Total Stress -Pressure Cells MN/m^2			Hydrostatic Pressure -Piezometer MPa		
	13	14	15	16	17	18
14	0.42	0.45	0.34	0	0	0
220	0.73	0.75	0.66	0.50	0.57	0.57
310	2.52	1.56	2.08	2.36	2.40	2.43
385	2.64	1.94	2.09	2.20	2.43	2.43

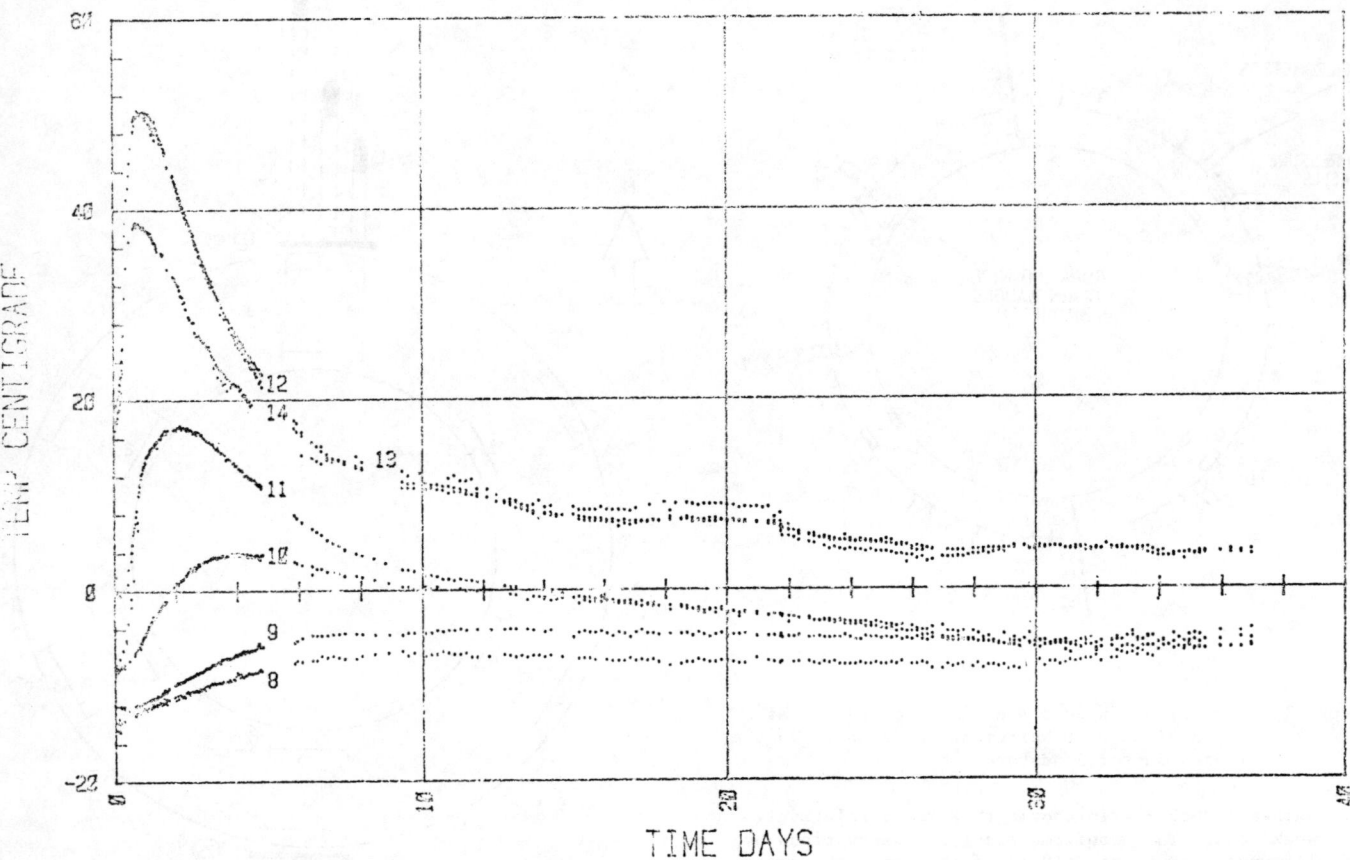

Figure 3. Plots of temperature against time for temperature gauges 8 - 13 (see Figure 1) at Whitemoor No. 2 Shaft. Note the positions of the gauges are:- No. 11, 50mm into the rock surface; No. 10, 270mm into the rock surface; No. 9, 720mm into the rock surface; No. 8, 1120mm into the rock surface; No. 12, is 290mm from the rock surface; No. 13, is 555mm from the rock surface; No. 14 is 930mm from the rock surface, and 115mm from the shaft wall. The lining was 1045mm thick at this point from 45 MN/m^2 (BRE Digest, 174) concrete. Note the points represent individually scanned data plots from the microprocessor.

6. RICCALL NO 2 SHAFT

Riccall No. 2 shaft, sunk to an inset level depth of 771m passes through 25m of glacial drift and 225m of Bunter Sandstone aquifer rock where groundwater was controlled by freezing (see Wild and Forrest, 1981). This was followed in the Permo-Trias succession by 40m of Upper Permian Marl, 25m of Upper Magnesian Limestone and 40m of Middle Permian Marls including evaporites, which were water free, overlying 80m of the Lower Magnesian Limestone aquifer. This contained water in fissures and was grouted using cement grouts. The underlying Basal Permian sands, about 7m thick were grouted with chemical grouts. The Coal Measures succession starting at about 430m contained one major aquifer rock - the Wooley Edge Sandstone - 12.5m thick at a depth of 660m. Laboratory samples gave the following average index data:

Compressive strength	19.7 MN/m^2
Tensile strength (indirect)	2.9 MN/m^2
Deformation modulus (secant at 50% strength)	8.3 GN/m^2
Poissons ratio (at 50% strength)	0.24
Dry unit weight	21.00 kN/m^3
Saturated unit weight	22.5 kN/m^3
Porosity	0.13
Voids ratio	0.15
Permeability	3.2 x 10^{-6} m/s
Potential water make	108 l/min

Drainage was chosen as the preferred method of groundwater control because of the relatively low permeability of the sandstone. The general dewatering layout is illustrated in Figure 4. Water was collected from 18 No. 25m long holes drilled at an angle of 22° from a level about 8.5m above the sandstone. Initially water flowed into the boreholes at a rate of 60 l/min, reducing to a rate of 18 l/min following excavation, the remaining water seeping into the excavation.

During lining construction, shaft wall surface water was kept from contact with the fresh concrete by a 'flexipane' sheeting. A secondary water collection system was installed behind this sheeting to prevent pressure build up until the lining had developed sufficient strength. Grout seals were then installed below and the lining was extensively backwall grouted. Following this the dewatering holes were grouted.

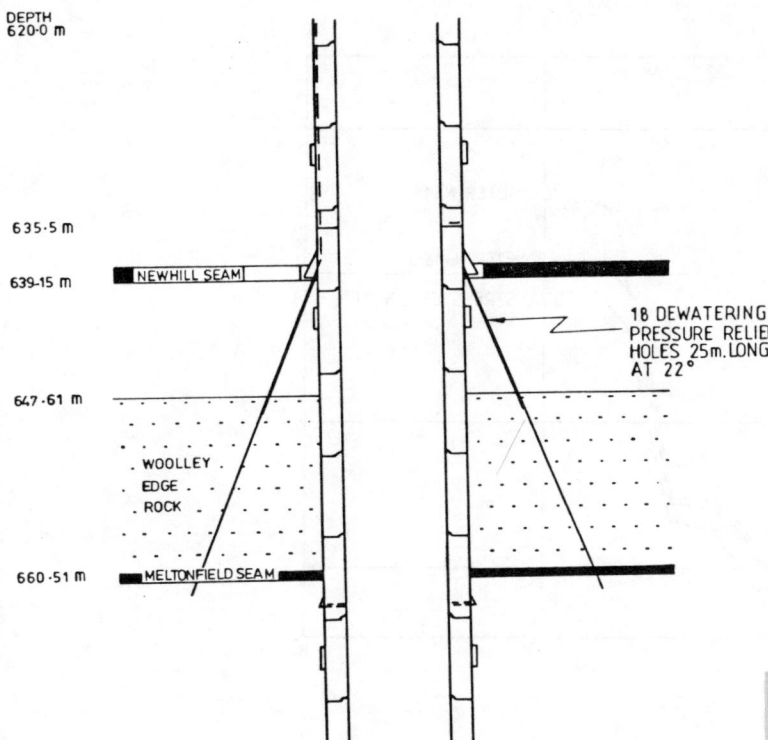

Figure 4. Dewatering layout planned for the Wooley Edge sandstone at Riccall No. 2 shaft.

Figure 7 includes strain measurements over the same period and temperature changes with time over a period of 39 days.

The temperature measurements (Figure 8) show a rapid rise to about 68° at the centre of the lining after approximately 40 hours, showing a similar pattern to those in Figure 3. These are currently being analysed in detail, but occasional occurrence of thermal cracking in high strength concrete can cause problems in shaft sinking.

It is interesting to note that the increase in temperature, whilst causing fluctuations in the vibrating wire transducers also caused a high negative reading in the mercury filled stress cells (see cell 14 in Figure 6) which did not completely recover.

These cells and the accompanying piezometers produced the most interesting and most significant results. Following grouting at 42 days there was a general increase in pressure of 0.8 MPa which reached a steady state when the grouting was completed, when a flow of 48 1/min was observed from the drain holes. Sealing of the grout holes raised the pressure level to between 4.2 and 5.2 MPa a magnitude equal to between 62 and 80% of the full hydrostatic head. It is interesting to note that at this point a total water flow of less than 6 1/min was estimated to be leaking through the lining over the 12.5 length of the shaft in the Wooley Edge Sandstone.

During this period hoop strains and temperatures in the concrete (a 50 MN/m^2 mix comprising 400 kg sulphate resisting cement; 775 kg Zone 2 sand 985 kg aggregate and 180 kg water was used) and total and hydrostatic stresses on the concrete lining interface were monitored.

The instrumentation layout at a depth of 665m is illustrated in Figure 2. It comprised six pairs of vibrating wire strain gauges (1-12), briquetted and positioned to measure near surface and backwall hoop strains, and two sets of vibrating wire temperature gauges (21-28) to measure temperature gradients through the lining and into the shaft wall. In addition two total stress cells (13,14) were placed at the lining interface and six piezometers (15-20) were placed at depths up to 3m into the sidewall rock. Figure 5 shows a selection of the instrumentation (4, 10, 14, 18-20, 25-28) placed above the previously poured concrete lining and behind the shuttering ready for pouring of the next concrete lining section. The concrete had a 1.2m design thickness but it can be seen from the data below Figure 2 that this was increased by overbreak to as much as 1.9m in places. Because of the large amount of concrete involved; pours were reduced from a lift of 6m to 3m in order to avoid overloading the kerb ring.

Following casting of the lining, a grout stop was installed and grouting commenced after 42 days, continuing until 49 days when the dewatering holes were also grouted. The results of piezometric pressure and total stress measurements in Figure 6 cover this period and the succeeding period up to 56 days. The data is typical of the readouts obtained and is based on 20 minute scans.

Figure 5. Installation of instrumentation from top of previously concreted lining sections at Riccall No. 2 Shaft.

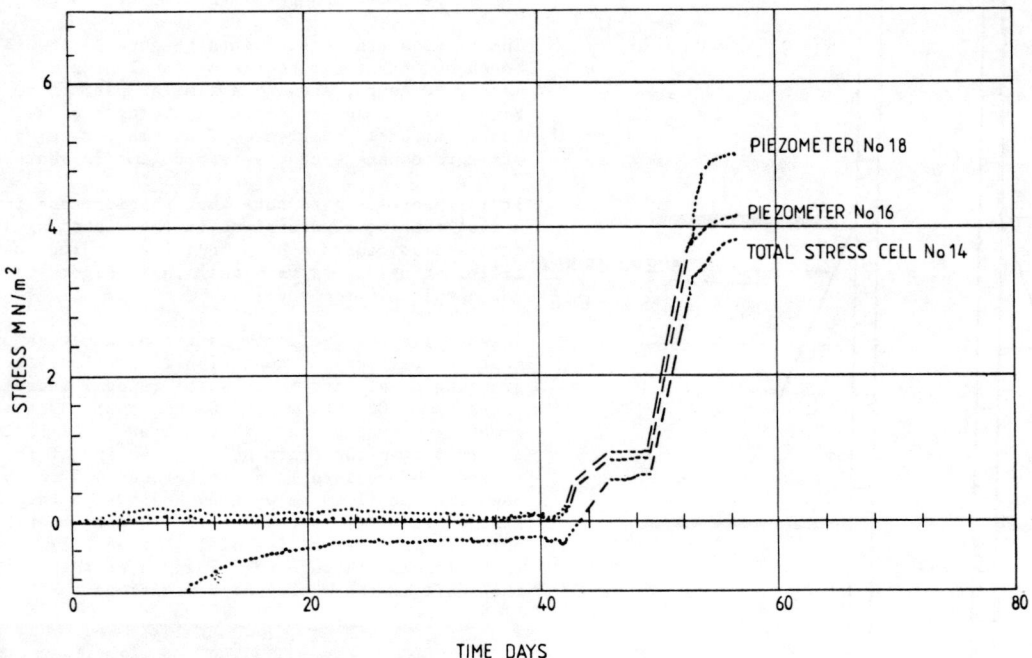

Figure 6. Changes in piezometric and total pressure up to 56 days at Riccall No. 2 Shaft. Over the period to 120 days pressures rose to levels roughly equivalent to full hydrostatic pressure. Gauge locations are shown in Figure 2.

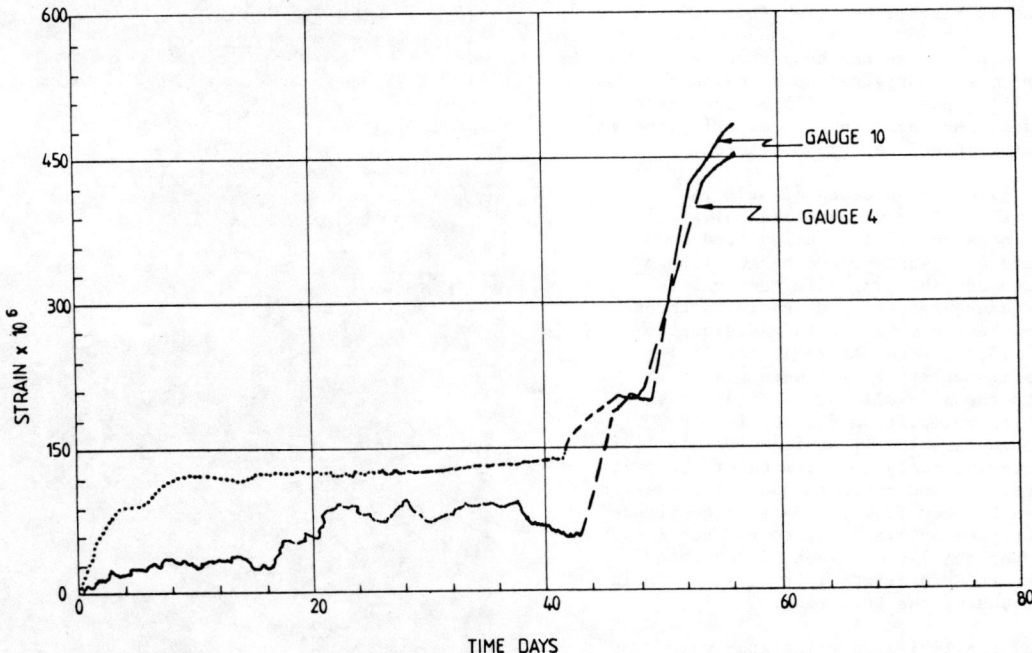

Figure 7. Changes in strain at the inside[10] and outside[4] surface of the shaft lining at Riccall No. 2 Shaft. Gauge locations are shown in Figure 2.

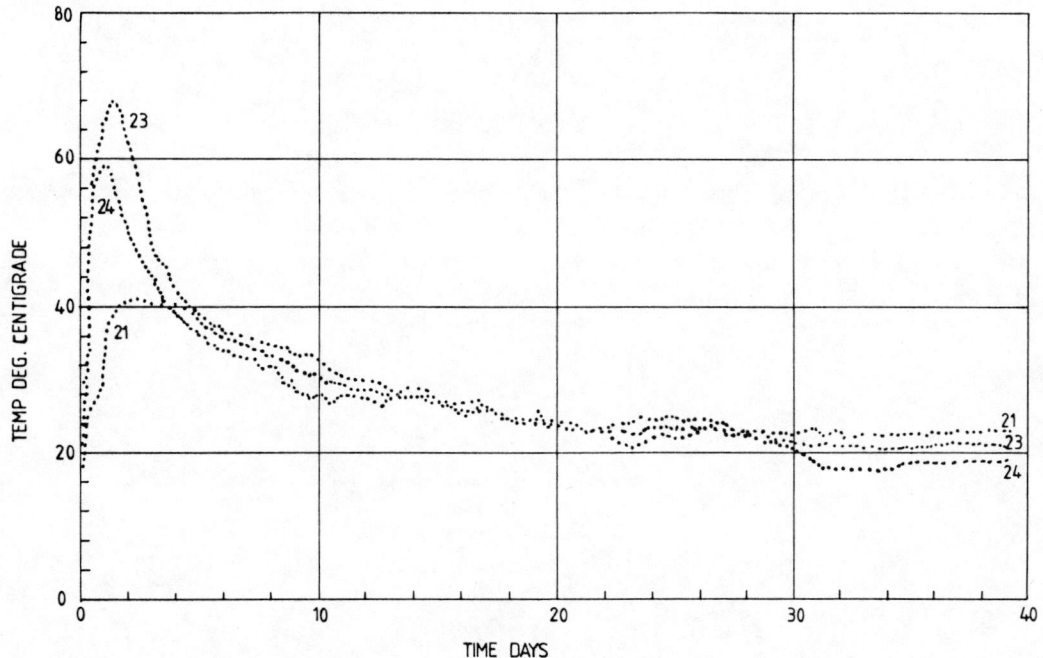

Figure 8. Changes in temperature with time at the inside[24] and centre[23] of the shaft lining and 50mm[21] into the rock wall at Riccall No. 2 Shaft. Gauge locations are shown in Figure 2.

During the succeeding 60 days, whilst continuous readings were not taken, both piezometric and hydrostatic pressures increased to levels close to hydrostatic pressure levels. Lining strains showed close correlation and there were no indications of any contribution to radial lining stresses from the rock wall.

The break in slope of the curves in Figure 6 cannot easily be explained. The most likely reason for the differential recharge is that over the first 50 days partial drawdown occurred through leakage into the adjacent shaft. Reduction of leakage in this shaft then allowed accelerated recharge. Since the outcrop of the Wooley Edge confined aquifer rock is about 20 miles to the West, it is likely that continual adjustments will occur for some time.

CONCLUSIONS

Examination of the results shows:

(a) Water pressures were the major source of radial lining pressure, having magnitudes at or close to hydrostatic levels, unless the lining has sufficiently high permeability to allow large scale drainage.

(b) Effective rock wall geostatic radial stresses were low or non-existent, even in weak rocks. The only indications of significant effective stress occurred during refreezing of porewater in frozen rock after it had thawed due to hydration temperatures in the cast in place lining.

(c) Concrete lining strains were related to radial lining pressures.

ACKNOWLEDGEMENTS

The work was carried out as part of a research project supported by the National Coal Board and the Research Fund of the European Coal and Steel Community (ref. 7220-AC/84). The views expressed are solely those of the authors.

REFERENCES

ALTOUNYAN, P.F.R. (1982) Measurement of stresses, strain and temperature in concrete in shafts and insets. Strata Mechanics (ed. I.W. Farmer) Vol 32, Developments in geotechnical engineering, Amsterdam, Elsevier, pp 154-149.

ALTOUNYAN, P.F.R. and FARMER, I.W. (1981) Tunnel lining pressures during groundwater freezing and thawing. Proc. 5th Rapid Excavation and Tunnelling Conf., San Francisco, 784-800.

ALTOUNYAN, P.F.R., BELL, M.J., FARMER, I.W. and HAPPER, C.J., (1982) Temperature, stress and strain measurements during and after construction of concrete lining in fozen sandstone. Proc. 3rd. Int. Symp. Ground Freezing, Hanover, N.H. pp

AULD, F.A. (1979) Design of concrete shaft linings. Prc. Inst. Civ. Engrs, 67, 817-832.

AULD, F.A. (1982) Ultimate strength of concrete shaft linings and its influence on design. Strata Mechanics (ed I.W. Farmer) Vol. 32, Developments in geotechnical engineering, Amsterdam, Elsevier, pp 134-140.

BELL, M.J. (1982) The design of shaft linings in Coal Measures rocks, ibid pp 160-166.

WILD, W.M. and FORREST, W. (1981) Application of the freezing process to ten shafts and two drifts at the Selby project. Mining Engineer, 140, 895-904.

LINING DESIGN FOR PERMANENT WORKINGS AND TUNNELS
Calcul des excavations définitives et des galeries
Dimensionierung des Ausbaues in permanenten Hohlraumbauten und Tunneln

N. S. Bulychev
N. N. Fotiyeva
Department Heads, Tula Polytechnic Institute, Tula

Yu. A. Veksler
S. K. Tutanov
Karaganda Polytechnic Institute, Karaganda

G. A. Katkov
D. I. Kolin
E. L. Kokosadze
S. A. Chesnokov
USSR Mining Institute, USSR Institute of Transportation Construction, Orgenergostroi,
Moscow, USSR

SYNOPSIS

This paper deals with certain methods of lining design for underground workings, based on investigations of the joint deformation of the lining and the surrounding rock massif. Discussed are a method for designing a system of anchors around the working in a linearly deformable medium, a method for designing the linings of a set of parallel circular tunnels of interacting behaviour, subject to static loads and the seismic effects of earthquakes. Also dealt with is the lining design in a massif subject to creep strain and failure.

RESUME

En se basant sur l'étude des déformations communes aux soutènements et aux massifs rocheux, les auteurs présentent certaines méthodes de calcul de soutènements et d'excavations. On discute une méthode de calcul d'un système d'ancrages autour d'excavations dans une zone susceptible de déformation linéaire ainsi qu'une méthode de calcul des revêtements d'un ensemble de galeries circulaires parallèles s'influençant réciproquement et soumises aux charges statiques et à l'activité séismique. D'autre part, on traite du calcul des soutènements dans un massif soumis aux déformations de fluage et à la rupture.

ZUSAMMENFASSUNG

Es werden einige Berechnungsverfahren zur Dimensionierung des Ausbaus unterirdischer Hohlraumbauten behandelt, die auf der Untersuchung des gegenseitigen Verformungsverhaltens von Ausbau und Gebirge beruhen. Das Berechnungsverfahren von Systemankerung um den Hohlraum herum in einem linear deformierbaren Medium sowie das Berechnungsverfahren von Auskleidungen aufeinander einwirkender, parallel verlaufender Tunnelrohre auf statische und Erdbebenbeanspruchungen werden dargelegt. Die Berechnung des Ausbaus in einem von Kriechverformung und Bruch gefährdeten Gebirge wird ebenfalls betrachtet.

INTRODUCTION

A new scientific trend, the mechanics of underground structures, is being successfully developed at the present time. It is based on the concept that the lining and the rock massif, in designing underground structures, are to be considered as elements of a single deformable system in contact with each other. This approach enables various kinds of static effects (underground pressure, tectonic stresses, external hydraulic pressure, internal pressure head, etc.) and dynamic effects to be dealt with from a single point of view. It also enables the behaviour of various kinds of lining (monolithic, precast, sprayed concrete, stayed, or anchored, etc.) to be investigated, and the mutual effects of closely located structures to be taken into account.
A special feature of this up-to-date approach is that the initial data in structure design does not include the so-called loads on the lining. These loads (stresses at the contact surface between the lining and the massif) are determined in the process of lining design, with the lining interacting with the rock massif.

On the basis of the above-mentioned principles, methods have been developed in the USSR in recent years for designing closed monolithic concrete and reinforced-concrete tunnel linings of arbitrary cross section, subject to static loads (Fotiyeva, 1974) and to the seismic effects of earthquakes (Fotiyeva, 1980); for designing multiple-layer and combination linings of tunnels and vertical shafts (Bulychev, 1982). New designing procedures take into account the linear hereditary creep of the rock, etc. The present paper sets forth the results of a further development of the above approach as applied to the lining design of a set, or complex, of parallel circular tunnels, taking into account their effects on one another, to the design of anchored linings for underground structures, located in a nonlinearly deformable massif subject to intensive rheological processes.

1. LINING DESIGN FOR A SET OF PARALLEL CIRCULAR TUNNELS

Procedures have been developed for designing the linings of an arbitrary number of parallel circular tunnels that mutually affect one another, that are of various radii and with any arrangement of their centres (including those in which they do not lie on a straight line), supported by monolithic linings of various thicknesses and made of various materials. The effects are considered of the main kinds of static loads: internal pressure head, which may differ in the tunnels, including the case in which some of the tunnels have been emptied; underground pressure and the pressure of subsoil waters, both uniform pressure and pressure that varies along the height of the linings. Also proposed is a procedure of lining design for a set of parallel tunnels subject to the seismic action of earthquakes.

The method of designing linings subject to static loads is based on analytical solutions of the corresponding plane contact problems of elasticity theory for a multiply connected piecewise homogeneous region, i.e. a medium weakened by reinforced circular holes, with boundary conditions corresponding to the kinds of loads being considered. The problems are solved by a procedure developed from the method proposed by D.I. Sherman (1951), making use of the theory of analytic functions of a complex variable, complex series and the properties of Cauchy-type integrals.

Design calculations for structures subject to the seismic action of earthquakes are based on finding the most unfavourable stressed state in each cross section of each lining of the set for the various combinations of joint action of long seismic longitudinal and transverse waves having any direction in the cross-sectional plane of the structure. In each radial section, for this purpose, the general expressions for the normal tangential stresses of each lining are investigated to find the extreme angle α of wave incidence. These expressions are obtained as the result of superposing the solutions of two quasi-static plane contact problems for the above-mentioned multiply connected region. The first problem concerns biaxial compression of the medium at infinity at the arbitrary angle α to the horizontal by loads depending on the wave parameters. Its solution enables the stresses in the linings caused by a long longitudinal wave of arbitrary direction to be obtained. The second problem concerns pure shear at infinity at the angle α. Its solution enables the stressed state of the lining to be determined upon the action of a long transverse wave of arbitrary direction. An examination of the sum and difference of the solutions of the second and first problems concerning the extreme angle α of wave incidence enables the combination of simultaneously arriving longitudinal and transverse waves (the least favourable case) to be found for each cross section, as well as their directions, at which the normal tangential stresses are a maximum in magnitude in the given cross section. This permits the analytic construction of the envelope of the normal tangential stress diagram for each lining. The forces in each cross section are determined for the combination and directions of the waves at which the normal tangential stresses are maximal in the given cross section. In the final analysis these stresses determine the strength of the design. The forces due to seismic effects are taken with + and - signs and are added to the forces applied by other types of loads.

The forces obtained in this manner represent, not a single kind of action, but the whole possible totality of loads in the most unfavourable version for each cross section of all the structures, taking into account their effects on each other. Such an approach increases the reliability and earthquakeproof quality of the linings.

Provision is also made to take into account the case in which the lining breaks away from the rock. For this purpose, the action of the longitudinal wave in the tensile phase is excluded from consideration. Then design calculations are based on two different envelopes, analytically plotted, of the normal tangential stress diagrams. The envelopes are plotted from the maximum values of compressive and tensile stresses due to transverse waves and to longitudinal waves in the compressive phase. This method is recommended for linings designed to permit crack formation.

The proposed methods of design calculations have been programmed for an ES series electronic computer and are being used in the design of parallel tunnels. Given in Fig.1 as an example are the results of calculations in designing two parallel tunnels subject to the external nonuniform pressure of subsoil waters. The initial data were: ratio of the moduli of deformation of the lining material to the modulus of deformation of the rock $E_1/E_0 = 6$, outside radii of the linings $R_1 = 11$ m and $R_2 = 5.5$ m, thicknesses of the linings $d_1 = 0.1 R_1$ and $d_2 = 0.1 R_2$, centre-to-centre distance of the tunnels $l = 22$ m, and the static head of subsoil waters, measured from a line passing through the centres of the tunnels, $h_1 = 16.5$ m.

2. DESIGN OF ANCHORED LININGS

The procedure for designing anchored linings is based on the solution of a contact problem on the interaction of a system of anchors, reinforcing the working, with the deforming massif (Bulychev, 1979). Displacement of the rock as the face advances and as the result of creep strain of the rock leads to deformation and loading of the anchors that prevent such displacements.

The massif surrounding the working is modeled by a linearly deformable medium. This enables, after replacing the effect of the anchors by a system of unknown forces, to deal with the resultant displacement field as the sum of

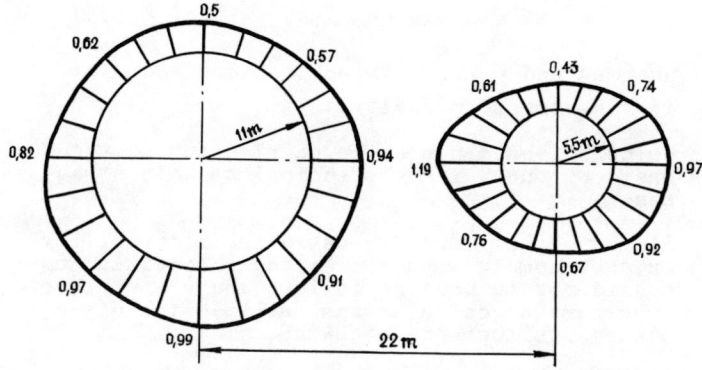

Fig.1. Diagrams of normal tangential stresses σ_θ at the inner outlines of the tunnel cross sections.

the displacement field of the ponderable massif with the unsupported working and the displacement field of the massif due to the action of the system of anchors. The forces acting in the anchors are determined from the condition of consistency of displacements of the anchors and the massif.

A closed system of equations has been derived for determining the relative displacements Δ_i of the ends of the i-th anchor in the process of its interaction with the massif. In matrix form, this system is as follows:

$$K\bar{\Delta} = \bar{\Delta}_H - c\bar{\Delta}_n, \qquad (2.1)$$

where K is a matrix whose elements take into consideration the relationship between the rigidities of the anchor rods and the rock and the effects of the anchors on one another, $\bar{\Delta}$ is the required vector with the components Δ_i, and $\bar{\Delta}_H$ and $\bar{\Delta}_n$ are vectors whose components take into account the displacement of the rock around the unreinforced working and the preliminary stresses of the anchors, respectively.

The required forces Q_i in the anchors are expressed in terms of the components of vector $\bar{\Delta}$ as follows:

$$Q_i = c(\Delta_i - \Delta_{in}) \qquad (2.2)$$

Taken into consideration in solving the problem is the creep of the rock within the framework of hereditary theory. This enables the effects of the lack of simultaneity in opening the face and in installing the anchors to be examined. Taking into account the creep of the rock and the distance from the face of the working to the place where the anchors are installed, the system of equations (2.1) is of the form

$$K(t)\bar{\Delta} = \bar{\Delta}_H[1 - \psi(z) + F(t + T_a) - F(T_a)] - c\bar{\Delta}_n[1 - \varphi(t)], \qquad (2.3)$$

where T_a is the time from the moment the face is opened to the time of installation of the anchors, and z is the distance from the face to the place where the anchors are installed.

Functions $K(t)$ and $\varphi(t)$ take into consideration the creep of the rock in the lock of the anchor and under its bearing element. Function $F(t)$ represents the creep of the rock massif, weakened by the unsupported working.

The design procedure takes into account the effect of the nonlinear dependence of displacement of the anchor lock on its resistance under conditions of simple loading of the system of anchors, i.e. the proportionality of the loads in the system of a single parameter.

The procedure for designing anchored linings has been programmed for an electronic computer.

An analysis of the results obtained in design calculations showed an essential dependence of the stressed state of the anchors on the time T_a that passes between the opening of a face and the installation of the anchored lining. If installation immediately follows the baring of the rock ($T_a \approx 0$), the anchor becomes loaded in the course of time and, after a definite length of time, its load is relieved, thereby reducing its efficiency. The effects of the lateral pressure in the massive and the shape of the cross section of the working were also investigated. At low values of the coefficient of lateral pressure, an anchor, installed in the roof of a working of circular cross section, is found to be overloaded, whereas those at the sides are underloaded (Fig.2) and, without prestressing, turn out to be ineffective.

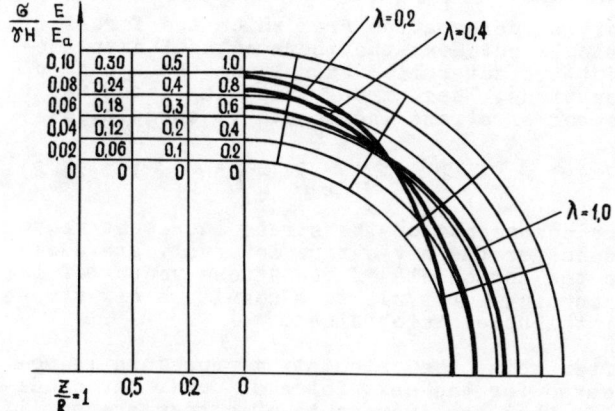

Fig.2. Stresses in the anchors ($l = 0.8 R$) as functions of their place of installment, coefficient of lateral pressure λ and the distance z to the face of the working.

The design procedure can be used to achieve optimum design parameters of the lining, taking into account its interaction with the massif.

3. LINING DESIGN FOR WORKINGS IN WEAK ROCK

The arrangement of workings in weak rock leads to intensive rheological processes in their vicinity.

Under consideration is a viscoelastic massif with an extended working. Deformation is assumed to be of the plane type. The physical equations are selected in the form of a Volterra integral equation. Calculations are carried out in running (current) coordinates.

It follows from the physical equations that the vector $\{\epsilon\}$ of total deformation is the vector sum of the vector $\{\epsilon^e\}$ of instantaneous elastic deformation and the vector $\{\epsilon^c\}$ of creep strain. In vector form the equation is

$$\{\epsilon\} = \{\epsilon^e\} + \{\epsilon_1^c\} + \ldots + \{\epsilon_i^c\} + \ldots , \quad (3.1)$$

where $\{\epsilon_i^c\}$ is the vector of creep strain in the chosen length of time.

With the equation written in this form it proves convenient to apply the method of initial deformations. The following algorithm is proposed for solving the problem.

For the instant of time $t = 0$, the elastic problem is solved by the finite-elements method (FEM). The elastic displacements and the new coordinates of the node points are determined. These displacements are employed to determine the stress vector of each element in the chosen region. The components of the stress vectors are determined and then used to calculate the vector of strain increment, and the running coordinates of the node points of the region are used to formulate a rigidity matrix of the system for the length of time t_i.

The quantity $\{\Delta\epsilon_1^c\}$ is considered to be the initial deformation from which the forces are calculated. Next the increments of displacement are determined by solving the system of equations. Then, from these displacements, the vector of stress increments is calculated:

$$\{\sigma_1\} = [D](\{\Delta\epsilon_1\} - \{\Delta\epsilon_1^c\}). \quad (3.2)$$

The components of the stress increment vector, found for the given time interval, are added to the components of the stress vector of the preceding interval. As a result, a new stress distribution is obtained.

After this, the calculation procedure is repeated for the next interval under the condition that the increments of stress are small.

Failure in the rock massif in the vicinity of the working is investigated by making use of the following strength conditions.

The durability of each finite element of the region is taken into account from the viewpoint of the thermal fluctuation concept of strength. The durability is calculated by the equation

$$T = A \exp(-\gamma\sigma_i), \quad (3.3)$$

where A and γ are constant factors and σ_i is the stress intensity.

Calculations can also be carried out on the basis of the maximum principal tensile stresses.

Elements that have exhausted their durability or that comply with the strength condition are considered to undergo failure and are excluded from further calculations. At this the displacements increase stepwise.

A programme has been worked out for carrying out calculations in practice in an electronic computer.

As an example for illustrating the potentialities of the method, the failure of a massif surrounding a working of trapezoidal cross section is considered (Fig.3). The basic design layout of the problem is a ponderable half-plane with a hole of the same shape, loaded by compressive stresses at infinity.

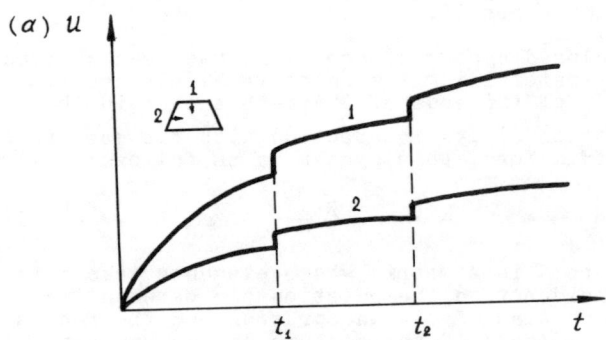

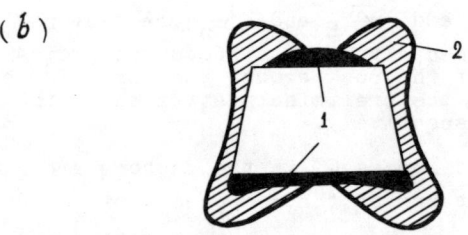

Fig.3. Development of displacements u in the roof and sides of a working with time (a), and the nature of rock failure (b): 1 -- at the instant t_1 of time, and (2) at the instant t_2.

It is established as a result of the solution that failure begins in the roof and bottom of the working.

REFERENCES

Bulychev N.S. (1982), Mechanics of Underground Structures, Nedra Publishers, Moscow (in Russian).

Bulychev N.S., Markin V.E. and Kolin D.I. (1979), Determining the Type and Optimum Design Parameters of Tentative Tunnel Linings, Journ. Transportnoye Stroitelstvo, No.6: p.47-49 (in Russian).

Fotiyeva N.N. (1974), Lining Design for Tunnels of Noncircular Cross Section, Stroiizdat Publishers, Moscow (in Russian).

Fotiyeva N.N. (1980), Lining Design for Underground Structures Located in Seismic Regions, Nedra Publishers, Moscow (in Russian).

Sherman D.I. (1951), On the Stresses in a Plane Ponderable Medium With Two Identical, Symmetrically Located Circular Holes, Journ. Prikladnaya Matematika i Mekhanika (Applied Mathematics and Mechanics) vol.15, issue 6, pp. 751-761 (in Russian).

TIME-DEPENDENT BEHAVIOUR OF TUNNELS IN HIGHLY STRESSED ROCK
Déformation en fonction du temps des tunnels en roche surchargée
Das zeitabhängige Verhalten von Tunneln im überbeanspruchten Gebirge

P. K. Kaiser
Associate Professor of Civil Engineering

S. Maloney
Research Associate

N. R. Morgenstern
Professor of Civil Engineering
Department of Civil Engineering, University of Alberta, Edmonton, T6G 2G7, Canada

SYNOPSIS

The time-dependent behaviour of small tunnels in highly stressed, jointed rock tested during laboratory tests is described after a brief description of the test apparatus and some typical test results. For the tunnels tested it was possible to determine unique tunnel convergence rate functions that separate three modes of tunnel behaviour: (a) pre-failure mode, (b) stable yield zone propagation mode, and (c) unstable or rupture mode. On the bases of these convergence rate functions the performance of tunnels excavated in the same rock material was predicted based on convergence measurements. The practical implications of these convergence rate functions are discussed.

RESUME

Après avoir donné une brève description des appareils d'expérimentation utilisés et présenté quelques résultats de tests typiques, on décrit le comportement, en fonction du temps, de petits tunnels excavés dans une roche fissurée soumise à de fortes contraintes et étudiés au cours d'expériences en laboratoire. On a pu déterminer des fonctions de convergence propres à trois modes de comportement des tunnels, à savoir: a) mode d'avant rupture; b) mode stable de propagation de la zone de rendement et c) mode instable ou de rupture. En se basant sur ces fonctions de taux de convergence on a pu prédire la performance de tunnels dans un même type de roche. Enfin, on examine les implications pratiques de ces fonctions de taux de convergence.

ZUSAMMENFASSUNG

Das zeitabhängige Verhalten von Tunneln in überbeanspruchtem, geklüftetem Gebirge wird beschrieben zusammen mit einer kurzen Erläuterung der Versuchsanlage und typischen Versuchsresultaten. Für die Tunnel, die im Labor geprüft wurden, konnten Konvergenz-Geschwindigkeits-Funktionen bestimmt werden, die es erlauben, drei Tunnelverhaltensgruppen zu differenzieren: (a) Verhalten vor dem Bruch, (b) Verhalten während beschränkter, und (c) während unbegrenzter Ausbreitung der Bruchzone. Mit Hilfe dieser Konvergenz-Geschwindigkeits-Funktionen war es möglich, das Verhalten von Tunneln in ähnlichen Gesteinsmaterialien vorherzusagen. Die praktische Bedeutung dieser Konvergenz-Geschwindigkeits-Funktionen wird kurz erläutert.

1. INTRODUCTION

The success of modern tunnelling technologies such as mechanized excavation by tunnel boring machines or the New Austrian Tunnelling Method (NATM) depends largely on the interpretation of the time-convergence relationship that is controlled by the time-dependent rock mass behaviour and the excavation - support history, particularly when excavating in incompetent rock. The stand-up-time and the rate of tunnel closure are dominated by the excavation method and sequence, the support system and the installation procedure. The time-dependent tunnel wall convergence is initially controlled by the rate of face advance and then by the time-dependent stress redistribution processes resulting from softening or yielding ahead of the tunnel face and near the tunnel walls. Far from the tunnel face the time-dependent behaviour is controlled largely by the creep properties of the rock mass.

For the purpose of studying these time-dependent processes controlling the performance of tunnels in highly stressed or overstressed rock masses, the tunnel convergence and the internal rock mass strains near the tunnel wall were monitored during process simulation tests (PST) in the laboratory. The instantaneous and time-dependent rock mass response was observed while a circular tunnel was excavated, as well as during creep stages following tunnel excavation, in a jointed rock mass with time - dependent strength and deformation properties.

The *total* accumulated tunnel closure or rock mass strain can seldom be determined accurately in the field and the *total* time-dependent displacements are difficult to determine because the initial strain rates are high when compared with the loading or excavation rate. Moreover, access for strain measurements ahead of the tunnel face is often limited or impossible. In practice, it has been found convenient to use the deformation rate rather than the displacements as an indicator to evaluate tunnel performance. For example, at the Arlberg Tunnel [John (1981)] the tunnel wall displacement rate was used to determine when additional rock bolting was required and what the optimum bolt length should be. Additional or longer bolts were installed when the deformation rate exceeded 50 mm/day, and reasonable development of loads on the final lining was observed if the residual tunnel wall deformation rate in a tunnel with a diameter of 11 m was equal to or less than 10 mm/month [John (1977)]. These values correspond approximately to tunnel closure rates of 37×10^{-3}%/hour and 0.25×10^{-3}%/hour respectively. Because of the practical significance of these closure or strain rates most of the following data interpretation will be based on the time-dependent observations presented in double logarithmic deformation rate vs time plots. This relationship was found to be nearly linear for the test material (coal) [Kaiser et al. (1981)] and a similar, linear relationship was observed from the tunnel closure and the radial rock mass strain observations [Kaiser et al. (1981)].

As discussed by Kaiser (1981) and Kaiser and Morgenstern (1982) tunnel closure measurements reflect the overall behaviour of the underground opening but must be combined with extensometer measurements to delineate the yield or failure mechanisms. Accordingly, the closure rate may be used to evaluate whether the behaviour of the opening is changing but a local radial strain rate is useful to determine whether and in what area stress redistribution is initiated or propagated.

The main objective of this paper is to evaluate recent test results in this perspective and to illustrate the applicability and limitations of tunnel closure rate measurements for the purpose of the evaluation of tunnel safety and for the design of temporary support systems.

2. TEST FACILITY

The experimental facility and typical test results are described in detail by Kaiser and Morgenstern (1981a) and Kaiser and Morgenstern (1981b) and are only reviewed briefly here. Certain aspects of both time-independent and time-dependent tunnel response have been discussed extensively by Kaiser and Morgenstern (1982).

The test facility was developed for physical modelling of the time-dependent processes that control the behaviour of underground openings, particularly tunnels. The tests were called process simulation tests (PST) because the intent was to simulate the physical process of stress redistribution rather than to model a specific prototype. Time-dependent stress redistribution near a tunnel excavated in a strain-weakening rock mass with time-dependent strength and deformation properties occurs during the propagation of a zone of softened rock material or while a yield or fracture zone develops when the rock mass strength is exceeded by the stresses near the opening wall.

The test frame was designed to allow for longterm tests of several weeks under constant pressures of up to 15 MPa at the sample boundary. The sectional view of the test frame presented in Figure 1 shows the longitudinal restraint and reaction frame that allows for maintaining plane strain conditions parallel to the axis of the tunnel. The sample is approximately 600x600x200 mm and is located in the center of the test frame. It is instrumented with radial extensometers installed from the outside of the sample and convergence gauges inside the tunnel. The convergence is measured in four directions (45° apart) and the extensometers record the average radial strain over a 35 mm distance near the tunnel wall (inside ring at $r/a=1.51$ to 1.67; where r = distance to center of the extensometer measuring range and a = tunnel radius) and further inside the rock mass (outside ring at $r/a=1.85$ to 2.13). The deformation of the sample boundaries were also recorded to determine the overall sample strains.

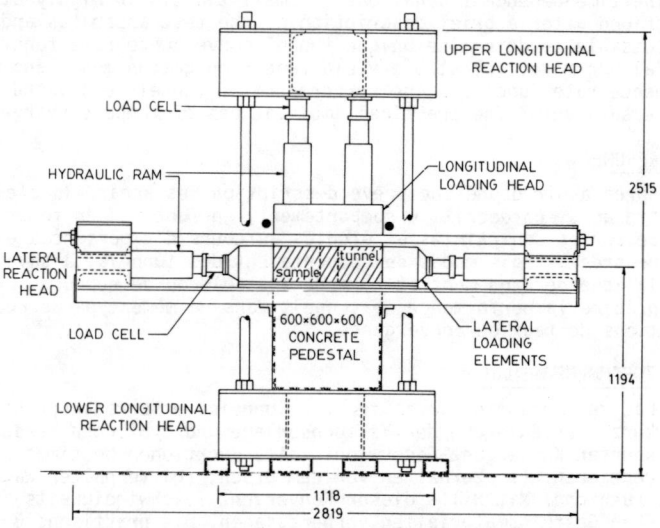

Figure 1. Process Simulation Test Frame; Side View.

The sample, with or without a pre-existing tunnel, was loaded with equal lateral pressures (stress ratio N=1) or unequal pressures (N<1 or N>1). This loading path (called external loading) does not correspond with the real loading condition in the field where a pre-stressed rock mass is unloaded internally during excavation. However, in an investigation designed to evaluate the effects of the different loading histories it was shown that the differences can be explained by superposition of load path increments [Kaiser et al. (1982)]. Most of the results presented in this paper originate from tests whereby the realistic loading sequence of internal unloading during excavation of a circular tunnel by drilling under load was used.

3. ROCK MASS PROPERTIES

The material for these process simulation

tests was selected to represent a rock mass with the following characteristics:

- jointed rock mass with regular pattern of joints or planes of weakness;
- frictional and cohesive strength components;
- strain-weakening stress strain curve;
- time-dependent peak strength;
- time-dependent post-peak strength behaviour; and
- time-dependent deformation properties.

Because it is extremely difficult or impossible to establish similitude requirements and to design an artificial material that satisfies all of these requirements, it was decided to use a natural material to model a rock mass as described above. A sub-bituminous coal found near Edmonton, Alberta, fulfils most of these requirements. It is regularly jointed with two orthogonal joint sets perpendicular to bedding (spacing between 10 and 30 mm) with one joint set being more dominant than the other. This coal also possesses all of the desired time-dependent characteristics and was described in some detail by Kaiser and Morgenstern (1981a). The pre-failure creep behaviour can be described by a power law resulting in a linear relationship between log-strain rate and log-time. The failure envelope is bilinear with negligible cohesion and an internal friction angle of 60 to 70° for normal stresses below 1 MPa and a cohesion c between 2 and 3 MPa with an ultimate friction angle of 30° at higher normal stresses. The Young's Modulus varies widely between 0.9 and 1.5 GPa as determined from triaxial tests and between 0.8 and 2.1 GPa measured during tests on large samples under hydrostatic pressure.

4. TYPICAL TIME-DEPENDENT RESPONSE

A typical plot of tunnel closure vs. time after loading, during an external loading test, is shown in Figure 2.a and typical radial strain development for four extensometers installed parallel to jointing is presented in Figure 2.b. In these figures the sample with a tunnel and two concentric rings of extensometers is shown schematically in the upper right corner. The direction of the major joint set is indicated by the lines extending from two corners of the sample. SIGMAV is the field stress applied to the upper and lower boundary of the sample and N x SIGMAV is the stress applied to the sides of the sample. TZERO gives the time from the beginning of the testing on the sample to the start of the load increment documented on the diagram. The location and orientation of the tunnel closure (convergence u divided by tunnel radius a) and radial strain measuring devices are indicated by the symbols or numbers on the curves and in the corner diagram.

The corresponding data in double logarithmic diagrams of closure and strain rate vs. time are given in Figures 3.a and 3.b respectively. The data corresponds to the results presented in Figure 2. The tunnel convergence is directed toward the tunnel while the radial strains are extensional. It can be seen that both the tunnel closure and the radial strain rate follow a power relationship similar to that for the test material itself. Deviation from this linear relationship was observed [Kaiser and Morgenstern (1982)] at elevated stress levels where stress redistribution near the opening walls occurred because of rock mass softening or yielding due to local or global overstressing of the rock. Figure 4 illustrates such a case where an external loading creep test was undertaken at increasing stress levels. The shaded area shows the narrow range of tunnel closure rates observed during a hydrostatic loading test (N=1) at stress levels below 13 MPa field stress. At 14 MPa a sudden rate increase at about one hour indicates that some stress redistribution caused a more rapid tunnel convergence. During a subsequent test, at 16 MPa field stress, yielding occurred and tunnel instability with rock fracture was observed visually. This yielding process consisted of several surges of rupture and failure zone propagation as indicated by the high and unsteady closure rate with several rate increases at approximately 0.2, 1.0 and 10 hours.

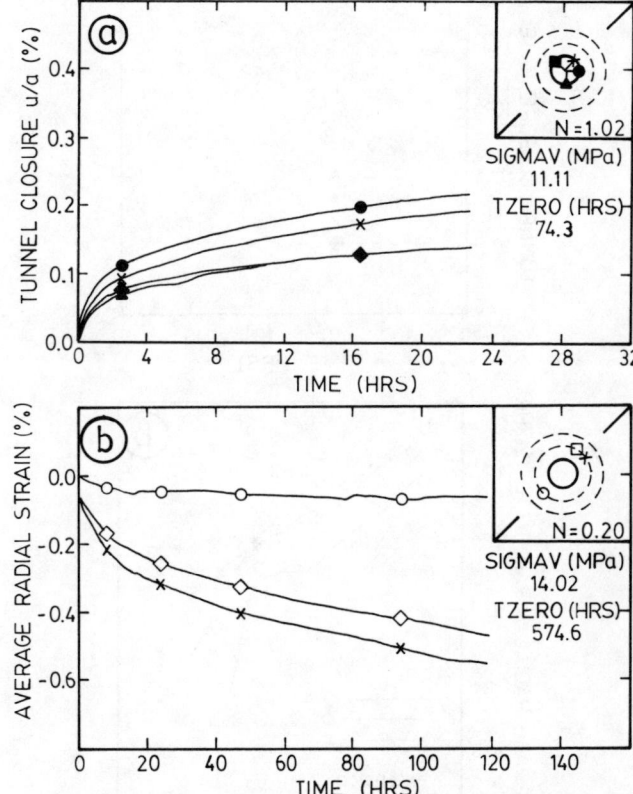

Figure 2. (a) Typical Tunnel Closure vs Time of Four Tunnel Diameters; (b) Typical Creep Curves Recorded by Three Extensometers near the Tunnel Wall. (negative=extension)

5. CRITICAL TUNNEL CLOSURE RATE

A tunnel is a highly statically indeterminate structure. Hence, initiation of yielding does not lead to immediate failure unless progressive strength loss results in uncontained propagation of the yield zone, e.g., to the ground surface. However, initiation of a yield zone changes the deformation behaviour of the opening and particularly the deformation rate. This can best be illustrated by the simple analogue model of a statically indeterminate beam of degree one presented in Figure 5. Under a load P_1, before yielding and plastic hinges are generated, the creep deformation rate at Point A depends on the creep properties of the material composing the beam. This creep rate will be below a critical creep rate that is only reached when yielding is initiated. The critical creep rate is a function of the time t measured from the time of load application and is called the Critical Creep Rate function CCR(t). It could also be called 'characteristic creep rate' because it depends mainly on the rheological properties of the rock mass. It is refer to as the critical creep rate because it also reflects the condition when yielding is initiated.

Post-yield Creep Rate PCR(t), the rate that is reached *after* propagation has terminated and a fully plastic hinge has been created at Point A. This post-yield creep rate function PCR(t) is not as uniquely defined as the critical creep rate function CCR(t). It depends on the yield mechanism or more accurately the final rock mass structure resulting from this yield mechanism, the post-peak stress strain behaviour of the rock mass, and the time-dependent post failure properties of the rock mass that themselves depend largely on the loading-history.

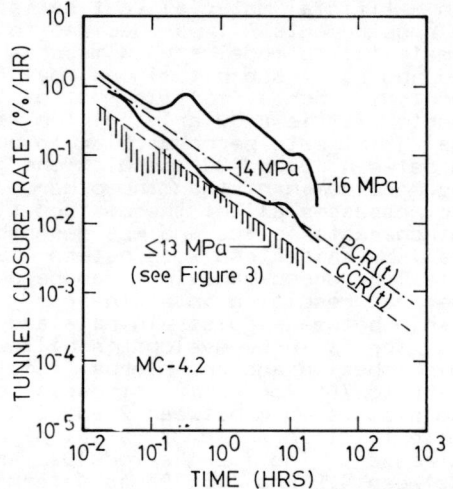

Figure 4. Tunnel Closure Rates recorded during Test MC-4.2.

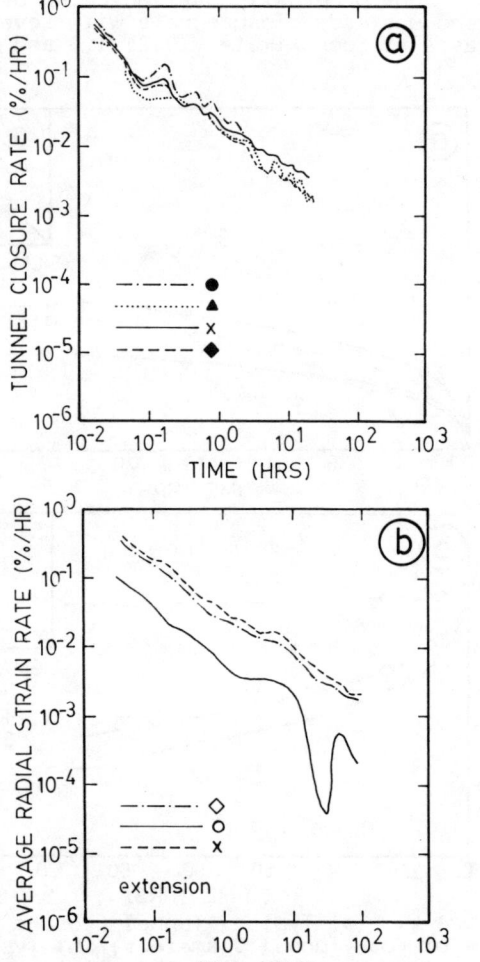

Figure 3. Same Data as Presented in Figure 2 Plotted in Double Logaritmic Diagrams of Closure Rate or Radial Strain Rate vs Time.

If the load is increased in magnitude to P_2 a plastic hinge is created, i.e. at point A, and as a result the deformation rate increases at the time of yielding. This is indicated by the sudden rate increase at Point A. After stress redistribution due to yielding has terminated a new statically changed but still stable system has been created and a new creep rate function reflecting the new structure and its creep properties is reached. This rate is called the

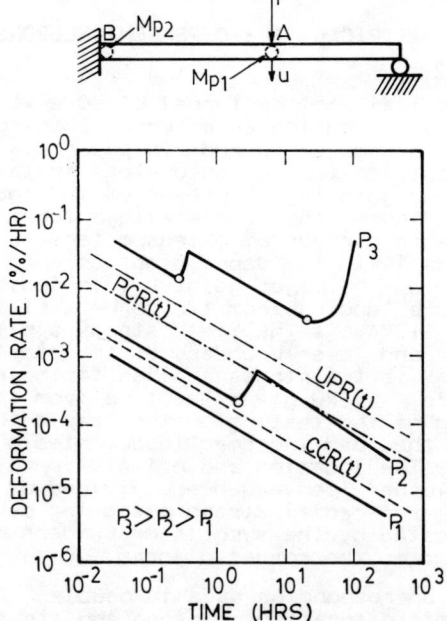

Figure 5. Schematic Diagram of a Structural System Illustrating the Three Creep Functions CCR(t), PCR(t) and UPR(t).

Under an even higher load P_3, yielding may also occur at Point B immediately after yielding at Point A or after some time-dependent stress redistribution as shown in Figure 5. At this time the system becomes unstable and collapses. The rate of deformation will increase beyond an

Ultimate Propagation Rate UPR(t). For an underground opening this limit can only be reached or exceeded if unstable failure occurs as can be expected in a cohesionless material or if propagation of a failure zone to the ground surface is possible.

The test results from an external loading test (MC-4.2; Figure 4) have been used to determine the CCR(t) and the PCR(t) for the small tunnels tested in the process simulation tests. The closure rates for all tests at stress levels below 13 MPa fell inside or below the shaded area shown in Figure 4 and no excessive yielding could be identified either from the extensometer readings or from visual inspection of the tunnel wall. At 14 MPa yielding in certain areas near the tunnel wall was evident from increased radial strain measurements and at 16 MPa extensive failure zones propagated from the tunnel wall in the direction perpendicular to the dominant joint set [Kaiser (1981)]. As a result of these observations the CCR(t) was defined by a straight line at the upper limit of the shaded range of closure rates and the PCR(t) was assumed as a line parallel to the CCR(t) touching the the highest rates measured during the test at 14 MPa. Similarly a PCR(t) for the test at 16 MPa could be determined. However, the PCR(t) as determined from the test at 14 MPa where first yielding was observed can be used to identify the range of closure rates that are indicative of initiation of yielding with terminating yield zone propagation. Closure rates in excess of this PCR(t) indicate that a yield zone is propagating further away from the tunnel wall. The ultimate propagation rate UPR(t) could not be determined in this test because no unstable propagation of yielding occurred.

At three specific times t=1, 5 and 20 hours the magnitude of these rate functions rate are:

- CCR (1) = 30 x 10^{-3} %/hr
- CCR (5) = 9 x 10^{-3} %/hr
- CCR (20) = 3 x 10^{-3} %/hr

and

- PCR (1) = 80 to 100 x 10^{-3} %/hr
- PCR (5) = 25 to 30 x 10^{-3} %/hr
- PCR (20) = 8 to 10 x 10^{-3} %/hr

These creep rate limits bounding yield initiation (I) and yield propagation (P) were then used to predict the initiation and propagation of stress redistribution near the tunnel by comparison with closure rate measurements made during fourteen subsequent tests on samples MC-5 and MC-6. In Table 1 the observed and predicted behaviour for only three of these tests (MC-6.02, 6.05o and 6.06) are summarized. Good agreement between prediction and actual behaviour was achieved in all other cases not documented in this paper [Kaiser and Morgenstern (1982)]. In column 3 and 4 of Table 1 are listed the stress levels where initiation (CCR(t)) and propagation (PCR(t)) was predicted from closure rate measurements. In column 5 the maximum radial creep strain increment measured at 12.4 MPa is tabulated. In all three cases the closure rate measurements indicate that yield initiation and minor propagation should occur at this stress level.

Table 1. Predicted and Observed Stress Levels where Initiation and Propagation of Yielding Occurs.

STRESS RATIO N	TEST NUMBER MC-	PREDICTED STRESS LEVEL (MPa)		OBSERVED MAXIMUM RADIAL CREEP STRAIN INCREMENT (extensometer)
		CCR(20) ( )Mean	PCR(20) ( )Mean	
1	6.02	≪12.6	<12.6	13-<0.1% at 29hrs <3.8% at 30hrs
1	6.05o	<12.5	<12.5-(>12.5)->12.5	13-<0.8%
1	6.06	<12.5-(<12.5)->12.5	>12.5	13-<0.4%

Figures 6.a to 6.c illustrate the incremental radial creep strain distribution measured during the creep stage at approximately 12.5 MPa for the three tests presented in Table 1. The observed creep strains have been contoured without consideration of the effects of the proximity of the tunnel wall. Hence, these plots do not reflect the actual distribution of radial strain but rather illustrate the variability of radial creep strains along concentric circles to radii corresponding with the radii of the two rings of centers of extensometers.

During Test MC-6.02 a 108 mm wide tunnel was excavated in an externally preloaded sample, widened to a diameter of 152 mm during Test MC-6.05o and then reloaded externally during Test MC-6.06. While during Test MC-6.02 creep strains just exceeded creep strains typical for the pre-failure behaviour of the tunnels (dashed lines), the radial creep strains recorded during the other tests clearly indicate that yielding had just occurred during Test MC-6.05o and that the yield zone propagated during Test MC-6.06 as indicated by the high creep strains measured by the extensometer on the left springline. During Test MC-6.02 large creep strains were recorded just before the end of the load increment. They are shown by the full lines on Figure 6.a. This indicates that a yield zone propagated as predicted by the tunnel closure rates. It can also been seen from Figure 6 that the spatial distribution of radial creep strains varies from test to test. For example, during Test MC-6.05o significant time-dependent yielding occurs near the spring line of the tunnel while during the subsequent reloading test (MC-6.06) this zone showed little additional creep but a new creep zone had developed in the right lower quadrant. A creep strain or total strain limit to identify initiation and to differentiate it from propagation of stress redistribution cannot be selected rationally at the present time because failure is generally defined by a critical stress rather than a critical strain criterion. If one selects creep strain rate limits at 0.1 %/20 hrs for initiation of failure and 0.4 %/20 hrs for propagation one would find that yielding was just initiated during Test MC-6.05o while the tunnel was widened, and that further propagation occurred during reloading of the enlarged tunnel (Test MC-6.06). These limits seem reasonable based on field observations and further support the predictions made by application of the critical creep rate functions. Considering that it is difficult to clearly identify the exact stress level at which yielding or stress redistribution is initiated and to differentiate between initiation and propagation, it can be concluded that the mode of behaviour could be predicted accurately.

In summary, the CCR(t) and PCR(t) have been shown to be useful means of evaluating the

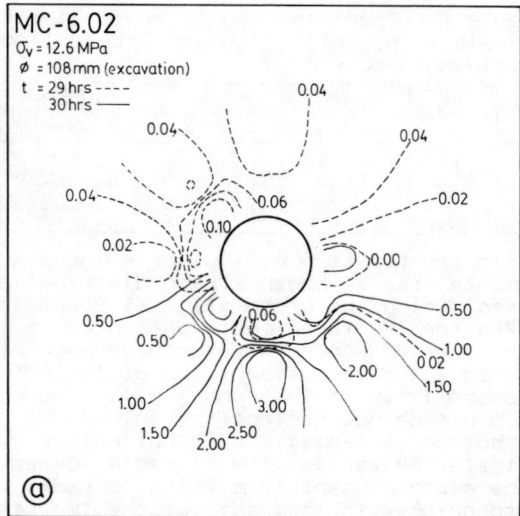
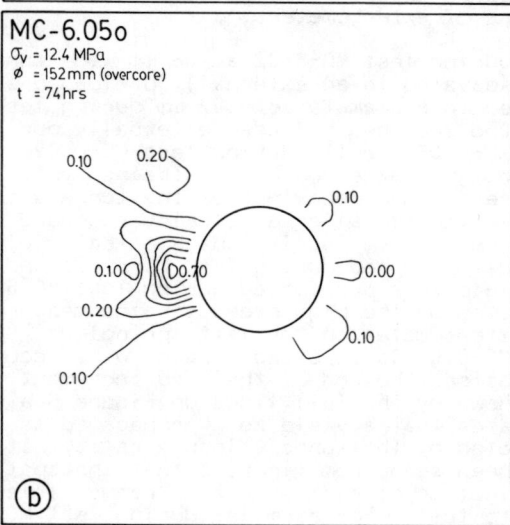
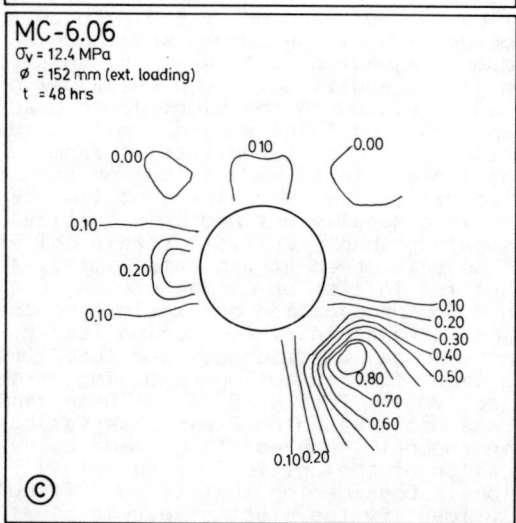

Figure 6. Radial Creep Strain Distribution Recorded at 12.4 MPa Field Stress: (a) Test MC-6.02; (b) Test MC-6.05o; and (c) Test MC-6.06. (Numbers in figures give % creep strain measured during given time increment; negative = extension)

stability of the small tunnels tested in the laboratory. Extrapolation from these laboratory tests to actual field conditions must be done with care because the magnitude of these limits depends on the time-dependent properties of the rock mass, its dependence on the loading history, the magnitude of the insitu stresses relative to the rock mass strength, and more generally the local geology and the failure modes controlled by the local geology or related planes of weakness. For practical tunnelling projects it will be necessary to establish these limits from field monitoring. A combination of convergence and extensometer measurements will be needed to establish when and where the processes of stress redistribution occur.

The convergence rates used at the Arlberg Tunnel (see Intoduction) to evaluate tunnel stability and support performance are suprisingly close to the critical rates established from the laboratory tests. Further research will show whether typical rates or classes of behaviour can be identified in pratice that can be applied to evaluate the stability of underground openings in a similar manner as proposed in this paper. Nevertheless, the results from our tests on small tunnels provide support for the approach adopted at the Arlberg Tunnel and indicate that critical or characteristic creep rate functions may provide a practical means of evaluating tunnel performance.

6. TIME-DEPENDENT BEHAVIOUR NEAR THE TUNNEL FACE

The time-dependent behaviour near the tunnel face was evaluated by measurements of the deformation pattern during the widening of a tunnel (Test MC-6.05o). The radial strain development measured by 16 extensometers during widening of the 108 mm wide tunnel to a tunnel with a diameter of 152 mm is presented in Figures 7.a and 7.b. The excavation progress is also shown at the bottom of these figures. The inner ring extensometer (full lines) were installed at Station 100 and the outer ring extensometers (dashed lines) at Station 120.

It can be detected from these figures that the radial strain increases rapidly during excavation and levels off during work interruptions. The closure of the tunnel is affected by the face advance in a similar manner and this must be considered if the closure rate is used to evaluate the tunnel performance by the method discussed earlier. By application of the equations proposed by Panet and Guenot (1982) it was possible to separate the effects of the tunnel advance rate from the time-dependent deformations related to progressive yielding and other time-dependent deformation processes.

It was found that the influence of the tunnel advance rate was negligible after the face had advanced by more than two to three tunnel diameters. During the laboratory tests the sample boundary was reached well before this limit was exceeded and the face effect was eliminated as soon as the sample was fully penetrated after 26 min. From the evaluation of the near face strain developments [Kaiser et al. (1982)] we concluded that the convergence rate

D 334

can only be used for the stability evaluation of tunnels once the influence of the face effect becomes insignificant, e.g., at a distance of more than two to three tunnel diameters from the face or, during multi-bench excavation, from the last bench. If this is not practical our data confirms the equations proposed by Panet and Guenot (1982) for the calculation of the closure rate as a function of the tunnel advance rate and hence these equations can be used to separate the convergence rate due to face advance from the total tunnel closure rate measured. Otherwise the near face behaviour cannot be evaluated on the bases of convergence measurements.

Nevertheless, for the tunnel tested, the radial strain measurements obtained by extensometers near the tunnel face provide sufficient information to determine that a yield zone started to propagate at the spring line of this tunnel shortly behind the tunnel face (see Figure 7.a: full line x-x). However, to successfully use the extensometer readings for the stability evaluation it is necessary to locate the extensometers at the critical location. Only one of the 16 extensometers provided, by comparison with the other extensometers, sufficient information to detect that a yield zone started to propagate and that propagation terminated shortly after excavation was completed.

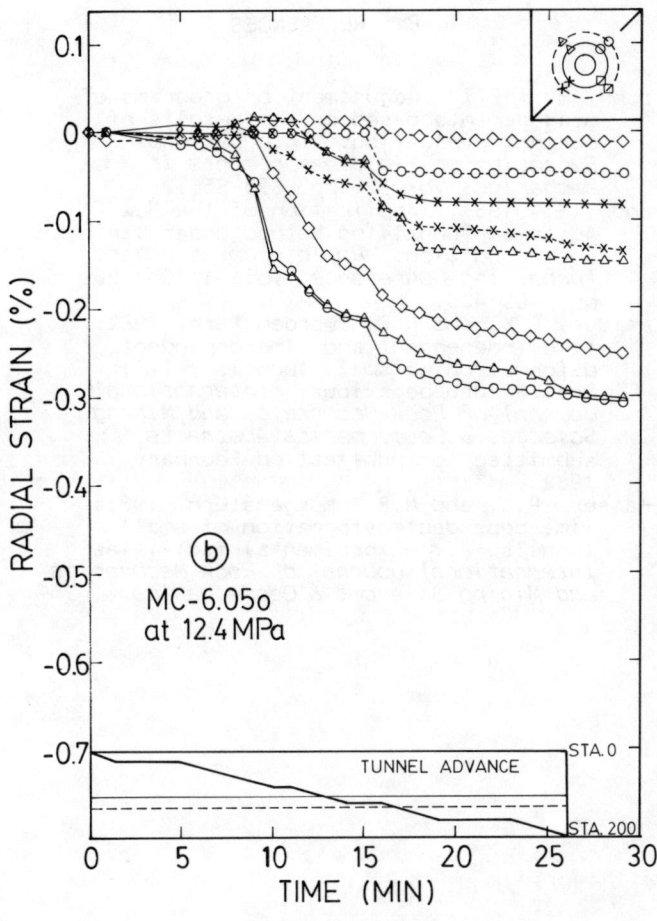

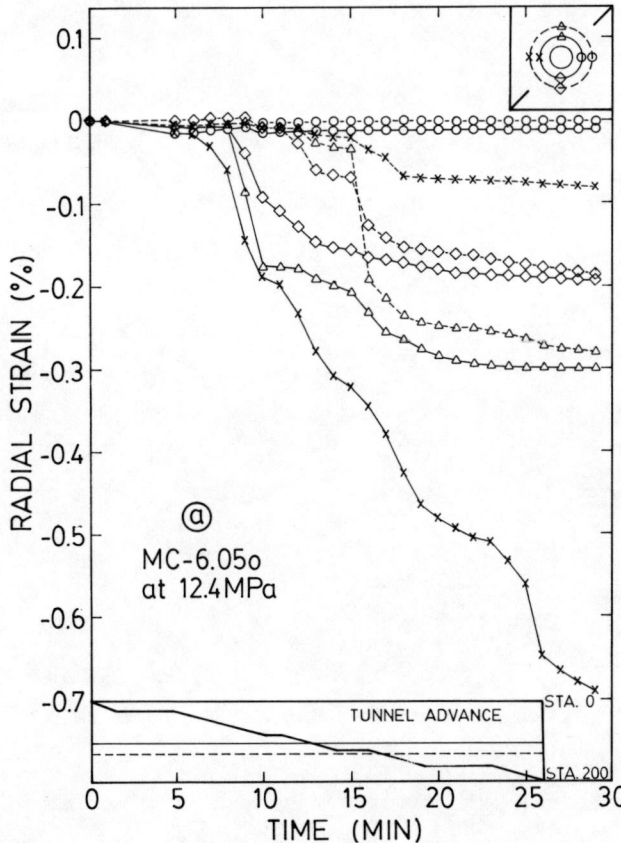

Figure 7. Radial Strain Development during Widening from a 108 mm to a 152 mm Wide Tunnel (including 4 stoppages of excavation).

7. CONCLUSIONS

Results from longterm tests of small tunnels in a rock mass with time-dependent deformation and strength properties were presented. The data from three tunnels excavated in similar but not identical rock mass conditions were used to evaluate the stability of the tunnels and particularly the mode of behaviour at the onset of propagation of a yield zone. It was shown that, for the tunnels tested, a critical creep or closure rate function can be determined that allows one to predict the mode of tunnel behaviour, the point of yield initiation and yield propagation. The convergence rate is, however, strongly influenced by the face advance rate and the rate of widening of the tunnel. This effect must be separated from the total convergence rate as long as the convergence is recorded within two to three tunnel diameters from the advancing face or widening operation. Only if it is possible to separate these two time-dependent effects can the convergence measurements be used for the stability evaluation. Otherwise, it is necessary to rely on extensometer measurements. Extensometers, however, only reflect the local rock mass behaviour and hence are only useful if installed at the appropriate location. The convergence measurements reflect the *overall* response of the rock mass and its response to local yield processes and thus are much better indicators of the integration of rock mass behaviour around the tunnel.

8. REFERENCES

John, M., 1977. Adjustment of programs of measurements based on the results of current evaluation. *International Symposium on Fieldmeasurements in Rock Mechanics*, Vol. 2, pp. 639-656.,

John, M., 1981. Application of the New Austrian Tunnelling Method under various rock condition. *Rapid Excavation and Tunnelling Conference*, Vol. 1, Ch. 26, pp. 409-426.,

Kaiser, P.K. and N.R. Morgenstern, 1982. Time-independent and time-dependent deformation of small tunnels - III. Pre-failure behaviour. *International Journal of Rock Mechnanics and Mining Sciences & Geomechanics Abstracts*, submitted for publication February 1982.,

Kaiser, P.K. and N.R. Morgenstern, 1981a. Time-dependent deformation of small tunnels - I. Experimental facilities. *International Journal of Rock Mechnanics and Mining Sciences & Geomechanics Abstracts*, Vol.18, pp. 129-140.,

Kaiser, P.K. and N.R. Morgenstern, 1981b. Time-dependent deformation of small tunnels - II. Typical test data. *International Journal of Rock Mechnanics and Mining Sciences & Geomechanics Abstracts*, Vol.18, pp. 141-152.,

Kaiser, P.K., S. Maloney and N.R. Morgenstern, 1981. Support Design for Underground Cavities in Weak Rock. Part I, *Energy Resources Research Fund*, Contract No. U-80-3, 98p.,

Kaiser, P.K., S. Maloney and N.R. Morgenstern, 1982. Support Design for Underground Cavities in Weak Rock. Part II, *Energy Resources Research Fund*, Contract No. U-80-3, 157p.,

Kaiser, P.K., 1981. Monitoring for the evaluation of the stability of underground openings. *1st Annual Conference on Ground Control in Mining*, West Virginia, pp. 90-97.,

Panet, M. and A. Guenot, 1982. Analysis of convergence behind the face of a tunnel. *Tunnelling'82*, The Institution of Mining and Metallurgy, pp. 197-204.,

CONTROLLED SINKING OF AN OPEN END CAISSON IN WEAK ROCK
Forage contrôlé d'un puits dans une roche fragile
Absenken eines grossen Brunnens in weichem Fels

Ivan Vrkljan, C.E., M.S.
Prof. Ervin Nonveiller, C.E., Ph.D.
Antun Szavits-Nossan, C.E., Ph.D., Sen. Lect.
Zvonimir Lisac, C.E.
Ivan Viseć, C.E.
Faculty of Civil Engineering, University of Zagreb — Yugoslavia

SYNOPSIS

A concret open-end caisson (depth 60 m, diameter 30 m) was sunk for the shaft of a reversible hydro-electric power plant, first through 10 m of clay and through weak rock (very hard marly clay) for the remaining 50 m. The sinking of such a heavy structure to the required depth calls for a very strict control of centre line inclination and penetration depth at every stage of undercutting the edge. Different failure modes of soil under the cutting edge have been studied of which the most adequate was implemented. Shaft behaviour during excavation and the experience acquired during the construction of this outstanding structure are described.

RESUME

Un puits en béton (profondeur 60 m, diamètre 30 m) a été excavé pour la salle de machines d'une centrale hydroéléctrique reversible. Les premiers 10 m du puits ont été creusés dans de l'argile, et les 50 m restants, dans de la roche fragile — une argile marneuse très dure. L'excavation d'une telle structure à la profondeur requise éxige un contrôle précis de l'inclinaison de l'axe et de la profondeur de pénétration à chaque stade d'excavation. Divers modes de rupture sous la trousse coupante ont été étudiés — le plus adéquat d'entre eux a été choisi et utilisé. Le comportement du puits au cours de son excavation ainsi que l'expérience acquise pendant la construction de cet ouvrage hors série sont décrits ci-après.

ZUSAMMENFASSUNG

Ein 60 m tiefer Stahlbeton-Brunnen mit 30 m Durchmesser für den Schacht einer Pumpenspeicher Wasserkraftanlage wurde durch 10 m weichen und 50 m sehr steifen mergeligen Ton abgesenkt. Die Absenkung eines so schweren Bauwerks durch weichen Fels erfordert eine genaue Kontrolle des Absenkvorganges und des Aushubs um die Brunnenschneide. Der Bruch des Materials unter der Schneide hängt von der angewandten Aushubsart ab. Anhand der Untersuchung einiger Varianten wurde der Aushub mit 1 m³ Tieflöffelbaggern durchgeführt. Das Verhalten des Schachtes während des Aushubs und die Erfahrungen während der Ausführung dieser außergewöhnlichen Bauaufgabe werden beschrieben.

1. INTRODUCTION

A shaft 60 m deep and having a diameter of 30 m was designed for a pumped storage hydroelectric power plant. The following options for the excavation of the foundation pit were considered:
- shaft excavation by mining technique, with sections approximately 3 m deep and simulataneous pouring of the concrete supporting lining wall of the shaft;
- excavation protected by diaphragm wall with struts at specific depths;
- excavation of foundation pit protected by a reinforced concrete shaft, sunk by excavating the bottom, and simultaneous concreting the shaft lining wall above ground level.

After a detailed consideration of these options, the third method was selected, as it offered the lowest cost and the shortest construction time.

The first concrete ring of the shaft, with the stell cutting edge, was poured on the ground surface. After removal of formwork, pouring the concrete lining was continued with a slipform (Fig. 1). A 20 cm gap filled with bentonite suspension was provided between the shaft lining and soil in order to eliminate friction. A 1 cu.m. bucket excavator was used for soil excavation and loading. The excavated material was hoisted by two 7 cu.m. containers and two tower cranes.

2. MATERIAL PROPERTIES

The shaft for the power station was excavated in a karst hollow filled down to 200 m with overconsolidated very stiff marly clay. Extensive preliminary explorations had shown two different beds of soil in the selected site
- a top 10 m bed of soft alluvial sediment and organic clay, and
- a bed of homogeneous, slightly stratified, very stiff marly-clayey soil extending to the bedrock.

The alluvial sediment bed was inadequate to support the first pours of the shaft, and it was accordingly replaced by compacted clay. The shearing strength parameters of the clay, compacted at optimum moisture (w = 14,4-24,2%) were $c = 6-13$ kNm^{-2}, $\varphi = 26,3-29,5°$. Most of the excavation was performed in the marly clay bed. The stiff clay was fully saturated and of low permeability, thus the penetration of the cutting edge would cause undrained failure.

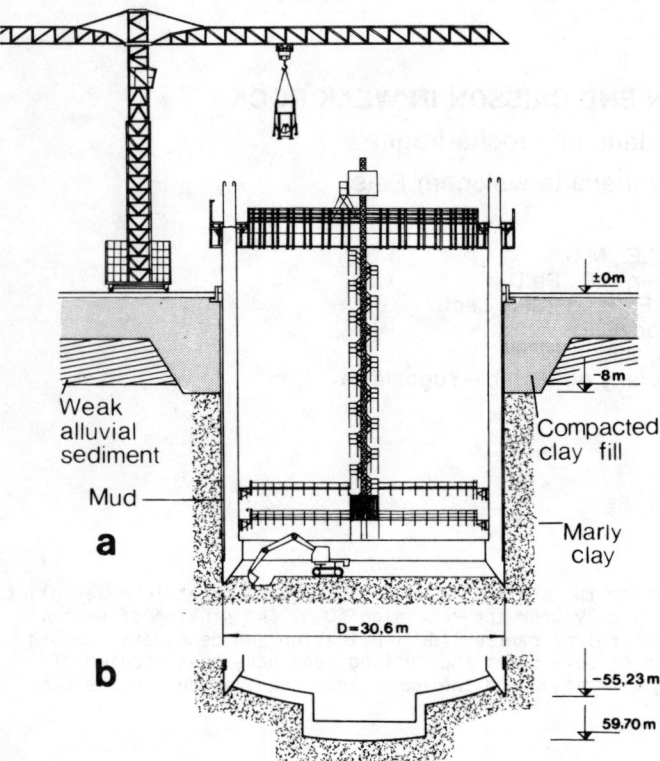

Fig. 1 a) Vertical cross section of shaft during construction
b) Reversed dome on bottom of shaft after sinking to design depth

Fig. 2 Pouring of concrete shaft lining, and container for material hoisting

Laboratory tests have shown strong strain softening character (Fig. 3), and it was assumed that soil failure under the cutting edge will be of progressive nature. The value of axial strength, q_a = 2000 kNm^{-2}, assumed as the design value for undrained failure, results in c_u = 1000 kNm^{-2} and φ_u = 0 as parameters for the analysis of the ultimate bearing capacity.

The change of stress caused by the pressure of the cutting edge causes a change of pore water pressure, which in turn influences the process of consolidation occuring in the material under consideration owing to the substantially decreased strength. In the case of 100% consolidation, parameters obtained by the drained test are valid for failure around the cutting edge. Values of φ = 26°, c = 100 kN/m^2, were assumed for the analysis of the "drained" failure around the shaft cutting edge. Shear test along the concrete-soil interface were also carried out, the strength parameters, were φ = 22,3 , c = 0.

Table 1. Marly clay properties

Natural moisture content	18-22%
Plastic limit	20-25%
Liquid limit	48-55%
Plasticity index	25-30%
Consistency index	1
Unit mass	2,0-2,1 Mg/m^3
Specific gravity	2,61-2,78
Activity (Skempton)	1,1-1,5
CaCO$_3$ content	23-28%
Axial strength	240-4000 kN/m^2
Cohesion	13-82 kN/m^2
Angle of internal friction CID	22-32°
Consolidation index	$2,5.10^{-3}$ cm^2/s

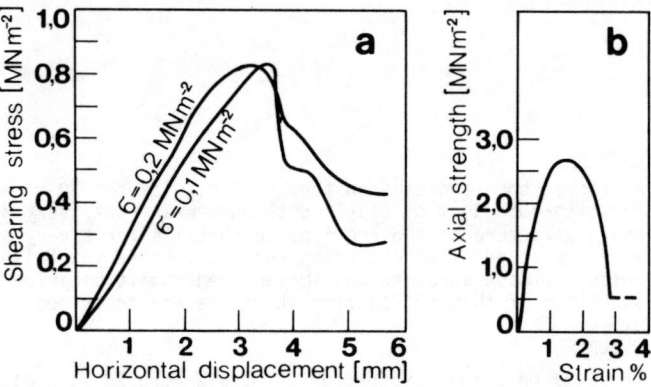

Fig. 3 Diagrams of a) direct shear, b) axial strength of marly clay

The large difference between peak and residual strength appears to suggest the occurence of progressive failure, meaning that failure occurs after a very small initial strain, and that the strength of material is rapidly reducted at failure. This property of the marly-clay influenced the choice of excavation technology, focused on avoiding rapid and uneven shaft penetration.

3. SOIL FAILURE ALONG THE CUTTING EDGE

The controlled sinking of a shaft requires plastic soil failure to be achieved by adequate excavation along the perimeter of its cutting edge.

3.1. Compacted clay layer

The penetration of the shaft cutting edge in the range of the clayey replacement was calculated according to G.G. Meyerhof, 1961.
The calculation was done for two extreme sets of shear strength parameters:

c = 6 kN/m^2 φ = 26,3°
c = 13 kN/m^2 φ = 29,5°

It was assumed that the penetration depth (d) is lower than the height (v) of the cutting edge and the depth member of Meyerhof's equation was omitted in the computation of the penetration depth (E_q.1).

After removal of the formwork on which the first ring with the cutting edge of the shaft was poured, the cutting edge sank by d = 1,5 m, which was within the expected range (as calculated, the depth of penetration was to be 1,0 < d < 1,8 m; Fig. 5). Fig. 6 illustrates the zone of plastic failure around the cutting edge after removal of the formwork. Determination of the area of soil plastification along the cutting edge was needed in order to define the boundary where the excavation does not cause sinking of the shaft.

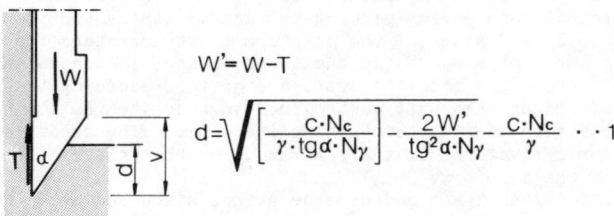

Fig. 4 Depth of cutting edge penetration in compacted clay layer.

Any excavation from this boundary to the cutting edge influences the stability of the cutting edge. An even and continous sinking of the shaft is achieved by excavation in horizontal layers uniformly around the perimeter of the cutting edge.

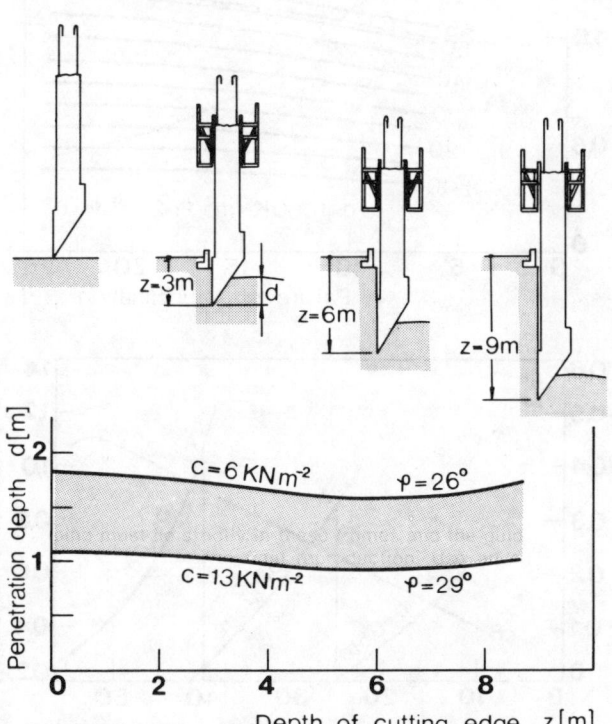

Fig. 5 Relation of the cutting edge penetration (d) to the depth of the cutting edge (z), according to equation 1.

3.2. Marly clay

The mechanical properties of the marly clay are substantially different. Accordingly, the appropriate technique of excavation along the cutting edge of the shaft was applied. The failure mode for the analysis of excavation geometry around the cutting edge is shown on Fig. 7. The active force, W (weight of the shaft and mud), is transferred to the resistance wedge formed by the excavation. The stability of the resistance wedge depends on the length of the potential sliding plane, and on its inclination β. Accordingly, the resistance wedge can be brought to a condition of limit equilibrium by various critical combination of length (l), and the angle β. By defining one of these two values the other which causes the failure of the resistance wedge can be computed. Thus failure under the cutting edge can be caused in two ways (Fig. 8).

Fig. 6 Cutting edge of the shaft after formwork removal

Excavation Type A maintains a constant width of the resistance wedge. The limit equilibrium condition is achieved with a critical value of the angle of inclination of the potential sliding plane, β_k, and undercutting depth, f_k.

In excavation Type B, the undercutting depth, f, is constant, while the width of the sliding body is reduced until the critical value, b_k, is achieved. For the adopted failure mode (Fig. 7), and from the conditions of equilibrium of forces acting on the resistance wedge, a simple expression was obtained for the safety factor (F) along the potential failure plane.

$$F = \frac{[T \cdot \sin(\gamma+\beta) + N \cdot \cos(\gamma+\beta)] \cdot tg\varphi_2 + c \cdot l}{N \cdot [\sin(\gamma+\beta) - \cos(\gamma+\beta) \, tg\varphi_1]} \quad \ldots \quad 2$$

$$N = \frac{W'}{\sin\alpha + \cos\alpha \cdot tg\varphi_1} \quad \ldots \quad 3$$

$$T = N \cdot tg\varphi_1 \quad \ldots \quad 4$$

$$\gamma = 90 - \alpha \quad \ldots \quad 5$$

Expressions for l_{kr} or b_{kr} were obtained provided a limit equilibrium condition ($F = 1$) is assumed.
As already mentioned, shear strength depends on the degree of consolidation. Accordingly, the dimensions of the resistance wedge at the moment of its failure also depend on the time elapsed since the failure of the preceding wedge. Intensive shaft sinking was assumed in some instances when no consolidation is possible (undrained failure).

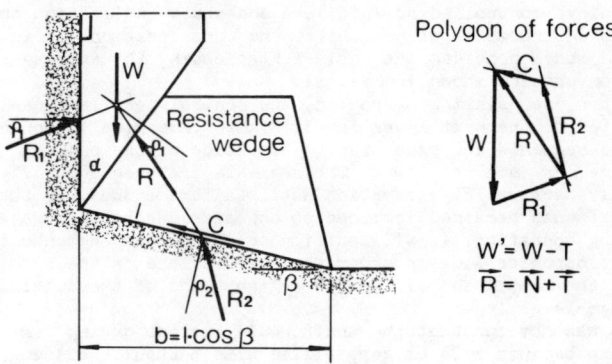

Fig. 7 Model of marly clay failure

In such a case, "undrained strength" parameters are valid for the calculation. When delays in sinking permit consolidation, "drained" strength parameters are valid. Practical experience proved this reasoning, because after sinking stopped for some days the resistance wedge dimensions were considerably larger than in periods of intensive shaft sinking.

Critical combinations of geometrical b-β magnitudes at which resistance wedge failure occurs (Fig. 9 a; Fig. 1oa) were computed for the adopted parameters of "drained" and "undrained strength" of marly clay. Because the active force W changes with depth of the cutting edge the relations between b and β are given for 5 m depth intervals (1o < Z < 55) on Figs 9 a and 1o a.

Theoretically, a resistance wedge of greater width can be brought to a state of limit equilibrium with high angles β. Deep undercutting may however result in sudden uncontrolled deep penetration of the shaft. Therefore the inclination of the failure plane was limited to β = 26°. The analysis of the diagram shown in Fig. 1o a shows that excavation Type A - in "undrained" conditions can be achieved by a step-wise reduction of the resistance wedge width b, while with excavation Type B, a constant value of undercutting depth, f = o,2 m (Fig. 9 b) should be maintained.

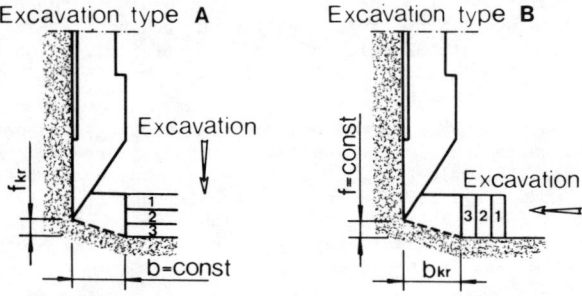

Fig. 8 Two ways of excavation along the cutting edge

The analysis of the diagram in Fig. 1o a shows that excavation Type A can be carried out with a constant width of the resistance wedge , b = 2,5 m (Fig. 11 b) throughout the whole excavation depth.

As shown, failure occurs with relatively small dimensions of the resistance wedge, meaning that excavation along a substantial part of the bottom surface does not cause failure along the cutting edge and corresponding shaft sinking. This fact makes management of the excavation and material loading much easier. Most of the excavation can be carried out in bulk without causing unwanted shaft penetration. It is only in the narrow annulus around the cutting edge, when the resistance wedge is formed, that excavation should be carried out carefully in order to achieve controlled soil failure and shaft inclination on the desired side of the shaft. The considerations reviewed above provided the basis for selecting the excavation pattern shown in Fig. 11.

After each sinking operation, the central part is excavated to the next lower level. Eight symmetrically arranged segments are excavated to elevation of the cutting edge, or somewhat lower. The segments are then successively narrowed by excavation until failure ensues. If the shaft has remained inclined to one side during the preceding operation, it will suffice to narrow the segments on the opposite side in order to cause failure on the side of the more intensely loaded segments and of the cutting edge levelling.

It was obvious that the shaft would incline during sinking because a 2o cm gap, filled with bentonite suspension was left between the shaft lining and soil. Several measuring systems were designed (hydraulic level with 8 indicator tubes; tables on the outer surface of the shaft on which a writing point recorded the wertical and horizontal path during sinking the shaft; lasers for the proper adjusting of the slipforms; extensometers for the control of stresses in the shaft lining; inclinometers incorporated into the cutting edge for recording the position of the shaft centreline) in order to monitor the position of the shaft centreline, this being essential for excavations in the shaft and for the proper lifting of the slipform.

All the analyses and considerations discussed above relate to the stage of design of excavation technology based on in situ and laboratory tests. No experience acquired on similar projects could be used because, to our knowledge, structures of such dimensions in similar materials have not been achieved so far.

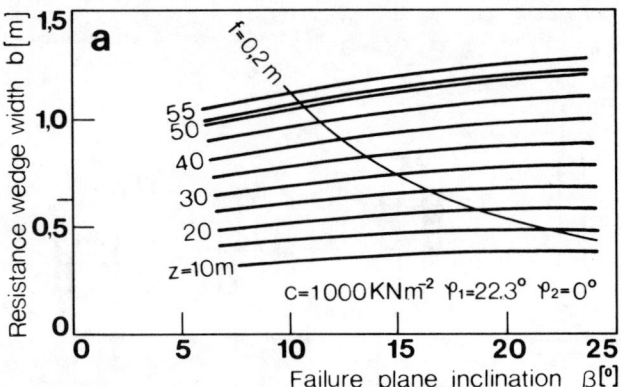

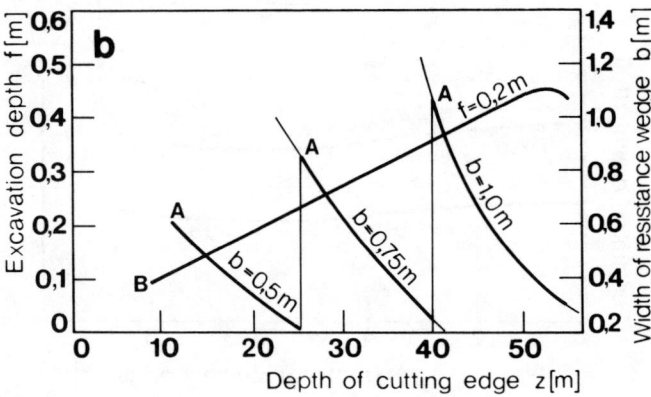

Fig. 9 Resistance wedge failure in undrained strength conditions

4. PROCESS OF SHAFT SINKING

The process of shaft sinking proved the predictions related to soil failure both in the compacted layer clay and in marly clay to be correct. In the compacted clay, the plastification zone was about 6 m wide, and the depth of penetration about 1,5 m. The shaft was sunk continuously with excavation of the plastification zone in horizontal layers.

Failure was entirely different during shaft penetration in the stiff marly clay. Because of its marked strain softening property, failure occured after only slight initial deformation, and the penetration was discontinuous - in steps. The average penetration of the shaft was about 6o cm in every operation.

Excavation in marly clay was carried out as shown in Fig. 11. However, nonsymmetrical excavation proved to be necessary, in order to adjust inclination on the desired side (Fig 12), in some parts where the shearing strength of the interlayer discontinuities was considerably reduced owing to the presence of a larger quantity of the sandy component. Fig. 13 shows the bedding planes in marly clay, and the characteristic forms of brittle material failure.

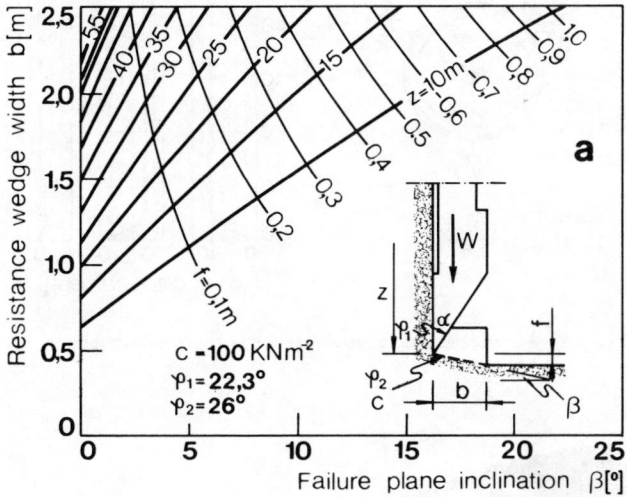

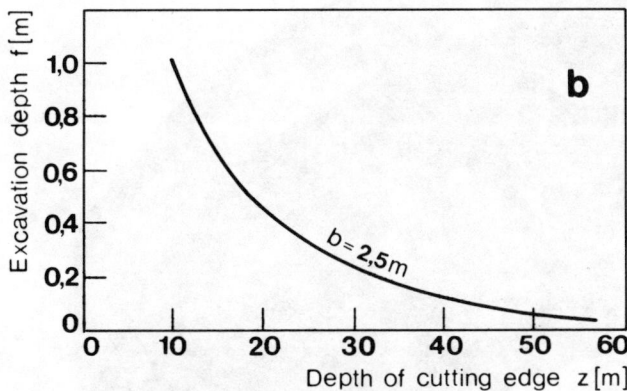

Fig. 1o Resistance wedge failure in drained strength conditions

Figs. 14 and 15 show the start of excavation for resistance wedge formation. In order to gain knowledge on shaft behaviour during sinking, three tables were mounted on the outer lining, and a recorder fixed to the protective wall, in order to record the sinking process in terms of depth and inclination, as shown in Fig. 16.
It was observed that the shaft, at incipient failure of the resistance wedges, showed a tendency to slight inclination towards the more weakened wedges. At the moment of failure of the most loaded resistance wedge, the load of the shaft was transferred to the adjacent wedges, causing their subsequent failure; than the shaft penetrated without any change in the position of its centreline, until a new condition of equilibrium was established. The conslusions derived from the sinking diagram shown in Fig. 16 were decisive for the selection of technology to be used in excavationg the reversed dome at the bottom of the shaft (Fig. 1) in order to prevent undesirable shaft subsidence during the excavation of the dome.
Any shaft inclination was prevented because, in this stage, the bentonite mud was replaced by cement mortar.

Accordingly, subsidence could occur only with the simultaneous attainment of the failure condition throughout the perimeter of the shaft. Namely, by preventing shaft inclination, the concentration of stress in a segment of the perimeter (which would subsequently cause the failure of other parts as well) was also made impossible. The excavation of the dome was achieved without any displacement.

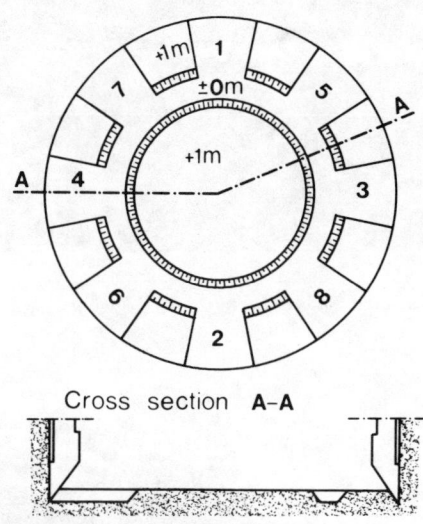

Fig. 11 Pattern of excavation along the cutting edge

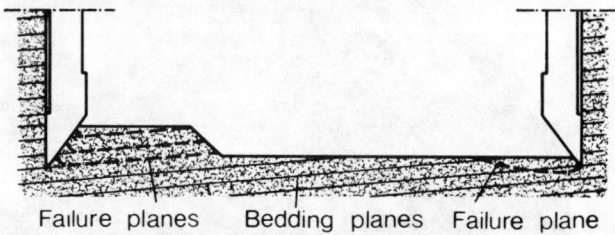

Fig. 12 Excavation in a zone of marked shearing strength anisotropy

Fig. 13 Marly clay stratification with inclined bedding planes

D 341

Fig. 14 Reistance wedge formation

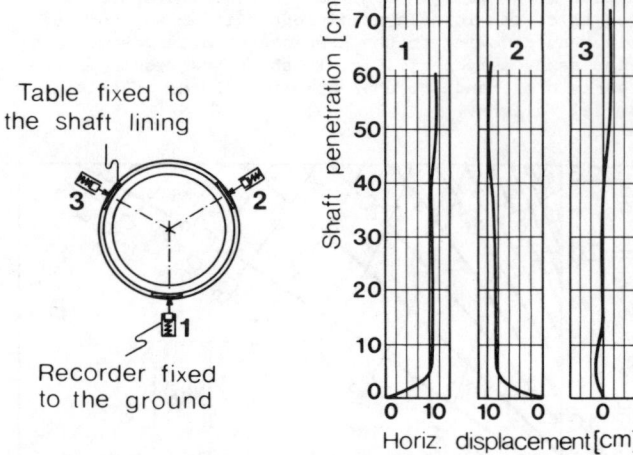

Fig. 16 A single shaft subsidence recorded on three tables

Fig. 15 Formation of resistance wedges along the cutting edge of the shaft

5. FINAL NOTES

The construction of a shaft having such unprecedented dimensions, can be achieved successfully in soft rock by sinking of the preeast lining as open end caisson. The shaft was sunk to the final depth on schedule after six months of operation, a feat that would prove impossible by any other method.

The experience suggests that the method can also be used for sinking larger shafts in a shorter time.

The construction of the powerhouse foundation as an open end caisson was suggested by the Industrogradnja Co. Zagreb. The same organization excavated the shaft, and was also awarded other civil works related to the constructtion of the power house.

References:

Meyerhof, G.G.: "The ultimate bearing capacity of wedge-shaped foundations", Int. conf. SMFE, Paris, 1961, 3B/16, pp. 1o5-1o9.

Jaroslav Černi Institute: Engine-Room of Hydroelectric Power Plant, Complex Explorations - Final Report, Belgrade, 1977.

Faculty of Civil Engineering University of Zagreb, and Geoexpert: Power house Shaft - Soil Mechanics Analyses, Design Report, Vol. I, 1979.

Elektroprojekt: Hydropower Plant - Power house, Stage I, Open end caisson - Power house; Zagreb, 1979.

Faculty of Civil Engineering University of Zagreb: Hydropower Plant - Engine-Room, Sliding Shaft - Technology of Shaft Sinking and Excavation; Zagreb, 198o.

INTERPRETATION DE LA STRUCTURE DU BASSIN HOUILLER DE ST-ELOY, D'APRES LES RESULTATS D'ESSAIS MECANIQUES ET LES MODELES TECTONIQUES

Structural interpretation of the St. Eloy coal basin from mechanical tests and tectonic models

Interpretierung der Geologie des St. Eloy Kohlebeckens mit Hilfe von Ergebnissen aus mechanischen Versuchen und von tektonischen Modellen

J. Bergues
Laboratoire de Mécanique des Solides — Ecole Polytechnique 91128 Palaiseau (France)

J. Grolier
Laboratoire de Pétrologie, Université d'Orléans 45046 Orléans (France)

J. C. Soula
Laboratoire de Géologie Pétrologie, Université de Toulouse 31000 Toulouse (France)

P. Travert
77, rue Jean Jaurès 63700 Saint-Eloy (France)

RESUME

La géométrie du bassin de St Eloy témoigne du fluage du charbon vers les charnières anticlinales. Se fondant sur les caractéristiques mécaniques des roches en présence, on précise le comportement rhéologique du charbon et l'on tente à partir de modèles analogiques d'expliquer la structure de ce bassin.

SYNOPSIS

The structural geology of the St-Eloy coal basin suggests coal creep towards anticlinal hinges. According to the mechanical properties of rocks, the rheological behaviour of coal is ascertained and the structure of this basin is tentatively explained by analogical tectonic models.

ZUSAMMENFASSUNG

Das rheologische Verhalten der Kohle wird bestimmt und die Geologie der St-Eloy Lagerstätte wird mit Hilfe analoger Modelle untersucht.

INTRODUCTION

Le bassin houiller de St-Eloy-les-Mines, Puy-de-Dôme, est l'une des principales accumulations de sédiments stéphaniens jalonnant le grand sillon houiller du Massif Central français (Letourneur 1953, Grolier & Letourneur 1968, Destieux 1980). Ce petit bassin a livré 42 MT de charbon, production cumulée depuis 1837 jusqu'à fin 1978 date de l'arrêt de l'exploitation.
Cité parfois en exemple de domaine à plis complexes fournis par coulissage entre deux moles rigides (Goguel 1965), le gîte de St-Eloy a été examiné du point de vue géologique par L. de Launay (1894) et J. Letourneur (1953) notamment et, du point de vue de l'exploitant, par R. Travert (1962). La structure du bassin semble témoigner d'un comportement extrêmement plastique du charbon en particulier dans les anticlinaux, obliques sur la direction générale du bassin (Figure 1a).
L'objectif envisagé dans cet article est la réalisation d'un modèle réduit, correct au point de vue dimensionnel et reflétant aussi exactement que possible la dynamique du charbon au cours de l'évolution tectonique du bassin de St-Eloy. Il s'agit en particulier de vérifier l'interprétation classique (Nodot, cité et repris par Letourneur 1953) selon laquelle les plis résulteraient d'un mouvement relatif sénestre des bordures rigides du bassin.
Les résultats présentés ici sont un premier pas vers une telle modélisation. En effet,
1) Il apparaît licite de s'inspirer pour une simulation approximative de St-Eloy de certains types de modèles déjà réalisés par Soula (1982) pour d'autres études ;
2) On voit mieux, comparant ces modèles à la structure réelle du bassin les particularités structurales locales qu'il conviendra d'introduire dans une simulation plus précise ;
3) enfin les essais mécaniques réalisés sur diverses roches de St-Eloy et des résultats d'essais pris dans la littérature permettront de choisir en toute connaissance de cause des matériaux ayant les propriétés rhéologiques appropriées.

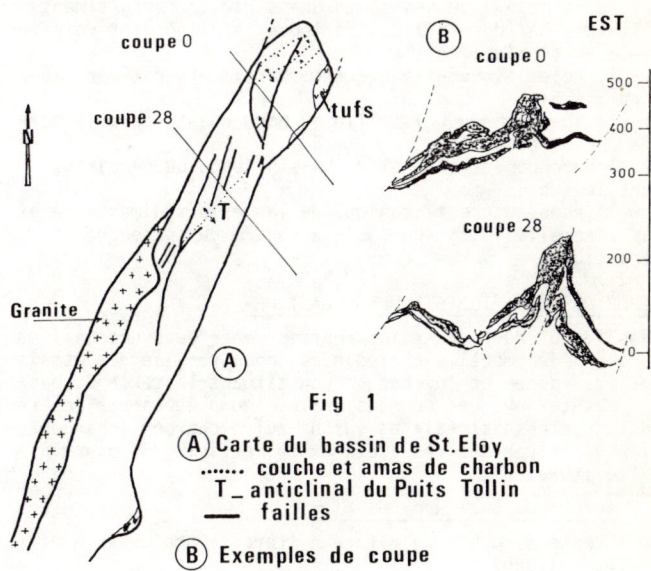

Fig 1
(A) Carte du bassin de St.Eloy
.... couche et amas de charbon
T _ anticlinal du Puits Tollin
— failles
(B) Exemples de coupe

DONNEES GEOLOGIQUES ET STRUCTURALES

La structure générale du bassin est relativement bien décrite par l'importante série des coupes verticales où les exploitants ont consigné la géométrie des couches de charbon (Figure 1b). Les faits essentiels qu'il faut en retenir sont les suivants :

E 1

1.- la disposition en échelon des plis multiples et variés qui affectent la série sédimentaire ; l'angle de l'axe des plis avec l'allongement du bassin étant de l'ordre de 20°;
2.- l'existence d'un remarquable amas de charbon anticlinal et à caractère diapirique connu sous l'appellation d'anticlinal du Puits Tollin (Figures 1 et 2).

TABLEAU I

Roche	σ_R (MPa)	E (MPa)	σ_0/σ_R	$\varepsilon_d/\varepsilon_i$
A	55	9000	0,73	0,15
B	48	7500	0,63	0,25
C	29	8500	0,69	0,12
D	40	13000	0,75	Rupture

avec
σ_R = résistance à la rupture (valeur moyennée sur 3 essais)
E = module élastique dans la partie linéaire des courbes "efforts déformations"
σ_0 = contrainte de fluage monoaxiale
ε_d = déformations différées sous σ_0
ε_i = déformations instantanées sous σ_0

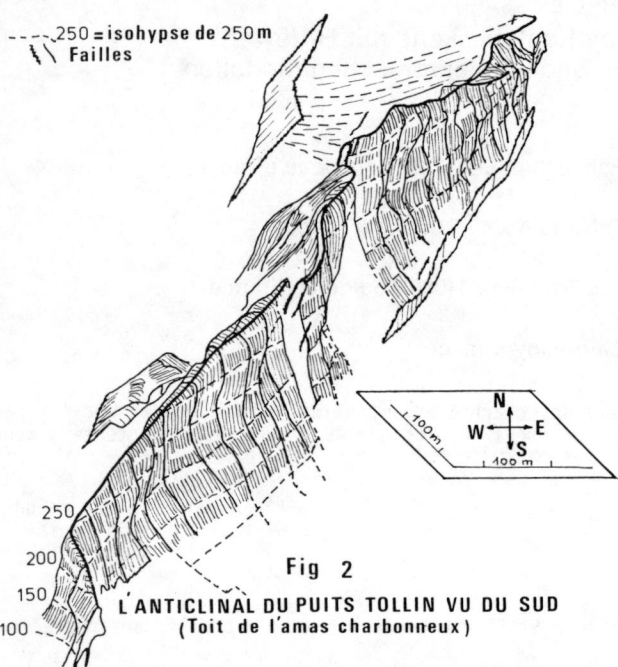

Fig 2
L'ANTICLINAL DU PUITS TOLLIN VU DU SUD
(Toit de l'amas charbonneux)

2 - *Caractéristiques mécaniques du charbon*
Ces caractéristiques ont été déterminées sur du charbon de la mine de Blanzy (13% en moyenne de matière volatile provenant du siège Darcy) et sur quelques échantillons du bassin de St-Eloy (35% en moyenne de matière volatile, bloc provenant du travers banc 192).
On distingue deux séries d'essais :
1ère série : Essais de chargement instantané et de fluage à la pression atmosphérique sur les charbons de Blanzy et St-Eloy.
2ème série : Fluage triaxial du charbon de Blanzy sous une pression latérale de 10 MPa.

a. Compression uniaxiale (1ère série)
Les principaux résultats sont consignés dans le tableau II

TABLEAU II

Charbon	σ_R (MPa)	E (MPa)	σ_0/σ_R	$\varepsilon_d/\varepsilon_i$
Blanzy	9	3800	0,70	0,1
			0,80	Rupture
St Eloy	8	2000	pas d'essai de fluage	

Les valeurs de σ_R et E sont moyennées sur trois essais.

b. Fluage triaxial du charbon (Blanzy)
Afin de se rapprocher des conditions "in situ" on a réalisé une série d'essais de fluage sous contraintes triaxiales classiques de compression ($\sigma_1 > P$) avec une pression latérale P de 10 MPa. Les résultats sont donnés par la figure 3 représentent les déformations différées en fonction du temps pour différents niveaux de chargement $(\sigma_1 - P)_0$.
Des essais à chargement instantané ont donné les valeurs suivantes :

$$P = 10 \text{ MPa} \begin{cases} (\sigma_1 - P)_R = 60 \text{ MPa} \\ E = 4000 \text{ MPa} \end{cases}$$

σ_1 = contrainte axiale, indice R = rupture

3.- le découpage du bassin en une série de compartiments distincts par des failles, fortement inclinées et voisines de la verticale ;
4.- le rejet systématique des compartiments est vers le nord ;
5.- le surplomb du Cristallin du bord ouest, dans la partie sud du bassin ;
6.- l'enfoncement général du bassin quand on se dirige vers le sud ;
7.- l'indépendance tectonique de la série sédimentaire et de la semelle volcanique qui la sépare du socle granitogneissique.

ETUDE DES CARACTERISTIQUES MECANIQUES
Dans le but de fournir les données mécaniques nécessaires au choix des modèles analogiques, on a réalisé des essais sur les roches et le charbon constituant la série du bassin de St-Eloy. Ces données ont été complétées par des résultats d'essais réalisés sur un autre charbon (charbon de Blanzy) et par des résultats expérimentaux pris dans la littérature.

1 - *Roches du bassin de St-Eloy*
On a testé 4 roches de nature différente prélevées à différents niveaux :
. Roche A - grès fin cote = 47 mètres
. Roche B - grès fin rubanné cote = 137 mètres
. Roche C - conglomérat cote = 200 mètres
. Roche D - conglomérat cote = 212 mètres

On a réalisé des essais de compression simple instantanée et des essais de fluage. Les principaux résultats sont consignés dans le tableau I.

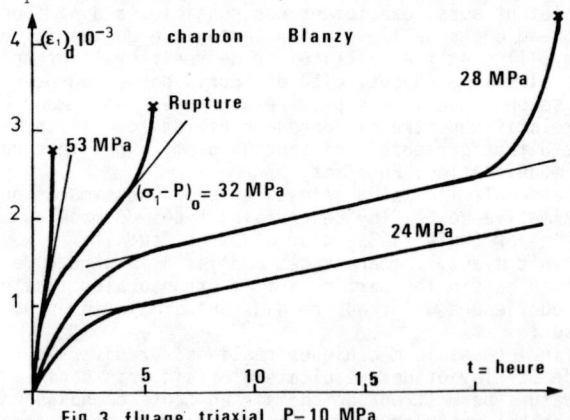

Fig 3 fluage triaxial P=10 MPa

3 - Interprétation

L'analyse des résultats d'essai montre qu'à la pression atmosphérique toutes les roches étudiées, y compris les charbons, présentent un comportement fragile sans fluage significatif. Comme cela a déjà été démontré, (Morlier 1964) ces types de roches sont surtout marqués avant la rupture par de faibles phénomènes différés avec un fluage asymptotique se stabilisant rapidement ou avec une rupture quasi instantanée ; la viscosité de ces roches se manifestant surtout après la rupture (Nguyen Minh D. et al.1981). Toutefois dans le cadre d'un problème relatif à la modélisation d'une série de roches, il est intéressant de remarquer en comparant les tableaux I et II que le charbon apparaît comme un matériau mécaniquement très différents des roches encaissantes, il est en particulier <u>beaucoup moins résistant et beaucoup plus déformable</u>. De plus, on a pu mettre en évidence sur le charbon de Blanzy qu'avec l'augmentation de la contrainte moyenne apparait très vite des phénomènes visqueux avec du fluage à trois phases même pour des niveaux de contraintes relativement faibles par rapport à la rupture.

Dans le tableau III, on a mentionné pour les différents essais les déformations différées de chaque phase avec leur durée respective.

ε_{dI} = déformations de la phase transitoire, temps correspondant = t_I

ε_{dII} = déformations de la phase stationnaire, temps correspondant = t_{II}

ε_{dIII} = déformations de la phase accélérée, temps correspondant = t_{III}

TABLEAU III

$(\sigma_1-P)_0/(\sigma_1-P)_R$	0,88	0,53	0,46	0,40
$\varepsilon_{dI} \cdot 10^{-3}$	1,1	1,2	1,6	0,9
t_I mn	13	60	300	180
$\varepsilon_{dII} \cdot 10^{-3}$	0,2	1	0,7	> 1
t_{II} mn	17	210	900	> 6000
$\varepsilon_{dIII} \cdot 10^{-3}$	0,3	0,9	2,2	-
t_{III} mn	12	90	420	-

A partir de ce tableau, on peut faire la remarque importante suivante : Comparés à ceux de la phase transitoire les temps des phénomènes différés de la phase stationnaire sont les plus longs et ce, d'autant plus que la contrainte de fluage est faible. Ainsi pour le problème qui nous intéresse ici qui consiste à modéliser le comportement rhéologique du charbon pour tenter de mettre en évidence des phénomènes dont les temps sont des temps d'ordre géologique, c'est surtout cette phase stationnaire du fluage qu'il faut prendre en considération. Le modèle le plus simple revient alors à considérer un modèle à seuil du type Bingham dont le fluage est linéaire en fonction du temps pour une charge donnée, et éventuellement d'introduire les déformations différées de la phase transitoire ε_{dI} dans la partie élastique (déformations instantanées). Notons que cet artifice a déjà été utilisé par Nguyen Minh D. (1982) pour interpréter des phénomènes de convergences différées dans les tunnels. Les caractéristiques du modèle sont déterminées à partir de la figure 4 représentant les vitesses de fluage de la phase stationnaire $\dot{\varepsilon}_{dII}$ en fonction de la contrainte de fluage $(\sigma_1 - P)_0$.

De cette figure on déduit les caractéristiques suivantes pour P = 10 MPa

$$\begin{cases} \text{seuil } S = (\sigma_1 - P) = 22 \text{ MPa} \\ \text{viscosité } \eta_C = 4,60 \text{ MPa.an} \end{cases}$$

On peut noter la faible valeur du seuil limite par rapport à la résistance instantanée.

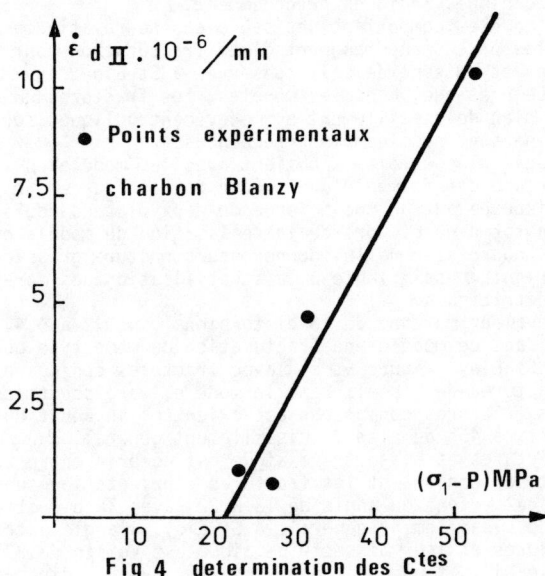

Fig 4 determination des C^{tes}

Il n'a malheureusement pas été possible d'effectuer ces mêmes mesures sur les roches constituant la série du bassin de St-Eloy. Or, pour avoir une idée du rapport des viscosités entre les différentes couches constituant cette série (paramètre important dans le choix du modèle analogique) on a calculé par la méthode qui vient d'être décrite les viscosités d'un grès et d'une argile consolidée à partir de résultats expérimentaux donnés par Ardeshir Afrouz et Jon M. Harvey (1974).

η_G(grès) = 978 MPa.an η_A(argile) = 64 MPa.an

Ces résultats montrent que ces roches sont plus visqueuses que le charbon avec des rapports de viscosité de :

η_G/η_C = 212 η_A/η_C = 14

En fait ce rapport peut en réalité être plus important, en effet l'expérience montre que la viscosité d'une roche augmente avec la contrainte moyenne or les valeurs mentionnées pour le grès et l'argile ont été déterminées dans un cas de chargement uniaxial.

APPORT DES MODELES ANALOGIQUES A L'INTERPRETATION MECANIQUE DU BASSIN DE ST-ELOY

D'une manière générale, le glissement favorise et contrôle le diapirisme associé et, réciproquement, l'instabilité diapirique accentue l'instabilité mécanique responsable du glissement (Ramberg 1967). On considérera donc ici que les structures produites dans les essais de glissement lié à un cisaillement et dans les essais de diapirisme sont associées dans la nature.

1 - Modèles en cisaillement simple

Pour modéliser des plis obliques tels que ceux de St-Eloy, il suffit de disposer une mince couverture horizontale sur la discontinuité verticale d'un socle glissant jouant en cisaillement simple (Vialon et al. 1982). Dans ce type de modèle, la couverture mince est formée d'un matériau fragile-ductile (plasticine, parafine ou gélatine). La présence d'une couche très plastique entre deux couches plus rigides ne modifie pas sensiblement la longeur d'onde, l'amplitude et la forme des plis. Les rapports de

viscosité entre les différentes couches sont pris de l'ordre de 100 à 1000 ce qui, dans le problème qui nous occupe ici, est tout à fait acceptable comme rapport de viscosité charbon/autres roches, compte tenu des résultats des essais mécaniques indiqués précédemment.
Dans les modèles comportant une seule couche relativement rigide, les plis s'accompagnent d'une fracturation dont la géométrie est voisine de celle observée à St-Eloy, à cette différence près que, dans les modèles, les fractures parallèles au plan de cisaillement ou modérément obliques sur celui-ci ne sont pas les plus fréquentes.
Une analogie plus étroite s'obtient avec les modèles où la couverture est formée d'une couche de talc ou de grès broyés disposée sur un socle formé de deux plaques coulissantes en forme de L. Lors de la réalisation du modèle, on tasse la poudre afin de lui donner une structure grossièrement orientée assimilable à la stratification des formations détritiques.
Pour une valeur moyenne de la distorsion (γ = 0,2 à 0,4) apparaît dans ce modèle une fracturation du même type que celle de St-Eloy (Figure 5 a et b) avec fractures conjuguées dites fractures de Riedel. Dans le modèle, la bissectrice aiguë des fractures conjuguées est orientée, au début du glissement, à 45° du plan de cisaillement général. L'angle entre fractures et bissectrice aiguë est compris entre 23 et 32° et par conséquent les fractures d'orientation sublongitudinale, font un angle de 13 à 22° avec la direction du plan de cisaillement général. A St-Eloy, l'angle entre les fractures et leur bissectrice aiguë est voisin de 30°, mais celle-ci fait un angle d'environ 40° avec la direction du plan de cisaillement; en conséquence les fractures d'orientation sublongitudinale font un angle de 10° en moyenne.

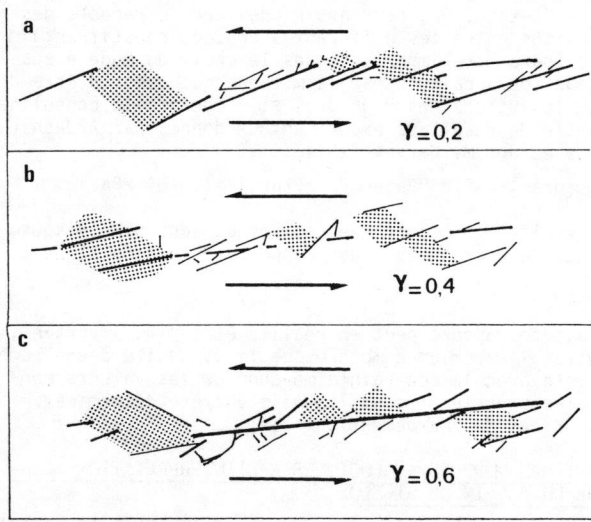

Fig.5 : Cisaillement simple d'une couverture de talc broyée (les zones anticlinales sont en pointillés)

Le point le plus important de ces modèles est qu'avec la fracturation, se développent des zones anticlinales formant avec le plan de cisaillement un angle compris entre 120 et 135°. Ces zones anticlinales sont parfois <u>limitées par des failles parallèles à leur axe et recoupées</u> par des fractures du type de celles décrites précédemment (figure 5 c.). Ces zones anticlinales semblent devoir être l'équivalent des anticlinaux obliques de St-Eloy, tels que l'anticlinal du Puits Tollin. Des essais avec une couche mince de plasticine située sous la couverture pulvérulente ont montré le développement de fractures parallèles aux axes dans la charnière des anticlinaux, analogues à celles observées localement dans les relevés.
Pour des valeurs de cisaillement plus fortes, les mêmes modèles (γ = 0,6), développent aussi des fractures de plus grandes dimensions formant un angle faible ou nul avec le plan de cisaillement général (inférieur ou égal à 5-6°). Certaines failles longitudinales de St-Eloy pourraient être dues à ce mécanisme.
Ainsi, la comparaison entre les modèles et le site géologique montre que l'hypothèse d'un cisaillement simple sénestre selon le sillon houiller est plausible. Toutefois, comme la bissectrice aiguë des fractures conjuguées fait un angle inférieur à 45° avec le plan de cisaillement, le processus de cisaillement simple ne saurait à lui seul expliquer la configuration observée et il faut admettre en plus que le rapprochement des deux bords du sillon a dû jouer un rôle, sans qu'on puisse préciser si ce serrage est contemporain du cisaillement ou postérieur. La faible déviation angulaire par rapport au cisaillement qui indique que la composante de raccourcissement transversal doit être relativement faible.

2 - *Modèles par centrifugation*

Les modèles précédents expliquent la genèse et la disposition des plis et fractures observées à St-Eloy, mais non la géométrie des accumulations de charbon. On se réfère donc à des modèles complémentaires par centrifugation.
Dans ces modèles, dont le principe est décrit par Ramberg (1967), on considère le diapirisme simultané de deux couches situées à deux niveaux différents d'une série stratifiée. La compression synchrone du diapirisme s'obtient par une construction particulière des modèles conduisant à un rapprochement des bords vers la partie centrale (Soula, 1982). Ce raccourcissement s'accompagne d'un chevauchement des bords (figure 6.a) analogue à celui observé dans le site géologique à la bordure ouest du bassin de St-Eloy où la lame bordière granitique surplombe (localement tout au moins) les formations sédimentaires. Dans ces modèles, le rapport de viscosité effective entre couches diapiriques et non diapiriques est de l'ordre de 100, ce qui semble raisonnable compte tenu des essais mécaniques.

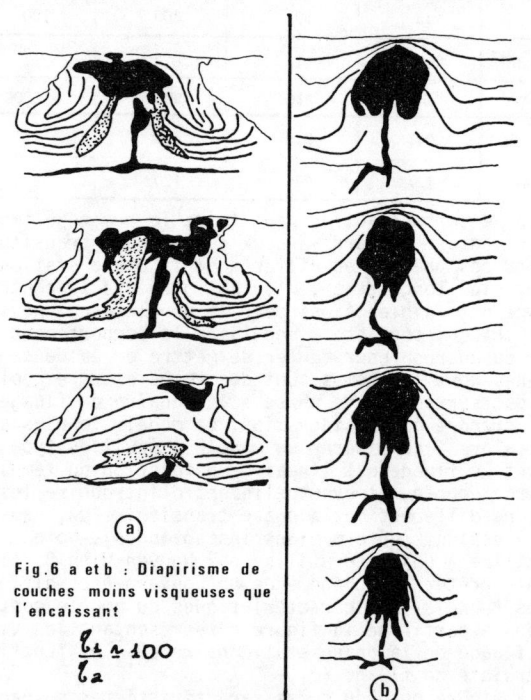

Fig.6 a et b : Diapirisme de couches moins visqueuses que l'encaissant

$\dfrac{\mu_1}{\mu_2} \simeq 100$

a) : Deux couches initialement situées à des profondeurs différentes : en noir la plus profonde (moins visqueuse) en pointillés la plus haute. Coupes sériées dans un même modèle. Diapirisme en compression avec chevauchement centripète des bords.
b) : Monocouche, même légende que a). Pas d'étalement dans les niveaux supérieurs.

E 4

Les figures données par les coupes sériées d'un même modèle centrifugé sont de même type que celles observées dans le charbon de St-Eloy : couches diapiriques partiellement discordantes, pouvant être séparées l'une de l'autre ou, au contraire, anastomosées selon la localisation, développement de fractures en réponse à la montée diapirique en bordure de zones anticlinales, etc. Dans le modèle représenté dans la figure 6.a on observe un étalement de la couche diapirique dans les niveaux supérieurs, ce matériel tendant à former une couche allochtone, intensive, intercalée dans les niveaux supérieurs. Ceci amène à reconsidérer l'interprétation structurale de l'extrêmité nord du bassin. Là, la disposition périsynclinale (coupes 25 à 32 des archives minières) est relativement simple. Mais certaines singularités du dessin des couches peuvent s'expliquer par l'étalement diapirique du charbon. Certains amas stratiformes de charbon en relation avec un anticlinal sous-jacent (coupes 2,1,0,1S à 5S) ont peut-être cette origine. Cependant, cet étalement n'est pas une constante des modèles diapiriques. En effet, on observe aussi dans ces modèles des figures (Figure 6.b) sans étalement dans les niveaux supérieurs, analogues aux coupes 23-29 et aux coupes 2-0. Seule l'analyse pétrographique et structurale de détail aurait permis, lors de l'exploitation de trancher entre les deux hypothèses.

CONCLUSION

Tous les éléments précédents suggèrent que la localisation, la fréquence et la géométrie d'ensemble des amas charbonneux est liée principalement au plissement, c'est-à-dire à des plis obliques formés lors d'une déformation avec forte composante de cisaillement simple en décrochement senestre parallèle au Sillon houiller. La géométrie de détail des accumulations semble, au contraire, être déterminée par le diapirisme du charbon lui-même, lié à ses caractéristiques physiques. Ce diapirisme a même conduit à des accumulations, de localisation aberrante par rapport aux anticlinaux, et peut-être aussi à une montée et un étalement du charbon dans les niveaux structuraux supérieurs ; ce qui dans les deux cas entraînait des disparitions brutales "imprévisibles" de la veine.
Ce travail, outre les informations nouvelles sur les propriétés intrinsèques de certaines roches de St-Eloy et sur le charbon de Blanzy, montre plus clairement la possibilité de réaliser un modèle réduit analogue spécifiquement représentatif du bassin de St-Eloy, mais il faudra :
a) tenir compte de l'asymétrie du site géologique (chevauchement seulement sur un bord du Sillon)
b) résoudre le problème technique consistant à réaliser un cisaillement en cours de centrifugation.

Le cas de St-Eloy dont la géométrie est connue en assez grand détail permettra de mettre au point une méthode de modélisation applicable à d'autres bassins ; en particulier ceux qui sont encore en cours d'exploitation.
Dans la mesure où les modèles permettent des interprétations structurales nouvelles, on peut espérer en déduire des indications pour la recherche et l'exploitation des gisements.

REFERENCES

A. AFROUZ and J.M. HARVEY (1974), *Rheology of rocks within the soft to medium strength range*, Int. Jour. Rock Mech. and Min. Sci., Vol. 11, n° 7, July 1974.

J. DERAMON, P. SIRIEYS et J.C. SOULA (1982), *Mécanismes de déformation de l'écorce terrestre : Structures et anisotropies induites*, 5ème Congrès I.S.R.M. Melbourne.

F. DESTHIEUX (1980), Carte géologique de la France, 1/50 000 feuille Montaigut en Combrailles.

J. GOGUEL (1965), Traité de Tectonique, 2ème édition, pp. 257-258.

J. GROLIER (1981), *Contribution de l'analyse structurale à l'interprétation génétique des bassins houillers (quelques exemples du Massif Central)*, Bull. Cent. Rech. Explor. Prod. Elf Aquitaine, 51,2,601,620.

J. GROLER et J. LETOURNEUR (1968), *L'évolution tectonique du grand Sillon Houiller du Massif Central*, XXIII Int. Geological Congress, Vol. 1, pp. 107-116.

L. de LAUNAY (1894), Carte géologique de la France, 1/80 000 (7ème édition) feuille Gannat.

J. LETOURNEUR (1953), *Le Grand Sillon Houiller du Plateau Central Français*, Bill. Serv. carte géol. Fr. n° 238, tome LI, pp. 1-236.

P. MORLIER (1964), *Etude expérimentale de la déformation des roches*, Thèse Faculté des Sciences de l'Université de Paris, 1964.

D. MINH NGUYEN, P. BEREST, J. BERGUES, P. HABIB (1981), *Softening behaviour of rock in engineering practice*, C.R. Congrès Int. Mech. Roches, Tokyo, Japon.

D. MINH NGUYEN, *The contraction of hollow circular cylinders with time dependent strain softening behaviour. Application to supported tunnels*, A paraître dans Int. J. Roch Mech. and Min. Sc. 1982.

F. ODONNE, J.F. GAMOND, GRATIER J.P., VIALON (1982), *Modélisation analogique de plis sur un décrochement de socle*, Réunion annuelle des Sciences de la terre, Paris, p. 479.

H. RAMBERG (1967), *Gravity deformation and the Earth's crust*, Academic Press, London-New-York, pp. 214.

J.C. SOULA (1982), *Characteristics and mode of emplacement of gneiss domes and plutonic domes in Central Eastern Pyrénées*, J. Struct. Geol., (sous presse) (34 pp).

R. TRAVERT (1962), *Les Houillères du Bassin d'Auvergne*, Revue "Le Mineur d'Auvergne", n°s 5,6 et 7.

STUDIES ON THE MECHANISM OF GAS AND COAL BURSTS IN JAPANESE COAL MINES

Etudes du mécanisme de coup de grisou dans les mines de houille au Japon

Untersuchung über den Mechanismus der plötzlichen Kohle- und Gasausbrüche in japanischen Kohlebergwerken

Y. Hiramatsu and T. Saito
Dept. of Mineral Science and Technology
Kyoto University, Kyoto, Japan

N. Oda
Coal Mining Research Center, Tokyo, Japan

SYNOPSIS

Considering the features of gas and coal bursts observed in Japanese coal mines, the authors studied the mechanism of the burst based on the results of stress analysis and several measurements. They found that the bursts could be explained as the mechanical fracture of a coal seam by the combined effect of rock stress and gas pressure, and suggested the basis for improved preventive measures.

RESUME

En tenant compte des caractéristiques des coups de grisou observés dans les houillères japonaises, les auteurs ont étudié le mécanisme de coup de grisou sur la base des résultats d'analyse de contraintes et de certaines mesures. Ils ont vu que les coups de grisou peuvent être considérés comme la fracture mécanique d'une couche de charbon provoquée sous l'effet combiné de la contrainte dans la roche et de la pression de gas. Ils ont suggéré des bases pour l'amélioration des mesures préventives.

ZUSAMMENFASSUNG

Unter Berücksichtigung der verschiedenen Faktoren bei plötzlichen Kohle- und Gasausbrüchen in japanischen Kohlenbergwerken wurde der Mechanismus dieser Ausbrüche, basierend auf den Ergebnissen von Kräftebestimmung und verschiedenen Messungen betrachtet. Daraus ergab sich, daß die Ausbrüche als mechanischer Bruch eines Kohlenflözes in Folge des kombinierten Effekts von Gebirgsspannung und Gasdruck erklärt werden kann. Diesbezüglich sollen die Grundlagen für eine verbesserte Vorbeugung umrissen werden.

1. INTRODUCTION

In several Japanese coal mines, mostly those in Hokkaido district, they have experienced frequently gas and coal bursts, though they have made efforts to prevent them. In recent years, gas and coal bursts have come to be of small scale and less and less frequent. But we had a grave disaster by a gas and coal burst in the New-Yubari coal mine on 16th Oct. 1981, and also a few minor ones around that time. Therefore improvement of the preventing measures for gas and coal bursts are strongly demanded. In order to contribute to find some better preventive measure for the bursts, the authors have investigated into the mechanism of them occurring in Japanese coal mines.

Note: Hereafter, a gas and coal burst will be simply expressed as "the burst".

2. FEATURES OF THE BURSTS IN JAPANESE COAL MINES

To provide meterials for discussing the mechanism of the bursts in Japanese coal mines, the features of them were investigated.

The bursts are apt to occur in various cases, such as in geologically disturbed areas, in the front areas of development, on driving gateroads, at the time when crosscuts meet coal seams, in layers of poor or weak coal, on driving raises compared with driving winzes, when some change in the state of the face is found, in the cases that, on boring, either the rate of gas outflow or of production of cuttings increases, when the boring rod is arrested by jamming, and just after blasting. (But in some cases bursts occur a few minutes after blasting or at any time, independent of blasting.) It is confirmed that the frequency as well as magnitude of the bursts increases with the mining depth, that the pre-mining a coal seam not prone to bursts is effective to prevent the bursts of other coal seams, and that the bursts occur successively over some period of time and are never accompanied by earthquakes, whereas rock bursts are always accompanied by earthquakes. It is noted that the amount of burst gas is very large, sometimes larger than 100m^3 per ton of burst coal, that burst coal is fractured into small fragments, and that the burst area has a flat and irregular shape.

3. THE GAS CONTAINED IN COAL SEAMS AND ITS OUTFLOW

3.1 Quantity of gas contained in coal seams

The quantity of gas that can be put into coal was tested by placing each coal specimen in a vacuum vessel and then introducing methane gas under varied pressures into the vessel (Hiramatsu et al. 1961). It was found from this test that the amount of gas which could be put into coal was not large even under a high pressure, so to say 17–25m^3 per ton of coal under the pressure of 10MN/m^2, and that the greater part of the gas was adsorbed.

However the amount of burst gas is generally large in Japanese coal mines, frequently larger than 100m^3 per ton of burst coal. Moreover in every coal mine that has ever experienced the burst, a large amount of gas is flowing out of the underground continually. A part of the gas is the drained gas, another part the freed gas from the mined coal and the rest is the gas flowed out through the numerous cracks in the strata created by all kinds of mining activities. The gas flowed out by the latter two ways is exhausted by the ventilated air current. Table 1 shows the quantities of gas outflow in several coal mines, which are expressed by the volume of gas per ton of mined coal. The reason for it is that in coal mines working steadily the gas outflow will be roughly proportional to the tonnage of mined coal although some quantity of gas may come from other coal seams than those being mined.

From these data it is concluded that coal seams prone to burst contain a large amount of gas of the order of 50–100m^3 per ton of coal, though

Table 1 Volumes of gas drained / exhausted by ventilated air current in m³/t of mined coal

Coal Mine	Volume of gas		Total gas outflow	Occurrence of bursts
	drained	exhausted by air current		
New-Yubari	50.6	50.0	100.6	Gas & coal bursts
Akabira	22.2	55.2	77.4	"
Minami-Ooyubari	45.4	27.7	73.1	"
Sunagawa	26.5	44.5	71.0	"
Ashibetsu	31.7	44.2	75.9	"
Horonai	26.4	15.8	42.2	Gas & rock bursts
Miike	0	3.5	3.5	none

the authors did not succeed to put such a large amount of gas into coal experimentally. The reason for it is suspected that the coal specimen once taken from the original coal seam may be changed in its physico-chemical properties.

3.2 Gas pressure
It is imagined that the gas pressure in virgin coal seams containing such a large amount of gas must be very high. However this gas pressure will be the pressure of gas contained in pores, therefore it will be hard to measure it by boring a borehole into a coal seam. At the stage of the geological survey before developing the New-Yubari coal mine, some of the boreholes bored from the surface blowed out the mud water, which proved the existence of high pressure gas, about $10MN/m^2$ at the depth of 850m from the surface.

3.3 Permeability of gas in coal seams
It is considered that the virgin coal seams are generally impervious for gas, but that they will become pervious only by initiation of cracks in them. These cracks will be created by the various mining activities, such as boring, driving galleries and gateroads, shaft sinking, longwall working and so on. Some geologically disturbed areas may be pervious and some may be impervious.

3.4 Rate of gas outflow
It is supposed that the rate of gas outflow through the cracks will naturally be proportional to the total area of them and depends on the time elapsed from the crack initiation. Fig. 1 shows the decrease in the rate of gas outflow from boreholes in the New-Yubari coal mine. (The gas outflow from boreholes will also be that through cracks.) These boreholes were bored into the Jusshaku seam from the galleries driven in rock for the purpose of gas drainage. Each curve shows the mean rate of gas outflow from a borehole in each of three districts, W1, W3 and N3.

The rate of gas outflow from a borehole can be approximately expressed by:

$$q(t) = q_0 \exp(-t/\lambda), \qquad (1)$$

where q_0 is the initial rate of gas outflow (=691m³/d on an average), $q(t)$ the rate on the t-th day from the completion of the borehole, λ the time constant (=about 100 days). Denote the period of time from the initiation of a borehole to the time when the rate of gas outflow decreases down to a nigligible small value by t_0 (=about 500 days).

Then the total gas outflow from a borehole, q_T, is given by:

$$q_T = \int_0^{t_0} q_0 \exp(-t/\lambda) dt. \qquad (2)$$

Now let us assume that the rate of gas outflow through any kind of crack can be also expressed by the similar equations as (1) and (2), and denote the initial daily gas outflow from all the cracks created every day by Q_0 and the gas outflow on the t-th day from the same cracks by $Q(t)$. Then we have:

$$Q(t) = Q_0 \exp(-t/\lambda). \qquad (3)$$

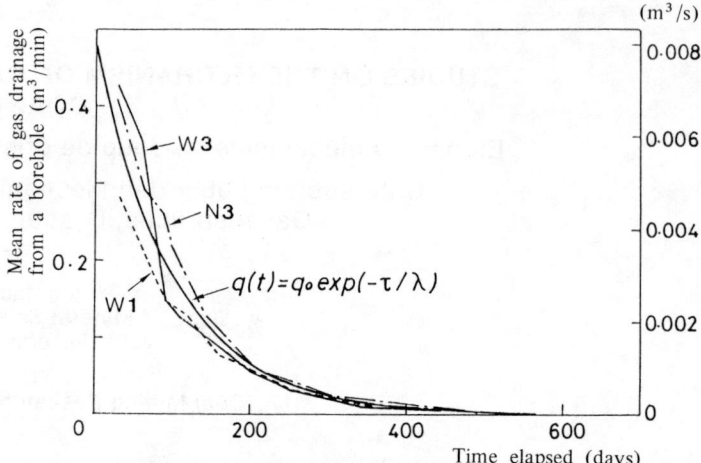

Fig. 1 Decrease in mean rate of gas drainage from a borehole bored through the virgin coal seam (W1, W3 and N3 denote the mining district)

In coal mines working steadily, the daily quantity of mined coal, the length of daily bored boreholes and the cracks created every day will be even. Since today's gas outflow concerns all the cracks created in the period of time from t_0 days ago to today as well as the amount of coal mined today, the daily total gas outflow from a coal mine, Q, will be given by:

$$Q = \int_0^{t_0} Q_0 \exp(-t/\lambda) dt + Q', \qquad (4)$$

where Q' is the volume of gas freed from the mined coal every day.

If all the mining activities are suspended, the gas outflow will decrease gradually, and the gas outflow on the n-th day from the suspension of works, $Q^{(n)}$, is given by:

$$Q^{(n)} = \int_n^{t_0} Q_0 \exp(-t/\lambda) dt. \qquad (5)$$

The theory above mentioned was examined by comparing the decrease in the rate of gas outflow calculated by eq. (5) with that measured in the New-Yubari coal mine. In this mine all works had been suspended from the accident of 16th Oct. 1981 for 103 days. Fig. 2 shows the decrease in the rate of gas outflow. The average rates of gas outflow in six months before the accident were, 88.6m³/min by ventilation and 89.7m³/min by gas drainage. Since these two valuales are near with each other, the value of Q was assumed as 89m³/min for the both gas outflows. The calculation of $Q^{(n)}$ was carried out by taking $\lambda = 100$ days, the same time constant as that of the gas outflow from boreholes shown in Fig. 1.

At a glance at this figure, it may be thought that the theory does not coincide with the results of measurement. But by a further investigation it is considered that since a large amount of gas (about 600km³) had burst in a few hours, the rate of gas outflow decreased after the burst, but became about the same as the theoretical values after eighty days from the accident. The theoretical total gas outflow for 100 days after the accident is, by calculation, 16.3Mm³, while the measured total gas outflow for the same period plus the burst gas is 15.2Mm³. Between these two values there is no large difference. Therefore the theory on the gas outflow may be accepted.

To lower the gas pressure in coal seams, we have to extract gas from the coal seams. It is only done through the numerous cracks created by the boring and the various mining activities. It should be noted that the gas outflow through the cracks is very slow but lasts over a long period of time.

4. CRITERION ON THE FRACTURE OF COAL CONTAINING HIGH PRESSURE GAS

It is supposed that the coal containing high pressure gas will be fractured under lower stress level than the coal with low pressure gas. However the criterion on the fracture of coal containing gas has not yet been studied. Accordingly the authors wished to obtain a clue for this criterion. Dr. Y. Mizuta, Yamaguchi University, has kindly analyzed the stress concentration around penny-shaped cracks by the theory of elasticity and suggested a criterion for the present problem as follows.

Assume that there are numerous minute pores of a penny shape, in a coal seam, which contain gas of a pressure p_0. Let the principal rock stresses of the coal seam be σ_1, σ_2 and σ_3, σ_3 being the major compressive stress. By an analytical investigation, it was found that the condition of fracture of coal, caused by the concentrated tensile stress sppearing around the tips of the pores, was given by:

$$(1+b/a)(\overset{*}{\sigma}_1 - \overset{*}{\sigma}_3)^2 - 2(b/a)^2(\overset{*}{\sigma}_1 + \overset{*}{\sigma}_3 - 2p)^2 T$$
$$+ 2b(\overset{*}{\sigma}_1 + \overset{*}{\sigma}_3 - 2p)[1 - (b/a)T] - 4b = 0 \quad (6)$$

in the region of compressive fratures and

$$\overset{*}{\sigma}_1 = 1 + p \quad (7)$$

in the region of tensile fractures, where,

$$T = 1 - \sqrt{1 - \frac{a}{b}\left(\frac{\overset{*}{\sigma}_1 - \overset{*}{\sigma}_3}{\overset{*}{\sigma}_1 + \overset{*}{\sigma}_3 - 2p}\right)^2}$$

$$\overset{*}{\sigma}_1 = \sigma_1/\sigma_t, \quad \overset{*}{\sigma}_3 = \sigma_3/\sigma_t, \quad p = p_0/\sigma_t$$

$$a = \nu(\nu - 3)(4 - 3\nu + \nu^2), \quad b = (1-\nu)^2(2-\nu)^2$$

ν; Poisson's ratio σ_t; tensile strength.

The condition of fracture of coal in case of $\nu = 0.2$ is shown in Fig. 3 for various gas pressures. Any stress state is shown by plotting a point in the area above the chain line in this figure. If the point is on the right hand side of the corresponding gas pressure curve, the coal seam will be in a stable state. It is seen that the higher the gas pressure is, the nearer is the coal seam to the critical state of burst on the assumption of the same stress state. When a coal seam is fractured by a combined action of rock stress and gas pressure, gas and finely crashed coal will burst.

5. MECHANISM OF THE BURSTS

In consideration of the features of the bursts and the investigation mentioned so far, the authors have attempted to form a new conception on the mechanism of bursts which can explain the phenomena of the bursts more reasonably than before.

5.1 Assumption
Let us assume tentatively that the bursts are pure mechanical fracture of a coal seam under the combined effect of rock stress and gas pressure, for which an appropriate criterion will be found in future. (See the preceding section.) Then some of the features of the bursts are easily understood.
(1) The bursts are apt to occur in the geologically disturbed areas as well as in the layers of poor coal or weak coal.
(2) The bursts are apt to occur in the areas where the gas extraction in advance of driving gateroads and so on is insufficient.
(3) The burst coal has a large amount of finely crashed coal which indicates that the high pressure gas contained in the structure of coal plays a role in the burst.
(4) Pre-mining a coal seam which seems not to burst, if any, is effective to prevent the bursts of other coal seams.

There remain, however, several problems to be discussed further.

5.2 Discussions
If the gas pressure in the coal seam is low, a gateroad will be driven safely even if the concentrated stress if higher than the strength of the coal, whereas if the gas pressure is high, the coal seam can be fractured completely and the burst can occur. The reason for it will be as follows. When the gas pressure is low, the stress concentration in the coal seam behind the face will be as shown in Fig. 4(a), provided that the coal seam is sufficiently strong. Even if the coal seam is not strong, the stresses will distribute as shown in Fig. 4(b) owing to the mechanical property of coal which is illustrated by the stress-strain curves in Fig. 4(c), and no collapse of the coal seam will occur. Only loosening of coal will be seen near the face. Such distribution of stresses move step for step with the face advance.

When the coal behind the face contains high pressure gas, the coal can be fractured by the combined effect of rock stress and gas pressure, mostly at the moment of the blasting. Such a fracture of coal will take place instantaneously like the sudden fracture of a rock specimen tested by a soft testing machine, and the gas and coal will burst. Once a part of coal seam bursts, the coal behind it may also be able to burst provided that the state of stress and gas pressure is near the critical state for the burst. In this way the burst will take place successively. It is supposed that the change in gas pressure may be slower than the change in stress when the coal is blasted. This matter might assist the occurrence of the burst.

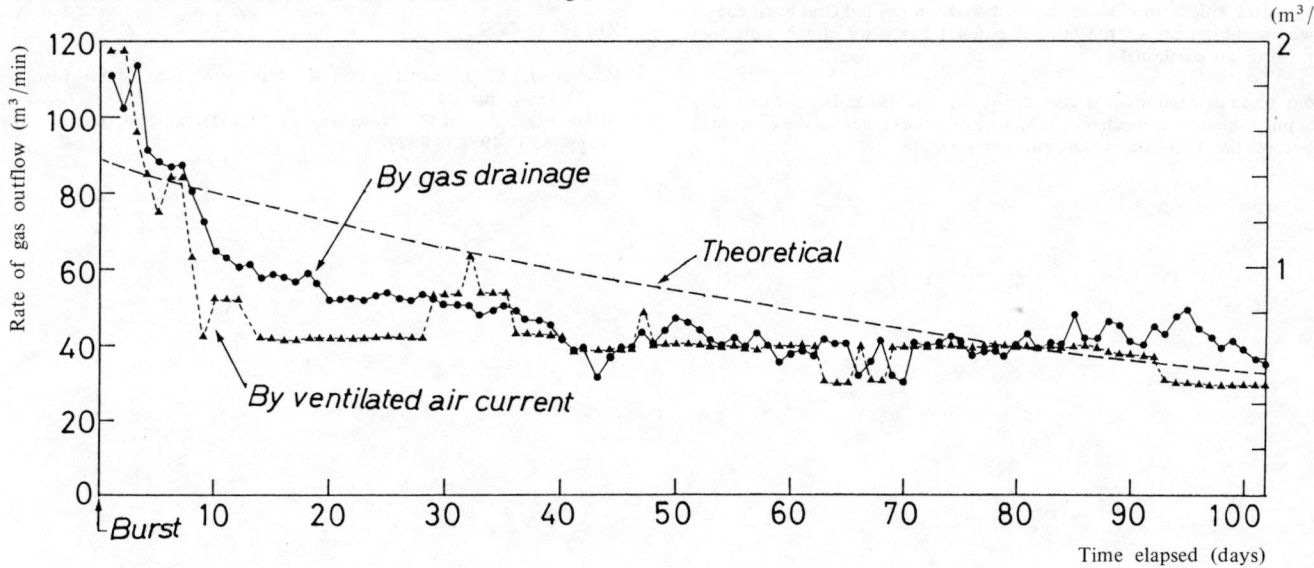

Fig. 2   Decrease in the rate of gas outflow with the time elapsed from the day when the gas and coal burst occurred on 16th Oct. 1981, in New-Yubari coal mine

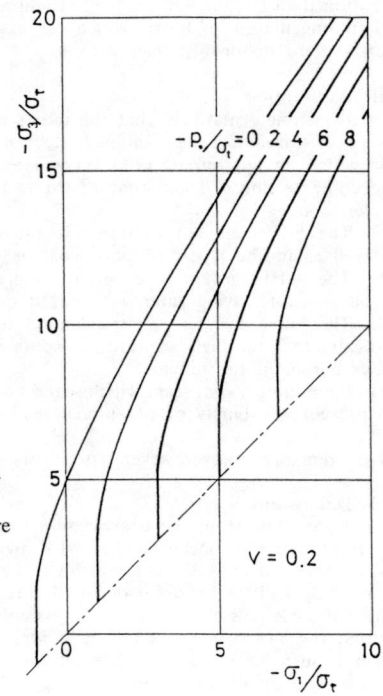

Fig. 3 Criterion for penny shaped crack containing high pressure gas

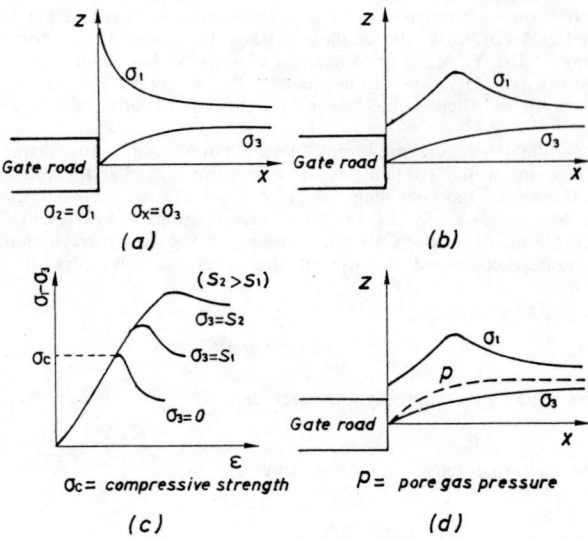

Fig. 4 Schematic diagrams showing distributions of rock stresses and gaspressure for three cases (1), (2) and (3) together with stress strain curves of a rock material

They have experienced the bursts which occurred several minutes after the blasting or at unexpected times. These bursts will be explained by the delayed fracture of rock materials. The authors found formerly by experiments that the greater part of the rock specimens, subjected to a uniaxial compressive stress of 80% of the compressive strength, broke in course of time, some of them instantaneously, some of them in a few minutes and the most delayed one after 183 hours. (Hiramatsu et al. 1957) The delayed bursts will be prevented by raising the safety factor, that is by placing the coal behind the face in a safe state, far apart from the critical state, by means of extracting the gas.

In case of a rock bursts an earthquake will usually take place, while a gas and coal burst is not accompanied by an earthquake. The reason for it may be that, while a rock burst causes a sudden subsidence of the ground which brings about an earthquake, a gas and coal burst causes indeed a subsidence of the ground but it takes place slowly and does not cause an earthquake.

From these discussions it is considered that the gas and coal burst may be a phenomenon of mechanical fracture of a coal seam by the combined effect of the rock stress and the gas pressure.

6. CONCLUSIONS

Taking account of the features of gas and coal bursts observed in Japanese coal mine, the authors studied the mechanism of the burst based on the results of stress analysis and several measurements. They found that the phenomena of gas and coal bursts could be explained by assuming them as the mechanical fracture of coal seam by the combined effect of rock stress and gas pressure.

It follows that the quantity of gas contained in a coal seam will be the major factor for occurrence of the burst, and extracting gas from a coal seam to lower the gas pressure will be the only measure to prevent the burst. Besides the rate of gas outflow for this purpose, the mechanism of the delayed burst and so on were also discussed.

The authors are very grateful to Dr. Mizuta for giving them the valuable results of analysis.

REFERENCES

Hiramatsu, Y., J. Kokado and M. Danno: J. Min, Metal. Inst. Japan, 77, 874, pp. 247–252 (1961)

Hiramatsu, Y. and M. Nishihara: J. Min. Metal. Inst. Japan, 73, 830, pp. 493–497, (1957)

STRATA CONTROL IN DEEP COAL MINES IN HOKKAIDO – IN-SITU MONITORING AND INTERPRETATION OF STRESS CHANGES IN COAL SEAMS

Contrôle des couches dans les mines souterraines profondes à Hokkaido – Surveillance in situ et interprétation du changement de contrainte dans les couches de charbon

Flözbeherrschung in den tiefen Kohlebergwerken in Hokkaido – Feldmessung und Analyse der Beanspruchungsänderungen in Kohleflözen

K. Fukuda
Graduate Student, Dep. of Resources Development Engineering, Hokkaido University, Japan

Y. Ishijima
Associate Prof., Dep. of Resources Development Engineering, Hokkaido University, Japan

S. Kinoshita
Professor, Dep. of Resources Development Engineering, Hokkaido University, Japan

SYNOPSIS

It has been shown from case studies that the occurrences of coal bumps in deep coal mines in Hokkaido, Japan, was restricted to the zone around coal pillars which had a sandstone roof measuring more than four times the thickness of the coal seam. In order to clarify the mechanism of coal bumps, laboratory tests, field measurements and numerical analyses on the coal seam behaviour were performed. As a result, it was ascertained that, when coal fails, a roof composed of sandstone supplies a large amount of energy to the coal seam compared to that composed of shale. Furthermore, both the mode of energy release and the patterns of stress changes are remarkably disturbed when an irregular distribution of mechanical strength of coal appears in the coal seam. These two conditions are the most important factors causing coal bumps.

RESUME

Les études relatives à de profondes mines souterraines de charbon à Hokkaido, au Japon, nous amènent à conclure que les éclatements du charbon (coal bumps) sont localisés autour des stots ayant pour toit des bancs de grès mesurant plus de 4 fois l'épaisseur de la couche du charbon. Afin de pouvoir expliquer le mécanisme de ces éclatements, on a procédé à l'exécution d'essais en laboratoire, au prélèvement de données sur le terrain et ensuite aux analyses numériques relatives au comportement de la couche de charbon. Comme résultat, il a été conclu qu'à la rupture, le toit composé de grès fournit beaucoup plus d'énergie que celui composé de schiste et que le mode de relâchement de l'énergie et celui du changement de la contrainte sont bien perturbés si la distribution de la résistance mécanique du charbon s'avère irrégulière dans la couche de charbon. Ce sont les deux facteurs les plus importants parmi ceux qui provoquent les éclatements du charbon.

ZUSAMMENFASSUNG

Es zeigte sich in Fallstudien, daß das Auftreten von Kohlebergschlägen in tiefen Kohlegruben Hokkaidos, Japan, auf Zonen um Kohlepfeiler beschränkt war, die ein Sandsteinhangendes mit mindestens der vierfachen Mächtigkeit des Flözes aufwiesen. Um den Mechanismus der Kohlebergschläge aufzuklären, wurden Laboratoriumsuntersuchungen, Feldmessungen und numerische Analysen durchgeführt. Es ergab sich, daß mit dem Bruch der Kohle ein Sandsteinhangendes eine im Vergleich zum Schiefertonhangenden vergleichsweise große Energiemenge an die Kohle abgibt. Ferner zeigte sich, daß die Art der Energiefreisetzung und die Änderung des Gebirgsdruckes stark von Unregelmäßigkeiten der Kohlefestigkeiten abhängen. Diese zwei Faktoren sind für Kohlebergschläge von entscheidender Bedeutung.

1. INTRODUCTION

In deep coal mines in Hokkaido there are frequent occurences of the brittle fracture of coal seam in or near working areas, being accompanied with fracture noise and seismic activity.

The present paper is the summary of the investigations which have been conducted as a part of the research program in order to clarify the cause and mechanism of coal bumps. Though most of the informations attained in this study concern brittle fractures, they will be available to understand the mechanism of coal bumps and rock burst.

2. COAL BUMPS OCCURED IN HOKKAIDO

The largest one of the nine fatal coal bumps occured in the past was that of Bibai Coal Mine in 1968. The coal bumps took place three times in different places of the same protecting pillars left in each side of the roadway along the upper gate of longwall pannel. The sites of the bumps are illustrated in Fig.1. The pillars situated at 650m below surface and had a dimension of about 25m in width and 3m in height.

The first bump occured at the part of the pillar of small and complex shape 25m ahead of working face as designated by ① in Fig.1. With further advancing of the face, the second and third bumps happened sequently at the place of ② and

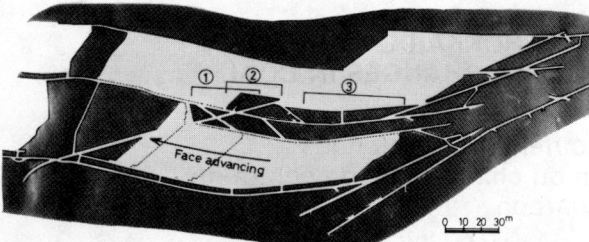

Fig.1 Sites of bumps in Bibai Coal Mine

③ behind the face front. All the three bumps were so severe that an earth tremor could be observed at the seismogram station of the meteorological observatory, 120km distant from the coal mine. The seismic shock was estimated at 2.2 in magnitude for the first bump, 2.7 for the third bump and less than 1.9 for the second bump. The damage of roadway attained to 100m long in the first bump, 170m in the third bump. In the severest case the part of roadway was completely closed due to the floor heave and the extrusion of wall.

The coal is hard and its uniaxial compression strength is 20 to 30MPa. The immediate hanging sandstone is 80-120MPa in compression strength and 13m in thickness, about four times as thick as the coal seam.

The presence of a strong and thick overlying sandstone measuring more than four times the thickness of the coal seam is a common factor found in the nine bumps experienced in Hokkaido.

3. DEFORMATION BEHAVIOR OF DOUBLE LAYERED MODEL

In order to investigate how sandstone roof exert influences on the deformation behavior of coal, uniaxial compression tests were performed using the double layered models of sandstone and coal connected in series.

The stress-strain diagrams for the three types of models are shown in Fig.2. The curve of ① is for the single coal model, ② is for the double layered model with the same height and ③ is for the model in which the sandstone is two times the thickness of the coal. The post-peak slope of the double layered models are steeper than that of the single coal model and it is ascertained that the thicker the layer of sandstone is the more violently the coal fails.

The ratio between the energy released from the sandstone and the energy consumed for the coal to deform in post-failure region is 0.12 for the case of ②, 0.44 for the case of ③. That is, the ratio becomes larger as the relative thickness of the sandstone increases. If the ratio exceeds 1, the specimen will break violently without any supply of energy to the system by the applied force.

This concept can be applied to practical mining problem. It will be possible to predict the proneness of unstable fracture in the working face, if the energy ratio described above is assessed for the underground structure concerned.

Furthermore, it must be noticed that there is a possibility of violent rupture for the coal, which does not manifest any property of unstable fracture in laboratory tests, because the brittle fracture of coal concerns structural characteristics rather than coal itself.

4. MONITORING OF STRESS CHANGES IN COAL SEAM

4.1 Examples in Horonai Coal Mine

A bump-like fracture occured in the pillar of 16m wide installed in the ribside of the upper gate of the advancing longwall face situated in a depth of 930m below surface when the face working just commenced. The geological condition and configuration of longwall pannel were nearly the same as in Bibai Coal Mine. After the incident, the face operation was continued while monitoring the stress changes using the hydraulic capsules of Brauner's type (Brauner,1973) installed in the pillar at 5m distant from the wall.

Fig.3 shows the results. No.1 capsule which was at 18m distant from the site of the bump-like fracture, did not indicate any changes. This is because this part of the pillar was under the influence of the bump and has been relaxed.

No.2 capsule put in the site of 80m ahead of No.1 showed a remarkable change immediately after installation. The change was so extraordinary that No.3 capsule was set additionally in the neighborhood of No.2 for confirmation. No.3 capsule gave almost the same result as No.2, then it was confirmed that this area was

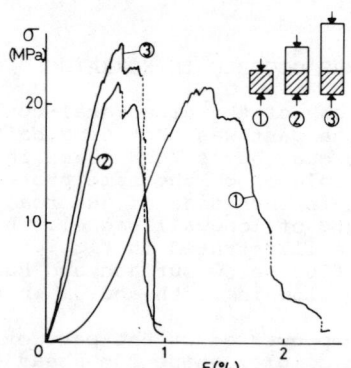

Fig.2 Stress-strain curves for double layered models of sandstone and coal

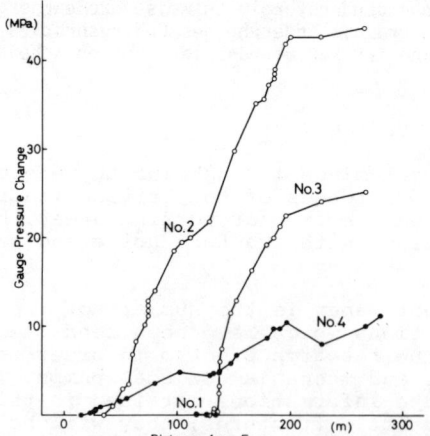

Fig.3 Stress changes in coal pillar (Horonai Coal Mine)

subjected to hard stress concentration. No.4 capsule located 25m ahead of No.3 showed a comparatively moderate rising of stress.

4.2 Examples in Minamioyubari Coal Mine

The monitoring of stress changes of coal seam was conducted in the tail gate of a depth of 610m below surface, where a series of bumps had been occured. Several hydraulic capsules were installed in the coal seam of tail gate ahead of the working face at intervals of 50m, starting from the vicinity of the area the bumps have taken place. The location of each hydraulic capsule was at the depth of 9m in the borehole drilled into the ribside coal of the tail gate.

The results of the measurement are shown in Fig.4. It is seen that stress changes appear suddenly and frequently while the face approaches measuring points. This itarate stress change corresponds to the stick-slip like behavior observed in the compression tests of coal under confining pressure, and means that the mother coal around the measuring point repeats the relaxation and compaction owing to the change of stress state resulting from the face advancing.

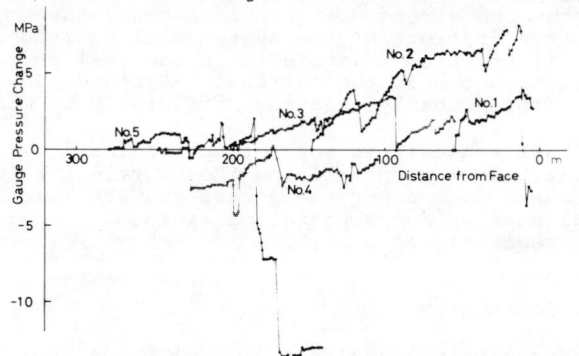

Fig.4 Stress changes in coal induced by face approaching (Minamioyubari Coal Mine)

Fig.5 shows the relation between the amount of stress change normal to bedding plane and the apparent lateral strain which is defined as the lateral displacement of wall divided by the distance of measuring point from the mouth of borehole. It is clear in comparison with No.2 and No.3 curves that the lateral strain undergoes a rapid increase every step of stress drop and that the steeper the slope of stress increment is, the more the amount of lateral strain increases.

As far as our experience concerns, the extrusion rate of coal is nearly constant so that the slope of the curve can be related to the increasing rate of rock pressure in the coal seam. Therefore it is supposed that the large amount of lateral strain was produced by the violent rupture of coal on account of the rapid stress rise. This is evidenced by the fact that the stress rise is the steepest at No.1 measuring point nearest to the area where the bumps had occured while the rate of stress rise diminishes moderately as being apart from the area.

5. NUMERICAL STUDIES ON BEHAVIOR OF COAL SEAM

5.1 Influence of overlying strata on coal seam fracture

As illustrated in the proceeding chapter 2, the coal bump is liable to occur in the case that the roof rock is hard sandstone and has not been experienced if the roof is consisted of weak rocks. Therefore in this chapter we tried an elasto-plastic analysis on the effect of roof rock property upon the yielding behavior of coal seam by means of F.E.M.

As the fracture criterion of both coal and rock, the following Coulomb's law and the maximum tensile strength theory were applied, and once the materials failed it assumed to drop immediately to the residual strength.

$$|\tau| = c + \sigma \tan\phi , \quad \sigma = \sigma_t$$

where τ : shear stress σ : normal stress
       c : cohesion ϕ : friction angle
       σ_t: tensile strength

The roof rock is taken as sandstone or shale and the constants necessary for computation are assumed as indicated in Fig.6. The results shown in the figure illustrate how the fracture zone develops when the goaf is enlarged to 54m behind the face.

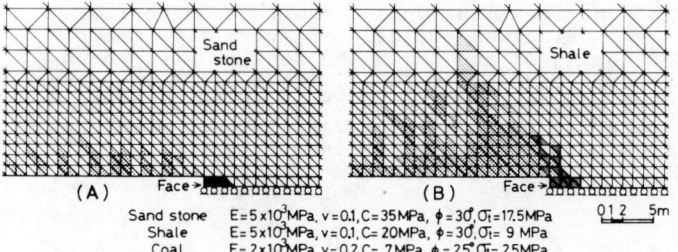

Fig.6 Plastic zone around face

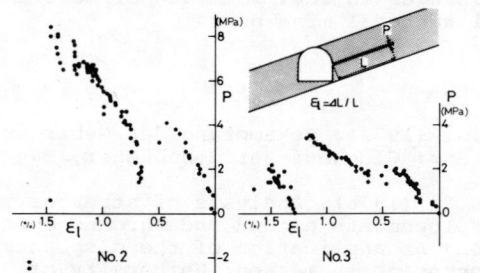

Fig.5 Relationships between normal stress change and apparent lateral strain

In case of sandstone roof, it is found that the tensile fracture of elements developed regularly in the goaf roof as shaded lightly in Fig.6-A. However, the roof rocks in front of the face are not ruptured but they recover elastically when the coal yields. On the other hand, in the case of shale roof, as seen in Fig.6-B, the yielded zone is fairly large as compared with sandstone roof and roof rocks in front of the face undergoes shear fracture as shaded deeply in the figure, which is probably caused at the same time or befor the excavation. The elastic recovery of deformation is hardly expected for shale roof, then its released energy may be small when the coal fails.

As an indicator which represents the violentness of fracture, that is, proneness of the brittle

E 13

fracture, the following index will be taken.

$$I = \frac{YRE}{TCE}$$

where YRE : Sum of the energy per unit advance of the longwall face, released from the failed elements while they transfer from an initial state of yield condition to the final state of yield condition through the unloading process and the energy released from unbroken elements which overlie the failed elements.
TCE : The energy per unit advance of the face required for failed elements to deform.

The value of I calculated for the model of sandstone roof shown in Fig.6 is four times larger than that of shale roof, and it is proved that the brittle fracture is more likely to occur in the case of sandstone roof.

5.2 Change of I value caused by mining

Assuming a numerical model of a coal bed sandwiched between two elastic rock beds, numerical computation with regard to I value was carried out to investigate the possibility of the brittle fracture of coal seam which might be caused by face advancing. In computing the two dimensional displacement discontinuity method (Crouch,1976) was applied. The four kinds of the complete stress-strain curves were assumed for the coal elements as illustrated in the right of Fig.7.

For the first, the computation was made under the assumption that all the coal elements lying in the range of five meter from the face, take ④ behavior, in the range of next five meter ③ behavior and so on. Then the coal lying more than 20m in front of the face is intact. Provided that the arrangement of the elements is kept identical during the face advancing, the I value was calculated every five meter advancing of the face. The results were shown by open circles in Fig.7. The I value increases very slowly from the onset of excavation and there are no apparent variation even when the face advances as far as 150m.

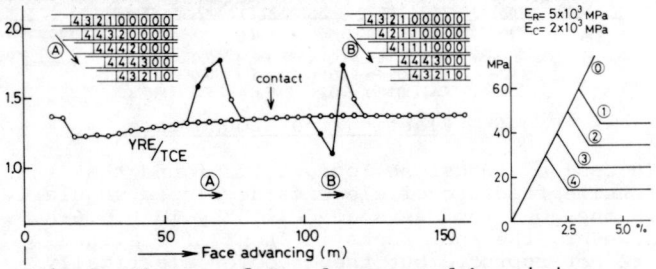

Fig.7 Change of I value caused by mining

For the second, the computation was made for the case where the arrangement of the elements was varied irregularly. The irregular arrangement of coal elements which are shown in the upper left and right of Fig.7, was given only at the places of A and B respectively. It is clear that I value increases sharply when the face arrived at the place of the irregular arrangement.

Fig.8 shows the peak normal stress in the front abutment of the face. The plotted point indicated by open circle corresponds to the first case mentioned above. In this case the curve of peak normal stress rises gradually as the face advances and becomes rather flat on and after the goaf begins to close. On the other hand, at the place of irregular arrangement, the distinguished increase of normal pressure takes place.

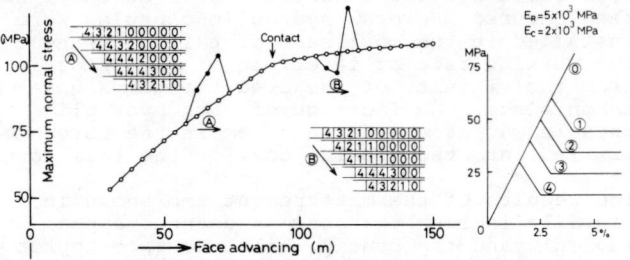

Fig.8 Peak normal stress in front abutment

It is assured from the results obtained that the variation of strength of coal seam elements is an important causative factor for the brittle fracture of coal and that if such variations are present in a mechanical system, the change of energy and stress sufficient to cause the brittle fracture of the system will be induced. The fluctuation of strength in the coal seam will be not only the intrinsic nature of coal but also probably created artificially by mining.

In this sense it is most effective for predicting and controlling the brittle fracture or bumps to monitor the stress changes in the coal seam in every pannel of mining as frequently as possible.

6. CONCLUSION

The statistical analysis of coal bumps in Hokkaido coal mines shows that most of them occured in pillars, particularly if the coal seam is overlain by thick sandstone over four times as thick as the coal seam.

In order to clarify the mechanism of coal bumps, laboratory tests, in-situ measurements and numerical analyses were performed. As a result, it is ascertained that if the roof is composed of a thick and hard sandstone, a large amount of energy is released from the structure when the coal fails and that released energy and normal stress increment in the coal seam increases markedly if the irregularities in strength are present in the coal seam.

Finaly, it is concluded that the monitoring of stress changes in coal seam are effective to control the coal mine bumps.

REFERENCES

Brauner,G. (1973). Bekampfung der Gebirgsschlaggefahr: Gluckauf-Betriebsbucher, Band 16.

Crouch,S.L. (1976). Analysis of stresses and displacements around underground excavation; An application of the displacement discontinuity method: University of Minnesota Geomechanics report.

APPLICATION OF GEOMECHANICAL CLASSIFICATION TO PREDICT THE CONVERGENCE OF COAL MINE GALLERIES AND TO DESIGN THEIR SUPPORTS

Applications des classifications géomécaniques pour prévoir les convergences des galeries de mines de charbon et pour calculer leur soutènement

Anwendung der geomechanischen Klassifizierung zur Vorabschätzung der Konvergenz und zum Entwurf des Ausbaus von Strecken in Kohlegruben

J. Abad
Dr. Mining Engineer, IGME

B. Celada
Dr. Mining Engineer, Geocontrol S.A.

E. Chacon
Prof. Dr. Mining Engineer, High School of Mines, Madrid

V. Gutierrez
Mining Engineer, Esboga Geotecnica S.A.

E. Hidalgo
Dr. Mining Engineer, IGME

SYNOPSIS

The geological conditions of the Asturias coal fields mined by HUNOSA are particularly difficult, due to the high seam inclination. The INSTITUTO GEOLOGICO Y MINERO DE ESPANA (IGME) is developing a research program to try to find scientific rules which make it possible to design the characteristics of the roof supports for the HUNOSA mines roads. A great amount of information about geometric, geological and geotechnical parameters relating to 187 roads from the HULLERA DEL NALON coal field (HUNOSA) has been analysed by means of the program BMPD-2R. This led to the definition of a convergence function providing for the design of the roof supports for these roads.

RESUME

Le fort pendage des couches rend l'exploitation du gisement de charbon d'HUNOSA particulièrement difficile. L'Institut Géologique et Minier d'Espagne y effectue un programme de recherches afin de trouver des indications scientifiques permettant de calculer au préalable le soutènement des galeries. On a enregistré une grande quantité de données sur les caractéristiques géologiques et géotechniques de 187 galeries au bassin de Nalon (HUNOSA), et l'analyse de ces données à l'aide d'un programme BMDP-2R a permis d'établir une fonction de convergence grâce à laquelle on peut dimensionner le soutènement des galeries.

ZUSAMMENFASSUNG

Die geologischen Verhältnisse der spanischen, von der HUNOSA abgebauten Kohlevorkommen sind wegen des starken Einfallens der Flöze besonders schwierig. Das INSTITUTO GEOLOGICO Y MINERO DE ESPANA (IGME) beschäftigt sich mit der Entwicklung eines Forschungsprogramms, das versucht, wissenschaftlich fundierte Regeln für den Firstausbau der Strecken aufzustellen. Eine große Datensammlung über die geologischen, geometrischen und geotechnischen Kennzeichen von insgesamt 187 Strecken der Hullera del Nalon (HUNOSA) wurde angelegt. Die Analyse dieser Daten mittels des EDV Programms BMPD-2R ermöglichte das Aufstellen einer Konvergenzfunktion, die einen rationellen Entwurf des Streckenausbaues ermöglicht.

1.- Introduction

About 3.960 million tons of coal have been evaluated in Spain the 85% of these must be exploited by underground mining. The conditions of this mining are not specially favourable in Spain, since 44% of the reserves are in seams with a slope of 35º or greater.

At this moment, the underground coal total output in Spain is about 33Mt including antracite and lignite per year. The coal production is 9,7Mt, and 41% of this is mined by the EMPRESA NACIONAL HULLERAS DEL NORTE (HUNOSA) in the Asturias basin in northern Spain.

In this area, the face depth varies from 400 to 800m, and the roads cross-section between 7 and 15 m^2, being 9 m^2 the most usual. The roof supports are mostly steel arches in TH section as shown in fig. 1. At this moment, the geotechnic conditions of the seams, are beggining to produce difficulties for the road conservation, and for

this reason the INSTITUTO GEOLOGICO Y MINERO DE ESPAÑA (IGME), considered necessary to develop a research program to stablish rules that allow sizing of the roof supports, as a function of the rocks geomechanic carachteristics and the face-imposed conditions.

2.- Research description

The project target is to obtain a scientific rules, based on the rocks geomechanic characteristics, that allow to solve two important questions in the mine roads design:
- Which must the initial road corss-section be, so that taking into account the deformations that will be produced, the conservation works be minimum?
- Which must be the roof supports to install to keep the stability of the road with a minimum cost?

To be able to solve these questions, the research was directed to define a convergence function that let to calculate a road convergence as a function of the goemetric characteristics of the road, the supports, the geotechnic parameters and the face influence.

To this, the program has been split in four steps :
1 - Actual conditions investigation
2 - Actual situation analysis
3 - Checking tests
4 - Final results obtention

The project began on July 1981 and in this moment the two first points have been completed and a massive road deformation data acquisition campaign has been initiated.
The project is expected to be finished at the end of 1983

2.1.- Definition of the object roads

The project is being developed at the HULLERA DEL NALON (HUNOSA) which has 8 pits.

As initial point, an technical enquiry has been done at each pit, to know the expected roads to be advanced till 1984. The purpose for this enquiry is to define the geometric carachteristics, methods and equipment that are going to be used in this work. A total of 275 roads where studied, 227 of which were gate roads in coal, and the rest main roads in rock.

2.2.- Geomechanic carachteristics

Once known the planned roads till 1984, the ones to be operatives during the duration of this study have been selected, and have been defined from the geomechanic point of view.

For the field data acquisition, the ISRM proposed methodology in the document BASIC GEOTECHNICAL DESCRIPTION (B. G.D.) has been used.

The geomechanical classifications from BARTON (1974) and BIENIAWSKI (1979) have been applied. Also the typical Laboratory tests as the concentrated load resistance with the Schmidt hammer, and the simple compression ones.

Convergence measurements have been done in each studied road and using the SCHMARTZ et alt. (1962) theory, the expected convergence for each road, once the face effects have finished, has been stablished.

2.3.- Data processins

In the geomechanic carachterisation term, 187 road heads have been studied "in situ", having been done more that 7000 Schmidt hammer tests, 1500 breaking under concentrated load tests, and 120 simple compression ones. About 7500 different data have been stored. To manage with this information, the program BANCO 1 has been developed. This program works as a high flexibility data storage, and operates in a 4431 IBM 370 system. Fig. 2 shows the program flow chart. Once the information has been compiled, the main program offers up to 28 outputs or "menus" that mean

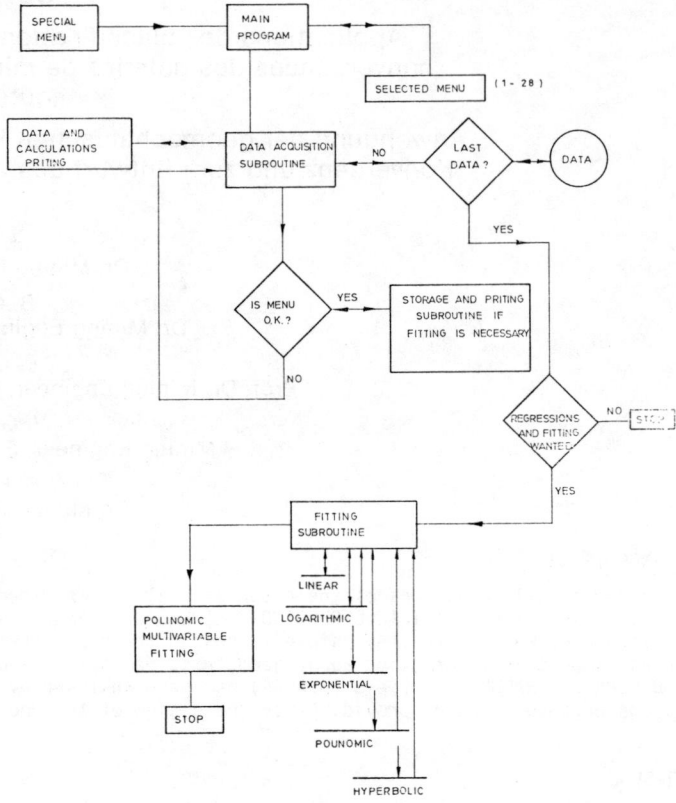

Fig. 2. Bancol Program Flow Chart

the same number of filterings.

The LECTUR subroutine reads the cards once the desired menu has been selected and process them to produce the wanted information.

If the selected menu includes the fittings betwen the studied parameters, the ESCRIT subroutine stores the results in a three dimension matrix. The three dimensions are the data number, the value of the considered data, and the pit where it comes from.

This makes easier the work of the IMPRIM subroutine that prints the stored data. Each time LECTUR reads, the menu is checked and ESCRIT stores de vector till reading the last data. In this moment LECTUR gets control again, calling AJUSTE which makes 12 differents bidimensional relations within the parameters, in six different modes : Linear, logarithmic, exponential, powers, hiperbolic, parabolic and a multiple lineal relation, to define the road convergence function.

An special menu allows direct access to the data storage, by means of the IMPRIM subroutine that calls LECTUR and processes the stored data, evaluating some parameters that are considered of interest for the rock characterisation :
- Road situation
- Rock type.
- Simple compression resistance in the pure rock
- Traction resistance

- Cut resistance
- Protodiaknov index
- Schmidt index
- Rock type percentage in the road cross-section
- Rock situation in respect to the coal seam.

Other printing facility of IMPRIM, provides the data from each B.G.D. with the origin pit name, road name, identification number, Barton & and Bieniawski R.M.R.

3.- Final results

From the geomechanical data, it has been done an statistical analysis to reach the project target, and the following results have been obtained :

3.1.- Relation between RMR and Q

This relation has been stablished by different authors :
BIENIAWSKI (1976) : $RMR = 9.\ln Q + 44$
RUTELEDGE (1978) : $RMR = 13,5 . \log Q + 43$
MORENO (1980) : $RMR = 12,5 . \log Q + 55,2$

with the results of this project, we have got the following :

$$RMR = 10,53 . \ln Q + 41,83$$

with the correlation coeficient $r = 0,934$. In Fig. 3, it is shown the correlation representation.

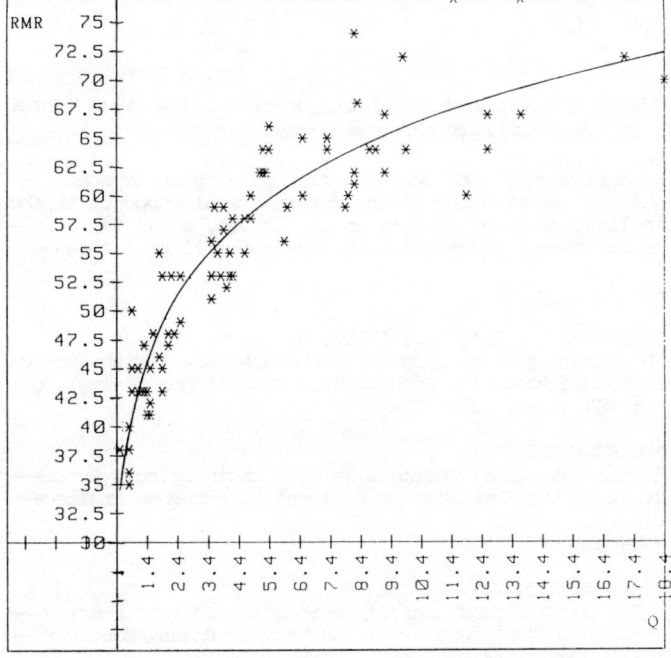

Fig. 3. Correlation between Q and RMR

It is easy to check that the expressions (1) to (4) produce similar results, although, obviously, the (4) expresions fits better the conditions of the HUNOSA basin.

3.2.- Convergence function

The following convergence function was initially defined
$$C = A_1 + A_2 . W + A_3 . (T_G) + A_4 . H + A_5 (DS) + A_6 . RMR \quad (5)$$
being

C = Final road convergence, in percentage of initial height
A_1 to A_5 = Parameters
W = Seam thickness (m)
TG = Road type. It varies from 1 to 5 depending of the rock quality.
H = Road depth (m)
DS = Supports density (steel Kg/mined m^3).
RMR = Bieniawski index.

Once the adjustment out fitting were made, the following relation was found ont; (IGME 1982) :
$$C = 38,45 - 56,33 W + 7,48 (TG) + 0,016 . H - 1,84 (DS) + 0,86 . RMR.$$

The total correlation coeficient, adjusting by the square minimums method, is $r = 0,798$

The partial correlations of convergence in respect to the other parameters, are :
Correlation C in respect to W............... 0,821
" C " " " DS............. 0,588
" C " " " TG............. 0,2
" C " " " H............... 0,047
" C " " " RMR............. 0,014

As expected, the biggest weight in the function corresponds to seam thickness, followed by supports density. It is not rare the little influence of road depth, because this parameter has not big variations in the studied area, and, in the other hand, many roads ramain in elastic state before of the faces influence. Nevertheless, it is difficult to accept the little weight of galery type (T_G) and RMR.

Finally, to make clear these doubts, a step by step regression was used instead of the multiple lineal correlation. For this the BMOP-2R program was used. This program has been developed by the HEALTH SCIENCES COMPUTING FACILITY in the California University (U.S.A.)

From the results pointed out by the BMOP-2R, the following conclusions have been stablished :
- In the concrete case of HUNOSA roads, it is more significative the cross-section convergence than the highness convergence. This is as a consequence of the high seams dip.

- with the analysed data, it results to be more significative to make a correlation of the Neperian logarithm of convergence, with the analysed parameters.
- In the HUNOSA gate roads in coal, the influence of Barton Q and RMR is little compared with the cross-section convergence.
- After the statistical analysis, it became clear that the most significative parameters are the supports density, joints state strata inclination and seam thickness.

According with the obtained results it was decided to stablish a new geomechanical classification to define an index, called IGME 82, that points the convergence function together with the supports density. To this, the parameters used for the road classification have been processed by the BMOP2R program, selecting the most significative data obtained from this, evaluating the influence of all them. Fig. nº 4 shows the results :

PARAMETER	DEFINITION	CALIFICATION
JOINTS STATE	FAULT POLISH OR OPEN JOINT.	0
	SLIGHTLY RUGGED JOINTS OR WITH SOFT EDGES	18
	RUGGED JOINTS OR WITH HARD EDGES	22
STRATA INCLINATION	20 - 40º	5
	45 - 60º	0
	60 - 90º	28
SEAM THICKNESS	0,3 m	50
	0,3 - 0,8 m	21
	0,8 - 1,8 m	14
	1,8 m	0

Fig. Nº 4. Weighing of parameters in the IGME 82 classification.

The IGME 82 index varies from 0 to 100. The less favourable case is the one of a road with a seam thickness greater than 1,8 m, a strata inclination steween 45 and 60º, and the discontinuities are fault polishes or open joints. The most favourable case is a road in which the seam thickness is less tham 0,3 m, the strata inclination is between 60 and 90º, and the present joints are rugged or with hard edges.

From this index and the supports density, in steel kg/mined m^3, a new correlation with the cross-section convergence Neperian logarithm, has been done, finaling the following expression:
$$\text{Ln } C = 4,2614 - 0,032 \text{ (DS)} - 0,0167 \text{ (IGME 82)}. \quad (7)$$

The multiple correlation index is $r = 0,572$ and the Fischer F, for the fitting made, is $F = 33,644$, what shows that the probability for the fitting to be random is less than 0,001 %.

3.3.- Support sizing

To display graphically the use of expression (7) the chart shown in fig. Nº 5 has been done. This chart can be used to foresee the sizing of the road roof supports in the HUNOSA collieries.

It is necessary to define the wished final convergence, in percentage, of the road at the end of its life. With this and the road heading calification after the IGME 82 index, the supports density is determined. From this, knowing the road cross-section (m^2), and the TH supports weight (kg/m), the distance between metalic arches is pointed out.

This chart was stablished after an statistical study of a great quantity of data from 187 roads in the HULLERA DEL NALON, HUNOSA, and it is going to be cheked in a wide road deformation measuremente campaign, in which also the loads actuating over the supports are going to be measured.

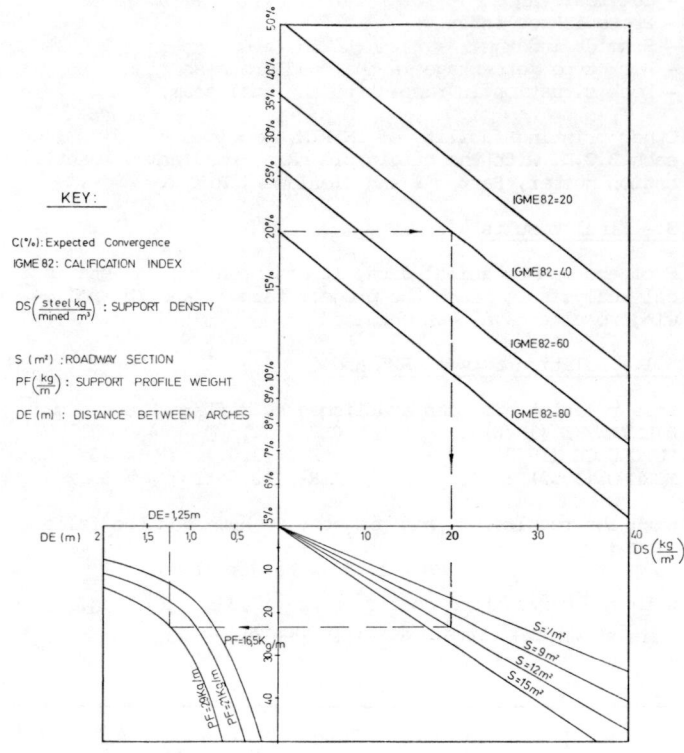

Fig. 5. Support prediction chart for the Hunosa Collieries roadways

In this moment, the chart in fig. Nº 5 is an important step in the previous sizing for road roof supports in the inclined seam collieries. In the future, after the referred checking, the name of this method will be increased.

4.- Bibliografy

* BARTON, N.; LIEU, R.; LUNDE, J.;
"Engineering Classification of Rock Masses for the Design of tunnel support". Rock Mechanics, Springer Verlag, vol 6 : 189 - 236. 1974.

* BIENIAWSKI, Z.T.
"Rock Mass Classifications in rock engineering". Procedings of the Symposium on exploration for rock engineering.
Johannesburg, 1976.

* BIENIAWSKI, Z.T.
" The Geomechanics Classifications in rock ingineering aplications" 4º congreso mt de Mer. de Rocas, Montreux 1979. Tomo 2, 41-48.

* INSTITUTO GEOLOGICO Y MINERO DE ESPAÑA

Optimizacion del sostenimiento en las galerías de las minas de hulla según las características geomecánicas de las rocas y de los factores de explotación. Informe anual de 1981.
Madrid, Diciembre 1981.

* MORENO TALLON, E.
Aplicación de las clasificaciones geomecánicas a los túneles de Pajares.
II Curso de Sostenimientos activos en Galerías y Túneles.
Fundación Gómez-Pardo. Madrid 1980.

*RUTLEDGE, J.C.; PRESTON R.L.
New Zealand Experience with Engineering Classifications of Rock for Prediction of Tunnel Support.
Tunnels Under Difficult Conditions. Tokyo, 1978.

*SCHWARTZ, B.; CHAMBON, C.; DECOMPS, J.; VIALLET, F.
Previsions des convergences dans les voies influences — par les tailles qu'elles desservent.
Revue de l'Industrie Minerale. Sep. 1962.

PLANMÄSSIGE ÜBERWACHUNG WICHTIGER GRUBENRÄUME IM STEINKOHLENBERGBAU MIT FERNÜBERTRAGUNG DER MESSWERTE

Systematic monitoring of important cavities in hard coal mines with remote data transmission

Contrôle régulier d'ouvrages souterrains importants dans les mines de charbon comprenant la télé-transmission de données de mesure

Dr.-Ing. Fritz Schuermann
Wissenschaftlicher Mitarbeiter, Bergbauforschung GmbH, 4300 Essen 13, West Germany

SYNOPSIS

The number of parameters to be measured increases continually, and the cost-effectiveness of measuring work decreases. A survey of the system comprising sensor, data transmission, recording, and evaluation yielded indications on possible improvements. This information was utilized. Accordingly, a processor-controlled monitoring and control technology without operator was developed. Transducers with 5 ... 15 Hz output are energized successively by a multiplexer with 96 connections, interrogated, the respective data are transmitted, recorded by a processor, and evaluated. The multiplexer is remote-controlled by the processor. Important transducers are interrogated more frequently. The processor can give instructions relative to necessary remote control or assume remote control directly. This technology is described by the example of a 1 600 m remote-monitored roadway exposed to the effects of a working coal face advancing at a rate of 8 to 14 m per day.

RESUME

Le nombre de mesures à effectuer va constamment en augmentant et le rapport coût/utilité se dégrade. L'étude du système composé de capteurs, l'installation de transmission de données, l'enregistrement et le dépouillement de données de mesure montre qu'il est possible d'y apporter des améliorations. Ces possibilités ont été utilisées. Ainsi une technique de contrôle et de commande au microprocesseur c.à.d. sans intervention humaine a été réalisée. Des capteurs de 5 à 15 Hz de sortie sont alimentés successivement par un multiplexeur à 96 connexions. On les interroge pour obtenir leurs données de mesure respectives qui sont télétransmises et enregistrées par un processeur et dépouillées. Le processeur assure la télécommande du multiplexeur. Il interroge plus souvent les capteurs importants. Le processeur peut donner des indications afférentes à des télécommandes nécessaires ou les exécuter lui-même. Cette technique est décrite pour une galerie télécontrôlée de 1 600 m soumise à l'effet d'exploitation d'une taille dont l'avancement journalier est de 8 à 14 m.

ZUSAMMENFASSUNG

Die Zahl der Meßaufgaben nimmt ständig zu. Das Aufwand/Nutzen-Verhältnis verschlechtert sich. Eine Untersuchung des Systems Meßwertaufnehmer-Datenübertragungsanlage-Meßwertregistrierung und -auswertung zeigte Verbesserungsmöglichkeiten. Sie wurden genutzt. So entstand eine unbemannte prozessorgesteuerte Überwachungs- und Steuerungstechnik. Meßgeber mit 5 ... 15 Hz Ausgang werden über einen Multiplexer mit 96 Anschlüssen nacheinander mit Strom versorgt, ihre Meßwerte abgefragt, diese fernübertragen, von einem Prozessor registriert und ausgewertet. Der Prozessor fernsteuert den Multiplexer. Er fragt wichtige Geber häufiger ab. Der Prozessor kann Hinweise für notwendige Fernsteuerung geben oder diese selbst ausführen. Am Beispiel einer 1 600 m langen fernüberwachten Strecke unter Abbaueinwirkung eines Strebes mit 8-14 m/d Fortschritt wird die Technik beschrieben.

1. Einleitung

Wesentliche Elemente der "Neuen Österreichischen Tunnelbauweise" (NÖT) sind das stetige Messen der Gebirgsverformungen direkt nach Freilegung des Gebirges und die frühzeitige Anpassung der Vortriebs- und Ausbautechnik an das gegebene Gebirgsverhalten. Die zugehörige Meßtechnik konnte in den vergangenen Jahren verbessert und ausgeweitet werden. Das hat den Aufwand für Messen und Auswerten ansteigen lassen. Die Meßaufgaben werden von Auftraggeber und Auftragnehmer häufig aus unterschiedlicher Sicht betrachtet. Den Auftraggeber interessieren mehr die Messungen für die Überwachung des Tunnelbauwerks, den Auftragnehmer mehr die Messungen über die Ausnutzung des eingesetzten Maschinenparks, den täglichen Fortschritt und die Abwehr aller Einflüsse, die die Arbeiten behindern können, wie z.B. die Gefahr von Gasausbrüchen oder Wasserzuflüssen.

Im Steinkohlenbergbau kommen zu den genannten Meßaufgaben noch die vom Abbau ausgelösten großflächigen Absenkungen des Gebirges mit ihren vielfältigen Einflüssen hinzu. Das führt zu einem so erheblichen Aufwand für die Meßtechnik, daß nach Wegen gesucht werden mußte, trotz steigender Zahl von Meßaufgaben, zumindest die spezifischen Kosten für die Meßtechnik entscheidend zu verringern. Dies ist möglich geworden, durch eine mannlose prozessorgesteuerte Vielstellenmeßtechnik, die bei Bedarf auch eine Fernsteuerung von Maschinen und Geräten einschließt.

2. Die Entwicklung der prozessorgesteuerten Vielstellenmeßtechnik

Die großen Entfernungen der Betriebspunkte von den Schächten und deren ständige Ortsveränderung haben die auf allen westdeutschen Bergwerken übliche Tonfrequenz-Multiplextechnik entstehen lassen. Mit je einem Adernpaar können in 20 und mehr Kanälen (je nach Typ und Hersteller) Signale im Bereich von 5 ..... 15 Hz ohne Verstärkung über 10 - 15 km Entfernung mit hoher Genauigkeit übertragen werden. Dabei wird für jeden Geber oder für Ja/Nein-Anzeigen usw. je ein Tonfrequenzkanal belegt. Die Kanalzahl engt ebenso wie der Stromverbrauch die Zahl der anschließbaren Meßgeber ein. Bei größerer Zahl an Gebern ist ein weiteres Adernpaar erforderlich. Diese Technik ist teuer, weil die Kapazität des Systems schlecht genutzt wird. Die meisten Meßwerte ändern sich nur geringfügig und langsam.

Das bisherige System kann aus Abb. 1 entnommen werden. Ein Geber wird mit Strom versorgt und gibt ein, seinem Meßwert proportionales, Signal zurück. Das wird im Eingang des Tonfrequenz-Netzes zunächst durch Umwandeln in ein Frequenzsignal 5 ..... 15 Hz übertragbar gemacht und in dem für den Meßwertgeber zur Verfügung gestellten Kanal mit Hilfe eines zugehörigen Senders auf den Weg gebracht. Übertage wird das ankommende -jetzt schwächere Signal- von einem auf den TF-Kanal eingestellten Empfänger aufgefangen und in ein Stromsignal umgewandelt, das den Arm eines Schreibers in seiner Höhe verstellt. Dort wird das Signal als Schreibstreifen über der Zeit registriert.

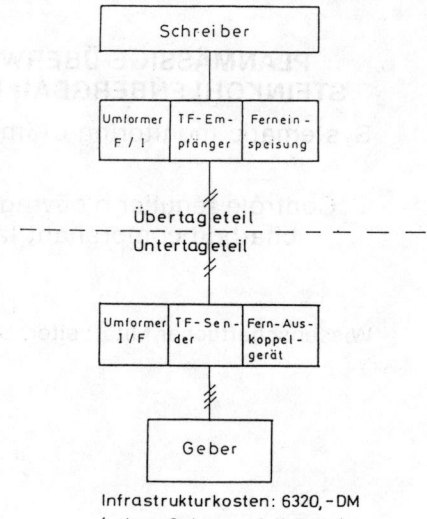

Infrastrukturkosten: 6320,-DM
(ohne Geber und Kabel)

Abb.1 Übliche Fernüberwachungstechnik

Durch mehrere Änderungen und Ergänzungen konnte das übliche System verbessert und wesentlich verbilligt werden. Das neue System kann aus Abb. 2 entnommen werden. Es enthält folgende Unterschiede:

1. Alle Meßgeber sind in der Elektronik so verändert worden, daß sie bei Erhalt von 12 V und 8 mA Strom ein dem Meßwert proportionales Frequenzsignal von 5 ..... 15 Hz abgeben. Das gibt zuverlässige Meßwerte (kein Temperatur- und Leitungslängeneinfluß, keine Meßwertverfälschung bei Kabelquetschungen).
2. Ein Multiplexer ist zwischen Gebern und Tonfrequenz-Anlage zwischengeschaltet. Mit diesem Gerät können 96 Meßgeber nacheinander mit Strom versorgt und ihre Anzeige abgefragt werden.
3. Das Übertragungssystem ist geblieben. Es ist jedoch ein zweiter Kanal (hier mit A bezeichnet) hinzugeschaltet worden. Er dient zur Fernsteuerung des Multiplexers.
4. Im Übertageteil (Grubenwarte z.B.) wird das Frequenzsignal von einem modernen Frequenzzähler mit BCD-Ausgang gelesen und digitalisiert direkt einem Tischrechner (Prozessor) zugeführt. Der prüft das Signal, registriert es und schaltet dann mit Hilfe des Kanals A den Multiplexer einen Meßgeber weiter. Um einen Geber von übertage her abzufragen und weiterzuschalten, ist ein Zeitbedarf von 1,2 bis 3 Sekunden erforderlich. Ein vollständiger Zyklus mit 96 Gebern erfordert 2 bis 5 Minuten. Von diesem ist aber nur ein Teil so wichtig, daß er ständig abgefragt werden muß. Der Prozessor kann nun aus der Entwicklung der Meßwerte und mit Hilfe einer vorgegebenen Gewichtung Dringlichkeitsklassen aufstellen:

Klasse 1 (alles entscheidend)	Meßwert wird bei angehaltenem Multiplexer auf Bildschirm graphisch aufgetragen.
Klasse 2 (sehr wichtig)	Abfrage alle 30 Sekunden erforderlich.

Klasse 3 (wichtig) Abfrage alle 5 bis 10 Minuten
Klasse 4 (möglicherweise wichtig) Abfrage alle 30 Minuten bis zu 24 Stunden

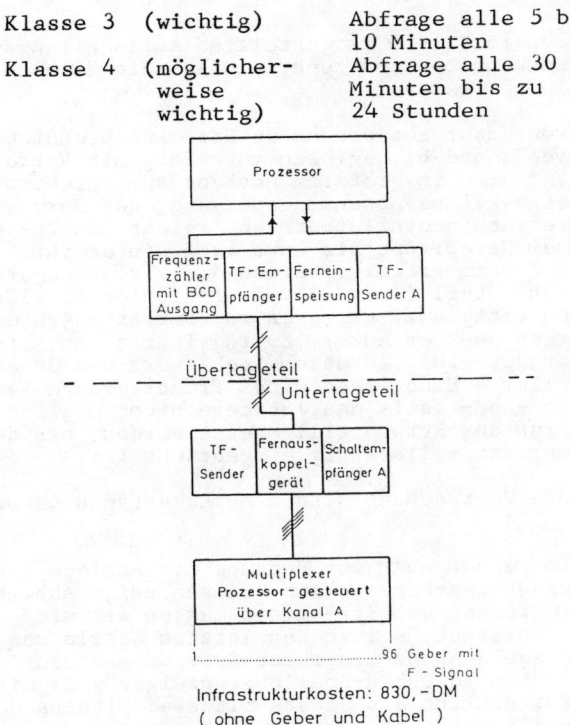

Abb. 2 Prozessorgesteuerte Vielstellenmeßtechnik

Dadurch, daß alle weniger wichtigen Geber zwischenzeitlich überschlagen werden können, ergeben sich Zykluszeiten von nur 30 Sekunden, im Vergleich zu 2 bis 5 Minuten bei einem vollständigen Zyklus. Trotz der Vielfachausnutzung der einzelnen Tonfrequenz-Kanäle ergibt sich durch die periphere Prozessorsteuerung eine vorzügliche Flexibilität. Der Prozessor kann mit geeigneten Hilfsgeräten, wie graphischem Bildschirm und Schnelldrucker die gesamte Auswertung und Berichterstattung übernehmen und durch Trendberechnung auf beginnende Gefahren hinweisen.

Mehrere solcher Vielstellenmeßanlagen sind bereits in Betrieb. Alle Meßgeber sind eigensicher und die Tonfrequenz-Fernübertragungsanlage explosionsgeschützt ausgeführt. Alle intelligenten Geräte und die Auswertung sind übertage in der Grubenwarte konzentriert. Der Multiplexer ist spritzwasserfest und kann auch vorübergehend Überflutungen vertragen.

3. Betriebsbeispiele

Auf dem Steinkohlenbergwerk Niederberg wird eine 1.600 m lange Abbaustrecke, die mit einem Anker-Maschendraht-Verbundausbau aufgefahren wurde und deren Zustand vor der Abbaueinwirkung Abb. 3 wiedergibt, durch einen an ihr auf einer Seite vorbeigeführten Abbau beansprucht. In dieser Strecke sind, annähernd gleichmäßig verteilt (in 20 m Abstand), sogenannte Überwachungsanker eingebaut. Sie dienen zur Überwachung der Auflockerung im Bereich des geankerten Gebirges. Zusätzlich sind 4, 6, 9 und 12 m lange Meßanker oder auch Mehrfachextensometer in größeren Abständen eingebaut worden und schließlich werden die Querschnittsverluste nach dem Verfahren der "Betrieblichen Streckenbeobachtung" der Forschungsstelle "Grubenausbau und Gebirgsmechanik" überwacht. An die Fernübertragung sind mit Hilfe von 79 elektronischen Wegmeßgebern jeder zweite Überwachungsanker, einige 6 m lange Meßanker, Mehrfachextensometer, Konvergenz-Meßstrecken und ein Temperaturmeßgeber angeschlossen worden.

Abb. 3 Bogenförmige Ankerstrecke

Die überwachte Strecke liegt in der Luftlinie rd. 7 km vom Schacht entfernt. Bei einer Leitungslänge von über 11 km kann der Multiplexer noch gut vom Prozessor übertage her gesteuert werden und die Meßwerte kommen exakt in der Grubenwarte an. Eine Reihe von Auswerteprogrammen mit der Möglichkeit der Früherkennung von Trends sind vorhanden und helfen der Leitung des Bergwerksbetriebes Gefahren rechtzeitig zu erkennen und abzuwenden.

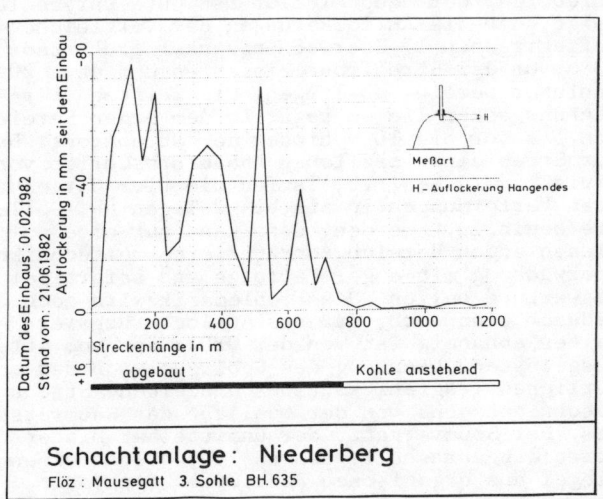

Abb. 4 Einfluß der Abbaueinwirkung auf die Auflockerung

Eine Übersicht über die bisherige Abbaueinwirkung des Strebs in bezug auf die in der Strecke gemessene Auflockerung gibt Abb. 4. Während diese Strecken ohne Abbaueinwirkung sehr gut stehen, wird der Einfluß des Strebs gravierend, sobald die vom 251 langen Streb freigelegte

Hangendfläche eine Breite von mehr als 20 m erreicht hat. Es wurden Auflockerungsbeträge hinter der Strebfront von 2 bis 75 mm gemessen. Nach Überschreiten von 60 mm Auflockerung platzen einzelne Muttern der auf der ganzen Länge eingeklebten 2,3 m langen Vollklebeanker mit 22 mm Ankerstangendurchmesser ab und es bilden sich "Beulen" aus aufgelockertem Gestein, die vom Maschendraht zusammengehalten werden.

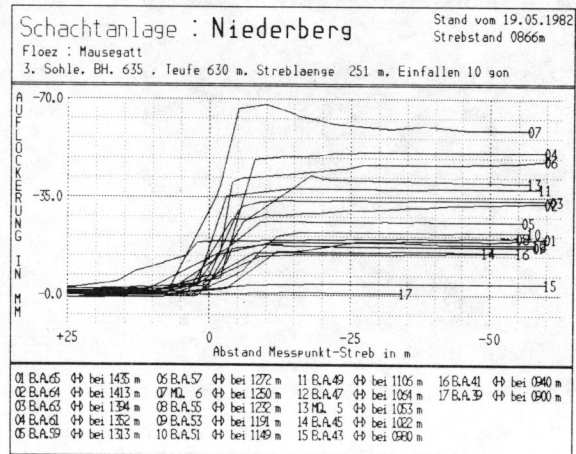

Abb. 5 Auflockerungsablauf in den einzelnen Abschnitten der Strecke

Zur Beurteilung des Auflockerungsverhaltens hat sich die vom Prozessor ausgedruckte Darstellung nach Bild 5 bewährt. Aufgetragen werden die gemessenen Auflockerungen über dem Abstand zum Streb, der der Haupteinfluß ist. Die Darstellung zeigt, daß die Auflockerungen der vergleichend aufgetragenen 17 Überwachungsanker frühestens 13 m vor der Strebfront einsetzen und etwa 20 m dahinter bereits abklingen. Die meisten Überwachungsanker liegen sogar in dem engen Bereich von 5 m vor bis 10 m hinter der Strebfront. Da der Streb einen täglichen Abbaufortschritt von 8 bis 14 m/d besitzt, laufen die hier dargestellten Verformungen in ein bis 2 Tagen ab. Sollen sie beeinflußt werden, dann kann nur eine auf diesen engen Bereich konzentrierte meßtechnische Überwachung mit enger Meßfolge und sofortiger Auswertung helfen. Die Problematik wird noch dadurch erschwert, daß das Auflockerungsverhalten abhängig ist von der Größe der vom Streb ausgelösten Absenkung des Gebirges, von der natürlichen Tragfähigkeit des einzelnen Streckenabschnittes und von der Qualität des Saumversatzes. Der Saumversatz, der unmittelbar hinter dem Streb eingebracht werden sollte, bildet einen Riegel aus organischen Baustoffen (Anhydrit, Zement, oder dergleichen), der in 1,5 m Breite parallel zur Strecke auf der Seite des Strebs diesem folgend eingebracht wird. Er ist notwendig, um die im Bruchfeld des Strebs eintretende große Absenkung in möglichst großem Umfang von der Strecke fernzuhalten. Von besonderem Einfluß sind das frühe Einbringen und das schnelle Erhärten des Dammbaustoffes. Treten Mängel auf, kommt es zu größeren Auflockerungsbewegungen. Das führt zur Überbeanspruchung des Gebirges und der Ausbauanker.

Aus den im Bild 5 dargestellten Auflockerungskurven kann bereits recht einfach die Ursache abgeleitet werden:

Die von Natur aus schwachen Streckenabschnitte (Kurven 2 und 6) beginnen vorzeitig mit Verformungen, also in größerem Abstand zur Strebfront. Der eigentliche Absenkungsvorgang, der durch den Saumversatz beeinflußbar ist, reicht von 2 - 4 m vor der Strebfront bis etwa 12 m hinter ihr. Ist er besonders kräftig, werden die Verformungskurven sehr steil (1 und 3). Treten beide Einflüsse gleichzeitig auf, kommt es zu Überlastungen des Gebirges und der Anker. Zu regeln ist das Auflockerungsverhalten durch Nachankern und durch schnelleres Einbringen eines frühtragenden Versatzes -und- falls das letztere nicht möglich ist, muß der Streb stillgesetzt werden, bis der Saumversatz vollständig eingebracht ist.

Für die Überwachung sind 2 Meßgebertypen notwendig:

Ein Längenmeßgeber zur Messung der Auflockerung an den Überwachungsankern. Diesen zeigt Abb. 6. Das Vorrücken der Strebfront messen wir mit einem Meßgeber, der in den letzten Schild des Strebregelausbaus eingebaut ist. Er hat den Namen "Schildausbau-Stellungsanzeiger". Er mißt den Fortschritt des Strebs mit zwei miteinander gekoppelten Potentiometern nach Meter und Zentimeter unter Verwendung eines verlorenen ablaufenden Seiles. Zusätzlich ist in dem Stellungsanzeiger auch ein Doppelneigungsgeber eingebaut. Er gibt die Neigung des Schildes in Streb- und Streckenrichtung an. Eine Übersichtszeichnung dieses Gebers ist in Abb. 7 beigefügt.

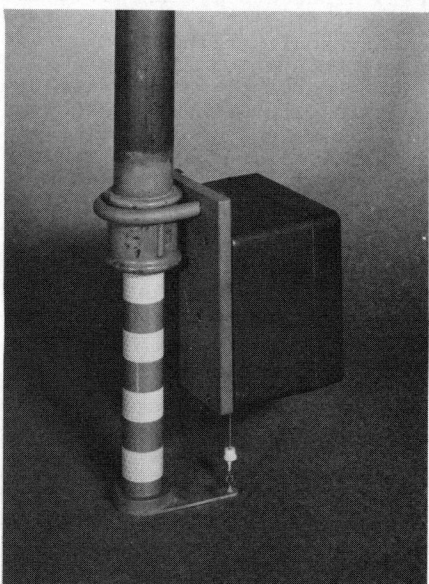

Abb. 6 Geber zur Messung der Auflockerung

Weitere Geber für andere physikalische Größen sind vorhanden oder in der Erprobung. Dazu gehören weitere Wegmeßgeber, Druck- und Temperaturgeber, Neigungsgeber, ein Staubmeßgerät, ein Geber zum Einbau in hydraulische Zylinder, um deren Stellung abzufragen, Kraftmeßdosen, usw.

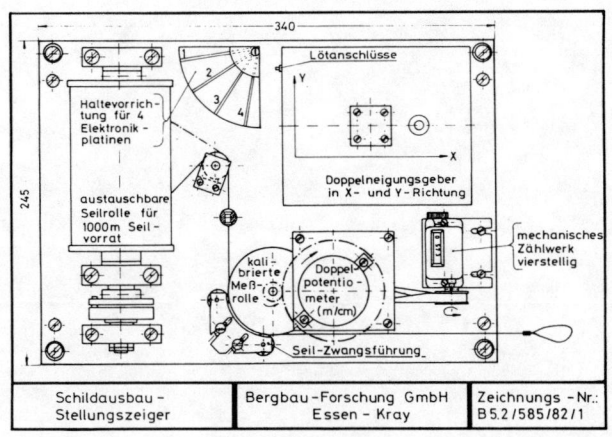

Abb. 7 Geber zur Messung des Abbaufortschrittes

4. Steuerungstechnik

Ebenso wie der Multiplexer von übertage her über weite Entfernungen über den TF-Kanal A (Abb. 2) ferngesteuert wird, können mehrere Maschinen, Vorgänge oder Regelkreise mit nur einem weiteren TF-Kanal ferngesteuert werden. Dazu sind die vom Multiplexer abzufragenden Geber und die diesem zuzuordnenden Schaltempfänger, wie Abbildung 8 zeigt, parallel zu schalten. Nur wenn der Meßgeber mit seinem Meßsignal auf dem Prozessor erscheint, hat der Schaltempfänger Strom. Geschaltet wird er durch ein Signal über den Kanal B. Dann zeigt der Meßgeber sofort das Einsetzen der Wirkung an. Die Steuerung erfolgt also kontrolliert.

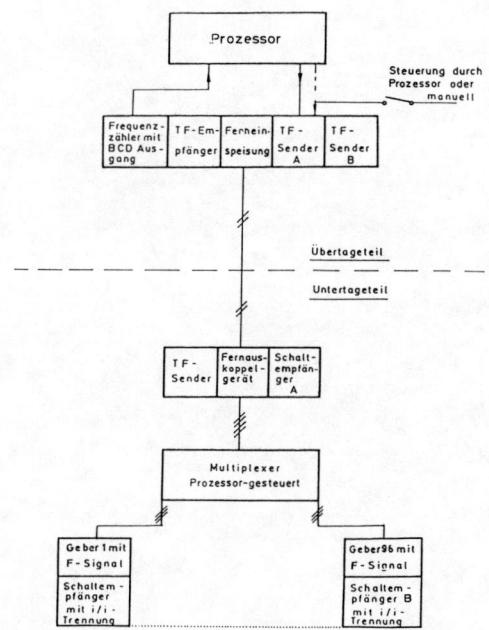

Abb. 8 Prozessorgesteuerte Vielstellen-Meß- und Steuerungstechnik

5. Zusammenfassung

Trotz zunehmender Zahl von Meßgebern und Meßaufgaben konnte die vorhandene Tonfrequenz-Multiplextechnik durch höhere Ausnutzung mit Hilfe eines Multiplexers an einem Ende und durch Anschluß intelligenter Prozessoren und Meßwertverarbeitungsgeräte am anderen Ende des TF-Netzes nicht nur verbessert, sondern auch entscheidend verbilligt werden.

Die gesamte Überwachung und Steuerung erfolgt durch ein 2-adriges Kabel über große Entfernungen. Es läßt eine Mehrfachausnutzung jedes Kanals für meßtechnische und ebenso für Steuerzwecke zu. Außerdem wird über dieses Kabel -wie bisher- telefoniert.

NEW DESIGN APPROACH FOR ROOM-AND-PILLAR COAL MINES IN THE U.S.A.

Une nouvelle méthode de dimensionnement pour abattage par chambres et dépilage dans les houillères aux Etats Unis

Eine neue Dimensionierungsmethode für Kammer- und Pfeilerbau im amerikanischen Kohlenbergbau

Z. T. Bieniawski
Professor of Mineral Engineering and Director Mining and Mineral Resources Research Institute, The Pennsylvania State University, University Park, Pennsylvania, U.S.A.

SYNOPSIS

A detailed study has been made of the U.S. practice and design needs for room-and-pillar coal mining. Although some 90% of underground coal mining in the United States is by the room-and-pillar method, no comprehensive design procedure is available for this purpose. Following a survey of over 200 coal mines, typical mining conditions and room-and-pillar configurations were analyzed during a three-year research project. Specific investigations were conducted to improve the methods of span selection, roof support and pillar design in U.S. coal mines. In particular, pillar strength formulae for coal mining were studied. Based on the results obtained, an improved procedure for room-and-pillar coal mining in the United States was proposed.

RESUME

La pratique américaine d'abattage par chambres et dépilage a été étudiée en détail en ce qui concerne la nécessité de développer des méthodes de dimensionnement. Bien que 90% de l'extraction souterraine de charbon aux Etats Unis soient sous forme d'abattage par chambres et dépilage aucun procédé général de dimensionnement n'existe. Dans le cadre d'un projet de recherche (d'une durée de trois années) plus de 200 houillères ont été inventoriées pour analyser les conditions typiques de minage et les configurations des chambres et piliers. Des recherches détaillées ont été effectuées pour améliorer les méthodes de dimensionnement pour les portées des chambres, pour le soutènement du toit et pour la grandeur des piliers. En particulier les "formules de résistance" pour les piliers ont été examinées. Les résultats de ces études permettent de proposer dès maintenant un procédé amélioré pour le dimensionnement de l'abattage par chambres et dépilage.

ZUSAMMENFASSUNG

Die Praxis des Kammer- und Pfeilerbaus im amerikanischen Kohlenbergbau wurde im Detail untersucht, vor allem bezüglich notwendiger Entwicklungen von Dimensionierungsmethoden. Obschon Raum- und Pfeilerbau ca. 90% des bergmännischen Kohlenabbaus in den USA ausmachen, besteht kein umfassendes Entwurfsverfahren für diese Abbaumethode. Im Rahmen eines dreijährigen Forschungsprojektes wurden deshalb Aufnahmen in über 200 Kohlengruben gemacht, und typische Abbauverhältnisse und Raum- und Pfeilerabmessungen aufgezeichnet. Spezifische Untersuchungen wurden durchgeführt, um Dimensionierungsmethoden für Kammer-Tragweiten, für Firstenausbau und für Pfeilerabmessungen zu verbessern. Vor allem wurden die "Pfeiler-Widerstandsformeln" untersucht. Die Ergebnisse dieser Forschung ermöglichen es nun, ein verbessertes Verfahren für Raum- und Pfeilerbau vorzuschlagen.

1.0 INTRODUCTION

The room-and-pillar mining method is used in the United States in some 90% of underground coal mining. Yet, systematic design guidelines are lacking both for roof control and pillar sizing. Since this method of mining will continue to be prominent in U.S. coal mining in the next two decades, in spite of an interest in longwall mining, it is obvious that rational room-and-pillar design guidelines are needed.

Detailed studies conducted at Penn State during the last few years have revealed that enough scientific and practical data are available to enable consolidating all this information together for the purpose of a design code for room-and-pillar mining. The aim of such a code would be to facilitate selection of roof spans in the coal mines and the sizing of coal pillars in accordance with the coal mining conditions in the United States. This means that the established mining practice and experience would be utilized in preparing the design guidelines. In addition, the objective would be to incorporate as much up-to-date scientific knowledge as is compatible with practical applications in coal mining to facilitate improved planning of new mines as well as optimizing the mining operations already in progress.

1.1. Room-and-Pillar Mining in the U.S.A.

The room-and-pillar method of coal mining as used in the United States involves a system known as retreat mining. In this method, the rooms (openings) and pillars are developed as mining advances but subsequently the pillars are recovered "on retreat" which can result in a high percentage of coal extraction, e.g., 70%.

However, room-and-pillar retreat mining techniques vary widely in the United States, depending on specific local conditions. The average coal extraction in the U.S. is just over 50% but the average percentage of mines practicing pillar extraction is only 26%, although in some areas (western Pennsylvania and northern West Virginia) this percentage is 60% to 70% (Kauffman et al., 1981). Clearly, there is much scope for improvement.

While pillar stability is of much importance in the room-and-pillar mining method, the stability of mine roofs is a crucial aspect of strata control in the United States. Hence, selection of roof spans and their reinforcement must be addressed in any design code. This may not be the situation in other countries using the room-and-pillar method. There, the pillars receive the main attention as they are generally not recovered on retreat but left permanently in place.

2.0 SURVEY OF U.S. MINING CONDITIONS

In 1979, the Pennsylvania State University initiated a national survey of room-and-pillar design practice and mining dimensions in the United States. The survey included a comprehensive study of such parameters as the depth below surface, seam thickness, roof spans, pillar height, pillar width, pillar length, width-to-height ratios, percentage extraction and method of design. A total of 174 cases were available for stability analyses of coal pillars (see Table 1) plus 58 cases of roof falls for analyzing roof spans. The pillar cases included only only three instances of pillar failure from the United States. However, 20 pillar failure cases were collected by the author from other countries. Nevertheless, it should be noted that pillar failure cases were not essential for establishing an improved pillar design for U.S. mining, contrary to an objective of a study by Salamon and Munro (1967) in South Africa. There, no pillar strength formula was available at the time while in the USA a number of formulae have been used since 1911, and it was the aim of this study to select one that would be the most economical yet safe for coal pillar design.

Table 1. Range of Room-and-Pillar Parameters Surveyed in U.S. Coal Mines (Penn State Survey of 174 cases)

	Range	Typical
Depth below surface	25-480 m	150 m
Seam or pillar height	1-4.5 m	2 m
Entry width	3.7-8.2 m	4.8-6.1 m
Pillar width	5-23 m	15 m
Pillar length	6.1-27 m	20 m
Width-to-height ratio	2-16	8.0
Length-to-width ratio	1-3	1.25
Percentage extraction	25%-85%	50%

The results obtained are depicted in Figure 1. The survey involved the coal seams commonly found in the United States. However, 38% of the data involved the Pittsburgh seam. As will be seen from Figure 1, the average depth of room-and-pillar mines in the United States is 150 m, with the range varying between 25 and 480 m. The typical seam thickness, which is synonymous with pillar height, is 2 m varying between 1.0 and 4.5 m. Roof spans, called entry widths in the U.S., are typically 4-6 m wide, varying from 3.6 to 8.2 m. The typical pillar width is 15 m varying from 5 to 23 m while the typical pillar length is 20 m. The average percentage extraction is 50% varying between 25 to 85 percent. An important observation from Figure 1 is that the typical pillar width-to-height ratio is 8.0 varying from 2 to 16.

The data base from the survey will be used in a later section to establish the safety factors for U.S. coal mining.

3.0 SELECTION OF ROOF SPANS

It is proposed that estimating safe roof spans in U.S. coal mines be achieved by means of an engineering rock mass classification system, specifically the Geomechanics Classification, as modified for coal mining applications. Reliance on beam theories is not recommended due to uncertainties in the assumptions as well as in the required input data (e.g., 'modulus of rupture').

Rock classifications are well known empirical methods for assessing the stability of underground openings in rock. Developed primarily for the purposes of tunnel and chamber design in civil engineering, they have received increasing attention in recent years as a means of roof stability appraisal. In mining, a detailed study of the application of rock classifications to U.S. hard rock mining was initiated by the Bureau of Mines in 1980 while in coal mining, Penn State researchers have been studying the rock classification approach systematically since 1978. As a result, 58 case histories of roof falls have been compiled, enabling estimation of safe roof spans in coal mines by means of the Geomechanics Classification (also known as the Rock Mass Rating, or RMR, System).

Full details of the step-by-step procedure for using the Geomechanics Classification are given elsewhere (Bieniawski, 1980), but reference should be made to Figure 2 which is the output from the Geomechanics Classification. This figure gives a relationship between the stand-up time and the roof span for various rock mass ratings. For example, a roof strata rating of RMR = 35 will mean that the maximum span possible in this rock is 5 m as read off from the interception of rating line 35 with the top line in Figure 2. On the other hand, if the roof span is only 1.5 m, it should last unsupported indefinitely (read off from the interception of rating line of 35 with the bottom line in Figure 2).

If the mining practice requires a roof span of, say, 4.0 m in the rock strata of RMR = 35, then it will be seen from Figure 2 that for a span of 4.0 m and a rating of 35, the stand-up time is about 8 hours. This means that such a span could be left completely unsupported for 8 hours but it will collapse after that period of time. What can be done to ensure long-term stability? The answer lies in appropriate support measures.

The Geomechanics Classification provides guidelines for the selection of roof support to ensure long-term stability of various rock mass classes. These guidelines are given elsewhere (Unal, 1982). It is recognized that by comparison with tunneling different objectives must be considered in coal mining. Hence, for coal mines, appropriate factors such as use and life of entries as well as the effect of intersections must be considered for support measures.

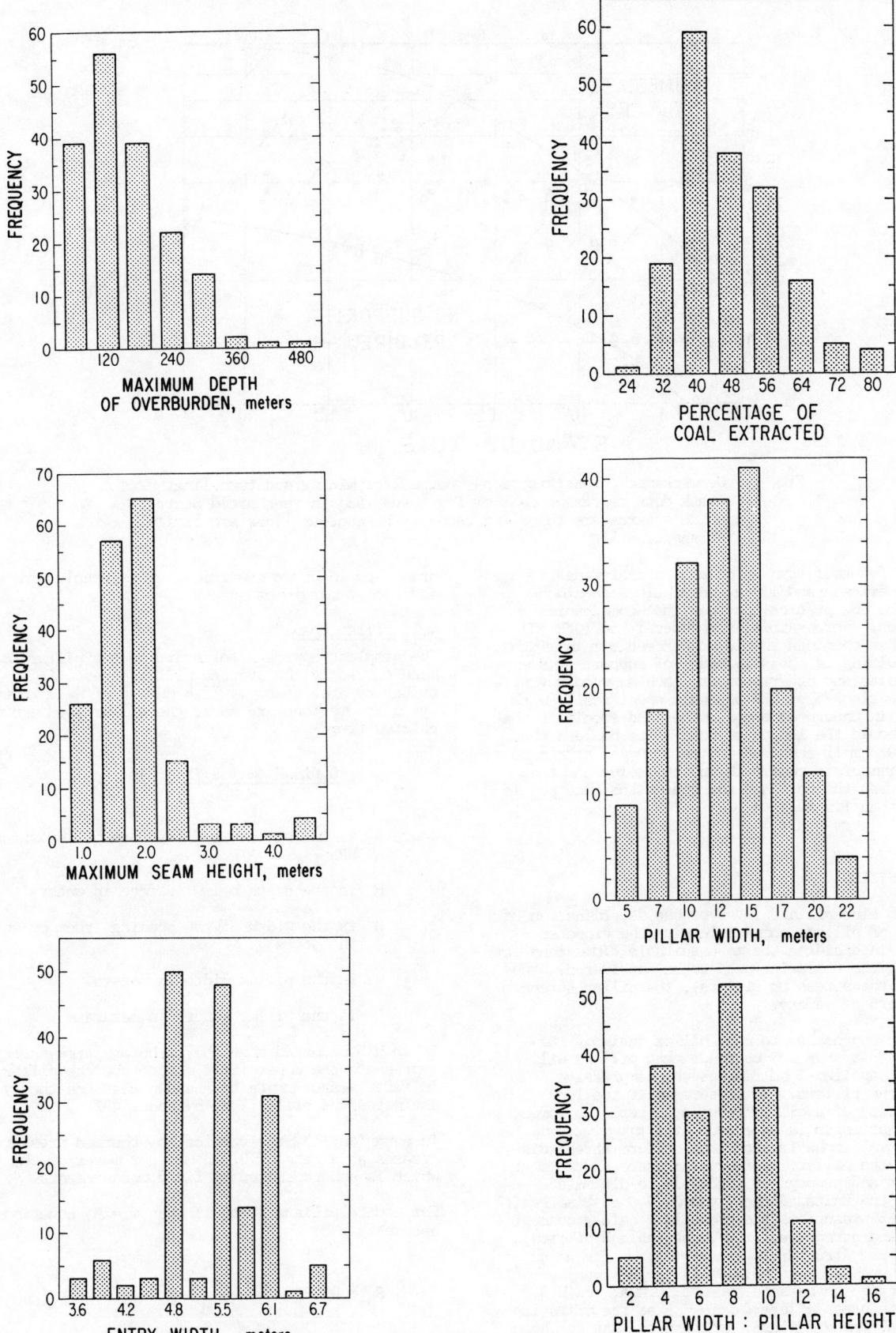

Figure 1. Penn State Survey of Room and Pillar Parameters in U.S. Coal Mining

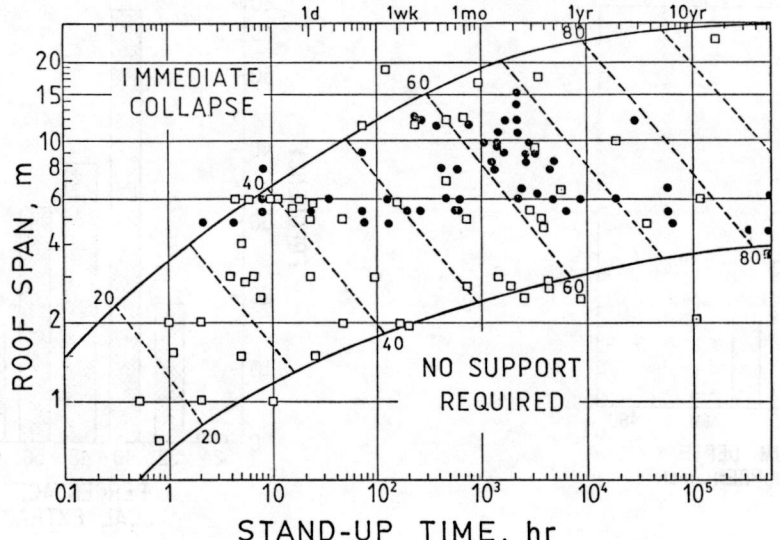

Fig. 2. Geomechanics Classification - output for mining and tunneling. Black dots represent coal mining cases (58) in the United States. White squares are tunneling cases. The contour lines are limits of applicability.

Furthermore, for roof span selection in coal mines, practical experience and mining regulations should be used to assess the predictions from the Geomechanics Classification. For example, the Federal Law (CFR 30) stipulates that openings should not exceed 6 m in width where roof bolting is the sole means of support or 9 m where roof bolts and other support, such as wood posts, are used. Section 75 of CFR 30 deals specifically with roof control in underground coal mines and specifies that in no case should the length of roof bolts be less than 0.75 m plus 0.3 m if anchored in the stronger strata to suspend the immediate roof. The bolt spacing and the distance between the bolt and the rib or the face should not be more than 1.5 m.

4.0 PILLAR SIZING

The design of mine pillars involves the determination of proper sizes of pillars compatible with the expected load. Thus, in deciding the most suitable dimensions for mine pillars, the following need to be considered: the pillar load (the stress on pillars), the pillar strength and the factors of safety.

There are two approaches to coal pillar design. *The ultimate strength approach* contends that pillars will fail when the applied load reaches the compressive strength of the pillars. It presumes that the load-bearing capacity of a pillar reduces to zero the moment its ultimate strength is exceeded. *The progressive failure approach* emphasizes the non-uniform stress distribution in the pillar. Failure initiates at the most crucial point and propagates gradually to ultimate failure. In the United States, the *1969 Coal Mine Health and Safety Act* prescribes the prevailing safety concept for underground structures (i.e., entries and pillars), namely, that *the structural elements (roof, floor and ribs) of the opening must be kept in perfect or nearly perfect condition to be safe for use*. Thus, the U.S. safety concept uses *failure initiation* as the criterion rather than ultimate failure, contrary to what has been practiced in other countries. This results in an additional margin of safety.

4.1 Pillar Load

The simplest approach to determine the pillar load, or more correctly the average pillar stress, is by the tributary area theory. If a number of well known simplifying assumptions are made, the pillar load can be calculated from:

$$S_p = \frac{0.025\ H\ (w + B)(L + B)}{w \times L} \qquad (1)$$

where S_p is the pillar load (average pillar stress) in MPa

H is the depth below surface in meters

B is the width of the opening (room or entry) in meters

w is the pillar width in meters

L is the pillar length in meters

It should be noted that the tributary area theory represents the upper limit of the average pillar stress. In fact, measurements have shown that the theory overestimates the pillar load by about 40%.

In equation (1), the vertical overburden pressure increases at a rate of 0.025 MPa per meter of the depth which is substantiated by field measurements.

For square pillars, that is when w = L, equation (1) becomes:

$$S_p = 0.025\ H \left[\frac{w + B}{w}\right]^2 \qquad (2)$$

If the term "extraction" e is introduced (100 e is percentage extraction) which is defined as the ratio of the mined out area to total area, then for square pillars:

$$e = 1 - \left[\frac{w}{w+B}\right]^2 \quad (3)$$

4.2 Pillar Strength

The strength of coal pillars, that is the ultimate load per unit area, is dependent upon three elements: (i) the size or volume effect (strength reduction from a small laboratory specimen of coal to a full size coal pillar); (ii) the effect of pillar geometry (shape effect); and (iii) the properties of the coal material. Although for years many pillar strength formulae have been proposed, two types of expressions are predominant:

$$\sigma_p = \sigma_1 \left(A + B \frac{w}{h}\right) \quad (4)$$

and

$$\sigma_p = K \frac{w^\alpha}{h^\beta} \quad (5)$$

where σ_p is the pillar strength, σ_1 is the strength of a cubical pillar at the *critical* specimen size (scaled strength incorporating the size effect), K is a constant characteristic of a coal seam while α and β are constants expressing the shape effect, w is the pillar width and h is the pillar height (usually the same as the seam height). A and B are constants.

The Size Effect

The concept of the 'critical size' (Bieniawski, 1968) is very important in practical strata control engineering. It means that for cubical specimens of coal, the strength decreases with increasing specimen size until it becomes constant from a "critical specimen size" onwards. This is depicted in Figure 3. The significance of this phenomenon is, of course, that the strength values at the critical size (about 1.0 m for coal) are directly applicalbe to full size pillars.

The size effect characterizes the difference in the strength between the small size spcimens tested in the laboratory and the large size coal pillars mined in situ. Research in the United States has shown (Hustrulid, 1976) that the scaling of coal properties from laboratory measured data to field values can be satisfactorily achieved by the following equations:

$$\sigma_1 = \frac{k}{\sqrt{36}} \quad (6)$$

applicable to cubical pillars if the height h > 36 inches (0.9 m), or

$$\sigma_1 = \frac{k}{\sqrt{h}} \quad (7)$$

applicable to cubical pillars if the height h is less than 36 inches (0.9 m).

In the above equations, the constant k must be determined for each coal seam and is obtained as shown by Gaddy (1956):

$$k = \sigma_c \sqrt{D} \quad (8)$$

where σ_c is the uniaxial compressive strength of coal specimens tested in the laboratory and having a diameter or cube side dimension D (in inches).

It should be noted that although there is a difference in laboratory results depending on whether cylindrical or cubical specimens are used, for practical engineering pruposes this difference is not significant within the range of D between 2 to 4 in (50 mm to 100 mm). (Hustrulid 1976).

Although the strength of coal at the critical size is most conveniently determined by the Hustrulid equations, a number of other approaches can also be used. These are: in situ large scale tests, "petite sismique" geophysical tests, the Protodyakonov method or the Hoek-Brown criterion for rock mass strength (Bieniawski, 1982).

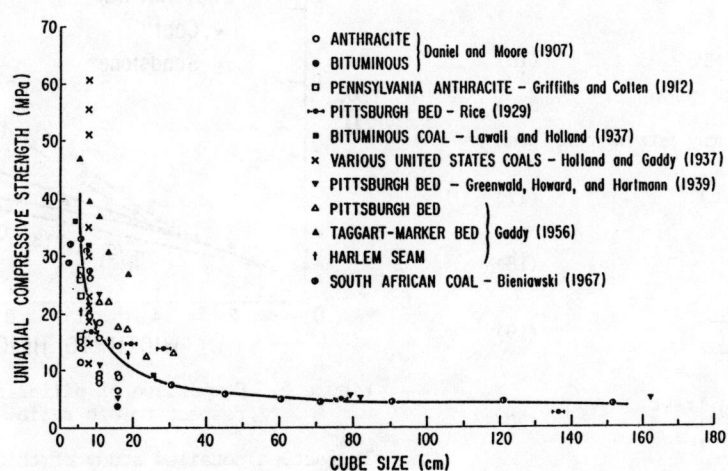

Figure 3. Specimen size effects in coal (Singh, 1981).

E 31

The Shape Effect

From all the available pillar strength formulae, the following five expressions are relevant for applications to U.S. room-and-pillar coal mining:

1. $\sigma_p = \sigma_1 (0.778 + 0.222\, w/h)$ [Obert & Duvall, 1967]   (9)

2. $\sigma_p = \sigma_1 \sqrt{w/h}$ [Holland, 1973]   (10)

3. $\sigma_p = \dfrac{k\sqrt{w}}{h}$ [Holland & Gaddy, 1964]   (11)

4. $\sigma_p = 1320 \dfrac{w^{0.46}}{h^{0.66}}$ [Salamon & Munro, 1967]   (12)

5. $\sigma_p = \sigma_1 (0.64 + 0.36\, w/h)$ [Bieniawski, 1969, 1977]   (13)

All these formulas have been applied extensively in coal mining. Using the Pittsburgh coal seam as an example, the above formulae may be compared. From the research by Gaddy (1956), Hustrulid (1976) and Bieniawski (1982), the following value is characteristic of the Pittsburgh coal:

$$k = 5580 \qquad (14)$$

which will be used for equation (11).

Thus

$$\sigma_1 = \frac{k}{\sqrt{36}} = 930 \text{ psi} = 6.4 \text{ MPa} \qquad (15)$$

The above value will be used for equations (9), (10), and (13).

For equation (12), the constant 1320 represents the strength in lb/sq in. of a one-foot cube specimen of coal determined in South Africa. For U.S. conditions the appropriate constant may be determined by the Hustrulid equation (7).

Thus, for Pittsburgh coal:

$$K = \frac{k}{\sqrt{12}} = 1610 \text{ psi} = 11.1 \text{ MPa} \qquad (16)$$

Consequently, the five formulae may now be rewritten as:

1. $\sigma_p = 930\,(0.778 + 0.222\, w/h)$ (17)

2. $\sigma_p = 930\,\sqrt{w/h}$ (18)

3. $\sigma_p = \dfrac{5580\sqrt{w}}{h} = \dfrac{5580}{\sqrt{h}}\sqrt{\dfrac{w}{h}}$ (19)

4. $\sigma_p = 1610\,\dfrac{w^{0.46}}{h^{0.66}} = \dfrac{1610}{h^{0.2}}\left[\dfrac{w}{h}\right]^{0.46}$ (20)

5. $\sigma_p = 930\,(0.64 + 0.36\, w/h)$ (21)

In Figure 4 the above five formulae are plotted as the strength versus the width-to-height ratio. It is apparent that for higher width-to-height ratios equation (19) predicts the lowest strength while equation (21) predicts the highest strength. At the same time, the form of equation (18) is such that it will become very conservative at high width-to-height ratios. The higher strength values predicted by equation (21) are consistent with the fact that for high width-to-height ratios there is a very rapid strength increase as shown by the experimental data plotted in Figure 5. In fact, pillars are thought to be almost indestructible for width-to-height ratios greater than 10 (Cook, 1978). This is further substantiated by the PSU survey of U.S. coal mines (described in a previous section) from which it is apparent that pillars are generally overdesigned in the U.S. for width-to-height ratios of five and over.

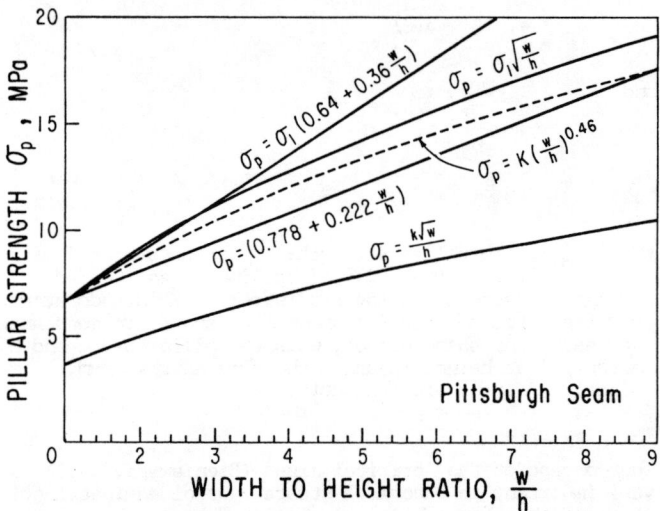

Figure 4. Pittsburgh seam: strength vs w/h ratio of pillars.

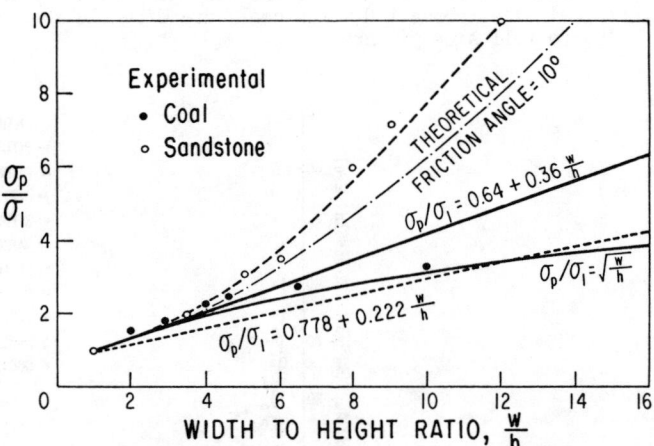

Figure 5. Comparison of pillar strength formulae with respect to w/h ratio.

In fact, a detailed study of this aspect conducted at Penn State has revealed that the theoretical strength (Bauer, 1980) of coal pillars is considerably higher than that predicted by equation (21). Bauer calculates the pillar strength based on rock mass properties such as: the cohesion and the angle of internal friction of coal, the cohesion and the angle of friction at the pillar-roof interface as well as the pillar dimensions of width, length and height.

It should nevertheless be noted that the results of underground tests on coal pillars performed by Wagner (1974) have shown that pillars of rectangular cross-section are about 40 percent stronger than square pillars of the same width and height. He measured the stress distribution across pillars at various stages of deformation and found that the perimeter of a pillar is capable of carrying relatively little stress, but this portion of the pillar provides lateral confinement which enhances the strength at the center of the pillar. He also concluded that long rectangular pillars having a width greater than 10 times their height are unlikely to fail except for punching into the roof or the floor.

Furthermore, it is important to observe that all of the five pillar strength formulae plotted in Figure 4 make use of the tributary area theory (for determination of factors of safety) which overestimates the pillar load (stress on pillars) by some 40%.

It must be also emphasized that for scaling of the laboratory determined results to field values by equation (6), it is imperative that the k values, which represent the properties of coal, are determined for each mine locality, as there is much variation in the coal strength from various U.S. seams.

Therefore although equation (21) predicts the highest strength of the three approaches, it is believed that because of the strength reserves its use is justified on the following counts:

(i) the tributary area theory overestimates the pillar load;
(ii) rectangular pillars are considerably stronger than the square pillars of the same width
(iii) high width-to-height ratios lead to rapid strength increase.

4.3 Improved Pillar Design

Since the aim of this research was to provide an improved design for room-and-pillar coal mining, it is believed that the pillar strength equation (21) makes the greatest contribution in this respect. This is so because the formula provides the highest pillar strength by comparison with the other formulae which will lead to a greater recovery of coal and hence will result in an improved and more economical, yet safe, mining operation.

Thus, for an improved design of coal pillars in the United States, the following formula is recommended:

$$\sigma_p = \sigma_1 (0.64 + 0.36 \, w/h) \qquad (21)$$

where σ_1 = 930 psi (6.4 MPa) for Pittsburgh coal and, for general use,

$$\sigma_1 = \frac{k}{\sqrt{36}}$$

Nevertheless, other strength formulae should be used as a cross-check to enable the designer to exercise his judgment based on the estimated pillar sizes and his own observations in the mine.

It should be noted that equation (21) is valid for pillar sizes greater than the critical size (which is about 1 m for coal) and for width-to-height ratios equal to and greater than unity. Other formulae are available for w/h <1 and for smaller specimen sizes (Hustrulid, 1975; Bieniawski and Van Heerden, 1975). Furthermore, for width-to-height ratios of 10 and greater, equation (21) will underestimate the strength of pillars.

5.0 FACTORS OF SAFETY

A factor of safety is defined as the ratio of the strength of a pillar to the load acting on it. Thus:

$$F = \frac{\sigma_p}{S_p} \qquad (22)$$

A factor of safety is necessary because in order to derive the load and the strength formulae presented in the previous section, certain assumptions had to be made.

A structure is generally defined as stable when F > 1. In engineering practice the factor of safety used is smaller when the conditions are well explored and understood. It is important to note that if the factor of safety is unity, the probability of failure is 50%. The factor of safety should be greater than unity to achieve a low probability of failure. What probability of failure is acceptable in coal mining?

5.1 Selection of Safety Factors

Using the PSU computer storage and retrieval system, the room-and-pillar dimensions from the PSU survey were used together with the various pillar strength formulae. A histogram of safety factors for the strength formula in equation (21) is given in Figure 6 for 171 case histories featuring stable pillars from the United States and 20 case histories involving failed pillars in other countries.

It is concluded from Figure 6 that factors of safety ranging from 1.5 to 2.0 are appropriate for U.S. coal mining conditions when using equation (21). It must be emphasized, however, that this range should be used as a guide only and the local mining experience should be taken into consideration.

6.0 DESIGN PROCEDURE

The following step-by-step procedure is recommended:

Step 1 - From geological data, borehole logs and rock and coal specimen testing (50 mm core or cubes) tabulate the following: uniaxial compressive strength of roof rock and coal, σ_c; spacing, condition and orientation of geological discontinuities; rock quality designation (RQD); and groundwater conditions.

Step 2 - Determine Rock Mass Rating (RMR) for roof rock in accordance with the published procedures for the Geomechanics Classification (Bieniawski 1980).

Step 3 - Select the roof span B from Figure 2 and decide on roof support.

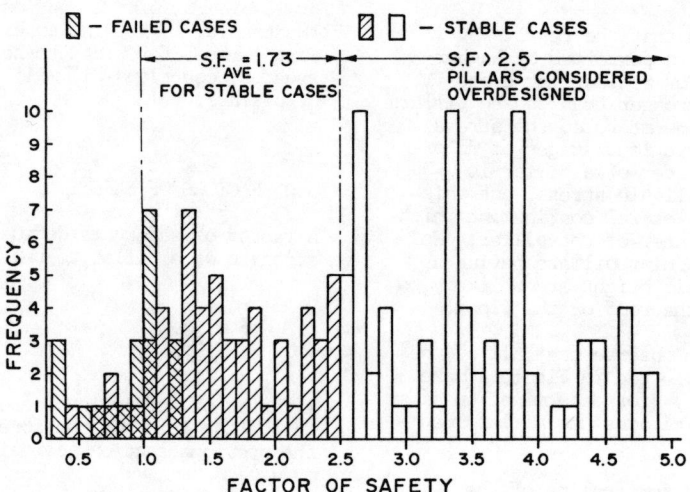

Figure 6. Histogram of safety factors for the strength formula given by equation (21).

Step 4 - Based on the uniaxial compressive strength of coal σ_c, determined either from 50-75 mm core or 50-75 mm cubes (both acceptable for practical purposes), determine the value of k for the mine locality:

$$k = \sigma_c \sqrt{D}$$

where D is the specimen diameter or cube size.

Step 5 - Determine the strength of cubical coal pillars at the critical size (no further size effect):

$$\sigma_1 = \frac{k}{\sqrt{36}} \quad \text{Formula 1}$$

if the seam height h is equal to or more than 0.9 m or

$$\sigma_1 = \frac{k}{\sqrt{h}} \quad \text{Formula 2}$$

if the seam height h is less than 0.9.

Step 6 - Write down the pillar strength formula which will be used to estimate the pillar width w, for a known seam height h:

$$\sigma_p = \sigma_1 (0.64 + 0.36 \, w/h) \quad \text{Strength Formula}$$

Note that the unknowns in this formula are the pillar strength σ_p and the pillar width w. To simplify the design procedure, assume that the pillars are of a square cross-section which will result in an underestimated pillar strength as shown in the previous section. This can be compensated for by using a smaller factor safety.

Step 7 - Write down the equation for the pillar load (the stress on the pillar) based on the tributary area theory:

$$S_p = 0.025 \, H \left[\frac{w + B}{w}\right]^2 \quad \ldots \quad \text{Load Formula}$$

where S_p is the pillar load in MPa, H is the depth below surface, B is the entry span from Step 3 and w is the pillar width. Note that the unknowns in the above formula are the pillar load S_p and the pillar width w.

Step 8 - From economic considerations, decide what percentage extraction (100e) is needed for a profitable mining venture. From the equation:

$$e = 1 - \left[\frac{w}{w + B}\right]^2 \quad \ldots \quad \text{Extraction Ratio}$$

select a pillar width w which would give the required coal extraction.

Step 9 - Test the selected pillar width w to determine whether it is acceptable from the mine stability point of view or whether the pillar width needs to be changed. This requires calculation of the factor of safety:

$$F = \sigma_p/S_p \quad \ldots \ldots \quad \text{Factor of Safety}$$

where σ_p is the pillar strength from Step 6 while S_p is the pillar load from Step 7.

Step 10 - The safety factors should be between 1.5 and 2.0. Cross-check the results by using other formulae. Exercise engineering judgment by considering a range of parameters, to assess the various options for mine planning.

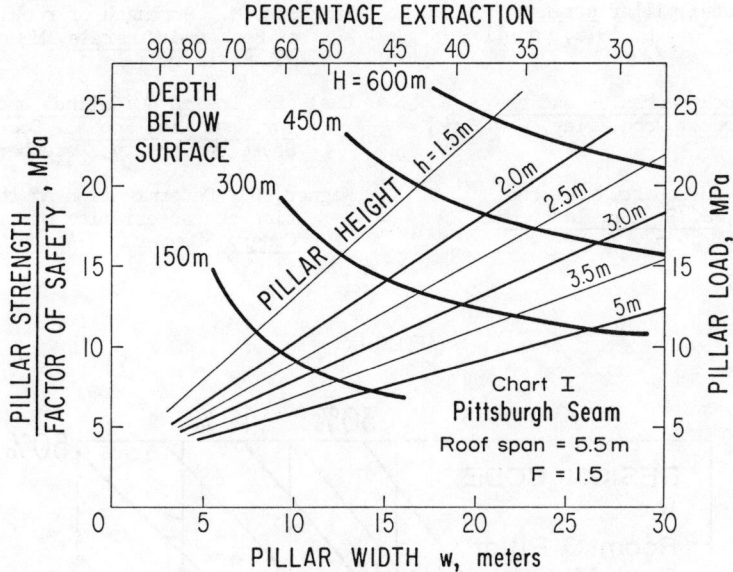

Figure 7. Pittsburgh seam: pillar strength vs pillar width for different pillar heights and different depths. (Span 5.5 m; PSU strength formula.)

In order to assess the stability of an existing mining operation, all the mining dimensions are known and the factor of safety can be determined for optimizing future design.

To facilitate this process, design Chart I is used, as given in Figure 7. Alternatively, design Chart II in the Appendix may be utilized.

7.0 CONCLUSION

A design approach for improved room-and-pillar coal mining has been developed and shown to be both economical and safe. Although valid for use in other countries, the proposed design procedure is particularly relevant to the United States where about 90% of underground coal mining is by the room-and-pillar method.

ACKNOWLEDGEMENTS

This paper was prepared with the support of the United States Department of Energy, Office of Advanced Research and Technology, under Grant No. DE-FG-01-78ET-11428, Distinguished Scientist/Engineer Grant Program. However, the opinion, findings, conclusions and recommendations expressed herein are those of the author and do not necessarily reflect the views of the Department of Energy or of the U.S. Government.

A Co-Director of this project was the late Dr. Robert Stefanko, professor of mining engineering. Dr. George Luxbacher, research assistant, was responsible for the analyses of the coal mine survey data. Three graduate assistants were also involved: Robert Belesky (coal pillar stability and 'petite sismique' measurements), Farzan Rafia (assessment of mine roof spans by the Geomechanics Classification) and David A. Newman (engineering geological mapping, integral sampling of floor strata and rock stress measurements). Drs. J. Bauer and D. Krzyszton, visiting post-doctoral fellows assisted with analytical studies.

REFERENCES

Bauer, J. A limit stress theory. Internal Report on Strata Control, The Pennsylvania State University, 1980.

Bieniawski, Z. T., Rafia, F. and Newman, D. A. Ground control investigations for assessment of roof conditions in coal mines. Proc. 21st Symp. on Rock Mech., Univ. of Missouri, Rolla, 1980, pp. 691-800.

Bieniawski, Z. T. Improved Design of Room-and-Pillar Coal Mining, Final Report, U.S. Department of Energy, Grant No. DE-FG01-78ET-11428, June 1982, 164 pp.

Bieniawski, Z. T. Discussion of Pillar Strength Formulas, Rock Mechanics, Vol. 10, 1977, pp. 107-110.

Bieniawski, Z. T. and van Heerden, W. L. The significance of in situ tests on large rock specimens. Int. J. of Rock Mech. & Min. Sci., Vol. 12, 1975, pp. 101-113.

Bieniawski, Z. T. In situ large scale testing of coal. Proc. Conf. In Situ Investigations, British Geotech. Soc., London, 1969, pp. 67-74.

Cook, N. G. W. and Hood, M. The stability of underground coal mine workings. Proc. Int. Symp. on Stability in Coal Mining, Vancouver, Canada, 1978, pp. 135-147.

Gaddy, F. L. A study of the ultimate strength of coal. Bulletin, Series, No. 112, August 1956, pp. 1-27.

Holland, C. T. The strength of coal in mine pillars. Proc. of 6th Symp. on Rock Mech., Univ. of Missouri, Rolla, 1964, pp. 450-466.

Holland, C. T. Pillar design for support of the overburden in coal mines. Proc. 9th Can. Symp. Rock Mech., McGill Univ., Montreal, 1973, pp. 114-139.

Hustrulid, W. A. A review of coal pillar strength formulas. Rock Mechanics, Vol. 8, 1976, pp. 115-145.

Obert, L. and Duvall, W. I. Rock Mechanics and the Design of Structures in Rock. John Wiley, New York, 1967, pp. 542-545.

Salamon, M. D. G. and Munro, A. H. A study of the strength of coal pillars. J. S. Afr. Inst. Min. Metall., Vol. 68, 1967, pp. 55-67.

Singh, M. M. Strength of rock. In: Physical Properties of Rock and Minerals, McGraw-Hill, New York, 1981, pp. 83-121.

Unal, E. Design Guidelines and Roof Control Standards for Coal Mine Roofs. Doctoral Thesis, Pennsylvania State University, December, 1982.

Wagner, H. Determination of the complete load deformation characteristics of coal pillars. Proc. 3rd Congr., Denver, 1974, Vol. IIB, pp. 1076-81.

APPENDIX

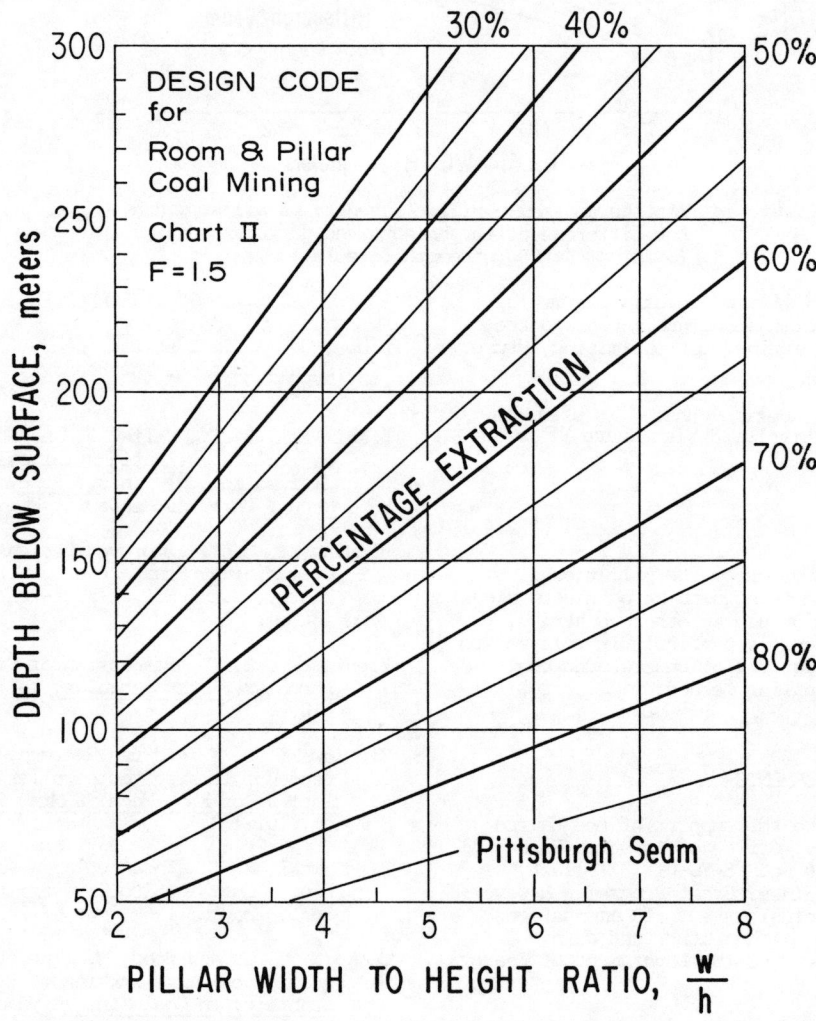

STABILITY EVALUATION OF RETREATING LONGWALL CHAIN PILLARS WITH REGRESSIVE INTEGRITY FACTORS

Evaluation de la stabilité des chaînes-piliers dans les longwalls au moyen de facteurs d'intégrité régressifs

Bewertung der Stabilität von Strebkettenpfeilern mittels regressiver Integritätsfaktoren

Paul H. Lu, Ph.D.
Mining Engineer, U.S. Bureau of Mines, Department of the Interior Denver, Colorado, U.S.A.

SYNOPSIS

Profiles of mining-induced loading and residual strength across a pillar vary with the position of the longwall face. The residual pillar-strength profile can be constructed on the basis of laboratory-determined triaxial compressive strength, in which the in situ measured horizontal pressure is considered as the constraint. The vertical-loading profile can be established with the measured vertical pressures. Vertical and horizontal pillar pressures can be measured with hydraulic borehole pressure cells. Defined as the ratio of the integrated total strength to the integrated total load under the profiles, the integrity factor is proposed here as a rational parameter for evaluating chain-pillar stability.

RESUME

Les profils des charges causées par l'exploitation minière, ainsi que les profils des résistances residuelles au travers des piliers varient avec la position du "longwall". Le profil de résistance residuelle au travers des piliers peut être construit à partir d'essais en laboratoire sur les résistances triaxiales en compression; dans ces essais la pression horizontale mesurée sur place est considerée être la contrainte. Le profil des charges verticales peut être établi à partir des pressions verticales mesurées. La pression verticale et la pression horizontale des piliers peuvent être mesurées au moyen de cellules de pression hydraulique en sondage. Le facteur d'intégrité, défini comme le quotient de la résistance totale intégrée sur la charge totale intégrée sous les profiles, est le paramètre rationnel pour évaluer la stabilité des chaînes-piliers.

ZUSAMMENFASSUNG

Sowohl die Profile des Lastzustandes, der während Bergbauarbeiten auftritt, als auch die Profile der verbleibenden Restfestigkeit quer über die Pfeiler, hängen von der Position der Strebfront ab. Das Profil der Restfestigkeit des Pfeilers kann in Laborexperimenten aus Messungen der dreiachsigen Druckfestigkeit konstruiert werden, wo der in situ gemessene horizontale Druck als Zwang betrachtet werden kann. Das vertikale Lastzustandsprofil kann über die gemessenen vertikalen Drücke festgesetzt werden. Der vertikale Druck und der horizontale Druck in einem Pfeiler können mittels hydraulischer Bohrungdruckzellen gemessen werden. Es wird ein Integritätsfaktor der gesamten integrierten Festigkeit zur gesamten integrierten Last unter Profilen definiert. Dieser Integritätsfaktor erweist sich als ein rationeller Parameter für die Stabilität von Kettenpfeilern.

1.0 INTRODUCTION

Since the beginning of this century, a number of investigators (Bunting, 1911; Greenwald, 1939; Gaddy, 1956; Holland, 1964; Salamon, 1967; Bieniawski, 1968; and many others) have tried to predict the strength of coal-mine pillars on the basis of unconfined compression tests on small samples. Hustrulid (1976) categorized these proposed formulas into two types: 1) compressive strength-size relationship, and 2) compressive strength-shape relationship. These formulas, however, have not been widely accepted and applied in stability analysis and design of coal-mine pillars, particularly in longwall operations. A great deal of effort has been given to establishing a rational approach, and among many investigators, Wilson (1972, 1977) finally arrived at the concept of stress balance: "Any rise in stress in rib sides and pillars must be offset by an equivalent stress reduction across roadways and other areas of extraction, and vice versa" (Wilson, 1981). His concept, however, still invites questions on the assumptions that he made. Moreover, some existing pillar design formulas do not take into account the time element corresponding to the progress of mining. Therefore, they are inadequate for evaluating the stability of pillars while the face is advanced.

In this report a set of field measurement data is presented, that may be helpful in stability analysis and hopefully lead to a rational design of chain pillars in retreating-longwall operations. The in situ strength of a full-size pillar, which is estimated from the measured horizontal pillar-confining pressure, is used instead of the laboratory-tested strength of a small sample. The integrated total actual loading on the coal pillar is used for analysis instead of the predicted average loading over the entire pillar based on the tributary principle, or of the hypothetical loading postulated from the stress balance principle. Also, instead of assuming that the entire pillar is solid, both the yielded and unyielded portions of the pillar are considered. The integrity factor is used instead of the safety factor because, although the latter is conventionally used for permanent or semi-permanent structures, the former better fits temporary structures such as mine pillars.

This paper briefly describes the basic concept and in-mine observations and presents two case studies of chain pillars.

2.0 BASIC CONCEPT AND IN-MINE OBSERVATIONS

For stability evaluation of the active long chain pillars in retreating longwall workings, the integrity factor, which decreases as mining progresses, may be a rational parameter. Barron (1978) hinted that the "pillar integrity index", defined as the ratio of pillar core area to total pillar cross-sectional area, might be a useful quantitative estimate of the relative structural and, in turn, the bearing capacity of a pillar. He identified the core area on the basis of the "fracture index" across a pillar, which was derived from the air injection tests. However, Barron's approach is indirect and insufficient as a reliable means of evaluating pillar stability. Therefore in this paper the pillar integrity factor, I, defined as the ratio A_S/A_ℓ, where A_S (respectively A_ℓ) is the integrated total residual pillar strength (respectively integrated total pillar load), which is represented by the area under the residual pillar strength (respectively pillar load) profile as shown in Figure 1, is introduced. The variation of this factor is dependent on the position of the face relative to the pillar as shown in Figures 2 and 3. Then, $I > 1$ is taken as the stable condition and $I \leq 1$ as the unstable condition.

Mine pillars are subjected to triaxial compression. The pre-mining vertical and horizontal pillar pressures can be easily measured with hydraulic borehole pressure cells. "The magnitude of the biaxial ground pressures existing in a rock mass can be determined by pressure convergence tests with a combination of one cylindrical and two flat hydraulic pressure cells installed in a single hole drilled into that rock mass" (Lu, 1981). From this same test, the response ratio between cell pressure and ground pressure also can be determined. For coal, this ratio is approximately one-to-one. With the knowledge of this response ratio, the mining-induced load transfer to the chain pillars can be monitored with the cement pre-encapsulated borehole pressure cells installed in a pillar (Lu, 1982). With the instrumentation shown in Figs. 4A and 5A, both vertical and horizontal pressures existing at the same point can be measured. Based on these measurements, the vertical and horizontal loading profiles for the selected phases of panel extraction can be constructed as shown in Figures 4B,C and 5B,C. Stresses at pillar edges are assumed to be zero. The residual compressive pillar-strength profiles can then be constructed based on the residual horizontal-pressure profiles, by using the laboratory-determined triaxial compressive strength interpolated from the strength versus confining-pressure plot. By comparing the integrated total areas under the strength profile, A_S, and under the vertical pressure profile, A_ℓ, the integrity factor, I, for each selected stage of panel extraction can be obtained as

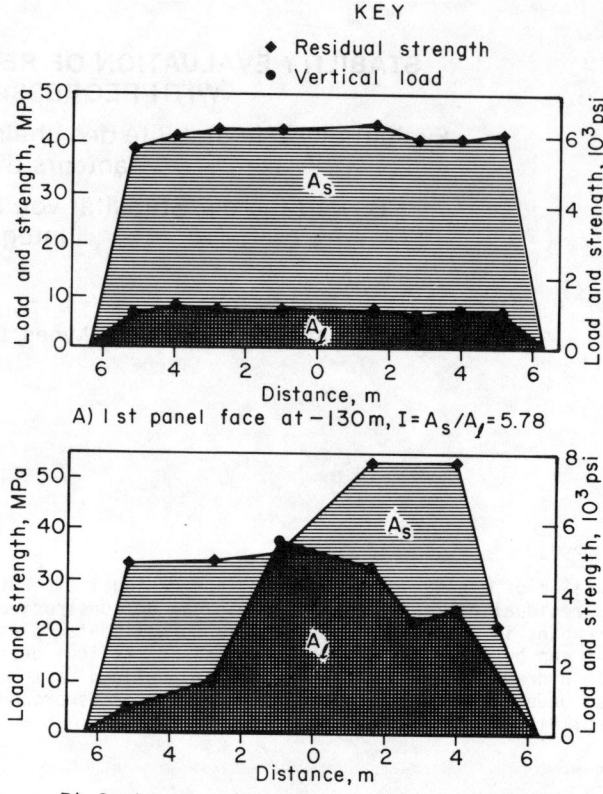

Figure 1 Profiles of progressive chain-pillar loading and residual strength (Mine A).

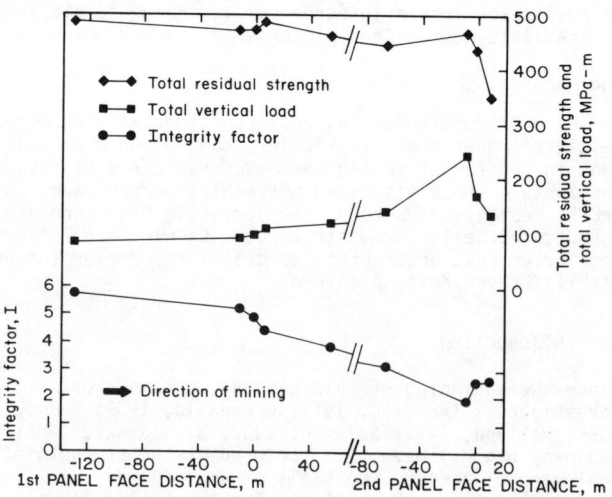

Figure 2 Total residual strength, total vertical load, and integrity factor versus face distance (Mine A).

shown in Figure 1. If these integrity factors are plotted against mining progress in terms of face distance, as shown in Figures 2 and 3, the plots will indicate regressive curves from the start of mining through the passage of the longwall face.

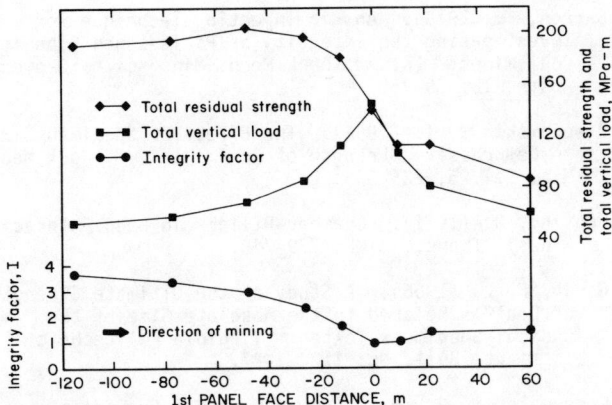

Figure 3 Total residual strength, total vertical load, and integrity factor versus face distance (Mine B).

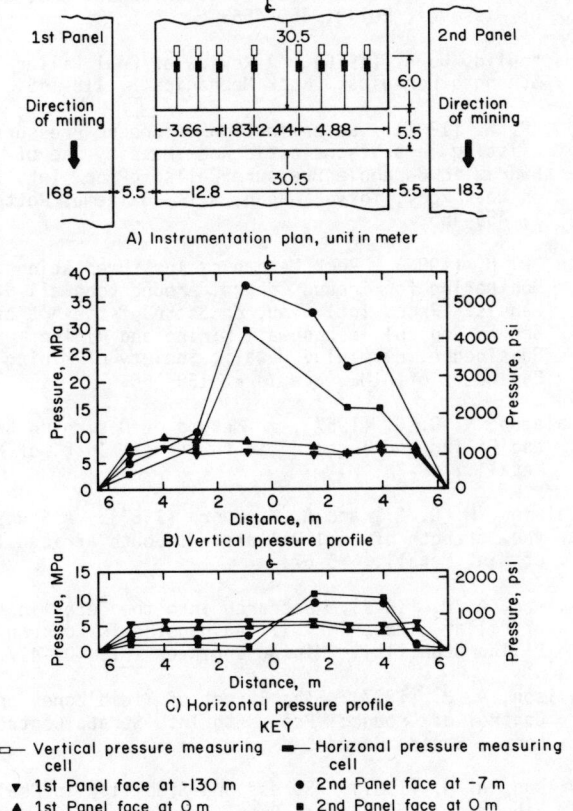

Figure 4 Progressive chain-pillar loading profiles induced by mining (Mine A).

3.0 CASE STUDIES

Case studies were conducted at two retreating-longwall mines. Both mines employed the two-entry system for their gate roads. The chain pillar studied at Mine A was situated under 140 m of cover, and that at Mine B under 442 m of cover. The extraction height at both mines was 2.4 meters. The pillar of Mine A was stable until the passage of the second longwall face, while the pillar ribs of Mine B had failed by the passage of the first longwall face.

3.1 Mine A

A single row of chain pillars, 30.5 m long by 12.8 m wide, was flanked by two adjacent panels. The widths of the 1st and 2nd panels are 168 and 183 m, respectively. A set of vertical-and horizontal-pressure measuring cells was installed in the same drillhole 6 m in length in one of the chain pillars as shown in Figure 4A, to measure the vertical and horizontal pressures at various positions across the pillar. Thus the pressure profiles at four selected stages of panel extraction --1st panel face at -130 m, 1st panel face at 0 m (passing), 2nd panel face at -7 m, and 2nd panel face at 0 m (passing)--are constructed as shown in Figures 4B and 4C. Negative face distance indicates the inby position and positive, the outby position.

The total residual-strength and total vertical-loading profiles across the pillar are integrated as shown in Figure 1 for each stage as mining progresses. Thus, the integrity factors at each stage are calculated and plotted in Figure 2. The integrity-factor plotted against the face distance is a curve with negative slopes. The I-value decreases gradually from 5.78 at -130 m face distance of the 1st panel to 1.90 at -7 m face distance of the 2nd panel. As indicated by Figure 2, in spite of decrease of the total pillar strength, the integrity factor improved after passage of the 2nd panel face due to a reduction in the vertical pillar-load because the gob started to take load in the caved area.

The fact that the lowest point on the I-curve was 1.90 reveals that the pillar was stable for the rest of its

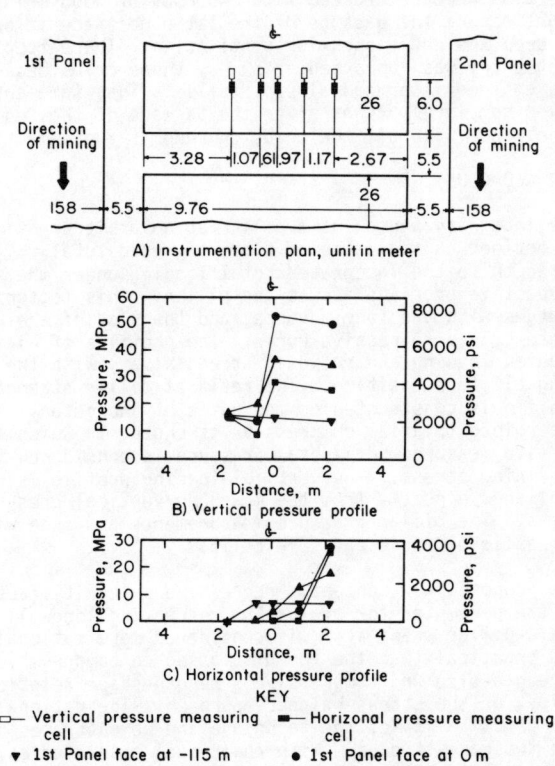

Figure 5 Progressive chain-pillar loading profiles induced by mining (Mine B).

life to protect the tailgate for ventilation, i.e., until completion of the 2nd panel extraction. This was verified by mine practice, in which two rows of wood cribs were supplemented to keep the tailgate entry open for ventilation only after the passage of the 2nd panel face. Before the passage of the 2nd panel face, merely the nearest 60 m interval of the tailgate entry was reinforced with steel beams and wooden posts for material transportation.

3.2 Mine B

In Mine B, a single row of chain pillars 260 m long by 9.76 m wide was flanked by two adjacent panels of the same width (158 m). Since these pillars were so narrow that their ribs had failed considerably by cracking and spalling prior to the 1st panel extraction, the pressure cells were installed only in the pillar core, as shown in Figure 5A, to evaluate the stability of the central portion, approximately 40 percent of the cross section of the entire pillar.

The profiles of both vertical and horizontal pressures at four selected stages of the 1st panel extraction--face at -115 m, face at -12 m, face at 0 m (face was passing), and face at 22 m--are shown in Figures 5B and 5C. Based on these pressure profiles, the strength- and vertical-loading-profiles diagrams were constructed and the integrity factors at each selected stage of the panel extraction were calculated. These pillar integrity factors are plotted in Figure 3 along with the total residual strengths and total vertical load.

As indicated in Figure 3, the integrity factor of pillar core decreased from 3.67 at -115 m face distance to 0.96 at 0 m face distance of the 1st panel. In practice, the tailgate entry supports for the 2nd panel extraction were reinforced with two rows of wood cribs right before the passage of the 1st panel face in order to keep the entry open for ventilation. Thereafter, that entry was supported mainly by those cribs because the entire pillar had already failed. That same entry caved completely right after the passage of the 2nd panel face with all the cribs crushed at the same time.

4.0 CONCLUSIONS

The integrity factor of a pillar at an arbitrary time is defined as the ratio of the integrated total strength to the integrated total loading under the respective profiles across the pillar. This factor changes with the face advance, and the plot of the factors is a regressive curve. The profiles of mining-induced loading and residual strength vary with the longwall face position. The residual pillar strength profile is constructed on the basis of laboratory-determined triaxial compressive strength, in which the in situ measured horizontal pressure is considered the confining stress. The vertical-loading profile is established directly from the measured vertical pressures. The pillar pressure measurements are made with hydraulic borehole pressure cells.

The proposed new concept, "regressive integrity factor as the parameter for evaluating retreating-longwall chain-pillar stability", is considered more rational and practical than the formulas based on compressive strength-size and compressive strength-shape relationships, or the stress-balance approach. The rationality and practicality of the technique are demonstrated by the two case studies. With additional case studies, a design criterion for retreating-longwall chain pillars may be established.

5.0 REFERENCES

Barron, K. (1978). An Air Injection Technique for Investigating the Integrity of Pillars and Ribs in Coal Mines: Int. J. Rock Mech. Min. Sci. & Geomech. Abstr. 15, 69-76.

Bieniawski, Z. T. (1968). The Effect of Specimen Size on Compressive Strength of Coal: Int. J. Rock Mech. Min. Sci. 5, 325-335.

Bunting, D. (1911). Chamber-Pillars in Deep Anthracite Mines: Trans., AIME, 739-748.

Gaddy, F. L. (1956). A Study of the Ultimate Strength of Coal as Related to the Absolute Size of the Cubical Specimens Tested: Virginia Polytechnic Institute Bull. No. 112, 1-27.

Greenwald, H. P., H. C. Howarth, and I. Hartmann (1939). Experiments on Strengths of Small Pillars of Coal in the Pittsburgh Bed: U.S.BuMines Technical Paper 605, 1-22.

Holland, C. T. (1964). The Strength of Coal in Mine Pillars: Proc. 6th U.S. Symp. on Rock Mech., Univ. of Missouri, Rolla, 450-466.

Hustrulid, W. A. (1976). A Review of Coal Pillar Strength Formulas: Rock Mechanics 8, 115-145.

Lu, P. H. (1981). Determination of Ground Pressure Existing in a Viscoelastic Rock Mass by Use of Hydraulic Borehole Pressure Cells: Proc. Int. Symp. on Weak Rock, Tokyo, Japan, A. A. Balkema, Rotterdam 1, 459-465.

Lu, P. H. (1982). Rock Mechanics Instrumentation and Monitoring for Ground Control Around Longwall Panels: Proc. Int. Symp. on State-of-the-Art of Ground Control in Longwall Mining and Mining Subsidence, Honolulu, Hawaii, Society of Mining Engineers of AIME, New York, 159-166.

Salamon, M. D. G. (1967). A Method of Designing Board and Pillar Workings: J. South African Inst. of Min. Metall., 68-78.

Salamon, M. D. G., and A. H. Munro (1967). A Study of the Strength of Coal Pillars: J. South African Inst. of Min. Metall., 55-67.

Wilson, A. H. (1972). Research into the Determination of Pillar Sizes, Part 1: An Hypothesis Concerning Pillar Stability: Mining Engineer 131, 409-417.

Wilson, A. H. (1977). The Effect of Yield Zones on the Control of Ground: Proc. 6th Int. Strata Control Conf., Banff, Canada.

Wilson, A. H. (1981). Stress and Stability in Coal Ribsides and Pillars: Proc. 1st Conf. on Ground Control in Mining, West Virginia Univ., Morgantown, W. Va., U. S. A., 1-12.

SEVERAL BASIC ASPECTS OF THE FORMING OF SUDDEN OUTBURSTS OF COAL (ROCK) AND GAS

Quelques questions fondamentales concernant la formation de dégagements instantanés dans les mines de charbon

Einige Grundfragen zum Entstehen plötzlicher Kohle- (Gesteins-) und Gasausbrüche

S. A. Khristianovich
Member of the USSR Academy of Sciences, Head of Laboratory Institute for Problems in Mechanics, the USSR Academy of Sciences, Moscow, USSR

R. L. Salganik
D.Sc., Professor, Institute for Problems in Mechanics, the USSR Academy of Sciences, Moscow, USSR

SYNOPSIS
Several aspects of the theory of the formation of outburst-prone situations and the estimations of outburst-preventing measures by means of directed unloading are considered. The important role of the development of oriented systems of gas-filled cracks in outburst-prone seams in the process of outburst and allied phenomena (outbursts of large masses of gas, squeezing, sudden caving, etc.) is shown. General possibilities of detecting such systems of cracks by means of active geophysical methods are discussed. The aspects of the dynamics of the phenomen of sudden outburst are considered: The propagation of crush, spalling, fracture and outburst waves.

RESUME
Cette communication analyse la théorie de l'apparition des situations se prêtant aux dégagements et celles de l'évaluation de l'efficacité des mesures de prévention des dégagements par les soulagements orientés de la pression de terrain. Les autres montrent le rôle important joué par la formation dans la couche recélant le danger de dégagements du système des fissures remplies de gaz au cours des phénomènes mentionnés et dans des phénomènes connexes (dégagements de grandes masses de gaz, essorages, apparitions instantanées, etc.). Les possibilités des principes du dépistage de ces systèmes de fissures à l'aide des méthodes géophysiques actives sont examinées ainsi que les questions de la dynamique du phénomène du dégagement instantané: diffusion des ondes de broyage, détachement, déstruction et dégagement.

ZUSAMMENFASSUNG
Es werden Fragen der Entstehungstheorie von ausbruchsgefährlichen Situationen und der Einschätzung der Effektivität der Maßnahmen zur Verhinderung von Ausbrüchen durch gerichtete Gebirgsdruckentlastungen behandelt. Es wird auf die Rolle hingewiesen, die bei solchen oder ähnlichen Erscheinungen, (Ausbrüche von großen Gasvolumina, Ausquetschen, plötzliche Zusammenbrüche u.a.) die Ausbildung eines orientierten Systems gasgefüllter Spalten in ausbruchgefährdeten Flözen hat. Erörtert werden prinzipielle Möglichkeiten der Ermittlung solcher Kluftsysteme mit Hilfe aktiver geophysikalischer Methoden. Des weiteren werden Fragen der Dynamik der Ausbrucherscheinungen – die Ausbreitung von Zerkleinerungs-, Abspaltungs-, Zerstörungs- und Ausbruchswellen – untersucht.

Coals, enclosing rock of coal deposits, sandstones and salt seams often contain high-pressured gas (methane, carbon dioxide). Under mining in such rock, sudden catastrophic outbursts may happen in the course of which the gas release is accompanied by coal (rock) fracture and outburst of extensive masses of coal (rock) and gas into mined-out space (tens, hundreds and thousands of tons of coal and sometimes many thousands of cubic metres of gas).

For many years the problems of sudden outbursts and protection measures have been investigated and significant practical experience has been gained of measures for preventing the outbursts and diminishing their harmful results. First of all, these are underhand and overhand mining in order to attain directed stress relief in outburst-prone seams. These problems have often been dealt with in the previous literature (see Orlova, 1981; Zabigailo, 1978; Chernov and Puzyriov, 1979; Nikolin et al., 1981; Petukhov et al., 1976; Novichikhin et al., 1977; Zorin et al., 1978; Proskuriakov, 1980).

Nevertheless, the outbursts still occur and

they often happen in the course of antioutburst actions. This calls for a profound investigation into the problems connected with the processes causing outbursts and with their essence.

This paper is an attempt of a theoretical study of the variety of effects leading to sudden outbursts of coal (rock) and gas, outbursts of considerable masses of gas, squeezing and other related processes, whose initiation and development take place under stress relief due to mining. The paper is based on the studies mainly performed at the Institute for Problems in Mechanics, Academy of Science, USSR, Moscow, their review being presented by Khristianovich and Salganik, 1980ab. It considers the effects connected with coal fracture due to the expansion of gas contained in pores and the dynamics of outburst.

1. ON THE PHENOMENON OF OUTBURST

At first sight the sudden outbursts seem random events as they depend on many factors such as the structure of coal and rock, geological disturbances, rock pressure, methods and rates of mining etc. However, the analysis reveals that the process of outburst initiation and development seems to obey to a relatively simple physical mechanism while a number of random factors only add to its intricate nature.

First of all let us specify the terminology involved. In the description of catastrophyc phenomena which can occur in the process of mining, a detailed classification is adopted. However, all these phenomena can be divided into two extensive classes according to the mechanism of rock failure.

Such dynamic processes as sudden roof fall, spalling, caving, inrush, etc. in which the failure occurs with no gas present are attributed to the first class. They can be ascribed to rock bursts.

The other class comprises the processes in which the fracture of coal (rock) is mainly caused by expansion of gas it contains. These are the outbursts proper and related phenomena mentioned above.

In reality even if there is some gas in pores, the process of fracture can resemble a rock burst greatly intensified by the energy of gas. On the contrary, in the coal (rock) and gas outburst processes occuring due to sudden change of the state of stresses, the bursts intermittent with coal (rock) crushing and effuse by expanding gas can play a considerable part.

What is the essence of sudden outburst in general?
In the course of mining, occasional dynamic fracture and crushing of coal (rock) occur suddenly and almost without any precursors. In the destructed coal mass, spalls of various sizes are present - ranging from the finest ("furious flour") up to rather massive ones - and all this mass of fractured and crushed coal (rock) is effused by expanding gas into the mined-out space. The process of fracture propagates in the form of a wave in seam, often up to distances of tens and even hundreds of metres (the waves of crushing, fracture and spalling). The proportion between the masses of outbursted coal and gas varies greatly. There can be observed both sudden outbursts of tremendous gas masses accompanied by small quantities of coal and outbursts with comparatively low degree of gas liberation.

Fig.1 schematizes an example of outburst into longwall (the arrows show the direction of gas propagation. It will be described in detail later).

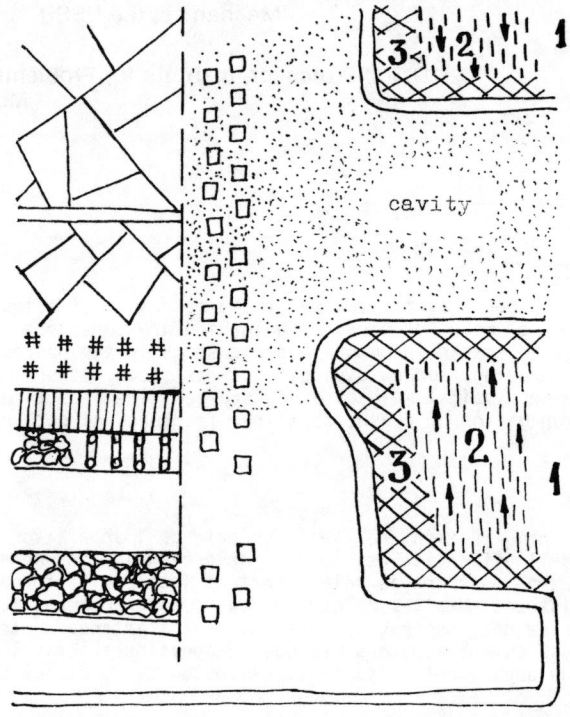

Fig.1. The outburst caused a cavity and then additional outburst followed. 40 t of coal and 8400 m³ of methane were outbursted; 1 - elastic zone; 2 - zone of splitting; 3 - edge plastic zone

2. GAS IN SEAMS

To understand the phenomenon of outburst one is to know the state of gas contained in undisturbed seam.

The typical content of gas in outburst-prone seams is estimated at average as 15-25 m³ per ton of coal (here and further the gas volume is given in normal cubic metres). A considerable part of the gas is contained in coal in dissolved (sorbed) state (Ettinger and Shulman, 1975; Ivanov et al., 1979). However, the analysis of observations leads to the conclusion that sudden outbursts are caused by the presence of free gas in coal. The analysis of overwhelming majority of outbursts reveals that they could not have originated without free gas.

To properly take measurements one is to prevent, wherever possible, fracture of coal and escape of gas from it prior to measurements (Kuznetsov, 1980). When gas pressure in outburst-prone coal seams undisturbed by mining is measured in such a way, it displays the values which are markedly above the critical saturation pressure of sorption.

The further theoretical discussion is based on some ideas of the way the free gas is kept in outburst-prone coal (rock) and liberated from it.

The outburst-prone coals and salts differ from gas-containing rock (sandstones, limestones) of gas fields in their plasticity which effects isolation of gas-filled pores so that the permeability of seams with such pores in undisturbed state is equal to zero. This accounts for the fact of gas conservation in outburst-prone seams during geological time even when such seams are overlapped by permeable rock, which is frequently the case.

The plasticity of coals, salts and other rock (their ability to flow under shear stresses) is expected to cause the levelling of the stresses in all directions for geological time. This results in compressive stresses at sufficient depth which are similar in all directions and equal to inherent local rock pressure.

We assume that the pressure of free gas in pores of outburst-prone seam is also approximately equal to rock pressure at the given depth, e.g. to the value of γH, where γ is average unit weight of overburden, H - the depth of the seam.

Therefore the outburst-prone seam is supposed to contain free gas. In calculations, the results of which will be presented later, the gas concentration in the seam was taken equal to 4 m^3/t. At the depth of 800 m, such quantity of gas can be accomodated provided that porosity is of about 2,5%. The question is: in what way will the gas be released out of the coal (rock) when relieved of compressive stresses?

3. GAS RELEASE OUT OF COAL UNDER RELIEF OF EXTERNAL COMPRESSIVE STRESSES

It is obvious that without filtration the gas cannot considerably increase its volume and lower its pressure without rock rupture.

In order to investigate general regularities of gas release out of coal, the following tests were performed in the Donetsk Physico-Technical Institute, Academy of Science of the Ukraine (Alekseev et al., 1980); they are of qualitative character as the value of stress relief was considerably influenced by friction in the elements of test equipment. Cubic coal specimens with 5 cm edges were subjected to triaxial compression by loading plates. The front plate had a hatch which could be instantaneously opened.

At first, non gas-saturated specimens were tested. They had been subjected to comparatively equidirectional compression, then the hatch was opened. No considerable change was noticed up to 100 MPa, further increase of compression led to squeezing of the coal out of the hatch.

Then the specimens preliminarily saturated with gas were tested. Again approximately equidirectional compression up to 30 MPa was attained. The hatch was opened and the outburst occurred. After that the same test was performed again, but the pressure on the front loading plate was considerably lowered in advance. And each hatch opening was accompanied by an outburst. In the next series of tests the pressure on the side plate was considerably lowered in advance. After that, no outbursts were observed when the hatch was opened.

How can this phenomenon be explained?
After the sample had been extructed from the seam, it already possessed permeability and due to this fact it had been saturating with gas. Then the gas became isolated in pores under all-round compression quite the same way as we understand it takes place in seam. Under a uniaxial stress relief, the gas can expand only by rupturing the coal (rock) and thus forming a system of cracks which gradually obtain a penny-shaped form and grow in planes orthogonal to the direction of the minor compression, i.e. parallel to the plate the pressure on which decreases.

Therefore the coal specimen subjected to de-stressing only in the direction of the front plate gradually splits forming a system of oriented cracks and loses its tensile strength in the direction orthogonal to cracks. This is why the opening of the hatch is accompanied by an outburst and formation of a cavity.

In the case of lateral de-stressing, a system of oriented cracks is formed in planes mainly parallel to the side plate. As a result, the rupturing action of gas in the direction of the front plate is very small, and the coal does not lose its strength in this direction. This is why the outburst does not occur.

The mentioned physical mechanism is the background for the most wide-spread and effective measure for outburst prevention - underhand and overhand mining (see Fig.2). Actually, the majority of effective outburst-preventing measures is based on the use of the same mechanism. This is why it is very important to study this mechanism and to construct an appropriate mechanical and computational model.

4. COAL (ROCK) WITH A SYSTEM OF ORIENTED GAS-FILLED CRACKS

The main task of the theoretical study of gas release out of solid material under de-stressing is the calculation of development of the system of gas-filled cracks, rupturing the material, and the evaluation of corresponding variations of effective deformation and filtration properties. As the result of these variations, initially isotropic material becomes transversely isotropic with an axis of isotropy orthogonal to the cracks.

Proceeding from the results of Vavakin and Salganik, 1978 (see also Salganik, 1982), an effective method for calculation of differential moduli (coefficients in differential relationship between stresses and strains) for material with a system of oriented gas-filled cracks under de-stressing was proposed in the case of elastic behavior of material, i.e. up to some value of unloading (Kovalenko, 1980b). The nonlinear stress-strain relationship is to be determined by integration in terms of known differential moduli (modified self-consistent method). The method is based on the solution for a single isolated crack (regarded as penny-shaped), and the calculation is performed in successive steps. At every step, the material already containing cracks is considered as one possessing the effective moduli corresponding to these cracks and unknown beforehand. A small portion of cracks, which are relatively sparse and assumingly non-interacting, is added to the ones existing before, and the corresponding moduli are calculated.

The method is applicable provided the crack distribution (subject to their size distribution if there is one) is sufficiently random and the concentration of cracks is not too close to limiting value so that the cracks could be described as isolated.

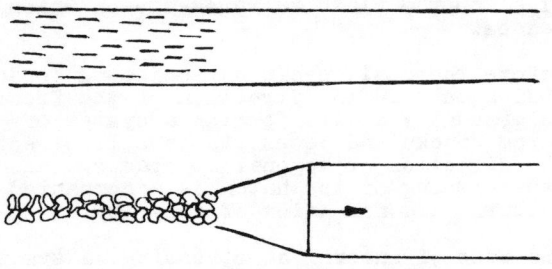

Fig.2. Scheme of underhand mining. The unloading of outburst-prone seam orthogonally to its roof and floor causes the formation of a system of oriented cracks (shown by strokes) which impedes the initiation of outburst of coal and gas; the direction of the face advance of underhand seam is indicated by arrow

A study of the growth of a single gas-filled penny-shaped crack (with constant gas content) required for calculation of effective moduli is performed by Kovalenko, 1980a. Fig.3 schematizes the limit equilibrium of a gas-filled penny-shaped crack corresponding to the highest possible stress intensity at the edge of the crack which is determined by cohesion modulus K or by critical stress intensity factor $K_{1C} = \sqrt{2/\pi}\, K$ (Liebovitz, 1968).

If the radius of initial crack is less then a_1, the limit state is not obtained by compressive stress relief orthogonally to the crack, and the radius of the crack does not change, only the volume increases. If the radius of crack a_0 slightly exceeds a_1, at first only the volume of the crack increases under sufficient stress relief (segment AB), then, having obtained the state of limit equilibrium, the radius increases in jump (segment BC; due to inertia the crack may stop when the radius is a little bit bigger than it is shown in Fig.3). After that, the radius gradually increases along the curve of limit equilibrium according to the de-stressing. The cracks, whose initial radius exceeds a_3, do not change their radius under stress relief and their volume increases exclusively due to increase of their opening. Finally, when the stress relief is less than some critical value (corresponding to the maximum in Fig.3), no cracks will change their initial radii, only their volume will increase.

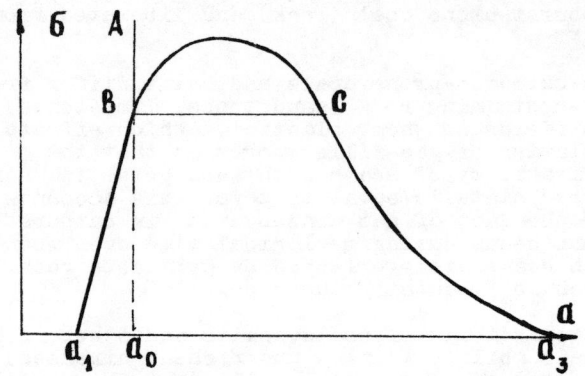

Fig.3. Schematized curve of limit equilibrium of gas-filled penny-shaped crack and the variation of crack radius under unloading; a - crack radius, δ - compressive stress

Fig.4 demonstrates the variation of parameters of gas-filled penny-shaped cracks under stress relief (Kovalenko, 1980a).

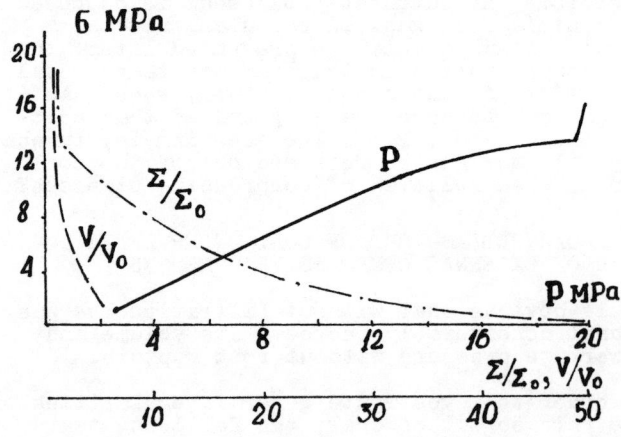

Fig.4. Variation of parameters of penny-shaped gas-filled crack under stress relief; $S_0 = 4.5$ MPa, $p^0 = \delta^0 = 20$ MPa; V - crack volume, Σ - crack area, p - pressure in the crack, δ - compressive stress (calculations were executed for coal); $S_0 = K/\sqrt{2a_0}$.

Fig.5 represents the results of calculation (Kovalenko, 1980b) of the variation in a differential modulus of elasticity in the direction orthogonal to cracks under stress relief in the case when initial cracks are small enough and their development starts with jump; the cohesion modulus determines the character of the curves only in combination $S_0 = K/\sqrt{2a_0}$, where a_0 is the radius of initial crack (the values of the parameters are chosen typical for coal).

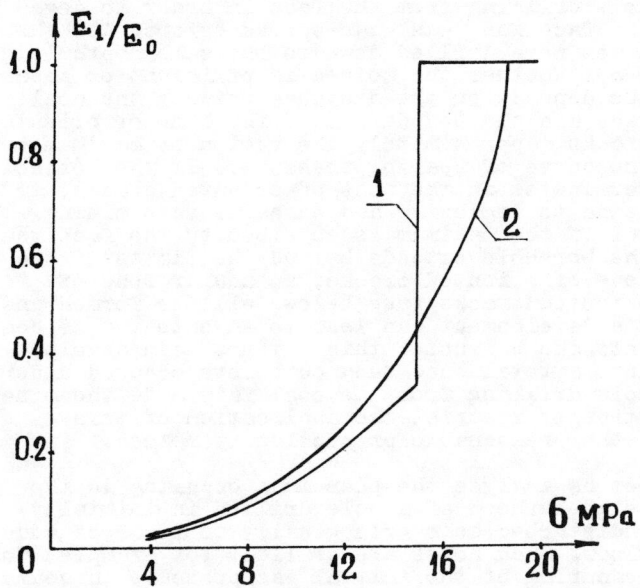

Fig.5. Variation of relative differential modulus of elasticity E/E_0 of coal with a system of oriented cracks under decrease of compressive stress σ (calculation). Initial gas pressure is equal to initial compressive stress of 20 MPa.
1 - all cracks are of the same size, $S_0=4.5$ MPa; 2 - cracks of three different sizes are present, $S_0=1$; 2.8; 4.5 MPa

After the jump of the modulus of elasticity or its sharp decrease, the compliance of cracked material becomes intermediate between its value in undisturbed state and the gas compliance. The material turns into a kind of anisotropic sponge with oriented cracks filled with compressed gas, whose pressure, though less than initial, acts over a considerably larger area. And so this sponge, like a piston, strongly presses upon the restricting walls parallel to the cracks (in other directions the crack growth insignificantly changes the sponge's action). As a result, the outburst initiation facilitates and can be induced (see below).

Under the development of internal process of coal fracturing and splitting, its cracks and pores become interconnecting. It results in non-zero permeability. However, the permeability remains rather low in the direction of acting pressure differential (orthogonal to the cracks). On the contrary, it becomes high in the directions parallel to the cracks when the stress relief is sufficient, so the zone of oriented cracks can instantaneously be emptied along the planes of cracks under considerable pressure differential in these directions. It accounts for the outbursts of large masses of gas out of seams in the direction parallel to coal face (see Fig.1, arrows).

5. OUTBURST-PRONE SITUATIONS

The best way of preventing outbursts would be pre-extraction of gas using hole drilling. However, the experience reveals that the gas practically does not enter these boreholes as non de-stressed coal does not filtrate. For degassing, the coal must be sufficiently unloaded in a vast area. It is realized in practice in the method of hydrorupture by means of durable infusion of large masses of water into the seam under pressure exceeding the rock pressure. Then the pressure is relieved (Chernov and Rozantsev, 1975; Nozhkin, 1979). The experience proved that this method works at comparatively shallow depths and for sufficiently strong coals. In the majority of cases, the mining is still conducted in non pre-degassed seams, many of them being outburst-prone.

An analysis of numerous outbursts shows that some of them are initiated by instantaneous dynamic loading (roof collapse, pillar crush, blasts) coasing the further self-sustaining (due to gas expansion energy) process of coal (rock) fracture. Nevertheless, most of the sudden outbursts occur as if without any external cause in the course of regular face advance. We suppose that the latter case results from the origination, in the adjacent neighbourhood of coal face, of a considerable zone of coal disrupted by a system of oriented cracks mostly parallel to the face. The situations when such system of oriented cracks developes under mining and causes a particular predisposal to outbursts will be referred to as outburst-prone situations.

For safe mining, it would be very important to work out geophysical methods for detection of the zones of oriented cracks (see below). It is also necessary to work out calculating methods for determining the mining process precluding outburst-prone situations.

In practice, the outburst-prone situations are formed, other things being equal, in the neighbourhood of the regions of geological disruptions in seams, where the coals are shaken and weak. It should be always borne in mind.

Let us regard several typical situations which occur under mining and measures used in practice to prevent outbursts. Let us begin with the central part of a longwall in a gently dipping seam (Fig.6). If there is no normal component of stress on coal face while a considerable vertical load acts at the face zone of the seam, considerable shear stresses occur and result in formation of a filtrating plastic zone. The shear stresses in this zone are limited by the curve of limit equilibrium and it leads to the increase of bearing pressure σ_y in the plastic zone in the direction opposite to the coal face, σ_y being different from the horisontal component of normal stress σ_x in the value of shear stresses. At the same time in the elastic zone, the compression of the seam increases in the direction from its undisturbed part to the face and, according to the laws of elasticity, the bearing pressure must increase in this direction too. As a result, the curve of bearing pressure upon the seam has a shape shown in Fig.6. Calculations of this curve are presented by Barenblatt and Khristianovich, 1955; Kuznetsov,

1968; Kuznetsov and Khapilova, 1975. The location of maximum on this curve approximately corresponds to the end of the plastic zone and is mainly determined by the cohesion of the seam with enclosing rock and by the angle of internal dry friction and the yield strength of the coal k obeying the law of limit equilibrium $\tau_n = \sigma_n \tg\varphi + k$, where τ_n and σ_n are the values of shear and normal stress on the element of surface of limit equilibrium.

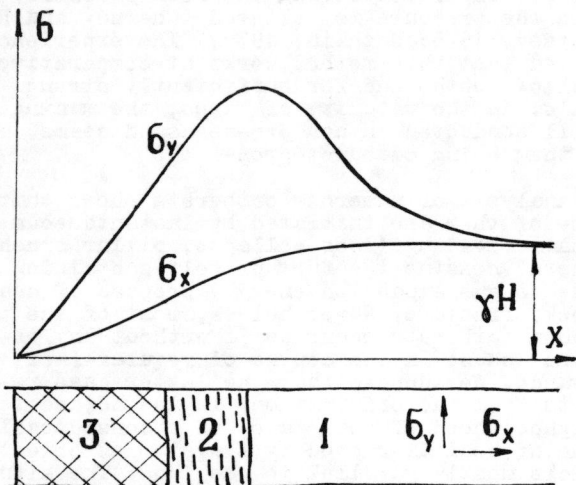

Fig.6. Scheme of stress-strain state at the coal face. 1 - elastic zone; 2 - zone of oriented cracks; 3 - plastic protecting zone

With weak cohesion (Kuznetsov, 1968), e.g. clay, and low angle of friction φ, this maximum moves away from the face. When the maximum is sufficiently close to the coal face, the longitudinal stress component $\sigma_x^{(p)}$ in the end of the plastic zone is less then the stress σ_{cr} which causes intensive development of the system of oriented cracks (see Fig.6). It results in formation of a zone of oriented cracks between elastic and plastic zones; the greater the difference between σ_{cr} and $\sigma_x^{(p)}$ the thicker this zone.

In the course of face advance, the undisturbed coal changes into splitted state in the zone of oriented cracks and therefrom into the state of plastic deformations in the plastic zone where interconnecting filtrating canals appear and degassing occurs. Therefore, the plastic zone acts as a protective screen which takes up the load of the anisotropic gas "sponge" (see Section 4). The thinner the screen, i.e. the closer the maximum of the curve of bearing pressure to the coal face, the greater is the probability of breaking down the screen and outburst initiation. The outburst shown in Fig.1 corresponds to this scheme. A large volume of evolved gas, many times exceeding its initial content in outbursted coal, entered from the zone of oriented cracks; the length of this zone far exceeded the size of newly generated cavity.

The presence of gas in the plastic zone influences the law of plastic deformations as the gas pressure changes the state of stresses. An attempt of taking this fact into consideration is presented by Protosenia et al., 1982. The gas pressure in plastic zone depends on the rate of face advance as the degassing rate depends on it.

The method of preventing outbursts by means of hole drilling from the face in order to degas the face zone was wide-spread before. All the holes were drilled down to the same depth (e.g. 5 m). Whether the method is efficient or hazardous depends on the distance between the coal face and the boundaries of the zone of oriented cracks (approximately the region of maximum on the curve of bearing pressure). If the borehole terminates in the zone of oriented cracks, it leads to degassing and enshures safe mining. But if the maximum is so close to the face that the borehole extends beyond the limits of the zone of oriented cracks, an anular zone of oriented cracks (see below) will be formed and its development can lead to an outburst of coal into the borehole; this outburst can develop into a total one. Many outbursts occured under hole drilling from the coal face made the miners strongly restrict the application of this rather reasonable precautionary method.

Let us analyse the phenomena occuring in the neighbourhood of a hole drilled in initially undisturbed seam orthogonally to its roof and floor. Such holes are drilled, for example, for measuring of the initial gas pressure in seams.

The method of measuring gas pressure in seams by means of the holes mentioned is usually based on the assumption that undisturbed coal filtrates the gas. The measurements revealed that the data scatter and the values of pressure differ very much, usually being very low. In reality, the gas pressure in borehole is determined by the volume of gas evolved from restricted de-stressed zone and depends on both the size of this zone (i.e. the coal strength) and the amount of culm carried away by drilling. Accounting for these factors reveals (Kovalenko, 1980a) that the pressure measured by this standart method is generally many times lower than actual initial pressure of free gas in pores.

Fig.7 shows the plastic zone and the zone of oriented cracks. If the plastic zone is not thick enough, it fails, breaks; it results in the origination of a wave of fracture and outburst into the borehole (Kovalenko, 1980a).

Preliminary underhand and overhand mining of outburst-prone seams are the most effective and wide-spread methods for preventing outbursts. In this case the directed unloading of the coal seam causes the development of a system of cracks oriented parallelly to the roof and floor of the seam (Fig.2). As a result, in spite of its splitting, the coal does not lose strength in the direction of mining; the gas practically does not affect rupture in this direction. The orientation of cracks promotes gas filtration in the direction towards the face. It also gives possibility of the pre-degassing of the seam by means of hole drilling into the zone subjected to unloading.

Fig.8 shows the variation of the discharge of a borehole drilled into a coal seam depending on the distance from gradually advancing face of the underhand seam. When there is no unloading, the coal does not filtrate and the gas does not enter the borehole. With the advance of the face of underhand seam, the discharge of the borehole increases suddenly because the cracks develop under unloading, and then the discharge decreases down to zero under rock pressure as the result of the closure of filtrating canals.

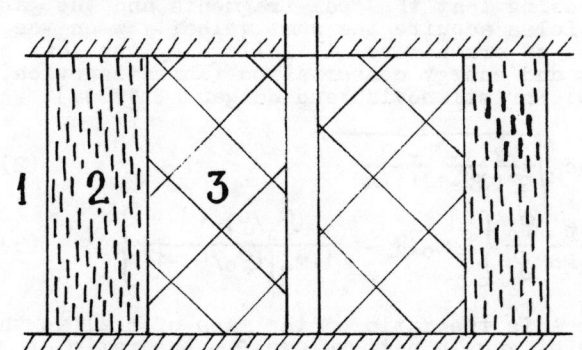

Fig.7. Scheme of a borehole drilled orthogonally to the seam in order to measure the gas pressure in seam. 1 - elastic zone; 2 - zone of oriented cracks; 3 - zone of plastic deformations

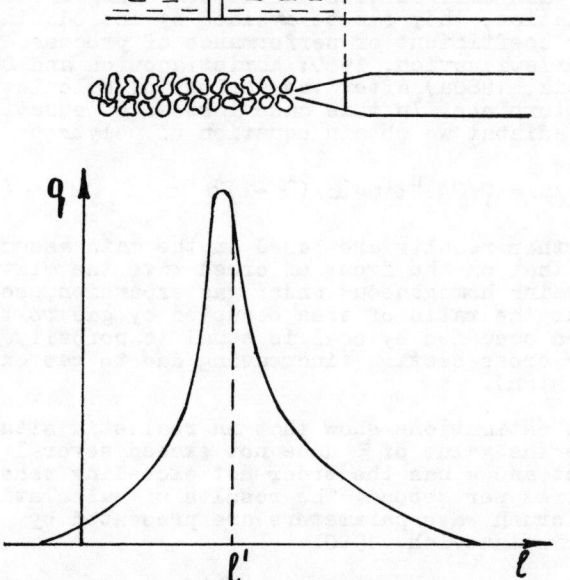

Fig.8. While the face of unloading seam approaches, the gas starts evolving out of the borehole and its discharge highly increases, then the discharge decreases as the face moves away

The majority of outbursts occur in the course of development mining (opening of seams, drift mining) and the stoping at the corners and benches of breakage face (longwall). This is due to abrupt change of the state of stresses occuring in the adjacent neighbourhood of the face. For example, in the neighbourhood of the drift face even as near as 1-1.5 diameters from it, initial rock pressure practically restores (Kurlaev, 1980a). This is why the protecting plastic zone is poorly developed in comparison with that in the central parts of the face (longwall). The zone of oriented cracks occurs in the adjacent neighbourhood of the stope face and the danger of outbursts strongly increases.

The most reasonable way of preventing outbursts in these situations is local directed stress relief due to pre-excavation of enclosing rock in order to produce in the neighbourhood of the face a system of oriented cracks orthogonal or nearly orthogonal to the face (Zorin, 1978).

The aspects of calculations and estimations of protective effects of underhand and overhand mining are discussed by Petukhov et al, 1976; Mokhel, 1980. The results of the actions preventing outbursts in gas-filled salts by means of directed unloading are described by Proskuriakov, 1980. The calculations of the effects of unloading are impeded by the lack of experimental data on strength, deformation and filtration characteristics of coal and enclosing rock under the loading conditions close to conditions in situ (high initial all-round compression and non-equiaxial stress relief of this state). This is why the most important problem is to construct presses which would allow us to test sufficiently large specimens and to determine the strength, deformation and filtration characteristics of coals, salts and rocks under independent stress relief in one, two or three directions.

6. SOME POSSIBILITIES OF GEOPHYSICAL DETECTION OF SYSTEMS OF ORIENTED CRACKS

The specialists apply many methods of geophysical control over state of rock in the neighbourhood of openings (Rzhevski and Yamshikov, 1968, Glushko et al, 1978). The passive methods of seismoacoustic estimation of the degree of outburst threat (Antsiferov et al, 1968; 1971) are now very popular. Recent development of these methods made it possible to determine the distance between the face and the seats of splitting. Let us briefly discuss some principal possibilities of using active geophysical methods to detect the systems of oriented cracks which allow us to judge upon the degree of outburst's risk and efficiency of antioutburst actions.

The development of systems of oriented cracks makes the material transversely isotropic. In the zone of oriented cracks the velocities of elastic wave propagation change. The calculations (with parameters typical for coal) show (Kovalenko, 1980b; Salganik, 1980) that under realistic de-stressing (up to 10 MPa) the velocities of elastic waves can decrease by tens of per cent due to the development of the systems of oriented cracks. And there exists essencial dependence of wave velocities on the direction of their propagation and their polarization. That makes it possible to detect by active acoustic methods both the degree of crackness and orientations of cracks.

Gas-filled cracks in electroconductive coal act

like screens which increase electric resistance under current transmission orthogonal to cracks and practically do not change it when the current is directed parallel to cracks. This effect is greater than the effect of elastic wave velocity variation under the same degree of crack development. Hence the measuring of the resistance in different directions can inform us of the degree of crackness of material and of the orientation of cracks.

The cracks start acting as electric capacitors under the influence of rapidly alternating electric fields in a coal seam. This leads to appearance of reactive resistance component depending on the crack opening, and by its variation one can judge of gas pressure variation in pores. A similar effect will be observed in the constant current field under the variation of crack opening with time which occurs with sufficient speed (e.g. under blasts).

As far as the determination of outburst threat requires investigations of the face zone of seam at the distances of several metres, non-contact electromagnetic methods can be applied for the crack diagnostics. These methods are effective within the depth of penetration of electromagnetic field into the coal seam. The estimations show that under realistic values of parameters this depth also stretches for several metres. A more detailed discussion is presented by Salganik, 1980; 1982.

7. THE WAVES OF SPALLING, CRUSHING, FRACTURING AND OUTBURST

Though the outburst-prone situation develops gradually during mining and dynamic effects practically do not influence its generation, the process of outburst itself is a dynamic one and similar to explosion.

In outburst-prone situations the origination of outbursts is greatly facilitated and it can be induced by very low disturbances. But sudden outbursts can occur even without outburst-prone situation if a considerable area of the surface even of almost untouched coal instantaneously exposes (e.g. as the result of a rock burst). It always takes place on the front of the process of coal fracture propagating into rock.

The resulting process bears a complicated character (see Section I), the propagation of the crush waves being the most significant part of this process. A rapid dynamic stress relief in the crush wave makes the total gas pressure differential concentrate in a very narrow near-face layer in coal. It causes layer-by-layer breaking away, coal crushing into very fine particles, practically complete evolution of free gas out of the coal. The crush wave propagation has been theoretically studied for a long time (Nikolski, 1953; Khristianovich, 1953). Recently, the theory of its propagation was essentially developed (Khristianovich, 1980; 1982; Kurlaev, 1980b).

If the expanding gas is taken as ideal neglecting the Joule's effect, and the heat capacities C_v, C_p are assumed to be independent on the temperature, then

$$e=(1/(\varkappa-1))p/\varrho + const; \quad c=\sqrt{\varkappa p/\varrho};$$
$$p/p_o=(\varrho/\varrho_o)^\varkappa; \quad \varkappa=C_p/C_v \qquad (1)$$

where e is internal energy of unit mass of gas, c - the sound velocity in it, p - pressure, ϱ - density; the third relationship is the equation of adiabat; index "o" marks the initial state.

Supposing that the coal fragments and the gas particles acquire the same velocity w on the front of crush wave, it follows by the laws of mass and energy conservation (Khristianovich, 1980; Khristianovich and Salganik, 1980a):

$$w=c_o\sqrt{\frac{2(1-\xi-B)}{\varkappa(\varkappa-1)(1+A)}} \qquad (2)$$

$$B=\frac{p}{p_o}\frac{\varrho_o}{\varrho}\left\{1+m_o(\varkappa-1)\frac{(\varrho_o/\varrho)-1}{1+m_o[(\varrho_o/\varrho)-1]}\right\} \qquad (3)$$

Here A is the ratio of the mass of coal to the mass of gas in unit volume, ξ - the ratio of crushing work to initial internal energy of gas, m_o - initial porosity. The velocity of propagation of the crush wave front N is determined by

$$N=-\frac{w}{m_o[(\varrho_o/\varrho)-1]} \qquad (4)$$

For a complete description of the process of crush wave propagation, it would be desirable to use the equation of impulse. There are certain difficulties in the composing of this equation; they can be avoided by introducing the coefficient of performance of process η (Khristianovich, 1980; Khristianovich and Salganik, 1980a) after the analogy of calculations of turbines. In this case instead of equation of adiabat we obtain equation of polytrope

$$p/p_o=(\varrho/\varrho_o)^n; \quad n=\varkappa/(\varkappa-1)\eta \qquad (5)$$

Further results are based on the main assumption that on the front of crush wave the mixture remains homogeneous under gas expansion, so that the ratio of area occupied by gas to the area occupied by coal is equal to porosity in any cross-section (increasing due to gas expansion).

The estimations show that in realistic situations the value of ξ does not exceed several per cent and w has the order not exceeding tens of metres per second. The results of calculations of crush wave parameters are presented by Khristianovich, 1980.

The investigations into outbursts are complicated due to a wide variety of ways of coal fracturing, and the way it does occur depends not only on the properties of the coal itself but also on the degree of outburst risk of the situation resulting from previous mining, on the size of coal and cohesion strength of coal blocks with each other and of coal with rock.

The ways of coal fracture due to outburst can be vividly exemplified with a unidimensional model of outburst development.

Imagine that the coal with pores containing gas under high pressure is enclosed in a semi-infinite tube closed at one end by a piston. The coal is subjected to all-round compressive stress, and the pressure of free gas in pores is equal to this stress. In some parts of the tube the cohesion of the coal with sidewalls is strong, in others - weak (it models the difference in the strength of cohesion of blocks). Let us suppose that the part of coal weakly cohered with the tube is adjacent to the piston, and the piston instantaneously moves out but so that the stress in this zone falls not lower than the threshold pressure of the initiation of intense crack development. Then the wave of elastic unloading will propagate along the part of coal adjacent to the piston (the coal slips against the walls of the tube). When the wave reaches the zone of strong cohesion of coal with tube it will be reflected with amplification (its amplitude nearly doubles); it can lead to spalling. A crush wave described above can start propagating from the exposed surface formed in coal by spalling. Having reached the zone of weak cohesion of coal with tube it can again provoke the process of spalling.

In nature this alternation of spalling process and crush wave propagation is of course more complicated. As a result, a cavity of intricate shape is formed and the coal is destructed into fragments of various sizes - from extremely fine ("furious flour") up to rather massive ones (being formed under spalling and following degassing). It should be emphasized that by defining the wave front as a layer whose thickness sufficiently exceeds the sizes of main mass of large spalls, we can regard this wave as extended crush wave and theoretically describe it in a way similar to the description of crush waves (see above).

Special attention should be paid to the development of outburst in outburst-prone situation which can be simulated in the example with a tube by preliminary gradual extension of the piston. It leads to splitting of coal by a system of cracks embedded in the planes parallel to the face of piston. Under further abrupt extension of the piston, the gas contained in pores will push the coal out, and a wave that can be called a fracture wave will start propagating over the splitted coal. In these situations the portion of energy transformed into kinetic energy strongly increases and therefore the outbursts can be more dangerous. Having passed the zone of splitting, the fracture wave can engender a crush wave propagating in coal and then the process described above can go on: alternating processes of crushing and spalling.

The gas evolves by filtration from the coal fragments originated by spalling, and it is the process of just a mere throttling when the gas does not work and hence changes neither its initial temperature nor its internal energy (Joule's effect is neglected). It allows us to consider that the energy of the mass of gas retained in expanded pores does not contribute into the acceleration of the coal fragments. It can be considered by representing in the following form (Khristianovich and Salganik, 1980a):

$$\xi = \xi_1 + \xi_2 \qquad (6)$$

where ξ_1 is actual relative crushing loss, ξ_2 - relative energy of gas which does not take part in the process of acceleration of coal fragments. Then it can be put down that

$$\xi_1 = (1-\mu)\xi_0 \qquad (7)$$

where ξ_0 is relative crush energy under complete crushing ("furious flour") of the total coal mass and liberation of the total mass of free gas contained in coal. It can be also put down that

$$\xi_2 = \mu m/m_0 \qquad (8)$$

where μ is the portion of coal mass outbursted in the form of spalls.

From (2), (6)-(8) it is seen that under high values of μ (many fragments) the energy of liberated free gas is not enough for the acceleration of coal. In this case the outburst degenerates into the process of crumbling and caving.

The waves of crushing, spalling and fracturing are accompanied by outburst wave in which the fractured coal is accelerated by compressed gas. The analysis reveals that the difference between the crush wave and the outburst wave diminishes and the crush wave continuously transforms into the outburst wave.

The investigations of Khristianovich, 1980; 1982, and Kurlaev, 1980b, revealed that the throttling effect must be taken into consideration in calculations of the outburst wave. It should not be supposed (as it had been done before) that under gas expansion practically all its energy is spent on converting into the kinetic energy of coal. In reality the most part of mechanical energy is lost in the process of throttling. Allowing for this circumstance leads to better accordance of calculated rethults with observed data.

REFERENCES

Alekseev, A.D., Nedodaev, N.V. and Starikov, G.P. (1980). Fracture of Gas-saturated Coal under Volumetric State of Stress in the Course of Unloading. Modelling of Outbursts of Coal and Gas. (In Russian). Moscow: Inst.Probl.Mech. Acad. Sci. USSR, Preprint 139.

Antsiferov, M.S. et al. (1968). Provisional Procedure of Forecast of Outburst Threat for Coal Seams by Use of Seismoacoustic Method. (In Russian). Moscow: The Skochinski Inst. of Mining.

Antsiferov, M.S., Antsiferova, N.G. and Kogan, Y.Y. (1971). Seismoacoustic Investigations and the Problems of Forecast of Dynamic Phenomena. (In Russian). Moscow: Nauka.

Barenblatt, G.I. and Khristianovich, S.A. (1955). On the roof caving in openings. (In Russian). Izv. Acad. Nauk SSSR, OTN, 11. (See also Khristianovich, S.A. Mechanics of Solids. (In Russian). Moscow: Nauka, 1981).

Chernov, O.I. and Puzyrev, V.N. (1979). The Forecast of Sudden Outbursts of Coal and Gas. (In Russian). Moscow: Nedra.

Chernov, O.I. and Rozantsev, E.S. (1975). Preparation of Mine Fields with Gas-Outburst Prone Seams. (In Russian). Moscow: Nedra.

Ettinger, I.L. and Shulman, N.V. (1975). Distribution of Methane in Pores of Fassil Coals. (In Russian). Moscow: Nauka.

Glushko, V.T., Yamshikov, V.S. and Yalanski, A.A. (1978). Geophysical Control in Coal Mines. (In Russian). Kiev: Naukova Dumka.

Ivanov, B.M., Feit, G.N. and Yanovskaya, M.F. (1979). The Mechanical and Physici-Chemical Properties of Coals in Outburst-Prone Seams. (in Russian). Moscow: Nauka.

Khristianovich, S.A. (1953). Distribution of gas pressure near advancing free coal face. On the outburst wave. On crushing wave. (in Russian). Izv. Acad Nauk SSSR, OTN, 12.

Khristianovich, S.A. (1980). Unrestricted Flow of soil Mass Caused by the Expansion of High-Pressured Gas Contained in Pores. Crush Wave. (In Russian). Moscow: Inst. Probl. Mech. Acad. Sci. USSR, Preprint 128.

Khristianovich, S.A. (1982). Unsteady flow of soil mass containing high-pressured gas in pores. Phys.-Techn. Probl. Razrabotki Poleznykh Iskopaemykh, 3.

Khristianovich, S.A. and Salganik, R.L. (1980a). Outburst-Prone Situations. Crushing. Outburst Wave. (In Russian). Moscow: Inst. Probl. Mech. Acad. Sci. USSR, Preprint 152.

Khristianovich, S.A. and Salganik, R.L. (1980b). Sudden Outbursts of Coal (Rock) and Gas. Stresses and Strains. (In Russian). Moscow: Inst. Probl. Mech. Acad. Sci. USSR, Preprint 153.

Kovalenko, Y.F. (1980a). Elementary Act of the Phenomenon of Sudden Outburst. Outburst Into Borehole. (In Russian). Moscow: Inst. Probl. Mech. Acad.Sci USSR, Preprint 145.

Kovalenko, Y.F. (1980b). Effective Characteristics of Bodies with Isolated Gas-filled Cracks. Fracture wave. (In Russian). Moscow: Inst. Probl. Mech. Acad. Sci. USSR, Preprint 155.

Kurlaev, A.R. (1980a). Stress-Strain State around the Face of Cylindrical Opening under All-round Compression far from the Face. (In Russian). Moscow: Inst. Probl. Mech. Acad. Sci. USSR, Preprint 158.

Kurlaev, A.R. (1980b). Estimation of the Effect of Degassing on Unrestricted Flow of Soil Mass containing Compressed Gas in Pores. (In Russian). Moscow: Inst. Probl. Mech. Acad. Sci.USSR, Preprint 163.

Kuznetsov, A.F. (1980). The Study and Working out of Apparatus and Methods of Seam Tests with Devices on Cable in Coal-Prospecting Boreholes. (In Russian). Moscow: Res. Inst. Geophysics.

Kuznetsov, S.V. (1968). The influence of shear stresses at the contact interface of seam and rock onthe state of stress of rock mass. (In Russian). Phys.-Techn. Probl. Razrabotki Poleznykh Iskopaemykh, 3.

Kuznetsov, S.V. and Khapilova, N.S. (1975). On sudden squeezing of coal seam under mining. Mech. Solids (10), 3.

Liebovitz, H. (Ed.). (1968). Fracture. An Advance Treatise. Vol.II. New-York, London: AP.

Mokhel, A.N. (1980). Theoretical Estimation of the Effect of Mining of Protecting Seam on Protected Seam. (In Russian). Moscow: Inst. Probl. Mech. Acad. Sci. USSR, Preprint 156.

Nikolski, A.A. (1953). On the waves of sudden outbursts of aerated rock. On the waves of fracture of aerated rock. (In Russian). Doklady AN SSSR (88), 4; (91), 5.

Novichikhin, I.A., Kuleshov, V.M. and Zaitsev, Y.A. (1977). Use of Protecting Flat-lying Seams in Donbass Mines. (In Russian). Donetsk: Donbass.

Nozhkin, N.V. (1979). Advance Degassing of Coal Deposits. (In Russian). Moscow: Nedra.

Orlova, A.V. (Comp.). (1981). List of Literature on the Problems of Fight against Coal, Rock and Gas Outbursts (1925-1979). Ed. A.P.Kulikov. Moscow: the Skochinski Inst.Mining.

Petukhov, I.M., Linkov, A.M., Sidorov, V.S. and Feldman, I.A. (1976). The Theory of Protecting Seams. (In Russian). Moscow: Nedra.

Nikolin, V.I., Balinichenko, I.I. and Simonov, A.A. (1981). Fight against Outbursts of Coal and Gas in Mines. (In Russian). Moscow: Nedra.

Proskuriakov, N.M. (1980). Sudden Outbursts of Rock and Gas in Kalium Mines. (In Russian). Moscow: Nedra.

Protosenia, A.G., Stavrogin, A.N., Chernikov, A.K and Shirkes, O.A. (1982). On the equations of limit state of gas-saturated homogeneous and non-homogeneous media and rock. (In Russian). Phys.-Techn. Probl. Razrabotki Poleznykh Iskopaemykh, 1.

Rzhevski, V.V. and Yamshikov, V.S. (1968). Ultrasonic Control and Investigations in Mining. (In Russian). Moscow: Nedra.

Salganik, R.L. (1980). On the Effective Characteristics of Material with a large Number of Cracks. Geophysical Determining of Crack Parameters with Respect to the Problem of Outburst Safety. (In Russian). Moscow:Inst. Probl. Mech. Acad. Sci. USSR, Preprint 154.

Salganik, R.L. (1982). Overall Effects due to cracks and crack-like defects: Proc. 1-st Internat. Symp. on Defects and Fracture; Tuchno, Poland, 1980. The Hagre/Boston/: Martinus Nijhoff Publ.

Vavakin, A.S. and Salganik, R.L. (1978). Effective elastic characteristics of bodies with isolated cracks, cavities and rigid inhomogeneities. Mech. Solids (13), 2.

Zabigailo, V.E. (1978). Geological Grounds for the theory of Forecast of Outburst Threat of Coal Seams and Rock.(In Russian). Kiev: Naukova Dumka.

Zorin, A.I. (1978). The Control of Dynamic Manifestations of Rock Pressure. (In Russian). Moscow: Nedra.

STRATA DEFORMATION AND SUPPORT PERFORMANCE AT A LONGWALL COAL FACE

Déformation des bancs du toit et rendement obtenu dans une longue taille de charbon

Gebirgsverformung und Verhalten des Schildes im Streb

R. N. Gupta
Lecturer in Mining Engineering, University College, Cardiff, Wales

I. W. Farmer
Reader in Mining Engineering, University of Newcastle upon Tyne, England

SYNOPSIS

During a series of detailed observations on three retreating longwall faces at Westoe Colliery, Tyne and Wear, U.K. the effect of setting pressure on support performance and strata deformation was investigated. Increased setting pressures, resulting in an increase in setting load density from .22 MN/m² were shown to reduce convergence and face spalling and roof flaking. This was shown to result mainly from the change in roof strata deformation from a wedge shaped compression zone with tension at the face edge to an even beam shaped compression zone over the face area.

RESUME

Lors d'une série d'observations détaillées effectuées sur trois longues tailles rabattantes à la mine Westoe, Tyne and Wear, R.U., on a étudié l'effet de la pression de pose du soutènement sur le rendement et la déformation des bancs du toit. Des pressions de pose accrues, se traduisant par une augmentation de la densité de charge de pose de 0,22 MN/m² à 0,45 MN/m², ont permis de réduire la convergence, le délavage du front de taille et l'écaillage du toit. Cela a semblé résulter principalement du changement de la déformation des bancs du toit passant d'une zone de compression en forme de coin, avec tension en bordure du front de taille, à une zone de compression en forme de poutre sur toute la surface de la taille.

ZUSAMMENFASSUNG

Im Laufe einer Reihe von Einzelbeobachtungen an drei Streben in der Grube Westoe, Tyne and Wear, G.B., wurden die Auswirkungen von Setzdruck auf Ausbauleistung und Gebirgsverformung untersucht. Gesteigerte Setzdrücke und daraus sich ergebende gesteigerte Setzlastdichten von 0,22 MN/m² führen zu verringerter Konvergenz, zu weniger häufigem Ausböschen des Kohlestoßes und zu geringeren Hangendausbrüchen. Es stellte sich heraus, daß diese Ergebnisse im wesentlichen auf andersartige Verformung der Hangendschichten zurückzuführen waren, d.h. anstelle einer keilförmigen Kompressionszone mit Spannungsauswirkung auf den Kohlestoß ergab sich eine homogene balkenförmige Kompressionszone über die gesamte Strebfläche.

1. INTRODUCTION

The majority of deep mined coal in the world is obtained from longwall faces - either advancing or retreating - supported by power operated hydraulic face supports. These can have frameworks based on chocks, shields, chock shields or shields incorporating lemniscate linkages, but the basic purpose is to support the roof strata of the working area. This is achieved initially by applying a positive resisting pressure normal to the plane of the roof, and subsequently by having available the capacity to resist roof lowering as the supported span increases temporarily during mining.

The general deformation regime to which the roof strata above a longwall face may be subjected is illustrated in Figure 1, which shows the vertical strain contours to a height of 70m above a retreating longwall face measured during an investigation at Lynemouth Colliery, Northumberland, and described by Hodkin (1978) and Farmer and Altounyan (1980). Strictly speaking these are not contours of engineering strain, but of deformation normalised in terms of the distance between anchors located in a borehole drilled vertically downwards from a tunnel above the face centre line. The deformation includes bed separation, loosening and fracture. Nevertheless the contours do show that ahead of the face there is a zone of compression which changes over the face area to a zone of tension where the unsupported roof

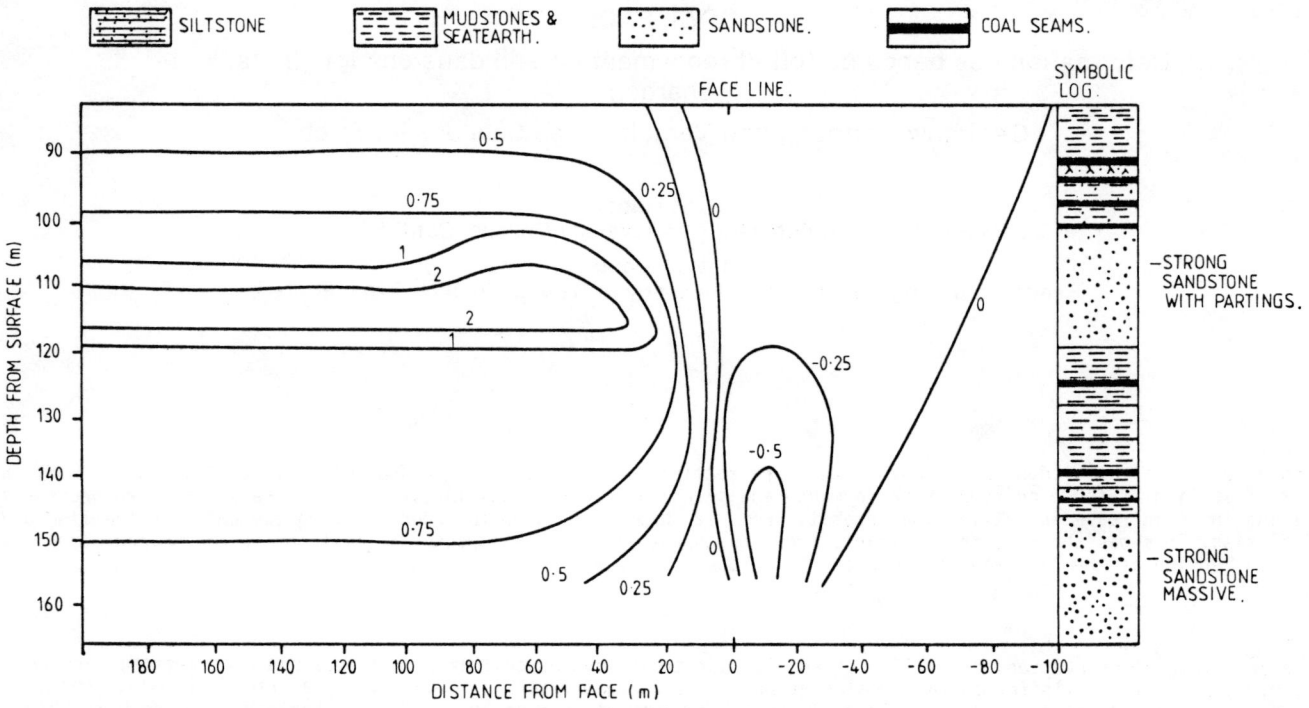

Figure 1. Contours of vertical strain computed from relative settlements between anchors in a vertical borehole above the centreline of a retreating longwall face at a depth of 165m (Farmer and Altounyan, 1980).

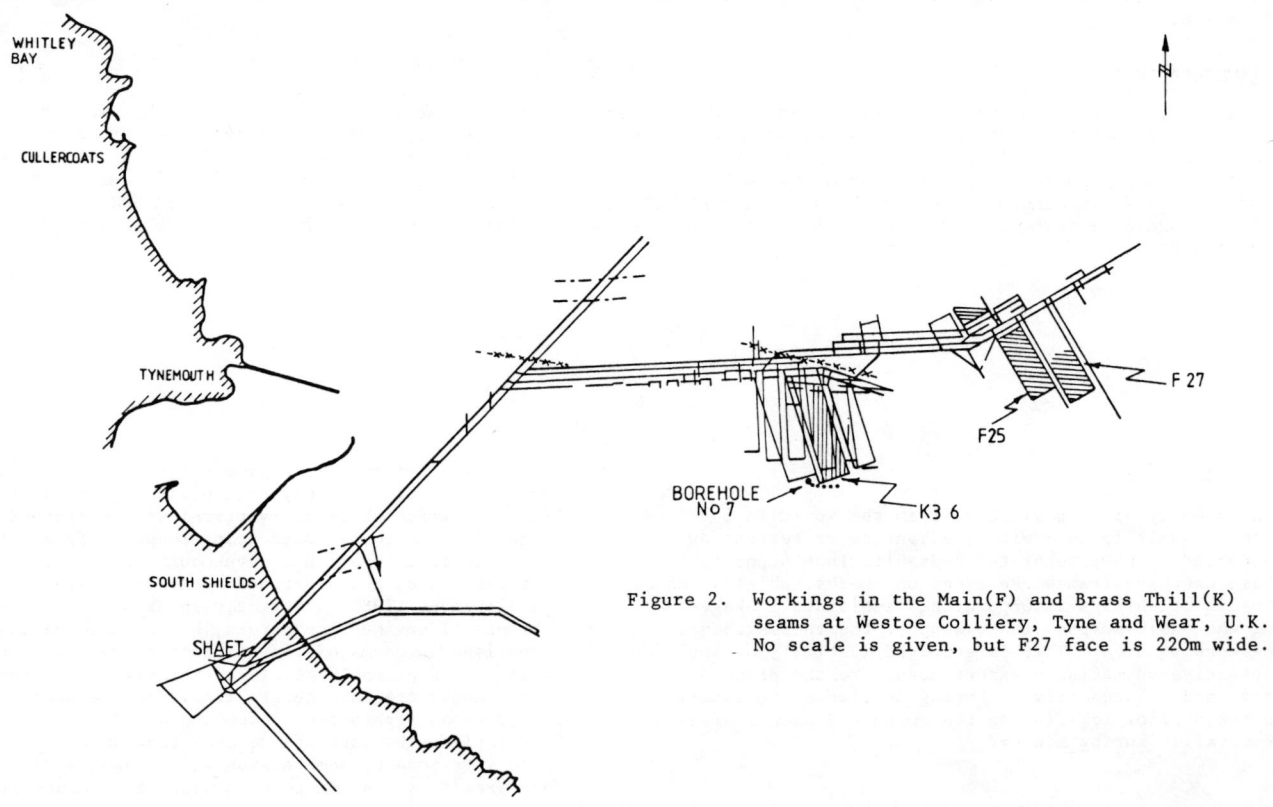

Figure 2. Workings in the Main(F) and Brass Thill(K) seams at Westoe Colliery, Tyne and Wear, U.K. No scale is given, but F27 face is 220m wide.

strata is allowed to cave. The compression zone has low strain magnitude and is spread over about 20m of the face abutment, due to yielding or fracturing of the coal seam. The tension zone reaches a stable level about 100m behind the face, demonstrating a capacity for long term dilation which is responsible for residual pillar stresses in coal mines. These are responsible for considerable stress interaction problems in multi-seam workings in mature coalfields.

The coal face lies in an intermediate zone between the compressed coal seam and the caving collapsed area, and is usually subjected to relatively minor stress and deformation. It has some similarities with the zone of rotational shear between active and passive stress zones in subsurface structural design.

The purpose of support in the face area can be explained by considering the basic mechanics of deformation at a typical longwall face line. The yielding of the front abutment, evident in Figure 1, under compression stresses redistributed from the unsupported caved area, results, not from plastic distortion, but from dilation of the coal seam in a direction normal to the face line. This results from induced vertical fractures (see Gupta 1982) parallel to the coal face, usually at intervals of 0.5 - 1m, which extend into the roof and floor. Gupta (1982) has shown that these can exist at distances greater than 10m into the coal face, but the greatest concentration is in the 3m of face abutment adjacent to the face.

The mechanism for crack formation can be explained in terms of Griffith's hypothesis. The redistributed compression stresses are predominantly vertical in the coal face abutment. Lateral restraint is reduced by removal of coal from the face and collapse of the caved roof. This allows the formation of fractures parallel to the major (vertical) stress direction and normal to the minor (lateral) stress direction, which may be imposed upon or exacerbate existing structural discontinuities such as cleat. The roof in the working area is partly supported as a cantilever by the coal at the coal face - although this support may be significantly reduced where coincidence of a major cleat direction parallel to the face with imposed fractures causes excessive spalling of the coal. The major objective of the supports is therefore to maintain the integrity of the roof working area - an advancing distance of up to 3m. The purpose of the paper which describes investigations at Westoe Colliery, South Shields, Tyne & Wear, is to examine the way in which the supports achieve this objective.

2. OBSERVATIONS

Westoe Colliery workings extend under the North Sea for about 8km roughly in line with the mouth of the River Tyne. The current workings (Figure 2) are in the Main(F) and Brass Thill(K) seams on either side of the 90 Fathom Fault. The faces chosen for investigation were F25 and F27 at an average depth of 277m below sea level and 232m below sea bed level, and K36 at an average depth of 345m below sea level. A section above K36 is given in Figure 3. The seam section worked on F25 and F27 face was 2.13m and on K36 face 2.0m. The supports used on F27 and K36 face were Gullick Dobson 6/300 tonne rigid base fully shielded chock supports at 1.2m centres. The supports on F25 face were Gullick Dobson 4/450 tonne chock shield supports with lemniscate linkages at 1.5m centres. All three faces were retreating longwall with face lengths of 219m (F25), 220m (F27) and 215m (K36). The face retreat directions were NW - roughly parallel to the major cleat direction.

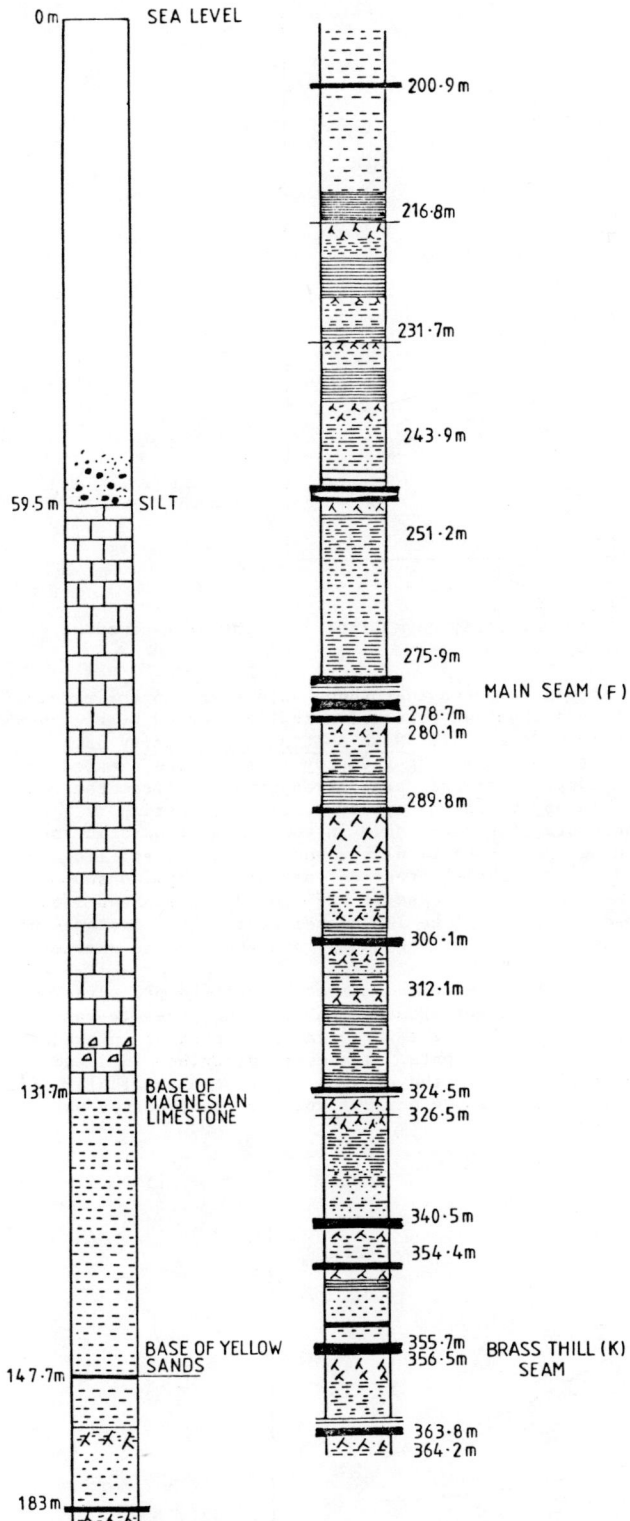

Figure 3. Symbolic borehole log from Borehole No 7 adjacent to K36 face in the Brass Thill seam. The position of the borehole is shown in Figure 2. Note that in the Coal Measures, coal seams are indicated by depth, shales by continuous lines, sandstones by broken lines and seatearths by root symbols.

E 53

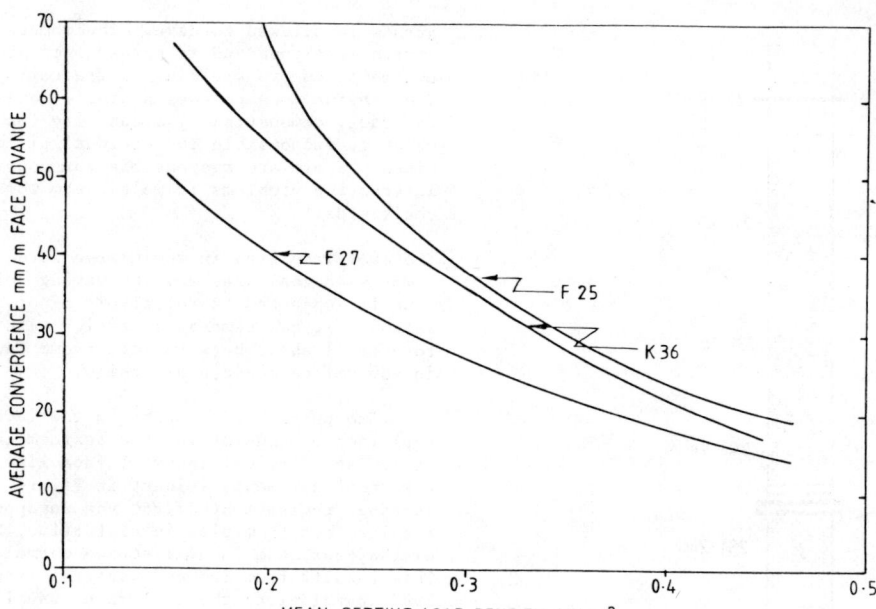

Figure 4. The effect of mean setting load density on roof to floor convergence, normalised in terms of face advance, over a single cutting cycle.

The ring main setting pressure for each of the face support systems was between 13.8 MPa and 16.8 MPa with positive set facilities capable of boosting setting pressures to 31 MPa. Variations in nominal values occurred during operations. Expressed in terms of support setting load density exerted on the roof these setting pressures gave the following mean setting load densities for F25 face:

Setting Pressure MPa	Mean Setting Load Density MN/m^2
13.8	0.208
17.2	0.260
20.7	0.312
24.1	0.364
27.5	0.416
31.0	0.468

It should be noted that these values are approximately 20% less than the pressure actually exerted by the support canopy on the roof. <u>Mean Setting Load Density</u> as defined in the mining industry is the force exerted by the support per unit area of exposed roof from the rear of the canopy to the face, at the time of setting. It allows, therefore, for the change in roof area exposed during the mining cycle - particularly where there is spalling, but may be an unsatisfactory definition of applied support pressure. In the case of F27 and K36 faces Mean Setting Load Density will be 10% higher because the distance between the front of the canopy and the face is shorter.

In order to examine the effect of setting pressures on roof and face performance, the setting pressure was varied on each of the three faces as part of an extensive series of experiments. These are described in detail by Gupta (1982). In the present paper, the particular effect of setting pressure on roof and face performance was monitored.

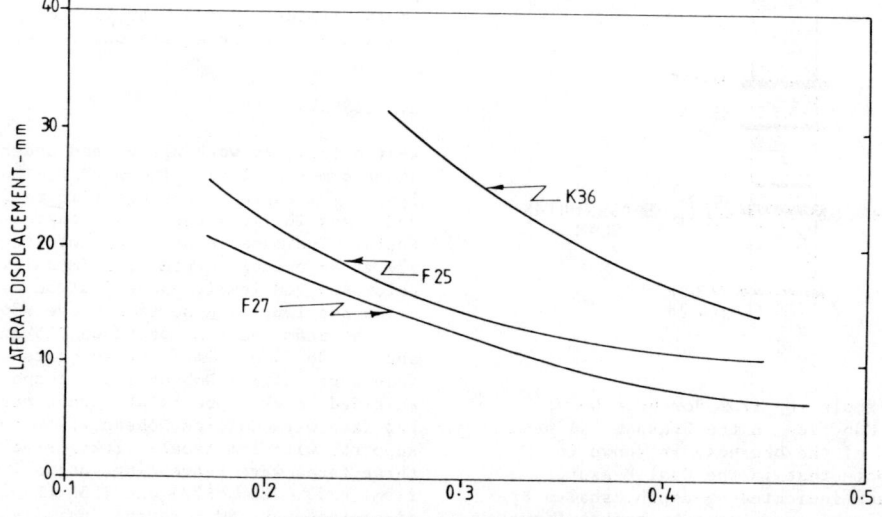

Figure 5. The effect of mean setting load density on lateral face movement over a 60 hour period.

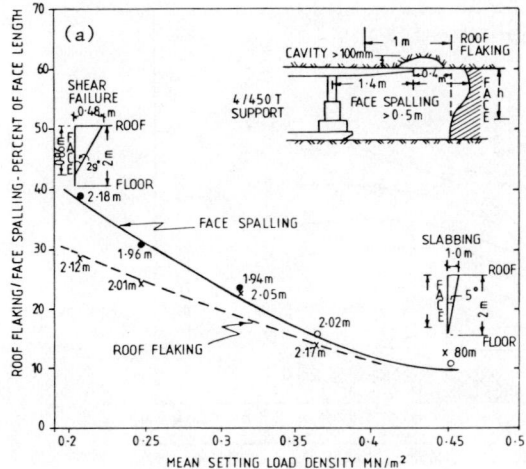
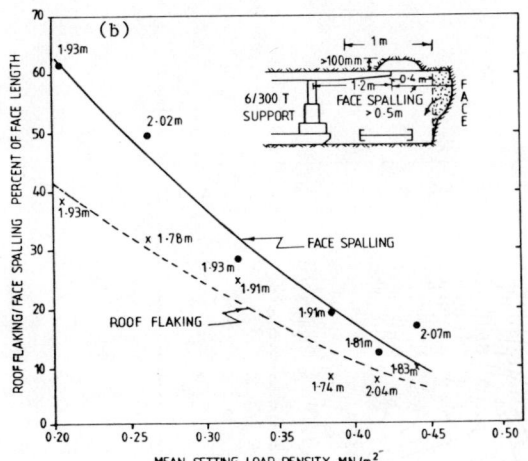
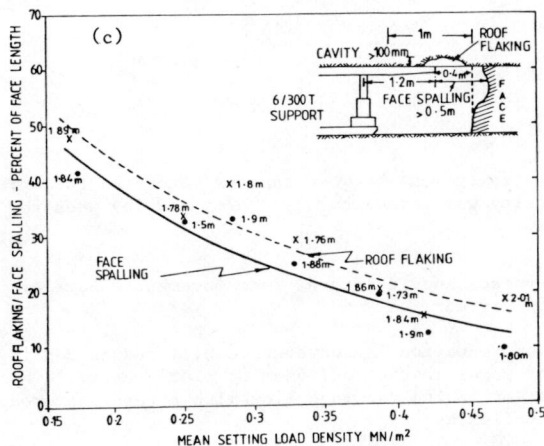

Figure 6. The effect of mean setting load density on roof flaking and face spalling on (a) F25 face, (b) K36 face and (c) F27 face. Note: flaking is defined as roof cavities greater than 100mm deep and 300mm long. Spalling is defined as a face distance greater than 0.5m from the support tip. Distances on the curves are prop free front distances.

In designing the experiment, the setting pressure, and the related mean setting load density, was selected as the independent variable since this is the feature of the face supports which immediately affects the lower roof strata. In the full programme the following observations were obtained at each of the setting pressures listed above:

(a) All leg pressures at each support at the time of setting and immediately before moreover.

(b) Leg closures at 10% of supports.

(c) Roof to floor convergence at the same location as leg closures.

(d) Support inclinations at these locations.

(e) Roof and floor debris compaction at these locations.

(f) Lateral movement of roof relative to floor at these locations.

(g) Roof flaking and cavity formation throughout the face. Cavities greater than 100mm high and 300mm wide were monitored.

(h) Face spalling throughout the face. This was considered significant and monitored when the distance from the tip of the support canopy to the face exceeded 500mm.

(i) Lateral movement of the coal face resulting from dilation behind the face.

Each of these observations was obtained over 4 cutting cycles and two weekend breaks at each pressure. The cutting cycles represented an average duration 2 hrs 30 mins during which the face and supports were advanced 0.45m on F25 and K36 face and 0.65m on F27 face. The weekend break was of average duration 60 hours from setting to first moreover.

In addition to the main observation programme, at one location at the centre of F25 and K36 face and 3 locations on F37 face, roof and floor deformations were observed using an anchor wire or magnetic ring extensometer system installed in a borehole vertically above the face supports. The detailed instrumentation has been described by Gupta and Farmer (1981).

3. CONVERGENCE, ROOF FLAKING AND FACE SPALLING

From the large amount of data (see Gupta 1982) collected, that concerning the effect of setting pressure on face stability is the most important, and the observations from (g) (h) and (i) of the programme above and from the extensometer observations illustrate both the effect of setting pressure, and more importantly the reason for that effect.

The effect of setting pressure on roof to floor convergence is illustrated in Figure 4 for the three faces. This is the average convergence which takes place over a single extraction cycle, normalised in terms of the distance advanced. It is interesting to note the reduced convergence of F27 face where a wider cut and a resultant increased advance was obtained. If the results were adjusted for this, the convergence per cut would be almost identical for the three faces. There is a similar relation between setting load density and lateral movement of the coal face, which is illustrated in Figure 5. This can probably be explained in terms of the reduced vertical compression transmitted to the coal closest to the face at higher setting loads.

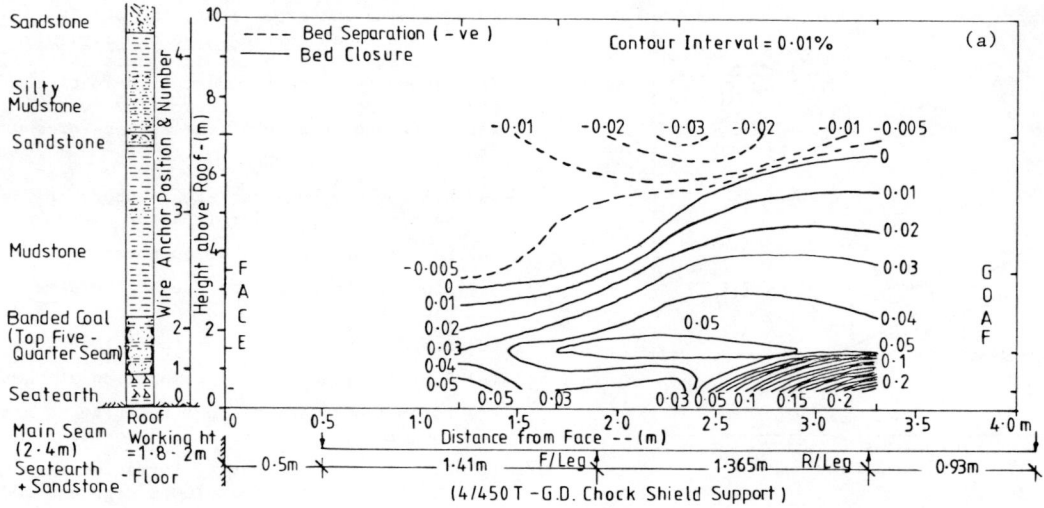

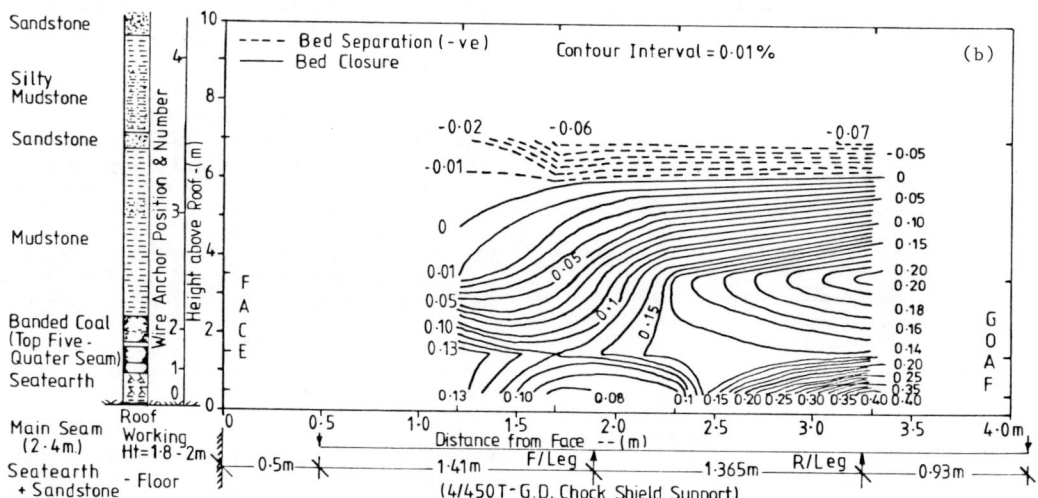

Figure 7. Contours of vertical "strain" computed from differential displacement between anchors connected to a wire extensometer above F25 face (a) when the support load density was raised to 0.26 MN/m^2 and (b) when the support load density was raised to 0.42 MN/m^2.

It is interesting to note here that whereas lateral displacement - measured in this case over a weekend period - is similar in the case of F25 and F27, the two Main Seam faces, it is much higher in the case of K36, in the Brass Thill seam. This is particularly the case at low setting pressures and it is accompanied (Figure 6b) by much greater face spalling. The explanation for both of these probably lies in the presence of 9 minor faults in the face and the presence of pillars and caved Main Seam workings about 70m above the face.

The most interesting data is in Figures 6 a, b, c which show the effect of setting pressures on roof flaking and face spalling, defined earlier and redefined in the figures. It can be seen that the frequency of roof cavities decreases significantly with increasing setting pressure. Similarly face spalling reduces by the same sort of magnitude. The reasons for this are probably interrelated and also related to the reduction in convergence and in lateral face movement. Some of these may be listed:

(a) The reduction in convergence will reduce bending stresses in the roof beam or plate where it "hinges" at or behind the face line with a resultant reduction in flaking.

(b) The reduction in lateral expansion of the face will reduce the tendency to spalling along expansion fractures.

(c) The reduction in spalling by giving greater support to the roof and reducing the canopy tip to face distance also reduces the unsupported roof area.

These and other points may be illustrated by examining the "strain" contours obtained from borehole measurements above the face in Figures 7 - 9. These have been selected to compare for each face the strain distribution

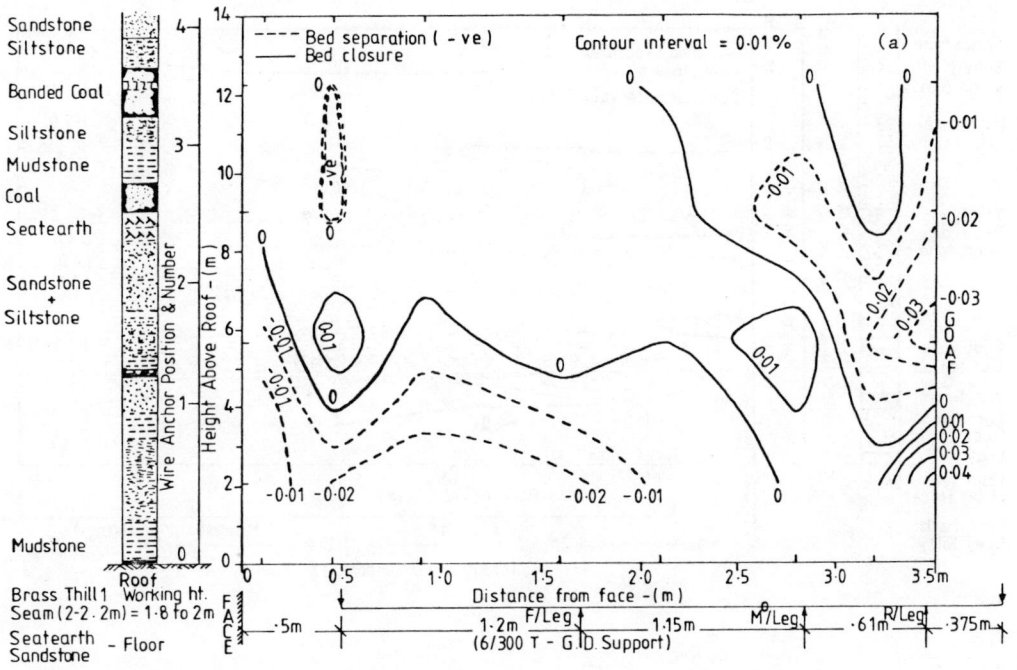

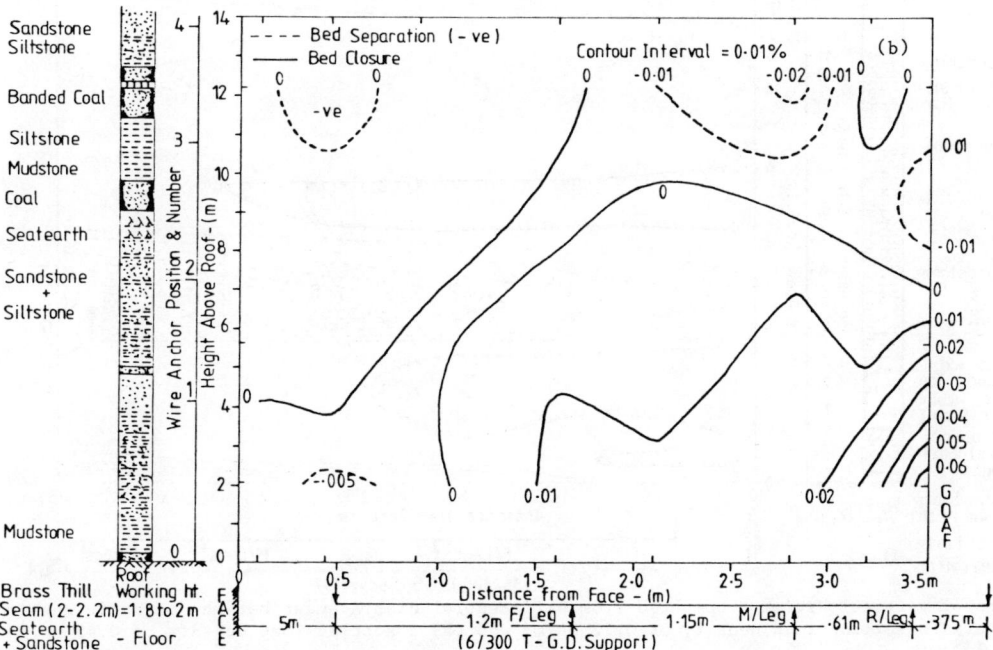

Figure 8. Contours of vertical "strain" computed from differential displacement between anchors connected to a wire extensometer above K36 face (a) when the support load density was raised to 0.22 MN/m² and (b) when the support load density was raised to 0.42 MN/m².

at two mean setting load densities, one at about 0.22 MN/m² equivalent to that exerted by the normal power support ring main pressure; the other at about 0.42 MN/m² equivalent to that exerted by the maximum boosted positive set pressure.

4. STRATA DEFORMATION

In Figures 7 - 9 "strain" has been computed from the relative deformation of anchor wire extensometers (Figures 7 and 8) and magnetic extensometers (Figure 9) installed in a borehole initially drilled in front of the support canopy tip. The experimental procedure varied. On F25 and K36 faces (Figures 7 and 8) following each advance of the face, support pressures for 10m each side of the

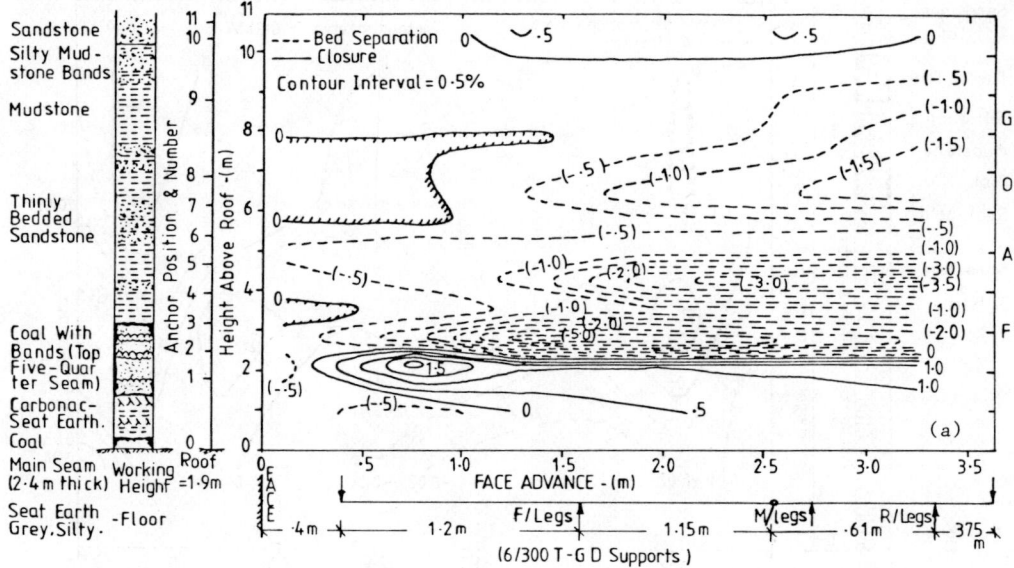

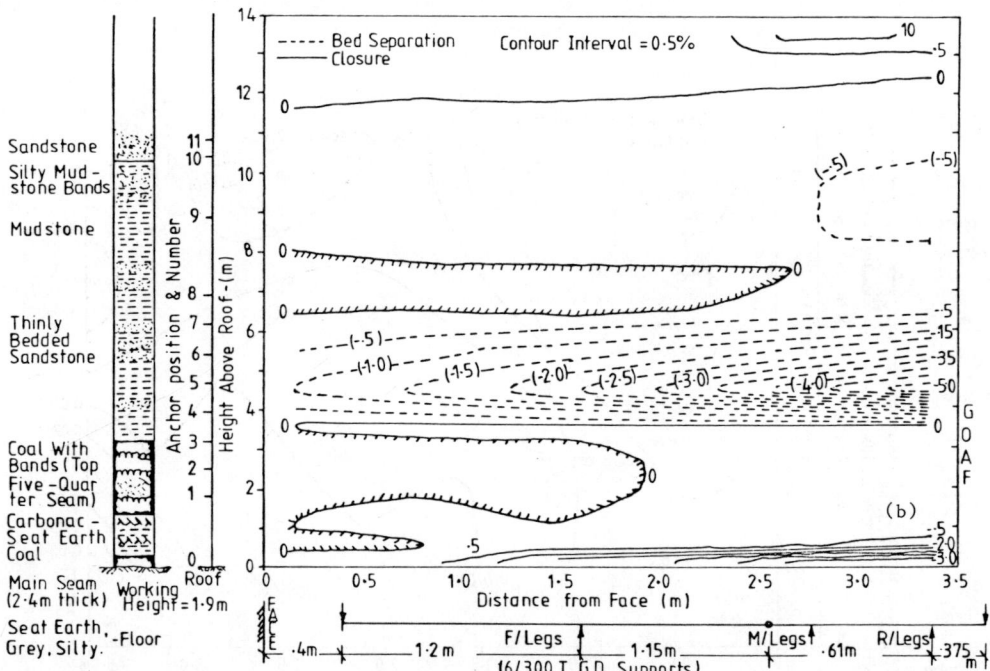

Figure 9. Contours of vertical "strain" computed from differential displacement between magnetic anchors above F27 face at (a) a setting load density of 0.22 MN/m^2 and (b) a setting load density of 0.45 MN/m^2.

borehole were raised through the full range of available pressures. The data in Figures 7 and 8 is selected from this range.

On F27 face two boreholes were drilled and all the supports on the face were maintained at a pressure sufficient to give the two setting load densities quoted in Figure 9. Although the techniques differ, the necessary act of support lowering during longwall mining is sufficient to allow valid comparisons to be made.

Divergent views have been expressed as to the height of the immediate roof which is affected by roof support (see forinstance Alder 1968, Habnicht, 1972, Wagner and Steijn, 1979). The thickness of lower strata which cave in the goaf after the removal of support has been quoted in the literature as varying from 1.5 times to 13 times the seam thickness (Ilstein, 1960). Few, if any, observations have been reported which explain in detail the mechanics of deformation of the immediate roof strata supported by the face supports.

The experiments on F-25 (Figure 7) and K-36 (Figure 8) faces confirmed that with a wire extensometer system the presence of wires limits the number of anchors which can be installed in a borehole to four and introduces many possible sources of error. Whilst useful information was obtained with the system, the higher resolution possible with a greater number of anchors was necessary in order to study in greater detail the influence of increasing support setting pressure on strata deformation.

It is appropriate to reiterate here that the plots of the percentage vertical strain contours against face advance, and height above the seam, in Figures 7 - 9, do not represent pure engineering strain, since the movements can consist of bed separation, rock failure, and elastic deformation. The magnitude of strain appears to be influenced by the distance between the anchors with higher anchor distances tending to give lower strain values.

The more important observations which can be made are:

(a) On F25 face the data is confined to the section above the support (Figure 7). The main change with increased pressure is from a wedge shaped compression zone above the support at the lower pressure to a beam shaped compression zone, with significantly increased compression. It is interesting to note that the change from compression to tension about 7m above the roof level, and indicating a possible zone of detatchment, coincides with the change in lithology from mudstone to sandstone.

(b) On K36 face the data is confused but at the lower load density there is evidence of considerable tension or roof sag at the front of the supports (Figure 8). It was on this face (see Figure 6b) that considerable roof flaking occurred at lower setting load densities. Even at higher pressures there is a limited tension zone at the front of the support. The "domed" shape of the strain contours at both high and low stress can be attributed to hinging of the chock shield support - particularly in the roof disturbed by flaking.

(c) The strain contours above F25 (Figure 9) face again demonstrate the existence of a tensile zone in front of the support canopy at lower pressures. At the higher setting load density, a compression "beam" 4m high and extending to the face line is created. This illustrates the basic mechanism of beam building in the roof above face supports at higher setting pressures.

5. CONCLUSIONS

(a) Increases in support setting load density on three retreating longwall faces at Westoe Colliery, from about 0.22 to 0.42 MN/m^2 were observed to:

    (i) reduce average roof to floor convergence over the supported face area

    (ii) reduce lateral expansion of the coal face into the working area

    (iii) reduce the extent of roof cavity formation

    (iv) reduce the extent of face spalling

(b) Observations of roof deformation showed that increases in support load density changed the vertical strain pattern immediately above the supports from an uneven compression zone with a tension zone ahead of the supports, to an evenly distributed compression zone over the supported area.

ACKNOWLEDGEMENTS

The work was carried out with the support of a grant from the National Coal Board extramural research fund and the E.C.S.C. research fund (Contract 7220 - AC/806). The authors wish to acknowledge the assistance of Mr. N.H. McLeod and Mr. D. Kelly of the North East Area Headquarters and Mr. I.W. Day and Mr. T.F. Burns of Westoe Colliery.

REFERENCES

Alder, L., (1968). Roof control in longwall mining. Mining Congress Jl., March, pp 58-67

Farmer, I.W., and Altounyan, P.F.R., (1980). The mechanics of ground deformation above a caving longwall face. Ground movements and structures (ed. J.D. Geddes) London, Pentech, pp 75-91.

Gupta, R.N., and Farmer I.W., (1981). A magnetic ring extensometer system for strata deformation measurement in coal mines. Mining Engineer, 141, pp 303-305.

Gupta, R.N., (1982). Influence of setting pressure on support performance and stata behaviour on longwall faces. Ph.D. Thesis, University of Newcastle upon Tyne.

Habernicht, H., (1972). Systematic development of powered supports for faces in weak rock. Proc. 5th Int. Conf. Strata Control, London, Paper 6.

Hodkin, D.L., (1978). Interaction between longwall and pillared workings at Lynemouth Colliery. Ph.D. Thesis, University of Newcastle upon Tyne.

Ilstein, A., (1960). Influence of the resistance of the support system on the manifestation of rock pressure in longwall faces. Proc. 3rd. Int. Conf. Strata Control, Paris pp 127-132.

Wagner, H., and Steijn, J.J., (1979). Effect of local strata conditions on the support requirements on longwall faces. Proc. 4th Cong. Int. Soc. Rock Mechs., Montreux, 1, 557-564.

APPLICATION OF THE DISPLACEMENT DISCONTINUITY METHOD TO THE PLANNING OF COAL MINE LAYOUTS

Application de la méthode de déplacement discontinu à la planification des traçages de houillères

Anwendung der Verschiebungs-Diskontinuitäts-Methode auf die Planung von Kohlegruben

L. J. Wardle
Senior Research Scientist, CSIRO, Division of Applied Geomechanics, Syndal, Melbourne, Australia

J. R. Enever
Principal Research Scientist
CSIRO, Division of Applied Geomechanics, Syndal, Melbourne, Australia

SYNOPSIS

A three-dimensional displacement discontinuity method for stress analysis of practical mine layouts in a layered, anisotropic rock mass is described. The methods used to determine reliable estimates of input parameters are illustrated by results from a coal mine currently being studied. Novel laboratory and field approaches used to obtain the stress-strain properties of the coal, the rock mass and caved waste material are described.

RESUME

On décrit une méthode de discontinuité de déplacement à trois dimensions pour l'analyse des efforts dans le cas de traçage d'accès dans un massif anisotropique stratifié. On démontre les méthodes employées pour parvenir à des appréciations sûres des paramètres d'entrée au moyen des résultats obtenus dans une mine à charbon actuellement en cours d'étude. On décrit de nouvelles méthodes d'étude, en laboratoire ainsi que sur le terrain, pour déterminer les propriétés d'efforts et de contraintes du charbon, du massif et des éboulis de foudroyage.

ZUSAMMENFASSUNG

Es wird eine dreidimensionale Verschiebungsunstetigkeits-Methode der Spannungsanalyse bei praktischen Grubenanordnungen in einem geschichteten, anisotropen Gebirgskörper beschrieben. Die zur Bestimmung sicherer Eingangsparameterabschätzungen verwendeten Methoden werden mittels aus einer zur Zeit untersuchten Kohlengrube erhaltener Ergebnisse veranschaulicht. Neue, im Labor und im Felde zur Bestimmung der Spannungs- bzw. Verformungseigenschaften der Kohle, des Gebirgskörpers sowie der Bruchberge verwendete Betrachtungsweisen werden beschrieben.

1. INTRODUCTION

In Australia, an increasing proportion of underground coal mining is being carried out by so-called 'total extraction' methods such as longwalling, shortwalling and pillar extraction. These mining systems involve caving of the immediate roof of the excavation and subsequent load carrying by the caved waste material. For these mining methods, numerical stress analysis allows assessment of the vertical load sharing arrangement at any stage in the excavation sequence between:

i) the coal left intact as pillars, excavation blocks etc.

ii) the area of caved waste that is in contact with the overlying competent strata, and

iii) any locally introduced support system, such as longwall face supports.

Reliable predictions of this relative load sharing allow conclusions to be drawn about the likely performance of any coal left either "permanently" or temporarily intact, and deformations induced in the "intact" strata overlying (and underlying) the mining area, particularly if seams in close proximity are to be subsequently extracted or if subsidence damage to surface structures is an important consideration.

Many geometrical features of total extraction systems, for instance near the gate ends of a longwall face, cannot be adequately modelled by two-dimensional methods.

Although in principle three-dimensional finite element or general boundary element models can be used for such geometries, the high cost and effort required for such analyses can seldom be justified. The displacement discontinuity method (a special form of boundary element method) is based on replacing actual excavation geometries by thin slits of the same plane area. This approach allows three-dimensional stress-analysis at a fraction of the cost of alternative methods.

Early displacement discontinuity models were somewhat restrictive in the range of problems that could be analysed. Extensions to the method to incorporate aspects necessary for realistic modelling of underground coal mine layouts, for example anisotropy and layering of the rock mass, are briefly described in the next Section.

The reliability of predictions obtained from a numerical model can only be commensurate with the reliability of the input parameters used. The methods used to determine the relevant input parameters are described by reference to a mine that is currently being investigated by the authors. A number of novel experimental approaches are described, including measurement of the deformability of the caved waste <u>in situ</u> by plate loading, and large scale triaxial testing of the coal. The numerical model has been used to back-calculate the anisotropic "elastic" constants of the rock mass from observed subsidence data. It is shown that the order of magnitude disparity between typical laboratory values of Young's modulus and values obtained by back-calculation using observed subsidence data and isotropic elastic models, can be removed by using appropriate anisotropic moduli.

Finally the three-dimensional capabilities of the numerical model are illustrated by some results from a detailed stress analysis of a typical panel extraction scheme.

2. DISPLACEMENT DISCONTINUITY METHOD

The idea of modelling the behaviour of tabular excavations by equivalent cracks or slits dates back to Hackett (1959). The closure of each pair of points in roof and floor is treated mathematically as a discontinuity in displacement at a single point. Berry (1960) and Berry and Sales (1961, 1962) used simple displacement discontinuity solutions for isotropic and anisotropic media to model surface subsidence profiles. All of this early work assumed that the mined panel could be modelled by a single uniform displacement discontinuity. The extension to allow more realistic boundary conditions and excavation geometries was made possible by dividing the plan area of the excavation into 'elements', each representing a uniform displacement discontinuity. (Plewman <u>et al</u>., 1969, Starfield and Crouch, 1973).

The aforementioned techniques have been restricted to a homogeneous continuum. Diering (1980) treated piecewise homogeneous isotropic continuum properties by using elements along the geological interfaces. However, the iterative equation solution would only converge if the equations were ordered in a particular way.

The authors have extended the displacement discontinuity method to a layered anisotropic rock mass (Wardle, in preparation). The rock mass can consist of an arbitrary number of layers parallel to the earth's surface. The elastic properties in any layer are cross-anisotropic (i.e. transversely isotropic) with a vertical symmetry axis and can be specified in terms of the Young's moduli E_h and E_v, Poisson's ratios ν_h, ν_{vh}, $\nu_{hv}(=\nu_{vh}E_h/E_v)$, and shear modulus G. The subscripts h and v denote horizontal and vertical directions.

The interfaces between the layers can be fully continuous or fully frictionless (i.e. zero shear stresses). The numerical solution procedure involves subdividing the plan area of interest into rectangular "elements". Each element is assigned a property code indicating whether it is mined out, intact coal or caved waste material. For elements transmitting load between roof and floor (i.e. intact coal or bulked waste), one-dimensional stress-strain properties are used. For the coal these are specified in terms of stiffness G_s/s and E_s/s where E_s, G_s and s are the Young's modulus, shear modulus and thickness of the coal. The bulked waste is assumed to have negligible shear stiffness and a compressive stiffness E_c/h_0 where E_c is the "modulus" and h_0 the initial height of the waste.

The resulting system of equations is solved iteratively. This allows non-linear coal and waste properties to be used without a substantial increase in computer time.

3. CASE STUDY

3.1 Background

The actual problem used to illustrate the capabilities of the method described is based on a specific study currently being made by the authors of some aspects of mine layout design at the Thiess Bros. Laleham No. 1 Colliery at South Blackwater, Queensland (Fig. 1). Mining at the colliery is being conducted in the lower or "C" seam of three seams that occur in the area. Figure 2 summarizes the typical stratigraphy. Early efforts at mining Laleham were concerned with resource proving and assessment of potential for total extraction. For the last few years effort has concentrated primarily on development of the lease for eventual longwall extraction of "C"-seam. Experience with a limited amount of pillar extraction by various versions of the "Wongawilli" system has emphasized the need for systematic geotechnical planning prior to the introduction of any comprehensive total extraction system. Particular interest has been focussed on the need for optimal sizing of coal pillars, both permanent and temporary, the sizing of openings and alternate solid coal developed during coal winning, and the possible effect of total extraction of "C"-seam coal on likely subsequent mining of the overlying "A"-seam.

Experience has shown that the "C"-seam coal is prone to spontaneous combustion when crushed sufficiently to allow short-circuit passage of ventilation air through pillars. This implies the need to design pillars that will not crush excessively during any part of the mining cycle, when critical ventilation conditions exist. On the other hand, successful total extraction may require pillars left as part of the mining process to yield at some stage, in order to avoid formation of uncontrollable stress concentrations. These could make mining conditions untenable in subsequent extraction. Careful attention must be paid to the likely performance of pillars at various stages of the mining process, to ensure they meet these potentially conflicting requirements.

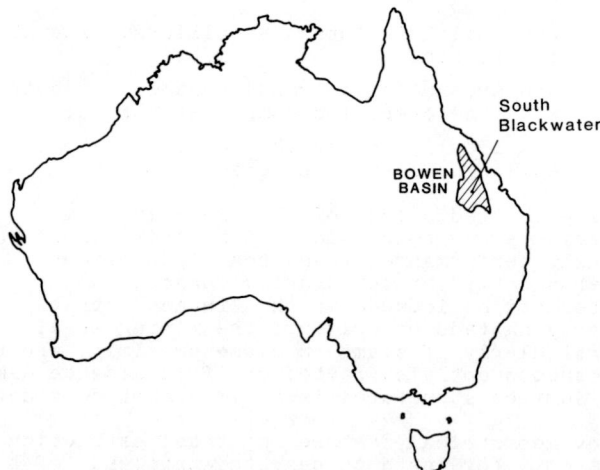

Figure 1. Location Map

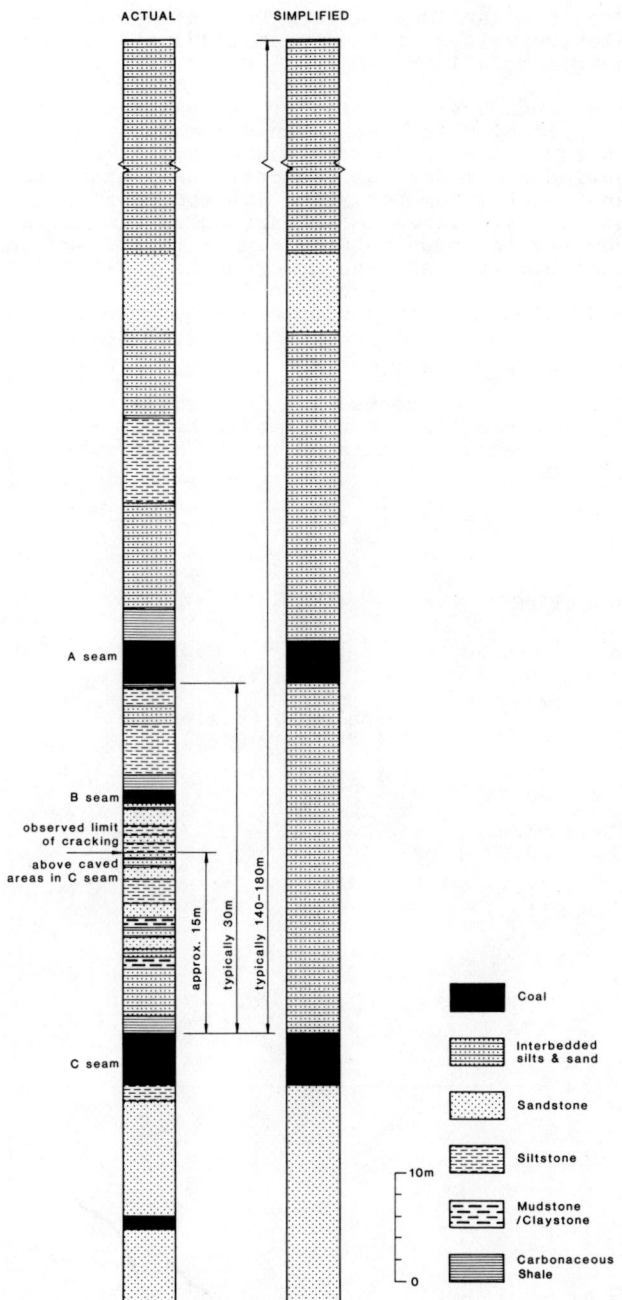

Figure 2. Stratigraphic sections

Operational considerations often dictate the practical limits of geometric layout that can be employed for any total extraction system. Experience at Laleham has, however, indicated the need for careful manipulation of the relative sizes and spacings of transient openings, within the limits set by operational considerations, to avoid problems such as premature crushing of blocks of coal which are still required to remain intact as an integral part of the mining system and the location of extraction openings in abutments highly stressed from prior extraction.

Surface drilling into the roof strata above extracted areas of "C"-seam has shown cracking to approximately 15 m above the top of the seam. This, in conjunction with the relatively thin and soft interburden between the "C"- and "A"-seams (Fig. 2) points to the likelihood of an adverse influence of "C"-seam extraction on extraction of "A"-seam. Careful planning is needed to ensure that development layouts proposed for "A"-seam extraction are located to minimize the influence of deformations at "A"-seam caused by prior mining of "C"-seam.

3.2 Determination of Input Parameters

Substantial effort has been devoted to the determination of meaningful input data for the numerical model. The authors have placed particular emphasis on this activity because reliable predictions can only be achieved if the in situ scale properties are approximated.

The particular input parameters requiring substantial effort to define are as follows:

i) deformational properties of caved waste

ii) elastic characteristics of layered continuum, and

iii) non-linear stress-strain properties of coal

3.2.1 Caved waste properties

Initial effort at determining the response of simulated caved waste piles in the laboratory proved quantitatively unsatisfactory because of the obvious problems associated with reproducing the spatial distribution of the component particles. A series of plate loading tests was performed on piles of caved material resulting from roof failures at Laleham Colliery. This was considered to provide the most realistic estimate of the behaviour of caved waste in a total extraction panel that could be practically achieved. The test arrangement is shown in Fig. 3. It was noted during the test that the loaded plate punched into the pile and that the visible surrounding material did not move substantially. This observation led to interpretation of the results on the basis of simple uniaxial loading of a column of the material, of a height determined by survey of the pile. The result of one test so interpreted is shown in Fig. 4. The near linear behaviour observed is consistent with the performance indicated during laboratory testing of simulated piles at relatively low levels of loading.

The maximum contact pressure reached in this test (1.3 MPa) represents approximately 35% of the overburden stress anticipated to occur at the depth of "C"-seam in the areas of interest for the study. A similar order of loading is predicted by the numerical modelling described later. (See Fig. 9).

3.2.2 Continuum properties

As a first step towards quantifying the continuum properties a series of unconfined compression tests was conducted on specimens prepared from core collected from a vertical hole drilled through to below C-seam. Tests were conducted on all the characteristic rock types. The properties determined were the Young's modulus in the vertical direction, (E_v), the vertical-to-horizontal Poisson's ratio (ν_{vh}) and the confined compressive strength (UCS). The results are summarised in Table 1. A simplified representation of the stratigraphic column with corresponding "lumped" properties is shown in Fig. 2. The arrangement was composed to incorporate material property contrasts thought significant in the overall performance of the layered medium.

The obvious shortcomings of relying entirely on laboratory values for the continuum properties led the authors to investigate the use of back-analysis of measured subsidence data. Using this, laboratory values were adjusted to give the most realistic estimates of these parameters, as input data for detailed modelling. This approach also allowed the relative importance of the various anisotropic parameters to be assessed in the light of the observable gross response of the rock mass.

Of particular importance were considered the relative values of the shear modulus (G), and Young's moduli (E_v, E_h).

As a first step a series of generalized two-dimensional models were run using the computer programs developed by the authors. A cross-section was modelled of a previously extracted panel at Laleham for which limited subsidence data was available. The particular simplified geometry is shown in Fig. 5. At this stage the continuum was taken as homogeneous and the coal as rigid. The surface subsidence profile predictions for this particular geometry, for a range of ratios of G/E_v, were calculated for a ratio of $E_v/E_h = 1.0$. The subsidence profiles in Figure 5 are expressed in the normalized form $\delta(x)/\bar{\delta}$ where $\delta(x)$ is the subsidence at a point x on the transverse axis, $\bar{\delta}$ is the mean subsidence defined by

$$\bar{\delta} = (\frac{1}{W}) \int_{-\infty}^{\infty} \delta(x)\ dx$$

TABLE 1 - ROCK PROPERTIES

Rock types	n	Laboratory values			Used for modelling		
		E_v (GPa)	ν_{vh}	UCS (MPa)	E_v (unscaled) (GPa)	E_v (scaled) (GPa)	ν_{vh}
Sandstones	16	14.6 (9.8)	.25 (.05)	71.2 (20.7)	12.0	3.4	.25
Interbedded silts & sands	6	5.6 (1.5)	.20 (.05)	51.0 (8.5)			
Siltstone, Mudstone Carbonaceous Shale	5	3.4 (1.4)	.19 (.03)	37.1 (11.1)	6.0	1.7	.25

(NB. Standard deviations in brackets)

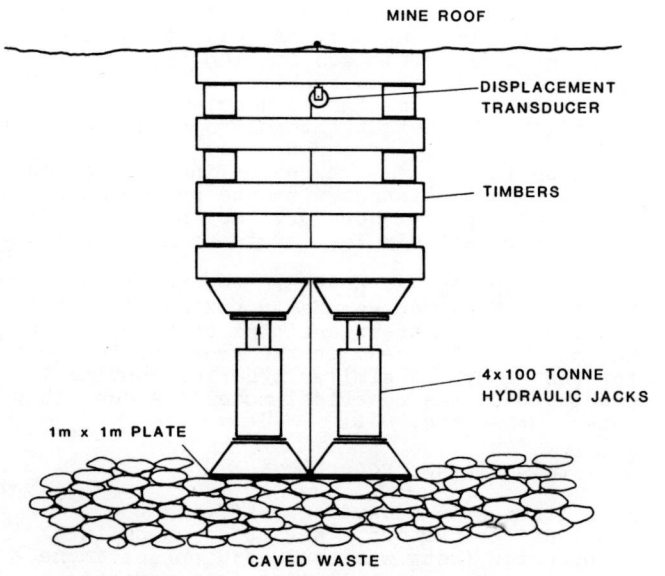

Figure 3. In situ plate loading apparatus

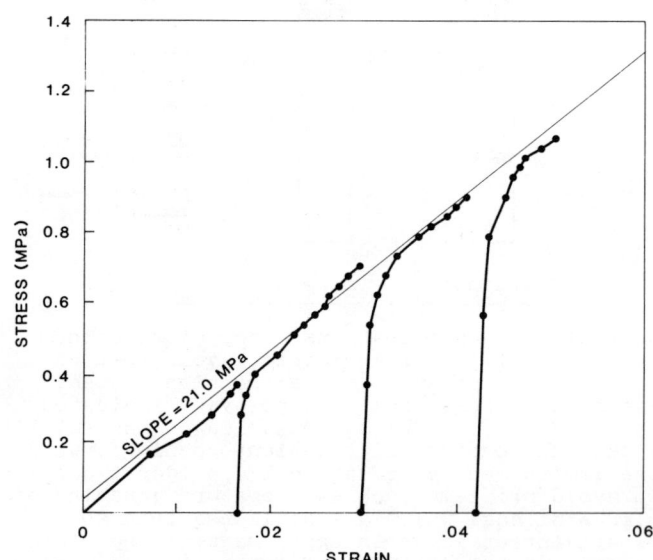

Figure 4. Results from in situ plate loading of caved waste

and w is the width of the panel. Using some results derived by Salamon (1963), $\bar{\delta}$ can be shown to equal the mean closure on the panel, if closure over intact coal is ignored. To graphical accuracy variations in Poisson's ratios for the continuum and stiffness of the caved waste had negligible effects on the subsidence profiles as plotted in Fig. 5. Superimposed on Fig. 5 is measured data for a single shortwall panel at Lambton colliery, Newcastle (Kapp, 1978). This particular case was geometrically very similar to the panel in question at Laleham. Evidence in this case points to the necessity of using anisotropic continuum properties (G/E_v much less than 0.5) to obtain reasonable agreement with observed subsidence profiles. This has been shown for U.K. conditions also (Crouch, 1976).

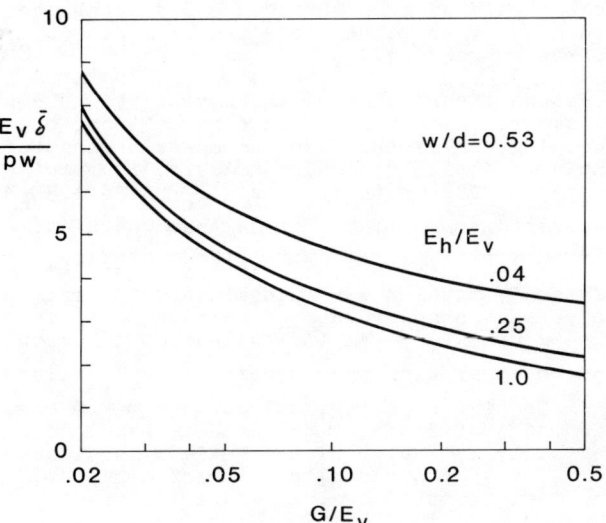

Figure 6. Effect of E_h/E_v and G/E_v on predicted subsidence

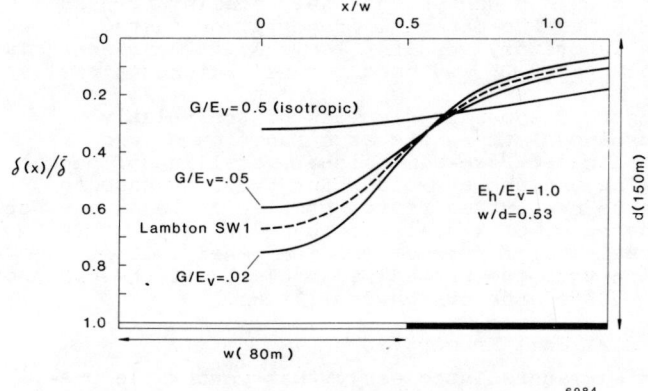

Figure 5. Effect of anisotropy of continuum properties predicted on subsidence profile

Figure 6 is based on results of a series of analyses for the same general conditions as previously described, but for a range of values of E_h/E_v. In this case the dimensionless variable $E_v\bar{\delta}/pw$, where p is the pre-mining vertical stress, is used to indicate the sensitivity of subsidence predictions to varying E_h/E_v for a range of G/E_v values. As can be seen, for higher degrees of anisotropy ($G/E_v = 0.05 \rightarrow 0.02$), the sensitivity to substantial variations in E_h/E_v is modest. Based on this observation and in the absence of any more detailed information on the in situ E_h/E_v value, the ratio was kept at 1.0 for all subsequent modelling. Although it is recognized that detailed predictions of stress changes will be sensitive to E_h/E_v, this effect was considered to be of second order importance at this time.

Figure 7 is derived from the previous analyses for $E_h/E_v = 1.0$. In this case the curves represent the inter-relationship between possible absolute values of E_v and the waste stiffness E_c/h_0 for given values of mean subsidence $\bar{\delta}$, overburden stress p and panel width w. Dimensionless variables have been used for generality.

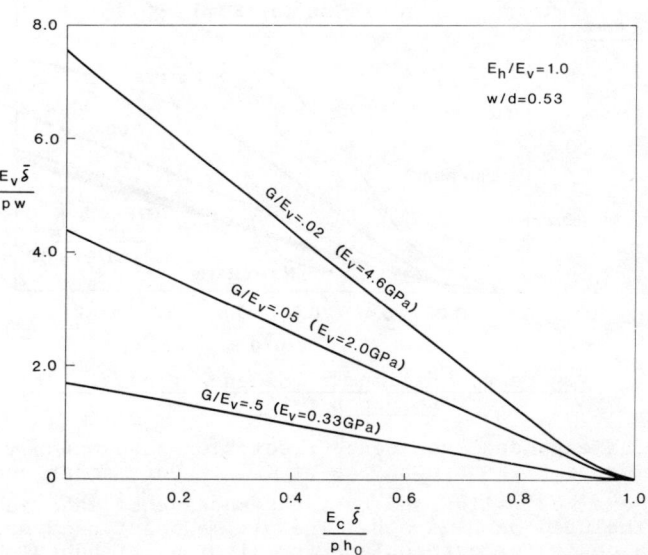

Figure 7. Inter-relationship between continuum modulus (E_v) and waste stiffness (E_c/h_0)

The values of E_v shown superimposed on Figure 7 for the various G/E_v ratios correspond to a value of E_c/h_0 = 20 MPa/15m = 1.33 MPa m which was considered appropriate for Laleham in general, and for values of $\bar{\delta}$, p (3.6 MPa) and w(80m) specific to the δ_{max} = 0.3/m panel at Laleham for which subsidence data was available. The value of E_c (20 MPa) is derived directly from the in situ plate loading tests. The value of h_0 (15m) was based on evidence available from drilling into the roof of extracted panels in "C"-seam. It is obvious from Figure 7 that as the degree of anisotropy of the continuum increases (i.e. G/E_v decreasing), values of E_v

back-calculated from subsidence data approach more closely to values obtained from direct laboratory testing.

To extend the range over which predictions could be compared with subsidence observations, the modelling was expanded to encompass a range of panel geometries typical for Australian mining practice. The waste stiffness E_c/h_0 (= 1.33 MPa). considered appropriate for Laleham was used. A constant value p = 3.6 MPa (=.024 MPa/m x 150m) consistent with the particular panel at Laleham previously studied was maintained while examining the effect on subsidence of various possible panel widths (w). Various values of G/E_v were used together with the corresponding E_v values shown in Figure 7 (i.e. selected to match the measured subsidence of the Laleham panel described previously). The predicted maximum subsidence (normalized by dividing by extracted height, s = 3m) is plotted against w/d in Figure 8.

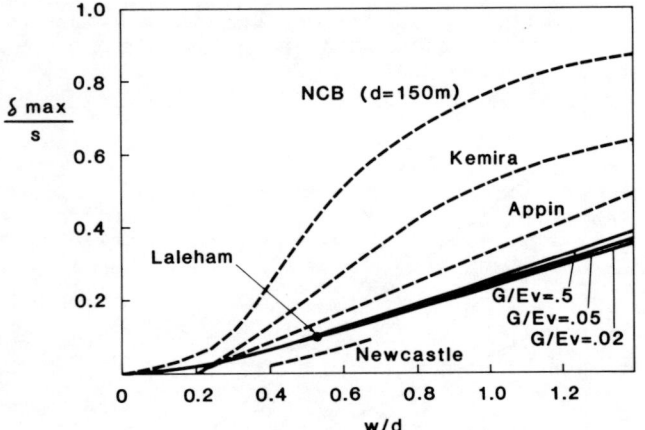

Figure 8. Maximum subsidence versus width to depth ratio

The expanded subsidence predictions are notably insensitive to a choice of G/E_v. For comparison some Australian and overseas subsidence data is included on Figure 8. The single point used as a basis for extrapolated predictions of behaviour at Laleham appears to be reasonably consistent with Australian experience, particularly in relation to the data for the Newcastle area which involves similar d and w values to the experience at Laleham. The general trend of the extrapolated predictions shows reasonable consistency with trends for other Australian experience. The paucity of data for Australian conditions mitigates against being any more conclusive than this. The apparent gross disparity between Australian data and that of the U.K. has been attributed to differences in typical rock types (Kapp, 1978). It is interesting to note, however, that the limiting value of maximum normalized subsidence (δ_{max}/s) predicted by the model for extreme panel widths is 0.9, equivalent to the limit of U.K. observations.

After consideration of the foregoing, the values E_v = 2.0 GPa, G/E_v = 0.05 and E_v/E_h = 1.0 were selected as representing the gross response of the rock mass at Laleham. This G/E_v value was preferred because it had been used previously to successfully model subsidence observations (Crouch, 1976). The corresponding value of E_v (= 2.0 GPa) was seen to be reasonably consistent with "average" laboratory values, given the well documented disparities between laboratory and field values of Young's modulus.

For more detailed modelling, the simplified layered system shown in Figure 2 was used with the chosen G/E_v, E_h/E_v values and with E_v values obtained by scaling the lumped laboratory values so that predicted subsidence for the Laleham back-analysis panel matched that obtained when using the simplified homogeneous model with E_v = 2.0 GPa. The scaled values are included in Table 1. It should be noted that this scaling process retained the material property contrasts. Based on laboratory results (Table 1), a representative value of 0.25 was used for all Poisson's ratios.

Figure 9 shows predictions of the relative loading of the waste as a function of w/d obtained by two-dimensional modelling of panels, using the chosen values for the continuum and waste properties representative of Laleham. For the range of w/d likely to be of interest for studies of future extraction panels, it can be seen that the waste load could be of the order of 20-50% of the overburden stress.

3.2.2 Coal Properties

It was appreciated early that worthwhile predictions relating to the performance of coal pillars and blocks would depend directly on the representativeness of the properties used to model the coal. An initial series of laboratory tests conducted on 50 mm diameter specimens to determine the axial stress-strain relationship at different levels of lateral confinement yielded results summarized in Figures 10 and 11. The specimens in this series of tests all exhibited brittle failure, with no discernable post-failure characteristic. Figure 10 indicates the significant effect of confinement pressure on peak stress. Figure 11 indicates no consistent

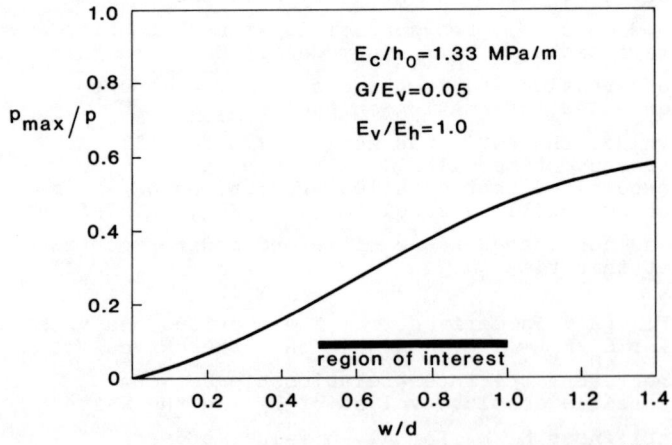

Figure 9. Relative load in waste versus width to depth ratio

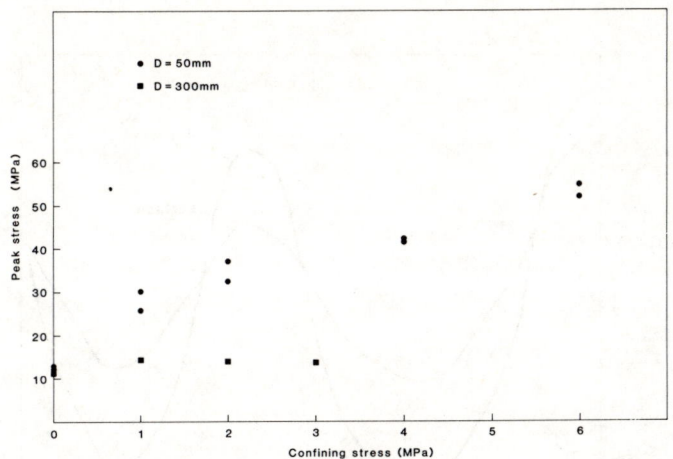

Figure 10. Peak axial stress versus confining stress for Laleham coal

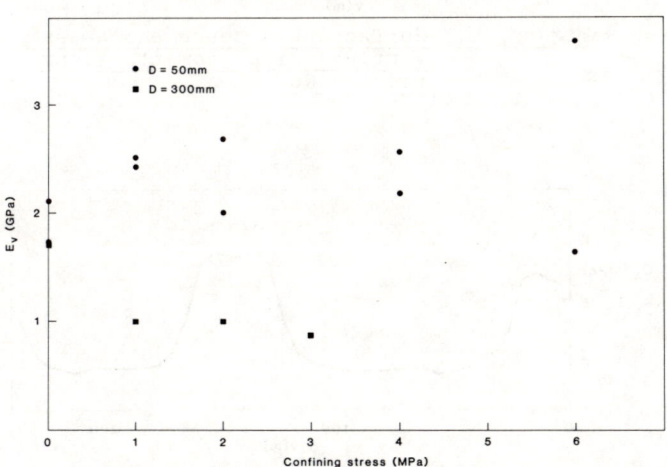

Figure 11. Young's Modulus versus confining stress for Laleham coal

influence of confining pressure on the Young's modulus (E_v) during loading. These trends are also reflected in the results of the first few of an ongoing series of tests conducted on much larger specimens, up to 300 mm diameter x 600 mm high, shown superimposed in Figures 10 and 11. In the case of the large specimens, the influence of confining pressure on peak stress appears tentatively much less dramatic than for the small specimens. Figures 10 and 11 clearly show the effect of scale on both peak stress and Young's modulus.

The large scale tests were conducted in a large triaxial cell with a capacity to apply 4 MPa confining pressure and a 400 tonne axial load. This level of confinement capability was consistent with estimates of likely horizontal stresses in the coal seam in the study area. Simple two-dimensional finite element studies were made of pillar cross-sections representative of those envisaged for Laleham to assess the likely extent to which the periphery of a coal results of these analyses were used to select the range of confining pressure used for testing. The design of the cell allows constant strain-rate testing to be approximated. This in turn meant that it was possible to obtain results in the post-failure range. Figure 12 is typical of results obtained to date.

For the detailed modelling described here, the stress-strain properties of the coal used were based on the large scale test results because of the obvious advantages in closer approximation to <u>in situ</u> scale. The relative lack of sensitivity of either peak stress or Young's modulus (on loading) to confining pressure for the large tests conducted to date led to the selection of a single representative stress-strain property for the coal as shown on Figure 12. This may be updated in the light of further test results.

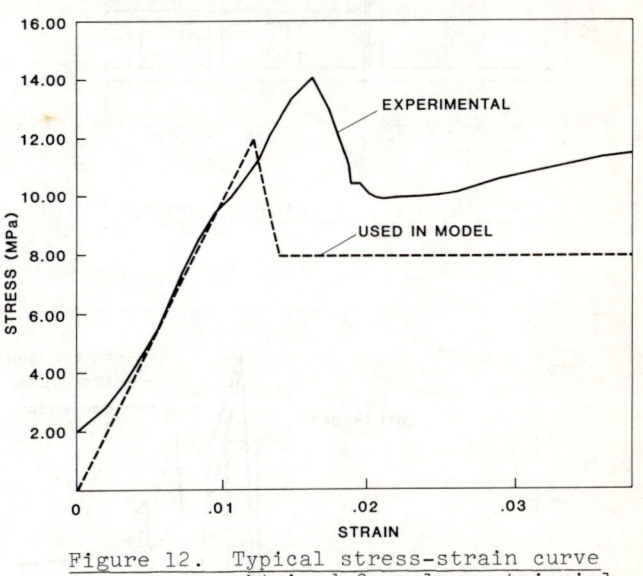

Figure 12. Typical stress-strain curve obtained from large triaxial test on Laleham coal

3.3 Sample Analysis

For the sake of illustration some initial results obtained for analysis of a generalized "Wongawilli" extraction system shown in Figure 13, are given in Figures 14 to 17. In this analysis the material properties summarized in the previous section were used. All other input data is summarized on Figure 13.

Figure 14 shows vertical stress at the level of the workings for the long-section AA shown in Figure 13, for fender widths of 6 m and 12 m as well as for the situation prior to fender creation. The caved waste stresses can be seen to be low relative to the initial overburden stress. Of the two fender widths considered, the 6 m width resulted in complete yielding of the fender, which in practical terms may be undesirable. The 12 m fender, while becoming relatively highly loaded, did not yield, at least on this section. Figure 15 shows the vertical stresses for cross-section BB for the 12 m fender width case. The results indicate that even for this width localized yielding can be expected at one end of the fender prior to commencement of fender

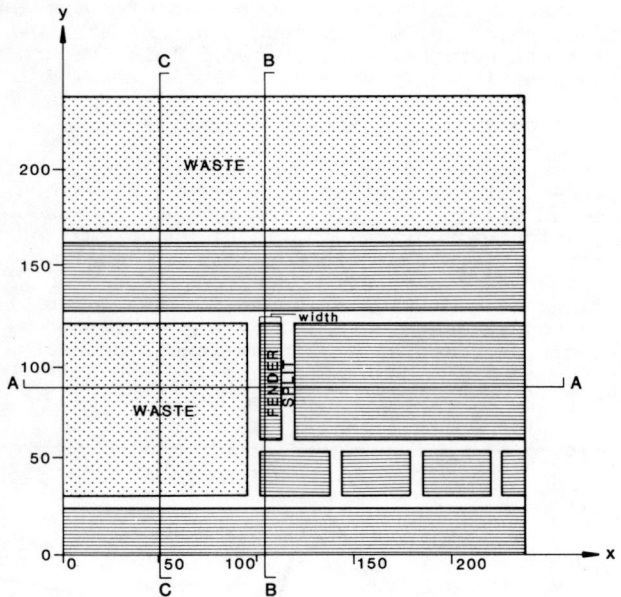

Figure 13. Layout for trial numerical analysis

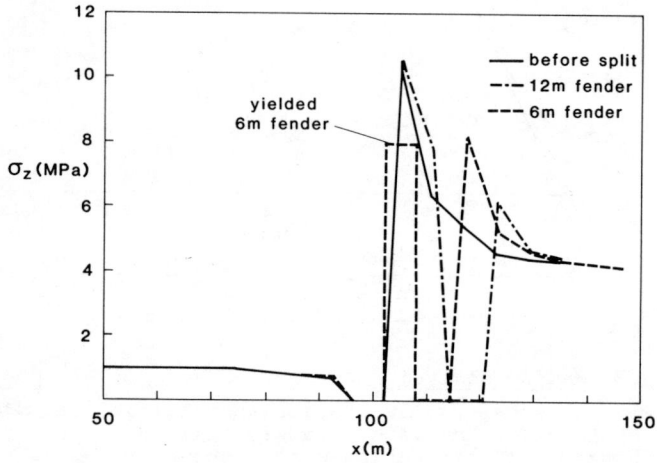

Figure 14. σ_z along section AA

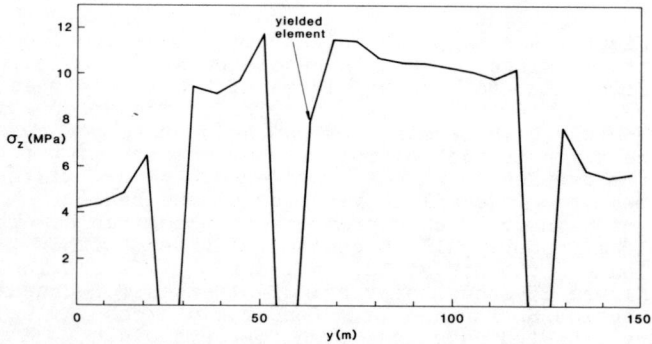

Figure 15. σ_z along section BB (12 m fender)

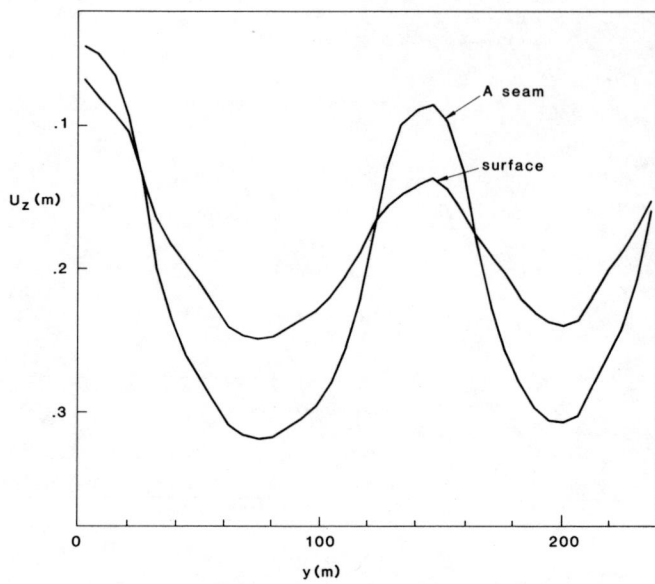

Figure 16. Surface subsidence and A-seam vertical displacements along section CC

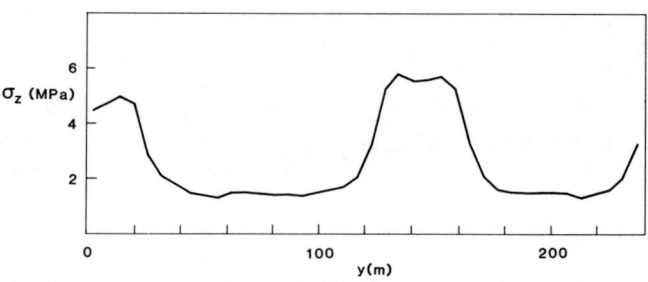

Figure 17. Vertical stresses along section CC

extraction. This is a result which could only be obtained using a three-dimensional model.

Figure 16 gives surface subsidence and vertical displacements for the level of "A" seam (approximately 30 m above the level of extraction), for the cross-section CC shown on Figure 13. These results are indicative of what may be expected after completion of a panel in C seam and are of relevance to the situation that may pertain at commencement of "A" seam development. The general level of vertical movements are in accord with the measured subsidence upon which the back-calculation of input parameters was based. Figure 17 shows the vertical stress at the level of 'A"-seam, again for cross-section CC. The effect of the panel and barrier pillar layout in "C" seam on the initial stresses for subsequent "A"-seam development are obvious. It is on results like these, along with those for various other stress- and displacement components at "A" seam level that decisions on the optimum locations for development of "A" seam could be based.

4. CONCLUSIONS

In the context of this paper no attempt has been made to give design-related answers to specific planning problems. Rather, an attempt has been made to focus attention on those aspects, both theoretical and experimental, that experience has shown needed to be refined in order to achieve any realistic predictive capability. Of particular importance are seen to be:

i) the need for anisotropic material response in the analysis technique

ii) the need for meaningful estimates of the stress-strain characteristics of the coal, caved waste and continuum

iii) the need for a three-dimensional modelling capability which may be critical in many practical cases.

In parallel with the numerical modelling for Laleham Colliery, an intensive programme of *in situ* performance monitoring has commenced. Measurement of pillar stresses and deformations, surface and sub-surface subsidence, caved waste loading and the vertical extent of caving are either under way or anticipated for the near future. The results of this monitoring will be used to update the input data for the modelling programme and to validate the predictive capability of the technique.

5. ACKNOWLEDGEMENTS

The valuable assistance of Malcolm Worswick and Alan White along with other members of CSIRO Division of Applied Geomechanics is gratefully appreciated. The willing assistance and cooperation of the staff and employees at Laleham is acknowledged.

6. REFERENCES

Berry, D.S. (1960). An elastic treatment of ground movement due to mining - Part I Isotropic ground. J. Mech. Phys. Solids, Vol. 8, pp. 280-292 (also Corrigendum Vol. 11, pp. 373-375).

Berry, D.S. and Sales, T.W. (1961). An elastic treatment of ground movement due to mining - Part II Transversely isotropic ground. J. Mech. Phys. Solids, Vol. 9, pp. 52-62 (also Corrigendum Vol. 11 pp. 372-375).

Berry, D.S. and Sales, T.W. (1962). An elastic treatment of ground movement due to mining Part III Three dimensional problem transversely isotropic ground. J. Mech. Phys. Solids, Vol. 10, pp. 73-83.

Crouch, S.L. (1976). Analysis of stresses and displacements around underground excavations: An application of the displacement discontinuity method. University of Minnesota, Geomechanics Report Dept. of Civil and Mineral Engineering, University of Minnesota, Minneapolis, Minnesota 55455.

Diering, J.A.C. (1980). Simulation of mining in non-homogeneous ground using the displacement discontinuity method. J.S. Afr. Inst. Min. Met., Vol. 80, pp. 225-228.

Hackett, P. (1959). An elastic analysis of rock movements caused by mining. Trans. Instn. Min. Engnrs. London, Vol. 118, pp. 421-433.

Kapp, W.A. (1978). Subsidence investigations in the Northern coalfields, New South Wales, and their application to the design of mine layouts in residential areas. Proc. Eleventh Commonwealth Min. & Metall. Congress Hong Kong, pp. 159-169.

Plewman, R.P., Deist, F.H. and Ortlepp, W.D. (1969). The development and application of a digital computer method for the solution of strata control problems. J.S. Afr. Inst. Min. Metall., Vol. 70, pp. 33-44.

Salamon, M.D.G. (1963). Elastic analysis of displacements and stresses induced by the mining seam or roof deposits - Part I. J.S. Afr. Inst. Min. Metall., Vol. 64, pp. 138-149.

Starfield, A.M. and Crouch, S.L. (1973). Elastic analysis of single seam extraction. In: New Horizon in Rock Mechanics (H.R. Hardy Jnr. & R. Stefanko (eds)), A.S.C.E. pp. 421-439. (proceedings 14th U.S. Symp. on Rock Mech., Penn. State Univ.).

Wardle, L.J. (In preparation). Displacement discontinuity method for three-dimensional stress analysis of tabular excavations in non-homogeneous ground. CSIRO Aust., IEER, Division of Applied Geomechanics Technical Report.

INVESTIGATIONS PRIOR TO THE INTRODUCTION OF LONGWALL MINING
Recherche avant la mise en oeuvre de l'exploitation minière par "Longwall"
Untersuchungen vor Einführung des Strebbaues

R. A. Yeates
Senior Mining Engineer, Peko-Wallsend Ltd: Coal Division, Australia

J. R. Enever
Principal Research Scientist, CSIRO, Australia

B. K. Hebblewhite
Manager-Mining Research, ACIRL, Australia

SYNOPSIS

Prior to the introduction of longwall mining to the Greta Seam in New South Wales, Australia, a comprehensive rock mechanics testing programme was carried out to assess the suitability of the longwall method for the particular seam, strata and geological conditions present. The information obtained was used to provide design guidelines for equipment supply and mine layout. The investigations commenced with the collection of basic data including the mechanical properties of the strata and the virgin stress levels. Various modelling techniques (finite element analysis, displacement discontinuity analysis and physical modelling) were used to provide information on roadway stability, strata caving characteristics, gate-pillar dimensions and longwall support density requirements.

RESUME

Avant d'introduire la technique "longwall" au gisement de Greta dans l'état de Nouvelle Galles du Sud (Australie) on a effectué un important programme d'études de mécanique des roches afin de savoir si cette technique y convient, compte tenu de sa géologie et de sa stratification. De plus, les données obtenues ont fourni des éléments pour le plan de la mine et pour le choix des équipements. On a commencé par rassembler des données de base telles que les propriétés mécaniques des couches et le niveau des contraintes avant travaux. On s'est servi de plusieurs techniques de traitement (analyse des éléments finis, analyse de la discontinuité du déplacement, modèles physiques) pour obtenir des indications sur la stabilité des voies de passage, sur les caractéristiques d'effondrement ou d'éboulement des couches, sur les dimensions entre piliers et sur les piliers de soutènement nécessaires au "longwall".

ZUSAMMENFASSUNG

Vor Einführung des Strebbaues in das Greta Flöz in Neu-Südwales, Australien, wurde ein umfassendes Untersuchungsprogramm ausgeführt, um die Eignung der Strebbaumethode für dieses Flöz, seine Hangend- und Liegendschichten und die geologischen Verhältnisse zu erfassen. Diese Untersuchung sollte ebenfalls Richtlinien für die Geräteauswahl und die Ausrichtung der Grube liefern. Die Untersuchungen begannen mit der Sammlung von grundlegenden Daten bezüglich der gesteinsmechanischen Kennwerte und der urspünglichen Spannungszustände. Verschiedene gesteinsmechanische Untersuchungsverfahren (Finite Elemente Analyse, Störungs- und Kluftanalysen und Modellversuche) wurden dann angewandt, um Kenntnisse über Streckensicherheit, Bruchbauwerte, Strecken-Pfeiler Dimensionen und die Ausbaudichte für die Strebbaue zu gewinnen.

INTRODUCTION

Ellalong Colliery is a new mine being developed to extract Greta Seam coal at depths of cover from 350 to 650m in the South Maitland Coalfield of New South Wales. This Coalfield is located 150 km north of Sydney.

The mine is being developed as an extension of Pelton Colliery, which has been mining the shallower Greta Seam coal (up to 350m) since 1916. Difficulties experienced mining the deeper sections of Pelton indicated that mining by the bord (room) and pillar system or its derivatives would not be viable at the depths present in Ellalong. Preliminary feasability studies pointed towards the longwall method of extraction.

The longwall method had not previously been used in the Greta Seam. This, coupled with the fact that all collieries which had attempted to mine the Greta Seam at depths greater than 350m had closed for a combination of economic and strata control reasons, led to the commencement of an extensive rock mechanics study of mining the

the Greta Seam at depth in Ellalong.

The three major aims of the study were:
- to assess, from a rock mechanics point of view, the applicability of the longwall method at depth in the Greta Seam at Ellalong
- to assist with the planning of the new mine
- to provide longwall equipment design guidelines

Each of these aims has been achieved to some extent. The mine has been developed in readiness for longwall extraction and a set of longwall equipment has been designed and is on order at the time of writing. When this equipment becomes operational it is intended to monitor its performance in order to assess the reliability of the predictions.

The investigations have been carried out as a co-operative effort by workers from the Australian Coal Industry Research Laboratories (A.C.I.R.L.), Commonwealth Scientific and Industrial Research Organization: Division of Applied Geomechanics (C.S.I.R.O.), the Geology Department of Newcastle University and the Coal Division of Peko-Wallsend Ltd. It is the aim of this paper to show the extent of the investigations and how the results have been used to aid mine planning and equipment design.

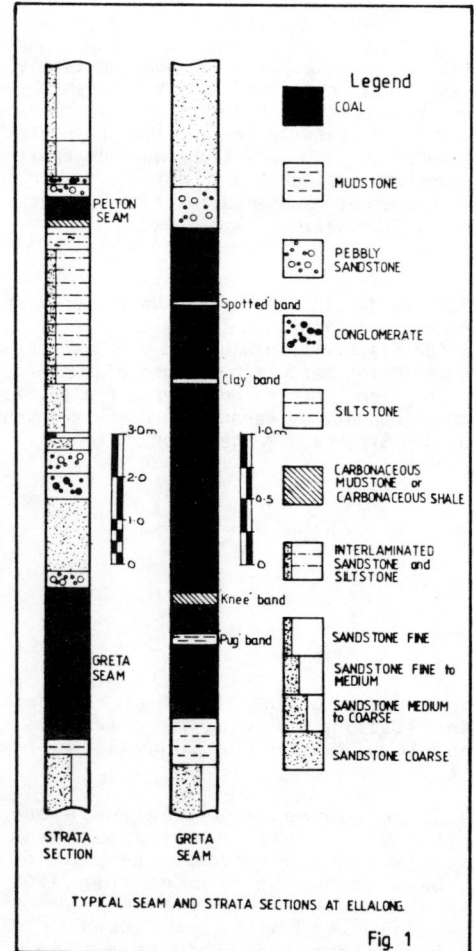

TYPICAL SEAM AND STRATA SECTIONS AT ELLALONG.

Fig. 1

COLLECTION OF BASIC DATA

Since little rock mechanics data was available on the Greta Seam and its adjacent strata the investigations began with the collection of basic data, as discussed below.

Sedimentology

The sedimentology of the Ellalong area was deduced by examining all available information from surface mapping, adjacent mines, borecores and from underground mapping as mining proceeded. Figure 1 shows typical sections of the Greta Seam and adjacent strata. The floor of the Seam is generally a thin mudstone unit. The Seam varies from 3 to 4.5m thick and may contain up to four distinct bands. The immediate roof is generally a coarse to pebbly sandstone, overlain by various bedded sandstone, conglomerate and laminite strata up to the 400mm thick Pelton Seam. The Pelton Seam is from 5 to 15m above the Greta Seam and is overlain by several hundred metres of sandstones and siltstones.

In some areas palaeochannels protrude into the Greta Seam from the roof strata. The palaeo-drainage system is interpreted as flowing from northwest to southeast in the area of the Ellalong Lease (Rawlings, 1982). Some of the palaeochannels protrude up to 1.5m into the Seam, but the majority protrude less than 0.5m. Material in the palaeochannels is generally a strong conglomerate or coarse sandstone and is difficult to cut with coal mining equipment. Dip of the seam is variable but averages 3°.

Strata Mechanical Properties

The mechanical properties of the strata present in the vicinity of the Greta Seam at Ellalong were measured by laboratory testing of numerous 50mm diameter samples taken by drilling from both the surface and underground. Properties measured included uniaxial and triaxial strengths, Young's Modulus and Poissons Ratio. Table 1 shows average values of properties measured for various identifiable strata.

ROCK TYPE	HORIZON	AVERAGE PROPERTIES			
		UNIAXIAL COMPRESSIVE STRENGTH (MPa)	UNIAXIAL TENSILE STRENGTH (MPa)	YOUNG'S MODULUS (MPa)	POISSONS RATIO (-)
SANDSTONE/SILTSTONE	ABOVE PELTON SEAM	74	5.8	16400	0.31
PELTON SEAM	5-15M ABOVE GRETA SEAM	30	1.2	-	-
LAMINITE	BELOW PELTON SEAM	66	4.7	6700	0.22
SANDSTONE	ROOF	64	4.1	11500	0.23
CONGLOMERATE	ROOF	48	4.3	13600	0.25
SANDSTONE	IMMEDIATE ROOF	37	2.6	12800	0.17
GRETA SEAM		37	1.6	3900	0.36
MUDSTONE	IMMEDIATE FLOOR	44	4.6	6000	0.26
SANDSTONE	FLOOR	48	4.2	10800	0.28

TABLE 1: AVERAGE MECHANICAL PROPERTIES OF STRATA AT ELLALONG COLLIERY (FROM HALL (1978), JAGGAR (1978) AND WALLMAN (1980))

Use was also made in the field of the point load method of testing to obtain further rock property data (Enever et al., 1979).

At several locations sufficient data was gathered to enable near continuous plots of strength along strata sections, as shown on Figure 2.

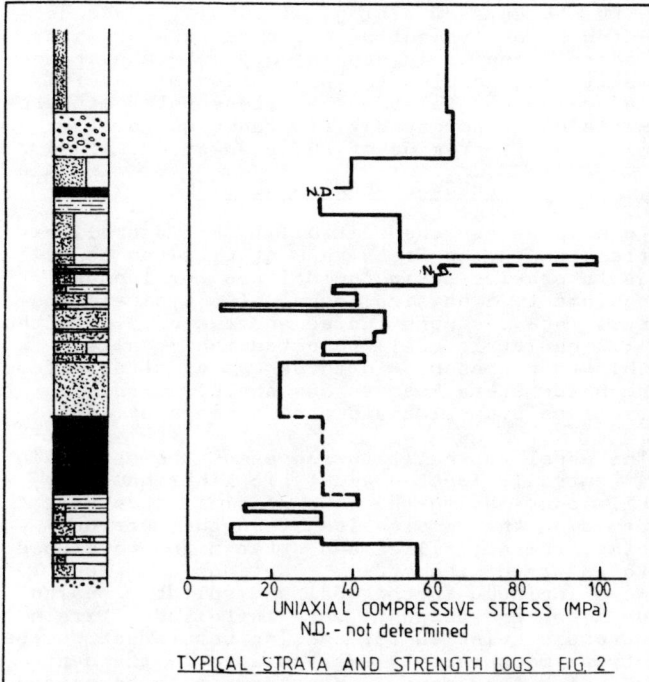

TYPICAL STRATA AND STRENGTH LOGS   FIG. 2

Large scale (300mm by 300mm by 600mm blocks) triaxial testing of coal was carried out. Figure 3 shows the results.

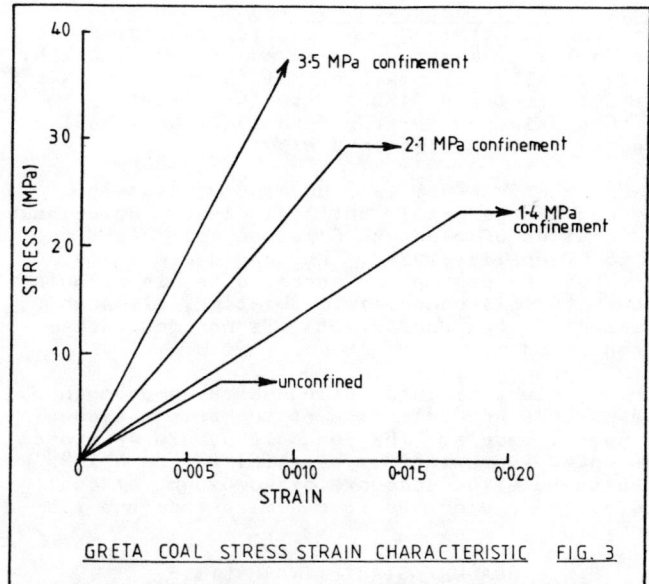

GRETA COAL STRESS STRAIN CHARACTERISTIC   FIG. 3

Virgin Stress Levels
Strata virgin stress levels were measured at three different locations in the Ellalong workings. Techniques used to measure the stress were the C.S.I.R.O. Hollow inclusion triaxial cell and the C.S.I.R. Triaxial cell. Pressiometer and hydrofracturing techniques were also attempted. All these techniques are described in Jaggar and Enever (1978). The results are given in Table 11.

From these results it has been concluded that the virgin stress field at Ellalong can vary from approximately hydrostatic, with the vertical component consistent with overburden depth to a situation with the maximum horizontal stress being approximately twice the vertical overburden stress.

LOCATION	METHOD	MAXIMUM STRESS			INTERMEDIATE STRESS			MINOR STRESS		
		STRESS (MPa)	BEARING (°mag)	ELEVATION (°)	STRESS (MPa)	BEARING (°mag)	ELEVATION (°)	STRESS (MPa)	BEARING (°mag)	ELEVATION (°)
Site 1	CSIR Cell No. 1	11	159	36	5	58	15	4	310	50
	CSIR Cell No. 2	16	208	34	13	103	17	8	351	51
Site 2	CSIR Cell No. 1	11	243	82	7	47	8	5	138	2
	CSIR Cell No. 2	6	196	87	5	37	3	3	127	1
	CSIRO Cell No. 1	9	260	70	8	20	12	5	115	17
	CSIRO Cell No. 2	6	210	76	3	347	10	1	80	9
Site 3	CSIRO Cell	10	82	0	8	172	0	7	82	90

TABLE 11:   VIRGIN STRESS MEASUREMENTS AT ELLALONG (from Jaggar (1978), Jaggar and Enever (1978) and Wallman (1980))

Geological Structural Features

Pelton and Ellalong contain two major directions of faulting, namely N 15°W and N 65°W (Moelle, 1979). All faults are normal faults. The N15°W faults are the dominant structural features with displacements from 5 to 25m and extending for many kilometres on strike.

The Greta Seam has well developed, closely spaced and generally vertical cleat. Principal directions of cleat at Ellalong are N 65°W and N 25°E (Moelle, 1979). Low angle cleat are noted to be present in zones, often in association with palaeochannels. Jointing, although present in the roof strata, is not generally pronounced.

This geological information has been used in the assessment and selection of the orientation of longwall panels. The longwall panels have been oriented approximately parallel to the N 15°W faults to allow adequate sized blocks of coal, to assist caving and to minimize face and rib spall.

ROADWAY STABILITY ANALYSIS

Utilizing the data collected, the finite element method was used to assess the potential stability of development roadways at different horizons in the seam. The finite element programme used was two dimensional and incorporated non-linear elasto-plastic deformation moduli. Calculated stresses were compared with the appropriate failure envelope and failure potential assessed. The analysis was cycled until no additional failed elements were detected.

Roadway development machinery in use at the mine drove 2.4m high rectangular roadways in the 3 to 4m seam. Analysis showed roadways driven in the bottom or intermediate sections of the seam would be relatively more stable than those driven in the top of the seam. Roadways driven in the top of the seam were predicted to suffer rib spall, a finding substantiated by practice. In the mine the rib spall tended to result in wide roof spans and consequent roof control problems. Except under exceptional circumstances, no mine roadways are now driven in the top of the seam at Ellalong.

Further finite element analyses indicated that some roadway stability problems might be expected at the stress levels extrapolated from measured values to those expected at the maximum depth of development anticipated. Some problems have been encountered with development headings driven to that depth.

The same analysis method was used to study the stability of an elliptical shaped roadway driven by a machine the Colliery was considering introducing. The analysis showed this roadway to be more stable at greater stresses than rectangular roadways as were in common use at the mine (Wallman, 1981). This study greatly assisted in the introduction of the machine (and as a consequence the single entry system) to the mine and the acceptance of the machine (and the system) by the work-force.

To check on the reliability of these predictions a programme of roof to floor convergence and borehole extensometer measurements was undertaken. The convergence measurements have been found to provide quick and relatively easily obtainable information on the stability of roof and floor and the relative stability of different roadways and support systems. The magnitude of roof to floor convergence measured was generally in the range 10 to 70mm. (Wallman unpublished data). Most convergence occurred within 8 days of mining. Instability was detected by assessment of the time rate of deformation. The early warning given by convergence monitoring enabled the prevention of several major roadway falls. The extensometer measurements have indicated yield zones up to 3.5m wide on either side of the roadways.

LONGWALL FACE ANALYSIS

To help assess the suitability of longwall extraction in the Greta Seam at Ellalong as well as to provide design guidelines for longwall equipment, a physical model of a typical longwall face was constructed and tested. With the 1:22 geometric scale selected the model was able to represent 41m thickness of strata (from 8m below Greta Seam to 30m above), a section of the face 8.8m wide and face advance of 175m.

The model was built to represent the strata in a centrally located and typical borehole at Ellalong. Using the laws of similitude, and based on the compressive strengths, various sand, cement, silica and water mixes were used to construct the model. Partings and cleat were included in the coal and suitably scaled down powered supports were installed. Virgin stresses selected for testing were based on the stress measurements extrapolated to the depth at which the longwall panels were to be extracted.

The face was cut and the supports advanced in a typical longwall mining sequence. Strata behaviour, face conditions, stress distribution and support performance were all monitored. Figure 4 shows typical goaf behaviour after an initial break had been formed.

The following summarises the major findings of the model (details given in Wallman and Hall, 1980):
- The longwall extraction technique was found to be suitable for Ellalong Colliery
- A support setting load of 360t (1.5m wide) gave good mining conditions and was sufficient to control the roof (this indicates a 450t yield support)
- Front abutment was detected 40m in advance of the face and peaked within 5m of the face at three times the virgin stress
- First load taken by the goaf was 40m behind the face, with a rear abutment of 1½ times virgin stress occurring 70m behind

An independent analytical analysis was carried out on the cantilever mode of failure observed in the model to attempt to generalise the information obtained on support density requirements (Enever and Crawford, 1982). Figure 5 shows the results. T, as defined in this analysis may be up to 9m, giving a maximum support setting density of 220t/m (330t support set load) and support yield load of approximately 600t.

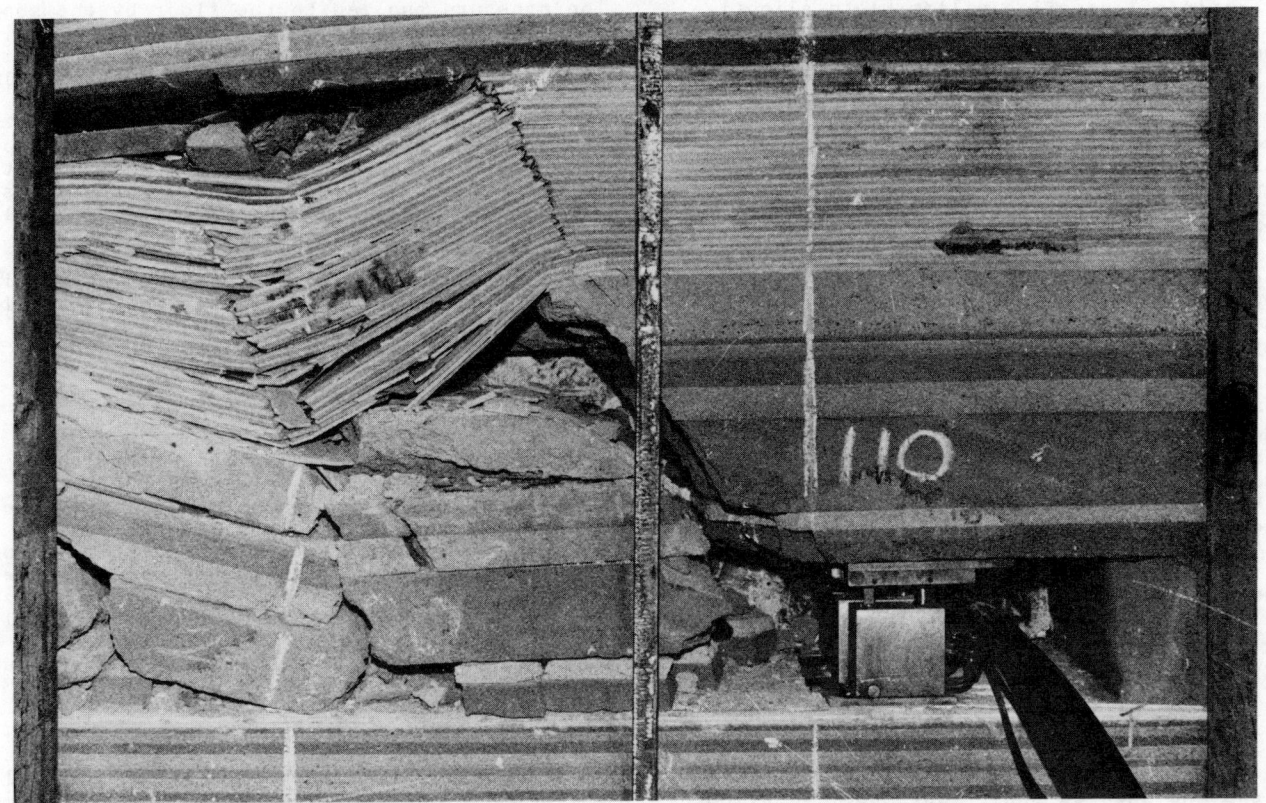

FIGURE 4: PHOTO OF ELLALONG COLLIERY LONGWALL MODEL TEST

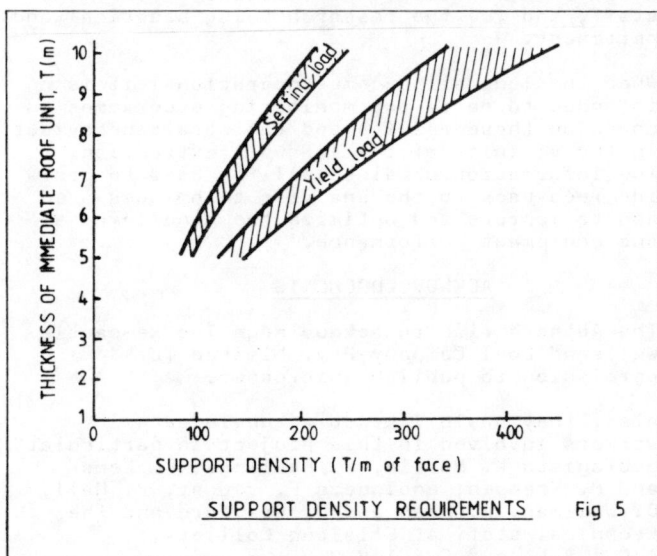

SUPPORT DENSITY REQUIREMENTS   Fig 5

These results greatly assisted the mine management in their decision to change to longwall mining.

The information obtained was utilized in the selection of the capacity of the longwall supports. A complete face of longwall supports of 600t yield capacity is on order for the mine and will be placed in operation shortly. The physical model was also a particularly useful demonstrative tool for directors, management and men.

GATEROAD PILLAR INVESTIGATION

An important parameter of two heading layout retreating longwall extraction, as planned for Ellalong, is the width of the gateroad pillars. Pillar width should be as narrow as possible to avoid loss of the coal resource but wide enough to ensure adequate stability of the roadways while in use. Pillar widths used on Australian longwalls have varied from 19m to 31m at depths comparable with Ellalong.

To investigate this pillar width under Ellalong conditions, use was made of a numerical method based on displacement discontinuity theory. The method can analyse the stability of underground, planar mining layouts (Hebblewhite et al., 1979). A 150m wide face with gatepillar widths of 15, 20, 25 and 30m was studied in separate analyses. A goaf caving criterion and caved element material property were incorporated as part of the analysis. Results of this work are summarised below:
- front abutment commenced 50m in advance of the face and peaked at twice the virgin stress about 8m in advance of the face
- rear abutment peak pressure occurred about 55m behind the face
- gateroad pillars from 14m to 19m wide were recommended as at this width they would be stable when the first face passes and yield after the second face has passed

For the first longwall a pillar width slightly in excess of that recommended by this study was chosen to ensure stability until the second face has passed.

When longwall extraction commences the stability of the gateroad pillars will be monitored to see if the width of subsequent pillars can be safely decreased.

FLOOR BEARING CAPACITY

Mining difficulties related to a soft floor horizon and early testing of the floor showing it to have a poor wet strength pointed to a potential problem for longwall equipment. Further testing to evaluate the bearing capacity of the floor was therefore conducted.

In the laboratory, the effect of moisture on the strength of samples of the floor strata was evaluated. A reduction in strength to about one third of the dry value was noted. Cyclic loading was also found to significantly reduce the strength.

In the mine a series of large scale in-situ plate bearing tests were conducted to measure the floor bearing capacity. Four 100t hydraulic jacks were used to load the various possible floor horizons through a 500mm diameter steel plate (Enever et al., 1979). The results are summarised in Figure 6.

These tests have identified the soft floor horizons and enabled their 'softness' (bearing capacity) to be quantified. On the basis of these results it is now planned to operate the longwall at Ellalong on either the mudstone floor (curve 1) or the siltstone floor (curve 3)

The pressure applied to the floor by the supports at set or yield will not cause the failure or excessive deformation of these floor horizons. The siltstone lens (curves 6 and 7) will not be used as a longwall floor.

Various devices and designs of longwall supports are available to enable longwall operation on soft floors. These generally carry a cost premium and cause other operational difficulties. As a result of this study it has been possible to omit any such devices from the Ellalong supports. Some minor and inexpensive changes to standard support base designs will ensure that support floor loadings are kept to acceptable levels without affecting general operational efficiency.

CONCLUSION

The major conclusion to be drawn from these investigations is that as a result of the investigations, plus others not covered here, it has been possible to proceed with the mine planning for Ellalong Colliery and the introduction of the longwall mining method to the mine with a far greater degree of confidence than would otherwise have been possible. Also, the investigations have provided guidelines for mine roadway layouts, longwall equipment design and longwall operating procedures.

The investigations have been a co-operative effort between a number of research organizations, consultants and the mine staff. At all times there has been good communication between the mine staff and the investigators resulting in (a) the results of the investigations being understood, accepted and utilized by the mine staff, and (b) the research being practical and pertinent.

When the longwall becomes operational it is intended to carry out monitoring programmes to check on these results and the strata behaviour in the vicinity of the longwall extraction. The information obtained will be used to provide feed-back to the analysis techniques used and to improve and optimize the mine layouts and equipment performance.

ACKNOWLEDGEMENTS

The Authors wish to acknowledge The Newcastle Wallsend Coal Company Pty. Limited for permission to publish this paper.

Also, they would like to acknowledge all workers involved in this project in particular geologists K. Moelle, C. Rawlings, R. Lees and P. Krempin, engineers F. Jaggar, R. Hall, D. Wallman, A. Dean and G. Crawford and the technical staff at Ellalong Colliery, A.C.I.R.L. and C.S.I.R.O.

Parts of the work described in this paper were funded by the Greta Seam Research Fund and the National Energy Research Development and Demonstration Council.

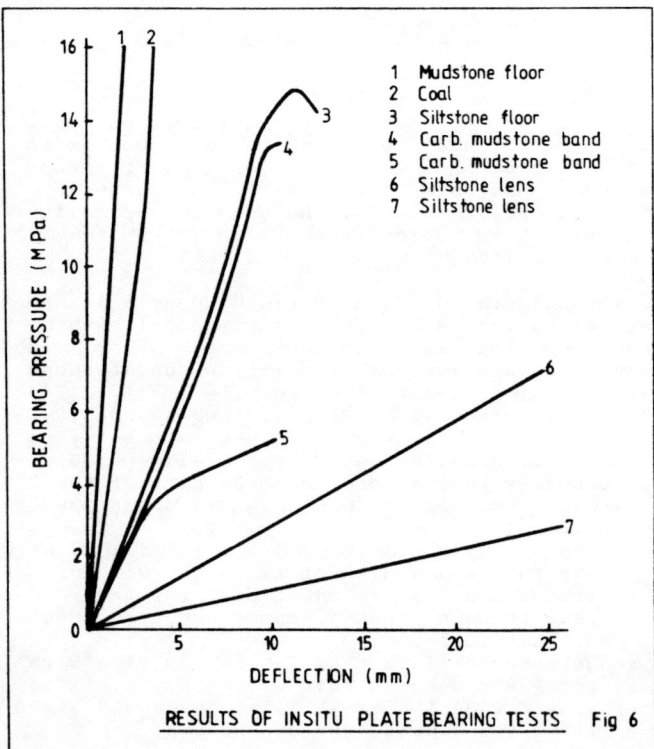

RESULTS OF INSITU PLATE BEARING TESTS   Fig 6

1 Mudstone floor
2 Coal
3 Siltstone floor
4 Carb. mudstone band
5 Carb. mudstone band
6 Siltstone lens
7 Siltstone lens

REFERENCES

Enever, J.R., Yeates, R.A., and Lees, R. (1979) An example of the use of point load testing and rock quality designation. C.S.I.R.O. Div. Appl. Geomech., G.C.M. Report No. 10.

Enever, J.R., Yeates, R.A., Dean, A.K. and Crawford, G.R. (1979). Geomechanical considerations related to the choice of a working floor for a proposed longwall development. C.S.I.R.O. Div. Appl. Geomech., G.C.M. Report No. 14.

Enever, J.R., and Crawford, G.R. (1982). An example of the use of a borehole penetrometer to investigate the caveability of a roof sequence. C.S.I.R.O. Div. Appl. Geomech., G.C.M. Report No. 40.

Hall, R.T. (1978). Rock property determinations in and adjacent to the Greta Seam. A.C.I.R.L. Reports 420 and 08/0007/1.

Hebblewhite, B.K. et al., (1979). Development and application of a computer modelling technique for assessment of the stability of underground mining layouts. A.C.I.R.L. Report PR 79-9.

Jaggar, F. (1978). Rock mechanics investigations of structural stability in the Greta Seam at Pelton Colliery. A.C.I.R.L. Report PR 78-1.

Jaggar, F., and Enever, J.R. (1978). An evaluation of stress measuring techniques conducted in the Southern Development Headings at Pelton Colliery. A.C.I.R.L. Report PR 78-6.

Moelle, K.H.R. (1979). Report on geological investigations at Ellalong Colliery. Unpublished report.

Rawlings, C.D. (1982). On structural analyses of the Lochinvar Anticline and their implications to the mining of the Greta Coal. Unpublished Ph.D thesis, Univ. of Newcastle.

Wallman, D. (1980). Geotechnical logging, stress measurement and rock mechanics testing of core from underground boreholes at Ellalong Colliery. A.C.I.R.L. Report 08/0083-A.

Wallman, D. (1981). Proposed single entry development at Pelton/Ellalong Colliery. A.C.I.R.L. Report PR 81-1.

Wallman, D., and Hall, R. (1980). A physical model to study longwall extraction at Ellalong Colliery. A.C.I.R.L. Report PR 80-1.

RESEARCH INTO THE PHENOMENON OF OUTBURSTS OF COAL AND GAS IN SOME AUSTRALIAN COLLIERIES

Recherche sur les dégagements instantanés dans les mines de charbon australiennes

Forschung über das Phänomen der Kohle- und Gasausbrüche in einigen australischen Kohlegruben

J. Hanes
Senior Geologist, Broken Hill Proprietary Co. Ltd., Minerals Division, Newcastle, N.S.W.

R. D. Lama
Manager Technology, Kembla Coal and Coke Pty. Ltd., Wollongong, N.S.W. Australia

J. Shepherd
Senior Structural Geologist, Australian Coal Industry Research Laboratories Ltd., North Ryde, N.S.W. Australia

SYNOPSIS

This paper presents results of research into frequently occurring outbursts in two Australian collieries. Investigations have been conducted in relation to geology, geomechanical parameters of coal, gas parameters including content, pressure, adsorption and desorption. Results show that outbursts are stress initiated and gas driven phenomena. At Leichhardt Colliery stress plays an important role as an initiating factor whereas at West Cliff Colliery, gas plays the dominant role. Outburst prediction techniques investigated include geological projections, gas adsorption-desorption, and gas pressure measurements. A successful method of alleviating outbursts at the two collieries involved drainage of gas from the solid coal and from shear zones.

RESUME

Cette communication présente les résultats de recherches sur les fréquents dégagements instantanés dans deux mines de charbon australiennes. Les recherches ont été conduites en termes de géologie et paramètres géomécaniques, y compris la teneur, la pression, l'adsorption et la désorption. Les résultats indiquent que les dégagements instantanés sont initiés par des contraintes et actionnés par le gaz. A la mine de charbon de Leichhardt les contraintes jouent un rôle important en tant que facteur initiateur, tandis qu'à la mine de charbon West Cliff la part dominante appartient aux gaz. Les techniques de prédiction de dégagements instantanés comprennent les prévisions géologiques, l'adsorption-désorption de gaz et les mesures des pressions. Une méthode efficace d'atténuation des dégagements instantanés dans les deux mines de charbon mentionnées se base sur le drainage des gaz du charbon solide et des zones cisaillées.

ZUSAMMENFASSUNG

Die Abhandlung präsentiert Forschungsresultate im Zusammenhang mit häufig vorkommenden Ausbrüchen in zwei australischen Kohlengruben. Untersuchungen mit Bezug auf die Geologie, geomechanischen Parameter der Kohle, Gasparameter einschließlich Inhalt, Adsorption und Desorption, wurden durchgeführt. Die Resultate zeigen, daß Ausbrüche durch Streß entstandene und von Gas angetriebene Erscheinungen sind. In der Leichhardt-Kohlengrube spielt Streß als einleitender Faktor eine wesentliche Rolle, während in der West Cliff-Grube Gas die dominierende Rolle spielt. Zu den in Betracht gezogenen Methoden, um Ausbrüche vorauszusagen, gehören geologische Projektionen, Gasadsorption-Desorption und Gasdruckmessungen. Eine erfolgreiche Methode, Ausbrüche in den zwei Kohlengruben zu vermindern, beinhaltet die Drainierung von Gas aus der kompakten Kohle und von Abbruchzonen.

INTRODUCTION

The problem of outbursts in Australian collieries is almost a century old. The first Australian outburst occurred at Metropolitan Colliery in 1895. Since then more than 450 outbursts have been recorded with the largest occurring at Collinsville in 1957 displacing over 1000 tonnes of material (Sheehy et al, 1956). The shallowest outburst occurred at Moura in 1982 at a depth of 130 m. The problem is on the increase because of the increasing depths of mining accompanied by increases in stress and gas. In the early nineteen seventies there were 8 mines facing this problem. Their number has increased to 14 in the last 10 years. Table 1 gives details of the occurrences of outbursts in Australian mines. Though the largest outbursts in Australia have been associated with carbon dioxide (CO_2), a majority are associated with methane and invariably occur with some form of geological discontinuity in the form of dykes or faults. Almost every underground mine in the Bowen Basin (Queensland) and most of the deeper mines in the Southern Coalfield (New South Wales) (Figure 1) are facing the problem.

Research into outbursts in Australia began during the nineteen fifties at Metropolitan Colliery where outbursts occur with mainly CO_2 and are associated with faults and dykes (Hargraves et al, 1964). An empirical gas desorption index, the Hargraves emission value, was developed for predicting outburst conditions at Metropolitan Colliery and was applied with some success at Collinsville and Leichhardt.

Current research is being conducted into three main areas:
1. The role of the geomechanical properties of the coal,
2. The role of the geological and stress environments,

and
3. The role of gas and filtration properties of the coal.

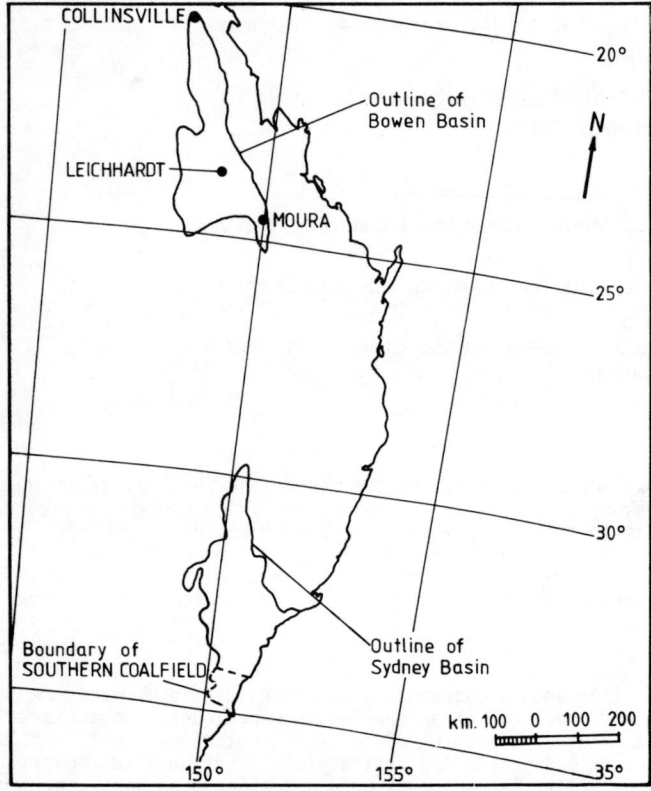

Figure 1. Location of collieries and coalfields prone to outbursts of coal and gas in Eastern Australia.

The manifestation of an outburst can be classified into 4 stages. Stage 1 is called the trigger stage where the coal face behaves as a triaxially stressed solid until it is suddenly exposed, when its resistance to pressure reduces producing extensive fracturing which extends into the solid coal. In the second stage, the adsorbed gas is released due to fracturing and a drop in gas pressure takes place with continued desorption and permeation of gas through the fractured coal. The critical stage is reached when the rate of desorption of gas exceeds its rate of permeation through the coal. This gives rise to the third stage at which mass movement takes place. The expanding gas serves as a transport medium and the fractured/pulverised coal is expelled. The final stage is the end of the process when all the fractured material has been ejected.

The above mechanism postulates that both stress and gas play a role in the occurrence of outbursts. The part played by each of these is important, though one may be more dominant than the other depending upon the local conditions.

At Leichhardt Colliery outbursts in general are not associated with prominent geological structures. They are initiated by stress concentrations at the face. At other collieries (for example West Cliff), the presence of strike-slip shear zones locally reduce the strength of the coal. The condition is thus ripe for an outburst when such an area is approached. High pressure gas in the weak coal initiates outbursts.

More than 80% of the total number of outbursts in Australia have occurred at these two collieries.

MECHANISM OF OUTBURSTS AT LEICHHARDT COLLIERY

Leichhardt Colliery closed in 1982 because of severe outbursting. It mined the 6 m thick Gemini Seam of the Upper Permian Rangal Coal Measures. The seam contains high frequency unidirectional systematic cleats with minor fault slip and gouge. Cleats with a strike length greater than 1 m occur at average spacings of 2 m with smaller parallel cleats at 10 mm spacing.

Approximately 250 outbursts ranging fromn 1 tonne to 500 tonnes occurred between 1975 and 1981. They were total or partial ejections of fractured coal leaving conical cavities. The axes of the cones were perpendicular to the most prominent discontinuity in the coal, usually the principal cleats; however some bursts occurred with axes perpendicular to coal bedding or to intense mining induced cleavage. Most bursts occurred as violent buckling of cleated coal from the rib or face on the side of the opening which first intersected the cleat indicating these to be a stress relaxation phenomenon.

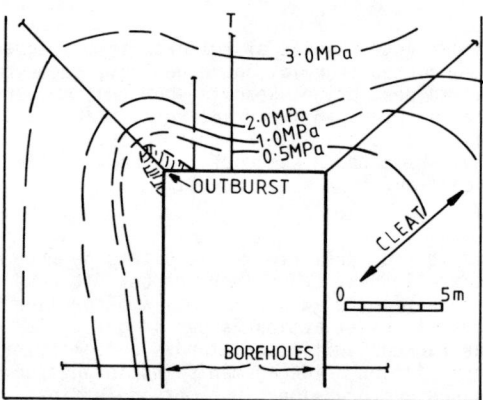

Figure 2. Gas pressures related to cleat and a blind-ended heading at Leichhardt Colliery.

However, the largest outburst of 500 tonnes occurred on a thrust fault of 7 m throw. Two occurrences of violent coal floor heave also occurred on a 3.5 m thrust fault. No outbursts occurred on intersection of two >10 m normal faults. A large outburst of 380 tonnes was not associated with any structures other than the well defined cleat.

GEOMECHANICAL PROPERTIES

Stress measurements were conducted in the stone roof of the Gemini Seam using an overcoring technique. The results of measurements are given below:

σ_1 = 29 MPa NNE near horizontal
σ_2 = 18 MPa ESE near horizontal
σ_3 = 10 MPa near vertical (overburden depth 400 m)

The major principal stress is near horizontal and about three times the vertical load. Horizontal/vertical stress ratios in coal would be a bit lower than in the stone roof and may vary from 0.75 - 1.15.

Tests were conducted to determine if any changes in the physical and mechanical properties of the coal are responsible for areal variations in outburst proneness. Samples were taken from four sites (Table 2). The results show that there is a reduction in strength and surface fracture energy of about 20% at site 3, where outbursts occurred. The ratio of apparent to true porosity for site 3 is relatively low indicating that site 3 is a region of comparatively lower permeability (Lama and Mitchell, 1981). Pore size distribution studies conducted using a mercury porosimeter show that medium pore (area) size of Gemini Seam is about 0.0038 μm compared to 0.0052 μm at West Cliff and other collieries in Australia and is equal to about 40% of the value for Canadian and European coals. Gas pressure measurements were conducted within the Gemini Seam using packers placed in boreholes. Figure 4 shows the effect of cleat on pressure gradient about an underground opening. In situ permeability tests conducted using filtration of gas from vertical boreholes indicated a permeability of approximately 0.02 mD. (Hemela, M., Hematite Petroleum Ltd., pers. comm.). Permeability of the Gemini Seam is equal to about 20% of other Australian coals and 0.2% of U.S. coals. Smaller pore size and smaller ratio of apparent to true porosity results in very high gas pressure gradients close to the face. High stress concentrations further accentuate the problem.

MINING STRAIN AND OUTBURSTS

Mining strain manifestations include low angle roof shear (roof guttering), crushing of ribs and intensely induced cleavages in the coal and stone (Moore and Hanes, 1980; Hanes and Shepherd, 1981). Induced cleavages form a fracture envelope which curves around the openings (Figure 3). These are most prominent in continuous miner driven openings where frequencies of 100 to 1000 cleavages per metre are noted, especially in a 0.5 m dull coal band at the top of the seam. Induced cleavage is absent from the upper dull coal in those drives where bursts did not occur.

Mining strain patterns at Leichhardt Colliery are dependent on the orientation of the opening relative to the orientations of maximum principal stress and systematic cleat. Figure 4 shows a schematic of heading orientations in relation to the principal stress and cleat directions. When headings are driven parallel to σ_1, face crush is prominent but cleavages and outbursts are rare. When the angle between cleat and direction of drivage is less than 30°, rib crush and cleavage are common and outbursts are rare. When headings are driven at an angle to the principal stress direction and at an angle greater than about 30° to the main cleat, face zone crush is reduced and it is here that outbursts occur frequently.

Around the advancing face, the effective stress level at which failure of the coal occurs is determined by the interaction of pore fluid pressure and coal mass strength anisotropy due to structure. The coal mass strength is lowest perpendicular to the well developed cleat. In situ gas pressure measurements show that gas pressure gradients are highest perpendicular to the cleat (Figure 2). The magnitudes of stress concentrations are highest in the corners which first intersect the maximum principal stress trajectories.

The interaction effect of high gas pressure gradients and high stress at the corners where a given cleat is first intersected forms the potential focus of an outburst. Where progressive crush of the face zone takes place, gas pressure and stress are relieved minimising the danger of an outburst. Where the face remains solid, high stress and high gas pressure gradients are maintained giving rise to conditions of outbursts.

Figure 3. Curviplanar, mining induced cleavage at Leichhardt Colliery.

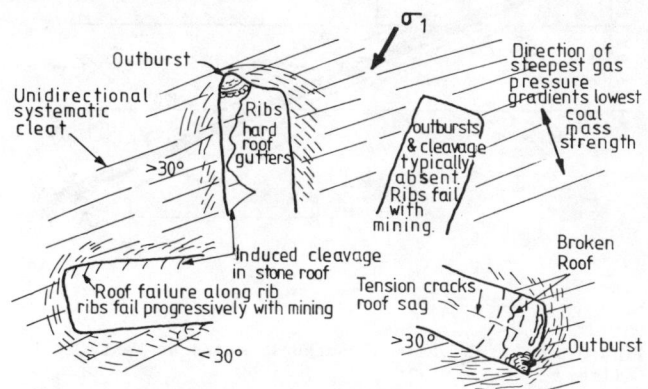

Figure 4. Mining strain related to orientations of drivage, cleat and stress at Leichhardt Colliery.

MECHANISM OF OUTBURSTS AT WEST CLIFF COLLIERY

The mechanism of outbursts at West Cliff and Collinsville No.2 Mines, and also at many other mines, is very similar and West Cliff Colliery has been chosen as an example to explain the mechanism. West Cliff Colliery mines the Bulli seam at a depth of 485 m. The thickness of the seam is 2.6 m and over 123 outbursts have occurred at the Colliery, invariably at strike-slip faults (Shepherd & Creasey, 1979, Marshall et al, 1980, Shepherd et al, in press.).

The faults are relatively minor structures with throws up to 0.1 m and horizontal displacement probably only 0.5 m. The zones are about 0.15 m wide in the stone roof widening in the coal to a maximum of 2.4 m of sub-vertical coplanar fractures and very soft coal gouge (mylonite). The coal ejected from the outbursts is highly pulverised and originates from the fault gouge.

The thickness of the shear zones is highest in the centre of the colliery holding where the main roof is predominantly sandstone. The shear zones diminish in frequency where the roof type changes to predominant shale/mudstone facies.

The average size of the outbursts is 32 tonnes with the largest reaching 140 tonnes. Where faults are intersected face-on, outbursts average 38 tonnes and where intersected end-on, outbursts average 51 tonnes.

The geomechanical properties of Bulli seam along with its filtration properties are given in Tables 2 and 3. The high strength of Bulli seam coal and its comparatively higher permeability result in lower gas pressure gradients and greater stability of the advancing face. Induced cleavage which is very prominent at Leichhardt Colliery is uncommon at West Cliff Colliery.

Figure 5 shows plans of two typical outbursts with a heading advancing "face on" and "end on" to shear zone along with associated features.

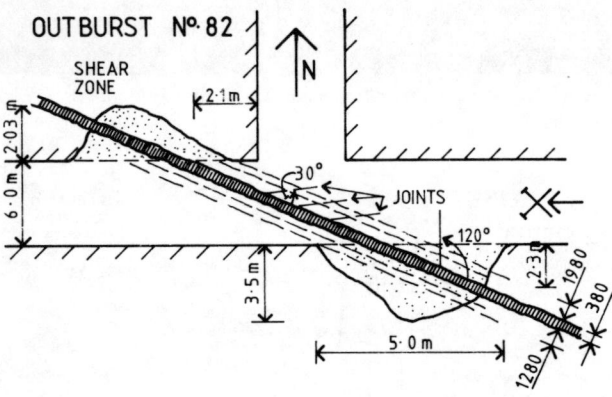

Figure 5 Sketch plans of outburst sites at West Cliff Colliery
(a) drivage at a low angle to a shear zone.

The amount of coal dislodged and the area affected by an outburst is determined by the width of the shear zone and associated joints and the angle of intersection of the drivage with the shear zone.

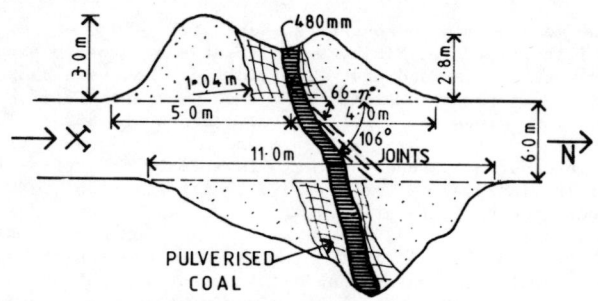

(b) drivage at a high angle to a shear zone.

The width of the zone of intense fractures ahead of an advancing face varies between 2 and 5 m. Outbursts occur when the solid rib of coal between a shear zone and an advancing face is reduced to less than 2 to 5 m.

The mechanism of outbursts at West Cliff Colliery is illustrated in Figure 6. While the shear zone is distant from the face, the face remains stable.

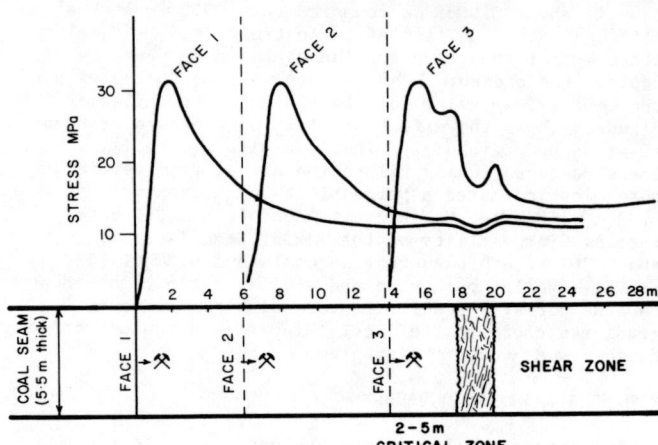

Figure 6. Mechanism of outburst occurrence at West Cliff Colliery.

As the distance between the shear zone and the face is reduced during advance, stress concentration builds up. The shear zone, because of its high horizontal stress on the coal surrounding it, and because of its inherent higher permeability, maintains a high gas pressure. The coal at the face softens due to fracturing as a result of high stress concentrations indicating an imminent outburst. A critical stage is reached when the rib between the shear zone and the face fractures and disintegrates and is then dislodged by the high gas pressure present in the shear zone.

PREDICTION OF OUTBURSTS

A number of approaches have been tried to predict outbursts at both Leichhardt and West Cliff Collieries. These include:
1. Geological and structural mapping,
2. Gas filtration properties of coal.

1. Geological-Structural Approach
At West Cliff Colliery, where outbursts are associated with small-throw strike-slip faults, a technique of geological mapping has been developed to predict the faults. Systematic coal joints are associated with the strike-slip faults. They are distinguished from cleats by their wider spacing and persistence through the entire seam section. They trend sub-parallel to the butt cleats and at a low angle to the faults. Results of scanline surveys show (Figure 7) that the joint frequency increases markedly close to strike-slip faults.

In some cases swarms of roof joints are present up to 50 m from a fault. With routine measurements at an advancing face, an outburst threshold can be defined where a sustained increase in joint frequency occurs. Strike-slip faulting occurs within 45 m of the outburst threshold.

At Collinsville, detailed geological mapping of advancing faces shows that as large, outburst prone faults are approached, there is an irregular but sustained increase in shear plane frequency. Fracture counting using a sample interval of 3 m showed that undistrubed coal has a fracture count of less than 1 per metre whereas extremely sheared coal adjacent to a thrust zone has a count of greater than

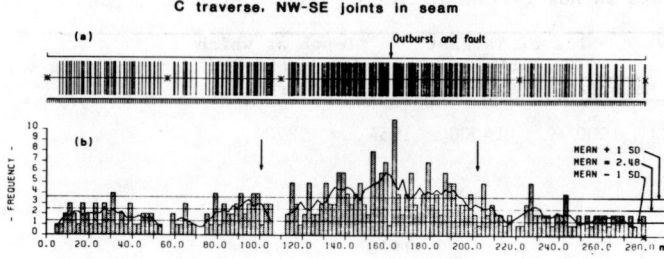

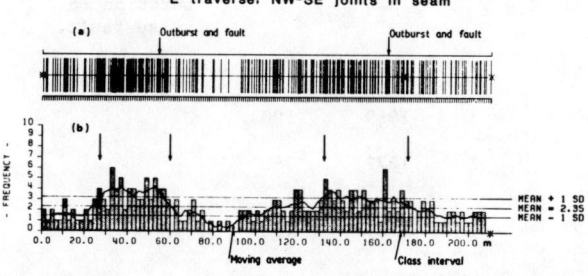

Figure 7. Joint spacing scan line traverses (C and E) across sites of outbursts and coincident strike-slip faults (shear zones) at West Cliff Colliery.
(a) Schematic profile along mine rib;
(b) frequency distribution of joints along traverse. Arrows indicate outburst threshold. 1SD = 1 standard deviation; moving average = 10m; class interval = 2 m; * = survey pin at intersection.

6 per metre. Thrust faults can be predicted using the fracture technique at a distance of about 120 m.

2. Gas filtration Properties of Coal
Techniques using gas filtration properties of coal are mainly based on gas pressure measurements and gas desorption. Gas pressure measurement at Leichhardt Colliery showed that outburst conditions displayed steep gas pressure gradients (Figure 4) with near virgin pressures existing within a few metres of the opening. At West Cliff Colliery, gas pressure testing shows that gas pressure gradients increase rapidly on approaching a shear zone (Figure 8). Using this technique, the presence of a shear zone could possibly be detected up to 50 - 100 m away from the shear zone.

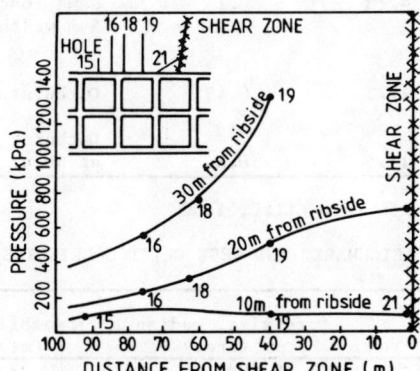

Figure 8. Gas pressure changes close to a strike-slip fault at West Cliff Colliery.

As gas pressure testing is too time consuming for inclusion in routine production schedules, measurement of gas desorption indices offer less interference with production and therefore a more attractive method of outburst prediction. Hargraves' emission testing at Leichhardt Colliery showed a dramatic increase in emission value (E.V.) as workings advanced from non-outbursting to outbursting conditions. In non-outbursting conditions the E.V. was typically less than 0.1 g/cc. It rose to greater than 1.0 g/cc within about 10 m of actual outbursting and values up to 2.8 g/cc were recorded after outbursts. An increase in free water content of the coal accompanied the onset of outbursting and wet coal reduced the validity of the test.

Certain other indices including K_T, ΔP, and ΔP express showed variations in the coal type and its properties, but because of large scatter of values and insensitivity to small changes they have not been successfully used for routine outburst predictions. A new index called L_2 index has showed promise and can detect changes in the rate of desorption of coal up to 10 m away from shear zones at West Cliff Colliery (Lama, 1980 (a)).

Advance drainage of gas as a method for alleviation of outbursts has been tried at West Cliff and Leichhardt Collieries. Advance drainage is successful only if the area is drained for about 120 days and if the gas pressures are lowered to under 150 kPa at the shear zones. Under such conditions outburst activity has been greatly reduced if not completely eliminated (Lama, 1980(b)).

CONCLUSIONS

At Leichhardt Colliery, stress concentrations at the corners of an advancing face produce failure of the structurally anisotropic coal. Where failure is abrupt, the concentrated stress and high gas pressure gradients interact to produce an outburst. At West Cliff Colliery, outbursts occur where concentrated stress causes disintegration of a 2 to 5 m thick rib of solid coal between an advancing face and a shear zone containing pulverised coal and gas under pressure.

It is possible to predict location of structures up to 100 m ahead of a face using geological fracture techniques or gas pressure measurement techniques. Other methods of prediction using Hargraves emission values, K_t, ΔP and ΔP express are not viable tools for use in prediction of outbursts on a regular basis. The only successful method of alleviation to date seems to be effective gas drainage.

ACKNOWLEDGEMENTS

Thanks are due to Broken Hill Proprietary Company Limited, Kembla Coal and Coke Pty. Limited and Collinsville Coal Company for permission to publish the data from Leichhardt, West Cliff and Collinsville Collieries. Financial support from The National Energy Research Development and Demonstration Council administered by the Department of National Development and Energy for conducting investigations at these mines is gratefully acknowledged. Two authors (RDL and JS) were formerly officers of CSIRO where most of their work was conducted. Thanks are due to the colleagues in the various organisations with which the authors are associated for support and encouragement.

TABLE 1. Occurrences of Coal and Gas Outbursts in Australian Collieries (June 1982)

Colliery* (Basin)	Seam	Outbursts experienced Period	Number	Gas type	Association with geological features	Size of largest Outburst Material(t)	Gas(m^3)	Year	Depth at which outbursts recurred (m)	Remarks
Clnsvlle (Bowen)	Bowen	1954-61	13	CO_2	Strike-slip fault	1000	14000	1954	220	
Clnsvlle (Bowen)	Bowen	1972	2	CO_2	Strike-slip fault	1	?	1972	230	
CCP No.2 (Bowen)	Bowen	1972-78	4	CO_2	Thrust	25	?	1979	260	
Lchdt (Bowen)	Gemini	1974-81	~250	CH_4	Normally none	500	12000	1978	380	Largest outburst on reverse fault.
Moura (Bowen)	'C'	1982	3	CH_4	Strike-slip fault	0.5	?	1982	130	
Appin (Sydney)	Bulli	1965-82	8	CH_4	Strike-slip fault	60	?	1969	500	
Bulli (Sydney)	Bulli	1972	1	CH_4		100	?	1972	380	
Coal Cliff (Sydney)	Bulli	1961	2	CH_4/CO_2		2	?	1961	450	
Corrimal (Sydney)	Bulli	1961-67	5	CH_4/CO_2	Shear zone	50	?	1967	360	
Mtrpltan (Sydney)	Bulli	1895-1957	+18	CO_2	Faults and dykes	300	?	1961	400	Several induced outbursts occurred after 1957
NthBulli (Sydney)	Bulli	1911	1	CH_4		1	?		370	
Tahmoor (Sydney)	Bulli	1980-82	29	CO_2/CH_4	Dykes and strike-slip faults	40	?	1982	400	
Tower (Sydney)	Bulli	1982	2	CH_4	Dykes and faults	100	?	1982	420	
WstClff (Sydney)	Bulli	1976-82	123	CH_4	Strike-slip faults	148	?	1976	470	

TABLE 2. GEOMECHANICAL PROPERTIES OF COAL FROM LEICHHARDT AND WEST CLIFF COLLIERIES

Colliery*	Site	Depth (m)	Uniaxial* Compressive Strength (MPa)	Strength / Vertical stress	Modulus** (GPa)	Fracture surface energy Ergs/$cm^2/10^4$	Apparent Porosity %	True Porosity %	Apparent / True Porosity	Remarks
Lchdt Gemini	I	380	12.8	1.35	2.8	15.87	1.43	2.98	0.48	Non-outburst area
Lchdt Gemini	II	380	11.7	1.23	2.4	14.02	2.25	3.34	0.67	Low activity area. Opposite to Site 4 in the same roadway 5 m width.
Lchdt Gemini	III	380	10.5	1.14	2.0	11.79	0.61	4.73	0.13	
Lchdt Gemini	IV	380	1.9	0.2	0.7	11.16	0.84	1.73	0.49	Outburst site.
WstClff (Bulli)	Coal	480	22.0	1.83	3.0	32.0	10.9	13.5	0.8	Outburst only at shear zones

** value determined from Rebound Index using Schmidt hammer.
vitrinite reflectance $R_o$max % averages: Leichhardt, 1.25 West Cliff, 1.30

TABLE 3. GAS FILTRATION AND GAS ADSORPTION INDICES OF COAL FROM LEICHHARDT AND WEST CLIFF COLLIERIES

Colliery	Gas Pressure kPa	Gas pressure / Hydrostatic* head	Gas pressure gradient kPa/metre	Gas Content m^3/tonne	Pore size, median μm (area)	Permeability (mD)
Leichhardt	3640	0.91	1100	15 - 16	0.0038	0.02
West Cliff	3000	0.60	150	12 - 13	0.0052	0.10

* Hydrostatic head is taken as equal to depth without accounting for depth of water table.

* Colliery abbreviations from Tables.

Collinsville	Clnsvlle
Collinsville Coal Pty. Ltd.	CCP
Leichhardt	Lchdt
Metropolitan	Mtrpltan
North Bulli	Nth.Bulli
West Cliff	Wst.Clff

REFERENCES

HANES, J., and SHEPHERD, J. (1981). Mining induced cleavage, cleats and instantaneous outbursts in the Gemini seam at Leichhardt Colliery, Blackwater, Queensland. Proc. Australas. Inst. Min. Metall., (No.277), 17-26.

HARGRAVES, A.J., et al. (1964). The control of instantaneous outbursts at Metropolitan Colliery, N.S.W. Proc. Australas. Inst.Min.Metall., (No.209), 133-166.

LAMA, R.D. (1980a). the use of adsorption/desorption isotherms in predicting outburst conditions. CSIRO, Division of Applied Geomechanics Coal Mining Report No. 21, 76 pp.

LAMA, R.D. (1980b). Drainage of gas from the solid, optimisation of drainage hole parameters. CSIRO, Division of Applied Geomechanics Coal Mining Report No. 18, 78 pp.

LAMA, R.D., and MITCHELL, G.W. (1981). Investigations on geomechanical parameters in relation to outbursts of gas and coal at Leichhardt Colliery, Queensland Coal Mining Co., Blackwater, Qld. CSIRO, Division of Applied Geomechanics Coal Mining Report No. 9, 49 pp.

MARSHALL, P. et al. (1980). Occurrence of outbursts at West Cliff Colliery. Symp. The Occurrence, Prediction and Control of Outbursts in Coal Mines. Brisbane 1980. Australas.Inst. Min. Metall., 19-39.

MOORE, R.D., and HANES, J. (1980). Bursts in coal at Leichhardt Colliery, Central Queensland and apparent benefits of mining by shotfiring. Symp. The Occurrence, Prediction and Control of Outbursts in Coal Mines. Brisbane 1980. Australas. Inst. Min. Metall., 71-83.

SHEEHY, J.A. et al. (1956). Report of the Royal Commission appointed to enquire into certain matters relating to the State Coal Mine, Collinsville, Queensland Government Printer, Brisbane.

SHEPHERD, J., and CREASEY, J.W. (1979). Forewarning of faults and outbursts of coal and gas at West Cliff Colliery, Australia. Colliery Guard. Coal Int. (227), 13-22.

SHEPHERD, J., et al. (1982). Instantaneous outbursts of coal and gas with reference to geological structures and lateral stresses in collieries. Symp. Seismicity in Mines, International Society of Rock Mechanics, Johannesburg, 1982 (in press).

PREDICTING MOISTURE-INDUCED DETERIORATION OF SHALES
Prédiction de la détérioration des schistes argileux provoquée par l'humidité
Die Vorhersage der feuchtigkeitsbedingten Entfestigung von Schiefer

M. M. Singh and R. A. Cummings
President and Section Head, Geology Engineers International, Inc. Westmont, IL 60559,
U.S.A.

SYNOPSIS

Certain shales have a tendency to deteriorate when exposed to humid atmospheres. This can have serious consequences for many mining and civil engineering structures. This investigation consisted of a laboratory study of a susceptible shale obtained from a coal mine. The laboratory study was correlated with field observations. To perform meaningful laboratory tests, specimens should be taken from fresh shale material and the tests conducted soon. Sampling history and technique strongly affect results. Large expansions were measured in the shale, for high atmospheric humidities, despite the absence of significant swelling clays in their mineralogy. The associated swelling strains exceeded the levels required to fail the rock.

RESUME

Exposés à une atmosphère humide certains schistes argileux ont tendance à se dégrader, ce qui peut entraîner des conséquences graves pour beaucoup d'ouvrages de mines et de génie civil. Cette étude comporte l'analyse en laboratoire d'un schiste argileux prélevé dans une mine de charbon, les résultats de l'analyse étant ensuite comparés aux observations relevées sur le terrain. Afin d'effectuer des essais en laboratoire valables les échantillons doivent être prélevés sur carottes fraîchement obtenues et les tests réalisés sans tarder. Les conditions et la technique d'échantillonage ont une forte influence sur les résultats. Dans des conditions de forte humidité et malgré l'absence d'argiles gonflantes dans leur minéralogie des dilatations importantes ont été observées dans les schistes argileux étudiés. Les déformations gonflantes s'y rapportant ont dépassé les taux nécessaires pour entraîner l'éboulement des roches.

ZUSAMMENFASSUNG

Manche Schiefertone tendieren dazu, zu verfallen, wenn sie der Luftfeuchtigkeit ausgesetzt sind. Dies kann für viele Bergbau- und Felsbauwerke schwerwiegende Konsequenzen haben. Es wird von Laboratoriumsversuchen an feuchtigkeitsempfindlichen Schiefertonen berichtet, die von einer Kohlengrube gewonnen wurden. Die Laboratoriumsuntersuchungen werden mit Geländebeobachtungen korrelliert. Um sinnvolle Laboratoriumsuntersuchungen durchführen zu können, sollten die Proben nur von frischem Schiefertonmaterial entnommen und die Versuche rasch ausgeführt werden. Ablauf und Technik der Probenahme haben einen starken Einfluß auf die Ergebnisse. In den Schiefertonen wurden bei großer Luftfeuchtigkeit substantielle Ausdehnungsbeträge gemessen, obwohl keine quellenden Tonminerale vorlagen. Die Ausdehnung überschritt den Dehnungsbetrag, der zum Bruch erforderlich ist.

INTRODUCTION

Mine roof shale deterioration results in difficulties in ventilation, increased costs for roof maintenance, reduced safety, production delays, and consequently increased costs for mining. Part of the reason for not being able to control this roof behavior is a lack of understanding of the process(es) involved.

To date, efforts to control shale deterioration in mine roofs have sought to establish and maintain stable humidity levels year round, at the rock surface. This is done by either treating the mine atmosphere with some sort of tempering or conditioning system, or treating the rock surface with a sealant to protect it from amospheric humidity fluctuations.

This study encompassed both field and laboratory investigations. It was conducted, in part, to gain some insight into the mechanism of shale failure.

The field program involved collection of data on the mine atmosphere (monitoring of airflow and atmospheric conditions) and on roof deterioration (measurement of roof strength, roof sag, mapping of roof conditions, photographic logs of roof behavior). This work continued for one full year. Most work was carried out in "conditioning entries" -- workings near the main intake where the air is brought to temperature and humidity equilibrium before being delivered to the working places.

At the same time, laboratory determinations of moisture uptake, natural moisture, static slaking, mineralogy, and expansion were conducted on samples of mine roof shale. This was done with the dual objectives of providing information on shale behavior in the mine roof, and of evaluating tests with the greatest potential for predicting moisture-induced disintegration.

At the mine studied, the roof fell into two categories. The most stable roof was formed by a hard, black, fissile shale in the immediate roof, up to 1.8 m (6 ft) thick, overlain by 0.6 m to 1.2 m (2 ft to 4 ft) of weak, highly slickensided claystone. Rashing roof consisted of a 0.3 m (1 ft) thick layer of banded black and dark gray shale with claystone, overlain by 0.6 m to 1.5 m (2 ft to 5 ft) of claystone. The claystone unit is only sparingly pyritic and contains layers of hard, gray silty clay or siltstone. Overlying both types of immediate roof is a pyritic, impure limestone that is very stable and can withstand roof spans of 6.1 m (20 ft) and more.

FIELD DATA

Baseline field data included continuous atmospheric monitoring at certain points along the air course, airflow measurements, and periodic determinations of temperature and humidity within the conditioning entries, by means of a psychrometer.

Methods for measuring roof deterioration included: concrete test hammer (Schmidt hammer) tests, photographic logs, collection of fallen material, roof convergence, reconnaissance inspection, and mapping of roof conditions.

Photographic Logs. In order to record the progress of roof deterioration, photos were taken in all three sets of conditioning entries during each mine visit.

Two types of photo logging were employed: reconnaissance and detail.

The reconnaissance photos were taken to document deterioration with respect to air residence time. No permanent installations were built, because the objective was to take a large number of photos, thus increasing the chance of documenting the development of new falls.

The detailed photo log was compiled by installing six (6) permanent photo (PH) stations, whereby the exact same photo could be taken each time, documenting progressive deterioration. These stations were set up near sites of active deterioration.

Fallen Material. Near each detailed photo logging station, a "deterioration monitoring" (DM) station was set up to collect fallen material and measure roof height increase. Eight (8) of these deterioration monitoring stations were set up.

These stations were strategically located so as to be easily correlated with other data elements. A large brattice curtain was spread out beneath visibly disintegrating roof to collect material falling from the roof. On each subsequent visit, fallen material was collected, its weight determined, and the curtain was cleared and spread out again.

In addition, at each DM station, a number of small strips of plastic surveyor's flagging were tacked to the half headers, extended straight up to the roof, and cut off. This provided a ready means for checking the increase of roof height. This was noted once each mine visit.

Roof Convergence. It was recognized early in the project that moisture-induced deterioration might be accomplished by dilation of the roof due to moisture uptake. A combination of absorption and crystal growth was thought to be the cause, affecting both roof coal and the overlying claystone top. This has the effect of hastening the loss of protective top coal, as well as promoting movements along discontinuities (chiefly small slickensides) within the claystone top. The convergence stations were designed to detect dilation due to both expansion and weakening, as a function of seasonality as well as air residence time.

Reconnaissance Inspection. The first day of each mine visit was spent in the mine inspecting and noting general roof conditions, evidence of atmospheric moisture conditions, and locating materials and places to perform any particular test. This helped to better organize activities planned during that particular visit.

Mapping Roof Conditions. Mapping of the roof conditions was done four times during the course of the project, during or immediately following the summer, autumn, winter and spring months. Joints, clay veins, slips, areas of shear over the rib, rib sloughage, and roof falls and deterioration were mapped at a scale of 1:1200 and detailed notes taken as to the height and rock type of each fall. In this way, progressive deterioration could be documented and the geology taken into account when evaluating other data elements.

Details of the field work may be found in Cummings, et al. (1981, 1982) and Engineers International, Inc. (1981) and are not discussed further here.

LABORATORY STUDIES

In order to measure shale response under conditions similar to those encountered in a mine, a laboratory program was undertaken. Based on the field visits and literature search, it was concluded that the test program should deal principally with the effects of moisture on shale. It was intended that lab results could be correlated with in-mine testing.

Throughout the program, a concerted effort was made to keep shale samples as "fresh" as possible. Both obtaining and storing fresh shale samples are difficult activities; nonetheless, the importance of great care must be stressed. Specimens used in the study were stored in wax-sealed plastic bags when not in use. Still, moisture changes almost certainly developed during storage (up to several months) and this probably contributed to data scatter.

The lab work included sampling of the roof rock, static slaking tests, moisture content determinations, moisture uptake tests, and shale expansion tests. Prior work done by others (Aughenbaugh, 1973, 1976; Chenevert, 1970; Harper et al., 1979; Adam, 1975; Curran, 1973, McNary, 1941), suggested that these types of tests best correlate with observed shale roof behavior.

Sampling. It was necessary to collect enough of the three types of shale to be able to run an adequate laboratory program. Furthermore, it was required that the method of sampling preserve the structure and moisture content of the roof rock. Accordingly, a 5-ft NX core barrel, that is adaptable to a roof bolter, was obtained. The core barrel is of the double-tube type, designed to minimize contact between rock and drilling medium (water).

Sampling of the roof rock at the mine was performed twice during the project. A combination of factors resulted in poor core recoveries. The major problem encountered was driller's breaks of the core, as the roof bolter operator had to regulate both rotation speed and advance rate by touch. Another problem was encountered in finding the correct amount of water to use while drilling. The claystone top, which is the most critical lithologic unit, readily absorbs water. If too much water was used, the claystone weakened and fell apart during coring, yet a certain minimum amount was needed for lubrication and cooling of the bit. Despite these difficulties, it was felt that the sampling procedure used was superior to other possible methods, such as bulk sampling of fallen material.

Samples were immediately put into specially designed plastic core bags, logged, and marked. At the end of the shift, the bags were sealed in wax and boxed. The samples were transported to the laboratory in EI's own truck.

Static Slaking. One aspect of the laboratory work was to determine the extent to which roof deterioration is a slaking problem. A procedure followed by Moriwaki (1974) formed the basis of these tests. It involved suspending a 1.5-cm (1/2-in.) thick disc of shale core vertically in a beaker of water by means of a holder specially designed to provide minimal restriction to the specimen. The specimens were suspended with monofilament line from beneath a triple-beam balance, so that weight changes could be measured. After an initial weight gain due to water absorption, any slaking would be evidenced by a weight loss. Sample condition was noted visually, as well.

Not surprisingly, specimens of the black shale and the hard gray silty shale showed no slaking tendencies during the tests. The samples seemed completely unaffected by immersion. Little slaking tendency was observed for the claystone. Although cracking was often observed, only surface chips that may have been loose prior to immersion fell off the specimens. This occurred shortly after the start of the test. In no instance did this activity persist for more than a few minutes, nor did continued crumbling and cracking, normally considered slaking behavior, occur. All the claystone specimens were greatly weakened by immersion, and could not withstand even the most delicate handling after the test.

It was concluded that static slaking would not be an objective measurement of roof deterioration susceptibility, because evaluation of the degree of cracking and weakening would require a trained eye. It was felt that other measures would be better suited to widespread practice.

Moisture Content. Natural moisture content determinations were run on the three shale types, and were the first tests performed. Natural moisture has been considered to be correlative with shale roof durability (Augenbaugh, 1973) and with observed roof conditions (Conroy, 1973).

Fresh, 1.5-cm (1/2-in.) thick discs of core were dried to constant weight at 49°C (120°F) to determine moisture content. It was desired that only readily-exchangeable moisture be determined; accordingly, the temperature selected was low enough to prevent dehydration of certain hydrated sulfate minerals that may possibly have been present. Samples placed in the 0% humidity dessicator (see Table I) provided additional moisture content determinations.

Table I. Saturated Salt Solutions Used in Moisture Uptake Determinations

Salt (dry)	Relative Humidity at 25° C (77° F)
P_2O_5	0
$MgCl_2 \cdot 6H_2O$	32.5
$Ca(NO_3)_2 \cdot 4H_2O$	50.5
$Na_2Cr_2O_7 \cdot H_2O$	53.0
NH_4NO_3	62.5
K Tartrate	75
KNa Tartrate	87
H_2O (distilled)	100

Ten samples were oven-dried or dried in dessicators. Samples which were initially placed in dessicators containing P_2O_5 (0% relative humidity) for moisture uptake tests were also considered natural moisture determinations.

For the hard gray silty shale, the average weight loss of 6 oven-dried specimens was 2.4%, compared with 3.9% for 3 dessicated specimens. For claystone, 2 oven-dried samples averaged 3.6% weight loss; 7 dessicator-dried samples averaged 4.5% weight loss. This last figure would have been 4.2% if one extraordinarily high value, 7.2%, was eliminated. Two black shale specimens dried in dessicators averaged 2.5% weight loss.

As expected, claystone moisture contents are highest, followed by hard gray silty shale and black shale. Detailed consideration of the data shows that moisture contents are higher for dessicated samples than for ovendried samples, suggesting that dessicators are more thorough in drying rocks, since P_2O_5 is a very effective dessicant. The use of dessicators for laboratory drying also removes any doubt about the possible adverse effects of elevated temperatures.

Sample history and sampling technique, especially for hydroscopic rocks such as the claystone, have a marked influence on natural moisture contents. Specimens should be taken from fresh material and determined as soon after sampling as is possible. These types of errors certainly resulted in scatter of the data from the tests.

Moisture Uptake. The manner in which roof shales react to various humidities was of concern in this study. Prior research (Aughenbaugh, 1973) indicated that a "null point" for shales exists; this is the humidity at which the rock will not take up or give off moisture. As a result of laboratory experiments, Aughenbaugh (1976) was unable to confirm an invariant null point, but found that it depends on sampling and experimental procedures. Because of the potential significance of the null point in conditioning entry design, the authors attempted to determine it for shale samples at natural moisture contents. Further, by testing for the null point, actual mine conditions were simulated in the lab, since shale samples were cycled through various humidities.

The experimental procedure was to place shale samples at natural moisture contents in standard aluminum dessicators in a controlled humidity environment.

The dessicators were prepared by placing saturated salt solutions (Table I) in shallow dishes at the bottom. The saturated solutions maintain a constant vapor pressure and thus regulate humidity. The containers were covered and a humidity detector probe affixed to the underside of the cover. The probes allowed for the humidities to be read remotely on an electronic readout. Readings were taken weekly to assure that the humidities did not vary significantly once a constant value was reached. The temperature in the lab was kept constant because some dessicant salt solutions are slightly temperature-sensitive. Further assurance of constant temperature was gained by placing the dessicators in an insulated enclosure. The dessicant solutions were selected because of low temperature influence on relative humidity.

Five humidities were achieved from six dessicators; these varied between 0 and 100%, with the least humidity spreads at the upper end of the scale. Two small samples of each of the three shale types were weighed and put into each dessicator. Once the humidity in the dessicator had stabilized at an acceptable level (about a month), the samples were removed and weighed. This process was repeated until no change in weight was recorded. At this point, all samples were switched to the second higher numbered dessicator; for example, a specimen equilibrated at 100% humidity was next exposed to 30%, and so on. In this way, the effect of humidity history on shale behavior could be indicated.

Null Point. Two rounds of testing were performed for the determination of the null point. In the first round, one dessicator switch was made. New samples were drilled at the mine, and a second round of tests conducted with fresh specimens, in which two dessicator switches were made. The sample distribution was as follows:

	Round 1	Round 2	Total
Hard Shale	8	10	18
Claystone	9	8	17
Black Shale	9	8	17
Lime	4	0	4
Coal	0	4	4
Total	30	30	60

Weight changes corresponding to the various relative humidities were plotted after equilibrium had been achieved. The results are shown in Figures 1, 2, 3, 4, and 5. These plots have been corrected to remove the effect of sample size, which imparted scatter to the data. Additional scatter could be generated by variations of initial moisture content away from the natural levels, time of sample exposure to air before weighing, weighing errors, and imprecision in relative humidity control. The characteristic line represents a least-squares fit.

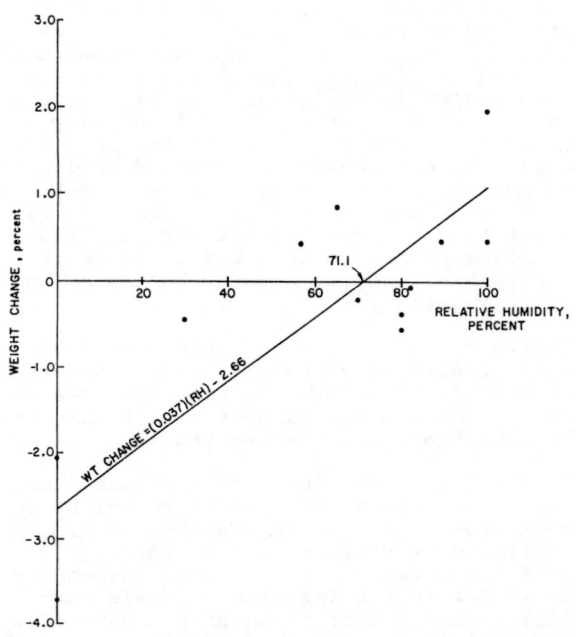

Fig. 1 Percent weight change versus relative humidity for gray shale. (Data corrected for influence of sample weight.)

There is a clear trend in all the material types toward moisture loss at low humidities, and toward net moisture gain at high humidities. The null points as determined by the regression lines are near 80% relative humidity for the Pittsburgh roof strata investigated. At 20°C (68°F) this corresponds to a specific humidity of 11.73 g/kg (82 grains/lb) of dry air. The null point may be assumed with confidence to be between 70% and 90% relative humidity. Further testing, under stringently controlled conditions, is needed to eliminate the sources of nonlithologic scatter, so that comparisons among various rock types may be obtained. It seems reasonable that the null points for Pittsburgh roof rocks are close, because the lithologic differences are not great (when compared to the entire spectrum of rock types).

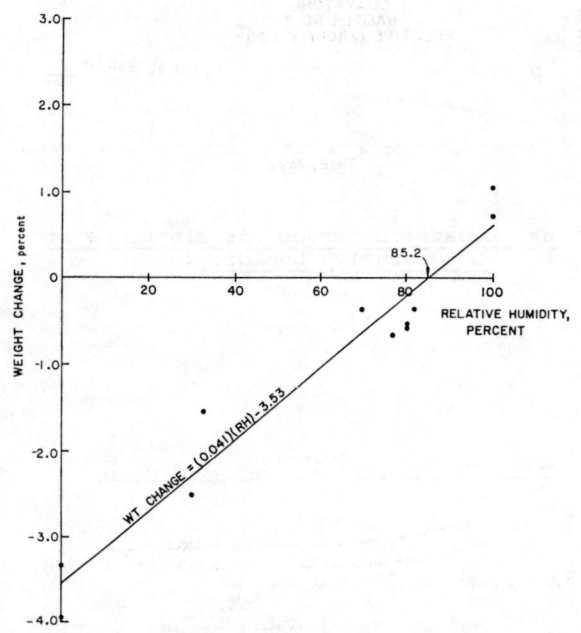

Fig. 2 Percent weight change versus relative humidity for claystone. (Data corrected for influence of sample weight.)

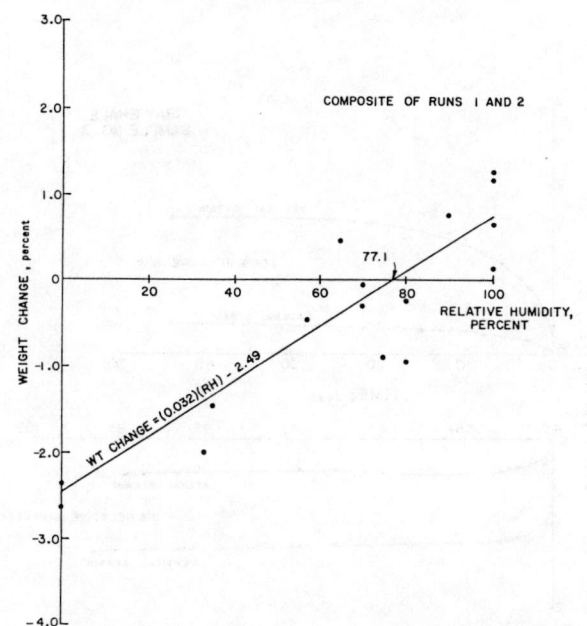

Fig. 3 Percent weight change versus relative humidity for black shale. (Data corrected for influence of sample weight.)

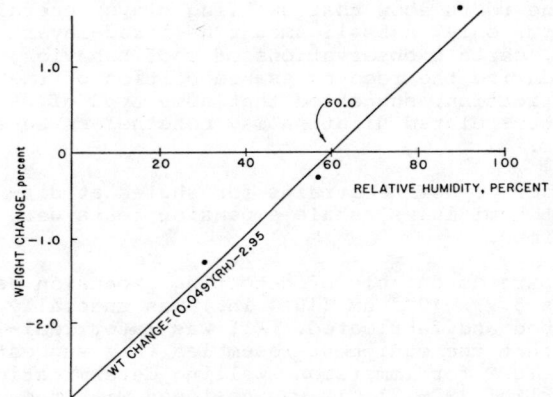

Fig. 4 Percent weight change versus relative humidity for top coal.

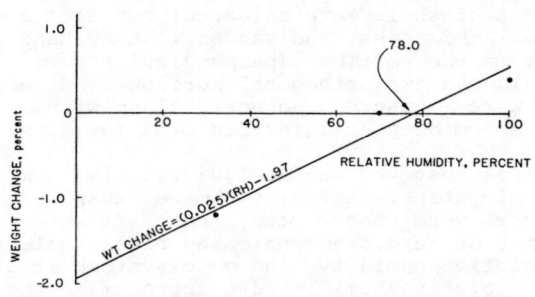

Fig. 5 Percent weight change versus relative humidity for lime.

In examining the moisture absorption trends, it is seen that the hard gray shale and claystone compare closely in moisture exchange capability -- they gain or lose similar amounts of moisture at equivalent relative humidities. The black shale is not as greatly affected by moisture changes. Limited data for coal and lime indicate a lower null point for coal (60%) and while the null point for lime is similar to that for the clastic sediments (78%), its overall response to moisture change is markedly lower. Table II summarizes the null points and regression data for the five lithologies tested. The slope of the regression line is a measure of the sensitivity of the material to moisture change.

Table II. Summary of Null Point and Regression Data

Material	Null Point	Regression Equation	No. of Data Points
Claystone	82.2%	=0.041(RH)-3.53%	11
Gray Shale	71.1%	=0.037(RH)-2.66%	13
Black Shale	77.1%	=0.032(RH)-2.49%	16
Coal	60.0%	=0.049(RH)-2.95%	3
Lime	78.0%	=0.025(RH)-1.97%	4

Shale Expansion. It may be concluded that swelling behavior of roof shale may have pronounced effects on roof stability. Although several survey papers of Pennsylvanian shales in mine roofs show that swelling clays generally occur only in small amount as mixed-layer clays, certain observations of roof behavior, made during the reconnaissance portion of the investigation, suggested that some type of moisture-related dilation may nonetheless be a factor.

In order to define strains for shales at different humidities, shale expansion tests were performed.

An apparatus capable of detecting expansion as low as 3.9×10^{-5} cm (10^{-4} in.) was specially designed and fabricated. (It was later realized that the equipment resembled that suggested by ISRM for immersion swelling determination (Franklin, 1979).) It was designed with the capability of holding a dessicant salt solution and a humidity detector probe for environment control.

In order to ensure representative behavior, specimens of shale were selected, cut to 1.5-cm (1/2-in.) thickness, and sanded flat by hand. Strains in the vertical (perpendicular to bedding), and two orthogonal horizontal directions, were measured and noted along with relative humidity at intervals of a few days.

An initial test was run at 100% relative humidity on claystone. Later, two more expansion enclosures were constructed, and tests were performed on hard gray shale and black shale at 100% relative humidity, and on claystone at 0% and 70% relative humidity (to approximate the null point).

A larger number of these tests would be desirable to minimize sampling and experimental effects. However, although these tests are simple and straightforward, they are time-consuming and there was not sufficient time to test more samples.

Despite the absence of swelling clays, large expansions were measured for claystone discs under 100% relative humidity. A change to a 0% relative humidity environment resulted in some contraction, but never enough to completely recover from the dilation.

Although the data base consisted only of two tests for claystone and one test for each rock type of gray shale and black shale, the claystone seemed the most expansive in a humid environment. After 153 days at 100%, just under 7% swelling strain perpendicular to bedding had been experienced (Figure 6a). The sample exhibited signs of severe distress when the test was terminated to change to a drying environment (Figure 6b). The weight gain was only 0.7%, in fair agreement with the moisture uptake determinations.

By comparison, gray shale equilibrated at 1.6% vertical swelling strain after 44 days at 100% relative humidity (Figure 7), and the black shale swelled 1.3% after 26 days at 96% relative humidity (Figure 8); its weight gain was 0.6%. The gray shale showed little damage and indeed recovered quite well upon drying, shrinking 1.3% after 49 additional days at 0% (Figure 7). The black shale was not dried but presumably would have behaved similarly.

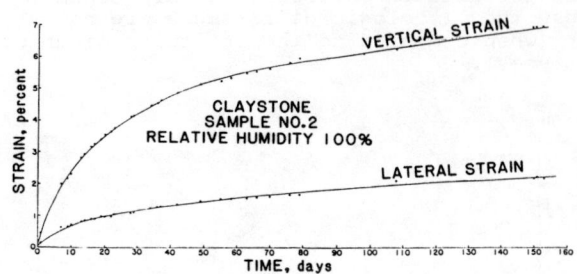

Fig. 6a Expansion curves for claystone at 100% relative humidity.

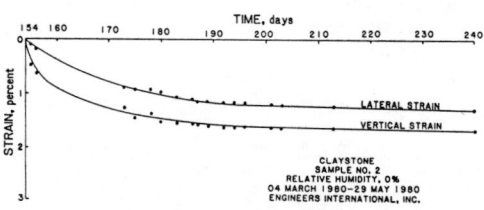

Fig. 6 b Expansion curves for claystone at 0% relative humidity.

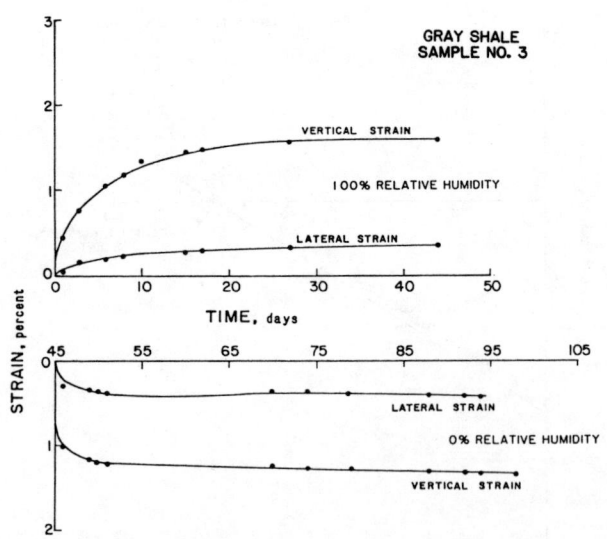

Fig. 7 Expansion curves for hard gray shale at 100% and 0% relative humidity.

E 92

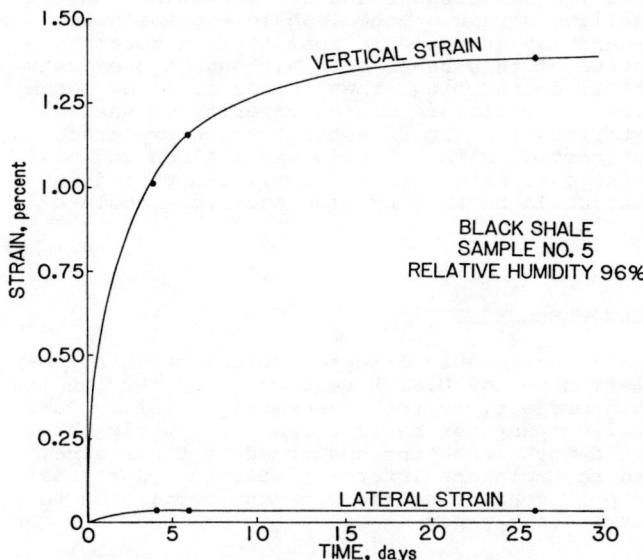

Fig. 8 Expansion curves for black shale at 96% relative humidity.

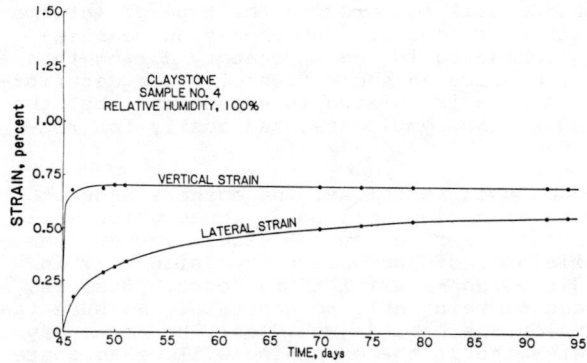

Fig. 9b Expansion curves for claystone at 100% relative humidity.

Lateral strains were about a third of vertical strains for claystone, a fifth of vertical strains for gray shale, and were extremely small for black shale, reflecting the degree of fissility exhibited by each rock type. Claystone, with its high density of slickensides at random orientations, has the most nearly isotropic swelling properties.

A claystone specimen placed in an expansion box at 70% relative humidity (believed at the time to be close to the null point, which had not yet been determined) exhibited far less swelling, 0.2%, at equilibrium, after 46 days (Figure 9a). The humidity was increased to 100%, and the swelling strain increased rapidly by 0.7% (Figure 9b). Swelling abruptly ceased, and it was found later that corrosion had caused a dial gage, responsible for measuring vertical swelling strain, to seize up. This unfortunately prevented measurement of the ultimate swelling strain, but the strain rate prior to the gage failure indicated that the ultimate strain would have been substantial.

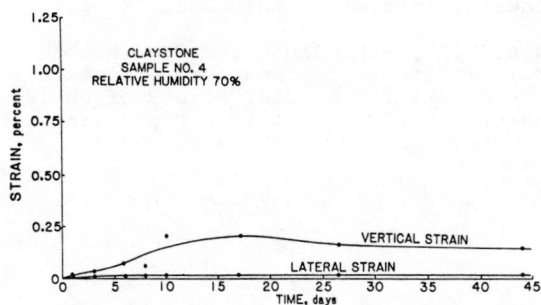

Fig. 9a Expansion curves for claystone at 70% relative humidity.

The tests showed that swelling strains in claystone surrounded by 100% relative humidity are far in excess of the levels required to fail the rock. Furthermore, expansion is not recovered during dry periods, and renders the rock increasingly susceptible to further humid cycles.

The mechanism for this expansion can only be speculated upon. While it is not due to adsorption of moisture by swelling clays such as montmorillonite, uptake of moisture is clearly associated. Probably a certain amount of the volume increase is merely a result of the absorption of water, in much the same fashion as a sponge. It is also possible that the claystone is overcompacted, as evidenced by the slickensides throughout the material. Addition of moisture would serve to weaken the rock fabric, releasing compaction strain energy, and causing swelling. This swelling would be self-enhancing, in that it would augment negative pore pressures, creating suction that increases absorption. Swelling that takes place along discontinuities (slickensides) is nonrecoverable, and loosens the claystone mass.

Gray shale and black shale are stronger, have less clay, and are not overcompacted. Thus, feedback mechanisms do not operate and high swelling strains are not achieved. Inasmuch as the total number of specimens is small, it is impossible to claim repeatability for the results. More tests of this type are definitely needed.

These data strongly support the field observations, which showed that measurable deterioration, roof convergence, and roof weakening are markedly higher in areas of temperature/humidity fluctuation, and that the claystone could not remain stable in such areas if exposed.

SUGGESTED TEST PROCEDURE FOR MINE OPERATORS

It would be desirable if a mine operator could ascertain if the roof shale in a mine is liable to deterioration upon exposure to variable humidity levels. It is evident from the investigations conducted by prior researchers that soaking a shale core in water, or some such technique will not provide the type of information that is sought. The expansion measurements conducted in the laboratory during this study, do give an indication of shale deterioration (which is related to expansion), but the method is too complicated and costly for routine use.

Based on this technique, the authors suggest that an expansion cell be designed which will accommodate a core (say NX size), rather than a parallelopiped, and detect expansion only in one direction -- axial to the core. Besides, instead of being able to accurately measure the expansion, perhaps it would only be necessary to determine if the expansion will be adequate to cause serious rashing of the roof. A much wider data base, with a number of shale types, would be necessary to establish what percentage expansion is tolerable. Probably, this value would be related to the strength properties of the rock. However, as an example, based on the results obtained in this investigation it may be suggested that an expansion of 1.5 to 2% is tolerable since this was observed in the gray and black shale. However, an expansion of about 3% and over becomes excessive, as noted with the claystone. These expansions are all at 100% relative humidity, for a period of 15 days. If it expands 3% during this period, it could make an electrical contact to light up a red bulb. This simple procedure would warn the operator that the shale expected to be encountered could cause problems during mining. It may be recommended that at least 10 tests be conducted on each shale prior to deciding whether the roof was liable to serious deterioration. It could further be suggested that samples of shales up to about 3 ft above the roof be examined.

It should be recognized that the scheme presented above is merely a concept for a test procedure that may be used by mine operators. Only the principle of the method is shown to be valid. The precise duration of the tests, expansion limits acceptable, minimum number of tests required, and depth of shale to be tested may vary considerably. A systematic program of tests for a number of shale types should be conducted to ascertain the figures that should be recommended in any test procedure. The numbers given are only illustrative, based on the scanty data obtained during this investigation.

CLOSURE

This study collaborates some prior work showing that shale deterioration in mine roofs is not a slaking problem in the conventional sense. It is more closely related to the stresses which result from moisture-induced weakening and swelling strain. Hence, shale expansion tests under high atmospheric humidity are more indicative of this behavior. Although these tests are time-consuming, they appear to offer potential as predictors of the severity of shale deterioration (roof rashing) to be expected, in conjunction with geologic examination and null point determination. A simplified procedure that could be used by mine operators could be developed.

ACKNOWLEDGEMENTS

The authors would like to express their appreciation to the U.S. Bureau of Mines for funding this project, to the management of Valley Camp Coal Company for their cooperation while the field work was being conducted at their mine, and to Engineers International, Inc. for staff help in conducting the work and permission to submit this paper.

REFERENCES

Adam, M.E. (1975). "Ground Control Evaluation of the Sahara No. 20 Mine, Marion, IL (Masters Thesis), University of Missouri at Rolla, 103 pp.

Aughenbaugh, N.B., and R.F. Bruzewski (1973). "Investigation of the Failure of Roofs in Coal Mines" (Final Report on U.S. Bureau of Mines Contract No. HO232057), University of Missouri at Rolla, 161 pp.

Bruzewski, R.F., and N.B. Aughenbaugh (1977). "Effects of Weather on Mine Air," Mining Congress Journal, September, p. 4.

Chenevert, M.E. (1970). "Shale Alteration by Water Absorption," Journal of Petroleum Technology, AIME, September, pp. 1142-1148.

Conroy, P.J. (1973). "Investigation of Roof Shales in Illinois Coal Mines" (Ph.D. Dissertation), University of Missouri at Rolla, 111 pp.

Cummings, R.A., M.M. Singh and N.N. Moebs (1981). "Effect of Atmospheric Moisture on the Deterioration of Coal Mine Roof Shales" (Preprint No. 81-159), Society of Mining Engineers, Littleton, Colorado, 14 pp.

Cummings, R.A., M.M. Singh and N.N. Moebs (1982). "Effect of Atmospheric Moisture on the Deterioration of Coal Mine Roof Shales," Transactions AIME, Technical Paper (in print).

Engineers International, Inc. (1981). <u>Control of Shale Roof Deterioration With Air Tempering, Vol. 1-Field and Laboratory Investigations</u> (Final Report on U.S. Bureau of Mines Contract J0188028), Westmont, Illinois, 162 pp.

Franklin, J.A. (1979). "Suggested Methods for Determining Water Content, Porosity, Density, Absorption and Related Properties, and Swelling and Slake-Durability Index Properties," Internat. J. Rock Mech. Min. Sci. Geomech. Abstr. Vol. 16, No. 2, April, pp. 153-154.

Harper, T.R., G. Appel, M.W. Pendleton, J.S. Szymanski, and R.K. Taylor (1979). "Swelling Strain Development in Sedimentary Rock in Northern New York," Internat. J. Rock Mech. Min. Sci. Geomech. Abstr. (16), pp. 271-292.

McNary, H.B. (1941). "Investigation of Mine Roof Deterioration," Mining Congress J., April, 3 pp.

Moriwaki, Y. (1974). "Causes of Slaking in Argillaceous Materials" (Ph.D. Dissertation), University of California at Berkeley, November, 291 pp.

INSTRUMENTATION SYSTEM FOR THE MEASUREMENT AND RECORDING OF TRANSIENT GEODYNAMIC PHENOMENA

Système d'enregistrement et de mesure des phénomènes géodynamiques transitoires

Instrumentierung zur Messung und Aufzeichnung kurzzeitiger geodynamischer Phänomene

J. Carrasco
Dr. Mining Engineering, Research Director, Asociación de Investigación tecnológica de equipos mineros (AITEMIN) Madrid, Spain

SYNOPSIS

This paper presents a measurement and recording instrumentation system for the transient geodynamic phenomena quantification in gassy mines. Its application in the Potasas de Navarra mines, where the stresses in powered supports due to roof blastings were measured, is also explained.

RESUME

Cet article présente des appareils de mesure et d'enregistrement développés pour mesurer des phénomènes transitoires dans des mines grisouteuses. On décrit aussi des expériences dans les mines de Potasas de Navarra, S.A. où on a mesuré les tensions produites par le tir dans les côtés arrière-taille du toit.

ZUSAMMENFASSUNG

Es wird die Instrumentierung zur Messung und Aufzeichnung kurzzeitiger geodynamischer Phänomene vorgestellt, die auch in gasausbruchgefährdeten Gruben eingesetzt werden kann. Ihre Anwendung in der Potasas de Navarra Grube wird aufgezeigt, in der die Lasten auf einen schreitenden Schildausbau infolge Sprengungen im überhängenden Hangenden gemessen wurden.

1.- INTRODUCTION

The presence of transient geodynamic phenomena in underground mining is a well known fact, and their effects commonly produce alterations in the faces operation or structures stability.

The most usual are roof bursts and load bursts in longwall faces, as well as stresses produced by blasting in the roof in the goaf side, or in road headings.

The measurement of these stresses is very difficult due to the almost random appearance and its great speed. Nevertheless, in some roof bursts, the action can be estimated to a minimum value, examining the produced effects, as can be to stablish the minimum load or pressure in the case of prop or powered support destructions.

From the technical point of view, it is necessary to know these stresses to have a basis for the design and calculations of the elements that will support these phenomena, and for the proper evaluation of the mining operations tending to lower their effects. For this reason, in the Asociación de Investigación Tecnológica de Equipos Mineros (AITEMIN), belonging to the Comisión Asesora de Investigación Científica y Técnica, a measurement and recording system has been developed, wich is designed specifically to the requirements of geodynamic phenomena quantifycation.

2.- FUNCTIONAL PRINCIPLES

The basis of the measurement system is an extensometry application by means of strain gange transducers.

The system must comply with the following requirements :
- Measurement, without attenuation, of all the pressures and stresses, what makes it to have a wide enough pass-band.

- Sampling and storing of the interesting periods, without considering the rest.

- Adequate protection for working in a coal mine, that is, in a potentially explosive atmosphere.

Al this makes the system complicated, with power supply, excitation, recorder...etc.

A possible general assembly is shown in Fig. 1 wich representes its installation on powered supports to measure pressure variations in the props. The pressure generated in the main cylinders of the support is measured with a strain gauge transducer fitted on them, that converts the pressure variations to resistance variations and is fed from a power supply situated in a flameproff enclosure togheter with a signal conditioner and the necessary set of barriers to make the external circuit be an intrinsically safe one.

The output signal is conected to the recording equipment which is situated in the gate end inside an internal over pressure enclosure.

The sistem mains is 220 V and is obtained from a common coal mining gate end box.

3.- COMPONENTS

The measurement system is composed by the following elements :

- Pressure transducer : Straing gauge type, Shown in fig. 2, with the following characteristics :
- Range (nominal) : 100 MPa
- Output, full scale : 2mV/V, $\pm$ 1 %

E 97

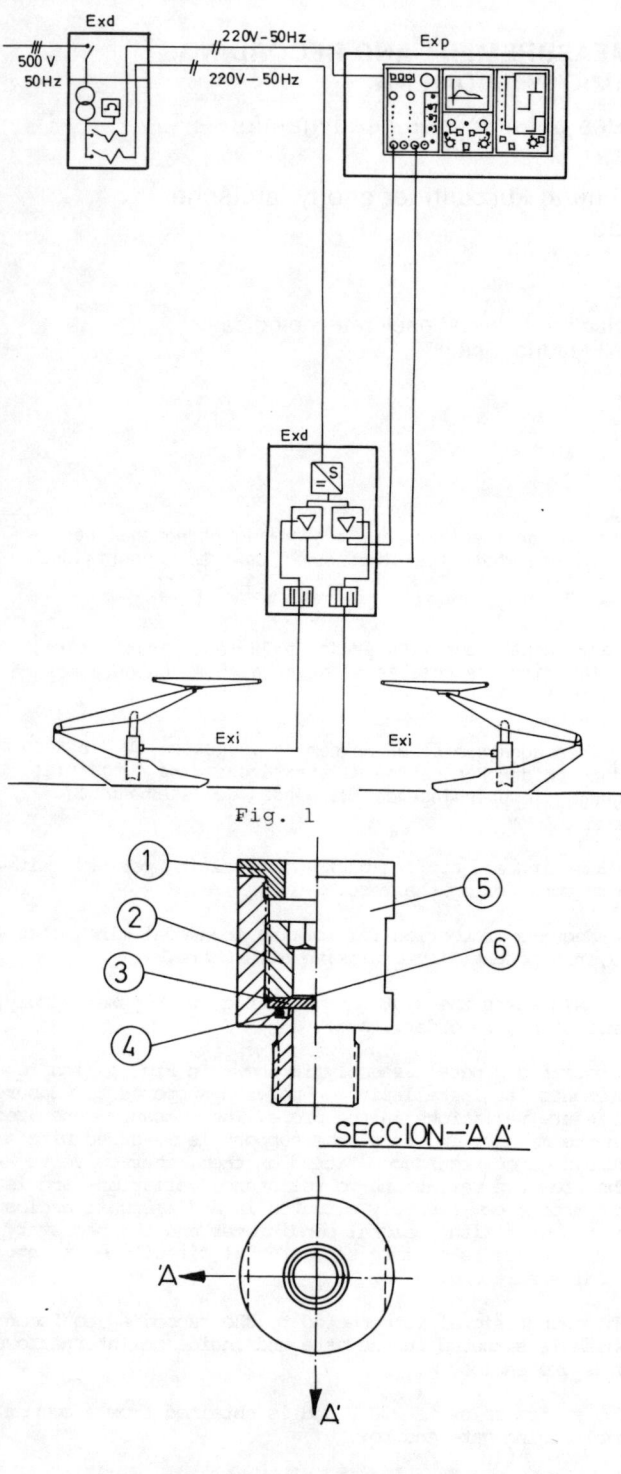

Fig. 1

SECCION-A A'

PIEZA Nº	DESIGNACION
1	TAPA DE CIERRE
2	TORNILLO DE APRIETE
3	PLACA SOPORTE
4	JUNTA TORICA
5	CUERPO DEL CAPTADOR
6	ROSETA DE GALGAS

Fig. 2

- Maximun excitation : 10 V, DC
- Input/Output resistance : 350
- Sensibility : 0,02 mV/V/MPa
- Initial de-balance : 0,01 mV/V
- Lineality : ± 0,2%
- Own frequency : 180 KHz
- Maximal Pressure : 200 MPa
- Admissible Pressure : 500 MPa

- Signal Conditioner. Its block diagram is shown in fig. 3. It is mounted in a flameproof enclosure together with the safety barriers for the external circuits to the transducers.

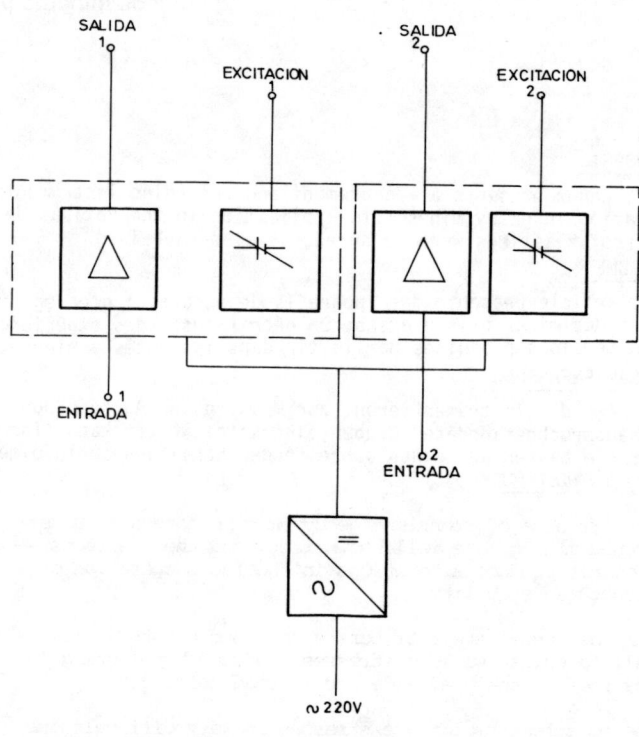

Fig. 3

- Transient signal recorder. It is an storage oscilloscope with auxiliar memory disks. Its caracteristics are the following :
- Memory capacity (points per record)....... 4 k bytes
- Byte length............................. 12 bits
- Resolution..............................0,024% F.S.
- Accuracy................................±0,25% F.S.
- Lineality...............................0,1%
- Maximal conversion speed................. 500 ns/conv
- Time between points............ 500 ns to 220s (27 steps).
- Input voltage ranges.......... ± 100 mV to ± 10 V F.S. with x1, x2 and x4 multipliers.

- Signal input............................. balanced
- Input impedance.......................... 1M , 50 pF
- Trigger.................................. level, slope manual or external

- Disk capacity............................ 32 K bytes (8 records)
- Analoy output............................ ± 10 V
- Output interface......................... DMA,IEEE488
- Power consumption........................ 150 W.

E 98

All the components are conected by special cables to defi
ne the supply, signal and transmission circuits.

4.- APPLICATIONS

The system was installed underground in the Potasas de Na
varra mine in the GN-25 face which carachteristics are :

- Geological conditions:
  - Mined seam   : Silvinite
  - Meam depth   : 740 m.
  - Seam thickness : 2,4 m.
  - Meam slope   : 12º
  - Roof   : Intermediate salt and carnalite.
  - Floor   : Salt.
  - Face length   : 270 m.
- Mining conditions :
  - Longwall face with integral roof falling
  - Retreat face
  - Marrel 14-27 Shield supports
  - Setting pressure 30 MPa
  - Anderson AB-16 shearer
  - Westfalia PF-IV A.F.C. with 2 x 90 KW drives
  - Monorail service to the lower gate end.

The equipment was installed in two sets: in the face, the
pressure transducers were fitted in the nº 56 and 63 supports, with the signal conditioner. In the gate end, the
register unit and power supply were situated.

For safety reasons, a BM1 Sieger methanometer was installed with this equipment, the alarm level being adjusted
to 1%, since on first stage the equipment should be opera
ted with the enclosure being open and in presence of vol
tage, for testing and adjusment purposes.

The aimed object of this installation was to record the
pressures that were generated inside the hidraulic cylinders, due to the blastings made periodically behind the
supports to blow the roof. The obtained records are shown
in fig. 4 and 5.

In support number 56, the pressure increase is 37 MPa
with a rising time of 7 ms. The total duration is 30 ms
and a pressure rise of 4 MPa remains permanenthy.

The pressure was over the nominal release value of the
valves, without them to limit the peak value, for that
reason it is supposed that their working has been similar
to the one perfomed in the impact testings.

In the nº 63 support, the pressure rise was 42 MPa with a
5 ms rising time. The phenomena duration was 20 ms and a
permanent overpressure of 25 MPa remained. In this case
the pressure was also over the valve release value, wi -
thout any limiting effect being observed.

In latter trials, the blastings were made between the
shields in the area between the monitored supports, so
that the distance from the blasting to he shield be shorter.

The drilling length was 4 m and 12 dynamite cardtridges
were used. In the registered effects, a considerable pressure rise is observed, since the conditions are harder
than usual.

5.- CONCLUSIONS

A measurement and recording system has been developed,
with the following carachteristics:
- Distance between transducer and recording
  unit... 0 to 500 m.
- Pressure range................................... ± 0,3 %

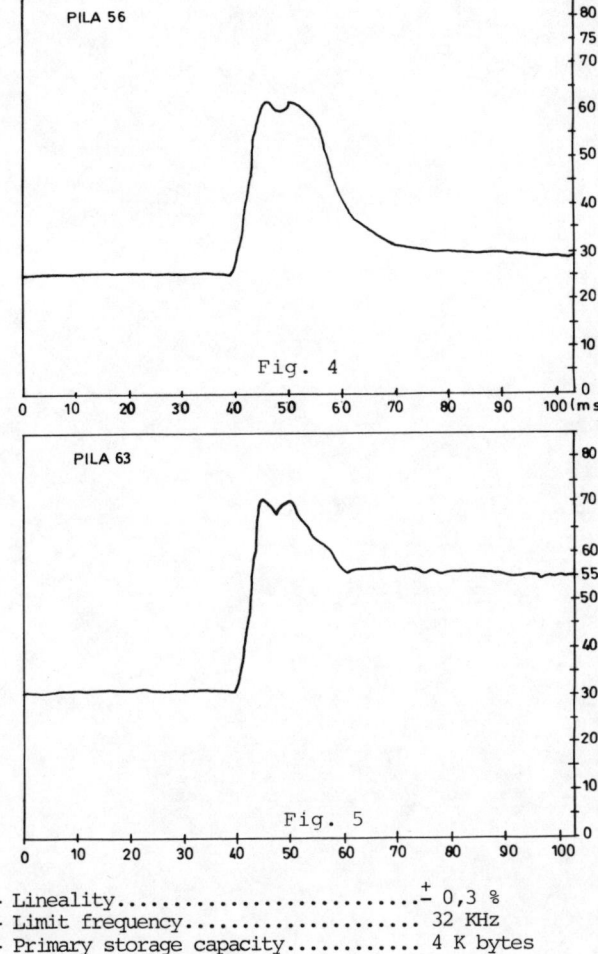

Fig. 4

Fig. 5

- Lineality....................... ± 0,3 %
- Limit frequency................. 32 KHz
- Primary storage capacity........ 4 K bytes
- Auxiliar storage capacity....... 32 K bytes
- Level, slope, manual or external
  trigger.
- Transducer power supply......... 10 V DC
- Protection method............... Ex d, p, i.

The system allows to meet the requirement of geodynamic
pressure measurements in potentially explosive atmosphe
res in underground mining. It is a basic instrument to
define geodynamic events and analyse, from the rock mechanics technology, the sudden actions produced in under
ground mining.

6.- BIBLIOGRAPHY

- Organo Permanente de Luxemburgo (1.980)
Mesures practiques destinées a reduire les risques des
coup de terrain.
- Josien - Poirrot (1.976)
Etude du comportement de verins de soutenement marchant
sous de sollicitations dynamiques.
CERCHAR
- Carrasco, J. (1.982)
Dinamica de un sistema de sostenimiento frente a fenóme
nos geodinámicos transitorios.
Escuela Técnica Superior de Ingenieros de Minas de Ovie-
do. España.

MAJOR DEVELOPMENT ON AN UNDERMINED SITE IN A CENTRAL CITY AREA

Constructions importantes sur d'anciennes mines situées sous un centre urbain

Grossbauten in einem innerstädtischen, vom Untertagebau beeinflussten Baugebiet

T. R. Stacey, H. A. D. Kirsten and B. L. Wiid
Steffen, Robertson and Kirsten, Consulting Engineers, Johannesburg, South Africa

SYNOPSIS

As a result of the extensive mining of gold reefs through to surface in the Johannesburg urban area, severe restrictions have been placed on building development over undermined land. The investigation of an undermined site and the subsequent consideration of stability and design of remedial stabilising measures are described. These formed the basis of a sound technical motivation for the relaxation of building restrictions, which has permitted the construction of a major development across the mined out reef outcrops.

RESUME

En raison de l'exploitation extensive des filons aurifères sous-jacents dans la zone urbaine de Johannesburg, il s'est avéré nécessaire d'imposer des règlements ayant trait à la construction d'immeubles sur ce terrain minier. Ces règles imposent l'examen des sols et déterminent les mesures à prendre afin d'assurer des critères de stabilité. Elles ont motivé un relâchement des règlements de la construction, ce qui a permis de bâtir un grand ensemble sur ces sites auparavant soumis à l'exploitation minière.

ZUSAMMENFASSUNG

Der umfassende Abbau von goldführenden Adern (Reefs) bis zur Erdoberfläche innerhalb des Stadtbezirkes von Johannesburg hatte die Folge, daß strenge Einschränkungen der Neubautätigkeit erhoben wurden. Die Untersuchung eines von Untertagebauten beeinflußten Baugeländes und die daraus folgenden Erwägungen zur Stabilität des Baugrundes und zum Entwurf von stabilisierenden Abhilfsmaßnahmen werden beschrieben. Diese Arbeit bildete die Grundlage für eine technisch motivierte Abminderung der Baueinschränkungen, die die Errichtung großen Neubaues auf abgebauten Reefs zuließ.

1 INTRODUCTION

The city of Johannesburg has grown up around the gold mining industry. The main gold bearing reefs of the central Witwatersrand area traverse the full length of the city in an east-west direction. These reefs have been extensively mined, which has resulted in an extensive area of undermined land very close to the central business district, as shown in Figure 1. The erection of buildings on undermined land is very strictly controlled by the Government Mining Engineer. The standard restrictions are summarized in Table 1.

TABLE I : STANDARD BUILDING RESTRICTIONS IMPOSED BY THE GOVERNMENT MINING ENGINEER

Depth of Mining (m)	Building permitted (storeys)
0 - 90	No building
90 - 120	One with one basement
120 - 150	Two with one basement
150 - 180	Three with one basement
180 - 210	Four with one basement
210 - 240	Five with one basement
240 -	No restrictions unless mining circumstances are unusual

Fig 1 : Mining outcrops adjacent to the central business district

There is a large area of property between the outcrops and a mining depth of 240 m which, being close to the central business district, is potentially valuable if satisfactory remedial works can be carried out to stabilise the old mine workings. Applications for relaxation of the standard building restrictions are considered very favourably by the Government Mining Engineer provided they are supported by sound technical reasoning and, where necessary, design of remedial measures to ensure stability of the mine workings. This paper describes a case study of the site investigation, and design and construction of the remedial works for a major development across the outcrop mine workings.

2 GEOLOGY AND MINING

The site lies on the Upper Witwatersrand gold bearing quartzites with a broad intrusion of diabase stretching across from east to west as illustrated in Figure 2. The quartzite as a whole is a coarse grained heavily jointed to massive rock dipping southwards at a steep angle of about 80°. Two tabular conglomerate reefs, the Main Reef Leader and the South Reef were mined on the site. The quartzite in the middling between the two reefs is distinctly weaker and more friable than either the hangingwall or footwall quarzites.

The diabase is an igneous intrusion in the form of a dyke sandwiched between the quartzite at approximately the same steep dip of 80° to the south. It is significantly weathered far more deeply than the quartzites.

The average stoping widths of the Main Reef Leader and South Reef were respectively 1,3 m (50 inches) and 1,4 m (54 inches). Mining in the area began in the late 19th century and before the turn of the century the Main Reef Leader had been opened up and extensively mined. Mining was initially from the surface down on dip, with crown pillars being left at surface, and other reef pillars being left underground, generally below each development level, to ensure stability. These pillars were often significantly reduced in size or even totally removed during the early years of uncontrolled mining. Since the shallow mining was completed some 80 years ago, no detailed records were available on the extent of mining. Many of the old stopes have been tip filled with waste rock, rubble and refuse from the surface. This filling continued until the stopes were choked. Over the years, however, aided often by inflow of rainwater, this loose fill has, in many cases ravelled away to greater depths in the stopes. This has resulted in significant open cavities in the stopes at shallow depth.

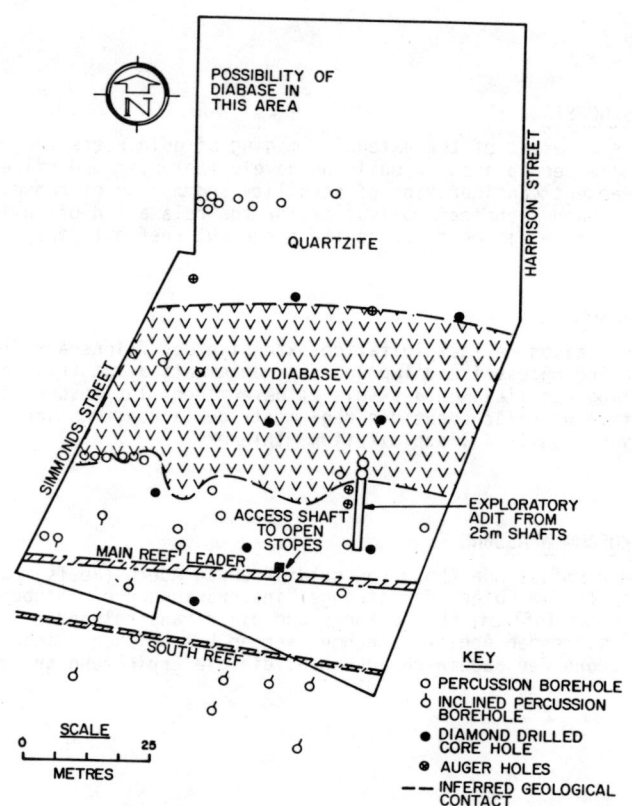

Fig 2 : Plan of site showing surface geology

3 SITE INVESTIGATIONS

The site has a considerable history of investigation as can be seen from the number of holes shown in Figure 2. This reflects the fact that subsurface conditions were involved and also the different requirements of several proposed developments over a period of 10 years.

In 1973, a comprehensive foundation investigation was carried out for a proposed multistorey building. The tower block for this development was to be located on the northern portion of the site, with the basement excavation crossing the Main Reef Leader. The investigation therefore dealt with both the undermining as well as conventional foundation conditions. The presence of gouge-filled joints to a depth of 25 m and slickensided joints to a depth of 50 m was identified in the footwall rocks. With the reefs extensively mined and stopes potentially open, or at least without substantial support, the location of the tower block is in fact on the edge of a "subsurface cliff". Local instability of the footwall was identified as a distinct possibility. It was recommended that the design of remedial measures should only be finalised after the Main Reef Leader had been adequately exposed to examine the true state of affairs in the old workings.

The proposed tower block project did not proceed and it was only in 1978 when further interest was revived in a development on the site. The recommendations of the 1973 investigation were then implemented. The objectives were:

(i) to investigate the condition of the Main Reef Leader stope to a depth of 50 m
(ii) to examine the condition of the footwall surface

A suitable location for access to the Main Reef Leader workings was chosen from the results of the previous investigation. A short shaft was sunk and found to correspond with what is thought to have been an old exploratory shaft. A strong updraft of air was encountered in the shaft indicating that there was connection to ventilated workings at greater depth. At the base of this first shaft there was a small opening connecting through into the Main Reef Leader stope. It was found that this reef had been mined out almost to the surface, leaving only a 3 m crown pillar of solid rock. The stope was almost vertical, and had been tip-filled with quartzite rubble, sand and ash from a crown hole close to the eastern boundary of the site.

Figure 3 shows the route followed during the investigation. Figure 4 shows the inclined timber walkway leading off in a westerly direction from the bottom of the short vertical shaft. It also shows, on the right of the figure, the condition of the footwall rock. Exploration of the old workings was terminated at a depth of 46 m below surface. At that depth, disposal of spoil had become difficult and the objectives of this aspect of the investigation had been accomplished - the stope was apparently filled with loose quartzite boulder backfill, though it was not ascertained what was supporting this fill; the footwall surface had been mapped, revealing the presence of joints with clayey filling, continuous in a strike direction and dipping into the stope at angles of up to 25°. In addition, the presence of a very significant solid pillar had been identified.

Access to the footwall rocks to examine the gouge-filled joints was gained by augering a 25 m deep shaft in the weathered diabase dyke (see Figure 2), and driving an adit back towards the stope. This approach also provided

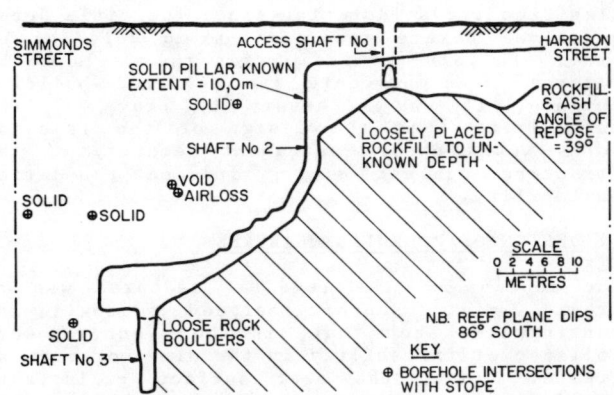

Fig 3 : Route of underground exploration of the Main Reef Leader Stope

Fig 4 : Underground access on inclined surface of rubble backfill

the opportunity of studying the in situ condition of the quartzite diabase contact. This zone proved to be very blocky with considerable evidence of slickensiding on joints. The adit followed one of the major joints which was found to be very significantly clay and gouge filled. However, the joints were very wavy and although dips of the order of 25° were measured, the average dip was approximately 5°. This shallow dip precluded the possibility of large scale instability of the footwall. Owing to this and the presence of the solid pillar it was considered that a remedial solution could be designed which would allow building development across both reef outcrops. This represented a significant consideration since the title deeds contained severe restrictions on building on site. To gain permission for the deeds to be altered to accommodate the required building heights, it was necessary to prove to the authorities that the design of the remedial works would ensure permanent security of the structure against mining induced foundation instability.

4 DESIGN OF REMEDIAL MEASURES

The concept of the remedial measures was to provide rigid support between footwall and hangingwall, such that, in the event of some collapse or instability in the mine workings at greater depth, the near surface rigid zone would form an arch. Concrete dip pillars were proposed to provide the rigid in-stope support. Two- and three-dimensional stress analysis techniques were used to analyse the stability and design requirements of the foundation and remedial measure system. These analyses took into account the following:

(i) the simulation of the removal of the solid rock crown pillars during the excavation of the basement;

(ii) the simulation of the construction of concrete pillars and the consequent design sizing and layout of the pillars;

(iii) the assessment of possible movements resulting from instability of the footwall rocks, associated with the presence of the dyke, after construction of the remedial measures;

(iv) the simulation of collapse in the old mine workings below the remedial measures, and the assessment of potential resulting surface movements.

Two-dimensional displacement discontinuity stress analyses (Crouch 1976) were used for the majority of the calculations. The geometry of the problem is shown diagrammatically in Figure 5, only one stope being illustrated, with the above four analysis steps indicated. All potential failure surfaces were modelled with displacement discontinuity fault elements. The stopes were modelled with displacement discontinuity seam elements and the concrete pillars with backfill elements. From the two-dimensional analyses the following results were obtained :

(i) the footwall is stable. The maximum calculated surface movement resulting from deformation of the footwall and stope system was 5 mm in the horizontal direction,

(ii) the maximum calculated surface movement resulting from collapse below the remedial works, assuming the presence of stable concrete pillars, was 1 mm.

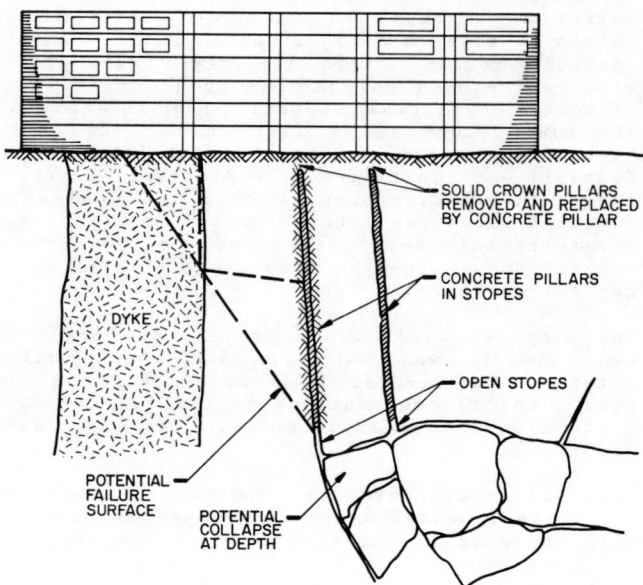

Fig 5 : Diagrammatic illustration of the subsurface geometry

Three-dimensional mining simulation techniques with displacement discontinuity elements (Diering 1980) were used to evaluate the sizing and layout of the concrete pillars. The final design was evolved from a series of analyses of a single stope as shown in Figure 6(a) and (b). The construction of regularly spaced pillars, all to a depth of 60 m below surface was simulated, followed by the removal of the solid rock crown pillar. This showed that the stress in the boundary pillar and in the lower third of the other pillars could potentially develop to unacceptably high levels as shown in Figure 6(a). In the upper section of the inner pillars the calculated stress levels were very low. The model was consequently modified to that shown in Figure 6(b) in which intermediate pillars were shortened, to extend to a depth of only 35 m, and a horizontal strike pillar inserted to join the bases of the deeper pillars. This layout resulted in much more acceptable levels of stress in the pillars as shown in Figure 6(b). Linear extrapolation and interpolation of these results allowed the effects of variations in pillar widths and and spacings to be considered.

5 CONSTRUCTION OF REMEDIAL MEASURES

Winzes, 2 m wide on strike and 8 m apart, were sunk in the old stopes as shown in Figure 7. Both stopes were found to be loosely filled with quartzite rubble, which could potentially have made the construction of the strike pillars a very hazardous operation. Fortunately, however, small solid reef pillars had been left just above the second level (200 feet) which corresponded to the required

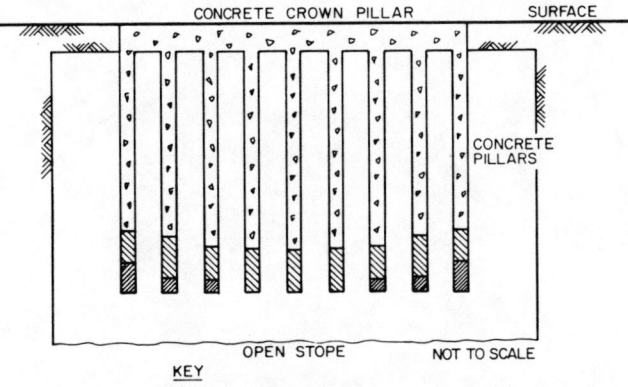

Fig 6(a) : Initial simulation of concrete dip pillars

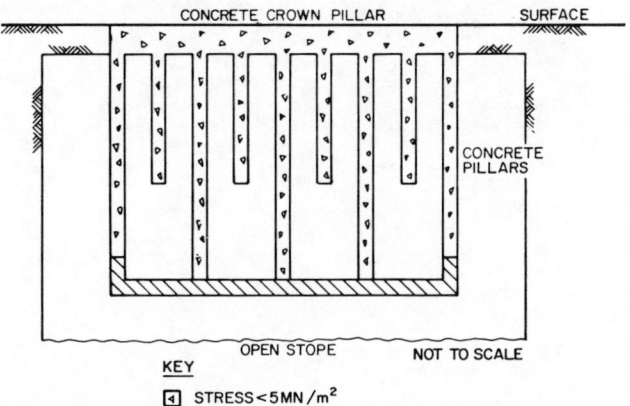

Fig 6(b) : Simulation of final configuration of concrete pillar support

depth of the deeper pillars. The stopes were found to be open along the full strike length just below this level. Some precariously supported rubble backfill formed the bases of these cavities which, after choking holes where some of the rubble had disappeared into the depths of the stopes below, could support the construction of the strike pillars by pouring in concrete grout in stages. Once the strike pillars had been completed, the winzes were backfilled with concrete. The immediate footwall and hangingwall surfaces at the pillar locations were checked to depths of 3 metres for open joints and fractures. If found, these were pressure grouted to ensure the integrity of the rock at the pillar locations. At the time of the writing, the building structure was almost complete as illustrated in Figure 8.

6 CONCLUSIONS

The procedure followed for geotechnical investigation and design on this project has illustrated that stable development over undermined land can proceed without risk provided that the correct technical approach is adopted. With this approach it can be expected that substantial development could take place in the future in Johannesburg in these undermined areas presently hampered by severe building restrictions.

Fig 7 : Sinking of winzes for construction of dip pillars

Fig 8 : Building development nearing completion

ACKNOWLEDGEMENTS

The permission of the Standard Bank of South Africa Ltd to publish this paper is acknowledged. The work described was carried out for Lillicrap, Wassenaar and Partners who are the structural engineering consultants for the project.

REFERENCES

CROUCH, S L (1976) Analysis of stresses and displacements around underground excavations : An application of the displacement discontinuity method, University of Minnesota Geomechanics Report.

DIERING, J A C (1980) An improved method for the determination, by a MINSIM type of analysis, of stresses and displacements around tabular excavations, Jl S Afr Inst Min Metall, Vol 80, No 12, December 1980, pp 425-430.

LINEAR MODELS FOR PREDICTING SURFACE SUBSIDENCE
Modèles linéaires pour prédiction d'affaissements de surface
Lineare Modelle zur Voraussage von Oberflächensetzungen

M. D. G. Salamon
Director General, Research Organization, Chamber of Mines of South Africa,
Johannesburg, South Africa

SYNOPSIS

First, the notion of a preliminary screening of models is introduced using critical measures of surface movement. Next, exact elastic media are examined with attention focussed on the modelling of stratified rock masses using an equivalent medium. The Monte Carlo technique is employed to estimate from the properties of individual layers, which are treated as independent random variables, the moduli of the equivalent transversely isotropic mass. Finally, semi-empirical models are discussed and their application is illustrated by an example.

RESUME

Cette communication commence par évoquer le concept du tri préalable des modèles d'affaissement en se basant sur des mesures critiques de déplacements de surface. Ensuite on examine les milieux élastiques exactes, surtout en ce qui concerne le modelage de masses rocheuses stratifiées, à l'aide d'un milieu équivalent. La technique Monte Carlo est utilisée pour l'évaluation des propriétés des couches discrètes envisagées comme variables indépendantes aléatoires — ce sont les coefficients de la masse transversalement isotrope. En conclusion on discute les modèles semi-empiriques et donne un exemple de leur application.

ZUSAMMENFASSUNG

Es wird einleitend eine Voruntersuchung zur Modellauswertung mit Hilfe kritischer Faktoren bezüglich Oberflächenbewegungen beschrieben. Danach folgen die Untersuchungsergebnisse von exakt-elastischen Materialien, wobei besonderes Augenmerk auf die Modellierung von geschichteten Felsmassen unter Verwendungnahme von maßstabgerechtem Material gerichtet wird. Zur Bewertung der Eigenschaften einzelner Schichten, welche als unabhängige, zufällig Veränderliche angenommen werden, sind die Moduli der äquivalenten transvers-isotropischen Massen mit Hilfe des "Monte Carlo" Verfahrens erarbeitet. Zum Abschluß werden semi-empirische Modelle und deren Verwendung besprochen und an Hand eines Beispieles veranschaulicht.

1. INTRODUCTION

There are essentially two procedures for the prediction of mining induced surface subsidence. The first of these is entirely empirical, being based on a large body of field observations and being without recourse to mechanistic ideas about ground movement. The second approach rests on the postulate that the deformation of the rock mass is governed, at least approximately, by linear laws which permit superposition.

The best known example of the empirical approach is summarized in the Subsidence Engineers' Handbook (1975) published by the National Coal Board of Great Britain. The methods described there offer reasonably accurate predictions of subsidence, tilt and horizontal strain under conditions similar to those prevailing in British coal fields. It is doubtful whether the results can be used elsewhere without extensive prior field observations to check their validity.

The theory of linear models, as applied to the prediction of surface subsidence due to mining of tabular deposits, has been advanced considerably since H Keinhorst proposed the 'zone calculation' method in the 1930's. In spite of the theoretical developments made since, the breadth of application of linear models to the solution of practical subsidence problems has been limited.

This paper attempts to pave the way for the wider employment of linear models by examining the practical applicability of specific examples. To do this, some critical aspects of observed surface movements are compared with the behaviour of the models.

2. A DIGEST OF SURFACE OBSERVATIONS

To simplify the discussion, the observations used are restricted to those obtained in

Britain. The aim is to identify a few critical measures of surface movements and to use these critical values for the preliminary screening of linear models. That is, models which are unable to reproduce reasonably closely the chosen values are rejected and those which are able deserve further, more detailed testing. Naturally, the conclusion regarding the applicability or otherwise of a model applies only to Britain and to those regions where conditions are similar to those prevalent in Britain. For the sake of simplicity the analysis is restricted to horizontal stratification and time dependency is neglected.

Experience in Britain has shown (Subsidence Engineers' Handbook (1975)) that:

(i) the radius of the 'critical area', that is, the area over the centre of which subsidence is virtually total, is $0,7H$, where H is the depth below surface,

(ii) the maximum tilt in a subsidence trough caused by working a super-critical panel is $2,75 (S_m/H)$, where S_m is the magnitude of maximum convergence,

(iii) the maximum tensile and compressive strains over a super-critical and parallel sided panel are $0,65 (S_m/H)$ and $-0,51 (S_m/H)$, respectively, and

(iv) it has been suggested that the maximum horizontal displacement is in the region of $0,2 S_m$, Wardell (1959). (The Handbook makes no mention of this relatively poorly substantiated result.)

3. BASIC POSTULATES OF LINEAR MODELS

Since fairly comprehensive reviews of the theory of linear models are available elsewhere, e.g. Salamon (1974, 1977), here only their basic postulates are summarized. The basic principle of the theory is that the mined-out region can be subdivided into small areas or 'face elements' and the total effect of mining equals to the union of the contributions of individual face elements.

The first step in the application of the theory is to define the displacements induced by a face element of area ΔA centred at a point defined by horizontal coordinates ξ, η. In general, displacements are induced by the convergence and ride of the roof and floor of the face element. In the case of horizontal stratification, however, ride components are small and can be neglected. The displacement distribution due to convergence $S(\xi,\eta)$ in the face element is axially symmetric, provided the behaviour of the rock mass is invariant with respect to rotation around a vertical axis. Under these circumstances the elementary vertical, ΔW, and horizontal radial, ΔU_r displacement components due to the volume of convergence $S(\xi,\eta)\Delta A$ in the face element are, Salamon (1962)

$$\Delta W = S(\xi,\eta) F(\rho^2) \Delta A^*/\pi, \quad \Delta U_r = -S(\xi,\eta)\rho G(\rho^2)\Delta A^*/\pi. \quad (1)$$

In these expressions

$$\rho = r/H = \{(x-\xi)^2 + (y-\eta)^2\}^{\frac{1}{2}}/H, \quad \Delta A^* = \Delta A/H^2, \quad (2)$$

where x, y are the horizontal coordinates of the surface point where ground movement is to be calculated. Note that both F and G, referred to as influence functions, are even, monotonic functions of radius ρ. Furthermore, function F must satisfy the condition that:

$$\int_0^\infty F(t) dt = 1 \quad (3)$$

and that both $\rho^2 dF/d\rho^2$ and $\rho^2 G(\rho^2)$ should approach zero as ρ approaches infinity, Salamon (1977).

There are two ways of obtaining specific forms of functions F and G. The first route relies on exact models of linear elastic media. In this case the solution of the boundary value problem involving a face element yields F and G and, if the mining layout is specified, the distribution of convergence also can be derived. Thus, the specification of the constitutive relations of an elastic medium leads, without any further assumption, to the full solution of the subsidence problem.

The second route is semi-empirical. This can be followed by selecting appropriate influence functions which satisfy the earlier mentioned conditions and by estimating the convergence distribution in the mined-out area. In many instances practical problems can be solved effectively by accepting limiting assumptions for convergence distributions.

4. CRITERIA FOR THE ACCEPTABILITY OF INFLUENCE FUNCTIONS

Having defined the basic concepts of linear models it is now possible to formulate criteria for testing the applicability of pairs of influence functions.

It has been noted that in Britain the critical radius is about $0,7H$. It follows from this that if a circular area of radius $0,7H$ would be mined out and would suffer uniform convergence of magnitude S_m then the vertical subsidence over the centre of this area W_o would be approaching S_m. This criterion can be expressed, on the basis of (1), in terms of the following inequality:

$$\frac{W_o}{S_m} = \frac{1}{\pi H^2} \int_0^{2\pi} \int_0^{0,7H} rF(\frac{r^2}{H^2})drd\psi = \int_0^{0,49} F(t)dt \geq 0,9. \quad (4)$$

The face element principle offers the opportunity of computing the theoretical <u>upper bounds</u> for tilt, I_m, horizontal displacements, U_m, and strain $\pm \varepsilon_m$ for a maximum convergence of S_m in a super-critical panel. To obtain the required upper bounds it is necessary to compute the surface movement due to an idealized mining geometry where the seam is undisturbed on the one side and totally extracted and uniformly closed on the other side of a straight face line.

The maximum tilt and horizontal displacement occur over the face line. The maximum tensile, ε_{tm}, and compressive, ε_{cm}, strains are attained at the same distance ℓ ahead of and behind the face, respectively, Salamon (1974). Naturally, for an acceptable pair of influence functions, these upper bounds should exceed the maximum values obtained through observations for a similar mining geometry, see Section 2.

5. EXACT MODELS

A number of well known elastic models exists which can be examined in the light of the tests proposed in Sections 2 and 4. In this paper only the homogeneous isotropic and transversely isotropic models will be analysed.

5.1 Homogeneous isotropic elastic model

Berry (1960) concluded "... that it is of little use to consider subsidence in British coal measure rocks as the deformation of homogeneous, isotropic, elastic medium." Nevertheless this model is included here, partly because of its relevance to the description of the deformation of hard rocks at depth, Cook et al. (1966) and partly because, being the simplest of elastic models, its behaviour is useful for comparison purposes.

The influence functions in this model are as follows,

$$F(\rho^2) = 3/2(1+\rho^2)^{5/2}, \quad G(\rho^2) = F(\rho^2) \qquad (5)$$

Perhaps the most interesting feature of these functions is that they are independent, as was noted by Berry (1960), of the mechanical properties of the medium.

Using procedures introduced in the early 60's, Salamon (1962), and refined later, Salamon (1964), the following results are obtained:

$$\frac{W_o}{S_m} = 1 - \frac{1}{(1+R^2/H^2)^{3/2}} = 1 - \frac{1}{\{1+(0,7)^2\}^{3/2}} \cong 0,45$$

$$I_m = \frac{2}{\pi} \frac{S_m}{H} \qquad (6)$$

$$U_m = \frac{S_m}{\pi} \qquad \varepsilon_m = \pm \frac{3\sqrt{3}}{8\pi} \frac{S_m}{H} = 0,2067 \frac{S_m}{H} \qquad (7)$$

These formulae confirm that when the results are compared with the criteria for the acceptability of influence functions, the homogeneous, isotropic, elastic model is unable to describe surface subsidence in British coal-measures.

5.2 Homogeneous transversely isotropic model

Berry and Sales (1961) suggested more than two decades ago that this model may have an application in subsidence engineering related to stratified rock masses. The stress-strain relations of this model are given by:

$$\varepsilon_x = \frac{1}{E_1}(\sigma_x - \upsilon_1\sigma_y - \upsilon_2\sigma_z), \quad \varepsilon_{xy} = \frac{\sigma_{xy}}{G_1}$$

$$\varepsilon_y = \frac{1}{E_1}(-\upsilon_1\sigma_x + \sigma_y - \upsilon_2\sigma_z), \quad \varepsilon_{xz} = \frac{\sigma_{xy}}{G_2}$$

$$\varepsilon_z = \frac{\upsilon_2}{E_1}(-\sigma_x - \sigma_y) + \frac{\sigma_z}{E_2}, \quad \varepsilon_{yz} = \frac{\sigma_{yz}}{G_2} \qquad (8)$$

where the z axis is perpendicular to the bedding planes and

$$G_1 = \frac{E_1}{2(1+\upsilon_1)} \qquad (9)$$

Thus, it is clear that this medium has five independent elastic moduli.

In the sequel, parameters α_1 and α_2 are used extensively. These constants are those two roots of

$$\alpha^4 - 2k_2\alpha^2 + k_1^2 = 0 \qquad (10)$$

which have positive real parts. Here, k_1 and k_2 are two dimensionless, real and independent combinations of the elastic moduli of the model:

$$k_1 = \left(\frac{1-\upsilon_1^2}{E_1/E_2-\upsilon_2^2}\right)^{\frac{1}{2}}, \quad k_2 = \frac{(1+\upsilon_1)(G_1/G_2-\upsilon_2)}{E_1/E_2-\upsilon_2^2} \qquad (11)$$

These notations were introduced by Berry and Sales (1961) who observed that:

$$\alpha_1 = \frac{1}{\sqrt{2}}\{(k_2+k_1)^{\frac{1}{2}} + (k_2-k_1)^{\frac{1}{2}}\},$$

$$\alpha_2 = \frac{1}{\sqrt{2}}\{(k_2+k_1)^{\frac{1}{2}} - (k_2-k_1)^{\frac{1}{2}}\}, \qquad (12)$$

which show that α_1 and α_2 are real when $k_2 \geq k_1 > 0$ and complex conjugate when $-k_1 < k_2 \leq k_1$.

The next step is to record the influence functions in the transversely isotropic model:

$$F(\rho^2) = \frac{1}{2(\alpha_1-\alpha_2)}\left\{\frac{\alpha_1^3}{(1+\alpha_1^2\rho^2)^{3/2}} - \frac{\alpha_2^3}{(1+\alpha_2^2\rho^2)^{3/2}}\right\},$$

$$G(\rho^2) = \alpha_1\alpha_2 F(\rho^2), \qquad (13)$$

from which the upper bounds, permitting the preliminary testing of these functions, follow:

$$\frac{W_o(R/H)}{S_m} = 1 - \frac{1}{\alpha_1-\alpha_2}\left(\frac{\alpha_1}{\sqrt{1+\alpha_1^2R^2/H^2}} - \frac{\alpha_2}{\sqrt{1+\alpha_2^2R^2/H^2}}\right) \quad (14)$$

$$I_m = \frac{\alpha_1+\alpha_2}{\pi}\frac{S_m}{H}, \quad U_m = \frac{\alpha_1\alpha_2 S_m}{\pi(\alpha_1-\alpha_2)}\log_e\left(\frac{\alpha_1}{\alpha_2}\right) \qquad (15)$$

The upper bound of the strain components is given by:

$$\varepsilon_m = \pm \frac{3(\alpha_1 + \alpha_2)}{4\pi} \frac{\alpha_1 \alpha_2 (\ell/H)}{1 + \frac{1}{2}(\alpha_1^2 + \alpha_2^2)(\ell/H)^2} \frac{S_m}{H} \quad (16)$$

where the distance ℓ is obtained from:

$$\frac{\ell}{H} = \frac{(\alpha_1^2 + \alpha_2^2)^{\frac{1}{2}}}{\sqrt{6}\,\alpha_1\alpha_2} \left\{ \sqrt{1 + 12\left(\frac{\alpha_1\alpha_2}{\alpha_1^2 + \alpha_2^2}\right)^2} - 1 \right\}^{\frac{1}{2}} \quad (17)$$

These formulae contain two parameters defined by the moduli of the medium. Thus, unlike in the case of the isotropic model, there are no unique upper bounds here. Moreover, there does not appear to be an independent way of estimating those values of parameters α_1 and α_2 which correspond to a particular region, for example, in this instance to the British coalfields. The alternative is to examine whether a pair of values can be found which permit the formulae in (14) to (17) to satisfy the criteria in Sections 2 and 4.

After considerable numerical work, values $\alpha_1 = 9,0$ and $\alpha_2 = 0,5$ have been selected for the purpose since they appear to represent an acceptable compromise. The substitution of these values into the expressions in (14) to (17) yields:

$$\alpha_1 = 9,0 \quad \frac{W_o(0,7)}{S_m} = 0,89, \quad I_m = 3,02 \frac{S_m}{H} \quad (18)$$
$$\alpha_2 = 0,5$$

$$U_m = 0,49\, S_m, \quad \varepsilon_m = \pm\, 0,75 \frac{S_m}{H}$$

According to these results, the tilt and strain values are acceptable, the subsidence basin is marginally too wide and the maximum horizontal displacement seems too large, although this latter parameter is the least reliable of the test data. On balance, the influence functions in (13) with $\alpha_1 = 9,0$ and $\alpha_2 = 0,5$ need not be rejected on account of this evidence. However, before any further investigation of the acceptability of this model is done, it is necessary to examine whether the particular values of the parameters can be supported on physical grounds.

5.3 Stratified elastic medium - an equivalent model

The rock mass surrounding coal seams is usually well stratified. The elastic properties of the layers may vary considerably. While each stratum can be regarded as homogeneous, the union of these, that is, the rock mass itself is clearly non-homogeneous. It has been suggested some time ago, Salamon (1968), that it might be possible to define an equivalent homogeneous medium, the behaviour of which will closely resemble that of the prototype. The discussion in this section is based on this proposition. The idea is to examine whether on the basis of the equivalent mass, which was shown to be a homogeneous transversely isotropic medium, physical support can be provided to the particular values of parameters α_1 and α_2 selected in the previous section.

In deriving the equivalent medium it was postulated, Salamon (1968), that all layers are homogeneous, transversely isotropic and that their thickness and elastic properties vary randomly with the depth below surface. Furthermore, it was assumed that the rock mass remains continuous after deformation due to mining. In this paper a less general set of assumptions seems satisfactory. It will be assumed that the individual rock beds are homogeneous, isotropic - there is no evidence to suggest that in coal measures individual layers depart markedly from isotropic behaviour.

The analysis was based on an examination of the behaviour of two cubes both having an edge dimension L. One of these cubes is cut from the rock mass and the other from the equivalent homogeneous medium. Assuming that the bedding planes are parallel, the rock cube is so cut that two of its sides are parallel with these planes. The cube must be sufficient in size to contain a large sample of the layers to constitute a representative sample of the rock mass.

Let the properties of the i-th layer be $\bar{E}_i$ and $\bar{\nu}_i$ and its thickness be h_i. Introduce the notation $\lambda_i = h_i/L$ and denote the number of layers in the cube by n. Now, clearly

$$\sum \lambda_i = 1, \quad (19)$$

where the summation is from 1 to n with respect to the subscript i. The same limits will be implied in the sequel whenever the summation sign occurs.

In terms of these notations and assumptions the equivalent medium is transversely isotropic and its stress-strain relationships are given in (8). The formulae, expressing the five moduli of the transversely isotropic model in terms of the moduli and thickness of the layers, are as follows:

$$\nu_1 = \frac{\sum \frac{\lambda_i \bar{\nu}_i \bar{G}_i}{1 - \bar{\nu}_i}}{\sum \frac{\lambda_i \bar{G}_i}{1 - \bar{\nu}_i}}, \quad \nu_2 = (1 - \nu_1) \sum \frac{\lambda_i \bar{\nu}_i}{1 - \bar{\nu}_i}$$

(20)

$$E_1 = 2(1 - \nu_1^2) \sum \frac{\lambda_i \bar{G}_i}{1 - \bar{\nu}_i}, \quad E_2 = \frac{1}{\frac{1}{2}\sum \frac{\lambda_i(1 - 2\bar{\nu}_i)}{(1 - \bar{\nu}_i)\bar{G}_i} + \frac{2\nu_2^2}{(1 - \nu_1)E_1}}$$

$$G_1 = \frac{E_1}{2(1 + \nu_1)} = \sum \lambda_i \bar{G}_i, \quad G_2 = \frac{1}{\sum \frac{\lambda_i}{\bar{G}_i}}$$

For the sake of brevity these expressions are formulated using the modulus of rigidity $\bar{G}_i$, instead of the Young's moduli $\bar{E}_i$ but, of course, this is of little practical significance because $\bar{G}_i = \bar{E}_i/2(1 - \bar{\nu}_i)$.

There are many sets of assumptions which could be proposed for the properties of individual rock beds. Each set would lead to different estimates for the properties of the equivalent medium. Perhaps the most obvious and fundamental is the proposition that the Young's modulus, $\bar{E}_i$, the Poisson's ratio, $\bar{\nu}_i$, and the thickness, h_i, of beds are independent random variables. There are no observations to support such a postulate but an analysis based on it is expected to give a useful insight into the relationship between the moduli of the equivalent medium and the parameters of the frequency distributions purporting to describe the variation in the properties of beds constituting the rock mass.

Postulate that the variations in the thickness of individual beds can be represented by an exponential distribution, that is, by

$$f(h, h_o) = \frac{1}{h_o}\exp(-h/h_o), \quad h, h_o > 0, \quad (21)$$

which has the cumulative distribution:

$$F(h, h_o) = 1 - \exp(-h/h_o) \quad (22)$$

The expected value, Ex(.), and standard deviation s.d.(.) of this distribution are:

$$Ex(h) = h_o, \quad s.d.(h) = \pm h_o \quad (23)$$

Assume next that the variation in Young's modulus can be described by the Weibull distribution:

$$f(E, E_o, a) = \frac{a}{E_o}\left(\frac{E}{E_o}\right)^{a-1}\exp\{-(E/E_o)^a\} \quad (24)$$

$$F(E, E_o, a) = 1 - \exp\{-(E/E_o)^a\}, \quad E, a > 0.$$

The expected value and standard deviation of this distribution are:

$$Ex(E) = E_o \Gamma\left(\frac{1}{a} + 1\right),$$

$$s.d.(E) = \pm E_o\{\Gamma(\frac{2}{a}+1) - [\Gamma(\frac{1}{a}+1)]^2\}^{\frac{1}{2}} \quad (25)$$

where $\Gamma(.)$ denotes the gamma function. Clearly, the Weibull distribution for $a = 1$ reduces to the exponential distribution.

Some examples of the numerical values of the properties of the Weibull distribution are given in Table I.

It is assumed that the variation in the Poisson's ratio of the beds can be represented by a particular form of the Beta distribution:

$$f(\nu, 2, 3, 0, \tfrac{1}{2}) = 24(2\nu)(1-2\nu)^2$$

$$F(\nu, 2, 3, 0, \tfrac{1}{2}) = 8\nu^2(3-8\nu+6\nu^2), \quad (26)$$

with

$$Ex(\nu) = 0,2, \quad s.d.(\nu) = \pm 0,1 \quad (27)$$

Further details concerning the statistical distributions employed in this section can be obtained from Hahn and Shapiro (1968).

The assumptions regarding the density functions of the various distributions having been made, the Monte Carlo technique can be employed to compute the properties of the equivalent medium. This is done by drawing repeated random samples from the three distributions and substituting these into the formulae in (20). The number of sample sets is determined so as to have $\Sigma h_i = L$. In practice distances are normalized with respect to h_o in (23), that is, $h_i^* = h_i/h_o$ and $L^* = L/h_o$ are used.

A detailed numerical study has been performed on the basis of these assumptions. The purpose of this investigation was twofold. Firstly, examples of the numerical values of the moduli of the equivalent medium were computed for cubes of different sizes and secondly, from these cube size-related values, an attempt was made to determine the size of the cube which yields reasonably representative values for the moduli. To obtain estimates of moduli corresponding to a given size, the calculations were repeated with fresh sets of random samples several times and from these sets the mean and the standard deviation were computed. The number of repetitions was 100 for $L^* = 5$ and 50 for $L^* = 10, 20, 40$ and 80, respectively. These results are given in Table II.

The results in Table II and III reveal firstly, that the Monte Carlo estimates appear to shift with cube size and secondly, that their standard deviations decrease as the expected number of layers in the cube, that is the ratio L/h_o, increases. The latter trend is a result of the increase in sample size associated with the growing ratio L/h_o. The presence of these features in the results makes the determination of a critical cube size difficult. Nevertheless, it is obvious that the assumptions accepted in this Section do not lead to a physical justification of the particular values of α_1 and α_2 used in (18).

TABLE I.

a	1,25	1,50	2,00	2,50	3,00	4,00
$Ex(E)/E_o$	0,931	0,903	0,886	0,887	0,893	0,906
$s.d.(E)/E_o$	± 0,750	± 0,613	± 0,463	± 0,380	± 0,325	± 0,254

TABLE II.

L/h_o		5	10	20	40	80
ν_1	mean s.d.	0,204 ± 0,067	0,204 ± 0,056	0,204 ± 0,046	0,197 ± 0,024	0,209 ± 0,024
ν_2	mean s.d.	0,214 ± 0,061	0,218 ± 0,047	0,215 ± 0,040	0,213 ± 0,024	0,219 ± 0,017
E_1/E_o	mean s.d.	0,984 ± 0,482	0,996 ± 0,349	0,937 ± 0,230	0,919 ± 0,144	0,940 ± 0,128
E_2/E_o	mean s.d.	0,599 ± 0,382	0,559 ± 0,275	0,426 ± 0,172	0,426 ± 0,144	0,393 ± 0,134
G_2/E_o	mean s.d.	0,229 ± 0,148	0,213 ± 0,104	0,162 ± 0,067	0,159 ± 0,056	0,147 ± 0,052

a = 1,25

In addition to the moduli in Table II, the Monte Carlo technique also provided the values of parameters α_1 and α_2. These are summarized in Table III.

TABLE III

L/h_o		5	10	20	40	80
α_1	mean s.d.	1,435 ± 0,179	1,451 ± 0,146	1,498 ± 0,096	1,513 ± 0,097	1,530 ± 0,083
α_2	mean s.d.	0,543 ± 0,169	0,516 ± 0,146	0,447 ± 0,110	0,445 ± 0,092	0,414 ± 0,079

a = 1,25

On the basis of Table III and the formulae in (14) to (17) the following results are obtained:

$$\alpha_1 = 1,53 \quad \frac{W_o(0,7)}{S_m} = 0,42 \quad I_m = 0,62 \frac{S_m}{H}$$
$$\alpha_2 = 0,41 \quad U_m = 0,23 S_m, \quad \varepsilon_m = \pm 0,119 \frac{S_m}{H} \quad (28)$$

which are drastically different from the test values in Sections 2 and 4.

6. EMPIRICAL INFLUENCE FUNCTIONS

Nothing prevents an investigator from testing various forms of influence functions. If a pair of functions $F(\rho^2)$ and $G(\rho^2)$ results in acceptable ground movement predictions then their use is justified in those geological regions where their validity has been proved.

It seems prudent not to choose independently the influence function for horizontal displacement and that for vertical displacement. Over the years two assumptions relating these two functions have been proposed by various investigators. Some of them have noted that the tilt and horizontal displacement profiles over a rectangular panel are similar. In the light of this similarity they postulated that these two quantities are proportional at all points on the ground surface. This assumption leads to the following relationship, Salamon (1964):

$$G(\rho^2) = -2K_1 \frac{dF(\rho^2)}{d\rho^2} \qquad (29)$$

Other investigators have noted that the displacement vector everywhere on the surface appears to point towards the centre of gravity of the mined-out area. This observation leads to the hypothesis that the displacement vector is directed towards the centre of gravity of the convergence volume in the mined-out area. This will be so if the elementary displacement vector induced by a face element falls into the straight line joining the surface point in question to the centre of a face element. This means that the radial horizontal and the vertical components of the elementary displacement vector should obey the following relationship, Salamon (1974):

$$\left| \frac{\Delta U_r}{\Delta W} \right| = \frac{r}{H} = \rho.$$

It is of interest to note from (5) that the homogeneous isotropic model does, in fact, fulfill the requirements of this formula.

For practical purposes it will be preferable to generalize this relationship and postulate that

$$\left| \frac{\Delta U_r}{\Delta W} \right| = K_2 \frac{r}{H} = K_2 \rho \qquad (30)$$

Constant K_1 in (29) and K_2 here are dimensionless parameters. The relationship in (30) leads to:

$$G(\rho^2) = K_2 F(\rho^2) \qquad (31)$$

which, as seen in (13) is obeyed by the homogeneous transversely isotropic model with $K_2 = \alpha_1 \alpha_2$.

There appears to be no theoretical justification for accepting the relationship in either (29) or in (31). However, it should be noted that both the isotropic and transversely isotropic models behave according to (31). This fact does lend some credibility to this formula. In the remainder of this Section it is assumed that the formula in (31) is valid.

To illustrate the employment of empirical influence functions let $F(\rho^2)$ be given by

$$F(\rho^2) = \frac{(2m-1)b^2}{2(1+b^2\rho^2)^{(2m+1)/2}}$$

$$m = 2,3,4,\ldots \qquad (32)$$

where b is a dimensionless constant. This function obeys (3).

It is possible to evaluate the quantities which are required to test the acceptability of this model:

$$I_m = \frac{b\Gamma(m)}{\sqrt{\pi}\,\Gamma(m-\tfrac{1}{2})} \frac{S_m}{H}, \quad \frac{W_0}{S_m} = 1 - \frac{1}{\{(bR/H)^2+1\}^{(2m-1)/2}}$$

$$U_m = \frac{K_2 \Gamma(m) S_m}{2(m-1)b\sqrt{\pi}\,\Gamma(m-\tfrac{1}{2})}, \quad \varepsilon_m = \pm \frac{K_2 \Gamma(m)(2m-1)^{m-\tfrac{1}{2}}}{\sqrt{\pi}\,\Gamma(m-\tfrac{1}{2})(2m)^m} \qquad (33)$$

Some numerical work indicates that this model might be acceptable if $m = 3$ and $K_2 = b = 3,5$. In this case the test quantities are as follows:

$$I_m = 2,97 \frac{S_m}{H}, \qquad \frac{W_0}{S_m} = 0,992,$$

$$\frac{U_m}{S_m} = 0,21, \qquad \varepsilon_m \pm 0,77 \frac{S_m}{H}, \qquad (34)$$

and which fit well the requirements posed earlier.

7. CONCLUSIONS

This review of linear models indicates their flexibility and adaptability. Although it has not been possible to find an exact elastic model which provides acceptable surface subsidence predictions in British coalfields, the models do simulate ground movement qualitatively.

In the circumstances there is a reasonable chance to find an empirical model which will yield reasonable predictions provided $S_m/H < 0,01$, that is, when the depth of mining is not unreasonably small.

Furthermore, a simple method of preliminary screening of proposed models is introduced. The principles discussed can, of course, be broadened and applied to any geological region where subsidence observations are available.

REFERENCES

Berry, D.S., (1960). An elastic treatment of movement due to mining - Part I. Isotropic ground: J. Mech. Phys. Solids, (8) 280-292.

Berry, D.S., and Sales, T.W. (1961). An elastic treatment of ground movement due to mining - Part II. Transversely isotropic ground: J. Mech. Phys. Solids, (9), 52-62.

Cook, N.G.W., Hoek, E., Pretorius, J.P.G., Ortlepp, W.D., and Salamon M.D.G., (1966). Rock mechanics applied to the study of rockbursts: J.S. Afr. Inst. Min. Metall., (66), 435-528.

Hahn, G.J., and Shapiro, S.S. (1968). Statistical models in engineering. 355 pp. New York: Wiley.

Salamon, M.D.G. (1962). The influence of strata movement and control on mining development and design: Thesis, University of Durham.

Salamon, M.D.G. (1964). Elastic analysis of displacements and stresses induced by the mining of seam or reef deposits - Part II. An application of the elastic theory: Protection of surface installations by underground pillars: J.S. Afr. Inst. Min. Metall., (64), 197-218.

Salamon, M.D.G. (1968). Elastic moduli of a stratified rock mass: Inst. J. Rock Mech. Min. Sci., (5), 519-527.

Salamon, M.D.G. (1974). Rock mechanics of underground excavations: Proc. 3rd Congr. Int. Soc. Rock Mech. (1), Part B, 951-1099, Denver.

Salamon, M.D.G. (1977). The role of linear models in the estimation of surface ground movements induced by mining tabular deposits: Large ground movements and structures, Proc. of the Congr., Cardiff, July 1977, 187-208, Pentech Press, London, 1978.

Subsidence Engineers' Handbook (1975) 2nd Edition, 111 pp. National Coal Board, Mining Department, London.

Wardell, K. (1959). The problem of analysing and interpreting observed ground movements: Colliery Engineering.

EXPLOITATIONS DE SURFACE AU-DESSUS D'EXPLOITATIONS SOUTERRAINES — PROBLEMES D'INTERACTION

Open-pit mining over underground extraction — Interactive problems

Tagebau über Untertagebauten — Probleme der gegenseitigen Beeinflussung

J. Arcamone and R. Poirot
Ingénieurs au Centre d'Etudes et Recherches, des Charbonnages de France, Verneuil-en-Halatte (France)

RESUME

Dans certains cas, pour résoudre des problèmes liés à l'exploitation et à la préparation du minerai (conditions d'exploitation, répartition des teneurs, problèmes minéralurgiques), le mineur peut avoir à conduire ses travaux de surface au-dessus d'exploitations souterraines anciennes ou en cours. De tels problèmes nécessitent de définir une pente évolutive en considérant la roche de surface comme un milieu rocheux ou un sol, fracturé et affecté par les affaissements. Pour des raisons économiques ou purement techniques, il est parfois nécessaire de modifier la répartition tres travaux souterrains. Après un examen des notions essentielles pour l'estimation des affaissements, différentes solutions adaptées aux problème étudié sont proposées. Elles sont présentées à partir du cas d'une mine à ciel ouvert et associées à une méthodologie d'étude générale.

SYNOPSIS

To solve problems connected with the exploitation and extraction of minerals (conditions of mining, distribution of minerals, mineral processing problems), open-pit mine operators may have to work out ore bodies overlying active or old underground workings. Such problems make it necessary to define an evolute slope by considering surface rock as a rock-body or soil subjected to fracturing and affected by subsidence. For economical or purely technical reasons it may be necessary to redefine the underground mine area. After reviewing the general principles of subsidence estimation, different solutions adapted to the problem under study are proposed. They are presented in the light of an example of an open-pit mine with reference to a general methodology.

ZUSAMMENFASSUNG

In Beziehung mit den die Ausbeutung und Gewinnung von Mineralvorkommen betreffenden Problemen (Abbaubedingungen, Mineralverteilung und -aufbereitung) wird in einigen Fällen der Tagebau über aktive oder verlassene unterliegende Untertagebauten geführt. Solche Probleme machen es notwendig, eine sich ändernde Böschung zu definieren, indem man das Oberflächengebirge als einen infolge von Bodenbewegungen verklüfteten und geänderten Gebirgskörper betrachtet. Aus wirtschaftlichen oder rein technischen Gründen kann es sich als notwendig erweisen, den Untertagegrubenbau zu modifizieren. Nach einem Überblick über die Grundlagen der Senkungsberechnungen über Abbaufeldern werden einige dem gestellten Problem gerecht werdende Lösungen vorgeschlagen. Solche Lösungen werden im Lichte des Beispiels Tagebau mit Bezug auf eine Methodik vorgelegt.

Les exploitations souterraines par tailles foudroyées ou par petits piliers torpillés entraînent en général des modifications de la topographie. Il peut en résulter des désordres pour les pentes naturelles ou artificielles, talus de découverte par exemple, ou les bâtiments.

Cet article est consacré plus particulièrement au cas des exploitations de découverte et souterraines simultanées. Ce type d'exploitations combinées est parfois retenu pour résoudre des problèmes de mise en valeur ou d'exploitabilité du minerai (conditions d'exploitation, répartition des teneurs, problèmes minéralurgiques).

Après un examen des problèmes posés, affaissements, stabilité des pentes, combinaison des exploitations, une méthodologie d'étude sera proposée et appliquée sur un cas.

I - EXAMEN DES PROBLEMES POSES

I.1 - Les exploitations souterraines et les affaissements

Dès qu'elle a atteint une surface suffisante, une exploitation avec foudroyage entraîne un déplacement d'ensemble de la surface au-dessus des travaux du fond. Le vecteur déplacement peut être décomposé en un vecteur déplacement horizontal et un vecteur déplacement vertical, l'affaissement proprement dit. Ce dernier est de l'ordre de 70 à 90 pour cent de l'ouverture exploitée et peut atteindre plusieurs mètres (ARCAMONE, 1980). Les mouvements différentiels associés sont respectivement la déformation et la pente (cf figure 1). L'ordre de grandeur des déformations, extension ou compression est très variable et peut aller jusqu'à plusieurs dizaines de millimètres par mètre.

Ce niveau de déformation entraîne en général une rupture du matériau avec fracturation (matériau élasto-fragile) ou passage dans le domaine plastique (matériau ductile). Dans la mesure où le front d'exploitation se déplace dans le temps, il faut considérer que la plus grande partie des terrains influencés par les travaux du fond aura subi des déplacements importants et, dans son ensemble, aura quitté le domaine élastique.

En effet, les mouvements transitoires atteignent 50 à 80 % des mouvements finaux et sont souvent suffisants pour provoquer des ruptures.

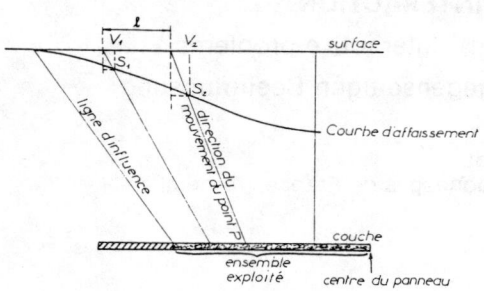

Figure 1 : Relation affaissement/déplacement-schéma
V : déplacement (composante horizontale)
S : affaissement (composante verticale)
$V_1 - V_2$: déformation $S_1 - S_2$: pente

Ces phénomènes ne sont pas toujours visibles in situ. Ainsi, par exemple, les matériaux argileux et pulvérulents ont tendance à se réorganiser à l'échelle des grains. Ils existent cependant et SHADBOLT (1975) rapporte des cas d'ouverture de fissures d'extension dans des roches préalablement diaclasées ou non et dans des sols. Ces fissures sont associées à des boursouflures dans les zones en compression. L'ouverture de telles fissures, dans un massif rocheux essentiellement, tend à annuler la résistance à la traction et diminue les autres résistances. De même, les déplacements relatifs sur les surfaces de discontinuité diminuent la cohésion qu'elle soit due à l'argile ou à l'imbrication des aspérités et des ondulations.

Le remaniement est associé à une perte de résistance du matériau et à une augmentation concomitante de la teneur en eau. Cette augmentation entraîne une baisse sensible de la résistance au cisaillement de la matrice dans les sols ou des discontinuités dans les roches.

Des méthodes de calcul permettent de prévoir les déplacements continus, absolus ou différentiels, de la surface (PROUST 1964). Mais on constate souvent que les déformations sont concentrées en certains points : les déplacements sont discontinus. De nombreux auteurs ont étudié les facteurs qui influencent les mouvements de la surface et des ouvrages implantés (SHADBOLT 1975, ARCAMONE 1980, FORRESTER 1976).
- un facteur <u>minier</u> lié à l'exploitation proprement dite, géométrie et <u>type</u>,
- un facteur <u>recouvrement</u> lié à la nature stratigraphique et à l'intensité des phénomènes tectoniques au niveau du recouvrement,
- un facteur <u>de site</u> lié à la nature des terrains de surface qui conditionnent de plus près l'évolution géomorphologique,
- un facteur <u>de structure</u> lié à la nature des fondations de l'ouvrage et de l'interface ouvrage-terrain naturel.

Une étude de ces différents facteurs associée à un calcul des mouvements doit permettre de prévoir assez précisément tout problème potentiel en surface.

I.2 - <u>Les exploitations de surface et la stabilité des pentes</u>

Une exploitation par découverte est soumise à deux problèmes essentiels :
- la stabilité des flancs de la fosse et des pentes des terrils.
- l'organisation des travaux et plus particulièrement les problèmes de transport du matériau exploité et des déblais.

Les matériaux sont le plus souvent sortis de la fosse par convoyeur ou par camions (cf figure 2) (BASTID - 1978).

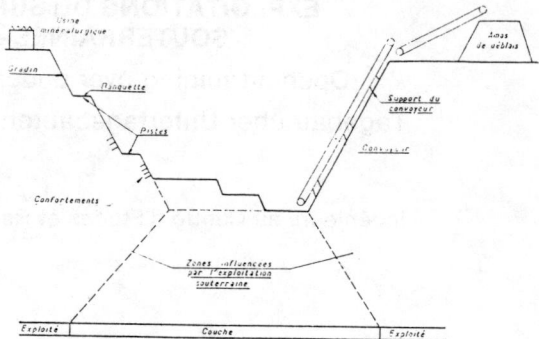

Figure 2 : Exploitation en découverte - Schéma de principe

Des impératif existent :
- les camions ne peuvent rouler que sur des pistes présentant une pente compatible avec les caractéristiques du camion, la qualité de la piste et le coût du transport. L'écart toléré autour de la valeur de la pente calculée dans le projet est très faible en général.
- les convoyeurs ont des pentes limites à partir desquelles les matériaux se mettent à glisser sur la bande.

Parallèlement, les flancs de la découverte doivent être stables au cours de l'ensemble du creusement jusqu'au stade ultime pour permettre une récupération totale du minerai en fond de fosse (LILLICO - 1974).

L'analyse de la stabilité d'une pente définie dans un massif hétérogène est couramment conduite à l'aide de méthodes à la rupture. De telles méthodes ont été proposées par de nombreux auteurs. Elles sont basées généralement sur la connaissance des caractéristiques à la rupture des différents matériaux (cohésion et angle de frottement). Elles ont été synthétisées par RAULIN et al. (1974) Utilisables dans les sols et les roches, elles présentent l'inconvénient majeur de supposer la rupture simultanée en tous les points de la ligne de rupture. Cette hypothèse habituelle peut devenir inacceptable dans un massif soumis aux affaissements miniers. Dans ce cas, en effet, les caractéristiques mécaniques de la matrice et/ou des discontinuités évoluent dans le temps. On est donc conduit à faire différentes hypothèses sur l'état du matériau en fonction de l'avancement des travaux souterrains.

Non loin de la crête des flancs de découverte, il est courant de trouver le terril et/ou l'usine minéralurgique. Ces surcharges ont une action néfaste sur la stabilité. Sous l'action des déformations, les terrils peuvent devenir brutalement instables (GEDDES 1977, FORRESTER 1976).

II - METHODOLOGIE D'ETUDE

L'approche proposée résulte de la synthèse de méthodologies d'étude appliquées aux affaissements miniers et aux découvertes (HOEK et BRAY, 1981). Elle s'appuie sur le dualisme des informations naturalistes et mécaniques et doit permettre de résoudre les problèmes posés au chapitre précédent. Elle sera exposée à partir du cas d'une mine du Sud de la France.

Trois grandes étapes peuvent être distinguées :
- l'analyse préalable, détermination des principales phases de l'exploitation,
- l'analyse des affaissements et de la stabilité des pentes pour chacune de ces phases,
- le contrôle des mouvements au cours des travaux et les modifications éventuelles du schéma général.

II.1 - L'analyse préalable comprend l'étude des éléments nécessaires pour établir la décomposition par phases de l'exploitation et pour conduire les calculs qui s'y rapportent.

Pour la mine étudiée (POIROT, 1981), en voie d'épuisement la découverte a pour objet d'exploiter le stot de protection des puits. Les travaux souterrains conduits à une profondeur maximale de 290 m, seront concomitants du creusement de la fosse dans des argiles et sables du Tertiaire jusqu'au niveau des terrains houillers (composés d'alternance de schistes, de grès et de couches plus ou moins épaisses de charbon) (cf figures 3 et 4). Seules les couches I et J d'une épaisseur totale de 14 m environ seront exploitées en souterrain.

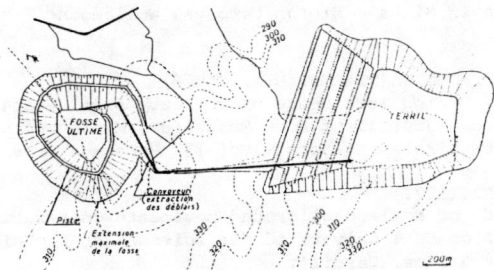

Figure 3 : Plan d'ensemble de la découverte et du terril

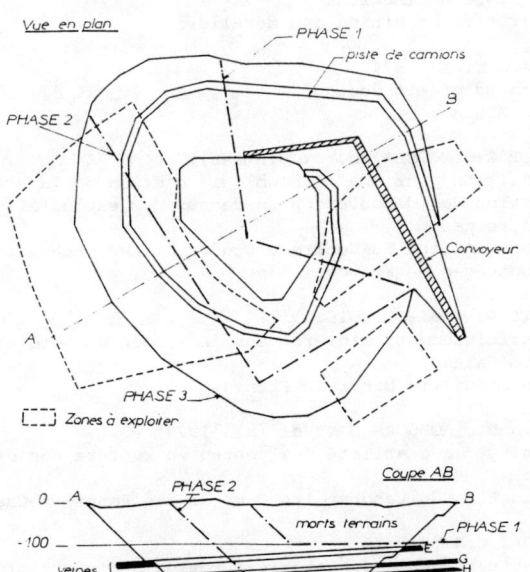

Figure 4 : Schéma d'exploitation de la découverte

Une étude structurale du houiller réalisée à partir de la surface dans les zones d'affleurement a permis de distinguer les principales familles de discontinuités ; leur répartition est plutôt aléatoire. Les terrains du Tertiaire sont globalement homogènes.

Une campagne d'essais a été réalisée sur les différents niveaux stratigraphiques. L'effort a porté sur les essais de cisaillement sur des éprouvettes de dimensions différentes (5 et 40 cm), pour préciser la cohésion et l'angle de frottement moyen.

Dans les argiles et sables du Tertiaire certaines zones plus sableuses se sont révélées aquifères. Les terrains houillers sont pratiquement imperméables.

L'exploitation souterraine devrait s'approcher jusqu'à 100 m du versant de la découverte. En fonction de cet ensemble, trois phases de développement ont été distinguées:
1) décapage des morts-terrains du Tertiaire
2) exploitation du houiller jusqu'à mi-hauteur
3) état final.

II.2 - Pour chacune de ces phases, il est nécessaire de prévoir différentes géométries du fond liées aux hypothèses sur les vitesses d'avancement relatives des chantiers. C'est une difficulté importante inhérente au problème des exploitations simultanées.

Sur ces bases, il est possible de calculer la répartition des affaissements, des pentes et des déformations au niveau des flancs de la découverte et d'analyser la stabilité de ces derniers.

Dans le cas de la mine étudiée, les travaux souterrains influencent surtout les travaux de surface dans les phases II et III. Le cône d'influence maximal de l'exploitation par foudroyage a été défini avec un angle de 35° par rapport à la verticale. Pour les calculs de stabilité effectués avec la méthode de Bishop, les propriétés mécaniques des terrains affaissés ont été modifiées en annulant la cohésion en en conservant l'angle de frottement moyen obtenu lors des essais.

La méthode de Bishop se justifie dans les terrains houillers en raison de l'intense fracturation présente dans ces terrains et de sa répartition aléatoire. Un exemple de modélisation schématique des terrains est proposée figure 5 (phase III).

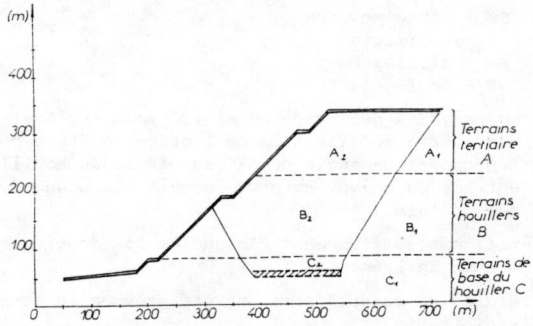

Figure 5 : Analyse de stabilité. Modélisation d'un flanc de découverte dans un cas d'interférence.

Pour chacun des terrains (A,B,C) les propriétés mécaniques sont différentes dans les zones soumises aux affaissements (A_2, B_2, C_2) et dans les autres (A_1, B_1, C_1).

Les calculs de stabilité ont montré que les angles de talus stables sont de 37° dans les argiles du Tertiaire et de 45° dans le houiller en dehors de toute exploitation souterraine.

En présence d'une exploitation souterraine, la stabilité des talus dépend de l'extension des zones sous-cavées. Ainsi, lorsque l'exploitation souterraine arrive à une distance horizontale de 100 m du versant de la découverte, le coefficient de sécurité est insuffisant. Pour une distance de 250 m, on obtient un coefficient de sécurité raisonnable.

Cette analyse de la stabilité globale a été associée à une analyse plus fine des banquettes.

Après la stabilisation des affaissements, deux hypothèses peuvent être formulées :
- une hypothèse pessimiste : la cohésion est définitivement annulée dans le volume influencé
- une hypothèse optimiste : l'exploitation ayant produit un déplacement d'ensemble, seul le volume influencé directement par la taille, de l'ordre de 3 fois l'ouverture de la veine, est perturbé durablement.

En supposant que le comportement réel est entre ces deux extrêmes, il est intéressant d'organiser l'exploitation en modifiant certaines pentes à la fin des travaux souterrains :
- un angle de 37° est nécessaire dans les terrains tertiaires et le houiller durant la mise en place des affaissements
- ultérieurement, une fois les affaissements définitivement achevés, le houiller pourrait être repris avec un angle de 45°.

Pour chaque phase, les accès et de déblocage sont analysés. Les pistes constituées par les banquettes sont étudiées avec ces dernières. La pente initiale du convoyeur doit être prévue pour intégrer les modifications de pente consécutives aux affaissements. La veine J est seule à influencer le convoyeur. Pour le calcul on a considéré deux parties dans le tracé : une partie superficielle et une partie profonde.

D'après les travaux de DEJEAN (1973), les affaissements en différents points du recouvrement sont fonction de l'affaissement calculé en surface. Le calcul de ces derniers à différentes profondeurs, au niveau du convoyeur permet de prévoir la pente moyenne suivant la formule :

$$E_m = k \frac{A_m}{P}$$

avec E_m : pente moyenne
   k : coefficient
   A_m : affaissement
   P : profondeur

Dans ce cas, des pentes de 60 et 100 mm/m ont été obtenues. Cette modification de l'ordre de 5°, a conduit à proposer un angle de 37° au niveau du houiller sur le tracé du convoyeur pour garantir la bonne évacuation des déblais.

Le terril est suffisamment éloigné de la découverte pour ne pas être influencé.

Des mesures de surveillance ont été prévues au cours de l'exploitation.

CONCLUSION

Dans tout projet d'exploitation simultanée en souterrain et en découverte, la coordination des travaux entre le fond et le jour est un élément essentiel de l'étude géotechnique préalable.

La méthodologie utilisée pour résoudre le problème a déjà été appliquée dans plusieurs mines. Pour la conduite des calculs de stabilité, on a recours à des simplifications importantes, car il est actuellement difficile de chiffrer précisément les modifications des propriétés mécaniques des terrains soumis aux affaissements miniers et aux déformations associées. Ces problèmes ouvrent des perspectives de recherche d'autant plus intéressantes que ces exploitations simultanées ont tendance à se développer.

BIBLIOGRAPHIE

ARCAMONE J. (Juin 1980)
Méthodologie d'études des affaissements miniers en exploitation totale et partielle.
Thèse INPL France

BASTID P.H. (Avril 1978)
Elaboration de projets d'exploitation minière à ciel ouvert.
Revue de l'Industrie Minérale.

DEJEAN M., MARTIN F. (1973)
Amplitude of subsidence of underground openings subject to the influence of mining adjacent and below.
Subsidence in Mines - Mining Congress Wollongong Australie.

FORRESTER D.J., WHITTACKER B.N. (1976)
Effects of Mining Subsidence on colliery spoil Heaps
International Journal of Rock Mechanics, Mining sciences and Geomechanical Abstracts - Vol 13. Pergamon Press.

GEDDES J.F. (1977)
The effect of horizontal ground movements on structures
Communication de l'Université des sciences et techniques du Pays de Galles. Cardiff.

HOEK E., BRAY J.W. (1981)
Rock slope Engineering
Institution of Mining and Metallurgy

LILLICO T.M. (1974
Economics of pit design
World Mining.

POIROT R., MUTEBA Ph. (Juin 1981)
Proposition pour une méthodologie d'étude de la stabilité des talus des découvertes au-dessus des exploitations souterraines.
Rapport interne du Centre d'Etudes et Recherches des Charbonnages de France.

PROUST A. (Mai-Juin 1964)
Les affaissements miniers dans le Bassin du Nord et du Pas-de-Calais
Revue Industrie Minérale France.

RAULIN P., ROUQUES G et al (Mai 1974)
Calcul de la stabilité des pentes en rupture non circulaire.
Publication du Laboratoire Central des Ponts-et-Chaussées.

SHABOLT C.H. (1975)
Mining subsidence ; historical review and state of the art.
Midlands region. Subsidence Engineering
Publication du National Coal Board.

PREDICTION OF MINE SUBSIDENCE IN EASTERN AUSTRALIA BY MATHEMATICAL MODELLING

La prédiction par modélisation mathématique des affaissements provoqués par l'extraction de charbon en Australie orientale

Über die mathematische Vorraussagung von Bodensenkung infolge Kohlegewinnung in Ostaustralien

P. A. Mikula
Senior Geotechnical Engineer, Australian Coal Industry Research Laboratories Ltd., Sydney, Australia

G. E. Holt
Senior Engineering Geologist, Australian Coal Industry Research Laboratories Ltd., Sydney, Australia

SYNOPSIS

Finite element modelling of subsidence due to coal extraction in Eastern Australia is described under certain limiting conditions. A systematic means of data acquisition and handling was developed to provide realistic input for the geotechnical model. The constant strain finite element program requires large, carefully designed meshes and empirical reduction of laboratory strength properties. Anisotropy needs to be considered for coal measures strata. The inclusion of joint elements improves subsidence simulation but is not essential for generalised prediction. Examples of successful modelling at shallow depths are discussed.

RESUME

La modélisation d'éléments finis des affaissements causés par l'extraction de charbon en Australie Orientale est décrite selon certaines conditions limitatives. Une méthode systématique d'acquisition et de traitement des données a été développée pour fournir des entrées réalistes pour le modèle géotechnique. Le programme d'éléments finis d'effort constant requiert de grandes mailles soigneusement conçues et une limitation empirique des propriétés de force étudiées en laboratoire. L'anisotropie doit être examinée pour les couches de mesures de charbon. L'inclusion d'éléments "joints" améliore la simulation de l'affaissement, mais n'est pas essentielle pour une précision générale. Des exemples de modélisation effectuée à de faibles profondeurs sont examinés.

ZUSAMMENFASSUNG

Die Methode endlicher Elemente zur mathematischen Darstellung von Bodensenkung infolge Kohlegewinnung in Ostaustralien wird unter gewissen einschränkenden Bedingungen beschrieben. Ein systematischer Weg zur Datenerfassung und -verarbeitung wurde entwickelt, um das geotechnische Modell mit genauen Eingaben zu versorgen. Ein Programm nach der Methode endlicher Elemente mit konstanter Dehnung erfordert ein großes, sorgfältig gewähltes Elementennetz und empirische Vereinfachung von im Labor gemessenen Festigkeitseigenschaften. Anisotropie muß für Kohleflözschichten berücksichtigt werden. Die Einbeziehung von Kluftelementen verbessert die Nachbildung von Bodensenkung, ist aber für verallgemeinerte Voraussagen nicht unbedingt erforderlich. Beispiele von erfolgreichen Modelldarstellungen bei geringen Teufen werden besprochen.

1. INTRODUCTION

British and Continental coal industries over the last 30 years have developed acceptable empirical methods for subsidence prediction, based on mathematical best fit of years of detailed survey data. The variety of prediction methods are satisfactory when employed in the appropriate area.

In Australia there is no general purpose prediction method. There has been monitoring of the surface effects of subsidence for over 17 years (Kapp 1973, 1982) in attempts to develop graphical procedures similar to those for the United Kingdom (NCB 1975). Empirical methods developed and successfully applied overseas give erroneous results when used under Australian conditions. Differences in strata behaviour are the main reasons for reduced subsidence effects (Frankham and Mould 1980, Kapp 1982).

Simple mechanistic models have proven incapable of simulating the complex strata behaviour involved in subsidence and computer based mathematical modelling appears the only practical alternative. Finite element methods utilising both two and three dimensional models have been trialled in Australia and the United States in recent years (Germanis and Valliappan 1975, Daemen and Hood 1982).

In this paper a close examination of particular problems in using finite element methods to simulate subsidence is undertaken. Emphasis was given to the development of accurate subsidence profiles and angles of draw since other properties such as surface strain and curvature follow if these two parameters are correctly determined. Angle of draw is defined herein by the vertical line drawn at the goaf edge and the line from the goaf edge to the 5 mm subsidence point.

2. THE GEOTECHNICAL MODEL

The complexity of strata behaviour during near total extraction of a coal seam imposes special problems for finite element modelling. Rock strata can behave elastically and plastically. Mass movements can be several metres as beds cantilever, crack, fall as blocks and rubble, then recompact. Time dependent movement occurs. Pre-existing discontinuities can open, close and shear, while new ones are formed. Anisotropic strength properties and stress fields add further constraints to subsidence modelling.

The first stage in analysing a rock mass is the development of an accurate geotechnical model since the results obtained from any mathematical modelling are no more reliable than the data on which the model has been based. The complexity of modern computer technology and voluminous output data often overshadow this fact.

In many subsidence modelling exercises overburden has been regarded as homogeneous, or simplified to a few major lithologies with average or convenient properties selected. However, if a geotechnical model does not make a fair attempt to represent reality, doubts on the validity of subsequent analyses always remain.

The information which should be considered in the formulation of a geotechnical model includes:

- regional and local geology
- structural geology
- strata geomechanical properties
- regional and local stress fields
- hydrology
- proposed mining geometry and sequence
- environmental restraints.

2.1 Data Acquisition

Acquisition of representative data is of paramount importance, but this objective is extremely difficult to achieve in the case of strata geomechanical properties. Sampling is usually limited to core retrieved by diamond drilling, and frequently only core from coal seams is recovered.

Assuming careful testing procedures to I.S.R.M. standards, maximising the amount of rock testing in each lithological unit under examination is the most practical way of improving reliability of data. Increasing the number of sampling points is the next most desirable course. Even with a large rock testing budget the number of test results from a stratum is almost invariably below a statistically desirable number. This is unfortunate but there is no real alternative.

Rock properties testing of exploration borecore is a cost effective way of obtaining a maximum number of sampling points.

The geotechnical properties of borecore of interest include
(1) The mechanical properties of the 'intact' core.
(2) The degree to which the core is fractured.
(3) The susceptibility of the various rock types represented in the core to strength loss with time.

If a comprehensive evaluation of all these properties is carried out by study of the core then a sound basis is available upon which to quantify the geotechnical properties of the in situ rock mass from which the core has been drilled.

The field logging procedures consist of various methods of fracture logging, lithological and discontinuity description plus point load strength testing for indication of strength anisotropy. Prior to point load testing (in the case of core logged in the field and not subjected to laboratory testing), core photography is undertaken to give accurately scaled and colour controlled prints for mounting on the graphic geotechnical log of each borehole. This provides a direct visual display of the core adjacent to the geotechnical properties, and provides a useful check against geological descriptions.

Results from the following tests, carried out to I.S.R.M. standards (I.S.R.M. 1981) are incorporated in the geotechnical model.

(a) Uniaxial compressive strength
(b) Indirect tensile strength
(c) Triaxial compressive strengths (gives Mohr Envelope)
(d) Point load strength estimate
(e) Elastic Modulus
(f) Poisson's Ratio
(g) Slake durability index
(h) Normal and shear stiffness of joints.

Wherever possible stress levels in Eastern Australia are determined underground by the overcoring strain relief technique. Two types of strain cells are used, the CSIR and the CSIRO cells. When strain measurements are not possible vertical stress is calculated by $\sigma_v = \rho g h$ and horizontal stress commonly by $\sigma_h = 2\sigma_v$.

2.2 Data handling system

In order to handle the large quantities of data involved in this type of geotechnical investigation a series of computer programs have been written (Pauncz and Holt, 1982). Figure 1 is a schematic flowchart of the system. All strata logging and fracture data is entered directly on to special coding sheets in the field. This data is loaded into the system via programs STRLOD and CINDEX. Laboratory test results are similarly entered via the STONES program so that all relevant information is recorded within the general purpose geotechnical database according to borehole and depth.

The major graphical output from the system is via two programs. TEKLOG produces a scaled graphical plot of all data obtained for a nominated borehole according to depth. The computer drawn log which covers all fracture data, point load test results, discontinuity and lithology descriptions plus laboratory test results located according to the section of core from which they were determined. The centre of the log is reserved for the scaled colour photographic strip of the actual core.

TEKLIS presents lithology, laboratory test results and limited fracture logging data as A4 size columnated tables.

Data in these two formats can be readily incorporated in the geotechnical model.

3. FINITE ELEMENT COMPUTER PROGRAM

The ACIRL finite element computer program used in this work is of the "constant strain" type. The program represents a vertical section through the strata by a large number of triangular and quadrilateral-shaped elements, with the size of the elements varied according

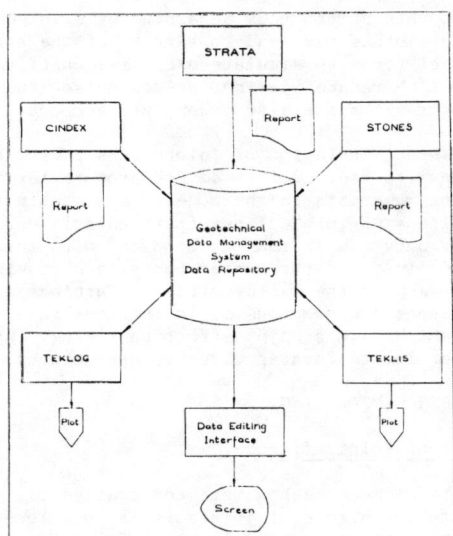

Figure 1. The geotechnical logging system.

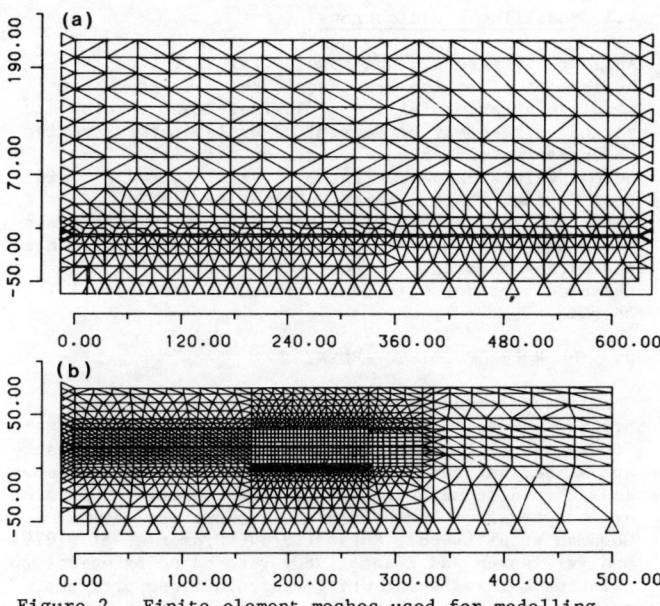

Figure 2. Finite element meshes used for modelling.
(a) Mesh for Model No. 1 (dimensions in metres)
(b) Mesh for Model No. 7 (dimensions in metres)

to the detail required. Joints, faults, bedding planes and other discontinuities are represented by Goodman joint elements. When the finite element mesh extends up to the ground surface, subsidence can be modelled. The program is operated in "cycles", where each cycle consists of the redistribution of any non-equilibrium forces left over from the previous cycle. Non-equilibrium forces arise from the creation of an excavation and from the yielding of material. Different rock properties, primitive stresses or excavation dimensions can be optionally specified at the appropriate stage of each cycle.

3.1 Mesh Design

A well-designed finite element mesh should be capable of accurately simulating structural variation and material failure at any location in the model. Design requirements include: small elements in area of interest; grading of element sizes; and where movement of the strata between the excavation and the surface is important, a fairly constant density of elements in that strata zone.

Two of the meshes used in this research are shown in Figure 2, for Models No. 1 (Hartley Main No. 4, N.S.W.) and 7 (United Collieries, N.S.W.).

Boundary conditions are imposed on the finite element meshes. Firstly, no movement occurs at a large enough distance from the excavation. Therefore boundaries outside the area affected by the excavation are fixed appropriately. Secondly no movement is possible across an axis of symmetry. Therefore one side of the mesh representing the centreline of the goaf in Figures 3, 4 and 6 is given horizontal fixity, while vertical movement is unaffected. Thirdly, the ground surface is free to move.

4. SELECTION OF STRATA PROPERTIES

Finite element modelling has great potential for the accurate simulation of nonelastic material behaviour, as it can model the effect of confinement realistically. However, post-failure material property data are necessary and this is not generated in the usual rock mechanics testing programs. Ideally, the complete stress-strain characteristic curve for all the material types that undergo failure will be available from laboratory testing results; otherwise estimation according to the typical behaviour of the rock type is necessary.

4.1 In situ material properties

For realistic modelling, laboratory test results should be modified to correlate with the in situ rock mass conditions. This is achieved using one or more of the following approaches.

(i) Use of the C-factor. Ideally, the C-factor when applied to laboratory results accounts for visible breaks in the rock core (most breaks are horizontal) and gives in situ properties in the vertical direction.

(ii) Comparison of predicted and measured stress and displacements around the excavation. This requires underground access and monitoring equipment.

(iii) Comparison of predicted and measured subsidence data, specifically maximum subsidence and the shape of the subsidence profile.

Methods (i) and (iii) were used for all modelling described in this paper.

4.2 Modelling of anisotropy

Anisotropic strength and modulus values in the horizontal direction are used in the F.E. program to simulate shearing failure of laminated and interbedded strata. These may be determined by laboratory testing of appropriately oriented core, or calculated from normal values using the ratio of axial to radial Point Load Strength Estimations (Hebblewhite, 1982). This latter method has been less successful with subsidence modelling, and an empirical approach of adjusting normal properties according to best fit of calculated and measured subsidence is demonstrated in Section 6.4 below.

5. THE ROLE OF JOINT ELEMENTS

The joint elements employed in the Finite Element program are based on the joint model originated by Goodman et al (1968). Joints are able to open, close and slide, and represent faults, joints, bedding planes and other discontinuities in the rock mass. Properties are adapted from Lackey (1974). Comparison between Goodman et al (1968), Rosso (1976), Cramer et al (1979) and Valliappan and Evans (1980) yielded no agreement on joint normal and shear stiffness values, despite the authors' apparent success at modelling. There was no available current test data; thus properties known to have been successful in previous mine stability modelling were selected, and are shown in Table II.

5.1 Joint behaviour in the model

In these simulations the joints behaved well, at least to the extent that the program was mathematically capable of modelling. A primary result from the modelling is that joints were unlikely to move more than minimally in shear, except close to an excavation. The finite element model is of an area bounded by fixities restricting movement, and in particular preventing slip at the end of a joint. A further result indicated that vertical joints generally were prevented from opening by the surrounding compressive stress field. Only after extensive failure (i.e. close to an excavation or goaf), or at the ground surface, was the joint normal stress relieved to the point where opening followed in the model. The following points arise from these observations:

(i) Joints generally contribute very little to rock mass slipping and shearing, except local to an excavation or goaf.

(ii) Joint opening is also generally confined to the immediate vicinity of excavation or goaf areas.

5.2 Influence of joints on caving simulation

The finite element model has the capability of automatically determining the extent of the caved area forming the goaf over an underground excavation. Given a bulking factor, the program calculates the amount of bulking of caved material, and allows the goaf to take load when reconsolidation begins. Caved areas are detected following tensile or compressive failure of strata material over an excavated area. Blocks of material isolated by open joints or joints failed in shear are caved, and therefore inclusion of joints in the model facilitates caving. Failed material does not necessarily cave as material higher above the goaf is often fractured but still in place.

A model of thin strata beds separated by a large number of parallel joints may lead to sagging of the strata. Such a model tends to simulate extensive cantilevering of strata with reduced caving, unless anisotropic material strengths are also taken into account.

It appears that inclusion of joints aids caving of goaf up to a massive bed. Joints do not promote large shear and opening movements in the model. A possible reason is that extensive joint planes (particularly vertical) are rather uncommon in reality compared with the volume of rock involved and the number and size of caving-induced breaks in the failed strata. Partings along bedding planes are numerous in laminated strata, and it is these which have a major effect on caving. Exception is made for joints located close to excavations.

6. SUBSIDENCE MODELLING

6.1 Model of joint effects

Four finite element meshes were constructed to investigate the effect of joints in various locations and orientations on the development of goafing and surface subsidence over Hartley Main No. 4 Colliery. The meshes were identical in all respects except for the joint element arrangements, which were as follows:

Model No. 1   No joints, as in Figure 2(a).
Model No. 2   Horizontal joints, located at distances of 2, 22, 70, 108 and 165 metres above the base of the seam.
Model No. 3   Vertical joints, located 70 metres apart.
Model No. 4   Combination of horizontal and vertical joints from meshes 2 and 3.

Each mesh contains from 661 to 992 nodes, 1020 solid elements and up to 314 joint elements. The material properties used for the meshes are given in Table I, and the joint properties in Table II. Joints detected as continuous from surface to seam were found to be variably spaced from 35 metres to several hundred metres apart (Shepherd et al, 1981). As a compromise 70 metre spacings were adopted for modelling. Stress levels were determined from overcoring measurements in Hartley Vale No. 4 colliery. Horizontal stress was 8 MPa, vertical stress 4 MPa approximately, in good agreement with the general formulae adopted in 2.1.

TABLE I - MATERIAL PROPERTIES, MODELS NO. 1-4

Rock unit	UCS MPa	E1 MPa	E2 MPa	PR	TS MPa	COH MPa	T10 MPa	T20 MPa	T40 MPa	T80 MPa	Bulking factor
1 sandstone	10.5	3570	2600	0.47	0.8	2	15	22	33	50	1.025
2 claystone	11.5	5011	2778	0.40	0.7	2	14	20	29	42	1.075
3 sandstone/ claystone	8.1	8032	5688	0.44	0.6	2	11	17	26	38	1.150
4 shale	47.1	11477	7897	0.24	3.0	8	21	29	41	53	1.200
5 coal	8.0	1004	1004	0.34	0.4	2	11	16	23	26	1.200
6 sandstone	123.5	2100	790	0.50	8.0	20	35	45	55	65	1.200

UCS Uniaxial compressive strength
E1 Isotropic elastic modulus
E2 Anisotropic elastic modulus
PR Poisson's ratio
TS Tensile strength
COH Cohesion
T10, T20, T40, T80 Mohr envelope coordinates for normal stress of 10, 20, 40 and 80 MPa.

TABLE II - JOINT PROPERTIES, ALL MODELS

Model No.	Joint orientation and type	N/STF MPa/m	CONS/STF MPa/m	SHR MPa/m	RES MPa/m	COH MPa	I/F Rad
3,4	vertical-all strata	20000	10000000	90	40	0.50	0.45
2,4	horizontal-sandstone	26000	10000000	1000	500	0.27	0.50
2,4	horizontal-shale	15000	10000000	1000	300	0.75	0.40
2,4	horizontal-coal	15000	7500000	800	300	0.75	0.40
5,6	horizontal-shale	1100000	750000000	1500	200	0.50	0.65
7	horizontal-sandstone	1200000	1000000000	2000	200	1.70	0.55

N/STF Normal stiffness
CONS/STF Consolidated stiffness
SHR Shear stiffness
RES Residual stiffness
COH Cohesion
I/F Angle of internal friction

The overall results of the four joint-pattern simulation runs are shown in Figures 3(a)(b)(c)(d), with details as follows, and relevant data given in Table III.

(i) <u>Subsidence</u>. The greatest maximum subsidence level, 590 mm (28% of extracted seam thickness) was obtained in Model No. 4 (horizontal and vertical joints), followed by Model No. 1 (no joints), Model No. 2 (horizontal joints) and Model No. 3 (vertical joints). Although the maximum subsidence was less than observed, the simulated angle of draw was reasonable in each case.

(ii) <u>Goaf formation</u>. Perhaps paradoxically, Model No. 3 simulated the greatest extent of strata disturbance and caving above the goaf, but the smallest maximum surface subsidence. It is probable here that the joint pattern rather than the goaf formation was the dominant influence on subsidence simulation. Model No. 2 shows clearly the cantilevering of horizontally-jointed strata beds in preference to caving. The provision of vertical jointing as well as horizontal (Model No. 4) prevents excessive cantilevering. The Models also demonstrate the effect of the joints on the extent of caving.

(iii) <u>Joint Behaviour</u>. Nowhere in these test Models was the opening of a joint recorded. The maximum shear along joints ranged from 1 mm in Model No. 2 to 16 mm in Model No. 3.

TABLE III - SUMMARY OF MODEL SUBSIDENCE RESULTS

Model No.	Joint pattern	Depth of cover m	Seam thickness m	Width of extraction m	Maximum subsidence m	Angle of draw degr
1	none	220	2.11	332	0.56	29
2	horizontal	220	2.11	332	0.49	28
3	vertical	220	2.11	332	0.47	28
4	horizontal & vertical	220	2.11	332	0.59	29
5	horizontal	80	2.60	150	0.64	16
6	horizontal	170	2.60	150	0.25	43
7	horizontal	74	3.11	412	2.18	42

6.2 Model of depth effect

Finite element modelling of subsidence over Angus Place colliery in New South Wales was done for two depths of cover, 100 metres (Model No. 5) and 170 metres (Model No. 6). The models were identical apart from the thickness of the uppermost rock strata unit, and horizontal joints were located in the vicinity of the seam only. All the strata above the seam was shale, with material properties similar to those given for shale in Table I. Joint properties were as listed in Table II.

The results for these two simulations are shown in Figure 4, with relevant data given in Table III. Maximum subsidence of 1.55 metres and angle of draw up to 14° was recorded at Angus Place (Schaller and Hebblewhite 1981), and a similar phenomenon was recorded by Galvin (1982) with regard to massive dolerite sills over South African mines. The shallow depth model, where strata caving breaks through to the surface, gave almost half this maximum subsidence and a 16° angle of draw, much improved on Model No. 6 with a subsidence of only 0.25 metres and a profile typical of elastic finite element models (Germanis and Valliappan 1975, Girrens et al 1982). The subsidence profiles for the simulations are plotted in Figure 5.

6.3 Model of large width of extraction

A proposed longwall operation for United Collieries, N.S.W., with depth of cover 74 metres and with two panels each of width 200 metres separated by a 12 metre chain pillar, was modelled using a 1608-node, 2471-element mesh [Figure 2(b), Model No. 7]. Geotechnical material properties were similar to those given for claystone in Table I. The joints had properties as listed in Table II.

The result of the simulation is presented in Figure 6, with the associated surface subsidence profiles in Figure 5. Maximum surface subsidence is 2.18 metres, or 70.1% of the thickness of extraction. This is in accord with maximum subsidence values recorded in New South Wales. The angle of draw of 42° is also in reasonable agreement with the usually expected value of 35° (Frankham and Mould 1980).

This result indicates that the finite element model is able to predict subsidence levels ahead of mining with accuracy for a supercritical width of extraction.

6.4 Effect of anisotropy

Using Model No. 5 the width of extraction was doubled to ensure that by all empirical criteria, maximum subsidence would result. The ratios of anisotropic moduli and strengths to normal values were varied as shown in Table IV, and subsidence profiles for each run plotted (Figure 7). The measured profile is shown in its correct position. Modelled and measured maximum subsidence levels coincided with an anisotropic ratio of 0.038:1. The profile noticeably steepened, when compared with Figure 4(a).

This was an improvement on earlier work undertaken at Bowen No. 2 colliery, Queensland where smaller variations in anisotropic properties had indicated low to negligible influence on subsidence (Mikula 1981). At the time stress magnitudes appeared to have greatest effect on subsidence. While they are important this work suggests they are but one factor in subsidence modelling.

TABLE IV - EFFECT OF ANISOTROPY

Run	COAL (Isotropic) UCS MPa	COAL (Isotropic) MODULUS MPa	COAL (Anisotropic) UCS MPa	COAL (Anisotropic) MODULUS MPa	Anisotropic Ratio	Maximum Subsidence m
Measured	8.70	1700.0	2.18	425.0	0.250	1.550
5a	8.70	1700.0	0.87	170.0	0.100	1.087
5b	8.70	1700.0	0.39	77.0	0.045	1.327
5c	8.70	1700.0	0.33	64.0	0.038	1.551
5d	8.70	1700.0	0.22	42.5	0.025	1.814

7. DISCUSSION

The results shown above highlight some of the problems encountered in mathematical modelling of subsidence using the finite element method.

Of the first four Models, it is encouraging to note that the most realistic model (No. 4), with horizontal and vertical joints, simulated the greatest maximum subsidence. However, the next best model (No. 1) was that with no joints at all, yielding 95% of the maximum subsidence of the best model. The difference between the two is minimal, and considering the extra costs involved in mesh generation and computer solution time

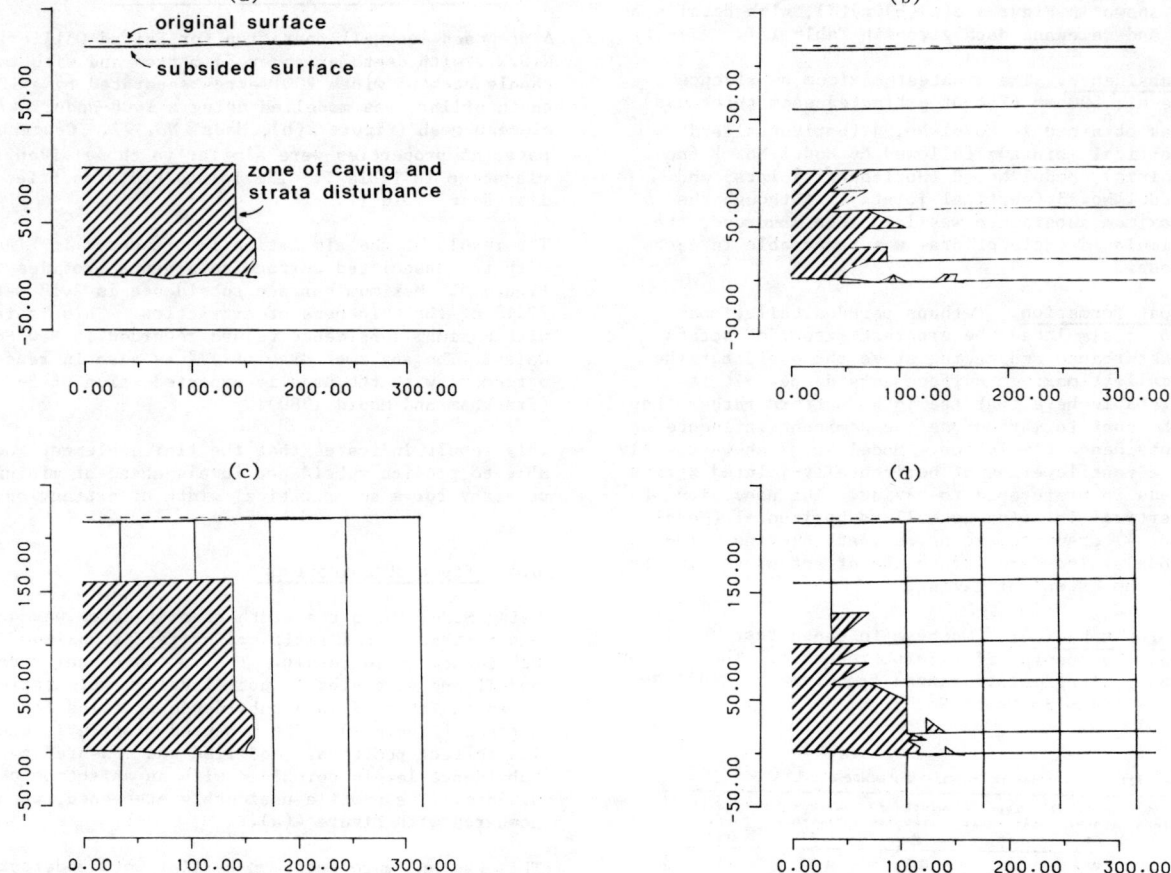

Figure 3. Model results for the joint-pattern simulations (dimensions in metres).
(a) Model No. 1 - no joints (b) Model No. 2 - horizontal joints
(c) Model No. 3 - vertical joints (d) Model No. 4 - horizontal and vertical joints.

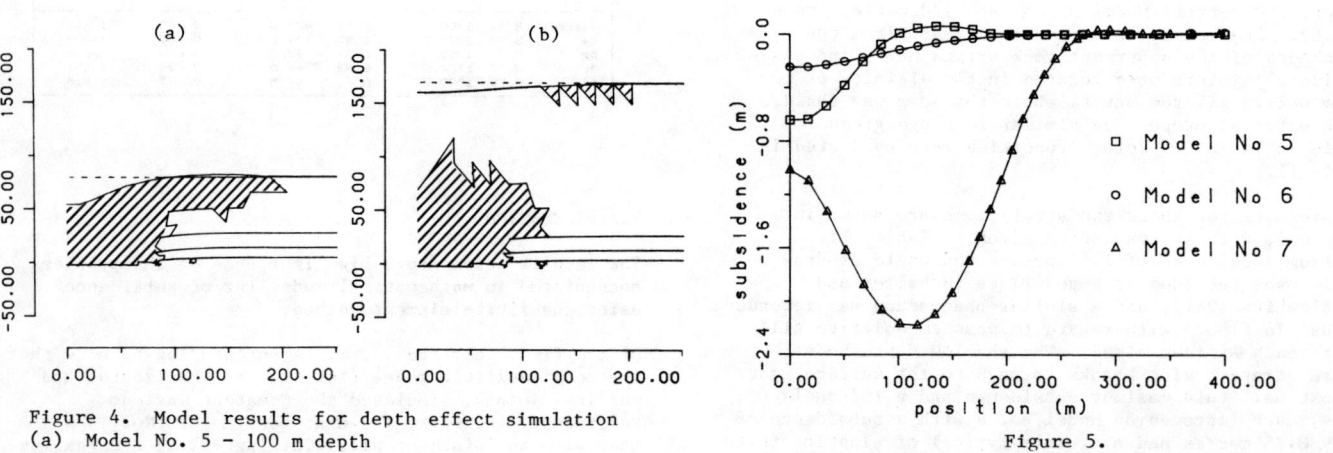

Figure 4. Model results for depth effect simulation
(a) Model No. 5 - 100 m depth
(b) Model No. 6 - 170 m depth

Figure 5.
Surface subsidence profiles for Models No. 5 and 6.

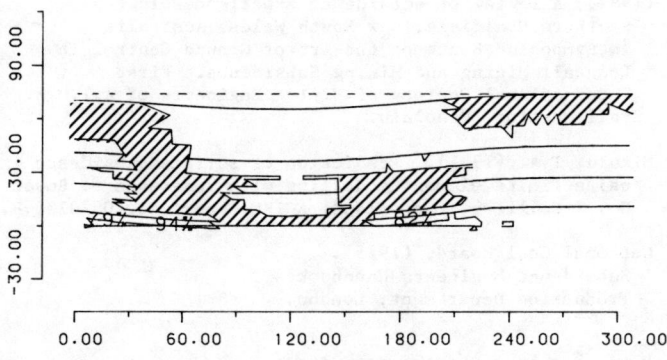

Figure 6
Model result for large width of excavation (Model No. 7)

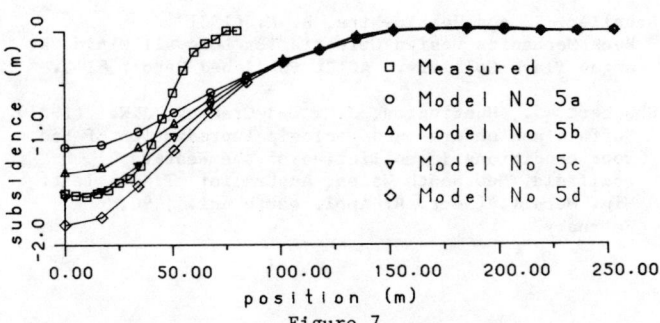

Figure 7
Surface subsidence profiles showing anisotropic influence

for the jointed mesh, it does not seem worthwhile to construct and solve a relatively complex mesh model where a much simpler model will perform almost as well. The poor performance of the horizontal joint model can be attributed to the tendency for strata beds to cantilever excessively, while the vertical joint only model can be considered to be less realistic. However, the best maximum subsidence result obtained (Model No. 4) was less than half that measured at the colliery, and indicates that there are other factors besides those modelled which influence the shape of the subsidence profile.

The remaining Models provide an insight into some of these factors. The poor performance of Model No. 6, simulating only 33% of the maximum subsidence obtained by Model No. 5, can be attributed more to the larger extent of strata disturbance indicated in the shallower model, rather than to the increase in depth of cover in Model No. 6.

Model No. 7 shows that maximum subsidence can be achieved over a very wide, shallow excavation. Although the extent of caving and strata disturbance in the model is vast, maximum subsidence is nevertheless obtained over only a small area. Empirical reduction of anisotropic properties improves the failure behaviour of the program as shown in Figure 7.

Finite element methods were developed to simulate small displacements over relative small mesh areas. Subsidence modelling simulates large, complex strata movement over large areas. One economic method to satisfy boundary conditions is to use large meshes as described in this paper.

Incorporation of all the features discussed above has not resulted in accurate simulation of subsidence. However, there are no alternative general purpose techniques, not the least because of the sheer impractibility of obtaining sufficient subsidence survey data over all coalfields in Eastern Australia. Improvements to finite element modelling including interfacing with other mathematical techniques does appear from the results of our research to date to be a practical proposition.

8. CONCLUSIONS

Due to the difficulties of developing reliable empirical or mechanistic models for the prediction of surface subsidence, attention has been given to the finite element technique. An essential requirement for realistic modelling is the collection and processing of geotechnical data in an accurate and reproducible manner. Material properties should be representative of in situ conditions, and data on anisotropy are important, whether or not joint elements are included in the finite element model.

Joint elements appear to facilitate caving and goafing simulation, rather than promote shear and opening movements in the strata. Since similar results were obtained from both jointed and joint-free models, it would appear that the increased accuracy afforded by inclusion of joints is not warranted in general subsidence modelling. However, all factors should be considered where other engineered structures may be affected.

It is also apparent that caving and reconsolidation simulation are crucial in determining the subsidence generated in the models. Modelling of maximum subsidence over a subcritical or critical excavation width is still difficult. The model, however, of a supercritical excavation width, demonstrated the development of maximum subsidence of about two-thirds of the extracted thickness, something not fully simulated in the earlier models. Empirical reduction of anisotropic properties does assist. This is an encouraging result, showing that, with continuing research, finite element modelling of subsidence can be successful.

ACKNOWLEDGEMENTS

Support for this project is provided by the Australian Coal Association. Permission to use subsidence and modelling data provided by Austen & Butta Ltd. and United Collieries Pty. Ltd. is also acknowledged.

REFERENCES

Daemen, J.J.K., and Hood, M. (1982). Subsidence profile functions derived from mechanistic rock models: Proc. Workshop on Surface Subsidence Due to Underground Mining, Ed Peng, S.S. and Harthill, M. Morgantown, U.S.A. pp124-140.

Frankham, B.S., and Mould, C.R. (1980). Mining subsidence in N.S.W. - Recent developments: Proc. A.I.M.M. Symposium, New Zealand, May, 167-179.

Galvin, J.M., (1982). Total extraction of coal seams - the significance and behaviour of massive dolerite sills: Chamber of Mines of South Africa, Res. Rep. 19/82.

Germanis, E. and Valliappan, S. (1975). Mining subsidence at the graving dock site, Newcastle: Symposium on Recent Developments in the Analysis of Soil Behaviour and their Application to Geotechnical Structures, Univ. of N.S.W., July, 173 - 185.

Girrens, S.P., Anderson, C.A., Bennett, J. G. and Kramer, M. (1982) Numerical prediction of subsidence with coupled geomechanical-hydrological modeling: Proc. Workshop on Surface Subsidence Due to Underground Mining, Ed Peng, S. S. and Harthill, M., Morgantown, U.S.A. pp63-70.

Goodman, R.E., Taylor, R.L. and Brekke, T.L. (1968) A model for the mechanics of jointed rock: Jour. of the Soil Mechs and Foundn Divn, ASCE, May, 1968, 637-659 (SM3).

Hansagi, I. (1973). A method of determining the degree of fissuration in rock: Int. J. Rock Mech. Min. Sci. & Geomech. Abstracts, Vol. 11, 379-388.

Hebblewhite, B.K. (1982). The use of modelling techniques for mine design: Aust. Jour. Coal Min. Tech. and Res. No. 1.

I.S.R.M. (1981). Rock Characterization Testing and Monitoring.: Ed E.T. Brown. Pergamon Press, Oxford.

Kapp, W., (1973). Mine subsidence: Proc. A.I.M.M. Symp. Subsidence in Mines Paper I.

(1982), A review of subsidence experiences in the Southern Coalfield, New South Wales, Australia in Symposium State of the Art of Ground Control in Longwall Mining and Mining Subsidence. First International Society of Mining Engineers of A.I.M.E. Fall Meeting, Honolulu.

Mikula, P.A. (1981). Prediction of surface subsidence using finite element modelling with reference to Bowen No. 2 Colliery, Queensland. ACIRL Report No. 08/0225-B.

National Coal Board, (1975). Subsidence Engineers Handbook: Production Department, London.

Pauncz, I. and Holt, G.E., (1982). Computer based geotechnical data management system for coal exploration and mine planning: Proc. Symp. Coal Resources, Origin Exploration and Utilization in Australia, Melbourne, November 15-19, 1982.

Schaller, S. and Hebblewhite, B. K. (1981) Rock Mechanics Design Criteria for Longwall Mining at Angus Place Colliery. ACIRL Published Report 81-3.

Shepherd, J., Huntington, J.F. and Creasey, J.W. (1981) Surface and underground geological prediction of bad roof conditions in collieries of the Western coalfield, New South Wales, Australia: Trans. Inst. Min. Metall. (Sect. B: Appl. earth sci.), 90, February.

THE DEVELOPMENT OF THE THEORY OF DOLERITE SILL BEHAVIOUR
Le développement de la théorie du comportement d'un sill de dolerite
Entwicklung einer Theorie über das Verhalten von Doleritschichten

J. M. Galvin, Ph.D.
Head, Coal Strata Control Section, Chamber of Mines of South Africa

SYNOPSIS

The presence of massive (>30 m thick) dolerite sills in the superincumbent strata of many South African coal mine workings has an overriding influence on the manner in which stresses are distributed around the mine workings. The success of total extraction methods depends to a large extent on inducing failure of overlying dolerite sills soon after the commencement of mining operations. Elastic thin plate theory has formed the basis for developing an equation for calculating the minimum panel dimension required to induce dolerite failure. Field observations and measurements have been utilized periodically to refine the equation, and critical panel dimensions can now be calculated very accurately.

RESUME

La présence de filons-couches massifs de dolérite (>30 m d'épaisseur) dans les strates superposées de beaucoup de mines de charbon en Afrique du Sud exerce une influence prépondérante sur la répartition des contraintes à travers l'ensemble de la mine. La réussite des méthodes de défruitement total dépend en large mesure de la capacité de provoquer la fracture des sills de dolérite superposés peu de temps après le début des travaux d'exploitation. La théorie des plaques élastiques minces a servi de base d'une équation pour calculer les dimensions minimales de panneaux servant à provoquer la fracture de la dolérite. On s'est servi de mesures et observations effectuées périodiquement sur le terrain pour raffiner l'équation de sorte qu'il est maintenant possible de calculer les dimensions critiques des panneaux avec une grande précision.

ZUSAMMENFASSUNG

Mächtige horizontale Doleritlagen haven einen überragenden Einfluß auf die Spannungsverteilung um Abbauhohlräume in südafrikanischen Kohlenlagerstätten. Ein erfolgreicher Totalabbau der Flöze hängt vom frühzeitigen Brechen dieser Doleritlagen ab. Die Theorie dünner elastischer Platten liefert die Grundlage von Berechnungsverfahren für die Vorhersage von Abbaudimensionen, die für das Brechen der Doleritbänke notwendig sind. Feldbeobachtungen und Messungen wurden periodisch durchgeführt, um das Bemessungsverfahren zu verbessern. Kritische Abbaudimensionen können mit großer Genauigkeit bestimmt werden.

1. INTRODUCTION

The presence of massive (>30 m thick) dolerite sills in the superincumbent strata of South African coal mine workings constitutes a specific, almost unique, problem. The strength and elastic modulus of such sills are an order of magnitude greater than that of normal coal measure rocks. This feature and the shallow depth (<250 m) of coal mine workings result in massive dolerite sills having an over-riding influence on the manner in which stresses are distributed around mine workings.

2. THE SIGNIFICANCE OF DOLERITE SILLS

2.1 General Significance

Investigations into the influence of massive dolerite sills on mine workings were initiated following the disastrous collapse of bord and pillar workings overlain by a 36 m thick dolerite sill at Coalbrook Collieries in 1960. The investigations were extended in 1964 to include the monitoring of massive dolerite sills overlying 'total' extraction panels. Two important points emerged from these investigations, namely:

1. The strength and elastic modulus of dolerite are an order of magnitude greater than that of normal coal measure rocks, Table 1.

TABLE 1 Typical range in strength and elastic modulus of South African coal measure rocks, as determined on 25,4 mm diameter, 76,2 mm long specimens

Stratum	Strength (MPa)		Elastic Modulus (GPa)	
	Range	Typical	Range	Typical
Shale	60- 80	70	0,5-13,0	3
Sandstone	40-100	70	4,0-15,2	7
Coal	15- 40	30	0,4- 4,1	2
Dolerite	250-390	300	50,0-98,5	70

2. Dolerite sills more than 30 m in thickness, henceforth referred to as massive, have a significant influence on mining operations. Such sills are capable of bridging over spans exceeding 150 m and thus, over whole panels if the minimum panel dimensions are too small to induce failure of the sill. The high abutment stresses induced by the weight of the uncaved strata cause extensive fracturing of the panel abutments, resulting in serious face and gateroad support problems. Air blasts, water inflow and high gas concentrations are hazards associated with the failure of a massive dolerite sill.

2.2 Bord and Pillar Workings

It is a well-known fact that the load acting on individual pillars in bord and pillar workings depends on the deformation characteristics, or stiffness, of both the surrounding strata and the pillars themselves. A simple qualitative proof of this fact has been provided by Salamon and Oravecz (1975). Briefly, it may be stated that pillar stiffness decreases with decreasing pillar width, decreasing material modulus and increasing pillar height. The stiffness of the surrounding strata decreases with increasing panel width, decreasing material modulus and decreasing strata thickness. As the ratio of surrounding strata stiffness:pillar stiffness increases the load acting on a panel pillar decreases. Furthermore, should pillar load exceed pillar strength, the possibility of a sudden, or uncontrollable, pillar collapse occurring is reduced.

It follows, therefore, that because of the high material modulus of dolerite, the stiffness of the superincumbent strata is greater when a massive dolerite sill is present within this strata. Thus, the stability of the workings will be improved. An accurate quantitative assessment of this improvement can be obtained utilizing analytical and numerical techniques, such as the analogue computer and the boundary element program 'MINAP' (Crouch, 1976). However, well established pillar design and panel layout procedures have been developed in South Africa since the Coalbrook collapse (Salamon and Oravecz, 1975) and it is not necessary to utilize these techniques when designing most bord and pillar layouts.

2.3 'Total' Extraction Workings

Total extraction workings refers to either pillar extraction operations or longwall mining operations, both of which are practised extensively beneath massive dolerite sills in South African coalfields.

Early experiences with total extraction methods revealed that initially, caving of the roof strata only extends up to the base of a massive dolerite sill and that a sill could bridge over spans exceeding 150 m, thus inducing high abutment stresses.

The magnitude of these stresses depends to a large extent on the location of the dolerite sill within the superincumbent strata, and the magnitude of the minimum panel dimension. An indication of the actual magnitude of these abutment stresses under typical South African conditions is given in Figure 1. Stresses have been calculated using the boundary element program 'DINCL' (Stephansen, 1981).

Experience has shown that once a massive dolerite sill has failed, caving of the roof strata extends to surface and then continues to keep pace with face advance. Consequently, abutment stresses are reduced considerably. When a sill does not fail Galvin et al. (1981) have shown that the maximum stress acting in the base of the sill changes insignificantly once the face advance exceeds twice the panel width. Similarly, the abutment stresses ahead of the working face also reach a maximum at a face advance of about twice the panel width.

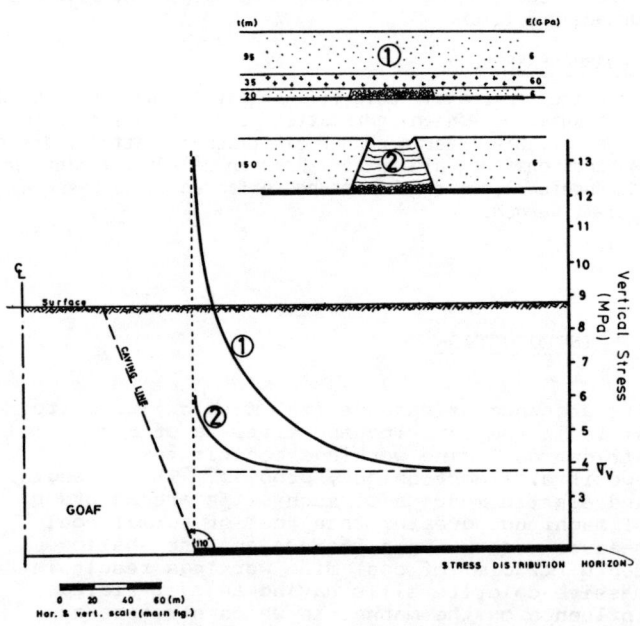

FIGURE 1 Influence of a massive dolerite sill on abutment stresses around a 'total' extraction panel

If panel dimensions are sub-critical, whereby dolerite failure is imminent but never actually occurs, high abutment stresses exist throughout the mining life of a panel. These stresses have an adverse effect on mining operations, especially in regard to face support and roadway support and maintenance. Abutment stresses may be limited by restricting panel width but this solution results in large coal losses in the form of wide interpanel pillars. Consequently, wherever practical panel width is designed to ensure failure of massive dolerite sills very soon after the commencement of mining operations.

Under South African economic conditions it is essential that panel widths are designed accurately. The price of steam coal locally is only of the order of R6 to R10 per tonne (R1=US$0,95), about one-fifth of that in other countries where longwall mining is utilized. As such, the economic viability of longwall mining in this country is marginal. If panel widths are sub-critical the cost of additional support in highly stressed gateroads and possibly on the longwall face make longwall mining an uneconomic proposition. For each metre that the panel width is overdesigned an additional R35 000 or more in capital must be outlayed. Little benefit in terms of rate of production and productivity is gained from this capital outlay since panel widths required to induce failure of a massive dolerite sill usually exceed the optimum economic panel width.

3. SIMPLE MODELS OF DOLERITE BEHAVIOUR

First attempts to calculate critical panel widths were made by Salamon et al. (1972) on the basis of elastic thin plate theory. According to this theory, the maximum stress in a sill is proportional to the factor:

$$\frac{D_D S^2}{t_D^2}$$

where D_D = depth to base of sill
t_D = thickness of sill
S = critical panel span.

Having initially determined the critical panel span, S, from experience, the critical panel span for any geometry can be determined by forming the dimensionless factor, ϕ, such that

$$\phi = \left(\frac{S_1 t_{D_2}}{S_2 t_{D_1}}\right)^2 \frac{D_{D_1}}{D_{D_2}} \qquad (1)$$

With the introduction of a number of new longwall installations in South Africa in the mid seventies it became apparent that the elastic thin plate model was unreliable under many conditions. A review of the model was undertaken, with particular attention being given to the critical stress, σ_c, required to induce dolerite failure and defined as:

$$\sigma_c = K_\gamma \frac{D_D S^2}{t_D^2} , \text{(MPa)} \qquad (2)$$

where K_γ is a (specific weight) constant (MN/m^3)

If the model was reasonably correct, the critical stress associated with each case where a sill had failed should be near constant. A review of data, Figure 2, clearly indicated that this condition did not hold. However, the review highlighted that the three high points in Figure 2, namely DNC 410, Sigma 4 and Coalbrook 1, represented situations where a dolerite sill occurred at depth. The depth: thickness ratio, D_D/t_D, of the sill for these situations was 2,0, 2,19 and 2,79 respectively. All other failed or transitional cases (refer Figure 2) had a D_D/t_D ratio of 1,5 or less. Therefore, the critical stress, σ_c, was plotted against the D_D/t_D ratio and the following relationship was derived:

$$S = \sqrt{1400 t_D - \frac{800 t_D^2}{D_D}} \quad (m) \qquad (3)$$

where the numerical values 1400 and 800 have units of metres.

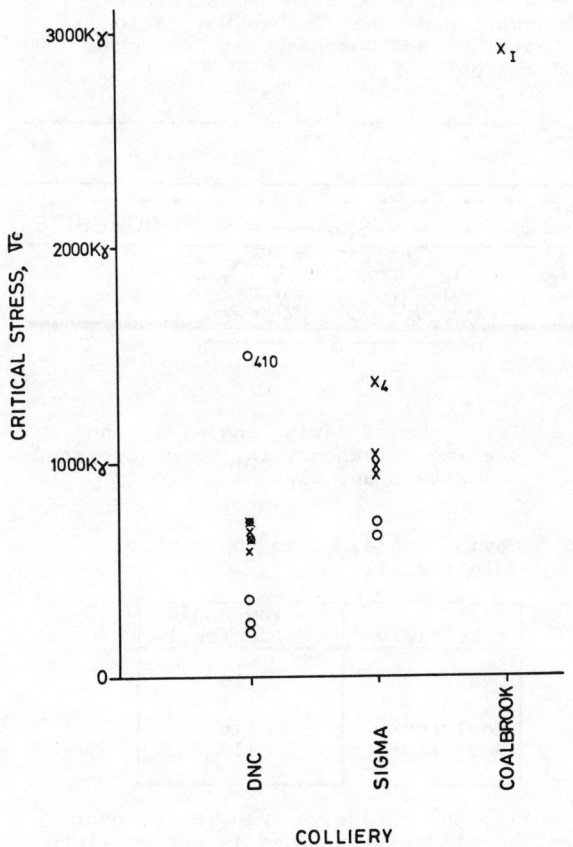

FIGURE 2 Review of elastic thin plate data

It was concluded that the term D_D/t_D accounted for the effects of weathering on the strength of a dolerite sill. At greater depth a sill would be less weathered and, thus, more competent than a sill near surface. Consequently, a greater stress would be required to induce failure of the deeper sill.

4. ADVANCED MODELS OF DOLERITE BEHAVIOUR

4.1 Approach to Developing Advanced Models

During the last three years, renewed investigations have been undertaken into strata behaviour associated with total extraction operations. The approach adopted in advancing the theory was that of viewing the mining environment as an in situ testing machine. In some instances, strata behaviour was monitored as longwall panel widths were increased until dolerite failure was induced.

4.2 Effective Span Elastic Thin Plate Model

One of the more important points to emerge from the new investigations was that the effective unsupported span of a sill is influenced significantly by the caving angle, β, and the parting thickness, t_p, of the strata between a seam and a sill. By definition, the caving angle, β, is measured relative to the plane of the seam ahead of the face, Figure 3. The average caving angle associated with the strata overlying longwall panels at Durban Navigation, Sigma, Coalbrook and Bosjesspruit Collieries is recorded in Table 2.

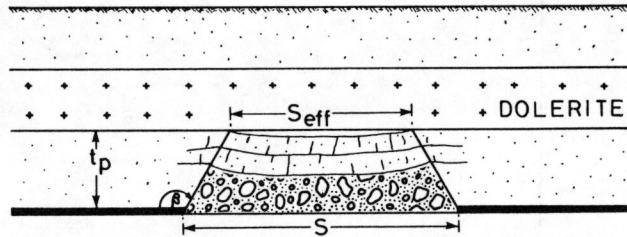

FIGURE 3 Influence of caving angle, β, and parting thickness, t_p, on unsupported dolerite span, S_{eff}

TABLE 2 Caving angle, β, to the base of a dolerite sill

Colliery	Caving Angle β (degrees)
DNC	110
Sigma	117
Coalbrook	110
Bosjesspruit	119

Consequently, the effective span, S_{eff}, over which an unfailed sill bridges is not equal to the panel dimension, S, but is defined by the equation

$$S_{eff} = S - 2t_p \mathrm{Tan}(\beta - 90) \quad (m) \quad (4)$$

The influence of the effective span, S_{eff}, and the caving angle, β, on dolerite behaviour is well illustrated by subsidence observations made above longwall panel 411 at DNC, Figure 4. The face width of this panel was increased in a number of steps during the non-simultaneous extraction of both an upper and lower seam separated by about a 1 m thick stone parting.

During extraction of the upper seam an effective span of at least 130 m was required to induce substantial subsidence of the sill. However, even an effective span of over 140 m was not sufficient to induce total collapse of the sill. Total collapse only occurred some two years later when caving was reactivated during the extraction of the lower seam.

Experience has shown that the caving angle steepens with time. Thus, reactivation of caving during lower seam mining results in an increase in the effective unsupported span of the dolerite sill.

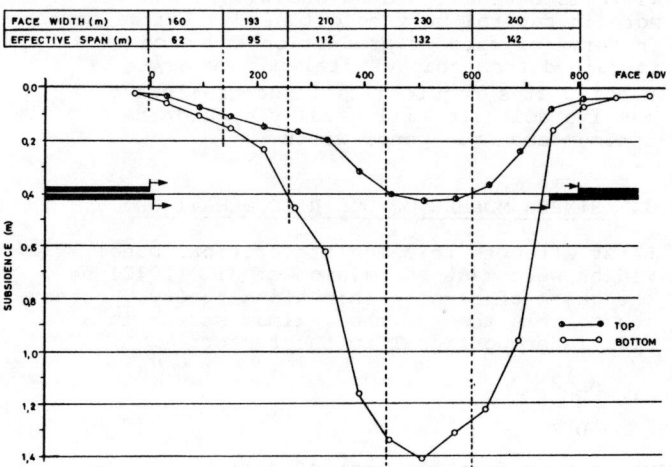

FIGURE 4 Influence of effective span on surface (dolerite) subsidence at DNC

In the light of this finding and additional longwall information, Equation (4) has been updated (Galvin, 1981) to the Effective Span Elastic Thin Plate Model, defined by the equation

$$S = \sqrt{1165 t_D - 935 \frac{t_D^2}{D_D}} + 2 t_p \mathrm{Tan}(\beta - 90) \quad (m) \quad (5)$$

where the numerical values 1165 and 935 have units of metres.

The difference between the actual span at which a sill failed and that as calculated from this new model ranges from one under-estimation of only 3 m to an over-estimation of 33 m.

4.3 Loading of a Sill

Although the Effective Span Elastic Thin Plate Model provides an accurate means for calculating the critical panel width, it does not model accurately the loading on a dolerite sill. The elastic thin plate models assume that a sill is loaded uniformly by a load equal to γD_D, where γ is the specific weight of the strata. As such, this loading condition takes no account of the stiffness of the dolerite and the overlying strata, which serves to transfer load onto the abutment areas. Furthermore, even if deadweight loading were the case, a load equal to γD_D would only act at the base of the sill and not throughout the sill. Thus, the load acting on a dolerite sill is overestimated considerably by the elastic thin plate models.

Further investigations into the stiffness of the strata loading a dolerite sill have revealed that the D_D/t_D ratio is, in fact, primarily a measure of this stiffness and not of the degree of weathering of the strata. Analytical analysis has shown that for a D_D/t_D ratio between 1 and 1,6 , a negligible error results if the external load is equated to the deadweight of the overlying strata. But, for an increasing D_D/t_D ratio greater than 1,6 , the external load becomes an increasingly smaller fraction of the deadweight load until at $D_D/t_D=4$, the external load is almost zero.

5. LIMITATIONS OF ELASTIC THIN PLATE MODELS

The elastic thin plate model was derived on the basis of limited data and was intended as a quick and simple 'analytical' solution to a rather complex problem. Although the model has been refined and developed in the light of further information, it still represents a simple solution. Therefore, it is not surprising that a number of anomalies are associated with the model, one of the most important being the failure mode of a sill..

On the basis of elastic plate theory, caving (of the base) of a sill, initiated by inducing the effective critical stress, σ_c , in the sill, results in an increase in the stress acting in the remainder of the sill. Thus caving should propagate rapidly through the sill. However, recent measurements and borehole camera observations have shown that a dolerite sill fails as a number of distinct plates, each plate subsidence being separated by a considerable period of time, Figure 5. Bulking occurs at the interface between plates but is negligible within a plate.

Furthermore, plates may fail not only in bending at the midspan of the undermined sill, as prescribed by elastic plate theory, but also in shear at the abutments of the undermined sill. The expression of a shear failure on the surface topography is shown in Figure 6.

Another interesting point is the area over which a thin dolerite plate can span. For example, the plate of dolerite containing anchors No. 5 and No. 6, Figure 5, was only between 7,6 m and 15,5 m in thickness yet this thin plate supported itself and the weight of 36 m of overlying strata

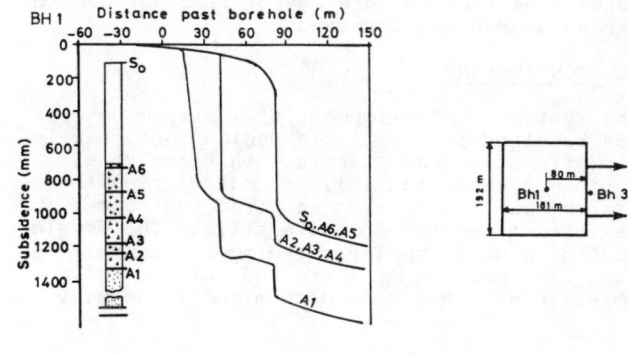

FIGURE 5 Progression of failure through a massive dolerite sill, Sigma Colliery

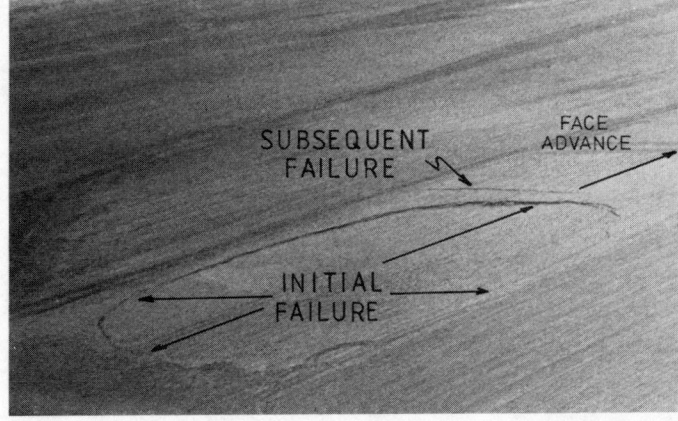

FIGURE 6 Shear failure of a massive dolerite sill

over spans in excess of 150 m. This aspect is being researched currently and early results indicate that this behaviour may be associated with the redistribution of horizontal stresses that occurs after each plate failure.

Dolerite sills contain a high density of well defined near vertical joints and poorly defined near horizontal joints. This jointing appears to control both the thickness of a dolerite plate and the manner in which it fails, that is, either in bending or shear. Following a plate failure, horizontal stresses are redistributed in the remaining portion of a sill, thereby increasing the clamping forces acting across joint planes. Thus, the stability of the remaining portion of the sill is improved. Further mining must take place before the weight of the undermined strata is sufficient to overcome the horizontal clamping forces within the sill and result in failure.

Obviously, the presence of well defined near vertical and near horizontal joints will have an over-riding influence on the occurrence of failure. Should such joints be absent over a large area then failure may still occur due to excessive bending stresses.

6. CONCLUSIONS

The systematic development of a simple concept has resulted in a reliable design tool, namely, the Effective Span Elastic Thin Plate Model defined by Equation (5), for calculating the minimum panel dimension required to induce failure of massive dolerite sills. The development of a more sophisticated model can only be justified once the failure mode of a massive dolerite sill has been determined in detail.

REFERENCES

CROUCH, S.L., 1976: Analysis of Stresses and Displacements Around Underground Excavations: An Application of the Displacement Discontinuity Method. Dept. of Civil and Mineral Engineering, University of Minnesota, November 1976.

GALVIN, J.M., STEIJN, J.J. and WAGNER, H., 1981: Rock Mechanics of Total Extraction. Lecture Notes for the South African Institute of Mining and Metallurgy Vacation School, 1981, entitled 'Increased Underground Extraction of Coal'.

GALVIN, J.M., 1981: The Mining of South African Thick Coal Seams - Rock Mechanics and Mining Considerations. Ph.D. thesis. University of the Witwatersrand.

SALAMON, M.D.G. and ORAVECZ, K.I., 1975: Rock Mechanics in Coal Mining. Chamber of Mines of South Africa. P.R.D. series No. 198, 1975.

SALAMON, M.D.G., ORAVECZ, K.I. and HARDMAN, D.R., 1972: Rock Mechanics Problems Associated with Longwall Trials in South Africa. Chamber of Mines of South Africa, Research Report RR.6/72.

STEPHANSEN, S., 1981: Stress Analysis in Inhomogeneously Layered Rock. M.Sc. thesis to be submitted to the University of the Witwatersrand, Johannesburg.

ANALYSIS OF SURFACE SUBSIDENCES CAUSED BY TWIN TUNNELS BUILT IN COHESIVE SOILS

Analyse des affaissements de surface au-dessus des tunnels doubles construits dans un sol cohérent

Analyse der Oberflächensenkungen, hervorgerufen durch in bindigen Böden gebaute Doppeltunnel

Dr. G. Petrasovits
Professor, Department for Geotechnique, Technical University of Budapest, Hungary

SYNOPSIS
The paper analyses the major factors influencing the amount and occurrence of surface subsidences observed above twin tunnels built by the shield method. The role of soil conditions, construction technology and tunnel structure as well as their interaction are demonstrated.

RESUME
Cette communication analyse l'effet des éléments principaux influençant la grandeur et le développement des affaissements de surface au-dessus des tunnels doubles construtis par la technique du bouclier. L'étude révèle le rôle des conditions de sol, de la technique de construction et de la structure des tunnels dans le cas des affaissements de surface au cours de la réalisation de structures souterraines ainsi que leurs effets les uns sur les autres.

ZUSAMMENFASSUNG
Die Abhandlung analysiert die wichtigeren Faktoren, die die Größe und Entstehung der Oberflächensetzungen bei über mittels Schildvortrieb gebauten Doppeltunneln beeinflussen. Die Rolle der Bodenverhältnisse, der Ausführungstechnologie und deren Zusammenwirken werden aufgezeigt.

Introduction

At the tunnels built in urban areas, the basic task is to keep the surface subsidences between such limits prevent any damage to surface or underground structures in the vicinity of the tunnel beside the ensurance of the stability and allowable deformations of the tunnelstructure. The selection of a tunnel structure which gives the best solution from the safety and aconomical aspects, suits the given soil conditions and corresponds to the allowable settlements is a very complex task which can be solved only by the possibly separate analyses - but combined evoluation - of the above mentioned three factors. One of the considerable conditions of the proper settlement analysis is the scheduled settlement measurement of a benchmark independent of any surface facilities combined with the accurate location of the shield and the knowledge of soil conditions. In the generalization of the obtained results and relations the numerical calculation methods play an important part. They allow the independent analysis of the individual factors, the determination of the amount and distribution of stresses developed in the tunnelstructure and the soil and the observation of deformations.

Beside the empirical relation to forecast the surface subsidences above twin tunnels the article also presents a method based on the finite elements considering the characteristics of the construction technology too.

1. Review of soil conditions, construction technology and structure

The investigated section of the underground line lies on the eastern side of the Danube at Budapest where the overburden consist of sand and gravelly sand deposits to a depth of 8 to 12 m. The groundwater is encountered 3-5 m from the surface. The sandy deposits are underlain by soft clay strata with sand laminae changing to a hard clay layer with depth. Its natural water content decreases with depth from 25 % to 15 %. Its index of platicity is in the order of I_p=20-30 % near the sandy deposits while it ranges from I_p=30 to 40 % at a depth of 20 m. The unconfined compression strength of the soil is σ =10-30 kN/m2 immediately below the sandy overburden but it increases to σ =30-45 kN/m2 15 m deeper. The modulus of compressibility of the cohesive soil increases from 10 Mpa with depth and reaches a value of 30 MPa at a depth of 30 m. The tunnels were built by means of shield method in the clay layer by keeping a minimum clay cover of 3 m above the tunnel.

The outside diameter of the shield is D_s=5,62 m, and the length L_s=4,9 m. The construction was carried out under compressed air from 0,7 to 1,4 bar. The two tunnels have on outside diameter of D=5,5 m each consisting of seven uniform pieces of 20 cm thick precast r/c segments and one smaller closing key element connected by hinges. The c-c distance of the two tunnels ranges between 20 and 22 m.

2. Analysis of surface subsidences along the longitudinal axis of the twin tunnels

The relationship between shield construction technology and surface subsidence can be investigated in the longitudinal profile of the tunnel. The surface subsidence in the shield technology is composed of the intrusion of soil at the tunnelface and at the shield tail. The influence of construction technology in the total settlement can be definately proven in the elastic or plastic-elastic movement of the tunnelface. In the longitudinal profile of the tunnel, surface subsidences can be observed already when the cutting edge of the shield is considerably away. This distance increases by the depth of tunnel below the surface.

It should be mentioned the surface movement can be in some cases even heaving, mainly at thin covering layer. This phenomenon was observed in a section of some metres in front of cutting edge at the metro construction of Budapest too. This can be explained by the increase of the resistance of jacks supporting the face and by the increase of mantle friction which are due to twisting and change in direction of the shield. When the shield moves ahead soil is intruded into the gap between the opening and the tunnel structure. The amount of soil is influenced by the type of soil; the thickness of covering layer above the tunnel, quality of the grouting, the rigidity of the tunnel structure and stress condition around the tunnel.

It will be indicated later that at the traditional shield construction method the surface subsidences can not be decreased effectively by the increase of the rigidity of tunnel lining.

The tunnel section built by shield technology runs at a depth of 10-30 m below the surface. The thickness of covering clay layer above the tunnel ranged from 10 to 20 m.

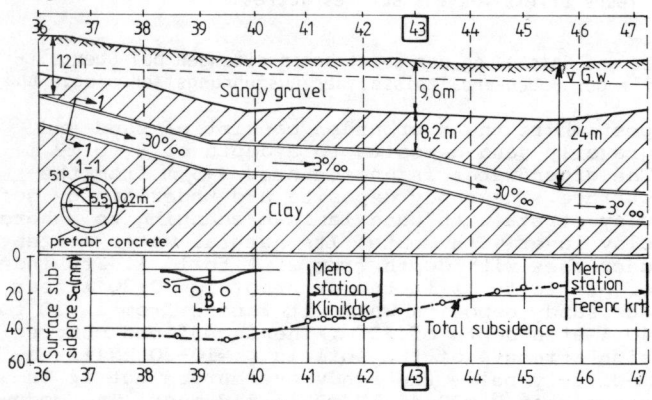

Fig.1.

Fig.1. presents one section of the South-North metroline showing the alignment and the stratigraphy and the final surface subsidences. Attention is called to the correlation between the subsidences and the thickness of clay cover.
It should be mentioned that no correlation was found between the thickness of the granular layer and the amount of surface subsidence.
The detailed analysis of surface subsidence is demonstrated on Fig.2. for Sta.43+00. Here the granular layer is 9,6 m and the cohesive layer is 8,2 m thick. Fig.2/1 presents the process of surface subsidence above the "L" left tunnel axis versus the distance from the investigated chainage and also the additional settlements above the "L" tunnel axis caused by the "R" right tunnel. After passing of both shields at the investigated section the $S_1 = 16,0$ mm subsidence occured.
The figure clearly shows that the surface subsidence commences when the "L" shield approaches the investigated section to a distance of 30 m and that this value increases to 3 mm about one third of the total subsidence at the passing of the first shield $/S_1^L = 9$ mm/. The maximum intensity of

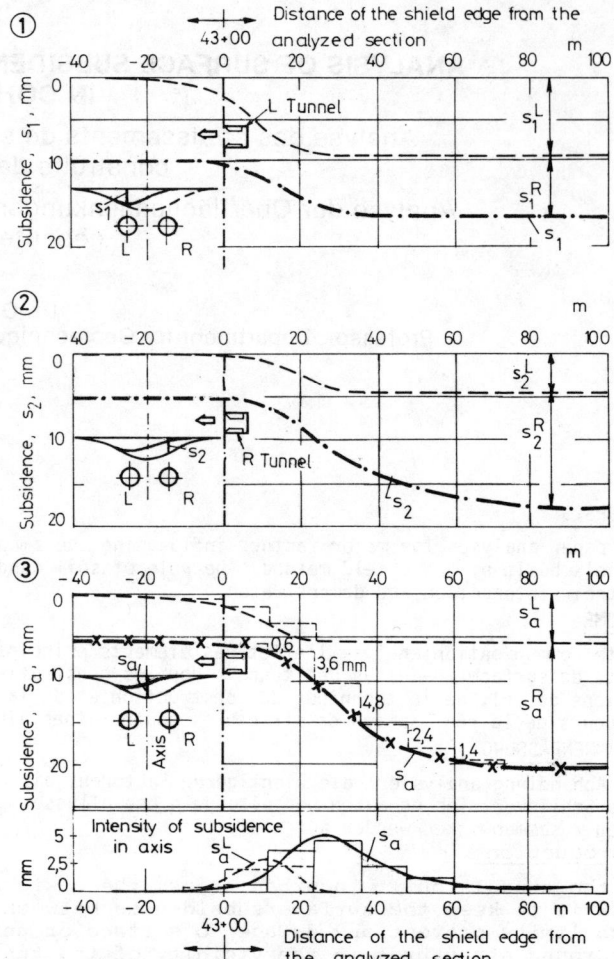

Fig.2.

the surface subsidence above the tunnel is reached when the shield edge is at a distance of 1-2 times its length. Fig.2/2 shows the same phenomenon above the "R" right tunnel axis with the calculated settlement values plotted on the $/S_2/$ curve. Fig.2/3 presents the change of surface subsidence and its intensity observed in the centerline of the twin tunnel in function of the shield distance from the analysed section. As seen in the figure the intensity of surface subsidence in the common centerline accelerates at a greater distance following the passing of the shield and the maximum intensity /4,8 mm/ is reached when the shield is at a distance of 4-5 times its length. The lower part of the figure presents the change and amount of settlement intensity. The delated surface subsidence is due to the considerable distance /22 m/ of the tunnels.

3. **Comparative analysis of the extension and shape of settlement trough**

The process of development of settlement trough in the centerline of the twin tunnel is presented in Fig.3. for the selected sta.43+00. Fig.3/1 shows the extension of settlement trough caused by the passing of "L" shield. Fig.3/3 indicates the amount and extension of surface subsidence due to the passing of both "L" and "R" shields. The difference of the two settlement trough represents the surface subsidence caused by the "R" shield /Fig. 3/2/.

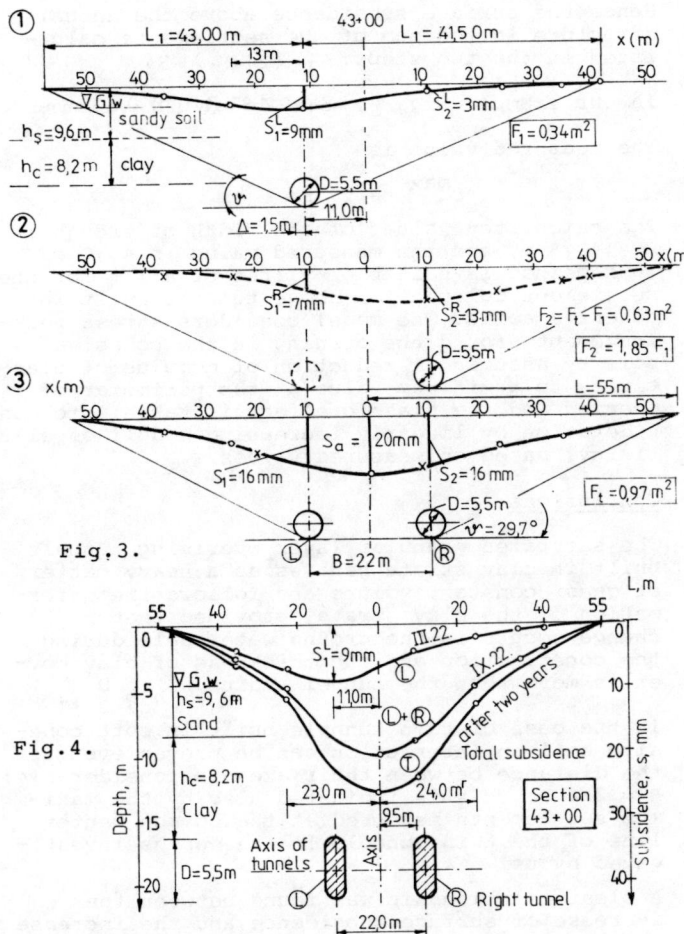

Fig.3.

Fig.4.

Fig.4. presents the final surface subsidence measured after the consolidation, i.e. two years after the completion of construction.
Evaluation of the great many settlement data shows /Fig.5./ that the settlement curve can be conformed closely by the Gaussian probability distribution function. The average curve of the surface subsidence measured and evaluated on the two metrolines can be closely conformed by the following equation

$$\frac{S}{S_a} = e^{-c\left(\frac{x^2}{L^2-x^2}\right)^n}$$
where $c = 2.135$
$n = 0.712$

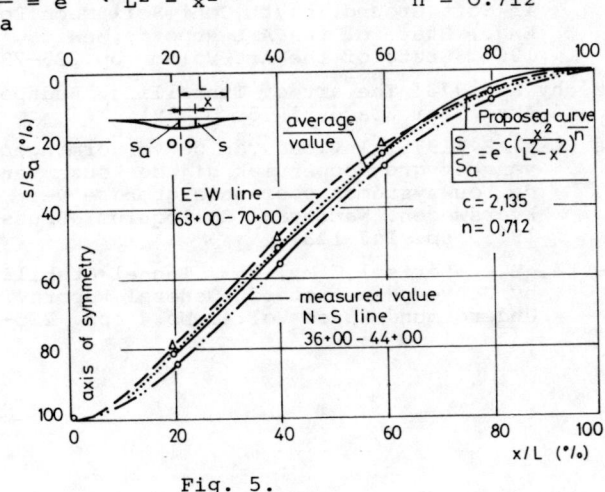

Fig. 5.

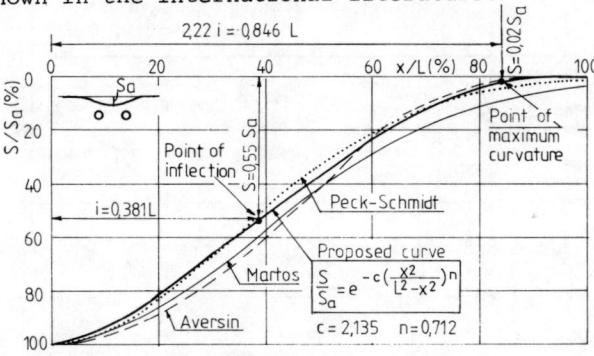

Fig.6.

Fig.6 presents beside the characteristic points of the curve suggested by Author, also the curves proposed by Peck-Schmidt, Aversin and Martos, well known in the international literature.

4. Prediction of surface subsidences

Empirical approach

Based on the statistical evaluation of a considerable number of cross-section settlement data, the curve presented on Fig.7 can be easily used to predict the surface subsidences to be expected above twin tunnels built in medium clay layers in function of the thickness of clay cover. It can be obtained by the following relation in the case of 3-20 m thick clay cover:

$$S_a^{mm} = \frac{271}{h_{clay}} /m/$$
where h_c = thickness of clay layer in meter.

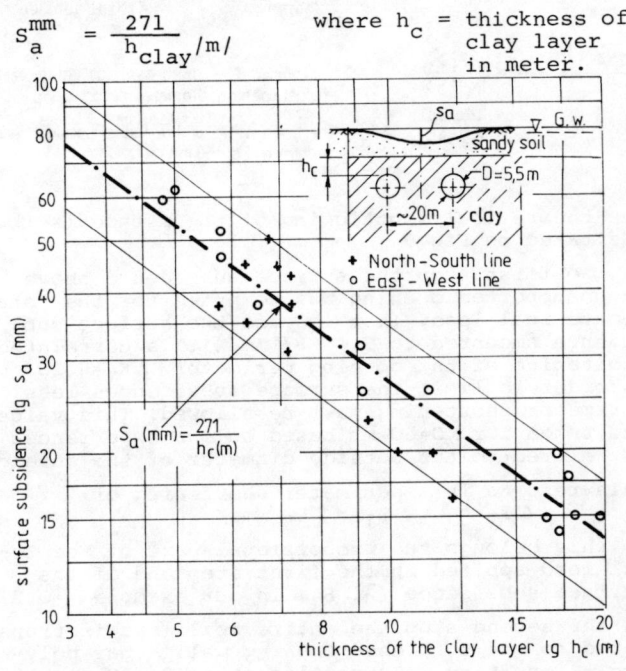

Fig.7.

Theoretical approach

Finite element method is suitable for the determination of deformations and stresses at any place selected. Although the model characterises the soil as elastic medium by its two elastic parameters /E/, their modification from element allows the consideration of its inhomogenity. Fig.8. shows the case of twin tunnels. At the assumption of the network two limitation problems occur: 1. the width of the region must be so great that the movements at the edge of region should be negligeable /12 D=60 m/, 2. the thickness of layer under tunnel is limited. Greater values than 3 D have no influence on the subsidences.

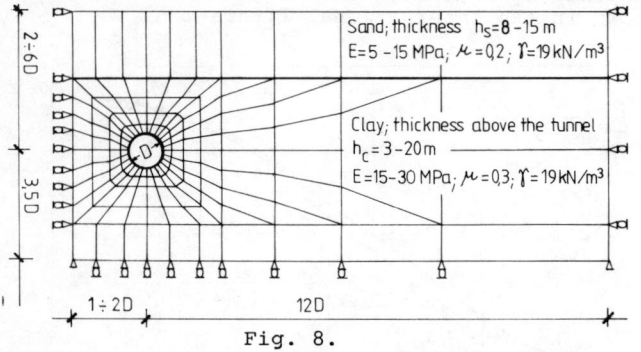
Fig. 8.

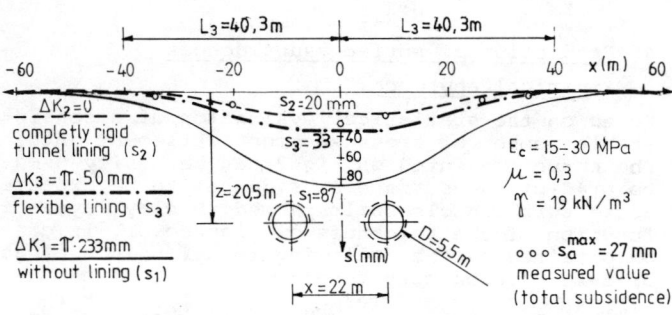
Fig. 9.

The theoretical functioning of the method is illustrated on Fig.9.

In the first step the surface subsidence above an unsupported opening was computed for the total geostatical load. In this case the surface subsidence amounted to $S_1 = 87$ mm with a correlated shortening of the opening perimeter $\Delta K_1 = \pi \cdot 233$ mm. In order to limit the surface subsidences less perimeter shortening must be allowed. This value was taken for $\Delta D = 50$ mm based on own experiences as related to the outside diameter of the shield.

This results in a perimeter shortening of
$\Delta K_3 = \lambda \cdot \Delta K_1 = \pi \cdot 50$ mm.

To this belongs the proportional part of the total load applied in the first step and of the surface subsidence /$\lambda \cdot S_1$/ in our example $\lambda = 0,21$.

In the second step the entire soil load is transfered to the stiffened opening wall - bar polygon - resulting in a settlement of $S_2 = 20$ mm without any further perimeter shortening /$\Delta K_2 = 0$/.

Since only the load remaining from the first step /$1-\lambda$/. p - in our example it is 79 % - acts on the opening with the stiffened wall, therefore the surface subsidence occuring in the second step appears only as the proportionally decreased part /$1-\lambda$/ . S_2 of S_2.

Hence the surface subsidence above the actual structure is the sum of the settlements calculated in the two steps: $S_3 = \lambda \cdot S_1 + (1-\lambda) \cdot S_2$

In our example Fig. 9. $S_3 = 0,21 \cdot 87 + (1-0,21) \cdot 20 = 33$ mm

The measured value is
$$S_3^{max} = 27 \text{ mm}$$
The calculated values of the width of trough is $L = 40,3$ m, and the measured value is 43,0 m. This method is suitable - first of all - for the settlement calculation above tunnels built in cohesive soils. The model considers stress rearrangement around the opening in the cohesive soil by assuming a reduction of modulus of elasticity to a minimum value at the perimeter of opening and the plastic effect is taken into consideration by linearly increasing moduli of elasticity, based on measured values.

Conclusions

The saturated granular layer overlying tunnels built in clay strata behaves as a heavy matter of quasi constant volume and follows the deformation of the clay strata, provided that no change occurs in the ground watertable during the construction and the thickness of clay cover is more than the tunnel radius /$h_c \geq \frac{D}{2}$/.

In the case of twin tunnels built in soft cohesive soil an interaction can be proven even if the distance between their axes is considerable; B = 3-4 D. This is indicated also by the maximum settlements measured at the common centerline of the twin tunnels in the depths investigated by us.

A clear relationship was found between the decrease of surface subsidence and the increase of clay cover thickness which can be readily applied in the prediction of surface subsidences to be anticipated /Fig. 7./.

The method of finite elements can be reasonably used for the prediction of surface subsidences based on the reliable determination of soil parameters.

References

Peck, R.B./1969/ Deep Excavations and Tunnelling in Soft Ground.In:7th Conf.Soil Mech.Found. Eng./ State of the Art Report/,Mexico, 1969. State of the Art Volume pp.225-290.

Széchy,K./1973/ The art of Tunnelling. Budapest, Akadémiai Kiadó, 1973. p. 891.

De Beer, E./1975/ Invloed van de vervormingen van het grondoppervlak bij de keuze van de bouwsystemen voor ondergrondse verkeerswegen. Nationaal Colloquium, Brussel, 1975. pp. 107-123.

Gesta, M.P.-Kerisel,J.M./1980/ Tunnel Stability by Convergence method /General Report/ Underground Space Vol.4. No.4. pp. 225-232.

THERMAL PROPERTIES OF STRESSED ROCKS
Propriétés thermiques des roches sous compression
Thermische Eigenschaften von belasteten Gesteinen

S. Ehara, M. Terada and T. Yanagidani
Faculty of Engineering, Kyoto University, Kyoto, Japan

SYNOPSIS

The thermal cyclic heating and cooling of both stressed and unstressed rocks were conducted in order to see the relation between thermal expansion and thermal cracking. The effects of the thermal cycling cracks on the thermal expansion were apparent under atmospheric pressure. The anisotropic expansion observed in stressed granites is explained by the suppression of thermal cycling cracks perpendicular to the loading direction. Temperature-independence of the thermal expansion coefficient in the loading direction is also caused by the uniaxial stress.

RESUME

On a procédé à un échauffement et à un refroidissement thermiques cycliques des roches sous pression atmosphérique ou sous compression afin de déterminer la relation entre l'expansion thermique et la fissuration thermique. Les effets des fissures de cycle thermique sur l'expansion thermique sont apparus à la pression atmosphérique. L'expansion anisotropique observée dans les granites sous compression s'explique par la suppression des fissures de cycle thermique perpendiculaires à la direction de la charge. L'indépendance à la température du coefficient d'expansion thermique dans la direction de la charge est également provoquée par l'effort uniaxial.

ZUSAMMENFASSUNG

Bei diesem Versuch wurden in einem Zyklus thermische Erhitzung und Abkühlung von belastetem und unbelastetem Gestein durchgeführt, um die Beziehung zwischen thermischer Dehnung und thermischer Rißbildung zu ermitteln. Dabei waren die Auswirkungen der thermischen Zyklusrisse auf die thermische Dehnung unter Luftdruck zu erkennen. Die in belastetem Granitgestein beobachtete richtungsabhängige Dehnung läßt sich auf die Unterdrückung von thermischen Zyklusrissen senkrecht zur Belastungsrichtung zurückführen. Die Temperaturunabhängigkeit des thermalen Dehnungskoeffizienten in der Belastungsrichtung ist ebenfalls durch die lineare Belastung bedingt.

1. INTRODUCTION

It is important to know the mechanical and thermal properties of rocks in order to design a rock cavern for the storage of LNG and to exploit geothermal energy.
When a rock is subjected to temperature change, formation and extention of crack can occur and the intrinsic properties may change even if temperature gradient within it is very low. This thermally induced cracking is produced by inhomogeneous strain due to the differences of the thermal expansion coefficients among the rock forming minerals and anisotropy of the thermal expansion inherent in each mineral (Simmons et al., 1978). There are two methods to measure thermal cracking: one is to measure the occurrence of acoustic emission (AE) directly (Johnson et al., 1978, Yong et al., 1980), and the other is to monitor the thermal expansion behaviour (Cooper et al., 1977). In previous investigations concerning thermal cracking, the thermal expansion and AE activities for the rock specimens subjected to a heating cycle, were mainly observed under atmospheric pressure. In this investigation, we carried out the thermal cyclic experiments with slow heating and cooling rate to study the relation between thermal expansion and AE activities of stressed and unstressed rocks.

2. EXPERIMENTAL PROCEDURES AND APPARATUS

The rocks studied were Westerly and Oshima granites and Murata basalt. Cylindrical samples (30mm in diameter and 75mm long) with ends grounded parallel within ±0.01mm were used. They were air dried for more than a month.
The schematic diagram of this testing system is shown in Fig. 1. Two extension rods could be moved into the chamber for pressing the sample. Temperature was measured by the thermocouples mounted on the sample and on the standard. Stress of the sample was measured by a load cell insulated thermally from the chanber using water circulation around the extension rods. A PZT transducer (diameter : 6mm, resonant

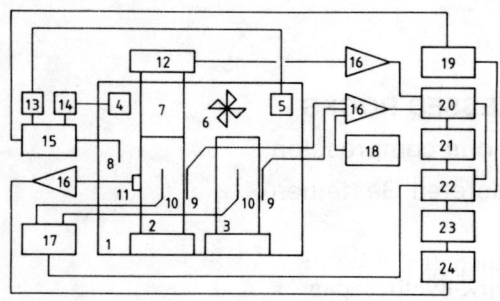

Fig. 1. Experimental apparatus

1: chamber, 2: rock sample, 3: standard sample, 4: heater, 5: nozzle, 6: fan, 7: extension rod, 8: platinum resistance thermometer, 9: thermocouple, 10: strain gage, 11: AE transducer, 12: load cell, 13: valve, 14: switch for heater, 15: temperature controller, 16: amplifier, 17: digital strainmeter, 18: ice bath, 19: DA converter, 20: switch, 21: digital voltmeter, 22: microcomputer, 23: counter, 24: discriminator

frequency : 1MHz) was attached to the sample to detect AE signals, that were amplified (40dB), then band-pass filtered (10kHz-1MHz). AE events were counted using a discriminator (threshold level 25mV) and a universal counter. Thermal expansion was measured by using the resistance strain gages (Thirumalai et al., 1970, Wong et al., 1979). In this experiment, we must eliminate all temperature-induced apparent strains except for the thermal expansion of the sample. The net temperature-induced apparent strains of the sample and the standard (Poore et al.,1978) are

$$\varepsilon_a = (\alpha - \alpha_g + \beta/F)(T_1 - T_0) \quad (1),$$

$$\varepsilon_a' = (\alpha' - \alpha_g + \beta/F)(T_1' - T_0') \quad (2),$$

where, ε_a: the apparent strain of the gage on the sample, α: coefficient of thermal expansion of the sample, α_g: coefficient of thermal expansion of the gage and gage base including adhesive material, β: temperature coefficient of resistivity of the grid conductor, F: gage factor, T_0: initial temperature of the sample, T_1: temperature of subsequent stage, and primes refer to the standard.

If the temperature of the sample is nearly equal to that of the standard or α and α' are independent of temperature from T_0 to T_1, the following expression for α can be obtained:

$$\alpha = \varepsilon_a/(T_1-T_0) - \varepsilon_a'/(T_1'-T_0') + \alpha' \quad (3).$$

To reduce error, fused silica was employed as the standard material because of its low-expansion. And its thermal conductivity is nearly the same value as those of rocks (Clark, 1966).

To study the relation between thermal expansion and thermal cracking, we carried out the following three sets of experiment. The rock samples were independently 1) heated cyclically (300-370K) with a slow heating rate (within 0.5K/min) and 2) cooled (300-210K), and 3) subjected to cyclic heating under uniaxial compressive stress. The experimental procedure was to heat or cool the rock sample and the standard materials in steps, allowing thermal equilibrium to occur at each intervening temperature; a complete thermal cycle consisted of stabilizing at five different temperature. The heating and cooling rates during the cycle were approximately 0.5K/min, not including the stabilization period (typically 120min). The accuracy of α was determined within 5% and 15% for the heating experiment, and the cooling experiment, respectively.

3. RESULTS

Relation between AE activities and hysteresis

α were calculated by using Eq. (3) over each of the four temperature intervals and plotted at the midpoint of them. In Fig. 2, averaged value for α in the axial direction of Westerly granite (No. 104) are plotted against temperature. Although the thermal expansion coefficients are independent of a previous thermal exposure below 330K, α is temperature dependent; it increases in proportion to temperature. There was a clear temperature dependence of α in the circumferential direction as well as in the axial direction and were hysteresis in the first cycle and residual strain (about 4×10^{-5}) at the end of the cycle. Accumulated events of AE versus temperature plot are presented in Fig. 3. Above about 330K, AE activities gradually increase with temperature in the first heating cycle. But in the second cycle, minor activities exist. Very few events occurred in the cooling part of each cycle. Clear correspondence between AE activities and hysteresis was observed for Westerly granite in the heating experiment. Above 330K, thermal cracking initiated in the first cycle and hysteresis appeared.

Thermal cracking in the low temperature

Figure 4 shows the relationship between α and temperature for Westerly granite (No. 105) in the cooling experiment. α decreases in proportion to temperature in the cooling part of the first cycle and increase with temperature in the subsequent cycles except for

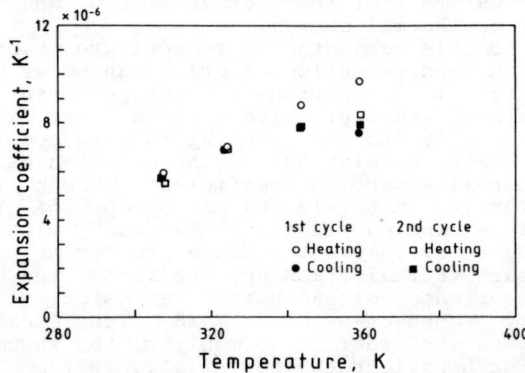

Fig. 2. Thermal expansion coefficient in axial direction vs. temperature in heating experiment for Westerly granite (No. 104).

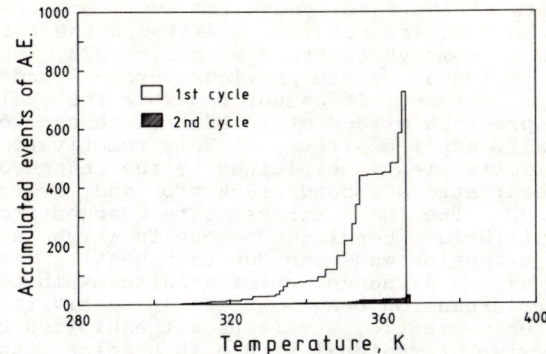

Fig. 3. Accumulated AE events of heating phase vs. temperature in each cycle of heating experiment for Westerly granite (No. 104).

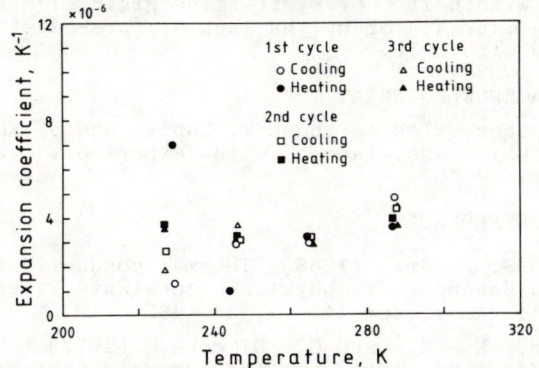

Fig. 4. Thermal expansion coefficient in axial direction vs. temperature in cooling experiment for Westerly granite (No. 105).

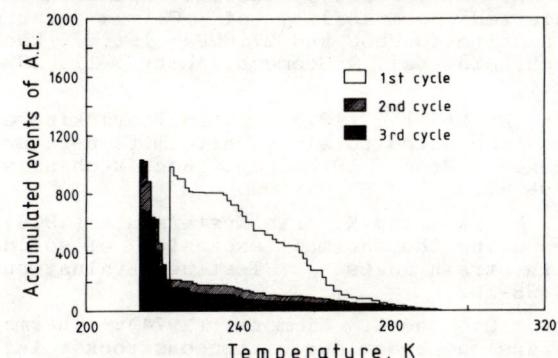

Fig. 5. Accumulated AE events of cooling phase vs. temperature in each cycle of cooling experiment for westerly granite (No. 105).

in the range below 250K. In this temperature range, thermal strain showed small hysteresis in the subsequent cycles. Figure 5 shows the relation between the accumulated AE events and temperature. Below about 270K, AE activities progressively increases in the first cooling cycle. Very few events occurred in the heating part of this cycle as was seen in cooling part of the heating cycle. Correlation between AE activities and hysteresis was observed but this correlation was less clear than that in the heating experiment. In the second and third cycle, although hysteresis was observed whole the cycle, AE activities was restricted about bottom of each cycle.

Stress effect on the thermal cracking

Figure 6 shows the relation between α and temperature in the axial (loading) direction under constant stress of 73MPa (about a third of its strength) for Westerly granite (No. 103). α is less dependent on temperature and α of the first heating cycle is the lowest of all. The increase in α in the circumferential direction to temperature was similar to that without external stress, but a smaller residual strain (about 1×10^{-5}) was observed. Accumulated events of AE in the first heating cycle under uniaxial stress and in the first reheating cycle after the stress was removed are plotted against temperature in Fig. 7. A large number of events occurred under external stress as well as in intact rock in the first heating cycle, and very few events occurred in the second cycle. But in the reheating cycles after the stress was removed, AE activities again increased in number in the first cycle and then it became smaller in the second cycle. After the stress was removed, α during the first reheating cycle is the highest among those of the four cycles. The similar temperature dependencies of α as observed in intact rock was investigated. Residual strains after the first cycle of reheating were about 7.6×10^{-5} in the axial direction and 2.3×10^{-5} in the circumferential direction, respectively.

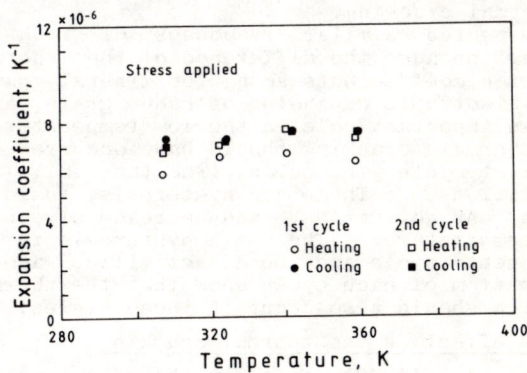

Fig. 6. Thermal expansion coefficient in axial direction vs. temperature in heating experiment under uniaxial compression for Westerly granite (No. 103).

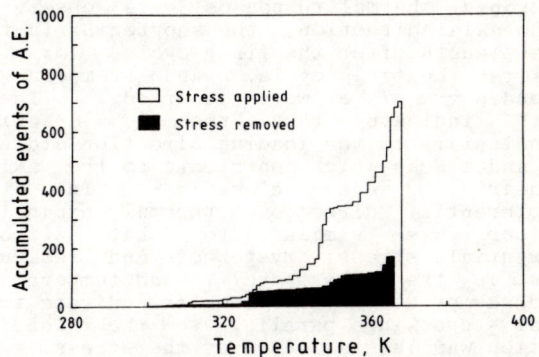

Fig. 7. Accumulated AE events of heating phase vs. temperature in 1st cycle when the stress was applied and removed in the heating experiment for Westerly granite (No. 103).

4. DISCUSSION

Thermal cracking and thermal expansion

When the temperature is raised in the first cycle in heating experiment, an intergranular stress due to non-uniform expansion of each grain (Simmons et al., 1978) increases in magnitude. If it exceeds the local strength at the grain-boundary, it causes thermal cracking. High α in the first heating cycle, hysteresis, and residual strain were attributed to the thermal cracking, because it increases crack porosity (Richter et al., 1974). In the second cycle of the same temperature range, the presence of cracks reduces the local stresses and decreases α by allowing some mineral grains to expand into the cracks (Cooper et al., 1977). The occurrence of thermal cracking and the irreversible change of the rock properties were apparent from hysteresis and residual strain in the first cycle and high α (above 330K) in the first heating cycle observed in two granites in heating without differential stress. Gradual increase of AE activities initiated around above 330K in the first cycle was also the result of thermal cracking, for it is a phenomenon related to crack extension (Yong et al., 1980). Correlation among these results was excellent and AE activities and the change of α seemed to be a good indicator of thermal cracking.

We expected similar responses of rock in cooling because the difference of the thermal expansion coefficients among the mineral grains and anisotropic expansion of each grain also play an inportant role in the low temperature. The thermal cracking should have occurred in the first cycle in cooling from the following observations. These are hysteresis, residual strain, AE activities, and decrease of α near room temperature. The small hysteresis in the subsequent cycle and the AE activities around the bottom of each cycle show that the thermal cracking should also occur in these cycles.

Stress effect on the thermal cracking

Wong et al., (1979), in their thermal expansion measurement at high confining pressures, observed that the thermal strain-temperature curves were perfectly linear at high pressures. Stress effects on the thermal expansion behaviour were also expected in the rock under uniaxial compressive stress (Heard, 1980).

In the sample with a constant stress, anisotropic thermal responses were observed. In the axial direction, the shortening in the sample length after the first cycle, low α in the first heating cycle, and temperature independency of α were observed. These results indicate that thermal cracking perpendicular to the loading direction did not occur and cracks which contribute to the axial nonlinearity were closed. In the circumferential direction, thermal expansion behaviour was similar to that without differential stress; hysteresis and residual strain in the first cycle, and temperature dependency of α. These results indicate that thermal cracking parallel to the loading direction was less affected by the stress.

A large number of AE events in the first heating cycle and a few events in the second cycle was observed; these were similar to those observed without stress and a comparable number of events at the first reheating cycle occurred after the load was removed. Although the rock retains a "memory" (Johnson et al., 1978, Yong et al., 1980) of the previous cycle under constant stress, it cannot remember the cycle of the previous stress state after a change of externally applied stress. This reactivation of AE activities is explained by the change of the local stress around crack tip and grain-boundary. The local stress state changed from an equilibrium condition to one in which new crack extension was made due to reheating and unloading. Although Oshima granite exhibited similar expansion behaviour and AE activities as Westerly granite, Murata basalt exhibited no hysteresis or residual strain in heating and showed small hysteresis after the first cooling experiment. The behaviour of Murata basalt is probably explained either by the lack of quartz grain within it, or by its fine grain size and glassy matrix, or by the lack of pre-existing cracks within it.

ACKNOWLEDGEMENT

The authors wish to thank T. Oshita and S. Kaku for their assistance with experiment and analysis.

REFERENCES

Clark, S. P. Jr., (1966). Thermal conductivity: in Handbook of physical constants, Geol. Soc. Amer. Memoir, 97, 459-482.

Cooper, H. W. and G. Simmons, (1977). The effect of cracks on the thermal expansion of rocks: Earth Planet. Sci. Lett., 36, 404-412.

Heard, H. C., (1980). Thermal expansion and inferred permeability of Climax quartz monzonite to 300C and 27.6MPa: Int. J. Rock Mech. Min. Sci. & Geomech. Abstr., 17, 289-296.

Johnson, B. et al., (1978). Thermal cracking of rock subjected to slow, uniform temperature change: Proc. 19th Symp. Rock Mechanics, 259-267.

Poore, M. W. and K. F. Kesterson, (1978). Measuring the thermal expansion of solids with strain gages: J. Testing Evaluation, 6, 98-102.

Richter, D. and G. Simmons, (1974). Thermal expansion behaviour of igneous rocks: Int. J. Rock Mech. Min. Sci. & Geomech. Abstr., 11, 403-411.

Simmons, G. and H. W. Cooper, (1978). Thermal cycling cracks in three igneous rocks: Int. J. Rock Mech. Min. Sci. & Geomech. Abstr., 15, 145-148.

Thirumalai, K. and S. G. Demou, (1970). Effect of reduced pressure on thermal-expansion behaviour of rocks and its significance to thermal fragmentation: J. Appl. Phys. 41, 5147-5151.

Wong, T. F. and W. F. Brace, (1979). Thermal expansion of rocks: Some measurements at high pressure: Tectonophysics, 57, 95-117.

Yong, C. and C. Wang, (1980). Thermally induced acoustic emission in Westerly granite: Geophys. Res. Lett., 7, 1089-1092.

HEAT TRANSFER IN THE ROCK MASS AROUND MINE OPENINGS

Diffusion de la chaleur dans la roche autour des vides miniers

Wärmeaustausch in der Umgebung bergmännischer Felshohlräume

G. Barla and N. Innaurato
Polytechnic of Torino, Italy

G. Pantaleoni
Solmine S.p.A., Boccheggiano (Grosseto), Italy

SYNOPSIS

The heat transfer in the rock mass around mine openings is studied with reference to the Campiano Mine (Grosseto, Italy), where the geothermal gradient was measured to be approximately 0.08°C per meter depth. The following problems are considered: (1) Laboratory tests carried out in order to determine the thermo-physical properties of the rock. (2) In situ tests performed to measure the temperature distribution in the rock mass around mine openings and to evaluate the most important thermo-physical parameters of the rock mass. (3) Analytical and numerical modelling performed in order to interpret the in situ tests and to predict the temperature distribution in the rock mass around mine openings.

RESUME

On étudie la diffusion de la chaleur dans la roche autour des vides miniers dans le cas de haut gradient de température (0,08°C/m) dans la mine de Campiano (Grosseto, Italie). L'étude a été réalisée par: 1) essais en laboratoire pour déduire les propriétés thermo-physiques de la roche; 2) essais in situ pour mesurer la distribution de la température dans la roche autour des vides miniers et d'évaluer les plus importants paramètres thermiques au niveau de la roche in situ; 3) mise au point des modèles analytiques et numériques ayant pour but d'interpréter les essais in situ et de prédire la distribution de la température dans la roche.

ZUSAMMENFASSUNG

Der Wärmeaustausch im Gebirge um bergmännische Hohlräume wird in Bezug auf die Campiano Grube (Grosseto, Italien) untersucht, wo der geothermische Gradient einer Temperaturzunahme von 0.08°C je Meter Tiefe — nach eigenen Messungen — zu entsprechen scheint. Man hat deswegen die folgenden Untersuchungen durchgeführt: 1) Laborversuche, um die Wärmeeigenschaften des Felsens zu bestimmen. 2) In situ Messungen, um die Wärmeverteilung in der Umgebung von bergmännischen Hohlräumen festzustellen und die wichtigsten Wärmeparameter des Gebirges abzuschätzen. 3) Analytische und numerische Modelle wurden hergestellt, um die in situ Messungen wissenschaftlich zu erklären und um die Temperaturverteilung um Grubenhohlräume vorhersagen zu können.

INTRODUCTION

A number of very important engineering problems relate to heat transfer in rock masses. Among the most interesting examples, the following may be recalled: 1. the stabilization of rock by artificial freezing, for the purpose of excavation of wells and tunnels; 2. the industrial use of geothermal reservoirs; 3. the choice of underground spaces for radioactive waste storage; 4. the exploitation of minerals in deep mines, with high temperature conditions.

With reference to the problem of evaluating the thermomechanical behavior of rock masses (i.e. stress analysis and heat transfer of rock masses), the present state of knowledge is limited. For example, rock thermo-physical properties (e.g. thermal diffusivity and/or thermal conductivity; specific heat) are less known than mechanical properties (e.g. stress-strain laws, failure criteria, etc.). The need for obtaining additional data on rock thermo-physical properties in the laboratory and in situ is therefore well recognized. At the same time, appropriate analytical and numerical methods and solutions are to be developed for the study of heat transfer in rock masses around underground openings.

This paper describes a number of problems relating to the analysis of heat transfer around underground openings in a deep mine near to Grosseto (Italy). The purpose is to obtain the basic data needed for predicting the temperature distribution in the rock mass and in the air, where a mine opening is being excavated. The following problems are considered: (1) Laboratory tests carried out in order to determine the rock thermo-physical properties. (2) In situ tests performed to measure the temperature distribution in the rock mass around mine openings and to evaluate the rock mass thermo-physi cal parameters. (3) Analytical and numerical modelling performed in order to interpret the in situ tests and to predict the temperature distribution in the rock mass around mine openings.

THE MINE UNDER STUDY

The study was carried out with reference to the Campiano mine, near to Grosseto (Italy), Fig. 1.

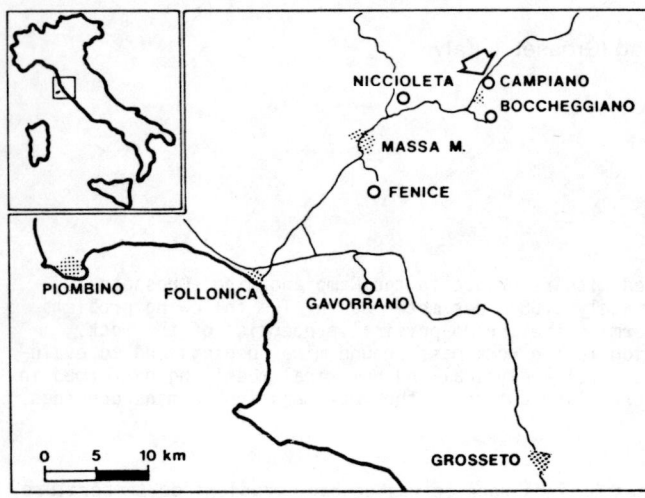

Fig. 1 - Campiano mine location

The mineral deposit is formed of pyrite, mixed (Pb, Zn, Cu) sulphurs, and pyrite and magnetite, with a total of 25 Mt presently estimated reserve. An evaporitic series, made mostly of anhydrite with dolomites, is above the deposit, with phyllitic rocks being below. A very high geothermal gradient (approximately 0.08 °C per meter depth) is present. It is to be noted that Campiano is located approximately 20 km away from the well known Larderello geothermal reservoirs.

In addition to the high temperature of the virgin rock mass (75° C at 500 m depth below sea level, near to 800 m below ground surface), a strong influence on the temperature conditions underground is that due to the heavy mechanized systems used for excavation purposes. The experience up to the present shows a certain difficulty in keeping the air temperature and underground climate acceptable for the workers, according to the safety and hygienic working conditions required by law.

For the purpose of the present study, the heat transfer problem in situ was examined, mainly by considering the rock mass thermal behavior around a drift (spiral decline) of a 20 m^2 cross section, created in order to reach the orebody, Fig. 2. However, in a few cases, also the openings located in the ore and neighbouring rocks were considered.

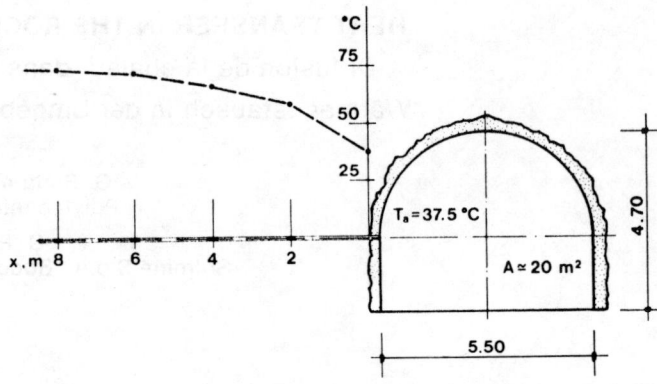

Fig. 2 - Cross section of the access drift (spiral decline) and simplified representation of temperature distribution in the rock mass

LABORATORY TESTING

The measurement of thermophysical properties of rocks may be carried out according to the methods commonly used in solid physics. However, a number of changes are to be introduced in order to account for the type and nature of the materials now being tested, which usually do not exhibit homogeneity, isotropy, and continuity. Definitely, the specimens to be used will be at least of the same size as adopted in conventional Rock Mechanics tests for design parameters.

A number of experimental methods are known to be adopted in solid physics for determining the physical properties relating to thermaly activity (Carslaw and Jaeger, 1962). In the present study, two different techniques were used for thermal conductivity, depending on the materials being tested, Fig. 3. For high conductivity materials, such as pyrite, the cylindrical specimens were 50 mm in diameter and 100 mm in length. For low conductivity materials, such as phyllite, the specimens were again cylindrical, with the same diameter, but only 20 mm in length.

For high conductivity rock materials, as shown on Fig. 3a, the specimen is insulated on the lateral surface; the heat source, at a constant temperature T_o, is on the upper end, with a thin layer of colloidal silver being interposed between the specimen and the same heat source. Temperatures T_1 and T_2 are measured (at time t_1) along the cylinder axis, at points 1 and 2 located at distances x_1 and x_2 with respect to the heat source.

Under the assumption that no heat exchange will take place (in the transient phenomenon being considered) between the source and the opposite specimen end, the theory of heat transfer in a semi-infinite homogeneous and isotropic space, with the surface at a constant temperature T_o, can be used for interpretation purposes (Carslaw and Jaeger, 1962)

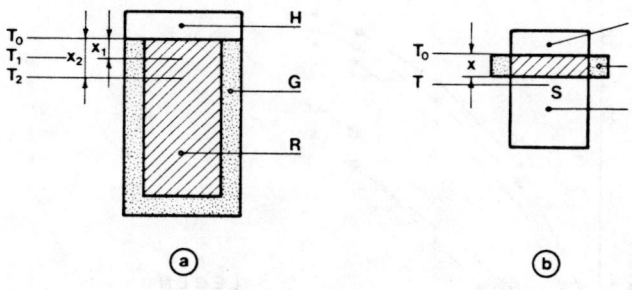

Fig. 3 - Experimental techniques used for determining thermal conductivity in the laboratory

$$T_1 - T_i = (T_o - T_i)\,\mathrm{erfc}\,\frac{x_1}{2\sqrt{a\,t_1}} \qquad (1)$$

$$T_2 - T_i = (T_o - T_i)\,\mathrm{erfc}\,\frac{x_2}{2\sqrt{a\,t_1}} \qquad (2)$$

where: T_i = initial specimen temperature;
T_o = heat source temperature;
a = thermal diffusivity;
erfc = 1 - erf;
erf = error function.

Equations (1) and (2) can be written

$$\frac{T_1 - T_i}{T_2 - T_i} = \frac{\mathrm{erfc}\,\dfrac{x_1}{2\sqrt{a\,t_1}}}{\mathrm{erfc}\,\dfrac{x_2}{2\sqrt{a\,t_2}}} \qquad (3)$$

so as not to account for the temperature T_o of the heat source. It is therefore sufficient to know temperatures T_1 and T_2, at t_1, to determining, by a trial and error procedure, the thermal diffusivity a. Since the thermal conductivity λ is a direct function of thermal diffusivity a

$$\lambda = \rho\,c\,a \qquad (4)$$

where: c = specific heat;
ρ = density;
the thermal conductivity λ is also evaluated.

For low conductivity rock materials, the thermal conductivity λ can be determined directly, as shown on Fig. 3b. The cylindrical specimen is again insulated on the lateral surfaces; the upper end is kept at a constant temperatura T_o as the lower end is in contact with a copper plate. The temperature T_1 at time t_1 of the copper plate is measured. The thermal conductivity λ is given by (Carslaw and Jaeger, 1962)

$$\lambda = -\frac{cx}{s}\,\frac{\ln\dfrac{T_o - T_i}{T_o - T_a}}{t_1} \qquad (5)$$

where: c = heat capacity of copper plate;
x = specimen thickness;
s = specimen cross section;
T_o = heat source temperature;
T_a = specimen and air temperature.

Tests were also performed in the laboratory to determine the *specific heat* for the same rock materials previously subjected to thermal conductivity (diffusivity) measurements. The calorimeter used is a standard Regnault type. The specific heat c is evaluated as follows

$$c = \frac{C}{m}\,\frac{(T - T_o)}{(T_1 - T)} \qquad (6)$$

where: C = calorimeter capacity;
m = specimen mass;
T_o = calorimeter initial temperature;
T_1 = specimen initial temperature;
T = overall system final equilibrium temperature.

In order that the rock specimen be kept dry during testing, a special varnish cover was used. Additionally, each specimen was weighted before and after each measurement. As for the conductivity tests, the specific heat was determined in the range 70°-25° C, according to the temperature limit values found in the rock mass at depth. The temperatures were measured with thermocouples with a sensitivity of 0.1°C.

Tests were carried out on a total of sixty samples of the following lithotypes: clay-shale; evaporites; fault breccias; skarn; pyrite; mixed sulphurs with magnetite; phyllites. The mean experimental values obtained for conductivity, density, specific heat and percentage of sulphures are reported in Table I.

Table I - Summary of experimental data

Lithotype	Specific weight kg/m³	Conductivity W/(m°C)	Sulphur Percentage %	Specific Heat kJ/(kg°C)
Clay-shale	2890	2.10	–	0.690
Evaporites	3100	3.22	–	0.686
Fault-Breccia	–	2.67	–	0.750
Skarn	3450	1.74	–	0.920
Phyllites	2820	1.88	–	0.837
Pyrite	4770	13.00	94.4	0.552
Mixed sulphurs and magnetite	4800	9.88	31.3	0.669

IN SITU TESTING

As for the mechanical properties (strength and deformability), also for the thermal parameters the representativeness of values determined in the laboratory is often questioned, when solving problems relating to the thermomechanical behavior of rock masses. It was therefore felt advisable to determine rock mass diffusivity by in situ tests. Among the most commonly used experimental techniques
• needle probe method
• rock surface radiation
• heat diffusion during a transient state

the latest one was adopted.

The heat transfer following the excavation of the access drift (Fig. 2), at various locations below the ground surface, in *phyllite*, was studied in detail, with the most important results being reported below (°). Boreholes (35 mm in diameter and up to 11 m in length) were drilled from the opening contour into the rock mass. By means of thermal transducers positioned along each borehole at 2,4 and 6 m depths and at the borehole end, the temperatures were measured.

In a few cases boreholes (30 mm in diameter) were also drilled at a depth ranging between 5 and 100 cm, so as to obtain the temperature distribution in the near proximity of the opening, where the temperature gradient is expected to be the highest. At the same time as the rock mass temperature was determined in boreholes, temperature measurements were carried out at the wall of the opening and in the air (respectively, at the center of the opening and at a distance equal to 20 cm from the wall).

The following data were therefore made available along the access drift, at various depths below the ground surface (measurements were carried out at eleven different cross sections, at a depth ranging from 165 m to 700 m below the ground surface):
  • temperature at the center and wall of the opening
  • temperature distribution in the rock mass, from the opening contour to 11 m distance from it.

Typical results obtained for the access drift, in *phyllite*, at a depth of 700 m below the ground surface, are depicted in Fig. 4 and 5. Also reported in Fig. 6 are the values of the mean temperature at various depths below the ground surface, in the air and in the rock mass, as inferred according to the measurements carried out up to the present.

ANALYTICAL AND NUMERICAL MODELLING

In order to determine the thermal parameters by means of the heat transfer experiments in situ, analytical and/or numerical models were developed for interpretative purposes. The following were considered (Barla et al., 1982):

   (1) Heat flow in a half space (Fig. 7a)
   (2) Heat flow in an infinitely extended medium containing a circular hole (Fig. 7b)
   (3) Heat flow in an infinitely extended medium containing an arbitrary shape hole (Fig. 7c).

A transient solution for temperatures at different points located in the rock mass at known distances from the boundary representing the

_____

(°) *Measurements were also carried out in the access drift in clay shales. At present, the same methods are being applied in mine openings excavated in pyrite. However, the available results refer to a very short time duration after excavation.*

opening contour was needed.

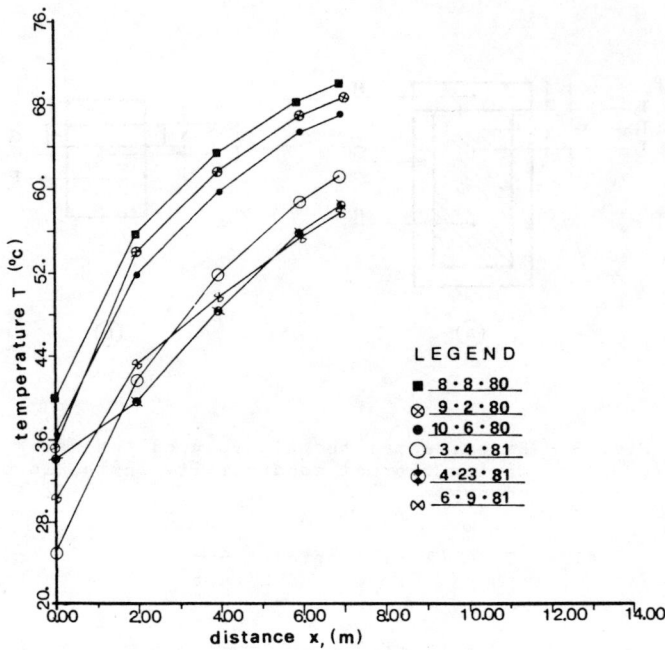

Fig. 4 - Temperature distribution around the access drift in phyllite at a depth of 700 m below the ground surface

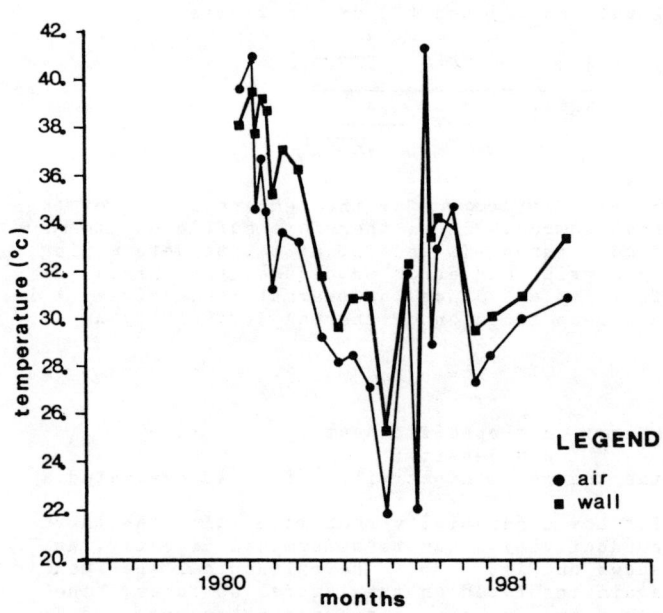

Fig. 5 - Temperature in the wall and in the air - Access drift in phyllite, 700 m below the ground surface

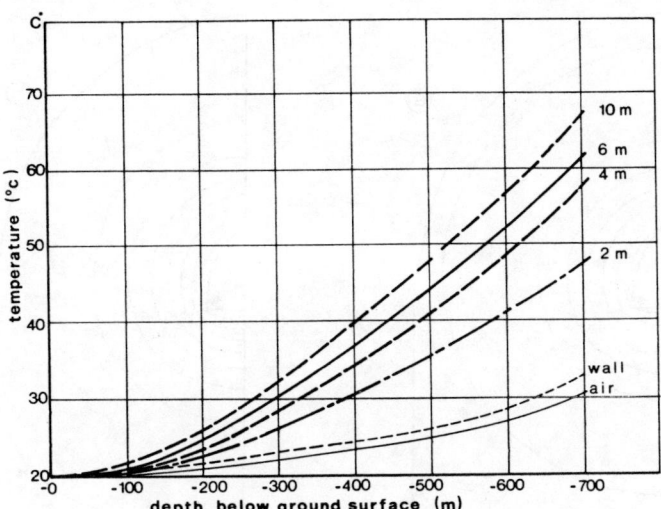

Fig. 6 - Mean temperature distribution below ground surface as inferred on the basis of measurements carried out up to the present

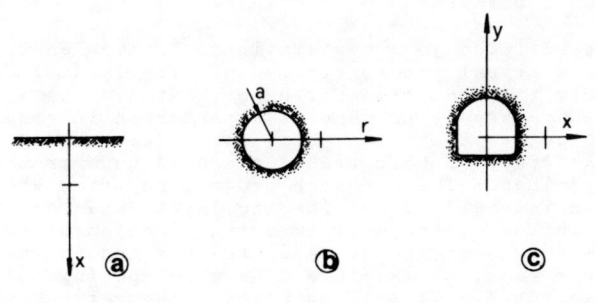

Fig. 7 - Theoretical models adopted for interpreting the heat transfer experiments in situ

Analytical solutions were developed for (1) and (2). The Finite Element Method (FEM) was used for (3). In all cases, problem oriented computer codes were written (Barla et al., 1982) so as to allow one to carry out sensitivity analyses to obtain a good representation of the results of observations and measurements in situ.

The effects on temperature distribution in the rock mass around mine openings due to changes in thermal conductivity were evaluated. Also, differing values were assumed in the calculations for the convective heat transfer coefficient at the opening contour, the air temperatures being known at various times.

The results obtained for the access drift in phyllite, at a depth of 700 m below the ground surface, are briefly discussed below, with consideration being given also to the temperature distribution in the rock mass around the opening. A two dimensional finite element model, formed of 462 quadrilateral isoparametric elements and 510 nodal points, was created in order to reproduce the cross section being considered.

The finite element model extends for 40 m (approximately eight times the equivalent radius of the opening) from the contour into the rock mass, so as to neglect the influence of the outer boundaries on the temperature distribution being evaluated. No heat flow is assumed to occur along the axis of the drift, orthogonally to the plane containing the model.

Accordingly, the numerical calculations were carried out by using both the analytical and numerical solution schemes. In all cases, an attempt was made to reproduce as closely as possible the time history of the heat transfer phenomenon taking place subsequently to the excavation of the access drift in phyllite.

Sensitivity analyses have shown that the values chosen for the rock mass thermal conductivity are very relevant for correct prediction of the temperature distribution around the opening. On the contrary, the values assumed for the convective heat transfer coefficient are found to influence the results of the transient phenomenon only in the first time steps following excavation.

The results obtained for the temperature distribution in the rock mass around the access drift, for a known set of thermal properties, are depicted in Fig. 8 and 9, where analytical and numerical values are compared with the experimental data. Also shown in Fig. 10 are the temperature contour lines around the same opening, as evaluated by the finite element method at two different time steps following excavation.

The analytical and numerical predictions of temperature distribution around underground openings are shown to compare well with experimental data, allowing one to define the most appropriate values for the thermal parameters to be adopted when solving problems on the thermomechanical behavior of rock masses. On the basis of the work carried out up to the present, the following parameters are suggested for phyllites and evaporites.
- Thermal conductivity . . . . $\lambda = 3.0-3.5$ W/(m°C)
- Convective heat transfer coefficient . . . . $\alpha = 25-30$ W/(m²°C).

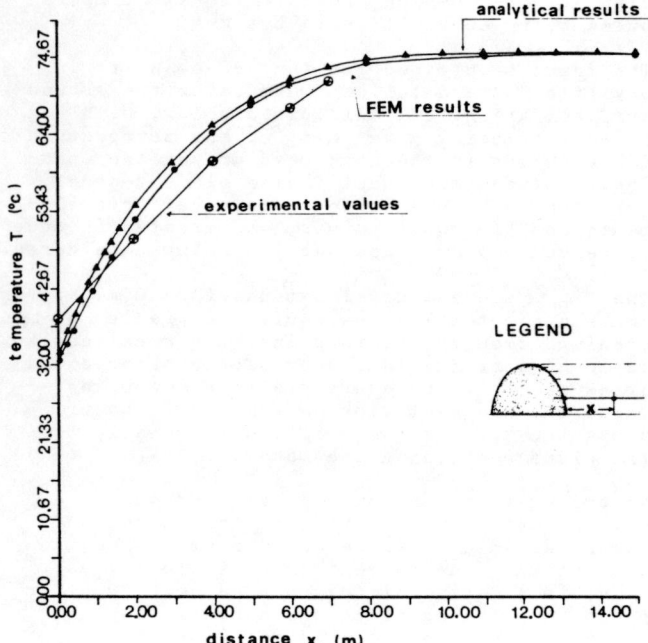

Fig. 8 - Temperature distribution around the access drift in phyllite, 700 m below the ground surface (Time = 53 days subsequent to excavation)

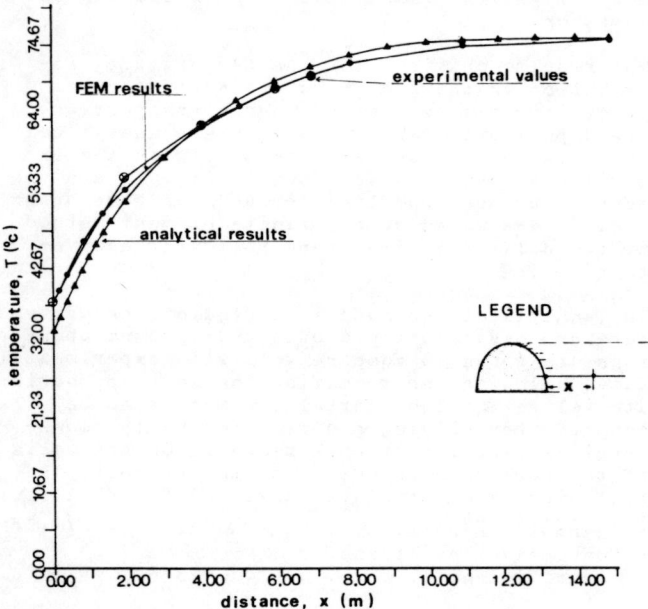

Fig. 9 - Temperature distribution around the access drift in phyllite, 700 m below the ground surface (Time = 73 days subsequent to excavation)

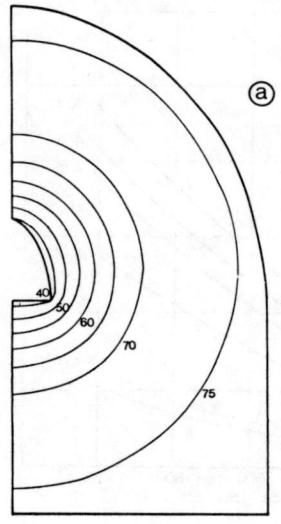 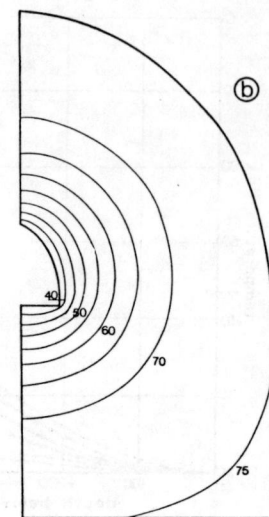

Fig. 10 - Temperature (°C) distribution around the access drift in phyllite, 700 m below the ground surface - Finite Element Results at (a) 69 days and (b) 98 days subsequent to excavation

CONCLUDING REMARKS

In addition to a systematic collection and theoretical interpretation of experimental data from the heat transfer experiments in the mine, laboratory tests were also performed in order to determine the material parameters needed for the study of heat transfer around underground openings. The research program reported above was intended to provide the input data for the solution of the most important problem of evaluating the temperature in the air and in the rock mass, at the face of a mine opening being excavated. To this purpose, a theoretical model based upon the Finite Element Method has already been developed and in situ testing is being carried out with the specific objective of checking its predictive capability.

ACKNOWLEDGEMENTS

The work reported in the present paper is supported by the Commission of the European Communities and Solmine S.p.A., as a part of the European Mining Technology Program, Contract No. 098-79-MPP-I. The help of the following persons, who have contributed at various times to the research program, is acknowledged: G. Carosso, G.P. Giani, P. Jarre, A. Visetti, and G. Zasso.

REFERENCES

Barla G. at al. (1982). Studio della diffusione del calore in coltivazioni minerarie profonde. Final report on Contract No. 189. Politecnico di Torino.

Carslaw H.S. and Jaeger J.C. (1962). Conduction of heat in solids. Oxford, Clarendon Press.

AN IN SITU DETERMINATION OF THE THERMAL CONDUCTIVITY OF GRANITIC ROCK

Détermination in situ de la conductivité thermique de la roche granitique

In situ Bestimmung der Wärmeleitfähigkeit eines granitischen Gneises

Michio Kuriyagawa, Isao Matsunaga and Tsutomu Yamaguchi
Researchers, National Research Institute for Pollution and Resources, Tsukuba, Japan

SYNOPSIS

A field experiment was carried out to measure the temperature profile produced in a granitic rock by electrical heating, and the thermal conductivity as a function of temperature was obtained from these results. For this purpose, a cylindrical heater of 60 cm length was placed in the rock at a depth of 3 metres and kept at 440°C for 2,000 hours. Temperature was measured with thermocouples set at various distances from the heater to determine an in situ temperature profile. The temperature profile was then calculated using the Finite Element Method for various assumed values of the conductivity, and these results were compared with those measured. From these comparisons, the best fit of the thermal conductivity k(W/mK) to the data is expressed as a function of temperature T(K): $k = 5.319 \times 10^3 T^{-1.5507}$.

RESUME

Une expérience in situ à été réalisée pour déterminer le profil thermique induit dans une roche granitique par chauffage électrique. On a dérivé de ces résultats la conductivité thermique en fonction de la température. Dans ce but un élément chauffant cylindrique 60 cm de long a été placé dans la roche à 3 mètres de profondeur et maintenu pendant 2000 heures à 440°C. La température en a été mesurée à l'aide des thermocouples à des distances allant de 0,5 à 11 mètres de l'élément. Pour obtenir la conductivité de la roche le profil thermique mesuré a été comparé avec le profil calculé par une méthode d'éléments finis pour plusieurs valeurs assumées de la conductivité. A partir de ces comparaisons la conductivité thermique k(W/mK) est représentée comme fonction de la température T(K) par $k = 5319(T^{-1.5507})$.

ZUSAMMENFASSUNG

Ein in situ Versuch zur Ermittlung der Wärmeleitfähigkeit eines Felses (granitischer Gneis) als Funktion der Temperatur wird beschrieben. Zu diesem Zweck wurde eine zylindrische Wärmequelle im Gebirge eingebracht und 2 000 Stunden lang auf 440°C gehalten. Die Gebirgstemperaturen wurden mit nach der Finite Element Methode berechneten Modellen verglichen, wobei die Wärmeleitfähigkeit und ihre Änderung mit der Temperatur variiert wurden. Die beste Anpassung an die gemessenen Temperaturen ergab sich, wenn für die Abhängigkeit der Wärmeleitfähigkeit k(W/mK) von der Temperatur T(K) die Funktion $k = 5,319 \times 10^3 T^{-1.5507}$ angenommen wurde.

1. INTRODUCTION

In designing a Hot Dry Rock geothermal reservoir, an underground waste isolation plant, or developing mines and tunnels where the control of temperature is critical, it is important to know the thermal conductivity of the rock encountered. Methods of measuring conductivities using a small specimen have been developed and used widely. But there remains a question of how the thermal conductivity measured in small specimens correlates with that of the in-situ rock mass. From this point of view, the field measurement of the thermal conductivity is necessary. Murphy, H.D. [1] obtained estimates of thermal conductivities from borehole measurements in a geothermal reservoir. This is a useful method for determining the in-situ thermal conductivity. This paper describes a field test using a line heater in the rock to determine the in-situ rock thermal conductivity and its dependence on temperature. A cylindrical electric heater was placed at the depth of 3 m inside the borehole and kept at high temperature for about three months. The temperature around the heater was measured during the experiments. The temperature profile was also calculated using Finite Element Method (FEM) by assuming several thermal conductivities. These results were compared with the experimental data and the thermal conductivity of the rock at the test site was determined. The thermal conductivity was also measured with specimens sampled from the test site and a comparison was made between the values obtained from the field test and in the laboratory.

2. LAYOUT OF THE FIELD TEST

The field experiment was carried out at Tochibora pit of Kamioka Mine, Mitsui Mining & Smelting Co.. The location of the field is shown in Fig.1. Kamioka Mine is the one of the largest metals mines in Japan and produces primarily lead, zinc and silver.

The test site selected is a roadway with the cross-section of approximately 3 m x 3 m and the rock is formed of granitic gneiss. A cylindrical heater, 60 cm in length and 6 cm in diameter, was placed at the depth of 3 m inside a 10 cm diameter borehole drilled into the sidewall of the roadway. The gap between the heater and borehole was filled with fire-proof materials. Thermocouples for temperature measurement of the rock arranged at distances of 0.5, 1.0, 2.0, 3.9 and 5.1 m from the heater. Fig.2 shows a schematic of the test layout.

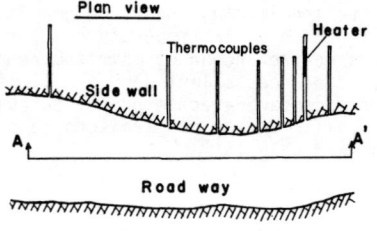

Fig.1 The location of the test site

These were placed at a depth of 2.9 to 3.0 m. In the measurement boreholes, No.4 and 5 located 0.5 and 1.0 m from the heater, thermocouples with the length of 1 m were placed to clarify the heat flow toward the free surface of the sidewall. Three other measurement boreholes, No.1, 2 and 3 in Fig.2 were drilled at 1 m from the

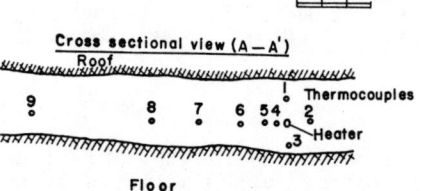

Fig.2 Schematic of the test layout

heater to obtain the anisotropy of the thermal conductivity of the rock. The anisotropy could not be determined because the relative position of the heater and thermocouples could not be measured exactly. This was due to the fact that the boreholes were drilled to a depth of 3 m from the sidewall which is not in a plane, as shown in Fig.2, and boreholes were slightly deviated. Also accuracy of the temperature measurement devices was not high enough to determine the temperature difference between boreholes which were located at the same distance from the heater. The thermocouple placed in borehole No.9, which is 11 m from the heater, was used to get the background temperature of the rock.

3. RESULTS OF TEMPERATURE MEASUREMENT

The temperature of the heater rose linearly at the rate of 25 °C per hour until a temperature of 440 °C was attained, after which it was held constant for 2,000 hours. The temperature of the heater and surrounding rock was recoreded continuously. Fig.3 shows the temperature of the rock at distances of 0.5, 1.0, 2.0, 3.9 and 5.1 m from the heater at 10, 20, 50, 100, 200 and 700 hours.

According to this figure, the temperature of the rock dropped quickly as the distance from the heater increased and had changed only 3 °C at 3 m from the heater after 700 hours. Fig.4 shows the temperature history of the rock at 0.5 m from the heater. Also

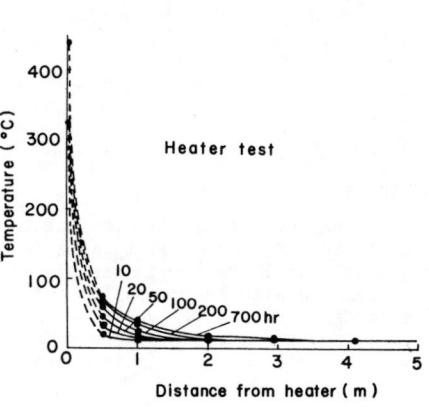

Fig.3 Temperature profile of the rock during heating

shown are temperature profiles calculted by the FEM which will be discussed later. This figure shows that the temperature had risen to 68 °C from the original temperature of 13 °C after 200 hours, but rose only 4 °C more to 72 °C after 700 hours. No heat flow toward the free

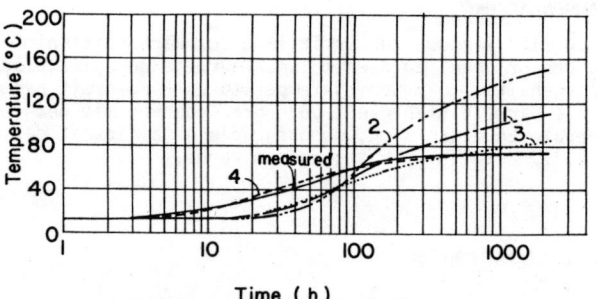

Fig.4 Temperature-time curve measured and calculated with various thermal conductivities at a distance of 0.5 m from the heater.

surface was observed with the measurement of the temperature using the No.4 and 5 boreholes. Similar results were obtained by the FEM calculations.

4. DETERMINATION OF THERMAL CONDUCTIVITY WITH FEM

E 148

This problem can be approximately considered as axi-symmetric by taking the Z-axis as the axial coordinate through the center of the heater as shown in Fig.5. The equation for the thermal conductivity can then be written in radial coordinate (r,z) as

$$\frac{\partial T}{\partial t} = \frac{1}{r}\left[\frac{\partial}{\partial r}\left(r\alpha\frac{\partial T}{\partial r}\right) + \frac{\partial}{\partial z}\left(r\alpha\frac{\partial T}{\partial z}\right)\right] \quad (1)$$

in which T is the temperature, t is the time and α is the thermal diffusivity which is expressed using the thermal conductivity k, density ρ, and specific heat c as $k/\rho c$. If α is a function of the temperature only, Eq. (1) can be written as

$$\frac{\partial T}{\partial t} = \alpha\left[\frac{\partial^2 T}{\partial r^2} + \frac{1}{r}\frac{\partial T}{\partial r} + \frac{\partial^2 T}{\partial z^2}\right] + \frac{\partial \alpha}{\partial T}\left[\left(\frac{\partial T}{\partial r}\right)^2 + \left(\frac{\partial T}{\partial z}\right)^2\right] \quad (2)$$

Eq.(2) is nonlinear and is difficult to solve analytically with the boundary conditions described later, so FEM was used. The code used here is AYER developed at Los Alamos National Laboratory in the U.S.. The FEM mesh was shown in Fig. 5. For boundary conditions, the sidewall denoted by the R-axis was kept at the ambient rock temperature of 13 °C, and zero heat flow was assumed at the other boundaries. The temperature change with time at 0.5 m from the heater was calculated assuming a temperature dependent thermal conductivity in various forms. These results were compared with the field measurements. According to Murphy,D. J.[2] the behavior of the thermal conductivity with temperature is dependent on the rock type. This dependence is shown in Fig.6. Rocks are classified as ones in which the thermal conductivity increases with temperature such as obsidian, ones in which it remains constant such as basalt, and ones in which it decreases with temperature rise such as granite. Several equations for the thermal conductivity k as a function of temperature were used for the numerical calculations. These equations ① k = 1, ② k = 2.5×10⁻³T−0.25, and ③ k = −2.5×10⁻³T+2.25 are shown in Fig.

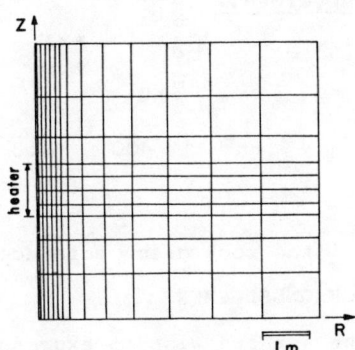

Fig.5 Coordinate system to calculate temerature in the rock.

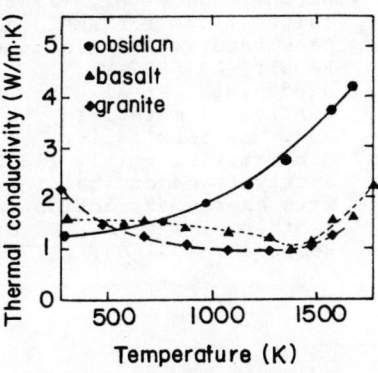

Fig.6 Change of thermal conductivity of different type of rocks.

7. In the calculations, the desity and specific heat of rock were assumed to be independent of temperature and location, and were set to be ρ = 3000 kg/m³ and c = 1000 J/kg K. As an example, the temperature profile calculated with the thermal conductivity equations ① and ② are shown in Fig.8 and Fig.9 respectively. As is clear from these figures, the heat diffused widely and the temperature change showed different behavior from that of the field measurement shown in Fig.3. The temperature changes with time at 0.5 m from the heater were also calculated using equations ①, ② and ③ and are shown in Fig. 5. It is clear from this figure that the temperature history gets closer to that of the field measurements when the thermal conductivity is assumed to decrease with temperature rise, so the thermal conductivity equation of the form k = aT^b is assumed, because the thermal conductivity of rocks changes nonlinearly with temperature[3] and this equation is simplest form to describe this relation. Here a and b are constants. A trial and error method was used to determine the best temperature history match. The match was close to the field data when a =5.319 ×10³ and b = −1.5507 were used. The relative error of the temperature calculated and measured is 11.1 % at the maximum and 4.4 % at the average. The cur-

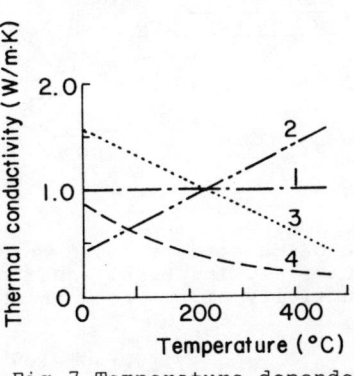

Fig.7 Temperature dependence of thermal conductivities predicted.

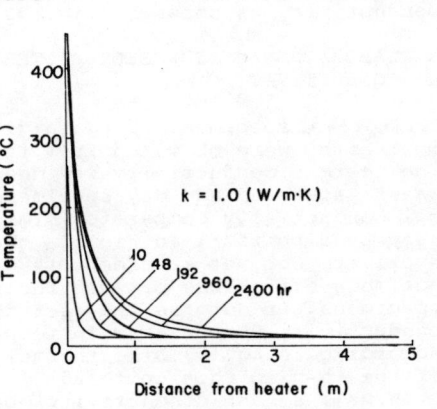

Fig.8 Temperature change of rock when k = 1.0 is assumed.

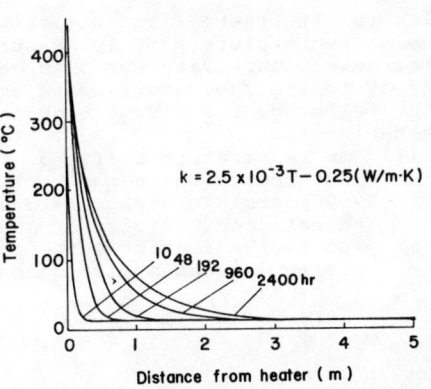

Fig.8 Temperature change of rock when k = 2.5×10⁻³ xT−0.25 is assumed.

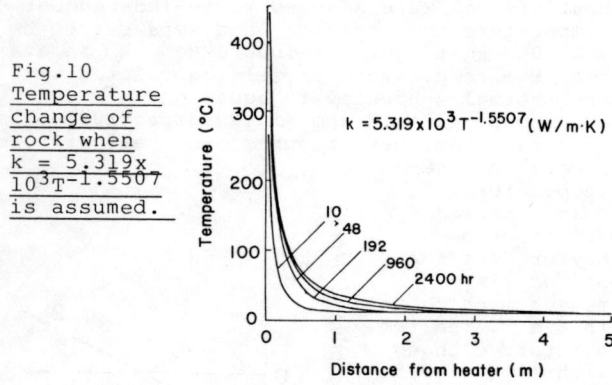

Fig.10 Temperature change of rock when $k = 5.319 \times 10^3 T^{-1.5507}$ is assumed.

ve calculated is shown in Fig.5 by the curve number 4. The final equation for thermal conductivity of the rock in the test site is given by

$$k = 5.319 \times 10^3 T^{-1.5507} \qquad (3)$$

The temperature profile of the rock along the R-axis at different time using above thermal conductivity is shown in Fig.10.

5. LABORATORY MEASUREMENT OF THERMAL CONDUCTIVITY

Cylindrical specimens, 5 cm in diameter and 3 cm in length, were obtained from the test site. The thermal conductivity was measured for both water-saturated and dry specimens with a thermal conductivity comparator similar to that used by Sibbitt[4]. In Fig.11, the thermal conductivity for wet specimens is given. There was not much difference between the wet and dry specimens. Of note is the fact that the thermal conductivity decreases with temperature. According to Eq.(3), the thermal conductivity of the in-situ rock up to 35 °C is less than 1 (W/mK), so the conductivity obtained from the laboratory samples was twice as high as that determined from the field measurements.

6. CONCLUSIONS

The in-situ thermal conductivity was determined using temperature profile measured in rock. The thermal conductivity was also measured up to 35 °C in the laboratory using small specimens. The following results were obtained from these experiments.

(1) The temperature diffused only 3 m from the heat source which was kept at 440 °C after 700 hours. This fact is due to the low thermal conductivity of granitic gneiss and to the fact that the thermal conductivity decreases with temperature.

(2) The best fit of the thermal conductivity of the in-situ rock is expressed as $k = 5.319 \times 10^3 T^{-1.5507}$.

(3) The thermal conductivity obtained with specimens up to 35 °C is more than twice as high as that obtained by the in-situ test. This difference is due to the size effect of the rock and the low accuracy of the in-situ experimental system. The authors are continuing the in-situ test to evaluate the thermal conductivity of

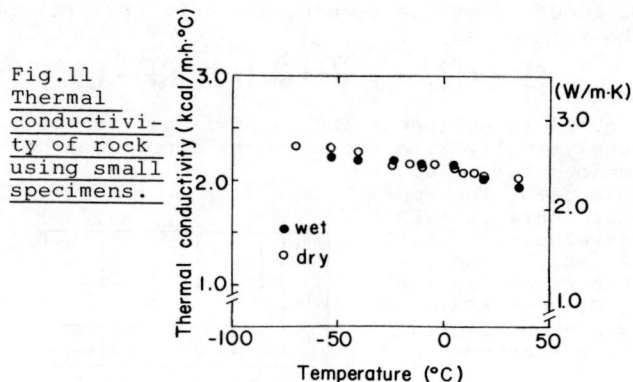

Fig.11 Thermal conductivity of rock using small specimens.

the rock mass more accurately.

ACKNOWLEDGEMENT

The authors wish to express their thanks to the Mitsui Mining & Smelting Co. which allowed them to do the field test and helped in gathering of the field data. Help on the AYER code by G. Bennett and the comments on this paper by G. Zyvolski and Z. Dash at the Los Alamos National Laboratory in the U.S. are greatfully acknowledged.

REFERENCE

1) Murphy, H.D. et al. (1977). Downhole measurements of the thermal conductivity in geothermal resources. J. of Pressure Vessel. Trans AIME. (99) 607-611.
2) Murphy, D.J. et al. (1973). Heat loss calculations for small diameter subterrene penetrators. Los Alamos National Lab. Report, LA-5207-MS.
3) Hayden, H.W. et al. The structure and properties of materials, vol.3. (1967) 201-202. New York, Wiley.
4) Sibbitt, W.L. et al. (1979). Thermal conductivity of crystalline rocks associated with energy extraction from Hot Dry Rock geothermal systems, J. of Geophysical Research. (84) B3, 1117-1124.

CLAB – AN INTERMEDIATE STORAGE FOR SPENT NUCLEAR FUEL IN SWEDEN

Le CLAB – Un stockage intermédiaire de combustibe nucléaire usé en Suède
CLAB – Vorläufige Verwahrung von verbrauchtem Kernbrennstoff in Schweden

K. Röshoff, Ph.D.
Division of Rock Mechanics, University of Luleå, Luleå, Sweden

O. Stephansson
Professor, Division of Rock Mechanics, University of Luleå, Luleå, Sweden

H. Larsson, Civ.Eng.
Swedish State Power Board, Vällingby, Sweden

R. Stanfors, Ph.D.
Division of Geology, University of Lund, Lund, Sweden

K. Eriksson, Civ.Eng.
VIAK AB, Vällingby, Sweden

SYNOPSIS
CLAB is the first licenced underground storage of spent nuclear fuel in the western world. The storage consists of water-filled pools constructed in a 117 m long and 27 m high cavern with a span of 21 m. The cavern is situated in granitic vulcanite 30 m below ground surface. Site investigations, design and construction are described. Extensive reinforcements and rock support were carried out. To ensure rock stability a thorough control program was undertaken.

RESUME
Le CLAB est la première installation souterraine agréée dans le monde occidental pour le stockage de combustible nucléaire usé. Cette installation se compose de bassins remplis d'eau, creusés dans une grotte longue de 117 m et haute de 27 m dont la voûte a une portée de 21 m. La grotte en question est située dans une volcanite acide à 30 m au-dessous du niveau du sol. On décrit les phases de prospection géologique et de construction. La roche a fait l'objet d'importants renforcements et, pour assurer un maximum de tenue à l'ensemble, un programme de contrôle minutieux a été réalisé.

ZUSAMMENFASSUNG
CLAB ist die erste zugelassene unterirdische Verwahrungsweise für verbrauchten Kernbrennstoff im Westen. Die Verwahrung erfolgt in Wasserbecken, die in eine 117 m lange und 27 m hohe Kaverne von 21 m Breite eingebaut sind. Die Kaverne liegt in einem granitartigen Tiefengestein 30 m unter der Erdoberfläche. Die Untersuchungen an Ort und Stelle, der Entwurf und die Bauausführung werden beschrieben. Umfangreicher Ausbau wurde ausgeführt. Zur Sicherung der Standfestigkeit des Gebirges wird ein sorgfältiges Kontrollprogramm durchgeführt.

1 INTRODUCTION

An intermediate storage for spent nuclear fuel - CLAB - is in the final stage of completion in southeastern Sweden. The underground storage is located in hard crystalline bedrock adjacent to the nuclear power plant Simpevarp. Spent fuel will be received from this plant and from Ringhals, Barsebäck and Forsmark. The storage capacity is about 3000 tons of uranium. The fuel will be stored in the cavern until the final deposition when it is reprocessed or deposit without reprocessing, Gustavsson et al., 1980.

The storage is owned by the Swedish Nuclear Fuel Supply Company (SKBF). OKG Aktiebolag (OKG) is responsible for the over all project management and construction work. The obligation of the civil engineering design is held by the Swedish State Power Board (SSPB). The contractor is a consortium comprising of ABV, Skånska Cementgjuteriet AB and WP-System AB. The Swedish Nuclear Power Inspectorate (SKI) possesses the control of the storage for the Swedish Government.

A general description of CLAB is given by Gustavsson et al., 1980. This paper discusses the realization phase of the storage from the site investigation to the completion of the underground chamber.

2 LAY-OUT OF THE STORAGE FACILITIES

CLAB storage facilities comprise three main units on the ground, namely fuel reception, auxiliaries and office, Fig. 1. The storage chamber is connected to the fuel reception by a fuel shaft and a service shaft.

The cavern has the dimension 117 m in length, 27 m in heith and a span of 21 m, Fig. 2. The cavern contains four water-filled pools for storage of spent fuel and one central pool. The pools are of concrete with an inside cover of stainless steel.

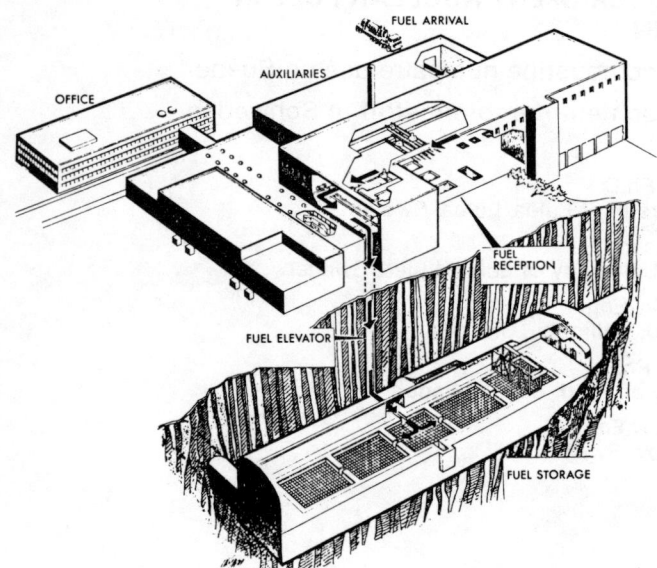

Fig. 1 Perspective view of CLAB (after Gustavsson et al., 1980).

A transept, 70 m in length, is located perpendicular to the long axis of the cavern. This enables an expansion of the storage capacity in the future, Fig. 4.

3 SECURITY REQUIREMENTS

The fuel will be stored in water-filled pools in the cavern. Extensive security is required in order to protect the fuel against any mechanical damage or outside forces.

3.1 Seismicity and ground movements

The cavern and the pool are designed and constructed to resist an earthquake which generates a ground acceleration of 0.1 g. CLAB is located in an area of Fennoscandia, which is characterized by very low seismicity, Kulhanek and Wahlström, 1977. The rate of glacial uplift in the area is of the order of 2 mm/year.

3.2 Stability of the cavern

The requirements for the stability of roof and walls of the cavern is extremely high as no rock burst, rock falls or collapse are allowed. Therefore the site investigation for location and orientation of the cavern in relation to tectonic structures has been more thourough than is normal for caverns in crystalline bedrock. Both the transport tunnel and cavern have been mapped in detail during the excavation phase in order to analyses the rock structures and stability.

The displacement of both walls and roof have been measured. Theoretical calculations of stresses and deformations were performed using a two-dimensional linear elastic finite element model. Larger safety factors then normally is required in this type of rock has been used. Smooth blasting and contour blasting has minimized the damage of the rock mass. The rock support consists of systematic rock bolting and shotcrete lining.

Fig. 2 CLAB storage room seen from the southern gable. Concrete walls and water pools foundation under construction

3.3 The environment in the cavern

High demands are required for the environment inside the cavern, e.g. humidity, temperatur, dust, and contamination. For the fulfil of these requirements the facilities in the rock cavern are surrounded by free-standing walls and roof. The walls are made of concrete and the roof is a light-gauge steel construction, Fig. 2 and 3. Roof and walls are air- and water-proof. The open space between the steel construction and the roof of the cavern will be ventilated and drained separately.

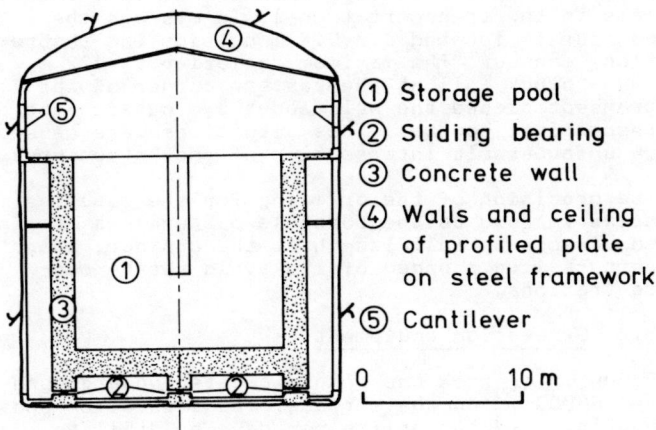

1. Storage pool
2. Sliding bearing
3. Concrete wall
4. Walls and ceiling of profiled plate on steel framework
5. Cantilever

Fig. 3 Cross section of storage room

3.4 Safety against outside impact

The cavern must be safe against forces introduced from the outside, i.e. sabotage, bombs, etc. Therefore, the room has been placed at a depth of 30 m below the ground surface.

4 SITE INVESTIGATIONS

The site investigation started in 1978 with a geological surface mapping, a few seismic profiles and one 80 m deep vertical borehole, Fig. 4. The borehole was analysed by TV-inspection and hydraulic tests were conducted, Moberg, M., 1979.

Later the same year this investigation was followed by a more detail investigation and the subsurface was investigated by eleven diamond boreholes (D1-D11) with a core diameter of 32 mm and a total length of 750 m. Water pump tests were performed in each borehole.

The virgin stresses were measured and boreholes were drilled for investigation of the geohydrological situation around the cavern. The elastic parameters of the intact rock were tested.

4.1 General geology

The topography of the area is flat and raises about 10 m above sea level. The bedrock consists of crystalline Precambrian granitic and volcanic rocks of post Svecofennian age, i.e. 1700-1800 m.ys. The granite is fine- and medium grained consisting of feldspar, quartz and biotite. The volcanic rocks are fine grained and normally

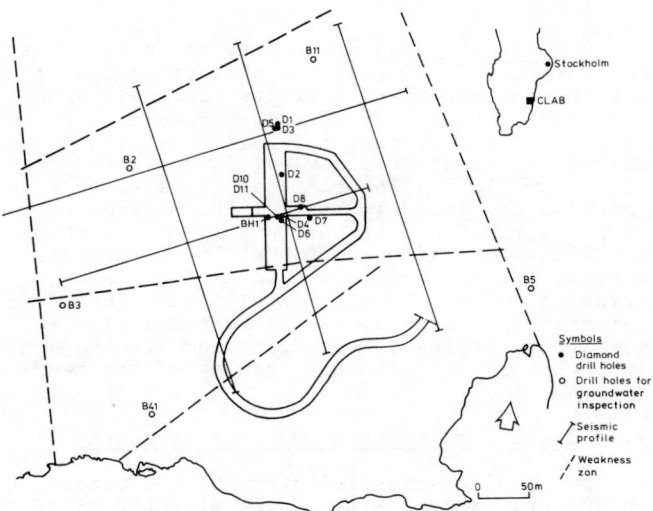

Fig. 4 General map of the site

greyish black in colour and the texture is porphyritic with feldspar megacrysts in a glassy matrix. No sharp contacts exists between the two rock types. The dominant rock type for the storage site is a granitic volcanite cut by dykes of aplite and pegmatite. The dykes are mostly striking in N65°E and are steeply dipping. The contact between aplite dykes and surrounding rock is often coated with chlorite.

The main tectonic weak zones are detected from the seismic profiles and later confirmed by diamond drilling, Fig. 4. The weak zone north of the cavern is 30 m wide and dips steeply towards south. The mean joint frequency of the zone is 16 joints per meter. Soil was observed in the tectonic weak zone. The zone west of the cavern is 6 m wide and dips westwards. The zone to the south is 10 m wide, dips towards south and cuts the cavern and the transport tunnel.

The orientation of the joints belongs to three sets, namely N10°W-10°E/80°W, E-W to N45°E, vertical and a third group with varying strike dipping 20-30°. The joint frequency is rather high with a mean joint frequency of 7.2 joints per meter. Some parts are intensly fractured. 47 % of the cores have a frequency of more than 6 joints per meter and only 15 % are known to have a low frequency. Joint fillings are rare but a coating of chlorite on the joint surfaces is often observed.

4.2 Water pump test

Water pump tests were performed after drilling of each borehole. A double packer unit, with test section of 3 m length, was used. Some holes were analysed with single packer system. The water pressure during a test was 5 bars except for the uppermost parts of the borehole. The result of the testing show that the bedrock is very tight as 94 % of the tested boreholes have a leakage which is less than 0.1 l/min·m·bar.

4.3 Virgin stresses

The in situ stresses were measured in the vertical borehole D1, Fig. 4. An overcoring technique for determining the 3-D stress field was used. Hitscher et al., 1979. The measurements were taken at three different levels, all within the volume of the planned cavern, Ingevald and Strindell, 1979,. The residual stresses were measured and found to be low. The maximum principle stress, σ_1, is horizontal and oriented E-W with a magnitude of 6 MPa. The intermedium stress, σ_2, has the magnitude of 4 MPa and is oriented N-S in the horizontal plane. The minimum principle stress, σ_3, is vertical with a magnitude varying between zero and 2.5 MPa and is in accordance with the loading of the overburden.

4.4 Elastic parameters and rock strength

The uniaxial compressive strength was determined on 30 core samples with a core diameter of 32 mm and 42 mm. Mean uniaxial compressive strength of the granitic volcanite is 200 MPa ± 40 MPa. Young's Modulus (E_{lab}) and Poisson's ratio (ν) were determined from a compressive test and a bending test. The mean value for Young's Modulus of the granitic volcanite is 85 GPa ± 5 GPa and a Poisson's ratio equal to 0.25.

4.5 Geohydrology

The groundwater table has been observed in six boreholes in the rock mass and one hole in the soil. The holes were located in the vicinity of the cavern, Fig. 4, and observations started 8 months prior the excavation.

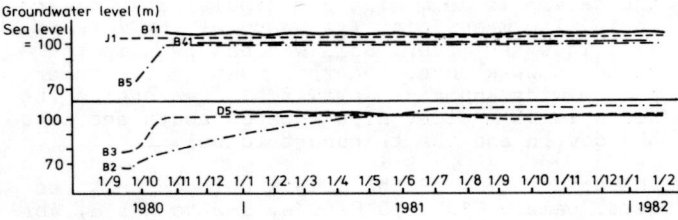

Fig. 5 Variation of groundwater level during excavation

The groundwater level was kept constant during the excavation phase, Fig. 5. From simultaneous registrations of temperature and rain fall hardly any seasonal variation of the level can be found. A raise of the water table is shown in borehole B2 which is due to air blown out after completion of the drilling. Since that time the water table has raised 40 m to attain a constant level. Borehole D5, located 22 m from the cavern, has been affected by the excavation as the water table sunk 2.7 m.

5 CONSTRUCTION WORK

5.1 Smoth blasting

The cavern was blasted with a gallery and three benches. The gallery section is 125 m^2 and each bench is 6.3 m high. The safety requirements demanded smooth blasting and contour blasting. The contour holes were spaced 0.6 m and loaded with Gurit 0.25 kg/m borehole. The two next holes in the row demanded limited concentration and were loaded with 0.52 kg/m and 0.91 kg/m borehole explosives to avoid damage of the rock mass. The floor of the cavern is smoothly blasted to limit the damage, as the water pools are founded directly on the intact rock mass.

The contour of the cavern was measured at 5 m-invervals in the storage room and at 10 m-intervals in the transport tunnel. In average the contour is located 0.3-0.4 m outside the theoretical contour. The maximum deviation is 1.2 m and a block fall at the eastern corner of the transept placed the wall about 3 m outside the theoretical contour. This situations were caused by unfavourable intersection of rock structures.

The precision of the drilling for the blasting holes is good as the borehole pattern has remained visible in full length at the contour. Sharp corners were rounded off to avoid stress concentrations.

5.2 Excavation equipment

98000 m^3 of rock has been excavated underground and 65000 m^3 on the surface. The excavation phase has involved two shifts per day for drilling, blasting, loading and transportation. Rock support was conducted at the night shift.

Drilling for blasting were made by a Bromec T470 with 3 boms drill rig. Loading was made by a CAT 980B and transport by three Kockums 32 tons trucks.

6. CONTROL OF ROCK MASS STABILITY

To ensure rock stability during the excavation phase and in the future the rock mass has been inspected in detail by extensive mapping of the geology and measurements of rock mass deformation. Finite element analyses were undertaken prior to the excavation for planning of field measurements, analyses of stress concentrations and deformations around the cavern. Data from the geological mapping has assisted the designing of the rock support.

6.1 Geological situation

6.1.1 Ground surface

The detailed geological mapping on the surface took place at the foundations of the auxiliary, Fig. 1. Here the rock is dominated by a granitic volcanite cut by aplite and pegmatite dykes, Eriksson, K., 1982,. One weakness zone cuts the foundation in N60°E/75°S. Two joint sets are identified with orientation N60-70°E/30°S and N60-70°E/80°S. A few joints have fillings but no swelling clay was detacted.

6.1.2 Transport tunnel

The rock observed in the transport tunnel between sections 1/140 and 1/150 is granitic volcanite, Fig. 6. Further along the bedrock is dominated by vulcanite. Aplite dykes are mainly found in the southern parts of the tunnel, while pegmatites are frequent in the northern end. During the excavation two weakness zones with clay seams were transected at section 1/275 and 1/400. The joints are grouped into three sets oriented in N50°E/90°, N65°W/80°S and N25°W/90°. Subhorizontal joints are met within section 1/400-1/500.

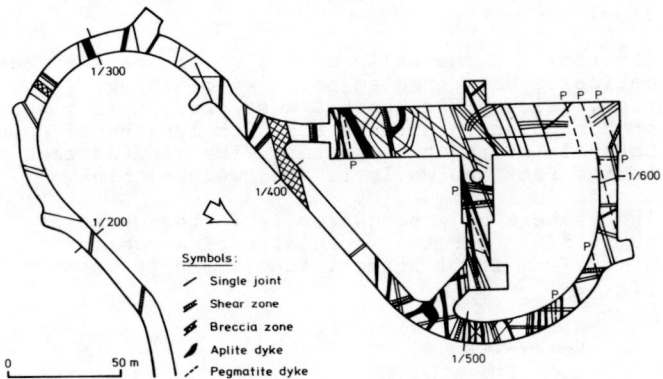

Fig. 6 Simplified geological and structural map of transport tunnel and storage room.

6.1.3 Rock cavern

The bedrock in the rock cavern is dominated by granitic volcanite. Several aplite and pegmatite dykes cut the cavern at high angles. Usually, the rock mass is more fractured in the volcanic rocks. Three sets of joints are observed, N60°E/80°S, N30°E/80°NW, and N20°W/80°E-80°W. The joints belonging to the last set often are coated with chlorite. As this set is almost subparallel to the length of the cavern, several rock falls have occured but non have caused severe problems. Locally a fourth set of joints is observed in the southern part of the cavern and oriented about E-W/30°S. In the northern part another set is observed in N45°W/80-90°N.

Joints oriented in N30°E/80°W have caused stability problems in the southeastern and northwestern corners at the crosscut of the cavern and the transept.

6.2 Deformation measurements

Movements of the rock mass around the cavern have been measured in two sections of the walls and along two vertical boreholes in the roof. The convergency of the contour of the cavern was measured by the Distometer ISETH and the displacement of the rock mass in the walls were recorded by the Sliding Micrometer ISETH, Kovari et al., 1979. The accuracy of the Sliding Micrometer is ± 0.003 mm within a distance of 1 m, and ± 0.02 mm at a measuring length below 20 m for the Distometer.

The measuring program involves 10 boreholes, M1-M10, of which M1 and M2 are vertically drilled from the surface, Fig. 7. M3-M6 are drilled in the gallery and M7-M10 in bench 1. These boreholes are 23 m in length and oriented horisontally. Convergency measuring points are located at each borehole in the walls and in the roof below boreholes M1 and M2. Fourteen measuring lines were selected all together. M1 and M5 were however both damaged during the excavation.

The total deformation of the cavern in relation to the excavation stages is presented in Fig. 7. The result is based on convergence measurements of the roof and bench 2 and a mean value of data from the sliding micrometer and the distometer for the gallery and bench 1.

The roof is heaved to a maximum value of 2 mm in the northern part of the cavern when bench 2 and 3 were passed. Later the roof has sagged and almost recovered to 0.4-0.5 mm above original level after completion of the cavern. The roof in the northern part of the cavern is likely to behave similar. However, data are lost because of damage of measuring point M1.

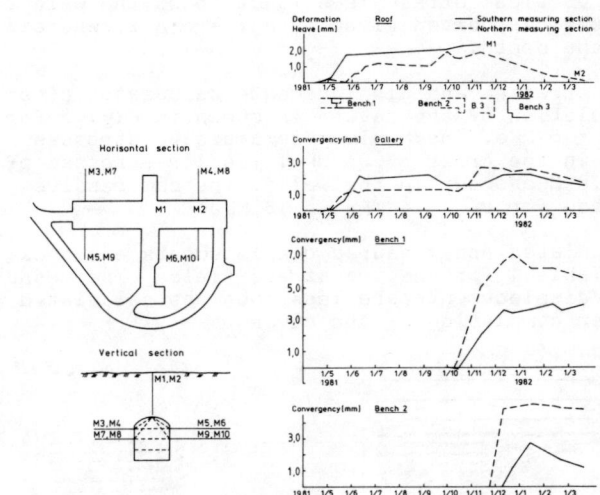

Fig. 7 Measured rock mass movements during excavation

The walls in the gallery show the maximum deformation when benches 1 and 2 are passing. For the southern section the deformation increases when benches 2 and 3 are passed. The finite convergency is small and is found to be 1.7 mm and 2.1 mm in the southern and the northern sections. The largest convergency occur for bench 1 when excavating bench 2 and 3. Then the finite deformation in the northern section is 3.5 mm and 6 mm in the southern. In conclusion it can be stated that in general the displacements are less in the southern section. This is probably due to a lower joint frequency.

6.3 Finite element analyses

A two-dimensional linear finite element analyses was performed to evaluate the stresses and displacements around the cavern, Jonasson, P., 1980. The model contains 1083 elements with 1150 node points.

The material parameters were chozen as E_{mass} = 30 GPa, which is 1/3 of Young's Modulus for intact rock, Poisson's ratio = 0.27 and the density of rock (ρ) to 2.6 t/m^3.

Table 1. Measured and calculated displacements for σ_{hi} = 4 and 8 MPa

	Displacements when excavating bench 1 (mm)				Displacements when excavating bench 3 (mm)			
	Calculated		Measured		Calculated		Measured	
	4 MPa	8 MPa	South. sect.	North. sect.	4 MPa	8 MPa	South. sect.	North. sect.
Roof	0.37	0.78	-	1.0	1.24	2.59	-	0.4
Gallery	1.01	2.04	2.2	1.2	4.8	9.7	1.7	2.1
Bench 1	-	-	-	-	5.2	10.4	3.5	6.2
Bench 2	-	-	-	-	7.7	15.8	1.5	3.8

The excavation was simulated for three stages, namely excavation of gallery bench 1 and complete cavern. As the virgin horisontal stresses varies between 4 and 8 MPa two calculations were performed, where σ_{hi} was taken as 4 MPa and 8 MPa. The vertical stress is a function of the weight of the overburden given by $\sigma_{vi} = \rho \cdot g \cdot z$, where z is the depth.

The stresses and displacements calculated after completion of the cavern is shown in Fig. 8 for σ_{hi} = 8 MPa. The maximum compressive stresses are in the order of 25 MPa. Tensile stresses of - 5.8 MPa occur in the walls. The compressive stress for σ_{hi} = 4 MPa is 13 MPa.

Calculated and measured displacements are shown in Table 1 for the two stress fields. The measured displacements are less then the calculated after completion of the cavern

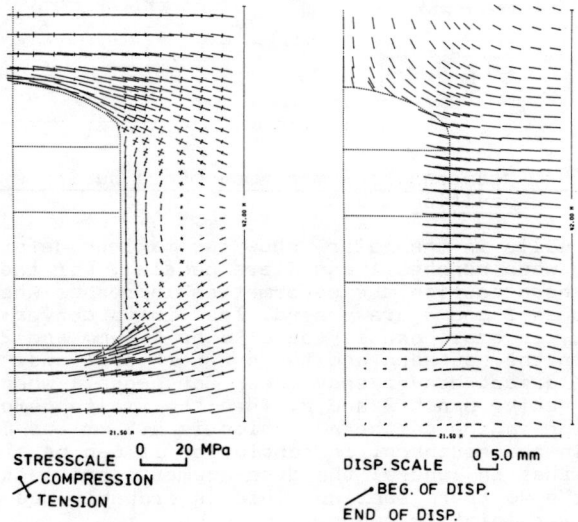

Fig. 8 FEM analyses showing calculated stresses (8a) and displacements (8b) at a virgin stress of 8 MPa

7 ROCK SUPPORT

7.1 Support program

The rock support for the CLAB cavern comprices grouting, rock bolting, anchoring, and shotcreting. Concrete arches and columns have been constructed in areas of bad rock quality.

7.2 Grouting

To minimize leakage of groundwater into the cavern and to stabilize the rock mass injection of cement grout into open joints have been conducted. Boreholes for rock bolting and ancoring have also been grouted in order to eleminate corrosion.

The roof and the walls of the cavern have systematically been grouted prior to blasting. The roof, gallery and bench 1 were grouted from seven boreholes in a fan shape with lengths of each borehole of about 6.0 meters. The grouted zone in the roof and walls is about 4.5 m thick.

The leakage of groundwater is in the order of 40-50 l/min for the whole site of which most comes from the transport tunnel and the shafts, Fig. 9.

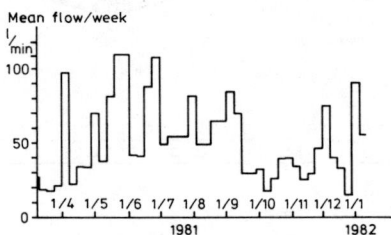

Fig. 9 Inflow of water in transport tunnel, storage room and shafts.

7.3 Rockbolts, anchors and shotcrete

The rock bolts are of steel (Ks 40) with diameter 25 mm and the length varying between 3 and 8 m. All bolts are fully grouted. The anchors are of type Dywidag with maximum lengths of 8 m and diameter 35 mm. The shotcrete is a mixture of cement, water and 8 mm sand.

7.3.1 Transport tunnel

The permanent support of the roof of the transport tunnel is 50 mm shotcrete without reinforcement. Individual rock bolts were placed selectivly along the tunnel in order to prevent single rock falls. Extra support were needed at three sections of weak rocks located at 1/274-1/287, 1/376-1/435 and 1/460-1/550, Fig. 6. Here the thickness of the shotcrete was increased to 100 mm with reinforcement and a systematic rock bolting of 1 bolt/4 m^2.

7.3.2 Rock cavern

The permanent support of the cavern consists of 100 mm reinforced shotcrete in the roof and 50 mm shotcrete on the walls. At the transept 50 mm of shotcrete covers both roof and walls. The thickness of the shotcrete has been controlled at each 200 m^2 of emplaced shotcrete. Ten samples were taken from each test area. The strength of the shotcrete is tested for each 500 m^2 of shotcreting. More than 50 % of the samples have a tensile strength > 1 MPa.

The roof has been systematically anchored with two anchors per m². The length of the bolts vary between 4 and 6 m. Bench 1 is systematically bolted with slake bolts 6.4 m in length and a density of one bolt per 2 m². The bolting of bench 2 and 3 is made selectively with bolts 6 m in length to prevent local rock fall, Fig. 10.

The transept tunnel and the fuel transport shaft has sytematically been reinforced with 2 bolts per m² in lengths of 4 to 6 meter.

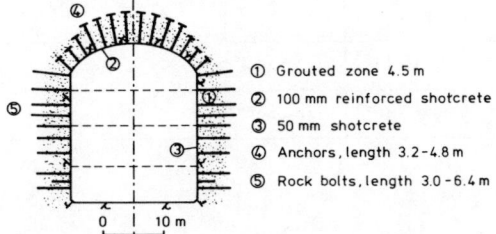

① Grouted zone 4.5 m
② 100 mm reinforced shotcrete
③ 50 mm shotcrete
④ Anchors, length 3.2-4.8 m
⑤ Rock bolts, length 3.0-6.4 m

Fig. 10 Permanent support of the storage room

7.3.3 Special support

The roof at the crosscut between the cavern and the transept tunnel is supported by six strongly reinforced concrete arches. Two of the arches are placed diagonally and four perpendicular to the cavern and the transept. The arches rest on reinforced concrete coloumns placed in the corners of the crosscut. Their thickness is 0.5-1.0 m and the length varies between 4 and 8 m. The rock mass at the corners are bolted selectively with slake fully grouted bolts 8 m in length.

The travers rest on concrete cantilevers. Each cantilever is anchord to the rock wall by one rock anchor type Dywiday ST110/125 GWS, diameter 36 mm and stressed to 83 tons. Five 8 m long slake fully grouted bolts of diameter 25 mm are also supporting the cantilever.

8 FOUNDATION OF THE WATER POOLS

The water pools form a single reinforced concrete monolithic unit with dimensions 96 x 17 x 15 m. The walls and the bottom are 1.5 and 1.7 m in thickness respectively. The central pool, 17 x 17 m in dimension, is founded directly on the unlined rock, Fig. 1.

The pressure on the rock surface from the water filled monolithics is calculated to give a mean value of 0.6-0.7 MPa and a maximum value of 2-3 MPa. The construction is very stiff and sensitive for unequal subsidence of the rock. The FEM calculations indicate a total subsidence of 0.5-2 mm in the floor due to this weight. Therefore an acceptable tolerance of the even subsidence was set to 2 mm over a length of 40 m of the concrete foundation.

8.1 Rock mass characteristics

The rock mass forming the floor of the cavern was investigated in detail to evaluate the rock characteristics and the rock mass deformability.

Twelve vertical diamond boreholes, equally spaced on the floor, were drilled to a depth of 10 m. In addition four short holes were drilled to depths between 2 m and 4 m. Rock type, joint frequency and RQD were determined from the cores and on the floor. The damage caused by blasting was evaluated.

A volcanic rock dominates the northwestern part of the floor, while more granitic rocks occur in the southern and eastern parts. The joints end their frequency is the same as for other parts of the cavern. The RQD value has a mean value of 71 for the twelve long holes and 45.9 for the shorter. The lower value is due to higher frequency of joints close to the floor, where the rock is damaged by blasting.

Water pump tests were performed in all boreholes, using a single packer unit placed at the depths of 0.3 m, 1.0 m and 3.0 m. The pressure was 3.6 and 10 bars during 10 minutes. The time was for one borehole increased to 5 hours to test the long term leakage. No leakage was detected in eleven of the sixteen boreholes. The maximum leakage, $2.7 \cdot 10^{-2}$ l/min·m·bar, occured close to the floor. The conclusion is that the rock mass is fairly tight.

8.2 Rock mass deformability

The rock mass deformability was evaluated by back calculation from measured movements in the wall and by using the reduction factor $E_{rock\,mass}/E_{lab}$ versus RQD, according to Deer et al., 1966.

8.2.1 Back calculatins

The calculation of the rock mass deformability is based on measured virgin stresses, FEM-calculations and the measured movements in the walls. The assumption is that

$$\varepsilon_{fem} \cdot E_{fem} = \varepsilon_{measured} \cdot E_{rock\,mass} \qquad (1)$$

The value of E_{fem} is chosen to 30 GPa, i.e. 1/3 of the value for intact rock. The calculations were made for two magnitudes of initial horizontal stresses, namely 8 and 4 MPa.

The measured values of the movements in the walls have been evaluated from the first 11 meter of the extensometers as this part of the rock mass is most likely to represent the behaviour of the rock mass in the floor. The single readings from every one meter intervall along the extensometer clearly indicate that the major movements occur within the first 10 to 12 meters from the contour, with one exception for measuring point M4. Table 2 gives the result of the calculated rock mass modulus from the gallery and the bench 1 during excavation of bench 1 and 3. The eastern and western walls behaved differently. The western wall has a modulus almost three times higher than the eastern wall as observed in the gallery. This difference is less pronounced when comparing the situation for bench 1.

8.2.2 Reduction factor E_{mass}/E_{lab}

The other approach was based on the reduction factor ratio E_{mass}/E_{lab} versus RQD, according to Deer et al., 1966.

Table 2. Measured and calculated displacements

Excavation stage		Bench 1					Bench 3				
Measuring point		Measured displacement ε (mm)	Calculated displacement (mm)		Calculated $E_{rock\ mass}$ (GPa)		Measured displacement ε (mm)	Calculated displacement (mm)		Calculated $E_{rock\ mass}$ (GPa)	
			σ_1 = 8 MPa	σ_1 = 4 MPa	σ_1 = 8 MPa	σ_1 = 4 MPa		σ_1 = 8 MPa	σ_1 = 4 MPa	σ_1 = 8 MPa	σ_1 = 4 MPa
Gallery	Wall E	$0.81 \cdot 10^{-4}$	$0.55 \cdot 10^{-4}$	$0.29 \cdot 10^{-4}$	20	11	$0.13 \cdot 10^{-3}$	$0.34 \cdot 10^{-4}$	$0.18 \cdot 10^{-4}$	12	7
	W	$0.33 \cdot 10^{-4}$	$0.55 \cdot 10^{-4}$	$0.29 \cdot 10^{-4}$	50	26	$0.23 \cdot 10^{-4}$	$0.34 \cdot 10^{-4}$	$0.18 \cdot 10^{-4}$	57	30
Bench 1	E	—					$0.24 \cdot 10^{-3}$	$0.91 \cdot 10^{-4}$	$0.48 \cdot 10^{-4}$	11	6
	W	—					$0.14 \cdot 10^{-3}$	$0.91 \cdot 10^{-4}$	$0.48 \cdot 10^{-4}$	20	10

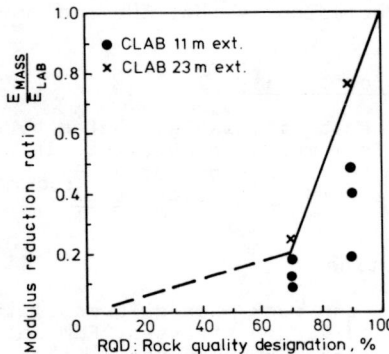

Fig. 11 Reduction ratio E_{mass}/E_{lab} in relation to RQD. Regression line after Bieniawski, 1978.

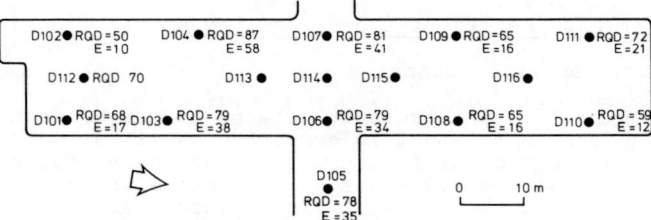

Fig. 12 Variation of modulus of deformability (E) and RQD in the floor

For the ratio E_{mass}/E_{lab} the value of E_{lab} is determined from uniaxial compressive tests, which gave a mean value of 85 GPa. A mean value of RQD has been calculated for each diamond drilled hole. The value of the modulus of deformability is a mean value estimated for each borehole. In order to get the relationship between RQD and the reduction factor, values were taken from the measurements performed in the walls, Table 2, where the RQD-value for the eastern wall at the measuring point is 70 % and 95 % for the western wall. The regression line with correlation coefficient 0.554 in Fig. 11 is taken from Bieniawski, 1978. The modulus of deformability in the floor varies between 10 GPa to 50 GPa, Fig. 12, with a mean value of 26 GPa. This value is almost double that of the results from the back calculation for fully excavated cavern. For design perposes a value of 15-25 GPa has been recommended. This value is also in agreement with practical results from the water pump tests.

9 CONCLUSIONS

1) The cavern for storage of spent nuclear fuel has been successfully excavated and the safety requirements has been fulfilled.

2) Through thorough geological mapping, rock stress measurements and theoretical analysis rock movements and rock stability have been controlled during excavation.

3) The excavation has been performed without any severe problems despite a rather fractured rock mass. This favourable situation is probably due to a combination of moderate rock stress, small leakage of water and very few clay filled joints. A second factor is the blasting technique by using smooth and contour blasting in order to minimize rock damage.

4) Extensive rock support has been installed and the cost of the reinforcement is about 50 % more than installations for conventional underground caverns under similar rock conditions.

5) The blasting started in May 1980 and the cavern was excavated in January 1982. Two waterfilled pools for spent fuel are in construction and the first elements of spent fuel will be stored in 1985.

6) The total final cost for CLAB will be 1600 million SEK. Excavation of tunnels, shafts and cavern is 14 million and cost for support is in order of 13 million. The cost of site investigation is 0.65 million and for rock control 1.4 million or 7.6 % of the cost of the underground work.

10 ACKNOWLEDGEMENTS

The authors wish to acknowledge Mr Kurt Angéus, SKBF for giving the permission to publish this article. Special thanks are given to Mr Hans Norrby and Mr Sigge Halvarsson SSPB and Mr Tommy Hedman OKG, for their contributions.

11 REFERENCES

Bieniawski, Z. T. 1978. Determining rock mass deformability: Experience from case histories. Int. J. Rock Mech. Min. Sci & Geomech. Abstr (15) 237-247.

Deer, P. U., Hendron, A. J., Patton, F. D., and Cording, E. J. 1966. Design of surface and near surface construction in rock. Proc. 8th Symp. Rock Mech., 237-302, Minneapolis.

Eriksson, K. 1982. Geological documentation of transport-tunnel and rock cavern at CLAB. VIAK, Falun. Report No 15.1099. 30 pp. (In Swedish)

Gustavsson, B., Hedman, T., and Larsson, H. 1980. An underground storage for spent nuclear fuel in Sweden. Proc. Int. Symp. Subsurface space. (2), 881-888. Stockholm.

Hiltscher, R., Martna, J., and Strindell, L. 1979. The measurement of triaxial rock tresses in deep boreholes and the use of the rock stress measurements in the design and construction of rock openings. Proc. Int. Cong. on Rock Mech. (2), 227-234. Montreux.

Ingevald, K., and Strindell, L. 1979. Simpevarp Project CLAB, rock stress measurements. Swedish State Power Board, Stockholm. Report No BTH 3710/L-583. 11 pp. (In Swedish)

Jonasson, P. 1981. CLAB-FEM. Linear elastic calculations. Björsta Ingenjörskonsult HB, Ösmo. 9 pp. (In Swedish)

Kovari, K., Amstad, C., and Köppel, J. 1979. New developments in the instrumentation of underground openings. Proc. 4th Rapid Excavation and Tunneling Conf., 817-834. Atlanta.

Kulhanek, O., and Wahlström, R. 1977. Earthquakes of Sweden 1891-1957. KBS Teknisk Rapport No. 21. Stockholm.

Moberg, M. 1979. CLAB; Simpevarp geological conditions. Swedish State Power Board, Stockholm. Report No BSU3-MM/MB-3532. 6 pp. (In Swedish)

UK ROCK MECHANICS RESEARCH FOR RADIOACTIVE WASTE DISPOSAL

Recherches en cours au Royaume Uni sur la mécanique des roches en ce qui concerne l'entreposage des déchets radioactifs

Felsmechanikforschungen in Grossbritannien für die Endlagerung radioaktiver Abfallstoffe

J. A. Hudson
Principal Scientific Officer, Geotechnics Division, Building Research Establishment,
Garston, Watford, WD2 7JR, UK

SYNOPSIS

The design of a repository in rock for disposal of radioactive waste requires a knowledge of many rock mechanics factors which includes the rock structure, properties associated with the migration of radionuclides and the in situ stress field. It is also necessary to understand how the factors influence the excavation, support and backfilling techniques that could be used in construction. These subjects and the interrelations between them are discussed within the context of the rock mechanics projects being co-ordinated by the Building Research Establishment as part of the UK research programme for radioactive waste disposal.

RESUME

L'élaboration de projet d'une galerie rocheuse destinée à recevoir des déchets radioactifs nécessite la connaissance de beaucoup de facteurs de la mécanique des roches, y compris la structure des roches, les propriétés associées à la migration des radionuclides et les contraintes rocheuses. Il faut aussi comprendre la façcon dont ces facteurs influent sur les techniques d'excavation, de support et de remblayage à employer au cours de construction. Les sujets sus-mentionnés et les rapports entre eux se présentent dans le contexte des projets de mécanique des roches coordonnés par le Building Research Establishment et qui font partie du programme de recherches du Royaume Uni concernant la gestion des déchets radioactifs.

ZUSAMMENFASSUNG

Die Planung einer Felskavernenanlage für die Endlagerung radioaktiver Abfallstoffe verlangt, daß man eine Kenntnis vieler Felsmechanikfaktoren besitzt — darunter die Struktur des Felsens, die Eigenschaften, die sich mit der Wanderung von Radionukliden assoziieren, und der Spannungszustand. Man muß verstehen, wie diese Faktoren die Aushub-, Abstütz- und Auffüllungsmethoden beeinflussen, die in der Konstruktion verwendet werden können. Diese Themen und ihre Beziehungen zueinander werden im Kontext der Felsmechanikprojekte diskutiert, die als Teil des britischen Forschungsprogramms für die Endlagerung radioaktiver Abfallstoffe durch das Building Research Establishment koordiniert werden.

INTRODUCTION

Because the rock mechanics factors in the design of a land based radioactive waste repository are both complex and interactive, the development of an adequate research programme naturally focuses attention on the fundamental knowledge that is required. Previously, the UK research programme included a wide range of studies on high-level waste disposal, and reports produced for the Department of the Environment (Mott, Hay and Anderson, 1981; Sir William Halcrow and Partners, 1981; Principia Mechanica, 1981) on high-level waste repository design and instrumentation have drawn attention to a variety of rock mechanics subjects. As a result of the decision to store high-level waste for many years, research is now concentrated on disposal of low and intermediate level wastes. The rock mechanics aspects of intermediate level waste disposal are similar to those of high-level waste apart from the thermal aspects of high-level waste. In fact, with the exception of rock properties directly associated with sorption of radionuclides on to rock surfaces, the rock mechanics problems are fundamental in nature and common to most civil and mining engineering projects.

This is illustrated by a report on rock mechanics research requirements related to energy and mineral resource development, civil and defence construction, and earthquake hazard reduction (US National Committee for Rock Mechanics, 1981) which has provided advice on the requirements and priorities for rock mechanics research in general. The major subjects covered by this report are permeability, in-situ stress, fractures, excavation, in-situ rock mass characteristics, thermal properties and numerical modelling. All of these subjects are relevant to the design of a radioactive waste repository and consequently many, if not most, of the key research areas identified are also relevant.

In the UK radioactive waste disposal research programme, which is managed by the Department of the Environment and includes all aspects of disposal, several rock mechanics projects are underway. In this paper, the projects that are either being conducted or co-ordinated by the Building Research Establishment are outlined. Other work with a rock mechanics connection, mainly being undertaken by the UK Atomic Energy Authority and the Institute of Geological Sciences has been reported elsewhere (eg P J Bourke et al, 1979; McEwen and Lintern, 1980).

The overall repository design concept is explained in the next section and followed by separate discussions of rock mass properties, permeability and in-situ stress. The special considerations for excavation and support are also explained.

DESIGN CONSIDERATIONS

It is useful to summarise briefly the fundamental objective of radioactive waste disposal in order to consider the special context within which the rock mechanics information is required. The fundamental objective is to ensure that unacceptable quantities of radionuclides do not migrate from the repository to the biosphere. The rock mechanics information is therefore required for:

(a) conventional aspects of excavation and support, and
(b) studies of radionuclide migration and safety analyses.

Thus, the importance of a given rock mechanics parameter depends not only on conventional aspects of underground rock engineering but also on its importance in the safety analysis. Any anisotropy in a rock of relatively low permeability, for example, would be of little concern in mining design but could be of paramount importance for a radioactive waste repository.

For modelling the mechanical behaviour and radionuclide migration, there are two main requirements: firstly, the modelling must accurately reflect the physical processes; and, secondly, the properties and boundary conditions of the rock mass must be known. There are considerable rock mechanics problems associated with both these requirements. The modelling difficulty, certainly in the near-field, is establishing whether a continuum approach can be valid and, if so, under what circumstances. The rock property and boundary condition difficulty relates to parameters under three headings:

(i) Basic properties of the rock mass (fracturing, stiffness, etc).
(ii) Migration associated parameters (permeability, sorption characteristics).
(iii) Boundary conditions (ie the in-situ stress field, hydrogeological regime).

It is important to note that although these headings provide a guideline for considering the requirements, there are inter-relations between them. For example, the jointing characteristics will generally dictate the permeability, the permeability is affected by the in-situ stress and a stronger rock mass can sustain a higher stress. Moreover, the design life of the repository is extremely long so the modelling, rock properties and boundary conditions assumed should be valid not only at the time of site investigation and directly after construction but for hundreds of years into the future (and for thousands of years in the case of high-level waste).

Within this context, the rock mechanics research is concentrated on understanding rock mass structure, modelling of water flow through rock masses and in-situ stress.

ROCK MASS PROPERTIES

The single most important aspect of in-situ rock masses for radioactive waste disposal is the properties of the discontinuities. This is because they govern both the mechanical and fluid flow properties of most rock masses. Additionally, and because the fracturing in rock is neither uniform in orientation nor location, the discontinuities cause anisotropy and inhomogeneity of these properties. All the discontinuity characteristics of a given rock mass cannot be obtained specifically and a statistical characterisation is therefore necessary. Boreholes, shafts and tunnels provide the access for specific information at restricted locations; further assessment is possible via indirect techniques such as measurement of seismic velocities and wave attenuation. From these data, it is possible to provide the statistical base for generating a synthetic but equivalent rock structure for modelling. It is ironic perhaps that, for radioactive waste disposal where a maximum amount of information is required about the rock mass, the number of boreholes may have to be minimised in order to avoid potential preferential migration paths for the radionuclides.

The difficulty in characterising the geometry of rock mass discontinuities is illustrated in Figs 1 and 2.

Fig 1 Discontinuities in Carboniferous sandstone (the scale is indicated by the 1 m long white strips)

For the rock face in Fig 1, the intensity and orientation of the discontinuities can readily be estimated but the distribution of trace length and aperture, even in this relatively simple case, are not determined simply, yet these four characteristics are an absolute minimum for defining the discontinuity geometry.

In Fig 2, representing a larger area of rock, it can be seen that the combination of long trace length and large aperture will cause some of the discontinuities to have far greater mechanical and fluid flow significance than others, thus underlining the importance of these two characteristics.

Fig 2 Discontinuities in Cambrian sandstone (the white strip is 1 m long)

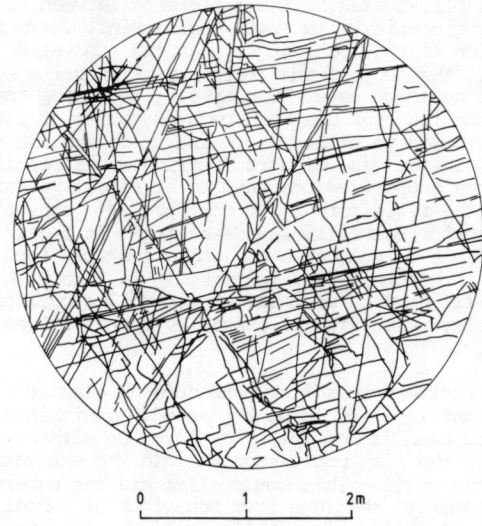

Fig 3 Line drawing of the discontinuities in the central part of the rock face illustrated in Fig 2

The line drawing in Fig 3 illustrates another feature of the discontinuity pattern - that of the connectivity of the discontinuities. The characteristics of intensity, orientation, trace length and aperture can be established from a systematic study of the individual discontinuities but those of the connectivity cannot. The connectivity is a high level feature, being a function of the pattern as a whole rather than as a summation of the individual components. This subject will be discussed in the next section on permeability.

A joint research programme involving the Building Research Establishment and Imperial College is concerned with the problems of analysing discontinuity arrays and assessing the discontinuity characteristics in three dimensions with optimal effectiveness. Completed work has resulted in a method for estimating fracture frequency in any direction through a given rock mass, including algorithms for establishing the directions and magnitudes of maximum and minimum fracture frequency (Hudson and Priest, 1982). Further work is reported in another paper to this Congress (Priest and Samaniego, 1983). The results of these theoretical studies will assist in both the modelling of discontinuous rock masses and the development of site investigation programmes. It is expected that future theoretical work will assist in the understanding of the anisotropy of rock stiffness and other properties as a function of the discontinuity geometry, together with the ability to establish through sensitivity analyses when the rock can validly be modelled as a continuum.

A programme of field work has been commenced in association with Golder Associates on the methodology of site assessment with specific reference to the adequacy of existing discontinuity survey techniques. It is hoped that improved computer programs will be developed for the reduction and analysis of site data. The use of seismic techniques to evaluate rock fracture characteristics mentioned earlier is being studied by the Transport and Road Research Laboratory as part of the field work.

PERMEABILITY

Table I illustrates the wide variation in the in-situ permeability of rocks (expressed with reference to water flow as hydraulic conductivity in cm/s).

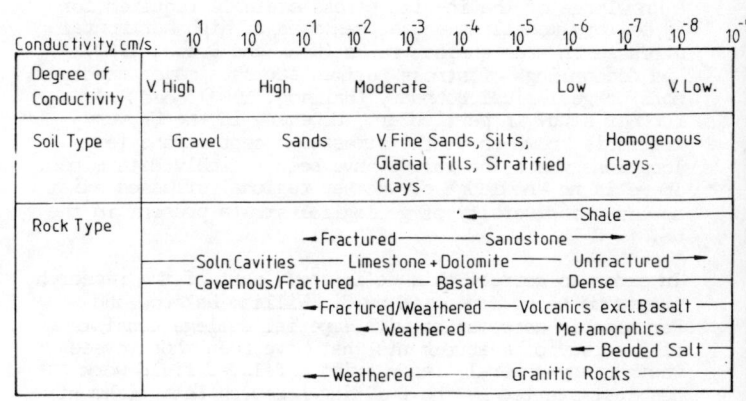

Table I In-situ values of hydraulic conductivity (from Garritty and Farmer, 1982)

The wide variation for a particular rock type is to a large extent caused by differences in discontinuity characteristics. This raised the question of whether water flow modelling can be successful using only a continuum approach. Also, and as Garritty and Farmer, have pointed out, the in-situ values are generally obtained from vertical boreholes with a resultant bias towards the horizontal conductivity of the rock mass. A detailed report on the modelling and sampling problems has been produced by Rogiers et al, 1979.

The modelling of discontinuity arrays both as a discontinuum and as an equivalent continuum is being studied in the joint research programme of the Building Research Establishment and Imperial College, and by Geognosis. Early work involved the development of printed circuits as analogues for rock jointing patterns in the two-dimensional case (Hudson, 1981). Measurements of resistance through printed circuits made from computer graphics gives an indication of the anisotropy in equivalent path length through a variety of arrays. Subsequent work has concentrated on numerical modelling of jointing patterns and water flow for both the two and three dimensional cases.

The objective of the work is to make a comparison of the discontinuum versus continuum approaches and hence to study the sensitivity of any differences to the discontinuity characteristics. In this way, it will be possible to state under what circumstances the continuum approach is valid, with the subsequent ramifications for the requirements of a site investigation programme. Furthermore, the results will assist in the interpretation of tracer tests which are likely to prove of great value in in-situ testing (Bourke, 1979; Tester and Potter, 1979).

A numerical modelling technique with great potential for this subject is Monte Carlo simulation. Having established a certain discontinuity pattern on the computer, water flow can be studied by 'firing' particles through the network of planes and compiling a distribution of exit times. This represents a direct modelling of water flow and tracer tests with the added advantage that radionuclide interaction with rock surfaces through sorption and desorption could be included simply.

IN-SITU STRESS

A knowledge of the in-situ stress state is required for repository modelling. The phenomena of high horizontal stresses in near surface rocks (Hoek and Brown, 1980) and decoupling of stress regimes across a major fracture zone or geological boundary (Haimson, 1980) need further study in particular. However, in the UK very little is known about the stresses except at the few locations where the values have been reliably determined. There is no knowledge of whether regional stresses exist across the diversity of geological strata present in the UK.

In order to correct this deficiency, part of the research programme being conducted by Sir William Halcrow and Partners in association with Imperial College consists of a survey of measurements that have been made to see whether any general trends exist. Related field work has just started by the Building Research Establishment and Golder Associates to evaluate different stress measurement techniques in order that recommendations can be made for future site evaluation instrumentation. Hydraulic fracturing and two types of overcoring systems will be used, one of the latter being a system developed at Newcastle University.

EXCAVATION

A research project related to permeability and stresses concerns the damage to the rock mass which was mentioned earlier. It would be inappropriate to model radionuclide migration using values determined from the site investigation before construction if the values were significantly altered as a result of excavation. This consideration serves as a good example of the interactive nature of the rock mechanics problems with the whole repository design, the site investigation, the numerical modelling, the safety analysis and the construction itself. The layout of the repository can be designed in relation to the rock mass structure to minimise water flow but the influence of excavation induced fractures must also be considered. Research at the University of Newcastle is aimed at qualitatively and quantitatively establishing the effect of excavation on near-field permeability through studies of the extent of the yield or fracture zone, the damage occurring as a result of stress redistribution and the excavation process itself.

It is likely that excavation techniques using explosives will have to be modified to minimise the damage caused by the stress waves and gas pressure. Alternatively, a tunnel boring machine could be used; then the damage caused by the excavation process itself would be much less. But until adequate models of radionuclide migration are developed it is not clear

(a) whether the whole repository layout being designed to minimise intersection with discontinuities is most important,
(b) to what extent damage should be avoided,
(c) whether the near-field parameters need to be determined after excavation, or
(d) whether any of these are significant compared to the integrity of the far-field geological environment.

SUPPORT AND BACKFILLING

A difficulty in considering the rock mechanics aspects of the support requirements is the fundamental design philosophy of the support, bearing in mind the very long design life of the repository. The two main philosophies are well illustrated by the difference between civil and mining engineering. In civil engineering, where the excavation is required rather than the excavated material, the support must retain its integrity so the excavation can be used. Conversely, in mining engineering, once the rock has been removed, parts of the mine can be allowed to collapse, depending on the mining method. In the case of a radioactive waste repository, should the support be designed to be effective for the life of the repository or to collapse in a controlled way following backfilling and sealing? In fact, it is not possible to guarantee any support system for hundreds or thousands of years so it is likely that the design will have to be based on controlled collapse, thus replacing artificial stability by natural geological stability.

For a repository in soft ground such as clay or some shales, the waste could be emplaced from within a pipe which had been jacked into the clay. On withdrawing the pipe, the clay would squeeze into the excavation around the waste. This would eliminate the uncertainty associated with the long term behaviour of tunnel linings. A review of stability ratios and their relation to construction techniques in possible repository candidate clay formations has been compiled by Hudson and Boden (1982).

In harder rocks, such collapse could not be easily achieved without significantly affecting the near-field permeability so backfilling with a material having some load-bearing capability will probably be necessary, as is the case with the stopes in certain mining methods. The backfilling material will have to be chosen very carefully because it has to serve the additional function of providing a barrier to radionuclide migration.

CONCLUSIONS

The rock mechanics components of the UK radioactive waste disposal research programme have been discussed with specific reference to the intra- and extra-mural work of the Building Research Establishment. The subjects on which most emphasis is being placed are

(i) the rock mass discontinuities,
(ii) discontinuum versus continuum methods for modelling water flow, and
(iii) in-situ stress.

The discontinuities are crucially important because they govern not only the mechanical properties of rock masses but also the permeability, and much more work is required on the validity of modelling radionuclide migration using continuum models. In-situ stress is being emphasised because a knowledge of the stress state is required for design, yet currently very little is known about the in-situ state of stress in UK rocks. The work is being conducted to develop a fundamental understanding of the influence of discontinuities and stress on rock mass behaviour and permeability so that the appropriate site assessment methodologies and instrumentation can be developed.

ACKNOWLEDGEMENTS

The research described in this paper is guided and supported by the Radioactive Waste Professional Division (Head, Dr F S Feates) of the Department of the Environment. The projects are being co-ordinated by the Geotechnics Division of the Geotechnics and the Structures Department at the Building Research Establishment and the paper is published with the permission of the Director of the Building Research Establishment.

REFERENCES

Bourke, P J (1979). Tests of porous permeable medium hypothesis for flow over long distances in fractured deep hard rock: Report AERE-R 9487, Atomic Energy Research Establishment, Harwell, UK

Garritty, P and Farmer, I W (1982). The permeability of near field rocks: Unpublished report submitted to the UK Department of the Environment, 14 pp

Haimson, B C (1981). Confirmation of hydrofracturing results through comparisons with other stress measurements: Proc 22nd US Symposium on Rock Mechanics, MIT Press

Hoek, E and Brown, E T (1980). Underground excavations in rock. 527 pp. London: Institution of Mining and Metallurgy

Hudson, J A (1981). Computer graphics and printed circuits for studying rock jointing patterns: Paper presented to the Computer Applications in Geology III meeting held on 26 February 1981 at the Geological Society of London

Hudson, J A and Boden, J B (1982). Geotechnical and tunnelling aspects of radioactive waste disposal: Proceedings of the Symposium Tunnelling '82, pp 271-281, Institution of Mining and Metallurgy, London

Hudson, J A and Priest, S D (1982). Discontinuity frequency in rock masses: Submitted to the Int J Rock Mech, Min Sci and Geomech Absts

McEwen, T J and Lintern, B C (1980). Fracture analysis of the rocks of the Altnabreac area: Report ENPU 80-8, Institute of Geological Sciences, Environmental Protection Unit, Harwell, UK, 28 pp

Mott, Hay and Anderson (1981). A review and synthesis of international proposals for the disposal of high-level radioactive wastes into crystalline rock formations: Report to the United Kingdom Department of the Environment, 204 pp

Panel on Rock Mechanical Research Requirements of the US National Committee for Rock Mechanics (1981). Rock Mechanics Research Requirements. 222 pp. National Academy Press, Washington, DC

Priest, S D and Samaniego, A (1983). A model for the analysis of discontinuity characteristics in two dimensions. Paper presented to the 5th International Congress on Rock Mechanics, Melbourne, Australia

Principia Mechanica (1981). Field instrumentation and testing needs for a high-level waste repository: Report to the United Kingdom Department of the Environment

Rogiers, J C, Curran, J H and Bawden, W F (1979). Numerical modelling of flow in fractured rock masses - discontinuous versus continuous approaches. Report produced for the Earth Sciences Branch, Energy Mines and Resources, Ottawa, Canada

Sir William Halcrow and Partners (1981). Repository schemes for high-level radioactive waste disposal - review of schemes for argillaceous and saliferous formations: Report to the United Kingdom Department of the Environment, 122 pp

Tester, J W and Potter, R M (1979). Interwell tracer analyses of a hydraulically fractured granitic geothermal reservoir: Paper SPE 8270 presented at the 54th Annual Conference of the SME of AIME held at Las Vegas in September 1979, 7 pp

THERMOMECHANICAL ROOM REGION ANALYSIS OF FOUR POTENTIAL NUCLEAR WASTE REPOSITORY SITES IN SALT

Analyse thermomécanique de la zone entourant une chambre à quatre sites éventuels pour les résidus nucléaires en gisements de sel

Thermomechanical Analyse der Einlagerungsstrecke für radioaktiven Abfall an vier potentiellen Endlagerstandorten im Salz

R. A. Wagner, M. C. Loken and H. Y. Tammemagi
RE/SPEC Inc., One Concourse Drive, P.O. Box 725, Rapid City, South Dakota 57709, U.S.A.

SYNOPSIS

This paper describes a thermomechanical numerical analysis of four potential nuclear waste repositories in salt in the Unite States: Paradox Basin, Utah; Permian Basin, Texas; Richton Dome, Mississippi; and Vacherie Dome, Louisiana. The analysis considers the region encompassing the disposal room. Thermomechanical responses are compared to help assess the suitability of each site as a nuclear waste repository.

RESUME

Cette communication décrit une analyse numérique thermomécanique de quatre dépôts éventuels pour les résidus nucléaires dans des formations salifères aux Etats-Unis, à savoir: Paradox Basin, Utah; Permian Basin, Texas; Richton Dome, Mississippi; et Vacherie Dome, Louisiana: L'analyse concerne la zone qui entoure la chambre d'isolation. On compare les réponses thermomécaniques afin d'établir l'ordre de choix relatif de chaque emplacement comme dépôt pour les résidus nucléaires.

ZUSAMMENFASSUNG

Die Studie beschreibt eine thermomechanische numerische Analyse der vier potentiellen Endlager für radioaktiven Abfall im Salz in den U.S.A.: Paradox Basin, Utah; Permian Basin, Texas; Richton Dome, Mississippi; and Vacherie Dome, Louisiana. Die Analyse bezieht sich auf die nähere Umgebung der Einlagerungsstrecke. Die thermomechanischen Auswirkungen werden verglichen, um die Eignung jedes Standortes für ein Endlager für radioaktiven Abfall beurteilen zu können.

1. INTRODUCTION

1.1 Background

The development of nuclear energy has created a need for the safe disposal of radioactive wastes. In the United States, the National Waste Terminal Storage (NWTS) Program has been established to determine suitable repository sites in deep geologic formations that will provide a safe barrier between the disposed commercial nuclear wastes and the surface environment. Salt is a primary candidate to host a nuclear waste repository because it has a relatively high thermal conductivity, is relatively free of water, and is a ductile material that can undergo large deformations without failing. Currently, four salt formations are being considered as potential sites for a nuclear waste repository. Two of these formations are domal salts (Richton Dome in Mississippi and Vacherie Dome in Louisiana) and two are bedded salts (Paradox Basin in Utah and the Permian Basin in Texas).

1.2 Objectives

The objectives of this study are two fold, viz:

- Compare the thermal and thermomechanical responses of the four potential nuclear waste repository sites in salt.

- Provide predictions of thermal and thermomechanical responses at each of the four sites which will assist in the repository design process.

The first and primary objective pertains to thermal and thermomechanical considerations which constitute a part of the selection process for a repository site. In the comparison of the four sites, the predicted temperatures and deformations are considered at strategic locations within the disposal room region.

The secondary objective of this study provides a unique opportunity to evaluate "site-specific" repository conditions. A complementary study was done previously to provide data for this analysis that is unique to each site (Tammemagi et al., 1981). This extensive data report discusses the exploration and laboratory testing of core samples that have been done for each of the sites to provide information for this analysis. Most of the previous similar studies associated with nuclear waste isolation have relied on "generic" properties and/or geometries. Consequently, the results of this study should provide pertinent information for repository design because the data considered are site specific to the extent possible.

RSI PUBL. NO. 82-11

Table I. Characteristics of a baseline repository

*NUCLEAR WASTE:

Type	Commercial high-level waste from reprocessing of a 3:1 mix of fresh UO_2 and MOX fuels. Borosilicate glass in a stainless steel canister.
Areal Thermal Loading	25 W/m^2
Age	10 Years Out-of-Reactor
Canister: Initial Power	2160 W
Length	3.0 m
Diameter	0.324 m
Drillhole Depth	5.5 m
Pitch	3.66 m

*ROOM DIMENSIONS:

Width	5.5 m
Height	5.5 m
Length	Unspecified
Pillar Width	18.3 m
Extraction Ratio	23%

REPOSITORY DEPTH (m):

Paradox	= 850 (plus mesa effect)
Permian	= 750
Richton	= 580
Vacherie	= 790

INITIAL REPOSITORY TEMPERATURE (°C):

Paradox	= 30
Permian	= 30
Richton	= 38
Vacherie	= 57

INITIAL LITHOSTATIC STRESS AT THE REPOSITORY LEVEL (MPa):

Paradox	= 23.0 (includes mesa effect)
Permian	= 15.0
Richton	= 12.5
Vacherie	= 10.5

* Repository Conditions-Interface Working Group, 1980

1.3 Methodology

This study represents the first phase of a series of planned thermomechanical analyses of potential nuclear waste repositories in salt. This first phase involves the use of a baseline repository, which is summarized in Table I, to make a comparison of the thermal and thermomechanical responses for each of the four sites. The baseline repository concept simplifies the comparison of the four sites by fixing many parameters.

The primary purpose of the planned second phase is to help optimize the repository design by performing a sensitivity or parametric study of the more important and controlling parameters. The parameters to be considered in this future sensitivity analysis will include room and pillar dimensions, areal thermal loading, and repository depth. These parameters are considered more influential than others since variations in their magnitudes should have a substantial effect on the thermal and thermomechanical responses.

The room region encompasses the excavated repository facility. This includes one room height of rock mass extending above the facility, below the bottom of the waste canisters, and horizontally beyond the repository edge.

This analysis has been performed numerically with the finite element method using eight-noded isoparametric elements. The finite element programs are from the SPECTROM series of computer programs, which have been specifically developed to analyze rock mechanics problems. The thermal analyses employed SPECTROM-41 which has been documented by Svalstad (1981). The thermomechanical analyses were performed with SPECTROM-21 (Fossum et al., 1982).

The scenario used in this study assumes the disposal rooms will be sealed, but not backfilled, during the initial 25 years after waste emplacement. Subsequently, the rooms are backfilled with crushed salt for an additional 25 years. After these 50 years of operation, the repository will be decommissioned and abandoned. The heat-generating nuclear waste is ten-year-old commercial high-level waste that is emplaced at an areal thermal loading of 25 W/m^2. The age of the waste represents the time between the removal of the waste from reprocessing to the emplacement in the disposal rooms.

2. MATERIAL PROPERTIES

The primary material considered in the room region analysis is salt. Thermomechanical properties of salt used in this study are listed in Table II. These properties were taken from a data report assembled by Tammemagi et al., (1981). The thermal conductivity of the salt at each of the four sites is essentially the same and is a function of temperature:

$$k(T) = 6.02 - 1.84 \times 10^{-2} T + 3.20 \times 10^{-5} T^2 \quad (1)$$

where k is thermal conductivity (W/m-K) and T is temperature (°C).

Table II. Thermomechanical properties of salt

Property	Paradox	Permian	Richton	Vacherie
Density (kg/m^3)	2180	2180	2180	2180
Specific Heat Capacity (J/kg-K)	909	909	909	909
Thermal Conductivity (W/m-K)	Eqn. 1	Eqn. 1	Eqn. 1	Eqn. 1
Coefficient of Thermal Expansion (10^{-6}/K)	41.0	41.0	41.0	41.0
Modulus of Elasticity (GPa)	31.0	26.6	31.5	31.1
Poisson's Ratio	0.36	0.33	0.36	0.34

The creep behavior of salt for each of the four sites has been determined from laboratory tests (Pfeifle et al., 1981). Although the creep law parameters derived from these tests are unique to each site, they are applied to a common constitutive relation for salt creep. This baseline creep law has recently been selected to be used in numerical analyses that assess the performance of a potential nuclear waste repository

(Senseny, 1981). This particular form, known as an exponential-time law, was found to be more desirable than seven other common forms of constitutive relations for salt creep.

3. SUPPLEMENTARY STUDIES

An initial lithostatic state of stress was assumed prior to the instantaneous excavation of the disposal room (Table I), with the stress state at each site proportional to the repository depth and overburden density of each site. Because of this assumption, the initial state of stress at the repository level is easily defined at three of the four sites. However, it is difficult to assess at the Paradox site because the topography consists of valleys and mesas which can have as much as a 450 m differential in ground elevation. Therefore, a supplementary study was performed in which three effective repository depths at the Paradox site were considered. Based on that study, an effective repository depth of 850 m plus an additional 150 m (mesa effect) was chosen for the Paradox site. Also, a secondary value of depth at the Permian site was evaluated because another suitable salt formation may exist at a shallower depth.

The assumption of an initial lithostatic state of stress may not be applicable to the Paradox site based on hydrofracture tests conducted by Woodward-Clyde Consultants (Schnapp et al., 1981). Therefore, the influence of initial stress state before excavation was investigated.

At the initiation of this study, only salt core from cycle 7 of the Paradox site was available for laboratory testing. Consequently, the elastic moduli and parameter values for the creep law for the Paradox site do not correspond to salt from the proposed repository horizon (i.e., cycle 6). After the completion of the room region analysis, the laboratory test results of salt from cycle 7 became available. Therefore, a supplementary study was performed for the Paradox site to determine the thermomechanical influence of elastic moduli and creep law parameters obtained from cycles 6 and 7.

As mentioned, the room was considered to be open and unventilated during the initial 25 years after waste emplacement and then backfilled for an additional 25 years. Since the time at which the room is backfilled may change, two alternate situations were considered: (1) the room was backfilled 5 years after waste emplacement, and (2) the room was not backfilled.

These brief discussions are intended only to introduce the various types of supplementary or supportive studies considered in this analysis. A detailed presentation of these studies is given by Wagner et al., 1982.

4. RESULTS

4.1 Thermal analysis

The temperature distributions in the room region are influenced largely by the initial ambient temperature associated with each site. These temperatures, which are 57°C (Vacherie), 38°C (Richton), and 30°C (Paradox and Permian), are significant because all other thermal considerations are nearly identical for the four sites (Table II). Despite the nearly identical parameters, a single thermal analysis representative of all sites was not possible because the assumed thermal conductivity expression for salt is temperature dependent (Equation 1).

The 50-year temperature distributions at four different locations in the room region (centerline of the roof and floor, midpoint of the pillar, and midheight of the canister/salt interface) are shown for each of the four sites in Figure 1. The temperatures are the highest for the Vacherie site and lowest for the two bedded sites (Paradox and Permian). The difference in these temperature distributions corresponds closely to the differences in the initial ambient temperatures. This implies that the thermal conductivity expression is not very sensitive to the temperature difference of approximately 30 K observed in these analyses.

The abrupt change in temperature after 25 years is caused by backfilling the disposal room with crushed salt. The heat transfer through the crushed salt is assumed to be considerably less than in the air that exists in the disposal room during the initial 25 years (Fossum and Callahan, 1981). The modes of heat transfer through the air in the disposal room include both conduction and radiation; only conduction is assumed through the crushed-salt backfill which has a thermal conductivity one-tenth that of salt. This reduced thermal conductivity perturbs the temperature distribution and is most noticeable at locations within the room (i.e., the centerline of the floor and roof).

The temperature distributions indicate the relation between the time of peak temperature and the proximity of a particular location to the heat source. The maximum temperature within the room and at the center of the pillar occurs between 22 and 32 years after waste emplacement.

4.2 Thermomechanical analysis

The primary indicator of the thermomechanical response in this room region analysis is room deformation. The degree of room deformation is affected mostly by the creep behavior of salt which is influenced significantly by the initial in situ temperature and the repository depth. These factors (creep behavior of salt, initial in situ temperatures, and repository depth) are the most prominent because other seemingly influential parameters, such as thermal loading and room-and-pillar geometry, were identical for each of the four sites (Table I). Roof-to-floor closures along the centerline of the disposal room have been calculated for each of the four sites. These analyses were limited to a maximum of 10 percent centerline roof-to-floor closure. The analyses did not proceed beyond this limit because additional closure may require remining of the room periphery and support of the roof and rib to provide adequate space and safety for operation of retrieval equipment. Room closures of less than the limit imply that the repository should remain structurally stable throughout the retrieval period, although some local failure may occur (Russell, 1979).

The comparison of the roof-to-floor closures at the four sites (Figure 2) indicates the influence of the aforementioned factors. The Vacherie domal salt site, which has the greatest room closure, has a substantially greater initial in situ temperature (57°C) and involves a relatively deep repository horizon (790 m). The Permian site exhibits a relatively great amount of room closure because of its high creep rate (Pfeifle et al., 1981) even though the initial temperature and repository depth (30°C and 750 m, respectively) are near or below the average values for the four sites. Similar room deformation at the Paradox and Richton sites suggests that tradeoffs exist in the governing factors. Although the Richton site is located at a considerably shallower repository depth (580 m), the Paradox salt creeps at a slower rate (Pfeifle et al., 1981).

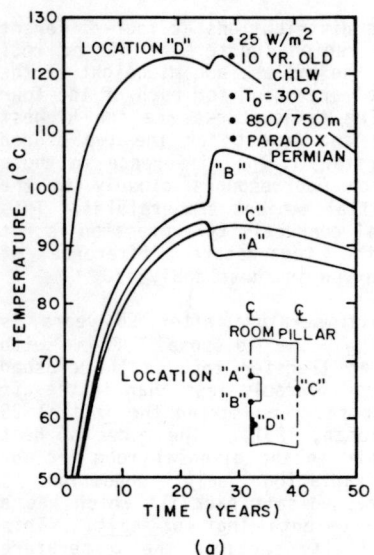 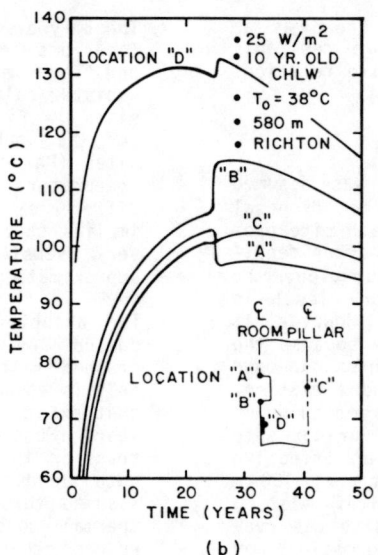 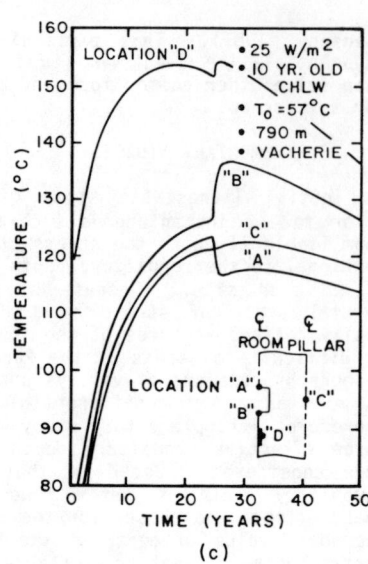

Figure 1. Temperature Distributions in the Disposal Room Region

4.3 Summary of results

A quantitative summary of the thermomechanical results is shown in Table III. These results will assist the NWTS site selection process. However, it is important to emphasize that these results are preliminary and only represent one of many considerations in the overall nuclear waste repository program.

Table III. Summary of results

Description	Paradox	Permian	Richton	Vacherie
Maximum Room Floor Temperature (°C)	106	106	115	136
Maximum Room Roof Temperature (°C)	94	94	103	123
Maximum Pillar Center Temperature (°C)	94	94	102	122
10% Roof-to-Floor Closure (yrs)	5.6	3.1	6.4	2.0

Variations in certain parameters affect the amount of room deformation to varying degrees. Thus, supplementary studies (see Section 3) were performed to investigate the influence on room deformation of repository depth, initial stress state, laboratory testing of salt from various salt cycles, and time of room backfill. The results of the supplementary studies are summarized below:

- Repository depth has a significantly greater influence on the Permian site than Paradox site.

- Nonlithostatic initial stress fields have little significance at the Paradox site.

- The predicted room deformation is conservative for the Paradox site when creep law parameters and elastic moduli derived from laboratory tests on cycle 7, instead of cycle 6, salt specimens are considered.

- The time at which the disposal room is backfilled should not perturb the temperature more than 10°C for the times and locations considered.

5. CONCLUSIONS

The relative comparison of the four sites, based on the temperatures in the room region, corresponds to the initial temperature along the repository horizon. The two bedded salt sites have the lowest temperatures and the Vacherie site has the highest temperatures. The time at which the maximum temperatures are reached (25 to 28 years) along the room periphery corresponds closely to the time at which room backfilling takes place (25 years).

The temperatures in the room region will have an impact on the working conditions, which relates to the ventilation design. The maximum temperature that exists before backfilling must be considered. Therefore, the ranges in room temperature that should be used for ventilation design vary from 96 to 126°C at the four sites. A ventilation study performed by Svalstad (1982) used a maximum rock surface temperature of 113°C for his baseline case. This study indicated that an acceptable working environment characterized by a maximum rock surface temperature of 49°C (Mine Design - Working Group, 1980) and a maximum wet bulb globe air temperature of 26°C (National Institute for Occupational Safety and Health, 1972) may be possible at all four sites. The correlation between these two studies is possible because of similarities in the room surface temperatures, nuclear waste type and age, material properties, and model geometry. Of the four sites being considered, Vacherie will be more difficult to cool than the other three sites and will require a higher volumetric flow rate and/or a lower cooling air temperature.

Comparison of the four sites based on room deformation is straightforward. The predicted room deformation is the least for the Richton and Paradox sites and the greatest for the Vacherie site. These rankings do not change for the initial five years after waste emplacement unless alternate repository depths and creep law parameters are considered (Wagner et al., 1982).

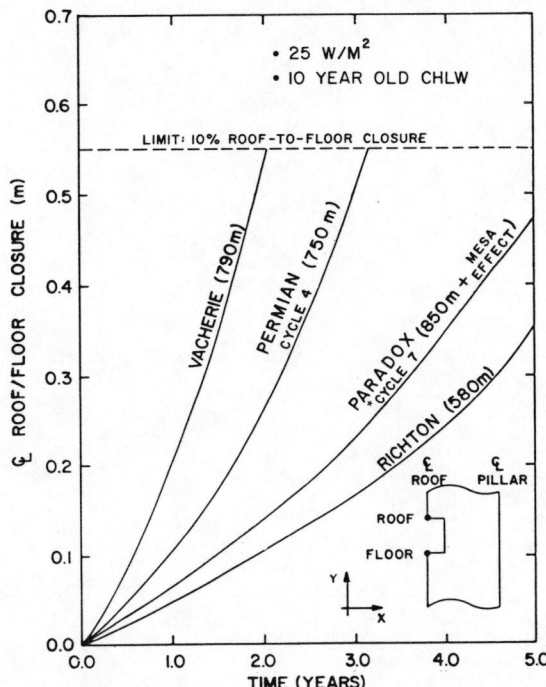

Figure 2. Comparison of Thermo/viscoelastic Centerline Roof-to-Floor Closures. (* Material Properties from Salt Cycle 7).

Excessive room closure will (1) have a deteriorating effect on room stability, (2) impair the ability to carry out retrieval operations if they become necessary, and (3) affect the ability to place backfill. From this point of view, the Richton and Paradox sites appear to have a significant advantage over the Vacherie and Permian sites for the specific parameters used in this study. The significance of room closure would be reduced by backfilling the repository rooms earlier than 25 years.

Based on these thermomechanical analyses and additional studies by Wagner et al., 1982, it is recommended that the Vacherie site be deferred as a potential nuclear waste repository. The primary reasons for the deferment would be the comparatively high temperatures and room deformation rate. Also, the size of the repository would be restricted at Vacherie because of the small domal salt structure.

7. ACKNOWLEDGEMENTS

The opportunity to perform this work under contract with the Office of Nuclear Waste Isolation (ONWI) of Battelle Memorial Institute is most appreciated. The technical contents of this report have been reviewed by Dr. Gary D. Callahan and Dr. Joe L. Ratigan. The neat appearance of this document can be attributed to Mr. Dan P. Nelson (drafting) and Ms. Jean M. Wilson (typing).

8. REFERENCES

Fossum, A. F., G. D. Callahan, and D. K. Svalstad (1982). User's Manual for SPECTROM-21: A Finite Element Thermo/viscoelastic Stress Analysis Program. RE/SPEC Report RSI-0201 prepared for Office of Nuclear Waste Isolation, Battelle Memorial Institute, Columbus, OH, (In Preparation).

Fossum, A. F. and G. D. Callahan (1981). Material Properties Testing and Analysis for the National Waste Terminal Storage Program. RE/SPEC Inc. Report RSI-0138 prepared for Office of Nuclear Waste Isolation, Battelle Memorial Institute, Columbus, OH, ONWI-

MIDES-WG (Mine Design - Working Group) (1980). Blast Cooling and Continuous Cooling, Activity Work Package, Sandia National Laboratories, Albuquerque, NM.

National Institute for Occupational Safety and Health (1972). Criteria for Recommended Standards - Occupational Exposure to Hot Environments. U. S. Department of Health, Education, and Welfare Publication No. HSM 72-10269.

Pfeifle, T. W., K. D. Mellegard, and P. E. Senseny (1981). Constitutive Properties of Salt from Four Sites. RE/SPEC Inc. Report RSI-0165 prepared for Office of Nuclear Waste Isolation, Battelle Memorial Institute, Columbus, OH, ONWI-314.

Pfeifle, T. W., K. D. Mellegard, and P. E. Senseny (1982). Preliminary Constitutive Properties for Salt and Nonsalt Rocks from Four Potential Repository Sites. RE/SPEC Inc. Report RSI-0191 prepared for Office of Nuclear Waste Isolation, Battelle Memorial Institute, Columbus, OH, ONWI- .

Russell, J. E. (1979). Areal Thermal Loading Recommendations for Nuclear Waste Repositories in Salt. Prepared for Office of Waste Isolation, Union Carbide Corp., Nuclear Division, Oak Ridge, TN, Y/OWI/TM-37.

Schnapp, M., J. G. Kocherhans, and R. A. Nelson (1981). Stress Measurements in Salt in a Deep Borehole, Southeastern Utah. Prepared for the Workshop on Hydraulic Fracturing and Stress Measurements, U. S. Geological Survey, Monterey, CA.

Senseny, P. E., K. D. Mellegard, and N. M. Eslinger (1981). Statistical Analysis of Transient Creep Data from Laboratory Tests on Avery Island, Dome Salt. RE/SPEC Inc. Report RSI-0141 prepared for Office of Nuclear Waste Isolation, Battelle Memorial Institute, Columbus, OH, ONWI-236.

Svalstad, D. K. (1981). User's Manual for SPECTROM-41 A Finite Element Heat Transfer Program. RE/SPEC Inc. Report RSI-0152 prepared for Office of Nuclear Waste Isolation, Battelle Memorial Institute, Columbus, OH, ONWI-326.

Svalstad, D. K. (1982). Forced Ventilation Analysis of a Commercial High-Level Nuclear Waste Repository in Salt. RE/SPEC Inc. Report RSI-0179 prepared for Office of Nuclear Waste Isolation, Battelle Memorial Institute, Columbus, OH, ONWI- .

Tammemagi, H. Y., M. C. Loken, J. D. Osnes, and R. A. Wagner (1981). Data Requirements for Site Characterization Repository Analyses. RE/SPEC Inc. Report RSI-0181 prepared for Office of Nuclear Waste Isolation, Battelle Memorial Institute, Columbus, OH, ONWI-364.

Wagner, R. A., M. C. Loken, and H. Y. Tammemagi (1982). Thermomechanical Analyses of a Baseline Repository at Four Potential Salt Sites. RE/SPEC Inc. Report RSI-0186 prepared for Office of Nuclear Waste Isolation, Battelle Memorial Institute, Columbus, OH, ONWI- .

STABILITY OF A ROCK OPENING SUBJECTED TO PULSATING TEMPERATURE

Stabilité d'une caverne rocheuse soumise à des températures pulsatiles

Die Stabilität eines Bergraumes, der pulsierenden Temperaturen ausgesetzt ist

Sten Bjurström, Dr.Eng.
Swedish Rock Mechanics Research Foundation, Stockholm, Sweden

Juri Martna
Chief Eng. Geologist, Swedish State Power Board, Vällingby, Sweden

Göran Rehbinder, Dr. Eng.
Swedish Rock Mechanics Research Foundation, Stockholm, Sweden

Kennert Röshoff, Ph.D.
Division of Rock Mechanics, University of Luleå, Sweden

SYNOPSIS

Large scale storage of heat is a vital part of the energy saving program being carried out in Sweden. In order to study the possibilities of storing heated water underground, a research rock cavern with a volume of 15 000 m³ has been built in Sweden. This cavern is filled with hot water and the temperature varies between 40 and 115°C. A research program will simulate the situation of seasonal heat storage and short term heat storage as well. In the cavern and in the surrounding rock a comprehensive instrumentation has been installed for the registration of, among other things, the stabilization process in the heated rock. This paper describes briefly the geological situation at the site and the selected method to measure the displacement in the heated rock. It also describes the preliminary results of the displacements in the rock due to the initial heating.

RESUME

Le stockage de chaleur à grande échelle constitue une partie vitale du programme d'économie d'énergie en Suède où une caverne expérimentale de 15 000 m³ a été construite en roche pour étudier les possibilités de stockage souterrain d'eau chaude. Cette caverne a été remplie d'eau chaude, la température y variant entre 40 et 115°C. Un programme de recherches permettra de simuler le stockage de chaleur sur une base saisonnière et à court terme. Un ensemble complet d'instruments a été installé dans la caverne et dans les roches environnantes pour enregistrer, entre autre, les processus de stabilisation intervenant dans la roche sous l'effet de la chaleur. Le présent document donne une description brève des conditions géologiques du site ainsi que de la méthode retenue pour mesurer le déplacement de la roche sous l'effet de la chaleur. Il donne aussi une description des résultats préliminaires relatifs au déplacement dans la roche dû à l'echauffement initial.

ZUSAMMENFASSUNG

Die Großraumspeicherung von Wärme ist ein äußerst wichtiger Bestandteil des schwedischen Energiesparprogramms. Mit dem Ziel, die Möglichkeiten der Untertagespeicherung von erhitztem Wasser zu erkunden, wurde in Schweden eine unausgekleidete Felskaverne im Ausmaß von 15 000 m³ gebaut. Diese Kaverne wurde mit Heißwasser im Temperaturbereich 40-115°C gefüllt. Das Untersuchungsprogramm sieht sowohl eine Simulierung der saisonbedingten als auch eine der kurzfristigen Wärmespeicherung vor. Innerhalb der Kaverne und im umgebenden Gebirge wurden Instrumente installiert, die u.a. Stabilitätsveränderungen im erhitzten Gebirge registrieren sollen. Dieser Aufsatz beschreibt kurz die geologische Situation und die ausgewählte Methode, um Verformungen im erhitzten Gebirge messen zu können. Er enthält auch die ersten vorläufigen Resultate der anfänglichen Erhitzung.

1. INTRODUCTION

A large part of world energy consumption is used for heating or cooling. In this sector, storage of energy can make substantial contributions to energy savings by levelling out the discrepancy between energy available and the need. This is particularly true for the economical development of new alternative energy sources and for the use of today often wasted or not used energy.

Such energy is often of relatively low quality (has low temperature). The economical use of this energy requires low cost and often large scale heat storage.

The utilization of the subsurface space for heat storage has great potential for economical storage of large volumes. This has been increasingly recognized and confirmed in the Swedish energy programme in recent years.

Sweden has particular reasons to develop all the possibilities for better energy efficiency and to find and develop new energy sources. Sweden is among the largest oil consumers per capita and totally dependent on imported oil. Sweden has also taken the political decision to end its nuclear era in 2010, which means that there may be a large gap between supply and demand of energy at that time.

Coal is expected to be main substitute for oil and nuclear energy. Besides great efforts to introduce coal, several hundred millions of Swedish Crowns are annually spent on solar and wind energy, biomass and peat, but also to a great extent on systems for heat extraction from the ground and on heat storage. Underground heat storage has proved to be essential for this and for the use of surplus heat from many industrial processes.

The concepts for subsurface storage of heat being introduced in Sweden today are judged to be of relevance also for most thermal accumulation problems outside of Sweden, for example storage of cold in hot parts of our world.

Storage of heat in water, rock or soil can be achieved in a number of ways: This has been described by Jansson, Bjurström & Hultin, 1980. One of the possibilities is to use manmade rock caverns for storage of heated water. In a suitable geological setting this method has several advantages. By using largely conventional civil engineering techniques, compact storages of desired size can be built close to either the producer or the consumer of heat, even in densely populated areas. Furthermore, heated oil has been stored in large rock caverns for many years in Sweden, providing experience in construction and operation of this type of storage.

A rock cavern can be pressurized comparatively easily, thus increasing the "energy density" of the storage and the principle of storage allows rapid inputs and outputs, increasing the versatility of the system. Rock formations suitable for this purpose are common in Sweden.

The storage of heated water in rock chambers has been discussed and studied in several Swedish projects (Bjurström, 1977, Bjurström et al., 1974, 1977). Laboratory studies on certain interesting questions have also been made (Claesson and Ronge, 1980, Martna, 1977).

This development means an introduction of new technologies. There are quite a number of not well known questions regarding thermal mechanisms, geohydrology, rock mechanics etc. Some questions are qualitatively relatively well understood, but there is a serious lack of quantitative information.

Full-scale tests are necessary for verifying the results of laboratory investigations and theoretical calculations. For this purpose, and also for demonstrating the technique under realistic conditions, a test cavern has been built in the town of Avesta. In the town of Uppsala another and larger cavern of 100 000 m³ for storage of solar heated water is under construction. A similar but somewhat limited research programme will be carried out also in this cavern.

2. THE AVESTA PROJECT

The research cavern of the Avesta project is connected to a district heating thermal plant (Fig.2). After the completion of the test programme the cavern will be used in Avesta's district heating system.

The Avesta cavern is, however, mainly a research project. It is considered to be of a suitable size for short-time storage in the Avesta district heating system, but it is too small to be optimal for seasonal storage of heated water.

The research programme of the Avesta project covers the whole range of problems connected to engineering geology and rock mechanics as well as the thermal problems and water chemistry.

A short review of the programme is given below Of this programme, only the aspects of the stability of the storage cavern due to pulsating thermal load will be discussed in the present paper.

3. RESEARCH PROGRAMME

By means of a full-scale rock cavern the research programme is expected to give a better quantitative evaluation of a number of problems than would be possible to achieve from laboratory experiments only, and also lead to a better understanding and evaluation of design and costs. The purpose is to obtain design criteria for rock caverns considerably larger than the present one.

Most of the data will be obtained from instruments placed in boreholes made from the surface or from the special research tunnel (Fig. 1). The data are collected and stored in a microcomputer on the location and are transmitted to a centre for calculations.

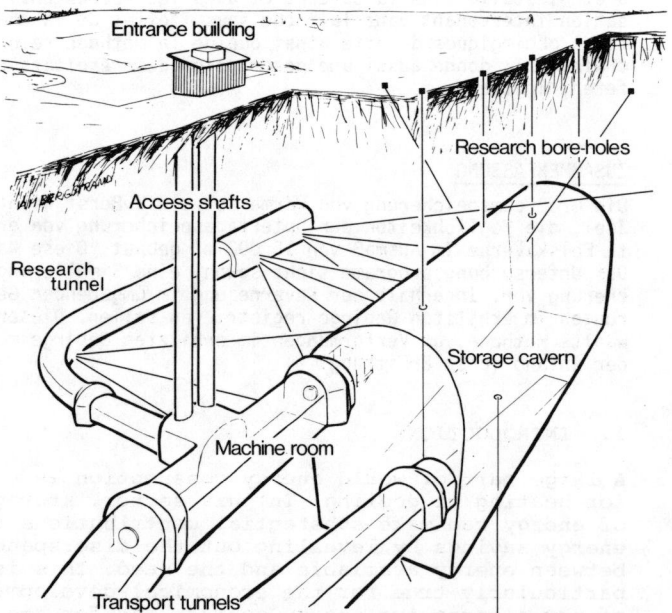

Fig. 1 A sketch of the hot water storage

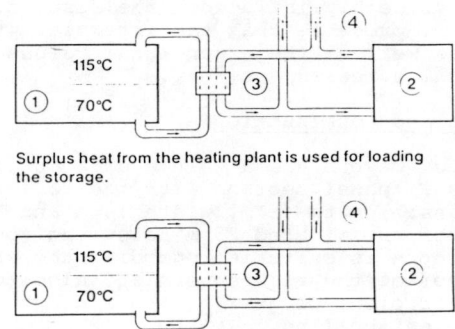

Surplus heat from the heating plant is used for loading the storage.

Heating plant out of operation. Unloading of the storage cavern.

Fig. 2   Operation principle of the hot water storage. 1. Storage cavern. 2. Thermal plant. 3. Heat exchanger. 4. District heating network

Besides the main topic of the present paper, the stability of a rock cavern exposed to pulsating thermal loads, a number of other problems are covered by the research programme.

Heat losses

No values based on data exist for total thermal losses for rock caverns. Although the losses can be calculated in parts (e.g. for the rock, the piping etc.), better precision is necessary for the evaluation of the thermal economy.

Thermal layering in the storage water

No adequate data concerning the stability of the thermal layer structure in the storage water exist for full-scale rock caverns. The factors having a considerable importance in this respect and thus also on the design and economy of the storage are for instance convection currents along the rock walls and the effect of unsymmetric placement of inlet and outlet of water.

Water chemistry, heat exchangers and construction materials

The construction materials are exposed to a rather unique environment and we do not have much knowledge of their behaviour. Laboratory studies have shown that significant amounts of mineral substances can be dissolved from crystalline rocks in hot water. These substances can be deposited in the system and could also accelerate corrosion.

Influence of the storage on the environment

Since there will be thermal losses from the storage cavern, the environment will be thermally influenced by heat transfer through the rock mass surrounding the cavern and by warming the ground water in the rock joints. The storage water may also have a tendency towards a partial exchange with the surrounding ground water.

4. WORKING PRINCIPLES FOR THE ACCUMULATING AND HEATING SYSTEMS OF AVESTA

The thermal plant of the Avesta district heating network to which the accumulator will be connected is at present fuelled by refuse and, in the future, probably partly with peat. The district heating system will be connected to the rock cavern water via heat exchangers (Fig. 2) since it is considered unsuitable to use water from the rock cavern directly in the district heating network due to possible risks for deposits of mineral substances dissolved from the cavern walls.

The thermal effect of the plant is 11 MW. During the warmer part of the year the heat load in the district heating network will, however, be less than 11 MW. The surplus heat can be used to load the heat storage. The heating plant will be in operation only on working days and the surplus heat stored during these days can be used during weekends. Thus the final, commercial operation of the test plant is based on short-time accumulation.

Since the test plant with components will also have a thermal effect of 11 MW, the fully loaded accumulator of 15.000 m^3 can replace the heating for almost three days.

During the research period, heated water from the thermal plant will be used to run one or two long-term cycles simulating the usage of solar energy, prior to an experimental study of the weekly cycles (Fig. 3).

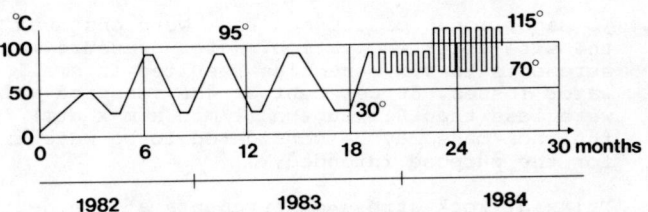

Fig. 3   Planned time-temperature curve for the research period

5. GEOLOGY AND TECTONICS

The purpose of the project is to study the behaviour of the "average" Swedish Precambrian rock. Thus, no exceptional rock qualities have been required. The absence of major zones of crushed or very fractured rock is, however, desirable to prevent either excessive leakage or grouting.

The soil cover at the site is thin, not exceeding a few metres. The ground water level is situated a few metres below the surface. The site was investigated by means of seismic profiles. Based on these, six vertical and inclined diamond drill holes were placed and bored. The holes were water-pressured tested and the cores logged regarding the lithology and the jointing. During the excavation a detailed geological mapping was carried out.

The dominating rock at the site is a Precambrian gneiss of sedimentary origin. This fine-grained grey rock contains mainly quartz, biotite and hornblend. The proportions of these minerals vary considerably in different layers. The gneiss is intersected by a number of pegmatite dikes with a maximum thickness of about 4 m.

The gneiss shows a schistosity as well as a linearity. The schistosity is folded and has therefore a varying orientation.

The number of joints in the rock mass is moderate, on the average about 2 per metre. Four joint sets have been distinguished. Three of them have an approximately vertical dip whereas the fourth one is flat-lying.

The first of the steep-dipping sets coincides with the above-mentioned schistosity of the gneiss. This joint set is distinctly wavy with a large amplitude and has a considerable variation in strike and dip. The long axis of the storage cavern is oriented about 15-20° from the average strike of this joint set.

The two other joint sets with a vertical dip strike on the average 65 and 90° from the first set. These two sets together comprise about 60 per cent of all observed joints.

The fourth set of joints is flat-lying and has a low frequency. The dip is on the average about 15° and does not exceed 50°.

The above joint sets occur in the gneiss. The pegmatite dikes have only a few joints with erratic directions. As a general rule, the joints in the gneiss do not continue into the pegmatite dikes.

No major zones of crushed rock were observed on the site after excavation. Water-pressure measurements in six boreholes resulted in small water losses. 86 per cent of the measured values were less than 0.1 Lugeon (l/min x m x atm), i.e., the rock mass may be considered to be watertight for the purpose intended.

Triaxial rock stress measurements at the depth of the storage cavern were performed prior to excavation in a 51 m deep borehole using the Hiltscher-probe of the State Power Board (Hiltscher et al., 1979). The stresses are mostly small, less than 2-3 MPa, markedly oblique to the horizontal plane and show a rather large scattering in respect of both magnitude and direction.

The stresses in the horizontal plane are consequently also variable, but can be said to be on the average about 2 MPa both longitudinal and transversally to the storage cavern. The mean value of the vertical stress is at the uppermost measurement level, 24 m below the rock surface, 6.5 MPa. This is about ten times the theoretical value, 0.6 MPa. Further down the vertical stress diminishes markedly.

The reliability of the gauges and the linearity of the stress-strain relation has for all measurements been carefully checked by calibration with the obtained tubular rock cores.

It is possible that, because of the low absolute value of the tectonic stresses, the intergranular stresses, which in this case have been estimated to be about 1 MPa, and the measurement errors have noticeably influenced the results. It is, however, probable that the irregular stress pattern as well as the exceptional values of the vertical stresses are real.

Design and Construction

The rock cavern has a volume of about 15.000 m^3. It has a tunnel section with the following dimensions: length 45 m, width 18 m and height 22 m. Its roof is situated 25 m below the rock surface. This depth is suitable regarding the stability of the roof of the rock cavern and also for obtaining sufficient counter-pressure of ground water to prevent boiling.

The cavern contains no temperature insulation. Parts of the roof have a shotcrete lining, the main purpose of which was to ensure the safety of the personnel during construction.

An underground machine room in the upper transport tunnel contains heat exchangers, circulation pumps for rock cavern water and district heating water as well as water treatment and electrical equipment. The machine room is separated from the rock cavern by a plug of concrete.

The cavern will be top-filled with water and put under pressure (115°). Model tests have been carried out for the design of the equipment for in- and outlet of water in the cavern. An unsymmetrical placing in one end of the cavern has been chosen for the in- and outlet equipment to minimize the length of pipes inside the cavern.

The excavation procedures followed in general the conventional Swedish practice for caverns of this size. A temporary access tunnel was used for transports during the construction. The storage cavern was excavated by means of a top heading and two benches. Close to the cavern, the access tunnel parted into two, an upper one for the top heading and a lower one for the benches. Subsequently, an underground machine room was installed in the upper access tunnel. Smooth blasting was prescribed for this part of the tunnel and the storage cavern.

The rock was secured with rock-bolts whenever necessary. Subsequently, the stability of the cavern was calculated on the basis of obtained joint data and the necessary additional rock bolts installed. All rock-bolts have been individually placed considering the jointing and the nature of the rock face. All rock-bolts are cement-grouted rebars with a diameter of 25 mm and all, except a number in the roof, have a standard length of 4 m. Thus no prestressed rock anchors have been used.

In the walls of the cavern a total of about 280 rock-bolts have been installed. In the roof, 140 bolts with a length of 4 m and 90 bolts with a length of 2.4 m were installed.

An observation gallery has been built for a direct visual inspection of rock surfaces and reinforcements in the storage cavern.

6. POSSIBLE PROBLEMS WITH STRESSES AND DEFORMATIONS

It is well known that heating of rock is the oldest destruction method used in mining. It is also known that fire in tunnels can cause damage. The question is then, of course, if the heating of the rock around an unlined hot water storage will induce stresses that can jeopardize the stability of the cavern.

Extensive investigations of this kind, both theoretical and experimental, have been performed within the Swedish-American cooperative program on radioactive waste storage in mined caverns in crystalline rock. A thorough theoretical study is presented by Tin Chan, 1979. A simple theoretical estimation, taken from some elementary text book on elasticity, i.e., Timoshenko and Grodier indicates that the temperature will induce a compressive tangential stress around the cavern which has its maximum at the wall whereas it induces a compressive normal stress around the cavern which has its maximum at some distance from the wall. This is interesting since probable tensile stresses in the walls due to the excavation will then be counteracted by the thermally induced stresses.

So far the discussion of thermal stresses have been confined to a static or a monotonic temperature load. This is also the case in the investigation on radioactive waste depository. This has also been found by Bjurström, Cederberg et al. concerning storage of hot water. In a hot water storage, however, the temperature varies periodically with time. This is certainly a complication.

If, which is most likely, the rock mass behaves in different ways in loading and in unloading a permanent deformation will remain after every load cycle and it is extremely important that such possible deformations tend to zero with number of load cycles. We can thus say that the stress level and to some extent even the magnitude of the deformations in the vicinity of the cavern during the initial heating are of less interest in comparison with the accumulation of even small deformations over a long period.

A laboratory test of this kind has recently been reported by Barton and Lingle, 1982. An 8 m^3 block of rock has been exposed to normal pressures and different temperatures according to a prescribed load cycle. The results clearly show that permanent deformations occur after mechanical loadings and after thermal loadings as well.

7. MEASUREMENTS OF DISPLACEMENTS AND STRESSES

Regarding the risk of accumulated deformations due to cyclic loading it is of great importance to measure to deformations of the rock during a long time, i.e., during several years. Measurements of this kind are difficult. A stationary gauge like an extensiometer has to stand a tough environmental impact. The gauge is located below the ground water table, the temperature is high and it is exposed to all this for a long time. Besides this there is another serious problem. Since the gauge has the same temperature as the rock the gauge as well as the rock have been thermally expanded and it is difficult to separate the rock component of the output signal from the gauge component. Such a separation requires a careful calibration of the instrument in the whole temperature range.

At the planning stage of this project experiences from the aforementioned project on radioactive waste depository indicated that the temperature influence on stationary extensiometers can cause problems which, if possible, should be avoided.

The most attractive solution to these problems is to have a loose gauge which is introduced and cooled in its measuring position during the very measuring procedure.

The idea is then of course that the cooling is such that the instrument is kept at a low, i.e. unaltered, temperature, whereas the rock keeps its high temperature. This is possible since the measuring takes short time and since the thermal inertia of the rock is great and the heat conduction of the rock is low.

A measuring device which satisfies the requirements mentioned above is the SLIDING MICROMETER-ISETH which has been installed. This instrument is presented by Kovari et al., 1979.

The aim has been to measure the displacements of the rock mainly normal to the contour of the cavern. To this end 10 holes, 6 from the surface above the cavity and 4 from the research tunnel beside the cavity, have been drilled towards the cavity. The drilling of the holes has been stopped about 1 m from the wall of the cavity since direct connection between the hole and the hot water cannot be accepted. There are two reasons for this. Firstly, the bottom of the plastic tube with the measuring marks is connected to a hose running parallel with the tube and grouted in the hole together with the tube. Through this hose the tube can be rinsed and the instrument cooled during the measuring procedure. If a minor rock fall in the cavity spoils this connection between the flushing hose and the plastic tube the whole measuring hole is spoilt. For this reason the bottom at the hole is kept at some distance from the wall. Secondly, there are safety reasons for avoiding any risk of contact between the cavern and the measuring holes coming from the research tunnel. The research tunnel is located below the top of the cavern and has open air pressure in it. This means that if a measuring hole from this tunnel becomes connected with the cavern the tunnel will immediately be filled with overheated steam.

Thus the introduction of a stationary flushing hose implies that the immediate vicinity of the cavern cannot be investigated with respect to displacements. This is the price one has to pay in order to avoid the difficulties with heating of the instrument.

The measuring holes from the ground were drilled and the tubes installed before the cavity was excavated. In this way it has been possible to measure the displacements of the rock induced by the very excavation. The importance of these figures is mainly to serve as reference values to be compared with the values of the temperature induced displacements.

It is of interest to measure the stress variations in the vicinity of the cavern. Firstly since the real situation can be expected to differ from a simple estimation and secondly since the hysteresis effects can be followed.

Therefore Glötzl cells have been introduced in boreholes drilled from the surface normal to the top of the cavern. The cells have been introduced such that they can measure the tangential stress as well as the axial stress in the vicinity of the cavern.

8. PRELIMINARY RESULTS OF DISPLACEMENT MEASUREMENTS

Up to now a limited set of deformation data are available. They are presented in Fig. 4., which shows the displacements along three holes in a cross section at the center of the cavern. The temperature-rise that has induced these displacements is 50°C. One interesting feature is striking. At the top, the cavern has expanded whereas at the bottom it has been compressed. The reason for this is most likely the fact that, according to the operation principle of the storage, the temperature of the water is stratified. This implies that the top of the cavern has been exposed to a higher temperature and for a longer time than the bottom.

9. CONCLUSIONS

Simple theoretical estimations have shown that the thermally induced stress probably is a problem of less magnitude than the possible accumulation of deformation due to cyclic temperature loading.

Preliminary displacement measurements has shown that for a temperature rise "half way" to the operating temperature, the top of the cavern has risen approximately 1 mm whereas the bottom of the cavern has "shrunk" 0.5 mm.

Acknowledgement

The investigation, which has been presented here, is supported by The Swedish Council for Building Research, The Research Council for Energy Resources and by The Swedish State Power Board.

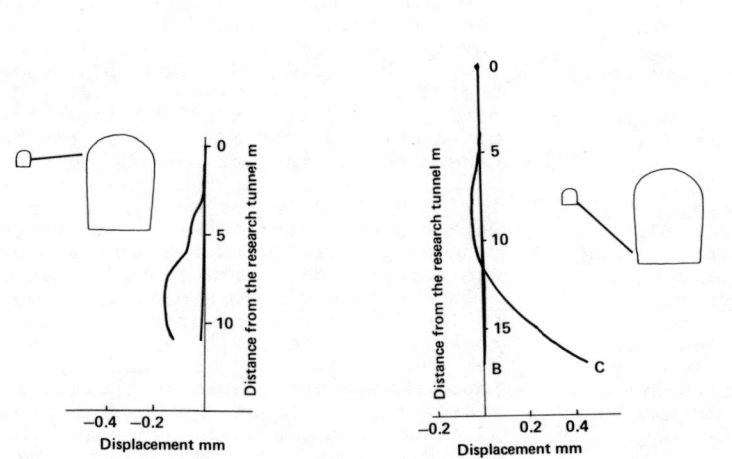
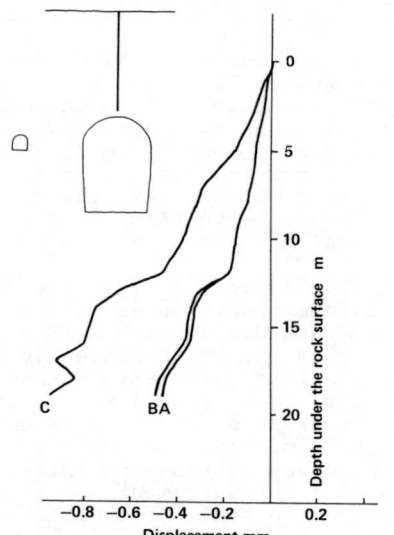

Fig. 4 Relative axial displacements measured in 3 boreholes, one from the surface and 2 from the research tunnel. Positive values of the displacement means axial extension of the hole.
A: Before excavation of the cavern. B: After excavation but before heating.
C: After heating 50°C.

REFERENCES

BARTON, W. & LINGLE, R.: Rock mass characterization methods for nuclear waste repositories in jointed rock. ISRM-symposium Aachen 1982.

BJURSTRÖM, S. 1977. Transport and storage of heated water in unlined rock openings. Proc. 1st Int. Symp. on Storage in Excavated Rock Caverns, Rockstore 77, vol. 2:212-218. Stockholm.

BJURSTRÖM, S., CEDERBERG, H., HANSSON, T., LINDSKOG, R. and MARTNA, J. 1974. Hetvattenlagring i bergrum. (Storage of hot water in rock caverns. In Swedish.) Swedish Rock Mechanics Research Foundation (BeFo). Pp. 1-78. Stockholm.

BJURSTRÖM, S., CEDERBERG, H., GÖRANSSON, A., HANSSON, T., LINDSKOG, R. and MARTNA, J. 1977. Lagring och transport av vattenburen värme i berg - fältförsök. (Storage and distribution of hot water in unlined rock caverns and tunnels - field tests. In Swedish.) Swedish Rock Mechanics Research Foundation (BeFo). Report No. 11:14/77, pp. 1-61. Stockholm.

BJURSTRÖM, S., KARLSSON, P.O. and MARTNA, J. 1980. The Avesta project. A test plant for storage of heated water in rock caverns. Proc. Int. Symp. on Subsurface Space, Rockstore 80, vol. 2:571-578. Stockholm.

CLAESSON, T. and RONGE, B. 1980. Water-rock interaction problems when storing and distributing hot water in unlined rock tunnels and caverns. Proc. Int. Symp. on Subsurface Space, Rockstore 80, vol. 2:587-592. Stockholm.

HILTSCHER, R., MARTNA, J. and STRINDELL, L. 1979. The measurement of triaxial rock stresses in deep boreholes and the use of rock stress measurements in the design and construction of rock openings. Proc. 4. Int. Congr. Rock Mechanics (2), 227-234. Montreux.

JANSSON, B., BJURSTRÖM, S. & HULTIN, S.A. Subsurface use for Energy Savings. Annex to the Swedish state of the art report for the Rockstore 80 symposium. BeFo-report Stockholm 1980.

KARLSSON, P.O. and REHBINDER, G. 1981. Avesta-projektet. Beskrivning av anläggning och forskningsprogram. (The Avesta project, Description of the plant and the research programme. In Swedish.) Avesta-projektet. Försöksanläggning för hetvattenlagring i bergrum 1981:1 Stockholm.

KOVARI, K., AMSTAD, C. & KÖPPEL, J. New Developments in the Instrumentation of Underground Openings. Proc. of the 4th Rapid Excavation and Tunneling Conference, Atlanta, USA, 1979.

MARTNA, J. 1977. Concrete for use in rock caverns for storing hot water. Proc. 1st Int. Symp. on Storage in Excavated Rock caverns, Rockstore 77, vol. 2:271-276. Stockholm.

TIN CHAN & COOK, N.G.W. Calculated thermally induced displacements and stresses for water experiments at Stripa, Sweden. Swed.-Am.coop. progr. on radioactive waste storage in mineral caverns in cryst. rock. Rep. SAC-22 Lawrence Berkely Lab. 1979.

Fig. 5 A photograph showing the wall of the hot water cavern. The length of the measuring rod is 4 meters.

Photo: Göran Hansson/N

KLASSIFIZIERUNG DER BOHRBARKEIT DES GEBIRGES DURCH IN SITU- UND LABORUNTERSUCHUNGEN
Classification of the drillability of a rock mass by in situ and laboratory testing
Classification de la forabilité des massifs rocheux par des essais in situ et en laboratoire

E. Mikura
Baugeologe, Universale Hoch- und Tiefbau A.G., Vienna 1, Austria

ZUSAMMENFASSUNG
Für die Vorhersage der Leistung von Vortrieben mit Vollschnitt-Tunnelbohrmaschinen (TBM) ist die quantifizierbare Erfassung der Gesteins- und Gebirgseigenschaften notwendig. Durch eine Kombination bekannter Labor- und in situ-Versuche soll nun dieses Parametergruppenpaar erfaßt werden. Im Labor wird hauptsächlich das Parameterspektrum der Gesteinseigenschaften ermittelt. Als anwendungsfreundlichstes technisches Hilfsmittel wird dabei ein für die Betonprüfung entwickelter Schmidthammer verwendet. Durch die Kürze des Beitrages kann jedoch in diesem Bericht nur kursorisch auf die Grenzen der Anwendbarkeit und möglichen Fehler der Aussage sowie die Vorteile des Verfahrens eingegangen werden. Auch bei der Beschreibung der Durchführungsdetails muß auf ältere Veröffentlichungen verwiesen werden.

RESUME
Pour le pronostic des performances de tunneliers il est nécessaire de pouvoir quantifier les caractéristiques des roches des masses rocheuses. Il faut enregistrer les deux groupes de paramètres d'essais en laboratoire et in situ et les combiner entre eux. Au laboratoire on détermine surtout la gamme de paramètres pour des roches tandis que les essais in situ servent à connaître les caractéristiques de masses rocheuses qui se montrent avant tout au niveau des surfaces séparatrices. L'outil le plus adapté en est le marteau Schmidt, développé pour les essais de béton. Le rapport est assez bref et ébauche simplement les limitations et les défauts possibles de cette théorie ainsi que les avantages du procédé. Il faut également se référer à des publications antérieures pour la description et les détails d'exécution.

SYNOPSIS
In order to be able to forecast the drilling performance of tunnel boring machines it is necessary to establish a quantifiable register of the characteristics of rocks and rock masses. The parameters of known laboratory as well as in situ tests are to be combined and registered. The laboratory tests mainly serve to establish the parameters of the characteristics of rocks whereas the in situ tests yield information on the characteristics of rock masses that chiefly show in the separating surfaces. The most convenient mechanical aid in this is the Schmidt-hammer which was developed for the testing of concrete. The present report, being rather short, can only give a cursory survey on limitations and possible shortcomings of this thesis as well as on the advantages of the procedure. For a description and details of its execution we have to refer to earlier publications on this subject.

1. ALLGEMEINES

Zur Beurteilung der Trennflächen in ihrer Häufigkeit und Wirkung zur quantifizierbaren Klärung der Bohrbarkeit und Standsicherheit mußten für den Vortrieb mit Vollschnitt-Tunnelbohrmaschinen rasche und billige Versuchsverfahren entwickelt werden. Neben den aufwendigen Labormethoden und Großversuchen wurde, aufbauend auf langjährigen Erfahrungen mit Geologenhammer und Stollenseismik, unter Verwendung des von der Betonprüfung kommenden Schmidthammers für einige Fälle ein in situ-Testverfahren entwickelt und erprobt.

2. AUSGANGSBASIS

2.1 Bisherige Methoden

Zur Beurteilung der Bohrbarkeit werden meist aus dem Verständnis der Metallurgen, Materialeigenschaften des Gesteins ermittelt. Dabei wird die Abriebsfestigkeit entweder aus dem Mineralaufbau oder aus einem Versuch abgeleitet und ein Druck- oder Zugversuch oder eine Kombination aus beiden (Brasil-Test, Point Load Index) durchgeführt und bei anisotropem Gestein die Lage zum Vortrieb berücksichtigt (Anstellwinkel). All dies erfolgt an Hand von in Laboratorien eingesandter Gesteinsstücke und geologischer Karten.

2.2 Beurteilung der bisherigen Methoden

Dieses Vorgehen ist als erster Schritt zur Ermittlung der gesteinsbedingten Basisbohrbarkeit grundsätzlich richtig. Vernachlässigt wird der, dem erfahrenen Praktiker bekannte Unterschied zwischen Gesteinseigenschaften und Gebirgseigenschaften, wobei als

Gesteinseigenschaften die physikalischen und mineralogischen Eigenschaften des Einzelstückes des Felsens,

als

Gebirgseigenschaften das Zusammenspiel der Gesteinseigenschaften mit aus Trennflächen verursachten Eigenschaften verstanden wird.

Die Gebirgseigenschaften, als Gesteins-Verbandsschwächung, fördern die Bohrbarkeit aus den Gesteinseigenschaften um ein Vielfaches. Diese leistungsfördernden Faktoren werden meist durch allgemeine, gleichmäßige Erhöhung der Leistungsangabe berücksichtigt. Dies ist jedoch ohne Quantifizierung und Qualifizierung der Trennflächen nicht möglich.

Als Methoden der Quantifizierung und Qualifizierung von Trennflächen werden angewandt: Augenschein im Gelände, ev. quantifiziert nach Pacher, 1963, (Durchtrennungsgrad), qualifiziert durch Abklopfen mit Geologenhammer, Bohrkernauswertungen nach Habenicht, 1979, seismische Versuche. Die Quantifizierung durch Augenschein oder nach Pacher, 1963, ist bei Fräsvortrieben meist, mangels Sichtbarkeit aller Trennflächen, nicht möglich, und damit fehlt auch die Qualifizierung in Richtung Bohrbarkeit oder Standsicherheitsverhalten.

Die seismischen Verfahren "Laufzeiten" erlauben bei bekannten Randbedingungen Aussagen, bedürfen dazu jedoch neben erfahrenem Personal größerer techn. Anlagen für Messung und Auswertung.

3. NEUE KOMBINATION VON IN-SITU- UND LABOR-TEST-VERFAHREN

3.1 Allgemeines

Lange Beobachtungen von erfahrenen Geologen bei ihrer "klassischen Hammerarbeit" zur Quantifizierung und Qualifizierung von Trennflächen und eigene "Versuche" führten, angeregt von Häusler, 1963, und Young, 1978, zu Versuchen der Beurteilung der Trennflächen durch den Schmidthammer.

3.2 Grenzen des Verfahrens

Das zu untersuchende Festgestein muß eine Würfeldruckfestigkeit deutlich größer 6 kN/cm2 haben.

Die Rauhigkeit der Oberfläche muß durch Erfahrungsbeiwerte berücksichtigt werden.

Es ist nur der Faktor zur Ermittlung der Druckfestigkeit, des Zusammendrückungsmoduls und der Bohrbarkeitsförderung der Trennflächen des Gebirges gegenüber dem Gestein ableitbar.

3.3 Vorteile des Verfahrens

Auch nicht sichtbare Trennflächen werden, wenn sie wirksam sind, "erkannt".

Die Ausstrahlung der gebirgsschwächenden Eigenschaften einer Trennfläche durch Fiederklüfte, bzw. Mikrorisse wird erfaßbar.

Es ist wenig Zeit erforderlich.

Mit vorauseilender Erkundung der Gesteinsparameter (Petrographie, Druckfestigkeit, E-Modul, etc.) lassen sich für Bereiche gleicher Petrographie die Gebirgsparameter ableiten bzw. kann man auf die Bearbeitbarkeit des Gebirges (z.B. Bohrbarkeit) schließen. Zur Vermeidung gröberer Fehlbeurteilungen ist jedoch ein erfahrener Geologe notwendig.

3.4 Durchführung des Verfahrens

Ähnlich dem Durchtrennungsgrad wird entlang bestimmter Linien oder Flächen eine Vielzahl von Versuchen in bestimmten Abständen getätigt, offensichtliche Fehlmessungen ausgeschieden und aus den verbleibenden Ergebnissen ein Mittel gebildet.

Die Abweichung dieses Mittels von der Höchstmarke (72 bei Schmidthammer N = 6 kN/cm2) ist direkt proportional der Schwächung des Gesteinsverbandes zum Gebirgsverband (bezüglich Druckfestigkeit und E-Modul).

Dabei muß die Oberfläche, die Petrographie und die Lagerung gleich bleiben oder durch Faktoren berücksichtigt werden.

4. ANWENDUNG DES VERFAHRENS

4.1 Statistisch gesicherte Anwendungen

4.11 Fräsleistungsvorhersage

Eine extrapolative Beurteilung der Durchtrennung (bzw. vorweggenommenen Brüche) erlaubt die Feststellung der Fräsleistungskomponente "Gebirgszerlegung", Mikura, 1980.

Nicht zu verwechseln mit dem Anstellwinkel der Anisotropieflächen zur Vortriebsrichtung.

4.2 Mögliche, statistisch noch nicht gesicherte Anwendungen

4.21 Standsicherheitsbeurteilung bei TBM-Vortrieben

Bei Vollschnittfräsvortrieben, in sprödem, geklüftetem Gebirge, können Trennflächen kurz hinter dem Bohrkopf nicht rechtzeitig erkannt werden.

Der Schmidthammer reagiert auch durch die Schmutz- oder dünne Spritzbetonschicht auf Schwächungen des Gesteinsverbandes. So können Gefährdungen durch Nachbrüche erkannt und wirtschaftliche Stützungsmittel eingebaut werden.

4.22 Dimensionierung der Auskleidung

Bei unterirdischen Hohlraumbauwerken mit statisch notwendiger Auskleidung kann bei standfesten Gebirgen die unterschiedliche "Qualität", bedingt durch Trennflächenanzahlen, genauer beurteilt, und es können dadurch Einsparungen getroffen werden.

References:

Habenicht, H., (1979)
Zur Beschreibung des Trennflächengefüges aus Bohrkernen, 3 Abb., Rock Mechanics 11, S.217-242, Wien/Springer

Häusler, H., (1963)
Zur Anschätzung der mechanischen Eigenschaften des Gesteinsverbandes bei Fundierungen. Mitt. d. Inst. f. Grundbau und Bodenmechanik TH Wien, Heft 5, S.35-44

Mikura, E., (1980)
Schnelle u. verläßliche Verfahren zur Prognostizierung der Fräsleistung, Rock Mechanics 12, S. 221-230, Wien

Pacher, F., (1963)
Durchtrennungsgrad u. Kluftflächenanteile, Der Felsbau von Dr.Ing.L.Müller, Bd.1 mit 307 Abb. u. 22 Tafeln, S 232-237, Stuttgart.

Young, R.P., (1978)
Assessing Rock Discontinuities, Tunnels & Tunnelling, S.45-48, Juni 1978

PRE-SPLIT BLAST DESIGN FOR OPEN-PIT AND UNDERGROUND MINES
Tirage en deux temps pour l'exploitation à ciel ouvert et exploitation au fond
Entwurf zum pre-split Sprengen für Tage- und Untertagebau

P. N. Calder
Professor and Head, Department of Mining Engineering, Queen's University, Kingston, Ontario, Canada

A. Bauer
Professor, Department of Mining Engineering, Queen's University, Kingston, Ontario, Canada

SYNOPSIS

Perimeter blasting techniques limit damage to the final walls of excavations. This is accomplished by lowering the explosive energy concentration of the excavation boundary. The most common method is pre-splitting which involves the formation of a crack coincident with the excavation boundary prior to firing the main blast. This crack serves as a vent for the explosive gases generated by the production blast holes. A theory of perimeter blast design based on static borehole pressure is described. The effect of rock strength parameters and structure is included, as well as the influence of the in situ stress conditions.

RESUME

Le tir de charge extérieur limite les dommages aux murs d'excavation. Ceci est obtenu en réduisant la concentration de charge des trous extérieurs. La méthode courante est le tirage à deux temps qui provoque la formation de fissures coincidentes avec la ligne des murs avant le tir principal. Cette fissure permet aux gaz générés par l'explosion lors du tir des charges principales de s'échapper. La théorie du comportement des charges extérieures basée sur la pression statique dans le trou foré est décrite. Les effets de force et de structure de la roche sont inclus, ainsi que l'influence des contraintes in situ.

ZUSAMMENFASSUNG

Umkreissprengarbeitsmethoden mindern Schäden der Aushubswandungen. Dieses kommt die Abnahme der Sprengstoffenergiekonzentration bei der Aushöhlungsgrenze zustande. Das Einbruchschießen, das aus einer Rissbildung zusammenfallend mit der Aushöhlungsgrenze vor dem Zünden des Hauptsprengschusses besteht, ist die gewöhnlichste Methode. Dieser Riss dient als eine Entlüftungsöffnung für die Sprenggase, die von den Produktionsbohrlöchern erzeugt werden. Eine Umkreisschießplantheorie, die auf den Bohrlochdruck begründet ist, wird beschrieben. Die Wirkung der Gesteinsfestigkeitsparameter, des Gefüges und der Einfluß der in situ Spannungsverhältnisse sind eingeschlossen.

1. PRE-SPLIT DESIGN THEORY

Various theories have been put forward over the years regarding the mechanism by which a pre-split line is formed. Most of these theories involve the interaction of the ground vibrations resulting from the detonation of the explosives in the pre-split blast holes. These approaches are highly theoretical and in the author's opinion do not provide a practical method for pre-split design. Another controversial issue is the mechanism by which the pre-split line functions to protect the final wall of the excavation. Here again most theories assume that the pre-split fracture acts as a boundary from which the ground vibrations generated from the production blast holes are reflected. Work performed by the authors (Calder and Bauer, 1977) indicate that the presplit line has virtually no effect on the magnitude of ground vibrations transmitted to the final excavation boundary. Similar conclusions have been reached in studies undertaken by the U.S. Bureau of Mines.

Our research has indicated that the pre-split line is actually formed by the borehole pressures generated within the pre-split holes causing the rock to fail in tension along that line. Borehole pressures act for a significantly long period of time to be regarded as static pressures. The main force that must be overcome in forming a pre-split line is the tensile strength of the rock along the surface area of the pre-slit line. If ground stresses of signficant magnitudes are present these must also be overcome. Our experience has also indicated that the presplit line protects the final excavation boundary by acting as a vent along which gases driving fractures from the main production blastholes are released. When a natural fracture is present the pre-split line is sometimes unnecessary (Calder and Tuomi, 1980) providing the loading of blastholes is controlled at the perimeter.

The methods developed have been extensively field tested in open pit mines (Calder and Morash, 1971), (Calder and Tuomi, 1980), (Crosby and Bauer, 1982).

2. BOREHOLE PRESSURE

Each explosive type has a characteristic borehole pressure which is generated when the explosive completely fills the borehole. Methods of determining borehole pressure have been extensively described by the authors, (Calder and Bauer, 1977), (Calder and Jackson, 1981), (Calder and Tuomi, 1980). The general relationship is given in the following equation:

$$(Pb)_c = N \rho D^2 \qquad (1)$$

where:

- $(Pb)_c$ = Borehole pressure of a fully coupled charge (i.e. charge diameter = blasthole diameter).
- D = Velocity of detonation for a confined explosive. The velocity of detonation for an unconfined or decoupled explosive is usually 85% of that for the same explosive when confined if detonating non ideally.
- ρ = Specific gravity of the explosive.
- N = Explosive constant related to specific gravity.

Note that the borehole pressure is not a function of the blasthole diameter.

Borehole pressure, and hence backbreak, can be reduced by decoupling. Charges are decoupled when they have a diameter smaller than the borehole in which they are loaded. The ratio of the charge radius to the hole radius is a measure of the degree of decoupling. Another method of decoupling is to separate the individual charges along the axis of the borehole. This is not a commonly used method at the present time due to the practical difficulties of loading the blastholes in this manner. The desired degree of decoupling can normally be achieved by selecting the correct charge diameter with respect to the hole diameter. However if a space is left between individual charges in the blasthole this can also be taken into account. The following equation indicates the method by which the coupling ratio is determined.

$$C.R. = \frac{r_c}{r_h} \times C^{1/2} \qquad (2)$$

where
- C = Percentage of explosive column that is loaded.
- r_c = Radius of charge.
- r_h = Radius of borehole.

The borehole pressure is drastically reduced by decoupling, as follows:

$$(Pb)_{dc} = 0.72 \, (Pb)_c \times (C.R.)^{2.4} \qquad (3)$$

$(Pb)_{dc}$ = Borehole pressure for a decoupled charge.

3. DESIREABLE BOREHOLE PRESSURE

In designing a perimeter blast high borehole pressures are desireable in the sense that they will minimize the amount of drilling required. There is a maximum value of borehole pressure however beyond which damage will occur to the final excavation wall. Our initial approach was to limit borehole pressures to the compressive strength of the rock mass. Our more recent experiences indicated that, although this approach produced excellent results, it was possible to increase the borehole pressure beyond this limit and still achieve acceptable results. The amount by which the borehole pressure exceeds the compressive strength of rock mass depends on the conditions. In most mining applications cosmetic results are not necessary and the borehole pressure is often in the range of two to five times the compressive strength. In other applications, such as excavations involving building foundations and in hydro construction projects, the borehole pressure would be limited to a value close to the compressive strength of the rock to eliminate any possible damage to the final wall.

4. PRE-SPLIT DESIGN

4.1 For insignificant in-situ field stresses

It is assumed that the borehole pressures generated by the detonation of the charges in the pre-split holes act over a finite period of time and are regarded as static. This is analogous to a situation in which the borehole collars were sealed and a borehole pressure created by injecting water under high pressure in the hole. Under these conditions the borehole pressures attempt to overcome the tensile strength of the rock thus forming a split along the line of the blastholes. Figure 1 is a sketch indicating a typical pre-split blast layout.

The pre-split line is fired instantaneously normally several days, but a minimum of fifty milli-seconds, prior to the firing of the main production blastholes. The situation is as illustrated in Figure 2.

The borehole pressure acts on an area equivalent to the diameter of the hole. On a per unit area basis it is resisted by the tensile strength of the rock. The necessary borehole

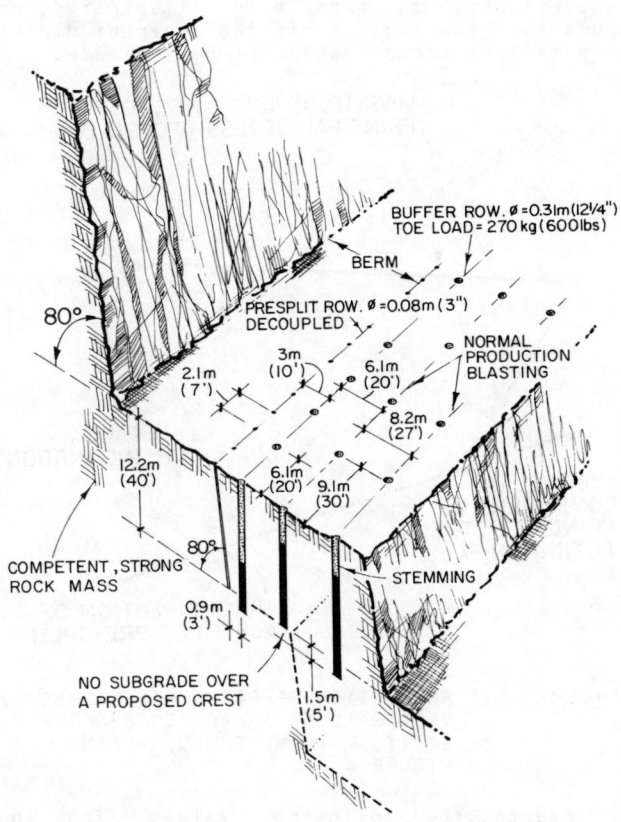

FIGURE 1: TYPICAL PRE-SPLIT BLAST LAYOUT IN AN OPEN PIT MINE

pressure to overcome the tensile strength of the rock is determined as follows:

$$Pb_{dc} D = (S-D)T \qquad (4)$$

$$S = \frac{D(Pb_{(dc)} + T)}{T} \qquad (5)$$

where:

- S = Hole spacing.
- D = Hole diameter.
- $(Pd)_{dc}$ = Decoupled borehole pressure of the explosive charge from Equation 3.
- T = In-situ tensile strength of the rock.

Note the hole spacing will be expressed in the same units in which the borehole diameter is specified.

4.2 Including in-situ field stresses

If significant values of compressive in-situ field stresses are present, the normal component of the principal stresses will attempt to prevent the pre-split line from forming and must be overcome in addition to the tensile strength of the rock mass. This is illustrated in Figure 2. Having measured and determined the direction cosines for the principal in-situ stresses, (Obert and Duvall, 1967) the stress vectors can be plotted on the lower hemisphere of a stereonet. Using a coordinate system such as that indicated in Figure 3 the three

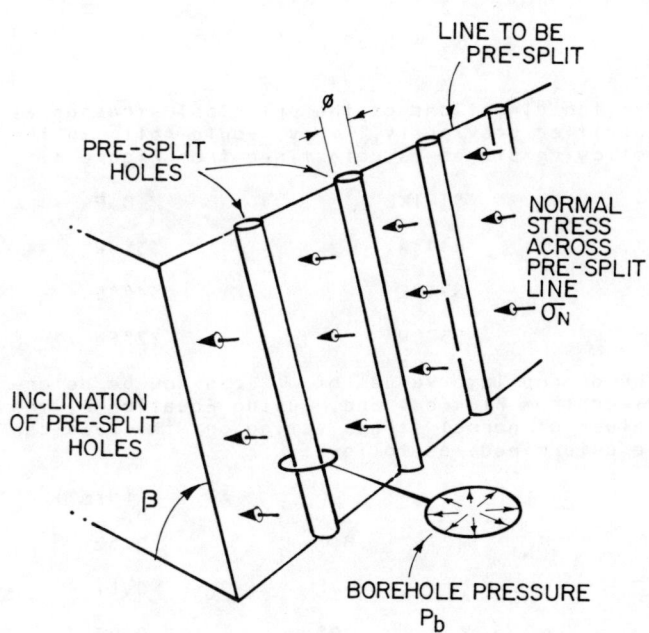

FIGURE 2: ILLUSTRATION OF PRE-SPLIT MECHANISM

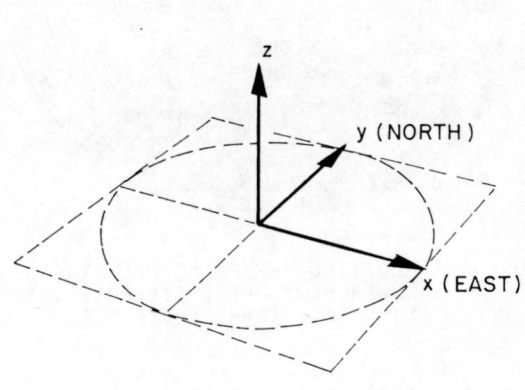

FIGURE 3: COORDINATE SYSTEM FOR PRINCIPAL STRESSES

principal stresses may be plotted as in the following example.

4.3 Example including field stresses in pre-split design

Assume the following directions for the principal stresses have been determined:

	X	Y	Z
σ_1	54°	127°	57°
σ_2	123°	79°	36°
σ_3	54°	39°	78°

Note that since the angles the stresses make with the z axis are less than 90°, the vectors as specified would appear in the upper hemisphere. In order to plot them on the lower hemisphere as desired, it is necessary to measure from the negative x and y axes. Having plotted the three principal stress vectors on a lower hemisphere stereonet as in Figure 4, the next step is to plot the orientation of the face which is to be pre-split. In this example the face is assumed to have an orientation of north 70° East dipping 50° to the southeast.

The values of θ can readily be determined from the stereonet, the example is illustrated on Figure 4. The angle θ is the apparent dip of the principal stress vector along the face.

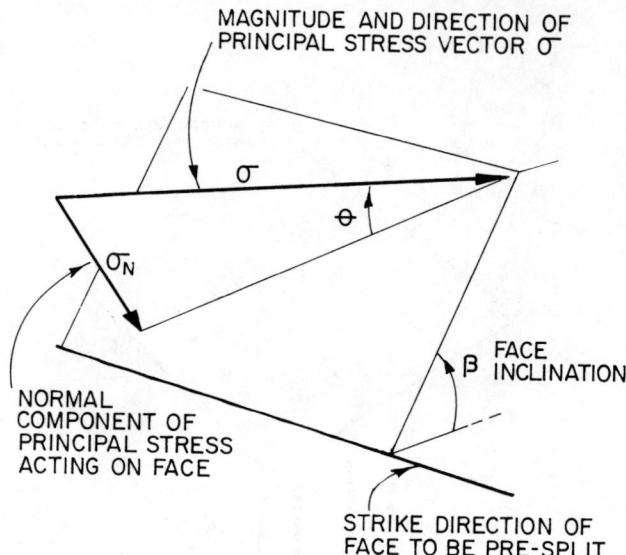

FIGURE 5: RESOLUTION OF PRINCIPAL STRESS VECTOR ALONG A FACE TO BE PRE-SPLIT, θ IS MEASURED AS IN FIGURE 4.

Assume the following values for the principal stresses.

$$\sigma_1 = 0.53 \text{ N/m}^2$$
$$\sigma_2 = 0.39 \text{ N/m}^2$$
$$\sigma_3 = 0.22 \text{ N/m}^2$$

For the directions of the principal stresses as specified previously, they would fall in the following planes as determined from Figure 4.

	STRIKE	DIP
σ_1	N44°W	33°NW
σ_2	S71°E	54°SE
σ_3	S37°W	12°SW

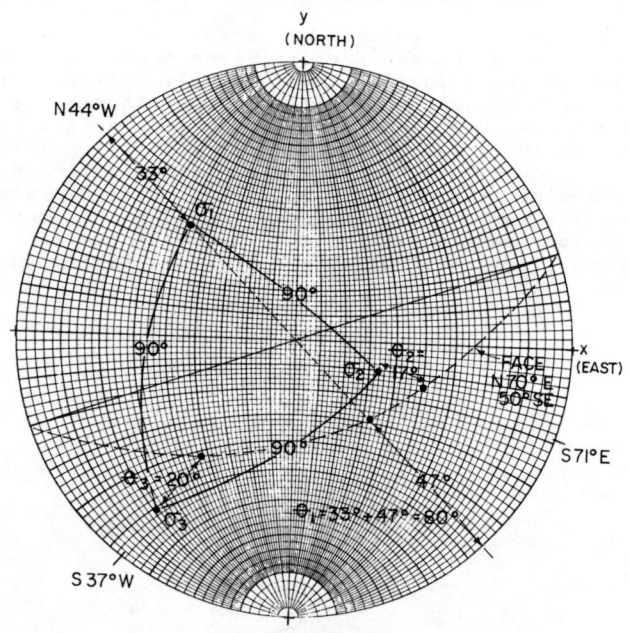

FIGURE 4: PLOTTING THE PRINCIPAL STRESSES ON A LOWER HEMISPHERE STEREONET, TO MEASURE θ, (See Figure 5)

Figure 5 illustrates a principal stress vector intersecting a face which is to be pre-split. The angle θ is a measure of the orientation of the principal stress vector in a plane perpendicular to the face to be pre-split. The normal value of stress, σ_N, acting on the face is defined as follows:

$$\sigma_N = \sigma \sin \theta \qquad (6)$$

The appropriate values of θ can now be determined from Figure 4 and, using Equation 6, the values of normal stress acting on the face can be determined, as follows:

	θ	σ_N (N/m^2)
σ_{N1}	80°	0.52
σ_{N2}	17°	0.11
σ_{N3}	20°	0.075

Equation 5 may now be modified to incorporate the effect of the normal stresses acting across the pre-split plane as follows:

$$Pb_{dc}D - \sigma_{NT}(S-D) = (S-D)T \qquad (7)$$

where:

$$\sigma_{NT} = \sigma_{N1} + \sigma_{N2} + \sigma_{N3}$$

or rearranging:

$$S = \frac{D(Pb + T + \sigma_{NT})}{T + \sigma_{NT}} \qquad (8)$$

5. ALTERNATE SPLITTING METHOD FOR HIGH FIELD STRESSES

In situations where the in-situ stresses are particularly high, presplitting may be impractical due to the very high borehole pressures needed to overcome these stresses. In this case a common technique would be to excavate in two stages as illustrated in Figure 6.

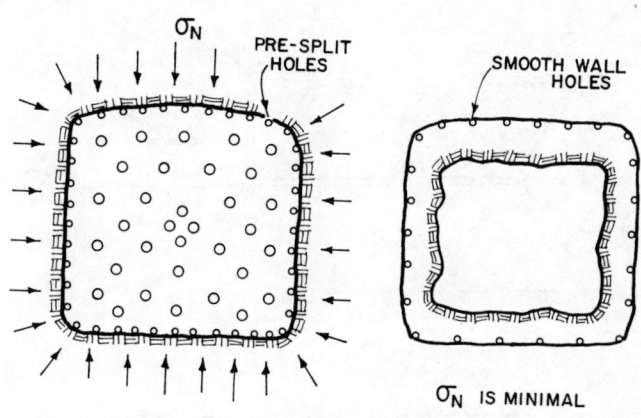

FIGURE 6: ELIMINATION OF NORMAL STRESSES BY CREATING A PILOT EXCAVATION.

This technique is generally referred to as smoothwall blasting or slashing. By removing the central core of the excavation the stresses normal to the faces to be split are generally removed, and the splitting technique as described for low field stresses would be used.

6. BUFFER ROW DESIGN

The distance between the buffer row and the pre-split line must be great enough to ensure that the stresses due to the static borehole pressure do not crush the rock which is to form the final wall. On the other hand, the charge must be sufficiently close so that the toe in front of the pre-split line is broken.

The stress generated in the rock by the borehole pressure is given by the following expression:

$$r = P_b \frac{r_h^2}{r^2} \qquad (9)$$

where:

r = radial (compressive) stress in psi

P_b = the borehole pressure in psi

r_h = the borehole radius in feet

r = distance from hole to point of interest in feet.

The charge in the buffer row must be sufficient to adequately fragment the rock between the pre-split rows. Because of its low centre of gravity, the charge in the buffer row acts as a spherical energy source and cube root scaling applies. Spherical cratering tests (Bauer, 1978) in hard jointed rock have indicated that the onset of fracturing occurs at a critical depth d_c defined as follows:

$$d_c = A\,W^{1/3} \qquad (10)$$

where:

d_c = distance in feet from the centre of gravity of the charge to the upper surface.

W = explosive charge in lbs.

A = a factor dependent on the nature of the ground.

Using metric measurements of metres and kilograms; A varies from 1.11 from high strength brittle rocks to 1.79 for soft plastic rocks. Using imperial measurements of feet and pounds A varies from 2.80 to 4.50 respectively.

Once the value of A has been corectly selected for the particular rock type the relationship in Equation 10 provides a valid guideline for determining the minimum buffer charge which will provide adequate fragmentation to the upper surface.

7. PRE-SPLIT BLASTING FOR DRAGLINES

In order to throw blast effectively it is necessary to have good control on the burden of the front row of blastholes. (Crosby and Bauer, 1982). This can be best achieved by presplitting the face ahead of drilling off the throw blast, as illustrated in Figure 7. This allows accurate control on the burden for the full length of the front row blastholes. This permits the throwing of a much higher percentage of material than would normally be achieved without the presplit.

The pre-splitting is somewhat different than that described earlier for normal pit operations. Horizontal bedding and a very pronounced weakness plane at the coal/overburden interface means that a single charge fully coupled at the bottom of the presplit hole is adequate to give excellent results. Figure 8 is a plan view of a typical and main blast tie-in at a dragline operation.

The pre-split hole spacing for a 25 cm diameter blasthole varies from 3 to 4 metres depending on the rock strength and competency. Highly fractured softer materials require the 3 metre spacing while this can be extended towards 4 metres for more competent higher strength sandstones.

Figure 9 illustrates the relationship between hole diameter and spacing, based on published field data. The pre-split holes are drilled to the coal and a charging of 2.2 to 3.7 kg/m of hole is used. The charge is placed at the bottom of the hole or suspended several feet from the bottom in a borehole liner. No stemming is used and it is preferrable to dewater if possible. The pre-split row is fired instantaneously prior to the main blast.

REFERENCES

Bauer, A., (1978). Trends in Drilling and Blasting, C.I.M.M. Bulletin, September.

Calder, P.N., Morash, B.J., (1971). Pit Wall Control at Adams Mine, Mining Congress Mining Congress Journal, Vol. 57, No. 8, p. 34-42.

Calder, P.N., (1977). Perimeter Blasting, Chapter 7, CANMET Pit Slopes Manual, Queen's Printer, Ottawa.

Calder, P.N., Tuomi, J., (1980). Control Blasting at Sherman Mine, 6th Annual Conference of the Society of Explosive Engineers, Tampa, Florida.

Calder, P.N., Jackson, R.J., (1981). Revised Perimeter Blasting Chapter Pit Slope Manual, CANMET, Ottawa.

Crosby, W., Bauer, A., (1982). Wall Control Blasting in Open Pits. Presented at the AIME meeting Tueson, Arizona, December, 1981. Published in Mining Engineering, February, 1982.

Obert, L., Duvall, W.I., (1967). Rock Mechanics and the Design of Structures in Rock. John Wiley and Sons, Inc., New York.

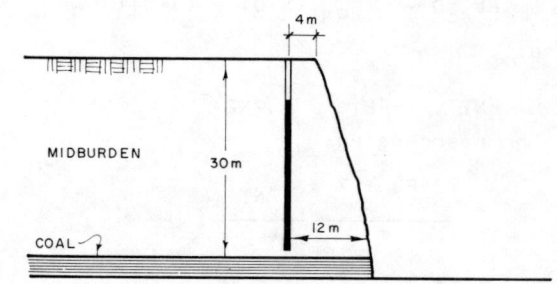

(b) MIDBURDEN BLAST FRONT ROW DESIGN SHOWING POOR CONTROL BECAUSE OF BACKBREAK RESULTING IN VARYING BURDENS WHEN PRESPLIT IS NOT USED.

FIGURE 7: COMPARISON BETWEEN FRONT ROW BURDEN WITH AND WITHOUT PRE-SPLIT

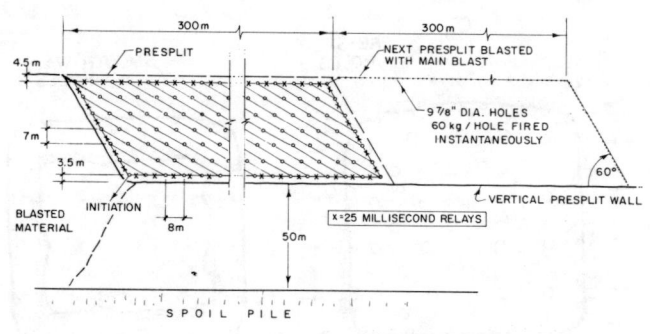

FIGURE 8: PLAN VIEW OF TYPICAL PRE-SPLIT AND MAIN BLAST TIE-IN AT A DRAGLINE OPERATION

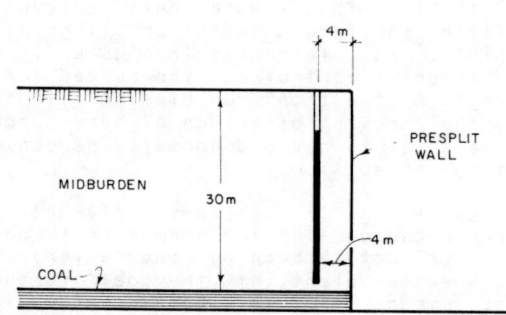

(a) MIDBURDEN BLAST FRONT ROW DESIGN SHOWING ACCURATE CONTROL ON THE BURDEN WHEN PRESPLIT IS USED

FIGURE 7: COMPARISON BETWEEN FRONT ROW BURDEN WITH AND WITHOUT PRE-SPLIT

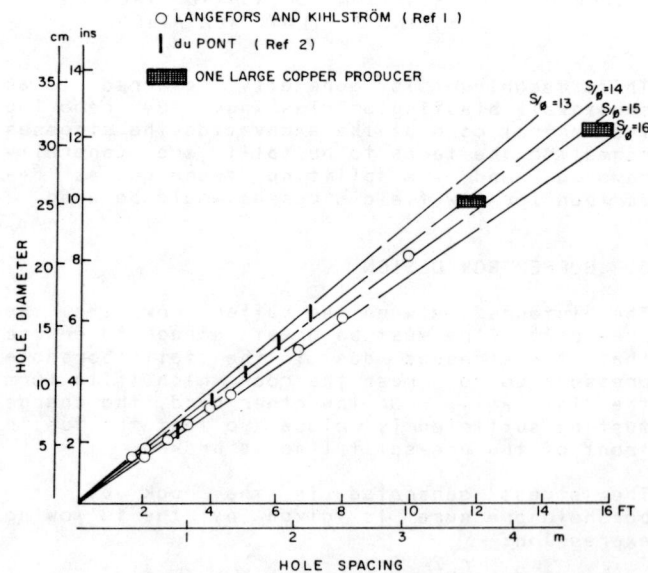

FIGURE 9: RELATIONSHIPS BETWEEN HOLE DIAMETER AND HOLE SPACING FOR TRIM BLASTING

ANALYTICAL CONTRIBUTION TO TUNNEL BEHAVIOUR CAUSED BY BLASTING

Contribution analytique sur le comportement des tunnels suite à l'abattage

Ein analytischer Beitrag zum Verhalten eines Tunnels beim Sprengen

M. Hisatake
Research Assoc. of Civil Eng., Osaka University, Osaka, Japan

S. Sakurai
Prof. of Civil Eng., Kobe University, Kobe, Japan

T. Ito
Prof. of Civil Eng., Osaka University, Osaka, Japan

Y. Kobayashi
Research Engr. of Sumitomo Metal Industries, Ibaraki, Japan

SYNOPSIS

In this paper an analytical approach to forecasting the dynamic behaviour of an existing tunnel due to adjacent blasting is proposed, taking detonation velocity, weight of charge and three-dimensional blasting effects into account. High accuracy of the proposed approach is assured through field measurements.

RESUME

Dans cet article, une approche analytique sur la prévision du comportement dynamique des tunnels existants suite aux abattages à proximité est proposée, en tenant compte de la vitesse des détonations, du poids de la charge et des effets tridimensionnels de l'abattage. L'exactitude de l'approche ainsi proposée est vérifiée par des mesures en réel.

ZUSAMMENFASSUNG

Es wird eine analytische Methode beschrieben, die das Verhalten eines Tunnels bei in der Nähe vor sich gehenden Sprengarbeiten voraussagt. Sie berücksichtigt die Explosionsgeschwindigkeit, das Gewicht der Sprengladung und den räumlichen Sprengeffekt. Die große Genauigkeit der vorgeschlagenen Methode wird durch Geländemessungen bestätigt.

1. INTRODUCTION

When construction works are conducted by blasting near an existing old tunnel, much attention should be paid not to cause any serious damages on the existing tunnel. For this purpose, an observational approach has been employed, in which the weight of charge and the method of blasting are controlled by monitoring the particle velocity of tunnel linings. This approach, however, is difficult to apply to such a special tunnel as water supply tunnel in which the monitoring is unable to operate. And also, in this approach, an allowable value of the particle velocity for the safety margin of the lining is not reasonably determined, so that the sufficient planning of blasting operation can not be made prior to blasting.

The objective of this study is to propose an analytical approach to forecast the dynamic behavior of existing tunnel linings due to an adjacent blast operation. In this approach, a finite element method (FEM) is employed, in which such executive conditions as weight and detonation velocity of charge, distance between existing tunnel and blasting point, and three dimensional blasting effects are taken into consideration. Then, the results obtained by this approach are compared with those of field measurements.

2. ANALYTICAL APPROACH

In this study, the dynamic behavior of existing tunnel linings is analyzed by FEM, in which θ-method is employed (Wilson et al.,1973), and the time length Δt for one step in numerical integration is taken to be $15/10^6$ Sec..
A plane strain condition is assumed so that two-dimensional analysis is done in a plane perpendicular to the tunnel axis, which passes through blasting points. In the two-dimensional analysis, however, unrealistically greater energy would be applied to the ground, because in reality the explosive energy propagates and attenuates three-dimensionally in the ground. To avoid this shortcoming the following method is proposed introducing the equivalent radius $\bar{r}$ of blasting hole, instead of using the actual radius of borehole in which blasting operates.

Firstly, field investigations are carried out through a trial blasting to establish the relation between the weight of charge W and the maximum particle velocity V_{max} of the ground.

$$V_{max} = f_1(W) \qquad (1)$$

Secondly, the finite element analysis is performed for the trial blasting conditions to obtain the relation between the analytical maximum particle velocity V_{max} and the radius $\bar{r}$ of blasting hole.

$$V_{max} = f_2(\bar{r}) \qquad (2)$$

By eliminating V_{max} in Eqs.(1) and (2), the radius $\bar{r}$ can be easily expressed in relation with W as follows,

$$\bar{r} = f(W) \qquad (3)$$

Therefore, the dynamic behavior of the tunnel linings can be easily analyzed by FEM in which the equivalent radius $\bar{r}$ obtained by Eq.(3) is used. According to this method, accurate results can be expected, because the geometrical and geological conditions near the tunnels are easily taken into account, and the equivalent radius of the blasting hole corresponding to the weight of charge is also considered.

3. DETERMINATION OF IMPUT DATA

3.1 Mechanical properties of ground and lining

Mechanical properties of rock specimen in general are quantitatively very different from those of rock masses, because they are very much affected by the specimen size as well as the arrangement of microscopic particles which compose the rock, and, on the other hand, the mechanical properties of rock masses depend mainly on macroscopic structural characteristics such as joints and faults. Therefore, the mechanical properties of rock masses should be directly evaluated by in situ tests. In this study, Young's modulus E of the ground is determined with the longitudinal wave velocity C_p by the following equation,

$$E = \frac{(1+\nu)(1-2\nu)}{1-\nu} \rho C_p \qquad (4)$$

where ρ and ν are, respectively, density and Poisson's ratio of the ground.

On the other hand, damping characteristics are determined as follows. Fig.1 shows the experimental results for several different types of rock indicating the relationship between the frequency f_t of harmonic applied stress and coefficient of viscosity η which is evaluated by assuming the ground to be Voigt type viscoelastic material (Hayashi et al.,1973). This relationship may be approximated by a straight line as shown in the figure and may be expressed by the following equations,

$$\eta = a_0 f_t^{-\lambda} \quad (s \cdot kgf/cm^2) \qquad (5)$$
$$a_0 = 675, \quad \lambda = 0.93$$

After determination of η, a damping factor h is easily evaluated and a damping matrix $[c]_q$ of each element q in FEM can be calculated by the following equations (Idriss et al.,1974),

$$[c]_q = \alpha_q [m]_q + \beta_q [k]_q \qquad (6)$$

$\alpha_q = h\omega, \quad \beta_q = h/\omega,$
$\omega = 2\pi f$: Circular frequency
$[m]_q$: Consistent mass matrix
$[k]_q$: Stiffness matrix

Vibrations caused by blasting are limitted in a local area of the ground and different from those of earthquake in which the ground vibrates in a widely spread region. The frequency of vibration measured in rock due to blasting of dynamite ranges from about 10 to 1000Hz, and its spectrum shows the peak value at about 200Hz (Ito et al.,1971). Therefore, in determining the frequencies in Eqs.(5) and (6), it is unfavorable to use the minimum value of natural circular frequency of the ground, which is usually used in earthquake problems. Owing to the above reasons, in the following analysis, the frequencies in Eqs.(5) and (6) are determined as 200Hz, and also this condition is extended to the concrete lining because of its similar mechanical properties to rock.

3.2 Dynamic pressure acting on rocks

The study on detonation pressure indicates that the maximum detonation pressure P_d may be approximated by the following equation presented by Jones, providing that detonation velocity v is more than 4,000m/s (Sassa et al.,1971),

$$P_d = 0.000424 v^2 \bar{\rho} (1 - 0.543\bar{\rho} + 0.193\bar{\rho}^2), \qquad (7)$$

$\bar{\rho}$: Density of charge, (C.G.S unit)

However, the maximum pressure P_{max} acting on rocks, which is different from the detonation pressure, is determined as follows,

$$P_{max} = \frac{2\rho C_p}{\rho C_p + v\bar{\rho}} P_d \qquad (8)$$

On the other hand, the time length of impact action of detonation pressure is about $10^{-6} \sim 10^{-4}$ sec., and that of gas pressure is about $10^{-3} \sim 10^{-1}$ sec.. Considering the above relations, the following dynamic pressure $P(t)$ acting on rocks is employed in the analysis (Starfield et al.1968),

$$P(t) = 4P_{max}\{exp(-Bt/\sqrt{2}) - exp(-\sqrt{2}Bt)\} \qquad (9)$$

$B = 16338$, t: Second

The above equation is shown in Fig.2.

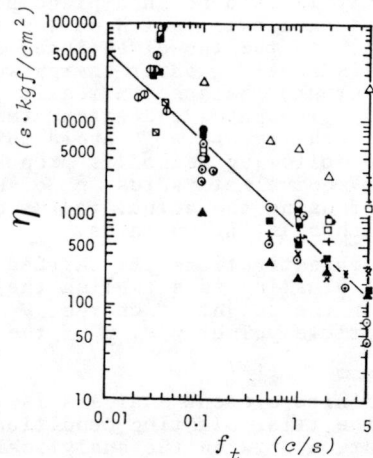

Fig.1 Relationship between frequency f_t of harmonic applied stress and coefficient of viscosity η for several different types of rock

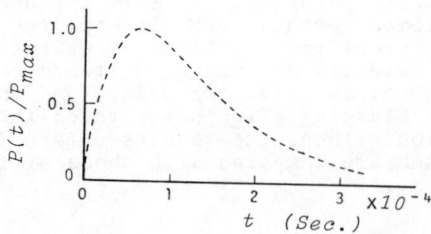

Fig.2 Dynamic pressure $P(t)$ acting on rocks

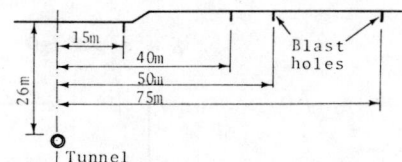

Fig.3 Site of trial blasting

Table 1
Material constants

	ρ g/cm³	ν	C_p km/s
Lining	2.30	0.16	4.5
Ground	2.67	0.21	1.7

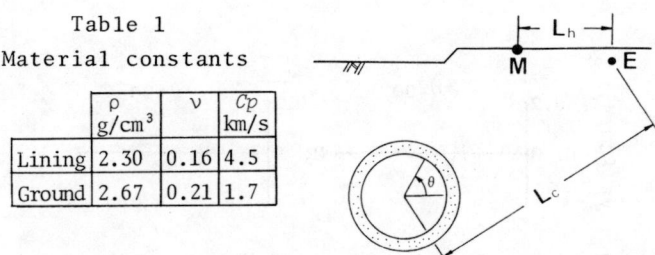

Fig.4 Geometrical condition

4. APPLICATION OF THE PROPOSED METHOD

To verify the accuracy of the proposed method, analytical results are compared with field data.

4.1 Test site

The test site where trial blasting has been conducted is shown in Fig.3. No surface soil layer exists and the propagation velocity of the longitudinal wave is approximately 1.7km/s. The charge (v=6,000m/s, $\bar{\rho}$=1.43g/cm³) with constant weight of W=400gf is set at each blasting point of 2.7m deep from the ground surface and is fired from a farther blasting point, so as not to damage the ground through which blasting waves travel to the measuring points. The diameter of tunnel and the thickness of its lining are 2.3m and 25cm, respectively.

The measurement of the vibration is taken at the ground surface and the inner surface of the tunnel lining as well. The geometrical relationship among the blasting point E, the location of tunnel and the measuring point M on the ground surface is shown in Fig.4, and the material constants of both the ground and the lining are shown in Table 1.

4.2 Relationship between radius $\bar{r}$ of blasting hole and weight of charge W

The analytical results of the horizontal and vertical components ($V_{max,h}$ and $V_{max,v}$, respectively) of the maximum particle velocity at the ground surface are shown in Fig.5 in relation with the distance L_h. These results can be represented by the following equations

$$V_{max,h} = C_1 \bar{r}^{\alpha_1}, \quad V_{max,v} = C_2 \bar{r}^{\alpha_2} \quad (10)$$

where C_1, C_2, α_1 and α_2 are constants determined from Fig.5.

On the other hand, Fig.6 shows field measurement results indicating a relationship between the distance L_h and the measured values of the maximum particle velocities at the ground surface. Substituting the measured values of particle velocity for L_h=15m and 28m, shown in Fig.6, into

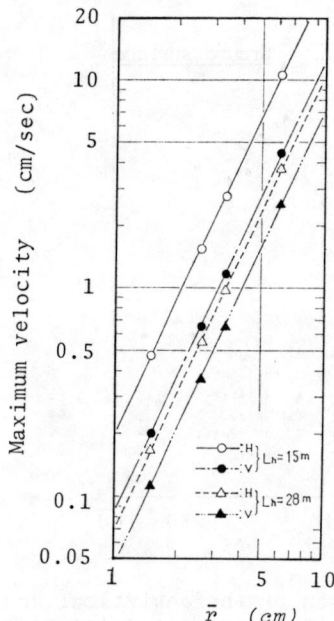

Fig.5 Analytical results for relationship between the maximum particle velocity and the radius $\bar{r}$ of blasting hole in L_h=15m and 28m (H:Horizontal, V:Vertical)

the Eq.(10) yields the result that the average value of the equivalent radius $\bar{r}$ of blasting hole, corresponding to W=400gf, is 2.2cm. The analytical and the experimental values of the maximum particle velocity obtained at several measuring points on the ground surface are compared in Fig.7. It is seen from the figure that both the results show a good agreement. This means that the value of the equivalent radius $\bar{r}$ of blasting hole could reasonably be evaluated.

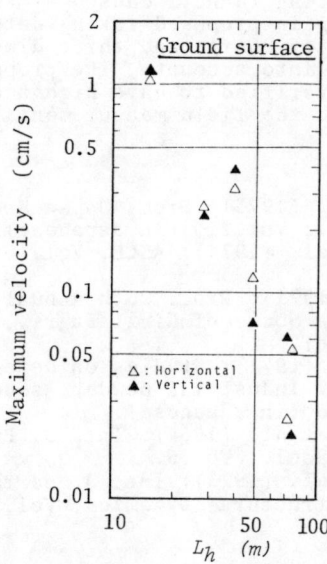

Fig.6 Experimental results for relationship between the maximum particle velocity and the distance L_h

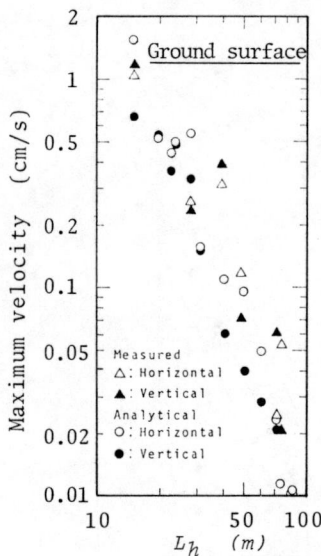

Fig.7 Comparison of the analytical and the experimental particle velocities

4.3 Comparison of measured and analytical results obtained at the tunnel lining

Fig.8 shows the comparison of seismograms of the analytical and measured particle velocities at the inner surface of the tunnel lining ($\theta=0°$ and $90°$), and the maximum particle velocities are shown in Fig.9. From these figures, it may be seen that there is a good agreement between the analytical and the measured results. This means that the analytical approach proposed here may be sufficiently applicable to practical problems.

5. CONCLUSIONS

An analytical approach to forecast the dynamic behavior of existing tunnels caused by an adjacent blasting is proposed taking detonation velocity, weight of charge and three dimensional blasting effects into account. The proposed method has been verified to have high accuracy by comparing with the field measurements.

REFERENCES

Hayashi, M. et al. (1973): Proc. Japan Soci. of Civil Engrs., Vol.217 (in Japanese).
Idriss, I.M. et al. (1974): ASCE, Vol.100, Gt 1, Jan..
Ito, I. et al. (1971): Proc. 26th Annual Conf. of the Japan Soci. of Civil Engrs., Vol.3 (in Japanese).
Sassa, K. et al. (1971): Studies on detonation pressure, J. Industrial powder Assoc., Vol.32, No.6 (in Japanese).
Starfield, A.M. et al. (1968): Int. J. Rock Mech. Min. Soci., Vol.5.
Wilson, E.L. et al. (1973): Int. J. Earthquake Engg. and Structural Dynamics, Vol.2, No.2.

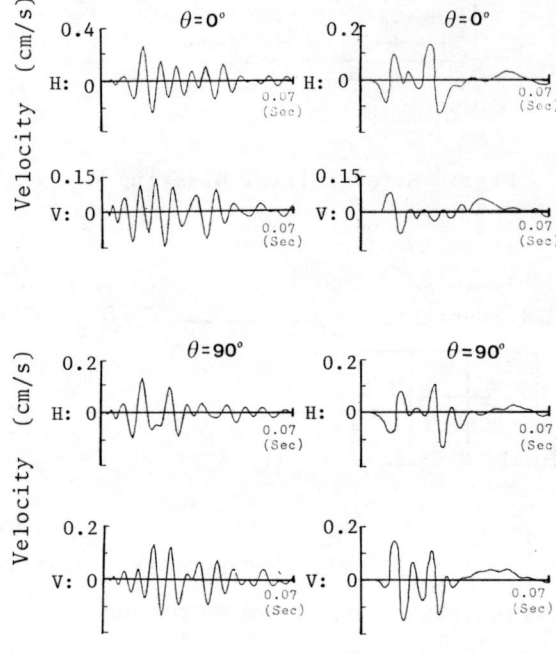

Fig.8 Comparison of seismograms of the analytical and the measured particle velocities (H:Horizontal, V:Vertical, $L_C=28m$)

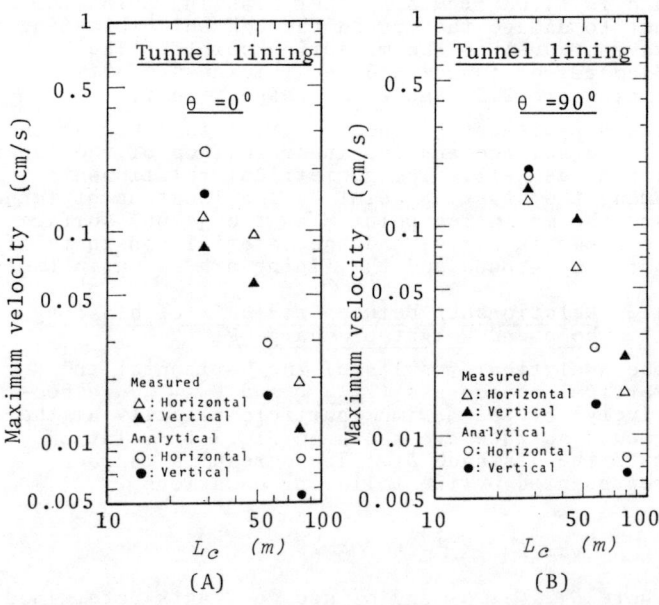

Fig.9 Comparison of the analytical and the measured maximum particle velocities ($L_C=28m$)

DRILLABILITY AND DRILLING METHODS
Forabilité et méthodes de forage
Bohrbarkeit und Bohrmethoden

A. Mouraz Miranda
Assistant Professor, Universidade Técnica de Lisboa, Portugal

F. Mello Mendes
Chairman Professor, Universidade Técnica de Lisboa, Portugal

SYNOPSIS

Some comments concerning the concept of the "drillability" of rock are followed by an investigation aimed at determining this concept. The investigation was carried out on the basis of two parameters, Vickers microhardness and specific energy, with the aim of determining the most suitable drilling method for a given type of rock on the basis of previous laboratory tests. The study was carried out using eight different rocks which had been drilled industrially by various methods and tested in the laboratory. The results obtained so far suggest that the research be continued.

RESUME

L'article commence par des commentaires sur la définition de forabilité des roches et a pour but de clarifier cette définition en faisant intervenir deux paramètres, à savoir: la microdureté Vickers et l'énergie spécifique, en choisissant la méthode de forage la plus adéquate pour chaque roche lors des essais en laboratoire. L'étude a été faite sur huit types de roches bien différentes. Elles ont été forées par diverses méthodes et testées en laboratoire. Les résultats obtenus sont encourageants.

ZUSAMMENFASSUNG

Nach einigen Betrachtungen über den Begriff der Gesteinsbohrbarkeit wird eine Untersuchung zur Bestimmung dieses Begriffs beschrieben. Diese Untersuchung wurde auf der Basis zweier Parameter, der Vickers Mikrohärte und der Spezifischen Energie, durchgeführt, mit der Absicht, daß aufgrund der Laborprüfungen vorausgesagt werden kann, welche die für ein gegebenes Gestein ratsamste Bohrmethode ist. Die Untersuchung wurde an acht verschiedenen Gesteinsarten durchgeführt, die mittels verschiedener industrieller Verfahren durchbohrt und dann im Labor geprüft wurden. Die erzielten Ergebnisse zeigten sich als ermutigend für den Fortgang der Untersuchung.

1. INTRODUCTION

The drillability of a rock is one facet of its workability. Meaning the easy or difficult way to drill an hole in the rock. However this common sense definition is not enoughly accurate being dependent of rock itself and the used drilling method.

For practical purposes, normally an economic one, two objectives are always requested to the drilling operations: drill the hole as fast as possible and get the minimum wear on the tool.

Bearing in mind this two objectives one tried to take them always into account for a drillability definition. An example of this way to define drillability was followed by SIEVERS (1952): using the rate of penetration and the corresponding wear of drilling tools, both determined by standard laboratory testing.

Plotting points for a set of typical rocks in a drillability diagram defined by these variables, and using the empirical knowledge to choose the most adequate drilling method (rotary or percussion) enables the outline of a boundary between the fields for these largely known drilling methods (Fig. 1). This same diagram also gives information on drillability of a rock only laboratory tested MELLO MENDES (1961). The parameters determined under laboratory procedure normally give good correlations for rate of penetration and corresponding wear for various rocks and different industrial equipment.

Meanwhile the split of the drillability diagram in only two domains (one for rotary and another for percussion) for selection of the best drilling method, is oversimplified. As a matter of fact, trying to be more specific with nowadays available drilling methods, it is rather difficult to present them by a biparametric scheme, such

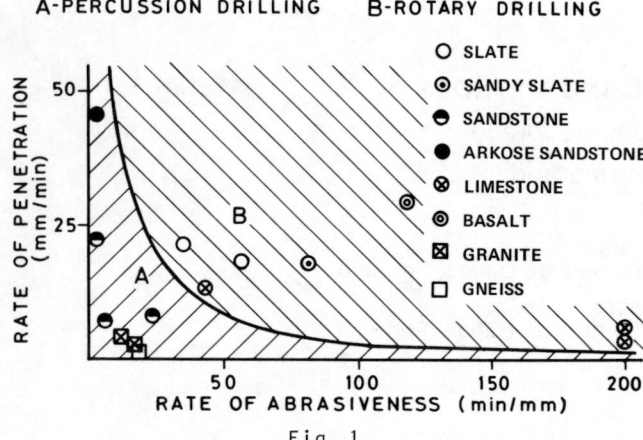

Fig. 1

as the above mentionned.

Reasoning about the phenomenous nature, conclusion can be achieved that both parameters are partial features of the same reality: the energy globally required by the drilling process, which must be linked with the drillability concept.

A more deep analysis about the rock drilling mechanism, whichever the drilling method used, and besides the relative predominant feature, allows to point the following main actions: indentations, of quasi-static type predominantly associated with plastic deformations; chipping, with brittle fracture, essentially dynamic; vibration, or percussion, whose frequencies and intensities may mobilize action of both referred types; longitunal thrust actions of drilling tools with more or less constant character.

No doubts arise about a selection method interest for the most adequate drilling technique to be used for a given rock, based on petrographic analysis, associated with physical parameters easily obtained by laboratory testing.

Among laboratory tests Vickers microhardness, has been elsewhere reported (MELLO MENDES et al. 1979), to give good correlation with industrial rate of penetration, using the same type of drilling equipment. Besides this, Vickers microhardness has shown very interesting as a rock workability index (AIRES-BARROS et al. 1974).

Correlations obtained between Vickers microhardness and rate of penetration for different types of equipment does not process the same meaning, despite some outlined overlaps (Fig. 2). This is not a very uncommon fact, having in mind the introducing comments about the various terms which make up global energy required for drilling.

Correlation between rate of penetration and specific energy only gives overlapped zones, each one for each type of equipment (Fig. 3).

This sum of evident facts took us to go on with the research hereby reported. The main purpose was to establish a drillability concept, giving step for a selection scheme of drilling equipment, made through a petrographic analysis of a given new rock. In the study we tried a biparametric definition for drillability using Vickers microhardness and specific energy used for drilling, and calculated according TEALE's (1965) proposal and later BULLOCK (1976) development.

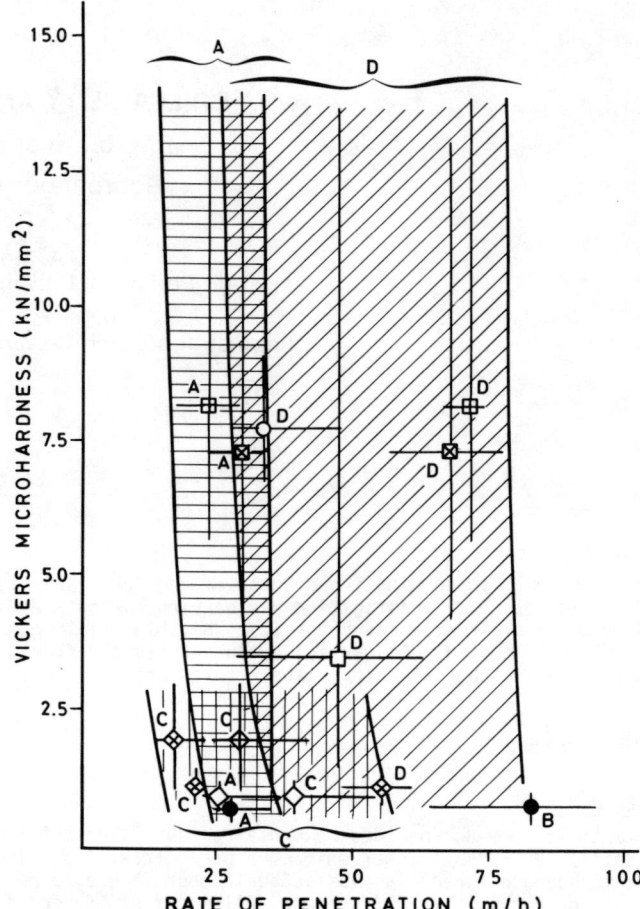

Fig. 2

2. STUDY DESCRIPTION

The study was conducted in production quarries and mines, using different drilling methods and over the following rocks:

Maceira Marl - A soft greyish brown, medium lime content marl.

Borba crystalline Limestone - Almost made by medium granular calcite, and very rare epidote which would be better named white granular marble.

Cantanhede Limestone - A cretaceous reciffal limestone with a massif character.

Ferreiras (Algarve) dolomitic Limestone - A brownish - grey, dolomitic clastic limestone.

Aljustrel Piritite - A pyrite ore with a fine to medium grain character. Pyrite is dominant with sphalerite and calcopirite with acessory character.

Aguieira Graywacke - A fine greyish grained pellitic - metaclastic rock, with alkalic feldspar,

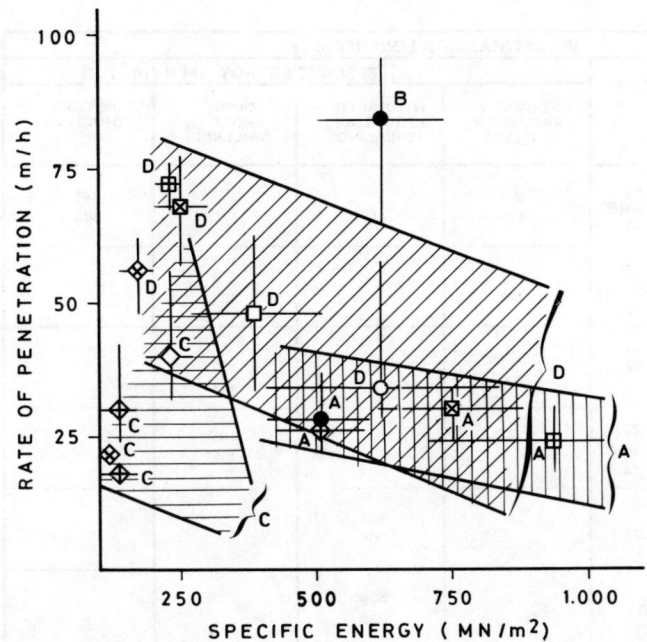

- ● — MACEIRA MARL
- ○ — ALJUSTREL PYRITITE
- ◇ — BORBA CRYSTALLINE LIMESTONE
- □ — AGUIEIRA GRAYWACKE
- ◈ — CANTANHEDE LIMESTONE
- ⊠ — AROUCA GRANITE
- ⊕ — FERREIRAS (ALGARVE) DOLOMITIC
- ⊞ — PENAFIEL GRANITE

A — PNEUMATIC PERCUSSION DRILLING C — DOWN HOLE DRILLING
B — ROTARY DRILLING D — HYDRAULIC PERCUSSION DRILLING

Fig. 3

some amphibole and fine sub-angular quartz grains, biotite, chlorite and same clayish material between grains.

Penafiel Granite - It is a medium to coarse grain, alkalic feldspar rich granite, some quartz, biotite and alcaline amphibole.

Arouca Granite - A porphyroid granite with a K-feldspar associated with an intermediate feldspar, large grains of quartz, and large amounts of two types of dark mica.

From all rocks, and as close as possible to the borehole place, samples were colleted, with aproximately 30x30x30 cm^3, for laboratory tests. From each sample, the polished surfaces were prepared for Vickers microhardness testing. Petrographic slides were prepared. A core drill, with 46 mm diameter, were also done for ultimate compressive strength determination.

Microhardness was determined using equipment and a tecnique described elsewhere (MELLO MENDES et al. 1979). Mean value, together with the highest and the lowest values are reported in this paper, due to the interest of the dispersion of values.

The ultimate compressive tests, done for a better geomechanic characterisation of all rocks, were requested to different laboratories for safety reasons. Average, maximum and minimum values are reported.

For each rock and for each drilling method meam, maximum and minimum values of rate of penetration are presented due to the importance of the values dispersion. The reported figures are the result of dozens of tests for each rock and for each drilling method taking care to use always the same operating conditions (rpm, thrust, air pressure, air delivery, shape of cutters, etc.). For drifter tests using extendable steel only the first rod was taken into account. In each operating site, only the results with the same operator and its equipment were considered. The data collection was carried during two years and an half of field work.

Specific energies were determined based on the collected data and using BULLOCK (1976) formula, which is made by three components for the required energy to drill an unit volume of rock: the first term is for percussive mechanism (if percussion hammers are used); the second component accounts for rotation; and finally the third for thrust or applied pulldown.

For each rock and each drilling method, maximum and minimum values were considered, derived from experimental data.

Table 1 shows the major figures for collected data as well as calculated values.

With all values put together, we tried first to establish correlation between Vickers microhardness and rate of penetration, searching for a

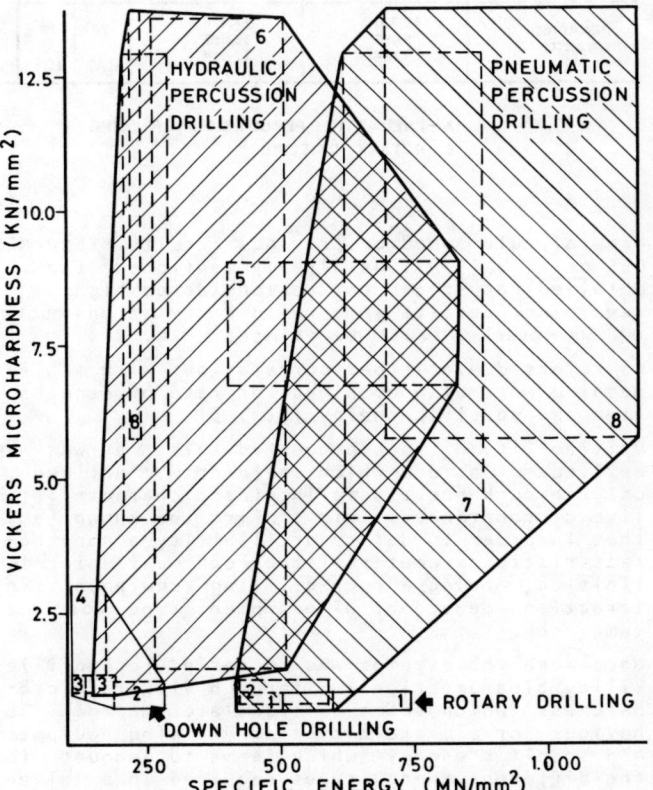

1 — MACEIRA MARL 5 — ALJUSTREL PYRITITE
2 — BORBA CRYSTALLINE LIMESTONE 6 — AGUIEIRA GRAYWACKE
3 — CANTANHEDE LIMESTONE 7 — AROUCA GRANITE
4 — FERREIRAS (ALGARVE) DOLOMITIC 8 — PENAFIEL GRANITE

Fig. 4

ROCK TYPE	LABORATORY TESTS		INDUSTRIAL DRILLING RESULTS					
	VICKERS MICRO HARDNESS (N/mm2)	ULTIMATE COMPRESSIVE STRENGTH (N/m2)	BOREHOLE DIAMETER (mm)		PENETRATION RATE (m/h)		SPECIFIC ENERGY (MN/m2)			
							PNEUMATIC PERCUSSION DRILLING	HYDRAULIC PERCUSSION DRILLING	DOWN HOLE DRILLING	ROTARY DRILLING
MACEIRA MARL.	1,049 / 764 / 627	2,058 / 1,666 / 1,078	102 (A)	89 (B)	35 / 28(A) / 24	96 / 84(B) / 64	598 / 410			745 / 503
BORBA CRYSTALLINE LIMESTONE	1,235 / 941 / 764	3,136 / 2,940 / 2,695	102 (A)	165 (C)	34 / 26(A) / 24	56 / 40(C) / 38	598 / 423		270 / 183	
CANTANHEDE LIMESTONE	1,382 / 1,049 / 990	4,547 / 3,724 / 3,214	114 (C)	102 (D)	24 / 22(C) / 20	62 / 56(D) / 48		190 / 148	130 / 108	
FERREIRAS (Algarve) DOLOMITIC LIMESTONE	2,999 / 1,940 / 1,078	5,743 / 4,361 / 3,430	165 (C)	114 (C)	42 / 30(C) / 24	22 / 18(C) / 14			165 / 105	
ALJUSTREL PYRITITE	9,016 / 7,683 / 6,684	39,200 / 37,240 / 23,128	64 (D)		58 / 34 (D) / 28			835 / 404		
AGUIEIRA GRAYWACKE	13,524 / 3,430 / 1,421	7,703 / 6,135 / 4,136	76 (D)		62 / 48 (D) / 32			508 / 263		
AROUCA GRANITE	12,936 / 7,252 / 4,214	12,201 / 11,711 / 10,584	76 (A)	76 (D)	34 / 30(A) / 24	78 / 68(D) / 56	874 / 618	290 / 209		
PENAFIEL GRANITE	13,720 / 8,095 / 5,684	17,444 / 16,562 / 12,054	76 (A)	76 (D)	30 / 24(A) / 18	74 / 72(D) / 68	1,165 / 699	239 / 220		

A – PNEUMATIC PERCUSSION DRILLING
B – ROTARY DRILLING
C – DOWN HOLE DRILLING
D – HYDRAULIC PERCUSSION DRILLING

Table 1

general rule based on previous results (MELLO MENDES et al. 1979). As already mentioned the obtained result can not be considered highly rewarding, despite some valid correlations outlined. Some overlaps must be pointed (Fig. 2).

As referred overlapped correlations with similar meaning were also obtained for rate of penetration and specific energy (Fig. 3).

Drillability of rock presented this way was an aptitude of a rock which could not be defined only through one single intrinsic characteristic, like microhardness. This confirm the known fact that in order to define drillability a non-characteristic parameter of the rock must enter definition, being a rock-drilling equipment interaction, dependent of equipment type for the same rock.

Once more the attempt was to define rock drillability biparametrically: through Vickers microhardness (which seems to translate the rock behaviour for a certain type of drilling equipment) and specific energy (which seems to account for the different energy types required in a given rock per unit of volume, for different types of drilling equipment).

Plotting each rock representing point on a microhardness - specific energy diagram, specifiying the different drilling equipments, it allowed to outline the domain boundaries for those equipments suitability. Those domains are drillability fields for, these equipments defined by microhardness.

The knowledge of those drillability domains allow to antecipate for a given rock, knowing its microhardness, which is the most suitable and convenient drilling method among those available. This method will correspond to the smallest specific energy.

Drillability fields outlined in Fig.4 were obtained using a very limited number of rocks. The shape and dimension of areas of these fields is primary resulting from the intrinsic characteristics dispersion. This expected fact is easily explained by the dispersed behaviours for the used drilling methods.

It should be pointed that the dispersion of values for the used drillability parameters can be justified by heterogeneity and anisotropy of rocks. This could also have been forecasted by the petrographical analysis and the mechanical tests such as the ultimate compression test, which emphasizes the fact better.

This fact is quickly remarked in Fig. 5, plotting along one axe microhardness and another the ultimate compression, on a X-Y diagram and marking all representing points for the rocks under analysis with respective dispersion of values. It is also remarkable the increase of both characteristics for the areas of dispersion of plotted points with the average microhardness. For each rock, dispersion is greater for those values obtained in such a scale that rock is more hete

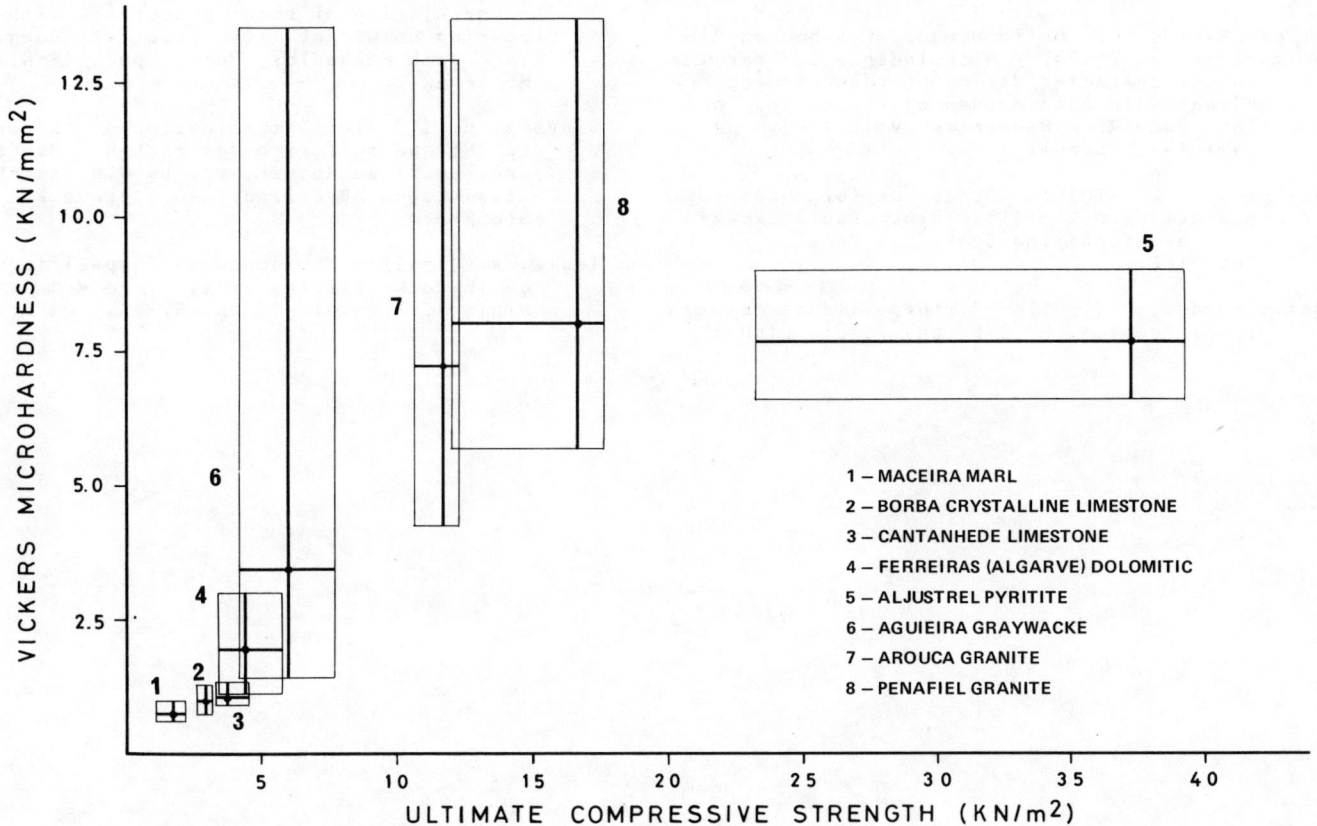

Fig. 5

rogeneous.

Starting with lowest microhardness, which is the same as to begin with marl and limestones with almost homogeneous composition, it is remarkable the continuous trend to increase the dispersion of values for ultimate compression. This fact is possibly due to rock anisotropy not taken into account during sampling procedures.

Graywacke presents an highly dispersed microhardness possibly due to matrix heterogeneity at microhardness test scale, as ultimate compressive strength is less dispersed. The same feature is observed in the granitic rocks, being more evident with the fine grain granite which is more homogeneous at the ultimate compressive strength test scale. At last, piritite which is an highly homogeneous rock material, showing a large dispersion for the ultimate compression values, probably resulting from major anisotropies at massif scale not taken into account.

The similarity between specific energy and compression dispersion is an expected fact. Drilling operations are obviously actions considered for specific energy definition work at a very close scale to compression tests. Being so, the knowledge of ultimate compressive strength and its dispersion is a precious tool to predict rock drillability dispersion in the described biparametric system.

3. CONCLUSIONS

An accurate definition for a rock drillability field for the various available drilling equipment requires a larger number of tests than those used. The presented results are by no means interesting. Vickers microhardness, as previously mentionned (AIRES-BARROS et al. 1974), is a very easy to obtain characteristic in the laboratory, giving a good guidance to identify the relative importance of indentation phenomena to the drilling mechanism.

Specific energy being highly influenced by rock--reaction characteristic to the used drilling equipment, can be obtained by standard laboratory tests. This is are possible way to carry research.

Finally, rock drillability definition based on these two parameters seems to point to a logical selection scheme for the most adequate rock drilling equipment based only on rock laboratory tests. We hope that continued research allow to confirm this assumption.

4. REFERENCES

Aires-Barros, L., Mello Mendes, F. & Mouraz Miranda, A. (1974) - Microindentation hardness in the characterization of rock sawing. Advances in Rock Mechanics. Proc. 3rd Cong. Int. Soc. Rock Mechanics, Vol. II-2, pp. 1471-1475. Denver.

Bullock, R. L. (1976) - Actual performances of hydraulic rock drills. Proc. Rapid Excavation and Tunneling Conf. (Las Vegas, Flo.) SME/AIME.

Mello Mendes, F. (1961) - Perfurabilidade de rochas. Técnica, Nº 316, pp. 75-86. Lisboa.

Mello Mendes, F. & Mouraz Miranda, A. (1979) - Correlation of rock properties with a bearing on workability. Proc. 4th Cong. Int. Soc. Rock Mechanics, Vol.2, pp. 429-431. Montreux.

Sievers, H. (1952) - Détermination de la caractéristique de forage des roches. Manuel de Creusement au Rocher, ed. by K.H.Fraenkel. Atlas Copco AB & Sandvikens Jernverks AB. Stockholm.

Teale, R. (1965) - The concept of specific energy in rock drilling. Int. J. Rock Mechanics Mining Sci., Vol. 2, pp. 57-73. Oxford.

UTILISATION DES JETS D'EAU A HAUTE PRESSION DANS LES TRAVAUX MINIERS
The use of high-pressure water jets in mining operations
Einsatz von Hochdruckwasserstrahlen im Grubenbetrieb

F. Pechalat
Chef du Groupe "MACHINES", Cerchar, BP n° 2,60550 Verneuil-en-Halatte
Y. Lefin
Ingénieur au Groupe "MACHINES", Cerchar, BP n° 2,60550 Verneuil-en-Halatte

RESUME

La communication présente les travaux en cours au CERCHAR dans le domaine de la destruction des roches par jets d'eau à haute pression. Dans une première partie, les auteurs décrivent les installations d'essais et résument brièvement les premiers résultats généraux obtenues à la suite d'essais de coupe linéaire avec un seul jet d'eau dans la roche en mettant l'accent sur certains d'entre eux. La deuxième partie est consacrée aux études liées à l'utilisation des jets d'eau à haute pression dans les travaux miniers. Sont notamment présentés:
- Un jumbo expérimental de sciage-foration par jets d'eau seuls (pression d'eau 350 MPa) destiné au prédécoupage des filons de minerais suivi d'un abattage sélectif.
- Un matériel de foration rotative par taillant mécanique assisté de jets d'eau (pression d'eau jusqu'à 250 MPa).
Ces deux engins sont destinés à être essayés dans des mines françaises dans un avenir proche.

SYNOPSIS

This paper presents the work in progress performed by Cerchar on rock cutting by means of high-pressure water jets. In the first part of the paper, the authors describe the test bench used and summarize the initial results of tests consisting of linear rock cutting by means of a single water jet, special emphasis being placed on some of the results. In the second part studies on the application of high pressure water jets in mines are presented. The equipment described includes:
- An experimental jumbo for drilling and kerfing only by means of water jets (350 MPa water pressure), intended for precutting ore deposits prior to selective mining.
- A machine for rotary drilling by means of a mechanical tool assisted by water jets (up to 200 MPa water pressure).
It is planned to test both machines in Frerch mines at an early date.

ZUSAMMENFASSUNG

Es wird über die Arbeiten des Cerchar über das Lösen von Gestein mit Hochdruckwasserstrahlen berichtet. Im ersten Teil beschreiben die Verfasser den Versuchsstand und fassen ihre ersten Versuchsergebnisse beim geradlinigen Schneiden mit einem einzigen Wasserstrahl zusammen, wobei auf gewisse Ergebnisse hingewiesen wird. Im zweiten Teil wird über den Einsatz dieser Technik im Bergbau berichtet. Beschrieben werden insbesondere:
- Ein experimenteller, nur mit Hochdruckwasserstrahlen (bis 350 MPa) arbeitender Kerb und Bohrwagen zum Vorschneiden von Erzgängen zwecks selektiver Gewinnung.
- Eine Drehbohrmaschine bei der das mechanische Bohrwerkzeug durch Hochdruckwasserstrahlen (Wasserdruck bis 200 MPa) unterstützt wird.
Diese Maschinen sollen in Kürze in französischen Gruben erprobt werden.

Pendant des millénaires, le métier de mineur a consisté à "griffer" roches et minerais au moyen d'outils métalliques ne transmettant au matériau à détruire qu'une faible énergie. Les machines puissantes utilisées actuellement et équipées de pics ou de molettes ne sont que des perfectionnements de ce principe ancien. Depuis plusieurs années, de nombreuses recherches ont été entreprises dans quelques pays avancés, visant à substituer à cette technique, un mode de destruction dans lequel une quantité d'énergie importante est concentrée sur une petite surface du massif à abattre. C'est ainsi que des expérimentations ont été faites -notamment en France- mettant en oeuvre des lasers et des jets d'eau à haute pression.

Le passage des expériences de laboratoires aux applications effectives en mine implique que l'Industrie mette à la disposition des mineurs des sources d'énergie puissantes qui, au point de vue des dimensions, de la fiabilité et de la sécurité soient compatibles avec l'environnement du fond. Si cette exigence n'est pas encore actuellement satisfaite pour les lasers, il n'en est pas de même pour les jets d'eau à haute pression : dans ce domaine des matériels existent -à coup sûr perfectibles- et des applications ont commencé à voir le jour.

Nous donnerons, dans ce qui suit, un aperçu des recherches et des réalisations entreprises en France par le Cerchar. Après une rapide présentation des expérimentations à caractère fondamental faites pour déterminer les lois de la coupe des roches par des jets d'eau, nous décrirons deux applications mettant en oeuvre :
- l'une, des "jets coupants", c'est-à-dire des jets d'eau à très haute pression attaquant la roche, seuls,
- l'autre, des "jets assistants", c'est-à-dire des jets d'eau à haute pression travaillant en association avec des outils de coupe conventionnels : ici, des taillants de foration.

I. ETUDES SUR LES LOIS DE LA COUPE DES ROCHES PAR UN JET D'EAU

Des lois régissant la coupe d'une roche par un jet -c'est-à-dire essentiellement l'influence de la pression, du débit et de la distance jet/roche sur la profondeur de la découpe- ont été établies depuis plusieurs années à la suite des travaux effectués notamment aux U.S.A. et en Grande-Bretagne. Il nous a paru cependant nécessaire de reprendre l'étude de ces lois, d'une part pour nous familiariser avec l'emploi d'un matériel sophistiqué -préliminaire indispensable à une utilisation de ce matériel dans un chantier de production- et, d'autre part pour approfondir certains points particuliers qui nous semblaient encore insuffisamment connus- en particulier l'influence de la vitesse de déplacement du jet par rapport à la roche et celle des caractéristiques géomécaniques du matériau à abattre. Nous ne traiterons dans ce chapitre que de ces points particuliers, supposant les lois générales connues.

I.1 Le banc d'essai

Il se compose d'une table support d'échantillons, mobile suivant deux axes perpendiculaires dans le plan horizontal et d'un portique fixe supportant la tête d'éjection du jet d'eau. Cette tête est mobile suivant un axe vertical, ce qui permet le réglage de la distance buse-matériau. La vitesse de déplacement de la table peut-être sélectionnée dans une plage allant de 0,5 à 15 m/min. La tête, munie de sa buse d'éjection en saphir synthétique de diamètre variable est alimentée en eau par une centrale dont les caractéristiques principales sont :
- puissance électrique installée : 250 kw
- pression d'eau maximale : 360 MPa
- débit d'eau maximal à 360 MPa : 27 l/min.

Cette centrale, construite par FLOW-INDUSTRIES a été notablement modifiée par le Cerchar, de façon, d'une part à la rendre compatible avec la règlementation minière, et d'autre part à réduire ses dimensions dans l'optique d'une descente au fond (Fig. 1).

Fig. 1 <u>Centrale haute pression</u>

Quant au mode opératoire des essais, il consiste à faire varier un par un les paramètres -pression, débit, vitesse, roche etc...- et à mesurer de façon soignée la profondeur de la saignée réalisée (Fig. 2).

Fig. 2 <u>Saignées dans un bloc de roche</u>

I.2 Quelques résultats remarquables

Nous ne reviendrons pas, ainsi qu'il a déjà été indiqué sur les lois générales.

I.2.1. Influence de la puissance hydraulique fournie P

A chaque couple de valeurs pression-débit correspond une puissance hydraulique P passant par la buse d'éjection que l'on recherche par la formule

$$P = \frac{k\pi d^2}{4} \left(\frac{2}{\rho}\right)^{\frac{1}{2}} p^{3/2}$$

d = diamètre de la buse
p = pression
ρ = masse volumique du fluide
k = coefficient sans dimension lié à la buse

Si l'on porte sur un graphique la profondeur de saignée h en fonction de cette puissance hydraulique pour différentes valeurs du couple pression-diamètre, on observe que les points se répartissent en première approximation sur une droite. Un examen plus fin du graphique (Fig. 3) montre que cette droite peut, en fait, se décomposer en segments de droite correspondant chacun à un diamètre de buse (c'est-à-dire à un débit) et que la conclusion pratique à en tirer est que, pour une puissance hydraulique donnée, on a intérêt à choisir le diamètre de buse minimal et la pression maximale correspondant à cette puissance. L'importance du gain sur h retiré d'un choix optimal du couple pression-débit dépend fortement de la nature de la roche à abattre.

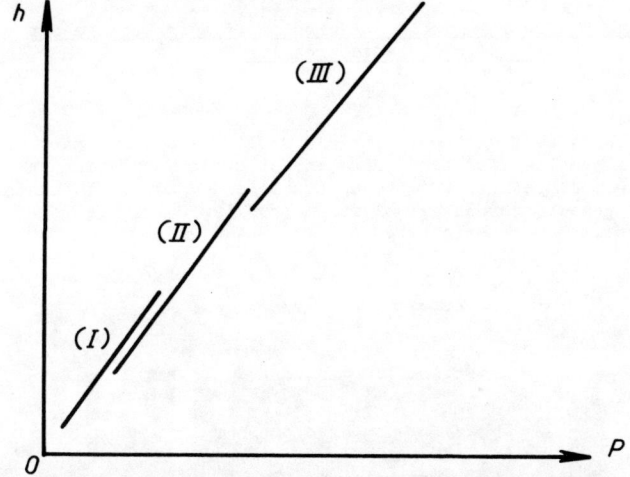

Fig. 3 Influence de la puissance hydraulique fournie P sur la profondeur de la saignée h, à vitesse de déplacement constante.
Les trois segments de droite (I), (II) et (III) correspondent à des diamètres de buse $\emptyset 1$, $\emptyset 2$, $\emptyset 3$ tels que : $\emptyset 3 > \emptyset 2 > \emptyset 1$.

I.2.2. Influence de la vitesse de déplacement

Les jets d'eau à haute pression ne sont pas destinés à effectuer de l'abattage intégral : toutes leurs utilisations potentielles dans le domaine de la destruction des roches se ramènent à une opération de découpe de saignées étroites. Il faut donc s'intéresser non plus au poids de produits abattu et à l'énergie spécifique qui en découle -ce qui n'aurait pas de sens- mais à la vitesse de sciage qui est égale à :

$$h \times V_T = \text{Profondeur de saignée} \times \text{vitesse de déplacement}$$

et à l'énergie consommée par unité de surface sciée que nous appellerons par commodité Energie Spécifique de Sciage

$$E_{ss} = \frac{P}{hV_T} = \frac{\text{Puissance fournie}}{\text{Vitesse de sciage}}$$

La relation la plus remarquable en ce qui concerne la vitesse de sciage est celle qui la lie à la vitesse de translation V_T. La figure 4 fait apparaître sur les courbes une valeur de V_T pour laquelle la vitesse de sciage est maximale et l'énergie spécifique de sciage minimale. On a donc intérêt, lors de la mise en oeuvre de jets coupants, à se placer au voisinage de cette vitesse de déplacement optimale. Celle-ci devra être déterminée expérimentalement et elle dépendra de la pression et du diamètre de buse utilisés. Il sera donc nécessaire de disposer d'un banc d'essai autorisant de plus grandes vitesses de déplacement jet/roche. Les essais à plus grande vitesse qui sont inscrits au programme de recherches du Cerchar, permettront de préciser l'évolution de la puissance hydraulique consommée. Au vu d'un petit nombre de résultats, il semblerait que les valeurs de l'énergie spécifique minimales seraient pratiquement indépendantes de la puissance consommée et seraient donc caractéristiques d'une roche vis-à-vis d'un jet.

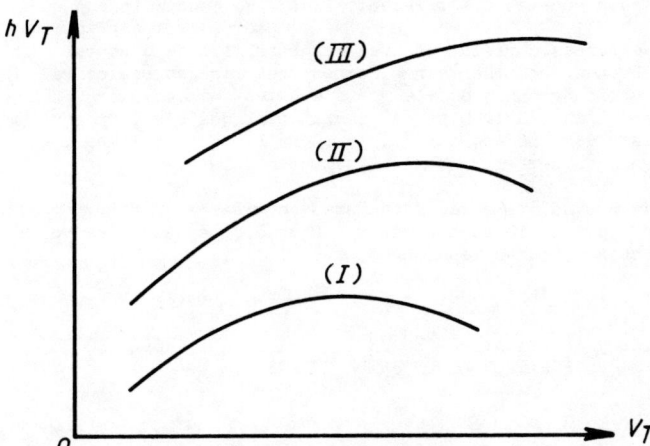

Fig. 4 Influence de la vitesse de déplacement V_T sur la vitesse de sciage mesurée par le produit $h V_T$.
Les trois courbes (I), (II), (III) correspondent à des pressions d'eau P_1, P_2, P_3 telles que : $P_3 > P_2 > P_1$

I.3 Influence des caractéristiques géomécaniques de la roche

Pour pouvoir déterminer et prévoir les valeurs optimales des paramètres de découpe au jet d'eau à haute pression (diamètre de buse, vitesse de translation, pression, angle, etc...), il est nécessaire de comprendre les mécanismes de destruction des roches par le jet d'eau, et, de là, de construire des modèles de prévision établissant des relations entre les performances de découpe et certaines propriétés géomécaniques de la roche.

La destruction d'une roche par un jet d'eau à haute pression est un phénomène complexe, mettant en jeu en général plusieurs mécanismes : déformation élasto-plastique, arrachement des grains, fracturation par initiation et propagation des fissures préexistantes, rupture fragile due aux ondes de traction et de cisaillement se propageant lors de l'impact du jet.

Le mécanisme dominant diffère selon :
- le type de chargement appliqué : le jet d'eau peut être continu, pulsé ou modulé ; il peut se déplacer au-dessus de la roche ou rester fixe ; la buse peut être soit proche de la surface de la roche, soit éloignée. Dans chaque cas, la découpe met en jeu des mécanismes différents.
- la nature du matériau à découper : pour les roches à texture granulaire et poreuse (calcaire, grès) le jet d'eau agit par érosion et arrachement des grains individuellement, du fait des contraintes de cisaillement qui apparaissent au niveau des grains à la surface de la roche par l'écoulement du jet sous pression, et, ceci, dans le cas de découpe de saignées. La porosité, la perméabilité, la granulométrie sont ici des facteurs en bonne corrélation avec les performances de découpe. Pour les roches cristallines (calcaire cristallisé, granit), l'eau sous pression pénètre dans les fissures préexistantes dans la roche, les élargit et les fait éclater. L'état de fissuration de la roche (intergranulaire, mais aussi au niveau des joints entre cristaux) influe dans ce cas énormément sur les performances de découpe. Les paramètres quantifiant cette fissuration (vitesse du son, résistance au cisaillement, indice de continuité) permettent d'obtenir de bonnes corrélation avec les résultats de découpe obtenus. Pour certaines roches très rigides (calcaire sublithographique), le jet d'eau les fait éclater par les ondes de traction et de cisaillement qui s'y propagent lors de l'impact du jet, la roche restituant l'énergie emmagasinée -lors de l'impact- par éclatement.

Néanmoins, tous ces mécanismes coexistent en fait, et il est difficile de construire un modèle mathématique applicable à toutes les roches.

II. APPLICATION DES JETS COUPANTS AUX PROBLEMES DE FORATION ET DE PREDECOUPAGE

Parmi les problèmes qui se posent au mineur, l'un des plus importants est celui de l'augmentation des vitesses de creusement de galeries en roches dures et abrasives. Il n'existe pas, pour le moment, de machine à attaque ponctuelle capable de traverser économiquement ce genre de terrains, et l'exploitant doit le plus souvent utiliser l'explosif. Avec cette technique l'accroissement de la vitesse d'avancement passe alors par :
- l'augmentation de la vitesse de foration des trous de mines,
- la réduction des hors-profils pour réduire le garnissage au minimum possible.

D'autre part, certaines mines métalliques souhaitent, pour réduire leur prix de revient, mettre en oeuvre un abattage sélectif conduisant à une extraction séparée du filon et de ses épontes.
L'utilisation de jets d'eau à haute pression permet d'envisager une réponse à ces problèmes.

Le Cerchar et la Sté Pénarroya ont, dans cette optique, étudié et réalisé en commun un engin expérimental mettant en oeuvre des jets d'eau à très haute pression et destiné à :
- forer des trous de mines,
- créer dans ces trous des saignées permettant un découpage du massif selon des orientations déterminées.

L'objectif initial consistant à effectuer un découpage complet autour du filon, par confection de saignées continues, a dû être abandonné au cours des essais préliminaires sur banc : si cette opération est techniquement possible, elle s'est avérée économiquement non rentable en raison des vitesses de sciage insuffisantes obtenues.
L'engin expérimental se compose :
- de la centrale de production d'eau sous pression (Fig. 1),
- d'un véhicule tractant la centrale et portant le bras de foration (Fig. 5).

Fig. 5 Vue d'ensemble du jumbo de foration-sciage par jets d'eau à haute pression

Le bras de foration supporte une glissière permettant l'avance de l'outil de foration et son alimentation en eau ; il permet, grâce à un ensemble de joints tournants, de positionner la glissière en tous les points de la section de galerie (Fig. 6).

Fig. 6 Vue du jumbo en cours de foration

L'étude de l'outil de foration a été menée en collaboration avec la Société Flow-Industries. Des essais préliminaires sur échantillons ont permis de déterminer la pression et le débit d'eau nécessaire pour les jets. Des essais sur banc ont ensuite permis d'aboutir à une "tête de coupe" constituée de 3 jets d'eau d'orientation déterminée (Fig. 7). La tête de coupe est placée à l'extrémité d'une tige d'amenée d'eau animée d'un mouvement de rotation au moyen d'un moteur hydraulique à vitesse réglable. La vitesse d'avance, elle-même réglable, permet ainsi d'obtenir le diamètre de trou voulu.

Fig. 7 Tête de coupe par jets d'eau à haute pression

La figure 8 montre les essais préliminaires de sciage intégral finalement abandonné au profit du découpage de saignées, de part et d'autre, de ces trous, visibles sur la figure 9.

Fig. 8 Outil de sciage-foration par jets d'eau à haute pression en fonctionnement

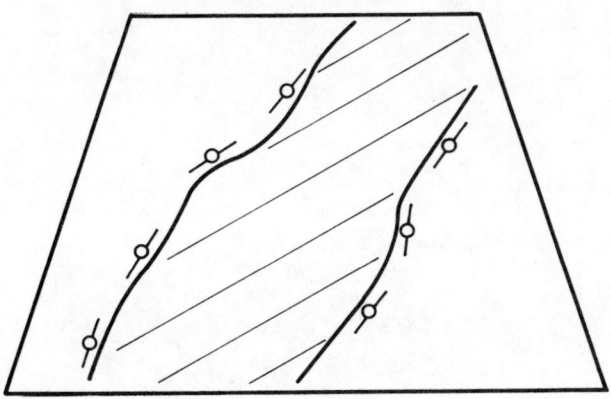

Fig. 9 Principe de l'abattage sélectif par foration et saignées

Au moment où ces lignes sont rédigées, cet engin de foration et prédécoupage est prêt à descendre dans une mine pour expérimentations in-situ. Il est attendu que ces expérimentations mettent en évidence des problèmes technologiques qui devront être résolus au fur et à mesure.

III. ASSISTANCE PAR JETS D'EAU DE LA FORATION ROTATIVE CLASSIQUE

Dans les terrains résistants et abrasifs, la foration rotative est totalement insuffisante et doit céder la place à la foration roto-percutante dont on connait les inconvénients majeurs :
- sollicitations importantes sur le matériel,
- niveau de bruit difficilement supportable.

L'extension du domaine d'utilisation de la foration est donc recherchée dans différents pays, la solution de principe retenue étant celle de l'assistance du taillant à plaquettes de carbures par des jets d'eau à haute pression.

La concrétisation de ce principe se heurte à de nombreuses difficultés :
- nécessité de disposer, dans la perforatrice, d'un joint tournant fiable capable de supporter une pression élevée et une grande vitesse de rotation.
- conception et réalisation de taillants percés d'orifices de très petits diamètres, judicieusement disposés de façon à amener l'eau sous pression au contact de la roche à l'endroit où elle sera la plus efficace.

A ce sujet, le mode d'action de l'eau est également mal connu et diverses hypothèses sont avancées :
- formation d'un film d'eau sous pression protégeant les plaquettes de carbure contre l'usure,
- pénétration de l'eau dans les fissures engendrées par les plaquettes de carbure et départ d'éclats sous l'action de la pression d'eau, évitant ainsi un broyage inutile des débris.

Le Cerchar et un constructeur français -la Sté CMM- se sont associés pour mener à bien la mise au point d'un matériel susceptible d'être introduit dans les mines. La perforatrice et son joint tournant, fournis par le constructeur, s'étant révélés immédiatement fiables et performants, la mise au point des barres de foration ayant été elle-même rapidement faite, l'étude s'est portée sur le taillant.

Un banc d'essai a été édifié (Fig. 10) et des expérimentations systématiques ont été faites mettant en oeuvre des taillants munis d'orifices positionnés de façons très diverses. La figure 11 montre un petit échantillonnage non exhaustif des dispositions utilisées. La figure 12 montre l'un de ces taillants en position de travail.

Au moment où cette communication est rédigée, un certain nombre de résultats ont déjà pû être obtenus concernant le positionnement des jets, les valeurs de la poussée et de la vitesse de rotation. Un grand nombre d'expérimentations seront cependant encore nécessaires avant d'aboutir à une réalisation définitive.

Fig. 10 Vue générale du banc d'essais de foration assistée par jets d'eau

Fig. 12 Taillant pour la foration assistée par jets d'eau

CONCLUSION

Les recherches menées dans un petit nombre de pays avancés dans le domaine des jets d'eau à haute pression appliqués à la destruction des roches, semblent indiquer qu'il s'agit là d'une voie pleine de promesses. Elles font apparaître cependant des difficultés technologiques nombreuses qui nécessiteront, pour être surmontées, du temps, des moyens et une collaboration étroite entre les chercheurs, les constructeurs et les utilisateurs.

En ce qui concerne ces derniers, on peut considérer qu'ils ne se limitent pas aux exploitants miniers : les matériels en cours d'étude en France et qui ont été présentés dans ce qui précède, peuvent, par exemple, être adaptés aux travaux en carrière et dans les travaux publics.

Fig. 11 Taillants divers utilisés pour la foration assistée par jets d'eau

STUDIES IN WATER JET ASSISTED DRAG TOOL ROCK EXCAVATION
Etudes sur l'excavation par outils à dragline secondés par jet d'eau
Untersuchungen über das Zerspanen von Gesteinen mit Radialmeissel und
Hochdruck—Wasserstrahl

R. J. Fowell
Lecturer, Department of Geotechnical Engineering, University of Newcastle upon Tyne,
Great Britain

O. Tecen
Former Research Student, Department of Geotechnical Engineering, University of
Newcastle upon Tyne, Great Britain

SYNOPSIS

Due to the limitations of cutting picks in terms of rock strength and abrasivity, experiments with water jet assisted drag tools have been undertaken in a range of rock materials to determine the advantageous influence of water jets on excavation performance. It was found that the depth of penetration of the water jet influences the mode of failure of the rock and an expression which includes rock toughness, porosity, cementing and microfractures best predicts the cutting forces experienced by the tools.

RESUME

Etant donné que la performance des pics de taille se trouve freinée par des facteurs tels que la force et l'abrasivité rocheuses, des essais ont été effectués avec des outils à dragline secondés par jet d'eau, sur une gamme de matériaux rocheux, de manière à déterminer l'effet salutaire des jets d'eau sur les travaux d'excavation. Ces essais ont montré que la profondeur de pénétration du jet d'eau influe sur la manière dont la roche cède et le meilleur moyen de prévoir les contraintes de taille auxquelles sont soumis les outils est de considérer la dureté de la roche, sa porosité, la cimentation et les microfractures.

ZUSAMMENFASSUNG

Wegen der Beschränkungen von Schneidezähnen, die durch Gesteinsfestigkeit und Abschleifen bedingt sind, wurden Untersuchungen mit Meißel und Hochdruck-Wasserstrahl an verschiedenen Gesteinen durchgeführt, um die vorteilhaften Einflüsse des Wasserstrahles zu erforschen. Es wurde festgestellt, daß die Tiefe des Eindringens des Wasserstrahles einen besonderen Einfluß auf die Bruchweise des Bruchspanes hat. Die entwickelte Formel gibt die Andruckkraft des Meißels mit Beziehung auf Zähigkeit, Porosität, Bindemittel und Mikro-Brüche wieder.

INTRODUCTION

Drag picks are the principal mechanical excavation tool employed by most coal and rock excavation machines used in mines. Unfortunately, they are limited by the abrasivity and strength of the rock materials they can excavate economically and safely. It is considered that the use of high pressure water jets with drag tools would yield the following advantages:

1. Significantly reduce cutting forces acting on the tool and reduce torque fluctuations on the cutting head.

2. Provide cooling for the tungsten carbide tipped cutting tool, extending its economical life.

3. Cool the rock in the track behind the tool eliminating hot spots that could cause ignitions.

4. Suppress dust at source.

5. Provide lubrication at the tool/rock interface.

6. Assist with debris removal.

However, little is known of the mechanics rock failure under the action of hybrid water jet cutting with drag tools and the interaction of the tool and water jets.

PREVIOUS RESEARCH

Several research projects have been carried out in Japan, U.S.A. and W. Germany into water jet assisted disc cutting and in South Africa and the U.K. into water jet assisted drag tool cutting. Hood (1976) found that the force acting on a drag bit, when cutting strong rock, can be reduced by directing a high pressure water jet immediately ahead of the bit, and reported a twofold increase in depth of cut when the jet position was optimised for the same cutting force. Wang et al. (1978) and Henneke et al. (1978) experimented with water jet assisted disc cutting when mounted on a full-face tunnelling machine with jets positioned at various locations. A 2-3 times increase in the penetration rate with water jet assisted cutting has been reported by Wang et al. (1978).

Plumpton et al. (1982) augmented a Dosco roadheader with high pressure water jets and reported 50% reduction in normal forces and 30% reduction in cutting forces. Although promising results have been reported by these researchers, they have placed great emphasis on the cutting performance of machines in individual rock types rather than a programme which included a range of rock

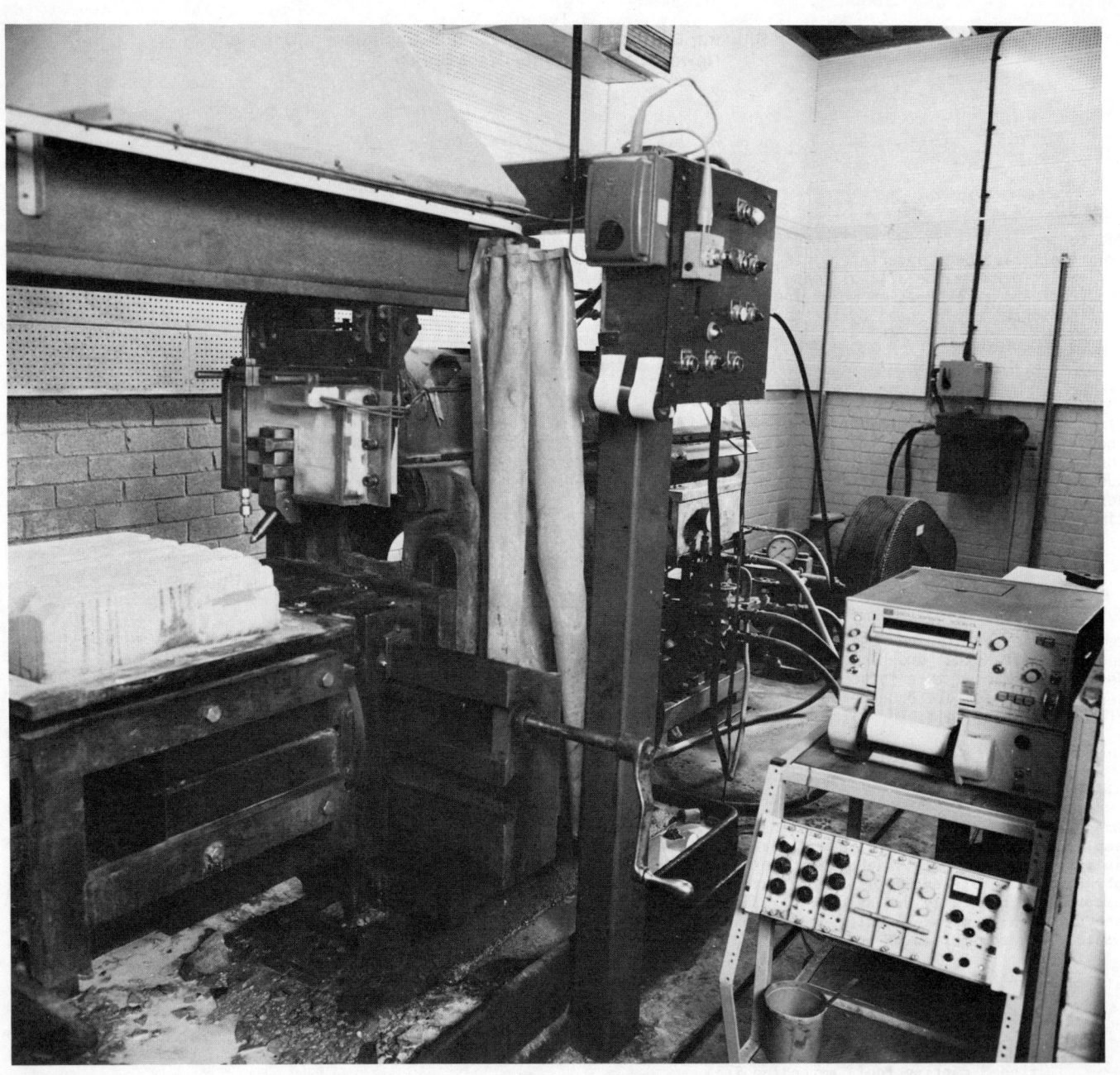

Fig.1 View of Water Jet Cutting Room.

materials with varying mechanical and physical properties, which was one of the objectives of the work reported in later sections of this paper.

ROCK CUTTING RIG

The tests were conducted on a modified shaping machine having a stroke of 800mm and capable of providing 5 tonnes in-line thrust force (Fig.1). Rock specimens up to 500mm x 500mm x 300mm dimension can be accommodated on the machine table and lowered and laterally traversed with respect to the cutting tool. The cutting tool is mounted in a triaxial dynamometer on the machine crosshead. A Uraca three-piston, positive displacement pump was used in jet assisted cutting and it was powered by a continuously rated 30 horsepower electric motor, and delivered 13.0 litre/min at a pressure of 62 MN/m^2. Nikonov type nozzles, which have a 13° contraction angle followed by a straight section of 3 times the nozzle exit diameter, were used. These were made of silver steel and oil hardened to prevent brittleness.

EXPERIMENTAL PROGRAMME

A series of laboratory tests were carried out to determine various rock properties (Table I). Experiments were conducted at 10mm depth of cut with the mechanical tool to determine the corresponding tool forces, quantities of debris and mechanical specific energies for the mechanical tool and hybrid cutting systems (Figs. 2a, 2b). Hydraulic variables are given in Table II (Fig.3).

Each cutting test was repeated four times and carried out in random order to minimise any changes that may have occurred in the rock samples and operating conditions during the cutting process. Computer curve-fitting analysis was performed on the experimental output and best-fit functions are as listed in Tables III and IV.

TABLE I

Rock Type	Compressive Strength (MPa)	Tensile Strength (MPa)	Bulk Density (g/cm^3)	Apparent Porosity (%)	Scleroscope Hardness Number	Plasticity (%)	Schmidt Hammer Hardness Number	NCB Cone Indenter Hardness Number	Grain Density (g/cm^3)	Dynamic Modulus (GPax10)
Sandstone A	43.2	3.00	2.21	16.40	36.7	42.3	52.0	1.98	2.60	1.79
Sandstone B	64.5	4.40	2.18	8.50	35.3	30.6	43.4	2.53	2.65	1.00
Sandstone C	57.5	3.70	2.18	7.90	58.4	27.3	44.8	2.48	2.67	0.98
Sandstone D	71.3	4.40	2.20	10.00	53.6	31.9	35.2	2.65	2.85	2.80
Limestone A	71.7	6.10	2.33	6.10	42.5	30.1	36.1	3.14	2.71	1.88
Limestone B	149.1	12.3	2.38	3.40	47.3	15.6	-	3.83	2.75	5.54

Fig.2a Hybrid system : before cutting.

Fig. 2b Hybrid system : during cutting.

TABLE II

Experimental Conditions

Depth of cut	...	10mm
Nozzle diameter	...	0.85mm
Water Jet Pressure	...	55.2MPa
Stand-off distance	...	15mm
Lead-on distance	...	1mm
Side-off distance	...	0
Cutting Speed	...	165mm/s
Offset Angle	...	6.5°
Tip Angle ...	...	87°
Angle of Attack	...	45°

TABLE III

Mechanical Tool Alone*

Rock Property	Curve-fit Equation	Index of Determination
Compressive Strength	MCF=20.7CS -27.23	0.85
Tensile Strength	MCF=1.82TS -3.38	0.83
Compressive/Tensile	MCF=1/ 0.06+0.0025(CS/TS)	0.17
Bulk Density	MCF=2.03+0.0043BD	0.69
NCB Cone Indenter	MCF=0.34NCB+1.07	0.78
Apparent Porosity	MCF=1/(0.04 x Porosity-0.06)	0.68
Shore Rebound Hardness	MCF=1/(0.03-7.7x10^{-4}Shore)	0.08
Plasticity	MCF=1/(2.9x10^{-3}+6.82x10^{-3}P)	0.66
Dynamic Modulus	MCF=0.99DM-2.61	0.93

* For cutting conditions given in Table II.

TABLE IV

Mechanical Tool with Water Jet Assistance*

Rock Property	Empirical Curve-Fit Equation	Index of Determination
Compressive Strength	WMCF=26.71CS-19.91	0.92
Tensile Strength	WMCF=2.25TS-2.36	0.82
Compressive/Tensile	WMCF=1/ 6.78x10^{-2}+1.54x10^{-3}(CS/TS)	0.04
Bulk Density	WMCF=4.72x10^{-2}BD+2.08	0.54
Apparent Porosity	WMCF=1/ 5.35z10^{-2}Porosity-0.048)	0.79
NCB Cone Indenter	WMCF=0.435NCB+1.204	0.83
Shore Rebound Hardness	WMCF=Shore/(2.69x10^{-2} +1.44x10^{-2}Shore)	0.37
Plasticity	WMCF=1/(9.53x10^{3}P+2.67x10^{3}	0.83
Dynamic Modulus	WMCF=1.167DM-1.863	0.84

*For cutting conditions given in Table II.

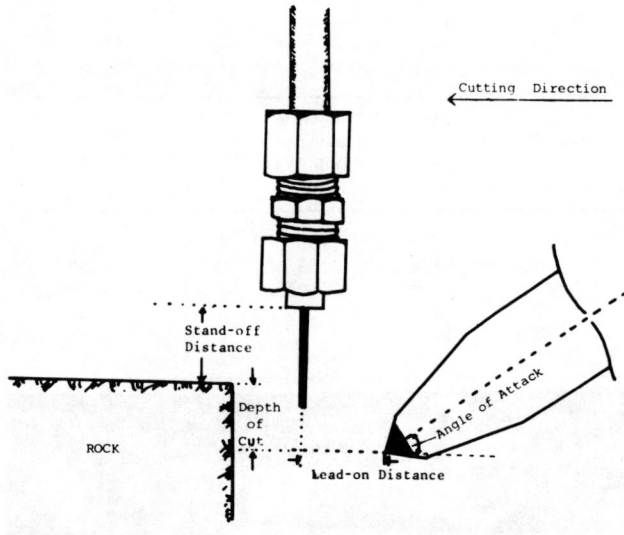
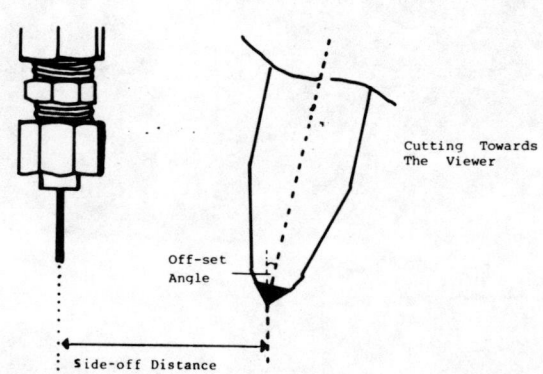

Fig.3 Diagram showing the variables.

DISCUSSION OF RESULTS

As can be seen from Figure 4, increasing water jet pressure causes increase in depth of penetration. The

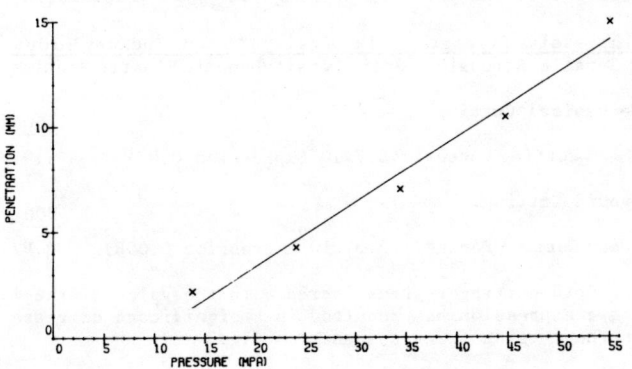

Fig.4 Sandstone B : Increase in pressure.

relationship between the variables is of linear form. When the curve is extrapolated the point at which it intersects the horizontal gives the threshold pressure required to initiate cutting. Some experiments were carried out on Sandstone B at a constant 7mm depth of cut and the water jet pressure was increased in constant increments. The results have shown that increasing jet pressure and, consequently, its depth of penetration, has a strong influence on the operating efficiency of the mechanical tool. Mean Cutting Force decreased exponentially with increase in jet penetration depth (Fig.5). Sandstone D cutting graphs indicate similar

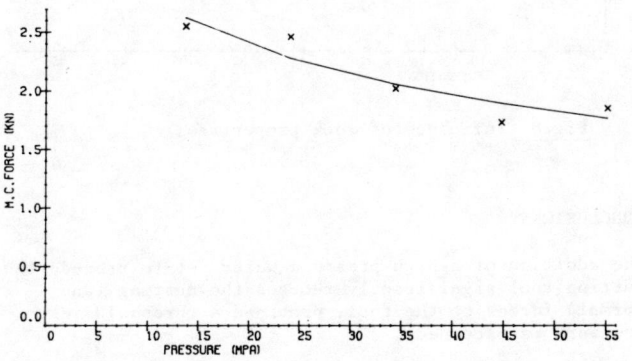

Fig.5 Effect of pressure on Mean Cutting Force.

reductions in forces and mechanical specific energies that may be achieved with high pressure water jet assistance (Figs. 6a, 6b).

The cutting action of a point attack tool may be examined in two stages. Initially, when a point attack tool penetrates a brittle rock, stresses are created in the vicinity of the tool/rock interface. While the tool is pushed through the rock, a critical force level, varying with rock type, is reached which initiates tensile fracturing at the tool tip followed by final breakage of rock ahead of the tool. The rock face is

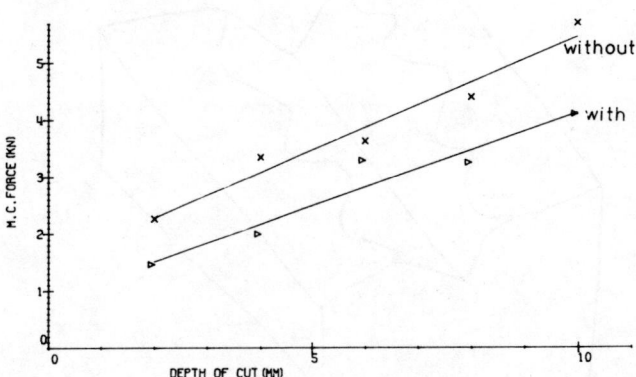

Fig.6a Cutting with and without jets, MCF

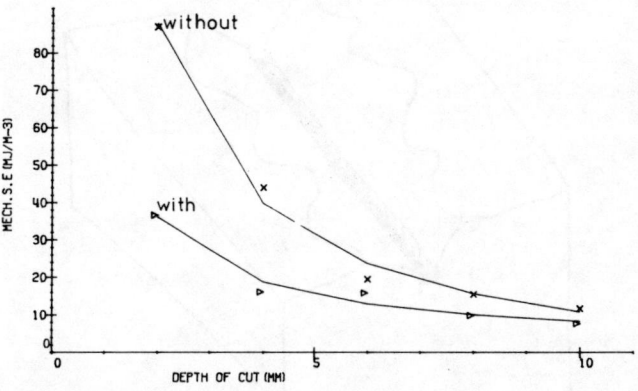

Fig.6b Cutting with and without jets, S.E.

left with V-shaped sloping surfaces after the initial action of the tip. Due to the geometry, the tip makes a rubbing contact with the sides of the cut and causes further rock failure. The second action of the tip is therefore causing a shearing failure due to the profiling action of the tool body along the cut length (Fig.7).

With the hybrid system the cutting action of the point attack tool changes due to jet action, dependent on whether the jet penetrates below the depth of cut taken by the mechanical tool. If the jet penetration depth is less than that of the mechanical tool, the pick tip comes into contact with the rock face and initiates cracks ahead of the tool. High pressure water is introduced into these newly formed cracks and exerts pressure at the walls, which assists fracture propagation and final removal of the material. If jet penetration is greater than the depth taken by the mechanical tool, the sides of the point attack tool come into contact with the rock. The tip of the tool does not initiate any cracking ahead of the tool, since its tip is not in contact with the rock. The rock in the immediate vicinity of the tool is crushed and this leads to tensile fracturing of the surrounding material with a bursting action.

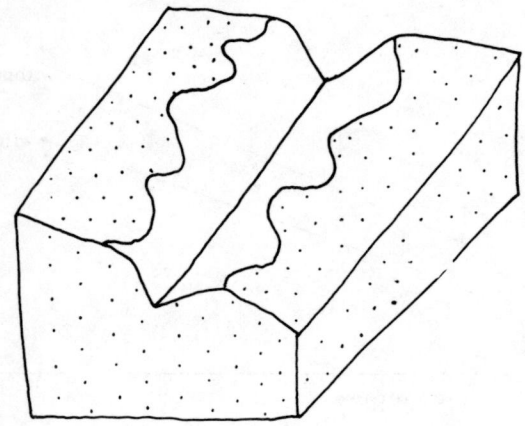

Rock Surface after the Initial Action of the Tool Tip
(Idealized view)

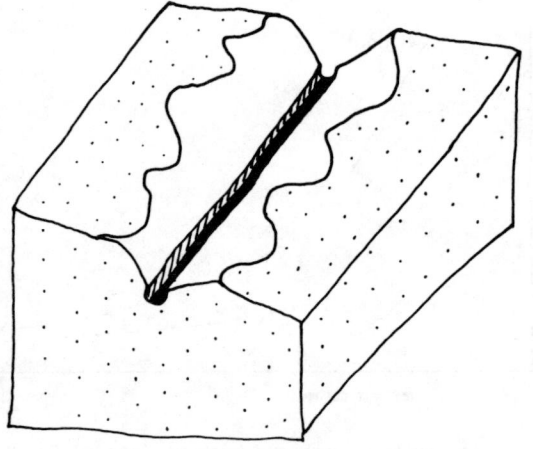

Rock Surface after the Shearing Action of the Tool Tip
(Actual)

Fig.7 Illustration of Cutting.

When the influence of rock properties on cutting performance was investigated it was found that the uniaxial compressive strength and indirect tensile strength of rocks exhibited good correlations with all the tool forces (Tables III and IV). The higher the rock strength, the more difficult it is to break it in both cutting situations. However, rocks of similar strength but differing composition and structure have shown variations in forces. The NCB Cone Indenter, which was used to test the hardness of the cementing material, has also shown a direct relationship with the forces. Increasing cone indenter hardness resulted in higher tool forces, showing that rocks with a closely intergrown fabric are more difficult to break than those in which the mineral grains are separated by a weak matrix. Dynamic Elastic Modulus, which is related to microfracturing in rock, gave good correlation with the tool forces. Apparent porosity has shown a trend in which it is easier to cut porous material, all other properties being equal.

The individual influences of some rock properties were combined by regression analysis to give the overall effect. The highest Index of Determination (IOD) between tool forces and properties was given by the expression:

$$\frac{\text{Compressive Strength}}{\text{Tensile Strength}} \times \frac{\text{Bulk Density}}{\text{Grain Density}} \times \frac{\text{NCB Cone Indenter Hardness}}{\text{Dynamic Elastic Modulus}}$$

Mechanical Cutting	IOD
Mean Cutting Force $=1/(2.7 \times 10^{-2} \text{Expression} - 0.07)$	0.97

Hybrid Cutting	IOD
Mean Cutting Force $=1/(3.32 \times 10^{-2} \text{Expression} - 0.058)$	0.97

For both cutting systems increase in the value of the above expression has resulted in a significant decrease in the cutting force components (Fig.8)

However, when a percentage reduction in tool forces, as a consequence of water jet assistance, was considered, it was found that no single rock property exhibited any significant influence.

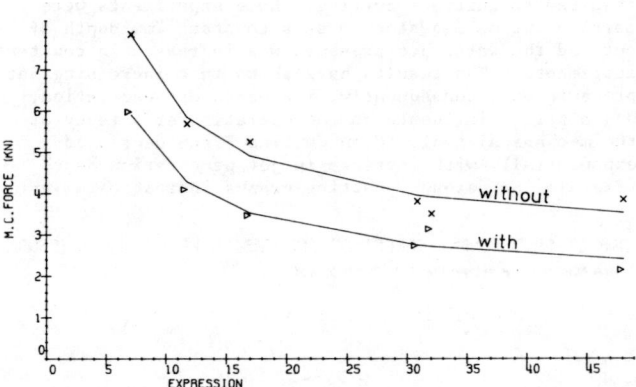

Fig.8 Influence of rock properties.

CONCLUSIONS

The addition of a high pressure water jet to precede the cutting tool significantly reduces the cutting (and normal) forces on the tool, provided a threshold jet pressure is exceeded.

There is greater advantage in limiting the penetration depth of the water jet to less than the depth of cut taken by the mechanical tool.

The parameter which gave the best prediction of the cutting force component experienced by the pencil point tool used was found to be a function of rock toughness as measured by the ratio of compressive to tensile strengths; porosity as measured by the ratio of bulk density to grain densities, cementing material hardness and microfractures.

The results reported in this paper were obtained from an initial study of water jet assisted drag tool cutting. Further studies are being undertaken on a wider range of rock lithologies and experimental conditions to establish with greater confidence the relationship derived. The investigations are also being extended to provide information for optimising the many variables involved to increase the range of rock materials that may be economically excavated by drag tool machines.

REFERENCES

Henneke, J., Baumann L. (1978). Jet assisted tunnel-boring in coal-measure strata: Proc. 4th Int. Symp. on Jet Cutting Technol. (Canterbury, U.K.: Apr. 12-14), vol.1, Cranfield, U.K. BHRA Fluid Engng. Paper J1,1-J1,12.

Hood, M. (1976). Cutting strong rock with a drag bit assisted by high-pressure water jets: J. South African Inst. Min. & Metall., vol. 77, no. 4, Nov., pp. 79-90.

Plumpton, N.A. and Tomlin M.G. (1982). The development of a water jet system to improve the performance of a boom type roadheader: Proc. 6th Int. Symp. on Jet Cutting Technol., Paper G1, pp.267-82. Guildford, U.K., 1982.

Tecen, O. (1978). M.Sc. Dissertation. University of Newcastle upon Tyne.

Tecen, O. (1982). High Pressure Water Jet Assisted Drag Tool Cutting of Rock Materials. Ph.D. Thesis, University of Newcastle upon Tyne.

Wang, F.D. et al. (1978). Water jet assisted tunnel boring: Proc. 3rd Int. Symp. on Jet Cutting Technol., Paper E6, pp x63-x71. Chicago, 1978.

LIMIT BLAST DESIGN EVALUATION
Evaluation du modèle de charge explosive limite
Überlegungen zum Entwurf von begrenztem Sprengen

C. K. McKenzie
Julius Kruttschnitt Mineral Research Centre, University of Queensland, Brisbane, Australia
P. D. Forbes
Julius Kruttschnitt Mineral Research Centre, University of Queensland, Brisbane, Australia
G. E. LeJuge
Julius Kruttschnitt Mineral Research Centre, University of Queensland, Brisbane, Australia
I. H. Lewis
Hamersley Iron Pty. Ltd., Tom Price, Australia
P. A. Lilly
Hamersley Iron Pty. Ltd., Paraburdoo, Australia
J. D. Lilly
Hamersley Iron Pty. Ltd., Tom Price, Australia

SYNOPSIS

The evaluation of blast effectiveness in limit blasting is concerned primarily with the minimisation of damage behind the blast zone. Techniques to assess the extent, degree and possible mechanisms of damage are presented and include vibration monitoring, blast timing, cross-hole seismic surveying and structural assessment. Research indicates that for large hole limit blasting in fractured rock: 1) Effective delays in excess of 70 ms are required in single hole firing; 2) Delays in excess of 100 ms are required in row firing; 3) Reduced explosive density reduces the vibration level but not the vibration period; 4) The vibration level is confirmed as one mechanism of blast damage.

RESUME

L'évaluation de l'efficacité d'explosion dans le cas de charge explosive limite a pour intérêt principal de diminuer les dommages à l'arrière de la zone minée. Des techniques pour évaluer au maximum les dommages possibles sont présentées et comprennent la régularisation de vibration, le contrôle de la durée de l'explosion, la surveillance séismique et l'étude de la structure du sol. Dans le cas de charge explosive limite pour larges trous en roches fracturées les études indiquent que: 1) des délais effectifs de plus de 70 ms sont nécessaires pour un simple trou de tir; 2) des délais de plus de 100 ms sont nécessaires dans le tir en rangées; 3) une plus faible densité d'explosifs réduit le niveau mais non pas le temps de vibration; 4) le niveau de vibration est confirmé comme l'un des mécanismes de dommages causés par l'explosion.

ZUSAMMENFASSUNG

Die Abschätzung der Sprengwirksamkeit beim begrenzten Sprengen beschäftigt sich in erster Linie mit der Verringerung der Beschädigung hinter der Sprengzone. Verfahren zur Feststellung des Ausmaßes, des Grades und des möglichen Mechanismus des Schadens sind dargestellt und enthalten: Überwachung der Vibration, Sprengzeitmessung, seismische Vermessung zwischen Bohrlöchern und strukturelle Einschätzung. Forschung zeigt uns, daß für begrenztes Sprengen in großen Löchern in brüchigem Fels: 1) Es erforderlich ist, das Sprengen im Einzelloch effektiv um über 70 ms zu verzögern; 2) Verzögerungen von über 100 ms bei Reihenschüssen erforderlich sind; 3) Die verminderte Explosionsdichte zwar das Vibrationsniveau vermindert, aber nicht die Vibrationsdauer; 4) Das Vibrationsniveau als ein Mechanismus des Sprengschadens bestätigt worden ist.

INTRODUCTION

Without careful control, blasting close to final slopes in open cut mines may have detrimental effects on the rock mass, jeopardising the stability of the slope so that final design angles cannot be safely attained. Providing that blast damage can be quantitatively assessed, limit blast designs can be modified to minimise destructive effects on the planned final wall while maintaining adequate fragmentation and movement of the blasted material.

Research conducted at the Hamersley Iron operations in the North West of Western Australia has demonstrated the ability to assess blast damage and evaluate the effectiveness of blast designs. The research programme has been carried out by the Julius Kruttschnitt Mineral Research Centre (JKMRC) and the Rock Mechanics Section of Hamersley Iron at the Tom Price and Paraburdoo mines (Figure 1).

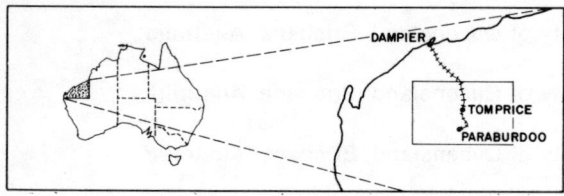

Figure 1 Locality plan showing Pilbara iron ore area and Hamersley Iron Tom Price and Paraburdoo mines.

Conventional open pit mining methods are used at both mines, with ore being drilled and blasted from 15 metre (Tom Price) or 14 metre (Paraburdoo) benches, shovel loaded and hauled by trucks to crushers or dumps.

The Hamersley Iron Province of Western Australia contains a group of Lower Proterozoic age sediments known as the Hamersley Group. These deposits are predominantly of chemical origin and occur in a large but discrete sedimentary basin covering an area of some 78,000 square kilometres. Banded iron formations (BIFs), cherts, dolomites and lutites are the principal components.(Bourn and Jackson, 1979; Baldwin, 1975)

The thickest BIF unit, the Brockman Iron Formation, is the most important unit at the Hamersley Iron Pty. Ltd. mine sites. At Tom Price, the Dales Gorge Member provides the bulk of the mineralisation, while Paraburdoo mines both the Dales Gorge Member and the Joffre Member. The Mt. McRae Shale forms the footwall at both sites.

SLOPE DESIGN AND ACHIEVEMENT

The Rock Mechanics Section of Hamersley Iron's Mining Engineering Department has gradually redesigned most of the Tom Price and Paraburdoo pits in an effort to maximise the economic benefits of mining, and to minimise pit slope instability. The conventional pit designs used arbitrary 45 degree slope angles. As a result, large scale failures were possible in structurally unfavourable areas and the opportunity to steepen slopes in structurally favourable areas was lost.

New procedures called for modified blasting and excavation techniques, which have now been successfully implemented over six bench levels at Tom Price. Rill accumulation is visibly less than on normal, shovel dug faces in Mt. McRae Shale.

Where hard material such as BIF, hematite or goethite occurs in the final faces, an overall slope angle of some 52 degrees is being attempted where structure allows. This desired angle demands that the damage to the final pit walls and benches be limited as much as possible. In order to achieve the designs it will be necessary to reduce the amount of backbreak, crest fracture and loose face rock to a minimum.

BLASTING AND DRILLING BACKGROUND

Blast holes, varying from 311mm (Paraburdoo) to 381mm diameter (Tom Price) are drilled vertically and sub-drilling to a depth of 2-4 metres is incorporated to give a clean break on the target bench.

Ammonium nitrate fuel oil (ANFO) is bulk loaded into the blast holes. Charged holes are connected by detonating cord with a blasting sequence delineated by detonating relay connectors (D.R.C.s) between subsequent breakage zones. Most blasts are fired in a 'V1' equilateral pattern.

Hagan (1977) has suggested a delay period of 8ms/m in row blasting with long stemming columns, low powder factors and soft, heavily fractured strata. This delay period was considered the optimum which allows good fragmentation and displacement of each burden without the presence of cut-offs. Using this guideline 35, 45, and 60ms D.R.C.'s are used between successive rows in production blasts. The basis of production blast design has therefore been the achievement of acceptable fragmentation and burden movement by maximising surface delay intervals without causing cut-offs. No down-hole delay systems are used.

Present final design blasting practice at Tom Price and Paraburdoo incorporates:

1. single hole delay blasting;
2. 'V1' equilateral row blasts;
3. control of bulk strength of ANFO;
4. sub-drilling where necessary in the front two holes of each row;*
5. surface delaying only.

The JKMRC and Hamersley Iron Rock Mechanics Section have made a study of each of these features in production and final limit blasting

In all final limit blast designs air decks are incorporated in the charge column to reduce effective bulk strength of explosive and to increase the rate of decay of blasthole pressure. This reduces the intensity of all mechanisms which can cause cracking (Harries, 1981). Air decks reduce the charge density from 0.8 g/cm^3 (normal ANFO) to approximately 0.2 g/cm^3 in the back holes of all rows. Single hole firing is common practice on final design blasts to minimise or reduce the effects of blast induced damage (Harries and Hagan, 1979) resulting from the enhancement of vibration (strain) waves. Current effective delay intervals are in the range 5ms to 10ms with delays greater than this being difficult to achieve using surface delays only.

---
* the term 'row' is defined by the initiation line rather than the direction of the free face.

The current research suggests that neither row-delayed nor single hole delayed blasting has satisfactorily reduced blast damage resulting from enhancement, as evidenced by:

1. visible back-cracking up to 15-20 metres behind blasts;
2. current effective delay intervals which are insufficient to produce separation of individual wave packets.

EXPERIMENTAL PROGRAMME

The applied blasting research programme aims to:

1. characterise rock types in terms of their response to blasting;
2. detect and assess the extent of blast induced damage behind blasts (in particular, limit blasts) in a variety of rock types;
3. reduce the damage profiles by modification of blast design parameters as assessed through an experimental blasting programme.

Cross Hole Seismic Assessment

The technique relates changes in the acoustic properties of signals propagated between the same holes before and after a blast to changes in the state of competence of the rock. Seismic techniques have been used in operating mines to study 'cavability' (Obert et.al, 1976) in block caving operations and also to study pillar stability (Kaneko et.al, (1979). The structural dependence of compressional wave velocity, the most readily obtained seismic parameter, has been illustrated in studies using both seismic (Broadbent, 1974; Murphy, 1972; Stacy, 1976) and ultrasonic (Grujic, 1974; Meister, 1974; McKenzie et.al, 1982) signals. However, in general it is considered that compressional velocity alone is a relatively insensitive indicator of structural or geological condition.

Full waveform analysis has been conducted in more recent studies by Young and Coffey (1979) and Albright, Pearson and Fehler (1980) in order to define in-situ rock condition more completely. First arrival pulse broadening has been used as a measure of rock quality (Gladwin and Stacey, 1974; Ramana and Rao, 1974; Ricker, 1953) but requires a high signal to noise ratio and an undistorted first arrival. The current study considers both pulse broadening (defined by the first arrival rise time) and spectral analysis to quantify changes in acoustic properties.

The acoustic properties evaluated in the current study include:

1. p-wave velocity;
2. amplitude - first arrival, peak and root-mean-square (RMS);
3. first arrival rise time;
4. spectral content.

The measured changes are related totally to a change in the fracture characteristics of the rock mass, i.e., increased fracture frequency or increased fracture aperture. In either case the result is expected to be decreased rock mass strength, and an increased tendency to collapse or slump if exposed as a final slope. The quantification of these effects requires a mathematical model incorporating fracture patterns and the strength and stiffness of joint surfaces. All seismic testing is performed under dry rock conditions.

Blast Vibration Monitoring

Recognised mechanisms of blast induced damage include release-of-load failure, fracture opening and extension by high pressure gas penetration, and fracture initiation and extension by vibration (Hagan and Morriss, 1981). Vibration monitoring in conjunction with seismic assessment of damage aims to investigate the third mechanism of damage. Further, it represents a means of assessing the interaction of blast parameters including delays, explosive type and density, initiation sequence, etc., in production or final limit blasting. Particle velocity is measured using geophones and taken to be proportional to dynamic strain produced by the blast. Constraints on vibration levels assume a strain 'threshhold', related to dynamic tensile strength, below which structural damage will not occur. The vibration limit can therefore be expected to vary markedly in accordance with in-situ rock strength and needs to be established for the specific rock types encountered in the particular mining environment.

Blast Timing

Multi-hole and even single hole firing can produce complicated vibration waveforms. When investigating the accuracy of delays, or the extent of enhancement, it is necessary to know the precise moment of initiation of blast holes or decks. The timing data are unambiguous and can be accurately transposed to the vibration record once the compressional wave velocity of the rock is known, together with the blast hole/monitor spacing. The technique also allows analysis of the influence of single holes, or rows, in a multi-row blast.

Instrumentation

The experimental configuration is shown in Figure 2, with geophone elements, used for vibration monitoring, fully grouted at various depths behind the blast zone. The geophone arrangement may be a single, radially-oriented transducer or a triaxial configuration, with transducers shunted in both cases to remove resonant distortion. As a check on the accuracy of the geophones, combination geophone/accelerometer arrangements have been installed. Vibration signals are recorded on a seven channel FM recorder.

Electronic timing is performed using a 10 channel timing device built at the JKMRC. Wires are tied to the downlines of holes, to the tie lines of rows, or to the primer if down hole

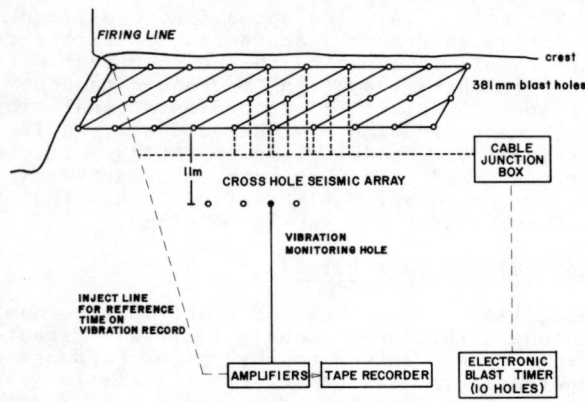

Figure 2 Typical experimental arrangement for blast monitoring showing electronic timing, cross hole seismic array and vibration monitoring.

delays are used. The moment of initiation is determined (± 0.1ms) as the wire is broken. A pulse, corresponding to the moment of initiation of the first hole or row, is also injected on to the vibration record to assist vibration analysis.

Cross hole seismic analysis incorporates a number 8 electric detonator in a water filled hole as a seismic source. The detonator may be an instantaneous, delay, or seismic detonator since electronic circuitry is also used to determine the moment (±2µs) of detonation. Pulse injection onto the seismic signal is also undertaken to assist velocity calculations. Seismic detectors are hydrophones with a flat frequency response to 100kHz. Water is used to couple both detectors and source to the rock mass. Only two detectors are usually used with the cross hole spacings dependent on rock competence and rarely exceeding 25 metres. Signals are captured using a Norland digital waveform analyser mounted in a recording vehicle. Full analysis in terms of velocity, rise time, amplitude, spectral content and signal averaging is performed on the analyser and the data stored on floppy disk.

DISCUSSION

Effect of Delay Interval

Figure 3 shows the effects of delaying in single hole blasting using 381mm diameter blast holes in heavily fractured mineralised material. The holes were drilled as normal face holes approximately 2 metres back from the crest, spaced at 8m intervals and drilled to 18.5 metres. The holes featured effective charge densities varying progressively from 0.8 g/cm^3 to approximately 0.13 g/cm^3 by use of air spacers in a constant length charge column. The holes were fired 'individually' with 45 ms delays.

Recording geophones were fully grouted in a triaxial configuration at a depth of 15 metres to minimise interference from surface waves, and approximately 30 metres behind the blast to

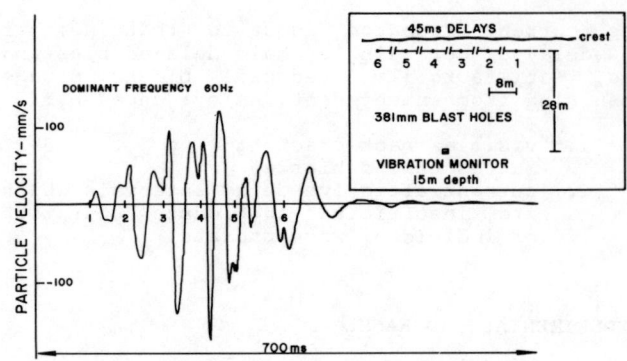

Figure 3 Vibration recording from single hole firing using 45ms delays and 381mm diameter blast holes.

minimise the path length difference between blast holes and recording station.

Superimposed on the vibration record are the initiation times for each hole as measured by an electronic timing device. The recording shows no separation of individual vibration packets, with clear indication of enhancement occurring. The delay intervals are inadequate to achieve minimisation of peak vibration levels and blast induced damage with large diameter blast holes, which are the main aims of single hole firing in limit blasting.

A second experimental blast was fired using a normal face hole charge (3-3.5 metres subdrill, 6 metre charge length) but incorporating a 240ms delay. Figure 4 shows the experimental arrangements with a single hole delayed to detonate separately from a group of three 'simultaneous' holes. The vibration record indicates separation of vibration packets after approximately 70ms. An analysis of the required delays for maximum effectiveness of single hole blasting can be undertaken from knowledge of local p-wave velocity, Vp, and the elemental vibration period T. This period is expected to vary with charge diameter and charge length and has been observed to range from 40ms to approximately 100ms (80Hz-30Hz) in the current research programme.

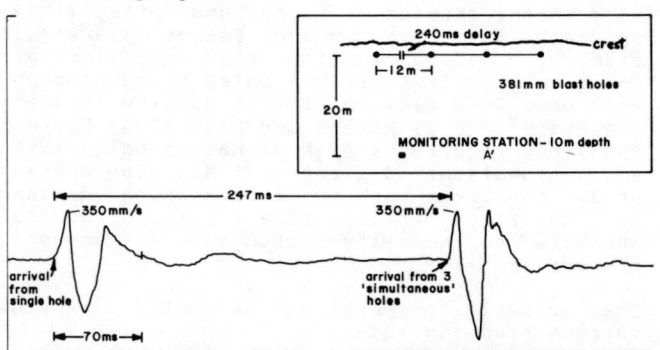

Figure 4 Vibration recording of single hole firing and 'simultaneous' firing of three holes as detailed in INSET.

Considering the case of monitoring behind a multi-row blast at a point on the planned final wall (Figure 5), the vibrations from a hole in the ith row arrive at the detector at a time dependent on the inter-hole delay (including cord burn time) and the p-wave velocity of the rock mass. Successive vibration packets arrive within a period of $(D_h - \Delta S/V_p)$ where D_h is the inter-hole delay period and ΔS is the path length difference. For minimum enhancement, this time must exceed the vibration period, T, of the preceeding vibration packet. Hence

$$(D_h - \Delta S/V_p) > T.$$

For a 6 metre equilateral pattern, a p-wave velocity of 1000m/s and a vibration period of 70ms, the interhole delay must exceed 75ms. This delay will approach 70ms as the local p-wave velocity increases.

The vibrations from the first hole of row (i+1) must also arrive at a time, T, after those from the last hole of the ith row. Hence, the minimum inter-row delay, D_r can be calculated for a given number, n, of blast holes per row:

$$D_r > (n-1) * (T - \Delta S/V_p)$$

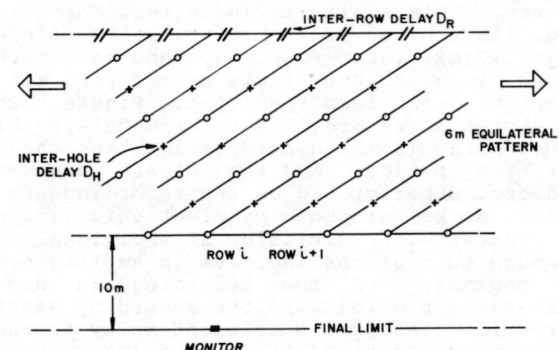

Figure 5   Portion of a long limit blast, fired staggered V1, showing inter-hole and inter-row delays and monitoring stations on final wall.

For 4 holes per row and a p-wave velocity of 1000m/s, the inter-row delay must exceed approximately 190ms.

The inter-hole and inter-row delays will change for different initiation sequences but the effective separation between successive holes must produce a wave packet separation of T at the final wall location in order to minimise vibration enhancement.

In practice it has been observed that single hole firing in limit blasts, using effective inter-hole delays in the range 5 to 10ms, has resulted in no discernable improvement in final wall condition over that obtained with row-delayed blasting. Minimum inter-row delays for row-delayed blasting can be determined from knowledge of the elemental vibration shape and by applying the principle of superposition (assuming linearity). Successive vibration packets from holes with a single row arrive within a period of $(\Delta S/V_p - T_b)$, where T_b is the burn time of the inter-hole detonating cord, with packets arriving at the monitor in reverse sequence to the initiation sequence under normal practice of front initiation first. To eliminate enhancement from the following row the inter-row delay must exceed the vibration period T for the composite waveform which will have a period increased by the sum of the terms $(\Delta S/V_p - T_b)$ for each hole in the row. In the case being considered, four holes per row would increase the vibration period from 70ms to approximately 90 ms. To evaluate the degree of enhancement of particle velocity, the attenuation factor for each vibration packet must be determined. It has been observed from spectral analysis of vibration data, particularly in the hard ore, that increasing row delays produces a distinct bimodal frequency distribution, with spectral peaks around 30Hz and 80Hz. The lower peak reflects the delay period between rows. This feature can be seen from waveform shape analysis to be the partial separation of discrete waveform 'packets' corresponding to the elemental wave shapes for individual rows.

The above analysis indicates the difficulties involved in obtaining complete separation of vibrations in single hole firing with large diameter holes. This procedure is considered preferable to row blasting but practical problems associated with tie-in probably prevent its routine use. Row-delayed blasting is also seen to require longer delays than are currently being used and will almost certainly require the application of down-hole delaying.

It is interesting to note that similar studies conducted in underground hard rock environments with 100mm and 165mm diameter blast holes have produced vibration periods of approximately 10ms. A comparison of these two environments, using identical detection and analysis systems is shown in Figure 6. Single hole firing in underground mines is usually associated with the firing of a single ring, making the practice considerably less complicated than in large hole, open-cut mining.

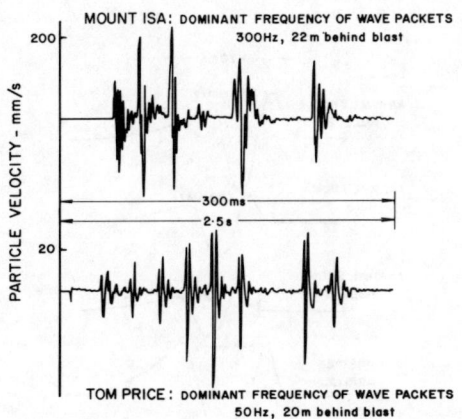

Figure 6   Comparison of single hole firing in underground and open-cut blasting using identical detection systems.

E 219

Effect of Charge Weight Per Delay

Figure 4 indicates no increase in peak vibration level as a result of the 'simultaneous' detonation of three holes, though the shape of the waveform has changed. This absence of enhancement can be explained by considering the geometry of the experimental arrangement and by using the principle of superposition. Figure 7 considers the superposition of three elemental wave shapes (closely resembling that produced by the single hole in Figure 4) delayed according to the difference in path length to the recording station and the burn time of the connecting detonating cord, and attenuated due to the difference in travel time. The composite waveform produced this way closely resembles the second waveform in Figure 4, displaying a distortion between the compressional and tensile peaks and an increase in the relative magnitude of the second compressional peak, with no significant increase in the magnitude of the first compressional peak.

The same procedure predicts that at point A' (Figure 4) an increase in peak particle velocity of approximately 150% would be observed. This feature of geometry is not expected to apply to such a significant effect in the case of monitoring behind a long blast where enhancement similar to that predicted at A' is expected. The degree of enhancement is dependent on the amplitude of the individual components and the delay between successive components. The amplitude of the vibrations is expected to decrease according to the equation

$$A(r) = A_o * (r/r_o)^{-\alpha}$$

(Harries, 1981)

where $A(r)$ is the amplitude at radial distance r from the blasthole of radius r_o, A_o is the amplitude at the borehole wall, and α is the strain absorption coefficient ranging from 1.5 to 2.0 for the material studied in the Pilbara mining areas.

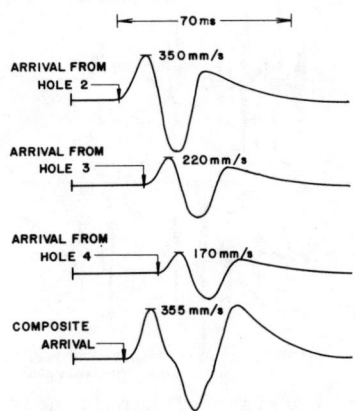

Figure 7   Superposition of three elemental waveforms from three holes fired 'simultaneously'.

Assuming a value of 1.87 for the attenuation coefficient, the 'instantaneous' initiation of

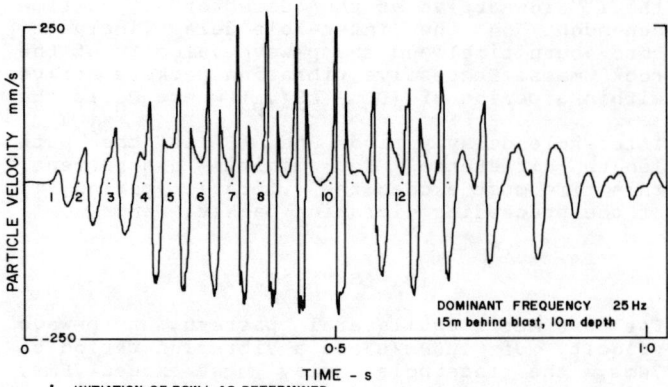

Figure 8   Vibration record for 45ms row-delayed blast indicating inadequate separation of waveforms produced by successive rows.

row i (Figure 5) would result in an increased vibration level of approximately 60% over that recorded at the monitoring station from the detonation of the single perimeter hole in the ith row. This analysis indicates that for a long limit blast it is preferable to fire with large (>70ms) inter-row delays and no interhole delays than to use single hole firing with an effective delay less than 15ms. Figure 8 shows the vibration recording for a row-delayed blast with a delay (45ms) considerably less than the vibration period, and the clear enhancement produced. Superimposed on the recording are the timing marks produced by electronic timing of the moment of initiation of individual rows. Although much of the increase in amplitude from the beginning of the recording is due to increasing proximity to the recording station, there is clearly insufficient delay to enable complete separation of successive vibration packets. There were between 7 and 9 holes per row in the blast of Figure 8.

Cross Hole Seismic Testing

Four cross hole seismic surveys were conducted behind a limit blast parallel to the last row of blast holes at distances varying from 10 to 40 metres. Cross hole seismic recordings were taken at 1 metre intervals down to full bench depth (15 metres). Diamond drilling was undertaken at two of these sites. In general, velocity increased with depth and variations between pre-blast and post-blast profiles were small. Figure 9 shows the pre- and post-blast rise time profiles for the three arrays closest to the blast zone, from which the following features can be observed:

1. decreased pulse rise time with depth;
2. significant local variations in rise time;
3. increased rise times in post-blast survey;
4. upper zones exhibit greatest changes in rise time;
5. the depth and severity of changes decrease with increasing distance behind the blast zone.

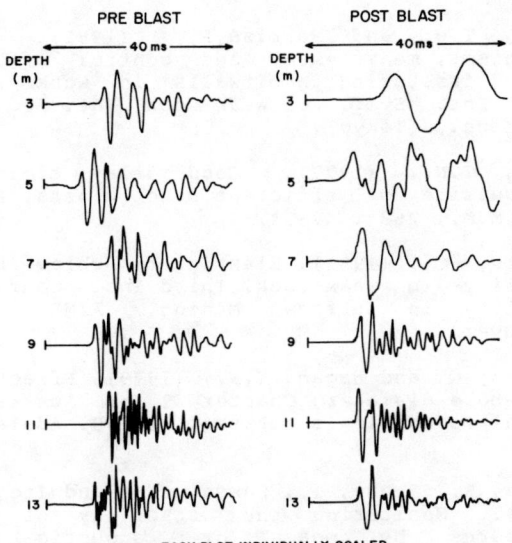

Figure 9   Change in cross hole acoustic wave shape from pre-blast to post-blast condition.

Figure 10 compares the change in rise time with depth from the pre-blast and post-blast condition for the three arrays, and indicates marked zones of damage in arrays 1 and 2 (10m and 15m behind blast) but no apparent damage in array 3 (20m behind blast) below 3m. Neither core logging nor face mapping could give self-consistent, reliable data representative of a large volume of rock for comparison. However, an overall analysis of fracture frequency from all cores showed no detectable difference between pre- and post-blast condition as shown in Figure 11. It is therefore apparent that the observed blast damage (1 metre heave between arrays 1 and 2) and the response of the acoustic survey are the result of the increased aperture of existing fractures rather than the creation of new fractures.

Where pulse rise times cannot be accurately measured, spectral analysis (consisting of a simple description of the shape of the frequency spectrum) has been shown to reveal the same boundaries of damage. The techniques have been applied to assess the damage zone behind six blasts with considerable variation in extent and degree being observed.

Two zones of damage are evident from the studies:

1. an upper zone of varying depth indicating major damage. The depth of this zone decreases with increasing distance behind the blast and corresponds to the limit of observed back-cracking;
2. a lower zone of relatively minor damage which may extend beyond the zone of back-cracking and to full bench depth. Because the zones are located behind final limits, direct logging techniques cannot be used to determine the significance of this damage.

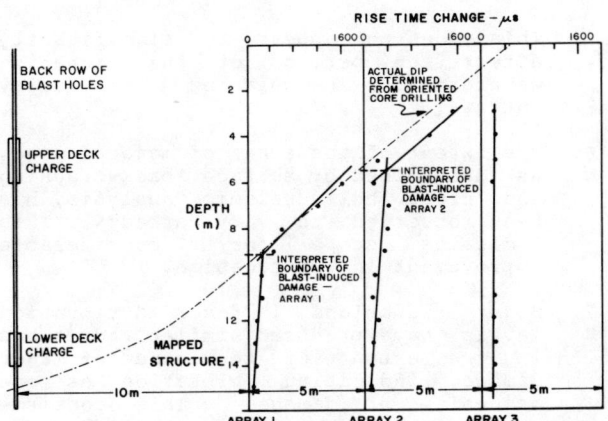

Figure 10   Change in rise time from pre-blast to post-blast condition.

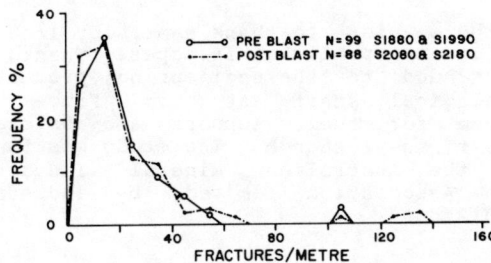

Figure 11   Fracture frequency distributions pre- and post-blast.

The research programme aims to minimise the degree of damage in both zones. A reduction in the extent of the zone of major damage is expected to have a similar reduction in the extent of the zone of minor damage.

CONCLUSIONS

1. Delay intervals to achieve effective separation of successive wave packets in delay blasting can be calculated from knowledge of vibration periods of individual charges (rows or holes).

2. For large hole blasting in fractured rock effective delays in excess of 70ms, required for complete waveform separation, make single hole blasting considerably more difficult than at present.

3. Delays in excess of 100ms are required to separate successive wave packets in row-delayed blasting.

4. Down-hole delays will be required in blasts with long delays to avoid cut-offs.

5. Preliminary analysis indicates a significant reduction of vibration level when the back holes of each row incorporate a reduced explosive density.

This does not appear to significantly affect the period of the vibration waveform for calculation of delay interval.

6. The extent of the zone of major damage, as determined by surface back-cracking and cross hole seismic analysis, has been observed to vary markedly. This indicates scope for considerable improvement in blast design.

7. Major reductions in recorded vibration levels have produced similar reductions in the extent of the zone of major damage, indicating vibration as one mechanism of damage. This confirms previous work.

ACKNOWLEDGEMENTS

The authors wish to thank Hamersley Iron for permission to publish this paper. Thanks are also extended to the engineering, production and geological staff at Tom Price and Paraburdoo for their support and assistance throughout the research. The study was funded through the Australian Mineral Industries Research Association Limited by industrial research grants.

REFERENCES

Albright, J.N., Pearson, C.F. and Fehler, M.C., (1980). Transmission of acoustic signals through hydraulic fractures, S.P.W.L.A. Twenty-first Annual Logging Symposium : 1-17.

Baldwin, J.T., (1975). Paraburdoo and Koodaideri iron ore deposits, and comparisons with Tom Price iron ore deposits, Hamersley Iron province, Economic Geology of Australia and Papua New Guinea (ed. C.L. Knight), 898-905.

Bourn, R. and Jackson, D.G., (1979). A generalised account of the Paraburdoo iron orebodies, Western Australian Conf., Aus.I.M.M., 187-201.

Broadbent, C.D., (1974). Predictable blasting with in-situ seismic surveys, Min. Engng., 37.

Gladwin, M.T. and Stacey, F.D., (1974). Anelastic degradation of acoustic pulses in rock, Phys. Earth Planet. Interiors, 8: 332-336.

Grujic, N., (1974). Ultrasonic testing of foundation rock, Proc. 3rd Congr. Int. Soc. for Rock Mech., Denver, Vol. 2A, 404-409.

Hagan, T.N. and Morriss, P., (1981). The mechanisms, measurement and control of blast induced fracturing in pitwalls in weak rock, Proc. Int. Symp. on Weak Rock, Int. Soc. for Rock Mech., (Tokyo).

Hagan, T.N., (1977). Good delay timing - prerequisite of efficient bench blasts, Proc. Aus.I.M.M., 263 : 47-54.

Harries, G., (1981). Blasting to achieve slope stability in weak rock. Third Inc. Conf. on Stability in Surface Mining, AIME, SME, Vancouver.

Harries, G. and Hagan, T.N., (1979). Effects of blast-hole diameter, Chapter 1 of Australian Mineral Foundation's Workshop 120/79, Adelaide, June.

Kaneko, K., Inoue, I., Sassa, K. and Ito, K., (1979). Monitoring the stability of rock structures by means of acoustic wave attenuation, Int. Soc. Rock Mech., 4th Int. Cong. Rock Mech., 2 : 287-292 (Montreux).

McKenzie, C.K., Stacey, G.P. and Gladwin, M.T., (1982). Ultrasonic characteristics of a rock mass, Int. J. Rock Mech. Min. Sci. and Geomech. Abstr., 19 : 25-30.

Meister, D., (1974). A new ultrasonic borehole meter for measuring the geotechnical properties of intact rock (in German), Proc. 3rd Congr. Int. Soc. for Rock Mech. , Denver, Vol. 2A, 410-417.

Murphy, V.J., (1972). Seismic velocity measurements for moduli determinations in tunnels, Proc. North American Rapid Excavation and Tunnelling Conf., 1 : 209-216.

Obert, L., Munson, R. and Rich, C., (1976). Caving properties of the Climax orebody, Trans. Soc. Min. Eng., AIME, 260 : 129-133.

Ramana, Y.V. and Rao, M.V.M.S., (1974). Q by pulse broadening in rock under pressure, Phys. Earth Planet. Interiors, 8 : 337-341.

Ricker, N., (1953). The form and laws of propagation of seismic wavelets, Geophysics, 18 : 10-39.

Stacy, T.R., (1976). Seismic assessment of rock masses, Proc. Symp. on Exploration for Rock Engineering, 2 : 113-117 (Johannesburg).

Young, R.P., Coffey, J.R., and Hill, J.J., (1979). The application of spectral analysis to rock quality evaluation for mapping purposes, Bull. Int. Assoc. Eng. Geol. Symp., 19 : 268-274 (Newcastle-upon-Tyne).

DYNAMIC SHEAR MODULUS AND DAMPING RATIO OF ROCKS FOR A WIDE CONFINING PRESSURE RANGE

Modules de cisaillement dynamique et rapports d'amortissement de roches soumis à une grande variation de pression reserrants de confinement

Dynamischer Schermodulus und Dämpfungsfaktor für dreiachsig-beanspruchte Gesteine

K. Nishi, T. Kokusho and Y. Esashi
Central Research Institute of Electric Power Industry, Abiko, Japan

SYNOPSIS
A cyclic triaxial testing machine for rocks was developed, making it possible to investigate the dynamic deformation properties of rock specimens at small as well as large strain levels over a wide range of confining pressures. This paper describes the results obtained in dynamic deformation tests for shear modulus and damping ratio performed on ten kinds of rock materials, paying particular attention to the dependence of these two characteristics on the frequency, initial stress, and amplitude of strain.

RESUME
Une machine d'essais cycliques à trois axes a été développée. Elle permet l'investigation des propriétés de déformation dynamiques de spécimens de roches soumises à une grande variation de pression de confinement et aussi des magnitudes de déformation. Les auteurs décrivent les modules de cisaillements et les rapports d'amortissement obtenus pour une dizaine de matériaux rocheux différents examinés dans une série d'essais sur la déformation dynamique, en prêtant une attention particulière à la fréquence, la charge initiale, et l'amplitude de déformation dont dépendent ces deux propriétés.

ZUSAMMENFASSUNG
Ein Dreiachsialgerät mit zyklischer Lastaufbringung wurde entwickelt, das die Bestimmung der dynamischen Verformungseigenschaften von Gesteinen bei kleinen und großen Dehnungen, über einen großen Bereich von Manteldrücken erlaubt. Es werden Ergebnisse dynamischer Deformationsversuche vorgestellt mit Werten der Schermoduli und Dämpfungsfaktoren von 10 verschiedenen Gesteinen. Besondere Aufmerksamkeit wird auf die Abhängigkeit dieser zwei Parameter von der Frequenz, Anfangsspannung und Amplitude der Dehnung gerichtet.

1. INTRODUCTION

While the shear modulus and the damping ratio of rocks under dynamic loading conditions have usually been measured in the laboratory using the resonant column device and ultrasonic methods, it is necessary to examine the influence of the confining pressure, and the deviatoric stress and strain applied to rock specimens on the

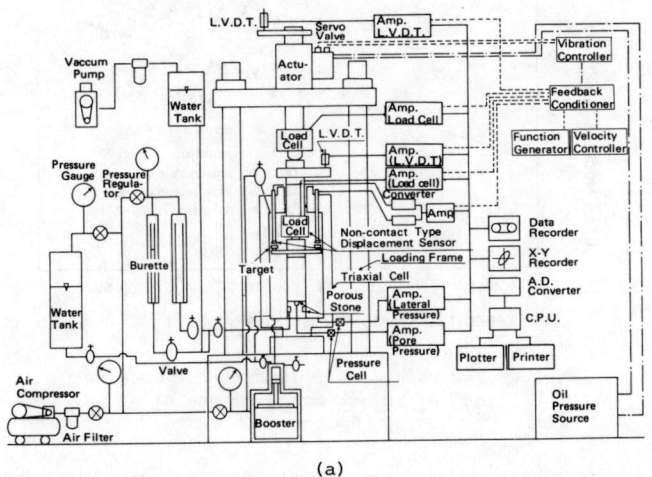
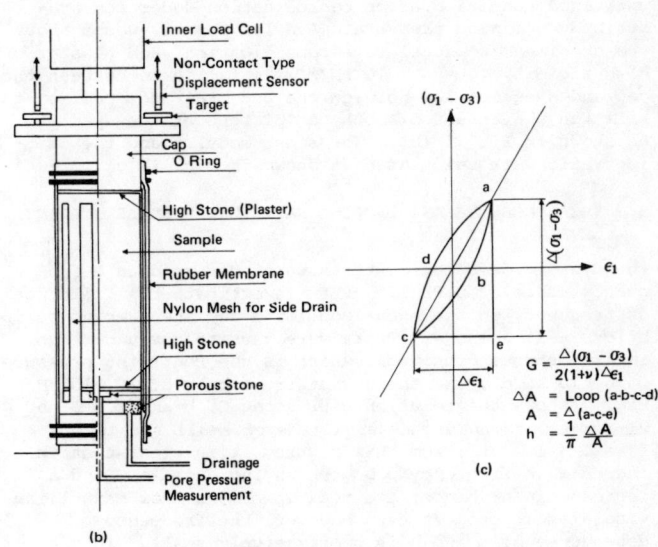

Fig. 1 (a) The outline of the cyclic triaxial apparatus, (b) set of specimen and (c) the calculation method for G and h

Table 1 Physical properties of rock materials

	dry density γ_d (tf/m³)	wet density γ_t (tf/m³)	unconfined strength q_u (10^2 × KN/m²)	effective porosity n (%)
pumiceous tuff (Tf-1)	1.14	1.70	35.3	46.5
tuff (Tf-2)	1.60	1.99	193.0	40.3
tuff (Tf-3)	1.55	1.96	42.7	41.3
tuff (Tf-4)	1.78	2.08	117.0	30.3
mud stone (Ms-1)	1.64	1.97	121.4	39.0
mud stone (Ms-2)	1.52	1.91	89.0	32.9
tuffaceous m. stone (Ms-3)	1.40	1.87	83.9	47.9
mud stone (Ms-4)	1.09	1.67	25.0	59.8
sand stone (Sa)	2.25	2.39	200.0	15.8
shale (Sh)	—	2.52	149.0	12.2

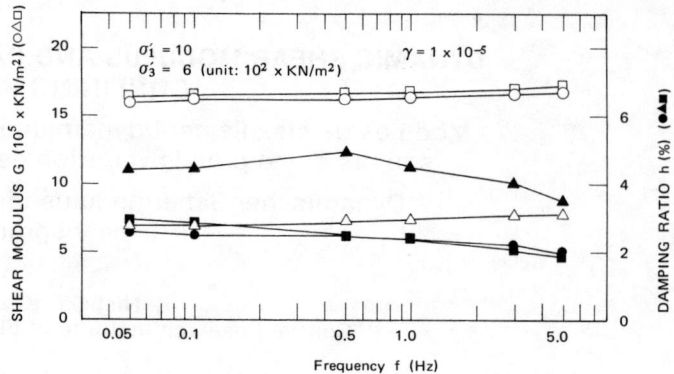

Fig. 2 The dependency of shear modulus G and damping ratio h on frequency
(○●: Tf-2, △▲: Tf-1, □■: Tf-4)

dynamic deformation characteristics of comparatively soft rock. We developed a cyclic triaxial testing machine in order to be able to investigate the shear modulus and the damping ratio at small as well as large strain levels over a wide range of confining pressures. This paper describes the dynamic deformation characteristics obtained for ten kinds of rock materials having unconfined compressive strength of $30-200 \times 10^2$ KN/m², with particular emphasis on the dependency of the shear modulus and the damping ratio on the frequency, initial stress and nonlinear deformation behavior.

2. APPARATUS AND MATERIALS

The newly developed cyclic triaxial testing machine is shown in Fig. 1 (a). This apparatus features non-contact type's displacement sensors (capacity: 2mm) and a load cell (capacity: 3 ton) set up in the triaxial cell to remove the influence of mechanical friction on the deformation characteristics of the specimen (Kokusho, 1980). The maximum applied confining pressure and axial load are 100×10^2 KN/m² and 5 ton, respectively. The rock specimens used (diameter, 50mm; height, 100mm) are the four kinds of tuffs, the four kinds of mudstones, sandstone and shale having the physical properties shown in Table 1. These specimens are in fully saturated states. The cyclic axial load was applied under the undrained condition after consolidation under the prescribed confining pressure. A nylon mesh was wound about the specimens to accelerate consolidation, and plaster of high rigidity ($G \fallingdotseq 4 \times 10^6$ KN/m²) is filled in between the cap and specimen and between the pedestal and specimen to kill the unevenness of both end platens of the specimen as shown in Fig. 1 (b). The shear modulus and the damping ratio were calculated as shown in Fig. 1 (c).

3. SHEAR MODULUS AND DAMPING RATIO UNDER SMALL STRAIN LEVEL

The dynamic deformation tests were performed in a frequency range of 0.05~5.0 Hz to investigate the influence of frequency on the shear modulus G and the damping ratio h for Tf-1, 2 and 4. Tests were carried out under the initial stress condition, which is the confining pressure of 6×10^2 KN/m², and the deviatoric stress of 10×10^2 KN/m². Fig. 2 shows the relationships among G, h and f. G and h were obtained under the amplitude of small strain ($\gamma = 1 \times 10^{-5}$). From this figure, it is clear that an increase in frequency produces an increase in G and a decrease in h; hence, the rock specimens show Maxwellian viscoelastic properties. However, the frequency-dependence of G and h is comparatively small. In the case of Tf-2, G and h at 0.05 Hz are 16×10^5 KN/m² and 2.6%, respectively. These are nearly identical to

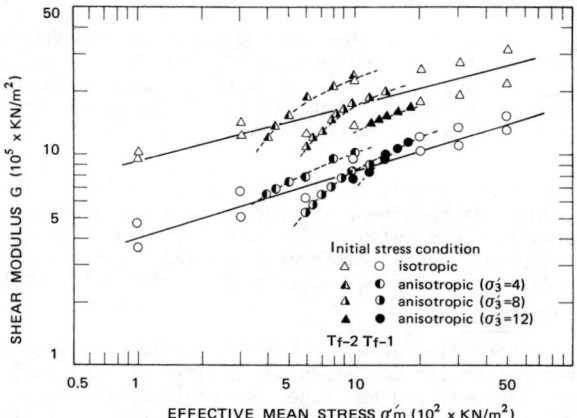

Fig. 3 The relations between shear modulus G and effective mean stress σ_m'

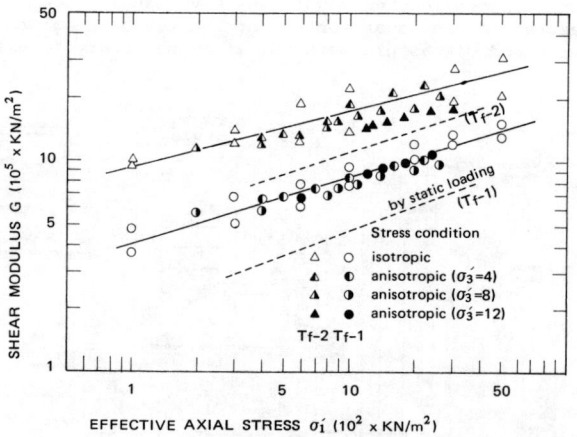

Fig. 4 The relations between shear modulus G and effective axial stress σ_1'

G($= 16.8 \times 10^5$ KN/m²) and h($= 2\%$) under 5.0 Hz. The same tendency was seen for other rock materials as well.

E 224

Fig. 3 shows the relationship between the shear modulus under $\gamma = 1 \times 10^{-5}$ and the effective mean stress $\sigma_m{'}$ for T_f-1 and T_f-2 on log-log scale paper, which were obtained under initial stress conditions of isotropic ($\sigma_1 = \sigma_3$) and the anisotropic stress ($\sigma_1 \neq \sigma_3$). The figure shows that the relationship between G and $\sigma_m{'}$ under isotropic stress is approximately linear, and that G increases with increasing $\sigma_m{'}$. Namely, the following equation is proposed.

$$G = A \cdot (\sigma_m{'})^B \qquad (1)$$

where A and B are the material constants. On the other hand, the relationship between G and $\sigma_m{'}$ under anisotropic stress are not linear. These tendencies were also seen in other rock materials.

Fig. 4 shows the relationship between G and the initial effective axial stress $\sigma_1{'}$, which is equivalent to the effective overburden pressure. The relationship between G and $\sigma_m{'}$ obtained by cyclic loading under isotropic stress conditions, and the same relationship obtained by static loading and represented in this figure by a solid and a broken line, respectively. G given by the static loading was calculated as the tangent modulus of the stress-strain curve. It is apparent from this figure that the relationship between G and $\sigma_1{'}$ are uniquely determined irrespective of differences in the isotropic or anisotropic stress conditions (Nishi et al., 1981). The relationship between G at small strain level and $\sigma_1{'}$ is given by the following equation.

$$G = A \cdot (\sigma_1{'})^B \qquad (2)$$

where the material constants A and B is obtained from the cyclic loading tests performed under isotropic stress conditions. Fig. 5 gives the relationship between G and $\sigma_1{'}$ for the ten kinds of rock materials. Some rock materials, Ms-1, Sh, T_f-4, and Sa, show a constant shear modulus to a stress level, but the relationship between G and $\sigma_1{'}$ over its stress level is given by Eq. (2). The material constant B in Eq. (2) is given as 0.306, 0.253, 0.249 and 0.440 (average value $\bar{B}$ is 0.312) for T_f-1, 2, 3 and 4; 0.187, 0.269, 0.162, and 0.09 ($\bar{B}$ is 0.177) for Ms-1, 2, 3 and 4; and 0.377 and 0.362 for Sa and Sh, respectively. From these test results, the degree of dependency on the axial stress of G of tuffs is larger than that of mudstones.

Fig. 6 shows the relationship between G under an initial axial stress of $5 \times 10^2 KN/m^2$ and the unconfined compressive strength q_u. The relationship between G during static loading under the confining pressure of $5 \times 10^2 KN/m^2$ and q_u is also shown in this figure. It is clear from the figure that (1) rock materials with a larger unconfined compressive strength have a larger shear modulus, (2) compared with the relationship between G and q_u under cyclic and static loading, the increasing rate of G for q_u is nearly identical, but the G obtained by static loading is about 70% of the G obtained by cyclic loading under an unconfined compressive strength of $10 \times 10^2 KN/m^2$.

Fig. 7 shows the average relationship between the damping ratio at $\gamma = 1 \times 10^{-5}$ and the initial effective axial stress for the ten kinds of rock materials. The damping ratio shows a tendency to decrease with increasing $\sigma_1{'}$ for all rock specimens. The lower and upper limit values of the damping ratios of the rock materials used were 1.2% under $\sigma_1{'} = 50 \times 10^2 KN/m^2$ and 6% under $\sigma_1{'} = 1 \times 10^2 KN/m^2$, respectively. Such dependency of h on the initial axial stress is remarkable for tuffs and mudstones in comparison with that of Sa and Sh. Also, Ms-4, which has the smallest unconfined compressive strength, shows the smallest damping ratio. Furthermore, looking at the damping ratio for sand and clay at $\gamma = 1 \times 10^{-5}$ (Seed and Idriss, 1972), the h of rock materials under low stress levels is somewhat larger than that of soils. This is

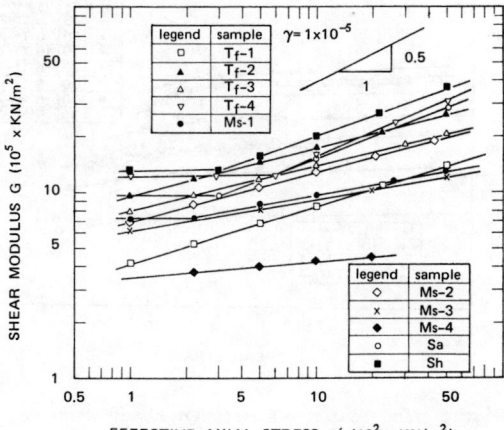

Fig. 5 *The relations between shear modulus G and effective axial stress $\sigma_1{'}$ average line*

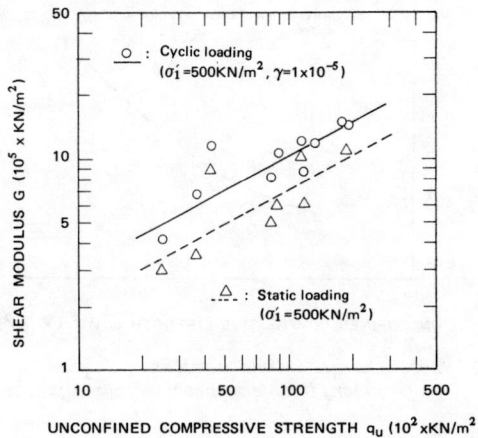

Fig. 6 *The relations between shear modulus G and unconfined compressive strength q_u*

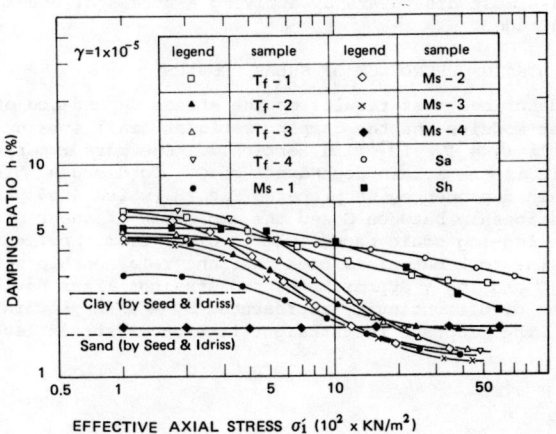

Fig. 7 *The relations between damping ratio h and effective axial stress $\sigma_1{'}$ (average curve)*

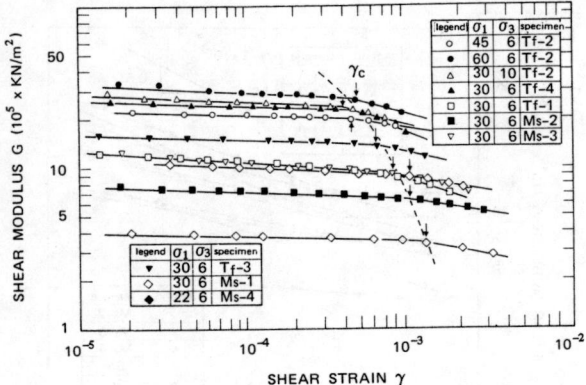

Fig. 8 *The relations between shear modulus G and amplitude of shear strain* γ

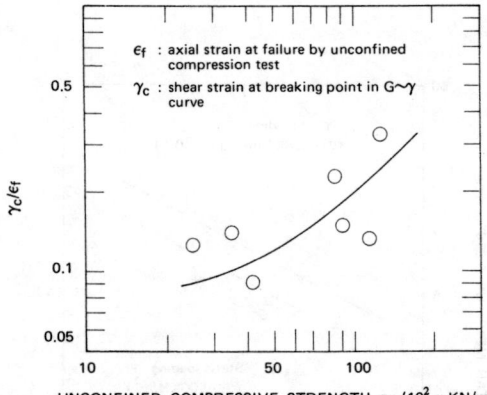

Fig. 9 *The relations between* γ_c/ε_f *and unconfined compressive strength* q_u

due to the opening of hidden cracks accompanying stress release during sampling. The dependence of h on the initial axial stress becomes noticeably smaller at larger stress levels than a σ_1' of $20 \times 10^2 \text{KN/m}^2$. Consequently, it may be concluded that the influence of stress release on h almost disappears by applying a stress of equal magnitude.

4. STRAIN DEPENDENCE OF SHEAR MODULUS

We described test results on the stress dependence of the shear modulus and the damping ratio at small strain levels ($\gamma = 1 \times 10^{-5}$) in Section 3. One more important topic is the strain dependence of G. To discuss this strain dependence, we prepared Fig. 8, which shows the relationship between G and the amplitude of shear strain γ on log-log scale paper. It is clear from the figure for the rock materials used that the relationship between log G and log γ approximates two straight lines having a point of discontinuity represented by a shear strain of γ_c. The slope of the straight line under strain levels smaller than γ_c is very small and so the strain dependency of G can be essentially neglected. However, the influence of the strain magnitude on the shear modulus is large under strain levels larger than γ_c. We therefore conclude that γ_c provides one index for expressing the non-linear characteristics of G. While γ_c ranges from 5×10^{-4} to 1.2×10^{-3}, Fig. 8 also shows that stiffer rock materials have a smaller γ_c. Furthermore, examining the test results for T_f-2, it is clear that γ_c does not depend on the magnitude of the initial stress under the range of stresses applied in this tests. Fig. 9 shows the relationship between γ_c/ε_f, when ε_f is the axial strain at failure obtained by unconfined compression tests, and the unconfined compressive strength q_u. The value of γ_c/ε_f ranges from 0.1 to 0.36. It can be pointed out from the figure that rock specimens having larger unconfined compressive strengths generally show larger values of γ_c/ε_f. This means that the rock specimens with larger strains at failure show nonlinear dynamic deformation behavior at smaller strain levels in comparison with strain at failure.

5. CONCLUSIONS

Dynamic deformation tests for ten kinds of rocks were performed using a newly developed cyclic triaxial testing machine, which makes it possible to investigate the dynamic deformation characteristics over a wide range of strain levels and confining pressures. The main conclusions obtained through this research are as follows.

1) The shear modulus and the damping ratio are not almost dependent on the frequency.

2) The shear modulus at small strain levels can be given as the function of the initial effective axial stress applied to the specimen, and this relationship approximated by straight lines on log-log scale paper.

3) The damping ratio is dependent on the magnitude of the initial effective axial stress. Namely, the damping ratio decreases with increase of the axial stress.

4) The strain dependence of G becomes pronounced under an amplitude of strain larger than $\gamma_c = 5 \times 10^{-4} \sim 1.2 \times 10^{-3}$. The ratio of γ_c to the axial strain ε_f at failure under unconfined stress conditions is in the range of about $0.1 \sim 0.3$.

References

Kokusho, T. (1980)
Cyclic triaxial test of dynamic soil properties for wide strain range: Soils and Found., Vol. 10, No.2, 45-60.

Nishi, K. and Uno, H. (1981)
Dynamic shear modulus of tuffs for wide confining pressure range: Proc. 17th JNSMFE, 657-660, Naha, in Japanese.

Seed, H.B. and Idriss, I.M. (1970)
Soil moduli and damping factors for dynamic response analysis: Report No. EERC 70-10, University of California.

A NEW CRUSHING TEST FOR ROCK EXCAVATION PROPERTIES

Un nouvel essai de concassage appliqué aux propriétés d'excavation en roches

Eine neue Quetschprüfung zur Ermittlung der Aushubeigenschaften von Gesteinen

Takashige Haga and Jiro Saito
Senior Research Engineers, Technical Research Institute, Ohbayashi-Gumi Ltd., Tokyo, Japan

SYNOPSIS

The Protodyakonov type crushing test involves some problematic points, and the authors have made efforts to solve these problems and to propose an improved test method. This paper describes details of the proposed testing method, the method of its utilization and the mechanical significance of specific energy as obtained by this test.

RESUME

L'essai de concassage du type Protodyakonov entraîne des problèmes pratiques: Les auteurs se sont efforcés de les résoudre et d'aboutir à une méthode d'essai améliorée. Ce rapport décrit les détails de la méthode d'essais proposée, son mode d'emploi et les caractéristiques mécaniques de l'énergie spécifique résultant de l'essai.

ZUSAMMENFASSUNG

In dem Quetschprüfungstest nach Protodyakonov gibt es gewisse praktische Probleme. Die Verfasser haben sich bemüht, diese Probleme zu lösen und schlagen ein verbessertes Verfahren vor. In diesem Referat werden die Einzelheiten der vorgeschlagenen Prüfungsmethode, dessen Anwendung, sowie die mechanische Bedeutung der spezifischen Leistung, welche von der Prüfung erhalten wird, beschrieben.

1. INTRODUCTION

Excavation properties of bedrock are judged by geological surveys, field rock excavation tests and laboratory tests of rock samples. With regard to the latter, it is desirable for the tests to be performed using simple apparatus, with impact crushing strengths of rock samples obtained rapidly.

It is thought the Protodyakonov type test well-known as an impact crushing test is the most suitable for this purpose. However, this testing method is difficult to apply to very hard and very soft rocks, while the mechanical significances of test values obtained are not distinct. The authors therefore endeavored to dissolve these problematic points and made improvements on the testing methods.

This paper describes details of the improved testing method, and the utilization method and mechanical significance of specific energy as obtained by this test.

2. PROBLEMATIC POINTS OF PROTODYAKONOV TYPE TEST

Problems about this Protodyakonov type test are such as the following:
(i) Samples collected from muck in actual excavation work and from core boring often cannot be utilized for this test because of the aspects of quantity and grain size.
(ii) Measuring the volume of the sample is difficult and since the size of the sample used in testing would differ due to personal equations, errors are produced in test results.
(iii) Differences in test results are produced depending on the number of impacts, while application of the test is difficult when impact crushing strength of the sample is very low, or very high.
(iv) The relationship between the Protodyakonov coefficients f and the energy required for excavation is not clarified and working efficiency cannot be rationally computed.

3. THEORETICAL DISCUSSION OF IMPACT CRUSHING TEST

Fig. 1 shows a case of the relationship between impact energy using the testing apparatus of Fig. 2 and the increased surface area of sample. Based on these test results, if it is assumed that energy required for impact crushing is proportional to the n-th power of increased surface area ΔS, Eq. (1) will hold true.

$$E \propto (\Delta S)^n \qquad (1)$$

Table 1 shows values of n obtained for various rocks performing tests as indicated in Fig. 1.

When a rock of density ρ, particle size D_f and

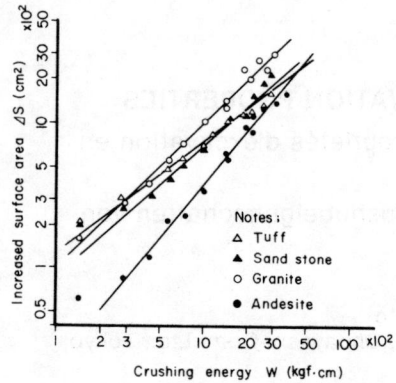

Fig. 1 Relationship between crushing energy and increased surface area of various rocks

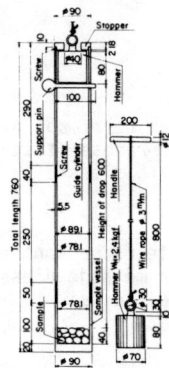

Fig. 2 Impact crushing test apparatus (Protodyakonov type)

weight M is crushed to any grain size D_i and a reference grain size D_o, the respective increased surfaces ΔS_i and ΔS_o will be given by Eq. (2).

$$\left.\begin{array}{l} \Delta S_i = \dfrac{6M}{\rho}\left(\dfrac{1}{D_i} - \dfrac{1}{D_f}\right) \\ \Delta S_o = \dfrac{6M}{\rho}\left(\dfrac{1}{D_o} - \dfrac{1}{D_f}\right) \end{array}\right\} \quad (2)$$

With energy required to crush rock to D_i and D_o as E_i and E_o, Eq. (3) is deduced from Eq. (1).

$$E_i = E_o\left(\dfrac{\Delta S_i}{\Delta S_o}\right)^n \quad (3)$$

Substituting Eq. (2) into Eq. (3), and further, assuming D_f to be infinite,

$$E_i = E_o\left(\dfrac{D_o}{D_i}\right)^n \quad (4)$$

Therefore, the energy ΔE_i required for crushing from grain size D_s of the sample to any grain size D_i will be

$$\Delta E_i = E_o D_o^n\left(\dfrac{1}{D_i^n} - \dfrac{1}{D_s^n}\right) \quad (5)$$

where, with the respective grain size and volumes of particles produced by crushing of samples as D_i and V_i, the energy required for crushing will be indicated by Eq. (6)

$$W = \Sigma \Delta E_i V_i \quad (6)$$

From Eqs. (5) and (6),

$$W = E_o D_o^n\left\{\Sigma\left(\dfrac{V_i}{D_i^n}\right) - \dfrac{\Sigma V_i}{D_s}\right\} \quad (7)$$

In the crushing test, with hammer weight as W_H, drop height as H, number of impacts as N, density of sample as ρ, weight of crushed sample in any grain size range as M_i, and further, considering $n = 1$ from Table 1, the energy required for crushing rock to D_o, the so-called specific energy E_o, will be according to Eq. (8) deduced from Eq. (7).

$$E_o = \dfrac{N \cdot W_H \cdot H \cdot \rho}{D_o\left\{\Sigma\left(\dfrac{M_i}{D_i}\right) - \dfrac{\Sigma M_i}{D_s}\right\}} \quad (8)$$

where, if $D_o = 0.1$ cm, $W_H = 2.4$ kgf, and $H = 60$ cm,

$$E_o = \dfrac{1440 \cdot N \cdot \rho}{\Sigma\left(\dfrac{M_i}{D_i}\right) - \dfrac{\Sigma M_i}{D_s}} \quad (9)$$

Table 1. Index of crushing of various rocks

Classification		Index of crushing n
		0.7 0.8 0.9 1.0 1.1 1.2 1.3 1.4 1.5
Igneous rock	Volcanic rock	
	Plutonic rock	
Sedimentary rock and tuff	Breccia / Tuff breccia	
	Sand stone / Coarse tuff	
	Mud stone / Fine tuff	
Metamorphic rock	Igneous rock	
	Sedimentary rock	

4. NEW CRUSHING TEST

This testing method basically is similar to the Protodyakonov type test, but differs in that specific energy is obtained from the sample used and the test results, and this is employed as an index of crushing strength.

The testing method is as described below.

i) Testing Apparatus

   a) Crushing test apparatus: The testing apparatus shown in Fig. 2 is used.

   b) Sieves: Sieves of the 7 sizes of 15.9 mm, 5.6 mm, 2.0 mm, 1.0 mm, 0.5 mm, 0.25 mm, and a pan are employed.

c) Scales: Scales of sensitivity 0.1 g are used.

ii) Testing Procedure

a) Preparation of sample: After drying the rock collected in a drying furnace at 60°C for 24 hours (*1), crushing is done by hammer, and a 50-gf sample passing the 15.9-mm sieve and retained on the 9.5-mm sieve is made (*2).

b) Crushing of sample: The 50-gf sample is put into the crushing test apparatus and crushing is done dropping the hammer. The number of impacts by the hammer is to be about 2 to 20 times (*3) depending on the condition of crushing of the sample.

c) Sieve analysis: After crushing, the sample is taken out from the crushing test apparatus, sieve analysis is performed, and the weights of the crushed sample retained on the various sieves and the pan are measured.

d) The average grain size D_i and weight M_i of the crushed samples retained on the various sieves, the number of impacts N, and the sample density ρ are substituted into Eq. (9) to obtain the specific energy E_O.

*1: The values of E_O differ between cases of the sample being dry and being wet. Therefore, material dried beforehand by drying furnace is made the sample in order to unify test results.

*2: The scatter of the test data obtained from 50gf of ϕ95-15.9 mm samples was observed to be smaller than those of ϕ20-40 mm samples in the Protodyakonov type test.

*3: In the cases of this testing method the value of E_O varies depending on the number of impacts. Fig. 3 shows an example of the relationship between number of impacts and E_O. At the number of impacts in a certain range, it may be seen that the variation in values are small. With rock of ordinary hardness, 10 times is suitable as the number of impacts.

5. UTILIZATION OF TEST RESULTS

i) Estimation of Protodyakonov Coefficient

There have been many studies reported in the past regarding the relationship between Protodyakonov coefficient f and excavation properties, and it will be significant to estimate the value of f from the results of the new impact crushing test.

Fig. 4 shows the relationship between E_O and f. Based on these figures the value of f may be estimated by Eq. (10).

$$f = 0.02 E_O^{1.2} \qquad (10)$$

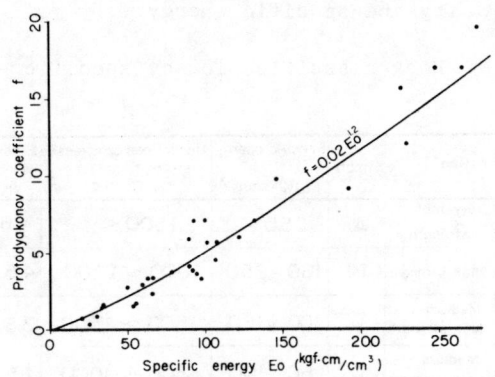

Fig. 4 Relationship between specific energy and Protodyakonov coefficient

ii) Engineering Classification of Rock by E_O

Fig. 5 and Fig. 6 show the six cateogories of rock obtained from the relationship between E_O, unconfined compressive strength σ_c of rock, and velocity V_p. Table 2 is a compilation of the above. If data on excavation properties of bedrock composed of these rocks were to be gathered according to each of the categories given in Table 2, it will be possible to obtain E_O of rock and rationally and readily determine the excavation properties of the bedrock.

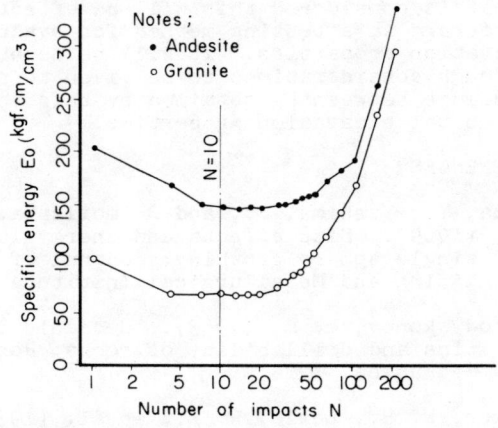

Fig. 3 Influence of number of impacts on specific energy

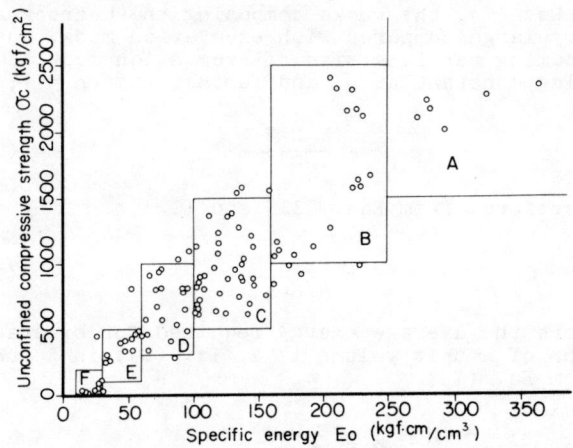

Fig. 5 Relationship between unconfined compressive strength and specific energy

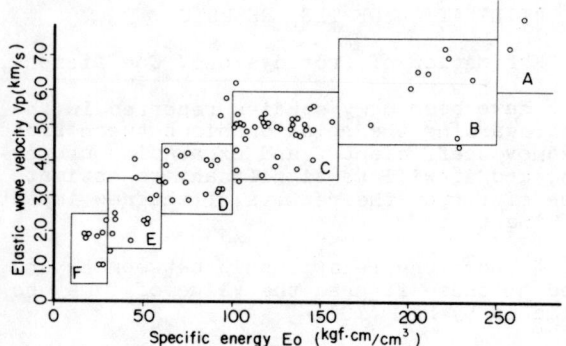

Fig. 6 Relationship between elastic wave velocity and specific energy

Table 2 Rock classification by specific energy

Classification		Symbol	Specific energy E_0 (kgf·cm/cm³)	Unconfined compressive strength σ_c (kgf/cm²)	Elastic wave velocity V_p (km/s)
Hard rock	Very high strength	A	250 <	1500 <	6.0 <
	High strength	B	160~250	1000~2500	4.5~7.5
Medium rock	Medium high strength	C	100~160	500~1500	3.5~6.0
	Medium low strength	D	60~100	300~1000	2.5~4.5
Soft rock	Low strength	E	30~60	100~500	1.5~3.5
	Very low strength	F	10~30	10~200	0.5~2.5

iii) Computation of Drilling Rate and Modulus of Machine

With P_0 as output of excavation equipment, e as modulus of the machine, drilling rate as U, borehole sectional area as A, and excavation volume as V, the mechanical equivalent W is given by Eq. (11).

$$W = P_0 \cdot e \cdot \frac{V}{AU} \quad (11)$$

In Eq. (7), the rocks composing the bedrock are very large compared with excavation muck, and assuming particle size of excavation muck as being constant at D, and further that n = 1,

$$W = \frac{E_0 D_0 V}{D} \quad (12)$$

Therefore, from Eqs. (11) and (12),

$$U = \frac{P_0 \cdot e \cdot D}{A E_0 D_0} \quad (13)$$

Where the average energy required for excavation of a unit volume is E, Eq. (14) is deduced from Eq. (13).

$$e = \frac{E_0 D_0}{E D} \quad (14)$$

6. APPLICATION TO ACTUAL EXCAVATION WORK

The project to which application was made was drilling of a granite bedrock to a depth of approximately 7 m at a diameter of 1.3 m using a rotary drilling machine of gear cutter type. Fig. 7 shows the properties of the ground at the location to be drilled, specific energy E_0 determined from samples collected from each depth, the actual excavating time and the specific energy E required for drilling a unit volume, and the modulus of the machine e obtained by Eq. (14).

In the graph giving the excavating time the values estimated prior to excavation from E_0 and Eq. (13) are shown and they are values which are roughly approximate. In making the estimates, speed of 15 rpm, torque of 800 kgf, thrust of 15 tf and e = 0.2 were assumed.

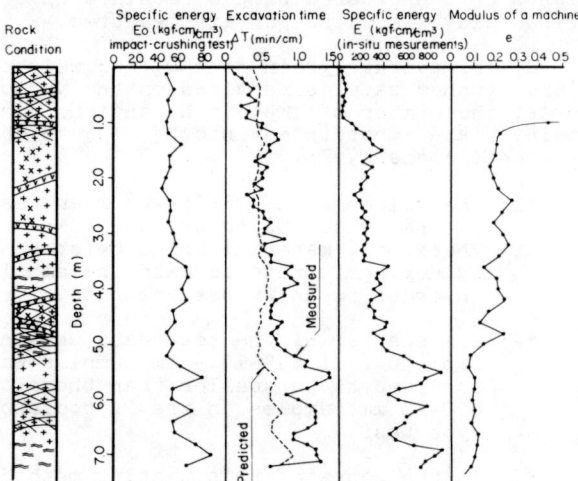

Fig. 7 Comparison of specific energy and actual excavation time and modulas of a machine

7. AFTERWORD

The impact crushing test proposed in this report requires a smaller quantity of sample compared with the conventional method, while the scatter in measured values is also smaller. Accordingly, it is considered this will be effective in the future as a testing method for evaluating excavation properties. It will be necessary for thorough considerations to be given to correspondence between E_0 obtained by this test and the actual excavation properties.

REFERENCES

Kanda, Y., Yashima, S., and Shimoiizaka, J. (1969). Size effects and energy laws of single sphere crushing: Journal of the Mining and Metallurgical Institute of Japan.

Protodyakonov, M. M. (1963). Mechanical properties and drillability of rocks: Rock Mechanics.

Walker, W. H., Lewis, W. K., et al. (1937). Principles of Chemical Engineering. McGraw-Hill.

BLAST DAMAGE AND STRESS MEASUREMENTS IN THE LKAB-MALMBERGET FABIAN OREBODY

Dommages causés par une charge explosive et détermination de contraintes dans l'amas métallifère "Fabian" de LKAB-Malmberget

Sprengauflockerung und Spannungsmessungen im Erzkörper "Fabian" der LKAB-Malmberget Grube

R. Holmberg
Swedish Detonic Research Foundation (SveDeFo), Stockholm, Sweden

K. Mäki
Swedish Detonic Research Foundation (SveDeFo), Stockholm, Sweden

W. Hustrulid
Department of Mining Engineering, Colorado School of Mines, Golden, Colorado, USA

H. Sellden
LKAB, Malmberget, Sweden

SYNOPSIS

The CSM-cell was used to determine the extent of blast-induced damage during the excavation of an opening slot in a sublevel stope. In some cases the deformation modulus decreased with decreasing distance as might be expected due to induced damage. However, in other cases it remained constant or increased. The explanation offered is that the modulus is stress-dependent. Thus as the slot extends, the high stresses around the end of the slot tend to increase the modulus. The CSM-cell system by itself cannot distinguish between these two opposing driving forces. An analysis of the expected stress fields and the modulus changes suggested a close connection between the two.

RESUME

La cellule de CSM a été employée pour déterminer l'importance des dommages causés par une charge explosive pendant l'excavation d'une rainure dans une exploitation souterraine. Dans certains cas le module de déformation a diminué avec la distance, ce qui concerde bien avec l'importance des dommages provoqués. Dans d'autres cas, cependant, le module de déformation est resté constant ou bien a augmenté. L'explication suggérée en est que le module est sensible aux contraintes. Par conséquent, au fur et à mesure que la rainure s'élargit les contraintes élevées autour de son extrêmité tendent à augmenter le module. Le dispositif de la cellule de CSM seul n'est past capable de distinguer entre ces deux forces opposées, mais l'analyse des contraintes attendues et les changements du module indique une relation très proche entre les deux.

ZUSAMMENFASSUNG

Die CSM Zelle wurde zur Bestimmung der Ausdehnung der Sprengauflockerung verwendet, die beim Absprengen eines Schlitzes in einem Untergrundabbaustoß zustandekommt. In einigen Fällen nahm der Deformationsmodul mit dem Abstand ab, was infolge der hervorgebrachten Beschädigung auch zu erwarten war. In anderen Fällen dagegen blieb der Deformationsmodul unverändert oder nahm zu, was damit zu erklären wäre, daß der Deformationsmodul spannungsabhängig ist. Infolge der hohen Spannungen am Ende des Schlitzes neigt der Modul mit der Verlängerung des Schlitzes zur Steigerung. Das CSM Zelle Gerät an sich vermag nicht diese beiden entgegengesetzten Kräfte voneinander zu unterscheiden. Die Analyse der bestimmbaren Spannungen und die Veränderungen des Moduls aber scheinen auf eine enge Relation zwischen den beiden hinzuweisen.

INTRODUCTION

The minimization of unwanted damage from blasting has always been a concern in mining and civil engineering structures in rock.
The relatively recent advent of large hole stope blasting in underground mining has made the quantification of the extent of damage and means of preventing or at least minimizing the damage of high priority.
Of particular concern is

(a) potential damage to the hanging wall, footwall, and pillars adjacent to the stope which might present stability problems.

(b) potential damage to the remaining holes in the round which have been drilled but not shot. This results in costly delays in which holes must be redrilled and/or cleaned.
Fragmentation and subsequent ore recovery may suffer.

This paper presents the results of blast damage and stress measurements performed in the Fabian Orebody, LKAB, Malmberget, Sweden. The blast damage measurements are reported in detail by Hustrulid, Holmberg and Mäki, 1981.

MINING PROCEDURE, Olsson, 1979.

The Fabian orebody is about 400 m in length and 30-40 m wide. It has a strike of N 50° E and dips at 75°. The average iron content is about 54%. Previous mining of the eastern end of the orebody has resulted in a room having approximate dimensions of 180 m long, 40 m wide and 120 m high. Although some filling of the room has been done, the unsupported area of the hanging wall is about 18 000 m^2. As seen in Figure 1, a pillar with an average width of 20 m will be left between the existing stope and the new workings.

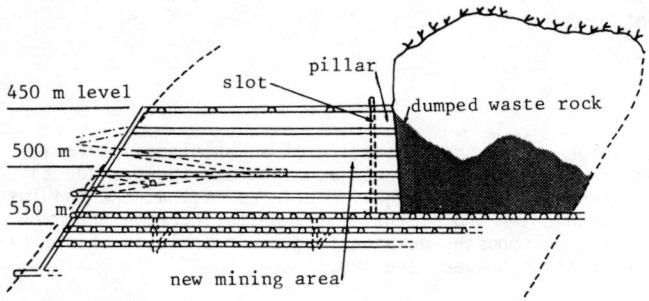

Figure 1. Longitudinal vertical section through the Fabian orebody.

The maintenance of the integrity of pillar is of importance and therefore special procedures were taken in creating the opening slot. This slot which is approximately 6 m in width is created by first driving a raise in the ore along the hanging wall contact. The full width of the orebody is then opened.
Either one or two rows of holes were blasted at a time. The most common practice during the period of the measurements June - Sept 1980 were two rows.

Table I. Data about loading of the slot holes between the 494 and 530 levels.
Holes per row:	5
Hole diameter:	64 mm
Explosive:	ANFO
Spacing × burden:	1 m × 1,5 m
Charge concentration:	2.9 kg/m
Nominal hole length:	29 m with 20 m in the last two rows.
Stemming:	1.5 m
Initiation:	Detonating cord (10 gr/m) along the whole charge
Delay:	25 ms
Delay pattern:	3 delays per row and one sometimes two rows per round.

The blasting of the raise between the 494 and 530 m levels was reported by Holmberg et al,1980. An attempt to investigate the damage zone around the raise was made by core drilling prior to and after blasting. The results of these investigations are reported by Mäki and Holmberg, 1980.

EXPERIMENTAL PROGRAM

Introduction

There are number of techniques which can be used for determination of the damage to the surrounding rock from blasting. These can be the examination of cores taken prior to and after blasting, geophysical surveys (cross hole and down hole) made from boreholes or from the rock surface. In all cases the parameter being measured reflects changes in some material properties. For example strength, elastic properties, number of fractures present, wave velocity, density, etc. In this series of experiments, the objective was to perform modulus measurements in boreholes as the progress of the slot blasting approached the measurement holes. The modulus would be expected to decrease with increasing damage. A quantitative measure of the degree of modulus reduction with distance could then be achieved.

Of the several bore hole tools available for use, the CSM cell was selected for use.

Description of the CSM cell, Hustrulid,1972/73.

The CSM (Colorado School of Mines) cell was developed as a tool for obtaining the deformation modulus of the rock surrounding a 38 mm diameter borehole. The basic system shown in Figure 2 consists of the CSM cell, a pressure gage and a screw type pressure generator with a vernier indicator.

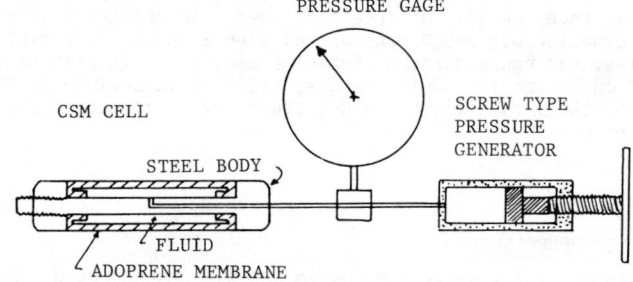

Figure 2. Diagrammatic representation of the CSM cell system.

As fluid is transferred from the pressure generator to the CSM cell by advancing the piston, the adoprene membrane inflates and pressurizes the wall of the borehole. By monitoring the transfer of fluid (turns of the screw type piston) and the resulting system pressure change one can obtain the pressure - volume relationship for the rock plus the pressurization system. By performing a similar test in a metal cylinder of known elastic properties (and pressure - volume relationship, M_c) one can determine the pressure-volume relationship for the pressurization system, M_s, alone. By subtracting the now known effect of the pressurization system from the total pressure - volume curve (borehole + pressurization system) one can calculate the desired pressure - volume relationship for the rock alone. The modulus of rigidity follows directly and if the

Poisson's ratio for the rock is known or can be estimated then the elastic modulus can be found. Since the pressure readings will be made at the half turn intervals of the screw type piston, it is convenient to describe the pressure - volume curves in terms of pressure change per half turn.

Location of the test holes

Two diamond drillholes each approximately 10 m in length were drilled in the floor of the slot crosscut on the 494 m level at the locations shown in Figure 3. During the drilling of the first hole problems were encountered near the surface due to rather heavy fracturing. To overcome this, a concrete pad was poured at the collar location and allowed to set. A hole 3" in diameter and approximately 2 m deep was drilled and a pipe was installed. The EX (nominal 38 mm) hole was then continued through the pipe. For hole 1, a bit having an outer diameter of 38.4 mm followed by a 38.5 mm reamer shell was used. This should have been satisfactory however initial testing with the CSM cell revealed that the hole diameter was too large (approximately 42 mm) for use. The second hole (42/80) was drilled using a 36.5 mm bit followed by a 38.5 mm reamer shell. With the exception of a couple of locations along the hole, this yielded the correct hole size.

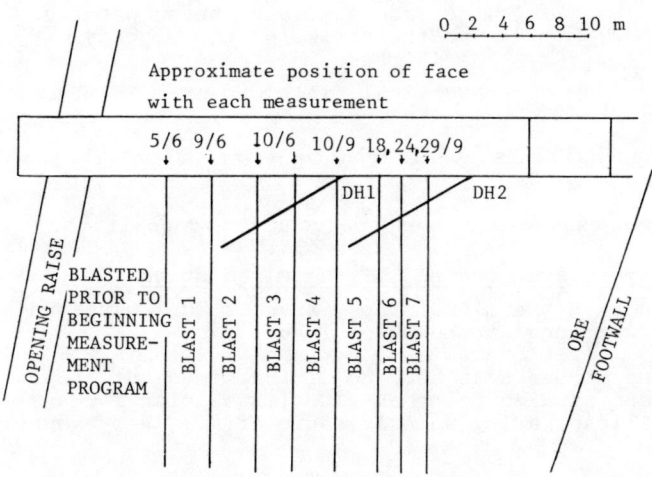

Figure 3. Vertical cross section showing the location of the measurement hole and blasting progress.

At the time the holes were drilled, the slot was approximately 6 m and 15 m away from the bottoms of holes 1 and 2 respectively. The cores obtained from both holes were carefully logged with respect to segment length. An attempt was made to identify both natural and boring induced fractures. Calculations of RQD (rock quality designation) were made based upon a minimum core length of 4.4 cm. The interval used was about 1 m. The results are given in Figure 4.

As can be seen the RQD values for hole 2 are considerably higher than those for hole 1. It is not known whether this reflects natural variability or possible blast damage to the rock at the end of hole 1. From the plots, it would suggest that significant damage to the floor of the drift due to its excavation extends to a depth (below surface) of between 0.5 and 1 m. It should be noted that because of the drill hole being drilled at an angle of 30°, the depth below surface is just one half the distance along the hole as measured from the collar. For hole 2 (the site of the CSM cell evaluation) the rock is of good quality in the zone 2 m to 5 m below the surface.

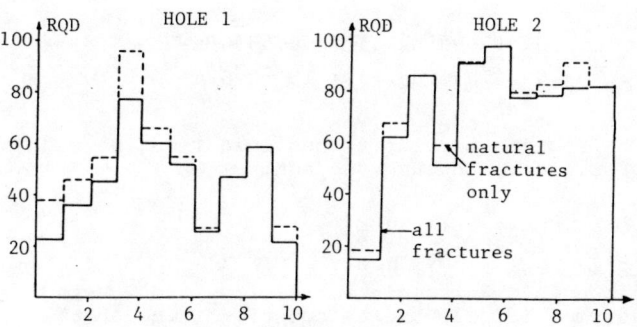

Figure 4. RQD as a function of distance along holes No. 1 and No. 2.

Experimental procedure

Prior to borehole use the CSM cell system was first calibrated by inserting it into a metal cylinder of known elastic properties. It is most desireable that the stiffness (pressure-volume curve) of the cylinder be similar to that expected for the rock in-situ. The stiffness of the cylinder can be varied by changing metals and outer diameter.

The decision to evaluate the modulus over a pressure range of 7 to 14 MPa, was made based upon (a) an initial evaluation of the rock strength and the stressfield (b) the expected linearity of the pressure-volume curve, and (c) the potential to minimize equipment problems.

After calibration the cell was inserted to the desired depth in the hole. Generally the best practice is to start at the bottom of the hole and proceed towards the collar. Testing was planned for depth increments of 0.5 m. The same procedure used in calibrating the cell was followed in the borehole tests. That is, the cell is pressurized to about 15 MPa (without taking readings). The pressure is reduced to 4-5 MPa and the first loading-unloading cycle performed. Second and sometimes third loading cycles were used.

Typical evaluation of CSM cell data

The first and second loading curves for the CSM cell located at a distance of 4.5 m along the hole No. 2 are shown in Figure 5a and 5b respectively.

For loading 1, the pressure-volume relationship (M_R) for the rock surrounding a borehole was determined to M_R=10.21 MPa/half turn and for loading 2 M_R was determined to M_R=19.00 MPa/half turn. The values for modulus of rigidity (G_R) are calculated using the equation

E 233

$$G_R = M_R \frac{\pi L r^2}{\gamma} \qquad (1)$$

where L = 16.5 cm (effective length of pressurization = 16.5 cm)

r = 1.905 cm (radius of the borehole)

γ = 0.1803 cm³/half turn (volume change per one half turn)

therefore

G_R (loading 1) = 10.65 GPa

G_R (loading 2) = 19.82 GPa

The Young's modulus is obtained from the modulus of rigidity through the eqn.

$$E_R = 2(1+\nu_R)G_R \qquad (2)$$

Although a Poisson's ratio of ν_R = 0.25 was assumed for all the calculations, the actual value appears to be closer to 0.30. The error introduced is about 4%.

Thus

E_R(loading 1) = 27 GPa

E_R(loading 2) = 50 GPa

These values differ somewhat from those given in Table II for June 10, 1980 and a distance of 4.5 m due to an averaged (as opposed to a single) CSM cell system stiffness being used.

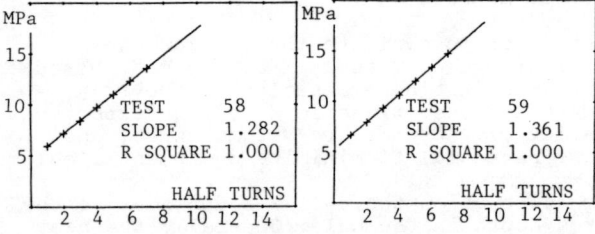

Figure 5a,b. Pressure-volume curves for the CSM cell in hole No. 2, slot drift, 494 m level. Distance along the hole = 4.5 m. June 10, 1980.

DISCUSSIONS OF THE RESULTS

Introduction

Before beginning a detailed examination of the modulii as obtained using the CSM cell, certain background information regarding (a) the elastic modulii for the Fabian ore body obtained in the laboratory and b) the expected stress state around the openings will be presented.

Laboratory determination of the elastic modulus

Wadood and Chitombo,1981,prepared cores from ore samples collected from the vicinity of hole 2. The average static elastic modulus is 84 GPa whereas the dynamic modulus is approximately twice this value (162 GPa). The average statically and dynamically determined values for Poisson's ratio are respectively 0.26 and 0.29. Average uniaxial compressive strength was determined to 249 MPa. Leijon,1981 as a part of a stress measurement program using the modified Leeman triaxial gage, obtained rock cylinders having an inner diameter of 38 mm, and an outer diameter of 71.5 mm. These cores together with their Leeman gages were placed in a cell capable of applying a radial pressure to the outer surface. Secant modulii were determined for the applied pressure range of 0 to 20 MPa. Although the range was large (40 - 127 GPa) the average value 79 and 73 GPa from the two holes (at 460 m level and 494 m level) respectively were similar to the values obtained from uniaxial compression tests. The values for Poisson's ratio 0.31 and 0.30 were also in good agreement.

Expected stress state around the openings

In situ stress field

Determination of the in situ stress have recently been made by Leijon,1981,on the 460 m, 494m, and 512 m levels in Fabian. An approximation of the stress field on the 494 m level made by the present authors considered to be reasonable for the purposes of this analysis is

σ_{Fy} = Horizontal field stress perpendicular to the strike of the orebody = 30 MPa

σ_{Fx} = Horizontal field stress parallel to the strike of the orebody = 15 MPa

σ_{Fz} = Vertical field stress = 15 MPa

These will be considered to be principal stresses.

Compressive stresses are considered positive.

Stresses at the end of the mined stope

In plan the mined stope is about 30 m wide and 180 m long at the 494 m level.

The stresses at the end of the mined out stope were studied using an elastic solution for an ellipse having an axis ratio of 4:1, Obert and Duvall, 1967.

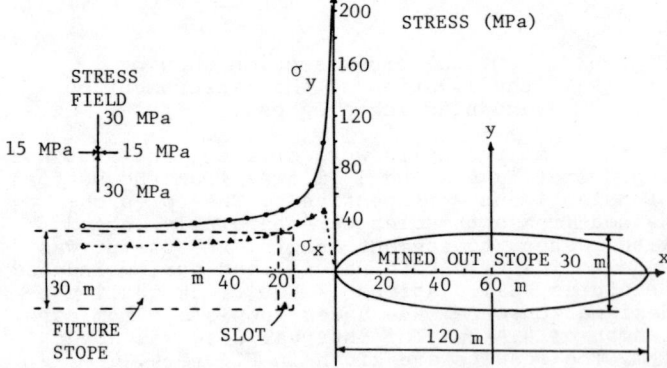

Figure 6. Diagrammatic plan representation of the mined out stope as an ellipse having an axis ratio of 4:1. Biaxial stress field imposed.

Stress in the floor of the slot drift.

An idealized section of the slot drift is shown in Figure 7. In the absence of the mined out stope, the vertical and horisontal stress in the plane of the cross section are $\sigma_{Fx} = \sigma_{Fz} = 15$ MPa

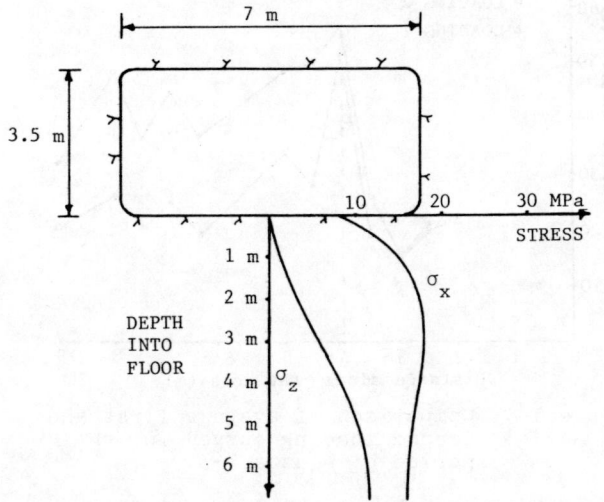

Figure 7. Stress distribution in the floor of the slot drift along the centerline.

Figure 7 provides an indication as to the potential stress distribution in the floor of the slot drift prior to being under the influence of the slot. It may be noticed that stress σ_x reaches a maximum at a depth of 3 m below the surface although it is relatively flat over the range from 2 to 4 m. The vertical stress σ_z increases almost linearly with depth below surface over the range of 0 to 5 m. The CSM-cell measurements were conducted at depths ranging from 1 to 4.5 m below the surface. The actual stress distribution will be somewhat different than that indicated due to the expected blast damage zone near the surface.

Stress changes in the floor of the slot drift due to the enlargement of the slot.

The slot is formed by enlarging the raise driven between levels at the ore-hangingwall contact. The final slot is about 6 m wide and 30 m long with its long axis perpendicular to the strike of the orebody. Two stages of slot formation have been represented in plan by the ellipses shown in Figure 8. The stress distribution along the major axis for these two ellipses are plotted in Figure 9. At the initiation of the CSM-cell measurement program, the situation is similar to that in Figure 9a whereas at the end of the program it is similar to Figure 9b.

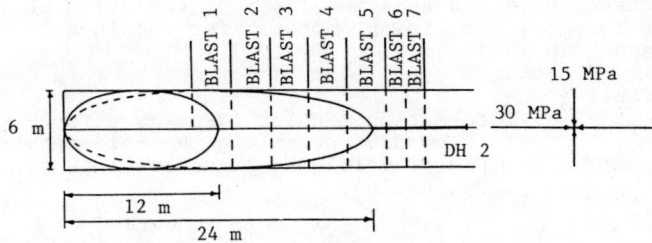

Figure 8. Approximation of two stages of the slot formation by ellipses having major axis ratios of 2:1 and 4:1. Plan view with the blasting sequence superimposed.

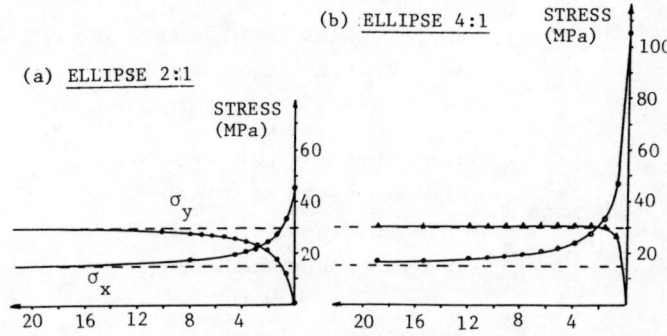

Figure 9 a,b. Stress distribution along the major axis for elliptical openings having a major to minor axis ratio of 2:1 and 4:1.

Conclusions regarding the influence of stress

The stress situation in the vicinity of drill hole 2 which is the site of the CSM-cell measurements is very complicated. However from the simple models which have been presented one would expect that

- For the slot drift alone, the stresses in the near vicinity of the floor (2 m depth below surface and less) are low. For depths between 2 and 4.5 m (the deepest extent of the CSM measurements), the σ_x stress is relatively constant and similar to the insitu stress. Over the depth range of 0 to 5 m the vertical stress σ_z increases linearly.

- With the extension of the slot from the raise, small stress changes are expected at the location of the CSM measurement hole for blasts 1 through 3. Major changes are expected between blasts 3 and 4.

E 235

CSM-cell testing in hole 2

The results of the CSM cell surveys performed in hole 2 are given as a function of distance along the hole (based upon the floor of the drift as reference) and mining geometry for loadings 1 and 2 in Table II.

Table II Deformation modulus (GPa) as a function of distance along the hole determined from loading 1 and loading 2 (in paranthesis).

Deformation modulus (GPa)

Distance (m)	Before blast	After blast 1	blast 2	blast 3	blast 4	blast 5	blast 6	blast 7
2.0			23(23)	(22)				
2.27	14							
2.5			21(21)	19(22)	17(17)	18(19)		14(15)
2.77	8							
3.0			14(15)	16(16)	16(18)	12(14)	15(15)	11(11)
3.5			19(27)	19(25)				
3.77	16							
4.0			48(42)	56(71)	31(45)	23(25)	27(24)	22(24)
4.27	20							
4.5		25	30(52)	29(48)	15(27)	21(36)	25(43)	26(27)
4.77	27							
5.0		34	37(47)	31(50)	32(69)	34(57)	76(80)	
5.27	17							
5.5					19)21)	21(21)	25(20)	
6.0					15(24)		26(24)	
6.27	15							
6.5			16(38)	20(33)	22(39)	13(28)	19(24)	30(27)
6.77	22							
7.0		25	24(40)	26(43)	24(33)	18(28)		
7.27	18							
7.5			17(33)	20(36)	25(40)	15(31)	26(28)	
7.77	20							
8.0			18(44)	30(51)	31(50)	29(41)	45(54)	
8.27	22							
8.5					24(44)			
8.77	19							
9.0			19(33)	24(36)	19(26)	29(37)		

The major variation in modulus with respect to hole position and mining geometry occurs in a region located between 4 and 5 m along the hole. This corresponds to a depth of 2 to 2.5 m below the floor of the drift. As can be seen from Figure 4, the ROD is the highest in this region and thus the higher modulus and greatest degree of change may be because of this. As has been discussed previously, this region is also one of higher expected stress and, as long as the strength is not exceeded, a higher modulus would be expected. As mining geometry in the near vicinity of the hole is changed one would expect a change in modulus.

A comparison of the average first and second loading curves for blasts 1 - 3 (during which a baseline or steady state case is assumed to exist) is shown in Figure 10. As can be seen, over the distance of 2 to 3 m along the hole (corresponding to a depth below surface of 1 to 1.5 m), the first and second loading modulii are essentially the same. This may be due to (a) blast damage and (b) the fact that the confining stress even under no blast damage conditions is low because of the presence of the drift. Although it is probably a combination of both, it is suggested that the primary effect is blast damage.

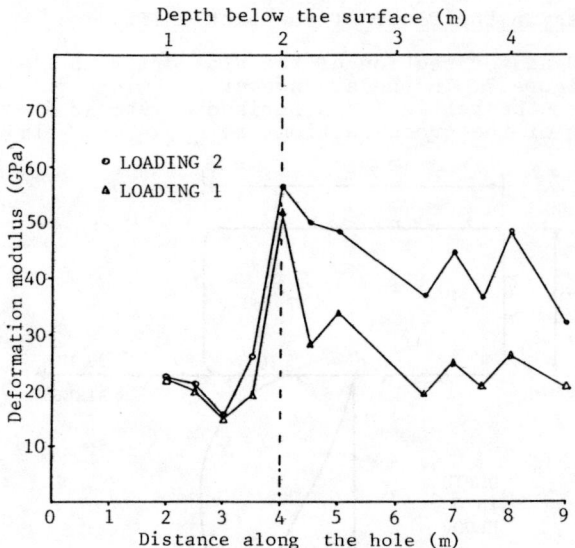

Figure 10. Comparison of average first and second loading curves for the period of blasts 1-3.

The rock is probably not able to sustain the high stress concentrations predicted based upon elastic theory (see Figure 9b). Rather the stresses are transferred further out away from the opening. Based upon the results of Figure 9b it is suggested that the high stresses produce the apparent high modulus zone (to a distance of 6.5 m or depth of 3.25 m). The insitu elastic modulus of the undisturbed ore under normal confinement would appear to be of the order of 35 GPa. The shape of the deformation modulus - depth below surface curve for loading 2 (Figure 10) is very similar to that expected for the σ_x stress.

As was indicated, a major stress change was expected to occur in the hole region between blasts 3 and 4. As can be seen from Table II, major changes in modulii occur in the zone of 4 to 8 m. The peak value shifted from a distance of 4 m to 5 m but the magnitude remained about the same (70 GPa). The modulii in the zone from 5.5 to 8 m were reduced by about 10 GPa. The modulus at 9 m increased.

The <u>general</u> trends for modulus behaviour during blasts 4 through 7 are:

Distance (m)	Modulus trend
4, 4.5	Decrease
5.0	Increase
5.5, 6.0	No change
6.5, 7.0	Decrease
7.5	Slight decrease
8.0, 8.5, 9.0	No change

Considered alone, one would expect blast damage to produce the greatest effect at the points closest to the blast. This is not demonstrated in this testing series. However, it is felt that one must consider the changes in the stress field as well. The greatest stress increases occur closest to the slot and the blasting. The combination of these two effects

- modulus decrease due to blast damage
- modulus increase due to increased stress

may in fact work to cancel each other.

Another way of evaluating the time at which blasting damage occurs (and to estimate its extent) is to examine the ratio R of the CSM cell determined modulii (E_D) for loadings 1 and 2

$$R = \frac{E_D \text{ (loading 2)}}{E_D \text{ (loading 1)}}$$

The results are presented in Table III. The point R where R decreases to about 1 is used as the criterion indicating significant blast damage.

Table III

Deformation modulus ratio as a function of distance along the hole and mining geometry.

Deformation modulus ratio (R)

Distance (m)	Before blast 1	After blast 2	blast 3	blast 4	blast 5	blast 6	blast 7
2.0		1.0	1.0				
2.5		1.0	1.2	1.0	1.1		1.1
3.0		1.1	1.0	1.1	1.2	1.0	1.0
3.5		1.4	1.3				
4.0		0.9	1.3	1.5	1.1	0.9	1.1
4.5		1.7	1.7	1.8	1.7	1.7	1.0
5.0		1.3	1.6		2.2	1.7	1.1
5.5				1.1		1.0	0.8
6.0				1.6			0.9
6.5	2.4	2.1	1.8	2.2		1.3	0.9
7.0		1.9	1.7	1.4		1.6	
7.5	1.9	1.8	1.6	2.1		1.1	
8.0	2.4	1.7	1.6	1.4		1.2	
8.5				1.8			
9.0	1.7	1.5	1.4	1.3			

Then one obtains the results presented in Table IV.

These results are far from conclusive but suggest that the damage zone is probably of the order of 0.5 to 1 m.

Table IV Estimate of extent of damage from large hole blasts.

Distance along hole (m)	Distance at which damage occurred (m)
2.0 - 3.0	Damage occurred during drifting. Distance 1 - 1.5 m from lifters.
3.5	-
4.0	3.0
4.5	-0.5 x)
5.0	0.5
5.5	4.5
6.0	1.5
6.5	1.5
7.0	-1.0 x)
7.5	0.5
8.0	0
8.5	0
9.0	-1.0 x)

x) The minus (-) sign indicates that the rock was still in place after the holes for that row were blasted. This is due to the fact that the holes were not loaded with explosive completely of the collar and an overhang developed for rounds 5 - 7.

STRESS MEASUREMENTS IN PILLAR

The stability of the pillar between the two stopes is crucial to successful sublevel stoping in the Fabian orebody. Therefore means of monitoring this stability were investigated. Due to limited access to the pillar it was found that the only practical way was to monitor stress changes in boreholes drilled from 548 m level upwards through the middle of the pillar, Figure 11.

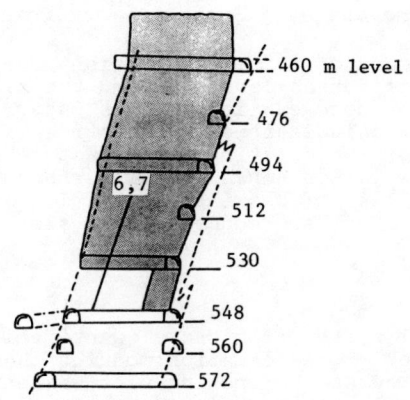

Figure 11. Side view showing IRAD gauge positions.

The IRAD gauge Vibrating Wire Stressmeter, described by Hawkes and Bailey, 1973, was chosen for the stress monitoring. The gauges yield information on stress changes only.

Originally, seven gauges were installed. They were installed before slot blasting was started but well after the already mined out stope had reached its final shape. Three more gauges were added after the slot was excavated.

Only the outputs from gauges 6 and 7 are considered in this paper since they were installed in the part of the pillar adjacent to the damage measurements at 494 m level, Figure 11. In Figure 12 gauge readings are plotted as a function of time. Recorded dates for slot blasts and sublevel stoping blasts are marked by short vertical lines.

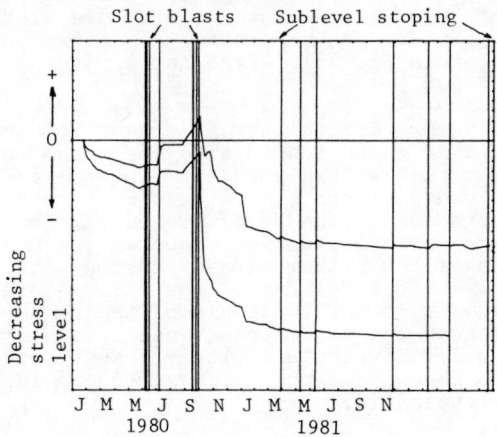

Figure 12. Readings for gauges 6 and 7.

As the slot is extended the readings increase for gauges 6 and 7. When the slot is excavated (Oct. 1980) the gauge readings start to decrease. The pillar has been exposed to stresses higher than its strength and they are now decreasing to a level where the remaining strength of the pillar can withstand them. When sublevel stoping is started after the CSM cell measurements was carried out the gauge readings show very small changes. A steady, but very slow, decrease can be observed and the readings increase a few units right after each blast, but no dramatic drops are occurring.

The consequence of these observations for the estimation of stress levels around the slot, that has been done earlier in this paper, is as follows: The calculations of the stress field at the end of the existing stope most likely give too low values since the pillar has a reduced capability of bearing load. This capability is even further reduced during slot blasting.

CONCLUSIONS

The CSM cell system provided a quantitative evaluation of the deformation modulus changes which occurred as the opening slot was extended toward the measurement hole. In some cases the modulus decreased with decreasing blasting distance as might be expected due to induced damage. However in other cases it remained constant or increased. The explanation offered is that the modulus of the Fabian ore is stress dependent. Thus as the slot extends toward the hole, the high stresses around the end of the slot tend to increase the modulus. The CSM cell system by itself cannot distinguish between these two opposing driving forces. An analysis of the expected stress fields and the modulus changes suggested a close connection between the two. Further work is required however before definitive conclusions can be reached. Great care must be exercised in evaluating blast damage zones when stress field as well as the rock are disturbed during excavation. The measurement suggested that a zone of reduced modulus extended to a depth of 1.5 m below the floor of the slot drift. This is probably due both to blast damage and lower confining stresses on the rock. Away from the opening the modulus increased. Both increased stress and less blast damage could be used to explain this. An evaluation of the extent of damage from the slot blasting based upon the ratio R of first and second loading modulii suggested a range of the order of 0.5 - 1 m.

The following conclusions are drawn from the review of IRAD gauge readings: - The gauges are sensing actual stress changes - The pillar was highly loaded before slot blasting was started - The stresses in the parts of the pillar, where the gauges were installed, exceeded the maximum strength of the pillar already during slot blasting - Total failure of the pillare did not occur and it is still capable of taking a reduced load - The estimated stress levels around the slot are likely to be lower than the real stresses because of the pillars reduced load bearing capability.

ACKNOWLEDGEMENT

The authors would like to express their appreciation to LKAB, Malmberget for providing the location for the research. The financial support for the project was through BeFo and LKAB.

REFERENCES

Hawkes, I. and Bailey, W.V. (1973). Design, Develop, Fabricate, Test and Demonstrate Low Cost Cylindrical Stress Gauges and Associated Components Capable of Measuring Change of Stress as a Function of Time in Underground Coal Mines: U.S. Bureau of Mines Contract, Report 4022050, USA.

Holmberg, R., Rustan, A., Naarttijärvi, T. and Mäki, K. (1980). Kratersprängning av stigort i LKABs gruva i Malmberget: SveDeFo Report DS 1980:12, Stockholm, Sweden.

Hustrulid, W., Holmberg, R. and Mäki, K. (1981). Damage Zone Adjacent to Large Hole Blasts at LKAB Malmberget Mine as Evaluated Using the CSM Cell: SveDeFo Report DS 1981:3, Stockholm, Sweden.

Hustrulid, A. and Hustrulid, W. (1972). Development of a Borehole Device to Measure the Modulus of Rigidity of Coal Measure Rocks: Final report on U.S.B.M. Contract HO 101705, July 6, USA.

Hustrulid, W. and Hustrulid, A. (1973). The CSM-cell - A Borehole Device for Determining the Deformation Modulus of Rock: 15th U.S. Symposium on Rock Mechanics, Rapid City, South Dakota, USA.

Leijon, B. (1981). Resultat från bergspänningsmätningar i Fabian: Report from Division of Rock Mechanics, University of Luleå, April 10, Luleå, Sweden.

Mäki, K. and Holmberg, R. (1980). Sprängskador i omgivande berg vid kratersprängning av stigort: SveDeFo Report DS 1980:8, Stockholm, Sweden.

Obert, L. and Duvall, W.I. (1967). Rock Mechanics and the Design of Structures in Rock: Wiley, New York, USA.

Olsson, M. (1979). Brytning av Fabian-malmkroppen: Internal Report, LKAB, Malmberget Sweden.

Wadood, A. and Chitombo, G. (1981). Elastic Property Determinations on Fabian Ore: Report, Coloardo School of Mines, Golden, Colorado, USA.

EFFECT OF PARTICLE SIZE AND STRAIN ON THE STRENGTH OF CRUSHED ROCK

Influence de la dimension et de la déformation des particules sur la résistance des roches broyées

Einfluss der Partikelgrösse und der Dehnung auf die Festigkeit zerkleinerten Gesteins

Mossaid Al-Hussaini
Associate Professor in Civil Engineering, University of Kuwait, Kuwait

SYNOPSIS

A number of consolidated drained triaxial compression and plane strain tests were performed on crushed basalt to study the influence of particle size and strain conditions on the strength and compressibility of the material. Results of the tests showed that the increase in particle size increases the strength and decreases the axial strain at failure and crushing of particles. The study also showed that crushed basalt, sheared under plane strain conditions, indicated higher strength than that tested under triaxial compression.

RESUME

Plusieurs essais de compression triaxiale et de déformation plane ont été effectués sur du basalt broyé, drainé et consolidé pour étudier l'influence des dimensions des particules et des conditions de déformation sur la résistance et la compressibilité du matériau. Les résultats des essais ont montré que l'augmentation des dimensions des particules augmente la résistance et diminue la déformation axiale à la rupture et à l'écrasement des particules. L'étude a montré aussi que le cisaillement du basalt broyé dans les conditions de déformation plane indique une résistance plus grande que celle obtenue dans un essai de compression triaxiale.

ZUSAMMENFASSUNG

Zur Beurteilung des Einflusses der Partikelgröße und der Dehnungen auf die Festigkeit und die Verdichtbarkeit wurden Versuchsreihen an zerkleinertem Basalt durchgeführt und zwei Prüfverfahren, der konsolidierte entwässerte dreiachsige Druck-, sowie der einachsige Scherversuch angewandt. Versuchsergebnisse zeigten, daß, je gröber die Körner der Mineralmischung, desto höher ihre Festigkeit und desto geringer ihre Scherverschiebungs- und Zerkleinerungswerte beim Bruch sind. Es ist erkennbar, daß die Scherfestigkeit des gebrochenen Basalts bei einachsiger Scherverschiebung höher ist als beim dreiachsigen Druckversuch.

1. INTRODUCTION

The problem of the effect of particle shape, particle size and confining pressure on the deformation characteristics of cohesionless rockfill material has been the subject of investigation in the past and will be investigated for many years to come. The reason for such interest stems from the experimental evidence which indicates that the strength and compressibility of granular and rockfill materials are greatly influenced by geometrical characteristics and density of particles, and confining pressure. As early as 1956, Holtz and Gibbs reported numerous drained triaxial compression tests on a mixture of sand and gravel from river deposits. Their test results indicated that the strength of the mixture increased with increasing gravel content until an optimum value is reached, then the shear strength either remained constant or decreased thereafter. The influence of particle size on the strength of granular material was studied by Vallerga et al., 1957. Using vacuum triaxial compression, they found that particle size has no effect on the angle of internal friction. Marchi, 1969, studied the strength and deformation of rockfill material using drained triaxial compression tests. The results of this study showed that the angle of internal friction showed a decrease with increasing particle size.

The effect of particle size on compressibility of granular and rockfill material also received a significant attention by researchers. Among the early studies concerning the compressibility of granular material, the one reported by Terzaghi and Peck, 1948, showed that the compressibility using triaxial compression tests increased with increasing particle size. Another study, conducted by Kolbuszewski and Fredrick, 1963, also showed that the compressibility of granular material increases with increasing particle size. Extensive isotropic and anisotropic triaxial compression tests were conducted by Lee and Farhoomand, 1967, to study the effect of particle size on the compressibility of granular material. The material used in the test varied from 6.5 mm to finer than

No. 200 U.S. sieve. Results of these tests indicated that coarse material compressed more than fine materials; also that uniformly graded material compressed and crushed more than well graded material of the same maximum particle size. The effect of particle size on compressibility of rock fill material was reported by several investigators; Fumagalli, 1969; Marchi, 1969; who showed that the axial strain and compressibility increased with increasing particle size of the rockfill materials.

From the previous studies it is apparent that while there is a general consensus that the increase in particle size increases the compressibility, there is a disagreement among investigators regarding the effect of particle size on the strength of granular and rockfill material.

The purpose of this study is to present results of test program on the effect of particle size, density, confining pressure and strain conditions on the strength and compressiblity of crushed basalt.

1.1. Material

The crushed basalt used in the test program was obtained from the basalt company, Blue Rock Quarry, Napa, California. The mineral composition of the crushed basalt, as determined by X-ray diffraction, consists of plagioclase, diopside, augite, and trace of montmorillonite clay minerals.

Different size fractions of the crushed basalt were combined to obtain straight line grain size distribution curves with maximum sizes ranging from 76.2, 50.8, 25.4, 12.7 and 6.35 mm, and minimum size of No. 30 U.S. standard sieve, to form uniformity coefficients of 11.6, 9.44, 6.67 4.68 and 3.31 respectively. Other physical properties as obtained by the U.S. Corps of Engineer procedure outlined in EM 1110-2-1906 are presented in Table I. The specific gravity of the material was 2.89 and the unconfined compressive strength of the intact rock was 172.2 MN/m^2

Table 1. Physical properties of crushed basalt

Limit of Particles Gradation mm	Coefficient of Uniformity	Void ratio	
		e_{max}	e_{min}
6.4 - 0.6	3.31	0.95	0.53
12.7 - 0.6	4.68	0.84	0.45
25.4 - 0.6	6.67	0.74	0.41
50.8 - 0.6	9.44	0.69	0.37
76.2 - 0.6	11.6	0.65	0.36

1.2 Testing Equipment

Two types of testing devices were used in this study. The first type consists of two sizes of conventional triaxial compression cells, while the second type is a plane strain device. The smallest triaxial cell can accomodate 154 mm diameter specimen used to test crushed basalt with maximum particle size up to 12.7 mm; the largest cell can accomodate 305 mm diameter specimens and used to test crushed basalt with maximum particle size up to 76 mm. All triaxial compression specimens tested had a height to diameter ratio of more than 2.

The second type of testing device used is a plane strain apparatus which can accomodate prismatic specimens 50.8 mm wide, 127 mm high, and 406 mm long, for testing the crushed basalt with maximum particle size of 6.3 mm. Detailed description of the plane strain apparatus is presented elsewhere; Al-Hussaini, 1981.

1.3 Preparation of specimens and test procedure

A total weight of dry material was first calculated according to the relative density desired for given test. The weight of different fractions of the crushed basalt, as dictated by the desired gradation, were measured and mixed thoroughly before saturating the material. The saturation procedure consists of flooding the material in a pan with water, boiling it for about 10 min. and then cooling it to room temperature. Water was allowed to flow into the specimen's forming mold from the base of the triaxial cell in order to free any air bubbles that might be trapped in the hydraulic system. The material was then transferred to the specimen mold in small quantities, and spread evenly under water in layers to insure uniform density. High relative density was achieved by vigorously vibrating each layer with a hand vibrator, while medium density was obtained by vibrating the side of specimen former when all material was placed in the mold. After the crushed basalt was placed in the specimen mold the upper platen was placed on the top of it and the rubber membrane was rolled around the platen and sealed. A vacuum of about 30 KN/m^2 was applied to make the specimen self supporting.

After the pressure chamber was assembled around the specimen, the pressure chamber was filled with silicon oil. The chamber pressure was raised slowly to replace the vacuum in the specimen. A back pressure of about 140 KN/m^2 was used to maintain a high degree of saturation. The confining pressure is gradually increased until the maximum confining pressure is reached and the specimen is consolidated. The specimen was then sheared under drained conditions at an axial deformation rate of 0.076 mm per minute.

2. TEST PROCEDURE AND DATA

All specimens were prepared at either a medium dense or dense state; they were consolidated isotropically to maximum consolidation pressures of 413, 861, 2067 and 3100 KN/m^2 and were sheared under drained conditions. A summary of the test results is presented in Table II for dense material and Table III for medium dense material.

The consolidation procedure used is to raise the cell pressure by small increments until the desired confining pressure is reached. Measurements of volume change, cell pressure and axial deformation at the end of each pressure increment were recorded.

The loading procedure during shear was to increase the axial load at a constant rate of

Table II. Triaxial compression test results for dense crushed basalt

Test No	START OF TEST		AT FAILURE			
	Max. Particle size mm	Relative Density %	σ_3 KN/m²	$\sigma_1 - \sigma_3$ KN/m²	ϵ_1 %	ϕ Deg.
1	6.3	95	413	2032	9.1	45.3
2	6.3	100	861	3479	13.8	42
3	6.3	96	2067	6359	18.5	37.3
4	6.3	98	3100	8840	20.1	36.0
5	12.7	96	413	2212	8.2	46.7
6	12.7	98	861	3734	11.6	43.2
7	12.7	100	2067	6856	15.8	38.6
8	12.7	100	3100	9480	17.5	37.2
9	25.4	100	413	2618	7.2	49.4
10	25.4	100	861	4306	10	45.6
11	25.4	95	2067	7613	13.6	40.4
12	25.4	95	3100	10011	15.5	38.1
13	50.8	98	413	2825	5.9	50.7
14	50.8	100	861	4547	8.2	46.5
15	50.8	97	2067	7979	12.3	41.2
16	50.8	99	3100	10645	14.8	39.2
17	76.2	99	413	2976	5.2	51.6
18	76.2	98	861	4844	7.0	47.6
19	76.2	97	2067	8454	12.1	42.2
20	76.2	100	3100	11375	14.3	40.1

Table III Triaxial compression tests results for medium dense crushed basalt

Test No	START OF TEST		AT FAILURE			
	Max. Particle size mm	Relative Density %	σ_3 KN/m²	$\sigma_1 - \sigma_3$ KN/m²	ϵ_1 %	ϕ Deg.
21	6.3	73	413	1660	12.5	41.9
22	6.3	73	861	2921	16.2	39
23	6.3	72	2067	6070	21.1	36.5
24	6.3	73	3100	8537	24.3	35.4
25	12.7	73	413	1895	11.7	44.1
26	12.7	73	861	3225	14.8	40.7
27	12.7	73	2067	6428	19.4	37.5
28	12.7	73	3100	9047	21.9	36.4
29	25.4	79	413	2184	9.2	46.5
30	25.4	79	861	3659	12.6	42.7
31	25.4	78	2067	7028	16.7	39
32	25.4	78	3100	9481	18.2	37.2
33	50.8	74	413	2287	7.5	47.2
34	50.8	76	861	3893	10.9	43.9
35	50.8	75	2067	7262	15.3	39.6
36	50.8	74	3100	9935	16.8	38.0
37	76.2	75	413	2522	7.0	48.8
38	76.2	76	861	4168	10.2	45.1
39	76.2	74	2067	7634	14.2	40.6
40	76.2	74	3100	10452	16.1	38.9

strain while keeping the confining pressure constant. The stress-strain and volume change relationships during shear were plotted for each specimen, but were not presented because of space limitations. The test was terminated when the axial load remained constant, or started to decline with increasing axial deformation.

3. RESULTS AND DISCUSSION

3.1. Drained compressive strength

The conventional Mohr-Coulomb criterion was used to calculate the angle of internal friction which is expressed as

$$\phi = \sin^{-1}(\sigma_1 - \sigma_3) / (\sigma_1 + \sigma_3) \quad (1)$$

The values of ϕ, as expressed in Equation (1) were plotted against confining pressure and the results are presented in Fig. 1 for medium dense specimens and Fig. 2 for dense specimens. Both figures show that the angle of internal friction increases with increasing particle size but the influence of particle size on ϕ tends to decrease with increasing confining pressure. It is clear from Figs. 1 and 2 that dense material shows a higher angle of internal friction than that for the medium dense material and the influence of density on ϕ decreases with increasing confining pressure for all particle sizes. The figures also show that the angle of internal friction decreases significantly with

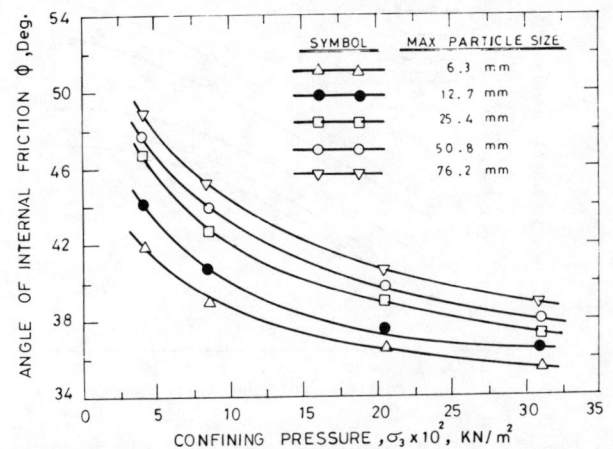

Fig. 1 Variation of ϕ with confining pressure for medium dense crushed basalt

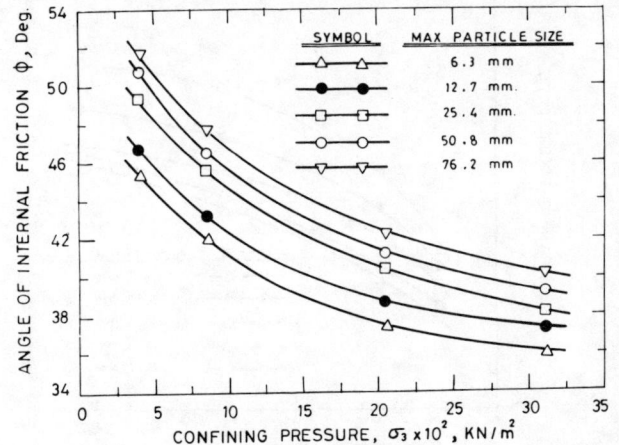

Fig. 2 Variation of ϕ with confining pressure for dense crushed basalt

increasing confining pressure regardless of maximum particle size.

3.2 Axial Strain at failure

The compressibility of crushed basalt is greatly influenced by the axial strain. In this study, the axial strain at failure $(\epsilon_1)_f$ is considered as a significant indicator of compressibility. Conventionally, the strain at failure is defined as the axial strain that the crushed basalt exhibits at maximum stress difference. The relationship between the failure strain and the consolidation pressure is presented in Fig. 3 for medium dense material and Fig. 4 for dense material. Both figures show that the failure strain decreases with increasing particle size and decreasing confining pressure. The figures also show that the failure strain decreases with increasing density of the material, however, the influence of density on the failure strain becomes less significant with increasing particle size.

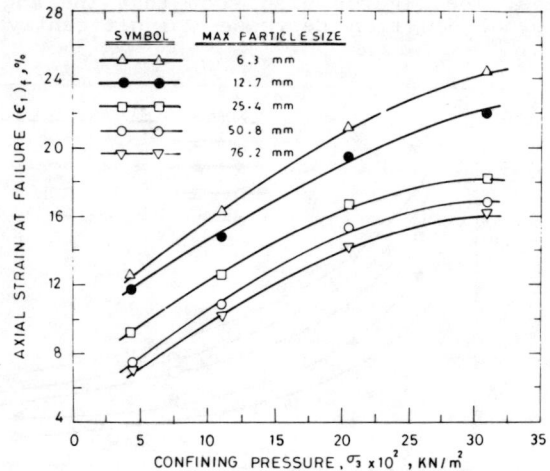

Fig. 3 Variation of failure strain with confining pressure for medium dense crushed basalt

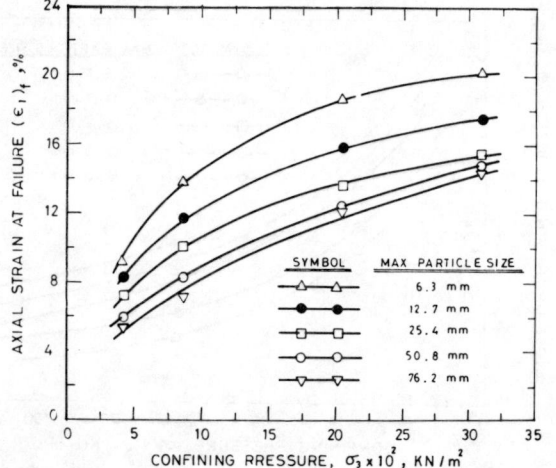

Fig. 4 Variation of failure strain with confining pressure for dense crushed basalt

4. INFLUENCE OF STRAIN CONDITION

Several consolidated plane strain tests were conducted on dense specimens of the same type of crushed basalt tested previously in the triaxial compression apparatus with maximum particle size of 6.3 mm. Results of the plane strain test series is presented in Table IV. Comparison between results presented in Table IV and those presented in Table II indicated that crushed basalt specimens sheared under plane strain condition exhibit higher angles of internal friction and less axial strain at failure than a comparable one tested under triaxial compression. This is significant since it implies that design based on triaxial compression test data for plane strain problem will always have an inherent margin of safety.

Table IV Plane strain test results for dense crushed basalt

START OF TEST		σ_3	σ_2	$\sigma_1 - \sigma_3$	ϵ_1	ϕ
Max. Particle size mm	Relative Density %	KN/m^2	KN/m^2	KN/m^2	%	Deg.
6.3	99	413	909	2749	8.1	50.1
6.3	98	861	1798	4458	12.2	46.2
6.3	98	2067	3528	8654	15.8	42.6

5. CRUSHING OF PARTICLES

Crushing of basalt grains occurs both during isotropic consolidation of the specimen and also during shear. However, test results show that crushing of particles during consolidation is very small compared to that which occurs during shear. Also that the major portion of particle crushing within the failure zone. Compression between grain size distribution curves before testing and after shear (not shown for space limitation) indicate that particle crushing increases with decreasing particle size and increasing confining pressure. The comparison also shows that the initial density has no significant effect on particle crushing although medium dense specimen tends to undergo more particle crushing than dense specimen. Particle crushing in this study is defined as the change in the effective diameter D_{10}, of the specimen before and after testing. The relationship between the changing in the effective diameter D_{10} with respect to the pressure during shear for specimens with maximum particle size of 76.2 mm and 6.3 mm is shown in Fig. 5. The figure clearly indicates that specimens with small particles exhibit more particle crushing than similar ones but with larger particle sizes.

6. CONCLUSIONS

Based on results of consolidated drained triaxial compression and plane strain tests conducted in this study, it is found that the strength of crushed basalt increases with increasing particle size and density and decreases with increasing confining pressure. Results of test also showed that specimens sheared under plane strain conditions showed significantly higher strength and less failure strain than the com-

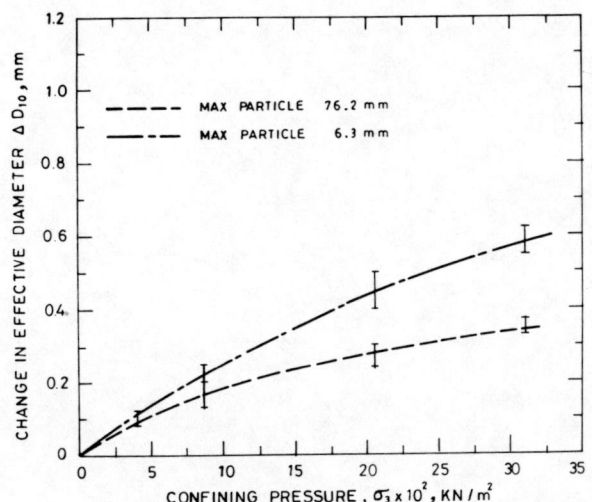

Fig. 5 Change in effective diameter during shear for crushed basalt

parable specimens tested under axially symmetric conditions. The study also showed that particle crushing decreases with increasing particle size of the crushed basalt tested.

7. ACKNOWLEDGEMENT

The test data presented herein were obtained from research conducted at the USAE Waterways Experiment Station, Vicksburg, Mississippi.

8. REFERENCES

Al-Hussaini, M. (1981). Deformation of crushed rock under plane strain condition, Int. Symposium on Weak Rock, Tokyo, Japan.

Fumagalli, E. (1969). Tests on cohesionless material for rockfill dams. Proc. JSMFD, ASCE, Vol. 95. No. SMI, pp. 313-329.

Holtz, W.G., and Gibbs, H.J. (1956). Triaxial shear tests on previous gravelly soils. Proc. JSMFD, ASCE, SMI, pp. 1-22.

Kolbuszewski, J., and Fredrick, M.R. (1963). The significance of particle shape and size on the mechanical behaviour of granular materials. European Conf. SMFE. Vol. 1, Wiesbaden, pp. 253-263.

Lee, K.L., and Farhoomand, I. (1967). Compressibility and crushing of granular soil in anisotropic triaxial compression. Canadian Geot. J., Vol. IV, No. 1, pp. 68-86.

Marchi, N.D. (1969). Strength and deformation characteristics of rockfill material. PhD thesis, University of California, Berkeley.

Terzaghi, K., and Peck, R. (1948). Soil mechanics in engineering practice, John Wiley, New York.

Vallegra, B.A., et al. (1957). Effect of shape size and surface roughness of aggregate particles on the strength of granular materail. ASTM, STP No. 212, pp. 63-74.

A NEW METHOD FOR PARTICLE SHAPE DETERMINATION
Une nouvelle technique pour la détermination de la forme des particules
Ein neues Verfahren zur Kornformbestimmung

Lineu Azuaga Ayres Da Silva, M.Eng. and Wildor Theodoro Hennies, Prof.Dr.Eng.
Mining Engineering Department, Politechnical School, The University of São Paulo, Brazil

SYNOPSIS

For a perfect geotechnical characterization of rocks and a better understanding of their behaviour, the shape of their natural constituting grains or the artificially induced fragments are very important and a new method for this determination is presented. An analysis of the existing particle shape determination methods shows that the majority of processes now used are very laborious and are associated with the operator's subjectivity if the instrumentation employed is simple, or, if more specific instruments are used, the resulting data are only of a qualitative nature. The simple, rapid and exact new method aims at eliminating the operator's subjectivity and furnishes data of a quantitative nature.

RESUME

Pour une parfaite définition géotechnique des roches et une meilleure compréhension de leur comportement il est très important de connaître la forme des particules naturelles ou des fragments produits artificiellement. Une nouvelle technique pour cette détermination est présentée ici. L'analyse des techniques de détermination existentes montre que les procédés utilisés sont très laborieux et associés à la subjectivité de l'opérateur si l'instrumentation employée est simple, ou, si les appareils sont plus spécifiques, le résultat est seulement d'une nature qualitative. Cette nouvelle technique simple, rapide et exacte cherche à éliminer la subjectivité de l'opérateur et à fournir des données d'ordre quantitatif.

ZUSAMMENFASSUNG

Zu einer sicheren Auswertung der geotechnischen Eigenschaften der Gesteine und einer besseren Kenntnis ihres Verhaltens ist die Form der natürlichen Körner oder der künstlich hergestellten Fragmente sehr wichtig, und ein neues Verfahren zu dieser Bestimmung wird hier beschrieben. Eine Untersuchung der aktuellen Verfahren zur Kornformbestimmung zeigt, daß der größte Teil dieser Verfahren sehr zeitraubend ist und sich der Subjektivität des Auswerters unterstellt, wenn die verwendeten Meßgeräte einfach sind. Wenn jedoch spezifischere Geräte verwendet werden, ist das Versuchsergebnis nur qualitativer Natur. Mit dem einfachen, schnellen und genauen neuen Verfahren beabsichtigt man, die Subjektivität des Auswerters auszuschalten und Angaben quantitativer Natur zu erzeugen.

1. INTRODUCTION

1.1. Statement of the object of the paper

Technological test must satisfy certain basic factors of which the most important are: precision, simplicity, reproductibility, facility of operation, elimination of operator's subjectivity and representativity of the sampled universe. As these enumerated factors are not independent the search for the right technological test for a specific determination is a relatively complex problem.

In searching for the right method for particle shape determination as rock fragments, the authors were led to develop a new method of technological test which is simple, rapid and accurate, eliminates the operator's subjectivity and furnishes data of a quantitative nature. The statement of the object of the paper is to describe this new method whose data is very important for rock behaviour not only in civil engineering projects but also in the mining industry.

1.2. The existing methods compared with the new method

A careful analysis of the existing particle shape determination methods shows that they can be considered from a qualitative or quantitative nature and also considering the procedure used to make this verification direct, semi-direct and indirect.

The methods of direct verification use measuring apparatus or mechanical devices or others, which, of any form, give directly the shape name.

The resulting data are generally of a qualitative nature and the particles or rock fragments are named rounded, irregular, angular, lamelar, alongated or alongated and lamelar.

In the semi-direct methods of verification, the measured data are processed for the determination of a volume number or designated as alongated, lamelar, cubic, alongated and lamelar respectively as quantitative or qualitative nature.

Finally, the indirect verification methods, which are all of a quantitative nature, are those in which a comparison is made of the measured data with mathematical models where values are know, tabulated or possible to calculate, defining the studied parameter as a numerical shape index.

In table 1, are presented some of the most used methods of particle shape determination. In this table, are presented the method number, country, granulometric range, verification process, nature scope, necessary equipment, shape classification and similar methods.

In the new proposed method the resulting indexes give the possibility to make a total shape classification and condense the data of qualitative and quantitative nature in only one nomogram

In the following sections we make some considerations about the used nomenclature, the mathematical model of the new method, the description of the apparatus, the principle of the operation, the operation and the resulting data.

2. NOMENCLATURE

As some terms of the used nomenclature are not conventional the following characterization is made:

rock fragment - granular material produced by rock comminution with natural or artificial processes.

rock - strictly, any naturally formed aggregate or mass of mineral matter, whether or not coherent, constituting an essential and appreciable part of the earth's crust.

shape of rock fragment - external aspect which rock presents with respect to its dimensions, faces, arests and vertices.

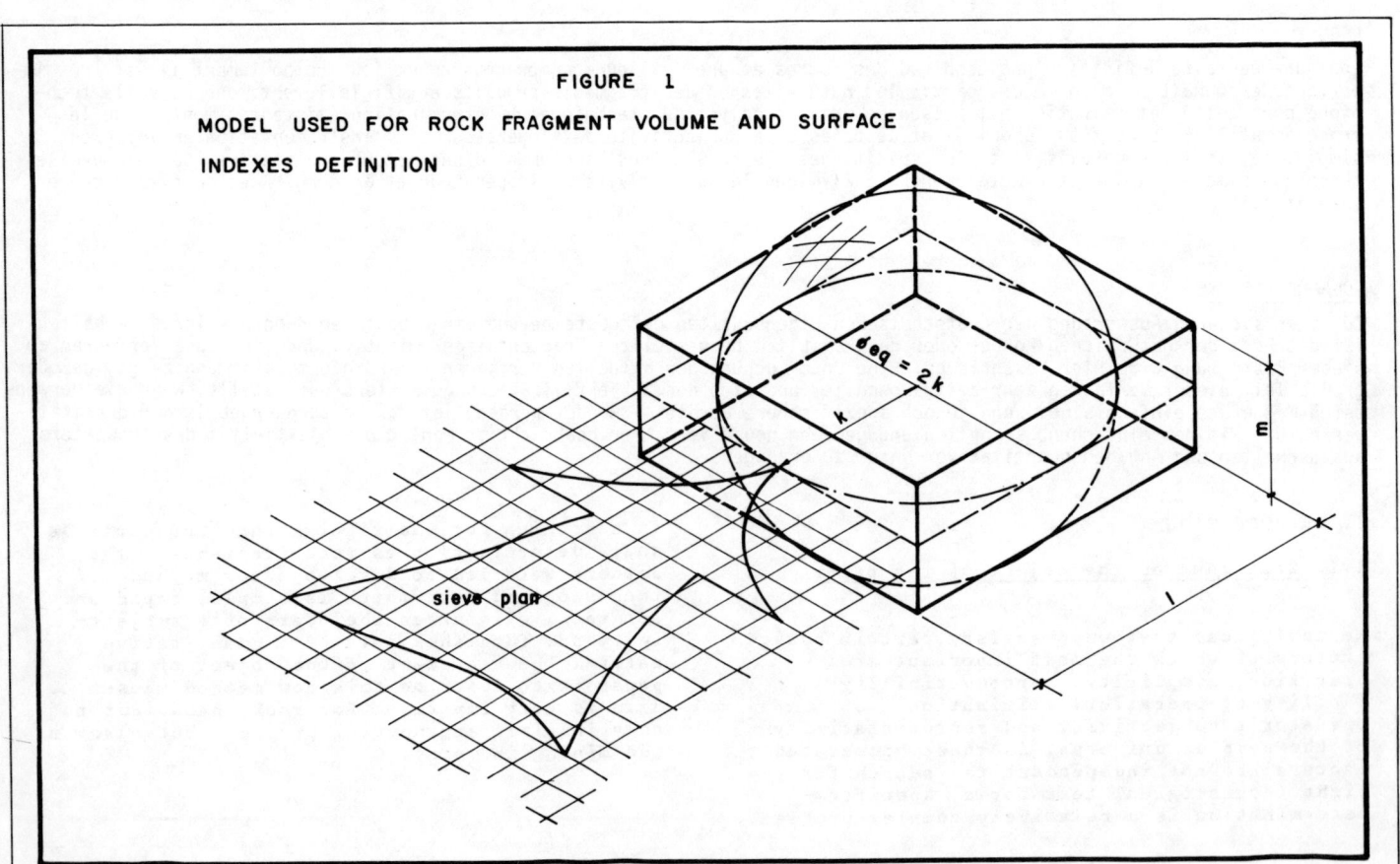

FIGURE 1
MODEL USED FOR ROCK FRAGMENT VOLUME AND SURFACE INDEXES DEFINITION

Method	CRD.C 119-53 (USA)	BS-812 : 1975 (ENGLAND)	Specification LNEC Method E.223-1968 (PORTUGAL)	DIN-DVN-1991 (GERMANY)
Granulometric range	2.54-0.95 (mm)	From Sample	50.8-4.76 (mm)	From Sample
Verification Nature	direct qualitative	direct qualitative	semi-direct qualitative	semi-direct qualitative
Scope	particle shape determination by measure of the three principal dimensions of rock fragments by mechanical device	particle shape determination by measure of the three principal dimensions of rock fragments by standard devices and tables	of a volume index definition by the relation between real volume of fragment and volume of spheres with same diameter as the lengths of rock fragments	definition of particle shape by the relation between the principal dimension of rock fragments
Equipment	-set of standard sieves -proportional calibrator -balance	-set of standard sieves -standard devices for alongment and lamerity determination	-set of standard sieves -paquimeter -graduated bottle	-paquimeter
Shape classification	-lamelar -alongate (and by exclusion) -cubic	-rounded -irregular -angular -lamelar -alongated-lamelar and alongated	-volume number	-alongated -lamelar -cubic -alongated lamelar
Similar Methods			French Norm AFNOR NFP 18-301	Brazilian Norm ABNT-MB 894 IPT and M-49

TABLE 1 — Principal methods of particle shape determination

Method	GOST-8269-64 (SOVIET UNION)	BERTHIER & CALTAUX (FRANCE)	IRAM 1681 (ARGENTINA)	Proposed Method (BRAZIL)
Granulometric range	Greater than 5 mm	From Sample	From Sample	50.8 - 4.76 mm
Verification Nature	semi-direct quantitative	indirect quantitative	indirect quantitative	indirect quantitative
Scope	percentual lamelarity index determination having as fundamental parameter thickness and particle length	shape coefficient determination from rock fragments measuring the thickness and volume with standard devices	shape index determination with standard device by measure of the rock fragments	particle spherical index determination which is related in the relation between surface and volume indexes of rock fragments
Equipment	-set of standard sieves -paquimeter	-set of standard sieves -parallel bar device -balance -counting machine	-set of standard sieves -balance -standards devices from circular and retangular openings	-set of standard sieves -special device (here described) -balance
Shape classification	-percentual lamelar index	numerical shape index	numerical shape index	numerical shape index
Similar Methods			similar to brazilian DER DPT M-86/64	The method can correlate all others of qualitative or quantitative nature

equivalent diameter - the diameter of a hypothetical sphere which passes through the sieve mesh of the screen through which passes the rock fragment.

correspondent sphere - the sphere which has the equivalent diameter of the rock fragment.

3. MATHEMATICAL MODEL

The developed mathematical model is based on the comparison of each rock fragment with a sphere of equivalent diameter named the correspondent sphere.

For this we suppose that the rock fragment is an ortogonal prism with a regular basal section, that has a side of dimension 1 and has n sides and a high m. These rock fragments pass through a sieve mesh whose dimension is 2K. In this case the correspondent sphere has an equivalent diameter of 2K (fig. 1).

The prism volume (rock fragment) is Vr and the correspondent sphere volume Ve. The relation Ve/Vr is named the volume index Iv.

The prism surface is Sr and of the surface of the correspondent sphere is Se. The relation of Se/Sr is named the surface index Is.

Now, if we take every prism and make a variation of the dimension m, the equivalent diameter of the sphere change for $2K \gtreqless m$ and is a function of the number of sides from the basal section.

If m is greater than 2K, the equivalent diameter is constant and equal to 2K for every prism.

As these indexes where defined by comparison rock fragment/sphere one by one, we can generalize this for comparison between identical granulometric distributions or rock fragments and corresponding spheres and:

$$Iv(n) = \frac{\Sigma Ve}{\Sigma Vr} \quad (1)$$

$$Is(n) = \frac{\Sigma Se}{\Sigma Sr} \quad (2)$$

for any granulometric fraction

From these models, if we take n, l and m for the prism of figure 1, we can make a tabulation of the indexes Iv and Is which led us to the following conclusions:

1. $Iv = Is = 1$ (3)

The rock fragment or particle is spherical if

$Iv(n) = Is(n) = 1$ (4)

The granulometric distribution is a set of spheres

2. for $m \lesseqgtr 2K$, when n tends to a very great number

$$Iv = 1,333 K/m \quad (5)$$

$$Is = \frac{4\pi K}{2\pi K + 2\pi m} \quad (6)$$

and if we make

2.1.
$$\lim_{m \to 0} Iv = \infty$$
$$\lim_{m \to 0} Is = 2$$

if we make

2.2.
$$\lim_{m \to \infty} Iv = 0$$
$$\lim_{m \to \infty} Is = 0$$

From these data we can construct the "nomogram of figure 2" which correlates the indexes Is, Iv and Ie.

The spherical index Ie, is a function of the other two defined indexes and gives a quantitative number of the shape so we can establish that:

$$Ie = f(Iv, Is) \quad (7)$$

which characterizes the quantitative nature of shape determination.

Additionally this nomogram presents regions in which the rock fragments shapes are named as spherical, cubic, quadratic, alongated and lamelar characteristic of the classifications of qualitative nature.

In conclusion if we know two of the three indexes Ie, Is and Iv the third can be determined. In our method the determination of Is and Iv gives a point in the nomogram which characterizes Ie as a quantitative number, and also, a qualitative designation of the shape of rock fragments.

4. DESCRIPTION OF THE SPECIAL DEVICE

The apparatus for the determination of the above indexes Is and Iv, consists of a glass bottle protected by metallic parts for its rigidity (see fig. 3).

A metallic cover permits a sufficient stoppage of the apparatus so that they can be manipulated in every direction without pouring out the liquid. This apparatus can be divided into three parts, which are:

1. The principal part of the apparatus has a cylindrical shape with an upper conical termination. At the bottom of the cylindrical part between this and the conical termination there is a sieve mesh which permits the passing of the water but not of the rock fragments. At the side of the cylindrical part there is a graduated translucid scale.

2. The next part of the apparatus is the bottle neck which connects the upper conical termination to the bottle. This is basically a graduated tube of a sensibility of 0.705ml.

3. The third part is a semispheric bottle and serves as liquid deposit, where the water is put.

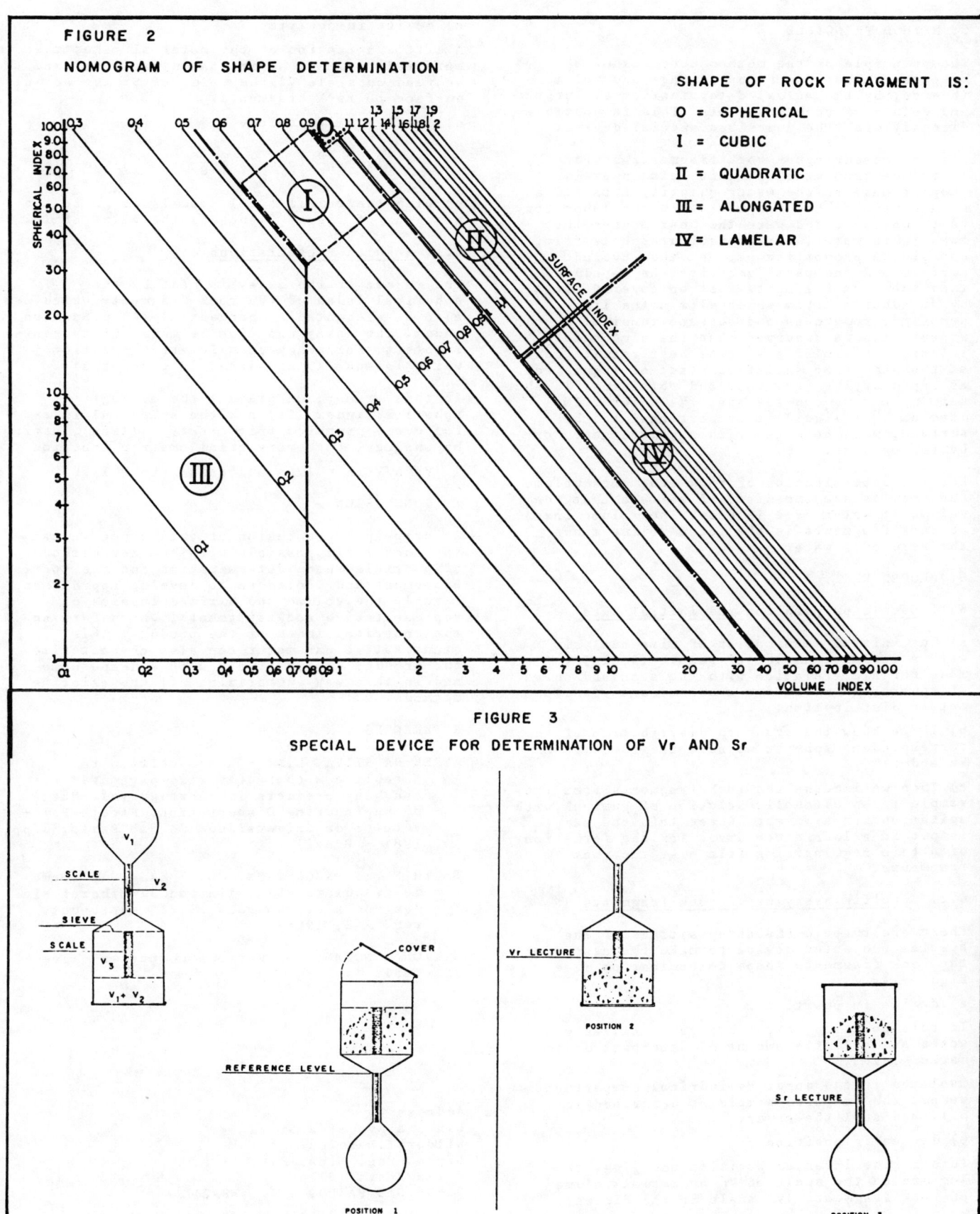

5. METHOD PRINCIPLE

The principle of the method consists of a precise evaluation of the surface and volume indexes, by the actual determination of surface and volume of rock fragments. This is possible directly with the described special device.

For the measuring of rock fragments surface index, we can establish that this index is proportional to the water quantity attached as a thin film to the fragments. In the laboratory tests we conducted were the best manner that the liquid water, that constitutes this film, can give a proportion between the covered surface and the water quantity. We concluded that the rock fragments must be covered first by an adherent film which eliminates its porosity, roughness and surface tension. In our experiments we observed that the alcoholic solution of gumlac gives the best results and at the same time satisfies a set of requisites as rapid drying, easy use and obtaining. In the same manner we can see that the water film used as best liquid is proportional to the surface, must be mixed with any detergent for better results.

For the determination of rock fragments volume the results are immediate if we take that the volume of water used is know. The volume index of rock fragments is the total volume minus the volume of water.

6. PROCEDURE

6.1. Sample preparation and initial data

a) The initial examination of size analysis with a definite amount of sample (5.0kg) is made for a correlation with the standard curve of spheres in which are a coincident granulometric distribution.

b) If we know the standard distribution of the correspondent spheres we note their values of Ve and Se.

c) Then we immerse the rock fragments from the sample in an alcoholic solution of gumlack, with agitation for a moment. After this the sample is put in a laboratory dryer for the formation of a thin regularizing film over the rock fragments.

6.2. Specific assaying of rock fragments

There are three different positions of the special projected device to make the assay for the rock fragments shape determination (see figure 3).

a) device in position 1

In this position, first bottle is filled with water and a little amount of detergent up to a marked level.

Isolated in the upper cylindrical compartment we put the prepared sample as described in 6.1. and seal the cover.

b) device in position 2

This is the inverted position and gives the lecture on the scale of Vr or actual volume of rock fragments (V3 scale in the figure).

c) device in position 3

A little agitation of the material makes the material assume a horizontal upper level and we read on scale V2 the value of Sr the actual surface of rock fragments.

d) Now we can calculate

$$Iv = \frac{Ve}{Vr} \quad \text{and} \quad Is = \frac{Se}{Sr}$$

e) with the calculated Iv and Is we go to the nomogram of figure 2 and has the value of Ie.

6.3. Results Presentations

The result of the assay is made by the spherical index of the rock fragments which can also be presented as percentual (%) of a mean index. The assay can also be made, if there is interest, for every granulometric fraction which is equally presented in percentual.

In this manner, we present the average spherical index (Ie) and the spherical index for every granulometric fraction (Iefn), having by analogy, for every granulometric fraction

$$Vefn/Vrfn = Ivfn \quad \text{and} \quad Sefn/Srfn = Isfn$$

7. CONCLUSION

As principal conclusion of this paper we can say that it is possible with the new method of particle shape determination and the special equipment and procedure to develop rapidly and exactly the volume and surface indexes of representative rock fragments, which furnish the spherical index of the product. This quantitative parameter can also characterize the product qualitatively with a shape name and in this way summarizing all the existing methods.

REFERENCES

AYRES DA SILVA, L.A. - A proposition to establish a characteristic parameter for crushing products (in portuguese). Master of Engineering Dissertation, Escola Politécnica da Universidade de São Paulo, 67pp, 1981 - Brazil

BERTHIER, J. & CALTAUX, CL. - Essai de forme de granulats. Bul. liaison du Laboratorie des Ponts et chausses nº 21, sept.-oct. ref. 342, 1966

NATIONAL NORMS - of various countries - see table 1.

Address:
Lineu A. Ayres da Silva
Wildor T. Hennies
Depto. Eng. Minas E.P.U.S.P.
Caixa Postal 8174
05508 SÃO PAULO, S.P., BRAZIL

ROCK MECHANICS APPLIED TO THE REGION NEAR A WELLBORE
Mécanique des roches appliqué à proximité d'un puits
Gesteinsmechanik, angewandt auf die Bohrstellenumgebung

Rolf K. Bratli,
Per Horsrud
Scientists, Continental Shelf Institute, Tr.heim, Norway

Rasmus Risnes
Sen.Petr.Eng. Norsk Agip A/S, Sandnes, Norway

SYNOPSIS

Analysis of stresses around a sand arch shows that a failure criterion exists. When a critical flow rate is reached the arch will collapse, leaving behind a greater cavity. Field test data can consistently be described by this theory. Extending the stress analysis to cylindrical wellbores shows that a plastically strained zone develops in poorly cemented rocks, when the well is drilled. During production this zone increases with the flow rate until the entire layer is fluidized. When injecting, fracture conditions may be reached before the material returns to an elastic state of stress. The stress analysis can be used to estimate the strength of the rock near the wellbore.

RESUME

L'analyse des contraintes autour d'une arche de sable montre l'existence d'une condition de stabilité critique. Quand le débit de fluide atteint un niveau maximum, l'arche s'écroule, provoquant une plus grande cavité. Cette théorie est vérifiée par des résultats d'essais effectués sur un puits réel. L'extension de cette analyse à un trou cylindrique montre que l'on aura une zone plastiquement déformé autour du puits quand on fore dans une formation mal cimentée. En production cette zone croît avec le débit jusqu'à ce que la formation entière s'écoule comme un fluide. En injection les conditions de fracturation peuvent être atteintes avant que la formation ne soit revenue à l'état élastique. Cette théorie peut aussi être utilisée pour obtenir indirectement une valeur de la résistance de la roche autour du puits.

ZUSAMMENFASSUNG

Die Analyse von Spannungen um ein Gewölbe aus Sand zeigt, daß ein Bruchkriterium existiert. Wenn eine kritische Strömungsrate erreicht ist, bricht das Gewölbe und hinterläßt einen größeren Hohlraum. Feldtestdaten können mit dieser Theorie widerspruchsfrei beschrieben werden. Wendet man die Spannungsanalyse auf zylindrische Bohrlöcher an, dann zeigt sich, daß in schlecht zementiertem Gestein eine plastisch deformierte Zone entsteht, wenn der Brunnen gebohrt wird. Während der Produktionsphase erweitert sich diese Zone entsprechend der Strömungsrate, bis die gesamte Schicht verflüssigt ist. Bei Injektion können die Bruchbedingungen erreicht werden, bevor das Material wieder zu einem elastischen Spannungszustand zurückgekehrt ist. Die Spannungsanalyse kann genutzt werden, um die Festigkeit des Gesteins in Bohrlochnähe abzuschätzen.

INTRODUCTION

The purpose of drilling a well in petroleum exploration is to locate and produce hydrocarbons on a commercial basis. Before the drill penetrates the downhole strata, there will be horizontal and vertical stresses, caused by the weight of the overlying strata and tectonic activity. As the drilling bit makes the hole, the vertical and horizontal stresses will be changed around the wellbore. This paper describes stress analyses performed on a poorly consolidated layer of rock around a well. The main purpose of this work, when it started several years ago, was to investigate the sand problem occurring in poorly consolidated sandstones when oil and gas is produced. Recently, a more general stability analysis has been made and this analysis has also served as a basis for the investigation of hydraulic fracture initiation pressures. The sand problem is normally experienced as sand influx from the formation into the production string. Flow of formation fines together with the produced fluids is often experienced during normal production. Such particle flow will not be a serious problem because the particles are small and small in quantity. Particle flow starts to be a real problem only when load bearing grains are removed from the formation, which will result in a reduction in the load carrying capacity and a potential failure of the formation. Production usually takes place through perforations which is shot through a protective casing. The analysis of the sand problem was therefore started by considering the stability of the sand behind perforation openings.

Formation of sand arches behind perforation openings is a mechanism that can stabilize a poorly consolidated sand and prevent it from flowing into the well.
Sand stability by arching was first treated by Terzaghi (1936) in his trap door experiment, demonstrating that arching was a real and stable phenomenon. Later Hall and Harrisberger (1970) made an experimental study on arches in relation to maximum sand free production rates. Stein et al. (1974) and Stein (1977) assumed that the maximum flow rate an arch can withstand is proportional to the shear modulus G for the sand. The G-modulus they obtained from accoustic and density logging data. Tippie and Kohlhaas (1974) and Cleary et al. (1979) made laboratory studies of the arching phenomenon and its relation to flow rate and confining stress levels. Tixier et al. (1974) treated the prediction of sanding from an interpretation of the mechanical properties log. Coates and Denoo (1981) introduced a refinement of the method presented by Tixier et al. Typical for these works is that they are based on empirical relations derived from experimental data and field operations.

A theoretical study of the stresses around a sand arch, supported by laboratory work, has been made by Bratli and Risnes (1981). The laboratory investigation showed there existed a limit to the load imposed by the fluid drag forces that a given arch can susstain. In the theoretical study a stability criterion relating the critical flow rate to the strength parameters of the sand was derived. Therefore, the corresponding sand control method essentially consists in keeping the production rates lower than a critical flow rate that will cause continuous sand influx. Knowledge of this critical flow rate is the key element in successful application of this method.

The theoretical stress analysis was extended to cylindrical wellbores in poorly consolidated sandstones, Risnes et.al (1982). The main conclusion from this work was that a plasticly strained zone develops around the wellbore in poorly consolidated rocks. Also in this case a fundamental stability criterion was found to exist when fluid flowed radially into the well.

Fracture initiation pressures in poorly consolidated sands were studied by simply reversing the fluid flow and letting fluid flow radially into the formation, Horsrud et al. (1982). The stresses around the wellbore could be followed as a function of increasing wellbore pressure until finally the fracture initiation pressure was reached. The analysis showed that, dependent on the rock properties, the initially plastic zone would either increase or decrease as the injection pressure was increased. In both cases fractures will, however, be initiated in a material which has been plastically strained. This analysis also showed that fracture initiation pressures calculated by this model can be significantly lower than values predicted by standard elastic theory, again this depends on the rock properties.

LABORATORY WORK ON ARCHING

The laboratory model is shown in fig. 1. The set-up consisted of a steel cylinder with a central hole in the bottom to simulate a perforation. This cylinder was filled with sand and compressed vertically by a piston to simulate overburden.

In this set-up the flow rate q was increased in steps until a small amount of sand was produced through the simulated perforation in the bottom of the cell. The flow rate then could be further increased until a new amount of sand was produced. The sand expulsion repeated itself several (5 to 10) times before the sand pack was fluidized and started to flow out of the container as a heavy liquid. The sequences are shown in fig. 1. There are thus registrated two modes of failure: a total failure and failure of thin inner shells.

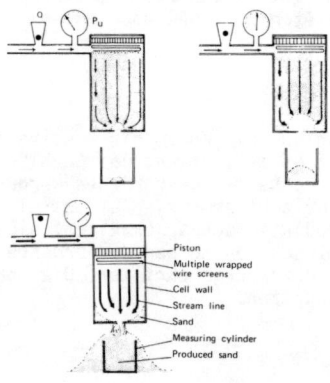

Fig. 1 Arching set-up and experimental sequence.

Some of the experiments were stopped before total collapse, and gypsum was injected to get an impression of the cavities formed behind the perforation opening. Examples of some of the casts are shown in fig. 2.

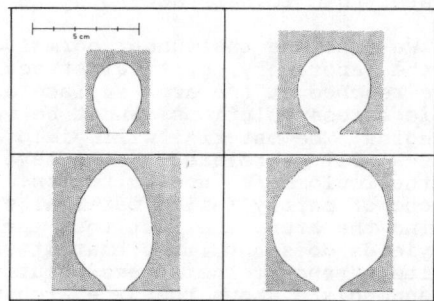

Fig. 2   Examples of cavities formed behind perforation.

It can be seen that the cavity grows behind the opening both in width and depth. The flow lines in the sand will converge towards the upper part of the cavity, while there will be no fluid flow in the sand around the lower part of the cavity.

The cavity may be regarded as consisting of two parts. The upper part, which is subject to the fluid drag forces, forms an approximately hemispherical arch, while the lower part acts as a conduit to the opening. Because the volumes of the upper and the lower parts are approximately equal, the cavity size (arch radius, r_1) can be estimated from the weight of the produced sand by assuming a spherical cavity model.

Typical arch radii from the experimental data were in the order of magnitude of 2 to 3 cm.

In the experiments the sand pack was sedimented under water and afterwards drained to irreducible water content by using a small air pressure. The cohesive strenght of the sand pack thus was given by the binding forces from water bridges between the grains. The cohesive strenght c can be determined by extrapolation of the tangent to the Mohr circle to zero stress, assuming the internal friction angle to be equal with the one of a dry sand. This is shown in fig. 3. For rock poorly cemented or not, a reduction in the cohesive strength will take place when the rock starts to fail. This is mainly due to breaking of cementation bonds. With only water bridges giving the strenght, the effect on c will be small.

STABILITY THEORY

Introduction

This stability analysis contains an investigation of the stability of the sand behind a perforation opening as well as an analysis of the stresses around an open hole. During both these analyses the rock is assumed isotropic and homogenous, obeying the linear Coulomb failure criterion, and the pores are completely filled with fluid. The total stress concept is used, with contributions both from the forces in the solid material and the pressure in the pore fluid. Compressive stresses and strains are considered positive. Deformations are considered to be small, and plain strain is assumed.

Arch stability

An idealized theoretical model for the experiments described in the previous chapter is a spherical arch. This arch is situated in an isotropic and homogenous material subjected to hydrostatic stresses at great distance. The model is shown in fig. 4. This model assumes spherical symmetry with σ_θ and σ_r as the greatest respective smallest principal stress using polar coordinates. The vertical stress component is excluded from this analysis, which reduces it to a two-dimensional problem.

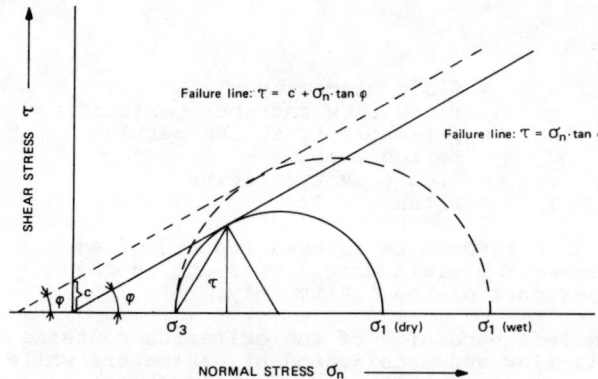

Fig. 3   Mohr-Coulomb plot with illustration of the method used to estimate the cohesive strength c.

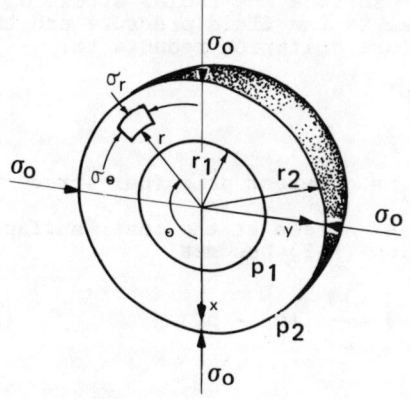

Fig. 4   Arch model.

The elastic stress solutions for this model is shown in fig. 5. The elastic stress equations is given in appendix A and the development of these equations can be found in Bratli and Risnes (1981).

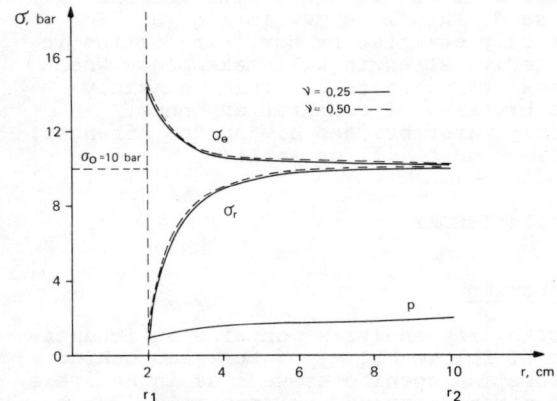

Fig. 5 Elastic stress solutions.

From figure 5 it turns out that the stress solutions are almost independent of the Poisson's ratio ν.
The greatest difference between the stresses is at the inner surface, so it is here the material eventually would fail. The linear Coulomb criterion is used to describe plastic flow of the sand. This is given as:

$$(\sigma_1 - p) = \sigma_c + (\sigma_3 - p)\tan^2 \alpha \qquad (1)$$

where:

- σ_1 = greatest principal stress
- σ_3 = smallest principal stress
- σ_c = uniaxial compressive strength = $2c\tan\alpha$
- c = cohesive strength
- α = failure angle = $\pi/4 + \dfrac{\phi}{2}$
- p = fluid pressure (given by Darcy's law)
- ϕ = internal friction angle

At the inner surface the radial stress σ_r will be equal to the fluid pressure and the Coulomb failure criterion reduces to:

$$(\sigma_\theta - p) = \sigma_c \qquad (2)$$

where:

$\sigma_\theta = \sigma_1$ = greatest principal stress

The tangential stress at the inner surface using equation (A-3) becomes:

$$(\sigma_\theta - p) = \frac{3}{2}(\sigma_o - p_1) \qquad (3)$$

where:

- σ_o = hydrostatic far field stress
- p_1 = fluid pressure at the arch surface

The cohesive strenght c for the sand in our laboratory set-up was less than 1 bar. The failure angle was typically around 65°. The calculated uniaxial compressive strength σ_c will thus have values less than 4 bar, while the left hand side of the criterion has a value greater than 10 bars according to fig. 5.
From this we conclude that under normal conditions in the laboratory, critical stress values are reached at the arch surface and the elastic stress solutions cannot be valid in this region. The material will yield and the stresses will be relaxed to the level given by the Coulomb failure criterion. A plastic zone of partly failed material will exist behind the arch. The fact that the material yields does not imply that it will lose all its strenght. The stress solutions given in appendix B shows that the arch will be stable. A schematic illustration of the stresses in the failed zone, combined with the elastic solutions is shown in fig. 6.

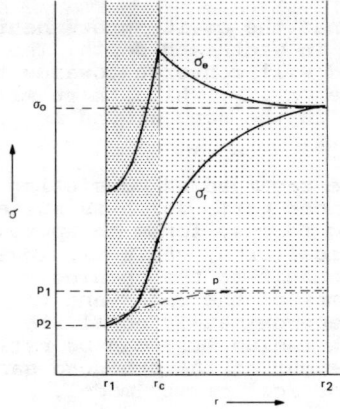

Fig. 6 Combined stress solutions

When fluid flow is introduced, the effect of the drag forces is to increase the depth of the failed zone. For a given critical flow rate the plastic zone will extend through the whole material, making combined stress solutions impossible. This leads to a fundamental stability criterion:

$$\frac{\mu q}{2\pi k r_1} = \frac{T+1}{T} \cdot 2\sigma_c \qquad (4)$$

where:

- μ = fluid viscocity
- q = fluid flow rate per perforation
- k = permeability in the partly failed zone
- r_1 = arch (cavity) radius
- T = $2(\tan \alpha - 1)$

In this formula 2π is used instead of 4π because the fluid drag forces act on the upper part of the cavity only.

The left hand side of the criterion contains only flow and arch dependent parameters while the right hand side depends only on the strenght of the material.

The fundamental stability criterion (eq. 4) predicts only total collapse. The observed behavior with a succession of shell collapses is caused by another mechanism. When the flow term on the left hand side in the criterion exceeds the level $2\sigma_c$ a region of tensile effective stress in the radial direction will be developed behind the arch. As the fluid pressure and the radial stresses are always equal at the arch surface, and as the fluid pressure will be smaller than the radial stress at some distance from the arch, the region of tensile stress will be limited, and the maximum value will be found a short distance behind the arch. This gives a possibility for internal fracturing causing a shell to collapse and a new arch to be established. This situation is illustrated in fig. 7.

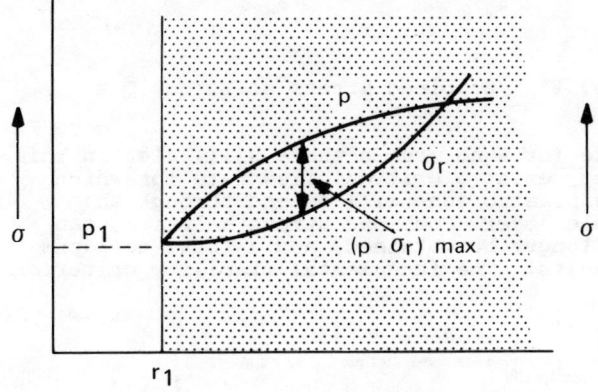

Fig. 7 Fracturing situation.

The condition for this mechanism to be possible is that:

$$\sigma_T < \frac{2\sigma_c}{T} \quad (5)$$

where:

σ_T = uniaxial tensile strength

For poorly consolidated sands this will normally be fulfilled.

The criterion for shell collapses then reads, Bratli & Risnes (1981):

$$\frac{\mu q}{2\pi k r_1} = \left(\frac{T+1}{T}\right) \cdot \frac{2\sigma_c}{1 + \frac{1}{T}\left(\frac{r_1}{r_s}\right)^{T+1}} \quad (6)$$

where:

r_s = radius of new arch at shell collapse

The strength parameter T will normally have values of five or more. The radius ratio (r_1/r_s) depends on the strength parameters and is approximately given by:

$$\frac{r_s}{r_1} = \frac{T+1}{T} \cdot \frac{1}{1 - \frac{T \cdot \sigma_T}{2\sigma_c}} \quad (7)$$

Equation (6) shows that the flow rate causing shell collapse will be only slightly below the value predicted by equation (4) for total collapse. Therefore, the fundamental stability criterion can be used also for estimating the critical values of flow rate causing shell collapses. The radius r_c of the failed zone, when shell collapses occur can be estimated from the formula:

$$\left(\frac{r_c}{r_1}\right)^T = \frac{3T}{T+3} \cdot \frac{(\sigma_o - p_1)\left[1 + T\left(\frac{r_s}{r_1}\right)^{T+1}\right]}{2\sigma_c} \quad (8)$$

The failure criteria also show that there can not be any transition between the two modes of failure as the flow dependent terms appear in the same way in both criteria. If thin shells start to fall in, this fracturing mechanism will continue. However, in all real systems there will be geometrical limitations, and when the failed zone interferes with these, conditions for total collapse may occur.

Wellbore stability

When a well is drilled, the material around the wellbore will lose radial support.

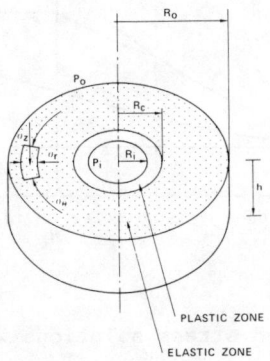

Fig. 8 Wellbore model.

The model for analysing this situation is a vertical cylindrical open hole through a horizontal layer of porous and permeable rock. Axial symmetry around the well axis is assumed. The configuration of the problem is shown in fig. 8, Risnes et al. (1982).

The material is subject to stresses in three dimensions, the principal stresses being radial, tangential and vertical, with the vertical stress parallel to the borehole axis. The vertical strain component will contain only deformation caused by initial overburden loading of the material, i.e. plain strain. The elastic stress solutions for this model are shown in fig. 9, and the stress solutions are given in appendix C.

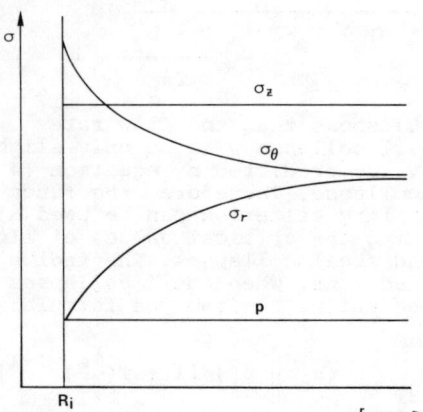

Fig. 9   Elastic stress solutions.

For poorly cemented materials failure conditions will normally be reached and a zone of critically stressed material will be formed around the well. Fig. 10 shows examples of the combined plastic/elastic stress solutions when there is no fluid flow. Appendix D gives the plastic solutions.

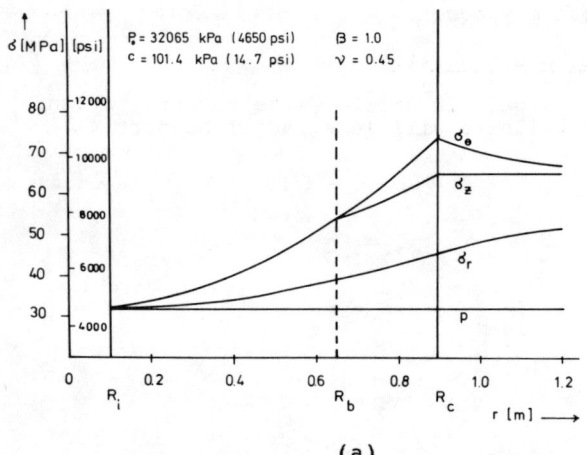

(a)

Fig. 10   Combined stress solutions with no fluid flow.
(a) $\nu = 0.45$ (b) $\nu = 0.3$

The depth of such a plastic zone may typically be in the order of magnitude of one meter. The depth of the failed zone seem also to be fairly independent of Poisson's ratio ν. When fluid is flowing into the well, this zone will be extended due to the fluid drag forces. In fig. 11 is shown the example from fig. 10a with a fluid flow of Q = 200 cc/s into the wellbore. This increases the plastic zone depth from approximately one meter to approximately three meters.

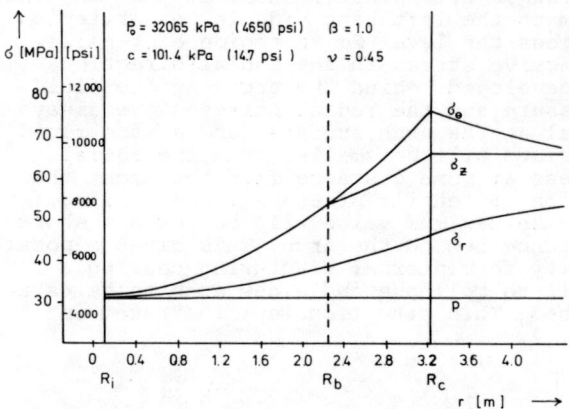

Fig. 11   Combined stress solutions Q = 200 cc/s.

Like for sand arches there will, also in this case, exist a critical flow rate for which the plastic zone will extend through the whole layer. Combined stress solutions can no longer be obtained, and a total collapse results. The fundamental stability criterion

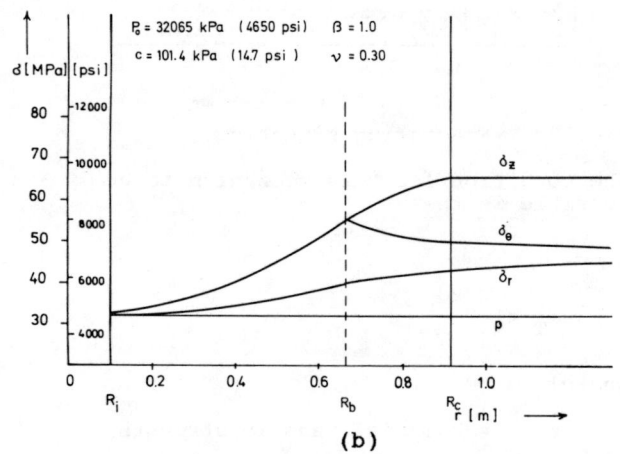

(b)

for an open hole is given as:

$$\frac{\mu Q}{2\pi h k} = \sigma_c \qquad (9)$$

where:

Q = fluid flow rate per well
h = height of producing layer

In a cased hole the formation will be supported by the cement and the casing. The critical flow rate that will be able to fluidize the whole layer will be considerably higher and is given as:

$$\frac{\mu Q}{2\pi h k} = \sigma_c + \frac{T}{2}(\sigma_{R_i} - P_i) \qquad (10)$$

where:

σ_{R_i} = radial stress from cement column at the wellbore wall
P_i = fluid pressure at the wellbore wall

For a cased well producing through perforations the sum of flow rates from each single perforation should be less than the critical flow rate given by equation (10). Normally this is fulfilled and the stability of the sand is governed by the stability of the arches. If, however, too much sand is removed around the well the total critical rate as given by equation (10) may be reduced towards the value predicted for an open hole, equation (9). When this is the case, the total collapse criterion, equation (9), may come into play even when we have shell collapses at the individual arches. Reasonable estimates for strenght, arch dimensions and perforation density show that the total critical rate given by eq. (9) may become comparable with the sum of critical flow rates for arch failure given by eq. (6).

Fracturing of wellbores

When a liquid is pumped into a borehole at high enough pressure, the stress at the borehole surface will become tensile. If this stress exceeds the tensile strength of the material, a fracture will occur. This technique, commonly known as hydraulic fracturing, is a method used in the oil industry to improve the productivity of a formation. However, for flooding projects it is essential to avoid initiating fractures. Determination of fracture initiation pressures is therefore important.

One way of estimating the fracture initiation pressures is to purposely initiate a fracture by increasing the wellbore pressure. This test, commonly called formation integrity test or leak-off test, is not always straight forward to interprete, especially in permeable formations.

Unintended fracturing around injection wells can have severe consequences to a flooding project. Such fractures can penetrate deep into the formation. Although this will lead to improved injectivity, it can jeopardize the whole injection project because of the reduction in sweep efficiency. Another result from such unintended fracturing could be influx of particles into the injectors at a system shut down. This could eventually plug the injection wells.

When killing a blow-out by creating a water bank around the well, it is essential that no fractures are created between the injection well and the blowing well. Such fractures would effectively short-circuit the flow between the wells, and make it impossible to block the blowing fluid by a water bank.

When studying fracture initiation pressures the same model as described under the previous paragraph and illustrated in fig. 8 was used. The starting point for our analysis was the stress solutions for no fluid flow shown in fig. 10. By increasing the fluid pressure in the well, fluid will flow radially into the formation. By keeping each fluid pressure increase small enough, steady state conditions could be ensured. As the injection pressure is increased, several different stress distributions will appear as illustrated in fig. 12. In this figure only the stress solution in the plastic zone is shown, and the plastic zone radius is kept constant for simplicity. Each stress distribution is composed of one or more different stress solutions linked together. This is solved by numerical iteration. The complete analysis is given in Horsrud et al. (1982).

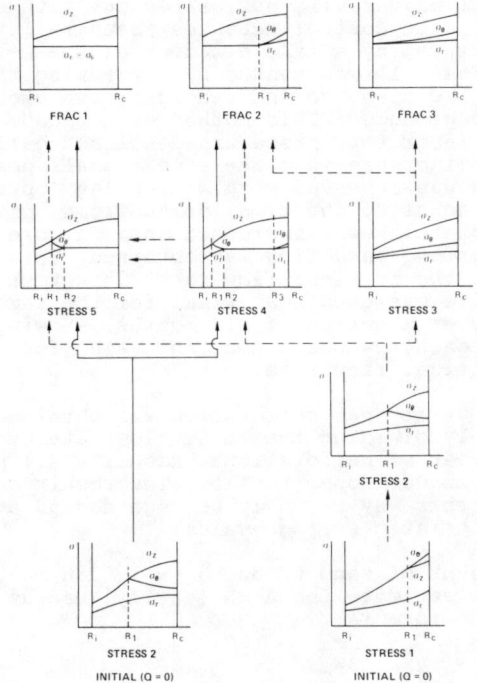

Fig. 12 Fracturing situation.

As injection pressure is increased the plastic zone developed when the well was drilled may increase or decrease, depending on the rock properties. But even if it decreases, and possibly disappears, the material cannot return to its original state. In a poorly consolidated sand fractures will therefore be initiated in a material that has been plastically strained.
When fracture initiation pressures are calculated using this model it can be signifi-

cantly lower than if calculated by elastic theory. The difference depends on the initial state of stress (stress ratio) and the rock properties (Poisson's ratio and uniaxial compressive strength) in the plastic zone. Typically with a Poisson's ratio of 0.2 and a horizontal to vertical effective stress ratio of 0.4, the difference is in the order of magnitude of 10 percent.

APPLICATION

Sand Control

The stability criterion for sand arches makes it possible to predict maximum allowable flow rates before sanding occurs. This theoretical result has been supported by experimental data, Bratli and Risnes (1981), and field data, Risnes et al. (1982).

In the latter work is shown an analysis of well test data in view of the stability theory for sand arches. The test refers to a well completed in poorly cemented sandstone. The well was heavily perforated but with no special sand control measures. Sand influx was measured by a sand detector at the wellhead. The well was tested by increasing the choke size stepwise and recording the amount of sand produced. This method corresponds to the laboratory procedure described earlier. At each increase in choke size a small peak of sand was observed with a time lag approximately equal to the time for "bottoms up". The choke size was increased until a more long lasting sand flow was obtained. This defined the critical flow rate. This test procedure was used four times for the same well, over a period of six months, showing an approach towards a constant value for the critical flow rate.

The flow rate per perforation was obtained by simply dividing the total flow rate by the number of perforations, assuming all perforations were open to flow. The result obtained this way may only be regarded as a characteristic average value.

The amount of sand produced (W_s), can be used to estimate the arch (cavity) radius from the formula:

$$r_1 = \left(\frac{3W_s}{4\pi(1-f)\rho_s n}\right)^{1/3} \quad (11)$$

where:

- r_1 = arch radius (cavity size)
- W_s = amount of sand produced
- f = fractional porosity; assumed 0.3
- ρ_s = density of sand; assumed 2.64 g/cc
- n = number of arches

The amount of sand produced can also be used to get an idea of how many arches (n) are formed, by comparing the arch radius calculated from eq (11) with the distance between the perforations. The studied well was heavily perforated with 8 shots per foot, 4 shots vertical in line and 4 shots in one plane. The distance between the vertical line shots is approximately 10 cm while the distance between the circumferential shots is approximately 20 cm.

Making the assumption that the sand is produced from specially weak zones, say 2 ft corresponding to 16 perforations, we obtain for the two last tests arch radii of about 5 cm. Compared with the distance between the perforations such large cavities would probably result in interference and instabilities. As this is not observed, we may reasonably assume that arches are formed behind many, maybe most, of the perforations. The fact that the critical rate does not change between the two last tests also suggests that the maximum number of arches is reached. On the other hand, if a typical sand peak is assumed to correspond to shell collapses at all the perforations, small values for the shell thickness are obtained. The most reasonable explanation of the single sand peaks therefore seems to be that they correspond to shell collapses and clean up at some of the perforations.

When the maximum flow rate is observed, the situation can be interpreted as collective shell collapses from many arches and the shell collapse criterion in eq (6) should apply. Since no details are available for the individual arches, eq. (6) can for estimates be replaced by the simpler fundamental stability criterion in eq. (4). When the field data were analysed we used the following typical parameters:

μ = 1.2 cp
B_o = 1.2 BBL/STB
α = 60°
k = 2.0 Darcy

The uniaxial compressive strength σ_c can then be calculated from the criterion, provided we assume a certain number of arches to determine the radius. For each test we have calculated the strenght as a function of the number of arches, and the results are given in Table 1.

Table 1.

CALCULATED COMPRESSIVE STRENGTH

TEST NO	1	2	3	4
NUMBER OF ARCHES n		σ_c (atm)		
16	5.7	4.4	3.6	3.4
32	7.1	5.5	4.5	3.9
64	9.0	7.0	5.9	4.9
128	11.4	8.8	7.1	6.2
176	12.7	9.8	8.0	6.9

From this table we can draw a number of conclusions. First, by the end of the 4th test we would assume all or nearly all perforations to be involved, i.e. n = 176. This gives a uniaxial compressive strength σ_c of 7.0 atm, which is a typical value for a poorly cemented sandstone. Thus the theory seems to apply without any special adjustment.

The strength of a sandstone would be assumed to be constant during a 6 months period. From Table 1 it can be seen that an assumption of a fixed number of arches will lead to contradictory results. There is no reason to believe the strength should have been reduced to approximately one half during this period.

From inspection of Table 1 it is obvious that constant values for σ_c are found along the diagonal from the lower right corner towards the upper left corner. This is consistent with the picture that in the first tests, arches are formed only at some of the perforations. By successive tests the number of arches increases until it reaches the maximum value.

The most important conclusion to be drawn from this case study is that the field data can consistently be described by the arch theory. This means that the arch theory can be used to predict maximum allowable flow rates in poorly cemented sandstones. These values can then be used by the production engineer as design criteria when he shall decide whether or not to use any sand control measures in the particular well.

To be able to use the sand stability theory in such a predictive manner, the in-situ strength of the rocks close to the wellbore should be known. Methods for determination of the mechanical parameters under in-situ conditions are therefore important.

Indirect methods for in-situ determination of strength

Many methods for direct measurement of in-situ strength have been investigated, but to date no single method has proved successful. It has therefore been decided to investigate the possibilities of using this theoretical stress analysis as an indirect measure of the in-situ strength for the material close to the wellbore.

From Risnes et al. (1982) it was found that a plastic zone developed around a well drilled through a poorly cemented sandstone. The depth of the plastic zone was mainly determined by the uniaxial compressive strength and the failure angle of the material near the wellbore. An indirect method for in-situ compressive strength measurement then could be to get a measurement of the plastic zone depth, and use this value to estimate the in-situ strength. The failure angle α has to be assumed if this method should apply. Normally α will vary between $55°$ and $65°$, indicating the limits which the strength must fall within.

To determine the depth of the plastic zone we are, together with Schlumberger Doll Research (SDR) investigating the use of a new Sonic Tool. A schematic configuration of such a tool is given in fig. 13.

The tool records a full wave train and by using standard refraction (time average method) on these readings, it should be possible to detect the depth of the plastic zone. The compressional wave velocity in the plastic zone has to be lower than the compressional wave velocity in the elastic material further away from the well if this should be possible.

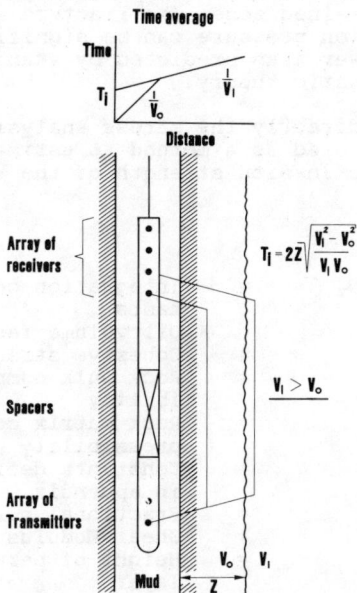

Fig. 13 Configuration for Expandable Sonic Tool.

Another indirect method could be using the fracture initiation pressure analysis. Once the fracture initiation pressure is determined, the strength can be estimated by back-calculating from the theory.

The third indirect method, which is already described, is use of production or production test data. The strength obtained from this analysis may be used to the same strata in nearby wells.

CONCLUSIONS

The major conclusions from this work can be summarized as follows:

1: For a single sand arch a failure criterion has been developed. This criterion relates the strength parameters with fluid flow forces.

2: Field test data has successfully

been described by the arching theory.

3: When a well is drilled through a layer of poorly cemented sandstone, a plastically strained zone may be formed around the wellbore.

4: When fluid flows into an open hole there is an upper limit for the flowrate before the entire layer will be fluidized. This upper limit is significantly increased when the hole is cased.

5: When fracturing a poorly cemented rock, the fracture initiation will take place in the plasticly strained zone. The fracture initiation pressure can be significantly lower than predicted by standard elastic theory.

6: Indirectly the stress analysis can be used as a method to estimate the in-situ strength of the rock.

NOMENCLATURE

A_1, B_1, D_1, D_2	=	Integration constants
B_o	=	Oil volume factor
C_o	=	Cohesive strength
C_b	=	Rock bulk compressibility
C_{ma}	=	Rock matrix compressibility
D_a, D_b	=	Constants defined in appendix D
f	=	Fractional porosity
G	=	Shear Modulus
h	=	Height of permeable layer
k	=	Permeability
p	=	Fluid pressure
p_1	=	Fluid pressure at r_1
p_2	=	Fluid pressure at r_2
P_i	=	Fluid pressure at R_i
P_c	=	Fluid pressure at plastic-elastic boundary
P_o	=	Fluid pressure at R_o
q	=	Fluid flow rate per perforation
Q	=	Fluid flow rate per well
r	=	Radial distance from center of arch/wellbore
r_1	=	Arch radius
r_2	=	Outer boundary radius, arch model
R_i	=	Wellbore radius
R_o	=	Outer boundary radius, wellbore model
R_b	=	Radius of inner plastic zone
R_1, R_2, R_3	=	Radii of different plastic zones, fracturing model
R_c	=	Radius of total plastic zone, wellbore model
r_c	=	Radius of plastic zone, arch model
t	=	Constant defined in appendix D
T	=	Constant defined in appendix B
Z	=	Depth
α	=	$\pi/4 + \Phi/2$ = Failure angle
β	=	$1 - C_{ma}/C_b$ = Biot constant
γ	=	Constant defined in appendix D
μ	=	Fluid viscosity
ϕ	=	Internal friction angle
ν	=	Poisson's ratio
σ_o	=	Far field hydrostatic stress, arch model
σ_c	=	$2c \tan\alpha$ = Uniaxial compressive strength
$\sigma_1, \sigma_2, \sigma_3$	=	Principal stress components
$\sigma_r, \sigma_\theta, \sigma_z$	=	Stress components in cylindrical coordinates
σ_{ri}, σ_r	=	Radial stress at: R_i, r_1
σ_{rc}	=	Radial stress at R_c, r_c
σ_{ro}, σ_{zo}	=	Radial and vertical stress at R_o
$\sigma_h (=\sigma_{ro})$	=	Horizontal stress at R_o
σ_T	=	Uniaxial tensile strength

REFERENCES

BIOT, M.A., (1941): General theory of three-dimensional consolidation. J. Appl. Phys. 12, 155 - 164.

BIOT, M.A., (1955): Theory of Elasticity and Consolidation for a Porous Anisotropic Solid: J. Appl. Phys. 26. 182 - 185.

BIOT, M.A., (1974): Exact simplified non-linear stress and fracture analysis around cavities in rock: Int. J. Rock Mech. Min. Sci. & Geom. Abstr. 11, 261 - 266.

BRADLEY, W.B., (1978): Failure of inclined boreholes: ASME paper presented at the Energy Technology Conference and Exhibition, Nov. 5 - 9.

BRADLEY, W.B., (1979): Mathematical Concept - Stress Cloud can predict Borehole Failure: Oil and Gas Journal 77, No. 8, 92 - 102.

BRATLI, R.K. and RISNES, R., (1981): Stability and Failure of Sand Arches: SPEJ April, 236 - 248.

BRECKELS, I.M. and van EEKELEN, H.A.M., (1981): Relationship between horizontal stress and depth of sedimentary basins: paper SPE 10336, 56th Annual Fall Technical Conference and Exhibition.

CLEARY et al. (1979): The Effect of Confining Stress and Fluid Properties on Arch Stability in Unconsolidated Sands: paper SPE 8426 presented at the 54th Annual Technical Conference and Exhibition, Sept.23 - 26.

COATES, G.R. and DENOO, S.A., (1981): Mechanical Properties Program Using Borehole Analysis and Mohr's Circle: SPWLA 22nd Annual Logging Symposium, June 23 - 26.

EATON, B.A., (1969): Fracture gradient prediction and its application in oil field operations: JPT 1353 - 1360.

FLORENCE, A.L. and SCHWER, L.E., (1978): Axisymmetric compression of a Mohr-Coulomb medium around a circular hole: Int. J. for Numerical and Analytical Methods in Geomechanics, Vol. 2, 367 - 379.

GEERTSMA, J., (1957): Effect of Fluid Pressure Decline on Volumetric Changes of Porous Rocks: Trans. AIME 210, 331 - 340.

GEERTSMA, J., (1966): Problems of rock mechanics in petroleum production engineering: Proc. First Congress Int. Soc. Rock Mech. 1, 585 - 594.

GEERTSMA, J., (1979): Some rock mechanical aspects of oil and gas well completions: Proc. European Off-shore Petr. Conf. & Exhibition 1, 301 -310.

GNIRK, P.F., (1972): The mechanical behaviour of uncased wellbores situated in elastic/plastic media under hydrostatic stress: SPEJ, Feb. 49 - 59.

HAIMSON, B. and FAIRHURST, C., (1967): Initiation and extension of hydraulic fractures in rocks: SPEJ, 310 - 318.

HALL, C.D. and HARRISBERGER, W.H., (1970): Stability of Sand Arches: A Key to Sand Control: JPT, July, 821 - 829.

HORSRUD, P., RISNES, R. and BRATLI, R.K., (1982): Fracture initiation pressures in permeable poorly consolidated sands: To be published in Int. J. Rock Mech. Min. Sci.

HUBBERT, M.K., and WILLIS, D.G.: Mechanics of hydraulic fracturing: Trans. AIME 210, 153 - 163.

JAEGER, J.C. and COOK, N.G.W., (1976): Fundamentals of Rock Mechanics: Chapman and Hall, London.

LOVE, A.E.H., (1944): A Treatise on the Mathematical Theory of Elasticity: Dover Publications, New York.

MENDLIN, W.L. and MASSÉ, L., (1979): Laboratory investigation of fracture initiation pressure and orientation, SPEJ 19, 129 - 144.

PASLAY, P.R. and CHEATHAM, J.B., (1963): Rock Stresses Induced by Flow of Fluids into Boreholes: SPEJ March, 85 - 94, Trans. AIME, 228.

RISNES, R., HORSRUD, P. and BRATLI, R.K., (1982): Sand Stresses Around a Wellbore: SPEJ December.

RISNES, R., BRATLI, R.K. and HORSRUD, P. (1982): Sand Arching - A Case Study: paper EUR 310 presented at the European Petr. Conf. London 25 - 28 Oct.

STEIN, N. et al. (1974): Estimating Maximum Sand-Free Production Rates From Friable Sands for Different Well Completion Geometries: JPT, Oct. 1156 - 1158; Trans. AIME, 257.

STEIN, N., (1976): Mechanical Properties of Friable Sands From Conventional Log Data: JPT July, 757 - 763.

TERZAGHI, K.V (1936): Stress Distribution in Dry and in Saturated Sand Above a Yielding Trap-Door: Proc. First Int. Conf. on Soil Mech. and Foundation Eng. 307 - 311.

TIPPIE, D.B. and KOHLHAAS, C.A., (1974): Variation of Skin Damage with Flowrate Associated with Sand Reservoirs: SPE 4866, SPE - 44th Annual California Meeting, April 4 - 5.

TIXIER, M.P. et al. (1973): Estimation of Formation Strength from the Mechanical Properties Log: SPE 4532, 48th Annual Meeting, Sept. 30 - Oct. 1.

APPENDIX A

ELASTIC STRESS SOLUTIONS, ARCH MODEL.

The final stress solutions can be given as:

$$\sigma_r = p_1 + \beta \left(\frac{1-2v}{1-v}\right)(p_2 - p_1)\frac{r_1 r_2}{r_2 - r_1} \cdot \left(\frac{1}{r_1} - \frac{1}{r}\right)$$

$$+ \left[\sigma_o - p_1 - \beta \left(\frac{1-2v}{1-v}\right)(p_2 - p_1)\right] \cdot \frac{r_1^3 \cdot r_2^3}{r_2^3 - r_1^3} \cdot \left(\frac{1}{r_1^3} - \frac{1}{r^3}\right) \quad (A1)$$

and

$$\sigma_\theta = p_1 + \beta \left(\frac{1-2v}{1-v}\right)(p_2 - p_1)\frac{r_1 r_2}{r_2 - r_1} \cdot \left(\frac{1}{r_1} - \frac{1}{2r}\right)$$

$$+ \left[\sigma_o - p_1 + \beta \left(\frac{1-2v}{1-v}\right)(p_2 - p_1)\right]\frac{r_1^3 \cdot r_2^3}{r_2^3 - r_1^3} \cdot \left(\frac{1}{r_1^3} + \frac{1}{2r^3}\right) \quad (A2)$$

Assuming no fluid flow, which is equivalent to assuming $p_2 = p_1$, these expressions are reduced to:

$$\sigma_r = p_1 + (\sigma_o - p_1)\frac{r_1^3 r_2^3}{r_2^3 - r_1^3} \cdot \left(\frac{1}{r_1^3} - \frac{1}{r^3}\right) \quad (A3)$$

and

$$\sigma_\theta = p_1 + (\sigma_o - p_1) \cdot \frac{r_1^3 \cdot r_2^3}{r_2^3 - r_1^3} \cdot \left(\frac{1}{r_1^3} + \frac{1}{2r^3}\right) \quad (A4)$$

For further details see Bratli & Risnes (1981).

APPENDIX B

STRESS SOLUTIONS IN PLASTIC ZONE, ARCH MODEL.

The final stress solutions in this zone are given as:

$$\sigma_r = p_1 - \frac{2\sigma_c}{T} + \frac{\mu q}{4\pi k} \left(\frac{1}{r_1} - \frac{1}{T+1} \cdot \frac{1}{r} \right) + C \cdot r^T \tag{B1}$$

and

$$\sigma_\theta = p_1 - \frac{2\sigma_c}{T} + \frac{\mu q}{4\pi k} \left(\frac{1}{r_1} - \frac{T}{T+1} \cdot \frac{1}{2r} \right)$$

$$+ \left(1 + \frac{T}{2} \right) \cdot C r^T \tag{B2}$$

where

$$T = 2(\tan^2 \alpha - 1) \tag{B3}$$

$$C = \left(\frac{2\sigma_c}{T} - \frac{1}{T+1} \cdot \frac{\mu q}{4\pi k} \right) \cdot r_1^{-T} \tag{B4}$$

The radius for the plastic zone can be found from the approximate solution:

$$\left(\frac{r_c}{r_1} \right)^T = \frac{3 \cdot \frac{T+1}{T+3} \cdot (\sigma_o - p_1)}{\frac{2 \cdot (T+1) \cdot \sigma_c}{T} - \frac{\mu q}{4\mu k r_1}} \tag{B5}$$

For further details see Bratli & Risnes (1981).

APPENDIX C

ELASTIC STRESS SOLUTIONS, WELLBORE MODEL

The final elastic stress solutions can be given as:

$$\sigma_r = \sigma_{ro} + (\sigma_{ro} - \sigma_{ri}) \cdot \frac{R_i^2}{R_o^2 - R_i^2} \cdot \left[1 - \left(\frac{R_o}{r}\right)^2\right]$$

$$- (P_o - P_i) \frac{1 - 2\nu}{2(1 - \nu)} \beta \cdot \left\{ \frac{R_i^2}{R_o^2 - R_i^2} \cdot \left[1 - \left(\frac{R_o}{r}\right)^2\right] \right.$$

$$\left. + \frac{\ln(R_o/r)}{\ln(R_o/R_i)} \right\} \quad (C1)$$

$$\sigma_\theta = \sigma_{ro} + (\sigma_{ro} - \sigma_{rc}) \frac{R_i^2}{R_o^2 - R_i^2} \cdot \left[1 + \left(\frac{R_o}{r}\right)^2\right] - (P_o - P_i)$$

$$\frac{1 - 2\nu}{2(1 - \nu)} \cdot \beta \left\{ \frac{R_i^2}{R_o^2 - R_i^2} \cdot \left[1 + \left(\frac{R_o}{r}\right)^2\right] \right.$$

$$\left. + \frac{1}{\ln(R_o/R_i)} \left[\ln(R_o/r) - 1\right] \right\} \quad (C2)$$

$$\sigma_z = \sigma_{zo} + 2\nu(\sigma_{ro} - \sigma_{ri}) \cdot \frac{R_i^2}{R_o^2 - R_i^2} - (P_o - P_i) \frac{1 - 2\nu}{2(1 - \nu)} \beta \cdot$$

$$\left\{ \nu \cdot \frac{2R_i^2}{R_o^2 - R_i^2} + \frac{2}{\ln(R_o/R_i)} \cdot \left[\ln(R_o/r) - \frac{\nu}{2}\right] \right\} \quad (C3)$$

For further details see Risnes et al. (1982).

APPENDIX D

PLASTIC STRESS SOLUTIONS, WELLBORE MODEL

The general solutions are:

i) $\underline{\sigma_r < \sigma_z < \sigma_\theta}$

$$\sigma_r = p - \frac{1}{t}\left[\sigma_c - \frac{\mu Q}{2\pi hk}\right] + 2(t+1)\frac{A_1}{V} r^t$$

$$\sigma_\theta = p - \frac{1}{t}\left[\sigma_c - (t+1)\frac{\mu Q}{2\pi hk}\right] + 2(t+1)^2 \frac{A_1}{V} r^t \qquad (D1)$$

$$\sigma_z = [2\nu(1-\beta) + \beta]\, p - \frac{\nu}{t}\left[2\sigma_c - (t+2)\frac{\mu Q}{2\pi hk}\right]$$

$$\qquad + \frac{(1+\nu)(1-2\nu)}{1-\nu}(\sigma_{zo} - \beta P_o) + 2\nu(t+2)(t+1)\frac{A_1}{V} r^t$$

where

$$A_1 = \text{integration constant}$$
$$t = \tan^2\alpha - 1$$
$$V = (t+1)^2 + 1 - \nu(t+2)^2$$

ii) $\underline{\sigma_r < \sigma_\theta = \sigma_z}$

$$\sigma_r = p - \frac{1}{t}\left[\sigma_c - \frac{\mu Q}{2\pi hk}\right] + 2(t+1)(1+\nu) B_1 r^t$$

$$\sigma_\theta = \sigma_z = p - \frac{1}{t}\left[\sigma_c - (t+1)\frac{\mu Q}{2\pi hk}\right] + 2(t+1)^2 (1+\nu) B_1 r^t$$

where

B_1 = integration constant

iii) $\underline{\sigma_r < \sigma_\theta < \sigma_z}$

$$\sigma_r = \frac{1}{V}[t(t+1) - vt(t+2) + (1-2v)(t+2)\beta]\,p + [v(t+2)+1]\frac{D_a}{V} + \frac{D_b}{V}$$

$$+ [v(t+2) + 1]\frac{D_b}{V}\ln r - \frac{1}{V}[(1-v)(t+1) - v]\,\sigma_c$$

$$+ \frac{1}{V}\frac{1-2v}{1-v}(t+1)(\sigma_{zo} - \beta P_o) + \frac{1}{V}[v(t+2) + \gamma]\,D_1 r^{\gamma-1} +$$

$$\frac{1}{V}[v(t+2) - \gamma]\,D_2 r^{-\gamma-1}$$

$$\sigma_\theta = \frac{1}{V}[vt^2 + (1-2v)((t+1)^2 + 1)\beta]\,p \qquad\qquad (D3)$$

$$+ [(v(t+2) + \gamma^2]\frac{D_a}{V} + v(t+2)\frac{D_b}{V} + [v(t+2) + \gamma^2]\frac{D_b}{V}\ln r$$

$$- \frac{1}{V}vt\sigma_c + \frac{1}{V}v\frac{1-2v}{1-v}(t+1)(t+2)(\sigma_{zo} - \beta P_o) +$$

$$\frac{\gamma}{V}[v(t+2) + \gamma]\,D_1 r^{\gamma-1} - \frac{\gamma}{V}[v(t+2) - \gamma]\,D_2\,r^{-\gamma-1}$$

$$\sigma_z = \frac{t+1}{V}[t(t+1) - vt(t+2) + (1-2v)(t+2)\beta - \frac{V}{t+1}t]\,p$$

$$+ (t+1)[v(t+2) + 1]\frac{D_a}{V} + (t+1)\frac{D_b}{V} + (t+1)$$

$$[v(t+2) + 1]\frac{D_b}{V}\ln r$$

$$-\frac{(t+1)}{V}\left[(1-v)(t+1) - v - \frac{V}{t+1}\right]\sigma_c +$$

$$\frac{1}{V}\frac{1-2v}{1-v}(t+1)^2(\sigma_{zo} - \beta P_o)$$

$$+\frac{t+1}{V}[v(t+2) + \gamma]\, D_1 r^{\gamma-1} + \frac{t+1}{V}[v(t+2) - \gamma]\, D_2 r^{-\gamma-1}$$

where

D_1 = Integration constant

D_2 = Integration constant

$\gamma^2 = (t+1)^2 - 2v(t+1) + 1$

$$D_a = \left[\frac{2(1-2v)t(1-\beta)}{(t+1)(t+1-2v)^2} + \frac{t(t+1) - vt(t+2) + (1-2v)(t+2)\beta}{\gamma^2 - 1}\right]\frac{\mu Q}{2\pi hk}$$

$$+ \frac{(1-2v)\,t(1-\beta)}{t+1-2v}\left[P_i - \frac{\mu Q}{2\pi hk}\ln R_i\right] - \frac{1-2v}{t+1-2v}\sigma_c$$

$$- \frac{(1-2v)(v(t+2)-1)}{(1-v)(t+1-2v)}(\sigma_{zo} - \beta P_o)$$

$$D_b = \frac{(1-2v)t(1-\beta)}{t+1-2v}\frac{\mu Q}{2\pi hk}$$

COMMUNICATING ELEMENTS FOR THE SIMULATION OF CRACK GROWING PROCESSES IN ROCKMASSES

Eléments en action réciproque pour simulation de la propagation des fissures dans la roche

Kommunizierende Elemente zur Simulation von Risswachstumsprozessen im Gebirge

H. Gross
Lehrstuhl für Felsmechanik, Universität Karlsruhe, Bundesrepublik Deutschland

SYNOPSIS

The knowledge of joint openings and crack propagations is important for the exploitation of natural resources out of deep rock. The Finite-Element-Method in standard form either leads to too inexact solutions or requires too large systems of equations. Therefore an iteration procedure based on the Finite-Element-Method will be introduced in this article. At critical positions a net refinement is carried out automatically. Although the system of equations retains its original size, the same solution is nevertheless obtained as by the use of the appropriate fine net. A proof of convergence follows.

RESUME

Pour l'exploitation des richesses naturelles du sous-sol il est très important de connaître le degré d'ouverture des failles et la propagation des fissures. La méthode des éléments finis dans sa forme standard fournit des résultats trop imprécis, ou bien exige d'utiliser des systèmes d'équations trop importants. C'est pourquoi on présente dans cet article un procédé d'itération basé sur la méthode des éléments finis. Dans ce procédé le quadrillage est raffiné automatiquement aux endroits critiques. Bien que le système d'équations conserve dans ce procédé sa taille habituelle, on obtient la même solution que celle que l'on obtiendrait avec un quadrillage convenable plus fin. Une démonstration de la convergence du procédé le prouve.

ZUSAMMENFASSUNG

Für die Gewinnung von Bodenschätzen aus dem Tiefengestein ist die Kenntnis von Kluftöffnungen und Rißausbreitungen von beträchtlichem Interesse. Die Finite-Element-Methode in Standardform liefert entweder zu ungenaue oder erfordert zu große Gleichungssysteme. Deshalb wird in dieser Arbeit ein Iterationsverfahren auf der Basis der FE-Methode vorgestellt, bei dem automatisch eine Netzverfeinerung an kritischen Stellen durchgeführt wird. Obwohl bei diesem Verfahren das Gleichungssystem seine ursprüngliche Größe beibehält, gewinnt man dennoch dieselbe Lösung wie für das zugehörige feine Netz. Ein Konvergenzbeweis verdeutlicht dies.

1. ENVIRONS OF THE PROBLEM

More and more it becomes necessary to mine the natural resouces and supports of energy out of the deep rock. Even if many of these can be opened by open-cast or by gallery, e. g. oil, gas and heat of the earth can't be gained by those methods.

Therefore it can be necessary to produce areas of joints or to extend existing joints inside the rock by high inner pressure. Along this joints oil and gas can flow to the bore-hole or the water can be conducted through them to profit the heat of the earth. Besides it is of important interest for the projecting engineer to be able to decide the positions of the openings of disjoining areas to know the developed field.

This knowledge gives information about the possibility or the necessity of further boreholes.

Therefore the propagation of the joint openings and the cracks influences the further planning and the carrying-through of the project.

About crack calculation in brittle material you find extensive literature, e. g. Hahn (1970) and Ingraffea, Heuze (1980). Analytical solutions exist only for very special geometries and boundary value problems. Numerical procedures yield partly unsatisfying results or use statements by which the solution is already partly given. Nevertheless certain process of crack propagations can be described thereby.

As well special formulations of the displacements of the elements at the crack tip as a suitable refined net are used in the Finite-Element-Method (FEM). Independent on the criterion decisive for the rise of a crack and his propagation the FEM is also described by smeared cracks, this means the properties of the cracks are valid for a whole element at the crack tip (Bazant, Cedolin, 1979).

A combination between net refinement and smeared cracks in scope of the FEM will be presented in this thesis. As the refinement is controlled automatically by the computer knowledge about the crack directions isn't provided in advance.

2. THE FE-METHOD: ITERATIVE USE

In this part the FEM is shortly described and the possibility to refine interactivly the used elementnet is shown. Hereby it is started from the displacement method. The process is available analogical for mixed and other procedures. Hence it is to minimize the functional ϕ for statical problems relative to the displacement field.

$$\phi = \int_V \tfrac{1}{2}\sigma_{ij}\varepsilon_{ij}dV + \int_V \gamma_i u_i dV - \int_F t_i u_i dF \quad (1)$$

Herein mean

- V = whole volume
- σ_{ij} = stress
- ε_{ij} = strain
- γ_i = unit weight
- u_i = displacement
- t_i = boundary stress
- F = surface of V

Unfortunately this optimization problem is as a rule neither numerical nor analytical soluble. Therefore the whole interesting area is divided in elements, which may have 4 to 8 nodes for twodimensional problems (Bathe, Wilson, 1976), fig. 1. Only the displacement components u_i^k at every node are supposed as unknown. By the choice of suitable shape functions it is attained that the displacements in an element depend only on the node displacements of this element and the displacements of one element edge depend only on the nodes of this edge of this element.

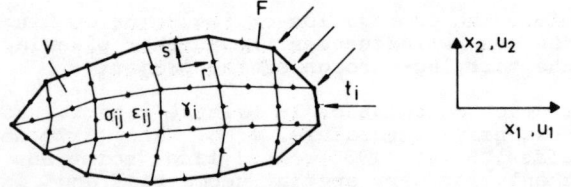

Fig. 1 Finite Element Net

$$u_i = \sum_k h_k(r,s,t) u_i^k \qquad \bar{u} = \underline{H}\,\bar{u}^k \quad (2)$$

with

$\bar{u}^k$ = vector of all node displacement components

where

r, s, t runs from -1. to 1. in every element. For special structure of the shape functions see Bathe, Wilson (1976).

Are x_i^k the node coordinates and is

$$x_i = \sum_k h_k(r,s,t)\, x_i^k \qquad \bar{x} = \underline{H}\bar{x}^k \quad (3)$$

then the elements are called isoparametric. The functions h_k are selected in a way that they take on the value 1. only at the node k and 0. at other nodes. As the functions h_k in (2) and (3) are identical combatibility of the displacements at the element edges is achieved without any additional condition. Thereby the strain field can be calculated

$$\underline{\varepsilon} := \underline{D}\bar{u} := \tfrac{1}{2}(\text{Grad }\bar{u} + \text{Grad}^T \bar{u}) \quad (4)$$

and by the material law the stress field

$$\underline{\sigma} := \underline{C}\,\underline{\varepsilon}. \quad (5)$$

Although the formula (2) is the same for all elements the differentiation in (4) in connection with (3) leads to elementwise different integrands in (1) the more so since the material matrix $\underline{C}$ isn't the same everywhere. Therefore it must be summed over all elements e respectively boundaries r.

$$\phi = \tfrac{1}{2}\sum_e \int_{V_e} \underline{\sigma}^T\cdot\underline{\varepsilon}\,dV + \sum_e \int_{V_e} \bar{\gamma}\cdot\bar{u}\,dV - \sum_r \int_{F_r} \bar{t}\cdot\bar{u}\,dF \quad (6)$$

The statement functions inserted, yields

$$\phi = \tfrac{1}{2}\sum_e \int_{V_e} \bar{u}^{kT}\underline{H}^T\underline{D}^T\underline{C}\,\underline{D}\,\underline{H}\,\bar{u}^k dV + \sum_e \int_{V_e} \bar{\gamma}^T\underline{H}\bar{u}^k dV \quad (7)$$

$$- \sum_r \int_{F_r} \bar{t}^T\underline{H}\bar{u}^k dF$$

This function ϕ has to become minimal with regard to the node displacement components. The variation of ϕ guides to the linear equation system for u_i^k:

$$\sum_e \int_{V_e} \underline{H}^T\underline{D}^T\underline{C}\,\underline{D}\,\underline{H}\,dV\bar{u}^k + \sum_e \int_{V_e} \underline{H}^T\bar{\gamma}\,dV - \sum_r \int_{F_r} \underline{H}^T\bar{t}\,dF = \bar{0} \quad (8)$$

Thus

$$\underline{K}\bar{u}^k = \bar{P}$$

where

$$\underline{K} = \sum_e \int_{V_e} \underline{H}^T\underline{D}^T\underline{C}\,\underline{D}\,\underline{H}\,dV \quad (10)$$

and

$$\bar{P} = \sum_r \int_{F_r} \underline{H}^T\bar{t}\,dF - \sum_e \int_{V_e} \underline{H}^T\bar{\gamma}\,dV.$$

Unfortunately the material law is frequently nonlinear. The insertion of the solution $\bar{u}^k$ in (2), (4) and (5) guides to a new matrix $\underline{C}$. $\underline{C}$ set in (8) delivers $P_c - P$ which generally isn't equal $\bar{0}$ where

$$\bar{P}_c := \sum_e \int_{V_e} \underline{H}^T \underline{D}^T \underline{\sigma}(\bar{u}^k) dV. \quad (11)$$

Nevertheless a solution is attained by the equation system

$$\underline{K} \Delta \bar{u}^k = \bar{P} - \bar{P}_c \quad (12)$$

where

$$\bar{u}^k := \bar{u}^k + \Delta \bar{u}^k \quad (13)$$

is put iteratively until the resultant $\bar{P}_c$ is equal to $\bar{P}$ (Zienkiewicz, 1975).

It is easy to see that the number of equations increases in according to the number of the nodes which is tied with the fineness of the net. Nevertheless a band structure of the matrix is attained with the help of a suited numbering of the nodes, because every displacement component of any node can only be related with displacements of the nodes of the adjoining elements because of the shape functions. To avoid too large equation systems you try to select a most practicable wide meshed net at areas with low displacement gradients and a most feasible, fine net at the ranges with high gradients.

Therefore it is necessary to choose a sufficient fine net close by the crack tip, but in some distance from it wide elements. Unfortunately as a rule the position of the crack tip is perhaps known at the beginning but not the behaviour under forces. Therefore it could be necessary to select a fine sharing by elements in sufficient wide surroundings. This lead to a large linear equation system.

A second possibility is an interactive net refinement. Based on the results of a Finite-Element calculation with a relatively wide meshed net, this net is refined for the consequence calculation only there where the conditions of stresses require it. The new system has more nodes and consequently more degrees of freedom. To retain the band structure of the stiffness matrix, it is necessary to carry out either a programinternal bandwideoptimization or a complete renumbering of the nodes. The second is only sensible if a proper netgenerator exists. So the crack course can be found out successively without using excessively large equation systems.

Since this procedure is workintensiv the automatisation of this progress is explained in the following. In the next section the influence of the net refinement on the original problem will be researched.

3. INFLUENCE OF A SUBSTRUCTURE

Two net are given for the same problem. In fig. 2a a wide meshed net is plotted, in fig. 2b a similar net with the element p further divided.

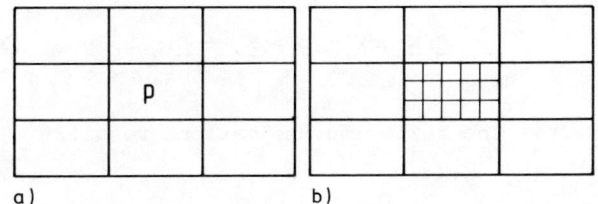

Fig. 2a wide meshed net, 2b refined net

The index 0 means the wide meshed net, 1 the fine one.

With

$$\underline{N} := \underline{D} \; \underline{H} \quad (14)$$

it is

$$\underline{K}_i \bar{u}_i = \bar{b}_i \qquad i = 0,1 \quad (15)$$

where

$$\underline{K}_i = \sum_{e_i} \int_{V_{e_i}} \underline{N}_i^T \underline{C} \; \underline{N}_i dV \qquad i = 0,1 \quad (16)$$

and

$$\bar{b}_i = \sum_r \int_{F_r} \underline{H}_i^T \bar{t} \; dF - \sum_{e_i} \int_{V_{e_i}} \underline{H}_i^T \bar{\gamma} \; dV \; i = 0,1. \quad (17)$$

If the unit weight is neglectable and the element p does not lie at the boundary then is

$$\bar{b} := \bar{b}_o = \bar{b}_1 \quad (18)$$

As the number of degrees of freedom in system 1 is greater than in 0, the dimension of the matrix $\underline{K}_o$ is smaller than that of $\underline{K}_1$. To be able to compare both, $\underline{K}_o$ is filled up by zerocolums and-rows. Only the belonging diagonal elements are set 1.

$$\underline{K}_o := \begin{pmatrix} \underline{K}_o & \underline{0} \\ \underline{0} & \underline{1} \end{pmatrix} \quad (19)$$

$\bar{b}_o$ is filled up by zero

$$\bar{b}_o := \begin{pmatrix} \bar{b}_o \\ \bar{0} \end{pmatrix} \quad (20)$$

it is

$$\underline{K}_1 = \sum_{e_1} \int_{V_{e_1}} \underline{N}_1^T \underline{C} \; \underline{N}_1 \; dV$$

$$= \sum_{e_o - p} \int_{V_{e_o}} \underline{N}_o^T \underline{C} \; \underline{N}_o \; dV + \int_{V_p} \underline{N}_1^T \underline{C} \; \underline{N}_1 \; dV. \quad (21)$$

With

$$\underline{B} := \int_{V_p} (\underline{N}_1 - \underline{N}_o)^T \underline{C} \; \underline{N}_o + \underline{N}_1^T \underline{C} \; (\underline{N}_1 - \underline{N}_o) \; dV \quad (22)$$

it supplies

$$\underline{K}_1 = \sum_{e_o} \int_{V_{e_o}} \underline{N}_o^T \underline{C} \underline{N}_o \, dV + \underline{B} = \underline{K}_o + \underline{B}. \quad (23)$$

Therefrom the following estimations result:

$$\bar{u}_o = \underline{K}_o^{-1} \bar{b}_o$$

$$\bar{u}_1 = \underline{K}_1^{-1} \bar{b}_1$$

Thus with (18)

$$\|\bar{u}_o - \bar{u}_1\| = \|\underline{K}_o^{-1}\bar{b}_o - \underline{K}_1^{-1}\bar{b}_1\| \leq \|\underline{K}_o^{-1} - \underline{K}_1^{-1}\| \, \|\bar{b}\|$$

or (24)

$$\underline{K}_o \bar{u}_o = \bar{b} = \underline{K}_1 \bar{u}_1 = \underline{K}_o \bar{u}_1 + \underline{B} \bar{u}_1$$

$$\frac{\|\bar{u}_o - \bar{u}_1\|}{\|\bar{u}_1\|} \leq \|\underline{K}_o^{-1} \underline{B}\| = \|\underline{K}_o^{-1}\underline{K}_1 - \underline{1}\| \quad (25)$$

analogiously

$$\frac{\|\bar{u}_o - \bar{u}_1\|}{\|\bar{u}_o\|} \leq \|\underline{K}_1^{-1}\underline{B}\| = \|\underline{1} - \underline{K}_1^{-1}\underline{K}_o\| \quad (26)$$

and

$$(\underline{K}_o + \underline{K}_1)(\bar{u}_o - \bar{u}_1) = \underline{K}_o\bar{u}_o + \underline{K}_1\bar{u}_o - \underline{K}_o\bar{u}_1 - \underline{K}_1\bar{u}_1$$

$$= (\underline{K}_o + \underline{B})\bar{u}_o - (\underline{K}_1 - \underline{B})\bar{u}_1$$

$$= \underline{B}(\bar{u}_o + \bar{u}_1)$$

hence

$$\frac{\|\bar{u}_o - \bar{u}_1\|}{\|\bar{u}_o + \bar{u}_1\|} \leq \|(\underline{K}_o + \underline{K}_1)^{-1}\underline{B}\| = \|(\underline{K}_o + \underline{K}_1)^{-1}(\underline{K}_1 - \underline{K}_o)\| \quad (27)$$

The knowledge of the stiffness matrices for both nets is sufficient to the determination of the relative difference between both solutions.

If the net refinement does not take place like plotted in fig. 2b because this leads to displacement incompatibilities, but by continuous transition of the degree of freedom like plotted in fig. 3, then not only the integral over the element p but also over all changed elements must be determinated again for the calculation of the stiffness matrix $\underline{K}_1$. The result affects only the matrix $\underline{B}$. The estimations (24) to (27) remain valid.

4. ELIMINATION OF A SUBSTRUCTURE BY STATIC CONDENSATION

For a good result it is necessary to have degrees of freedom (nodes) as most as possible in the displacement statement. Otherwise this leads to a large equation system. In contrast the size of the system does not change if added nodes can be eliminated.

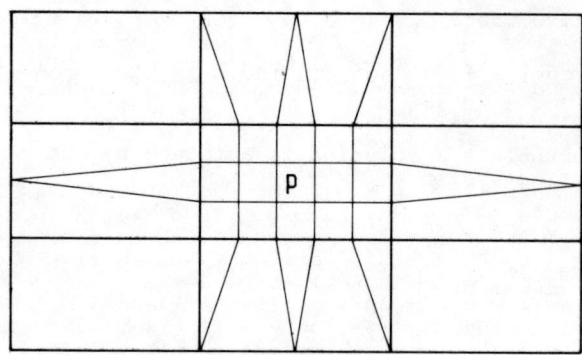

Fig. 3 Continuous net refinement

Suppose $\bar{u}_o$ is the displacement vector of the nodes of the wide meshed net $\bar{u}_1$ the displacement vector of the added nodes and

$$\bar{u} := \begin{pmatrix} \bar{u}_o \\ \bar{u}_1 \end{pmatrix} \text{ the total displacement vector} \quad (28)$$

The stiffness matrix is divided then in three parts.

$$\underline{K} := \begin{pmatrix} \underline{K}_{oo} & \underline{K}_{o1} \\ \underline{K}_{o1}^T & \underline{K}_{11} \end{pmatrix} \quad (29)$$

Where $\underline{K}_{oo}$ joins the nodes of the wide meshed system, $\underline{K}_{o1}$ the nodes of the wide meshed system with the added nodes and $\underline{K}_{11}$ the added nodes mutually only. $\underline{K}_{oo}$ is not the same as $\underline{K}_o$ of chapter 3. Also the node force vector can be divided into the parts of the original nodes and the added nodes.

$$\bar{b} := \begin{pmatrix} \bar{b}_o \\ \bar{b}_1 \end{pmatrix} \quad (30)$$

Thus

$$\underline{K} \bar{u} = \bar{b}. \quad (31)$$

Extended written:

$$\underline{K}_{oo} \bar{u}_o + \underline{K}_{o1} \bar{u}_1 = \bar{b}_o$$

$$\underline{K}_{o1}^T \bar{u}_o + \underline{K}_{11} \bar{u}_1 = \bar{b}_1 \quad (32)$$

To obtain the original size of the equation system in spite of the greater number of the degrees of freedom, $\bar{u}_1$ is eliminated from (32) by:

$$\bar{u}_1 = \underline{K}_{11}^{-1}(\bar{b}_1 - \underline{K}_{o1}^T \bar{u}_o). \quad (33)$$

Inserted in (32), this yields

$$(\underline{K}_{oo} - \underline{K}_{o1}\underline{K}_{11}^{-1}\underline{K}_{o1}^T)\bar{u}_o = \bar{b}_o - \underline{K}_{o1}\underline{K}_{11}^{-1}\bar{b}_1 \quad (34)$$

Frequently $\bar{b}_1 = \bar{o}$ is valid.

If this static condensation is carried through after the complete construction of the stiffness matrix for the refined net, then the static condensation corresponds to the solution of the linear equation system where first the coefficients of $\underline{K}_{11}$ are selected as pivot elements. Therefore the static condensation in this form does not yield any advantage in relation to the system size, exept any iteration of the balance is eventually required. If the static condensation succeeds on element base - this means during the construction of the element stiffness matrices - then you get the same equation system (34) but without building the equation system (32) completely. This is possible, if all added nodes correspond only with nodes of a single element of the original net, thus they do not fall on an element edge. Otherwise you have to include in the static condensation several elements but at most all elements containing added nodes.

5. SEPARATED CALCULATION OF THE MAIN- AND SUBSTRUCTURE

As not only at the general Finite-Element-Method but also at the method of static condensation the addition of one or more nodes influences all matrix components belonging to the nodes of the same element, thus e. g. in $\underline{K}_{oo}$, the stiffness matrices have to be suitable changed respectively to be put up again. To avoid this the following procedure is proposed. First a displacement field $\bar{u}_o$ for the base system (Fig. 2a) is calculated by the previously usual FEM.

$$\underline{K}_o \bar{u}_o = \bar{b}_o \tag{35}$$

If in element p the crack criterion is exceeded, then this element will be covered with a refined net. As boundary conditions at the original nodes the adjoining displacement components of $\bar{u}_o$ are valid and at the new nodes lying on an element edge the new node displacements which can be computed by (2) for the base net. You obtain for the divided element p the linear equation system (D = Double-Technic)

$$\underline{K}_D \bar{u}_D = \bar{b}_D. \tag{36}$$

$\underline{K}_{D_o}$ denotes the stiffness matrix to $\underline{K}_D$ without consideration of the displacement boundary conditions.

It is built

$$\bar{b}_{D_o} := \underline{K}_{D_o} \bar{u}_D \tag{37}$$

You attain among other things a force vector at the base nodes of the element p.

As a rule this force vector will be different from the vector

$$\bar{b}_p := \underline{K}_p \bar{u}_o \tag{38}$$

where $\underline{K}_p$ = element stiffness matrix of p.

The diffence force vector is used for the correction of the vector $\bar{b}_o$. The following equation system supplies the proof.

$$\underline{K}_o \bar{u}_o - \underline{K}_p \bar{u}_o + \underline{K}_{oo} \bar{u}_o + \underline{K}_{o1} \bar{u}_1 = \bar{b}_{o1} \tag{39}$$

and

$$\underline{K}_{o1}^T \bar{u}_o + \underline{K}_{11} \bar{u}_1 = \bar{b}_{11} \tag{40}$$

where

$$\underline{K}_{D_o} = \begin{pmatrix} \underline{K}_{oo} & \underline{K}_{o1} \\ \underline{K}_{o1}^T & \underline{K}_{11} \end{pmatrix} \tag{41}$$

$\bar{u}_1$ the displacement components at the added nodes
$\bar{b}_{o1}$ the force vector at the original nodes
$\bar{b}_{11}$ the force vector at the added nodes

This equation system is identical with the complete system of the refined net.

The algorithmus of iteration for (39) under the condition (40) reads:

$$\underline{K}_o \bar{u}_o^{n+1} = \bar{b}_{o1}^n + \underline{K}_p \bar{u}_o^n - \underline{K}_{oo} \bar{u}_o^n - \underline{K}_{o1} \bar{u}_1^n \tag{42}$$

with

$$\bar{b}_{o1}^n = \begin{cases} \bar{b}_o & \text{for } n = 0 \\ \bar{b}_{o1} & \text{for } n > 0 \end{cases}$$

n = iteration step

Thus with (37) and 38)

$$\underline{K}_o \bar{u}_o^{n+1} = \bar{b}_{o1}^n + \bar{b}_p - \bar{b}_{D_o}\big|_A$$

$\bar{b}_{D_o}\big|_A$:= components of $\bar{b}_{D_o}$ at nodes of the original net.

Here the correction vector is $\bar{b}_p - \bar{b}_{D_o}\big|_A$ like mentioned before.

For the consideration of the convergence $\bar{u}_1$ in (42) is eliminated by (40).

$$\underline{K}_o \bar{u}_o^{n+1} = \bar{b}_{o1}^n - \underline{K}_{o1} \underline{K}_{11}^{-1} \bar{b}_{11} + (\underline{K}_p - \underline{K}_{oo} + \underline{K}_{o1} \underline{K}_{11}^{-1} \underline{K}_{o1}^T) \bar{u}_o \tag{43}$$

With

$$\bar{f} := \underline{K}_o^{-1} (\bar{b}_{o1}^n - \underline{K}_{o1} \underline{K}_{11}^{-1} \bar{b}_{11}) \quad n > 0 \tag{44}$$

$$\underline{F} := \underline{K}_o^{-1} (\underline{K}_p - \underline{K}_{oo} + \underline{K}_{o1} \underline{K}_{11}^{-1} \underline{K}_{o1}^T) \quad n > 0$$

you obtain

$$\bar{u}_o^1 = \underline{K}_o^{-1} \bar{b}_o \tag{45}$$

and

$$\bar{u}_o^{n+1} = \bar{f} + \underline{F} \bar{u}_o^n = \sum_{i=0}^{n-1} \underline{F}^i \bar{f} + \underline{F}^n \underline{K}_o^{-1} \bar{b}_o \quad n > 0 \tag{46}$$

$$\bar{u}_o^{n+1} = (\underline{I} - \underline{F})^{-1} (\underline{I} - \underline{F}^n) \bar{f} + \underline{F}^n \underline{K}_o^{-1} \bar{b}_o \quad n > 0 \tag{47}$$

in case

$$\|\underline{F}\| < 1 \tag{48}$$

$$\bar{u}_o := \lim_{n\to\infty}(\underline{I} - \underline{F})^{-1}(\underline{I} - \underline{F}^n)\bar{f} + \underline{F}^n \underline{K}_o^{-1}\bar{b}_o \quad (49)$$

$$\bar{u}_o = (\underline{I} - \underline{F})^{-1}\bar{f}$$

$$\bar{u}_o = (\underline{K}_o - \underline{K}_p + \underline{K}_{oo} - \underline{K}_{o1}\underline{K}_{11}^{-1}\underline{K}_{o1}^T)^{-1}$$
$$(\bar{b}_{o1} - \underline{K}_{o1}\underline{K}_{11}^{-1}\bar{b}_{11}) \quad (50)$$

If in contrast the equation system (39) and (40) is considered as a complete Finite-Element-Problem, meaning the boundary value problem, which is to be calculated, is only solved with the refined net, and if $\bar{u}_1$ is eliminated from (39) by the static condensation, then you attain

$$(\underline{K}_o - \underline{K}_p + \underline{K}_{oo} - \underline{K}_{o1}\underline{K}_{11}^{-1}\underline{K}_{o1}^T)\bar{u}_o = \bar{b}_{o1} - \underline{K}_{o1}\underline{K}_{11}^{-1}\bar{b}_{11}$$
$$(51)$$

As this direct procedure delivers the same result like the iterative method (50), both are equivalent. But whereas the fine net must be given in advance for the direct way, the fineness can be made dependent on the result of the iterative method. So you avoid to get the stiffness matrix unnecessaryly too large.

6. CONSEQUENCES

Both the choice of a fine Finite-Element net in advance and an additional refinement have advantages and disadvantages.

If the positions with high stress gradients or critical stresses are known a fine net can be advantageous because on the on hand you do not need the iteration like described under 5., on the other hand only those areas are covered with small elements which lie near critical points. By this method an excessive increase of the number of the unknowns is avoided.

If on the contrary these critical points are not known or only inexactly, or if they even change their positions during the calculation, then the iterative method is recommendable, because only the critical area is refined and thus the number of unknowns can be kept as small as possible. The proof of convergence in chapter 5. demonstrates that there is not any loss of accuracy in comparison with the refined net.

7. CONCLUSION

The Finite-Element-Method is a suitable tool for calculations of crack propagation processes for any given boundary value problem. As crack elements with special geometrie and displacement statements can be set in, in few cases so-called smeared elements are used. This requires a sufficient fine net at the environs of the crack tip.

To avoid to inlarge excessively the existing linear equation system, a method is proposed, where a net refinement automatically takes place at critical positions. With the aid of a suitable iterative method the same solution is attained as by the standard Finite-Element-Method. The proof of convergence shows this. Thereby stress situations and displacement fields can be determined sufficient enough in the environs of cracks without being forced to return to large equation systems.

8. REFERENCES

Bathe, K.-J., Wilson, E. (1976). Numerical Methods in Finite Element Analysis: Prentice Hall, Inc.

Bazant, Z., Cedolin, L. (1979). Blunt Crack Band Propagation in Finite Element Analysis: J. of the Engineering Mech. Div, ASCE, Vol. 105, No. EM2, 297 - 315.

Hahn, H. G. (1970). Spannungsverteilung an Rissen in festen Körpern: VDI-Forschungsheft 542, VDI-Verlag Düsseldorf.

Ingraffea, A., Heuze, F. (1980). Finite Element Models for Rock Fracture Mechanics: Int. J. for Num. and Anal. Meth. in Geom., Vol. 4, 25 - 43.

Zienkiewicz, O. C. (1975). Methode der finiten Elemente; Carl Hanser Verlag München Wien.

9. ACKNOWLEDGEMENT

This work was supported by Deutsche Forschungsgemeinschaft, Grant NA 69/11 and NA 69/14-2. The author gratefully acknowledges the admitted funds.

Address of the author:

Groß, H. Dipl.-Math.
Lehrstuhl für Felsmechanik
Universität Karlsruhe
7500 Karlsruhe 1
Federal Republic of Germany

SIMULATION NUMERIQUE DE L'ECOULEMENT DE L'EAU DANS UNE VEINE DE CHARBON SUITE A DES ESSAIS DE FRACTURE ENTRE DEUX FORAGES
Numerical simulation of water flow in a coal vein by cracking tests between two boreholes
Numerische Simulation des Wasserflusses in einer Kohlenschicht auf Grundlage von Bruchversuchen zwischen zwei Bohrungen

J. P. Bruggeman
Ingénieur, Université Catholique de Louvain, Louvain-la-Neuve, Belgique

J. P. Coyette
Assistant, Université Catholique de Louvain, Louvain-la-Neuve, Belgique

E. Lousberg
Professeur, Université Catholique de Louvain, Louvain-la-Neuve, Belgique

J. F. Thimus
Premier Assistant, Université Catholique de Louvain, Louvain-la-Neuve, Belgique

RESUME

On applique la méthode des éléments finis à la simulation d'un processus de "linking" dans une veine de charbon profonde. Malgré la polyvalence de la technique des éléments finis appliquée avec expansion polynome localement définie elle est sujette à certains inconvénients. De tels inconvénients se manifestent parce que le domaine étudié est infini. Dans ce contexte on a appliqué le concept des éléments finis à l'analyse non-linéaire de l'écoulement de l'eau dans une veine de charbon. La concordance entre les données expérimentales et numériques est bonne. On attribue cette performance satisfaisante d'une part à l'incorporation de ces éléments spécifiques et d'autre part à un modèlage correct des variations de la perméabilité en fonction de la pression.

SYNOPSIS

The Finite Element Method is applied to the simulation of a linking process in a deep coal-layer. Despite its versatility, the Finite Element Technique implemented with locally defined polynomial expansion presents certain difficulties. Such difficulties are manifested because the field of study is infinite. In this context, the concept of infinite elements is applied to the non-linear analysis of the water flow through a coal layer. Comparisons between experimental and numerical data show a high rate of agreement. This good performance is attributed to the use of these special elements and a correct modelling of permeability variations with pressure.

ZUSAMMENFASSUNG

Die Finite Element Methode wird zur Simulierung eines Verbindungsvorgangs in einem tiefen Kohlenflöz herangezogen. Trotz ihrer vielseitigen Anwendbarkeit bietet diese Methode jedoch gewisse Schwierigkeiten, wenn sie mit örtlich definierten polynomialen Auflösungen angewandt wird. Solche Schwierigkeiten erweisen sich, weil das Studiengebiet unendlich ist. In diesem Zusammenhang wird der Begriff der unendlichen Elemente auf die nicht-lineare Analyse des Wasserflusses durch eine Kohlenschicht angewandt. Vergleiche zwischen experimentellen und numerischen Daten weisen gute Übereinstimmung auf. Diese guten Resultate schreibt man der Anwendung dieser Sonder-Elemente und der korrekten Modellierung von Durchdringunsvermögen und Druck zu.

1. INTRODUCTION

Plusieurs pays européens, dont la Belgique, possèdent d'énormes réserves de charbon situées à grande profondeur (vers 1000 m) et constituées de veines de faible puissance (de l'ordre du m).

Le coût de l'exploitation directe de ces veines est prohibitif et dans divers pays les techniciens ont tenté d'utiliser l'énergie calorifique en gazéifiant le charbon.

Ce procédé a été essayé en Belgique, mais le rendement s'est avéré insuffisant vu la très faible perméabilité du charbon (k $\simeq$ 0.6 10^{-8} m/sec) due à sa grande compacité.

Divers autres procédés ont été mis au point pour augmenter la perméabilité du charbon, à savoir : l'injection d'eau et de gaz sous forte pression (appelé le linking hydraulique ou pneumatique), l'explosion, la rétrocombustion, les forages dans les veines, etc...

En Belgique, l'Institution pour le Développement de la Gazéification Souterraine a mené des recherches de linking hydraulique et pneumatique dans divers sites miniers du pays. Ces essais consistent à forer deux ou plusieurs puits à une distance variant de 35 à 100 m, d'injecter dans l'un d'eux un fluide (eau ou gaz) sous des pressions de l'ordre de 13 à 50 MPa en vue de provoquer la fissuration du charbon et de recueillir le gaz par les puits exutoires [Ledent et Chandelle, 1978]. Ces méthodes se sont avérées infructueuses en Belgique parce que la zone fissurée de charbon ne s'est développée que dans un rayon limité autour du puits d'injection.

Disposant des résultats des essais effectués in situ, les auteurs de la présente contribution ont cherché à établir une simulation numérique de l'écoulement, compte tenu notamment de la variation de la perméabilité du charbon en fonction de la contrainte qui lui est appliquée. Cette simulation numérique a utilisé la méthode des éléments finis et des éléments infinis.

2. DE CERTAINES CARACTERISTIQUES DU CHARBON BELGE

Le charbon se rencontre, en Belgique, en veines de faible épaisseur (0.5 à 1.5 m) situées à grande profondeur.

Il présente une microfissuration intense, de l'ordre de 2 à 120 fissures par mm [Ledent et Chandelle, 1978]. L'écoulement de l'eau à travers les veines de charbon peut donc être simulé comme l'écoulement dans un matériau granulaire. Néanmoins, la structure complexe du charbon montre qu'il existe un double réseau de fissures : un réseau de fissures ouvertes dans lesquelles l'effet d'un gradient peut se faire sentir rapidement à assez grande distance (30 m) et un réseau de fissures très fines capables d'absorber lors d'un processus d'injection un volume d'eau égal à 1 % du volume du charbon. Il y a donc dans tout essai de linking une phase transitoire durant laquelle il y a absorption suivie d'une phase d'écoulement permanent.

La perméabilité du charbon belge a été étudiée par plusieurs auteurs [Brych et Defourny, 1981; Defourny, 1980; Ledent, 1981]. Elle dépend des contraintes naturelles et de la pression de linking.

La courbe donnant la valeur de la perméabilité k en fonction de la pression d'injection de l'eau lors d'un essai de linking est donnée à la figure 1. Lorsque la pression d'injection p_{inj} est inférieure à la pression hydrostatique p_h mesurée dans le terrain (p_h = 3 MPa), k reste constant (0.64 10^{-8} m/sec); pour p_{inj} compris entre p_h et la pression de fracking p_f (p_f = 10.7 MPa), la valeur de k augmente linéairement et pour $p_{inj} > p_f$, la perméabilité croît rapidement.

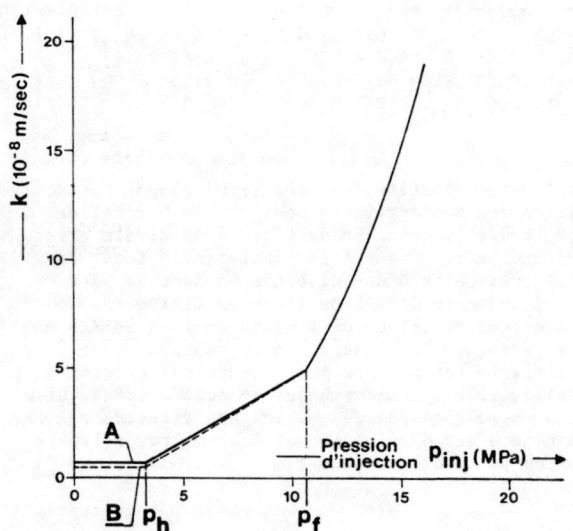

Fig. 1.

3. DONNEES EXPERIMENTALES ET HYPOTHESES

La simulation numérique décrite ci-après est basée sur les hypothèses et données suivantes (Site de Helchteren-Zolder) :

- Veine de charbon d'une épaisseur de 1.10 m;
- Ecoulement plan laminaire (Loi de Darcy);
- Perméabilité du charbon donnée à la figure 1;

  la courbe A correspond au cas normal tandis que la prise en compte du colmatage conduit à utiliser la courbe B caractérisée par une perméabilité de 0.42 x 10^{-8} m/sec lorsque la pression d'injection p_{inj} est inférieure à la pression hydrostatique p_h mesurée (3 MPa);

- Deux puits l'un d'injection, l'autre exutoire, distants de 80 m;
- Hypothèse du régime permanent;

  - pression d'injection 130 MPa
  - pression au puits exutoire 0 MPa,

  - débit d'injection Q_{inj} = 3.6 m^3/jour,
  - débit recueilli Q_{rec} = 0.504 m^3/jour.

Le débit recueilli Q_{rec} représente 14 % du débit injecté Q_{inj} : la différence représente la perte dans la veine (supposée avoir une extension infinie).

4. ANALYSE NUMERIQUE DE L'ECOULEMENT

4.1 Eléments finis et infinis

Equations de base

L'équation aux dérivées partielles gouvernant l'écoulement bidimensionnel étudié peut s'écrire sous la forme

$$k_x \frac{\partial^2 \phi}{\partial x^2} + k_y \frac{\partial^2 \phi}{\partial y^2} + q = 0 \quad \text{sur } \Omega \qquad (1)$$

où ϕ désigne le potentiel,

k_x et k_y sont les perméabilités selon x et y,

q représente le débit,

et Ω est le domaine géométrique sur lequel le problème est étudié.

Les conditions frontières concernent le potentiel ϕ et sa dérivée normale sur certaines zones frontières Γ :

$$\phi = \phi_o$$
$$\quad \text{sur } \Gamma \qquad (2)$$
$$\frac{\partial \phi}{\partial n} = 0$$

L'écoulement laminaire considéré est régi par des considérations de minimisation de l'énergie [Zienkiewicz, 1977]. La formulation énergétique équivalent à (1) consiste en la minimisation de l'intégrale

$$\iint_\Omega (\frac{1}{2} k_x (\frac{\partial \emptyset}{\partial x})^2 + \frac{1}{2} k_y (\frac{\partial \emptyset}{\partial y})^2 - q \emptyset) \, dx \, dy \qquad (3)$$

pour toute fonction $\emptyset (x, y)$.

Si on utilise le procédé habituel de discrétisation du domaine en éléments finis, le potentiel $\emptyset$ peut s'exprimer en fonction des potentiels aux noeuds

$$\emptyset = N.\emptyset^e \qquad (4)$$

où $\emptyset^e$ désigne les potentiels nodaux et N la fonction d'interpolation adoptée.

Dans ces conditions, le procédé de minimisation conduit à un système linéaire de la forme

$$K. \emptyset^e = Q \qquad (5)$$

à partir duquel sont calculés les potentiels aux noeuds du maillage.

Approches alternatives

En dépit de sa puissance et de son énorme champ d'application, la technique des éléments finis (implantée avec de classiques fonctions d'interpolation polynomiales) se heurte parfois à de sérieuses difficultés.

Ainsi, celles-ci peuvent se manifester lorsque le domaine géométrique devient infini.

Ce cas est classiquement rencontré en mécanique des sols ou des roches.

Dès lors, les conditions limites imposées sur la frontière du domaine (fini) discrétisé peuvent, dans une mesure plus ou moins importante, affecter la réponse du système.

Quoique, dans ce contexte, le recours à la méthode des intégrales frontières [Brebbia, 1980] puisse se révéler fort attractif, il a été jugé plus rentable de recourir ici au concept d'élément infini introduit par Bettess [Bettess, 1977].

Ce type d'élément utilisé conjointement aux classiques éléments finis est appliqué à l'analyse non linéaire de l'écoulement d'eau dans une veine de charbon.

Les comparaisons entre résultats de calcul et mesures in situ se révèlent très bonnes.

Cet excellent résultat est essentiellement attribuable à l'utilisation simultanée d'éléments finis et infinis ainsi qu'à une modélisation correcte des variations de perméabilité avec la pression.

4.2 Approche par éléments finis

Une analyse par éléments finis d'un tel problème impose à l'utilisateur de délimiter strictement le domaine géométrique où l'écoulement est supposé confiné.

Compte tenu de l'entredistance d entre puits (80 m), il semble acceptable de limiter la région étudiée au rectangle $S_1 S_2 S_3 S_4$ de la figure 2.

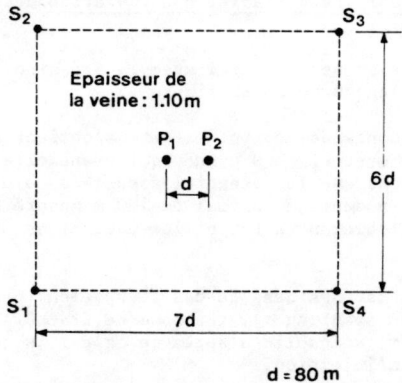

Fig. 2.

En vertu de la symétrie, seule une demi veine est analysée (figure 3) et AD est considéré comme ligne de courant

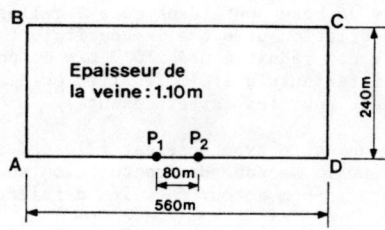

Fig. 3.

Sur les autres bords, diverses possibilités peuvent être envisagées, soit que l'on considère le contour ABCD comme ligne de courant (ce qui implique dès lors l'absence de débit de fuite) ou comme équipotentielle. Les deux possibilités sont examinées avec ou sans prise en compte des variations de perméabilité avec la pression.

Dans tous les cas présentés, la pression d'injection (puits P1) vaut $p_{inj} = 130$ MPa alors qu'à l'exutoire (puits P2), la pression p_{rec} est nulle.

Cas 1 : Perméabilité constante, absence de débit de fuite

La veine de charbon est supposée homogène et isotrope et le contour ABCD est assimilé à une ligne de courant.

Le calcul a été réalisé par deux valeurs de perméabilité (cas 1a : $k = 5 \times 10^{-8}$ m/sec, cas 1b : $k = 0.6 \times 10^{-8}$ m/sec.). Les résultats obtenus sont repris dans le tableau 1.

Par comparaison aux mesures faites, ceux-ci n'apparaissent pas particulièrement réalistes. Les valeurs de perméabilité affectent en effet considérablement le temps de transit et le faible taux de récupération mesuré est tributaire de l'existence d'importants débits de fuite non considérés dans cette approche.

Cas 2 : Perméabilité variable, considération des débits de fuite

Les variations de perméabilité avec la pression sont décrites par la courbe de la figure 1.

La prise en compte de débits de fuite s'obtient en assimilant le contour ABCD à une équipotentielle. Nous supposerons que le potentiel associé à celui-ci correspond au potentiel initial de 300 m mesuré dans la veine préalablement à toutes les opérations d'injection.

Les résultats obtenus dans ce cas se révèlent beaucoup plus réalistes (tableau 1). La zone de fracturation ($\emptyset >$ 1092 m) s'établit d'après ce calcul jusqu'à 6 m du puits d'injection.

Un examen plus détaillé des résultats du calcul semble indiquer que la perméabilité utilisée est trop élevée à proximité du puits exutoire alors qu'elle se révèle trop faible ailleurs.

Ce fait est semble-t-il dû au colmatage du puits d'extraction. Aussi un troisième cas a-t-il été traité de façon à simuler cet effet.

Cas 3 : Prise en compte du colmatage

Les hypothèses de base sont identiques à celles du cas 2 à la restriction près que le coefficient de perméabilité k est réduit à 0.44×10^{-8} m/sec. pour les potentiels $\emptyset$ inférieurs à 316 m (en lieu et place de 0.64×10^{-8} m/sec. pour les cas précédents).

Le bien fondé de cette hypothèse est illustré par les résultats obtenus. La zone de fracturation se maintient dans un rayon de 5-6 m autour du puits d'injection.

TABLEAU 1

	Cas	Analyse	Perméabilité k (x 10^{-8} m/sec)	Débit injecté Q_{inj} (m³/j)	Débit recueilli Q_{rec} m³/j)	Taux de récupération (%)	Temps de transit (j)
Calculs	1a	EF	5.0	3.456	3.456	100	19.3
	1b	EF	0.64	3.456	3.456	100	143.6
	2	EF	Courbe A ✱	3.428	0.790	24	30.5
	3	EF	Courbe B ✱	3.146	0.566	18	34.3
	4	EF+EI	Courbe A ✱	3.707	0.533	14.3	35.4
	Mesures			3.600	0.504	14	35

EF = Eléments Finis
EI = Eléments Infinis

✱ Les courbes A et B sont représentées à la Fig. 1

4.3 Mise en oeuvre des éléments infinis

Un doute subsistant sur l'influence réelle de la condition imposée sur le contour ABCD dans l'approche par éléments finis, il a été jugé intéressant d'obtenir une meilleure représentation géométrique du domaine étudié par la mise en oeuvre simultanée d'éléments finis et infinis.

Les éléments finis seront utilisés dans la zone où les gradients de pression sont les plus importants (zone voisine du puits) tandis que les éléments infinis seront localisés à l'extérieur de cette zone.

Elément quadrilatère infini

Dans cette étude, l'élément choisi est un élément quadrilatère à 9 noeuds de la famille de Lagrange dont l'extension dans la direction locale ξ est infinie (figure 4).

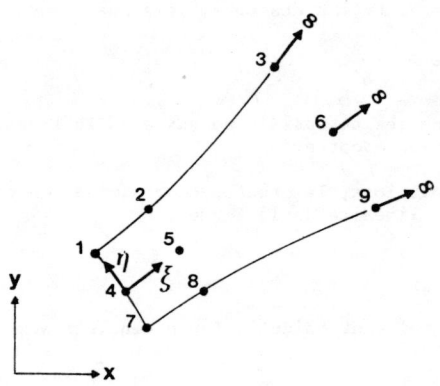

Fig. 4.

Les fonctions d'interpolation associées à un tel élément doivent respecter deux exigences. D'une part, elles doivent être "réalistes" et d'autre part, elles doivent impérativement conduire à des intégrales finies sur le domaine de l'élément.

Un choix acceptable pour la direction infinie est constitué par des fonctions basées sur les polynomes de Lagrange multipliées par une exponentielle décroissante avec la distance.

En limitant notre attention aux noeuds 1, 2 et 3 de l'élément considéré, nous pouvons écrire les fonctions d'interpolation N_1, N_2 associées aux noeuds 1 et 2 sous la forme

$$N_1 = e^{(\xi_1 - \xi)/L} \cdot \left(\frac{\xi_2 - \xi}{\xi_2 - \xi_1}\right) \qquad (6)$$

$$N_2 = e^{(\xi_2 - \xi)/L} \cdot \left(\frac{\xi_2 - \xi}{\xi_2 - \xi_1}\right) \qquad (7)$$

où ξ_1, ξ_2 désignent les coordonnées locales des noeuds 1 et 2 et L est un facteur qui contrôle la décroissance exponentielle.

Quant à la fonction N_3 associée au noeud 3 rejeté à l'infini elle s'obtient par

$$N_3 = 1 - N_1 - N_2 \qquad (8)$$

Ces expressions satisfont aux conditions usuellement imposées aux fonctions de forme. [Zienkiewicz, 1977].

Dans la direction locale η, des polynomes conventionnels de Lagrange sont utilisés.

La combinaison des deux types de fonction fournit les fonctions d'interpolation associées à l'élément quadrilatère infini.

Représentation paramétrique

Le procédé de calcul par éléments finis exige non seulement le choix de fonctions d'interpolation acceptables mais également le respect de la configuration géométrique du domaine étudié.

A cet égard, les éléments finis paramétriques [Zienkiewicz, 1977] se révèlent particulièrement bien adaptés.

Les polynomes d'interpolation vus plus haut ne sont toutefois pas satisfaisants comme base paramétrique de l'élément [Bettess, 1977]. Il est préférable de leur substituer les polynomes conventionnels de Lagrange définis sur un quadrilatère dont les noeuds 3, 6 et 9 sont placés, non plus à l'infini mais bien à une grande distance (finie) de l'origine du système local ξ, η.

Calcul des matrices élémentaires

La dérivation des matrices de "raideur" élémentaires des éléments infinis suit la même filière qu'habituellement.

La principale différence résulte du choix particulier des fonctions d'interpolation et de forme de l'élément infini.

L'adoption du terme exponentiel dans les fonctions d'interpolation associées à la direction locale ξ implique l'utilisation des règles intégration de Gauss-Laguerre [Rabinowitz, Weiss, 1959] tandis que dans la direction η, la technique habituelle de Gauss-Legendre est utilisée.

Cas 4 : Modélisation par éléments infinis

Pour réaliser cette analyse, le domaine étudié est partitionné en 2 zones (figure 5), l'une (voisine des puits) discrétisée en éléments finis tandis que l'autre (extérieure) est décomposée en éléments infinis.

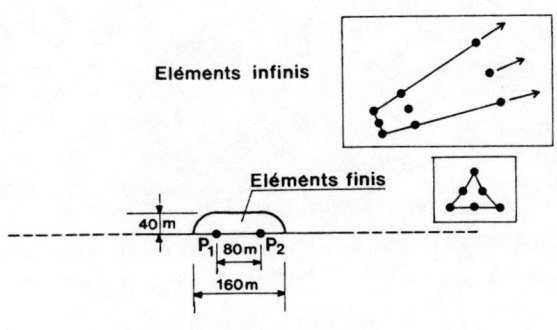

Fig. 5.

Les potentiels aux puits d'injection et exutoire demeurent égaux à 1300 m et 0 m respectivement. A distance infinie, le potentiel est supposé s'identifier au potentiel préexistant de 300 m.

Les variations de perméabilité avec la pression sont également intégrées au calcul.
Enfin, le facteur L de décroissance exponentielle a été choisi égal à 2.
Comme le montre le tableau 1, les résultats obtenus dans ce cas se comparent très favorablement aux résultats expérimentaux.

5. CONCLUSIONS

L'écoulement étudié montre une forte dépendance vis-à-vis des conditions frontières imposées. Celle-ci est illustrée par les résultats de calcul par éléments finis.
La mise en oeuvre simultanée d'éléments finis et infinis permet de s'affranchir d'hypothèses relatives aux débits de fuite.

La prise en compte supplémentaire des importantes variations de perméabilité avec la pression fournit des résultats numériques en très bonne corrélation avec les mesures effectuées in situ.

6. REMERCIEMENTS

Les auteurs tiennent à remercier tout particulièrement l'Institut National des Industries Extractives (INIEX) pour avoir mis à leur disposition les résultats des essais effectués sur le site de Helchteren-Zolder.

7. BIBLIOGRAPHIE

BETTESS P.,[1977] "Infinite Elements", Int. J. Num. Meth. Engng, 11, 53-64

BREBBIA C.A., [1980] "The Boundary Element Method for Engineers", Second Edition", Pentech Press

BRYCH J. et DEFOURNY P., [1981]. "Considérations sur les propriétés mécaniques du charbon". Faculté Polytechnique de Mons, rapport INIEX n° 8

DEFOURNY P., [1980]. "Résultats d'une série d'essais de perméabilité réalisés avec la cellule 700/3-200 sur des échantillons de charbon". Faculté Polytechnique de Mons, rapport INIEX n° 3

LEDENT P. et CHANDELLE V. [1978]. "Rapport final sur les expériences de linking dans les charbonnages belges". INIEX, Liège

LEDENT P., [1981]. "Etude de la perméabilité du charbon autour d'un sondage d'injection d'eau". INIEX, Liège

RABINOWITZ P., WEISS G., [1959] "Tables of abscissas and weights for numerical evaluation of integrals of the form $\int_0^\infty e^{-x} f(x) dx$", Math. Tables and other Aids to Computation, 13, 285-294.

ZIENKIEWICZ O.C., [1977] "The Finite Element Method", Third Edition, Mac Graw Hill.

ROCK STRESS INFLUENCE ON WATER FLOW IN FRACTURES
Effect de la contrainte des roches sur l'écoulement des eaux dans les fractures
Der Einfluss von Spannungen im Felsen auf die Wasserströmung in Klüften

Anders Carlsson
Ph.D., Associate Professor, Swedish State Power Board, Stockholm, Sweden
Tommy Olsson
Ph.D., Associate Professor, K-Konsult, Stockholm, Sweden

SYNOPSIS

A field test has been carried out to study the relationship between hydraulic conductivity in fractured bedrock and the state of stress in the rock mass. The test was carried out in situ, by applying a load of 3.6 MN to a 1 m^2 area of the rock and injecting water under pressure into a central borehole. The results show that the conductivity through the fractures changes noticeably under load. Fractures at right angles to the direction of load close, thus reducing their conductivity to 60-70% of the original value. But the conductivity of fractures more or less parallel to the load increases by up to 110% of the original value.

RESUME

On a effectué un essai sur le terrain afin d'étudier le rapport entre la conductivité hydraulique dans la roche mère fracturée et l'état de contrainte dans la masse rocheuse. L'essai a été effectué in situ en appliquant une charge de 3,6 MN à une superficie de 1 m^2 de la roche et en injectant de l'eau sous pression dans un trou de sonde central. Les résultats montrent que la conductivité à travers les fractures varie sensiblement sous charge. Les fractures à angle droit de l'axe de charge se ferment, diminuant ainsi leur conductivité jusqu'à 60-70% de la valeur primitive. Par contre, la conductivité des fractures se trouvant plus ou moins parallèles à l'axe de charge augmente jusqu'à 110% de la valeur primitive.

ZUSAMMENFASSUNG

Eine Feldstudie wurde durchgeführt, um das Verhältnis zwischen hydraulischer Leitfähigkeit in geklüftetem Gestein und dem Spannungszustand in der Felsmasse zu untersuchen. Die Prüfung wurde auf folgende Weise in situ durchgeführt: 1 m^2 des Gesteins wurde einer 3,6 MN Last unterworfen, während Hochdruckwasser in ein Bohrloch eingeführt wurde. Es wurde herausgefunden, daß sich die Leitfähigkeit durch Klüfte unter einer Last wesentlich ändert. Klüfte im rechten Winkel zur Lastrichtung schließen sich und vermindern dadurch ihre Leitfähigkeit auf 60-70% des Ausgangswertes. Jedoch vergrößert sich die Leitfähigkeit von Klüften, die mehr oder weniger parallel zur Last liegen, bis auf 110% des Ausgangswertes.

INTRODUCTION

In different geological contexts in conjunction with construction work, it has been shown that there are clear tendencies towards a characteristic interaction between variations in the load to which the rock mass is subjected and the hydraulic conductivity of the fractures (Carlsson and Olsson 1978 and 1979, Olsson 1979). Under certain circumstances, this effect can be extremely important to the performance of an installation, such as a dam, if the applied load has a negative effect on the tightness of the bedrock under the foundations. But the redistribution of stresses around tunnels and rock caverns also affects the rate of leakage into the structures.

A field test has been carried out under controlled and natural conditions, to study the effect of stress on the hydraulic properties of the fractures. In-situ tests have earlier been carried out in granitic rock masses by Pratt et al. (1977). But the test configuration used on that occasion was quite different to the current one, and involved horizontal loading of a vertical joint system.

The site for the present test was at Forsmark power station, to the north of Stockholm in Sweden. Among the reasons for selecting this site were:

- The geological conditions on site were well known.
- The superficial rock mass is dominated by a horizontal joint set, which facilitates tests employing vertical loading.
- The availability of electricity, water and compressed air.
- The availability of equipment for rock and construction works.

A central hole offers a possibility of installing either measuring equipment, to determine

Figure 1. The test arrangement, with six jacks located on a bedrock surface which has been levelled using concrete. The borehole visible in the foreground was used to measure the resulting water pressure. Photograph by G. Hansson/N.

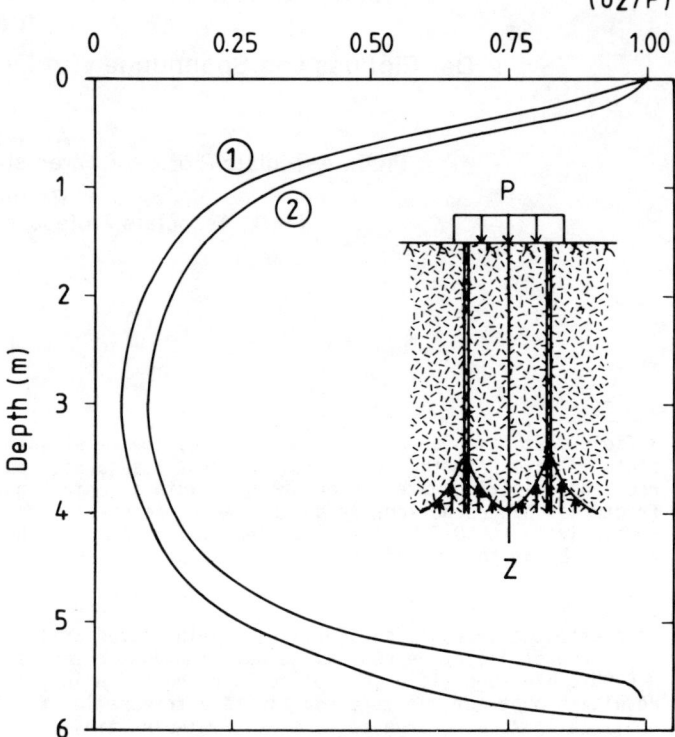

Figure 2. Stress distribution as a function of depth, calculated using a finite element method (1) and in accordance with an analytical model (2).

TEST CONFIGURATION — EQUIPMENT

To carry out a worthwhile in-situ test of this type, it must be possible to control both the loading of the rock mass and the flow of water in the fractures. This has been achieved by the combination of two methods for in-situ testing:

- Determination of the modulus of deformation, using a method developed by the Swedish State Power Board.
- Injection of water into boreholes.

The first method has been used to load the rock mass. The load was applied by six jacks connected in parallel, which each act against anchor bars grouted into 9 m deep holes in the rock mass. These jacks permit the application of a total of approximately 3.6 MPa (60 tonnes per jack) which is uniformly distributed over the 1 m^2 area levelled with concrete, as shown in Figure 1. Of course this test configuration does not achieve a uniform stress throughout the whole volume of rock under load; the stress diminishes with increasing distance from the jacks and anchors. The minimum value is recorded at a depth of 3 m, where the stress is only 5 - 10 % of the applied load, as shown in Figure 2.

the deformation properties of the rock mass, or a system of packers for water injection. The latter system is illustrated in Figure 3. It consists of double packers at 0.5 m centres. The water pressure used during measurements could be raised to about 1 MPa, with the equipment used.

During the test, it is possible to vary the load on the rock mass (using jacks) and the water pressure applied to the fractures (by water injection), permitting measurement of the following parameters:

- The load.
- The water pressure in the borehole and the water supply lines.
- The flow of water.

Figure 4 shows the interior of the test station, indicating different components for control and recording. In addition to the central borehole, there are four further peripheral holes which are used for making hydraulic measurements. The loacations of those holes are shown in Figure 1. The system of packers may also be installed in these holes, to record the water pressure outside the loaded area resulting from the water injection.

Figure 3. The system of packers for water injection tests. Photograph by G. Hansson/N.

GEOLOGY

The bedrock in the Forsmark area consists mainly of gneiss granite, which has a well developed granitic system of joints. The upper part of the rock mass is heavily fractured with open, continuous horizontal fractures. For the test, bedrock containing both horizontal and vertical fractures was sought; a condition that was well satisfied by the site selected.

RESULTS

The test configuration selected permits water to flow from the central hole, in a radial direction along the existing pattern of fractures. The flow of water is affected by double boundary conditions, partly in the boreholes where the anchor bars are installed, and partly in the measurement holes located around the periphery. The resulting water pressure at the outer boundary can be measured directly by the installed pressure gauges, whereas the inner anchor-bar holes only act as drains, with a given maximum head.

The relatively short distance between the injection hole and the controlling boundary conditions, together with the high hydraulic conductivity of the rock mass result in quick achievment of steady-state conditions of flow. This is shown in Figure 5, where the pressure build-up and fall-off are shown to a linear scale. In the measurement section considered, where the conductivity is $5 \cdot 10^{-7}$ m/s, the internal water pressure build-up took place very quickly. Steady-state conditions are reached after approximately 5 seconds. This fast and short-lived course of events makes it more difficult to evaluate the results and reduces the accuracy. The course of events is slower for recovery after the completion of injection, an effect which is attributed to a reservoir-filling effect which is equalized slowly. Three different evaluation methods have been used, one for each stage of the curve in Figure 5. Jacob's method of evaluation has been used for the pressure build-up and fall-off, i.e. the injection anf the recovery stages. For the stationary stage a simple technique based on Lugeon (1933) has been used. Despite these differences of response, the evaluation methods gave relatively similar results.

Figure 6 shows two curves with the relative hydraulic conductivity as a function of the external loading. In section 0.835-1.335 m, there is a clear tendency to reduced hydraulic conductivity with increased load. At maximum load, the value of the hydraulic conductivity has fallen to about 60%. This measurement section is dominated by horizontal fractures at right angles to the direction of the action of the load. The other example reported is for measurement section 2.90-3.40 m, a measurement section which contains steeply dipping fractures. In this section, the tendency is that the hydraulic conductivity increases with increasing load. But the increase is not as significant as the reduction obtained in the upper measurement section, and amounts to only about 10% at maximum load.

REFERENCES

Carlsson, A. and Olsson, T. (1978). Hydrogeological aspects of groundwater inflow in the Juktan tunnels, Sweden. Proc. Int. Symp. on Water in Min. and Underground Works, (1), 373-390, Granada.

Carlsson, A. and Olsson, T. (1979). Hydraulic conductivity and its stress dependence. Proc. Workshop on Low-Flow, Low-Permeability Measurements in Largely Impermeable Rocks, OECD, 249-259, Paris.

Lugeon, M. (1933). Barrages et géologie. 139 pp, Paris, Dunod.

Figure 4. *The interior of the test station. 1. The test site, with jacks on the concrete slab. 2. Water tank. 3. Panel with flow meter and controls. 4. Panel for inflation. 5. Pressure gauge. 6. Power supply and pressure gauge signal converter. 7. Printer, with jack control equipment. 8. Water pressure recorder. Photograph by G. Hansson/N.*

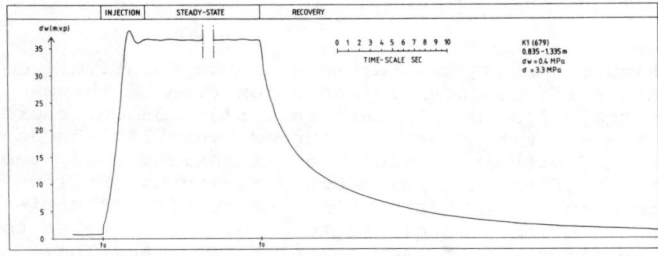

Figure 5. *Pressure change during the course of injection. The points t_0 represent the time for starting water injection and the time for starting recovery.*

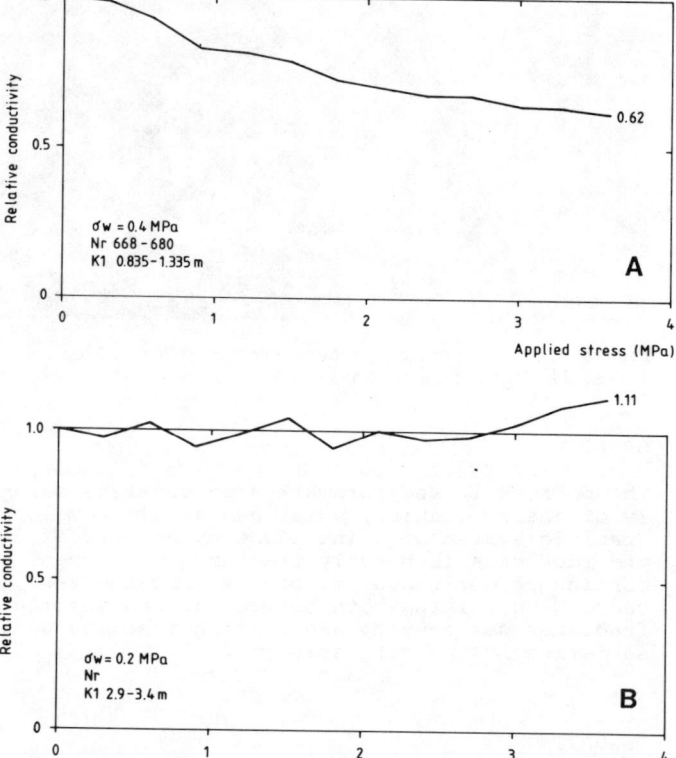

Figure 6. *The relative hydraulic conductivity as a function of the applied load. Figure 6A is taken from a section with a horizontal joint set and Figure 6B is taken from a section containing mainly a steeply dipping set.*

Olsson, T. (1979). Hydraulic properties and groundwater balance in a soil-rock aquifer system in the Juktan area, northern Sweden. Striae, (12), 72 pp. Uppsala.

Pratt, H.R., Swolfs, H.S., Bruce, W.F., Black, A.D. and Handin, J.W. (1977). Elastic and transport properties of in-situ jointed granite. Int. J. Rock Mech., Min. Sci. & Geomech. Abstr., (14), 34-45, Oxford.

THE TIME-DEPENDENT BEHAVIOUR OF RESERVOIR ROCKS IN RELATION TO FLUID PRODUCTION

Comportement des roches d'un gisement géothermique par rapport à la production des fluides, et en fonction du temps

Das zeitliche Verhalten von Speichergesteinen bei fortschreitender Förderung

T. W. Thompson
Associate Director, Center for Earth Sciences and Engineering, The University of Texas at Austin, Austin, Texas, U.S.A.

K. E. Gray
Director, Center for Earth Sciences and Engineering, The University of Texas at Austin, Austin, Texas, U.S.A.

P. N. Jogi
Research Engineer, Center for Earth Sciences and Engineering, The University of Texas at Austin, Austin, Texas, U.S.A.

SYNOPSIS

A series of creep tests have been conducted on cores from a deep geopressured-geothermal reservoir. The rocks exhibit time-dependent volume and deviatoric strains under applied hydrostatic stress and on pore pressure reduction. The results are discussed, and the effects of this behaviour on fluid production, compaction and surface subsidence are illustrated.

RESUME

Une série de tests sur le fluage à été réalisée sur des carottes ayant été extraites d'un gisement géothermique profond sous pression géostatique. Sous l'effet de l'application de contraintes hydrostatiques et de la réduction de la pression des pores, l'allongement déviatorique et le volume de la roche deviennent une fonction du temps. On présente une discussion des résultats, et les effets d'un tel comportement sur la production de fluide et sur la compaction et la subsidence de surface sont illustrés.

ZUSAMMENFASSUNG

Eine Reihe von Kriechversuchen wurde an Kernen aus tiefen, sich durch ungewöhnlich hohe Drücke und hohe Temperaturen auszeichnende Speichergesteinen durchgeführt. Unter hydrostatischem Druck und bei Absenkung des Porendruckes zeigen die Gesteine zeitabhängige Volumenänderungen und Versetzungen. Die Ergebnisse werden erläutert und ihre Auswirkungen auf Förderverlauf, Gesteinsverdichtung und Bodenabsenkung beschrieben.

INTRODUCTION

The mechanical response of reservoir rocks to fluid production can have a significant effect on this production and can lead to major surface subsidence. For example, Finol and Farouq Ali (1975) and, more recently, Rago et al. (1982) have illustrated how increases in rock compressibility can change the pore pressure decline history, and thus the total production, of a fluid reservoir. Geertsma (1966, 1973) has also examined this problem and has shown how reservoir compaction can lead to surface subsidence. This subsidence effect is well known in many areas. Examples may be quoted of large effects caused by oil production near Long Beach, California (Poland and Davis, 1962) and those caused by ground water and oil production near Houston, Texas (Gabrysch and Bonnet, 1975).

Until very recently, this rock response has been treated as a purely linear elastic problem. While this simple field behavior gives a reasonable approximation under some conditions, recent work has indicated that the deformation behavior of reservoir rocks can show significant stress dependence and can be time dependent (e.g., Jogi et al., 1981; Thompson et al., 1981; Sinha et al., 1981).

The study reported here had as its aim the determination of time-dependent mechanisms of rocks retrieved from a geopressured-geothermal reservoir on the Gulf Coast of Texas, and the investigation of the effects of this behavior on fluid production and the potential for subsidence. The reservoirs in question are deep sandstones occurring in deltaic sand and shale sequences. The contained fluids are high salinity brines occurring at high temperature and pressure. Varying quantities of methane are dissolved in these brines (Bebout and Gutierrez, 1981). For economic development these fluids must be produced at very high rates over a number of years, a situation which makes an understanding of reservoir compaction necessary. While this study is specifically aimed at this resource, the discussion of rock deformation given here has applications to all fluid reservoirs.

THE GEOPRESSURED-GEOTHERMAL RESOURCE

The Resource

Investigations of the geopressured-geothermal resources of Texas have been concentrated primarily on the Frio Formation of Oligocene age. This formation is composed of a number of shale-sandstone cycles consisting of prodelta shale at the base, delta front shale and sand in the middle, and shale at the top (Bebout et al., 1978). The existence of a blanketing shale combined with active sedimentation and subsidence has led to abnormally high fluid pressures and temperatures in the formation.

The formation fluids are primarily brines with high salinities, up to 120,000 ppm NaCl. At depths in excess of about 3000 m (10,000 ft), the fluids can have pressures of between 10.5 and 22.2 kPa/m depth (0.465 and 0.98 psi/ft depth) and temperatures in the range of 2.6 to 4.4°C/m depth (1.4 to 2.4°F/ft depth). Under these conditions considerable quantities of methane can be dissolved in the brines, forming the major economic incentive for investigating these formations.

Activities in these formations have been concentrated on the "Brazoria Fairway," located about 97 km (60 miles) southeast of Houston, Texas. Two test wells have been drilled in this area. General Crude Oil — Department of Energy Pleasant Bayou No. 1 was drilled to a depth of 4778 m (15,675 ft), but was plugged back to 2615 m (8581 ft) because of hole stability problems. Pleasant Bayou No. 2 was drilled to 5029 m (16,500 ft) and completed. Potentially productive sands were established between 4663 m (14,644 ft) and 4892 m (16,050 ft) (Bebout et al., 1978). Cores taken from these wells provided material for the current study.

Potential Production Problems

The geopressured-geothermal reservoirs have a number of potential problems which can be related to the mechanical properties of the rocks. For economic production, fluid withdrawal rates of 3200 to 6400 m^3/day (20,000 to 40,000 bbls/day) are needed. At these high rates reservoir compaction is required if production is to continue for any length of time. However, with this compaction comes the possibility of adversely affecting nearby growth faults (Bebout and Guttierez, 1981) and the potential of inducing surface subsidence (Okoye, 1980).

The present study was initiated to evaluate the mechanical response of the formation to fluid withdrawal, and hence to assess the potential for reservoir compaction.

THE TEST PROGRAM

A comprehensive rock testing program has been conducted using cores retrieved from the Pleasant Bayou wells. A large number of short-term tests were conducted (Jogi et al., 1981), during which it was found that several cores showed time-dependent deformation under constant stress conditions. As a result, a series of medium-term, constant stress creep tests have been conducted to evaluate this behavior.

Specimen Description

A total of 16 samples have been tested from various horizons. In general, the rocks may be described as very fine to fine-grained silty sands. The rocks are indurated and have porosity of 9 to 19 percent and permeability of 1×10^{-10} to 5×10^{-9} mm^2 (.1 to 5 millidarcies). The specimens show varying quantities of clay (0 to 40 percent) which occurs in thin laminae to more substantial streaks (Richardson, 1981). Samples with higher clay contents tend to show more pronounced creep behavior.

All samples were prepared by recoring to a 51 mm (2 in.) diameter and end finishing to a 102 mm (4 in.) length. Tolerances were to ISRM recommendations (ISRM, 1977). All samples were prepared with their long axes approximately perpendicular to bedding. After preparation the samples were saturated with a 6 percent saline solution to inhibit clay swelling.

Apparatus and Procedure

The samples were tested in a high-pressure triaxial cell modified to allow constant stress conditions to be maintained by using a series of pressure intensifiers and regulator controlled nitrogen cylinders (Thompson et al., 1981). All tests were run at room temperature. Axial and radial strains were monitored using strain gaged cantilevers mounted on the specimen inside the triaxial cell (Richardson, 1981).

Three main types of tests have been run. The first type was fully drained tests with an initial hydrostatic loading ($\sigma_1 = \sigma_2 = \sigma_3$) followed by a period under non-hydrostatic load ($\sigma_1 > \sigma_2 = \sigma_3$). In the second type, the same external loading path was followed, but pore pressure was allowed to build up by closing off the pore lines. In both of these tests the external loads were maintained until creep was completed or had reached an apparently constant rate. Unloading was achieved by first removing the incremental axial stress, then by removing all stresses. In the third type of test, an initial hydrostatic state of stress was imposed, with pore pressure buildup. After the sample had stabilized, the pore pressure was reduced to a new constant value. The various tests are summarized in Table 1.

THE TEST RESULTS

All of the specimens show, to a greater or lesser degree, time-dependent deformation in both the axial and radial directions. For the purposes of analysis, these deformations have been treated as dilational or volume strains, e, and axial deviatoric strains, e_z. These strains are defined by:

$$e = \varepsilon_z + 2\varepsilon_r$$

and

$$e_z = \varepsilon_z - \frac{1}{3} e = \frac{2}{3}(\varepsilon_z - \varepsilon_r)$$

where ε_z and ε_r are the axial and radial strains, respectively.

The results of the various tests are summarized in Figures 1 through 6.

Table 1. Summary of Tests Conducted under Drained, Undrained, and Pore Pressure Decline Conditions

Test #	Depth (m)	Time (Hours)	Axial Stress (MPa)	Confining Pressure (MPa)	Test Conditions*
1	4297	0-25	48.3	48.3	D
		25-122	96.6	48.3	
		122-138	48.3	48.3	
		138-186	0	0	
2	4296	0-960	48.3	48.3	U
		960-2161	96.6	96.6	
		2161-2808	48.3	48.3	
		2808-2974	0	0	
3	4284	0-497	96.6	48.3	U
		497-546	0	0	
6	4293	0-71	69.0	48.3	U
		71-80	96.6	48.3	
7	4292	0-5	69.0	48.3	U
		5-46	96.6	48.3	
8	4296	0-95	69.0	48.3	U
		95-100	96.6	48.3	
10	3126	0-51	41.4	34.5	U
		51-166	48.3	34.5	
		166-222	58.7	34.5	
		222-294	62.1	34.5	
11	3128	0-5	48.3	34.5	U
		5-338	58.7	34.5	
		338-360	0	0	
12	3128	0-333	34.5	34.5	D
		333-719	58.7	34.5	
		719-840	34.5	34.5	
		840-1006	0	0	
13A	3127	9-145	34.5	34.5	U
		145-315	58.7	34.5	
		315-341	34.5	34.5	
		341-454	0	0	
13B	3127	0-71	34.5	34.5	D
		71-240	58.7	34.5	
		240-264	34.5	34.5	
		264-343	0	0	
14A	3128	0-142	34.5	34.5	U
		142-240	58.7	34.5	
		240-362	34.5	34.5	
		362-481	0	0	
14B	3128	0-167	34.5	34.5	D
		167-474	58.7	34.5	
		474-553	34.5	34.5	
		553-690	0	0	

Table 1 (cont.)

Test #	Depth (m)	Time (Hours)	Axial Stress (MPa)	Confining Pressure (MPa)	Test Conditions*
15	3124	0-93	34.5	34.5	D
		93-240	58.7	34.5	
		240-358	34.5	34.5	
		358-383	0	0	
16	3124	0-420	34.5	34.5	PPD
		420-1776	34.5	20.7	
		1776-2256	34.5	13.8	
		2256-2400	0	0	
17	4499	0-168	34.5	34.5	PPD
		168-960	34.5	20.7	

* D = Drained, U = Undrained, PPD = Pore Pressure Decline

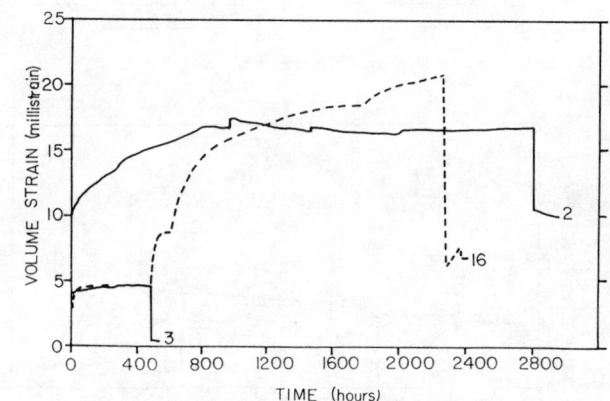

Fig. 1 Volume strain results for longer undrained tests. Solid lines (Tests 2 and 3) are fully undrained tests. Dashed line (Test 16) is a pore pressure reduction test.

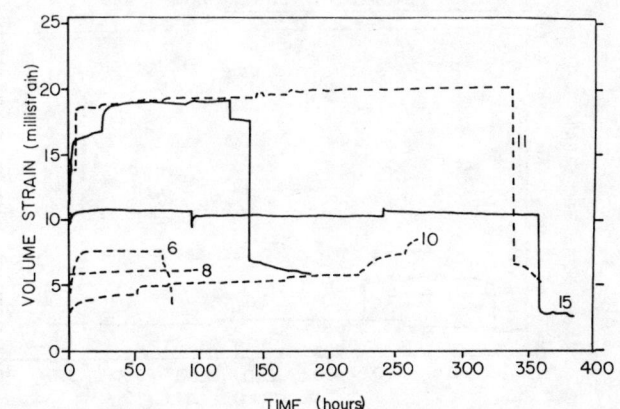

Fig. 2 Volume strain results for shorter tests. Solid lines (Tests 1 and 15) are drained tests. Dashed lines (Tests 6, 8, 10, and 11) are undrained tests.

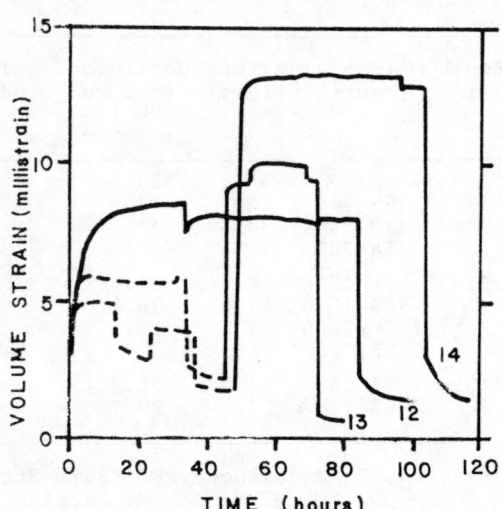

Fig. 3 Volume strain results for shorter tests. Solid lines (Tests 12, 13B, and 14B) are drained tests. Dashed lines (Tests 13A and 14A) are undrained tests.

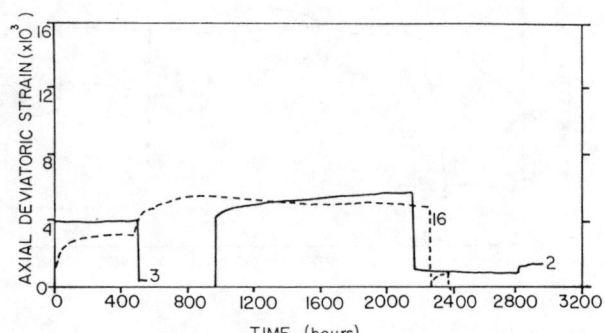

Fig. 4 Axial deviatoric strain results for longer tests. Solid lines (Tests 2 and 3) are fully undrained tests. Dashed line (Test 16) is a pore pressure reduction test.

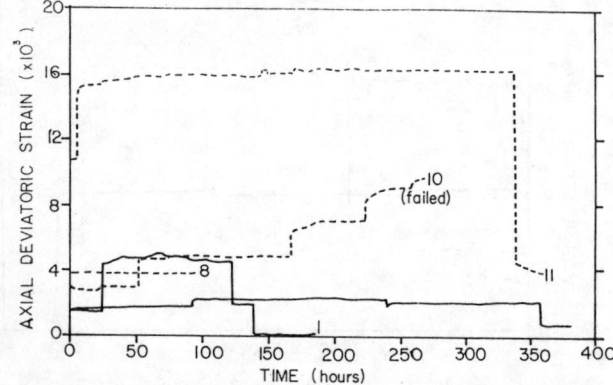

Fig. 5 Axial deviatoric strain results for shorter tests. Solid lines (Tests 1 and 15) are drained tests. Dashed lines (Tests 8, 10, and 11) are undrained tests.

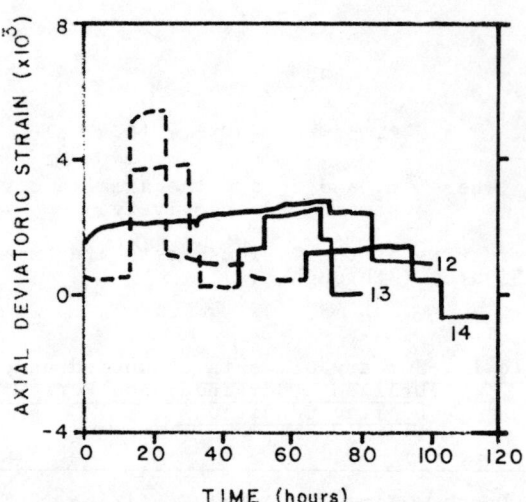

Fig. 6 Axial deviatoric strain results for shorter tests. Solid lines (Tests 12, 13B and 14B) are drained tests. Dashed lines (Tests 13A and 14A) are undrained tests.

In the remainder of this section emphasis will be placed on the dilational strains because they are of more immediate significance to the volume compaction of a reservoir than are deviatoric strains, and because their analysis is further progressed. However, it is worth noting that some deviatoric strain is seen under hydrostatic conditions, a result of the well-developed anisotropy of the samples.

Certain general conclusions may be reached from the test results, as follows:

1. Creep strain rates are high immediately after reaching full external load and in most cases reduce fairly rapidly. Final strains at a particular stress condition are nearly constant in time.

2. In undrained tests (Tests 2, 3, 6, 8, 10, 11, 13A, 14A) pore pressures do not build up instantaneously (Figure 7). This is probably due to imperfect saturation of these rather tight sands. However, over a period of time these pressures do increase, accompanying an increase in volume reduction. It is likely that the resulting decrease in "effective stress" causes some reduction in strain rates.

3. With increasing external stress, additional instantaneous and creep strains occur, but the samples behave in a "stiffer" manner than under initial loading. Pore pressures do not increase much with increasing load, which suggests a significant effect of matrix compressibility. This matrix compressibility is also seen in short-term tests (Jogi et al., 1981).

4. With increasing axial stress, radial strains tend to decrease instantaneously, which reflects radial expansion due to the applied load increment. However, in time the radial strains increase, due to continued radial strain under the constant confining pressure (Figure 8).

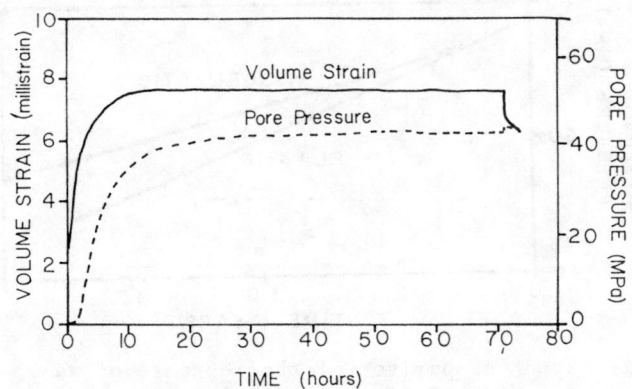

Fig. 7 Typical pore pressure response on an undrained test (Test 6).

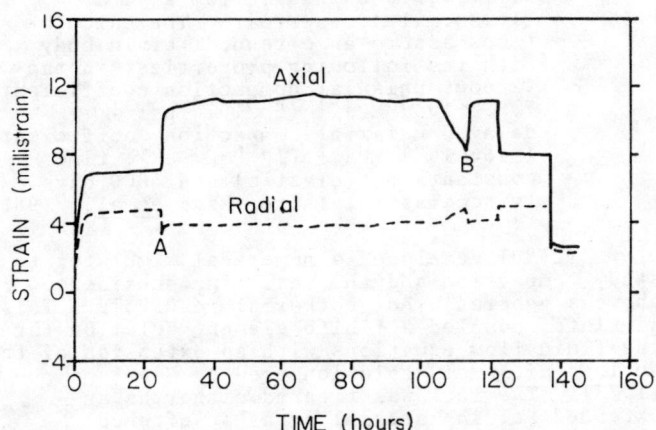

Fig. 8 Typical axial and radial strain results (Test 1). Note the instantaneous loss of radial strain on load increase followed by creep recovery (Point A). Point B represents a temporary loss of load.

5. Creep strain is accompanied by a reduction in ultimate strength. Thus, in Test 10, for example, failure occurred after about 10 hours at a differential axial stress of 27.6 MPa (4000 psi).

6. All unfailed specimens show considerable permanent set, both axially and radially, which is reflected as a permanent loss of volume. However, little permanent loss of volume is noticed on returning to a hydrostatic stress state at the start of unloading.

7. Creep can be initiated under non-deviatoric applied stress. Thus, all the tests show some creep under the initial hydrostatic stress. In particular, it will be noted that in Test 16 creep is initiated by pore pressure reduction.

A simple comparison of the test results may be made by examining their overall compressibilities. Thus, if the simple effective stress is defined as

$$\hat{p} = \sigma_m - p_p$$

where p_p is pore fluid pressure
and σ_m is the mean external stress,
then an instantaneous compressibility may be defined as:

$$C_i = \frac{\Delta e}{\hat{p}_2 - \hat{p}_1}$$

where Δe is the volume strain change occurring between effective stress $\hat{p}_1$ and $\hat{p}_2$.

By a simple extension, a "creep compressibility" may be defined as:

$$C_c = \frac{\Delta e_c}{\hat{p}_2 - \hat{p}_1}$$

where Δe_c is the total strain developed both instantaneously and over time by changing the effective stress from $\hat{p}_1$ to $\hat{p}_2$. For comparison purposes, Δe_c is taken after creep strain rates approach zero.

Table 2 compares the instantaneous and creep compressibilities for the loading parts of the different tests. An analysis of this data indicates that although there are some anomalies, there appears to be a well developed tendency for decreasing compressibility with effective stress.

Table 2. Instantaneous and Creep Compressibilities

TEST	INITIAL			AT END OF FIRST CREEP PERIOD		
	Mean Effective Stress (MPa)	Volume Strain (mε)	Compressibility (x 10^{-4} MPa^{-1})	Mean Effective Stress (MPa)	Volume Strain (mε)	Compressibility (x 10^{-4} MPa^{-1})
1	48.3	3.68	0.77	48.3	16.76	3.47
2	39.0	3.97	1.02	26.9	16.82	7.19
3	19.7	4.25	2.16	25.9	4.63	1.78
6	55.2	2.55	0.46	12.6	7.67	6.05
7	55.2	3.48	0.88	55.2	3.23	0.83
8	23.9	5.22	2.18	9.0	6.22	6.93
10	36.8	1.22	0.33	1.9	4.38	22.4
11	12.3	13.00	10.56	3.6	16.10	45.3
12	34.5	1.49	0.43	34.5	8.48	2.46
13A	8.6	4.51	5.37	5.6	5.76	10.4
13B	34.5	3.75	1.09	34.5	7.14	2.07
14A	32.3	2.86	0.88	1.2	4.65	36.2
14B	34.5	0.20	0.06	34.5	13.03	3.30
15	34.5	7.82	2.26	34.5	10.71	3.09

MODELING

A generally applicable model for volumetric creep strain of these materials has not been fully developed at the time of writing. However, it is interesting to consider the general form that this model might take. In choosing a model we have been guided by certain principles. Thus, we feel that the observed volumetric creep behavior is likely to be controlled by the action of local deviatoric stresses set up in the heterogeneous porous matrix. We also feel that the reduction of creep strain with time under load and the increasing stiffness at increasing load are reflections of the increased compaction of the material. Thus, we favor a strain hardening law.

Fossum (1977) proposed a strain hardening law to describe halite creep under deviatoric stress of the form:

$$(\dot{e}_{ij})_{cr} = B_o \, \hat{\sigma}^{n-1} \, (\hat{\varepsilon})_{cr}^{m} \, s_{ij}$$

where $(\dot{e}_{ij})_{cr}$ is the deviatoric creep strain rate,
s_{ij} is the deviatoric stress,
B_o, n and m are constants,
and $\hat{\sigma}$ is the effective creep stress defined by

$$\hat{\sigma} = \sqrt{3J_2}$$

where J_2 is the second invariant of deviatoric stress
and $(\hat{\varepsilon})_{cr}$ is the effective creep strain, defined by

$$(\hat{\varepsilon})_{cr} = \frac{1}{2} (e_{ij})_{cr} (e_{ji})_{cr} .$$

If one treats the porous rock as having single-size spherical pores in the way described by Bhatt et al. (1975) and Carroll and Holt (1975), and reported by Schatz (1975), it can be shown that Fossum's deviatoric law can be written as a volumetric law for hydrostatic stress in the form (Thompson, 1982):

$$(\dot{e})_{cr} = c_o \, \hat{p}^{n} \, (e)_{cr}^{m}$$

where $(\dot{e})_{cr}$ is the volumetric creep strain rate,
$(e)_{cr}$ is the volumetric creep strain,
c_o is a new constant,
and $\hat{p}$ is the simple effective stress

$$\hat{p} = p_c - p_p$$

where p_c and p_p are the confining and pore pressures, respectively.

Early results suggest that this law may be useful in describing the volume strain data.

APPLICATIONS

Recent work reported by Rago et al. (1982) has illustrated the effects of time-dependent deformation on reservoir production. In this work a finite difference reservoir-simulator was modified to allow the bulk rock to deform as a linear extended Kelvin material. Figure 9 shows the result of one set of simulations and illustrates how an increase in creep deformation leads to pore pressure maintenance, and can lead to increased total production.

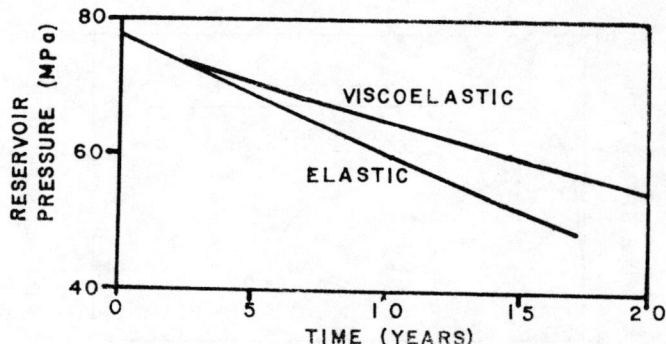

Fig. 9 A comparison of the response of an elastic and viscoelastic reservoir. Note the decrease of pore pressure reduction for the viscoelastic case. Results are for a production of 1600 m^3/day (10,000 bbl/day) from a 41 km^2 (16 sq.mile) reservoir. The calculations assume an extended Kelvin body with the following properties: instantaneous uniaxial compaction coefficient 7.25×10^{-5} MPa^{-1} (5×10^{-7} psi^{-1}), delayed uniaxial compaction coefficient 1.45×10^{-4} MPa^{-1} (10^{-6} psi^{-1}), time constants ∞ (elastic) and 3000 days (viscoelastic) (after Rago et al., 1982).

Okoye (1980) developed a numerical simulator to study compaction and the resulting subsidence above a geopressured-geothermal reservoir. This simulator coupled a finite element solution for the fluid flow equations with an extension of the nucleus of strain solution used by Geertsma (1973). The rock was treated either as an extended Kelvin material or as a softened elastic material, that is, one whose compressibility includes the effect of creep. Figure 10 shows the compaction and subsidence calculated for one set of conditions.

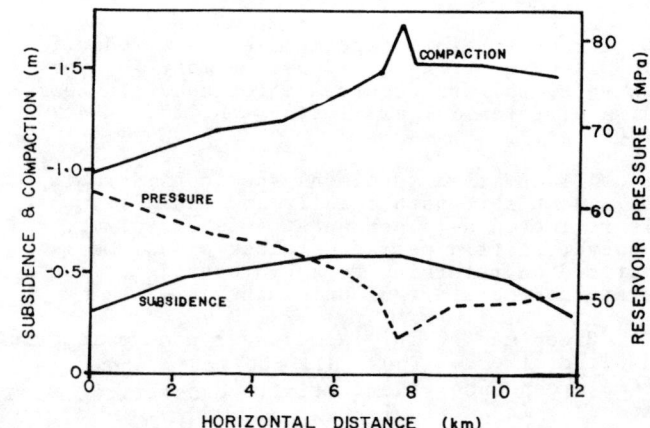

Fig. 10 Reservoir pressure, compaction and surface subsidence computed for a linear elastic reservoir. The results are for a production of 24,000 m^3/day (150,000 bbl/day) from 6 wells for a 61 m (200 ft) thick reservoir of 73 km^2 (28 sq.mile) areal extend (after Okoye, 1980).

CONCLUSIONS

The following conclusions may be drawn from this study:

1. Reservoir rocks can show appreciable time-dependent deformation within the time frame associated with fluid production operations. This creep behavior is seen in both the volumetric and deviatoric components of strain. It is initiated either by an increase in external stress or by a reduction in pore pressure.

2. It appears that the total compressibility of the rocks decreases with increasing effective stress or with increasing strain.

3. Early indications suggest that volume creep strains may be modeled using a strain hardening law (developed from a deviatoric creep law) relating volume strain rate and effective stress.

4. Creep of reservoir rocks reduces the rate of pore pressure decline and can increase total production.

5. Time-dependent reservoir compaction can increase surface subsidence effects.

ACKNOWLEDGEMENTS

The work reported here forms part of a major research effort at the University of Texas. Thanks are due to many students and faculty of the Department of Petroleum Engineering for their input and help. Particular thanks are due to the staff of the Center for Earth Sciences and Engineering for their help in the testing program.

This work was sponsored by the United States Department of Energy, under Contract No. DE-AC08-79-ET-27112.

REFERENCES

Bhatt, J.J., Carroll, M.M., and Schatz, J.F. (1975). A spherical model calculation for volumetric response of porous rocks. Journal Applied Mechanics, 42 363-368.

Bebout, D.G., Loucks, R.G., and Gregory, A.R. (1978). Frio sandstone reservoirs in the deep subsurface along the Texas Gulf Coast: University of Texas, Austin, Bureau of Economic Geology, Rept. Inv. 91, 92 pp.

Bebout, D.G., Gutierrez, D.R. (1981). Geopressured-geothermal resource in Texas and Louisiana — geological constraints: Proc. 5th Conf. Geopressured-Geothermal Energy, Baton Rouge, Louisiana, 13-24.

Carroll, M.M. and Holt, A.C. (1972). Static and dynamic pore-collapse relation for ductile porous materials: Journal Applied Physics, 43, 1626-1636.

Finol, A. and Farouq-Ali, S.M. (1975). Numerical simulation of oil production with simultaneous ground subsidence: Soc. Pet. Eng. Jour.

Fossum, A. (1977). Viscoplastic behavior during the excavation phase of a salt cavity: Int. J. Num. and Anal. Methods Geomech., 1, 1.

Gabrysch, R.K. and Bonnet, C.W. (1975). Land surface subsidence in the Houston-Galveston region: Texas Water Development Board Rept. 188, Austin, Texas.

Geertsma, J. (1966). Problems of rock mechanics in petroleum production engineering: Proc. 1st Int. Cong. Rock Mech., Lisbon.

Geertsma, J. (1973). A basic theory of subsidence due to reservoir compaction: The homogeneous case: Verhandelingen, Kon Ned. Geol. Mijnbouwk. Gen, 28.

International Society for Rock Mechanics (1977). Suggested methods for determining the strength of rock materials in triaxial compression.

Jogi, P.N., Gray, K.E., Ashman, T.R., and Thompson, T.W. (1981). Compaction measurements on cores from the Pleasant Bayou wells: Proc. 5th Conf. Geopressured-Geothermal Energy, Baton Rouge, Louisiana, 75-82.

Morton, R.A. and Ewing, T.E. (1981). Geometry and reservoir heterogeneity of tertiary sandstones — A guide to reservoir continuity and geothermal resource development: Proc. 5th Conf. Geopressured-Geothermal Energy, Baton Rouge, Louisiana, 7-12.

Okoye, D. (1980). Use of finite element methods to simulate the performance of geopressured-geothermal aquifers undergoing elastic and linear viscoelastic deformation: Ph.D. Dissertation, University of Texas, Austin, Texas.

Poland, J.F. and Davis, C.H. (1962). Land subsidence due to withdrawal of fluids: Geol. Soc. Am., Reviews in Eng. Geol., 2, 187.

Rago, F.M., Ohkuma, H., Sepehrnoori, K., and Thompson, T.W. (1982). Reservoir performance in viscoelastic porous media: Center for Energy Studies, University of Texas, Austin, Texas, 32 pp.

Richardson, J.W. (1981). A preliminary creep analysis of geopressured-geothermal reservoir rocks: M.S. thesis, University of Texas, Austin, Texas.

Schatz, J.F. (1975). Models of inelastic volume deformation for porous geologic materials: in The Effects of Voids on Material Deformation, ASME, 141-170.

Sinha, K.P., Borschel, T.F., Holland, M.J., and Schatz, J.F. (1981). Laboratory determination of mechanical properties of rocks from the Parcperdue geopressured-geothermal site: Proc. 5th Conf. Geopressured-Geothermal Energy, Baton Rouge, Louisiana, 279.

Thompson, T.W., Jogi, P.N., Gray, K.E., and Richardson, J. (1981). The time-dependent behavior of cores from the Pleasant Bayou wells: Proc. 5th Conf. Geopressured-Geothermal Energy, Baton Rouge, Louisiana, 83-86.

Thompson, T.W. (1982). Petroleum engineering research: in Edgar, T.F., Thompson, T.W., Britton, L.N., Humenick, M.J. Support research in chemical, mechanical, and environmental factors of underground coal gasification: University of Texas, Austin, Texas.

IN-SITU STRESS VARIATIONS AND HYDRAULIC FRACTURE PROPAGATION IN LAYERED ROCK — OBSERVATIONS FROM A MINEBACK EXPERIMENT

Variations des contraintes in situ et propagation des fractures hydrauliques dans une roche litée: Observations effectuées à partir d'une galerie creusée après coup

Imhomogene in situ Spannungszustände und die Ausdehnung eines hydraulisch erzeugten Bruches in geschichteten Gebirge — Beobachtungen eines Versuches vor Ort

L. W. Teufel and N. R. Warpinski
Members of the Technical Staff, Sandia National Laboratories, Albuquerque, New Mexico, U.S.A.

SYNOPSIS

In situ experiments, which were accessible for direct observations by mineback, have been conducted to determine the effect that material property interfaces and in situ stress differences have on hydraulic fracture propagation and the resultant overall fracture geometry. Fractures were observed to terminate only in regions of high minimum principal in situ stress. Fracture growth into a higher (by a factor of 5 to 15) elastic modulus region was preferred to propagation into a region of higher (by a factor of 2) stress. Determination of the in situ stress in the reservoir rock and adjacent layers can be used to assess the economical success of hydraulic fracture treatments, a priori, by predicting whether containment of the fracture within the reservoir or out-of-zone propagation and failure of the treatment will occur.

RESUME

On a procédé à des expériences in situ, suivies du creusement d'une galerie d'accès permettant des observations directes, afin de déterminer les propriétés caractéristiques des interfaces des matériaux et des différences de contraintes in situ sur la propagation des fractures hydrauliques et la géométrie d'ensemble qui en résulte. On a observé que toutes les fractures aboutissaient à des régions où la contrainte principale minimale in situ se trouvait être élevée. La propagation des fractures se faisait vers des régions à module plus élévé (de 5 à 15 fois plus élevés) plutôt que vers des régions à plus forte contrainte (2 fois plus forte). Il est donc essentiel de déterminer l'importance des contraintes in situ dans la roche réservoir et dans les couches adjacentes lorsqu'on évalue des traitements susceptibles de stimuler les fractures hydrauliques, car on peut ainsi prédire si la fracture restera confinée au réservoir ou, au contraire, se propagera hors de la zone réservoir, causant ainsi l'échec du traitement.

ZUSAMMENFASSUNG

Es wurden in situ Experimente durchgeführt, die direkter Beobachtung vor Ort (mineback) zugänglich waren, um den Einfluß verschiedener Materialgrenzflächen bzw. Änderungen des in situ Spannungsfeldes auf das Fortschreiten von hydraulischer Spaltung und auf die daraus entstehende geometrische Anordnung zu bestimmen. Es wurde beobachtet, daß Risse nur in einer Region mit hohem in situ σ_3 aufhörten. Fortschreiten der Spalten in eine Region mit höherem (5-15 fach) E-Modul wurde dem Fortschreiten in eine Region mit bis zu zweifach höheren Spannungen vorgezogen. Daraus ergibt sich, daß die Ermittlung der in situ Spannungsgrößen im Speichergestein und dessen Umgebung für die Beurteilung etwaiger hydraulischer Spaltanregungsverfahren wichtig ist, um damit voraussagen zu können, ob die Spaltung ausschließlich innerhalb der Lagerstätte stattfindet oder auch außerhalb der Zone, was ein Versagen des Verfahrens bedeuten würde.

1. INTRODUCTION

Massive hydraulic fracturing is the most promising technique for stimulation of low-permeability gas reservoirs at the present time. This technique is at least an order of magnitude scale-up from conventional hydraulic fracture technology and it is designed to create long penetrating fractures which contact large areas of the reservoir. However, the results to date have often been disappointing and the general applicability of these treatments for unconventional gas resources is uncertain.

Although there are several possible causes for the lack of success, one of the most likely reasons is the failure of the fracture to contact a sufficiently large area of the reservoir due to unfavorable vertical propagation out of the reservoir into formations lying above and below the producing zone. When a treatment is designed, the height of the fracture is the parameter about which the least is known a priori, yet this influences all aspects of the design (Perkins and Kern, 1961). Therefore, it is extremely important to recognize and understand the mechanisms which may influence the height of a fracture by restricting vertical fracture propagation (containment). Of course, these same features would be available to influence lateral propagation also.

Several parameters have been suggested as being important for hydraulic fracture containment. A difference in elastic modulus between the reservoir rock and the barrier rock is often singled out as a primary mechanism controlling containment. In their work on composite materials, Cook and Erdogan (1972) calculated the stress intensity factor for a two-dimensional crack approaching an interface between two materials with different elastic moduli. Simonson et al (1978) applied these results to hydraulic fracturing and observed that since the stress intensity factor, K, at the tip approaches zero as a fracture in a lower modulus material propagates toward a higher modulus material, the fracture will tend to be arrested. Conversely, for a fracture propagating in a higher modulus material toward a lower modulus material, they observed that K becomes large as the interface is approached, and the fracture should accelerate through the interface.

Daneshy (1978) conducted laboratory experiments and found that differences in rock properties were insufficient to stop fracture growth at an interface. He suggested that barriers may need to be defined as formations that reduce vertical fracture growth rather than prevent it. Daneshy further suggested that fracture containment may be more a result of the nature of the interface itself rather than any difference in material properties, but he thought that this would be most often the case at shallow depths where the bonding is likely to be weaker. Teufel (1979) and Anderson (1981) performed laboratory experiments to study crack growth near both bonded and unbonded interfaces. They showed that, for unbonded surfaces, the stress normal to the interface (thus the friction along the interface and the interfacial shear strength) was the determining parameter for crack arrest or continued propagation.

Perkins and Kern (1961) suggested that in-situ stress differences in different rock strata may limit fracture height. Simonson et al. (1978) also studied the effect that stress variations have on fracture propagation. They showed that a layer of greater in-situ stress would provide an effective barrier because of the increase in the fracturing pressure necessary to continue propagating a fracture in this layer. The effectiveness of the barrier would, of course, depend on the difference in stress between the two layers. They also showed that the upward or downward migration of the fractures can be influenced by the hydrostatic gradient of the frac-fluid relative to the vertical gradient of the minimum horizontal in-situ stress.

While all of these theoretical studies and laboratory experiments offer insight to the problem of containment, it is clear that such idealized results are not easily applied directly. This paper presents the results of realistic in-situ experiments which have been conducted in an existing tunnel complex at the U. S. Department of Energy's Nevada Test Site to examine hydraulic fracture behavior under many different conditions. These facilities are ideal for hydraulic fracturing experiments because they provide a realistic in-situ medium with the appropriate boundary conditions (in-situ stresses, no free surfaces) yet still allow for detailed examination of the created fractures and geological features through mineback (physical excavation of the rock to observe the fracture directly). A detailed physical description can be obtained through photography and mapping, and this can be correlated with measured geologic properties, in-situ stress distributions, fluid behavior and the operational parameters of the test. Although the various volcanic tuffs in which these fractures are propagated are not the sandstones and shales usually encountered in gas reservoirs, proper application of rock mechanics principles allows the extrapolation of these results to gas well conditions.

2. EXPERIMENT

The hydraulic fracture experiments were conducted in a tunnel complex driven into volcanic tuffs which comprise Rainer Mesa at the Nevada Test Site. At tunnel level there is approximately 400 m of overburden which provides a realistic in-situ stress distribution. The experiments are conducted in ash-fall tuffs, which are soft, low modulus, high porosity, low permeability tuffs which allow for easy excavation with a continuous mining machine. Overlying the ash-fall tuff is an ash-flow tuff which is much denser, higher modulus and lower porosity than the ash-fall tuff. The ash-flow tuff grades upward from an unwelded basal ash-flow tuff into a densely welded ash-flow tuff. The elastic properties of the region near the ash-fall tuff/ash-flow tuff interface are given in Table 1. The elastic modulus of the ash-flow tuff is a factor of 5 to 15 greater than the ash-fall tuff. The permeability of the tuffs are very low, ranging from 10^{-17} to 10^{-18} m^2.

Table 1
Elastic Properties of Volcanic Tuffs Used in This Experiment

Lithology	Elastic Modulus (GPa)	Poisson's Ratio
Welded ash-flow	20.5 - 25.3	0.12 - 0.21
Basal ash-flow	8.6 - 12.7	0.21 - 0.28
Ash-fall	0.5 - 2.7	0.26 - 0.43

The experiments utilize the immediate contact between the ash-fall tuff and ash-flow tuff as a material-property interface. As shown in Figure 1, inclined boreholes are cored either below or through the interface from an existing nearby tunnel. The holes are 0.1 m in diameter and are uncased. The core is examined and several unfractured intervals in each hole are chosen for hydraulic fracturing. These zones are usually 3 to 5 m apart.

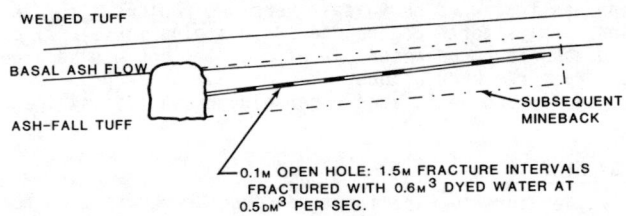

Fig. 1 Schematic of experiment

For fracturing, a mobile pump system is brought into the tunnel and situated near the collar of the hole. The pump system is capable of pumping dyed water at a maximum flow rate of 0.15 m^3/sec at 35 MPa. Straddle packers are inserted in the open hole, centered over the zone and inflated. The length of the remaining open hole zone is approximately 1.5 m. The length of the packer elements is 1.7 m. Packers are typically inflated to 5 to 10 MPa.

The fracturing procedure is as follows. After the packers are situated and inflated, the frac job begins with a very small breakdown pump at maximum flow rate to determine the breakdown pressure and the instantaneous shut in pressure (ISIP). This is usually on the order of 0.6 m^3 of dyed water, just sufficient an amount for the formation to break down and the fracturing pressure to stabilize. In this way the fracture has propagated out away from borehole effects, yet the fracture is still small enough so that the ISIP, which is equivalent to the minimum principal in-situ stress, is measured over a small area. After the pressure decays to a small value, another 0.2 to 0.6 m³ of fluid are injected into the fracture to obtain a repeat value of the ISIP. Finally, when the pressure again decays to a low value, the remaining volume to be pumped is injected into the formation at the required flow rate. These experiments are conducted with volumes of fluid from 5 m³ to 15 m³. Afterwards, the packers are moved to the next zone and the procedure is repeated.

3. RESULTS

3.1 Mineback observations

Hydraulic fractures were initiated in two inclined boreholes, CFE-1 and CFE-2, at different stratigraphic horizons in the ash-fall tuff. The instantaneous-shut-in-pressure (ISIP) obtained after the initial breakdown of each fracture zone was used to establish the vertical distribution of the magnitude of the minimum principal in-situ stress below the ash-flow tuff/ash-fall tuff interface. The magnitude of the minimum principal in-situ stress varied from 2.2 MPa immediately below the interface to 8 MPa at 2 m below the interface (Figure 2). Multiple ISIP stress determinations within the same stratigraphic horizon showed differences of less than 0.4 MPa.

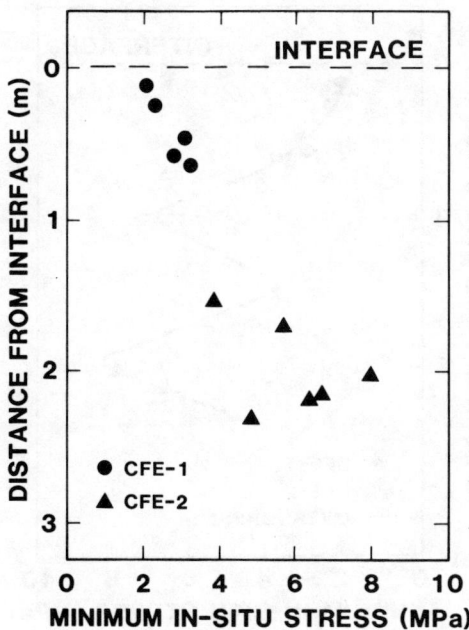

Fig. 2 Plot of minimum principal in-situ stress in ash-fall tuff below interface. Minimum stress is determined from ISIP of fractures

After completion of all of the hydraulic fractures in the CFE-1 and CFE-2 holes, a mineback was conducted of each hole to observe directly the behavior of hydraulic fractures near a major material-property interface and in regions of varying in-situ stress. During mineback it was found that each fracture initiated nearly perpendicular or at high angles to the borehole, usually near one of the packers, and propagated essentially in a vertical plane which consistently had a trend of N20°E to N30°E.

In the CFE-1 drift, all of the fractures preferentially propagated horizontally outwards and upwards through the interface into the higher modulus ash-flow tuff (Figure 3). The orientation of the fractures did not change during propagation across the interface, but remained in a vertical plane. There were, however, significant changes in the width of the fractures in regions of differing modulus with the larger width found in the low modulus region. The upper extent of these fractures was always several meters into the densely welded ash-flow tuff. The higher modulus welded tuff was clearly not a containment barrier to fracture propagation as predicted by current fracture models (Simonson et al., 1978).

Fig. 3 Photograph of partially exposed fracture which is nearly parallel to the mineback face. The fracture clearly breaks through the interface

In sharp contrast to the uncontained upward propagation of the CFE-1 fractures, downward propagation in the ash-fall tuff was severely restricted. All of the hydraulic fractures were arrested at the same stratigraphic horizon about 1 m below the interface (Figure 4). Near the location of the fracture termination, the fractures were generally impregnated with more dye than at any other location. This indicates that the fractures probably propagated to this point at an early time and remained inflated during pumping even though fracture propagation had terminated in that direction. Fluid continued to leak off, resulting in the heavy dye residue along the fracture surface at the termination point.

3.2 Stress variations and fracture containment

A possible explanation of why the lower termination of all the CFE-1 hydraulic fractures occurred abruptly about 1 m below the interface was that the minimum in-situ stress was significantly greater in that horizon than in overlying horizons where the fractures had been initiated. Previous ISIP data clearly showed that high stress regions can occur in the ash-fall tuff (Figure 2). In order to obtain additional data on the vertical distribution of the magnitude of the minimum

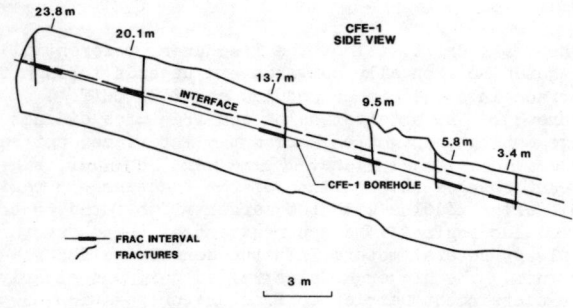

Fig. 4 Side view schematic of CFE-1 mineback drift showing vertical extent of fractures. All of the fractures cross the interface but terminate 1 m below interface

principal in-situ stress, particularly in the region of fracture termination, overcoring stress measurements were made in the CFE-1 drift at three stratigraphic horizons (Figure 5). The horizons were 0.2 m, 0.6 m, and 1.0 m below the interface. The upper two horizons corresponded to horizons where hydraulic fractures had been initiated and thus a direct comparison could be made between stresses calculated from overcoring strain-relief measurements and ISIP data. The lower horizon was the horizon of fracture termination.

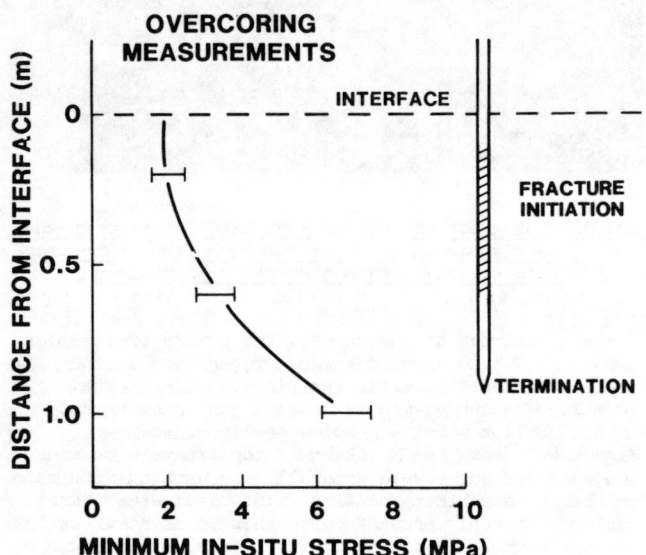

Fig. 5 Plot of minimum principal in-situ stress in ash-fall tuff below interface and extent of fractures in CFE-1. Stress determined from overcoring measurements

All of the overcoring stress determinations were made with the well-documented USBM overcoring technique (Hooker and Bickel, 1974). Elastic moduli for overcore samples were determined with a biaxial loading device and the USBM three-component borehole deformation gage (Fitzpatrick, 1962). The overcoring measurements were at drill-hole depths outside the zone of significant influence of the mineback drift. As such, the overcoring stress data should be representative of free-field in-situ stresses. The drill-holes were parallel to the general azimuth of the vertical hydraulic fracture planes, and had an orientation of N25°E. Three overcoring measurements were made at each horizon.

At the abrupt lower termination of all the CFE-1 hydraulic fractures, 1 m below the interface, the magnitude of the minimum principal in-situ stress was 6.8 ± 0.6 MPa, which was a factor of 2 greater than the two overlying horizons where these fractures were initiated (Figure 5). The 3 MPa increase in the minimum in-situ stress was sufficient to arrest the fractures and act as a containment barrier. Fracture pressure was apparently not sufficient to drive any of these fractures through the high stress horizon. Instead, fracture growth after initiation was always outward and upward into the higher modulus and lower stress regions.

Variations in the minimum in-situ stress magnitude also influenced the propagation of fractures initiated in the CFE-2 hole. Combining ISIP data and overcoring measurements provides a detailed plot of the vertical distribution of the minimum in-situ stress magnitude (Figure 6). At 1 m and 2 m below the interface, stress peaks occurred and the minimum stress was greater than in surrounding horizons by up to a factor of 3. In the CFE-2 hole, fractures were initiated between the two

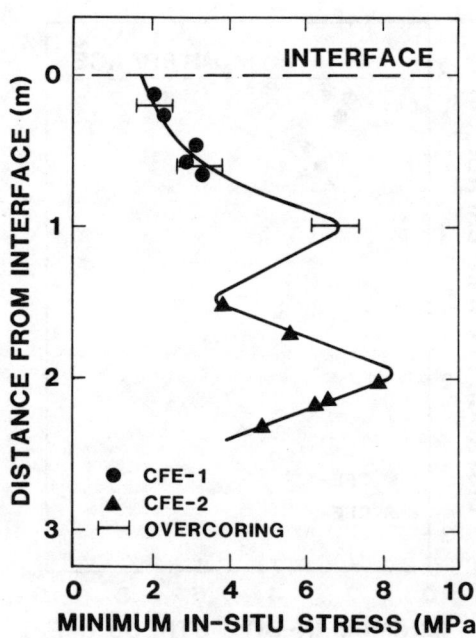

Fig. 6 Plot of minimum principal in situ stress in ash-fall tuff below interface from ISIP data and overcoring measurements. Two stress peaks occur at 1 m and 2 m

stress peaks, in the lower stress peak, and below the lower stress peak. In the CFE-1 hole, fractures were initiated only above the upper stress peak. The effect of these two stress peaks on fracture propagation is shown schematically in Figure 7. Fractures which were initiated above or below the stress peaks propagated away from the high in-situ stress regions when the stress difference was greater than 2 MPa. Fractures initiated between the stress peaks were totally contained vertically in that region. These contained fractures had rectangular fracture geometries and the fracture length was considerably greater than the height. Fractures in the lower stress peak region (ISIP's of fractures were 6.5 to 8.0 MPa) had uncontained downward propagation, but upward propagation was contained at the upper stress peak after the fractures had propagated through the intervening lower stress region. These results demonstrate that fracture growth in tuff can be contained by a region of high stress, but only when the resulting stress difference ahead of the propagating fracture is greater than 2 MPa.

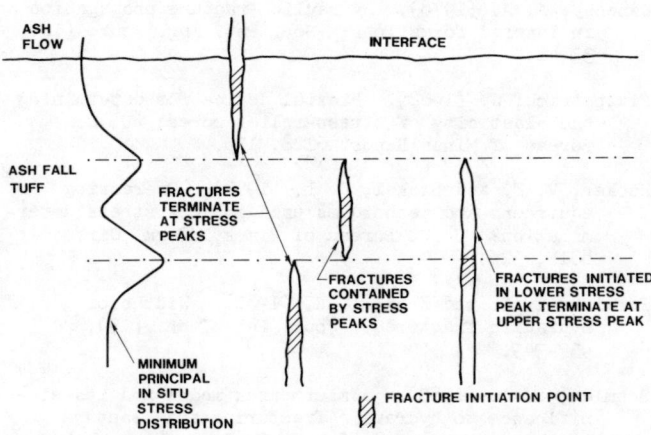

Fig. 7 Schematic of results showing effect of stress peaks on fracture containment

4. DISCUSSION

Hydraulic fractures were initiated below a material property interface in a region with large variations in the minimum principal in-situ stress magnitude. Each fracture was injected with 5 to 10 m^3 of dyed water. The location of fracture initiation varied from 0.15 m below the interface (higher modulus material above) to 2.3 m below the interface. None of the fractures had their growth restricted in any apparent way by the higher modulus layer (factor of 5 to 15 difference). These results are consistent with those of a previous large scale experiment (Warpinski, et al., 1982) which was also conducted near the same interface. Laboratory experiments also show that fractures will not be contained by a change in properties across an interface (Teufel, 1979). These results do not agree with the present two-dimensional analyses of crack behavior at an interface which would predict containment (Simonson et al., 1978). The failure of these analyses apparently lies in modeling the behavior at the interface using only the stress intensity factor at the tip of the crack for the failure criterion. In order to properly model the mechanisms operating in this problem, it may be necessary to link a more realistic failure criterion with the stress analysis of a propagating crack (Schmidt, 1980).

The ISIP obtained during breakdown of each fracture test and overcoring measurements were used to establish the vertical distribution of the minimum principal in-situ stress magnitude (Figure 6). At two horizons (0.2 m and 0.6 m below the interface) the ISIP data and overcoring measurements overlapped and were in good agreement. Within the ash-fall tuff the minimum stress was not constant nor a linear gradient with depth, but exhibited large variations in magnitude. At 1 m and 2 m below the interface the minimum stress increased by a factor of 2 to 3 or 2 to 4 MPa relative to the surrounding horizons. These variations in stress magnitude had a significant and dominant influence on the propagation and overall geometry of the hydraulic fractures. Fractures were arrested whenever the fractures propagated from a low stress to a high stress region in which the stress difference of the minimum in-situ stress was greater than 2 MPa. Fractures initiated in a low stress layer which was bounded by high stress layers were totally contained vertically and had rectangular fracture geometries.

The high stress layers provide an effective containment barrier because fracture propagation into these layers requires an additional increase in fracture pressure (Simonson et al., 1978). If the fracture pressure does not increase sufficiently, propagation terminates at the high stress boundary but continues in other directions. The effectiveness of the stress barrier depends on the difference in the minimum in-situ stress between the bounding layer and the layer with the propagating fracture. In these experiments with volcanic tuff, at a depth of 400 m, the critical stress difference is 2 to 3 MPa. In deeper wells and other lithologies, the critical stress difference for containment may be higher. Laboratory experiments have shown that the critical stress difference to contain a hydraulic fracture ranges from 2 to 5 MPa, depending on the boundary conditions and lithologies used in the test (Teufel and Clark; 1982 and Warpinski, et al., 1982). We are presently conducting a field experiment in low-permeability, gas-bearing sandstones and bounding shales to evaluate the magnitude of the critical stress difference required for fracture containment in this type of reservoir at a depth to 2500 m.

The cause of the in-situ stress variations found in this experiment are thought to be due to a combination of gravitational loading applied to the layered rock mass, which had varying elastic properties and boundary relief at the sides of the mesa. For a layered rock mass subjected only to gravitational loading the magnitude of the horizontal stresses in individual layers increases with increasing depth. The relative difference in the horizontal stress from one layer to another will be a function of the relative difference in elastic properties of the layers (Teufel and Clark, 1982). A compressional increase in the minimum horizontal stress will occur in going from a layer with a high shear modulus (due to a high elastic modulus and/or low Poisson's ratio) into a layer with a low shear modulus (due to a low elastic modulus and/or high Poisson's ratio). In this experiment the stress peaks occurred in horizons where the shear modulus was the lowest (Figure 8). This indicates that differences in elastic properties may be important, not as a containment barrier per se, but because of the manner in which elastic property variations influence the vertical distribution of the minimum horizontal in-situ stress magnitude. It is expected that the most significant changes in in-situ stress will occur in the vicinity of a change in rock properties, typically due to layering.

The results of this study have important implications to the design of massive hydraulic fracture treatments for stimulation and enhanced recovery of low-permeability gas reservoirs where it is economically imperative that the fracture be contained within the reservoir

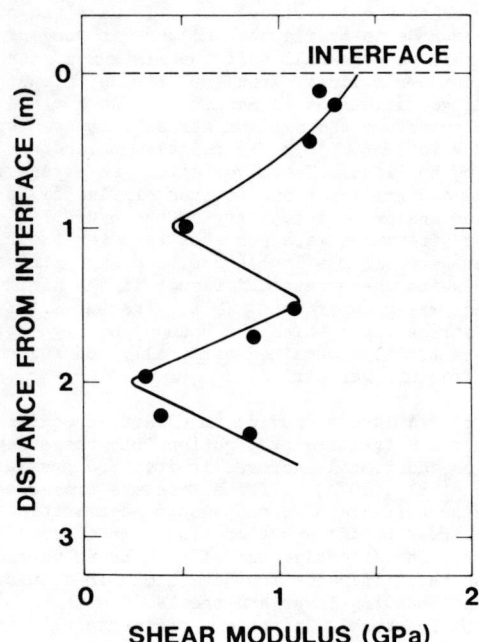

Fig. 8 Plot of shear modulus in ash-fall tuff below interface. Horizons with lowest modulus correlate with stress peaks (Figure 6)

formation. The predominant influence of in-situ differences in the minimum stress on hydraulic fracture containment clearly indicates that an evaluation of the economic success of potential fracture treatments, prior to stimulation, can be made by determining the in-situ stress state in the reservoir-fracture interval as well as in the bounding layers. This will enable selection, a priori, of zones that are bounded by high stress layers and that will most likely provide economic production due to favorable growth and containment conditions.

5. CONCLUSIONS

Mineback experiments have provided insight into the mechanisms which are responsible for controlling fracture geometry. It has been demonstrated that a factor of 5 difference in elastic modulus between adjacent layers is insufficient to arrest a fracture at an interface, as predicted by present fracture models. These experiments have shown that the minimum principal in-situ stress is the predominant influence on fracture containment. Fractures were contained whenever there was a sufficiently large increase in stress in the bounding rock layers. In layered volcanic tuff the critical stress difference was 2 to 3 MPa. These results indicate that an assessment of the potential success of fracture treatments of low-permeability gas reservoirs in which containment is desired can be made by determining the in-situ stress in the fracture interval and bounding layers.

6. ACKNOWLEDGEMENTS

This work performed at Sandia National Laboratories supported by the U. S. Department of Energy under Contract Number DE-AC04-76-DP00789.

7. REFERENCES

Anderson, G. D. (1981). Effects of friction on hydraulic fracture growth near unbonded interfaces in rocks: Soc. Pet. Eng. Jour., 21, 21-29.

Cook, T. S. and Erdogan, F. (1972). Stresses in bonded materials with a crack perpendicular to the interface: Int'l. Jour. Eng. Sci. (10), 677-697.

Daneshy, A. A. (1978). Hydraulic fracture propagation in layered formations: Soc. Pet. Eng. Jour. (18), 33-41.

Fitzpatrick, J. (1962). Biaxial device for determining the elasticity of stress-relief cores: U. S. Bureau of Mines Report 6128, 13p.

Hooker, V. E. and Bickel, D. L. (1974). Overcoring equipment and techniques used in rock stress determination: U. S. Bureau of Mines Inform. Cir. 8618, 32p.

Perkins, T. K. and Kern, L. R. (1961). Widths of hydraulic fractures: Jour. Pet. Tech. (13), 937-949.

Schmidt, R. A. (1980). A microcrack model and its significance to hydraulic fracturing and fracture toughness testing: 21st U. S. Symp. Rock Mech., Rolla, MO, 581-590.

Simonson, E. R., Abou-Sayed, A. S., and Clifton, R. J. (1978). Containment of massive hydraulic fractures: Soc. Pet. Eng. Jour. (18), 27-32.

Teufel, L. W. (1979). An experimental study of hydraulic fracture propagation in layered rock: Ph.D. Dissertation, Texas A&M University, College Station, TX.

Teufel, L. W. and Clark, J. A. (1982). Hydraulic fracture propagation in layered rock: experimental studies of fracture containment: Soc. Pet. Eng. Jour., Paper No. 9834, in press.

Warpinski, N. R., Northrop, D. A., and Schmidt, R. A. (1978). Direct observation of hydraulic fractures: behavior at a formation interface: Sandia National Laboratories Report, SAND78-1935.

Warpinski, N. R., Clark, J. A., Schmidt, R. A., and Huddle, C. W. (1982). Laboratory investigation on the effect of in-situ stresses on hydraulic fracture containment: Soc. Pet. Eng. Jour., 22, 333-340.

Warpinski, N. R., Schmidt, R. A., and Northrop, D. A. (1982). In situ stresses: the predominant influence on hydraulic fracture containment: Jour. Pet. Tech., (34), 653-664.

DEVELOPMENT AND ESTIMATION OF ROCK BREAKING METHODS
Le développement de la technologie et du calcul d'abattage des roches
Die Entwicklung der Verfahren und der Berechnung zum Lösen des Gebirges

Y. I. Protasov
I. F. Oksanich
P. S. Mironov
VIOGEM Institute, Belgorod, USSR; Moscow Mining Institute, Moscow, USSR

SYNOPSIS
This paper deals with questions of solving the intensity problem of rock breaking by blasting and other practices except blasting, and methods of blasted mass lumpiness prediction. Factors affecting rock breaking extent are considered. The breakability constant is determined. It was established that the rock breaking mechanism is estimated by the value of the blast pulse and its duration. The estimation methods for blasting parameters and blast control are stated.

RESUME
Cet exposé concerne le processus d'abattage des roches par l'explosion ou sans explosion ainsi que les procédés de prévision de granulométrie des masses abattues. On a envisagé les facteurs qui influencent le degré de fragmentation des roches et on a déterminé la constante d'aptitude au fractionnement. On a établi qu'on peut déterminer le mécanisme de fragmentation non seulement par l'intensité, mais aussi par la durée de l'impulsion de l'explosion. On a exposé les méthodes de calcul des paramètres des tirs de mine et de contrôle de l'explosion.

ZUSAMMENFASSUNG
Im Beitrag werden die Hauptrichtungen der Lösung des Intensivierungsproblems der Gesteinszerstörung mit Hilfe der Spreng- und Nichtsprengverfahren, sowie die Methoden der Voraussage der Zusammensetzung des gesprengten Haufwerks erläutert. Es werden die Faktoren, die den Grad der Gesteinszerstörung beeinflussen, erörtet und die Konstante der Zerkleinerung festgestellt. Es wird festgestellt, daß der Mechanismus der Gesteinszerstörung nicht nur durch die Größe, sondern auch durch die Dauer des Sprengimpulses festgestellt wird. Es sind die Berechnungsmethoden der Parameter der Bohrsprengarbeiten und der Sprengsteuerung dargelegt.

The question of rock breaking intensity increase and blasted massif lumpiness prediction acquires specific actuality in USSR of late because in broad sense of the word the notion "breaking" must cover the following meanings: blast breaking in pit (mine) and mechanical crushing at crushing and dressing plants. It should be noted that all kinds of breaking through the full cycle beginning from rock separation from massif up to the third stage, i.e. crushing at crushing plants, occupy 35-45% in mining and ferruginous quartzite processing cost as concerns mining and ore dressing integrated works of Krivbass.

The importance of this problem is conditioned by necessity of on-line or cyclic and on-line mining in pits characterized by high production rate. This problem solution depends on possibility of predetermined lumpiness prediction on while rock blasting. Thus mining flowsheet can't be now adopted if efficiency of rock breaking and lumpiness prediction is not taken into account. Lumpiness prediction while blasting is based on regularities of rock blasting. Therefore during last years in our country statistical method of description of rock blasting regularities and lumpiness prediction is developed as it proved to be the most effective one. Below main principles of the method developed referring to rock breaking regularities and lumpiness prediction are stated. This method has been already put into practice. It can be used at mine design stage or at design stage of some separate systems and types of mining. While applying statistical approach the authors used energy conservation law, i.e. energy of breaking is equal to useful energy, which explosives deliver to rocks. The authors also used the law of conservation of form i.e. the kind of distribution law.

As a criterien of rock mass breakage efficiency the fuller usage of blast energy on useful work is assumed. The authors used a law which can be called as the law of conservation of form, i.e. the law of distribution of lumps in broken massif corresponds to lump distribution in massif before blasting. Lumps in massif may be bonded with each other and that's why breaking occurs along the weakness planes. Lump exponential distribution in broken massif based on rock specimen statistical estimation meets practical requirements. In order to get closed solution it is necessary to state geometric relationship between lump surface and one of the distribution parameters. In the given set of equations the factor for useful energy transfer from explosive to rock and energy expenditure on lump surface unit formation are not known. In principle these parameters can be easily estimated theoretically. The easier way of their achieving is the experimental one. A relationship between lump surface unit value with energy specific unit expenditure and rock strength and mean distance between fractures are achieved on the base of industrial pit and mine blasting data processing. The three mentioned relationships and relationships for breakability constant values determine law of breaking. The law of conservation of energy is written in the following form

$$K_2 S = K_1 W \quad (1)$$

where
  W - blast energy;
  K_1 - factor of useful energy transfer from explosive to rock;
  S - lump surface;
  K_2 - energy needed for surface unit formation.

The law of conservation of form is written as follows:

$$p_o = 1 - e^{-\alpha x^{\gamma}} \quad (2)$$

where p_o - probability of getting of lumps, sizes from Δx up to x, α and γ are distribution parameters.

For most rocks γ is equal to 1. For monomineral rocks of gorblendite type which is highly unbreakable it is equal to 1,5.

According to the given lump distribution law lump surface in broken massif depends on distribution parameters α as follows:

$$S = K_\gamma \alpha^{\frac{1}{\gamma}} V \quad (3)$$

where V - bulk of massif being broken;
  K_γ - factor depending on parameter γ in distribution.

After set of equations (1), (2), (3) are solved the law breaking will be written as follows:

$$p_o = 1 - e^{-(\varphi q x)^{\gamma}}; \quad q = \frac{W}{V}; \quad \varphi = \frac{K_1}{K_2 K_\gamma} \quad (4)$$

The equation (4) is the law of lump distribution in broken massif. The lump mean diameter in broken massif for rocks characterized by normal breakability is equal to

$$D = \frac{1}{\varphi q} \quad (5)$$

and for highly breakable rocks

$$D = 0.9 (\varphi q)^{-\frac{2}{3}} \quad (6)$$

The authors call the parameter φ a breakability constant.

Thus for achievement of breakage to the given extent as an estimation base a breakability constant is assumed which depends on many factors; the main of them are physical and mechanical properties of massif being blasted; type of applied explosives, explosive charge design and diameter, blast practice and patterns, detonation direction; blast conditions (free exposure plane availability, blasting in closed space) parameters of drilling and blasting operations etc.

Breakability constant is detrmined by industrial blasts data for standard type explosive (ammonite 6 B), cylindric charge standard diameter 0,225m and while blast benching in pits. Basing on test data a set of curves of relationship of reference breakability constant against rocks compression strength and mean distance between fractures was plotted (Fig.1).

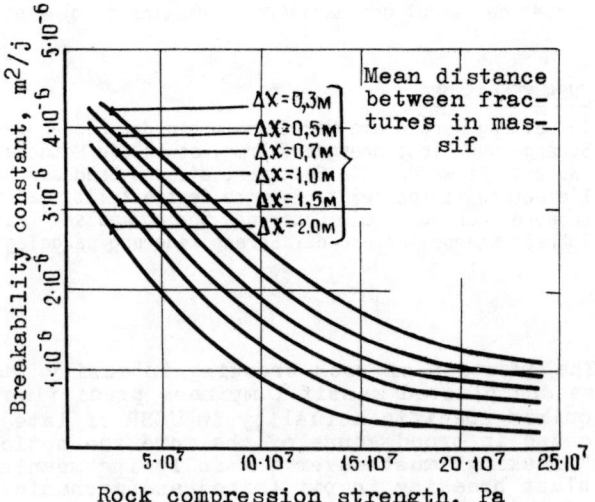

Fig.1.Relationship of breakability constant against rock ultimate uniaxial compression strength and fracturing

Under different conditions of blasting design breakability constant is found by way of correction to reference constant in the form of coefficients for detonation impedance, well (blast-hole) diameter, exposed surface number, well (blast-hole) nonparallelelism, row number, excavation cross-section.

Using the law of breaking (formulas 4,5,6) and relationships for breakability constants it is possible to predict lumpiness and estimate parameters of blasting operations according to predicted data for driving and mining in pits and mines.

For lumpiness prediction it is necessary to determine rock compression strength of specimen and mean distance between fractures in massif. One must choose cylindric charge diameter, explosive energy requirements, explosive type (estimate explosive detonation impedance), specify blasting conditions which determine ψ corrections to breakability constant. Then one must estimate this constant taking into account blasting conditions, explosive type, well (hole) diameter. Further lump mean diameter is estimated according to formula (5) on account of energy expenditures on rock volume unit or probability of predetermined lumpiness yielding is determined according to formula (4).

As a base of breaking intensity increase methods as concerns rock blasting and breaking uniformity achievement the authors took results of research of Soviet and foreign scientists (O.E.Vlasov, M.V.Machinskii, M.A.Sadovskii, T.J.Pokrovskii, A.N.Khanukaev, Kuamo Khino, White et al), who had developed the theory of blast action in rocks. The reveal of mechanism of rock breaking and crushing by blasting, searching new methods of blast control are the main trends of technical progress development in the field of rock breaking intensity increase while blasting. It was stated that the most important characteristics of rock breaking by blasting are stress field parameters depending on blast pulse value and shape.

Changing value and shape of blast pulse resulted in predetermined lumpiness of rock. A laboratory experiment with results of blast action simulation and achieved results checking as concerns block blasting under industrial conditions were assumed as a base of breaking mechanism research. The most important parameter of breaking process is the fracturing velocity; its estimation for various modes of blasting made it possible to consider the process of blasting with taking time factor into account and to evaluate the optimal delay period for short-delay blasting under specific conditions and time of beginning of breaking process and displacements in massif as well. As a result one of the valid relationships between breaking intensity and uniformity and energy action duration increase while blasting in massif was established.

This permitted to increase blast energy efficiency while rock breaking. It was proved that blast impulse value determined by explosion gases pressure and detonation velocity are not so important as they were prescribed earlier. The proper chosen detonation direction, initiating charge embedding position and blasting conditions (short-delay blasting on broken massif etc.) that favour time increase of blast energy effect are of particular importance. Time delay for multiple-shot blasting is 15-25m/s (in a group of charges) and 25-50m/s (between the groups of them).

Some other factors which influence blast pulse shape and mode were determined. Explosive type and specific charge for explosives, charge shape and diameter, the sequence and conditions of their blasting, short-delay blasting patterns and techniques etc. are determined. The resulted relationships made it possible to control blasting process and its performance. Rock breaking mechanism while single charge and multiple short-delay blasting, i.e. sequence of rock breaking with various systems of fractures, has been studied in order to develop blast control technique. Breakability depends mainly on characteristics and intensity of stress fields which move at a velocity of 2000-5000m/s. Massif breaking intensity is determined by fracture system density.

Breaking mechanism is governed mainly not only by the value of blast pulse but by its duration as well, i.e. for rock breaking stress in massif and its duration in accordance with kinetic theory of strength are significant factors. This is one of the main conclusions resulted from rock breaking mechanism studies.

Specific charge increase according to formulas (5) and (6) results in rock lump mean size reduction after blasting. Besides the greater axplosive requirements the lower rock strength in broken lumps and rock strength decreases mainly due to disturbance of boundaries among minerals. This results in less expenditures of energy on the further ore dressing because crushing and grinding power intensity reduces, besides, concentrate quality is improved due to better mineral grains opening. Thus the increased specific charge for rock breaking proved to be correct, as truck capacity together with crusher and mill production rate increases greatly resulting in mentioned operations cost cut down. Specific charge increase is due to well pattern dimension decrease, well diameter widening, pocket formation in wells. In each situation one of the methods of specific charge increase must be chosen. Well widening by means of thermal rock breacking is widely used. As a result of these studies the following technical progress trends in the field of breaking efficiency increase and rock blasting can be formulated.

As a basis for developing new types of explosives and charge design the increase of duration of blast action on massif is taken. This duration increase is achieved owing to using granulated explosives, decked and composite charges, detonating points optimal position, pocket and parallel closed spaced charges. In order to achieve more uniform rock mass breaking, diminish blast detrimental effects on enviromental and desintegration in broken rock mass there was decided to introduce short-delay blasting.

Buffer blasting and blasting in blasted earlier massif meet these requirements. The most effective patterns of multiple short-delay blasting are as follows: row by row blasting, row by row cut-hole blasting, perpendicular cut-hole blasting, diagonal, herring-bone like, radial, wave-like, combined and other types of blasting.

One of the main trends of technical progress in the field of rock breaking pattern improvement is the perfection of blasted rock mass lumpiness prediction. Simultaneously with development and improvement of rock blasting reseaches and attacking a problem of rock breaking by mechanical, thermal and electrical processes are being done.

As per two last methods rock comes forward as working body, i.e. it transforms the applied energy, thermal, for instance, into breaking. In the above-mentioned practice rock comes forward as a tool which breaks rock. This practice is characterized by the absence of needs for special tools and by the fact that thermal and electric energy break rocks.

Let's consider this practice from the power-producing point of view. Breaking power production means that energy Q being introduced into rock is transformed in power of breaking. The general form of power equation is as follows:

$$dA = dA_c - \frac{V_o P}{E_o} dP \qquad (7)$$

where V_o - rock bulk which energy is introduced into;
dA_c - power in bulk V_o as a result of energy of bulk V_o transforming into power;
P - rock massif response to bulk V_o;
E_o - bulk modulus.

In equation (7) the second term represents deformation power of bulk V_o by specific force P which is response of massif being broken to dilation of bulk V_o. The relationship of two terms of equation (7) gives maximum value of breaking power and energy transformation.

Energy transformation by bulk V_o occurs at the expense of its dilation. For thermal practice value dA_c is written as:

$$dA_T = \alpha P V_o dT \qquad (8)$$

where α - is the heat-expansion factor;
dT - temperature increment in bulk V_o in comparison with the ambient temperature.

For electricity-applied practice of rock breaking

$$dA_э = \alpha P \frac{UI dt}{cm} \qquad (9)$$

where U - voltage;
I - current in bulk V_o;
dt - energy release time.

For mechanical breaking the value dA_c is written as follows:

$$dA_m = \frac{M V_o P}{SE} dF \qquad (10)$$

where M - Poisson's ratio;
S - area of contact of tool with rock;
E - Young's modulus for rock;
dF - force F increment affecting bulk V_o.

In equation (7) P value is the process parameter and that's why it can be changed if necessary. But if changed P value may result in being equal $P = P_{opt}$ maximum value of power for one and the same energy value introduced in bulk V_o of rock.

Changing dA_T (8), $dA_э$ (9), dA_m (10) and introducing them into equation (7) one can get for thermal, electric and mechanic energy, respectively, the following equations:

$$A_T = \alpha P V T - \frac{V_o P^2}{2 E_o} \qquad (11)$$

$$A_э = \frac{\alpha P U I t}{cm} - \frac{V_o P^2}{2 E_o} \qquad (12)$$

$$A_m = \frac{2 V_o M P F}{S E} - \frac{V_o P^2}{2 E_o} \qquad (13)$$

Using formulas (11), (12), (13) from equations $\frac{\partial A}{\partial P} = 0$ we get the value

$$P_{mT} = \alpha E_o T$$

$$P_{mэ} = \frac{\alpha U I E_o t}{cm V_o} \qquad (14)$$

$$P_{mm} = \frac{2 M F}{3 S (1 - 2M)} \qquad (15)$$

Using P_m from equations (13), (14), (15) we get power maximum value:

$$A_{mT} = \frac{1}{2} \alpha^2 E_o V_o T^2 \qquad (16)$$

$$A_{mэ} = \frac{\alpha^2 U^2 I^2 E_o t^2}{6 c^2 m^2 (1 - 2M) V_o} \qquad (17)$$

$$A_{mm} = \frac{2 M^2 V_o F^2}{3 (1 - 2M) S^2 E} \qquad (18)$$

Using values of maximum power from equations (17), (18), (16) we can get maximum efficiency of thermal, electrical and mechanical energy transformation into power of breaking.

$$\eta_T = \frac{\alpha^2 E T}{6 cm (1 - 2M)} \qquad (19)$$

$$\eta_э = \frac{\alpha^2 U I E t}{6 c^2 m^2 (1 - 2M) V_o} \qquad (20)$$

$$\eta = \frac{2 M^2}{3 (1 - 2M)} \qquad (21)$$

Thus if working body is affected by load P_m we get maximum power and maximum efficiency from it. The value of maximum efficiency for thermal and electrical energy varies in range of 2-3% (for solid) - 50-60% (for plasma); as concerns mechanical breaking this value varies in the range of 3-5 - 15%.

Estimation of rock breaking processes being considered in this paper is done basing on law of energy conservation which for mechanical breaking is written as follows:

$$4h\mu P\alpha F - 3(1-2M)V_o P\alpha P = 2K\sigma^2 dV \qquad (22)$$

From equations like equation (22) the maximum value of bulk being blasted is determined:

$$V_m = \frac{2M^2 F^3}{3B\sigma^2 \sigma_o S(1-2M)K} \qquad (23)$$

and rock breaking minimum energy expenditure:

$$q_{min} = \frac{3\sigma^2 K(1-2M)\sigma_o S}{2M^2 EF} \qquad (24)$$

where σ - rock ultimate tensile strength;
B - tool blade width;
σ_o - rock bulk strength.

Comparing the effectiveness of mechanical, electrical and thermal practices of rock breaking one can state that the most effective is the electrical practice which is characterized by energy expenditure 0,1-0,3 kWhr per ton. The advantage of electrical and thermal practices lies in that fact that tools applied actually can't be worn out unlike tools applied in mechanical practice.

The thermal practice is used for well drilling in rocks characterized by low energy expenditure for breaking with temperature of rock breaking surface not more than 250°C. As rock massif is characterized by the presence of hard-broken seams the application of thermal practice of drilling is limited by few deposits, sites and rock massives. The most wide application of thermal breaking is typical for well expansion including mechanically drilled wells.

The field of application of thermal breaking depends on area of its affecting: if rock surface subjected to thermal breaking is increased the deformation of heated seam progresses as well together with rock breaking temperature; rock elastic properties are almost the same, as to efficiency of breaking it increases sharply. Well expansion permits well pattern to be widen that results in breaking performance rise and cost reduction.

Electrical breaking is used for ore mining, mining excavation driving, well expansion in underground mines. Mode of breaking provides for massif large area simultaneous treating that results in energy expenditure reduction for breaking up to 40-60kWhr per m^3. High-strength rock driving rate is 0,3-0,5m/h. It should be noted that mining excavations driven with the help of electricity are more stable than ones driven with drilling and blasting operations because electrical practice of breaking doesn't disturb massif beyond the excavation borders and at the same time it permits to preserve massif strength.

Underground well expansion rate upon the diameter is 0,3m/h. It should be noted that by means of one working tool it is possible to expand well diameter up to 0,5-0,7m.

Electrical practice is used for oversize rock lump breaking in open pits and in underground mines. It is ease of handling as it doesn't require main process stopping and doesn't affect mechanisms' workability. Hard ore breaking from massif while mining isn't widely used. Mining machine production rate is 30-50t/h. In principle there is a possibility to increase it for an order greater than the one mentioned and though there are some engineering difficulties there they are to be overcome. Electric ore grinding makes it possible to increase mill production by 20-30% and reduce grinding energy expenditure by 1,5-2kWhr if mill power is increased by 10-12%.

The combined practice of breaking is evidently of the greatest interest. The thermal and electrical practice is combined with the mechanical one; the electrical practice is combined with blasting etc. The combined practice if proper combinations and optimum conditions are kept to makes it possible to avoid drawbacks of practices being combined and to maintain and intensify their advantages. For instance, combining of electrical and mechanical practices makes it possible to reduce breaking energy expenditure by 30-40% as compared with each practice separately and to increase machine production in 2-3 times. Combining being applied mechanical tool wear reduces 1,5-2 times, machine mass is lowered, its price and breaking production cost reduce.

Rock massif electrical treating by blasting makes it possible to reduce specific charge and oversize output nearly to zero, and to widen well pattern. For such treating electrical and thermal break-down is used.

Such method applied the predetermined lumpiness may be achieved with high reliability.

Rock breaking done with the help of some other practices including the combined ones excepting blasting makes it possible to develop new mechanisms and methods of rock breaking, reduce losses and mineral impoverishment. In the end the application of these practices will yield high economic efficiency.

Evidently it is combined practice that is reasonable to develop further as the base for new mining machines and processes creation.

Literature

Оксанич И.Ф., Миронов П.С. Закономерности дробления горных пород взрывом и прогнозирование гранулометрического состава. М., Недра, 1982, 155с.

THE APPLICATION OF THE FINITE ELEMENT MODEL OF THE NÄSLIDEN MINE TO THE PREDICTION OF FUTURE MINING CONDITIONS

Application du modèle en éléments finis établi pour la mine de Näsliden à la prévision des conditions futures d'exploitation à cette mine

Anwendung eines Modells mit finiten Elementen für das Bergwerk Näsliden zur Vorhersage künftiger Bergbaubedingungen

T. Borg
Division of Rock Mechanics, University of Luleå, S-951 87 Luleå, Sweden

N. Krauland
Boliden Mineral Inc., S-936 00 Boliden, Sweden

SYNOPSIS

Finite element models of the Näsliden Mine, Sweden, were developed and their agreement with mine behaviour was checked in the Näsliden Project. This paper describes the application of an elastic FEM-model to the prediction of mining conditions to the end of the mine's life. The extension strain failure criterion is used for the prediction of potential failure zones. The criterion is calibrated against actual mine behaviour in one stope. Stages of failure are predicted for the remaining stopes and the predictions are checked against available evidence.

RESUME

Des modèles en éléments finis concernant la mine de Näsliden, en Suède, ont été réalisés et leur concordance avec le fonctionnement effectif d'une mine a été contrôlée dans le cadre du projet Näsliden. Cet exposé décrit l'application d'un modèle élastique en éléments finis à la prévision des conditions de l'exploitation minière jusqu'à la fermeture définitive de la mine concernée. Le critère de rupture par tensions d'allongement est utilisé pour prévoir les zones de défaillance potentielles, ce critère étant basé sur la situation réellement constatée pour un gradin. Le degré de rique de défaillance est évalué en ce qui concerne les autres gradins et ces évaluations sont vérifiées par rapport aux données concrètes disponibles.

ZUSAMMENFASSUNG

Modelle mit finiten Elementen für das schwedische Bergwerk Näsliden sind entwickelt und hinsichtlich der Übereinstimmung mit den wirklichen Grubenverhältnissen für das Projekt Näsliden geprüft worden. Dieser Aufsatz beschreibt die Anwendung eines elastischen Modells mit finiten Elementen zur Vorhersage der Abbauverhältnisse bis zum Ende der Lebensdauer des Bergwerks. Das Kriterium des Dehnungsbruches wird für die Vorhersage von potentiellen Bruchzonen benutzt. Das Kriterium wird auf das wirkliche Gebirgsverhalten einer Kammer abgestimmt. Bruchzustände werden für die restlichen Kammern vorhergesagt, und die Vorhersagen werden mit vorhandenen Berichten verglichen.

1 INTRODUCTION

The present work is a continuation of the Näsliden project, which aimed at evaluating the suitability and reliability of finite element models as a rock mechanics planning instrument in mining. In the project, a linearly elastic model and a model, in which joint elements were included, simulating the weak alteration zones adjacent to the orebody, were developed. The degree of agreement with actual mine behaviour was established by comparison with rock mechanics observations and measurements carried out in the Näsliden Mine from the start of mining in 1970. A comprehensive presentation of the Näsliden project has been given at the Conference on the Application of Rock Mechanics to Cut and Fill Mining, Luleå, June 1980, Stephansson and Jones, 1981.

For the purpose of the present study the main results of the Näsliden project were

- the development of a linearly elastic model with a known degree of agreement between model and mine behaviour

- a joint element model giving better qualitative information on the rock behaviour in the immediate vicinity of an excavation when the rock mass approaches a state of failure. However, introduction of joint elements increased quantitative uncertainty mainly due to difficulties in determining the necessary mechanical parameters of joints. Complexity and cost of model work increased significantly.

The Näsliden project covered the early stages of mining. Only one stope was approaching the final stage of excavation and experiencing stability problems.

The aim of the present study is to develop a method for the prediction of future mining conditions. From the miners' point of view it is important to obtain a prediction of those stages of mining, at which mining has to be stopped or the mining process has to be changed either by changing the mining method or introducing new stabilization measures. The study is therefore concentrated on the prediction of those types of failure that are due to excess loading of the rock mass.

2 CUT AND FILL MINING IN THE NÄSLIDEN MINE

The geology and structures of the Näsliden mine have been described by Stephansson, 1981, Fig. 1. Cut and fill mining is used and the ore body is mined in slices 3.5 m high. The open stope is back filled with either sand or mill tailings.

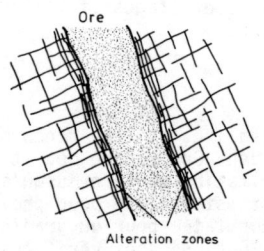

Fig. 1 Vertical section across ore body, alteration zones and side wall of the Näsliden mine, Stephansson, 1981

After back filling mining of the next slice starts with entries from ramps in the foot-wall, Fig. 2.

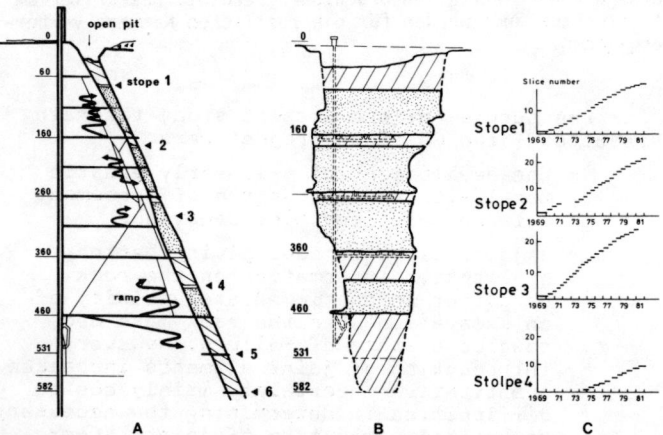

Fig. 2 Vertical sections of the Näsliden mine
A) cross section
B) section along the strike
C) mining sequences

At present mining is carried out in four stopes, starting at 460 m, 355 m, 255 m and 155 m levels. Ramps are being driven for the opening of stopes 5 and 6 in years 1983 to 1984. An open pit was mined to the 40 m level during 1972-1973. The annual production in 1975 was 250,000 tonnes of ore.

3 METHOD

In this study it was decided to try an approach comprising the following steps:

a) Calculate the elastic response of the rock mass to mining in terms of stresses and/or strains.

b) Determine a failure criterion for the rock mass.

c) Compare elastic stresses/strains with limits of the failure criterion. The loci of points where the stresses/strains equal the limits of the failure criteria defines the boundary of the potential failure zone.

d) Critical levels in the development of failure are defined from observations in the mine:

 - fracturing, often violent

 - rock fall problems; mining can still be continued

 - ultimate failure; mining has to be stopped or changed

 The criterion for these levels are quantified by comparing observations from stope 3 in the mine with calculated values of stresses/strains for the same stage of mining.

e) Prediction of the critical levels for all mine stopes up to the end of mining. Validity of this prediction is checked by comparing predicted critical levels for the stopes with practical experience from the mine.

4 OBSERVATIONS OF FAILURES IN THE MINE

An important part of the follow up program in the Näsliden mine is the observation of failure phenomena which were presented by Nilsson and Krauland, 1981.

4.1 Types of failure

Fig. 3 shows the common types of failure in cut and fill mining as observed in Näsliden and other Boliden Mineral Inc. mines.

4.2 Sequence of events

Considering roof failure due to high, nearly horizontal stresses the following critical levels in the development of failure can be distinguished in the stopes:

Stage 1: Brittle fracture of the roof occurs, often violently; nearly horizontal fracture surfaces are thereby created. Stability of the stopes is maintained by fairly sparse rock reinforcement, mainly rock bolting.

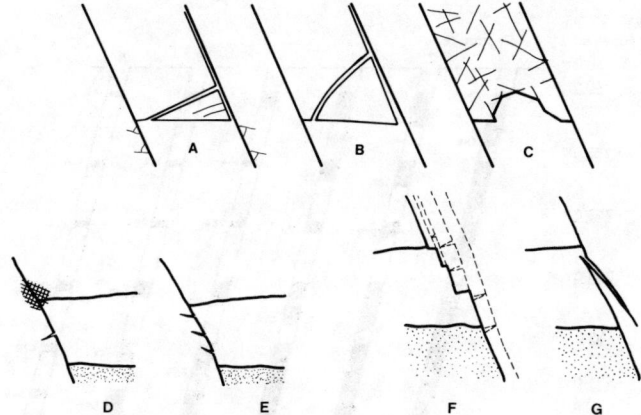

Fig. 3 Types of failure
A) wedge formation in roof, often violent, B) roof failure when hanging wall is very weak, C) roof failure due to structural features in the roof, D, E) foot wall failures, F, G) hanging wall failures, Nilsson et al., 1981

Stage 2: Failure of the face by means of subsidence of large wedges and some roof falls. Mining can still be continued but frequent scaling and an increased amount of support are necessary.

Stage 3: Ultimate failure, continuous roof failure including crushing, also at a great distance from the face. Stable roof conditions cannot be achieved by scaling and roof bolting. Termination of mining or change over to another mining method/stabilizing measures.

A graphical presentation of the field observations defining stage 1-3 is given in Fig. 4.

4.3 Variations in failure phenomena and stability conditions

There is, of course, a fairly wide variation in geological conditions and hence in mechanical properties of the rock mass, resulting in variations in rock behaviour. One significant difference is the composition of the foot wall alteration zone. In the southern part of stope 3 it consists of hard sericitic quartzite. In this area of the mine violent failure of the roof occurred, whereas the foot wall remained intact. In the northern part of the ore body the foot wall alteration zone consists of soft chloritic quartzite with low strength and a low modulus of elasticity. Here the roof in the ore body remained intact, whereas the foot wall was more or less fractured.

5 FAILURE PREDICTION

The extension strain failure criterion (ESFC) suggested by Stacey, 1981, was used in this study for the prediction of failure. The ESFC is stated by Stacey as follows: "Fracture of brittle rock will initiate when the total extension strain in the rock exceeds a critical value which is characteristic of that rock type. This may be expressed as follows: Fracture initiates when

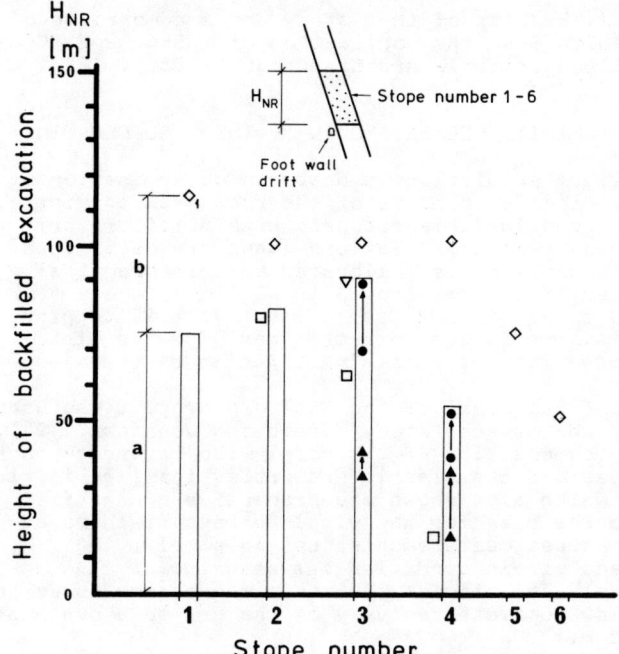

NR stope number
a height of backfilled excavation
b height of remaining ore up to bottom of nearest backfilled excavation
▲ fractures
● roof falls
▽ ultimate failure
◇ bottom of nearest backfilled excavation above stope
◊ open pit bottom
⊓ present level of roof (1982)
☐ onset of failure in foot wall drift above stope

Fig. 4 Summary of failure observations in Näsliden up to 1982

$$\varepsilon_3 \geq \varepsilon_c \qquad (1)$$

where ε_c is the critical value of extension strain and ε_3 is the minimum principal strain.

The fractures will form in planes normal to the direction of the extension strain, which corresponds with the direction of the minimum principal strain. For a material which shows ideal linear elastic deformation behaviour, the strain in this direction is related to the three principal stresses by the following equation:

$$\varepsilon_3 = \frac{1}{E} \left[\sigma_3 - \nu (\sigma_1 + \sigma_2) \right] \qquad (2)$$

where σ_1, σ_2 and σ_3 are the principal stresses, E is the modulus of elasticity, and ν is Poisson's ratio."

Inspection of failure surfaces in the mine indicated that the extension strain criterion could be applied. However, the Coulomb failure criterion (CFC) has also been applied to the Näsliden mine. Results of that study and a comparison of results from the application of (ESFC) and (CFC) failure criteria are presented by Borg, 1982.

6 FINITE ELEMENT MODEL OF THE NÄSLIDEN MINE

Failure prediction is based on determination of the elastic response of the rock mass to mining. By comparing this response with a failure criterion, potential failure zones are determined. The criterion is calibrated against practical experience from stope 3 in Näsliden, where mining had to be stopped due to severe stability problems. Predictions are then made for the other stopes for the remaining lifetime of the mine.

The final model of the Näsliden project was used for the present study, Groth and Jonasson, 1981. The composition of the model with regard to rock types and the elastic properties is given in Fig. 5, which also shows the properties of the fill and the data for the virgin stress field based on stress measurements, Leijon et al., 1981. Plane strain condition was assumed for the model. The size of the model was chosen so as to allow complete recovery of the ore to a depth of 582 m.

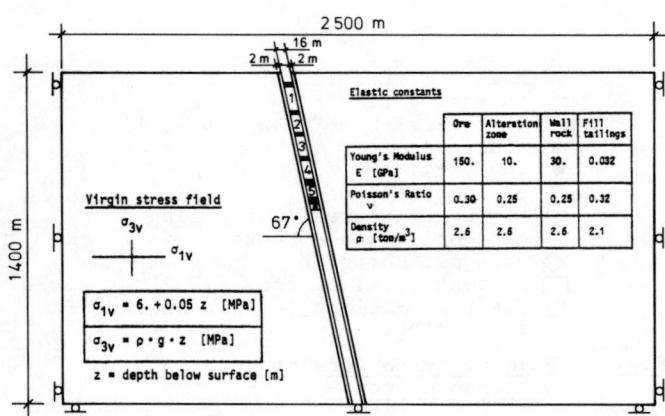

Fig. 5 Virgin stress field, material data, model geometry, and boundary conditions of finite element model

6.1 Mining sequences

The mining stages were chosen as follows, Fig. 6:

- Three historic mining sequences, 1-3, were selected in which failures important for the calibration of the failure criterion occurred
- On the basis of long term mining plans, three future mining situations, 4-6, were selected, at which significant changes in mine behaviour were expected from the rock mechanics point of view.

7 APPLICATION OF THE EXTENSION STRAIN FAILURE CRITERIA (ESFC)

The ESFC have been calibrated against important events in stope 3 of the Näsliden mine, defined as the three critical stages shown in Fig. 4,

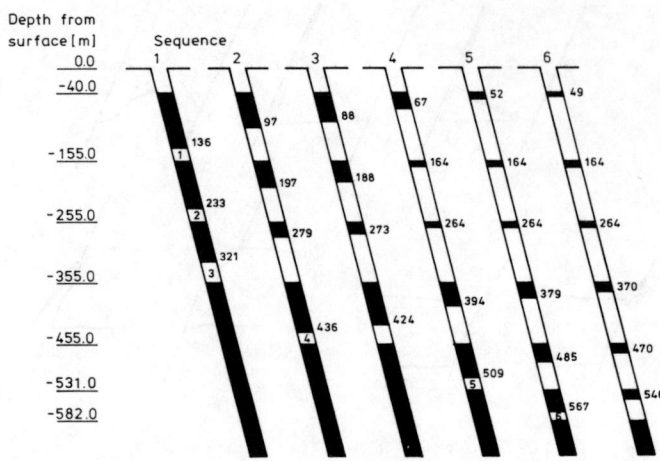

Fig. 6 Mining sequences simulated in the finite element model

and summarized in Table 1. The position in the model where the maximal extension strain occurs is used for quantification of the failure criterion. The minimum principal strain ε_3 in the lowest row of finite elements across the roof of the stope is plotted in Fig. 7a for mining sequences 1-4 as shown in Fig. 6. The maximum values of ε_3 at the early mining stages occur at a distance of about 3 m from the hanging wall. This point is referred to as Q_1. At a late stage in mining, sequence 4, and when the sill pillar condition is reached, the position of the maximum value shifts to the proximity of the foot wall, Q_2.

Table 1. Stages of failure as observed in stope 3 and extension strain levels in model

Stage of failure	Year	Height of back-filled excavation [m]	Height of ore between stope 2 and 3 [m]	Elevation of roof [m]	Extension strain $\varepsilon_3 \cdot 10^{-6}$
1	1973	34.	66.	321.	150.
2	1979	70.	30.	285.	210.
3	1981	89.	11.	266.	300.

Fig. 7b shows ε_3 as a function of the height above the roof. Only a narrow zone of the roof is subjected to high extension strain until the sill pillar condition is reached when high ε_3 values occur over a large part of the ore body. Development of ε_3 at point Q_1 for all six stopes as a function of the height of the ore body between adjacent stopes is shown in Fig. 7c. Determining ε_3 for stope 3 for the three defined stages of Table 1 gives the ε_3 values, shown in the same table.

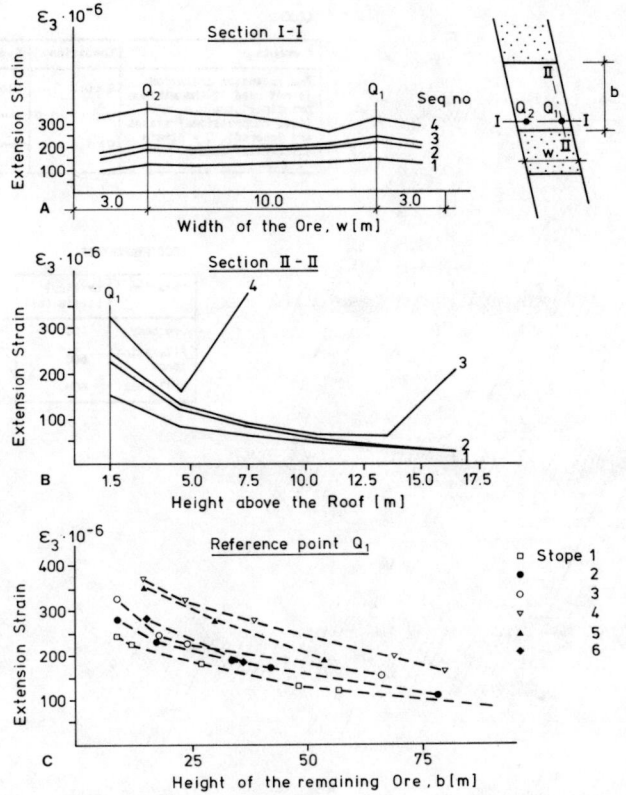

Fig. 7 Development of extension strains in the ore

Stress measurements were performed in the roof of stope 3 at roof elevation 295 m, at a distance of about 5 m from the foot wall, Leijon et al., 1981. In the ore adjacent to the roof surface, violent failure occurred at the time of the measurements. Core discing was observed during the rock stress measurements and only measurements with doorstoppers were sucessful. At a depth of 2.25 m from the surface of the roof the first successful measurement with a three-dimensional Leeman cell was obtained, giving strain values of 160. µs. This result is in good agreement with the estimated critical strain value 150. µs, Table 1, earlier estimated from model results.

7.1 Failure prediction with application of ESFC

When the critical strain has been determined, critical failure zones can also be located. These zones are evaluated as the quotient between the actual strain and the critical strain corresponding to mining, stage 1, i.e. 150. µs. Thus the development of critical extensions around the stopes for mining sequence 1-6 can be plotted, Fig. 8.

In a similar way as for the ore, the critical extensions were determined for the alterations zones, ε_c = 600. µs, and the side wall rock ε_c = 400. µs. The results indicate a narrow critical zone in the roof of the stopes in the early stages of mining. Critical strains also occur at the hanging wall side in the alteration zone above and immediately beneath the roof level and on the foot wall side beneath the roof surface.

At a late stage in mining, when the sill pillar condition is reached, critical strains also occur in the side wall at the levels of the pillar and within large parts of the pillars. This indicates that failure of the pillars or punching of the pillars into the side wall are possible modes of failure.

Finally the development of extensions at the reference point Q_1 in Fig. 7c is used to determine the critical stages for stope 1, 2, 4, 5, and 6, as shown in Fig. 9.

7.2 Comparison of the prediction and in situ observations

If observed failures shown in Fig. 4 are compared with the prediction, Fig. 9 this gives:

Stope 1: Roof failures in few locations have occurred. Good agreement.

Stope 2: Fracturing has been observed only during mining of the two latest slices.

Stope 3: Not applicable to this analysis as the observations are used for the calibration of the model.

Stope 4: Violent failures started somewhat earlier than predicted. Serious failures in the hanging wall and in the roof of the mining face are being experienced since the height of the backfilled excavation was 36 m. An unfavourable joint orientation affects the situation.

Foot wall drifts along the ore body exists at 160 m, 260 m and 360 m levels. They are situated mainly in the alteration zone at a distance of 0 to 4 m from the ore body. Extensive damage to these drifts is observed as mining approaches from below and as the height of the ore body between the stopes above and below the drift decreases. Previous investigations have shown that onset of failure in the drifts is well correlated to the maximum principal stress at approximately 40 MPa at the position of the drift, Krauland et al., 1981.

8 DISCUSSION

Three restrictions for the presented prediction are discussed. Firstly the elastic model is applied to quantify failures due to excess loading of the rock mass. Secondly, the elastic model is applicable strictly to the stage of fracture initiation. After that redistribution of stresses will occur. Thirdly, the failure observations in stope 3 of the mine are used for the prediction of failure for the other stopes. Thus, the validity of the results depends on the assumption that the properties of stope 3 is relevant for the other stopes.

The failure type which has been used as the main prediction event, is that of brittle failure of the roof due to excess loading. From mining experience brittle failure is known to cause the most serious problems in the Näsliden mine. However, other types of failures do occur, as described in chapter 4. Failure or rock fall from

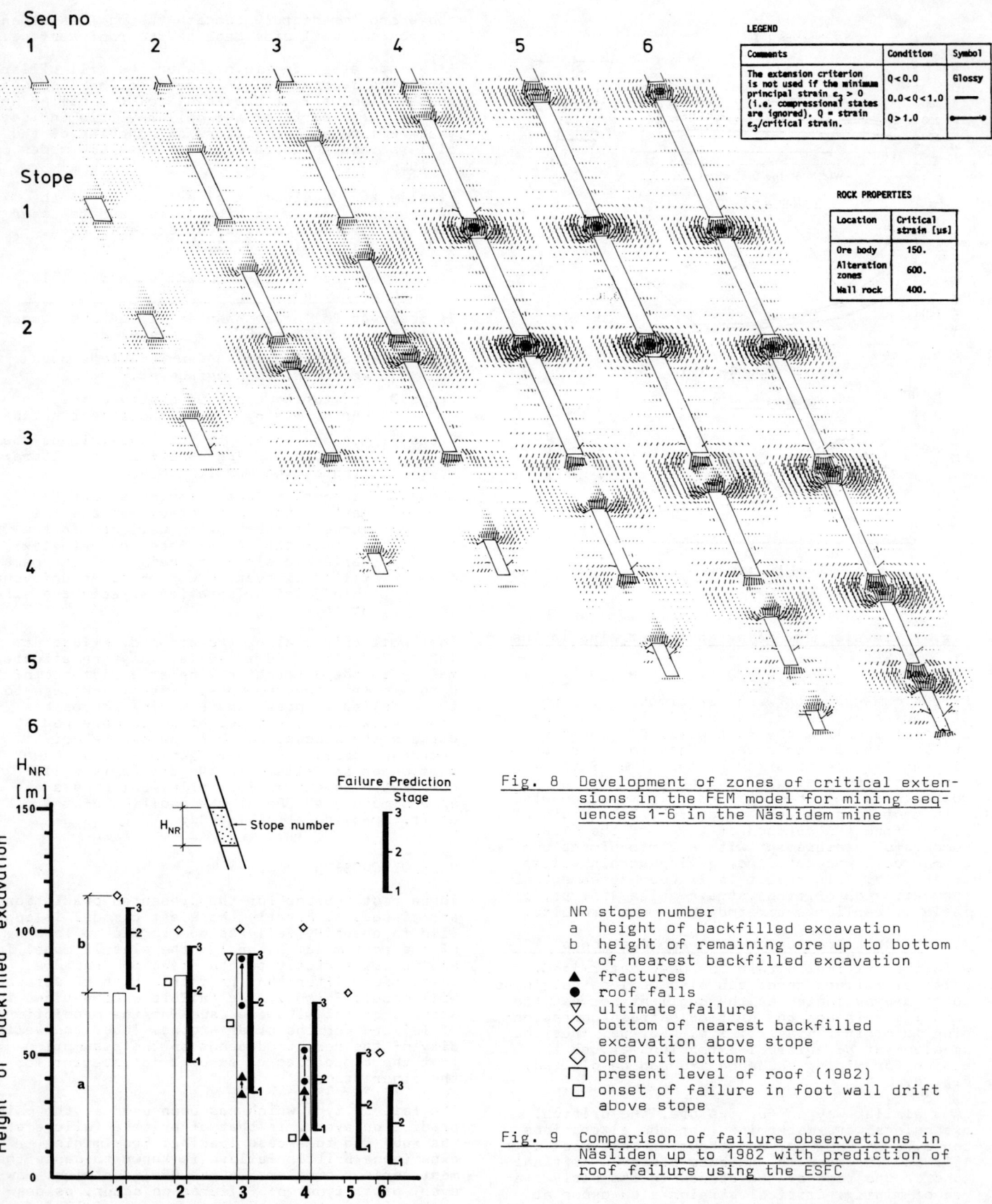

Fig. 8 Development of zones of critical extensions in the FEM model for mining sequences 1-6 in the Näslidem mine

NR stope number
a height of backfilled excavation
b height of remaining ore up to bottom of nearest backfilled excavation
▲ fractures
● roof falls
▽ ultimate failure
◇ bottom of nearest backfilled excavation above stope
◇ open pit bottom
▭ present level of roof (1982)
☐ onset of failure in foot wall drift above stope

Fig. 9 Comparison of failure observations in Näsliden up to 1982 with prediction of roof failure using the ESFC

one or both side walls leads to downward movements of ore slices along the dip of the alteration zone. This induces movements in the ore body, which results in additional extension strains deeper into the ore body. Falls of slices up to 3-5 m in thickness have been observed in other Boliden Mineral Inc. mines.

The joint model presented by Groth and Jonasson, 1981, provides a better insight into the strain and stress distributions in the vicinity of the stope. However, the tested joint model, is a slightly improved approximation of the real conditions, as failures of many different types are present. Detailed modelling of all failure modes is a very difficult problem. The determination of relevant joint properties and the development of model techniques for complex rock masses are at present not applicable to mine planning.

The applied elastic model has only been evaluated for mining stages up to 1979. At later stages of mining the failure process becomes more dominant and deviations from the elastic model are likely to increase, especially in the vicinity of the stopes. For these late stages of mining dilatation will lead to redistributions of stresses and strains. From uniaxial compression we know that axial strain often shows linearity up to peak strength. At the same time the lateral strain, ε_3, increases at the stage of fracture initiation, Jaeger and Cook, 1981.

However, the application of EFSC indicates that the excess of extension strains, ε_3, is concentrated to a rather narrow zone close to the roof until a late stage of mining, Fig. 7b. At a stage of failure the redistribution of stress along the strike of the ore will influence the fracture development in the roof. This study has demonstrated that the elastic model can be applied beyond the stage of fracture initiation.

The approach quantifying the critical levels of failures from mine observations is expected to have the following advantages:

- The critical stages 2 and 3 in Fig. 4 are dependent on the mining method and the support methods used.
- The size dependancy of the rock strength is often very pronounced. Therefore it is better to use rock strength criteria that are determined from rock volumes of the same size as the rock volumes in the mine.

However, it is important to remember that the properties in the vicinity of stope 3 are assumed to be relevant for all stopes.

The quantification of stage 2 and 3 of the model is based on the loci where the highest extensions occur. Another possibility is to base prediction on the development and the extent of potential failure zones, as will be demonstrated by Borg, 1982.

9 CONCLUSIONS

1) An elastic finite element model and the extension strain criterion of Stacey, 1981, have successfully been applied to predict brittle failures in the Näsliden mine.

2) The model of critical stages in the mine can be used to determine:

   - the duration of mining with minor disturbances
   - the stage of mining when an increased amount of conventional support is required
   - the stage when a change to other support systems or other mining methods is necessary
   - the time for close-down of a stope and the start of new production areas.

An analysis of the same mining sequences and with the application of the Coulomb failure criterion is in progress.

10 ACKNOWLEDGEMENT

This study forms part of a doctoral thesis by the first author. Financial support by the Swedish Rock Engineering Research Foundation is acknowledged. Thanks are due to the Boliden Mineral Inc. for permission to use data from the Näsliden mine. The authors would like to thank Dr. Sten G. A. Bergman for his great assistance during the project and Professor Ove Stephansson for proposing improvements in the manuscript.

11 REFERENCES

Borg, T. 1982. The evaluation and application of a finite element model of the Näsliden mine to the prediction of future mining conditions. University of Luleå, Div. of Rock Mechanics (Doctoral thesis) To be published.

Groth, T., and Jonasson, P. 1981. Application of the BEFEM code to the Näsliden mine models. Proc. Application of Rock Mechanics to cut and fill mining. The Institution of Mining and Metallurgy, London, 226-232.

Jaeger, J. C., and Cook, N. G. W. 1979. Fundamentals of Rock Mechanics. 3rd Edition. 593 pp. London, Chapman and Hall.

Krauland, N., Nilsson, G., and Jonasson, P. 1981. Comparison of rock mechanics observations and measurements with FEM calculations. Proc. Application of Rock Mechanics to cut and fill mining. The Institution of Mining and Metallurgy, London, 250-260.

Leijon, B., Carlsson, H., and Myrvang, A. 1981. Stress measurements in Näsliden mine. Proc. Application of Rock Mechanics to cut and fill mining. The Institution of Mining and Metallurgy, London, 162-168.

Nilsson, G., and Krauland, N. 1981. Rock mechanics observations and measurements in Näsliden mine. Proc. Application of Rock Mechanics to cut and fill mining. The Institution of Mining and Metallurgy, London, 233-249.

Stacey, T. R. 1981. A simple extension strain criterion for fracture of brittle rock. Int. J. Rock Mech. Min. Sci. & Geomech. Abstr. (18), 6, 469-474.

Stephansson, O., and Jones, M. 1981. Proc. Application of Rock Mechanics to cut and fill mining. 376 pp. London, The Institution of Mining and Metallurgy.

Stephansson, O. 1981. The Näsliden project - rock mass investigations. Proc. Application of Rock Mechanics to cut and fill mining. The Institution of Mining and Metallurgy, London, 145-161.

UNTERSUCHUNG MÖGLICHER AUSWERTUNGSFEHLER BEI DER INTERPRETATION VON IN-SITU SPANNUNGSMESSUNGEN IN ANISOTROPEN GESTEINEN

Analysis of Potential Errors of Interpretation of In-Situ Stress Measurements in Anisotropic Rocks

Etude des erreurs potentielles d'interprétation des contraintes primaires en roches anisotropes

W. Rahn
Ehem. Wissenschaftlicher Mitarbeiter, Arbeitsgruppe Felsmechanik der, Ruhr-Universität
Bochum, Bundesrepublik Deutschland

ZUSAMMENFASSUNG

Die mechanischen Eigenschaften vieler Gesteine der Erdkruste sind anisotrop. Wird die elastische Gesteinsanisotropie bei der Auswertung von in situ-Spannungsmessungen nicht berücksichtigt, kann dies zu beachtlichen Fehlbeurteilungen hinsichtlich der Größe und Richtung der Hauptspannungen führen. Dies wird für die "Doorstopper"-Methode anhand von Parameterstudien unter Verwendung anisotroper Spannungskonzentrationsfaktoren gezeigt.

SYNOPSIS

Anisotropy of mechanical behaviour is a common feature of many crustal rocks. If this is neglected in the analysis of in situ stress measurements, considerable misinterpretations with respect to the magnitudes and directions of the principal stresses may result. This is demonstrated for the "doorstopper"-method by means of parametric studies, in which anisotropic stress concentration factors are used.

RESUME

De nombreuses roches de la lithosphère sont anisotropes. Si l'anisotropie élastique n'est pas pris en considération par l'évaluation de mesures des contraintes primaires, des erreurs importantes se produisent quant à l'amplitude et la direction des contraintes principales. Ceci est démontré pour la méthode de "doorstopper" à l'aide de calculs paramétriques et par l'application de facteurs de concentration de contraintes anisotropes.

1. EINLEITUNG

Ursache und Ergebnis des in quantitativer und quanlitativer Hinsicht beachtlichen Fortschritts des Felshohlraumbaus der letzten zwei Jahrzehnte ist nicht zuletzt die Entwicklung anpassungsfähiger Tunnelbaumethoden sowie der Einsatz effektiver, computerorientierter Berechnungsverfahren. Gleichermaßen erhöhten sich auch die Ansprüche bezüglich des Umfangs und der Qualität der ingenieurgeologischen und gebirgsmechanischen Untersuchungen zur Ermittlung der für die Planung maßgebenden Gebirgskenngrößen. Hierzu gehören vor allem Parameter zur Beschreibung der Festigkeit und Verformbarkeit sowie des primären Beanspruchungszustandes des Gebirges.

Neben richtungsabhängigen Festigkeitseigenschaften besitzen die meisten Gesteine der Erdkruste auch eine oftmals ausgeprägte elastische Anisotropie (Gerrard,1975;Peres Rodrigues,1979). Eine Vielzahl von Untersuchungen zeigt, daß vor allem Sedimentgesteine häufig transversal anisotrope elastische Eigenschaften besitzen.

Die Kenntnis der Primärspannungen des Gebirges ist gerade bei der Anwendung numerischer Berechnungsmethoden von großer Bedeutung. Daß die Ergebnisse von in-situ-Spannungsmessungen trotz zahlreicher gerätetechnischer Verfeinerungen und Neuentwicklungen immer noch mit großen Unsicherheiten behaftet sind, liegt u.a. auch an den i.a. zu stark vereinfachten Ansätzen für die Auswertung der Messungen. Ein Schritt zur realitätsnäheren Datenauswertung, insbesondere bei Messungen durch Bohrlochentspannung, ist die Berücksichtigung der elastischen Anisotropie der Gesteine. Für eine der Entspannungsmeßmethoden, nämlich das "Doorstopper"-Verfahren (Leeman,1971) werden im folgenden die Auswertungsfehler untersucht, die sich aus der Vernachlässigung der elastischen Gesteinsanisotropie bei der Interpretation von in-situ-Meßdaten ergeben können.

2. ANWENDUNG DER DOORSTOPPER-METHODE IN TRANSVERSAL ANISOTROPEN GESTEINEN

Die Einzelheiten der Doorstopper-Meßzelle, des Meßverfahrens und der Auswertung der Meßergebnisse (für den Fall der elastischen Gesteinsisotropie) sind bei Leeman,1971 beschrieben. Wichtig für die Anwendung dieses Meßverfahrens ist, daß die Messungen in mindestens 3 Bohrungen durchgeführt werden und daß die Werte der Spannungskonzentrationsfaktoren bekannt sind, also jener Koeffizienten, die den Einfluß der 6 Gebirgsspannungskomponenten auf die 3 lokalen Sekundärspannungskomponenten am Bohrlochboden im Bereich der Meßzelle quantifizieren. Die Spannungskonzentrationsfaktoren müssen experimentell bzw. mit numerischen Verfahren bestimmt werden. Es gilt:

$$(\sigma) = [K] \cdot (S') = [D] \cdot (\varepsilon)$$
$$(\sigma) = (\sigma_{77}, \sigma_{55}, \sigma_{75})^t$$

$$(S') = (S'_{\xi\xi}, S'_{\eta\eta}, S'_{\zeta\zeta}, S'_{\xi\eta}, S'_{\eta\zeta}, S'_{\zeta\xi})^t$$

$$(\varepsilon) = (\varepsilon_{\eta\eta}, \varepsilon_{\zeta\zeta}, \tfrac{1}{2}\gamma_{\eta\zeta})^t$$

$$= (e_A, e_B, (e_C - \tfrac{1}{2}(e_A + e_B)))^t$$

(vgl. Abb. 1)

(σ) : Lokale Spannungen am Bohrlochboden
(S') : Gebirgsspannungen (lokales K.O.-System)
[D] : Matrix des elastischen Stoffgesetzes
[K] : Matrix der Spannungskonzentrationsfaktoren
(ε) : Dehnungen am Bohrlochboden

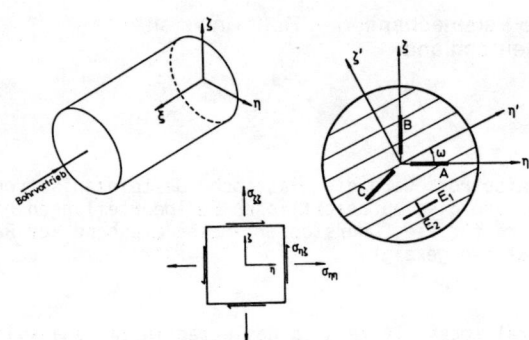

Abb. 1 Doorstopper-Meßrosette in einer Bohrung parallel zur Schichtung; lokales Koordinatensystem

Die theoretischen Grundlagen für die Auswertung von Doorstopper-Messungen im anisotropen Gestein ergeben sich aus der Verallgemeinerung der Ansätze für isotropes Gestein. Der Unterschied zur isotropen Form der Berechnungsformeln besteht im wesentlichen darin, daß ein anisotropes Stoffgesetz [D] und eine anisotrope, unsymmetrische Koeffizientenmatrix [K] verwendet wird. Der Aufbau der Flexibilitätsmatrix [C] = [D]$^{-1}$ für die verschiedenen Stufen elastischer Symmetrie ist aus der Kontinuumsmechanik bekannt. Für den Fall der transversalen Anisotropie ergibt sich folgende Form:

$$[C] = \begin{bmatrix} \frac{1}{E_2} & -\frac{\nu_2}{E_2} & -\frac{\nu_2}{E_2} & 0 & 0 & 0 \\ -\frac{\nu_2}{E_2} & \frac{1}{E_1} & -\frac{\nu_1}{E_1} & 0 & 0 & 0 \\ -\frac{\nu_2}{E_2} & -\frac{\nu_1}{E_1} & \frac{1}{E_1} & 0 & 0 & 0 \\ 0 & 0 & 0 & \frac{1}{2G_{12}} & 0 & 0 \\ 0 & 0 & 0 & 0 & \frac{1+\nu_1}{E_1} & 0 \\ 0 & 0 & 0 & 0 & 0 & \frac{1}{2G_{12}} \end{bmatrix}$$

Der Aufbau der Koeffizientenmatrix [K] in Abhängigkeit von der Symmetrieordnung des elastischen Stoffgesetzes und von der jeweiligen Orientierung des Bohrlochs in bezug auf die Symmetrieachsen des Stoffgesetzes wird u.a. bei Ribacchi,1977 diskutiert. Für den Fall, daß die Meßbohrung in der Isotropieebene eines transversal anisotropen Gesteins liegt, ergibt sich die folgende Form (vgl. Abb. 1):

$$[K]_{\parallel} = \begin{bmatrix} c_y & a_y & b_y & 0 & 0 & 0 \\ c_z & b_z & a_z & 0 & 0 & 0 \\ 0 & 0 & 0 & 0 & 0 & d \end{bmatrix}$$

Verläuft die Meßbohrung normal zur Isotropieebene so gilt wie im isotropen Gestein:

$$[K]_{\perp} = \begin{bmatrix} c & a & b & 0 & 0 & 0 \\ c & b & a & 0 & 0 & 0 \\ 0 & 0 & 0 & 0 & (a-b) & 0 \end{bmatrix}$$

jedoch sind hier die Werte a,b,c vom Grad der Materialanisotropie abhängig.

Für eine Reihe unterschiedlicher anisotroper Materialkennwerte sind die einzelnen Koeffizienten der Matrizen [K]$_{\parallel}$ und [K]$_{\perp}$ bekannt (Rahn,1981). Die mit Hilfe von Finite-Elemente-Berechnungen bestimmten Spannungskonzentrationsfaktoren für Bohrungen parallel und senkrecht zur Isotropieebene (z.B. Schichtung) sind für verschiedene Anisotropiegrade $n_E = E_1:E_2$ in Tab. 1 zusammengestellt.

n_E	ν	senkrecht			parallel zur Schichtung					
n_G		a	b	c	a_y	a_z	b_y	b_z	c_y	c_z	d
1.5	0.0	1.323	-0.128	-0.441	1.333	1.291	-0.133	-0.093	-0.392	-0.278	1.460
	0.1	1.332	-0.104	-0.549	1.346	1.300	-0.117	-0.077	-0.470	-0.346	1.439
	0.2	1.343	-0.072	-0.665	1.359	1.310	-0.087	-0.055	-0.547	-0.415	1.412
	.25	1.350	-0.054	-0.726	1.366	1.316	-0.069	-0.042	-0.586	-0.449	1.396
	0.3	1.358	-0.034	-0.789	1.373	1.322	-0.047	-0.027	-0.625	-0.484	1.378
	0.4	1.375	+0.012	-0.922	1.386	1.335	-0.010	+0.008	-0.702	-0.553	1.339
2.0	0.0	1.328	-0.130	-0.503	1.365	1.285	-0.152	-0.076	-0.427	-0.216	1.418
	0.1	1.336	-0.104	-0.616	1.383	1.295	-0.139	-0.065	-0.493	-0.265	1.400
	0.2	1.348	-0.072	-0.740	1.399	1.305	-0.111	-0.049	-0.558	-0.314	1.379
	.25	1.356	-0.054	-0.806	1.407	1.311	-0.092	-0.040	-0.590	-0.338	1.367
	0.3	1.365	-0.034	-0.876	1.414	1.317	-0.070	-0.030	-0.622	-0.362	1.354
	0.4	1.386	+0.011	-1.019	1.427	1.329	-0.041	-0.011	-0.684	-0.409	1.325
3.0	0.0	1.337	-0.133	-0.607	1.420	1.275	-0.183	-0.052	-0.481	-0.130	1.357
	0.1	1.344	-0.105	-0.722	1.449	1.289	-0.174	-0.049	-0.533	-0.153	1.342
	0.2	1.357	-0.072	-0.858	1.472	1.300	-0.149	-0.039	-0.581	-0.175	1.328
	.25	1.366	-0.053	-0.930	1.481	1.306	-0.131	-0.034	-0.604	-0.185	1.321
	0.3	1.376	-0.033	-1.006	1.488	1.311	-0.110	-0.032	-0.626	-0.195	1.315
	0.4	1.400	+0.011	-1.167	1.499	1.320	-0.090	-0.022	-0.666	-0.214	1.302
5.0	0.0	1.350	-0.138	-0.740	1.495	1.270	-0.222	-0.041	-0.527	-0.109	1.334
	0.1	1.357	-0.107	-0.875	1.553	1.297	-0.208	-0.059	-0.586	-0.150	1.304
	0.2	1.370	-0.072	-1.025	1.590	1.313	-0.189	-0.027	-0.635	-0.149	1.284
	.25	1.379	-0.052	-1.105	1.601	1.318	-0.178	-0.026	-0.656	-0.159	1.278
	0.3	1.390	-0.031	-1.189	1.606	1.319	-0.165	-0.019	-0.674	-0.170	1.276
	0.4	1.416	+0.014	-1.369	1.602	1.316	-0.136	-0.010	-0.704	-0.192	1.278

Tab. 1 Anisotrope Spannungskonzentrationsfaktoren für Bohrungen senkrecht und parallel zur Schichtung oder Schieferung (Rahn,1981)

Bei der Bestimmung der anisotropen Konzentrationsfaktoren wurde der Sonderfall der Isotropie ebenfalls erfaßt. Die Spannungskonzentrationsfaktoren für isotropes Gestein sind in Tab. 2 zusammengestellt. Diese Werte zeigen gute Übereinstimmung mit den Angaben anderer Autoren (z.B. Hocking,1976).

n_E	n_G	ν	a	b	c
1.0	1.0	0.0	1.316	-0.125	-0.366
		0.1	1.325	-0.101	-0.466
		0.2	1.335	-0.070	-0.571
		0.25	1.340	-0.053	-0.625
		0.3	1.346	-0.033	-0.681
		0.4	1.358	+0.016	-0.796
		0.425	1.362	0.031	-0.824
		0.45	1.367	0.046	-0.849
		0.475	1.372	0.060	-0.863
		0.49	1.361	0.051	-0.838

Tab. 2 Spannungskonzentrationsfaktoren für isotropes Gestein (Rahn,1981)

3. KONZEPTION DER PARAMETERSTUDIEN

Mit Hilfe eines Auswertungsprogramms für Doorstopper-Messungen in elastisch transversal anisotropen Gesteinen und der numerisch ermittelten Spannungskonzentrationsfaktoren wurden in Parameterstudien die Auswertungsfehler untersucht, die sich aus der Verwendung inadäquater Materialparameter und Spannungskonzentrationsfaktoren bei der Berechnung der Gebirgsspannungen ergeben. Zu diesem Zweck wurden zunächst der "wahre" Gebirgsspannungszustand und das "wahre" Materialverhalten vorgegeben und die Richtungen der Meßbohrungen festgelegt. Im zweiten Schritt wurde der lokale Spannungszustand am Boden der Bohrungen mit Hilfe der Spannungskonzentrationsfaktoren berechnet, die den vorgegebenen Materialeigenschaften entsprechen. Aus den lokalen Spannungen ließen sich so die für den Doorstopper relevanten Dehnungen e_A, e_B, e_C (vgl. Abb. 1) berechnen; hierfür wurden die vorgegebenen "wahren" Materialkennwerte verwendet.

Die auf diese Weise vorgegebenen fiktiven Meßdaten (Dehnungen) wurden als Eingabegrößen für das o.g. Auswertungsprogramm verwendet und bilden die Ausgangsbasis für die Untersuchung der oben beschriebenen Auswertungsfehler in 2 Parameterstudien. Der Einfachheit halber wurde eine söhlige Schichtlagerung vorausgesetzt. Die Bohrungen liegen schichtparallel bzw. schichtnormal. Die Raumstellung aller berücksichtigten Bohrungen ist in den Abbildungen 2a und 3a dargestellt.

In Studie 1 basieren die fiktiven Meßdaten (Dehnungen) auf den folgenden Gebirgsspannungswerten und Materialkenndaten:

$S_{xx} = 30$ $S_{yy} = 20$ $S_{zz} = 10$
$S_{xy} = 5$ $S_{yz} = -5$ $S_{zx} = -5$

x = Nord y = Ost z = Lotrichtung

$E_2 = 1000$ $\nu_1 = \nu_2 = \nu$
$E_1 = n_E \cdot E_2$ $G_{12} = G_1/n_G$ $n_G = n_E$
$G_1 = E_1/(2(1+\nu_1))$.

Der Übersichtlichkeit wegen wurde im Rahmen dieser Studien immer vorausgesetzt, daß die "wahre" Poisson-Zahl ν des Gesteins bekannt ist. Da die untersuchten Fehler sowohl durch die Verwendung falscher Konzentrationsfaktoren als auch falscher Materialkennwerte in den Berechnungen verursacht werden können, wurde in einer Teilserie die Bedeutung der anisotropen Konzentrationsfaktoren allein und in einer zweiten Teilserie der kombinierte Einfluß der Anisotropie des Materialgesetzes und der Konzentrationsfaktoren untersucht. Zu diesem Zweck wurden in der ersten Teilserie zwar die "wahren" anisotropen Materialkennwerte in die Berechnungsformeln eingesetzt (d.h. die lokalen Spannungen am Bohrlochboden wurden aus den vorgegebenen Dehnungen richtig berechnet) aber die Spannungskonzentrationsfaktoren verwendet, die einem isotropen Material mit dem vorgegebenen ν-Wert entsprechen. In der zweiten Teilserie wurden zusätzlich noch gemittelte, isotrope Materialkennwerte $\bar{E}$, $\bar{\nu}$, $\bar{G}$ für die Berechnungen verwendet.

$\bar{E} = \frac{1}{2}(E_1+E_2)$, $\bar{\nu} = \frac{1}{2}(\nu_1+\nu_2)$, $\bar{G} = \frac{\bar{E}}{2(1+\bar{\nu})}$.

Die einzelnen Fälle sind in Tab. 3 zusammengestellt.

Die zweite Parameterstudie verdeutlicht exemplarisch für einen ideellen Primärspannungszustand die jeweilige Auswirkung der Vernachlässigung der Anisotropie der E-Moduli und der Schubmoduli auf die Berechnungsergebnisse. Die fiktiven Meßdaten (e_A, e_B, e_C) basieren auf folgenden Gebirgsspannungen und Materialkennwerten:

$S_{xx} = S_{yy} = S_{zz} = S_{xy} = S_{yz} = S_{zx} = 1.0$

Für die Materialkennwerte gelten dieselben Angaben wie bei Studie 1, mit der Einschränkung:

$\nu_1 = \nu_2 = \nu = 0.25 = $ const.

Die einzelnen Fälle sind der Tabelle 4 zu entnehmen.

Fall	$n_E = n_G$	$\nu_1 = \nu_2$	Material-kennwerte	Konzentr. faktoren
F0	2	0.0	anisotrop	isotrop
F2	2	0.2	anisotrop	isotrop
F4	2	0.4	anisotrop	isotrop
G0	5	0.0	anisotrop	isotrop
G2	5	0.2	anisotrop	isotrop
G4	5	0.4	anisotrop	isotrop
H0	2	0.0	isotrop	isotrop
H2	2	0.2	isotrop	isotrop
H4	2	0.4	isotrop	isotrop
I0	5	0.0	isotrop	isotrop
I2	5	0.2	isotrop	isotrop
I4	5	0.4	isotrop	isotrop

Tab. 3 Zusammenstellung der Fälle in Studie 1

Fall	n_E	n_G	Material-Kennwerte	Konzentr. faktoren
J0	1	1	isotrop	isotrop
J1	2	2	isotrop	isotrop
J2	5	5	isotrop	isotrop
J3	1	5	isotrop	isotrop
J4	5	1	isotrop	isotrop

Tab. 4 Zusammenstellung der Fälle in Studie 2

4. ERGEBNISSE

Abb. 2 und Abb. 3 zeigen die Ergebnisse der in Tab. 2 bzw. Tab. 3 erläuterten Berechnungen. Die jeweilige Orientierung der wahren und scheinbaren Hauptspannungen ist in Abb. 2a und Abb. 3a durch die Achsendurchstoßpunkte in der unteren Lagenkugelhälfte dargestellt. Die scheinbaren Änderungen der Hauptachsenrichtungen als Folge der beschriebenen Eingabedatenfehler ist durch Buchstaben an den jeweiligen Achsendurchstoßpunkten bzw. durch unterschiedliche Signaturen gekennzeichnet.

4.1 Studie 1

Als Folge der Vernachlässigung der Materialanisotropie und/oder der Anisotropie der Konzentrationsfaktoren ergeben sich in der ersten Studie Bestimmungsfehler für die Hauptachsenrichtungen von bis zu 30° (für $n_E = 5$). Das sind Größenordnungen von unmittelbar praktischer Bedeutung. Die Lage der scheinbaren Rotationsachsen und die scheinbare Drehrichtung der Hauptspannungsachsen, die sich durch die einzelnen Parametervariationen ergeben, sind verschieden, je nachdem, ob sich die Variation auf die Verwendung isotroper Konzentrationsfaktoren allein oder in Verbindung mit der Verwendung quasi-isotroper, gemittelter Materialkennwerte bei der Spannungsberechnung bezieht. Dies ist aus dem Vergleich der Ergebnisse der Teilserien F,G und H,I zu erkennen. Es wird deutlich, daß die Vernachlässigung der Anisotropie durch die Verwendung quasi-isotroper, gemittelter Materialkennwerte die stärkere scheinbare Hauptachsenrotation bewirkt. Diese ist zusätzlich vom ν-Wert abhängig, besonders stark bei den Fällen H und I (vgl. Tab. 3). Bemerkenswert ist die Ermittlung scheinbarer Zugspannungen im Fall I.

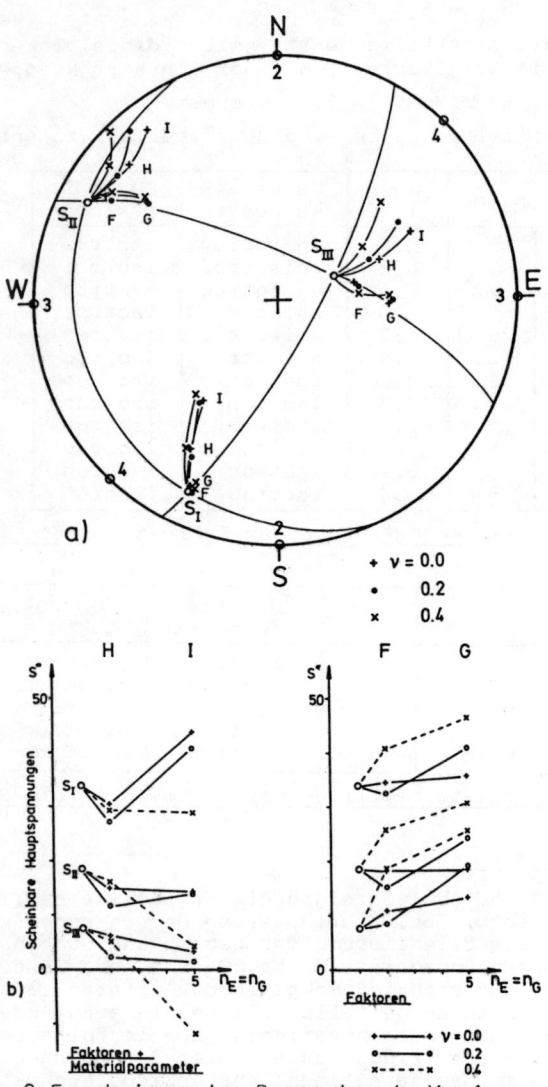

Abb. 2 Ergebnisse der Parameterstudie 1

Die in der Abb. 2b dargestellte Abhängigkeit der scheinbaren Hauptspannungswerte vom Anisotropiefaktor n folgt keinem einheitlichen Trend. Die Größe des Berechnungsfehlers ist auf komplexe Weise von den "wahren" Gebirgsspannungen, der Bohrlochanordnung in bezug auf die Hauptspannungsrichtungen und von der Größe und Richtung der Gebirgsanisotropie abhängig. Eine Interpretation der Zusammenhänge auf theoretischer Basis im Hinblick auf eine nachträgliche Korrektur falscher Berechnungsergebnisse ist deshalb nicht praktikabel. Allgemein zeichnet sich jedoch gemäß Diagramm 2b die Tendenz ab, daß insbesondere die "Verschmierung" der Materialanisotropie durch die Mittelwertbildung bei der Materialparameterbestimmung zu einer starken Polarisierung der scheinbaren Hauptspannungen führen kann. Dadurch werden evtl. hohe Deviatorspannungen im Gebirge vorgetäuscht (im Fall I(a,2) ergibt sich für S_I eine scheinbare Erhöhung um ca. 25% für S_{III} eine scheinbare Verminderung um ca. 85%). Da hohe primäre Deviatorspannungen i.a. einen ungünstigen Einfluß auf die Standsicherheit von Felshohlräumen haben, könnten hieraus möglicherweise ungerechtfertigt aufwendige Stütz- und Sicherungsmaßnahmen abgeleitet werden.

Demgegenüber verursacht allein die Eingabe isotroper Konzentrationsfaktoren in die Doorstopper-Auswertungsformeln tendenziell eine Anhebung des allgemeinen scheinbaren Spannungsniveaus; dabei werden zugleich die wahren Deviatorspannungen scheinbar reduziert. Allgemein werden die hier beschriebenen Tendenzen mit zunehmenden ν-Werten verstärkt.

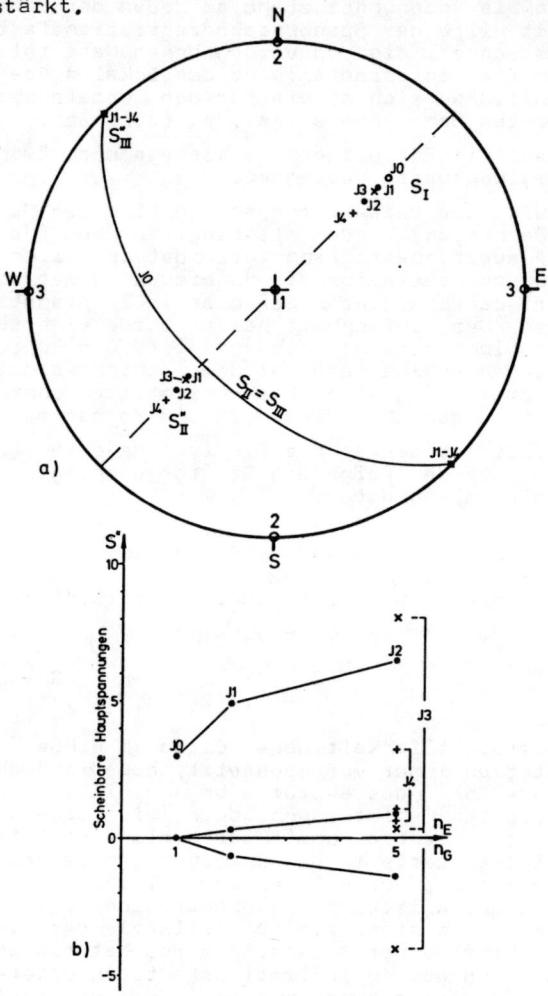

Abb. 3 Ergebnisse der Parameterstudie 2

4.2 Studie 2

Für die zweite Studie wurde ein einachsiger Spannungszustand ($S_I=3$, $S_{II}=S_{III}=0$) vorgegeben.

Durch Einsetzen isotroper Materialparameter und Spannungskonzentrationsfaktoren in die Berechnungsformeln wird ein scheinbar dreiachsiger Primärspannungszustand im Gebirge vorgetäuscht. Die Richtungen der scheinbaren Hauptspannungen liegen auf einem senkrecht stehenden, NE-SW-streichenden Großkreis. Dies ist auf die besondere geometrische Konfiguration zwischen den Bohrungen, der horizontalen Schichtlagerung und der Richtung der wahren Hauptspannung S_I als Raumdiagonale im ersten Oktanten zurückzuführen. Aufgrund dieser Zusammenhänge sind die lokalen, fiktiven Meßdaten (Dehnungen) in den horizontalen Bohrungen symmetrisch zueinander; eine Variation der Materialparameter oder Spannungskonzentrationsfaktoren bei der Primärspannungsberechnung ändert nichts an dieser Symmetrie.

Die Vernachlässigung der Anisotropie bei der Auswertung der Meßdaten verursacht (wie in Studie 1) generell eine mit zunehmender Anisotropie wachsende Polarisierung des scheinbaren Primärspannungszustandes, wobei i.a. die Hauptspannungen S_I und S_{II} überschätzt und die kleinste Hauptspannung S_{III} unterschätzt werden. Sehr unterschiedliche Ergebnisse ergeben sich für den (unrealistischen) Fall, daß die Messungen in einem bezüglich der E-Moduli isotropen, aber bezüglich der Schubmoduli anisotropen Gebirge durchgeführt werden (vgl. Abb. 3). Hier liegen die Auswertungsfehler im Bereich von ca. +165% bzw. ca -135% des "wahren" Wertes von S_I. Im durchaus realistischen Fall J2 liegen die Auswertungsfehler im Bereich von ca. +116% bzw. ca. -45% des "wahren" S_I-Wertes.

5. ZUSAMMENFASSUNG

Es wurde anhand einiger Parameterstudien aufgezeigt, daß bei der Anwendung der "Doorstopper"-Methode in anisotropen Gesteinen die Vernachlässigung der Anisotropie bei der Auswertung der Meßdaten zu beachtlichen Fehlbeurteilungen führen kann. So kann in Gesteinen mit einem Anisotropieverhältnis $n_E = n_G = 5$ die maximale Hauptspannungsdifferenz $(S_I - S_{III})$ um ca. 50% des wahren Wertes überschätzt werden, wobei Richtungsfehler für die Hauptspannungen von ca. $30°$ auftreten können!

Abschließend ist jedoch hervorzuheben, daß die aufgezeigten Zusammenhänge nur exemplarischen Charakter haben und längst nicht alle Parameter- und Bohrlochkonfigurationen berücksichtigt sind. So verdeutlichen einige, hier nicht beschriebene Zusatzuntersuchungen, daß z.B. der vorgegebene wahre Gebirgsdruck um so stärker (scheinbar) verfälscht wird, je geringer die deviatorischen Primärspannungsanteile sind. Das heißt, daß bei kleinen Hauptspannungsdifferenzen $(S_I - S_{III})$ bzw. $(S_{II} - S_{III})$ durch die Verwendung quasi-isotroper Materialkennwerte und/oder Konzentrationsfaktoren große Fehler - insbesondere hinsichtlich der Hauptspannungsrichtungen - auftreten können.

6. LITERATUR

Gerrard, C.M.(1975).
Background to mathematical modeling in geomechanics-The roles of fabric and stress history: Proc. Int.Symp. on Num.Meth., Karlsruhe, 33-120, Karlsruhe.

Hocking, G.(1976).
Three-dimensional elastic stress distribution around the flat end of a cylindrical cavity: Int.J.Rock Mech.Min.Sci.&Geomech. Abstr.,13,331-337,London.

Leeman, E.R.(1971).
The C.S.I.R. "Doorstopper" and triaxial rock stress measuring instruments: Rock Mechanics ,3,25-50,Wien.

Peres Rodrigues, F.(1979).
The anisotropy of the moduli of elasticity and of the ultimate stresses in rocks and rock masses: Proc. 4th Cong. I.S.R.M.,(2), 517-523,Rotterdam.

Rahn, W.(1981).
Zum Einfluß der Gesteinsanisotropie und des bruchbedingten nichtlinearen Materialverhaltens auf die Ergebnisse von Spannungsmessungen im Bohrloch: Bochumer geol. u. geotechn. Arb.,5,1-209,Bochum.

Ribacchi, R.(1977).
Rock stress measurements in anisotropic masses: Proc. Int.Symp. on Field Measurements in Rock Mechanics,Zürich,(1),183-196,Clausthal.

DIE BESTIMMUNG DES VOLLSTÄNDIGEN SPANNUNGSZUSTANDES IM GEBIRGE MIT DER KOMPENSATIONSMETHODE

The determination of the complete state of stress in rock with the flat jack method

La détermination des contraintes dans les massifs rocheux par la méthode du vérin plat

K. Balthasar
E. Wenz
Lehrstuhl für Felsmechanik, Universität Karlsruhe, Bundesrepublik Deutschland

ZUSAMMENFASSUNG

Trotz des hohen Entwicklungsstandes der Kompensationsmeßtechnik sind die Messungen noch immer auf die Gebirgsoberfläche beschränkt. Am Beispiel eines Meßeinsatzes in einem Wasserüberleitstollen im Harz wird die Anwendung der Kompensationsmethode zur Bestimmung des vollständigen räumlichen Spannungszustandes im Gebirge beschrieben. Es werden die dazu erforderlichen theoretischen und technischen Voraussetzungen erläutert und der Bezug der Meßergebnisse zu andernorts ermittelten Meßdaten hergestellt.

SYNOPSIS

In spite of the high state of development of the flat jack method, the measurements are still restricted on the surface of the rock. Giving an example by a measurement which was carried out in a water overflow gallery in the Harz Mountains in Germany, the application of the flat jack method for the determination of the complete state of stress in rock is described. The necessary technical equipment and the theoretical assumptions for this method are explained. Further, the results are compared with the results of stress measurements in other areas of Central Europe.

RESUME

Malgré le stade avancé du développement de la technique de mesures par la méthode du verrin plat les résultats sont toujours restreints à la surface des massifs rocheux. L'emploi de cette méthode pour la détermination du champ naturel des contraintes dans les massifs rocheux est décrit dans le cas d'une galerie d'écoulement d'eau dans le massif du Harz (Allemagne). On explique les impératifs théoriques et techniques et compare les résultats avec ceux obtenus dans d'autres zones en Europe Centrale.

1. EINLEITUNG

Seit der ersten Messung von Gebirgsspannungen im Bereich des Boulder-Dammes durch Lieurance im Jahre 1932 ist eine Vielzahl von neuen, auf verschiedensten Meßprinzipien basierenden Spannungsmeßverfahren entwickelt worden. Trotzdem gibt es noch immer sehr wenige Verfahren, die es ermöglichen, den vollständigen räumlichen Spannungszustand mit einem einzigen Meßzyklus zu ermitteln. Solche Verfahren gehören bisher ausschließlich der Gruppe der Entlastungsmethoden bei bekannten Moduli an. Hier werden im Regelfall geeignete Verformungsgeber in Bohrlöcher eingebracht. Anschließend werden diese Meßbohrlöcher überbohrt, die dabei auftretenden Entlastungsverformungen gemessen und nach der Elastizitätstheorie in Spannungen umgerechnet.

Obwohl diese Verfahren teilweise einen hohen Entwicklungsstand erreicht haben, stellen sie jedoch hohe Anforderungen an die Qualität des Gebirges hinsichtlich Homogenität, Klüftigkeit und Kristallkorngrößen. Auch bei feuchtem Gebirge sind diese Verfahren oft nicht mehr sinnvoll einsetzbar. Zudem müssen die Verformungsmoduli des Gebirges noch zusätzlich bestimmt werden.

Im Gegensatz zu den Entlastungsmethoden ist bei den Kompensationsmethoden eine Kenntnis der elastischen Konstanten des an der Meßstelle anstehenden Gesteins nicht notwendig. Dieses Meßverfahren besteht darin, daß die während einer künstlichen Entspannung des Gebirges auftretenden Verformungen durch einen Kompensationsdruck, der mit geeigneten Belastungseinrichtungen aufgebracht wird, wieder rückgängig gemacht werden. Aus diesem Kompensationsdruck läßt sich die senkrecht zur Kompensationsrichtung wirkende Spannungskomponente ermitteln.

Das heute weitaus am häufigsten angewendete Kompensationsverfahren benutzt als Kompensationselement ein flaches hydraulisches Druckkissen (flat jack), das kraftschlüssig in einen vorher hergestellten Entspannungsschlitz im Gebirge eingebaut wird. In dieser Form ist die Kompensationsmethode auch erstmals durch Mayer, Habib und

Marchand (1951) sowie durch Tincelin (1951) bekannt geworden. Aber auch andere Kompensationseinrichtungen wurden entwickelt und erprobt, so z. B. ein zweiachsig arbeitender Bohrlochdruckgeber durch Natau (1967).

Wie bereits erwähnt, müssen bei der Kompensationsmethode die elastischen Konstanten des Gebirges nicht mehr bekannt sein. Auch wirken sich wegen der relativ großflächigen Meßweise kleinere Inhomogenitäten an der Meßstelle kaum auf das Meßergebnis aus. Bedingt durch das Meßprinzip können Kompensationsmessungen jedoch nur an der Gebirgsoberfläche durchgeführt werden. Dies hat zwar den für die Praxis nicht zu unterschätzenden Vorteil, daß die Meßstelle direkt zugänglich und sichtbar ist, andererseits ist dadurch aber die unmittelbare Bestimmung des vollständigen räumlichen Spannungszustandes im Gebirge nicht möglich.

Vorschläge zur Lösung dieses Problems sind schon vor längerer Zeit bekannt geworden. Sie bestehen grundsätzlich darin, die Spannungen in der Wandung eines untertägigen Hohlraumes an mehreren Punkten in mehreren definierten Richtungen zu messen und unter Verwendung von Spannungskonzentrationsfaktoren, die für den jeweiligen Hohlraum gesondert bestimmt werden müssen, auf die Primärspannungen zurückzurechnen. Auf diese Weise konnte Alexander (1960) die Primärspannungen im Bereich einer geplanten untertägigen Maschinenhalle für ein Wasserkraftwerk rechnerisch abschätzen. Die Spannungskonzentrationsfaktoren für den Erkundungsstollen, in dessen Wandung er die Messungen durchführte, wurden von ihm mit Hilfe spannungsoptischer Modelluntersuchungen ermittelt. Bei einfacherer geometrischer Form des untertägigen Hohlraumes, so z. B. bei Tunneln mit Kreisprofil, können die Spannungskonzentrationsfaktoren aber auch, wie Jaeger und Cook (1969) dies vorschlagen, mit analytischen Methoden berechnet werden.

Die immer zahlreicher werdenden Einsätze von Vollschnitt-Vortriebsmaschinen im Tunnel- und Schachtbau haben Natau (1972) dazu angeregt, für die Anwendung der Kompensationsmethode in langgestreckten untertägigen Hohlräumen mit Kreisprofil einfache Berechnungsformeln zur Bestimmung des Primärspannungszustandes zu erarbeiten. Dieses Verfahren konnte mittlerweile erstmals praktisch erprobt werden, worüber im folgenden noch berichtet wird.

2. THEORIE ZUR VOLLSTÄNDIGEN BESTIMMUNG DES PRIMÄRSPANNUNGSZUSTANDES MIT HILFE DER KOMPENSATIONSMETHODE IN GEBOHRTEN STOLLEN ODER SCHÄCHTEN

Die theoretische Voraussetzung zu diesem Verfahren ist die Kenntnis des Zusammenhangs zwischen dem Spannungszustand im unverritzten Gebirge und dem gestörten Spannungszustand im Nahbereich eines Bohrstollens. Dieser Zusammenhang wird durch die Gleichungen zur Berechnung der Spannungen um ein langes, kreisrundes Loch beschrieben, das sich in einem elastischen und isotropen Medium befindet, in dem ein homogener Spannungszustand herrscht. Die Gleichungen sind aus der bekannten Lösung von Kirsch abgeleitet und wurden bereits von mehreren Autoren, so z. B. Hiramatsu und Oka (1962) oder Leeman (1968) veröffentlicht.

Unmittelbar am Bohrlochrand kann ein zweiachsiger Spannungszustand vorausgesetzt werden. Dann ergibt sich unter Zugrundelegung des in Abb. 1 dargestellten Koordinatensystems mit den Primärspannungskomponenten σ_x, σ_y, σ_z, sowie τ_{xy}, τ_{yz}

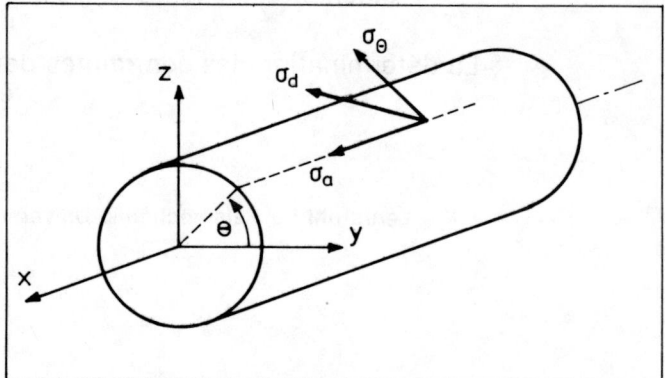

Abb. 1 Koordinatensystem für die Berechnung der Spannungen in der Bohrlochwandung

und τ_{zx} für die Spannungen an einem beliebigen Punkt der Bohrlochwandung:

$$\sigma_\theta = \sigma_y + \sigma_z - 2(\sigma_y - \sigma_z)\cos 2\theta - 4\tau_{yz}\sin 2\theta$$
$$\sigma_a = \sigma_x - \nu\left[2(\sigma_y - \sigma_z)\cos 2\theta + 4\tau_{yx}\sin 2\theta\right]$$
$$\tau_{\theta a} = 2\tau_{zx}\cos\theta - 2\tau_{xy}\sin\theta \qquad (1)$$

Während die Tangentialspannungskomponente σ_θ und die Axialspannungskomponente σ_a an jedem, durch den Winkel θ gegebenen Punkt der Bohrlochwandung (die Spannungen in der Bohrlochwandung sind von der Axiallage unabhängig) mit der Kompensationsmethode direkt gemessen werden kann, ist eine direkte Messung der zugehörigen Schubspannungskomponente $\tau_{\theta a}$ mit diesem Verfahren nicht möglich. Sie läßt sich aber dadurch einbeziehen, daß an diesem Punkt eine dritte Normalspannungskomponente, beispielsweise in diagonaler Richtung, d. h. unter einem Winkel von 45° zu den Richtungen von σ_θ und σ_a gemessen wird (siehe Abb. 1). Für diese zusätzlich gemessene Spannungskomponente σ_d gilt:

$$\sigma_d = \tau_{\theta a} + \frac{1}{2}(\sigma_\theta + \sigma_a) \qquad (2)$$

Die o. a. Bestimmungsgleichung (1) für $\tau_{\theta a}$ läßt sich mit Gleichung (2) ersetzen und man erhält für die Spannungen an einem beliebigen Punkt der Bohrlochwandung die folgenden drei Ausgangsgleichungen:

$$\sigma_\theta = \sigma_y + \sigma_z - 2(\sigma_y - \sigma_z)\cos 2\theta - 4\tau_{yz}\sin 2\theta$$
$$\sigma_a = \sigma_x - \nu\left[2(\sigma_y - \sigma_z)\cos 2\theta + 4\tau_{yz}\sin 2\theta\right]$$
$$\sigma_d = \frac{1}{2}(\sigma_x + \sigma_y + \sigma_z)$$
$$\quad - (1+\nu)\left[(\sigma_y - \sigma_z)\cos 2\theta + 2\tau_{yz}\sin 2\theta\right]$$
$$\quad - 2\tau_{xy}\sin\theta + 2\tau_{zx}\cos\theta \qquad (3-5)$$

Neben den gesuchten sechs Komponenten des Primärspannungszustandes enthalten diese Gleichungen als zusätzliche unbekannte Größe die Poissonzahl ν, weshalb, will man diesen Materialparameter nicht experimentell ermitteln, für eine vollständige Bestimmung des Primärspannungszustandes mindestens sieben verschiedene Spannungskomponenten an mindestens drei Punkten der Bohrlochwandung gemessen werden müssen. Aus den Gleichungen (3 - 5) erhält man durch Einsetzen der Winkelfunktionswerte für die gewählten Meßpunkte die Bestimmungsgleichungen für die sechs unbekannten Tensorkomponenten. Die siebente Unbekannte, die Poisonzahl ν, läßt sich unter Verwendung der Tangential- und Axialspannungskomponenten für zwei beliebige Meßpunkte 1 und 2 aus folgender Gleichung bestimmen:

$$\nu = (\sigma_{a1} - \sigma_{a2})/(\sigma_{\theta 1} - \sigma_{\theta 2}).$$

Durch eine gezielte Wahl der Meßpunkte erhält man aufgrund der Winkelfunktionswerte besonders einfache Bestimmungsgleichungen, wofür hier nur zwei Beispiele gezeigt werden sollen.

Das erste Beispiel entspricht dem bereits erwähnten, von Natau (1972) erarbeiteten Vorschlag. Die Anordnung der drei Meßpunkte auf dem Bohrstollenumfang geht aus Abb. 2 hervor. Es werden folgende Spannungskomponenten gemessen:

An Meßpunkt 1 bei $\theta = 0°$:
- die Tangentialkomponente $\sigma_{\theta 0}$
- die Axialkomponente σ_{a0}
- die Diagonalkomponente σ_{d0}

An Meßpunkt 2 bei $\theta = 90°$:
- die Tangentialkomponente $\sigma_{\theta 90}$
- die Axialkomponente σ_{a90}
- die Diagonalkomponente σ_{d90}

An Meßpunkt 3 bei $\theta = 225°$:
- die Tangentialkomponente $\sigma_{\theta 225}$

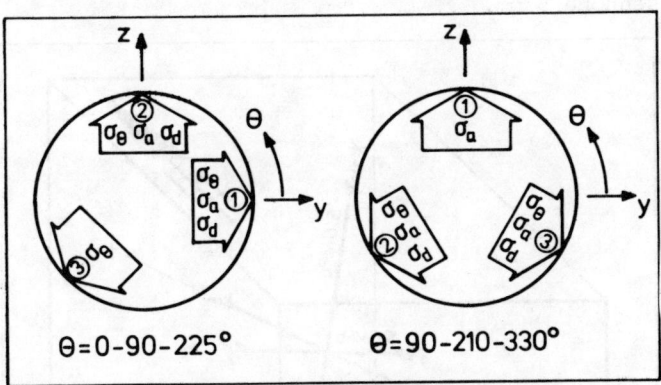

Abb. 2 Zwei Beispiele für Meßpunktanordnungen zur Kompensationsmessung im Bohrstollen

Für diese Konfiguration ergeben sich folgende Bestimmungsgleichungen für die sechs Tensorkomponenten des Primärspannungszustandes:

$$\sigma_x = \frac{1}{2} (\sigma_{a0} + \sigma_{a90})$$
$$\sigma_y = \frac{1}{8} (3\sigma_{\theta 90} + \sigma_{\theta 0})$$
$$\sigma_z = \frac{1}{8} (3\sigma_{\theta 0} + \sigma_{\theta 90})$$

$$\tau_{xy} = -\frac{1}{4} (2\sigma_{d90} - \sigma_{a90} - \sigma_{\theta 90})$$
$$\tau_{yz} = \frac{1}{8} (\sigma_{\theta 0} + \sigma_{\theta 90} - 2\sigma_{\theta 225})$$
$$\tau_{zx} = \frac{1}{4} (2\sigma_{d0} - \sigma_{a0} - \sigma_{\theta 0})$$

Diese Bestimmungsgleichungen gelten auch dann, wenn für Meßpunkt 3 anstelle des Positonswinkels von $\theta = 225°$ ein Winkel von $\theta = 45°$ gewählt wird.

Das zweite Beispiel weist eine äquidistante Anordnung der Meßpunkte auf dem Bohrstollenumfang auf, wie Abb. 2 zeigt. Hier ist die Messung folgender Spannungskomponenten notwendig:

An Meßpunkt 1 bei $\theta = 90°$:
- die Axialkomponente σ_{a90}

An Meßpunkt 2 bei $\theta = 210°$:
- die Tangentialkomponente $\sigma_{\theta 210}$
- die Axialkomponente σ_{a210}
- die Diagonalkomponente σ_{d210}

An Meßpunkt 3 bei $\theta = 330°$:
- Tangentialkomponente $\sigma_{\theta 330}$
- die Axialkomponente σ_{a330}
- die Diagonalkomponente σ_{d330}

Bei dieser Meßpunktanordnung muß in der Firste (Meßpunkt 1, $\theta = 90°$) nur die Axialkomponente gemessen werden. Dies wird einer an dieser Stelle ebenfalls möglichen Messung der Tangentialkomponente vorgezogen, da in der Firste ein erhöhtes Risiko des Auftretens tangentialer Zugspannungen besteht, die Kompensationsmethode jedoch nicht für die Ermittlung von Zugspannungen geeignet ist. Wie bereits beim ersten Beispiel, wird auch hier die Anlage von Meßpunkten im Sohlenbereich des Bohrstollens vermieden, da dort fast immer Wasseransammlungen auftreten, die eine Messung sehr erschweren können. Die Bestimmungsgleichungen für die sechs Tensorkomponenten des Primärspannungszustandes sind bei dieser Meßpunktanordnung nur geringfügig aufwendiger. Sie lauten:

$$\sigma_x = \frac{1}{3} (\sigma_{a90} + \sigma_{a210} + \sigma_{a330})$$

$$\sigma_y = \frac{1}{4} (\sigma_{\theta 210} + \sigma_{\theta 330})$$
$$\quad + \frac{1}{6\nu} (2\sigma_{a90} - \sigma_{a210} - \sigma_{a330})$$

mit: $\nu = (\sigma_{a330} - \sigma_{a210})/(\sigma_{\theta 330} - \sigma_{\theta 210})$

$$\sigma_z = \frac{1}{4} (\sigma_{\theta 210} + \sigma_{\theta 330})$$

$$\tau_{xy} = \frac{1}{2} (\sigma_{d330} + \sigma_{d210})$$
$$\quad - \frac{1}{4} (\sigma_{a330} + \sigma_{a210} + \sigma_{\theta 330} + \sigma_{\theta 210})$$

$$\tau_{yz} = \frac{1}{4\sqrt{3}} (\sigma_{\theta 330} - \sigma_{\theta 210})$$

$$\tau_{zx} = \frac{1}{2\sqrt{3}} (\sigma_{d330} - \sigma_{d210})$$
$$\quad - \frac{1}{4\sqrt{3}} (\sigma_{a330} - \sigma_{a210} + \sigma_{\theta 330} - \sigma_{\theta 210})$$

Aus den mit Hilfe dieser Gleichungen erhaltenen Tensorkomponenten lassen sich die Hauptspannungen und die zugehörigen Richtungswinkel unter Verwendung bekannter Gleichungen der Elastizitätstheorie berechnen. Hierzu wird auf die umfangreiche Fachliteratur verwiesen, so z. B. Jaeger und Cook (1969).

Aus den theoretischen Grundlagen folgt eine wichtige Einschränkung hinsichtlich der Anwendbarkeit der Kompensationsmethode im Bohrstollen: Das Verfahren verlangt ein elastisches und isotropes Gebirge an der Meßstelle. Dies ist beim Kompensationsverfahren als solchem nicht unbedingt erforderlich, sondern ergibt sich erst aus seiner Anwendung bei der Tensorbestimmung in gebohrten Stollen oder Schächten. Besonders auch im Hinblick auf ein mögliches Auftreten plastischer Bruchzonen um den Hohlraum ist diese Einschränkung von Bedeutung.

3. PRAKTISCHE DURCHFÜHRUNG EINER SPANNUNGSMESSUNG MIT DEM KOMPENSATIONSVERFAHREN IN EINEM GEBOHRTEN WASSER-ÜBERLEITSTOLLEN

Wie schon erwähnt, konnte das beschriebene Verfahren mittlerweile praktisch erprobt werden. Die Messung wurde im Zuge der Auffahrung des Radau-Stollens durchgeführt, eines Wasser-Überleitstollens im Harz/West-Deutschland. Er hat eine Länge von 4800 m und verbindet, nahezu ost-westlich verlaufend, das Radau-Tal mit dem Oker-Tal. Der Vortrieb erfolgte größtenteils mit Hilfe einer Vollschnitt-Vortriebsmaschine, die ein Kreisprofil von 2,3 m Durchmesser herstellte. Aufgrund des überwiegend standfesten Gebirges waren nur in wenigen Streckenabschnitten Ausbaumaßnahmen notwendig. Die geologischen Gegebenheiten im Bereich der Stollentrasse und auffahrungstechnischen Einzelheiten wurden von Lange (1979) beschrieben.

Als Meßort wurde ein etwa 6 m langer, trockener und ungeklüfteter Streckenabschnitt, 3525 m vom westlichen Stollenmundloch entfernt, gewählt. Die Gebirgsüberlagerung betrug hier 188 m. Der an der Meßstelle anstehende kontaktmetamorphe Hornfels zeigte, wie Laboruntersuchungen an mehreren Bohrkernen ergaben, neben einer hohen Druckfestigkeit von 180 MPa ein linear-elastisches Spannungs-Verformungs-Verhalten mit einem Elastizitätsmodul von 68 000 MPa und einer Poissonzahl von 0,22. Anisotrope Verformungseigenschaften waren nicht erkennbar.

Zur Messung der einzelnen Spannungskomponenten wurde, mit wenigen Änderungen, das von Rocha, Lopes und Silva (1966) erstmals beschriebene Verfahren der Kompensation im Sägeschlitz angewendet. Die Lage der Meßpunkte auf dem Stollenumfang mit den dort jeweils zu messenden Normalspannungskomponenten entsprach im wesentlichen dem von Natau (1972) veröffentlichten Vorschlag, wie er im vorangegangenen Abschnitt als Beispiel 1 erläutert wurde. Wegen verschiedener Rohr- und Kabelinstallationen an der nördlichen Streckenulme wurde jedoch für den Meßpunkt 3 anstelle eines Positionswinkels von 225° ein Winkel von 45° gewählt. Änderungen an den Bestimmungsgleichungen waren deswegen nicht notwendig. Abb. 3 zeigt die Lage der Meßschlitze in der Tunnelwandung. Die beiden Diagonalmeßschlitze wurden, wie sich erkennen läßt, versehentlich in der falschen, um 90° verdrehten Diagonalrichtung geschnitten. Die dabei ermittelten Meßwerte ließen sich jedoch mit Hilfe einfacher elastizitätstheoretischer Beziehungen

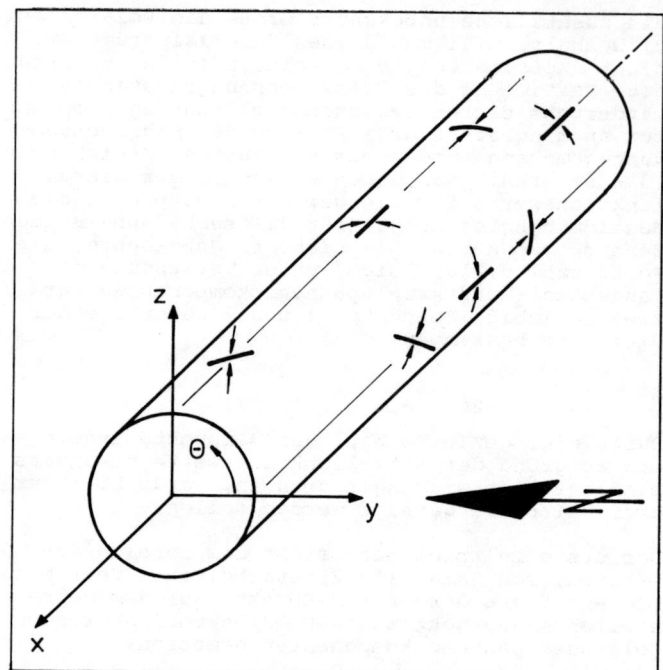

Abb. 3 Lage der Kompensationsmeßschlitze bei einer Spannungsmessung im Radau-Stollen/Harz

hungen in die richtigen Diagonalspannungskomponenten umrechnen.

Zur Herstellung der Kompensations-Meßschlitze wurde eine druckluftbetriebene Steinsäge benutzt, die mit einem 5 mm starken Diamantsägeblatt ausgestattet war. Die halbmondförmigen Druckkissen bestanden aus zwei an ihren Rändern aufeinander verschweißten, 2 mm starken Stahlblech-Kreissegmenten von 622 mm Breite und 135 mm Tiefe. Der Sägeschlitzweite angepaßt, betrug die Druckkissenhöhe 5 mm.

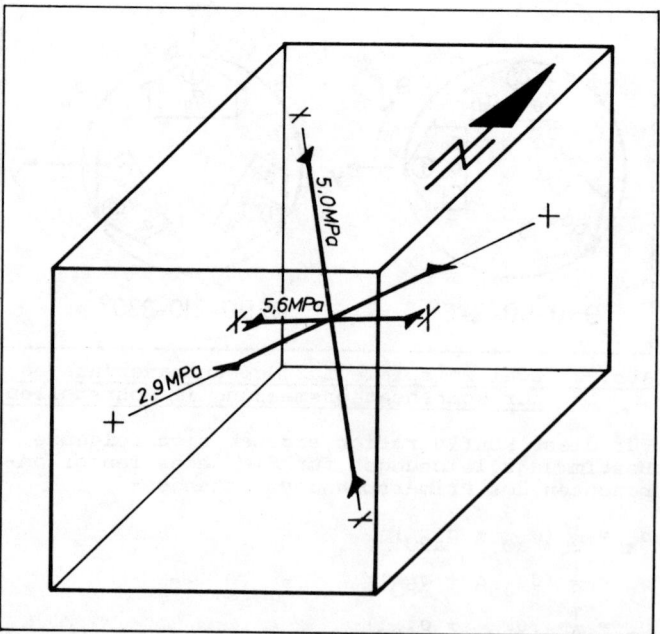

Abb. 4 Ergebnis der im Radau-Stollen/Harz durchgeführten Spannungsmessung

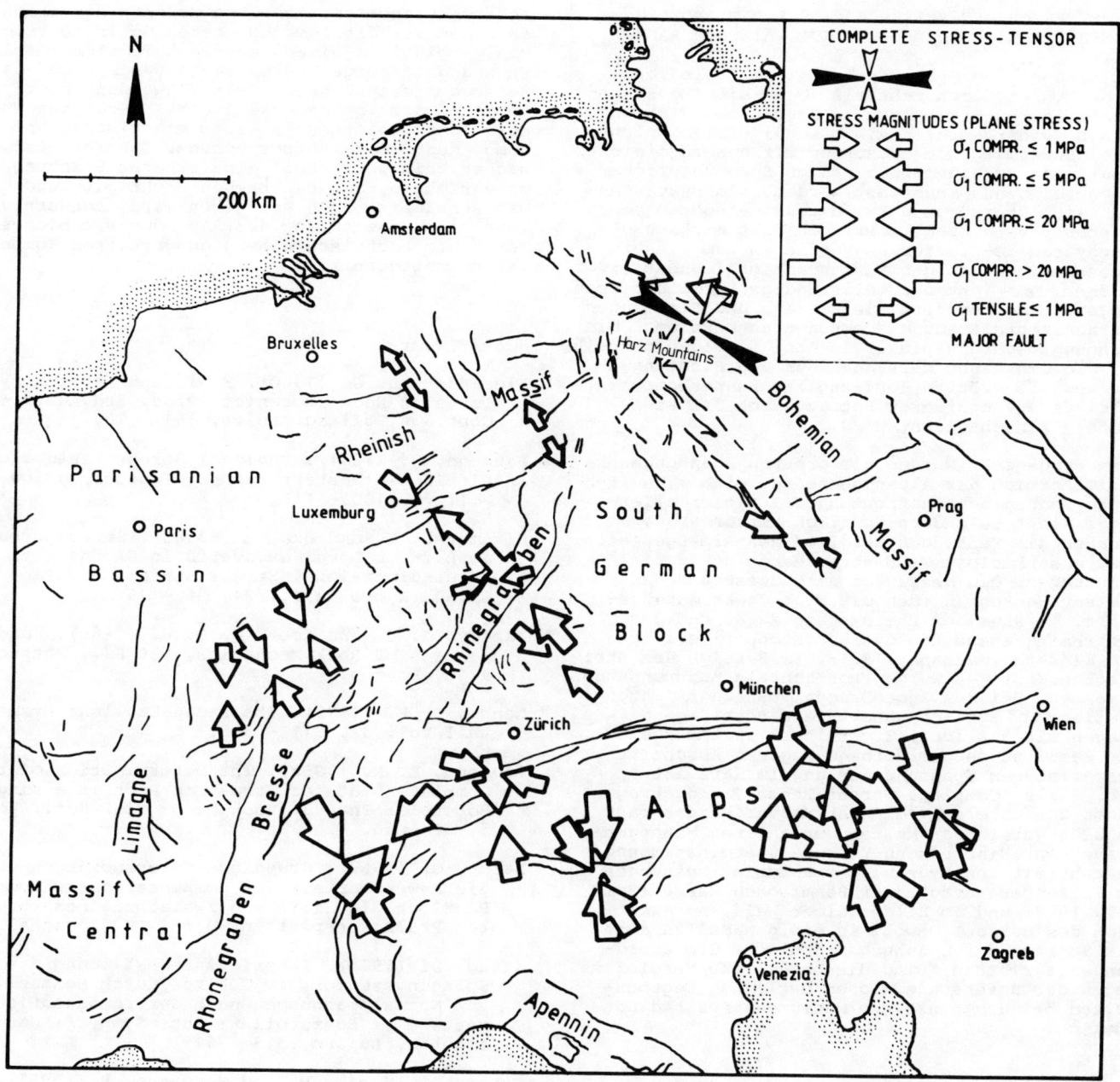

Abb. 5 Tektonisches Spannungsfeld in Mitteleuropa

Da bei Verwendung der für Flat-Jack-Messungen meistens benutzten Deformeter wegen der gekrümmten Bohrstollenwandung Probleme zu erwarten waren, kamen zur Messung der Gebirgsverformungen am Meßschlitz Dehnungsmeßstreifen zum Einsatz. Sie wurden ca. 12 cm vom Sägeschnitt entfernt (= ca. 20 % der Meßschlitzbreite), mittig auf die vorher mit einem Handschleifgerät geglättete Gebirgsoberfläche geklebt und mit Silikon-Abdeckmittel gegen Feuchtigkeitseinwirkung geschützt. Zur Sicherheit waren alle Dehnungsmeßstreifen doppelt vorhanden. Die deshalb jeweils zweifach gemessenen Kompensationsdrücke stimmten in allen Fällen gut überein. Wegen der durch die verschiedenen Schnittrichtungen entstandenen unterschiedlichen Meßschlitzgeometrien, aber auch durch die aus konstruktiven Gründen im Randbereich inaktive Druckkissenfläche war eine vollflächige Einleitung des Kompensationsdruckes im Meßschlitz nicht möglich. Die gemessenen Kompensationsdrücke wurden deshalb, diesen Flächenunterschieden entsprechend, abgemindert.

Das Ergebnis der Messung ist in Abb. 4 dargestellt: Die größte der drei Hauptspannungen beträgt 5,6 MPa und fällt unter einem Winkel von 23° nach NW (301°) ein. Die mittlere Hauptspannung von 5,0 MPa fällt steil mit 66° nach SE (128°) ein. Die kleinste Hauptspannung schließlich mißt 2,9 MPa und fällt mit 3° nur sehr flach in NE-Richtung (32°) ein. Der Betrag der Vertikalspannungskomponente liegt nur wenig über dem aus der Gebirgsüberlagerung errechenbaren Wert.

4. VERGLEICH DES MESSERGEBNISSES MIT ANDEREN MESSDATEN AUS DEM MITTELEUROPÄISCHEN RAUM

Das tektonische Spannungsfeld in Mitteleuropa wurde bisher durch mehr als 40 in-situ-Messungen erfaßt, deren Ergebnisse in Betrag und Richtung in Abb. 5 dargestellt sind, wobei örtlich dicht beieinanderliegende Messungen mit nahezu gleichen Resultaten zu einem gemeinsamen Spannungsrichtungspfeil zusammengefaßt wurden. Als Meßverfahren kamen überwiegend Entlastungsmethoden (Doorstopper), vereinzelt aber auch Flat Jacks und das hydraulic-fracturing-Verfahren zum Einsatz. Abgesehen von einigen Messungen im Alpengebiet wurden meist nicht der vollständige räumliche Spannungszustand, sondern die Beträge und Richtungen der Horizontalspannungskomponenten bestimmt. Die Spannungsrichtungspfeile in Abb. 5 zeigen deshalb für die einzelnen Messungen nur die Richtungen der jeweils größten Horizontalspannungskomponente, wobei deren ungefährer Betrag durch die Pfeilgröße veranschaulicht wird.

Erwartungsgemäß wurden die größten Spannungswerte im Bereich der Alpen gemessen, eine einheitliche Beanspruchungsrichtung tritt hier allerdings nicht besonders deutlich hervor. Dagegen ergaben die Messungen nördlich des Alpengebietes fast ausnahmslos NW-SE streichende Horizontalspannungen. Die Messungen am Südwestrand des Pariser Beckens wurden mit Flat Jacks durchgeführt, im Raum der Rheinischen Masse und des Oberrheingrabens kam das Doorstopper-Verfahren zum Einsatz (Baumann, 1981). Im Bereich des Rheinischen Massives wurden horizontale Zugspannungen gemessen, bei den zugehörigen Spannungsrichtungspfeilen sind örtlich nahe beieinanderliegende Messungen mit ähnlichen Ergebnissen zusammengefaßt. Das Resultat der im vorangegangenen Abschnitt beschriebenen Spannungsmessung im Harz ist in Abb. 5 als "Complete Stress-Tensor" eingetragen. Trotz des unterschiedlichen Meßverfahrens paßt es sich gut in das Muster der anderen Meßergebnisse ein. Eine besonders gute Übereinstimmung besteht mit den hydraulic-fracturing-Messungen im Falkenberg-Granit der Böhmischen Masse (Rummel, 1979) und im Hildesheimer Wald, nordwestlich des Harzes. Obwohl in einigen Fällen auch die Beträge der Spannungen ähnliche Größenordnungen aufweisen, sind diesbezügliche Vergleiche wegen der unterschiedlichen örtlichen Gegebenheiten bei den einzelnen Messungen nur bedingt möglich.

5. SCHLUSSBETRACHTUNG

Das Kompensationsmeßverfahren weist neben vielen Vorzügen auch einen durch das Meßprinzip bedingten Nachteil auf: Es ermöglicht nur Spannungsmessungen an der Gebirgsoberfläche, weshalb die Ermittlung von Spannungen im Gebirgsinneren mit dieser Methode bisher nicht möglich war. Ein erfolgreich verlaufender Meßeinsatz zur Bestimmung des vollständigen räumlichen Spannungszustandes im Gebirge durch Anwendung der Kompensationsmeßtechnik in einem gebohrten Wasserstollen im Harz hat die praktische Durchführbarkeit dieses in seinen theoretischen Grundlagen schon länger bekannten Verfahrens bewiesen. Das erzielte Meßergebnis stimmt mit den Resultaten anderer in Mitteleuropa durchgeführter Messungen überein. Das Verfahren ist gut beherrschbar, der technische Aufwand ist relativ gering. Im Gegensatz zu anderen bisher entwickelten Meßmethoden erfaßt es einen verhältnismäßig großen Gebirgsbereich von m- bis 10 m-Dimension, weshalb sich lokale Spannungsinhomogenitäten weniger stark auf das Meßergebnis auswirken. Seine Anwendung setzt allerdings das Vorhandensein eines gebohrten Tunnels oder Schachtes in einem elastischen und homogenen Gebirgskörper voraus. Insofern ist es sicher kein universell einsetzbares Spannungsmeßverfahren, das es ohnehin nicht gibt und wahrscheinlich auch nie geben wird, sondern vielmehr eine Erweiterung des Anwendungsbereiches der heute technisch nahezu ausgereiften Kompensationsmeßtechnik.

SCHRIFTTUM:

Alexander, L. G. (1960). Field and Laboratory Tests in Rock Mechanics: Proc. 3rd Austr.-N.Z. Conf. on Soil Mechanics, 161 - 168.

Baumann, H. (1981). Regional Stress Field and Rifting in Western Europe: Tectonophysics Vol. 73, 105 - 111.

Hiramatsu, Y. and Oka, Y. (1962). Stress around a Shaft or Level excavated in Ground with a three-dimensional Stress State: Mem. Fac. Engng. Kyoto Univ., 24, 56 - 76.

Jaeger, J. C. and Cook, N. G. W. (1969). Fundamentals of Rock Mechanics, 1st ed., Methuen & Co, London.

Lange, R. P. (1979). The Radau-Stollen: Erzmetall, Vol. 32, 311 - 315.

Leemann, E. R. (1968). The Determination of the Complete State of Stress in Rock in a single Borehole: Int. J. Rock Mech. Min. Sci., Vol. 5, 31 - 56.

Natau, O. (1967). Grundlagenuntersuchungen über die Anwendbarkeit der Kompensationsmethode im Bohrloch mit Hilfe von zweiachsig beanspruchten Prüfkörperplatten: Diss. TH Clausthal.

Natau, O. (1972). Theorie zur Bestimmung des Spannungstensors im Gebirge durch Messung von Normalspannungen nach der Kompensationsmethode im Bohrstollen: Int. Symp. f. Untertagebau, Luzern, 513 - 519.

Mayer, A., Habib, P. and Marchand, R. (1951). Conf. int. sur les pressions de terrains et le soutènement dans les chantiers d'exploration, Liège, 217 - 221.

Rocha, M., Lopes, J. and Silva, J. (1966). A new Technique for Applying the Method of the Flat Jack in the Determination of Stresses inside Rock Masses: Proc. 1st Congr. Int. Soc. Rock Mech., Lisbon, Vol. 2, 57 - 66.

Rummel, F. and Alheid, H. J. (1979). Hydraulic Fracturing Stress Measurements in SE-Germany and Tectonic Stress Pattern in Central Europe: Proc. Int. Res. Conf. on Intra-Continental Earthquakes, Lake Ohrid, Yugoslavia, 33 - 65.

Tincelin, M. E. (1951). Conf. int. sur les pressions de terrains et le soutènement dans les chantiers d'exploration, Liège, 158 - 175.

FAULT MECHANISM IN THE TOLEDO SHEAR ZONE IN SPAIN
Mécanisme des failles dans la zone de cisaillement de Tolédo en Espagne
Mechanismus einer geologischen Störung in der Scherzone von Toledo in Spanien

J. L. Hernandez Enrile
Profesor Adjunto de Geodinámica Interna, Universidad Complutense, Madrid, España

SYNOPSIS

The processes dominant in the recrystallization of quartz in the mylonite of Toledo are the "dynamic recovery" development and dynamic recrystallization. The quasi-plastic shear zone of Toledo has developed under steady strain rates of aseismic shear.

RESUME

Les phénomènes dominants de la recristallisation du quartz dans la mylonite de Tolède sont le développement du "rééquilibrage dynamique" et la recristallisation dynamique. La zone de cisaillement quasi-plastique de Tolède s'est développée à des taux de déformation constante dans des conditions de cisaillement non-sismique.

ZUSAMMENFASSUNG

Die dominanten Prozesse bei der Quarz-Rekristallisation im Mylonit von Toledo sind "dynamic recovery"-Entwicklung und dynamische Rekristallisation. Die quasi-plastische Scherzone von Toledo hat sich unter stetiger, aseismischer Scher-Deformation entwickelt.

1. INTRODUCTION

A fundamental understanding of seismic and aseismic faulting in the earth, must be based in part on a kowledge of fault rocks which are useful source of information on processes operating within major fault zones. By studying their textures, related grain scale deformation mechanism and their distribution within dislocation zones, we can to establish criteria by which the rock products of seismic and aseismic faulting can be distinguished.

In the present study, attention is given to fault rocks from ancient deeply eroded shear zone developed in cristalline quartzo-feldespathic crust. Correlation their textures and microstructure result of increasing strain with mineral deformation processes fundamentally of quartz, has been subjet of study in order to establish the macroscopic fracture mechanism developed in Toledo Shear zone. The enviromental parameters during shearing are briefly also discussed in relation to progressive development of the formation induced microstructures observed within dislocation zone. Thus, can be expected to yield information about the change in mineral deformation processes and fault mechanism with depth.

2. GEOLOGICAL SETTING

The Toledo shear zone is situated in the Toledo Cristalline Massif belonging to the axial zone of the Hesperic Belt of the Iberian Peninsula. Polymetamorphic precambrian rocks, large amounts of metasediments subjected to a polyphase hercynian tectonic history and hercynian granites, are characteristic in this extensive area.

This shear belt occur in a narrow planar zone wich extends for a length of 50 Kilometers with general trend east-west. Both ends are covered by tertiary sediments. Is a example of a postorogenic shearing structure commonly found in the cristalline basement rocks, which separates quartz-feldspathic rocks (migmatites) from lower palaezoic cover and granitic rocks showed in the Fig. 1. This explains the abrup jump in metamorphic intensity between complex amphibolite facies and lower Palaezoic metasediments greenschist facies, Aparicio, (1.971). This author interpreted the dislocation zone as a normal fault with 30° 45° dips towards the south.

3. STYLE OF FAULTING AND ASSOCIATED FAULT ROCKS

The study of fault rock textures make evident that the rock type which develops within the Toledo shear zone, are mylonites. In fact, on the north side of the shearing zone, amphibolite facies migmatites have undergone ductile deformation resulting ultimately in mylonites.

The mylonite zone has generally a constant oucrop width of the one kilometer a true thickness of between 300 and 400 meters. For high values of strain, mylonitic foliation may be expected to lie subparallel to the shear zone walls. Moreover a penetrative mineral lineation is present in mylonite, marked by quartz crystals show much grater elongation than individual feldspar porphyroclasts wich are mainly fragmen-

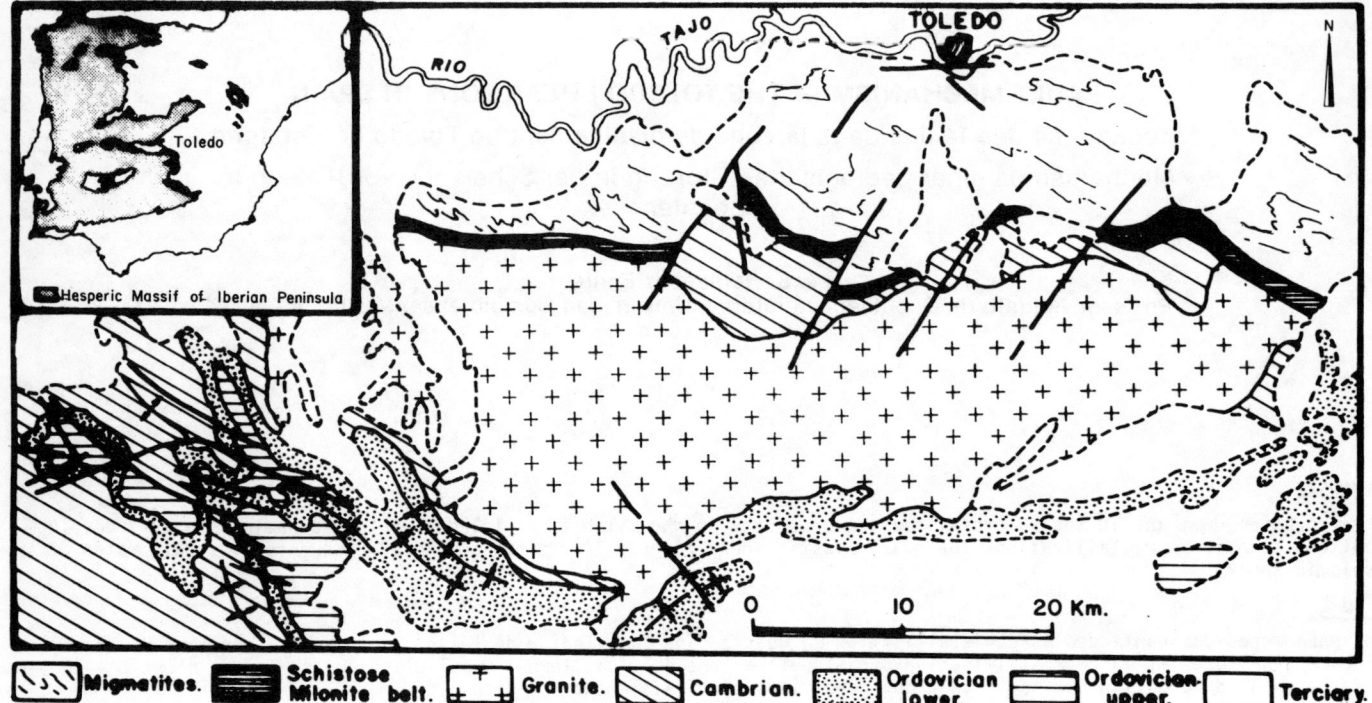

Fig. 1 Geological setting of Toledo Shear Zone

ted and slightly flattened.

From north to south across the fault zone, there are progressive textures changes result of increasing strain for a style of faulting. Thus, from the north margin country rock with incipient preferred orientation (protomylonite) may be observed gradually passing to mylonite and ultramylonite towards the interior of ductile shear zone.

3.1 Proto-mylonite

In intact rock, slip surfaces are defined by grain boundary sliding accompanied by rotation of grains as a result of shear planes along or near and parallel to the boundary between grains A preferred orientation of porphyroclasts is developed together with a small increase in the matrix that surrounds them, showed in the Fig. 2-b. The fault rocks are composed of feldspar porphyroclasts, wich make up more than 50 percent of the rock. In addition, the matrix constitutes 10 to 15 percent of the rock, Sibson (1.977).

3.2 Mylonite

Mylonitized quarzo-feldespathic rocks containing strained porphyroclasts within a fine-grained mica-quartz matrix, which has increased to 50-80 percent of the total rock. Recrystallized quartz grains have been reduced in gran sice. Moreover, large feldspar porphyroclasts show the effect of deformation through fragmentation and rotation followed by marginal granulation. Thus, the initial fragments of feldspar are reduced in size with increasing deformation. This is represented in the Fig. 2-c, as augen mylonite texture.

3.3 Ultramylonite

In the mylonite zone as cohesive rock occurs in which a very fine grain sized matrix constitutes more than 95 percent of the bulk of the rock, showed in the Fig. 2-d. Most of the porphyroclasts have been reduced to sizes smaller than one millimiter. Ultramylonites are found interspersed within foliated mylonite in thin and discontinous zones. They are rocks of black colour, generally homogeneous but sometimes having compositional layering. They represent the most intense mylonitization and are localised along the southern limit of the mylonite belt. Therefore the mylonite zone is clearly asymmetric as a result of the intensity of deformation increasing from north to south.

4. MICROSTRUCTURAL DEVELOPMENT

Different stages of deformation have been established from the microstructural evolution of the different minerals constituents of the quarzo-feldspathic rocks (migmatites) on the Toledo shear zone, Hernández-Enrile (1.982). Grain refinement in these rocks, is the result of stress intensification during ductile flow.

The effect of increasing strain on the quartz grain size is represented in the Fig. 3 were a marked reduction in new recristallized sice, occurs from protomylonite-to mylonite. Deformed quartz are common around the grain boundaries of the old grains which have subgrains at the edges and undeformed cores. The marked grain size reduction of the old grains, is accompanied by the development of subgrains and new grains. The quartz new grains are distinguished by increasing misorientation across subgrains. Hence, recovery and recrystallization processes has taken place in the most highly strained grains, produ-

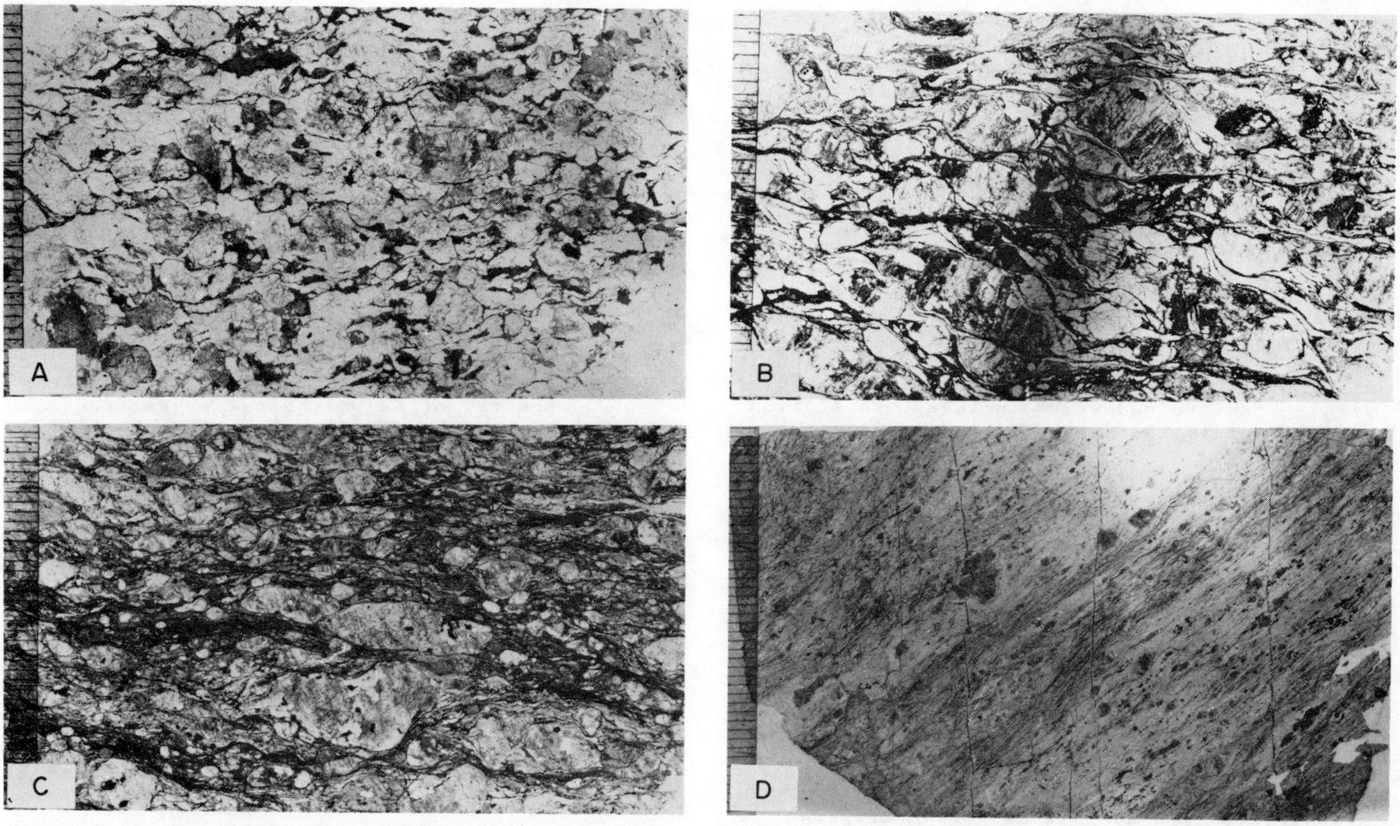

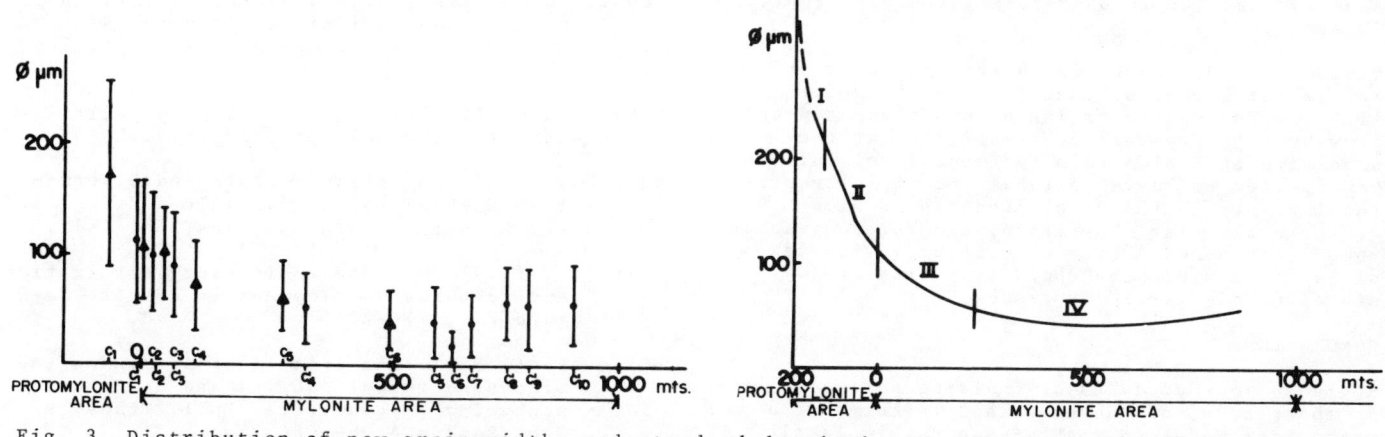

Fig. 2. Mylonite series from migmatites. a) Intact rock passing to incipient protomylonite. b) Protomylonite. c) Mylonite. d) Ultramylonite

Fig. 3. Distribution of new grain widths and standard desviations and summary of grain size distribution from all three traverses. Deformations Stages I-IV are based on changes in grain size

cing sub-grains and new flattened grains along and near their margins, forming an incipient schistosity.

For high values of strain within ductile shear zone, the old quartz grains have completely recrystallized into narrow aggregates of equant new grains marking the development of a quasi-stable microstructure. Thus, the constant size of the new grains constituting the quasi-steady microstructure, is mantained by continous grain boundary sliding or by dinamic recrystallization

Thus, the ultimate product of mylonitisation is the development of an ultramylonite which is homogeneous on the microscale. Originali quartz grains are defined by small relict aggregates of equidimensional undeformed new grains. The small new grains (60 μm) have straight grain boundaries giving rise to polygonal shapes.

In mylonites derived from a quartzo-feldspathic rocks, it has been show that quartz by virtue of its ductility, is essential for the formation of mylonites. Bell and Etheridge (1.973). White

(1.973, 1.976, 1.979), Bouchez (1.977), Bossiere and Vauchez (1.978). It is therefore concluded, that cristal plastic processes in the Toledo shear zone were controlled by the mechanical behaviour of quartz mineral.

On the other hand, in the Toledo shear zone the resistence of feldspars to deformation is in direct contrast to the behaviour of quartz under the same conditions, Hernández-Enrile (1.982). The feldspars are essentially affected by a process of rotational deformation involving the comminution of feldspars in a ductile matrix of quartz and biotite. Microstructural evolution of feldspars involves at low strain states, crushing, granulation and intergranular sliding. At higher strain states, intragranular gliding is involved which includes translation (slip), deformation twinning and partial recrystallization.

5. CRUSTAL CONDITIONS AND FAULT MECHANISM.

The ancient Toledo fault zone is now exposed at erosion level wich correspond to considerable depth when the fault was active. Processes of rock deformation within fault zone, must be influence with depth in the crust as a result of varying, temperature, confining pressure, variables such as strain rate and differential stress. Depth is likely to be 6-8 kilometers for mylonites in Toledo fault zone, Hernández-Enrile (1.982) assuming anormal geothermal gradients of 50°C/Kmt, gives a temperature between 300-400°C. Therefore, the migmatites have been mylonitized in a ductile manner under greenschist facies metamorphic conditions. On this basis fault mechanism can accomodate shearing, where the quartz first beginning to deform by crystal plasticity. It is under these and higher grades of metamorphism that quasi-plastic shear zone may develop in quartzo-feldspathic crust, Sibson (1.977).

Dislocation processes in quartz dominante at strain rates of less than 10^{-9} sec^{-1}, White (1.976). The quartz grains become elongated and flow by intracrystalline plasticity, at the same time optical strain features form. This can be expected to leave textural and microstructural imprints on the rocks products of slow aseismic faulting. Therefore, the Toledo shearing zone corresponds aseismic fault creep and takes place at strain rates compatible with steady state, quasi-plastic processes.

CONCLUSIONS

Style an rock products of faulting across Toledo shear zone, can be explained in terms of a mechanical model in wich a zone quasi-plastic behaviour, where mylonite series rocks are developed. Strain rates compatible with flow by intracrystalline plasticity are associated with steady aseismic faulting.

REFERENCES

Aparicio, A. (1.971). Estudio geológico del Macizo Cristalino de Toledo. Estudios Geol. 27, 369-414.

Bell, T.H. and Etheridge, M.A. (1.976). The deformation and recrystallization of quartz in a mylonite zone, central Australia, Tectonophysics, 32, 235-267.

Bossiere, G. and Vauchez, A. (1.978). Deformation naturelle par cisaillement ductile d'un granite de Grande Kabile occidentale (Algerie). Tectonophysics, 51, 57-81.

Brune, J.N. (1.970). Tectonic stress and the spectra of seismic shear waves from earthquakes. J. Geophys. Res. 75, 4997-5009.

Bouchez, J.L. (1.977). Plastic deformation of quartzites at low temperature in an area of natural strain gradient. Tectonophysics, 39, 25-50.

Griggs, D.T. (1.974). A model of hydrolytic wakening in quartz. J. Geophys. Res. 79, 1653-1661.

Hernández-Enrile, J.L. (1.982). Evolución microestructural como resultado del aumento de la deformación en la milonita de Toledo (España). Cuadernos de Geología Ibérica, 7.

Mitra, G. (1.978). Ductile deformation zones and mylonites: The mechanical processes involved in the deformation of crystalline basement rocks. Am. J. Sci. 278, 1057-1084.

Scholz, C.H., Wyss, M. and Smith, S.W. (1.969). Seismic and aseismic slop on the San Andreas Fault. J. Geophys. Res. 74, 2049-69.

Sibson, R.H. (1.977). Fault rocks and fault mechanism. J. Geol. Soc. 133, 191-213.

Vauchez, A. (1.980).Ribbon Texture and Deformation Mechanisms of Quartz In a Mylonitized Granite. Tectonophysics, 67, 1-12.

White, S. (1.973). Syntectonic recrystallization and textures development in quartz. Nature 244, 276-278.

White, S. (1.976). The effects of strain on the microstructures, fabrics and deformation mechanism in quartzite. Phil. Trans. R. Soc. London A 069-086.

White, S. (1.979). Subgrain and grainsize variations across a Shearzone. Contrib. Mineral. Petrol. 70, 193-202.

White, S. et al. (1.980).Milonites in Shearzones. J. Struc. Geol. 2, 175-187.

DETERMINATION OF THE STATE OF STRESS OF ROCK MASSES BY THE SMALL FLAT JACK (SFJ) METHOD

Détermination de l'état de contrainte des massifs rocheux par la méthode des petits vérins plats (SFJ)

Bestimmung des Gebirgsspannungszustandes mit der Methode der kleinen Druckkissen (SFJ)

José Loureiro Pinto
José Gabriel Charrua-Graça
Research Officers, Rock Foundation Division — Dams Department — LNEC, Lisboa, Portugal

SYNOPSIS

The application of the SFJ method to the determination of the state of stress of rock masses is analysed for circular, square and elliptical galleries. Different configurations are discussed. Solutions are presented and procedures are suggested with a view to optimizing the application of the method.

RESUME

On étudie l'application de la méthode SFJ à la détermination de l'état de contrainte des massifs rocheux pour les galeries circulaires, carrées et elliptiques et on discute des différentes configurations. On présente des solutions et on suggère des procédés ayant pour but d'optimiser l'application de la méthode.

ZUSAMMENFASSUNG

Die Anwendung der SFJ Methode zur Bestimmung des Gebirgsspannungszustandes wird für kreisförmige, quadratische und elliptische Stollen betrachtet. Verschiedene Stellungen werden erörtert. Es werden Lösungen vorgestellt und Verfahren im Hinblick auf die bestmögliche Anwendung der Methode vorgeschlagen.

1. INTRODUCTION

The importance of knowing the state of stress of the rock masses where works are to be carried out has been amply illustrated in the literature on the subject. However, it is only recently that the different entities responsible for these works have shown they are sufficiently aware of that importance to concern themselves with the determination of such stresses.

LNEC developed a method using slots and another by overcoring. The article concerns the slot method, which has the advantage of allowing direct measurement of the stresses in the rock mass. It is also easier to execute, although it is rather lengthy and makes it necessary to open galleries or shafts inside the rock mass.

2. TECHNIQUE OF THE METHOD

The SFJ method technique, ROCHA (1966), can be briefly described as follows:

- On the surface to be studied, duly made even, are placed pairs of measuring bases and the distances between them are measured.

- A slot is cut between these bases with a diamond-edge disk thus relieving the normal stress in the plane of the slot. As a consequence, the distance between the measuring bases varies, usually by decreasing.

- A flat jact of suitable shape is placed inside the slot, fitting it closely, oil under pressure is introduced in the jack, and the distance between the bases is measured until the initial position is obtained.

- The pressure introduced in the jack for obtaining again the initial pressure (cancelling pressure) is, apart from small correcting factors, the normal stress in the area of the slot.

Figs. 1, 2 and 3 show the cutting machine with the motor for operating it, flat jacks with differ

Fig. 1 - Cutting machine.

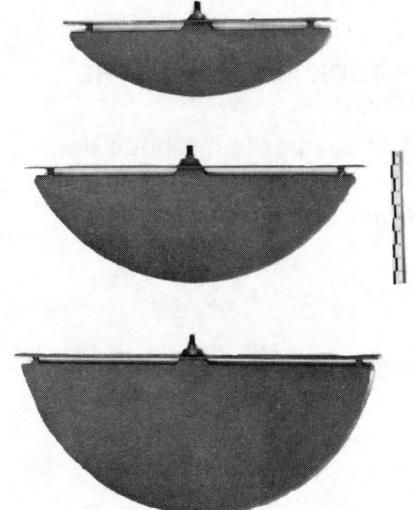

Fig. 2 - Small Flat Jacks (SFJ)

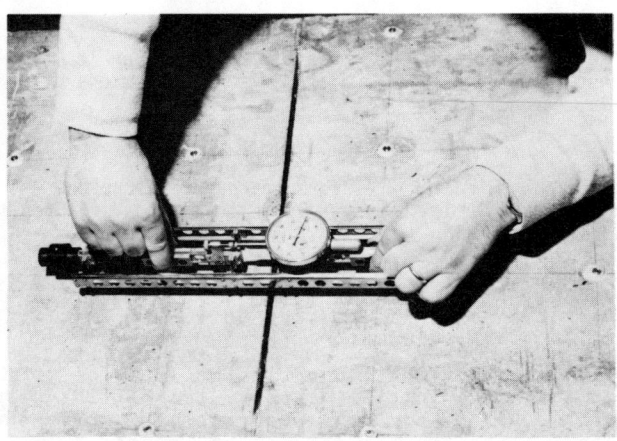

Fig. 3 - Measurement of the distance between the bases.

ent rise to be introduced in the slot according to the depth attained, and the measurement of the distance between the bases.

The carrying out of three tests of the type described, with the slots forming a rosette, makes it possible to determine the state of stress at the point of the plane in question. Four slots forming a 45° rosette are the usual procedure because this makes it possible, by means of the superabundant test, to check the results obtained.

If the study of the plane referred to is repeated in two more planes with a different orientation from the first this will make it possible to determine the whole state of stress at the point considered.

3. THE STATE OF STRESS AROUND A CIRCULAR SECTION GALLERY

From what has been said it can be seen that it is easy to find out the state of stress in gallery walls. The problem lies in the fact that these stresses do not correspond to those of the rock mass owing to the disturbance caused by the opening of the gallery.

In fact, the opening of the gallery brings about two local effects in the field of stresses of the rock mass. The first causes the stresses to concentrate around the gallery, this effect decreasing as depth increases. The second, which is very much dependent on the kind of material, on the excavation process and on several other factors, leads to a decompression of the rock mass and to a decrease in the value of the stresses, which diminishes with depth. Unlike the first it is extremely difficult to assess it without resorting to methods that will not be dealt with in this paper and therefore it shall not be considered.

The opening of a circular hole in a body subject to the state of stress defined by stresses P_x, P_y, P_z, P_{yz}, P_{zx}, P_{xy}, gives rise to the appearance, around the edge of the slot, of strains with values, LNEC (1974):

$$\varepsilon_r = -\nu \frac{P_x + P_y}{E} + 2\nu(1+\nu)\frac{P_x - P_y}{E}\cos 2\theta - \nu\frac{P_z}{E} - 4\nu(1+\nu)\frac{P_{xy}}{E}\sen 2\theta$$

$$\varepsilon_\theta = \frac{P_x + P_y}{E} - 2(1-\nu^2)\frac{P_x - P_y}{E} - \nu\frac{P_z}{E} + 4(1-\nu^2)\frac{P_{xy}}{E}\sen 2\theta$$

$$\varepsilon_z = -\nu\frac{P_x + P_y}{E} + \frac{P_z}{E} \qquad \qquad 1)$$

$$\gamma_{r\theta} = 0$$

$$\gamma_{\theta z} = 4(1+\nu)\left(\frac{P_{yz}}{E}\cos\theta - \frac{P_{zx}}{E}\sen\theta\right)$$

$$\gamma_{zr} = 0$$

where r, θ and z are the radial, perimetral and longitudinal directions of the gallery; x, y and z the horizontal and vertical directions transverse to the gallery and longitudinal to the gallery; P_i and P_{ik} are the normal and tangential stresses of the rock mass, in the already defined system of axes xyz, before the opening of the gallery; ε and γ are the strains and distorsions at the surface of the gallery; E and ν the modulus of deformability and the Poisson ratio of the rock mass; θ the angle that defines the position of the point in which the stresses indicated in Fig. 4 are to be found.

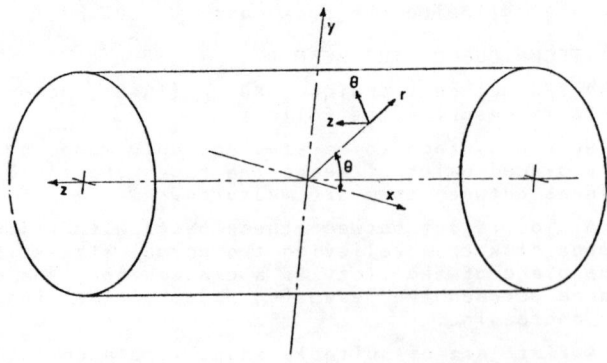

Fig. 4 - Systems of axes xyz and rθz.

From the system of equations (1) can be calculated the general expressions of the state of stress around a point of the contour of the gallery, which is given by:

$$\sigma_\theta = (1-2\cos2\theta)P_x + (1+2\cos2\theta)P_y + 4P_{xy}\sen2\theta$$

$$\sigma_z = 2\nu\cos2\theta(P_y-P_x) + P_z + 4\nu P_{xy}\sen2\theta \qquad 2)$$

$$\tau_{\theta z} = -2P_{zx}\sen\theta + 2P_{yz}\cos\theta$$

4. DETERMINATION OF THE STATE OF STRESS OF A ROCK MASS BY MEANS OF TESTS CARRIED OUT IN A CIRCULAR GALLERY

By executing, in a circular section gallery, a rosette of four slots at 45° halfway along the side walls, another with the same characteristics on the ground, and still another equal to these, at 45° and with the vertical as indicated in Fig. 5, it is possible to deduce from expressions, LNEC (1974), the stresses that should

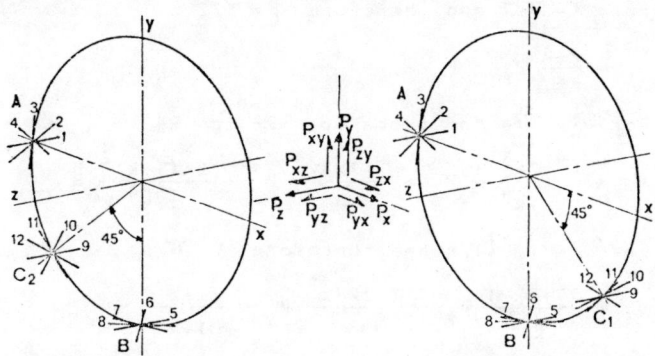

Fig. 5 - Location of the rosettes and of the measuring surface elements.

be measured along the 12 directions defined by the three plane rosettes, thus obtaining:

Rosette A

$$\sigma_1 = -P_x + 3P_y$$

$$\sigma_2 = -(0,5+\nu)P_x + (1,5+\nu)P_y + 0,5P_z + P_{yz}$$

$$\sigma_3 = -2\nu P_x + 2\nu P_y + P_z$$

$$\sigma_4 = -(0,5+\nu)P_x + (1,5+\nu)P_y + 0,5P_z - P_{yz}$$

Rosette B

$$\sigma_5 = 3P_x - P_y$$

$$\sigma_6 = (1,5+\nu)P_x - (0,5+\nu)P_y + 0,5P_z - P_{zx} \qquad 3)$$

$$\sigma_7 = 2\nu P_x - 2\nu P_y + P_z$$

$$\sigma_8 = (1,5+\nu)P_x - (0,5+\nu)P_y + 0,5P_z + P_{zx}$$

Rosette C

$$\sigma_9 = -P_x + P_y + 4P_{xy}$$

$$\sigma_{10} = 0,5P_x + 0,5P_y + 0,5P_z + \sqrt{2}/2P_{yz} \mp \sqrt{2}/2P_{zx} + 2(1+\nu)P_{xy}$$

$$\sigma_{11} = P_z + 4\nu P_{xy}$$

$$\sigma_{12} = 0,5P_x + 0,5P_y + 0,5P_z - \sqrt{2}/2P_{yz} \mp \sqrt{2}/2P_{zx} + 2(1+\nu)P_{xy}$$

As the stresses measured do not meet the above expressions the method of the least squares must be resorted to for determining the unknowns $P_x \ldots P_{xy}$ that define the state of stress in the rock mass before opening of the gallery.

This solution is an exact solution and it is easy to apply in galleries of perfectly circular section, like those made with a tunnel boring machine, when the section is not circular it is sometimes difficult to form rosette C and it is then preferable to determine the stresses at the top of the gallery. The execution of rosette C at the top leads to the changing of stresses σ_9, σ_{10}, σ_{11} and σ_{12}; the expressions of the other stresses remain the same.

4.1 - The use of the top of the gallery

These seems to be no exact solution for this case but studies, Neuber (1946), of the stress concentration around a spherical cavity are available.

Although it may seem forced and lacking in verisimilitude to consider the performance of the top of the gallery as that of a spherical cavity, several authors have accepted it as the solution closest to reality available so far. The acceptance of this criterion is based on the hypotheses:

a) That the top of the gallery actually has a behaviour similar to that of a semi-spherical calotte with a radius equal to that of the gallery. This hypothesis is based on the consideration that the fire, by inciding particularly on the central zone, and the concentration at the contours give rise to this type of concentration. In case of doubt and to find out the state of stress of the rock mass one may take the precaution of making it as close as possible to the shape considered by means of manual breaking.

b) The behaviour of a semi-spherical cavity connected to a cylindrical cavity is not very unlike the behaviour of a spherical cavity. Although there is of course different behaviour it may be admitted that the differences will be essentially in the zone next to the contour of the top of the gallery and that they diminish towards the central zone of the callote. Tests should therefore be carried out as close as possible to this zone to minimize that effect.

The general equations that define the state of stress at the surface of a spheric cavity can be deduced from those indicated by Neuber (1964) and they are as follows:

$$\sigma_\theta = \frac{1}{14-10\nu} \sum_i P_i (30\sen^2\theta_i - 3 - 15\nu)$$

$$\sigma_\phi = \frac{1}{14-10\nu} \sum_i P_i (30\sen^2\theta_i - 3 - 15\nu) \qquad 4)$$

where θ_i are the angles of stresses P_x, P_y and P_z, measured on planes longitudinal to the gallery, with the vector normal to the gallery wall at the point where stresses σ_θ and σ_ϕ normal to one another and to the longitudinal direction and parallel to directions x and occur.

By Considering the angles θ corresponding to the slots we have:

$$\sigma_9 = \frac{1}{14-10\nu}\left[P_x(-3+15\nu)P_y(27-15\nu)-P_z(3+15\nu)\right]$$

$$\sigma_{10} = \frac{1}{14-10\nu}\left[12(P_x+P_y)-P_z(3+15\nu)-30P_{xy}(1-\nu)\right]$$

$$\sigma_{11} = \frac{1}{14-10\nu}\left[P_x(27-15\nu)+P_y(-3+15\nu)-P_z(3+15\nu)\right] \quad 5)$$

$$\sigma_{12} = \frac{1}{14-10\nu}\left[12(P_x+P_y)-P_z(3+15\nu)+30P_{xy}(1-\nu)\right]$$

The solution of system (3) in which the four last equations are replaced by those indicated in (5) according to the least square method leads to the solution of the problem.

5. NON-CIRCULAR GALLERIES

Galleries, except those made with a tunelling machine, which have circular sections, usually have very irregular sections and only in some cases may they be considered as approximately circular. One hypothesis for determining the state of stress is the execution of test chambers with sections close to the circular. In some cases it is possible to obtain a more or less exact solution or to have an idea of the errors made.

5.1 - Square - section galleries

In these galeries the stress concentration halfway along the side walls or the ground is the same as in circular galleries, Timoshenko (1934). In the case of approximately square - section galleries equations (5) may thus be used, which are suitable for these galleries.

5.2 - Elliptic - section galleries

In the case of galleries with elliptic section the expressions deduced by Greenspan (1944) may be written as follows:

Only $P_x \neq 0$; $\sigma_t = \dfrac{(p+q)^2 \mathrm{sen}^2\beta - q^2}{(p^2-q^2)\mathrm{sen}^2\beta + q^2}$

Only $P_y \neq 0$; $\sigma_t = \dfrac{-(p+q)^2 \mathrm{sen}^2\beta + q(q+2p)}{(p^2-q^2)\mathrm{sen}^2\beta + q^2} P_y \quad 6)$

Only $P_{xy} \neq 0$; $\sigma_t = \dfrac{(p+q)^2 \mathrm{sen}2\beta}{(p^2+q^2)\mathrm{sen}^2\beta + q^2} P_{xy}$

where p and q are the minor and major axis of the section. The minor axis coincides with the axis of the xx and the major axis with that of the yy; β is an angle that defines the position of the point considered and σ_t the stress normal to the surface element that passes through that point and normal to the contour of the gallery.

Considering another angle that defines the position of the point of the contour under study as represented in Fig. 6, stresses σ_t corresponding to those considered in the previous paragraphs will be, taking $k = p/q \leq 1$:

For $\alpha = 0$ and therefore $\beta = 0$

$\sigma_t = -P_x + (1+2k)P_y$

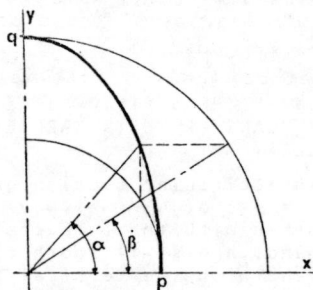

Fig. 6 - Angles α and β defining the contour

For $\alpha = \pi/2$ and therefore $\beta = \pi/2$

$\sigma_t = \dfrac{2+k}{k} P_x - P_y$

For $\alpha = \pi/4$ and therefore $\beta = tg^{-1}k$

$\sigma_t = \dfrac{k^4+2k^3-1}{1+k^4} P_x + \dfrac{1+2k-k^4}{1+k^4} P_y + \dfrac{2k(1+k)^2}{1+k^4} P_{xy}$

For $\alpha = tg^{-1}1/k$ and therefore $\beta = \pi/4$

$\sigma_t = \dfrac{k^2+2k-1}{1+k^2} P_x + \dfrac{1+2k-k^2}{1+k^2} P_y + \dfrac{(1+k)^2}{1+k^2} P_{xy}$

Bearing in mind these expressions, the stresses that should be measured in rosettes A, B and C of Fig. 5, this time considering an elliptic contour instead of a circular one would be:

Rosette A

$\sigma_1 = -P_x + 2aP_y$

$\sigma_2 = -(0,5+b\nu)P_x + (a+kb\nu)P_y + 0,5P_z + P_{yz}$

$\sigma_3 = -2b\nu P_x + 2kb\nu P_y + P_z$

$\sigma_4 = -(0,5+b\nu)P_x + (a+kb\nu)P_y + 0,5P_z - P_{yz}$

Rosette B

$\sigma_5 = 2cP_x - P_y$

$\sigma_6 = (c+d\nu)P_x - (0,5+kd\nu)P_y + 0,5P_z - P_{zx} \quad 7)$

$\sigma_7 = 2d\nu P_x - 2kd\nu P_y + P_z$

$\sigma_8 = (c+d\nu)P_x - (0,5+kd\nu)P_y + 0,5P_z + P_{zx}$

Rosette C

$\sigma_9 = 2eP_x + 2fP_y + 2gP_{xy}$

$\sigma_{10} = -(e-kh\nu)P_x+(f+h\nu)P_y+0,5P_z+iP_{yz}+kiP_{zx}+(1+\nu)P_{xy}$

$\sigma_{11} = -2kh\nu P_x + 2h\nu P_y + P_z + 2gP_{xy}$

$\sigma_{12} = (e-kh\nu)P_x+(f+h\nu)P_y+0,5P_z-iP_{yz}\mp kiP_{zx}+(1+\nu)P_{xy}$

In these expressions parameters a, b, c, d, e, f, g, h and i are parameters that depend on k and therefore on the relation between the minor and major semi-axes of the ellipse and $K(\varepsilon)$ and $E(\varepsilon)$ are the complete elliptical integrals of the 1st and 2nd kind as regards excentricity ε of the ellipse. The variation of these parameters with k is indicated in Fig. 7 which also gives the analytical expressions of these parameters.

The solving of this system of equations by the least squares method leads to the solution of the problem.

5.3 - Galleries with sections having two symmetry axes

The study presented for elliptical sections can be made in the same way for any section with two symmetry axes by using the expressions of Greenspan (1944), it shall however not be described here because circular section, elliptic section and square section cover the usual range of gallery sections. Anyway in the determination of the state of stress of a rock mass by the SFJ it may be necessary to open galleries with the forms studied here.

Bibliography

BONNECHERE, F. (1971) - Contribution à la determination de l'état de contrainte des massifs rocheux. Thèse, Université de Liége.

GREENSPAN, M. (1944) - Effect of a Small Hole on the Stresses in a Uniformly Loaded Plate. Quarterly Appl. Math. 2 pp 60-71.

LNEC, (1974) - Estudo das deformações de um cilindro de plástico colocado no interior de um maciço rochoso. LNEC Jan, 1974.

NEUBER, H. (1946) - Theory of Noch Stresses. Edwards Brothers, Ann Arbos, Michigan (cit. Obert, L. and Duwall, W. - Rock Mechanics and the Design of Structures in Rock. John Wiley and Sons, USA 1966).

ROCHA, M., BAPTISTA LOPES, J.J. & NEVES DA SILVA, J. (1966) - A new technique for applying the method of the flat jack in the determination of stresses inside rock masses. 1st Cong. of IRMS.

TIMOSHENKO, S. (1934) - Theory of Elasticity. Mcgraw Hill (cit. Mello Mendes - Mecânica das Rochas. I.S.T. (1967).

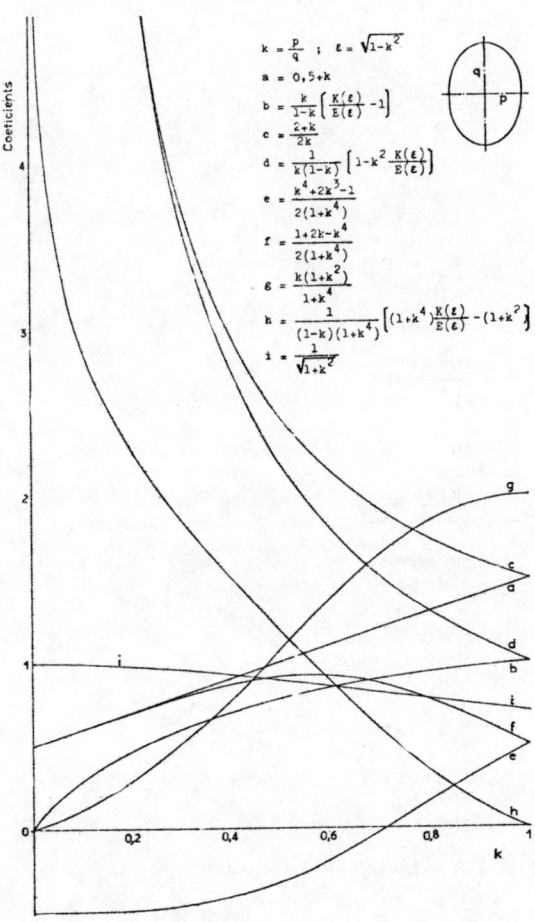

Fig. 7 - Parameters that define the state of stress in an elliptic gallery.

MEASUREMENTS OF TECTONIC STRESSES, STRAIN RATES RELATED TO ACTIVE FAULTS AND OBSERVED EARTHQUAKES AROUND LARGE CAVERNS

Mesure de contraintes tectoniques, progression de cisaillement géodésique des failles actives, et observations de tremblements de terre aux alentours d'une centrale électrique souterraine

Messungen tektonischer Spannungen, Dehnungsgeschwindigkeiten an aktiven Störungen und von Erdbeben nahe grosser Kraftwerkskavernen

T. Kanagawa
Senior Researcher, Central Research Institute of Electric Power industry, Abiko, Abiko City, Japan

H. Komada
Dr. Eng., Senior Researcher, Central Research Institute of Electric Power Industry, Abiko, Abiko City, Japan

M. Hayashi
Dr. Eng., Associate Director, Central Research Institute of Electric Power Industry, Abiko, Abiko City, Japan

SYNOPSIS
On the basis of approx. 20 experiments of tectonic stress measurements in large underground excavations and active faults the suitability of employing the acoustic emission method and a miniaturization of the overcoring method as well as the effects of tectonics, topography and faults are examined. Moreover, the connection between the geodetically ascertained shear strain rate in Japan and the range of historically observed earthquakes over the past 1 200 years in the surroundings of underground power stations is discussed.

RESUME
Cette communication examine l'efficacité de la méthode d'émission acoustique, la miniaturisation de la méthode de surcarottage et les influences tectoniques et topographiques en se basant sur les données acquises à partir d'une vingtaine de cas de mesures de la géocontrainte en rapport avec des fouilles en sous-sol, près de failles actives. De plus, on donne le taux de progression de la contrainte de cisaillement géodésique ainsi que l'influence des circonstances régionales de failles actives au cours des 1 200 années passées relativement aux tremblements de terre historiques. Enfin, on examine les mesures et les observations dynamiques effectuées aux alentours d'une centrale électrique souterraine lors d'un tremblement de terre.

ZUSAMMENFASSUNG
Auf der Grundlage von etwa 20 tektonischen Spannungsmessungen in Bezug auf große unterirdische Hohlräume und aktive Verwerfungen wurden Betrachtungen in Bezug auf die Verwendbarkeit der akustischen Emissionsmethode, Verkleinerung der "Overcoring"-Methode, tektonische Einflüsse, Einfluß der Topographie und der Erdschichten usw. angestellt. Weiterhin wurde der Zusammenhang zwischen der geodätisch festgestellten Dehnungsrate in Japan, dem Bereichszustand der geschichtlichen Erdbeben über 1 200 Jahre, aktiven Verwerfungsgruppen usw. diskutiert. Abschließend werden Messungen bei Erdbeben in der Umgebung unterirdischer Kraftwerke und dynamische Betrachtungen angestellt.

1. TWENTY EXPERIMENTS OF OVERCORING GEOSTRESSES AND CATEGORIZATION

Japan Islands are characterized by tectonic strain obtained by Geodetic survey[2],[6] in Fig. 1 [7] This geodetic result might be influenced by the rather deep bed rock deformation than overcoring method or Acoustic Emission Method in shallow ground.

Measured overcoring stresses in Table 1 are principally affected by overburden, but they might be affected more or less by other rock mechanics factors:

1) Effect of Topography is characterized by the direction of maximum compression σ_1 toward the summit.

2) Effect of Geological Structure is featured by the direction of σ_1 toward dip or strike of fault, fractured zone or strata.

3) Effect of Tectonics might often be recognized as rather large ratios σ_H max/σ_V. The maximum ratio were around 2.4 ~ 2.6.

4) Multiplicated Effect is of course met as illustrated in Fig. 2, obtained at No.12 in Chugoku district. There is recognized σ_1 toward the summit and σ_3 is normal to the steep slope. There is recognized a coincidence between the direction of σ_1 and strike of fault in Fig. 3, too. In addition to these features, the ratio σ_H max/σ_V is rather large and overcoring feature is almost consistent to the independent survey by geodetic method.

5) Tectonic Effect is illustrated representatively by Fig. 4.[3] This shows the coincidence between measured σ_1 direction and joint strike. Measured direction of σ_H max N 87° E is almost consistent to the other measurement N88° E and N89° W in Mining tunnel[4] (EL. 1220 - 1550) located at 65 km far from the site. This measured direction is also confirmed with the another seismic report[5] which has analysed the compressive

direction deduced from the earthquake wave analysis of the shallow earthquakes with depth 20 km, 1975 - 1976 in Fig. 4.

6) Stress ratios σ_H max/σ_V were plotted in Fig. 5. for three groupes based on Table 1.

7) It should be noticed that result of overcoring boring are affected by the depth of boring, change of rididity of rock heterogeneity of rock or change of P wave velocity, as shown in Fig. 6.

2. Zonation of Maximum Shear Strain Distribution

It might be interested in possible zone of maximum shear strain which might be mobilized at most. Rough scope of this resolution is illustrated in Fig. 1.[8] Based on this illustration, geodetic strike shear along the Median Tectonic Line is shown in Fig. 7. This procedure will be performed to another smaller ambiguious active faults.

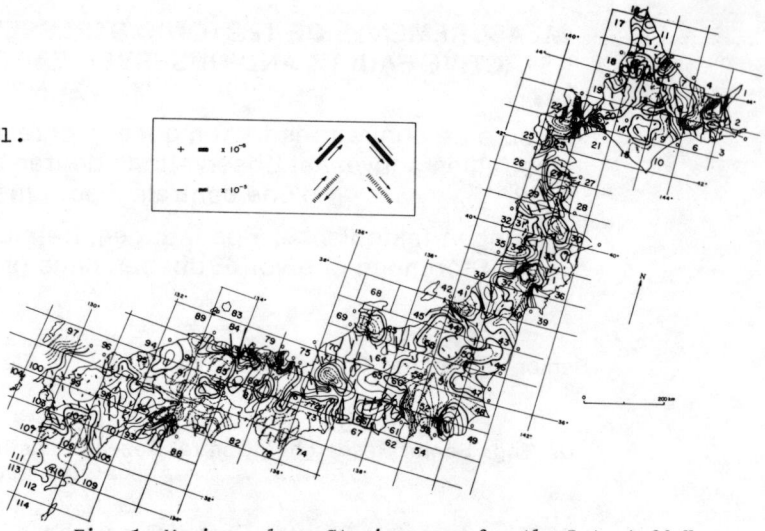

Fig. 1 Maximum shear Strain γ max for the Latest 60 Years by Geodetic Survey
(Vectors were presented by T. Harada and A. Kasai, 1971, and strain contours are illustrated by M. Hayashi, 1981)

3. Earthquake Behavior in the Mountain with Large Cavern

Quantitative observation by Means of accelerometers and dynamic strain meters has been continued at Shiroyama Underground Power Station[9] since 1976 as shown in Fig. 8.

1) It is easily recognized that the underground siting is seismically feasible than the on ground siting as shown in Fig.9. And also it is confirmed that the top of mountain is not so earthquaked compared with foot of mountain. Accelarations around the cavern were approximately about 1/2 in

Table 1. The results of the in-situ stress measurements and topographic, geologic and tectonic effects on them.
{ A: high sensitivity
  B: moderate sensitivity
  C: low sensitivity

Site No.	Location					Principal Stress (MPa)			Horizontal Stress (MPa)		Vertical Stress (MPa)	σ_{Hmax}/σ_v	Sensitivity		
	Longitude	Latitude	Rock	Elevation(m)	Depth(m)	σ_1	σ_2	σ_3	σ_{Hmax}	σ_{Hmin}	σ_v		Topography	Geologic Structure	Tectonic Environment
1	142°30'E	42°30'N	schalstein	140	210	6.2	4.8	4.7	6.2	4.8	4.8	1.29	B	A	C
2	139°30'E	37°30'N	rhyolite	260	165	4.2	3.3	2.5	4.0	3.0	2.9	1.37	B	A	B
3	138°35'E	37°20'N	mudstone	-40	70	1.24	1.08	1.07	1.16	1.08	1.17	0.99	B	A	C
4	139°40'E	36°50'N	siliceous sandstone	540	420	15.7	10.5	7.8	14.7	8.8	10.6	1.39	C	A	C
5	139°40'E	36°50'N	breccia	540	395	12.1	8.5	7.6	11.4	7.6	9.1	1.25	A	C	C
6	138°55'E	36°45'N	conglomerate	601	270	8.2	5.5	4.9	8.1	5.0	5.5	1.47	B	B	B
7*	139°15'E	35°25'N	quartz diorite	550	15	—	—	—	7.4	2.6	2.8	2.64	—	—	A
8*	138°10'E	34°35'N	mudstone	-20	30	—	—	—	0.49	0.45	0.55	0.89	—	B	C
9	137°20'E	34°35'N	granite	320	280	9.6	7.5	4.9	7.9	4.9	9.2	0.86	C	A	C
10	136°40'E	35°40'N	rhyolite	460	335	9.6	6.2	4.6	8.9	5.6	5.2	1.71	B	B	A
11*	135°50'E	34°00'N	shale	386	214	—	—	—	9.0	4.6	7.3	1.23	—	C	B
12	133°20'E	35°05'N	granite	185	370	23.4	13.2	7.2	20.2	11.1	12.5	1.62	B	A	A
13	133°20'E	33°40'N	black schist	580	270	11.1	5.4	3.7	8.7	4.4	7.2	1.21	A	C	C
14	132°50'E	34°00'N	granite	-47	71	5.5	4.6	4.1	5.5	4.5	4.2	1.31	B	C	A
15	132°20'E	33°30'N	greenschist	5	30	0.89	0.66	0.46	0.77	0.53	0.71	1.08	A	C	C
16	130°10'E	33°25'N	granite	130	510	15.8	11.1	6.3	15.5	6.4	11.2	1.38	C	A	B
17	130°10'E	31°50'N	conglomerate	-16	22	1.06	0.72	0.41	0.92	0.71	0.55	1.67	C	A	C
18	139°50'E	37°00'N	rhyolite	663	192	5.1	4.3	1.7	4.4	1.7	5.0	0.88	A	C	C
19	139°50'E	37°00'N	tuff breccia	664	241	5.0	3.7	2.9	4.1	2.9	4.6	0.89	A	C	C
20	137°40'E	36°30'N	granite	1054	250	13.7	4.6	0.4	12.3	1.2	5.1	2.41	B	—	A

* The horizontal and vertical normal stress components are clarified, but three demensional principal stresses can not be determined in these sites.

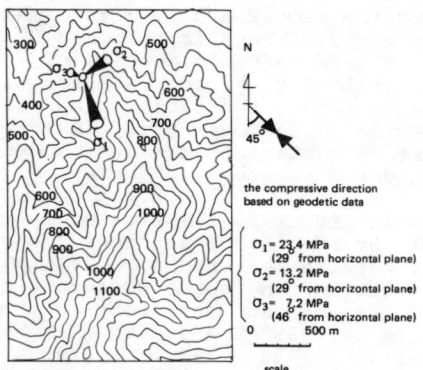

Fig. 2 Topography and the result of in-situ stress measurements (No. 12 Site)

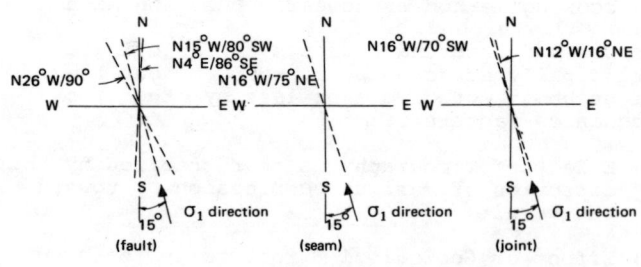

Fig. 3 Relation of the results of in-situ stress measurement to topography and geologic structure at No. 12 site

horizontal directions and about 2/3 in vertical direction compared with top of mountain or ground surface as summarized in Fig. 10. Dynamic analysis will be discussed in the near future.

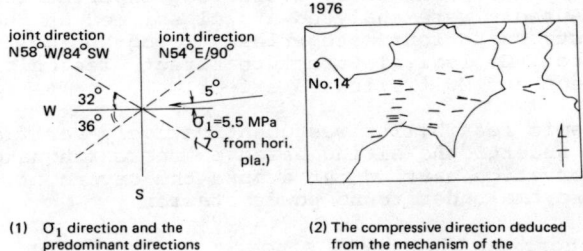

(1) σ_1 direction and the predominant directions of the joints

(2) The compressive direction deduced from the mechanism of the shallow earthquakes (Okano, 1980)

Fig. 4 Relation between the result of the in-situ stress measurement and the compressive directions deduced from joints and earthquakes mechanism at No. 14 site

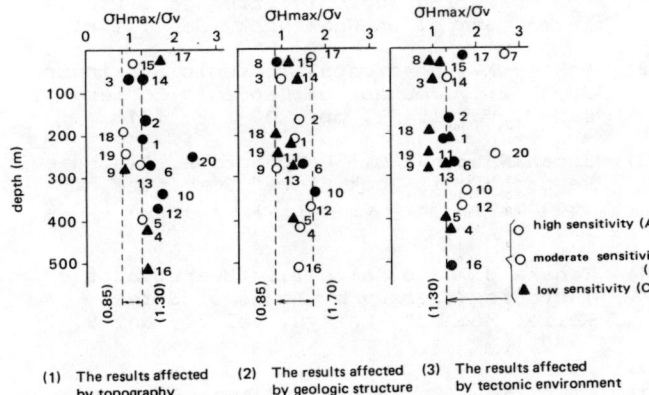

(1) The results affected by topography
(2) The results affected by geologic structure
(3) The results affected by tectonic environment

Fig. 5 Relation of σ_H MAX/σ_v to topography, geological structure and tectonic environment

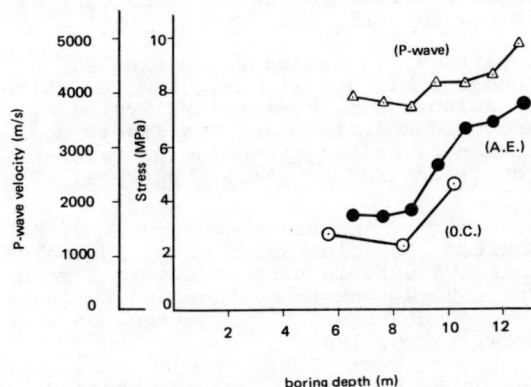

Fig. 6 Dependences of boring depth on A. E. and Overcoring results

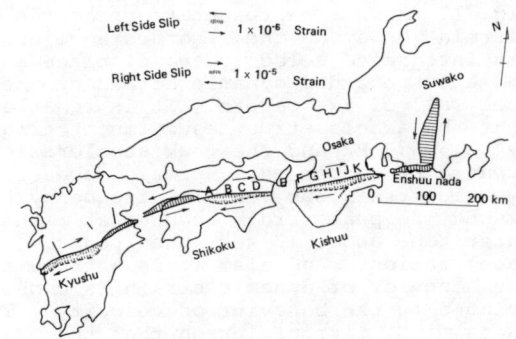

Fig. 7 Geodetic Strike Slip along the Median Tectonic Line during the Latest 60 Years

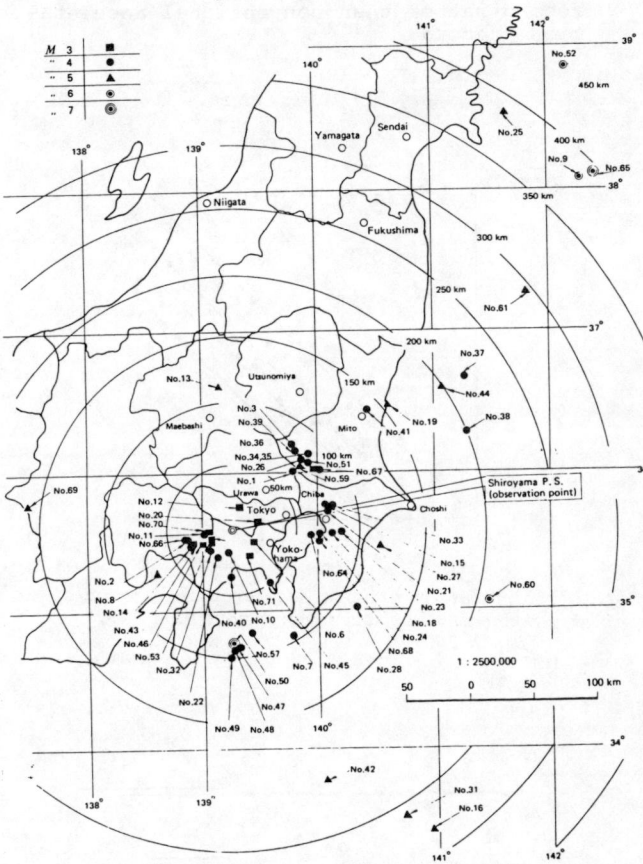

Fig. 8 Epicenters and Magnitudes of Recorded Earthquakes

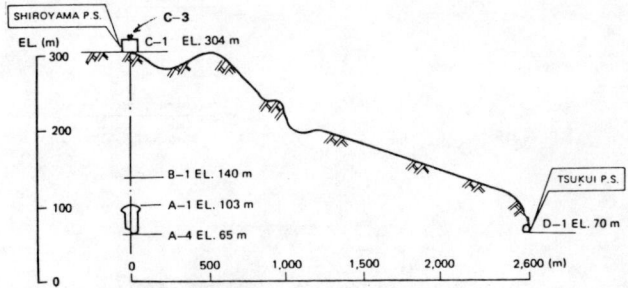

Fig. 9 Distribution of max acceleration and max dynamic strain.

2) Fig. 11 shows a representative records of correlation among the recorded acceleration, its integrated velocity and displacement and also recorded dynamic strain on the rock wall of the cavern. It is remarked that the acceleration has dominant frequency 3.5~6.5 Hz and the peak acceleration occured at rather early time. However, the dynamic strain has very low frequency 0.1 Hz and the peak strain occured at rather later time about 12 sec. than peak acceleration. And also it is noticed that the behavior of dynamic strain is closely related to the behavior of velocity. These facts might give us lesson that the engineering knowledge related to the failure or strain of rock masses during earthquake would be obtained from velocity record rather than conventional acceleration record.

Conclusion

Microzonation of maximum shear strain vectors was illustrated, using the previously published national geodetic survey in Japan and geodetic zonation was shown along the Median Tectonic Line.

Twenty Measurements of overcoring experiments by Authors were analyzed and classified on the Geostress to four categories: Topographical effect, Geological Structure Effect, Tectonic Effect and Multiplicated effect.

Seismic feasibility was quantitatively varified for underground siting based on the earthquake observation carried out around the cavern of Shiroyama underground power station.

Authors would express sincere gratitude for colleagues in lavoratory and related engineers.

References

1) Kanagawa T., Hayashi M., and Kitahara Y., Acoustic Emission and Overcoring Methods for Measuring Tectonic Stresses", Proc. Inter. Sympo. on Weak Rock, Sept. 1981

2) Nakane K., Horizontal Tectomic Strain in Japan(2)., Journal of Geodetic Society of Japan, Vol 19, 3, pp. 200-208, 1973

3) Japan National Oil Corporation, "Technical Report of Oil Rock Store" Committee of Technology Development of Oil Rock Store, 1981

4) Tanaka U., and Saito T., "Measurement of Tectonic Stresses by Means of Stress Relief Method", Chikyu, Vol 21, No. 9, 1980

5) Okano K., Kimura S., and Konomi T. and Nakamura M., "Focal Mechanism in Shikoku, Japan Inferred from Microearthquake Observations." Mem. Fac. Sci. Koochi Univ., 1, Ser. B, 1980.

6) Harada T. and Kasai A., "Horizontal Strain of the Crust in Japan for the Latest 60 years", Journal of Geodetic Society of Japan, Vol, 17, I-2, pp 4-7, 1971

7) Hayashi M., Kanagawa T., Hibino S., Motojima M., and Kitahara Y. "Detection of Anisotropic Geo-stresses Trying by Acoustic Emission, and Non linear Rock Mechanics on Large Excavating Caverns", Proc. 4th ISRM Congress, Montreux, 1979.

8) Hayashi M., "Engineering Aspects of Active Faults", Relation among Geomorphological Active Faults in Land, Tectonic Strain, Epicenters, Dating Methods and Engineering Assessment, Proc. Inter. Sympo. on Weak Rock, Tokyo, 1981

9) Komada H. and Hayashi M. and Hotta M., "Earthquake Observation around Cavern of Underground Power Station", Proc. Japan Society of Civil Engineers, No. 309, 1981

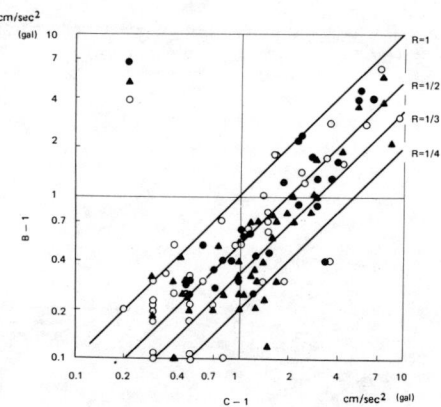

Fig. 10 Correlation between Accelerations at Top of Mountain C-1, and Underground Rock B-1. (Depth 160 m)

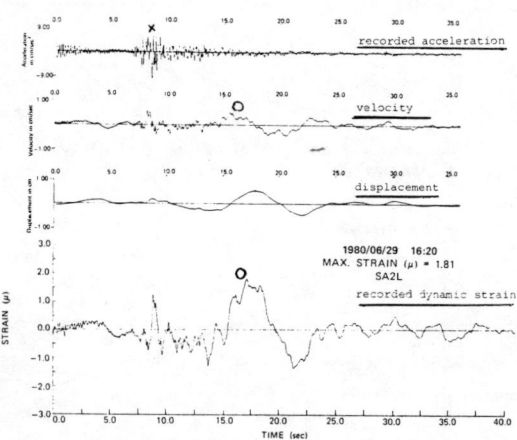

Fig. 11 Correlation among recorded acceleration, velocity and displacement and dynamic strain on the rock wall of the cavern during earthquake.

MECANISMES DE DEFORMATION DE L'ECORCE TERRESTRE – STRUCTURES ET ANISOTROPIE INDUITES

Mechanisms of the deformation of the Earth's crust – Induced structures and anisotropy

Mechanismen der Deformation der Erdkruste – Induzierte Strukturen und Anisotropie

J. Deramond
P. Sirieys
J. C. Soula
Laboratoire de Tectonophysique, Université Paul Sabatier, Toulouse, France

RESUME

La déformation de l'écorce terrestre se manifeste de façon continue (écoulement) ou résulte de discontinuités (fractures). Ses mécanismes sont analysés en fonction de la structure du milieu (structure induite par la déformation finie elle-même ou structure préexistante qui peut être active ou passive). Des essais en laboratoire sur échantillons et sur modèles sont interprétés et appliqués à des exemples naturels, pris dans les Pyrénées et le Sud du Massif Central (France), relatifs à l'ouverture des bassins sédimentaires et à l'écoulement-glissement des nappes.

SYNOPSIS

The deformation of the earth's crust appears as continuous (viscous or plastic flow) or discontinuous fracturing. Its mechanisms have been investigated with emphasis on the role of the structure of the deformed medium (strain induced or pre-existing structure) which can be active or passive. Laboratory experiments on rock samples and analogical models are interpreted and applied to natural examples from the Pyrenees and the southern French Massif Central, relating to the opening of sedimentary basins and the sliding-spreading of nappes.

ZUSAMMENFASSUNG

Die Deformation der Erdkruste zeigt sich in kontinuierlicher Weise oder ist das Ergebnis von Unstetigkeiten. Ihre Mechanismen werden je nach der Milieustruktur analysiert (induzierte Struktur), die aktiv oder passiv sein kann. Laboruntersuchungen an Proben oder Modellen werden interpretiert und auf natürliche Beispiele angewendet, die aus den Pyrenäen und dem südlichen Zentralmassif (Frankreich) stammen, und die die Öffnung der Sedimentbecken und das Fließen-Gleiten der Schichten betreffen.

INTRODUCTION

Les mécanismes de déformation de l'écorce terrestre sont liés à sa structure, qui peut être induite par la déformation (finie) elle-même ou préexistante (c'est-à-dire résultant du mode de formation des roches et de déformations antérieures). Les déformations de tels milieux peuvent être continues et/ou résulter de déplacements discontinus. Ces divers mécanismes sont analysés ici à l'aide d'essais de laboratoire sur échantillons et modèles analogiques et appliqués à des exemples géologiques concrets dans les Pyrénées et le Sud du Massif Central (France).

1 - DEFORMATIONS ET STRUCTURES

Aux structures génétiques résultant de la formation des roches, telles que la stratification pour les roches sédimentaires, se superposent des structures, dites induites, dues à la déformation tectonique, de deux types, continus et/ou discontinus. Les surfaces structurales ainsi créées (la schistosité S, les fentes extensives Fe et fractures cisaillantes Fc) sont orientées par rapport aux axes des tenseurs de déformations et contraintes tectoniques.
Dans le cas particulier d'une déformation pure (coaxialité de la déformation finie D, infinitésimale d et des contraintes σ) de directions principales XYZ, la schistosité S est caractérisée par le plan XY, les fentes cisaillantes Fc par des plans XL_1 (dextre) et YL_2 (senestre) faisant $\pm\mu$ avec Z.
Dans le cas général d'une déformation rotationnelle ces surfaces tournent au cours de la déformation. Dans l'exemple du glissement simple de direction g, de taux de glissement γ (ici choisi négatif) en déformation plane (fig.1):

La schistosité S est repérée par la direction X' de la déformation principale majeure finie après rotation, c'est-à-dire par $\phi' = (g,X')$ tel que : $\tan 2\phi' = 2/\gamma$ (1)

Les fentes d'extension, initiées sensiblement à $\pi/4$ de g (c'est-à-dire Y' de la déformation infinitésimale) subissent rotation et ouverture au cours de la déformation progressive.

Les fentes de cisaillement conjuguées (dites de Riedel) L1 et L2, initiées sensiblement avec bissectrice aigüe à $\pi/4$ de g, évoluent au cours de D en cisaillement, la famille L1 de façon synthétique (de même sens que celui qu'il a dans l'infinitésimal), la famille L2 de façon antithétique (sens inverse).

L'apparition préférentielle de ces structures dépend du matériel et des conditions pression-température de la déformation. Que ces structures soient préexistantes ou induites par la déformation progressive, la déformation d'un tel milieu est celle d'un matériau anisotrope structuré, elle s'effectue, soit indépendamment des surfaces structurales (structure passive) soit en faisant intervenir ces surfaces structurales (structure active), selon l'orientation de la déformation, c'est-à-dire des contraintes appliquées (tectoniques ou non) et les caractéristiques structurales. Les caractéristiques structurales, qui conditionnent éventuellement la déformation du milieu, font intervenir essentiellement deux éléments relatifs aux surfaces structurales : leur fréquence et leurs caractéristiques mécaniques propres.

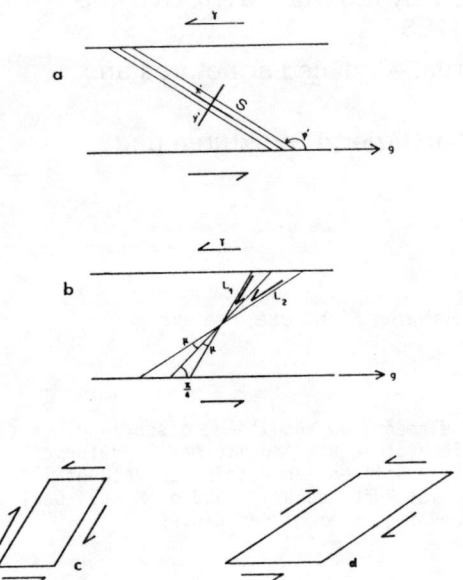

Fig.1 - Structures induites. a) Schistosité; b) Fentes extensives et fractures cisaillantes; c_1) Famille L1 synthétique; c_2) Famille L2 antithétique.

a) **Fréquence des "discontinuités".**

La fréquence des surfaces structurales est définie en fonction de l'échelle considérée (suivant le projet technique envisagé, ou l'étude tectonique considérée) par le nombre n de surfaces éventuellement actives) par unité de longueur L dans la direction perpendiculaire, c'est-à-dire : f = n/L. f a pour dimensions L^{-1}. On définit ainsi la fréquence métrique, décamétrique, hectométrique, kilométrique etc... selon l'échelle. Une structure discrète est caractérisée par un petit nombre de surfaces structurales (f petit), une structure continue (pénétrative) par un grand nombre de surfaces structurales (f grand). Ce second cas est celui d'une structure schisteuse, pour laquelle on considère qu'en tout point du milieu passe une surface structurale ($f \infty$).

b) **Caractéristiques mécaniques structurales.**

Les déformations discontinues apparaissent lorsque sur une surface structurale une composante (au moins) du déplacement U (u_n, u_t) dans le référentiel (n,t) lié à la structure, subit une discontinuité (Fig. 2).
Pour $u_t^1 \neq u_t^2$, avec $u_n^1 = u_n^2$, apparaît une discontinuité de cisaillement (caractérisée par le rejet $u_t^1 - u_t^2$), lorsque est vérifiée la loi du cisaillement structural,

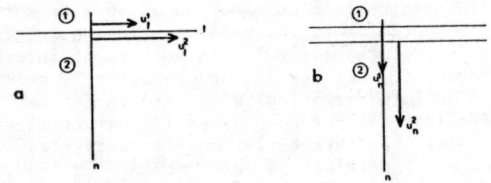

Fig.2 - Discontinuité des déplacements (n= normale à la surface de discontinuité plane t). a) Discontinuité tangentielle; b) Discontinuité normale.

qui s'exprime par une loi linéaire de type Coulomb :
$$|\tau_{nt}| = \bar{c} + \sigma_n \tan\bar{\phi} \qquad (2)$$
où $\bar{c}$ et $\bar{\phi}$, cohésion et angle de frottement structuraux, sont uniquement valables sur les surfaces structurales.
Pour $u_n^1 \neq u_n^2$ apparaît une discontinuité extensive caractérisée par l'ouverture $u_n^1 - u_n^2$), lorsqu'est vérifiée une loi de rupture extensive telle que la loi d'extension maximale :
$$\varepsilon_n = \bar{\varepsilon}_o \qquad (3)$$
où $\bar{\varepsilon}_o$ est l'extension limite sur la surface structurale.

2 - RESULTATS EXPERIMENTAUX

2.1. Sur échantillons (Caractéristiques structurales schisteuses)

Les caractéristiques de cisaillement sur la schistosité ont été déterminées par compressions monoaxiales sur échantillons de schistes ardoisiers de Lacaune (Tarn, France).
Ces schistes présentent un fort feuilletage, visible sur échantillons et en lame mince, coïncidant avec les plans axiaux de plis à longueur d'onde déca-à hectométrique. En lame mince apparaissent dans les phyllites (composant essentiel de la roche, de l'ordre de 60%) des microplis d'une longueur d'onde de 10 à 20 m, dont les plans axiaux sont les plans de schistosité S, à fréquence très élevée. En outre, le milieu est affecté par deux systèmes de joints conjugués à fréquence décamétrique de l'ordre de 2, d'angles 2μ égaux à 25 et 35° dont les bissectrices aiguës sont sensiblement orthogonales.

Sur les plans S ont été obtenues (Saint-Leu et al.) les valeurs $\bar{c}$ = 16 MPa et $\bar{\phi}$= 22 degrés, alors que les essais en compression normale à la schistosité, induisent des fractures astructurales, fournissent les valeurs astructurales égales à c = 80 MPa et ϕ= 33 degrés.

2.2. Essais analogiques

Les mécanismes de cisaillement analysés jusqu'ici en petite déformation sont étudiés expérimentalement en déformation finie par glissement simple (avec des taux ≤ 1) sur des matériaux fragiles-ductiles, puis comparés à des cas naturels. Le modèle, notamment destiné à simuler l'effet d'un cisaillement profond (horizontal le long d'une surface Fe verticale) sur une couverture sédimentaire, est constitué par deux demi-plaques en forme de L, induisant un confinement de la couverture parallèle à g, comme c'est le cas dans la nature. Les matériaux utilisés ont été de deux types : cohésifs (plasticine) et pulvérulents à très faible cohésion, tels que talc et grès broyés à granulométrie variable.

Deux types d'essais ont été réalisés : les essais monocouches et les essais bicouches (effectués à l'aide d'une plaque intercalaire structurée).

A - Essais monocouches.

1) Milieu initial pulvérulent.

Aux faibles $|\gamma|$ apparaissent les fractures conjuguées Fc de type Riedel. Leur angle aigü 2μ (dont la bissectrice est à 45° de g) varie avec le matériau du modèle (23 à 32°). Leur fréquence dépend de la granulométrie, les matériaux plus grossiers ayant des périodes plus élevées.
Ces fractures sont parfois accompagnées de fractures extensives Fe (à 45° de g) en échelon et/ou de fractures conjuguées et fractures extensives F'c secondaires (L'1 et L'2)(Fig. 3a).

Pour $|\gamma|$ supérieur à 0,2-0,3 apparaissent des discontinuités de la composante normale (et tangentielle) du déplacement, entraînant un écartement des lèvres des fractures

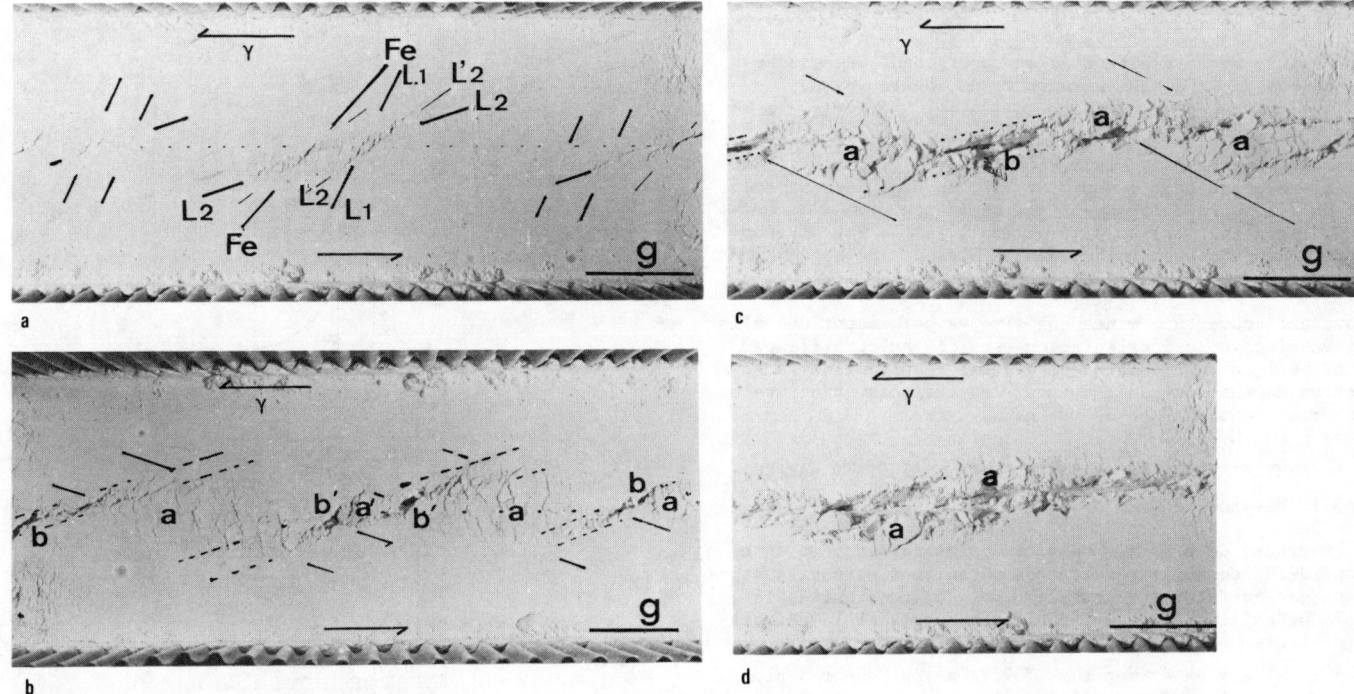

Fig.3 - Evolution du modèle monocouche. a) $\gamma = 0,2$; b) $\gamma = 0,4$, (zones anticlinales (a), bassins (b), zones anticlinales secondaires (a') et bassins secondaires (b'); c) $\gamma = 0,6$; d) $\gamma = 1$

L et Fe qui subissent une rotation (faible mais mesurable) au cours de la déformation progressive (Fig. 3b).
Il s'effectue une individualisation d'un certain nombre de fractures, celles extensives (à 45° de g) et celles de la famille L2 (antithétique) qui subissent une ouverture, de direction L2 (Fig. 3b, 3c).

Entre les zones d'ouverture se marquent des zones " en contraction " qui se manifestent par un bombement et/ou des chevauchements à faible inclinaison de direction voisine de la normale à g. Le développement de ces zones est systématique, quel que soit le matériau. Les bombements ont une forme allongée de direction (135-160°) avec g. Ces contractions compensent les extensions (structure géologique en "pull-apart") (Fig. 4) en sorte que la déformation globale est isoaire. Au cours de la déformation progressive peuvent se produire des figures de même type et de dimensions plus faibles (fractures secondaires).

Pour $|\gamma|$ voisin de 1, enfin, s'individualise une fracture en décrochement, verticale, parallèle à g, séparant en deux les zones de creux et bombement (Fig. 3d). Par coalescence on aboutit à la formation d'un "bassin" central, sur la trace du décrochement, bordé par deux rides latérales également parallèles à g.

2) Milieu initial cohésif.

a) - Couverture mince
Les essais avec couche mince d'épaisseur h déformable conduisent à la formation de plis en échelon dont la longueur d'onde dépend de h en accord avec les théories de Biot (1961) et Ramberg (1962). Ces plis naissent avec un axe dont l'angle à 135° de g, décroît lorsque $|\gamma|$ croit tandis que le plissement s'accentue et que plusieurs générations de fractures conjuguées naissent avec une bissectrice aigüe à 45° de g, mais tournant dans le sens

de g lorsque $|\gamma|$ croit, et en même temps que des fentes extensives initialement à 45° de g.

b) Couverture épaisse structurée
Les essais avec couche épaisse cohésive structurée (constituée par une plaque déformable comportant des discontinuités) induisent des mécanismes anisotropes. Les structures dans le champ contractif (c'est-à-dire à normale dans le champ extensif) sont actives et l'objet de discontinuités (normale et tangentielle) de type ouverture avec cisaillement antithétique, c'est-à-dire inverse de g., accompagné d'une contraction normale pour les surfaces structurales dans le champ extensif (dont la normale est dans le champ contractif) qui se manifeste parfois par un chevauchement (pour des surfaces structurales inclinées (Soula et al. 1979).

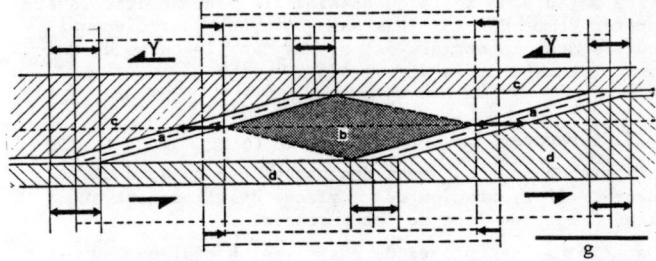

Fig.4 - Schéma du fonctionnement en "pull-apart" après la phase d'initiation et de rotation des fractures induites. Le mouvement relatif senestre parallèle à g des compartiments supérieurs (c) et inférieurs (d) entraîne à la fois une ouverture suivant les fractures (ici L2) (a) et un raccourcissement dans la zone comprise entre deux fractures adjacentes. La zone de raccourcissement maximum (b) est due à la combinaison de ce raccourcissement local et du raccourcissement lié au glissement général suivant g.

B - Essais bicouches.

Des essais ont été effectués en superposant une couche au-dessus de la plaque structurée du modèle 2b.

Pour une couche pulvérulente, des figures analogues à celles des essais monocouches se développent au-dessus des discontinuités actives, formant avec g un angle compris, théoriquement entre 0 et 90°, en pratique entre 15 et 80°. Il apparaît souvent une ouverture selon chacune de ces discontinuités; la déformation s'initie par développement de fractures conjuguées, à bissectrice aigüe parallèle à la direction de la discontinuité.

Pour une couverture mince cohésive, apparaissent des plis en échelons dont l'orientation est liée au cisaillement antithétique du cisaillement général résultant de la rotation des discontinuités, pour les surfaces structurales comprises entre 15 et 80° de g.

3 - EXEMPLES NATURELS (BASSINS SEDIMENTAIRES ET NAPPES)

3.1. Bassins sédimentaires

L'ouverture de nombreux bassins sédimentaires peut être considérée comme le résultat de mécanismes extensifs suivant des surfaces structurales (des "discontinuités") soit préexistantes, soit induites au cours de la déformation finie.

Les résultats expérimentaux (§.2.2) s'appliquent à plusieurs bassins sédimentaires naturels, analysés à la fois du point de vue structural et sédimentologique en particulier les bassins Permiens et Triasiques des Pyrénées (Soula et al. 1979) et les bassins houillers stéphaniens des Pyrénées et du Sud du Massif Central.

Le bassin houiller de Carmaux, encore en exploitation, peut être pris comme exemple. Il semble résulter d'un cisaillement dextre suivant une fracture majeure du socle hercynien orientée N 20-30 dans un matériel gneissique et micaschisteux fortement anisotrope, à schistosité sensiblement verticale N 130-160. Le bassin apparaît formé de quatre sous-bassins, séparés par des zones hautes, où les axes de drainage, mis en évidence par l'étude sédimentologique (Delsahut, 1981), correspondent aux directions d'ouvertures obtenues dans les modèles.

Des cônes de débris latéraux, révélés par l'analyse sédimentologique, semblent également correspondre aux cônes d'éboulis obtenus par Soula et al. (1979). De même, des décalages horizontaux et verticaux, sans incidence sur la sédimentation mais contemporains de celle-ci, s'observent également suivant des accidents parallèles à la schistosité majeure du socle du bassin. Ils peuvent être interprétés comme dûs aux jeux sans ouverture des discontinuités dont la normale est située dans le champ de raccourcissement de la déformation.

3.2. Nappes

L'aptitude au glissement des nappes dépend de deux facteurs : l'orientation des surfaces de glissement et l'adhérence le long de ces surfaces.

Dans le cas des nappes de charriage, à déplacement sensiblement horizontal, des surfaces structurales peu pentées peuvent jouer le rôle de surfaces de glissement. Dans les formations peu métamorphiques où la schistosité est du type crénulation avec des surfaces peu individualisées, l'aptitude au glissement, est faible. Toutefois si la déformation s'accompagne d'une intense recristallisation, la présence de minéraux en feuillets, tels que les micas, favorise le glissement sur la schistosité de flux. Dans la plupart des nappes d'origines tégumentaires ("thin-skinned sheets"), le déplacement s'effectue par

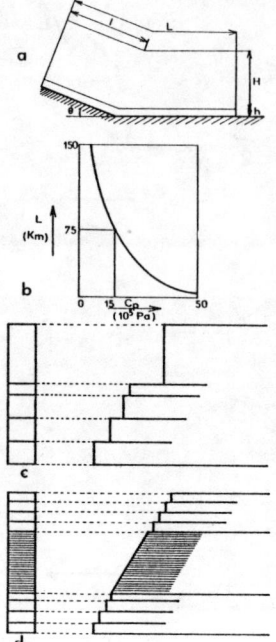

Fig. 5 - Nappes. a) Déplacement de l'unité Crétacée-Eocène; b) Valeurs de L en fonction de la cohésion c_p du Permo-Trias (assimilé à un corps de Bingham); c) Discontinuités tangentielles structurales (calculées avec $c_q = 30 \times 10^5$ Pa); d) Nappes du Minervois (schématisation) : alternance de zones à forte et à faible f_s.

glissement discontinu le plus souvent sur les surfaces de stratification et éventuellement les surfaces de failles). Lorsque la fréquence f_s des surfaces de stratification, qui dépend des conditions de sédimentation, est élevée (séries rythmiques) la nappe se comporte comme un milieu continu fluant, à l'échelle de la structure géologique. Lorsque, par contre, f_s est faible, les déplacements en bloc de ces formations géologiques sont discontinus, ils peuvent en outre s'accompagner d'une déformation continue dans les couches les plus déformables. Ces mécanismes sont illustrés par des exemples des Pyrénées et du Sud du Massif Central (France).

a) Nappe de Gavarnie (Pyrénées).

La nappe de Gavarnie comporte trois unités fondamentales (J. Deramond et al. 1980) : la zone de racine au N, la zone centrale constituée de terrains paléozoïques (N.P.) et la zone méridionale crétacée-éocène reposant sur quelques mètres de trias.

La fréquence hectométrique des surfaces structurales de stratification f_s est supérieure à 10 dans les formations paléozoïques (séries rythmiques) et inférieure à 5 dans les formations crétacées (massives).

Le contact Paléozoïque-Crétacé est constitué de formations peu épaisses (quelques mètres) du Permo-Trias constitué d'argilites à comportement viscoplastique avec des constantes mécaniques sans doute peu élevées.

1) Glissement Crétacé sur Permo-Trias.

Ce glissement est, schématiquement celui d'une couche

coudée d'épaisseur H + h (Fig. 5). L'assimilation du Permo-Trias à un corps de Bingham à seuil de cisaillement Cp (Rambach et Deramond, 1979) conduit à la longueur de chevauchement L égale à :

$$L = 1 \frac{\gamma h \sin\theta}{Cp} \qquad (4)$$

(où γ est le poids volumique, L l'épaisseur du Permo-Trias, θ l'inclinaison de la partie coudée).

Les valeurs de L en fonction de Cp, sont reproduites (Fig. 5b) pour $l = 8$ km, $\theta = 12°$ et Cp compris entre 2 et 50×10^5 Pa. Pour Cp = 15.10^5 Pa (valeur moyenne admise pour ce type de formations) L = 75 km, valeur qui correspond à la distance de Gavarnie au front de la nappe.

2) Glissement du Crétacé.

La fréquence f_s des surfaces structurales étant faible (f_s hecto < 5), le déplacement du Crétacé s'effectue selon un schéma de blocs (Fig. 4c). Dans cet ensemble calcaire, une cohésion structurale de Ca = 30.10^5 Pa fournit un déplacement schématisé dans la figure 5c. Ces résultats conduisent à un taux de glissement moyen (pour le Crétacé) de 2,3.

3) Ecoulement du Paléozoïque.

La fréquence f_s dans le Paléozoïque est élevée, le milieu est assimilable à un milieu continu, la déformation s'effectuant sans discontinuité. L'écoulement s'effectue, lorsque le seuil de plasticité est atteint, de façon analogue à celui des glaciers, il peut être assimilé à un modèle de Prandtl. Les lignes de cisaillement maximum, astructurales, forment un réseau de cycloïdes orthogonales

Les failles ou les zones de cisaillement observées dans le Paléozoïque ont une allure curviligne et matérialisent les lignes de cisaillement maximum cycloïdales du modèle de Prandtl (pour lequel les bordures rigides sont représentées par les formations massives métamorphiques). Le sens de l'écoulement est vertical dans la zone de racine, puis sensiblement horizontal pour la nappe proprement dite.

b) Nappes du Minervois.

Les nappes du Minervois sont des ensembles paléozoïques chevauchants (du Nord vers le Sud) situés au Sud de la Montagne Noire (S du Massif Central français), hétérogènes constitués d'une superposition de zones à forte f_s alternant avec des zones à faibles f_s de même puissance (plusieurs centaines de mètres).

Le milieu peut donc être assimilé à une succession de zones à écoulement plastique et de zones à glissement structural sur quelques surfaces parallèles aux surfaces de stratification (Fig. 5d).

REFERENCES

- Biot M.A. (1961)
  Theory of folding of stratified viscoelastic media and its implications in tectonics and orogenesis.
  Geol. Soc. Am. Bull. 72 - p. 1595-1620.

- Delsahut B. (1981)
  Dynamique du bassin de Carmaux (Tarn) et géologie du Stéphano-Permien des environs entre Réalmont et Najac.
  Th. spec. Toulouse - 232 p.

- Deramond J. et P. Sirieys (1980).
  Mécanisme de déplacement et déformations des nappes, exemple de la nappe de Gavarnie (Pyrénées Centrales).
  Rev. Fr. de Géotechnique n° 16 p. 41-54.

- Rambach J.M. et Deramond J. (1979).
  Constant thickness overthrust on variable slope support
  Tectonophysics, 60, p. 7-16.

- Ramberg H. (1962).
  Contact strain and folding instability.
  Geol. Rundschau, 51-P. 405-439.

- Saint-Leu C, Lerau J. et Sirieys P. (1978).
  Mécanismes de ruptures des schistes de Lacaune (Tarn). Influence de la pression isotrope.
  Bull. Soc. Fr. Mineral - 101 -(p. 437-441).

- Soula J.C., Lucas C. et Bessière G. (1979)
  Genesis and evolution of Permian and Triasias basin in the Pyrenees by regional simple shear acting on older variscan structures : field evidence and experimental models.
  Tectonophysics, 58, T1-T9.

TIME-DEPENDENT DILATANCY PRIOR TO ROCK FAILURE AND EARTHQUAKES

Dilatation en fonction du temps avant la fracture de roches et les tremblements de terre

Zeitabhängige Dilatanz vor dem Felsbruch und vor Erdbeben

Prof. Dr. ir TAN Tjong Kie
Institute of Geophysics, Academia Sinica, Beijing, China
KANG Wen Fa, Eng.
Institute of Geophysics, Academia Sinica, Beijing, China

SYNOPSIS

A brief review is presented of the importance of dilatancy prior to rock failure and especially of its fundamental role in earthquake prediction. Constitutive equations for the visco-elastic-dilatant behaviour of rocks are presented. The dilatancy parameters are determined from various triaxial tests. The process of dilatancy prior to earthquakes is discussed on the basis of this hypothetical model.

RESUME

Cette communication examine le rôle important des dilatations survenant avant la rupture de roches et plus particulièrement leur rôle fondamental dans la prédiction des tremblements de terre. On présente des équations pour l'étude du comportement visco-élastique des roches et on détermine les paramètres de dilatation pour divers essais triaxiaux. C'est sur la base de ce modèle hypothétique qu'on analyse le processus de dilatation avant les tremblements de terre.

ZUSAMMENFASSUNG

Die Arbeit gibt einen kurzen Überblick über die Bedeutung der Dilatanz vor dem Bruch des Gesteins und besonders ihre grundlegende Rolle in der Voraussage der Erdbeben. Die Ausgangsgleichungen für das elasto-visko-dilatante Verhalten der Gesteine werden aufgestellt und die Dilatations-Parameter für verschiedene Triaxial-Versuche bestimmt. Die Dilatations-Vorgänge vor den Erdbeben werden aufgrund dieses hypothetischen Modells behandelt.

The concept of dilatancy is well known in rheology and soil mechanics. Its origin is in the observation that wet sands dilates under the action of shearing stresses (Reynolds 1885). Dilatancy in granular media is associated with the overall decrease of packing density due to relative movements of groupes of grains; it is a geometrical necessity in the deformation process.

Dilatancy of rocks was first observed by Bridgman (1949); Handin et al (1963) measured dilatancy of rock samples of Berea sandstone at low confining pressures, while decrease in porosity (volume hardening) at high confining pressures. Dilatancy under deviatoric stresses is observed as a time dependent volume increase; a dilatancy model is shown in Fig.1; later many experimental results and their analysis have been published from laboratory testing by Paterson (1963), Edmond and Paterson (1972), Brace, Paulding and Scholz (1966), Bienawski (1967), Crouch (1970), Zoback and Byerlee (1975), Perkins, Green and Friedman (1970), Rummel (1974), Mogi (1977), Tan (1964) and others which are not listed here.

The results of these investigators will be summarised

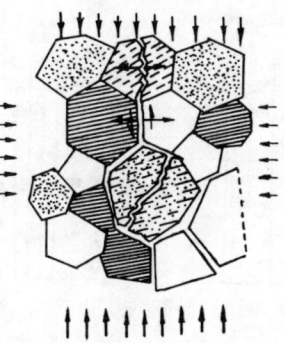

Fig.1 Dilatancy model

and analysed in the next paragraphs.

In Rock-engineering the phenomenon of dilatancy in rocks is often observed; although it is a fundamental factor in the stability of underground structures it has not yet got the attention it deserves. In another paper to this Congress (Tan 1983) it is emphasized that dilatant volume increase in rock masses plays a crucial role in the stability of potentially swelling rocks.

Rock dilatancy has gained worldwide attention since it has been observed that the earthcrust dilates prior to earthquakes; direct evidence of dilatancy have been observed in the San Andreas Fault near Parkfield (Cherry and Savage 1972); a very clear evidence have been reported from the bulging of the earthcrust prior to the large Haycheng Earthquake (1975). This has been one of the crucial scientific materials for the prediction of this earthquake. The Haycheng earthquake was the first earthquake, which was predicted correctly in site, magnitude and time (Fig.7A,B,C).

Indirect evidence for crustal dilatancy has been reported from the lowering of deep water wells situated in a radius of 100 km around the epicentre of the large Tangshan Earthquake (Tan, He 1982); a further indirect indication is the steady decrease in earth resistivity of the upper layer of the crust (Fig.2a-b). The changes in v_p/v_s ratio (velocities of normal wave to shear waves) which is considered to be directly related to dilatancy, have been frequently measured in our country (Feng et al 1976; Duan et al 1976). Such changes in this v_p/v_s ratios have been earlier reported by Semenov (1969), Aggarwal (1973). Some U.S. scientists belief that fluid inflow during dilatancy is a crucial factor leading to earthquakes as it is accompanied by the generation of waterpressure (i.e. decrease of the effective normal stress, thus strength) and the lubrication of fissures.

A physico rheological model for Earthquake fore runners has been recently suggested (Tan 1982); it is based on the fundamental assumption that the earthcrust is a rheological dilatant body traversed by a network of planes of easier glide, the seismic belts.

Time dependent dilatancy is an important problem, which has not yet been explored extensively. The inwards motion of rock cavities with the time shows a remarkable time dependency.

It is generally believed that this is due to "creep" but it is not clearly specified what is understood under "creep". As it will be discussed next, creep is due to the continuous compatible straining of grains which is increasing with the time with decreasing rate. This is the case as far as the deviatoric stresses remain below an upper yield limit $f_3(f_3$ for shear and $f^*=\sqrt{3}f_3$ for compression).

As soon as this upper limit is exceeded, then non compatible anelastic deformations will occur leading to void and crack formation and opening of inborn cracks. In this range the creep is thus accompanied by time dependent volume increase due to dilatancy (Fig.1 and 3).

MATHEMATICAL MODELLING OF TIME-DEPENDENT DILATANCY

The testresults of the investigators mentionned above may be summarised in the following points (Fig.3):

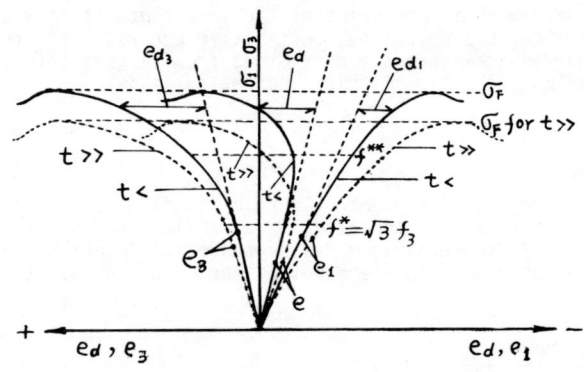

Fig.3 Stress-Strain relationship for dilatant rocks.

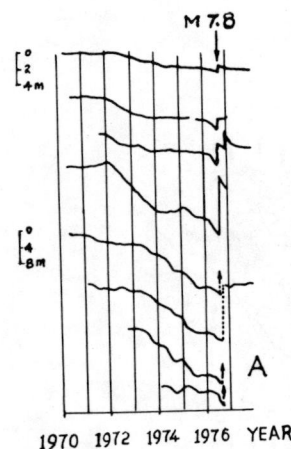

Fig.2A Waterlevel changes in wells (Wang, 1979)

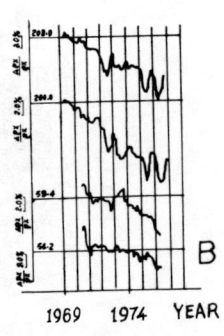

Fig.2B Changes in electric resistivity (Zhao and Zian, 1979)

1. the deviatoric stress-strain relationships are nearly linear in the beginning; in this range the deviatoric stress-volume strain relations are also linear; the rock assumes visco-elastic behaviour.

2. when the deviatoric stresses are increased, the above relationships become non linear and the material dilates (onset of dilatancy) as soon as the stress differences exceeds a certain limit, which is denoted by $f^*=\sqrt{3}f_3$; obviously dilation is observed at the turning point f^{**}.
For many rocks $f^{**} >$ one-half of the fracture strength σ_F.
The dilation increases with increasing stress differences.

3. after reaching the maximum strength σ_F the stresses in constant strain rate tests decrease for low confining pressures but the dilation continues to increase steadily.

4. the radial strain e_3 grows faster than the axial strain e_1 in uni-and triaxial tests.

5. dilation decreases when the confining pressure is increased.

6. dilatancy causes a change in the v_p/v_s ratio.

7. dilatant volume deformations show an instantaneous part, followed by a time dependent part, increasing with

the time.

8. strain induced anisotropic changes may occur.

Our main purpose is to model the rock in great lines and to apply it to Earthquake Prediction and Rock-engineering. Hence in the mathematical modelling of creep dilatancy behaviour, we must keep the equations as simply as possible. So the following assumptions are made:

1. the rock remains a continuum and the laws of continuum mechanics are valid; in this paper a tension is positive;
2. cataclastic sliding and fracturing after the stresses reach their maximum σ_F is beyond the scope of this modelling. So the equations can only describe the deformation process up to the maximal stress difference $(\sigma_1 - \sigma_3)_{max} = \sigma_F$.

On the basis of these eight points the following constitutive equations have been derived; it is a modest attempt for the mathematical modelling of this complicated process.

These equations have been published already in tensor notation (Tan et al 1980); this paper presents these equations in more practical corrected form, better suitable for experimental analysis. For instance the parameter CD in the first paper is now replaced by D^*. For the total strain $e(t)$, we can write

$$e(t)_{ij} = e_{e_{ij}} + e_c(t)_{ij} + e_{d_{ij}} + e_d(t)_{ij} \quad (1)$$

whereby

$e(t)_{ij}$ = time dependent total strain

$e_{e_{ij}}$ = elastic strain

$e_{d_{ij}}$ = **instantaneous dilatant** strain

$e_d(t)_{ij}$ = time-dependent dilatant strain

we now write:

$$e_x(t) = \frac{s_x}{2G} + \frac{1-2\nu}{E}p + \int_{-T}^{t} \psi(t-\theta) \, d\frac{s_x(\theta)}{d\theta} \, d\theta +$$
$$\int_{-T}^{t} \phi(t-\theta) \frac{dp(\theta)}{d\theta} d\theta + D^*\langle F_1(\frac{\sigma_{oct}}{f_3})\rangle +$$
$$C\langle F_2(\frac{\sigma_{oct}}{f_3})\rangle \frac{s_x}{\sigma_{oct}} +$$
$$\varsigma \int_{-T}^{t} \left[D^*\langle F_1(\frac{\sigma_{oct}}{f_3})\rangle + C\langle F_2(\frac{\sigma_{oct}}{f_3})\rangle \frac{s_x}{\sigma_{oct}} \right] dt \quad (2)$$

$$\tfrac{1}{2}\gamma_{xy}(t) = e_{xy}(t) = \frac{\tau_{xy}}{2G} + \int_{-T}^{t}\psi(t-\theta)d\frac{\tau_{xy}(\theta)}{d\theta} d\theta +$$
$$C\langle F_2(\frac{\sigma_{oct}}{f_3})\rangle \frac{\tau_{xy}}{\sigma_{oct}} +$$
$$C\varsigma \int_{-T}^{t}\langle F_2(\frac{\sigma_{oct}}{f_3})\rangle \frac{\tau_{xy}}{\sigma_{oct}} dt \quad (3)$$

and similar expressions for $e_y(t)$, $e_z(t)$, $\tau_{yz}(t)$, $\tau_{zx}(t)$

$$e(t) = 3\frac{1-2\nu}{E}p + 3\int_{-T}^{t}\phi(t-\theta)\frac{dp(\theta)}{d\theta} d\theta +$$
$$3D^*\langle F_1(\frac{\sigma_{oct}}{f_3})\rangle + 3D^*\varsigma\int_{-T}^{t}\langle F_1(\frac{\sigma_{oct}}{f_3})\rangle dt \quad (4)$$

whereby $s_x = \sigma_x - p$; $p = (\sigma_1 + \sigma_2 + \sigma_3)/3$,

$e(t) = e_x(t) + e_y(t) + e_z(t)$ = volume strain

$$\sigma_{oct} = \sqrt{J_2} = [1/6\{(\sigma_x - \sigma_y)^2 + (\sigma_y - \sigma_z)^2 + (\sigma_z - \sigma_x)^2 + 6\tau_{xy}^2 + 6\tau_{yz}^2 + 6\tau_{zx}^2\}]^{\tfrac{1}{2}}.$$

The first two terms in (2) and the first term in (3) and (4) represent the elastic part of the deformation e_e. The creep deformation is represented by the third and fourth terms in (2) and the second terms in (3) and (4);

$\psi(t) = \beta_s \log(1+\alpha t)$ = creep compliance for unit stress difference;

$\phi(t) = \beta_v(1-e^{-t/r})$ = volume creep compliance for unit hydrostatic stress,

whereby β_s, β_v, α and r are constants.

The instantaneous dilatant deformations $e_{d_{ij}}$ are represented by the fifth and sixth terms in (2) and the third terms in (3) and (4). The time dependent dilatant deformation is described by the seventh term in (2) and the fourth term in (3) and (4).

It is understood that the onset of dilatancy occurs for
$$\sigma_{oct} > f_3$$

thus $\langle F_1(\frac{\sigma_{oct}}{f_3})\rangle = F_1(\frac{\sigma_{oct}}{f_3})$
$= 0$
$\langle F_2(\frac{\sigma_{oct}}{f_3})\rangle = F_2(\frac{\sigma_{oct}}{f_3})$
$= 0$ } for $\frac{\sigma_{oct}}{f_3} \begin{cases} > 1 \\ \leq 1 \end{cases}$.

In this paper a cone is assumed for the yield surface; A coaxial cone is assumed for the failure surface. In $\sigma_1 - \sigma_2 - \sigma_3$ space (Fig.4) they are described by

$$\sqrt{J_2} = f_3 = f_{3_0} - m\,p \quad , \quad (5)$$
$$\sqrt{J_2} = \sigma_F = \sigma_{F_0} - m''\,p \quad . \quad (6)$$

The space between these cones is the zone of dilatancy. In our first paper it is tentatively taken (Tan 1980)

$$F_1(\frac{\sigma_{oct}}{f_3}) = F_2(\frac{\sigma_{oct}}{f_3}) = (\frac{\sigma_{oct}}{f_3} - 1)^n.$$

From many testresults, however, we come to learn that the following relationship:

$$F_1\left(\frac{\sigma_{oct}}{f_3}\right) = F_2\left(\frac{\sigma_{oct}}{f_3}\right) = \left(\frac{\sigma_{oct}}{f_3}\right)^n \quad (7)$$

is more suitable for the rocks studied.

Of course all the parameters $G, E, \beta_s, \beta_v, D^*, C, m, n, \zeta$ in the above equations must be regarded as scalar functions of the stress invariants I_1, I_2, I_3 and Temperature T.

In potentially swelling rocks moreover they will be strongly dependent on **the degree of swelling (Tan 1983)**.

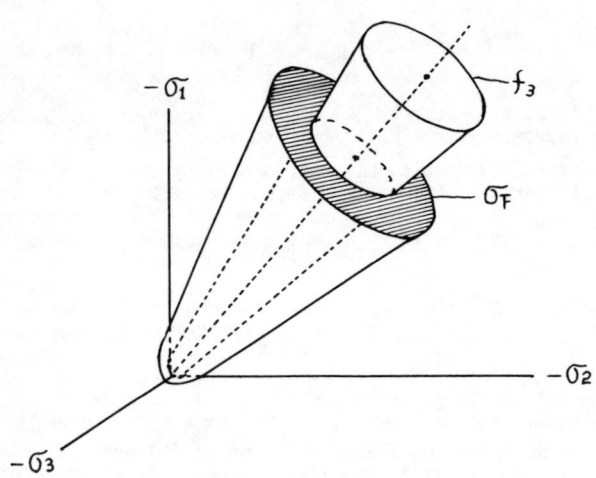

Fig.4 Yield and failure Surfaces

DETERMINATION OF THE DILATANCY PARAMETERS

The elasto-viscoous parts of the deformation will not be studied here as no adequate data are available. We will focus our attention on the determinations of the dilatancy parameters D^*, C, m, n and ζ for many rocksamples and some crustal upheavals prior to Earthquakes.

For triaxial conditions we get:

$$\sigma_{oct} = \sqrt{\tfrac{2}{3}}(\sigma_1 - \sigma_3) \quad ; \quad \sigma_1 > \sigma_2 = \sigma_3 ;$$

$$s_x = -\tfrac{2}{3}(\sigma_1 - \sigma_3);$$

then from (2):

$$e_{d1} = \left(\frac{\sigma_1 - \sigma_3}{f^*}\right)^n \left(D^* - \frac{2}{\sqrt{3}}C\right) ; \quad (8)$$

$$e_{d3} = \left(\frac{\sigma_1 - \sigma_3}{f^*}\right)^n \left(D^* + \frac{1}{\sqrt{3}}C\right) ; \quad (9)$$

$$e_d = 3 D^* \left(\frac{\sigma_1 - \sigma_3}{f^*}\right)^n . \quad (10)$$

The stress differences $(\sigma_1 - \sigma_3)$ are now plotted against the dilatant volume strains e_d in double logarithmic coordinate axis:

$$\log e_{d1} = n \log\left(\frac{\sigma_1 - \sigma_3}{f^*}\right) + \log\left(D^* - \frac{2}{\sqrt{3}}C\right) ; \quad (11)$$

$$\log e_{d3} = n \log\left(\frac{\sigma_1 - \sigma_3}{f^*}\right) + \log\left(D^* + \frac{1}{\sqrt{3}}C\right) ; \quad (12)$$

$$\log e_d = n \log\left(\frac{\sigma_1 - \sigma_3}{f^*}\right) + \log 3 + \log D^* ; \quad (13)$$

Further: $\quad f^* = f_o^* - m \, p \quad (14)$

$$\sigma_F = \sigma_{F_o} - m'' \, p \quad (15)$$

The results for samples of various rocks are shown in Fig.5

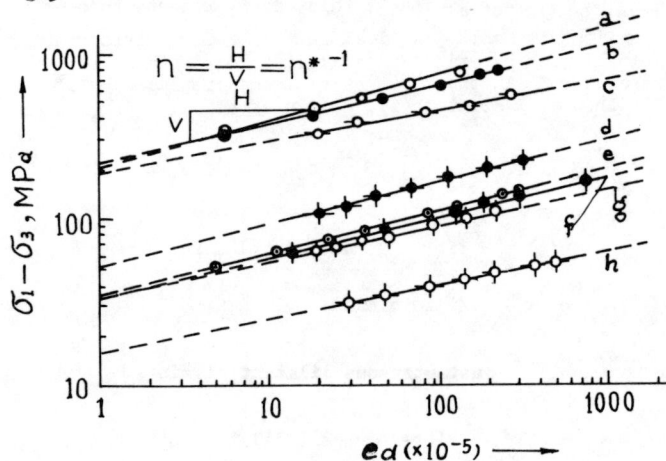

Fig.5 Relationship between $\sigma_1 - \sigma_3$ and e_d for various rocks.
a. o Mogi,1977; b. • Hadley,1975; c. ⊙ Wu,1980; d. ◆ Makoto,1979; e. ⊙ Perkins,1970; f. ◆ Hiroyuki,1978; g. ◊ Tan,1964; h. ◇ Toshiyo,1979.

The yield values f^* for compression are taken for $e_d = 10^{-4}$. These results are further presented in Table I. It must be remarked that these data have been obtained from figures published by the various authors; as these figures are rather small the above determined values can not be very accurate; nevertheless they can serve for reference.

A special analysis will now be given of the testresults by Rummel on Ruhr sandstone. As the tests covered a range of lateral stresses, we can measure the f^*, f^{**} and σ_F also the m, m' and m'' values in formula (14) and (15).

Further we can determine n, D^* and C. The testresults are reproduced in Fig.6A, Fig.6B and Fig.6C give the double logarithmic plot for $\sigma_1 - \sigma_3$ vs e_d and $(\sigma_1 - \sigma_3)$ vs e_{d1}. The results are shown on Table II.

Noteworthy is that n decreases with the pressure σ_3; on the contrary D^* and C, f^*, f^{**} and σ_F increase with the pressure σ_3. The f^* and f^{**}, σ_F vs p relationships are shown in Fig.6D; we find m=0.62 ; m'=1.09; m''=1.10 . The ($\sigma_1-\sigma_3$) vs e_{dl} and ($\sigma_1-\sigma_3$) vs. e_d are computed using the values in Table II. The results are plotted in Fig.6E. The agreement with the experiments is satisfactory,

TABLE I

No.	AUTHOR	ROCK	DIM(cm)	σ_3 (Mpa)	σ_2 (Mpa)	n	$D^*(10^{-6})$	f^* (Mpa)	f^{**} (Mpa)	σ_F (Mpa)	f^*/σ_F	f^{**}/σ_F
a	MOGI (1977)	GRANITE	1.5x1.5x3.0	70	253	3.9	35.2	340	640	790	0.43	0.81
b	KATE HADLEY (1975)	WESTERLY GRANITE	φ3x6	50	50	4.3	2.96	200	300	400	0.50	0.75
c	WU JING NONG (1980)	MIGMATITE	φ9x18	120	120	5.0	55.1	300	475	553	0.54	0.86
d	MAKOTO TERADA (1979)	GRANODIORITE	φ4.76x9.5	0	0	3.9	24.6	81	146	200	0.41	0.73
e	PERKINS (1970)	PORPHYRITIC TONALITE	φ1.27x2.54	0	0	4.0	3.25	33	95	150	0.22	0.63
f	HIROYUKI INOUE et al (1978)	GRANITE	6x6x12	0	0	4.6	46.3	56	95	130	0.43	0.73
g	TAN TJONG KIE (1964)	LIMESTONE	40x40x80	0	0	4.0	14.9	17	30	40.5	0.42	0.74
h	TOSHIYO ISOBE YUUSAKU TOMINAGA (1979)	SANDSTONE	φ3x6	0	0	3.9	26.2	50	120	175	0.29	0.69

TABLE II

AUTHOR	ROCK	DIM	σ_3 (Mpa)	n	$D^*(10^{-5})$	$c(10^{-5})$	$f^*=\sqrt{3}f_3$ (Mpa)	f^{**} (Mpa)	σ_F (Mpa)	f^*/σ_F	f^{**}/σ_F
F. RUMMEL (1974)	Ruhr-sandstone	l=6cm φ=3cm	80	3.4	6.74	4.93	200	530	570	0.35	0.93
			50	4.2	4.84	3.83	120	250	315	0.38	0.79
			30	4.3	5.83		120	230	300	0.40	0.77
			20	5.0	3.25	1.12	108	180	220	0.49	0.82
			10	5.9	3.81		90	140	185	0.49	0.76

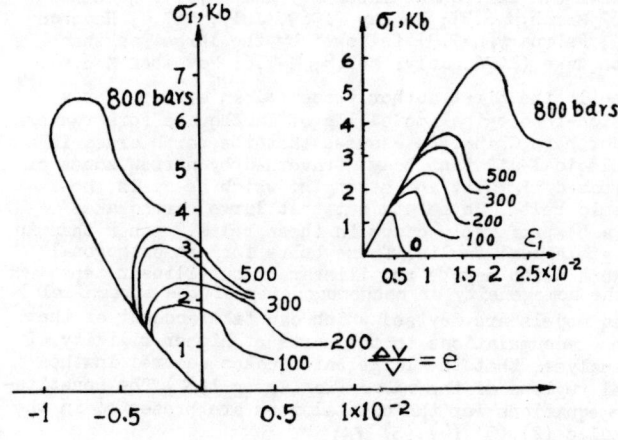

Fig.6A Test results on Ruhr-Sandstone (Rummel, 1974).

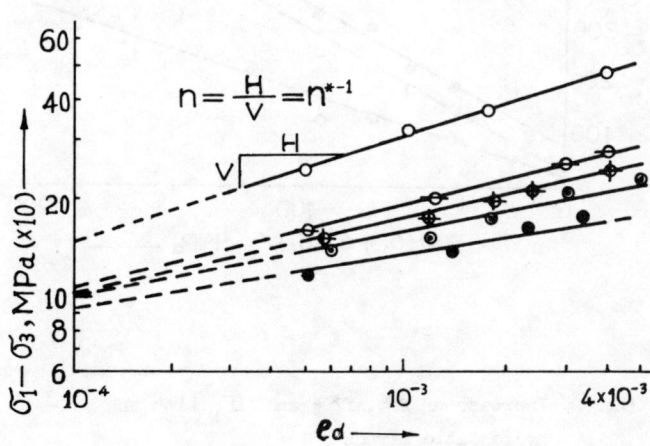

Fig.6B Influence of confining pressure on volume dilatancy.
σ_3=10MPa, n=5.9; σ_3=20MPa, n=5.0;
σ_3=30MPa, n=4.3; σ_3=50MPa, n=4.2;
σ_3=80MPa, n=3.4;

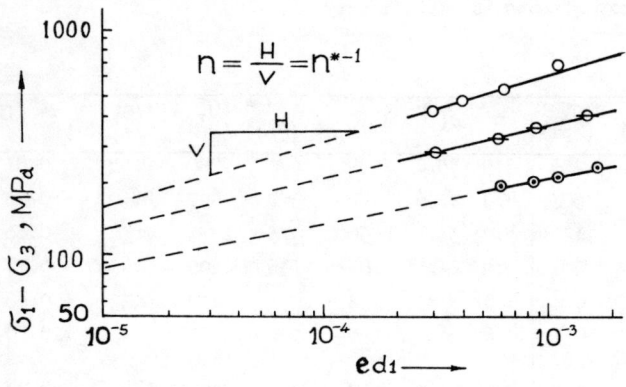

Fig.6C Relationship $\sigma_1 - \sigma_3$ vs e_{d1}
σ_3=20MPa, n=5.1;
σ_3=50MPa, n=4.3;
σ_3=80MPa, n=3.5;

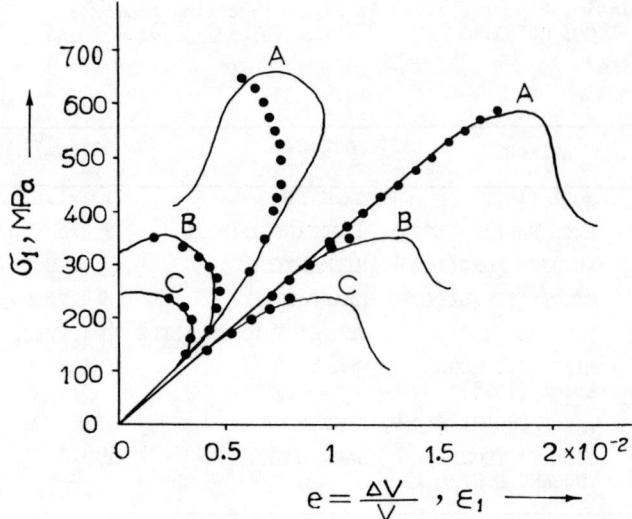

Fig.6E Measured (——, Rummel) and Computed (• •, Tan) stress-strain relationship.
A. σ_3=80MPa; B. σ_3=50MPa;
C. σ_3=20MPa;

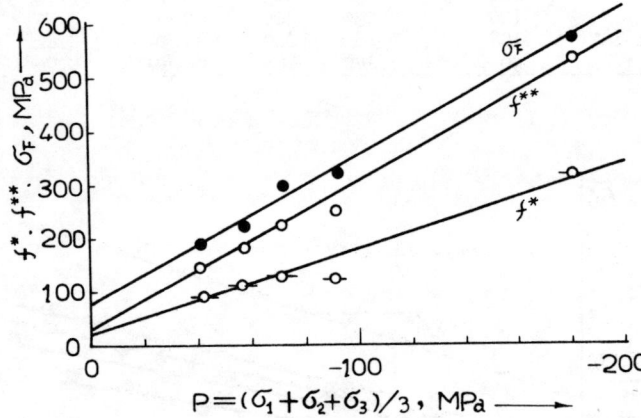

Fig.6D Increase of f^*, f^{**} and σ_F with the hydrostatic stress p
$f^* = f_o^* - mp$; $f^{**} = f_o^{**} - m'p$; $\sigma_F = \sigma_{Fo} - m''p$;
f_o^*=25MPa; f_o^{**}=38MPa ; σ_{Fo}=75MPa ;
m = 0.62;   m' = 1.09 ;   m'' = 1.10.

CRUSTAL UPHEAVAL AND DILATANCY

In Northern China the upheavals of the crust have been observed for twenty to thirty years. In this regions many large earthquakes has occurrured: Xing Tay (1966, March, M=6.8 Earthquake Intensity and M=7.2); Hochian (1967 March, M=6.3); Po Hay (1969, July, M=7.4; Haycheng 1975, February, M=7.3) followed by the large Tangshan Earthquake (1976, July: M=7.8; M=7.6; November M=6.9).

Recently the first author presented an attenpt for a physico-rheological modelling of Earthquake fore runners in Northern China. He assumes that the earth crust is a rheological-dilatant body, traversed by narrow zones of decreased rigidity and strength, which he calls the seismic belts. He points out that large earthquakes (M ≥ 6) preferably occur in these belts, rather than in the geological faults. These belts form an orthogonal network which may be rectilinear or ourvilinear dependent on the homogeneity or nonhomogoneity of the stressfield. Three models are devised which can take account of the stress accumulations to the parts of higher rigidity. It is analysed that the large earthquakes occured in the nodal regions of the **belts** (Tan,Cheng 1982). The constitutive equations for the crustal rock are presented in the formulas (2),(3),(4),(5) ,**(6).**

Well documented are the crustal upheavals of the Hai-Cheng, Tangshan and Xing tai earthquakes (Geod. Surv. 1975 1977). The isocontours of constant upheavals relative to the Pan Shen observation point are shown in Fig.7A, and a crossection AB in Fig.7B. It can be seen that the dilatancy process covers an area of some thousands of km^2 and an extension of a few hundreds of km. A typical example of vertical upheavals increasing with the time is shown in Fig.7C. The total upheaval u_{max} wihh relation to a point outside this dilating region and their average rate are given in **the following table III (Geod. Surv.1977).** when the stress distribution and the rate of stress increase are known, then hypothetically we can estimate the crustal horizontal deformations and uppeavals with the help of the constitutive equations presented above. The inverse problem however is very difficult and some rigorous simplifications must be made: shallow earthquakes occur at a depth of 10-25 km with a maximum frequency at ca 15 km. The earthquake is generated in the dilatant region and then the crack propagates up-down- and side-wards.

Let us now for example analyse the Hay-cheng earthquake and make use of equation (2) but rewritten in the y and z directions then we will have to deal with the following unknowns:
(1) depth of the dilating region H; (2) n; (3) σ_{oct}/f_3
(4) D^*; (5) C; (6) $(\sigma_z - p)/\sigma_{oct}$; (7) ξ.

Now we will venture in attempting to give a crude estimation of the possible limits between which some of the above quantities may vary.

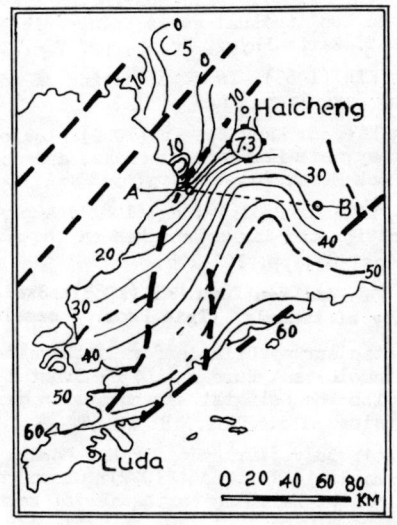

Fig.7A Isocontours of Landupheaval for Haycheng earthquake (National Seismological Bureau, 1977).

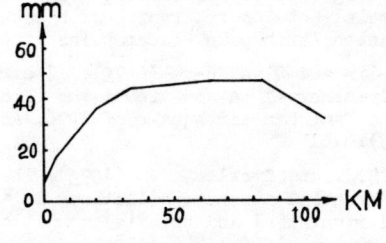

Fig.7B Cross-Section AB (National seismological Bureau, 1977).

The only quantity which can be directly determined is the average value of ξ_{av}; From (1) or (3) we find.

$$u = u_o + \dot{u}_o \xi \Delta t$$

From Fig.7C and Table III we find u_o = 115 mm;

TABLE III

Earthq.Loc.	Xing Tai	Po Hay	Hay cheng
M	7.2	7.7	7.3
Period	1920-1955	1958-1969	1937-1975
u_{max} (mm)	350	80	120
$\dot{u}$ (mm/y)	10	8	3-5
Focal Depth km	10	35	12

$\dot{u}_{av}$ = 5 mm/y then we get ξ_{av} = 2.9x10^{-2} pro year.

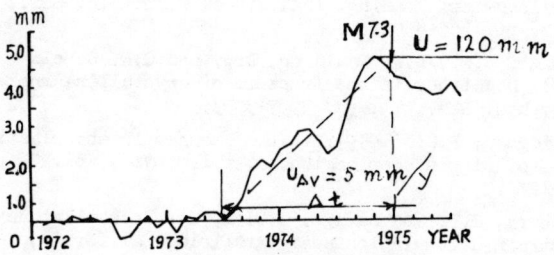

Fig.7C Landupheaval increasing with time (National Seismological Bureau, 1977).

Next we assume that volume dilatancy plays a major role thus we can take equation (4) instead of (2). The dilatancy zone may be extended as deep as ten to twenty km; however its contribution to the upheaval will decrease sharply with the increase in f_3 (see form (5); apparently only the upper 3 to 10 km of the crust contributed to the upheaval, so the average strain will be e_{dav} = 11.5 to 38x10^{-6}; further assume D^*=10^{-6} (Table II) then we find for the average ratio

$$(\sigma_{oct}/f^*)_{av.} = \begin{cases} 1.40 \text{ to } 1.89 \text{ for } n=4 ; \\ 1.31 \text{ to } 1.63 \text{ for } n=5 ; \end{cases}$$

The maximal ratio in the region, where the earthquake crack was generated must be higher than these values; these average values are plausible since we can find in Table I that for granites $(\sigma_{oct}/f^*) > 2$.

CONCLUSION

In this paper the dilatancy parameters have been estimated from small samples for various rocks in the range of 0 - 2,50 Mpa lateral pressure. Engineering rocks always contain a large variety and number of cracks, far more than in small samples. Hence it is known that large samples have lower rigidities and strenghts than small samples. Hence for engineering purposes it can be expected that the D^* and C values will be much higher. In earthquakes the hydrostatic stress may amount to hundreds to thousands Mpa and it can be expected, that the D^* and C values may be much lower due to the higher densities of the rock.

The important role of time **dependent dilatancy** prior to rock failure and earthquakes is discussed and it is hoped that more readers will be interested in it.

REFERENCES

Aggarwal, Y.P., Sykes, L.R., Armbruster, J. and Sbar, M.L. (1973): Premonitory changes in seismic velocities and prediction of earthquakes. Nature, Vol. 241. Jan. 12, pp. 101-104.

Bieniawski, Z.T. (1967):Mechanism of brittle fracture of rock. Part I-Theory of the fracture process; Part II-Experimental studies; Part III-Fracture in tension and under long-term loading. Int. J.Rock Mech. Min. Sci., Vol. 4, pp. 365-430.

Brace, W.F., B.W. Paulding, Jr., and C.H. Scholz (1966): Dilatancy in the fracture of crystalline rocks J. Geophys. Res., 71, pp. 3939-3953.

Bridgman, P.W. (1949): Volume changes in the plastic stages of simple compression. J.Appl. phys., Vol.20, pp. 1241-1251.

Cherry, J.T. and Savage, J.C.(1972): Rock dilatancy and strain accumulation near Parkfield, California, Bull. Seism. Soc. Am., Vol. 62, pp. 1343-1347.

Crouch, S.L. (1970): Experimental determination of volumetric strains in failed rock, Int. J. Rock Mech. Min. Sci., Vol. 7, pp. 589-603.

Duan Xing-bei, Zheng Jian-zhong, Zhou Zhi-gun, Yan Shou-min and Sun Ci-chang (1976): Variations in teleseismic P wave residuals before the Haicheng earthquake. Acta Geophysica Sinica V. 19, No. 4, pp. 286-294.

Edmond, J.M. and Paterson, M.S.(1972): Volume changes during the deformation of rocks at high pressures. Int. J.Rock Mech. Min. Sci., Vol. 9,pp. 161-182.

Feng Rui, Pang Zing-yan, Fu zheng-xiang, Zheng Jianzhong, Sun Ci-chang and Li Bao-xuabg (1976): Variations of v_p/v_s before and after the Haicheng earthquake of 1975. Acta Geophysica Sinica, V. 19, No. 4, pp. 295-304.

Hadley, K. (1975): Azimuthal variation of dilatancy. J. Geophys. Res., Vol. 80, pp. 4845-4850.

Handin, J., Hager, R.V., Friedman, M. and Feather, J.N. (1963): Experimental deformation of sedimentary rocks underconfining pressure: Pore pressure tests. Bull Am. Assoc. Pet. Geol., Vol. 47, No. 5, May, pp. 717-755.

Hiroyuki, I., Katsuhiko, K., Koichi, S. and Ichiro. I. (1978): The attenuation of the wave due to progress of breakage in rocks (Original prints in Japanese), Journal of Japanese Mining society, Vol. 94, No. 1083, pp. 7-11.

John, M. (1974): Time dependence of fracture processes of rock mechanics (in German). Proc. 3rd cong. ISRM (Denver), Vol. 2A, pp. 330-335.

Makoto, T., Takashi, Y., Osamu, S. and Ichiro, I. (1979): The fracture behaviour of the granodiorite under uniaxial compression. Rock Mech. in Japan, Vol. III, pp. 50-52.

Mogi, K. (1977): Dilatancy of rocks general triaxial stress with special reference to earthquake precursors. J.Phys. Earth. 25, Suppl., 5203-5217.

Paterson, M.S. (1963): Secondary changes in length with pressure in experimentally deformed rocks. Proc. Roy. Soc. London, Series A, V. 271, pp. 57-87.

Perkins, R.D., Green, S. J. and Friedman, M. (1970): Uniaxial stress behaviour of porphyritic tonalite at strain rates to 10^{-3}/second. Int. J.Rock Mech. Min. Sci., Vol. 7, pp. 527-535.

Reynolds, O. (1885): On the dilatancy of media composed of rigid particles in contact, Phylosophical Magazine, Vol. 20, pp. 469-481.

Rummel, F. (1974): Changes in the P-wave velocity with increasing inelastic deformation in rock specimens under conpression, Reprinted from advances in rock mechanics. ISBN 0-309-02246-0 National Academy of Sciences, Washington, pp. 517-523.

Semenov, A.H. (1969): Variation in the travel time of transverse and longitudinal waves before violent earthquakes. Invest. Earth Phys., No. 4, pp. 72-77.

Tan Tjong Kie (1964): In situ testing of rocks on behalf of the Daye iron mine.

Tan Tjong Kie and Kang Wen-fa (1980): Locked in stresses, creep and dilatancy of rocks, and constitutive equations. Rock Mech, Vol. 13, pp. 5-22.

Tan Tjong Kie and He Zhi-tong (1982): A physico-Rheological model for the large Tangshan earthquake. Tectonophysics, 85, pp. 123-148.

Tan Tjong Kie **and Wen Xuan Mei (1983): Swelling rocks and stability of tunnels. (This Proc. session).**

The Geodetic survey Brigade for Earthquake Research, National Seismological Bureau (1975): Crustal deformation associated with the Hsingtai earthquake in March, 1966. Acta Geophysica Sinica Vol. 18, No. 4, pp. 153-163.

The Geodetic Survey Brigade for Earthquake Research, National Seismological Bureau (1977): Ground surface deformation of the Haicheng earthquake of magnitude 7.3. Acta Geophysica Sinica, Vol. 20, pp. 251-263.

Toshiro, I. and Yunsaky, T.(1979): Effects formerly acted stress on stress-strain behaviour of rock and its anisotropical property, Proc. 4th congress ISRM, Vol.1, pp. 181-186. Montreux.

Wang Chen-ming et al. (1979): Characteristies of deepwell water level variation before and after the Tangshan earthquake. Symp. Earthquake Prediction, UNESCO, Paris.

Wu Jing-nong (1980): Rocksoftening and reservoir earthquakes, Special note on reservoir earthquake. pp. 84-95. Reports: Canton Earthquake Research Institute.

Zhao Yu-lin and Qian Fu-ye (1979): Electrical resistivity anomaly observed in and around the epicentral area prior to the Tangshan earthquake of 1976. Acta Geophys. Sin., 22 (3): 181.

Zoback, M.D. and Byerlee, J.D.(1975): The effect of cyclic differential stress on dilatancy in Westerly granite under uniaxial and triaxial condition. J. Geophys, Res., Vol. 80, pp. 1526-1530.

EVALUATION DES CONTRAINTES NATURELLES DANS LES COUCHES SUPERIEURES D'UN MASSIF SCHISTEUX

Assessment of in situ stresses in the upper layers of a schistous bedrock
Abschätzung der natürlichen Spannungen in den oberen Schichten eines Schieferuntergrundes

P. J. Huergo
Chef de Travaux, Université Libre de Bruxelles, Bruxelles, Belgique

RESUME

L'évaluation du comportement mécanique réel d'un massif rocheux et notamment la réponse déformationnelle, ne dépend pas seulement des sollicitations externes qui lui sont appliquées mais aussi — et ce à un dégré significatif — de l'état des contraintes naturelles du massif dont celles dues à la pesanteur et/ou aux forces tectoniques associées au passé géologique de la roche. La détermination du champ de contraintes résultant de la superposition des contraintes induites et naturelles est donc essentielle pour la connaissance du comportement mécanique du massif et ce notamment pour les couches supérieures de celui-ci qui sont le plus fortement sollicitées par les charges appliquées. Dans le but d'évaluer ce comportement on a procédé, lors de l'étude d'un système de fondation sur un massif constitué d'une succession de couches de schistes siluriens, à une série d'essais in situ — notamment des essais pressiométriques — ainsi que des essais en laboratoire dont les résultats ont permis de mesurer les contraintes naturelles.

SYNOPSIS

The evaluation of the actual mechanical behaviour of a rock mass, and in particular its deformational reaction, does not depend solely on external forces to which it is subjected but also — and to a significant degree — on the state of the natural stresses of the rock mass, some of which are due to its weight and/or to tectonic forces associated with the geological past of the rock. The determination of the stress field resulting from the superposition of induced and natural stresses is, after all, essential in order to know the mechanical behaviour of the rock mass. This applies particularly to its upper layers which are most strongly affected by the forces applied to it. The assessment of this behaviour proceeds from the study of a foundation on a rock mass, consisting of a series of layers of silurian schist, to a sequence of in situ tests whose results made it possible to measure the natural stresses.

ZUSAMMENFASSUNG

Die Bewertung des wirklichen mechanischen Verhaltens einer Gesteinsmasse, und insbesondere ihre Verformungsreaktion, hängt nicht nur von den auf sie einwirkenden Kräften ab, sondern auch in hohem Maße von dem Zustand der in der Gesteinsmasse herrschenden natürlichen Spannungen. Einige dieser Spannungen hängen mit dem Gewicht der Gesteinsmasse zusammen und/oder beruhen auf tektonischen Kräften, die mit der geologischen Vergangenheit des Felses in Zusammenhang stehen. Die Bestimmung des durch die Überlagerung von induzierten und natürlichen Spannungen hervorgerufenen Spannungsfeldes ist ja notwendig, um das mechanische Verhalten der Gesteinsmasse zu verstehen. Dies trifft insbesondere auf die oberen Schichten zu, die am stärksten von den auf sie einwirkenden Kräften betroffen werden. Die Ermittlung dieses Verhaltens beginnt mit dem Studium eines Fundamentes, welches auf einer Gesteinsmasse ruht, die auf einer Reihe aus Silur-Schiefer bestehenden Schichten aufgebaut ist, und führt dann zu in situ Untersuchungen, aus deren Resultaten die natürlichen Spannungen gemessen werden können.

1. Introduction

Toute analyse géomécanique tendant à évaluer la réponse mécanique d'un massif rocheux sous l'effet des sollicitations externes doit tenir compte de l'état des contraintes internes qui y existent du fait de la genèse et de l'âge des roches constituantes ainsi que de la situation du massif dans son environnement géologique.

En effet, à l'opposé des sols - formés le plus souvent par des dépôts successifs plus ou moins uniformes conduisant à une organisation granulaire faiblement liée - les massifs rocheux,

de par leur nature même, sont ou ont été soumis au cours de leur histoire géologique à des efforts considérables dépendant des processus endogènes (gravifiques, tectoniques, thermiques etc.) et exogènes (érosion, sédimentation et autres phénomènes de surface) qui ont lieu en rapport avec l'évolution continue de l'écorce terrestre.

Dans le cas des travaux de génie civil, travaux qui en général se trouvent soit à la surface, soit à faible profondeur par rapport à l'épaisseur de l'écorce, l'existence dans le massif rocheux d'un champ de contraintes internes ou naturelles, auquel vient se superposer le champ de contraintes induit par les surcharges de l'ouvrage, peut bouleverser la distribution des contraintes qui en résultent et modifier significativement le comportement mécanique du massif rocheux. Il en découle la nécessité, pour la bonne conception des ouvrages, de mener une campagne expérimentale de mesures tendant à déterminer, de la façon la plus précise possible, la gamme des contraintes liées à la géométrie et à l'histoire de l'assise rocheuse.

La présente communication fait état de l'approche expérimentale utilisée dans le cadre du projet d'un système de fondation, pour mettre en évidence, à l'aide d'une campagne de mesures in situ et en laboratoire, le champ des contraintes naturelles et pour en évaluer l'influence sur le comportement de la fondation envisagée.

2. Description géologique du site

Le site de construction, situé en bordure de la Meuse au S.E. de la Belgique, se présente sous la forme d'un socle rocheux à surface subhorizontale, recouvert par des couches alluvionnaires constituées essentiellement d'horizons sablo-graveleux de 7 à 11 m d'épaisseur. Ces horizons sablo-graveleux gisent sous une couverture superficielle de limons fluviatiles d'une épaisseur allant de 2 à 6 m selon les endroits.

Le socle rocheux est constitué d'une succession de couches schisteuses - schistes siluriens - dont l'allure générale est orientée E.O. avec des pendages divers selon les plissements de la structure. Le massif révèle un passé tectonique caractérisé par l'existence de cassures et de rejets ayant comme résultat la formation d'une structure à géométrie fort complexe. Du point de vue lithologique, on décèle des couches à schistes micacés compacts ou à schistes tendres facilement délitables ainsi que des couches à schistes argileux caractérisant souvent les zones tectonisées ou altérées. A l'échelle microscopique, les schistes se présentent sous des aspects très divers allant de l'anamorphisme complet à la micro-straticulation.

Les horizons supérieurs du massif schisteux et notamment ceux qui se trouvent au contact des terrains alluvionnaires de couverture, sont en état d'altération dont le degré varie en fonction de la tectonisation et de la plus ou moins grande sensibilité des minéraux constituants à l'agression des divers agents physico-chimiques, essentiellement ceux liés à la percolation des eaux souterraines.

En ce qui concerne les contraintes naturelles autres que les contraintes tectoniques, il y a lieu de prendre en considération les contraintes géostatiques liées à l'évolution de la vallée de la Meuse à la verticale du site étudié. Si l'on se réfère aux niveaux des terrasses laissées par le fleuve au cours de son évolution pendant le quaternaire (Clairbois, 1959), on peut supposer que la contrainte historique maximale, au moins en ce qui concerne cette époque géologique, devrait être en rapport avec le niveau de l'assise de la plus haute terrasse décelée dans les versants de la vallée. Ce niveau se situerait aux environs de la cote + 118 alors que le sommet actuel de l'assise schisteuse se situe à la cote + 60 environ. Par conséquent, il y aurait eu, à ce niveau une contrainte géostatique due au poids des couches érodées d'une épaisseur maximale d'environ 58 m.

L'existence de ces contraintes et leur disparition ultérieure sont à mettre en rapport, au moins partiellement, avec la distribution actuelle de zones consolidées et déconsolidées que l'on peut déceler dans le massif rocheux.

3. Mesures in situ

La campagne de reconnaissance géotechnique du site a été réalisée principalement à l'aide d'essais pressiométriques effectués avec une sonde Ménard GC de 60 mm de diamètre et dont la plage des pressions varie de 0 à 4000 kPa. Les cellules de garde sont pressurisées au CO_2 et la cellule de mesure à l'eau. L'entredistance des essais est d'un mètre et les lectures ont été effectuées par paliers de 400 kPa. La profondeur maximale atteinte est d'environ 40 m.

La grille des forages pressiométriques au nombre de 41 a été doublée d'une série de 26 sondages de même profondeur. On y a prélevé des échantillons non remaniés (ϕ = 63 mm) à l'aide d'un double carottier à câble. Les relevés et mesures effectués sur les échantillons ont permis d'évaluer d'une part l'indice de qualité de la roche R.Q.D. et le taux de carottage t.c. et, d'autre part, les divers paramètres physiques tels que la teneur en eau w et le poids spécifique γ ainsi que l'altérabilité et la sensibilité au gonflement.

En outre, afin de mieux préciser la valeur des contraintes géostatiques, on a effectué des mesures de densité in situ à partir de fouilles dans les terrains de couverture et au moyen d'une sonde gamma-gamma alimentée au caesium 137 dans les couches du substratum schisteux. Le poids spécifique moyen des schistes en place est de γ = 26 kN/m^3.

4. Essais en laboratoire

Afin de cerner statistiquement la valeur la plus probable de la résistance à la compression σ_c et de la résistance à la traction σ_t de la roche, le programme d'essais en laboratoire a été basé essentiellement sur la réalisation du nombre le plus grand possible d'essais de compression et d'essais de traction par compression diamétrale. Ceci a permis d'avoir une connaissance globale du comportement mécanique du massif schisteux.

La plupart des éprouvettes d'essais ont été façonnées à partir des échantillons prélevés dans les zones à schistes compacts et peu fracturés (R.Q.D. > 0.5). Dans les zones à schistes foliés, il s'est avéré difficile d'obtenir des éprouvettes d'élancement adéquat ($l/\phi \geqslant 2$) du fait que les plus sveltes subissaient au cours des manipulations, et ce malgré les précautions prises, des cassures par délitage le long des plans de foliation. Il en a été de même pour les éprouvettes des schistes altérés des couches supérieures du substratum ou des zones fortement tectonisées, caractérisées toutes les deux par des taux de carottage faibles et des indices R.Q.D. bas (R.Q.D. < 0.25). Dans ces schistes, les éprouvettes pouvaient s'effriter, selon des plans de rupture quelconques, sous l'effet de contraintes même minimes. Tant pour les schistes altérés que pour les schistes tectonisés, le pourcentage de déchet a été très important (>50%) et les manipulations expérimentales se sont révélés extrêmement délicates.

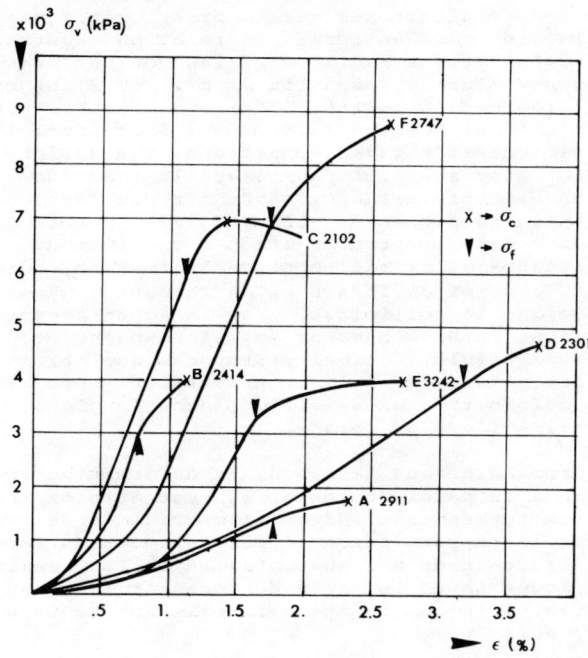

Fig. 3.1

Courbes caractéristiques contraintes σ - déformations ε obtenues en compression uniaxiale

Les essais de compression uniaxiale ont été réalisés par cycles progressifs de charge et de décharge appliqués par paliers de contraintes allant de 100 à 1000 kPa selon la valeur de la résistance à la compression uniaxiale σ_c connue par des essais préalables ou évaluée selon le type de schiste. La courbe contraintes σ - déformations ε utilisée est la courbe enveloppe des diagrammes cycliques dont l'allure dépend des caractéristiques de la structure planaire de la roche ainsi que des propriétés mécaniques de la matrice interfoliaire. Selon la valeur de l'angle θ formé par l'axe vertical d'application de la charge et le plan de foliation, les courbes contraintes-déformations mettent en évidence des comportements élasto-plastiques (anamorphisme ou $\theta \to 0$, courbe C, fig. 4.1) ou plus fréquemment des comportements plasto-élasto-plastiques ($\theta \to \pi/2$).

La figure 4.1 montre quelques courbes σ-ε caractéristiques obtenues à partir d'essais réalisés sur éprouvettes prélevées à différents endroits et profondeurs du massif schisteux. On a considéré que le seuil de fluage σ_f correspond au seuil de fissuration caractérisé par un accroissement non linéaire de la déformation volumique. Ceci situe souvent le seuil de fluage légèrement en-deça du début apparent de la phase plastique telle qu'elle se manifeste par la fin de la linéarité des relations contraintes-déformations.

D'après les valeurs de la résistance à la compression uniaxiale σ_c et du module d'élasticité E, module évalué d'après la pente de la tangente à la courbe σ-ε pour une contrainte $\sigma = 0.5\ \sigma_c$, il s'avère que l'ensemble des schistes du site tombe, selon la classification Deere-Miller, dans les catégories ML et EL.

La courbe A ($z = 20.7$ m, $\alpha = 78°$) illustre le comportement d'un schiste altéré de faible résistance σ_c alors que les courbes B et E mettent en évidence respectivement la réponse mécanique différente d'un schiste compact peu déformable et celle d'un schiste tectonisé d'une résistance analogue mais de plus grande déformabilité. Dans ce cas, les relations déformations-contraintes dépendent essentiellement des propriétés mécaniques intrinsèques de la roche et sont peu ou pas influencées par ses caractéristiques géométriques ($\alpha(B) = 59°$, $\alpha(E) = 65°$).

Les paramètres caractéristiques correspondant aux schistes types représentés dans la figure 4.1 sont repris dans le tableau 4.I. Y figurent par ailleurs les valeurs de la résistance à la traction σ_t ainsi que les valeurs de la résistance au cisaillement calculées par la méthode d'Horibe (Vutukuri et al., 1974), elle-même basée sur la théorie de Griffith. Cette méthode permet de déterminer la cohésion intrinsèque τ_o de la roche à partir des couples de valeurs expérimentales (σ_t, σ_c). La valeur τ_o est déduite d'après l'intersection de l'axe τ d'ordonnées avec la tangente commune au cercle de rayon $4\sigma_t(-\sigma_t, 3\sigma_t)$ et au cercle de Mohr correspondant à la résistance à la compression uniaxiale σ_c. L'expression théorique de la cohésion intrinsèque est

$$\tau_o = \frac{\sigma_c\ \sigma_t}{\sqrt{2\sigma_t(\sigma_c - 3\sigma_t)}} \qquad (4.1)$$

Tableau 4.I

Sondage n°	Profondeur z (m)	Pendage α (°)	E(0.5 σ_c) (kPa)	ν	σ_c (kPa)	σ_f (kPa)	σ_t (kPa)	τ_o (kPa)
2911	20.7	78	126732	0.29	1765	1373	243	427
2414	29.4	59	743242	0.19	3942	2942	543	955
2102	33.5	90	538404	0.15	6864	5884	839	1508
2301	35.1	38	165704	0.38	4668	3923	643	1181
3242	35.2	65	396028	0.29	3962	3432	547	962
2747	36.6	89	595404	0.25	8826	6865	1999	3710

La figure 4.2 montre les cercles σ_t, $4\sigma_t$ et σ_c correspondant aux courbes C et F de la figure 4.1 ainsi que les paraboles de Griffith délimitant la plage des relations $\tau - \sigma_{1,3}$ associées au début de la désorganisation de la roche.

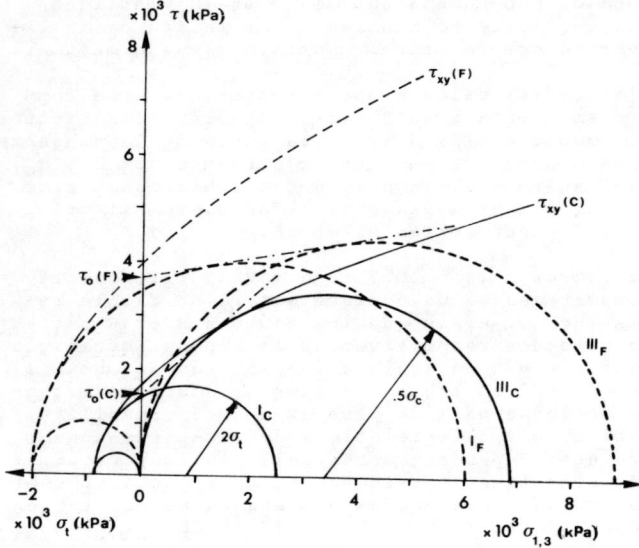

Fig. 4.2
Détermination de la résistance au cisaillement τ_o à partir des relations σ_t, $\sigma_{1,3}$, τ

5. Evaluation des contraintes naturelles à partir des résultats expérimentaux

La réalisation de mesures des contraintes in situ se heurte dans le cas des massifs schisteux du type décrit à des problèmes expérimentaux difficiles à résoudre.

En effet, dans le cas du massif étudié situé en bordure d'un fleuve, recouvert d'une épaisseur importante d'alluvions gorgés d'eau, l'utilisation de méthodes de mesure basées soit sur la mise en place en fond de fouille de vérins plats, soit à l'aide de dilatomètres ou de jauges nécessitant la libération des contraintes par surcarottage, s'est avérée pratiquement impossible du fait de la rupture fragile de la roche ou de la mauvaise tenue des parois du forage dans les passées altérées ou tectonisées.

En conséquence, il a été décidé d'évaluer indirectement l'état naturel des contraintes à partir des résultats des essais pressiométriques, en interprétant en termes de relations contraintes-déformations radiales ε_r les courbes pression p- volume V (Baguelin et al., 1972) et ce à l'aide des paramètres déduits des essais en laboratoire. Pour ce faire, on a fait appel à la méthode itérative de Marsland et Randolph (1976) mise au point dans l'argile raide de Londres. Cette méthode, basée sur des développements théoriques de Palmer (1972), permet de déduire dans un premier stade les relations résistance au cisaillement τ-déformation radiale à la paroi du forage ε_r en faisant l'hypothèse que la sollicitation de la sonde pressiométrique insérée dans un massif homogène et incompressible - ce qui présuppose que celui-ci est saturé ($S_r = 1$) et non drainé - produit une déformation axisymétrique dans le plan normal à l'axe du forage.

D'après Palmer et Marsland, la déformation radiale à la paroi du forage ε_r peut être exprimée en termes de ε_r', déformation radiale définie comme le rapport entre l'accroissement du rayon du forage Δr dû à l'augmentation de la pression appliquée Δp et le rayon du forage r_o correspondant à la contrainte naturelle horizontale p_{oh}, soit

$$\varepsilon_r = \varepsilon_r' = \frac{\Delta r}{r_o} \qquad (5.1)$$

où $\Delta r = f(\Delta p) \rightarrow \Delta p = p - p_{oh}$ et
$r_o = f(p_{oh})$

On considère donc qu'après application de la contrainte p_{oh}, le massif encaissant retrouve l'état d'équilibre existant avant l'exécution du forage et on prend celui-ci comme état de référence pour les calculs.

La déformation radiale ε_r' est exprimée en termes de variation ΔV du volume de la sonde par rapport à celui correspondant à l'état de référence et du volume V de la sonde à la pression mesurée p_i, soit

Tableau 5.I

Essai n°	Profondeur z(m)	E_p (kPa)	$P\ell e$ (kPa)	P_{oh} (kPa)	c_u (kPa)	p'/σ_f (kPa)	c_u/τ_o (kPa)
2931	21	368534	35990	741	633	1.001	1.482
4016	21	172695	18142	1424	537	1.000	0.720
2702	24	968407	87083	1855	891	1.129	1.377
3123	27	228887	21476	1450	1296	0.933	1.090
3208	28	153964	13337	1753	995	0.934	0.837
"	29	236340	24320	1894	1247	0.801	1.090
"	35	987530	79728	2923	607	1.029	0.631
2702	37	686760	51289	2354	1813	0.944	1.487
2421	38	333132	39521	1961	406	1.006	1.041

$$\varepsilon'_r = \frac{1}{\sqrt{1-\Delta V/V}} - 1 \qquad (5.2)$$

et la résistance au cisaillement

$$\tau = \frac{1}{2} \varepsilon'_r (1 + \varepsilon'_r)(2 + \varepsilon'_r) \frac{dp}{d\varepsilon'_r} \qquad (5.3)$$

d'où

$$\tau \cong \varepsilon'_r \frac{dp}{d\varepsilon'_r} \text{ en petites déformations} \qquad (5.4)$$

et

$$\tau = \frac{dp}{d(\ln \Delta V/V)} \text{ en grandes déformations} (5.5)$$

Cette dernière expression permet de déduire la résistance au cisaillement d'après la pente de la courbe p-ln $\Delta V/V$ et la valeur de la cohésion non drainée $c_u = \tau_{max}$.

Afin de diminuer les erreurs systématiques liées à la mesure du volume de la sonde (Baguelin et al., 1973), on a comparé les courbes pressiométriques obtenues à égale profondeur et dans le même type de schistes pour vérifier l'existence en phase pseudo-plastique des variations ΔV anormales non attribuables à un changement de nature de la roche.

D'après la méthode de Marsland et Randolph, la contrainte géostatique horizontale p_{oh} peut être déterminée par itérations successives sur les couples de valeurs p_{oh}, $c_u(p_{oh})$ en variant la contrainte p_{oh} considérée, jusqu'à ce que la somme $p_{oh} + c_u$ soit égale la valeur p' (correspondant dans le cas d'un matériau élasto-plastique parfait, au seuil de fluage) matérialisée par la fin de la linéarité des relations contraintes p - déformations radiales ε'_r.

L'application de cette méthode au cas étudié a conduit à l'obtention d'un grand nombre de courbes p-ε'_r dont quelques exemples caractéristiques sont représentés dans la figure 5.1 ci-après.

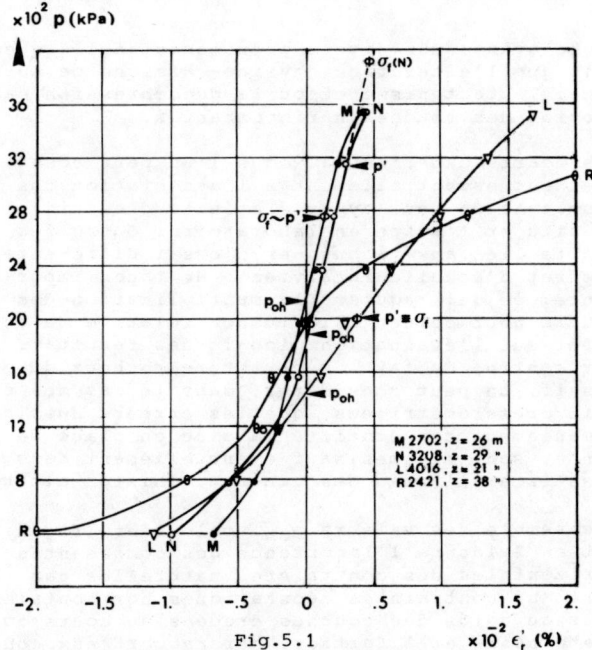

Fig.5.1
Exemples de courbes pression p - déformation ε'_r déterminées d'après la méthode itérative de Marsland et Randolph

Les courbes M et N correspondent à des essais réalisés dans des zones à schistes compacts peu fracturés. Les relations p-V montrent un palier pseudo-élastique s'étendant sur une large gamme de pressions avant d'atteindre la pression de fluage p_f. Ceci se traduit par des courbes p-ε'_r essentiellement linéaires dans l'intervalle de contraintes correspondant grosso modo à ladite phase pseudo-élastique de l'essai. Il apparaît, par ailleurs qu'en général la pression naturelle horizontale p_{oh} déterminée au point p-ε'_r ($\varepsilon'_r = 0$) tend à se rapprocher de la pression p_o (début de la phase pseudo-élastique) au fur et à mesure que la profondeur z de l'essai augmente. L'essai correspondant à une profondeur z = 38 m montre l'allure des courbes p-ε'_r obtenues dans les schistes tectonisés ou très fracturés avec fortes déformations dans la gamme des pressions appliquées.

Le tableau 5.I donne quelques résultats significatifs pris dans l'ensemble des mesures pressiométriques effectuées. On peut y remarquer que les composantes gravifiques des contraintes naturelles ont une influence certaine sur les valeurs des contraintes horizontales dont la tendance générale est d'augmenter avec la profondeur et ce même dans le cas des schistes altérés ou tectonisés. Ce tableau donne aussi les valeurs des rapports entre la contrainte p'= $p_o + c_u$ obtenues d'après la méthode itérative employée et la contrainte correspondant au seuil de fluage σ_f déterminée en compression uniaxiale. On peut relever que pour l'ensemble des mesures effectuées, la plupart des valeurs p'/σ_f sont comprises entre 0.7 et 1.1. Ce fait a permis, dans certains cas de supposer que la contrainte p' est associée au seuil de fluage σ_f. Il s'agit des cas où la pression de fluage p_f - à fortiori la pression limite p_ℓ - n'a pu être atteinte ou n'est pas décelable, la transition entre les domaines pseudo-élastique et plastique s'effectuant sans changement apparent de linéarité des relations p-V obtenues. En prédéterminant la valeur p' sur la droite p-ε'_r, le calcul itératif conduit à des valeurs "apparentes" de la contrainte naturelle horizontale p'_{oh} liées à l'incertitude de la transposition des caractéristiques intrinsèques du matériau découlant du type d'essai utilisé et de l'effet d'échelle (Londe, 1973). Toutefois, on peut admettre que ces valeurs apparentes sont proches de la valeur p_{oh} réelle telle qu'elle aurait pu être déterminée s'il n'y avait pas eu d'ambiguité expérimentale sur la valeur du seuil de fluage.

Le tableau 5.I donne en outre la valeur de la cohésion non drainée c_u - contrainte de cisaillement maximale τ_{max} définie par Palmer - déterminée d'après l'équation (5.5).

Ces valeurs c_u sont comparées à la résistance au cisaillement τ_o calculée selon le critère d'Horibe décrit précédemment. La dispersion des valeurs c_u/τ_o est plus grande que pour les rapports p'/σ_f tout en restant dans une gamme allant de 0.5 à 1.5. Ces résultats sont évidemment le fait des écarts résultant de l'anisotropie des schistes, à la variation de la teneur en eau au niveau des facettes cisaillées ainsi qu'à l'effet d'échelle. On peut constater, par ailleurs, qu'en valeur absolue, la cohésion non drainée semble être indépendante des contraintes gravifiques ou tectoniques mais être fonction de la présence des passées argileuses (schistes altérés ou tectonisés) ou encore de la nature des contacts au niveau des plans de schistosité.

Finalement, il y a lieu d'indiquer que les pressions limites reprises dans la 3ème colonne du tableau 5.I sont les pressions limites élastiques $p_{\ell e}$ calculées par extrapolation de la partie linéaire de la courbe pressiométrique exprimée en termes de pression p - inverses des volumes 1/V (Van Wambeke et d'Hemricourt, 1975).

La figure 5.2 montre l'ensemble de points correspondant aux relations p_{oh} - z obtenues. On peut constater qu'il existe deux populations de points distinctes, l'une associée aux contraintes p_{oh} inférieures et moyennes, déterminées dans les schistes altérés ou décomprimés des couches supérieures du massif ou encore dans les schistes tectonisés, et l'autre associée aux contraintes maximales des zones de schistes compacts peu ou pas fracturés. Les droites de régression α - α et β - β des deux populations sont à peu près parallèles, ce qui montrerait que, bien que l'influence des phénomènes de surface se réduise progressivement en profondeur, l'influence des phénomènes liés au tectonisme se fait sentir bien au-delà de la zone reconnue.

On remarque que l'ensemble des valeurs des contraintes naturelles horizontales p_{oh} tend à augmenter avec la profondeur et ce beaucoup plus rapidement que les contraintes géostatiques historiques. En effet, dans la figure 5.2, on a représenté la droite σ_{hpc} correspondant aux contraintes géostatiques horizontales calculées d'une part sur base des coefficients de Poisson ν obtenus en laboratoire et, d'autre part, en tenant compte de l'épaisseur des terrains érodés au-dessus du niveau actuel du substratum rocheux. L'influence des efforts tectoniques est donc mise en évidence par l'écart entre les points p_{oh} déterminés et la droite σ_{hpc}.

Cet écart ne devrait pas augmenter au-delà d'une profondeur limite h_f à laquelle on atteindrait une contrainte seuil en rapport avec la contrainte de fluage σ'_f de la roche à grande profondeur (Horvath, 1965).

6. Conclusions

La mise en évidence de l'état naturel des contraintes est essentielle pour la connaissance du comportement mécanique du substratum rocheux et ce notamment en ce qui concerne sa déformabilité. Dans le cas du massif de fondation étudié, l'évaluation des contraintes naturelles a été réalisée à partir des résultats d'essais pressiométriques à l'aide de la méthode itérative de Marsland et Randolph. Du fait des incertitudes inhérentes à la précision expérimentale, l'interprétation des courbes pressions p - déformations radiales ε'_r a été effectuée sur base des paramètres caractéristiques, notamment la pression du fluage σ_f et la résistance au cisaillement τ_o, déterminés par des essais en laboratoire. Ces essais, pour la plupart des essais de compression uniaxiale et de compression diamétrale (essais brésiliens) ont été quasiment les seuls à pouvoir être réalisés en grand nombre par suite de la nature particulière de la roche étudiée - schistes siluriens -, c'est à dire

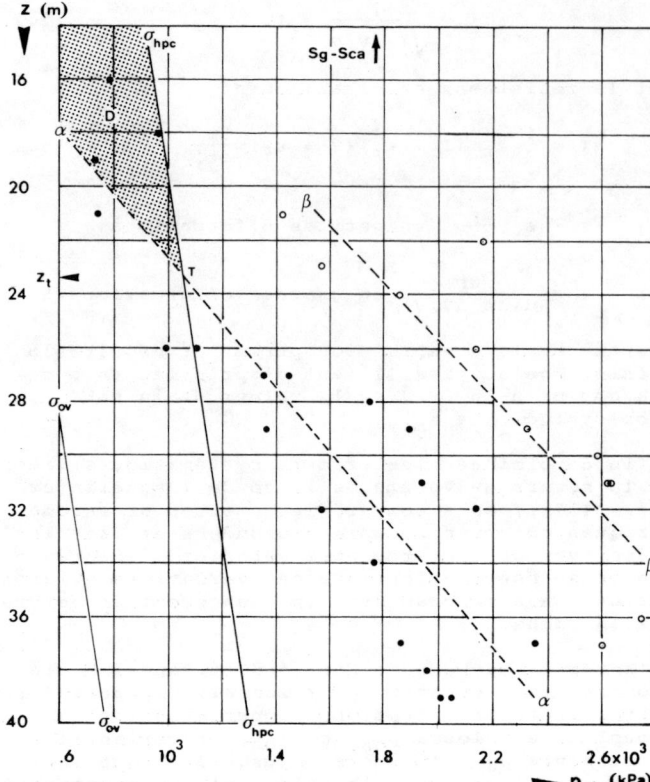

Sg-Sca : couches de couverture

α - α p_{oh} (kPa) = - 869.33 + 82.334 z(m) , r = 0.883

β - β p_{oh} (kPa) = - 387.853 + 92.781 z(m) , r = 0.913

D : zone décomprimée

Fig. 5.2
Plage de variation des contraintes naturelles horizontales p_{oh}, p'_{oh} en fonction de la profondeur

roche ayant subi divers phénomènes géologiques tels que l'altération physico-chimique de surface, le tectonisme et/ou la décompression par érosion des couches surincombantes.

Les écarts expérimentaux que l'on peut constater tiennent essentiellement à l'association des résultats de deux types distincts d'essais, l'un in situ et l'autre en laboratoire. Outre les écarts liés aux techniques d'essai différentes, l'effet d'échelle intervient de façon importante, ce qui requiert la multiplication des essais pour cerner l'influence relative de cet effet sur l'évaluation globale des résultats. Par contre, du fait du caractère rocheux du massif, on peut considérer, dans le cas des essais pressiométriques, que les erreurs dues au remaniement consécutif à la mise en place de la sonde, sont minimes, sauf éventuellement dans les passées argileuses des zones à schistes altérés.

L'ensemble des valeurs p_{oh} calculées (fig.5.2) met en évidence l'importance des composantes horizontales des contraintes naturelles par rapport aux contraintes géostatiques horizontales dues au poids des couches érodées au cours du quaternaire et à fortiori par rapport aux contraintes géostatiques σ_{ov} correspondant à l'épaisseur actuelle des terrains.

Si l'on considère qu'à partir d'une certaine profondeur z_t (point T), les efforts tectoniques augmentent sensiblement les contraintes de confinement $\sigma_{1,3}$ qui accroissent à leur tour les contraintes déviatoriques nécessaires à l'obtention d'un même pourcentage de déformation, on peut admettre que la partie essentielle des déformations qui ont eu lieu sous l'effet de sollicitations de la fondation, se développe dans les couches supérieures du massif schisteux au-dessus de la profondeur seuil z_t.

En conclusion, on peut dire que les méthodes expérimentales et théoriques utilisées permettent d'évaluer les composantes horizontales des contraintes naturelles existant au sein du massif rocheux et de déterminer, dans la mesure où la connaissance géotechnique du massif est suffisamment étendue, l'influence des différents facteurs géologique intervenus dans la genèse de la roche pour en déduire son comportement mécanique.

Bibliographie

Baguelin, F., Jézéquel, J., Le Mée, E. et Le Méhauté, A. (1972) "Expansion des sondes cylindriques dans les sols cohérents", Bulletin de Liaison des Laboratoires des Ponts et Chaussées, n° 61, Paris

Baguelin, F. et Jézéquel, J. (1973), Discussion, Géotechnique, volume 23, n° 2, Londres

Clairbois, A.-M. (1959) "L'évolution de la Meuse entre Liège et Anseremme au cours du Quaternaire", Annales de la Société Géologique de Belgique, Tome LXXXII, n° 1, Bruxelles

Horvath, J. (1965) "A new approach to the determination of stresses in the earth's crust and strata pressure on tunnel linings", International Journal on Rock Mechanics and Mining Sciences, volume 2, pp. 327-340, Pergamon Press, Londres

Londe, P. (1973) "The role of rock mechanics in the reconnaissance of rock foundations", Quarterly Journal Engineering Geology, volume 6, pp. 57-74, Londres

Marsland, A. et Randolph, M.F. (1976) "Comparisons of the results from pressuremeter tests and large in situ test in London clay", Géotechnique, volume 27, n° 2, Londres

Palmer, A.C. (1972) "Undrained plane-strain expression of a cylindrical cavity in clay: a simple interpretation of the pressuremeter text", Géotechnique, volume 22, n° 3, Londres

Van Wambeke, A. et d'Henricourt, J. (1975), "Courbes pressiométriques inverses", Sols-Soils, tome VII, n° 25, Paris

Vutukuri, V.S., Lama, R.D. et Saluja, S.S. (1974), "Handbook on mechanical properties of rocks", volume I, § 4.5., Trans-tech Publications

GEOLOGY AND ROCK STRESSES IN DEEP BOREHOLES AT FORSMARK IN SWEDEN

Géologie et contraintes des roches dans les trous de forage en profondeur à Forsmark en Suède

Geologie und Gebirgsspannungen in tiefen Bohrungen in Forsmark in Schweden

J. Martna
Chief Engineering Geologist, Swedish State Power Board, S-162 87 Vällingby, Stockholm, Sweden

R. Hiltscher
Dr.-Ing., Swedish State Power Board, S-162 87 Vällingby, Stockholm, Sweden

K. Ingevald
Research Engineer, Swedish State Power Board, S-162 87 Vällingby, Stockholm, Sweden

SYNOPSIS

Rock stress measurements made in two research boreholes, 503 and 250 m deep respectively, are given together with a description of the geology and the tectonic features of the area. The boreholes are located in a Precambrian gneiss granite with dikes of dolerite and pegmatite. The principal stresses are almost horizontal and vertical. Down to 320 m the horizontal stresses remain more or less unchanged with average values of about 15-20 MPa. The vertical stress increases on the whole with the weight of the overburden. At a depth of 320 m a fractured zone of rock is penetrated. Below this zone the horizontal stresses increase abruptly to 65 MPa and possibly over this value. The results demonstrate that the initial rock stresses may change abruptly and considerably even in a seemingly homogeneous rock and need not necessarily follow any particular law. Therefore comprehensive local measurements are necessary in every case for the knowledge of the stresses in a sizable rock mass.

RESUME

Les mesures de contrainte des roches obtenues dans les trous de forage de recherche, respectivement à 503 et 250 m de profondeur, sont fournis en même temps qu'une description de la géologie et des caractéristiques tectoniques du terrain. Les trous de forage sont situés dans un granite gneissique pré-cambrien avec des digues de diabase et de pegmatite. Les contraintes principales sont pratiquement horizontales et verticales. A 320 m de profondeur les contraintes horizontales demeurent plus ou moins inchangées avec des valeurs moyennes de 15-20 MPa. La contrainte verticale augmente très approximativement avec le poids du terrain de couverture. A 320 m de profondeur une zone de roches fissurée est pénétrée. En dessous de cette zone les contraintes horizontales augmentent abruptement jusqu'à 65 MPa, et vraisemblablement au dessus de 65 MPa. Les résultats montrent que les contraintes initiales de la roche peuvent changer abruptement et considérablement même dans une roche en apparence homogène et ne pas nécessairement suivre une loi spéciale. Ainsi les mesures locales sont nécessaires dans chaque cas pour connaître les contraintes dans les grandes masses rocheuses.

ZUSAMMENFASSUNG

Es wird über Gebirgsspannungsmessungen in zwei 503 bzw. 250 m tiefen Forschungs-Bohrungen berichtet, zusammen mit einer Beschreibung der Geologie und der tektonischen Charakteristik des Gebiets. Die Bohrungen liegen in einem präkambrischen Gneis-Granit mit Gängen von Diabas und Pegmatit. Die drei Hauptnormalspannungen im Gebirge sind nahezu horizontal bzw. vertikal. Bis in eine Tiefe von 320 m halten sich die beiden horizontalen Hauptnormalspannungen mehr oder minder unverändert auf einem Mittelwert von etwa 15-20 MPa. Die vertikale Druckspannung stimmt im wesentlichen mit dem Gewicht des überlagerten Gebirges überein. In 320 m Tiefe findet sich eine Bruchzone. Unterhalb dieser Zone steigen die Spannungen sprunghaft auf 65 MPa und vermutlich sogar noch höhere Werte an. Die Ergebnisse zeigen, daß die initialen Gebirgsspannungen selbst im augenscheinlich homogenen Gebirge sprunghaften und beträchtlichen Änderungen unterworfen sein können und keinswegs einer bestimmten Gesetzmässigkeit zu folgen haben. Um die Gebirgsspannungsverhältnisse in einer größeren Gebirgsmasse zu erfassen, sind deshalb jeweils umfassende lokale Messungen nötig.

1. INTRODUCTION

If the stress situation in a rock mass is to be used in the lay-out and design of rock chambers, the rock stresses must be measured in good time before the start of the construction work, i.e. in connection with the geological survey of the site. For this purpose a special probe and procedure for triaxial rock stress measurements in deep boreholes has been developed by Hiltscher and collaborators (Hiltscher et al. 1979) at the Swedish State Power Board.

To test the probe, but also to obtain a better knowledge of the state of stresses in the upper part of the earth's crust, stress measurements were performed in two deep boreholes DBT-1(503 m) and DBT-3 (250 m) drilled for research and development of the methods of characterization of deepseated rock masses. Both holes are vertical, have a diameter of 76 mm and are situated about 120 m from each other.

For practical reasons these research boreholes were drilled at the building site of the Forsmark nuclear power plant. The site is situated on the east coast of Sweden about 130 km north of Stockholm. The geology of the area is quite well known owing to the extensive investigations before and during the conctruction of the three nuclear power units.

2. OUTLINE OF GEOLOGY

The Precambrian bedrock (age 1700 - 1800 M years) of the Forsmark area is chiefly composed of a gneiss granite. Also other rocks occur, such as mica gneiss, mica schist and diorite. The gneiss granite has a rather well developed foliation with a NW strike and a steep dip towards NE. Numerous dolerites or amphibolitic greenstones and pegmatites occur as dikes and small massifs constituting about 10 to 15 per cent of the rock mass. Within the site of the boreholes DBT-1 and DBT-3, a fine-to-medium-grained gneiss granite with relatively minor inclusions of dolerite and pegmatite is the main rock type.

The topography of the area is extraordinarily flat. The ground level lies between one and five metres above sea level and the rock surface a few metres lower. Also, the general slope of the surface is negligible and the area may be considered as being horizontal for the purpose of the present paper. The ground water level is at the surface or a few metres below it. The boreholes are thus waterfilled.

3. TECTONIC FEATURES

The folded Precambrian strata of the Forsmark area have subsequently been subject to tectonic fracturing. The area is situated between two major, or first order, fracture zones which are situated about 5 km apart. These zones have a NW strike, a steep dip and a width of some hundred metres. A large part of this width is sealed by mainly quartz and calcite mineralizations. The intervening rock mass is intersected by a number of second order steeply dipping fracture zones which have a generally N-S or E-W strike and a width of some tens of metres.

The rock mass is thus divided into irregular polygonal blocks, each with a surface of, say, from less than one to several square kilometres. These blocks are dissected by minor, up to a width of some metres, fracture zones and contain numerous mostly tectonic joints. Three vertical and one horizontal joint set have been described (Carlsson 1979). The jointing of the rock mass is illustrated by Fig. 1.

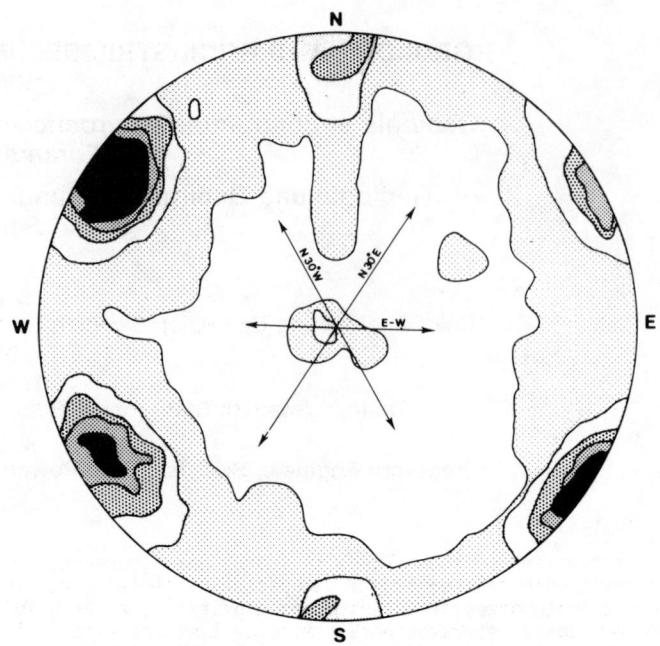

Fig.1. Fractures in the rock mass of Unit 3. Stereographic projection of the poles of fractures. Lower hemisphere, Schmidt's net. Contour density 3-6-9-12-15%. Arrows indicate main directions of sets of joints. From Carlsson 1979.

A remarkable feature of the area is the occurrence of wide, more or less horizontal joints, filled with Pleistocene sedimentary silt which is sometimes varved. These joints are common in the uppermost 5 m of the bedrock. Their maximum observed width is 820 mm (Carlsson 1979). No such silt-filled joints have been observed deeper than 13 m below the rock surface.

4. LITHOLOGY OF THE CORES

The cores are dominated by a grey foliated rock, traditionally called gneiss granite. It is fine-to-medium-grained and has a usually rather well defined foliation of dark and light minerals. The foliation has a steep NE dip, 70-90°, and strikes NW. This so-called gneiss granite has an approximate granodioritic composition, as shown in Table I.

Table I. Mineral composition of the gneiss granite. Median values of three samples. Carlsson and Olsson, 1982.

Mineral	%
Quartz	23
Plagioclase	30
Potassium feldspar	32
Biotite	11
Other minerals	4

A red variety of gneiss granite and rocks of leptite and aplite type also occur in the cores.

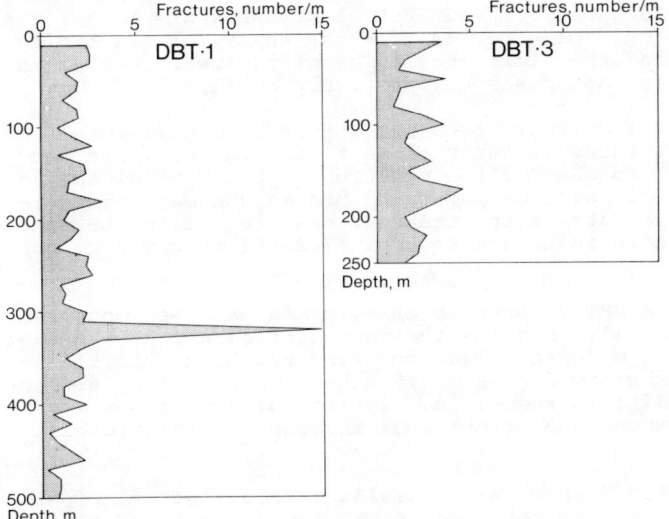

Fig. 2. Frequencies of coated and weathered fractures in the boreholes DBT-1 and DBT-3. From Carlsson and Olsson 1982.

They have foliations parallel to that of the grey gneiss granite.

Dikes and veins of dolerite or amphibolitic greenstone and pegmatite are common but constitute a minor part of the core. The contacts are, in general, parallel to the foliation of the gneiss granite.

The thickness of the dolerites varies from a few centimetres to about 10 metres. The dolerite is fine-to-medium-grained, slightly foliated and has a more or less preserved ophitic texture.

The thickness of the pegmatites varies from a few centimetres to a few metres. The grain size is about 5-50 mm and the crystals are often cracked. Mineralogically the pegmatites are dominated by feldspars.

No essential changes with depth can be discerned in the lithology of cores.

5. FRACTURE FREQUENCIES

In logging the cores, open fractures with fresh, clean and uneven surfaces were considered to be induced by drilling operations and discarded.

The fracture frequencies discussed below refer to open fractures with weathered or coated surfaces. Of these, all fractures with weathered surfaces are considered to be open in rock mass. About half of the fractures with coated surfaces are considered to be open in the rock. The remainder, here also defined as open, was initially probably at least partially sealed. Owing to their inherent weakness they have been broken up during the drilling.

The fractures appearing as sealed in the cores have mostly a strength comparable to that of the rock and they are not acting as fractures in the physical sense in the rock mass any more. They are not considered in the fracture frequencies.

In DBT-1, the number of sealed fractures (1046 fractures) is slightly higher than that of open fractures (989 fractures). There is no pronounced difference in fracturing between the gneiss granite and dolerite or pegmatite.

Borehole DBT-3 shows a higher fracture frequency than DBT-1 (Fig. 2). The mean fracture frequency in DBT-3 is 2.3 fractures per metre and that in DBT-1 down to 250 m level is 1.9 per metre. The value for the whole borehole DBT-1 is slightly higher, 2.0 fractures per metre, owing to the fractured zone at 320 m. Below 400 m depth, the average fracture frequency is very low, 1.0-1.1 per metre. Considering, on the other hand, only the uppermost 30 m of bedrock, the frequency is 2.5 fractures per metre in both boreholes.

An earlier study of horizontal fracturing in 10 boreholes in the area of Unit 3 has shown that for the levels 0-10 m, 10-20 m and 20-30 m the mean fracture frequencies for the respective depth levels were 3.7, 3.4 and 2.0 per metre (Carlsson 1979).

Thus, on the average, the frequency of fractures decreases rapidly from the surface to about 20 m depth and thereafter rather slowly. However, as is shown in Fig. 2, there is an obvious variation throughout the boreholes in this respect. The rock mass is divided into parts with varying fracture frequencies. It seems also as if the zones with higher fracture frequencies show less tendency of frequency decrease with depth than the zones with lower frequencies.

The fracture zone at 320 m depth in DBT-1, where the rock stresses (as shown in sections 6 and 7) abruptly increase, can be described as a zone of fractured red gneiss granite situated between 319.2 and 322.5 m. It contains two thin layers of crushed rock with a total width of 40 cm. The dip of this zone is uncertain but probably comparatively low.

6. ROCK STRESS MEASUREMENTS

The measurement of rock stresses was performed with the Hiltscher-probe of the Power Board using the technique described by Hiltscher et al.1979. Thus the reliability of the gauges and the linearity of the stress-strain relation has for all measurements been checked by calibration with the obtained tubular rock cores. The reliability of the measurements is therefore very good.

The error of measurement due to the instrumentation, creep of rock and temperature effects has been calculated to amount to maximum ± 2 MPa for a modulus of elasticity of 75 GPa, an average value for the present rock. To this error, the error caused by the variation of the modulus of elasticity and Poisson´s ratio have to be added, which is estimated to be maximally about ± 10 per cent of the respective stress value.

The stress measurements were performed at different measuring levels along the boreholes with a distance of about 30 to 40 m. At each of these levels one, two or three consecutive triaxial measurements (the number depending on various circumstances) were made with a short vertical distance between these measurement points.

Above the 320 m level, 92 per cent of the prepared measurements were successful. At 320 m, borehole DBT-1 penetrated a fracture zone. Below this zone the horizontal stresses became very high and instead of tubular rock cores 12-18 mm thick annulae were obtained making the measurements impossible. This phenomenon of discing owing to excessive rock pressure seems to occur when the largest principal stress exceeds some 65 MPa and constitutes the uppermost limit for the application of method in the present case. The compressive strength of the gneiss granite is on the average about 280 MPa. For rocks with other physical properties and also for other dimensions of cores, this limit of applicability will be different.

Owing to the successive failure of nine attempts, there are no stress values available from the nearest 50 m below the 320 m fracture zone. It is estimated that the maximum principal stress, σ_1, between 320 and 370 m depth is at least 65 MPa.

Between 370 and 500 m, 8 out of 11 attempts were successful, i.e. it was possible to measure the stresses at points which, by chance, were situated at levels where the varying stress field locally showed lower values. In general, the maximum stresses below 320 m are probably higher than 65 MPa.

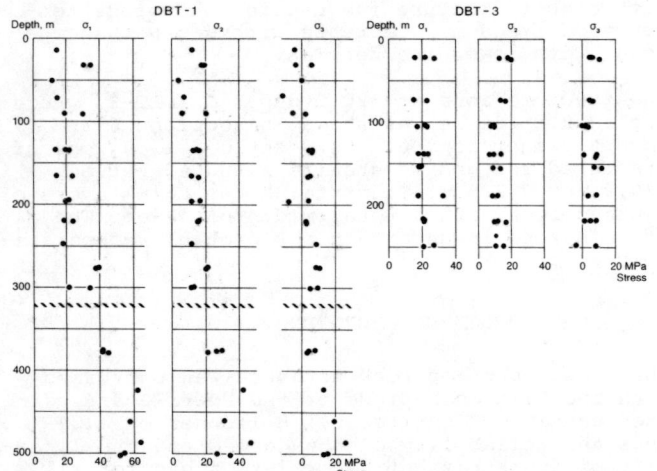

Fig.3. *All obtained individual values of principal stresses in DBT-1 and DBT-3. In DBT-1, the fractured zone of rock at 320 m depth is shown.*

7. DISCUSSION

The measurements show that the minimum principal stress, σ_3 is approximately vertical whilst σ_2 and σ_1 lie more or less in the horizontal plane. σ_3 corresponds on the average more or less to the load of the superimposed rock. There are, however, numerous local exceptions resulting in a notable scattering of values which to the main part are attributed to the influence of varying geological structures.

As usual both horizontal stresses are larger than could be expected from the theory of elasticity and the difference in magnitude between them is in most cases comparatively small.

The Forsmark area has rather high horizontal stresses already close to the surface. In the borehole D 358 a measurement 6.5 m below the rock surface shows 6.7 MPa as the major principal stress and the next measurement in the same hole shows a stress of 17.2 MPa at 8.6 m depth.

In DBT-1, a value of 14.0 MPa has been measured at 13.9 m below the rock surface and 35.5 MPa at 32 m depth. Rock bursting has been observed at shallow depths (5-15 m below surface) under conditions suggesting significant variations in the magnitude of the rock stresses at this level.

Fig.3 shows all individual values of principal stresses obtained in DBT-1 and DBT-3. The diagrams show a certain scattering at all levels of measurement. Considering the magnitude of the measurement errors involved it seems probable that this scattering essentially reflects the variation of stress magnitudes in the rock mass.

Fig.4 and 5 represent the three-dimensional stress tensors showing the magnitude of the principal stresses, their directions projected to the horizontal plane and their angles β_1, β_2 and β_3 downwards from the horizontal plane. These figures show also the vertical stresses and the maximum and minimum horizontal stresses.

In DBT-1, the variation in the magnitude of the stresses is considerable between 0 and 100 m depth. In the borehole DBT-3 this variation is not evident, possibly due to a different location of measuring points. As mentioned above (section 5), the joint frequency measurements indicate that the surface rock with an increased joint frequency is stretched to a depth of about 20 m. It could be possible that, in connection with stress release in the uppermost part of the bedrock, local increase of stresses may occur. An increased number of measurements is necessary to fully understand the stress distribution at these levels.

Between the depths of 100 and 300 m in DBT-1 (250 m in DBT-3), where the scattering of measurements within the levels is rather moderate, already the individual measurements are quite characteristic of the state of stresses in the rock mass and it is possible to calculate quite reliable mean values for the levels of measurement by combining two or more stress tensors to a new one. The stress diagrams in Figs. 4 and 5 are based on such mean values. The number of successful measurements (n) constituting each mean value is also indicated.

Between 50 and 250 m depth there is no appreciable increase in the magnitude of the two almost horizontal principal stresses. In both DBT-1 and DBT-3, σ_1 is about 20 MPa and σ_2 about 10-15 MPa. There follows (only in DBT-1 since DBT-3 ends at 250 m) an increase of σ_1 to about 40 MPa at 275 m, followed by a decrease. This might be a significant change in connection with the fractured zone of rock at 320m depth. However, although these latter changes in stress magnitude conside-

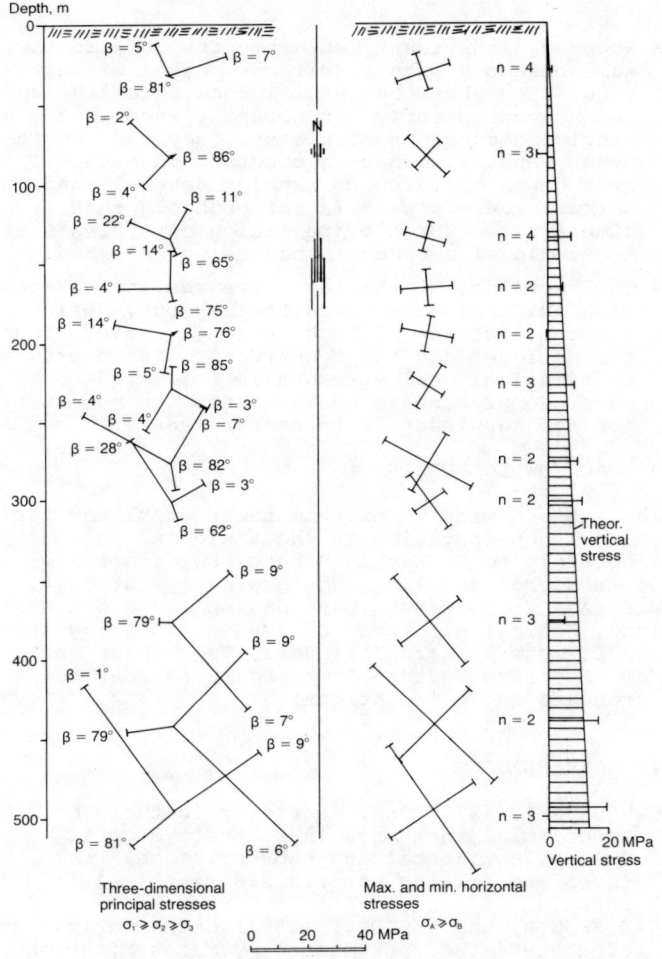

Fig.4. *Three-dimensional principal stresses, and horizontal and vertical stresses versus depth in DBT-1. The mean values of n measurements are shown. The three-dimensional stresses are drawn with their true bearing and according to scale magnitude. The angle β downwards from the horizontal plane is given for each stress.*

rably exceed the probable error of measurement, they are noted only in a few measurements and may therefore be considered as an indication rather than confirmation of such changes.

Perhaps the most interesting fact disclosed by the borehole DBT-1 is the abrupt change of the stress level when the borehole passes through a zone of fractured rock at 320 m depth without any corresponding changes in lithology or geology. The leap is evidently still larger than shown by the measurements, since the values from the successful measurements are lower than the average as the highest values could not be measured.

Thus the borehole DBT-1 shows the stress situation in two separate blocks, i.e. an upper block which is 320 m thick and a lower one of unknown thickness.

From many construction sites from all over the world a large number of observations of rock bursting and tunnel compression exists, implicating

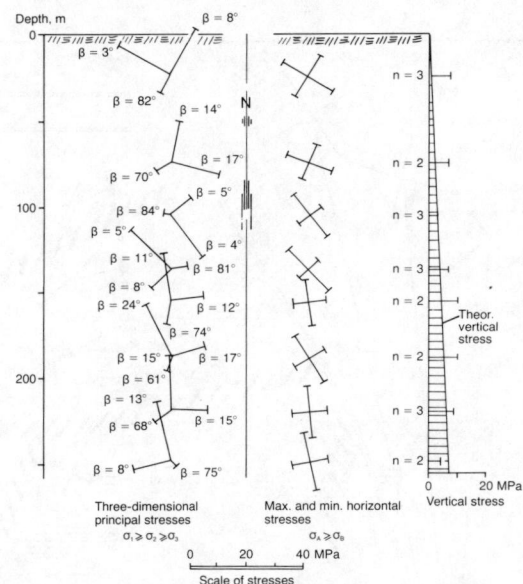

Fig.5. *Three-dimensional principal stresses, and horizontal and vertical stresses versus depth in DBT-3. The mean values of n measurements are shown. The three-dimensional stresses are drawn with their true bearing and according to scale magnitude. The angle β downwards from the horizontal plane is given for each stress.*

more or less sudden changes in stress conditions in limited rock masses. Some of these changes in stress effects are explained by changing external conditions such as depth below surface or proceeding excavations or lithological changes. However, in a number of cases, as for instance the present one, none of these explanations is applicable.

Phenomena similar to the present case, i.e. abrupt changes in the stress level at seemingly constant geological conditions have been observed in several tunnels in Sweden. Sometimes there is a connection with faults or fractured zones of rock and sometimes no such connection can be shown to exist. Large changes in the magnitude of the stresses under similar rock conditions and small differences in depth have been verified by measurements in the Suorva-Vietas headrace tunnel (Hiltscher 1972, Martna 1970, 1971).

These observations and measurements are an indication of the fact that the upper part of the earth's crust may be divided into blocks and zones with different, sometimes widely different stress conditions. Their boundaries can be quite sharp. Such blocks and zones would be parts of larger stress fields and there might be no local explanations to their existence. This applies not only to active mountain ranges and earthquake zones but seems to occur also in such tectonically quiet areas as the Fennoscandian shield.

In the Forsmark area in-situ measurements of rock stresses have been made in seven boreholes. In addition to the research boreholes DBT-1 and DBT-3, triaxial measurements have been made in the 32 m deep D 358.

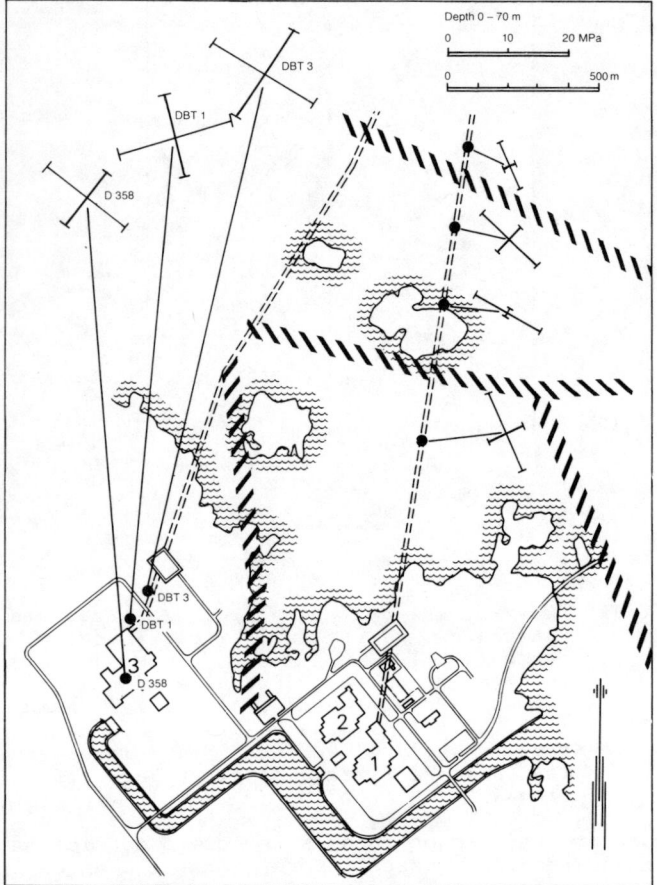

Fig.6. *Average horizontal stresses in the uppermost 70 m of bedrock and the main fracture zones in the Forsmark area.*

There are also four boreholes in the roof of the discharge tunnel No.1 for measurement of horizontal stresses with the "doorstopper"-method. These holes are situated about 50-60 m below the rock surface.

As an attempt to visualize the distribution of the horizontal stress conditions in the area, the situation of these boreholes and the magnitude and direction of the horizontal stresses down to 70 m depth below surface is shown in Fig.6 together with the main tectonic features of the area. It seems as if, although there exists a general NW-SE direction of the maximum horizontal stresses, each of the large tectonic bedrock blocks tends to have its own variant of stress magnitudes and directions.

8. CONCLUSIONS

The present study confirms similar opinions expressed earlier by the authors (Hiltscher et al, 1979, Scherman et al. 1980).

- As seems to be the usual case near the rock surface, the horizontal stresses in the Forsmark area are in all directions markedly higher than the values corresponding to the state of equilibrium according to the theory of elasticity. As a general tendency, the magnitude of these stresses increases with depth below surface.

- However, individual measurements, made in the same borehole with a reliable method of measuring, show that the initial rock stresses can change considerably and abruptly even in a seemingly homogeneous rock mass. They can, on the other hand, also have a constant magnitude over depth differences ranging several hundred metres. Therefore it is not probable that simple rules for the prediction of stresses at a certain site or depth can be established.

- The knowledge of the real stresses in the rock of a building site makes the lay-out, design and construction of rock openings easier. Thus the designer and the constructor are interested in existing local stresses in a limited rock mass. Comprehensive measurements are necessary for the knowledge of these stresses.

9. ACKNOWLEDGEMENTS

The authors wish to express their gratitude for valuable co-operation in the field and in the laboratory to Mr.Martin Moberg, Swedish State Power Board (drilling and logging operations), Miss Åsa Noro, Messrs.Lars Hansen and Björn Westlund, Geological Survey of Sweden (core-logging) and to Messrs. Lars Strindell, Tage Öhman and Mats Andersson, State Power Board (stress measurements and calculations).

10. REFERENCES

CARLSSON,A.(1979).Characteristic features of a superficial rock mass in southern central Sweden. Horizontal and sub-horizontal fractures and filling materials:Striae 11,1-79. Uppsala.

CARLSSON,A. and OLSSON,T.(1982).Characterization of deep-seated rock masses by means of borehole investigations. In situ rock stress measurements, hydraulic testing and core-logging: Swedish State Power Board, Research and Development Rep. $\underline{5:1}$, 1-155. Stockholm.

HILTSCHER,R.(1972).Anwendung der Gebirgsspannungsmessung bei der Schwedischen Staatlichen Kraftwerksverwaltung: Int.Symp. Underground Openings, 555-560.Lucerne.

HILTSCHER,R.,MARTNA,J.and STRINDELL,L.(1979).The measurement of triaxial rock stresses in deep boreholes and the use of rock stress measurements in the design and construction of rock openings: 4.Int.Cong.Rock Mech.(2),227-234 Montreux.

MARTNA,J.(1970).Rock bursting in the Suorva-Vietas headrace tunnel: 1.Int. Cong. Int.Ass. Engineering Geology, (2), 1134-1139. Paris.

MARTNA,J.(1971).Geological aspects on rock bursting in the Suorva-Vietas tunnel(In Swedish, English summary):Swed.Acad.Eng.Sci.Rep.$\underline{38}$., 141-151 Stockholm.

SCHERMAN,K-A.,HILTSCHER,R.and MARTNA,J.(1980).Experiences from rock stress measurements made by the Swedish State Power Board during the last fifteen years (In Swedish, English summary): Swedish Rock Eng.Res.Foundation (BeFo),Rock Mech. Meeting 1980,77-95. Stockholm.

DIFFERENTIAL STRAIN CURVE ANALYSIS – A NEW METHOD FOR DETERMINING THE PRE-EXISTING IN-SITU STRESS STATE FROM ROCK CORE MEASUREMENTS

Analyse de la courbe des déformations différentielles – Une nouvelle méthode de détermination des contraintes in-situ à partir de mesures sur carottes

Differentialanalyse von Dehnungskurven – Eine neue Methode zur Bestimmung des Spannungszustandes in situ durch Messungen an Gesteinskernen

N-K. Ren
J.-C. Roegiers
Dowell Division of Dow Chemical U.S.A., P.O. Box 2710, Tulsa, Oklahoma 74101

SYNOPSIS

Extensive laboratory and field investigations have led to validation of the Differential Strain Curve Analysis (DSCA) as a technique to determine the in situ stress field from measurements performed on available core. The method is based on the assumption that a rock specimen, retrieved from its downhole environment, will expand due to the generation of randomly oriented microcracks. Their density is proportional to the stress magnitude differential. Careful monitoring of a rock specimen's behaviour upon reloading reflects the past stress history. Statistical analyses compensate for the localized inherent inhomogeneities. DSCA results are compared with data generated by other techniques for several geological formations.

RESUME

Un grand nombre d'investigations en laboratoire et sur le terrain ont mené à la validation de l'Analyse de la Courbe des Déformations Différentielles en tant que technique pour déterminer le champ de contraintes in situ à partir de mesures effectuées sur carottes. La méthode est basée sur l'hypothèse qu'un échantillon de roche, détaché de l'environnement souterrain, se dilatera par microfissurations dont la densité est proportionelle à la différence de contrainte. Des mesures précises du comportement de cette roche lors de son rechargement permettront donc d'obtenir une idée de son chargement initial. L'introduction de méthodes statistiques permettront, en outre, d'écarter les problèmes des non-homogénéités locales. Des résultats obtenus par cette méthode ont été comparés avec les données obtenues par d'autres techniques. Plusieurs formations géologiques ont été considérées et l'Analyse de la Courbe des Déformations Différentielles semble donner de bons résultats.

ZUSAMMENFASSUNG

Ausgedehnte Labor- und Felduntersuchungen haben bestätigt, daß die Differentialanalyse von Dehnungskurven eine geeignete Methode ist, das Spannungsfeld vor Ort durch Messungen an vorhandenen Kernen zu bestimmen. Die Methode beruht auf der Annahme, daß eine Gesteinsprobe, wenn sie von den in der Bohrlochsohle herrschenden Bedingungen entfernt wird, sich durch zufällig orientierte Mikrorisse ausdehnt, deren Dichte proportional zu der Gröse der differentialen Spannung ist. Ein sorgfältiges Verfolgen des Verhaltens einer Gesteinsprobe, wenn die Spannung wieder erhöht wird, zeigt daher den Verlauf der Spannung in der Vergangenheit. Eine statistische Analyse ermittelt den Einfluß der unvermeidlichen lokalen Inhomogeneitäten. Ergebnisse der Differentialanalyse von Dehnungskurven werden für mehrere geologische Formationen verglichen mit Ergebnissen, die durch andere Methoden gewonnen wurden. Die Methode erscheint vielversprechend.

1. INTRODUCTION

Rock formations are frequently fractured to increase production in the oil and gas fields. The attitude of these induced hydraulic fractures is governed by the magnitude and orientation of the pre-existing stress field.

As sophisticated stimulation techniques become available, more "difficult" reservoir developments are contemplated. However, the constantly increasing well-completion costs make it increasingly vital to anticipate the geometry of these man-made features as their pattern and spacing influence the recovery efficiency. This is especially true for low-permeability formations in which the drainage area is of limited extent; hence, massive hydraulic fracture treatments are usually required. Similarly, an appreciation of the stress tensor is important in the case of solution mining and geothermal hot-dry rock energy extraction, as the man-made fracture constitutes a crucial link between the injection and withdrawal wells.

Although the pre-existing in-situ stress field is only one of the parameters controlling the geometry of such fractures, it generally plays the most important role. A similar argument may easily be developed when considering large underground openings for mining, civil or military purposes, as their relative orientation with respect to the stress tensor will affect the overall stability of the structure.

Several techniques to determine the in-situ stress field are presently available. Although involving a great variety of tools, they can be classified in four main categories:

(i)   techniques relying on complete strain relief;

(ii) techniques relying on partial strain relief;

(iii) techniques relying on rock flow or fractures; and

(iv) techniques based on correlations existing between rock properties and applied stresses.

The method described in this paper relies upon complete strain relief because it considers the strains a rock core undergoes after removal from its underground environment.

2. PRINCIPLE

As early as 1923, Adams and Williamson noted that rocks containing cracks are more compressible than the same rocks would be without cracks. Based on this basic observation, and assuming that the microcracks present in rock samples are related to the applied stresses, several authors predicted that the crack spectrum could be used to evaluate the past history of a particular formation (Simmons and Richter, 1976; Richter et al., 1976; Batzle and Simmons, 1976). Correlation of the time-dependent strain induced in rocks after removal from the ground and the pre-existing stress is not new (Emery, 1962; Voight and St. Pierre, 1974). Appreciation of these concepts led to initial attempts at stress prediction using Differential Strain Analysis resulting solely from core relaxation (Siegfried and Simmons, 1978; Simmons and Richter, 1976; Simmons et al., 1974; Simmons et al., 1978a; Strickland et al., 1979). However, the data generated from such strain relaxation analyses were far from convincing, even when stringent efforts were made to monitor strain immediately upon core recovery (Strickland, personal communication, 1980). The major drawback of such an approach was that strains had predominantly relaxed during coring and recovery. To overcome this difficulty, a modified testing philosophy was adopted.

When a core is removed from its underground environment, the drilling or sawing operation constitutes a stress relief mechanism. The physical detachment of the specimen from the rock mass allows it to undergo differential relaxation. In most rock types, the mechanism involved is the creation of randomly oriented microcracks (Strickland et al., 1979). If one assumes that the density of such microscopic discontinuities is allegedly proportional to the stress change, it is expected that the induced crack spectrum will reflect the stress history. However, stress-controlled crack generation is also accompanied by changes in rock properties as softening links are introduced, leading to anisotropic behavior. Upon reloading, these microcracks are extremely sensitive to even the lowest externally applied load. This is due to their low stiffness. Their progressive closure under increasing pressure will transform the rock specimen from a fissured to a semi-continuous state, hence modifying the stress-strain response. A relationship, such as the following (which tacitly assumes that the induced microcracks are noninteractive), can be developed to reflect sample behavior under the application of hydrostatic stresses (Walsh, 1965):

$$\Delta V = \frac{3(1-2\nu)}{E} p + \eta(p) \qquad (1)$$

where

ΔV = change in volume of rock specimen,

ν = Poisson's ratio,

E = Young's modulus,

p = hydrostatic, externally applied pressure, and

$\eta(p)$ = fracture porosity.

The first term in equation (1) corresponds to a linear elastic approximation, while the second term reflects the contribution of the fractures. Figure 1 schematically represents this relationship. The zero-pressure intercept, η_o, corresponds to the strain at zero pressure due to all the cracks that closed completely once the critical pressure, p_c, had been reached.

FIGURE 1

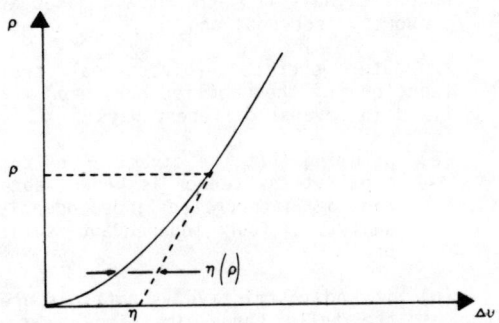

Figure 1. Schematic Pressure - Volumetric Strain Behavior.

Such concepts of crack porosity were further developed by Walsh (1965) who proposed an expression relating the number of cracks which will be closed at each pressure increment to their aspect ratio, α (i.e., crack width divided by crack length):

$$p = \frac{\pi E \alpha}{4(1-\nu^2)}, \qquad (2)$$

for the case of ellipsoidal cracks.

In 1971, Morlier attempted to quantify the crack aspect ratio distribution, predicting it is proportional to the second derivative of the strains. This implied that extremely sensitive measuring devices would be required to detect the appropriate strain levels. A further corollary is that specimen preparation must be conducted with extreme care to minimize the induced damage. Moreover, efforts should be made to avoid experimental errors due to secondary effects from pressure and temperature on the instrumentation used. For this reason, Feves et al. (1976) proposed the use of fused silica for correction purposes. The simultaneous pressurization of this reference (the fused silica standard) under similar conditions eliminates all experimental nonlinearities (Simmons et al., 1974, Simmons et al., 1978).

Consequently, if linear crack closure is assumed (verified by Siegfried and Simmons, 1978), the strain due to the presence of a crack can be obtained by subtracting the average matrix strain from the magnitude of the observed total strain. Specifically,

$$d\eta_{ij} = \delta\varepsilon_{ij} - \delta\varepsilon^m_{ij} \qquad (3)$$

or

$$\eta_{ij}(p) = \varepsilon_{ij}(p) + \eta_0(p_c) - \varepsilon^m_{ij}(p) \qquad (4)$$

$$= \varepsilon_{ij}(p) + \eta_0(p_c) - p\beta \qquad (5)$$

where

β = matrix compressibility,

$\eta_{ij}(p)$ = directional fracture porosity, and

$\varepsilon_{ij}(p)$ = strain tensor.

The foregoing equations are the basis of Differential Strain Curve Analysis (DSCA).

A similar approach is used to compute the principal stress ratios, i.e.,

2-D transverse isotropic situation (Perry and Lissner, 1979):

$$\frac{\sigma_{MAX}}{\sigma_{MIN}} = \frac{\varepsilon_{MAX} + \nu\varepsilon_{MIN}}{\varepsilon_{MIN} + \nu\varepsilon_{MAX}} \qquad (6)$$

where

ε_{MIN} = minimum, in-plane, principal strain,

ε_{MAX} = maximum, in-plane, principal strain,

σ_{MIN} = minimum, in-plane, principal stress, and

σ_{MAX} = maximum, in-plane, principal stress.

3-D, transversely isotropic situation (Duvall, 1965):

$$\sigma_1 : \sigma_2 : \sigma_3 = (C_{11}\varepsilon_1 + C_{12}\varepsilon_2 + C_{13}\varepsilon_3) :$$
$$(C_{12}\varepsilon_1 + C_{11}\varepsilon_2 + C_{13}\varepsilon_3) :$$
$$(C_{13}\varepsilon_1 + C_{23}\varepsilon_2 + C_{33}\varepsilon_3) \qquad (7)$$

where

$$C_{11} = \frac{E_1(1-\nu_{31}\nu_{13})}{(1+\nu_{12})(1-\nu_{12}-2\nu_{31}\nu_{13})},$$

$$C_{12} = \frac{E_1(\nu_{12}+\nu_{31}\nu_{13})}{(1+\nu_{12})(1-\nu_{12}-2\nu_{31}\nu_{13})},$$

$$C_{13} = \frac{E_3\nu_{13}}{1-\nu_{12}-2\nu_{31}\nu_{13}},$$

$$C_{33} = \frac{E_3(1-\nu_{12})}{1-\nu_{12}-2\nu_{31}\nu_{13}},$$

ε_i = principal strain acting in direction i,

σ_i = principal stress acting in direction i,

E_i = Young's modulus measured in direction i,

ν_{ij} = Poisson's ratio measured in direction i, due to a load acting along direction j, and

i,j = 1, 2, 3.

Results of three case histories will be discussed in a following section.

3. PROCEDURE

The standard procedure, which evolved through extensive experimentation, can be summarized as follows:

(i) a cubical sample is extracted from the center of the core, thus avoiding the zone of drilling damage;

(ii) the specimen is carefully hand-lapped to avoid further crack generations;

(iii) the specimen is instrumented with 12 strain gages, four on three orthogonal faces;

(iv) the specimen is vacuum-dried and then vented with nitrogen prior to encapsulation in clear, flexible and impermeable epoxy;

(v) after curing, the specimen along with the fused silica standard is subjected to increasing hydrostatic pressure during which output signals from the strain gages are continuously recorded; and

(vi) the data generated (directional strain as a function of the applied pressure) are analyzed in several different ways:

(a) assuming that the direction of the principal stress tensor is known, each plane can be interpreted independently (2-D analysis, four independent solutions); or

(b) if no restrictive assumptions are made, the fully three-dimensional case (3-D) leads to 64 combinations.

This statistical approach leads to the determination of a "confidence level" in the data generated.

4. CASE HISTORIES

To date, DSCA has been used to predict the stress field in nine different formations. These sites were selected since information from other techniques would be available, allowing the data to be compared and the value of DSCA assessed. Overall the results have been consistent and encouraging. Three of the cases will be discussed. (Information from the others remains proprietary.)

Table 1 summarizes the formations, locations, rock types and depth ranges for the three test cases.

TABLE 1. SUMMARY OF CASE HISTORIES

Case	Location	Formation	Rock Type	Depths
1	Grayson County, Texas	S - sand[1]	Vugular, fine-grained sandstone	3,770 to 5,880 ft
2	Cold Spring, Minnesota	Charcoal gray[2] granite	Fine-grained granite	Quarry-Maximum 50 ft
3	Garfield County, Colorado	Cozzette[3]	Porous, medium-grained sandstone	5,000 to 8,000 ft

---

[1] Extremely fine-grained sandstone with clay particles dispersed throughout.

[2] Composed of a variety of mineral types with major constituents of pyroxenes and amphiboles which attribute the dark gray color. The nonoriented mineral grains are angular to subangular in shape and are coarse to medium-fine in size.

[3] Dirty, fine-grained sandstone formed in a marine environment which accounts for the presence of intermixed gray-to-black limy and nonlimy shale particles.

4.1 Case History 1

Downhole seismic surveys, surface electrical potential system surveys, tiltmeter surveys, temperature surveys and focused gamma ray logs were run in an attempt to predict the azimuth of a hydraulic fracture created for production purposes. (The completion being inside casing, it was not possible to run impression packers.) Eight pieces of oriented cores were available for DSCA investigations. Table 2 summarizes all of the results.

The data interpretation from DSCA seems to show a good correlation with the other techniques used in this case history. The discrepancy in the tiltmeter results is mainly due to the erraticism in the acquired field data, leading to questionable interpretation. Furthermore, it should be emphasized that DSCA provides the full three-dimensional stress tensor.

TABLE 2. DATA SUMMARY

Technique	Predicted Stress Orientation and Ratios	Nomenclature
Downhole Seismic	σ_{MAX} : N 35° E ± 15°	σ_{MAX} = Maximum Principal Stress
Surface Electrical Potential System	σ_{MAX} : N 50° E ± 10°	σ_{MIN} = Minimum Principal Stress
		σ_{INT} = Intermediate Principal Stress
Focused Gamma Ray	σ_{MAX} : N 20° E ± 10°	σ_{HMAX} = Maximum Horizontal Stress Component
Tiltmeter	σ_{MAX} : N 117° E^(†)	σ_{HMIN} = Minimum Horizontal Stress Component
Microhydraulic Fracturing	$\dfrac{\sigma_{HMAX}}{\sigma_{HMIN}} = 1.5$	
Differential Strain Curve Analysis	2-D σ_{HMAX} : N 25° E ± 15° σ_{HMIN} : N 55° W ± 15° $\dfrac{\sigma_{HMAX}}{\sigma_{HMIN}} = 1.3 \pm 0.2$ 3-D σ_{HMAX} = Vertical ± 20° σ_{INT} = N 30° E ± 15° σ_{MIN} = N 60° W ± 15° $\sigma_1 : \sigma_2 : \sigma_3 = (1.8 \pm 0.3):(1.4 \pm 0.2):1.0$	

(†) The apparent discrepancy in the tiltmeter data interpretation led to further analysis of that data. At that time, the analysts indicated the possibility of a "secondary" fracturing system, suggesting a principal stress acting at N27°E.

4.2 Case History 2

In this particular case, oriented block samples were obtained from a quarry floor in northern Minnesota. The selection of this test case was made on the basis of reputedly high stress contrasts that were thought to be an ideal correlation for a well-defined prediction. A total of 19 samples was tested. Figure 2 represents a typical stress-strain plot of a particular plane. Table 3 is an example of the data interpretation for the same test and Figure 3 is the associated stereonet. Table 4 summarizes the comparison with other techniques (Crouch, 1967; Bonnechere, 1969; von Schoenfeldt, 1970).

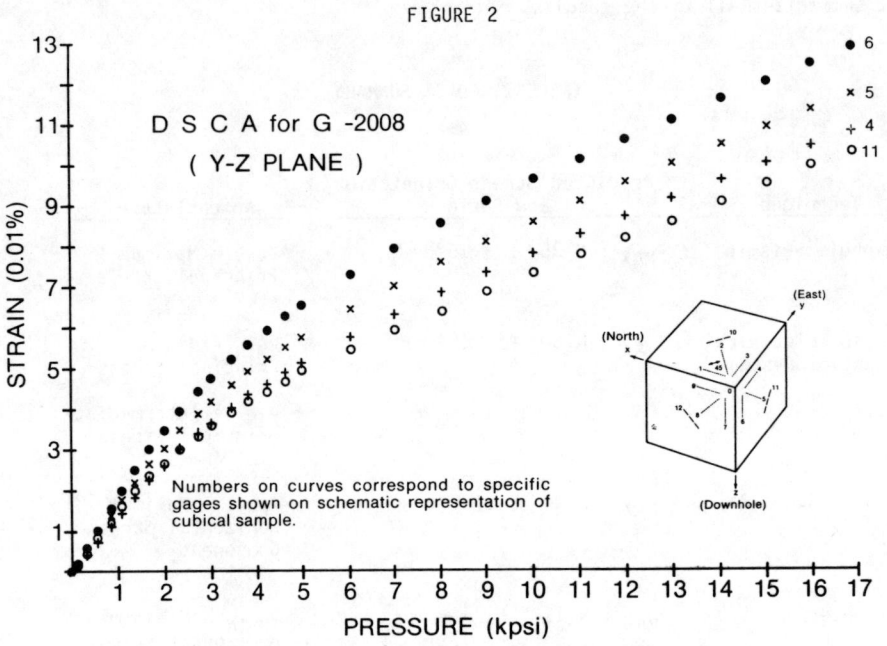

Figure 2. Typical Pressure/Strain Output.

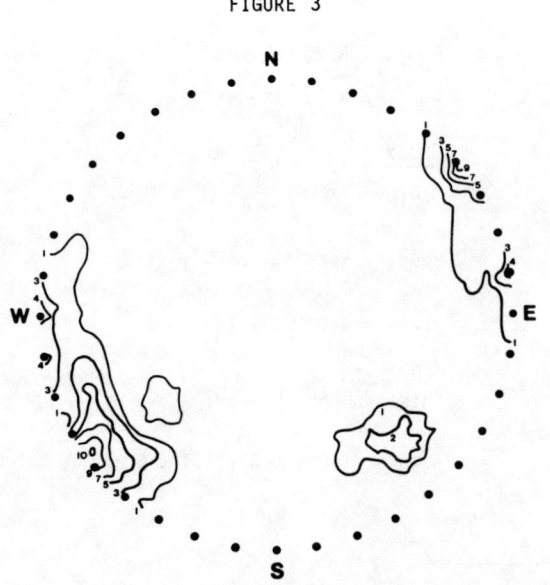

Figure 3. Stereonet of Stress Azimuth Determination.

TABLE 3. TWO-DIMENSIONAL AND THREE-DIMENSIONAL DIFFERENTIAL STRAIN CURVE ANALYSIS

SAMPLE NO. G 2008

TWO-DIMENSIONAL ANALYSIS

	X Y PLANE				X Z PLANE				YZ PLANE			
Gage	1	2	3	10	9	8	7	12	4	5	6	11
Strains	ε_X	ε_{XY}	ε_Y	ε_{XY}	ε_X	ε_{XZ}	ε_Z	ε_{XZ}	ε_Y	ε_{YZ}	ε_Z	ε_{YZ}
θ	0.1482	0.1202	0.1269	0.1412	0.1412	0.1532	0.2132	0.1524	0.1372	0.1634	0.1863	0.1515
β	0.0438	0.0409	0.0383	0.0437	0.0418	0.0335	0.0506	0.0410	0.0442	0.0448	0.0477	0.0432
$\varepsilon' = \theta-\beta$	0.1044	0.0793	0.0886	0.0975	0.0995	0.1197	0.1626	0.1114	0.0930	0.1186	0.1386	0.1083

Ratio	$-24° \pm 16°$ (**)	$-94° \pm 9°$	$84° \pm 2°$
	1.4 ± 0.2	1.8 ± 0.4	1.5 ± 0.1
Stress(*) Ratio	1.2 ± 0.2	---	---

(**) Maximum Horizontal Stress at N 24° W
(*) Poisson's Ratio = 0.18

THREE-DIMENSIONAL ANALYSIS

	ε_X	ε_Y	ε_Z	ε_{XY}	ε_{XZ}	ε_{YZ}
1	0.1044	0.0886	0.1626	0.0793	0.1197	0.1186
2	0.0995	0.0930	0.1386	0.0975	0.1114	0.1083

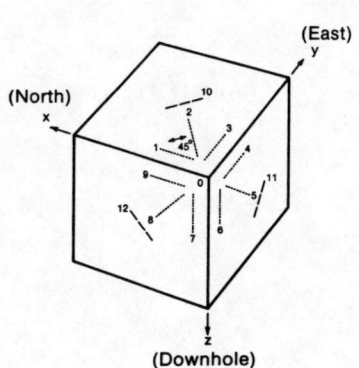

DATA INTERPRETATION

Strain Ratio $\varepsilon_1 : \varepsilon_2 : \varepsilon_3 = (1.9:1.3:1.0) \pm 0.2$

Stress Ratio $\sigma_1:\sigma_2:\sigma_3 = (1.7:1.2: 1.0) \pm 0.2$

Principal Stress Directions   Max.: N 40° W Horizontal
                                                    Int.: N 50° E Horizontal
                                                    Min.: Vertical

TABLE 4. DATA INTERPRETATION COMPARISON

Technique	Predicted Stress Orientations and Ratios
Microhydraulic Fracturing	σ_{MIN} : Vertical, 50 psi σ_{INT} : N 50° W, Horizontal, 810 psi σ_{MAX} : M 40° E, Horizontal, 2,390 psi $\dfrac{\sigma_{MAX}}{\sigma_{INT}} = 2.9$
Overcoring	σ_{MIN} : Vertical, 0 psi σ_{INT} : N 14° W, Horizontal, 2,200 psi σ_{MAX} : N 76° E, Horizontal, 4,100 psi $\dfrac{\sigma_{MAX}}{\sigma_{INT}} = 1.9$
Joint Survey	Direction of prominent fracturing: N 50°E
Differential Strain Curve Analysis	2-D σ_{HMAX} : N 32° W ± 10° σ_{HMIN} : N 58° E ± 10° $\dfrac{\sigma_{HMAX}}{\sigma_{HMIN}} = 1.2 \pm 0.1$ 3-D σ_{MIN} : Horizontal, N 38° E ± 5° σ_{INT} : Horizontal, N 45° W ± 12° σ_{MAX} : Vertical ± 5° Stress Ratios: 1.9:1.2:1.0
Velocity Profiling	2-D $V_{c,max}$: N 44° E ± 7° $\dfrac{V_{c,max}}{V_{c,min}} = 1.1 \pm 0.1$ 3-D $V_{c,max}$: N 40° E ± 10° $V_{c,int}$: N 50° W ± 10° $V_{c,min}$: Vertical Ratios = (1.1 ± 0.01):(1.06 ± 0.01):1.0

It should be pointed out that the overcoring was performed very close to the surface of the quarry floor. This could easily explain the high stress magnitudes, when considering the additional thermal stress regime induced during the summer months when the testing was performed. These thermal stresses rapidly disappear with depth and probably were not a factor in the hydraulic fracturing measurements.

As can be seen from the previous table, the DSCA data interpretation led to a 90 degree shift in the stress tensor as compared to the other techniques. The authors credit this discrepancy to the fact that a dominant fracture spectrum was already present in the formation, caused by past high loading history. This inherent microfracturing spectrum overshadows the fracturing pattern induced upon unloading. Such a phase shift was also discovered by others (Engleder and Plumb, 1981) when attempting to correlate ultrasonic wave velocity anisotropies with a pre-existing stress field. Although the maximum stress direction always corresponded to the minimum compressional wave velocity, the opposite prevailed in granites.

Three-dimensional velocity profiles were determined on the rock specimens prior to encapsulation, using the same statistical approaches (refer to Table 3). In order to correct for anisotropy, it is recommended that velocities be measured under similar progressive loading conditions. The authors are further investigating whether a 90 degree phase shift is characteristic or coincidental for granitic rocks.

4.3 Case History 3

This case is part of the Multiwell Experiment sponsored by the U.S. Department of Energy. This formation was selected because oriented core over a large depth interval was available. In addition, a regional study by the U.S. Geological Survey had shown consistent stress orientation in this area. Table 5 summarizes the results obtained using various techniques.

TABLE 5. DATA INTERPRETATION COMPARISON

Technique	Predicted Stress Orientations and Ratio
Microhydraulic Fracturing[†]	σ_{MAX} : Horizontal, N 70° W ± 5°
Surface Fracture Mapping	σ_{MAX} : Horizontal, N 79° W ± 10°
Overcoring (surface outcrops)	σ_{MAX} : Horizontal, N 90° W ± 15°
Viscoelastic Relaxation	σ_{MAX} : N 81° W
Vertical Fractures in Cores	σ_{MAX} : N 75° W
Differential Strain Curve Analysis	2-D σ_{HMAX} : N 75° W ± 5°
	$\dfrac{\sigma_{HMAX}}{\sigma_{HMIN}} = 1.4 \pm 0.1$
	3-D σ_{MAX} : Horizontal, N 80° W ± 10°
	σ_{INT} : Vertical
	σ_{MIN} : Horizontal, N 10° E ± 10°
	Stress ratios (1.4 ± 0.2):(1.2 ± 0.1):1.0
Velocity Profiling	3-D Ratios = (1.13 ± 0.01):(1.04 ± 0.01):1.0
	$V_{c,MIN}$ = N 72° W ± 10°
	$V_{c,INT}$ = Vertical
	$V_{c,MAX}$ = N 15° E ± 10°

[†] At the time of writing, the pressure data had not been analyzed; hence, the stress magnitudes were unknown.

Again, the DSCA gave results which are in good agreement with other predictions. The three-dimensional analysis showed that the vertical component was the intermediate stress, a situation which had been found by others (Haimson, 1973; Bredehoeft et al., 1976). Also, the magnitudes which can easily be obtained by assuming the overburden to be 1.1 psi/ft seem to be reasonable when compared with data generated by others (Lindner et al., 1978; Zoback et al., 1980).

5. DISCUSSION AND CONCLUSIONS

The three different geological formations which are described provided additional data to assess the Differential Strain Curve Analysis as a technique to determine the pre-existing stress field. Although the method gives reasonable results in the case of sedimentary formations, it has not been validated in the following cases:

(i) where the rock has been subjected, in its past geological history, to very high stress magnitudes which have induced a predominant microcrack spectrum (and, therefore, the fracture pattern induced by the relaxation becomes secondary); and

(ii) where the rock, due to large cohesion, has difficulty relaxing and the differential strains induced by relaxation result in additional microstresses locked in the various constituent grains.

Further work is in progress in an effort to separate these effects.

However, particularly for sedimentary sequences, the DSCA procedure has been well developed and is a cost-effective measurement technique allowing stress prediction a priori. Another advantage is that no assumptions are needed regarding the relative orientation of the borehole with respect to the in-situ stress tensor. Finally, the introduction of statistical approaches has increased the level of confidence by eliminating the inherent local stress inhomogeneities.

It should, however, be emphasized that the quality and reliability of the data generated by the DSCA procedure are extremely dependent upon experience.

6. REFERENCES

Adams, L. H., and E. D. Williamson (1923). "The Compressibility of Minerals and Rocks at High Pressure," J. Franklin Inst. 195, 475-529.

Batzle, M. L., and G. Simmons (1976). "Microfractures in Rocks from Two Geothermal Areas," Earth Planet. Sci. Lett. 30, 71-93.

Bonnechère, F. (1969). "A Comparative Field Study of Rock Stress Determination Techniques" (Technical Report 1-69) U.S. Army Corps of Engineers, Missouri River Division, Omaha, Nebraska.

Bredehoeft, J. D., R. G. Wolff, W. S. Keys, and E. Shuter (1976). "Hydraulic Fracturing to Determine the Regional In-situ Stress Field, Piceance Basin, Colorado," Geol. Soc. Amer. Bull. 87, 250-258.

Crouch, S. L. (1967). "Development of an Instrument for the Experimental Determination of Stresses in Surface Rock" (Masters Thesis), University of Minnesota.

Duvall, W. I. (1965). "The Effect of Anisotropy on the Determination of Dynamic Elastic Constants of Rock," Soc. Min. Eng. Bull. 309-315.

Emery, C. L. (1962). "The Measurement of Strains in Mine Rock," Proc. Int. Symp. on Mining Research, Pergamon Press, Vol. 2, pp. 541-554.

Engelder, T., and R. Plumb (1981). "The Relation Between In-situ Ultrasonic Properties of Rock and In-situ Stress," Proc. USGS Workshop on Hydraulic Fracturing Stress Measurement, Monterey, California (to be published).

Feves, M., and G. Simmons (1976b). "Effects of Stress on Cracks in Westerly Granite," Bull. Seism. Soc. Am. 66, 1755-1765.

Feves, M., G. Simmons, and R. W. Siegfried (1976). "Microcracks in Crustal Igneous Rocks: Physical Properties," The Earth's Crust (J. G. Heacock, ed.) (Geoph. Monograph No. 20), Am. Geoph. Union, pp. 95-117.

Haimson, B. C. (1973). "Earthquake Related Stresses at Rangely, Colorado," New Horizons in Rock Mechanics (Proceedings, 14th Symposium on Rock Mechanics), American Society of Civil Engineers, New York, pp. 689-708.

Lindner, E. N., and J. A. Halpern (1978). "In-Situ Stress in North, A Compilation," Int. J. Rock Mech. Min. Sci. Geomech. Abstr. 15, 183-203.

Montgomery, C. T., and N-K. Ren (1981). "Differential Strain Curve Analysis -- Does It Work?" Proc. USGS Workshop on Hydraulic Fracturing Stress Measurement, Monterey, CA (to be published).

Morlier, P. (1971). "Description de l'état de Fissuration d'une Roche à Partir d'essais Non-Destructifs Simples," Rock Mech. 3, 135-138.

O'Connell, R. J., and B. Budiansky (1974). "Seismic Velocities in Dry and Saturated Cracked Solids," J. Geoph. Res. 79, 5412-5426.

Perry, C. C., and H. R. Lissner (1979). The Strain Gage Primer, McGraw-Hill, New York.

Richter, D., G. Simmons, and R. Siegfried (1976). "Microcracks, Micropores, and Their Petrologic Interpretation for 72415 and 15418," Proc. 7th Lunar Sci. Conf., pp. 1901-1923.

Siegfried, R. W. (1977). "Differential Strain Analysis: Application to Shock Induced Microfractures" (Doctorate Dissertation), Massachusetts Institute of Technology.

Siegfried, R. W., and G. Simmons (1978). "Characterization of Oriented Cracks with Differential Strain Analysis," J. Geoph. Res. 83(B3), 1269-1278.

Simmons, G., R. W. Siegfried, and M. L. Feves (1974). "Differential Strain Analysis: A New Method for Examining Cracks in Rocks," J. Geoph. Res. 79(29), 4383-4385.

Simmons, G., and D. Richter, (1976). "Microcracks in Rocks," The Physics and Chemistry of Minerals and Rocks (R. G. J. Strens, ed.), Interscience, New York, pp. 105-137.

Simmons, G., M. Batzle, H. Cooper, R. Siegfried, and M. L. Feves (1978). "Characterization of Microcracks," Proc. Conf. on Mechanics of Deformation and Faulting, Lulea, Sweden.

Simmons, G., and H. Cooper (1978). "Thermal Cycling Cracks in Three Igneous Rocks," Int. J. Rock Mech. Sci. Geomech. Abstr. 15, 145-148.

Smith, M. B. (1979). "Effect of Fracture Azimuth on Production with Application to the Wattenberg Gas Field" (Preprint Paper 8298), Society of Petroleum Engineers, Dallas, Texas.

Strickland, F. G., M. L. Feves, and D. Sorrells, (1979). "Microstructural Damage in Cotton Valley Formation Cores" (Preprint Paper 8303), Society of Petroleum Engineers, Dallas, Texas.

Strickland, F. G., and N-K. Ren (1980). "Use of Differential Strain Curve Analysis in Predicting In-Situ Stress State for Deep Wells," Proc. 21st U.S. Rock Mechanics Symposium, Rolla, Missouri.

Voight, B. (1960). "Determination of the Virgin State of Stress in the Vicinity of a Borehole from Measurements of a Partial Anelastic Strain Tensor in Drill Holes," Felsmechanik und Ingenieurgeologie 6, 201-215.

Voight, B., and B. H. P. St. Pierre (1974). "Stress History and Rock Stress," Proc. 3rd Cong. ISRM, Denver, Vol. 2A, pp. 580-582.

von Schoenfeldt, H. (1970). "An Experimental Study of Open-Hole Hydraulic Fracturing as a Stress Management Method with Particular Emphasis on Field Tests" (Technical Report MRD 3-70) U.S. Army Corps of Engineers, Missouri River Division, Omaha, Nebraska, 204 pp.

Walsh, J. B. (1965). "The Effects of Cracks on the Compressibility of Rock," J. Geoph. Res. 70, 381-389.

Zoback, M. L., and M. Zoback (1980). "State of Stress in the Conterminous United States in the Magnitude of Deviatoric Stresses in the Earth's Crust," Special Issue of J. Geoph. Res. (T. Hanks and B. Raleigh, eds.).

ROCK STRESS MEASUREMENT BY SLEEVE FRACTURING

Mesure par la technique du manchon fissurant des contraintes subies par une roche

Gebirgsdrucksmessung durch Manschetten-Bruchbelastung

O. Stephansson
Professor in Rock Mechanics, University of Luleå, Luleå, Sweden

SYNOPSIS

By the sleeve fracturing technique rock deformability and rock stresses are determined in one and the same borehole test. Axial borehole fractures are induced without fluid interaction with existing fractures and joints. Sleeve fracturing of blocks and thick wall cylinders are described. Stress determination by means of sleeve fracturing in the Pomona Basalt of the Hanford Test Site are found to be of the same magnitude and direction as stresses determined by by hydrofracturing and overcoring techniques.

RESUME

La technique du manchon fissurant permet de déterminer au niveau d'un seul et même trou de forage la tendance à la déformation et les contraintes subies par la roche. Des fissures axiales du trou de forage sont provoquées sans interaction par voie fluide avec des fissures ou jointures déjà existantes. La fissuration de blocs et de cylindres à parois épaisses au moyen de manchons est décrite. Les mesures de contraintes par l'intermédiaire de manchons fissurants dans le cas du basalte de Pomona sur le site expérimental de Ranford se sont révélées être du même ordre de grandeur et aller dans le même sens que les contraintes déterminées par les techniques d'hydrofissuration et de sur-carottage.

ZUSAMMENFASSUNG

Mit Hilfe der Manschetten-Bruchbelastung können das Gebirgsverhalten und der Gebirgsdruck durch Prüfung in einem einzigen Bohrloch ermittelt werden. Axiale Bohrlochbrüche werden hervorgerufen, ohne daß eine Flüssigkeit mit vorhandenen Brüchen und Rissen zusammenwirkt. Manschetten-Bruchbelastung von Blöcken und dickwandigen Zylindern wird beschrieben. Es zeigt sich, daß der mit Hilfe von Manschetten-Bruchbelastung festgestellte Gebirgsdruck von Pomona Basalt in der Versuchsstrecke von Hanford die gleiche Größe und Richtung hat wie die durch hydraulische Bruchbelastung und Untersuchung von Bohrkernen ermittelten Gebirgsdrücke.

1. INTRODUCTION

Hubbert and Willis, 1957, developed the now classical equation

$$P_b = 3\sigma_2 - \sigma_1 + T - P_p \qquad (1)$$

relating the breakdown pressure of hydraulic fracture formation, P_b to the maximum and minimum principal stresses σ_1 and σ_2 in a plane perpendicular to the axis of the borehole. T is the tensile strength and P_p is the pore pressure in the rock mass. Hydraulic fracturing is now widely used for in situ rock stress measurement in a number of countries; Haimson, 1974, Zoback and Zoback, 1980 and Enever and Wooltorton, 1981.

Stress measurements by hydraulic fracturing technique are usually performed in sections of the borehole which seem to be free of joints or have a low frequency of joints and fractures. Hydraulic fracturing in the laboratory of cylindrical samples of jointed igneous rocks loaded under tri-axial state of stress and fractured by pressurization water in a central borehole of the cylinder bring out some of the problems related to hydrofracturing of jointed rocks; Stephansson, 1982. Out of twenty samples tested only six show the expected hydraulic fracture type. The penetration of water and the build up of pore pressure in the vicinity of the borehole was found to have a strong effect on the break down pressure.

This paper describes the development of a technique called *sleeve fracturing* which provides a method of inducing axial borehole fractures at any depth without introducing fluids into the fracturing process. The sleeve breakdown pressure would even be more representative of stresses than hydraulic fracture breakdown pressure in the same rock as the sleeve prevents fluid interactions with flaws, microfractures and joints in the borehole. Another difference is the distance of fracture propagation. Hydraulically induced fractures can propagate long distances from the borehole; whereas sleeve induced fractures are limited by a couple of borehole radii from the borehole wall.

2. DESIGN OF THE SLEEVE FRACTURING SYSTEM

The pressuremeter is probably the most versatile in-situ testing device available at present for investigating static and cyclic strength and deformation properties of soils, Baguelin et al., 1978. The application of pressuremeter technique to rocks is less developed.

Panek et al., 1964, developed a borehole cell called the Cylindrical Pressure Cell (CPC) for determining the modulus of rigidity of rock. The CPC system had some problems, so later Hustrulid and Hustrulid, 1975, improved the system, the calibration and the data reduction procedure and called the new system the Colorado School of Mines (CSM) cell. Ozdemir and Wang, 1981, further improved the CSM cell by application of strain gauges to the sleeve. For testing in-situ modulus of rigidity in NX holes of the Experimental mine of CSM, El Rabaa, 1981, increased the fluid capacity of the system by adding another pressure generator. Ladanyi and Gill, 1981, determined creep parameters of rock salt by means of the CSM cell.

The sleeve fracturing system used in this study consists of the following items, see Fig. 1.

1. High pressure generator with vernier indicator rated at a pressure capacity of 70 MPa and a fluid capacity of 30 cm^3.
2. High pressure generator with vernier indicator rated at a pressure capacity of 35 MPa and a fluid capacity of 60 cm^3.
3. Three-way high pressure valve.
4. CSM borehole cell.
5. Pressure gauge rated at 0 to 140 MPa.
6. High line differential pressure transducer of diaphragm type ranges from 35 kPa to 85 MPa and a readout unit.
7. Linear displacement transducer, resolution 0.05-0.08 mm.
8. X-Y-recorder.

The heart of the sleeve fracturing system is the borehole cell. The CSM cell that has been used in this study has a diameter that fits an EX borehole with diameter 38 mm. It consists of three parts:

(i) a membrane made of Adiprene (L-100, L-167)

(ii) a central steel mandril with an end cup

(iii) a removable steel cup

The membrane has a self-sealing construction so that when the pressurizing fluid enters the cavity between the mandril and the inner membrane wall; the pressure seals the flange of the membrane against the mandril and the ends against the steel end cups, Hustrulid and Hustrulid, 1975.

Once the system has been assembled and filled with fluid it is bled to get rid of the air,

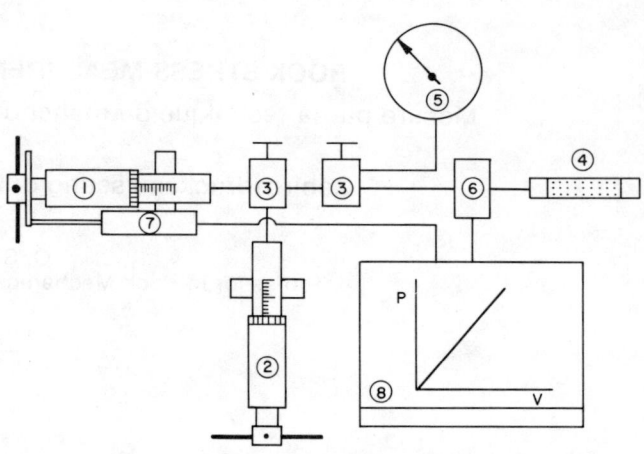

Fig. 1 System for sleeve fracturing.

otherwise the pressure-volume curve will be nonlinear. In this study a minimum of tubing has been applied and water was used as the pressurizing fluid in order to obtain a stiff system.

3. DETERMINATION OF ROCK MODULUS AND ROCK STRESS FROM SLEEVE FRACTURING

The sleeve fracturing system offers the possibility to determine two important rock parameters in one and the same borehole test. The system is first pressurized to a level where no fractures are initiated and the modulus of rigidity of the rock mass is determined. If the Poisson's ratio (ν) of the rock is known or can be estimated, the elastic modulus (E) can also be found. By increasing the pressure further a borehole fracture is induced and a sleeve breakdown pressure is recorded. The direction of the fracture indicates the direction of maximum stress in the plane perpendicular to the axis of the borehole. By recording the pressure for re-opening the fracture during a second pressurization and knowing the tensile strength of the rock mass the magnitude of principal stresses can be determined from the classical equation for hydraulic fracturing, Equation (1).

3.1 Determination of rock modulus

The complete mathematical analysis of rock modulus determination according to the CSM cell system is presented by Hustrulid and Hustrulid, 1975. A summary of the major steps in rock modulus determination is presented here.

The first step in the procedure of using the sleeve for rock modulus determination is to determine the stiffness of the entire system (pressure generators, fluid, valves, membrane and pressure gauges). Stiffness calibration is accomplished by pressurizing the cell in a metal cylinder of known geometry and elastic properties and recording the slope of the pressure-volume curve, M_m. The slope of the curve (M_m) includes the stiffness contribution from the calibration cylinder, M_c, which is calculated plus the stiffness from the complete pressurization system, M_s. The stiffness M_s

can be calculated from

$$M_s = \frac{M_c M_m}{M_c - M_m} \quad (2)$$

where

M_c = calculated stiffness of the calibration cylinder (MPa/turn)

M_m = measured stiffness of the pressurizing system plus the calibration cylinder (MPa/turn)

M_s = stiffness for the pressurization system alone (MPa/turn)

After calibration of the sleeve it is inserted into a borehole or rock cylinder of unknown elastic properties. The slope of the curve for the pressurizing system and the rock mass is determined, M_T, and the pressure-volume relationship for the rock mass alone is then calculated from

$$M_R = \frac{M_s M_T}{M_s - M_T} \quad (3)$$

where

M_T = stiffness for the pressurization system plus the unknown rock mass (MPa/turn)

M_s = stiffness for the pressurization system (MPa/turn)

M_R = stiffness for the rock mass (MPa/turn)

For testing of a borehole in a rock mass with an infinite outer radius the modulus of rigidity, G_R, can be calculated from

$$G_R = \frac{M_R \pi L r_i^2}{\gamma} \quad (4)$$

where L is the length of the sleeve, r_i radius of the borehole and γ fluid volume ejected per turn of the pressure generator.

If the value of Poisson's ratio of the rock mass, ν_R is known then the modulus of elasticity, E can be calculated from

$$E = 2(1 + \nu_R)G_R \quad (5)$$

3.2 Determination of rock stresses

In principal, fractures produced by the sleeve are formed by similar boundary conditions as fractures produced by conventional hydraulic fracturing. Let us consider a rock mass subjected to a uniform stress system $\sigma_1 > \sigma_2$ in the far field and the borehole surface subjected to a uniform pressure, P_O from the sleeve. Prior to fracturing the circumferential stress at the borehole is

$$\sigma_{\theta\theta}(\theta) = -P_O + \sigma_1 + \sigma_2 - 2(\sigma_1 - \sigma_2)\cos 2\theta \quad (6)$$

As the sleeve pressure increases the tangential stress at the borehole wall will decrease until it becomes equal to the tensile strength of the rock, T. At the critical pressure P_c a fracture will form at an angle $\theta = 0°$ and Equation (6) results in

$$P_c = 3\sigma_2 - \sigma_1 + T \quad (7)$$

This equation is valid for the assumption that the critical pressure P_c is equal to the contact stress, σ_{rr}, at the interface between the borehole wall and the sleeve. However, the contact stress is lower than the internal pressure of the sleeve and needs to be calculated for the particular geometry and material properties of the sleeve. As the fracture has been formed at a particular critical pressure the deviatoric stress can be determined from Equation (7) for known values of tensile strength.

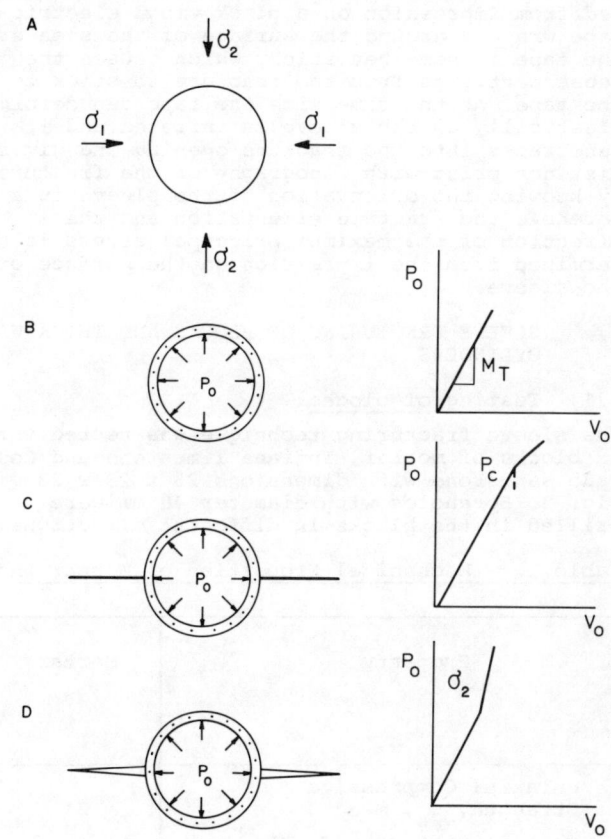

Fig. 2 Principals of sleeve fracturing of boreholes. A, virgin state of stress. B, pressurization and determination of stiffness M_T for the system and rock mass. C, sleeve fracturing and determination of breakdown pressure P_c. D, repressurization and determination of least principal stress σ_2.

The direction of the maximum principal stress is also known from the orientation of the axial fracture plane in the borehole. In hydraulic fracturing the magnitude of the least principal stress is determined from the so-called shut-in pressure, i.e. the fluid pressure that is needed to open the fracture and let the fluid flow into the fracture. In sleeve fracturing, fracture opening of the borehole is determined from the inflexion point of the pressure-volume curve. This is based on the assumption that the rate of volume change for the pressurized sleeve is different prior and after opening of the fracture.

The principal of sleeve fracturing of a borehole and determination of stiffness for calculation of rock mass deformability and determination of critical pressure, P_c and least principal stress, σ_2 are shown in Fig. 2. Inserting the value of σ_2 and P_c into Equation (7) gives the maximum principal stress, σ_1 for known values of tensile strength.

Fracture orientation in the borehole is determined from impression on a black vinyl electrical tape wrapped around the surface of the sleeve. The tape is somewhat sticky which causes the loose particles from the fracture to stick to the tape. At the same time the tape can deform plastically as the sleeve is inflated and it penetrates into the fracture opening and gives a distinct print with topography of the fracture. By knowing the orientation of the sleeve in a borehole the fracture orientation and the direction of the maximum principal stress is determined from the impression on the surface of the sleeve.

4. SLEEVE FRACTURING OF BLOCKS AND THICK WALL CYLINDERS

4.1 Testing of blocks

The sleeve fracturing technique was tested first in blocks of mortar, Indiana limestone and Colorado sandstone with dimensions 28 x 28 x 23 cm, Fig. 3. Boreholes with diameter 40 mm were drilled in the blocks in different directions.

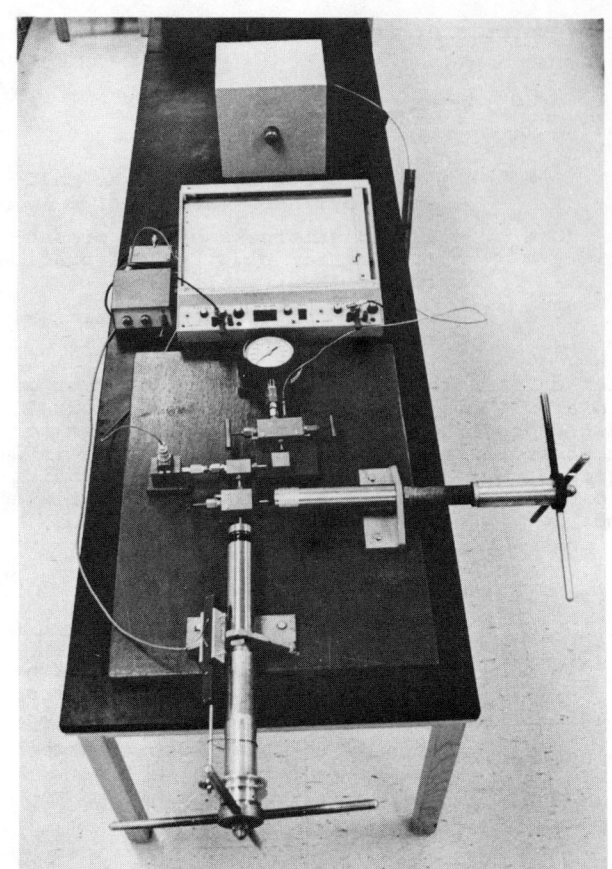

Fig. 3 Sleeve fracturing of blocks.

Table I. Mechanical Properties of Mortar and Rocks

Property	Mortar	Indiana* Limestone	Colorado* Sandstone	
			Parallel Bedding	Perpendicular Bedding
Uniaxial Compressive Strength, σ_c, MPa	21	34.33	156.3	135.8
Tensile Strength, T, MPa		2.76	8.41	3.10
Cohesion, from Triaxial Tests, C, MPa		6.82	22.61	
Internal Friction, $\phi°$		33.9	49.8	
Shear Strength, τ, MPa		9.69	25.86	31.02
Shear Strength $\tau=K(\sigma+T)^m$, MPa		0.63 $\tau=59.5(\sigma+2.8)$	0.74 $\tau=16.6(\sigma+8.4)$	
Young's Modulus, E, GPa	4.7	31.7	49.6	33.1
Poisson's Ratio, ν	0.15	0.28	0.15	0.15
Shear Modulus, G, GPa	2.0	12.4	21.6	14.4

* After Moreno-Martinez, 1981.

The mortar and rock types were tested in accordance with conventional rock mechanics testing procedures, and the properties are shown in Table I.

Pressure-volume curves for four mortar blocks with different hole configurations are shown in Fig. 4. For low pressure the stiffness of the mortar is highly non-linear. As breakdown pressure P_C is reached the fracture initiates and propagates followed by an instantaneous pressure drop. Fractures always propagate in the direction of least resistance and the fracture initiations at the wall of the borehole are in most cases 180 degrees opposite to each other. After fractures were created the effective fracture surface was determined, cf. Table II. The results indicate that a higher value of the breakdown pressure, P_C, is followed by a larger effective fracture surface for the mortar.

The breakdown pressure and the slope of the pressure-volume curve are similar for the three blocks of Indiana limestone. The direction of fractures of the unloaded blocks varied and the effective fracture surfaces are listed in Table II. A higher value of the breakdown pressure, P_C, is followed by a larger effective surface.

Two blocks of Colorado sandstone were tested with the borehole oriented parallel and perpendicular to bedding respectively.

The tensile strength and Young's Modulus of the blocks can be calculated approximately from the elastic theory of thick wall cylinder. Consider a thick wall cylinder incircled in the block. The tangential stress reach the maximum value at the inner wall of the thick wall cylinder and decreases as the wall thickness increases. Stress concentration σ_θ/P_o, at the inner wall of the cylinder is used to determine the tensile strength of the material. For block A1 in Table II the stress concentration $\sigma_\theta/P_o=1.04$. As the

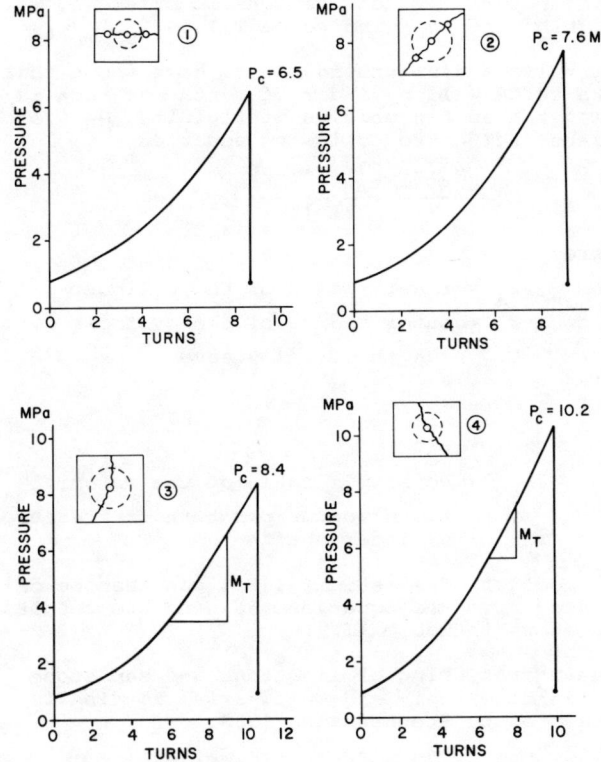

Fig. 4 Pressure-volume curves, breakdown pressure and residual pressure for sleeve fracturing of mortar blocks with different borehole configurations.

pressure at failure is known from the breakdown pressure, P_C, and the material is assumed to

Table II. Deformation and Fracturing of Mortar and Rock Blocks.

Material		Sample	Breakdown Pressure P_C {MPa}	Effective Frac. Surface A_E {cm^2}	Radii Ratio $\frac{r_o}{r_i}$	Tensile Strength T {MPa}	Pressure Volume Ratio M_T {MPa/turn}	Young's Modulus E_R {GPa}
Mortar		1	6.5	401	3	8.2	–	
		2	7.6	560	4	8.6	–	
		3	8.4	570	7	8.7	1.15	1.5
		4	10.2	645	6	10.7	1.33	2.0
Indiana Limestone		A1	13.8	570	7	14.3	1.57	7.2
		A2	13.5	560	7	14.1	1.53	6.7
		A3	13.9	575	7	14.4	1.60	7.6
Colo. Sandstone	Perp. Bedding	B1*	13.4	575	7	13.9	1.38	5.1
	Parallel Bedding	B2	14.3	682	7	14.9	1.72	9.8

* Block contained joints

fail in pure tension the tensile strength, T, for block A1 is found to be $T=1.04 \cdot P_c$.

Hustrulid and Hustrulid, 1975, have shown that if a thick wall cylinder of mortar or rock is being tested the modulus of rigidity, G_R, is calculated using the following equation

$$G_R = \frac{M_R \cdot \pi \cdot L r_i^2}{\gamma} \frac{1+\beta-2\nu_M}{1-\beta} \qquad (8)$$

where

r_i = inner radius of the cylinder
r_o = outer radius of the cylinder
L = length of the sleeve
$\beta = \dfrac{r_i^2}{r_o^2}$
ν_R = Poisson's ratio of the material
γ = fluid volume per turn of pressure cylinder shaft

The modulus of elasticity, E_R, can then be calculated from the Equation (5), and the results are shown in Table II.

Sleeve fracturing of limestone and sandstone blocks under uni- and multi-axial loading is presented by Stephansson, 1983.

4.2 Testing of Thich Wall Cylinders

Sleeve fracturing technique was applied to thick wall cylinders for determination of tensile strength. This is the most favorable testing technique as it provides the same stress distribution as field testing of sleeve fracturing in boreholes. Thick wall cylinders from four test sites are available from previous stress measurements with overcoring technique.

Values of tensile strength determined by sleeve fracturing of thick wall cylinders are listed in Table III and are later applied in the classical equation for hydraulic fracturing, Equation (1), to solve the secondary principal stresses in the geological formations of the test sites.

5. SLEEVE FRACTURING IN GEOLOGICAL FORMATIONS

Field testing of sleeve fracturing has been conducted in the following geological formations and test sites:

a) Migmatic gneiss, Colorado School of Mines Experimental Mine at Idaho Springs, Colorado.

b) Latitic lava flow, U.S. Geological Survey test site, South Table Mountain, Golden, Colorado.

c) Climax Granite, Spent Fuel Test-Climax, Nevada Test Site, Nevada.

d) Pomona Basalt, Near Surface Test Facility, Hanford Test Site, Washington.

These test sites are all related to the U.S. program of storage of radioactive waste in geological formations and they have been chosen primarily due to existing rock mechanical information at the sites, such as rock mass deformation modulus, rock stresses, strength and elastic parameters. For this study results from d) will be presented.

Sleeve fracturing has been conducted in three boreholes at Near-Surface Test Facility (NSTF) on the Hanford Site, Washington. A number of rock mechanics field and laboratory tests have been conducted by the Basalt Waste Isolation.

Table III. Sleeve Fracturing of Thick Wall Cylinders.

Rock type	Bore hole No	Depth {m}	Length of Cylinder {cm}	Radii r_i {cm}	Radii r_o {cm}	Stress Concentration Factor σ_θ/P_o	Breakdown Pressure P_c {MPa}	Tensile Strength T {MPa}
Migmatite Gneiss Idaho Springs	3	2.6	28	2.92	7.18	1.41*	6.9*	9.7*
	1	4.6	30	2.93	7.19	1.40	9.0	12.6
	1	3.3	30	2.92	7.18	1.41	16.9	23.8
Lava Flow South Table Mountain Golden			25	2.94	7.18	1.40	22.8	31.9
Climax Granite Nevada Test Site	Hor 2	5.4	35	2.9	7.41	1.36	18.3	24.9
	Hor 2	4.0	37	2.9	7.33	1.37	23.8	32.6
	Hor 3	6.2	60	2.9	7.28	1.38	24.5	33.8
	Hor 3	6.8	60	2.9	7.28	1.38	24.5	33.8
Average							22.8	31.3

* Failure Parallel Foliation

Project to enhance the capability of predicting the response of Columbia River basalt to thermomechanical loading resulting from storage of radioactive waste, Gregory and Kim, 1981.

Three vertical boreholes were inspected with a boroscope prior to testing. Suitable sections with less jointing were selected for testing and sections showing vertical joints were omitted. Four tests were conducted under water. Two testing procedures were used. In the first, pressure was increased until the first fracture appeared followed by a pressure drop and a continues increase of pressure to the range of 35-70 MPa and unloading. Later this was followed by cyclic loading and unloading.

The other test procedure consisted of loading to the breakdown pressure where the fracture appeared and a pressure drop was recorded. After unloading the system a second pressurization cycle was imposed till a pressure drop was recorded for the second time. Repeated loading and unloading indicated minor changes in rock mass stiffness as indicated in Fig. 5 for borehole 8T3. The fracture initiation was also recorded from the clear and in most cases loud sound transmitted from the fracture via the extension rod to the pressure equipment.

The breakdown pressure has been recorded in all tests except two and the pressure drop due to fracture initiation varies between a few megapascals to tenths of megapascals. Breakdown pressures, P_c, listed in Table IV, have been applied to Equation (7) for in situ stress determination.

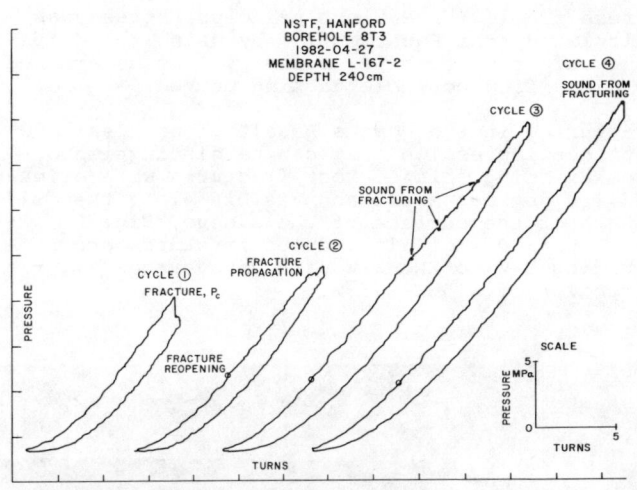

Fig 5. Pressure-volume relationship for Pomona Basalt, NSTF, Hanford Test Site, Washington.

The pressure for reopening the fracture in subsequent pressurizations is defined as the minimum principal stress. This point is determined by overlapping the pressure-volume curves from the first and later pressurizations. The point where the curves diverge indicates the change in stiffness and hence the reopening of the fracture due to exceeding minimum principal,

Table IV. Rock mass Modulus and In Situ Rock Stresses by Sleeve Fracturing Near Surface Test Facility, Hanford Test Site, Hanford, Washington.

Borehole	Depth {cm}	Rock Mass Modulus Prior to Fracturing E_R^1 {cm}	Rock Mass Modulus After Fracturing E_R^{11} {GPa}	Breakdown Pressure P_c {MPa}	Tensile Strength T {MPa}	Maximum Principal Stress σ_1 {MPa}	Minimum Principal Stress σ_2	Ratio σ_1/σ_2	Direction of Impression on Sleeve	Note
3T8	147	17.4	13.4	14.5	10.9	4.8	2.8	1.9	N60°E-S60°W	
	241	9.8	14.7	16.9	10.9	4.2	3.4	1.3	N80°E-S80°W	
3T12	368	14.6	11.7	11.7	10.9	5.5	2.1	2.6	N51°E-S66°W	
	480	14.7	9.1	23.8	10.9	-8.1	1.6	3.0	E-W	*
	717	2.7	4.6	16.2	10.9	15.4	6.9	2.2	N80°E-S85°W	**
	841	21.2	15.9	18.3	10.9	4.9	4.1	1.2	N42°E-S78°W	
	915	20.2	4.0	27.5	10.9	-4.6	4.0	1.7	E-W	
	950	22.1	35.5	37.9	10.9	-15.0	4.0	2.5	N75°E-S81°W	
	970	26.5	31.1	25.1	10.9	-	-	-	N63°E-S78°W	**
	1024	17.8	4.4	35.2	10.9	-	-	-	E-W	**
8T3	240	4.1	3.4	11.7	17.4	21.2	5.2	4.1	N70°W-S75°E	
	420	4.8	4.9	28.3	17.4	5.7	5.5	1.0		***
	520	3.7	4.3	14.3	17.4	18.4	5.1	3.6	E-W	
	627	3.7	2.9	10.3	17.4	23.6	5.5	4.3	N80°E-S80°W	

* Fracture Section
** Large Joint Intersection
*** Three Fractures Formed

stress, Table IV. Maximum principal stress was calculated from Equation (7) by using two different values of tensile strength, T as determined by a modified borehole jacking method.

Fracturing in the Pomona Basalt gives clear and distinct impression that can be distinguished from existing joints. Most fractures were oriented 180 degrees apart and visible along the full length of the surface of the sleeve, Fig. 6. More than 70 % of all induced fractures are striking E-W to ENE-WSW with an average direction of N75°E-S75°W.

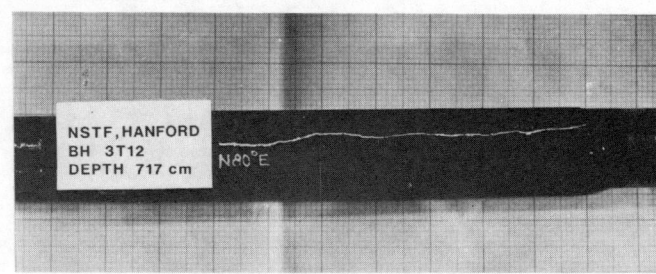

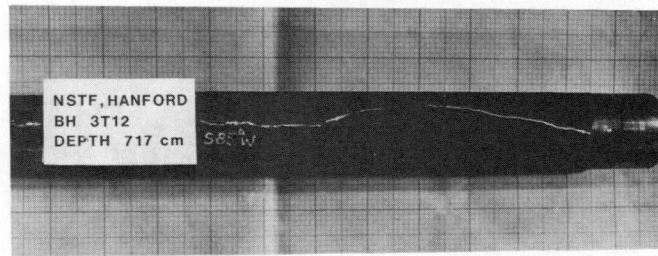

Fig. 6 Impression from sleeve fracturing.

Comparison of rock mass results from borehole jacking experiments and results from sleeve inflating technique of this study show that the modulus are about the same despite variations in level of pressure, orientation of the jacks and test location. Rock stress measurements by hydraulic fracturing and overcoring stress relief methods were conducted at NSTF Hanford, Kim, 1980. Results obtained by these methods and sleeve fracturing are shown in Table V.

Examining the results in Table V we notice an acceptable agreement between the methods considering the difficulties encountered in the test conditions for jointed basalts. The results also find support in the structural geological situation at NSTF, where a northerly trending fault is located near the test site and reflects an east-west direction compressiona force, Kim, 1980.

6. CONCLUSION

1. The borehole sleeve system used in this study is a development of the Colorado School of Mines (CSM) technique proposed in this paper both rock mass deformability and rock stress are determined in one and the same test.

2. Sleeve fracturing has been tested in blocks and thick wall cylinders and with loading on the periphery. Tensile strength determined from sleeve fracturing is higher than results from Brazilian tests.

3. Sleeve fracturing has been conducted in various geological formations at four test sites related to the U.S. program for storage of radioactive waste.

4. Sleeve fracturing was tested in three vertical boreholes of jointed basalt at the Near Surface Test Facility, Hanford Test Site. The stresses are found to be of the same magnitude as stresses determined by hydrofracturing and USBM overcoring. The direction of maximum principal stress from sleeve impressions are found to be very consistent and in agreement with tectonic stresses in the area.

7. ACKNOWLEDGEMENTS

The author is most grateful to W. Hustrulid for valuable support and stimulating discussions.

H. Swolfs, W. Patrick and C. Gregory gave valuable suggestions in interpretation of pressure- volume curves.

This work was supported by Battelle Memorial Institute, Project Management Division, contract E512-04800 under contract Ey-76-C-06-1830 with the Department of Energy. The University of Luleå, Sweden and the Swedish Board for

Table V. Comparison of stress determinations at NSTF, Hanford Test Site.

Method	Principal stress					
	σ_1		σ_2		σ_3	
	Magnitude {MPa}	Orientation	Magnitude {MPa}	Orientation	Magnitude {MPa}	Orientation
Overcoring* USBM-gauge	6.9	N81°W 11 deg. up	2.1	N8°E 5 deg. up	2.1	N65°W 77 deg. up
Hydrofracturing*	13.6	N70°W	1.5	N20°E	1.4	vertical
Sleeve fracturing	7.5	N72°E	3.6	N18°W	1.4	vertical

* Data from Kim, 1980.

Technical Development under contract 79-5104 also supported the work.

8. REFERENCES

Baguelin, F., Jézéquel, J.F. and Shields, D.H. (1978). The pressuremeter and foundation engineering. Trans. Tech. Publications, Clausthal, 617 pp.

El Rabaa, A.W.M.A. (1981). Measurements and modeling of rock mass response to underground excavation. M.S. Thesis Department of Mining Engineering, Colorado School of Mines, T2470.

Enever, J.R. and Wooltorton, B.A. (1982). Experimence with hydraulic fracturing as a means of estimating in situ stress in Australian coal basin sediments. Proceeding from Workshop on Hydraulic Fracturing Stress Measurements, Monterey, Dec 3-5, 1981 (in press).

Gregory, C. and Kim, K. (1981). Preliminary results from the full-scale heater tests at the near-surface test facility, 22nd U.S. Symposium on Rock Mechanics, Cambridge, Massachusettes, 137-142.

Haimson, B.C. (1974). Determination of stresses in deep holes and around tunnels by hydraulic fracturing. Proceedings of the 1974 Rapid Excavation and Tunneling Conference, Soc. Mining Engineers of AIME, (2), 1539-1560.

Hubbert, M.K. and Willis, D.G. (1957). Mechanics of hydraulic fracturing, Trans. A.I.M.E., (210), 153-168.

Hustrulid, W. and Hustrulid, A. (1975). The CSM cell - a borehole device for determining the modulus of rigidigy of rock. In E.R. Hoskins Jr. (editor) Applications of Rock Mechanics. Proceedings 15th Symposium on Rock Mechanics, Sept 17-19, 1973. American Society of Civil Engineers, New York, 181-125.

Kim, K. (1980). Basalt waste isolation project, annual report - fiscal year 1980. Section on rock mechanics field test result to date, Rockwell Hanford Operations, RHO-BWI80-100, V26-V36.

Ladanyi, B. and Gill, D.E. (1981). Determination of creep parameters of rock salt by means of a borehole dilatometer. First Conference on the Mechanical Behaviour of Salt. Pennsylvania State University (in press).

Moreno-Martinez, M.R. (1981). An investigation of fracture initiation and propagation near interfaces in coal measure rocks. M.S. thesis T-2381. Department of Mining Engineering, Colorado School of Mines, 256 pp.

Ozdemir, L. and Wang, F.D. (1981). Development of an inclusiongaged sleeve for use on a culindrical pressure cell. Excavation Engineering and Earth Mechanics Institute of CSM. U.S. Bureau of Mines Contract No. H0282020. 66 pp.

Panek, L.A., Hornsey, E.E. and Lappi, R.L. (1964). Determination of the modulus of rigidity of rock by expanding a cylindrical pressure cell in a drillhole. 6th Symposium on Rock Mechanics, Rolla, Missouri, 427-449.

Stephansson, O. (1982). State of the art and future plans about hydraulic fracturing stress measurements in Sweden. Proceedings from Workshop on Hydraulic Fracturing Stress Measurements, Monterey, Dec. 3-5, 1981 (in press).

Stephansson, O. (1983). In situ rock stress measurement by means of sleeve fracturing. International Symposium on Soil and Rock Investigations by In-situ Testing, Paris May 18-20, 1983, (in preparation).

Zoback, M.L. and Zoback, M. (1980). State of stress in the conterminous United States. Journal of Geophysical Research, vol 85, 6113-6156 pp.

THE STATE OF STRESS IN ROCK AND METHODS OF ITS DETERMINATION

L'état de contrainte dans les masses rocheuses et méthodes de sa détermination

Der Spannungszustand des Gebirges und die Methoden seiner Bestimmung

S. V. Kuznetsov
D.Sc., Professor, Head of Laboratory, Institute of Complex Development of Mineral Resources, the USSR Academy of Sciences, Moscow, USSR

D. M. Bronnikov
Corresponding Member of the USSR Academy of Sciences, Director of Institute of Complex Development of Mineral Resources, the USSR Academy of Sciences, Moscow, USSR

I. A. Parabuchev
Dr., Chief Geologist, "Hydroproject" Institute, Moscow, USSR

V. D. Parphenov
Dr., Chief Specialist, "Hydroproject" Institute, Moscow, USSR

I. T. Aitmatov
Dr., Director of Rock Physics and Mechanics Institute, the Kirghiz SSR Academy of Sciences, Frunze, USSR

G. A. Markov
D.Sc., First Deputy of Chairman of the Presidium of the Kola Branch, The USSR Academy of Sciences, Apatity, USSR

SYNOPSIS

A proof is given for the systems of current control and forecast of the stresses and behaviour of rock masses during exploitation of solid mineral deposits. The methods to be used for the estimation of stress in a rock mass are described. A method of stress determination is proposed, whose application is based on measurements of compensatory pressure which keeps the initial axial strain of the core constant under its overcoring in the rock mass.

RESUME

On donne une démonstration des systèmes courantes de contrôle et de prédiction de contraintes et du comportement des masses rocheuses lors de l'exploitation de gisements de minerais solides et on décrit les méthodes à utiliser pour l'évaluation des contraintes dans les masses rocheuses. Ensuite, on propose une méthode d'évaluation des contraintes dont l'application se base sur des mesures de la pression compensatoire qui maintient la contrainte axiale primitive de la carotte dans un état constant lors du surcarottage dans la masse rocheuse.

ZUSAMMENFASSUNG

Es wird die Schaffung von Systemen für die kontinuierliche Kontrolle und die Prognose des Spannungszustandes und des Verhaltens von Gesteinsmassiven beim Abbau von Lagerstätten fester Bodenschätze begründet. Es wird eine Beurteilungsmethode für Spannungen in Gesteinsmassiven und eine Möglichkeit der Bestimmung von Spannungen beschrieben, die auf der Messung des Ausgleichsdrucks beruht, der die Ausgangsdeformierung des Bohrkerns bei seinem Herausbohren aus dem Massiv aufrechterhält.

1. PRINCIPLES OF FORECASTING STRESSES AND BEHAVIOR OF ROCK MASS FOR EXPLOITATION OF SOLID MINERAL DEPOSITS

At present, the countries which possess well-developed mining industry exploit their mineral deposits persistently raising the intensity of work and penetrating to progressively greater depth of mining. It results in higher rock pressure and richer variety of its manifestations in the forms of rock bursts, sudden caving, outbursts of coal and gas, etc. Similar manifestations take place, as a rule, at different depths in different regions. This is due to the mining geological features of deposits (or their certain regions) including the stress-strain state and physico-mechanical properties of rock as well as to the technology and parameters of mining systems.

Geological and mining technical conditions of rock pressure manifestations, state of stress in rock mass, physico-mechanical properties of rock, strains and displacements in the influence zones of openings, acoustic emission and velocities of elastic wave propagation in rock mass, etc. are investigated to ensure safe mining and increased efficiency of underground mining. Proceeding from rich experience of controlling harmful effects of rock pressure in coal and ore mines, new facilities and methods have been worked out and widely used for in situ and laboratorial measurements of various physical quantities reflecting certain properties of rock or the state and behavior of rock mass.

Attempts of using different mining systems which should preclude (either independently or together with suitable engineering precautions) considerable dynamic phenomena are made in coal and ore mines prone to dynamic manifestations of rock pressure. Normally the degree of potential risk of such dynamic manifestations and the necessity of appropriate prevention measures in regions under mining are determined using preliminary studies and observations. However, the experience of controlling sudden caving, rock bursts and rock, coal and gas outbursts, which has been gained in all countries, reveals that technically feasible and economically reasonable precautions sometimes fail to ensure sufficient control of rock mass behavior and to avoid dangerous dynamic manifestations of rock pressure.

The search for means of higher labour productivity in ore and coal mines in countries with well-developed mining industries is aimed, in particular, at extensive use of high-productive high-power machinery, including self-propelled equipment, which favours a sharp increase in intensity of recovery of mineral resources. This equipment requires comparatively large operating space and haulageways which could ensure uniform breaking and haulage of large masses of rock and mineral resources while using underground mining. The formation of large operating space and haulageways in rock mass is known to lead to considerable increase in stress concentration, in particular in constructional elements of mining systems, while high rates of mining, as a rule, do not allow any relaxation of these stresses and thus add to the hazards due to manifestations of rock pressure.

All this calls for a special approach to the problem of rock pressure manifestations, the one in which the mining systems are chosen subject to compulsory accomplishment of a package of investigations aimed at forecasting of the sites of dynamic rock pressure manifestations and at the control of the state of rock mass in the course of mining. In these investigations, apart from estimations of the degree of hazard for some part of rock mass and precautions to be taken, a possibility must invariably be provided for obtaining information on location, time of genesis of the sites of dynamic phenomena, their development and intensity at the stage of corresponding self-preparation of rock mass conditioned by natural and mining-technical situation.

As a result, the present approach consisting of two basic elements (mining system - prevention measures) is changed for that of three elements (mining system - forecast and control - prevention measures). The efforts on control and forecast of rock mass state, subject to general logical scheme, together with necessary scientific and engineering means are included into systems of current control and forecast of rock mass state and behavior. The application of this approach is now possible thanks to a number of methods developed for modelling rock mass behavior taking into consideration its deformation properties and geological structure. Thus, even at the stage of mine designing, one can use geological prospecting data for preliminary zoning of the mine field as to the degree of risk of dynamic phenomena and outline precautions necessary in the course of development working.

Besides, the dynamic phenomena in rock mass are associated with some preparation processes attended with characteristic symptoms (so called precursors) which bear information on rock mass behavior assiciated with rock bursts, caving, coal and gas outbursts, etc. The recording of these precursors, estimation of their statistical distribution, establishing of criteria for rock mass state using these precursors, comparison of current factors with the criteria found will help us to study the mechanism of the phenomena mentioned and to correct rock mass parameters used for model forecasting. Current instrumental industrial control of rock mass state and behavior exercised throughout the mine field regions under mining is to become the basis for these operations.

Therefore, the implementation of this approach requires a number of operations of both scienti-

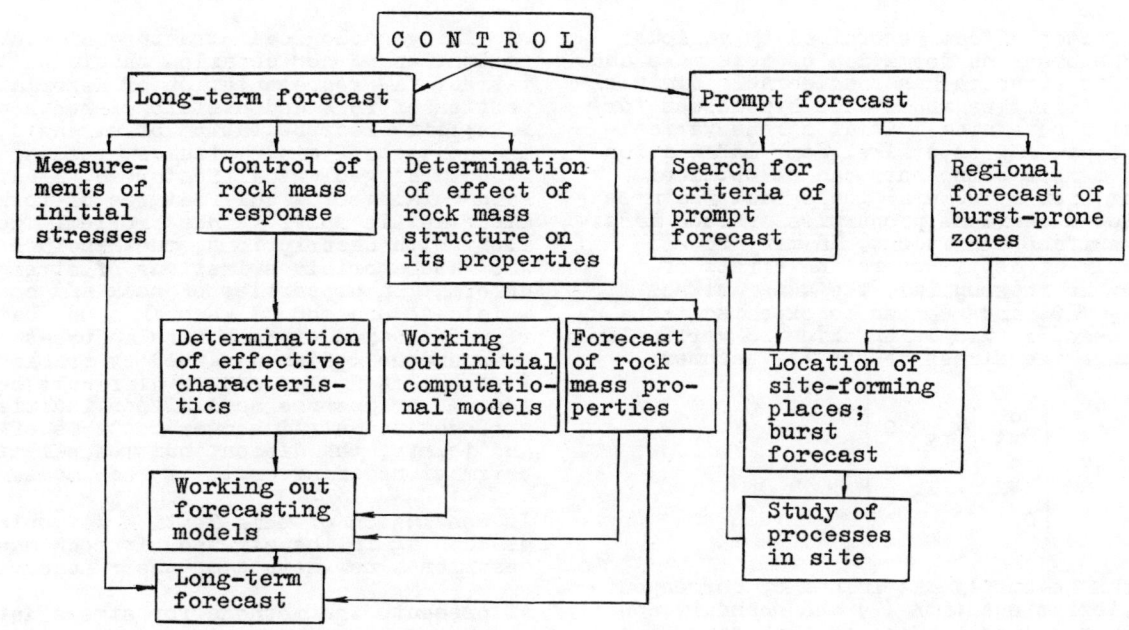

Fig.1. Block diagram of geomechanical control system

fic and practical nature which must be included in a common program of organizational and other actions. All this **emphasizes** the necessity of systems of current geomechanical control of rock mass state and behavior to prevent the hazard of dynamic manifestations of rock pressure. Fig.1 is a block diagram of such a system.

No control and forecasting system can be put into practice unless a number of problems are solved to obtain necessary a priori information on the object controlled and on the working conditions of the control system.

By present, extensive experimental data have been accumulated displaying the complicated character of stress distribution in the Earth's crust. The regularities in variability of stress fields are not yet studied enough, and the forecast required for engineering geological estimations of mineral deposits is often insufficiently grounded. The experience gained from investigations shows that detection and forecast of regularities in stress variations should be based on the analysis of geological structures and direct measurements of stresses in rock mass.

The forecasting estimations of stresses in the course of engineering geological studies of mineral deposits depend on both regional stress values, characteristic of different tectonic structures and tectonic regions, and the stresses due to relatively small tectonic structures (faults, overlaps, folds, etc.) and petrogenetic features of rock mass.

In prospecting practice, when there are no openings and no direct stress measurements are possible, the stresses can be evaluated by means of a complex of engineering geological, seismo-acoustic and computational methods. When selecting the parameters of mining systems for mineral deposits at considerable depth, the question always arises: what are the present state of stress and deformation properties of rock at this depth? The answers to this question are often contradictory due to both the imperfection of measuring methods and the erroneous estimations of properties and strains in rock.

In accordance with this, consider stress tensor (Bronnikov et al., 1977)

$$T_\sigma^o = \begin{vmatrix} \sigma_{xx}^o & \sigma_{xy}^o & \sigma_{xz}^o \\ \sigma_{xy}^o & \sigma_{yy}^o & \sigma_{yz}^o \\ \sigma_{xz}^o & \sigma_{yz}^o & \sigma_{zz}^o \end{vmatrix} \quad (1)$$

which completely determines the present state of stress in rock mass. In general case, it can be only asserted that all the stress components depend on coordinates and satisfy equilibrium equations

$$\frac{\partial \sigma_{xx}^o}{\partial x} + \frac{\partial \sigma_{xy}^o}{\partial y} + \frac{\partial \sigma_{xz}^o}{\partial z} + \rho g_x = 0$$

$$\frac{\partial \sigma_{xy}^o}{\partial x} + \frac{\partial \sigma_{yy}^o}{\partial y} + \frac{\partial \sigma_{yz}^o}{\partial z} + \rho g_y = 0 \quad (2)$$

$$\frac{\partial \sigma_{xz}^o}{\partial x} + \frac{\partial \sigma_{yz}^o}{\partial y} + \frac{\partial \sigma_{zz}^o}{\partial z} + \rho g_z = 0$$

where ρ is density of rock; g_x, g_y and g_z are components of gravity acceleration vector.

The multifactor effect exercised by various natural phenomena on formation of rock mass and hence on its stresses has considerably diminished the possibilities to advance hypotheses for regularities governing initial stress variations with depth and on strike. Some information on stress tensor components can be obtained proceeding from structural geological features and physico-mechanical properties of rock mass and from surface conditions. In many rocks, particularly at depths where the effect of topography is inessential, the shear stress components σ^o_{xz} and σ^o_{yz} can be expected to be equal to zero, z-axis being oriented vertically. In this case the stress tensor (1) becomes

$$T^o_6 = \begin{vmatrix} \sigma^o_{xx} & \sigma^o_{xy} & 0 \\ \sigma^o_{xy} & \sigma^o_{yy} & 0 \\ 0 & 0 & \sigma^o_z \end{vmatrix}$$

Here σ^o_z can be easily determined by corresponding equilibrium equation (2) and boundary conditions on the free surface of rock mass, i.e. for z=0:

$$\sigma^o_z = \sigma^o_{xz} = \sigma^o_{yz} = 0 \qquad (3)$$

As $\sigma^o_{xz} = \sigma^o_{yz} = 0$ in rock mass considered, the following relationship results from equation (2) allowing for conditions (3):

$$\sigma^o_z = -\gamma z \qquad (4)$$

where γ is average unit weight of overburden at the depth z.

Therefore, in rock mass with no shear stresses on horizontal planes the vertical stress component σ^o_z continuously varies with depth, its absolute value being equal to the weight of overlying rock column with unit cross-section. The other two principal stress components σ^o_x, σ^o_y with corresponding orientations may be related to tectonic processes and many other nature's phenomena affecting the formation of rock mass. Their absolute value can be more, equal to, or less than, σ^o_z and even possess discontinuities when changing from one rock layer to another. All the rest of information on the present stress tensor can only be obtained by means of experiments in natural conditions of a certain rock mass.

In a rock mass of complicated structure and distinct tectonic disruptions, the information on stress tensor (1) should be obtained by means of experiments. But it can only be done in certain points of rock mass. This is why, along with stress measurements, selection of measuring points and stress field interpolation over the total area of investigation in terms of measured results becomes paramount.

Another important problem of rock pressure at considerable depths is the model representation of rock mass in terms of general principles of the mechanics of solids (Kuznetsov et al., 1981; Kuznetsov and Slonim, 1981). This problem includes three basic aspects: i) representation of mining geological structure of rock mass and estimation of how detailed should be the elaboration; ii) representation of deformation properties of rock and certain elements of rock mass; iii) representation of strength of rock and contacts (joints, faults, bedding and other structural geological features). Studies of structural geological features of rock mass are important in solving this problem. These studies, along with descriptions, must include in situ and laboratorial estimations of strength and deformation properties of rock and contacts. In this case one should keep in mind that model representation of rock mass as pre-stressed inhomogeneous medium weakened by cracks and joints (open, closed, sealed) is difficult because the medium may possess some discontinuities of stresses and strains over contacts of layers and joints, the discontinuities effecting the estimations of strength of rock mass.

In conclusion of this Section we would like to discuss again the stresses in rock mass and to describe a new method of their measuring.

At present, the methods for stress determination are based on rock strain measurements under complete or partial stress relief of the section to be studied. The formulae known for calculating stress components σ_i have, in the simplest case, the following form:

$$\sigma_i = E \cdot f(\varepsilon_{ij}, \nu)$$

where E is the modulus of elasticity, ν is Poisson's ratio, ε_{ij} are measured strains.

Function $f(\varepsilon_{ij}, \nu)$ is usually found by solving a problem of continuum mechanics corresponding to the method of stress determination. The stress-strain relations are generally nonlinear for many rocks, especially for shales, coals, ores of non-ferrous metals, cracked rocks. Moreover, there is no one-to-one correspondence between stresses and strains under repeated loading and unloading while plastic deformations develop in the process. All this largely restricts the use of these methods and makes stress determination less reliable.

Therefore, to determine stresses in nonlinearly deforming rock, one has to find a method of obtaining information on the state of stress which preserves, completely or partially, the initial state of strain. The calculation formulae must not contain the modulus of elasticity, which is used, as a rule, as one of the main parameters.

In this connection we have developed a new method for stress measurements in rock mass by means of strain compensation of the section to be studied under overcoring (Kuznetsov and Savostianov, 1981). The method is based on measuring the pressure applied to the surface of the section under study to preclude its deformations in the course ov overcoring, at least in the direction of applied pressure.

A way to use this method is as follows (Fig.2). At first, a large-diameter research hole (e.g. 150 mm) is to be drilled in rock mass toward the zone of measurements. Then from its face a coaxial small-diameter measuring hole is drilled in this zone (e.g. 36 mm). This borehole is used to grapple a rod by means of an anchor lock and to place the gages for axial strain control (see Fig.1). Jack-type compensator is set up in the large-diameter research hole, its travelling pivot being joined to its rod and retained against the hole face. The compensator casing is rigidly attached to the rod. When pressure is developed in the compensator, it is transmitted via the travelling pivot onto the face of the research hole because the rod joined to rock mass at the end of the measuring hole grips the compensator casing. The initial pressure in the compensator should ensure transmission of load p_o of about $0.4 - 0.5 \gamma H$ via the travelling pivot onto the face. Here H is the depth of the point of stress measurements below the surface. When the load is p_o, the overcoring of the core starts.

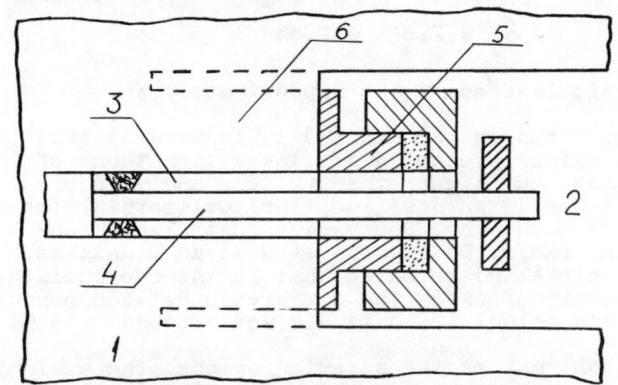

Fig.2. Method for measuring compensatory pressure to determine stresses in rock

1 - rock mass; 2 - research hole; 3 - measuring hole; 4 - rod with anchor lock; 5 - travelling pivot; 6 - overcored core

If the core develops axial strains, i.e. $\Delta\varepsilon_x^o \neq 0$, the pressure in compensator is changed so as to keep initial axial strains of the overcored part of the core. The face load on the core p_x^o, which maintains initial axial strain ($\Delta\varepsilon_x^o = 0$) of the overcored part, can be called compensatory pressure. It is the pressure to be detected. Since this pressure is apparently connected with chosen direction Ox, it is necessary to chose two other mutually perpendicular directions Oy and Oz in a plane orthogonal to direction Ox. The research and measuring holes are drilled in these directions and compensatory pressures p_y^o and p_z^o are measured in the way described above. Using the measured values of compensatory pressures, the normal stress components σ_x^o, σ_y^o, σ_z^o in rock mass are determined, allowing for Poisson's ratio ν, from calculation formulae

$$\sigma_x^o = -\frac{(1-\nu)p_x + (p_y + p_z)\nu}{(1+\nu)(1-2\nu)}$$

$$\sigma_y^o = -\frac{(1-\nu)p_y + (p_z + p_x)\nu}{(1+\nu)(1-2\nu)}$$

$$\sigma_z^o = -\frac{(1-\nu)p_z + (p_x + p_y)\nu}{(1+\nu)(1-2\nu)}$$

To determine the complete stress tensor, additional measurements of compensatory pressure must be taken in directions non-parallel to Ox, Oy, Oz.

The proposed method of stress measuring offers an important advantage of avoiding the use of Young's modulus and obviating the measurements of core strains in the course of its overcoring.

2. GEOLOGICAL-KINEMATIC METHOD OF ESTIMATING THE STATE OF STRESS IN ROCK MASS,

At present, the geological-kinematic method is widely used in studies of the state of stress in rock foundations of hydrotechnical structures. It allows us to determine the orientations of principal normal stresses σ_1, σ_2, σ_3 in fields of various scale and age and to judge upon the values of differential stresses proceeding from investigations into the frequency of deformational elements in minerals possessing different shear strength under twinning.

The essence of the geological-kinematic method is a statistical analysis of spatial distribution of directions of displacements on shear faults and joints in rock and of displacement vectors on twin planes and rink bands in deformed crystals of calcite, sulphate (Parphenov, 1974; 1981), mica, etc.

The kinematic analysis of stress orientations is based on well-known mechanical relations between orientations of principal stresses and the location of shear faults (twins) and the directions of displacements on the disruptions mentioned. On measuring the elements of shear fault bedding and slip sfriations and determining the character of displacements on the fault, the orientation of quasi-principal axes σ_1', σ_2', σ_3' is calculated for shear fault which is conventionally taken as action surface of maximum shear stresses. Graphical method is the simplest way of determining this orientation; the slip traction of the fault l and its pole P are projected on stereographic equiangular net. On the fault plane axis σ_2' is directed orthogonally to l while on the lP plane axes σ_1', σ_3' are bisectors of lP right angles. To distinguish between these axes the displacement sign is considered: axis σ_1' is oriented at an acute angle to the direction of displacement on fault. The mean statistical position of quasi-principal stress axes with same indices corresponds to the orientations of the axes of principal normal stresses σ_1, σ_2, σ_3 (The mean is calculated over directions of quasi-principal stress axes for all recorded shear faults).

The reconstruction of quasi-principal axes at a micro-level is managed using the method described for microsections under a microscope on the Fedorov's stage. These axes are plotted for the planes of known twin systems and kink bands where displacement vectors are fixed in the crystalline structure of minerals relative to crystallographic and optical axes.

The geological-kinematic analysis amplifies mechanical and geophysical in situ stress measurements. Investigations applied for rock foundation of Rogun hydroelectric station under construction in the Southern Tadjikistan (Parabuchev and Parphenov, 1982) is an example of successful combination of different methods to determine the state of stress. This foundation is a tectonic block confined by deep upthrust and feathering thrust actively developing at recent tectonic stage.

The kinematic analysis of the Alpine shear jointing and microstructure of anhydride and calcite tectonites has shown that rhe axis of maximum compression σ_3 gently sloped in the NW direction and was approximately orthogonal to the plane of the Ionashski fault and the stratification, the latter steeply dipping in the SE azimuth. Axis σ_2 was gently sloping at small angle towards the fault strike; axis σ_1 steeply sloped. These data are in a good accordance with the results of measuring stresses by means of stress relief method (Kolichko et al., 1982) and seismologic measurements at earthquake focuses. These results display the inherited nature of tectonic regime at the given region. This conclusion is of principal importance for seismic zoning of given territory.

So, owing to simplicity and low labour input, the geological-kinematic method can be used for preliminary prompt estimation of stress-strain state in rock mass and choice of grounds and directions of observations for in situ mechanical and geophysical measurements.

3. NATURAL STRESSES OF ROCK MASS IN DIFFERENT TECTONIC STRUCTURES

Extensive stress measurements in rock mass were performed which revealed that global stress fields in the upper part of Earth's crust, on the whole, are inhomogeneous and closely associated with large tectonic structures. It was proved by investigations into in situ stresses performed at the Rock Physics and Mechanics Institute, Acad. Sci. Kirghizia (USSR), in particular in the folded seismic area of the Central Asia and S.-E. Kazakhstan (Aitmatov, 1981).

By present, a number of investigators considered the Earth's crust, including its uppermost parts, to be subjected to the state of relatively high stress. However, in situ measurements performed in different parts of Central Asia down to the depth of 1 km revealed that lower stresses compared with those in ancient stable shields (e.g. Baltic, Canadian, Australian, etc.) exist in the upper parts of Earth's crust in seismic orogenic regions.

To estimate the variation of stresses with depth N.Hast, 1967, proposed a well-known empiric formula

$$\sigma_x + \sigma_y = 19 + 0.098 H, \text{MPa} \quad (5)$$

where H is measured in metres.

However, our measurements in the folded seismic regions of Central Asia and S.-E. Kazakhstan (Tien Shan, Pamir, Dzhungaria) displayed the following stress variation with depth. For rocks with Young's modulus ranging from 60,000 to 110,000 MPa this variation is

$$\sigma_x = 4.5 + 0.045H$$
$$\sigma_y = 5.0 + 0.03\gamma H \quad (6)$$

For other rocks of this region, with Young's modulus ranging from 30,000 to 70,000 MPa the dependence is

$$\sigma_x = 3.0 + 0.03\gamma H$$
$$\sigma_y = 2.0 + 0.028H \quad (7)$$

Vertical stress σ_z is approximately γH.

From formulae (6) and (7) it is evident that the values presented are lower than those of N.Hast, but higher than γH. The analysis of data for 42 points (the Alps, California, the Rocky Mountains, South-East Australia, Tien-Shan, Pamir, Dzhungaria, the Sayan Mountains, Sikhote-Alin) revealed that in the overwhelming majority of cases the difference between measured and calculated data did not exceed 10-15 %.

In contrast to the seismic regions, the aseismic folded regions (i.e. the Urals, the Appalachian Mountains, the Scandinavian Caledonides) possess more inhomogeneous stress distribution. For example, at the Urals in 6 of 17 measurement points $(\sigma_x + \sigma_y)$ is close to value calculated by formula (5), in 5 points - by formulae (6)-(7), in 3 points - by hydrostatic law.(Vlokh et al., 1979).

So called mobil shields and platforms differ from stable ones in their seismicity and higher position above the sea level. These mobil platforms consist of large isolated stable blocks (cratons) separated from each other by mobil zones including fold systems and a number of abyssal fractures.

The stresses in cratons can be estimated as

$$\sigma_x = \sigma_y = \frac{\nu}{1-\nu} \sigma_z, \sigma_z = \gamma H \quad (8)$$

Stresses in mobil zones can be evaluated by either formula (5) or formulae (6)-(7) in accordance with the disturbances of rock mass.

In rift zones, grabens and within island arcs the stresses are close to those calculated by hydrostatic law. Sedimentary covers of platforms are also characterized by relatively high degree of inhomogeneity.

Thus, the structures in upper parts of the Earth's crust can be divided into 6 types according to inherent stress conditions. These types are:
1. Stable ancient shields and basements of steady platforms.
2. Folded seismic belts and regions.

3. Mobil shields and platforms.
4. Aseismic folded regions.
5. Grabens, reefs, island arcs.
6. Sedimentary covers of platforms.

Certainly, within these tectonic structures there may exist isolated local areas with stresses differing from the basic ones in the regions considered.

4. TECTONIC STRESSES IN THE UPPER PART OF THE EARTH'S CRUST AND A CHOICE OF RATIONAL LOCATION OF UNDERGROUND WORKINGS

A desire to construct important underground workings in the most strong and monolithic rock is traditional. However, investigations show that the most strong and monolithic rocks are often subjected to increased tectonic stresses. In accordance with this, the choise of design and location of underground workings requires a special approach.

Thus, in the process of construction of vertical shafts of oreshutes 600 m in depth and 6 m in diameter in one of mines at the Kola Peninsula, unusual conditions of their destruction took place. The oreshutes were constructed in strong monolithic isotropic rock (compressive strength - 150-250 MPa, tensile strength - 10-20 MPa, modulus of elasticity - 10,000-100,000 MPa). The oreshute sinking was accompanied by rock bursts and burstings. In due course, the oreshutes were intensively destructed and their cross-section, initially circular, was transformed into ellipse, its major axis being equal to 40-60 m (Markov et al., 1978). No destructions have been observed in the direction of its minor axis for more than ten years.

In situ investigations in the rocks mentioned displayed tectonic stresses 4-5 times exceeding γH. The walls of oreshute did not fail until the stresses on the shute contour exceeded 1/3 of the compressive strength of rock. The rate of wall destruction was experimentally determined under stresses constituting 0.5-2.5 of the rock strength.

Rational form of oreshutes cross-section and design of adjoining workings (parallel to, and intersecting with, oreshute's shaft) allows us to preclude the destruction or to reduce its rate (Markov et al., 1980). So, the forecast of initial stresses becomes paramount (see Turchaninov et al., 1974).

At present, the regularities in tectonic stresses are determined (Markov, 1980), in accordance with which the forecast is executed with successive stages. At the primary stage, the zones of present tectonic lifts of the Earth's crust are determined. In these zones higher tectonic stresses are the most probable events.

Then the differentiation of stress conditions is carried out depending on the structure of geological blocks and terrain where underground constructions are suggested to be placed. The highest tectonic stresses are in central parts of lifting blocks. The rocks situated at the tops of mountains are releived from tectonic stresses. At the same time, massifs situated below the bottom ov valleys and possessing the same lithology are the most stressed ones even in the close proximity of the Earth's surface.

At the next stage, the regions of the most elastic, monolithic and strong rocks are determined. At the final stages instrumental in situ determination of stress tensor components is exercised. In accordance with the data obtained, a preliminary determination, forecast and differentiation of expected stress conditions in rock mass are made and a decision is adopted for the way of experimental studies and a choice of rational construction of underground structure and its location.

The experience shows that the described methodological approach allows us to prevent dangerous manifestations of rock pressure in the process of construction and to prolong the terms of underground structure's service.

REFERENCES

Aitmatov, I.T. (1981). The stress state of rock mass in the upper part of the Earth's crust in seismic regions of Central Asia and South-East Kazakhstan. (In Russian). In: Geomekhanich. Uslovia i Dinamich. Proyavleniya Gornogo Davlenia na Rudnikakh Sredney Azii. Frunze: Ilim, 3-20.

Aitmatov, I.T. (1981). The state of stress in the upper part of the Earth's crust in seismic regions. (In Russian). Izv. Acad. Nauk Kirghiz. SSR, 1, 39-45.

Bronnikov, D.M., Kuznetsov, S.V. and Zamesov, N.F. (1977). Problems of rock pressure in mining at considerable depths. (In Russian). Prikladn. Zadachi Mekh. Gorn. Porod. Moscow: Nauka, 3-7.

Hast, N. (1967). The state of stresses in the upper part of the Earth's crust. Engng. Geol. (2), 1, 5-17.

Kolichko, A.V., Parabuchev, I.A. and Stepanov, V.I. (1982). Effect of geological stressed conditions of rock mass in designing undergroumd power house: Proc. Intern. Symp. "Rock Mechanics Related to Cavems and Pressure Shafts", FRG.

Kuznetsov, S.V, Odintsev, V.N., Slonim, M.E. and Trofimov, V.A. (1981). Methodology of Rock Pressure Calculation. Moscow: Nauka.

Kuznetsov, S.V. and Savostianov, E.V. (1981). Method for measuring mechanical stresses in rock mass. (In Russian). Ofits. Bull. Gos.Kom.SSSR po Delam Izobreteniy i Otkrytiy, 26, 156 (No 846730).

Kuznetsov, S.V. and Slonim, M.E. (1981). A mathematical model for cracked weak rock with non-linear deformative and collector properties: Proc. Internat. Symp. on Weak Rock, Tokyo, 731-736.

Markov, G.A. (1980). On the distribution of horizontal tectonic stresses near the surface of zones of liftings of the Earth's crust. (In Russian). Inzhenernaya Geologiya, 1, 20-31.

Markov, G.A., Demidov Y.V., Sazonov, G.V. and Sizov, Y.P. (1978). The influence of tectonoc forces on the stability of deep oreshutes. (In Russian). Gorny Zhurnal, 7, 63-66.

Markov, G.A., Lovchikov, A.V. and Eryomin, V.I. (1980). The increase of the stability of workings in conditions of tectonic forces influence. (In Russian). Tsvetnaya Metallurgiya, 16, 5-8.

Parabuchev, I.A. and Parphenov, V.D. (1982). On importance of geological methods to determine the orientation of stresses in measuring the state of stress of rock mass in situ. (In Russian). Trudy Seminara po Izmereniyu napryazheniy v Massive Gornykh Porod. Novosibirsk.

Parphenov, V.D. (1974). On the possibility to use microstructural analysis of barite tectonites to reconstruct paleostressed state in rock. (In Russian). Izv. AN SSSR, Ser. Geolog., 1, 122-129.

Parphenov, V.D. (1981). Analysis of the state of stress in anhydrite tectonites. (In Russian). Doklady AN SSSR, (260), 3, 695-698.

Turchaninov, I.A., Markov, G.A., Panin, V.I. and Ivanov, V.I. (1974). Complex analysis and experimental determination of complete stress tensor in the rock mass: Advances in Rock Mechanics. Proc. 3-rd Congress ISRM, 575-579, Denver, Colorado.

Vlokh, N.P., Aleynikov, A.L., Zubkov, A.V. and Lipin, Y.I. (1979). Some particularities of regional elastic stress field in the Earth's crust in the Urals. (In Russian). Gornoye Davlenie, Metody Upravlenia i Kontrola: Proc. VI All-Union Conference on Rock Mechanics. Frunze: Ilim, 60-69.

A NEW JOINT ELEMENT FOR THE ANALYSIS OF FRACTURED ROCK
Un nouvel élément sur les joints pour l'analyse des massifs rocheux fissurés
Ein neues Kluftelement für die Analyse von geklüftetem Fels

I. Carol
E. E. Alonso
E.T.S. Ingenieros de Caminos, Barcelona, Spain

SYNOPSIS

A joint element which is believed to improve previous models has been developed and adapted for implementation in more general Finite Element Codes. It incorporates all the relevant behaviour of joints in shear and compression. Two examples of markedly different character have been solved to illustrate the capabilities of the model.

RESUME

Un élément joint qui améliore des modèles précédents a été développé et adapté pour l'implantation dans des programmes modernes, plus généraux, d'éléments finis. Il emmagasine toute l'information rélévant du comportement de joints sous compression et sous l'effort tranchant. Deux exemples de caractère très différent ont été résolus pour montrer la capacité du modèle.

ZUSAMMENFASSUNG

Ein verbessertes Kluftelement wurde entwickelt und zur Anwendung in mehr generelle Finite-Element-Kodes vorbereitet. Dieses Element erfaßt alle relevanten Vorgänge beim Scheren und Schließen von Klüften. Zwei grundverschiedene Beispiele sind angeführt, um die Fähigkeiten des Modelles zu illustrieren.

1. INTRODUCTION

In order to investigate the influence of a jointed abutment in the overall behaviour of an arch dam several previously developed rock joint models were reviewed. Their capability to fit coherently into modern Finite Element codes as well as to represent comprehensively the main features of rock joint stress-strain behaviour was analyzed.

The joint element developed by Goodman (in Goodman et al., 1968; Goodman, 1975; Goodman an St.John, 1977) presents several difficulties in order to be associated to eight-nodes cuadrangular or six-nodes triangular isoparametric elements since it only has four nodes. This implies lineal displacement distributions within the element (against the quadratic formulation of the isoparametric elements mentioned). The Goodman's element does not use shape functions. Instead, the element displacement variables are referred to the mid point of the element and are characterized by two relative displacements and a relative rotation which attempts to reflect the variation of normal displacements along the joint (considered as a rigid body).

Other developments (Wilson, 1977; Heuze, 1980) introduce several modifications to the basic Goodman element. In particular Heuze includes joint dilatancy through a fictitious stiffness of the remaining of the body for each iteration. Ghaboussi et al (1973) propose a saw-teeth model for the joint and derive the relative stress-displacement relationships for a point within the joint.

None of the above joint elements was considered to be appropiate for the intended use and the structure and capabilities of the available Finite Element codes. accordingly an isoparametric rock joint element was developed and used to analyze the foundation of an arch dam. The basic formulation of this new element is described in this paper. It is then applied to solve a few simple cases which show its capabilities.

2. BASIC FORMULATION

The element which has six nodes and twelve degrees of freedom is represented in Fig. 1. It is accepted that its thickness is at least one order of magnitude smaller than its length. The joint surface is represented by the three mid points MP_i whose relative displacements are the difference between the displacements of each pair of associated nodes. The geometry of the joint is interpolated between the coordinates of the midplane nodes through the appropriate quadratic shape functions $\left(N_1 = \tfrac{1}{2}s(s-1);\ N_2 = 1-s^2;\ N_3 = \tfrac{1}{2}s(s+1)\right)$:

$$x(s) = \sum_{i=1}^{3} N_i(s)\, x_{MP_i} \quad;\quad Y(s) = \sum_{i=1}^{3} N_i(s)\, Y_{MP_i} \quad (1a,b)$$

The relative tangential, u_t, and normal, u_n, displacements along the joint can be expressed as a function of relative cartesian displacements of midplane joints u_x and u_y through the angle α (fig. 1).

$$u_t = u_x \cos\alpha + u_y \operatorname{sen}\alpha;$$
$$u_n = -u_x \operatorname{sen}\alpha + u_y \cos\alpha \qquad (2a,b)$$

where

$$\operatorname{sen}\alpha = \frac{1}{J}\frac{dy}{ds} \quad;\quad \cos\alpha = \frac{1}{J}\frac{dx}{ds} \qquad (3a,b)$$

Can be obtained from equations (1). J is the Jacobian of this transformation. Equations (2) expressed in matrix terms become

$$\underline{u}_{nt}(y) = \underline{G}\,\underline{u}_{xy}(s) \qquad (4)$$

Where

$$\underset{\sim}{G} = \begin{bmatrix} \cos \alpha & \sen \alpha \\ -\sen \alpha & \cos \alpha \end{bmatrix} \quad (5)$$

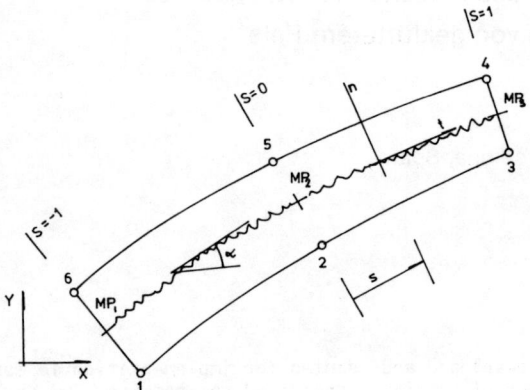

Fig. 1. Isoparametric joint element.

The relative cartesian displacements u_{xy} are easily obtained from the twelve nodal displacements $\underset{\sim}{\delta}$ through a geometric matrix $\underset{\sim}{M}$:

$$\underset{\sim}{u}_{xy} = \underset{\sim}{M} \underset{\sim}{\delta} \quad (6)$$

In addition the cartesian displacements $u_{xy}(s)$ within the joint are also expressed through expressions similar to equation (1a,b) and in matrix form

$$\underset{\sim}{u}_{xy}(s) = \underset{\sim}{N} \underset{\sim}{u}_{xy} = \underset{\sim}{N} \underset{\sim}{M} \underset{\sim}{\delta} \quad (7)$$

where $\underset{\sim}{N}$ is the matrix of shape functions. Finally, the relative normal and tangential displacements within the joint can be expressed, in term of nodal cartesian displacements, through (4) and (7) as

$$\underset{\sim}{u}_{nt}(s) = \underset{\sim}{G} \underset{\sim}{N} \underset{\sim}{M} \underset{\sim}{\delta} = \underset{\sim}{B} \underset{\sim}{\delta} \quad (8)$$

where $\underset{\sim}{B}$ has the following structure:

$$\underset{\sim}{B} = \left| \underset{\sim}{B}_1 \mid \underset{\sim}{B}_2 \mid \underset{\sim}{B}_3 \mid -\underset{\sim}{B}_3 \mid -\underset{\sim}{B}_2 \mid -\underset{\sim}{B}_1 \right| ;$$

$$\underset{\sim}{B}_i = \begin{vmatrix} -N_i \cos \alpha & -N_i \sen \alpha \\ N_i \sen \alpha & -N_i \cos \alpha \end{vmatrix} \quad (9)$$

The nonlinear stress-strain behaviour of the joint is represented by the "tangent" stiffness matrix $\underset{\sim}{D}$ (developed in the next section) given by

$$d\underset{\sim}{\sigma}(s) = \begin{Bmatrix} d\tau \\ d\sigma \end{Bmatrix} = \underset{\sim}{D} d \underset{\sim}{u}_{nt} = \underset{\sim}{D} \underset{\sim}{B} d \underset{\sim}{\delta} \quad (10)$$

The overall joint stiffness will be derived through the principle of virtual work applied to a tensional state I (forces and stresses in equilibrium $\underset{\sim}{F}^I$ and $\underset{\sim}{\sigma}^I$) and a deformational state II (displacements and compatible deformations $\Delta\underset{\sim}{\delta}^{II}$ and $\Delta\underset{\sim}{\varepsilon}^{II}$). The principle expresses

$$(\Delta\underset{\sim}{\delta}^{II})^t \underset{\sim}{F}^I = \int_{JOINT} (\Delta\underset{\sim}{\varepsilon}^{II})^t \underset{\sim}{\sigma}^I \, ds \quad (11)$$

If δ^I represents the actual displacements caused by the forces F^I and equations (8) and (10) are taken into account,

$$(\Delta\underset{\sim}{\delta}^{II})^t \underset{\sim}{F}^I = \int_{JOINT} (\underset{\sim}{B}\Delta\underset{\sim}{\delta}^{II})^t \underset{\sim}{D} \underset{\sim}{B} \underset{\sim}{\delta}^I \, ds \quad (12)$$

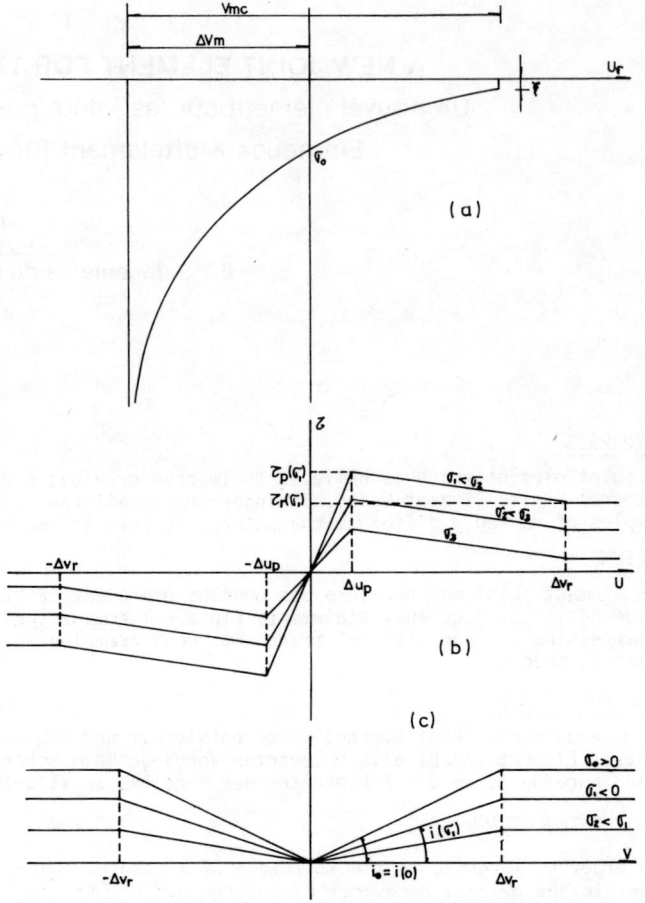

Fig. 2. Stress behaviour of joints: (a) Compression behaviour. b) Shear behaviour. c) Dilatancy.

Therefore

$$\underset{\sim}{F}^I = \left| \int_{JOINT} \underset{\sim}{B}^t \underset{\sim}{D} \underset{\sim}{B} \, ds \right| \cdot \underset{\sim}{\delta}^I = \underset{\sim}{K} \underset{\sim}{\delta}^I \quad (13)$$

which defines the stiffness matrix $\underset{\sim}{K}$ of the joint.

3. STRESS-STRAIN BEHAVIOUR OF JOINTS AND THE FORMULATION OF MATRIX D

The compression behaviour (normal stress, σ_n versus normal displacement, u) adopted follows the relationship proposed by Goodman and St. Jhon (1977) (Fig. 2a):

$$\sigma_n = -\xi + c \left(\frac{v_{mc} - \Delta v_m - u}{\Delta v_m} \right)^t \quad (14)$$

where ξ is a (small) tension cutoff, v_{mc} is the maximum shortening of the joint, Δv_m is the maximum shortening when the origin is at the initial compression stress, σ_o, of the joint, t is a coefficient and c is a constant which can be expressed in terms of σ_o. A family of curves, controlled by σ_n, for the shear stress-shear displacement behaviour has been adopted following the Fig. 2b. Among the different expressions for the peak shear strength, τ_p, the following relationship proposed by Barton (1971) has been adopted in the present analysis for its simplicity:

$$\tau_p = -\sigma_n \, tg \left(R \log \frac{q_u}{\sigma_n} + \phi_\mu \right) \quad (15)$$

F 148

Where R is a parameter describing the rugosity of the joint, q_u is the compression strength and ϕ_μ the residual angle of friction. The relationships shown in Fig. 2b are completed by other necessary parameters: Δu_p the displacement at the peak shear strength, Δu_r the residual displacement and the residual strength τ_r which is usually expressed as a fraction $\alpha = \tau_r/\tau_p$.

In order to model dilatancy the following description of normal displacement is proposed:

$$u = u_{DIL} + u_\sigma \quad (16)$$

where u_σ is the deformation associated to normal stresses, u_{DIL} is a purely geometrical term associated to dilatancy and u is the total normal displacement, selected as a nodal variable.

Following Ladanyi and Archambault (1970) (see also Fig.2c) the dilatancy is characterized, below the residual shear deformation, by an angle i such that

$$tg\, i = tg\, i_0 \left(1 - \frac{\sigma_n}{q_u}\right)^4 \quad (17)$$

where i_0 is the dilatancy angle for $\sigma_n = 0$.

Summarizing, 11 experimental parameters are needed, according to the previous relationships, to characterize the joint behaviour. They are used subsequently to formulate the D matrix.

From equation (18) it can be written, $du_\sigma = du - du_{DIL}$ and therefore

$$d\sigma = \frac{\partial \sigma}{\partial u_\sigma}(du - du_{DIL}) = \frac{\partial \sigma}{\partial u_\sigma} du - \frac{\partial \sigma}{\partial u_\sigma} \frac{\partial u_{DIL}}{\partial v} dv \ldots (18)$$

On the other hand, taking eq. (18) into account

$$d\tau = \frac{\partial \tau}{\partial \sigma} d\sigma + \frac{\partial \tau}{\partial v} dv =$$
$$\frac{\partial \tau}{\partial \sigma} \frac{\partial \sigma}{\partial u_\sigma} du + \left(\frac{\partial \tau}{\partial v} - \frac{\partial \tau}{\partial \sigma} \frac{\partial \sigma}{\partial u_\sigma} \frac{\partial u_{DIL}}{\partial v}\right) dv \quad (19)$$

If the matrix D is expressed as

$$\underset{\sim}{D}_S = \begin{vmatrix} D_S & D_{SN} \\ D_{NS} & D_N \end{vmatrix}, \quad (20)$$

its terms can be expressed as

$$D_S = \frac{\partial \tau}{\partial v} - \frac{\partial \tau}{\partial \sigma} \frac{\partial \sigma}{\partial u_\sigma} \frac{\partial u_{DIL}}{\partial v};$$
$$D_{SN} = \frac{\partial \tau}{\partial \sigma} \frac{\partial \sigma}{\partial u_\sigma};\quad (21\ a,b,c,d)$$
$$D_{NS} = -\frac{\partial \sigma}{\partial u_\sigma}\frac{\partial u_{DIL}}{\partial v}; \quad D_N = \frac{\partial \sigma}{\partial u_\sigma}$$

Each one of the derivatives can easily be calculated as a function of the above mentioned laws. The matrix $\underset{\sim}{D}$ is non symmetric and this presents some inconveniences for the numerical solution of the problem. Since the method of solution adopted relies in an iteration procedure based in a combination of the initial stress and initial displacements methods, the tangent stiffness matrix has been made symmetric maintaining only the diagonal terms. This assumption does not affect the final answer, simplifies the computations and may only affect the speed of convergence.

4. SOLUTION PROCEDURE

The solution procedure must ensure, simultaneously, convergence for the σ-u, τ-v and dilatancy laws. In addition the opening of the joint should be distinguished from the joint in full contact (compression).

For each iteration residual forces are computed for each joint element and "relaxed" throughout the structure. If, for iteration k, a given joint is in tension ($\sigma_k > 0$) the $\underset{\sim}{D}_{k+1}$ tangent matrix of the joint is set to zero and all the stresses (σ_k, τ_k) are converted to residual stresses, $\sigma_k^{res} = \sigma_k$; $\tau_k^{res} = \tau_k$. The displacements are accumulated.

If the joint is in full contact the iterative procedure has been schematized in Fig. 3. The convergence for the τ-v relationship follows standard and well known procedures. However the σ_n-u law presents some difficulties due to the dilatancy effect and the rapid rigidification of the joint in compression. Once the point 1 (Fig. 3) has been reached starting at the initial stress point, the actual u displacement u_k^* (point 1') is found with the aid of equation (14). The new stiffness matrix is found with this particular value. For a given "initial" state for the kth iteration the normal displacement due to dilatancy can be computed. Then the "accumulated" σ_k^{ac} value is found as

$$\sigma_k^{ac} = \sigma_k - (u_k - u_k^*)K_N - (u_{DILk} - u_{DILk-1})K_N \quad (22)$$

and the corresponding normal displacement u_k^{ac},

$$u_k^{ac} = u_k - (u_{DILk} - u_{DILk-1}) \quad (23)$$

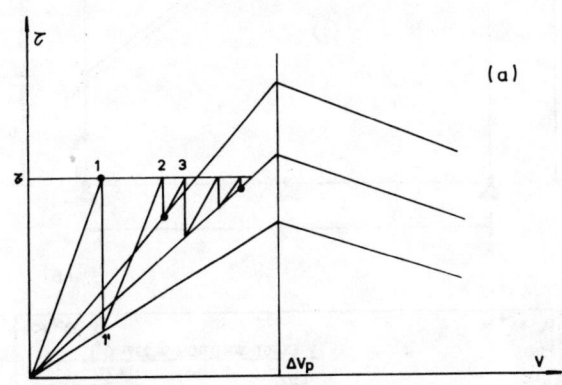

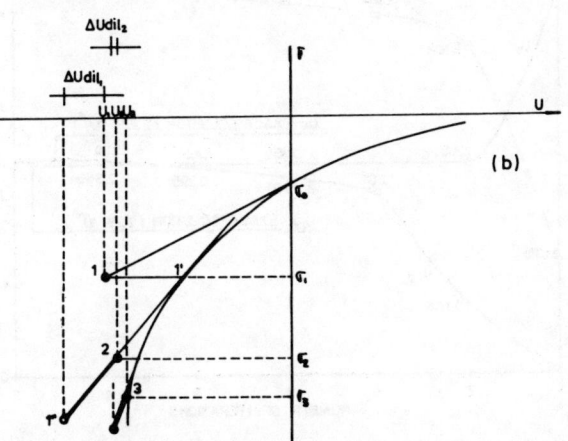

Fig.3. Iterative process for: a) The shear behaviour. b) The compression behaviour.

The point 1" in Fig.3b has been reached. The sequence of events is then the following: New nodal forces and new global stiffness matrices are computed. The system is solved and a set of increment of nodal displacements, $\Delta\delta_k$, for the whole structure and a set of values Δu_k, Δv_k, $\Delta\sigma_k$, $\Delta\tau_k$ found for the joints. A new stress-displacement state
$$u_{k+1} = u_k^{ac} + \Delta u_k; \quad v_{k+1} = v_k^{ac} + \Delta v_k; \quad \sigma_{k+1} = \sigma_k^{ac} + \Delta\sigma_k;$$
$$\tau_{k+1} = \tau_k^{ac} + \Delta\tau_k \qquad (24)$$
is found. The joint is now in point 2 of Fig. 3. The process continues until convergence.

5. SOLVED CASES

The first case solved is the isostatic compression of an inclined joint (Fig. 4a). Stresses and displacements can be computed directly and compared with the model.

The following joint properties have been adopted:
$\Delta v_m = 0.4$ mm., $v_{mc} = 0.8$ mm., $\xi = 0$; $t = 0.01$; $i_0 = 9°$; $R = 10$; $q_u = -10$ MN/m² $\phi_\mu = 20°$; $\Delta u_p = 5$ mm.;
$\Delta v_r = 40$ mm; $\tau_r/\tau_p = 0.7$. In addition, the initial stresses were taken as $\sigma_0 = -1$ MN/m² ; $\tau_0 = 0$. Two Gauss points were used for numerical integration in the joint element and the convergence of successive iterations towards the exact solution (Fig. 4b) corresponds to one of the Gauss points. The convergence of displacements is rather rapid in this case. It can be observed that the normal joint displacement due to normal stress is almost identical (and opposite sign) to the dilatant effect. Both effects are clearly distinguished in this model.

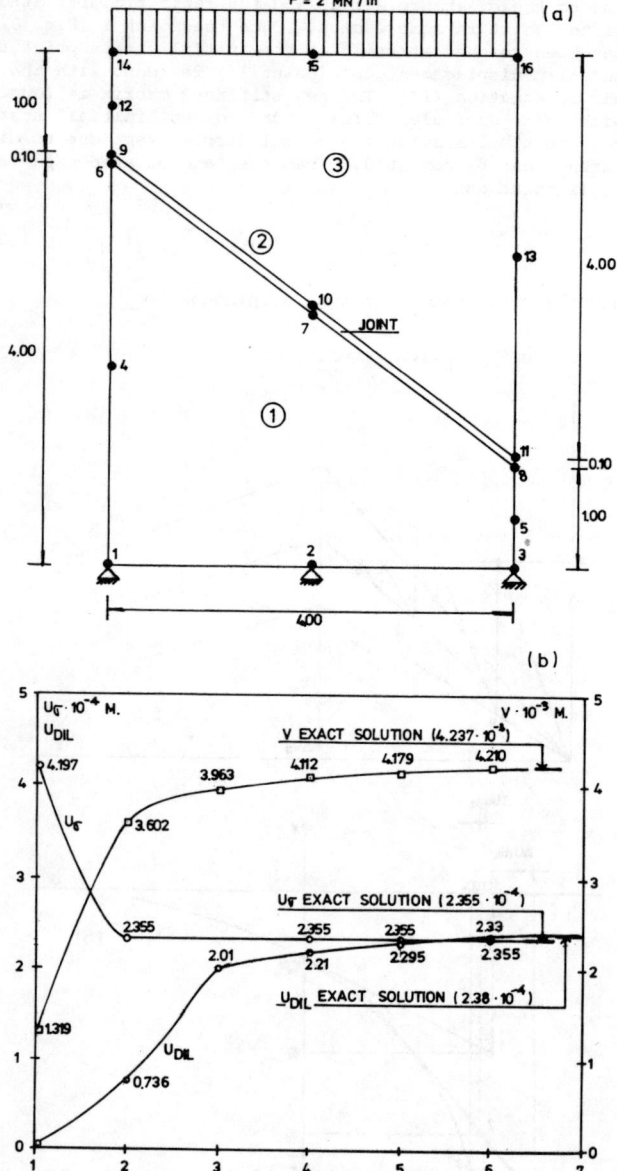

Fig. 4. Case No.1.: a) Geometry and loading conditions. b) Convergence towards exact solution of displacements.

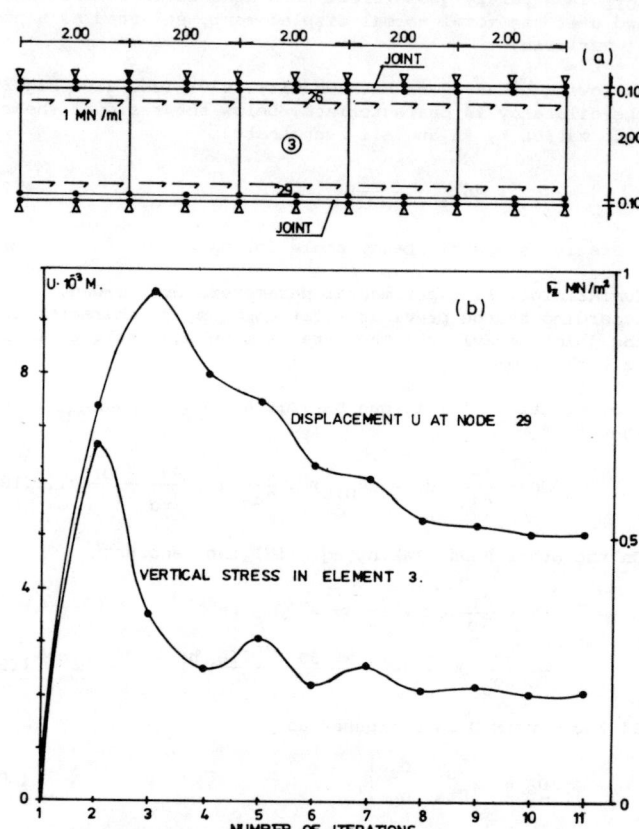

Fig. 5. Solved case No. 2.:(a) Geometry and loading conditions.(b) Convergence of vertical displacement u at node 29 and vertical stress σ_z in element 3.

A second case (Fig. 5a) refers to a non-isostatic shear behaviour of a joint (with the same properties as above). The elongated geometry and the number of elements were selected to get uniform conditions in the midelement ③. The vertical stress of the midelement and the normal displacement u of the joint at the mid-point (node 29) have been represented in Fig. 5b as a function of number of iterations. In this case the interaction effects due to dilatancy are stronger and this is reflected in the slower rate towards convergence. It can be observed in this case that a "residual" vertical stress of 0'2MN/m² has been generated. Furthermore it was checked that the converged stress and displacement values belonged to the stress-displacement laws with small error.

6. CONCLUSIONS

A general isoparametric joint element which takes into

account the relevant behaviour of rock joints has been developed. It is compatible with modern F.E.M. codes and is presently used in the analysis of dam foundations on fractured rock. Due to the strong non linear behaviour of joints in compression and shear a combined initial stress-initial strain solution method has been implemented. Normal displacements across the joint due to normal stress and dilatancy are properly combined in the solution procedure. Two simple problems have been solved to illustrate the capabilities of the element developed and the typical rate of convergence towards the correct solution.

7. REFERENCES

Barton, N.R. (1971). A relationship between joint roughness and joint shear strength. Proc. Int-Symp. Rock Fracture. ISRM. Nancy.

Ghaboussi, J.E.L. Wilson and J. Isenberg (1973). Finite element for rock joints and interfaces. J. Soil. Mech. and Found. Div. ASCE, vol. 99, nº SM10.pp.833-848.

Goodman, R.E., R.L. Taylor and T.L. Brekke (1968). A Model for the mechanics of jointed rock. J. Soil Mech. and Found. Div. ASCE. Vol. 94. nº SM3 pp. 637-659.

Goodman, R.E. (1975). Methods of Geological Engineering in Discontinous Rocks. West Publishing Company. St. Paul.

Goodman, R.E. and C. St. John. (1977). Finite element analysis for discontinuous rocks. In Numerical Methods in Geotechnical Engineering. Ed. by C.S. Desai and U.T. Christian. McGraw-Hill.

Heuze, F.E. (1980). New models for joints and interfaces. VC RL-8522-Preprint- Lawrence Livermore Laboratory.

Ladanyi, B. and G. Archambault. (1970). Simulation of shear behaviour of a jointed rock mass. Proc. 11th. Symp. Rock Mech. AIME, pp 105-125.

Wilson, E.L. (1977). Finite elements for foundations, joints and fluids. In Finite Elements in Geomechanics Ed. by Gudehus. Wiley.

ANALYSIS OF ADVANCING TUNNELS IN ROCK
Comportement des tunnels au voisinage du front de taille
Analytische Untersuchungen zum Gebirgsverhalten beim Tunnelvortrieb

Antonio P. Cunha
Research Officer of the Underground Construction Division, Laboratorio Nacional de
Engenharia Civil (LNEC) Lisbon, Portugal

SYNOPSIS

A Finite Element study of axisymmetric tunnels, advancing in linear continuous rock masses is described, isotropic, anisotropic, homogeneous, and non-homogeneous situations being considered. Advancing in the neighbourhood of a contact between two geological formations and the crossing of a geological fault are studied with some detail.

RESUME

On présente une étude par éléments finis de l'avancement de tunnels axissymétriques en massifs rocheux simulés comme moyen continu de comportement linéaire. Des situations d'isotropie, anisotropie, homogénéité et hétérogénéité sont considérées. En particulier, les traversées d'une faille et d'un contact entre deux formations rocheuses sont étudiées en détail.

ZUSAMMENFASSUNG

Es wird eine Untersuchung mit Hilfe der finiten Elemente über den Vortrieb von achsen-symmetrischen Tunneln in Felsen beschrieben, wobei das Gebirge als lineares Kontinuum betrachtet wird, und isotrope, anisotrope, homogene und nichthomogene Situationen berücksichtigt werden. Es wird der Vortrieb in der Nähe eines Kontaktes zwischen zwei geologischen Formationen und das Durchschreiten einer Störung im Einzelnen untersucht.

1. INTRODUCTION

Advancing of tunnels — whatever the excavation and lining systems, the initial state of stress, the mechanical properties of the surrounding rock and the shape, dimensions and depth of the underground structure — is a paradigmatic example of a three-dimensional equilibrium, whose solution may be obtained by the finite element method. The accuracy of such an analysis is, of course, strictly dependent both on the modelling approaches and the adequacy of the parameter values considered in calculus.

If the rock mass is either scarcely or intensively fractured, a continuous approach is very often suitable for the analysis. Besides, if the tunnel has a circular cross-section and is located deeply enough to allow the consideration of an hydrostatic and uniform initial state of stress, the equilibrium is of the axisymmetric type, provided any excavation, lining, heterogeneity or anisotropy fits the same symmetry condition.

In the paper a study is described of axisymmetric advancing tunnels, in linear continuous modelled rock mass. Isotropic and anisotropic situations, as well as homogeneous and non-homogeneous ones are considered. Advancing in the neighbourhood of a contact between two geological formations and the crossing of a geological fault are described with some detail.

2. ANALYSIS OF ADVANCING TUNNELS

2.1. Axisymmetric model

A finite element axisymmetric model was developed for the special purpose of studying the phenomenology of advancing tunnels (Cunha, 1981). The computing method allows a step-by-step analysis in which, by deactivating ground elements or activating support ones, at appropriate stages, during the calculations, excavation phases or liner installation can be simulated. Whenever new elements are activated or deactivated, a geometrically non-linear problem arises, as a new structure appears, in which there is no equilibrium between forces corresponding to the previous stage loads and the stress field. The problem is solved, based on a procedure proposed by Ghaboussi and Ranken (1974): a displacement increment is calculated at each stage, corresponding to the non-balanced forces plus the incremental loads of the stage. As regards non-linear material problems, solution is achieved by using iterative techniques together with stress-transfer methods.

A more thorough description of the applied model - including the basic geometrical and mechanical assumptions of the axissymetric approach, its scope and limitations, the displacement formulation of the FEM study of axissymetric equilibria, the mechanical behaviour of materials considered (isotropic and transverse-isotropic linear-elastic materials or non-linear isotropic, both elastoplastic and with residual strenght ones), the technique used for the simulation of the excavation, and the computer program with its flow-chart and finite elements available — — can be found in the paper (Cunha,1981) referred above.

For the research reported hereafter, a FEM mesh involving three-nodal and four-nodal isoparametric elements (1492 finite elements, 1441 nodal points) was used (fig. 1) and unsupported tunnels were considered in a multi-step advance through elastic rock formations.

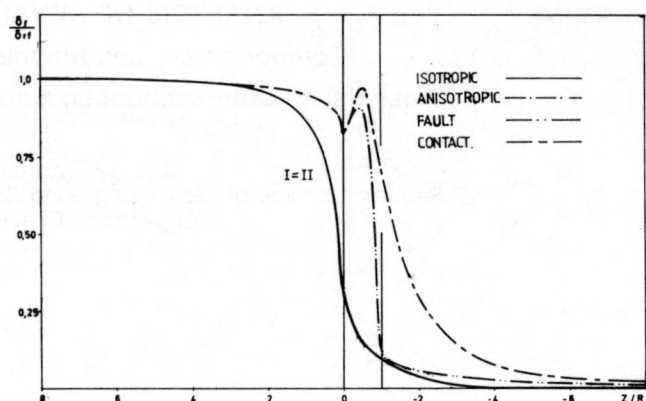

Fig. 2. Radial displacements along the wall

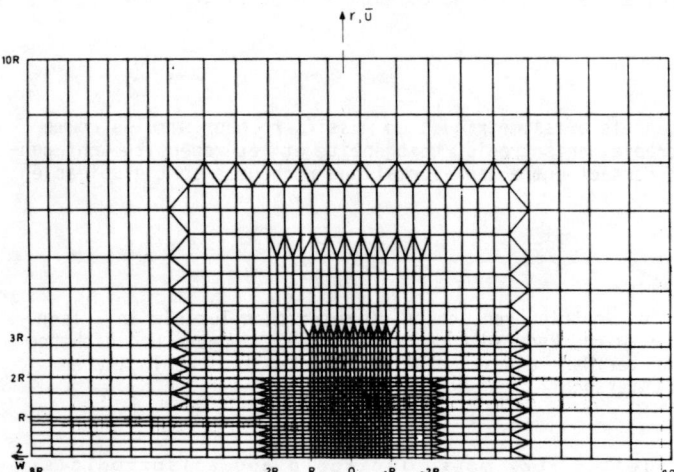

Fig. 1. Finite element mesh

2.2. Advancing in homogeneous media

With the aim of obtaining a clear reference situation for the forthcoming research, a complete graphic description of the displacement and stress fields around a non-lined tunnel in a homogeneous and isotropic rock mass (E, ν) was made, which was reported in detail elsewhere (Cunha, 1981). Radial displacements of the tunnel wall and longitudinal displacements of the face are presented here in fig. 2 and 3, respectively, for comparative purposes with the results presented hereafter.

A tunnel driven in a homogeneous but transverse-isotropic rock mass (two elasticity moduli E_1, E_2, two Poisson's ratio ν_1, ν_2 and a shear modulus G) was studied, the anisotropy axis being coincident with the tunnel axis. The change in calculation parameters was reduced to a minimum, relatively to the previous isotropic situation. Thus, it was assumed that the rock mass kept, in the cross section, the same mechanical properties, but the longitudinal elasticity modulus was reduced to 0.25 of its previous value, and the correspondent Poisson's ratio corrected accordingly. The distorsion at plane ORZ was also maintained, by using the same independent elastic constant G. As transverse deformability has been kept unchanged, wall radial displacements are practically identical to isotropic ones (fig. 2) but face longitudinal displacements increase significantly (fig.3) due to the increase in the longitudinal deformability. Such difference will depend directly on the anisotropy coefficient of the rock mass.

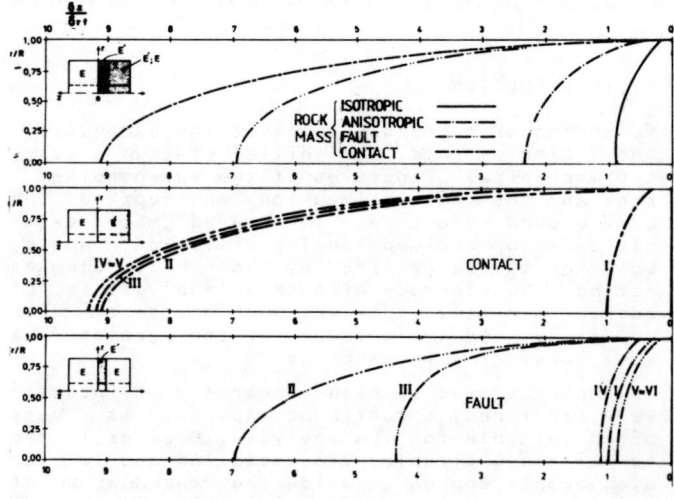

Fig. 3 - Longitudinal displacements of the face

F 154

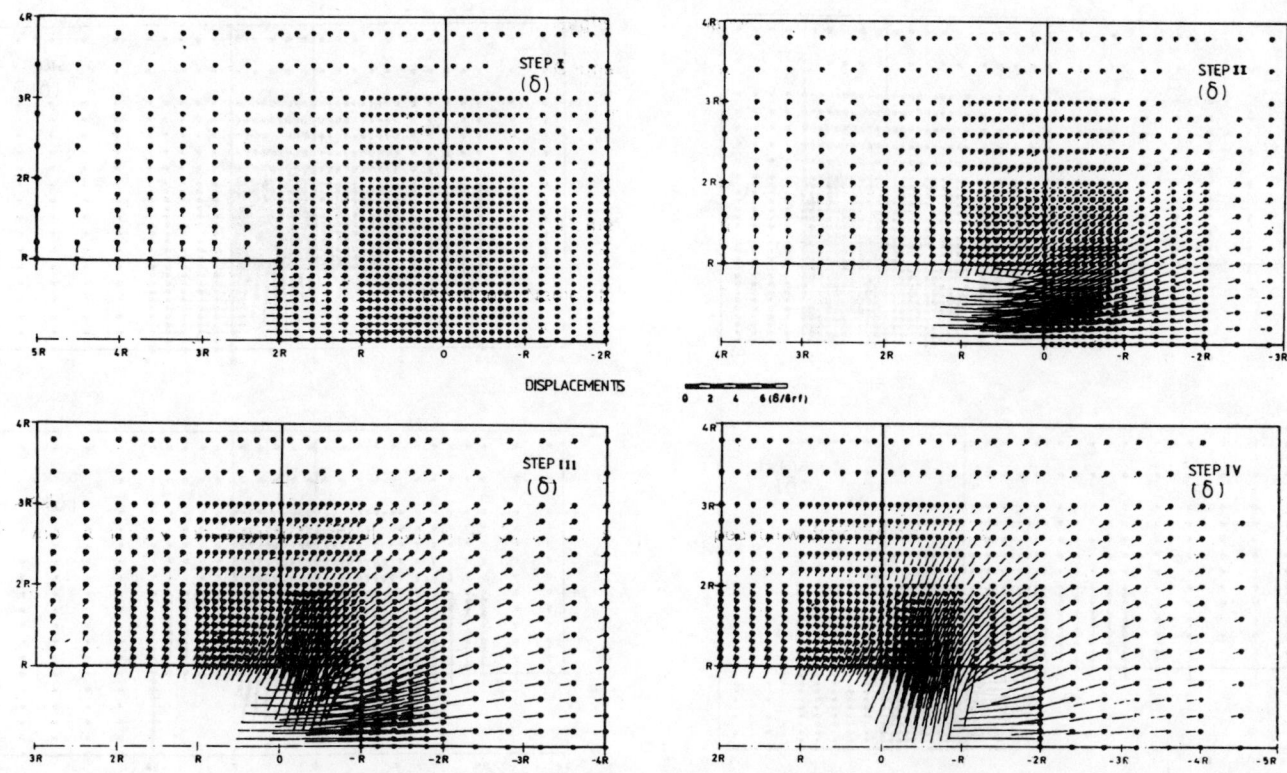

Fig.4. Displacement field during the step-by-step advance through a contact

2.3. Advancing in heterogeneous media

To enlighten the influence of rock heterogeneity on tunnel behaviour, near the face zone, namely when a weaker formation is encountered, the advance in the neighbourhood of a vertical contact between two rock masses with different mechanical characteristics, was considered. The tunnel was leaving a rock formation, with an elasticity modulus E (identical to the isotropic medium considered in section 2.2) and entering a soft rock mass with $E' = 0.1\ E$. Fig. 4 illustrates the displacement fields corresponding to four successive positions of the face relative to the rock contact. In step one, the face is one diameter before the contact, and the displacement pattern remains almost similar to the initial homogeneous medium; in step II (face on the contact) an explosion can be observed in the displacement field, at the face and in the non--excavated zone, corresponding to a ten times greater deformability of the second medium. Excavation rounds are then reduced and steps III and IV show that, for the second rock mass, the contact acts like a stiff wall, preventing the expression of longitudinal displacements in its vicinity. Thus, in the second medium the displacements of the wall zone near the contact have a predominantly radial pattern. In fig. 2, radial wall displacements for step II are drawn, clearly increasing behind the face, relative to the homogeneous situations reported in section 2.2. As regards fig. 3 (intermediate graph) the step-by-step longitudinal face displacements are shown, virtually similar to the homogeneous medium of section 2.2 in step I, varying with $1/E$ in step II-V.

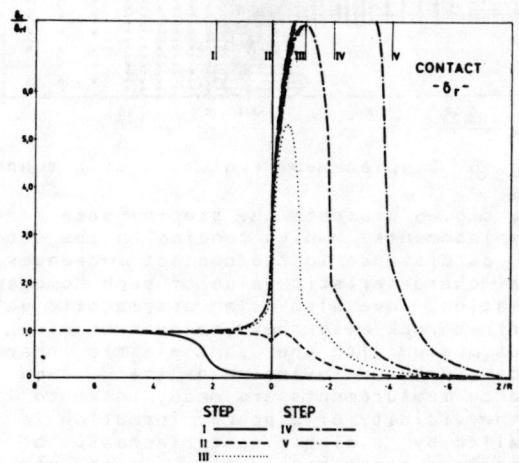

Fig. 5. Radial wall displacements during the advance through a geological contact

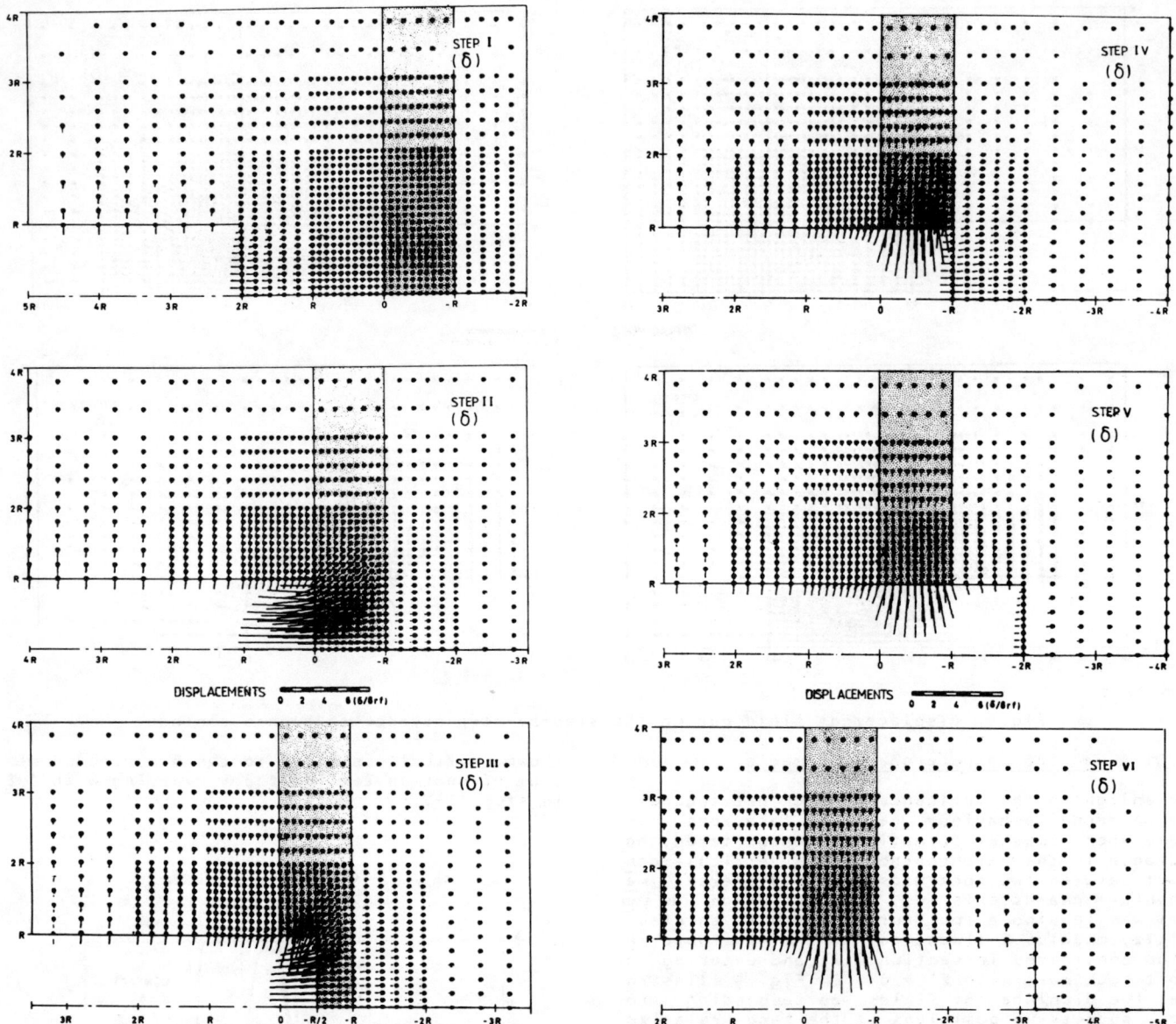

Fig. 6. Displacement evolution of a tunnel during the step-by-step advance through a fault

Finally, fig. 5 presents the step-by-step radial wall displacements, which tending in the second medium, as distance to the contact increases, to the final characteristic value of such homogeneous formation, have also as an assymptotic value in the first rock mass the displacement of a homogeneous medium with the same elastic characteristics. Such conclusion means that, even if convergence measurements are made, next to the face, the vicinity of a poorer formation is only signalled by a significant increase of the convergencies after penetrating in the new medium, that is, the rock mass gives no previous clear warning of the next sudden deterioration of its characteristics when convergence measurements are carried out inside the excavation, during tunnel advance.

Another heterogeneity situation analysed was the advance of a circular tunnel through a geological accident, perpendicular to the tunnel axis. The fault was filled by soft material, with precarious mechanical properties ($E' = 0.1\ E$) and had a thickness equal to the tunnel radius. Fig. 6 (displacement field for 6 advance steps, with a lenght per round in connection with the geological characteristics encountered) stresses a fan-shaped expression of the fault displacements, as the distance to the face increases, suggest - ing the image of a deformable beam supported at the two adjoining stiff zones. The face and wall displacements are obviously conditioned by the characteristics of the media in which they are located, at each advancing stage.

In fig.2, wall radial displacements for step II are also presented. It can be shown that behind the face, they are similar to the homogeneous medium E, until the fault boundary is reached; a sudden increase arises inside the fault, owing to its much greater deformability, but the assymptotic displacement, far from the face, in the excavated zone, as the wall is placed in the first homogeneous medium E, corresponds again to the characteristic value of such medium.

As regards the face deformation, fig. 3 (lower graph) shows that a maximum is reached at stage II (all the fault is then contributing to the face displacement) and it reassumes the characteristic displacement values of the homogeneous stiffer medium, at small distances from the fault. The stiffness of the rock formation conditions the deformation of the fault as well, since face displacements never reach, during the crossing of the fault, the 10 times higher displacement value which would correspond to the deformability of the fault material. The same happens with the radial displacements of the wall (fig. 7), which reassume, on either side of the geological accident, the expression and amplitude corresponding to the surrounding rock mass, without clearly signalling the existence of the fault, in terms of deformations, in the adjoining zones. Again it is found that evidence of the fault is not clearly marked by an increase of convergence, except during the crossing of the accident itself.

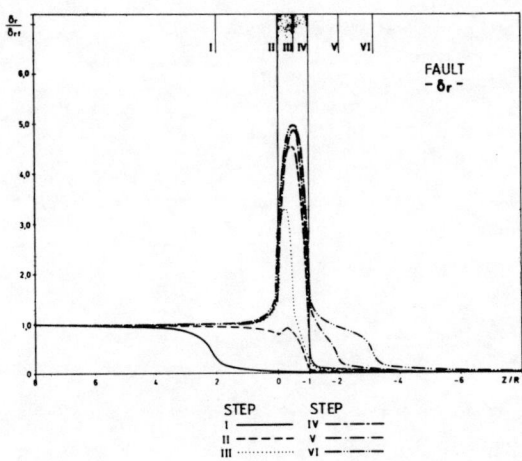

Fig. 7. Radial wall displacements during the advance through a geological accident

4. CONCLUSIONS

A study was described of the influence of anisotropy and heterogeneity in tunnel advance. Both wall and face displacements are, at each stage of the excavation, directly dependent on the deformability characteristics of the media in which they are located. Thus, the rock mass doesn't allow, by a previous neat increase of convergency, an anticipated clear warning of the next sudden deterioration of its characteristics, even if convergence measurements are carried out near the face.

5. REFERENCES

Cunha, A.P. (1981). Aplicação de modelos matemáticos ao estudo de túneis em maciços rochosos (Mathematical modelling of rock tunnels). Research Engineer Thesis at LNEC 342 pp, Lisbon.

Ghaboussi, J., Ranken, R.E. (1974). Tunnel design considerations - analysis of medium support interaction. Univ. Illinois, Urbana.

Pedro, J.O. (1973). Finite element stress analysis of plates, shells and massive structures: Int. Course on Struct. Concrete CEB, Lisbon.

Rocha, M. (1976). Estruturas Subterrâneas. LNEC, Lisboa.

Sousa, L., Teles, M. (1980). Modelo de cálculo para estudo de túneis pelo método dos elementos finitos. Internal Report, 545 pp, LNEC, Lisbon.

EXTENDED BOUNDARY ELEMENT METHODS IN THE MODELLING OF BRITTLE ROCK BEHAVIOUR

Extension des méthodes d'éléments limite pour modélisation du comportement de roches dures et fragiles

Erweiterte "boundary element" Methode zur Modellierung von Sprödbruchverhalten von Gebirge

A. P. Peirce
J. A. Ryder
Chamber of Mines of South Africa Research Organization, Johannesburg, South Africa

SYNOPSIS

A plain strain boundary element formulation is described for modelling inelastic behaviour of a brittle rock continuum in which mining excavations are embedded. Efficient iteration, kernel pre-calculation and lumping procedures were devised to enhance computational efficiency. A novel interpretation of interacting element self-effects in terms of load lines further facilitates the modelling of the non-linear zones of the rock mass. The hard brittle quartzites of South African gold mines exhibit extreme load-shedding, strengthening with confinement according to a Mohr-Coulomb law, and strong post-failure volume dilatancy. The behaviour of such rock in the fracture zone around square tunnels is modelled as a case study in the paper.

RESUME

On décrit un plan de déformation d'allongement des éléments limite pour modéliser le comportement inélastique d'une roche dure et fragile renfermant des travaux miniers. Des itérations efficaces du coefficient d'influences pré-calculées et le procédé de groupement ont été établis pour améliorer l'efficacité des traitements de données. Une interprétation originale des éléments interactifs "effet indépendant" au niveau des charges a l'avantage de faciliter la modélisation des zones non-linéaires des masses de roche. Les quartzites durs et fragiles des mines d'or Sud-Africaines montrent que le déslestage extrême augmente la compression d'après la loi de Mohr-Coulomb et l'élargissement rapide des volumes apparaît "post rupture". On modélise le comportement d'une telle roche dans la zone de fracture autour de tunnels de section carrée pour en donner un exemple.

ZUSAMMENFASSUNG

Das inelastische Verhalten eines bergmännischen Abbaues in sprödem Gebirge wird mit Hilfe von "boundary elements" des ebenen Dehnungszustandes modelliert. Wirtschaftliche Iteration, Vorherberechnung von Einflußkoeffizienten und Zusammenfassung der Einflußbereiche wurden zur Verbesserung der Berechnungswirtschaftlichkeit entwickelt. Ein neue Interpretation der Wechselwirkung von Einflußelementen in Bezug auf Lastlinien, vereinfachen weiterhin das Modellieren von nichtlinearen Zonen in dem Gebirge. Die harten und spröden Quartzite der südafrikanischen Goldminen weisen extremen Lastabfall nach dem Bruch, sowie eine Verfestigung mit steigenden Umschlingungsdrucken, entsprechend des Mohr-Coulomb Gesetzes und stark nachbrüchige Volumensvergrößerungen nach dem Bruch auf. Das Verhalten eines derartigen Gebirges in der Bruchzone eines quadratischen Tunnelprofiles wird als typischer Fall beschrieben.

1. INTRODUCTION

Since the late 1960's, the gold mining industry in South Africa has made increasing use of numerical rock stress analysis programs as an integral part of the mine design process. "MINSIM" (Plewman, Deist and Ortlepp, 1969) and its successors currently account for nearly 1500 problem runs per year in South Africa, where rock mechanics engineers are interested in designing deep level tabular mining layouts and sequences aimed at minimizing service excavation stresses and face energy release rates. "MINAP" (Crouch, 1976) and its extensions is another popular Boundary Element program; restricted to 2 dimensions but otherwise more general in that excavations need not all be tabular, and inhomogeneities, fault-sliding and other quasi-nonlinearities can easily be studied. These programs base their applicability on the observation that remote from excavations, South African rock strata commonly appear to behave in a strictly linear-elastic fashion (for example, Ryder and Officer, 1964). It is clear however that for designing support to control instabilities of pillar, haulage or ore-pass sidewalls or stope hangingwalls, to understand the vagaries of face pre-fracturing - a necessity for economic non-explosive stoping in hard rock (Joughin, 1976) - or to obtain any measure of quantitative understanding of the rockburst problem, some form of modelling of the inelastic behaviour of the skin of fractured material surrounding deep-level excavations must be undertaken (Figure 1). Two pioneering workers in the field (Deist, 1966; Crouch, 1970) recognized this need by taking it as the theme in their respective doctoral theses.

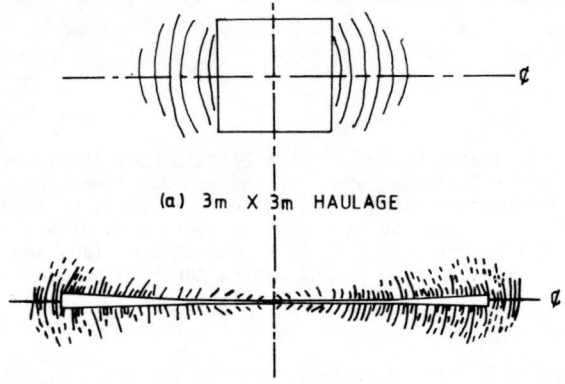

FIGURE 1: Typical South African Gold Mine Excavations at a depth of more than 2000m in hard brittle quartzite, showing the extent of fractured ground (after Adams et al., 1981)

Few aspects of the problem yield easily to analysis. In-situ studies of the detailed development, extent and mobility of the fracture zone are difficult, though some progress has been made: Fig. 1b depicts the shape of the fracture zone surrounding a deep stope. The extension-type fractures ahead of such stope faces are subtly banded and occur together with less common inclined fractures along which displacements of up to 140mm have been observed (Adams, et al., 1981). Laboratory studies of the full triaxial behaviour of brittle rock required the development of stiff testing machines (Cook and Hojem, 1966) (Figure 2), but the dependence of particularly the detailed post-failure characteristics on such aspects as specimen size and geometry, end effects, strain-rate, loading path, intermediate principal stress, lateral and axial stiffness of the testing apparatus and the like, remain still very much open questions. However, the qualitative aspects of Figure 2 can be taken as established: the rapid increase in peak strength with lateral confinement according to an approximately linear Mohr-Coulomb law, the pronounced negative-slope or load-shedding post-failure characteristic

with equally pronounced lateral dilatation, tending finally into a regime of heavy damage where behaviour may perhaps be described in terms of simple plasticity or soil mechanics theory.

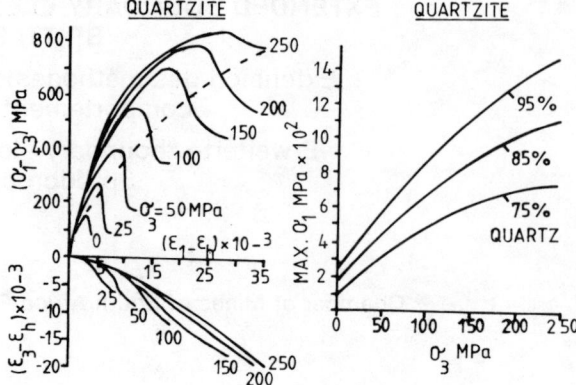

FIGURE 2: Triaxial behaviour of quartzite (after Stavropoulou, 1982)

In this paper a novel boundary element methodology for modelling the kind of behaviour shown in Figure 2 is introduced. Some key points in the particular modelling philosophy are reviewed. In addition simplifications chosen to reduce the complexities of the real world suggested by Figure 1b and 2 to more reasonable bounds are discussed. The simplifying assumptions marked (x) in Section 2 can in fact be relaxed relatively easily, and at the cost of some increase in computational effort, permit more general configurations to be modelled.

2. MODELLING APPROACH

2.1 Plane-strain

Many important mining situations such as shown in Figures 1a and 1b are essentially two dimensional in character, and the same appears to be true of the associated fracture zones. Plane-strain modelling is therefore followed throughout and the constitutive laws of brittle material (Figure 2) are assumed to be available directly in plane-strain form, obtained either by direct measurement or by specialization of more general polyaxial laws of behaviour.

2.2 Four-way symmetry (x) (Figures 1a and 1b)

The assumption of four-way symmetry reduces the scale of the modelling by a factor of four; though certain, hopefully non-essential, asymmetries between hanging and footwalls, such as organized shifts in material properties or bedding-plane separations, can no longer be modelled explicitly.

2.3 Homogeneity of behaviour (x)

All rock both inside and outside the fracture zone is assumed to follow the same constitutive law of Figure 2. This assumption is not well substantiated in the field, where fracture bands are often seen to terminate or drastically change in density at parting planes or shale layers. This may be mainly a question of detail: the existence and general extent of fracture bands and zones does not seem to demand the existence of inhomogeneities of this type.

2.4 Static analysis

Laboratory studies have ruled out strong pre-failure time dependencies of brittle rocks such as quartzite or norite. Post-failure, the situation may be different and field data of closure in stopes subsequent to blasting suggest time-constants in the fracture zone in the order of hours or days (Kersten, 1981). Provided Figure 2 represents conditions of appropriately slow loading rates, it could be argued (as in the present paper) that given small increments in loading, static analyses of fracture zone development should suffice. If, however, instabilities in the form of small strain bursts or full-scale rockbursts manifest themselves, Deist, (1966) may well have been right in asserting that full dynamic modelling of the propagation of stress-strain waves and the dynamic build-up of any newly-failed regions is imperative because of the probable strong path-dependence and localization phenomena involved. Time wave-propagations add a full dimension to the computational modelling effort involved, and so, for the time being at least, this complication is ignored.

2.5 Continuum modelling

The high density of fracturing in an otherwise strongly-knit fabric, seen even in the exposed stope hangingwall, suggests that it would be inappropriate to attempt to model discrete blocks (per Cundall, 1974) or discrete fractures in the sense of limiting equilibrium or fracture-mechanics studies. Rather a continuum view is pursued. an arbitrary sub-division of the fractured zone into elements, each possibly containing numerous fractures, but which follow phenomenological stress-strain laws of the nature of Figure 2, is made.

2.6 Boundary element (BE) vs Finite element (FE)

Given a continuum formulation in which at least the fractured inelastic zone must be divided up into a large number of individual elements, it might be thought that FE would be the preferred method of modelling. FE codes have reached a high degree of sophistication over the past two decades. FE have been successfully used in modelling not only linear inhomogeneities but also a variety of non-linear problems involving plastic flow in metals or consolidation of clay-like materials (Zienkiewicz, 1977). On the other hand, there are some potential drawbacks:

a) FE requires the entire continuum - unfractured as well as fractured - to be divided up into elements. Mining excavations are embedded in very large masses of undisturbed rock and huge numbers of elements are ordinarily required - whence the historical popularity of BE over FE for modelling elastic mining problems. This disadvantage has to a large extent fallen away in recent years. One solution (Crouch, 1970; Brady and Wassyng, 1981) has been to surround those FE's involved in modelling non-linear behaviour with a skin of BE's which effectively 'see' the infinite intact rock mass: this is however a hybrid marriage of disjoint philosophies with concomitant numerical complications. A potentially more elegant and simple solution may lie in the recent invention of 'infinite elements' (Bettes, 1977; Beer and Meek, 1981) which do not complicate the basic FE code and require only a fairly thin overcoat of normal FE's to model the nearby elastic portions of the problem geometry.

b) Operational experience in the early 1970's with a simple FE program in modelling slot-like excavations with extreme aspect ratios (Figure 1b) had suggested matrix ill-conditioning problems; whereas suitable BE formulations can certainly model this geometry in elastic ground without particular difficulty (Figure 3).

c) FE normally uses direct system matrix solution techniques, and while these have been made very efficient, considerable computing resources are required if a large number of elements have to be included as in the present problem. BE on the other hand normally use iterative system solution techniques very successfully and this fact alone could swing the computational balance strongly in favour of BE.

For these reasons, and also admittedly very largely because of the authors' greater familiarity with BE philosophy and methodologies, the latter was chosen to provide the required framework for modelling not only the boundaries of the mining geometry but also the inelastic behaviour of the fractured zone itself. The name "SIMPLER" has been given to the computer code involved.

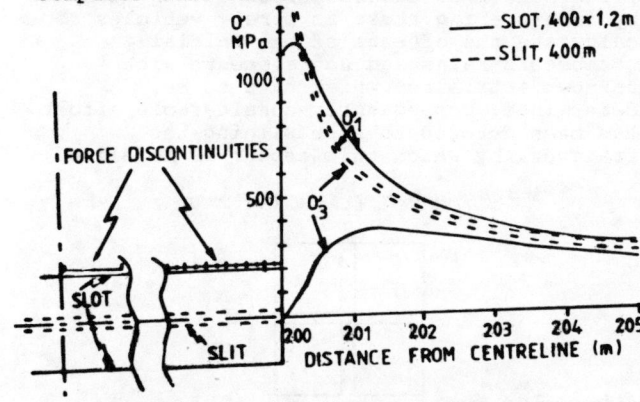

FIGURE 3: Stress distribution along the reef plane in front of the face of a 400x1,2m stope in purely elastic unfractured ground, modelled by B.E. analysis

3. THE BOUNDARY ELEMENT FORMULATION

3.1 Mining Boundaries

The methodology is similar to that used in MINAP (Crouch, 1976) though SIMPLER is in fact a totally new code. An indirect (Banerjee and Butterfield, 1981) formulation is set up for the boundaries using force discontinuities rather than Crouch's displacement discontinuities to avoid numerical difficulties associated with large aspect-ratio stopes and to give smoother stress profiles near the skin of the excavation. Induced traction boundary conditions are used and numerical stability appears to be satisfactory (Figure 3). A direct or 'boundary integral equation' approach (Banerjee and Cathie, 1979) would presumably have worked at least as well, but would have required double the number of remote influence coefficients and was rejected on these grounds.

3.2 Non-linear parts of the continuum

Since any BE formulation explicity assumes the presence of an infinite matrix of linear elastic material, what has to be modelled here are just those components of stress or strain that deviate from pure linear elastic behaviour. Use of BE for this purpose is not unheard of. Swedlow and Cruise, (1971) first formulated an "initial strain" approach which was successfully implemented by Riccardella, (1972). In Figure 5, for a given stress level σ active at some interior point in the body, an excess or 'initial' strain BC has to be invoked. Banerjee and Cathie, (1980) formulated an 'initial stress' approach, whereby corresponding to a given level ϵ of strain, a stress drop DC is involved. Banerjee and Mustoe, (1978) also formulated a 'fictitious body force' approach which generated the required amounts of either initial stress or
initial strain. These authors (Banerjee and Cathie, 1980)(Banerjee and Mustoe, 1978) have tended to gloss over the computational burden imposed by the presence of an in general very large number of excess stress/strain elements by interpreting these as merely vehicles to calculate the effects of the initial stresses/strains and not elements with unknown attributes which have to be determined. Consequently considerable effort has been devoted to streamlining the framework by which this aspect is modelled.

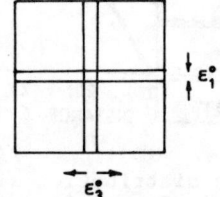

FIGURE 4: Cruciform injection of excess strains

In an initial strain approach, the simplest conceptual vehicle for injecting excess strains into an otherwise elastic element would be to set up a network of cruciform displacement discontinuities (Figure 4) which embody the amounts of contraction ϵ_1° and lateral dilatation ϵ_3° required to simulate the stress/strain behaviour shown in Figure 2. A program such as MINAP could handle such a formulation with minimal modification. Inclined strain fields could be modelled in a fixed rectilinear framework by simply including suitable ride discontinuities (Clifton, 1979). This approach could, indeed in principle, model any required constitutive behaviour, but in practice the sharp stress gradients generated by these discrete discontinuities cause numerical impracticalities. SIMPLER embodies the most obvious refinement of Figure 4 - smearing of the cruciform discontinuities across the full area of the square element, creating thereby what is essentially a discretized piecewise constant, excess strain field. Because a square element is not symmetric about every line through its centroid, the introduction of an excess strain field whose principal axes do not line up with an axis of symmetry of the element generates small spurious second order shear stresses along these principal axes. A case could be made for 'smearing' these discontinuities into a circular outline (Starfield, 1982) thereby obtaining an element with complete isotropy with regard to the injection of strain, but this possiblity has not been pursued as yet.

3.3 Discretization, system matrix and lumping

SIMPLER at present embodies a 4-way symmetric plane-strain geometry, discretized into up to 21x21 square elements (later to be enlarged to about 200x40 square elements of unequal size) each capable of carrying four non-zero excess strain components ϵ°. Up to 42 mining boundary elements, each carrying two piecewise-constant force discontinuity components τ, can be invoked oriented collinear with parts of the square grid. In its present form, SIMPLER is capable therefore of modelling geometries such as that of Figure 1a but not stopes of the extreme aspect ratio of Figure 1b. In outline (further details in the Appendix) the system matrix takes on the following form:

$$\begin{bmatrix} B & B' \\ \hline D' & D \end{bmatrix} \begin{bmatrix} \tau \\ \hline \epsilon^\circ \end{bmatrix} = \begin{bmatrix} t \\ \hline \sigma \end{bmatrix} \quad \ldots \ldots (1a)$$
$$\quad \ldots \ldots (1b)$$

The discretization into square elements permits the following considerable savings of computing time and memory resources:

a) Pre-calculated tables of influence coefficients can be held in memory, so that submatrices D, D' and B' of equ. (1) can be indexed rapidly by the iteration scheme whenever required.

b) Remote influences can be 'lumped' (Plewman et al., 1969). Specifically, in D of equ. (1), any 3 x 3 group of elements further than 10 grid distances from the 'receiving' elements on the main diagonal are lumped into one large element whose influence coefficient has been pre-calculated and stored. This technique alone reduces computational effort about ten-fold, and offsets to a great extent the disadvantage of a BE over a FE formulation of not enjoying a sparse system matrix.

3.4 Iterative solution technique

Submatrix B in equ. (1) is strictly linear and can thus be pre-inverted by LU decomposition, permitting an efficient block-relaxation scheme for solving the (1a) and (1b) segments of equ. (1) as a whole. The more formidable problem of solving (1b) is non-linear in character. A Gauss-Seidel iteration approach is the natural one, which in SIMPLER is generalized in a novel and also physically meaningful manner which seems to unify the 'initial stress' and 'initial-strain' approaches. Consider the stress vector at a typical receiving element in equ. (1b) centered at $\underline{p}'$:

$$\underline{\sigma}^{ext}(\underline{p}') + D'\underline{\varepsilon}^o(\underline{p}') = \underline{\sigma}(\underline{p}') \quad \ldots\ldots\ldots \quad (2)$$

Here, $\underline{\sigma}^{ext}(\underline{p}')$ comprises those stress components imposed at the centroid of the receiving element by all external influences including virgin stresses; $D'\underline{\varepsilon}^o(\underline{p}')$ represents the large self-effect stresses generated by the excess strain $\underline{\varepsilon}^o$ in the element itself, and $\underline{\sigma}(\underline{p}')$ is the resultant state of absolute stress which must be compatible with $\underline{\varepsilon}^o$ in terms of the constitutive law of Figure 2.

Given a fixed and known $\underline{\sigma}^{ext}$, equ. (2) implies the presence of a kind of 'load line' whose intersection with the constitutive curves defines the point of solution – Figure 5 illustrates one component only.

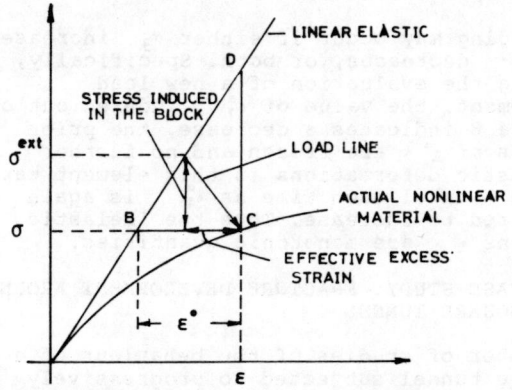

FIGURE 5: Representation of the initial strain, initial stress and load-line approaches

This generalized load-line is an improvement over either initial stress or initial strain: the solution point obtained thereby is fixed and requires no further iteration if $\underline{\sigma}^{ext}$ is itself fixed. In practice $\underline{\sigma}^{ext}$ does change slowly during the iteration procedure until final convergence is obtained; however the procedure is a complete analogue of how Gauss Seidel iteration works in a purely linear environment. Convergence should be somewhat enhanced, and a further advantage of the approach is that the strongly interacting horizontal and vertical components symbolized in matrix form in equ. (2) can be solved by algebraic or numeric means to obtain explicit relationships for each component of $\underline{\varepsilon}^o$ in terms of each component of $\underline{\sigma}^{ext}$ (Figure 8).

The physical background to Figure 5 is sketched in Figure 6. As already pointed out, any BE formulation assumes the presence of an infinite matrix of linear elastic material into which each element is inextricably embedded. When initial strains $\underline{\varepsilon}^o$ are injected into the element under consideration, the elastic surroundings react with stiffness D', corresponding to the slope of the load line. In particular, lateral dilatation will generate horizontal compression which, because of Mohr-Coulomb strengthening, leads to autogeneous stiffening of the material as suggested by the dotted load path in Figure 2. Of course this stiffening is overstated: the enveloping body actually contains free surfaces and inelastic blocks of ground, but as the iteration proceeds the effect of these is brought fully into play on to each element in turn. Temporarily however the extreme instabilities suggested by the solid curves of Figure 2 are not manifested and this may enhance to some extent the numerical stability of practical studies.

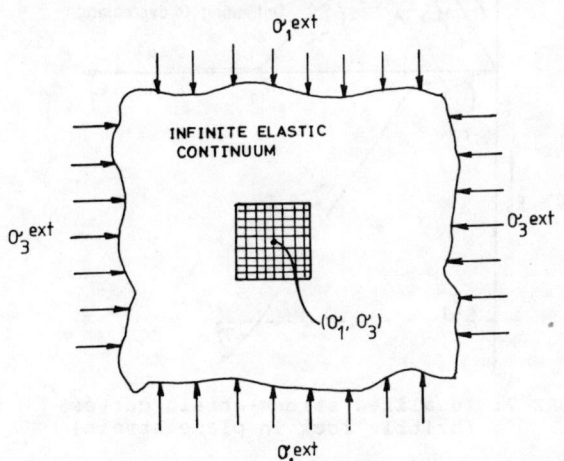

FIGURE 6: Physical interpretation of Gauss-Siedel iteration in SIMPLER

4. IDEALIZED CONSTITUTIVE LAW

4.1 Absolute vs Incremental Formulation

In plasticity theory, constitutive laws can be cast in one of two forms: a deformation formulation which is essentially a nonlinear theory of elasticity and an incremental formulation which is designed to cater for materials for which unique curves of the type of figure 2 do not hold for load paths likely to be encountered. Edelman (1951) has demonstrated the coincidence of these formulations for cases of proportional boundary loading similar to those we are considering. For brittle rocks, path dependency in pre-failure loading appears to be small (Stavropoulou, 1982), and certainly laboratory data required to fix the intricacies of path-dependent post-failure behaviour is woefully lacking at present.

While SIMPLER could easily handle incremental constitutive laws, a type of deformation formulation in absolute terms has initially been chosen: Figure 2 is assumed to represent a universal and unique set of curves, defining stress/strain behaviour irrespective of loading history or loading path.

4.2 Idealized curves

SIMPLER could be programmed to handle Figure 2 numerically as a set of arbitrary curves. Initially again, these have been simplified into a more tractable set (Figure 7); one which can be described in terms of a total of 6 parameters only: E, ν (elastic constants); k, σ_c (linear Mohr-Coulomb strength parameters) and $\bar{E}, \gamma$ (post-failure negative modulus and dilatation constant).

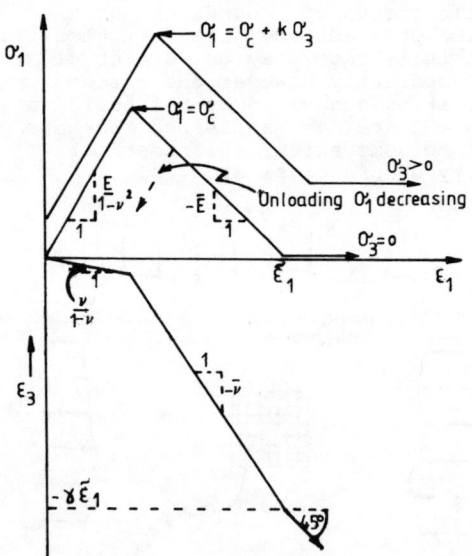

FIGURE 7: Idealized stress-strain curves (brittle rock in plane strain)

The idealizations include:

a) Pure elastic behaviour persists right up to peak strength point.

b) Peak strength follows the linear law
$\hat{\sigma}_1 = \sigma_c + k \sigma_3$

c) Post-failure load-shedding follows a constant negative modulus $\bar{E}$ in plane strain (and constant "Poissons ratio" $\bar{\nu}$).

d) Maximum stress drop is σ_c for all values of confinement.

e) Zero incremental dilatation ($d\epsilon_1 = -d\epsilon_3$) in the heavy-damage 'plastic' regime.

It can be seen from Figure 2 that while considerable liberties have been taken with quantitative aspects, hopefully the main qualitative aspects of brittle rock behaviour have been retained.

Setting up and solving equ. (2) using these assumptions gives Figure 8, which can be used directly in Gauss-Seidel iteration and which again clearly illustrates the self-strengthening behaviour of rock embedded in an elastic matrix.

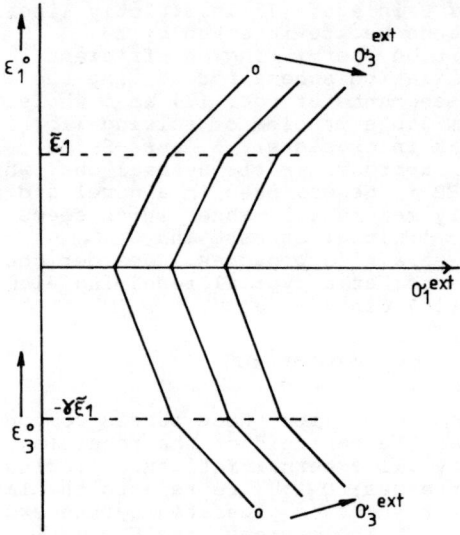

FIGURE 8: Idealized constitutive law for direct use in Gauss-Seidel iteration

4.3 Unloading

If failed rock is 'unloaded', then it is assumed to undergo strain increments which are purely elastic (Figure 7) - a mild idealization of laboratory evidence (Stavropoulou, 1982).

Unloading may occur if either σ_3 increases or σ_1 decreases, or both. Specifically, if during the evaluation of a new load increment, the value of ϵ_1^o indexed out of Figure 8 indicates a decrease, the prior values of $\underline{\epsilon}^o$ are frozen and no further inelastic deformations in this element take place - until such time as ϵ_1^o is again required to increase. Thus the inelastic strains $\underline{\epsilon}^o$ are monotonic quantities.

5. CASE STUDY: FRACTURE DEVELOPMENT AROUND A SQUARE TUNNEL

A number of studies of the behaviour of a square tunnel subjected to progressively increasing field stresses q_y, q_x were carried out. One quadrant of the tunnel is depicted in Figure 9(a); the 5x5 gridsize corresponds for a 3x3m tunnel to a blocksize of 0,3m. Parameters used were:

a) Horizontal: Vertical field stress ratio = 0,5.

b) Elastic moduli: $E = 70$ GPa,
   $\nu = 0,20$.

c) Unconfined plane-strain strength:
   $\sigma_c = 200$ MPa.

d) Negative slope modulus:
   $\bar{E} = 10\text{-}40$ GPa.

e) Mohr-Coulomb stiffening parameters:
   $k = 6\text{-}10$.

f) Dilatation parameter (Figure 7):
   $\gamma = 1,5\text{-}3$.

Typical results showing the onset and buildup of fracture are depicted in Figure 9(b). The extent of fracturing significantly exceeds the zone that would be predicted by simplistic application of elastic Mohr-Coulomb failure criteria.

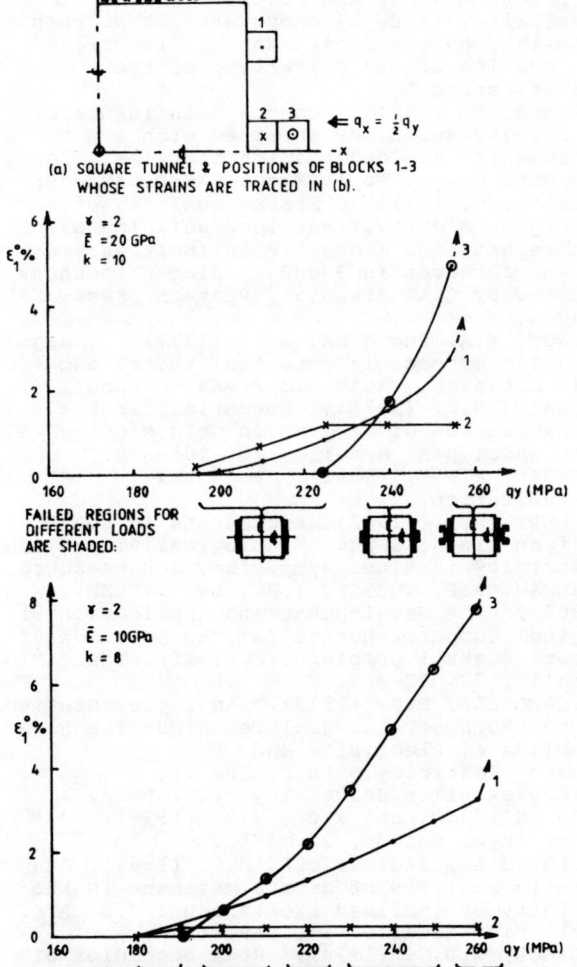

FIGURE 9: Fracturing around a square tunnel for various field loads and material properties

The detailed behaviour depicted in Figure 9(b) must be viewed as tentative at this stage, but some general observations may be of interest:

1) First onset of failure occurs at q_y values around 190 MPa. In actual fact tunnels in South African quartzites already show severe damage at field stresses of 100 MPa. This large discrepancy between strength in the laboratory and in the field is well known (e.g. Wilson and More O' Ferrall, 1970); the explanation may be partly size effect (large blocks contain more flaws than laboratory specimens), partly the effect of blast fractures dating from the creation of the tunnel, and partly the influence of geological inhomogeneities such as moduli changes which generate tensile stresses thereby reducing the effective strength of rocks in situ. To compensate, a case could be made for our reducing the effective strength parameter σ_c to around 100 MPa.

2) As the field load is increased, fracturing procedes in an initially stable pattern. At some load (260 MPa in Figure 9(b)) an unstable process initiates: a small increment in load leads to a large increase in the extent and degree of fracturing. The physics of this situation is akin to the well-known analysis of stiff vs soft testing machines (Cook and Hojem 1966; Salamon, 1974). It may have some connection with the phenomenon known as 'strain bursting' (Ortlepp, 1982); certainly it seems that such potential instabilities characterize any tunnel modelled with parameters appropriate to brittle rock. While SIMPLER could in principle be used to track the progress of such a 'strain burst', it is doubtful whether this would be physically meaningful: a full dynamic treatment such as that of Deist (1966) becomes imperative here.

3) As might be expected the main effect of an increase in the negative slope modulus $\bar{E}$ is to enhance the magnitude of fracturing and hasten the onset of instability.

4) The effects of increasing k or γ, thereby increasing the relative 'brittleness' of the rock modelled, are more subtle: there is a tendency for groups of fracturing blocks to line up vertically rather than in diagonal patterns. This correlates with some characteristics of fracture patterns of 'brittle' and 'plastic' rocks seen in the field.

5) Strong localizations of inelastic deformation tend to occur: fracturing concentrates and intensifies in one but not both of any two neighbouring blocks - (eg curves 2 and 3 in Figure 9(b)). Such localization phenomena have been discussed by Rudnicki and Rice (1975), and are certainly seen in the field in the bands of stope fractures reported by Adams et al., (1981); but it is unclear whether SIMPLER in its present form is correctly modelling this important and complex aspect.

6. CONCLUSIONS

A vehicle for modelling inelastic rock behaviour utilizing boundary elements has been described. Whether this has computational advantage over conventional FE modelling has yet to be established and a great deal of further evaluation and development is outstanding, but it is believed at this stage that at least two potentially important insights have flowed from the approach:

1) Because brittle rock dilates under load and shows strong strength increase with confinement, the concept of 'lateral stiffness' should in fact play a more dominant role than that of classical 'axial stiffness' - both in characterizing fracture and instabilities in the field, and in governing the design of appropriate testing in the laboratory.

2) The tendency for modelled blocks of brittle rock to form vertical zones of failure - a behaviour predictable from direct examiniation of the BE kernels in equ. (1) - seems to throw light on the mechanism whereby extension fractures form up to 40m above or below deep stopes.

ACKNOWLEDGEMENTS

The development of SIMPLER was carried out as part of a research programme of the Chamber of Mines of South Africa.

REFERENCES

ADAMS, G.R., JAGER, A.J., and ROERING, C. (1981). Investigations of rock fracture around deep-level gold mine stopes: Proc. 22nd U.S. Symposium on rock mechanics, Massachusetts Institute of Technology, 213-218.

BANERJEE, P.K., and MUSTOE, G.G.W. (1978). The boundary element method for two-dimensional problems of Elastoplasticity: Proc. Int. Conf. Rec. Adv. in boundary element meth., 283-300, Pentech Press, London.

BANERJEE, P.K. and CATHIE, D.N. (1979). A direct formulation and numerical implementation of the boundary element method for two dimensional problems of elasto-plasticity: Int. J. Mech. Sci., 22, 238-245.

BANERJEE, P.K. and BUTTERFIELD, R. (1981). Boundary element methods in engineering science. McGraw Hill, London.

BETTES, P. (1977). Infinite elements: Int. J. Num. Meth. Engng., 11, 53-54.

BEER, G., and MEEK, J.L. (1981). Infinite domain elements: Int. J. Num. Meth. Engng, 17, 43-52.

BRADY, B.H.G., and BRAY, J.W. (1978). The boundary element method for elastic analysis of tabular orebody extraction, assuming complete plain strain: Int. Rock Mech. Min. Sci. and Geomech. Abstr., 15, 29-37.

BRADY, B.H.G., and WASSYNG, A. (1981). A coupled finite element-boundary element method of stress analysis: Int. J. Rock Mech. Min. Sci. and Geomech. Abstr., 18, 475-485.

CLIFTON, R. (1979). Personal communication.

COOK, N.G.W., and HOJEM, J.P.M. (1966). A rigid 50-ton compression and tension testing machine: S. Afr. Mech. Engr., 16, 89-92.

CROUCH, S.L. (1970). The influence of failed rock on the mechanical behaviour of underground excavations. PhD. Thesis submitted to the faculty of the graduate school of the University of Minnesota.

CROUCH, S.L., (1976). Analysis of stress and displacements around underground excavations. An application of the displacement discontinuity method. Dept. of Civil and Mineral Engineering, University of Minnesota.

CUNDALL, P.A. (1974). Rational design of tunnel supports: A computer model for rock mass behaviour using interactive graphics for the input and output of geometrical data. Final report to the department of the U.S. Army. Contract No. DACW45-74-C-0066.

DEIST, F.H. (1966). The development of a nonlinear continuum approach to the problem of fracture zones and rockbursts and feasibility study by computer. Ph.D. thesis submitted to the Department of Electrical Engineering of the University of the Witwatersrand.

EDELMAN, F. (1951). On the coincidence of plasticity solutions obtained with incremental and deformation theories: Proc. 1st Nat. Congr. Appl. Mech., ASME, 493-498.

HOCKING, G. (1978). Stress analysis of underground excavations incorporating slip and separations along discontinuities: In: Recent advances in Boundary Element methods (edited by C.A. Brebbia), Pentech press, London.

JASWON, M.A. and SYMM, G.T. (1977). Integral equation methods in potential theory and elastostatics. Academic Press, London.

JOUGHIN, N.C. (1976). Potential for the Mechanization of Stoping in Gold Mines: J.S. Afr. Inst. Min. Metall., 76, 285-300.

KERSTEN, R.W.O. (1981). Personal communication.

ORTLEPP, W.D. (1982). Rockbursts in South African mines: a phenomenological view: ISRM Seismicity in Mines symposium, Johannesburg.

PLEWMAN, R.P., DEIST, F.H., and ORTLEPP, W.D. (1969). The Development and Application of a Digital Computer Method for the Solution of Strata Control problems: J.S. Afr. Inst. Min. Metall., 70, 33-34.

RICCARDELLA, P.C. (1973). An Implementation of the Boundary Integral Technique for Plane Problems of Elasticity and Elasto-plasticity. Ph.D. Thesis, Carnegie-Mellon University, Pittsburg.

RUDNICKI, J.W. and RICE, J.R. (1978). J. Mech. Phys. Solids, 23, 371.

RYDER, J.A., and OFFICER, N.C. (1964). An elastic analysis of strata movement in the vicinity of inclined excavations: J.S. Afr. Inst. Min. Metall., 64, 219-44.

SALAMON, M.D.G. (1974). Rock mechanics of underground excavations: Proc. third congress of Int. Soc. for Rock. Mech., 1, part B. Denver, Col.

STAVROPOULOU, V. (1982). Constitutive laws for brittle rock. PhD thesis submitted to Department of Mining Engineering, University of the Witwatersrand.

SWEDLOW, J.L., and CRUISE, T.A. (1971). Formulation of Boundary Integral Equations for Three-dimensional Elasto-Plastic Flow: Int. J. Solids and Structs, 7, 144-151.

STARFIELD, A.J. (1982). Personal Communication.

WILSON, J.W. and MORE O'FERRALL (1970). The application of the electrical resistance analogue to mining operations: J.S. Afr. Inst. Min. Metall., 64, 219-44.

ZIENKIEWICZ, O.C. (1977). The Fininte Element Method. 3rd Edition, McGraw-Hill, London.

APPENDIX

DETAILS OF INDIRECT B.E. FORMULATION

The integral equation expressing the traction vector $t_r(\underline{P})$ at a point $\underline{P}$ on a smooth part of the boundary S due to the force discontinuity distribution τ_i along the boundary S and the initial strain distribution ϵ^o_{kl} within the body V is:

$$t_r(\underline{P}) = -\tfrac{1}{2}\tau_r(\underline{P}) + \int_S T(\underline{P}_r,\underline{Q}_i)\tau_i(\underline{Q})ds(\underline{Q}) + \int_V t(\underline{P}_r,\underline{q}_{kl})\epsilon^o_{kl}(\underline{q})dv(\underline{q}) \ldots \text{(A1)}$$

Here the integral over the boundary S has to be considered in a Cauchy Principal value sense.

The integral equation expressing the stress tensor σ_{ij} at an interior point $\underline{p}$ due to a force discontinuity distribution τ_k along the boundary S and the initial strain distribution ϵ^o_{kl} within the body V is:

$$\sigma_{ij}(\underline{p}) = \int_S \Sigma(\underline{p}_{ij},\underline{Q}_k)\tau_k(\underline{Q})ds(\underline{Q}) + \int_V \sigma(\underline{p}_{ij},\underline{q}_{kl})\epsilon^o_{kl}(\underline{q})dv(\underline{q}) \ldots \text{(A2)}$$

In this equation the integral over V has to be considered in a Cauchy Principal value sense.

The discretization of (A1) and (A2) into piecewise constant traction discontinuities over line-segments along mine boundaries and piece-wise constant initial-strains within square blocks is described in Section 3.3.

The integrated kernels for the remote stress and traction components due to a constant traction discontinuity vector along a line segment can be found in (Hocking, 1978). The ij th stress component at a point $\underline{p}$ due to a constant unit initial strain in the kl direction distributed within a square block of width 2a centered at the origin is denoted by:

$$D_{ijkl} = \int_\square \sigma(\underline{p}_{ij},\underline{q}_{kl})dv(\underline{q})$$

(width $2a$)

The expressions for these integrated kernels which are used in the discretization of (A2) are as follows:

$$D_{1112} = \frac{G}{2\pi(1-\nu)}\left[-\frac{(p_1-\xi)^2}{r^2} + \log_e r\right]_{\xi=-a,\ \eta=-a}^{\xi=a,\ \eta=a}$$

$$D_{1122} - \beta_o = \frac{G}{2\pi(1-\nu)}\left[-\frac{(p_1-\xi)(p_2-\eta)}{r^2}\right]_{\xi=-a,\ \eta=-a}^{\xi=a,\ \eta=a} = D_{1212} - \beta_1$$

$$\ldots\text{(A3)}$$

$$D_{1222} = \frac{G}{2\pi(1-\nu)}\left[\frac{(p_1-\xi)^2}{r^2} + \log_e r\right]_{\xi=-a,\ \eta=-a}^{\xi=a,\ \eta=a} = D_{2212}$$

$$D_{2222} = \frac{G}{2\pi(1-\nu)}\left[-2\tan^{-1}\left(\frac{p_2-\eta}{p_1-\xi}\right) + \frac{(p_1-\xi)(p_2-\eta)}{r^2}\right]_{\xi=-a,\ \eta=-a}^{\xi=a,\ \eta=a} + \beta_2$$

WHERE: $r = \sqrt{(p_1-\xi)^2 + (p_2-\eta)^2}$; G = Shear modulus; ν = Poisson's ratio

Here compressive stresses are assumed to be positive. The remaining stress components D_{1111} (etc) can be obtained by simple rotation of the stresses quoted here and have thus been omitted for the sake of brevity. The corresponding integrated traction kernels used in the discretization of (A1) can be obtained by taking the product of the stress components in (A3) and those of the appropriate unit normal to the surface S at the receiving point $\underline{P}$.

The constants β_o, β_1 & β_2 have been introduced in (A3) to represent different branches of the arctan function. Two branches which have been found to have interesting physical significance are:

(I) $\beta_o = \beta_1 = \beta_2 = 0$ For all $\underline{p}$.

(II) $\beta_o = \frac{2G}{(1-2\nu)}$; $\beta_1 = 2G$ & $\beta_2 = \frac{2G(1-\nu)}{(1-2\nu)}$ when $\underline{p} \in \square$.

$\beta_o = \beta_1 = \beta_2 = 0$ when $\underline{p} \notin \square$

These branches have the paradoxical property of generating very different stresses (of opposite sign) within the block itself but identical stresses outside. Branch I has in fact been used in SIMPLER, but a discussion of the significance of branch II may be of interest. It can be shown (Jaswon and Symm, 1977) that any linear differential operator transforms an elastostatic displacement field into another elastostatic displacement field except at a sending point. In particular Brady and Bray (1978) demonstrated how, by differentiating the Kelvin displacement due to a point force, the force dipole displacement field can be obtained. They also showed how the displacement field due to a force quadrupole can be generated. It is interesting to note that the point 'initial stress' kernel is identical to that for a force dipole and that the point 'initial strain' or point displacement discontinuity kernel only differs from the force quadrupole by a multiplicative constant. One of the branches (I) of the integrated kernels of equ. (A3) can be approximated by distributing a large but finite number of displacement discontinuities within a square block and the other (II) branch by distributing two pairs of opposing line segments of force along the periphery of the square element thus forming a 'finite force quadrupole'. That one can represent the initial strain kernels for a square element by a 'finite force quadrupole' is evocative of a scheme using a rectilinear lattice of force discontinuities in which only the nett force on the interface of two neighbouring blocks is used. This would enable one to use the same set of kernels for mining boundaries as for the representation of interior inelastic behaviour. A disadvantage of this force lattice approach is that there is no natural way of determining the magnitude of the nett force which represents a given material behaviour without actually referring to the pairs of adjacent constituent blocks. It is also interesting to note that this is an alternative way of establishing that inelastic behaviour within a body can be represented by a force field similar to the 'fictitious body force' field of Banerjee and Mustoe (1978).

THREE-DIMENSIONAL ANALYSIS OF LARGE UNDERGROUND POWER STATIONS

Analyse tridimensionelle de grandes cavernes pour des usines hydroélectriques

Dreidimensionale Analyse grosser unterirdischer Kraftwerke

L. R. Sousa
Research Officer of the Underground Construction Division, LNEC, Lisbon, Portugal

SYNOPSIS

Stability analysis of large openings may be carried out using 3-D numerical models, namely those based on finite elements and boundary element methods. Studies are made for an underground powerhouse by using finite element models, and the results are compared with those obtained on an experimental model and with the monitored displacements.

RESUME

La stabilité des grandes cavernes peut être analysée par des modèles numériques tridimensionels, en particulier en employant la méthode des élément finis et celle des éléments de frontière. Des études sont effectuées pour une centrale souterraine et les résultats sont comparés avec ceux d'un modèle expérimental et avec des déplacements mésurés in situ.

ZUSAMMENFASSUNG

Die Stabilitätsanalyse großer Hohlräume kann mit Hilfe dreidimensionaler numerischer Modelle durchgeführt werden, namentlich jener der Methoden der Finiten Elemente und der Grenzelemente. Es werden Untersuchungen für ein in Mozambique gelegenes unterirdisches Kraftwerk aufgeführt, und die Ergebnisse mit denjenigen, die an einem experimentellen Modell erhalten wurden, und mit den in situ gemessenen Verschiebungen verglichen.

1. INTRODUCTION

In the development of underground power plants, the construction of large underground openings plays an important role in the design of such undertakings. Concerning stability and safety problems, the conceptual models used for these underground structures, based on the approach of continuous or discontinuous media will be analysed. As the static calculations for the stability analysis are carried out generally with the help of numerical models, especial emphasis will be given to the use of three-dimensional numerical models based on finite element methods and also on boundary element methods.

A 3-D finite element model was developed, with which an analysis can be made of the influence of the excavation sequence and of the introduction of supports. For a continuous approach, it assumes an anisotropic elastic body, while for a discontinuous approach, it assumes inability to sustain tension and non-linear behaviour during shear deformations.

Studies are made with this 3-D model for an underground hydroelectric powerhouse. The characteristics of the main cavern are 216.7 m in length, 28.9 m in width and 57.0 m in maximum height. In the numerical model, a block was considered, which contains part of the machine hall and one of the surge chambers. Calculations of the excavation sequence were carried out and the results are compared with those obtained on a 3-D experimental model, developed at LNEC, and on a 2-D non linear finite element model. The monitored displacements obtained are also presented.

2. DESIGN CONSIDERATIONS FOR LARGE UNDERGROUND STRUCTURES

2.1 General

In the development of underground power plants the construction of large underground openings performs a relevant part in the design of such undertaking. The design stages of these underground structures imply the knowledge of the geological and geotechnical conditions of the rock masses where the underground space should be located. It should include the different types of rocks and their mechanical properties, the geometry and nature of the discontinuities, the weakness zones, the initial state of stress and the hydraulic and thermal properties of the rock masses.

To the design of such underground openings, a procedure may consist of the following stages (Olsen and Broch, 1977): i) selection of the location which shows optimal conditions from a stability point of view; ii) definition of the orientation of the openings axis that will be suitable with respect to the in situ stress field and to major discontinuities; iii) shaping of the underground openings taking into account mechanical properties of rock masses, in situ stresses and jointing; iv) dimensioning of the different underground structures followed by detailed static calculations concerning stability and safety problems. This procedure may be used, in general, iteratively until we get the best optimal economic solution.

Dimensioning of large underground structures, as underground power plants, is essentially based on empirical methods, using the precedent practice and observation results of other works, on stability studies using results of stress analysis, on assuming continuous and discontinuous approaches, and on analysis with limit equilibrium techniques applied to specific failure mechanisms (Richards et al, 1977). The shaping of supports should generally be a function of the average rock mass block size, of the span dimension and of the space between rock bolts.

Moreover, as regards the calculation methods for analysing underground works, one question is frequently put: what kind of idealization or idealizations of the rock masses are adequate to the different situations to be found in practice, that are reproduced in the conceptual models, and particularly which is the meaning of continuous and discontinuous approaches (Rocha, 1976). It should be emphasized that the insufficient knowledge of the properties of rock masses does not, in general, justify complex idealizations.

The continuous approach is justified in the case of sound rock masses, without relevant discontinuities, and also in the case of jointed rock masses with small ratio between the distance of joints and dimensions of the openings. In the latter case, the ubiquitous joint method is generally used. The discontinuous approach should be applied to rock masses in which the distance of discontinuities in each set is of the same order as the opening dimensions. Nevertheless, the discontinuous approach has only been applied in the case of continuous joints.

2.2 Numerical models

The use of models has been of special interest in the design of large underground structures. It is possible to use physical models, in which the prototype will be duplicated at a convenient scale with a minimum of distorsion, and to use numerical models with a spectacular progress. The principal numerical models are essentially based on continuum mechanics principles and the more important models are based on the finite element method and on the boundary element method. The limit equilibrium methods are also important and can be used to check the stability of blocks formed in rock masses in relation to the shear strength characteristics of the joints.

The finite element models are suitable for a very wide range of applications, because fairly sophisticated techniques are available for modelling the stress-strain behaviour of the rock mass as continuous or discontinuous media, different construction sequences and the in situ stress field. In structural analyses, these models are intrinsically correlated with the idealization adopted (continuous or discontinuous). In discontinuous models, special finite elements have been developed, joint elements, which can simulate the discontinuities of rock masses and the interfaces between rock mass and supports. The finite element models are also important to analyse loadings in underground openings. In the case of water percolation through the rock masses calculation models are developed in order to solve its effects. Also in the case of thermal effects, more important in storage caverns and underground nuclear power plants, finite element models considering the time effect have been developed.

In large underground openings, we have in general tridimensional equilibria. Sometimes the 3-D finite element models are inefficient (large amount of data and impracticable computer times) and so the boundary element models can be efficient in the analysis of underground structures particularly in 3-D problems. The boundary element method is an integral method that uses approximating functions that satisfy the governing equations of the problem in the domain, but are approximated in the boundaries by using the standard finite element techniques. Unfortunately, the models using such a method are not sufficiently adequate to the analysis of anisotropic bodies and to the analysis of complex non-linear behaviour which generally appear in the construction sequences of underground openings. Nevertheless, they permit to consider problems with infinite boundaries and so in underground openings it is only necessary to discretize the underground surface boundaries.

3. THREE-DIMENSIONAL NUMERICAL MODEL

3.1 General description

A general finite element model was developed at LNEC that can consider the tridimensional nature of the equilibrium in the vicinity of large underground openings (Ocampo, 1982). The model can analyse rock mass as a continuous medium, or a continuous medium intercepted by given discontinuities, assuming non-linear behaviour in these surfaces. It also simulates the excavation sequences with or without supports by using appropriate techniques.

3.2 Construction sequences and non-linear analysis

In order to simulate the excavation sequences and the possibility to install supports, the model permits a step-by-step analysis, each step of which means a given stage of construction, by deactivating ground elements and activating support elements at the appropriate stage of the calculations.

Such analysis means a geometrically non-linearity, because when the elements are desactivated in each stage a new structure appears without equilibrium between forces corresponding to the previous stage loads and the field of stresses. In each stage, these resulting forces more the incremental loads at this stage imply an increment in the displacements calculated by an equation given in the work of Cunha (1981).

Together with this geometrically non-linearity, non-linear material problems are considered in the model and correlated with the discontinuities. The solution is achieved by using iterative techniques and stress-transfer methods, which permit a constant stiffness matrix at each stage.

3.3 Continuous approach

Following the continuous approach, rock mass behaves linearly elastically with transverse anisotropic stress-strain behaviour, which can be described by five independent elastic constants: the Youngs' moduli E_1 and E_2 for loads perpendicular and parallel to the planes of schistosity, the Poisson's ratios ν_1 and ν_2 and the shear modulus G_1.

The domain is discretized by several kinds of finite elements, namely isoparametric prismatic elements of 8 and 6 nodes, tetrahedrons of 4 nodes and elements of 5 nodes composed by two tetrahedrons. The simulation of the concrete supports is made by subparametric elements of 8 nodes (Pedro and Pina, 1982).

3.4 Discontinuous approach

Following the discontinuous approach, the joints were idealized as continuous, non-dilatant and with linear elastic or non-linear behaviour. In the case of linear behaviour, the discontinuities are defined by values of shear stiffness and normal stiffness. In the case of non-linear behaviour, the model is based on the assumption of no tension in the direction normal to the discontinuities; in the direction of the discontinuities the model follows the Coulomb failure criteria. In the model two types of superficial joint elements are considered (Fig. 1).

Fig. 1 Superficial joint elements

4. ANALYSIS OF AN UNDERGROUND POWER STATION

4.1 General description

To evaluate the capabilities of the 3-D model developed and to illustrate the ieas exposed in this work, studies were made for a large underground power station, located in Mozambique (Figueiredo and Bastos, 1975). Fig. 2 gives a schematic perspective of the powerhouse and annexes. The underground openings are situated at a depth that, in the case of the powerhouse, varies from about 130 m at one of the ends to about 230 m at the other end, the dimensions characteristic of this cavern being 216.7 m in length, 28.9 m in width and 24 and 57 m in minimum and maximum height. The two surge chambers are located in parallel to the powerhouse and the dimensions characteristic of the caverns are as follows: 82.5 and 87.7 m in length for the north and the south chambers; 19.0 m in width; 72.0 m and 70.3 m in height for the north and the south chambers.

The rock mass consists of granitic gneiss rock with several dyke intrusions. The existence of lamprophyric dykes becomes important in the behaviour of such undertaking (Rocha, 1976). Three principal discontinuity sets are detected and geological accidents like faults are scarce.

The geotechnical structure of the site was analysed with particularly emphasis by LNEC (1969, 1970). The in situ tests performed consist essentially in the determination of deformability of the rock mass and of the shear strength of discontinuities and of rock masses. The initial stresses are also determined in tests carried out using the small flat jack and the stress tensor gauge methods developed at LNEC. Some significant results are summarized in Table 1.

TABLE 1 - SUMMARY OF IN SITU TEST PARAMETERS

PARAMETER	VALUE
Rock mass modulus	
mean	65×10^3 MPa
maximum	116×10^3 MPa
minimum	23×10^3 MPa
Shear strength	
gneissic discontinuities	$C=0.3$ MPa; $\emptyset=33$ to $41°$
lamprophyric dykes	$C=0.2$ MPa; $\emptyset = 20°$
In situ stresses	
major principal stress	13 to 18 MPa
minor principal stress	9 to 12 MPa

4.2 Finite element models used

For the purpose of the stability analysis of the underground openings of the power station, a 3-D finite element idealization was selected, considering a block that contains part of the cavern corresponding to the powerhouse, and half of the south surge chamber and respective diffuser (Fig. 3). Because the power station is situated at a great depth, it can be assumed that the presence of the ground surface has a negligible effect on the boundaries of the excavation. For this reason and by introducing some simplifications, the boundaries of the finite element model were assumed fixed with reference to normal movements. The finite element mesh contains 1034 solid elements, with a total of 756 nodes.

Following the continuous approach, the rock mass was supposed to behave linearly elastic and isotropic, described by a Young's modulus equal to 70×10^3 MPa and a Poisson's ratio equal to 0.2. The initial state of stress was supposed equal to a vertical stress $\sigma_v = 15$ MPa, a horizontal stress $\sigma_h = 10$ MPa and a shear stress $\tau_{vh} = 0$. The concrete supports used in the principal caverns, we

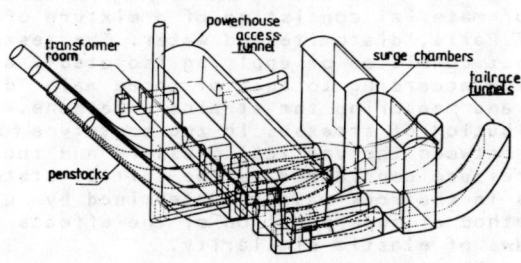

Fig.2 Schematic perspective of the underground openings

former room, penstocks and tailrace tunnels (Fig. 5). This model uses a complex computer program, which includes construction sequences, non-linear

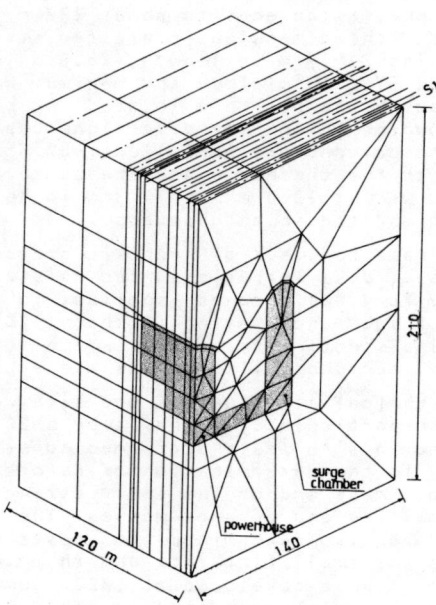

Fig.3 3-D finite element mesh

re assumed to present a Young's mudulus equal to 40×10^3 MPa and a Poisson's ratio equal to 0.2, taking into account results of tests.

To study the behaviour of the openings, two hypotheses were made: i) a one-stage excavation sequence analysis deactivating finite elements in zones corresponding to openings and to concrete supports (fig. 3); ii) a four-stage construction and excavation sequence as shown in Fig. 4, section S_1, was simulated.

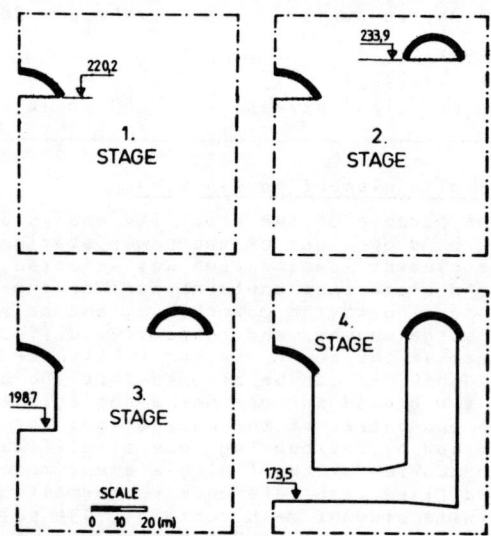

Fig.4 Construction sequence simulated by 3-D model at section S_1

Finally, a plane finite element model containing a cross-section through section S_1 was built for the purpose of comparing results with the 3-D idealization, and to analyse more correctly the influence of other openings, like the trans

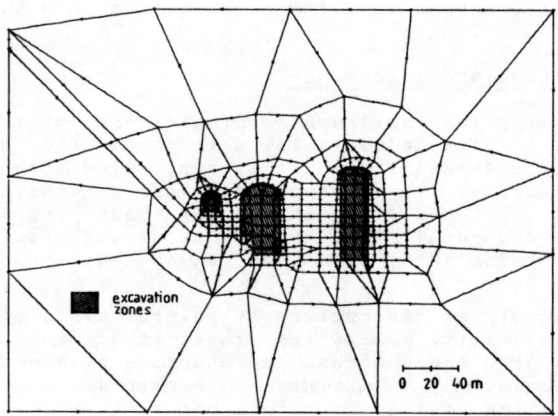

Fig. 5 2-D finite element mesh

characteristics for the continuum elements and for curved joint elements and special lagrangean elements to concrete supports (Sousa and Teles, 1980). For simplicity a one-stage excavation sequence was adopted in this model.

4.3 Laboratory model

LNEC (Silveira et al, 1974) conducted a series of tests in a 3-D experimental model trying to know what stresses would set up in the boundaries of the powerhouse and of the surge chambers. The model used had the form a cube with 1.75 m edge, inside which were reproduced the underground openings (Fig. 6).

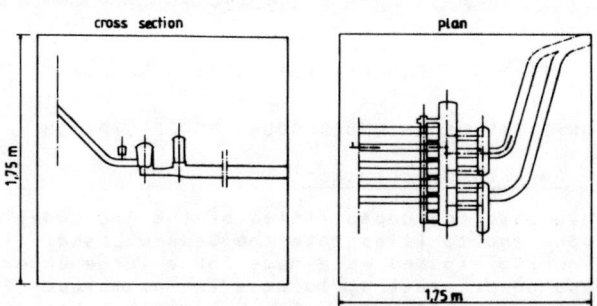

Fig.6 Schematic drawing of the experimental model

The model was obtained by assemblying various blocks made of material consisting of a mixture of plaster of Paris, diatomite and water. The tests carried out consisted of applying isolatedly an compression according to each of three main directions and measuring the strains undergone. The distribution of stresses in the prototype for any ratio between the vertical pressure and the lateral pressure considered as the initial state of stress in the rock mass is determined by using the method of superposition of the effects and the laws of elastic similarity.

The initial stage of stress was supposed to equal that considered in the numerical model.

4.4 Comparison between experimental and numerical values

The laboratory results show that in the main cavern the field of stresses is practically compressive since the few tensile stresses recorded are of little importance and the highest concentrations of stresses occur in the powerhouse, in the springing of the ceiling arch, by the upstream wall (Silveira et al, 1974). The highest compression recorded was 55 MPa and as regard tensions a maximum of about 3 MPa was observed.

These results were compared with those obtained in the first hypothesis used in the 3-D finite element model. In Fig. 7 are shown the main stres

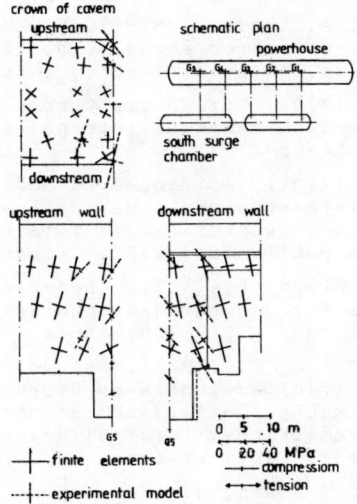

Fig.7 Main stresses in the powerhouse cavern

ses in the powerhouse cavern, whereas Fig.8 shows the main stresses in the cavern of the south surge chamber.

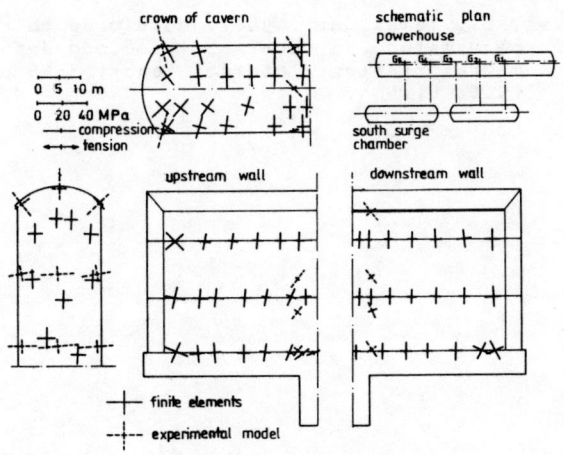

Fig.8 Main stresses in the caverns of the south surge chamber

There is a reasonable agreement between these two kind of values, taking into account that the finite element mesh was not refined enough for computation difficulties. Nevertheless there is a better agreement in the walls.

4.5 Comparisons between numerical and monitored displacements

The results of the four-stage construction sequence presented in Fig. 9 show the displacements calculated at section S_1 by using the 3-D model. It can be noted that there is a great in

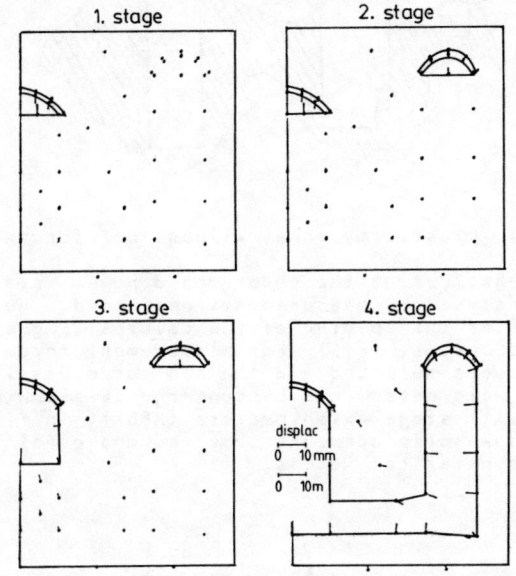

Fig.9 Displacement vectors by 3-D model at Section S_1

fluence of the excavation sequence. In Fig.10 are

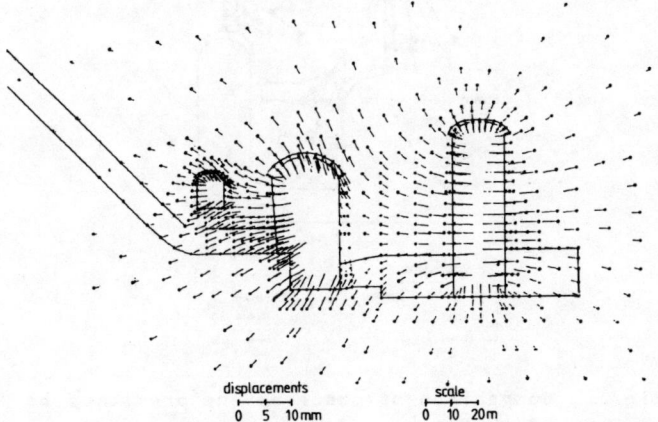

Fig.10 Displacement vectors by 2-D model

shown displacements obtained at the same section by using the 2-D finite element model. In this case we can see that displacements are oriented towards the surge chamber which shows the large influence of this large cavern over results.

With the aim to complement these results and because of the limitations of the elastic-linear models used, non-linear behaviour was considered in the bidimensional model, characterized by

Coulomb criteria, using a cohesion of 0.3 MPa and an angle of friction of 41° (Table 1). As Fig. 11 shows, the failure zones around the cavities are very important, but it is worth mentioning that the presence of reinforcements produces a considerable reduction in the extent of the plasticized zones.

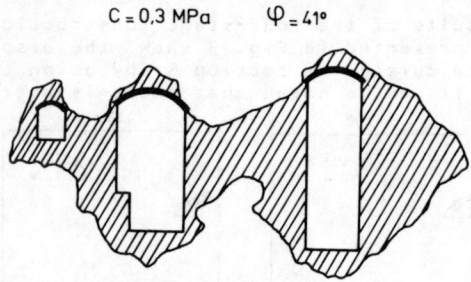

Fig.11 Plasticity zones without reinforcements

The behaviour of the underground power station was analysed by measurements performed during and after the opening of the caverns (Figueiredo and Bastos, 1975). For measurement, three sections were selected and the monitored displacements were determined by convergence measurements and multi-stage extensometers (MPBX). In Fig.12 are shown displacements observed and predicted by the numerical models.

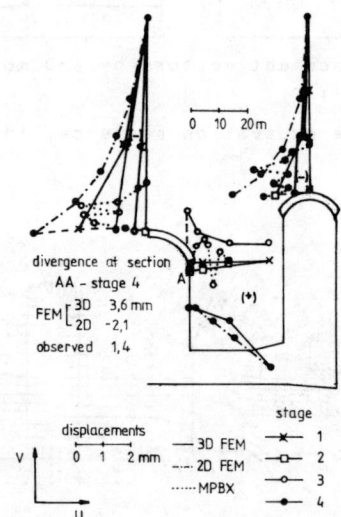

Fig.12 Comparison of observed and predicted behaviour

5. CONCLUSIONS

The discussion in this paper as regards the problem of using calculation models do analyse large underground openings, and particularly underground power stations, clearly shows the importance of numerical models.

A general 3-D finite element model is presented and this model has been applied successfully to an underground hydroelectric powerhouse. The results are compared with those calculated on a 2-D finite element model and on a 3-D experimental model developed at LNEC and also with the monitored displacements obtained.

6. ACKNOWLEDGEMENTS

The author wishes to acknowledge the collaboration of Mr. Ocampo, unfortunately disappeared, and of Mr. Abel Mascarenhas.

REFERENCES

Cunha, A. (1981). Aplicação de modelos matemáticos ao estudo de túneis em maciços rochosos. Research Engineer Thesis at LNEC, 342 pp, Lisbon.

Figueiredo, A., Bastos, M. (1975). Observação das obras subterrâneas da barragem de Cahora-Bassa. Internal Report, LNEC, Lisbon.

LNEC (1969,1970). Estudo das fundações da barragem de Cahora-Bassa. Internal Reports (4), LNEC, Lisbon.

Ocampo, F. (1982). Aplicação de modelos de cálculo tridimensionais na análise de grandes estruturas subterrâneas. Internal Report (to be published), LNEC, Lisbon.

Olsen, R., Broch, E. (1977). General design procedure for underground openings in Norway: Rockstore 77, 219-226, Stockholm.

Pedro, J., Pina, C. (1982). Estudo de equilíbrios tridimensionais em estruturas laminares e maciças utilizando elementos finitos subparamétricos. Internal Report (to be published), LNEC, Lisbon.

Richards, L., Sharp, J., Pine, R. (1977). Design considerations for large unlined caverns at shallow depths in jointed rock. Rockstore 77, 239-246, Stockholm.

Rocha, M. (1976). Estruturas subterrâneas. LNEC, Lisbon.

Silveira A., Azevedo, M., Costa, P. (1974). Contribuição para o estudo da central subterrânea de Cahora-Bassa. Memória nº 430, 14 pp, LNEC, Lisbon.

Sousa, L., Teles, M. (1980). Modelo de cálculo para estudo de túneis pelo método dos elementos finitos. Internal Report, 545 pp, LNEC, Lisbon.

MODELES NUMERIQUES D'ANALYSE A LA RUPTURE AVEC DILATANCE
Numerical models for failure analysis with dilatancy
Numerische Modelle zur Bruchanalyse bei Dilatanz

L. Rochet
Ingénieur Civil des Mînes, Chef du Groupe de Mecanique des Roches, Laboratoire des
Ponts et Chaussées de Lyon, Bron, France

RESUME

Les propriétés mécaniques les plus importantes des discontinuités dans les roches résultent de leur comportement au cisaillement. Les phénomènes de dilatance ont une importance fondamentale dans le comportement à la rupture qui demeure lié à un effet d'échelle. L'étude du comportement à la rupture et post-rupture à l'échelle des masses rocheuses et des ouvrages nécessite l'analyse sur modèles numériques des mécanismes de broyage et de dilatance qui se développent au cours du cisaillement.

SYNOPSIS

The most important mechanical properties of discontinuities in rocks are those which result from their shear behaviour. The dilatancy phenomena exert a basic influence on the shear failure behaviour which is related to the scale effect. The study of the failure and post-failure behaviour in the scale of the rock masses and of the structures requires the numerical model analysis of grinding and dilatancy mechanisms which develop during shearing.

ZUSAMMENFASSUNG

Die bedeutendsten mechanischen Eigenschaften der Felsdiskontinuitäten hängen mit dem Scherverhalten zusammen. Dilatanzphänomene besitzen eine grundlegende Bedeutung im Bruchverhalten, das mit einem Größenordnungseffekt verbunden ist. Die Untersuchung des Bruch- und Nachbruchverhaltens von Gebirge und Felsbauwerken erfordert die Analyse mit Hilfe numerischer Modelle der Bruch- und Dilatanzmechanismen, die sich während des Scherprozesses entwickeln.

1 - INFLUENCE DES DISCONTINUITES SUR LE COMPORTEMENT A LA RUPTURE DES MASSES ROCHEUSES

L'existence dans les roches et les massifs rocheux de discontinuités naturelles joue un rôle déterminant dans leurs propriétés mécaniques et leur comportement. Présentes, aux différentes échelles (matrice, échantillon massif), de nature, forme, fréquence, extension variables leur influence est essentielle dans les problèmes de stabilité des massifs rocheux et des ouvrages.

Chaque discontinuité ne peut être prise en compte individuellement. Leur influence est saisie à travers la définition de milieux homogènes et continus mécaniquement "équivalents" permettant de définir des caractéristiques moyennes, représentatives, à l'échelle de l'échantillon pour les essais de Laboratoire, du bloc pour les essais in situ, de l'ouvrage pour l'analyse du comportement des ouvrages. Seules sont individualisées les discontinuités qui existent à l'échelle du problème considéré.

Le comportement réel des masses rocheuses apparaît complexe et traduit l'influence des discontinuités aux différentes échelles. Cette interférence est également sensible dans les essais de mécanique des roches et traduit l'effet d'echelle qui affecte de nombreux essais.

L'existence des discontinuités à l'échelle de l'ouvrage ou du site conditionne le comportement du massif rocheux en définissant des surfaces de faiblesse et des modes de rupture préférentiels (glissement, basculement, arrachement...). L'analyse de stabilité nécessite que soient prises en compte les caractéristiques de ces surfaces particulières dont le rôle est déterminant.

Les propriétés mécaniques les plus importantes des discontinuités dans les roches résultent de leur comportement au cisaillement. Leur résistance à la traction normalement à leur surface est généralement très faible et souvent nulle dans le cas des massifs rocheux superficiels décomprimés et soumis à une altération active. La cohésion suivant les discontinuités est un paramètre dont le poids est généralement très important dans tous les problèmes de stabilité. La cohésion est très hétérogène à toutes échelles et soumise à un effet d'échelle très important. En laboratoire cette hétérogénéité se traduit par une dispersion importante des valeurs mesurées pour une même discontinuité. La détermination de valeurs de la cohésion à l'échelle des ouvrages n'est généralement pas possible. Par ailleurs, les phénomènes d'altération entrainent une destruction progressive de la cohésion dans les discontinuités. La cohésion est en outre profondément altérée par le processus de rupture entrainant ainsi une perte de résistance importante en particulier dans le cas des phénomènes de rupture progressive.

L'introduction de la cohésion dans les modèles d'analyse de stabilité des masses rocheuses n'apparaît pas possible actuellement. Par ailleurs, elle n'est pas souhaitable dans le cas des massifs rocheux superficiels soumis aux mécanismes de décompression et d'altération (talus rocheux, fondations d'ouvrages, falaises).

Divers auteurs ont souligné l'influence fondamentale des phénomènes de dilatance dans le comportement à la rupture par cisaillement des discontinuités dans les roches. Le caractère statistique et global des critères proposés apparaît mal adapté à l'analyse de la rupture à l'échelle des ouvrages pour laquelle les surfaces de rupture potentielle doivent être prises en compte de manière individualisée, avec leurs caractéristiques propres. De même l'étude des mécanismes de rupture progressive observables à grande échelle nécessite une analyse plus fine du comportement à la rupture et post-rupture.

2 - DETERMINATION DES CARACTERISTIQUES DE CISAILLEMENT DANS LES ROCHES

La détermination des caractéristiques de cisaillement des discontinuités dans les roches est effectuée en laboratoire par des essais de cisaillement direct sur échantillons. Ces essais sont couramment effectués à l'aide d'une machine de cisaillement équipée d'une boîte type CASAGRANDE (fig.1). Chaque essai est effectué à contrainte normale constante. L'effort de cisaillement et la dilatance sont enregistrés pendant toute la durée de l'essai à l'aide d'un enregistreur bitrace qui permet de tracer en fonction du déplacement de cisaillement δl les deux courbes caractéristiques de l'essai (fig.2).
- la courbe de cisaillement : $\tau(\delta l)$ pour la valeur de la contrainte normale σ_n considérée
- la courbe de dilatance : $\delta h (\delta l)$

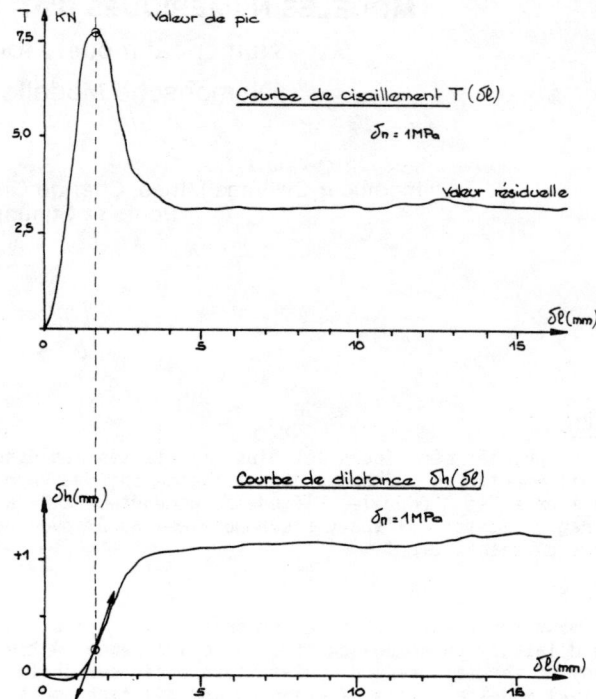

Fig:2 - Cisaillement d'une discontinuité naturelle d'une roche Courbes de cisaillement et de dilatance enregistrées au cours de l'essai (σ_n constant)

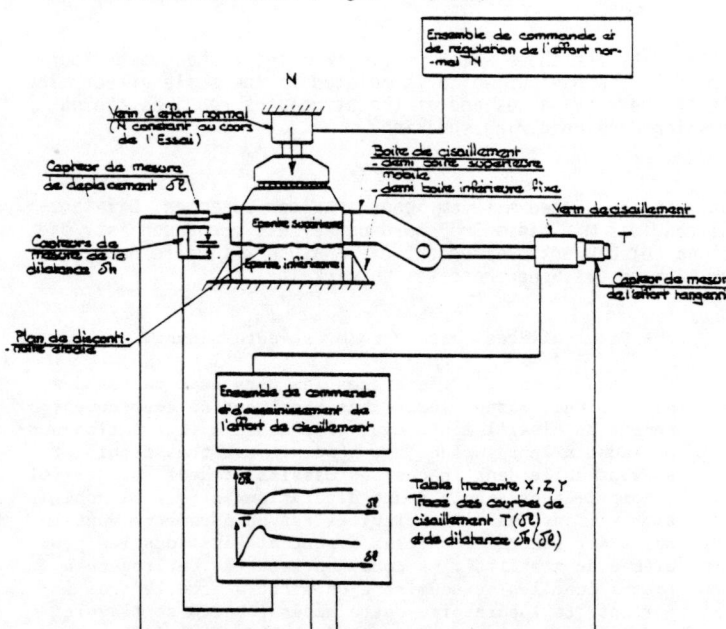

Fig:1. Schéma de principe de l'essai de cisaillement direct

De ces courbes qui constituent le véritable résultat de l'essai on déduit généralement certaines valeurs caractéristiques de la contrainte de cisaillement :
- la résistance de pic $\tau_p(\sigma_n)$ qui correspond au maximum de la valeur de cisaillement enregistrée.
- la résistance résiduelle $\tau_r(\sigma_n)$ qui est déterminée à partir du palier de la courbe de cisaillement et correspond à un processus de frottement des épontes.

Le tracé des courbes $\tau_p(\sigma_n)$ et $\tau_r(\sigma_n)$ permet de faire apparaître des caractéristiques de rupture (courbe de résistance maximum et courbe de résistance ultime (fig. 3). Toutefois ces courbes sont insuffisantes à elles seules pour rendre compte du processus de rupture. Elles doivent être complétées des données relatives à la dilatance.

La détermination de la courbe de résistance de pic est rarement possible compte tenu de la dispersion généralement importante de la valeur de la résistance maximum. Par contre le tracé de la courbe de résistance résiduelle est généralement bien défini, cette caractéristique, contrairement à la résistance de pic, n'apparaissant pas affectée par l'effet d'échelle de manière significative.

Le tracé des courbes de dilatance traduisant les déformations au cours de la rupture et au-delà de la rupture pendant le cisaillement est lié à la géométrie des épontes et soumis à l'effet d'échelle.

Les essais de cisaillement sont effectués en laboratoire sur des échantillons de dimension limitée, dont la surface n'excède pas quelques décimètres carrés. La réalisation d'essais de cisaillement sur des surfaces plus importantes atteignant plusieurs mètres carrés ne peut être pratiquement envisagée, si l'on exclut les rares essais de cisaillement réalisés in situ dans le cadre des projets de barrages importants. Ces essais ne peuvent constituer un moyen d'investigation applicable dans la pratique à la détermination des caractéristiques de cisaillement des discontinuités à l'échelle des ouvrages (surfaces de plusieurs dizaines à plusieurs centaines de mètres carrés).

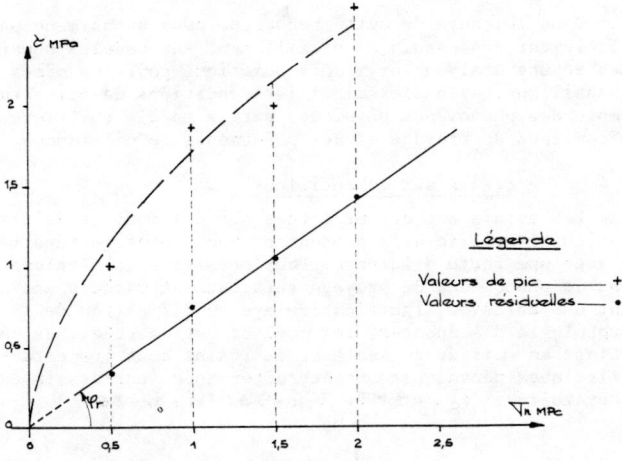

Fig. 3 - Courbes caractéristiques. $\tau (\sigma_n)$ obtenues par cisaillement direct

Les recherches développées actuellement sont orientées vers la mise au point de modèles numériques d'analyse du comportement au cisaillement des discontinuités réelles avec prise en compte des phénomènes de dilatance et d'abrasion, à partir de la détermination de paramètres mécaniques mesurés en laboratoire et géométriques mesurés sur échantillon et in situ.

3 - ETUDE DE LA MODELISATION DES DISCONTINUITES NATURELLES DANS LES ROCHES

Les propriétés mécaniques des discontinuités font intervenir de nombreux paramètres : caractéristiques mécaniques et géométriques des épontes, existence de remplissage présence d'eau ... Un cas fréquent est cependant constitué par des discontinuités pratiquement sans remplissage et sans sous-pressions sur les épontes (celles-ci pouvant être introduites séparément dans le calcul. L'étude du modèle de discontinuité conduit à prendre en compte deux types de paramètres essentiels :
- la géométrie des surfaces en contact (morphologie et extension des épontes considérée à l'échelle de l'ouvrage ou du site)
- les caractéristiques mécaniques des épontes (cisaillement et abrasion).

L'exploitation du modèle numérique est effectuée sur ordinateur et permet ainsi de prendre en compte l'étude des discontinuités de grande dimension.

3 - 1 - Constitution du modèle

On considère une discontinuité naturelle sans remplissage et dépourvue de résistance à la traction entre les épontes (fig.4). Le modèle étant orienté vers l'analyse de stabilité et le dimensionnement des confortements, les hypothèses retenues sont celles dont la détermination est la plus fiable et qui présentent par ailleurs une certaine stabilité ou une évolution progressive au cours du cisaillement le comportement peut être représenté par un mécanisme de frottement avec broyage des aspérités caractérisé par :
- une fonction de cisaillement Φ (frottement avec broyage) égale en première analyse au frottement résiduel déterminé sur échantillon : $\Phi = tg\, \varphi_r$
- une caractéristique de résistance des épontes R (rupture et broyage local des épontes). Le critère de rupture a été relié à la valeur σ_c déterminé sur échantillon.

- la morphologie des épontes déterminée par un ensemble de profils concordants des deux épontes, parallèlement à la direction X' X du cisaillement (description à trois dimensions des surfaces en contact sous la forme de courbes $Z_i (Y_i, X_i)$

3 - 2 - Conditions du déplacement

Les deux épontes en contact sont supposées rigides et présentant des déformations d'ensemble négligeables à l'échelle des irrégularités de surface. Le modèle étant destiné à analyser le comportement au cisaillement de discontinuités d'extension importante, le déplacement relatif des épontes est effectué parallèlement à lui-même sans basculement appréciable. (En fait pour des surfaces courbes il est possible d'imposer des conditions de déplacement différentes tenant compte de la courbure générale des surfaces en contact).

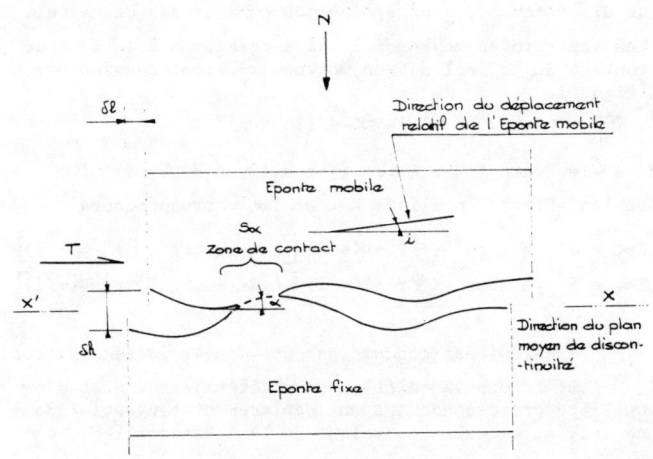

Fig. 4 - Discontinuité naturelle au voisinage d'une zone de contact des deux épontes

3 - 3 - Mécanisme de cisaillement

La partie mobile du modèle (éponte supérieure fig.4) est soumise à un effort normal N. Cet effort est constant pendant tout le cisaillement. L'effort de cisaillement T est parallèle au plan moyen. Le cisaillement du modèle s'accompagne d'un déplacement relatif des épontes δl, d'un mouvement de dilatance δh et d'une variation de l'effort de cisaillement T. Le modèle permet de déterminer en fonction du déplacement tangentiel δl :
- la valeur de l'effort de cisaillement T (δl)
- la valeur de la dilatance δh (δl)
- l'altération de la géométrie des épontes au cours du cisaillement par suite des phénomènes de broyage qui se développent aux points de contact.

Ces éléments sont déterminés à partir de l'analyse des conditions d'équilibre relatif des épontes sous l'action des efforts extérieurs N et T.(fig. 5). Les conditions de contact sont déterminées, pour chaque zone de contact, par :

- la géométrie locale des deux épontes
- la considération du critère de rupture pris en compte

Le calcul permet de déterminer les contraintes moyennes σ_n et τ normale et tangentielle relatives à chaque aire de contact S, et les forces locales correspondantes dans le système général N, T lié au plan moyen de la discontinuité.

L'état de contrainte limite dans chaque zone de contact est déterminé par le critère de rupture et l'angle de dilatance $i_{j,k}$ déterminé pour chaque pas de calcul. Les contraintes moyennes locales relatives à la zone de contact S_α d'inclinaison moyenne α sont données par (fig. 6).

$$\sigma_n = R.\cos^2(\varphi_r - \alpha + i)$$

$$\tau = R\sin(\varphi_r - \alpha + i)\cos(\varphi_r - \alpha + i)$$

et les efforts localisés N_α et T_α correspondants :

$$N_\alpha = S'.R.\cos^2(\varphi_r - \alpha + i)\left[1 - \mathrm{tg}\alpha\,\mathrm{tg}(\varphi_r - \alpha + i)\right]$$

$$T_\alpha = S'.R.\cos^2(\varphi_r - \alpha + i)\left[\mathrm{tg}\alpha + \mathrm{tg}(\varphi_r - \alpha + i)\right]$$

Le calcul est conduit par une double incrémentation:

- sur le pas le calcul k pour déterminer l'état d'équilibre correspondant à un déplacement tangentiel donné $\delta 1_j$ et calculer la valeur de la dilatance δh_j, la valeur de l'angle de dilatance i_j, l'effort de cisaillement T_j, la forme résultante des épontes par suite du broyage dans les zones de contact.

- sur le pas de calcul j pour déterminer l'évolution de la dilatance δh et de l'effort de cisaillement T au cours de l'essai et tracer les courbes correspondantes $T(\delta 1)$ et $\delta h(\delta 1)$.

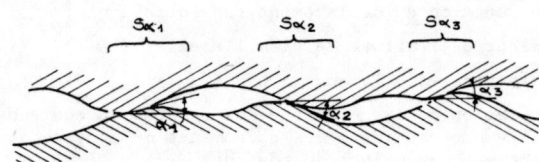

Fig: 5. Distribution des zones de contact entre deux épontes naturelles avec broyage

4 - EXPERIMENTATION EN LABORATOIRE

Dans le cadre de cette recherche nous avons mené parallèlement des essais de cisaillement sur modèles physiques et une analyse sur modèle numérique pour les mêmes échantillons, afin d'examiner les conditions de prise en compte des phénomènes physiques par le modèle numérique (mécanismes de broyage et des phénomènes de dilatance).

4 - 1 - Essais sur échantillons

Les essais ont été effectués sur des modèles de discontinuité artificielle présentant une géométrie régulière avec une forte dilatance. Pour permettre le développement de phénomènes de broyage suffisamment marqués entraînant une abrasion significative avec modification de la morphologie des épontes, les modèles ont été réalisés par moulage en mortier de sable et de résine dont les caractéristiques mécaniques ont été déterminées par essais en laboratoire : $\sigma_c = 48$ MPa ; $\sigma_{TB} = 9,3$ MPa ; $\varphi_r = 36°$

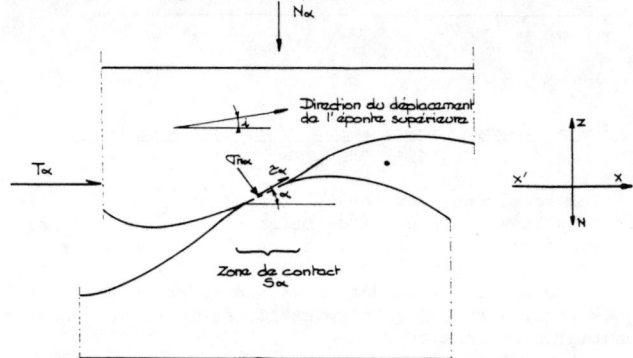

Fig: 6. Contraintes appliquées à une zone de contact S_α

L'angle de frottement résiduel a été déterminé par cisaillement direct.

Les épontes présentaient un profil régulier en forme d'onde sinusoïdale de 0,8mm d'amplitude et 6,2mm de longueur d'onde. Les modèles de dimension 10cm X 5cm ont été obtenus par moulage et contre moulage simultanés dans un moule métallique (fig. 7).

Les essais de cisaillement direct ont été effectués sur une machine de cisaillement rigide à vitesse courante et déplacement parallèlement à elle-même de la partie mobile du modèle. L'effort de cisaillement T et la dilatance δh étaient enregistrés par un enregistreur bitrace permettant le tracé direct des courbes $T(\delta 1)$ et $\delta h(\delta 1)$.

4 - 2 - Analyse sur modèle numérique

Parallèlement l'étude du comportement au cisaillement de la même discontinuité a été effectuée sur modèle numérique. Les caractéristiques géométriques initiales des deux épontes ont été déterminées par enregistrement à l'aide d'un lecteur de profil. Les profils ont été levés avec un pas de mesure de 0,1mm. Les caractéristiques mécaniques ont été déterminées sur échantillon par des essais de laboratoire et définies au paragraphe précédent.

Le déplacement tangentiel $\delta 1$ au cours du cisaillement était le même que dans le cas du modèle physique soumis au cisaillement direct, soit : $\delta 1 = 18,5$mm et correspondait à la longueur de trois ondulations successives.

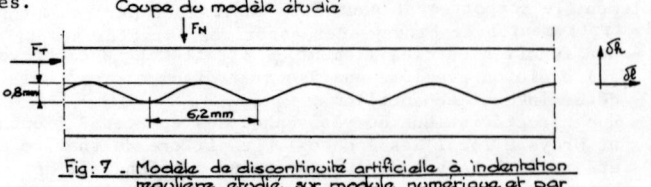

Fig: 7. Modèle de discontinuité artificielle à indentation régulière étudié sur modèle numérique et par cisaillement direct.

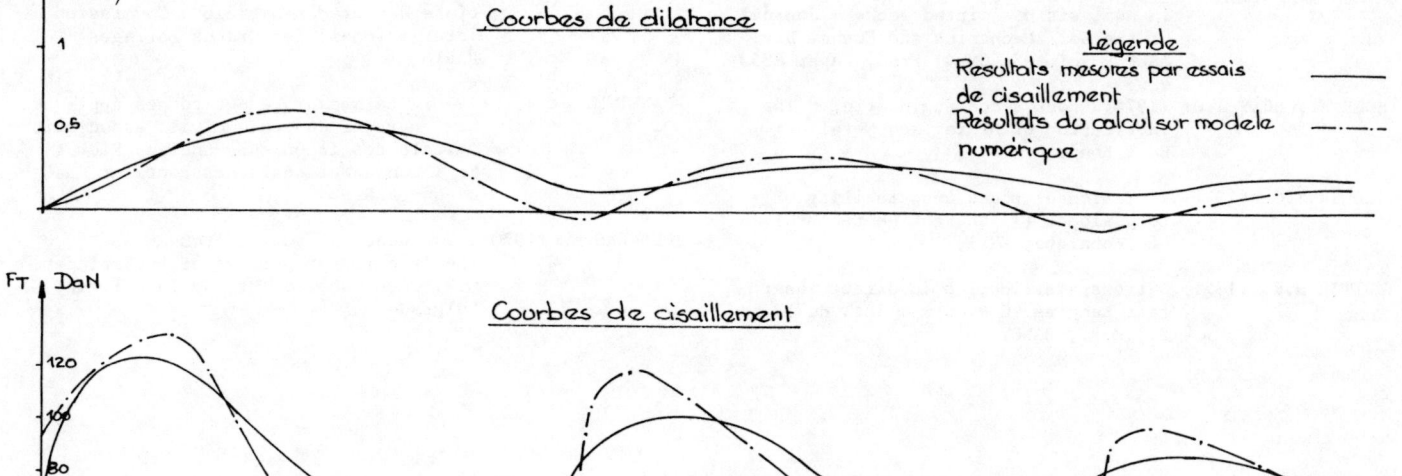

Fig: 8 _ Comparaison des résultats de l'étude sur modèle numérique et de l'essai de cisaillement direct effectué sur un même échantillon

4 - 3 - Comparaison des résultats du modèle numérique et de l'essai de cisaillement direct

Les courbes de cisaillement T (δ1) et de dilatance δh (δ1) obtenues dans les deux cas sont reportées sur le graphique fig. 8. L'examen des courbes montre une bonne représentativité du modèle numérique qui traduit bien le comportement observé lors de l'essai de cisaillement direct. Les courbes mettent en évidence de manière très nette les phénomènes de dilatance et d'abrasion qui accompagnent le cisaillement.

Les mécanismes de broyage et d'abrasion qui se développent dans les zones de contact modifient progressivement la morphologie des épontes ainsi que le montrent les courbes de dilatance dont l'amplitude et les fluctuations diminuent au cours du cisaillement. Ce phénomène apparaît également très nettement sur les courbes de cisaillement T (δ1) qui présentent une dissymétrie très marquée et une atténuation progressive des fluctuations de l'effort de cisaillement qui tend vers la valeur du frottement résiduel du matériau.

On note d'une manière générale que les courbes expérimentales présentent un aspect plus arrondi que celles qui ont été déterminées par le modèle numérique par suite de la présence entre les épontes des matériaux broyés au cours du cisaillement.

5 - CONCLUSION

La mise au point des modèles numériques permettant l'analyse sur ordinateur du comportement au cisaillement de discontinuités de grande extension présente un grand intérêt pour l'étude des problèmes de stabilité des masses rocheuses. Les modèles numériques permettent de s'affranchir dans une certaine mesure de l'effet d'échelle qui interdit généralement l'extrapolation des résultats des essais de cisaillement en laboratoire ou in situ à l'échelle des masses rocheuses ou des ouvrages.

Les résultats des recherches entreprises dans le domaine de la modélisation des discontinuités dans les roches permettent une analyse fine des mécanismes qui se développent au cours du cisaillement au contact des deux épontes, par la prise en compte des mécanismes de dilatance et de broyage dont le rôle est déterminant dans le comportement et les caractéristiques mécaniques des discontinuités dans les roches.

Ces recherches se développent activement en vue de l'application de ces modèles numériques à l'étude des problèmes de stabilité des masses rocheuses et des ouvrages. Parallèlement à la mise au point des modèles numériques, des études sont développées dans le domaine de l'analyse et la caractérisation de la morphologie des surfaces de discontinuité naturelles.

BIBLIOGRAPHIE

BARTON N.R. (1973). Review of a new shear-strength criterion for rocks joints Engineering geology, 7

BERNAIX J. (1974). Propriétés des roches et des massifs rocheux. Rapport général (Thème 1, 3ème Congrès Int. de la S.I.M.R. Denver

FECKER E., RENGER N. (1971). Measurement of large scale roughnesses of rock planes by means of profilograph and geological compass C.R. Symp. Int. de la S.I.M.R., NANCY

GOODMAN R.E., DUBOIS J. (1972). Duplication of dilatancy in analysis of jointed rocks - Journal of the Soil Mechanics and Found. Div. ASCE - Vol. 98, SM.4, Proc. Paper 8853

HOEK E., BRAY J.W. (1974). Rock slope engineering - The Institution of Mining and Metallurgy Ed., London

JAEGER J.C. (1971). Friction of rocks and stability of rock slopes (11 th Rankine Lecture) Géotechnique, XXII,2

KUTTER H.K. (1971). Stress distribution in direct shear test samples. C.R. Symp. Int. de la S.I.M.R., NANCY

LONDE P. (1974). La mécanique des roches et les fondations des grands barrages. Commission Internationale des Grands Barrages - PARIS

PANET M. et al. (1976). La mécanique des roches appliquée aux ouvrages du génie civil. Association Amicale des Ingénieurs Anciens Elèves de l'Ecole Nationale des Ponts et Chaussées - PARIS

RENGERS N. (1970). Influence of surface roughnesses on the friction properties of rock planes. C.R. 2ème Congrès Int. de la S.I.M.R. Belgrade

ETUDE NUMERIQUE D'OUVRAGE EN MASSIF ROCHEUX A STRUCTURE PLANAIRE

Numerical analysis of rock masses with planar structure

Numerische Analyse von geschichtetem Gebirge

A. Guenot
M. Panet
Laboratoire Central des Ponts et Chaussées, Paris (France)

RESUME

L'analyse numérique de masse rocheuse à structure planaire est effectuée à l'aide d'un modèle de calcul par éléments finis, élastoplastique, avec deux critères de rupture superposés. Dans le cas particulier d'un talus chargé en tête, on montre qu'on peut obtenir, outre les déplacements et les contraintes, une bonne estimation du mode de rupture probable et de la capacité portante.

SYNOPSIS

An elastoplastic finite element model with two superimposed criteria is used to study the behaviour of stratified rock masses. In the particular case of a footing resting on the crest of a slope, it is shown that, beyond the classical stress, and displacement fields, the model provides a good evaluation of the rupture mechanism and the bearing capacity.

ZUSAMMENFASSUNG

Ein elastoplastisches Gesetz mit zwei sich überlagernden Bruchkriterien und die Methode der Finiten Elemente werden angewandt, um die numerische Analyse von geschichtetem Gebirge durchzuführen. An Hand einer an der Oberkante belasteten Böschung wird gezeigt, daß nicht nur die Verformungen und Spannungen erkannt werden können, sondern daß auch eine gute Bewertung des vermutlichen Bruchmechanismus und der Tragfähigkeit erreicht wird.

1 - INTRODUCTION -

La pratique conduit souvent les géotechniciens à devoir évaluer la perturbation qu'apporte à un massif rocheux à structure planaire (stratification, foliation, schistosité) un ouvrage de génie civil, qu'il s'agisse d'une culée de barrage ou d'une semelle de fondation. On lui demande le plus souvent de déterminer une valeur de la capacité portante et/ou une estimation des déplacements. La capacité portante d'un tel massif rocheux est évaluée selon les cas par deux approches différentes :

- celle des milieux continus utilisée en mécanique des sols à l'aide d'un modèle rigide plastique.

- celle des équilibres de blocs limités par des surfaces de discontinuités.

Ces méthodes ne fournissent pas une évaluation acceptable du champ de déplacements. L'objet de l'étude présentée dans cet article, est d'évaluer la possibilité d'utiliser un modèle numérique de comportement élastoplastique pour obtenir simultanément les deux informations.

Le modèle numérique permettant de représenter des massifs rocheux stratifiés ou schisteux, et son application à des modèles tridimensionnels, ont déjà été présentés précédemment par Franck, Guenot et Humbert, 1980, 1982. Afin d'apprécier ses possibilités quant à l'évaluation des modes de ruptures et d'une capacité portante, on présente le cas simple du talus chargé en tête. Ce cas est intéressant car on peut comparer les résultats à une solution analytique présentée par Rambach et Sirieys, 1979. La solution décrite par ces auteurs a cependant des limites :

- d'une part géométrique : le modèle est à deux dimensions, ce qui impose le même azimuth aux directions horizontales du talus et des plans de discontinuités.

- d'autre part sur les propriétés mécaniques du matériau : la résolution simple du problème impose une condition supplémentaire sur les propriétés mécaniques rappelées plus loin, qui n'est pas forcément réaliste.

Une concordance entre les résultats des calculs analytique et numérique sur ce cas particulier, permettrait de conclure favorablement, quant à l'utilisation de cette méthode numérique pour des cas plus généraux.

2 - EQUILIBRE LIMITE D'UNE PENTE CHARGEE EN TETE DANS UN MASSIF A STRUCTURE PLANAIRE -

Ce paragraphe reprend essentiellement certains des résultats présentés par Rambach et Sirieys, 1979. Nous avons conservé les mêmes notations que ces auteurs.

Soit une pente d'inclinaison β dans un massif rocheux présentant des plans de discontinuités (S) inclinés d'un angle α sur l'horizontale (cf fig. 1). Elle est considérée en déformation plane. Il existe deux critères de rupture :

- l'un relatif à la roche
$$|\tau| = C + \sigma_n \, tg \, \phi \qquad (1)$$
- l'autre est un critère orienté relatif aux ruptures sur les plans de discontinuités :
$$|\tau| = \overline{C} + \sigma_n \, tg \, \overline{\phi} \qquad (2)$$

où τ et σ_n sont les contraintes tangentielles et normale sur la facette de rupture, de direction quelconque dans le premier cas, et parallèle aux plans de discontinuités dans le second. Les plans (S) sont des plans de résistance plus faibles, en particulier :
$$\overline{C} \ll C \qquad (3)$$

La rupture peut alors prendre 4 formes principales reprises sur la figure 2 et dans le cas où l'on impose la relation supplémentaire suivante :
$$C \; \text{cotg} \; \phi = \overline{C} \; \text{cotg} \; \overline{\phi} \qquad (4)$$

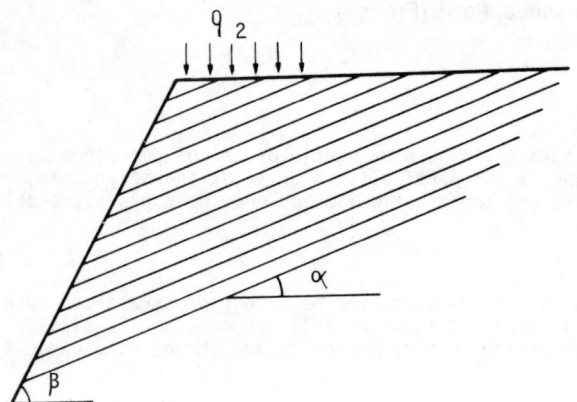

Fig. 1 - Schéma de principe
Slope with stratified rock mass.

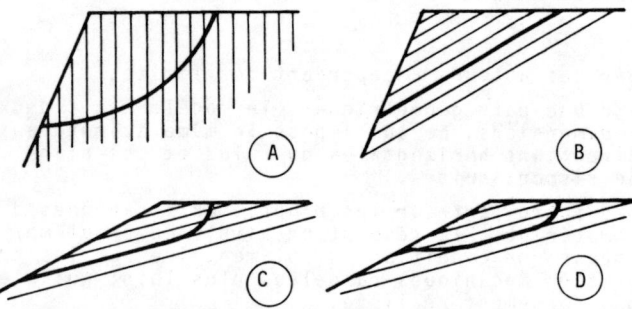

Fig. 2 - Différents types de rupture.
Several rupture mechanisms.

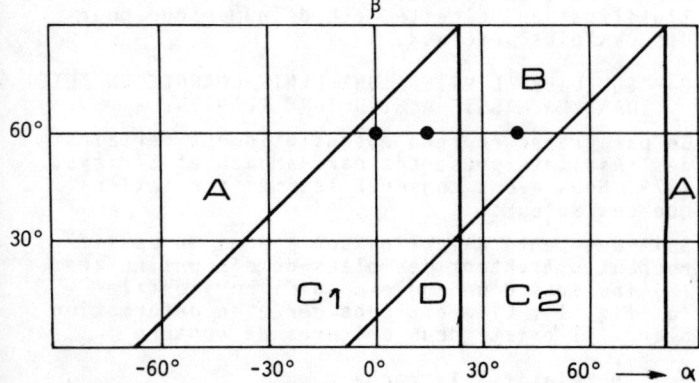

● Etude par la méthode des éléments finis.
Fig. 3 - Répartition des mécanismes de rupture en poussée dans le cas où $\phi = 30°$, $\overline{\phi} = 15°$.
Chart giving the rupture mechanism, knowing α and β.

On peut distinguer facilement ces modes de rupture connaissant α et β. C'est l'objet de la figure 3. Les modes de rupture sont les suivants :

A/ rupture "astructurale" où seule la résistance au cisaillement de la matrice rocheuse intervient.

B/ rupture "structurale" par glissement sur un plan de discontinuité.

C/ rupture mixte avec une seule surface de cisaillement de la matrice.

D/ rupture mixte avec deux surfaces de cisaillement de la matrice.

Les phénomènes de fauchage, courants dans ce type de massif rocheux, ne sont évidemment pas pris en compte par ce modèle.

3 - ETUDE NUMERIQUE -

3.1 - Modèle numérique -

Les deux critères de rupture décrits dans le paragraphe précédent sont utilisés dans un calcul élastoplastique, effectué à l'aide d'un programme de calcul par éléments finis. Ils sont superposés, le critère orienté étant prioritaire. Les paramètres définissant la résistance du massif sont donc la cohésion C et l'angle de frottement ϕ du critère (1), la cohésion $\overline{C}$, l'angle de frottement $\overline{\phi}$ du second critère, ainsi que l'inclinaison α des discontinuités. Pour traiter le problème non linéaire, nous utilisons l'algorithme des contraintes initiales à rigidité constante. Son intérêt principal est de ne nécessiter qu'une seule résolution complète du système linéaire pour étudier plusieurs valeurs de paramètres rappelés ci-dessus y compris l'angle d'inclinaison des discontinuités.

L'utilisation d'un massif continu avec un critère de rupture orienté revient à faire l'hypothèse d'un nombre très grand de plans de discontinuités vis à vis de l'échelle du problème. Dans l'étude d'un cas réel, il sera sans doute nécessaire de prendre en compte un autre comportement que le comportement parfaitement plastique après la rupture et d'échelonner suffisamment le chargement. Dans l'étude de ce cas théorique, ces améliorations ont été omises, sciemment, dans un but de simplification.

3.2 - Etude réalisée -

Nous présentons le cas d'un talus incliné à 60° ($\beta = 60°$); trois valeurs différentes de l'angle α permettent d'effectuer la comparaison ($\alpha = 0°$, 15°, 40°). Les points correspondants à ces cas de figures sont portés sur la figure 3 et indiquent les types de rupture prévisible en l'occurence C_1 et B.

Les valeurs des paramètres retenues pour le calcul sont indiquées ci-dessous :

$E = 50.000$ MPa
$\nu = 0.2$
$C = 5$ MPa $\overline{C} = 2.32$ MPa
$\phi = 30°$ $\overline{\phi} = 15°$

Ces valeurs ont été choisies de manière à satisfaire la relation (4) qui impose une cohésion élevée pour les surfaces de discontinuités.

Le comportement est parfaitement plastique tant pour la rupture de la matrice que pour la rupture des discontinuités. Le maillage est représenté sur la figure 4 avec ses caractéristiques et conditions

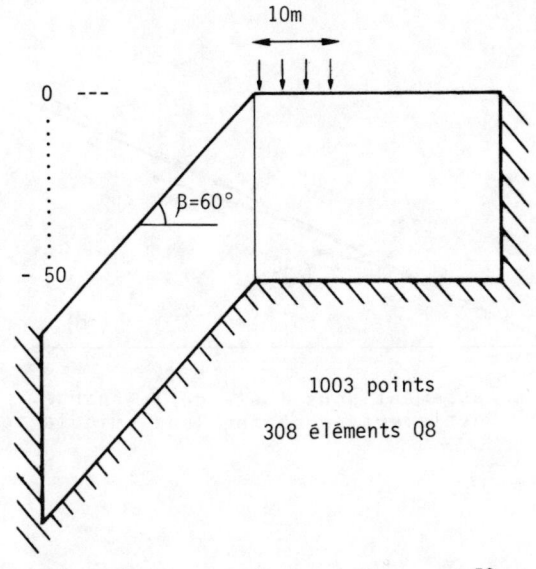

Fig. 4 - Schéma de principe du maillage.
Limits of the Mesh and Boundary conditions.

aux limites. Le chargement de pression normale uniforme est appliquée en tête du talus par incréments successifs. Le calcul numérique a été effectué par Bayle, 1981, à l'aide du système de calcul ROSALIE du Laboratoire Central des Ponts et Chaussées.

3.3 - Evolution des zones plastiques -

L'évolution des zones plastiques est représentée sur les figures 5 à 7 pour les trois calculs réalisés. On peut distinguer les régions foncées qui correspondent à la rupture de la matrice rocheuse, les autres parties correspondant à une rupture selon les plans de discontinuités. Une surface de rupture probable est également indiquée. On s'aperçoit que la rupture a lieu conformément aux modes prévus dans la carte de la figure 3. Ce type de modélisation est donc capable de représenter convenablement ces modes de rupture plane ou mixte. La progressivité de la rupture est également apparente, notamment dans le cas ou $\alpha = 15°$ (figure 6) où l'on voit que les ruptures, selon les deux modes, apparaissent dans deux régions différentes, puis évoluent l'une vers l'autre. Finalement une zone plastifiée se développe jusqu'à ce qu'elle ne soit plus confinée par des zones élastiques, "traverse" la pente et permette ensuite un écoulement libre qui ne peut pas être étudié avec un modèle de ce type. On obtient alors une divergence de l'algorithme de traitement du problème non linéaire.

3.4 - capacité portante -

On peut étudier la capacité portante du massif à l'aide d'une courbe présentant les variations des déplacements verticaux dans l'axe de la charge en fonction de la valeur de la charge appliquée (figure 8). On s'aperçoit que même pour des valeurs importantes de la charge, où il existe des zones plastiques non confirmées, on atteint difficilement une tangente horizontale. C'est un résultat classique rapportée encore par Smith,

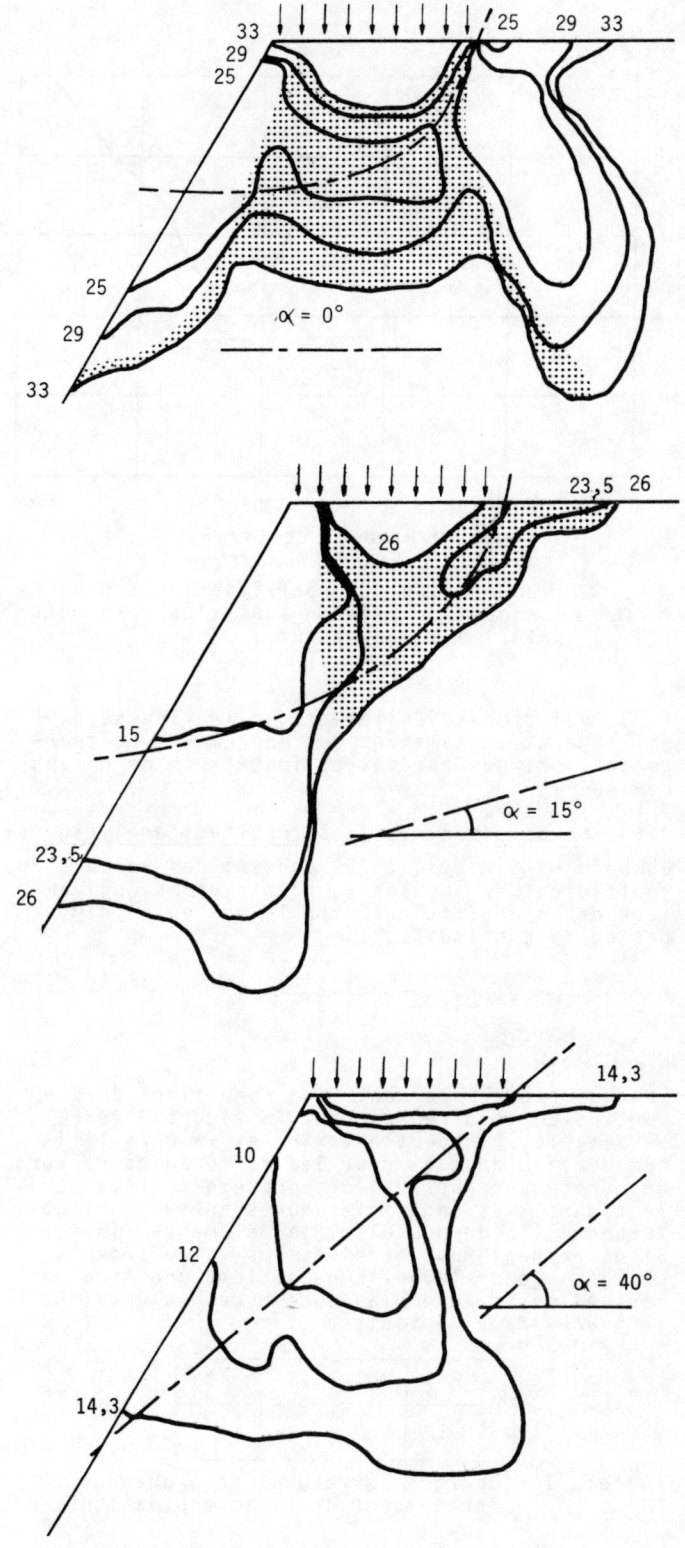

Fig. 5 - 6 - 7 -
Répartition des zones plastiques.
Evolution of plastic zones with the applied load.

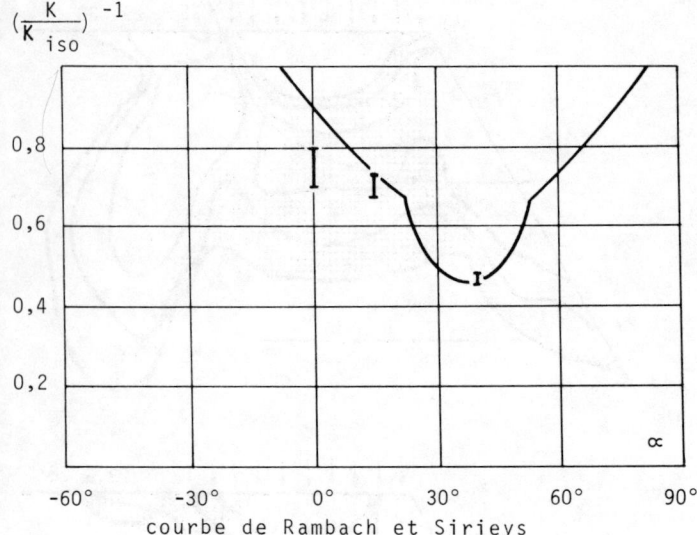

courbe de Rambach et Sirieys
résultats méthode numérique

Fig. 8 - Comparaison des coefficients de poussée. Comparison between analytical and numerical active pressure.

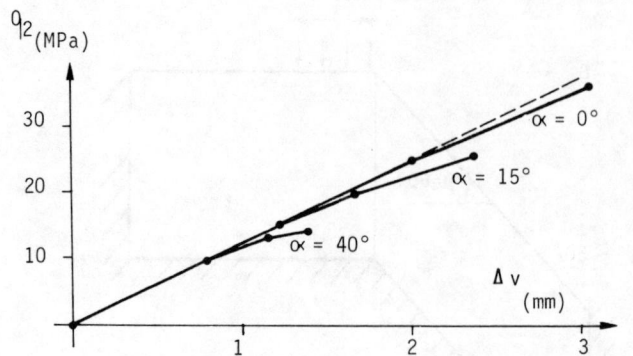

Fig. 9 - Tassement sous l'axe de la charge. Settlement under the load (middle point)

1982, qui précise également que la finesse du maillage et l'étalement des incréments de chargement sont des paramètres importants de ce phénomène.

3.5 - Comparaison avec les résultats analytiques -

Rambach et Sirieys, 1979, donnent des valeurs du coefficient de poussée K, normée par K_{iso}, valeur de ce coefficient dans le cas où il n'y a pas de discontinuités :

$$K = \frac{H}{q_2 + H} \text{ et } K_{iso} = \frac{1 - \sin \phi}{1 + \sin \phi} e^{-(\pi - 2\beta)tg\phi}$$

$$H = C \cot g \phi \qquad (5)$$

La courbe correspondant aux conditions du présent calcul est reprise sur la figure 9 ($\beta = 60°$, α variable). On peut obtenir les valeurs limites de la charge q_2 pour les différentes valeurs de α retenues. Ces valeurs ont été portées sur la figure 8 et sont inférieures aux valeurs obtenues à l'aide du calcul par éléments finis. Si maintenant nous étudions les valeurs de la charge q_2 pour laquelle on obtient une "traversée" de la zone plastique : ces valeurs sont indiquées dans le tableau 1.

α	0	15	40
q_2 MPa	29	26	14.4

Tableau 1 - charges correspondant à une "traversée" de la zone plastique.

Les valeurs correspondantes du coefficient de poussée normé sont portées sur la figure 9. Les segments de droite indiquent la variation de charge entre la charge indiquée dans le tableau 1 et l'incrément de charge précédent où la zone plastique était encore confinée. On observe alors un bon accord entre les valeurs obtenues analytiquement et numériquement.

Le développement de zones plastiques non confinées par des zones élastiques constitue un critère permettant d'obtenir un minorant de la capacité portante que l'on peut obtenir numériquement; il correspond dans ce cas à la valeur de la capacité portante obtenue analytiquement.

3.6 - Autres résultats -

Les calculs donnent une image complète des champs de déplacement et de contrainte. Dans le cadre limité de cette communication, nous ne présenterons que les déplacements différentiels sous la charge, en plus du déplacement vertical présenté précédemment.

La figure 10 montre le tassement différentiel sous la charge appliquée sur une largeur de 10m. On observe un résultat intéressant en accord avec les modes de rupture. Dans les cas $\alpha = 15°$ et $\alpha = 40°$, l'état de plastification immédiatement sous la charge est important (cf figures 6 et 7), et le tassement différentiel est plus important que dans le cas élastique. Par contre dans le cas où l'angle α est nul, on obtient une zone de rupture quasi-circulaire sous la charge qui provoque une rotation de la surface dans le sens opposé à la rotation obtenue élastiquement.

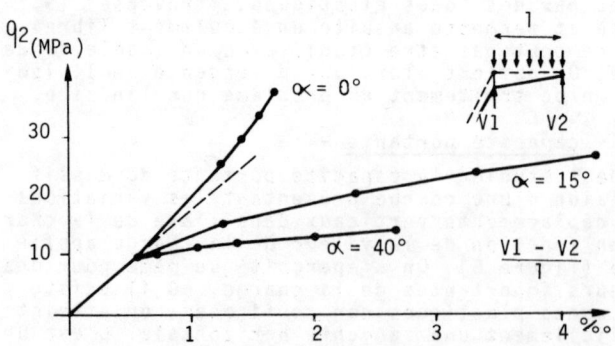

Fig. 10 - Tassement différentiel sous la charge. Differential settlement under the load.

4 - CONCLUSION -

On a montré qu'un modèle de calcul par éléments finis, avec un comportement élastoplastique particulier, pouvait permettre de déterminer les modes de rupture et les charges limites d'une structure mettant en jeu une masse rocheuse à structure planaire. Il fournit bien sûr en plus des informations sur les déplacements et les contraintes et peut être également utilisé pour le dimensionnement d'un renforcement éventuel.

Ce modèle fournit certainement plus d'informations que le modèle élastique orthotrope, utilisé quelquefois pour la modélisation d'une telle structure, sans pour cela être plus compliqué. De plus, il a été montré que le modèle orthotrope était très dépendant du module de cisaillement équivalent, de la masse rocheuse, qu'il est difficile de déterminer expérimentalement.

Il serait illusoire de vouloir tirer plus de conclusions de ces quelques calculs effectués sur un cas particulier. Le modèle nécessite maintenant la sanction d'une application à des cas réels.

REFERENCES -

Bayle A. (1981) -
Introduction d'une famille de discontinuités dans un programme d'éléments finis.
TFE, Ecole Nationale des Ponts et Chaussées.
RDUP, Laboratoire Central des Ponts et Chaussées, Paris.

Comité Français de Mécanique des Roches (1979) -
Rôle de l'auscultation dans la conception et l'exécution d'ouvrages souterrains.
Proc. 4e Congr. SIMR (1), 773-786, Montreux.

Franck R., Guenot A., Humbert P. (1982) -
Numerical analysis of contacts in geomecanics.
Proc. 4th Iconmig. (1), 37-45, Edmonton.

Franck R., Guenot A., Humbert P. (1980) -
Etude par éléments finis de quelques critères de plasticité orientés.
Proc. 2nd Int. Congr. Num. Meth. Eng. Gamni2.
Dunod Ed. France (2), 765-775.

Rambach J.M., Sirieys P. (1979) -
Stabilité des massifs rocheux à structure planaire.
Proc. 4e Congr. SIMR (1), 279-285, Montreux.

Smith I.M. (1982) -
Programming the Finite element method.
351 pp. Chichester : Wiley.

NON-LINEAR ANALYSIS OF ROCK FOUNDATIONS WITH SOFT INTERFACES

L'analyse non-linéaire de fondations en roche avec interfaces molles

Nichtlineare Analyse von Felsgründungen mit weichen Zwischenschichten

Jia-Shou Zhuo
Yin-Tang Wang
Dept. of Mechanics, East China Technical University of Water Resources, Nanjing, China

SYNOPSIS

In this paper, in the light of experimental results, three models for rock and concrete are proposed: 1) "Low-tension elasto-plastic model" 2) "Fracture model" 3) Combined model". For soft interstitial layers, the "no-tension elasto-plastic model" is proposed. Their nonlinear constitutive relations are established respectively. In the paper, the formula of F E M and "The Iteration with Changeable Kp" are introduced. Some examples of practical engineering problems are analysed.

RESUME

A la lumière des résultats expérimentaux on présente trois modèles pour roche et béton à savoir: 1) modèle élasto-plastique à tension faible; 2) modèle de rupture; 3) modèle composé. Pour la couche interstitielle molle, le modèle élasto-plastique sans tension est proposé. Leurs relations constitutives non-linéaires sont respectivement établies. On introduit la formule de F.E.M. et l'itération à Kp variés. Des exemples pratiques des travaux sont analysés.

ZUSAMMENFASSUNG

Der Artikel schlägt, auf der Grundlage experimenteller Ergebnisse, für Fels und Beton drei Modelle vor: 1) Elasto-plastisches Modell bei niedrigen Druckspannungen, 2) Bruchmodell, 3) Kombinations-Modell; für die weiche Zwischenschicht wird ein zugspannungsfreies, elasto-plastisches Modell vorgeschlagen. Dabei wird auch für sie ein nichtlineares Verhältnis hergestellt. Außerdem hat der Artikel noch die Formel der F E M und die "Iteration mit geändertem Kp" dargestellt. Am Schluß wird ein Beispiel für die Ingenieurwesensberechnung dargeboten.

1. THE IDEALIZED MATERIAL MODEL OF ROCK AND CONCRETE AND THE CONSTITUTIVE RELATIONS

1-1. The idealized material model

It is well known that a material behaves differently in complex stress states. For instance, rocks show plastic properties at higher confining pressure while brittle properties at lower confining pressure. The experimental results (stress-strain curves for Carrara marble at various confining pressures) in Fig. 1 show that when confining pressure P=0, the curve designates typical brittle properties, when P=50 MN/m², it shows ideal plastic properties, and when P=165 MN/m², hardening character appears. On the other hand, the tensile strength in rocks, especially with growing fissures in it, is very low. Therefore, from the viewpoint of elasto-plastic and fracture theories, it is considered that the failure of the material may occur due to either plastic yield or brittle destruction under external loads. Three models of material failure are proposed for different cases.

1). Low tension elasto-plastic model
From the results of the shear test of rock, we can obtain the simplified curves shown in Fig.2 and Fig.3. Here, rock is considered as an elasto-plastic material with strain hardening and softening. A and B refer to the yield point and the failure point respectively. They are determined by φ_A, C_A and φ_B, C_B separately. D. C. Drucker's criterion $F = \alpha I_1 + \sqrt{J_2} - k = 0$ is used as yield criterion, and $F_i = \sigma_i - \sigma_t = 0$ are used as criteria for failure, where σ_i are principal stresses, $i = 1, 2, 3$ and σ_t is the tensile strength of the material.

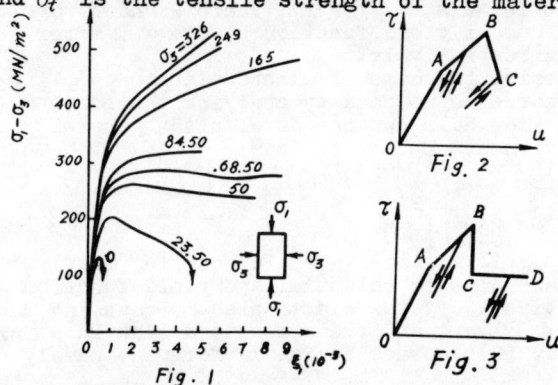

Fig. 1

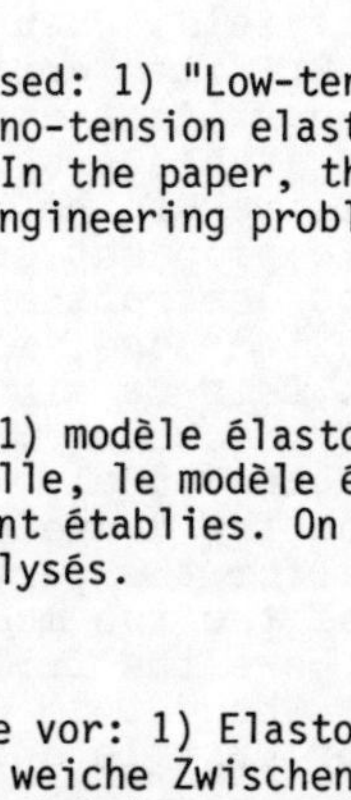

Fig. 2

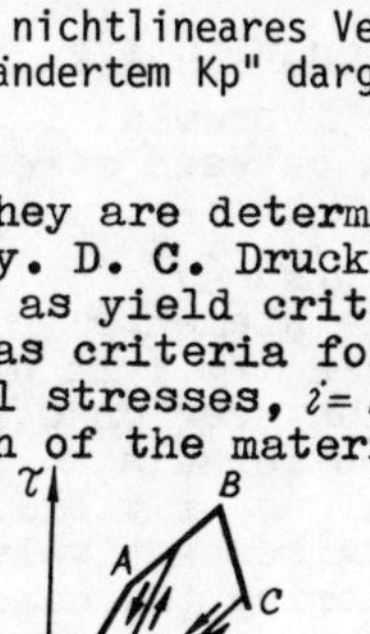

Fig. 3

2). Fracture model
In most rocks and concrete blocks, there are always minute cracks of different sizes and with random orientations, all of which may be considered as Griffith Crack. The stress-concentration at the tip of minute cracks will cause them to extend and lead to brittle destruction. In

the light of different stress states the brittle destruction can be divided into three types, i.e. Pure-Tensile failure, Tensile-Torsional failure and Compressive-Torsional failure, called for short failures of P-T, T-T and C-T respectively. And the following equations are used as criteria for failure. Griffith criteria:

P-T type: $F = \sigma_1 - \sigma_t = 0$ when $3\sigma_1 + \sigma_3 \geq 0$

T-T type: $F = (\sigma_1 - \sigma_3)^2 + 8\sigma_t(\sigma_1 + \sigma_3) = 0$
when $3\sigma_1 + \sigma_3 < 0$ and $\sigma_1 + (3-2\sqrt{2})\sigma_3 \geq 0$

Modified Griffith criterion:

C-T type: $F = (\sqrt{1+f^2} + f)\sigma_1 - (\sqrt{1+f^2} - f)\sigma_3 - 4\sigma_t = 0$
when $\sigma_1 + (\sqrt{1+f^2} - f)^2 \sigma_3 \leq 0$

The suitable regions of the above three models are shown in Fig. 4, where the slope of line ③ varies with friction coefficient f, hence, sometimes the suitable regions of T-T and C-T types may overlap and sometimes may separate from each other. When the stresses satisfy Griffith criteria, P-T or T-T type destruction will take place and the cracks will extend rapidly and automatically. It is similar to the phenomenon of the failure of ideal elasto-plastic material due to plastic yield. When the stresses satisfy the Modified Griffith criterion, C-T type destruction will occur. At that time, crack propagation becomes a stable process, which does not lead to the failure of the material so quickly as the unstable propagation of Griffith crack in P-T or T-T type destruction. Only when the load continues to increase, will the cracks extend incessantly. This is similar to the phenomenon of failure of hardening material due to plastic yield.

3) Combind model:

Based on the "envelope" conception in structural design (for the purpose of safety), the combined model of the two models mentioned above is proposed. Here the failure of the material like rock is thought to occur due to either plastic yield or brittle destruction in different conditions.

1-2. Constitutive Relation:

In the incremental theory of plasticity, the relation between stress and strain is expressed as: $\{d\sigma\} = ([D] - (1-r)[D_p])\{d\varepsilon\} = [D_{ep}]\{d\varepsilon\}$ (1)

where $0 \leq r \leq 1$.

For the elements in elastic region or in elastic unloading r=1; in plastic region r=0; and in transient region 0<r<1. Here, we approximately use the formula, r=-F/(F'-F), where F and F' are the values of yield function before and after loading respectively.

Considering the case of elasto-plastic coupling and bizarre point, and by applying the nonassociated Flow Rule, we can develop the general expression of [Dp] as follows:

$$[D_p] = ([D] - [Q])\left[\frac{\partial\{G\}}{\partial\{\sigma\}}\right]\left(A + \left[\frac{\partial\{F\}}{\partial\{\sigma\}}\right]^T ([D]-[Q])\left[\frac{\partial\{G\}}{\partial\{\sigma\}}\right]\right)^{-1}\left[\frac{\partial\{F\}}{\partial\{\sigma\}}\right]^T [D]$$ (2)

where [Q] is a elasto-plastic coupling matrix, G and F are plastic potential and yield function respectively. If the elasto-plastic coupling is ignored ([Q]=0), and the associated Flow Rule G=F is used, (2) becomes Y. Yamada formula, namely

$$[D_p] = [D]\left\{\frac{\partial F}{\partial\sigma}\right\}\left(A + \left\{\frac{\partial F}{\partial\sigma}\right\}^T[D]\left\{\frac{\partial F}{\partial\sigma}\right\}\right)^{-1}\left\{\frac{\partial F}{\partial\sigma}\right\}^T[D]$$ (3)

As there are some similarities between the two physical phenomena, plastic yield and Griffith brittle destruction, the constitutive relation similar to (1) can be established by analogy when the stresses satisfy Griffith or Modified Griffith criteria. And hence a united computational formula can be obtained. Substituting F and [D] into (3), we can obtain [Dp] for each criterion. The hardening effect is represented by parameter A, which has to be determined from the curve obtained in uniaxial tests.

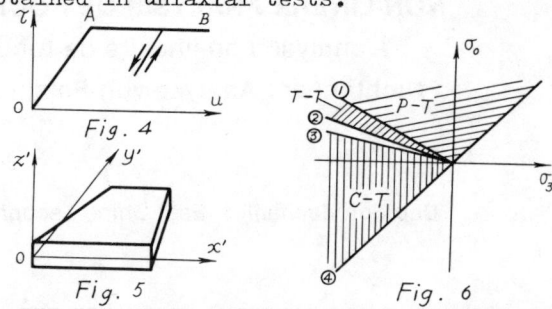

Fig. 4

Fig. 5

Fig. 6

2. THE IDEALIZED MATERIAL MODEL OF SOFT INTERSTITIAL LAYER AND ITS CONSTITUTIVE RELATION

For soft interstitial layers, no-tension elasto-plastic model is used. It is assumed that the material may fail due to plastic yield, or crack by tension. In view of the fact that the thickness of the layer is much less than the dimension of its surface, it can be assumed that $\varepsilon_{x'} = \varepsilon_{y'} = \gamma_{x'y'} = 0$ in the local coordinates x', y', z', shown in Fig. 5. Therefore, the computational formulae can be simplified in some respects. The stress-displacement curve obtained from shear test of the layer may be simplified as shown in Fig. 6. The Drucker criterion is used as yield criterion. Here, its simplified form $F = 3\alpha\sigma_{z'} + \sqrt{(1+12\alpha^2)(\tau_{y'z'}^2 + \tau_{z'x'}^2)} - k = 0$ has been derived. $F = \sigma_{z'} = 0$ is used as no-tension criterion. Moreover the constitutive relation of the layer will be established as we have done for the model of rock and concrete, as follows:

$\{d\sigma_{z'}, d\tau_{y'z'}, d\tau_{z'x'}\}^T = ([D'] - (1-\gamma)[D'_p])\{d\varepsilon_{z'}, d\gamma_{y'z'}, d\gamma_{z'x'}\}^T$
$= [D'_{ep}]\{d\varepsilon_{z'}, d\gamma_{y'z'}, d\gamma_{z'x'}\}^T$

where $[D'] = \begin{bmatrix} \lambda+2G & 0 & 0 \\ 0 & G & 0 \\ 0 & 0 & G \end{bmatrix}$ is the elastic matrix of the layer.

3. THE FORMULA OF FEM AND "ITERATION WITH CHANGEABLE Kp"

For the purpose of both the computational precision and economical benifits, the spacial isoparametric element with eight nodes was used. For the element of interstitial layer, some simplifications were made. When applying the incremental method to computation, the expression becomes $[K]\{\Delta\delta\} = \{\Delta R\}$. Here, $[K]$ varies with the stress states. By means of transforming the formula, its iterative algorithm can be obtained as follows:

$\{[K_e]\{\Delta\bar{\delta}\} = \{\Delta R_p\}$
$\{\Delta R_p\} = [K_p](\{\Delta\delta^*\} + \{\Delta\bar{\delta}\})$

where $[K_e]$ is the elastic stiffness matrix of the structure, $\{\Delta\delta^*\}$ is the elastic displacement increment of the nodes, $\{\Delta\bar{\delta}\}$ is the additional displacement increment of the nodes, and $[K_p]$ is the plastic stiffness matrix when some elements enter plastic stage.

The computational results show that due to the revision of the values of $[K_p]$ and γ after each step of iteration, the iteration is successful and precision can be obtained. "The Iteration with Changeable Kp" method not only has the advantages of both the direct method with variable stiffness and the iterative method with constant stiffness, but also is good in stability and convergence.

4. COMPUTATIONAL EXAMPLES:

According to the above idealized material models, applying the incremental theory and "Iteration with Changeable Kp", and using spacial isoparametric element with eight nodes, we have worked out a three dimensional nonlinear finite element analysis program TNFP. For different rock foundations, different models of rock or interstitial layer can be chosen to make nonlinear calculations. The following are some test examples and computational results of practical projects. (There is not the example of the "Combined model")

4-1. Solution of a thick wall cylinder subjected to inner pressure

Making use of symmetry, we took a quarter of it to compute. It was divided into fifty elements. $E=98$ MN/m², $\mu=0.2$, $C=565.82$ KN/m², $\varphi=0$, $A=0$. The computational results show that both the values of the displacements and stresses are fairly approximate to the exact values (see Tables. 1 and 2). Therefore, program TNFP has been proved reliable.

The values with asterisk * in the two tables are the values in plastic region. The exact values were computed for $\mu=0.5$. (There are not theoretical solution for $\mu=0.2$) In the tables, EV are exact values and CV are computing values.

Table 1   Radial Displacement Ur

r(m)	$q=q_1=314.38$ KN/m²		$q=q_2=54.98$ KN/m²	
	EV (cm)	CV (cm)	EV (cm)	EV (cm)
10	0.87771	0.87601	0.15349*	0.17952
11	0.83319	0.83174	0.14570*	0.16588
12	0.79917	0.79792	0.16761	0.15914
13	0.77323	0.77214	0.16217	0.15400
14	0.75304	0.75267	0.15806	0.15012
15	0.73912	0.73824	0.15501	0.14724

Table 2   Radial Stress

r(m)	$q=q_1=314.38$ KN/m²		$q=q_1+q_2=369.36$ KN/m²	
	EV(KN/m)	CV(KN/m)	EV(KN/m)	CV(KN/m)
10.5	-261.77	-260.54	-314.15*	-308.72
11.5	-176.39	-175.30	-207.23	-210.24
12.5	-110.66	-109.75	-130.02	-131.63
13.5	-59.00	-58.24	-69.31	-69.85
14.5	-17.64	-17.01	-20.73	-20.41

4-2. The stability analysis of resisting slide of GEZHOU BA ERJIANG Sluice Dam on the rock foundation with soft interstitial layers

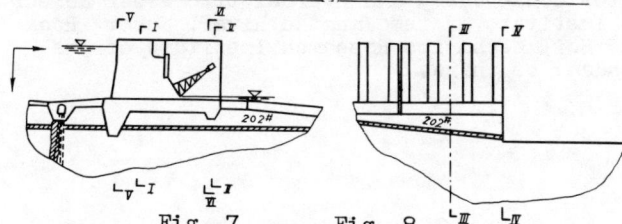

Fig. 7   Fig. 8

The scheme of the dam is shown in Figs. 7 and 8. In the rock foundation under the dam there is a soft interstitial layer ($E=9.8$ MN/m², $\mu=0.4$ $f=0.2$, $C=0$) along the river. One of the key problems in the project is whether the dam will slide along the river under the water pressure. The dam and rock foundation as a whole were divided into 220 elements, in which 32 elements were interstitial layer elements. The computational results show that under normal design water level, the stability of the dam against sliding can be guaranteed. From the shapes of envelopes of the stresses $f\sigma_{z'} \sim \tau_{y'z'}$ in the interstitial layer, shown in Fig. 9, it is seen that the elastic solution indicates that at the downstream part of the footing plate the curve of $\tau_{y'z'}^{e}$ is not enclosed in that of $f\sigma_{z'}^{e}$, while considering the nonlinearity of the layer, the curve of $\tau_{y'z'}^{ep}$ is always enclosed in that of $f\sigma_{z'}^{ep}$. The factor of safety against sliding for the whole dam along the stream and in the traverse directions are 3.02 and 3.01 respectively.

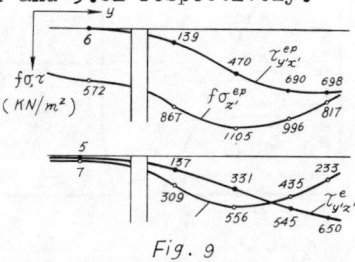

Fig. 9

4-3. The strength and stability analysis of HENAN GUXIAN DAM and its foundation with faults:

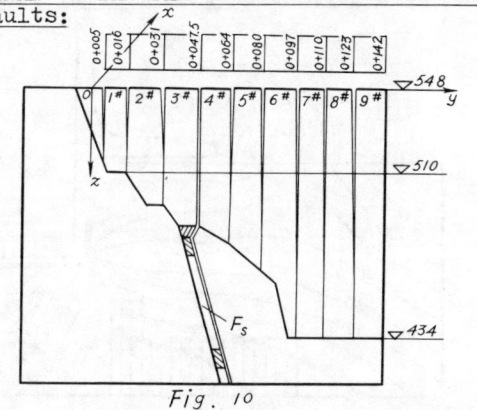

Fig. 10

The scheme of the dam is shown in Fig. 10. The problem worrying people is how the existing F_5 Fault in the rock foundation ($E=14.7$ MN/m², $\mu=0.4$ $f=0.25$, $C=0$) will affect the strength and stability of the dam. The computations were done in the light of three programmes: with and without concrete offsets put into the fault, and assuming the fault inexistent. (The total number of elements is 504, in which the number of interstitial layer elements is 33). The results show:
1). the existence of the F_5 Fault makes a notable impact on the strength and stability of the dam. In the 35th element between the third and fourth sections of the dam, the difference of τ_{yz} and σ_1 computed in the three programmes is great. (see Table 3)

Table 3

stress \ programmes	I	II	III
τ_{yz} (MN/m²)	1.35	0.81	0.37
σ_1 (MN/m²)	1.86	0.88	0.39

After the treatment the strength and stability of the dam are guaranteed foundamentally. 2). The concrete offsets located in deep layers have to be strengthened. 3). For stress results (see Fig. 11), the distribution of σ_z and τ_{zx} in the middle section of the dam approximates to the sphenoidal body solution of elastic theory in most regions. It is especially satisfactory that the normal stress σ_N near the upstream face of the dam tallies with the strength of water pressure fairly well, unless there is bizarre phenomenon at the heel of the dam.

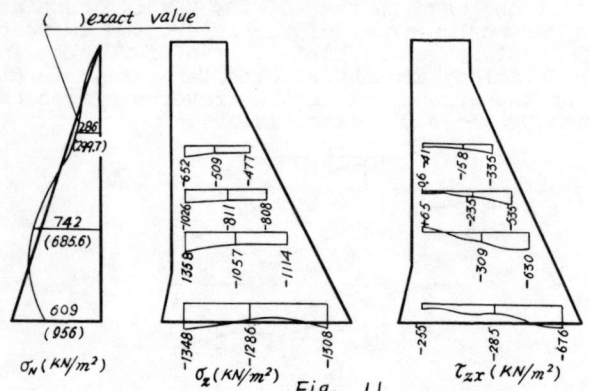

Fig. 11

4-4. The stress analysis of a certain lock on the rock foundation with growing fissures:

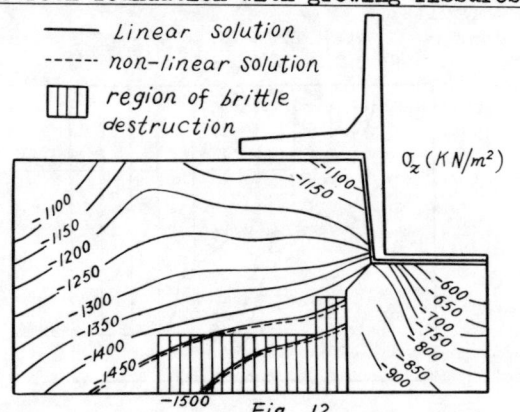

Fig. 12

The lock is built on a rock foundation with growing fissures. The problem is how the rock foundation will behave and how it will influence the stress and displacement of the upper structure under the weight of the lock body and the earth presure. The results show that under such loads there are six elements which will fail due to brittle destruction of the C-T type in the bottom of the rock foundation. The stress isogram in Fig. 12 indicates that $|\sigma_x|$ increases and $|\sigma_z|$ decreases in the elements which failed. But their influence upon the stress of the rest elements of rock and the lock elements is small. In the lock $\sigma_{max}=0.47$ MN/m², while $\sigma_t=0.98$ MN/m². Thus, the strength of the lock is enough.

5. CONCLUSION

5-1. The computational results of practical projects show that the idealized material models, the computing method of FEM and Program TNFP proposed in this paper can be used in making analysis of strength and stability of structures on complex rock foundations, so as to provide useful suggestions and data, for engineering units concerned.They have won admirations of design units.

5-2. Nonlinear analysis of the rock foundation is very complex. The work done in the paper is still elementary. The reasonal extent and suitable region of many assumptions remain to be further studied.

5-3. Some parameters in the paper, such as initial stress, and the hardening parameter A of material, have great influence on the computing result and hence, it is necessary to obtain correct data from practical meassurements or experiments.

REFERENCES

Drucker, D. C. Prager, W. (1952). Soil Mechanics and Plastic Analysis on Limit Design. Quart. of APPL. Math. Vol. 10 No. 2.

Goodman, R. E., Taylor, R. L., and Brekke, T. L. (1968). A model for the Mechanics of Jointed Rock. Journal of the Soil Mechanics and and Foundations Division A. S. C. E. Vol.94 No. SM3

Jaeger, J. C. and Cook, N. G. W. (1976) Fundamentals of Rock Mechanics. John Wiley & Sons, Inc, New York

Mcclintock, F. A., Walsh, J. B. (1962) Friction on Griffith Cracks in Rock under Pressure. Proc, of the Fourth U. S. National Congress of Applied Mechanics, Vol. 2

Prager, W. Hodge, P. G. (1951). Theory of Perfectly Plastic Solids. Johu wiley.

Sandhu, R. S. Huang, S. W. (1975). Application of Griffith's Theory to Analysis of Progressive Fracture Inter. Jour, of Fracture, Vol. 11. No.1.

Yamada, Y. Recent Janpanese Developments in Matrix Displacement Method for Elastic-Plastic Problems. Recent Advances

Yin, Y. Q., Qu. S. N. (1981). Constitutive Relation in the Finite Element Analysis of Structures Made of Concrete, Rock, etc.

Zhuo, J. S. (1979). Three Dimensional Nonlinear Finite Element Method Program AAA3 and the comments.

Zienkiewicz, O. C. (1971). The Finite Element Method in Engineering Science.

Experimental study of the Shearing Strength of the Rock Foundation at Erjiang and Sanjiang of the GEZHOU BA Project on the Yangtze River. Water Conservancy and Hydroeletric Power Research Institute of the Yangtze River, Huber Rock and Soil Mechanics Research Institute of the Acadamy of China.

MECANIQUE DES ROCHES ET MODELES MATHEMATIQUES APPLIQUES A LA GEOLOGIE

Rock mechanics and mathematical models applied to geology

Gesteinsmechanik und auf Geologie angewandte mathematische Modelle

J. A. Quiblier
K. Ngokwey
Institut Français du Pétrole, 1 & 4 avenue de Bois Préau, B.P. 311,92506 Rueil Malmaison, Cedex, France

RESUME

Un modèle mathématique visco-élastique en grandes déformations appliqué au chevauchement de deux plaques de l'écorce terrestre rend bien compte de la déformation de la couverture sédimentaire et permet d'étudier certains paramètres. Un modèle élastique appliqué à une faille permet de déterminer la poussée qui cause sa propagation.

SYNOPSIS

A large-deformation viscoelastic mathematical model applied to the overthrusting of two plates of the earth's crust clearly illustrates the deformation of the sedimentary cover and can be used to analyse various parameters. An elastic model applied to a fault can be used to determine the thrust that caused the fault to be propagated.

ZUSAMMENFASSUNG

Ein mathematisches Modell einer weiträumigen, zähflüssig-elastischen Verformung, das auf die Überschiebung zweier Platten der Erdkruste angewandt wird, gibt sehr gut die Verformung der sedimentären Überdeckung wieder, und macht es möglich, verschiedene Parameter zu studieren. Ein elastisches Modell, das auf eine Verwerfung angewandt wird, ermöglicht die Bestimmung der Schubgröße, die ihr Fortschreiten verursacht.

1 - GENERALITES

De 1970 à 1982, l'Institut Français du Pétrole a développé un modèle de calcul par éléments finis spécifiquement adapté aux problèmes géologiques. Ce modèle a, jusqu'ici été utilisé pour trois objectifs :

- Essayer de prévoir la localisation des zones les plus fracturées d'un gisement d'huile ou de gaz, à partir de la connaissance de sa forme actuelle et des phases tectoniques qui l'ont conduit à cette forme.

- Déterminer les phénomènes géologiques (poussées tectoniques, mouvements du socle etc.) qui peuvent rendre compte des observations géologiques (plis, fractures, stries fossiles déformés, stylolithes etc.) dans une zone donnée.

- Déterminer, pour des mécanismes géologiques dont on connaît l'allure de façon qualitative, l'influence plus particulière de tel ou tel paramètre (glissement entre couches, vitesse de la déformation, épaisseur et rigidité des couches etc.)

Le modèle inclut des milieux stratifiés bi- ou tridimensionnels dont chaque couche est visco-élastique et orthotrope. Les discontinuités sont modélisées par de fines couches de mailles. Le modèle prend en compte les grandes déformations. Les conditions aux limites sont des contraintes aux frontières, des vitesses de déplacement imposées, des déplacements imposés, des glissements sur une surface de forme imposée. Les calculs sont effectués sur CDC 7600 ou sur CRAY 1.

Le modèle fournit, en fonction du temps, les déplacements, les contraintes et une large gamme de représentations graphiques. La fracturation est étudiée, à partir des contraintes par la méthode dite du "risque relatif de fracturation". Rappelons brièvement cette technique :

On ne sait pas vraiment déterminer les paramètres mécaniques des roches pour des déformations aussi lentes que les déformations géologiques. On va donc leur attribuer des valeurs arbitraires qui n'auront pour seule ambition que de classer les roches du modèle l'une par rapport à l'autre. Si l'on a deux points A et B, comme ceux de la fig.1, qui représentent les contraintes en deux points du modèle, calculées avec ces valeurs arbitraires des paramètres, il est impossible de comparer directement ces points à un critère expérimental de rupture comme le critère (C1) de la fig.1 ou à un seuil (S1) d'entrée en domaine pseudoplastique.

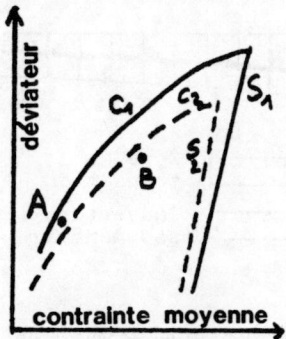

Fig. 1 - Détermination du "risque relatif de fracturation" entre A et B

Par des transformations appropriées (affinités) on détermine à partir du critère expérimental (C1)-(S1) une famille de courbes (C2)-(S2), (C3)-(S3) etc. et on classe A et B d'après leur position entre ces courbes. En conclusion, on ne dit pas qu'il y a fracturation en A ou en B mais simplement que le risque de fracturation est plus élevé en A qu'en B. Pour compléter ce critère, on conviendra que le seuil de rupture en traction est quasi-nul.

Comme il s'agit de modèles géologiques, l'interprétation doit parfois être plus globale que locale. L'expérience nous a, par exemple, montré que lorsqu'on détermine par la méthode ci-dessus des zones d'égal risque de fracturation, des zones en traction et des zones pseudo-plastiques, les frontières entre ces zones peuvent elles-mêmes être interprétées comme des méga-fractures (failles) qui ont alors une orientation tout à fait différente de celle que fournit l'interprétation ponctuelle des contraintes. Nous attachons pour notre part une grande importance à cette notion.

2 - EXEMPLE N° 1 : CHEVAUCHEMENT DE DEUX PLAQUES.

2 - 1 - Généralités

On connaît aujourd'hui assez bien le concept de "lithosphère", plaque superficielle rigide et mobile à la surface de la terre. Les "zones de subduction" sont des limites convergentes où une plaque de lithosphère océanique plonge sous une autre de même type (Arc Insulaire) ou de type continental (Marge Continentale Active). L'observation de ces marges actives montre que, par le jeu des déformations plus ou moins importantes, les sédiments convoyés par la plaque plongeante acquièrent des morphologies et des structures fort variées pour former ce que l'on appelle "prisme d'accrétion". Si la subduction est le moteur principal nécessaire à cette structuration, elle ne suffit cependant pas à rendre compte de la diversité et encore moins de l'amplitude des structures ainsi édifiées. Il faut compter avec des paramètres tels que la nature et l'épaisseur des sédiments en cause, le taux de convergence, la présence éventuelle de niveaux de décollement etc. Cette situation rend nécessaire l'emploi de modèles mathématiques déterministes qui peuvent permettre de connaître la contribution de chaque facteur dans les processus impliqués. L'exemple que nous présentons traite de l'influence d'un niveau de décollement, les sédiments étant supposés avoir un comportement visco-élastique.

2 - 2 - Description du modèle.

Maillage et caractéristiques mécaniques.

La fig. 2 représente le maillage du modèle visant à reproduire les traits généraux d'une zone de subduction. Ce modèle est inspiré de la Ride de la Barbade (Arc Insulaire des Petites Antilles) étudiée depuis 1973 dans le cadre du projet IPOD et d'un projet propre à l'IFP (Géodynamique des Marges Actives).

TABLEAU I. Caractéristiques mécaniques adoptées

	$E1$ 10^6bar	$E2$ 10^6bar	$G2$ 10^6bar	ν_1	ν_2	$EV1$ 10^6bar	$EV2$ 10^6bar	$\dfrac{\eta}{E1}$
1	0.8	1.	0.3	0.3	0.3	0.8	1.	0.8
2	1.5	2.	0.5	0.3	0.3	1.5	2.	1.5
3	0.02	0.05	0.01	0.4	0.3	0.02	0.05	0.02
4	0.1	0.05	0.025	0.4	0.3	0.1	0.05	0.1

L'indice 1 se réfère aux caractéristiques dans le plan de la stratification tandis que l'indice 2 se réfère aux caractéristiques dans la direction normale à ce plan. $EV1$ et $EV2$ sont les modules d'Young pour l'élasticité de l'élément de Kelvin-Voigt.

Conditions aux limites.

La frontière gauche (AE) est supposée bloquée, aucun déplacement horizontal n'y est admis. La frontière droite (CD) est soumise au chargement géostatique (poussée simulant le poids d'un massif rocheux infini et au repos situé à droite de cette frontière). La limite inférieure de la plaque plongeante (BC), dont les noeuds sont astreints à se déplacer sur la courbe (BC) elle-même, est chargée par une pression tangentielle de 1500 bar. La limite inférieure de la plaque chevauchante (AA') glisse librement sur elle-même sans décollement. La portion (BB') de la plaque plongeante est entraînée vers le bas suivant l'angle de plongement du point B. La face supérieure est libre.

2 - 3 - Résultats.

Le modèle DEC1 (fig. 3a) présente un couplage mécanique entre la couverture sédimentaire et les deux plaques. Dans le modèle DEC2 (fig. 3b), un mince niveau moins rigide assure la transition entre la couverture et les deux plaques, tout en maintenant le couplage au-delà d'une certaine distance de la jonction des plaques. Les répartitions des contraintes indiquent, dans les deux cas, un régime global compressif. Au voisinage immédiat de la jonction des plaques, les trajectoires des contraintes principales de compression maximale forment des arcs dont les rayons de courbure augmentent quand on s'éloigne de la jonction.

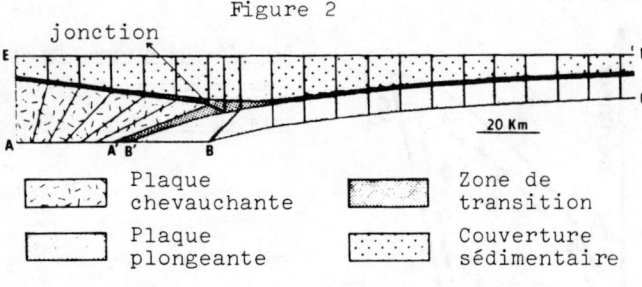

Figure 2

- Plaque chevauchante
- Plaque plongeante
- Zone de transition
- Couverture sédimentaire

Ce modèle représente une zone de 400 km de longueur et est composé de quatre unités mécaniques : une plaque chevauchante, une plaque plongeante, un niveau de transition possible et une couverture sédimentaire. Chacune de ces unités est orthotrope pour prendre en compte la stratification. Les caractéristiques mécaniques adoptées sont présentées dans le tableau I

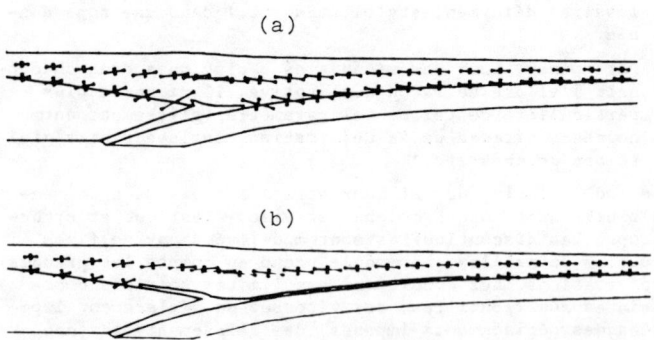

Figure 3. (a) Modèle avec couplage DEC1
(b) Modèle avec découplage partiel DEC2

Ces mêmes contraintes sont quasi horizontales partout ailleurs. La traduction des contraintes en termes de ruptures, par la méthode du risque relatif de fracturation exposée dans les généralités, conduit à délimiter des zones du modèle en fonction du plus ou moins grand risque de fracturation qu'elles présentent. Les fig. 4a et 4b représentent cette analyse. Si l'on admet, comme cela a été souligné, que les frontières entre zones dont les risques présentent de forts contrastes peuvent être interprétées comme des failles, les fig. 4a et 4b donnent directement les types de fracturation et de déformation possibles dans la couverture. Les fractures de type F1, F2 et F3 sont effectivement observées dans la Ride de la Barbade.

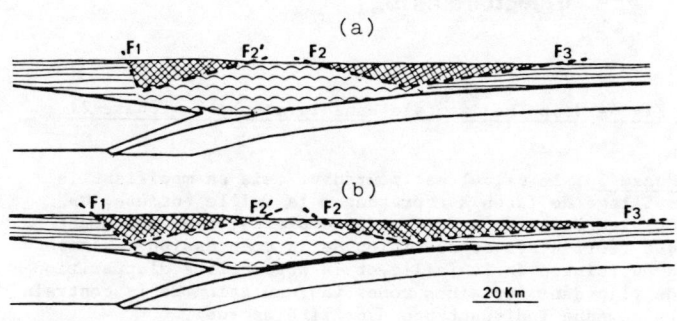

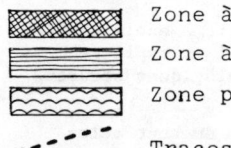
Zone à grand risque de rupture
Zone à faible risque de rupture
Zone pseudo-plastique
--- Traces des grandes fractures possibles

Fig. 4   Interprétation des contraintes

La fig. 5 donne les déformées de la surface libre rapportées à un même niveau zéro pour les deux cas. Si on considère la distance entre les fractures de type F1 et F3, on note que les deux déformées diffèrent à la fois en amplitude et en extension latérale.

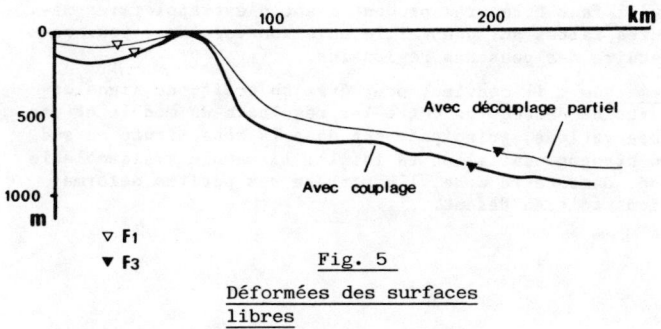

Fig. 5
Déformées des surfaces libres

La fig. 6 présente les déplacements verticaux (v) de la surface libre en fonction du temps, relevés à l'intersection de cette surface et de la verticale indiquée sur la fig. 2.

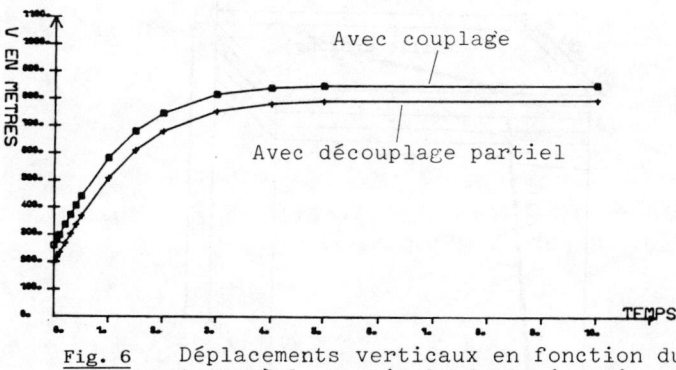

Fig. 6   Déplacements verticaux en fonction du temps à la verticale de la jonction.

Ces deux figures montrent que la présence d'une zone de transition moins rigide a pour effet de diminuer l'amplitude de la déformation tout en accentuant son extension latérale. Il semble qu'un niveau de décollement confère une certaine indépendance à la couverture sédimentaire vis à vis des réactions de la plaque plongeante aux contraintes.

Le modèle rejoint les observations qualitatives opérées dans cette zone. Il est impossible d'entrer ici dans le détail des résultats obtenus mais on peut conclure à l'intérêt de ce type d'approche qui permet de bien isoler l'influence de chacun des paramètres.

3 - EXEMPLE N° 2 : PROPAGATION D'UNE FAILLE.

La fig. 7 localise le modèle. La zone choisie est traversée par un faisceau de failles. On s'intéresse à la principale d'entre elles (faille de Saint Michel-Vissec). Les géologues démontrent que sa partie Sud (Faille de Saint Michel) est beaucoup plus ancienne que sa partie Nord.

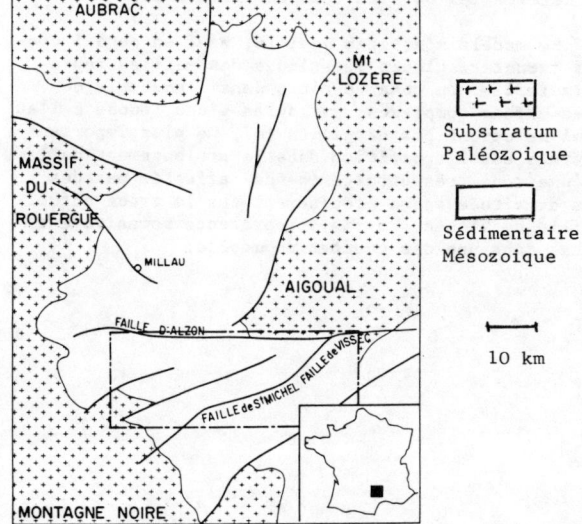

Fig. 7 - Localisation et contexte géologique du modèle.

La fig. 8 schématise l'état initial du massif, servant à la construction du modèle. Cet état initial ne comprend que l'amorce de la faille. Le modèle ignore bien entendu tout du prolongement de cette faille, c'est à dire du tracé actuel de la faille de Vissec.

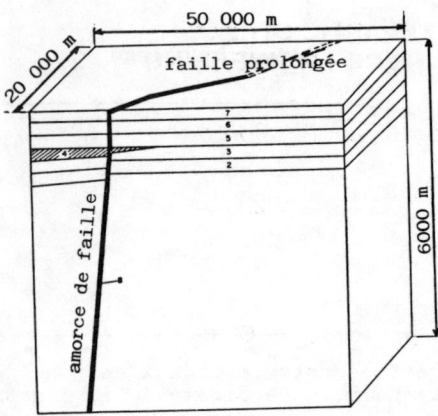

Fig. 8 - Schéma du modèle.

Les géologues souhaitaient savoir si la récente poussée tectonique Nord-Sud (poussée d'âge pyrénéen) qui affecte cette zone suffit à expliquer la propagation de la faille ou s'il est nécessaire d'introduire d'autres mécanismes plus complexes. Le calcul est fait dans l'hypothèse élastique et en petites déformations (cette dernière hypothèse semblant autorisée par les observations dans cette région)

Le calcul se déroule en trois phases :

Phase 1 : L'amorce de faille est très consolidée et moins déformable que le massif qui l'entoure. On applique la poussée Nord-Sud. On retrouve partout dans le massif une compression Nord-Sud, sauf dans la faille où des concentrations de contraintes sont calculées. On note aussi des tractions contenues dans le "plan" de la faille. On en conclut que cette phase n'affecte pas le massif mais est de nature à affaiblir mécaniquement la faille par des ruptures internes. A l'issue de cette phase, la faille devient donc mobile et sera modélisée par une série de mailles plus déformables que les couches traversées.

Phase 2 : Le modèle n'est pas modifié, sauf en ce qui concerne les caractéristiques mécaniques des mailles représentant la faille. On observe alors dans l'ensemble du massif des effets complexes, variables d'une couche à l'autre et qui ne seront pas détaillés ici. Le plus important de ces effets est l'apparition dans le prolongement de la faille d'une zone très étroite (bande) affectée par des tractions et située assez précisément sur le tracé actuel de la faille de Vissec. La fig. 9 représente sommairement ce résultat dans une des couches du modèle.

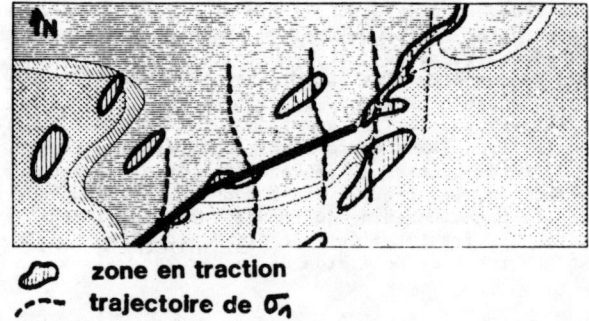

Fig. 9- Coupe horizontale dans la couche 3 (phase 2)

Phase 3 : le calcul est poursuivi mais en modifiant le maillage de façon à représenter la faille entière. On observe des effets également complexes, avec notamment des réorientations sensibles de la compression maximale au voisinage de la faille et la possibilité d'apparition de plis dans certaines zones (augmentation de la contrainte moyenne indiquant une ductilité accrue).

Conclusions de ce calcul :

(a) La poussée tectonique Nord-Sud suffit à expliquer la propagation de la faille de Saint Michel et explique également de nombreux autres traits géologiques structuraux observés sur le terrain.

(b) Cette propagation ne s'effectue pas du tout selon l'analogie classique qui assimile une faille à une grande fracture se propageant en milieu continu. La bande en traction qui apparaît en avant de la faille dans la phase 2 du calcul est occupée par des tractions Est-Ouest. Elle va donc d'abord être le siège d'une fracturation locale Nord-Sud qui va l'affaiblir mécaniquement. C'est cet affaiblissement localisé dans une bande étroite qui va transformer la bande en faille. Ce mécanisme requiert moins de puissance que le mécanisme de propagation directe.

(c) Les observations sur des modèles de ce type montrent qu'il faut être très prudent avant d'extrapoler les mesures faites sur une faille ou à son voisinage pour en déduire des poussées régionales.

remarque : il convient, pour être objectif, de signaler certains désaccords entre les résultats du modèle et les observations, principalement dans la zone située au sud du tronçon central de la faille. Il semble vraisemblable que, dans cette zone, l'hypothèse des petites déformations soit en défaut.

BOUNDARY ELEMENT METHOD FOR LINEAR VISCO-ELASTIC STRESS ANALYSIS IN A ROCK MASS AND ITS APPLICATION IN ROCK ENGINEERING

Méthode des éléments limites appiiquée à des massifs rocheux viscoélastiques linéaires — analyse des contraintes et son application en géotechnique

Granzelementmethode zur Spannungsanalyse für lineare Viskoelastizität des Gebirges und ihre Anwendung im Felsbau

Lin Dezhang
Research Assistant, Research Institute of Mining and Metallurgy, Changsha, China
Liu Baoshen, Dr. Eng.
Associate Professor, Research Institute of Mining and Metallurgy, Changsha, China

SYNOPSIS

This paper presents the boundary element method for the analysis of linear viscoelastic deformation of rock engineering. The illustrative problem was chosen to demonstrate the interaction characteristics between rockmass and support, regarded as systems influencing each other's deformation. An example involving interaction in the binary system of rock and support was discussed and the results obtained for this problem by the BIM were in agreement with in situ observations.

RESUME

Cette communication présente la méthode des éléments limite de déformation viscoélastique linéaire avec interaction entre la roche et le support comme un système de déformation à influence réciproque. On discute un exemple de l'interaction entre une roche et un support. Les résultats sont conformes à l'observation in situ.

ZUSAMMENFASSUNG

In vorliegender Arbeit wird eine Deformations-Grenzelementsmethode für lineare Viskoelastizität des Felsgebirges vorgestellt. Es wird der Interaktionscharakter zwischen Fels und Ausbau hervorgehoben. Es werden binäre Systeme, einschließlich Interaktion zwischen Fels und Ausbau, diskutiert. Das mit der Grenzelementsmethode erhaltene Ergebnis ist mit in situ Beobachtungen identisch.

1. Introduction

The observations of rock engineering practice so far and the laboratory tests adduced as a factual basis of theoretical research suggest that the consequences of excavating caves, tunnels, underground spaces etc, must be regarded as time-dependent phenomena. This viscoelastic constitutive equations are described in the form of differential or integral operators. The complicated constitutive equations proposed for representing the reological behavior and widely variety of irregular domain shapes lead to the fact that the analytical solutions can not be used directly for such varied rock engineering projects as tunnels, caverns, and other underground spaces.
Remarkable progress has been made in recent years in the application of the boundary element method to stress analysis. The principle advantages of this method are that it effectively reduces the dimensions of a problem by one, can be used relatively easily to obtain displacements and stresses in multiply connected bodies of arbitary shape, treats uniform and mixed boundary valur problems with equal ease, and requires only specification data on the boundary of the region being considered, as a consequence of this, the number of simultaneous equations to be solved is relatively small.
In this paper we present the boundary element method for the analysis of linear viscoelatic deformation. The approach accommodates a broad class of linear constitute equations of differential and integral type. The essence of our me-
thod is the use of the associated elastic procedure to obtain the time histories of stress, strain and displacement based on the correspondence principle. The illustrative problem was chosen to demonstrate the interaction characteristics between the rockmass and the support, regaded as systems influencing each other's deformation. In this procedure we made use of B. Brady's complete plane stress concept and elastic results from the boundary element analysis in the Laplace transform space, the linear viscoelastic results are obtained via the numerical transform inversion. Most important is the feasibility of this approach that only one dimensional elements on the boundary need to be defined, and the time-dependent three dimensional deformation behavior of surrounding rockmass can be obtained.

2. Boundary Element Method Formulation and Correspondence Principle

The basic equations for linear viscoelastic deformation, as elastic deformation, are as follows:

2.1 Strain-displacement relation

$$\varepsilon = \frac{1}{2}(\nabla u + \nabla u^T) \qquad (1)$$

2.2 Constitutive equations

The stress at a particle is a linear function of

the infinitesimal strain history at the particle is usually expressed in the following form

$$\sigma = L[\varepsilon(s)\big|_{-\infty}^{t}] \qquad (2)$$

Specific representations for L lead to well known models for material behavior. The representations appropriate to linear elasticity and linear viscoelasticity are

$$L[\varepsilon(s)\big|_{-\infty}^{t}] = E\varepsilon(t) \qquad (3)$$

and

$$L[\varepsilon(s)\big|_{-\infty}^{t}] = G(0)\varepsilon(t) - \int_{0}^{t}\varepsilon(t-s)\,dG(s) \qquad (4)$$

where E is constant and G(s) is a tensor-valued function of boundary variation on every subinterval. For linear, homogeneous, isotropic material, the constitutive equation can be simplified into: for linear elasic material

$$\sigma_{ij} = \lambda\Theta\delta_{ij} + 2G\varepsilon_{ij} \qquad (5)$$

where λ, G are the Lame constants, $\Theta = \varepsilon_{ii}$.
for linear viscoelastic material
a) integral equation

$$\sigma_{ij} = [\lambda_{0}\Theta(t) - \int_{0}^{t}\lambda(t-\tau)\Theta(\tau)]\delta_{ij} + 2[G_{0}\varepsilon_{ij}(t) - \int_{0}^{t}G(t-\tau)\varepsilon_{ij}(\tau)d\tau \qquad (6)$$

b) differential equation

$$\sigma_{ij} = K(D)\delta_{ij} + 2G(D)(\varepsilon_{ij}(t) - \frac{1}{3}\Theta(t)\delta_{ij}) \qquad (7)$$

where

$$3K(D) \equiv \frac{Q_{k}(D)}{P_{k}(D)}, \quad 2G(D) \equiv \frac{Q_{g}(D)}{P_{g}(D)} \qquad (8)$$

$$G(D) \equiv (Q_{k}(D)/P_{k}(D) - Q_{g}(D)/P_{g}(D))/(2Q_{k}(D)/P_{k}(D) + Q_{g}(D)/P_{g}(D))$$

and $Q_{\lambda}(D), P_{\lambda}(D)$ $(\lambda=k,g)$ are polynomials with constant voefficients in the operator.

2.3 Equilibrium equation

$$\text{div}\,\sigma + b = 0 \qquad (9)$$

we consider the mixed boundary value problem in a finite, multiply connected open region Ω

$$u = \bar{u}, \quad u \in \Gamma_{1}$$
$$t = \sigma n = \bar{t}, \quad u \in \Gamma_{2} \quad \Gamma_{1} \cup \Gamma_{2} = \Gamma = \partial\Omega \qquad (10)$$

2.4 Principle of correspondence

The components of vector field and tensor field in elasticity and viscoelasticity are the elements in Hilbert space, Let R be A H-space belonging to rheology and $\tilde{\omega}_j$ is a element in this space, $\{\tilde{\omega}_j\} \subset R$ and E a H-space belonging to elasticity. If the star operator $*: R \to E$ is a linear mapping at $\{\omega_j\}$ the inverse mapping of the star operator $*^{-1}$ is an one-one mapping:

$$\tilde{\omega}_j \to \omega_j = *^{-1}(\tilde{\omega}_j) \qquad (11)$$

The solution for a quasi-static problem for viscoelastic material can be obtained from the solution for the corresponding problem for an elastic solid by applying the star operator $*$ to the elastic solution and inverting the mapping. It is understood that the boundary conditions for the two problems are identical. The star operator $*$ used in this paper is Laplace transform.

$$f^{*}(t) = \mathcal{L}(f(t)) = \int_{-\infty}^{+\infty}f(t)\,e^{-st}\,dt \qquad (12)$$

The remember function $\lambda(t-\tau)$ and $G(t-\tau)$ can be given by the combination of idealized models or the rheological tests. Taking Maxwell's model for example

$$\lambda(t-\tau) \equiv \frac{\lambda_0}{t_\lambda}\exp[-(t-\tau)/t_\lambda]$$
$$G(t-\tau) \equiv \frac{G_0}{t_u}\exp[-(t-\tau)/t_u] \qquad (13)$$

generally we introduce

$$\lambda(t-\tau) = \int_{0}^{\infty}S_{\lambda}(\tau)\exp[-(t-\tau)/\tau_\lambda]\,d\tau$$
$$G(t-\tau) = \int_{0}^{\infty}S_{g}(\tau)\exp[-(t-\tau)/\tau_g]\,d\tau \qquad (14)$$

where $S(\tau)$ is relaxation spectrum.

2.5 Direct formulation

In Laplace transform space, we use of the Kelvin singular state due to unit load at point Y, $[u_Y(l), \sigma_Y(l)]$, the elastic state $[u, \sigma]$ and Betti's reciprocal theorem, the solution known as Somigliana's identity can be arrived at, namely

$$u(Y)l + \int_{\Gamma}\sigma_Y[l]\,n u\,d\Gamma = \int_{\Gamma}u_Y[l]\sigma n\,d\Gamma + \int_{\Omega}b\,u_Y[l]\,d\Omega \qquad (15)$$

The displacement at any interier point X can be obtained

$$u_Y[l](X) = \frac{1}{cr}[\frac{P \otimes P}{r^2} + (3-4G)\,1]l \qquad (16)$$

where $c = 16\pi G(1+G)$, $P = X-Y$, $r = |X-Y|$, 1 is unit tensor.
The stress field is given by

$$\sigma_Y[l](X) = \frac{-2G}{cr^3}\{\frac{3(P\cdot l)P\otimes P}{r^2} + (1-2G)[P\otimes l + l\otimes P - (P\,l)1]\} \qquad (17)$$

For an arbitrary point i at the discreted boundary we obtain the equation for plane problem

$$C^{i}\bar{u}_i + \sum_{j=1}^{n}[\int_{\Gamma_j}t_j^{T}\Phi_j\,d\Gamma]\bar{u}_j = \sum_{j=1}^{n}[\int_{\Gamma_j}u_j^{T}\psi_j\,d\Gamma]\bar{t}_j + \sum_{K=1}^{n}[\int_{e_K}u_j^{T}\,b\,d\Omega] \qquad (18)$$

where

$$u_j = [u_{j1}, u_{j2}]^T, \quad \bar{u}_j = [\bar{u}_{j,1}, \bar{u}_{j,2}, \bar{u}_{j+1,1}, \bar{u}_{j+1,2}]^T$$
$$t_j = [t_{j1}, t_{j2}]^T, \quad \bar{t}_j = [\bar{t}_{j,1}, \bar{t}_{j,2}, \bar{t}_{j+1,1}, \bar{t}_{j+1,2}]^T$$
$$\Phi_j = \begin{bmatrix}\phi_{j1} & 0 & \phi_{j2} & 0 \\ 0 & \phi_{j1} & 0 & \phi_{j2}\end{bmatrix}, \quad \psi_j = \begin{bmatrix}\psi_{j1} & 0 & \psi_{j2} & 0 \\ 0 & \psi_{j1} & 0 & \psi_{j2}\end{bmatrix} \qquad (19)$$

The integration is taken over the corresponding intervals and two square $2N \times 2N$ matrices may be evaluated. Once t and u are known everywhere on Γ, the desired field quantities in Laplace transform space are immediately obtainable by means of (16) and (17).

2.6 Indirect formulation

In indirect method the discreted integral equations are formulated in terms of fictitious distributions ϕ, ψ of the singular solutions $f_{\alpha\beta}, g_{\alpha\beta}, M_{\alpha\beta r}$ over the problem boundaries. In the presence of a known body force the displacements and the tractions over the surface can be obtained from

$$u_\alpha(Y) = \int \phi_\beta(X)f_{\alpha\beta}(Y,X)\,d\Gamma(X) + \int \psi_\beta(Z)f_{\alpha\beta}(Y,Z)\,d\Omega(Z)$$
$$t_\alpha(Y) = \int \phi_\beta(X)g_{\alpha\beta}(Y,X)\,d\Gamma(X) + \int \psi_\beta(Z)g_{\alpha\beta}(Y,Z)\,d\Omega(Z) + \frac{1}{2}\phi_\alpha(Y)$$
$$\sigma_{\alpha\beta r}(Z) = \int M_{\alpha\beta r}(Z,X)\phi_\gamma(X)\,d\Gamma(X) \qquad (\alpha,\beta,r=1,2,3) \qquad (20)$$

Onve the fictitious density functions ϕ, ψ have been determined the stresses and displacements at all in the Laplace transform space. In this elastic solution replace the elastic moduli by the moduli which is the Laplace transfoum of the corresponding differential operators. Inversion of the expressions so obtained for the transforms of stresses and displacements gives the vicoelastic solutions.

3. Illustrative problem

The illustrative problem was chosen to show the feasibility of the present approach and accuracy of the approximations to the associated analytical solution, Problems involving more irregular geometry and more complicated constitutive equations and boundary conditions are being investigated in order to lend confidence to the associated results in the absence of an analytical solution.

The concept of complete plane strain makes it possible to regard the mechanical state about a roadway driven in a primary field of arbitrary orientation as the superposition of two simple states, plane state and anti-plane state.

To obtain the viscoelastic solution for the problem of interaction in the binary system of rock and support, first determine the stresses and induced displacements around the opening excavated in homogeneous state which is equivalent to the sum of the case of the primary stress field p_x, p_y, p_z, p_{xy}, p_{yz}, p_{xz} and imposed tractions t_l, t_m, t_n on the boundary of excavation. The system of equations are written for the midpoint of the boundary elements and can be given as folows:

$$\begin{bmatrix} a_{nn} & b_{nn} \\ a_{nl} & b_{nl} \end{bmatrix} \begin{bmatrix} \varphi_n^* \\ \varphi_l^* \end{bmatrix} = \begin{bmatrix} t_n^* \\ t_l^* \end{bmatrix} \quad , \quad [C_{mn}^*][\varphi_m^*] = [t_m^*] \quad (21.1)$$

$$\begin{bmatrix} d_{nn} & e_{nn} \\ d_{nl} & e_{nl} \end{bmatrix} \begin{bmatrix} \varphi_n^* \\ \varphi_l^* \end{bmatrix} = \begin{bmatrix} u_n^* \\ u_l^* \end{bmatrix} \quad , \quad [f_{mn}^*][\varphi_m^*] = [u_m^*] \quad (21.2)$$

assuming the support characteristic to be a linear one, that is, the load on the support is proportional to the displacement of the peripheral points. This assumption can be expressed in terms of the rigidity matrix

$$\begin{Bmatrix} t_l^* \\ t_m^* \\ t_n^* \end{Bmatrix} = \begin{bmatrix} K_l & 0 & 0 \\ 0 & K_m & 0 \\ 0 & 0 & K_n \end{bmatrix} \begin{Bmatrix} u_l^* \\ u_m^* \\ u_n^* \end{Bmatrix} \quad (2.2)$$

Since the radial displacements play the decisive role, we shall adopt the simplification of considering the load on the support depends exclusicely on the radial displacement, introducing $t_n^* = k u_n^*$ and $t_l^* = t_m^* = 0$ into (21.1) and (21.2). Next invert the coefficient matrix to get the solutions for fictitious forces φ_l, φ_m, φ_n.

As an example, in north-western China a testing cavern with circular cross section was built in underlying 19.5 m of loess overburden. The bulk density of loess is about 1.8 t/m^3, the radius of the testing cavern R=3.0 m, $p_x = p_y = 28.08$ t/m, $p_z = 35.1$ t/m^2 $p_{xy} = p_{yz} = p_{xz} = 0$, and the deviatoric component for rock may be regarded as Kelvin's body, $G_r = 0.04 \times 10^5$ t/m^2 $\eta_r = 2.04 \times 10^6$ day·ton/m^2 and the dilatational as Hooke's body, $\nu_r = 0.25$

$$P_{rg}/Q_{rg} = 2G_r + 2\eta_r \frac{\partial}{\partial t} \quad , \quad P_{rK}/Q_{rK} = 3K = 5G_r \quad (23)$$

the support behaves as a linearly elastic Hookean solid $G_s = 5.03 \times 10^5$ t/m^2 and incompressible, $\nu_s = 0.5$.

$$P_{sg}/Q_{sg} = 2G_s \quad , \quad K = K_n = 2G_s(R^2-(R-b)^2)/(R(R-b)^2) = 4.1 \times 10^4 \quad (24)$$

By analytical method, the radial displacements about a opening of circular cross section are of arbitrary direction in a primary field of unstricted orientation and triaxial stress distribution take the form

$$u_r = \frac{1}{2}(P_z + P_x)\left(\frac{R}{r}\right)\frac{R}{kR+2G_r}(1-\exp(-(kR+2G_r)/2\eta_r \cdot t)) \quad (25)$$

Fig.1. shows the radial displscement vs. time of the point at $\theta = 45°$, r=4.5m from boundary element method analysis and analytical solution in the supportless case and the comparision between the calculated results and observed data. Thirty six boundary elements of equal length were disposed around the complete circumference of the cavern. From Fig.1 it is seen the boundary element method for linear viscoelasticity can be applied to study the interaction in the binary system of rock and support. Only one dimentional elements with its interaction factor of rock and support need to be defined, and the time-dependent three dimensional deformations behavior of surrounding rockmass can be obtained.

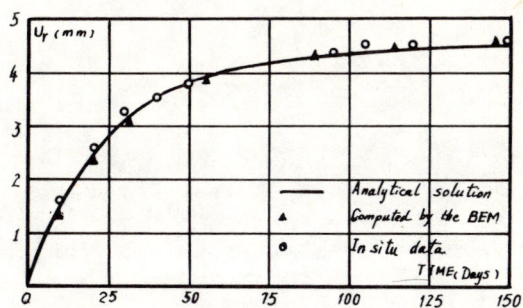

Fig.1.　Radial displacements vs. time of the point at $\theta = 45°$, r=4.5 m

Conclusion

In this paper we have presented the boundary element method for the analysis of time-dependent stress-strain field of rock engineering. The approach accomodates a broad class of boundary value problems of arbitrarily shaped domain, connectivity, and different material constitutions in the form of differential or integral operators. The approximate numerical solution procedure is capable of efficiently and accurately treating difficult problems within the framework of cooresponcence principle. An illustrative example involving interaction in the binary system of rock and support is discussed. The numerical results of displacement and stress obtained for this problem by the BEM agree with expressions obtained earlier by the present authors directly from the differential equation.

References

(1) Brady,B.H.G., Bray,J.W. (1978). The Boundary Element Method for Determining Stresses and Displacements Around Long Openings in A Triaxial Stress Field: Int.J. Rock Mech Min. Sci & Geomech. Abstr. Vol.15,1-28.
(2) Riggo,F., Shippy, D. (1971). An Application of the Coreespondence Principle of Linear Viscoelasticity Theory: SIAM Appl.Math. Vol.21 No.2. 321-330.
(3) Bland,D.R.(1960). The Theory of Viscoelasticity. 124 pp. Pergamon Press.
(4) Brebbia,C.A. (1978). Recant Advances in Boundary Element Methods. 424 pp. Pentech Press London; Plymouth.
(5) Assonyi,Cs., Richter, R. The Continuum Theory of Rock Mechanics.First Edition (1979) Trans Tech Publications.
(6) Brady,B.H.G., and Bray, J.W. (1978). The Boundary Element Method for Elastic Analysis of Tabular Orebody Extraction, Assuming Complete Plane Strain. Int. J. Rock Mech. & Min. Sci., Vol.15. 29-37.
(7) Rizzo,F.J. (1967) An Integral Equation Approach to Boundary Value Broblems of Classical Elastostatics. Quart. Appl. Math., Vol.25, 83-95.
(8) Jaeger, J.C. and Cook, N.G.W. (1971) Fundamental of Rock Mechanics. Chapman and Hall, London.

A MODEL FOR THE ANALYSIS OF DISCONTINUITY CHARACTERISTICS IN TWO DIMENSIONS

L'analyse des caractéristiques de discontinuités en deux dimensions

Ein ebenes Computermodell für die Untersuchung der Klufteigenschaften

S. D. Priest
Lecturer in Rock Mechanics, Imperial College of Science & Technology, London, U.K.

A. Samaniego
Research Student in Rock Mechanics, Imperial College of Science & Technology, London, U.K.

SYNOPSIS

A computer model was developed, capable of generating in two dimensions a random discontinuity fabric which can be presented numerically or graphically. The input data are obtained from surveys of the rock mass. The resulting fabric can be interrogated numerically using scanlines of any orientation, to give the values of discontinuity frequency, spacing and trace length at this orientation. The model shows that there is a simple linear relation between mean frequency and the ratio of discontinuity density and mean trace length. Theoretical predictions concerning the anisotropy of discontinuity frequency were validated for impersistent discontinuities of variable orientation. Further use of the model in the analysis of rock mass permeability is described.

RESUME

Au cours de l'analyse des caractéristiques de discontinuités en deux dimensions, nous avons développé une méthode de génération automatique sur ordinateur d'un réseau de discontinuités de longueur et d'orientation aléatoires. La représentation de ce réseau peut se faire numériquement ou graphiquement. Les données d'entrée sont obtenues à partir des observations d'échantillons de la masse rocheuse. Nous pouvons superposer sur le réseau des lignes de balayage d'orientation quelconque, afin de calculer numériquement la fréquence des discontinuités, leurs espacements et longueurs de trace. Nous démontrons qu'il existe une relation linéaire entre la fréquence moyenne et la rapport de la densité des discontinuités à la longueur de trace moyenne. Nous trouvons que la théorie de l'anisotropie de la fréquence des discontinuités de longueur infinie est encore valable pour le cas des discontinuités de longueurs finies et d'orientation variables. Nous présentons également une analyse plus évoluée qui permet le calcul de la perméabilité de la masse rocheuse, à partir des perméabilités des discontinuités elles-mêmes.

ZUSAMMENFASSUNG

Ein ebenes Computermodell wurde entwickelt, das die Fähigkeit hat, ein zufällig diskontinuierliches Gefüge zu erzeugen. Diese Gefüge können numerisch oder zeichnerisch dargestellt werden. Die Eingabedaten können aus Messungen auf dem Felsgebirge erhalten werden. Die resultierenden Gefüge können mit den Werten der Klufthäufigkeit, des Kluftabstandes und der Ausbißlänge verifiziert werden. Diese Werte können mit Milfe der Abtastlinien (scanline-Methode) ermittelt werden. Das Modell zeigt, daß eine einfache lineare Beziehung zwischen der Durchschnittshäufigkeit, den Werten der Kluftdichtigkeit und der mittleren Ausbißlänge existiert. Die Theorie der Anisotropie der Klufthäufigkeit ist damit für nichtdurchgehende Klüfte mit veränderlicher Richtung überprüft. Eine weitere Anwendung des Modells für die Untersuchung der Wasserdurchlässigkeit der Felsgebirge wird beschrieben.

1. INTRODUCTION

It is well known that discontinuities such as joints, fractures, fissures, faults and bedding planes can have a major influence on the engineering properties of a rock mass, in particular its deformability, strength and permeability. The following discontinuity characteristics are recognised as controlling the nature and extent of this influence on the engineering properties: (1) orientation, (2) frequency, (3) size, (4) surface geometry and aperture, (5) nature of fill material.

In many cases, discontinuities have a generally planar geometry and are orientated in sub-parallel groups, or sets. The frequency of discontinuities is usually expressed in terms of the average number per unit length intersecting a line passing through, or on the surface of, the rock mass. The frequency may be recorded for individual sets or for all sets aggregated. A corollary to the concept of frequency is that of discontinuity spacing. An examination of the relevant literature by Merritt and Baecher (1981) has revealed that the individual spacings between discontinuities, when plotted in histogram form, are usually found to follow an exponential distribution. Mean discontinuity spacing is, of course, the reciprocal of mean frequency. A more fundamental measure of discontinuity frequency is the areal or volumetric density, expressed in terms of the average number of discontinuity centres per unit area, or unit volume of the rock mass respectively.

The size and shape of discontinuities can rarely be examined directly since this would involve excavation of the rock mass. The most practical measure of discontinuity size is the length of the trace produced where the discontinuity intersects a rock face. Measurement of this trace length can be very difficult where the rock face is very irregular or of limited area, as is often the case underground. As with spacings, the individual trace lengths can be plotted in histogram form. In general, authors have found, or assumed, that discontinuity trace lengths follow either an exponential or log-normal distribution. The discontinuity shape is usually assumed to be circular, elliptical or rectangular (Merritt and Baecher, 1981). Sampling problems make the verification of these assumptions a difficult task.

Surface geometry, aperture and the fill material are important in determining both the shear strength and fluid conductivity of a given discontinuity. Surface geometry is difficult to express in quantitative terms and is therefore usually described using indices, related to shear strength on an empirical basis (International Society for Rock Mechanics, 1978). Aperture can be measured directly using a feeler-gauge or indirectly from water flow measurements. Snow (1970) has suggested that discontinuity apertures follow a log-normal distribution.

Finding the relation between the discontinuity characteristics described above and the engineering properties of the rock mass is a complex and difficult task, for the following reasons:

1. Discontinuity characteristics are a three-dimensional property. Sampling of these characteristics can only be carried out in a one-dimensional, or at best two-dimensional, framework.

2. The problem of sampling variable properties and predicting the values of these properties at depth within the rock mass requires the use of statistical and probabilistic methods that are unfamiliar to most rock mechanics engineers.

3. The verification of any link between discontinuity characteristics and rock mass properties requires the use of large scale, and usually expensive, in-situ tests.

As a first step in this task it is desirable to be able to reproduce, in numerical form, a given discontinuity fabric observed in the field. This model can then be interrogated to investigate how the various discontinuities interact and influence other discontinuity characteristics that cannot readily be measured in the field, such as discontinuity density, maximum and minimum discontinuity frequency, and also connectivity. The model can then be used to investigate the deformability, strength and permeability of the fabric and to study how these properties are dependent upon the input and derived discontinuity characteristics. It is the aim of this paper to describe how a simple, two-dimensional model can be used to study the discontinuity characteristics and permeability of a rock mass in this way.

The two-dimensional model is outlined in section 2. This is followed in section 3 by a study of the relation between discontinuity density, trace length and frequency, and in section 4 by a study of the anisotropy of discontinuity frequency. The final section of the paper explains, using a simple example, how the model can be used to examine connectivity and water flow in a discontinuous rock mass.

2. DISCONTINUITY FABRIC GENERATION

The characteristics of discontinuities exposed at a planar rock face can be determined using the scanline survey techniques explained by Piteau (1970), Priest and Hudson (1976), Priest and Hudson (1981) and Hudson and Priest (1982). During a scanline survey the characteristics of only those discontinuities intersecting a line set up on the rock exposure are measured systematically. Quantitative properties such as discontinuity spacing, orientation, trace length and aperture are measured, whilst at the same time recording a description of discontinuity surface geometry, fill material and other qualitative data. In collecting and processing the quantitative data it is important to ensure that (i) the sample size is sufficiently large (ii) that the bias imposed by a linear survey is allowed for and (iii) that the restrictions imposed by an exposure of limited extent are overcome (Priest and Hudson, 1981). The numerical results of a typical scanline survey usually contain the following data:

1. Number of sets, n.

2. The trend α_i and plunge β_i of the mean normal to the ith set.

3. Mean discontinuity frequency along the mean normal to the ith set, λ_i.

4. Mean trace length for the ith set, $1/\mu_i$

5. Distributions and associated parameters for discontinuity orientation, spacing and trace length.

A computer program, DICHA, has been developed to generate, in two dimensions, a random discontinuity fabric using scanline data of the form listed above. The resulting fabrics, which can be presented both numerically and graphically, are similar to those presented by Hudson and La Pointe (1980) but differ in that they are interrogated numerically rather than by the electrical analogue methods adopted by Hudson and La Pointe.

The main assumptions of the program are as follows:

1. The spatial distribution of discontinuity centres within the generation area is assumed to obey a statistically homogeneous, two-dimensional Poisson process.

2. The orientation of the lines representing each discontinuity set are assumed to be normally distributed about a given mean value

Fig. 1 Typical graphical output of the discontinuity fabric generation program.

3. The discontinuity trace length for each set is assumed to obey one of the following probability density distributions:

    i) Negative exponential
    ii) Uniform
    iii) Log-normal

4. The number of discontinuities for each set is defined in terms of the discontinuity density, N: the number of discontinuity centres per unit area

The program can typically handle up to 4 sets, each containing up to 3000 discontinuities. Random values from the appropriate density distributions are generated in order to produce randomly located discontinuities from each set within a given generation area. A typical example of the graphical output is shown in Fig. 1. In order to avoid edge effects a sample area, smaller than the generation area, is interrogated using scanlines of any desired orientation. The following data are obtained for the discontinuities intersected by each scanline:

1. The coordinates of the discontinuity ends and the intersection point.

2. The distributions and mean values of total trace length and also semi-trace length for each set and for all sets aggregated. The semi-trace length is that portion of the discontinuity trace between the scanline intersection point and the end of the trace.

3. The distributions and mean values of orientation for each set and for all sets aggregated.

4. The distributions and mean values of

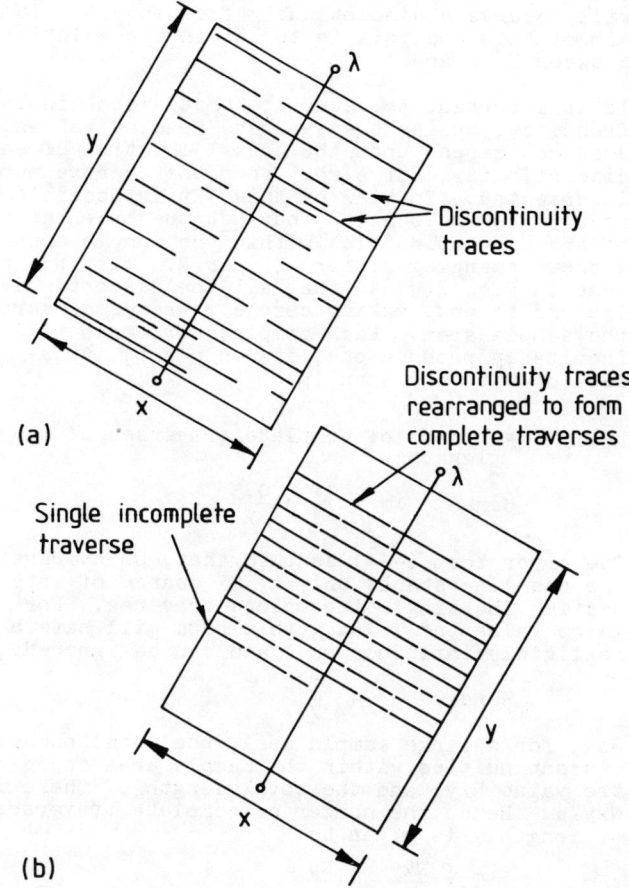

Fig. 2 Illustration of the geometrical analysis used to determine the relation between discontinuity density, trace length and frequency.

spacing for each set and for all sets aggregated. Mean discontinuity frequency, λ, is the reciprocal of mean spacing.

5. The distributions and mean values of censored semi-trace length for up to 20 censoring levels for each scanline.

Although the graphical output provides a useful visual impression of the discontinuity fabric, it is the numerical output that is of most value when studying the properties of this fabric. It is to this latter area that the remaining sections of this paper are devoted.

3. THE RELATION BETWEEN DISCONTINUITY DENSITY, TRACE LENGTH AND FREQUENCY

Consider a single set of parallel discontinuities with a density of N discontinuity centres per unit area and a mean trace length $1/\mu$. Fig. 2a shows a rectangular sample area, of side lengths x by y orientated with the sides of length y normal to the discontinuity traces. It is assumed that x and y are both very large compared with $1/\mu$, such that edge effects are negligible. A scanline, of length y, across the sample area, normal to the discontinuity set will observe a discontinuity frequency λ. The aim of this analysis is to obtain the relation between N, μ and λ.

It is important to recognize that discontinuity frequency, unlike the separate spacing values, does not depend upon the actual location of each discontinuity, but simply upon the average number intersected. In view of this, it is possible to re-arrange the location (but not the orientation) of the discontinuities within the sample area without changing either λ, μ or N. This has been done in Fig. 2(b) so that all the discontinuities lie end to end, making complete traverses across the sample area. Each complete traverse will inevitably produce one, and only one, intersection with the scanline.

Let T = number of complete traverses of length x

$$\lambda = \frac{T}{y} + \frac{0.5}{y}$$

The error term $(\frac{0.5}{y})$ assumes that, on average, the scanline stands only a 50% chance of intersecting the single incomplete traverse. For large values of T and y this term will have a negligible influence on λ and can be ignored.

$$\text{Hence} \quad \lambda = \frac{T}{y}$$

Now, for a large sample area, the total number of discontinuities within the sample area approaches the value Nxy, and the total length of these is Nxy/μ. Hence the number of complete traverses of length x is given by

$$T = \frac{Nxy}{\mu x} = \frac{Ny}{\mu}$$

$$\text{Hence} \quad \lambda = \frac{N}{\mu} \quad (1)$$

The relation between the number of intersections and the average line length has been considered from a stereological point of view by Underwood (1967). He found that for a totally random line (i.e. discontinuity) orientation, an estimate of the average total length, L, of discontinuity traces per unit area is given by $\pi\lambda/2$.

For a large sample size $L = \frac{\pi\lambda}{2}$

but $\mu = \frac{N}{L}$ hence $\frac{N}{\mu} = \frac{\pi\lambda}{2}$

or $\lambda = \frac{2N}{\pi\mu}$ (2)

Equation (1) is applicable for perfectly parallel discontinuities, and Equation (2) for totally random orientation. It is reasonable to expect, therefore, that discontinuities showing some variability about a preferred orientation will have a discontinuity frequency given by

$$\frac{2N}{\pi\mu} \leq \lambda \leq \frac{N}{\mu}$$

The hypotheses presented in Equations (1) and (2) were tested by means of a series of numerical simulations, using the computer program referred to in the previous section. For each test, a given value of μ was selected and the discontinuities were generated, according to pre-defined distributions of discontinuity trace length and orientation, to achieve the required density, N. The resulting fabric was sampled, using a scanline normal to the mean discontinuity orientation, to determine the discontinuity frequency λ. The process was repeated for different values of N and μ, and also different types of trace length distribution. Samples of the results, presented graphically as plots of N/μ against λ, are given in Figs. 3(a) to (c) for negative exponential, uniform and log-normal distributions of trace length respectively. Each figure contains data for a single, parallel discontinuity set and also for totally random discontinuity orientation. Where the hypothesis in Equation (1) is applicable, the best straight line through the points in these figures should have a slope of unity, and zero intercept. Where Equation (2) is applicable, the best straight line should have a slope of $\pi/2$ (i.e. 1.571) and also zero intercept. The complete set of simulation results indicate that, in all cases, there is an approximately linear relation between N/μ and λ. These results showed that for a single set of parallel discontinuities, the slope of the best straight line ranged between 0.812 and 1.415 with a mean of 1.113, and that the intercept ranged between $0.100m^{-1}$ and $0.469m^{-1}$ with a mean of $0.284m^{-1}$. For totally random discontinuity geometry the slope ranged between 1.064 and 1.779 with a mean of 1.421, while the intercept ranged between $0.114m^{-1}$ and $0.765m^{-1}$ with a mean of $0.439m^{-1}$. In all cases the slopes of the best straight lines tended to increase with increasing mean trace length. Further tests are required to assess the validity of Equations (1) and (2) in describing the N,μ,λ relation for multiple discontinuity sets, where each set has some variable orientation about a mean value. The results of this work have important implications for the measurement of discontinuity characteristics using linear surveys such as scanlines or boreholes. In particular, the simple linear relation between N/μ and λ provides a basis for analysing the complex relation between discontinuity frequency and discontinuity size in three dimensions.

4. ANISOTROPY OF DISCONTINUITY FREQUENCY

Consider n sets of vertical, planar discontinuities intersecting a horizontal plane. Each set is assumed to consist of parallel, persistent discontinuities. The normal to the ith discontinuity set has an azimuth direction α_i; the discontinuity frequency for the ith set along this normal is λ_i. Hudson and Priest (1982) have shown that the total discontinuity frequency, λ_s, along a scanline of azimuth α, is given by

$$\lambda_s = \sum_{i=1}^{n} \lambda_i \cos|\alpha - \alpha_i| \quad (3)$$

This equation can be used, by incrementing α in the range $0°$ to $360°$, to generate the locus of λ_s in two dimensions. Hudson and Priest have shown that this locus is composed of 2n circular arcs that intersect to define cusps. In particular, there will be a direction of maximum frequency and also, but not necessarily at right angles to it, a direction of minimum frequency.

In practice, discontinuities are not persistent but have some finite value of mean trace length.

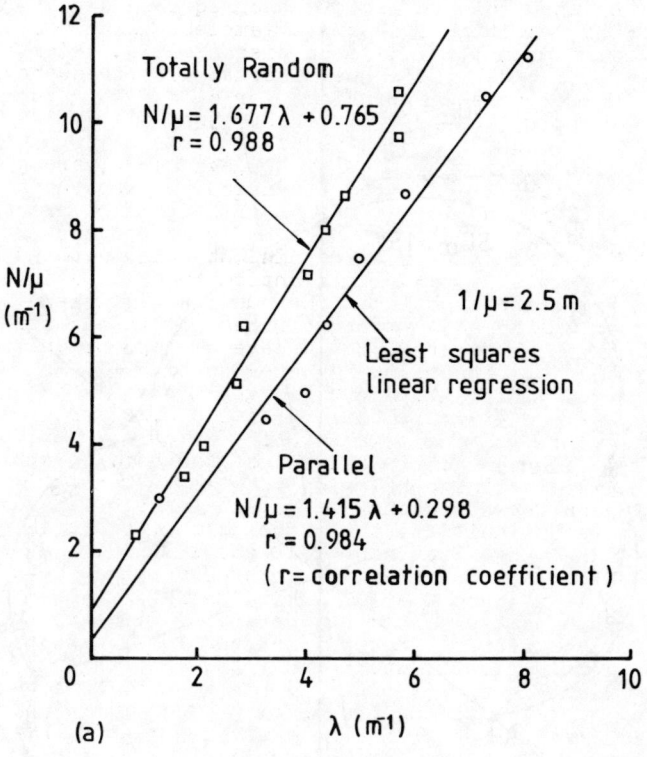

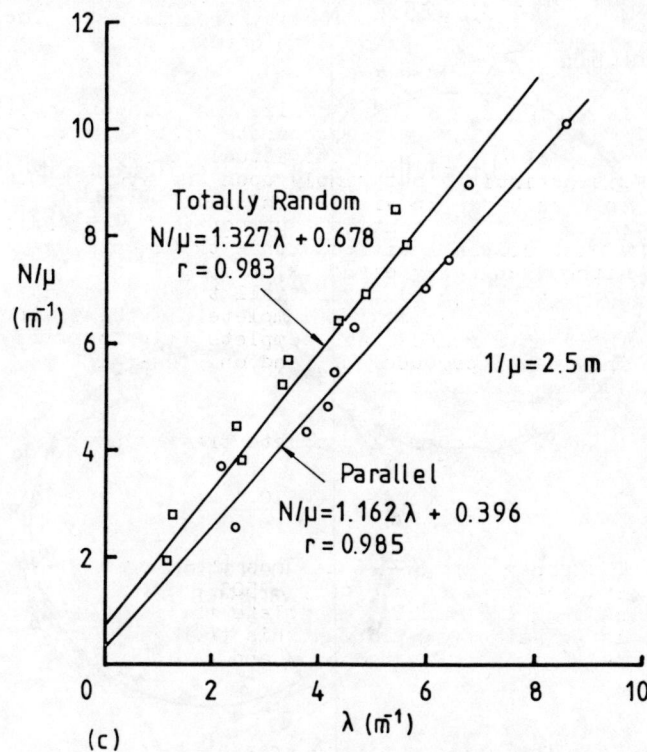

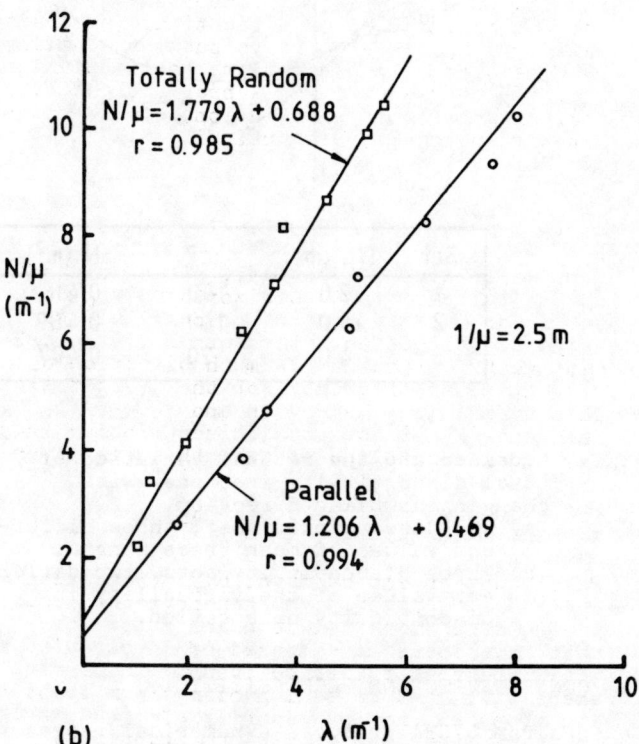

Fig. 3 The relation between discontinuity density (N), mean trace length ($1/\mu$) and mean frequency (λ) for totally random and also parallel discontinuities with a mean trace length of 2.5m.
(a) Negative exponential distribution of trace lengths.
(b) Uniform distribution of trace lengths
(c) Log-normal distribution of trace lengths.

Moreover, all the values of trace length will never be exactly the same but will follow a distribution of some form. Similarly, the discontinuities will never be perfectly parallel, but will be distributed about a mean orientation. The program DICHA was used to generate the λ_s locus for up to 3 sets of 4000 discontinuities having various distributions of trace length and orientation. Each set was defined using the following:

1. The mean orientation of the set normal, α_i.
2. The standard deviation of discontinuity orientation, SDα, assuming a normal distribution about the mean orientation, α_i.
3. The form of the trace length distribution (negative exponential or log-normal).
4. The mean trace length, $1/\mu$.
5. The number of discontinuity centres per unit area, N.

A sample of the results of this work is shown in Fig. 4(a) for one discontinuity set, and Fig. 4(b) for three sets of various orientations. In each case, the trace lengths were assumed to obey a negative exponential distribution. The solid line in each figure shows the theoretical

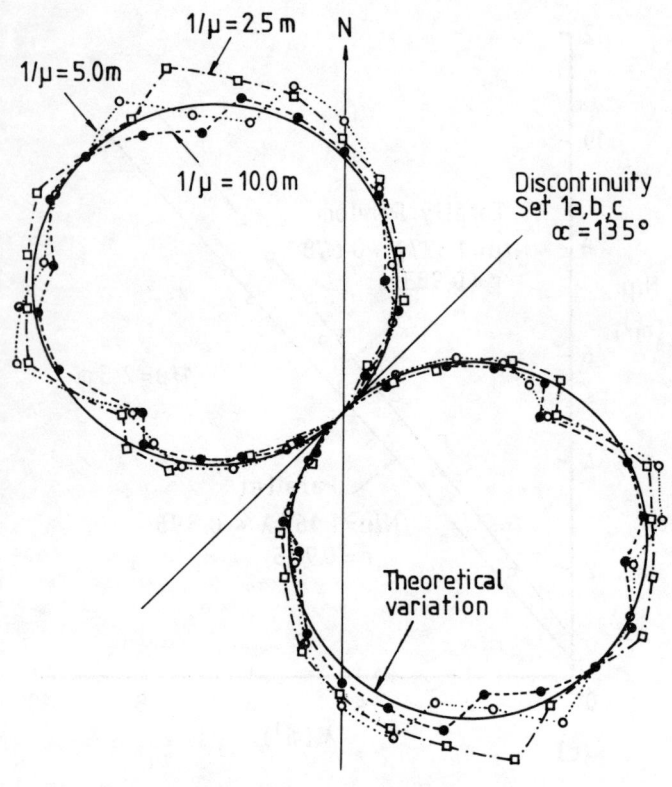

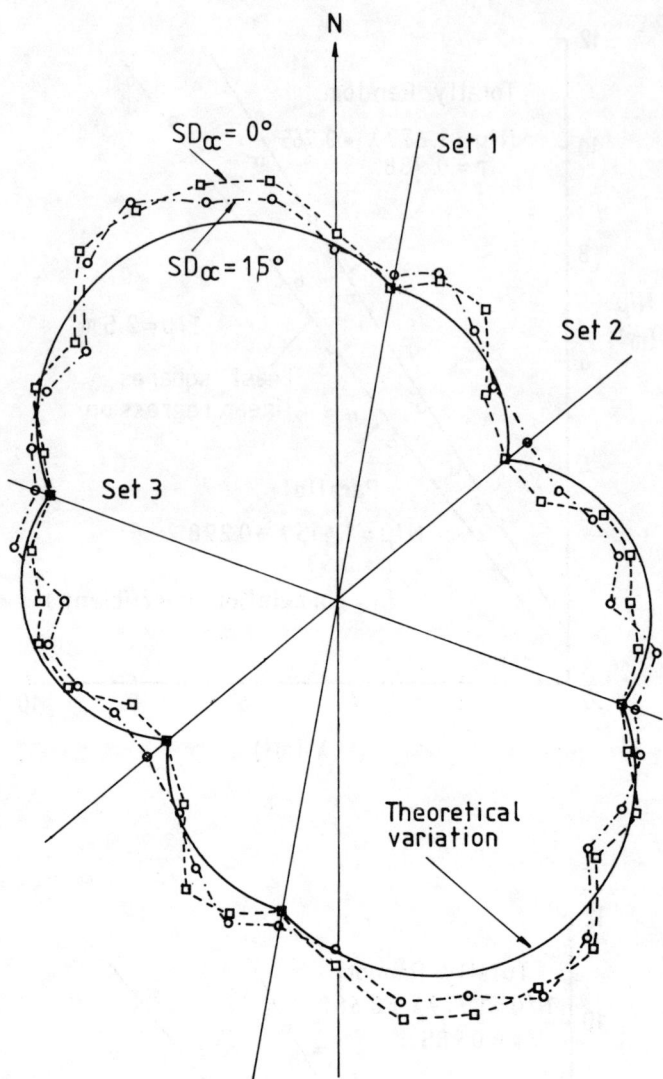

variation of total discontinuity frequency, λ_s, with α as defined by Equation (3), assuming there to be one or more sets of parallel persistent discontinuities. The broken lines in each figure give values of λ_s, measured, at 9° intervals of α, from a random discontinuity mesh generated by DICHA. The input frequency parameter(s) for each set λ_i, used in Equation (3), were obtained from scanlines in the direction(s) α_i across the generated mesh. This enabled the theoretical values, and each group of modelled values, of λ_s to be plotted at the same scale on the same diagram.

Figs. 4(a) and (b) show that the theory presented in Equation (3) is valid even where the discontinuity traces have a finite mean length and follow a negative exponential distribution. The modelled variation of λ_s also agrees with the theory when there is some variability in discontinuity orientation. These results not only serve to validate the theory presented by Hudson and Priest (1982) but also suggest that the theory will be applicable to real rock masses containing impersistent discontinuities of variable orientation.

Fig.4 Modelled and theoretical variation of total discontinuity frequency with changing sampling direction.
(a) One discontinuity set, three different values of mean trace length.
(b) Three discontinuity sets, two different values of the variability in discontinuity orientation.

5. CONNECTIVITY AND PERMEABILITY

The program DICHA produces a numerical representation of a two-dimensional discontinuity mesh, recorded as the coordinates of the end points of each trace within the sample area. The mesh is therefore a two-dimensional representation of

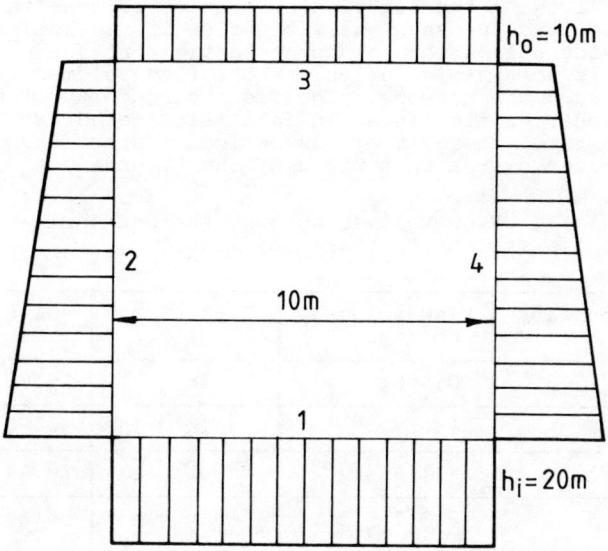

Fig. 5 Boundary conditions of the network.

one possible random discontinuity geometry, generated according to the various input parameters. The flow of water through the mesh is controlled both by the connectivity and also conductance of the discontinuity traces. Two programs, DICONN and DIFLOW, have been written to analyse connectivity and water flow through the mesh generated by DICHA.

In order to begin an analysis of connectivity it is necessary to define the boundary conditions of the zone under investigation. The approach used here is to define a square zone at some given orientation within the sample area. Two opposite sides of the square act as input and output lines; the other two opposite sides represent permeable boundaries through which there can be input or output. The potentials (i.e. heads) applied to these boundaries are shown diagrammatically in Fig.5. Sides 1 and 3, which are the input and output lines respectively, carry a constant potential h_i and h_o respectively. The potentials on sides 2 and 4 are assumed to vary linearly between h_i and h_o. The first step in the analysis of connectivity is to identify the subset of discontinuities that are connected either directly, or through other discontinuities, to one or more of the boundary lines. This is achieved by carrying out a systematic search of all discontinuities and all intersection points. For the purposes of the search the four boundary lines are treated as discontinuities. During the search, each of the m discontinuities is assigned an index number; this leads to the generation of a list of discontinuities that intersect the ith discontinuity, for i = 1 to m. The distances $\{S_i\}$, from the lower end of the ith discontinuity, to each of the intersection points are then calculated and listed. Any discontinuities that do not intersect any other, will be automatically excluded from this list. Isolated, and also "dead-end", closed loops of discontinuities within the interconnected sample area may still remain at this stage and must be removed by tracing pathways from each discontinuity to the boundary lines.

The next step in the analysis is to sort the values $\{S_i\}$ into ascending length for the ith discontinuity, and the jth intersection, so that $S_{i,j} > S_{i,(j-1)}$. Each of the intersection points, or nodes, is then assigned an index number, called the node number. The intersection between discontinuities i and k must of course, have the same number as the intersection between discontinuities k and i. A path is defined as the portion of a discontinuity that lies between two consecutive nodes. The sorted lists of intersection distances and node numbers can be examined to determine the length of each path. Each path is then characterised by the node numbers at each end, the length of the path, and the discontinuity on which it lies. The list of paths will automatically exclude portions of single discontinuities terminating at a "dead end".

The analysis of connectivity is illustrated, using a simple example, in Fig.6 which shows three discontinuity sets in an area 10m square. The characteristics of these sets are listed in Table 1. Fig. 6(a) shows the initial mesh and Fig.6(b) the interconnected network along which water can flow between the boundaries.

Table I Characteristics of discontinuities in Fig. 6.

	Set 1	Set 2	Set 3
Orientation α	150°	60°	110°
Standard deviation of orientation* SDα	2.5°	2.5°	20°
Mean trace length, $1/\mu$ (m)	2.5	2.8	3.0
Standard deviation of trace length† (m)	0.8	0.9	1.0
Discontinuity density N (m^{-2})	0.52	0.60	0.28
Aperture e (mm)	0.03	0.02	0.025
Standard deviation of aperture† (mm)	0.01	0.01	0.01

* Normal distribution   † Log-normal distribution

The flow of water through the network depends upon the conductance of the various discontinuities which, in turn, is critically dependent upon their aperture. The conductance c of a smooth, parallel-walled, planar discontinuity of length L, aperture e and width w is, for laminar flow, given by the standard equation

$$c = \frac{ge^3 w}{12 \nu L} \quad (4)$$

where g is the acceleration due to gravity
and ν is the kinematic viscosity of the fluid

If there is a head loss Δh over the length L, the quantity, Q, of fluid flowing in unit time is given by

$$Q = c \, \Delta h \quad (5)$$

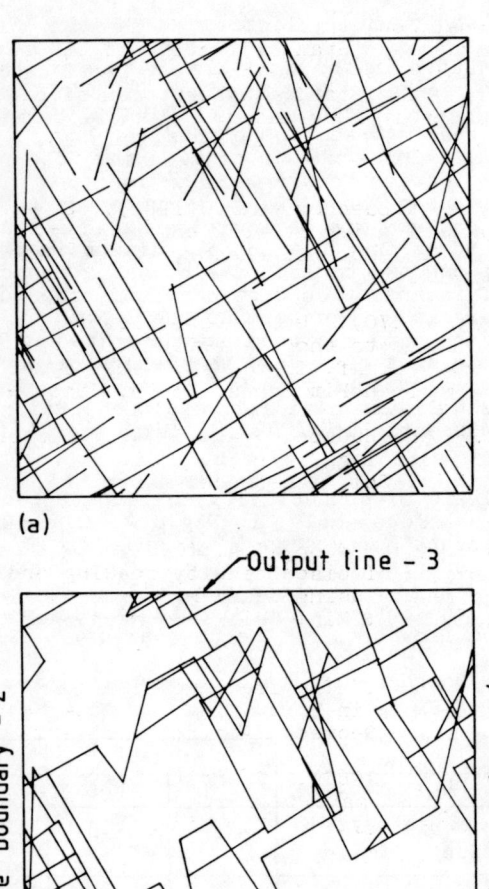

Fig. 6 Analysis of connectivity

For simplicity in this analysis, each discontinuity was assumed to have unit width normal to the plane of the mesh. Following Snow (1970) the apertures of each discontinuity set were assumed to be log-normally distributed, having the parameters given in Table I. The aperture of each discontinuity was generated as a random value from the appropriate log-normal distribution. Equation (4) was then used to calculate the conductance of each discontinuity path. The flow volumes along each path are given by Equation (5), assuming that there is no head loss due to turbulence at the node, and imposing the constraint that the flow into any given node should be equal to the flow out. This constraint generates a system of linear simultaneous equations, in which the nodal potentials are unknown except at the boundaries. Solution of these equations was carried out using a relaxation method incorporating an over-relaxation factor to accelerate convergence. This approach, although relatively expensive, had the advantage of allowing the solution of systems of non-linear equations in cases where the flow is non-laminar. Having solved the unknown potentials at the nodes, it is possible to calculate the flow volumes through the network. The flows through the four boundaries are listed in Table II together with the total area, A, of discontinuity apertures exposed over a unit width of each boundary.

Table II Flow volumes through the boundaries of the network in Fig.6.

Boundary	Flow volume cm^3/sec	Direction	A cm^2
1	0.540	in	15.65
2	1.287	in	36.26
3	4.144	out	76.94
4	2.317	in	56.26

The results in Table II confirm that, to a large extent, the flow across a given boundary is determined by the total discontinuity aperture at that boundary. The relatively small number of discontinuities in the interconnected network means that, at the 10m scale, there is considerable inhomogeneity in the frequency of conducting discontinuities. This in turn has caused the inhomogeneity in water flow evident in Table II. The flow of water through the network shown in Fig.6 cannot, therefore, be represented satisfactorily in terms of the permeability of an equivalent continuum. In general, as the number of discontinuities in the interconnected network becomes larger, the continuum representation will become more valid. The analysis of water flows through dense networks of this type, coupled with an investigation of the relation between discontinuity characteristics and the anisotropy and inhomogeneity of flow patterns, will form the subject of future work.

6. CONCLUSIONS

1. Using a two-dimensional model it has been shown that, for a single discontinuity set, there is a linear relation between mean discontinuity frequency (λ) sampled along a line, and the values of discontinuity density (N) and mean trace length ($1/\mu$). This linear relation is of the form

$$\frac{N}{\mu} = b\lambda + d$$

where b is a constant generally in the range $1.0 \leq b \leq \pi/2$ and d is close to zero.

2. Theoretical predictions of the variation of total discontinuity frequency with scanline direction have been validated for impersistent discontinuities of variable orientation and whose trace lengths follow a negative exponential or log-normal distribution.

3. The model can be used to determine the connectivity and flow volumes through any two-

dimensional discontinuity fabric in any chosen direction.

7. ACKNOWLEDGEMENTS

The work described in this paper was carried out in the Department of Mineral Resources Engineering at the Imperial College of Science and Technology under a research contract sponsored by the Building Research Establishment of the Department of the Environment as part of the DOE programme of radioactive waste management research. The authors thank Dr. John Bray and Dr. John Watson, both of Imperial College, for their assistance during this work.

8. REFERENCES

Hudson, J.A. and La Pointe, P.R. (1980). Printed circuits for studying rock mass peremeability. Technical Note. Int. J. Rock Mech. Min.Sci. & Geomech. Abstr., 17, 297-301.

Hudson, J.A. and Priest, S.D. (1982). Discontinuity frequency in rock masses. To be published in Int. J. Rock Mech.Min.Sc. & Geomech.Abstr.

International Society for Rock Mechanics, Commission on Standardisation of Laboratory and Field Tests, (1978). Suggested methods for the quantitative description of discontinuities in rock masses. Int. J. Rock. Mech. Min.Sci. & Geomech. Abstr., 15, 319-368.

Merritt, A.H. & Baecher, G.B. (1981). Site characterization in rock engineering. 22nd U.S. Symp. Rock Mech., 47-63, Cambridge, Massachusetts.

Piteau, D.R. (1970). Geological factors significant to the stability of slopes cut in rock. S. Af. Inst. Min. Met., Symp. Planning Open Pit Mines, 33-53, Johannesburg.

Priest, S.D. and Hudson, J.A. (1976). Discontinuity spacings in rock. Int. J. Rock Mech. Min. Sci. & Geomech. Abstr. 13, 135-148.

Priest, S.D. and Hudson, J.A. (1981). Estimation of discontinuity spacing and trace length using scanline surveys. Int. J. Rock Mech. Min. Sci. & Geomech. Abstr. 18, 183-197.

Snow, D.T. (1970). The frequency and apertures of fractures in rock. Int. J. Rock Mech. Min. Sci. & Geomech. Abstr. 7, 23-40.

Underwood, E.E. (1967). Quantitative evaluation of sectioned material. Proc. 22nd Int. Congr. for Stereology, 49-60, Chicago.

AN APPLICATION OF FINITE ELEMENT PROCEDURE FOR UNDERGROUND STRUCTURES WITH NON-LINEAR MATERIALS AND JOINTS

Application de la méthode des éléments finis à des structures souterraines avec des matériaux non-linéaires et des joints

Die Anwendung einer Finiten Element Prozedur im Hohlraumbau mit nichtlinearen Materialien und Klüften

C. S. Desai
Dept. of Civil Eng. and Eng. Mech., University of Arizona, Tucson, AZ, USA

I. M. Eitani
Dept. of Mining Eng., King Abdul Aziz University, Jeddah

C. Haycocks
Dept. of Mining Eng., Virginia Tech, Blacksburg, VA, USA

SYNOPSIS

A Finite Element procedure is applied for the prediction of observed field behaviour of underground excavations related to tunneling and mining. The procedure includes a number of improvements and features such as considerations of constitutive behaviour of rock and joints including laboratory testing, development and use of a thin-layer element for joints and interfaces, and a scheme for creating stress-free excavated surfaces. It permits use of various elasticity and plasticity based stress-strain models and the simulation of various construction sequences.

RESUME

Un procédé à éléments finis est appliqué pour la détermination des résultats mesurés in situ dans des excavations souterraines associées à des constructions de tunnels et de mines. Ce procédé est un modèle perfectionné par de nombreuses caractéristiques, par exemple la considération de comportement des roches et des joints par une loi comprenant des expériences en laboratoire, le développement et l'application d'un élément disposé par lits étroits pour des joints et interfaces, et un schéma pour réaliser une excavation sans contrainte. Ainsi, le procédé permet l'application des divers modèles de contraintes basés sur la théorie d'élasticité et plasticité, et sur les simulations de différentes séquences de la construction.

ZUSAMMENFASSUNG

Eine Finite Element Prozedur wurde zur Voraussage beobachteter Auswirkungen im Felde bei Ausbruchsarbeiten im Tunnel- und Bergbau entwickelt. Die Prozedur enthält eine Anzahl von Verbesserungen und Besonderheiten wie die Betrachtung des konstitutiven Verhaltens von Fels und von Klüften einschließlich der Laborversuche, sowie die Entwicklung und Anwendung dünnschichtiger Elemente für Klüfte und Grenzflächen, und ein Schema zur Schaffung einer spannungsfreien Ausbruchsfläche. Es erlaubt die Anwendung verschiedener elastizitäts- und plastizitätstheoretisch begründeter Spannungs-Verformungs-Modelle und die Simulation der Abfolge verschiedener Ausbruchsphasen.

1. INTRODUCTION

The finite element procedure has been applied for analysis and design of underground structures such as tunnels, caverns, and cavities (Agbabian Associates, 1973; Azzouz et al., 1979; Desai and Christian, 1979; Ghaboussi and Ranken, 1974; Kovari et al., 1976; Kulhawy, 1977; Rose et al., 1979; Wittke, 1977). The objective of this study has been application of a finite element displacement procedure for simulation of sequential construction, including underground excavation. Its contributions include (1) a scheme for simulation of excavation, (2) use of a thin-layer element for joints and interfaces, (3) use of various constitutive models for rock and joints based on laboratory testing, and (4) application to typical field problems in tunneling.

2. FINITE ELEMENT PROCEDURE

The displacement finite element procedure is commonly used for underground works. Mixed and hybrid procedures that can allow improved simulation of excavation have also been used recently for geomechanics problems (Desai et al., 1982a). Since the main objective herein is application of a two-dimensional displacement procedure, brief details of that procedure are first presented.

Use of a variational principle leads to the following incremental finite element equations:

$$[k] \{\Delta q\} = \{\Delta Q\} + \{\Delta Q_o\} \qquad (1)$$

where $[k]$ = tangent stiffness matrix, $\{\Delta q\}$ = vector of incremental nodal displacements, $\{\Delta Q\}$ = vector of incremental nodal loads, and $\{\Delta Q_o\}$ = incremental residual or initial load vector. Once the displacements are found, the strains, $\{\Delta\varepsilon\}$, and stresses, $\{\Delta\sigma\}$ are computed as

$$\{\Delta\varepsilon\} = [B] \{\Delta q\} \qquad (2)$$

$$\{\Delta\sigma\} = [C] \{\varepsilon\} \qquad (3)$$

where [B] = transformation matrix and [C] = tangent constitutive matrix. An incremental-iterative technique is used for solution of the nonlinear Eqs. (1).

An eight-node isoparametric element (Figure 1) is used and the components of displacement, u and v, in the x and y directions are given by

$$u = \sum_{i=1}^{8} N_i u_i$$
$$v = \sum_{i=1}^{8} N_i v_i \quad (4)$$

where u_i, v_i (i = 1, 2, ...8) are nodal displacements and N_i = (i = 1, 2, ...8) are interpolation functions.

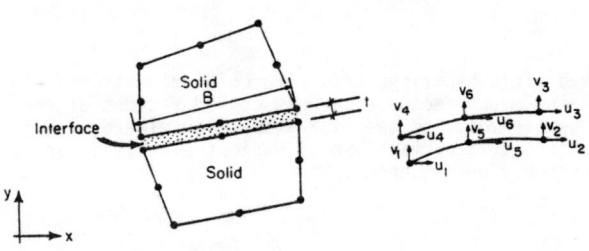

Figure 1 Two-dimensional solid and thin-layer joint/interface elements

3. CONSTRUCTION SEQUENCES

Before the analysis starts, the in situ conditions are established on the basis of known coefficient of earth pressure, K. The sequences of construction that can be simulated are (1) dewatering, (2) excavation, (3) deposition or embankment, and (4) support systems such as anchors and tie-backs. Details of incorporation of these sequences are available elsewhere (Lightner and Desai, 1979). Here, details of excavation simulation are given.

Simulation of a stress-free surface (Figure 2) for a given stage of excavation is obtained through an iterative scheme in which a residual load vector, $\{Q_o\}$, is applied successively. The residual load vector is found from

$$\{Q_o\} = \int_V [B]^T \{\sigma_o\} \, dV \quad (5)$$

where $\{\sigma_o\}$ = state of stress in the elements near the excavated surface (Figure 2) caused by removal of a portion of the geological mass. This load vector represents the degree to which the assemblage is out of equilibrium due to the excavation step.

4. JOINT/INTERFACE ELEMENT

A rather new and simple thin-layer element is used to simulate joints in rock masses and interfaces between structural and geological media (Desai, 1981; Desai et al., 1981). Figure 1 shows a schematic of the thin-layer element between two solid elements for the two-dimensional problems considered here. This element has a "small" finite thickness and is treated essentially the same as the solid geological elements. A parametric study indicated that such an element provides equally as good or improved simulation of interface behavior as the conventional relative displacement element (Goodman et al., 1968) if the following condition is fulfilled:

$$t/B = 0.01 \text{ to } 0.1 \quad (6)$$

where t = thickness of the element and B = (mean) width of the adjoining elements (Figure 1).

The constitutive matrix [C] for the thin-layer element is expressed as

$$[C]_j = \begin{bmatrix} [C_{nn}] & [C_{ns}] \\ [C_{sn}] & [C_{ss}] \end{bmatrix} \quad (7)$$

Since it is difficult to determine the coupling terms $[C_{ns}]$ and $[C_{sn}]$ at this time, they are deleted. However, the normal and the shear components, $[C_{nn}]$ and $[C_{ss}]$, are defined in a special way. $[C_{nn}]$ is expressed in terms of the properties and state of stress in the thin-layer element itself as well as the adjoining geological elements as

$$[C_{nn}] = [C_{nn} (\alpha_m^j, \beta_m^g)] \quad (8)$$

where α_m^j and β_m^g (m = 1, 2, ..) denote the properties of the interface and geological elements, respectively. Thus the normal stiffness during the deformation of the joint can be dependent upon the state of stress in the joint and in the surrounding elements.

The shear component is expressed by defining shear modulus, G_j, as

$$G_j (\sigma_{nn}, \tau, u_r) = \left. \frac{\partial (\tau, \sigma_r, u_r)}{\partial u_r} \times t \right|_{\sigma_n} \quad (9)$$

where u_r = relative displacement. Figure 3 shows a schematic of a direct shear test from which linear or nonlinear values of G_j are determined.

5. CONSTITUTIVE MODELS AND TESTING

The finite-element procedure includes the option for use of various stress-strain models: elasticity based-linear elastic, nonlinear elastic (hyperbolic), and plasticity based-Drucker-Prager. Details of these models are available in DiMaggio and Sandler (1971), Desai and Christian (1977) and Desai and Siriwardane (1982). These models can be used for both the geological media and joints.

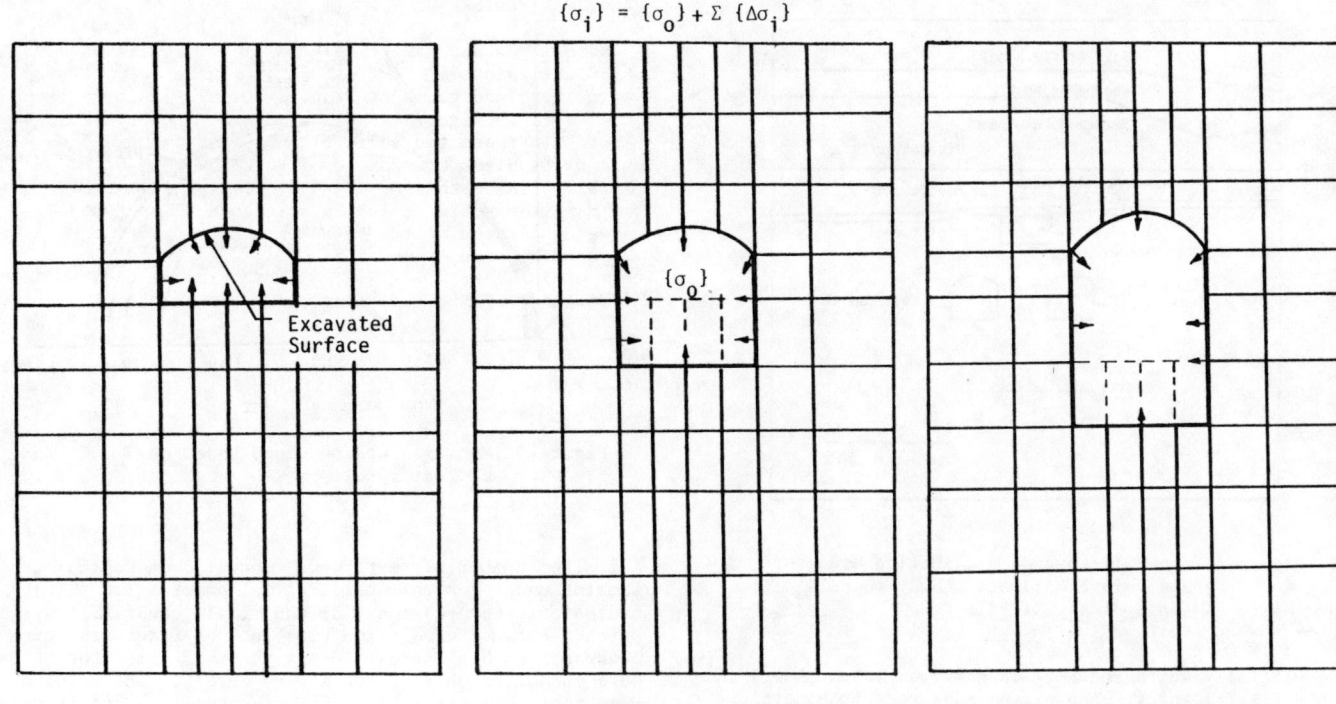

Figure 2 Schematic for sequential excavation.

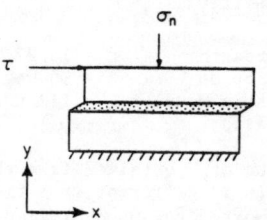

(a) Schematic of Direct Shear Test

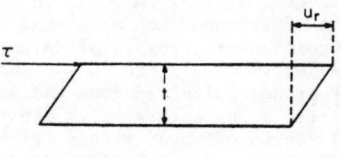

(b) Deformations at the Interface

Figure 3 Behavior at the interface.

In the incremental nonlinear analysis, the constitutive law is expressed as

$$\{d\sigma\} = [C^{ep}] \{d\varepsilon\} = ([C^e] - [C^p])\{d\varepsilon\}) \qquad (10)$$

where $[C^e]$ = (tangent) elasticity matrix and $[C^p]$ = (tangent) plasticity matrix.

5.1 Site and Materials

Appropriate characterization of geological materials and joints plays a significant role in obtaining realistic predictions. Hence, laboratory tests were performed by using conventional cylindrical triaxial and truly triaxial devices for (a part of) rocks and direct shear for the discontinuities or joints. Since the rock and joints tested are relevant to the subsequent application problem, a description of the site and materials is presented below; these are relevant to the Research Chamber at the Peachtree Station, Atlanta Subway, and are adopted from Law Engineering and Testing Co. (1976, 1977), Rose et al. (1979), and Tudor Engineering Company (1977). Figure 4 shows a section through the subway with the Research Chamber, Pilot Tunnel, and the Twin Tunnels. The sub-surface profile in this area includes a fill, residual soil, partially weathered rock and intact rock. The fill covers only a limited area and its depth varies from 2.13 m (7 ft) to 5.18 m (17 ft). The residual soil exists beneath the fill or beneath the surface pavement where fill is not present. The major portion of the residual soil consists of firm and micaceous silty sands whose denseness increases with depth.

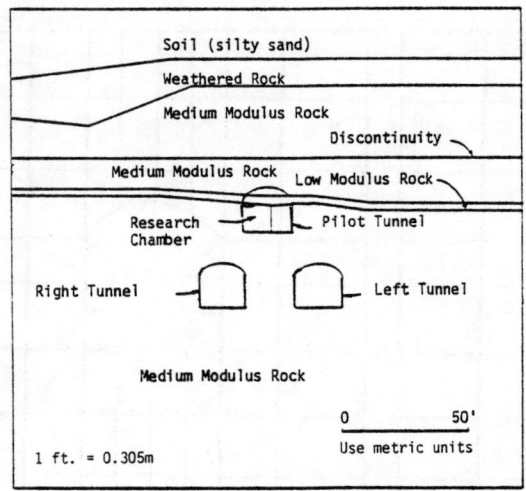

Figure 4 Generalized geologic section used for analysis of Atlanta subway tunnels (Rose et al., 1979).

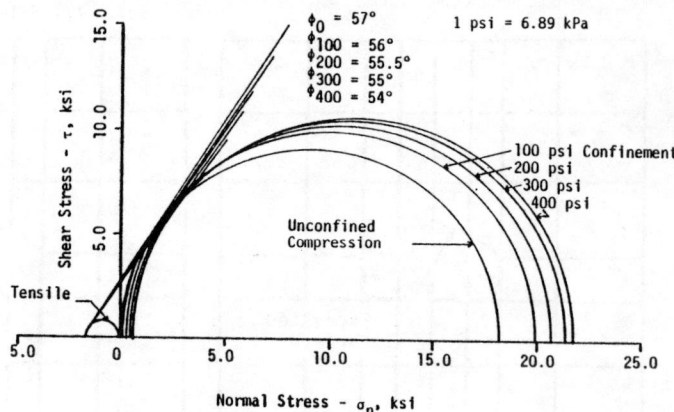

Figure 5 Mohr failure envelope based on T_o, C_o and C_p test results.

The partially weathered rock ranges from 1.5 m (5 ft) to 4.5 m (15 ft) thick. The predominant rock is biotite gneiss and biotite amphibole gneiss. The upper 3.0 m (10 ft) to 6.0 m (20 ft) of the rock is characterized by a high degree of weathering. At greater depth, the rock is generally of excellent quality with isolated zones of fracturing and weathering.

5.2 Material Properties

Material properties for all materials except the weathered rock and the discontinuity (joint) were adopted from those reported and used by Kulhawy (1977) and Rose et al. (1979). They were derived from conventional triaxial laboratory and in situ (flat jack and overcore) tests (Law Engineering Testing Company, 1976, 1977).

Under these studies, a series of cylindrical or conventional triaxial and three-dimensional triaxial tests were performed for the weathered rock; the latter involved use of a truly triaxial device (Desai et al., 1982b). The cylindrical specimens were prepared from the core samples provided by Law Engineering Testing Company, Atlanta, Georgia. The average diameter of the samples was 5.00 cm (2.0 in.), with L/D ratio equal to 2.0. The cubical samples, 10 x 10 x 10 cm (4 x 4 x 4 inch), were cut from large size pieces of the rock collected from the site.

The testing for cylindrical samples included unconfined compression (C_o), confined compression (C_p) and indirect tensile (Brazilian) (T_o) tests. An MTS testing device with a capacity of 4.45×10^3 KN (10^6 lbs) in compression and 27×10^2 KN (6×10^5 lbs) in tension was used. The triaxial loading cell used was Hoek's cell. Tests conducted were 16 Brazilian, 6 unconfined compression, and 4 confined compression. For the latter, the confining pressures were 690 KPa, 1.38 MPa, 2.07 MPa and 2.76 MPa (100, 200, 300, 400 psi). The Mohr-Coulomb diagram based on the C_o, C_p and T_o tests is shown in Figure 5.

A limited number of tests were conducted on cubical samples under conventional triaxial compression (CTC), hydrostatic compression (HC) and simple shear (SS) stress paths. Because of difficulties such as fracturing of membranes at high stress levels, at this time, the tests were conducted only at low stress levels. The results from the truly triaxial tests were generally consistent with those from the cylindrical triaxial tests at low stress levels.

The following properties for the weathered rock were derived on the basis of the foregoing tests and used in the finite element analysis:

Cohesive Strength   c = 18.27 MPa (2650 psi)
Angle of Friction ϕ = 56 Deg.
Young's Elastic Modulus E = 118 MPa (2460 Ksf)
Poisson's Ratio ν = 0.19

Note that the value of E obtained from the laboratory test results herein is different from the value of 21,600 Ksf (1035-MPa) reported and used previously for the weathered rock (Kulhawy, 1977; Rose et al., 1979).

The value of the Poisson's ratio of 0.19 was obtained by measuring radial deformations in the unconfined tests; it compares closely with that of 0.17 reported previously.

5.3 Joints

A series of direct shear tests were performed by Kane (1981) on 10 x 10 cm (4 x 4 in.) samples of granite gneiss with thickness of 1.27 cm (0.50 in.). These included variation of factors such as normal stress, roughness (natural and artificial) of joint, and moisture at the junction. In the case of natural roughness, rock pieces collected from the site were carefully separated along natural cleavages. Figure 6 shows results in terms of shear stress (τ) vs (relative) horizontal displacement, u_r, for three different normal stresses. Based on these results, a value of G_j = 20.00 MPa was used for discontinuity (see Figure 7), with thickness of the thin-layer element equal to 6.1 cm (0.20 ft).

In view of the laboratory observation that the rock behavior was essentially linear in the range of loadings encountered, linear elastic stress-strain behavior was assumed.

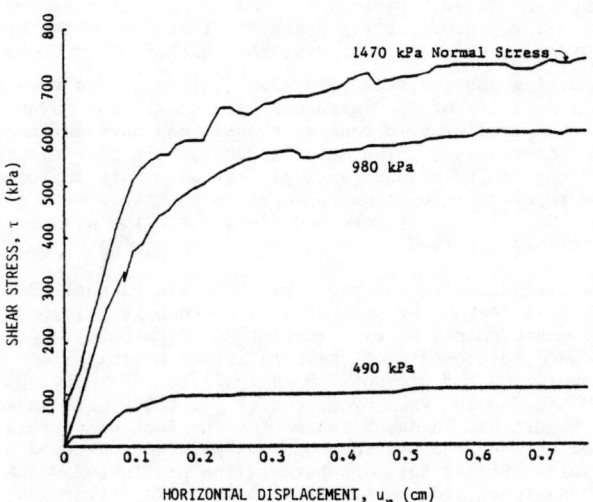

Figure 6 Shear behavior of natural cleavage at different normal stresses (Kane, 1981).

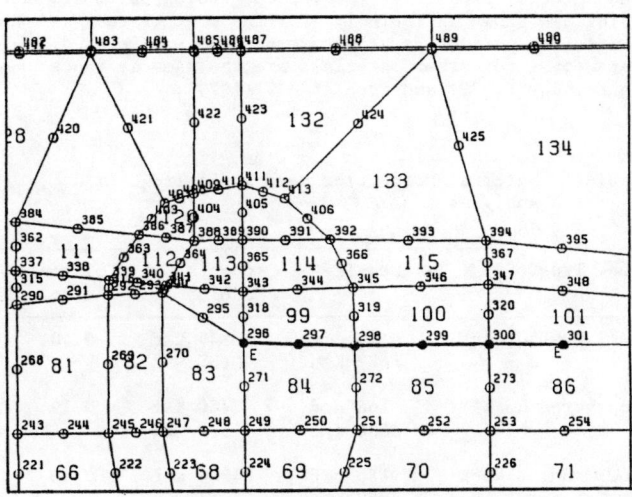

(b) Details of mesh near Research Chamber with location of extensometer

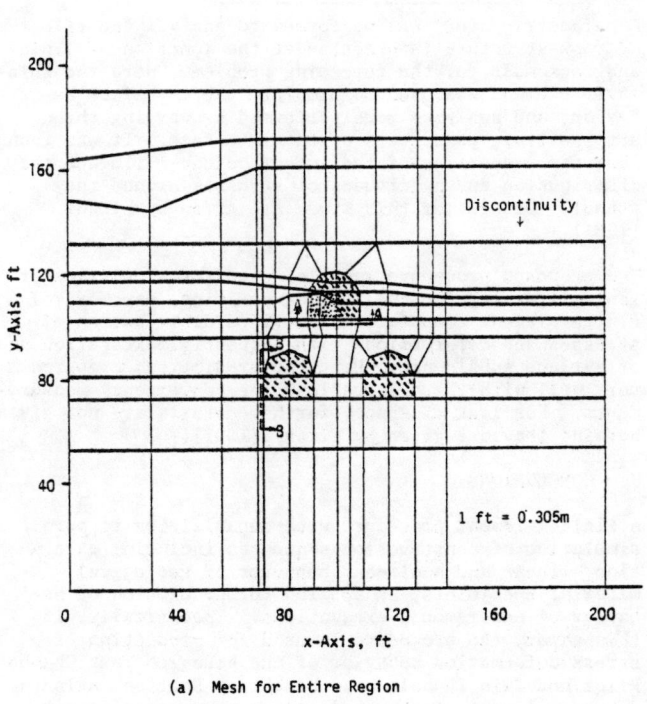

(a) Mesh for Entire Region

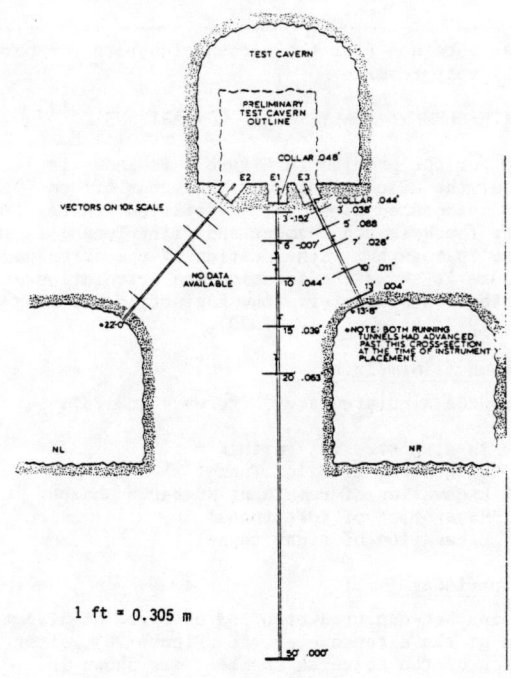

(c) Extensometers E_1, E_2, E_3

Figure 7 Finite element meshes and location of extensometers for Atlanta subway problem.

F 213

The material properties for various geological media and joint are shown in Table 1. For the weather rock and the joint, the parameters as determined above were used. Parameters for other materials are the same as those used in Kulhawy (1977) and Rose et al. (1979).

Table 1 Material properties used in Finite Element Analyses

Rock Type	Density	Young's Modulus	Poisson's Ratio
Soil Overburden	105 pcf 1682 Kgm^{-3}	847 Ksf 40.55 MPa	0.40
Weathered Rock*	168 pcf 2688 Kgm^{-3}	2460 Ksf 118 MPa	0.19
Low Modulus Rock	180 pcf 2880 Kgm^{-3}	64800 Ksf 3 GPa	0.10
Medium Modulus Rock	180 pcf 2880 Kgm^{-3}	108000 Ksf 5 GPa	0.17
Discontinuity	162 pcf 2592 Kgm^{-3}	21600 Ksf 1 GPa	0.17
	G = 20 MPa	t = 6.1 cm (0.2 ft)	

* Values obtained from the present laboratory experimental test results.

6. FINITE ELEMENT ANALYSIS AND COMPARISONS

The mesh for the problem in Figure 4 is shown in Figure 7a; the discontinuity of thickness 6.1 cm (0.20 ft.) was introduced as shown. A detailed finite-element mesh near the Research Chamber and Pilot Tunnel is shown in Figure 7b together with location of the extensometer EE. Figure 7c shows the location and orientation of two of the other extensometers (Law Engineering Testing Company, 1977; Rose et al., 1979).

6.1 Sequences Simulated

The sequences simulated are (Figures 4 and 7a):

1. In situ stresses (with K = 1.0)
2. Excavation of Pilot Tunnel
3. Excavation of remaining Research Chamber
4. Excavation of left tunnel
5. Excavation of right tunnel

6.2 Comparisons

Comparisons between predicted and observed displacements recorded at the extensometer, EE (Figure 7b), after the excavation of the Research Chamber, are shown in Figure 8a. The correlation appears to be satisfactory. Table 2 shows comparisons between predictions and observations at extensometers E_1 and E_3 (Figure 7c) at the end of excavation of the Right Tunnel.

Table 2 Observed and calculated displacements.

Point	Displacement	
	Observed	Calculated
E_1	0.12 cm	0.03 cm
E_3	0.11 cm	0.03 cm

The calculated displacements are much smaller than the observed values. Various reasons can be advanced for the discrepancy. It was suggested that the collar station of E_1 was initially disturbed with further damage occurring subsequently (Wier-Jones, 1979). The excavation activity of the large underground station adjacent to the parallel twin running tunnels may have influenced the observations. The material characteristics adopted for the finite-element analysis may also have influenced the results; it will be appropriate to have access to more detailed laboratory and field tests for all the materials involved.

The prediction from a previous three-dimensional (3-D) analysis (Azzouz et al., 1979) are shown in Figure 8a for measurements at extensometer EE (Figure 7b). Computed displacements of the deformation of the pilot tunnel from the present 2-D analysis and the previous 2-D and 3-D analyses by Azzouz et al. (1979) are shown in Figure 8b. A major reason for the lack of correlation between the previous 3-D analysis and observed results (Figure 8a), and between the previous 2-D and 3-D analyses and the present 2-D analysis (Figure 8b), can be due to the use of uniform material properties for the entire mass in the previous 2-D and 3-D analyses by Azzouz et al. (1979).

Figure 9a, b shows distribution of typical computed displacements and principal stresses, respectively, after the excavation of the right tunnel. These results show realistic trends under the sequential excavation.

6.3 Additional Finite-Element Studies

A parametric study was performed to analyze the effect of rock-structure interaction at the junction of lining and rock mass for the foregoing problem. Here the thin-layer element was used to simulate the interface behavior, and analyses were performed by varying the stress-strain parameters of the interface. It was found that the properties of the interface can influence the distribution and nagnitudes of stresses around the tunnel. Details of this study are given by Eitani (1981).

The proposed procedure was also used for simulation of longwall mining for the York Canyon Mine, New Mexico. Finite-element computations provided distribution of stresses and deformations with sequential excavation and of surface subsidence. Here the predictions compared very well with the observations for convergency measurements. For lack of space, further details are not given herein; they are given by Eitani (1981).

7. CONCLUSIONS

A finite element procedure with capabilities to permit simulation of construction sequences including excavation, linear and nonlinear behavior of geological material and joints, is applied for prediction of behavior of underground excavations. Specifically, in this paper, the procedure is used for prediction of stress deformation behavior of the Research Test Chamber, Pilot and Twin Tunnels near Peachtree Station, Atlanta Subway. A part of the predictions show satisfactory comparison with observations, whereas the other part shows a discrepancy; possible reasons for the latter are discussed. Based on these results and other results for York Canyon Mine problem, it can be concluded that the proposed procedure can provide satisfactory simulation of the behavior of structures at surface and underground. It is necessary to have appropriate stress-strain models for geological material, joints and interfaces for realistic predictions from the numerical procedure.

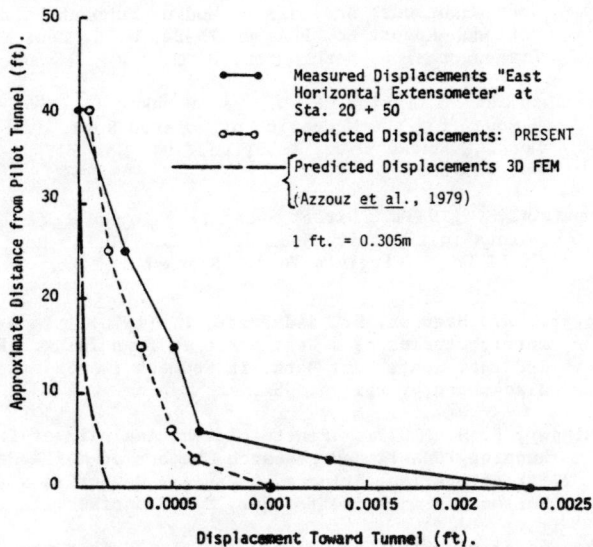

(a) Comparison between calculated and observed displacements at EE.

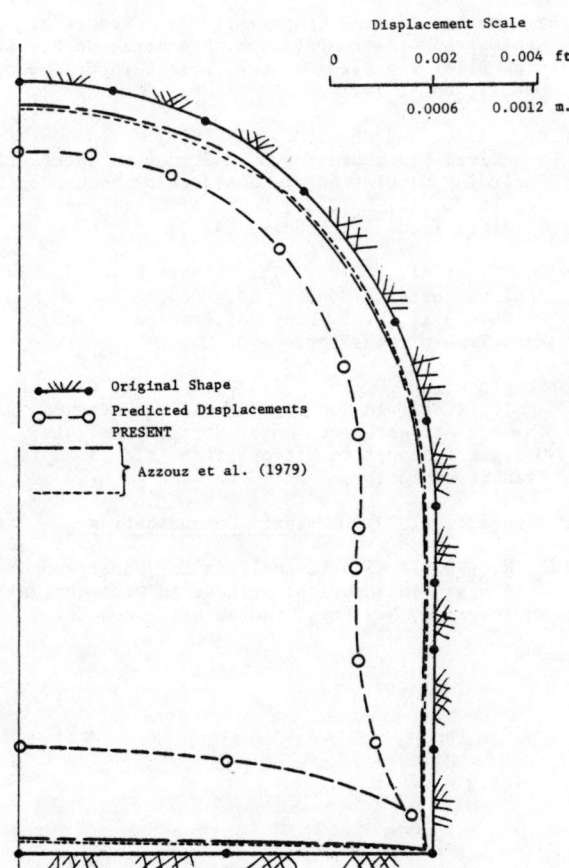

(b) Comparison between the pilot tunnel deformations obtained by the present analysis, and 3-D and 2-D analysis by Azzouz et al. (1979).

Figure 8 Comparison of predictions and observations.

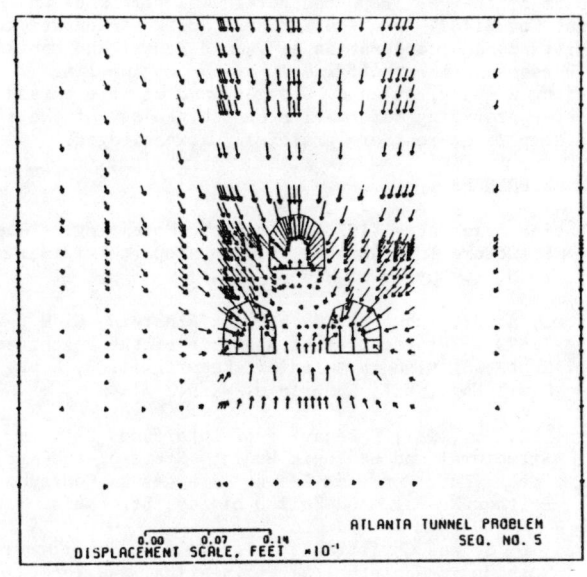

(a) Displacements

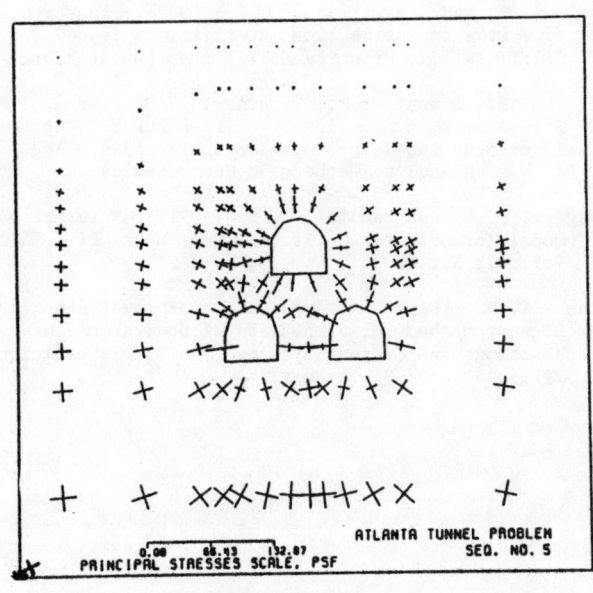

(b) Principal Stresses   1 psf = 47.9 Pa

Figure 9 Typical results after excavation of the right tunnel.

8. ACKNOWLEDGMENTS

A part of the work reported herein was performed under Grant No. 814844 from the National Science Foundation. Assistance and cooperation of Thomas Keusel and Don Rose of Parsons, Brinkerhoff/Tudor, of Law Engineering Testing Company, Atlanta, Georgia, and of site personnel towards providing information on the geology of the site and samples or rock are gratefully acknowledged.

9. REFERENCES

Agbabian Associates (1973). Analytic Modelling of Rock-Structure Interaction, Vol. 1: Report R-7215-1-2701 to U. S. Bureau of Mines.

Azzouz, A. S., Schwartz, C. W., and Einstein, H. H. (1979). Finite Element Analysis of the Peachtree Center Station in Atlanta: Report Vol. 3, Dept. of Civil Eng., MIT, Cambridge, Mass.

Desai, C. S. (1981). Behavior of Interfaces Between Structural and Geologic Media: State-of-the-Art Paper, Int. Conf. on Recent Advances in Geotech. Earthquake Eng. and Soil Dynamics, St. Louis, Mo.

Desai, C. S. and Christian, J. T. (Eds.) (1979). Numerical Methods in Geotechnical Engineering, New York, McGraw-Hill.

Desai, C. S., Lightner, J. G., and Sargand, S. M. (1982a). Mixed and Hybrid Procedures for Nonlinear Problems in Geomechanics: Proc., 4th Int. Conf. on Num. Meth. in Geomechanics, Edmonton, Alberta, Canada.

Desai, C. S., Janardhanam, R., and Sture, S. (1982b). High Capacity Truly Triaxial Device: J. Geotech. Testing, ASTM. March.

Desai, C. S. and Siriwardane, H. J. (1982). Constitutive Laws of Engineering Materials, Englewood Cliffs, N. J., Prentice-Hall, under publication.

Desai, C. S., Zaman, M. M., Lightner, J. G., and Siriwardane, H. J. (1981). Thin-Layer Element for Interfaces and Joints: Under publication, Int. J. Num. & Analyt. Methods in Geomechanics.

DiMaggio, F. L. and Sandler, I. S. (1971). Material Model for Granular Soils. J. Eng. Mech. Div., ASCE, Vol. 97, No. EM 3.

Eitani, I. M. (1981). An Application of the Finite Element Method for Simulation of Underground Excavations and Support Systems: Ph.D. Dissertation, Virginia Tech, Blacksburg, VA.

Ghaboussi, J. and Ranken, R. E. (1974). Tunnel Design Considerations; Analysis of Medium-Support Interaction: Report No. FRA-ORD 75-24, U. S. Dept. of Transportation, Washington, D. C.

Goodman, R. E.; Taylor, R. L.; and Brekke, T. (1968). A Model for the Mechanics of Jointed Rock, J. Soil Mech. & Found. Div., ASCE, Vol. 94, No. SM 3, pp. 637-659.

Kane, W. F. (1981). Direct Shear Behavior of Rock Joints in a Granite Gneiss: M. S. Thesis, Dept. of Civil Eng., Virginia Tech., Blacksburg, VA.

Kovari, K., Hagedon, H., and Fritz, P. (1976). Parametric Studies as a Design Aid in Tunnelling: Proc. 2nd Int. Conf. Num. Meth. in Geomech., ASCE, Blacksburg, Virginia, USA.

Kulhawy, F. H. (1977). Finite Element Analysis of Twin Running Tunnels and Research Chamber at MARTA Cain Street Station, Atlanta, Georgia: Report to Parsons, Brinckerhoff/Tudor, San Francisco.

Law Engineering Testing Company (1976). Report of Subsurface Investigation - Final Design, Construction Unit DN-11/Tunnelling Alternatives: Vol. 1-3, to Parsons, Brinckerhoff/Tudor, San Francisco.

Law Engineering Testing Company (1977). Report of Geology and Instrumentation, Peachtree Center Station Pilot Tunnel: Parsons, Brinckerhoff/Tudor, DN-124, Vols. I-III.

Lightner, J. G. and Desai, C. S. (1979). Improved Numerical Procedure for Soil-Structure Interaction Including Simulation of Construction Sequences: Report No. VPI-E-79.32, Dept. of Civil Eng., Virginia Tech, Blacksburg, VA.

Rose, D. C. et al. (1979). The Atlanta Research Chamber, Applied Research Monographs. Report No. UMTA-GA-06.007-79-1: U. S. Dept. of Transportation, Urban Mass Transp., Washington, D. C.

Tudor Engineering Company. (1977). Finite Element Analysis of Twin Running Tunnels and Research Chamber at Peachtree Center Station, Atlanta, Georgia: Report to Metropolitan Atlanta Rapid Transit Authority.

Weir, Jones, I. (1979). Private Communication.

Wittke, W. (1977). Static Analysis for Underground Openings: in Numerical Methods in Geotechnical Engineering, New York, McGraw-Hill Book Co.

THE BEHAVIOUR OF REINFORCED JOINTED ROCK MASSES UNDER VARIOUS SIMPLE LOADING STATES

Le comportement de masses rocheuses jointes et armées sous divers chargements simples

Das Verhalten eines armierten, geklüfteten Felsgebirges unter verschiedenen einfachen Belastungszuständen

G. N. Pande
Lecturer in Civil Engineering, University College of Swansea, Swansea, U.K.

C. M. Gerrard
Senior Research Scientist, CSIRO Division of Building Research, Victoria, Australia

SYNOPSIS

A model for the behaviour of reinforced jointed rock masses is proposed. The rock material and the joint sets are represented by elasto-visco-plastic rheological units connected in series to model the behaviour of jointed rock mass. The model is extended to account for sets of fully grouted reinforcement by considering each set of reinforcement represented by an elasto-visco-plastic unit to be acting in parallel with the jointed rock mass. The response of the model to various simple loading configurations of uniaxial compression, uniaxial tension and simple shear is computed for various orientations of a singel set of reinforcement. The model can be readily incorporated in a Finite Element Code to assess the optimum quantity and direction of reinforcement required for any complex geometry, load path and rock fabric.

RESUME

Un modèle est proposé pour le comportement des roches jointes et armées. La roche et les séries d'armatures sont représentées par des éléments rhéologiques élasto-visco-plastiques. Le modèle est appliqué à plusieurs séries d'armatures scellées: chaque ensemble d'armatures est représenté par un élément élasto-visco-plastique agissant en parallèle avec le matériau rocheux. La réponse du modèle à diverses configurations de chargements simples — compression uniaxiale, tension uniaxiale et cisaillement simple — est calculée pour diverses orientations des armatures. Le modèle peut être aisément incorporé dans un programme aux éléments finis qui évalue la direction et la quantité optimale d'armatures nécessaires selon la géométrie du problème, l'histoire du chargement et le matériau rocheux.

ZUSAMMENFASSUNG

Ein Modell für das Verhalten von bewehrtem, geklüftetem Gebirge wird vorgeschlagen. Das Gestein und die Kluftscharen werden jeweils durch hintereinandergeschaltete elasto-plasto-viskose Elemente dargestellt. Hinzu kommt ein Satz von mit dem Gebirge verbundenen Bewehrungen, die ebenfalls durch ein elasto-plasto-viskoses Element dargestellt werden und die im Verbund mit dem Gebirge wirken. Das Verhalten des Modells auf verschiedene einfache Belastungszustände, wie z.B einachsigen Druck, einachsigen Zug oder einfache Scherung, wird für verschiedene Orientierungen von einzelnen Bewehrungsscharen untersucht. Das Modell kann ohne weiteres in ein Finite-Element-Programm eingefügt werden, um die optimale Anzahl, und Richtung der Bewehrungen zu ermitteln, die für beliebige Geometrien, Belastungszustände und Kluftgefüge erforderlich sind.

1. INTRODUCTION

In many civil engineering and mining structures, the engineers have to accept the rock mass as encountered during planning or construction and devise methods of improving its properties to obtain a safe and an economical structure. Rock reinforcing has proved to be an effective technique and has been widely used. The design of rock reinforcement, however, is largely based on empirical rules and it has not been hitherto possible to make a series of engineering calculations to decide the optimum orientation, spacing and length of rock reinforcement. With the advent of high speed computers, the situation is rapidly changing. Various numerical methods particularly the Finite Element Method have reached a stage of development that they can readily be used for the solution of practical engineering problems. In the recent years the research has been concentrated on developing models of non-linear behaviour of materials which can be incorporated in the Finite Element codes.

A time dependent multi-laminate model of jointed rock masses based on Wittke (1977) has been proposed by Zienkiewicz and Pande (1977). It accounts for the multiple planes of weakness by defining a yield or failure criterion for each of them. Assumptions of elasto visco-plasticity were invoked, (though it is now clear that it is not necessary and model can be implemented within elasto plastic framework (Pande and Pietruszczak, to be published)) to obtain the rates of plastic strains as a sum of contributions made by each of the sets of weakness planes (joints) and the rock material. This model which has been used extensively in design and research for jointed rock masses, is extended in this paper for reinforced jointed rock masses. The reinforcement sets are assumed to be fully grouted and evenly distributed throughout the jointed rock mass. Numerical calculation for stress-strain response in various simple loading situations has been made. The fact that different material characteristics can be readily assigned to various elements in the Finite Element method, enables the model to have an important role in the parametric analyis of reinforced rock structures to obtain optimum spacing, orientation and length of reinforcement.

2. RHEOLOGICAL MODEL FOR REINFORCED JOINTED ROCK MASSES

Following previous work [Zienkiewicz and Pande, 1977], relating to jointed rock masses, the simple rheological model of an elasto/visco-plastic material, as shown in Fig.1, is suggested as the basic building block to construct a model for reinforced jointed rock masses. The simple model, consisting of a spring connected in series with a parallel coupling of a dashpot and slider has the properties that an applied load produces elastic strain and, if the yield value of the slider is exceeded, a visco-plastic strain rate. This behaviour represents the following three important features commonly observed in jointed rock masses,

a) deformations are not fully recoverable and hence contain non-elastic components,

b) deformations are time dependent, and

c) the apparent 'stiffness' under rapid loading is much higher than the apparent 'stiffness' under quasi-static loading.

For problems where only quasi-static response is required to be modelled the elasto/visco-plastic algorithm is still employed since it offers a number of computational advantages. In applying the simple model to processes involving incremental changes of the strain vector, referred to global coordinates, it is postulated that such changes consist of two parts, elastic and visco-plastic, i.e.

$$\Delta \varepsilon = \Delta \varepsilon^e + \Delta \varepsilon^{vp} \qquad (1)$$

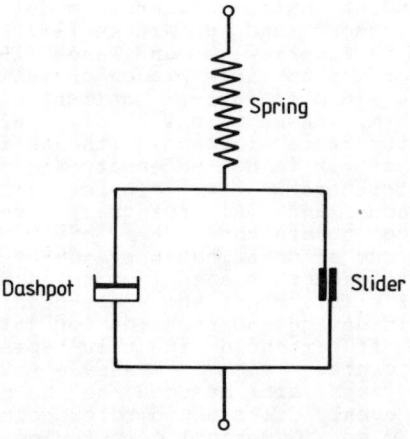

FIGURE 1 SIMPLE RHEOLOGICAL MODEL OF ELASTO/VISCO-PLASTIC BEHAVIOUR

In the idealized jointed rock mass the joints within each set are assumed to be parallel, continuous, and evenly spaced. Since it is further assumed that the joints contain no fill material and occupy a negligible proportion of the volume of the rock mass, it follows that when an increment of stress, $\Delta \sigma$, referred to global coordinates, is applied to a representative sample of the rock mass the resultant distribution of stress will be homogeneous throughout the mass. As discussed previously [Zienkiewicz and Pande, 1977, Gerrard, 1982], this means that in rheological terms the rock material and each of the joints sets can be considered to be connected in series so that, referred to global coordinates, they experience the same increment of stress and their resultant incremental strains are additive to produce the total incremental strain of the jointed rock mass.

Reinforcement for the jointed rock mass is provided by the insertion of several sets of parallel, fully grouted, and evenly spaced reinforcing bars. Because of the effectiveness of the grouting and the negligible proportion of the total volume occupied by the bars, it follows that in any representative sample the incremental strain of the jointed rock mass, $\Delta \varepsilon$, in terms of global coordinates, will also be the incremental strain of each of the sets of reinforcement, and hence that for the reinforced jointed rock mass. In rheological terms the jointed rock mass and each of the reinforcement sets can be considered to be connected in parallel (Gerrard, 1982). This is illustrated in Fig.2 with the jointed rock mass being represented on the left hand side by a series connection of the rock material and the joint sets. Consideration of the model of reinforced jointed rock masses, shown in Fig.2, indicates that the following four relationships, in terms of global coordinates, must always apply during incremental changes of stress and/or strain,

a) the stress increment in the rock material will be equal to that in each of the joint sets and thus can be considered as the stress increment for the jointed rock mass, i.e.

$$\Delta \sigma_J = \Delta \sigma_M = \Delta \sigma_{J1} = \Delta \sigma_{Ji} = \Delta \sigma_{Jm} \qquad (2)$$

where M represents the rock material,

Ji represents the i:th set of joints in a total of m sets,

and J represents the jointed rock mass,

b) the strain increment for the jointed rock mass is given by summing contributions from the rock material and each of the joint sets, i.e.

$$\Delta \varepsilon_J = \Delta \varepsilon_M + \sum_{i}^{m} \Delta \varepsilon_{Ji} \qquad (3)$$

c) the strain increment for the jointed rock mass is equal to each of those for the various reinforcement sets and hence can be regarded as the strain increment for the reinforced jointed rock mass, i.e.

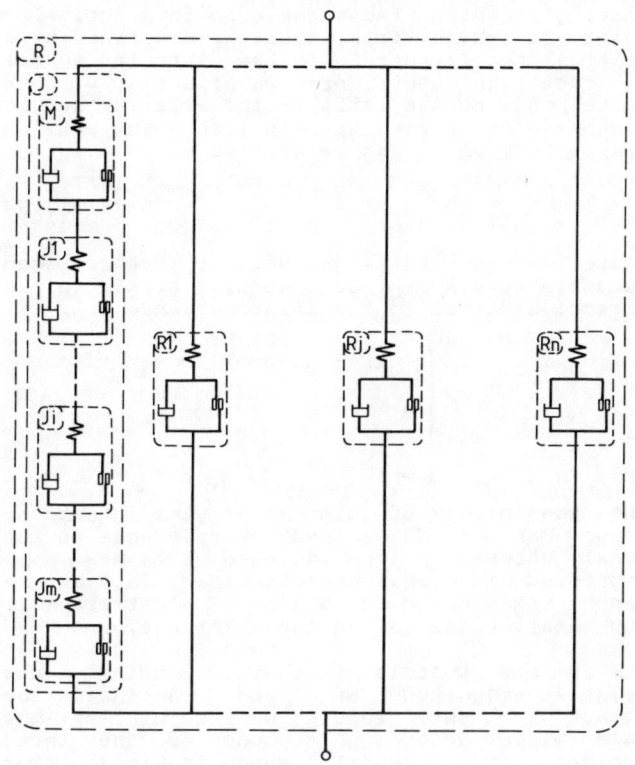

FIGURE 2 RHEOLOGICAL ANALOGUE FOR REINFORCED JOINTED ROCK MASS

$$\Delta \varepsilon_R = \Delta \varepsilon_J = \Delta \varepsilon_{R1} = \Delta \varepsilon_{Rj} = \Delta \varepsilon_{Rn} \quad (4)$$

where Rj represents the j:th set of reinforcement in a total of n sets,

and R represents the reinforced jointed rock mass,

d) the stress increment for the reinforced jointed rock mass is given by summing the products of the volumetric proportions and the stress increments for the jointed rock mass and each of the reinforcement sets, i.e.,

$$\Delta \sigma_R = P_J \Delta \sigma_J + \sum_{j}^{n} P_{Rj} \Delta \sigma_{Rj} \quad (5)$$

where, as indicated by subscripts, P is the volumetric proportion of the jointed rock mass or a reinforcement set to the total such that $\Sigma P = 1$.

Before using equations 1 to 5 to describe, in detail, the behaviour of a reinforced jointed rock mass it is necessary to consider the properties of the components of the system, i.e. the rock material, the rock joint sets, and the reinforcement sets.

3. PROPERTIES OF ROCK MATERIAL, ROCK JOINT SETS, AND REINFORCEMENT SETS.

As shown in Fig.2, each component in the system is assumed to behave in an elasto/visco-plastic fashion. This means that the rock material, the joint sets, and the reinforcement sets are assumed to respond elastically to instantaneous loading. After such loading, if their stress states are such as to cause yielding, visco-plastic strain rates will develop. As discussed previously [Zienkiewicz and Pande, 1977] the behaviour of a typical component, in general terms, can be described by three relationships, involving firstly the matrix of elastic constants, D,

$$\Delta \sigma = D \Delta \varepsilon^e, \text{ secondly} \quad (6)$$

the yield function, F, defined by

$$F(\sigma, \varepsilon^{vp}) = 0, \text{ and finally} \quad (7)$$

the visco-plastic flow law involving the plastic potential function, Q,

$$\dot{\varepsilon}^{vp} = \gamma <\Psi(F)> \frac{\partial Q}{\partial \sigma} \quad (8)$$

where the dot indicates a time rate, γ is a fluidity parameter, and the bracket, $<>$, indicates,

$$<\Psi(F)> = \Psi(F) \text{ when } F > 0,$$
$$<\Psi(F)> = 0 \text{ when } F < 0.$$

$\Psi(F)$ represents a monotonic function of F. In line with other applications of visco-plastic theory to the problems of soil and rock mechanics problems, in subsequent development it is assumed that,

$$\Psi(F) = F \quad (9)$$

Thus, equation (8) reduces to

$$\dot{\varepsilon}^{vp} = \gamma <F> \frac{\partial Q}{\partial \sigma} \quad (10)$$

The form of elasticity matrix, yield and plastic potential functions is different for the rock material, the rock joint sets, and the reinforcement sets, as discussed below.

a) Rock material: Assuming rock material to be isotropic elastic, its elasticity matrix, D can be easily written in terms of Young's modulus, E and Poissons' ratio ν. If the intact rock is transversely anisotropic as in case of many sedimentary rocks, 5 independent material constants ($E_1, E_2, \nu_1, \nu_2, G_2$) are required to describe the intact rock behaviour. The D matrix is set up in a local system of coordinates corresponding to the orientation of orthotropy and a simple transformation gives the global elasticity matrix.

The failure of intact rock is a complex phenomenon. Various yield criteria have been advanced in the past [Griffith, 1921, Cook, 1965, Hoek and Brown, 1980, Gerogiannopoulous and Brown, 1978]. It is fortunate that the choice of yield criterion of intact rock is hardly ever critical in failure of jointed rock masses as failure of joints preceeds the failure of intact rock. Detailed discussion on this aspect is beyond the scope of this paper. Nevertheless, the frame work of the proposed model enables any yield criterion to be incorporated in a modular fashion.

b) Joint sets: The mechanical response of a typical joint set is conveniently described in terms of a set of local cartesian coordinates adopting dip and strike of the joint and normal to the joint plane as three axes. If the spacing of the i:th joint set is d_{Ji}, the elasticity matrix of the joint set in global coordinates can be written as

$$D_{Ji} = d_{Ji} \, T^T \, k \, T \qquad (11)$$

where T represents an appropriate transformation matrix and k represents a matrix of elastic 'joint stiffness', [Goodman et al, 1968]. The coefficients of k matrix can be determined experimentally and a large body of data on joint stiffness is already available.

An alternative to computing elasticity matrices of rock material and joint sets individually as discussed above is to estimate averaged values of Young's modulus and Poisson's ratio based on in-situ tests on rock mass. This approach was adopted by Zienkiewicz and Pande (1977) and will be used in the numerical examples presented later in section 5. However if sufficient data are available on stiffnesses of joint sets, a corresponding elasticity matrix can be computed.

The joints in the i:th set are assumed to be capable of yielding in both shear and tension. The shear failure criteria of Coulomb and empirical relationship of Barton and Choubey (1977) can be put in the form,

$$F_{Ji} = f_n (\tau_c, \sigma_n) = 0 \qquad (12)$$

where τ_c is the critical shear stress, and σ_n the normal stress, on the joint plane.

A tensile cut-off can be simply defined by,

$$F_{Ji} = - \sigma_n = 0 \qquad (13)$$

The failure criterion proposed by Barton and Choubey, (1978) has been discussed and implemented by Pande and Xiong, (1982). In this work, further development will be restricted to discussion of the Coulomb criteria described by the friction angle of the joint plane, ϕ, and the cohesion, C_o, that is available when the joints are fully closed. In general C_o and ϕ can be assumed to be functions of relative orientation of the asperities on the joint plane with respect to the direction of maximum shear stress.

The tensile cut-off implies that joint opening is simulated in the model by tensile visco-plastic straining normal to the joint plane. It is assumed that when this straining takes place the effective cohesion, C_e, of the joint set decreases linearly from a maximum, when the joints are closed, to zero when the 'mating' joint surfaces cease to interact. The point at which this occurs is defined by a critical tensile strain in the direction normal to the joint set, of magnitude ε_c, that corresponds to the ratio of the asperity height to the joint spacing for the i:th joint set. A simple form of C_e can be written as

$$C_e = C_o \frac{\langle \varepsilon_c - \varepsilon_t \rangle}{\varepsilon_c} \qquad (14)$$

where ε_t is the magnitude of the computed accumulated visco-plastic tensile strain in the direction normal to the joint set, and,

$$\langle \varepsilon_c - \varepsilon_t \rangle \begin{cases} = \varepsilon_c - \varepsilon_t & \text{if } \varepsilon_c - \varepsilon_t > 0 \\ = 0 & \text{if } \varepsilon_c - \varepsilon_t < 0 \end{cases}$$

The possibility of joint set i yielding at any stage can be determined by reference to the total stress in the jointed rock mass, σ_j, expressed in global coordinates. This stress can be transformed to obtain normal stress (σ_n) and shear stress (τ) on the joint set.

It is now postulated that depending on the relative magnitude of σ_n and τ acting on the i:th joint set, yielding on the joint set may take place according to one of the three regimes, I, II and III shown in Fig.3. For regime I the yielding is in shear according to the Coulomb criterion i.e.

$$F_{Ji} = |\tau| - \sigma_n \tan\phi - C_e \qquad (15)$$

with non-associated flow described by the plastic potential function,

$$Q_{Ji} = |\tau| - \sigma_n \tan\delta - \text{constant} = 0 \qquad (16)$$

where δ is the dilatancy angle for joint set i. The region occupied by regime I is bounded on the left in Fig.3 by the line,

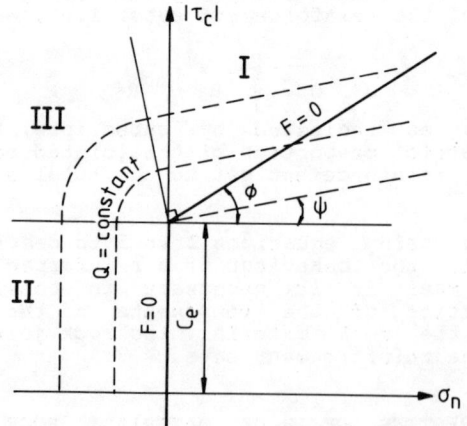

FIGURE 3 TENSILE AND SHEAR YIELD REGIMES FOR A TYPICAL JOINT SET

$$|\tau| = C_e - \sigma_n \cot\delta \quad (17)$$

Regime II corresponds to yielding in tension according to the criterion given by eqn.(13), with associated flow being assumed. The boundaries of the regime are,

$$\sigma_n < 0, \quad |\tau_c| < C_e \quad (18)$$

The remaining regime III is a transition between yielding in shear and yielding in tension. Any stress condition initially lying in this regime is forced to ultimately adopt the stress state,

$$\sigma_n = 0, \tau_c = C_e \quad (19)$$

This is achieved by adopting the yield criterion,

$$F_{Ji} = \{(|\tau_c| - C_e)^2 + \sigma_n^2\}^{\frac{1}{2}} \quad (20)$$

with associated flow.

Depending on whether joint set i is following regime I, II or III in yielding, the relevant yield criterion and plastic potential function can be selected and used to determine the visco-plastic strain.

c) Reinforcement.

For a typical set of reinforcement, the j:th, the elastic response is determined by the Young's modulus (E_a) of the reinforcement material along the reinforcement axis, this axis being defined with respect to the global coordinates by the direction cosines $\underline{\ell}$. E can be regarded as a 1 x 1 matrix and transformed into global coordinates to give the stiffness of the j:th set of reinforcement, D_{Rj}, as,

$$D_{Rj} = T_{Rj}^T E_a T_{Rj} \quad (21)$$

where T_{Rj} is a 1 x 6 transformation matrix that is a function of $\underline{\ell}$ and whose elements are given by:

$$(T_{Rj})_s = \ell_m \ell_n, \text{ where} \quad (22)$$

$$m = s \text{ Int } \tfrac{1}{3}(6-s) + \text{Int } \tfrac{1}{3}(s-1) \text{ Int } \tfrac{1}{2}(s-2)$$

$$n = s \text{ Int } \tfrac{1}{3}(6-s) + \text{Int } \tfrac{1}{3}(s-1) \text{ Int } \tfrac{1}{2}(s+1)$$

Yielding of the j:th set of reinforcement is assumed to be governed by the simple strain hardening type of yield function,

$$F_{Rj} = |\sigma_a| - \sigma_y (\varepsilon_{Rj}^{vp}) \quad (23)$$

where σ_y is the magnitude of the uni-axial yield stress of the reinforcement material, ε_{Rj}^{vp} is the plastic strain in the reinforcement and σ_a is stress in reinforcement.

4. FORMULATION FOR REINFORCED JOINTED ROCK MASS

The properties of the individual components of the reinforced jointed rock mass, i.e. the rock material, the rock joint sets, and the reinforcement sets can be used to define quantities that relate to aggregates of components in the system. These are,

a) the stiffness of the jointed rock mass,

$$D_J^{-1} = D_M^{-1} + \sum_i^m D_{Ji}^{-1} \quad (24)$$

b) the stiffness of the reinforced jointed rock mass,

$$D_R = P_J D_J + \sum_j^n P_{Rj} D_{Rj} \quad (25)$$

c) the visco-plastic strain rate for the jointed rock mass,

$$\dot{\chi}_J = \dot{\chi}_M + \sum_i^m \dot{\chi}_{ji} \quad (26)$$

and
d) the stress rate for the reinforced jointed rock mass,

$$\dot{S}_R = P_J D_J (\dot{\varepsilon}_J - \dot{\chi}_J) + \sum_j^n P_{Rj} D_{Rj} (\dot{\varepsilon}_{Rj} - \dot{\chi}_{Rj}) \quad (27)$$

The above relations are quite general and apply to any portion of a particular loading path.

The solution of practical problems is approached by considering an increment of load to be applied instantaneously. This increment may involve only stress components, only strain components, or a self consistent combination of both. At time t = 0 the entire strains produced are elastic in nature. The computed incremental stresses are added to those that pre-existed to give the revised total stress Mtates in the rock material, the rock joint sets, and the reinforcement sets. A series of time increments are now allowed to elapse, during which visco-plastic yielding can occur. For problems in which quasi-static behaviour is being modelled the number of time increments allowed after the application of a load increment needs to be sufficient to enable steady-state conditions to be approached. During a time increment the required stress and /or strain conditions can be specified in the form that,

a) the reinforced jointed rock mass does not change stress i.e. $\Delta \underline{S}_R = 0$,

b) there is no change in strain, i.e. $\Delta \underline{\varepsilon}_R = 0$,

or
c) a self consistent combination of restraints on particular stress and strain components.

Each of these possibilities will have a different impact on the incremental stress-strain relations, i.e. eqns. (27).

In general, during a time increment the components of the reinforced jointed rock mass, i.e. the rock material, the rock joint sets, and the reinforcement sets will undergo both elastic and visco-plastic straining. The former strain will be reflected in a change in the stress acting on the particular component. At the conclusion of the time increment the total stress state and the accumulated elastic and visco-plastic strains for each of the components can hence be updated. The required loading path may then dictate the application of a further increment of load, the elapse of a further increment of time, or the termination of the loading and deformation processes.

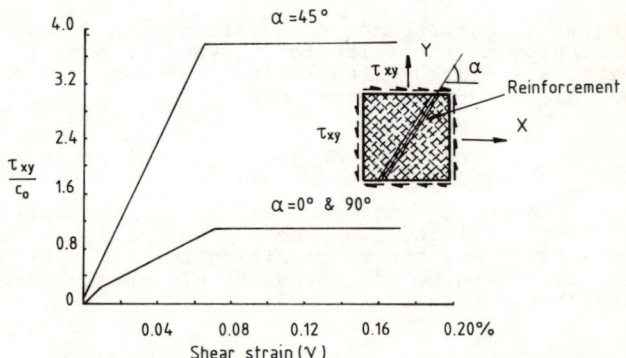

FIGURE 6 STRESS-STRAIN RESPONSE OF REINFORCED JOINTED ROCK MASS IN SIMPLE SHEAR

5. NUMERICAL EXAMPLES

Numerical examples of stress-strain response of the proposed model to various simple configurations of loading are presented in this section. For these examples isotropic elastic properties, relevant to jointed rock mass have been assigned (Young's modulus = 7×10^6 KN/m^2, Poisson's ratio = 0.15) without making an attempt to compute them from the elastic joint stiffnesses and joint spacing. The rock fabric consists of two sets of joints perpendicular to each other and a single set of reinforcement is provided. The initial value of C_0 is assumed as 70 KN/m^2 and has been used to normalize the applied compressive, tensile or shear stress. Plane strain conditions are assumed for all loading cases. Figure 4(a) shows the stress-strain relationship in uniaxial compression when the reinforcement is placed along the direction of uniaxial stress as shown in the inset. Curves for $\phi = 40°, 35°, 30°$ and two values of volumetric proportion (P) are plotted. The stress strain relationship is seen to be tri-linear. The two kinks correspond to the situations when sliding on the joint sets commences and reinforcement yields to give unlimited strain. Figure 4(b) and 4(c) show the curves for P = 0.001% when reinforcement is placed normal and at 45° to the direction of compressive loading. Tri-linear stress-strain characteristic is again noticed in figure 4(b). For reinforcement at 45° (coinciding with the orientation of the joint sets), the stress-strain relationship is elasto ideally plastic or visco-plastic. The reason is that the reinforcement so placed has no contributing effect on the sliding strength of the other joint set.

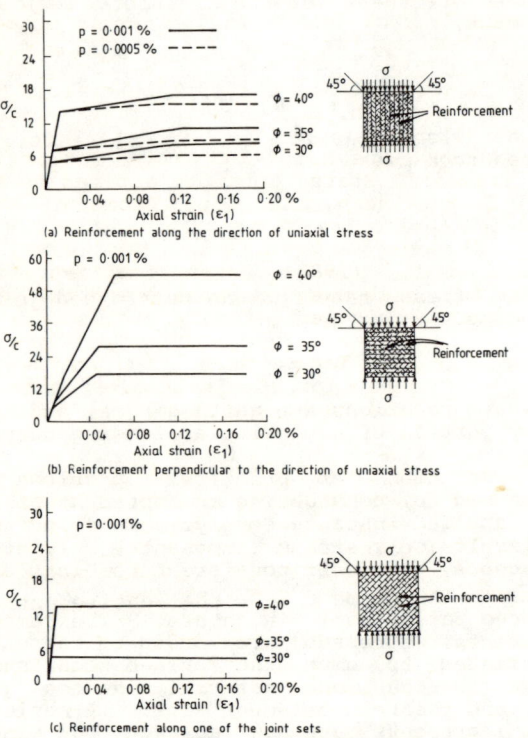

FIGURE 4 STRESS-STRAIN RESPONSE OF REINFORCED JOINTED ROCK MASS IN UNIAXIAL COMPRESSION

Figure 5 shows response in uniaxial tension when reinforcement is placed at angle of $\alpha = 0°$, 30°, 60° and 90° to the x-axis (see inset on figure 5). With $\alpha = 45°$, no tensile stress can be withstood as one of the joint set yields in tension and reinforcement along the second set can not contribute any strength. Fig. 6 shows shear stress-strain response when a simple shear stress is applied. Here $\alpha = 45°$ is the best direction to reinforce. However if the sign of shearing stress was changed, $\alpha = 135°$ would give the best direction.

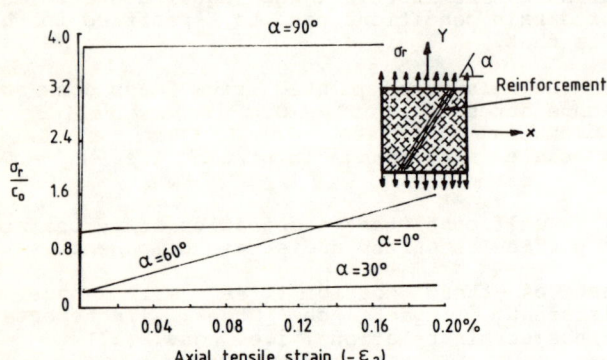

FIGURE 5 STRESS-STRAIN RESPONSE OF REINFORCED JOINTED ROCK MASS IN UNIAXIAL TENSION

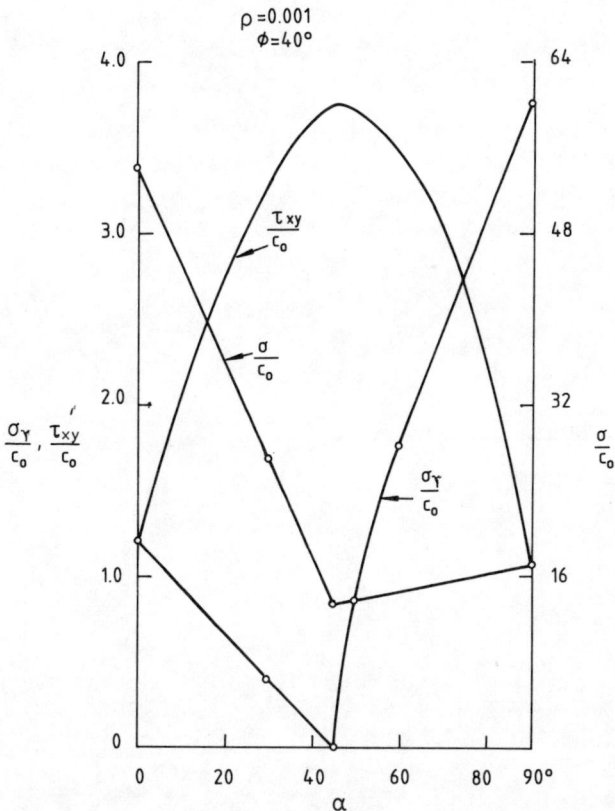

FIGURE 7 VARIATION OF UNIAXIAL STRENGTH WITH THE DIRECTION OF REINFORCEMENT

Results of figures 4,5 and 6 are summarized in Figure 7 where the peak strength developed for the three loading situations is plotted against α. It is seen that for uniaxial compression the most effective direction is one perpendicular to the applied stress, for uniaxial tension the best direction is along the applied stress while for simple shear the best direction is along the diagonal which undergoes extension. In a multi-axial situation, the optimum reinforcement orientation would depend on the stress path followed by an element of jointed rock mass.

6 CONCLUSION

A model of the behaviour of reinforced jointed rock masses has been presented. The model is formulated by representing rock material, joint and reinforcement sets by elasto visco-plastic units. Numerical tests on simple loading configurations indicate that the best orientation of the reinforcing steel is dependent on the type of loading. The model proposed is quite general and is applicable to multi-axial stress state. It can be readily incorporated in a Finite Element code. For any boundary value problem, depending on the stress path followed by various zones, an optimum direction of reinforcing requiring minimum volume of steel can be found through parametric studies. The results will have to be, however, modeified to take into account the practicability of providing reinforcement in the optimum direction. Further studies on the solution of practical problems are planned to be undertaken by the authors.

7 REFERENCES

Barton, N. and Choubey, V. (1977). The shear strength of rock joints in theory and practice, Rock Mechanics, 10, 1-54.

Cook, N.G.W. (1965). The failure of rock, Int. J. Rock Mech. Min. Sci., 2, 389-403.

Gerogiannopoulous, N.G. and Brown, E.T. (1978) The Critical State Concept applied to Rock, Int. J. Rock Mech. Min. Sci. Geomech. Abstr. 15. 1-10.

Gerrard, C.M. (1982) Joint Compliances as a basis for rock mass properties and the design of support requirements (to be published).

Goodman, R.E., Taylor, R.L. and Brekke, T.L. (1968) A Model for the mechanics of jointed rock, J. of Soil Mech. and Found Div. ASCE, 94, 3, 637-659.

Griffith, A.A., (1921). The phenomenon of rupture and flow in solids, Phil. Trans. Roy Soc. London, Sec. A, 221, 163-198.

Hock, E. and Brown, E.T. (1980) Underground excavations in rock, I.M.M. London.

Pande, G.N. and Pietruszczak, S (1982) A plasticity formulation of multi-laminate model for rocks and soils (To be published).

Pande, G.N. and Xiong, W. (1982). An improved multi-laminate model of rocks. Numerical Models in Geomechanics, Eds., Dungar, Pande & Studer, A.A. Balkema, Rotterdam, 218-226.

Wittke, W. (1977) New design concept for underground openings in rock. Chapter 13, Finite Elements in Geomechanics, Ed. G. Gudehus, Wiley.

Zienkiewicz, O.C. and Pande, G.N. (1977) Time dependent multi-laminate model of rocks - a numerical study of deformation and failure of jointed rock masses. Int.J. Num. and Anal. Methods in Geomech. 1, 219 - 247.

APPLICATIONS OF A NEW COMPUTER MODEL FOR RECONSTRUCTING BLOCKY ROCK GEOMETRY — ANALYSING SINGLE BLOCK STABILITY AND IDENTIFYING KEYSTONES

Applications d'un nouveau modèle informatique pour la réconstitution de la géométrie des rochers en blocs — l'analyse de la stabilité des blocs simples et la détermination des clefs de voûtes

Anwendungen eines neuen Rechnermodells zur Nachahmung der Blockgesteinsgeometrie, zur Analyse der Stabilität einfacher Blöcke sowie zur Identifizierung von Schluss-Steinen

P. M. Warburton
Senior Research Scientist, CSIRO, Division of Applied Geomechanics, Syndal, Victoria, Australia

SYNOPSIS

The paper introduces a new computer model for blocky rock and describes some of its applications. The current program can reconstruct the hidden three-dimensional block structure around a complex excavation, eliminate selected blocks, and analyse the stabilities of individual surface blocks. Detailed perspective views can readily be obtained. Recent theoretical developments permit the analysis of simple systems of interacting blocks and support, including keystone systems.

RESUME

On présente un nouveau modèle informatique pour les roches en blocs et on décrit quelques-unes de ses applications. Le programme courant peut reconstruire la structure cachée en blocs à trois dimensions autour d'une excavation complexe, éliminer des blocs choisis et analyser la stabilité individuelle des blocs en surface. Il est possible d'obtenir sans difficulté des vues détaillées en perspective. De récents développements théoriques permettent d'analyser les systèmes simples de blocs qui réagissent réciproquement ainsi que les systèmes de support, y compris les systèmes de clefs de voûte.

ZUSAMMENFASSUNG

Ein neues Rechnermodell für Blockgesteine wird vorgestellt und einige dessen Anwendungen werden beschrieben. Das vorliegende Programm vermag es, den verborgenen, dreidimensionalen, um eine komplexe Ausgrabung vorkommenden Blockaufbau wiederzugeben, ausgewählte Blöcke auszuschließen und die Stabilität der einzelnen oberflächlichen Blöcke zu analysieren. Ausführliche Perspektivansichten können ohne Schwierigkeiten erhalten werden. Neue theoretische Entwicklungen gestatten es, einfache Systeme wechselwirkender Blöcke sowie Stützsysteme, einschließlich Schlußstein-Systeme, zu analysieren.

INTRODUCTION

Blocky rock is usually associated with the combination of extensive discontinuities, hard rock and low stress fields. Since the discontinuities are extensive, they frequently intersect each other, dividing the rock mass into the characteristic blocky pattern. Clearly, individual blocks in such an assemblage need not be restricted to particular shapes or sizes. The discontinuities dominate not only the structure of the system but also its behaviour, because the stress fields are too low to have much effect on the integrity of the hard blocks of rock. Thus deformations take place almost exclusively at the discontinuities, and the blocks themselves move essentially as rigid bodies with virtually no deformation or cracking.

In mining and civil engineering, the most familiar occurrences of blocky rock are in excavations at or near the earth's surface, where stress fields are generally low. Rock slopes and tunnels are common examples. But the conditions for blocky rock may also be approximated in some excavations at greater depths if local stress relief has occurred.

Since rigid block assumptions are possible and gravitational driving forces predominate, the stabilities of individual blocks are often analysed by comparatively simple methods. The problem is generally not one for which detailed computation of stress distributions is needed to achieve a satisfactory solution. In any case, current stress analysis computer programs, such as those based on finite elements or boundary elements, do not provide a practicable means of coping with the very complex three-dimensional geometry of a typical blocky system, including the discontinuities. The distinct element method of Cundall (1980), though designed specifically for interacting blocks, lacks the realism of three dimensions.

Despite the success of the simple methods of stability analysis, their scope too has been limited by the complexities of real block geometries. Some of these methods use stereographic projection and are essentially graphical, whereas others are based on vector techniques and may be called analytical. Until recently, however, the most complex geometry that either type could handle consisted of a tetrahedral block with one free face and the potential to slide on one of its other faces or on two simultaneously. Shih (1978) and Priest (1980) greatly extended the range of block geometries available in graphical methods, but all methods of this type share the disadvantage of being unsuitable for computers. Fortunately the opposite is true of the vector techniques in analytical solutions, and their natural computational advantages have been exploited in a new procedure derived by Warburton (1981). The stability analysis on which that procedure is based generalizes the block geometry to an arbitrary polyhedron with any number of free faces. The block can have various re-entrant surface features, such as notches and cavities, but its possible movements are assumed to be limited to translation only, as in most other solutions for three-dimensional blocks.

Although the block geometries that can be handled by the new stability analysis are much more general than in previous vector solutions, the generality carries a penalty. Previously, the permissible geometries were so simple that it was possible to visualize them easily and to calculate their properties by hand. After all, the block was never allowed more than three (or at most four) bounding discontinuities and a single free face. But an arbitrary polyhedral block with any number of free faces cannot be visualized easily and definition of its geometry would be impracticable without appropriate computer routines.

Up till now, no routines with the required generality have been available in rock mechanics. The nearest substitute has been the program KINWEG written by Croney et al. (1978). Although KINWEG could take into account the boundary of an excavation and all the dominant discontinuities in its vicinity, the excavation itself was restricted to being a tunnel with a standard idealized shape, and the program could identify and analyse only those blocks that were bounded by three or four discontinuities and an exposure on the tunnel surface. Apparently, nested combinations of blocks were computed as additional single blocks. Another drawback is that KINWEG was written for a major consulting firm and is not freely available. Many of the sophisticated geometric modelling systems that have been developed for computer-aided design and might conceivably be adapted to rock mechanics are constrained by similar proprietary restrictions.

New algorithms have now been developed to compute the complex geometry of a three-dimensional blocky system. The algorithms have been specially designed to complement the stability analysis in the previous paper by Warburton (1981), and the combined procedures have been implemented in a single FORTRAN program called BLOCKS. Tests on CSIRO's CDC CYBER 76 computer have confirmed the validity of the FORTRAN routines.

Full technical details of the computer program BLOCKS and its new algorithms will be published elsewhere. So too will recent theoretical developments that permit the analysis of simple systems of interacting blocks and support. The present paper is intended to give a more descriptive introduction to the new model for the benefit of practitioners who might wish to apply it.

CAPABILITIES AND APPLICATIONS OF MODEL

The description of the new model will be accompanied by ten figures showing perspective views of blocky systems. It should be pointed out in advance that all these figures were produced by the computer. Detailed perspective views are indispensable in an actual analysis, and one of the features of program BLOCKS is that they can be obtained with a minimum of manual intervention. At appropriate stages in the program, BLOCKS automatically computes pertinent data and stores them on a special file for subsequent transfer to an auxiliary display package. The particular package in use at CSIRO is MOVIE, which was obtained from Brigham Young University, Utah.

Before carrying out an analysis with the new model, the user must first select a local three-dimensional region of interest in the rock mass. Parts of the boundary of such a region will be free surfaces exposed at adjacent excavations, but the parts inside the rock are fictitious and can be defined artificially so as to provide a sufficient working volume for the subsequent block reconstructions. It is also necessary to design an imaginary structure of convex polyhedral components occupying the selected region. This is done purely for mathematical convenience to facilitate later subdivisions along arbitrary planes. Although the components must be convex, any number of them may be used, so it is clear that exceedingly complex excavations can be represented.

Figure 1 shows an example from underground mining, in which the selected region of interest consists of a pillar, together with portions of the roof and floor. As can be seen from the view in Figure 2, the shape of the roof is constructed from five convex components; five more are used for the floor, and a single seven-faced component represents the irregular pillar. It must be emphasized that these eleven components are merely an artifice for defining the shape of the selected region in terms acceptable to the program. The components do not represent real blocks of rock, nor do the interfaces between them correspond to actual discontinuities.

Data on the shape of the selected region are submitted to the program in an ascending hierarchy of vertices, edges, faces and polyhedral components, together with the planes containing the faces. The input for each face includes a code denoting whether that face is free, or abutting on the rock mass outside the region, or shared by two components. Another code is available for discontinuities. Although the internal operation of the program requires a data structure complying with strict mathematical conventions, conversion to the correct form is done automatically and does not concern the user. One of the few restrictions on data preparation is that the positive x, y and z axes must point north, west and vertically upwards respectively.

The only other geometrical input needed is data on the dominant discontinuities. Each discon-

FIGURES 1 TO 4.   UNDERGROUND MINING

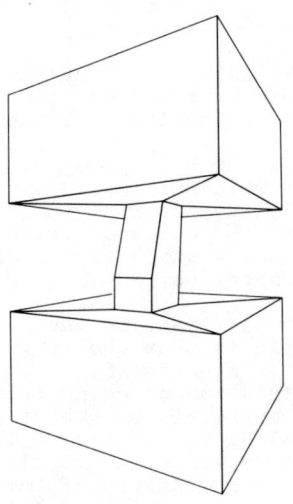

Figure 1.   Selected region of interest

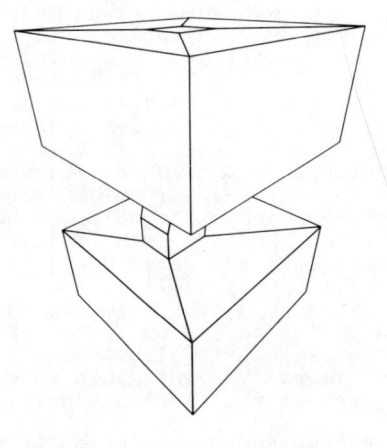

Figure 2.   Selected region of interest

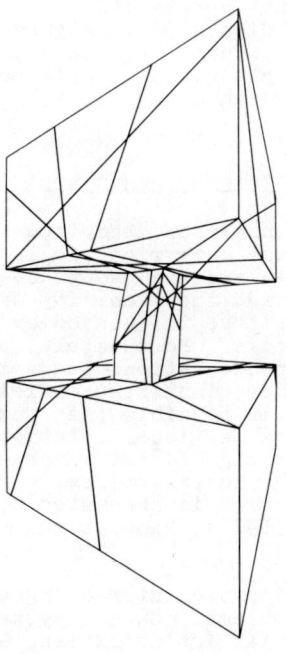

Figure 3.   Block structure produced by five dominant discontinuities

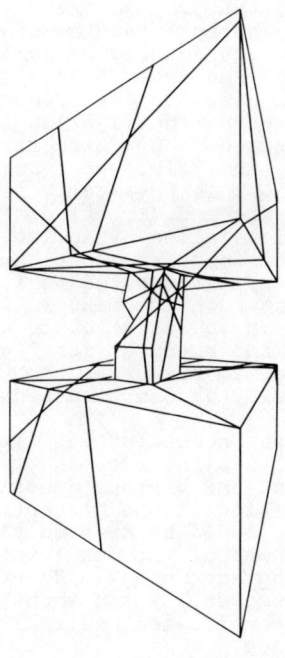

Figure 4.   Block structure after elimination of one block

tinuity is assumed to be an infinite plane, which is defined, in the same way as the other planes, either by three points or by dip, dip direction and one point. It will be shown shortly how finite persistence could be allowed for.

The program reconstructs the hidden block structure in the selected region by effectively slicing through the region along successive discontinuity planes. Vertices, edges, faces and blocks are modified or generated automatically as required. The initial convex components must be treated as blocks during this process, but at the end of it a supplementary routine effectively restores the continuity across all internal interfaces that do not represent actual discontinuities. Thus blocks in the amended set correspond directly to those in the real structure. Nevertheless, traces of the initial components are retained in the face data and still appear as lines when surfaces are displayed. Figure 3, for example, shows the block structure produced by five particular discontinuities, but it also includes all the lines from Figure 1, even though some of them are incidental to the final blocks.

The routine that restores continuity across specified interfaces could also provide one means of reducing the persistence of a discontinuity, though some modifications to the program would be necessary. This approach might be taken if the rock bridges were expected to stay intact. Alternatively, if rock bridge failure were possible, it could be allowed for by retaining the infinite discontinuity in the model but using effective shear strengths, such as those derived by Lajtai (1969).

Once the block structure of the region has been reconstructed, the program examines every block in succession and analyses the stabilities of all those that are bounded by combinations of discontinuities and free faces. This part of the program is based on the procedure derived previously by Warburton (1981). Adjacent blocks are assumed to be fixed at the start of each stability analysis, which begins by finding whether the block is physically free to move. If so, the nature and direction of the attempted movement are determined, associated supporting fixed faces (if any) are identified, and a factor of safety is computed to indicate the likelihood that such movement will be prevented by friction. The block's volume and weight, together with the areas of its faces, are always supplied to the user, and centres of mass could easily be computed too if required.

Additional input needed at the time of the stability analyses includes the weight density of the rock and the various cohesions and angles of friction of the discontinuities. Water pressure could be treated simply as another component of the resultant driving force acting on each block. However, the influence of rock bolts and anchors on stability is much more complicated, as will be seen in the next section.

The program includes facilities for eliminating particular blocks and repeating the stability analyses with the new configuration. The user specifies which blocks are to be eliminated and how many are to be eliminated simultaneously.

A typical result is shown in Figure 4, where the pillar has been reshaped by the elimination of one of the twelve blocks that were found to be potentially unstable in Figure 3. Another option available to the user is to modify the selected region of interest, either to expose some internal section in the original region, or else to investigate a proposed variant or extension of the excavation.

Figures 5 to 10 illustrate the application of the model to open-pit mining. The region selected is shown in Figure 5, though the views in Figures 6 and 7 will probably help to clarify the geometry of the horizontal bench and the slopes, all of which dip at 70°. It can be inferred fairly readily from the traces on the top surface and the exposed vertical surface of the region that its shape is constructed from nine convex components.

An interesting sidelight to the data preparation for this example is that the program itself was used to compute the geometry of the nine components. This was done in preliminary runs by setting up appropriate cuboids and slicing through them along the planes of the bench, the slopes and the two internal vertical sections. Alternatively, the geometry of the components could have been visualized and calculated manually by trigonometry.

Figure 8 shows the block structure produced by five particular discontinuities, which happen, incidentally, to be identical to those in Figure 3. This time the stability analyses found only one potentially unstable block, whose elimination would leave the configuration shown in Figure 9. As a further demonstration of the program, the block reconstructions were repeated with the same discontinuities but without three of the components that represented the original region. The result is displayed in Figure 10, which exposes three internal sections in Figure 8 and shows how the pit would look if the rock below the bench had been excavated too.

RECENT THEORETICAL DEVELOPMENTS

The stability analyses described in the previous section can deal only with single blocks, and adjacent blocks are assumed to be fixed. In reality, however, the stability of a surface block might well be influenced by interactions with other blocks. Fortunately, stress relief is common near excavation boundaries in blocky rock, and the original stress fields are low anyhow, so it is possible that one would have to consider only a few local interactions to obtain all the significant external influences on a particular surface block. The validity of such an assumption is supported by observations made by Hoek (1970), Hammett and Hoek (1981) and Cartney (1977).

Analysis of even two interacting blocks poses formidable problems. Such a system can have various potential instabilities, which might be dealt with in practice by installing rock bolts to provide a specified factor of safety. The analysis should begin by identifying the type of potential instability and indicating whether rock bolts are needed in either or both of the blocks. Sometimes the stability of one block

FIGURES 5 TO 10. OPEN-PIT MINING

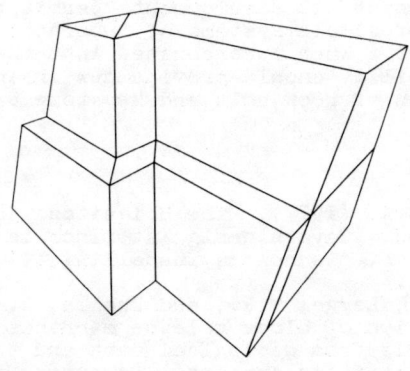

Figure 5. Selected region of interest

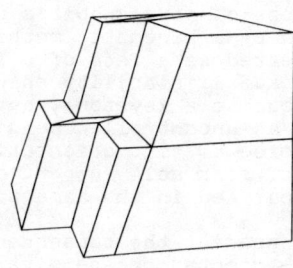

Figure 6. Selected region of interest

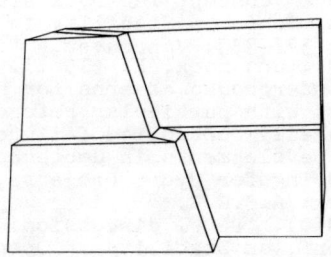

Figure 7. Selected region of interest

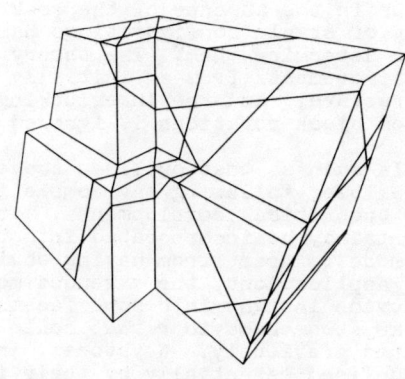

Figure 8. Block structure produced by five dominant discontinuities

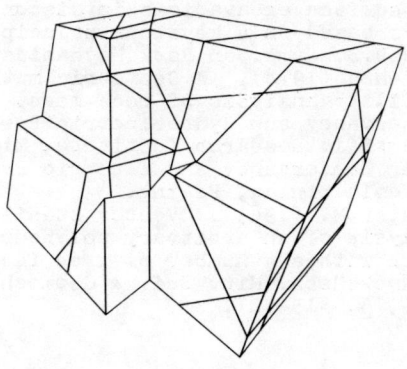

Figure 9. Block structure after elimination of one block

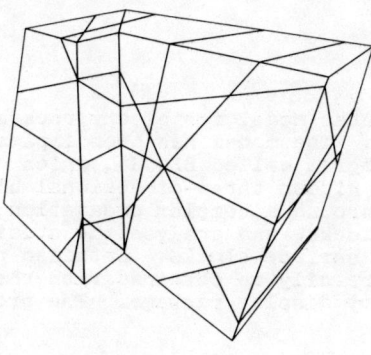

Figure 10. Block structure of modified region

will depend on that of the other, and stabilization of the complete system will be possible by supporting the keystone alone.

Rock bolts introduce new interacting components into the system, and their influence on the blocks is analogous to the action of one block on another. It turns out, for instance, that the effectiveness of a rock bolt's orientation can be analysed by a kinematic method in which the bolt is treated as a face of a fictitious additional block. If stability ensues, then the bolt is analogous to a keystone, but instability is interpreted as uncontrolled bending resulting from a poor choice of bolt orientation. The effectiveness of rock bolt support of a single block can be analysed in the same way.

The ability to predict the consequences of different bolt orientations permits poor choices to be improved until a satisfactory combination is found. Theoretically, there is then enough information to calculate the bolt tensions for any given factor of safety. In order to carry out this calculation, however, the analysis must first determine the movements that would have to be resisted if the blocks began to give way. Since these movements can differ from those that would occur in the absence of the rock bolts, a new theory of stable configurations has been derived to determine them. The theory considers only small movements from an initially close-packed structure, and the interlocking that can result from block rotations is ignored.

An analysis such as that outlined above has now become feasible, following the completion of necessary theoretical developments. It will almost certainly be incorporated into the new computer model. Apart from having obvious practical applications, the extended model should provide insight into some facets of rock bolt and keystone behaviour that could not be investigated previously. Keystones, in particular, are defined essentially by their interactions with their surroundings. A block that tends to be pushed out by its neighbour might cease to be a keystone, and might even attempt to push out its neighbour, if some detail of the nearby blocks or rock bolts were different. Consequently the keystones that occur in systems of interacting blocks and rock bolts are not necessarily the same as the 'keyblocks' that would be identified by Goodman and Shi (1981) and Shi and Goodman (1981).

CONCLUSION

A new computer model for blocky rock has been introduced. The model has been implemented in a FORTRAN program called BLOCKS, which can reconstruct the hidden three-dimensional block structure around a complex excavation, eliminate selected blocks, and analyse the stabilities of individual surface blocks. Detailed perspective views can readily be obtained with the help of an auxiliary display package. The program also enables the user to examine internal sections and to analyse the stabilities of the blocks that would be exposed if the excavation were modified or extended.

Recent theoretical developments permit the analysis of simple systems of interacting blocks and support. When incorporated into the model, this capability should provide new insight into some facets of rock bolt and keystone behaviour.

REFERENCES

Cartney, S.A. (1977). The Ubiquitous joint method. Cavern design at Dinorwic Power Station. Tunnels & Tunnelling (9), 3, 54-57.

Croney, P., Legge, T.F., and Dhalla, A. (1978). Location of block release mechanisms in tunnels from geological data and the design of associated support. Computer Methods in Tunnel Design, 97-119. London: Institution of Civil Engineers.

Cundall, P. (1980). UDEC - a generalized distinct element program for modelling jointed rock. Final Technical Report. European Research Office, U.S. Army, London.

Goodman, R.E., and Shi, Gen-Hua (1981). Geology and rock slope stability - application of a 'keyblock' concept for rock slopes: Proc. 3rd Int. Conf. on Stability in Surface Mining, 347-373, Vancouver.

Hammett, R.D., and Hoek, E. (1981). Design of large underground caverns for hydroelectric projects with particular reference to structurally controlled failure mechanisms. Recent Developments in Geotechnical Engineering for Hydro Projects, 192-206. New York: A.S.C.E.

Hoek, E. (1970). Panel discussion: Proc. 1st Int. Conf. on Stability in Open Pit Mining, 235, Vancouver.

Lajtai, E.Z. (1969). Strength of discontinuous rocks in direct shear. Géotechnique (19), 2, 218-233.

Priest, S.D. (1980). The Use of inclined hemisphere projection methods for the determination of kinematic feasibility, slide direction and volume of rock blocks. Int. J. Rock Mech. Min. Sci. & Geomech. Abstr. (17), 1, 1-23.

Shi, Gen-Hua, and Goodman, R.E. (1981). A New concept for support of underground and surface excavations in discontinuous rocks based on a keystone principle: Proc. 22nd U.S. Symp. on Rock Mechanics, M.I.T.

Shih, Ken-hua (1978). A Geometric method of stability analysis of rock mass. Water Conservancy and Hydroelectric Power Scientific Research Institute, Ministry of Water Conservancy and Electric Power, Academia Sinica, Peking.

Warburton, P.M. (1981). Vector stability analysis of an arbitrary polyhedral rock block with any number of free faces. Int. J. Rock Mech. Min. Sci. & Geomech. Abstr. (18), 5, 415-427.

THREE-DIMENSIONAL ANALYSIS OF NO TENSION MATERIALS
Analyse tridimensionelle des matériaux "no traction"
Dreidimensionale Analyse für Materialien ohne Zugfestigkeit

Manuel Casteleiro
Eugenio Oñate
Antonio Huerba
Jordi Roig
Eduardo Alonso
Escuela Técnica Superior de Ingenieros de Caminos, Canales y Puertos, Universidad
Politécnica de Barcelona

SYNOPSIS

A technique to dissipate tensile stresses in no-tension materials has been developed and applied to the linear elastic analysis of an arch dam to be built in the north of Spain. Convergence proved to be satisfactory, and the technique could easily be extended to non-linear analysis of structures involving this type of material.

RESUME

On a développé une technique pour disperser les tensions de traction dans des matériaux qui ne peuvent pas les admettre, et on l'a appliquée dans l'analyse élastique linéaire d'un barrage-voûte qui sera bâti dans le Nord de l'Espagne. La convergence de la méthode itérative a été satisfaisante.

ZUSAMMENFASSUNG

Es wurde eine Technik zur Umlagerung von Spannungen in Materialien ohne Zugfestigkeit entwickelt und in ein linear-elastisches Modell zur Berechnung einer Bogenstaumauer in Nordspanien eingebaut. Das Konvergenzverhalten war befriedigend, und es wäre leicht, die entwickelte Technik auf nicht-lineare Probleme auszudehnen.

INTRODUCTION

Linear elastic analysis is usually accepted to obtain a global overlook of the behavior of homogeneous materials structures under moderate loads, far away from a failure situation of the structure or of any of its parts.

However, certain materials can no resist tensile stresses or well, the maximum tension than they can withstand is well below the maximum compressive stress. Granular soils, rock, concrete, etc., fall into this class of materials. They can transmit compression loads to other elements of the structure in a linear elastic way.

Tensile stresses which are officially allowed (from Normative of different countries) range from cero for non cohesive granular soils to $5 \times 10^5 - 7 \times 10^5$ N/m^2 for concrete (Bonaldi et. al., 1975). In the case of rock masses, these maximum values vary from $.5 \times 10^5$ to 2×10^5 N/m^2, depending on the rock type and on its degree of cracking and fissuration.

To be able to calculate structures composed in whole or part by these no tension materials, it is necessary, in order to evaluate the global strength and stability under external loads, to consider that these materials can not support tensile stresses farther than a certain threshold, and therefore, the structure should be able to redistribute the excess of stresses.

If these tensions can be "relaxed" to a point which is below the maximum threshold and the resulting compressive stresses are also allowable, the global structure will be stable under the imposed external loads.

Zienkiewicz et. al. (1968) have developed a technique to redistribute tensile stresses by means of the finite element method. This technique is exposed in this paper, slightly modified to be implemented in three-dimensional programs, and applied to the study of arch dams foundations.

METHOD OF SOLUTION

The iterative technique to eliminate part - or the totality - of tensile stresses from "no tension" materials, by means of a finite element analysis, can be resumed as follows:

a) A linear elastic finite element analysis (Zienkiewicz, 1971) is performed, and displacements and principal stresses in each element, $\underline{u}_o^e$ and $\underline{\sigma}_o^e$ respectively, are calculated.

b) If the particular element corresponds to a "no tension" material, the greatest tensile stress throughout the element is compared with the permissible threshold impossed to this material.

c) If the maximum tensile stresses are greater than the corresponding threshold, these stresses are eliminated and, to maintain equilibrium, nodal compression forces are introduced. In particular, if

$$\underline{\sigma}_o^e = \underline{\sigma}_c^e + \underline{\sigma}_t^e \qquad (1)$$

where σ_c^e are principal compressions stresses and σ_t^e principal tensions, the nodal forces F_o that should be imposed to substitute the cancelled tensile stresses are

$$F_o = \int_V B \, \sigma_t^e \, dV \qquad (2)$$

where B is the strain-displacement relationship matrix and integration is extended to the volume of the element.

d) Once nodal forces F_o have been calculated, the problem is once more solved in a linear elastic hypothesis and displacements $\bar{u}_o^e$ and stresses $\bar{\sigma}_o^e$ are obtained. By the principle of superposition, the displacements and stresses after this iteration are, thus

$$\begin{aligned} u_1^e &= u_o^e + \bar{u}_o^e \\ \sigma_1^e &= \sigma_o^e + \bar{\sigma}_o^e \end{aligned} \qquad (3)$$

e) The process is repeated from point b) until the greatest tensile stress in the no tension material is less than the corresponding stress threshold.

This iterative technique can lead to two different situations: if all tensions can be relaxed to a point below the material stress threshold, the structure is stable and the solution obtained provides a lower bound on the stability of the structure (Zienkiewicz et. al. 1968). On the other hand, if it is not possible to transfer all tensile stresses, the particular structure is not stable under this load configuration, and therefore, at least a part of it will fail.

It is worthwhile to repeat that, even if this model can not represent exactly the stress-strain behavior of no tension materials, it allows a reasonable study of the global stability of the structure.

From the operational point of view it is convenient to realize that the corresponding stiffness matrix does not change in the different iterations (if a linear elastic analysis is been performed), and therefore, once the load vector (2) has been calculated, it is only necessary to solve the corresponding system of algebraic equations. It could be, therefore, convenient a partial inversion of the stiffness matrix in the computer analysis, solving the different systems by simple matrix computations.

APPLICATION TO ARCH DAMS ROCK FOUNDATIONS

The technique described above has been applied to the linear elastic analysis of Isil dam, to be built at the Noguera-Pallaresa river in the North of Spain.

This dam is a 70 m. high three centers arch, with spillway in the right buttres, and its finite element discretization is shown in Figure 1.

Foundation rock is a good quality shale, with relatively few fractures. Material characteristics are presented in Table I.

	Young Modulus (N/m^2)	Poisson Ratio
Concrete	3×10^{10}	0.2
Shale	1×10^{10}	0.2

Table I. Material Characteristics

A finite element program with isoparametric 20 nodes (60 d.o.f.) elements has been used to solve the three dimensional problem under linear elastic hypothesis, with a total of 98 elements and 756 nodes.

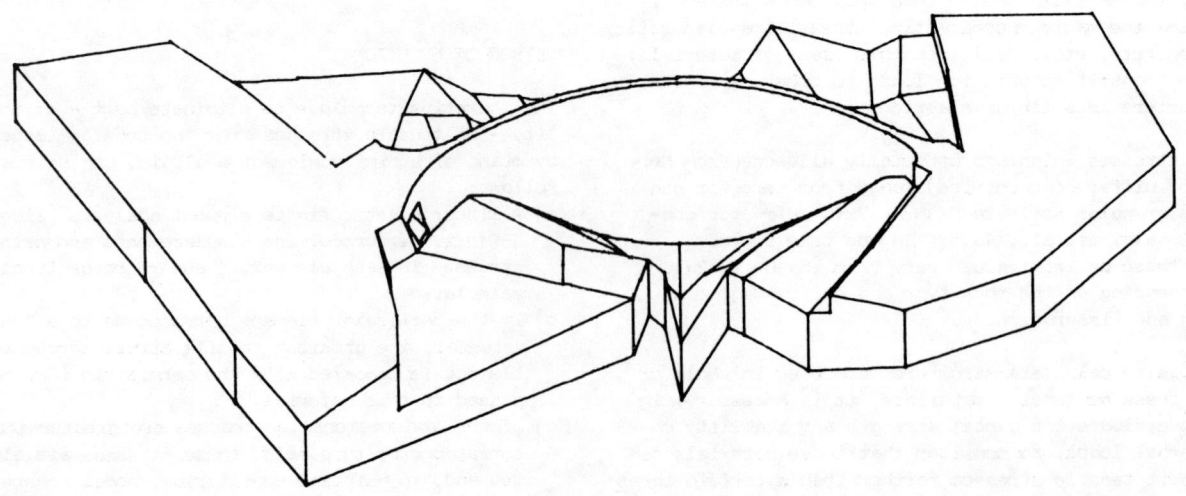

Figure 1. Finite Element Representation of Isil Dam

The calculus has been performed under 21 different load cases, including all the officially contemplated in the Spanish Normative (1967) plus some more originated by extreme conditions. To simplify this presentation, only one load situation will be analyzed. It is considered to be a representative state of solicitations.

This loading case is as follows:
1. Weight of dam (concrete density 2400 Kg/m^3).
2. Water pressure at flood level. The maximum foreseeable elevation is 2.5 m. below dam crest.
3. Sediments (submerged density 1400 Kg/m^3). Maximum deposits height is 10 m. above the base of the dam, where low level outlets are placed.
4. Thermal dilatation. From the assumed end of construction temperature (10º C) a maximum increment of 12º C has been considered on the downstream face. On the upstream face a linear temperature variation has been supposed, from + 12º C of increment at the surface of the water to –6º C of decrement at a depth of 20 m. Below this level, temperature is constant and equal to 4º C. The temperature varies linearly through the depth of the dam, and the thermal dilatation coefficient of concrete is 0.7x10^{-5} ºC^{-1}

Once principal stresses have been calculated in the dam and in the foundation rock mass, it was supposed that the shale behaved as a no tension material, and tensile stresses were relaxed by means of the iterative technique described above.

With only 4 iterations, 98% of rock nodes presented principal tensions of less than 10^5 N/m^2, and the maximum tensile stress in the foundation was 1.52x10^5 N/m^2.

Table II shows the decrement of principal tensions in the foundation rock nodes with the greatest initial tensile stresses.

Principal Tension Elastic Solution x10^5 N/m^2	Principal Tension After 4 Iterations x10^5 N/m^2
2.93	0.85
5.85	0.51
7.38	0.41
3.11	0.31
5.03	0.84
3.40	0.72
4.45	1.14
3.31	0.71
4.37	0.49
3.33	0.82

Table II. Tension Relaxation at Rock Mass Nodes

Figure 2 represents the sum of all principal tensions in the rock mass as a function of the number of iterations performed. This graphic gives an idea of the speed of convergence of the iterative technique, but does not imply that a stable solution is achieved at each point of the rock. Convergence must be studied at each node in order to assure the impossibility of a local failure.

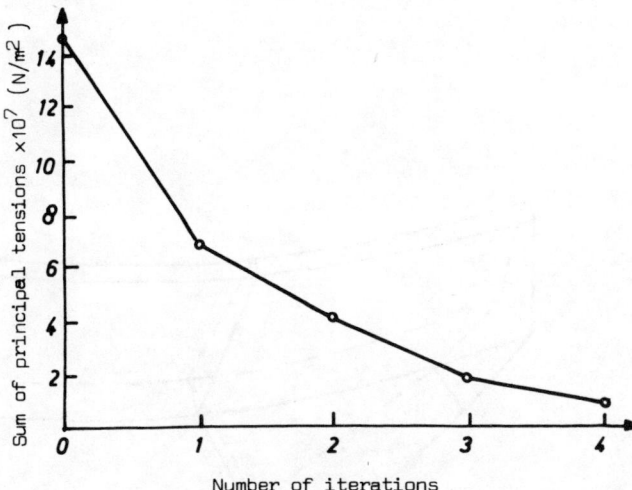

Figure 2. Convergence of the Iterative Process

It is obvious that this relaxation technique must generate a variation in the dam stress field. Tables III and IV show compression and tension changes in some representative nodes of the dam, which are situated in the arch finite element discretization represented in Figure 3. It is clear from the data that in this particular structure, increments of stress generated by the relaxation of tensions throughout the rock mass is not important and can be supported by the concrete.

Node	Maximum Compression Elastic Solution x10^5 N/m^2	Maximum Compression After 4 Iterations x10^5 N/m^2
28	16.48	16.72
84	28.39	29.27
18	20.00	20.10
38	19.16	19.21

Table III. Compression Changes in the Dam due to Tension Relaxation in the Rock

Node	Maximum Tension Elastic Solution x10^5 N/m^2	Maximum Tension After 4 Iterations x10^5 N/m^2
26	7.33	0.41
28	0.56	- -
83	6.75	7.14
18	0.43	0.86
38	1.03	0.84

Table IV. Tension Changes in the Dam due to Tension Relaxation in the Rock

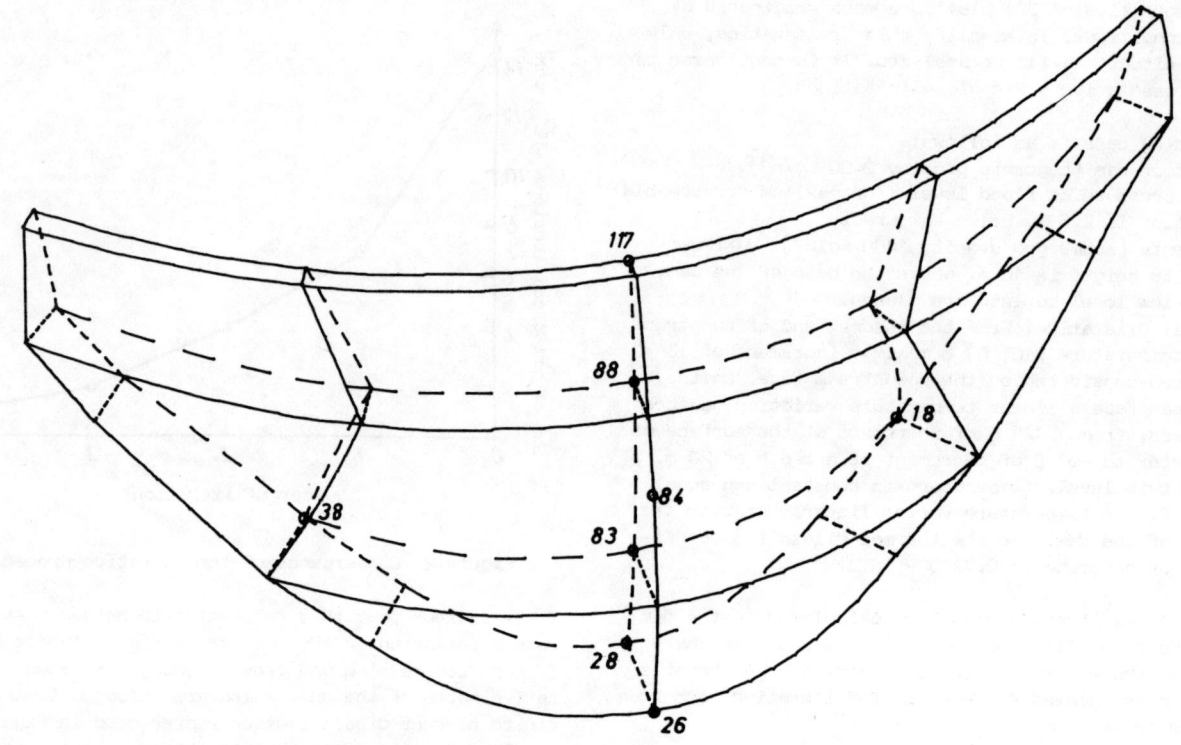

Figure 3. Situation of Nodes Referred to in Tables III, IV and V

Displacement changes in the nodes with maximum movements in the linear elastic solution are represented in Table V (x horizontal, positive downstream. z vertical, positive upward). Variations are again very small and never greater than 0.4×10^{-3} m. in x direction and 0.06×10^{-3} m. in z direction.

Node	Maximum Displacement Elastic Solution $\times 10^{-3}$ m	Maximum Displacement After 4 Iterations $\times 10^{-3}$ m
117	Ux = 4.72 Uz = -0.88	Ux = 4.72 Uz = -0.85
88	Ux = 6.08 Uz = -1.18	Ux = 6.15 Uz = -1.15
83	Ux = 4.04 Uz = -1.73	Ux = 4.44 Uz = -1.67
28	Ux = 0.67 Uz = -1.21	Ux = 0.99 Uz = -1.20

Table V. Displacement Changes due to Tension Relaxation in the Foundation Rock

CONCLUDING REMARKS

The program was impletented on a Digital's Vax 780/11 with 2.5 Mbytes of central memory. Typical run time for elastic solution was 8 to 10 minutes, and the iterative solution time, up to five iterations, was always less than three times the elastic solution time.

CONCLUSIONS

An iterative process to dissipate tensile stresses in no tension materials has been developed. This technique has been applied to the elastic analysis of arch dam foundations.

Consideration of the results leads to the following conclusions:
1. The iterative technique described throughout this paper represents a powerful and economic way to study the relaxation of tensile stresses in no tension materials and the resulting global stability of the structure.
2. In the particular case studied, the convergence of the solution was satisfactory, and the computer time necesary for convergence was never superior to three times the elastic solution time.
3. The method can be easily extended to non linear studies, as well as to other problems outside the scope

of rock mechanics (earth dams, earth supported structures concrete, etc.)

AKNOWLEDGEMENTS

The work described in this paper was supported in part by Hidroeléctrica de Cataluña, S.A. and by the Institut Català d'Enginyería Civil. The authors would like to acknowledge the assistance received by the Computer Centers of the Universidad Politécnica of Barcelona and of Universidad Autónoma of Barcelona.

REFERENCES

Bonaldi, P., di Monaco, A., Fanelli, M., Giusepetti, G. and Ricioni, R. (1975). Concrete Dam Problems: An outline of the role, potential and limitations of numerical analysis, in Criteria and Assumptions for Numerical Analysis of Dams. University College of Swansea, U.K.

Ministerio de Obras Públicas (1967). Instruccion para el Proyecto, Construcción y Explotación de Grandes Presas. Madrid, Spain

Zienkiewicz, O. C. (1971). The Finite Element Method in Engineering Science. 521 pp. McGraw-Hill, London, U.K.

Zienkiewicz, O. C., Valliappan, S. and King, I. P. (1968). Stress Analysis of Rock as a No tension Material. Geotechnique, 18, 56–66

ELEMENT METHODS IN PLANNING OF MINE OPENINGS IN HIGHLY STRESSED PRECAMBRIAN BEDROCK

Die Element-Methoden in der Planung von Grubenhohlraumen in stark beanspruchten präkambrischen Gesteinen

Methodes a elements dans la planification d'excavations minieres en roches precambriennes soumises a de fortes contraintes

P. Lover
Research Engineer, Helsinki University of Technology, Finland

J. Oksanen
Research Engineer, Helsinki University of Technology, Finland

K. Äikäs
Research Geologist, Helsinki University of Technology, Finland

SYNOPSIS

Element methods (FEM and BEM) have been used to prepare a stoping plan for the Kotalahti nickel mine. The rock mass between levels +600 and +800 is of poor to fair quality. Maximum horizontal stress is about 50 MPa acting in an unfavourable direction and the vertical stress corresponds to the overburden pressure. The mining method applied is sublevel stoping with preplaced cable bolting. Element methods have been utilized in the dimensioning of the crown pillar and the planning of preplaced cable bolting. A monitoring system has ben installed to measure deformations and changes of stresses. Stoping has recently been started. Final feedback is expected in three years when the stoping area will have been exhausted.

RESUME

Des méthodes à éléments F.E.M. et B.E.M. ont été utilisées pour la planification d'abattage à la mine de nickel de Katalahti. La qualité de la masse rocheuse entre les niveaux +600 et +800 mètres varie de mauvaise à moyenne. La contrainte horizontale maximum est environ 50 MPa et en direction défavorable, et la contrainte verticale est de la même grandeur que celle du terrain de couverture. La méthode d'exploitation employée est celle d'abattage par gradins avec boulonnage de câbles pré-positionnés. Des méthodes à éléments ont été utilisés pour le dimensionnement des piliers en couronne et le boulonnage des câbles. On a installé un système de contrôle pour mesurer les déformations et les changements de contrainte, et on vient de commencer l'abattage. On compte disposer des résultats définitifs dans trois ans lorsque la zone d'abattage sera défruitée.

ZUSAMMENFASSUNG

Die Element Methoden (FEM und GEM) wurden zur Bearbeitung eines Abbauplanes für die Nickelgrube Kotalahti angewandt. Die Gebirge zwischen den +600- und +800-Sohlen sind von schwacher bis guter Qualität. Die höchste Horizontalspannung in ungünstiger Richtung liegt bei etwa 50 MPa. Die Vertikalspannung ist gleich mit dem Deckgebirgsdruck. Als Abbaumethode wurde der Weitungsbau mit voreingezogener Kabelankerung angewandt. Die Element-Methoden wurden bei der Dimensionierung des Firstenpfeilers und bei der Planung der voreingezogenen Kabelankerung ausgenützt. Zur Messung von Deformationen und Spannungsänderungen wurde ein Oberwachungssystem installiert. Mit dem Abbau wurde vor kurzem begonnen. Endgültige Resultate sind nach erfolgtem Abbau des Gebietes innerhalb von drei Jahren zu erwarten.

1. INTRODUCTION

Numerical methods, mostly the finite-element method, have been occasionally used in the Finnish mining industry for ten years. A two and a half years project for applying element calculation technology to practical mine planning was started on the initiation of the Finnish mining industry in June 1980. The project has been carried out in close cooperation between the Finnish mining industry and different state research institutes. The total budget was 300´000 US$ of which one half was financed by the Ministry for Commerce and Industry and the rest by the mining industry.

The practical problems of the Rautuvaara iron mine of Rautaruukki Oy and the Kotalahti nickel mine of Outokumpu Oy formed the base material for this project.

At present the Kotalahti mine extends down to a depth of 800 m. Annual production has been about 500´000 tons of ore. The Huuhtijärvi orebody is the largest one of the five orebodies. The ore is stoped mainly by sublevel stoping. Stoping between levels +600 and +800 has been planned during the project. This paper describes how element calculations have been applied to the stoping plan of the Huuhtijärvi orebody of Kotalahti mine. The determination of input data for element calculations is described, too.

2. GEOLOGY

The Kotalahti mine is situated in central Finland. The area is built up by gneisses, schists and plutonic rocks, metamorphosed under the conditions of high grade metamorphism. Five deformation phases, including polyphase folding, intrusion tectonics and shear tectonics, have been distinguished by structural analysis (Gaal, 1980).

The host rock of the Kotalahti nickel ore is an ultramafic - mafic intrusion. It is located in a brachysynform structure close to a granite gneiss dome. The ore is built up by five orebodies. One of them, the Huuhtijärvi orebody, is the project test area.

The Huuhtijärvi orebody is a pipelike formation (fig.1) dipping $80°$ NNW and reaching at least to the depth of one km. It has intruded along a subvertical fold axis and in horizontal section it forms a lens-like body (fig.2). Surrounding rocks are mica gneisses and amphibolites.

3. ROCK MECHANICAL CONDITIONS

3.1. Stress

The rocks of the Kotalahti area are under high horizontal compression. The direction of the major horizontal stress σ_{H1} is usually N-NNW and intensity some 50 MPa. σ_{H2} is 25 MPa and vertical stress corresponds to the overburden pressure. At Huuhtijärvi, +800 level, the stress field has turned some $50°$ clockwise. In western part of the Huuhtijärvi mafic formation the major stress is thus acting perpendicular to the long walls of the planned stopes (fig.2).

3.2. Jointing

In Huuhtijärvi the following joint maximums with steep dips are found: WNW and NE (the directions of the fold limbs), NNW (the direction of the axial plane). In addition there exists a subhorizontal maximum and some random jointing.

In the mafic formation the average joint frequence, counted from diamond drill cores, is ten and in the surrounding mica gneisses and amphibolites six. Joint fillings with weak residual shear strength (chlorite, talc and tremolite) are common in the host rock. In the surrounding rocks joint friction is higher. The host rock contacts are somewhat fractured at the SW and SE edges.

3.3. Rock mass parameters

The following rock mass parameters were determined: rock quality designation index (RQD), rock mass quality (Q) and dynamic rock mass modulus of deformation (E_d). RQD and Q are distinctly lower at the west end and SE contact of the mafic formation (fig.2) than elsewhere, where the rock quality is fair.

Dynamic rock mass modulus of deformation (E_d) was determined by hammer seismograph in sublevel and bolting drifts. The phase velocities of the compressional (P) wave and shear (S) wave were measured in the rock mass and E_d was calculated from:

$$E_d = \frac{D \, v_s^2 \left[3(\frac{v_p}{v_s}) - 4\right]}{(\frac{v_p}{v_s})^2 - 1} \text{ GPa} \quad (1)$$

where D = rock mass density
v_p = P-wave velocity
v_s = S-wave velocity

The following E_d values were calculated: mafic-ultramafic formation 100 GPa, amphibolite 80 GPa and mica gneiss 50 GPa.

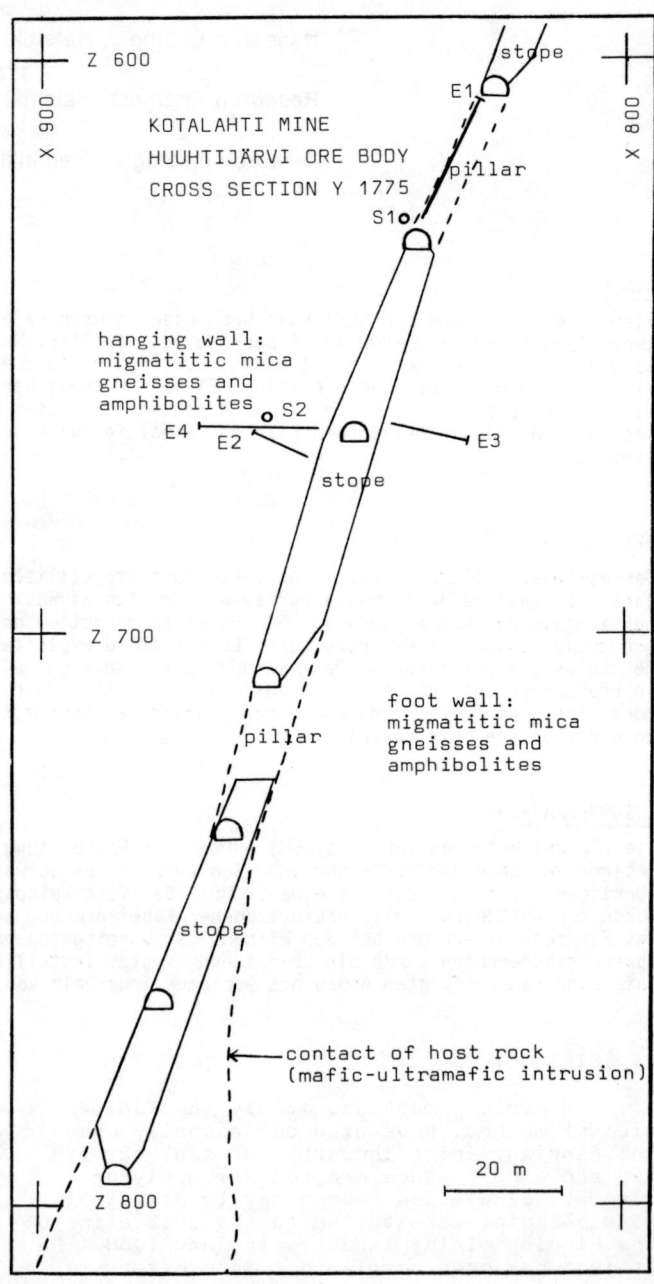

Fig.1. Cross section of the Huuhtijärvi mafic-formation with planned stopes. E represents extensometer and S represents stress-cell.

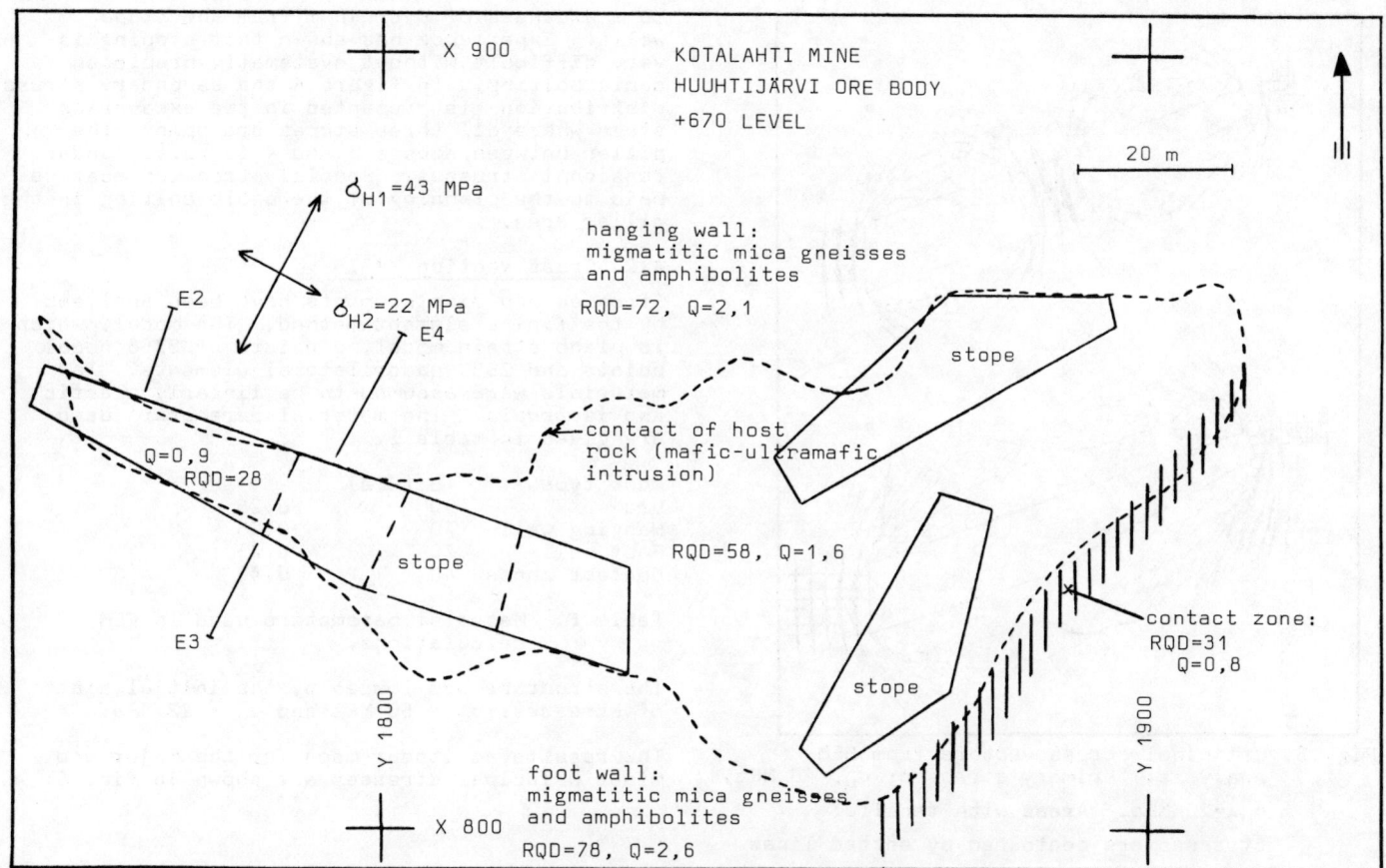

Fig.2. Horizontal section of the Huuhtijärvi mafic-ultramafic formation with planned stopes. E represents extensometer.

4. ELEMENT CALCULATIONS

To get information about the distribution of secondary stresses in the surroundings of the Huuhtijärvi stoping area, stress analysis have been carried out by using finite-element (FEM) and boundary element (BEM) methods. All numerical models were linearly elastic.

The boundary element program used is based on indirect formulation and the elements are of constant type. It can be used to calculate stresses and displacements for the following conditions: 1) the material is homogeneous, isotropic and linearly elastic; 2) the medium is infinite or closed by a finite external boundary; 3) the medium may contain a number of holes of arbitrary shape; 4) the conditions are those of plane strain; 5) the loading may consist of any combination of uniform field stresses or uniformly distributed loads on the boundaries.

The FEM calculations were made with a program using the two-dimensional finite-element method. The materials are assumed to be linearly elastic, though the elasticity parameters may be different in orthogonal directions. The loading may consist of gravity loads, pressure loads and the initial state of stress. The basic element is isoparametric quadrilateral. Two types of joint elements and a bar element can be used, too.

4.1. Horizontal section +670

The stress analysis were carried out by boundary element method. The model contains 65 to 100 boundary elements depending on the excavation stage. The modulus of elasticity (E) was 60GPa and Poisson's ratio (ν) 0.3. The structure was loaded by field stresses σ_{H1}=40 MPa and σ_{H2}=20 MPa. The direction of σ_{H1} is perpendicular to the long wall of the orebody.

Element calculations give the stresses at the center point of each boundary element and in the specified interior points. In figures 3 and 4 the directions of the secondary principal stresses are plotted, lengths indicating the magnitude of the stresses. The square in the centers of the stress vectors indicate the shear stress exceeding 50 MPa.

Two alternatives to stope the ore from the stope 1 are presented in figure 3. In the first case (fig. 3a) the tension zones are quite small but according to the calculations the average shear stress in the pillar is 37 MPa and the average strength/stress-ratio is about 0.9. The pillar contains about 50 000 tons of nickel ore. The later stoping of the pillar may be difficult because of the fragmentation. The secondary stress distribution in the case where the whole stope is open is presented in fig. 3b. Areas with tensional stresses extend

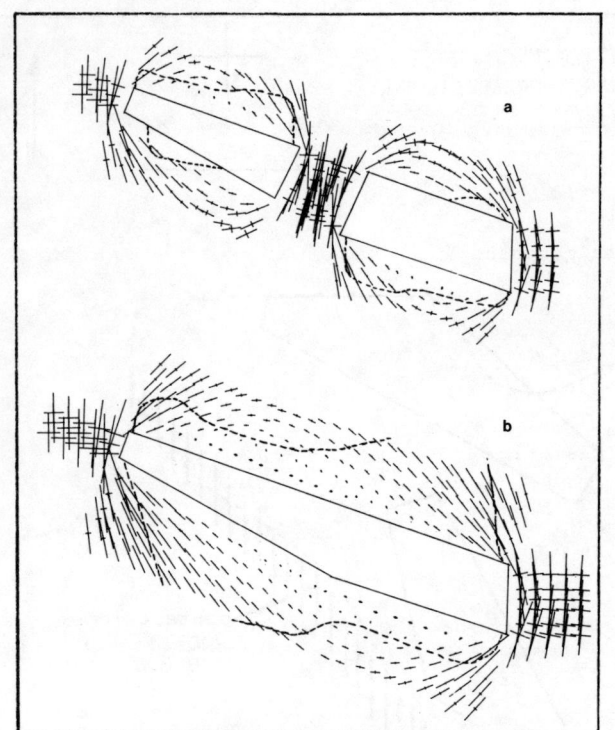

Fig. 3. Principal stress vectors from BEM analysis. Primary stresses: σ_{H1}=40 MPa, σ_{H2}=20 MPa. Areas with tensile stresses are contoured by dotted lines.

to a wideness of over 18 m from the stope walls. Experience has shown that stoping is very difficult without systematic preplaced cable bolting. In figure 4 the secondary stress distribution is presented in the excavation stage where all three stopes are open. The pillar between stopes 3 and 4 is partly under tensional stresses. Special attention must be paid to the planning of the cable bolting in the pillar area.

4.2. Cross section Y1775

Stresses and displacements have been analysed by the finite-element method. The model, which is plane strain model, consists of 2750 nodal points and 2652 quadrilateral elements. The materials were assumed to be linearly elastic and isotropic. The material parameters used are given in table I.

Rock type	E (GPa)	ν
Ore	90	0.25
Hanging wall	70	0.25
Foot wall	70	0.25
Contact zones	40	0.40

Table I. Material parameters used in FEM calculations.

The structure was loaded by the initial state of stresses: σ_H = 50 MPa and σ_V = 22 MPa.

The results as isobar-maps for the major and minor principal stresses are shown in fig. 5.

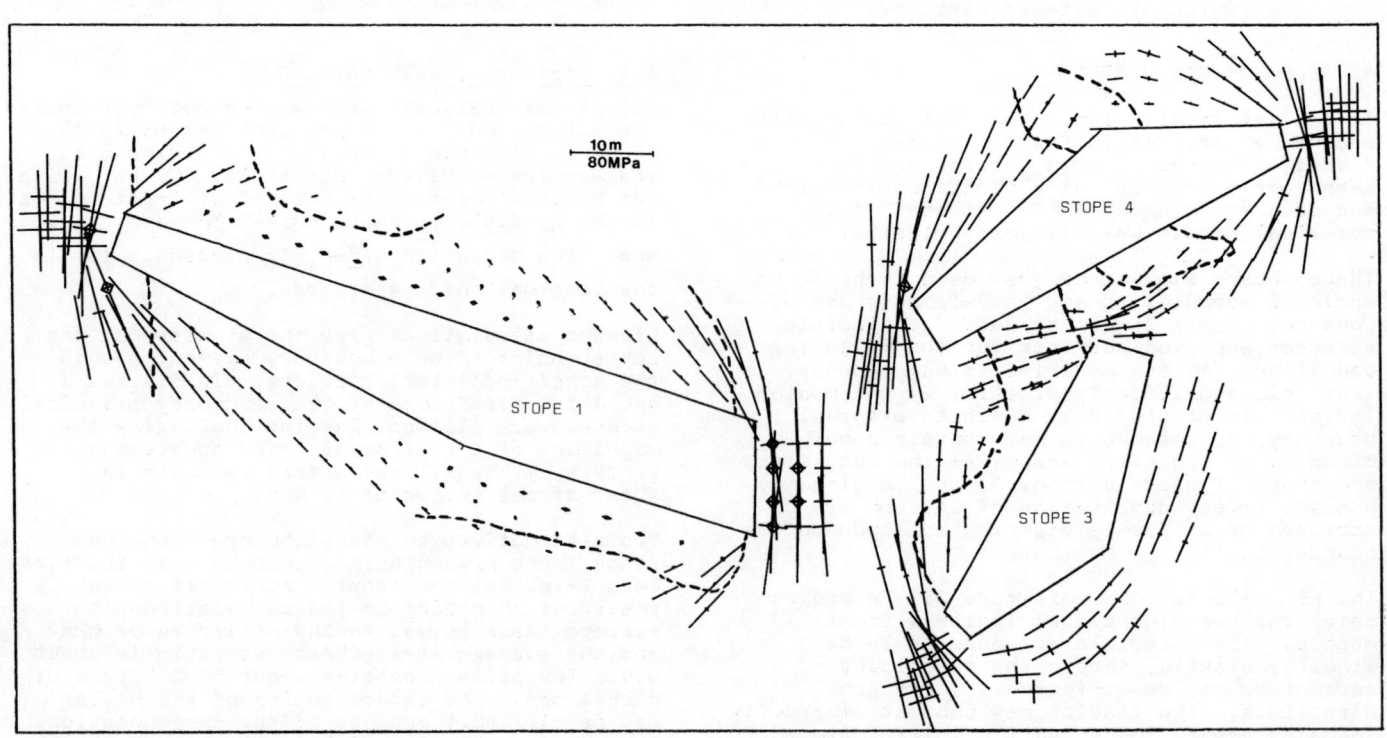

Fig. 4. Principal stress vectors from BEM analysis. Primary stresses: σ_{H1}=40 MPa, σ_{H2}=20 MPa. tensile stresses are contoured by dotted lines.

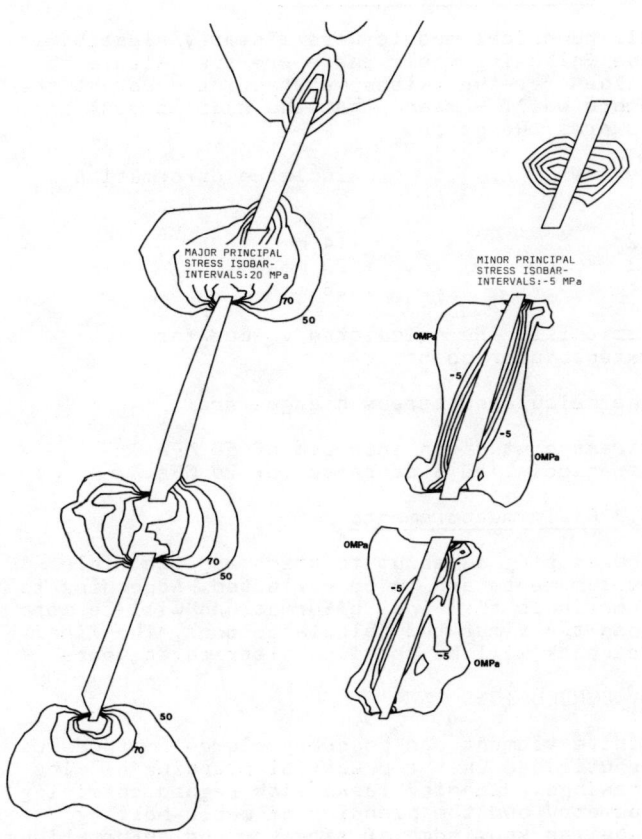

Fig. 5. Results from the FEM calculation. The distribution of major and minor principal stresses as isobar-maps.

4.3. Pre-reinforcement by cable bolting

Experience has shown that it is possible to use open stoping methods instead of the expensive cut and fill method in poor rock/stress conditions by pre-reinforcing stope walls, roof and pillars by cable bolting.

Preplaced cable bolting is done before stoping. That is why tensioning of fully grouted cables is not needed. Under high horizontal compression rock itself will stress the cables and makes cable bolts active after stoping has started.

In sublevel stoping with preplaced cable bolting bolt spacing $1/25$ m^2 ... $1/50$ m^2 will generally be enough when using fully grouted steel strands (2x⌀ 15.2 mm/hole) or steel cables (⌀25 ... ⌀35 mm). Cable bolt lengths vary from 15 m to 50 m and the are in fan shape (Lappalainen and Pulkkinen, 1982).

Scientific design in preplased cable bolting is difficult because of too many unknown factors. Element methods can still be utilized successfully. The most critical tension and shear zones are to be bolted and bolts are to be directed perpendicularly to the highest compression (Lappalainen and Pulkkinen, 1982).

The secondary principal stresses from FEM calculation and corresponding cable bolting plan for The Huuhtijärvi orebody is shown in fig. 6. Areas with tensional stresses are contoured by dotted lines. Shear stresses exceeding 50 MPa are marked by squares in the centroids of elements. The spacing of bolts is about $1/25$ m^2 on the hanging wall side.

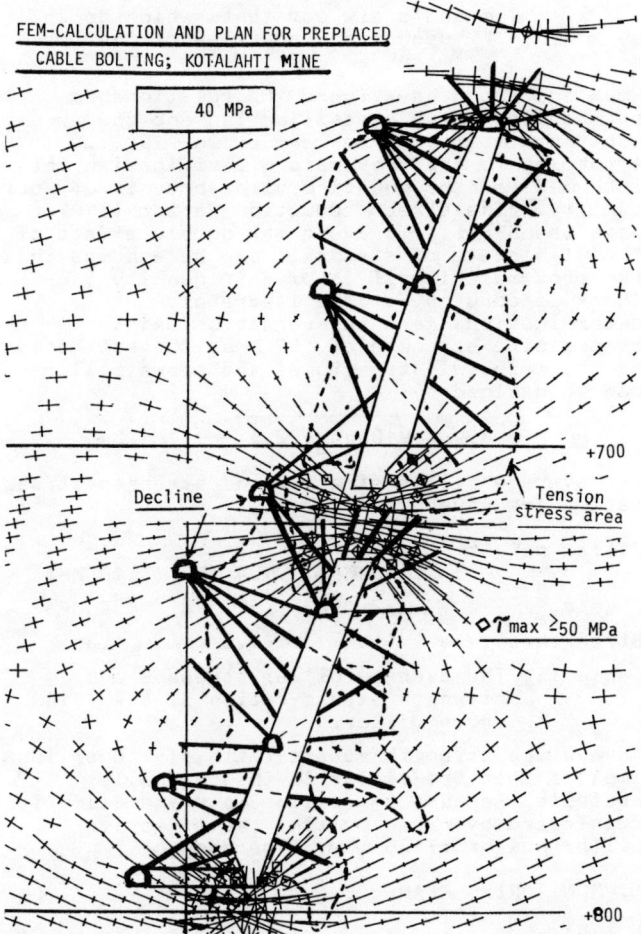

Fig. 6. FEM calculation and cable bolting plan, Kotalahti mine, Huuhtijärvi orebody. Primary stresses: σ_H=50 MPa, σ_V=22 MPa.
Areas with tensional stresses are contoured by dotted lines.

4.4. The dimensioning of the +710 crown pillar

The stability analysis of the +710 crown pillar is based on the empirical failure criterion by Hoek and Brown (Hoek and Brown, 1980). The strength/stress-ratio was determined for three different pillar heights (15m, 20m, 25m). Pillar width is 10m. Stress calculations were made by boundary element method. The material was assumed to be isotropic, homogeneous and linearly elastic. The modulus of elasticity was 60 GPa and the field stresses: σ_H=50 MPa and σ_V=22 MPa.

The triaxial strength of rock mass can be defined by equation 2 (Hoek and Brown, 1980).

$$\sigma_1 = \sigma_3 + \sqrt{m \times \sigma_c \times \sigma_3 + s \times \sigma_c^2} \quad (2)$$

where
- σ_1 is the major principal stress
- σ_3 is the minor principal stress
- σ_c is the uniaxial compressive strength of the intact rock material
- m and s are constants which depend upon the properties of the rock mass

Hoek and Brown have found the relationship between rock mass classification and the constants m and s (Hoek and Brown, 1980). According to the rock mass classification the rock mass of the Huuhtijärvi orebody is of poor quality. The stress reduction factor (SRF) used was SRF=5. To avoid the double effect of the high state of stress it can be assumed that the rock mass itself is of fair quality (Q-index is about 5.). The laboratory determinations gave an average uniaxial compressive strength of 150 MPa. The estimation of the triaxial strength of the crown pillar can be defined by:

$$\sigma_1 = \sigma_3 + \sqrt{188 \times \sigma_3 + 22.5} \quad (3)$$

The average pillar stresses and strength/stress ratios are given in table II.

Pillar height	15m	20m	25m
σ_{1a}	160 MPa	134 MPa	118 MPa
σ_{3a}	59 MPa	59 MPa	54 MPa
Strength/stress	1.0	1.2	1.3

Table II. The average pillar stresses and strength/stress-ratios of the +710 crown pillar.

An average strength/stress-ratio of 1.0 or less implies that the pillar is theoretically unstable. Because the pillar is systematically reinforced by cable bolting the effective pillar height of 20 m will be enough.

5. MONITORING PLAN

A monitoring plan was made to measure the real action of the rock mass during the excavation. The selection of the installation places was done after the calculations and interesting places were chosen. In the figures 1 and 2 the measuring places are roughly marked. It should be noted, that all the extensometers are not exactly situated in that cross section. The most critical places are the upper pillar and the hanging wall.

5.1 Measurement devices

The most used monitoring devices in Finland are the normal rod-extensometers to measure deformations and the stress-cells to measure stress changes. A convergense tape is now for the first time used in Finland, it will be tested in the tunnels in pillars. The list of the measuring devices is as follows:
- 2 three-point rod-extensometers
- 2 two-point rod-extensometers
- 2 stress-cells
- 1 convergense tape

5.2 Calculated deformations

All numerical models were linearly elastic. In the following table there are the calculated values for the extensometer points nearest the stope wall. + means that the displacement is towards the stope.

device	calculated deformation
E1	- 3 mm
E2	+ 14 mm
E3	+ 29 mm
E4	+ 26 mm

Table III. The calculated values for extensometer points.

The calculated stress changes are:

stresspoint σ1 an increase of 50 MPa
stresspoint σ2 a decrease of 20 MPa

5.3 Field measurements

The stoping is about to start and the field measurements are to be collected. According to experience the final deformations will be more than two times the calculated ones. The final feedback will be received after three years.

6. CONCLUSIONS

Finite-element and boundary element methods can be utilized in the practical planning of mine openings. Planning tasks with regard to mining geometry and the planning of cable bolting requires knowledge of stresses and deformations over a wide area. This information can, with sufficient accuracy be obtained with an elastic model.

ACKNOWLEDGEMENTS

The personnel of Outokumpu Oy, especially rock mechanics engineer, Mr. Pekka Lappalainen, MSc is acknowledged his co-operation in the preparation of this paper.

We are indebted to Mr. A. Öhberg, MSc for correcting the English text.

REFERENCES

Gáal, G.(1980) Geological setting and intrusion of the Kotalahti nickel-copper deposit, Finland. Bull. Geol. Soc. Finland 52 pp. 101-128.

Hoek, E., Brown, E.T.(1980) Underground Excavations in Rock. 527 pp. London: The Institute of Mining and Metallurgy.

Lappalainen, P., Pulkkinen, J.(1982) Pre-reinforcement by cable bolting at Outokumpu Oy mines. manuscript 15 pp.

GEGENWÄRTIGER STAND UND TENDENZEN IN DER ENTWICKLUNG DER GESTEINSMECHANIK IN DER TSCHECHOSLOWAKEI
The present state and tendencies in the development of rock mechanics in Czechoslovakia
L'état actuel et tendance du développement futur de la mécanique des roches en Tchécoslovaquie

Lubomir Siska
Prof.Ing.Dr.Sc. Mitglied-Korrespondent, Bergbauinstitut Akademie der Wissenschaften, CSSR, Direktor

ZUSAMMENFASSUNG

Die Bergbaugeomechanik in der ČSSR löst Aufgaben, die mit Bewertung der physikalischen und technischen Eigenschaften der Gesteine, der Gebirgsmassive und der Spannungszustände im Massiv verbunden sind und durch die Bergwerktätigkeit entstehen. In dem Beitrag wird der zeitgenössische Zustand und die Aussicht beschrieben und die Realisation an Beispielen aus dem Tage- und Untertagekohlenbergbau gezeigt.

SYNOPSIS

Mining geomechanics in ČSSR is solving tasks connected with the evaluation of physical and technical properties of rocks, rockmassif and stresses in the massif, originating in the process of mining activities. The present state, perspectives and realization are demonstrated in the report by examples from the open-pit and underground mining.

RESUME

La géomécanique minière en Tchécoslovaquie s'occupe de la solution des problèmes des propriétés physiques et techniques des roches dans les massifs rocheux ainsi que des tensions dans les massifs dues à l'activité minière. Ce compte rendu présente la situation actuelle et la perspective future et donne des exemples de l'exploitation de la houille en galeries et à ciel ouvert.

1. EINLEITUNG

Die Entwicklung und Zielsetzung der Bergbaugeomechanik in der Tschechoslowakischen Sozialistischen Republik wird durch immer tieferes Abteufen und den damit zusammenhängenden Druck- und Verformungserscheinungen in den Grubenbauen auf dem Gebiet des Lagerstättentiefbaus bestimmt, auf dem des Tagebaus durch die Forderung nach Sicherung der Stabilität der Hänge in tiefen Tagebauen, hohen Abraumkippen und Korridoren mit Ingenieursnetzen.

In beiden Fällen taucht auch das Problem des effektiven Lostrennens der Kohle und der Gesteine verschiedenartiger und hoher Festigkeit durch Bohren und Schneiden, in Zukunft auch auf nichtklassische Weisen auf.

Im allgemeinen betrifft daher die Forschung auf dem Gebiet der Bergbaugesteinsmechanik in der Tschechoslowakei vordringlich folgende Problemkreise:
1. Untersuchung und Bewertung der natürlichen Verhältnisse, insbesondere der physikalischen und mechanischen Eigenschaften der Gesteine und der technologischen Eigenschaften (Trennbarkeit, Abrasionsfähigkeit u.a.).
2. Untersuchung der Spannungs- und Verformungszustände in dem durch die bergmännische Tätigkeit beeinflussten Massiv, Methoden der Voraussagung und Messungen in situ. Besondere Aufmerksamkeit wird in diesem Zusammenhang Anomalitätserscheinungen der Gebirgsdrücke, -erschütterungen und Kohlen- und Gasausbrüchen gewidmet.

2. BEWERTUNG DER NATÜRLICHEN VERHÄLTNISSE

Die Bewertung der naturgegebenen Verhältnisse geht von der geologischen Beurteilung des Milieus aus, worauf unmittelbar die geomechanische Bewertung folgt.

Die Vollständige Bestimmung der physikalischen, mechanischen und technologischen Eigenschaften der Gesteine erfordert Messungen von über hundert Parametern für jedes einzelne Gestein (Probestück), was praktisch nicht zu bewältigen ist. (Konečný, 1977).

Aus diesem Grunde wurde in der Tschechoslowakei das System von drei Bewertungskategorien entwickelt: 1. das komplexe, 2. das Grundsystem und 3. das Betriebssystem. Dieses System zieht die technischen und ökonomischen Möglichkeiten des Bergbaus in Betracht und entspricht zugleich den Anforderungen der Entwicklung von Theorie und Praxis (Šiška, Konečný, 1982). Die Komplexbewertung umfasst die breiteste Messungsskala und wird nur an den Repräsentanten einzelner Gesteine vorgenommen, in der Regel unter Heranziehung wissenschaftlicher

Forschungsstellen. Die Grundbewertung repräsentiert zumeist fast ausschliesslich Labormessungen im durch die Foschung gegebenen Niveau und auch durch die Forderungen der zu lösenden konkreten Aufgaben. Sie dienen als Stützpunkte für die nachfolgenden in breitestem Ausmass realisierten Betriebsmessungen. Mit diesen werden nur die Grundeigenschaften (Festigkeit, Verformungsfähigkeit, technologische Eigenschaften) unter Zuhilfenahme rascher und einfacher Methoden bestimmt (Konečný, 1978). Diese Festigkeitsprüfungen werden in der Regel schon in den Grubenbetrieben selbst vorgenommen.

Das angeführte Bewertungssystem der naturegebenen Verhältnisse ermöglicht die Bildung eines geomechanischen Milieumodells, an das die weitere praktische und Forschungstätigkeit anknüpft.

Die Untersuchung der Methoden und Methodiken zur Messung der physikalischen, mechanischen und technologischen Eigenschaften zielt auf dem Gebiet der physikalischen und mechanischen Eigenschaften auf die Verbesserung und Vervollkommnung der bestehenden Metnoden (z.B. Messungen des Prozesses der spröden Gesteinszerstörung), die Einführung neuer Methoden in der Bergbaugeomechanik (z.B. die Resonanzmethoden zur Messung der elastischen Parameter). Auf dem Gebiet der technologischen Eigenschaften wird die Forschung durch die Forderungen der technologischen Prozesse hervorgerufen, insbesondere durch das Loslösen.

Im Rahmen der angeführten Forschungsproblematik wurde beispielsweise im Institut "Banícky ústav ČSAV" systematisch der Verfolgung der Funktionsbindungen zwischen den Input- und Outputparametern mit den Grössen des Prozesses des mechanischen Loslösens der Gesteine, die gesetzmässigen Charakter tragen, gewidmet.

Unter Zuhilfenahme einer speziellen Forschungseinrichtung wurden die energetischen Transformationsprozesse beim Bohren und Mahlen der Gesteine verfolgt. Es wurde eine lineare Abhängigkeit zwischen der verbrauchten Arbeit und der irreversibel akkumulierten Energie im Gestein und im Instrument während des Bohrprozesses festgestellt. Die im Mahlprozess angeführten Grössen weisen einetangenshyperbolische Abhängigkeit auf (Sekula, Kupka, 1970, Tkáčová, Sekula, 1970).

Aufgrund der Erkenntnisse aus den energetischen Transformationsprozessen beim Loslösen der Gesteine durch Rotationsbohren, sowie auch mit Hilfe direkter Experimentalmethoden wurde die Funktionsbindung zwischen der Intensität der Abnutzung der Arbeitselemente des Instruments und der spezifischen Umfangsarbeit des Loslösens nachgewiesen (Sekula, Bejda, Dunay, 1975).

Es wurde ein Indikator konstruiert und gefertigt, der es ermöglicht, aufgrund der Aufnahme der Inputgrundgrössen die spezifische Umfangsarbeit des Loslösens der Gesteine beim Bohrprozess zu bewerten und zu registrieren. Die Einrichtung wurde erfolgreich unter Betriebsbedingungen beim Bohren von 76-mm-Bohrlöchern in einer Teufe bis zu 100 m erprobt. Die Anwendungsmöglichkeiten sind jedoch breiter, besonders beim Bohren von Grossprofillöchern und Tunnelmaschinen.

Mit Hilfe der experimentiellen Standuntersuchung wurde die Funktionsabhängigkeit der Bohrgeschwindigkeit, des Torsionsmoments und der spezifischen Umfangsarbeit in Abhängigkeit von den veränderlichen Inputgrössen (Andruck und Umdrehungen) und den Parametern (Gestein und Instrumentengeometrie) bestimmt. Die Funktionen stellen die Lösung von partiellen Diffenrentialgleichungen mit konstanten Koeffizienten dar (Sekula, Bejda, 1977).

Diese mathematischen Modelle ermöglichen es, aufgrund einfacher technischer Grössen den Entwurf der Technologie im Vorbereitungs und Projektionsstadium, sowie auch beim eigentlichen Bohrprozess zu optimalisieren. Dabei kann die Leitung im Bohrprozess direkt und indirekt unter Zuhilfenahme automatisierter Systeme durch Abtasten der Eintrittsgrössen realisiert werden.

Manche der allgemeinen Prinzipien der angeführten Ergebnisse können auch auf die Bohrung von Grossprofilbohrlöcher und Tunnelierungsmaschinen angewendet werden. Im weiteren Forschungsstadium wird es sich um die Quantifizierung der angeführten Relationen auf diese Systeme handeln, insbesondere vom Blickpunkt der Energieersparnis und dem der superharten Sparmaterialien.

Die Lösung der Problematik des Prozesses der Loslösung fester abrasiver Gesteine beruht im Wissenschaftlichen Forschungsinstitut für Kohle in Ostrava-Radvanice auf der geometrischen Analyse der bestehenden Schneidewerkzeuge und auf der Labor- und Grubenforschung. Es wurden neue Erkenntnisse über die negativen Faktoren erzielt, die eine Herabsetzung der Loslösungsleistung hervorrufen, sowie auch eine Herabsetzung der Lebensdauer der Schneidewerkzeuge (Vašek, Matušek, Dlouhý, 1976, 1978).

Ausser den bestätigten, früher bekannten und diskutierten Erkenntnissen über den Einfluss des Freiwinkels γ und des Schnittwinkels δ wurde festgestellt, dass nach Überschreiten des Schnittwinkels bei über 110° beim Loslösen der Gesteine ein heftiges Anwachsen der Schneidewiderstände entsteht, so dass der Loslösungsprozess zum Zerkleinern führt. (Abb. 1). Es stellte sich auch heraus, dass es in der Praxis auch grössere Schnittwinkel als 110° existieren, und zwar an allen mehr oder weniger abgestumpften Schneideinstrumenten (Abb. 2). Der Zerkleinerungsprozess führt zu einem grossen Energieverbrauch, bildet eine grosse Verstaubungsquelle; in Kohlengruben erhöht er die Entzündungsgefahr des Gemisches von Methan und Luft durch den Zündfunken, er ist mit einem Temperaturanwachs verbunden, der den Arbeitswiderstand der Hartmetallmaterialien herabsetzt. Die Analyse der Entwicklung der Schneideinstrumente, die parallel durchgeführt wurde, ergab, dass an der Vervollkommnung der Schneideinstrumente für feste, abrasive Gesteine folgende Entwicklungsgesteine Anteil hatten (Abb. 3):
- Hartmetallverstiefung des Schneide (a),
- Vergrösserung des Schneidewinkels (b),
- Rotationslängsachse (c),

- Austauschbarkeit der Schneide (d).

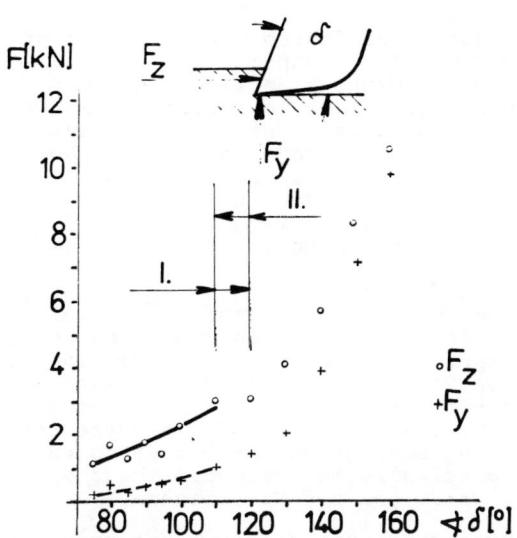

Abb. 1 Abhängigkeit des Schnittwinkels vom Schnittwiderstand F. - I. - Schnittgebiet, II. - Zerkleinerungsgebiet.

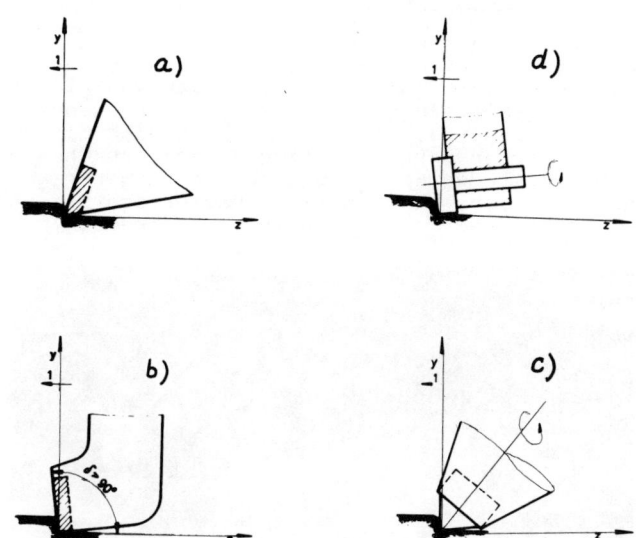

Abb. 3 Entwicklungselemente der Schneidewerkzeuge. 1- Bewegungsrichtung, O - Beginn des rechtwinkligen Koordinatensystems, δ - Schnittwinkel

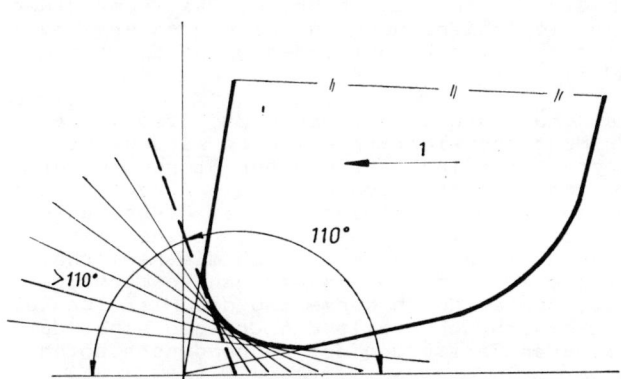

Abb. 2 Darstellung der Geometrie des amgestumpften Schneidewerkzeugs. 1 -Bewegungsrichtung des Schneidewerkzeugs

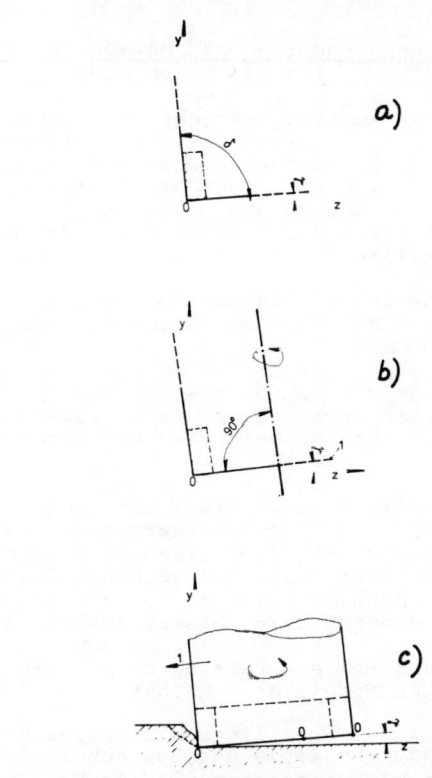

Abb. 4 Schematische Darstellung der Entstehung des Schneidewerzeugs mit drehbarer Rundschneide, yz -Achsen des rechtwinkligen Koordinatensystems, 1- Bewegungsrichtung des Schneidewerkzeugs, O-Schneidenpunkte, α - Freiwinkel.

Die neuen Forschungsarbeiten des Laboratoriums des Wissenschaftlichen Forschungsisnstituts für Kohle in Ostrava-Radvanice führten zum Entwurf einer Geometrie des Schneidewerkzeugs für die Loslösung fester Gesteine, die durch den negativen Freiwinkel in der Grössenordnung von 5° bis zu 8° und dem Schnittwinkel oberhalb von 90° definiert ist. Das Schneideinstrument ist mit einer Hartmetallschneide versehen (Abb. 4a). Eine Besonderheit stellt die Tatsache dar, dass die Rotationsachse nicht durch die Schneide hindurchgeht (Abb. 4b), wobei das Schneidewerkzeug die Form eines Zylinders oder Kegelstumpfes hat (Abb. 4c). (Patente USA, Grossbritannien). Das Schneideinstrument dieser Konstruktion wurde an einem

- Schnittwinkel

Laborstand geprüft und die erste Serie der Prototypschneiden an Loslösungsorganen von Kohlenkombinen (Abb. 5) beim Abbau von Flözen mit Bergemitteln und Nachnahme von Hangenden- und Liegendengesteinen. Beim Vergleich mit den gewöhnlichen Schneideinstrumenten tschechoslowakischer Produktion wurde ein geringerer Energieverbrauch, ein höherer Abbaufortschritt und ein ruhigerer Maschinengang verzeichnet.

Abb. 5 <u>Lostrennungsorgan der Abbaucombine mit drehbaren, kreisförmigen Schneiden</u>

Derzeit werden weitere Betriebsversuche vorgenommen, wobei zu erwarten ist, dass der angeführte Typ des Schneidewerkzeugs auch einen Beitrag zur Verminderung des Staubanfalls, des Risikos der Entzündung des Methan-Luft-Gemenges und zur Verlängerung der Lebensdauer der Hartmetallmaterialien der Schneide leisten kann.

3. SPANNUNGS- UND VERFORMUNGSZUSTÄNDE IM DURCH DIE BERGMÄNNISCHE TÄTIGKEIT ANGEGRIFFENEN MASSIV

Die Voraussage der Spannungs- und Verformungszustände in der Umgebung von Grubenbauen wird ebenfalls in mehreren methodischen Ebenen gelöst.

Für die operativen Bedürfnisse der Geomechaniker werden die sog. Ingenieurmethoden benutzt, die zwar eine orientierende, jedoch in der Praxis einfacher Fälle hinreichende Begrenzung der spannungsexponierten Gebiete im Massiv ermöglichen Konečný 1978, Šiška, Konečný 1982).

Kompliziertere Fälle werden durch physikalischmathematisches Modellieren gelöst.

Das physikalische Modellieren ist am meisten im Institut für Geologie und Geotechnik der Akademie der Wissenschaften ČSAV in Prag und im Wissenschaftlichen Forschungsinstitut für Kohle in Ostrava-Radvanice verbreitet. Die erreichten Ergebnisse sind sehr gut, den Hindernis bildet der Arbeitsaufwand und der hohe Zeitanspruch dieser Methode.

Das mathematische Modellieren wird am meisten in der Bergbau-Geomechanik im Bergbauinstitut der Akademie der Wissenschaften in Ostrava entwickelt. Zur Zeit befindet sich das Originalsystem GEM 3 in Betrieb, die Lösung von Raumaufgaben. Das System beruht auf einer ursprünglichen Philosophie, die von der Anwendung der Methode endlicher Elemente bei der durch Darstellung im regelmässigen Netz gegebenen Diskritisierung ausgeht. Dies ermöglicht eine überraschend leichte Aufgabenstellung auch bei verhältnismässig komplizierten Gebieten und eine rasche Lösung. Derzeit können Aufgaben vom Umfang bis zu 32 667 Freiheitsstufen (am Computer IBM 370/145) gelöst werden.

Die Prognosen der Spannungszustände und ihrer Erscheinungen werden durch Messungen in situ verifiziert. Für die Bestimmung der Spannung und ihrer Veränderungen im Massiv werden hauptsächlich Verformungsmethoden, empirische und geophysikalische Methoden benutzt (Konečný, 1980). Von immer grösserer Bedeutung werden die geophysikalischen Methoden, besonders die Seismoakustik und die Seismologie.

4. BEISPIELE DER REALISIERUNG DER ERGEBNISSE DER GEOMECHANIK

Deutliche Probleme auf dem Gebiet der Gesteinsmechanik entstehen in unseren Braunkohle-Tagbauen. Die jährliche Bewegung der Abraummassen beträgt cca. 190 Mio m^3 Erden. In der Weiterentwicklung ist ein Anwachsen bis zum Doppelten dieser Menge zu erwarten. Das kommt daher, dass die Kohlengruben in immer grössere Teufen vordringen (Mächtigkeit des Hangenden bis zu 200 m).

Die Kohäsion und damit auch die Stabilität der Hangenden-Tertiärsedimente wird nicht nur durch eine Reihe tektonischer Störungen unterbrochen, sondern auch durch die Abbautätigkeit der ehemaligen Tiefbauschächte. Diese bauten ein cca 40 m mächtiges Flöz in seinem mittleren höchstqualitativen Teil durch die Kammerbaumethode ab, und zwar in der ganzen Mächtigkeit, später durch Kammerbau der Mittelbänke. Der Kammerbruch stellt die Ursache einer deutlicheren Zerstörung der Hangendenschichten dar.

Durch postgenetische orogene Prozesse, durch die Feuchtigkeitsabnahme der Hangendentone und durch eine Reihe weiterer Einflüsse entstanden in den Hangendengesteinen prädestinierte Störflächen, die beim Abbau zu einem plötzlichen Abrutschen des abzubauenden Kohlenbolcks führen. Die Rissbildung infolge der Feuchtigkeitsabnahme führt zu Rissen und Flächen mit verminderter Festigkeit, die wiederum eine Abrutschgefahr hervorrufen. Zugleich sind diese Tone beim Abbau mit Schaufelradbaggern zum Quidrieren geneigt. Das bedeutet, dass das Loslösen nicht entlang der Trajektorie des Eimers vor sich geht, sondern nach prädestinierten Loslösungsflächen. Auf diese Weise bilden sich scharfkantige harte Tonstücke, die die Förderbänder beschädigen. Sie weisen eine Masse bis zu 1 t auf. Die Bildung und den Einfluss der Schlechte zeigt Abb. 6.

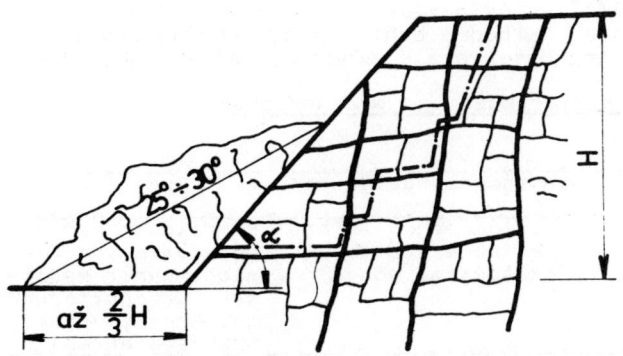

Abb. 6 Einfluss der Risse in den Hangendentonen

Die Stabilitätsberechnung der Hänge durch die gewohnten analytischen Methoden, z.B nach der Verdeyenschen Gleichung muss durch einen erhöhten Sicherheitskoeffizienten korrigiert werden.

Ein technisch ganz vereinzlter Fall wird im Grubenfeld der Grossgruben ČSA und J. Šverma gelöst. Im abgebauten Raum zwischen beiden Gruben ist eine innere Abraumkippe entstanden. Diese wird von beiden Seiten von den Grossgruben auf die gewohnte Art und Weise durch Schreitbagger und Schienenbandabsetzer aufgeschüttet. Derart wurde ein mächtiges Erdmassiv mit einem Inhalt von cca. 440 Mio m^3 Abraum gebildet, das stellenweise bis zu 160 m hoch ist. Der vereinfachte Durchschnitt durch beide Gruben (Abb. 7) zeigt auch das Profil des erwähnten Erdmassivs.

mit durchaus vrschiedenartigen Erden sind unzuverlässig. Während des Aufschüttens entstanden mehrere Flöze, auch Teilflöze, z.B. zwischen zwei Etagen, die höhengemäss im mittleren Teil der Abraumkippe liegen. Bei den Abgleitungen wurde wiederum der Spannungszustand des Massivs während ihrer Bildung bestimmt. Mi Methoden der mathematischen Statistik wurden die gegenseitigen Zusammenhängeder Stabilitätsgrundparameter bewertet.

Regelmässig wird auch der Verformungsverlauf des gesamten Massivs nachgemessen. Wir setzen voraus, dass es in der zweiten Etappe möglich sein wird, die einzelnen Faktoren so zu verallgemeinern, dass sie auch beim Ausrichten von Kohlengruben mit einer Teufe bis zu 400 m angewendet werden können.

Im Zusammenhang mit dem Vortrieb der Grossgrube ČSA am Fuss des Erzgebirges, wo das Flöz stellenweise steil mit einem Winkel bis zu 50° ausgeht, entstand eine Reihe Probleme mit der Stabilität der Hänge der Abbaufronten am Fuss des Gebirges, sowie auch mit der des gesamten Gebirgsmassivs des Erzgebirges. Diese Probleme bilden derzeit den Inhalt einer umfangreichen geologischen Untersuchung, von Stabilitätsberechnungen und Modellaborversuchen mit äquivalenten Materialien.

Zu den bedeutendsten Beispielen der praktischen Realisierung der Forschungsergebnisse der geomechanischen Probleme in der Tschechoslowakei gehört zweifellos der Kampf gegen Durchbrüche und Erschutterungen im Steinkohlenrevier von Ostrava-Karviná. Zur Zeit stammen cca. 35% des Abbaus in diesem Revier aus Flözen die durch Erschütterungen bedroht sind oder durch Kohlen- und gasdurchbruch, wobei die Durchbruche ausschliesslich im Westteil und die Erschütte-

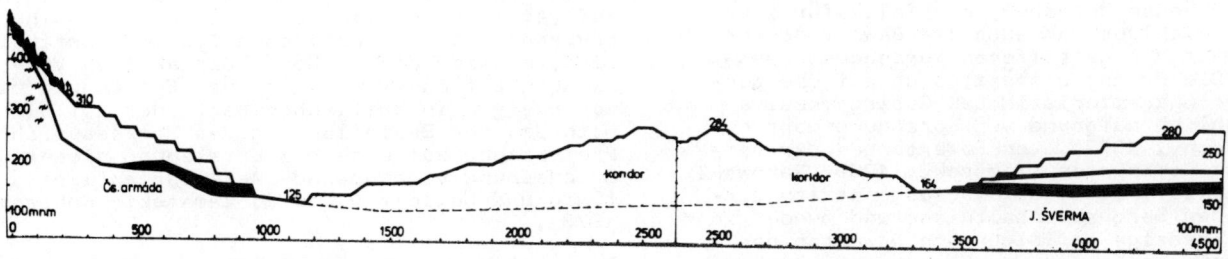

Abb. 7 Lage der inneren Abraumkippe der Gruben ČSA - Šverma

Auf das frisch aufgetragene Massiv wird in einer Länge von 4 km die viergleisige Staasbahn verlegt, die Fernstrasse und der Bach aus dem Vorfled der Grossgruben.

Die Sicherung der Stäbilität der Hänge dieses Massivs erforderte eine besonderes Entwässerungssystem des Liegenden, den Ausschluss der übermässug feuchten Tone aus dem Massiv und eine Neugestaltung der Abraumhänge mit einer Neigung von 1 : 12. Die herkömmlichen Berechnungsmethoden der Stabilität bei einem so hohen Massiv

rungen vor allem im Ostteil des Reviers auftreten, was eine der Erscheinungen der ausserordentlich grossen Variabilität der naturgegebenen geologischen Bedingungen in diesem Teil des Oberschlesischen Kohlenreviers darstellt. Den Umfang der angeführten Anwendung möge beispielsweise die Tatsache bestätigen, dass diese Arbeiten das Bohren von cca. 1400 km Bohrungen verschiedener Art im Laufe eines Jahres bilden. Insgesamt ist dies ein grosses System verschiedener systematisch angeordneter Tätigkeiten mit Benutzung einer

Reihe Methoden in der Prognose und mit der Vorbeugung von Erschütterungen und Durchbrüchen (Rakowski, Trávníček, 1982). Die Hauptelemente dieses Systems und der angewendeten Methode sind übersichtlich in Tabelle I angeführt.

den Erscheinung, die Beeinflussung der auf den Gefahrenzustand einwirkenden Faktoren und die Aufhebung oder wenigstens Einschränkung ihrer Einflüsse orientiert sind. Das Prinzip all dieser Methoden geht von der Auffassung des Gebirgsmassivs als einem System energetischer

Tabelle I: Grundelemente des Prognosensystems der Erschütterungen und Durchbrüche im Steinkohlenrevier von Ostrava-Karviná

Prognosenstufe	Durchführungszeit	Arbeitsmethode	Gebirgserschütterungen	Austritt
Makroregional	im Stadium der Projektierung neuer Gruben, neuer Sohlen usw.	- geologischer Aufbau der Lagerstätte - Analogiemethode - Paläontologische Rekonstruktion	- geologischer Aufbau der Lagerstätte - Analogiemethode	Einreihung der Lagerstätte vom Gesichtspunkt möglicher Erschütterungsgefahren (Durchbruchsgefahren)
Regional	In der Phase der Projektierung neuer Sohlen, der Ausarbeitung von Ausrichtungsplänen, der Vorbereitung und des Abbaus	Intensität der tektonischen Störungen Abgrenzung der gasführenden Schichten physikalische Eigenschaften der Kohlenmasse	Lagerstättengeologie Festigkeit der Gesteine im Hangenden und Liegenden Elastizitätsstufe der Kohle Teufe	Einreihung des Flözes in die Kategorie der Erschütterungs- oder Durchbruchsgefahr
Lokal	vor Beginn des Grubenbaus, in der Projektionsphase	Abgrenzung der gasführenden Schichten Abgrenzung der tektonischen Störungen in der Kohle Abgrenzung des Einflusses der Ausbeutekanten Teste der Lokalprognose	geometrische Abgrenzung der zusätzlichen Spannungen mathematisch-physikalisches Modellieren Seismik	Einreihung des Baus in die Stufe der Erschütterungs- oder Durchbruchsgefahr
Kontinuierlich (temporal und lokal)	während der Durchführung des Grubenbaus	Bestimmung der Zerstörungsstufe der Kohle Messen der Gasparameter	Bohrteste SA-Teste Seismoakustik, stationär Seismologie Deformometrie	Feststellung des aktuellen Stands der Erschütterungs- oder Durchbruchsgefahr

Die Methoden der Prognose sind auf glaubwürdige Zeit- und Raumbegrenzungen orientiert, in denen Bedingungen für die Entstehung von Gefahrenzuständen bestehen, ebenfalls für ihre Weiterentwicklung und auch die Bewertung der Wirksamkeit der getroffenen Vorbeugungsmassnahmen. Die Prognose stützt sich auf die geometrische Charakteristik des Gebirgsmassivs hauptsächlich aufgrund von Forschungsbohrungen (Konečný, 1977), von Bewertungen der geologischen Bedingungen (Zamarský, 1969, Rakowski und Kollektiv, 1982) der vorausgesetzten Einflusses der Bergbauverhältnisse und mündet in das regelmässige Verfolgen und Bewerten des Gefahrenstands in der Umgebung der Grubenbaue (Rakowski, Trávníček, 1982, Trávníček, Zamarski, 1982).

Im allgemeinen wird der Begriff der Vorbeugung als Tätigkeitsmenge aufgefasst, deren Ziel dahin gerichtet ist, die Bildung von gefährlichen Erscheinungen zu vermeiden oder wenigstens die Folgen derjenigen Erscheinungen einzuschränken, die beim derzeitigen Stand der Erkenntnisse und Technologien nicht zu vermeiden sind. Daraus geht sondann die Grundgliederung der Vorbeugungsmassnahmen in aktive und passive Methoden hervor. Vom Geschichtspunkt der Sicherheit, der Rationalität und der Kontinuität des Grubenbetriebs wird eindeutig den Methoden der aktiven Vorbeugung der Vorzug gegeben, die auf die Beseitigung der Ursachen der betreffen-

Systeme mit veränderlicher Entwicklung in Raum und Zeit aus. Ist es Aufgabe der Prognose, die Bildungsmöglichkeiten oder des tatsächlichen Auftretens eines eine gefährliche Erscheinung hrvorrufenden energetischen Systems festzustellen, so sind die Methoden der aktiven Vorbeugung auf das Verhindern der Entstehung dieser Energie im kritischen Wert oder auf die Methoden der Beeinflussung des Prozesses ihrer Freisetzung auf eine die Entstehung dieser Erscheinung verhindernde Weise orientiert (Šmíd und Kollektiv, 1978; Zamarski, Rakowski 1978).

Die praktischer Realisierung des gesamten Systems sichert ein besonders zu diesem Zweck gebildeter und geschulter, sowie auch technische ausgestatteter Dienst, der in allen Gruben des Reviers existiert, von einem speziellen geomechanischen Gremium geleitet wird und über moderne Laboratorien verfügt, die es ermöglichen, die erforderlichen Daten über die Beschaffenheit der Gesteine, der Kohle, des Wassers und des Gases zu gewinnen. Das System knüpft eng die wissenschaftliche Forschungsbasis an, wodurch es ermöglicht wird, die neuen Erkenntnisse operativ auf die Praxis und die erforderliche Rückkopplung anzuwenden.

Literatur:

Konečný P.: Geomechanische Bewertung des Gebirgsmassivs in Forschungsbohrungen. (Geomechanické hodnocení horninového masívu z průzkumných vrtů). Kandidaten-Dissertation des VVUÚ Ostrava-Radvanice 1977, 133 Seiten

Konečný P. u.Kollektiv: Untersuchung und Vervollkommnung der Grubenbetriebsmethoden zur Feststellung der mechanischen Parameter des Gebirgsmassivs. Bericht des UGG ČSAV, Ostrava, 1978

Konečný P.: Möglichkeiten zur Bestimmung des primären Spannungs- und Verformungszustands des Gebirgsmassivs des Steinkohlenreviers von Ostrava-Karviná. Konferenz "Ausbau von Grubenbauen unter anormalen Bedingungen", Gottwaldov, 1980

Konečný P.: Eine einfache Methode zur Bestimmung des Bereichs erhöhter Spannungen im Hangenden stehengelassener Pfeiler und deren Ausnutzung unter den Bedingungen des Steinkohlenreviers von Ostrava-Karviná. Sammelband "Neue Erkenntnisse der Wissenschaft, Forschung und Praxis auf dem Gebiet der Gesteinsmechanik", DTVTS Košice, 1978

Rakowski Z.; Lát J., Hruzík B., Dvořáček J. (1982): Neue Erkenntnisse auf dem Gebiet der Kohlen-und Gasdruchbrüche im Steinkohlengebiet von Ostrava-Karviná". SNTL, Praha, 206 Seiten

Rakowski Z., Trávníček L.(1982): Prognosen und Vorbeugungsmethoden, VŠB Ostrava, 203 Seiten

Sekula F., Bejda J., Dunay G.: Der Werkzeugverschluss beim drehenden Gesteinsbohren in Abhängigkeit von den Bohrbedingungen. Glückauf Forschungshefte J. 36, Heft 5, 1975

Sekula F., Bejda J., Koči M., Krajecová O.: Bohrgeschwindigkeit und spezifische Volumenenergie der Zerstörung in Abhängigkeit von den Bohrmethoden. Sammelband der Internationalen Konferenz der RWG-Länder "Intergeotechnika", 1977 (russisch).

Sekula F., Kupka J.: Physikomechanische Studie über die Felszerspanung. Internationale Gesellschaft für Felsmechanik, 2. Internationaler Kongress, Beograd 5-5, 1970

Šiška L., Konečný P.: Forschungsergebnisse der Gebirgserschütterungen in Beziehung zu den Zonen der zusätzlichen Spannungen, die durch die stehengelassenen Pfeiler im Hangenden hervorgerufen werden. Internationale Konferenz "Gebirgserschütterungen in Kohlengruben", Ostrava, 1982

Šiška L., Konečný P.: System geomechanischer Auswertung für bergmännische Zwecke. Sammelband ISRM, Montreaux, 1979

Šmíd M., Jelen B., Složil J., Rakowski Z.(1978): Anleitung zur Durchführung von Prognosen und Vorbeugungen von Kohlen- und Gasdurchbruchgefährdeten Kohlengruben im Steinkohlenrevier von Ostrava-Karviná. VVUÚ, Ostrava-Radvanice, 72 Seiten

Tkáčová K., Sekula F., Hocmanová, Krupa V.: Die Bestimmung von irreversibel gespeicherter Energie beim Mahlprozess (englisch). XIII. International Processing Congress, Warschau, 1979

Trávníček L., Zamarski B.(1982): Methode des seismoakustischen Testens. Internationale Konferenz "Gebirgserschütterungen in Kohlengruben", Ostrava

Vašek J.: Untersuchung des Einflusses der Geometrie des Schneidewerkzeugs auf den Loslösungsprozess der Kohle. Abschlussbericht des VVUÚ, 1978

Matušek Z., Vašek J.: Untersuchungen über den Einsatz von Streckenvortriebsmaschinen beim Nachreissen von Gestein". Glückauf-Forschungshefte, Oktober 1976

Vašek J., Dlouhý J.: Beurteilung der Leistung der Streckenvortriebsmaschine AM 50. Glückauf-Forschungshefte April 1978

Zamarski B. (1969): Beherrschung des Hangenden in Strebabbauen des Steinkohlenreviers von Ostrava-Karviná. VVUÚ Ostrava-Radvanice, 166 Seiten

Zamarski B., Rakowski Z., Franěk J.(1978): Katalog der Mittel und Massnahmen zum Kampf gegen Erschütterungen im Steinkohlenrevier von Ostrava-Karviná. DT ČSVTS Ostrava, 17 Seiten

Čs 10 Nr. 201 186 USA-Patent Nr.4 222 446, GB-Patent Nr. 2 009 287 B.

THE DEVELOPMENT AND CURRENT STATE OF ROCK MECHANICS IN CHINA

Développement et situation actuelle de la mécanique des roches en Chine

Die Entwicklung und der gegenwärtige Stand der Felsmechanik in China

Mei Jian-Yun
Research Engineer, Secretary General for China NG, ISRM. Deputy Director of Department
Institute of Geophysics, Academia Sinica, Beijing, China

Fu Bing-Jun
Engineer, Secretary for China NG, ISRM. Water Conservancy and Hydroelectric Power
Scientific Research Institute, Beijing, China

Kang Wen-Fa
Assistant Professor, Geodynamic Department, Institute of Geophysics, Academia Sinica,
Beijing, China

SYNOPSIS

This paper describes briefly the development and current state of rock mechanics since the founding of the People's Republic of China in 1949. The application of rock mechanics in dam foundation, underground structures, rock slopes and the improvement of rock masses were introduced. Some theoretical aspects and current lines of research are discussed briefly. Finally, some main points about the development of rock mechanics in China are presented.

RESUME

Cette communication présente une description précise de l'évaluation et de l'état actuel de la mécanique des roches dans la République Populaire de Chine depuis 1949. On présente les applications de la mécanique des roches dans les domaines de fondations de barrages, la construction souterraine, les talus de roches et la stabilité des roches massives. On décrit brièvement certains aspects théoriques et les grandes lignes de la recherche actuelle. En conclusion, on présente quelques points majeurs du développement de la mécanique des roches en Chine.

ZUSAMMENFASSUNG

Es wird eine kurze Beschreibung der Entwicklung und des heutigen Standes der Felsmechanik in der Volksrepublik China seit 1949 gegeben. Die Anwendung der Felsmechanik bei Dammgründungen, in Untertage-Hohlräumen, bei Felsböschungen und bei der Gebirgsvergütung wird erläutert. Es folgt eine kurze Diskussion über einige theoretische Aspekte und gegenwärtige Forschungskonzepte. Abschließend werden einige Hauptgesichtspunkte in der Entwicklung der Felsmechanik in China zusammengefaßt.

1. INTRODUCTION

The Chinese people have a long history in dealing with the problems of rock Mechanics, but the beginning of the scientific research work is considered after the founding of the People's Republic of China. In 1958 in order to fulfil the requirement then raised by the design and construction of the large dam on Yangtze River-Three Gorges project, under the leadership of the National Commission of Science and Technology, " Three Gorges Research Group of Rock Mechanics" was formed. At that time more than 100 technicians coming from 18 institutes all over the country was concentrated in this group. Under the scientific leadership of Prof. Tan Tjong-Kie, an extensive and systematic scientific research on rock foundation, underground structures, rock slopes, the dynamic properties of rocks and grouting techniques both in laboratory and in-situ was carried out. The first theories on the rheological properties, deformation characteristics and stress wave propagation in rock masses were started when the first complete apparatuses also appeared. In this manner, the basic technical staff for rock mechanics in China has been trained and the foundation in the development of rock mechanics all over China was laid successfully. Up to now, according to incomplete statistics there are more than 1200 engineers, associate professors and professors for rock mechanics working for 130 units, institutes or universities.

With the developing of rock mechanics in our country, various special committees or organizations on rock mechanics have been established in different Societies concerning the Water Conservancy and Hydroelectric Power, Railway, Coal Mine, Architecture, Metallurgy, Geology and so on. Lately under the leadership of Prof. Tan, two scientific organizations have been established, namely the China National Group, ISRM. (1978) and the Preparative Committee of Chinese Rock Mechanics and Engineering Society (1981). Up to now, more than 10 journals related to rock mechanics in our country have been published. All of these activities are of help to enhance the development of rock mechanics further.

At present, the movement of four modernizations in industries, agricultures, national defence, sciences and technology of our country is in full swing. Hence, a lot of projects in or on rocks are being constructed. In the respect of underground works, tunnels with the length up to several ten kilometers, mining excavations in the depth of up to 1000-2000 m, underground caverns with the span of 30-40 m, height of 70-80 m, high prussure hydraulic tunnel with the water head of 300-500 m and high-velocity flow will be constructed in near future. As for the surface works, a series of large dams will be built on complex foundations. Some of the rock slopes in mining industry have 400-500 m in height. And besides, China is a country with rich deposite resources and with frequent occurrence of earthquakes, so the research works on the rock mechanics for geodynamic purposes have started recently. All of these will certainly enhance the development of research on rock mechanics further, and will provide great prospects for this new discipline. We are eager to learn the advanced technology and experiences of foriegn countries and make great efforts to develope the friendship and cooperation with them.

2. ROCK MECHANICS IN ENGINEERING PRACTICE

2.1 Dam construction

China is very rich in water resources. The total theoretical capacity of water power reaches 680 million kw, but only 2.36% of it has been developed. After the founding of PRC in 1949, according to incomplete statistics, more than 84,000 dams of different types have been built including 80 larg concrete dams built on various rock foundations. They include gravity dams, double curved arch dams, buttress dams, gravity arch dams and multiple arch dams. The maximum heights of them are 147 m, 80 m, 128 m, 165 m and 88 m respectively. Besides, more than 400 stone masonry dams of various types have also been constructed.

With the aim of studying the mechanical properties of rock, early in the fifties, the routine experiments has been carried out. In 1959, the first triaxial testing machine with samples of 9cm in diameter and 20cm in height and the maximum vertical load of 500 T, maximum confined pressure up to 1500 kg/cm² was manufactured by "Three Georges Research Group of Rock Mechanics". In 1977, the first true triaxial testing machine with the maximum load of 200T and maximum confined pressure of 750 kg/cm² was provided by Gezhouba Project Construction Bureau. In the field of insitu triaxial tests, after 1965 some true triaxial test on rock mass blocks were performed. The dimensions of test blocks are generally 50x50x50 cm³ and 70x70x70 cm³. In 1978, true triaxial tests on samples with length of 100 cm, width of 100 cm, height of 150 cm were carried out at Gezhouba Project. From these tests, the failure mechanism, strength and deformation characteristics as well as the affect of intermediate principal stresses were studied.

As for the stability analysis of dam against sliding, there are two different cases: (1) sliding occurred along the contact zone between the concrete and rock foundation-plane shear failure; (2) sliding occurred within the rock foundation along the discontinuities-deep shear failure. A statistical evaluation of the shear tests from 148 sets at 56 dam sites with concrete blocks cast on rock foundation proved that, for the intact rock masses the coefficient of friction f=1.25-1.52 c=12.04-15.33 kg/cm² with the confidence probability of 95%; whereas the test results on various shear zones or weak seams from 113 sets showed that f=0.22-0.35, c=0.22-0.72 kg/cm² with the same confidence (Mei 1978 1981) A large amount of engineering practice has proved that previous parameters can easily meet the requirements for the stability of dam against sliding for the intact rock foundation. Thus we consider the crucial point that affect the dam stability is the weak seams underlying the dam foundation.

According to imcomplete statistics among the large dams that have been or are being built, about 90 of them have such weak seams, and 30 of them have been obliged to change design scheme in the following manner, such as decreasing the height of dam, increasing the amount of construction or making additional reinforcement due to shear zones or weak seams existed underlying the dam foundation. Since the seventies, we have made great efforts to the study of the affect of weak seams, in connection with the research on geological exploration, experimental study, design method and foundation treatment.

Among the complex dam foundation with shear zones, Gezhouba project may be taken as a typical example. Gezhouba is the first key project on the middle reaches of Yangtze River under construction with the maximum dam height of 47 m, the aggregate length of 2,561m, and the installed capacity of the power plant of 2,715 MW. It is based on a very complex foundation containing nearly 50 gentle dip shear zones. In order to investigate its genesis, type, occurrence and distribution, according to incomplete statistics, we have made a total of 1101 small boreholes with depth amounting to 35,489 m, 33 large boreholes (the diameter being more than 1 m) with an aggregate depth of 1,037 m. And besides, more than 10 exploration adits and shafts were performed. A large amount of geophysical prospecting and natural stress measurements have also been undertaken.

As to the experimental research done with the shear zones, various topics, such as the structural characteristics, mineral composition, physico-chemical properties, mechanical properties as well as its possible variation and their evolution, the permeability stability after the impounding of the reservoir have been finished. Meanwhile the geomechanical model tests on the stability against sliding have also been carried out.

For justifying the deformation characteristics and the long-term strength of the shear zones, both laboratory and in-situ shear creep tests, relaxation tests were conducted. For the test duration studied it is possible to estimate the strength after 100 years (Tan 1981) Residual strength tests with the applied reversed shear loading for 300 times have also performed . The test results of previous three methods are shown on Fig.1. By comparing test results it showed that the $\tan\varphi$ obtained from creep test is 0.79 of that obtained from quick shear.

For justifying the resistance of downstream rock masses against shear loading, in-situ large scale rock resistance block tests were made. As shown on Fig.2, their dimensions are 11.65x1.70x2.53 m³ and 9.45x1.70x2.30 m³ (length x width x height respectively). From the test results not only the various mechanical parameters about

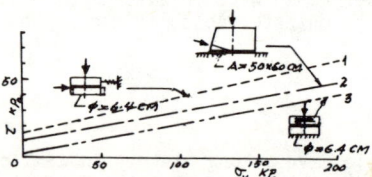

Fig.1 Long term strength: relaxation .
Creep and residual strength (Tan 1981)
1-Numb. of samples:9, $\tan\varphi_{rel}$=0.23
2-Field, $\tan\varphi_{cr}$=0.19
3-Numb. of samples:18, $\tan\varphi_{cycl}$=0.19

resistance force etc were obtained, but also the failure patterns and process of rock masses were presented vividly. The failure mode of horizontal laminated weak rocks was displayed as the dislocation between various layers and buckling of individual layers. Based on these analyses, it showed that the design method according to traditional Coulomb failure plane does not tally with the facts for the downstream resistance rock masses (Ren 1974) On the basis of research works of weak shear zone, rational design and construction principles were proposed (Tan 1981) For justifying the abutment stability of arch dams we commonly use the ultimate equilibrium method and F E M. Recently we have developed the three dimentional geomechanical model tests with model blocks simulating the rock masses (Zhou 1982). By means of these tests the sliding.

stability, the capacity against overloading, failure mechanism and the scheme about foundation treatment were studied.

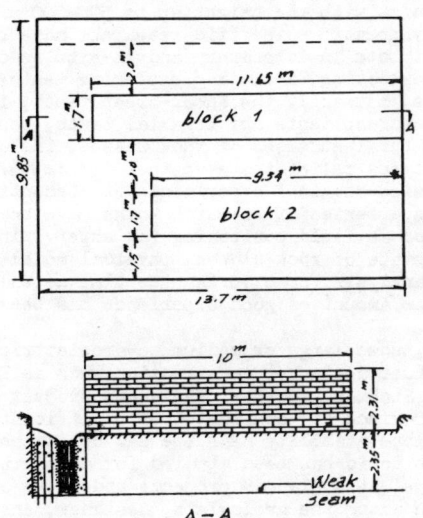

Fig.2 The arrangement of test blocks (Ren 1974)

2.2 Underground works

2.2.1 Underground hydropower plant and tunnel

According to the incomplete statistics, since the founding of our country we have built or are building 27 underground hydro-power plants. Among them the Liujiaxia underground hydropower plant built at the beginning of the sixties has its span of 30 M, height of 60 M. Additionally, we have constructed hydraulic tunnels with a total length exceeding 100 km. Among them, there are 28 tunnels with the length of more than 2000m each. The longest ones are more than 8 km. The designed head of the high pressure pipe line reaches 630 M.

It is known that the regularity of the distribution of crustal stress is one of the most important factors, affecting the stability of the tunnel which must be considered in determining the alignment, shapes, demensions and rational spacing of tunnels. For many of the underground power plants and tunnels in our country were constructed in the area with high crustal stress, early in the sixties, the task of measurement and research of the crustal stress had already begun. By the study of the test data, it shows 1/4-1/3 of them with its verticle components σ_v higher than that caused by the overburden, about 60% with its average horizontal stress components exceeding verticle stress. The coefficient of lateral pressure N ranges from 0.28 to 0.30. Horizontal stresses distributing on the plane are usually non-uniform (Liu 1979),(Lu 1982).

As for the deformation test in tunnels, early in 1958, the "Three Gorges Research Group of Rock Mechanics" started the water loading test. Later on, in some other projects, the water loading test as well as radial flat jack tests were carried out in lined or not lined, grouted or not grouted tunnels with the diameters up to 2 or 3 M.

A large amount of tests demonstrate that for the pressure tunnels where rocks are above medium strength, 70% of the internal water pressure is supported by the surrounding rocks. In case of discontinuities which influence the stability of the surrounding rock, large scale rheological shear test along the discontinuities was carried out, and based on these data the necessary bolting force to reinforce the unstable rock mass had been calculated.

On the evaluation of the stability of the surrounding rocks of the tunnel, according to different rock behavior, various theories and methods had been adopted such as the caved arch theory had been used for loose media, block equilibrium analysis for sound and jointed rock, and further the viscoelastic analysis, model test as well as FEM were also applied.

2.2.2 Railway Tunnel

The construction of railway tunnel started in 1887. Before the founding of our country in 1949, there were 427 tunnels with total length of 112 km. However, after 1949, within only 30 years from 1950-1979, there were 3,957 railway tunnels built, the total length of which is 1,897 km. Half of them distributed in the mountaineous region of southwest and northwest parts of our country. By the end of 1979, 4,386 railway tunnels were built with its total length of 2,009 km. Among them 154 tunnels had each length exceeding 2 km, 10 tunnels with each length exceeding 5 km, the longest Yimaling tunnel being more than 7 km. Dayaoshan double-line tunnel is preparing to construct, the length of which is designed to be more than 14 km (Liang 1981). With respect to the geological exploration for tunnels, multiple methods were applied. For the deep lying tunnels boring machine with its drilling capacity more than 1000 m had been used. Since 1970 the methods of geophysical exploration, bore hole television and remote sensing interpretations etc have been developed in the geological investigations. Simultaneously the tests of the physical and mechanical properties of rocks were also carried out.

In the evaluation of the stability of railway tunnels, the method by means of the classification of the surrounding rock had been adoped. With regard to this work, three stages could be divided. In the fifties, according to Protodyakonov coefficient of rock strength, the rock was classified into 9 grades.

In the sixties, the rock mass was classified into 6 grades. At the beginning of 1975, the surrounding rock was classified based on the stabilities. According to the degree of soundness (intactness), i.e. the fracture state and strength of rock mass, it was classified into four kinds: sufficient stable, stable, temporary stable and unstable. Besides, there are two subdivisions, altogether making six kinds.

The tendency of the classification of rock masses is to consider the multiple factors and comprehensive index instead of monofactor which was usually used in the past.

2.2.3 Coal mine tunnel

Our country is rich in coal resources, the productions of which ranks third in the world. Coal mine is the main energy resource in our country. According to incomplete statistics, in 1962-1979, the total length of the coal mining tunnels excavated were 11,535 km. In 1954-1979 the total depth of coal mining shaft was 118,130 M. The deepest one reaches 1058 M.

At present the outstanding problems related to the coal mine rock mechanics are: (Niu 1982)
(1) Roof control and support problems;
(2) The prevention of impact rock pressure;
(3) The safety problems of mining under constructions, railway and reservoir;

Serious damage is usually caused by the abrupt large area falling down of the roof of underground coal mine caverns. In the studying and solving this problem, the research procedure we used is firstly the investigation of the in-situ state of residual stress and the regularity of the movement of the overlying strata; secondly the study of the practical working conditions of supports and their affecting factors, especially the roof control of weak rocks and the supporting method; thirdly basing

upon the above research results, the treatment of caved regions, design of mining procedure, selection of support patterns and supporting methods, and the way as well as procedure of the calculations of support strength. The key point is to strengthen the observation of the regularity of mine pressure. In our country many mines got the successful experiences in predicting the large area roof falling problems and safty was ensured.
Some of the mines in our country have their excavation depth exceeding 1000 m. The deep shaft is always characterized by high rock pressure and temperature, difficulties in supporting, high frequency of impact rock pressure, especially the big amount of deformation of surrounding rock, some of them reaching or exceeding 50% of the demension of the tunnel cross section.
By researching the factors causing the occurrence of impact rock pressure are as follows:

(1) Physical and mechanical properties of the coal seam. According to investigations, where the coal seam with elastic deformation lower than 50%, generally impact rock pressure will not occur. Several coal seams in our country with impact rock pressure possess the water contents generally less than 1-3%.

(2) Geological factors including tectonic stress field, depth of coal seam and the properties of coal seam roof. In our country the impact rock pressure usually occurs in the syclinal and anticlinal zones, and the roofs of large majority of coal seam are rather sound with the compressive strength ranging from 900 to 1300 kg/cm^2

(3) The factors regarding the mining techniques including method of coal mining, sequence of mining, method of roof control, state of pillar, etc.
On the basis of the above mentioned factors, the ways to prevent the impact rock pressure are as the following:

(1) Rational arrangement of the tunnel locations and mining area;

(2) Improving the physical and mechanical properties of the coal seam, for example, adopting the method of weakening of the coal seam by high water-pressure;

(3) Lowering the stress in the coal seam by: forced explosion, elimination of coal pillar, adoption of rational mining sequence;

(4) Release the elastic energy of the coal seam, such as drilling the big bore hole or shallow bore hole with plenty explosives installed for the purpose to release the elastic potential energy.

(5) Prediction of impact rock pressure by observing the rock flour.

According to incomplete statistics, the coal resource lying under the constructions, railway and reservoir in our country is extimated to be 10 thousands million tons. Now we are just focusing the research work on the regularities of moving behavior of the earth surface and rock strata in such conditions, seeking the resonable method of mining and back filling, as to guarantee the safety of the surface constructions and underground works. (Niu 1982).

2.3 Rock slopes

There are two kinds of problems of rock slopes with important significance in engineering practice, namely:
(1) The rock slope stability related to large dams, reservoir banks and dam abutments. (2) The rock slope stability related to open pit minings.
The dam sites in our country are mostly located in narrow gorges between high mountains with complex geological conditions. Some large hydropower stations which are under design or construction often have troubles with the unstability of rock slopes nearby the dam sites or in the reservoirs. Obviously, we should pay great attention to the rock slope stability as we do to the stability of dam foundation.
As regards to the stability analysis, we prefer to take the research works on rock slope stability at Taye open pit mining as a typical example. In the early sixties, we applied for the first time the rheological theory in connection with the structural and tectonical stresses theory of rock masses to study the stability of rock slopes with the height up to 500-600 m. Extensive and systematic scientific researchs have been carried out both in laboratory and in-situ, such as dynamic prospecting, including explosive tests; static experiments, primarily the shear-creep tests, in-situ compression-creep tests and triaxial tests, the measurements of natural stresses of rock masses, including the borehole stress relief tests up to 100 m depth; as well as the physico-chemical experiments of discontinuities. And besides, a series of research works have been performed, such as field monitoring for supervizing the stability state of rock slopes, physical model tests, numerical analysis (considering the time dependence). Thus a large amount of good experience has been gained (Tan 1964).
At present, some large or medium hydroelectric power stations under design or construction such as Er-tan project in Sichuan province, Wujiangdu project in Gueizhou province are confronted with defficulties about the rock slope stability near the dam or at the reservoir banks. This topic has been studied for a long time. For instance, at Wujiangdu project, under the close cooperation among the geologists, designer, engineers of rock mechanics and constructors, a large amount of valuable information, such as the field monitoring materials (for a time period of 4-8 years), data about the main factors affecting the stability, mechanism of deformation and the possible failure modes have been obtained. With the aid of studies mentioned above, correct treatment such as removing the rock masses on the top, installation of reinforced concrete piles against sliding has been adopted. On this subject, in recent years, besides routine experimental researches we have emphasized following works: the field investigation of the natural slopes (their distrbution, frequency of occurrence, failure patterns and process, the internal and external causes of failure) and the field observation of the rock slopes.

2.4 The construction technique and reinforcing measures

2.4.1 Excavation and explosion

In the field of rock breaking, we have also obtained a series of achievements. For the underground construction we have used fully mechanized drilling and blasting method for the large cross sectional tunnel and full face tunnel boring machine (TBM).
In rock drilling we have made progress in the respect of strengthen percussion drill, hydraulic perussion drill, down hole drill, roller cone bit, the lengthening of service time of drill steels, manufacture of drill bit, the basic properties of rock drilling and wave mechanics of percussive drilling.
In the respect of blasting technique, large scale blasting, a wide range of millisecond delay blasting, compression blasting, smooth blasting, presplit blasting, adit and gallery bulding by blasting, damming by blasting technique and so on, have been widely used and got successful results in our country.
As to the physical breaking of rock, we have carried out a series of research on the jet drill, plasma jet, water jet cutting with high pressure, arc cutting, aluminothermic breaking and thrust hole by blasting (Xu 1979).
A large-scale blasting have been conducted in the caved area at Panchihua mining industry with the charges of blasting explosive of 1,380 tons, the volume of exploded ore reached 2.095 million m^3, thus resulted in anticipated consquences.
In 1959, the first dam for a power plant was built by means of directional explosion in Zhejiang province,

In 1960, in Guangdung province, another large dam with the average height of 65 m was constructed which was the largest in the world at that time ever built with directional explosion. The charges used in the project reached 1,395 tons, whereas the quarried rock was 1.67 million m^3. Since 1970, this method has been widely used in Water Conservancy and Hydroelectric power, Coal Mining, Metallurgical and Railway engineering for building of earth-rock dam, storage dam, tailing dam, permeable dike and embankment dam with a total number of more than 50, which ranks first in the world.

Moreover, some experiences have been obtained in the respect to rock plug blast under water. For instance, the intake heading of Jingpoh powerstation in Northeast China was successfully constructed by this method with the depth of submergence up to 25 m.

2.4.2 Rock bolting and shotcrete technique

Early in the fifties, the rock bolting and shotcrete technique was used in engineering practice which have got new advances in our country since the sixties. According to statistics in the Railway construction systems, more than 70 tunnels supported by rock bolts and shotcrete with an aggregate length of about 23 km were built. According to investigation result after completion, 89.2% of them have been operating in good conditions whereas 6.4% of them are in basically good conditions.

Statistics in the Water Conservancy and Hydroelectric Power units showed that in the time period before 1978, tunnels or underground caverns supported by rock bolting or shotcrete with an amount of areas 320×10^3 m^2 had been constructed. If they are converted into tunnels with cross section of 50 m^2, the corresponding length would be 12.7 km. Compared with traditional tunnelling method the rock bolting and shotcrete techniques have many advantages, such as saving cement by 1/3, saving manpower and investment by 1/2, shortening the construction period by 1/2-2/3 and nearly saving all of the wood (Yun 1982).

In the aspects of coal mine and metallurgy, the rock bolting and shotcrete techniques are used more widely and effectively. According to investigation data about the galleries supported by rockbolts and shotcrete with the total length of 4515 m in Kailuan Coal Mine after Tang-Shan catastrophic earthquake, 95% of them were in good condition after earthquake. That is to say, the aseismatic properties of rock bolting and shotcrete system are excellent.

At present, the rock bolting and shotcrete techniques are not only used for temporary supporting, but also for permanent supporting. Their uses have been extented from supporting the tunnels with smaller cross section to the tunnels or caverns with large spans and high side walls; from supporting the tunnels without water flow to the pressure tunnels. Additionally, they are used successfully to support underground works within fractured, even loosened rock masses. Besides, a large amount of new achievement concerning the study on mechanism of rock bolting and shotcrete techniques, their design method as well as construction technique have been obtained.

As for the reinforcement of rock masses, in 1973 two kinds of prestressed rock bolts with the length of 24 m, tension forces 60 T were provided, which have been successfully used in reinforcement to tunnel linings, rock slides treatment, reinforcement to surrounding rocks, anti-knock engineerings and so on, and are suitable for various types of rock masses.

2.4.3 Grouting

The grouting methods are widely used in our country to consolidate various rock masses, which are also used in anti-seepage, stopping up the leakage and etc. In the last twenty years, the chemical grouting have got good application.

In dam construction, we commonly used cement grouting as a main measure for consolidation or seepage prevention. Recently, we have conducted the high pressure cement grouting tests in karstic formation at Wujiangdu hydroelectric project (the maximum height of gravity-arch dam being 165 m, the rock foundation being limestone) and obtained valuable information. Test results showed that effective grout curtains can be made in the karstic formation by adopting high pressure (60-80 kg/cm^2) cement grouting. After completion of grouting, the unit water absorption could be reduced by smaller than 0.001 $liter/min.m^2$.

As regards chemical grouting, a series of problems have been solved where the traditional cement grouting are helpless. In the mean time in respect to study of various kinds of chemical agents, their prescription, control techniques, and construction procedures, we have gained nice experiences.

3. SOME THEORETICAL RESEARCHES OF ROCK MECHANICS

3.1 The rheological properties of rocks

In discussing problems dealing with the rheological properties of rocks, we have to mention the name of the brilliant scientist Prof. Tan Tjong-Kie who has made outstanding contribution in this field. Early in 1959 he introduced the conception of rheology to rock mechanics and instructed the rheological tests of surrounding rock in exploration tunnel at Three Gorges dam site (Tan 1959). The test results are shown on Fig.3.

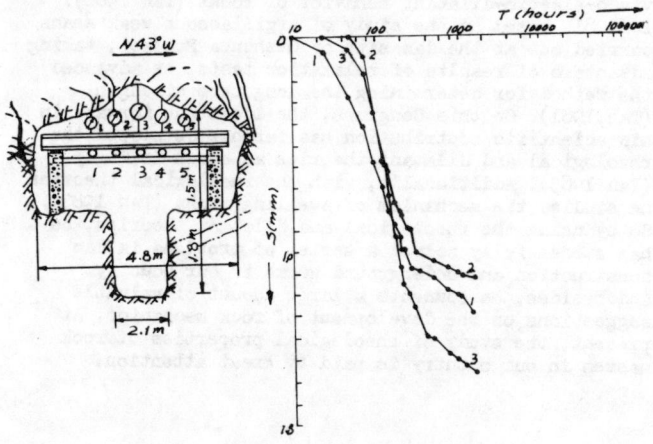

Fig.3 Creep of a tunnel (Tan 1980)

In 1961 he extended the rheological theory of rocks and made it applicable to anisotropic rock masses. In 1966 based on the experimental research on jointed rock formation he solved successfully the theoretical solution about the stress distribution of surrounding rocks within the laminated rock masses. At that time he proposed two basic conception: (1) the stress field of surrounding rocks may rotate with the lapse of time as shown in Fig.4; (2) the applied pressure to the lining will increase as time goes on due to the rheology and recovery of rock masses. Meanwhile he indicated that the widely used at that time the Protodyakonov's theory is not suitable (Tan 1966).

Starting from the study of microstructural properties of rocks, considering their genesis and history of evolution, Prof.Tan indicates that there must be existing disocations, cracks and internal stresses hence leading to rheological charateristic in rock masses. Further

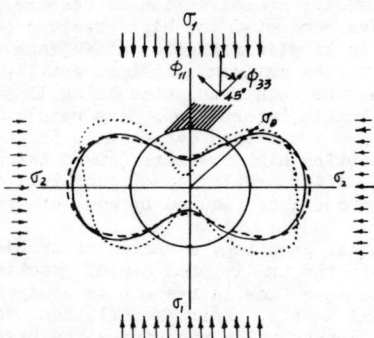

Fig.4 Tangential stress distribution σ_θ in polar diagram;— isotropic elastic case;____ transversely isotropic visco-elastic case for t=0; ...... the same for $t \to \infty$; layer thickness assumed very small in comparison with diameter tunnel; degree of anisotropy increases from 1.3 at t=0 to 2.3 at $t \to \infty$. In isotropic case $\sigma_{\theta\,max} = 2.5\,\sigma_1$; $\sigma_{\theta\,min} = 0.5\,\sigma_1$. (Tan 1966)

more based on experimental research results and physical reasoning, he set up the constitutive equations for visco-elastic-dilatant behavior of rocks (Tan 1980). In 1981, based on the study of argillaceous weak seams carried out at the dam site of Gezhouba Project, taking advantage of results of relaxation tests, he advanced the method for determining the long term strength (Tan 1981). On this Congress, the ideas put forward in his scientific contribution has further developed the rheological and dilatant theories as shown in Fig.5 (Tan 1983). Additionally, with the rheological theories he studied the mechanism of swelling rocks (Tan 1983). So by using the rheological and dilatant theories, he has successfully solved a series of problems in dam construction and underground works in our country. And besides, he comments a large amount of valuable suggestions on the development of rock mechanics. At present, the study of rheological properties of rock masses in our country is paid by great attention.

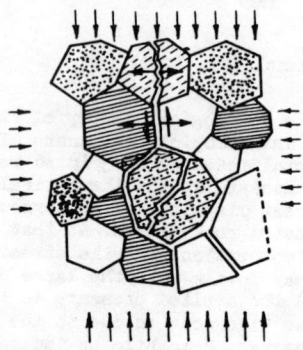

a. Dilatancy model

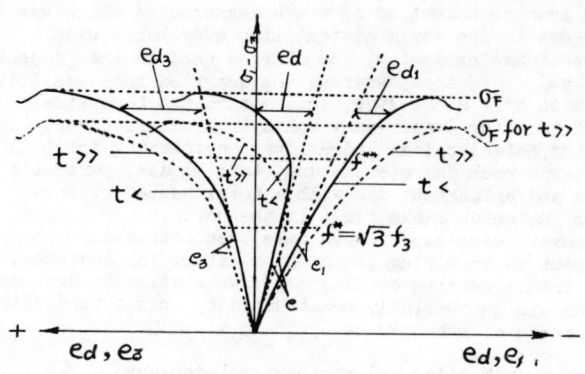

b. Stress-strain relationship for dilatant rocks

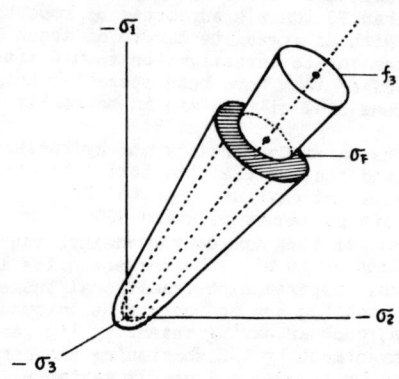

c. Yield and failure surfaces

Fig.5 The theory of rocks dilatancy (Tan 1983).

3.2 On the developement of rock mechanics in the field of geodynamics.

At present, the main points of study in the field of geodynamics are earth crust and upper mantle. Owing to the strain rate of tectonic movement of earth crust is rather low (less than 10^{-6} /sec). The more deeper, the more higher of the temperature and the more greater of the confined pressures. Hence, in the field of research on geodynamics the time factor, high temperature and high pressure are considered as the crucial points. In order to devolope the research works in this field, three sets of high pressure and high temperature triaxial apparatuses have been manufactured by Institute of Geophysics, Academia Sinica: (1) the maximum vertical load being 800 T with the confined pressure of 10,000 bars, temperature of $400^\circ C$, for the purpose of studying the rocks of earth crust; (2) the maximum vertical load being 300 T with the confined pressure 30-50 kb, temperature up to $1500^\circ C$, for the purpose of studying the lower part of earth crust and upper mantle; (3) the maximum vertical load being 15 T with the confined pressures of 500 b, temperature up to $200^\circ C$ for the purpose of studying the rock properties of shallow earth crust. All of these three apparatuses may be used to carry out rheological test with very low strain rate. Although this work has started not long age, it has a great prospect in developing the rock mechanics.

Lately, based on the conception about the accumulation of natural stresses, the delatancy of rock and the strength envelopes of rocks becoming gradually parallel to the axis of normal stresses in the depth of 10 km or more under the surface owing to high comfined pressures of 3-8 kb and high temperatures of 400-800°C a new hypotheses was advanced about the orthogonality of earthquake belt in North China and the physical rheological modes of catastrophic earthquake in Tangshan, which has been supported by practical data (Tan 1982).

3.3 The study on the fracture mechanics of rock mechanics

In the last few years, the research of fracture mechanics in the field of rock mechanics has been started in our country. A series of experiments and research works dealing with the propagation of cracks in rock and failure criteria have been conducted. It has been used to research the earthquake prediction, the stability of large dams and the mechanism of shallow earthquake of water reservoir (Zhou 1979).

3.4 On the theories concerning deformation, stress and strength

Early in the end of fifties, we started to research the deformation theory and rheological properties of rocks (Tan 1959). Lately, based on the structural characteristics of rock masses the constitutive equation of rock mass deformation were studied (Sun 1982). With respect to the strength theory of rocks, the affect of intermediate principal stresses on the strength was researched with true triaxial testing machines. And besides, the mechanical model of jointed rocks was presented through the test with stiff machines (Liu 1981). Recently the research works on the yield criteria of rocks have also been performed. As for the natural stresses of rock masses, the new conception of "locked in" stress was advanced which can be used to research the rock bursting and the mechanism of earthquake (Tan 1983).

4. FOUNTAMENTAL CONCEPTION ON THE DEVELOPMENT OF ROCK MECHANIS (Tan 1982)

4.1 Rock masses should be considered as a geological medium which has undergone a long period tectonic movement in history. Under the long term consistent action of the tectonic stress field, the rock mass is not homogeneous, in which a large amount of sliding planes between the cryctals, cracks, joints, fractures, bedding planes, weak seams and faults exist. So the rock mass assumes complex structure with characteristics of nonhomogeneity, anisotropy and discontinuity, within which various rock elements with different mechanical properties were contained. Therefore, each rock element will be also nonhomogenuous, anisotropic and discontinuous. Hence, its mechanical properties of discontinuities must be studied first, from which the mechanical properties of all elements and even the whole structure of rock mass could be deduced.

4.2 Natural stress exists in rock masses, that makes them somewhat different from other materials. Natural stresses can be classified as:
 (1) Continential stress field under the compression force of tectonic plate movement;
 (2) Continential stress field owing to the thermal convection of earth mantle under the plate;
 (3) Internal stress field originating from the internal stress of earth crust or palaeo-stress field;
 (4) Gravitational stress field due to gravity;
 (5) Local stress field including the local stress field due to stress accumulation at local zone, the stress field caused by the nonuniform distribution of temperature as well as the stress field caused by the gradient of underground water.
All of these stress fields must be considered as a great important factor in the studying of stability of underground structures, rock burst and the mechanism of earthquake.

4.3 Under the action of stresses, such phenomena, as sliding, dislocation, deformation may occur within the micro or macro structures of rock with obviously time-dependent properties, therefore rock must be regarded as a rheological body. The study on long term stability of structures must be carried out after this principle.

4.4 The researches on weak seams and discontinuities are often the crucial points that affect the stubility of structures. Hence, in engineering practice, their physico-mechanical properties and their effect on structure must be studied emphatically.

4.5 In view of rock mechanics is a new discipline which is closely related to engineering practice, especially its theory is still in developing stage. In engineering practice a large number of complicated problems concern rock mechanics, modern experimental techniques, the method of anological computation, construction techniques and other specialities. Hence, in order to solve these problems successfally, it is very important that close collaboration among geologists, scientific workers, designers and constructors must be insisted upon.

4.6 Crucial is a correct concept. It is well known that rock is a geological medium with very complicated properties. Thorough understanding of it is impossible, no matter how careful study we have done. Undoubtedly, the computer is a very useful tool, but it still has some limitation for us. Therefore a correct diagnosis and concept is essential in engineering practice which will give us some guidelines for further study in the field of engineering geology, experiments, design, construction and so on. Field monitoring is important to check our basic ideas and modify our design or construction scheme constantly so as our ideas, concept, scheme and plan can be in comformity with the actual situation. Only by these means, new development of rock mechanics theory could be brought forth creatively.

ACKNOWLEDGEMENT

Our warm thanks are due to Niu Xi-Zhuo, Gao Qu-Qing, Wang Cheng-Shu, Wang Wu-Lin, Chen Dan-Xi, Chen Cheng-Zeng, Zhu Zhi-Jie, Zhou Si-Meng, Liu Bao-Shen, Yu Yong-Hao, Zheng Yu-Tian, Lin Yun-Mei, Liang Guo-Qiang, Wang Zheng-Hong, Xu Xiao-He, Hua An-Zeng for their kindness in providing the reference materials.

REFERENCES:

Cheng Liang-Kui, Feng Shen-Duo (1978): The practice and cognition on the rockbolting and shotecrete support within fractured weak surrounding rocks. Architectural Research Institute of Ministry of Metallurgy.

Li Xi-Yun, Lin Yun-Mei, Zheng Yu-Tian (1981) Research on the mechanism of rockbolting support. Journal of Northeast Technological Institute, sum No.29

Liang Wu-Tao, Chen Cheng-Zeng, Cheng Sheng-Gao (1981): A general survey of tunnel construction in China. Journal of Civil Engineering Vol. 14 No.1 pp.51-55

Liu Bao-Shen, Yan Roungui (1981): Mechanical models of fractured rock. Proc. Int. Symp. Weak Rock. Tokyo, Vol.2 pp.619-624.

Liu Bao-Shen, Lin De-Zhang (1979):
Rheological study on the mechanism of supporting. Institute of Metallurgy and Mining, Ministry of Metallurgy

Liu Yong-Xue (1979):
Natural stresses of rock masses and their testing method. The theory and practice on rock mechanics, pp. 105-108

Lu Bu-Fan (1982):
The measurement of natural stresses of rock mass in mining construction of China. Research Institute of Coal Mine, Cheng-Qing Branch.

Mei Jian-Yun (1978):
On determination of shear failure criteria and the selection of calculation parameters of shear strength. Technical Information of Water conservancy and Hydroelectric Power, No.1, pp.1-32

Mei Jian-Yun (1981):
On the shear strength of dam foundations with interbedded weak seams. Proc. Int. Symp. Weak Rock Tokyo, Vol.1 pp.381-385

Niu Xi-Zhuo (1982):
Several problems on rock mechanics threatening the safty of coal mine. Journal of rock mechanics and engineering, 1982, No.1, (in press)

Ren Fang (1974):
The analysis on the large scale rock mechanics testing. Scientific Selected Works of Yangtze River Research Institute, No.1, pp.1-21

Sun Guang-Zhong (1982):
The constitutive equation of rock mass deformation. Institute of Geology , Academia Sinica.

Tan Tjong Kie (1959):
The partial differential equation of rock deformation.

Tan Tjong-Kie (1964):
In situ testing of rocks on behalf the Daye iron mine.

Tan Tjong-Kie (1980):
Locked in stresses, creep and dilatancy of rocks and constitutive equations. Rock Mech. Vol.13, pp. 5-22.

Tan Tjong-Kie and Li Ke-Ri (1981):
Relaxation and creep properties of thin interbedded clayey seams and their fundamental role in the stability of dams. Proc. Int. Symp. Weak Rock, Tokyo, Vol.1, pp. 369-374.

Tan Tjong-Kie (1966):
Eine Untersuchung der Verformung einiger. Arten Von. Zerklüfteten Gesteinsformationen und die Bestimmung ihrer rheologischen Eigenschaften in situ. Abhandlungen der deutschen akademie der wissenschaften zu berlin, klasse fur bergbau, Huttenwesen und Montangeologie, Nr. 1, pp. 214-223.

Tan Tjong-Kie and He Zhi-Tong (1982):
A physico-rheological model for the large Tangshan earthquake, Tectonophysics, 85, pp.123-148.

Tan Tjong-Kie (1982):
On the geodynamics of China plate. Acta mechanica Sinica. (in prss)

Tan Tjong-Kie and Wen Xuan Mei (1983):
Swelling rocks and stability of tunnels. (This Proc. session).

Tan Tjong-Kie and Kang Wen-Fa (1983):
Time dependent dilatancy prior to rockfailure and earthquakes (This Proc. session).

Tan Tjong-Kie (1982):
Foreword to the " Joural of rock Mechanics and Engineering",No.1 (in press)

Tan Tjong-Kie (1982):
The crucial is a correct concept. Hydrogeology and Engineering Geology, No.2, pp. 5-10

Xu Xiao-He (1979):
A priliminary research on the new discipline in mining industry-rock breaking. Rock breaking research department, Northeast Technological Institute.

Yan Ke-Qiang (1982):
The current state and development of the rockbolting and shotecrete technique in hydraulic underground construction, Rock Mechanics, No.5, pp.1-9

Zhou Wei-Yuan, Yang Ruo-Qiong (1982):
The stability analysis of arch dams with FEM and model test. Qing-hua University.

Zhou Qun-Li (1979):
Fracture Mechanics of rock and its application. The throry and practice of rock mechanics, pp.213-230.

RESULTS OF ROCK MECHANICS IN HUNGARY – AN ENGINEERING GEOLOGICAL MODEL OF ROCKS

Résultats dans la mécanique des roches en Hongrie – Un modèle géotechnique des roches

Fortschritte der Felsmechanik in Ungarn – Ein ingenieur – geologisches Gesteinsmodell

M. Gálos
Senior Research Officer, Technical University, Budapest, Hungary

P. Kertész
Associate Professor, Technical University, Budapest, Hungary

SYNOPSIS

A rock model based on geological-petrophysical-engineering geological characteristics, suitable for the solution of rock mechanics problems is presented and its elements such as geological formation, rock body, rock block, discontinuity are defined. The ways of designation of the model elements are described and the variation in time of the properties of the basic elements (rock body, rock block) and the main types of the spatial distribution of these characteristics are analysed.

RESUME

Cette communication présente un modèle de roche fondé sur des caractéristiques géologiques, pétrophysiques et géotechniques permettant la solution des problèmes de la mécanique des roches, et elle en définit les éléments du modèle (formation géologique, corps de roche, bloc de roche, discontinuité). On traite des modes de distinction des éléments du modèle et on analyse la variation en temps des propriétés des éléments fondamentaux (corps de roche, bloc de roche) ainsi que les types principaux de leur distribution spatiale.

ZUSAMMENFASSUNG

Ein zur Lösung felsmechanischer Probleme geeignetes Felsmodell, welches auf den Charakteristiken der Geologie, der Petrophysik und der Ingenieurgeologie beruht, wird hier vorgestellt und seine Bestandteile wie z.B. die geologische Formation, die Gesteinsmasse, Gesteinsblöcke und Diskontinuität werden definiert. Die Beschreibungsmethode der Bestandteile des Modells werden auch beschrieben, sowohl als auch die zeitabhängigen Veränderungen der Eigenschaften der Grundbestandteile (Gesteinsmasse, Gesteinsblock), während die Haupttypen der räumlichen Verteilung dieser Charakteristiken analysiert werden.

INTRODUCTION

Rock mechanics fuses the theoretical and practical knowledge on petrography and mechanics with the aim of enabling man to truly understand and forecast, by conscious efforts such as the opening of underground workings, location of engineering structures, etc., the disrupted stress state of the earth crust and the resulting field of strain, deformation and dislocation. Changes in the immediate or the broader neighbourhood of a structure may be only of a character or size not threattening the safety of people, nor of goods, i.e. not involving any inadmissible risk.

The changes cannot be traced, nor forecasted in their tangible reality, as the earth crust is a system of material composition and structural pattern too complex to allow such an approach. For this reason, we have to replace parts of the earth crust important for the problems to be dealt with by such a model which behaves in some of its – arbitrarily selected – variation processes as a real element of space, but which has a simple structure and composition, carrying in its elements all the properties and property-changes that we wish to take into consideration in designing the size characteristic of an engineering structure.

Modelling is based on our characterizing a concrete, tangible space element circumscribed as required for the problem in question in terms of a set of characteristics. The less important characteristics are neglected, and the more important ones are considered to be components of the model to be constructed. Both the material characteristics and the spatial shape of the crustal elements are modelled: our model has material characteristics and spatial extension.

Consequently, it is always a spatial element (V) of definite volum (and shape) and its selected characteristics $(T_1, \ldots, T_i)$ that participate in constituting a rock model; $V(T_1, T_2, \ldots T_i, \ldots, T_n)$. The smaller the volum, V, the simpler the model and the better it approaches to reality. However, the smaller the number of the properties taken into consideration, the simpler the model and the less the reality is approximated.

The assemblage of the spatial elements thus selected is the rock mechanical (engineering geological) model.

THE COMPOSITION OF A ROCK MODEL

A rock mechanical, engineering geological model should satisfy the requirements of continuum mechanics: its any space element should be filled continuously by material. From the viewpoint of the degree of stacking of the material any possible transition from the most rigorous stacking, i.e. from the crystalline state up to the gaseous or vapour-like phase, in other words, to the least fixed structure of the material may be represented. The resulting complex behaviour of these is studied in the rock model.

A rock model is engineering structure-centred. It is primarily the effect of the structure that must be taken into consideration while distinguishing model elements.

For the solution of our problem we have split up that part of the earth crust to be studied rock mechanically, i.e. the bedrock, into model elements that are superimposed to one another, but easy to handle each separately. Of these the largest, geologically separable unit is a <u>geological</u> formation. The principal genetic conditions within a formation are the same, the petrographic, geochemical and geophysical properties are definite and interdependent. Rocks of varied facies, but of similar origin, occur within a formation. Different formations either grade one into the other or are interconnected with a sharp boundary.

A formation is homogeneous in the geological sense, being characterized by one predominant lithofacies or the circumstances of one determinant geological process. From the rock mechanical viewpoint it does not behave homogeneously, but it can be split up into homogeneous rock bodies.

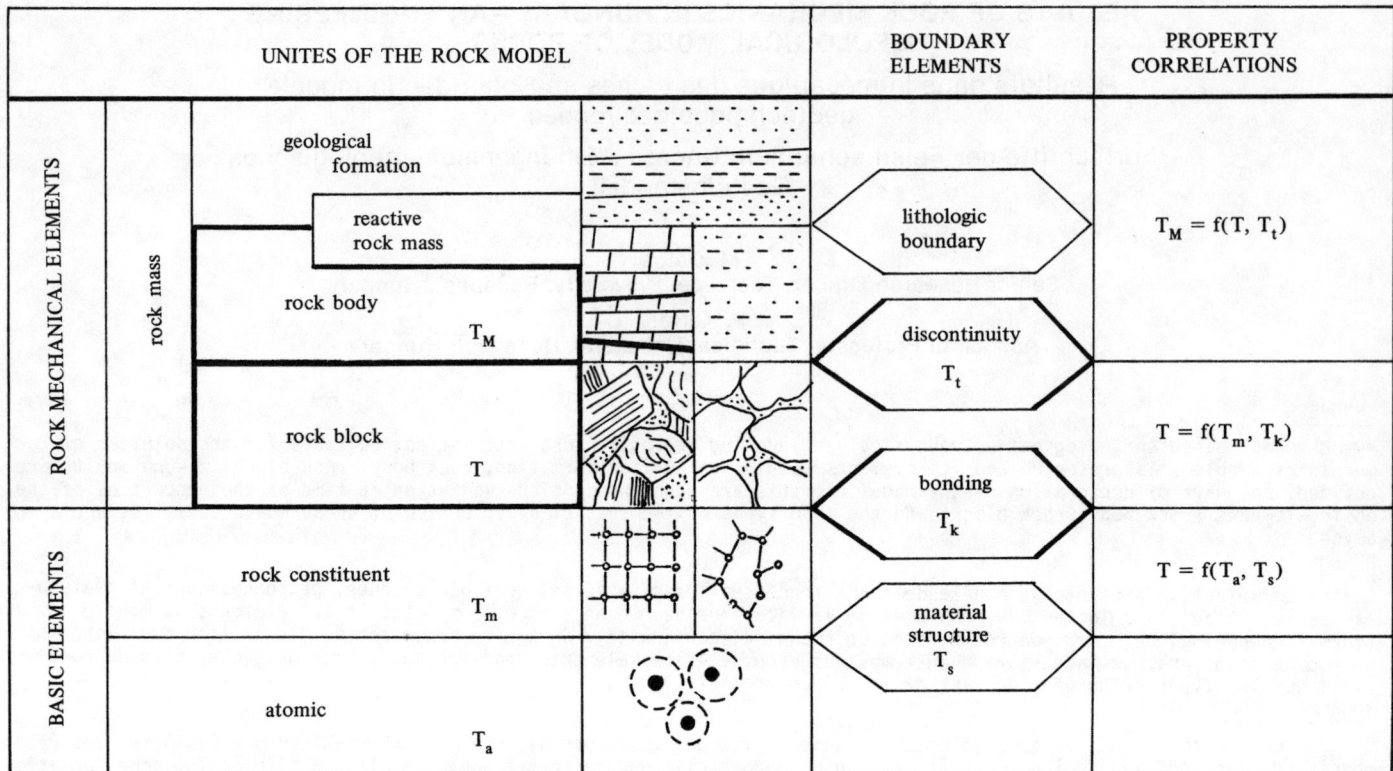

Table 1. Elements of the rock model

In general, the spatial unit affected by the engineering structure in question and called the reactive rock mass should be selected from one formation. The selection will be the result of the conscious engineering efforts already referred to. Less frequently, a reactive rock mass may encompass not one, but two or more formations.

The rock mass is composed of such rock bodies that can be regarded, from the viewpoint of the problem in question, to be homogeneous, of uniform behaviour. The rock mass is either uniform, continuous, or it consists of such homogeneous rock slabs of uniform quality which are not directly interconnected along their boundary surfaces, inasmuch as the very structure of their material is concerned. The dissection into rock slabs may be due either to the process of lithogenesis or to postgenetic effects.

A rock body always belongs to one genetic unit and, in general, uniform change-effects are supposed to be manifested in it. The anisotropy of a rock mass may be quite considerable, the homogeneity in it is interpreted by admitting that on any element on which the whole system of the rock body can be observed (e.g. its being discontinuous) the same characteristics are observable.

The unit surrounded by dividing surfaces of a rock body is a rock block. This is a continuous element of space not intersected by a dividing face. A rock block is the basic unit of rock mechanics. Laboratory rock mechanical test results practically always concern a rock block and they can be transferred to the relevant rock body only in the case if the system of discontinuity is known. The properties of a rock block include petrographic homogeneity, the identical composition of all spatial elements and the presence of one and the same system of material structure.

The properties of a rock block and particularly the changes in these properties cannot be analyzed nor evaluated, unless the internal composition, constitution of the rock-mechanically homogeneous block is examined with additional scrutiny. From this viewpoint the rock block directly corresponds to the notion „rock" in petrology and it can be said to represent — in a directly observable way — a loose or solid aggregate of some kinds of rock constituents.

Formed in terms of size and shape according to their genetic circumstances, the rock components constitute the rocks in a definite and typical spatial system, being connected with one another by material-structural or temporary bonds. This spatial system is the texture of the rock.

The design of a rock mechanical model is summarized in Table 1, according to which the characteristics of any model element can be traced back to the properties of the lower-rank units.

INTERPRETATION OF PROPERTY AND EFFECT IN THE ROCK MODEL

The property of a rock body (T_M) is composed of the properties of the rock block (T) and its discontinuity pattern (T_t)

$$T_M = f(T, T_t).$$

It also follows from this that any change in the properties of a rock body (ΔT_M) due to various effects such as the changes in energy level are, themselves, composed of the changes in the components.

$$\Delta T_M = f''(\Delta T, \Delta T_t).$$

The birth and variation of a property are time-dependent. Diagenetic effects will change the properties of a rock towards consolidation, weathering effects towards loosening. Fig. 1. 2. and 3. illustrate the character of the changes in rock properties.

A rock (rock mass, rock body, rock block) is affected during the time to

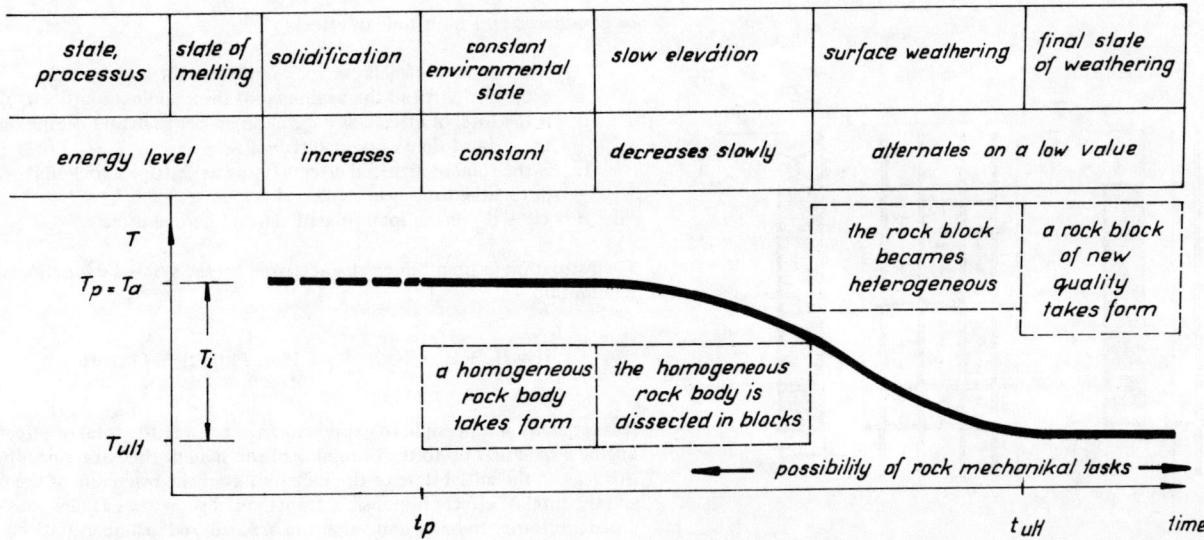

Fig. 1. Changes of rock properties in an igneous rocks. T_p – property of the just formed rock, T_{ult} – property in the final state of weathering, T_i – extension of T property, t_p – time of rock formation, t_{ult} – final state of weathering, T_a – property corresponding to maximal energy level

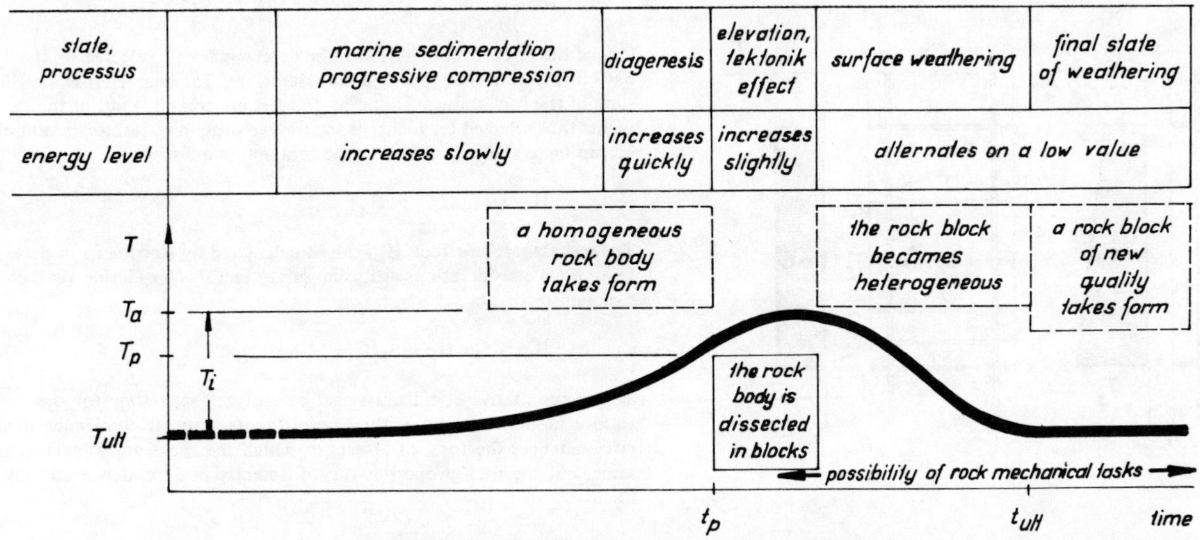

Fig. 2 Changes of rock properties in a sedimentary rock

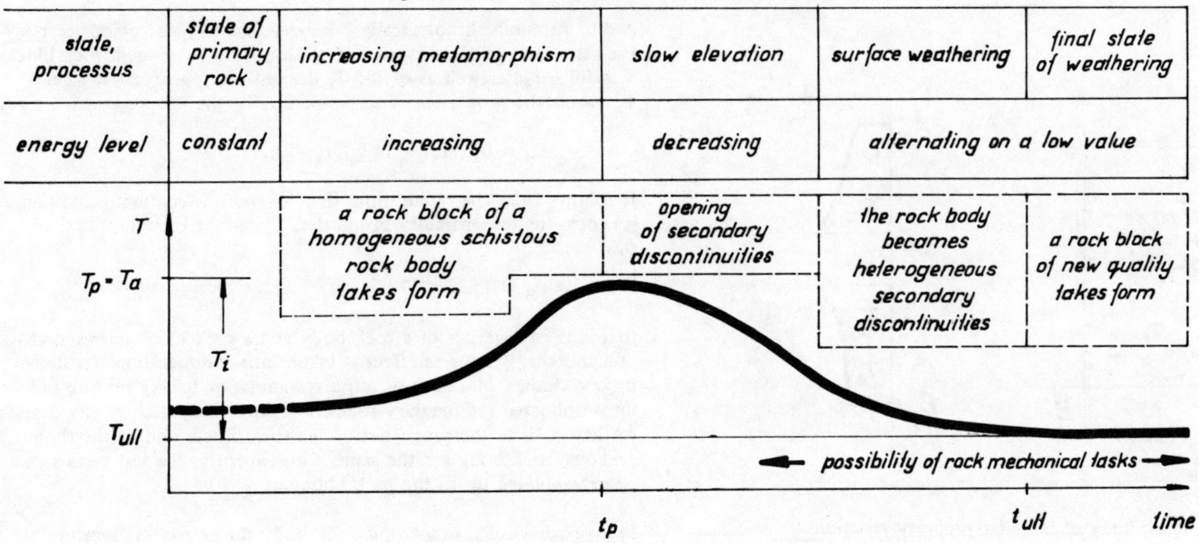

Fig. 3 Changes of rock properties in a metamorphic rock

be considered by a multitude of effects:

H_0 is the total of effects per rock unit from the moment of geological birth to the beginning of the examination $(0 - t_0)$,
H_t is the total of effects of natural origin per rock unit during the the span of time examined $(t_0 - t_i)$,
H_m is the total of artificial effects (impacts) hitting a rock unit during the time-span examined $(t_0 - t_i)$.
$H = H_t + H_m$ is the total of artificial and natural effects.

Their variation in time can be characterized by the symbolic functional relationship

$$H = H_0 + H_t + H_m = H_0 + H = \int_0^{t_0} h(t)dt + \int_{t_0}^{t_i} h(t)dt.$$

It is generally not possible to express in an exact way the total of effects hitting a rock unit up to the beginning of the span of time examined. For this reason, the initial state of the rock unit as of the beginning of the test and the total of effects responsible for its initial properties T_0 are considered arbitrarily to be of unit value and are analyzed just quantitatively.

$$H_0 = \int_0^{t_0} h(t)dt = 1.$$

Out of the effects those judged to be determinant are selected on the basis of theoretical or practical considerations. The characteristics of the state of the rock units as found by the test are compiled out of the elements thus selected (e. g. change in stress, change in water, temperature) according to the very nature of the problem in question.

$$H = f(H_1, H_2, \ldots, H_n).$$

The representative effect, H_M, can be calculated by starting from these theoretical considerations and using safety factors considering virtual experiences $(b_1, b_2, \ldots, b_n)$.

$$H_M = f(b_1 H_1 + b_2 H_2 + \ldots b_n H_n)$$

The representative effect must also be analyzed separately for the various model elements or their constituents, respectively. Under identical circumstances the total of effects to which the rocks are subject is the same, but the rock properties vary differently in dependence on rock quality.

MAIN TYPES OF ROCK BODIES

A discontinuous, lithologically homogeneous and isotropic <u>rock body</u> has properties (T_M) composed of those of the constituent rock blocks $T = f(T_a, T_k)$ as well as of the T_t discontinuity properties, where T_a means the properties of the minerals, T_k the properties of the bond.

$$T_M = f(T, T_t) = f'(T_a, T_k, T_t)$$

It follows from the relationship that if a rock is continous and homogeneous, the discontinuity properties, T_t cannot be interpreted and thus

$$T_M = T,$$

that is the properties of a rock body and a rock block correspond to one another, if the size effect is taken into consideration. As shown by our studies, in a rock of loosely granular and flaky texture (a loose, unconsolidated sedimentary rock) no virtual discontinuity can develop, so that in these the properties of the rock blocks and of the (homogeneous) rock body are the same. Consequently, the test results can be extrapolated up to the rock boundary.

In a <u>petrologically anisotropic</u> rock body the above relationships are

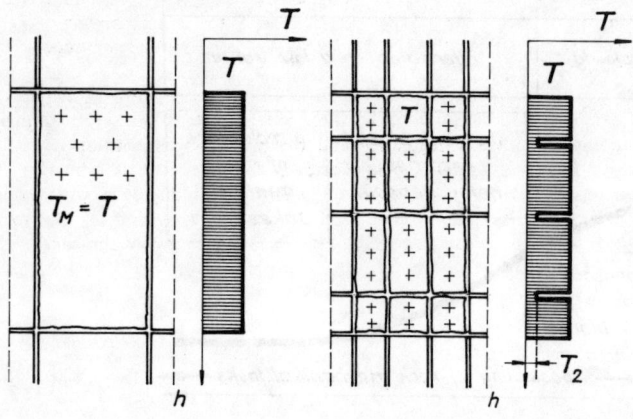

Fig. 4 Homogeneous rock body

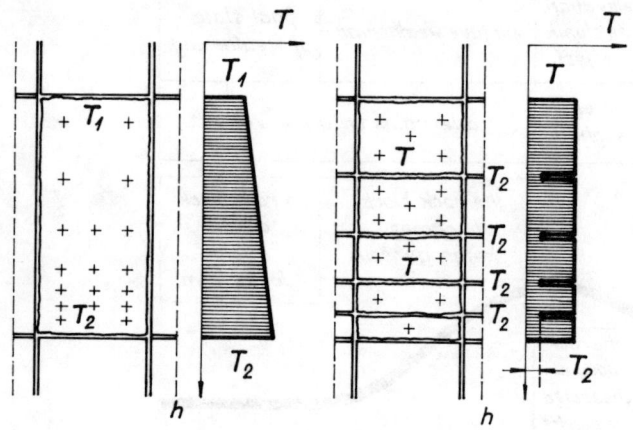

Fig. 5 Rock body of constant property-variation

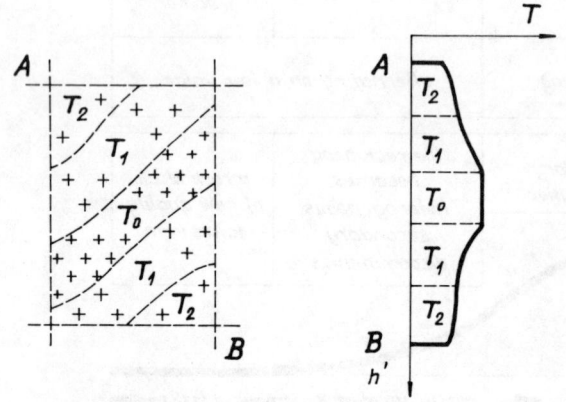

Fig. 6 Rock body of irregular property-variation

different only to the extent that they have to be interpreted separately in the characteristic directions of space (x, y, z) (T_{Mx}, T_{My}, T_{Mz}). This applies with the same character to the anisotropy resulting from discontinuity as well.

In the construction of a rock body comprising blocks and discontinuity the following principal cases of distribution of properties are distinguished.

Rock body with homogeneous (identical and constant) properties. A property is subconstant in any point of space, the curve of property distribution corresponds to the normal distribution pattern. So continuous or uniformly discontinuous is a considerable part of the rock bodies (Fig. 4).

The properties of a rock body with regular property variation vary in definite space directions to a determinable extent, continuously and monotonously (Fig. 5). The property distribution curve is asymmetric, divergent. A rock body of this kind is e.g. that of a clay on the way of compaction, a weathering granite or a granite being fractured.

The trend of property variation in a rock body with irregular property variation cannot be determined: the property values vary from point to point without any regularity and the property distribution is uneven and irregular, too. The change in the properties is primarily provoked by effects manifesting themselves irregularly (e.g. weathering, postvolcanic effects, etc.). A rock of this kind can be split up into rock blocks that can be handled separately for testing only by scrutinized analysis (Fig. 6).

RAYMOND H. FOGLER LIBRARY
DATE DUE

BOOKS ARE
RECALL AFT